To
JOHN VONDELING
visionary, innovator, editor without equal, and
a unique and cherished friend

With deepest heartfelt thanks for the many memories —
May the sun shine and the fishing be grand

			3A (13)	4A (14)	5A (15)	6A (16)	7A (17)	8A (18)	
								2 **He** Helium 4.0026	1
			5 **B** Boron 10.811	6 **C** Carbon 12.011	7 **N** Nitrogen 14.0067	8 **O** Oxygen 15.9994	9 **F** Fluorine 18.9984	10 **Ne** Neon 20.1797	2
8B (10)	1B (11)	2B (12)	13 **Al** Aluminum 26.9815	14 **Si** Silicon 28.0855	15 **P** Phosphorus 30.9738	16 **S** Sulfur 32.066	17 **Cl** Chlorine 35.4527	18 **Ar** Argon 39.948	3
28 **Ni** Nickel 58.693	29 **Cu** Copper 63.546	30 **Zn** Zinc 65.39	31 **Ga** Gallium 69.723	32 **Ge** Germanium 72.61	33 **As** Arsenic 74.9216	34 **Se** Selenium 78.96	35 **Br** Bromine 79.904	36 **Kr** Krypton 83.80	4
46 **Pd** Palladium 106.42	47 **Ag** Silver 107.8682	48 **Cd** Cadmium 112.411	49 **In** Indium 114.82	50 **Sn** Tin 118.710	51 **Sb** Antimony 121.757	52 **Te** Tellurium 127.60	53 **I** Iodine 126.9045	54 **Xe** Xenon 131.29	5
78 **Pt** Platinum 195.08	79 **Au** Gold 196.9665	80 **Hg** Mercury 200.59	81 **Tl** Thallium 204.3833	82 **Pb** Lead 207.2	83 **Bi** Bismuth 208.9804	84 **Po** Polonium (209)	85 **At** Astatine (210)	86 **Rn** Radon (222)	6
110 — — (269)	111 — — (272)	112 — — (277)		114 — — (285)		116 — — (289)		118 — — (293)	7

63 **Eu** Europium 151.965	64 **Gd** Gadolium 157.25	65 **Tb** Terbium 158.9253	66 **Dy** Dysprosium 162.50	67 **Ho** Holmium 164.9303	68 **Er** Erbium 167.26	69 **Tm** Thulium 168.9342	70 **Yb** Ytterbium 173.04	71 **Lu** Lutetium 174.967	6
95 **Am** Americium (243)	96 **Cm** Curium (247)	97 **Bk** Berkelium (247)	98 **Cf** Californium (251)	99 **Es** Einsteinium (252)	100 **Fm** Fermium (257)	101 **Md** Mendelevium (258)	102 **No** Nobelium (259)	103 **Lr** Lawrencium (260)	7

HERE'S WHAT YOUR COLLEAGUES ARE SAYING ABOUT MOORE, STANITSKI & JURS:
CHEMISTRY: THE MOLECULAR SCIENCE...

LAYING THE CHEMICAL FOUNDATION

"As a group of scientists we have usually made the first chapter of many texts incredibly boring and students quickly lose interest. Chapter 1 of Moore et al. is a nice diversion and shows how chemistry solved a societal problem. This is a strength of the book."

Rick White, Sam Houston State University

COVERAGE OF ORGANIC AND BIOCHEMISTRY

"I like the way that topics in organic and biochemistry are introduced in the text. I feel this is an effective way to engage students as well as to present principles. . . . I find the writing to be clear, concise, and accurate."

David Miller, California State University–Northridge

"I am quite supportive of this text for two principal reasons: It treats the subject in a fresh way and is written in a style that should appeal to the youthful customer it would serve. The integration of organic and biochemical topics . . . is an excellent idea that is well executed. The writing is remarkably uniform with no indication of disconnected joint efforts."

Richard Thompson, University of Missouri–Columbia

"I like the fact that this book integrates biochemistry and organic chemistry, two sub-disciplines of chemistry that are barely touched by other textbooks."

Jose Vites, Eastern Michigan University

"The section on Gibbs Energy and Biological Systems [Chapter 18] is great."

William P. Reinhardt, University of Washington

PROBLEMS AND EXERCISES

"I think the Conceptual Exercises are a good complement to the material provided. There were plenty of them and they had the equivalent rigor of the examples in the text."

Marcy McDonald, University of Alabama–Tuscaloosa

"The discussion on reasonableness for most sample problems is wonderful. It clearly shows students how to think critically."

Mapi M. Cuevas, Santa Fe Community College

"The conceptual rather than the plug-and-chug problems are outstanding and should be encouraged."

Richard Thompson, University of Missouri–Columbia

KEY TOPICS: THERMODYNAMICS AND KINETICS

"The chapter on Kinetics is well written, and the nanoscale view to understand elementary reactions is introduced earlier in the chapter compared to most textbooks. Consequently, the student should find it less difficult to understand reaction mechanisms."

Margaret R. Asirvatham, University of Colorado–Boulder

"The second chapter on thermodynamics [Chapter 18] is very well crafted; there is definitely a unique view of concepts that would distinguish this text. The level is good for the introduction of free energy and the definition of biochemical terms, exergonic and endergonic, is also useful. . . ."

Kathleen Murphy, Daemen College

OTHER TOPICS AND FEATURES . . .

"The most innovative feature that this textbook offers that goes beyond material presented in traditional general chemistry texts is *Chemistry You Can Do*. This is an excellent strategy to engage students in hands-on chemistry activities at home."

Cheryl Dammann, University of North Carolina–Charlotte

"The chapter on Gases covers the typical concepts of gases, but also does a great job of integrating environmental chemistry into it."

John DeKorte, Glendale Community College

Harcourt College Publishers

Where Learning Comes to Life

TECHNOLOGY

Technology is changing the learning experience, by increasing the power of your textbook and other learning materials; by allowing you to access more information, more quickly; and by bringing a wider array of choices in your course and content information sources.

Harcourt College Publishers has developed the most comprehensive Web sites, e-books, and electronic learning materials on the market to help you use technology to achieve your goals.

PARTNERS IN LEARNING

Harcourt partners with other companies to make technology work for you and to supply the learning resources you want and need. More importantly, Harcourt and its partners provide avenues to help you reduce your research time of numerous information sources.

Harcourt College Publishers and its partners offer increased opportunities to enhance your learning resources and address your learning style. With quick access to chapter-specific Web sites and e-books . . . from interactive study materials to quizzing, testing, and career advice . . . Harcourt and its partners bring learning to life.

Harcourt's partnership with Digital:Convergence™ brings :CRQ™ technology and the :CueCat™ reader to you and allows Harcourt to provide you with a complete and dynamic list of resources designed to help you achieve your learning goals. Just swipe the cue to view a list of Harcourt's partners and Harcourt's print and electronic learning solutions.

http://www.harcourtcollege.com/partners/

CHEMISTRY
The Molecular Science

JOHN W. MOORE
University of Wisconsin — Madison

CONRAD L. STANITSKI
University of Central Arkansas

PETER C. JURS
Pennsylvania State University

HARCOURT COLLEGE PUBLISHERS

FORT WORTH   PHILADELPHIA   SAN DIEGO   NEW YORK   ORLANDO   AUSTIN
SAN ANTONIO   TORONTO   MONTREAL   LONDON   SYDNEY   TOKYO

Publisher: Emily Barrosse
Publisher/Acquisitions Editor: John Vondeling
Marketing Strategist: Pauline Mula
Contributing Editor: Mary Castellion
Associate Editor: Marc Sherman
Project Editor: Robin C. Bonner
Production Manager: Charlene Catlett Squibb
Art Director and Text Designer: Caroline McGowan

Cover Credit: © Charles Krebs/Stone Images

Conceptual Challenge Problems © 1997 H. Graden Kirksey

Chemistry: The Molecular Science, First Edition
ISBN: 0-03-032011-9
Library of Congress Catalog Control Number: 2001089370

Copyright © 2002 by Harcourt, Inc.

All rights reserved. No part of this publication may be reproduced or transmitted in any form or by any means, electronic or mechanical, including photocopy, recording, or any information storage and retrieval system, without permission in writing from the publisher.

Requests for permission to make copies of any part of the work should be mailed to the following address: Permissions Department, Harcourt, Inc., 6277 Sea Harbor Drive, Orlando, FL 32887-6777.

Address for domestic orders:
Harcourt College Publishers,
6277 Sea Harbor Drive, Orlando, FL 32887-6777
1-800-782-4479
e-mail   collegesales@harcourt.com

Address for international orders:
International Customer Service, Harcourt, Inc.
6277 Sea Harbor Drive, Orlando FL 32887-6777
(407) 345-3800
Fax   (407)345-4060
e-mail   hbintl@harcourt.com

Address for editorial correspondence:
Harcourt College Publishers, Public Ledger Building, Suite 1250,
150 S. Independence Mall West, Philadelphia, PA 19106-3412

Web Site Address
http://www.harcourtcollege.com

Printed in the United States of America

1234567890   048   10 987654321

CONTENTS OVERVIEW

1 THE NATURE OF CHEMISTRY *1*

2 ATOMS AND ELEMENTS *41*

3 CHEMICAL COMPOUNDS *74*

4 QUANTITIES OF REACTANTS AND PRODUCTS *121*

5 CHEMICAL REACTIONS *162*

6 ENERGY AND CHEMICAL REACTIONS *210*

7 ELECTRON CONFIGURATIONS AND THE PERIODIC TABLE *265*

8 COVALENT BONDING *315*

9 MOLECULAR STRUCTURES *359*

10 GASES AND THE ATMOSPHERE *407*

11 LIQUIDS, SOLIDS, AND MATERIALS *459*

12 FUELS, ORGANIC CHEMICALS, AND POLYMERS *513*

13 CHEMICAL KINETICS: RATES OF REACTIONS *571*

14 CHEMICAL EQUILIBRIUM *632*

15 THE CHEMISTRY OF SOLUTES AND SOLUTIONS *680*

16 ACIDS AND BASES *727*

17 ADDITIONAL AQUEOUS EQUILIBRIA *776*

18 THERMODYNAMICS: DIRECTIONALITY OF CHEMICAL REACTIONS *812*

19 ELECTROCHEMISTRY AND ITS APPLICATIONS *863*

20 NUCLEAR CHEMISTRY *918*

21 CHEMISTRY OF SELECTED MAIN GROUP ELEMENTS *954*

22 CHEMISTRY OF SELECTED TRANSITION ELEMENTS AND COORDINATION COMPOUNDS *983*

APPENDICES *A.1*

ANSWERS TO PROBLEM-SOLVING PRACTICE PROBLEMS *A.49*

ANSWERS TO EXERCISES *A.69*

ANSWERS TO SELECTED QUESTIONS FOR REVIEW AND THOUGHT *A.93*

FIGURE CREDITS *C.1*

GLOSSARY *G.1*

INDEX *I.1*

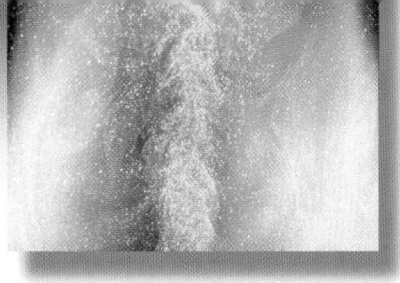

PREFACE

Chemistry is a mature science, yet new chemistry and new ways to apply chemical principles to the world around us are reported every day. Chemical research is helping to solve problems as diverse as how to make electronic circuits on the molecular scale; how to design and synthesize new, more effective drugs; and how to create metals, plastics, and other materials that have just the properties we want. All of these problems require the chemist's unique, molecular-scale viewpoint — a perspective with a value that has been proved many times over during the past century. Because it is so broadly applicable, much of today's cutting-edge chemical research involves collaborations with biochemists, biologists, pharmacologists, physicians, geologists, materials scientists, engineers, and others. We believe that it is crucial that students in first-year chemistry courses recognize our discipline's ability to solve important problems and its important contributions to other disciplines, and we have acted on that belief by writing this textbook.

GOALS

Our overarching goal is to involve science and engineering students in active study of what modern chemistry is, how it applies to a broad range of disciplines, and what effects it has on their own lives. We maintain a high level of rigor so that students in the mainstream general chemistry course will learn the concepts and develop the problem-solving skills essential to their future ability to use chemical ideas effectively. We have selected and carefully refined the book's features in support of this goal.

More specifically, we intend that this textbook will help students develop

- a broad overview of chemistry and chemical reactions,
- an understanding of the most important concepts and models used by chemists and those in chemistry-related fields,
- the ability to apply the facts, concepts, and models of chemistry appropriately to new situations in chemistry, other sciences and engineering, and to other disciplines,
- knowledge of the many practical applications of chemistry in other sciences, in engineering, and in other fields,
- an appreciation of the many ways that chemistry impacts the daily lives of everyone, students included, and
- motivation to study in ways that help them achieve long-term retention of facts and concepts.

Because modern chemistry is inextricably entwined with so many other disciplines, we have integrated organic chemistry, biochemistry, environmental chemistry, industrial chemistry, and materials chemistry into the chapters that deal with chemical principles and facts. Applications in these areas are discussed together with the principles on which they are based. This serves to motivate students whose interests lie in these related disciplines and also gives a more accurate picture of the multidisciplinary collaborations so prevalent in modern chemical research and modern industrial chemistry. Organic and biochemistry are especially important to a large fraction of the students who take mainstream general chemistry courses, and we have identified organic chemistry and biochemistry material by means of icons.

 Organic chemistry icon

 Biochemistry icon

AUDIENCE

Chemistry: The Molecular Science is intended for mainstream general chemistry courses for students who expect to pursue further study in science or science-related disciplines. Those planning to major in chemistry, biochemistry, biological sciences, engineering, geological sciences, agricultural sciences, materials science, physics, and many related areas will benefit from this book and its approach. We assume that the students who use this book have had a basic foundation in mathematics (algebra and geometry) and in general science. Most will also have had a chemistry course before coming to college.

FEATURES

We strongly encourage students to understand concepts and learn to apply them to problem solving. We believe that such understanding is essential if students are to be able to use what they learn in subsequent courses and in their future careers. Too often we hear from professors in courses for which general chemistry is prerequisite that students have not retained what we have taught them. This book's wide range of features has been carefully chosen to address this issue and to help students achieve long-term retention of the material.

Problem Solving

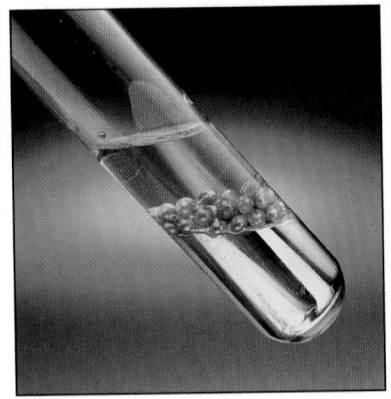

Problem solving is introduced in Chapter 1 and a framework is built there that is followed throughout the book. In each chapter there are many worked-out **Problem-Solving Examples** — a total of 236 in the book as a whole. Most consist of five parts: a Question (problem); an Answer, stated briefly; an Explanation that provides significant help for students whose answer did not agree with ours; a section marked with a ✔ that indicates how a student could check whether a result is reasonable, and an additional **Problem-Solving Practice** that provides similar questions with answers appearing only in an appendix. We explicitly encourage students first to define the problem, develop a plan, and work out an answer without looking at either the Answer or the Explanation, and only then to compare their answer with ours. If their answer did not agree with ours, students are asked to repeat their work. Only after that do we suggest that they look at the Explanation, which is couched in conceptual as well as numeric terms so that it will improve students' understanding, not just their ability to answer an identical question on an exam. The ✔ section helps students learn how to use estimated results and other criteria to decide whether an answer is reasonable, an ability that will serve them well in the future. By providing similar practice problems that are answered in the back of the book, we encourage students immediately to consolidate their thinking and improve their ability to apply their new understanding to related problems.

Enhancing students' abilities to estimate results is the goal of **Estimation** boxes in most chapters. Each Estimation poses a problem that relates to the content of the chapter in which it appears and for which a rough calculation suffices. Students gain knowledge of various means of approximation, such as back-of-the-envelope calculations and graphing, and diverse sources of information, such as encyclopedias, handbooks, and the Internet.

To further ensure that students are not merely memorizing algorithmic solutions to specific problems, we provide more than 300 **Exercises,** which are placed immediately following new concepts as they are introduced. Often the results students obtain from a numeric Exercise provide insights into the concepts they are learning. Most Exercises are thought-provoking and require that students apply conceptual thinking. Exercises and Problem-Solving Examples that are conceptual are designated by a special icon.

Conceptual icon

Examples, Practice problems, Estimation boxes, and Exercises are all intended to stimulate active thinking and participation by students as they read the text and to help students hone their understanding of concepts. The grand total of 561 of these **active-learning items** exceeds the number in any similar textbook.

Conceptual Understanding

We believe that a sound conceptual foundation is the best means by which students can approach and solve a wide variety of real-world problems. This is supported by considerable evidence in the literature: Students learn better and retain what they learn longer when they have mastered fundamental concepts. Chemistry requires familiarity with three conceptual levels:

- **Macroscale** (laboratory and real-world phenomena)
- **Nanoscale** (models involving particles: atoms, molecules, and ions)
- **Symbolic** (chemical formulas and equations).

These three conceptual levels are explicitly defined in Chapter 1. This chapter emphasizes the value of the chemist's unique nanoscale perspective on science and the world with a specific example of how chemical thinking can help solve a real-world problem — that of designing painkillers that do not produce chronic side effects. This theme of conceptual understanding and its application to problems continues throughout the book. Many of the problem-solving features already mentioned have been specifically designed to support it.

Units are introduced on a need-to-know basis at the first point in the book where they contribute to the discussion. Units for length and mass are defined in Chapter 2, in conjunction with the discussion of the sizes and masses of atoms and subatomic particles. Energy units are defined in Chapter 6, where they are first needed to deal with kinetic and potential energy, work, and heat. In each case, defining units at the time when the need for them can be made clear allows definitions that would otherwise appear pointless and arbitrary to support the development of closely related concepts.

Most important of all, we provide **clear, direct, thorough, and understandable explanations** of all topics, including those such as kinetics, thermodynamics, and electrochemistry that many students find daunting. The scientific method and concepts such as chemical and physical properties, purification and separation, the relation of macroscale, nanoscale, and symbolic representations, elements and compounds, and kinetic-molecular theory are introduced in Chapter 1 so that they can be used throughout the later discussion. Rather than being bogged down with discussions of units and nomenclature, students begin this book with an overview of what real chemistry is about — together with fundamental ideas they will need to understand it.

Visualization for Understanding

Illustrations in *Chemistry: The Molecular Science* have been designed to engage today's visually oriented students. They help students to visualize atoms and molecules and to make connections among macroscale observations, nanoscale models, and symbolic representations of chemistry. Excellent color photographs of substances and reactions, many by Charles D. Winters, are presented together with greatly magnified illustrations of the atoms, molecules, and/or ions involved that have been created by J/B Woolsey Associates LLC. Often these are accompanied by the symbolic formula for a substance or equation for a reaction, as in the example shown on page xii. These **nanoscale views of atoms, molecules, and ions** have been generated with molecular modeling software and then combined by a skilled artist with the photographs and formulas or equations. Similar illustrations appear in exercises, examples, and end-of-chapter problems, to make certain that students are tested on the ideas they represent. The result, we think, provides an

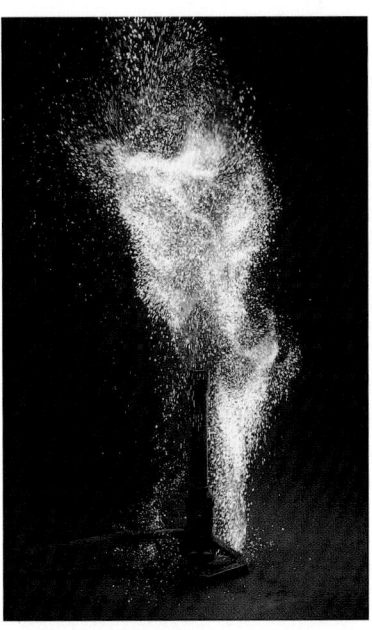

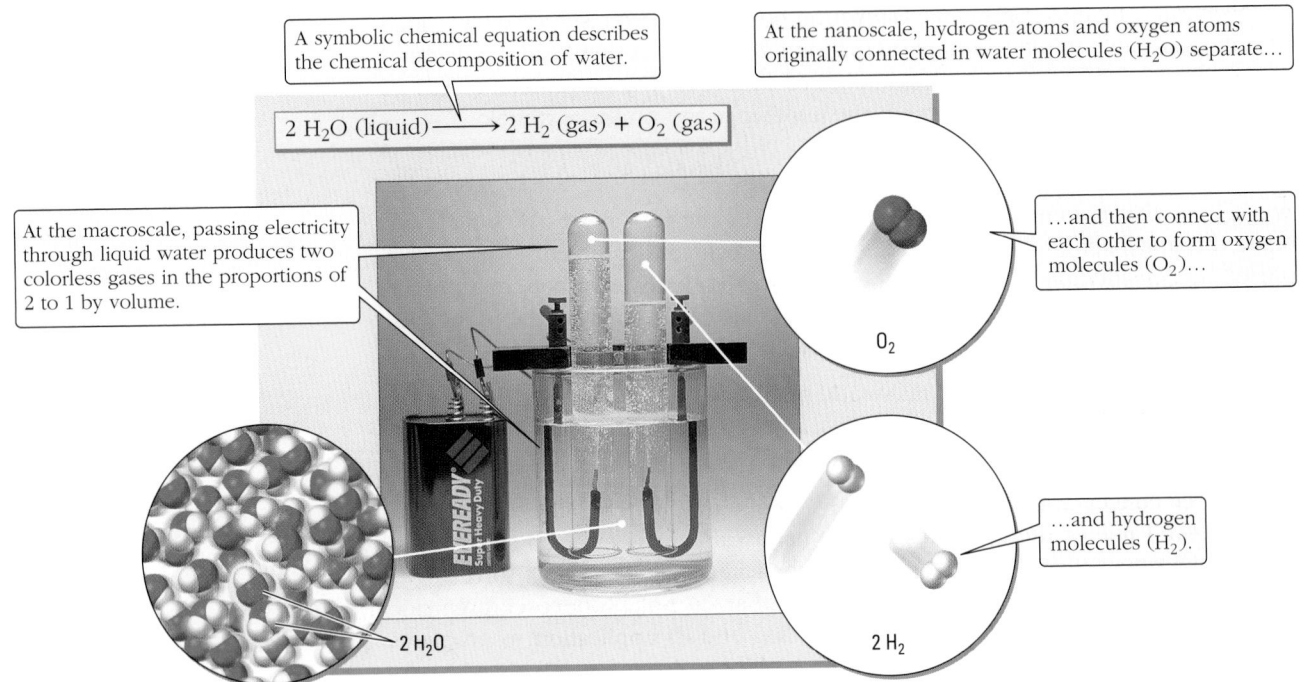

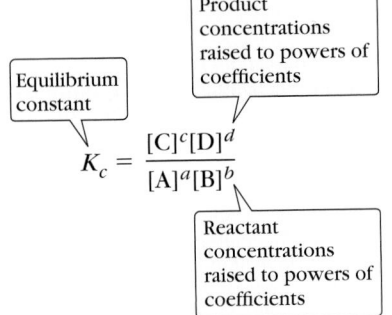

exceptionally effective way for students to learn how chemists think about the nanoscale world of atoms, molecules, and ions.

Often the story is carried solely by the illustrations and accompanying text that points out the most important parts of a figure. An example is the visual story of molecular structure shown below. In other cases, text in balloons is used to explain the operation of instruments, apparatus, and experiments, to clarify the development of a mathematical derivation, or to point out salient features of graphs or nanoscale pictures. Throughout the book visual interest is high, and visualizations of many kinds are used in support of conceptual development.

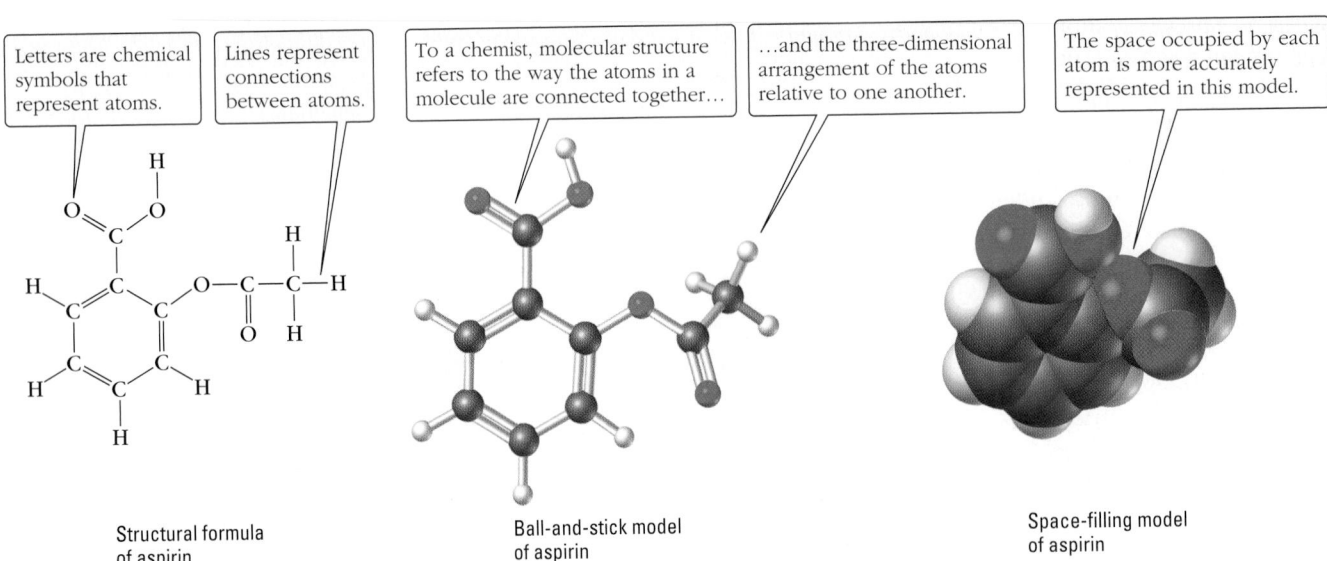

STYLE KEY

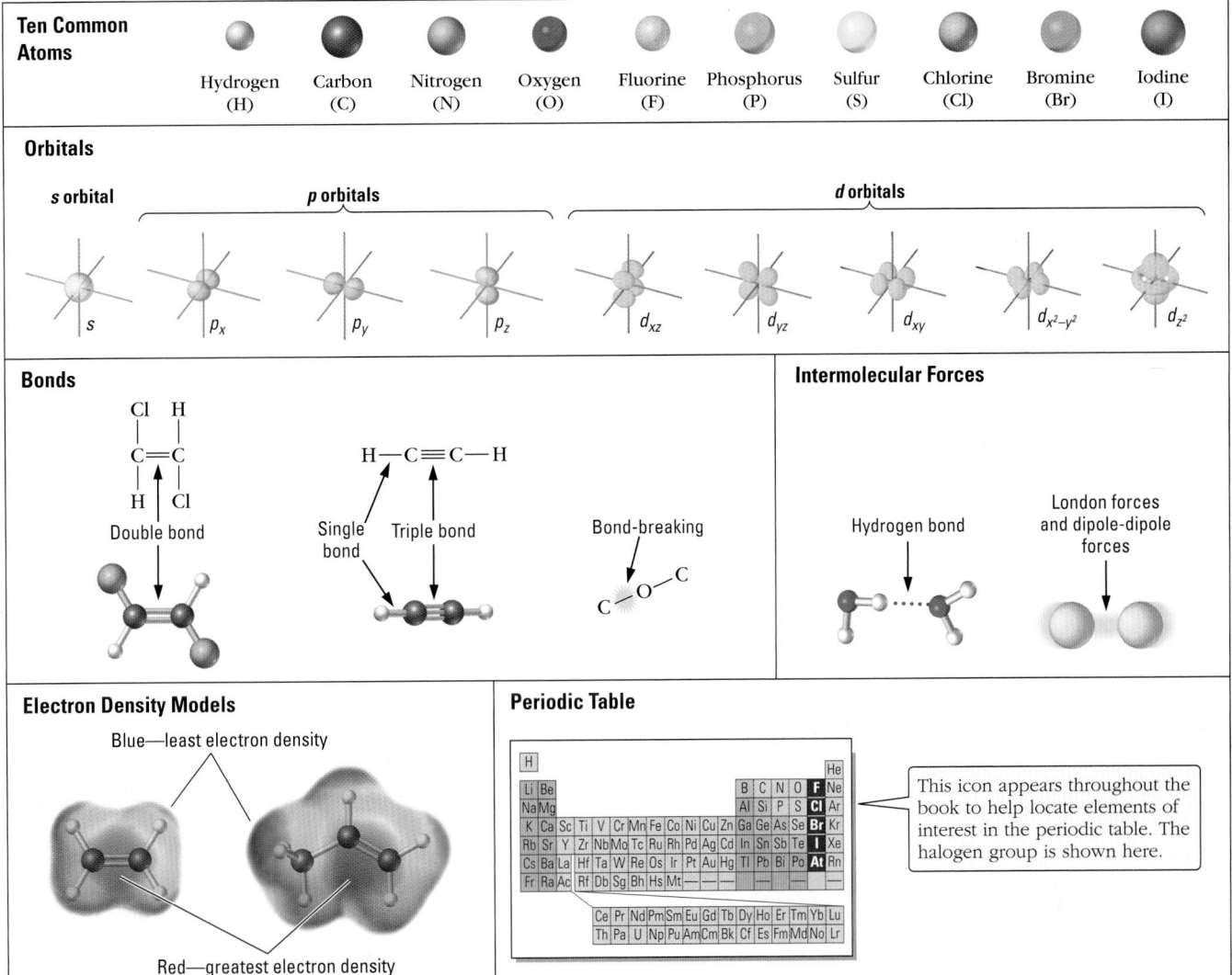

Interdisciplinary Applications

Whenever possible we include practical applications, especially those applications that students will revisit when they study other natural science and engineering disciplines. Applications have been integrated where they are relevant, not relegated to isolated chapters and separated from the principles and facts on which they are based. We intend that students should see that chemistry is a lively, relevant subject that is fundamental to a broad range of disciplines and that can help solve important, real-world problems.

We have especially emphasized the **integration of organic chemistry and biochemistry** throughout the book. In many areas, such as stoichiometry and molecular formulas, organic compounds provide excellent examples. To take advantage of this we have incorporated basic organic topics into the text beginning with Chapter 3 and used them wherever they are appropriate. In the discussion of molecules and the properties of molecular compounds, the concepts of structural formulas, functional groups, and isomers are developed naturally and effectively. Many of the principles that students encounter in general chemistry are directly applicable to biochemistry, and a large percentage of the students in most general chemistry courses are planning careers in biological or medical areas that make

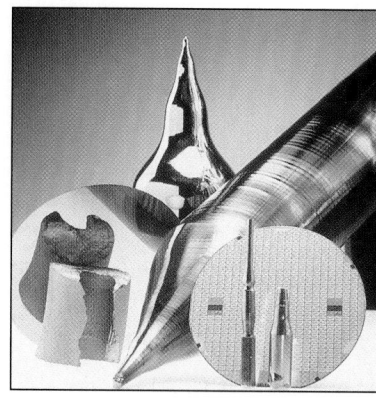

constant use of biochemistry. Therefore we have chosen to deal with biochemical topics in juxtaposition with the general chemistry principles that underlie them.

Here are some examples of integration of organic and biochemistry; the book contains many more.

- Section 3.3, *Hydrocarbons and Alcohols,* and Section 3.4, *Alkanes and Their Isomers,* introduce simple hydrocarbons, one type of functional group, and the concept of isomerism as a natural part of the discussion of molecular compounds.

- Section 6.12, *Foods: Fuels for Our Bodies,* applies thermochemical and calorimetric principles learned earlier in the chapter to the caloric values of proteins, fats, and carbohydrates in food.

- Section 9.6, *Biomolecules: DNA and the Importance of Molecular Structure,* describes structure and function of DNA in terms of noncovalent forces, a topic developed in the preceding section and one of paramount importance in DNA.

- Section 9.7, *Chiral Molecules,* introduces chirality as a natural aspect of molecular structure.

- Chapter 12, *Fuels, Organic Chemicals, and Polymers,* builds on principles and facts introduced earlier, applying them to organic molecules and functional groups selected for their relevance to synthetic and natural polymers. Proteins and polysaccharides illustrate the importance of biopolymers.

- Section 13.9, *Enzymes: Biological Catalysts,* applies kinetic principles developed earlier in the chapter and ideas about molecular structure from earlier chapters to enzyme catalysis and how it is influenced by protein structure.

- Section 18.9, *Gibbs Free Energy and Biological Systems,* discusses the role of Gibbs free energy and coupling of thermodynamic systems in metabolism, making clear the fact that metabolic pathways are governed by the rules of thermodynamics.

- Section 19.8, *Neuron Cells,* applies electrochemical principles to transmission of nerve impulses from one neuron to another, showing that changes in concentrations of ions result in changes in voltage and hence electrical signals.

Environmental and industrial chemistry are also integrated. In Chapter 6, *Energy and Chemical Reactions,* thermochemical principles are used to evaluate the energy densities of fuels. In Chapter 10, *Gases and the Atmosphere,* a discussion of gas-phase chemical reactions leads into the stories of stratospheric ozone depletion and air pollution. Chapter 12, *Fuels, Organic Chemicals, and Polymers,* deals with the consequences of combustion in a section on global warming. In Chapter 13, *Chemical Kinetics,* the importance of catalysts is illustrated by several industrial processes and exhaust-emission control on automobiles. In Chapter 16, *Acids and Bases,* practical acid-base chemistry illustrates many of the principles developed in the same chapter. In Chapter 21, *Chemistry of Selected Main Group Elements,* and Chapter 22, *Chemistry of Selected Transition Metals,* principles developed in earlier chapters are applied to extraction of elements from their ores. Students in a variety of disciplines will discover that chemistry is fundamental to their other studies.

Special Features

Another important aspect of our approach is that students should be involved in doing chemistry, and they ought to learn that common household materials are also chemicals. Most chapters include a ***Chemistry You Can Do*** experiment that can be done in a kitchen or dorm room and illustrates a topic included in the chapter. *Chemistry You Can Do* experiments require only simple equipment and familiar chemicals available at home or on a college campus.

Chemistry in the News boxes bring the latest discoveries in chemistry and applications of chemistry to the attention of students, making clear that chemistry is continually changing and developing—it is not just a static compendium of items to memorize. ***Tools of Chemistry*** boxes provide examples of how chemists use modern instrumentation to solve challenging problems. Like any other human pursuit, chemistry depends on people, and so we include in nearly every chapter **biographical sketches** of men and women who have advanced our understanding or applied chemistry imaginatively to important problems.

End-of-Chapter Study Aids

At the end of each chapter students will find many ways to test and consolidate their learning. Every chapter except the first ends with a **Summary Problem** that brings together concepts and problem-solving skills from throughout the chapter. Students are challenged to answer a multifaceted question that builds on and is relevant to the chapter's content. A section titled **In Closing** lists learning goals for the chapter and provides references to the sections in the chapter that address each goal. **Key Terms** are listed, again with reference to the section where they are defined.

A broad range of **Questions for Review and Thought** is provided to serve as a basis for homework or in-class problem solving. These begin with **Conceptual Challenge Problems,** most of them written by H. Graden Kirksey, University of Memphis. These are especially important in helping students assess and improve conceptual thinking ability. Designed for group work, the Conceptual Challenge Problems are rigorous and thought provoking. Much effective learning can be induced by dividing a class into groups of three or four and assigning the groups to work collaboratively on these problems. **Review Questions,** which are not answered in the back of the book, test vocabulary and simple concepts. These are followed by groups of questions that are keyed to each section in the chapter. These are often accompanied by photographs, graphs, and diagrams that make the situations described in the questions more concrete and realistic. Usually a question that is answered in an appendix is paired with a similar one that is not. There are many **General Questions** that are not keyed to chapter sections. Often these require students to integrate several concepts and many are more challenging than typical chapter-end exercises. **Applying Concepts** includes questions specifically designed to test conceptual learning. Often these involve diagrams of atoms, molecules, or ions and require students to relate macroscopic observations, atomic-scale models, and symbolic formulas and equations. Each chapter also has media questions, most of them related to the **General Chemistry CD-ROM.** To answer these questions students visit electronic media that provide a real-world flavor and additional realistic assessments of how well they understand the chapter's content.

General Chemistry CD-Rom Icon

ORGANIZATION

The order of chapters reflects the most common division of content between the first and second semesters of a typical general chemistry course. The first few chapters briefly review material that most students should have encountered in high school. Then we develop the ideas of chemical reactions, stoichiometry, and energy transfers during reactions. Throughout these early chapters organic chemistry, biochemistry, and applications of chemistry are integrated. Next we deal with electronic structure of atoms, bonding and molecular structures, and how structure affects properties. To finish up a first-semester course, there are adjacent chapters on gases, and on liquids and solids.

Next, we extend our integration of organic chemistry in a chapter that describes the role of organic chemicals in fuels, polymers, and biopolymers. Chapters on kinetics and equilibrium establish fundamental understanding of how fast reactions will go and what concentrations of reactants and products will remain when equilibrium is reached. Then these ideas are applied to solutions, and to acid-base and solubility equilibria in aqueous solutions. A chapter on thermodynamics and Gibbs free energy is followed by one on electrochemistry that makes use of thermodynamic ideas. Finally we deal with nuclear chemistry and descriptive chemistry of selected main group and transition elements.

To help students connect chemical ideas that are closely related, but are presented in different chapters, we have included **numerous cross references (indicated by the ⇐ symbol).** These will help students link a concept being developed in the chapter they are reading with an earlier, related principle or fact. They also provide many opportunities for students to review material learned earlier.

Chemistry: The Molecular Science can be divided into a number of sections, each of which treats an important aspect of chemistry.

Fundamental Ideas of Chemistry

Chapter 1, The Nature of Chemistry, is designed to capture students' interest from the start by concentrating on chemistry (not on math, units, and significant figures). It opens with a story of modern drug development that illustrates interdisciplinary chemical research. It also introduces major concepts that bear on all of chemistry, emphasizing the three conceptual levels that students must be familiar with: macroscale, nanoscale, and symbolic.

Chapter 2, Atoms and Elements, introduces units and dimensional analysis on a need-to-know basis in the context of the sizes of atoms. It concentrates on thorough, understandable treatment of the concepts of atomic structure, atomic weight, and moles of elements, making the connections among them clear. It concludes by introducing the periodic table and periodicity of properties of elements.

Chapter 3, Chemical Compounds, distinguishes ionic compounds from molecular compounds and illustrates molecular compounds with the simplest alkanes and alcohols. The important theme of structure is reinforced by showing several ways that organic structures can be written. Charges of monatomic ions are related to the periodic table, which is also used to show elements that are important in living systems. Molar masses of compounds and determining formulas fit logically into the chapter's structure.

Chemical Reactions

Chapter 4, Quantities of Reactants and Products, begins a three-chapter sequence that treats chemical reactions qualitatively and quantitatively. Students learn how to balance equations and to use typical inorganic reaction patterns to predict products. A single stepwise method is provided for solving all stoichiometry problems, and eleven examples demonstrate a broad range of stoichiometry problems.

Chapter 5, Chemical Reactions, has a strong descriptive chemistry focus, dealing with exchange reactions, acid-base reactions, and redox reactions in aqueous solutions. It includes real-world occurrences of each type of reaction. Students learn how to recognize a redox reaction from the chemical nature of the reactants (not just by using oxidation numbers) and how to do titration calculations.

Chapter 6, Energy and Chemical Reactions, begins with a thorough and straightforward introduction to forms of energy, conservation of energy, heat and work, system and surroundings, and exothermic and endothermic processes. Carefully designed figures help students to understand thermodynamic principles. Heat capacity, heats of changes of state, and heats of reactions are clearly explained, as are calorimetry and standard enthalpy changes. These ideas are then applied to fossil fuel combustion and to metabolism of biochemical fuels (proteins, carbohydrates, and fats) is described.

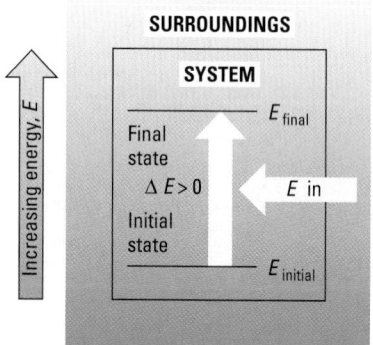

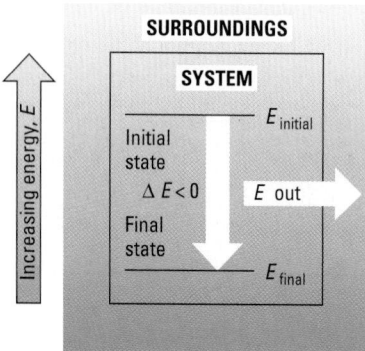

Electrons, Bonding, and Structure

Chapter 7, Electron Configurations and the Periodic Table, introduces spectra, quantum theory, and quantum numbers, using color-coded illustrations to visualize the different energy levels of s, p, d, and f orbitals. The s-, p-, d-, and f-block locations in the periodic table are used to predict electron configurations.

Chapter 8, Covalent Bonding, provides simple stepwise guidelines for writing Lewis structures, with many examples of how to use them. The role of single and multiple bonds in hydrocarbons is smoothly integrated with the introduction to covalent bonding. The discussion of polar bonds is enhanced by molecular models that show variations in electron density. A discussion of *cis* and *trans* fatty acids illustrates how small variations in molecular structure can have important consequences.

Chapter 9, Molecular Structures, provides a thorough presentation of VSEPR theory and orbital hybridization. Molecular geometry and polarity are extensively illustrated with computer-generated models, and the relation of structure, polarity, and hydrogen bonding to attractions among molecules is clearly developed and illustrated in solved problems. The importance of noncovalent interactions is emphasized by describing how they determine the structure of DNA.

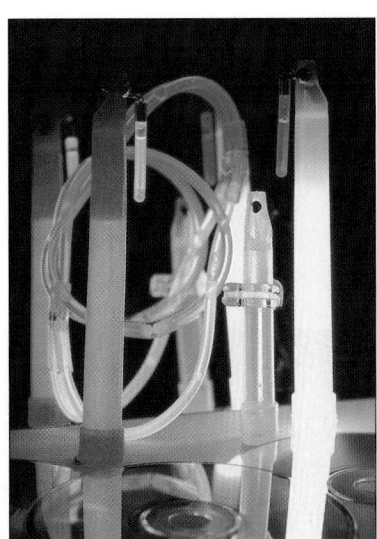

States of Matter

Chapter 10, Gases and the Atmosphere, uses kinetic-molecular theory to interpret the behavior of gases and then describes each of the individual gas laws. Mathematical problem solving focuses on the ideal gas law or the combined gas law, and many conceptual Exercises throughout the chapter emphasize qualitative understanding of gas properties. Gas stoichiometry is presented in a uniquely concise and clear manner. The properties of gases are then applied to chemical reactions in the atmosphere, the role of ozone in both the troposphere and stratosphere, and industrial and photochemical smog.

Chapter 11, Liquids, Solids, and Materials, begins by discussing the properties of liquids and the nature of phase changes. The unique and vitally important properties of water are covered thoroughly. The principles of crystal structure are introduced using cubic unit cells only. The fact that much current chemical research involves materials is illustrated by discussion of metals, n- and p-type semiconductors, insulators, superconductors, network solids, cement, ceramics and ceramic composites, and glasses, including optical fibers.

Important Industrial, Environmental, and Biological Molecules

Chapter 12, Fuels, Organic Chemicals, and Polymers, offers a distinctive combination of topics of major relevance to industrial, energy, and environmental concerns. Petroleum, natural gas, and coal are discussed as resources for energy and chemical materials, and in the context of global warming. Enough organic functional groups are introduced so that students can understand polymer formation, and the idea of condensation polymerization is extended to carbohydrates and proteins, which are compared with synthetic polymers.

Reactions: How Fast and How Far?

Chapter 13, Chemical Kinetics, presents one of the most difficult topics in the course with extraordinary clarity. Definition of reaction rate, finding rate laws from initial rates and integrated rate laws, and using the Arrhenius equation are thoroughly developed. How molecular changes during unimolecular and bimolecular elementary reactions relate to activation energy initiates the treatment of reaction mechanisms (including those with an initial fast equilibrium). Catalysis is shown to involve changing a reaction mechanism. Both enzymes and industrial catalysts are covered using concepts developed earlier in the chapter.

Chapter 14, Chemical Equilibrium, emphasizes equally a qualitative understanding of the nature of equilibrium and solving mathematical problems. That equilibrium results from equal but opposite reaction rates is fully explained. Both Le Chatelier's principle and the reaction quotient, Q, are used to predict shifts in equilibria. A unique section on equilibrium at the nanoscale introduces briefly and qualitatively how enthalpy changes and entropy changes affect equilibria. Optimizing the yield of the Haber-Bosch ammonia synthesis elegantly illustrates how kinetics, equilibrium, and enthalpy and entropy changes control the outcome of a chemical reaction.

Reactions in Aqueous Solution

Chapter 15, The Chemistry of Solutes and Solutions, builds on principles previously introduced, showing the influence of enthalpy and entropy on solution properties. Understanding of solubility, Henry's law, concentration units (including ppm and ppb), and colligative properties (including osmosis) is reinforced by applying these ideas to water as a resource, hard water, municipal water treatment, and the role of household waste in water pollution.

Chapter 16, Acids and Bases, concentrates initially on the Brønsted-Lowry acid-base concept, clearly delineating proton transfers using color coding and molecular models. In addition to a full exploration of pH and the meaning and use of K_a and K_b, acid strength is related to molecular structure, and the acid-base properties of carboxylic acids, amines, and amino acids are introduced. Student interest is enhanced by discussion of everyday use of acids and bases. Lewis acids and bases are defined and illustrated using examples.

Chapter 17, Additional Aqueous Equilibria, extends the treatment of acid-base and solubility equilibria to buffers, titration, and precipitation. The Henderson-Hasselbalch equation, which is widely used in biochemistry, is applied to buffer pH. Calculations of points on titration curves are shown, and the interpretation of several types of titration curves provides conceptual understanding. Acid-base concepts are applied to formation of acid rain. Factors that affect solubility (pH, common ions, complex ions, and amphoterism) are dealt with in the final section.

Thermodynamics and Electrochemistry

Chapter 18, Thermodynamics: Directionality of Chemical Reactions, explores the nature and significance of entropy, both qualitatively and quantitatively. The signs of Gibbs free energy changes are related to the easily understood classification of reactions as reactant- or product-favored, avoiding the term spontaneous, which is easily misinterpreted. The thermodynamic significance of coupling one reaction with another is illustrated using industrial, metabolic, and photosynthetic examples. Energy conservation is defined thermodynamically. A closing section reinforces the important distinction between thermodynamic and kinetic stability.

Chapter 19, Electrochemistry, defines redox reactions and uses half-reactions to balance redox equations. Electrochemical cells, cell voltage, standard cell potentials, the relation of cell potential to Gibbs free energy, and the effect of concentrations on cell potential are all explored. These ideas are then applied to the transmission of nerve impulses. Practical applications include batteries, fuel cells, electrolysis, and corrosion.

Nuclear Chemistry

Chapter 20, Nuclear Chemistry, deals with radioactivity, nuclear reactions, nuclear stability, and rates of disintegration reactions. There is a thorough description of nuclear fission and nuclear fusion and a thorough discussion of nuclear radiation and applications of radioactivity.

More Descriptive Chemistry

Chapter 21, Chemistry of Selected Main Group Elements, tells the interesting story of how the elements were formed and which are most important on earth. Physical separation of nitrogen, oxygen, and sulfur from natural sources, and extraction of sodium, chlorine, magnesium, and aluminum by electrolysis provide important industrial examples as well as an opportunity for students to apply principles learned earlier in the book.

Chapter 22, Chemistry of Selected Transition Metals and Coordination Compounds, continues the trend set in Chapter 21 of treating a few important elements in depth and integrating the review of principles learned earlier. Iron, copper, chromium, silver, and gold provide an interesting, motivating medium for transmitting principles of transition metal and coordination chemistry.

SUPPORTING MATERIALS

For the Student

The **Student Solutions Manual** contains fully worked-out solutions to end-of-chapter questions that have boldface numbers. Solutions match the problem-solving strategies used in the main text.

The **Study Guide** contains these learning tools: review of chapter objectives, review of key terms by a fill-in-the-blank exercise, and Practice Tests with Answers provided.

The **Interactive General Chemistry CD-ROM.** This upgraded version of the most popular general chemistry CD-ROM runs on major platforms including Windows 95™, Windows 98™, Windows 2000™, and Macintosh®.

OWL (On-Line Web-Based Learning System). OWL is an on-line homework, quizzing, and testing tool. On-line learning systems have been developed and used for more than 15 years at the University of Massachusetts, Amherst, Chemistry Department. Find out more about OWL at the Web site that accompanies this text or at **http://www.harcourtcollege.com/tech.**

The **Periodic Table Live! CD-ROM** (available separately from *JCE Software*) provides in html format a broad range of textual, numeric, and visual information about the chemical elements, including videos of reactions with air, water, acids, and bases.

The **General Chemistry Collection CD-ROM** (available separately from *JCE Software*) contains many software programs, animations, and videos that correlate with the content of this book. Arrangements can be made to make this item available to students at very low cost. Call 1(800)991-5534 for more information about JCE products.

For the Instructor

OWL OWL is an on-line homework, quizzing, and testing tool (see above description).

The **Instructor's Solutions Manual** contains fully worked-out solutions to the end-of-chapter questions (including Conceptual Challenge Problems). Solutions match the problem-solving strategies used in the main text.

A **Test Bank** contains more than 1100 questions, each one carefully matched to the text sections. Computerized versions of the Test Bank are available in Windows and Macintosh versions.

Overhead transparencies consist of 150 acetates of key illustrations.

The **Instructor CD-ROM** contains most of the illustrations from the text and makes them available in four formats: print-ready pdfs, Internet-ready pdfs, jpegs, and gif images.

The **site at http://www.harcourtcollege.com/chem** includes suggested Course Outlines, Syllabus Builder, PowerPoint™ presentation, online version of the Workbook to accompany the Interactive General Chemistry CD-ROM, the OWL learning system described above, correlations to laboratory experiments, and other resources.

The **Chemistry Comes Alive! series of five CD-ROMs** (available separately from *JCE Software*) includes in html format a broad range of videos and animations suitable for use in lecture presentations, for independent study, or for incorporation into an instructor's own tutorials.

For the Laboratory

The three manuals listed below have proven themselves as the most reliable on the market. As an extra aid, each manual's experiments are correlated to appropriate sections of the main text on the text Web site.

Chemical Principles in the Laboratory, **7/e,** by Emil J. Slowinski, Wayne C. Wolsey, and William L. Masterson.

Standard and Microscale Experiments in General Chemistry, **4/e,** by Carl B. Bishop, Muriel B. Bishop, and Kenneth W. Whitten.

Laboratory Experiments for General Chemistry, **3/e,** by Toby F. Block, George M. McKelvy, and Harold R. Hunt.

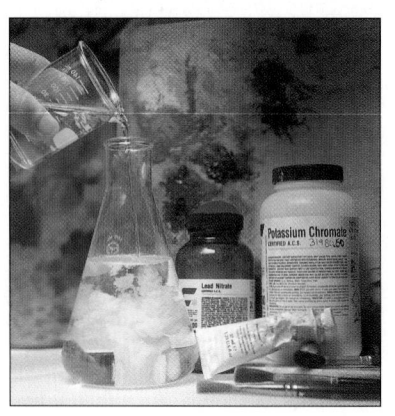

Harcourt College Publishers may provide complimentary instructional aids and supplements or supplement packages to those adopters qualified under our adoption policy. Please contact your sales representative for more information. If as an adopter or potential user you receive supplements you do not need, please return them to your sales representative or send them to

> Attn: Returns Department
> Troy Warehouse
> 465 South Lincoln Drive
> Troy, MO 63379

Note: Unless otherwise noted, the Website Domain Names (URLs) provided are not published by Harcourt, Inc. and the Publisher can accept no responsibility or liability for the content of these sites. Because of the dynamic nature of the Internet, Harcourt cannot in any case guarantee the continued availability of third-party web sites.

REVIEWERS

Reviewers have played a critical role in the preparation of this textbook. The individuals listed below helped to shape this text into one that is not only accurate and up to date, but is a practical resource for teaching and testing students.

Margaret Asirvatham, *University of Colorado - Boulder*

Donald Berry, *University of Pennsylvania*

Barbara Burke, *California State Polytechnic University, Pomona*

Dana Chatellier, *University of Delaware*

Mapi Cuevas, *Sante Fe Community College*

Cheryl Dammann, *University of North Carolina - Charlotte*

John DeKorte, *Glendale Community College*

Russ Geanangel, *University of Houston*

Peter Gold, *Pennsylvania State University*

Albert Martin, *Moravian College*

Marcy McDonald, *University of Alabama - Tuscaloosa*

Charles W. McLaughlin, *University of Nebraska*

David Metcalf, *University of Virginia*

David Miller, *California State University - Northridge*

Kathleen Murphy, *Daemen College*

William Reinhardt, *University of Washington*

Eugene Rochow, *Fort Myers, Florida*

Steven Socol, *McHenry County College*

Richard Thompson, *University of Missouri - Columbia*

Sheryl Tucker, *University of Missouri - Columbia*

Jose Vites, *Eastern Michigan University*

Sarah West, *University of Notre Dame*

Rick White, *Sam Houston State University*

We would also like to thank the following team of people who were dedicated to checking accuracy of various components of the text and art:

John DeKorte, *Glendale Community College*

Larry Fishel, *East Lansing, Michigan*

Leslie Kinsland, *Cornell University*

Judy Ozment-Payne, *Pennsylvania State University–Abington*

Gary Riley, *St. Louis School of Pharmacy*

ACKNOWLEDGMENTS

No textbook like this one is produced by its authors alone. We have had outstanding assistance in all aspects of production of this book, and we extend our thanks to everyone concerned.

First and foremost we thank John Vondeling, editor extraordinaire. John brought together the authors and encouraged us to produce a new book with just the right characteristics to suit today's rigorous, mainstream general chemistry course. John also was the driving force in assembling the exceptionally well-qualified team in Philadelphia who supported our every effort. John invariably provided appropriate leadership and guidance whenever we needed it.

Our Contributing Editor, Mary E. Castellion, has played a key role throughout the project. She has helped us establish goals, organize content, and get the best possible words onto paper. In addition she has kept tabs on what needed to be done and prodded, cajoled, and coerced us to meet deadlines and communicate effectively. Mary's unique combination of experience as an author and an editor have contributed greatly to making this book as good as it could possibly be. We thank Mary for her hard work and perseverance, and for her unfailing help exactly when it was needed most.

Senior Associate Editor Marc Sherman of Harcourt College Publishers has brought to the project a wide range of skills. His efforts in finding reviewers and encouraging them to provide feedback have helped us enormously. In addition he has provided advice on many issues and, based on his experience with textbooks in many different areas of science, helped ensure that interdisciplinary content is appropriate and correct.

This book is beautiful to look at, and its illustrations support its content magnificently. This extraordinary art program has been supervised by Caroline McGowan, Senior Art Director, who spent incredible time and effort to make it so. The many photographs by Charles D. Winters of Oneonta, New York, provide students with close-up views of chemistry in action. Other photos were obtained in timely fashion and with a keen sense of what we needed by Jane Sanders. George Kelvin, a fine scientific illustrator, contributed several of the chapter-opening figures. We are especially indebted to John Woolsey and the artists of J/B Woolsey Associates LLC. John's careful attention to detail and his many good ideas for illustrations that help students learn have helped and inspired us to create many new ways to impart chemistry pictorially. The drawings of molecular structures and of nanoscale models produced in John's studio are particularly noteworthy.

Charlene Squibb, Senior Production Manager, supervised the entire production process and has played a central role in keeping the project on time and on budget. Senior Project Editor Robin Bonner has cheerfully but firmly set deadlines and provided the encouragement we needed to meet them and still get things just right. Her considerable talents as an editor are greatly appreciated. Copy editor Janice Moore has kept the text consistent, caught many mistakes, and prompted us for things that we had omitted. Louise Robinson did an excellent job as the text proofreader.

We have received much feedback from Harcourt's excellent marketing team. Pauline Mula, Marketing Strategist, has summarized for us the information that many field representatives have transmitted to her. Her help in deciding what characteristics of a book are most useful and most wanted by those who will use it is greatly appreciated, as is the experience and enthusiasm of the sales force she supervises.

Many of the take-home *Chemistry You Can Do* experiments in this book were adapted from activities published by the Institute for Chemical Education as "Fun with Chemistry: Volumes I and II," by Mickey and Jerry Sarquis of Miami University (Ohio). Some were adapted from Classroom Activities published in the *Journal of Chemical Education*. Conceptual Challenge Problems at the end of most chapters were written by H. Graden Kirksey, University of Memphis, and we very much appreciate his contribution. The active-learning, conceptual approach of this book has been greatly influenced by the systemic curriculum project, *Establishing New Traditions: Revitalizing the Curriculum*, funded by the National Science Foundation, Directorate for Education and Human Resources, Division of Undergraduate Education, grant DUE-9455928.

We thank the many teachers, colleagues, students, and others who have contributed to our knowledge of chemistry and how best to help others learn it. Collectively we authors have many years of experience teaching and learning, and we have tried to incorporate as much of that as possible into what follows.

Finally we thank our families and friends who have supported all of our efforts—and who can reasonably expect more of our time and attention now that this project is complete.

We hope that using this book results in a lively and productive experience for both faculty and students.

John W. Moore   **Conrad L. Stanitski**   **Peter C. Jurs**
Madison, Wisconsin   *Conway, Arkansas*   *State College, Pennsylvania*

ABOUT THE AUTHORS

JOHN W. MOORE received an A.B. magna cum laude from Franklin and Marshall College and a Ph.D. from Northwestern University. He held a National Science Foundation postdoctoral fellowship at the University of Copenhagen, and taught at Indiana University–Bloomington and Eastern Michigan University before joining the faculty of the University of Wisconsin–Madison in 1989. At the University of Wisconsin Dr. Moore is Professor of Chemistry, Director of the Institute for Chemical Education, and Chair of the General Chemistry Division. In 2000 he was appointed to a named chair, W. T. Lippincott Professor of Chemistry, in honor of his excellence in research in chemical education. Dr. Moore has been Editor of *The Journal of Chemical Education* since 1996. He is very active in the American Chemical Society Division of Chemical Education and was its chair during 1996.

CONRAD L. STANITSKI is Professor of Chemistry and Department Chair at the University of Central Arkansas. He received his B.S. in Science Education from Bloomsburg State College, an M.A. in Chemical Education from the University of Northern Iowa, and Ph.D. in Inorganic Chemistry from the University of Connecticut. Professor Stanitski is a prolific author and has won many teaching awards, including the CMA CATALYST National Award for Excellence in Chemistry Teaching, the Thomas Branch Award for Teaching Excellence, and the Samuel Nelson Gray Distinguished Professor Award from Randolph-Macon College. Dr. Stanitski is Chair of the American Chemical Society Division of Chemical Education during 2001.

PETER C. JURS is Professor of Chemistry at the Pennsylvania State University. Dr. Jurs earned his B.S. in Chemistry from Stanford University in 1965 and his Ph.D. in Chemistry from the University of Washington in 1969. He then joined the faculty of Pennsylvania State University, where he has been Professor of Chemistry since 1978. In 1995 he was awarded the C.I. Noll Award for Outstanding Undergraduate Teaching. Jurs's research interests have included the application of computer methods to chemical and biological problems. A recent project in which Dr. Jurs participates is aimed at developing an artificial nose using fiber-optic sensor arrays for detection of volatile organic analytes. Dr. Jurs serves as an elected Councilor for the American Chemical Society Computer Division.

CONTENTS

1 THE NATURE OF CHEMISTRY *1*

1.1 Molecular Medicine *2*

1.2 How Science Is Done *5*

1.3 Physical Properties of Matter *6*

1.4 Chemical Changes and Chemical Properties *11*

1.5 Substances, Mixtures, and Separations *13*

1.6 Elements and Compounds *16*

1.7 Nanoscale Models: Solids, Liquids, and Gases *18*

1.8 Atomic Theory *22*

1.9 The Chemical Elements *24*

1.10 Chemical Symbolism *28*

1.11 Risks and Benefits *31*

1.12 Why Care About Chemistry? *32*

Chemistry You Can Do: Separation of Dyes *15*

Chemistry in the News: Nanobes: The Smallest Form of Life? *21*

Portrait of a Scientist: John Dalton *22*

Estimation: How Tiny Are Atoms and Molecules? *24*

Portrait of a Scientist: Richard E. Smalley *29*

2 ATOMS AND ELEMENTS *41*

2.1 Atomic Structure and Subatomic Particles *42*

2.2 The Nuclear Atom *44*

2.3 The Sizes of Atoms and the Units Used to Represent Them *46*

2.4 Atomic Numbers and Mass Numbers *52*

2.5 Isotopes and Atomic Weight *55*

2.6 Amounts of Substances — The Mole *57*

2.7 Molar Mass and Problem Solving *59*

2.8 The Periodic Table *60*

Portrait of a Scientist: Ernest Rutherford *46*

Tools of Chemistry: Scanning Tunneling Microscopy *48*

Tools of Chemistry: Mass Spectrometer *54*

Estimation: The Size of Avogadro's Number *58*

Portrait of a Scientist: Dmitri Mendeleev *61*

Chemistry in the News: Newest, Heaviest Elements *65*

Chemistry You Can Do: Preparing a Pure Sample of an Element *66*

3 CHEMICAL COMPOUNDS *74*

3.1 Molecular Compounds *75*

3.2 Naming Binary Molecular Compounds *78*

3.3 Hydrocarbons and Alcohols *79*

3.4 Alkanes and Their Isomers *83*

3.5 Ions and Ionic Compounds *86*

3.6 Naming Ions and Ionic Compounds *92*

3.7 Properties of Ionic Compounds *95*

3.8 Moles of Compounds *97*

3.9 Percent Composition *101*

3.10 Determining Empirical and Molecular Formulas *103*

3.11 The Biological Periodic Table *105*

3.12 Biomolecules: Carbohydrates and Fats *107*

Estimation: Number of Alkane Isomers *86*

Chemistry You Can Do: Pumping Iron: How Strong Is Your Breakfast Cereal? *106*

Chemistry in the News: Dietary Selenium *107*

4 QUANTITIES OF REACTANTS AND PRODUCTS *121*

4.1 Chemical Equations *122*

4.2 Patterns of Chemical Reactions *124*

4.3 Balancing Chemical Equations *132*

4.4 The Mole and Chemical Reactions: The Macro-Nano Connection *134*

4.5 Reactions with One Reactant in Limited Supply *140*

4.6 Evaluating the Success of a Synthesis: Percent Yield *145*

4.7 Percent Composition and Empirical Formulas *148*

Portrait of a Scientist: Antoine Lavoisier *123*

Portrait of a Scientist: Alfred Nobel *128*

Chemistry in the News: The Carbon Cycle: Calcium Carbonate, CO_2, and Global Warming *130*

Estimation: How Much CO_2 Is Produced by Your Car? *140*

Chemistry You Can Do: Vinegar and Baking Soda: A Stoichiometry Experiment *145*

5 CHEMICAL REACTIONS *162*

5.1 Exchange Reactions: Precipitation and Net Ionic Equations *163*

5.2 Acids, Bases, and Acid-Base Exchange Reactions *170*

5.3 Oxidation-Reduction Reactions *178*

5.4 Oxidation Numbers and Redox Reactions *182*

5.5   Displacement Reactions, Redox, and the Activity Series   *186*

5.6   Solution Concentration   *190*

5.7   Molarity and Reactions in Aqueous Solutions   *195*

5.8   Aqueous Solution Titrations   *199*

Chemistry You Can Do: Pennies, Redox, and the Activity Series of Metals   *189*

Chemistry in the News: Solving a Problem in Industrial Chemistry   *196*

6   ENERGY AND CHEMICAL REACTIONS   *210*

6.1   The Nature of Energy   *211*

6.2   Conservation of Energy   *214*

6.3   Heat Capacity   *218*

6.4   Energy and Enthalpy   *223*

6.5   Thermochemical Equations   *229*

6.6   Enthalpy Changes for Chemical Reactions   *231*

6.7   Where Does the Energy Come From?   *236*

6.8   Measuring Enthalpy Changes: Calorimetry   *238*

6.9   Hess's Law   *241*

6.10   Standard Molar Enthalpies of Formation   *243*

6.11   Chemical Fuels for Home and Industry   *247*

6.12   Foods: Fuels for Our Bodies   *249*

Portrait of a Scientist: James P. Joule   *213*

Estimation: Earth's Kinetic Energy   *214*

Chemistry in the News: Capturing the Sun's Energy   *225*

Chemistry You Can Do: Work and Volume Change   *228*

Chemistry You Can Do: Rusting and Heating   *235*

7   ELECTRON CONFIGURATIONS AND THE PERIODIC TABLE   *265*

7.1   Electromagnetic Radiation and Matter   *264*

7.2   Planck's Quantum Theory   *266*

7.3   The Bohr Model of the Hydrogen Atom   *269*

7.4   Beyond the Bohr Model: The Quantum Mechanical Model of the Atom   *275*

7.5   Quantum Numbers, Energy Levels, and Orbitals   *278*

7.6   Atom Electron Configurations   *284*

7.7   Ion Electron Configurations   *291*

7.8   Periodic Trends: Atomic Radii   *294*

7.9   Periodic Trends: Ionic Radii   *299*

7.10   Periodic Trends: Ionization Energies   *300*

7.11   Periodic Trends: Electron Affinities   *303*

Estimation: Seeing Red   *272*

Chemistry You Can Do: Using a Compact Disc (CD) as a Diffraction Grating   *274*

Portrait of a Scientist: Niels Bohr   *277*

Chemistry in the News: Keeping Time by Pumping Atoms   *278*

Tools of Chemistry: Nuclear Magnetic Resonance and Its Applications   *298*

8   COVALENT BONDING   *315*

8.1   Covalent Bonding   *316*

8.2   Single Covalent Bonds and Lewis Structures   *318*

8.3   Single Covalent Bonds in Hydrocarbons   *322*

8.4   Multiple Covalent Bonds   *325*

8.5   Multiple Covalent Bonds in Hydrocarbons   *328*

8.6   Bond Properties: Bond Length and Bond Energy   *333*

8.7   Bond Properties: Bond Polarity and Electronegativity   *338*

8.8   Formal Charge   *341*

8.9   Lewis Structures and Resonance   *343*

8.10   Exceptions to the Octet Rule   *346*

8.11   Aromatic Compounds   *349*

Portrait of a Scientist: Gilbert Newton Lewis   *317*

Chemistry in the News: Keeping Cool, But Safely   *325*

Chemistry in the News: *Trans* Fatty Acids   *335*

Portrait of a Scientist: Linus Pauling   *340*

Chemistry in the News: Biologically Reactive Oxygen and Antioxidants   *349*

9   MOLECULAR STRUCTURES   *359*

9.1   Using Molecular Models   *360*

9.2   Predicting Molecular Shapes: VSEPR   *361*

9.3   Orbitals Consistent with Molecular Shapes: Hybridization   *371*

9.4   Molecular Polarity   *379*

9.5   Noncovalent Interactions and Forces Between Molecules   *383*

9.6   Biomolecules: DNA and the Importance of Molecular Structure   *392*

9.7   Chiral Molecules   *396*

Tools of Chemistry: Infrared Spectroscopy   *362*

Chemistry in the News: A Molecule with a Bang   *372*

Tools of Chemistry: Ultraviolet-Visible Spectroscopy   *380*

Chemistry in the News: Sticking to the Task at Hand   *391*

Chemistry You Can Do: Molecular Structure and Biological Activity   *392*

10   GASES AND THE ATMOSPHERE   *407*

10.1   Properties of Gases   *408*

10.2   The Atmosphere   *412*

10.3   Kinetic-Molecular Theory   *414*

10.4   The Behavior of Ideal Gases   *417*

10.5   The Ideal Gas Law   *422*

10.6   Quantities of Gases in Chemical Reactions   *426*

10.7   Gas Density and Molar Masses   *429*

10.8   Gas Mixtures and Partial Pressures   *432*

10.9   The Behavior of Real Gases   *437*

10.10   Chemical Reactions in the Atmosphere   *439*

10.11   Ozone and Ozone Depletion   *441*

10.12   Chemistry and Pollution in the Troposphere   *444*

Estimation: Thickness of Earth's Atmosphere   *415*

Portrait of a Scientist: Jacques Alexandre Cesar Charles   *420*

Estimation: Helium Balloon Buoyancy   *431*

Portrait of a Scientist: F. Sherwood Rowland   *442*

Portrait of a Scientist: Susan Solomon   *443*

Chemistry in the News: Mexico City Air Pollution   *445*

Chemistry in the News: Particle Size and Visibility   *446*

11   LIQUIDS, SOLIDS, AND MATERIALS   *459*

11.1   The Liquid State   *460*

11.2   Vapor Pressure   *463*

11.3   Phase Changes: Solids, Liquids, and Gases   *465*

11.4   Water: An Important Liquid with Unusual Properties   *475*

11.5   Types of Solids   *479*

11.6   Crystalline Solids   *481*

11.7   Materials Science   *487*

11.8   Metals, Semiconductors, and Insulators   *491*

11.9   Silicon and the Chip   *496*

11.10   Network Solids   *500*

11.11   Cement, Ceramics, and Glass   *501*

Chemistry You Can Do: Melting Ice with Pressure   *475*

Chemistry in the News: Melting Below Zero   *476*

Chemistry You Can Do: Closest Packing of Spheres   *485*

Portrait of a Scientist: Dorothy Crowfoot Hodgkin   *487*

Tools of Chemistry: X-Ray Crystallography   *488*

12   FUELS, ORGANIC CHEMICALS, AND POLYMERS   *513*

12.1   Petroleum   *514*

12.2   Natural Gas and Coal   *521*

12.3   Energy Conversions   *523*

12.4   Atmospheric Carbon Dioxide, the Greenhouse Effect, and Global Warming   *525*

12.5   Organic Chemicals   *529*

12.6   Alcohols and Their Oxidation Products   *531*

12.7   Carboxylic Acids and Esters   *538*

12.8   Synthetic Organic Polymers   *542*

12.9   Biopolymers: Proteins and Polysaccharides   *555*

Estimation: Burning Coal   *524*

Chemistry in the News: Green Chemistry: Being Critical About Grease   *530*

Portrait of a Scientist: Percy Lavon Julian   *538*

Chemistry You Can Do: Making "Gluep"   *547*

Portrait of a Scientist: Stephanie Louise Kwolek   *554*

13   CHEMICAL KINETICS: RATES OF REACTIONS   *571*

13.1   Reaction Rate   *572*

13.2   Effect of Concentration on Reaction Rate   *577*

13.3   Rate Law and Order of Reaction   *581*

13.4   A Nanoscale View: Elementary Reactions   *587*

13.5   Temperature and Reaction Rate: The Arrhenius Equation   *593*

13.6   Rate Laws for Elementary Reactions   *597*

13.7   Reaction Mechanisms   *599*

13.8   Catalysts and Reaction Rate   *604*

13.9   Enzymes: Biological Catalysts   *607*

13.10 Catalysis in Industry   *613*

Estimation: Pesticide Decay   *587*

Chemistry You Can Do: Kinetics and Vision   *592*

Portrait of a Scientist: Ahmed H. Zewail   *593*

Chemistry You Can Do: Enzymes: Biological Catalysts   *608*

Chemistry in the News: Protease Inhibitors and AIDS   *614*

14   CHEMICAL EQUILIBRIUM   *632*

14.1   Characteristics of Chemical Equilibrium   *633*

14.2   The Equilibrium Constant   *637*

14.3   Determining Equilibrium Constants   *643*

14.4   The Meaning of the Equilibrium Constant   *646*

14.5   Using Equilibrium Constants   *650*

14.6   Shifting a Chemical Equilibrium: Le Chatelier's Principle   *655*

14.7   Equilibrium at the Nanoscale   *663*

14.8   Controlling Chemical Reactions: The Haber-Bosch Process   *666*

Chemistry in the News: Equilibrium and Alzheimer's Disease   *636*

Chemistry You Can Do: Growing Crystals by Shifting Equilibria   *658*

Estimation: Generating Gaseous Fuel   *663*

Portrait of a Scientist: Fritz Haber   *667*

15   THE CHEMISTRY OF SOLUTES AND SOLUTIONS   *680*

15.1   Solubility and Noncovalent Forces   *681*

15.2   Enthalpy, Entropy, and Dissolving Solutes   *687*

15.3   Solubility and Equilibrium   *688*

15.4   Temperature and Solubility   *692*

15.5   Pressure and Dissolving Gases in Liquids: Henry's Law   *693*

15.6   Solution Concentration: Keeping Track of Units   *694*

15.7   Vapor Pressures, Boiling Points, and Freezing Points of Solutions   *700*

15.8   Osmotic Pressure of Solutions   *708*

15.9   Colloids   *712*

15.10 Surfactants   *714*

15.11 Water: Natural, Clean, and Otherwise   *716*

Chemistry in the News: Bubbling Away: Delicate and Stout   *695*

Portrait of a Scientist: Jacobus Henricus van't Hoff   *707*

Chemistry You Can Do: Curdled Colloids   *715*

16   ACIDS AND BASES   *727*

16.1   The Brønsted-Lowry Concept of Acids and Bases   *728*

16.2   Carboxylic Acids and Amines   *734*

16.3   The Autoionization of Water   *736*

16.4   The pH Scale   *737*

16.5   Ionization Constants of Acids and Bases   *741*

16.6   Problem Solving Using K_a and K_b *747*

16.7   Molecular Structure and Acid Strength   *751*

16.8   Acid-Base Reactions of Salts   *756*

16.9   Practical Acid-Base Chemistry   *761*

16.10 Lewis Acids and Bases   *766*

Portrait of a Scientist: Arnold Beckman   *739*

Chemistry in the News: How Low Can You Go? Ultraacidic Water   *740*

Estimation: Using an Antacid   *762*

Chemistry You Can Do: Aspirin and Digestion   *767*

17   ADDITIONAL AQUEOUS EQUILIBRIA   *776*

17.1   Buffer Solutions   *777*

17.2   Acid-Base Titrations   *786*

17.3   Acid Rain   *793*

17.4   Solubility Equilibria and the Solubility Product Constant, K_{sp} *794*

17.5   Factors Affecting Solubility   *797*

17.6   Precipitation: Will It Occur?   *802*

Chemistry You Can Do: Making an Acid-Base Indicator   *787*

18   **THERMODYNAMICS: DIRECTIONALITY OF CHEMICAL REACTIONS**   *812*

18.1   Reactant-Favored and Product-Favored Processes   *813*

18.2   Probability and Chemical Reactions   *814*

18.3   Measuring Dispersal or Disorder: Entropy   *817*

18.4   Calculating Entropy Changes   *822*

18.5   Entropy and the Second Law of Thermodynamics   *822*

18.6   Gibbs Free Energy   *827*

18.7   Gibbs Free Energy Changes and Equilibrium Constants   *831*

18.8   Gibbs Free Energy, Maximum Work, and Energy Resources   *837*

18.9   Gibbs Free Energy and Biological Systems   *839*

18.10   Conservation of Gibbs Free Energy   *847*

18.11   Thermodynamic and Kinetic Stability   *849*

Portrait of a Scientist: Ludwig Boltzmann   *817*

Chemistry You Can Do: Rubber Bands and Thermodynamics   *824*

Chemistry in the News: The Hindenburg Disaster — Hydrogen or Thermite?   *840*

Estimation: Gibbs Free Energy and Automobile Travel   *849*

19   **ELECTROCHEMISTRY AND ITS APPLICATIONS**   *863*

19.1   Redox Reactions   *864*

19.2   Using Half-Reactions to Understand Redox Reactions   *866*

19.3   Electrochemical Cells   *871*

19.4   Electrochemical Cells and Voltage   *875*

19.5   Using Standard Cell Potentials   *879*

19.6 $E°$ and Gibbs Free Energy   *884*

19.7   Effect of Concentration on Cell Potential   *887*

19.8   Neuron Cells   *892*

19.9   Common Batteries   *894*

19.10   Fuel Cells   *898*

19.11   Electrolysis — Forcing Reactant-Favored Reactions to Occur   *899*

19.12   Counting Electrons   *904*

19.13   Corrosion—Product-Favored Reactions   *907*

Chemistry You Can Do: Remove Tarnish the Easy Way   *880*

Portrait of a Scientist: Michael Faraday   *885*

Portrait of a Scientist: Wilson Greatbatch   *896*

Chemistry in the News: Batteries for Electric Cars   *900*

Estimation: The Amount of Aluminum in a Soda Can   *907*

20   **NUCLEAR CHEMISTRY**   *918*

20.1   The Nature of Radioactivity   *919*

20.2   Nuclear Reactions   *920*

20.3   Stability of Atomic Nuclei   *924*

20.4   Rates of Disintegration Reactions   *928*

20.5   Artificial Transmutations   *935*

20.6   Nuclear Fission   *936*

20.7   Nuclear Fusion   *940*

20.8   Nuclear Radiation: Effects and Units   *942*

20.9   Applications of Radioactivity   *945*

Portrait of a Scientist: Glenn Seaborg   *935*

Portrait of a Scientist: Darleane C. Hoffmann   *936*

Chemistry in the News: Where to Put High-Level Nuclear Waste for Long-Term Storage?   *940*

Chemistry You Can Do: Counting Millirems: Your Radiation Exposure   *943*

21   **THE CHEMISTRY OF SELECTED MAIN GROUP ELEMENTS**   *954*

21.1   Formation of the Elements   *955*

21.2   Terrestrial Elements   *957*

21.3   Some Elements Extracted by Physical Methods: Nitrogen, Oxygen, and Sulfur   *961*

21.4   Some Main Group Elements Extracted by Electrolysis: Sodium, Chlorine, Magnesium, and Aluminum   *969*

21.5   Some Main Group Elements Extracted by
        Chemical Oxidation-Reduction: Phosphorus,
        Bromine, and Iodine   *976*

Chemistry in the News: HArF: Getting Together
        with Argon   *964*

Portrait of a Scientist: Charles Martin Hall   *976*

Portrait of a Scientist: Herbert H. Dow   *979*

**22   CHEMISTRY OF SELECTED TRANSITION
        ELEMENTS AND COORDINATION
        COMPOUNDS   *983*

22.1   Properties of the Transition (*d*-Block)
        Elements   *984*

22.2   Iron and Steel: The Use of Pyrometallurgy   *989*

22.3   Copper: A Coinage Metal   *993*

22.4   Silver and Gold: The Other Coinage
        Metals   *998*

22.5   Chromium   *1000*

22.6   Coordinate Covalent Bonds: Complex Ions and
        Coordination Compounds   *1002*

Estimation: Up and Down the East Coast   *990*

Chemistry in the News: The Sacagawea Dollar Coin: In Search
        of Gold   *997*

Chemistry You Can Do: A Penny for Your Thoughts   *1007*

Portrait of a Scientist: Alfred Werner   *1009*

APPENDICES   *A.1*

A   Problem Solving and Mathematical Operations
     A.2

B   Units, Equivalences, and Conversion
     Factors   *A.17*

C   Physical Constants   *A.22*

D   Molecular Orbitals   *A.23*

E   Naming Simple Organic Compounds   *A.28*

F   Ionization Constants for Weak Acids at
     25 °C   *A.34*

G   Ionization Constants for Weak Bases at
     25 °C   *A.35*

H   Solubility Product Constants for Some Inorganic
     Compounds at 25 °C   *A.36*

I   Standard Reduction Potentials in Aqueous Solution
     at 25 °C   *A.40*

J   Selected Thermodynamic Values   *A.41*

**ANSWERS TO PROBLEM-SOLVING
PRACTICE PROBLEMS   *A.49***

ANSWERS TO EXERCISES   *A.69*

**ANSWERS TO SELECTED QUESTIONS FOR
REVIEW AND THOUGHT   *A.93***

FIGURE CREDITS   *C.1*

GLOSSARY   *G.1*

INDEX   *I.1*

SPECIAL FEATURES

CHEMISTRY IN THE NEWS

Nanobes: The Smallest Form of Life?   21
Newest, Heaviest Elements   65
Dietary Selenium   107
The Carbon Cycle: Calcium Carbonate, CO_2, and Global Warming   130
Solving a Problem in Industrial Chemistry   196
Capturing the Sun's Energy   225
Keeping Time By Pumping Atoms   278
Keeping Cool, But Safely   325
Trans Fatty Acids   335
A Molecule with a Bang   372
Sticking to the Task at Hand   391
Mexico City Air Pollution   445
Melting Below Zero   476
Green Chemistry: Being Critical About Grease   530
Protease Inhibitors and AIDS   614
Equilibrium and Alzheimer's Disease   636
Bubbling Away: Delicate and Stout   695
How Low Can You Go? Ultraacidic Water   740
The Hindenburg Disaster: Hydrogen or Thermite?   840
Batteries for Electric Cars   900
Where to Put High-Level Nuclear Waste for Long-Term Storage?   940
HArF: Getting Together with Argon   964
The Sacagawea Dollar Coin: In Search of Gold   997

CHEMISTRY YOU CAN DO

Separation of Dyes   15
Preparing a Pure Sample of an Element   66
Pumping Iron: How Strong Is Your Breakfast Cereal?   106
Vinegar and Baking Soda: A Stoichiometry Experiment   145
Pennies, Redox, and the Activity Series of Metals   189
Work and Volume Change   228
Rusting and Heating   235
Using a Compact Disc (CD) as a Diffraction Grating   274
Molecular Structure and Biological Activity   392
Particle Size and Visibility   446
Melting Ice with Pressure   475
Closest Packing of Spheres   485
Making "Gluep"   547
Kinetics and Vision   592
Enzymes: Biological Catalysts   608
Growing Crystals by Shifting Equilibria   658
Curdled Colloids   715
Aspirin and Digestion   767
Making an Acid-Base Indicator   787
Rubber Bands and Thermodynamics   824
Remove Tarnish the Easy Way   880
Counting Millirems: Your Radiation Exposure   943
A Penny for Your Thoughts   1007

ESTIMATION

How Tiny Are Atoms and Molecules?   24
The Size of Avogadro's Number   58

(Estimation continued)

Number of Alkane Isomers   86
How Much CO_2 Is Produced by Your Car?   140
Earth's Kinetic Energy   214
Seeing Red   272
Thickness of Earth's Atmosphere   415
Helium Balloon Buoyancy   431
Burning Coal   524
Pesticide Decay   587
Generating Gaseous Fuel   663
Using an Antacid   762
Gibbs Free Energy and Automobile Travel   849
The Amount of Aluminum in a Soda Can   907
Up and Down the East Coast   990

TOOLS OF CHEMISTRY

Scanning Tunneling Microscopy   48
Mass Spectrometer   54
Nuclear Magnetic Resonance and Its Applications   298
Infrared Spectroscopy   362
Ultraviolet-Visible Spectroscopy   380
X-Ray Crystallography   488

PORTRAIT OF A SCIENTIST

Arnold Beckman   739
Niels Bohr   277
Ludwig Boltzmann   817
Jacques Alexandre Cesar Charles   420
John Dalton   22
Herbert H. Dow   979
Michael Faraday   885
Wilson Greatbatch   896
Fritz Haber   667
Charles Martin Hall   976
Dorothy Crowfoot Hodgkin   487
Darleane C. Hoffman   936
James P. Joule   213
Percy Lavon Julian   538
Stephanie Louise Kwolek   554
Antoine Lavoisier   123
Gilbert Newton Lewis   317
Dmitri Mendeleev   61
Alfred Nobel   128
Linus Pauling   340
F. Sherwood Rowland   442
Ernest Rutherford   46
Glenn Seaborg   935
Richard E. Smalley   29
Susan Solomon   443
Jacobus Henricus van't Hoff   707
Alfred Werner   1009
Ahmed H. Zewail   593

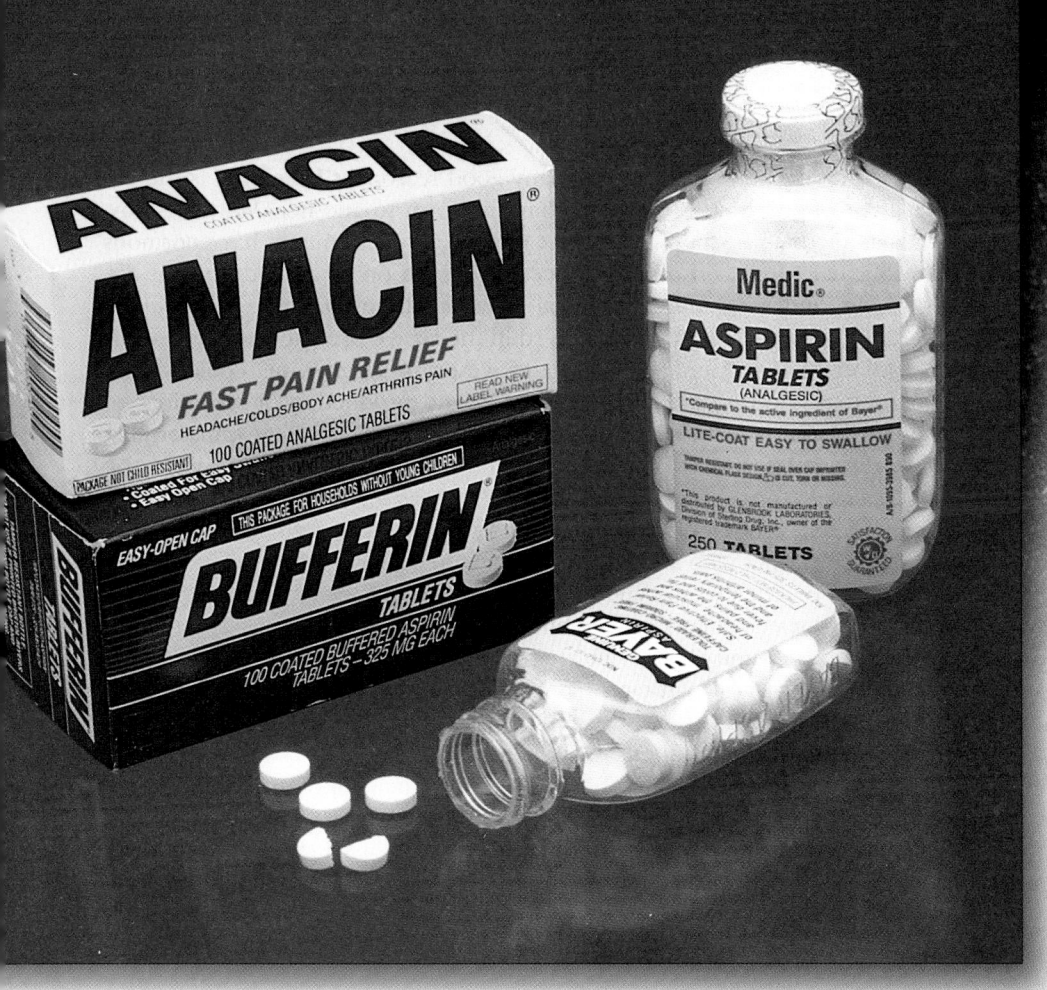

A product of chemical synthesis, aspirin is the most commonly used nonprescription painkiller; more than 30 billion tablets are consumed annually in the United States. In addition to relieving pain and inflammation of tissues, aspirin also causes unwanted side effects. By learning the molecular basis for how aspirin works, scientists can search more effectively for aspirin substitutes that have fewer harmful side effects.

THE NATURE OF CHEMISTRY

1.1 Molecular Medicine

1.2 How Science Is Done

1.3 Physical Properties of Matter

1.4 Chemical Changes and Chemical Properties

1.5 Substances, Mixtures, and Separations

1.6 Elements and Compounds

1.7 Nanoscale Models: Solids, Liquids, Gases

1.8 Atomic Theory

1.9 The Chemical Elements

1.10 Chemical Symbolism

1.11 Risks and Benefits

1.12 Why Care About Chemistry?

W hy study chemistry? There are many good reasons. **Chemistry** is the science of matter and its transformations from one form to another. **Matter** is anything that has mass and occupies space. Consequently, chemistry has enormous impact on our daily lives, on other sciences, and even on areas as diverse as art, music, cooking, and recreation. Chemical transformations happen all the time, everywhere. Chemistry is intimately involved in the air we breathe and the reasons we need to breathe it; in purifying the water we drink; in growing, cooking, and digesting the food we eat; and in the discovery and production of medicines to help maintain health. Chemistry continually provides new ways of transforming matter into different forms with useful properties. Examples include the plastic disks in CD players; the microchips and batteries in calculators or computers; and the steel, aluminum, rubber, plastic, and other components of automobiles.

Chemists are people who are fascinated by matter and its transformations—as you are likely to be after seeing and experiencing chemistry in action. Chemists have a unique and spectacularly successful way of thinking about and interpreting the material world around them—an atomic and molecular perspective. Atoms and molecules, the smallest particles that embody the chemical properties of substances, are fundamental to chemistry, and it is impossible to write or think about chemistry without referring to them.

Knowledge and understanding of chemistry are crucial in biology, pharmacology, medicine, geology, materials science, many branches of engineering, and other sciences. Modern research is often done by teams of scientists whose members represent several of these different disciplines. In such teams, ability to communicate and collaborate is just as important as knowledge in a single field. Studying chemistry can help you learn how chemists think about the world and solve problems, which in turn can lead to effective collaborations. Such knowledge will be useful in many career paths and will help you become a better informed citizen in a world that is becoming technologically more and more complex—and interesting.

This chapter illustrates and discusses some of the most fundamental ideas that chemists use every day. These ideas are extremely important and will be applied over and over throughout your study of chemistry and of many other sciences.

Very human accounts of how fascinating— even romantic—chemistry can be are provided by Primo Levi in his autobiography, *The Periodic Table* (New York: Schocken Books, 1984), and by Oliver Sacks in "Brilliant Light: A Chemical Boyhood" (*The New Yorker,* December 20, 1999). Levi was sentenced to a death camp during World War II but survived because the Nazis found his chemistry skills useful; those same skills made him a special kind of writer. Sacks describes how his mother and other relatives encouraged his interest in metals, diamonds, magnets, medicines, and other chemicals and how he learned that "science is a territory of freedom and friendship in the midst of tyranny and hatred."

  ## 1.1 MOLECULAR MEDICINE

How modern science works, and why the chemist's unique perspective is so valuable, can be seen through an example. From many possibilities, we have chosen the story of aspirin, a drug whose chemical name is acetylsalicylic acid. You have probably taken aspirin—lots of people use it to combat headache pain or fever. Others use it to reduce inflammation and pain caused by arthritis or sprains, or to make clotting of blood more difficult, thereby combating heart disease. Aspirin is recommended more often than any other drug, and annual production in the United States is more than 25 million kilograms (25,000 metric tons). This corresponds to about one tablet every three days for every man, woman and child in the country.

Substances similar to aspirin have been used to relieve pain since antiquity. Nearly 2500 years ago, the Greek physician Hippocrates recommended extracts of the bark of willow trees to alleviate the pain of childbirth. In 1763 the first scientific study of the effects of willow bark was reported in England, but it was not until 1827 that anyone was able to separate and identify the substance that was responsible for willow bark's salutary effects. By the 1860s chemists had found a way to make a similar substance, sodium salicylate, that also reduced fever, pain,

Willow tree. It has been known since antiquity that chewing willow bark can alleviate pain and inflammation.

and inflammation. Soon a factory was built for large-scale production of sodium salicylate, making it available at low cost for widespread use.

In 1898 the chemist Felix Hoffmann, observing that his father was experiencing significant stomach irritation as a result of taking six to eight grams per day of sodium salicylate to combat the pain of arthritis, found a way to convert salicylic acid into a less irritating form. He synthesized a related substance, acetylsalicylic acid, which his father found more palatable and more effective. The compound was named aspirin ("a" for acetyl and "spirin" for *Spirsaüre,* the German word for salicylic acid) by the director of research at the Bayer Company, where Hoffmann worked. Bayer began marketing aspirin in 1899, and it has become the most common nonprescription painkiller. The molecular structure of aspirin is shown here in three different ways.

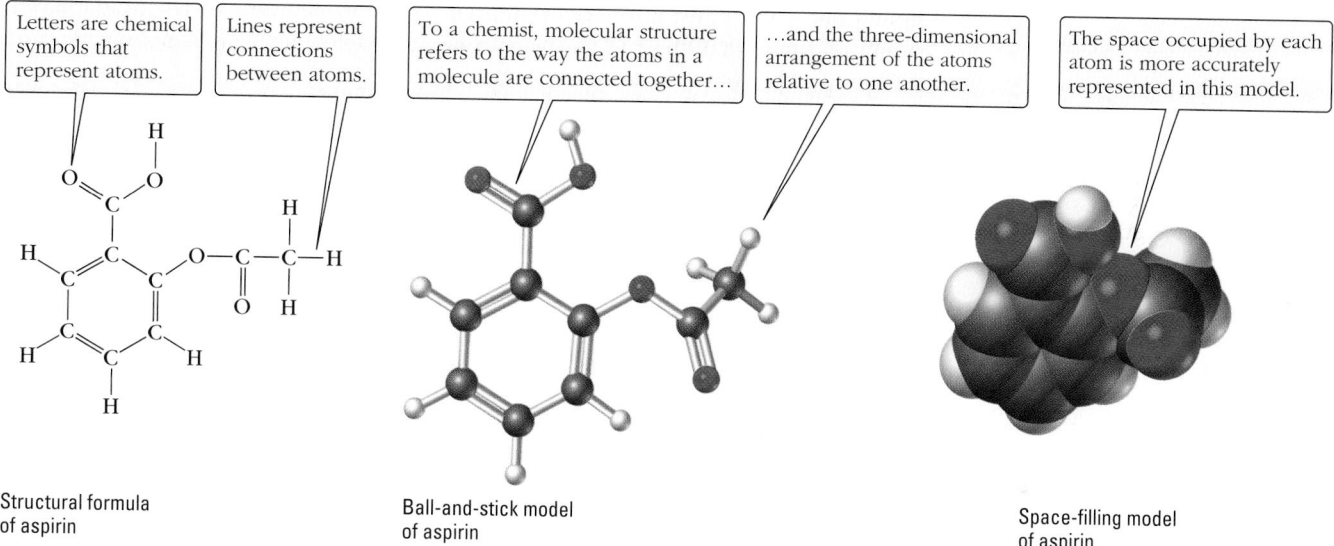

Letters are chemical symbols that represent atoms.

Lines represent connections between atoms.

To a chemist, molecular structure refers to the way the atoms in a molecule are connected together…

…and the three-dimensional arrangement of the atoms relative to one another.

The space occupied by each atom is more accurately represented in this model.

Structural formula of aspirin

Ball-and-stick model of aspirin

Space-filling model of aspirin

Aspirin, acetylsalicylic acid.

Despite aspirin's popularity and usefulness, it has only been during the past few decades that scientists have begun to understand how it works. (For a chemist, such understanding means having a mental picture of how aspirin molecules interact with molecules in the body to alleviate inflammation, pain, and fever.) In the 1970s it was noted that many forms of tissue injury are followed by the release of prostaglandins—hormone-like substances that produce pain, redness, and fever. In the absence of injury, prostaglandins protect the stomach walls and maintain chemical balance in the kidneys. Aspirin and other similar analgesics work by inhibiting production of prostaglandins. This alleviates inflammation, pain, and fever, but lack of prostaglandins also causes damage to the stomach and kidneys.

Production of prostaglandins is facilitated by two different cyclooxygenase (COX) enzymes. Enzyme molecules speed up the production of other molecules in living organisms. COX-1 accelerates reactions that produce "good" prostaglandins—those that protect stomach walls and kidneys. COX-2 facilitates production of "bad" prostaglandins—those associated with pain and inflammation. Aspirin has both good and bad effects because it interferes with the action of both COX-2 and COX-1 enzyme molecules.

This understanding at the atomic/molecular level of how prostaglandins are produced and of the action of aspirin opened the door to a search for a painkiller that would have the positive effects of aspirin without the negative side effects.

Throughout this book, computer-generated models of molecular structures will be used to help you visualize chemistry at the atomic and molecular levels.

An enzyme inhibitor is a molecule that reduces (inhibits) the enzyme's ability to speed up a reaction. An inhibitor can retard an enzyme-facilitated reaction so that fewer product molecules are produced in a given time. An inhibitor for COX-2 slows production of bad prostaglandins.

Because there are two enzymes, scientists realized that research based on molecular structures ought to be able to find a "superaspirin" that would inhibit only COX-2, leaving COX-1 to produce the prostaglandins necessary for good health. Searching for a molecule with exactly the right shape and composition to interact with and inhibit the action of a specific enzyme is often the goal of modern drug design.

There have been two main approaches to finding a superaspirin. In one, a large number of compounds that were known to inhibit prostaglandin production were tested separately for their effects on COX-1 and on COX-2. Once some compounds were found that inhibited COX-2 much better than COX-1, their molecular structures were examined to try to figure out why they affected only one of the two enzymes. Then other molecules with closely related molecular structures were synthesized and tested with both enzymes. Two such studies reported in 1997 illustrate how this works. One involved 20 scientists who studied more than 140 compounds. The other involved 23 collaborators who studied more than 50 compounds. Each group found a potential aspirin substitute that is a much better inhibitor for COX-2 than COX-1. The two molecules are called meloxicam and celecoxib. They have the structures shown here.

meloxicam

celecoxib

celecoxib

The second approach to finding a superaspirin begins with determining the molecular structures of the enzymes COX-1 and COX-2. Comparing the two enzyme structures should enable chemists to see how molecules that inhibit the enzymes fit into each enzyme structure. This in turn should allow molecules to be designed for optimum fit with COX-2 and poor fit with COX-1, thereby resulting in blocking of only the undesirable reactions. In 1994 the molecular structure of COX-1 was determined, and during the next two years two independent research groups determined the structure of COX-2. The COX-2 structure is shown in Figure 1.1. When researchers compared the two enzyme structures, they found

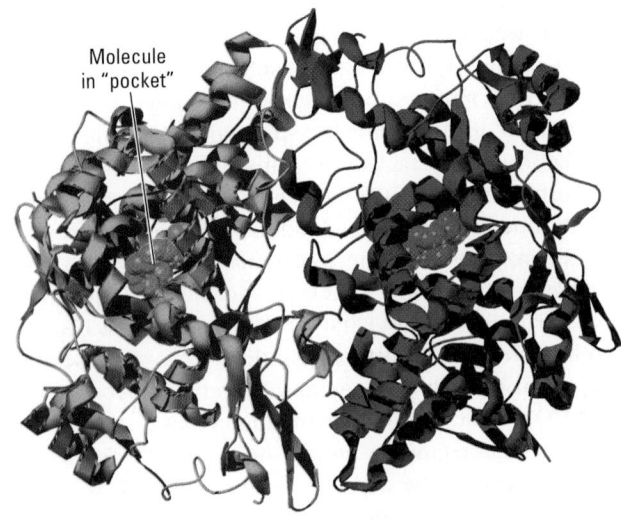

Molecule in "pocket"

COX-2 protein

Figure 1.1 Model of COX-2. The enzyme COX-2 facilitates release of prostaglandins that cause pain and inflammation. Two molecules are shown occupying "pockets" within the much larger enzyme. Because the enzyme molecule is so large, its individual atoms are not shown. The blue and green ribbon-like structures represent the "backbone" of the enzyme molecule, giving a rough indication of where the atoms are located.

that COX-2 had a somewhat larger "pocket" where an inhibitor molecule could fit. By designing inhibitor molecules with appropriate shapes to fit this pocket, chemists should be able to develop drugs that are even better than meloxicam and celecoxib in combating pain and inflammation without harmful side effects.

By 1999 both meloxicam and celecoxib had proved their worth in clinical trials and were on the market. Meloxicam is available from Boehringer Ingelheim Pharmaceuticals and Abbot Laboratories under the trademark Mobic, and celecoxib is marketed by Searle under the trademark Celebrex. A third drug, VIOXX, was developed based on the enzyme-structure approach. It is marketed by Merck & Co., Inc.

1.2 How Science Is Done

The story of aspirin and superaspirins illustrates many aspects of how people do science and how scientific knowledge changes and improves over time. In antiquity it was known that the bark of willow trees had medicinal properties. Probably this involved a chance observation that led to the hypothesis that willow bark was effective in relieving pain. A **hypothesis** is an idea that is tentatively proposed as an explanation for some observation and provides a basis for experimentation. Repeated administration of willow bark to those suffering from pain or inflammation provided a test for the hypothesis and verified that it was correct.

Testing a hypothesis may involve collecting qualitative or quantitative data. The observation that chewing willow bark alleviated pain was **qualitative:** It did not involve numeric data. Physicians recognized that their patients had less pain, but did not attempt to determine how much less. **Quantitative** information is obtained from measurements that produce numeric data. In studies attempting to identify superaspirins, a measured mass of each substance was mixed with COX-1 and COX-2 enzymes, and the speeds of production of prostaglandins were measured. These quantitative data allowed the researchers to say that celecoxib was more than 350 times as effective at inhibiting COX-2 as it was for COX-1.

A scientific **law** is a statement that summarizes and explains a wide range of experimental results and has not been contradicted by experiments. A law is capable of predicting unknown results and also of being disproved or falsified by new experiments. When the results of a new experiment contradict a law, that's exciting to a scientist. If several scientists repeat a contradictory experiment and get the same result, then the law must be modified to account for the new results — or even discarded altogether.

A successful hypothesis is often designated as a **theory** — a unifying principle that explains a body of facts and the laws based on them. A theory usually suggests new hypotheses and experiments, and, like a law, it may have to be modified or even discarded if contradicted by new experimental results. A **model** makes a theory more concrete, often in a physical or a mathematical form. Models of molecules, for example, provided the basis for recent efforts to find a superaspirin. Molecular models can be constructed by using spheres to represent atoms and sticks to represent the connections between the atoms. Or a computer can be used to calculate the locations of the atoms and display model molecular structures on a screen (as was done to create Figure 1.1). The theories that matter is made of atoms and molecules, that atoms are arranged in specific molecular structures, and that the properties of matter depend on those structures are fundamental to chemists' unique atomic/molecular perspective on the world and to nearly everything modern chemists do. Clearly it is important that you become as familiar as you can with these theories and with models based on them.

Another important aspect of the way science is done involves communication. Science is based on experiments and on hypotheses, laws, and theories that can be

"The whole of science is nothing more than a refinement of everyday thinking." —Albert Einstein.

 CD-Rom Screen Intro. 2: Science and Its Methods

How science is done is dealt with in *Oxygen,* a play written by chemists Carl Djerassi and Roald Hoffmann that premiered in 2001. By revisiting the discovery of oxygen, the play provides many insights regarding the process of science and the people who make science their life's work.

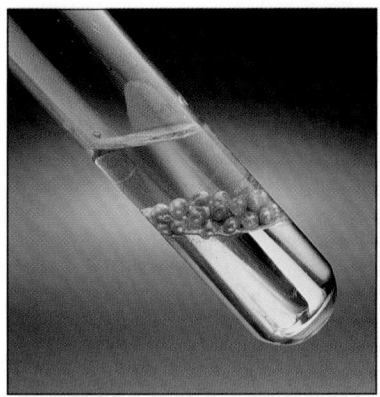

Figure 1.2 Substances have characteristic physical properties. The test tube contains silvery liquid mercury in the bottom, small spheres of solid orange copper in the middle, and colorless liquid water above the copper. Each of these substances has characteristic properties that differentiate it from the others.

CD-Rom Screen 1.2: Physical Properties of Matter

Some Physical Properties

Temperature
Pressure
Mass
Volume
State (solid, liquid, gas)
Melting point
Boiling point
Density
Color
Shape of solid crystals
Hardness, brittleness
Heat capacity
Thermal conductivity
Electrical conductivity

CD-Rom Screen 1.10: Temperature

contradicted by experiments. Therefore it is essential that experimental results be communicated to all scientists working in any specific area of research as quickly and accurately as possible. Scientific communication allows contributions to be made by scientists in different parts of the world and greatly enhances the rapidity with which science can develop. In addition, communication among members of scientific research teams, such as the groups of roughly 20 persons involved in the research that developed aspirin substitutes, is crucial to the success of such teams. The importance of scientific communication is emphasized by the fact that the Internet and World Wide Web were created not by commercial interests, but by scientists who saw their great potential for communicating scientific information.

1.3 PHYSICAL PROPERTIES OF MATTER

A **substance** is a type of matter that has specific properties and a particular composition. Each substance has characteristic properties that are different from the properties of any other substance (Figure 1.2). In addition, one sample of a substance has the same composition as every other sample of that substance—it consists of the same stuff in the same proportions.

Your friends can recognize you by your physical appearance: the shape of your face, the color of your eyes and hair, your height and weight. A substance's physical appearance also helps to identify it. You can distinguish sugar from water because you know that sugar consists of small white particles of solid, while water is a colorless liquid. Corn syrup is also a liquid, but it comes in light and dark colors and is much more viscous (pours more slowly) than water. Metals can be recognized because they usually are solids, have high densities, feel cold to the touch, and have shiny surfaces.

Properties such as these, that can be observed and measured without changing the composition of a substance, are called **physical properties.** Physical properties help us to identify substances of the material world, to classify them into groups of similar substances, such as metals, and to tell the difference between single substances and mixtures of two or more substances.

Physical Change

As a substance's temperature or pressure changes, or if it is mechanically manipulated, some of its physical properties may change. Changes in the physical properties of a substance are called **physical changes.** The same substance is present before and after a physical change, but the substance's physical state or the gross size and shape of its pieces may have changed. Examples are melting a solid (Figure 1.3), boiling a liquid, hammering a copper wire into a flat shape, and grinding sugar into a fine powder.

Temperature

The property of matter that determines whether there can be heat energy transfer from one object to another is called **temperature** and is represented by the symbol T. Energy transfers spontaneously from an object at a higher temperature to a cooler object. In the United States, everyday temperatures are reported using the Fahrenheit temperature scale. On this scale the freezing point of water is by definition 32 °F and the boiling point is 212 °F. The **Celsius temperature scale** is used in most countries of the world and in science. On this scale 0 °C is the freezing point and 100 °C is the boiling point of pure water at a pressure of one atmosphere. The number of units between the freezing and boiling points of water is 180 Fahrenheit degrees and 100 Celsius degrees. This means that the Celsius degree is almost twice as large as the Fahrenheit degree. It takes only 5 Celsius degrees to cover the same temperature range as 9 Fahrenheit degrees, and this relationship can be used to convert a temperature on one scale to a temperature on the other (see Appendix B.2).

Because temperatures in scientific studies are usually measured in Celsius units, there is little need to make conversions to and from the Fahrenheit scale, but it is quite useful to be familiar with how large various Celsius temperatures are. For example, it is useful to know that water freezes at 0 °C and boils at 100 °C, a comfortable room temperature is about 22 °C, your body temperature is 37 °C, and the hottest water you could put your hand into without serious burns is about 60 °C.

A substance's temperature cannot help us identify it, but two physical properties based on temperature can. These are the temperatures at which a solid melts (a substance's **melting point**) and a liquid boils (its **boiling point**). If two or more substances are mixed together, the melting point depends on how much of each is present, but for a single substance the melting point is always the same. This is also true of the boiling point (as long as the pressure on the boiling liquid is the same). In addition, the melting point of a pure sample of a substance is sharp — melting occurs with almost no change in temperature. When a mixture of two or more substances melts, the temperature when the first liquid appears can be quite different from the temperature when all of the solid is gone. In the studies of possible aspirin substitutes described in Section 1.1, the melting point of each solid was reported to help other scientists judge whether pure samples of the correct substances had actually been synthesized and used.

Figure 1.3 Physical change. When ice melts it changes — physically — from a solid to a liquid, but it is still water.

Exercise 1.1 Temperature

(a) Which is the higher temperature, 37 °C or 85 °F?
(b) Which is the lower temperature, 20 °F or 0 °C?
(c) The boiling point of a substance is 15 °C. If you hold a sample of the substance in your hand, will it boil?

The special icon at the beginning of this exercise indicates that it is designed to test your understanding of a concept.

Density

If you have ten pounds of sugar, it takes up ten times the volume that one pound of sugar does. In mathematical terms, a substance's volume is directly proportional to its mass. This means that a substance's **density,** the ratio of mass to volume, has the same value regardless of how big the sample is.

$$\text{Density} = \frac{\text{mass}}{\text{volume}} \qquad d = \frac{m}{V}$$

Density is a characteristic property that can help identify a solid or liquid sample. Even if they look the same, you can tell a sample of aluminum from a sample of lead by picking each up. Your brain will automatically estimate which sample has greater mass for the same volume, telling you which is the lead. Aluminum has a density of 2.70 g/mL, placing it among the least dense metals. Lead's density is

CD-Rom 1.8: Density; 1.9: Density on the Submicroscopic Scale

The density of a substance varies depending on the temperature and the pressure. Densities of liquids and solids change very little as pressure changes, and they change less with temperature than do densities of gases. Because the volume of a gas varies significantly with temperature and pressure, the density of a gas can help identify the gas only if the temperature and pressure are specified.

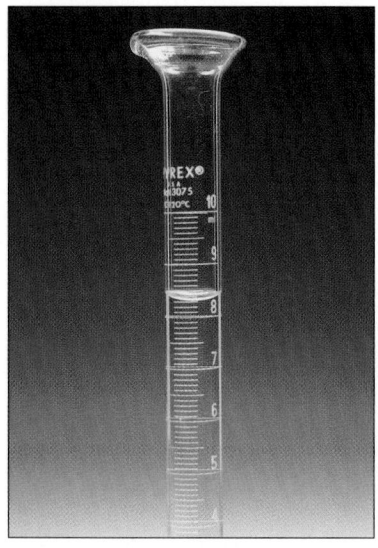

Graduated cylinder containing 8.30 mL of liquid.

11.34 g/mL, so a sample of lead is much heavier than a sample of aluminum of the same size.

Suppose that you are trying to identify a liquid that you think might be ethanol (ethyl alcohol), and you want to determine its density. You could weigh a clean, dry graduated cylinder and then add some of the liquid to it. Suppose that, from the markings on the cylinder, you read the volume of liquid to be 8.30 mL. You could then weigh the cylinder with the liquid and subtract the mass of the empty cylinder to obtain the mass of liquid. Suppose the liquid mass is 6.544 g. The density can then be calculated as

$$d = \frac{m}{V} = \frac{6.544\ \text{g}}{8.30\ \text{mL}} = 0.788\ \text{g/mL}$$

From a table of physical properties of various substances you find that the density of ethanol is 0.789 g/mL, which helps confirm your suspicion that the substance is ethanol.

Exercise 1.2 Density of Liquids

When 5.0 mL each of vegetable oil, water, and kerosene are put into a large test tube, they form three layers, as shown in the photo on the next page.
(a) Which is the most dense liquid?
(b) Which is the least dense liquid?
(c) If an additional 5.0 mL of vegetable oil is added to the test tube, will there be a permanent change in the order of liquids from bottom to top of the test tube?
(d) Describe the contents of the test tube after the addition of 5.0 mL of vegetable oil.

Exercise 1.3 Physical Properties and Changes

Identify each physical property and physical change in each of the following statements. Also identify the qualitative and the quantitative information given in each statement.
(a) The blue chemical compound azulene melts at 99 °C.
(b) The white crystals of table salt are cubic.
(c) A sample of lead has a mass of 0.123 g and melts at 327 °C.
(d) Ethanol is a colorless liquid that vaporizes easily; it boils at 78 °C and its density is 0.789 g/mL.

A useful source of data on densities and other physical properties of substances is the *Handbook of Chemistry and Physics,* published by the CRC Press. Information is also available via the Internet—for example, the National Institute for Standards and Technology's Webbook at **http://webbook.nist.gov.**

CD-Rom Screen 1.17: Using Numerical Information

In this book, units and dimensional analysis techniques are introduced at the first point where you need to know them. Appendices A and B provide all of this information in one place.

Dimensional Analysis and Problem Solving

The results of scientific measurements, such as 6.544 g or 8.30 mL, usually consist of a number and a unit. Both the number and the unit should be included in calculations. For example, the densities in Table 1.1 have units of grams per milliliter (g/mL), because density is defined as the mass of a sample divided by its volume. When a mass is divided by a volume, the units (g for the mass and mL for the volume) are also divided. The result is grams divided by milliliters (g/mL). That is, both numbers *and units* follow the rules of algebra. This is an example of **dimensional analysis,** a method of using units in calculations to check for correctness. More detailed descriptions of dimensional analysis are given in Section 2.3 and Appendix A.2. We will use this technique for problem solving throughout the book.

TABLE 1.1	Densities of Some Substances at 20 °C		
Substance	Density (g/mL)	Substance	Density (g/mL)
Butane	0.579	Titanium	4.50
Ethanol	0.789	Zinc	7.14
Benzene	0.880	Iron	7.86
Water	0.998	Nickel	8.90
Bromobenzene	1.49	Copper	8.93
Magnesium	1.74	Lead	11.34
Sodium chloride	2.16	Mercury	13.55
Aluminum	2.70	Gold	19.32

Liquid densities. Kerosene (*top layer*), vegetable oil (*middle layer*), and water (*bottom layer*) have different densities.

Suppose that you want to know whether you could lift a gallon (3784 mL) of the liquid metal mercury. To answer the question, calculate the mass of the mercury using the density, 13.55 g/mL, obtained from Table 1.1. One way to do this is to use the equation that defines density, $d = m/V$. Then solve algebraically for m, and calculate the result:

$$m = V \times d = 3784 \text{ mL} \times \frac{13.55 \text{ g}}{1 \text{ mL}} = 51{,}270 \text{ g} \tag{1.1}$$

This equation emphasizes the fact that mass is proportional to volume, because the volume is multiplied by a proportionality constant, the density. Notice also that the units of volume (mL) appeared once in the denominator of a fraction and once in the numerator, thereby dividing out and leaving only mass units (g). (The result, 51,270 g, is more than 100 pounds, so you could probably lift the mercury, but not easily.)

In Equation 1.1 a known quantity (the volume) was multiplied by a proportionality factor (the density), and the units canceled, giving an answer (the mass) with appropriate units. A general approach to this kind of problem is to recognize that the quantity you want to calculate (the mass) is proportional to a quantity whose value you know (the volume). Then use a proportionality factor that relates the two quantities, setting things up so that the units cancel.

Because mercury and mercury vapor are poisonous, carrying a gallon of it around is not a good idea unless it is in an appropriate closed container.

$$\text{known quantity units} \times \frac{\text{desired quantity units}}{\text{known quantity units}} = \text{desired quantity units}$$

proportionality (conversion) factor

$$3784 \text{ mL} \quad \times \quad \frac{13.55 \text{ g}}{1 \text{ mL}} \quad = \quad 51{,}270 \text{ g}$$

A **proportionality factor** is a ratio (fraction) whose numerator and denominator have different units but refer to the same thing. In the preceding example, the proportionality factor is the density, which relates the mass and volume of the same sample of mercury. A proportionality factor is often called a **conversion factor** because it enables us to convert from one kind of units to a different kind of units.

Because a conversion factor is a fraction, every conversion factor can be expressed in two ways. The conversion factor in the example just given could be expressed either as the density or as its reciprocal:

$$\frac{13.55 \text{ g}}{1 \text{ mL}} \quad \text{or} \quad \frac{1 \text{ mL}}{13.55 \text{ g}}$$

The first fraction enables conversion from volume (mL) to mass (g) units. The second allows mass units to be converted to volume units. Which conversion factor to use depends on which units are in the known quantity and which units are in the quantity that we want to calculate. Setting up the calculation so that the units cancel ensures that we are using the appropriate conversion factor. (See Appendix A.2 for more examples.)

Problem-Solving Example 1.1   Density

Find the volume occupied by a 4.33-g sample of benzene.

Answer   4.92 mL.

Explanation   A good approach to problem solving is to (1) define the problem, (2) develop a plan, (3) execute the plan, and (4) check your result to see whether it is reasonable. (These four steps are described in more detail in Appendix A.1.)

Step 1: Define the problem.   You are asked to find the volume of the sample, and you know the mass.

Step 2: Develop a plan.   Density relates mass and volume and is the appropriate conversion factor, so look up the density in a table. Volume is proportional to mass, so the mass either has to be multiplied by the density or multiplied by the reciprocal of the density. Use the units to decide which.

Step 3: Execute the plan.   According to Table 1.1, the density of benzene is 0.880 g/mL. Setting up the calculation so that the unit (grams) cancels gives

$$4.33 \text{ g} \times \frac{1 \text{ mL}}{0.880 \text{ g}} = 4.92 \text{ mL}$$

Notice that the result is expressed to three significant figures, because both the mass and the density had three significant figures. To refresh your memory regarding the rules for significant figures, read Appendix A.3.

Step 4. Check your answer.   Because the density is a little less than 1.00 g/mL, the volume in mL should be a little larger than the mass in g. The calculated answer, 4.92 mL, is a little larger than the mass, 4.33 g.

Problem-Solving Practice 1.1

Given that there are four quarts in a gallon, 453.59 g in exactly 1 pound, and 946.3 mL in exactly 1 quart, calculate the mass of one gallon of mercury, hence verifying the statement in the text that a gallon of mercury weighs more than 100 pounds.

This book includes many examples like Problem-Solving Example 1.1. These illustrate general problem-solving techniques and ways to approach specific types of problems. Usually, each of these examples states a problem, gives the answer, explains one way to analyze the problem, plan a solution, and execute the plan, and describes a way to check that the result is reasonable. We urge you to first try to solve the problem on your own. Then check to see whether your answer matches the one given. If it does not match, try again before reading the explanation. After you have tried twice, read the explanation to find out why your reasoning differs from that given. If your answer is correct, but your reasoning differs from the explanation, you may have discovered an alternative solution to the problem. Finally, work out the Problem-Solving Practice that accompanies the example. This relates to the same concept and allows you to improve your problem-solving skills.

Sucrose (Reactant) ——changes to——→ Carbon + Water (Products)

Figure 1.4 Chemical change: caramelizing sugar.

Water

Carbon

① When table sugar (sucrose) is heated . . .

② . . . it caramelizes, turning brown.

③ Heating to a higher temperature causes further decomposition (charring) to carbon and water vapor.

1.4 CHEMICAL CHANGES AND CHEMICAL PROPERTIES

If you take ordinary table sugar (sucrose) and heat it carefully on a stove, it will caramelize. That is, it will turn brown as the sucrose decomposes to give water and other new substances. If you heat it very hot, it will char, leaving behind a black residue that is mainly carbon—and is hard to clean up (Figure 1.4). If you drip some water onto a sample of sodium metal, the sodium will react violently with the water, producing a solution of lye (sodium hydroxide) and a flammable gas, hydrogen (Figure 1.5). These are examples of **chemical changes** or **chemical reactions.** In a chemical reaction, one or more substances (the **reactants**) are transformed into one or more different substances (the **products**). Reactant substances are replaced by product substances as the reaction occurs. This is indicated by writing the reactants, an arrow, and then the products:

 CD-Rom Screen 1.11: Chemical Change; 1.12: Chemical Change on the Molecular Scale

Sucrose ——→ carbon + water

Reactant   changes to   Products

(a)

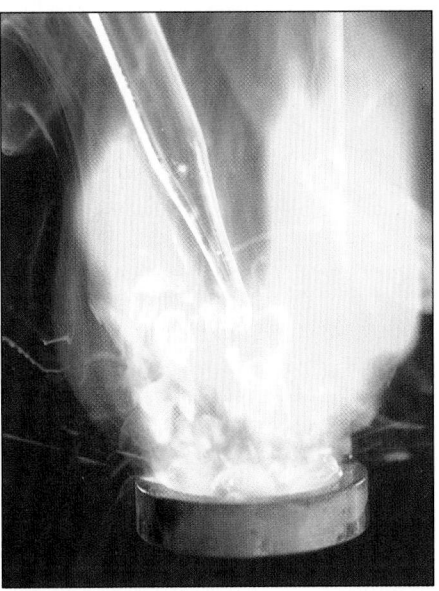

(b)

Figure 1.5 Reaction of sodium with water. When a drop of water (a) hits a piece of sodium, there is a violent reaction (b) that produces flammable hydrogen gas and a solution of sodium hydroxide (lye). Production of motion, heat, and light when substances are mixed is evidence that a chemical reaction is occurring.

Figure 1.6 Chemical change. An acid has been added to the green solid (nickel(II) carbonate), causing colorless carbon dioxide gas to bubble out. Production of gas bubbles when substances are mixed is evidence that a chemical reaction is occurring.

Chemical reactions make chemistry interesting, exciting, and valuable. If you know how, you can make a medicine from coal, clothing from crude petroleum, or even a silk purse from a sow's ear (it has been done). This is a very empowering idea, and human society has gained a great deal from it. Our way of life is greatly enhanced by our ability to use and control chemical reactions.

Chemical Properties

A substance's **chemical properties** describe the kinds of chemical reactions the substance can undergo. One chemical property of metallic sodium is that it reacts rapidly with water to produce hydrogen and a solution of sodium hydroxide (Figure 1.5). Because it also reacts rapidly with air and a number of other substances, sodium is also said to have a more general chemical property: It is highly reactive. A chemical property of substances known as metal carbonates is that they produce carbon dioxide when treated with an acid (Figure 1.6). Fuels are substances that have the chemical property of reacting with oxygen or air and at the same time transferring large quantities of energy to their surroundings. An example is natural gas (mainly methane), which is shown reacting with oxygen from the air in a gas stove in Figure 1.7. A substance's chemical properties tell us how it will behave when it contacts air or water, when it is heated or cooled, when it is exposed to sunlight, or when it is mixed with another substance. Such knowledge is very useful to chemists, biochemists, geologists, chemical engineers, and many other kinds of scientists.

Energy

Chemical reactions (and to a lesser extent physical changes) are usually accompanied by transfers of energy. **Energy** is defined as the capacity to do work, that is, to make something happen. Combustion of a fuel, as in Figure 1.7, transforms energy stored in chemical bonds in the fuel molecules into motion of the product molecules and of other nearby molecules. This corresponds to a higher temperature in the vicinity of the flame. The chemical reaction in a "light stick" transforms energy stored in chemical bonds into light energy, with only a little heat transfer

Figure 1.7 Combustion of natural gas. Natural gas, which consists mostly of methane, burns in air, transferring energy that raises the temperature of its surroundings.

(Figure 1.8). A chemical reaction in a battery makes a calculator work by forcing electrons to flow through an electric circuit.

Energy supplied from somewhere else can cause chemical reactions to occur. For example, photosynthesis takes place when sunlight illuminates green plants. Some of the sunlight's energy is stored in carbohydrate molecules that are produced from carbon dioxide and water by photosynthesis. Aluminum foil that you may have used to wrap and store food was produced by passing electricity through a molten sample of an aluminum-containing ore. You consume and metabolize food, using the energy stored in food molecules to cause chemical reactions to occur in the cells of your body. The relation between chemical changes and energy is an important theme of this book.

 Exercise 1.4 Chemical and Physical Changes

Identify the chemical and physical changes that are described in this statement: Propane gas burns, and the heat of the combustion reaction is used to hard boil an egg.

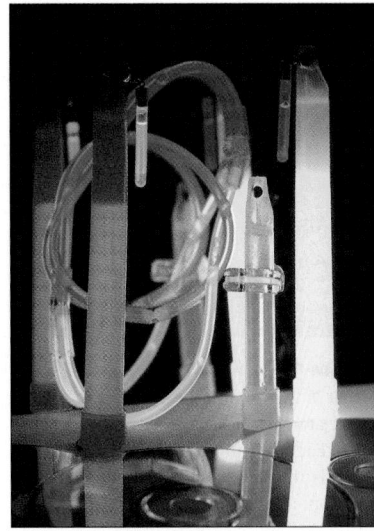

Figure 1.8 **Light sticks.** In each of these light sticks a chemical reaction transforms energy stored in molecules into light. Unlike an electric light bulb, the light sticks do not get hot.

1.5 SUBSTANCES, MIXTURES, AND SEPARATIONS

Most materials we encounter every day are like willow bark, concrete, or the carbon fiber composite frame of a high-tech bicycle — they are not uniform throughout. There are variations in color, hardness, and other properties from one part of the material to another. This makes these materials complicated, but also interesting. A major advance in chemistry was made when it was realized that it was possible to separate from such nonuniform samples several component substances. For example, in 1827 appropriate treatment of willow bark produced a substance, salicin, which was shown to be responsible for reducing pain.

Often, as in the case of the bark of a tree, we can easily see that one part of a sample is different from another part. In other cases a sample may appear completely uniform to the unaided eye, but a microscope can reveal that it is not. For example, blood appears smooth in texture, but magnification reveals particles within the liquid (Figure 1.9). The same is true of milk. A mixture in which the uneven texture of the material can be seen without magnification or with a microscope is called a **heterogeneous mixture.** Properties in one region are different from the properties in another region.

A **homogeneous mixture,** or **solution,** is completely uniform and consists of two or more substances in the same phase — solid, liquid, or gas (Figure 1.10). No amount of optical magnification will reveal different properties in one region of a solution compared with those in another. Heterogeneity exists in a solution only at the scale of atoms and molecules, which are too small to be seen with visible light. Examples of solutions are clear air (mostly a mixture of nitrogen and oxygen gases), sugar water, and some brass alloys (which are homogeneous mixtures of copper and zinc). The properties of a homogeneous mixture are the same everywhere in any particular sample, but they can vary from one sample to another depending on how much of one component is present relative to another component.

 CD-Rom Screen 1.13: Mixtures and Pure Substances; 1.14: Separation of Mixtures

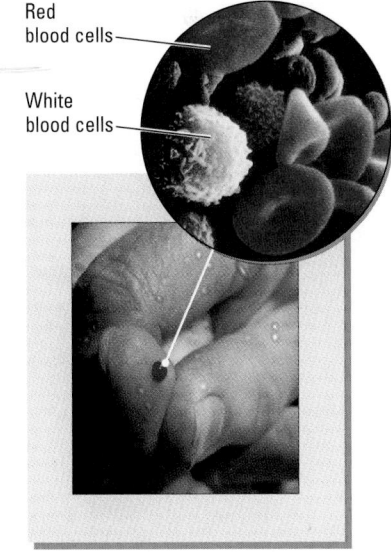

Red blood cells

White blood cells

Figure 1.9 **Blood cells.** Blood appears to be uniform to the unaided eye, but a microscope reveals that it is not homogeneous. The properties of red blood cells differ from the properties of the surrounding blood plasma, for example.

Separation and Purification

Earlier in this chapter we stated that a substance has characteristic properties that distinguish it from all other substances. However, for those characteristic properties to be observed, the substance must be separated from all other substances;

Figure 1.10 A solution. When solid sugar is stirred into liquid water it dissolves to form a homogeneous liquid mixture. Each portion of the solution has exactly the same sweetness as every other portion, and other properties are also the same throughout the solution.

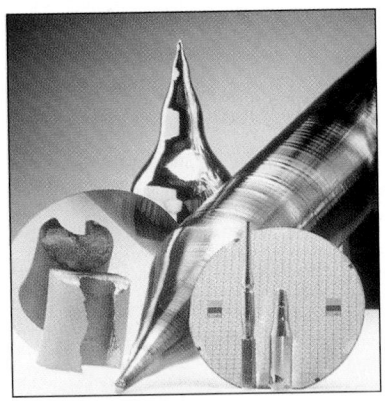

High-purity silicon.

that is, it must be purified. The melting point of an impure substance is different from that of the purified substance. Color and appearance of a mixture may also differ from those of a pure substance. Therefore, when we talk about the properties of a substance, it is assumed that we are referring to a pure substance — one from which all other substances have been separated.

Purification usually has to be done in several repeated steps and monitored by observing some property of the substance being purified. For example, the iron can be separated from a heterogeneous mixture of iron and sulfur with a magnet as shown in Figure 1.11. In this example, color, which depends on the relative quantities of iron and sulfur, indicates purity. The bright yellow color of sulfur is assumed to indicate that all the iron has been removed.

Concluding that a substance is pure on the basis of a single property of the mixture could be misleading, because other methods of purification might change some other properties of the sample. It is safe to call sulfur pure when a variety of methods of purification fail to change its physical and chemical properties. Purification is important because it allows us to attribute properties (such as alleviation of pain) to specific substances and then to study systematically which kinds of substances have properties that we find useful. There have been cases in which insufficient purification of a substance has led scientists to attribute to it properties that were actually due to a tiny trace of another substance.

Only a few substances occur in nature in pure form. Gold, diamonds, and sulfur are examples. We live in a world of mixtures; all living things, the air and food we depend on, and many products of technology are mixtures. Much of what we know about chemistry, however, is based on separating and purifying the components of those mixtures and then determining their properties. To date, more than 22 million substances have been reported, and many more are being discovered or synthesized by chemists every year. When pure, each of these substances has its own particular composition and its own characteristic properties.

A good example of the importance of purification is the high-purity silicon needed to produce transistors and computer chips. In a billion grams (about 1000 tons) of highly pure silicon there has to be less than one gram of impurity. Once the silicon has been purified, small but accurately known quantities of specific substances, such as boron or arsenic, can be introduced to give the electronic chip the desired properties. (See Sections 11.8 and 11.9.)

❹ Repeated stirrings eventually leave a bright yellow sample of sulfur that cannot be purified further by this technique.

❶ Iron and sulfur can be separated by stirring with a magnet.

❷ The first time that the magnet is removed, much of the iron is removed with it.

❸ The sulfur still looks dirty because a small quantity of iron remains.

Figure 1.11 Separating a mixture: iron and sulfur.

CHEMISTRY *You Can Do...*

Separation of Dyes

Find a piece of absorbent paper such as a coffee filter or paper towel. Cut it into a circular disk and fold it in half and then in half again. Put the paper in front of you with the point toward you. Make a horizontal mark with a water-soluble marker pen about 2 cm from the point.

Pour about an inch of water into a glass. Poke a pencil through the paper near the curved top so that you can rest the pencil on the top of the glass and suspend the paper inside the glass. The pointed tip should just enter the water so that the lowest 1 cm is below the surface. Make careful observations every minute or two, depending on how fast things are happening. What do you observe? Before the water reaches and wets the top of the paper, remove the paper from the glass and stop the experiment.

Find more pens of different colors and try them. Do one experiment in which you allow the paper to remain in the glass overnight. What other variations besides pens and time can you try?

Consider these questions and indicate how your observations apply to answering them:

1. Is the ink in your pen a single substance or a mixture?
2. Assuming that water is made of molecules, do you think these molecules are "sticky"? That is, is there some attraction of one water molecule for another? Why or why not?
3. Is there an attraction between water molecules and the molecules in the paper you used?
4. Draw a picture of what the molecules might be doing as the experiment proceeds. Label the ink, water, and paper molecules. Try to make your nanoscale picture correspond with and explain the macroscale observations you made.
5. What are some possible applications of the separation technique you have just used?

Separation of dyes

Analysis and Detection

Once we know that a given substance is either valuable or harmful, it becomes important to know whether that substance is present in a sample and to be able to find out how much of it is there. Does an ore contain enough of a valuable metal to make it worthwhile to mine the ore? Is there enough mercury in a sample of fish to make it unsafe for humans to eat the fish?

Answering questions like these is the job of *analytical chemists*, and they improve their methods every year. For example, in 1960 mercury could be detected at a concentration of one part per million, in 1970 the detection limit was one part per billion, and by 1980 the limit had dropped to one part per trillion. Thus, in 20 years the ability to detect small concentrations of mercury had increased by a factor of a million. This has an important effect. Because we can find smaller and smaller concentrations of contaminants, such contaminants can be found in many more samples. A few decades ago, toxic substances were usually not found when food, air, or water was tested, but that did not mean they were not there. It just meant that our analytical methods were unable to detect them. Today, with much better methods, toxic substances can be detected in most samples, which prompts demands that concentrations of such substances should be reduced to zero.

One part per million (ppm) means we can find one gram of a substance in one million grams of total sample. That corresponds to a tenth of a drop in a bucket. One part per billion corresponds to a drop in a swimming pool, and one part per trillion to a drop in a large supermarket.

Absence of evidence is not evidence of absence.

Although we expect that chemistry will push detection limits lower and lower, they will never get to zero. Proving that there are no contaminants in a sample will never be possible. This is a specific instance of the general rule that it is impossible to prove a negative. To put this another way, it will never be possible to prove that we have produced a completely pure sample of a substance, and therefore it is unproductive to legislate that there should be zero contamination in food or other substances. It is more important to use chemical analysis to determine a safe level of a toxin than to try to prove that it is completely absent. In some cases very small concentrations of a substance are beneficial but larger concentrations are toxic. Analytical chemistry can help us to find what ranges of concentration are optimal.

1.6 ELEMENTS AND COMPOUNDS

In 1661 Robert Boyle was the first to propose that elements could be defined by the fact that they could not be decomposed into two or more simpler substances.

Most of the substances separated from mixtures can be converted to two or more simpler substances by chemical reactions—a process called *decomposition.* Substances are often decomposed by heating them, illuminating them with sunlight, or passing electricity through them. For example, table sugar (sucrose) can be separated from sugar cane and purified. When heated it decomposes via a complex series of chemical changes (caramelization—shown earlier in Figure 1.4) that produces the brown color and flavor of caramel candy. If heated for a longer time at a high enough temperature, sucrose is converted completely to two other substances, carbon and water. Furthermore, if the water is collected, it can be decomposed still further to pure hydrogen and oxygen by passing an electric current through it. However, nobody has found a way to decompose carbon, hydrogen, or oxygen.

 CD-Rom Screen 1.5: Elements and Atoms; 1.6: Compounds and Molecules

Substances like carbon, hydrogen, and oxygen that cannot be changed by chemical reactions into two or more new substances are called **chemical elements** (or just elements). Substances that can be decomposed, like sucrose and water, are **chemical compounds,** or just compounds. When elements are chemically combined in a compound, their original characteristic properties—such as color, hardness, and melting point—are replaced by the characteristic properties of the compound. For example, sucrose is composed of the following three elements:

- Carbon, which is usually a black powder but is also commonly seen in the form of diamonds
- Hydrogen, a colorless, flammable gas with the lowest density known
- Oxygen, a colorless gas necessary for human and animal respiration

As you know from experience, sucrose is a white, crystalline powder that is completely unlike any of these three elements (Figure 1.12).

If a compound consists of two or more different elements, how is it different from a mixture? There are two ways: (1) a compound has specific composition and (2) a compound has specific properties. Both the composition and the properties of a mixture can vary. A solution of sugar in water can be very sweet or only a little sweet, depending on how much sugar has been dissolved. Other properties, such as viscosity (resistance to flow), can also vary. Corn syrup is much thicker and harder to pour than water with only a little sugar in it. There is no particular composition of a sugar solution that is favored over any other, and each different composition has its own set of properties. On the other hand, 100.0 grams of pure water always contains 11.2 grams of hydrogen and 88.8 grams of oxygen. Pure wa-

Figure 1.12 A compound and its elements. Table sugar, sucrose (a), is composed of the elements carbon (b), oxygen (c), and hydrogen (d). When elements are combined in a compound the properties of the elements are no longer evident. Only the properties of the compound can be observed.

ter always melts at 0.0 °C and boils at 100 °C (at one atmosphere pressure), and it is always a colorless liquid at room temperature.

Problem-Solving Example 1.2   Elements and Compounds

A black powder is placed in a long glass tube. Hydrogen gas is passed into the tube so that it sweeps out all other gases. The powder is then heated with a Bunsen burner. It turns red-orange, and water vapor can be seen condensing at the unheated far end of the tube. The red-orange color remains after the tube cools. (a) Was the original black substance an element? (b) Is the new red-orange substance an element?

Answer   (a) No. (b) Maybe.

Explanation   (a) Hydrogen is known to be an element. Hydrogen must have reacted chemically with the black substance, because the color changed and water was produced. Water contains hydrogen and oxygen, so there must have been oxygen in the original black substance. The red-orange substance must also have been derived from the black substance, so the black substance could not have been an element.

(b) There is not enough information to decide whether the red-orange substance is an element. If there were three or more elements in the black substance, then the red-orange substance must be a compound. However, the red-orange substance sounds suspiciously like copper, which has a black oxide. More experiments would have to be done to confirm this.

Problem-Solving Practice 1.2

A finely divided black substance is placed in a glass tube filled with air. When the tube is heated with a Bunsen burner, the black substance turns red-orange. The total mass of the red-orange substance is greater than that of the black substance. Can you conclude that the black substance is an element? Explain.

Classifying Matter

What we have just said about separating mixtures to obtain elements or compounds and decomposing compounds to obtain elements leads to a useful way to classify matter (Figure 1.13). Heterogeneous mixtures such as iron and sulfur can be separated using simple manipulation—such as a magnet. Homogeneous mixtures are somewhat more difficult to separate, but physical processes will serve. For example, salt water can be purified for drinking by distilling: heating to evaporate the water, and condensing the water vapor back to liquid. When enough water has evaporated away, salt crystals will form and they can be separated from the solution. Most difficult of all is separation of the elements that are combined in a compound. This requires a chemical change and may involve sizable inputs of energy or reactions with other substances.

Desalinization of water (removing salt) could provide drinking water for large numbers of people who live in dry climates near the ocean. However, distillation requires a lot of energy resources and therefore is expensive. When solar energy can be used to evaporate the water, desalinization is less costly.

Exercise 1.5   Classifying Matter

Classify each of the following with regard to the type of matter described:
(a)   Salt dissolved in water
(b)   Concrete (such as that used to make a road)
(c)   Water from a muddy river
(d)   The diamond in a piece of jewelry
(e)   A penny
(f)   Table salt

Figure 1.13 **A scheme for classifying matter.**

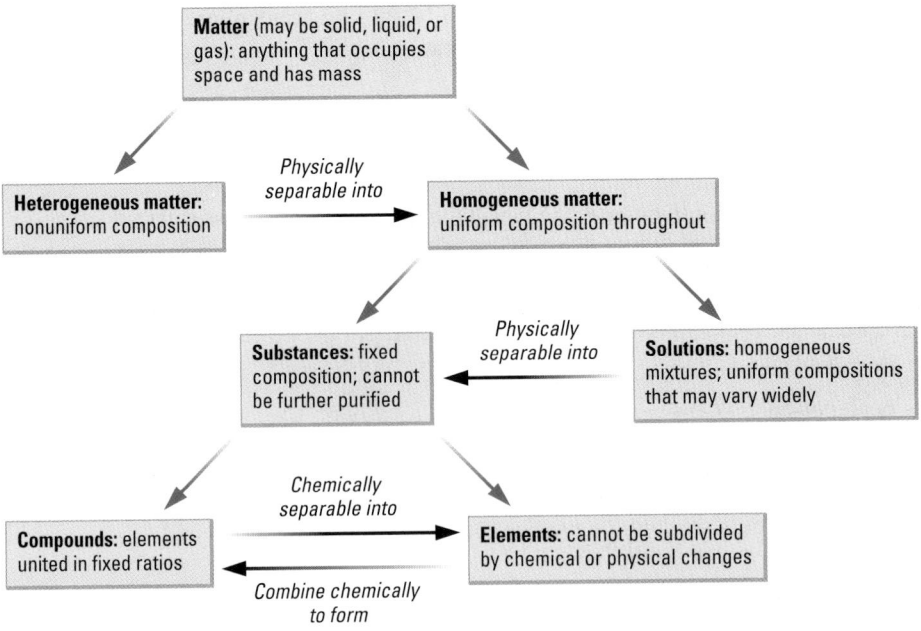

1.7 NANOSCALE MODELS: SOLIDS, LIQUIDS, GASES

An easily observed and very useful property of matter is its physical state. Is it a solid, liquid, or gas? A **solid** can be recognized because it has a rigid shape and a fixed volume that changes very little as temperature and pressure change (Figure 1.14). Like a solid, a **liquid** has a fixed volume, but a liquid is fluid—it takes on the shape of its container and has no definite form of its own. **Gases** are also fluid, but gases expand to fill whatever containers they occupy, and their volumes vary considerably with temperature and pressure. For most substances, when compared at the same conditions, the volume of the solid is slightly less than the volume of the same mass of liquid, but the volume of the same mass of gas is much, much larger. As the temperature is raised, most solids melt to form liquids; eventually, if the temperature is raised enough, most liquids boil to form gases.

All the physical and chemical properties we have already described can be observed by the unaided human senses and refer to samples of matter large enough to be seen, measured, and handled. Such samples are macroscopic; their size places them at the **macroscale.** By contrast, samples of matter so small that they have to be viewed with a microscope are **microscale** samples. Viruses and bacteria, for example, are matter at the microscale. The matter that really interests chemists, however, is at the **nanoscale.** The term is based on the prefix "nano," which comes from the International System of Units (SI units) and indicates something one billion times smaller than something else. (See Table 1.2 for

Figure 1.14 **Quartz crystal.** Quartz, like any solid, has a rigid shape. Its volume changes very little with changes in temperature or pressure.

The International System of Units is the modern version of the metric system. It is described in more detail in Appendix B.

CD-Rom Screen 1.15: Elements and Atoms; 1.6: Compounds and Molecules

TABLE 1.2	SI (Metric) Units for Length		
Prefix	**Abbreviation**	**Meaning**	**Example**
kilo	k	10^3	1 kilometer (km) $= 1 \times 10^3$ meter (m)
deci	d	10^{-1}	1 decimeter (dm) $= 1 \times 10^{-1}$ m $= 0.1$ m
centi	c	10^{-2}	1 centimeter (cm) $= 1 \times 10^{-2}$ m $= 0.01$ m
milli	m	10^{-3}	1 millimeter (mm) $= 1 \times 10^{-3}$ m $= 0.001$ m
micro	μ	10^{-6}	1 micrometer (μm) $= 1 \times 10^{-6}$ m
nano	n	10^{-9}	1 nanometer (nm) $= 1 \times 10^{-9}$ m

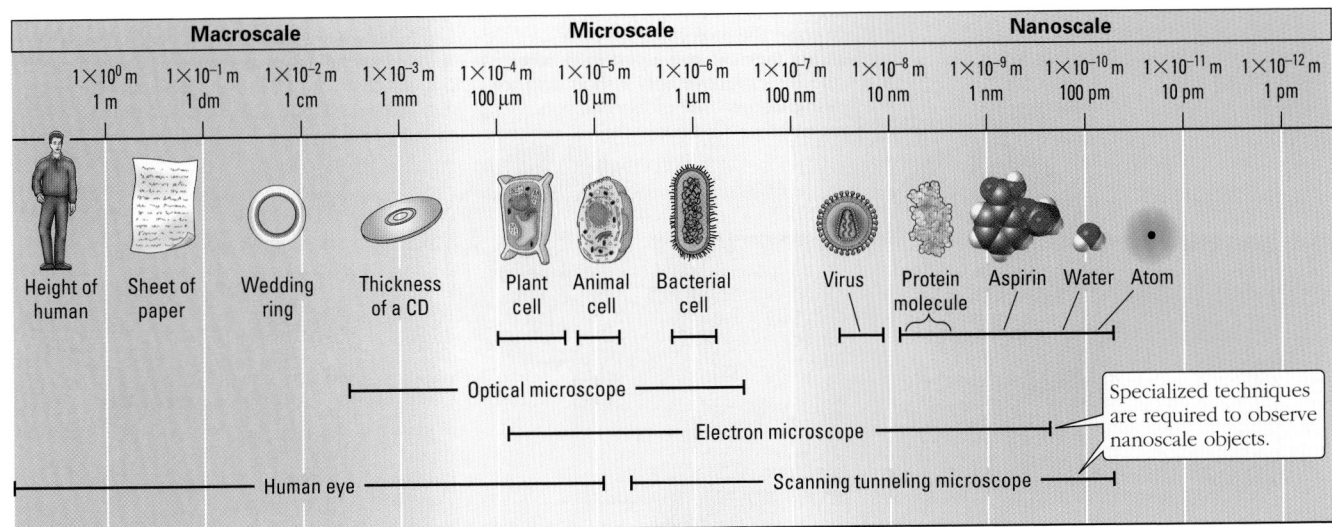

Figure 1.15 **Macroscale, microscale, and nanoscale.**

some important SI prefixes and length units.) For example, a line that is one billion (10^9) times shorter than 1 meter is 1 *nano*meter (1×10^{-9} m) long. The sizes of atoms and molecules are at the nanoscale. An average-sized atom such as a sulfur atom has a diameter of two-tenths of a nanometer (0.2 nm = 2×10^{-10} m), a water molecule is about the same size, and an aspirin molecule is about three quarters of a nanometer (0.75 nm = 7.5×10^{-10} m) across. Figure 1.15 indicates the relative sizes of various objects at the macroscale, the microscale, and the nanoscale.

We described earlier the chemist's unique atomic and molecular perspectives. It is a fundamental idea of chemistry that matter is the way it is because of the nature of its constituent atoms and molecules. Those atoms and molecules are very, very tiny. Therefore, we need to use imagination creatively to discover useful theories that connect the behavior of tiny nanoscale constituents to the observed behavior of chemical substances at the macroscale. Chemistry enables you to "see" in the things all around you nanoscale structure that cannot be seen with your eyes.

Using 10^9 to represent 1,000,000,000 or 1 billion is called scientific notation. It is explained in Appendix A.5.

Jacob Bronowski, in a television series and book titled The *Ascent of Man,* had this to say about the importance of imagination: "There are many gifts that are unique in man; but at the center of them all, the root from which all knowledge grows, lies the ability to draw conclusions from what we see to what we do not see."

Kinetic-Molecular Theory

A theory that deals with matter at the nanoscale is the **kinetic-molecular theory.** It states that all matter consists of extremely tiny particles (atoms or molecules) that are in constant motion. In a solid these particles are packed closely together in a regular array, as shown in Figure 1.16a. The particles vibrate back and forth about their average positions, but seldom does a particle in a solid squeeze past its immediate neighbors to come into contact with a new set of particles. Because the particles are packed so tightly and in such a regular arrangement, a solid is rigid, its volume is fixed, and the volume of a given mass is small. The external shape of a solid often reflects the internal arrangement of its particles. This relation between the observable structure of the solid and the arrangement of the particles from which it is made is one reason scientists have long been fascinated by the shapes of crystals and minerals (Figure 1.17).

The kinetic-molecular theory of matter can also interpret the properties of liquids, as shown in Figure 1.16b. Liquids are fluid because the atoms or molecules are arranged more randomly than in solids. Particles are not confined to specific locations but rather can move past one another. No particle goes very far without

 CD-Rom Screen 1.7: The Kinetic Molecular Theory

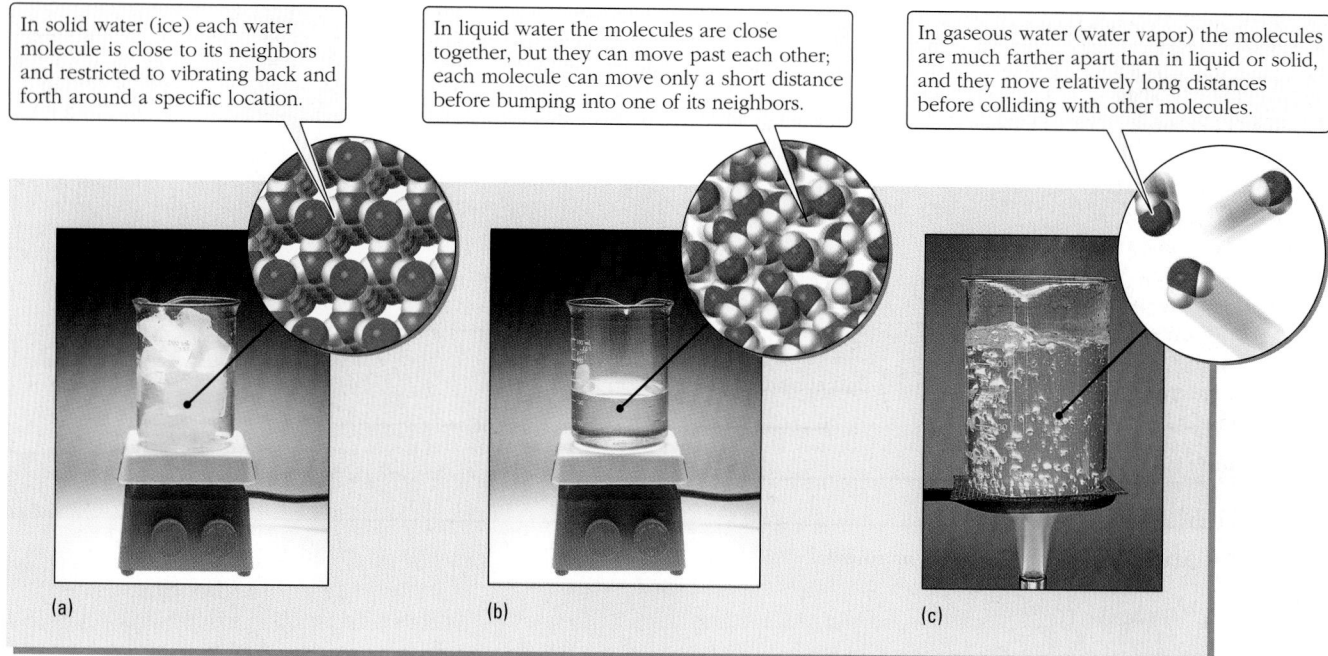

In solid water (ice) each water molecule is close to its neighbors and restricted to vibrating back and forth around a specific location.

In liquid water the molecules are close together, but they can move past each other; each molecule can move only a short distance before bumping into one of its neighbors.

In gaseous water (water vapor) the molecules are much farther apart than in liquid or solid, and they move relatively long distances before colliding with other molecules.

(a)   (b)   (c)

Figure 1.16 **Nanoscale representation of three states of matter.**

Because the nature of the particles is relatively unimportant in determining the behavior of gases, all gases can be described fairly accurately by the ideal gas law, which is introduced in Chapter 10.

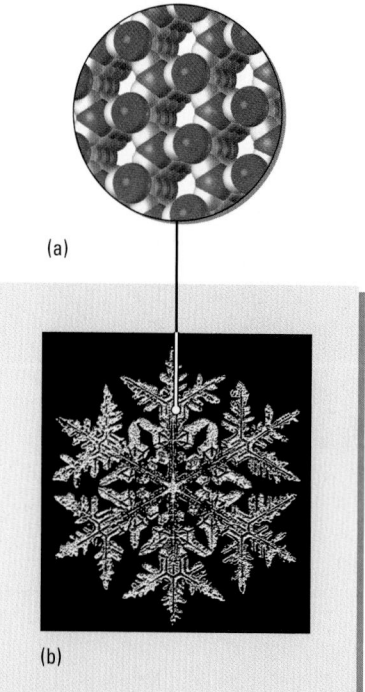

(a)

(b)

Figure 1.17 **Structure and form.**
(a) In the nanoscale structure of ice, each water molecule occupies a position in a regular array or lattice. (b) The form of a snowflake reflects the hexagonal symmetry of the nanoscale structure of ice.

bumping into another — the particles in a liquid interact with their neighbors continually. Because the particles are usually a little farther apart in a liquid than in the corresponding solid, the volume is usually a little bigger. (Ice and liquid water, which are shown in Figure 1.16, are an important exception to this last generality. As you can see from the figure, the water molecules in ice are arranged so that there are empty hexagonal channels. When ice melts, these channels become partially filled by water molecules, accounting for the slightly smaller volume of the same mass of liquid water.)

Like liquids, gases are fluid because their nanoscale particles can easily move past one another. As shown in Figure 1.16c, the particles fly about to fill any container they are in; hence, a gas has no fixed shape or volume. In a gas the particles are much farther apart than in a solid or a liquid. They move significant distances before hitting other particles or the walls of the container. The particles also move quite rapidly. In air at room temperature, for example, the average molecule is going faster than 1000 miles per hour. A particle hits another particle every so often, but most of the time each is quite far away from all the others. Consequently, the nature of the particles is much less important in determining the properties of a gas.

Temperature can also be interpreted using the kinetic-molecular theory. The higher the temperature is, the more active the nanoscale particles are. A solid melts when its temperature is raised to the point where the particles vibrate fast enough and far enough to push each other out of the way and move out of their regularly spaced positions. The substance becomes a liquid because the particles are now behaving as they do in a liquid, bumping into one another and pushing past their neighbors. As the temperature goes higher, the particles move even faster until finally they can escape the clutches of their comrades and become independent; the substance becomes a gas. Increasing temperature corresponds to faster and faster motions of atoms and molecules, a general rule that you will find useful in many future discussions of chemistry (Figure 1.18).

Using the kinetic-molecular theory to interpret the properties of solids, liquids, and gases and the effect of changing temperature provides a very simple example

Chemistry in the News

NANOBES: THE SMALLEST FORM OF LIFE?

Some scientists think there are living things even smaller than bacteria—small enough to come very close to the size of large molecules. Nanobes, which were discovered in 1998, are tiny filaments that grow on sandstone three miles beneath the surface of the sea off Australia. They range from 20 to 150 nm (nanometers) long and are about 10 nm in diameter. This is similar to the size of viruses, which are considered to be nonliving parasites because they need hosts to reproduce. Based on experimental evidence, Dr. Phillipa J. R. Uwins, a scientist at the Center for Microscopy and Microanalysis at the University of Queensland, Australia, has concluded that nanobes are living things. Nanobe colonies grow spontaneously, contain DNA (deoxyribonucleic acid) molecules, are rich in biological elements such as carbon, nitrogen, and oxygen, have some inner structure, and may have an area, like the nucleus of a cell, where the DNA is stored.

Not all scientists agree that nanobes are living organisms, and there has been a lot of debate over the issue of how small a living organism can be. Much of this debate was initiated in 1996 when tiny structures that appeared to be fossil microbes were found in the 4.5-billion-year-old Martian meteorite ALH 84001, which had fallen to earth in Antarctica long ago. These objects in the meteorite were 20 to 200 nm in size, and their small size led many scientists to question whether they really represented a life form that had originated outside the earth's biosphere. Whether or not nanobes are living things, they have certainly stirred up a lot of controversy in the scientific community.

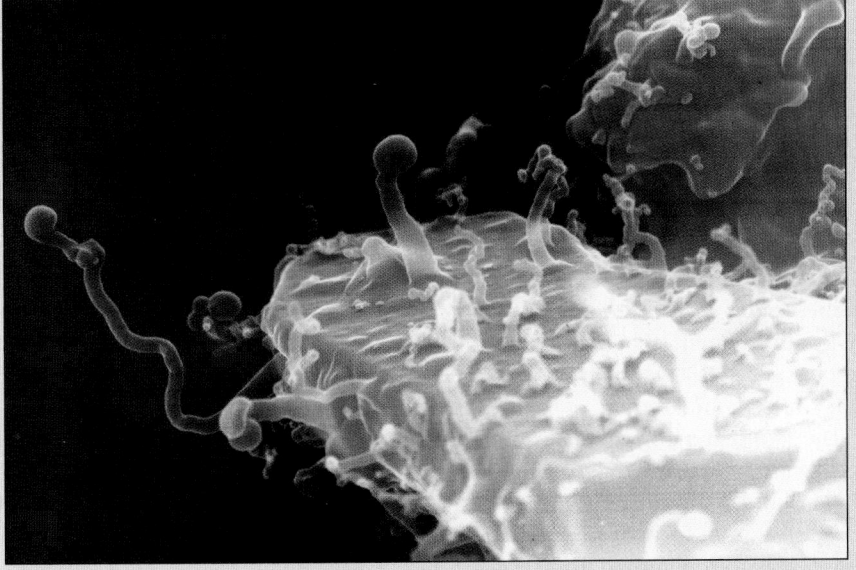

Electron micrograph of nanobe filaments projecting from sandstone.

Source:

The New York Times, January 18, 2000, p. D1.

of how chemists use nanoscale theories and models to interpret and explain macroscale observations. In the remainder of this chapter and throughout your study of chemistry, you should try to imagine how the atoms and molecules are arranged and what they are doing whenever you consider a macroscale sample of

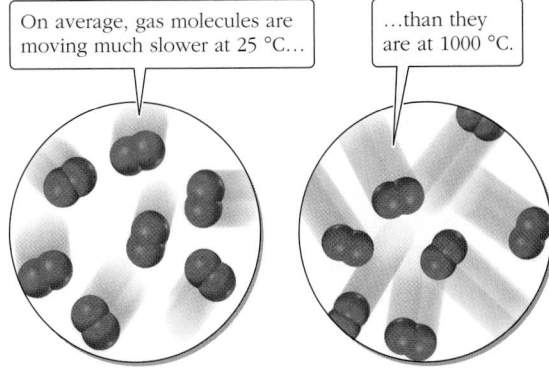

On average, gas molecules are moving much slower at 25 °C...

...than they are at 1000 °C.

Figure 1.18   **Molecular speed and temperature.**

matter. That is, you should try to develop the chemist's special perspective on the relation of nanoscale structure to macroscale behavior.

 Exercise 1.6 Kinetic-Molecular Theory

Use the idea that matter consists of tiny particles in motion to interpret each observation.
(a) An ice cube sitting in the sun slowly melts, and the liquid water eventually evaporates.
(b) Wet clothes hung on a line eventually dry.
(c) Moisture appears on the outside of a glass of ice water.
(d) Evaporation of a solution of sugar in water forms crystals.

1.8 ATOMIC THEORY

CD-Rom Screen 2.2:
Introduction to Atoms; 2.5:
The Dalton Atomic Theory

The existence of elements can be explained by a nanoscale model involving particles, just as the properties of solids, liquids, and gases can be. This model, which is closely related to the kinetic-molecular theory, is called the atomic theory. It was proposed in 1803 by John Dalton. According to Dalton's theory, an element cannot be decomposed into two or more new substances because at the nanoscale it consists of one and only one kind of atom and because atoms are indivisible under the conditions of chemical reactions. An **atom** is the smallest particle of an element that embodies the chemical properties of that element. An element, such as the sample of copper in Figure 1.19, is made up entirely of atoms of the same kind.

The fact that a compound can be decomposed into two or more different substances can be explained by saying that each compound must contain two or more different kinds of atoms. The process of decomposition involves separating at least one type of atom from atoms of the other kind(s). For example, charring of sugar corresponds to separating atoms of carbon from atoms of oxygen and atoms of hydrogen.

JOHN DALTON
(1766–1844)

John Dalton came from a poor background and ended formal schooling at age 11. About a year later he became a teacher in Manchester, England, where he spent most of his life. His lifelong interest in weather observations led to curiosity about the behavior of gases and then to chemistry. In 1803 he introduced his atomic theory to the Literary and Philosophical Society of Manchester, and in 1810 he became a member of the Royal Society, the most prestigious scientific society in Britain.

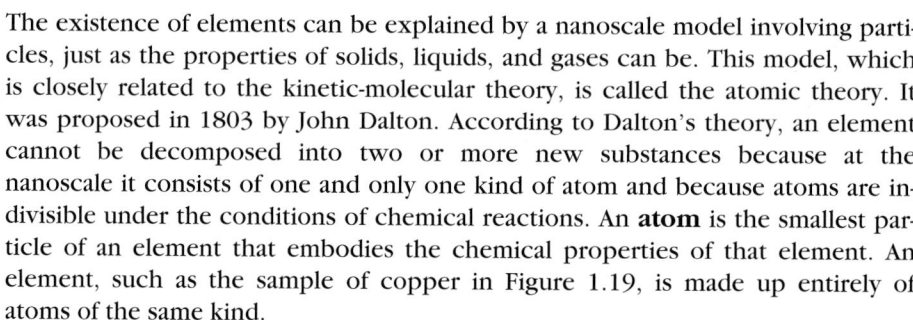
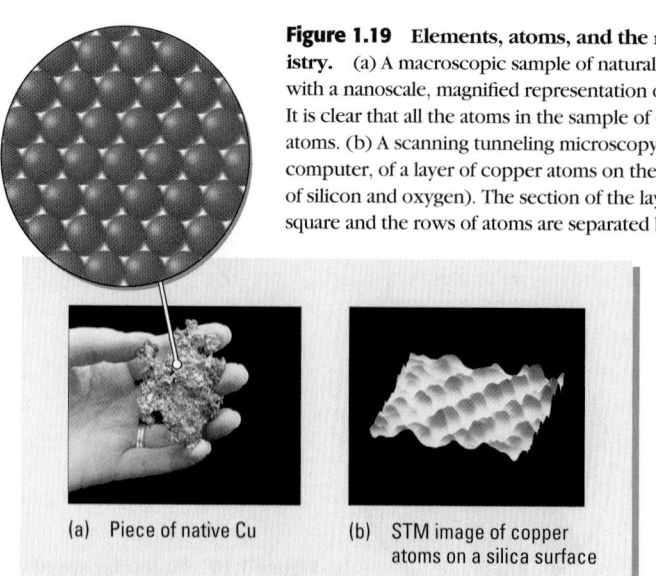

Figure 1.19 Elements, atoms, and the nanoscale world of chemistry. (a) A macroscopic sample of naturally occurring copper metal with a nanoscale, magnified representation of a tiny portion of its surface. It is clear that all the atoms in the sample of copper are the same kind of atoms. (b) A scanning tunneling microscopy (STM) image, enhanced by a computer, of a layer of copper atoms on the surface of silica (a compound of silicon and oxygen). The section of the layer shown is 1.7 nanometers square and the rows of atoms are separated by about 0.44 nanometer.

(a) Piece of native Cu   (b) STM image of copper atoms on a silica surface

Dalton also said that each kind of atom must have its own properties — in particular, a characteristic mass. This idea allowed his theory to account for the masses of different elements that combine in chemical reactions to form compounds. An important success of Dalton's ideas was that they could be used to interpret known chemical facts quantitatively.

Two laws known in Dalton's time could be explained by the atomic theory. One was based on experiments in which the reactants were carefully weighed before a chemical reaction, and the reaction products were carefully collected and weighed afterward. The results led to the **law of conservation of mass** (also called the law of conservation of matter): *There is no detectable change in mass during an ordinary chemical reaction.* The atomic theory says that mass is conserved because the same number of atoms of each kind is present before and after a reaction, and each of those kinds of atoms has its same characteristic mass before and after the reaction.

The other law was based on the observation that in a chemical compound the proportions of the elements by mass are always the same. Water always contains 1 gram of hydrogen for every 8 grams of oxygen, and carbon monoxide always contains 4 grams of oxygen for every 3 grams of carbon. The **law of constant composition** summarizes such observations: *A chemical compound always contains the same elements in the same proportions by mass.* The atomic theory explains this observation by saying that atoms of different elements always combine in the same ratio in a compound. For example, in carbon monoxide there is always one carbon atom for each oxygen atom. If the mass of an oxygen atom is $\frac{4}{3}$ times the mass of a carbon atom, then the ratio of mass of oxygen to mass of carbon in carbon monoxide will always be 4:3.

Dalton's theory has been modified to account for discoveries since his time. The *modern* atomic theory is based on the following assumptions:

All matter is composed of atoms, which are extremely tiny. Interactions among atoms account for the properties of matter.

All atoms of a given element have the same chemical properties. Atoms of different elements have different chemical properties.

Compounds are formed by the chemical combination of two or more different kinds of atoms. Atoms usually combine in the ratio of small whole numbers. For example, in a carbon monoxide molecule there is one carbon atom and one oxygen atom; a carbon dioxide molecule consists of one carbon atom and two oxygen atoms.

A chemical reaction involves joining, separating, or rearranging atoms. Atoms in the reactant substances form new combinations in the product substances. Atoms are not created, destroyed, or converted into other kinds of atoms during a chemical reaction.

The hallmark of a good theory is that it suggests new experiments, and this was true of the atomic theory. Dalton realized that it predicted a law that had not yet been discovered. If compounds are formed by combining atoms of different elements on the nanoscale, then in some cases there might be more than a single combination. An example is carbon monoxide and carbon dioxide. In one case there is one oxygen atom for each carbon atom, while in the other there are two oxygen atoms per carbon atom. Therefore, in carbon dioxide the mass of oxygen per gram of carbon ought to be twice as great as it is in carbon monoxide (because twice as many oxygen atoms will weigh twice as much). Dalton called this the **law of multiple proportions,** and he carried out quantitative experiments seeking data to confirm or deny it. Dalton and others obtained data consistent with the law of multiple proportions, thereby enhancing acceptance of the atomic theory.

Our bodies are made up of atoms from the distant past, atoms from other people and other things. Some of the carbon, hydrogen, and oxygen atoms in our carbohydrates have come from the breaths (first and last) of famous and ordinary persons of the past.

According to the modern theory, atoms of the same element have the same chemical properties but are not necessarily identical in all respects. The discussion of isotopes in Chapter 2 shows how atoms of the same element can differ in mass.

oxygen atom   carbon atom

carbon monoxide   carbon dioxide

ESTIMATION   **How Tiny Are Atoms and Molecules?**

It is often useful to estimate an approximate value for something. Usually this can be done quickly, and often it can be done without a calculator. The idea is to pick round numbers that you can work with in your head, or to use some other method that allows a quick estimate. If you really need an accurate value, an estimate is still useful to check whether the accurate value is in the right ball park. Often an estimate is referred to as a "back-of-the-envelope" calculation, because estimates might be done over lunch on any piece of paper that is at hand. Some estimates are referred to as "order-of-magnitude calculations" because only the power of ten (the order of magnitude) in the answer is obtained. To help you develop estimation skills, each chapter in this book will provide you with an example of estimating something.

To get a more intuitive feeling for how small atoms and molecules are, estimate how many hydrogen atoms could fit inside a 12-oz (240-mL) soft drink can. Make the same estimate for protein molecules. Use the approximate sizes given in Figure 1.15.

Because 1 mL is the same volume as a cube 1 cm on each side (1 cm^3), the volume of the can is the same as the volume of 240 cubes 1 cm on each side. Therefore we can first estimate how many atoms would fit into a 1-cm cube and then multiply that number by 240.

According to Figure 1.15, a typical atom has a diameter slightly less than 100 pm. Because this is an estimate, and to make the numbers easy to handle, assume that we are dealing with an atom that is 100 pm in diameter. Then the atom's diameter is 100×10^{-12} m $= 1 \times 10^{-10}$ m, and it will require

10^{10} of these atoms lined up in a row to make a length of 1 m. Since 1 cm is $\frac{1}{100}$ (10^{-2}) of a meter, only $10^{-2} \times 10^{10} = 10^8$ atoms would fit in 1 cm.

In three dimensions, there could be 10^8 atoms along each of the three perpendicular edges of a 1-cm cube (the x, y, and z directions). The one row along the x-axis could be repeated 10^8 times along the y-axis, and then that layer of atoms could be repeated 10^8 times along the z-axis. Therefore, the number of atoms that we estimate would fit inside the cube is $10^8 \times 10^8 \times 10^8 = 10^{24}$ atoms. Multiplying this by 240 gives $240 \times 10^{24} = 2.4 \times 10^{26}$ atoms in the soft drink can.

This estimate is a bit low. A hydrogen atom's diameter is less than 100 pm, so more hydrogen atoms would fit inside the can. Also, atoms are usually thought of as spheres, and so they could pack together more closely than they would if just lined up in rows. Therefore, an even larger number of atoms than 2.4×10^{26} could fit inside the can.

For a typical protein molecule, Figure 1.15 indicates a diameter on the order of 5 nm = 5000 pm. That is 50 times bigger than the 100-pm diameter we used for the hydrogen atom. This means that there would be 50 times fewer protein molecules in the x direction, 50 times fewer in the y direction, and 50 times fewer in the z direction. Therefore, the number of protein molecules would be fewer by $50 \times 50 \times 50 = 125,000$. The number of protein molecules can thus be estimated as $(2.4 \times 10^{26})/(1.25 \times 10^5)$. Because 2.4 is roughly twice 1.25, and because we are estimating, not calculating accurately, we can take the result to be 2×10^{21} protein molecules. That's still a whole lot of molecules!

1.9   THE CHEMICAL ELEMENTS

CD-Rom Screen 2.16: The Periodic Table

Every element has been given a unique *name* and a *symbol* derived from the name. These names and symbols are listed in the periodic table inside the front cover of the book. The first letter of each symbol is capitalized; the second letter, if there is one, is lower case, as in He, the symbol for helium. Elements discovered a long time ago have names and symbols with Latin or other origins, such as Au for gold (from *aurum,* meaning "bright dawn") and Fe for iron (from *ferrum*). The names of more recently discovered elements are derived from their place of discovery or from a person or place of significance (Table 1.3).

Ancient people knew of nine elements—gold (Au), silver (Ag), copper (Cu), tin (Sn), lead (Pb), mercury (Hg), iron (Fe), sulfur (S), and carbon (C). Most of the other elements were discovered during the 1800s, as one by one they were separated from minerals in the earth's crust or from the earth's oceans or atmosphere. Currently, more than 110 elements are known, but only 90 occur in nature. Elements such as technetium (Tc), neptunium (Np), mendeleevium (Md), seaborgium (Sg), and meitnerium (Mt) have been made using nuclear reactions (see Chapter 20), beginning in the 1930s.

Elements are being discovered even now. Elements 114, 116, and 118 were discovered in 1999, but only a few atoms of each were observed.

Types of Elements

The vast majority of the elements are **metals**—only 25 are not. You are probably familiar with many properties of metals. They are solids (except for mercury, which is a liquid), they conduct electricity (and conduct better as the temperature decreases), they are ductile (can be drawn into wires), they are malleable

TABLE 1.3   The Names of Some Chemical Elements

Element	Symbol	Date Discovered	Discoverer	Derivation of Name/Symbol
Carbon	C	Ancient	Ancient	Latin, *carbo* (charcoal)
Curium	Cm	1944	G. Seaborg, et al.	Honoring Marie and Pierre Curie, Nobel Prize winners for discovery of radioactive elements
Hydrogen	H	1766	H. Cavendish	Greek, *hydro* (water) and *genes* (generator)
Meitnerium	Mt	1982	P. Armbruster, et al.	Honoring Lise Meitner, codiscoverer of nuclear fission
Mendelevium	Md	1955	G. Seaborg, et al.	Honoring Dmitri Mendeleev, who devised the periodic table
Mercury	Hg	Ancient	Ancient	For Mercury, messenger of the gods, because it flows quickly; symbol from Greek *hydrargyrum*, liquid silver
Polonium	Po	1898	M. Curie and P. Curie	In honor of Poland, Marie Curie's native country
Seaborgium	Sg	1974	G. Seaborg, et al.	Honoring Glenn Seaborg, Nobel prize winner for synthesis of new elements
Sodium	Na	1807	H. Davy	Latin, *soda* (sodium carbonate); symbol from Latin *natrium*
Tin	Sn	Ancient	Ancient	German, *Zinn*; symbol from Latin, *stannum*

(can be rolled into sheets), and they can form alloys (solutions of one or more metals in another metal). In a solid metal, individual metal atoms are packed close to each other, and so metals usually have fairly high densities. Figure 1.20 shows some representative metals. Iron (Fe) and aluminum (Al) are used in automobile parts because of their ductility, malleability, and relatively low cost. Copper (Cu) is used in electrical wiring because it conducts electricity better than most metals. Gold (Au) is used for the vital electrical contacts in automobile air bags and in some computers because it does not corrode and is an excellent electrical conductor.

In contrast, **nonmetals** do not conduct electricity (with a few exceptions, such as graphite, one form of carbon). Nonmetals are more diverse in their physical properties than are metals (Figure 1.21). At room temperature some nonmetals are solids (such as phosphorus, sulfur, and iodine); bromine is a liquid; and others are gases (such as hydrogen, nitrogen, and oxygen). The nonmetals helium (He), neon (Ne), argon (Ar), krypton (Kr), xenon (Xe), and radon (Rn) are gases that consist of individual atoms.

Figure 1.20 Some metallic elements—iron, aluminum, copper, and gold. The steel ball bearing is principally iron. The rod is made of aluminum. The inner coil is gold and the other one copper. Metals are malleable, ductile, and conduct electricity.

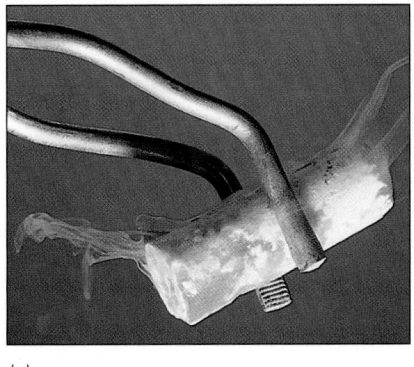

(a)

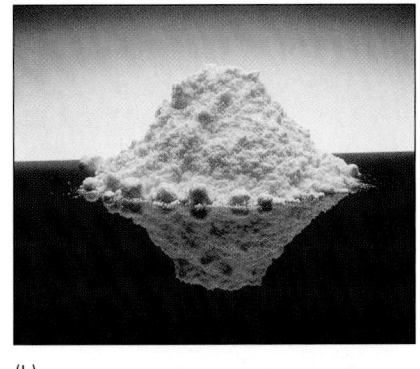

(b)

(c)   (d)

Figure 1.21 Some nonmetallic elements—(a) white phosphorus, (b) sulfur, (c) bromine, (d) iodine. Nonmetals do not conduct electricity. Bromine is the only nonmetal that is a liquid at room temperature.

On the periodic table inside the front cover of this book, the metals, nonmetals, and metalloids are color-coded: gray and blue for metals, lavender for nonmetals, and orange for metalloids.

A few elements — boron, silicon, germanium, arsenic, antimony, and tellurium — are classified as **metalloids,** elements that have some typically metallic properties and other properties that are characteristic of nonmetals. For example, some metalloids are shiny like metals, but they do not conduct electricity as well as metals. Many of them are semiconductors and are essential for the electronics industry. (See Sections 11.8 and 11.9.)

Exercise 1.7   Elements

Use Table 1.3, the periodic table in the inside front cover, and/or the list of elements inside the back cover to answer these questions.
(a) Three elements are named for planets in our solar system. Give their names and symbols.
(b) One element is named for a state in the United States. Name the element and give its symbol.
(c) Two elements are named in honor of women. What are their names and symbols?
(d) Several elements are named for countries or regions of the world. Find at least four of these and give names and symbols.
(e) List the symbols of all elements that are nonmetals.

CD-Rom Screen 3.2: Elements that Exist as Molecules

Space-filling model of Cl_2

Elements that Consist of Molecules

Most elements that are nonmetals consist of molecules on the nanoscale. A **molecule** is a unit of matter in which two or more atoms are chemically bonded together. For example, a chlorine molecule contains two chlorine atoms and can be represented by the chemical formula Cl_2. A **chemical formula** uses the symbols for the elements to represent the atomic composition of a substance. In gaseous chlorine, Cl_2 molecules are the particles that fly about and collide with each other and the container walls. Molecules like Cl_2, that consist of two atoms, are called **diatomic molecules.** Oxygen and nitrogen also exist as diatomic molecules, as do hydrogen (H_2), fluorine (F_2), bromine (Br_2), and iodine (I_2).

You have already seen (in Figure 1.1) that for really big molecules, such as the COX enzymes, molecules may be represented by ribbons or sticks, without showing individual atoms at all. Often such representations are drawn by computers, which help chemists to manipulate and understand the molecular structures.

Elements that consist of diatomic molecules are H_2, N_2, O_2, F_2, Cl_2, Br_2, and I_2. You need to remember that these elements consist of diatomic molecules, because they will be encountered frequently. Most of these elements are close together in the periodic table, which makes it easier to remember them.

Exercise 1.8   Elements that Consist of Diatomic Molecules or Are Metalloids

On a copy of the periodic table, circle the symbols of the elements that
(a) Consist of diatomic molecules;
(b) Are metalloids.

Devise rules related to the periodic table that will help you to remember which elements these are.

oxygen molecule (O_2)

ozone molecule (O_3)

Allotropes

Oxygen and carbon are among the elements that exist as **allotropes,** different forms of the same element that exist in the same physical state at the same temperature and pressure. Allotropes are possible because the same kind of atoms can be connected in different ways when they form molecules. For example, the allotropes of oxygen are O_2, sometimes called dioxygen, and O_3, ozone. Dioxygen, a major component of earth's atmosphere, is by far the more common allotropic form. Ozone is a highly reactive pale blue gas first detected by its characteristic pungent odor. Its name comes from *ozein,* a Greek word meaning "to smell."

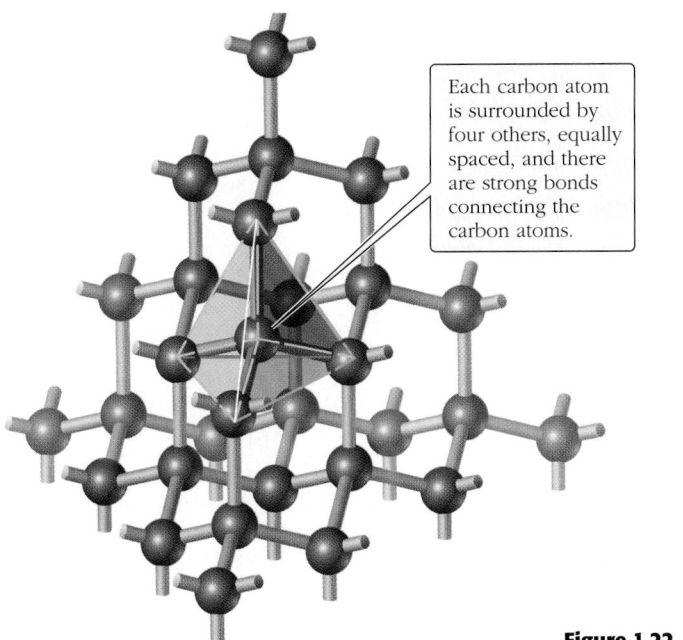

Figure 1.22 Structure of diamond.

Each carbon atom is surrounded by four others, equally spaced, and there are strong bonds connecting the carbon atoms.

Diamond and graphite, known for centuries, are two allotropes of carbon containing extended networks of carbon atoms. In diamond (Figure 1.22) each carbon atom is surrounded by four other carbon atoms and connected to them by strong bonds. In graphite each carbon atom is surrounded by three others, all strongly bonded, in a flat layer or sheet. The sheets are packed one over the other to form a layered structure, as shown in Figure 1.23.

Diamond and graphite had long been considered the only allotropes of carbon with well-defined structures. Therefore, it was a surprise in the 1980s when another carbon allotrope was discovered in soot produced when carbon-containing materials are burned with very little oxygen. The new allotrope consists of 60-carbon atom cages and represents a new class of molecules. The C_{60} molecule resembles a soccer ball with a carbon atom at each corner of each of the black

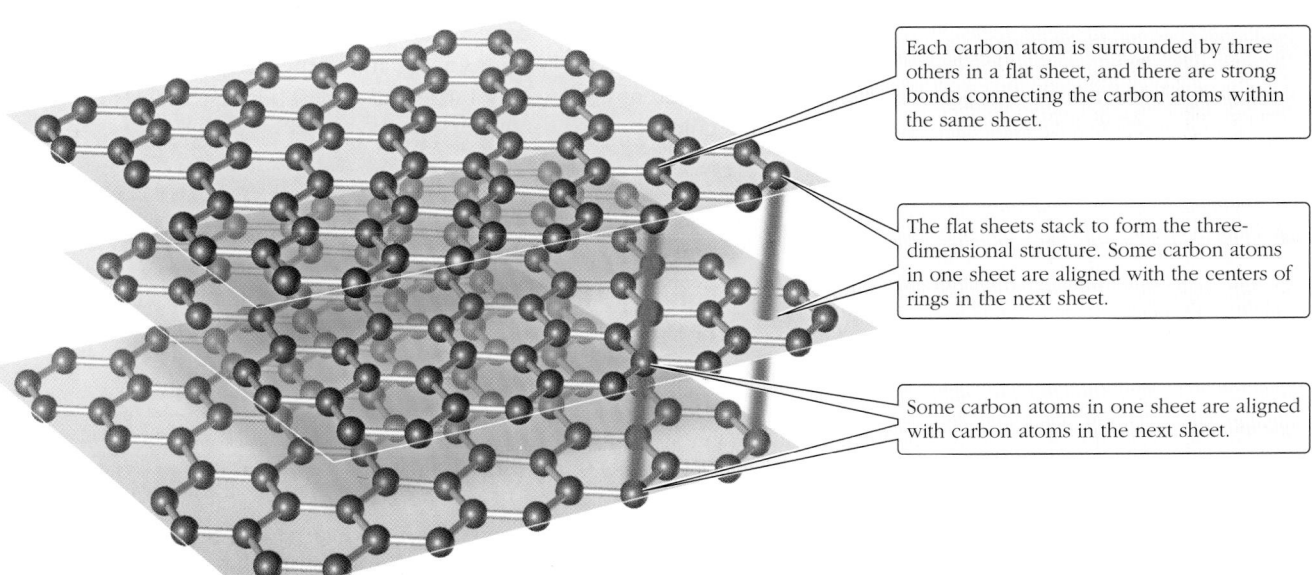

Each carbon atom is surrounded by three others in a flat sheet, and there are strong bonds connecting the carbon atoms within the same sheet.

The flat sheets stack to form the three-dimensional structure. Some carbon atoms in one sheet are aligned with the centers of rings in the next sheet.

Some carbon atoms in one sheet are aligned with carbon atoms in the next sheet.

Figure 1.23 Structure of graphite.

(a)

(b)

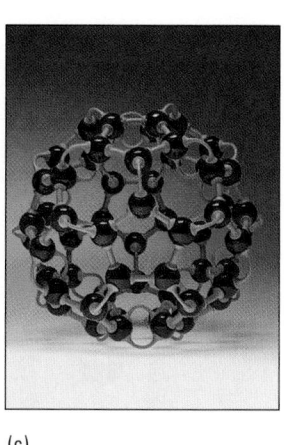
(c)

Figure 1.24 **Models for fullerenes.** (a) Geodesic domes at Elmira College, Elmira, NY. Geodesic domes, such as those designed originally by R. Buckminster Fuller, contain linked hexagons and pentagons. (b) A soccer ball is a model for the C_{60} structure. (c) The C_{60} fullerene molecule, which is made up of five-membered rings (black rings on the soccer ball) and six-membered rings (white rings on the ball).

RICHARD E. SMALLEY
(1943–)

Richard E. Smalley, Robert Curl, and Harry Kroto received the Nobel Prize for Chemistry in 1996 for discovering fullerenes. To model the C_{60} molecule, Smalley assembled paper hexagons and pentagons to make what looked like a soccer ball. According to Smalley, ". . . I asked Harry (Kroto) . . . who was the architect who worked with big domes. . . . He said it was Buckminster Fuller. Within a few moments we drew a ball on the blackboard and shouted, with rather Monty Pythonesque humor, 'IT'S BUCKMINSTER FULLER . . . ENE!'"

For an up-to-date description of research on nanotubes, see "Nanotubes for Electronics," *Scientific American*, Dec. 2000, p. 62.

CD-Rom Screen 3.4: Representing Compounds

pentagons in Figure 1.24b. Each five-membered ring of carbon atoms is surrounded by five six-membered rings. This molecular structure of carbon pentagons and hexagons reminded its discoverers of a geodesic dome, a structure popularized years ago by the innovative American philosopher and engineer R. Buckminster Fuller (Figure 1.24a). Therefore, the official name of the C_{60} allotrope is buckminsterfullerene. Chemists often call it simply a "buckyball." C_{60} buckyballs belong to a larger molecular family of even-numbered carbon cages that is collectively called fullerenes.

Carbon atoms can also form concentric tubes that resemble rolled-up chicken wire. These single- and multi-walled *nanotubes* of only carbon atoms are excellent electrical conductors and extremely strong. Imagine the exciting applications for such properties, including making buckyfibers that could substitute for the metal wires now used to transmit electrical power. Dozens of uses have been proposed for fullerenes, buckytubes, and buckyfibers, among them microscopic ball bearings, lightweight batteries, new lubricants, nanoscale electric switches, new plastics, and antitumor therapy for cancer patients (by enclosing a radioactive atom within the cage). All these applications await an inexpensive way of making buckyballs and other fullerenes. Currently buckyballs, the cheapest fullerene, are more expensive than gold.

Exercise 1.9 Allotropes

A student says that tin and lead are allotropes because they are both dull gray metals. Why is the statement wrong?

1.10 CHEMICAL SYMBOLISM

Chemical symbols, such as Na, I, or Mt, are a shorthand way of indicating what kind of atoms we are talking about. Chemical formulas tell us how many atoms of an element are combined in a molecule and in what ratios atoms are com-

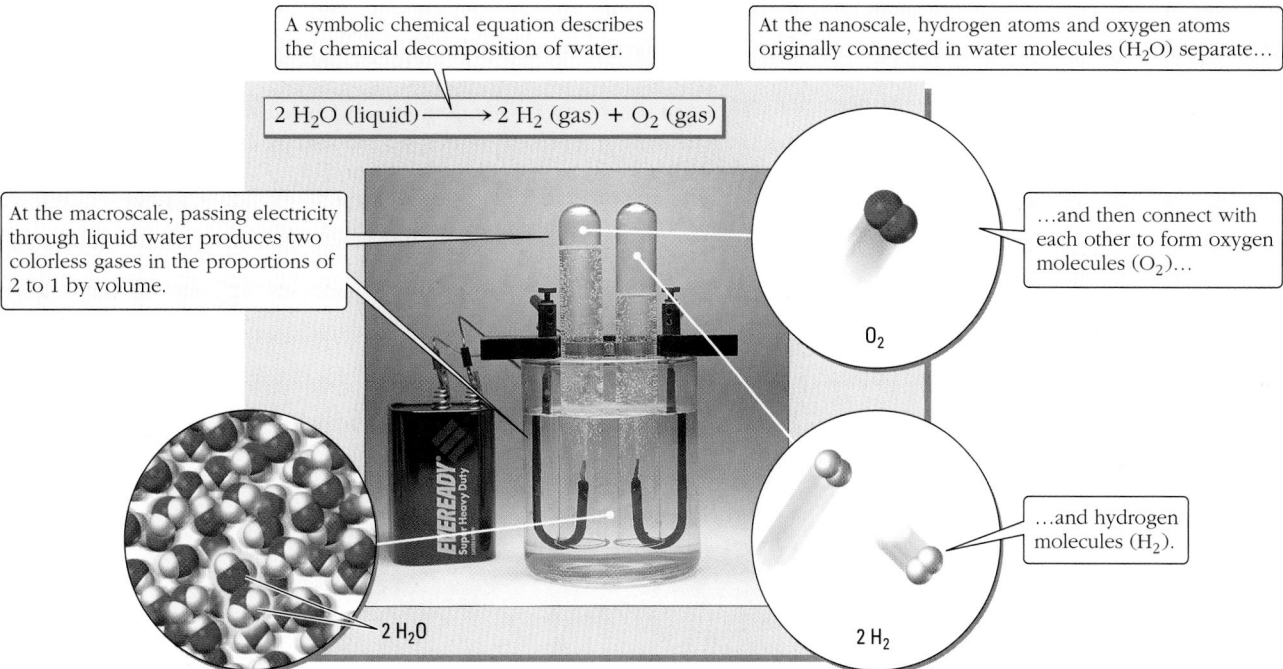

A symbolic chemical equation describes the chemical decomposition of water.

At the nanoscale, hydrogen atoms and oxygen atoms originally connected in water molecules (H_2O) separate...

$$2 \, H_2O \, \text{(liquid)} \longrightarrow 2 \, H_2 \, \text{(gas)} + O_2 \, \text{(gas)}$$

At the macroscale, passing electricity through liquid water produces two colorless gases in the proportions of 2 to 1 by volume.

...and then connect with each other to form oxygen molecules (O_2)...

O_2

...and hydrogen molecules (H_2).

$2 \, H_2O$

$2 \, H_2$

Figure 1.25 Symbolic, macroscale, and nanoscale representations of a chemical reaction.

bined in compounds. For example, the formula Cl_2 tells us that there are two chlorine atoms in a chlorine molecule. The formulas CO and CO_2 tell us that carbon and oxygen form two different compounds — one that has equal numbers of C and O atoms and one that has twice as many O atoms as C atoms. In other words, chemical symbols and formulas symbolize the nanoscale composition of each substance.

Chemical symbols and formulas also represent the macroscale properties of elements and compounds. That is, the symbol Na brings to mind a highly reactive metal, and the formula H_2O represents a colorless liquid that freezes at 0 °C, boils at 100 °C, and reacts violently with Na. Because chemists are familiar with both the nanoscale and macroscale characteristics of substances, they usually use symbols to abbreviate their representations of both. Symbols are also useful to represent chemical reactions. For example, the charring of sucrose mentioned earlier is represented by

Sucrose	$\longrightarrow$	carbon	+	water
$C_{12}H_{22}O_{11}$	$\longrightarrow$	12 C	+	11 H_2O
Reactant	changes to			Products

The symbolic aspect of chemistry is the third part of the chemist's special view of the world. It is important that you become familiar and comfortable with using chemical symbols and formulas to represent chemical substances and their reactions. Figure 1.25 shows how chemical symbolism can be applied to the process of decomposing water with electricity (electrolysis of water).

Problem-Solving Example 1.3 Macroscale, Nanoscale, and Symbolic Representations

The figure shows a sample of water boiling. In spaces labeled A, indicate whether the macroscale or the nanoscale is represented. In spaces labeled B, draw the molecules that would be present with appropriate distances between them. One of the circles represents a bubble of gas within the liquid. The other represents the liquid. In space C, write a symbolic representation of the boiling process.

C

B

A

B

Answer

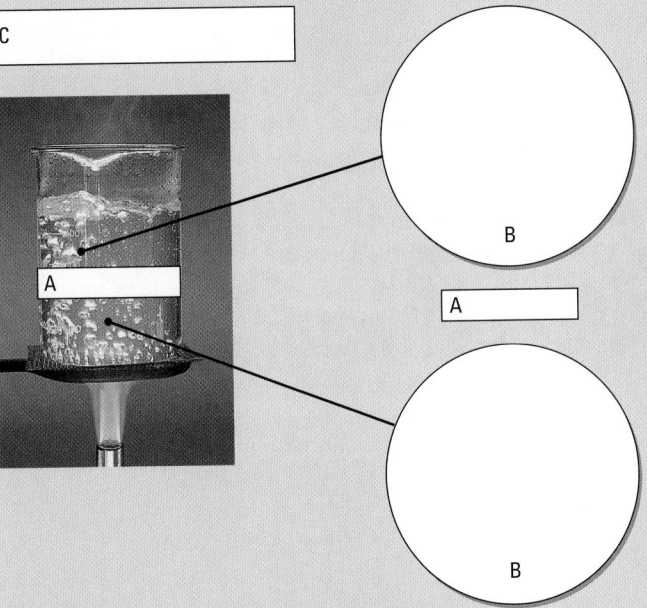

C H_2O (liquid) $\longrightarrow$ H_2O (gas)

A   Macroscale

B

A   Nanoscale

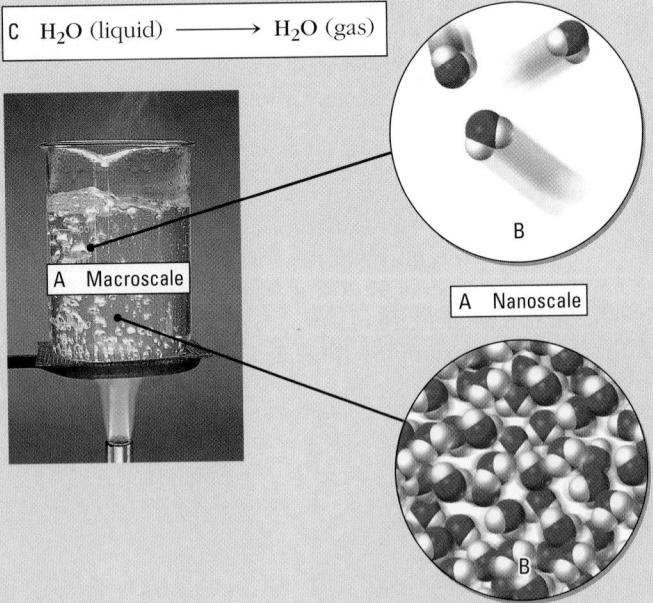

B

Explanation Each water molecule consists of two hydrogen atoms and one oxygen atom. In liquid water the molecules are close together and oriented in various directions. In a bubble of gaseous water the molecules are much farther apart, and there are fewer of them per unit volume. The symbolic representation is the equation

$$H_2O(\text{liquid}) \longrightarrow H_2O(\text{gas})$$

Problem-Solving Practice 1.3

Draw a nanoscale representation and a symbolic representation for both allotropes of oxygen. Describe the properties of each allotrope at the macroscale.

1.11 RISKS AND BENEFITS

We sometimes hear that a chemical is bad or toxic and should be banned. What is actually being said is that the risks of using the chemical seem to outweigh its benefits. Any scientific discovery carries with it both benefits and risks. It is unlikely that someone would suggest that aspirin be banned, but chronic use of this drug can cause bleeding in the stomach or intestines and kidney damage. Some people die as a result of such side effects. Aspirin benefits many, many people throughout the world, but it nevertheless does harm a small fraction of the population. You could say the same about automobiles or airplanes. They are risky, but for most of us the benefits outweigh the risks, and we use them. A very few chemicals are hazardous, but most are beneficial. It is important to know how to assess the benefits and risks, and science provides the necessary background.

Risk depends on both the nature of a substance and on the quantity of that substance to which you are exposed. In some cases the risk can be attributed to a specific substance, such as aspirin. In other cases, a particular element, such as mercury or selenium, may be targeted, regardless of what compound it may be in. Greater exposure means greater risk. Drinking the occasional beer is usually not harmful, but it is possible to kill yourself by drinking too much alcohol in too short a time. Assessing risk involves collecting quantitative data about the hazards of specific substances by doing animal tests or by examining records of human exposure. It also involves determining how people are exposed to a substance and to how much they are exposed.

People often conclude that anything manmade is bad, whereas anything natural is good, but often this is not confirmed by careful risk assessment. You may be surprised by some of the results in Table 1.4. Some risks that many people readily

TABLE 1.4	Estimates of Risk: Activities That Produce One Additional Death per One Million People Exposed to the Risk
Activity	**Cause of Death**
Smoking 1.4 cigarettes	Cancer, lung disease
Living 2 months with a cigarette smoker	Cancer, lung disease
Eating 40 tablespoons of peanut butter	Liver cancer caused by the natural carcinogen aflatoxin B
Drinking 40 cans of saccharin-sweetened soda	Cancer
Eating 100 charcoal-broiled steaks	Cancer
Traveling 6 minutes by canoe	Accident
Traveling 10 minutes by bicycle	Accident
Traveling 300 miles by car	Accident
Traveling 1000 miles by jet aircraft	Accident
Drinking Miami drinking water for 1 year	Cancer from chloroform
Living 2 months in Denver	Cancer caused by cosmic radiation
1 chest x-ray in a good hospital	Cancer
Living 5 years at the boundary of a typical nuclear power plant	Cancer

From Gough, L., and Gough, M. "Risky Business." *Chem Matters*, December 1993, pp 10–12.

accept are significantly higher than others that are more highly publicized in the media. Analytical chemists' skills have revealed that there are many substances in the environment, and even in our bodies, that might be harmful. Those skills also help to identify just how serious is the risk from each of these substances. A wealth of information is available, and analyzing it scientifically provides our best means to identify and avoid risks.

> **Exercise 1.10 Risks**
>
> Name five risks you have taken today and rank them in relative order of their danger. Compare your list with the list of a friend. Are any of the items on your list the same as your friend's?

1.12 WHY CARE ABOUT CHEMISTRY?

We hope that this chapter has made clear many of the reasons you should care about chemistry. Chemistry is fundamental to understanding many other sciences and to understanding how the material world around us works. Chemistry provides a unique, nanoscale perspective that has been highly successful in stimulating scientific inquiry and in the development of high-tech materials, modern medicines, and many other things that benefit humankind. Chemistry is happening all around us and within us. Knowledge of chemistry is key to understanding and making the most of our internal and external environments. Chemistry and chemical knowledge can help us to make better decisions about the balance between risks and benefits—decisions that are important ones for all informed citizens. Finally, chemistry—the properties of elements and compounds, the nanoscale theories and models that interpret those properties, and the changes of one kind of substance into another—is just plain interesting and fun.

Chemistry, and the chemist's way of thinking, can help answer a broad range of questions. Here are some that have occurred to us.

- How can a disease be caused or cured by a tiny change in a molecule?
- Why does rain fall as drops instead of cubes or cylinders?
- Why can some animals walk on ceilings or walls?
- Why does salt help to clear snow and ice from roads?
- Why do droplets of water form on the outside of a cold soft drink can?
- Where does the energy come from to make my muscles work?
- What are the molecules in my eyes doing when I watch a movie?
- Why does frost form on top of a parked car in winter, but not on the sides?
- Why is the sky blue?
- Why can some insects walk on water?
- Why is ozone depletion bad? I thought too much ozone was bad for your lungs.
- How does soap help to get clothes clean?
- What happens when I hard-boil an egg?
- How are plastics made, and why are there so many different kinds?

There are probably many more that you have thought of. We encourage you to add them to the list and think about them as you study this book.

You will be called upon to make many decisions in your life for your own good, or for the good of those in your community—whether that is your local community or the global community. An understanding of the nature of science in general, and of chemistry in particular, can only serve to help in these decisions. We believe that knowledge of chemistry and its methods can help you begin to analyze the risks and benefits from your own decisions and from the actions of business, industry, and government.

IN CLOSING

Having studied this chapter, you should be able to . . .

- Describe the approach used by scientists in solving problems (Sections 1.1, 1.2).
- Understand the differences among a hypothesis, a theory, and a law (Section 1.2).
- Define quantitative and qualitative observations (Section 1.2).
- Identify the physical properties of matter or physical changes occurring in a sample of matter (Section 1.3).
- Estimate Celsius temperatures for commonly encountered situations (Section 1.3).
- Calculate mass, volume, or density, given any two of the three (Section 1.3).
- Identify the chemical properties of matter or chemical changes occurring in a sample of matter (Section 1.4).
- Explain the difference between homogeneous and heterogeneous mixtures (Section 1.5).
- Describe the importance of separation, purification, and analysis (Section 1.5).
- Understand the difference between a chemical element and a chemical compound (Sections 1.6 and 1.8).
- Classify matter (Figure 1.13).
- Describe characteristic properties of the three states of matter—gases, liquids, and solids (Section 1.7).
- Identify relative sizes at the macroscale, microscale, and nanoscale levels (Section 1.7).
- Describe the kinetic-molecular theory at the nanoscale level (Section 1.7).
- Use the postulates of modern atomic theory to explain macroscopic observations about elements, compounds, conservation of mass, constant composition, and multiple proportions (Section 1.8)
- Distinguish metals, nonmetals, and metalloids according to their properties (Section 1.9).
- Identify elements that consist of molecules, and define allotropes (Section 1.9).
- Distinguish among macroscale, nanoscale, and symbolic representations of substances and chemical processes (Section 1.10).
- Appreciate the balance of benefits and risks in our technological world (Section 1.11).

KEY TERMS

The following terms were defined and given in boldface type in this chapter. Be sure to understand each of these terms and the concepts with which they are associated. (The number of the section where each term is introduced is given in parentheses.)

allotrope *(1.9)*

atom *(1.8)*

boiling point *(1.3)*

Celsius temperature scale *(1.3)*

chemical change *(1.4)*

chemical compounds *(1.6)*

chemical element *(1.6)*

chemical formula *(1.9)*

chemical property *(1.4)*

chemical reaction *(1.4)*

chemistry *(Introduction)*

conservation of mass, law of *(1.8)*

constant composition, law of *(1.8)*

conversion factor *(1.3)*

density *(1.3)*

diatomic molecule *(1.9)*

dimensional analysis *(1.3)*

energy *(1.4)*

gas *(1.7)*

heterogeneous mixture *(1.5)*

homogeneous mixture *(1.5)*

hypothesis *(1.2)*

kinetic-molecular theory *(1.7)*

law *(1.2)*

liquid *(1.7)*

macroscale *(1.7)*
matter *(Introduction)*
melting point *(1.3)*
metal *(1.9)*
metalloid *(1.9)*
microscale *(1.7)*
model *(1.2)*
molecule *(1.9)*
multiple proportions,
 law of *(1.8)*

nanoscale *(1.7)*
nonmetal *(1.9)*
physical changes *(1.3)*
physical properties
 (1.3)
product *(1.4)*
proportionality factor
 (1.3)
qualitative *(1.2)*
quantitative *(1.2)*

reactant *(1.4)*
solid *(1.7)*
solution *(1.5)*
substance *(1.3)*
temperature *(1.3)*
theory *(1.2)*

QUESTIONS FOR REVIEW AND THOUGHT

Conceptual Challenge Problems

CP-1.A (Section 1.2) Some people use expressions such as "a rolling stone gathers no moss" and "where there is no light there is no life." Why do you believe these are "laws of nature"?

CP-1.B (Section 1.2) Parents teach their children to wash their hands before eating. (a) Do all parents accept the germ theory of disease? (b) Are all diseases caused by germs?

CP-1.C (Section 1.7) In Section 1.7 you read that, on an atomic scale, all matter is in constant motion. (For example, the average speed of a molecule of nitrogen or oxygen in the air is greater than 1000 miles per hour at room temperature.) (a) What evidence can you put forward that supports the kinetic-molecular theory? (b) Suppose you accept the notion that molecules of air are moving at speeds near 1000 miles per hour. What can you then reason about the paths that these molecules take when moving at this speed?

CP-1.D (Section 1.7) One approach to determining whether the strange structures called nanobes (⬅ *p. 21*) are living is to estimate how many atoms and molecules could make up a nanobe. If the number is too small, then there would not be enough DNA, protein, and other biological molecules to carry out life processes. To test this method, estimate an upper limit for the number of atoms that could be in a nanobe. (Use a small atom, such as hydrogen.). Also estimate how many protein molecules could fit inside a nanobe. Do your estimates rule out the possibility that a nanobe could be living? Explain why or why not.

CP-1.E (Section 1.11) The life expectancy of United States citizens in 1992 was 76 years. In 1916 the life expectancy was only 52 years. This is an increase of 46% in a lifetime. (a) Could this astonishing increase occur again? (b) To what single source would you attribute this noteworthy increase in life expectancy? Why did you identify this one source as being most influential?

Answers to questions in **bold** can be found in the back of the book.

Review Questions

1. The category of drugs called NSAIDs (nonsteroidal anti-inflammatory drugs) includes aspirin, ibuprofen, and naproxen. All have side effects similar to those of aspirin (damage to the stomach lining and kidneys). Give a nanoscale explanation of why the effects of all of these substances might be similar. Speculate about whether these molecules are likely to be greatly different in size.

2. What is meant by the structure of a molecule? Why are molecular structures important?

3. Why is it often important to know the structure of an enzyme? How can knowledge of enzyme structures be useful in medicine?

4. Choose an object in your room, such as a CD player or television set. Write down five qualitative observations and five quantitative observations regarding the object you chose.

5. What are three important characteristics of a scientific law? Name two laws that were mentioned in this chapter. State each of the laws that you named.

6. How does a scientific theory differ from a law? How are theories and models related?

7. When scientists first suggested that aspirin works by blocking the release of prostaglandins, were they stating a theory or a hypothesis?

8. List three specific examples of how chemistry is used in your major area of study in college.

9. What is the unique perspective that chemists use to make sense out of the material world? Give at least one example of how that perspective can be applied to a significant problem.

10. While camping in the mountains you build a small fire out of tree limbs you find on the ground near your campsite. The dry wood crackles and burns brightly and warms you. Before slipping into your sleeping bag for the night, you put the fire out by dousing it with cold water from a nearby stream. Steam rises when the water hits the hot coals. Describe the physical and chemical changes in this scene.

11. Give two examples of situations in which purity of a chemical substance is important.

12. Methods of chemical analysis continually improve so that smaller concentrations of contaminants can be detected. Why might this sometimes mislead us into thinking we are at risk from toxic substances when we really are not?

Is it reasonable to expect zero contamination or zero risk?

How Science Is Done

13. Identify the information in each sentence as qualitative or quantitative.
    (a) The element gallium melts at 29.8 °C.
    (b) A chemical compound containing cobalt and chlorine is blue.
    (c) Aluminum metal is a conductor of electricity.
    (d) The chemical compound ethanol boils at 79 °C.
    (e) A chemical compound containing lead and sulfur forms shiny plate-like yellow crystals.
14. In the photograph, make as many qualitative and quantitative observations as you can.

15. Which of these statements are qualitative? Which are quantitative? Explain your choice in each case.
    (a) Sodium is a silvery-white metal.
    (b) Aluminum melts at 660 °C.
    (c) Carbon makes up about 23% of the human body by mass.
    (d) Pure carbon occurs in different forms: graphite, diamond, and fullerenes.
16. Which of the these statements are qualitative? Which are quantitative? Explain your choice in each case.
    (a) The atomic mass of carbon is 12.011 amu (atomic mass units).
    (b) Pure aluminum is a silvery-white metal that is non-magnetic, has a low density, and does not produce sparks when struck.
    (c) Sodium has a density of 0.968 g/mL.
    (d) In animals the sodium cation is the main extracellular cation and is important for nerve function.

Physical Properties of Matter

17. In the accompanying photo, you see a crystal of the mineral calcite surrounded by piles of calcium and carbon, two of the elements that combine to make the mineral. (The other element combined in calcite is oxygen.) Based on the photo, describe some of the physical properties of the elements and the mineral. Are any properties the same? Are any properties different?

Calcite (the clear crystal) and two of its constituent elements, calcium (chips) and carbon (black grains). The calcium chips are covered with a thin film of calcium oxide.

18. Solid gallium has a melting point of 29.8 °C. If you hold this metal in your hand, what will be its physical state? That is, will it be a solid or a liquid? Explain briefly.
19. Which temperature is higher?
    (a) 20 °C or 20 °F
    (b) 100 °C or 180 °F
    (c) 60 °C or 100 °F
    (d) −12 °C or 20 °F
20. Which has the higher temperature, a sample of water at 65 °C or a sample of iron at 65 °F?
21. The following temperatures are measured at various locations during the winter in North America: −10 °C at Montreal, 28 °F at Chicago, 20 °C at Charlotte, and 40 °F at Philadelphia. Which city is the warmest? Which city is the coldest?
22. A 105.5-g sample of a metal was placed into water in a graduated cylinder, and it completely submerged. The water level rose from 25.4 mL to 37.2 mL. Use data in Table 1.1 to identify the metal.
23. An irregularly shaped piece of lead weighs 10.0 g. It is carefully lowered into a graduated cylinder containing 30.0 mL of ethanol, and it sinks to the bottom of the cylinder. To what volume reading does the ethanol rise?
24. An unknown sample of a metal is 1.0 cm thick, 2.0 cm wide, and 10.0 cm long. Its mass is 54.0 g. Use data in Table 1.1 to identify the metal. (Remember that 1 cm³ = 1 mL.)
25. Calculate the volume of a 23.4-g sample of bromobenzene.
26. Calculate the mass of the sodium chloride crystal in the photo that accompanies Question 43 if the dimensions of the crystal are 10 cm thick by 12 cm long by 15 cm wide. (Remember that 1 cm³ = 1 mL.)

Chemical Changes and Chemical Properties

27. In each case, identify the underlined property as a physical or chemical property. Give a reason for your choice.

(a) The normal <u>color</u> of the element bromine is red-orange.

(b) Iron is <u>transformed into rust</u> in the presence of air and water.

(c) Dynamite can <u>explode.</u>

(d) Aluminum metal, the <u>shiny</u> "foil" you use in the kitchen, <u>melts</u> at 660 °C.

28. In each case, identify the underlined property as a physical or a chemical property. Give a reason for your choice.

(a) Dry Ice <u>sublimes</u> (changes directly from a solid to a gas) at −78 °C.

(b) Methanol (methyl alcohol) <u>burns in air</u> with a colorless flame.

(c) Sugar is <u>soluble in water.</u>

(d) Hydrogen peroxide, H_2O_2, <u>decomposes to form oxygen, O_2, and water, H_2O.</u>

29. In each case, describe the change as a chemical or physical change. Give a reason for your choice.

(a) A cup of household bleach changes the color of your favorite T-shirt from purple to pink.

(b) The fuels in the space shuttle (hydrogen and oxygen) combine to give water and provide the energy to lift the shuttle into space.

(c) An ice cube in your glass of lemonade melts.

30. In each case, describe the change as a chemical or physical change. Give a reason for your choice.

(a) Salt dissolves when you add it to water.

(b) Food is digested and metabolized in your body.

(c) Crystalline sugar is ground into a fine powder.

(d) When potassium is added to water there is a purplish-pink flame and the water becomes basic (alkaline).

31. In each situation, decide whether a chemical reaction is releasing energy and causing work to be done, or whether an outside source of energy is forcing a chemical reaction to occur.

(a) Your body converts excess intake of food into fat molecules.

(b) Sodium reacts with water as shown in Figure 1.5.

(c) Sodium azide in an automobile air bag decomposes, causing the bag to inflate.

(d) An egg is hard-boiled on your kitchen stove.

Substances, Mixtures, and Separation

32. Small chips of iron are mixed with sand (see photo). Is this a homogeneous or heterogeneous mixture? Suggest a way to separate the iron and sand from each other.

Layers of sand, iron, and sand.

33. Suppose that you have a solution of sugar in water. Is this a homogeneous or heterogeneous mixture? Describe an experimental procedure by which you can separate the two substances.

34. Identify each of the following as a homogeneous or a heterogeneous mixture.

(a) Vodka

(b) Blood

(c) Cowhide

(d) Bread

35. Identify each of the following as a homogeneous or a heterogeneous mixture.

(a) An asphalt (blacktop) road

(b) Clear ocean water

(c) Iced tea with ice cubes

(d) Filtered apple cider

36. The black ink in some pens is a mixture of several different colors. Describe an experiment by which you could determine whether black ink is a single substance or a mixture of two or more.

Elements and Compounds

37. For each of the changes described, decide whether two or more elements formed a compound or if a compound decomposed (to form elements or other compounds). Explain your reasoning in each case.

(a) Upon heating, a blue powder turned white and lost mass.

(b) A white solid forms three different gases when heated. The total mass of the gases is the same as that of the solid.

38. For each of the changes described, decide whether two or more elements formed a compound or if a compound decomposed (to form elements or other compounds). Explain your reasoning in each case.

(a) After a reddish-colored metal is placed in a flame, it turns black and has a higher mass.

(b) A white solid is heated in oxygen and forms two gases. The mass of gases is the same as the masses of the solid and the oxygen.

39. Classify each of the following with regard to the type of matter (element, compound, heterogeneous mixture, or homogeneous mixture). Explain your choice in each case.

(a) A piece of newspaper

(b) Solid, granulated sugar

(c) Freshly squeezed orange juice

(d) Gold jewelry

40. Classify each of the following with regard to the type of matter (element, compound, heterogeneous mixture, or homogeneous mixture). Explain your choice in each case.

(a) A cup of coffee

(b) A soft drink such as a Coke or Pepsi

(c) A piece of Dry Ice (a solid form of carbon dioxide)

41. Classify each of the following as an element, a compound, a heterogeneous mixture, or a homogeneous mixture. Explain your choice in each case.

(a) Chunky peanut butter   (b) Distilled water

(c) Platinum   (d) Air

42. Classify each of the following as an element, a compound, a heterogeneous mixture, or a homogeneous mixture. Explain your choice in each case.
    (a) Table salt (sodium chloride)
    (b) Methane (which burns in pure oxygen to form only carbon dioxide and water).
    (c) Chocolate chip cookie
    (d) Silicon

Nanoscale Models: Solids, Liquids, Gases

43. The accompanying photo shows a crystal of the mineral halite, a form of ordinary salt. Are these crystals in the macroscale, microscale, or nanoscale world? How would you describe the shape of these crystals? What might this tell you about the arrangement of the atoms deep inside the crystal?

A halite (sodium chloride) crystal.

44. Galena, shown in the photo, is a black mineral that contains lead and sulfur. It shares its name with a number of towns in the United States; they are located in Alaska, Illinois, Kansas, Maryland, Missouri, and Ohio. How would you describe the shape of the galena crystals? What might this tell you about the arrangement of the atoms deep inside the crystal?

A galena (lead sulfide) crystal.

45. The photograph shows a bacterium that is approximately 1 μm from top to bottom. Is this at the macroscale, the microscale, or the nanoscale? (1 μ = 1 μm)

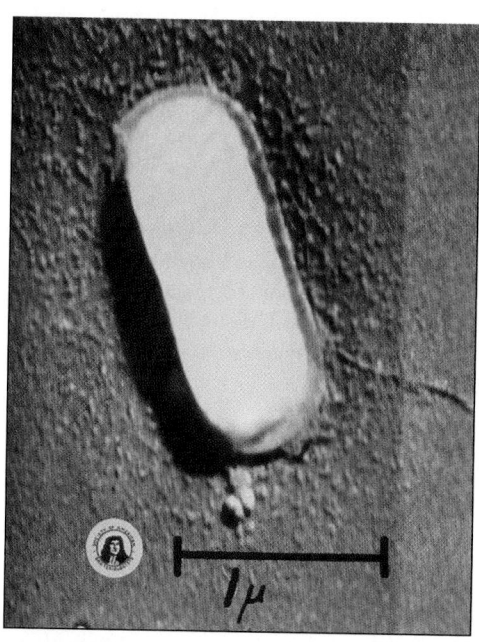

A bacterium.

46. What are the states of matter, and how do they differ from one another? Describe the characteristic properties of each state of matter.

47. When you open a can of soft drink, the carbon dioxide gas inside expands rapidly as it rushes from the can. Describe this process in terms of the kinetic-molecular theory.

48. After you wash your clothes, you hang them on a line in the sun to dry. Describe the change or changes that occur in terms of the kinetic-molecular theory. Are the changes that occur physical or chemical changes?

49. Sucrose has to be heated to a high temperature before it caramelizes. Use the kinetic-molecular theory to explain why sugar caramelizes only at high temperatures.

Atomic Theory

50. Explain in your own words, by writing a short paragraph, how the atomic theory explains conservation of mass during a chemical reaction and during a physical change.

51. Explain in your own words, by writing a short paragraph, how the atomic theory explains constant composition of chemical compounds.

52. State the four postulates of the modern atomic theory.

53. Explain in your own words, by writing a short paragraph, how the atomic theory predicts the law of multiple proportions.

54. State the law of multiple proportions in your own words.

55. The element chromium forms three different oxides (that contain only chromium and oxygen). The percentage of chromium (number of grams of chromium in 100 g of oxide) in these compounds is 52.0%, 68.4%, and 76.5%. Do

these data conform to the law of multiple proportions? Explain why or why not.

The Chemical Elements

56. Name and give the symbols for two elements that
(a) Are metals
(b) Are nonmetals
(c) Are metalloids
(d) Consist of diatomic molecules

57. Name and give the symbols for two elements that
(a) Are gases at room temperature
(b) Are solids at room temperature
(c) Do not consist of molecules
(d) Have different allotropic forms

Chemical Symbolism

58. Write a chemical formula for each substance, and draw a nanoscale picture of how the molecules are arranged at room temperature.
(a) Water, a liquid whose molecules contain two hydrogen atoms and one oxygen atom each
(b) Nitrogen, a gas that consists of diatomic molecules
(c) Neon
(d) Chlorine

59. Write a chemical formula for each substance and draw a nanoscale picture of how the molecules are arranged at room temperature.
(a) Iodine, a solid that consists of diatomic molecules
(b) Ozone
(c) Helium
(d) Carbon dioxide

60. Write a nanoscale representation and a symbolic representation and describe what happens at the macroscale when hydrogen reacts chemically with oxygen to form water vapor.

61. Write a nanoscale representation and a symbolic representation and describe what happens at the macroscale when carbon monoxide reacts with oxygen to form carbon dioxide.

62. Write a nanoscale representation and a symbolic representation and describe what happens at the macroscale when iodine sublimes (passes directly from solid to gas with no liquid formation) to form iodine vapor.

63. Write a nanoscale representation and a symbolic representation and describe what happens at the macroscale when bromine evaporates to form bromine vapor.

Risks and Benefits

64. Consider each of the following statements and decide (1) whether the benefits outweigh the risks or the risks outweigh the benefits and (2) who benefits and who is at risk. Give several reasons for your choice.
(a) Several new anti-AIDS drugs, which could prolong the lives of many individuals who are HIV-positive, have been discovered. The FDA, however, will not approve their use until they have been tested thoroughly.
(b) Potato chips cooked in Olestra, a fat substitute, have fewer calories and zero grams of fat compared with regular potato chips. They are also known to cause gastrointestinal side effects in many people.

65. Consider each of the following statements and decide (1) whether the benefits outweigh the risks or the risks outweigh the benefits and (2) who benefits and who is at risk. Give several reasons for your choice.
(a) We live in an age of plastics, but plastics are causing a major landfill problem in the United States. We should do away with plastics and return to the "good old days" of natural materials.
(b) Much of the country gets its electrical energy from coal-fired power plants. Coal-fired plants are a major contributor to acid rain and global warming.
(c) As the world population grows, so does the demand for food. The increased use of pesticides has helped farmers get more crops per acre. This also increases the chance of pesticides leaching out of the soil and ending up in our rivers, lakes, and aquifers.

Why Care About Chemistry?

66. Make a list of at least four questions you have wondered about that may involve chemistry. Compare your list with a list from another student taking the same chemistry course. Evaluate the quality of each other's questions and how "chemical" they are.

67. Make a list of at least four issues faced by our society that require scientific studies and scientific data before a democratic society can make informed, rational decisions. Exchange lists with another student and evaluate the quality of each other's choices.

General Questions

68. Classify the information in each of the following statements as quantitative or qualitative and as relating to a physical or chemical property.
(a) A white chemical compound has a mass of 1.456 grams. When placed in water containing a dye, it causes the red color of the dye to fade to colorless.
(b) A sample of lithium metal, with a mass of 0.6 gram, was placed in water. The metal reacted with the water to produce the compound lithium hydroxide and the element hydrogen.

69. Classify the information in each of the following statements as quantitative or qualitative and as relating to a physical or chemical property.
(a) A liter of water, colored with a purple dye, was passed through a charcoal filter. The charcoal adsorbed the dye, and colorless water came through. Later, the purple dye was removed from the charcoal and retained its color.

(b) When a white powder dissolved in a test tube of water, the test tube felt cold. Hydrochloric acid was then added, and a white solid formed.

70. The density of solid potassium is 0.86 g/mL. The density of solid calcium is 1.55 g/mL, almost twice as great. However, the mass of a potassium atom is only slightly less than the mass of a calcium atom. Provide a nanoscale explanation of these facts.

71. Describe in your own words how different allotropic forms of an element are different at the nanoscale.

72. Describe the structures of diamond, graphite, and buckyball.

Applying Concepts

73. Using Table 1.1, but without using your calculator, decide which has the larger mass:
(a) 20. mL of butane or 20. mL of bromobenzene
(b) 10. mL of benzene or 1.0 mL of gold
(c) 0.732 mL of copper or 0.732 mL of lead

74. Using Table 1.1, but without using your calculator, decide which has the larger volume:
(a) 1.0 g of ethanol or 1.0 g of bromobenzene
(b) 10. g of aluminum or 12. g of water
(c) 20 g of gold or 40 g of magnesium

75. At 25 °C the density of water is 0.997 g/mL, whereas the density of ice at −10 °C is 0.917 g/mL.
(a) If a plastic soft-drink bottle (volume = 250 mL) is completely filled with pure water, capped, and then frozen at −10 °C, what volume will the solid occupy?
(b) What will the bottle look like when you take it out of the freezer?

76. Of the substances listed in Table 1.1, which would not float on liquid mercury? (Assume that none of the substances would dissolve in the mercury.)

77. Which of the substances in Figure 1.2 has the greatest density? Which has the lowest density?

78. Water does not dissolve in bromobenzene.
(a) If you pour 2 mL of water into a test tube that contains 2 mL of bromobenzene, which liquid will be on top?
(b) If you pour 2 mL of benzene carefully into the test tube with bromobenzene and water described in part (a) without shaking or mixing the liquids, what will happen?
(c) If you pour 2 mL of ethanol carefully into the test tube with bromobenzene and water described in part (a) without shaking or mixing the liquids, what will happen?
(d) What will happen if you thoroughly stir the mixture in part (c)?

79. Give a nanoscale interpretation of the fact that at the melting point the density of solid mercury is greater than the density of liquid mercury, and at the boiling point the density of liquid mercury is greater than the density of gaseous mercury.

80. At the top of the next column is a nanoscale view of the atoms of mercury in a thermometer registering 10 °C.

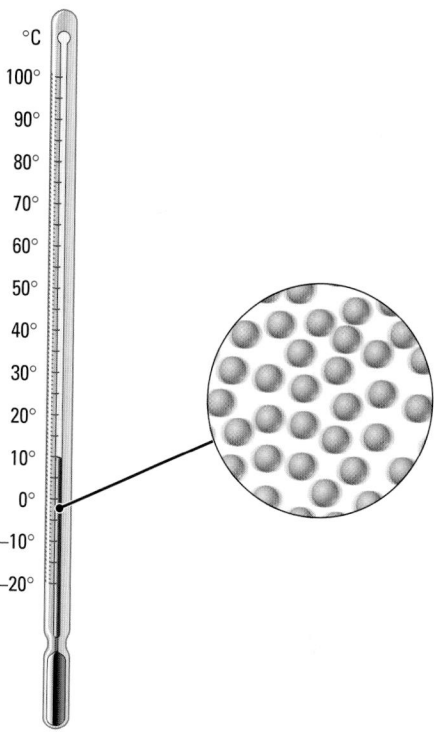

Which nanoscale drawing best represents the atoms in the liquid in this same thermometer at 90 °C? (Assume that the same volume of liquid is shown in each nanoscale drawing.)

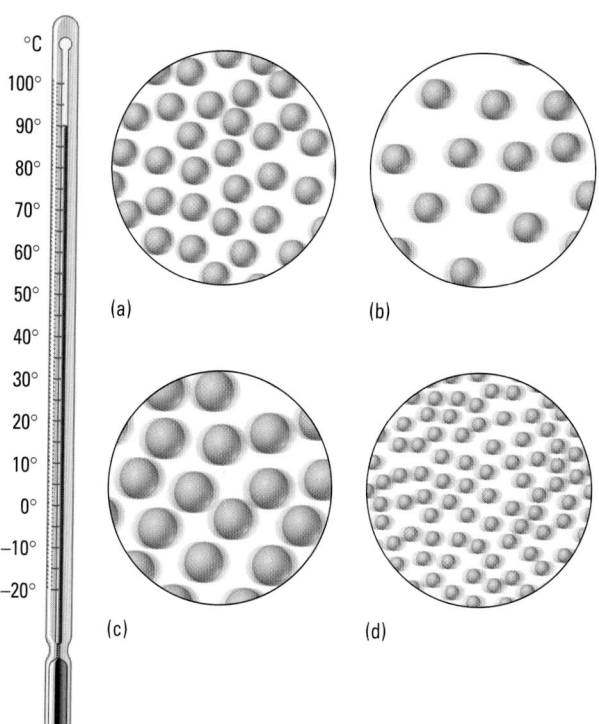

81. Answer the following questions using figures (a) to (i). (Each question may have more than one answer.)

(a) Which represents nanoscale particles in a sample of solid?

(b) Which represents nanoscale particles in a sample of liquid?

(c) Which represents nanoscale particles in a sample of gas?

(d) Which represents nanoscale particles in a sample of an element?

(e) Which represents nanoscale particles in a sample of a compound?

(f) Which represents nanoscale particles in a sample of a pure substance?

(g) Which represents nanoscale particles in a sample of a mixture?

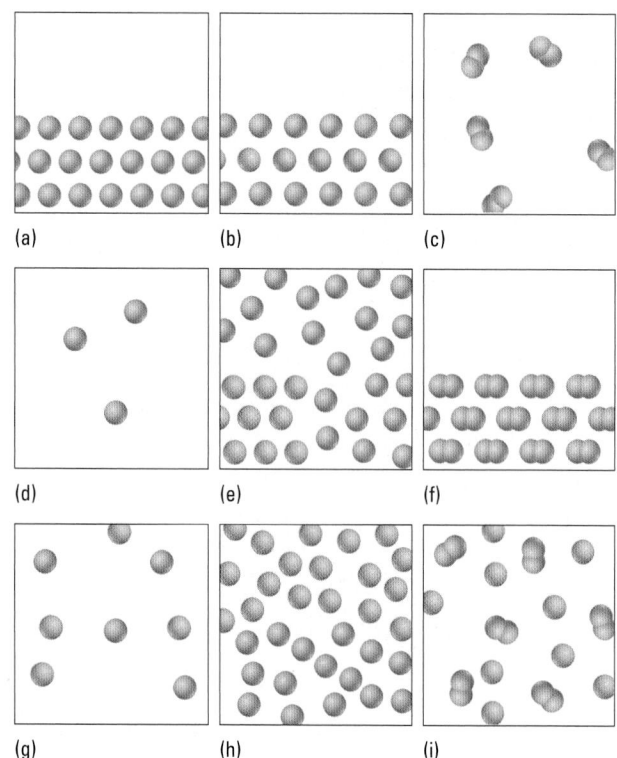

(a) (b) (c)

(d) (e) (f)

(g) (h) (i)

Media Questions

82. Find four articles, advertisements, or cartoons in a newspaper or magazine that are directly related to chemistry.

83. Find four Internet sites that are directly related to chemistry.

84. From one or more of the Internet sites listed below, obtain eight facts about aspirin or aspirin substitutes that were not given in this chapter.

http://www.bris.ac.uk/Depts/Chemistry/MOTM/ aspirin/aspirin1.htm

http://advance.byu.edu/BYM/1999/99spring/ images/superaspirin.html

http://arthritis.miningco.com/health/arthritis/ library/weekly/aa062398.htm

http://www.sciam.com/1999/0599issue/ 0599working.html

85. From one or more of the Internet sites listed below, obtain four facts about celecoxib or meloxicam that were not given in this chapter.

http://www.searlehealthnet.com/pi/celebrex/ celebrex.html

http://www.infomed.org/pharma-kritik-e/ abs0918.html

http://www.boehringer-ingelheim.com/ corporate/products/products.htm (search for meloxicam)

http://www.abbott.com/ (choose pain management, then mobic)

86. Search the Internet for sites related to nanoscale or nanotechnology. How many sites are available? Based on your search, is this a hot business area?

General Chemistry CD-ROM

CD1.1 Screen 1.8: Density. Explain why there is a difference in the density of the brick and the Styrofoam block shown on this screen, considering their structures on the molecular scale.

CD1.2 Screen 1.8: Density. The densities of the elements are listed in the *Periodic Table* database. (Click on the "Tools" button.)

(a) What is the density of aluminum at 25 °C?

(b) What is the density of uranium at 25 °C?

(c) Look up the densities of lead, platinum, and mercury at 25 °C. Which is the densest? Which is the least dense?

CD1.3 Screen 1.12: Chemical Changes on the Molecular Scale. Examine the video of the reaction of elemental phosphorus and chlorine and the animation of this reaction. Note that the animation is not meant to illustrate *how* the reaction occurs, but only that the number of P and Cl atoms is the same before and after reaction.

(a) How many P_4 molecules are there in the beginning?

(b) How many Cl_2 molecules are there in the beginning?

(c) How many PCl_3 molecules are formed?

CD1.4 Screen 1.5: Chemical Elements.

(a) How many elements are presently known?

(b) How many elements exist in nature?

(c) Five different elements are pictured on this screen. Under "normal" conditions, which ones are solid? Which ones are liquid? Which ones are gases?

CD1.5 Screen 1.15: Units of Measurement. The speedometer in your car is probably marked off in two scales (such as the one on this screen). One scale is in units of miles per hour (white numerals on this screen) and the other is kilometers per hour (yellow numerals on this screen).

(a) Does either scale use SI units? How should the speedometer be marked if SI units are used?

(b) Use the speedometer to arrive at the approximate relation between kilometers and miles. That is, how many kilometers are there, approximately, per mile?

An image of a strand of DNA, the molecule that carries genetic information in living organisms. The image was generated by a scanning tunneling microscope (STM), which can detect individual atoms or molecules, allowing us to make images of nanoscale atomic arrangements. The peaks in the curves correspond to individual atoms, which are 3.5 nm apart in the DNA molecule.

ATOMS AND ELEMENTS

2.1 Atomic Structure and Subatomic Particles

2.2 The Nuclear Atom

2.3 The Sizes of Atoms and the Units Used to Represent Them

2.4 Atomic Numbers and Mass Numbers

2.5 Isotopes and Atomic Weight

2.6 Amounts of Substances — The Mole

2.7 Molar Mass and Problem Solving

2.8 The Periodic Table

Early theories of the atom considered atoms to be indivisible, but we know now this is wrong. Elements are different from one another because of differences in the internal structure of their atoms. Under the right conditions, smaller particles within atoms — known as *subatomic particles* — can be removed or rearranged. The term **atomic structure** refers to the identity and arrangement of these subatomic particles in the atom. An understanding of the details of atomic structure aids in the understanding of how atoms combine to form compounds, are rearranged in chemical reactions, and account for the properties of materials. The next few sections describe how experiments support the idea that atoms are composed of smaller (subatomic) particles.

2.1 ATOMIC STRUCTURE AND SUBATOMIC PARTICLES

Electricity played an important role in many of the experiments from which the theory of atomic structure was derived. Two types of electrical charge exist — positive and negative. *Electrical charges of the same type repel one another, and charges of the opposite type attract one another.* A positively charged particle repels another positively charged particle. Likewise, two negatively charged particles repel each other. In contrast, two particles with opposite signs attract each other.

CD-ROM Screen 2.6: Electricity and Electric Charge: Attraction and Repulsion

> **Exercise 2.1   Electric Charge**
>
> When you comb your hair on a dry day, your hair sticks to the comb. How could you explain this behavior in terms of a nanoscale model in which atoms contain positive and negative charges?

Radioactivity

In 1896 Henri Becquerel discovered that a uranium ore emitted rays that exposed a photographic plate, even though the plate was covered by a protective black paper. In 1898 Marie and Pierre Curie isolated the new elements polonium and radium, which emitted the same kind of rays. Marie suggested that atoms of such elements spontaneously emit these rays and named the phenomenon **radioactivity.**

Radioactive elements can emit three types of radiation: alpha (α), beta (β), and gamma (γ) rays. These radiations behave differently when passed between electrically charged plates (Figure 2.1). Alpha and beta rays are deflected, but gamma rays are not. These observations can be explained by assuming that alpha rays and beta rays are composed of charged particles that come from within the radioactive atom. Alpha rays have a $+2$ charge, and beta rays have a -1 charge. Alpha rays and beta rays are particles because they have mass — they are matter. In the experiment shown in Figure 2.1, alpha particles are deflected less and so must be heavier than beta particles. Gamma rays have no detectable charge or mass — they behave like light rays. If radioactive atoms can break apart to produce subatomic alpha and beta particles, then there must be something smaller inside the atom.

CD-ROM Screen 2.7: Evidence of Subatomic Particles: Radioactivity

Electrons

CD-ROM Screen 2.8: Electrons

Further evidence that atoms are composed of subatomic particles came from experiments with specially constructed glass tubes called cathode-ray tubes. Most of the air has been removed from these tubes and a metal electrode sealed into each

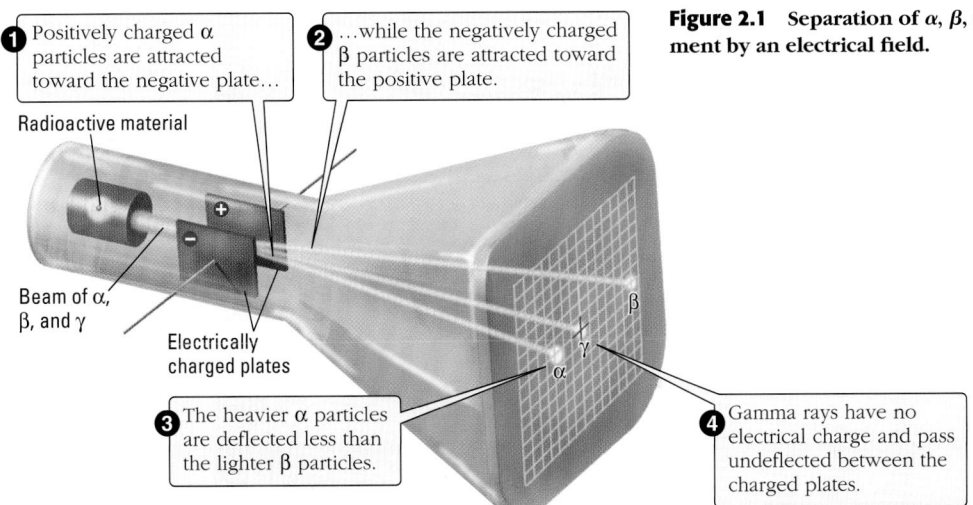

① Positively charged α particles are attracted toward the negative plate...

② ...while the negatively charged β particles are attracted toward the positive plate.

Radioactive material

Beam of α, β, and γ

Electrically charged plates

③ The heavier α particles are deflected less than the lighter β particles.

④ Gamma rays have no electrical charge and pass undeflected between the charged plates.

Figure 2.1 Separation of α, β, and γ rays from a radioactive element by an electrical field.

end. When a sufficiently high voltage is applied to the electrodes, a beam of rays flows from the negatively charged electrode (the *cathode*) to the positively charged electrode (the *anode*). These rays, known as cathode rays, are coming directly from the metal atoms of the cathode. The cathode rays travel in straight lines, are attracted toward positively charged plates, can be deflected by a magnetic field, cast sharp shadows, can heat metal objects red hot, and cause gases and fluorescent materials to glow. When cathode rays strike a fluorescent screen, the energy transferred causes light to be given off as tiny flashes. Thus, the properties of a cathode ray are those of a beam of negatively charged particles, each of which produces a light flash when it hits a fluorescent screen. Sir Joseph John Thomson suggested that cathode rays consist of the same particles that had earlier been named **electrons** and had been suggested to be the carriers of electricity. He also observed that cathode rays were produced from electrodes made of many different metals. This implied that electrons are constituents of the atoms of many elements.

In 1897 Thomson used a specially designed cathode-ray tube to simultaneously apply electric and magnetic fields to a beam of cathode rays. By balancing the electric field against the magnetic field and using the basic laws of electricity and magnetism, Thomson calculated the *ratio* of mass to charge for the electrons in the cathode-ray beam: 5.60×10^{-9} grams per coulomb (g/C). (The coulomb, C, is a fundamental unit of electrical charge.)

Fourteen years later, Robert Millikan used a cleverly devised experiment to measure the charge of an electron (Figure 2.2). Tiny oil droplets were sprayed into a chamber. As they settled slowly through the air, the droplets were exposed to x-rays, which caused electrons to be transferred from gas molecules in the air to the droplets. Using a small telescope to observe individual droplets, Millikan adjusted the electric charge of plates above and below the droplets so that the electrostatic attraction just balanced the gravitational attraction. In this way he could suspend a single droplet motionless. From equations describing these forces, Millikan calculated the charge on the droplet. Different droplets had different charges, but Millikan found that each was an integer multiple of the smallest charge. The smallest charge was 1.60×10^{-19} C. Millikan assumed this to be the fundamental quantity of charge, the charge on an electron. Given this value and the mass-to-charge ratio determined by Thomson, the mass of an electron could be computed: $(1.60 \times 10^{-19} \text{ C})(5.60 \times 10^{-9} \text{ g/C}) = 8.96 \times 10^{-28}$ g. The currently accepted most accurate value for the electron's mass is $9.10938188 \times 10^{-28}$ g, and the currently accepted most accurate value for the electron's charge is $-1.602176462 \times 10^{-19}$ C. This value is defined as the standard *electron charge*

The deflection of cathode rays by charged plates is used to create the picture on a television screen or a computer cathode-ray tube screen.

 CD-ROM Screen 2.9: Mass of the Electron

See Appendix A.2 for a review of scientific notation, which is used to represent very small or very large numbers as powers of 10. For example, 0.000001 is 1×10^{-6}, and 2,000,000 is 2×10^{6}.

Figure 2.2 Millikan oil-drop experiment. From the known mass of the droplets and the applied voltage at which the charged droplets were held stationary, Millikan could calculate the charges on the droplets.

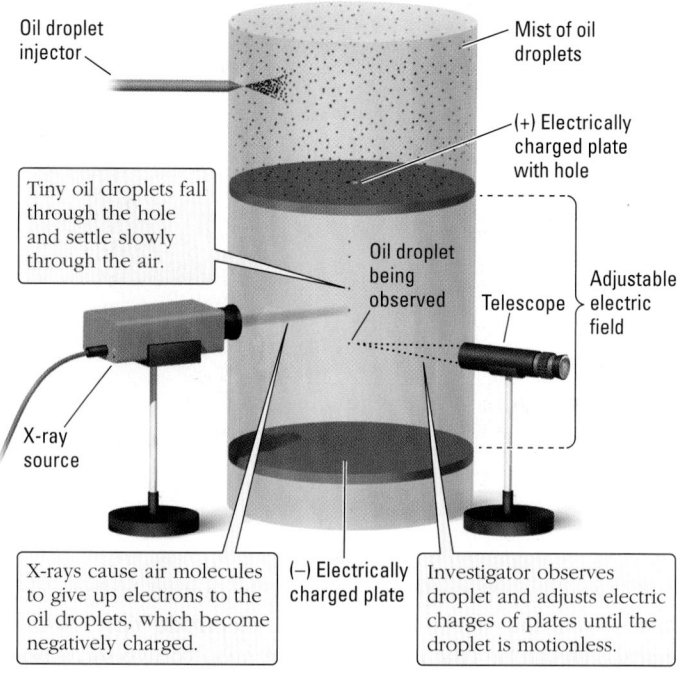

Oil droplet injector

Mist of oil droplets

(+) Electrically charged plate with hole

Tiny oil droplets fall through the hole and settle slowly through the air.

Oil droplet being observed

Telescope

Adjustable electric field

X-ray source

X-rays cause air molecules to give up electrons to the oil droplets, which become negatively charged.

(–) Electrically charged plate

Investigator observes droplet and adjusts electric charges of plates until the droplet is motionless.

Electron
Charge = –1
Mass = 9.1094×10^{-28} g

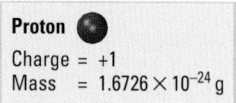

CD-ROM Screen 2.10: Protons

As mass increases, mass-to-charge ratio increases for a given amount of charge. For a fixed charge, doubling the mass will double the mass-to-charge ratio. For a fixed mass, doubling the charge will halve the mass-to-charge ratio.

Proton
Charge = +1
Mass = 1.6726×10^{-24} g

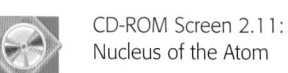

CD-ROM Screen 2.11: The Nucleus of the Atom

and is represented by –1. For convenience, the charges of atomic and subatomic particles are then given in multiples of –1 or +1.

Other experiments provided further evidence that the electron is a *fundamental* particle of matter—that is, it is present in *all* matter. The beta particles emitted by radioactive elements were found to have the same properties as cathode rays, which are streams of electrons.

Protons

When atoms lose electrons, they become positively charged, and when they gain electrons they become negatively charged. Such atoms, or similarly charged groups of atoms, are known as **ions.** From experiments with positive ions, formed by knocking electrons out of atoms, the existence of a positively charged, fundamental particle was deduced. Positively charged particles with different mass-to-charge ratios were formed by atoms of different elements. The variation in masses showed that atoms of different elements must contain different numbers of positive particles. Those from hydrogen atoms had the smallest mass-to-charge ratio, indicating that they are the fundamental positively charged particles of atomic structure. Such particles are called **protons.** The mass of a proton is known from experiment to be $1.67262158 \times 10^{-24}$ g, which is about 1800 times the mass of an electron. The charge on a proton at $1.602176462 \times 10^{-19}$ C is equal in size, but opposite in sign, to the charge on an electron. The proton's charge is +1 in units of standard electronic charge.

2.2 THE NUCLEAR ATOM

The Nucleus

Once it was known that there were subatomic particles, the next question scientists wanted to answer was, how are these particles arranged in an atom? In about 1910 Ernest Rutherford devised an experiment (Figure 2.3) that led to a better understanding of atomic structure. Alpha particles (which have the same mass as he-

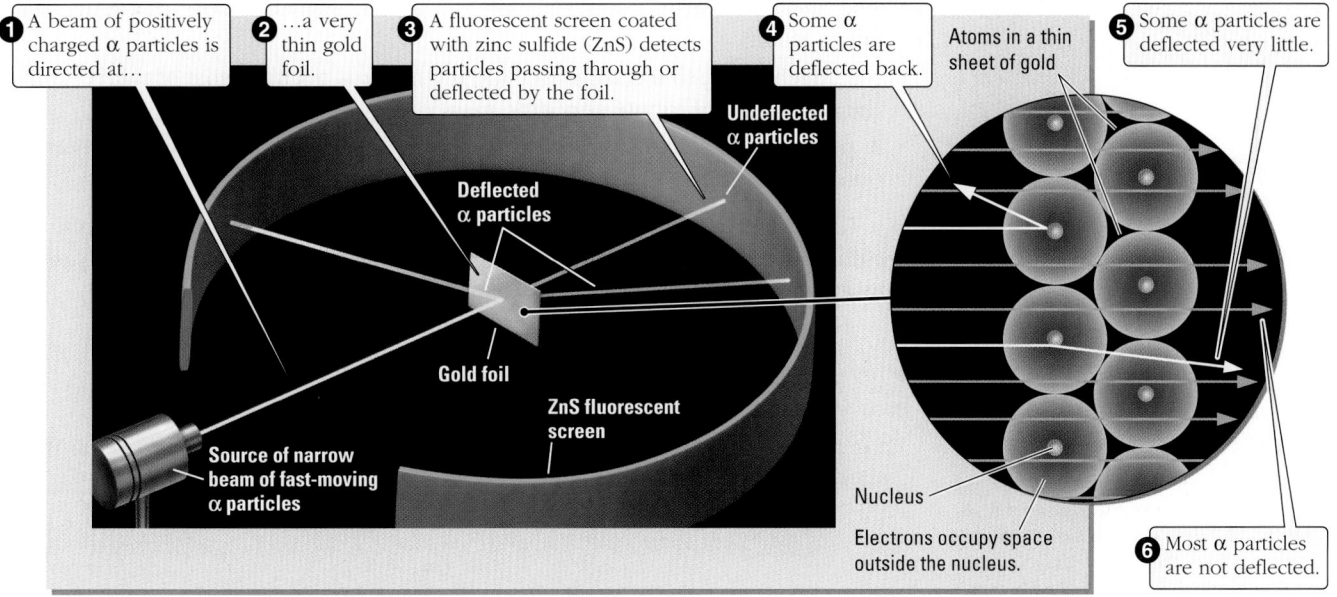

1 A beam of positively charged α particles is directed at...

2 ...a very thin gold foil.

3 A fluorescent screen coated with zinc sulfide (ZnS) detects particles passing through or deflected by the foil.

4 Some α particles are deflected back.

Atoms in a thin sheet of gold

5 Some α particles are deflected very little.

Undeflected α particles

Deflected α particles

Gold foil

ZnS fluorescent screen

Source of narrow beam of fast-moving α particles

Nucleus

Electrons occupy space outside the nucleus.

6 Most α particles are not deflected.

Figure 2.3   **The Rutherford experiment and its interpretation.**

lium atoms and a +2 charge) were allowed to hit a very thin sheet of gold foil. Almost all the alpha particles passed through undeflected. However, a very few alpha particles were deflected through large angles, and some came almost straight back toward the source. Rutherford described this unexpected result by saying, "It was about as credible as if you had fired a 15-inch [artillery] shell at a piece of paper and it came back and hit you."

The only way to account for the observations was to conclude that all of the positive charge and most of the mass of the atom are concentrated in a very small region (Figure 2.3). Rutherford called this tiny atomic core the **nucleus.** Only such a region could be sufficiently dense and highly charged to repel an alpha particle. From their results, Rutherford and his associates calculated values for the charge and radius of the gold nucleus. The currently accepted values are a charge of +79 and a radius of approximately 1×10^{-13} cm. This makes the nucleus about 10,000 times *smaller* than the atom. Most of the volume of the atom is occupied by the electrons. Somehow, the negative electrons occupy the space outside the nucleus, but their arrangement was unknown to Rutherford and scientists of the time. The arrangement is now well understood and is the subject of Chapter 7.

Alpha particles are four times heavier than the lightest atoms, which are hydrogen atoms.

Neutrons

Atoms are electrically neutral, so they must contain equal numbers of protons and electrons. However, most neutral atoms have masses greater than the sum of the mass of their protons and electrons. The additional mass indicates that subatomic particles with mass but no charge must also be present. Because they have no charge, these particles are more difficult to detect experimentally. In 1932 James Chadwick devised a clever experiment that detected the neutral particles by having them knock protons out of atoms and then detecting the protons. The neutral subatomic particles are called **neutrons.** They have no electric charge and a mass of $1.674928716 \times 10^{-24}$ g, nearly the same as the mass of a proton.

In summary, there are three primary subatomic particles: protons, neutrons, and electrons.

- Protons and neutrons make up the nucleus, providing most of the atom's mass; the protons provide all of its positive charge.

 CD-ROM Screen 2.12: Neutrons

In 1920 Ernest Rutherford proposed that the nucleus might contain an uncharged particle whose mass approximated that of a proton.

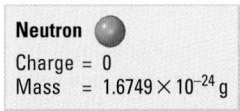

Neutron

Charge = 0
Mass   = 1.6749×10^{-24} g

ERNEST RUTHERFORD
(1871 – 1937)

Ernest Rutherford, born on a farm in New Zealand, earned his Ph.D. in physics from Cambridge University in 1895. He discovered alpha and beta radiation and coined the term half-life. For proving that alpha radiation is composed of helium nuclei and that beta radiation consists of electrons, he received the Nobel Prize in chemistry in 1908. As a professor at Cambridge University, he guided the work of no fewer than ten future Nobel Prize recipients. Element 104 is named in his honor.

CD-ROM Screen 1.16: The Metric System

Strictly speaking, the pound is a unit of weight rather than mass. The weight of an object depends on the local force of gravity. For measurements made at the earth's surface the distinction between mass and weight is not generally useful.

- The nuclear radius is approximately 10,000 times smaller than the radius of the entire atom.
- Negatively charged electrons outside the nucleus occupy most of the volume of the atom, but contribute very little mass.
- A neutral atom has no net electric charge, because the number of electrons outside the nucleus equals the number of protons inside the nucleus.

To chemists, the electrons are the important subatomic particles because they are the first part of the atom to contact another atom. The electrons largely dictate how atoms combine to form chemical compounds.

 Exercise 2.2   Describing Atoms

If an atom had a radius of 100 m, it would approximately fill a football stadium. (a) What would the approximate radius of the nucleus of such an atom be? (b) What common object is about that size?

2.3   THE SIZES OF ATOMS AND THE UNITS USED TO REPRESENT THEM

Atoms are extremely small. One teaspoon of water contains about three times as many atoms as the Atlantic Ocean contains teaspoons of water. It is important to understand the units used to express the sizes of very large and very small quantities.

To state the size of an object on the macroscale in the United States (for example, yourself) we would give your weight in pounds and your height in feet and inches. Pounds, feet, and inches are part of the measurement system used in the United States, but almost nowhere else in the world. Most of the world uses the **metric system** of units for recording and reporting measurements. The metric system is a decimal system that adjusts the size of its basic units by multiplying or dividing them by multiples of 10.

In the metric system, your weight (really, your mass) would be given in kilograms. The **mass** of an object is a fundamental measure of the quantity of matter in that object. The metric units for mass are *grams* or multiples or fractions of grams. The prefixes listed in Table 2.1 are used with all metric units. A *kilo*gram, for example, is equal to 1000 grams and is a convenient size for measuring the mass of a person.

For objects much smaller than people, prefixes that represent negative powers of 10 are used. For example, 1 *milli*gram equals 1×10^{-3} g.

$$1 \text{ } milli\text{gram (mg)} = \frac{1}{1000} \times 1 \text{ g} = 0.001 \text{ g} = 1 \times 10^{-3} \text{ g}$$

Individual atoms are too small to be weighed directly; their masses can be measured only by indirect experiments. An atom's mass is on the order of 1×10^{-22} g. For example, a sample of copper that weighs one *nano*gram (1 ng = 1×10^{-9} g) contains about 9×10^{12} copper atoms. The most sensitive laboratory balances can weigh samples of about 0.0000001 g (1×10^{-7} g = 0.1 μg).

Your height in metric units would be given in *meters,* the metric unit for length. Six feet is equivalent to 1.83 m. Atoms aren't nearly this big. Their sizes are reported in *pico*meters (1 pm = 1×10^{-12} m), and the radius of a typical atom is only between 30 and 300 pm. For example, the radius of a copper atom is 128 pm.

CD-ROM Screen 1.17: Using Numerical Information

TABLE 2.1 Prefixes Used in the SI and Metric Systems

Prefix	Abbreviation	Meaning	Example
Mega	M	10^6	1 megaton = 1×10^6 tons
Kilo	k	10^3	1 kilometer (km) = 1×10^3 meter (m)
			1 kilogram (kg) = 1×10^3 gram (g)
Deci	d	10^{-1}	1 decimeter (dm) = 1×10^{-1} meter (m)
			1 deciliter (dL) = 1×10^{-1} liter (L)
Centi	c	10^{-2}	1 centimeter (cm) = 1×10^{-2} meter (m)
Milli	m	10^{-3}	1 milligram (mg) = 1×10^{-3} gram (g)
Micro	μ	10^{-6}	1 micrometer (μm) = 1×10^{-6} m
Nano	n	10^{-9}	1 nanometer (nm) = 1×10^{-9} m
			1 nanogram (ng) = 1×10^{-9} g
Pico	p	10^{-12}	1 picometer (pm) = 1×10^{-12} m
Femto	f	10^{-15}	1 femtogram (fg) = 1×10^{-15} g

The International System of units (or SI units) is the officially recognized measurement system of science. It is derived from the metric system and is described in Appendix B. The units for mass, length, and volume are introduced here. Other units are introduced as they are needed in later chapters.

To get a feeling for these dimensions, consider how many copper atoms it would take to form a single file of copper atoms across a U.S. penny with a diameter of 1.90×10^{-2} m. This distance can be converted to picometers by using a conversion factor (⇐ *p. 9*) based on 1 pm = 1×10^{-12} m.

Conversion factors are the basis for dimensional analysis, a commonly used problem-solving technique. It is covered in detail in Appendix A.2.

$$1.90 \times 10^{-2} \, \cancel{m} \times \frac{1 \, pm}{1 \times 10^{-12} \, \cancel{m}} = 1.90 \times 10^{10} \, pm$$

conversion factor

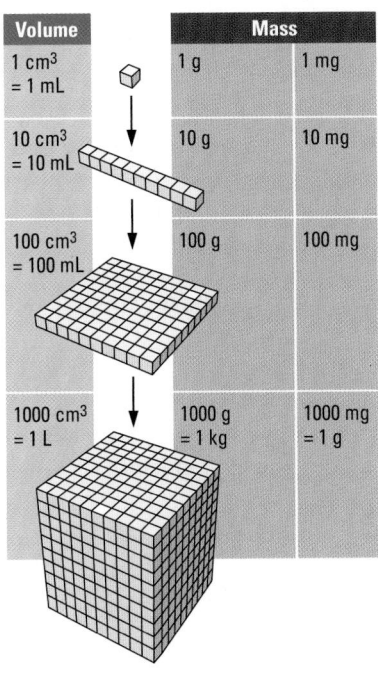

Relative sizes of mass and volume units.

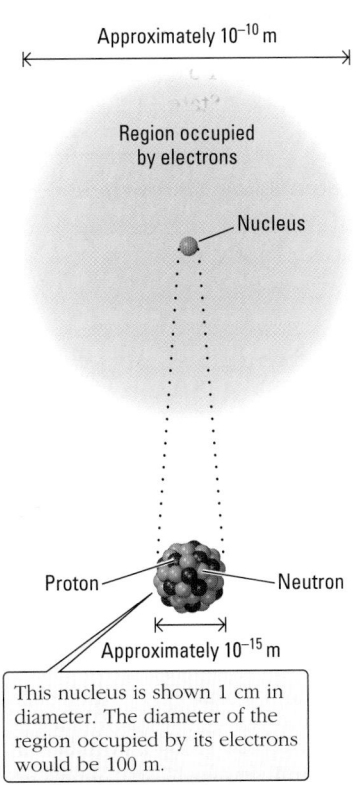

Relative sizes of the atomic nucleus and an atom.

Tools of Chemistry   Scanning Tunneling Microscopy (STM)

The ability to image individual atoms directly has long been a dream of chemists, and instrumental advances have now fulfilled this dream. The **scanning tunneling microscope** (STM) is an exciting analytical instrument that provides images of individual atoms or molecules on a surface. To do this, a metal probe in the shape of a needle with an extremely fine point (a nanoscale tip) is brought extremely close (within one or two atomic diameters, a few tenths of a nanometer) to the sample surface being examined. A voltage is applied between the probe and sample surface, and when the tip is close enough to the sample, electrons

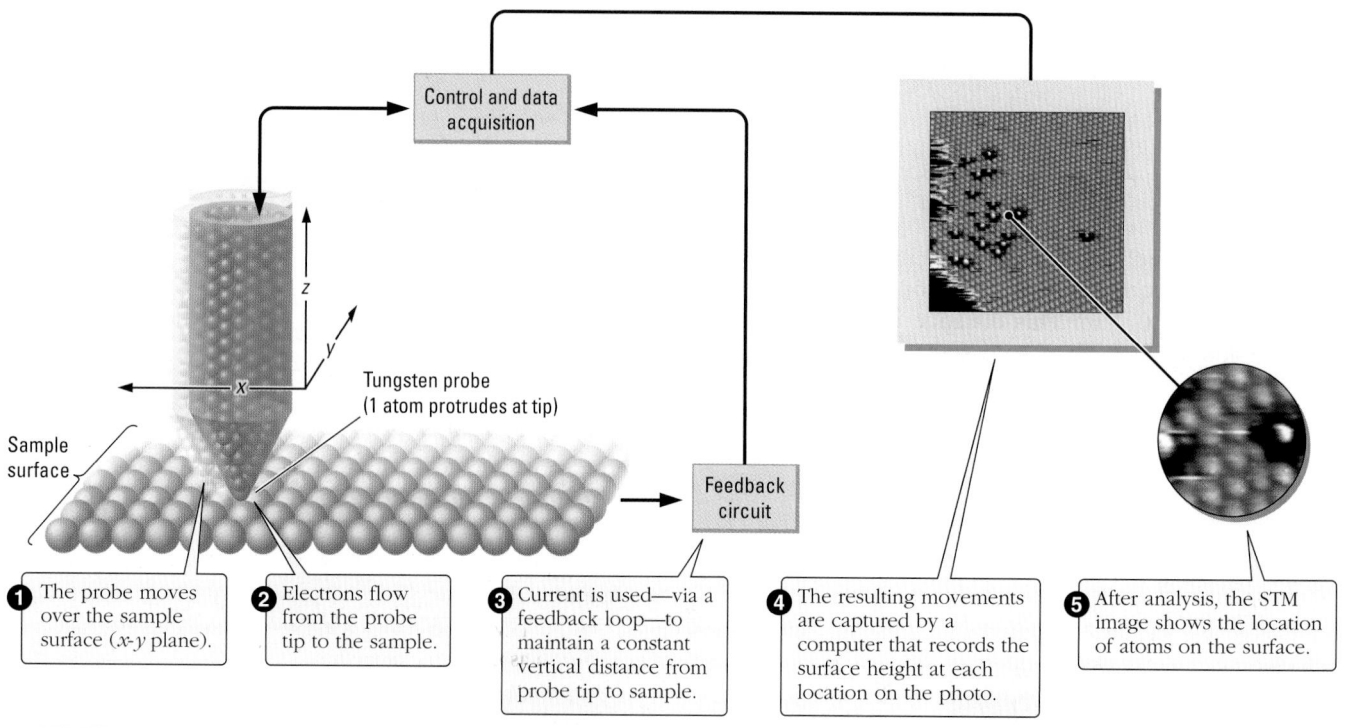

① The probe moves over the sample surface (*x-y* plane).

② Electrons flow from the probe tip to the sample.

③ Current is used—via a feedback loop—to maintain a constant vertical distance from probe tip to sample.

④ The resulting movements are captured by a computer that records the surface height at each location on the photo.

⑤ After analysis, the STM image shows the location of atoms on the surface.

A schematic diagram of a scanning tunneling microscope.

Note that the units m (for meters) cancel, leaving the answer in the units we want (pm). A penny is 1.90×10^{10} pm in diameter.

Every conversion factor can be used in two ways. We just converted meters to picometers by using

$$\frac{1 \text{ pm}}{1 \times 10^{-12} \text{ m}}$$

Picometers can be converted to meters by inverting the conversion factor:

$$8.70 \times 10^{10} \text{ pm} \times \frac{1 \times 10^{-12} \text{ m}}{1 \text{ pm}} = 8.70 \times 10^{-2} \text{ m} = 0.087 \text{ m}$$

The number of copper atoms needed to stretch across a penny can be calculated by using a conversion factor linking the penny's diameter in picometers with the diameter of a single copper atom. The diameter of a Cu atom is twice the radius, 2×128 pm $= 256$ pm. Therefore, the conversion factor is 1 Cu atom per 256 pm, and

jump between the probe and the sample in a process called tunneling (see figure at left). The size and direction of this electron flow (the current) depend on the applied voltage, the distance between probe tip and sample, and the identity and location of the nearest sample atom on the surface and its closest neighboring atoms.

The probe tip is attached to a control mechanism that maintains a constant distance between the tip and the sample at atomic scale resolution by keeping the current constant. The current decreases exponentially as the distance between the probe tip and the sample increases, so the current provides an extremely sensitive measure of the interatomic separation between probe tip and sample. The probe tip is scanned laterally across the sample surface, and as necessary it moves toward or away from the sample to maintain the constant current. These movements toward or away from the sample are recorded to capture the surface height on an atomic scale.

The probe tip is moved systematically across the surface to form a complete topographic map of that part of the surface. The computer controlling the STM probe records the surface height at each location on the surface. These resulting topographic data are processed by software to form the final images that depict the surface contours. The STM image, which appears much like a photographic image, shows the locations of atoms on the surface being investigated. (The image is actually of the electrons on the atoms.) The height of each point on the surface is represented by the brightness of the image at that point. The figure on the right shows an STM image of a copper surface with a well-ordered array of iron atoms on it.

The STM has been applied to a wide variety of problems throughout science and engineering. The greatest number of studies have focused on the properties of clean surfaces that have been modified. Very high vacuums in the range of 1×10^{-10} mm Hg are necessary to allow study over a period of

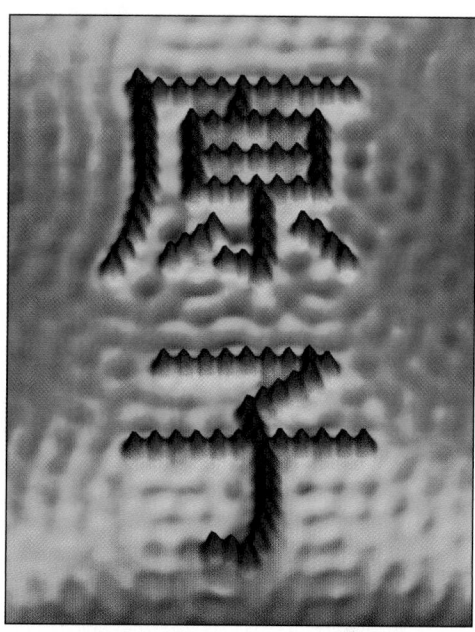

The Kanji characters for "atom" formed with iron atoms on a copper surface.

hours if contamination of the surface is to be avoided. The STM can also be used to study electrode surfaces in a liquid. Applications of STM and closely related techniques to biological molecules on surfaces is a growing scientific area of research as well. The STM can be used to move atoms on a surface, and researchers have used this capability to generate spectacular images such as the Kanji characters in the figure.

The scanning tunneling microscope was invented by Gerd Binnig and Heinrich Rohrer at IBM, Zurich, in 1981, and they shared a Nobel Prize in physics in 1986 for this work.

$$1.90 \times 10^{10} \text{ pm} \times \frac{1 \text{ Cu atom}}{256 \text{ pm}} = 7.42 \times 10^7 \text{ Cu atoms or } 74,200,000 \text{ Cu atoms}$$

Notice how "Cu atom" is included in the conversion so that cancellation of units gives the desired answer.

It takes 74 million copper atoms to reach across the penny's diameter. Atoms are indeed tiny.

In chemistry, the most commonly used length units are the centimeter, the millimeter, the nanometer, and the picometer. The most commonly used mass units are the kilogram, the gram, and the milligram. The relationships among these units and some other units are given in Table 2.2.

Examples 2.1 and 2.2 illustrate the use of dimensional analysis in unit conversion problems. Notice that in these examples, and throughout the book, the answers are given before the explanation of how the answers are found. We urge you to first try to answer the problem on your own. Then check to see if your answer is correct. If it does not match, try again. Finally, then read the explanation and find out why your reasoning differs from that given. If your answer is correct, but your reasoning differs from the explanation, you might have discovered an alternative way to solve the problem.

On September 23, 1999, the NASA Mars Climate Orbiter spacecraft approached too close and burned up in Mars' atmosphere because of a navigational error due to a failed translation of English units into metric units by the spacecraft's software.

|← 2 nm →|

DNA double helix.

Problem-Solving Example 2.1   Conversion of Units

A student is 6 feet 2 inches tall. What is the student's height in centimeters?

Answer   188 cm

Explanation   We use conversion factors derived from Table 2.2 to calculate the answer. One inch equals 2.54 cm. We first find the student's height in inches.

$$\left(6\ \text{ft} \times \frac{12\ \text{in}}{1\ \text{ft}}\right) + 2\ \text{in} = 74\ \text{in}$$

Having the height in inches, we can use the conversion factor to convert to cm

$$74\ \text{in} \times \frac{2.54\ \text{cm}}{1\ \text{in}} = 187.96\ \text{cm}$$

which we can consider to be 188 cm.

✔   There are about 2.5 cm per inch, and six feet is 72 inches, so 2.5×72 is about $72 + 72 + 36 = 180$, which is close to our more accurate answer.

Problem-Solving Practice 2.1

(a) If a car requires 10 gallons to fill its gas tank, how many liters would this be?
(b) A football field is 100 yards long. How many meters is this?

Problem-Solving Example 2.2   Nanoscale Distances

The double helix of the DNA molecule has a diameter of 2.0 nm. How many DNA molecules could be laid side by side to stretch a distance of 1.00 mm?

Answer 5.0×10^5 DNA molecules

Explanation   A nanometer is 1×10^{-9} m and a millimeter is 1×10^{-3} m. The diameter of the DNA in meters is

$$2.0\ \text{nm} \times \frac{1 \times 10^{-9}\ \text{m}}{1\ \text{nm}} = 2.0 \times 10^{-9}\ \text{m}$$

One millimeter is 1×10^{-3} m. Therefore, the number of DNA molecules of diameter 2.0 nm that could fit into 1.00 mm is

$$1 \times 10^{-3}\ \text{m} \times \frac{1\ \text{DNA molecule}}{2.0 \times 10^{-9}\ \text{m}} = 5.0 \times 10^5\ \text{DNA molecules}$$

✔   If DNA molecules are 2 nm in diameter, then one half of 10^9 could fit into a meter. One mm is 1/1000 of a meter, so multiplying these two estimates together gives $(0.5 \times 10^9)(1/1000) = 0.5 \times 10^6$, which is the same as the answer above.

Problem-Solving Practice 2.2

Do the following conversions using factors based on the equalities in Table 2.2.

(a) How many grams of sugar are in a 5-lb bag of sugar?
(b) Over a lengthy period of time, a donor gives 3 pints of blood. How many milliliters (mL) has the donor given?
(c) The same donor's 160-lb body contains approximately 5 L of blood. Considering that 1 L is nearly equal to 1 quart, estimate the percentage of the donor's blood that has been donated in all.

Table 2.2 also lists the liter (L) and milliliter (mL), which are the most common volume units of chemistry. There are 1000 mL in 1 L. One liter is a bit larger than a quart, and a teaspoon of water has a volume of about 5 mL. Chemists often use the terms milliliter and cubic centimeter (or "cc") interchangeably because they are equivalent (1 mL = 1 cm^3).

Problem-Solving Example 2.3   Volume Units

In a chemical analysis, a chemist uses 30 μL (microliters) of sample for an analysis (1 μL = 1×10^{-6} L). What is the volume in mL ? In cm^3?

Answer 3.0×10^{-2} mL ; 3.0×10^{-2} cm^3

Explanation   We need to use conversion factors based on two equalities to convert microliters to milliliters: 1 μL = 1×10^{-6} L and 1 L = 1000 mL. We multiply the conversion factors to cancel μL and L, leaving only mL.

$$30 \; \cancel{mL} \times \frac{1 \times 10^{-6} \; \cancel{L}}{1 \; \cancel{\mu L}} \times \frac{100 \; mL}{1 \; \cancel{L}} = 3.0 \times 10^{-2} \; mL$$

Because 1 mL and 1 cm^3 are equivalent, the sample size can also be expressed as 3.0×10^{-2} cm^3.

✔   There is a conversion factor of 10^3 between μL and mL, and the final answer is a factor of 10^3 larger than the original volume, so the answer is reasonable.

Problem-Solving Practice 2.3

A patient's blood cholesterol level measures 165 mg/dL. Express this value in g/L (1 deciliter (dL) = 1×10^{-1} L).

As illustrated in Problem-Solving Example 2.3, two or more steps of a calculation using dimensional analysis are best written in a single setup and entered into a calculator as a single calculation.

TABLE 2.2	Some Common Unit Equalities*
Length	
1 kilometer	= 1000 meters = 0.62137 mile
1 meter	= 100 centimeters
1 centimeter	= 10 millimeters
1 nanometer	= 1×10^{-9} meter
1 picometer	= 1×10^{-12} meter
1 inch	= 2.54 centimeters (exactly†)
1 angstrom (Å)	= 1×10^{-10} meter
Volume	
1 liter (L)	= 1×10^{-3} m^3 = 1000 mL
	= 1000 cm^3
	= 1.056710 quarts
1 gallon	= 4 quarts
1 quart	= 2 pints
1 cubic meter (m^3)	= 1×10^3 liter (L)
Mass	
1 kilogram	= 1000 grams
1 gram	= 1000 milligrams
1 amu	= 1.6606×10^{-24} grams
1 pound	= 453.59237 grams = 16 ounces
1 ton (metric)	= 1000 kilograms
1 ton (American)	= 2000 pounds

*See Appendix B for other unit equalities.

†This exact equality is important for length conversions.

Devices that measure the volume of liquids. (a) Beaker, (b) graduated cylinder, (c) pipette, (d) volumetric flask, (e) burette.

2.4 ATOMIC NUMBERS AND MASS NUMBERS

CD-ROM Screen 2.14:
Summary of Atomic
Composition

The periodic table is organized by atomic
number; it is discussed in Section 2.8.

1 atomic mass unit (amu) = $\frac{1}{12}$ the mass
of a carbon atom having 6 protons and 6
neutrons in the nucleus.

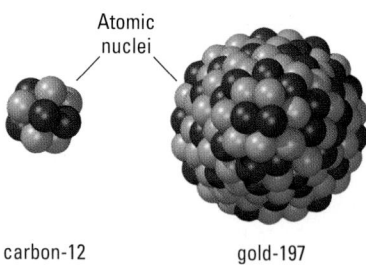

Atomic
nuclei

carbon-12   gold-197

Experiments done in the early years of the 20th century found that *atoms of the same element have the same numbers of protons in their nuclei.* This number is called the **atomic number** and is given the symbol Z. In the periodic table at the front of this book, the atomic number for each element is written above the element's symbol. For example, a copper atom has a nucleus containing 29 protons, so its atomic number is 29 (Z = 29). A lead atom (Pb) has 82 protons in its nucleus, so the atomic number for lead is 82.

The scale of atomic masses is defined relative to a standard. This standard is the mass of a carbon atom that has 6 protons and 6 neutrons in its nucleus. The masses of atoms of every other element are established relative to the mass of this carbon atom, which is defined as having a mass of 12 atomic mass units. In terms of macroscale mass units, 1 **atomic mass unit** (amu) = 1.66054×10^{-24} g. Thus, for example, when an experiment shows that a gold atom, on average, is 16.4 times as massive as the standard carbon atom, we then know its mass in amu and grams.

$$16.4 \times 12 \text{ amu} = 197 \text{ amu}$$

$$197 \text{ amu} \times \frac{1.66054 \times 10^{-24} \text{ g}}{1 \text{ amu}} = 3.27 \times 10^{-22} \text{ g}$$

The masses of the fundamental subatomic particles in atomic mass units have been determined experimentally. The proton and the neutron have masses very close to 1 amu, whereas the electron's mass is approximately 1800 times less.

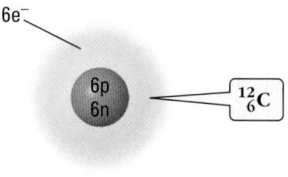

carbon-12

Particle	Mass (grams)	Mass (atomic mass units)	Charge
Electron	$9.10938188 \times 10^{-28}$	0.000548579	−1
Proton	$1.67262158 \times 10^{-24}$	1.00728	+1
Neutron	$1.67492716 \times 10^{-24}$	1.00866	0

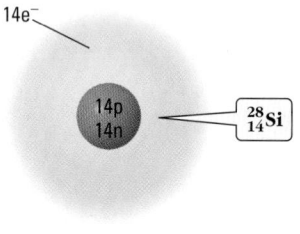

silicon-28

Once a relative scale of atomic masses has been established, we can estimate the mass of any atom whose nuclear composition is known. The proton and neutron have masses so close to 1 amu that the difference can essentially be ignored. Electrons are much, much lighter than protons or neutrons. Because the number of electrons in an atom must equal the number of protons, there are never enough electrons in an atom to significantly alter its mass, so their mass need not be considered. To estimate an atom's mass, we add up its number of protons and neutrons. This sum, called the **mass number** of that particular atom, is given the symbol A. For example, a copper atom that has 29 protons and 34 neutrons in its nucleus has a mass number, A, of 63. A lead atom that has 82 protons and 126 neutrons has A = 208.

With this information, an atom of known composition, such as a lead atom, can be represented as follows:

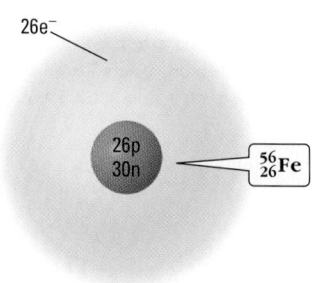

iron-56

$$\begin{array}{c} \text{A} \\ \text{X} \\ \text{Z} \end{array} \quad \begin{array}{l} \text{Mass number} \\ \text{Element symbol} \\ \text{Atomic number} \end{array} \quad {}^{208}_{82}\text{Pb}$$

Each element has its own unique one- or two-letter symbol. Because each element is defined by the number of protons it contains, knowing which element you are dealing with means you automatically know the number of protons it has. Thus, the Z part of the notation is superfluous. For example, the lead atom might be represented by the symbol ^{208}Pb because the Pb tells us the element is lead, and lead by definition always contains 82 protons. Another notation you may sometimes see is lead-208. We would simply say "lead-208" in any case.

Each element has a unique atomic number.

Problem-Solving Example 2.4   Atomic Nuclei

Iodine-127 is important in diagnostic medicine because it concentrates in the thyroid gland, where its quantity can be detected and used to measure thyroid gland activity. How many neutrons are there in an iodine atom of mass number 127?

Answer   74 neutrons

Explanation   The periodic table inside the front cover shows that the atomic number of iodine (I) is 53. Therefore, the atom has 53 protons in the nucleus. Because the mass number of the atom is the sum of the numbers of protons and neutrons in the nucleus,

$$\text{Mass number} = \text{number of protons} + \text{number of neutrons}$$

$$127 = 53 + \text{number of neutrons}$$

$$\text{Number of neutrons} = 127 - 53 = 74$$

The actual mass of an atom is slightly less than the sum of the masses of its protons, neutrons, and electrons. The difference, known as the mass defect, is related to the energy binding nuclear particles together, a topic discussed in Chapter 20.

Problem-Solving Practice 2.4

(a) What is the mass number of a phosphorus atom with 16 neutrons?
(b) How many protons, neutrons, and electrons are there in a neon-22 atom?
(c) Write the symbol for the atom with 82 protons and 125 neutrons.

Although an atom's mass approximately equals its mass number, the actual mass is not an integral number. For example, the actual mass of a gold-196 atom is 195.9231 amu, slightly less than the mass number 196. The masses of atoms are determined experimentally using mass spectrometers. The *Tools of Chemistry* discussion illustrates the use of a mass spectrometer to determine the atomic masses of neon atoms.

Mass spectrometric analysis of most naturally occurring elements reveals that not all atoms of an element have the same mass. For example, all silicon atoms have 14 protons, but some silicon nuclei have 14 neutrons, others have 15, and others have 16. Thus, naturally occurring silicon (atomic number 14) is always a mixture of silicon-28, silicon-29, and silicon-30 atoms. Such different atoms of the *same* element are called isotopes. **Isotopes** are atoms with the *same* atomic number (Z) but different mass numbers (A). All the silicon atoms have 14 protons — that is what makes them silicon atoms. But these isotopes differ in their mass numbers because they have different numbers of neutrons.

Two elements can't have the same atomic number. If two atoms differ in their number of protons, they are atoms of different elements. If only their number of neutrons differs, they are isotopes of a single element.

 CD-ROM Screen 2.13: Isotopes

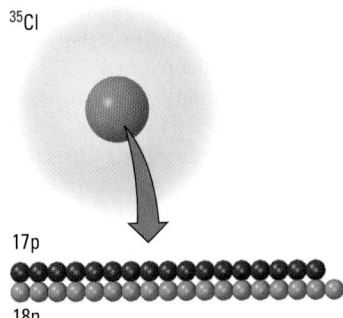

^{35}Cl

17p

18n

Problem-Solving Example 2.5   Isotopes

Copper has two isotopes, one with 34 neutrons and the other with 36 neutrons. What are the mass numbers and symbols of these isotopes?

Answer   The mass numbers are 63 and 65. The symbols are $^{63}_{29}$Cu and $^{65}_{29}$Cu.

Explanation   Copper has an atomic number of 29, so it has 29 protons in its nucleus. Therefore, the mass numbers of the two isotopes are

$$\text{Isotope 1: Cu} = 29 \text{ protons} + 34 \text{ neutrons} = 63 \text{ (copper-63)}$$

$$\text{Isotope 2: Cu} = 29 \text{ protons} + 36 \text{ neutrons} = 65 \text{ (copper-65)}$$

Placing the atomic number at the bottom left and the mass number at the top left gives the symbols $^{63}_{29}$Cu and $^{65}_{29}$Cu.

Problem-Solving Practice 2.5

Naturally occurring magnesium has three isotopes with 12, 13, and 14 neutrons. What are the mass numbers and notations of these three isotopes?

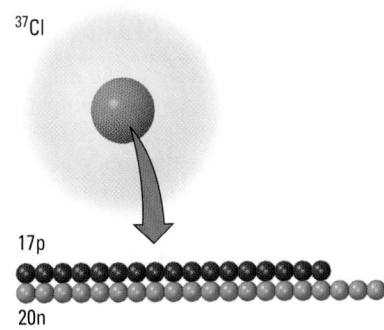

^{37}Cl

17p

20n

Chlorine isotopes.   Chlorine-35 and chlorine-37 atoms each contain 17 protons; chlorine-35 atoms have 18 neutrons, and chlorine-37 atoms contain 20 neutrons.

Tools of Chemistry Mass Spectrometer

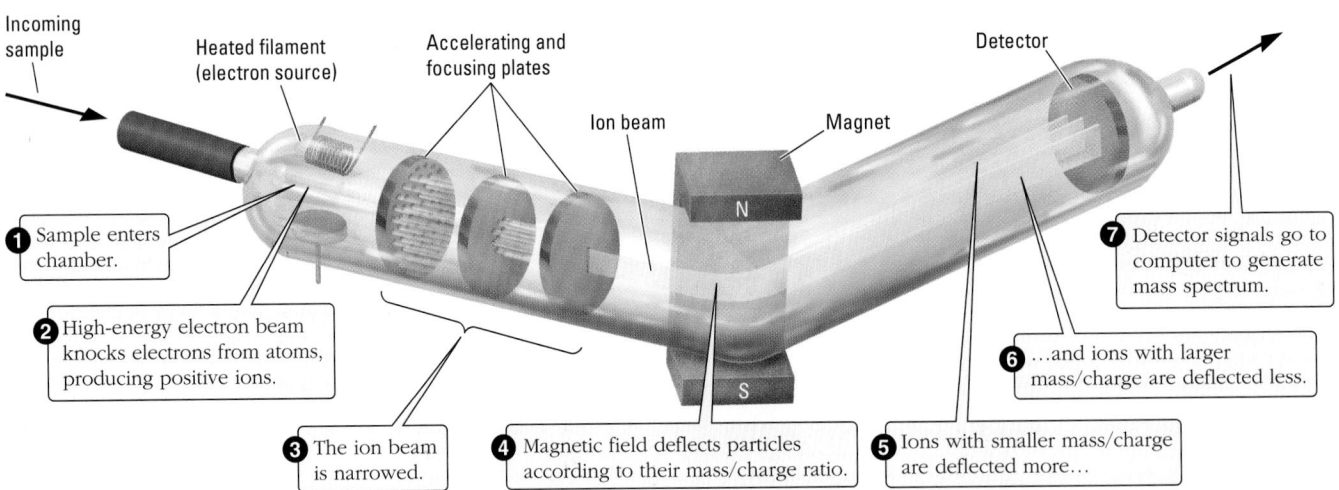

Incoming sample

Heated filament (electron source)

Accelerating and focusing plates

Ion beam

Magnet

Detector

1 Sample enters chamber.

2 High-energy electron beam knocks electrons from atoms, producing positive ions.

3 The ion beam is narrowed.

4 Magnetic field deflects particles according to their mass/charge ratio.

5 Ions with smaller mass/charge are deflected more…

6 …and ions with larger mass/charge are deflected less.

7 Detector signals go to computer to generate mass spectrum.

A **mass spectrometer** is an analytical instrument used to measure atomic and molecular masses directly. A gaseous sample of the substance being analyzed is bombarded by high-energy electrons. Collisions between the electrons and the sample's atoms (or molecules) produce positive ions, mostly with $+1$ charge, which are attracted to a negatively charged grid. The beam of ions is narrowed and passed through a magnetic field, which deflects the ions, with the amount of deflection depending on their mass and charge. Larger mass ions are deflected less; smaller mass ions are deflected more. This deflection amounts to sorting the ions by mass since most of them have the same $+1$ charge. The deflected ions pass through to a detector, which measures the ions as a current that is directly proportional to the number of ions. This allows the determination of the relative abundance of the various ions in the sample.

In practice, the mass spectrometer settings are varied to focus ions of different masses on the stationary detector at different times. The mass spectrometer records the current of ions (the ion abundance) as the magnetic field is varied systematically. After processing by software, the data are plotted as a graph of the ion abundance versus the mass of the ions, which is a **mass spectrum.**

The measurement of the mass spectrum of neon and the resulting spectrum are shown in the figures. The beam of Ne^{+} ions passing through the mass spectrometer is divided into three segments because three isotopes are present: ^{20}Ne, with an atomic mass of 19.9924 amu and an abundance of 90.92%, ^{21}Ne, with an atomic mass of 20.9940 and an abundance of only 0.26%, and ^{22}Ne, with an atomic mass of 21.9914 amu and an abundance of 8.82%.

The mass spectrometer described here is a simple one based on magnetic field deflection of the ions, and it is similar to those used in early experiments to determine isotopic abundances. Modern mass spectrometers operate on quite different principles, although they generate similar mass spectra. These instruments are used to measure the masses of chemical compounds as well as atoms. Mass spectrometers are also used to investigate details of molecular structure of compounds ranging in complexity from simple organic and inorganic compounds to biomolecules such as proteins. Mass spectrometry is an important and rapidly developing field of chemistry.

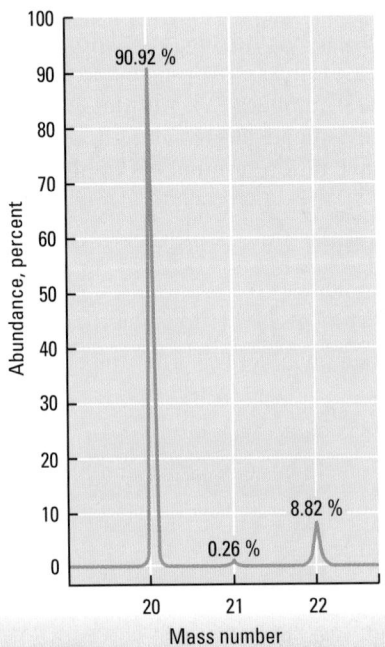

Mass spectrum of neon. The principal peak corresponds to the most abundant isotope, neon-20. The height of each peak indicates the percent relative abundance of each isotope.

We usually refer to a particular isotope by giving its mass number. For example, $^{238}_{92}U$ is referred to as uranium-238 or U-238. But a few isotopes have distinctive names and symbols because of their importance, such as the isotopes of hydrogen, all with just one proton. When the single proton is the only nuclear particle, the element is simply called hydrogen. With one neutron as well as one proton present, the isotope $^{2}_{1}H$ is called either deuterium or heavy hydrogen (symbol D). When two neutrons are present, the isotope $^{3}_{1}H$ is called tritium (symbol T).

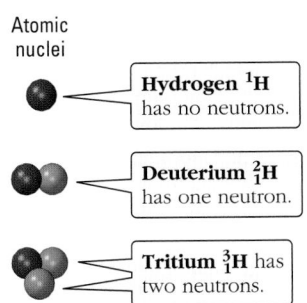

Atomic nuclei

Hydrogen ^{1}H has no neutrons.

Deuterium $^{2}_{1}H$ has one neutron.

Tritium $^{3}_{1}H$ has two neutrons.

Hydrogen isotopes. Hydrogen, deuterium, and tritium each contain one proton. Hydrogen has no neutrons; deuterium and tritium have one and two neutrons, respectively.

Exercise 2.3   Isotopes

A student in your chemistry class tells you that nitrogen-14 and nitrogen-15 are not isotopes because they have the same number of protons. How would you refute this statement?

2.5   ISOTOPES AND ATOMIC WEIGHT

Copper has two naturally occurring isotopes, copper-63 and copper-65, with atomic masses of 62.9296 amu and 64.9278 amu, respectively. In a macroscopic collection of naturally occurring copper atoms, the average mass of the atoms is neither 63 (all copper-63) nor 65 (all copper-65). Rather, the average atomic mass will fall between 63 and 65, with its exact value depending on the proportion of each isotope in the mixture. The proportion of atoms of each isotope in a natural sample of an element is called the **percent abundance,** the percentage of atoms of a *particular* isotope.

The concept of percent is widely used in chemistry, and it is worth briefly reviewing here. For example, earth's atmosphere contains approximately 78% nitrogen, 21% oxygen, and 1% argon. U.S. pennies minted after 1982 contain 2.4% copper; the remainder is zinc.

CD-ROM Screen 2.15: Atomic Mass

The sum of the percents in a sample must be 100.

Problem-Solving Example 2.6   Applying Percent

Currently, U.S. pennies are 2.40% copper and 97.60% zinc. What mass of each element is in a penny weighing 2.485 g ?

Answer   0.0596 g Cu; 2.425 g Zn

Explanation   Let's start by calculating the mass of copper in the penny. Its percentage, 2.40%, means that every 100 g of penny contains 2.40 g of copper.

$$2.485 \text{ g penny} \times \frac{2.40 \text{ g copper}}{100 \text{ g penny}} = 0.0596 \text{ g copper}$$

The mass of zinc is found the same way, using the conversion factor of 97.60 g of zinc per 100 g of penny.

$$2.485 \text{ g penny} \times \frac{97.60 \text{ g zinc}}{100 \text{ g penny}} = 2.425 \text{ g zinc}$$

We could also have directly obtained this value by recognizing that the masses of zinc and copper must add up to the mass of the penny. Therefore,

$$2.485 \text{ g penny} = 0.0596 \text{ g copper} + x \text{ g zinc}$$

Solving for x, the mass of zinc, gives 2.485 g − 0.0596 g = 2.425 g of zinc.

✔   The ratio of zinc to copper is about 40 : 1, and the ratios of the masses calculated are also about 40 : 1 (2.4/0.06 = 40), so the answer is reasonable.

U.S. pennies.   Part of the surface of the pre-1982 penny on the left has been removed to show that the coin is solid copper. The post-1982 penny on the right shows the inner zinc filling, covered by a thin coating of copper.

Problem-Solving Practice 2.6

Many heating devices such as hair dryers contain nichrome wire, an alloy containing 80.% nickel and 20.% chromium, which gets hot when an electric current passes through it. If a heating device contains 75 g of nichrome wire, how many grams of nickel and how many grams of chromium does it contain?

The percent abundance of each isotope in a sample of an element is given as follows:

$$\text{Percent abundance} = \frac{\text{number of atoms of a given isotope}}{\text{total number of atoms of all isotopes of that element}} \times 100\%$$

Table 2.3 gives information about the percent abundance for naturally occurring isotopes of hydrogen, boron, and bromine. The percent abundance and isotopic mass of each isotope can be used to find the average mass of atoms of that element, and this average mass is called the atomic weight of the element. The **atomic weight** of an element is the average mass of a representative sample of atoms of the element, expressed in atomic mass units.

Boron, for example, is a relatively rare element present in compounds used in laundry detergents, mild antiseptics, and Pyrex cookware. It has two naturally occurring isotopes: boron-10, with a mass of 10.0129 amu and 19.91% abundance, and boron-11, with a mass of 11.0093 amu and 80.09% abundance. Since the abundances are approximately 20% and 80%, respectively, you can estimate the atomic weight of boron: 20 atoms out of every 100, or 2 atoms out of every 10, are boron-10. If you then add up the masses of 10 atoms, you have 2 atoms with a mass of about 10 amu and 8 atoms with a mass of about 11 amu, so the sum is 108 amu, and the average is 108 amu/10 = 10.8 amu. This approximation is about right when you consider that the mass numbers of the boron isotopes are 10 and 11 and that boron is 80% boron-11 and only 20% boron-10. Therefore, the atomic weight should be about two tenths of the way down from 11 to 10, or 10.8. Thus, each atomic weight is a weighted average that accounts for the proportion of each isotope, not just the usual arithmetic average, in which the values are simply summed and divided by the number of values.

In general, the atomic weight of an element is found from the percent abundance data as shown by the following more exact calculation for boron. The mass of each isotope is multiplied by its percent abundance expressed as a decimal fraction to calculate the weighted average:

Atomic weight = [(% abundance B-10)(isotopic mass B-10)
$$+ \text{(% abundance B-11)(isotopic mass B-11)}]$$

$$= [(19.91\%)(10.0129 \text{ amu}) + (80.09\%)(11.0093 \text{ amu})]$$

$$= (0.1991)(10.0129 \text{ amu}) + (0.8009)(11.0093 \text{ amu})$$

$$= 10.81 \text{ amu}$$

Our earlier estimate was quite close to the more exact result. The arithmetic average of the isotopic masses of boron is (10.0129 + 11.093)/2 = 10.55, which is quite different from the actual atomic weight.

The atomic weight of each stable (nonradioactive) element has been determined; these values appear in the periodic table in the front inside cover of this book. For most elements, the abundances of the isotopes are the same no matter where a sample is collected. Therefore, the atomic weights in the periodic table are used whenever an atomic weight is needed. In the periodic table, each ele-

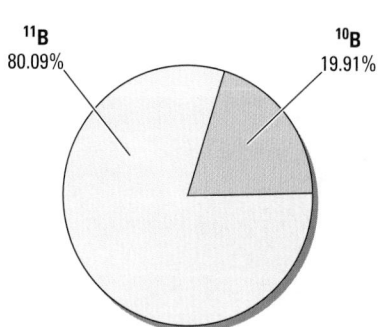

^{11}B
80.09%

^{10}B
19.91%

Percent abundance of boron-10 and boron-11.

1 amu = $\frac{1}{12}$ mass of a carbon-12 atom

TABLE 2.3	Isotopic Masses of the Stable Isotopes of Hydrogen, Boron, and Bromine				
Element	Symbol	Atomic Weight (amu)	Mass Number	Isotopic Mass (amu)	Percent Abundance
Hydrogen	H	1.00794	1	1.007825	99.9855
	D		2	2.0141022	0.0145
Boron	B	10.811	10	10.012939	19.91
			11	11.009305	80.09
Bromine	Br	79.904	79	78.918336	50.69
			81	80.916289	49.31

ment's box contains the atomic number, the symbol, and the atomic weight. For example, the periodic table entry for copper is

$$
\begin{array}{ll}
29 & \longleftarrow \text{Atomic number} \\
\textbf{Cu} & \longleftarrow \text{Symbol} \\
63.546 & \longleftarrow \text{Atomic weight}
\end{array}
$$

The term "atomic weight" is so commonly used that it has become accepted, even though it is really a mass rather than a weight.

Exercise 2.4   Atomic Weight

Verify that the atomic weight of lithium is 6.941 amu, given the following information:
^6_3Li mass = 6.015121 amu and percent abundance = 7.500%
^7_3Li mass = 7.016003 amu and percent abundance = 92.50%

 ### Exercise 2.5   Isotopic Abundance

Naturally occurring magnesium contains three isotopes: ^{24}Mg (78.70%), ^{25}Mg (10.13%), and ^{26}Mg (11.17%). Estimate the atomic weight of Mg and compare your estimate with the atomic weight calculated by finding the arithmetic average of the atomic masses. Which value is larger? Why is it larger?

 ### Exercise 2.6   Percent Abundance

Gallium has two abundant isotopes, and its atomic weight is 69.72 amu. If you knew only this value and not the percent abundances, make the case that the percent abundance of each of the two gallium isotopes cannot be 50%.

2.6   AMOUNTS OF SUBSTANCES—THE MOLE

Atoms are much too small to be seen directly or weighed individually on the most sensitive balance. However, when working with chemicals it is essential to know how many atoms, molecules, or other nanoscale units of an element or compound you are working with. To connect the macroscale world, where chemicals can be measured, weighed, and manipulated, to the nanoscale world of individual atoms or molecules, chemists have defined a convenient unit of matter that contains a known number of particles. This chemical-counting unit is the **mole (mol)**, defined as the amount of substance that contains as many atoms, molecules, ions, or other nanoscale units as there are atoms in *exactly* 12 g of carbon-12.

 CD-ROM Screen 2.18: The Mole

One mole of carbon has a mass of 12.01 g, not exactly 12 g, because naturally occurring carbon contains both carbon-12 (98.89%) and carbon-13 (1.11%). By definition, one mole of carbon-12 has a mass of exactly 12 g.

The essential point to understand about moles is that *one mole always contains the same number of particles,* no matter what substance or what kind of particles we are talking about. The number of particles in a mole is

$$1 \text{ mol} = 6.02214199 \times 10^{23} \text{ particles}$$

The mole is the connection between the macroscale and nanoscale worlds, the visible and the not directly visible.

The number of particles in a mole is known as **Avogadro's number** after Amadeo Avogadro (1776–1856), an Italian physicist who conceived the basic idea, but never experimentally determined the number, which came later. It is important to realize that the value of Avogadro's number is a definition tied to the values of 12 g of carbon-12.

$$\text{Avogadro's number} = 6.02214199 \times 10^{23} \text{ per mole} = 6.02214199 \times 10^{23} \text{ mol}^{-1}$$

One difficulty in comprehending Avogadro's number is its sheer size. Writing it out in full yields

$$6.02214199 \times 10^{23} = 602{,}214{,}199{,}000{,}000{,}000{,}000{,}000$$

or 602,214.199 × 1 million × 1 million × 1 million. There are many analogies used to try to give a feeling for the size of this number. If you poured Avogadro's number of marshmallows over the continental United States, the marshmallows would cover the country to a depth of 650 miles. Or, if one mole of pennies were divided evenly among every man, woman, and child in the United States, your share alone would pay off the national debt (about $3 trillion, or 3×10^{12}), and you would still have about $20 trillion left over to buy pizza for youself and your friends.

Thus, chemists heap up particles by the mole; one mole contains 6.022×10^{23} particles. You can think of the mole simply as a counting unit, analogous to the counting units we use for ordinary items such as doughnuts or bagels by the dozen, shoes by the pair, or sheets of paper by the ream (500 sheets). Atoms, molecules, and other particles in chemistry are counted by the mole. The different masses of the elements shown in Figure 2.4 each contain one mole of atoms. For

The term mole is derived from the Latin word *moles* meaning a "heap" or a "pile."

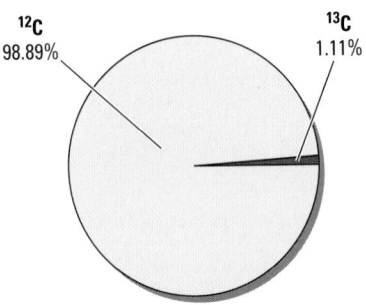

^{12}C
98.89%

^{13}C
1.11%

Percent abundance of carbon-12 and carbon-13.

Although Avogadro's number is known to eight significant figures, we will usually use it rounded to 6.022×10^{23}.

CD-ROM Screen 2.19: Moles and Molar Masses of the Elements

ESTIMATION   The Size of Avogadro's Number

Chemists and other scientists often use estimates in place of exact calculations when they want to know the approximate value of a quantity. Analogies to help us understand the extremely large value of Avogadro's number are an example.

If 1 mol of green peas were spread evenly over the continental United States, how deep would the layer of peas be? The surface area of the continental United States is about 3.0×10^6 square miles (mi^2). There are 5280 feet per mile.

Let's start with an estimate of a green pea's size: $\frac{1}{4}$-inch diameter. Then 4 peas would fit along a 1-inch line, and 48 would fit along a 1-foot line, and $48^3 = 110{,}592$ would fit into 1 cubic foot (ft^3). Since we are estimating, we will approximate by saying 1×10^5.

Now, let's estimate how many cubic feet of peas are in one mole of peas:

$$\frac{6.022 \times 10^{23} \text{ peas}}{1 \text{ mol peas}} \times \frac{1 \text{ ft}^3}{1 \times 10^5 \text{ peas}} \cong \frac{6.0 \times 10^{18} \text{ ft}^3}{1 \text{ mol peas}}$$

(The "approximately equal sign," $\cong$, is an indicator of these approximations.) The surface area of the continental United States is 3×10^6 square miles (mi^2), which is about

$$3.0 \times 10^6 \text{ mi}^2 \times \left(\frac{5280 \text{ ft}}{1 \text{ mi}} \right)^2 \cong 8.4 \times 10^{13} \text{ ft}^2$$

so 1 mol of peas spread evenly over this area will have a depth of

$$\frac{6.0 \times 10^{18} \text{ ft}^3}{1 \text{ mol peas}} \times \frac{1}{8.4 \times 10^{13} \text{ ft}^2} \cong \frac{7.2 \times 10^4 \text{ ft}}{1 \text{ mol peas}}$$

or

$$\frac{7.2 \times 10^4 \text{ ft}}{1 \text{ mol peas}} \times \frac{1 \text{ mi}}{5280 \text{ ft}} \cong \frac{14 \text{ mi}}{1 \text{ mol peas}}$$

Note that in many parts of the estimate, we rounded off or used fewer significant figures than we could have used. Our purpose was to *estimate* the final answer, not to compute it exactly. The final answer, 14 miles, is not particularly accurate, but it is a valid estimate. The depth would be more than 10 miles but less than 20 miles. It would not be 6 inches or even 6 feet. Estimating served the overall purpose of developing the analogy for understanding the size of Avogadro's number.

each element in the figure, the mass in grams (the macroscale) is numerically equal to the atomic weight in atomic mass units (the nanoscale). The **molar mass** of any substance is the mass, in grams, of one mole of that substance. Molar mass has the units of grams per mole (g/mol).

For example,

$$\text{molar mass of copper (Cu)} = \text{mass of 1 mol of Cu atoms}$$

$$= \text{mass of } 6.022 \times 10^{23} \text{ Cu atoms}$$

$$= 63.546 \text{ g/mol}$$

$$\text{molar mass of aluminum (Al)} = \text{mass of 1 mol of Al atoms}$$

$$= \text{mass of } 6.022 \times 10^{23} \text{ Al atoms}$$

$$= 26.9815 \text{ g/mol}$$

Each molar mass of copper or aluminum contains Avogadro's number of atoms. Molar mass differs from one element to the next because the atoms of different elements have different masses. Think of a mole as analogous to a dozen. We could have a dozen golf balls, a dozen baseballs, or a dozen bowling balls — in each case, 12 items. The dozen items do not weigh the same, however, because the individual items do not weigh the same — 45 g per golf ball, 134 g per baseball, and 7200 g per bowling ball. In a similar way, the mass of Avogadro's number of atoms of one element is different from the mass of Avogadro's number of atoms of another element.

Atomic weights are given in the periodic table of elements in the inside front cover of this book.

Figure 2.4 One-mole quantities of several elements. Top row (left to right): Cu, Al, Pb. Bottom row (left to right): S, Cr, Mg.

2.7 MOLAR MASS AND PROBLEM SOLVING

Understanding the idea of a mole and applying it properly are *essential* to doing quantitative chemistry. In particular, it is *absolutely necessary* to be able to make two basic conversions: *moles → mass* and *mass → moles*. To do these and many other calculations in chemistry, it is most helpful to use dimensional analysis in the same way it is used in unit conversions. Along with calculating the final answer, write the units with all quantities in a calculation and cancel the units. If the problem is set up properly, the answer will have the desired units.

Let's see how these concepts apply to converting mass to moles or moles to mass. *In either case, the conversion factor is provided by the molar mass of the substance, the number of grams in one mole, that is, grams per mole (g/mol).*

The mass of one dozen items. The mass of one dozen golf balls is 540 g (12 baseballs = 1608 g; 12 bowling balls = 86,400 g). Similarly, the mass of one mole of an element depends on the atomic weight of its atoms.

$$\text{Mass} \rightleftharpoons \text{Moles conversions for substance A}$$

Mass A $\longrightarrow$ moles A	Moles A $\longrightarrow$ mass A

$$\text{Grams A} \times \frac{1 \text{ mol A}}{\text{grams A}} = \text{moles A} \qquad \text{Moles A} \times \frac{\text{grams A}}{1 \text{ mol A}} = \text{grams A}$$

$$\underbrace{\frac{1}{\text{molar mass}}}_{} \qquad\qquad\qquad \underbrace{\frac{}{\text{molar mass}}}_{}$$

Suppose you need 0.250 mol of Cu for an experiment. How many grams of Cu should you use? The atomic weight of Cu is 63.546 amu, so the molar mass of Cu is 63.546 g/mol. To calculate the mass of 0.250 mol of Cu, you need the conversion factor 63.546 g Cu/1 mol Cu.

$$0.250 \text{ mol Cu} \times \frac{63.55 \text{ g Cu}}{1 \text{ mol Cu}} = 15.9 \text{ g Cu}$$

In this book we will, when possible, *use one more significant figure in the molar mass than in any of the other data in the problem* (see Appendix A). In the problem just completed, note that four significant figures were used in the

Items can be counted by weighing—a 5-pound box of nails.

Appendix A reviews the use of significant figures. If you are unfamiliar with how to use significant figures, read Appendix A now.

molar mass of Cu, when three were used in the number of moles. Using one more significant figure in the molar mass guarantees that its precision is greater than that of the other numbers and does not limit the precision of the result of the computation.

Exercise 2.7   Grams, Moles, and Avogadro's Number

You have a 10.00-g sample of lithium and a 10.00-g sample of iridium. How many atoms are in each sample, and how many more atoms are in the lithium sample than in the iridium sample?

Frequently, a problem requires converting a mass to the equivalent number of moles, such as calculating the number of moles of bromine in 10.00 g of bromine. Because bromine is a diatomic element, it consists of Br_2 molecules. Therefore, there are 2 moles of Br atoms in 1 mole of Br_2 molecules. The molar mass of bromine is twice its atomic mass, 2×79.904 g/mol $= 159.81$ g/mol. To calculate the moles of bromine in 10.00 g of Br_2, use the conversion factor 1 mol Br_2/159.81 g Br_2.

$$10.00 \text{ g } Br_2 \times \frac{1 \text{mol } Br_2}{159.81 \text{ g } Br_2} = 6.257 \times 10^{-2} \text{ mol } Br_2$$

Problem-Solving Example 2.7   Mass and Moles

(a) Titanium (Ti) is a metal used to build airplanes. How many moles of Ti are in a 100-g sample of the pure metal?
(b) Aluminum is also used in airplane manufacturing. A piece of Al contains 2.16 mol of Al. Is the mass of Al greater or less than the mass of Ti in Part (a)?

Answer   (a) 2.09 mol Ti (b) 58.3 g Al, which is less than the mass of Ti

Explanation
(a) This is a mass-to-moles conversion that is solved using the conversion factor 1 mol Ti / 47.88 g Ti.

$$100 \text{ g } Ti \times \frac{1 \text{ mol Ti}}{47.88 \text{ g } Ti} = 2.09 \text{ mol Ti}$$

(b) This is a moles-to-mass conversion that requires the conversion factor 26.98 g Al/ 1 mol Al.

$$2.16 \text{ mol } Al \times \frac{26.98 \text{ g Al}}{1 \text{ mol } Al} = 58.3 \text{ g Al}$$

This mass is less than the 100-g mass of titanium.

Problem-Solving Practice 2.7

Calculate (a) the number of moles in 1.00 mg of molybdenum (Mo) and (b) the number of grams in 5.00×10^{-3} mol of gold (Au).

Titanium and aluminum are metals used in modern airplane manufacturing.

2.8   THE PERIODIC TABLE

CD-ROM Screen 2.16: The Periodic Table

You have already used the periodic table inside the front cover of this book to obtain atomic numbers and atomic weights of elements. But it is much more valuable than this. The periodic table is an exceptionally useful tool in chemistry. It allows us to organize and interrelate the chemical and physical properties of the elements. For example, the periodic table can be used to classify elements as metals, nonmetals, or metalloids by their positions in the table. You should become familiar with its main features and terminology.

Dmitri Mendeleev (1834–1907), while a professor at the University of St. Petersburg, realized that by ordering the elements by increasing atomic weight, there appeared to be a periodicity to the properties of the elements. He summarized his findings in the table that has come to be called the periodic table. By lining up the elements in horizontal rows in order of increasing atomic weight, and starting a new row when he came to an element with properties similar to one already in the row, he saw that the resulting columns contained elements with similar properties. In generating his periodic table, Mendeleev found that some positions in his table were not filled in, and he predicted that new elements would be found that filled in the gaps. Two of the missing elements—gallium (Ga) and germanium (Ge)—were soon discovered, with properties very close to those Mendeleev had predicted.

Later experiments by H. G. J. Moseley demonstrated that elements in the periodic table should be ordered by atomic numbers rather than atomic weights. Arranging the elements in order of increasing atomic number gives the **law of chemical periodicity:** *The properties of the elements are periodic functions of their atomic numbers (numbers of protons).*

It gives a useful perspective to realize that Mendeleev did this nearly a half-century before electrons, protons, and neutrons were known.

Periodicity of the elements means a recurrence of similar properties at regular intervals when the elements are arranged in the correct order.

Periodic Table Features

Elements in the **periodic table** are arranged according to atomic number so that *elements with similar properties occur in vertical columns* called **groups.** The table commonly used in the United States has groups numbered 1 through 8 (Figure 2.5), with each number followed by either an A or a B. The A groups (Groups 1A and 2A on the left of the table and Groups 3A through 8A at the right) are collectively known as **main group elements.** The B groups (in the middle of the table) are called **transition elements.**

The horizontal rows of the table are called **periods,** and they are numbered beginning with 1 for the period containing only H and He. Sodium (Na) is, for example, in Group 1A and is the first element in the third period. Silver (Ag) is in Group 1B and is in the fifth period.

The table helps us to recognize that most elements are metals (gray and blue colors on the periodic table inside the front cover), far fewer elements are nonmetals (lavender), and even fewer are metalloids (orange). Elements generally become less metallic from left to right across a period, and eventually one or more nonmetals are found in each period. The six metalloids (B, Si, Ge, As, Sb, Te) fall along a zigzag line passing between Al and Si, Ge and As, and Sb and Te.

An alternative convention for numbering the groups in the periodic table uses the numbers 1 through 18, with no letters.

Periodicity of piano keys.

DMITRI MENDELEEV
(1834–1907)

Originally from Siberia, Mendeleev spent most of his life in St. Petersburg. He taught at St. Petersburg University, where he wrote books and published his concept of chemical periodicity, which helped systematize inorganic chemistry. Later in life he moved on to other interests, including studying the natural resources of Russia and their commercial applications.

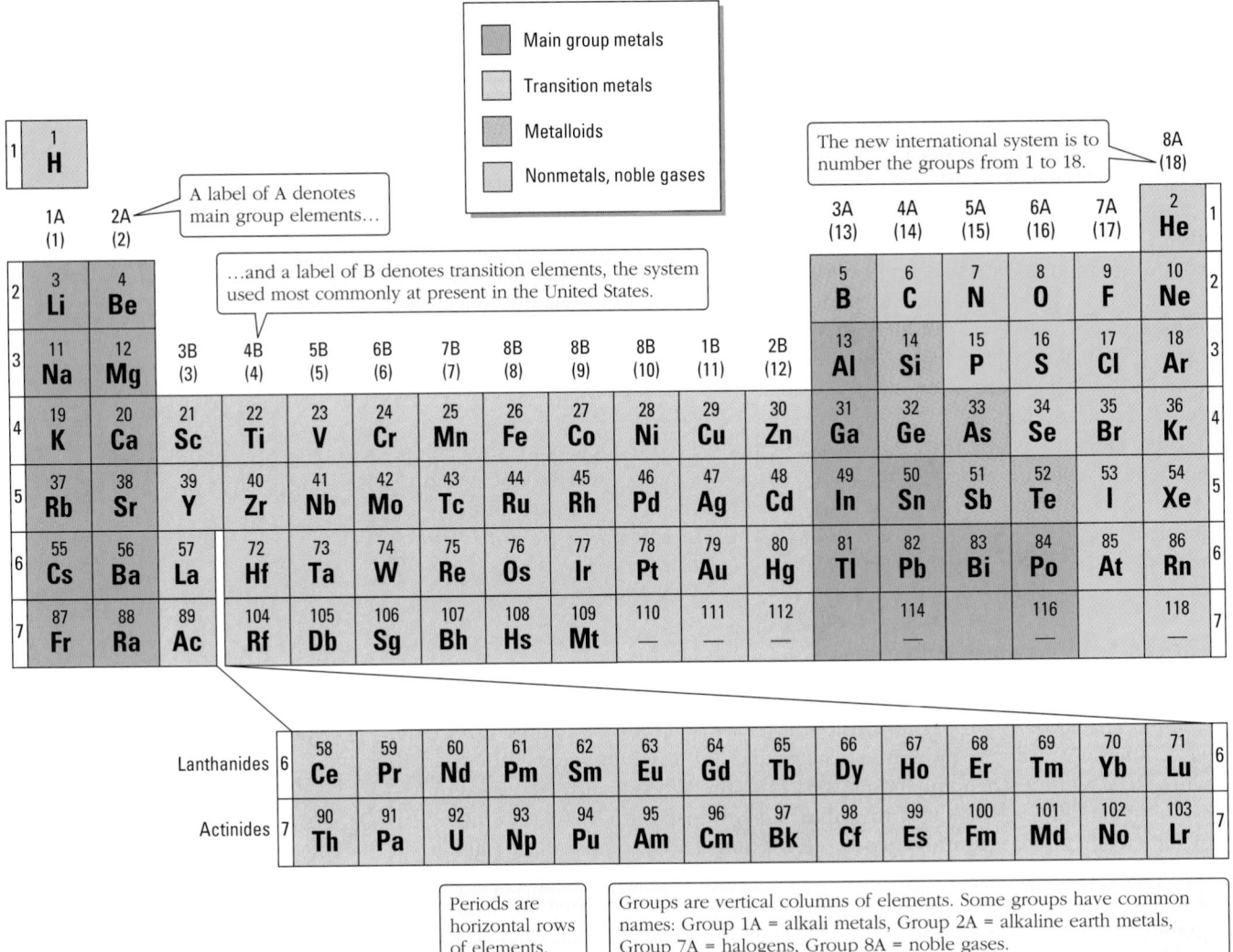

Figure 2.5 Modern periodic table of the elements. Elements are listed across the periods in ascending order of atomic number.

The Alkali Metals (Group 1A) and Alkaline Earth Metals (Group 2A)

"Alkali" comes from the Arabic language. Ancient Arabian chemists discovered that ashes of certain plants, which they called *al-qali,* produced water solutions that felt slippery and burned the skin.

Elements found (uncombined) in nature are sometimes called "free" elements.

Gold and silver as free metals in nature triggered the great gold and silver "rushes" of the 1800s in the United States.

Elements in the leftmost column (Group 1A) are called the **alkali metals** (except hydrogen) because their aqueous solutions are alkaline (basic). Elements in Group 2A, known as **alkaline earth metals,** are extracted from minerals (earths) and also produce alkaline aqueous solutions (except beryllium).

Alkali metals and alkaline earth metals are very reactive and are found in nature only combined with other elements in compounds, never as the free metallic elements. Their compounds are plentiful, and many are significant to human and plant life. Sodium (Na) in sodium chloride (table salt) is a fundamental part of human and animal diets, and throughout history civilizations have sought salt as a dietary necessity and a commercial commodity. Today, sodium chloride is commercially important as a source of two of the most important industrial chemicals — sodium hydroxide and chlorine. Magnesium (Mg) and calcium (Ca), the sixth and fifth most abundant elements in the earth's crust, are present in a vast array of chemical compounds.

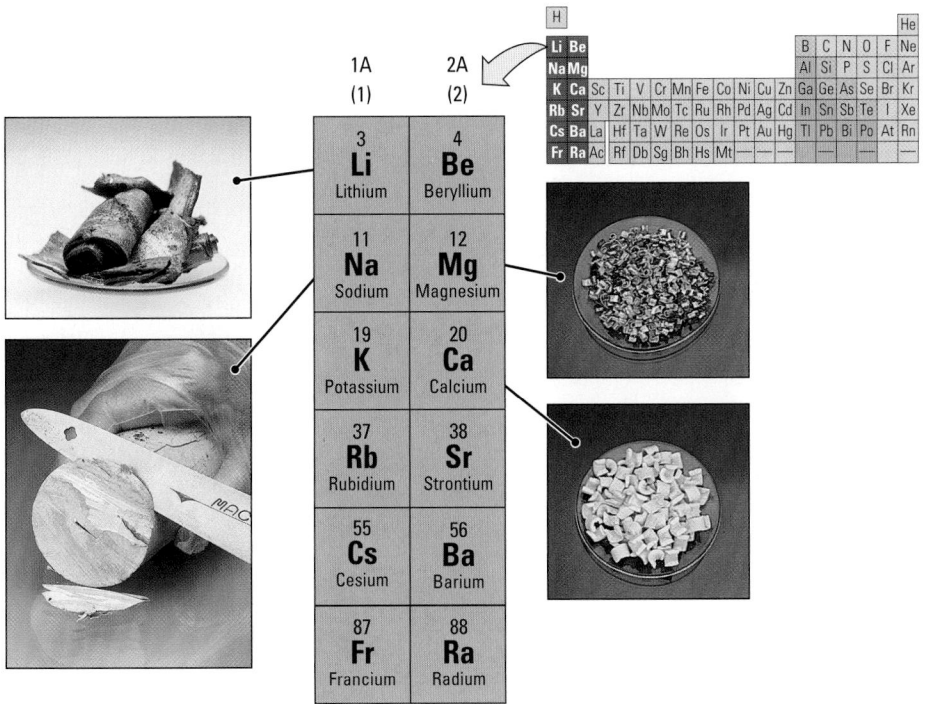

Transition Elements, Lanthanides, and Actinides

The transition elements (also known as the *transition metals*) fill the middle of the periodic table in periods 4 through 7, and most are found in nature only in compounds. The notable exceptions are gold, silver, platinum, copper, and liquid mercury, which can be found in elemental form. Iron, zinc, copper, and chromium are among the most important commercial metals. Because of their vivid colors, transition metal compounds are used for pigments.

The **lanthanides** and **actinides** are listed separately in two rows at the bottom of the periodic table. Using the extra, separate rows keeps the periodic table from becoming too wide and too cumbersome. These elements are relatively rare and not as commercially important as the transition elements.

Groups 3A to 6A

These four groups have no special names, but they contain the most abundant elements in the earth's crust and atmosphere (Table 2.4). They also contain the elements present in most of the important molecules in our bodies: carbon (C), nitrogen (N), and oxygen (O). Because of the ability of carbon atoms to bond extensively with each other, there are huge numbers of carbon compounds. Organic chemistry is the branch of chemistry devoted to the study of carbon compounds. Carbon atoms also provide the framework for the molecules essential to living things, which are the subject of the branch of chemistry known as biochemistry.

Groups 4A to 6A each begin with one or more nonmetals, include one or more metalloids, and end with a metal. Group 4A, for example, contains carbon, a nonmetal, includes two metalloids (Si and Ge), and finishes with two metals (Sn and Pb). Group 3A starts with boron (B), a metalloid.

The Halogens, Group 7A

The elements in this group consist of diatomic molecules and are highly reactive. The group name, **halogens,** comes from the Greek words *hals,* meaning "salt,"

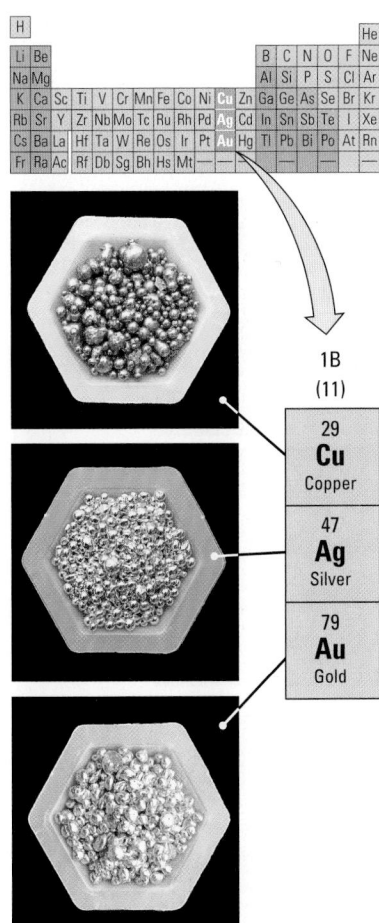

TABLE 2.4 Selected Group 3A–6A Elements

Group 3A

Aluminum: Most abundant *metal* in the earth's crust (7%). Always found combined naturally, especially with silicon and oxygen in clay minerals.

Group 4A

Carbon: Second most abundant element in living things. Provides the framework for organic and biochemical molecules.

Silicon: Second most abundant element in earth's crust (25%). Always found combined naturally, usually with oxygen in quartz and silicate minerals.

Group 5A

Nitrogen: Most abundant element in earth's *atmosphere* (78%) but not abundant in earth's crust because of relatively low chemical reactivity of N_2.

Group 6A

Oxygen: Most abundant element in earth's crust (47%) because of high chemical reactivity. Second most abundant element in earth's atmosphere (21%).

and *genes,* meaning "forming." The halogens all form salts — compounds similar to sodium chloride (NaCl) — by reacting vigorously with alkali metals and with other metals as well. Halogens also react with most nonmetals. Small carbon compounds containing chlorine and fluorine are relatively unreactive but are involved in seasonal ozone depletion in the upper atmosphere (Section 10.11).

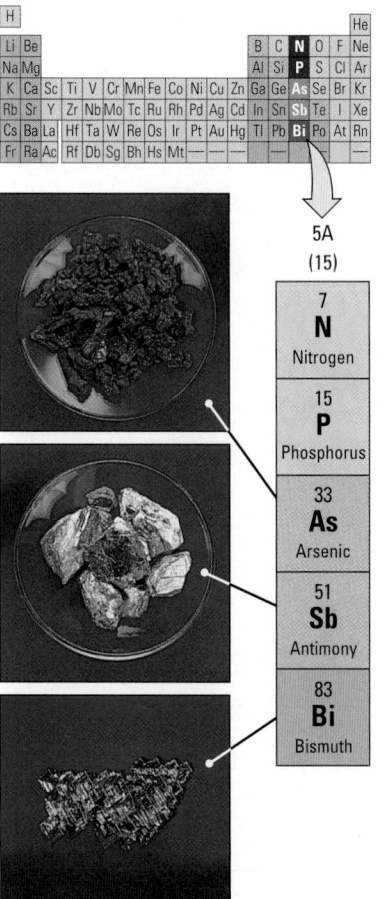

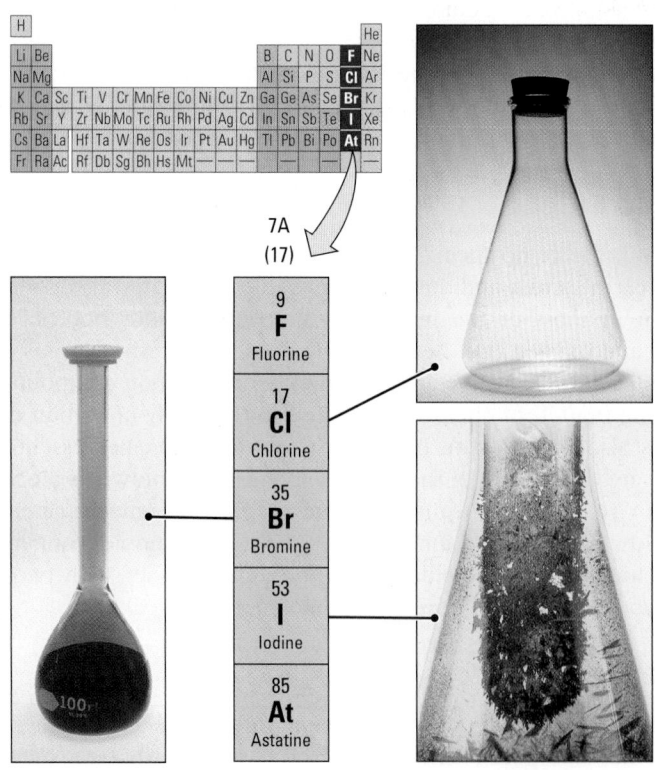

Chemistry in the News

NEWEST, HEAVIEST ELEMENTS

How many elements are known? The number depends on what year you ask the question, because scientists are continually trying to create new elements in the laboratory. Three new elements — 110, 111, and 112 — were synthesized between 1994 and 1996 at the Society for Heavy-Ion Research (GSI), Darmstadt, Germany. In 1999 reports of elements 114, 116, and 118 appeared: The Joint Institute for Nuclear Research, Dubna, Russia, reported observing element 114, and Lawrence Berkeley National Laboratory reported observing elements 116 and 118. The creation of these newest elements is a tale of cutting-edge science done by teams of international researchers.

GSI researchers synthesized elements 107 to 112 by bombarding lead-208 or bismuth-209 nuclei for 2 to 3 weeks with a beam of nuclei with atomic numbers from 24 (Cr) to 30 (Zn) (see table). The bombarding nuclei, because they carry a positive charge, must be accelerated to velocities that give them the correct energy to overcome the repulsion by the positive charge of the Pb or Bi target nuclei. The new elements are formed when the bombarding nuclei combine with the target nuclei, which then may emit one or more neutrons. The energy of the bombarding particles must be just right — if too energetic, the combination splits into two lighter nuclei; if there is not enough energy, the combination doesn't occur. When atoms of a new element are formed, they are separated from the beam by electric and magnetic fields along an 11-m heavy ion separator. Very sophisticated, ultrasensitive detectors are needed because very few of the new atoms are produced. A 2-week experiment, for example, produced just four atoms of element 110, and it took 18 days to create three atoms of element 111.

In spite of having so few atoms to work with, researchers can identify the new elements by the energies of the alpha particles they emit while undergoing radioactive decay to nuclei of lower atomic numbers.

The GSI team continues to seek even heavier elements by bombarding lead-208 with germanium-76 to get element 114, and bismuth-209 with zinc-70 or germanium-76 to form elements 113 and 115, respectively. The GSI team uses a somewhat different experimental approach from the Russian team, so searching for different pathways to synthesize a

The apparatus used at Lawrence Berkeley Laboratory for the discovery of superheavy element 118.

heavy element reported by others (here, element 114) is fruitful.

In 1999 the Joint Institute for Nuclear Research, in Dubna, Russia, in collaboration with the Lawrence Livermore National Laboratory, reported evidence for element 114 made by combining plutonium-244 and calcium-48. With an apparent 30-second lifetime (before emitting radioactive particles and thus changing to a different element), this would be an exceptionally long-lived heavy element, which has implications for nuclear theory and a postulated "island of stability" among heavy nuclei.

Also in 1999, an international team of scientists at Lawrence Berkeley National Laboratory discovered element 118 and element 116, which is formed from 118 by emission of an alpha particle, by bombarding a target of lead-208 atoms with a beam of krypton-86 in a cyclotron particle accelerator. Element 118 had a lifetime of less than a millisecond.

Sources:

Dagani, R. Two Superheavy Elements Created. *Chemical & Engineering News*, June 14, 1999, p. 6.

Freemantle, M. Heavy-Ion Research Institute Explores Limits of Periodic Table. *Chemical & Engineering News*, March 13, 1995, p. 37, and February 26, 1996, p. 6.

Rouhi, M. Element 112: New Element Made from Zinc and Lead. *Chemical & Engineering News*, February 26, 1996, p. 6.

Stone, R. Element 114 Lumbers into View. *Science*, January 22, 1999, p. 474.

Recently Discovered Elements

Bombarding Particle	Target Nucleus	Atomic Number of New Element	Mass Number of New Element	New Element Created
Cr-54	Bi-209	107	262	Feb. 1981
Fe-58	Pb-208	108	265	Mar. 1984
Fe-58	Bi-209	109	266	Sept. 1982
Ni-62	Pb-208	110	269	Nov. 1994
Ni-64	Bi-209	111	272	Dec. 1994
Zn-70	Pb-208	112	277	Feb. 1996
Ca-48	Pu-244	114	289	Jan. 1999
Kr-86	Pb-208	116	289	June 1999
Kr-86	Pb-208	118	293	June 1999

The Noble Gases, Group 8A

Once the arrangement of electrons in atoms was understood (Section 7.6), the place where the noble gases fit into the periodic table was obvious.

The **noble gases** at the far right of the periodic table are the least reactive elements. Their lack of chemical reactivity, as well as their rarity, prevented them from being discovered until about a century ago. Thus, they were not known when Mendeleev developed his periodic table. Until 1962 they were called the *inert* gases, because they were thought not to combine with any element. In 1962 this basic canon of chemistry was overturned when compounds of xenon with fluorine and with oxygen were synthesized. Since then other xenon compounds have been made, as well as compounds of fluorine with krypton and with radon.

Exercise 2.8 The Periodic Table

1. How many (a) metals, (b) nonmetals, and (c) metalloids are in the fourth period of the periodic table? Give the name and symbol for each element.
2. Which groups of the periodic table contain (a) only metals, (b) only nonmetals, (c) only metalloids?
3. Which period of the periodic table contains the most metals?

Exercise 2.9 Element Names

On Wednesday, June 14, 2000, a major daily American newspaper published the following paragraph:

> ABC's *Who Wants to Be a Millionaire* crowned its fourth million-dollar winner Tuesday night. Bob House . . . [answered] the final question: Which of the following men doesn't have a chemical compound named after him? (a) Enrico Fermi; (b) Albert Einstein, (c) Niels Bohr, (d) Isaac Newton

What is wrong with the question? What is the correct answer to the question after it is properly posed? (The question was properly posed and correctly answered on the TV show.)

CHEMISTRY You Can Do...   Preparing a Pure Sample of an Element

You will need the following items to do this experiment:

- Two glasses or plastic cups that will each hold about 250 mL of liquid
- Approximately 100 mL (about 3.5 oz) of vinegar
- Soap
- An iron nail, paper clip, or other similar-sized piece of iron
- Something abrasive, such as a piece of steel wool, Brillo, sandpaper, or nail file
- About 40 to 50 cm of thin string or thread
- Some table salt
- A magnifying glass (optional)
- 15 to 20 dull pennies (shiny pennies will not work)

Wash the piece of iron with soap, dry it, and clean the surface further with the steel wool or other abrasive until it is shiny. Tie one end of the string around one end of the piece of iron.

Place the pennies in one cup (labeled A) and pour in enough vinegar to cover them. Sprinkle on a little salt, swirl the liquid around so it contacts all the pennies, and observe what happens. When nothing more seems to be happening, pour the liquid into the second cup (labeled B), leaving the pennies in cup A (that is, pour off the liquid). Suspend the piece of iron from the thread so that it is half-submerged in the liquid in cup B.

Observe the piece of iron over a period of 10 minutes or so, and then use the thread to pull it out of the liquid. Observe it carefully, using a magnifying glass if you have one. Compare the part that was submerged with the part that remained above the surface of the liquid.

1. What did you observe happening to the pennies?
2. How could you account for what happened to the pennies in terms of a microscopic model? Cite observations that support your conclusion.
3. What did you observe happening to the piece of iron?
4. Interpret the experiment in terms of a microscopic model, citing observations that support your conclusions.
5. Would this method be of use in purifying copper? If so, can you suggest ways that it could be used effectively to obtain copper from ores?

SUMMARY PROBLEM

Atoms of one of the metallic elements contain 51 protons and 70 neutrons.

(a) Identify the element and give its symbol.

(b) What is this atom's (i) atomic number? (ii) mass number?

(c) This element has two naturally occurring isotopes.

Isotope	Mass Number	Percent Abundance	Isotopic Mass (amu)
1	121	57.25	120.9038
2	123	42.75	122.9041

Calculate the atomic weight of the element.

(d) This element is in which group of the periodic table? Is this element a metal, a metalloid, or a nonmetal? Explain your choice.

(e) A .357-caliber bullet contains 1.07×10^{-6} g of the element. (i) How many moles of the element are in this mass? (ii) How many atoms of the element are in the sample? (iii) Atoms of this element have an atomic diameter of 282 pm. If all the atoms of this element in the sample were laid side by side, how many meters long would the chain of atoms be?

IN CLOSING

Having studied this chapter, you should be able to . . .

- Describe radioactivity, electrons, protons, and neutrons and the general structure of the atom (Sections 2.1 – 2.2).
- Use conversion factors for the units for mass, volume, and length common in chemistry (Section 2.3).
- Define isotope and give the mass number and number of neutrons for a specific isotope (Section 2.4).
- Calculate the atomic weight of an element from isotopic abundances (Section 2.5).
- Explain the difference between the atomic number and the atomic weight of an element and find this information for any element (Sections 2.4 – 2.5).
- Relate masses of elements to the mole, Avogadro's number, and molar mass (Section 2.6).
- Do gram-mole and mole-gram conversions for elements (Section 2.7).
- Identify the periodic table location of groups, periods, alkali metals, alkaline earth metals, halogens, noble gases, transition elements, lanthanides, and actinides (Section 2.8).

KEY TERMS

actinides *(2.8)*

alkali metals *(2.8)*

alkaline earth metals *(2.8)*

atomic mass unit (amu) *(2.4)*

atomic number *(2.4)*

atomic structure *(Introduction)*

atomic weight *(2.5)*

Avogadro's number *(2.6)*

chemical periodicity, law of *(2.8)*

electron *(2.1)*

group *(2.8)*

halogens *(2.8)*

ion *(2.1)*

isotope *(2.4)*
lanthanides *(2.8)*
main group elements
  (2.8)
mass *(2.3)*
mass number *(2.4)*
mass spectrometer
  (p. 54)
mass spectrum
  (p. 54)

metric system *(2.3)*
molar mass *(2.6)*
mole (mol) *(2.6)*
neutron *(2.2)*
noble gases *(2.8)*
nucleus *(2.2)*
percent abundance
  (2.5)
period *(2.8)*
periodic table *(2.8)*

proton *(2.1)*
radioactivity *(2.1)*
scanning tunneling mi-
  croscope *(p. 48)*
transition elements
  (2.8)

QUESTIONS FOR REVIEW AND THOUGHT

Conceptual Challenge Problems

CP-2.A (Section 2.1)   Suppose that you are faced with a problem similar to the one faced by Robert Millikan when he analyzed data from his oil-drop experiment. Below are the masses of three stacks of dimes. What do you conclude to be the mass of a dime, and what is your argument?

Stack 1 = 9.12 g   Stack 2 = 15.96 g   Stack 3 = 27.36 g

CP-2.B (Section 2.3)   The age of the universe is unknown, but some conclude from measuring Hubble's constant that it is about 18 billion years old, which is about four times the age of the earth. If so, what is the age of the universe in seconds? If you had a sample of carbon with the same number of carbon atoms as there have been seconds since the universe began, could you measure this sample on a laboratory balance that can detect masses as small as 0.1 mg?

Answers to questions in **bold** can be found at the back of the book.

Review Questions

1. Two phenomena played an important role in the discovery of atomic structure.
   (a) The first, _____, was investigated by Benjamin Franklin using such experiments as the infamous key on a kite string in a thunderstorm. This phenomenon consists of two parts that are *conserved,* meaning you cannot find one part without finding one of its counterpart. These two parts are a _____ charge and a _____ charge. How is this phenomenon similar to the structure of an atom?
   (b) The second, _____, was investigated by many scientists including Marie Curie, who named it, but Henri Becquerel discovered it. Three types of this phenomenon can be observed coming from some elements. The types are _____, _____, and _____. What are the respective charges on each type? How does each type behave when passing through electrically charged plates?
2. (a) List the properties exhibited by cathode rays.
   (b) What is the charge on the particles making up cathode rays?
   (c) Which subatomic particles are cathode rays most like?
   (d) Cathode rays and beta particles consist of which subatomic particle?
3. What is the fundamental unit of electrical charge?
4. Beta particles consist of which subatomic particle?
5. Millikan was able to determine the charge on an electron using his famous oil-drop experiment. Describe the experiment and how Millikan was able to calculate the mass

of an electron using his results and the ratio discovered earlier by Thomson.
6. The positively charged particle in an atom is called the proton.
   (a) How much heavier is a proton than an electron?
   (b) What is the difference in the charge on a proton and an electron?
7. Ernest Rutherford's famous gold-foil experiment examined the structure of atoms.
   (a) What surprising result was observed?
   (b) The results of the gold-foil experiment enabled Rutherford to calculate that the nucleus is much smaller than the atom. How much smaller?
8. In any given *neutral* atom, how many protons are there compared with the number of electrons?

Units and Unit Conversions

9. If the nucleus of an atom were the size of a medium-sized orange (let us say with a diameter of about 6 cm), what would be the diameter of the atom?
10. The average lead pencil, new and unused, is 19 cm long. What is its length in millimeters? In meters? In inches?
11. The pole vault world record is 6.14 m. What is this in cm? In feet and inches?
12. The maximum speed limit in many states is 65 miles per hour. What is this speed in kilometers per hour?
13. Olympic stadium in Montreal has a center-field fence 404 ft from home plate. What is this distance in meters?
14. A Volkswagen engine has a displacement of 120 in.^{3}. What is this volume in cubic centimeters? In liters?
15. An automobile engine has a displacement of 250. in.^{3}. What is this volume in cubic centimeters? In liters?
16. Suppose your bedroom is 18 ft long, 15 ft wide, and the distance from floor to ceiling is 8 ft, 6 in. You need to

know the volume of the room in metric units for some scientific calculations. What is the room's volume in cubic meters? In liters?

17. A crystal of fluorite (a mineral that contains calcium and fluorine) has a mass of 2.83 g. What is this mass in kilograms? In pounds? Give the symbols for the elements in this crystal.

Percent

18. Silver jewelry is actually a mixture of silver and copper. If a bracelet with a mass of 17.6 g contains 14.1 g of silver, what is the percentage of silver? Of copper?

19. The solder once used by plumbers to fasten copper pipes together consists of 67% lead and 33% tin. What is the mass of lead (in grams) in a 1.00-lb block of solder? What is the mass of tin?

20. Automobile batteries are filled with sulfuric acid. What is the mass of the acid (in grams) in 500. mL of the battery acid solution if the density of the solution is 1.285 g/cm^3 and the solution is 38.08% sulfuric acid by mass?

21. When popcorn pops, it loses water explosively. If a kernel of corn weighing 0.125 g before popping weighs 0.106 g afterward, what percentage of its mass did it lose on popping?

22. A popular breakfast cereal contains 190 mg of sodium per 30-g serving. What percentage of the cereal is sodium?

Isotopes

23. What is wrong with the following statement? Atoms of the same element always have the same mass number, but atoms of the same element can have different atomic numbers due to the presence of *isotopes*.

24. Americium-241 is used in household smoke detectors and in bone mineral analysis. Give the number of electrons, protons, and neutrons in an atom of americium-241.

25. The artificial radioactive element technetium is used in many medical studies. Give the number of electrons, protons, and neutrons in an atom of technetium-99.

26. Atoms of the same element have the same number of protons in the nucleus, and therefore all the atoms for any given element would have the same atomic _____.

27. What is the definition of the atomic mass unit?

28. To estimate an atom's mass, one only need to add up the number of protons and neutrons in the nucleus. This estimate, a whole number, is referred to as the _____ number for a given atom. Why can we essentially ignore the mass of the electrons?

29. What is the difference between the mass number and the atomic number of an atom?

30. When you subtract the atomic number from the mass number for an atom, what do you obtain?

31. Give the mass number of each of the following atoms: (a) beryllium with 5 neutrons, (b) titanium with 26 neutrons, and (c) gallium with 39 neutrons.

32. Give the mass number of (a) an iron atom with 30 neutrons, (b) an americium atom with 148 neutrons, and (c) a tungsten atom with 110 neutrons.

33. Give the complete symbol $_Z^A X$ for each of the following atoms: (a) sodium with 12 neutrons, (b) argon with 21 neutrons, and (c) gallium with 38 neutrons

34. Give the complete symbol $_Z^A X$ for each of the following atoms: (a) nitrogen with 8 neutrons, (b) zinc with 34 neutrons, and (c) xenon with 75 neutrons

35. How many electrons, protons, and neutrons are there in an atom of (a) calcium-40, $_{20}^{40}Ca$, (b) tin-119, $_{50}^{119}Sn$, and (c) plutonium-244, $_{94}^{244}Pu$?

36. How many electrons, protons, and neutrons are there in an atom of (a) carbon-13, $_6^{13}C$, (b) chromium-50, $_{24}^{50}Cr$, and (c) bismuth-205, $_{83}^{205}Bi$?

37. Fill in the following table:

Z	A	Number of Neutrons	Element
35	81	___	___
___	___	62	Pd
77	___	115	
___	151	___	Eu

38. Fill in the following table:

Z	A	Number of Neutrons	Element
60	144	___	___
___	___	12	Mg
64	___	94	
___	37	___	Cl

39. Which of the following are isotopes of element X, whose atomic number is 9: $_9^{18}X$, $_9^{20}X$, $_4^9X$, $_9^{15}X$?

40. Cobalt has three radioactive isotopes used in medical studies. Atoms of these isotopes have 30, 31, and 33 neutrons, respectively. Give the symbol for each of these isotopes.

Atomic Weight

41. Verify that the atomic weight of lithium is 6.941 amu, given the following information:

^{6}Li, exact mass = 6.015121 amu

percent abundance = 7.500%

^{7}Li, exact mass = 7.016003 amu

percent abundance = 92.50%

42. Verify that the atomic weight of magnesium is 24.3050 amu, given the following information:

^{24}Mg, exact mass = 23.985042 amu

percent abundance = 78.99%

^{25}Mg, exact mass = 24.98537 amu

percent abundance = 10.00%

^{26}Mg, exact mass = 25.982593 amu

percent abundance = 11.01%

43. Gallium has two naturally occurring isotopes, ^{69}Ga and ^{71}Ga, with masses of 68.9257 amu and 70.9249 amu, respectively. Calculate the abundances of these isotopes of gallium.

44. Copper has two stable isotopes, ^{63}Cu and ^{65}Cu, with masses of 62.939598 amu and 64.927793 amu, respectively. Calculate the abundances of these isotopes of copper.

45. Lithium has two stable isotopes, ^{6}Li and ^{7}Li. Since the atomic weight of lithium is 6.941, which is the more abundant isotope?

46. Argon has three naturally occurring isotopes: 0.337% ^{36}Ar, 0.063% ^{38}Ar, and 99.60% ^{40}Ar. What would you estimate the atomic weight of argon to be? If the masses of the isotopes are 35.968, 37.963, and 39.962, respectively, what is the atomic weight of natural argon?

The Mole

47. The "mole" is simply a convenient unit for counting molecules and atoms. Name four "counting units" (such as a dozen for eggs and cookies) that you commonly encounter.

48. If you divide Avogadro's number of pennies among the 260 million men, women, and children in the United States, and if each person could count one penny each second every day of the year for eight hours a day, how long would it take to count all of the pennies?

49. Why do you think it is more convenient to use some chemical counting unit when doing calculations (chemists have adopted the unit of the mole, but it could have been something different) rather than using individual molecules?

50. Calculate the number of grams in
    (a) 2.5 mol of boron   (c) 1.25×10^{-3} mol of iron
    (b) 0.015 mol of O_2 (d) 653 mol of helium

51. Calculate the number of grams in
    (a) 6.03 mol of gold
    (b) 0.045 mol of uranium
    (c) 15.6 mol of Ne
    (d) 3.63×10^{-4} mol of plutonium

52. Calculate the number of moles represented by each of the following:
    (a) 127.08 g of Cu   (b) 20.0 g of calcium
    (c) 16.75 g of Al   (d) 0.012 g of potassium
    (e) 5.0 mg of americium

53. Calculate the number of moles represented by each of the following:
    (a) 16.0 g of Na   (b) 0.0034 g of platinum
    (c) 1.54 g of P   (d) 0.876 g of arsenic
    (e) 0.983 g of Xe

54. How many moles of Na are in 50.4 g?

55. Krypton really does not give Superman his strength. If you have 0.00789 g of the gaseous element, how many moles does this represent?

56. If you have 4.6×10^{-3} g of gaseous helium, how many moles do you have?

57. In an experiment, you need 0.125 mol of sodium metal. Sodium can be cut easily with a knife, so if you cut out a block of sodium, what should be the volume of the block in cubic centimeters? If you cut a perfect cube, what will be the length of the edge of the cube? (The density of sodium is 0.968 g/cm^3.) (*Caution:* Sodium is *very* reactive with water. The metal should be handled only by a knowledgeable chemist.)

58. If you have a 35.67-g piece of chromium metal on your car, how many atoms of chromium do you have?

59. If you have a ring that contains 1.94 g of gold, how many atoms of gold are in the ring?

60. What is the average mass in grams of one copper atom?

61. What is the average mass in grams of one atom of titanium?

The Periodic Table

62. What is the difference between a group and a period in the periodic table?

63. Name and give symbols for (a) three elements that are metals; (b) four elements that are nonmetals; and (c) two elements that are metalloids. In each case, also locate the element in the periodic table by giving the group and period in which the element is found.

64. Name and give symbols for three transition metals in the fourth period. Look up each of your choices in a dictionary or on the Internet, and make a list of their properties. Also list the uses of each element.

65. Name two halogens. Look up each of your choices in a dictionary, in a book such as *The Handbook of Chemistry and Physics*, or on the Internet, and make a list of their properties. Also list any uses of each element that are given by the source.

66. Name three transition elements, two halogens, and one alkali metal.

67. Name an alkali metal, an alkaline earth metal, and a halogen.

68. Name an element discovered by Madame Curie. Give its name, symbol, and atomic number. Use a dictionary or the Internet to find the origin of the name of this element.

69. How many elements are there in Group 4A of the periodic table? Give the name and symbol of each of these elements. Tell whether each is a metal, nonmetal, or metalloid.

70. How many elements are there in the fourth period of the periodic table? Give the name and symbol of each of these elements. Tell whether each is a metal, metalloid, or nonmetal.

71. Which single period in the periodic table contains the most metals, metalloids, *and* nonmetals?

72. How many periods of the periodic table have 8 elements, how many have 18 elements, and how many have 32 elements?

73. Use the periodic table to answer the following:
    (a) Name an element in Group 2A.
    (b) Name an element in the third period.
    (c) What element is in the second period in Group 4A?
    (d) What element is in the third period in Group 6A?

(e) What halogen is in the fifth period?

(f) What alkaline earth element is in the third period?

(g) What noble gas element is in the fourth period?

(h) What nonmetal is in Group 6A and the second period?

(i) Name a metalloid in the fourth period.

74. Use the periodic table to answer the following:

(a) Name an element in Group 2B.

(b) Name an element in the fifth period.

(c) What element is in the sixth period in Group 4A?

(d) What element is in the third period in Group 5A?

(e) What alkali metal is in the third period?

(f) What noble gas is in the fifth period?

(g) Name the element in Group 6A and the fourth period. Is it a metal, nonmetal, or metalloid?

(h) Name a metalloid in Group 5A.

75. The following chart is a plot of the logarithm of the relative abundances of elements 1 through 36 in the solar system. The abundances are given on a scale that assigns silicon a relative value of 1.00×10^6 (the logarithm of which is 6).

(a) What is the most abundant metal?

(b) What is the most abundant nonmetal?

(c) What is the most abundant metalloid?

(d) Which of the transition elements is most abundant?

(e) How many halogens are considered on this plot, and which is the most abundant?

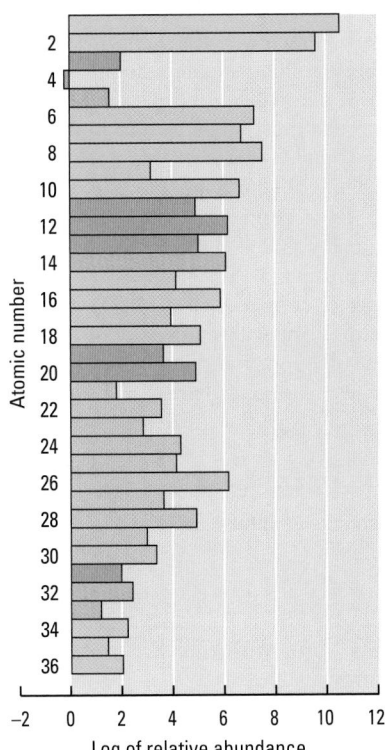

76. Consider the plot of relative abundance versus atomic number once again. Uncover any relation between abundance and atomic number. Is there any difference between elements of even atomic number and those of odd atomic number?

General Questions

77. In his beautifully written autobiography, *The Periodic Table,* Primo Levi says of zinc that "it is not an element which says much to the imagination; it is gray and its salts are colorless; it is not toxic, nor does it produce striking chromatic reactions; in short, it is a boring metal. It has been known to humanity for two or three centuries, so it is not a veteran covered with glory like copper, nor even one of these newly minted elements which are still surrounded with the glamour of their discovery." From this description, and from reading this chapter, make a list of the properties of zinc. For example, include in your list the position of the element in the periodic table, and tell how many electrons and protons an atom of zinc has. What are its atomic number and atomic weight? Zinc is important in our economy. Check your dictionary, a book such as *The Handbook of Chemistry and Physics*, or the Internet, and make a list of the uses of the element.

78. The density of a solution of sulfuric acid is 1.285 g/cm³, and it is 38.08% acid by mass. What volume of the acid solution (in mL) do you need to supply 125 g of sulfuric acid?

79. Molecular distances are usually given in nanometers (1 nm = 1×10^{-9} m) or in picometers (1 pm = 1×10^{-12} m). However, a commonly used unit has been the angstrom, where 1 Å = 1×10^{-10} m. If the distance between the Pt atom and the N atom in the cancer chemotherapy drug cisplatin is 1.97 Å, what is the distance in nm? In pm?

80. The separation between carbon atoms in diamond is 0.154 nm. (a) What is their separation in meters? (b) What is the carbon atom separation in angstroms (where 1 Å = 1×10^{-10} m)?

81. The smallest repeating unit of a crystal of common salt is a cube with an edge length of 0.563 nm. What is the volume of this cube in nm³? In cm³?

82. The cancer drug cisplatin contains 65.0% platinum. If you have 1.53 g of the compound, how many grams of platinum does this sample contain?

83. At 25 °C, the density of water is 0.997 g/cm³, whereas the density of ice at −10 °C is 0.917 g/cm³. (a) If a plastic soft drink bottle (volume = 250 mL) is filled with pure water, capped, and then frozen at −10 °C, what volume will the solid occupy? (b) Could the ice be contained within the bottle?

84. A common fertilizer used on lawns is designated as "16-4-8." These numbers mean that the fertilizer contains 16% nitrogen-containing compounds, 4.0% phosphorus-containing compounds, and 8.0% potassium-containing compounds. You buy a 40.0-lb bag of this fertilizer and use all of it on your lawn. How many grams of the phosphorus-containing compound are you putting on your lawn? If the phosphorus-containing compound consists of 43.64% phosphorus (the rest is oxygen), how many grams of phosphorus are there in 40.0 lb of fertilizer?

85. The fluoridation of city water supplies has been practiced in the United States for several decades because it is believed that fluoride prevents tooth decay, especially in young children. This is done by continuously adding sodium fluoride to water as it comes from a reservoir. Assume you live in a medium-sized city of 150,000 people

and that each person uses 175 gal of water per day. How many tons of sodium fluoride would you have to add to the water supply each year (365 days) in order to have the required fluoride concentration of 1 part per million (that is, 1 ton of fluoride per million tons of water)? (Sodium fluoride is 45.0% fluoride, and one U.S. gallon of water has a mass of 8.34 lb.)

86. Name three elements that you have encountered today. (Name only those that you have seen as elements, not those combined into compounds.) Give the location of each of these elements in the periodic table by specifying the group and period in which it is found.

87. Potassium has three stable isotopes, ^{39}K, ^{40}K, and ^{41}K, but ^{40}K has a very low natural abundance. Which of the other two is the more abundant?

88. The figure on p. 54 shows the mass spectrum of neon isotopes. What are the symbols of the isotopes? Which is the more abundant isotope? How many protons, neutrons, and electrons does this isotope have? Without looking at a periodic table, give the approximate atomic weight of neon.

89. When an athlete tears ligaments and tendons, they can be surgically attached to bone to keep them in place until they reattach themselves. A problem with current techniques, though, is that the screws and washers used are often too big to be positioned accurately or properly. Therefore, a titanium-containing device is used.
(a) What are the symbol, atomic number, and atomic weight of titanium?
(b) In what group and period is it found? Name the other elements of its group.
(c) What chemical properties do you suppose make titanium an excellent choice for this and other surgical applications?
(d) Use a dictionary, a book such as *The Handbook of Chemistry and Physics*, or the Internet to make a list of the properties of the element and its uses.

90. **Crossword puzzle:** In the 2×2 crossword puzzle shown, each letter must be correct four ways: horizontally, vertically, diagonally, and by itself. Instead of words, use symbols of elements. When the puzzle is complete, the four spaces will contain the overlapping symbols of ten elements. There is only one correct solution.*

1	2
3	4

*This puzzle appeared in *Chemical & Engineering News,* Dec. 14, 1987, p. 86 (submitted by S. J. Cyvin) and in *Chem Matters,* October 1988.

Horizontal

1–2: Two-letter symbol for a metal used in ancient times

3–4: Two-letter symbol for a metal that burns in air and is found in Group 5A

Vertical

1–3: Two-letter symbol for a metalloid

2–4: Two-letter symbol for a metal used in U.S. coins

Single squares (all one-letter symbols)

1: A colorful nonmetal

2: A colorless gaseous nonmetal

3: An element that makes fireworks green

4: An element that has medicinal uses

Diagonal

1–4: Two-letter symbol for an element used in electronics

2–3: Two-letter symbol for a metal used with Zr to make wires for superconducting magnets

91. Draw a picture showing the approximate positions of all protons, electrons, and neutrons in an atom of helium-4. Make certain that your diagram indicates both the number and position of each type of particle.

92. Gems and precious stones are measured in carats, a weight unit equivalent to 200 mg. If you have a 2.3-carat diamond in a ring, how many moles of carbon do you have?

93. The international markets in precious metals operate in the weight unit "troy ounce" (where 1 troy ounce is equivalent to 31.1 g). Platinum sells for $325 per troy ounce. (a) How many moles are there in 1 troy ounce? (b) If you have $5000 to spend, how many grams and how many moles of platinum can you purchase?

94. Gold prices fluctuate, depending on the international situation. If gold currently sells for $338.70 per troy ounce, how much must you spend to purchase 1.00 mol of gold (1 troy ounce is equivalent to 31.1 g)?

95. The Statue of Liberty in New York harbor is made of 2.00 $\times 10^5$ lb of copper sheets bolted to an iron framework. How many grams and how many moles of copper does this represent (1 lb = 454 g)?

96. A piece of copper wire is 25 ft long and has a diameter of 2.0 mm. Copper has a density of 8.92 g/cm^3. How many moles of copper and how many atoms of copper are there in the piece of wire?

Applying Concepts

97. Which sets of values are possible? Why are the others not possible?

	Mass Number	Atomic Number	Number of Protons	Number of Neutrons
(a)	19	42	19	23
(b)	235	92	92	143
(c)	53	131	131	79
(d)	32	15	15	15
(e)	14	7	7	7
(f)	40	18	18	40

98. Which member of the pair has the greater number of particles? Explain why.
(a) 1 mol of Cl or 1 mol of Cl_2
(b) 1 molecule of O_2 or 1 mol of O_2
(c) 1 nitrogen atom or 1 nitrogen molecule
(d) 6.022×10^{23} fluorine molecules or 1 mol of fluorine molecules
(e) 20.2 g of Ne or 1 mol of Ne
(f) 1 molecule of Br_2 or 159.8 g of Br_2
(g) 107.9 g of Ag or 6.9 g of Li
(h) 58.9 g of Co or 58.9 g of Cu
(i) 1 g of calcium or 6.022×10^{23} calcium atoms
(j) 1 g of chlorine atoms or 1 g of chlorine molecules

99. Which has the greater mass? Explain why.
(a) 1 mol of iron or 1 mol of aluminum
(b) 6.022×10^{23} lead atoms or 1 mol of lead
(c) 1 copper atom or 1 mol of copper
(d) 1 mol of Cl or 1 mol of Cl_2
(e) 1 g of oxygen atoms or 1 g of oxygen molecules
(f) 24.3 g of Mg or 1 mol of Mg
(g) 1 mol of Na or 1 g of Na
(h) 4.0 g of He or 6.022×10^{23} He atoms
(i) 1 molecule of I_2 or 1 mol of I_2
(j) 1 oxygen molecule or 1 oxygen atom

100. A group of astronauts in a spaceship accidentally encounters a space warp that traps them in an alternative universe where the chemical elements are quite different from the ones they are used to. The astronauts find these properties for the elements that they have discovered:

Atomic Symbol	Atomic Weight	State	Color	Electrical Conductivity	Electrical Reactivity
A	3.2	Solid	Silvery	High	Medium
D	13.5	Gas	Colorless	Very low	Very high
E	5.31	Solid	Golden	Very high	Medium
G	15.43	Solid	Silvery	High	Medium
J	27.89	Solid	Silvery	High	Medium
L	21.57	Liquid	Colorless	Very low	Medium
M	11.23	Gas	Colorless	Very low	Very low
Q	8.97	Liquid	Colorless	Very low	Medium
R	1.02	Gas	Colorless	Very low	Very high
T	33.85	Solid	Colorless	Very low	Medium
X	23.68	Gas	Colorless	Very low	Very low
Z	36.2	Gas	Colorless	Very low	Medium
Ab	29.85	Solid	Golden	Very high	Medium

(a) Arrange these elements into a periodic table.
(b) If a new element, X, with atomic weight 25.84 is discovered, what would its properties be? Where would it fit in the periodic table you constructed?
(c) Are there any elements that have not yet been discovered? If so, what would their properties be?

 General Chemistry CD-ROM

CD2.1 Screen 2.6: Electricity and Electric Charge.
(a) What is an important principle of operation of the electroscope?
(b) How does the attraction and repulsion of electric charges apply to the structure of the atom? Go back to Screen 2.2 and think again about the structure of the atom.

CD2.2 Screen 2.8: Electrons.
(a) How could you demonstrate that an electron is a negatively charged particle?
(b) What property of the electron was measured in Thomson's experiment?
(c) How are electrons related to the beta rays emitted by radioactive elements?

CD2.3 Screen 2.10: Protons.
(a) How are positively charged particles generated in the "canal-ray" experiment?
(b) Why does hydrogen have the largest charge-to-mass ratio of all the elements studied?

CD2.4 Screen 2.11: The Nucleus of the Atom.
(a) Why are so few particles deflected as they pass through the foil?
(b) Why are even fewer particles deflected back in the direction from which they came?

CD2.5 Screen 2.15: Atomic Mass. Magnesium has three stable isotopes, ^{24}Mg, ^{25}Mg, and ^{26}Mg.
(a) What is the average atomic weight of magnesium?
(b) One of these isotopes has an abundance of 78.99% and another is 10.00% abundant. What is the abundance of the third?
(c) What is the mass number of the most abundant isotope of the three?

3

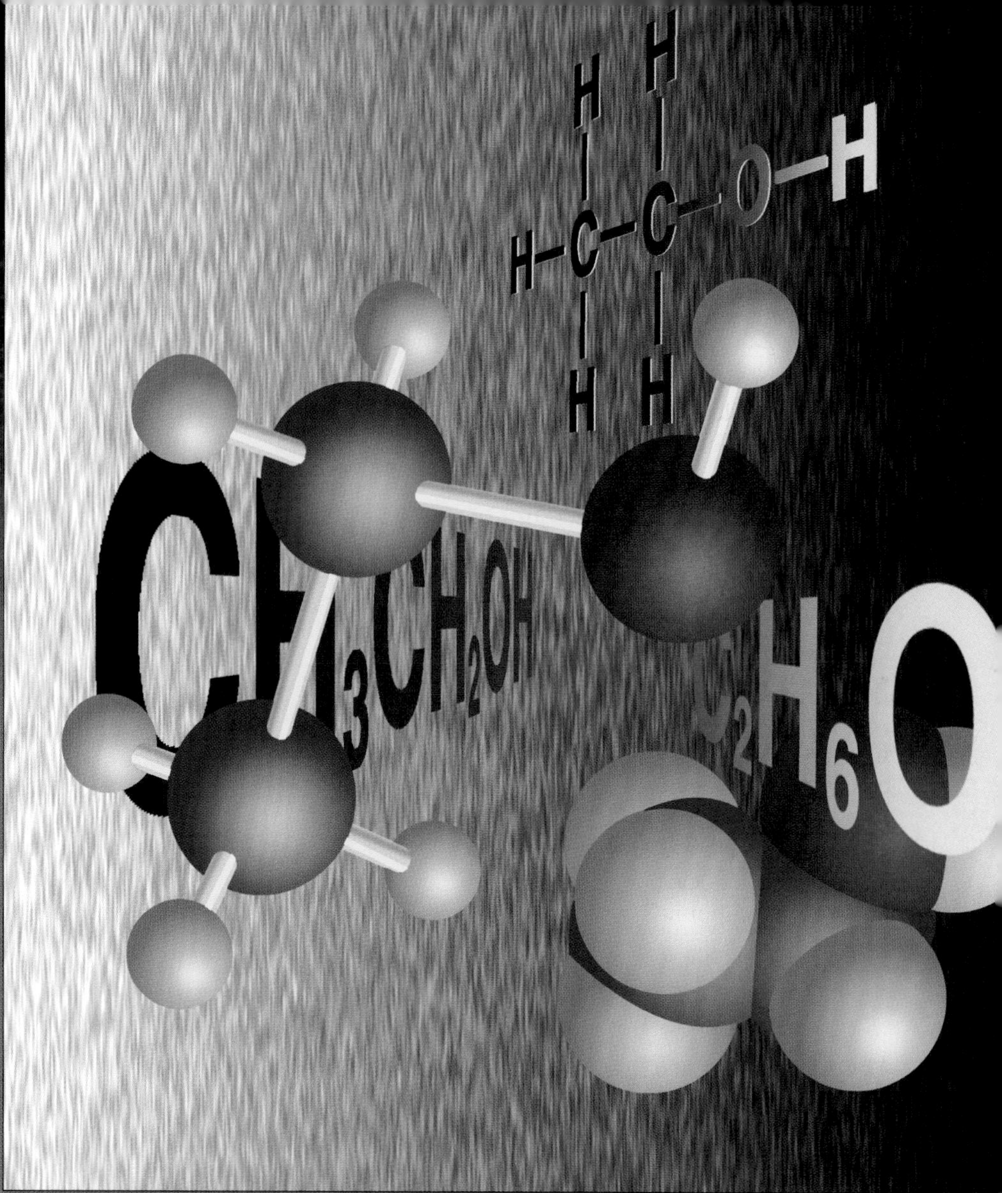

Ethanol, a simple molecular compound. There are many ways to represent a single molecular compound. The chemical formulas and the two molecular models shown here all represent ethanol.

CHEMICAL COMPOUNDS

3.1   Molecular Compounds

3.2   Naming Binary Molecular Compounds

3.3   Hydrocarbons and Alcohols

3.4   Alkanes and Their Isomers

3.5   Ions and Ionic Compounds

3.6   Naming Ions and Ionic Compounds

3.7   Properties of Ionic Compounds

3.8   Moles of Compounds

3.9   Percent Composition

3.10   Determining Empirical and Molecular Formulas

3.11   The Biological Periodic Table

3.12   Biomolecules: Carbohydrates and Fats

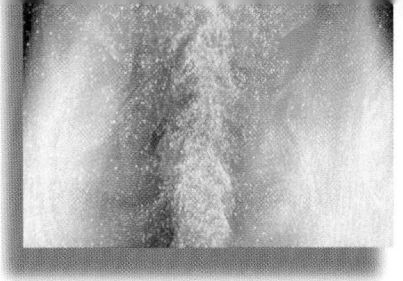

One of the most important things chemists do is make new chemical compounds, substances that on the nanoscale consist of new, unique combinations of atoms. These compounds may have properties similar to those of existing compounds, or they may be very different. Often chemists can custom-design a new compound to have desirable properties. All compounds contain at least two elements, and most compounds contain more than two elements. This chapter deals with two major, general types of chemical compounds—those consisting of individual molecules (⇐ *p. 26*) and those made of the positively and negatively charged atoms or groups of atoms called ions (⇐ *p. 44*). We will now examine how compounds are represented by symbols, formulas, and names and how formulas represent the macroscale masses and compositions of compounds. Finally, we will introduce the elements and two of the major classes of molecules that are essential to biochemistry.

3.1 MOLECULAR COMPOUNDS

In a **molecular compound** at the nanoscale level, atoms of two or more different elements are combined into the independent units known as molecules (⇐ *p. 26*). Every day we inhale, exhale, metabolize, and in other ways use thousands of molecular compounds. Water, carbon dioxide, sucrose (table sugar), and caffeine, as well as carbohydrates, proteins, and fats, are among the many common molecular compounds in our bodies.

 ### Molecular Formulas

The composition of a molecular compound is represented in writing by its **molecular formula,** in which the number and kinds of atoms combined to make one molecule of the compound are indicated by the subscripts and elemental symbols. For example, the molecular formula for water, H_2O, shows that there are three atoms per molecule—two hydrogen atoms and one oxygen atom. The subscript to the right of each element's symbol indicates the number of atoms of that element present in the molecule. If the subscript is omitted, it is understood to be one, as for the O in H_2O. These same principles apply to the molecular formulas of all molecules.

Some molecules are classified as **inorganic compounds** because they do not contain carbon or carbon and hydrogen—for example, sulfur dioxide, SO_2, an air pollutant, or ammonia, NH_3, which dissolved in water becomes a household cleaning agent. (Many inorganic compounds are ionic compounds, which are described in Section 3.5 of this chapter.) The majority of **organic compounds** are composed of molecules. Organic compounds invariably contain carbon, usually contain hydrogen, and may also contain oxygen, nitrogen, sulfur, phosphorus, or halogens. Such compounds are of great interest because they are the basis for the clothes we wear, the food we eat, the fuels we burn, and the living organisms in our environment. For example, ethanol (C_2H_6O) is the organic compound familiar as a component of "alcoholic" beverages, and methane (CH_4) is the organic compound that is the major component of natural gas.

The formula of a molecular compound, especially an organic compound, can be written in several different ways. The molecular formula given previously for ethanol, C_2H_6O, is one example. For an organic compound, the symbols of the elements other than carbon are frequently written in alphabetical order, and each has a subscript indicating the total number of atoms of that type in the molecule, as illustrated by C_2H_6O. Because of the huge number of organic compounds, this formula may not give sufficient information to indicate what compound is represented. This requires more information about how the atoms are connected

 CD-ROM Screen 3.3: Molecular Compounds

Metabolism is a general term for all of the chemical reactions that act to keep a living thing functioning. We *metabolize* food molecules to extract energy and produce other molecules needed by our bodies. Our *metabolic* reactions are controlled by enzymes (discussed in Section 13.9) and other kinds of molecules.

$$H_2O \qquad H\!-\!O\!-\!H$$

Space-filling model   Ball-and-stick model

 CD-ROM Screen 3.4: Representing Compounds

Remember that some elements are also composed of molecules. In oxygen, for example, two oxygen atoms are joined in an O_2 molecule (⇐ *p. 26*).

Atom colors in molecular models: H, light gray; C, dark gray; N, blue; O, red; S, yellow

TABLE 3.1	Examples of Simple Molecular Compounds		
Name	Molecular Formula	Number and Kind of Atoms	Molecular Model
Carbon dioxide	CO_2	3 total: 1 carbon, 2 oxygen	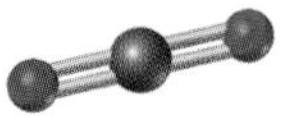
Ammonia	NH_3	4 total: 1 nitrogen, 3 hydrogen	
Nitrogen dioxide	NO_2	3 total: 1 nitrogen, 2 oxygen	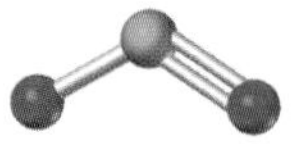
Carbon tetrachloride	CCl_4	5 total: 1 carbon, 4 chlorine	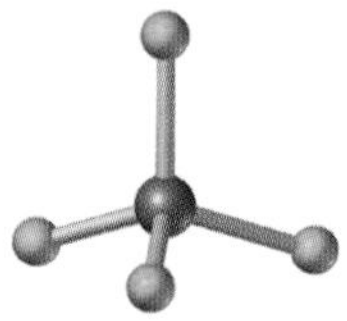
Octane	C_8H_{18}	26 total: 8 carbon, 18 hydrogen	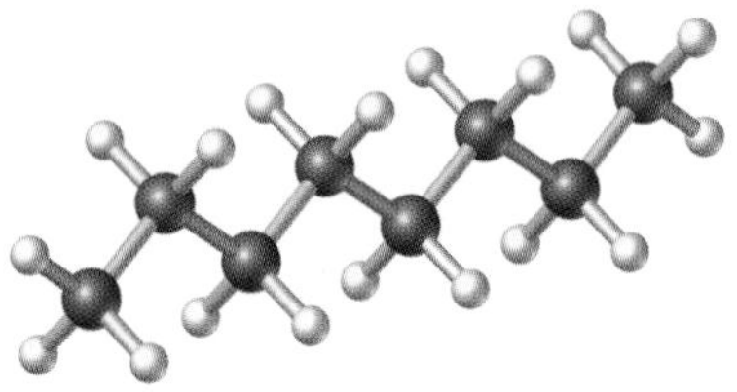

to each other. A **structural formula** shows exactly how atoms are connected. In ethanol, for example, the first carbon atom is connected to three hydrogen atoms, and the second carbon atom is connected to two hydrogen atoms and an —OH group.

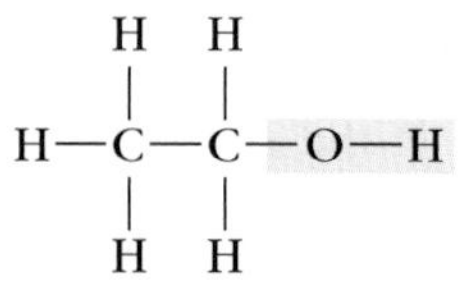

The formula can also be written in a modified form to show how the atoms are grouped together in the molecule. Such formulas, called **condensed formulas,** emphasize the atoms or groups of atoms connected to each carbon atom. For ethanol the condensed formula is CH_3CH_2OH. Compare this to the structural formula for ethanol and you will easily see that they represent the same structure. The —OH attached to the C atom is an example of a **functional group,** which is a distinctive group of atoms that, as part of an organic molecule, imparts specific properties to the molecule. A functional group is often the part of an organic molecule that changes when the molecule reacts with another molecule.

To summarize, there are three different ways of writing formulas, shown here for ethanol:

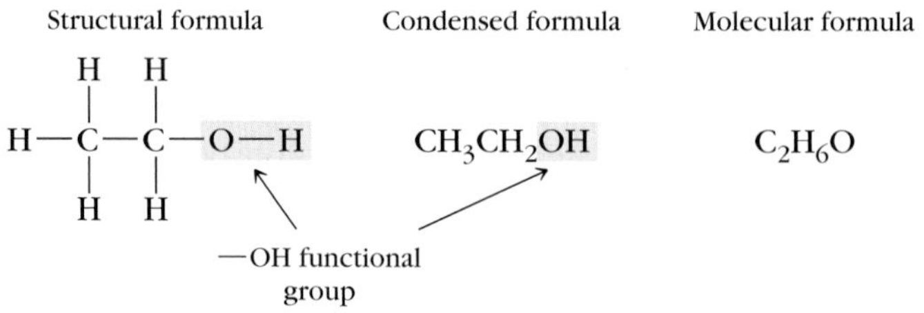

Functional groups are responsible for the characteristic chemical behavior of organic compounds. For example, the chemistry of alcohols depends on the presence of an —OH group attached to a C atom. Thus, ethanol (CH_3CH_2OH) is similar to propanol ($CH_3CH_2CH_2OH$) and other alcohols.

Organic compounds are divided into classes according to their functional groups. For example, compounds with —OH groups are classified as alcohols. A complete list of classes of organic compounds is given in Table E.2 in Appendix E.2, Functional Groups.

Structural formula   Condensed formula   Molecular formula

CH_3CH_2OH C_2H_6O

—OH functional group

As illustrated earlier for molecular elements (⇐ *p. 26*), molecular compounds can also be represented by ball-and-stick and space-filling models.

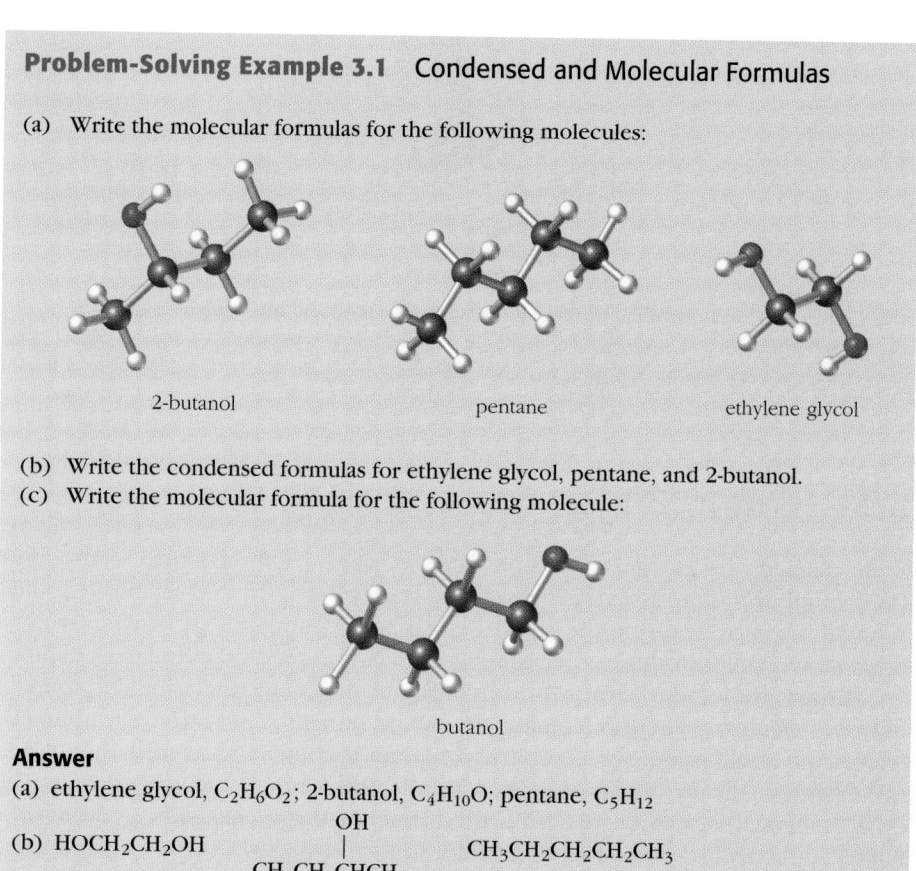

Structural
formula

Ball-and-stick
model

Space-filling
model

Some additional examples of ball-and-stick molecular models are given in Table 3.1.

Problem-Solving Example 3.1   Condensed and Molecular Formulas

(a) Write the molecular formulas for the following molecules:

2-butanol   pentane   ethylene glycol

(b) Write the condensed formulas for ethylene glycol, pentane, and 2-butanol.
(c) Write the molecular formula for the following molecule:

butanol

Answer

(a) ethylene glycol, $C_2H_6O_2$; 2-butanol, $C_4H_{10}O$; pentane, C_5H_{12}

(b) HOCH$_2$CH$_2$OH $\overset{\displaystyle OH}{\underset{\displaystyle |}{CH_3CH_2CHCH_3}}$ CH$_3$CH$_2$CH$_2$CH$_2$CH$_3$

ethylene glycol   2-butanol   pentane

(c) C_4H_9O

Explanation

(a) Simply count the atoms of each type in each compound to obtain the molecular formulas. Then write the symbols with their subscripts, putting C first and the others in alphabetical order.

(b) In condensed formulas, each carbon atom and its hydrogen atoms are written without connecting lines (CH$_3$, CH$_2$, or CH). Other groups are usually written on the same line with the carbon and hydrogen atoms if the groups are at the beginning or end of the molecule. Otherwise, they are connected above or below the line by straight lines to the respective carbon atoms. Condensed formulas emphasize important groups in molecules, such as the —OH groups in 2-butanol and ethylene glycol.

(c) Again, count the atoms of each type in the compound to obtain the molecular formula.

Problem-Solving Practice 3.1

Write the molecular formulas for the following compounds.

(a) Adenosine triphosphate (ATP), an energy source in biochemical reactions, which has 10 carbon, 11 hydrogen, 13 oxygen, 5 nitrogen, and 3 phosphorus atoms per molecule.
(b) Capsaicin, the active ingredient in chili peppers, which has 18 carbons, 27 hydrogens, 3 oxygens, and 1 nitrogen atom per molecule.
(c) Oxalic acid, found in rhubarb, which has the condensed formula HOOCCOOH.

Exercise 3.1   Structural, Condensed, and Molecular Formulas

A model of propylene glycol, used in some "environmentally friendly" antifreezes, looks like this:

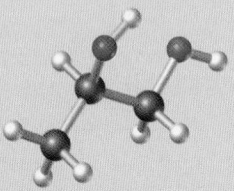

propylene glycol

Write the structural formula, the condensed formula, and the molecular formula for propylene glycol.

An automotive antifreeze that contains propylene glycol.

3.2 NAMING BINARY MOLECULAR COMPOUNDS

CD-ROM Screen 3.5: Binary Compounds of the Nonmetals

Hydrogen compounds with carbon are discussed in the next section.

Each element has a unique name and so does each chemical compound. Compound names are assigned in a systematic way based on well-established rules. We will begin by applying the rules used to name simple binary molecular compounds. We will introduce other naming rules as we need them in later chapters.

Binary molecular compounds consist of molecules that contain atoms of only two elements. There is a binary compound of hydrogen with every nonmetal except the noble gases. For hydrogen compounds containing oxygen, sulfur, and the halogens, the hydrogen is written first in the formula and named first. The other nonmetal is then named, with the element's name changed to end in *-ide*. For example, HCl is named hydrogen chloride.

Formula	Name
HCl	Hydrogen chloride
HBr	Hydrogen bromide
HI	Hydrogen iodide
H_2Se	Hydrogen selenide

hydrogen chloride   hydrogen bromide   hydrogen iodide

Many binary molecular compounds contain nonmetallic elements from Groups 4A, 5A, 6A, and 7A of the periodic table. In these compounds the elements are listed in formulas and names in the order of the group numbers, and prefixes are used to designate the number of a particular kind of atom. The prefixes are listed in Table 3.2. Table 3.3 illustrates how these prefixes are applied.

TABLE 3.2	Prefixes Used in Naming Chemical Compounds
Prefix	**Number**
Mono-	1
Di-	2
Tri-	3
Tetra-	4
Penta-	5
Hexa-	6
Hepta-	7
Octa-	8
Nona-	9
Deca-	10

TABLE 3.3	Examples of Binary Compounds	
Molecular Formula	**Name**	**Use**
CO	Carbon monoxide	Steel manufacturing
NO_2	Nitrogen dioxide	Preparation of nitric acid
N_2O	Dinitrogen oxide	Anesthetic; spray can propellant
N_2O_5	Dinitrogen pentoxide	Forms nitric acid
PBr_3	Phosphorus tribromide	Forms phosphorous acid
PBr_5	Phosphorus pentabromide	Forms phosphoric acid
SF_6	Sulfur hexafluoride	Transformer insulator
P_4O_{10}	Tetraphosphorus decoxide	Drying agent

A number of binary nonmetal compounds were discovered and named years ago, before systematic naming rules were developed. Such *common names* are still used today and must simply be learned.

Formula	Common Name	Formula	Common Name
H_2O	Water	NO	Nitric oxide
NH_3	Ammonia	N_2O	Nitrous oxide
N_2H_4	Hydrazine		("laughing gas")
PH_3	Phosphine		

H_2O is written with H before O, as are the hydrogen compounds of Groups 6A and 7A: H_2S, H_2Se, and HF, HCl, HBr, and HI. Other H-containing compounds are usually written with the H atom after the other atom.

Problem-Solving Example 3.2   Naming Binary Inorganic Compounds

Name the following compounds: (a) SO_2 (b) BF_3 (c) N_2O_4 (d) CCl_4

Answer   (a) sulfur dioxide (b) boron trifluoride (c) dinitrogen tetroxide (d) carbon tetrachloride

Explanation   These compounds consist entirely of nonmetals, so they are all molecular compounds. The prefixes in Table 3.2 are used as necessary. (a) Use *di-* to represent the two oxygen atoms. (b) Use *tri-* for the three fluorines. (c) Use *di-* for the two nitrogens and *tetra-* for the four oxygens. (d) Use *tetra-* for the four chlorines.

Problem-Solving Practice 3.2

Name the following compounds: (a) CS_2 (b) N_2F_4 (c) PCl_3

Exercise 3.2   Names and Formulas of Compounds

Give the formula for each of the following binary nonmetal compounds:
(a) carbon disulfide   (b) phosphorus trichloride   (c) sulfur dibromide
(d) selenium dioxide   (e) oxygen difluoride   (f) xenon trioxide

 # 3.3 HYDROCARBONS AND ALCOHOLS

Millions of organic compounds are known. They vary enormously in structure and function, ranging from the simple molecule methane (CH_4, the major constituent of natural gas) to large, complex biochemical molecules such as proteins, which often contain hundreds or thousands of atoms. Organic compounds are the main constituents of living matter. In organic compounds the carbon atoms are nearly

Pentane is an alkane used as a solvent.

 CD-ROM Screen 3.6: Alkanes

always bonded to other carbon atoms and to hydrogen atoms. Among the reasons for the enormous variety of organic compounds is the characteristic property of carbon atoms to form strong, stable bonds with up to four other carbon atoms. A **chemical bond** is an attractive force between two atoms holding them together. Thus, carbon atoms can form chains, branched chains, rings, and other more complicated structures. With such a large number of compounds, dividing them into classes is necessary to make organic chemistry manageable.

Hydrocarbons are organic compounds composed of only carbon and hydrogen atoms; they are the simplest class of organic compounds. The simplest major class of hydrocarbons is the **alkanes,** which are economically important fuels and lubricants. The simplest alkane is methane, CH_4, which has a central carbon atom with four bonds joining it to four H atoms. The general formula for alkanes is C_nH_{2n+2}, where n is an integer. Table 3.4 provides some information about the first ten alkanes. The first four (methane, ethane, propane, butane) have common names that must be memorized. For $n = 5$ or greater, the names are systematic: The prefixes of Table 3.2 indicate the number of carbon atoms in the molecule, and the ending *-ane* indicates that the compound is an alk*ane*. For example, the six-carbon alkane is *hexane*.

Methane, the simplest alkane, makes up about 85% of natural gas. It is also known to be one of the greenhouse gases (Section 12.4), meaning it is one of the chemicals involved in the problem of global warming. Ethane, propane, and butane are used as heating fuel for homes and in industry. In these simple alkanes, the carbon atoms are connected in unbranched chains, and each carbon atom is connected to either two or three hydrogen atoms:

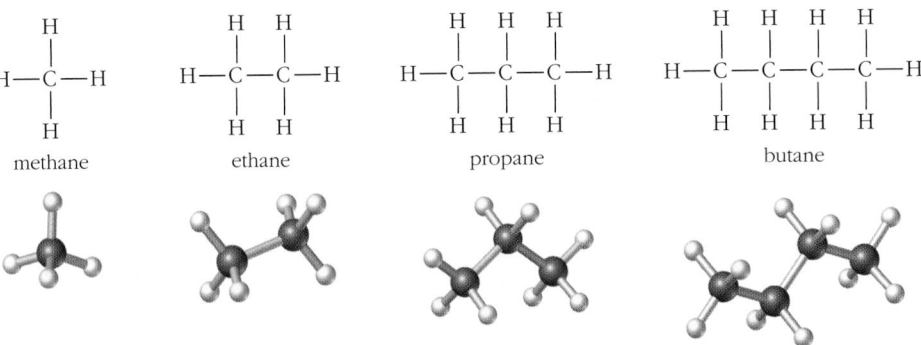

methane ethane propane butane

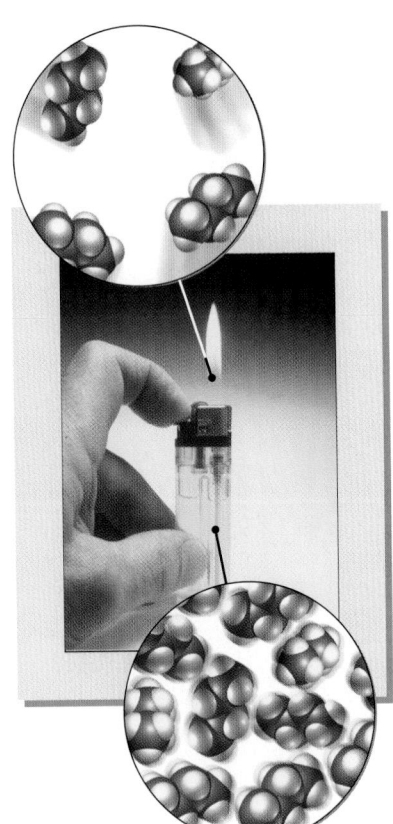

Butane ($CH_3CH_2CH_2CH_3$) is the fuel in this lighter. Butane molecules are present in the liquid and gaseous states in the lighter.

TABLE 3.4	The First Ten Alkane Hydrocarbons, C_nH_{2n+2}		
Molecular Formula	**Name**	**Boiling Point (°C)**	**Physical State at Room Temperature**
CH_4	Methane	−161.6	Gas
C_2H_6	Ethane	−88.6	Gas
C_3H_8	Propane	−42.1	Gas
C_4H_{10}	Butane	−0.5	Gas
C_5H_{12}	Pentane	36.1	Liquid
C_6H_{14}	Hexane	68.7	Liquid
C_7H_{16}	Heptane	98.4	Liquid
C_8H_{18}	Octane	125.7	Liquid
C_9H_{20}	Nonane	150.8	Liquid
$C_{10}H_{22}$	Decane	174.0	Liquid

Larger alkanes have longer chains of carbon atoms with hydrogens attached to each carbon. For example, heptane (C_7H_{16}) is found in petroleum, and eicosane ($C_{20}H_{42}$) is found in paraffin wax.

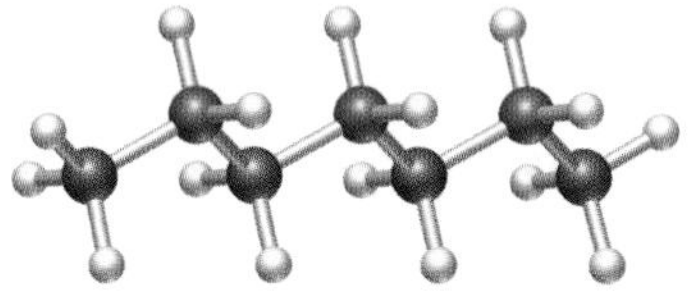

$$CH_3(CH_2)_5CH_3$$
heptane

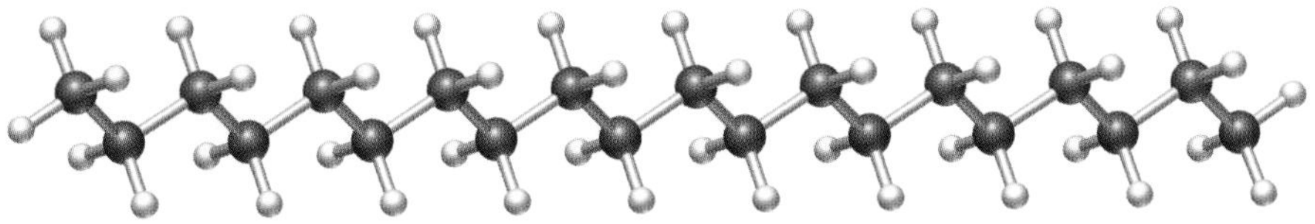

$$CH_3(CH_2)_{18}CH_3$$
eicosane

There are also cyclic hydrocarbons in which the carbon atoms are connected in rings, for example,

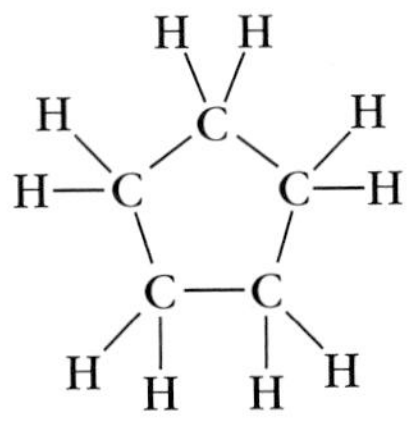

cyclopentane 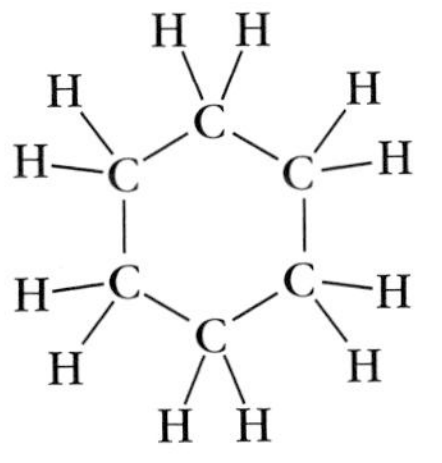   cyclohexane

---

Exercise 3.3   Alkane Molecular Formulas

(a) Using the general formula for alkanes, C_nH_{2n+2}, write the molecular formulas for the alkanes containing 16 and 28 carbons atoms.
(b) How many hydrogen atoms are present in tetradecane, which has 14 carbon atoms?

The molecular structures of hydrocarbons provide the framework for all other organic compounds. A few simple members of the class of organic compounds known as alcohols illustrate how this is possible. In **alcohols,** the functional group is the —OH group. A functional group such as this can take the place of one or more of the hydrogen atoms in a hydrocarbon. For example, the three simplest alcohols are methanol, ethanol, and propanol. In each case, one of the hydrogens of an alkane has been replaced by the —OH group.

The molecular structures of organic compounds determine their properties. For example, changing ethane, CH_3CH_3, to ethanol, CH_3CH_2OH, with the substitution of the —OH group for one of the hydrogens, changes the boiling point of the substance from $-88.6\ °C$ to $78.5\ °C$. This is due to the types of intermolecular interactions that are possible, and these effects will be fully explained in Section 9.5.

More organic functional groups will be introduced in later chapters.

Alkane		Alkane Boiling Point	Alcohol		Alcohol Boiling Point	Molecular Model
Methane	CH_4	−161.6 °C	Methanol	CH_3OH	65 °C	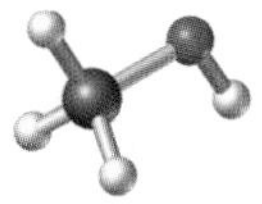
Ethane	CH_3CH_3	−88.6 °C	Ethanol	CH_3CH_2OH	78 °C	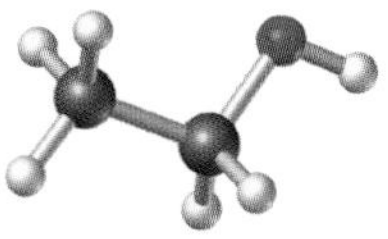
Propane	$CH_3CH_2CH_3$	−42.1 °C	Propanol	$CH_3CH_2CH_2OH$	97 °C	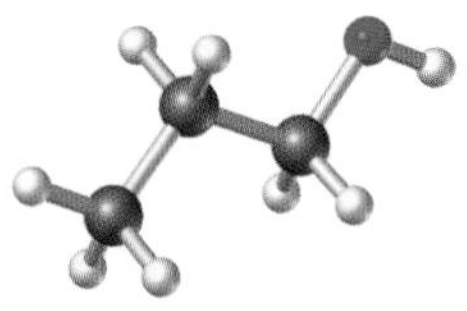

Exercise 3.4 Structural Formulas

Write the structural formula for propanol.

Exercise 3.5 Structural Formulas

Another alcohol has the same molecular formula as propanol, but the —OH group is connected to the middle carbon atom. Write the structural formula for this molecular relative of propanol.

Problem-Solving Example 3.3 Alkanes

Table 3.4 gives the boiling points for the first ten alkane hydrocarbons.

(a) Is the change in boiling point constant from one alkane to the next in the series?
(b) What do you propose as the explanation for the manner in which the boiling point changes from one alkane to the next?

Answer

(a) No. The increment is large between methane and ethane and gets progressively smaller as the compounds get larger. It is only 23 °C between nonane and decane.
(b) Larger molecules must interact more strongly and therefore require a higher temperature to move apart and to become gaseous.

Explanation

(a) The boiling point differences between successive alkanes are

	B.P.			B.P.	Change
Methane	(−162 °C)	to ethane		(−89 °C):	73 °C
Ethane	(−89 °C)	to propane		(−42 °C):	47 °C
Propane	(−42 °C)	to butane		(0 °C):	42 °C
Butane	(0 °C)	to pentane		(36 °C):	36 °C
Pentane	(36 °C)	to hexane		(69 °C):	33 °C

	B.P.		B.P.	Change
Hexane	(69 °C)	to heptane	(98 °C):	29 °C
Heptane	(98 °C)	to octane	(126 °C):	28 °C
Octane	(126 °C)	to nonane	(151 °C):	25 °C
Nonane	(151 °C)	to decane	(174 °C):	23 °C

The increments are getting smaller as the alkanes are getting larger.
(b) The larger molecules require a higher temperature in order to overcome their attraction to one another and to cease being liquid and become gaseous (⟵ *p. 20*, Section 1.7).

Problem-Solving Practice 3.3

Compare the boiling point of methane with that of methanol, ethane with that of ethanol, and propane with that of propanol. Comment on the trend(s) that you observe. (Alcohol boiling points in table on p. 82).

3.4 ALKANES AND THEIR ISOMERS

Two or more compounds that have the same molecular formula but different arrangements of atoms are called **isomers.** Isomers differ from one another in one or more physical or chemical properties such as boiling point, color, solubility, and reactivity. Several different types of isomerism are possible, particularly in organic compounds. **Constitutional isomers** (also called structural isomers) are compounds with the same formula that differ in the order in which their atoms are bonded together.

Straight-Chain and Branched-Chain Isomers of Alkanes

The first three alkanes—methane, ethane, and propane—have only one possible structural arrangement. When we come to the alkane with four carbon atoms, C_4H_{10}, there are two possible arrangements—a *straight* chain of four carbons or a *branched* chain of three carbons with the fourth carbon attached to the central atom of the chain of three.

In this context, "straight chain" means a chain of carbon atoms with no branches to other carbon atoms; the carbon atoms are in an unbranched sequence. As you can see from the molecular model of butane, the chain is not actually straight, but rather a zigzag.

Molecular Formula	Condensed Formula	Structural Formula	Molecular Model
Butane C_4H_{10}	$CH_3CH_2CH_2CH_3$ Melting point −138 °C Boiling point −0.5 °C	(structure)	(model)
Methylpropane	CH_3CHCH_3 CH_3 Boiling point −11.6 °C	(structure)	(model)

Historically, straight-chain hydrocarbons were once referred to as *normal* hydrocarbons, and *n*- was used as a prefix in their names. The current practice is not to use *n*-. If a hydrocarbon's name is given without indication that the compound is branched-chain, assume it is a straight-chain hydrocarbon.

Butane and methylpropane are constitutional isomers because they have the same molecular formula, but they are different compounds with different properties. Two constitutional isomers are different from each other in the same sense that two different structures built with identical Lego blocks are different from each other.

Methylpropane, the branched isomer of butane, has a *methyl* group (—CH_3) bonded to the central carbon atom. This is the simplest example of an **alkyl group,** the fragment of the molecule that remains when a hydrogen atom is removed from an alkane. In this case, removal of an H atom from methane gives the methyl group:

or CH_3—

methyl group

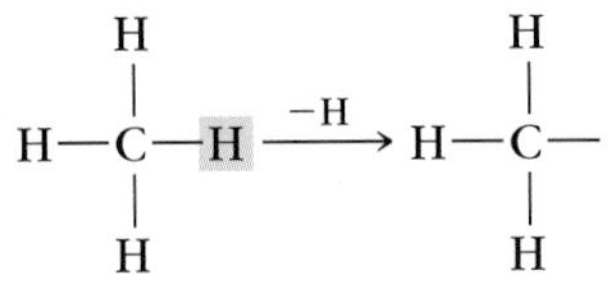

Removal of an H atom from ethane gives an ethyl group:

or CH_3CH_2— or —CH_2CH_3

ethyl group

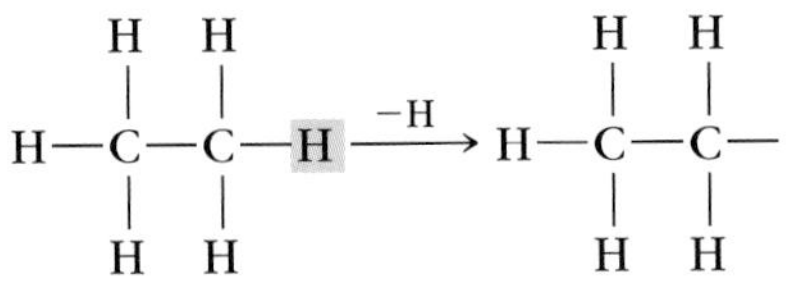

Notice that two different alkyl groups are possible when one H atom is removed from propane, C_3H_8.

or $CH_3CH_2CH_2$—

propyl group

or $(CH_3)_2CH$—

isopropyl group

Thus, although there is only one structure possible for propane, there are two for the propyl group. Alkyl groups are named by dropping *-ane* from the parent alkane name and adding *-yl*. Theoretically, any alkane can be converted to an alkyl group. Some of the more common examples of alkyl groups are given in Table 3.5.

The number of alkane constitutional isomers grows rapidly as the number of carbon atoms increases because of the possibility of chain branching. Table 3.6 shows the number of isomers for some alkanes. Chain branching is another reason for the extraordinary number of possible organic compounds.

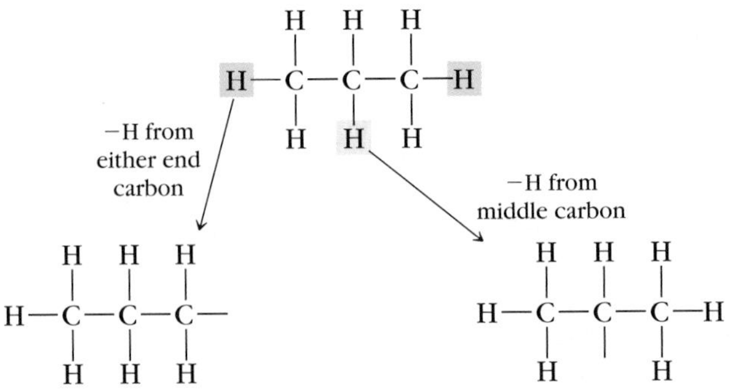

TABLE 3.5 Some Common Alkyl Groups

Name	Condensed Structural Representation
Methyl	CH_3—
Ethyl	CH_3CH_2—
Propyl	$CH_3CH_2CH_2$—
Isopropyl	CH_3CH— | CH_3 or $(CH_3)_2CH$—
Butyl	$CH_3CH_2CH_2CH_2$—
sec-butyl	CH_3CH_2CH— | CH_3
t-butyl	CH_3 | CH_3—C— | CH_3 or $(CH_3)_3C$—

Exercise 3.6 Straight-Chain and Branched-Chain Isomers

Three constitutional isomers are possible for pentane. Write structural and condensed formulas for these isomers.

TABLE 3.6 Alkane Isomers

Molecular Formula	Number of Isomers	Molecular Formula	Number of Isomers
CH_4	1	C_9H_{20}	35
C_2H_6	1	$C_{10}H_{22}$	75
C_3H_8	1	$C_{12}H_{26}$	355
C_4H_{10}	2	$C_{15}H_{32}$	4,347
C_5H_{12}	3	$C_{20}H_{42}$	366,319
C_6H_{14}	5	$C_{30}H_{62}$	4,111,846,763
C_7H_{16}	9	$C_{40}H_{82}$	62,491,178,805,831
C_8H_{18}	18		

Naming Branched-Chain Alkanes

Many alkanes and other organic compounds have both common names and systematic names. Usually the common name was assigned first and is widely known. Many consumer products are labeled with common names. When only a few isomers are possible, the common name adequately identifies the product for the consumer. However, common names quickly fail when several constitutional isomers are possible. For this reason, a formal system for naming organic compounds has been developed.

The rules for systematic names were formulated by the International Union of Pure and Applied Chemistry and are described in Appendix E.

For example, 2,2,4-trimethylpentane is the systematic name of the following branched-chain isomer of octane, $CH_3CH_2CH_2CH_2CH_2CH_2CH_2CH_3$.

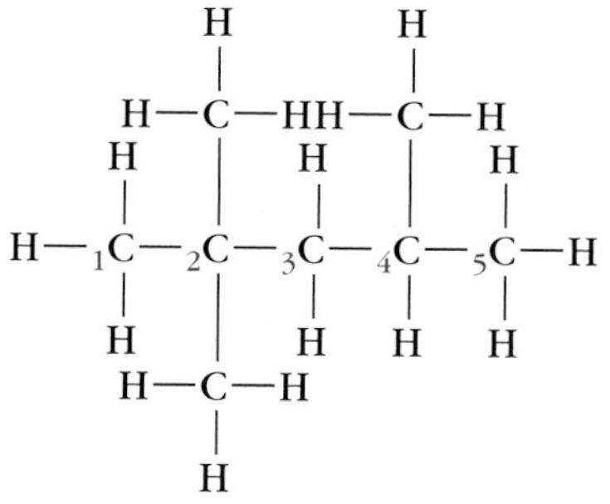

2,2,4-trimethylpentane

The two CH_3 groups of carbons 1 and 5 are not named because they are part of the pentane chain, not methyl groups substituted for hydrogen, as is the case at carbons 2 (twice) and 4.

The "pentane" part of the name, which means a straight (unbranched) five-carbon chain, identifies the longest chain in the molecule. The carbons in this chain are numbered from one end to the other, and the numbers "2,2,4-" indicate the locations of the three (*tri-*) methyl groups. This particular isomer of octane is used as a standard in assigning the *octane ratings* of various grades of gasoline (discussed in Section 12.1).

In this textbook we will not emphasize nor expect you to learn the extensive systematic methods for naming organic compounds. Such naming will be covered thoroughly if you take an organic chemistry course.

Exercise 3.7 Branched-Chain Alkanes

Draw the condensed formula for each of the following compounds: (a) 2-methylpentane, (b) 3-methylpentane, (c) 2,2-dimethylbutane, (d) 2,3-dimethylbutane

---

ESTIMATION **Number of Alkane Isomers**

The number of possible carbon compounds is truly enormous. Table 3.6 shows how the number of isomers of the simplest hydrocarbon compounds, alkanes, increases as the number of carbon atoms increases. How could these data be used to estimate the number of alkane isomers for a much larger number of carbon atoms? More specifically, let's estimate the number of alkane isomers for C_{40} and check the result against the last entry in the table.

To help get a picture of the growth rate, we could plot the number of alkane isomers versus the number of carbon atoms. If we made such a linear plot—that is, with the x-axis as the number of carbon atoms and the y-axis as the number of alkane isomers—we would see very little, because the plot would be rising so fast. In order to keep the final point on the plot, the y-axis would be so expanded that all the other points would be squashed toward the bottom of the plot. Therefore, to make these data easier to view, we plot the logarithm of the number of isomers, $\log(N_i)$, versus the number of carbon atoms (see figure). The points lie on a slightly concave upward curve, but a line fitted through them would be reasonably close to a straight line.

Now we are ready to make our estimate. To estimate how many isomers there are for C_{40}, we will extrapolate from the C_{20} and C_{30} points. The $\log(N_i)$ for C_{20} is 5.56, and the $\log(N_i)$ for C_{30} is 9.61; the difference is $9.61 - 5.56 = 4.05$. Our estimate of the $\log(N_i)$ at C_{40} will be this increment added to the value of $\log(N_i)$ for C_{30}:

$$\log(N_i) \text{ for } C_{40} = [\log(N_i) \text{ for } C_{30}] + \text{increment}$$

$$= 9.61 + 4.05 = 13.66$$

To calculate the number of isomers at C_{40} we take the antilog(13.66) = 4.57×10^{13}. Since the curve on the plot is concave upward, we know that our estimate will be a little too low, but it is still reasonable. The actual number of alkane isomers for C_{40} is 6.25×10^{13}. Our estimate is only

$$\frac{(6.25 - 4.57)}{6.25} \times 100\% = 27\% \text{ off.}$$

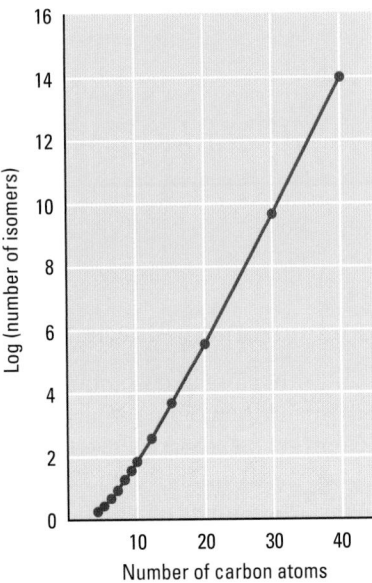

Semilog plot of the number of isomers versus the number of carbon atoms for alkanes.

3.5 IONS AND IONIC COMPOUNDS

CD-ROM Screen 3.7: Ions

Ionic compounds. Red iron(III) oxide, black copper(II) bromide, CaF_2 (front crystal), and NaCl (rear crystal).

The terms "cation" and "anion" are derived from the Greek words *ion* (traveling), *cat* (down), and *an* (up).

A compound whose nanoscale composition consists of positive and negative ions is classified as an **ionic compound.** Many common substances, such as table salt (NaCl), lime (CaO), lye (NaOH), and baking soda ($NaHCO_3$), are ionic compounds.

When metals react with nonmetals, the *metal atoms typically lose electrons to form positive ions.* Any positive ion is referred to as a **cation** (pronounced CAT-ion). Cations always have fewer electrons than protons. For example, Figure 3.1 shows how an electrically neutral sodium atom, which has 11 protons [11+] and 11 electrons [11−], loses one electron to become a sodium cation, which with 11 protons but only 10 electrons, has a net 1+ charge and is symbolized as Na^+. *The quantity of positive charge on a cation equals the number of electrons lost.* For example, when a neutral magnesium atom loses two electrons, it forms a 2+ magnesium ion, Mg^{2+}.

Conversely, when nonmetals react with metals, the nonmetal atoms typically gain electrons to form negatively charged ions. Any negative ion is an **anion** (pronounced ANN-ion). Anions always have more electrons than protons. Figure 3.1 shows how a neutral chlorine atom (17 protons, 17 electrons) can gain an electron to form a chlor*ide* ion, Cl^-. With 17 protons and 18 electrons, the chloride ion has a *net* 1− charge. *The quantity of negative charge on a nonmetal anion equals the number of electrons gained.* For example, a sulfur atom that gains two electrons forms a sulf*ide* ion, S^{2-}.

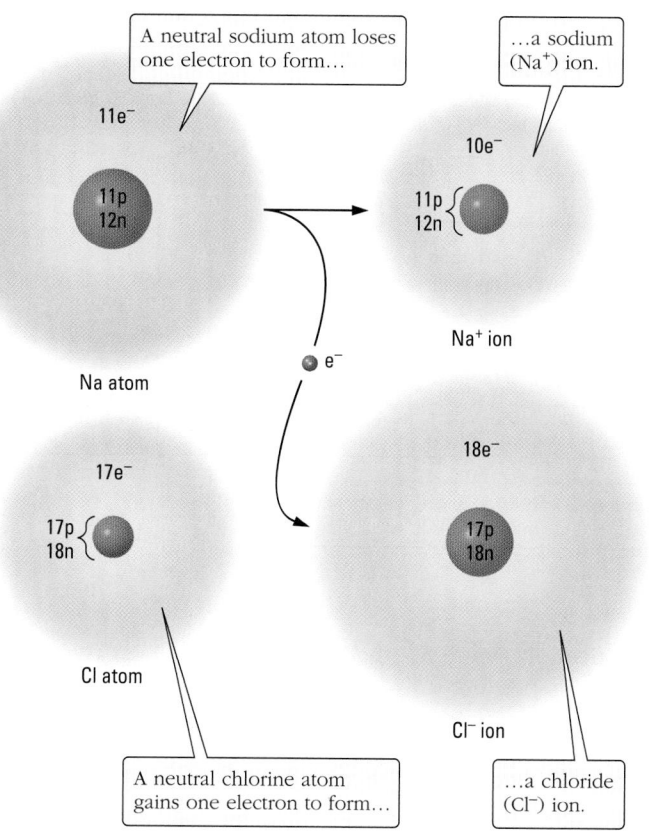

Figure 3.1 **Formation of the ionic compound NaCl.**

Monatomic Ions

A **monatomic ion** is a single atom that has lost or gained electrons. The charges of the common monatomic ions are given in Figure 3.2. Notice that metals of Groups 1A, 2A, and 3A form monatomic ions with charges equal to the group number. For example,

Group	Neutral Metal Atom	Electrons Lost	Metal Ion
1A	K (19 protons, 19 electrons)	1	K^+ (19 protons, 18 electrons)
2A	Mg (12 protons, 12 electrons)	2	Mg^{2+} (12 protons, 10 electrons)
3A	Al (13 protons, 13 electrons)	3	Al^{3+} (13 protons, 10 electrons)

It is extremely important that you know the ions commonly formed by the elements shown in Figure 3.2 so that you can recognize ionic compounds and their formulas and write their formulas as reaction products (Section 5.1).

Nonmetals of Groups 5A, 6A, and 7A form monatomic ions that have a negative charge usually equal to 8 minus the A group number. For example,

Group	Neutral Nonmetal Atom	Electrons Gained = 8 − Group No.	Nonmetal Ion
5A	N (7 protons, 7 electrons)	3 = (8 − 5)	N^{3-} (7 protons, 10 electrons)
6A	S (16 protons, 16 electrons)	2 = (8 − 6)	S^{2-} (16 protons, 18 electrons)
7A	F (9 protons, 9 electrons)	1 = (8 − 7)	F^- (9 protons, 10 electrons)

In Section 8.5 we explain the basis for the (8 − group number) relationship for nonmetals.

Figure 3.2 **Charges on some common monatomic cations and anions.** Note that metals generally form cations. The cation charge is given by the group number in the case of the main group elements of Groups 1A, 2A, and 3A *(gray).* For transition elements *(blue),* the positive charge is variable, and other ions in addition to those illustrated are possible. Nonmetals *(lavender)* generally form anions that have a charge equal to 8 minus the group number.

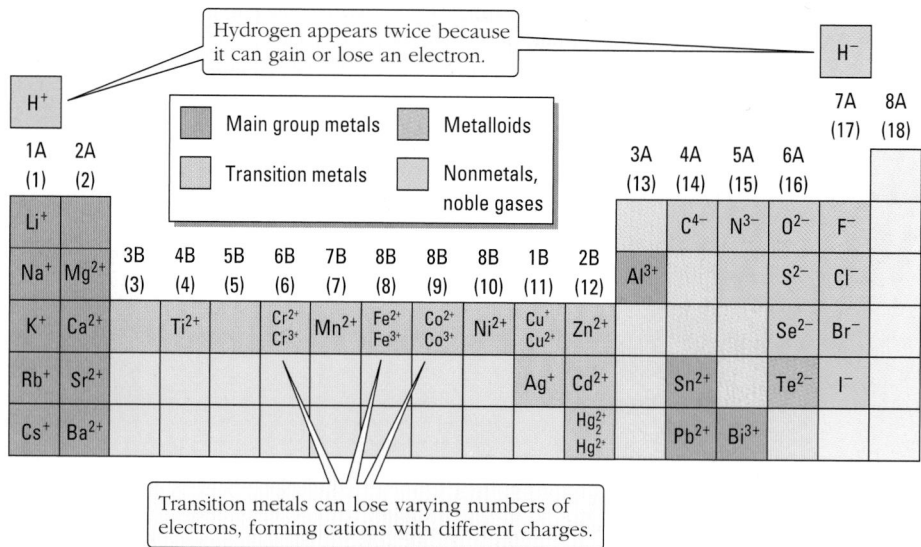

You might have noticed in Figure 3.2 that hydrogen appears at two locations in the periodic table. This is because a hydrogen atom can either gain or lose an electron. When it loses an electron, it forms a hydrogen ion, H^+ (1 proton, 0 electrons). When it gains an electron, it forms a hydride ion, H^- (1 proton, 2 electrons).

Noble gas atoms do not easily lose or gain electrons and have no common ions to list in Figure 3.2.

Transition metals form cations but can lose varying numbers of electrons, thus forming ions of different charges (Figure 3.2). Therefore, the group number is not an accurate guide to charges in these cases. You must learn which ions are formed most frequently by these metals. Many transition metals form 2+ and 3+ ions. For example, iron atoms can lose two or three electrons to form Fe^{2+} (26 protons, 24 electrons) or Fe^{3+} (26 protons, 23 electrons).

An older naming system for distinguishing between metal ions of different charges uses the ending *-ic* for the ion of higher charge and *-ous* for the ion of lower charge. These endings are combined with the element's name — for example, Fe^{2+} (ferrous) and Fe^{3+} (ferric) or Cu^+ (cuprous) and Cu^{2+} (cupric). We will not use these names in this book, but you may encounter them elsewhere.

Problem-Solving Example 3.4   Predicting Ion Charges

Predict the charges on ions of indium and of arsenic and write symbols for these ions.

Answer In^{3+} and As^{3-}

Explanation   Indium, a Group 3A metal, is predicted to lose three electrons to give the In^{3+} cation.

$$In \longrightarrow In^{3+} + 3\,e^-$$

Arsenic, a Group 5A metalloid, is predicted to gain $8 - 5 = 3$ electrons to give the As^{3-} anion.

$$As + 3\,e^- \longrightarrow As^{3-}$$

Problem-Solving Practice 3.4

For each of the ions listed below, explain whether it is likely to be found in an ionic compound.

(a) Ca^{4+} (b) Cr^{2+} (c) Sr^-

TABLE 3.7 Common Polyatomic Ions

Cation (1+)

NH_4^+	Ammonium

NH_4^+
ammonium ion

Anions (1−)

OH^-	Hydroxide	NO_2^-	Nitrite
HSO_4^-	Hydrogen sulfate	NO_3^-	Nitrate
CH_3COO^-	Acetate	MnO_4^-	Permanganate
ClO^-	Hypochlorite	$H_2PO_4^-$	Dihydrogen phosphate
ClO_2^-	Chlorite	CN^-	Cyanide
ClO_3^-	Chlorate	HCO_3^-	Hydrogen carbonate (bicarbonate)
ClO_4^-	Perchlorate		

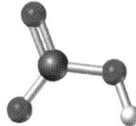

HCO_3^-
bicarbonate ion

Anions (2−)

CO_3^{2-}	Carbonate	SO_3^{2-}	Sulfite
HPO_4^{2-}	Hydrogen phosphate	SO_4^{2-}	Sulfate
$Cr_2O_7^{2-}$	Dichromate	$C_2O_4^{2-}$	Oxalate
$S_2O_3^{2-}$	Thiosulfate		

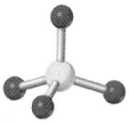

SO_4^{2-}
sulfate ion

Anion (3−)

PO_4^{3-}	Phosphate

Polyatomic Ions

A **polyatomic ion** is a *unit* of two or more atoms that bears a net electrical charge. Table 3.7 lists some common polyatomic ions. Polyatomic ions are found in many places—oceans, minerals, living cells, and foods. For example, hydrogen carbonate (bicarbonate) ion, HCO_3^-, is present in rain water, sea water, blood, and baking soda. It consists of one carbon atom, three oxygen atoms, and a hydrogen atom, with one unit of negative charge spread over the group of five atoms. The polyatomic sulfate ion, SO_4^{2-}, consists of one sulfur and four oxygen atoms and has an overall charge of 2−. One of the most common polyatomic cations is NH_4^+, the ammonium ion. In this case, four hydrogen atoms are connected to a nitrogen atom, and the group bears a net 1+ charge. We discuss the naming of polyatomic ions containing oxygen atoms in detail in the next section.

In many chemical reactions the polyatomic ion unit remains intact. *It is important to know the names, formulas, and charges of the common polyatomic ions listed in Table 3.7.*

CD-ROM Screen 3.8:
Polyatomic Ions

Ionic Compounds

In ionic compounds, cations and anions are attracted to each other by electrostatic forces—the forces of attraction between positive and negative charges. The strength of the electrostatic force dictates many of the properties of ionic compounds.

The attraction between oppositely charged ions increases with charge and decreases with the distance between the ions. Therefore, the attractive force between 2+ and 2− ions is greater than that between 1+ and 1− ions. The attractive force also increases as the distance between centers of the ions decreases. Thus, a small cation and a small anion will attract each other more strongly than will larger ions. We discuss the sizes of ions in detail in Section 7.8.

CD-ROM Screen 3.10: Ionic Compounds

Potassium dichromate K₂Cr₂O₇. This beautiful orange-red compound contains potassium ions (K^+) and dichromate ions ($Cr_2O_7^{2-}$).

Being able to classify compounds as ionic or molecular is very useful because, as you will see, these two types of compounds have quite different properties and thus different uses. The following generalizations will help you to predict whether a compound is ionic.

1. Metals (the elements in the gray and blue areas in Figure 3.2) generally form ionic compounds; metals lose electrons to form positive ions.
2. Nonmetals (the elements in the lavender area of Figure 3.2) can form both ionic and molecular compounds. They form monatomic negative ions *only* when combined with metals. Nonmetals form molecular compounds when combined with other nonmetals or with metalloids, or occasionally when several non-metal atoms are bonded to a metal atom.
3. It is difficult to predict when metalloids (the elements in the orange area of Figure 3.2) will form ions. These elements form both ionic and molecular compounds.

For example, the compounds formed when calcium combines with chlorine, sodium with sulfur, and strontium with bromine have the formulas $CaCl_2$, Na_2S, and $SrBr_2$, respectively. The metals form cations; the nonmetals form anions. These ionic compounds consist of Ca^{2+} and Cl^-, Na^+ and S^{2-}, and Sr^{2+} and Br^-, respectively. The nonmetal chlorine can also combine with the metalloid boron to form the molecular compound BCl_3 and with the nonmetal carbon to form the molecular compound CCl_4.

Sodium chloride, NaCl. This common ionic compound contains sodium ions (Na^+) and chloride ions (Cl^-).

Problem-Solving Example 3.5 Ionic and Molecular Compounds

Predict whether each compound is likely to be ionic or molecular:

(a) FeS   (b) $CoCl_2$ (c) CO_2 (d) $(NH_2)_2CO$ (e) $(NH_4)_2CO_3$ (f) C_8H_{18}

Answer
(a) ionic   (b) ionic   (c) molecular   (d) molecular   (e) ionic   (f) molecular

Explanation
(a) Iron is a metal, sulfur a nonmetal, so FeS is likely to be an ionic compound.
(b) Cobalt is a metal, chlorine a nonmetal, so they are likely to form an ionic compound.
(c) Carbon dioxide is formed from two nonmetals and so is molecular.
(d) Nitrogen, hydrogen, carbon, and oxygen are all nonmetals and NH_2 and CO are not listed as polyatomic ions in Table 3.7, so this is a molecular compound.
(e) The ammonium ion (NH_4^+) and the carbonate ion (CO_3^{2-}) form an ionic compound.
(f) This is octane, an alkane hydrocarbon composed entirely of nonmetal atoms, so it is molecular.

Problem-Solving Practice 3.5

Predict whether each of these compounds is likely to be ionic or molecular:

(a) CH_4 (b) $CaBr_2$ (c) $MgCl_2$ (d) PCl_3 (e) KCl

Writing Formulas for Ionic Compounds

All compounds are electrically neutral. Therefore, when cations and anions combine to form an ionic compound, there must be zero net charge. The total positive charge of all the cations must equal the total negative charge of all the anions. For example, consider the ionic compound formed when potassium reacts with sulfur. Potassium is a Group 1A metal, so a potassium atom loses one electron to become a K^+ ion. Sulfur is a Group 6A nonmetal, so a sulfur atom gains two electrons to become an S^{2-} ion. To make the compound electrically neutral, two K^+ ions (total charge is 2+) are needed for each S^{2-} ion. Consequently, the compound has the formula K_2S. The subscripts in an ionic compound show the numbers of ions included in the simplest formula unit. In this case, the subscript 2 indicates two K^+ ions for every S^{2-} ion.

Similarly, aluminum oxide, a combination of Al^{3+} and O^{2-} ions, has the formula Al_2O_3: $2\ Al^{3+} = 6+$; $3\ O^{2-} = 6-$; total charge $= 0$.

$$Al_2O_3$$
Two 3+ ions ↗   ↖ Three 2− ions

Notice that in writing the formulas for ionic compounds, *the cation symbol is written first, followed by the anion symbol.* The charges of the ions are *not* included in the formulas of ionic compounds.

Let's now consider several ionic compounds of magnesium, a Group 2A metal that forms Mg^{2+} ions.

As with the formulas for molecular compounds, a subscript of 1 in formulas of ionic compounds is understood to be there and is not written.

Combining Ions	Overall Charge	Formula
Mg^{2+} and Br^-	$(2+) + 2(1-) = 0$	$MgBr_2$
Mg^{2+} and SO_4^{2-}	$(2+) + (2-) = 0$	$MgSO_4$
Mg^{2+} and OH^-	$(2+) + 2(1-) = 0$	$Mg(OH)_2$
Mg^{2+} and PO_4^{3-}	$3(2+) + 2(3-) = 0$	$Mg_3(PO_4)_2$

Notice in the latter two cases that when a polyatomic ion occurs more than once in a formula, the polyatomic ion's formula is put in parentheses followed by the necessary subscript.

$$Mg_3(PO_4)_2$$
Three 2+ magnesium ions   Two 3− phosphate ions

Problem-Solving Example 3.6   Ions in Ionic Compounds

For each compound, give the symbol or formula of each ion present and indicate how many of each ion are represented in the formula: (a) Li_2S (b) Na_2SO_3 (c) $Ca(CH_3COO)_2$ (d) $Al_2(SO_4)_3$

Answer (a) two Li^+, one S^{2-} (b) two Na^+, one SO_3^{2-} (c) one Ca^{2+}, two CH_3COO^- (d) two Al^{3+}, three SO_4^{2-}

Explanation
(a) Lithium is a Group 1A element and *always* forms 1+ ions. The S^{2-} ion is formed from sulfur, a Group 6A element, by gaining two electrons ($8 - 6 = 2$). To maintain electrical neutrality, there must be two Li^+ ions for each S^{2-} ion.
(b) Sodium is also a Group 1A element and therefore forms Na^+. Two Na^+ ions offset the 2− charge of the single polyatomic sulfite ion, SO_3^{2-} (see Table 3.7).
(c) Calcium is a Group 2A element that always forms 2+ ions. In order to have electrical neutrality, the charge on each acetate ion must be 1− . Two acetate ions offset the 2+ charge on the Ca^{2+}.
(d) Aluminum, a Group 3A element, forms Al^{3+} ions. A 2:3 combination of two Al^{3+} ions ($2 \times 3+ = 6+$) with three 2− sulfate ions, SO_4^{2-}, ($3 \times 2- = 6-$) gives electrical neutrality.

Problem-Solving Practice 3.6

Determine how many ions and how many atoms there are in each formula.

(a) $In_2(SO_3)_3$ (b) $(NH_4)_3PO_4$

It is important to recognize that the subscript 2 in BaI_2 comes from the balancing of charges in this ionic compound. It is *not* because iodine is a diatomic molecule; the I_2 molecule is not involved here in any way.

Problem-Solving Example 3.7 Formulas of Ionic Compounds

Write formulas for ionic compounds composed of (a) barium and iodide ions, (b) lithium and carbonate ions, (c) Fe^{3+} and nitrate ions.

Answer (a) BaI_2 (b) Li_2CO_3 (c) $Fe(NO_3)_3$

Explanation

(a) Barium is a Group 2A metal, so it forms $2+$ ions. Iodine is a Group 7A nonmetal that forms $1-$ ions. Therefore, we need two iodide $1-$ ions for every Ba^{2+} ion.

(b) Carbonate is a $2-$ polyatomic ion that combines with two Li^+ ions to form Li_2CO_3.

(c) Because iron in this case has a $3+$ charge, it requires three $1-$ nitrate ions to balance the Fe^{3+}. Notice that the polyatomic ion is enclosed in parentheses followed by the subscript.

Problem-Solving Practice 3.7

Identify each constituent ion and tell how many there are in

(a) CaF_2 (b) $CoCl_2$ (c) $K_2(HPO_4)$

(d) Copper is a transition element that can form two compounds with bromine that contain either Cu^+ or Cu^{2+}. Write the formulas of these compounds.

(e) Write the formulas for the ionic compounds formed from the possible combinations of K^+ and Sr^{2+} with either O^{2-} or SO_4^{2-}.

3.6 NAMING IONS AND IONIC COMPOUNDS

CD-ROM Screen 3.13: Naming Ionic Compounds

Ionic compounds can be named unambiguously by using the rules given in the following paragraphs, rules you should learn thoroughly.

Naming Positive Ions

Don't confuse NH_4^+, ammonium ion, with NH_3, which is a molecular compound.

The Stock system is named after Alfred Stock (1876–1946), a German chemist famous for his work on the hydrogen compounds of boron and silicon.

An unusual cation that you will see on occasion is Hg_2^{2+}, the mercury(I) ion. The Roman numeral (I) is used to show that the ion is composed of two Hg^+ ions bonded together, giving a collective $2+$ charge.

Virtually all cations used in this book are metal ions that can be named by the rules given below. The ammonium ion (NH_4^+) is the major exception—a polyatomic ion composed of nonmetal atoms.

1a. *For metals that form only one kind of cation, the name is simply the name of the metal plus the word "ion."* For example, Mg^{2+} is the magnesium ion.

1b. *For metals that can form more than one kind of cation, the name of each ion must indicate its charge.* To do so, a Roman numeral in parentheses is given immediately following the ion's name (the Stock system). For example, Cu^{2+} is the copper(II) ion and Cu^+ is the copper(I) ion.

Naming Negative Ions

The following rules apply to naming anions.

2a. *A monatomic anion is named by adding* -ide *to the stem of the name of the nonmetal element from which the ion is derived.* For example, a *phosph*orus atom gives a *phosph*ide ion, and a *chlor*ine atom forms a *chlor*ide ion. Anions of Group 7A elements, the halogens, are collectively called **halide ions.**

2b. *The names of the most common polyatomic ions are given in Table 3.7 (p. 89).* Most must simply be memorized. There are, however, some guidelines that can help, especially for **oxoanions,** which are polyatomic ions that contain oxygen. For oxoanions with a nonmetal in addition to oxy-

Copper(I) oxide (*left*) and copper(II) oxide. The different copper ion charges result in different colors.

gen, the oxoanion with the greater number of oxygen atoms is given the suffix *-ate*. The oxoanion with the smaller number of oxygen atoms is given the suffix *-ite*.

Note that *-ate* and *-ite* suffixes do not relate to the ion's charge, but to the relative number of oxygen atoms.

For oxoanions with a nonmetal in addition to oxygen, the oxoanion with the greatest number of oxygen atoms is given the suffix **-ate**.

The oxoanion with the smaller number of oxygen atoms is given the suffix **-ite**.

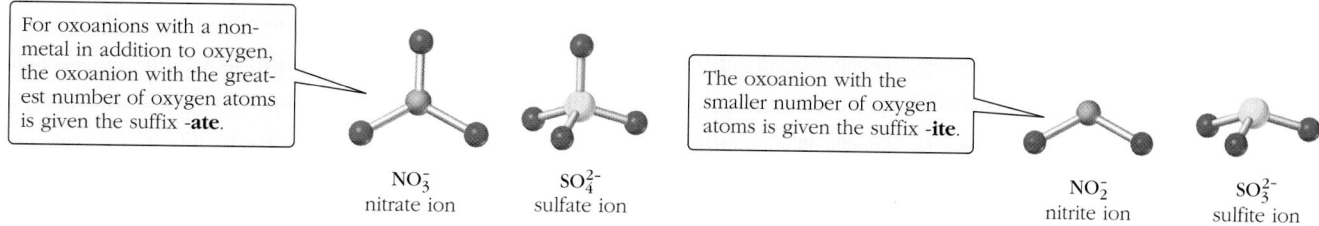

NO_3^-
nitrate ion

SO_4^{2-}
sulfate ion

NO_2^-
nitrite ion

SO_3^{2-}
sulfite ion

When more than two different oxoanions of a given nonmetal exist, a more extended naming scheme must be used. When there are four oxoanions involved, the two middle ones are named according to the *-ate* and *-ite* endings; then the ion containing the largest number of oxygen atoms is given the prefix *per-* and the suffix *-ate,* and the one containing the smallest number is given the prefix *hypo-* and the suffix *-ite*. The oxoanions of chlorine are good examples:

Oxoanions having one more oxygen aton than the **-ate** ion are named using the prefix **per-**.

Oxoanions having one less oxygen atom than the **-ite** ion are named using the prefix **hypo-**.

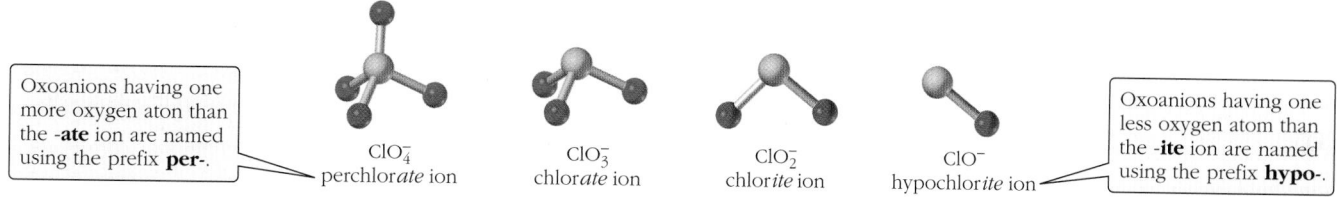

ClO_4^-
perchlor*ate* ion

ClO_3^-
chlor*ate* ion

ClO_2^-
chlor*ite* ion

ClO^-
hypochlor*ite* ion

The same naming rules apply to the oxoanions of bromine and iodine as well.

Oxoanions containing hydrogen are named simply by adding the word "hydrogen" before the name of the oxoanion, for example, hydrogen sulfate ion, HSO_4^-. When an oxoanion of a given nonmetal can combine with different numbers of hydrogen atoms, we must use prefixes to indicate which ion we are talking about: *di*hydrogen phosphate for $H_2PO_4^-$ and *mono*hydrogen phosphate for HPO_4^{2-}. Because many hydrogen-containing oxoanions have common names that are used often, you should know them. For example, the hydrogen carbonate ion, HCO_3^-, is often called the bicarbonate ion.

Anion	Systematic Name	Common Name
HCO_3^-	Hydrogen carbonate ion	Bicarbonate ion
HSO_4^-	Hydrogen sulfate ion	Bisulfate ion
HSO_3^-	Hydrogen sulfite ion	Bisulfite ion

Naming Ionic Compounds

Table 3.8 lists a number of common ionic compounds. We will use these compounds to demonstrate the rules for systematically naming ionic compounds. One basic naming rule is by now probably apparent—***the name of the cation comes first, then the name of the anion.*** Also, in naming a compound, the word "ion" is not used with the metal name.

Notice the following from Table 3.8:

- Calcium oxide, CaO, is named from calcium for Ca^{2+} (Rule 1a) and oxide for O^{2-} (Rule 2a). Likewise, sodium chloride is derived from sodium (Na^+, Rule 1a) and chloride (Cl^-, Rule 2a).

TABLE 3.8 Names of Some Common Ionic Compounds

Common Name	Systematic Name	Formula
Baking soda	Sodium hydrogen carbonate	$NaHCO_3$
Lime	Calcium oxide	CaO
Milk of magnesia	Magnesium hydroxide	$Mg(OH)_2$
Table salt	Sodium chloride	$NaCl$
Smelling salts	Ammonium carbonate	$(NH_4)_2CO_3$
Lye	Sodium hydroxide	$NaOH$

- Ammonium carbonate, $(NH_4)_2CO_3$, contains two polyatomic ions that are named in Table 3.7.
- In the name copper(II) sulfate, the (II) indicates that Cu^{2+} is present, not Cu^+, the other possibility.

Problem-Solving Example 3.8   Using Formulas to Name Ionic Compounds

Write the name for each of the following ionic compounds.
(a) Na_2SO_4 (b) $Ba(OH)_2$ (c) $AlCl_3$ (d) $Fe(ClO_3)_2$

Answer   (a) sodium sulfate, (b) barium hydroxide, (c) aluminum chloride, (d) iron(II) chlorate

Explanation
(a) The sodium ion Na^+ and the sulfate ion SO_4^{2-} are present, so the compound is named sodium sulfate.
(b) The Ba^{2+} ion and the hydroxide ion OH^- are present, so the compound is named barium hydroxide.
(c) The Al^{3+} ion and the chloride ion Cl^- combine to give aluminum chloride.
(d) The cation is the iron(II) ion, Fe^{2+}, and the anion is the chlorate ion ClO_3^-, so the compound is named iron(II) chlorate.

Problem-Solving Practice 3.8

Name the following ionic compounds: (a) KNO_2 (b) $NaHSO_3$ (c) $Mn(OH)_2$
(d) $Mn_2(SO_4)_3$ (e) Ba_3N_2 (f) LiH

Problem-Solving Example 3.9   Using Names to Write Formulas of Ionic Compounds

Write the formulas for the following ionic compounds: (a) ammonium hydrogen phosphate, (b) copper(II) chloride, (c) iron(III) nitrate, (d) potassium dichromate

Answer   (a) $(NH_4)_2HPO_4$ (b) $CuCl_2$ (c) $Fe(NO_3)_3$ (d) $K_2Cr_2O_7$

Explanation
(a) The ammonium cation is NH_4^+ and the hydrogen phosphate anion is HPO_4^{2-}, so two ammonium ions are needed for one hydrogen phosphate ion in the neutral compound.
(b) The copper(II) cation is Cu^{2+} and the anion is Cl^-, so two Cl^- ions are needed.
(c) The iron(III) ion is Fe^{3+} and the nitrate ion is NO_3^-, so three NO_3^- ions are needed.
(d) The potassium cation is K^+ and the dichromate anion is $Cr_2O_7^{2-}$, so two K^+ ions are needed.

Problem-Solving Practice 3.9

Write the correct formula for each of these ionic compounds.

(a) Potassium dihydrogen phosphate	(b) Copper(I) hydroxide
(c) Sodium hypochlorite	(d) Ammonium perchlorate
(e) Chromium(III) chloride	(f) Iron(II) sulfite

3.7 PROPERTIES OF IONIC COMPOUNDS

The properties of an ionic compound differ significantly from those of its component elements, as do the properties of all compounds. Consider the most ordinary of ionic compounds, table salt (sodium chloride, NaCl), composed of Na^+ and Cl^- ions. Sodium chloride is a white, crystalline, water-soluble solid, very different from its component elements, metallic sodium and gaseous chlorine. Metallic sodium reacts violently with water. Chlorine is a diatomic, toxic gas that reacts with water. Sodium *ions* and chloride *ions* do not undergo such reactions; NaCl dissolves uneventfully in water. You eat sodium ions and chloride ions daily in food, but you would not want to eat elemental sodium or inhale chlorine gas.

In ionic solids, cations and anions are held in an orderly array called a **crystal lattice,** in which each cation is surrounded by anions and each anion is surrounded by cations. This arrangement maximizes the attraction between cations and anions and minimizes the repulsion between ions of like charge. In sodium chloride, as shown in Figure 3.3, six chloride ions surround each sodium ion, and six sodium ions surround each chloride ion. As indicated in the formula, there is one sodium ion for each chloride ion.

The formula of an ionic compound indicates only the smallest whole-number ratio of the number of cations to the number of anions in the compound. In NaCl that ratio is 1:1. An Na^+Cl^- pair is referred to as a **formula unit** of sodium chloride. Note that the formula unit of an ionic compound has no independent existence, which is different from the molecules of a molecular compound.

The regular array of ions in a crystal lattice gives ionic compounds two characteristic properties—high melting points and distinctive crystalline shapes. The melting points are related to the charges and sizes of the ions. For ions of similar size, such as O^{2-} and F^-, *the higher the charges, the higher the melting point*, because of the greater attraction between ions of higher charge. For example, CaO (Ca^{2+} and O^{2-} ions) melts at 2572 °C, whereas NaF (Na^+ and F^-) melts at 993 °C.

The crystals of ionic solids have characteristic shapes because the ions are held rather rigidly in position by strong attractive forces. Such alignment creates planes of ions within the crystals. Ionic crystals can be cleaved by an outside force that causes the planes of ions to shift slightly, bringing ions of like charge closer together (Figure 3.4). The resulting repulsion causes the layers on opposite sides of the cleavage plane to separate, and the crystal splits apart.

 CD-ROM Screen 3.11: Properties of Ionic Compounds

Figure 3.3 **Two models of a sodium chloride crystal lattice.** (a) This ball-and-stick model illustrates clearly how the ions are arranged, although it shows the ions too far apart. (b) Although a space-filling model shows how the ions are packed, it is difficult to see the locations of ions other than those on the faces of the crystal lattice.

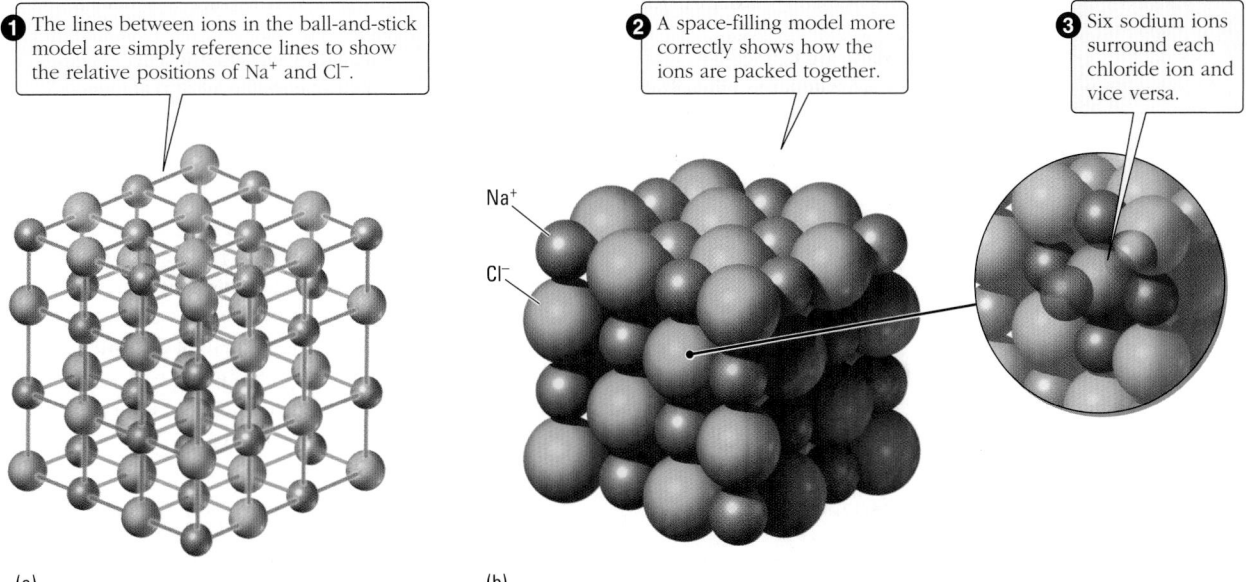

❶ The lines between ions in the ball-and-stick model are simply reference lines to show the relative positions of Na⁺ and Cl⁻.

❷ A space-filling model more correctly shows how the ions are packed together.

❸ Six sodium ions surround each chloride ion and vice versa.

Na⁺

Cl⁻

(a)   (b)

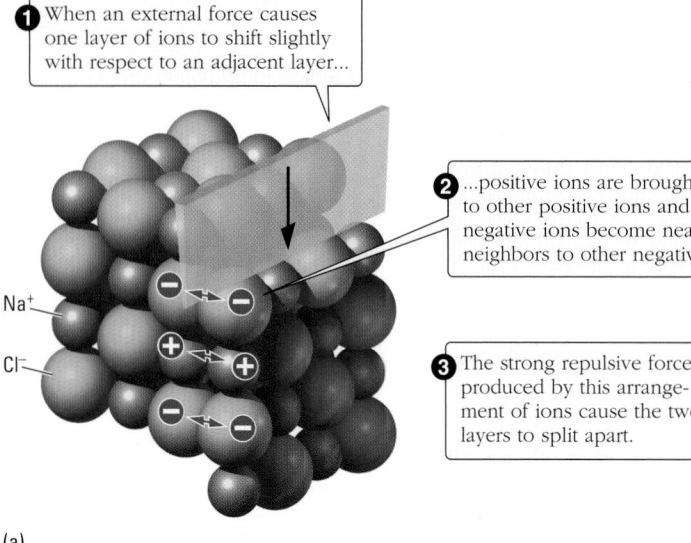

① When an external force causes one layer of ions to shift slightly with respect to an adjacent layer...

② ...positive ions are brought close to other positive ions and negative ions become nearest neighbors to other negative ions.

③ The strong repulsive forces produced by this arrangement of ions cause the two layers to split apart.

Na^+

Cl^-

(a)

(b)

Figure 3.4 Cleavage of an ionic crystal. (a) Diagram of forces involved in cleaving an ionic crystal. (b) A sharp blow on a knife edge lying along a plane of a salt crystal causes the crystal to split.

Because the ions in a crystal can only vibrate about fixed positions, ionic solids do not conduct electricity. However, when the solid melts, as shown in Figure 3.5, the ions are free to move and conduct an electric current. Cations move toward the negative electrode and anions move toward the positive electrode, which results in an electric current.

The general properties of molecular and ionic compounds are summarized in Table 3.9. In particular, note the differences in physical state, electrical conductivity, melting point, and water solubility.

An electric current is the movement of charged particles from one place to another. In a metal wire, the electric current is due to the movement of electrons through the wire.

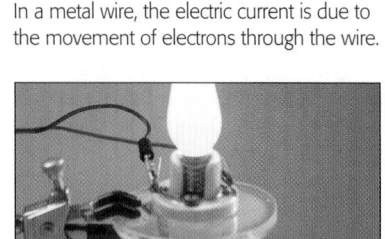

Figure 3.5 A molten ionic compound conducts an electric current. When an ionic compound is melted, ions are freed from the crystal lattice and are able to migrate to the electrodes dipping into the melt. The cations move toward the negative electrode, and the anions move toward the positive electrode. An electric current flows, and the light bulb illuminates, showing that the circuit is complete.

Exercise 3.8   Properties of Molecular and Ionic Compounds

Is a compound that is solid at room temperature and soluble in water likely to be a molecular or ionic compound? Why?

TABLE 3.9   Properties of Molecular and Ionic Compounds

Molecular Compounds	Ionic Compounds
Many are formed by combination of nonmetals with other nonmetals or with some metals.	Formed by combination of reactive metals with reactive nonmetals
Gases, liquids, solids	Crystalline solids
Brittle and weak or soft and waxy solids	Hard and brittle solids
Low melting points	High melting points
Low boiling points (−250 to 600 °C)	High boiling points (700 to 3500 °C)
Poor conductors of heat and electricity	Good conductors of electricity when molten; poor conductors of heat and electricity when solid
Many insoluble in water but soluble in organic solvents	Many soluble in water
Examples: hydrocarbons, H_2O, CO_2	Examples: NaCl, CaF_2

Ionic Compounds in Aqueous Solution: Electrolytes

Many ionic compounds are soluble in water. As a result, the oceans, rivers, lakes, and even the tap water in our residences contain many kinds of ions in solution. This makes the solubilities of ionic compounds and the properties of ions in solution of great practical interest.

When an ionic compound dissolves in water, it *dissociates*—the oppositely charged ions separate from one another. For example, when solid NaCl dissolves in water, it dissociates into Na^+ and Cl^- ions that become uniformly mixed with water molecules and dispersed throughout the solution.

Aqueous solutions of ionic compounds, like molten ionic compounds, conduct electricity because the ions are free to move about in the solution and conduct a current (Figure 3.6). All substances that conduct electricity when dissolved in water are called **electrolytes.**

Water-soluble ionic compounds are known as **strong electrolytes** because they are completely (100%) dissociated in solution—only ions are present. Most molecular compounds that are water soluble continue to exist as molecules in solution; table sugar (chemical name, sucrose) is an example of such a molecular compound. Such substances whose solutions do not conduct electricity are called **nonelectrolytes.** Be sure you understand the difference between these two important properties of a compound: its solubility and its ability to form ions in solution.

3.8 MOLES OF COMPOUNDS

Molar Mass of Molecular Compounds

The most recognizable molecular formula, H_2O, shows us that there are two H atoms for every O atom in a water molecule. In two water molecules, therefore, there are four H atoms and two O atoms; in a dozen water molecules there are two dozen H atoms and one dozen O atoms. We can extend this until we have one mole of water molecules (Avogadro's number of molecules, 6.022×10^{23}), each containing two hydrogen atoms and one oxygen atom ($\Leftarrow$ *p. 58*). We can also say that in 1.000 mol of water there are 2.000 mol of H atoms and 1.000 mol of O atoms:

H_2O	H	O
6.022×10^{23} water molecules	$2(6.022 \times 10^{23}$ H atoms)	6.022×10^{23} O atoms
1.000 mol H_2O molecules	2.000 mol H atoms	1.000 mol O atoms
18.0152 g H_2O	2(1.0079 g H) = 2.0158 g H atoms	15.9994 g O atoms

The mass of one mole of water molecules—the *molar mass*—is the sum of the masses of two moles of H atoms and one mole of O atoms: 2.0158 g of H + 15.9994 g of O = 18.0152 g of water in a mole of water. For chemical compounds, the molar mass, in grams per mole, is *numerically the same* as the **molecular weight,** the sum of the atomic weights (in amu) of *all* the atoms in the compound's formula. The molar masses of several molecular compounds are shown in this table:

CD-ROM Screen 4.5: Compounds in Aqueous Solution

We use the term "dissociate" for ionic compounds that separate into their constituent ions in water. The term "ionize" is used for molecular compounds whose molecules react with water to form ions.

Some substances, which are described in Section 5.2, are only partly converted into ions in solution and are known as weak electrolytes.

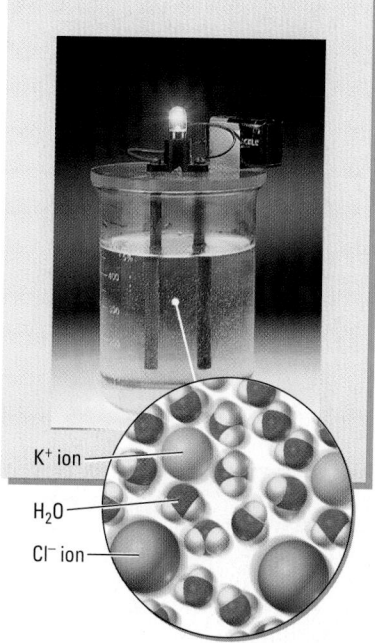

Figure 3.6 Electrical conductivity of an ionic compound solution. When an electrolyte, such as potassium chloride, KCl, is dissolved in water and provides ions that move about, the electrical circuit is completed between the electrodes, and the light bulb in the circuit glows. The ions of every KCl unit have dissociated: K^+ and Cl^-. The Cl^- ions move toward the positive electrode, and the K^+ ions move toward the negative electrode, transporting electric charge through the solution.

CD-ROM Screen 3.15: Compounds, Molecules, and the Mole

One mole of a compound does not mean one molecule. It means 6.022×10^{23} molecules.

Compound	Structural Formula	Molecular Weight	Molar Mass
Ammonia NH_3	H—N—H | H	14.01 amu, N + 3(1.01 amu, H) = 17.04 amu	17.04 g/mol
Trifluoromethane CHF_3	F | F—C—F | H	12.011 amu, C + 1.01 amu, H + 3(19.00 amu, F) = 70.02 amu	70.02 g/mol
Sulfur dioxide SO_2	O=S—O	32.07 amu, S + 2(16.00 amu, O) = 64.07 amu	64.07 g/mol
Glycerol $C_3H_8O_3$	CH_2OH | CHOH | CH_2OH	3(12.011 amu, C) + 8(1.01 amu, H) + 3(16.00 amu, O) = 92.11 amu	92.11 g/mol

Molar Mass of Ionic Compounds

Because ionic compounds do not contain individual molecules, the term "formula weight" is sometimes used for ionic compounds instead of "molecular weight." As with molecular weight, an ionic compound's **formula weight** is the sum of the atomic weights of all the atoms in the compound's formula. The molar mass of an ionic compound, expressed in grams per mole (g/mol), is numerically equivalent to its formula weight. The term "molar mass" is used for both molecular and ionic compounds.

Compound	Formula Weight	Molar Mass
Sodium chloride NaCl	22.99 amu, Na + 35.45 amu, Cl = 58.44 amu	58.44 g/mol
Magnesium oxide MgO	24.31 amu, Mg + 16.00 amu, O = 40.31 amu	40.31 g/mol
Potassium sulfide K_2S	2(39.10 amu, K) + 32.07 amu, S = 110.27 amu	110.27 g/mol
Calcium nitrate $Ca(NO_3)_2$	40.08 amu Ca + 2(14.01 amu, N) + 6(16.00 amu, O) = 164.10 amu	164.10 g/mol
Magnesium phosphate $Mg_3(PO_4)_2$	3(24.31 amu, Mg) + 2(30.97 amu, P) + 8(16.00 amu, O) = 262.87 amu	262.87 g/mol

Notice that $Mg_3(PO_4)_2$ has 2 P atoms and $2 \times 4 = 8$ O atoms because there are two PO_4^{3-} ions in the formula.

One-mole quantities of four compounds. H_2O (18.02 g/mol); small beaker NaCl (58.44 g/mol); large beaker aspirin (180.2 g/mol); green $NiCl_2 \cdot 6H_2O$ (237.7 g/mol).

Exercise 3.9 Molar Masses

Calculate the molar mass of each of the following compounds:
(a) K_2HPO_4 (b) $C_{27}H_{46}O$ (cholesterol)   (c) $Mn_2(SO_4)_3$ (d) $C_8H_{10}N_4O_2$ (caffeine)

Gram-Mole Conversions

CD-ROM Screen 3.16: Using Molar Mass

As you might expect, it is essential to be able to do gram-mole conversions for compounds, just as we did for elements (⇐ *p. 59*). Here also, the key to such conversions is using molar mass as a conversion factor.

Problem-Solving Example 3.10   Grams to Moles

Calcium phosphate, $Ca_3(PO_4)_2$, is an ionic compound that is the main constituent of bone. How many moles of $Ca_3(PO_4)_2$ are in 10.0 g of the compound?

Answer 3.22×10^{-2} mol $Ca_3(PO_4)_2$

Explanation

To convert from grams to moles, we first find calcium phosphate's molar mass, which is the sum of the molar masses of the atoms in the formula.

$$3(40.08 \text{ g/mol}) + 2(30.97 \text{ g/mol}) + 8(16.00 \text{ g/mol}) = 310.18 \text{ g/mol}$$

This is used to convert mass to moles:

$$10.0 \text{ g Ca}_3(\text{PO}_4)_2 \times \frac{1 \text{ mol Ca}_3(\text{PO}_4)_2}{310.18 \text{ g Ca}_3(\text{PO}_4)_2} = 3.22 \times 10^{-2} \text{ mol Ca}_3(\text{PO}_4)_2$$

✔   We started with 10.0 g of calcium phosphate, which has a molar mass of about 300, so 10/300 is about 1/30, which is close to the more accurate answer we calculated.

Problem-Solving Practice 3.10

Calculate the number of moles in 12.5 g of each of these ionic compounds: (a) $K_2Cr_2O_7$
(b) $KMnO_4$ (c) $(NH_4)_2CO_3$

Problem-Solving Example 3.11   Moles to Grams

Adrenocorticotropic hormone (ACTH) is a hormone of the anterior pituitary gland with a molar mass of approximately 4600 g. How many milligrams are in 3.0 micromoles (3.0×10^{-6} mol) of ACTH?

Answer 1.4×10^{-2} g of ACTH

Explanation

We use the molar mass to convert from moles to grams:

$$3.0 \times 10^{-6} \text{ mol ACTH} \times \frac{4600 \text{ g ACTH}}{1 \text{ mol ACTH}} = 1.4 \times 10^{-2} \text{ g ACTH}$$

or 140 mg ACTH.

Problem-Solving Practice 3.11

(a) How many grams are in 5.0 millimoles of sucrose, $C_{12}H_{22}O_{11}$?
(b) How many grams are in 5.0 millimoles of a protein with a molar mass of 7100 ?

Problem-Solving Example 3.12   Grams and Moles

Aspartame is a widely used artificial sweetener (NutraSweet) that is almost 200 times sweeter than sucrose. One sample of aspartame, $C_{14}H_{18}N_2O_5$, has a mass of 1.80 g; another contains 0.220 mol of aspartame. To see which sample is larger, answer these questions:

(a) What is the molar mass of aspartame?
(b) How many moles of aspartame are in the 1.80-g sample?
(c) How many grams of aspartame are in the 0.220-mol sample?

Answers   (a) 294.3 g/mol   (b) 6.12×10^{-3} mol   (c) 64.7 g. The 0.220-mol sample is larger.

Explanation

(a) The molar mass is numerically the same as the molecular weight, the sum of the atomic weights of all the atoms in the formula:

$$[14 \text{ C atoms} \times (12.011 \text{ amu/C atom})] + [18 \text{ H atoms} \times (1.008 \text{ amu/H atom})] +$$
$$[2 \text{ N atoms} \times (14.007 \text{ amu/N atom})] + [5 \text{ O atoms} \times$$
$$(15.999 \text{ amu/O atom})] = 294.3 \text{ amu}$$

Thus, the molar mass of aspartame is 294.3 g/mol.

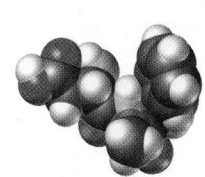

aspartame

(b) Use the molar mass to calculate moles from grams:

$$1.80 \text{ g aspartame} \times \frac{1 \text{ mol aspartame}}{294.3 \text{ g aspartame}} = 6.12 \times 10^{-3} \text{ mol aspartame}$$

(c) Converting moles to grams is the opposite of item (b):

$$0.220 \text{ mol aspartame} \times \frac{294.3 \text{ g aspartame}}{1 \text{ mol aspartame}} = 64.7 \text{ g aspartame}$$

The 0.220-mol sample of aspartame is the larger one.

Problem-Solving Practice 3.12

(a) Calculate the number of moles in 10.0 g of $C_{27}H_{46}O$ (cholesterol) and 10.0 g of $Mn_2(SO_4)_3$.

(b) Calculate the number of grams in 0.25 mol of K_2HPO_4 and 0.25 mol of $C_8H_{10}O_2N_4$ (caffeine).

Exercise 3.10   Moles and Formulas

Is the following statement true? "Two different compounds have the same formula. Therefore, 100 g of each compound contains the same number of moles." Justify your answer.

Moles of Ionic Hydrates

CD-ROM Screen 3.14: Hydrated Compounds

Many ionic compounds, known as **ionic hydrates** or hydrated compounds, have water molecules trapped within the crystal lattice. The associated water is called the **water of hydration.** For example, the formula for a beautiful deep-blue compound named copper(II) sulfate pentahydrate is $CuSO_4 \cdot 5H_2O$. The $\cdot 5H_2O$ and the term "pentahydrate" indicate five moles of water associated with every mole of copper(II) sulfate. The molar mass of a hydrate includes the mass of the water of hydration. Thus, the molar mass of $CuSO_4 \cdot 5H_2O$ is 249.7 g: 159.6 g $CuSO_4$ + 90.1 g (for 5 mol H_2O) = 249.7 g. There are many ionic hydrates, including the frequently encountered ones listed in Table 3.10.

Calcium sulfate *hemi*hydrate contains one water molecule per two $CaSO_4$ units. The prefix *hemi-* refers to $\frac{1}{2}$ just as in the familiar word "hemisphere."

One commonly used hydrate may well be in the walls of your room. Plasterboard (sometimes called wallboard or gypsum board) contains hydrated calcium sulfate, or gypsum, $CaSO_4 \cdot 2H_2O$, as well as unhydrated $CaSO_4$, sandwiched between two thicknesses of paper. Gypsum is a natural mineral that can be mined, but it is also obtained as a byproduct when sulfur dioxide is removed from exhaust gases in electric power plants.

$CuSO_4 \cdot 5H_2O$.

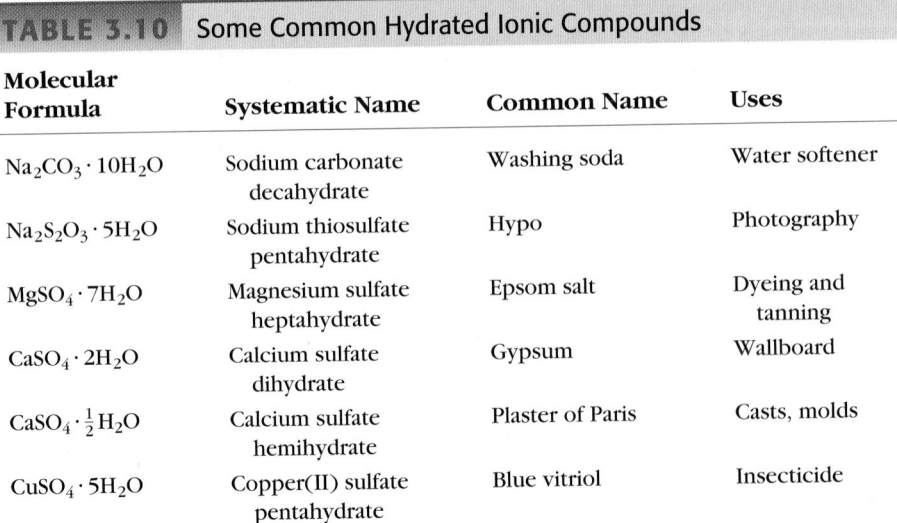

TABLE 3.10   Some Common Hydrated Ionic Compounds

Molecular Formula	Systematic Name	Common Name	Uses
$Na_2CO_3 \cdot 10H_2O$	Sodium carbonate decahydrate	Washing soda	Water softener
$Na_2S_2O_3 \cdot 5H_2O$	Sodium thiosulfate pentahydrate	Hypo	Photography
$MgSO_4 \cdot 7H_2O$	Magnesium sulfate heptahydrate	Epsom salt	Dyeing and tanning
$CaSO_4 \cdot 2H_2O$	Calcium sulfate dihydrate	Gypsum	Wallboard
$CaSO_4 \cdot \frac{1}{2}H_2O$	Calcium sulfate hemihydrate	Plaster of Paris	Casts, molds
$CuSO_4 \cdot 5H_2O$	Copper(II) sulfate pentahydrate	Blue vitriol	Insecticide

Heating gypsum to 180 °C drives off some of the water of hydration to form calcium sulfate hemihydrate, $CaSO_4 \cdot \frac{1}{2}H_2O$, commonly called "plaster of Paris." This compound is widely used in casts for broken limbs. When added to water, it forms a thick slurry that can be poured into a mold or spread out over a part of the body. As it hardens, it takes on additional water of hydration and its volume increases, forming a rigid protective cast.

Exercise 3.11 Moles of an Ionic Hydrate

A home remedy calls for 2 teaspoons (20 g) of Epsom salt (see Table 3.10). Calculate the number of moles of the hydrate this mass represents.

3.9 PERCENT COMPOSITION

You saw in the previous section that the composition of any compound can be expressed as either *(1) the number of atoms of each type per molecule or formula unit* or *(2) the mass of each element in a mole of the compound.* The latter relationship provides the information needed to find the **percent composition by mass** (also called the **mass percent**) of the compound.

Problem-Solving Example 3.13 Percent Composition by Mass

What are the percentages of carbon and oxygen in carbon dioxide?

Answer 27.29% carbon and 72.71% oxygen

Explanation
We calculate the percentage of carbon from the ratio of the mass of carbon in 1 mol of CO_2 to the total mass of 1 mol of CO_2:

$$\% \text{ C} = \frac{\text{mass of C in 1 mol } CO_2}{\text{mass of } CO_2 \text{ in 1 mol } CO_2} \times 100\%$$

$$= \frac{12.01 \text{ g C}}{44.01 \text{ g } CO_2} \times 100\%$$

$$= 27.29\% \text{ (also expressed as 27.29 g C per 100.0 g } CO_2)$$

We calculate the percentage of oxygen the same way:

$$\% \text{ O} = \frac{\text{mass of O in 1 mol } CO_2}{\text{mass of } CO_2 \text{ in 1 mol } CO_2} \times 100\%$$

$$= \frac{32.0 \text{ g O}}{44.01 \text{ g } CO_2} \times 100\%$$

$$= 72.71\% \text{ O (or 72.71 g O per 100.0 g } CO_2)$$

✔ There are two oxygen atoms for each carbon atom, so if the atomic weights of C and O were the same, CO_2 would be about $\frac{2}{3}$ oxygen and $\frac{1}{3}$ carbon by weight. Oxygen is 30% heavier than carbon, so the oxygen percentage should be somewhat larger than 67% and the carbon mass percentage somewhat less than 33%, and they are.

Problem-Solving Practice 3.13

What is the percentage of each element in silicon dioxide, SiO_2?

Note that the percentages calculated in Example 3.12 add up to 100%. Therefore, once we calculated the percentage of carbon, we also could have determined the percentage of oxygen simply by subtracting: $100\% - 27.29\%$ C = 72.71% O. Calculating all percentages and adding them to check that they give 100% is a good way to check for errors.

Gypsum in wallboard and in its crystalline form. Gypsum is hydrated calcium sulfate, $CaSO_4 \cdot 2H_2O$.

  CD-ROM Screen 3.17: Percent Composition

It is important to recognize that the percent composition of a compound by mass is independent of the quantity of the compound. The percent composition by mass remains the same whether a sample contains 1 mg, 1 g, or 1 kg of the compound.

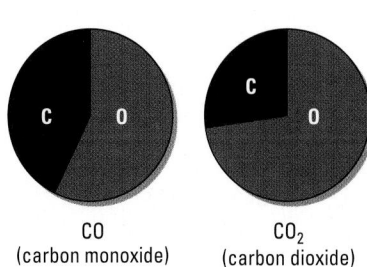

CO
(carbon monoxide)

CO_2
(carbon dioxide)

Mass percent carbon and oxygen in CO and CO_2.

In many circumstances, we do not need to know the percent composition of a compound with many significant figures, but rather we are interested in having an estimate of the percentage. For example, our objective could be to estimate whether the percentage of oxygen in a compound is about half or closer to two thirds. For example, the labels on garden fertilizers show the approximate percent composition of the product and allow the consumer to compare approximate compositions of products.

Plant fertilizers provide nitrogen, phosphorus, and potassium. The amounts of each are specified in the label.

Problem-Solving Example 3.14 Mass Percent of Fertilizer Components

The active ingredients in lawn and garden fertilizer are nitrogen, phosphorus, and potassium. Bags of fertilizer usually carry three numbers, as in 5-10-5 fertilizer (a typical fertilizer for flowers). The first number is the mass percent N; the second is the mass percent P_2O_5; the third is the mass percent K_2O. Thus, the active ingredients in a 5-10-5 product are equivalent to 5% N, 10% P_2O_5, and 5% K_2O, by weight. (The reporting of fertilizer ingredients in this way is a convention agreed upon by fertilizer manufacturers.) What is the mass in pounds of each of these three elements (N, P, K) in a 100-lb bag of fertilizer?

Answer   5 lb N, 4.4 lb P, 4.2 lb K

Explanation

The first number refers directly to the mass percent N, so 5% of 100 lb is 5 lb of N. The conversion from percent P_2O_5 to percent P is as follows:

$$\% \text{ P in } P_2O_5 = \frac{\text{mass of P in 1 mol of } P_2O_5}{\text{mass of } P_2O_5 \text{ in 1 mol of } P_2O_5} \times 100\%$$

$$= \frac{61.95 \text{ g P}}{141.95 \text{ g } P_2O_5} \times 100\% = 43.6\%$$

Ten percent (10%) P_2O_5 means that a 100-lb bag contains the equivalent of 10 lb of P_2O_5. But only 43% of this is phosphorus, so the 100-lb bag contains 10 lb $\times$ 0.436 = 4.4 lb of phosphorus.

The conversion from percent K_2O to percent K is as follows:

$$\% \text{ K in } K_2O = \frac{\text{mass of K in 1 mol } K_2O}{\text{mass of } K_2O \text{ in 1 mol of } K_2O} \times 100\%$$

$$= \frac{78.20 \text{ g K}}{94.2 \text{ g } K_2O} \times 100\% = 83.0\%$$

Five percent (5%) K_2O means that a 100-lb bag contains the equivalent of 5 lb of K_2O. But only 83% of this is potassium, so the 100-lb bag contains 5 lb $\times$ 0.830 = 4.2 lb of potassium.

Nitrogen is commonly present as ammonium nitrate, NH_4NO_3, or urea, $CO(NH_2)_2$. Although phosphorus content is reported as percent P_2O_5, this compound is not present in the product. Phosphorus is commonly present as $Ca(H_2PO_4)_2$ or $(NH_4)_2HPO_4$. Potassium is commonly present as KCl.

Problem-Solving Practice 3.14

How many pounds of urea are present in a 100-lb bag of 5-10-5 fertilizer if urea is the only nitrogen-containing compound present?

Exercise 3.12 Percent Composition

Express the composition of each compound first as the mass of each element in 1.000 mol of the compound, and then as the mass percent of each element:
(a) SF_6 (b) $C_{12}H_{22}O_{11}$ (c) $Al_2(SO_4)_3$ (d) $U(OTeF_5)_6$

3.10 DETERMINING EMPIRICAL AND MOLECULAR FORMULAS

A formula can be used to derive the percent composition by mass of a compound, and so the reverse process also works—we can determine the formula of a compound from mass percent data. In doing so, keep in mind that the subscripts in a formula indicate the relative numbers of moles of each element in one mole of that compound.

CD-ROM Screen 3.18: Determining Empirical Formulas

CD-ROM Screen 3.19: Determining Molecular Formulas

We can apply this method to finding the formula of diborane. Experiments show that diborane is 78.13% B and 21.87% H. Based on these percentages, a 100.0-g diborane sample contains 78.13 g of B and 21.87 g of H. From this information we can calculate the number of moles of each element in the sample:

$$78.13 \text{ g B} \times \frac{1 \text{ mol B}}{10.811 \text{ g B}} = 7.227 \text{ mol B}$$

$$21.87 \text{ g H} \times \frac{1 \text{ mol H}}{1.0079 \text{ g H}} = 21.70 \text{ mol H}$$

To determine the formula from these data, we next need to find the number of moles of each element *relative to the other,* in this case the ratio of moles of hydrogen to moles of boron. Looking at the numbers reveals that there are about three times as many moles of H atoms as there are moles of B atoms. To calculate the ratio exactly, divide the smaller number of moles into the larger number of moles. For diborane that ratio is

$$\frac{21.70 \text{ mol H}}{7.227 \text{ mol B}} = \frac{3.003 \text{ mol H}}{1.000 \text{ mol B}}$$

This confirms that there are three moles of H atoms for every one mole of B atoms and that there are three hydrogen atoms for each boron atom. This information gives the formula BH_3, which may or may not be correct.

For a molecular compound such as diborane, the molecular formula must also accurately reflect the *total number of atoms in a molecule of the compound.* The calculation we have done gives the *simplest possible ratio of atoms in the molecule,* and BH_3 is the simplest formula for diborane. A formula that reports the simplest possible ratio of atoms in the molecule is called an **empirical formula.** However, multiples of the simplest formula are possible, such as B_2H_6, and B_3H_9, and so on.

To determine the actual molecular formula from the empirical formula requires that we *experimentally determine* the molar mass of the compound and then compare our result with the molar mass predicted by the empirical formula. If the two molar masses are the same, the empirical and molecular formulas are the same. However, if the experimentally determined molar mass is some multiple of the value predicted by the empirical formula, the molecular formula is that multiple of the empirical formula. In the case of diborane, experiments indicate that the molar mass is 27.67 g . This compares with the molar mass of 13.84 g for BH_3, and so the molecular formula is a multiple of the empirical formula. That multiple is 27.67/13.84 = 2.00. Thus, the molecular formula of diborane is B_2H_6, two times BH_3.

Problem-Solving Example 3.15   Molecular Formula from Percent Composition by Mass Data

When oxygen reacts with phosphorus, two possible oxides can form. One contains 56.34% P and 43.66% O, and its experimentally determined molar mass is 219.90 g/mol. Determine its molecular formula.

Answer P_4O_6

Explanation The first step in finding a molecular formula from percent composition by mass and molar mass is to calculate the relative number of moles of each element and then determine the empirical formula. A 100.0-g sample of this phosphorus oxide contains 56.34 g of P and 43.66 g of O, and so you have

$$56.34 \text{ g P} \times \frac{1 \text{ mol P}}{30.97 \text{ g P}} = 1.819 \text{ mol P}$$

$$43.66 \text{ g O} \times \frac{1 \text{ mol O}}{16.00 \text{ g O}} = 2.729 \text{ mol O}$$

Thus, the mole ratio (and atom ratio) is

$$\frac{2.729 \text{ mol O}}{1.819 \text{ mol P}} = \frac{1.500 \text{ mol O}}{1.000 \text{ mol P}}$$

Because we can't have half-atoms, we double the numbers to convert to whole numbers, which gives us three oxygen atoms for every two phosphorus atoms. This gives an empirical formula of P_2O_3. The molar mass corresponding to this empirical formula is

$$\left(2 \text{ mol P} \times \frac{30.97 \text{ g P}}{1 \text{ mol P}}\right) + \left(3 \text{ mol O} \times \frac{16.00 \text{ g O}}{1 \text{ mol O}}\right) = 109.95 \text{ g } P_2O_3 \text{ per mole of } P_2O_3$$

compared with a known molar mass of 219.90 g/mol. Thus, the experimental molar mass is $\frac{219.90}{109.95} = 2$ times the molar mass predicted by the empirical formula, and so the molecular formula is P_4O_6, twice the empirical formula.

✔ The molar mass of P is about 31 g, so we should have 31 g × 4 = 124 g P in one mole of P_4O_6. This would give a P percent by mass of 124/220 = 56%, which is just about the right value.

Problem-Solving Practice 3.15

The other phosphorus oxide contains 43.64% P and 56.36% O, and its experimentally determined molar mass is 283.89 g/mol. Determine its empirical and molecular formulas.

 Problem-Solving Example 3.16 **Molecular Formula from Percent Composition By Mass Data**

Rubbing alcohol, also commonly called isopropyl alcohol, has a molar mass of 60.096 g/mol. It contains 59.96% C, 13.42% H, and the rest is oxygen. What are its empirical and molecular formulas?

Answer Both formulas are C_3H_8O.

Explanation First find the number of moles of each element in 100.0 g of the compound:

$$59.96 \text{ g C} \times \frac{1 \text{ mol C}}{12.01 \text{ g C}} = 4.992 \text{ mol C}$$

$$13.42 \text{ g H} \times \frac{1 \text{ mol H}}{1.0079 \text{ g H}} = 13.32 \text{ mol H}$$

That leaves moles of oxygen to be calculated. The mass of oxygen in the sample must be

$$100.0 \text{ g sample} - 59.96 \text{ g C} - 13.42 \text{ g H} = 26.62 \text{ g O}$$

Converting this to moles of oxygen gives

$$26.62 \text{ g O} \times \frac{1 \text{ mol O}}{15.9994 \text{ g O}} = 1.664 \text{ mol O}$$

Base the mole ratio on the smallest number of moles present, in this case, moles of oxygen:

$$\frac{4.992 \text{ mol C}}{1.664 \text{ mol O}} = \frac{3.000 \text{ mol C}}{1.000 \text{ mol O}} \qquad \frac{13.32 \text{ mol H}}{1.664 \text{ mol O}} = \frac{8.000 \text{ mol H}}{1.000 \text{ mol O}}$$

Therefore, the empirical formula has an atom ratio of 3 carbons to 8 hydrogens to 1 oxygen, C_3H_8O.

The molar mass predicted by this empirical formula is 60.096 g/mol, the same as the experimentally determined molar mass, indicating that the molecular formula is the same as the empirical formula.

✔ The molar mass of C is 12 g, so we should have 12 g × 3 = 36 g C in one mole of C_3H_8O. This would give a C percent by mass of 36/60 = 60%, which is just about the right value.

Problem-Solving Practice 3.16

Vitamin C (ascorbic acid) contains 40.9% C, 4.58% H, and 54.5% O and has an experimentally determined molar mass of 176.13 g/mol. Determine its empirical and molecular formulas.

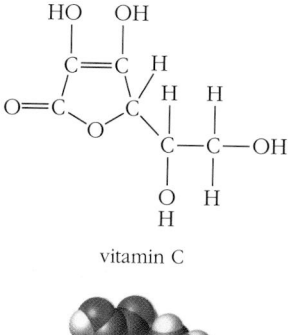

vitamin C

3.11 THE BIOLOGICAL PERIODIC TABLE

Most of the more than 100 known elements are not directly involved with our personal health and well-being. However, more than 30 of the elements, shown in Figure 3.7, are absolutely essential to us. Among these elements are metals, nonmetals, and metalloids from across the periodic table. All are part of a well-balanced diet.

Table 3.11 lists the building block elements and major minerals in the order of their relative abundances per million atoms in the body, showing the preeminence of four of the nonmetals—oxygen, carbon, hydrogen, and nitrogen. These four nonmetals, the building block elements, contribute most of the atoms in the biologically significant chemicals—the biochemicals—composing all plants and animals. With few exceptions, the biochemicals these nonmetals form are organic compounds.

Nonmetals are also present as anions in body fluids, including chloride ion (Cl^-), phosphorus in three ionic forms (PO_4^{3-}, HPO_4^{2-}, and $H_2PO_4^-$), and carbon as bicarbonate ion (HCO_3^-) and carbonate ion (CO_3^{2-}). Metals are present in the body

If you weigh 150 lb, about 90 lb (60%) is water, 30 lb is fat, and the remaining 30 lb is a combination of proteins, carbohydrates, and calcium, phosphorus, and other dietary minerals.

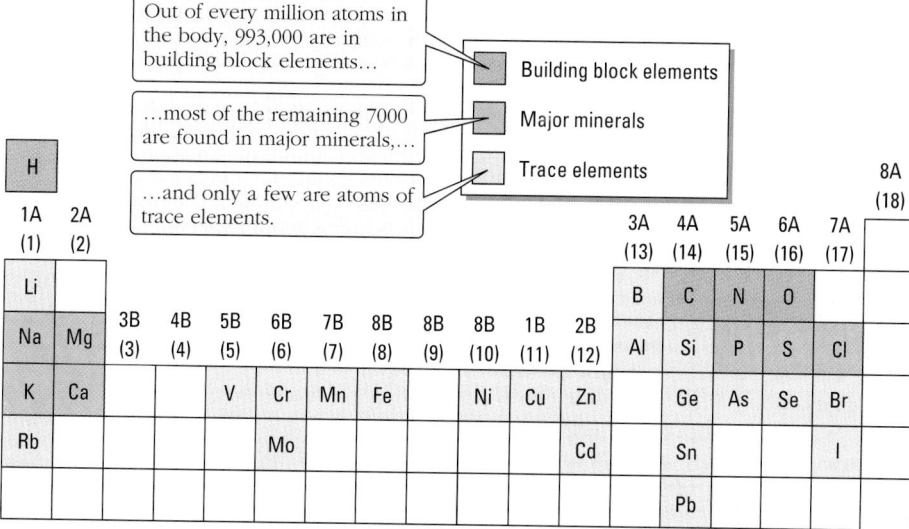

Figure 3.7 Elements essential to human health. Four elements, C, H, N, and O, form the many organic compounds that make up living organisms. The major minerals are required in relatively large amounts; trace elements are required in lesser amounts.

TABLE 3.11 Major Elements of the Human Body

Element	Symbol	Relative Abundance in Atoms/Million Atoms in the Body
Hydrogen	H	630,000
Oxygen	O	255,000
Carbon	C	94,500
Nitrogen	N	13,500
Calcium	Ca	3100
Phosphorus	P	2200
Chlorine	Cl	570
Sulfur	S	490
Sodium	Na	410
Potassium	K	260
Magnesium	Mg	130

as cations in solution (for example, Na^+, K^+) and in solids (Ca^{2+} in bones and teeth). They are also incorporated into large biomolecules (for example, Fe^{2+} in hemoglobin and Co^{3+} in vitamin B_{12}).

Exercise 3.13   Essential Elements

Using Figure 3.7, identify (a) the essential nonmetals, (b) the essential alkaline earth metals, (c) the essential halide ions, and (d) four essential transition metals.

The Dietary Minerals

The general term **dietary minerals** refers to the essential elements other than carbon, hydrogen, oxygen, or nitrogen. The dietary necessity and impact of these elements are far beyond their collective presence as only about 4% of our body

CHEMISTRY You Can Do... Pumping Iron: How Strong Is Your Breakfast Cereal?

Iron is an essential dietary mineral that enters our diets in many ways. To prevent dietary iron deficiency, the U.S. Food and Drug Administration (FDA) allows food manufacturers to "fortify" (add iron to) their products. One way of fortifying a cereal is to add an iron compound, such as iron(III) phosphate, that dissolves in the stomach acid (HCl). The iron in Special K cereal contains particles of pure iron metal baked into the flakes. It is listed on the ingredients label as "reduced iron." Can you detect metallic iron in a cereal? You can find out with the following experiment, for which you will need

• A reasonably strong magnet. You might find a good one holding things to your refrigerator door or in a toy store or hobby shop. Agricultural supply stores have "cow magnets," which will work well.

• A plastic freezer bag (quart size) and a rolling pin (or something else to crush the cereal).
• One serving of Special K (1 cup = about 30 g).

Put the Special K into the plastic bag and crush the cereal into small particles. Place the magnet into the bag and mix it well with the cereal for several minutes. Carefully remove the magnet from the bag and examine the magnet closely. A magnifying glass is helpful. Is anything clinging to the magnet that was not there before?

Think about the following questions:
1. Based on this experiment, is there metallic iron in the cereal?
2. What happens to the metallic iron after you swallow it?
3. Does this iron contribute to your daily iron requirement?

Chemistry in the News

DIETARY SELENIUM

It's easy to get confused by the claims being made about the value of dietary supplements. Many trace minerals are essential to good health, and some of those essential elements are found in our diets in extremely small amounts. Selenium is an example. Selenium supplements are marketed with claims that they will prevent cancer, improve immune function, and protect against environmental pollutants. The media have publicized studies recently showing that people who take supplemental selenium have lower rates for some types of cancers. But other studies show opposite results. Many of the benefits of taking selenium supplements were found for subjects in selenium-poor regions of the world where the people have very low blood levels of selenium. In addition, " . . . though the benefits of selenium supplements have been shown in people who are deficient, nobody is sure that supplemental selenium will do you any good if your diet is supplying adequate selenium."[1] A fur-

ther complication is the difficulty of finding out how much selenium you are getting from your foods. Finally, excess selenium is toxic. Thus, there is much confusion about the usefulness of selenium supplements.

In the body, selenium functions as part of an antioxidant enzyme system by interacting with vitamin E, so the health benefits of selenium and vitamin E are closely linked.

The Institute of Medicine of the National Academy of Sciences issues reports on Dietary Reference Intakes (DRIs). The DRIs have replaced the Recommended Dietary Allowances (RDAs) that the Academy issued for the past 60 years. The DRIs now include the old RDAs, but additionally set upper limits on dietary intake to help people avoid excessive intake of supplements. The latest thinking about selenium is that selenium deficiency is bad but that large supplementary intake can be harmful too. Men and women should get 55 μg

per day.[2] This amount of selenium is already acquired by most Americans from their food. Where does dietary selenium come from? Food sources of selenium include fish, meats, poultry, whole grains, and some kinds of nuts (especially Brazil nuts). The report also sets the upper intake level for selenium at 400 μg per day. More than this amount can cause nausea, hair loss, and tooth loss. Thus, selenium intake has a narrow range (55 to 400 μg per day) that is considered to be healthy. Too little can lead to severe chronic conditions, and too much can lead to other severe symptoms.

Sources:

[1]Wellness Letter, University of California, Berkeley, June 2000.

[2]*Dietary Reference Intakes for Vitamin C, Vitamin E, Selenium, and Carotenoids.* Washington, DC: National Academy of Science Press, 2000 (pamphlet).

weight. They exemplify the old saying that "good things come in small packages." Because the body uses them efficiently, recycling them through many reactions, dietary minerals are required in only small amounts, but their absence from your diet can cause significant health problems.

The 28 dietary minerals indicated in Figure 3.7 are classified into the relatively more abundant **major minerals** and the less plentiful **trace elements.** Major minerals are those present in quantities greater than 0.01% of body weight (100 mg per kg), which is, for example, more than 6 g for a 60-kg (132-lb) individual. Trace elements are present in smaller, sometimes far smaller, amounts. For example, the daily total intake of iodine is only 150 μg. In the context of nutrition, major minerals and trace elements usually refer to ions in soluble ionic compounds in the diet.

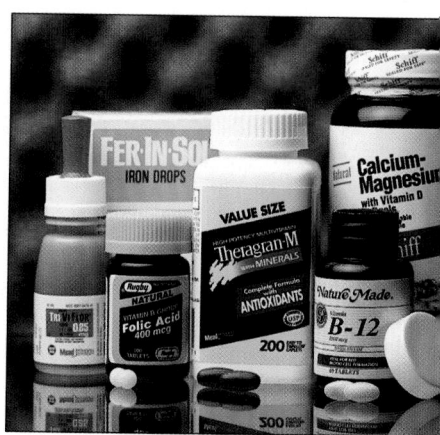

Vitamin and mineral supplements.

 ## 3.12 BIOMOLECULES: CARBOHYDRATES AND FATS

One benefit of understanding organic molecules is the application of this understanding to the molecules involved in life — organic molecules that we eat, that provide the structure of our bodies, and that allow our bodies to function. You have already learned enough chemistry to consider a few simple biomolecules. In this section, we're going to examine the structural formulas of two simple carbohydrates, glucose and sucrose, and tristearin, a typical solid fat.

Grains, breads, and legumes are sources of dietary carbohydrates.

CD-ROM Screen 11.5: Functional Groups

Some atoms in molecules are connected by a double bond (represented by ═), rather than a single bond (represented by —). The nature of these bonds is described in Chapter 8.

Like alcohols, aldehydes and ketones are major classes of organic compounds. The simplest aldehyde and ketone are formaldehyde and acetone, respectively.

$$\underset{\text{formaldehyde}}{\overset{\overset{\displaystyle O}{\|}}{H-C-H}} \qquad \underset{\text{acetone}}{\overset{\overset{\displaystyle O}{\|}}{CH_3CCH_3}}$$

As you proceed through the chapters in this book, you will be reading more about carbohydrates, fats, and other classes of biomolecules. For each class, there are two major kinds of information to be considered: the general structures of the molecules and how they function in living things.

At one time, biochemistry and chemistry were thought of as very different fields of study. Biochemists, after all, worked mainly with living material, and chemists almost never did. However, as time passes, the boundaries between chemistry and biochemistry are disappearing. Almost everything we know about the structure and shape of molecules, the behavior of ions, and the chemical properties of organic compounds applies to the chemistry of living things. Our goal in this text is to provide you with some insight into the application of chemical principles to biochemistry.

Carbohydrates

The word "carbohydrate" literally means "hydrate of carbon." **Carbohydrates** have the general formula $C_x(H_2O)_y$, in which x and y are whole numbers. The molecular formulas for glucose, $C_6H_{12}O_6$ ($x = 6$, $y = 6$), and sucrose (table sugar), $C_{12}H_{22}O_{11}$ ($x = 12$, $y = 11$), illustrate the general formula. Carbohydrates originally were thought to be simple combinations of carbon and water, but this is not the case. We now know that the carbon, hydrogen, and oxygen atoms in carbohydrates are arranged primarily in alcohol, aldehyde, and ketone functional groups (shown in color):

$$\underset{\substack{\text{alcohol}\\\text{group}}}{-C-O-H} \qquad \underset{\substack{\text{aldehyde}\\\text{group}}}{\overset{\overset{\displaystyle O}{\|}}{-C-C-H}} \qquad \underset{\substack{\text{ketone}\\\text{group}}}{\overset{\overset{\displaystyle O}{\|}}{-C-C-C-}}$$

Carbohydrates are divided into three classes. **Monosaccharides** (from the Latin *saccarum*, "sugar") contain one sugar unit and are the simplest carbohydrates; glucose is the most important member of this class. Its structural formula can be written either with the six carbon atoms in a straight chain or with the first carbon joined to the oxygen on the fifth carbon to form a ring. Note that each carbon atom in glucose is connected to an oxygen atom and that all but one of the oxygen atoms is in an —OH group. The other oxygen atom is part of an aldehyde group in the chain structure.

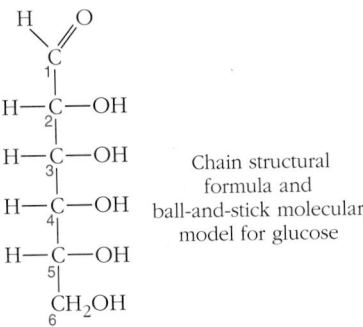

Chain structural formula and ball-and-stick molecular model for glucose

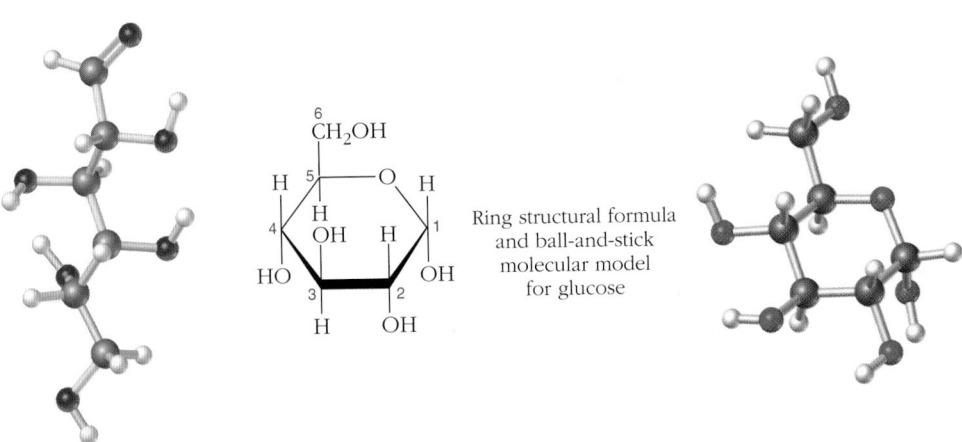

Ring structural formula and ball-and-stick molecular model for glucose

For simplicity, the ring structural formulas of glucose and other carbohydrates are drawn without showing the carbon atoms in the ring.

Carbohydrates make up about half the average human diet, most of them from fruit, vegetables, and grain products such as bread, pasta, and rice. Animals rely on glucose as a major energy source—we metabolize it to produce the energy needed to stay alive (Section 6.12), just as furnaces burn natural gas and fuel oil to heat where we live and work. Although glucose is present in fruit and honey, we don't have to get our daily supply from the foods we eat because chemical reactions in our bodies make glucose from molecules that are in our diets. Serious health problems arise when the glucose concentration in the blood is either too high or too low. For example, diabetes results from the body's failure to regulate glucose concentration at the proper level.

Sucrose is representative of the second class of carbohydrates, known as **disaccharides.** As you might guess from the name, disaccharides are molecules in which two monosaccharides are connected to each other. A sucrose molecule is made of two monosaccharide molecules in their ring forms—glucose and fructose.

Carbon atoms are represented by the junctions of lines in the plane of the ring.

This structure represents the five carbon atoms and the oxygen atom in the six-membered glucose ring.

Glucose is also known as dextrose or blood sugar.

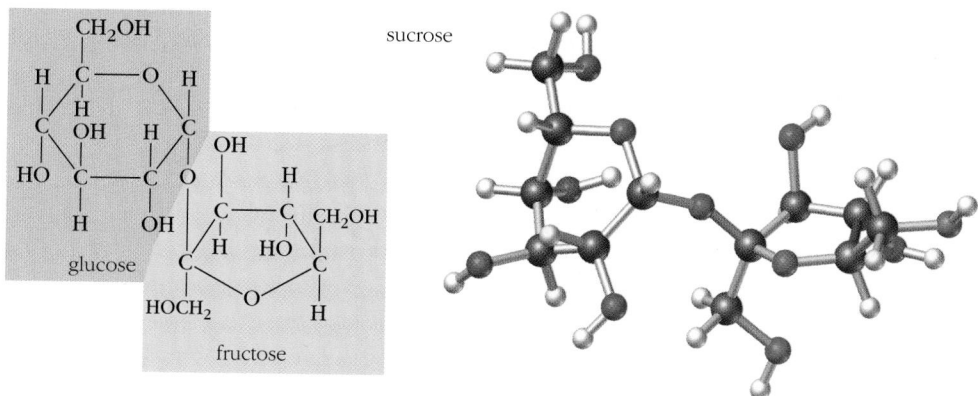

sucrose

Note that, as in sucrose, glucose, and fructose, the names of all mono- and disaccharides end in -ose.

Monosaccharides and disaccharides together are referred to as simple sugars. About half of the simple sugars in our diet come from milk and fresh fruit; the remainder come from sweeteners added to prepared foods. Table 3.12 compares the sweetness of some other simple sugars and some artificial sweeteners with that of sucrose.

The third class of carbohydrates comprises the **polysaccharides,** which consist of many monosaccharide molecules joined together, up to several thousand of them. Cellulose is a polysaccharide built from glucose units. Together with other carbohydrates, cellulose forms wood, cotton, and paper. It also acts as a dietary fiber in our foods because we cannot digest it. We can, however, digest starch, an-

The open-chain form of fructose contains a ketone group; the ring structure does not.

Polysaccharides are polymers of monosaccharides. The chemistry of natural and synthetic polymers is discussed in Chapter 12.

From Table 3.12 you can see that aspartame is nearly 200 times sweeter than sucrose. This means you need only a "pinch" of aspartame to be equivalent to a teaspoon of table sugar.

TABLE 3.12	Sweetness of Other Simple Sugars and Artificial Sweeteners Relative to Sucrose	
Substance	**Sweetness Relative to Sucrose (at 1.00)**
Lactose (milk sugar, a disaccharide)	0.16
Galactose (a monosaccharide in milk sugar)	0.32
Maltose (a disaccharide used in beer making)	0.33
Glucose (dextrose, a common monosaccharide)	0.74
Sucrose (table sugar, a disaccharide)	1.00
Fructose (fruit sugar, a monosaccharide)	1.74
Aspartame (artificial sweetener, NutraSweet)	180
Saccharin (artificial sweetener)	300

Foods high in dietary fiber. Dietary fiber is composed mostly of polysaccharides.

CD-ROM Screen 11.7: Functional Groups: Fats and Oils

The carboxylic acid group is present in all organic acids. Acids of all kinds are an important class of compounds, as is discussed in Chapter 16.

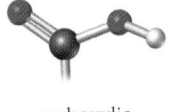

carboxylic acid group

other glucose-based polysaccharide. The arrangement of glucose units in cellulose is slightly different from the arrangement in starch. This relatively small difference in the arrangements of glucose units at the molecular level results in a major difference in how the body handles these two biomolecules.

> **Exercise 3.14 Carbohydrates**
>
> Which of the following four compounds might be a carbohydrate?
> (a) $C_9H_8O_4$ (b) $C_5H_{10}O_5$ (c) $C_5H_7NO_4$ (d) $C_6H_{10}O_5$

Fats and Oils

The structures of all fat and oil molecules are variations on the same structure. To illustrate how all fats and oils are similar, we will begin with tristearin, a very common animal fat, and separate the molecule into its component parts: glycerol and stearic acid.

Glycerol has three alcohol functional groups, the —OH groups. It is a syrupy liquid used in hydraulic fluid, in cosmetics, and as a sweetener. Stearic acid is representative of fatty acids, all of which are molecules with long chains of carbon atoms with their attached hydrogen atoms and, at the end of the carbon chain, a **carboxylic acid** functional group, —COOH. Stearic acid, a colorless, waxy solid obtained from animal fat, is used in soaps, lubricants, shoe polish, and other products.

Earlier we noted that the functional groups are where the chemical action is in organic molecules. All fats and oils have the same basic molecular structure— three long hydrocarbon chains from three fatty acids linked to glycerol. In tristearin formation, the carboxylic acid groups of three molecules of stearic acid react with the three —OH groups of a glycerol molecule. A water molecule is eliminated where each stearic acid molecule is joined to the glycerol molecule.

Within our bodies, fat is a form of stored energy. Fats and oils are the most concentrated energy source in our diets. One gram of fat releases more than twice the amount of energy available from 1 g of carbohydrate. The average American diet currently derives about 37% of its calories from fats and oils. Nutritionists urge that this be reduced to 30% because of mounting evidence that the risk of heart disease increases with the quantity of fats consumed.

Cooking oils are liquids at room temperature; butter is a solid. The difference between liquid oils and solid fats relates to how the carbon atoms in the fatty acids are connected to each other.

Cooking oils are liquid at room temperature, and fats are solid at room temperature, but their general stuctures are as illustrated for tristearin. The difference between liquid oils and solid fats relates to how the carbon atoms in the fatty acids are connected to one another.

SUMMARY PROBLEM

Part I

During each launch of the Space Shuttle, the booster rocket uses about 1.5×10^6 lb of ammonium perchlorate as fuel.

1. Write the chemical formula for (a) ammonium perchlorate, (b) ammonium chlorate, (c) ammonium chlorite.

2. (a) When ammonium perchlorate dissociates in water, what ions are dispersed in the solution?

(b) Would this aqueous solution conduct an electric current? Explain your answer.

3. How many moles of ammonium perchlorate are used in Space Shuttle booster rockets during a launch?

Part II

Chemical analysis of ibuprofen (Advil) indicates that it contains 75.69% carbon, 8.80% hydrogen, and the remainder, oxygen. The empirical formula is also the molecular formula.

1. Determine the molecular formula of ibuprofen.

2. Two 200-mg ibuprofen tablets were taken by a patient to relieve pain. Calculate the number of moles of ibuprofen contained in the two tablets.

IN CLOSING

Having studied this chapter, you should be able to . . .

- Interpret the meaning of molecular formulas, condensed formulas, and structural formulas (Section 3.1).
- Name binary molecular compounds, including straight-chain alkanes (Sections 3.2 and 3.3).
- Write structural formulas for and identify straight- and branched-chain alkane constitutional isomers (Section 3.4).

- Predict the charges on monatomic ions of metals and nonmetals (Section 3.5).
- Know the names and formulas of polyatomic ions (Section 3.5).
- Describe the properties of ionic compounds and compare them with the properties of molecular compounds (Sections 3.5 and 3.7).
- Write the formulas of ionic compounds (Section 3.5).
- Name ionic compounds (Section 3.6).
- Describe electrolytes in aqueous solution and summarize the differences between electrolytes and nonelectrolytes (Section 3.7).
- Thoroughly explain the use of the mole concept for chemical compounds (Section 3.8).
- Calculate the molar mass of a compound (Section 3.8).
- Calculate the number of moles of a compound given the mass, and vice versa (Section 3.8).
- Explain the formula of a hydrated ionic compound and calculate its molar mass (Section 3.8).
- Express molecular composition in terms of percent composition (Section 3.9).
- Use percent composition and molar mass to determine the empirical and molecular formulas of a compound (Section 3.10).
- Identify biologically important elements (Section 3.11).
- Identify the important functional groups in carbohydrates and fats (Section 3.12).

KEY TERMS

alcohol *(3.3)*
alkane *(3.3)*
alkyl group *(3.4)*
anion *(3.5)*
binary molecular compound *(3.2)*
carbohydrate *(3.12)*
carboxylic acid *(3.12)*
cation *(3.5)*
chemical bond *(3.3)*
condensed formula *(3.1)*
constitutional isomer *(3.4)*
crystal lattice *(3.7)*
dietary mineral *(3.11)*

disaccharide *(3.12)*
electrolyte *(3.7)*
empirical formula *(3.10)*
formula unit *(3.7)*
formula weight *(3.8)*
functional group *(3.1)*
halide ion *(3.6)*
hydrocarbon *(3.3)*
inorganic compound *(3.1)*
ionic compound *(3.5)*
ionic hydrate *(3.8)*
isomer *(3.4)*
major mineral *(3.11)*
molecular compound *(3.1)*

molecular formula *(3.1)*
molecular weight *(3.8)*
monatomic ion *(3.5)*
monosaccharide *(3.12)*
nonelectrolyte *(3.7)*
organic compound *(3.1)*
oxoanion *(3.6)*
percent composition by mass *(3.9)*
polyatomic ion *(3.5)*
polysaccharide *(3.12)*
strong electrolyte *(3.7)*
structural formula *(3.1)*
trace element *(3.12)*
water of hydration *(3.8)*

QUESTIONS FOR REVIEW AND THOUGHT

Conceptual Challenge Problems

CP-3.A (Section 3.9) A chemist analyzes three compounds and reports the following data for the percent by mass of the elements Ex, Ey, and Ez in each compound.

Compound	% Ex	% Ey	%Ez
A	37.485	12.583	49.931
B	40.002	6.7142	53.284
C	40.685	5.1216	54.193

Assume that you accept the notion that the numbers of atoms of the elements in compounds are in small whole-number ratios and that the number of atoms in a sample of any element is directly proportional to that sample's mass. What is possible for you to know about the empirical formulas for these three compounds?

CP-3.B (Section 3.10) The following table displays on each horizontal row an empirical formula for one of the three compounds noted in CP-3.A.

Compound A	Compound B	Compound C
	$Ex Ey_2 Ez$	
$Ex_6 Ey_8 Ez_3$		
		$Ex_3 Ey_2 Ez$
	$Ex_9 Ey_2 Ez_6$	
		$Ex Ey_2 Ez_3$
$Ex_3 Ey_8 Ez_3$		

Based only on what was learned in that problem, what is the empirical formula for the other two compounds in that row?

CP-3.C (Section 3.10) (a) Suppose that a chemist now determines that the ratio of the masses of equal numbers of atoms of Ez and Ex atoms is 1.3320 g Ez/1 g Ex. With this added information, what can now be known about the formulas for compounds A, B, and C in CP-3.A?

(b) Suppose that this chemist further determines that the ratio of the masses of equal numbers of atoms of Ex and Ey is 11.916 g Ex/1 g Ey. What is the ratio of the masses of equal numbers of Ez and Ey atoms?

(c) If the mass ratios of equal numbers of atoms of Ex, Ey, and Ez are known, what can be known about the formulas of the three compounds A, B, and C?

Answers to questions in **bold** can be found in the back of the book.

Review Questions

1. A dictionary defines the word "compound" as a "combination of two or more parts." What are the "parts" of a chemical compound? Identify three pure (or nearly pure) compounds you have encountered today. What is the difference between a compound and a mixture?

2. Nobel Prize–winning chemist Roald Hoffmann has said, "Today chemistry is the science of molecules and their transformations." What does that statement mean in the context of your own experience?

3. For each of the following structural formulas, write the molecular formulas and condensed formulas.

(a) (b) (c)

4. Given the following condensed formulas, write the structural and molecular formulas.
   (a) CH_3OH (b) $CH_3CH_2NH_2$
   (c) $CH_3CH_2SCH_2CH_3$

5. Give a molecular formula for each of these organic acids.

(a) pyruvic acid   (b) isocitric acid

6. Give a molecular formula for each of these molecules.

(a) valine   (b) 4-methyl-2-hexanol

7. Give the name for each of these binary nonmetal compounds.
   (a) NF_3 (b) HI   (c) BBr_3 (d) C_6H_{14}

8. Give the name for each of these binary nonmetal compounds.
   (a) C_8H_{18} (b) P_2S_3 (c) OF_2 (d) XeF_4

9. Give the formula for each of these nonmetal compounds.
   (a) Sulfur trioxide
   (b) Dinitrogen pentoxide
   (c) Phosphorous pentachloride
   (d) Silicon tetrachloride
   (e) Diboron trioxide (commonly called boric oxide)

Molecular and Structural Formulas

10. Give the formula for each of the following nonmetal compounds.
    (a) Bromine trifluoride   (b) Xenon difluoride
    (c) Diphosphorus tetrafluoride
    (d) Pentadecane   (e) Hydrazine

11. Write structural formulas for the following alkanes.
    (a) Butane   (b) Nonane   (c) Hexane
    (d) Octane   (e) Octadecane

12. Write the molecular, condensed, and structural formulas for the simplest alcohols derived from butane and pentane.

13. Octane is an alkane (Table 3.4). For the sake of this problem, we will assume that gasoline, a complex mixture of hydrocarbons, is represented by octane. If you fill the tank of your car with 18 gal of gasoline, how many grams and how many pounds of gasoline have you put into the car? Information you may need is (a) the density of octane, 0.692 g/cm^3, (b) the volume of 1 gal in milliliters (3790 mL).

14. Which of the following molecules contains more O atoms, and which contains more atoms of all kinds?
    (a) Sucrose, $C_{12}H_{22}O_{11}$
    (b) Glutathione, $C_{10}H_{17}N_3O_6S$ (the major low-molecular-weight sulfur-containing compound in plant or animal cells)

15. Write the molecular formula of each of the following compounds.
    (a) Benzene, a liquid hydrocarbon, with 6 carbon atoms and 6 hydrogen atoms per molecule
    (b) Vitamin C, with 6 carbon atoms, 8 hydrogen atoms, and 6 oxygen atoms per molecule

16. Write the formula for
    (a) A molecule of the organic compound heptane, which has 7 carbon atoms and 16 hydrogen atoms.
    (b) A molecule of acrylonitrile, the basis of Orlon and Acrilan fibers, which has 3 carbon atoms, 3 hydrogen atoms, and 1 nitrogen atom.
    (c) A molecule of Fenclorac, an antiinflammatory drug, which has 14 carbon atoms, 16 hydrogen atoms, 2 chlorine atoms, and 2 oxygen atoms.

17. Give the total number of atoms of each element in one formula unit of each of the following compounds.
    (a) CaC_2O_4 (b) $C_6H_5CHCH_2$ (c) $(NH_4)_2SO_4$
    (d) $Pt(NH_3)_2Cl_2$ (e) $K_4Fe(CN)_6$

18. Give the total number of atoms of each element in each of these molecules.
    (a) $C_6H_5COOC_2H_5$ (b) $HOOCCH_2CH_2COOH$
    (c) $NH_2CH_2CH_2COOH$ (d) $C_{10}H_9NH_2Fe$
    (e) $C_6H_2CH_3(NO_2)_3$

Constitutional Isomers

19. Draw condensed structures for the five constitutional isomers of C_6H_{14}.

20. Draw the structural formula and condensed formula of 2,2-dimethylpropane. Explain why there is no compound named 2,3-dimethylpropane.

Predicting Ion Charges

21. For each of the following metals, write the chemical symbol for the corresponding ion (with charge).
    (a) Lithium   (b) Strontium   (c) Aluminum
    (d) Calcium   (e) Zinc

22. For each of the following nonmetals, write the chemical symbol for the corresponding ion (with charge).
    (a) Nitrogen   (b) Sulfur   (c) Chlorine
    (d) Iodine   (e) Phosphorus

23. Predict the charges of the ions in an ionic compound composed of barium and bromine.

24. Predict the charges for ions of the following elements.
    (a) Magnesium   (b) Zinc
    (c) Iron   (d) Gallium

25. Predict the charges for ions of the following elements.
    (a) Selenium   (b) Fluorine
    (c) Nickel   (d) Nitrogen

26. Cobalt is a transition metal and so can form ions with at least two different charges. Write the formulas for the compounds formed between cobalt ions and oxide ion.

27. Although not a transition element, lead can also form two cations: Pb^{2+} and Pb^{4+}. Write the formulas for the compounds of these ions with the chloride ion.

28. Which of the following are the correct formulas of compounds? For those that are not, give the correct formula.
    (a) AlCl   (b) NaF_2 (c) Ga_2O_3 (d) MgS

29. Which of the following are the correct formulas of compounds? For those that are not, give the correct formula.
    (a) Ca_2O (b) $SrCl_2$ (c) Fe_2O_5 (d) K_2O

Polyatomic Ions

30. For each of the following compounds, tell what ions are present and how many there are per formula unit.
    (a) $Pb(NO_3)_2$ (b) $NiCO_3$ (c) $(NH_4)_3PO_4$ (d) K_2SO_4

31. For each of the following compounds, tell what ions are present and how many there are per formula unit.
    (a) $Ca(CH_3CO_2)_2$ (b) $Co_2(SO_4)_3$
    (c) $Al(OH)_3$ (d) $(NH_4)_2CO_3$

32. Determine the chemical formulas for barium sulfate, magnesium nitrate, and sodium acetate. Each compound contains a monatomic cation and a polyatomic anion. What are the names and electrical charges of these ions?

33. Write the chemical formulas for the following compounds:
    (a) Nickel(II) nitrate   (b) Sodium bicarbonate
    (c) Lithium hypochlorite   (d) Magnesium chlorate
    (e) Calcium sulfite

Ionic Compounds

34. Determine which of the following substances are ionic.
    (a) CF_4 (b) $SrBr_2$ (c) $Co(NO_3)_3$
    (d) SiO_2 (e) KCN   (f) SCl_2

35. Which of the following are ionic? Write the formula for each.
    (a) Methane   (b) Dinitrogen pentoxide
    (c) Ammonium sulfide   (d) Hydrogen selenide
    (e) Sodium perchlorate

36. Give the formula for each of the following ionic compounds.
    (a) Ammonium carbonate   (b) Calcium iodide
    (c) Copper(II) bromide   (d) Aluminum phosphate

37. Give the formula for each of the following ionic compounds.
    (a) Calcium hydrogen carbonate
    (b) Potassium permanganate
    (c) Magnesium perchlorate
    (d) Ammonium hydrogen phosphate

38. Name each of the following ionic compounds.
    (a) K_2S (b) $NiSO_4$ (c) $(NH_4)_3PO_4$

39. Name each of the following ionic compounds.
    (a) $Ca(CH_3CO_2)_2$ (b) $Co_2(SO_4)_3$ (c) $Al(OH)_3$

40. Name each of the following ionic compounds.
    (a) KH_2PO_4 (b) $CuSO_4$ (c) $CrCl_6$

41. Solid magnesium oxide melts at 2800 °C. This property, combined with the fact that magnesium oxide is not an electrical conductor, makes it an ideal heat insulator for electric wires in cooking ovens and toasters. In contrast, solid NaCl melts at the relatively low temperature of 801 °C. What is the formula of magnesium oxide? Suggest a reason that it has a melting temperature so much higher than that of NaCl.

42. Assume you have an unlabeled bottle containing a white crystalline powder. The powder melts at 310 °C . You are told that it could be NH_3, NO_2, or $NaNO_3$. What do you think it is and why?

Electrolytes

43. What is an electrolyte? How can we differentiate between a weak and a strong electrolyte? Give an example of each.

44. Epsom salt, $MgSO_4 \cdot 7H_2O$, is sold for various purposes over the counter in drug stores. Methanol, CH_3OH, is a small organic compound that is readily soluble in either water or gasoline. Which of these two compounds is an electrolyte and which is a nonelectrolyte?

45. Comment on the statement, "Molecular compounds are generally nonelectrolytes."

46. Comment on the statement, "Ionic compounds are generally electrolytes."

47. For each of the following electrolytes, what ions will be present in an aqueous solution?
    (a) KOH   (b) K_2SO_4 (c) $NaNO_3$ (d) NH_4Cl

48. For each of the following electrolytes, what ions will be present in an aqueous solution?
    (a) CaI_2 (b) $Mg_3(PO_4)_2$ (c) NiS   (d) $MgBr_2$

Moles of Compounds

49. Fill in the table for 1 mol of methanol, CH_3OH.

	CH_3OH	Carbon	Hydrogen	Oxygen
No. of moles	___	___	___	___
No. of molecules or atoms	___	___	___	___
Molar mass	___	___	___	___

50. Fill in the following table for 1 mol of glucose, $C_6H_{12}O_6$.

	$C_6H_{12}O_6$	Carbon	Hydrogen	Oxygen
No. of moles	___	___	___	___
No. of molecules or atoms	___	___	___	___
Molar mass	___	___	___	___

51. Calculate the molar mass of each of the following compounds.
    (a) Fe_2O_3, iron(III) oxide
    (b) BF_3, boron trifluoride
    (c) N_2O, dinitrogen oxide (laughing gas)
    (d) $MnCl_2 \cdot 4H_2O$, manganese(II) chloride tetrahydrate
    (e) $C_6H_8O_6$, ascorbic acid

52. Calculate the molar mass of each of the following compounds.
    (a) $B_{10}H_{14}$, a boron hydride once considered as a rocket fuel
    (b) $C_6H_2(CH_3)(NO_2)_3$, TNT, an explosive
    (c) $PtCl_2(NH_3)_2$, a cancer chemotherapy agent called cisplatin
    (d) $CH_3(CH_2)_3SH$, a compound that has a skunk-like odor
    (e) $C_{20}H_{24}N_2O_2$, quinine, used as an antimalarial drug

53. How many moles are represented by 1.00 g of each of the following compounds?
    (a) CH_3OH, methanol
    (b) Cl_2CO, phosgene, a poisonous gas
    (c) NH_4NO_3, ammonium nitrate
    (d) $MgSO_4 \cdot 7H_2O$, magnesium sulfate heptahydrate (Epsom salt)
    (e) $AgCH_3CO_2$, silver acetate

54. Assume you have 0.250 g of each of the following compounds. How many moles of each are represented?
    (a) $C_7H_5NO_3S$, saccharin, an artificial sweetener
    (b) $C_{13}H_{20}N_2O_2$, procaine, a painkiller used by dentists
    (c) $C_{20}H_{14}O_4$, phenolphthalein, a dye

55. (a) What is the molar mass of iron(II) nitrate, $Fe(NO_3)_2$?
    (b) What is the mass, in grams, of 0.200 mol of $Fe(NO_3)_2$?
    (c) How many moles of $Fe(NO_3)_2$ are present in 4.66 g?

56. Acetaminophen, an analgesic, has the molecular formula $C_8H_9O_2N$.
    (a) What is the molar mass of acetaminophen?
    (b) How many moles are present in 5.32 g of acetaminophen?
    (c) How many g are present in 0.166 mol of acetaminophen?

57. An Alka-Seltzer tablet contains 324 mg of aspirin ($C_9H_8O_4$), 1904 mg of $NaHCO_3$, and 1000 mg of citric acid ($C_6H_8O_7$). (The last two compounds react with each other to provide the "fizz," bubbles of CO_2, when the tablet is put into water.)
    (a) Calculate the number of moles of each substance in the tablet.
    (b) If you take one tablet, how many molecules of aspirin are you consuming?

58. The use of CFCs (chlorofluorocarbons) is being curtailed, because there is strong evidence that they cause environmental damage. If a spray can contains 250 g of one of these compounds, CCl_2F_2, how many molecules of this CFC are you releasing to the air when you empty the can?

59. Sulfur trioxide, SO_3, is made in enormous quantities by combining oxygen and sulfur dioxide, SO_2. The trioxide is not usually isolated but is converted to sulfuric acid.
    (a) If you have 1.00 lb (454 g) of sulfur trioxide, how many moles does this represent?
    (b) How many molecules?
    (c) How many sulfur atoms?
    (d) How many oxygen atoms?

60. CFCs (chlorofluorocarbons) are implicated in decreasing ozone in the stratosphere. A CFC substitute is CF_3CH_2F.
(a) If you have 25.5 g of this compound, how many moles does this represent?
(b) How many atoms of fluorine are contained in 25.5 g of the compound?

61. How many water molecules are in one drop of water? (One drop of water is $\frac{1}{20}$ of a mL or 0.050 mL, and the density of water is 1.0 g/mL.)

62. If the water from a well contains 0.10 ppb (parts per billion) of chloroform, $CHCl_3$, how many molecules of chloroform are present in one drop of the water? (One drop of water is $\frac{1}{20}$ of a mL or 0.050 mL, and the density of water is 1.0 g/mL.)

Percent Composition

63. Calculate the molar mass of each of these compounds and the weight percent of each element.
(a) PbS, lead(II) sulfide, galena
(b) C_2H_6, ethane, a hydrocarbon fuel
(c) CH_3CO_2H, acetic acid, an important ingredient in vinegar
(d) NH_4NO_3, ammonium nitrate, a fertilizer

64. Calculate the molar mass of each of these compounds and the weight percent of each element.
(a) $MgCO_3$, magnesium carbonate
(b) C_6H_5OH, phenol, an organic compound used in some cleaners
(c) $C_2H_3O_5N$, peroxyacetyl nitrate, an objectionable compound in photochemical smog
(d) $C_4H_{10}O_3NPS$, acephate, an insecticide

65. A certain metal, M, forms two oxides, M_2O and MO. If the percent by weight of M in M_2O is 73.4%, what is its percent by weight in MO?

66. The copper-containing compound $Cu(NH_3)_4SO_4 \cdot H_2O$ is a beautiful blue solid. Calculate the molar mass of the compound and the mass percent of each element.

67. Nitrogen fixation in the root nodules of peas and other legumes occurs with a reaction involving a molybdenum-containing enzyme named *nitrogenase*. This enzyme contains two Mo atoms per molecule and is 0.0872% Mo by mass. What is the molar mass of the enzyme?

68. If you heat Al with an element from Group 6A, an ionic compound is formed that contains 18.55% Al by mass.
(a) What is the likely charge on the nonmetal in the compound formed?
(b) Using X to represent the nonmetal, what is the empirical formula for this ionic compound?
(c) Which element in Group 6A has been combined with Al?

69. Disilane, Si_2H_x, contains 90.28% silicon by mass. What is the value of x in this compound?

70. Chalky, white crystals in mineral collections are often labeled borax, which has the molecular formula $Na_2B_4O_7 \cdot 10\ H_2O$, when actually they are partially dehydrated samples with the molecular formula $Na_2B_4O_7 \cdot 5\ H_2O$, which is more stable under the storage conditions. Real crystals of borax are clear, transparent crystals.
(a) What percent of the mass has the mineral lost when it partially dehydrates?

(b) Will the percent boron by mass be the same in the two compounds?

71. Chemists often express the composition of compounds in terms of the percentage of a particular element that is present. Look for some food product that gives the composition in terms of percentages. (a) What data are given? (b) Is percent by weight of an element or of a compound listed?

72. If a food label gives the percent composition of a particular compound in a food as 14.5%, what does that indicate about the compound in terms of the food?

Empirical and Molecular Formulas

73. What is the difference between an empirical formula and a molecular formula? Use the compound ethane, C_2H_6, to illustrate your answer.

74. The empirical formula of maleic acid is CHO. Its molar mass is 116.1 g/mol. What is its molecular formula?

75. A well-known reagent in analytical chemistry, dimethylglyoxime, has the empirical formula C_2H_4NO. If its molar mass is 116.1 g/mol, what is the molecular formula of the compound?

76. Acetylene is a colorless gas that is used as a fuel in welding torches, among other things. It is 92.26% C and 7.74% H. Its molar mass is 26.02 g/mol. Calculate the empirical and molecular formulas.

77. There is a large family of boron-hydrogen compounds called boron hydrides. All have the formula B_xH_y and almost all react with air and burn or explode. One member of this family contains 88.5% B; the remainder is hydrogen. Which of the following is its empirical formula: BH_3, B_2H_5, B_5H_7, B_5H_{11}, or BH_2?

78. Nitrogen and oxygen form an extensive series of at least seven oxides of general formula N_xO_y. One of them is a blue solid that comes apart, or "dissociates," reversibly, in the gas phase. It contains 36.84% N. What is the empirical formula of this oxide?

79. Cumene is a hydrocarbon, a compound composed only of C and H. It is 89.94% carbon, and the molar mass is 120.2 g/mol. What are the (a) empirical and (b) molecular formulas of cumene?

80. Acetic acid is the important ingredient in vinegar. It is composed of carbon (40.0%), hydrogen (6.71%), and oxygen (53.29%). Its molar mass is 60.0 g/mol. Determine the (a) empirical and (b) molecular formulas of the acid.

81. An analysis of nicotine, a poisonous compound found in tobacco leaves, shows that it is 74.0% C, 8.65% H, and 17.35% N. Its molar mass is 162 g/mol. What are the (a) empirical and (b) molecular formulas of nicotine?

82. Cacodyl, a compound containing arsenic, was reported in 1842 by the German chemist Bunsen. It has an almost intolerable garlic-like odor. Its molar mass is 210 g/mol, and it is 22.88% C, 5.76% H, and 71.36% As. Determine its (a) empirical and (b) molecular formulas.

83. The action of bacteria on meat and fish produces a poisonous compound called cadaverine. As its name and origin imply, it stinks! It is 58.77% C, 13.81% H, and 27.42% N. Its molar mass is 102.2 g/mol. Determine the molecular formula of cadaverine.

84. DDT is an insecticide with the following percent composition: 47.5% C, 2.54% H, and the remainder chlorine. What is the empirical formula of DDT?

85. If epsom salt, $MgSO_4 \cdot xH_2O$, is heated to 250 °C, all the water of hydration is lost. After a 1.687-g sample of the hydrate is heated, 0.824 g of $MgSO_4$ remains. How many molecules of water are there per formula unit of $MgSO_4$?

86. The alum used in cooking is potassium aluminum sulfate hydrate, $KAl(SO_4)_2 \cdot xH_2O$. To find the value of x, you can heat a sample of the compound to drive off all the water and leave only $KAl(SO_4)_2$. Assume that you heat 4.74 g of the hydrated compound and that it loses 2.16 g of water. What is the value of x?

Biological Periodic Table

87. Make a list of the top ten most abundant essential elements needed by the human body.

88. Which types of compounds contain the majority of the oxygen found in the human body?

89. (a) How are metals found in the body, as atoms or as ions? (b) What are two uses for metals in the human body?

90. Distinguish between macrominerals and microminerals.

91. Which minerals are essential at smaller concentrations but toxic at higher concentrations?

Carbohydrates and Fats

92. (a) Name the monosaccharide commonly found in fresh fruit. (b) What is the structural formula for this sugar?

93. (a) Which monosaccharide is the most important? (b) Give its name and structural formula. (c) How is it involved in plants and animals?

94. Why are fats and oils known as triglycerides?

General Questions

95. (a) Draw a diagram showing the crystal lattice of sodium chloride (NaCl). Show clearly why such a crystal can be cleaved easily by tapping on a knife blade properly aligned along the crystal. (b) Describe in words why the cleavage occurs as it does.

96. Give the molecular formula for each of the following molecules.

trinitrotoluene, TNT   serine, an essential amino acid

97. (a) Calculate the mass of one molecule of nitrogen. Now assume that someone decided to make Avogadro's number have a simpler value, say 1.000×10^{20}. (b) What would be the mass of a molecule of nitrogen in that case?

98. (a) Which of the following pairs of elements are likely to form ionic compounds? (b) Write appropriate formulas for the compounds you expect to form, and give the name of each.
   (a) Chlorine and bromine   (b) Lithium and tellurium
   (c) Sodium and argon   (d) Magnesium and fluorine
   (e) Nitrogen and bromine   (f) Indium and sulfur
   (g) Selenium and bromine

99. (a) Name each of the following compounds. (b) Tell which ones are best described as ionic.
   (a) $ClBr_3$ (b) NCl_3 (c) $CaSO_4$ (d) C_7H_{16}
   (e) XeF_4 (f) OF_2 (g) NaI (h) Al_2S_3
   (i) PCl_5 (j) K_3PO_4

100. (a) Write the formula for each of the following compounds. (b) Tell which ones are best described as ionic.
   (a) Sodium hypochlorite   (b) Aluminum perchlorate
   (c) Potassium permanganate   (d) Potassium dihydrogen phosphate
   (e) Chlorine trifluoride   (f) Boron tribromide
   (g) Calcium acetate   (h) Sodium sulfite
   (i) Disulfur tetrachloride   (j) Phosphorus trifluoride

101. Precious metals such as gold and platinum are sold in units of "troy ounces," where 1 troy ounce is equivalent to 31.1 g. (a) If you have a block of platinum with a mass of 15.0 troy ounces, how many moles of the metal do you have? (b) What is the size of the block in cubic centimeters? (The density of platinum is 21.45 g/cm^3 at 20 °C.)

102. "Dilithium" is the fuel for the Starship *Enterprise*. However, because its density is quite low, you will need a large space to store a large mass. As an estimate for the volume required, we shall use the element lithium. (a) If you want to have 256 mol for an interplanetary trip, what must the volume of a piece of lithium be? (b) If the piece of lithium is a cube, what is the dimension of an edge of the cube? (The density of lithium is 0.534 g/cm^3 at 20 °C.)

103. Fluorocarbonyl hypofluorite was recently isolated, and analysis showed it to be 14.6% C, 39.0% O, and 46.3% F. If the molar mass of the compound is 82 g/mol, determine the (a) empirical and (b) molecular formulas of the compound.

104. Azulene, a beautiful blue hydrocarbon, is 93.71% C and has a molar mass of 128.16 g/mol. What are the (a) empirical and (b) molecular formulas of azulene?

105. A major oil company has used a gasoline additive called MMT to boost the octane rating of its gasoline. What is the empirical formula of MMT if it is 49.5% C, 3.2% H, 22.0% O, and 25.2% Mn?

106. Direct reaction of iodine (I_2) and chlorine (Cl_2) produces an iodine chloride, I_xCl_y, a bright yellow solid. (a) If you completely use up 0.678 g of iodine and produce 1.246 g of I_xCl_y, what is the empirical formula of the compound? (b) A later experiment shows that the molar mass of I_xCl_y is 467 g/mol. What is the molecular formula of the compound?

107. Pepto-Bismol, which helps provide relief for an upset stomach, contains 300 mg of bismuth subsalicylate, $C_7H_5BiO_4$, per tablet. (a) If you take two tablets for your stomach distress, how many moles of the "active ingredient" are you taking? (b) How many grams of Bi are you consuming in two tablets?

108. Iron pyrite, often called "fool's gold," has the formula FeS_2. If you could convert 15.8 kg of iron pyrite to iron

metal and remove the sulfur, how many kilograms of the metal could you obtain?

109. Ilmenite is a mineral that is an oxide of iron and titanium, $FeTiO_3$. If an ore that contains ilmenite is 6.75% titanium, what is the mass (in grams) of ilmenite in 1.00 metric ton (exactly 1000 kg) of the ore?

110. Stibnite, Sb_2S_3, is a dark gray mineral from which antimony metal is obtained. If you have one pound of an ore that contains 10.6% antimony, what mass of Sb_2S_3 (in grams) is there in the ore?

111. Draw a diagram to indicate the arrangement of nanoscale particles of each substance. Consider each drawing to hold a very tiny portion of each substance. Each drawing should contain at least 16 particles, and it need not be three-dimensional.

$Br_2(\ell)$	LiF(s)

112. A piece of nickel foil, 0.550 mm thick and 1.25 cm square, was allowed to react with fluorine, F_2, to give a nickel fluoride. (a) How many moles of nickel foil were used? (b) If you isolate 1.261 g of the nickel fluoride, what is its formula? (c) What is its name? (The density of nickel is 8.908 g/cm^3.)

113. Uranium is used as a fuel, primarily in the form of uranium(IV) oxide, in nuclear power plants. This question considers some uranium chemistry.
(a) A small sample of uranium metal (0.169 g) is heated to 800 to 900 °C in air to give 0.199 g of a dark green oxide, U_xO_y. How many moles of uranium metal were used? What is the empirical formula of the oxide U_xO_y? What is the name of the oxide? How many moles of U_xO_y must have been obtained?
(b) The oxide U_xO_y is obtained if $UO_2(NO_3) \cdot nH_2O$ is heated to temperatures greater than 800 °C in the air. However, if you heat it gently, only the water of hydration is lost. If you have 0.865 g of $UO_2(NO_3) \cdot nH_2O$ and obtain 0.679 g of $UO_2(NO_3)$ on heating, how many molecules of water of hydration were there in each formula unit of the original compound?

114. Draw diagrams of each nanoscale situation. Represent atoms or monatomic ions as circles; represent molecules or polyatomic ions by overlapping circles for the atoms that make up the molecule or ion; and distinguish among different kinds of atoms by labeling or shading the circles. In each case draw representations of at least five nanoscale particles. Your diagrams can be two-dimensional.
(a) A crystal of solid sodium chloride
(b) The sodium chloride from part (a) after it has been melted
(c) A sample of molten aluminum oxide, Al_2O_3

115. Draw diagrams of each nanoscale situation. Represent atoms or monatomic ions as circles; represent molecules or polyatomic ions by overlapping circles for the atoms that make up the molecule or ion; and distinguish among different kinds of atoms by labeling or shading the circles. In each case draw representations of at least five nanoscale particles. Your diagrams can be two-dimensional.
(a) A sample of solid lithium nitrate, $LiNO_3$
(b) A sample of molten lithium nitrate
(c) The same sample of lithium nitrate after electrodes have been placed into it and a direct current applied to the electrodes
(d) A sample of solid lithium nitrate in contact with a solution of lithium nitrate in water

Applying Concepts

116. When asked to draw all the possible constitutional isomers for C_3H_8O, a student drew these structures. The student's instructor said some of the structures were identical. (a) How many actual isomers are there? (b) Which structures are identical?

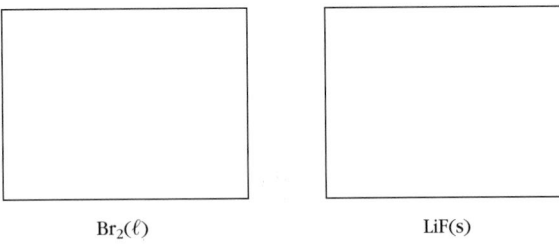

CH_3—CH_2—CH_2—OH CH_3—CH_2—O—CH_3

CH_3—O—CH_2—CH_3

HO—CH_2—CH_2 | CH_3

CH_3—CH—CH_3 | OH

HO—CH—CH_3 | CH_3

117. A student incorrectly named this structure as 2,4,4-trimethylbutane. What is the correct name?

$$CH_3-\overset{\overset{\displaystyle CH_3}{|}}{\underset{\underset{\displaystyle H}{|}}{C}}-CH_2-\overset{\overset{\displaystyle CH_3}{|}}{\underset{\underset{\displaystyle CH_3}{|}}{CH}}$$

118. The statement that "metals form positive ions by losing electrons" is difficult to grasp for some students because positive signs indicate a gain and negative signs a loss. How would you explain this contradiction to a classmate?

119. The formula for thallium nitrate is $TlNO_3$. Based on this information, what would be the formulas for thallium carbonate and thallium sulfate?

120. The name given with each of these formulas is incorrect. What are the correct names?
(a) CaF_2, calcium difluoride
(b) CuO, copper oxide
(c) $NaNO_3$, sodium nitroxide
(d) NI_3, nitrogen iodide
(e) $FeCl_3$, iron(I) chloride
(f) Li_2SO_4, dilithium sulfate

121. Based on the guidelines for naming oxyanions in a series, how would you name the following?
(a) BrO_4^-, BrO_3^-, BrO_2^-, BrO^-
(b) SeO_4^{2-}, SeO_3^{2-}

122. Which illustration best represents $CaCl_2$ dissolved in water?

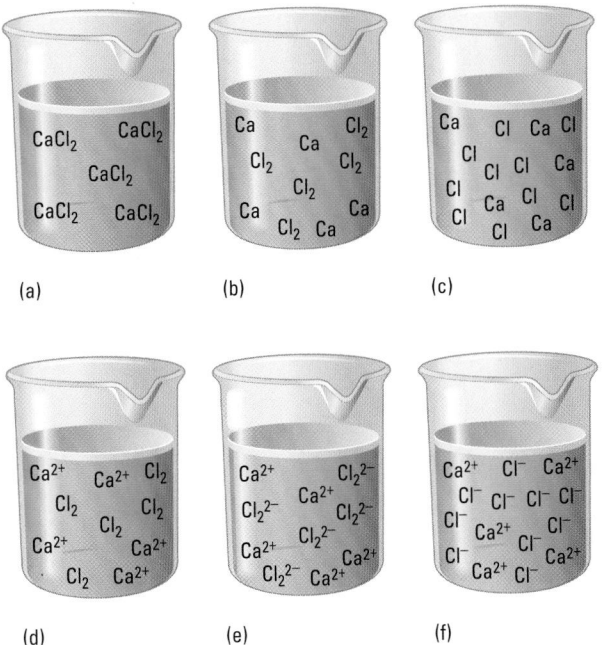

(a)　　　　　　　　(b)　　　　　　　　(c)

(d)　　　　　　　　(e)　　　　　　　　(f)

123. Which sample has the largest amount of NH_3?
    (a) 6.022×10^{24} molecules of NH_3
    (b) 0.1 mol of NH_3
    (c) 17.03 g of NH_3

124. One molecule of an unknown compound has a mass of 7.308×10^{-23} g, and 27.3% of that mass is due to carbon; the rest is oxygen. What is the compound?

 General Chemistry CD-ROM

CD3.1 Screen 3.3: Molecular Compounds. When examining the molecular models, recall that carbon atoms are gray, hydrogen atoms are white, oxygen atoms are red, and nitrogen atoms are blue.
(a) How many total atoms are there in the ethanol molecule?
(b) Is the ammonia molecule flat? How would you describe its structure?

CD3.2 Screen 3.7: Ions.
(a) When a magnesium atom forms an ion, are electrons gained or lost? How many? What happens to the size of the atom when it forms an ion?
(b) When a fluorine atom forms an ion, are electrons gained or lost? How many? What happens to the size of the atom when it forms an ion?

CD3.3 Screen 3.10: Ionic Compounds. Examine the video of the reaction of sodium with chlorine.
(a) The sodium metal for the reaction is removed from a beaker, where chips of the metal are clearly seen under a clear, colorless liquid. Is this liquid water? Explain. (Refer to Screen 2.17.)
(b) What happens to the sodium in the course of the reaction? Does it gain or lose one or more electrons? If so, how many electrons? What is the final form of the sodium in sodium chloride?
(c) Is energy involved in this reaction? Is there evidence for the evolution of energy? What is that evidence?

CD3.4 Screen 3.17: Percent Composition.
(a) Which has the greater percentage of N in 25.0 g of the compound: NO or NO_2?
(b) Which has the greatest percentage of H in 12 g of the compound: CH_4, C_2H_6, or C_5H_{12}?

CD3.5 Screen 3.18: Determining Empirical Formulas; Screen 3.19, Determining Molecular Formulas. Complete the table below.

Name	Molecular Formula	Molar Mass	Empirical Formula
Ethane	C_2H_6	30.1 g/mol	CH_3
_____	N_2O_4	_____	_____
Benzene	_____	78.1 g/mol	CH
Naphthalene	_____	128.2 g/mol	C_5H_4
Tartaric acid	$C_4H_6O_6$	_____	_____

4

A NASA space shuttle and its booster rockets are launched. The shuttle is powered by the reaction of liquid hydrogen and liquid oxygen. How do the scientists and engineers designing the engine know how much of each reactant to use? Using a balanced chemical equation allows precise calculations of the masses of reactants needed.

QUANTITIES OF REACTANTS AND PRODUCTS

4.1 Chemical Equations

4.2 Patterns of Chemical Reactions

4.3 Balancing Chemical Equations

4.4 The Mole and Chemical Reactions: The Macro-Nano Connection

4.5 Reactions with One Reactant in Limited Supply

4.6 Evaluating the Success of a Synthesis: Percent Yield

4.7 Percent Composition and Empirical Formulas

A major emphasis of chemistry is understanding chemical reactions. To work with chemical reactions requires writing down the exact composition of each reactant and product and the relative amounts of each species. This information is contained in balanced chemical equations—equations that are consistent with the law of conservation of mass and the existence of atoms. This chapter begins with a discussion of the nature of chemical equations. Then there is a brief introduction to some general types of chemical reactions. Many of the very large number of known chemical reactions can be assigned to a few categories: combination, decomposition, displacement, and exchange. Next is a description of how to write balanced chemical equations. After that, we will look at how the balanced chemical equation can be used to move from an understanding of what the reactants and products are to *how much* of each is involved under various conditions. Finally, we will introduce methods used to find the formulas of chemical compounds.

While the fundamentals of chemical reactions must be understood at the atomic and molecular levels (the nanoscale), the chemical reactions that we will describe are observable at the macroscale in the everyday world around us. Understanding and manipulating the quantitative relationships in chemical reactions are essential skills. You should know how to calculate the quantity of product a reaction will generate from a given amount of reactants. Facility with these quantitative calculations connects the nanoscale world of the chemical reaction to the macroscale world of laboratory and industrial manipulations of measurable quantities of chemicals.

4.1 CHEMICAL EQUATIONS

CD-Rom Screen 4.2: Chemical Equations

A candle flame can create a mood as well as provide light. It also is the result of a chemical reaction, a process in which reactants are converted into products (⟸ *p. 11)* The reactants and products can be elements, compounds, or both. In equation form we write

$$\text{Reactant(s)} \longrightarrow \text{product(s)}$$

where the arrow means "forms" or "yields" or "changes to."

In a burning candle, the reactants are hydrocarbons from the candle wax and oxygen from the air. Such reactions, in which an element or compound burns in

$$C_{25}H_{52}(s) \ + \ 38 \ O_2(g) \longrightarrow 25 \ CO_2(g) \ + \ 26 \ H_2O(\ell)$$

| a hydrocarbon in candle wax | diatomic oxygen | carbon dioxide | water |

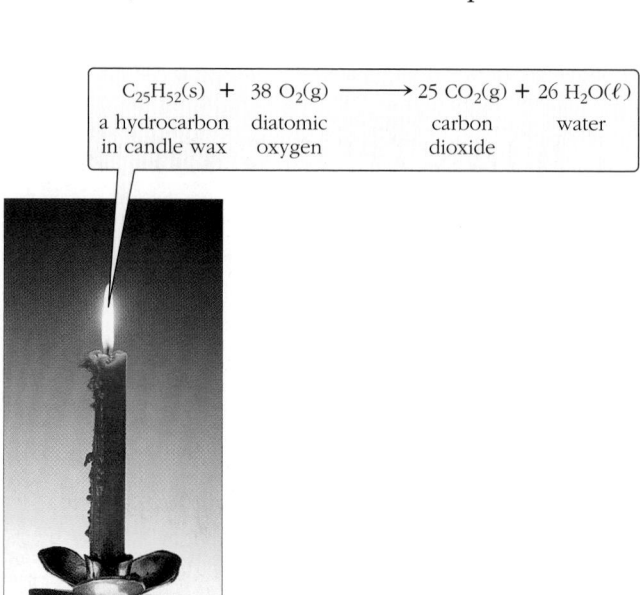

Burning hydrocarbons from candle wax create a candle flame.

air or oxygen, are called **combustion reactions.** The products of the complete combustion of hydrocarbons (⬅ *p. 80)* are always carbon dioxide and water.

$$C_{25}H_{52}(s) + 38 \, O_2(g) \longrightarrow 25 \, CO_2(g) + 26 \, H_2O \, (\ell)$$

a hydrocarbon   diatomic   carbon   water
in candle wax   oxygen   dioxide

This **balanced chemical equation** indicates the relative amounts of reactants and products so that the number of atoms of each element in the reactants equals the number of atoms of the same element in the products. In the next section we discuss how to write balanced equations.

Usually the physical states of the reactants and products are indicated in a chemical equation by placing one of the following symbols after each reactant and product: (s) for solid, (ℓ) for liquid, and (g) for gas. The symbol (aq) is used to represent a substance dissolved in water, an **aqueous solution.** This is illustrated by the equation for the reaction of zinc metal with hydrochloric acid, an aqueous solution of hydrogen chloride. The products are hydrogen gas and an aqueous solution of zinc chloride, a soluble ionic compound (⬅ *p. 86).*

$$Zn(s) + 2 \, HCl(aq) \longrightarrow H_2(g) + ZnCl_2(aq)$$

Reactants   Products

In the 18th century, the great French scientist Antoine Lavoisier introduced the law of conservation of mass, which later became part of Dalton's atomic theory. Lavoisier showed that mass is neither created nor destroyed in chemical reactions. Therefore, if you use 5 g of reactants they will form 5 g of products if the reaction is complete, and if you use 500 mg of reactants they will form 500 mg of products, and so on. Combined with Dalton's atomic theory, this also means that if there are 1000 atoms of a particular element in the reactants, then those 1000 atoms must appear in the products.

Consider, for example, the reaction between gaseous hydrogen and chlorine to produce hydrogen chloride gas:

$$H_2(g) + Cl_2(g) \longrightarrow 2 \, HCl(g)$$

When applied to this reaction, the conservation of mass means that one diatomic molecule of H_2 (two atoms of hydrogen) and one diatomic molecule of Cl_2 (two atoms of Cl) must produce *two* molecules of HCl. The numbers in front of formulas—the **coefficients**—in balanced equations show how matter is conserved. The 2 HCl indicates that two HCl molecules are formed, each containing one hydrogen atom and one chlorine atom. Note how the symbol of an element or the formula of a compound is multiplied through by the coefficient that precedes it. The equality of the number of atoms of each kind in the reactants and in the products is what makes the equation "balanced."

Multiplying all the coefficients by the same factor gives the relative amounts of reactants and products at any scale. For example, 4 H_2 molecules will react with 4 Cl_2 molecules to produce 8 HCl molecules (Figure 4.1). If we continue to scale up the reaction, we can use Avogadro's number as the common factor. Thus, 1 mol of H_2 reacting with 1 mol of Cl_2 molecules will produce 2 mol of HCl. As demanded by the conservation of mass, the number of atoms of each type in the reactants and the products is the same.

With the atoms balanced, the masses represented by the equation are also balanced. The molar masses show that 1.000 mol of H_2 is equivalent to 2.016 g of H_2 and that 1.000 mol of Cl_2 is equivalent to 70.90 g of Cl_2, so the total mass of reactants must be 2.016 g + 70.90 g = 72.92 g when 1.000 mol each of H_2 and Cl_2 are used. Conservation of mass demands that the same mass, 72.92 g of HCl, must result from the reaction, and it does.

$$2.000 \text{ mol HCl} \times \frac{36.45 \text{ g HCl}}{1 \text{ mol HCl}} = 72.92 \text{ g HCl}$$

ANTOINE LAVOISIER
(1743–1794)

Lavoisier was one of the first to recognize the importance of exact scientific measurements and of carefully planned experiments. He introduced principles for naming chemical substances that are still in use today. Further, he wrote a textbook, *Elements of Chemistry* (1789) in which he applied for the first time the principle of the conservation of mass to chemistry and used the idea to write early versions of chemical equations. His life was cut short during the Reign of Terror of the French Revolution.

 CD-Rom Screen 4.3: The Law of Conservation of Matter

There must be an accounting for *all* atoms in a chemical reaction.

Recall that there are 6.022×10^{23} atoms in a mole of any atomic element (⬅ *p. 58)* or 6.022×10^{23} molecules (⬅ *p. 97)* in a mole of any molecular element or compound.

Figure 4.1 Hydrogen (H_2) and chlorine (Cl_2) react to form hydrogen chloride (HCl). Two molecules of HCl are formed when one H_2 molecule reacts with one Cl_2 molecule. This ratio is maintained when the reaction is carried out on a larger scale.

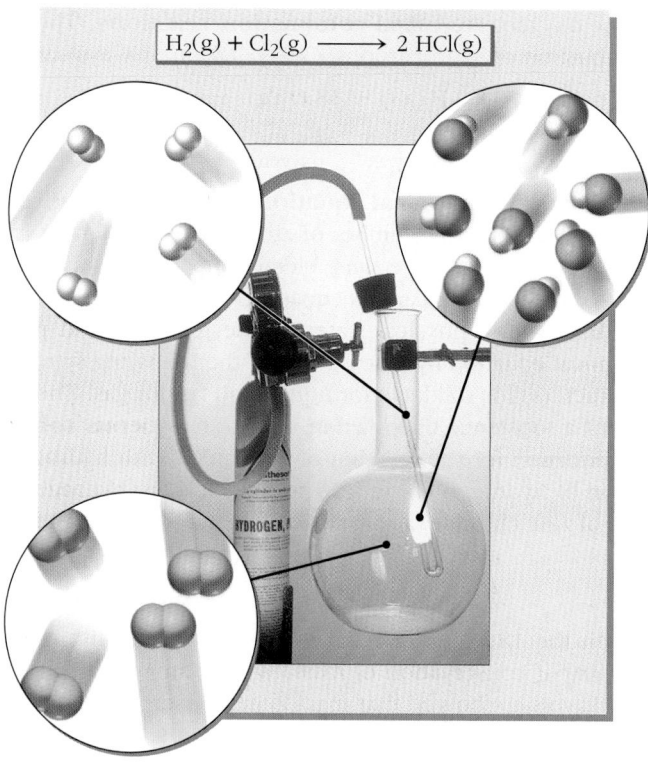

$$H_2(g) + Cl_2(g) \longrightarrow 2\ HCl(g)$$

Figure 4.2 Powdered iron burns in air to form the iron oxide Fe_2O_3. The flame and the energy released during the reaction heat the particles to incandescence.

The relationship between the masses of chemical reactants and products is called **stoichiometry** (stoy-key-AHM-uh-tree), and the coefficients (the multiplying numbers) in a balanced equation are the **stoichiometric coefficients.**

Exercise 4.1   Chemical Equations

When methane burns the following reaction is occurring:

$$CH_4(g) + 2\ O_2(g) \longrightarrow CO_2(g) + 2\ H_2O(g)$$

Write out in words the meaning of this chemical equation.

Exercise 4.2   Stoichiometric Coefficients

Heating iron metal in oxygen forms iron(III) oxide, Fe_2O_3 (Figure 4.2):

$$4\ Fe(s) + 3\ O_2(g) \longrightarrow 2\ Fe_2O_3(s)$$

(a) If 2.50 g of Fe_2O_3 is formed by this reaction, what is the maximum *total* mass of iron metal and oxygen that reacted?
(b) Identify the stoichiometric coefficients in this equation.
(c) If 10,000 oxygen atoms reacted, how many Fe atoms were needed to react with this amount of oxygen?

4.2   PATTERNS OF CHEMICAL REACTIONS

The classification of reaction patterns given in Figure 4.3 applies mainly to elements and inorganic compounds.

Many simple chemical reactions fall into one of the reaction patterns illustrated in Figure 4.3. Learning to recognize these reaction patterns is useful because they serve as a guide to predict what might happen when chemicals are mixed together or heated.

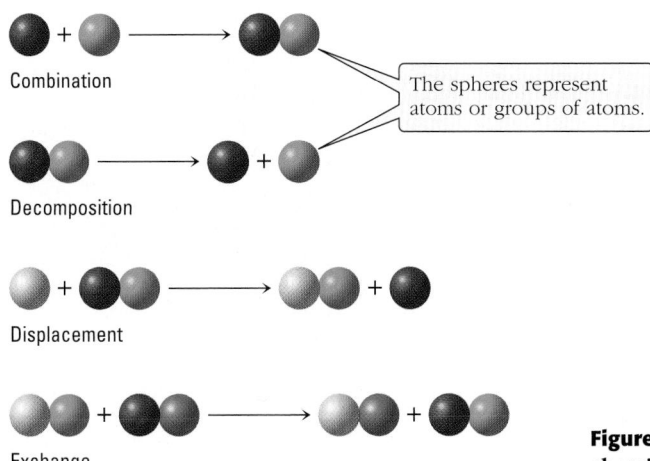

Combination

Decomposition

Displacement

Exchange

The spheres represent atoms or groups of atoms.

Figure 4.3 Four general types of chemical reactions.

Take a look at the equation below. What does it mean to you at this stage in your study of chemistry?

$$Cl_2(g) + 2\ KBr(aq) \longrightarrow 2\ KCl(aq) + Br_2(\ell)$$

It's easy to let your eye slide by an equation on the printed page, but *don't* do that; **read the equation.** In this case it shows you that gaseous diatomic chlorine mixed with an aqueous solution of potassium bromide reacts to produce an aqueous solution of potassium chloride plus liquid diatomic bromine. After you learn to recognize reaction patterns, you will see that this is a *displacement* reaction — chlorine has displaced bromine so that the resulting compound in solution is KCl instead of the KBr originally present. Because chlorine displaces bromine, chlorine is what chemists describe as "more active" than bromine. The occurrence of this reaction implies that when chlorine is mixed with a solution of a different ionic bromide compound, displacement might also take place.

Throughout the rest of this chapter and the rest of this book, you should read and interpret chemical equations as we have just illustrated. Note the physical states of reactants and products. Mentally classify the reactions as described in the following sections and look for what can be learned from each example. Most importantly, *don't* think that the equations are there to be memorized. There are far too many chemical reactions for that. Look instead for patterns, classes of reactions and the kinds of substances that undergo them, and for information that can be applied in other situations. Doing this will give you insight into how chemistry is used every day in a wide variety of applications.

Combination Reactions

In a **combination reaction,** two or more substances react to form a single product.

X   Z   XZ

Oxygen and the halogens (Group 7A) are such reactive elements that they undergo combination reactions with most other elements. Thus, if one of two possible reactants is oxygen or a halogen and the other is another element, it is reasonable to expect that a combination reaction will occur.

Recall that the halogens are F_2, Cl_2, Br_2, and I_2.

Figure 4.4 Combination of zinc and iodine. Gray powdered zinc metal (far left) reacts with grayish-purple iodine in a vigorous combination reaction. Zinc atoms react with diatomic iodine molecules to form zinc iodide, an ionic compound, and the heat of the reaction is great enough that excess iodine forms a purple vapor.

The combination reaction of a metal and oxygen produces an *ionic compound,* a metal oxide. Like any ionic compound, the metal oxide must be electrically neutral. Because you can predict a reasonable positive charge for the metal ion by knowing its position in the periodic table and by using the guidelines in Section 3.4, you can determine the formula of the metal oxide. For example, when aluminum (Al, Group 3A), which forms $3+$ ions, reacts with O_2, which forms the O^{2-} ion, the product must be aluminum oxide Al_2O_3 ($2\ Al^{3+}$ and $3\ O^{2-}$ ions), a compound also known as *alumina,* or *corundum.*

$$4\ Al(s) + 3\ O_2(g) \longrightarrow 2\ Al_2O_3(s)$$
$$\text{aluminum oxide}$$

The halogens also combine with metals to form ionic compounds with formulas that are predictable based on the charges of the ions formed. The halogens all form $1-$ ions in ionic compounds. For example, sodium combines with chlorine, and zinc combines with iodine, to form sodium chloride and zinc iodide, respectively (Figure 4.4).

$$2\ Na(s) + Cl_2(g) \longrightarrow 2\ NaCl(s)$$

$$Zn(s) + I_2(s) \longrightarrow ZnI_2(s)$$

When nonmetals combine with oxygen or chlorine, the compounds formed are not ionic but are composed of *molecules.* For example, sulfur, the Group 6A neighbor of oxygen, combines with oxygen to form two oxides, SO_2 and SO_3, in reactions of great environmental and industrial significance.

$$S_8(s) + 8\ O_2(g) \longrightarrow 8\ SO_2(g)$$
$$\text{sulfur dioxide}$$
$$\text{(colorless; choking odor)}$$

$$2\ SO_2(g) + O_2(g) \longrightarrow 2\ SO_3(g)$$
$$\text{sulfur trioxide}$$
$$\text{(colorless; even more choking odor)}$$

Sulfur dioxide enters the atmosphere from natural sources and from human activities. The eruption of Mount St. Helens in May 1980 injected millions of tons of SO_2 into the atmosphere, for example. But about 75% of the sulfur oxides in the atmosphere come from human activities, such as burning coal. All coal contains sulfur, usually from 1% to 4% by weight.

Mount St. Helens erupting.

Exercise 4.3 Combination Reactions

Indicate whether each equation for a combination reaction is balanced, and if it is not, why not.

(a) $Cu + O_2 \longrightarrow CuO$

(b) $Cr + Br_2 \longrightarrow CrBr_3$

(c) $S_8 + 3 F_2 \longrightarrow SF_6$

Exercise 4.4 Combination Reactions

(a) What information is needed to predict the product of a combination reaction between two elements? (b) Between calcium and fluorine? (c) What is the product formed by this reaction?

Decomposition Reactions

Decomposition reactions might be considered the opposite of combination reactions. In a **decomposition reaction,** one substance decomposes to form two or more products. The general reaction is

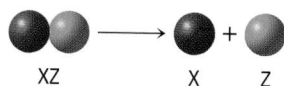

Many compounds that we would describe as "stable" because they exist without change under normal conditions of temperature and pressure undergo decomposition when the temperature is raised, a process known as *thermal decomposition.* For example, a few metal oxides decompose upon heating to give the metal and oxygen gas, the reverse of combination reactions. One of the best known metal oxide decomposition reactions is the reaction by which Joseph Priestley discovered oxygen in 1774.

$$2 \, HgO(s) \xrightarrow{\text{heat}} 2 \, Hg(\ell) + O_2(g)$$

A very common, and important, type of decomposition reaction is illustrated by the chemistry of *metal carbonates,* and calcium carbonate in particular. Carbonates decompose to give oxides plus carbon dioxide:

$$\underset{\substack{\text{calcium} \\ \text{carbonate}}}{CaCO_3(s)} \xrightarrow{\text{800–1000 °C}} \underset{\substack{\text{calcium} \\ \text{oxide}}}{CaO(s)} + CO_2(g)$$

Calcium is the fifth most abundant element in the earth's crust and is the third most abundant metal (after Al and Fe). Most naturally occurring calcium is in the form of calcium carbonate from the fossilized remains of early marine life. Limestone, a form of calcium carbonate, is one of the basic raw materials of industry. Lime (calcium oxide), made by the decomposition reaction just discussed, is a raw material needed for the manufacture of chemicals, in water treatment, and in the paper industry.

Some compounds are sufficiently unstable that their decomposition reactions are explosive. Nitroglycerine is so sensitive that it decomposes violently at the slightest shock.

$$4 \, C_3H_5(NO_3)_3(\ell) \longrightarrow 12 \, CO_2(g) + 10 \, H_2O(\ell) + 6 \, N_2(g) + O_2(g)$$

The formula for nitroglycerine contains parentheses around the NO_3 groups; they must be accounted for when balancing chemical equations. Each molecule of ni-

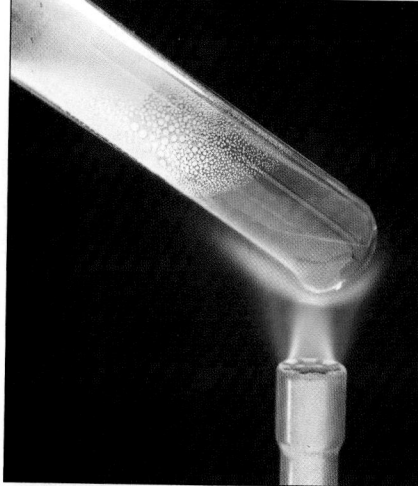

Decomposition of HgO. When heated, red mercury(II) oxide decomposes into liquid mercury and oxygen gas.

Sea shells are composed largely of calcium carbonate.

Dynamite contains nitroglycerine.

ALFRED NOBEL
(1833–1896)

Alfred Nobel was a Swedish chemist and engineer who discovered how to mix nitroglycerine (a liquid explosive that is extremely sensitive to light and heat) with diatomaceous earth to make dynamite, which could be handled and shipped safely. Nobel's talent as an entrepreneur combined with his many inventions (he held 355 patents) made him a very rich man. He never married and left his fortune to establish the Nobel Prizes, awarded annually to individuals who "have conferred the greatest benefits on mankind in the fields of physics, chemistry, physiology or medicine, literature and peace."

troglycerin contains 9 O atoms — 3 each for the 3 NO_3 groups. The NO_3 groups are not nitrate ions, NO_3^-. Nitroglycerine is a molecular organic compound with the structure shown below.

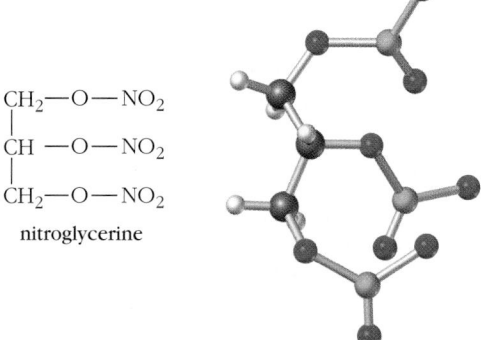

$$CH_2\!-\!O\!-\!NO_2$$
$$CH\ \,-\!O\!-\!NO_2$$
$$CH_2\!-\!O\!-\!NO_2$$

nitroglycerine

Water, by contrast, is such a stable compound that it can be decomposed to hydrogen and oxygen only at a very high temperature or by using an electric current, a process called electrolysis (Figure 4.5).

$$2\,H_2O(\ell) \xrightarrow{\text{direct current}} 2\,H_2(g) + O_2(g)$$

Problem-Solving Example 4.1   Combination and Decomposition Reactions

Predict the reaction type and the missing compound for each of the following reactions:

(a) $BaCO_3(s) \rightarrow BaO(s) +$ _____ (g)
(b) $2\,Mg(s) + O_2(g) \rightarrow 2$ _____ (s)
(c) $P_4(s) + 6$ _____ (g) $\rightarrow 4\,PCl_3(\ell)$
(d) $2\,KClO_3(s) \rightarrow 2$ _____ (s) $+ 3\,O_2(g)$

Answer   (a) CO_2 (b) MgO   (c) Cl_2 (d) KCl

Explanation
(a) Decomposition of $BaCO_3(s)$ will produce $CO_2(g)$ as well as BaO(s). (b) Combination of Mg(s) with $O_2(g)$ will produce the binary metal oxide MgO(s). (c) Combination of $P_4(s)$ with $Cl_2(g)$ will produce $PCl_3(\ell)$. (d) Decomposition of $KClO_3(s)$ will generate KCl(s) and $O_2(g)$.

Problem-Solving Practice 4.1

Predict the reaction type and the missing compound for each of the following reactions:

(a) _____ (g) $+ 2\,O_2(g) \rightarrow 2\,NO_2(g)$
(b) $4\,Fe(s) + 3$ _____ (g) $\rightarrow 2\,Fe_2O_3(s)$
(c) $2\,NaN_3(s) \rightarrow 2\,Na(s) + 3$ _____ (g)

Exercise 4.5   Combination and Decomposition Reactions

Predict the products formed by the following reactions:
(a) Magnesium with chlorine
(b) The thermal decomposition of magnesium carbonate

Displacement Reactions

Displacement reactions are those in which one element reacts with a compound to form a new compound and release a different element. The element released is said to have been displaced. The general equation for a **displacement reaction** is

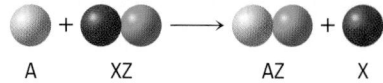

A   XZ   AZ   X

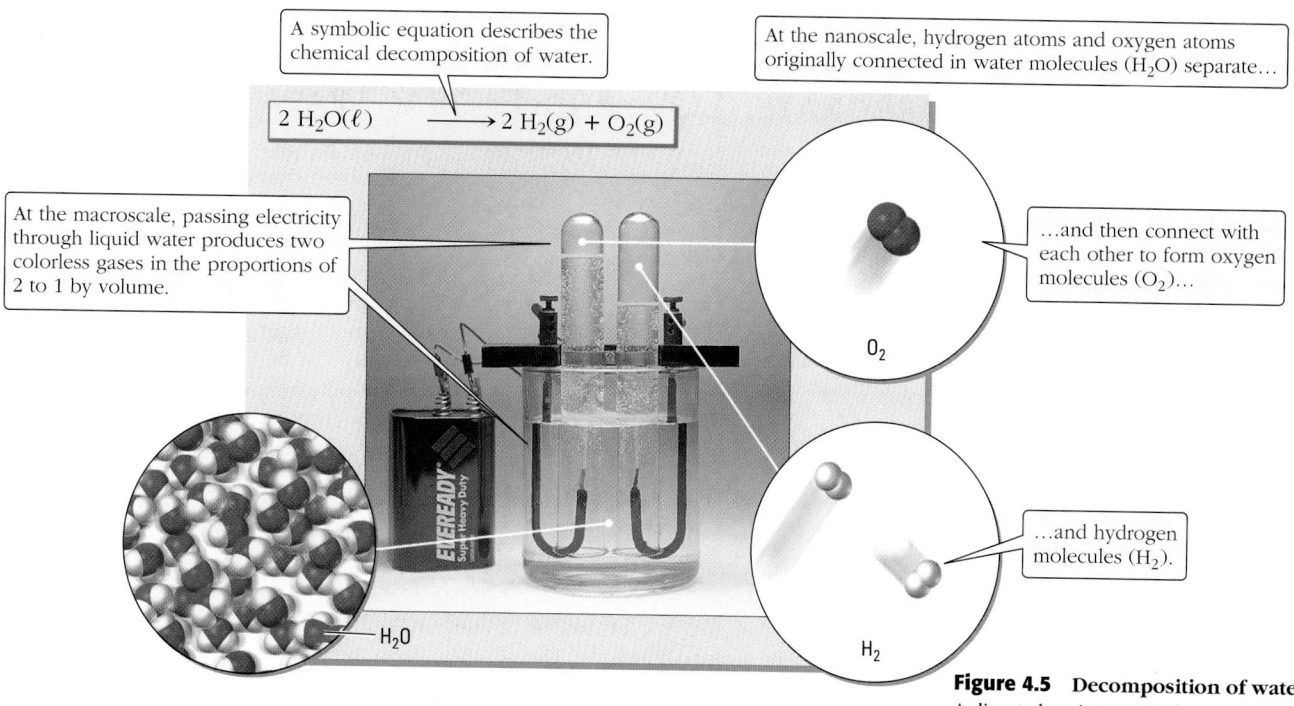

A symbolic equation describes the chemical decomposition of water.

$$2 H_2O(\ell) \longrightarrow 2 H_2(g) + O_2(g)$$

At the nanoscale, hydrogen atoms and oxygen atoms originally connected in water molecules (H_2O) separate...

At the macroscale, passing electricity through liquid water produces two colorless gases in the proportions of 2 to 1 by volume.

...and then connect with each other to form oxygen molecules (O_2)...

...and hydrogen molecules (H_2).

O_2

H_2

H_2O

Figure 4.5 **Decomposition of water.** A direct electric current decomposes water into gaseous hydrogen (H_2) and oxygen (O_2).

The reaction of metallic sodium with water is such a reaction.

$$2 Na(s) + 2 H_2O(\ell) \longrightarrow 2 NaOH(aq) + H_2(g)$$

Here sodium displaces hydrogen from water (Figure 4.6). All the alkali metals, which are very reactive elements, react in this way when exposed to water.

Another example is the reaction that occurs between metallic copper and a solution of silver nitrate.

$$Cu(s) + 2 AgNO_3(aq) \longrightarrow Cu(NO_3)_2(aq) + 2 Ag(s)$$

Here, one metal displaces another. As you will see in Chapter 5, the metals can be

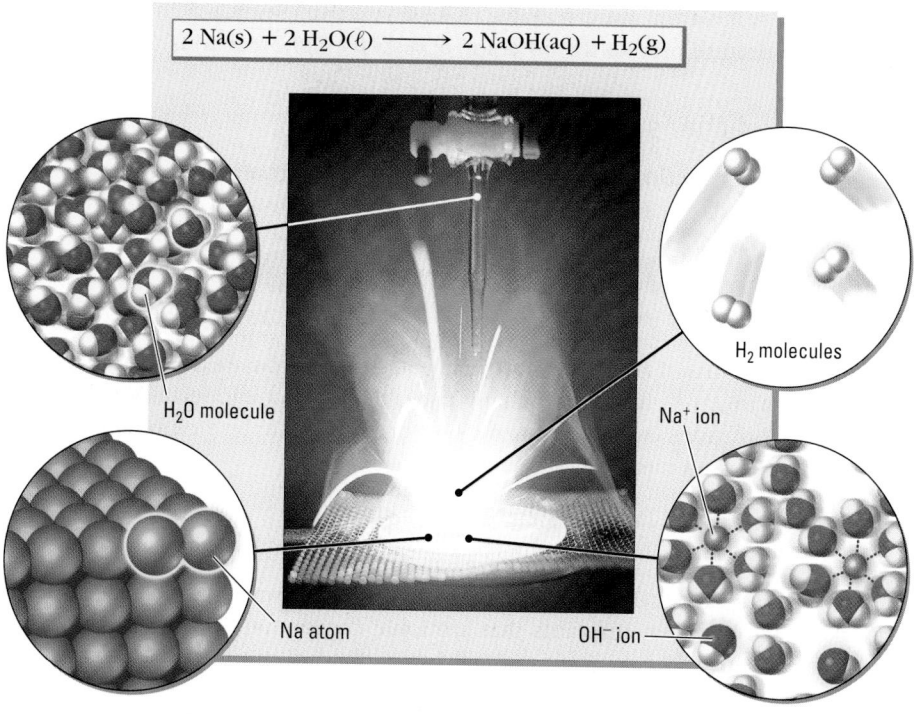

$$2 Na(s) + 2 H_2O(\ell) \longrightarrow 2 NaOH(aq) + H_2(g)$$

H_2 molecules

H_2O molecule

Na atom

Na^+ ion

OH^- ion

Figure 4.6 **A displacement reaction.** When liquid water drips from a buret onto a sample of solid sodium metal, the sodium displaces hydrogen gas from the water, and an aqueous solution of sodium hydroxide is formed. The hydrogen gas burns, producing the flame shown in the photograph. In the nanoscale pictures, the numbers of atoms, molecules, and ions that appear in the balanced equation are shown with yellow highlights.

THE CARBON CYCLE:
CALCIUM CARBONATE, CO_2, AND GLOBAL WARMING

The element carbon stands at the center of chemistry related to life and the ecosystem of earth that supports life. Carbon is found in nature in CO_2 gas in the atmosphere or in CO_2 dissolved in water, in HCO_3^- ions in water, in $CaCO_3$ and $MgCO_3$ in minerals, in many comp,ex forms in fossil fuels, and incorporated into all living things. The many chemical reactions that interrelate these forms of carbon illustrate important aspects of chemistry. These forms of carbon and their interrelationships also appear in the news regularly because of their extreme importance to the environment and the condition of the Earth's ecosystem. The most visible aspect of these questions in the media is the relationship between CO_2 concentration in the atmosphere and global warming.

Minerals, such as limestone, which is composed largely of calcium carbonate ($CaCO_3$), and dolomite, which is composed of $CaCO_3 \cdot MgCO_3$, form the single largest carbon reservoir on earth. Most limestone was formed by living organisms in the sea that generated their calcium carbonate skeletons using dissolved CO_2. When the organisms died,

Liquid carbon dioxide spills from a 4-L beaker on the ocean floor, 3650 m under water.

the skeletons fell to the sea floor and eventually formed minerals. The resulting limestone reservoir is connected to other carbon reservoirs on earth by the chemical reactions of the carbon cycle.

A small but very significant fraction of the carbon on earth is present as gaseous CO_2. Plants take in inorganic CO_2 from the atmosphere and through photosynthesis fix the carbon as biological carbon. Yet more carbon is incorporated into petroleum and natural gas and

other hydrocarbon materials such as coal.

As we enter the 21st century, earth faces a massive challenge in the form of global warming, which is related to the buildup of CO_2 in the atmosphere. The concentration of CO_2 has been increasing since the beginning of the industrial revolution and is linked to human activity, especially the burning of fossil fuels that necessarily releases CO_2.

$$\text{Fossil Fuel} + O_2 \longrightarrow CO_2 + H_2O$$

Many schemes have been advanced to slow the CO_2 buildup. They all revolve around the complex chemistry of CO_2, HCO_3^-, and $CaCO_3$. The control of CO_2 generation by decreasing our reliance on gasoline, coal, and other fossil fuels is one alternative. If we burn less fuel, we will generate less carbon dioxide.

Capture of CO_2 after generation is also feasible, a process known as sequestration of CO_2. The CO_2 would be stored in salt domes, petroleum fields, and the deep ocean—all natural reservoirs—by a pumping process that is the reverse of pumping oil or gas out of the ground. The oceans now hold about 50 times as

arranged in a series from most reactive to least reactive (Section 5.5). This activity series can be used to predict the outcome of displacement reactions.

Exchange Reactions

In an **exchange reaction**, there is an interchange of partners between two compounds. In general:

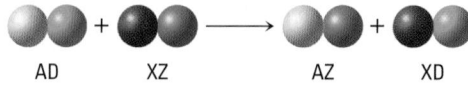

$$AD \qquad XZ \qquad AZ \qquad XD$$

Mixing solutions of lead(II) nitrate and potassium chromate, for example, illustrates an exchange reaction in which an insoluble product can form. The aqueous Pb^{2+} ions and K^+ ions exchange partners to form insoluble lead chromate and water-soluble potassium nitrate:

$$Pb(NO_3)_2(aq) + K_2CrO_4(aq) \longrightarrow PbCrO_4(s) + 2\ KNO_3(aq)$$

lead nitrate   potassium chromate   lead chromate   potassium nitrate

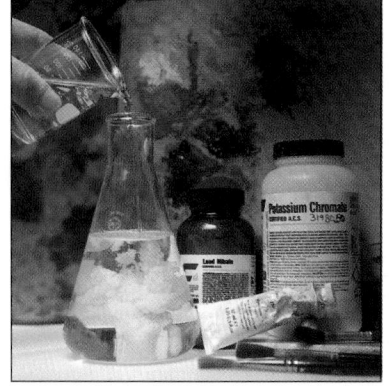

$$Pb(NO_3)_2 + K_2CrO_4 \longrightarrow$$
$$PbCrO_4 + 2\ KNO_3$$

Such reactions (further discussed in Chapter 5) include several kinds of reactions that take place between reactants that are ionic compounds and are dissolved in water. They occur when reactant ions are removed from solution by the formation

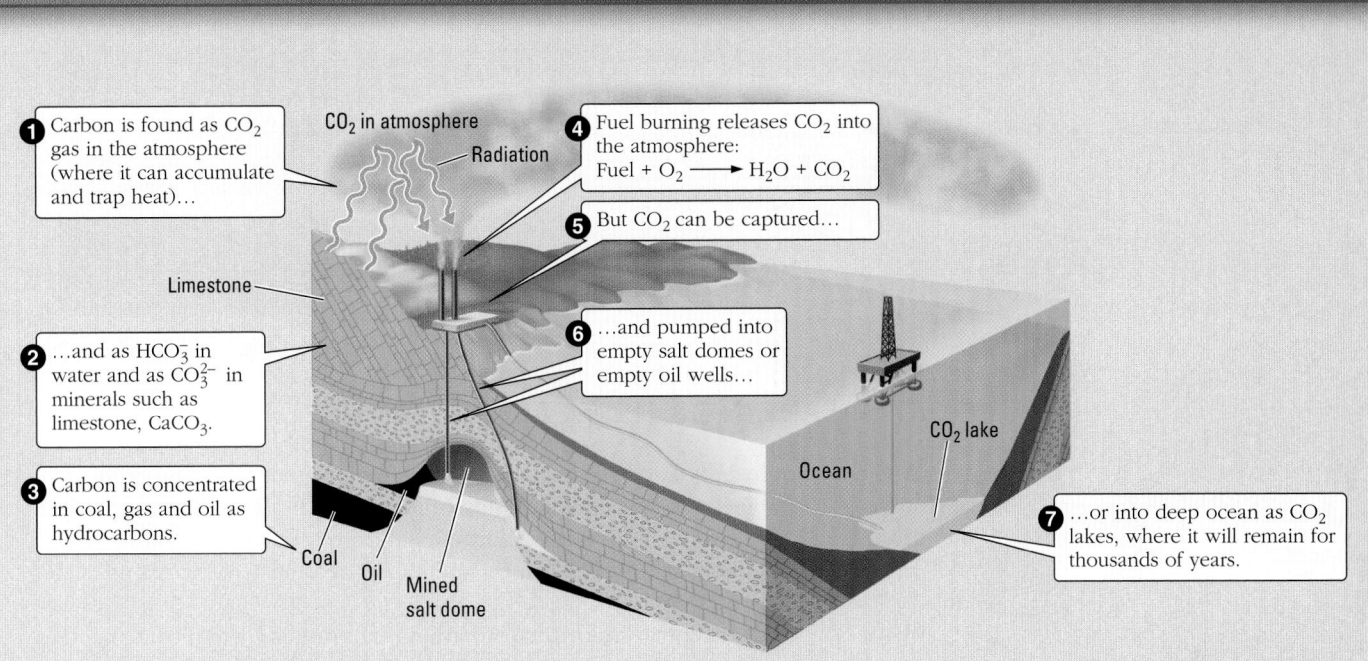

1 Carbon is found as CO_2 gas in the atmosphere (where it can accumulate and trap heat)...

CO_2 in atmosphere

Radiation

Limestone

2 ...and as HCO_3^- in water and as CO_3^{2-} in minerals such as limestone, $CaCO_3$.

3 Carbon is concentrated in coal, gas and oil as hydrocarbons.

Coal

Oil

Mined salt dome

4 Fuel burning releases CO_2 into the atmosphere:
Fuel + O_2 ⟶ H_2O + CO_2

5 But CO_2 can be captured...

6 ...and pumped into empty salt domes or empty oil wells...

CO_2 lake

Ocean

7 ...or into deep ocean as CO_2 lakes, where it will remain for thousands of years.

Carbon dioxide can be stored in the earth and the deep sea to keep it out of the atmosphere.

much CO_2 as the atmosphere, but their capacity is much larger, and CO_2 could be stored there without environmental damage. In general, storage of CO_2 seems to be one promising strategy for dealing with increasing CO_2 levels in the atmosphere.

Sources

Herzog, H., Eliasson, B., Kaarstad, O. Capturing Greenhouse Gases. *Scientific American,* February 2000, pp 72 – 79.

CO_2 Chemistry Deep in the Ocean. *Chemical & Engineering News,* May 10, 1999, p 6.

Energy Department Supports CO_2 Sequestration Research. *Chemical & Engineering News,* May 3, 1999, p 37.

Brewer, P.G., Friederich, G., Peltzer, E.T., Orr, F.M., Jr. Direct Experiments on the Ocean Disposal of Fossil Fuel CO_2. *Science,* Vol. 284, May 7, 1999, pp 943 – 945.

of one of three types of product: (1) a solid (Section 5.1), (2) a molecular compound (Section 5.2), or (3) a gas (Section 5.2).

Exchange reactions are also called metathesis or double-displacement reactions.

Problem-Solving Example 4.2 Classifying Reactions by Type

Classify each reaction as one of the four general types discussed in this section.

(a) $Pt(s) + 2 F_2(g \rightarrow PtF_4(\ell)$
(b) $3 Fe(s) + 4 H_2O(g) \rightarrow Fe_3O_4(s) + 4 H_2(g)$
(c) $2 H_3BO_3(s) \rightarrow B_2O_3(s) + 3 H_2O(\ell)$
(d) $BaCl_2(aq) + Na_2SO_4(aq) \rightarrow BaSO_4(s) + 2 KCl(aq)$

Answer (a) Combination (b) displacement (c) decomposition (d) exchange

Explanation

(a) With two reactants and a single product, this must be a combination reaction.
(b) The general equation for a displacement reaction, $A + XZ \rightarrow AZ + X$, matches what occurs in the given reaction: Iron (A) displaces hydrogen (X) from water (XZ) to form Fe_3O_4 (AZ) plus H_2 (X).
(c) In this reaction, a single substance, H_3BO_3 (boric acid), decomposes to form two products, B_2O_3 and H_2O.
(d) The reactants interchange partners in this exchange reaction. Applying the general equation for an exchange reaction ($AD + XZ \rightarrow AZ + XD$) to this case, we find $A = Ba^{2+}$, $D = Cl^-$, $X = Na^+$, and $Z = SO_4^{2-}$.

Problem-Solving Practice 4.2

Classify each of the following reactions as one of the four general reaction types described in Section 4.2.

(a) $2\,Al(OH)_3(s) \longrightarrow Al_2O_3(s) + 3\,H_2O(g)$
(b) $Na_2O(s) + H_2O(\ell) \longrightarrow 2\,NaOH(aq)$
(c) $S_8(s) + 24\,F_2(g) \longrightarrow 8\,SF_6(g)$
(d) $3\,NaOH(aq) + H_3PO_4(aq) \longrightarrow Na_3PO_4)_4(aq) + 3\,H_2O(\ell)$
(e) $3\,C(s) + Fe_2O_3(s) \longrightarrow 3\,CO(g) + 2\,Fe(\ell)$

4.3 BALANCING CHEMICAL EQUATIONS

 CD-Rom Screen 4.4: Balancing Chemical Equations

To balance a chemical equation means using coefficients so that the same number of atoms of each element appears on each side of the equation. We will begin with one of the general classes of reactions, the combination of reactants to produce a single product, to illustrate how to balance chemical equations by a largely trial-and-error process.

We will balance the equation for the formation of ammonia from nitrogen and hydrogen. Millions of tons of ammonia (NH_3) are manufactured worldwide annually by this reaction, using nitrogen extracted from air and hydrogen obtained from natural gas.

Step 1: ***Write an unbalanced equation containing the correct formulas of the reactants and products.***

$$\text{(unbalanced equation)} \qquad N_2 + H_2 \longrightarrow NH_3$$

Clearly, both nitrogen and hydrogen are unbalanced. There are two nitrogen atoms on the left and only one on the right, and two hydrogen atoms on the left and three on the right.

Step 2: ***Balance atoms of one of the elements.*** We start by using a coefficient of 2 on the right to balance the nitrogen atoms: $2\,NH_3$ indicates two ammonia molecules, each containing a nitrogen atom and three hydrogen atoms. On the right we now have two nitrogen atoms and six hydrogen atoms.

Balancing an equation involves changing the coefficients, but the subscripts in the formulas *cannot* be changed.

$$\text{(unbalanced equation)} \qquad N_2 + H_2 \longrightarrow 2\,NH_3$$

Step 3: ***Balance atoms of the remaining elements.*** To balance the six hydrogen atoms on the right, we use a coefficient of 3 for the H_2 on the left to furnish six hydrogen atoms.

$$\text{(balanced equation)} \qquad N_2 + 3\,H_2 \longrightarrow 2\,NH_3$$

Step 4: ***Verify that the number of atoms of each element is balanced.*** We do an atom count to check that the numbers of nitrogen and hydrogen atoms are the same on each side of the equation.

$$\text{(balanced equation)} \qquad N_2 + 3\,H_2 \longrightarrow 2\,NH_3$$

atom count:
$$2\,N + (3 \times 2)\,H = 2\,N + (2 \times 3)\,H$$
$$2\,N + 6\,H = 2\,N + 6\,H$$

The physical states of the reactants and products are usually also included in the balanced equation. Thus, the final equation for ammonia formation is

$$N_2(g) + 3\,H_2(g) \longrightarrow 2\,NH_3(g)$$

Problem-Solving Example 4.3 Balancing a Chemical Equation

Chromium metal combines with gaseous chlorine to form solid chromium(III) chloride. Write the balanced equation for the formation of $CrCl_3$.

Answer $2 Cr(s) + 3 Cl_2(g) \rightarrow 2 CrCl_3(s)$

Explanation

Step 1: *Write an initial unbalanced equation containing the correct formulas of all reactants and products.* The formula of chlorine is Cl_2 and the formula of chromium(III) chloride is $CrCl_3$. The unbalanced equation is

$$\text{(unbalanced equation)} \quad Cr + Cl_2 \longrightarrow CrCl_3$$

Step 2: *Start by balancing atoms of one of the elements.* Chlorine is unbalanced; there are two chlorine atoms on the left and three on the right. Whenever three and two atoms must be balanced, add coefficients to give six atoms in both places. Here, we use the coefficients of 3 on the left and 2 on the right to have six chlorine atoms on each side.

$$\text{(unbalanced equation)} \quad Cr + 3 Cl_2 \longrightarrow 2 CrCl_3$$

Note that the chromium atoms are now unbalanced.

Step 3: *Balance atoms of the remaining elements.* Chromium atoms are balanced by using a coefficient of 2 on the left, which balances the equation as well.

$$\text{(balanced equation)} \quad 2 Cr + 3 Cl_2 \longrightarrow 2 CrCl_3$$

Step 4: *Verify that the number of atoms of each element is balanced.*

$$2 Cr(s) + 3 Cl_2(g) \longrightarrow 2 CrCl_3(s)$$

$$
\begin{array}{ccc}
2 \, Cr & = & 2 \, Cr \\
6 \, Cl & = & 6 \, Cl
\end{array}
$$

Problem-Solving Practice 4.3

Balance these equations.

(a) $Xe(g) + F_2(g) \rightarrow XeF_4(g)$
(b) $As_2O_3(s) + H_2(g) \rightarrow As(s) + H_2O(\ell)$

We now turn to the combustion of octane (C_8H_{18}) to illustrate balancing a somewhat more complex chemical equation. We will assume that complete combustion occurs, meaning that the only products are carbon dioxide and water.

Problem-Solving Example 4.4 Balancing a Combustion Reaction Equation

Write a balanced equation for the complete combustion of C_8H_{18}, a gasoline component.

Answer $2 C_8H_{18}(\ell) + 25 O_2(g) \rightarrow 16 CO_2(g) + 18 H_2O(\ell)$

Explanation

Step 1: *Write an initial unbalanced equation.* The initial equation is

$$\text{(unbalanced equation)} \quad C_8H_{18} + O_2 \longrightarrow CO_2 + H_2O$$

Step 2: *Start by balancing the atoms of one of the elements.* None of the elements are balanced, so we could start with C, H, or O. We start with carbon because it appears in only one formula on each side of the equation. The eight C atoms in C_8H_{18} will produce eight CO_2 molecules.

$$\text{(unbalanced equation)} \quad C_8H_{18} + O_2 \longrightarrow 8 CO_2 + H_2O$$

Step 3: *Balance atoms of the remaining elements.* We next balance the hydrogen atoms. The 18 H atoms in the reactants will combine with oxygen to form 9 water molecules, each containing 2 H atoms.

$$\text{(unbalanced equation)} \quad C_8H_{18} + O_2 \longrightarrow 8 CO_2 + 9 H_2O$$

It usually works best to first balance the element that appears in the fewest formulas; balance the element that appears in the most formulas last.

Oxygen is the remaining element to balance. At this point, there are 25 oxygen atoms in the products (8×2 in 8 CO_2, plus 9×1 in 9 H_2O), but only 2 in the reactants. In this case, there is an even number of O atoms in the reactants and an odd number of them in the products. A stoichiometric coefficient of 25/2 for O_2 on the left can give us 25 oxygen atoms because 25/2 oxygen molecules each containing 2 oxygen atoms makes 25 oxygen atoms. Therefore the equation could be balanced as follows:

(balanced equation) $C_8H_{18} + 25/2\,O_2 \longrightarrow 8\,CO_2 + 9\,H_2O$

Except in special circumstances, however, it is customary to use only whole-number coefficients. Thus, multiplying every coefficient in the equation by 2 gives the balanced equation with whole-number coefficients.

(balanced equation)

$$2\,C_8H_{18} \;+\; 25\,O_2 \longrightarrow 16\,CO_2 \;+\; 18\,H_2O$$

Step 4: *Verify that the number of atoms of each element is balanced.*

$$2\,C_8H_{18}(\ell) + 25\,O_2(g) \longrightarrow 16\,CO_2(g) + 18\,H_2O(\ell)$$

16 C	=	16 C
36 H	=	36 H
50 O	=	32 O + 18 O

Problem-Solving Practice 4.4

Ethyl alcohol, C_2H_5OH, can be added to gasoline to create a cleaner burning fuel. Write the balanced equation for the combustion of ethyl alcohol.

(a) Complete combustion to produce carbon dioxide and water
(b) Incomplete combustion to produce carbon monoxide and water

Fractional coefficients are sometimes needed for one or more reactants or products when it is necessary to write an equation for a specific whole number of moles of another reactant or product.

When one or more polyatomic ions appears on both sides of a chemical equation, any such ion is treated as a whole during the balancing steps. Parentheses surround such an ion when it must have a subscript in the molecular formula. For example, in the equation for the reaction between sodium phosphate and barium nitrate to produce barium phosphate and sodium nitrate

$$2\,Na_3PO_4(aq) + 3\,Ba(NO_3)_2(aq) \longrightarrow Ba_3(PO_4)_2(s) + 6\,NaNO_3(aq)$$

nitrate ions and phosphate ions are kept together as units and are surrounded with parentheses when the polyatomic ion occurs more than once in the formula.

4.4 The Mole and Chemical Reactions: The Macro-Nano Connection

CD-Rom Screen 5.2: Weight Relations in Chemical Reactions

When the concept of the mole, which links numbers of atoms, molecules, or formula units with the mass of the atoms, molecules, or ionic compounds, is combined with a balanced chemical equation, the masses of the reactants and products can be calculated. In this way the nanoscale of chemical reactions is linked with the macroscale, at which we can measure masses of reactants and products by weighing.

We will explore these relationships using the combustion reaction between methane, CH_4, and oxygen, O_2, as an example (Figure 4.7). The balanced equation shows the number of molecules of the reactants, which are methane and oxygen, and of the products, which are carbon dioxide and water. The coefficients on each species can also be interpreted as the numbers of moles of each compound, and we can use the molar mass of each compound to calculate the mass of each reactant and product represented in the balanced equation.

$$CH_4 \quad + \quad 2\ O_2 \longrightarrow CO_2 \quad + \quad 2\ H_2O$$

1 CH$_4$ molecule	2 O$_2$ molecules	1 CO$_2$ molecule	2 H$_2$O molecules
1 mol CH$_4$	2 mol O$_2$	1 mol CO$_2$	2 mol H$_2$O
16.0 g CH$_4$	64.0 g O$_2$	44.0 g CO$_2$	36.0 g H$_2$O

80.0 g total   80.0 g total

Notice that the total mass of reactants (16.0 g CH$_4$ + 64.0 g O$_2$ = 80.0 g reactants) equals the total mass of products (44.0 g CO$_2$ + 36.0 g H$_2$O = 80.0 g products), as must always be the case for a balanced equation.

The stoichiometric coefficients in a balanced chemical equation provide the **mole ratios** that relate the numbers of moles of reactants and products to each other. These mole ratios are used in all quantitative calculations involving chemical reactions. Several of the mole ratios for the example equation are

$$\frac{2\ \text{mol O}_2}{1\ \text{mol CH}_4} \quad \text{or} \quad \frac{1\ \text{mol CO}_2}{1\ \text{mol CH}_4} \quad \text{or} \quad \frac{2\ \text{mol H}_2\text{O}}{2\ \text{mol O}_2}$$

These mole ratios express the following: 2 mol of O$_2$ reacts with every 1 mol of CH$_4$, or 1 mol of CO$_2$ is formed for each 1 mol of CH$_4$ that reacts, or 2 mol of H$_2$O is formed for each 2 mol of O$_2$ that react.

We can use the mole ratios to calculate moles of one reactant or product from moles of another reactant or product. For example, we can calculate the number of moles of H$_2$O produced when 0.40 mol of CH$_4$ is reacted fully with oxygen.

$$\boxed{\text{Mol of CH}_4} \longrightarrow \boxed{\text{Mol of H}_2\text{O}}$$

$$0.40\ \text{mol CH}_4 \times \frac{2\ \text{mol H}_2\text{O}}{1\ \text{mol CH}_4} = 0.80\ \text{mol H}_2\text{O}$$

Exercise 4.6   Mole Ratios

Write all the possible mole ratios that can be obtained from the balanced equation of the reaction between Al and Br$_2$ to form Al$_2$Br$_6$.

The molar mass and the mole ratio, as illustrated in Figure 4.8, provide the links between masses and moles of reactants and products.

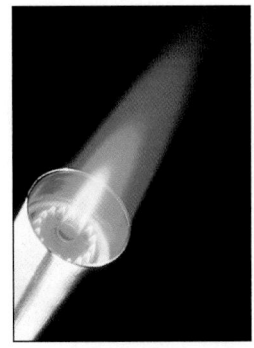

Figure 4.7 Combustion of methane with oxygen. Methane is the main component of natural gas, a primary fuel for industrial economies.

The mole ratio is also known as the stoichiometric factor.

CD-Rom Screen 5.3: Calculations in Stoichiometry

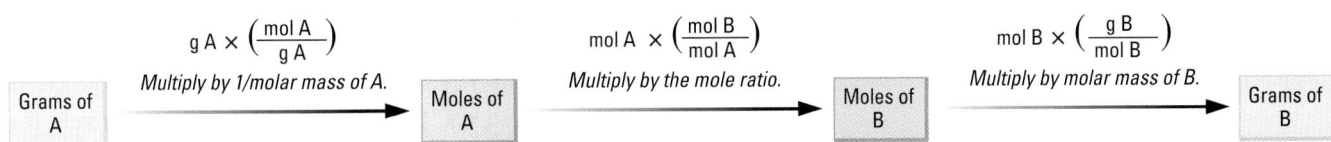

Figure 4.8 Stoichiometric relationships in a chemical reaction. The grams or moles of one reactant or product (A) are related to the grams or moles of another reactant or product (B) by the series of calculations shown.

Problem-Solving Example 4.5 Moles and Grams in Chemical Reactions

How many moles of water are formed from complete reaction of 4.32 g of methane with oxygen?

Answer 0.540 mol of H_2O

Explanation

We know the mass of methane, and we need to know the moles of methane in order to use the balanced equation linking methane with water. We convert mass of methane to moles of methane using the molar mass

$$\boxed{\text{Mass of CH}_4} \longrightarrow \boxed{\text{Mol of CH}_4}$$

Moles of CH_4 reacted: $4.32 \text{ g CH}_4 \times \dfrac{1 \text{ mol CH}_4}{16.0 \text{ g CH}_4} = 0.270 \text{ mol CH}_4$

We then use the mole ratio from the balanced equation

$$CH_4(g) + 2\,O_2(g) \longrightarrow CO_2(g) + 2\,H_2O(g)$$

to convert moles of methane reacted to moles of water produced.

$$\boxed{\text{Mol of CH}_4} \longrightarrow \boxed{\text{Mol of H}_2O}$$

Moles of H_2O formed: $0.270 \text{ mol CH}_4 \times \dfrac{2 \text{ mol H}_2O}{1 \text{ mol CH}_4} = 0.540 \text{ mol H}_2O$

✔ The balanced equation shows that for every 1 mol of methane that reacts, 2 mol of water is formed. Therefore, the reaction of approximately 0.25 mol of methane should produce twice that amount, approximately 0.5 mol, of water. The answer is reasonable.

Problem-Solving Practice 4.5

How many grams of oxygen are required to react completely with 0.55 mol of methane?

Solving stoichiometry problems relies on the relationships illustrated in Figure 4.8 to connect masses and molar amounts of reactants or products. We will illustrate using these relationships and a stepwise problem-solving method to answer the following question: How many grams of O_2 are needed to react completely with 5.0 g of CH_4?

Step 1: *Write the correct formulas for reactants and products and balance the chemical equation.* The balanced equation has been given.

$$CH_4(g) + 2\,O_2(g) \longrightarrow CO_2(g) + 2\,H_2O(g)$$

Step 2: *Decide what information about the problem is known and what is unknown. Map out a strategy for answering the question.* In this example, you know the mass of CH_4 and you want to calculate the mass of O_2. You also know that you can use molar mass to convert grams of CH_4 to moles of CH_4. Then you can use the mole ratio from the balanced equation (2 mol O_2/1 mol CH_4) to calculate the moles of O_2 needed. Finally, you can use the molar mass of O_2 to convert the moles of O_2 to grams of O_2.

Step 3: *Calculate moles from grams (if necessary).* The known mass of CH_4 must be converted to moles because the balanced equation expresses mole relationships.

$$5.0 \text{ g } CH_4 \times \frac{1 \text{ mol } CH_4}{16.0 \text{ g } CH_4} = 0.313 \text{ mol } CH_4$$

Step 4: *Use the mole ratio to calculate the unknown number of moles, and then convert the number of moles to number of grams (if necessary).* We calculate the number of moles and then grams of O_2.

$$0.313 \text{ mol } CH_4 \times \frac{2 \text{ mol } O_2}{1 \text{ mol } CH_4} = 0.626 \text{ mol } O_2$$

$$0.626 \text{ mol } O_2 \times \frac{32.0 \text{ g } O_2}{1 \text{ mol } O_2} = 20. \text{ g } O_2$$

In multistep calculations remember to carry one additional significant figure in intermediate steps before rounding to the final value.

Step 5: *Check the answer to see if it is reasonable.* The starting mass of CH_4 of 5.0 g is about one third of a mole of CH_4. One third of a mole of CH_4 should react with about two thirds of a mole of O_2 because the mole ratio is 1:2. The molar mass of O_2 is 32.0 g/mol, so two thirds of this is about 20. Therefore, the answer of 20. g of O_2 is reasonable.

Exercise 4.7   Moles and Grams in Chemical Reactions

Verify that 10.8 g of water is produced by the reaction of excess O_2 with 0.300 mol of CH_4.

Problem-Solving Examples 4.6 and 4.7 illustrate further the application of the steps for solving stoichiometry problems.

Problem-Solving Example 4.6   Moles and Grams in Chemical Reactions

Magnesium metal in fireworks reacts with oxygen in air to produce a brilliant white flash. The product of this combination reaction is magnesium oxide. How many grams of magnesium oxide will be formed by the complete reaction of 0.500 mol of magnesium with oxygen gas from the air?

Answer   20.2 g MgO

Explanation

Step 1: *Write the correct formulas for reactants and products and balance the chemical equation.* Magnesium metal and diatomic oxygen gas are the reactants; solid magnesium oxide is the product.

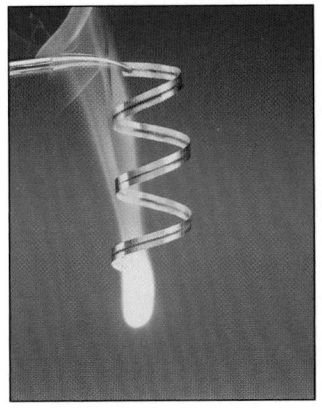

Magnesium oxide. MgO is produced when Mg metal burns in air — as a strip of metal and in fireworks.

$$2 \text{ Mg(s)} + \text{O}_2(\text{g}) \longrightarrow 2 \text{ MgO(s)}$$

Step 2: ***Decide what information about the problem is known and what is unknown. Map out a strategy for answering the question.*** We know how many moles of Mg are available. Oxygen from the air is unlimited. The moles of MgO and the equivalent mass of MgO must be calculated.

$$\boxed{\text{Moles of Mg}} \longrightarrow \boxed{\text{Moles of MgO}} \longrightarrow \boxed{\text{Grams of MgO}}$$

Step 3: ***Calculate moles from grams (if necessary).*** The moles of Mg are known.

Step 4: ***Use the mole ratio to calculate the unknown number of moles, and then convert the number of moles to number of grams (if necessary).*** Both steps are needed here to convert moles of Mg to grams of O_2. In a single setup the calculation is

$$0.500 \text{ mol Mg} \times \frac{2 \text{ mol MgO}}{2 \text{ mol Mg}} \times \frac{40.30 \text{ g MgO}}{1 \text{ mol MgO}} = 20.2 \text{ g MgO}$$

Step 5: ***Check the answer to see if it is reasonable.*** The answer of 20.2 g of MgO is reasonable because 1 mol of Mg would produce 1 mol of MgO (40.3 g); therefore, one-half mole of Mg will generate one-half mole of magnesium oxide (20.2 g).

Problem-Solving Practice 4.6

Tin is extracted from its ore cassiterite, SnO_2, by reaction with carbon from coal.

$$\text{SnO}_2(\text{s}) + 2 \text{ C(s)} \longrightarrow \text{Sn}(\ell) + 2 \text{ CO(g)}$$

(a) What mass of tin can be produced from 0.300 mol of cassiterite?
(b) How many grams of carbon are required to produce this much tin?

 Problem-Solving Example 4.7 Grams, Moles, and Grams

A popular candy bar contains 21.1 g of sucrose (cane sugar), $\text{C}_{12}\text{H}_{22}\text{O}_{11}$. When the candy bar is eaten, the sucrose is metabolized according to the overall equation

(unbalanced equation) $$\text{C}_{12}\text{H}_{22}\text{O}_{11}(\text{s}) + \text{O}_2(\text{g}) \longrightarrow \text{CO}_2(\text{g}) + \text{H}_2\text{O}(\ell)$$

Balance the chemical equation, and find the mass of O_2 consumed and the masses of CO_2 and H_2O produced by this reaction.

Answer $\text{C}_{12}\text{H}_{22}\text{O}_{11}(\text{s}) + 12 \text{ O}_2(\text{g}) \rightarrow 12 \text{ CO}_2(\text{g}) + 11 \text{ H}_2\text{O}(\ell)$
23.7 g of O_2 is consumed; 32.6 g of CO_2 and 12.2 g of H_2O are produced.

Explanation Coefficients of 12 and 11 for CO_2 and H_2O balance C and H, giving 35 oxygen atoms in the products. The oxygen atoms are balanced by 12 O_2 molecules plus the 11 oxygen atoms in sucrose.

$$\text{C}_{12}\text{H}_{22}\text{O}_{11}(\text{s}) + 12 \text{ O}_2(\text{g}) \longrightarrow 12 \text{ CO}_2(\text{g}) + 11 \text{ H}_2\text{O}(\ell)$$

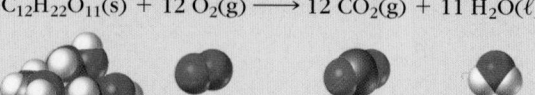

Use sucrose's molar mass (342.3 g/mol) to convert grams of sucrose to moles of sucrose.

$$21.1 \text{ g sucrose} \times \frac{1 \text{ mol sucrose}}{342.3 \text{ g sucrose}} = 0.06164 \text{ mol sucrose}$$

Next, use the mole ratios and molar masses to calculate the grams of O_2, CO_2, and H_2O.

$$0.06164 \text{ mol sucrose} \times \frac{12 \text{ mol } O_2}{1 \text{ mol sucrose}} \times \frac{31.99 \text{ g } O_2}{1 \text{ mol } O_2} = 23.7 \text{ g } O_2$$

$$0.06164 \text{ mol sucrose} \times \frac{12 \text{ mol } CO_2}{1 \text{ mol sucrose}} \times \frac{44.01 \text{ g } CO_2}{1 \text{ mol } CO_2} = 32.6 \text{ g } CO_2$$

$$0.06164 \text{ mol sucrose} \times \frac{11 \text{ mol } H_2O}{1 \text{ mol sucrose}} \times \frac{18.02 \text{ g } H_2O}{1 \text{ mol } H_2O} = 12.2 \text{ g } H_2O$$

The mass of water could have been found by using the conservation of mass.

Total mass of reactants = total mass of products

Total mass of reactants = 21.1 g sucrose + 23.7 g O_2 = 44.8 g

Total mass of products = g of CO_2 + ? g of H_2O = 44.8 g

Mass of H_2O = 44.8 g − 32.6 g = 12.2 g

✔ These are reasonable answers because approximately 0.05 mol of sucrose would require approximately 0.6 mol of O_2 and would produce approximately 0.6 mol of CO_2 and 0.5 mol of H_2O. Therefore, the calculated masses should be somewhat smaller than the molar masses, and they are.

Problem-Solving Practice 4.7

A lump of coke (carbon) weighs 57 g.

(a) What mass of oxygen is required to burn it to carbon monoxide?
(b) How many grams of CO are produced?

To this point we have used the methods of stoichiometry calculations to compute the amount of products given the amount of reactants. Now we turn to the reverse problem: Given the amount of products, what quantitative information can we deduce about the reactants? Questions such as these are often confronted by *analytical chemistry,* a field in which chemists creatively identify pure substances and measure the quantities of components of mixtures. Although analytical chemistry is now largely done by instrumental methods, classic chemical reactions and stoichiometry still play a central role. The analysis of mixtures is often challenging. It can take a great deal of imagination to figure out how to use chemistry to determine what, and how much, is there.

Problem-Solving Example 4.8   Evaluating an Ore

The mass percent of titanium dioxide (TiO_2) in an ore can be evaluated by carrying out the reaction of the ore with bromine trifluoride and measuring the mass of oxygen gas evolved.

$$3 \text{ TiO}_2(s) + 4 \text{ BrF}_3(\ell) \longrightarrow 3 \text{ TiF}_4(s) + 2 \text{ Br}_2(\ell) + 3 \text{ O}_2(g)$$

If 2.376 g of a TiO_2-containing ore generates 0.143 g of O_2, what is the mass percent of TiO_2 in the ore sample?

Answer   15.0% TiO_2

Explanation   The mass percent of TiO_2 is

$$\text{Mass percent TiO}_2 = \frac{\text{mass of TiO}_2}{\text{mass of ore sample}} \times 100\%$$

The mass of the sample is given, and the mass of TiO_2 can be determined from the known mass of oxygen and the balanced equation.

$$0.143 \text{ g } O_2 \times \frac{1 \text{ mol } O_2}{32.00 \text{ g } O_2} \times \frac{3 \text{ mol TiO}_2}{3 \text{ mol } O_2} \times \frac{79.88 \text{ g TiO}_2}{1 \text{ mol TiO}_2} = 0.3569 \text{ g TiO}_2$$

The mass percent of TiO_2 is

$$\frac{0.3569 \text{ g } TiO_2}{2.376 \text{ g sample}} \times 100\% = 15.0\%$$

Titanium dioxide is a valuable commercial product so widely used in paints and pigments that an ore, even with only 15% TiO_2, can be mined profitably.

✔ The molar mass of TiO_2 is about 80 g/mol and that of O_2 about 32 g/mol; the ratio is about 2.5 : 1. The balanced equation shows that equal numbers of moles of TiO_2 and O_2 are involved. So x g of TiO_2 should produce about $x/2.5$ g O_2; 2.4 g of TiO_2 should produce about 0.95 g of O_2. In fact about 0.14 g was produced, and this is about 15% of the 0.95 g, so the answer is reasonable.

Problem-Solving Practice 4.8

The purity of magnesium metal can be determined by reacting the metal with excess hydrochloric acid to form $MgCl_2$, evaporating the water from the resulting solution, and weighing the solid $MgCl_2$ formed.

$$Mg(s) + 2 \, HCl(aq) \longrightarrow MgCl_2(aq) + H_2(g)$$

Calculate the percentage of magnesium in a 1.72-g sample that produced 6.46 g of $MgCl_2$ when reacted with excess HCl.

4.5 REACTIONS WITH ONE REACTANT IN LIMITED SUPPLY

In the previous section, we assumed that exactly stoichiometric amounts of reactants were present; the reactants were entirely consumed when the reaction was over. However, this is rarely the case when chemists carry out an actual synthesis, whether for small quantities in a laboratory or on a large scale in an industrial process. Usually, one reactant is more expensive or less readily available than others. The cheaper or more available reactants are used in excess to ensure that the more expensive material is completely converted to product.

 CD-Rom Screen 5.4: Reactions Controlled by the Supply of One Reactant

The industrial production of methanol, CH_3OH, is such a case. Methanol, an important industrial product, is manufactured by the reaction of carbon monoxide and hydrogen.

ESTIMATION   How Much CO₂ Is Produced By Your Car?

Your car burns gasoline in a combustion reaction and produces water and carbon dioxide. CO_2 is involved in global warming because it is one of the major greenhouse gases, which we discuss in detail in Section 12.4. For each gallon of gasoline that you burn in your car, how much CO_2 is produced? How much CO_2 is produced by your car per year?

To proceed with the estimation, we need to write a balanced chemical equation with the stoichiometric relationship between the reactant, gasoline, and the product of interest, CO_2. To write the chemical equation we need to make an assumption about the composition of gasoline. We will assume that the gasoline is octane, C_8H_{18}, so the reaction of interest is

$$2 \, C_8H_{18} + 25 \, O_2 \longrightarrow 16 \, CO_2 + 18 \, H_2O$$

One gallon equals 4 quarts, which equals

$$4 \text{ quarts} \times \frac{1 \text{ L}}{1.057 \text{ quart}} = 3.78 \text{ L}.$$

Gasoline floats on water, so its density must be less than that of water. Assume it is 0.80 g/mL, so

$$3.78 \text{ L} \times 0.80 \text{ g/mL} \times 10^3 \text{ mL/L} = 3.02 \times 10^3 \text{ g}$$

We now convert the grams of octane to moles using the molar mass of octane.

$$3020 \text{ g octane} \times \frac{1 \text{ mol}}{114 \text{ g}} = 26.4 \text{ mol octane}$$

To convert to moles of CO_2 we use the balanced equation, which shows that for every mole of octane consumed, eight moles of CO_2 are produced, so we have $26.4 \times 8 = 211$ moles of CO_2. The molecular weight of CO_2 is 44 amu, so 211 mol $\times$ 44 g/mol = 9280 g of CO_2. Thus, for every gallon of gasoline burned, 9.3 kg (20.5 lb) of CO_2 is produced.

If you drive your car 10,000 miles per year and get an average of 25 miles per gallon, you use about 400 gallons of gasoline per year. Burning this quantity of gasoline produces 400×9.3 kg = 3720 kg of CO_2. That's 3720 kg $\times$ 2.2 lb/kg = 8200 lb, or more than 4 tons of CO_2.

$$CO(g) + 2 H_2(g) \longrightarrow CH_3OH(\ell)$$

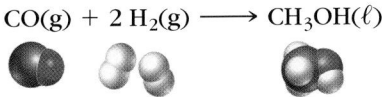

Carbon monoxide is manufactured cheaply by burning coke (which is mostly carbon) in a limited supply of air so that there is insufficient oxygen to form carbon dioxide. Hydrogen is more expensive to manufacture. Therefore, methanol synthesis uses an excess of carbon monoxide, and the amount of methanol produced is dictated by the amount of hydrogen available. Hydrogen acts as what is called a limiting reactant.

A **limiting reactant** is the reactant that is completely converted to products during a reaction. Once the limiting reactant has been used up, no more product can form. *The limiting reactant must be used as the basis for calculating the maximum possible amount of product(s)* because the limiting reactant limits the amount of product(s) that can be formed. ***The moles of product formed are always determined by the starting number of moles of the limiting reactant.***

We can make an analogy to a chemistry "limiting reactant" in the assembling of a pamphlet containing colored sheets of paper. Each pamphlet must have 5 yellow sheets, 3 blue sheets, and 2 pink sheets. Suppose we have in stock 400 yellow sheets, 300 blue sheets, and 200 pink sheets. How many complete pamphlets can be assembled?

Each pamphlet must have pages in the ratio of 5 yellow : 3 blue : 2 pink (analogous to coefficients in a balanced chemical equation). Using the stock on hand and the 5 : 3 : 2 requirement, only 80 complete pamphlets can be assembled.

Sheets in stock	Required for 1 pamphlet	Assembling		
Yellow: 400	5	400 yellow sheets $\times \dfrac{1 \text{ pamphlet}}{5 \text{ yellow sheets}}$	=	80 pamphlets
Blue: 300	3	400 blue sheets $\times \dfrac{1 \text{ pamphlet}}{3 \text{ blue sheets}}$	=	133 pamphlets
Pink: 400	2	400 pink sheets $\times \dfrac{1 \text{ pamphlet}}{2 \text{ pink sheets}}$	=	200 pamphlets

Yellow sheets are the "limiting reactant" because they yield the smallest number of pamphlets (Figure 4.9). Overall, the 80 pamphlets contain a total of 400 yellow, 240 blue, and 160 pink sheets; 60 blue sheets and 240 pink sheets are left over.

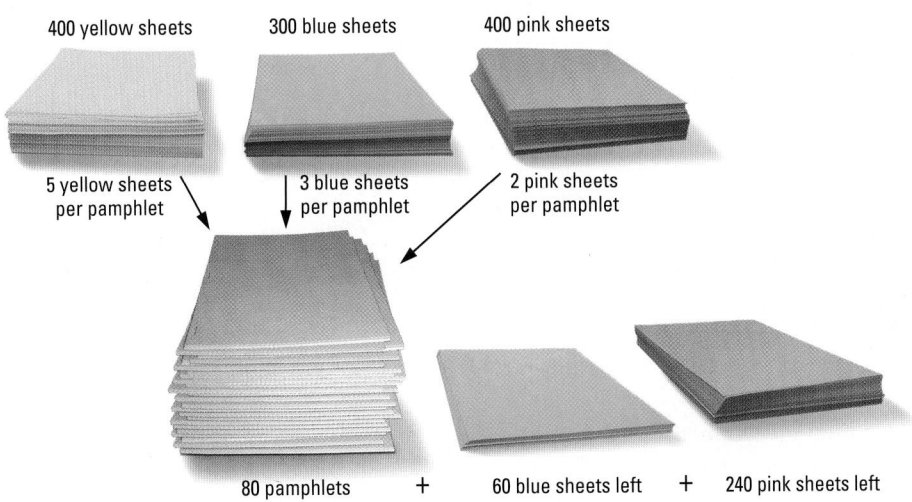

Figure 4.9 Limiting reactant in pamphlet assembly. Assembling a 10-page pamphlet requires 5 yellow sheets of paper, 3 blue sheets, and 2 pink sheets per pamphlet. The yellow sheets are the "limiting reactant" because all 400 yellow sheets are used when 80 pamphlets are produced, leaving 60 blue sheets and 240 pink sheets unused.

In determining how many pamphlets could be assembled, the "limiting reactant" was the number of yellow sheets. Similarly, the limiting reactant must be identified in a chemical reaction in order to determine how much product(s) will be produced if all the reactants are converted to the desired product(s).

If we know which one of a set of reactants is the limiting reactant, we can use that information to solve a quantitative problem directly, as illustrated in Problem-Solving Example 4.9.

Problem-Solving Example 4.9   Moles of Product from Limiting Reactant

Consider the following reaction between nitrogen and hydrogen to form ammonia.

$$N_2(g) + 3 H_2(g) \longrightarrow 2 NH_3(g)$$

If 0.60 mol of N_2 and 2.0 mol of H_2 are mixed, how many moles of NH_3 can be produced? N_2 is the limiting reactant.

Answer   1.2 mol of NH_3

Explanation   Start with the balanced equation and consider the stoichiometric coefficients. Concentrate on the N_2 since it is the limiting reactant. The coefficients show that for every 1 mol of N_2 reacted, 2 mol of NH_3 will be produced. Thus, we use this information to answer the question

$$0.60 \text{ mol } N_2 \times \frac{2 \text{ mol } NH_3}{1 \text{ mol } N_2} = 1.2 \text{ mol } NH_3$$

✔   The balanced equation shows that the number of moles of NH_3 produced must be twice the number of moles of N_2 that reacted, and it is.

Problem-Solving Practice 4.9

If we react 0.60 mol of N_2 with 1.7 mol of H_2 (and H_2 is now the limiting reactant), how many moles of NH_3 will be produced ?

CD-Rom Screen 5.5: Limiting Reactants: The Details

Next, we consider the case where the quantities of reactants are given, but the limiting reactant is not identified and must be determined. Problem-Solving Example 4.10 illustrates two methods for identifying a limiting reactant — the mole ratio method and the mass method. You can rely on whichever method works best for you.

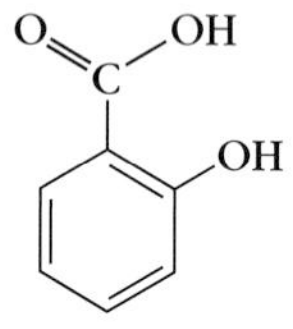

salicylic acid
$C_7H_6O_3$

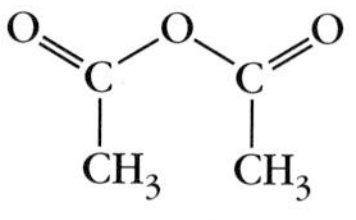

acetic anhydride
$C_4H_6O_3$

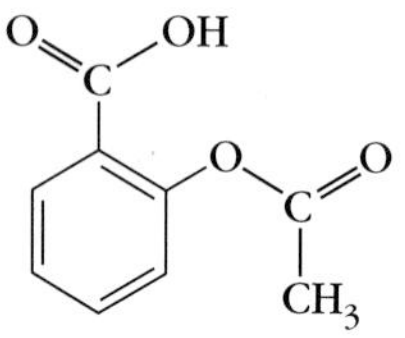

aspirin
$C_9H_8O_4$

Problem-Solving Example 4.10   Limiting Reactant

In a general chemistry laboratory experiment, a student prepares aspirin (acetylsalicylic acid, $C_9H_8O_4$) by the reaction of salicylic acid, $C_7H_6O_3$ (abbreviated SA), with acetic anhydride, $C_4H_6O_3$ (AA):

$$2 C_7H_6O_3(s) + C_4H_6O_3(\ell) \longrightarrow 2 C_9H_8O_4(s) + H_2O(\ell)$$

salicylic acid	acetic anhydride	aspirin
(SA)	(AA)	
138.1 g/mol	102.1 g/mol	180.2 g/mol

Her laboratory book directs her to use 2.0 g of salicylic acid and 5.4 g of acetic anhydride. (a) What is the limiting reactant? (b) What is the maximum mass of aspirin she can prepare?

Answer   (a) Salicylic acid is the limiting reactant. (b) 2.6 g aspirin

Explanation

   Method 1 (Mole Ratio Method)   The limiting reactant can be identified by finding the number of moles of each reactant available and then comparing their mole ratio with the mole ratio from the stoichiometric coefficients. The moles of each reactant are

$$2.0 \text{ g SA} \times \frac{1 \text{ mol SA}}{138.1 \text{ g SA}} = 0.0145 \text{ mol SA}$$

$$5.4 \text{ g AA} \times \frac{1 \text{ mol AA}}{102.1 \text{ g AA}} = 0.0529 \text{ mol AA}$$

From this information we can reason out which reactant is limiting. We refer to the stoichiometric coefficients of the balanced equation, which shows that for every two moles of SA reacted there is one mole of AA reacted.

$$\text{Mole ratio} = \frac{2 \text{ mol SA}}{1 \text{ mol AA}}$$

Consider two possibilities: (a) Salicylic acid is limiting, which means that 0.0145 mol of SA would require half that number of moles of AA, or 0.00725 mol of AA. There is 0.0529 mol of AA available, which is more than enough, so this possibility is supported by the facts. (b) Acetic anhydride is limiting, which means that 0.0529 mol of AA would require twice as many moles of SA, or 0.1058 mol of SA. There is not this much SA available, so AA cannot be the limiting reactant. Therefore, salicylic acid is the limiting reactant, and the quantity of aspirin that can be produced must be calculated from the 0.0145 mol of salicylic acid.

$$0.0145 \text{ mol SA} \times \frac{2 \text{ mol aspirin}}{2 \text{ mol SA}} \times \frac{180.2 \text{ g aspirin}}{1 \text{ mol aspirin}} = 2.6 \text{ g aspirin}$$

Method 2 (Mass Method)   Another way to identify salicylic acid as the limiting reactant is to calculate the mass of aspirin that would be produced from 0.0145 mol of salicylic acid and excess acetic anhydride, or from 0.0529 mol of acetic anhydride and excess salicylic acid. The limiting reactant is the one that produces the smaller mass of aspirin.

Mass of aspirin produced from 0.0145 mol of salicyclic acid and excess acetic anhydride:

$$0.0145 \text{ mol SA} \times \frac{2 \text{ mol aspirin}}{2 \text{ mol SA}} \times \frac{180.2 \text{ g aspirin}}{1 \text{ mol aspirin}} = 2.6 \text{ g aspirin}$$

Mass of aspirin produced from 0.0529 mol of acetic anhydride and excess salicylic acid:

$$0.0529 \text{ mol AA} \times \frac{2 \text{ mol aspirin}}{1 \text{ mol AA}} \times \frac{180.2 \text{ g aspirin}}{1 \text{ mol aspirin}} = 19. \text{ g aspirin}$$

This comparison shows that the amount of available salicylic acid would produce less aspirin than the amount of acetic anhydride available, providing further proof that salicylic acid is the limiting reactant. Using excess acetic anhydride makes economic sense as well as chemical sense because acetic anhydride costs only about half as much as salicylic acid.

✔   SA and aspirin have molar masses that are very roughly equal, and equal numbers of moles of SA and aspirin are in the balanced equation. So x g of SA should produce roughly x g aspirin, and it does.

Problem-Solving Practice 4.10

Carbon disulfide reacts with oxygen to form carbon dioxide and sulfur dioxide.

$$CS_2(\ell) + O_2(g) \longrightarrow CO_2(g) + SO_2(g)$$

A mixture of 3.5 g of CS_2 and 17.5 g O_2 is reacted.

(a) Balance the equation.
(b) What is the limiting reactant?
(c) What is the maximum number of grams of sulfur dioxide that can be formed?

Silicon carbide, SiC. The grinding wheel (left) is coated with SiC. Naturally occurring silicon carbide (right) is also known as carborundum. It is one of the hardest substances known, making it valuable as an abrasive.

Problem-Solving Example 4.11   Limiting Reactant

Silicon carbide, SiC, also known as carborundum, is an important industrial abrasive made by the high-temperature reaction of SiO_2 with carbon.

$$SiO_2(s) + 3\ C(s) \longrightarrow SiC(s) + 2\ CO(g)$$

(a) Determine the limiting reactant when a mixture of 5.00×10^3 g of SiO_2 and 5.00×10^3 g of carbon react. (b) What is the maximum number of grams of carborundum that can form? (c) How many grams of the excess reactant remain after the reaction is complete?

Answer   (a) SiO_2 is the limiting reactant. (b) 3.34×10^3 g SiC   (c) 2.00×10^3 g C remain.

Explanation   (a) We begin by determining how many moles of each reactant are available.

$$5.00 \times 10^3 \text{ g SiO}_2 \times \frac{1 \text{ mol SiO}_2}{60.1 \text{ g SiO}_2} = 83.19 \text{ mol SiO}_2$$

$$5.00 \times 10^3 \text{ g C} \times \frac{1 \text{ mol C}}{12.01 \text{ g C}} = 416.3 \text{ mol C}$$

The masses of carborundum produced, based on the masses available of each reactant, are

$$83.19 \text{ mol SiO}_2 \times \frac{1 \text{ mol SiC}}{1 \text{ mol SiO}_2} \times \frac{40.10 \text{ g SiC}}{1 \text{ mol SiC}} = 3.34 \times 10^3 \text{ g SiC}$$

$$416.3 \text{ mol C} \times \frac{1 \text{ mol SiC}}{3 \text{ mol C}} \times \frac{40.10 \text{ g SiC}}{1 \text{ mol SiC}} = 5.56 \times 10^3 \text{ g SiC}$$

Clearly, the quantity of carborundum that can be formed using the given quantities of carbon and silicon dioxide is controlled by the quantity of silicon dioxide; it is the limiting reactant.

(b) Thus, the maximum mass of carborundum that can form is that produced by reaction of the limiting reactant SiO_2 — that is, 3.34×10^3 g of SiC.

(c) We can find the moles of C that reacted from the moles of SiO_2 used and the mole ratio of C to SiO_2.

$$83.19 \text{ mol SiO}_2 \times \frac{3 \text{ mol C}}{1 \text{ mol SiO}_2} = 250. \text{ mol C}$$

By subtracting 250. mol of C from the initial amount of carbon (416.3 mol − 250. mol = 166. mol carbon), we find that 166. mol of carbon remains unreacted. The mass of unreacted carbon is therefore

$$166. \text{ mol C unreacted} \times \frac{12.01 \text{ g C}}{1.00 \text{ mol C}} = 2.00 \times 10^3 \text{ g C unreacted}$$

✔   The molar masses of SiC (40 g/mol) and SiO_2 (60 g/mol) are in the ratio of 2:3, and an equal number of moles of SiO_2 and SiC are in the balanced equation. So x g of SiO_2 should produce about $\frac{2}{3}x$ g SiC, and it does.

It is useful, although not necessary, to calculate the amount of excess reactant remaining to verify that the reactant in excess is not the limiting reactant.

Problem-Solving Practice 4.11

The pure silicon used in silicon chips involves the reaction between purified liquid silicon tetrachloride and magnesium.

$$SiCl_4(\ell) + 2\ Mg(s) \longrightarrow Si(s) + 2\ MgCl_2(s)$$

If the reaction were run with 100. g each of $SiCl_4$ and Mg, which reactant would be limiting, and what mass of Si would be produced?

Exercise 4.8   Limiting Reactant

Urea is used as a fertilizer because it can react with water to release ammonia, which provides nitrogen to plants.

$$(NH_2)_2CO(s) + H_2O(\ell) \longrightarrow 2\,NH_3(aq) + CO_2(g)$$

(a) Determine the limiting reactant when 300 g of urea and 100 g of water are combined.
(b) How many grams of ammonia and how many grams of carbon dioxide form?
(c) What mass of the excess reactant remains after reaction?

4.6   EVALUATING THE SUCCESS OF A SYNTHESIS: PERCENT YIELD

A reaction that forms the maximum possible quantity of product is said to have a 100 percent yield. This maximum possible quantity of product is called the **theoretical yield.** Often the **actual yield,** the quantity of desired product actually ob-

CD-Rom Screen 5.6: Percent Yield

CHEMISTRY
You Can Do...
Vinegar and Baking Soda: A Stoichiometry Experiment

This experiment focuses on the reactions of metal carbonates with acid. For example, limestone reacts with hydrochloric acid to give calcium chloride, carbon dioxide, and water:

$$CaCO_3(s) + 2\,HCl(aq) \longrightarrow CaCl_2(aq) + CO_2(g) + H_2O(\ell)$$

limestone

In a similar way, baking soda (sodium bicarbonate) and vinegar (aqueous acetic acid) react to give sodium acetate, carbon dioxide, and water:

$$NaHCO_3(s) + CH_3COOH(aq) \longrightarrow$$

baking soda   acetic acid

$$Na(CH_3COO)(aq) + CO_2(g) + H_2O(\ell)$$

sodium acetate

In this experiment we want to explore the relation between the quantity of acid or bicarbonate used and the quantity of carbon dioxide evolved. To do the experiment you need some baking soda, vinegar, a small balloon, and a bottle with a narrow neck and a volume of about 100 mL. The balloon should fit tightly but easily over the top of the bottle. (It may slip on more easily if the bottle and balloon are wet.) Inflate the balloon ahead of time to check that there are no leaks.

Place 1 level teaspoon of baking soda in the balloon. (You can make a funnel out of a rolled-up piece of paper to help get the baking soda into the balloon.) Add 3 teaspoons of vinegar to the bottle, and then slip the balloon over the neck of the bottle. Turn the balloon over so that the baking soda runs into the bottle, and then shake the bottle to make sure that the vinegar and baking soda are well mixed. What do you see? Does the balloon inflate? If so, why?

Now repeat the experiment several times using the following quantities of vinegar and baking soda:

Baking Soda	Vinegar
1 tsp	1 tsp
1 tsp	4 tsp
1 tsp	7 tsp
1 tsp	10 tsp

Be sure to use a new balloon each time and rinse out the bottle between tests. In each test, record how much the balloon inflates.

Is there a relationship between the quantity of vinegar and baking soda used and the extent to which the balloon inflates? If so, how can you explain this connection?

At which point does an increase in the quantity of vinegar not increase the volume of the balloon? Based on what we know about chemical reactions, how could increasing the amount of one reactant not have an effect on the balloon's size?

The setup for the study of the reaction of baking soda with acetic acid.

tained from a synthesis in a laboratory or industrial chemical plant, is less than the theoretical yield.

The efficiency of a particular synthesis method is evaluated by calculating the **percent yield,** which is defined as

$$\text{Percent yield} = \frac{\text{actual yield}}{\text{theoretical yield}} \times 100\%$$

Percent yield can be applied, for example, to the synthesis of aspirin as described in Problem-Solving Example 4.10. Suppose a student carried out the synthesis and obtained 2.2 g of aspirin rather than the theoretical yield of 2.6 g. What is the percent yield of this reaction?

$$\text{Percent yield} = \frac{\text{actual yield of product}}{\text{theoretical yield of product}} \times 100\% = \frac{2.2\ \text{g}}{2.6\ \text{g}} \times 100\% = 85\%$$

Although we hope to obtain as close to the theoretical yield as possible, few reactions or experimental manipulations are so efficient, despite controlled conditions and careful laboratory techniques. Side reactions can occur that form products other than the desired one, and during the isolation and purification of the desired product, some of it may be lost. When chemists report the synthesis of a new compound, or the development of a new synthesis, they state the percent yield of the reaction or the overall series of reactions. Other chemists who wish to repeat the synthesis then have an idea of how much product can be expected from a certain amount of reactants.

Popcorn yield. We began with 20 popcorn kernels, but only 16 of them popped. The percent yield of popcorn from this "reaction" was (16/20) × 100% = 80%.

 Problem-Solving Example 4.12 Calculating Percent Yield

Methanol, CH_3OH, is an excellent fuel, and it can be produced from carbon monoxide and hydrogen.

$$CO(g) + 2\ H_2(g) \longrightarrow CH_3OH(\ell)$$

If 500. g of CO reacts with excess H_2 and 485. g of CH_3OH is produced, what is the percent yield of the reaction?

Answer 85%

Explanation To solve the problem we need to calculate the theoretical yield of CH_3OH. We start by calculating the number of moles of reacting CO.

$$500.\ \text{g CO} \times \frac{1\ \text{mol CO}}{28.0\ \text{g CO}} = 17.86\ \text{mol CO}$$

The coefficients of the balanced equation show that for every 1 mol of CO reacted, 1 mol of CH_3OH will be produced. Therefore, the maximum number of moles of CH_3OH produced will be 17.86 mol of CH_3OH. We convert this to mass of CH_3OH.

$$17.86\ \text{mol } CH_3OH \times \frac{32.0\ \text{g } CH_3OH}{1\ \text{mol } CH_3OH} = 571.\ \text{g } CH_3OH$$

Thus, the theoretical yield is 571. g of CH_3OH. The problem states that 485. g of CH_3OH was produced. We calculate the percent yield.

$$\frac{485.\ \text{g } CH_3OH\ \text{(actual yield)}}{571.\ \text{g } CH_3OH\ \text{(theoretical yield)}} \times 100\% = 85.0\%$$

✔ The molar mass of CH_3OH is slightly greater than that of CO, and an equal number of moles of CO and CH_3OH are in the balanced equation. So x g of CO should produce somewhat more than x g of CH_3OH. Slightly less is actually produced, however, so a percent yield slightly less than 100% is about right.

Problem-Solving Practice 4.12

If this reaction is run with 85% percent yield, and you want to make 1.0 kg of CH_3OH, how many grams of H_2 should you use if CO is available in excess?

Problem-Solving Example 4.13   Percent Yield

Acetic acid (CH_3COOH) is produced industrially by the direct combination of methanol (CH_3OH) with carbon monoxide (CO).

$$CH_3OH(\ell) + CO(g) \longrightarrow CH_3COOH(\ell)$$

methanol   carbon   acetic
monoxide   acid

How many grams of methanol are needed to react with excess carbon monoxide to prepare 5.0×10^3 g of acetic acid if the expected yield is 88%?

Answer 3.0×10^3 g of methanol

Explanation   The first step is to calculate the theoretical yield. Expressing the 88% yield as a decimal fraction gives

$$\frac{\text{Actual yield}}{\text{Theoretical yield}} \times 100\% = 88\% \quad \text{so} \quad \frac{\text{Actual yield}}{\text{Theoretical yield}} = 0.88$$

$$\text{Theoretical yield} = \frac{5.0 \times 10^3 \text{ g acetic acid}}{0.88} = 5.68 \times 10^3 \text{ g acetic acid}$$

Carbon monoxide is in excess, so methanol, the limiting reactant, determines the maximum amount of acetic acid that can form. The mass of methanol needed can be calculated from the theoretical yield of acetic acid and the 1:1 mole ratio for methanol and acetic acid.

$$5.68 \times 10^3 \text{ g acetic acid} \times \frac{1 \text{ mol acetic acid}}{60.0 \text{ g acetic acid}} \times \frac{1 \text{ mol methanol}}{1 \text{ mol acetic acid}}$$

$$\times \frac{32.0 \text{ g methanol}}{1 \text{ mol methanol}} = 3.0 \times 10^3 \text{ g methanol}$$

✔   The molar mass of CH_3OH is about 32 g/mol and of CH_3COOH is about 60 g/mol, a ratio of about 1:2; an equal number of moles of each is in the balanced equation. So 5000 g of acetic acid would require about 2500 g of CH_3OH if the yield were 100%. But the yield is less, so a bit more than 2500 g is required. This agrees with the answer.

Problem-Solving Practice 4.13

You heat 2.50 g of copper with an excess of sulfur and synthesize 2.53 g of copper(I) sulfide, Cu_2S:

$$16 \text{ Cu(s)} + S_8(s) \longrightarrow 8 \text{ Cu}_2S(s)$$

Your laboratory instructor expects students to have at least a 60% yield for this reaction. Did your synthesis meet this standard?

Exercise 4.9   Percent Yield

Percent yield can be reduced by side reactions that produce undesired product(s) and by poor laboratory technique in isolating and purifying the desired product. Identify two other factors that could lead to a low percent yield.

Atom Economy—Another Approach to Tracing Starting Materials

The concept of **atom economy** focuses on the amount of starting materials that are incorporated into the desired final product. The higher the fraction of starting atoms incorporated into the final product, the fewer waste byproducts that are created. The objective is to devise syntheses that are as efficient as possible.

Whereas a high percent yield has often been the major goal of chemical synthesis, the concept of atom economy, quantified by the definition of percent atom utilization, is becoming important.

Percent atom economy

$$= \frac{\text{atomic weight of atoms in useful product}}{\text{atomic weight of all the reactants used in the reaction}} \times 100\%$$

Reactions for which all the atoms in the reactants are found in the desired product have a percent atom economy of 100%. As an example, consider the combination reaction discussed in Problem-Solving Example 4.12.

$$CO(g) + 2\,H_2(g) \longrightarrow CH_3OH(\ell)$$

The atomic weight of all the atoms in the reactants is $12.011 + 15.9994 + \{2 \times (2 \times 1.0079)\} = 32.042$ amu. The atomic weight of all the atoms in the product is $12.011 + \{3 \times (1.0079)\} + 15.9994 + 1.0079 = 32.042$ amu. The percent atom economy for this reaction is 100%.

Many other reactions in organic synthesis, however, involve the generation of other products in addition to the desired product. In such cases, the percent atom economy is far less than 100%. Devising strategies for synthesis of desired compounds with the least waste is a major goal of the current push toward "green chemistry."

4.7 PERCENT COMPOSITION AND EMPIRICAL FORMULAS

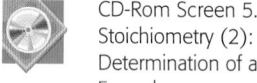
CD-Rom Screen 5.8: Using Stoichiometry (2): Determination of an Empirical Formula

In Section 3.10, percent composition data were used to derive empirical and molecular formulas, but nothing was mentioned about how such data are obtained. One way to obtain such data is combustion analysis, which is often used for organic compounds, most of which contain carbon and hydrogen. In **combustion analysis** a compound is burned in oxygen, which converts the carbon to carbon dioxide and the hydrogen to water. These combustion products are collected and weighed, and the weights are used to calculate the amounts of carbon and hydrogen in the original substance using the balanced combustion equation. The apparatus is shown in Figure 4.10. Many organic compounds also contain oxygen. In such cases, the mass of oxygen in the sample can be determined simply by difference.

$$\text{Mass of oxygen} = \text{mass of sample} - (\text{mass of C} + \text{mass of H})$$

As an example, we will consider the following problem. An analytical chemist used combustion analysis to determine the empirical formula of vitamin C, an organic compound containing only carbon, hydrogen, and oxygen. Combustion of 1.000 g of pure vitamin C produced 1.502 g of CO_2 and 0.409 g of H_2O. A different experiment determined that the molar mass of vitamin C is 176.12 g/mol.

The task is to determine the subscripts on C, H, and O in the empirical formula of vitamin C. Recall from Chapter 3 that the subscripts in a chemical formula

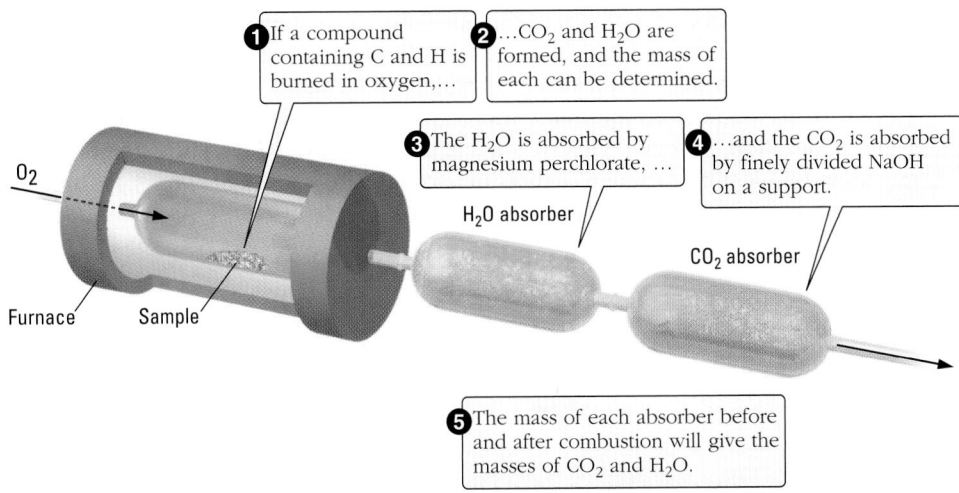

1 If a compound containing C and H is burned in oxygen,...

2 ...CO_2 and H_2O are formed, and the mass of each can be determined.

3 The H_2O is absorbed by magnesium perchlorate, ...

4 ...and the CO_2 is absorbed by finely divided NaOH on a support.

O_2

Furnace   Sample

H_2O absorber

CO_2 absorber

5 The mass of each absorber before and after combustion will give the masses of CO_2 and H_2O.

Figure 4.10   Combustion analysis. Only a few milligrams of a combustible compound are needed for analysis.

tell how many moles of atoms of each element are in 1 mol of the compound. We begin by determining how many grams of carbon, hydrogen, and oxygen were in the original sample. *All* of the carbon in the CO_2 and *all* of the hydrogen in the H_2O came from the vitamin C sample that was burned, so we can work backward to assess the composition of vitamin C.

First, we determine the masses of carbon and hydrogen in the original sample.

$$1.502 \text{ g CO}_2 \times \frac{1 \text{ mol CO}_2}{44.009 \text{ g CO}_2} \times \frac{1 \text{ mol C}}{1 \text{ mol CO}_2} \times \frac{12.011 \text{ g C}}{1 \text{ mol C}} = 0.4100 \text{ g C}$$

$$0.409 \text{ g H}_2\text{O} \times \frac{1 \text{ mol H}_2\text{O}}{18.015 \text{ g H}_2\text{O}} \times \frac{2 \text{ mol H}}{1 \text{ mol H}_2\text{O}} \times \frac{1.0079 \text{ g H}}{1 \text{ mol H}} = 0.04577 \text{ g H}$$

The mass of oxygen in the original sample can be calculated by difference.

1.000 g sample − (0.4100 g C in sample + 0.04577 g H in sample)

= 0.5442 g O in the original sample

From the mass data, we can now calculate how many moles of each element were in the sample.

$$0.4100 \text{ g C} \times \frac{1 \text{ mol C}}{12.011 \text{ g C}} = 0.03414 \text{ mol C}$$

$$0.04577 \text{ g H} \times \frac{1 \text{ mol H}}{1.0079 \text{ g H}} = 0.04541 \text{ mol H}$$

$$0.5442 \text{ g O} \times \frac{1 \text{ mol O}}{15.999 \text{ g O}} = 0.03401 \text{ mol O}$$

Next, we find the mole ratios of the elements in the compound by dividing by the smallest number of moles.

$$\frac{0.04541 \text{ mol H}}{0.03401 \text{ mol O}} = \frac{1.335 \text{ mol H}}{1.000 \text{ mol O}}$$

$$\frac{0.03414 \text{ mol O}}{0.03401 \text{ mol C}} = \frac{1.004 \text{ mol O}}{1.000 \text{ mol O}}$$

The ratios are very close to 1.33 mol of H to 1.00 mol of O and a one-to-one ratio of C to O, which on multiplication by 3 to get whole numbers gives the empirical

vitamin C

Carrying an extra digit during the intermediate parts of a multistep problem and rounding at the end is good practice.

If after dividing by the smallest number of moles, the ratios are not whole numbers, multiply each subscript by a number that converts the fractions to whole numbers. For example, multiplying $NO_{2.5}$ by 2 changes it to N_2O_5.

formula of vitamin C as $C_3H_4O_3$. From this we can calculate an empirical molar mass of 88.06 g. Because the calculated molar mass is one half the experimental molar mass, the molecular formula of vitamin C is $C_6H_8O_6$.

For many organic compounds, the empirical and molecular formulas are the same. In addition, several different organic compounds can have the identical empirical and molecular formulas—they are isomers (⬅ *p. 83, Section 3.4*). In such cases, you must know the structural formula to fully describe the compound.

$$CH_3CH_2OH \qquad CH_3OCH_3$$
ethanol dimethyl ether

Problem-Solving Example 4.14 Empirical Formula from Combustion Analysis

Suppose you have isolated a compound from clover leaves and want to know its empirical formula to help identify it. You know that the compound contains only carbon, hydrogen, and oxygen, and so you use combustion analysis. Burning 0.514 g of the compound produces 0.501 g of CO_2 and 0.103 g of H_2O. What is its empirical formula? Another experiment shows that the molar mass of the compound is 90.04 g/mol. What is its molecular formula?

Answer The empirical formula is CHO_2; the molecular formula is $C_2H_2O_4$.

Explanation All of the carbon and hydrogen in the compound are burned to carbon dioxide and water, respectively. Therefore, use the masses of CO_2 and H_2O to find how many grams of C and H, respectively, were in the unknown compound.

$$0.501 \text{ g } CO_2 \times \frac{1 \text{ mol } CO_2}{44.01 \text{ g } CO_2} \times \frac{1 \text{ mol C}}{1 \text{ mol } CO_2} \times \frac{12.01 \text{ g C}}{1 \text{ mol C}} = 0.1367 \text{ g C}$$

$$0.103 \text{ g } H_2O \times \frac{1 \text{ mol } H_2O}{18.02 \text{ g } H_2O} \times \frac{2 \text{ mol H}}{1 \text{ mol } H_2O} \times \frac{1.008 \text{ g H}}{1 \text{ mol H}}$$

$$= 0.1152 \text{ g H formerly in the sample burned}$$

These calculations show that the 0.514-g sample of the unknown compound contains 0.1367 g of C and 0.01152 g of H. The remaining mass, 0.3658 g, must be oxygen.

$$0.1367 \text{ g C} + 0.01152 \text{ g H} + 0.3658 \text{ g O} = 0.514\text{-g sample}$$

Finding the number of moles of each element in the unknown compound reveals its empirical formula (Section 3.10).

$$0.1367 \text{ g C} \times \frac{1 \text{ mol C}}{12.01 \text{ g C}} = 0.01138 \text{ mol C}$$

$$0.01152 \text{ g H} \times \frac{1 \text{ mol H}}{1.008 \text{ g H}} = 0.01142 \text{ mol H}$$

$$0.3658 \text{ g O} \times \frac{1 \text{ mol O}}{16.00 \text{ g O}} = 0.02286 \text{ mol O}$$

To find the mole ratios of the elements, we divide the moles of each element by the smallest number of moles.

$$\frac{0.01142 \text{ mol H}}{0.01138 \text{ mol C}} = \frac{1.004 \text{ mol H}}{1.00 \text{ mol C}} \quad \text{and} \quad \frac{0.02286 \text{ mol O}}{0.01138 \text{ mol C}} = \frac{2.009 \text{ mol O}}{1.000 \text{ mol C}}$$

The mole ratios indicate that for every C atom in the molecule, there are one H atom and two O atoms. Therefore, the empirical formula of the unknown compound is CHO_2 with an empirical formula molar mass of 45.02 g/mol. The experimental molar mass is known to be twice this value, so the molecular formula of the unknown compound is $C_2H_2O_4$, twice the empirical formula.

Problem-Solving Practice 4.14

Phenol is a compound of carbon, hydrogen, and oxygen that is used commonly as a disinfectant. Combustion analysis of a 175-mg sample of phenol yielded 491. mg of CO_2 and 46.5 mg of H_2O.

(a) Calculate the empirical formula of phenol.
(b) What other information is necessary to determine whether the empirical formula is the actual molecular formula?

 Exercise 4.10 **Formula from Combustion Analysis**

Nicotine, a compound found in cigarettes, contains C, H, and N. Outline a method by which you could use combustion analysis to determine the empirical formula for nicotine.

Determining Formulas from Experimental Data

One technique to determine the formula of a binary compound formed by direct combination of its two elements is to measure the mass of reactants that are converted to the product compound.

Problem-Solving Example 4.15 **Empirical Formula from Experimental Data**

Solid red phosphorus reacts with liquid bromine to produce a phosphorus bromide.

$$P_4(s) + Br_2(\ell) \longrightarrow P_x Br_y(\ell)$$

If 0.347 g of P_4 reacts with 0.860 mL of Br_2, what is the empirical formula of the product? The density of bromine is 3.12 g/mL.

Answer PBr_3

Explanation We start by calculating the moles of P_4 and Br_2 that combined.

$$0.347 \text{ g } P_4 \times \frac{1 \text{ mol } P_4}{123.90 \text{ g } P_4} = 2.801 \times 10^{-3} \text{ mol } P_4$$

To determine the moles of bromine, we first use its density to convert milliliters of bromine to grams

$$0.860 \text{ mL } Br_2 \times \frac{3.12 \text{ g } Br_2}{1 \text{ mL } Br_2} = 2.68 \text{ g } Br_2$$

and then convert from grams to moles

$$2.68 \text{ g } Br_2 \times \frac{1 \text{ mol } Br_2}{159.8 \text{ g } Br_2} = 1.677 \times 10^{-2} \text{ mol } Br_2$$

The mole ratio of bromine atoms to phosphorus atoms in a molecule of the product can be calculated from the moles of atoms of each element.

$$1.677 \times 10^{-2} \text{ mol } Br_2 \text{ molecules} \times \frac{2 \text{ mol Br atoms}}{1 \text{ mol } Br_2 \text{ molecule}} = 3.35 \times 10^{-2} \text{ mol Br atoms}$$

$$2.801 \times 10^{-3} \text{ mol } P_4 \text{ molecules} \times \frac{4 \text{ mol P atoms}}{1 \text{ mol } P_4 \text{ molecules}} = 1.12 \times 10^{-2} \text{ mol P atoms}$$

$$\frac{3.35 \times 10^{-2} \text{ mol Br atoms}}{1.12 \times 10^{-2} \text{ mol P atoms}} = \frac{2.99 \text{ mol Br atoms}}{1.00 \text{ mol P atoms}}$$

The mole ratio in the compound is 3.00 mol of bromine atoms for 1.00 mol of phosphorus atoms. The empirical formula is PBr_3. By other experimental methods, the known molar mass of this compound is found to be the same as the empirical formula molar mass. Given this additional information, we know that the molecular formula is also PBr_3.

Problem-Solving Practice 4.15

The complete reaction of 0.569 g of tin with 2.434 g of iodine formed Sn_xI_y. What is the empirical formula of this tin iodide?

SUMMARY PROBLEM

Iron can be smelted from iron oxide in ore via the following high-temperature reaction in a blast furnace.

$$Fe_2O_3(s) + 3\ CO(g) \longrightarrow 2\ Fe(\ell) + 3\ CO_2(g)$$

The iron product can be cooled and weighed.

(a) For 19.0 g of Fe_2O_3, what mass of CO is required to react completely?

(b) What mass of CO_2 is produced when the reaction runs to completion with 10.0 g of Fe_2O_3 as starting material?

When the reaction was run repeatedly with the same mass of iron oxide, 19.0 g of Fe_2O_3, but differing amounts of carbon monoxide, the following graph was obtained.

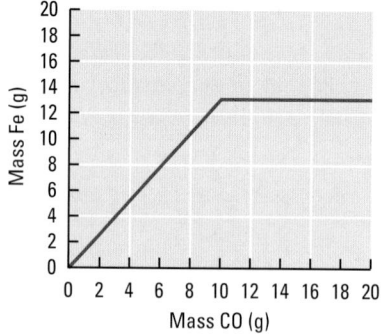

(c) Which reactant is limiting in the part of the graph where there is less than 10.0 g of CO available to react with 19.0 g of Fe_2O_3?

(d) Which reactant is limiting when more than 10.0 g of CO is available to react with 19.0 g of Fe_2O_3?

(e) If 24.0 g of Fe_2O_3 reacted with 20.0 g of CO and 15.9 g of Fe was produced, what was the percent yield of the reaction?

IN CLOSING

Having studied this chapter, you should be able to . . .

- Interpret the information conveyed by a balanced chemical equation (Section 4.1).
- Recognize the general reaction types: combination, decomposition, displacement, and exchange (Section 4.2).
- Balance simple chemical equations (Section 4.3).
- Use mole ratios to calculate the number of moles or mass of one reactant or product from the number of moles or mass of another reactant or product by using the balanced chemical equation (Section 4.4).
- Use stoichiometry principles in the chemical analysis of a mixture (Section 4.4).
- Determine which of two reactants is the limiting reactant (Section 4.5).
- Explain the differences among actual yield, theoretical yield, and percent yield, and calculate theoretical and percent yields (Section 4.6).
- Use stoichiometric principles to find the empirical formula of an unknown compound using combustion analysis and other mass data (Section 4.7).

KEY TERMS

actual yield *(4.6)*

aqueous solution *(4.1)*

atom economy *(4.6)*

balanced chemical equation *(4.1)*

coefficient *(4.1)*

combination reaction *(4.1)*

combustion analysis *(4.7)*

combustion reaction *(4.1)*

decomposition reaction *(4.2)*

displacement reaction *(4.2)*

exchange reaction *(4.2)*

limiting reactant *(4.5)*

mole ratio *(4.4)*

percent yield *(4.6)*

stoichiometric coefficient *(4.1)*

stoichiometry *(4.1)*

theoretical yield *(4.6)*

QUESTIONS FOR REVIEW AND THOUGHT

Conceptual Challenge Problems

CP-4.A (Section 4.4) In Example 4.7 it was not possible to find the mass of O_2 directly from a knowledge of the mass of sucrose. Are there chemical reactions in which the mass of a product or another reactant can be known directly if you know the mass of a reactant? Can you cite a couple of these reactions?

CP-4.B (Section 4.4) Glucose ($C_6H_{12}O_6$), a monosaccharide, and sucrose ($C_{12}H_{22}O_{11}$), a disaccharide, undergo complete combustion with O_2 (metabolic conversion) to produce H_2O and CO_2.

(a) How many moles of O_2 are needed per mole of each sugar for the reaction to proceed?

(b) How many grams of O_2 are needed per mole of each sugar for the reaction to proceed?

(c) Which combustion reaction produces more H_2O per gram of sugar? How many grams of H_2O are produced per gram of each sugar?

Answers to questions in **bold** can be found in the back of the book.

Review Questions

1. What information is provided by a balanced chemical equation?

2. This chapter introduced the quantitative aspects of chemical reactions. Which of these questions address quantitative issues?
   (a) Is this chemical process beneficial to society?
   (b) Does the yield justify the production costs?
   (c) How pure is the product?
   (d) Does the chemical reaction pollute the environment?
   (e) Will the reaction occur?
   (f) How much of the reactants are wasted in the synthesis?

3. Complete the table for the reaction

   $$3 H_2(g) + N_2(g) \longrightarrow 2 NH_3(g)$$

H$_2$	N$_2$	NH$_3$
_____ mol	1 mol	_____ mol
3 molecules	_____ molecules	_____ molecules
_____ g	_____ g	34.08 g

4. What is meant by the statement, "the reactants were present in stoichiometric amounts"?

5. Write all the possible stoichiometric factors for the reaction

   $$3 MgO(s) + 2 Fe(s) \longrightarrow Fe_2O_3(s) + 3 Mg(s)$$

6. If a 10.0-g mass of carbon is combined with an exact stoichiometric amount of oxygen (26.6 g) to make carbon dioxide, what mass in grams of CO_2 can be isolated?

7. Given the reaction

   $$2 Fe(s) + 3 Cl_2(g) \longrightarrow 2 FeCl_3(s)$$

   fill in the missing conversion factors for the scheme

   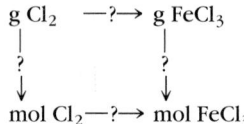

8. (a) If you are making cheeseburgers with two slices of cheese, one hamburger patty, and one bun, how many cheeseburgers can you make from eight buns, one-half dozen patties, and 20 slices of cheese? (b) What is the limiting "reactant"? (c) What "reactants" are in excess?

9. When an exam question asks, "What is the limiting reactant?", students often guess the reactant with the smallest mass. Why is this not a good strategy?

10. Why can't the product of a reaction ever be the limiting reactant?

11. A cookie recipe states that it makes a batch of 5 dozen cookies; however, you only get 4 dozen per batch. (a) What are your actual and theoretical yields? (b) What is your percent yield?

12. Does the limiting reactant determine the theoretical yield, actual yield, or both? Why?

Stoichiometry

13. For the following reaction, fill in the table.

    $$2 C_2H_6(g) + 7 O_2(g) \longrightarrow 4 CO_2(g) + 6 H_2O(g)$$

	C$_2$H$_6$(g)	O$_2$(g)	CO$_2$(g)	H$_2$O(g)
No. of molecules				
No. of atoms				
No. of moles of molecules				
Mass				
Total mass of reactants				
Total mass of products				

14. Magnesium metal burns brightly in the presence of oxygen to produce a white powdery substance, MgO.

    $$Mg(s) + O_2(g) \longrightarrow MgO(s)$$

    (a) If 1.00 g of MgO(s) is formed by this reaction, what is the maximum *total* mass of magnesium metal and oxygen that reacted?
    (b) Identify the stoichiometric coefficients in this equation.
    (c) If 50 *atoms* of oxygen reacted, how many magnesium *atoms* were needed to react with this much oxygen?

15. Balance the following *combination reaction* by adding coefficients as needed.

    $$Fe(s) + O_2(g) \longrightarrow Fe_2O_3(s)$$

16. The following diagram shows A (blue spheres) reacting with B (red spheres). Which equation best describes the stoichiometry of the reaction depicted in this diagram?

    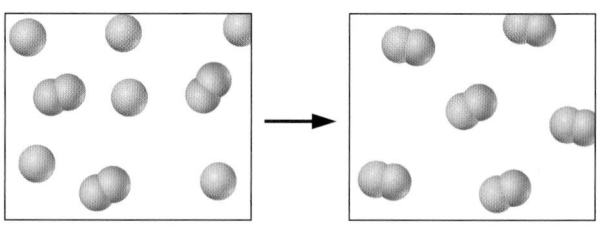

    (a) $3 A_2 + 6 B \rightarrow 6 AB$ (b) $A_2 + 2 B \rightarrow 2 AB$
    (c) $2 A + B \rightarrow AB$ (d) $3 A + 6 B \rightarrow 6 AB$

17. Given the following equation,

    $$4 A_2 + 3 B \longrightarrow B_3A_8$$

    use a diagram to illustrate the amount of reactant A and product (B_3A_8) that would be needed/produced from the reaction of six atoms of B.

18. Balance the following equation and determine which box represents reactants and which box represents products.

    $$Sb(g) + Cl_2(g) \longrightarrow SbCl_3(g)$$

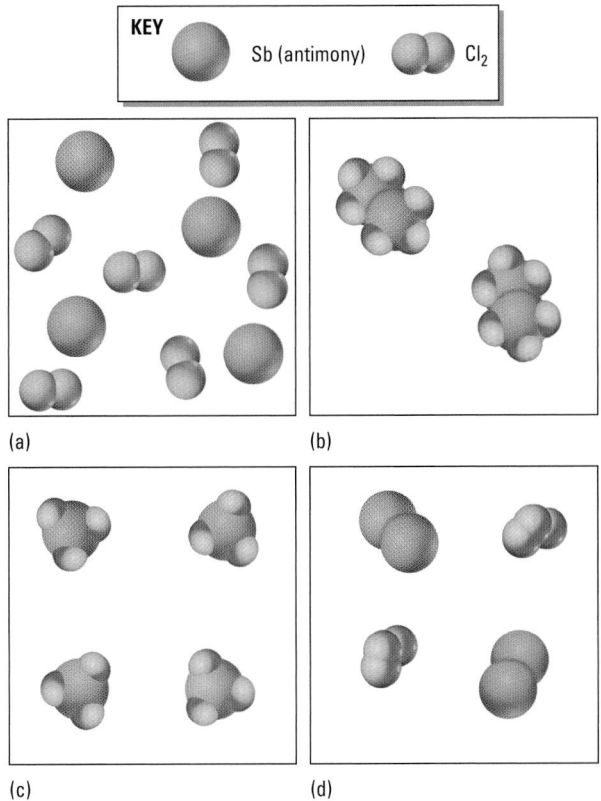

KEY

Sb (antimony)   Cl$_2$

(a)

(b)

(c)

(d)

Classification of Chemical Reactions

19. Write a balanced equation for the formation of each of the following compounds from the elements.
(a) Carbon monoxide
(b) Nickel(II) oxide
(c) Chromium(III) oxide

20. Write a balanced equation for the formation of each of the following compounds from the elements.
(a) Copper(I) oxide
(b) Arsenic(III) oxide
(c) Zinc oxide

21. Write a balanced equation for each of the following decomposition reactions. Name each product.
(a) BeCO$_3$(s) $\xrightarrow{\text{heat}}$
(b) NiCO$_3$(s) $\xrightarrow{\text{heat}}$
(c) Al$_2$(CO$_3$)$_3$(s) $\xrightarrow{\text{heat}}$

22. Write a balanced equation for each of the following decomposition reactions. Name each product.
(a) ZnCO$_3$(s) $\xrightarrow{\text{heat}}$
(b) MnCO$_3$(s) $\xrightarrow{\text{heat}}$
(c) PbCO$_3$(s) $\xrightarrow{\text{heat}}$

23. Write a balanced equation for the following combustion reactions.
(a) C$_4$H$_{10}$(g) + O$_2$(g) $\rightarrow$
(b) C$_6$H$_{12}$O$_6$(s) + O$_2$(g) $\rightarrow$
(c) C$_4$H$_8$O(ℓ) + O$_2$(g) $\rightarrow$

24. Complete and balance the following equations involving oxygen reacting with an element. Name the product in each case.
(a) Mg(s) + O$_2$(g) $\rightarrow$
(b) Ca(s) + O$_2$(g) $\rightarrow$
(c) In(s) + O$_2$(g) $\rightarrow$

25. Complete and balance the following equations involving oxygen reacting with an element.
(a) Ti(s) + O$_2$(g) $\rightarrow$ titanium(IV) oxide
(b) S$_8$(s) + O$_2$(g) $\rightarrow$ sulfur dioxide
(c) Se(s) + O$_2$(g) $\rightarrow$ selenium dioxide

26. Complete and balance the following equations involving the reaction of a halogen with a metal. Name the product in each case.
(a) K(s) + Cl$_2$(g) $\rightarrow$
(b) Mg(s) + Br$_2$(ℓ) $\rightarrow$
(c) Al(s) + F$_2$(g) $\rightarrow$

27. Complete and balance the following equations involving the reaction of a halogen with a metal.
(a) Cr(s) + Cl$_2$(g) $\rightarrow$ chromium(III) chloride
(b) Cu(s) + Br$_2$(ℓ) $\rightarrow$ copper(II) bromide
(c) Pt(s) + F$_2$(g) $\rightarrow$ platinum(IV) fluoride

Balancing Equations

28. Balance the following equations.
(a) Al(s) + O$_2$(g) $\rightarrow$ Al$_2$O$_3$(s)
(b) N$_2$(g) + H$_2$(g) $\rightarrow$ NH$_3$(g)
(c) C$_6$H$_6$(ℓ) + O$_2$(g) $\rightarrow$ H$_2$O(ℓ) + CO$_2$(g)

29. Balance the following equations.
(a) Fe(s) + Cl$_2$(g) $\rightarrow$ FeCl$_3$(s)
(b) SiO$_2$(s) + C(s) $\rightarrow$ Si(s) + CO(g)
(c) Fe(s) + H$_2$O(g) $\rightarrow$ Fe$_3$O$_4$(s) + H$_2$(g)

30. Balance the following equations.
(a) UO$_2$(s) + HF(ℓ) $\rightarrow$ UF$_4$(s) + H$_2$O(ℓ)
(b) B$_2$O$_3$(s) + HF(ℓ) $\rightarrow$ BF$_3$(g) + H$_2$O(ℓ)
(c) BF$_3$(g) + H$_2$O(ℓ) $\rightarrow$ HF(ℓ) + H$_3$BO$_3$(s)

31. Balance the following equations.
(a) MgO(s) + Fe(s) $\rightarrow$ Fe$_2$O$_3$(s) + Mg(s)
(b) H$_3$BO$_3$(s) $\rightarrow$ B$_2$O$_3$(s) + H$_2$O(ℓ)
(c) NaNO$_3$(s) + H$_2$SO$_4$(aq) $\rightarrow$ Na$_2$SO$_4$(aq) + HNO$_3$(g)

32. Balance the following equations.
(a) Reaction to produce hydrazine, N$_2$H$_4$, a good industrial reducing agent:

$$H_2NCl(aq) + NH_3(g) \longrightarrow NH_4Cl(aq) + N_2H_4(aq)$$

(b) Reaction of the fuel (dimethylhydrazine and dinitrogen tetraoxide) used in the moon lander and space shuttle:

$$(CH_3)_2N_2H_2(\ell) + N_2O_4(g) \longrightarrow N_2(g) + H_2O(g) + CO_2(g)$$

(c) Reaction of calcium carbide to produce acetylene, C$_2$H$_2$:

$$CaC_2(s) + H_2O(\ell) \longrightarrow Ca(OH)_2(s) + C_2H_2(g)$$

33. Balance the following equations.
(a) Reaction of calcium cyanamide to produce ammonia:

$$CaNCN(s) + H_2O(\ell) \longrightarrow CaCO_3(s) + NH_3(g)$$

(b) Reaction to produce diborane, B_2H_6:

$$NaBH_4(s) + H_2SO_4(aq) \longrightarrow$$
$$B_2H_6(g) + H_2(g) + Na_2SO_4(aq)$$

(c) Reaction to rid water of hydrogen sulfide, H_2S, a foul-smelling compound:

$$H_2S(aq) + Cl_2(aq) \longrightarrow S_8(s) + HCl(aq)$$

34. Balance the following combustion reactions:
(a) $C_6H_{12}O_6 + O_2 \rightarrow CO_2 + H_2O$
(b) $C_5H_{12} + O_2 \rightarrow CO_2 + H_2O$
(c) $C_7H_{14}O_2 + O_2 \rightarrow CO_2 + H_2O$
(d) $C_2H_4O_2 + O_2 \rightarrow CO_2 + H_2O$

35. Balance the following equations:
(a) $Mg + HNO_3 \rightarrow H_2 + Mg(NO_3)_2$
(b) $Al + Fe_2O_3 \rightarrow Al_2O_3 + Fe$
(c) $S + O_2 \rightarrow SO_3$
(d) $SO_3 + H_2O \rightarrow H_2SO_4$

The Mole and Chemical Reactions

36. Chlorine can be produced in the laboratory by the reaction of hydrochloric acid with manganese(IV) oxide.

$$4\,HCl(aq) + MnO_2(s) \longrightarrow$$
$$Cl_2(g) + 2\,H_2O(\ell) + MnCl_2(aq)$$

How many moles of HCl are needed to form 12.5 mol of Cl_2?

37. Methane, CH_4, is the major component of natural gas. How many moles of oxygen are needed to burn 16.5 mol of CH_4?

$$CH_4(g) + 2\,O_2(g) \longrightarrow CO_2(g) + 2\,H_2O(\ell)$$

38. Nitrogen monoxide is oxidized in air to give brown nitrogen dioxide.

$$2\,NO(g) + O_2(g) \longrightarrow 2\,NO_2(g)$$

Starting with 2.2 mol of NO, how many moles and how many grams of O_2 are required for complete reaction? What mass of NO_2, in grams, is produced?

39. Aluminum reacts with oxygen to give aluminum oxide.

$$4\,Al(s) + 3\,O_2(g) \longrightarrow 2\,Al_2O_3(s)$$

If you have 6.0 mol of Al, how many moles and how many grams of O_2 are needed for complete reaction? What mass of Al_2O_3, in grams, is produced?

40. Many metals react with halogens to give metal halides. For example, iron gives iron(II) chloride, $FeCl_2$.

$$Fe(s) + Cl_2(g) \longrightarrow FeCl_2(s)$$

Beginning with 10.0 g of iron, what mass of Cl_2, in grams, is required for complete reaction? What quantity of $FeCl_2$, in moles and in grams, is expected?

41. Like many metals, manganese reacts with a halogen to give a metal halide.

$$2\,Mn(s) + 3\,F_2(g) \longrightarrow 2\,MnF_3(s)$$

(a) If you begin with 5.12 g of Mn, what mass in grams of F_2 is required for complete reaction?
(b) What amount in moles and in grams of the red solid MnF_3 is expected?

42. The final step in the manufacture of platinum metal (for use in automotive catalytic converters and other products) is the reaction

$$3\,(NH_4)_2PtCl_6(s) \longrightarrow$$
$$3\,Pt(s) + 2\,NH_4Cl(s) + 2\,N_2(g) + 16\,HCl(g)$$

Complete this table of reaction quantities for the reaction of 12.35 g of $(NH_4)_2PtCl_6$.

$(NH_4)_2PtCl_6$	Pt	HCl
12.35 g	_____ g	_____ g
_____ mol	_____ mol	_____ mol

43. Disulfur dichloride, S_2Cl_2, is used to vulcanize rubber. It can be made by treating molten sulfur with gaseous chlorine.

$$S_8(\ell) + 4\,Cl_2(g) \longrightarrow 4\,S_2Cl_2(g)$$

Complete this table of reaction quantities for the production of 103.5 g of S_2Cl_2.

S_8	Cl_2	S_2Cl_2
_____ g	_____ g	103.5 g
_____ mol	_____ mol	_____ mol

44. Many metal halides react with water to produce the metal oxide (or hydroxide) and the appropriate hydrogen halide. For example,

$$TiCl_4(\ell) + 2\,H_2O(g) \longrightarrow TiO_2(s) + 4\,HCl(g)$$

(a) If you begin with 14.0 g of $TiCl_4$, how many moles of water are required for complete reaction? (b) How many grams of each product are expected?

Liquid titanium tetrachloride ($TiCl_4$). When exposed to air it forms a dense fog of titanium(IV) oxide, TiO_2.

45. Gaseous sulfur dioxide, SO_2, can be removed from smoke-stacks by treatment with limestone and oxygen.

$$2 SO_2(g) + 2 CaCO_3(s) + O_2(g) \longrightarrow$$
$$2 CaSO_4(s) + 2 CO_2(g)$$

(a) How many moles each of $CaCO_3$ and O_2 are required to remove 150. g of SO_2?
(b) What mass of $CaSO_4$ is formed when 150. g of SO_2 is consumed completely?

46. Tungsten oxide can be reduced to tungsten metal.

$$WO_3(s) + 3 H_2(g) \longrightarrow W(s) + 2 H_2O(\ell)$$

How many grams of tungsten are formed from 1.00 kg of WO_3?

47. If you want to synthesize 1.45 g of the semiconducting material GaAs, what masses of Ga and of As, in grams, are required?

48. Ammonium nitrate, NH_4NO_3, is a common fertilizer and explosive. It was used in the Oklahoma City bombing. When heated, it decomposes into gaseous products.

$$2 NH_4NO_3(s) \longrightarrow 2 N_2(g) + 4 H_2O(g) + O_2(g)$$

How many grams of each product are formed from 1.0 kg of NH_4NO_3?

49. Iron reacts with oxygen to give iron(III) oxide, Fe_2O_3.
(a) Write a balanced equation for this reaction. (b) If an ordinary iron nail (assumed to be pure iron) has a mass of 5.58 g, what mass in grams of Fe_2O_3 would be produced if the nail is converted completely to this oxide? (c) What mass of O_2 (in grams) is required for the reaction?

50. Nitroglycerin decomposes violently according to the equation

$$4 C_3H_5(NO_3)3(\ell) \longrightarrow$$
$$12 CO_2(g) + 10 H_2O(\ell) + 6 N_2(g) + O_2(g)$$

How many grams of each gaseous product are produced from 1.00 g of nitroglycerin?

51. Chlorinated fluorocarbons, such as CCl_2F_2, have been banned from use in automobile air conditioners because they are destructive to the ozone layer. Researchers at MIT have found an environmentally safe way to decompose these compounds by treating them with sodium oxalate, $Na_2C_2O_4$. The products of the reaction are carbon, carbon dioxide, sodium chloride, and sodium fluoride.
(a) Write a balanced equation for this reaction.
(b) What mass of $Na_2C_2O_4$ is needed to remove 76.8 g of CCl_2F_2?
(c) What mass of CO_2 is produced?

52. Careful decomposition of ammonium nitrate, NH_4NO_3, gives laughing gas (dinitrogen monoxide, N_2O) and water.
(a) Write a balanced equation for this reaction.
(b) Beginning with 10.0 g of NH_4NO_3, what masses of N_2O and water are expected?

53. In making iron from iron ore, the following reaction proceeds.

$$Fe_2O_3(s) + 3 CO(g) \longrightarrow 2 Fe(s) + 3 CO_2(g)$$

(a) How many grams of iron can be obtained from 1.00 kg of iron(III) oxide?
(b) How many grams of CO are required?

54. Cisplatin, $Pt(NH_3)_2Cl_2$, a drug used in the treatment of cancer, can be made by the reaction of K_2PtCl_4 with ammonia, NH_3. Besides cisplatin, the other product is KCl.
(a) Write a balanced equation for this reaction.
(b) In order to obtain 2.50 g of cisplatin, what masses in grams of K_2PtCl_4 and ammonia do you need?

Limiting Reactant

55. Aluminum chloride, Al_2Cl_6, is an inexpensive reagent used in many industrial processes. It is made by treating scrap aluminum with chlorine according to the balanced equation

$$2 Al(s) + 3 Cl_2(g) \longrightarrow Al_2Cl_6(s)$$

(a) Which reactant is limiting if 2.70 g of Al and 4.05 g of Cl_2 are mixed?
(b) What mass of $AlCl_3$ can be produced?
(c) What mass of the excess reactant will remain when the reaction is completed?

56. Hydrogen and oxygen react to form water.

$$2 H_2(g) + O2(g) \longrightarrow 2 H_2O$$

(a) If a mixture containing 100. g of each reactant is ignited, what is the limiting reactant?
(b) How much water is formed?

57. Methanol, CH_3OH, is a clean-burning, easily handled fuel. It can be made by the direct reaction of CO and H_2 (obtained from heating coal with steam).

$$CO(g) + 2 H_2(g) \longrightarrow CH_3OH(\ell)$$

(a) Starting with a mixture of 12.0 g of H_2 and 74.5 g of CO, which is the limiting reactant? (b) What mass of the excess reactant, in grams, is left after reaction is complete? (c) What mass of methanol can be obtained, in theory?

58. The reaction of methane and water is one way to prepare hydrogen.

$$CH_4(g) + 2 H_2O(g) \longrightarrow CO_2(g) + 4 H_2(g)$$

If 995 g of CH_4 reacts with 2510 g of water, how many moles of reactants and products are there when the reaction is finished?

59. Ammonia gas can be prepared by the following reaction.

$$CaO(s) + 2 NH_4Cl(s) \longrightarrow 2 NH_3(g) + H_2O(g) + CaCl_2(s)$$

If 112 g of CaO reacts with 224 g of NH_4Cl, how many moles of reactants and products are there when the reaction is finished?

60. The equation for one of the reactions in the process of turning iron ore into the metal is

$$Fe_2O_3(s) + 3 CO(g) \longrightarrow 2 Fe(s) + 3 CO_2(g)$$

If you start with 2.00 kg of each reactant, what is the maximum amount of iron you can produce?

61. Aspirin is produced by the reaction of salicylic acid and acetic anhydride.

$$2 C_7H_6O_3(s) + C_4H_6O_3(\ell) \longrightarrow 2 C_9H_8O_4(s) + H_2O(\ell)$$

| salicylic acid | acetic anhydride | aspirin |

If you mix 100 g of each of the reactants, what is the maximum mass of aspirin that can be obtained?

Percent Yield

62. Ammonia gas can be prepared by the reaction of calcium oxide with ammonium chloride.

$$CaO(s) + 2\ NH_4Cl(s) \longrightarrow 2\ NH_3(g) + H_2O(g) + CaCl_2(s)$$

If exactly 100 g of ammonia is isolated but the theoretical yield is 136 g, what is the percent yield of this gas?

63. Quicklime, CaO, is formed when calcium hydroxide is heated.

$$Ca(OH)_2(s) \longrightarrow CaO(s) + H_2O(\ell)$$

If the theoretical yield is 65.5 g but only 36.7 g of quicklime is produced, what is the percent yield?

64. Diborane, B_2H_6, is valuable for the synthesis of new organic compounds. The boron compound can be made by the reaction

$$2\ NaBH_4(s) + I_2(s) \longrightarrow B_2H_6(g) + 2\ NaI(s) + H_2(g)$$

Suppose you use 1.203 g of $NaBH_4$ and excess iodine, and you isolate 0.295 g of B_2H_6. What is the percent yield of B_2H_6?

65. Methanol, CH_3OH, is used in racing cars because it is a clean-burning fuel. It can be made by the reaction

$$CO(g) + 2\ H_2(g) \longrightarrow CH_3OH(\ell)$$

What is the percent yield if 5.0×10^3 g of H_2 reacts with excess CO to form 3.5×10^3 g of CH_3OH?

66. Disulfur dichloride, which has a revolting smell, can be prepared by directly combining S_8 and Cl_2, but it can also be made by the reaction

$$3\ SCl_2(\ell) + 4\ NaF(s) \longrightarrow SF_4(g) + S_2Cl_2(\ell) + 4\ NaCl(s)$$

What mass of SCl_2 is needed to react with excess NaF to prepare 1.19 g of S_2Cl_2, if the expected yield is 51%?

67. The ceramic silicon nitride, S_3N_4, is made by heating silicon and nitrogen at an elevated temperature.

$$3\ Si(s) + 2\ N_2(g) \longrightarrow Si_3N_4(s)$$

How many grams of silicon must combine with excess N_2 to produce 1.0 kg of Si_3N_4 if this process is 92% efficient?

Empirical Formulas

68. Styrene, the building block of polystyrene, is a hydrocarbon (a compound consisting only of C and H). If 0.438 g of the compound is burned and produces 1.481 g of CO_2 and 0.303 g of H_2O, what is the empirical formula of the compound?

69. Mesitylene is a liquid hydrocarbon. If 0.115 g of the compound is burned in pure O_2 to give 0.379 g of CO_2 and 0.1035 g of H_2O, what is the empirical formula of the compound?

70. Propionic acid, an organic acid, contains only C, H, and O. If 0.236 g of the acid burns completely in O_2 and gives 0.421 g of CO_2 and 0.172 g of H_2O, what is the empirical formula of the acid?

71. Quinone, which is used in the dye industry and in photography, is an organic compound containing only C, H, and O. What is the empirical formula of the compound if 0.105 g of the compound gives 0.257 g of CO_2 and 0.0350 g of H_2O when burned completely?

General Questions

72. Nitrogen gas can be prepared in the laboratory by the reaction of ammonia with copper(II) oxide according to the following unbalanced equation.

$$NH_3(g) + CuO(s) \longrightarrow N_2(g) + Cu(s) + H_2O(g)$$

If 26.3 g of gaseous NH_3 is passed over a bed of solid CuO (in stoichiometric excess), what mass, in grams, of N_2 can be isolated?

73. In an experiment, 1.056 g of a metal carbonate containing an unknown metal M was heated to give the metal oxide and 0.376 g CO_3.

$$MCO_3(s) \xrightarrow{\text{heat}} MO(s) + CO_2(g)$$

What is the identity of the metal M?
(a) Ni   (b) Cu   (c) Zn   (d) Ba

74. Aluminum bromide is a valuable laboratory chemical. What is the maximum theoretical yield, in grams, of Al_2Br_6 if 25.0 mL of liquid bromine (density = 3.1023 g/mL) and excess aluminum metal are used?

$$2\ Al(s) + 3\ Br_2(\ell) \longrightarrow Al_2Br_6(s)$$

75. Uranium(VI) oxide reacts with bromine trifluoride to give uranium(IV) fluoride, an important step in the purification of uranium ore.

$$6\ UO_3(s) + 8\ BrF_3(\ell) \longrightarrow 6\ UF_4(s) + 4\ Br_2(\ell) + 9\ O_2(g)$$

If you begin with 365 g each of UO_3 and BrF_3, what is the maximum yield, in grams, for UF_4?

76. The cancer chemotherapy agent cisplatin is made by the following reaction:

$$(NH_4)_2PtCl_4(s) + 2\ NH_3(aq) \longrightarrow$$
$$2\ NH_4Cl(aq) + Pt(NH_3)_2Cl_2(s)$$

Assume that 15.5 g of $(NH_4)_2PtCl_4$ is combined with 0.15 mol of aqueous NH_3 to make cisplatin. What is the theoretical mass, in grams, of cisplatin that can be formed?

77. Diborane, B_2H_6, can be produced by the following reaction:

$$2\ NaBH_4(aq) + H_2SO_4(aq) \longrightarrow$$
$$2\ H_2(g) + Na_2SO_4(aq) + B_2H_6(g)$$

What is the maximum yield, in grams, of B_2H_6 that can be prepared starting with 2.19×10^{-2} mol of H_2SO_4 and 1.55 g of $NaBH_4$?

78. Silicon and hydrogen form a series of interesting compounds, Si_xH_y. To find the formula of one of them, a 6.22-g sample of the compound is burned in oxygen. When this is done, all of the Si is converted to 11.64 g of SiO_2 and all of the H to 6.980 g of H_2O. What is the empirical formula of the silicon compound?

79. Boron forms an extensive series of compounds with hydrogen, all with the general formula B_xH_y. To analyze one of these compounds, you burn it in air and isolate the boron in the form of B_2O_3 and the hydrogen in the form of water. If 0.148 g of B_xH_y gives 0.422 g of B_2O_3 when burned in excess O_2, what is the empirical formula of B_xH_y?

80. What is the limiting reactant for the reaction

$$4\,KOH + 2\,MnO_2 + O_2 + Cl_2 \longrightarrow$$
$$2\,KMnO_4 + 2\,KCl + 2\,H_2O$$

if 5 mol of each reactant is present? What is the limiting reactant when 5 g of each reactant is present?

81. The Hargraves process is an industrial method for making sodium sulfate for use in papermaking.

$$4\,NaCl + 2\,SO_2 + 2\,H_2O + O_2 \longrightarrow 2\,Na_2SO_4 + 4\,HCl$$

(a) If you start with 10 mol of each reactant, which one will determine the amount of Na_2SO_4 produced?
(b) What if you start with 100 g of each reactant?

Applying Concepts

82. Chemical equations can be interpreted on either a particulate level (atoms, molecules, ions) or a mole level (moles of reactants and products). Write word statements to describe the combustion of butane on a particulate level and a mole level.

$$2\,C_4H_{10}(g) + 13\,O_2(g) \longrightarrow 8\,CO_2(g) + 10\,H_2O(\ell)$$

83. Write word statements to describe the following reaction on a particulate level and a mole level.

$$P_4(s) + 6\,Cl_2(g) \longrightarrow 4\,PCl_3(\ell)$$

84. What is the single product of this reaction?

$$4\,A_2 + AB_3 \longrightarrow 3 \,\underline{\hspace{2cm}}$$

85. What is the single product of this hypothetical reaction?

$$3\,A_2B_3 + B_3 \longrightarrow 6 \,\underline{\hspace{2cm}}$$

86. In a reaction, 1.2 g of element A reacts with exactly 3.2 grams of oxygen to form an oxide, AO_x; 2.4 g of element A reacts with exactly 3.2 g of oxygen to form a second oxide, AO_y.
(a) What is the ratio x/y?
(b) If $x = 2$, what is the identity of element A?

87. If 1.5 mol of Cu reacts with a solution containing 4.0 mol of $AgNO_3$, what ions will be present in the solution at the end of the reaction?

$$Cu(s) + 2\,AgNO_3(aq) \longrightarrow Cu(NO_3)_2(aq) + 2\,Ag(s)$$

88. Ammonia is formed in a direct reaction of nitrogen and hydrogen.

$$N_2(g) + 3\,H_2(g) \longrightarrow 2\,NH_3(g)$$

A tiny portion of the starting mixture is represented by the following diagram, where the blue circles represent N and the white circles represent H.

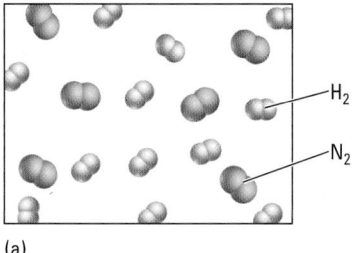

(a)

Which of the following represents the product mixture?

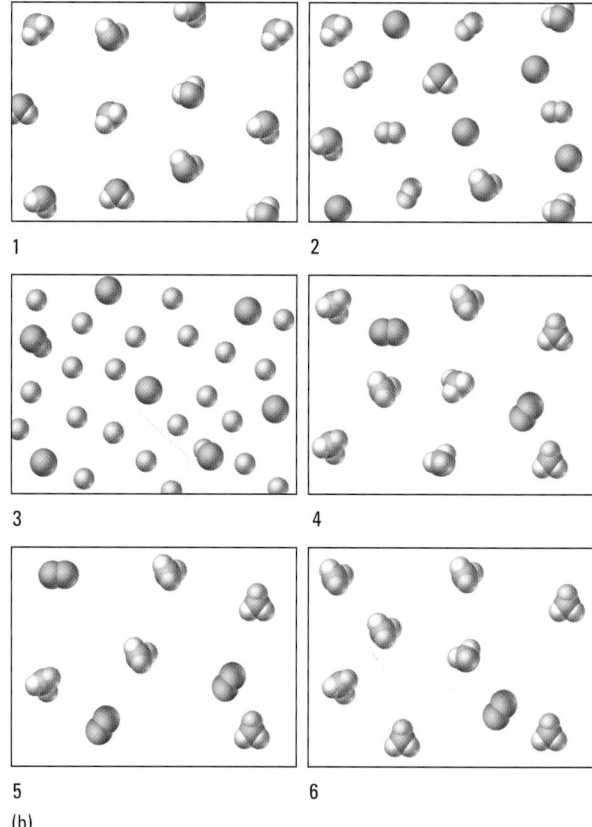

1   2

3   4

5   6

(b)

For the reaction of the given sample, which of the following is true?
(a) N_2 is the limiting reactant.
(b) H_2 is the limiting reactant.
(c) NH_3 is the limiting reactant.
(d) No reactant is limiting; they are present in the correct stoichiometric ratio.

89. Carbon monoxide burns readily in oxygen to form carbon dioxide.

$$2\,CO(g) + O_2(g) \longrightarrow 2\,CO_2(g)$$

The box on the left represents a tiny portion of a mixture of CO and O_2. If these molecules react to form CO_2, what should the contents of the box on the right look like?

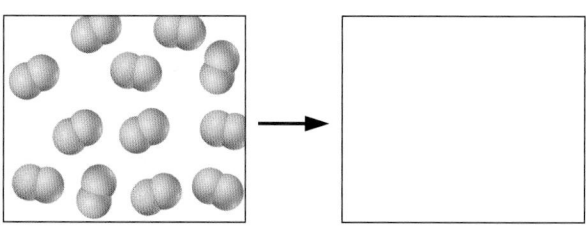

90. Which chemical equation best represents the reaction taking place in this illustration?
(a) $X_2 + Y_2 \rightarrow nXY_3$
(b) $X_2 + 3 Y_2 \rightarrow 2 XY_3$
(c) $6 X_2 + 6 Y_2 \rightarrow 4 XY_3 + 4 X_2$
(d) $6 X_2 + 6 Y_2 \rightarrow 4 X_3Y + 4 Y_2$

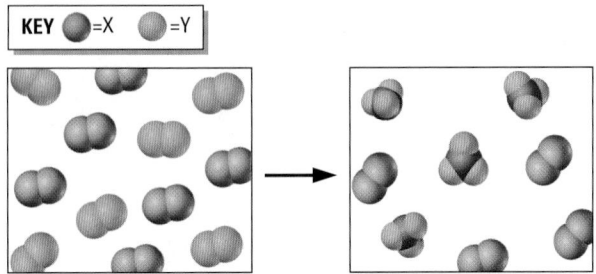

91. A student set up an experiment, like the one described in Chemistry You Can Do on page 000, for six different trials between acetic acid, CH_3COOH, and sodium bicarbonate, $NaHCO_3$ (see figure).

$$CH_3COOH(aq) + NaHCO_3(s) \longrightarrow$$
$$NaCH_3CO_2(aq) + CO_2(g) + H_2O(\ell)$$

The volume of acetic acid is kept constant, but the mass of sodium bicarbonate increased with each trial. The results of the tests are shown in the figure.
(a) In which trial(s) is the acetic acid the limiting reactant?
(b) In which trial(s) is sodium bicarbonate the limiting reactant?

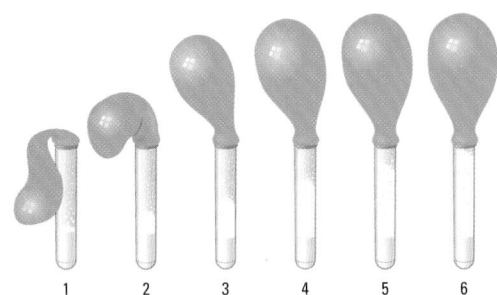

92. A weighed sample of a metal is added to liquid bromine and allowed to react completely. The product substance is then separated from any leftover reactants and weighed. This experiment is repeated with several masses of the metal but with the same volume of bromine. The following graph indicates the results. Explain why the graph has the shape that it does.

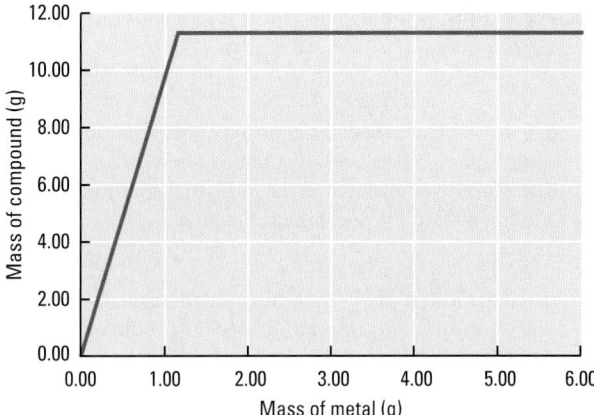

93. A series of experimental measurements like the ones described in Question 92 is carried out for iron reacting with bromine. The following graph is obtained. What is the empirical formula of the compound formed by iron and bromine? Write a balanced equation for the reaction between iron and bromine. Name the product.

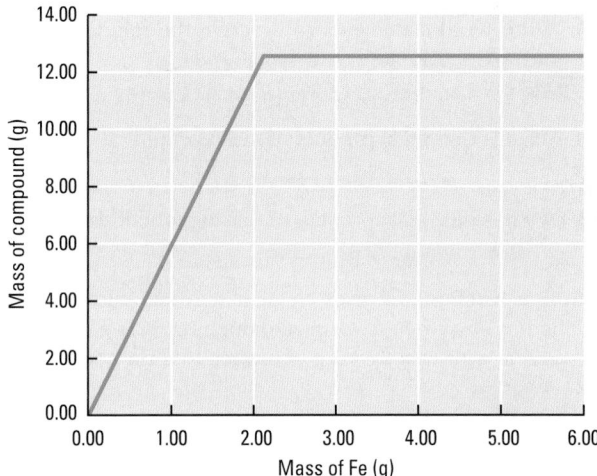

General Chemistry CD-ROM

CD4.1 Screen 4.3: The Law of Conservation of Matter. Observe the decomposition of mercury(II) oxide and the accompanying animation.
(a) What visual evidence is there for the decomposition of the oxide?

(b) What happens to the oxygen evolved in the decomposition?

(c) In the animation, how many molecules of oxygen are evolved? How many atoms of mercury? How many "molecules" of HgO must have decomposed? Does this agree with the balanced chemical equation for the decomposition?

CD4.2 Screen 4.4: Balancing Chemical Equations.

(a) *Reactions that form oxides:* Observe the video of the reaction of elemental phosphorus. Here the phosphorus is removed from a beaker full of water and placed on a laboratory spoon in the air. (The phosphorus used here is called "white" phosphorus, even though its color is often closer to yellow.)

   (1) Is elemental phosphorus a solid, liquid, or gas?

   (2) What can you conclude about the relative tendency of phosphorus to react with water and air? Why is the element stored under water?

(b) *Combustion reactions:* Observe the animation of the reaction of propane and oxygen. How does this animation illustrate the principle of the conservation of matter?

CD4.3 Screen 5.4: Reactions Controlled by the Supply of One Reactant.

(a) Explain how the video of methanol combustion illustrates the fact that methanol is the limiting reactant in this reaction.

(b) If 2.5 moles of methanol burn in oxygen, how many moles of O_2 are required?

(c) Examine the problem associated with this screen (Screen 5.4PR). Which of the diagrams at the right is correct? Why?

CD4.4 Screen 5.5: Limiting Reactants. Examine the series of photos of the reaction between varying amounts of Zn metal and a constant amount of hydrochloric acid.

(a) What gas inflates the balloons in the demonstration?

(b) Why are the volumes of gas in the first two balloons greater than the volume in the balloon on the right?

(c) In which flask is the Zn metal the limiting reactant and in which flask is the HCl(aq) the limiting reactant? Explain.

CD4.5 Screen 5.8: Using Stoichiometry (2). Assume you burn the white solid naphthalene, $C_{10}H_8$, in excess oxygen.

(a) Write the balanced equation for the combustion reaction.

(b) If you burn 0.100 g of naphthalene, what masses of H_2O and CO_2 are formed?

5

Steel wool burns vigorously in pure oxygen gas. In this reaction, two elements combine to form a compound, iron oxide:

$$4\ Fe + 3\ O_2 \longrightarrow 2\ Fe_2O_3$$

CHEMICAL REACTIONS

5.1 Exchange Reactions: Precipitation and Net Ionic Equations

5.2 Acids, Bases, and Acid-Base Exchange Reactions

5.3 Oxidation-Reduction Reactions

5.4 Oxidation Numbers and Redox Reactions

5.5 Displacement Reactions, Redox, and the Activity Series

5.6 Solution Concentration

5.7 Molarity and Reactions in Aqueous Solutions

5.8 Aqueous Solution Titrations

Chemistry is concerned with how substances react and what products are formed when they react. A chemical compound can consist of molecules or oppositely charged ions, and often the compound's properties can be deduced from the behavior of these molecules or ions. The *chemical* properties of a compound are the transformations that the molecules or ions can undergo when the substance reacts. A central focus of chemistry is providing answers to questions such as these: When two substances are mixed together, will a chemical reaction occur? If a chemical reaction is possible, what will the products be?

As you saw in Chapter 4 (⬅ *p. 125),* most chemical reactions can be assigned to a few general categories: combination, decomposition, displacement, and exchange. In this chapter we discuss chemical reactions in more detail, including oxidation-reduction reactions. The ability to recognize which type of reaction occurs for a particular set of reactants will allow you to predict the products.

Chemical reactions involving exchange of ions to form precipitates are discussed first, followed by net ionic equations, which focus on the active participants in such reactions. We move on to acid-base reactions, neutralization reactions, and reactions that form gases as a product. This is followed by a discussion of oxidation-reduction (redox) reactions, oxidation numbers as a means to organize our understanding of redox reactions, and the activity series of metals.

A great deal of chemistry, perhaps most, occurs in the solution phase, and we introduce the means for quantitatively describing solutions. This is followed by solution stoichiometry and finally by aqueous titration, an analytical technique that is used to measure solution concentrations.

5.1 EXCHANGE REACTIONS: PRECIPITATION AND NET IONIC EQUATIONS

Aqueous Solubility of Ionic Compounds

Many of the ionic compounds that you frequently encounter, such as table salt, baking soda, and household plant fertilizers, are soluble in water. It is therefore tempting to conclude that all ionic compounds are soluble in water, but such is not the case. Although many ionic compounds are water-soluble, some are only slightly soluble, and others dissolve hardly at all.

When ionic compounds dissolve in water, they dissociate into ions that are surrounded by water molecules, as illustrated in Figure 5.1a. By contrast, most

Figure 5.1 Dissolution of (a) an ionic compound and (b) a molecular compound (methanol, CH_3OH) in water.

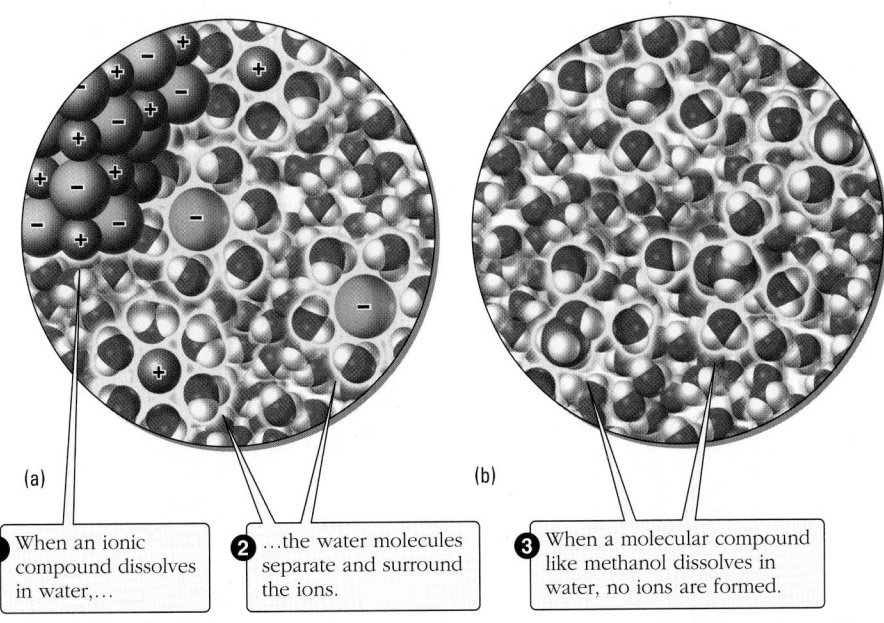

(a)

(b)

❶ When an ionic compound dissolves in water,...

❷ ...the water molecules separate and surround the ions.

❸ When a molecular compound like methanol dissolves in water, no ions are formed.

molecular compounds do not ionize when dissolved in water, as illustrated in Figure 5.1b.

 CD-ROM Screen 4.6: Solubility

The solubility rules given in Table 5.1 are general guidelines for predicting the water solubilities of ionic compounds based on the ions they contain. *If a compound contains at least one of the ions indicated for soluble compounds in Table 5.1, then the compound is at least moderately soluble.*

Figure 5.2 shows examples illustrating the solubility rules for a few nitrates, hydroxides, and sulfides. Suppose you want to know whether $NiSO_4$ is soluble in water. $NiSO_4$ contains Ni^{2+} and SO_4^{2-} ions. Although Ni^{2+} is not mentioned in Table 5.1, substances containing SO_4^{2-} are described as soluble (except for $SrSO_4$, $BaSO_4$, and $PbSO_4$). Because $NiSO_4$ contains an ion (SO_4^{2-}) that indicates solubility and $NiSO_4$ is not one of the sulfate exceptions, it is predicted to be soluble.

Problem-Solving Example 5.1 Using Solubility Rules

Indicate which ions are present in each of the following compounds. Then predict whether each compound is water-soluble.

(a) NH_4Cl (b) $Mg_3(PO_4)_2$ (c) $Ba(OH)_2$ (d) $Ca(NO_3)_2$

TABLE 5.1 Solubility Rules for Ionic Compounds

Ammonium, Group 1A, NH_4^+, Li^+, Na^+, K^+, Rb^+, Cs^+	All ammonium and Group 1A (alkali metal) salts are soluble.
Nitrates, NO_3^-	All nitrates are soluble.
Chlorides, bromides, iodides, Cl^-, Br^-, I^-	All common chlorides, bromides, and iodides are soluble except $AgCl$, Hg_2Cl_2, $PbCl_2$; $AgBr$, Hg_2Br_2, $PbBr_2$; AgI, Hg_2I_2; PbI_2.
Sulfates, SO_4^{2-}	Most sulfates are soluble; exceptions include $CaSO_4$, $SrSO_4$, $BaSO_4$, and $PbSO_4$.
Chlorates, ClO_3^-	All chlorates are soluble.
Perchlorates, ClO_4^-	All perchlorates are soluble.
Acetates, CH_3COO^-	All acetates are soluble.
Phosphates, PO_4^{3-}	All phosphates are insoluble except those of NH_4^+ and Group 1A elements (alkali metal cations).
Carbonates, CO_3^{2-}	All carbonates are insoluble except those of NH_4^+ and Group 1A elements (alkali metal cations).
Hydroxides, OH^-	All hydroxides are insoluble except those of NH_4^+ and Group 1A (alkali metal cations). $Sr(OH)_2$, $Ba(OH)_2$, and $Ca(OH)_2$ are slightly soluble.
Oxides, O^{2-}	All oxides are insoluble except those of Group 1A (alkali metal cations).
Oxalates, $C_2O_4^{2-}$	All oxalates are insoluble except those of NH_4^+ and Group 1A (alkali metal cations)
Sulfides, S^{2-}	All sulfides are insoluble except those of NH_4^+, Group 1A (alkali metal cations), and Group 2A (MgS, CaS, and BaS are sparingly soluble).

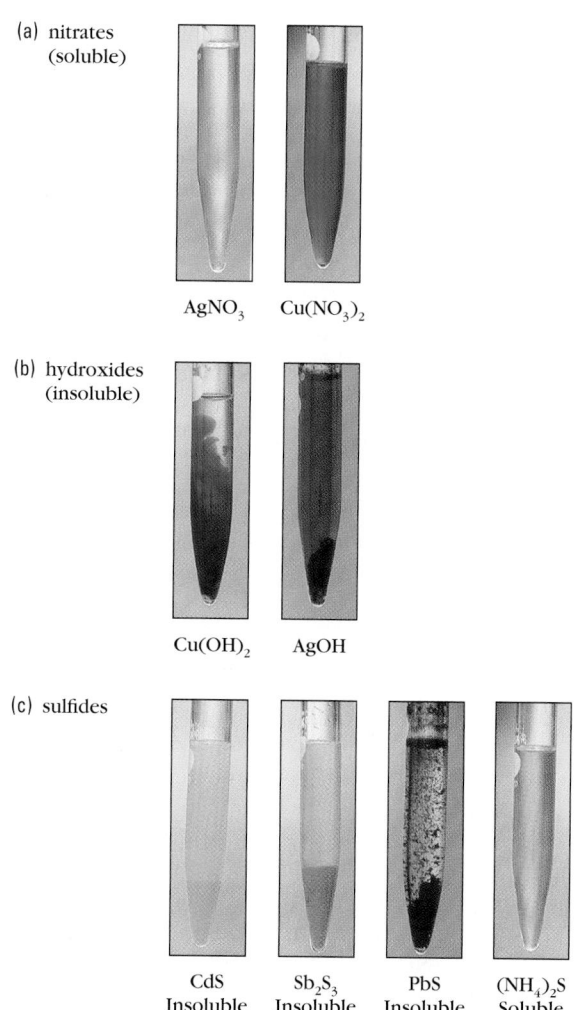

Figure 5.2 **Illustration of some solubility guidelines in Table 5.1.** $AgNO_3$ and $Cu(NO_3)_2$, like all nitrates, are soluble. $Cu(OH)_2$ and AgOH, like most hydroxides, are insoluble. CdS, Sb_2S_3, and PbS, like nearly all sulfides, are insoluble, but $(NH_4)_2S$ is an exception (it is soluble).

Answer (a) NH_4^+, Cl^-, (b) Mg^{2+}, PO_4^{3-}, (c) Ba^{2+}, OH^-, (d) Ca^{2+}, NO_3^-. NH_4Cl, $Ba(OH)_2$, and $Ca(NO_3)_2$ are soluble; $Mg_3(PO_4)_2$ is insoluble.

Explanation The use of solubility rules requires identifying the ions present and checking their effects on water solubility (Table 5.1).

(a) NH_4Cl contains NH_4^+ and Cl^- ions. The presence of NH_4^+ means that the compound is soluble.
(b) $Mg_3(PO_4)_2$ is composed of Mg^{2+} and PO_4^{3-} ions. Most phosphates are insoluble, with the exception of those containing ammonium ion or alkali metal cations. Because Mg^{2+} is not an alkali metal ion, $Mg_3(PO_4)_2$ is insoluble.
(c) Most hydroxides are insoluble, but not $Ba(OH)_2$. It is slightly soluble.
(d) $Ca(NO_3)_2$ contains Ca^{2+} and NO_3^- (nitrate) ions. Table 5.1 indicates that all nitrates are soluble, so $Ca(NO_3)_2$ is soluble.

Problem-Solving Practice 5.1

Predict whether each of the following compounds is likely to be water-soluble.
(a) NaF (b) $Ca(CH_3COO)_2$ (c) $SrCl_2$ (d) MgO (e) $PbCl_2$ (f) HgS

Recall that exchange reactions *(⇐ p. 130),* have the following reaction pattern:

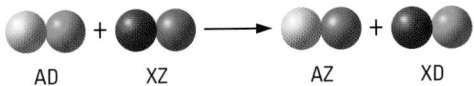

If both the reactants and the products of such a reaction are water-soluble ionic compounds, no overall reaction takes place. Mixing the solutions of AD and XZ just results in an aqueous solution containing the four ions. If, however, one of the potential products of the reaction removes ions from the solution, a reaction occurs. Three different kinds of products can cause an exchange reaction to occur in aqueous solution:

1. Formation of an *insoluble ionic compound:*

$$AgNO_3(aq) + KCl(aq) \longrightarrow KNO_3(aq) + \boxed{AgCl(s)}$$

2. Formation of a *molecular compound* that remains in solution. Most commonly this happens when water is produced in acid-base reactions:

$$H_2SO_4(aq) + 2\,NaOH(aq) \longrightarrow Na_2SO_4(aq) + \boxed{2\,H_2O(\ell)}$$

3. Formation of a *gaseous molecular compound* that escapes from the solution:

$$2\,HCl(aq) + Na_2S(aq) \longrightarrow 2\,NaCl(aq) + \boxed{H_2S(g)}$$

Precipitation Reactions

CD-ROM Screen 4.12: Precipitation Reactions

What might happen if two aqueous solutions are mixed together, one containing dissolved calcium nitrate, $Ca(NO_3)_2$, and the other containing dissolved sodium chloride, NaCl? These are both soluble ionic compounds (Table 5.1), so the resulting solution contains Ca^{2+}, NO_3^-, Na^+, and Cl^-. To decide whether a reaction will occur requires determining whether any of these ions can react with each other to form a new compound. For an exchange reaction to occur, the calcium ion and the chloride ion would have to form calcium chloride ($CaCl_2$), and the sodium ion and the nitrate ion would have to form sodium nitrate ($NaNO_3$). Is this a possible chemical reaction? The answer is yes if either of these compounds is insoluble. Checking the solubility rules shows that both of these compounds are water-soluble. No reaction to remove the ions from solution is possible; therefore, the answer is that there is no chemical reaction.

Consider the possibility of an exchange reaction between aqueous solutions of barium chloride and sodium sulfate.

$$BaCl_2(aq) + Na_2SO_4(aq) \longrightarrow ? + ?$$

If the barium ions and sodium ions exchange partners to form $BaSO_4$ and NaCl, the equation will be

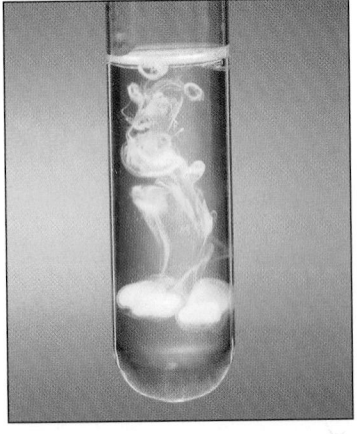

Figure 5.3 Precipitation of barium sulfate. The mixing of aqueous solutions of barium chloride ($BaCl_2$) and sodium sulfate (Na_2SO_4) forms a precipitate of barium sulfate ($BaSO_4$). Sodium chloride (NaCl), the other product of this exchange reaction, is water-soluble, and Na^+ and Cl^- ions remain in solution.

$$BaCl_2(aq) + Na_2SO_4(aq) \longrightarrow BaSO_4 + 2\,NaCl$$

| barium chloride | sodium sulfate | barium sulfate | sodium chloride |

Will a reaction occur? The answer is yes if an insoluble product — a **precipitate** — can form. Checking Table 5.1, we find that NaCl is soluble, but $BaSO_4$ is not soluble (sulfates of Sr^{2+}, Ba^{2+}, and Pb^{2+} are insoluble). Therefore, an exchange reaction will occur, and barium sulfate will precipitate from the solution. Precipitate formation is indicated by an (s) in the overall equation (Figure 5.3). Because it is soluble, NaCl remains dissolved in solution.

Ba^{2+} and Na^+ do not react with each other, and neither do Cl^- and SO_4^{2-}.

$$BaCl_2(aq) + Na_2SO_4(aq) \longrightarrow \boxed{BaSO_4(s)} + 2\,NaCl(aq)$$

Problem-Solving Example 5.2   Exchange Reactions

For each of the following pairs of ionic compounds, decide whether an exchange reaction will occur when their aqueous solutions are mixed and write a balanced equation for those reactions that will occur.

(a) $AgNO_3$ and $NaBr$ (b) $CaCl_2$ and $(NH_4)_3PO_4$ (c) KI and $MgCl_2$

Answer   Precipitates will form in (a), $AgBr$, and in (b), $Ca_3(PO_4)_2$.

(a) $AgNO_3(aq) + NaBr(aq) \longrightarrow AgBr(s) + NaNO_3(aq)$
(b) $3\ CaCl_2(aq) + 2\ (NH_4)_3PO_4(aq) \longrightarrow Ca_3(PO_4)_2(s) + 6\ NH_4Cl(aq)$
(c) No reaction occurs; both possible products KCl, and MgI_2, are soluble.

Explanation   In each case, consider which cation-anion combinations can be formed and whether the possible compounds will precipitate.

(a)   An exchange reaction between $AgNO_3$ and $NaBr$ forms $AgBr$ and $NaNO_3$. Table 5.1 indicates that $NaNO_3$ is soluble (all nitrates are soluble) and remains dissolved in solution, but $AgBr$ is not soluble and therefore precipitates.

$$AgNO_3(aq) + NaBr(aq) \longrightarrow AgBr(s) + NaNO_3(aq)$$

(b)   The exchange of Ca^{2+} for NH_4^+ forms insoluble $Ca_3(PO_4)_2$, leaving soluble NH_4Cl in solution.

$$3\ CaCl_2(aq) + 2\ (NH_4)_3PO_4(aq) \longrightarrow 6\ NH_4Cl(aq) + Ca_3(PO_4)_2(s)$$

(c)   No precipitate forms because each possible product, KCl and MgI_2, is soluble. Thus, all four of the ions (K^+, I^-, Mg^{2+}, and Cl^-) remain in the solution. No exchange reaction is possible because no product is formed that removes ions from the solution.

Problem-Solving Practice 5.2

Predict the products and write a balanced chemical equation for the exchange reaction in aqueous solution between the following ionic compounds. Use Table 5.1 to determine solubilities and indicate in the equation whether a precipitate forms.

(a) $NiCl_2$ and $NaOH$ (b) K_2CO_3 and $CaBr_2$

Net Ionic Equations

In writing equations for exchange reactions in the preceding section, we have used overall equations. But there is another way to represent what actually happens. In each case in which a precipitate formed, the product that did not precipitate remained in solution. Therefore, its ions were in solution as reactants and remained there after the reaction. Such ions are commonly called **spectator ions** because, like the spectators at a play or game, they are present but they are not involved in the real action. Consequently, the spectator ions can be left out of the equation that represents the chemical change that occurs. An equation that includes only the symbols or formulas of ions in solution or compounds that undergo change is called a **net ionic equation.** We will use the reaction of aqueous $NaCl$ with $AgNO_3$ to form $AgCl$ and $NaNO_3$ to illustrate the general steps for writing a net ionic equation.

CD-ROM Screen 4.10: Equations of Reactions in Aqueous Solution: Net Ionic Equations

Step 1: Write the overall balanced equation using the correct formulas for the reactants and products.

$$AgNO_3 + NaCl \longrightarrow AgCl + NaNO_3$$

silver nitrate   sodium chloride   silver chloride   sodium nitrate

Step 2: Use the general guidelines in Table 5.1 to determine the solubilities of reactants and products.   In this case, the guidelines indicate that nitrates are soluble, so $AgNO_3$ and $NaNO_3$ are soluble. NaCl is water-soluble because almost all chlorides are soluble. However, AgCl is one of the insoluble chlorides ($AgCl$, Hg_2Cl_2, and $PbCl_2$). Using this information we can write

$$AgNO_3(aq) + NaCl(aq) \longrightarrow AgCl(s) + NaNO_3(aq)$$

Step 3: *Recognize that all soluble ionic compounds dissociate into their component ions in aqueous solution.* Therefore we have

$$AgNO_3(aq) \longrightarrow Ag^+(aq) + NO_3^-(aq)$$

$$NaCl(aq) \longrightarrow Na^+(aq) + Cl^-(aq)$$

$$NaNO_3(aq) \longrightarrow Na^+(aq) + NO_3^-(aq)$$

Step 4: *Write a complete ionic equation with the ions in solution from each soluble compound shown separately.*

$$Ag^+(aq) + NO_3^-(aq) + Na^+(aq) + Cl^-(aq) \longrightarrow AgCl(s) + Na^+(aq) + NO_3^-(aq)$$

Note that the precipitate is represented by its complete formula.

Step 5: *Cancel out the spectator ions from each side of the complete ionic equation to create the net ionic equation.* Sodium ions and nitrate ions are the spectator ions in this example, and we cancel them from the complete ionic equation to give the net ionic equation (Figure 5.4).

Complete ionic equation:

$$Ag^+(aq) + \cancel{NO_3^-(aq)} + \cancel{Na^+(aq)} + Cl^-(aq) \longrightarrow$$
$$AgCl(s) + \cancel{Na^+(aq)} + \cancel{NO_3^-(aq)}$$

Net ionic equation:

$$Ag^+(aq) + Cl^-(aq) \longrightarrow AgCl(s)$$

Step 6: *Check to see that the sum of the charges is the same on each side of the net ionic equation.* For the equation in Step 5 the sum of

If the charge is not the same on both sides of a balanced equation, then electrons are being created or destroyed, which is impossible according to the law of conservation of mass.

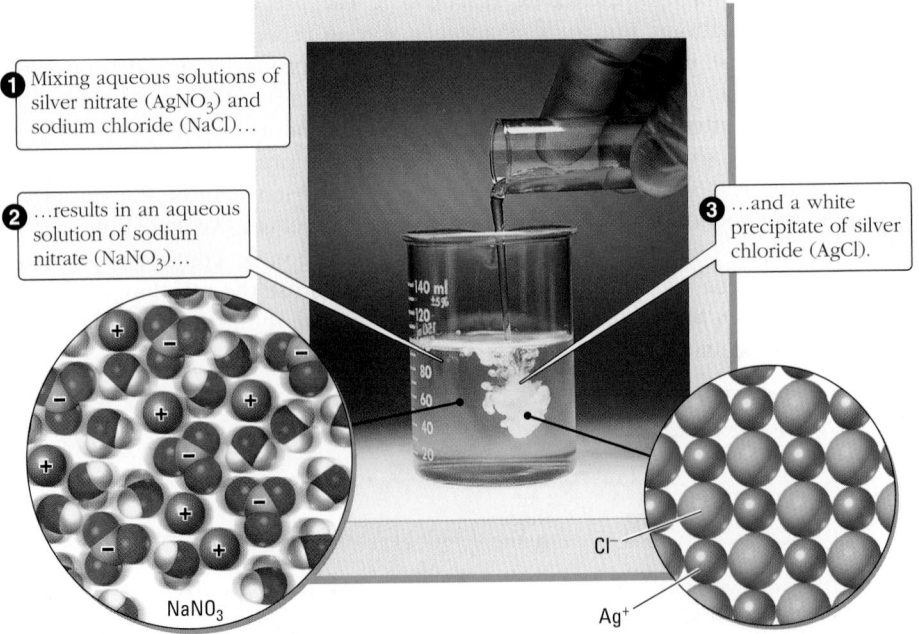

① Mixing aqueous solutions of silver nitrate ($AgNO_3$) and sodium chloride (NaCl)...

② ...results in an aqueous solution of sodium nitrate ($NaNO_3$)...

③ ...and a white precipitate of silver chloride (AgCl).

$NaNO_3$

Cl^-

Ag^+

Figure 5.4 Precipitation of silver chloride.

charges is zero on each side $(1+) + (1-) = 0$ on the left; AgCl is an ionic compound with no net charge on the right.

Problem-Solving Example 5.3   Net Ionic Equations

Write the net ionic equation for the reaction that occurs when aqueous solutions of magnesium sulfate ($MgSO_4$) and sodium oxalate ($Na_2C_2O_4$) are mixed.

Answer $Mg^{2+}(aq) + C_2O_4^{2-}(aq) \longrightarrow MgC_2O_4(s)$

Explanation

Step 1: *Write the overall balanced equation.* This is an exchange reaction.

$$MgSO_4 + Na_2C_2O_4 \longrightarrow MgC_2O_4 + Na_2SO_4$$

Step 2: *Determine the solubilities of reactants and products.* The solubility rules in Table 5.1 predict that all the reactants and products, except MgC_2O_4, are soluble.

$$MgSO_4(aq) + Na_2C_2O_4(aq) \longrightarrow MgC_2O_4(s) + Na_2SO_4(aq)$$

Step 3: *Write equations for the soluble compounds that dissociate in solution.*

$$MgSO_4(aq) \longrightarrow Mg^{2+}(aq) + SO_4^{2-}(aq)$$
$$Na_2C_2O_4(aq) \longrightarrow 2\,Na^+(aq) + C_2O_4^{2-}(aq)$$
$$Na_2SO_4(aq) \longrightarrow 2\,Na^+(aq) + SO_4^{2-}(aq)$$

Step 4: *Write the complete ionic equation.*

$$Mg^{2+}(aq) + SO_4^{2-}(aq) + 2\,Na^+(aq) + C_2O_4^{2-}(aq) \longrightarrow$$
$$MgC_2O_4(s) + 2\,Na^+(aq) + SO_4^{2-}(aq)$$

Steps 5 and 6: *Cancel spectator ions; check to see that charge is balanced.*

Net ionic equation: $Mg^{2+}(aq) + C_2O_4^{2-}(aq) \longrightarrow MgC_2O_4(s)$

Net charge $= (2+) + (2-) = 0$

Problem-Solving Practice 5.3

Write a balanced equation for the reaction (if any) for each of the following ionic compound pairs in aqueous solution. Then write their balanced net ionic equations.

(a) NaF and $Ca(CH_3COO)_2$ (CaF_2 is insoluble)   (b) $(NH_4)_2S$ and $FeCl_2$

Exercise 5.1   Net Ionic Equations

It is possible for an exchange reaction in which both products precipitate to occur in aqueous solution. Using Table 5.1, identify the reactants and products of an example of such a reaction.

If you live in an area with "hard water" you have probably noticed the scale that forms inside your teakettle or saucepans when you boil water in them. Hard water is mostly caused by the presence of the cations Ca^{2+}, Mg^{2+}, and also Fe^{2+} or Fe^{3+}. When the water also contains bicarbonate ion (HCO_3^-) the following reaction occurs upon heating:

$$2\,HCO_3^-(aq) \longrightarrow H_2O(\ell) + CO_2(g) + CO_3^{2-}(aq)$$

The carbon dioxide escapes from the hot water, and the bicarbonate is slowly converted to the carbonate. The carbonate ions can form precipitates with the calcium, magnesium, or iron ions to produce a scale that sticks to metal surfaces. In hot water heating systems in areas with high calcium ion concentrations, the buildup of such *boiler scale* can plug up the pipes.

$$Ca^{2+}(aq) + 2\,HCO_3^-(aq) \longrightarrow CaCO_3(s) + H_2O(\ell) + CO_2(g)$$

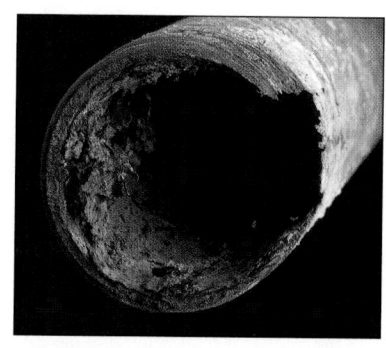

Boiler scale can form inside hot water pipes.

5.2 ACIDS, BASES, AND ACID-BASE EXCHANGE REACTIONS

Acids and bases are two important classes of compounds, so important that we include two chapters about them later (Chapters 16 and 17). Here, we focus on a few of their general properties and how acids and bases react with each other. Acids have a number of properties in common, and so do bases. Some properties of acids are related to properties of bases. Solutions of acids change the colors of pigments in specific ways; for example, all acids change the color of litmus from blue to red and cause the dye phenolphthalein to be colorless, but bases turn red litmus blue and make phenolphthalein pink. If an acid has made litmus red, adding a base will reverse the effect, making the litmus blue again. Thus, acids and bases seem to be opposites. A base can *neutralize* the effect of an acid, and an acid can neutralize the effect of a base.

Litmus is a dye derived from certain lichens. Phenolphthalein is a synthetic dye.

Acids have other characteristic properties. They taste sour, they produce bubbles of gas when reacting with limestone, and they dissolve many metals while producing a flammable gas. Although you should never taste substances in a chemistry laboratory, you have probably experienced the sour taste of at least one acid — vinegar, which is a solution of acetic acid in water. Bases, in contrast, have a bitter taste. Soap, for example, contains a base. Rather than dissolving metals, bases often cause metal ions to form insoluble compounds that precipitate from solution. Such precipitates can be made to dissolve by adding an acid, another case in which an acid counteracts a property of a base.

Acids

 CD-ROM Screen 4.7: Acids

The properties of acids can be explained by a common feature of acid molecules. An **acid** is any substance that, dissolved in water, increases the concentration of hydrogen ions, H^+. A "naked" H^+ ion, however, can't exist in water. Because it is just a proton, H^+ is the smallest possible ion and is strongly attracted to any negative charge in the vicinity. The oxygen part of a water molecule has a slight negative charge, so H^+ and H_2O combine to form H_3O^+, known as the **hydronium ion.** In Chapter 16 the importance of the hydronium ion to acid-base chemistry is explored. For now, we represent hydronium ion as $H^+(aq)$. The properties that acids have in common are those of hydrogen ions dissolved in water.

$$H^+ + H_2O \longrightarrow H_3O^+$$

Acids that are entirely converted to ions when dissolved in water are termed *strong electrolytes* (⇐ *p. 97*) and **strong acids.** One of the most common strong acids is hydrochloric acid, which ionizes completely in aqueous solution to form hydrogen ions and chloride ions (Figure 5.5a).

$$HCl(aq) \longrightarrow H^+(aq) + Cl^-(aq)$$

The more complete, and proper, way to write the reaction is

$$HCl(aq) + H_2O(\ell) \longrightarrow H_3O^+(aq) + Cl^-(aq)$$

which explicitly shows the hydronium ion, H_3O^+. Table 5.2 lists some other common acids.

In contrast, acids and other substances that form ions only to a limited extent are termed **weak electrolytes.** Acids that are only partially ionized in aqueous solution are termed **weak acids.** For example, when acetic acid dissolves in water,

 CD-ROM Screen 4.5: Compounds in Aqueous Solution

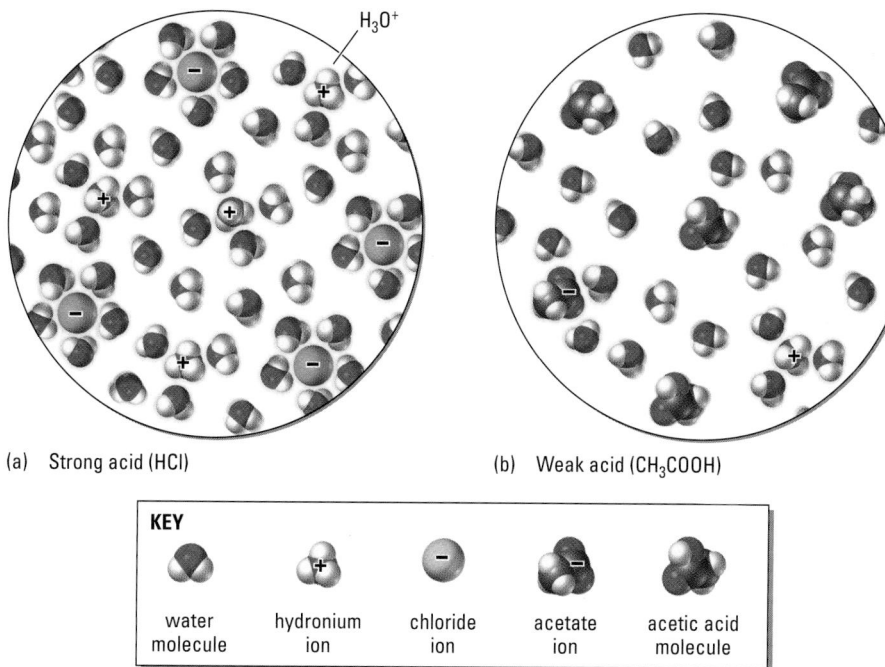

(a)   Strong acid (HCl)   (b)   Weak acid (CH₃COOH)

KEY

water molecule | hydronium ion | chloride ion | acetate ion | acetic acid molecule

Figure 5.5   The ionization of acids in water.   (a) A strong acid such as hydrochloric acid (HCl) is completely ionized in water; all the HCl molecules ionize to form $H_3O^+(aq)$ and $Cl^-(aq)$ ions. (b) Weak acids such as acetic acid (CH_3COOH) are only slightly ionized in water. Nonionized acetic acid molecules far outnumber aqueous H_3O^+ and CH_3COO^- ions formed by the ionization of acetic acid molecules.

usually less than 5% of the molecules are ionized at any time. The remainder of the acetic acid exists as nonionized molecules. Thus, because acetic acid is only partially ionized in aqueous solution, it is a weak electrolyte and classified as a weak acid (Figure 5.5b).

TABLE 5.2 Common Acids and Bases			
Strong acids (strong electrolytes)		**Strong bases (strong electrolytes)**	
HCl	Hydrochloric acid	LiOH	Lithium hydroxide
HNO_3	Nitric acid	NaOH	Sodium hydroxide
H_2SO_4	Sulfuric acid	KOH	Potassium hydroxide
$HClO_4$	Perchloric acid	$Ca(OH)_2$	Calcium hydroxide
HBr	Hydrobromic acid	$Ba(OH)_2$	Barium hydroxide
HI	Hydroiodic acid	$Sr(OH)_2$	Strontium hydroxide
Weak acids* (weak electrolytes)		**Weak bases† (weak electrolytes)**	
H_3PO_4	Phosphoric acid	NH_3	Ammonia
CH_3COOH	Acetic acid	CH_3NH_2	Methylamine
H_2CO_3	Carbonic acid		
HCN	Hydrocyanic acid		
HCOOH	Formic acid		
C_6H_5COOH	Benzoic acid		

* Many organic carboxylic acids are weak acids.

† Many organic amines (related to ammonia) are weak bases.

(a)

(b)

(c)

Acids and bases.   (a) Many common foods and household products are acidic or basic. Citrus fruits contain citric acid, and household ammonia and oven cleaner are basic. (b) The acid in lemon juice turns blue litmus paper red, whereas (c) household ammonia turns red litmus paper blue.

The organic functional group —COOH is present in all organic acids (Section 16.2). You saw it previously as part of the fatty acids that make up fats and oils (⇐ *p.110*).

$$CH_3COOH(aq) \rightleftharpoons H^+(aq) + CH_3COO^-(aq)$$

acetic acid   acetate ion

The double arrow in this equation for the ionization of acetic acid signifies a characteristic property of weak electrolytes. They establish a *dynamic equilibrium* in solution between the formation of the ions and their recombination. In acetic acid, hydrogen ions and acetate ions recombine to form CH_3COOH molecules.

Some common acids, such as sulfuric acid, can provide more than 1 mol of H^+ ions per mole of acid:

$$H_2SO_4(aq) \longrightarrow H^+(aq) + HSO_4^-(aq)$$

sulfuric acid   hydrogen sulfate ion

$$HSO_4^-(aq) \rightleftharpoons H^+(aq) + SO_4^{2-}(aq)$$

hydrogen sulfate ion   sulfate ion

The first ionization reaction is essentially complete, so sulfuric acid is considered a strong electrolyte (and a strong acid as well). However, the hydrogen sulfate ion, like acetic acid, is only partially ionized, so it is a weak electrolyte and also a weak acid.

 Exercise 5.2   Dissociation of Acids

Phosphoric acid, H_3PO_4, has three protons that can ionize. Write the equation for its three ionization reactions, each of which is a dynamic equilibrium.

Bases

 CD-ROM Screen 4.8: Bases

A **base** is a substance that increases the concentration of the **hydroxide ion, OH⁻**, when dissolved in pure water. The properties that bases have in common are the properties of $OH^-(aq)$. Compounds that contain hydroxide ions, such as sodium hydroxide or potassium hydroxide, are obvious bases. As ionic compounds they are strong electrolytes and **strong bases.**

$$NaOH(s) \xrightarrow{H_2O} Na^+(aq) + OH^-(aq)$$

A base that is not very water-soluble, such as $Ca(OH)_2$, can still be a strong electrolyte if the portion of the compound that does dissolve then completely dissociates into ions.

Ammonia, NH_3, is another very common base. Although the compound does not have an OH^- ion as part of its formula, it produces the ion by reaction with water.

$$NH_3(aq) + H_2O(\ell) \rightleftharpoons NH_4^+(aq) + OH^-(aq)$$

Acids and bases that are strong electrolytes are strong acids and bases. Acids and bases that are weak electrolytes are weak acids and bases.

In the equilibrium between NH_3 and the NH_4^+ and OH^- ions, only a small concentration of the ions is present, so ammonia is a weak electrolyte (<5% ionized) and a **weak base.**

Exercise 5.3   Acids and Bases

(a) What ions are produced when perchloric acid dissolves in water? (b) Calcium hydroxide is not very soluble in water. What little does dissolve, however, is dissociated. What ions are produced? Write an equation for the dissociation of calcium hydroxide.

Problem-Solving Example 5.4   Strong, Weak, and Nonelectrolytes

Identify whether each of the following dissolved substances is a strong electrolyte, a weak electrolyte, or a nonelectrolyte: HNO_3 (nitric acid); KOH (potassium hydroxide), $C_6H_{12}O_6$ (glucose), CH_3COOH (acetic acid).

Answer HNO_3 is a strong electrolyte; KOH is a strong electrolyte; $C_6H_{12}O_6$ is a non-electrolyte; CH_3COOH is a weak electrolyte.

Explanation   Nitric acid is a common strong acid and therefore is a strong electrolyte. Potassium hydroxide is a strong base and is completely dissociated into ions in aqueous solution, so it is a strong electrolyte. $C_6H_{12}O_6$ (glucose) is a molecular compound that does not dissociate into ions in aqueous solution, so it is a nonelectrolyte. Acetic acid, CH_3COOH is a weak acid because it is only about 5% dissociated in aqueous solution.

Problem-Solving Practice 5.4

Look back through the discussion of electrolytes and Table 5.2 and identify at least one additional strong electrolyte, one additional weak electrolyte, and one additional non-electrolyte beyond those discussed in Problem-Solving Example 5.4.

To summarize:

- Strong electrolytes are either ionic compounds or molecular compounds that are strong acids and ionize completely in aqueous solution.
- Weak electrolytes are molecular compounds that are weak acids or bases and establish equilibrium with water.
- Nonelectrolytes are molecular compounds that do not ionize in aqueous solution.

Neutralization Reactions

When solutions of a strong acid (HCl) and a strong base (NaOH) are mixed, the ions in solution are the hydrogen ion and the anion from the acid, the metal cation, and the hydroxide ion from the base:

CD-ROM Screen 4.13: Acid-Base Reactions

$$\text{From hydrochloric acid:}\quad H^+(aq),\ Cl^-(aq)$$

$$\text{From sodium hydroxide:}\quad Na^+(aq),\ OH^-(aq)$$

As in precipitation reactions, an exchange reaction will occur if two of these ions can react with each other to form a compound that removes ions from solution. In an acid-base reaction, that compound is water formed by the reaction of $H^+(aq)$ with $OH^-(aq)$.

When a strong acid and a strong base react, they neutralize each other. This happens because the hydrogen ions from the acid react with hydroxide ions from the base to form water. Water, the product, is a molecular compound that is neither acidic nor basic; it is neutral. The other ions remain in the solution, which is an aqueous solution of a **salt** (an ionic compound whose cation comes from a base and whose anion comes from an acid). If the water were evaporated the solid salt would remain. In the case of HCl plus NaOH, the salt is sodium chloride (NaCl).

$$\underset{\text{acid}}{HCl\,(aq)} + \underset{\text{base}}{NaOH\,(aq)} \longrightarrow \underset{\text{water}}{HOH\,(\ell)} + \underset{\text{salt}}{NaCl\,(aq)}$$

The overall neutralization reaction can be written in general as

$$\underset{\text{acid}}{HX\,(aq)} + \underset{\text{base}}{MOH\,(aq)} \longrightarrow \underset{\text{water}}{HOH\,(\ell)} + \underset{\text{salt}}{MX\,(aq)}$$

You should recognize this as an exchange reaction in which $H^+(aq)$ ions from the aqueous acid and $M^+(aq)$ ions from the metal hydroxide exchange partners.

The particular salt that forms depends on the acid and base used. Magnesium chloride, another salt, is formed when a commercial antacid containing magnesium hydroxide is swallowed to neutralize excess hydrochloric acid in the stomach.

Milk of magnesia consists of a suspension of finely divided particles of $Mg(OH)_2(s)$ in water.

$$2\ HCl(aq)\ +\ Mg(OH)_2(s)\ \longrightarrow\ 2\ H_2O(\ell)\ +\ MgCl_2(aq)$$

hydrochloric   magnesium   magnesium
acid   hydroxide   chloride

Organic acids, such as acetic acid and propionic acid, which contain the acid functional group —COOH, also neutralize bases to form salts. The H in the —COOH functional group is the acidic proton that is removed to generate the —COO$^-$ anion. The reaction of propionic acid, CH_3CH_2COOH, and sodium hydroxide produces the salt sodium propionate, $NaCH_3CH_2COO$, containing sodium ions (Na^+) and propionate ions ($CH_3CH_2COO^-$).

$$CH_3CH_2COOH(aq)\ +\ NaOH(aq)\ \longrightarrow\ H_2O(\ell)\ +\ NaCH_3CH_2COO(aq)$$

propionic acid   sodium propionate

Although the propionic acid molecule contains a number of H atoms, it is the H atom that is part of the acid functional group that is involved in this neutralization reaction. Sodium propionate is commonly used as a food preservative.

Problem-Solving Example 5.5   Balancing Neutralization Equations

Write a balanced chemical equation for the reaction of sulfuric acid, H_2SO_4, with iron(III) hydroxide, $Fe(OH)_3$, in aqueous solution.

Answer $3\ H_2SO_4(aq)\ +\ 2\ Fe(OH)_3(aq)\ \rightarrow\ Fe_2(SO_4)_3(aq)\ +\ 6\ H_2O(\ell)$

Explanation   This is a neutralization reaction between an acid and a base, so the products are a salt and water. We start by writing the unbalanced equation with all the species.

(unbalanced equation) $H_2SO_4(aq)\ +\ Fe(OH)_3(aq)\ \longrightarrow\ Fe_2(SO_4)_3(aq)\ +\ H_2O(\ell)$

It is generally a good idea to start with the ions and later balance the hydrogen and oxygen atoms. Two iron ions appear in the products, so at least two iron(III) hydroxides will be needed as reactants.

(unbalanced equation) $H_2SO_4(aq)\ +\ 2\ Fe(OH)_3(aq)\ \longrightarrow\ Fe_2(SO_4)_3(aq)\ +\ H_2O(\ell)$

The stoichiometric coefficient of 1 for the iron(III) sulfate is understood. Now we notice that at least three SO_4^{2-} ions will be needed for the one iron(III) sulfate in the products, so we use a 3 for the sulfuric acid coefficient.

(unbalanced equation) $3\ H_2SO_4(aq)\ +\ 2\ Fe(OH)_3(aq)\ \longrightarrow\ Fe_2(SO_4)_3(aq)\ +\ H_2O(\ell)$

Three of the four stoichiometric coefficients are determined. We must still find the correct coefficient for water. We count twelve hydrogen atoms in the reactants (6 from sulfuric acid and 6 from iron(III) hydroxide), so we must put a coefficient of 6 in front of the water.

$$3\ H_2SO_4(aq)\ +\ 2\ Fe(OH)_3(aq)\ \longrightarrow\ Fe_2(SO_4)_3(aq)\ +\ 6\ H_2O(\ell)$$

The equation is balanced.

Problem-Solving Practice 5.5

Write a balanced equation for the reaction of phosphoric acid, H_3PO_4, with sodium hydroxide, NaOH.

Problem-Solving Example 5.6   Acids, Bases, and Salts

Identify the base and acid that could be used to form each of the following salts: (a) $NaNO_3$, (b) Cs_3PO_4. Write the equations for the formation of these compounds.

Answer   (a) Sodium hydroxide ($NaOH$) and nitric acid (HNO_3), (b) cesium hydroxide ($CsOH$) and phosphoric acid (H_3PO_4).

Explanation   (a) A salt is formed from the cation of a base and the anion of an acid. $NaNO_3$ contains sodium and nitrate ions; Na^+ is derived from the base $NaOH$, and NO_3^- comes from nitric acid, HNO_3. The reaction of $NaOH$ and HNO_3 produces sodium nitrate (and water).

$$NaOH(aq) + HNO_3(aq) \longrightarrow NaNO_3(aq) + H_2O(\ell)$$

(b) Cesium phosphate is formed from cesium ions (Cs^+) and phosphate ions (PO_4^{3-}) that come from cesium hydroxide ($CsOH$) and phosphoric acid (H_3PO_4), respectively.

$$3\,CsOH(aq) + H_3PO_4(aq) \longrightarrow Cs_3PO_4(aq) + 3\,H_2O(\ell)$$

Problem-Solving Practice 5.6

Identify the acid and base that can react to form (a) $MgSO_4$ and (b) $SrCO_3$.

Net Ionic Equations for Acid-Base Reactions

As with precipitation reactions, net ionic equations can be written for acid-base reactions. This should not be surprising because precipitation and acid-base neutralization reactions are both exchange reactions.

Consider the reaction given earlier of magnesium hydroxide with hydrochloric acid to relieve excess stomach acid (HCl). The overall balanced equation is

$$2\,HCl(aq) + Mg(OH)_2(s) \longrightarrow 2\,H_2O(\ell) + MgCl_2(aq)$$

The acid and base furnish hydrogen ions and hydroxide ions, respectively.

$$2\,HCl(aq) \longrightarrow 2\,H^+(aq) + 2\,Cl^-(aq)$$

$$Mg(OH)_2(s) \longrightarrow Mg^{2+}(aq) + 2\,OH^-(aq)$$

Note that we retain the coefficients from the balanced overall equation (first step).

We now use this information to write a complete ionic equation. We use Table 5.1 to check the solubility of the product salt, $MgCl_2$. Magnesium chloride is soluble, so the Mg^{2+} and Cl^- ions remain in solution. The complete ionic equation is

$$Mg^{2+}(aq) + 2\,OH^-(aq) + 2\,H^+(aq) + 2\,\cancel{Cl^-(aq)} \longrightarrow$$
$$\cancel{Mg^{2+}(aq)} + 2\,\cancel{Cl^-(aq)} + 2\,H_2O(\ell)$$

Canceling spectator ions from each side of the complete ionic equation yields the net ionic equation. In this case magnesium ions and chloride ions are the spectator ions, and canceling them leaves us with the net ionic equation

$$2\,H^+(aq) + 2\,OH^-(aq) \longrightarrow 2\,H_2O(\ell)$$

or simply

$$H^+(aq) + OH^-(aq) \longrightarrow H_2O(\ell)$$

This is the net ionic equation for the neutralization reaction between a strong acid and a strong base that yields a soluble salt. Note that, as always, there is *conservation of charge* in the net ionic equation. On the left, $(1+) + (1-) = 0$; on the right, water has no net charge.

Although magnesium hydroxide is not very soluble, the little that dissolves is completely dissociated.

Next, consider a neutralization reaction between a weak acid, HCN, and a strong base, KOH.

$$HCN(aq) + KOH(aq) \longrightarrow KCN(aq) + H_2O(\ell)$$

The weak acid is not completely ionized, so we leave it in the molecular form, but KOH and KCN are strong electrolytes, so the complete ionic equation is

$$HCN(aq) + K^+(aq) + OH^-(aq) \longrightarrow K^+(aq) + CN^-(aq) + H_2O(\ell)$$

Canceling spectator ions yields

$$HCN(aq) + OH^-(aq) \longrightarrow CN^-(aq) + H_2O(\ell)$$

The net ionic equation for the neutralization of a weak acid with a strong base contains the molecular form of the acid and the anion of the salt. The net ionic equation shows that charge is conserved.

Problem-Solving Example 5.7 Neutralization Reaction with a Weak Acid

Write a balanced equation for the reaction of carbonic acid, H_2CO_3, with sodium hydroxide, NaOH. Then write the net ionic equation for this neutralization reaction.

Answer

Equation: $H_2CO_3(aq) + 2\,NaOH(aq) \rightarrow 2\,H_2O(\ell) + Na_2CO_3(aq)$

Net ionic equation: $H_2CO_3(aq) + 2\,OH^- \rightarrow 2\,H_2O(\ell) + CO_3^{2-}(aq)$

Explanation The information given is the formula of an acid and a base that will react. The products of the neutralization reaction will be water and the salt, Na_2CO_3, formed from the base's cation and the acid's anion. Table 5.1 shows that Na_2CO_3 is soluble.

We start with the unbalanced chemical equation

(unbalanced equation) $H_2CO_3(aq) + NaOH(aq) \longrightarrow Na_2CO_3(aq) + H_2O(\ell)$

To balance the equation, two hydroxide ions (OH^-) must be supplied to react with the two protons (H^+) from the carbonic acid. This also means that two water molecules will be produced. The balanced equation is

$$H_2CO_3(aq) + 2\,NaOH(aq) \longrightarrow Na_2CO_3(aq) + 2\,H_2O(\ell)$$

To write the net ionic equation we must know whether the three species involved in the reaction are strong or weak electrolytes. Carbonic acid is a weak electrolyte (it is a weak acid); NaOH is a strong electrolyte (strong base); sodium carbonate is a strong electrolyte (Table 5.1 shows this). We write the complete ionic equation

$$H_2CO_3(aq) + 2\,\cancel{Na^+(aq)} + 2\,OH^-(aq) \longrightarrow 2\,\cancel{Na^+(aq)} + CO_3^{2-}(aq) + 2\,H_2O(\ell)$$

The sodium ions are spectators and can be canceled to give the net ionic equation

$$H_2CO_3(aq) + 2\,OH^-(aq) \longrightarrow CO_3^{2-}(aq) + 2\,H_2O(\ell)$$

✔ As a check, notice that each side of the equation has the same charge (2−) and the same numbers and types of atoms.

Problem-Solving Practice 5.7

Write a balanced equation for the reaction of hydrocyanic acid, HCN, with calcium hydroxide, $Ca(OH)_2$. Then write the net ionic equation for this neutralization reaction.

Exercise 5.4 Neutralizations and Net Ionic Equations

Write balanced complete ionic equations and net ionic equations for the neutralization reactions of the following acids and bases: (a) HCl and KOH, (b) H_2SO_4 and $Ba(OH)_2$ (remember that sulfuric acid can provide 2 mol of $H^+(aq)$ per 1 mol of sulfuric acid), (c) CH_3COOH and $Ca(OH)_2$

Exercise 5.5   Net Ionic Equations and Antacids

The commercial antacids Maalox, Di-Gel tablets, and Mylanta contain aluminum or magnesium hydroxide that reacts with excess hydrochloric acid in the stomach. Write balanced complete ionic equations and net ionic equations for the soothing neutralization reactions of aluminum hydroxide with HCl.

Gas-Forming Exchange Reactions

The formation of a gas is the third factor causing exchange reactions to occur. Escape of the gas drives the reaction to form products from the reactants. Acids are involved in many gas-forming exchange reactions.

The reaction of metal carbonates with acids is an excellent example of a gas-forming exchange reaction (Figure 5.6).

$$CaCO_3(s) + 2\ HCl(aq) \longrightarrow CaCl_2(aq) + H_2CO_3(aq)$$

$$H_2CO_3(aq) \longrightarrow H_2O(\ell) + CO_2(g)$$

A salt and H_2CO_3 (carbonic acid) are *always* the products from an acid reacting with a metal carbonate, and their formation illustrates the exchange reaction pattern. Carbonic acid is unstable, however, and much of it is rapidly converted to water and CO_2 gas. If the reaction is done in an open container, most of the gas bubbles out of the solution.

Carbonates (which contain CO_3^{2-}) and hydrogen carbonates (which contain HCO_3^- and are sometimes called bicarbonates) are basic because they consume protons in neutralization reactions. Carbon dioxide is always released when acids react with a metal carbonate or a metal hydrogen carbonate. For example, excess hydrochloric acid in the stomach is neutralized by ingesting commercial antacids such as Alka-Seltzer ($NaHCO_3$), Tums ($CaCO_3$), or Di-Gel liquid ($MgCO_3$). Taking an Alka-Seltzer or a Tums to relieve excess stomach acid produces the following helpful reactions.

Alka-Seltzer: $NaHCO_3(aq) + HCl(aq) \longrightarrow NaCl(aq) + H_2O(\ell) + CO_2(g)$

Tums: $CaCO_3(aq) + 2\ HCl(aq) \longrightarrow CaCl_2(aq) + H_2O(\ell) + CO_2(g)$

The net ionic equations for these two reactions are

$$HCO_3^-(aq) + H^+(aq) \longrightarrow H_2O(\ell) + CO_2(g)$$

$$CO_3^{2-}(aq) + 2\ H^+(aq) \longrightarrow H_2O(\ell) + CO_2(g)$$

Acids also react by exchange reactions with metal sulfites or sulfides to produce foul-smelling gaseous SO_2 or H_2S, respectively. With sulfites, the initial product is sulfurous acid, which, like carbonic acid, quickly decomposes.

$$CaSO_3(aq) + 2\ HCl(aq) \longrightarrow CaCl_2(aq) + H_2SO_3(aq)$$

$$H_2SO_3(aq) \longrightarrow H_2O(\ell) + SO_2(g)$$

$$Na_2S(aq) + 2\ HCl(aq) \longrightarrow 2\ NaCl(aq) + H_2S(g)$$

Exercise 5.6   Gas-Forming Reactions

Predict the products and write the balanced overall equation and the net ionic equation for each of the following gas-generating reactions:
(a) $Na_2CO_3(aq) + H_2SO_4(aq) \rightarrow$
(b) $FeS(s) + HCl(aq) \rightarrow$
(c) $K_2SO_3(aq) + HCl(aq) \rightarrow$

Figure 5.6   Reaction of calcium carbonate with an acid.   A piece of coral that is largely calcium carbonate, $CaCO_3$, reacts readily with hydrochloric acid to give CO_2 gas and aqueous calcium chloride.

CD-ROM Screen 4.14: Gas-Forming Reactions

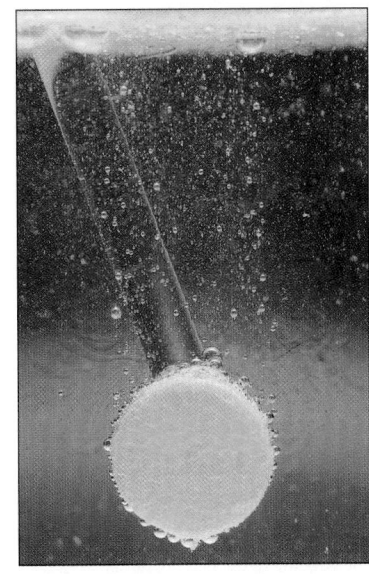

Antacid reacting with HCl.

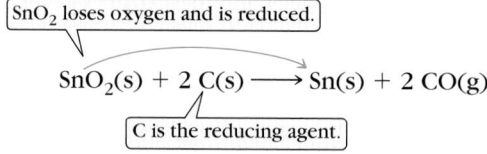

Exercise 5.7 Exchange Reaction Classification

Identify each of the following exchange reactions as a precipitation reaction, an acid-base reaction, or a gas-forming reaction. Predict the products of each reaction and write an overall balanced equation and net ionic equation for the reaction.
(a) $NiCO_3(s) + H_2SO_4(aq) \longrightarrow$
(b) $Sr(OH)_2(s) + HNO_3(aq) \longrightarrow$
(c) $BaCl_2(aq) + Na_2C_2O_4(aq) \longrightarrow$
(d) $PbCO_3(s) + H_2SO_4(aq) \longrightarrow$

5.3 OXIDATION-REDUCTION REACTIONS

CD-ROM Screen 4.15:
Oxidation-Reduction Reactions

Now we turn to oxidation-reduction reactions, which are classified not by any specific reaction pattern, but by what happens with electrons at the nanoscale level as a result of the reaction. Many of the reaction types already considered are oxidation-reduction reactions as well.

The terms "oxidation" and "reduction" come from reactions that have been known for centuries. Ancient civilizations learned how to change metal oxides and sulfides to the metal — that is, how to *reduce* ore to the metal. For example, cassiterite or tin(IV) oxide, SnO_2, is a tin ore discovered in Britain centuries ago. It is very easily reduced to tin by heating with carbon.

$$\boxed{SnO_2 \text{ loses oxygen and is reduced.}}$$
$$SnO_2(s) + 2\ C(s) \longrightarrow Sn(s) + 2\ CO(g)$$
$$\boxed{C \text{ is the reducing agent.}}$$

In this reaction, carbon brings about the reduction of tin ore to tin metal, so carbon is called the **reducing agent.**

When SnO_2 is reduced by carbon, oxygen is removed from the tin and added to the carbon, which is "oxidized" by the addition of oxygen. In fact, any process in which oxygen is added to another substance is an oxidation.

When magnesium burns in air, oxygen is the **oxidizing agent** because it is the agent that is responsible for the oxidation.

$$\boxed{Mg \text{ combines with oxygen and is oxidized.}}$$
$$2\ Mg(s) + O_2(g) \longrightarrow 2\ MgO(s)$$
$$\boxed{O_2 \text{ is the oxidizing agent.}}$$

The experimental observations we have just outlined point to several fundamental conclusions:

- If one substance is oxidized, another substance in the same reaction must simultaneously be reduced. For this reason, we refer to such reactions as **oxidation-reduction reactions,** or **redox reactions** for short.
- Oxidation is the reverse of reduction.
- The reducing agent is itself oxidized, and the oxidizing agent is reduced.

For example, the reactions we have just described show that addition of oxygen is oxidation and removal of oxygen is reduction.

Figure 5.7 Oxidation of copper metal by silver ion. Copper wire was formed into a "tree" by wrapping short lengths around a central "trunk." The tree was immersed in a solution of silver nitrate, $AgNO_3$. With time, the copper reduces Ag^+ ions to silver metal crystals, and the copper metal is oxidized to Cu^{2+} ions. The blue color of the solution is due to the presence of aqueous copper(II) ion.

All redox reactions contain both oxidizing and reducing agents as reactants.

Redox Reactions and Electron Transfer

Oxidation and reduction reactions involve transfer of electrons from one reactant to another. When a substance *accepts electrons,* it is said to be **reduced.** The language is descriptive because in a **reduction** there is a decrease (reduction) in the real or apparent electric charge on an atom. For example, in the following net ionic equation, Ag^+ ions are reduced to uncharged Ag atoms by accepting electrons from copper atoms. Since copper atoms supply the electrons and cause the Ag^+ ions to be reduced, Cu is the reducing agent (Figure 5.7).

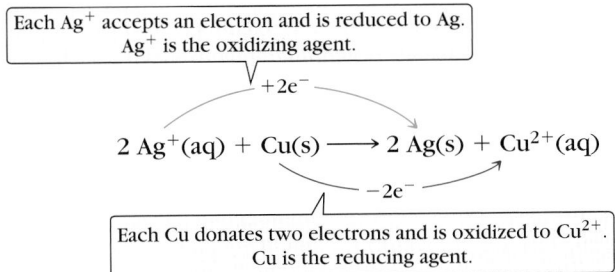

Each Ag^+ accepts an electron and is reduced to Ag. Ag^+ is the oxidizing agent.

$$+2e^-$$

$$2\,Ag^+(aq) + Cu(s) \longrightarrow 2\,Ag(s) + Cu^{2+}(aq)$$

$$-2e^-$$

Each Cu donates two electrons and is oxidized to Cu^{2+}. Cu is the reducing agent.

When a substance *loses electrons,* it is said to be **oxidized.** In **oxidation,** the real or apparent electric charge on an atom of the substance increases when it gives up electrons. In our example, copper metal releases electrons on forming Cu^{2+}; its electric charge has increased, and it is said to have been oxidized. In order for this to happen, there must be something available to take the electrons offered by the copper. In this case, Ag^+ is the electron acceptor, and therefore it is the oxidizing agent. *In every oxidation-reduction reaction, a reactant is reduced (and therefore is the oxidizing agent) and a reactant is oxidized (and therefore is the reducing agent).*

In the reaction of magnesium with oxygen (Figure 5.8), oxygen gains electrons when converted to the oxide ion. The charge of the O atoms changes from 0 to 2−.

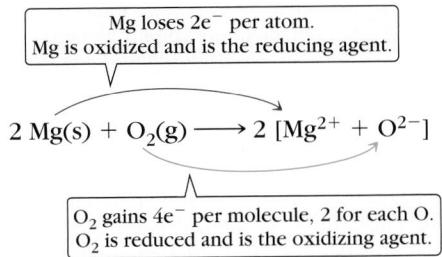

Mg loses $2e^-$ per atom. Mg is oxidized and is the reducing agent.

$$2\,Mg(s) + O_2(g) \longrightarrow 2\,[Mg^{2+} + O^{2-}]$$

O_2 gains $4e^-$ per molecule, 2 for each O. O_2 is reduced and is the oxidizing agent.

Magnesium is the reducing agent because it loses two electrons per atom to form the Mg^{2+} ion. All redox reactions can be analyzed in a similar manner.

Figure 5.9 provides some guidelines to determine which species involved in a redox reaction is the oxidizing agent and which is the reducing agent, and the relationships among the different ways of labeling the two species.

Common Oxidizing and Reducing Agents

Like oxygen, the halogens (F_2, Cl_2, Br_2, and I_2) are always oxidizing agents in their reactions with metals and most nonmetals. For example, consider the combination reaction of sodium metal with chlorine.

 CD-ROM Screen 4.16: Redox Reactions and Electron Transfer

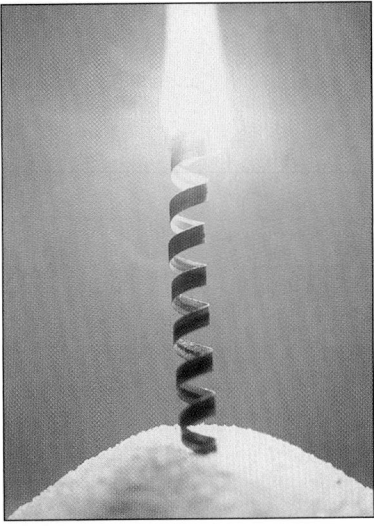

Figure 5.8 $Mg(s) + O_2(g)$. A piece of magnesium ribbon burns in air, oxidizing the metal to the white solid magnesium oxide, MgO.

Oxidation is the loss of electrons.

$$X \longrightarrow X^+ + e^-$$

X loses one or more electrons, is oxidized, and is the reducing agent (reduces something else).

Reduction is the gain of electrons.

$$Y + e^- \longrightarrow Y^-$$

Y gains one or more electrons, is reduced, and is the oxidizing agent (oxidizes something else).

A useful memory aid for keeping the oxidation and reduction definitions straight is OIL RIG (**O**xidation **i**s **l**oss; **R**eduction **i**s **g**ain).

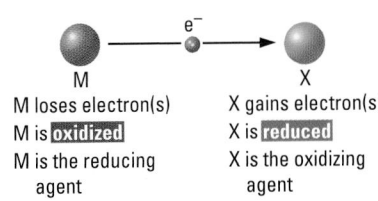

M loses electron(s)
M is `oxidized`
M is the reducing agent

X gains electron(s)
X is `reduced`
X is the oxidizing agent

Figure 5.9 **Oxidation-reduction relationships and electron transfer.**

 CD-ROM Screen 4.18: Recognizing Oxidation-Reduction Reactions

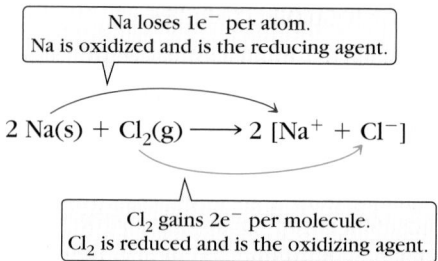

Here sodium begins as the metallic element, but it ends up as the Na^+ ion after combining with chlorine. Thus, sodium is oxidized and is the reducing agent. Chlorine ends up as Cl^-; it has been reduced and therefore is the oxidizing agent. The general reaction for halogen, X_2, reduction is

Reduction reaction: $X_2 + 2e^- \longrightarrow 2X^-$
                                      oxidizing
                                      agent

That is, a halogen will always oxidize a metal to give a metal halide, and the formula of the product can be predicted from the charge on the metal ion and the charge of the halide. The halogens in decreasing order of oxidizing ability are as follows:

Oxidizing agent	Usual reduction product
F_2	F^-
Cl_2	Cl^-
Br_2	Br^-
I_2	I^-

Exercise 5.8   Oxidizing and Reducing Agents

Identify which species is losing and which is gaining electrons, which is oxidized and which is reduced, and which is the oxidizing agent and which is the reducing agent in the reaction

$$2\,Ca(s) + O_2(g) \longrightarrow 2\,CaO(s)$$

Exercise 5.9   Redox Reactions

Write the chemical equation for chlorine gas undergoing a redox reaction with calcium metal. Which species is the oxidizing agent?

Chlorine is widely used as an oxidizing agent in water and sewage treatment. A common contaminant of water is hydrogen sulfide, H_2S, which gives a thoroughly unpleasant rotten egg odor to the water and may come from the decay of organic matter or from underground mineral deposits. Chlorine oxidizes H_2S to insoluble elemental sulfur, which is easily removed.

$$8\,Cl_2(g) + 8\,H_2S(aq) \longrightarrow S_8(s) + 16\,HCl(aq)$$

Oxidation and reduction occur readily when a strong oxidizing agent comes into contact with a strong reducing agent. Knowing the easily recognized oxidizing and reducing agents enables you to predict that a reaction will take place

TABLE 5.3	Common Oxidizing and Reducing Agents		
Oxidizing agent	Reaction product	Reducing agent	Reaction product
O_2 (oxygen)	O^{2-} (oxide ion)	H_2 (hydrogen)	H^+ (hydrogen ion) or H combined in H_2O
H_2O_2 (hydrogen peroxide)	$H_2O(\ell)$	C (carbon) used to reduce metal oxides	CO and CO_2
F_2, Cl_2, Br_2, or I_2 (halogens)	F^-, Cl^-, Br^-, or I^- (halide ions)	M, metals such as Na, K, Fe, or Al	M^{n+}, metal ions such as Na^+, K^+, Fe^{3+}, or Al^{3+}
HNO_3 (nitric acid)	Nitrogen oxides such as NO and NO_2		
$Cr_2O_7^{2-}$ (dichromate ion)	Cr^{3+} (chromium(III) ion), in acid solution		
MnO_4^- (permanganate ion)	Mn^{2+} (manganese(II) ion), in acid solution		

when they are combined and in some cases to predict what the products will be. Table 5.3 and the following points provide some guidelines.

• An element that has combined with oxygen has been oxidized. In the process the oxygen, O_2, is changed to the oxide ion, O^{2-} (as in a metal oxide), by adding electrons or is combined in a molecule such as CO_2 or H_2O (as occurs in the combustion reaction of a hydrocarbon). Therefore, the oxygen has been reduced. Since it must have taken on electrons, oxygen is the oxidizing agent.

• An element that has combined with a halogen has been oxidized. In the process the halogen, X_2, is changed to halide ions, X^-, by adding electrons, or it is combined in a molecule such as HCl. Therefore, the halogen has been reduced, and it is the oxidizing agent. Among the halogens, fluorine and chlorine are particularly strong oxidizing agents.

• If a metal combines with something, the metal has been oxidized. In the process, it has lost electrons, usually to form a positive ion.

Oxidation reaction: $\quad M \longrightarrow M^{n+} + ne^-$
$\qquad\qquad\qquad\qquad$ reducing
$\qquad\qquad\qquad\qquad$ agent

Therefore, the metal (an electron donor) has been oxidized and has functioned as a reducing agent. Most metals are reasonably good reducing agents, and metals such as sodium, magnesium, and aluminum from Groups 1A, 2A, and 3A are particularly good ones.

• Other common oxidizing and reducing agents are listed in Table 5.3, and some are described below. When one of these agents takes part in a reaction, it is reasonably certain that it is a redox reaction. (Nitric acid can be an exception. In addition to being a good oxidizing agent, it is also an acid and functions only as an acid in reactions such as the decomposition of a metal carbonate.)

Figure 5.10 illustrates the action of concentrated nitric acid, HNO_3, as an oxidizing agent. Here the acid oxidizes copper metal to give copper(II) nitrate, and

There are exceptions to the guideline that metals are always positively charged in compounds. However, you probably will not encounter these exceptions in introductory chemistry.

Figure 5.10 $\quad Cu(s) + HNO_3(aq)$.
Copper reacts vigorously with concentrated nitric acid to give brown NO_2 gas.

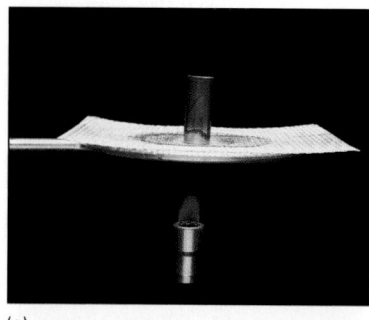

(a)

(b)

Figure 5.11 Reduction of copper oxide with hydrogen. (a) A piece of copper has been heated in air to form a film of black copper(II) oxide on the surface. (b) When the hot copper metal, with its film of CuO, is placed in a stream of hydrogen gas (from the yellow tank at the rear), the oxide is reduced to copper metal, and water forms as the byproduct.

the acid is reduced to the brown gas NO_2. The net ionic equation for the reaction is

$$Cu(s) + 4\,H^+(aq) + 2\,NO_3^-(aq) \longrightarrow Cu^{2+}(aq) + 2\,NO_2(g) + 2\,H_2O(\ell)$$

reducing agent   oxidizing agent

The metal is clearly the reducing agent, since it is the substance oxidized. In fact, the most common reducing agents in the laboratory are metals. Some metal ions such as Fe^{2+} can also be reducing agents because they can be oxidized to ions of higher charge. Aqueous Fe^{2+} ion reacts readily with the strong oxidizing agent MnO_4^-, the permanganate ion. The Fe^{2+} ion is oxidized to Fe^{3+}, and the MnO_4^- ion is reduced to the Mn^{2+} ion.

$$5\,Fe^{2+}(aq) + MnO_4^-(aq) + 8\,H^+(aq) \longrightarrow 5\,Fe^{3+}(aq) + Mn^{2+}(aq) + 4\,H_2O(\ell)$$

Carbon can reduce many metal oxides to metals, and it is widely used in the metals industry to free metals from their compounds in ores. For example, titanium is produced by treating a mineral containing titanium(IV) oxide with carbon and chlorine.

$$TiO_2(s) + C(s) + 2\,Cl_2(g) \longrightarrow TiCl_4(\ell) + CO_2(g)$$

In effect, the carbon reduces the oxide to titanium metal, and the chlorine then oxidizes it to titanium(IV) chloride. Because $TiCl_4$ is easily converted to a gas, it can be removed from the reaction mixture and recovered. The $TiCl_4$ is then reduced with another metal, such as magnesium, to give titanium metal.

$$TiCl_4(\ell) + 2\,Mg(s) \longrightarrow Ti(s) + 2\,MgCl_2(s)$$

Finally, H_2 gas is a common reducing agent, widely used in the laboratory and in industry. For example, it readily reduces copper(II) oxide to copper metal (Figure 5.11).

$$H_2(g) + CuO(s) \longrightarrow Cu(s) + H_2O(g)$$

reducing agent   oxidizing agent

It is important to be aware that it can be dangerous to mix a strong oxidizing agent with a strong reducing agent. A violent reaction, even an explosion, may take place. Chemicals are no longer stored on laboratory shelves in alphabetical order, because such an ordering may place a strong oxidizing agent next to a strong reducing agent. Swimming pool chemicals that contain chlorine and are strong oxidizing agents should not be stored in the hardware store or the garage next to easily oxidized materials.

Exercise 5.10   Oxidation-Reduction Reactions

Decide which of the following reactions are oxidation-reduction reactions. In each case explain your choice and identify the oxidizing and reducing agents.
(a) $NaOH(aq) + HNO_3(aq) \rightarrow NaNO_3(aq) + H_2O(\ell)$
(b) $4\,Cr(s) + 3\,O_2(g) \rightarrow 2\,Cr_2O_3(s)$
(c) $NiCO_3(s) + 2\,HCl(aq) \rightarrow NiCl_2(aq) + H_2O(\ell) + CO_2(g)$
(d) $Cu(s) + Cl_2(g) \rightarrow CuCl_2(s)$

5.4   OXIDATION NUMBERS AND REDOX REACTIONS

An arbitrary bookkeeping system has been devised for keeping track of electrons in redox reactions. It extends the obvious oxidation and reduction case when neutral atoms become ions in reactions in which the changes are less obvious. The

TABLE 5.4	Recognizing Oxidation-Reduction Reactions	
	Oxidation	**Reduction**
In terms of oxygen	Gain of oxygen	Loss of oxygen
In terms of halogen	Gain of halogen	Loss of halogen
In terms of hydrogen	Loss of hydrogen	Gain of hydrogen
In terms of electrons	Loss of electrons	Gain of electrons
In terms of oxidation numbers	Increase of oxidation number	Decrease of oxidation number

system is set up so that *oxidation numbers always change in redox reactions.* As a result, oxidation and reduction can be determined in the ways shown in Table 5.4.

An **oxidation number** compares the charge of an uncombined atom with its actual charge or its relative charge in a compound. All neutral atoms have an equal number of protons and electrons and thus have no net charge. When sodium metal atoms (0 net charge) combine with chlorine atoms (0 net charge) to form sodium chloride, each sodium atom loses an electron to form a sodium ion, Na^+, and each chlorine atom gains an electron, forming a chloride ion, Cl^-. Therefore, Na^+ has an oxidation number of $+1$ because it has one less electron than a sodium atom, and Cl^- has an oxidation number of -1 because it has one more electron than a chlorine atom. Oxidation numbers of atoms in molecular compounds are assigned as though electrons were completely transferred to form ions. In the molecular compound phosphorus trichloride (PCl_3), for example, chlorine is assigned an oxidation number of -1 even though it is not a Cl^- ion; the chlorine is directly bonded to the phosphorus. The chlorine atoms in PCl_3 are thought of as "possessing" more electrons than they have in Cl_2.

You can use the following set of rules to determine oxidation numbers.

Rule 1: ***The oxidation number of an atom of a pure element is 0.*** When the atoms are not combined with those of any other element (for example, oxygen in O_2, sulfur in S_8, iron in metallic Fe, or chlorine in Cl_2), the oxidation number is 0.

Rule 2: ***The oxidation number of a monatomic ion equals its charge.*** Thus, the oxidation number of Cu^{2+} is $+2$; that of S^{2-} is -2.

Rule 3: ***Some elements have the same oxidation number in almost all their compounds and can be used as references for oxidation numbers of other elements in compounds.***

- Hydrogen has an oxidation number of $+1$, unless it is combined with a metal, in which case its oxidation number is -1.
- Fluorine has an oxidation number of -1 in all its compounds.
- Oxygen has an oxidation number of -2, except in peroxides, such as hydrogen peroxide, H_2O_2, in which it has an oxidation number of -1 (and hydrogen is $+1$).
- In binary compounds (compounds of two elements) atoms of Group 6A elements (O, S, Se, Te) have an oxidation number of -2, except when combined with oxygen or halogens, where the Group 6A elements have positive oxidation numbers.

Rule 4: ***The sum of the oxidation numbers in a neutral compound is 0; the sum of the oxidation numbers in a polyatomic ion equals the charge on the ion.*** For example, in SO_2, the oxidation number of oxygen is -2, and with two O atoms, the total for oxygen is -4.

CD-ROM Screen 4.17: Oxidation Numbers

Oxidation numbers are written as +2, whereas charges on ions are written as 2+.

How electrons participate in bonding atoms in molecules is the subject of Chapter 8.

Oxidation numbers are also called oxidation states.

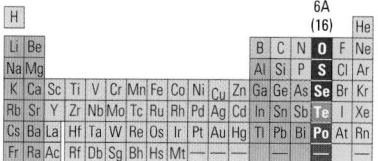

Because the sum of the oxidation numbers must equal zero, the oxidation number of sulfur must be $+4$: $(+4) + 2(-2) = 0$. In the sulfite ion, SO_3^{2-}, the net charge is $2-$. Because each oxygen is -2, the oxidation number of sulfur in sulfite must be $+4$: $(+4) + 3(-2) = 2-$.

$$\overset{+4\ \ -2}{SO_3^{2-}}$$

Now, let's apply these rules to the equations for simple combination and displacement reactions involving sulfur and oxygen.

Combination: $\overset{0}{S_8(s)} + 8\ \overset{0}{O_2(g)} \longrightarrow 8\ \overset{+4\,-2}{SO_2(g)}$

Combination: $\overset{+2\ -2}{ZnS(s)} + 2\ \overset{0}{O_2(aq)} \longrightarrow \overset{+2\,+6\,-2}{ZnSO_4(aq)}$

Displacement: $\overset{+1\ -2}{Cu_2S(s)} + \overset{0}{O_2(g)} \longrightarrow 2\ \overset{0}{Cu(s)} + \overset{+4\,-2}{SO_2(g)}$

These are all oxidation-reduction reactions, as shown by the fact that there has been a change in the oxidation numbers of atoms from reactants to products. ***Every reaction in which an element becomes combined in a compound is a redox reaction.*** The oxidation number of the element must increase or decrease from its original value of zero. Combination reactions and displacement reactions in which one element displaces another are all redox reactions.

Those decomposition reactions in which elemental gases are produced are also redox reactions. Millions of tons of ammonium nitrate, NH_4NO_3, are used as fertilizer to supply nitrogen to crops. Ammonium nitrate is also used as an explosive that is decomposed by heating.

$$2\ NH_4NO_3(s) \longrightarrow 2\ N_2(g) + 4\ H_2O(\ell) + O_2(g)$$

Like a number of other explosives, ammonium nitrate contains an element with two different oxidation numbers, in effect having an oxidizing and reducing agent in the same compound.

$$\overset{-3\,+1}{NH_4^+} \quad \overset{+5\ -2}{NO_3^-}$$

Note that nitrogen's oxidation number is -3 in the ammonium ion and $+5$ in the nitrate ion. Therefore, in the decomposition of ammonium nitrate to N_2, the N in the ammonium ion is oxidized from -3 to 0, and the ammonium ion is the reducing agent. The N in the nitrate ion is reduced from $+5$ to 0, and the nitrate ion is the oxidizing agent.

Ammonium nitrate was used in the 1995 bombing of the Federal Building in Oklahoma City, Oklahoma.

Problem-Solving Example 5.8 Applying Oxidation Numbers

Henry Cavendish discovered hydrogen in 1671 by dripping sulfuric acid down an iron gun barrel.

$$Fe(s) + H_2SO_4(aq) \longrightarrow FeSO_4(aq) + H_2(g)$$

Assign oxidation numbers for each atom in this equation. Identify what has been oxidized and what has been reduced.

Answer

$$\overset{0}{Fe(s)} + \overset{+2\ +6\,-2}{H_2SO_4(aq)} \longrightarrow \overset{+2\,+6\,-2}{FeSO_4(aq)} + \overset{0}{H_2(g)}$$

Fe is oxidized; H^+ is reduced.

Explanation Because iron is in its elemental state as a reactant, its oxidation number is 0 (Rule 1). The same rule applies for the product H_2: H in H_2 has a 0 oxidation number.

For sulfuric acid, we start by recognizing that it is a compound and has no net charge (Rule 4). Therefore, because the oxidation state of each oxygen is -2 (Rule 3c) for a total of -8, and the oxidation state of each hydrogen is $+1$ (Rule 3a), for a total of $+2$, the oxidation state of sulfur must be $+6$: $\quad 0 = 4(-2) + 2(+1) + (+6)$.

In the case of $FeSO_4$, we can assign an oxidation number to Fe by recognizing that it must be $+2$ to balance the charge of SO_4^{2-}. In the sulfate ion, given that oxygen is -2, the oxidation number of sulfur must be $+6$ to give the ion a net -2 charge. The iron has been oxidized, as shown by its oxidation number change from 0 to $+2$. The H^+ has been reduced, as indicated by the change in its oxidation number from $+1$ to 0.

Problem-Solving Practice 5.8

Determine the oxidation number for each atom in this equation:

$$Sb_2S_3(s) + 3\,Fe(s) \longrightarrow 3\,FeS(s) + 2\,Sb(s)$$

Cite the oxidation number rule(s) you used to choose your answers.

Problem-Solving Example 5.9   Oxidation-Reduction Reaction

Most of the metals that we use are found in nature as cations in ores. The metal ion must be reduced to its elemental form, which is done with an appropriate oxidation-reduction reaction. The copper ore chalcocite (Cu_2S) is reacted with oxygen in a process called roasting to release metallic copper.

$$Cu_2S(s) + O_2(g) \longrightarrow 2\,Cu(s) + SO_2(g)$$

Identify the atoms that are oxidized and reduced, and name the oxidizing and reducing agents.

Answer Cu and oxygen are reduced; sulfur is oxidized. O_2 is the oxidizing agent; Cu_2S is the reducing agent.

Explanation We first assign the oxidation states for all the atoms in the reaction according to Rules 1 through 4.

$$\overset{+1\ -2}{Cu_2S}(s) + \overset{0}{O_2}(g) \longrightarrow 2\,\overset{0}{Cu}(s) + \overset{+4\ -2}{SO_2}(g)$$

The oxidation state of the Cu^+ decreases from $+1$ to 0, so Cu^+ is reduced. The oxidation state of S increases from -2 to $+4$, so S is oxidized. The oxidation state of oxygen decreases from 0 to -2, so oxygen is reduced. The oxidizing agent is O_2, which accepts electrons. The reducing agent is S^{2-} in Cu_2S, which donates electrons.

Problem-Solving Practice 5.9

Which are the oxidizing and reducing agents, and which atoms are oxidized and reduced in the following reaction?

$$PbO(s) + CO(g) \longrightarrow Pb(s) + CO_2(g)$$

 ## Exercise 5.11   Redox in CFC Disposal

The following redox reaction is used for the disposal of chlorofluorocarbons (CFCs) by their reaction with sodium oxalate, $Na_2C_2O_4$:

$$CF_2Cl_2(g) + 2\,Na_2C_2O_4(s) \longrightarrow 2\,NaF(s) + 2\,NaCl(s) + C(s) + 4\,CO_2(g)$$

(a) What is oxidized in this reaction?
(b) What is reduced?

Exchange reactions of ionic compounds in aqueous solutions are not redox reactions because no change of oxidation numbers occurs. Consider, for example, the precipitation of barium sulfate when aqueous solutions of barium chloride and sulfuric acid are mixed.

$$Ba^{2+}(aq) + 2\,Cl^-(aq) + 2\,H^+(aq) + SO_4^{2-}(aq) \longrightarrow BaSO_4(s) + 2\,HCl(aq)$$

Net Ionic Equation: $Ba^{2+}(aq) + SO_4^{2-}(aq) \longrightarrow BaSO_4(s)$

The oxidation numbers of the reactant and product atoms remain unchanged.

5.5   DISPLACEMENT REACTIONS, REDOX, AND THE ACTIVITY SERIES

Recall that displacement reactions (⬅ *p. 127*) have the following reaction pattern:

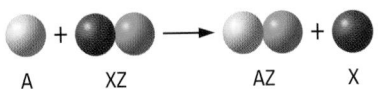

Displacement reactions, like combination reactions, are oxidation-reduction reactions. For example, in the reaction of hydrochloric acid with iron (Figure 5.12),

$$Fe(s) + 2\,HCl(aq) \longrightarrow FeCl_2(aq) + H_2(g)$$

metallic iron is the reducing agent; it is oxidized from an oxidation number of 0 in $Fe(s)$ to $+2$ in $FeCl_2$. Hydrogen ions, H^+, in hydrochloric acid are reduced to hydrogen gas (H_2), in which hydrogen has an oxidation number of 0.

Extensive studies with many metals have led to the development of a **metal activity series,** a ranking of relative reactivity of metals in displacement and other kinds of reactions (Table 5.5). The most reactive metals are at the top of the series,

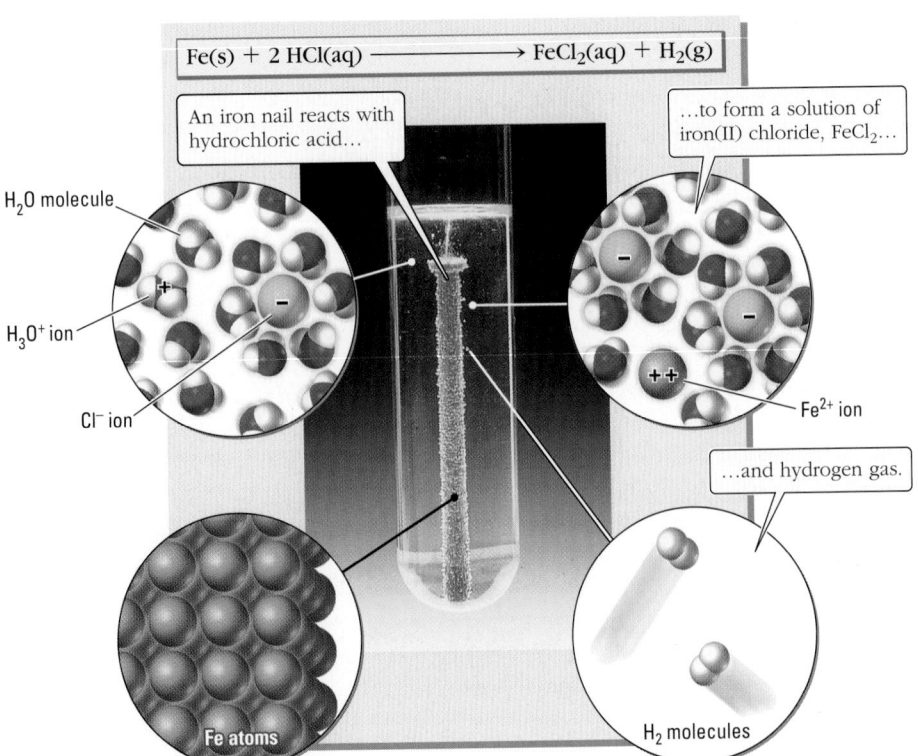

Figure 5.12   **Metal + acid displacement reaction.**

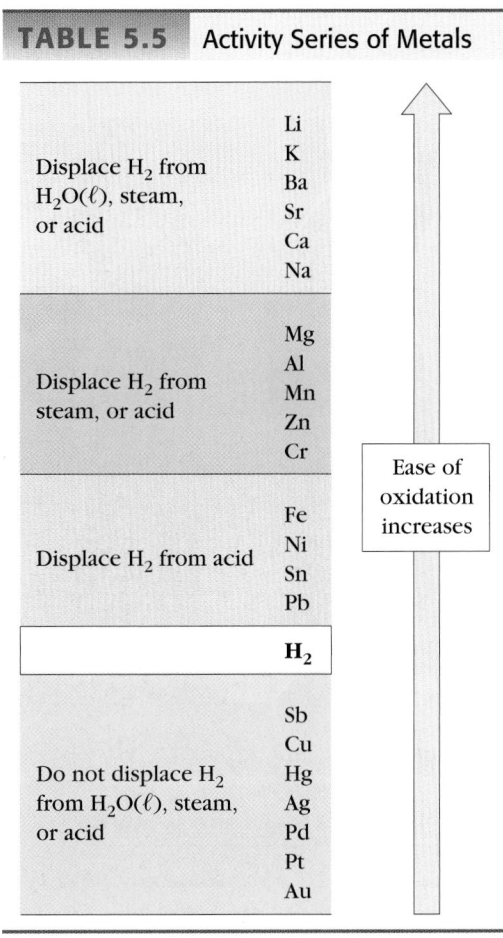

TABLE 5.5 Activity Series of Metals

Displace H_2 from $H_2O(\ell)$, steam, or acid	Li K Ba Sr Ca Na
Displace H_2 from steam, or acid	Mg Al Mn Zn Cr
Displace H_2 from acid	Fe Ni Sn Pb
	H_2
Do not displace H_2 from $H_2O(\ell)$, steam, or acid	Sb Cu Hg Ag Pd Pt Au

Ease of oxidation increases

and activity decreases down the series. Metals at the top are powerful reducing agents and readily form cations. The metals at the lower end of the series are poor reducing agents. However, their cations (Au^+, Ag^+) are powerful oxidizing agents that readily react to form the free metal.

A metal higher in the activity series will displace an element below it in the series from its compounds. For example, zinc displaces copper from copper(II) sulfate solution, and copper metal displaces silver from silver nitrate solution (Figure 5.13).

$$Zn(s) + CuSO_4(aq) \longrightarrow ZnSO_4(aq) + Cu(s)$$

$$Cu(s) + 2\,AgNO_3(aq) \longrightarrow Cu(NO_3)_2(aq) + 2\,Ag(s)$$

In each case, the elemental metal (Zn, Cu) is the reducing agent and is oxidized; Cu^{2+} ions and Ag^+ ions are oxidizing agents and are reduced to Cu(s) and Ag(s).

Metals above hydrogen in the series react with acids whose anions are not oxidizing agents, such as hydrochloric acid, to form hydrogen (H_2) and the metal salt containing the cation of the metal and the anion of the acid. For example, $FeCl_2$ is formed from iron and hydrochloric acid and $ZnBr_2$ from zinc and hydrobromic acid.

$$Fe(s) + 2\,HCl(aq) \longrightarrow FeCl_2(aq) + H_2(g)$$

$$Zn(s) + 2\,HBr(aq) \longrightarrow ZnBr_2(aq) + H_2(g)$$

Metals below hydrogen in the series do not displace hydrogen from acids in this way.

Very reactive metals—those at the top of the activity series, from lithium (Li) to sodium (Na)—can displace hydrogen from water. Some do so violently (Figure 5.14). Metals of intermediate activity (Mg through Cr) displace hydrogen from steam, but not from liquid water at room temperature.

When an acid such as HNO_3, whose nitrate ion is a strong oxidizing agent, reacts with a metal, the anion is also reduced and different products are formed.

Figure 5.13 Metal + aqueous metal salt displacement reaction. The oxidation of copper metal by silver ion. (Atoms or ions that take part in the reaction have been highlighted in the nanoscale pictures.)

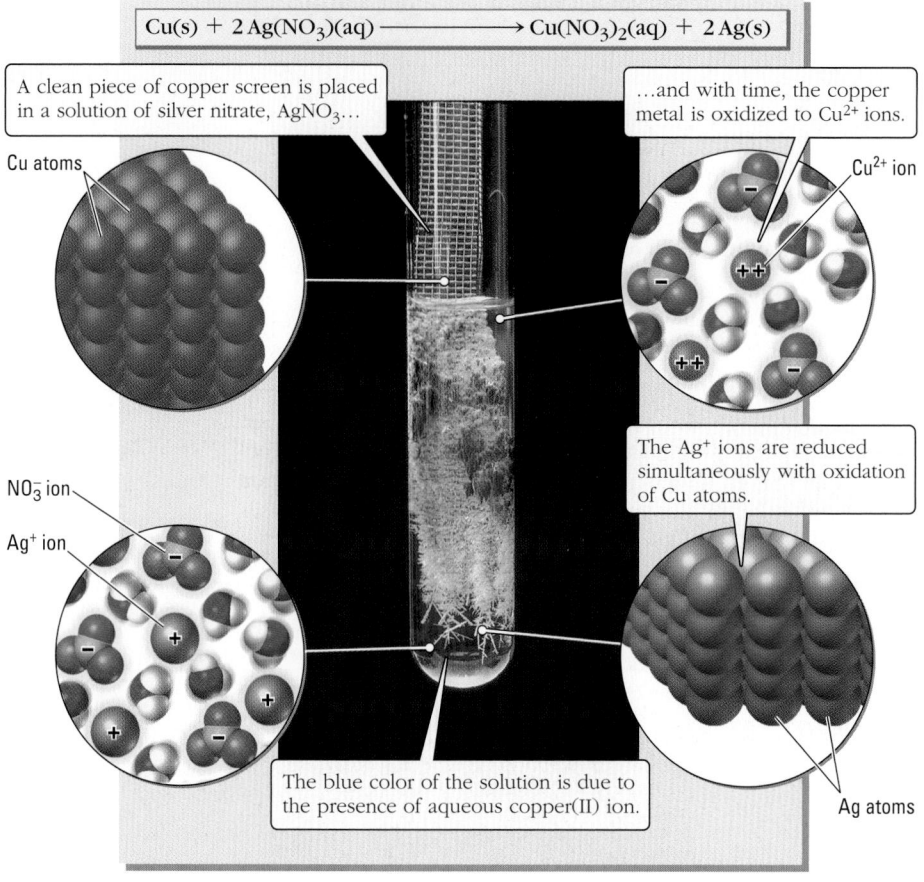

$$Cu(s) + 2\,Ag(NO_3)(aq) \longrightarrow Cu(NO_3)_2(aq) + 2\,Ag(s)$$

A clean piece of copper screen is placed in a solution of silver nitrate, $AgNO_3$...

Cu atoms

...and with time, the copper metal is oxidized to Cu^{2+} ions.

Cu^{2+} ion

NO_3^- ion

Ag^+ ion

The Ag^+ ions are reduced simultaneously with oxidation of Cu atoms.

The blue color of the solution is due to the presence of aqueous copper(II) ion.

Ag atoms

Elements very low in the activity series are unreactive. They are sometimes called noble metals (Au, Ag, Pt), and they are prized for their nonreactivity. It is no accident that gold and silver have been used extensively for coinage since antiquity. These metals do not react with air, water, or even common acids, thus maintaining their luster (and value) for many years. These metals are discussed in Chapter 22. Their low reactivity is also why they occur naturally as free metals and have been known as elements since antiquity.

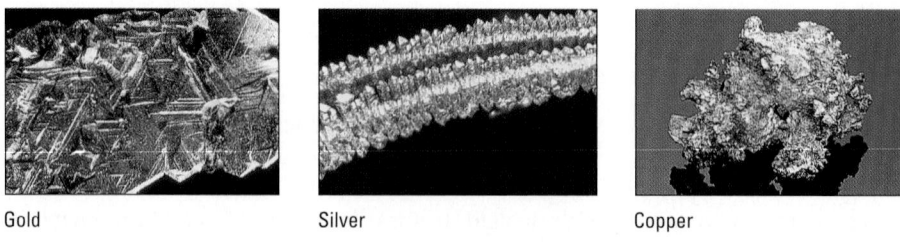

Gold   Silver   Copper

Figure 5.14 Potassium, an active metal. When a drop of water falls onto a sample of potassium metal, it reacts vigorously to give hydrogen gas and a solution of potassium hydroxide.

Problem-Solving Example 5.10 Activity Series of Metals

Use the activity series (Table 5.5) to predict whether any of the following reactions will occur. Complete and balance the equations for those reactions that will occur.
(a) $Fe(s) + CuCl(aq) \rightarrow$
(b) $Zn(s) + MgCl_2(aq) \rightarrow$
(c) Manganese + hydrochloric acid $\rightarrow$

Answer
(a) $Fe(s) + CuCl_2(aq) \rightarrow FeCl_2(aq) + Cu(s)$
(b) No reaction
(c) $Mn(s) + 2\,HCl(aq) \rightarrow MnCl_2(aq) + H_2(g)$

Explanation

(a) Iron is above copper in the activity series. Therefore, it will displace copper ions from a solution of copper(II) chloride to form metallic copper and Fe^{2+} ions.

$$Fe(s) + CuCl_2(aq) \longrightarrow FeCl_2(aq) + Cu(s)$$

(b) Zinc is a less active metal than magnesium. Therefore, it will not displace magnesium from magnesium chloride, and no reaction occurs.

(c) This reaction is analogous to the reaction of iron and sulfuric acid. Because manganese is above hydrogen in the activity series, it will displace hydrogen from HCl and form the metal salt ($MnCl_2$) plus hydrogen gas.

$$Mn(s) + 2 HCl(aq) \longrightarrow MnCl_2(aq) + H_2(g)$$

Problem-Solving Practice 5.10

Use Table 5.5 to predict whether each of the following reactions will occur. If a reaction occurs, identify what has been oxidized or reduced and what the oxidizing agent and the reducing agent are.

(a) $2 Al(s) + 3 CuSO_4(aq) \rightarrow Al_2(SO_4)_3(aq) + 3 Cu(s)$
(b) $2 Al(s) + Cr_2O_3(s) \rightarrow Al_2O_3(s) + 2 Cr(s)$
(c) $Pt(s) + 4 HCl(aq) \rightarrow PtCl_4(aq) + 2 H_2(g)$
(d) $Au(s) + 3 AgNO_3(aq) \rightarrow Au(NO_3)_3 + 3 Ag(s)$

Exercise 5.12 Reaction Product Prediction

For the following pairs of reactants, predict what kind of reaction would occur and what the products might be. Which reactions are redox reactions?
(a) CH_3CH_2OH (ethanol)$(\ell) + O_2(g) \rightarrow$?
(b) $Fe(s) + HNO_3(aq) \rightarrow$?
(c) $AgNO_3(aq) + KBr(aq) \rightarrow$?

CHEMISTRY *You Can Do...* Pennies, Redox, and the Activity Series of Metals

In this experiment, you will use pennies to test the reactivity of copper and zinc with acid. Post-1982 pennies are a copper and zinc "sandwich" with zinc in the middle covered by a layer of copper. Pre-1982 pennies do not have this composition.

To do this experiment you will need

- Two glasses or plastic cups that will each hold 50 mL (about 1.5 oz) of liquid
- About 100 mL of "pickling" vinegar, such as Heinz Ultrastrength brand. (Regular vinegar is only about 4–5% acetic acid.)
- An abrasive such as a piece of sandpaper, steel wool, or a Brillo pad
- A small file
- Four pennies—two pre-1982 and two post-1982

Clean the pennies with the abrasive until all the surfaces (including the edges) are shiny. Use the file to make two cuts into the edge of each penny, one across from the other. If you look carefully, you might observe a shiny metal where you cut into the post-1982 pennies.

Caution: Keep the vinegar away from your skin and clothes and especially your eyes. If vinegar spills on you, rinse it off with flowing water.

Place the two pre-1982 pennies into one of the cups and the post-1982 pennies into the other cup. Add the same volume of vinegar to each cup, making sure that the pennies are covered by the liquid. Let the pennies remain in the liquid for several hours (even overnight), and periodically observe any changes in the pennies. After several hours, pour off the vinegar and remove the pennies. Dry them carefully and observe any changes that have occurred.

1. What difference did you observe between the pre-1982 pennies and the post-1982 ones?
2. Which is the more reactive element—copper or zinc?
3. What happened to the zinc in the post-1982 pennies? Interpret the change in redox terms, and write a chemical equation to represent the reaction.
4. How could this experiment be modified to determine the percent zinc and percent copper in post-1982 pennies?

5.6 SOLUTION CONCENTRATION

CD-ROM Screen 5.9: Solutions

Many of the chemicals in your body or in a plant are dissolved in water; that is, they are in an aqueous solution. Just as a living system uses chemistry in solution, so do chemists, who do their work quantitatively. For example, intravenous fluids administered to patients contain many compounds (salts, nutrients, drugs, and so on), and the concentration of each must be known accurately. To accomplish this, we continue to use balanced equations and moles, but we measure volumes of solution rather than masses of solids, liquids, and gases. A solution is a homogeneous mixture of a **solute,** the substance that has been dissolved, and the **solvent,** the substance in which the solute has been dissolved. To know the quantity of solute in a given volume of a liquid solution requires knowing the **concentration** of the solution — the relative amounts of solute and solvent. Molarity, which relates the amount of solute in moles to the solution volume in liters, is the most useful of the many ways of expressing solution concentration for studying chemical reactions in solution.

Molarity

CD-ROM Screen 5.10: Solution Concentration: Molarity

The **molarity** of a solution is defined as moles of solute per liter (mol/L) of solution.

$$\text{Molarity} = \frac{\text{moles of solute}}{\text{liters of solution}}$$

Note that the volume term is liters of *solution,* not liters of solvent (such as water).

If, for example, 40.0 g (1.00 mol) of NaOH is dissolved in sufficient water to produce a solution with a total volume of 1.00 L, the solution has a concentration of 1.00 mol NaOH/1.00 L of solution, which is a 1.00 molar solution. The molarity of this solution is reported as 1.00 M, where the capital M stands for moles/liter. Molarity is also represented by square brackets around the formula of a compound or ion, such as [NaOH] or [OH$^-$]. The brackets indicate moles of the compound or ion per liter of solution.

A solution of known molarity can be made by adding the required amount of solute to a volumetric flask, adding some solvent to dissolve all the solute, and then adding sufficient solvent with continual mixing to fill the flask "to the mark." As shown in Figure 5.15, the etched marking indicates the liquid level equal to the specified volume of the flask.

Problem-Solving Example 5.11 Molarity

(a) Suppose 0.275 g of K_2CrO_4 is placed into a 500-mL volumetric flask and water is added until the solution volume is exactly 500 mL. What is the molarity of K_2CrO_4 in this solution?

(b) Suppose a 0.00375 M solution of K_2CrO_4 was needed for an experiment. How many grams of K_2CrO_4 should be added to a 1-L volumetric flask to give the desired molarity?

Answer (a) 0.00283 M (b) 0.728 g

Explanation

(a) To calculate the molarity, we need the solution volume in liters and the moles of solute. The volume has been given as 500 mL, which is 0.500 L. We use the molar mass of K_2CrO_4 (194.2 g/mol) to obtain moles of solute.

$$0.275 \text{ g K}_2\text{CrO}_4 \times \frac{1 \text{ mol K}_2\text{CrO}_4}{194.2 \text{ g K}_2\text{CrO}_4} = 1.416 \times 10^{-3} \text{ mol K}_2\text{CrO}_4$$

We can now calculate the molarity.

$$\text{Molarity of K}_2\text{CrO}_4 = \frac{1.416 \times 10^{-3} \text{ mol K}_2\text{CrO}_4}{0.500 \text{ L solution}} = 2.83 \times 10^{-3} \text{ mol/L}$$

This can be expressed as 0.00283 M, or in the notation $[K_2CrO_4] = 0.00283$ M.

(b) To make a 0.00375 M solution in a 1-L volumetric flask we will need 0.00375 mol of K_2CrO_4, which we convert to grams as follows.

$$0.00375 \text{ mol } K_2CrO_4 \times \frac{194.2 \text{ g } K_2CrO_4}{1 \text{ mol } K_2CrO_4} = 0.728 \text{ g } K_2CrO_4$$

✔ (a) We have about one quarter of a gram of K_2CrO_4 with a molar mass of about 200 g/mol, so we have about 0.25 g/(200 g/mol) or 0.00125 mol of K_2CrO_4. This made 0.5 L of solution, so the molarity must be 0.00125 mol/0.5 L = 0.0025 mol/L, which is about the answer we calculated.

(b) We need less than 1/200 of a mole of K_2CrO_4, with a molar mass of about 200, so we should need less than one gram, which is what we calculated.

Problem-Solving Practice 5.11

Calculate the molarity of sodium sulfate in a solution that contains 36.0 g of Na_2SO_4 in 750. mL of solution.

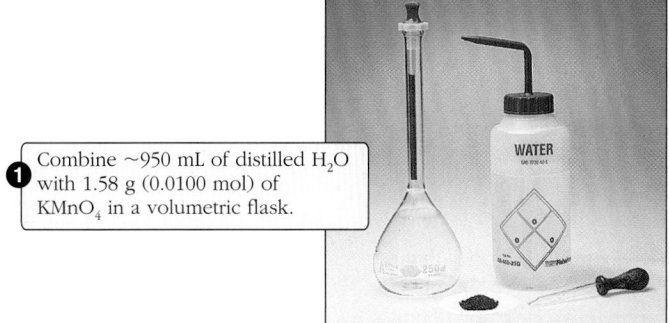

❶ Combine ~950 mL of distilled H_2O with 1.58 g (0.0100 mol) of $KMnO_4$ in a volumetric flask.

❷ Shake the flask to dissolve the $KMnO_4$.

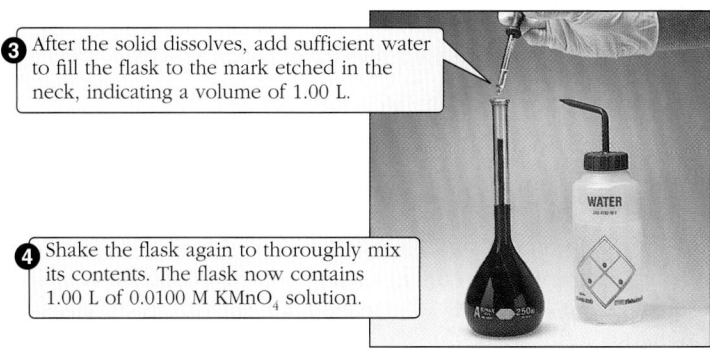

❸ After the solid dissolves, add sufficient water to fill the flask to the mark etched in the neck, indicating a volume of 1.00 L.

❹ Shake the flask again to thoroughly mix its contents. The flask now contains 1.00 L of 0.0100 M $KMnO_4$ solution.

Figure 5.15 **Solution preparation from a solid solute.**
Making a 0.0100 M aqueous solution of $KMnO_4$.

Exercise 5.13 Cholesterol Molarity

A blood serum cholesterol level greater than 240 mg of cholesterol per deciliter (0.100 L) of blood generally indicates the need for medical intervention. Calculate the serum cholesterol level in terms of molarity. Cholesterol's molecular formula is $C_{27}H_{46}O$.

Sometimes the molarity of a particular ion in a solution is required, a value that depends on the formula of the solute. For example, potassium chromate is a soluble ionic compound and a strong electrolyte that is completely dissociated in solution to form 2 mol of K^+ ions and 1 mol of CrO_4^{2-} ions for each mole of K_2CrO_4 that dissolves:

$$K_2CrO_4(aq) \longrightarrow 2\ K^+(aq) + CrO_4^{2-}(aq)$$

1 mol   2 mol   1 mol
100% dissociation

The K^+ concentration is twice the K_2CrO_4 concentration because each mole of K_2CrO_4 contains 2 mol of K^+. Therefore, the 0.00283 M K_2CrO_4 has a K^+ concentration of 2×0.00283 M = 0.00566 M and a CrO_4^{2-} concentration of 0.00283 M.

Exercise 5.14 Molarity

A student dissolves 6.37 g of aluminum nitrate in sufficient water to make 250 mL of solution. Calculate (a) the molarity of aluminum nitrate in this solution and (b) the molar concentration of aluminum ions and of nitrate ions in this solution.

Exercise 5.15 Molarity

When solutions are prepared, the final volume of solution can be different from the sum of the volumes of the solute and solvent because some expansion or contraction can occur. Why is it always better to describe solution preparation as "adding enough solvent" to make a certain volume of solution?

Preparing a Solution of Known Molarity by Diluting a More Concentrated One

CD-ROM Screen 5.12:
Preparing Solutions of Known Concentration (2): Solution by Dilution

Consider two cases: A teaspoonful of sugar ($C_{12}H_{22}O_{11}$) is dissolved in a glass of water and then the glass of sugar solution is poured into a swimming pool full of water. The swimming pool and the glass contain the same number of moles of sugar, but the concentration of sugar in the swimming pool is far less because the volume of water in the pool is much greater than that in the glass.

Frequently, a solution needs to be available in a wide variety of molarities. For example, hydrochloric acid is often used at concentrations of 6.0 M, 1.0 M, and 0.050 M. To make these solutions, chemists use a concentrated solution of known molarity and dilute samples of it with water to make solutions of lesser molarity. The number of moles of solute in the solution remains constant throughout the dilution operation. Diluting a solution does increase the volume, which lowers the molarity. Therefore, the moles of solute in the dilute solution must be the same as the number of moles of solute in the sample of the more concentrated solution.

The moles in each case are equal and a simple relationship applies:

$$Molarity(\text{conc}) \times V(\text{conc}) = Molarity(\text{dil}) \times V(\text{dil})$$

where *Molarity*(conc) and *V*(conc) represent the molarity and the volume (in liters) of the concentrated solution; *Molarity*(dil) and *V*(dil) represent the molarity and volume of the dilute solution. *Multiplying a volume in liters by a solution's molarity (moles/liter) yields moles of solute.*

We can calculate, for example, the concentration of a hydrochloric acid solution made by diluting 25.0 mL of 6.0 M HCl to 500 mL. In this case, we want to determine *Molarity*(dil); *Molarity*(conc) = 6.0 M, *V*(conc) = 0.0250 L, and

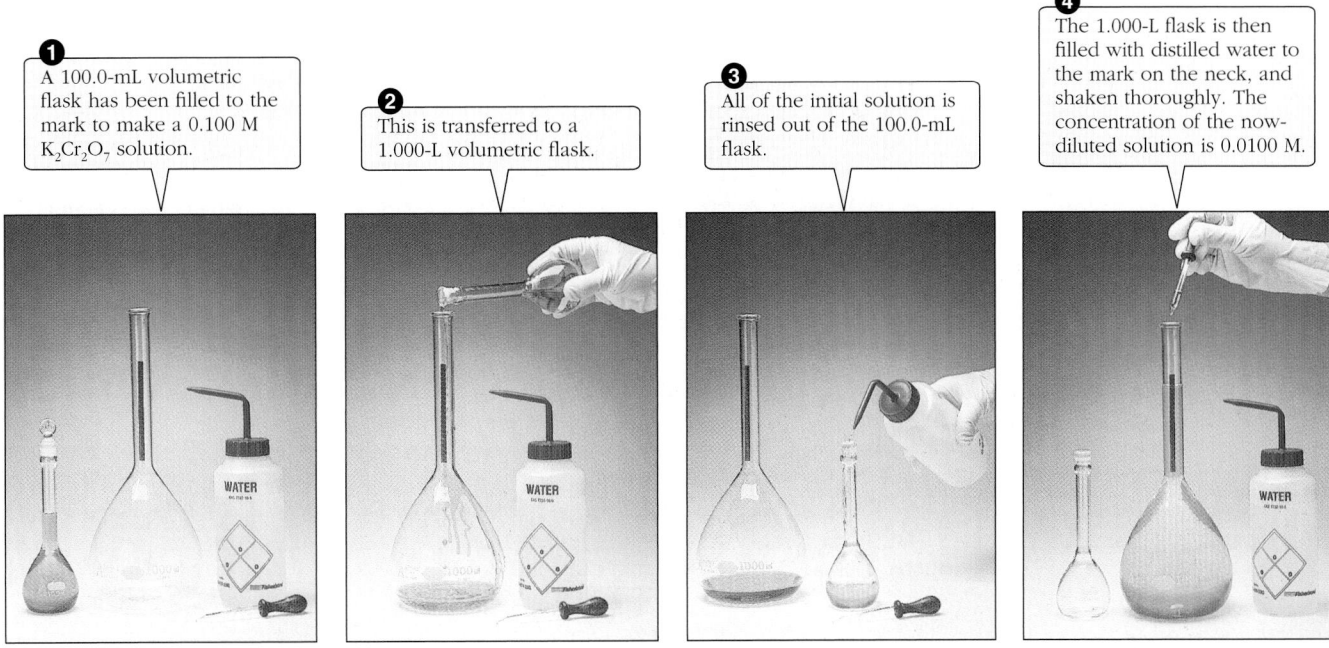

Figure 5.16   **Solution preparation by dilution.**

V(dil) = 0.500 L. We algebraically rearrange the relationship to get the concentration of the diluted HCl.

$$Molarity(\text{dil}) = \frac{Molarity(\text{conc}) \times V(\text{conc})}{V(\text{dil})}$$

$$= \frac{6.0 \text{ mol/L} \times 0.0250 \text{ L}}{0.500 \text{ L}} = 0.30 \text{ mol/L}$$

A diluted solution will always be less concentrated (lower molarity) than the more-concentrated stock solution (Figure 5.16).

Caution must be used when diluting a concentrated acid. The more-concentrated acid should be added slowly to the solvent (water) so that the heat generated during the dilution is slowly dissipated. If water is added to the acid, the heat generated by the dissolving could be sufficient to vaporize the solution, spraying the acid over you and those in the area.

A quick and useful check on a dilution calculation is to make certain that the molarity of the diluted solution is not larger than that of the original solution.

Exercise 5.16   Moles of solute in solutions

Consider 100. mL of 6.0 M HCl solution, which is diluted with water to yield 500. mL of 1.20 M HCl. Show that 100. mL of the more concentrated solution contains the same number of moles of HCl as 500. mL of the more dilute solution.

Problem-Solving Example 5.12   Solution Concentration and Dilution

How can 400 mL of 2.00 M HNO_3 be prepared from concentrated nitric acid, 16.0 M HNO_3?

Answer   Add 50.0 mL of the concentrated acid to enough water to make up a total volume of 400 mL of solution.

Explanation This is a dilution problem in which the concentrations of the concentrated (16.0 M) and less concentrated (2.00 M) solutions are given, as well as the volume of the diluted solution (0.400 L). The volume of the concentrated nitric acid, V(conc), to be diluted is needed, and can be calculated from the relationship

$$Molarity\text{(conc)} \times V\text{(conc)} = Molarity\text{(dil)} \times V\text{(dil)}$$

$$V\text{(conc)} = \frac{Molarity\text{(dil)} \times V\text{(dil)}}{Molarity\text{(conc)}} = \frac{2.00 \text{ mol/L} \times 0.400 \text{ L}}{16.0 \text{ mol/L}} = 0.0500 \text{ L} = 50.0 \text{ mL}$$

Therefore, 50.0 mL of concentrated nitric acid is added slowly, with stirring, to about 300 mL of pure water. When the solution has cooled to room temperature, sufficient water is added to bring the final volume to 400 mL, resulting in a 2.00 M nitric acid solution.

✔ The ratio of molarities is 1:8, so the ratio of volumes should be 1:8 as well, and it is.

Problem-Solving Practice 5.12

A laboratory procedure calls for 50.0 mL of 0.150 M NaOH. You have available 100 mL of 0.500 M NaOH. What volume of the more concentrated solution should be diluted to make the desired solution?

 Exercise 5.17 Solution Concentration

The molarity of a solution can be decreased by dilution. How could the molarity of a solution be increased without adding additional solute?

Preparing a Solution of Known Molarity from a Pure Solute

CD-ROM Screen 5.11:
Preparing Solutions of Known
Concentration (1): Direct
Addition

In Problem-Solving Example 5.11, we described finding the molarity of a K_2CrO_4 solution that was prepared from known amounts of solute and solution. More frequently, a solid or liquid solute (sometimes even a gas) must be used to make up a solution of known molarity. The problem becomes one of calculating what mass of solute to use to provide the proper number of moles.

Consider a laboratory experiment that requires 2.00 L of 0.750 M NH_4Cl solution. What mass of NH_4Cl must be dissolved in water to make 2.00 L of solution? The moles of NH_4Cl required can be calculated from the molarity:

$$0.750 \text{ mol/L } NH_4Cl \text{ solution} \times 2.00 \text{ L solution} = 1.500 \text{ mol } NH_4Cl$$

Then the molar mass can be used to calculate the number of grams of NH_4Cl needed.

$$1.500 \text{ mol } NH_4Cl \times 53.49 \text{ g/mol } NH_4Cl = 80.2 \text{ g } NH_4Cl$$

The solution is prepared by putting 80.2 g of NH_4Cl into a container and adding distilled water until the solution volume is 2.00 L, which results in a 0.750 M NH_4Cl solution.

Problem-Solving Example 5.13 Solute Mass and Molarity

How is 250 mL of 0.0150 M $Ce(SO_4)_2$ solution prepared from solid $Ce(SO_4)_2$?

Answer Dissolve 1.25 g of solid $Ce(SO_4)_2$ in water and add enough water to make 250 mL of solution.

Explanation First, find the number of moles of the solute, $Ce(SO_4)_2$, in 250 mL (0.250 L) of 0.0150 M $Ce(SO_4)_2$ solution. From this calculate the number of grams.

$$0.250 \text{ L solution} \times \frac{0.0150 \text{ mol } Ce(SO_4)_2}{1 \text{ L solution}} = 3.75 \times 10^{-3} \text{ mol } Ce(SO_4)_2$$

$$3.75 \times 10^{-3} \text{ mol } Ce(SO_4)_2 \times \frac{332.2 \text{ g } Ce(SO_4)_2}{1 \text{ mol } Ce(SO_4)_2} = 1.25 \text{ g } Ce(SO_4)_2$$

The solution is prepared by putting 1.25 g of $Ce(SO_4)_2$ into a container and adding distilled water until the solution volume is 250 mL, resulting in a 0.0150 M $Ce(SO_4)_2$ solution.

✔ The molar mass of $Ce(SO_4)_2$ is about 332 g/mol, and we want a 0.0150 M solution, so we need about 332 g/mol $\times$ 0.0150 mol/L ≈ 5 g/L. But only 250 mL of solution is needed, so one fourth of the 5. g, or 1.25 g of $Ce(SO_4)_2$, is needed, which checks with our answer.

Problem-Solving Practice 5.13

Describe how you would prepare

(a) 1.0 L of 0.125 M Na_2CO_3 from solid Na_2CO_3.
(b) 100 mL of 0.0500 M Na_2CO_3 from a 0.125 M Na_2CO_3 solution
(c) 500 mL of 0.0215 M $KMnO_4$ from solid $KMnO_4$
(d) 250 mL of 0.00450 M $KMnO_4$ from 0.0215 M $KMnO_4$

5.7 MOLARITY AND REACTIONS IN AQUEOUS SOLUTIONS

Many kinds of reactions — acid-base *(⇐ p. 171)*, precipitation *(⇐ p. 164)*, and redox *(⇐ p. 176)* —occur in aqueous solutions. In such reactions, molarity is the concentration unit of choice because it quantitatively relates a volume of one reactant and the moles of that reactant contained in solution to the volume and corresponding moles of another reactant or a product in solution. This allows us to make conversions between volumes of solutions and moles of reactants and products as given by the stoichiometric coefficients. Molarity is used to link mass, moles, and volume of solution (Figure 5.17).

 CD-ROM Screen 4.13:
Stoichiometry of Reactions in Solution

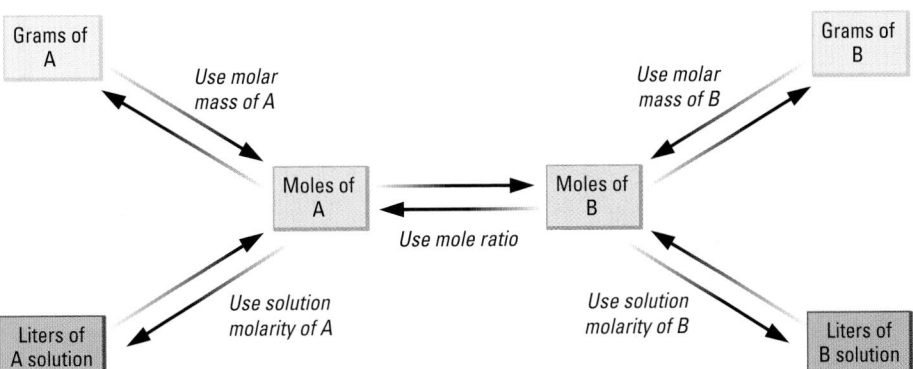

Figure 5.17 **Stoichiometric relationships for a chemical reaction in aqueous solution.** A mole ratio provides the connection between moles of a reactant or product and moles of another reactant or product.

Chemistry in the News

SOLVING A PROBLEM IN INDUSTRIAL CHEMISTRY

Many of the reactions you have seen in Chapter 4 and in this chapter involve acids. One reason for the emphasis on acid chemistry is that acids are some of the most important chemicals in our economy. Five acids are among the most important industrial chemicals produced in the United States, with quantities in the range of 5 to 100 billion pounds produced per year: sulfuric, phosphoric, nitric, hydrochloric, and acetic acids. More than 15,000 companies use these acids to make other chemicals, to clean and refinish metals, to plate metals onto other metals or onto plastics, and in many other applications. A problem faced by all of these industries is what to do with acid-containing waste. For example, when acids are used to wash a metal surface, the washings contain unused acid along with metal ions such as iron(III), copper(II), vanadium(II), silver(I), nickel(II), and lead(II). It is estimated that over 8 billion pounds of acid-containing wastes are generated annually, and they cannot simply be flushed into the nearest lake or river. Not only will the acid damage aquatic life, but many metals are toxic to plants and animals.

A process to significantly reduce the volume and toxicity of acid waste was developed at the U.S. Department of Energy's (DOE) Pacific Northwest Laboratory and is now a commercial system, WADR (Waste Acid Detoxification and Reclamation). The waste from a chemical or metallurgic operation can contain sulfuric, phosphoric, and nitric acids, along with metal ions dissolved in the aqueous acid. In the WADR process, the mixture is heated, the acids vaporize, and the metal ions remain in the liquid phase. The vapor is purified to yield a clean acid, in some cases cleaner than industrial-grade acids. The solution containing metal ions is collected in a tank, and the metal ions are removed by adding salts with anions that form precipitates with the heavy metal ions. The end products of the process are clean water that can be returned to the environment, acids that can be reused, and metal salts from which the metals can be reclaimed.

The process has received several awards for outstanding new technology and is commercially available for industrial uses. Savings of $1 to $5 per gallon in the cost of processing acid wastes are achievable. Furthermore, it is a technology that can be adapted to both large and small waste producers. A large steel company needs a system to process thousands of gallons of waste a week, whereas a small company plating metals or plastic may need to clean only a few hundred gallons. The process can be used to recover up to 90% of the spent acid as a clean product for reuse.

Source:

Chemical and Engineering News, Vol. 71, 1993; p. 27.

Viatec Recovery Systems, Richland, WA.

Waste Acid Detoxification and Reclamation installation. Used acids that contain dissolved metal ions are transformed in this equipment into reusable acid, reclaimable metal salts, and clean water.

Problem-Solving Example 5.14 Solution Reaction Stoichiometry

A major industrial use of hydrochloric acid is for "pickling," the removal of rust from metals, especially steel (largely iron) before it is used to fabricate steel products. The rust is removed by dipping the steel into very large baths of hydrochloric acid. The acid reacts in an exchange reaction with rust, which is essentially Fe_2O_3, leaving behind a clean steel surface.

$$Fe_2O_3(s) + 6\ HCl(aq) \longrightarrow 2\ FeCl_3(aq) + 3\ H_2O(\ell)$$

Once the rust is taken off, the steel is removed from the acid bath and rinsed before the acid reacts significantly with the iron in the steel.

How many pounds of rust can be removed when rust-covered steel reacts with 800. L of 12.0 M HCl? Assume that only the rust reacts with the HCl (1.000 lb = 453.6 g).

Answer   563 lb of Fe_2O_3

Explanation   First, calculate the number of moles of HCl available in 800 L.

$$800.\ L\ HCl \times \frac{12.0\ mol\ HCl}{1\ L\ solution} = 9.600 \times 10^3\ mol\ HCl$$

The mass of Fe_2O_3 can be determined using the remaining steps.

$$(9.600 \times 10^3\ mol\ HCl) \times \frac{1\ mol\ Fe_2O_3}{6\ mol\ HCl} = 1.600 \times 10^3\ mol\ Fe_2O_3$$

$$(1.600 \times 10^3\ mol\ Fe_2O_3) \times \frac{159.7\ g\ Fe_2O_3}{1\ mol\ Fe_2O_3} = 2.555 \times 10^5\ g\ Fe_2O_3$$

$$(2.555 \times 10^5\ g\ Fe_2O_3) \times \frac{1\ lb\ Fe_2O_3}{453.6\ g\ Fe_2O_3} = 563\ lb\ Fe_2O_3$$

Placing all the conversion factors in the same mathematical setup gives

$$800.\ L\ HCl \times \frac{12.0\ mol\ HCl}{1\ L\ solution} \times \frac{1\ mol\ Fe_2O_3}{6\ mol\ HCl}$$

$$\times \frac{159.7\ g\ Fe_2O_3}{1\ mol\ Fe_2O_3} \times \frac{1\ lb\ Fe_2O_3}{453.6\ g\ Fe_2O_3} = 563\ lb\ Fe_2O_3$$

The solution remaining in the acid bath presents a real disposal challenge because of the metal ions it contains. This is discussed in Chemistry in the News: Solving a Problem in Industrial Chemistry.

✔   We have about 10,000 mol of HCl. This will react with one-sixth as many moles of rust, or about 1600 mol of Fe_2O_3. Converting moles of Fe_2O_3 to pounds requires multiplying by the molar mass (160 g/mol) and then multiplying by the grams-to-pounds conversion factor (roughly 1 lb/500 g), which is equivalent to dividing by about 3. So 1600/3 ≈ 500, which checks with our more accurate answer.

Problem-Solving Practice 5.14

In 1995, 1.2×10^{10} kg of sodium hydroxide (NaOH) was produced in the United States by passing an electric current through brine, an aqueous solution of sodium chloride.

$$2\ NaCl(aq) + 2\ H_2O(\ell) \longrightarrow 2\ NaOH(aq) + Cl_2(g) + H_2(g)$$

What volume of brine is needed to produce this mass of NaOH? (1.0 L of brine contains 360 g of dissolved NaCl.)

The chemistry of photography provides another application of solution stoichiometry. When silver bromide in black-and-white photographic film is exposed to light, silver ions in the silver bromide are reduced to metallic silver. When the photograph is developed, this creates black regions on the negative. If left on the film, the unreacted silver bromide would ruin the picture because it would darken when exposed to light. AgBr(s) is dissolved away from the film with a solution of the "fixer," sodium thiosulfate ($Na_2S_2O_3$). Aqueous thiosulfate ions ($S_2O_3^{2-}$)

(a)   (b)

Figure 5.18   Dissolving silver bromide with thiosulfate.   (a) A precipitate of AgBr formed by adding $AgNO_3(aq)$ to KBr(aq). (b) When sodium thiosulfate, $Na_2S_2O_3(aq)$, is added, the solid AgBr dissolves.

combine with silver ions from AgBr to form a soluble product (Figure 5.18). The net ionic equation for this exchange reaction is

$$AgBr(s) + 2\,S_2O_3^{2-}(aq) \longrightarrow Ag(S_2O_3)_2^{3-}(aq) + Br^-(aq)$$

The following example is a quantitative look at this chemistry.

One mole of $Na_2S_2O_3$ dissociates into 2 mol of sodium ions and 1 mol of thiosulfate ions. Thus, the 0.0150 mol of $Na_2S_2O_3$ in 1 L of 0.0150 $Na_2S_2O_3$ solution dissociates into 0.0300 mol of Na^+ and 0.0150 mol of $S_2O_3^{2-}$ ions.

Sodium ions are spectator ions in this reaction.

Problem-Solving Example 5.15   Solution Reaction Stoichiometry

If you need to dissolve 50.0 mg (0.0500 g) of AgBr, how many milliliters of 0.0150 M $Na_2S_2O_3$ should you use?

Answer   35.5 mL of the 0.0150 M $Na_2S_2O_3$ solution is needed

Explanation   Use the diagram in Figure 5.17 and follow the appropriate path. We need to find the moles of $Na_2S_2O_3$ required, considering that 2 mol of $S_2O_3^{2-}$ are required for 1 mol AgBr, and that 1 mol of $Na_2S_2O_3$ contains 1 mol of of $S_2O_3^{2-}$ ions.

$$0.0500 \text{ g AgBr} \times \frac{1 \text{ mol AgBr}}{187.8 \text{ g AgBr}} \times \frac{2 \text{ mol } S_2O_3^{2-}}{1 \text{ mol AgBr}}$$

$$\times \frac{1 \text{ mol } Na_2S_2O_3}{1 \text{ mol } S_2O_3^{2-}} = 5.324 \times 10^{-4} \text{ mol } Na_2S_2O_3$$

The volume of 0.0150 M $Na_2S_2O_3$ required is obtained using the molarity of the solution.

$$5.324 \times 10^{-4} \text{ mol } Na_2S_2O_3 \times \frac{1 \text{ L solution}}{0.0150 \text{ mol } Na_2S_2O_3}$$

$$= 0.0355 \text{ L of } 0.0150 \text{ M } Na_2S_2O_3 \text{ solution, or } 35.5 \text{ mL}$$

✔   We have 0.050 g/190 g/mol ≈ 0.00026 mol AgBr. We have (0.0150mol/L)(0.0355 L) ≈ 0.00053 mol $S_2O_3^{2-}$. This is about the 2:1 ratio needed according to the balanced equation.

Problem-Solving Practice 5.15

If you had 125 mL of a 0.0200 M solution of $Na_2S_2O_3$ on hand, how many milligrams of AgBr could you dissolve?

Exercise 5.18 Molarity

Sodium chloride is used in intravenous solutions for medical applications. The NaCl concentration in such solutions must be accurately known and can be assessed by reacting the solution with an experimentally determined volume of $AgNO_3$ solution of known concentration. The net ionic equation is

$$Ag^+(aq) + Cl^-(aq) \longrightarrow AgCl(s)$$

Suppose that a chemical technician uses 19.3 mL of 0.200 M $AgNO_3$ to convert all the NaCl in a 25.0-mL sample of an intravenous solution to AgCl. Calculate the molarity of NaCl in the solution.

5.8 AQUEOUS SOLUTION TITRATIONS

One important use of aqueous solution reactions is to determine the unknown concentration of a reactant in a solution, such as the concentration of NaOH in a solution of NaOH. This is done by a titration using a **standard solution,** a solution whose concentration is known accurately. In a **titration,** a substance in the standard solution reacts with a known stoichiometry with the substance whose concentration is to be determined. When the stoichiometrically equivalent amount of standard solution has been added, the **equivalence point** is reached. At that point, the moles of reactant that have been added from the standard solution are exactly what are needed to react completely with the substance whose concentration is to be determined. The progress of the reaction is monitored by an indicator, a dye that changes color at the equivalence point, or through some other means with appropriate instruments. Phenolphthalein, for example, is commonly used as the indicator in strong acid-base titrations because it is colorless in acidic solutions and pink in basic solutions.

A common example of a titration is the determination of the molarity of an acid by titration of it with a standard solution of a base. We can use a standard solution of 0.100 M KOH to determine the concentration of an HCl solution. To carry out this titration, we use 50.0 mL of the HCl solution and slowly add the KOH solution until the equivalence point is reached (Figure 5.19). At that point, the moles of OH^- added to the HCl solution exactly match the moles of H^+ that were in the acid sample.

CD-ROM Screen 5.14: Titrations

Acid-base titrations are covered extensively in Chapter 17.

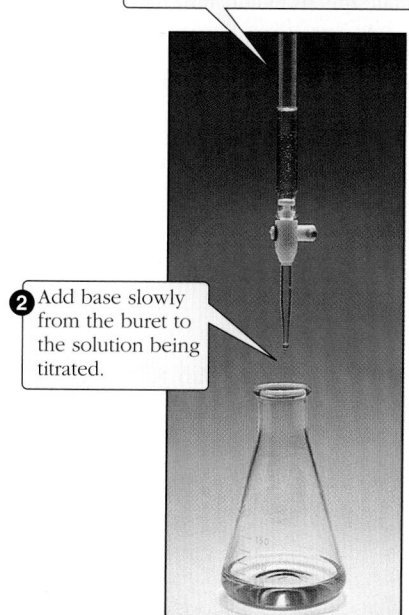

❶ A buret, a volumetric measuring device calibrated in divisions of 0.1 mL, holds an aqueous solution of a base of known concentration.

❷ Add base slowly from the buret to the solution being titrated.

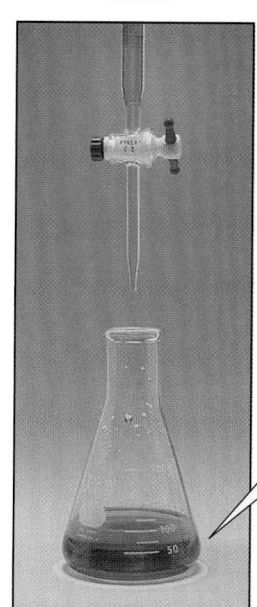

A change in the color of ❸ an indicator signals the equivalence point. (The indicator used here is phenolphthalein.)

Figure 5.19 Titration of an acid in aqueous solution with a standard solution of base.

CD-ROM Screen 5.15: Titration Simulation

Problem-Solving Example 5.16   Acid-Base Titration

A student has a solution of HCl that is approximately 0.2 M and must determine the exact concentration by titration. A 50.0 mL sample of the HCl solution is titrated with a 0.250 M KOH standard solution. The equivalence point is reached when 37.6 mL of the standard KOH solution has been added. What is the molarity of the HCl solution?

Answer   0.188 M HCl solution

Explanation   The balanced chemical equation for this neutralization reaction is

$$HCl(aq) + KOH(aq) \longrightarrow KCl(aq) + H_2O(\ell)$$

The net ionic equation is

$$H^+(aq) + OH^-(aq) \longrightarrow H_2O(\ell)$$

We first calculate the number of moles of KOH consumed in the titration.

$$0.0376 \text{ L KOH solution} \times \frac{0.250 \text{ mol KOH}}{1.00 \text{ L KOH solution}} = 9.41 \times 10^{-3} \text{ mol KOH}$$

From the balanced equation, we know that KOH and HCl react in a one-to-one mole ratio, so there must have been 9.41×10^{-3} mol of HCl in the initial solution. Knowing the number of moles of HCl, we can calculate the molarity of the HCl solution.

$$\frac{9.41 \times 10^{-3} \text{ mol HCl}}{0.0500 \text{ L HCl solution}} = 0.188 \text{ M HCl}$$

✔   At the equivalence point, the number of moles of $H^+(aq)$ and $OH^-(aq)$ must be equal. The number of moles of each reactant is the volume times the molarity. The volume of KOH is somewhat less than the volume of HCl, and the molarity of KOH is somewhat greater than the molarity of HCl. The two factors should about balance out, the answer should be quite close to 0.2 M, and it is.

Problem-Solving Practice 5.16

In a titration, a 20.0-mL sample of sulfuric acid (H_2SO_4) was titrated to the endpoint with 41.3 mL of 0.100 M NaOH. What is the molarity of the H_2SO_4 solution?

SUMMARY PROBLEM

Gold in its elemental state can be separated from gold-bearing rock by treating the ore with cyanide, CN^-, in the presence of oxygen via this reaction:

$$4 \text{ Au}(s) + 8 \text{ CN}^-(aq) + O_2(g) + 2 \text{ H}_2O(\ell) \longrightarrow 4 \text{ Au(CN)}_2^-(aq) + 4 \text{ OH}^-(aq)$$

The CN^- is supplied as NaCN, but Na^+ is a spectator ion and is left out of the net ionic equation.

(a) Which reactant is being oxidized? What are the oxidation numbers of this species as a reactant and as a product?

(b) Which reactant is being reduced? What are the oxidation numbers of this species as a reactant and as a product?

(c) What is the oxidizing agent?

(d) What is the reducing agent?

(e) What mass of NaCN would it take to prepare 1.0 L 0.075 M NaCN?

(f) If the ore contains 0.019% gold by weight, what mass of gold is found in one metric ton (1000 kg) of the ore?

(g) How many grams of NaCN would you require to extract the gold in one metric ton of this ore?

(h) How many liters of the 0.075 M NaCN solution would you require to extract the gold in one metric ton of this ore?

IN CLOSING

Having studied this chapter, you should be able to . . .

- Predict products of common types of chemical reactions: precipitation, acid-base, and gas-forming (Sections 5.1 – 5.2).
- Write a net ionic equation for a given reaction in aqueous solution (Section 5.1).
- Recognize common acids and bases and predict when neutralization reactions will occur (Section 5.2).
- Recognize oxidation-reduction reactions and common oxidizing and reducing agents (Section 5.3).
- Assign oxidation numbers to reactants and products in a redox reaction, identify what has been oxidized or reduced, and identify oxidizing agents and reducing agents (Section 5.4).
- Use the activity series to predict products of displacement redox reactions (Section 5.5).
- Define molarity and calculate molar concentrations (Section 5.6).
- Determine how to prepare a solution of a given molarity from the solute and water or by dilution of a more concentrated solution (Section 5.6).
- Solve stoichiometry problems using solution molarities (Section 5.7).
- Understand how aqueous solution titrations can be used to determine the concentration of an unknown solution (Section 5.8).

KEY TERMS

acid *(5.2)*

base *(5.2)*

concentration *(5.6)*

equivalence point *(5.8)*

hydronium ion *(5.2)*

hydroxide ion *(5.2)*

metal activity series *(5.5)*

molarity *(5.6)*

net ionic equation *(5.1)*

oxidation *(5.3)*

oxidation number *(5.4)*

oxidation-reduction reaction *(5.3)*

oxidized *(5.3)*

oxidizing agent *(5.3)*

precipitate *(5.1)*

redox reactions *(5.3)*

reduced *(5.3)*

reducing agent *(5.3)*

reduction *(5.3)*

salt *(5.2)*

solute *(5.6)*

solvent *(5.6)*

spectator ions *(5.1)*

standard solution *(5.8)*

strong acid *(5.2)*

strong base *(5.2)*

titration *(5.8)*

weak acid *(5.2)*

weak base *(5.2)*

weak electrolyte *(5.2)*

QUESTIONS FOR REVIEW AND THOUGHT

Conceptual Challenge Problems

CP-5.A (Section 5.3) There is a conservation of the number of electrons exchanged during redox reactions, which is tantamount to stating that electric charge is conserved during chemical reactions. The assignment of oxidation numbers is an arbitrary yet clever way to do the bookkeeping for these electrons. What makes it possible to assign the same oxidation number to all elements that are not bound to other elements not in chemical compounds?

CP-5.B (Section 5.4) Consider the following redox reactions:

$$HIO_3 + FeI_2 + HCl \longrightarrow FeCl_3 + ICl + H_2O$$

$$CuSCN + KIO_3 + HCl \longrightarrow$$

$$CuSO_4 + KCl + HCN + ICl + H_2O$$

(a) Identify the species that have been oxidized or reduced in each of the reactions.

(b) After you have correctly identified the species that have been oxidized or reduced in each equation, you might like to try using oxidation numbers to balance each equation. This will be a challenge because, as you have discovered, there is more than one kind of atom that is oxidized or reduced, although in all cases the product of the oxidation and reduction is unambiguous. Record the initial and final oxidation states of each kind of atom that is oxidized or reduced in each equation. Then decide on the coefficients that will equalize the oxidation number changes and satisfy any other atom balancing needed. Finally, balance the equation by adding the correct coefficients to it.

CP-5.C (Section 5.5) A student was given four metals (A, B, C, and D) and solutions of their corresponding salts (AZ, BZ, CZ, and DZ). The student was asked to determine the relative reactivity of the four metals by reacting the metals with the solutions. The student's laboratory observations are indicated in the table at the right. Arrange the four metals in order of decreasing activity.

CP-5.D (Section 5.6) How would you prepare 1 L of 1.00×10^{-6} M NaCl (molar mass = 58.44 g/mol) solution by using a balance that can measure mass only to 0.01 g?

CP-5.E (Section 5.8) How could you show that when baking soda reacts with the acetic acid, CH_3COOH, in vinegar, the carbon and oxygen atoms in the carbon dioxide produced all come from the baking soda alone and none from the acetic acid in vinegar?

Metal	AZ(aq)	BZ(aq)	CZ(aq)	DZ(aq)
A	No reaction	No reaction	No reaction	No reaction
B	Reaction	No reaction	Reaction	No reaction
C	Reaction	No reaction	No reaction	No reaction
D	Reaction	Reaction	Reaction	No reaction

Answers to questions in **bold** can be found in the back of the book.

Review Questions

1. Find in this chapter one example of each of the following reaction types, and write the balanced equation for the reaction: (a) combustion, (b) combination, (c) exchange, (d) decomposition, and (e) oxidation-reduction. Name the products of each reaction.
2. Classify each of the following reactions as a combination, decomposition, exchange, acid-base, or oxidation-reduction reaction.
   (a) $MgO(s) + 2\ HCl(aq) \xrightarrow{heat} MgCl_2(ag) + H_2O(\ell)$
   (b) $2\ NaHCO_3(s) \xrightarrow{heat} Na_2CO_3(s) + CO_2(g) + H_2O(g)$
   (c) $CaO(s) + SO_2(g) \rightarrow CaSO_3(s)$
   (d) $3\ Cu(s) + 8\ HNO_3(aq) \rightarrow 3\ Cu(NO_3)_2(aq) + 2\ NO(g) + 4\ H_2O(\ell)$
   (e) $2\ NO(g) + O_2(g) \rightarrow 2\ NO_2(g)$
3. Find two examples in this chapter of the reaction of a metal with a halogen, write a balanced equation for each example, and name the product.
4. Find two examples of acid-base reactions in this chapter. Write balanced equations for these reactions, and name the reactants and products.
5. Find two examples of precipitation reactions in this chapter. Write balanced equations for these reactions, and name the reactants and products.
6. Find an example of a gas-forming reaction in this chapter. Write a balanced equation for the reaction, and name the reactants and products.
7. Explain the difference between oxidation and reduction. Give an example of each.
8. For each of the following, does the oxidation number increase or decrease in the course of a redox reaction?
   (a) An oxidizing agent
   (b) A reducing agent
   (c) A substance undergoing oxidation
   (d) A substance undergoing reduction

9. Explain the difference between an oxidizing agent and a reducing agent. Give an example of each.

Solubility

10. Tell how to predict that $Ni(NO_3)_2$ is soluble in water, whereas $NiCO_3$ is not soluble in water.
11. Predict whether each of the following compounds is likely to be water-soluble. Indicate which ions are present for the water-soluble compounds.
   (a) $Fe(ClO_4)_2$ (b) Na_2SO_4
   (c) KBr   (d) Na_2CO_3
12. Predict whether each of the following compounds is likely to be water-soluble. Indicate which ions are present for the water-soluble compounds.
   (a) Potassium hydrogen phosphate
   (b) Sodium hypochlorite
   (c) Magnesium chloride
   (d) Calcium hydroxide
   (e) Aluminum bromide

Exchange Reactions

13. Write a balanced equation for the reaction of nitric acid with calcium hydroxide.
14. For each of the following pairs of ionic compounds, write a balanced equation reflecting whether precipitation will occur in aqueous solution. For those combinations that do not produce a precipitate, write "NR."
   (a) $MnCl_2 + Na_2S$
   (b) $HNO_3 + CuSO_4$
   (c) $NaOH + HClO_4$
   (d) $Hg(NO_3)_2 + Na_2S$
   (e) $Pb(NO_3)_2 + HCl$
   (f) $BaCl_2 + H_2SO_4$
15. Name the spectator ions in the reaction of nitric acid and magnesium hydroxide, and write the net ionic equation from the following complete ionic equation.

$$2\,H^+(aq) + 2\,NO_3^-(aq) + Mg(OH)_2(s) \longrightarrow$$
$$2\,H_2O(\ell) + Mg^{2+}(aq) + 2\,NO_3^-(aq)$$

What type of reaction is this?

16. Name the water-insoluble product in each reaction.
    (a) $CuCl_2(aq) + H_2S(aq) \rightarrow CuS + 2\,HCl$
    (b) $CaCl_2(aq) + K_2CO_3(aq) \rightarrow 2\,KCl + CaCO_3$
    (c) $AgNO_3(aq) + NaI(aq) \rightarrow AgI + NaNO_3$

17. Name the spectator ions in the reactions from Question 16, and write the net ionic equations for those reactions.

18. Balance each of the following equations, and then write the complete and net ionic equations.
    (a) $Zn(s) + HCl(aq) \rightarrow H_2(g) + ZnCl_2(aq)$
    (b) $Mg(OH)_2(s) + HCl(aq) \rightarrow MgCl_2(aq) + H_2O(\ell)$
    (c) $HNO_3(aq) + CaCO_3(s) \rightarrow Ca(NO_3)_2(aq) + H_2O(\ell) + CO_2(g)$
    (d) $HCl(aq) + MnO_2(s) \rightarrow MnCl_2(aq) + Cl_2(g) + H_2O(\ell)$

19. Balance each of the following equations, and then write the complete and net ionic equations.
    (a) $(NH_4)_2CO_3(aq) + Cu(NO_3)_2(aq) \rightarrow CuCO_3(s) + NH_4NO_3(aq)$
    (b) $Pb(NO_3)_2(aq) + HCl(aq) \rightarrow PbCl_2(s) + HNO_3(aq)$
    (c) $BaCO_3(s) + HCl(aq) \rightarrow BaCl_2(aq) + H_2O(\ell) + CO_2(g)$

20. Balance each of the following equations, and then write the complete and net ionic equations. Refer to Tables 5.1 and 5.2 for information on solubility and on acids and bases. Show states (s, ℓ, g, aq) for all reactants and products.
    (a) $Ca(OH)_2 + HNO_3 \rightarrow Ca(NO_3)_2 + H_2O$
    (b) $BaCl_2 + Na_2CO_3 \rightarrow BaCO_3 + NaCl$
    (c) $Na_3PO_4 + Ni(NO_3)_2 \rightarrow Ni_3(PO_4)_2 + NaNO_3$

21. Balance each of the following equations, and then write the complete and net ionic equations. Refer to Tables 5.1 and 5.2 for information on solubility and on acids and bases. Show states (s, ℓ, g, aq) for all reactants and products.
    (a) $ZnCl_2 + KOH \rightarrow KCl + Zn(OH)_2$
    (b) $AgNO_3 + KI \rightarrow AgI + KNO_3$
    (c) $NaOH + FeCl_2 \rightarrow Fe(OH)_2 + NaCl$

22. Barium hydroxide is used in lubricating oils and greases. Write a balanced equation for the reaction of this hydroxide with nitric acid to give barium nitrate, a compound used in pyrotechnics devices such as green flares.

23. Aluminum is obtained from bauxite, which is not a specific mineral but a name applied to a mixture of minerals. One of those minerals, which can dissolve in acids, is gibbsite, $Al(OH)_3$. Write a balanced equation for the reaction of gibbsite with sulfuric acid.

24. Balance the equation for the following precipitation reaction, and then write the complete and net ionic equations.

$$CdCl_2 + NaOH \longrightarrow Cd(OH)_2 + NaCl$$

25. Balance the equation for the following precipitation reaction, and then write the complete and net ionic equations.

$$Ni(NO_3)_2 + Na_2CO_3 \longrightarrow NiCO_3 + NaNO_3$$

26. Write an overall balanced equation for the precipitation reaction that occurs when aqueous lead(II) nitrate is mixed with an aqueous solution of potassium chloride. Name each reactant and product. Indicate the state of each substance (s, ℓ, g, or aq).

27. Write an overall balanced equation for the precipitation reaction that occurs when aqueous copper(II) nitrate is mixed with an aqueous solution of sodium carbonate. Name each reactant and product. Indicate the state of each substance (s, ℓ, g, or aq).

28. The beautiful mineral rhodochrosite is manganese(II) carbonate. Write an overall balanced equation for the reaction of the mineral with hydrochloric acid. Name each reactant and product.

Acids, Bases, and Salts

29. Classify each of the following as an acid or a base. What ions are produced when each is dissolved in water?
    (a) KOH   (b) $Mg(OH)_2$ (c) HClO
    (d) HBr   (e) LiOH   (f) H_2SO_3

30. For each acid and base in Question 29, which are strong and which are weak?

31. Identify the acid and base used to form the following salts.
    (a) $NaNO_2$ (b) $CaSO_4$ (c) NaI
    (d) $Mg_3(PO_4)_2$ (e) $NaCH_3COO$

32. For each salt in Question 31, write the overall neutralization reaction that formed each salt. Write the complete and net ionic equations for each neutralization reaction.

33. Classify each of the following exchange reactions as an acid-base reaction, a precipitation reaction, or a gas-forming reaction. Predict the products of the reaction, and then balance the completed equation.
    (a) $MnCl_2(aq) + Na_2S(aq) \rightarrow$
    (b) $Na_2CO_3(aq) + ZnCl_2(aq) \rightarrow$
    (c) $K_2CO_3(aq) + HClO_4(aq) \rightarrow$

34. Classify each of the following exchange reactions as an acid-base reaction, a precipitation reaction, or a gas-forming reaction. Predict the products of the reaction, and then balance the completed equation.
    (a) $Fe(OH)_3(s) + HNO_3(aq) \rightarrow$
    (b) $FeCO_3(s) + H_2SO_4(aq) \rightarrow$
    (c) $FeCl_2(aq) + (NH_4)_2S(aq) \rightarrow$
    (d) $Fe(NO_3)_2(aq) + Na_2CO_3(aq) \rightarrow$

Oxidation-Reduction Reactions

35. Assign oxidation numbers to each element in the following compounds.
    (a) SO_3 (b) HNO_3 (c) $KMnO_4$
    (d) H_2O (e) LiOH   (f) CH_2Cl_2

36. Assign oxidation numbers to each element in the following compounds.
    (a) $Fe(OH)_3$ (b) $HClO_3$ (c) $CuCl_2$
    (d) K_2CrO_4 (e) $Ni(OH)_2$ (f) N_2H_4

37. Assign oxidation numbers to each element in the following ions.
    (a) SO_4^{2-} (b) NO_3^- (c) MnO_4^-
    (d) $Cr(OH)_4^-$ (e) $H_2PO_4^-$ (f) $S_2O_3^-$

38. Which of the following reactions are oxidation-reduction reactions? Explain your answer briefly.

Classify the remaining reactions.
(a) $CdCl_2(aq) + Na_2S(aq) \longrightarrow CdS(s) + 2\ NaCl(aq)$
(b) $2\ Ca(s) + O_2(g) \longrightarrow 2\ CaO(s)$
(c) $Ca(OH)_2(s) + 2\ HCl(aq) \longrightarrow CaCl_2(aq) + 2\ H_2O(\ell)$

39. Which of the following reactions are oxidation-reduction reactions? Explain your answer in each case. Classify the remaining reactions.
(a) $Zn(s) + 2\ NO_3^-(aq) + 4\ H_3O^+(aq) \longrightarrow$
$Zn^{2+}(aq) + 2\ NO_2(g) + 6\ H_2O(\ell)$
(b) $Zn(OH)_2(s) + H_2SO_4(aq) \longrightarrow ZnSO_4(aq) + 2\ H_2O(\ell)$
(c) $Ca(s) + 2\ H_2O(\ell) \longrightarrow Ca(OH)_2(s) + H_2(g)$

40. Which region of the periodic table has the best reducing agents? The best oxidizing agents?

41. Which of the following substances are oxidizing agents?
(a) Zn (b) O_2 (c) HNO_3
(d) MnO_4^- (e) H_2 (f) H^+

42. Which of the following substances are reducing agents?
(a) Ca (b) Ca^{2+} (c) $Cr_2O_7^{2-}$
(d) Al (e) Br_2 (f) H_2

43. Identify the products of the following redox combination reactions.
(a) $C(s) + O_2(g) \longrightarrow$ (b) $P_4(s) + Cl_2(g) \longrightarrow$
(c) $Ti(s) + Cl_2(g) \longrightarrow$ (d) $Mg(s) + N_2(g) \longrightarrow$
(e) $FeO(s) + O_2(g) \longrightarrow$ (f) $NO(g) + O_2(g) \longrightarrow$

44. Complete and balance the following equations for redox displacement reactions.
(a) $K(s) + H_2O(\ell) \longrightarrow$ (b) $Mg(s) + HBr(aq) \longrightarrow$
(c) $NaBr(aq) + Cl_2(aq) \longrightarrow$ (d) $WO_3(s) + H_2(g) \longrightarrow$
(e) $H_2S(aq) + Cl_2(aq) \longrightarrow$

Activity Series

45. Give an example of a displacement reaction that is also a redox reaction and identify which species is (a) oxidized, (b) reduced, (c) the reducing agent, and (d) the oxidizing agent.

46. (a) In what groups of the periodic table are the most reactive metals found? Where do we find the least reactive metals?
(b) Silver (Ag) does not react with 1 M HCl solution. Will Ag react with a solution of aluminum nitrate, $Al(NO_3)_3$? If so, write a chemical equation for the reaction.
(c) Lead (Pb) will react very slowly with 1 M HCl solution. Aluminum will react with lead(II) sulfate solution $(PbSO_4)$. Will Pb react with an $AgNO_3$ solution? If so, write a chemical equation for the reaction.
(d) On the basis of the information obtained in answering parts (a), (b), and (c), arrange Ag, Al, and Pb in decreasing order of reactivity.

47. Use the activity series of metals (Table 5.5) to predict the outcome of each of the following reactions. If no reaction occurs, write N.R.
(a) $Na^+(aq) + Zn(s) \longrightarrow$ (b) $HCl(aq) + Pt(s) \longrightarrow$
(c) $Ag^+(aq) + Au(s) \longrightarrow$ (d) $Au^{3+}(aq) + Ag(s) \longrightarrow$

48. Using the activity series of metals (Table 5.5), predict whether the following reactions will occur in aqueous solution.
(a) $Mg(s) + Ca(s) \longrightarrow Mg^{2+} + Ca^{2+}$
(b) $2\ Al^{3+} + 3\ Pb^{2+} \longrightarrow 2\ Al(s) + 3\ Pb(s)$
(c) $H_2(g) + Zn^{2+} \longrightarrow 2\ H^+ + Zn(s)$
(d) $Mg(s) + Cu^{2+} \longrightarrow Mg^{2+} + Cu(s)$

(e) $Pb(s) + 2\ H^+ \longrightarrow H_2(g) + Pb^{2+}$
(f) $2\ Ag^+ + Cu(s) \longrightarrow 2\ Ag(s) + Cu^{2+}$
(g) $2\ Al^{3+} + 3\ Zn(s) \longrightarrow 3\ Zn^{2+} + 2\ Al(s)$

Halogens in Redox Reactions

49. Which halogen is the strongest oxidizing agent? Which is the strongest reducing agent?

50. Predict the products of the following halogen displacement reactions. If no reaction occurs, write N.R.
(a) $I_2(s) + NaBr(aq) \longrightarrow$ (b) $Br_2(\ell) + NaI(aq) \longrightarrow$
(c) $F_2(g) + NaCl(aq) \longrightarrow$ (d) $Cl_2(g) + NaBr(aq) \longrightarrow$
(e) $Br_2(\ell) + NaCl(aq) \longrightarrow$ (f) $Cl_2(g) + NaF(aq) \longrightarrow$

51. For the reactions in Question 50 that occurred, identify the species oxidized or reduced as well as the oxidizing and reducing agents.

52. For the reactions in Question 50 that do not occur, rewrite the equation so that a reaction does occur (consider the halogen activity series).

Solution Concentrations

53. You have a 0.12 M solution of $BaCl_2$. What ions exist in solution, and what are their concentrations?

54. A flask contains 0.25 M $(NH_4)_2SO_4$. What ions exist in the solution, and what are their concentrations?

55. Assume that 6.73 g of Na_2CO_3 is dissolved in enough water to make 250. mL of solution. (a) What is the molarity of the sodium carbonate? (b) What are the concentrations of the Na^+ and CO_3^- ions?

56. Some $K_2Cr_2O_7$, with a mass of 2.335 g, is dissolved in enough water to make 500. mL of solution. (a) What is the molarity of the potassium dichromate? (b) What are the concentrations of the K^+ and $Cr_2O_7^{2-}$ ions?

57. What is the mass, in grams, of solute in 250. mL of a 0.0125 M solution of $KMnO_4$?

58. What is the mass, in grams, of solute in 100. mL of a 1.023×10^{-3} M solution of Na_3PO_4?

59. What volume of 0.123 M NaOH, in milliliters, contains 25.0 g of NaOH?

60. What volume of 2.06 M $KMnO_4$, in liters, contains 322 g of solute?

61. If 6.00 mL of 0.0250 M $CuSO_4$ is diluted to 10.0 mL with pure water, what is the concentration of copper(II) sulfate in the diluted solution?

62. If you dilute 25.0 mL of 1.50 M HCl to 500. mL, what is the molar concentration of the diluted HCl?

63. If you need 1.00 L of 0.125 M H_2SO_4, which method would you use to prepare this solution?
(a) Dilute 36.0 mL of 1.25 M H_2SO_4 to a volume of 1.00 L.
(b) Dilute 20.8 mL of 6.00 M H_2SO_4 to a volume of 1.00 L.
(c) Add 950. mL of water to 50.0 mL of 3.00 M H_2SO_4.
(d) Add 500. mL of water to 500. mL of 0.500 M H_2SO_4.

64. If you need 300. mL of 0.500 M K_2CrO_7, which method would you use to prepare this solution?
(a) Dilute 250. mL of 0.600 M K_2CrO_7 to 300. mL.
(b) Add 50.0 mL of water to 250. mL of 0.250 M K_2CrO_7.
(c) Dilute 125 mL of 1.00 M K_2CrO_7 to 300. mL.
(d) Add 30.0 mL of 1.50 M K_2CrO_7 to 270. mL of water.

Calculations for Reactions in Solution

65. What mass in grams of Na_2CO_3 is required for complete reaction with 25.0 mL of 0.155 M HNO_3?

$$Na_2CO_3(aq) + 2\ HNO_3(aq) \longrightarrow$$
$$2\ NaNO_3(aq) + CO_2(g) + H_2O(\ell)$$

66. Hydrazine, N_2H_4, a base like ammonia, can react with an acid such as sulfuric acid.

$$2\ N_2H_4(aq) + H_2SO_4(aq) \longrightarrow 2\ N_2H_5^+(aq) + SO_4^{2-}(aq)$$

What mass of hydrazine can react with 250. mL of 0.225 M H_2SO_4?

67. What volume in milliliters of 0.125 M HNO_3 is required to react completely with 1.30 g of $Ba(OH)_2$?

$$2\ HNO_3(aq) + Ba(OH)_2(s) \longrightarrow Ba(NO_3)_2(aq) + 2\ H_2O(\ell)$$

68. Diborane, B_2H_6, can be produced by the following reaction.

$$2\ NaBH_4(aq) + H_2SO_4(aq) \longrightarrow$$
$$2\ H_2(g) + Na_2SO_4(aq) + B_2H_6(g)$$

What volume in milliliters of 0.0875 M H_2SO_4 should be used to completely consume 1.35 g of $NaBH_4$?

69. What volume in milliliters of 0.512 M NaOH is required to react completely with 25.0 mL of 0.234 M H_2SO_4?

70. What volume in milliliters of 0.812 M HCl would be required to neutralize 15.0 mL of 0.635 M NaOH?

71. What is the maximum mass, in grams, of AgCl that can be precipitated by mixing 50.0 mL of 0.025 M $AgNO_3$ solution with 100.0 mL of 0.025 M NaCl solution? Which reactant is in excess? What is the concentration of the excess reactant remaining in solution after the AgCl has been precipitated?

72. Suppose you mix 25.0 mL of 0.234 M $FeCl_3$ solution with 42.5 mL of 0.453 M NaOH. (a) What is the maximum mass, in grams, of $Fe(OH)_3$ that will precipitate? (b) Which reactant is in excess? (c) What is the concentration of the excess reactant remaining in solution after the maximum mass of $Fe(OH)_3$ has precipitated?

73. A soft drink contains an unknown amount of citric acid, $C_3H_5O(COOH)_3$. A volume of 10.0 mL of the soft drink requires 6.42 mL of 9.580×10^{-2} M NaOH to neutralize the citric acid.

$$C_3H_5O(COOH)_3(aq) + NaOH(aq) \longrightarrow$$
$$Na_3C_3H_5O(COO)_3(aq) + 3\ H_2O(\ell)$$

(a) Which step in the following calculations for mass of citric acid in 1 mL of soft drink is not correct?
(b) What is the correct answer?
  (i) Moles NaOH = (6.42 mL) (1 L/1000 mL) (9.580×10^{-2} mol/L)
  (ii) Moles citric acid = (6.15×10^{-4} mol NaOH) (3 mol citric acid/1 mol NaOH)
  (iii) Mass citric acid in sample = (1.85×10^{-3} mol citric acid) (192.12 g/mol citric acid)
  (iv) Mass citric acid in 1 mL soft drink = (0.354 g citric acid)/(10 mL soft drink)

74. Vitamin C is the compound $C_6H_8O_6$. Besides being an acid, it is also a reducing agent that reacts readily with bromine, Br_2, a good oxidizing agent.

$$C_6H_8O_6(aq) + Br_2(aq) \longrightarrow 2\ HBr(aq) + C_6H_6O_6(aq)$$

Suppose a 1.00-g chewable vitamin C tablet requires 27.85 mL of 0.102 M Br_2 to react completely.
(a) Which step in the following calculations for mass, in grams, of vitamin C in the tablet is incorrect?
(b) What is the correct answer?
  (i) Mole Br_2 = (27.85 mL) (0.102 mol/L)
  (ii) Moles $C_6H_8O_6$ = (2.84 mol Br_2) (1 mol $C_6H_8O_6$/1 mol Br_2)
  (iii) Mass $C_6H_8O_6$ = (2.84 mol $C_6H_8O_6$) (176 g/mol $C_6H_8O_6$)
  (iv) Mass $C_6H_8O_6$ = (500 g $C_6H_8O_6$)/(1 g tablet)

75. If a volume of 32.45 mL of HCl is used to completely neutralize 2.050 g of Na_2CO_3 according to the following equation, what is the molarity of the HCl?

$$Na_2CO_3(aq) + 2\ HCl(aq) \longrightarrow$$
$$2\ NaCl(aq) + CO_2(g) + H_2O(\ell)$$

76. Potassium acid phthalate, $KHC_8H_4O_4$, is used to standardize solutions of bases. The acidic anion reacts with bases according to the following net ionic equation.

$$HC_8H_4O_4^-(aq) + OH^-(aq) \longrightarrow H_2O(\ell) + C_8H_4O_4^{2-}(aq)$$

If a 0.902-g sample of potassium acid phthalate requires 26.45 mL of NaOH to react, what is the molarity of the NaOH?

77. Sodium thiosulfate, $Na_2S_2O_3$, is used as a "fixer" in black-and-white photography. Assume you have a bottle of sodium thiosulfate and want to determine its purity. The thiosulfate ion can be oxidized with I_2 according to the equation

$$I_2(aq) + 2\ S_2O_3^{2-}(aq) \longrightarrow 2\ I^-(aq) + S_4O_6^{2-}(aq)$$

If you use 40.21 mL of 0.246 M I_2 to completely react a 3.232-g sample of impure $Na_2S_2O_3$, what is the percent purity of the $Na_2S_2O_3$?

78. A sample of a mixture of oxalic acid, $H_2C_2O_4$, and sodium chloride, NaCl, has a mass of 4.554 g. If a volume of 29.58 mL of 0.550 M NaOH is required to neutralize all the $H_2C_2O_4$, what is the weight percent of oxalic acid in the mixture? Oxalic acid and NaOH react according to the equation

$$H_2C_2O_4(aq) + 2\ NaOH(aq) \longrightarrow Na_2C_2O_4(aq) + 2\ H_2O(\ell)$$

79. You are given 0.954 g of an unknown acid, H_2A, which reacts with NaOH according to the balanced equation

$$H_2A(aq) + 2\ NaOH(aq) \longrightarrow Na_2A(aq) + 2\ H_2O(\ell)$$

If a volume of 36.04 mL of 0.509 M NaOH is required to react with all of the acid, what is the molar mass of the acid?

80. You are given an acid and told only that it could be citric acid (molar mass = 192.1 g/mol) or tartaric acid (molar mass = 150.1 g/mol). To determine which acid you have, you react it with NaOH. The appropriate reactions are

Citric acid: $C_6H_8O_7 + 3\ NaOH \longrightarrow Na_3C_6H_5O_7 + 3\ H_2O$

Tartaric acid: $C_4H_6O_6 + 2\ NaOH \longrightarrow Na_2C_4H_4O_6 + 2\ H_2O$

You find that a 0.956-g sample requires 29.1 mL of 0.513 M NaOH to reach the equivalency point. What is the unknown acid?

General Questions

81. Name the spectator ions in the reaction of calcium carbonate and hydrochloric acid, and write the net ionic equation.

$$CaCO_3(s) + 2\,H^+(aq) + 2\,Cl^-(aq) \longrightarrow$$
$$CO_2(g) + Ca^{2+}(aq) + 2\,Cl^-(aq) + H_2O(\ell)$$

What type of reaction is this?

82. Magnesium metal reacts readily with HNO_3, as shown in the following equation.

$$Mg(s) + HNO_3(aq) \longrightarrow$$
$$Mg(NO_3)_2(aq) + NO_2(g) + H_2O(\ell)$$

(a) Balance the equation for the reaction.
(b) Name each reactant and product.
(c) Write the net ionic equation for the reaction.
(d) What type of reaction is this?

83. Aqueous solutions of $(NH_4)_2S$ and $Hg(NO_3)_2$ react to give HgS and NH_4NO_3.
(a) Write the overall balanced equation for the reaction. Indicate the state (s or aq) for each compound.
(b) Name each compound.
(c) Write the net ionic equation for the reaction.
(d) What type of reaction does this appear to be?

84. Classify the following reactions and predict the products formed.
(a) $Li(s) + H_2O(\ell) \longrightarrow$
(b) $Ag_2O(s) \xrightarrow{heat}$
(c) $Li_2O(s) + H_2O(\ell) \longrightarrow$
(d) $I_2(s) + Cl^-(aq) \longrightarrow$
(e) $Cu(s) + HCl(aq) \longrightarrow$
(f) $BaCO_3(s) \xrightarrow{heat}$

85. Classify the following reactions and predict the products formed:
(a) $SO_3(g) + H_2O(\ell) \longrightarrow$
(b) $Sr(s) + H_2(g) \longrightarrow$
(c) $Mg(s) + H_2SO_4(aq, dilute) \longrightarrow$
(d) $Na_3PO_4(aq) + AgNO_3(aq) \longrightarrow$
(e) $Ca(HCO_3)_2(s) \xrightarrow{heat}$
(f) $Fe^{3+}(aq) + Sn^{2+}(aq) \longrightarrow$

86. Azurite is a copper-containing mineral that often forms beautiful crystals. Its formula is $Cu_3(CO_3)_2(OH)_2$. Write a balanced equation for the reaction of this mineral with hydrochloric acid.

87. What species (atoms, molecules, ions) are present in an aqueous solution of each of the following compounds?
(a) NH_3 (b) CH_3COOH
(c) $NaOH$ (d) HBr

88. Use the activity series to predict whether the following reactions will occur.
(a) $Fe(s) + Mg^{2+}(aq) \longrightarrow Mg(s) + Fe^{2+}(aq)$
(b) $Ni(s) + Cu^{2+}(aq) \longrightarrow Ni^{2+}(aq) + Cu(s)$
(c) $Cu(s) + 2\,H^+(aq) \longrightarrow Cu^{2+}(aq) + H_2(g)$
(d) $Mg(s) + H_2O(g) \longrightarrow MgO(s) + H_2(g)$

89. Determine which of the following are redox reactions. Identify the oxidizing and reducing agents in each of the redox reactions.
(a) $NaOH(aq) + H_3PO_4(aq) \longrightarrow NaH_2PO_4(aq) + H_2O(\ell)$

(b) $NH_3(g) + CO_2(g) + H_2O(\ell) \longrightarrow NH_4HCO_3(aq)$
(c) $TiCl_4(g) + 2\,Mg(\ell) \xrightarrow{heat} Ti(s) + 2\,MgCl_2(\ell)$
(d) $NaCl(s) + NaHSO_4(aq) \xrightarrow{heat} HCl(g) + Na_2SO_4(s)$

90. Identify the substance oxidized, the substance reduced, the reducing agent, and the oxidizing agent in the equations in Question 89. For each oxidized or reduced substance, identify the change in its oxidation number.

91. Much has been written about chlorofluorocarbons and their impact on our environment. Their manufacture begins with the preparation of HF from the mineral fluorspar, CaF2, according to the following *unbalanced* equation.

$$CaF_2(s) + H_2SO_4(aq) \longrightarrow HF(g) + CaSO_4(s)$$

The HF is combined with, for example, CCl_4 in the presence of $SbCl_5$ to make CCl_2F_2, called dichlorodifluoromethane or CFC-12, and other chlorofluorocarbons.

$$2\,HF(g) + CCl_4(\ell) \longrightarrow CCl_2F_2(g) + 2\,HCl(g)$$

(a) Balance the first equation above and name each substance.
(b) Is the first reaction best classified as an acid-base reaction, an oxidation-reduction reaction, or a precipitation reaction?
(c) Give the names of the compounds CCl_4, $SbCl_5$, and HCl.
(d) Another chlorofluorocarbon produced in the reaction is composed of 8.74% C, 77.43% Cl, and 13.83% F. What is the empirical formula of the compound?

Applying Concepts

92. When the following pairs of reactants are combined in a beaker, (a) describe in words what the contents of the beaker would look like before and after any reaction occurs, (b) use different circles for atoms, molecules, and ions to draw a nanoscale (particulate-level) diagram of what the contents would look like, and (c) write a chemical equation to represent symbolically what the contents would look like.

$$LiCl(aq) \text{ and } AgNO_3(aq)$$
$$NaOH(aq) \text{ and } HCl(aq)$$

93. When the following pairs of reactants are combined in a beaker, (a) describe in words what the contents of the beaker would look like before and after any reaction occurs, (b) use different circles for atoms, molecules, and ions to draw a particulate-level diagram of what the contents would look like, and (c) write a chemical equation to represent symbolically what the contents would look like.

$$CaCO_3(s) \text{ and } HCl(aq)$$
$$NH_4NO_3(aq) \text{ and } KOH(aq)$$

94. Explain how you could prepare barium sulfate by (a) an acid-base reaction, (b) a precipitation reaction, and (c) a gas-forming reaction. The materials you have to start with are $BaCO_3$, $Ba(OH)_2$, Na_2SO_4, and H_2SO_4.

95. Students were asked to prepare nickel sulfate by reacting a nickel compound with a sulfate compound in water and

then evaporating the water. Three students chose these pairs of reactants.

Student 1	Ni(OH)$_2$ and H$_2$SO$_4$
Student 2	Ni(NO$_3$)$_2$ and Na$_2$SO$_4$
Student 3	NiCO$_3$ and H2SO$_4$

Comment on each student's choice of reactants and how successful you think each student will be at preparing nickel sulfate by the procedure indicated.

96. An unknown solution contains either lead ions or barium ions, but not both. Which one of the following aqueous solutions could you use to tell whether the ions present are Pb^{2+} or Ba^{2+}? Explain the reasoning behind the choice.

$$HCl(aq), H_2SO_4(aq), H_3PO_4(aq)$$

97. An unknown solution contains either calcium ions or strontium ions, but not both. Which one of the following solutions could you use to tell whether the ions present are Ca^{2+} or Sr^{2+}? Explain the reasoning behind the choice.

$$NaOH(aq), H_2SO_4(aq), H_2S(aq)$$

98. When given an oxidation-reduction reaction and asked what is oxidized or what is reduced, why should you never choose one of the products for your answer?

99. When given an oxidation-reduction reaction and asked what is the oxidizing agent or what is the reducing agent, why should you never choose one of the products for your answer?

100. You prepared a NaCl solution by adding 58.44 g of NaCl to a 1-L volumetric flask and then adding water to dissolve it. When finished, the final volume in your flask looked like this:

1.00-L flask

The solution you prepared is
(a) greater than 1 M because you added more solvent than necessary.
(b) less than 1 M because you added less solvent than necessary.
(c) greater than 1 M because you added less solvent than necessary.

(d) less than 1 M because you added more solvent than necessary.
(e) is 1 M because the amount of solute, not solvent, determines the concentration.

101. These drawings represent beakers of aqueous solutions. Each orange circle represents a dissolved solute particle.

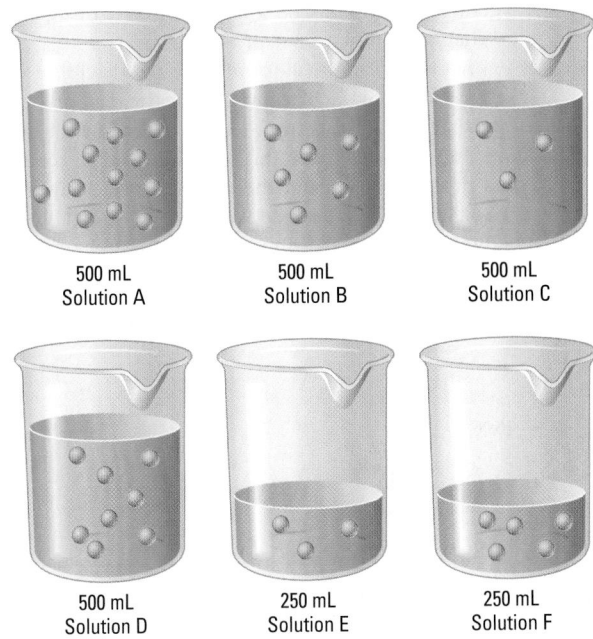

500 mL
Solution A

500 mL
Solution B

500 mL
Solution C

500 mL
Solution D

250 mL
Solution E

250 mL
Solution F

(a) Which solution is most concentrated?
(b) Which solution is least concentrated?
(c) Which two solutions have the same concentration?
(d) When solutions E and F are combined, the resulting solution has the same concentration as solution _____.
(e) When solutions C and E are combined, the resulting solution has the same concentration as solution _____.
(f) If you evaporate half of the water from solution B, the resulting solution will have the same concentration as solution _____.
(g) If you place half of solution A in another beaker and then add 250 mL of water, the resulting solution will have the same concentration as solution _____.

102. Ten milliliters of a solution of an acid is mixed with 10 mL of a solution of a base. When the mixture was tested with litmus paper, the blue litmus turned red, and the red litmus remained red. Which of the following interpretations is (are) correct?
(a) The mixture contains more hydrogen ions than hydroxide ions.
(b) The mixture contains more hydroxide ions than hydrogen ions.
(c) When an acid and a base react, water is formed, and so the mixture cannot be acidic or basic.
(d) If the acid was HCl and the base was NaOH, the concentration of HCl in the initial acidic solution must have been greater than the concentration of NaOH in the initial basic solution.
(e) If the acid was H$_2$SO$_4$ and the base was NaOH, the concentration of H$_2$SO$_4$ in the initial acidic solution must

have been greater than the concentration of NaOH in the initial basic solution.

103. A chemical company was interested in characterizing a competitor's organic acid (it consists of C, H, and O). After determining that it was a diacid, H_2X, a 0.1235-g sample was neutralized with 15.55 mL of 0.1087 M NaOH. Next, a 0.3469-g sample was burned completely in pure oxygen, producing 0.6268 g of CO_2 and 0.2138 g of H_2O.
    (a) What is the molar mass of H_2X?
    (b) What is the empirical formula for the diacid?
    (c) What is the molecular formula for the diacid?

104. Various masses of the three Group 2A elements magnesium, calcium, and strontium were allowed to react with liquid bromine, Br_2. After the reaction was complete, the reaction product was freed of excess reactant(s) and weighed. In each case, the mass of product (shown below) was plotted against the mass of metal used in the reaction.

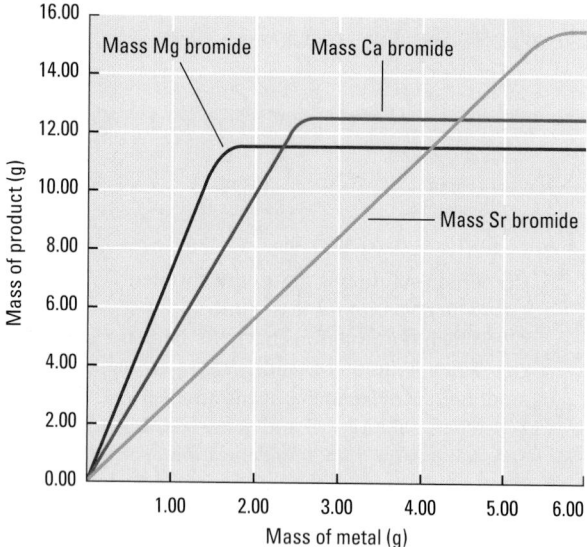

(a) Based on your knowledge of the reactions of metals with halogens, what product is predicted for each reaction? What are the name and formula for the reaction product in each case?
(b) Write a balanced equation for the reaction occurring in each case.
(c) What kind of reaction occurs between the metals and bromine — that is, is the reaction a gas-forming reaction, a precipitation reaction, or an oxidation-reduction reaction?
(d) Each plot shows that the mass of product increases with increasing mass of metal used, but the plot levels out at some point. Use these plots to verify your prediction of the formula of each product, and explain why the plots become level at different masses of metal and different masses of product.

105. Gold can be dissolved from gold-bearing rock by treating the rock with sodium cyanide in the presence of the oxygen in air.

$$4\ Au(s) + 8\ NaCN(aq) + O_2(g) + 2\ H_2O(\ell) \longrightarrow$$
$$4\ NaAu(CN)_2(aq) + 4\ NaOH(aq)$$

Once the gold is in solution in the form of the $Au(CN)_2^-$ ion, it can be precipitated as the metal according to the following unbalanced equation:

$$Au(CN)_2^-(aq) + Zn(s) \longrightarrow Zn^{2+}(aq) + Au(s) + CN^-(aq)$$

(a) Are the two reactions above acid-base or oxidation-reduction reactions? Briefly describe your reasoning.
(b) How many liters of 0.075 M NaCN will you need to extract the gold from 1000 kg of rock if the rock is 0.019% gold?
(c) How many kilograms of metallic zinc will you need to recover the gold from the $Au(CN)_2^-$ obtained from the gold in the rock?
(d) If the gold is recovered completely from the rock and the metal is made into a cylindrical rod 15.0 cm long, what is the diameter of the rod? (The density of gold is 19.3 g/cm^3.)

106. Four groups of students from an introductory chemistry laboratory are studying the reactions of solutions of alkali metal halides with aqueous silver nitrate, $AgNO_3$. They use the following salts.

Group A: NaCl   Group C: NaBr

Group B: KCl   Group D: KBr

Each of the four groups dissolves 0.004 mol of their salt in some water. Each then adds various masses of silver nitrate, $AgNO_3$, to their solutions. After each group collects the precipitated silver halide, the mass of this product is plotted versus the mass of $AgNO_3$ added. The results are given on the following graph.

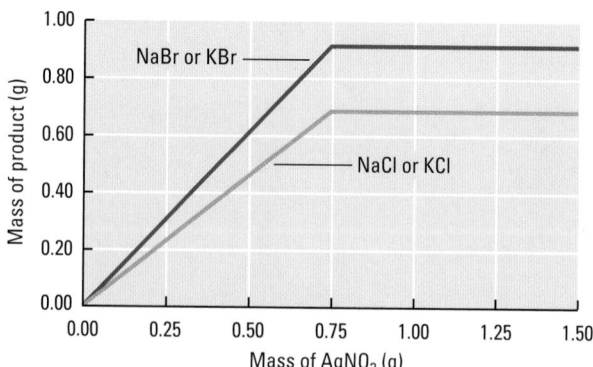

(a) Write the balanced net ionic equation for the reaction observed by each group.
(b) Explain why the data for groups A and B lie on the same line, whereas those for groups C and D lie on a different line.
(c) Explain the shape of the curve observed by each group. Why do they level off at the same mass of added $AgNO_3$ (0.75 g) but give different masses of product (0.6 g for groups A and B; 0.8 g for groups C and D)?

107. One way to determine the stoichiometric relationships among reactants is continuous variations. In this process, a series of reactions is carried out in which the reactants are varied systematically, while keeping the total volume of each reaction mixture constant. When the reactants combine stoichiometrically, they react completely; none is in excess. The following data were collected to determine the stoichiometric relationship for the reaction.

$$m\,X^{n+} + n\,Y^{m-} \longrightarrow X_m Y_n$$

Trial	A	B	C	D	E
0.10 M X^{n+}	7 mL	6 mL	5 mL	4 mL	3 mL
0.20 M Y^{m-}	3 mL	4 mL	5 mL	6 mL	7 mL
Excess X^{n+} present?	Yes	Yes	Yes	No	No
Excess Y^{m-} present?	No	No	No	No	Yes

(a) In which trial are the reactants present in stoichiometric amounts?
(b) How many moles of X^{n+} reacted in that trial?
(c) How many moles of Y^{m-} reacted in that trial?
(d) What is the whole-number mole ratio of X^{n+} to Y^{m-}?
(e) What is the chemical formula for the product $X_m Y_n$?

General Chemistry CD-ROM

CD5.1 Screen 4.10: Equations for Reactions in Aqueous Solution: Net Ionic Equations. Consider the following questions after watching the video of the addition of a colorless solution of lead(II) nitrate to a solution of yellow potassium chromate.

(a) What is the color of the precipitate in this reaction? Give its name and formula.
(b) Explain why K^+ is considered a "spectator ion" and can be eliminated from the chemical equation described on this screen.

(c) Is there another spectator ion in this equation? If so, what is its identity?

CD5.2 Screen 4.15: Oxidation-Reduction Reactions.
(a) Observe the photos of the reaction of magnesium with oxygen.
   (1) What evidence is there that a reaction has occurred?
   (2) What is the name of the product of this reaction? What is its color? Its physical state?
(b) Iron reacts with oxygen to give iron(III) oxide (rust).
   (1) Write a balanced chemical equation for this reaction.
   (2) What is the oxidizing agent in this reaction? What has been oxidized? What has been reduced?

CD5.3 Screen 5.3: Solutions.
(a) Describe what happens as $KMnO_4$ is added to water. Is the mixture homogeneous or heterogeneous?
(b) In the animation, the K^+ ions are shown as yellow spheres, whereas the MnO_4^- ions are a collection of blue and smaller red spheres.
   (1) Describe what you see as the animation proceeds. How does an ionic solid dissolve in water? What forces are at work that allow the solid to dissolve?
   (2) Water molecules are attached to both positive and negative ions. What does this tell you about the nature of water?
   (3) How does the orientation of the water molecules differ when they encounter a K^+ ion as compared with a MnO_4^- ion? What does this tell you about the nature of the water molecule? That is, describe how electric charge is distributed in the water molecule.

CD5.4 Screen 5.11: Preparing Solutions of Known Concentration (1).
(a) Sufficient water was added to the hydrated nickel(II) chloride to make exactly 250 mL of a 0.140 M solution. What is the concentration of Ni^{2+} ion? Of the Cl^- ion?
(b) Why is it important to shake the volumetric flask thoroughly before using the solution you have made?
(c) Examine the glassware on Screen 5.11SB. Which piece of glassware is least accurate?

Combustion of natural gas. Burning natural gas, which in the United States is mainly methane, CH_4, releases a great deal of energy to anything in contact with the reactant and product molecules. The energy released when a fossil fuel such as natural gas burns can be transformed to provide many of the benefits of our technology-intensive society.

ENERGY AND CHEMICAL REACTIONS

6.1 The Nature of Energy

6.2 Conservation of Energy

6.3 Heat Capacity

6.4 Energy and Enthalpy

6.5 Thermochemical Equations

6.6 Enthalpy Changes for Chemical Reactions

6.7 Where Does the Energy Come From?

6.8 Measuring Enthalpy Changes: Calorimetry

6.9 Hess's Law

6.10 Standard Molar Enthalpies of Formation

6.11 Chemical Fuels for Home and Industry

6.12 Foods: Fuels for our Bodies

In our industrialized, high-technology, appliance-oriented society, the average use of energy per person is at nearly its highest point in history. The United States, with only 5% of the world's population, consumes 30% of the world's energy resources. In every year since 1958 we have consumed more energy resources than have been produced within our borders. Most of the energy we use comes from chemical reactions: combustion of the fossil fuels coal, petroleum, and natural gas. The rest comes from hydroelectric power plants, nuclear power plants, solar energy and wind collectors, and burning wood and other plant material. Both U.S. and world energy use are growing rapidly.

Chemical reactions involve transfers of energy. When a fuel burns, the energy of the products is less than the energy of the reactants. The leftover energy shows up in anything that is in contact with the reactants and products. For example, when natural gas, which is mainly methane, burns in air, the carbon, hydrogen, and oxygen atoms that make up the reactant CH_4 and O_2 molecules rearrange to form CO_2 and H_2O molecules.

$$CH_4(g) \;+\; 2\,O_2(g) \longrightarrow CO_2(g) \;+\; 2\,H_2O(g)$$

Because of the way their atoms are bonded together, the CO_2 and H_2O molecules have less total energy than the reactant CH_4 and O_2 molecules did. After the reaction, some energy that was in the reactants is not contained in the product molecules. That energy heats everything that is close to where the reaction takes place. We say that the reaction transfers energy to its surroundings.

For the past hundred years or so, most of the energy society has used has come from combustion of fossil fuels, and this will continue to be true well into the future. Consequently, it is very important to understand how energy and chemical reactions are related and how chemistry might be used to alter our dependence on fossil fuels. This requires knowledge of **thermodynamics,** the science of heat, work, and transformations of one to the other. The fastest growing new industries in the new millennium may well be those that capitalize on such knowledge and the new chemistry and chemical industries it spawns.

6.1 THE NATURE OF ENERGY

What is energy? Where does the energy we use come from? And how can chemical reactions result in the transfer of energy to or from their surroundings? *Energy,* represented by E, was defined in Section 1.4 (⬅ *p. 12)* as the capacity to do work. If you climb a mountain or a staircase, you work against the force of gravity as you move upward, and your gravitational energy increases. The energy you use to do this work is released when food you have eaten is metabolized (undergoes chemical reactions) within your body. Energy from food enables you to work against the force of gravity as you climb, and it warms your body (climbing makes you hotter as well as higher). Therefore our study of the relations between energy and chemistry also needs to consider processes that involve work and processes that involve heat.

Energy can be assigned to one of two categories: kinetic or potential. **Kinetic energy** is energy that something has because it is moving (Figure 6.1):

- Energy of motion of a macroscale object, such as a moving baseball or automobile; this is often called *mechanical energy.*

Joseph Romm and Charles Curtis have clearly described the importance of energy in our daily lives, to our political system, and to our collective future. See The *Atlantic Monthly,* April 1996, pp. 59–74.

"A theory is the more impressive the greater the simplicity of its premises is, the more different kinds of things it relates, and the more extended is its area of applicability. Therefore, the deep impression which classical thermodynamics made upon me. It is the only physical theory of universal content concerning which I am convinced that, within the framework of the applicability of its basic concepts, it will never be overthrown." (Albert Einstein, quoted in Schlipp, P. A. [ed.] "Albert Einstein: Philosopher-Scientist." In *The Library of Living Philosophers,* Vol. VII. Autobiographical notes, 3rd ed. LaSalle, IL: Open Court Publishing, 1969; p. 33.)

Work and heat refer to the quantity of energy transferred from one object or sample to another by working or heating *processes.* However, we often talk about work and heat as if they were forms of energy. Working and heating processes transfer energy from one form or one place to another. To emphasize this, we often will use the words working and heating where many people would use work and heat.

 CD-ROM Screen 6.4: Energy

Figure 6.1 Kinetic energy. Race cars have kinetic energy that depends on their mass and velocity.

Rock climbing. (a) Climbing requires energy. (b) The higher the altitude, the greater the climber's gravitational energy.

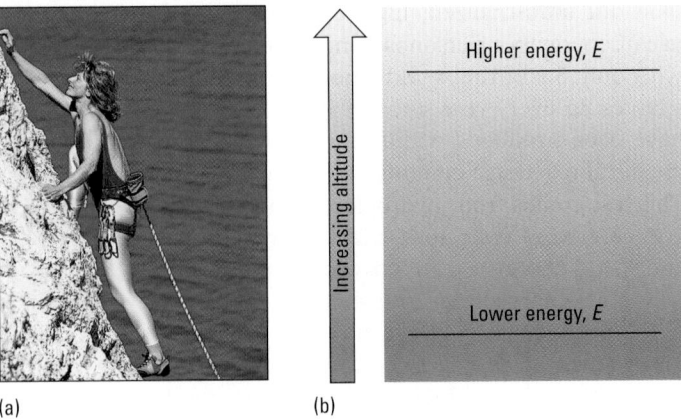

(a)   (b)

• Energy of motion of nanoscale objects such as atoms, molecules, or ions; this is often called *thermal energy.*
• Energy of motion of electrons through an electrical conductor; this is often called *electrical energy.*
• Energy of periodic motion of nanoscale particles when a macroscale sample is alternately compressed and expanded (as when a sound wave passes through air).

Kinetic energy E_k can be calculated as $E_k = \frac{1}{2}mv^2$, where m represents the mass and v represents the velocity of a moving object.

Potential energy is energy that something has as a result of its position and some force that is capable of changing that position. Examples are

• Energy that a ball held in your hand has because the force of gravity attracts it toward the floor; this is often called *gravitational energy.*
• Energy that charged particles have because they attract or repel each other; this is often called *electrostatic energy;* an example is potential energy of positive and negative ions a small distance apart.
• Energy resulting from attractions and repulsions among electrons and atomic nuclei in molecules; this is often called *chemical potential energy* and is the kind of energy stored in foods and fuels.

Figure 6.2 Gravitational potential energy. Water on the brink of a water-fall has potential energy (stored energy that could be used to do work) because of its position relative to the earth; that energy could be used to generate electricity, for example.

Potential energy can be calculated in different ways depending on the type of force that is involved. For example, near the surface of the earth, gravitational potential energy E_p can be calculated as $E_p = mgh$, where m is mass, g is the gravitational constant ($g = 9.8$ m/s^2), and h is the height above the surface.

Potential energy can be converted to kinetic energy, and vice versa. As droplets of water fall over a waterfall (Figure 6.2), the potential energy they had at the top is converted to kinetic energy — they move faster and faster. Conversely, the kinetic energy of falling water could drive a water wheel to pump water to an elevated reservoir, where its potential energy would be higher.

Energy Units

CD-ROM Screen 6.6: Energy Units

The joule is the unit of energy in the International System of units (SI units). SI units are described in Appendix B.

The SI unit of energy is the joule (rhymes with rule), symbol J. The joule is a derived unit, which means that it can be expressed as a combination of other units: 1 J = 1 kg m^2/s^2. If a 2.0-kg object (which weighs about $4\frac{1}{2}$ pounds) is moving with a velocity of 1.0 meter per second (roughly 2 miles per hour), its kinetic energy is

$$E_k = \frac{1}{2} mv^2 = \frac{1}{2} \times (2.0 \text{ kg})(1.0 \text{ m/s})^2 = 1.0 \text{ kg m}^2/\text{s}^2 = 1.0 \text{ J}$$

This is a relatively small quantity of energy, and because the joule is so small, we often use the kilojoule (1 kilojoule = 1 kJ = 1000 J).

Another energy unit is the calorie, symbol cal. By definition 1 cal = 4.184 J. A calorie is very close to the quantity of energy required to raise the temperature of one gram of water by 1 degree Celsius. The "calorie" that you hear about in connection with nutrition and dieting is actually a kilocalorie (kcal) and is usually represented with a capital C. Thus, a breakfast cereal that gives you 100 Calories of nutritional energy actually provides 100 kcal = 100×10^3 cal. In many countries food energy is reported in kilojoules rather than in Calories. For example, the label on the packet of nonsugar sweetener shown in Figure 6.3 indicates that it provides 16 kJ of nutritional energy.

A joule is approximately the quantity of energy required for one human heartbeat.

1 cal = 4.184 J exactly

1 Cal = 1 kcal = 1000 cal = 4.184 kJ = 4184 J

Figure 6.3   Food energy.   A packet of artificial sweetener from Australia. As its label shows, the sweetener in the packet supplies 16 kJ of nutritional energy. It is equivalent in sweetness to 2 level teaspoonfuls of sugar, which would supply 140 kJ of nutritional energy.

Problem-Solving Example 6.1   Energy Units

A single Cheetos snack chip has a food energy of 7.0 Cal. What is this energy in units of joules?

Answer 2.9×10^4 J

Explanation   To find the energy in joules, we use the fact that 1 Cal = 1 kcal, the definition of the prefix *kilo-* (= 1000), and the definition 1 cal = 4.184 J to generate appropriate proportionality factors (conversion factors).

$$E = 7.0 \text{ Cal} \times \frac{1 \text{ kcal}}{1 \text{ Cal}} \times \frac{1000 \text{ cal}}{1 \text{ kcal}} \times \frac{4.184 \text{ J}}{1 \text{ cal}} = 2.9 \times 10^4 \text{ J}$$

✔ 2.9×10^4 J is 29 kJ. Because 1 Cal = 1 kcal = 4.184 kJ, the result in kJ should be about four times the original 7 Cal — that is, about 28 kJ — which it is.

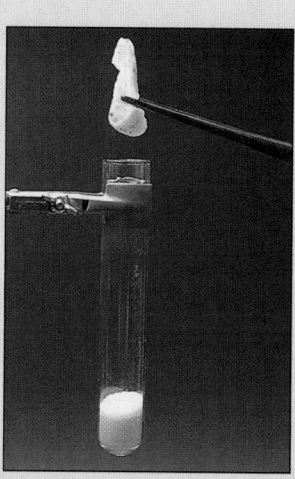

Food energy.   A single Cheetos chip burns in oxygen generated by thermal decomposition of potassium chlorate.

Problem-Solving Practice 6.1

(a) If you eat a hot dog, it will provide 160 Calories of energy. What is this energy in joules?

(b) A watt is a unit of power that corresponds to the transfer of one joule of energy in one second. The energy used by an x-watt light bulb operating for y seconds is $x \times y$ joules. If you turn on a 75-watt bulb for 3.0 hours, how many joules of electrical energy will be transformed into light and heat?

(c) The packet of nonsugar sweetener in Figure 6.3 provides 16 kJ of nutritional energy. What is this energy in kilocalories?

JAMES P. JOULE
(1818–1889)

The energy unit joule is named for James P. Joule. Joule was the son of a brewer in Manchester, England, and was a student of John Dalton (⬅ *p. 22*). Joule established the idea that working and heating are both processes by which energy can be transferred from one sample of matter to another.

ESTIMATION: Earth's Kinetic Energy

Estimate the earth's kinetic energy as it orbits the sun.

From an encyclopedia, a dictionary, or the Internet, you can obtain the facts that the earth's mass is about 3×10^{24} kg and its distance from the sun is about 150,000,000 km. Assume earth's orbit is a circle and calculate the distance traveled in a year as the circumference of the circle, $\pi d = 2\pi r = 2 \times 3.14 \times 1.5 \times 10^8$ km. Since $2 \times 3 = 6$, $1.5 \times 6 = 9$, and 3.14 is a bit more than 3, estimate the distance as 10×10^8 km. Earth's speed, then, is a bit less than 10×10^8 km/yr.

Because $1 \text{ J} = 1 \text{ kg m}^2/\text{s}^2$, convert the time unit from years to seconds. Estimate the number of seconds in a year as 60 s/min $\times$ 60 min/hr $\times$ 24 hr/d $\times$ 365 d/yr = 60 $\times$ 60 $\times$ 24 $\times$ 365 s/yr. To make the arithmetic easy, round 24 to 20 and 365 to 400, giving $60 \times 60 \times 20 \times 400$ s/yr = $6 \times 6 \times 2 \times 4 \times 10^5$ s/yr. This gives 288×10^5 s/yr, or 3×10^7 s/yr rounded to one significant figure. Therefore earth's speed is about $10/3 \times 10^8/10^7 \cong 30$ km/s or 3×10^4 m/s.

Now the equation for kinetic energy can be used.

$$E_k = \tfrac{1}{2} m v^2$$
$$= \tfrac{1}{2} \times (3 \times 10^{24} \text{ kg}) \times (3 \times 10^4 \text{ m/s})^2 \simeq 1.3 \times 10^{32} \text{ J}$$

Although the earth's speed is not high, its mass is very large. This results in an extraordinarily high kinetic energy.

6.2 CONSERVATION OF ENERGY

CD-ROM Screen 6.5: Forms of Energy

In Section 1.7 (→ *p. 19*) the kinetic-molecular theory was described qualitatively. A corollary to this theory is that *molecules move faster, on average, as the temperature increases.*

When you stand on a diving platform above a pool of water, poised to dive in, you have greater gravitational potential energy than you would at the surface of the water. Once you jump, that extra potential energy is converted progressively into kinetic energy (Figure 6.4). During the dive, the force of gravity accelerates your body to move faster and faster, so your velocity and kinetic energy increase. This happens at the expense of potential energy. At the moment you hit the water, your velocity is abruptly reduced. Much of your kinetic energy is converted to mechanical energy of the water, which splashes as your body does work on it (transfers energy to it) to move it aside. Eventually, you float on the surface and the water becomes still again. However, on average, the water molecules are moving a little faster in the vicinity of your point of impact; that is, the temperature of the water is now a little higher.

Your dive has caused a series of transformations of energy: from potential to kinetic and from macroscale kinetic to nanoscale kinetic (that is, thermal). However, the total quantity of energy, kinetic plus potential, is the same before and after the dive. In many, many experiments, the total energy has always been

Figure 6.4 Energy transformations. Potential and kinetic energy are interconverted when someone dives into water. These interconversions are governed by the law of conservation of energy. (a) The diver has greater gravitational potential energy on the platform than at the surface of the water, because the platform is higher above the earth. (b) Some of the potential energy has been converted into kinetic energy as the diver's altitude above the water decreases and velocity increases; maximum kinetic energy occurs just prior to impact with the water. (c) Upon impact, the diver works on the water, splashing it aside; eventually, the initial potential energy difference is converted into motion on the nanoscale — the temperature of the water has become slightly higher.

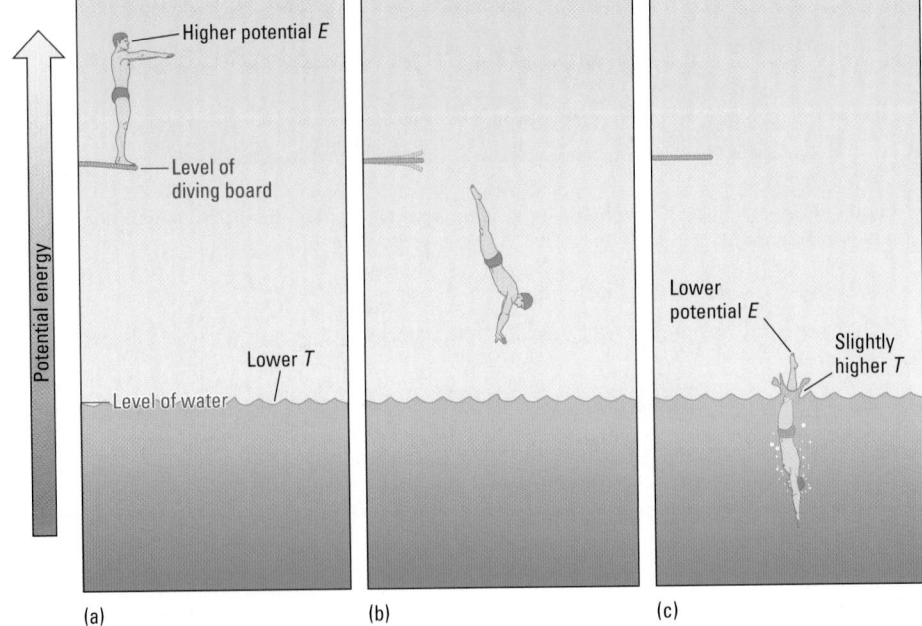

found to be the same before and after an event. These experiments are summarized by the **law of conservation of energy,** which states that *energy can neither be created nor destroyed — the total energy of the universe is constant.* This is also called the **first law of thermodynamics.**

CD-ROM Screen 6.13: The First Law of Thermodynamics

Exercise 6.1 Energy Transfers

You toss a rubber ball up into the air. It falls to the floor, bounces for a while, and eventually comes to rest. Several energy transfers are involved. Describe them and the changes they cause.

Scientific laws are discussed in Chapter 1 (⬅ *p. 5, Section 1.2*).

Energy and Working

When a force acts on an object and moves the object, the change in the object's kinetic energy is equal to the work done on the object. Work has to be done, for example, to accelerate a car from 0 to 60 miles per hour or to hit a baseball out of a stadium. Work is also required to increase the potential energy of an object. Thus, work has to be done to raise an object against the force of gravity (as in an elevator), to separate a sodium ion (Na^+) from a chloride ion (Cl^-), or to move an electron away from an atomic nucleus. The work done on an object corresponds to the quantity of energy transferred to that object; that is, doing **work** is *a process that transfers energy to an object.* Conversely, if an object does work on something else, the quantity of energy associated with the object must decrease.

Work is required to cause some chemical and biochemical processes to occur. Examples are moving ions across a cell membrane and synthesizing adenosine triphosphate (ATP) from adenosine diphosphate (ADP).

Energy, Temperature, and Heating

According to the kinetic-molecular theory (⬅ *p. 19*), all matter consists of nanoscale particles that are in constant motion (Figure 6.5). Therefore, all matter has thermal energy. For a given sample, the quantity of thermal energy is greater the higher the temperature is. Transferring energy to a sample of matter usually results in a temperature increase that can be measured with a thermometer. For example, when a mercury thermometer is placed into warm water (Figure 6.6), energy transfers from the water to the thermometer (the water heats the thermometer). The increased energy of the mercury atoms means that they move about more rapidly, which slightly increases the volume of the spaces between the

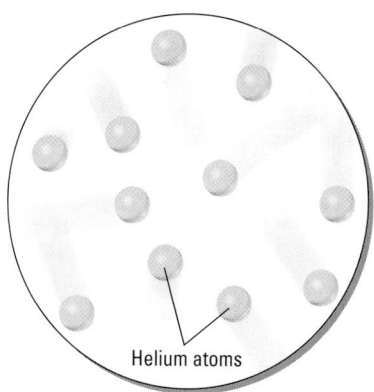

Helium atoms

Figure 6.5 Thermal energy. According to the kinetic-molecular theory, nanoscale particles (atoms, molecules, and/or ions) are in constant motion. Here, atoms of gaseous helium are shown. Each atom has kinetic energy that depends on how fast it is moving (as indicated by the lengths of the "tails," which show how far each atom travels per unit time). The thermal energy of the sample is the sum of the kinetic energies of all the helium atoms. The higher the temperature of the helium, the faster the average speed of the molecules, and therefore the greater the thermal energy.

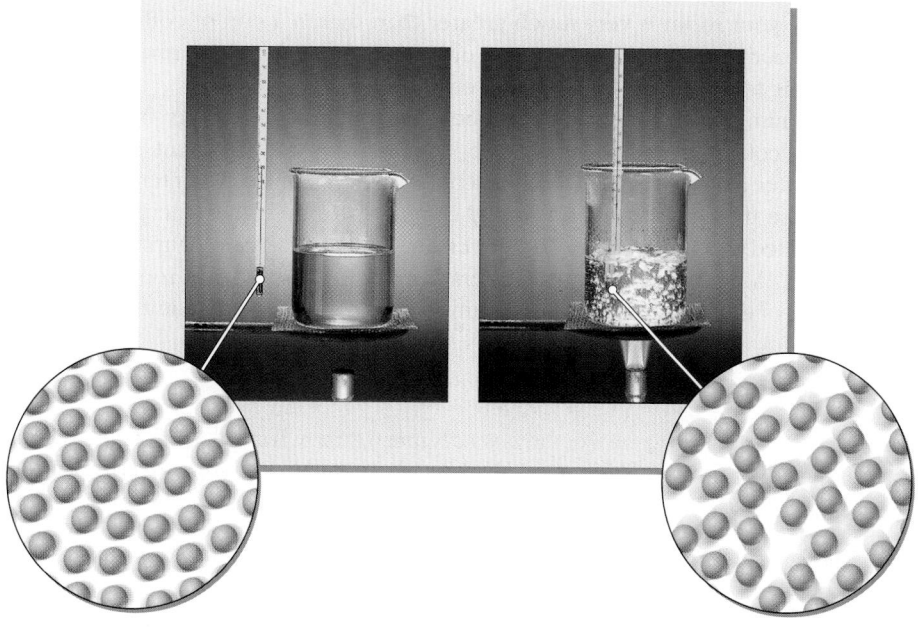

Figure 6.6 Measuring temperature. The mercury in a thermometer expands because the mercury atoms are moving faster (have more energy) after the boiling water transfers energy to (heats) the mercury; the temperature and the volume of the mercury have both increased.

 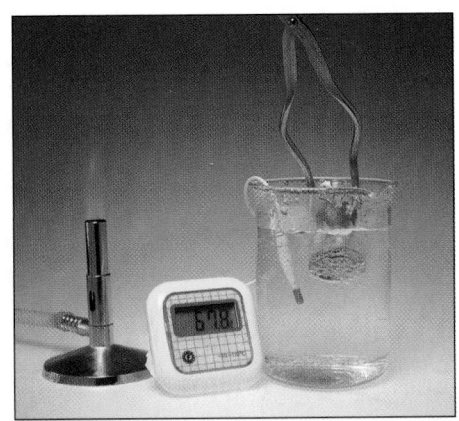

Figure 6.7 Energy transfer by heating. Water in a beaker is heated when a hotter sample (a brass bar) is plunged into the water. There is a transfer of energy from the hotter metal bar to the cooler water. Eventually, enough energy is transferred so that the bar and the water reach the same temperature. Thermal equilibrium is achieved.

 CD-ROM Screen 6.5SB: A Closer Look: Heat

Transferring energy by heating is a process, but it is common to talk about that process as if heat were a form of energy. It is often said that one sample transfers heat to another, when what is meant is that one sample transfers energy by heating the other.

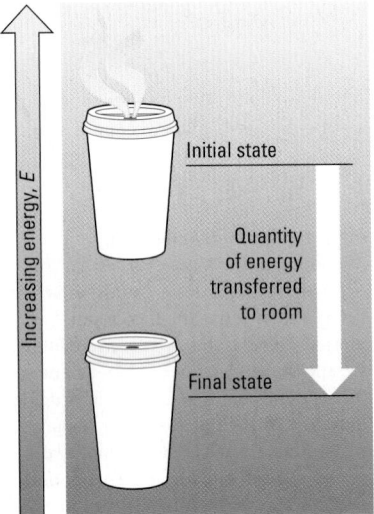

Figure 6.8 Energy diagram for a cup of hot coffee. The diagram compares the energy of a cup of hot coffee with the energy after the coffee has cooled to room temperature. The higher something is in the diagram, the more energy it has. As the coffee and cup cool to room temperature, energy is transferred to the surrounding matter in the room. According to the law of conservation of energy, the energy remaining in the coffee must be less after the change (in the final state) than it was before the change (in the initial state). The quantity of energy transferred is represented by the arrow from the initial to the final state.

atoms. Consequently, the mercury expands (as most substances do upon heating), and the column of mercury rises higher in the thermometer tube.

Heat (or heating) refers to *the energy transfer process that happens whenever two samples of matter at different temperatures are brought into contact. **Energy always transfers from the hotter to the cooler sample until both are at the same temperature.*** For example, a piece of metal at a high temperature in a Bunsen burner flame and a beaker of cold water (Figure 6.7a) are two samples of matter with different temperatures. When the hot metal is plunged into the cold water (Figure 6.7b), energy transfers from the metal to the water until the two samples reach the same temperature. Once that happens, the metal and water are said to be in **thermal equilibrium.** When thermal equilibrium is reached, the metal has heated the water (and the water has cooled the metal) to a common temperature.

Usually most objects in a given region, such as your room, are at about the same temperature — at thermal equilibrium. A fresh cup of coffee, which is hotter than room temperature, transfers energy by heating the rest of the room until the coffee cools off (and the rest of the room warms up a bit). A can of cold soda, which is much cooler than its surroundings, receives energy from everything else until it warms up (and your room cools off a little). Because the total quantity of material in your room is very much greater than that in a cup of coffee or a can of soda, the room temperature changes a only a tiny bit to reach thermal equilibrium, whereas the temperature of the coffee or the soda changes a lot.

A diagram such as Figure 6.8 can be used to show the energy transfer from a cup of hot coffee to your room. The upper horizontal line represents the energy of the hot coffee and the lower line represents the energy of the room temperature coffee. Because the coffee started at a higher temperature (higher energy), the upper line is labeled the initial state. The lower line is the final state. During the change from initial to final state, energy transfers from the coffee to your room. Therefore, the energy of the coffee is lower in the final state than it was in the initial state.

Exercise 6.2 Energy Diagrams

(a) Draw an energy diagram like the one in Figure 6.8 for warming a can of cold soda to room temperature. Label the initial and final states and use an arrow to represent the change in energy of the can of soda.

(b) Draw a second energy diagram, to the same scale, to show the change in energy of the room as the can of cold soda warms to room temperature.

System, Surroundings, and Internal Energy

In thermodynamics it is useful to define a *region of primary concern* as the **system.** Then we can decide whether energy transfers into or out of the system and keep an accounting of how much energy transfers in each direction. *Everything that can exchange energy with the system* is defined as the **surroundings.** A system may be delineated by an actual physical boundary, such as the inside surface of a flask or the membrane of a cell in your body. Or, the boundary may be indistinct, as in the case of the solar system within its surroundings, the rest of the galaxy. In the case of a hot cup of coffee in your room, the cup and the coffee might be the system, and your room would be the surroundings. For a chemical reaction, the system is usually defined to be all of the atoms that make up the reactants. These same atoms will be bonded in a different way in the products after the reaction, and it is their energy before and after reaction that we are most interested in.

The **internal energy** of a system is the *sum of the individual energies (kinetic and potential) of all nanoscale particles (atoms, molecules, or ions)* in that system. Increasing the temperature increases the internal energy because it increases the average speed of motion of nanoscale particles. ***The total internal energy of a sample of matter depends on temperature, the type of particles, and how many of them are in the sample.*** For a given substance, internal energy depends on temperature and the size of the sample. Thus, despite being at a higher temperature, a cupful of boiling water contains less energy than a bathtub full of warm water.

Calculating Thermodynamic Changes

If we represent a system's internal energy by E, then the change in internal energy during any process is calculated as $E_{final} - E_{initial}$. That is, from the internal energy after the process is over subtract the internal energy before it began. Such a calculation is designated by using a Greek letter Δ (capital delta) before the quantity that changes. Thus, $E_{final} - E_{initial} = \Delta E$. *Whenever a change is indicated by Δ, a positive value indicates an increase and a negative value a decrease.* Therefore, if the internal energy increases during a process, ΔE has a positive value ($\Delta E > 0$); if the internal energy decreases, ΔE is negative ($\Delta E < 0$).

A good analogy to this thermodynamic calculation is your bank account. Assume that in your account (the system) you have a balance B of $260 ($B_{initial}$), and you withdraw $60 in spending money. After the withdrawal the balance is $200 ($B_{final}$). The change in your balance is

$$\text{Change in balance} = \Delta B = B_{final} - B_{initial} = \$200 - \$260 = -\$60$$

The negative sign on the $60 indicates that money has been withdrawn from the system (account) and transferred to the surroundings (you, and wherever you spend the money). The cash itself is not negative, but during the process of withdrawing your money the balance in the bank went down, so ΔB was negative. Similarly, the magnitude of change of a thermodynamic quantity is a number with no sign. To indicate the direction of a change, we attach a negative sign (transferred out of the system) or a positive sign (transferred into the system).

Conservation of Energy and Chemical Reactions

For many chemical reactions the only energy transfer processes are heating and doing work. If no other energy transfers (such as emitting light) take place, the law of conservation of energy can be written as

$$\Delta E_{system} = q_{system} + w_{system} \qquad [6.1]$$

 CD-ROM Screen 6.12: Energy Changes in Chemical Processes

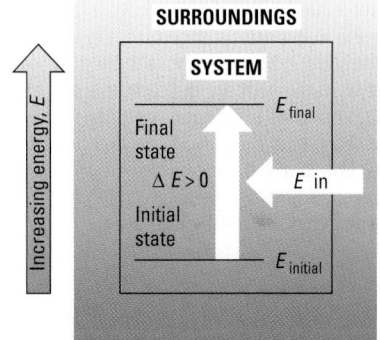

ΔE **positive: Internal energy increases.**

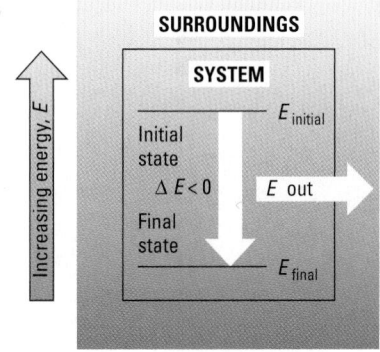

ΔE **negative: Internal energy decreases.**

Figure 6.9 **Internal energy, heat, and work.** Schematic diagram showing energy transfers between a thermodynamic system and its surroundings.

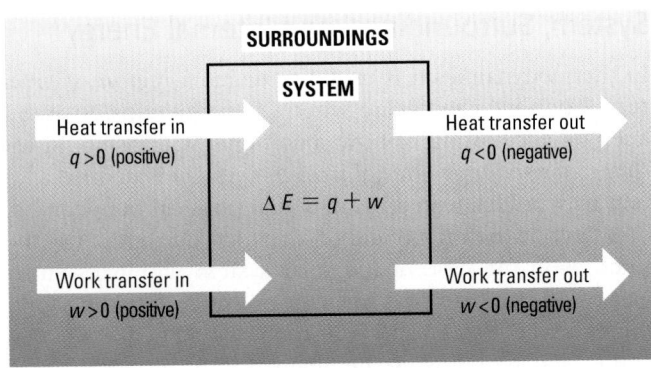

Sometimes Equation 6.1 is written as $\Delta E_{system} = q_{system} - w_{system}$. In such cases the meaning of the equation is the same, but w has been defined to have a positive value when work is *done by the system* on the surroundings. (Because the work transfer is in the opposite direction, the sign of w is opposite.) Engineers are often interested in how much work can be done by a system in which a fuel is burned, in which case it is helpful to define w to be positive when work is done by the system.

where q represents the quantity of energy transferred by heating the system, and w represents the quantity of energy transferred by doing work on the system. If energy is transferred into the system from the surroundings by heating, q is positive; if energy is transferred into the system because the surroundings do work on the system, w is positive. If energy is transferred out of the system by heating the surroundings, then q has a negative value; if energy is transferred out of the system because work is done on the surroundings, then w has a negative value. The *magnitudes* of q and w indicate the *quantities* of energy transferred, and the *signs* of q and w indicate the *direction* in which it is transferred. The relationships among ΔE, q, and w for a system are shown in Figure 6.9.

Problem-Solving Example 6.2   Internal Energy, Heat, and Work

Gasoline burns in the cylinder of an automobile engine. During the combustion reaction, the piston is forced to move, which requires 560 J. Also, the engine's cooling system has to carry away 2.2 kJ as a result of increased temperature of the coolant liquid. If the system is defined to be the gasoline and oxygen that react, what is the change in its internal energy?

Answer -2.7×10^3 J

Explanation   Because the system is the reactants and products, the piston and the coolant liquid are part of the surroundings. The problem states that the piston is forced to move, which requires that the system do work on the piston. Therefore w must be negative, and $w = -560$ J. Energy is transferred to the cooling system, so q must also be negative, and $q = -2.2$ kJ $= -2200$ J. Using Equation 6.1,

$$\Delta E = q + w = -2200\,\text{J} + (-560\,\text{J}) = -2.7 \times 10^3\,\text{J}$$

Thus, the internal energy of the products of the reaction, carbon dioxide and water, is 2700 J lower than the internal energy of the reactants, gasoline and oxygen.

✔ ΔE is negative, which is reasonable. The internal energy of the reaction products should be lower than for the reactants, because energy transferred from the reaction heats the surroundings and does work on the surroundings.

Problem-Solving Practice 6.2

Suppose that the internal energy decreases by 2400 J when a mixture of hydrogen and oxygen gases is ignited and burns. If the surroundings are heated by 1.89 kJ, how much work was done by this system on the surroundings?

6.3 HEAT CAPACITY

CD-ROM Screen 6.7: Heat Capacity

The quantity of energy transferred by doing work can be determined by measuring the height to which a weight can be raised, the change in speed of an object of known mass, or the distance through which a known force works. To determine the quantity of energy transferred by heating, we usually measure the increase in temperature of a substance whose heat capacity is known — often it is water. The **heat capacity** of a sample of matter is *the quantity of energy required to increase the temperature of that sample by one degree*. Heat capacity depends on the mass of the sample and the substance of which it is made (or substances, if it is not pure).

Specific Heat Capacity

To make useful comparisons among samples of different substances with different masses, the **specific heat capacity** (which is sometimes just called *specific heat*) is defined as the quantity of energy needed to increase the temperature of one gram of a substance by one degree. For water at 15 °C, the specific heat capacity is 1.00 cal g^{-1} °C^{-1} or 4.184 J g^{-1} °C^{-1}; for common window glass it is only about 0.8 J g^{-1} °C^{-1}. That is, it takes about five times as much heat to raise the temperature of a gram of water by 1 °C as it does for a gram of glass. Like density (⬅ *p. 7*), specific heat capacity is a property that can be used to distinguish one substance from another. It can also be used to distinguish a pure substance from a solution or mixture, because the specific heat capacity of a mixture will vary with the proportions of the mixture's components.

The specific heat capacity, *c*, of a substance can be determined experimentally by measuring the quantity of energy transferred to or from a known mass of the substance as its temperature rises or falls.

$$\text{Specific heat capacity} = \frac{\text{quantity of energy transferred by heating}}{\text{sample mass} \times \text{temperature change}}$$

or

$$c = \frac{q}{m \times \Delta T} \qquad [6.2]$$

Suppose that for a 25.0-g sample of ethylene glycol (a compound used as antifreeze in automobile engines) it takes 90.7 J to change the temperature from 23.7 °C to 25.2 °C. Thus,

$$\Delta T = (25.2\,°C - 23.7\,°C) = 1.5\,°C$$

and, from Equation 6.2, the specific heat capacity of ethylene glycol is

$$c = \frac{q}{m \times \Delta T} = \frac{90.7\,J}{25.0\,g \times 1.5\,°C} = 2.4\,J\,g^{-1}\,°C^{-1}$$

The specific heat capacities of many substances have been determined. A few values are listed in Table 6.1. Notice that water has one of the highest values. This is important because a high specific heat capacity means that a great deal of energy must be transferred to a large body of water to raise its temperature by just a degree. Conversely, a lot of energy must be transferred away from the water before its temperature falls by one degree. Thus, a lake or ocean can store an enormous quantity of energy and thereby moderate local temperatures. This has a profound influence on weather near lakes or oceans.

CD-ROM Screen 6.8: Heat Capacity of Pure Substances

The notation J g^{-1} °C^{-1} means that the units are joules divided by grams and divided by degrees Celsius; that is, $\frac{J}{g\,°C}$. Negative exponents will be used to show unambiguously which units are in the denominator whenever there are two or more units in the denominator.

The high specific heat capacity of water helps to keep your body temperature relatively constant. Water accounts for a large fraction of your body mass, and warming or cooling that water requires a lot of energy transfer.

Moderation of microclimate by water. In cities located near large bodies of water (such as Chicago, shown here), summertime temperatures are lower within a few hundred meters of the waterfront than they are a few kilometers away from the water. Wintertime temperatures are higher, unless the water freezes, in which case the moderating effect is less, because ice on the surface insulates the rest of the water from the air.

TABLE 6.1 Specific Heat Capacities for Some Elements, Compounds, and Common Solids

Symbol or formula	Name	Specific heat capacity $(J\,g^{-1}\,{}^{\circ}C^{-1})$
Elements		
Al	Aluminum	0.902
C	Carbon (graphite)	0.720
Fe	Iron	0.451
Cu	Copper	0.385
Au	Gold	0.128
Compounds		
$NH_3(\ell)$	Ammonia	4.70
$H_2O(\ell)$	Water (liquid)	4.184
$C_2H_5OH(\ell)$	Ethanol	2.46
$(CH_2OH)_2(\ell)$	Ethylene glycol (antifreeze)	2.42
$H_2O(s)$	Water (ice)	2.06
$CCl_4(\ell)$	Carbon tetrachloride	0.861
$CCl_2F_2(\ell)$	A chlorofluorocarbon (CFC)	0.598
Common solids		
	Wood	1.76
	Concrete	0.88
	Glass	0.84
	Granite	0.79

CD-ROM Screen 6.9:
Calculating Heat Transfer: Use
of Heat Capacity

When the specific heat capacity of a substance is known, you can calculate the temperature change that should occur when a given quantity of energy is transferred to or from a sample of known mass. More important, by measuring the temperature change and the mass of a substance, you can calculate q, the quantity of energy transferred to or from it by heating. For these calculations it is more convenient to rearrange Equation 6.2 algebraically as

$$\Delta T = \frac{q}{c \times m} \qquad \text{or} \qquad q = c \times m \times \Delta T \qquad [6.2]$$

Problem-Solving Example 6.3 Using Specific Heat Capacity

If 50.0 g of water is warmed from 25.3 °C to 33.7 °C, what quantity of energy has been transferred to heat the water?

Answer 1.76 kJ

Explanation The quantity of energy is proportional to the specific heat capacity of water (Table 6.1), the mass of water, and the change in temperature. This is summarized in Equation 6.2 as

$q = c \times m \times \Delta T$

$\quad = 4.184\,J\,g^{-1}\,{}^{\circ}C^{-1} \times 50.0\,g \times (33.7\,{}^{\circ}C - 25.3\,{}^{\circ}C) = 1.76 \times 10^3\,J = 1.76\,kJ$

✔ It requires about 4 J to heat 1 g of water by 1 °C. In this case the temperature change is not quite 10 °C and there is 50 g of water, so q should be about $4\,J\,g^{-1}\,{}^{\circ}C^{-1} \times 10\,{}^{\circ}C \times 50\,g = 2000\,J = 2\,kJ$, which it is.

Problem-Solving Practice 6.3

If 24.1 kJ is supplied to warm a piece of aluminum with a mass of 250. g from an initial temperature of 5.0 °C, what is the final temperature of the aluminum? Obtain the specific heat capacity of Al from Table 6.1.

Exercise 6.3 Specific Heat Capacity and Temperature Change

Suppose you use a hot plate to heat two 50-ml beakers at the same constant rate. If one beaker contains 10. g of pulverized granite rock and one contains 10. g of water, which has the higher temperature after 3 minutes of heating?

Molar Heat Capacity

It is often useful to know the heat capacity of a sample in terms of the same number of particles instead of the same mass. For this purpose we use the **molar heat capacity,** symbol c_m. This is *the quantity of energy that must be transferred to increase the temperature of one mole of a substance by 1 °C.* The molar heat capacity is easily calculated from the specific heat capacity by using the molar mass of the substance. For example, the specific heat capacity of liquid ethanol is given in Table 6.1 as 2.46 J g^{-1} °C^{-1}. The formula of ethanol is CH_3CH_2OH, and so its molar mass is 46.07 g/mol. The molar heat capacity can be calculated as

$$c_m = \frac{2.46\,J}{g\,°C} \times \frac{46.07\,g}{mol} = 113\,J\,mol^{-1}\,°C^{-1}$$

 Exercise 6.4 Molar Heat Capacity

Calculate the molar heat capacities of all the metals listed in Table 6.1. Compare these with the value just calculated for ethanol. Based on your results, suggest a way to predict the molar heat capacity of a metal. Can this same rule be applied to other kinds of substances?

As you should have found in Exercise 6.4, molar heat capacities of metals are very similar. This can be explained if we consider what happens on the nanoscale when a metal is heated. The energy transferred by heating a solid makes the atoms vibrate more extensively about their average positions in the solid crystal lattice. Every metal consists of many, many atoms, all of the same kind and packed close together; that is, the structures of all metals are very similar. Because of this, the ways that the metal atoms can vibrate (and therefore the ways that their energies can be increased) are very similar. Thus, no matter what the metal, nearly the same quantity of energy must be transferred per metal atom to increase the temperature by one degree. The quantity of energy per mole is therefore very similar for all metals.

Problem-Solving Example 6.4 Direction of Energy Transfer

Calculate the quantity of energy transferred from a 250.-mL cup of coffee to your body and the surrounding air when the temperature of the coffee drops from 60.0 °C to 37.0 °C (normal body temperature). Make reasonable assumptions to obtain the mass and specific heat capacity of the liquid.

Answer 24.1 kJ transferred out of the coffee

Explanation Assume the density of coffee, which is mostly water, is 1.00 g/mL, so the coffee has a mass of 250. g; also assume the specific heat capacity of coffee is the same as that of water, 4.184 J g^{-1} °C^{-1}. The initial temperature is 60.0 °C and the final temperature is 37.0 °C. Thus the quantity of energy transferred is

$$q = c \times m \times \Delta T$$
$$= 4.184\,J\,g^{-1}\,°C^{-1} \times 250.\,g \times (37.0 - 60.0)\,°C = -24.1 \times 10^3\,J = -24.1\,kJ$$

 CD-ROM Screen 6.10: Heat Transfer Between Substances

Usually the surroundings contain a great deal more matter than the system and hence have a much greater heat capacity. Consequently the change in temperature of the surroundings is often so small that it cannot be measured.

The negative sign of the result indicates that 24.1 kJ is transferred from the coffee (system) to the surroundings as the temperature of the coffee decreases.

✔ Estimate the heat transfer as a bit more than $(4 \times 20 \times 250)$ J = 20,000 J = 20 kJ, which it is. The transfer is from the coffee, so q should be negative, which it is.

Problem-Solving Practice 6.4

Assume that the same cup of coffee described in Problem-Solving Example 6.4 is warmed from 37 to 65 °C and there is no work done by the heating process. What is ΔE_{system} for this process?

Problem-Solving Example 6.5   Transfer of Energy between Samples by Heating

A 55.0-g piece of iron is heated to 425 °C and then plunged into a beaker of water. The beaker holds 600. mL of water, and its temperature is 25 °C before the iron is dropped in. What is the temperature of both the water and the piece of iron when thermal equilibrium is reached? (Assume that there is no energy transfer to the glass beaker or to the air or to anything else but the water. Assume that the density of water is 1.00 g/mL.)

Answer $T_{final} = 29$ °C

Explanation   Thermal equilibrium means that the water and the iron bar will have the same final temperature, which is what we want to calculate. The law of conservation of energy requires that the total quantity of energy be the same before and after the experiment. Because the energy transfer involves only the iron and the water, summing the transfers must give zero:

$$q_{iron} + q_{water} = 0 \qquad \text{or, by algebraic rearrangement} \qquad q_{water} = -q_{iron}$$

The quantity of energy transferred to the water and the quantity transferred from the iron are equal. They are opposite in algebraic sign because energy was transferred *from* the iron as its temperature dropped, and energy was transferred *to* the water to raise its temperature. Specific heat capacities of iron and water are in Table 6.1. The mass of water is 600. mL × 1.00 g/mL = 600. g. $T_{initial}$ for the iron is 425 °C and $T_{initial}$ for the water is 25 °C.

> The quantities of energy transferred have opposite signs because they take place from the iron (negative) to the water (positive).

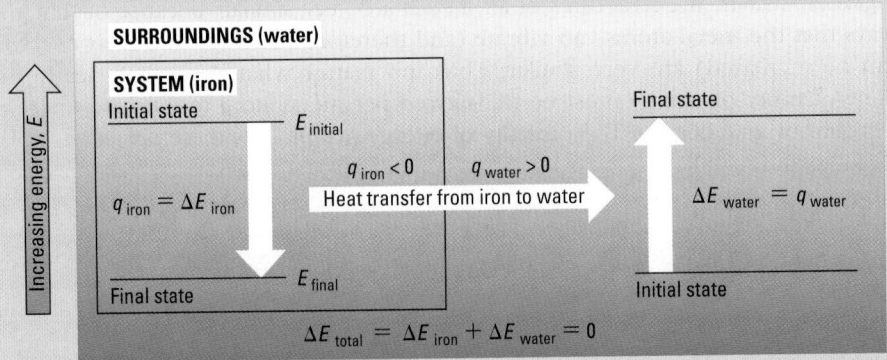

Heat transfer from iron to water.

$$q_{water} = -q_{iron}$$
$$(4.184 \text{ J g}^{-1}\,°C^{-1})(600.\text{ g})(T_{final} - 25\,°C) = -(0.451 \text{ J g}^{-1}\,°C^{-1})(55.0\text{ g})(T_{final} - 425\,°C)$$
$$(2510.\text{ J}\,°C^{-1})T_{final} - (6.276 \times 10^4 \text{ J}) = -(24.80 \text{ J}\,°C^{-1})T_{final} + (1.054 \times 10^4 \text{ J})$$
$$(2535 \text{ J}\,°C^{-1})T_{final} = 7.330 \times 10^4 \text{ J}$$

> $\Delta T_{iron} = 29\,°C - 425\,°C = -396\,°C$
>
> $\Delta T_{water} = 29\,°C - 25\,°C = 4\,°C$

Solving, we find $T_{final} = 29$ °C. The iron has cooled a lot ($\Delta T_{iron} = -396$ °C) and the water has warmed a little ($\Delta T_{water} = 4$ °C).

✔   As a check, note that the final temperature must be between the two initial values, which it is. Also, don't be concerned by the fact that transferring the same quantity of energy has resulted in two very different values of ΔT; this is because the specific heat capacities and masses of iron and water are different. There is much less iron and its specific heat capacity is smaller, so its temperature changes a lot more than for the water.

Problem-Solving Practice 6.5

A 400.-g piece of iron is heated in a flame and then immersed in 1000. g of water in a beaker. The initial temperature of the water was 20.0 °C, and both the iron and the water are at 32.8 °C at the end of the experiment. What was the original temperature of the hot iron bar? (Assume that all energy transfer is between the water and the iron.)

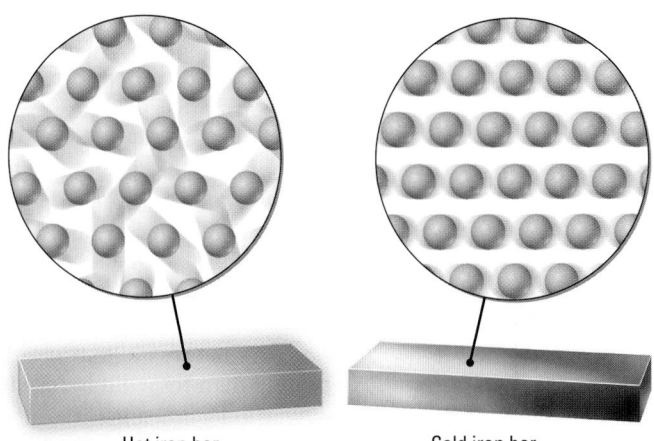

Hot iron bar   Cold iron bar

Hot and cold iron.   On the nanoscale the atoms in the sample of hot iron are vibrating much farther from their average positions than those in the sample of room temperature iron. The greater vibration of atoms in hot iron means harder collisions of iron atoms with water molecules. Such collisions transfer energy to the water molecules, heating the water.

6.4   ENERGY AND ENTHALPY

Using heat capacity we can account for transfers of energy between samples of matter as a result of temperature differences. But energy transfers also accompany physical or chemical changes, *even though there may be no change in temperature.* We will first consider the simpler case of physical change and then apply the same ideas to chemical changes.

 CD-ROM Screen 6.11: Heat Associated with Phase Changes

Changes of state (between solid and liquid, liquid and gas, or solid and gas) are described in more detail in Section 11.3. Because the temperature remains constant during a change of state, melting points and boiling points can be measured relatively easily and used to identify substances (⇐ *p. 7*).

Freezing and Melting (Fusion)

As an example, consider what happens when ice is heated at a slow, constant rate from −50 °C to +50 °C. A graph of temperature as a function of quantity of transferred energy is shown in Figure 6.10. When the temperature reaches 0 °C, it remains constant, despite the fact that the sample is still being heated. As long as ice is melting, thermal energy must be continually supplied to overcome forces that hold the water molecules in their regularly spaced positions in the nanoscale structure of solid ice. Overcoming these forces raises the potential energy of the water molecules and therefore requires transfer of energy into the system.

Figure 6.10 Heating graph. When a 1.0-g sample of ice is heated at a constant rate, the temperature does not always increase at a constant rate.

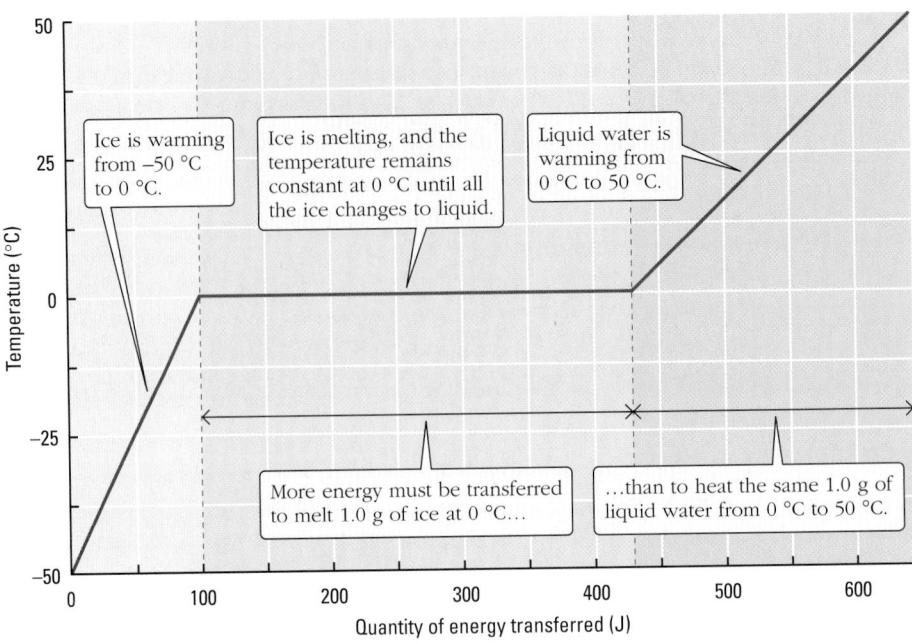

Ice is warming from −50 °C to 0 °C.

Ice is melting, and the temperature remains constant at 0 °C until all the ice changes to liquid.

Liquid water is warming from 0 °C to 50 °C.

More energy must be transferred to melt 1.0 g of ice at 0 °C…

…than to heat the same 1.0 g of liquid water from 0 °C to 50 °C.

Protecting crops from freezing. Because heat is transferred to the surroundings as water freezes, one way to protect plants from freezing if the temperature drops just below the freezing point is to spray water on them. As the water freezes, energy transfer to the leaves and stems keeps the plants themselves from freezing.

Melting a solid is an example of a **change of state** or **phase change,** a physical process in which one state of matter is transformed into another. During a change of state, the temperature remains constant, but energy must be continually transferred into or out of the system because the nanoscale particles have higher or lower potential energy after the phase change than they did before it. As shown in Figure 6.10, the quantity of energy transferred is significant.

The quantity of thermal energy that must be transferred to a solid as it melts is called the **heat of fusion.** For ice the heat of fusion is 333 J/g at 0 °C. This same quantity of energy could raise the temperature of a 1.00-g block of iron from 0 °C to 738 °C (red hot), or it could melt 0.50 g of ice and heat the liquid water from 0 °C to 80 °C. This is illustrated schematically in Figure 6.11. The opposite of melting is freezing. When water freezes, the quantity of energy transferred is the same as when water melts, but energy transfers in the opposite direction—from system to surroundings. Thus, under the same conditions of temperature and pressure,

$$q_{\text{fusion}} = -q_{\text{freezing}}.$$

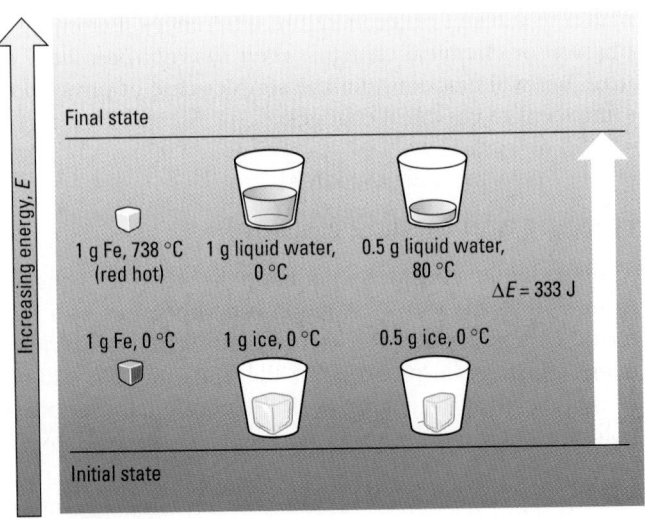

Figure 6.11 Heating, temperature change, and phase change. Heating a substance can cause a temperature change, a phase change, or both. Here, 333 J has been transferred to each of three samples: a 1-g block of iron at 0 °C; a 1-g block of ice at 0 °C; and a 0.5-g block of ice at 0 °C. The iron block becomes red hot; its temperature increases to 738 °C. The 1-g block of ice melts, resulting in 1 g of liquid water at 0 °C. The 0.5-g block of ice melts, and there is enough energy to heat the liquid water to 80 °C.

Chemistry in the News

CAPTURING THE SUN'S ENERGY

In the Mojave Desert near Daggett, California, a $39 million solar power plant called Solar Two was dedicated on June 5, 1996. Its experimental operation was completed in April 1999 and demonstrated successfully the latest technology for converting sunlight into electricity. Solar Two consists of 1926 mirrors surrounding a tower nearly 100 m high. Each mirror can be adjusted by two computer-controlled electric motors to reflect sunlight to the top of the tower. There, an array of pipes coated with special energy-absorbing paint collects the solar energy. The top of the tower glows because of the concentrated sunlight, which heats it to nearly 600 °C.

More than 1.3 million kg of molten sodium nitrate and potassium nitrate circulates through the pipes and is heated to 565 °C. Once heated, the mixture of salts is stored in specially insulated tanks. The large heat capacity of the molten salts allows solar energy to be collected during the day and stored for use at night when demand for electricity is high. The hot mixture is then used to boil water, which passes through steam generators similar to those used in coal-burning electric power plants.

Solar Two produced 10 MW of electricity, enough for a city of 10,000 homes, and delivered it to an electric power grid. Because it stores solar energy for use when utility customers want electricity, Solar Two is attractive as a means of converting solar energy for human use, and industry is planning to implement a commercial version of this pioneering solar energy experiment.

_____

Source

http://www.energylan.sandia.gov/sunlab/

Solar energy collection. Solar Two's central tower is surrounded by 1926 adjustable mirrors that reflect sunlight onto the tower, nearly 100 m above the ground.

Vaporization and Condensation

The quantity of energy that must be transferred to convert a liquid to vapor is called the **heat of vaporization.** For water it is 2260 J/g at 100 °C. This is considerably larger than the heat of fusion, because the water molecules become completely separated during the transition from liquid to vapor. As they separate, a great deal of energy is required to overcome the attractions among water molecules. Therefore, the potential energy of the vapor is considerably higher than for the liquid. Although 333 J can melt 1.00 g of ice at 0 °C, it will boil only 0.147 g of water at 100 °C.

$$333 \text{ J} \times \frac{1.00 \text{ g water vaporized}}{2260 \text{ J}} = 0.147 \text{ g water vaporized}$$

The opposite of vaporization is condensation. As you might expect, the heat transfer for condensation is in the opposite direction from that for vaporization. Therefore, under the same conditions, $q_{\text{vaporization}} = -q_{\text{condensation}}$.

Exercise 6.5 Heating and Cooling Graphs

(a) Assume that a 1.0-g sample of ice at −5 °C is heated at a uniform rate until the temperature is 105 °C. Draw a graph like the one in Figure 6.10 to show how temperature varies with energy transferred. Your graph should be to approximately the correct scale.

(b) Assume that a 0.50-g sample of water is cooled (at the same uniform rate as the heating in part a) from 105 °C to −5 °C. Draw a cooling curve to show how temperature varies with energy transferred. Your graph should be to the same scale as in part (a).

Exercise 6.6 Changes of State

Assume you have 1 cup of ice (237 g) at 0.0 °C. How much heating is required to melt the ice, warm the resulting water to 100.0 °C, and then boil the water to vapor at 100.0 °C? (Hint: Do three separate calculations and then add the results.)

Conservation of Energy and Changes of State

Consider a system that consists of water at its boiling temperature in a container with a balloon attached (Figure 6.12). The system is under a constant atmospheric pressure. If the water is heated, it will boil, the temperature will remain at 100 °C, and the steam produced by boiling the water will inflate the balloon (Figure 6.12b). If the heating stops, then the water will stop boiling, some of the steam will condense to liquid, and the volume of steam will decrease (Figure 6.12c). There will be heat transfer of energy to the surroundings. However, as long as there is steam condensing to liquid water, the temperature will remain at 100 °C. In summary, transferring energy *into* the system produces more steam; transferring energy *out of* the system results in less steam. Both the boiling and condensing processes occur at the same temperature — the boiling point.

The boiling process can be represented by the equation

$$H_2O(\ell) \longrightarrow H_2O(g) \qquad \text{endothermic}$$

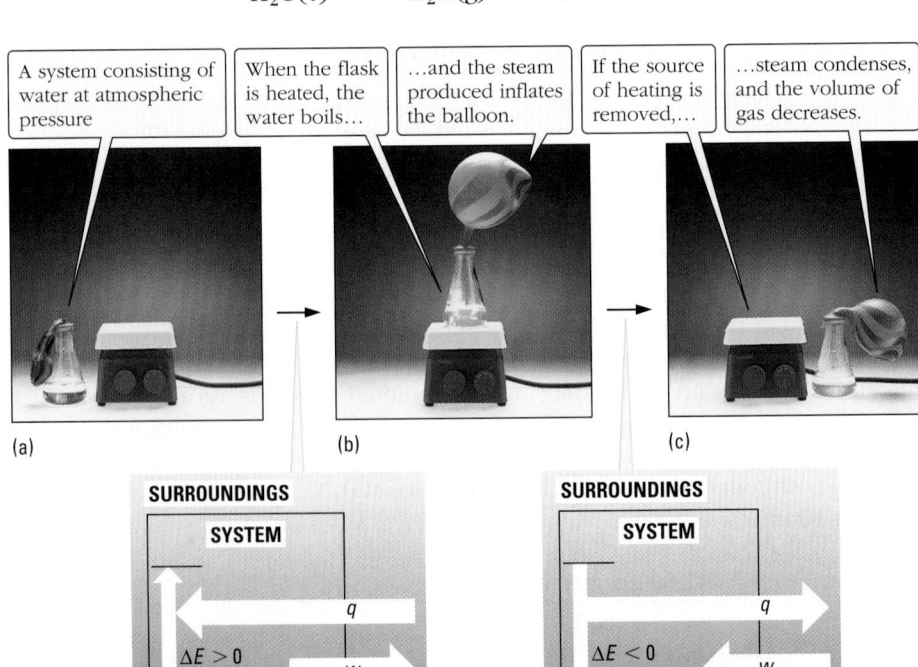

Figure 6.12 Boiling water at constant pressure. When water boils, the steam pushes against atmospheric pressure and does work on the atmosphere (which is part of the surroundings). The balloon is used so that the expansion of the steam can be seen; even if the balloon were not there, the steam would push back the surrounding air. In general, for any constant pressure process, if there is a change in volume, some work is done, either on the surroundings or on the system.

We call this process **endothermic** because, as it occurs, energy must be transferred *into* the system to maintain constant temperature. If there were no energy transfer, the liquid water would get cooler. Evaporation of water (perspiration) from your skin, which occurs at a lower temperature than boiling, is an endothermic process that you are certainly familiar with. Energy must be transferred from your skin to the evaporating water, and this transfer cools your skin.

The opposite of boiling is condensation. It can be represented by the opposite equation,

$$H_2O(g) \longrightarrow H_2O(\ell) \qquad \text{exothermic}$$

This process is said to be **exothermic** because energy must be transferred *out of* the system to maintain constant temperature. Because condensation of $H_2O(g)$ (steam) is exothermic, a burn from steam at 100 °C is much worse than a burn from liquid water at 100 °C. The steam heats the skin a lot more because there is a heat transfer due to the condensation as well as the difference in temperature between the water and your skin.

CD-ROM Screen 6.12: Energy Changes in Chemical Processes

"Thermic" or "thermo" comes from a Greek word *thermé*, which means heat. "Endo" comes from the Greek word *endon*, which means within or inside. "Endothermic" therefore indicates transfer of energy *into* the system. "Exo" comes from the Greek word *exō*, which means out of. "Exothermic" indicates transfer of energy *out of* the system.

Phase change	Direction of energy transfer	Sign of q_{system}	Type of change
$H_2O(\ell) \to H_2O(g)$	Surroundings $\to$ system	Positive ($q > 0$)	*Endo*thermic
$H_2O(g) \to H_2O(\ell)$	System $\to$ surroundings	Negative ($q < 0$)	*Exo*thermic

The system in Figure 6.12 can be analyzed by using the law of conservation of energy, $\Delta E = q + w$. Suppose that 1.0 g of water is vaporized. This requires heat transfer equal to the heat of vaporization, and so $q_{\text{system}} = 2260$ J (a positive value because the transfer is from the surroundings to the system). At the same time, the expansion of the steam pushes back the atmosphere, doing work. The quantity of work is more difficult to calculate, but it is clear that w must be negative, because the system does work on the surroundings. Therefore, the internal energy of the system is increased by the quantity of heating and is decreased by the quantity of work done.

Now suppose that the heating is stopped and the direction of heat transfer is reversed. The water stops boiling and some of the steam condenses to liquid water. The balloon deflates and the atmosphere pushes back the steam. If 1.0 g of steam condenses, then $q = -2260$ J. Because the surrounding atmosphere pushes on the system, the surroundings have done work on the system. This makes w positive. As long as steam is condensing to liquid, the temperature remains at 100 °C. The internal energy of the system is increased by the quantity of work done on it and decreased by the quantity of energy transfer to the surroundings.

The device described here is a crude example of a steam engine. Burning fuel boils water, and the steam does work. In a real steam engine the steam would drive a piston and then be allowed to escape, providing a means for the system to continually do work on its surroundings. Systems that convert heat into work are called heat engines. Another example is the engine in an automobile, which converts heat from the combustion of fuel into work to move the car.

Enthalpy: Heat Transfer at Constant Pressure

In the previous section we did not calculate how much work was done by the system when it expanded against the atmosphere, but it was clear that work was done because there was a change in volume. Work is done when a force moves an object through some distance. If nothing moves, no work is done. Therefore, in a closed container where the system's volume is constant, $w = 0$, and

$$\Delta E = q + w = q + 0 = q_V$$

where the subscript V indicates constant volume; that is, q_V is the heat transfer into a constant-volume system. This means that *if a process is carried out in a closed container and the heat transfer is measured, ΔE has been determined.*

CD-ROM Screen 6.14: Enthalpy Change and ΔH

CHEMISTRY You Can Do... Work and Volume Change

Obtain an empty aluminum soft drink can, a hot plate or electric stove that can boil water, tongs, a glove, or a potholder you can use to pick up the can when it is hot, and a container of cold water large enough so that you can immerse the soft drink in the water. Rinse out the can with clean water and then pour water into the can until it is about 1 cm deep. Put the can on the hot plate and heat it until the water starts to boil. Let the water boil until steam has been coming

out of the opening for at least 1 min. (**Caution:** Watch the can carefully while it is being heated. If it boils dry, the temperature will go way above 100 °C, the aluminum can will melt, your hot plate will be messed up, and you might start a fire.) While a steady stream of steam continues to come out of the can, pick it up with the tongs and in one smooth, quick motion turn it upside down and immerse the opening in the cold water. Be prepared for a surprise. What happens?

Now analyze what happened thermodynamically. The following questions may help your analysis.

1. Write a thermochemical equation for the process of boiling the water.
2. What energy transfers occur between the system and surroundings as the water boils?
3. What was in the can after the water had boiled for a minute or two? What happened to the air that was originally in the can?
4. What happened to the contents of the can as soon as it was immersed in the cold water?
5. Write a thermochemical equation for the process in Question 4.
6. Did the atmosphere do work on the can and its contents after the can was immersed in the water? Cite observations to support your answer.

In plants, animals, laboratories, and the environment, physical processes and chemical reactions seldom take place in closed containers. Instead they are carried out in contact with the atmosphere. For example, the vaporization of water shown in Figure 6.12 took place under conditions of constant atmospheric pressure, and the expanding steam had to push back the atmosphere. In such a case,

$$\Delta E = q_P + w_{\text{expansion}}$$

That is, ΔE differs from the heat transfer at constant pressure, q_P, by the work done to push back the atmosphere, $w_{\text{expansion}}$. For processes that do not involve gases, $w_{\text{expansion}}$ is very small. Even when gases are involved, $w_{\text{expansion}}$ is usually much smaller than q_P, and, for the same system and the same initial and final conditions, $w_{\text{expansion}}$ always has the same value. Therefore it is convenient to use q_P to characterize energy transfers at constant pressure. *The quantity of thermal energy transferred into a system at constant pressure, q_P, is called the* **enthalpy change** *of the system.* Enthalpy change is symbolized by ΔH; that is, $H_{\text{final}} - H_{\text{initial}}$. Thus, $\Delta H = q_P$, and

$$\Delta E = q_P + w_{\text{expansion}} = \Delta H + w_{\text{expansion}}$$

Because it is equal to the quantity of thermal energy transferred at constant pressure, and because most chemical reactions are carried out at atmospheric (constant) pressure, the enthalpy change for a process is often called the *heat of that process.* For example, the enthalpy change for melting (fusion) is also called the *heat of fusion.*

ΔH accounts for all the energy transferred except the quantity that does the work of pushing back the atmosphere, which is usually relatively small. That is, ΔH is closely related to the change in the internal energy of the system but is slightly different in magnitude. ***Whenever heat transfer is measured at constant pressure, it is ΔH that is determined.***

Problem-Solving Example 6.6   Changes of State, ΔH, and ΔE

When 1.0 g of methanol, CH_3OH, boils away at 65.0 °C and atmospheric pressure, the volume of gaseous methanol is 864 mL greater than the volume of the original liquid. This change in volume corresponds to expansion work of 88 J. The heat transfer is 1173 J, and the process is endothermic. Calculate ΔH and ΔE.

Answer $\Delta H = 1173\ \text{J}$; $\Delta E = 1085\ \text{J}$

Explanation   The process takes place at constant pressure. By definition, $\Delta H = q_P$. Because thermal energy is transferred *to* the system, the sign of q_P must be positive. Therefore $\Delta H = 1173$ J. Because the system expands and pushes back the atmosphere, work must be done by the system. Therefore the sign of w is negative, and $w_{\text{expansion}} = -88$ J. To calculate ΔE, add the expansion work to the enthalpy change.

$$\Delta E = \Delta H + w_{\text{expansion}} = 1173\ \text{J} - 88\ \text{J} = 1085\ \text{J}$$

✔   Boiling is an endothermic process, so ΔH must be positive. Because the system did work on the surroundings, the change in internal energy must be less than the enthalpy change, and it is.

Problem-Solving Practice 6.6

When potassium melts at atmospheric pressure, the heat transfer is 14.6 cal/g. The density of liquid potassium at its melting point is 0.82 g/mL, and that of solid potassium is 0.86 g/mL. Given that a volume change of 1.00 mL at atmospheric pressure corresponds to 0.10 J, calculate ΔH and ΔE for melting 1.00 g of potassium.

State Functions and Path Independence

Both energy and enthalpy are **state functions,** *properties whose values are invariably the same if a system is in the same state.* A system's state is defined by its temperature, pressure, volume, mass, and composition. For the same initial and final states, a change in a state function does not depend on the path by which the system changes from one state to another (Figure 6.13). Returning to the bank account analogy (⬅ *p. 217*), your bank balance is independent of the path by which you change it. If you have $1000 in the bank (initial state) and withdraw $100, your balance will go down to $900 (final state) and $\Delta B = -\$100$. If instead you had deposited $500 and withdrawn $600 you would have achieved the same change of $\Delta B = -\$100$ by a different pathway, and your final balance would still be $900.

The fact that changes in a state function are independent of the sequence of events by which change occurs is important, because it allows us to apply laboratory measurements to real-life situations. For example, if you measure in the lab the heat transfer when 1.0 g of glucose (dextrose sugar) burns in exactly the amount of oxygen required to convert it to carbon dioxide and water, you will find that $\Delta H = -15.5$ kJ. When you eat something that contains 1.0 g of glucose and your body metabolizes the glucose (producing the same products at the same temperature and pressure), there is the same change in enthalpy. This means that laboratory measurements can be used to determine how much energy you can get from a given quantity of food, which is the basis for the caloric values listed on the labels of the foods you eat.

6.5 THERMOCHEMICAL EQUATIONS

To indicate the heat transfer that occurs when either a physical or chemical process takes place, we write a **thermochemical equation.** This is *a balanced chemical equation together with the corresponding value of the enthalpy*

The results of this example show that ΔE differs by less than 10% from ΔH—that is, by 88 J out of 1173 J, which is 7.5%. It is true for most physical and chemical processes that the work of pushing back the atmosphere is only a small fraction of the heat transfer of energy.

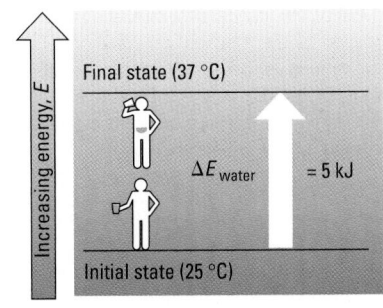

(a)

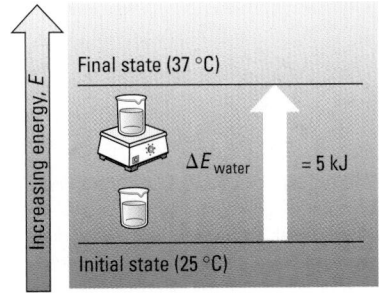

(b)

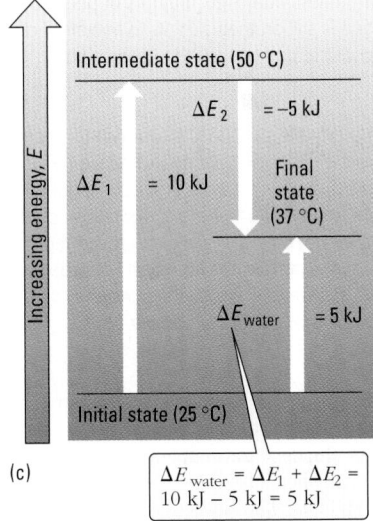

(c)

Figure 6.13   Energy change is path-independent.   If 100. g of water at 25 °C is warmed to 37 °C (body temperature) at atmospheric pressure, the change in energy of the water is the same whether (a) you drank the water and your body warmed it to 37 °C, or (b) you put the water in a beaker and heated it with a hot plate, or (c) you heated the water to 50. °C and then cooled it to 37 °C.

change. For evaporation of water near room temperature and at typical atmospheric pressure this thermochemical equation can be written:

$$H_2O(\ell) \longrightarrow H_2O(g) \qquad \Delta H° = +44.0 \text{ kJ} \qquad (25\ °C,\ 1\ bar)$$

The symbol $\Delta H°$ (pronounced "delta-aitch-standard") represents the **standard enthalpy change,** which is defined as the *enthalpy change at the standard pressure of 1 bar and a specified temperature.* Because the value of the enthalpy change depends on the pressure at which the process is carried out, all enthalpy changes are reported at the same standard pressure, 1 bar. (The bar is a unit of pressure that is very close to the pressure of the earth's atmosphere at sea level; you may have heard the bar used in a weather report.) The value of the enthalpy change also varies slightly with temperature. For thermochemical equations in this book, the temperature can be assumed to be 25 °C, unless some other temperature is specified.

In 1982 the International Union of Pure and Applied Chemistry chose a pressure of 1 bar as the standard for tabulating information for thermochemical equations. This pressure is very close to the standard atmosphere: 1 bar = 0.98692 atm = 1 × 10^5 kg m^{-1} s^{-2}. (Pressure units are discussed further in Section 10.1.)

Exercise 6.7 Interpreting Thermochemical Equations

What part of the thermochemical equation for vaporization of water indicates that energy is transferred from the surroundings to the system when the evaporation process occurs?

You experience cooling due to evaporation of water when you perspire. If you work up a real sweat, then lots of water evaporates from your skin, producing a much greater cooling effect. People who exercise in cool weather need to carry a sweatshirt or jacket. When they stop exercising they generate less body heat, but lots of perspiration is still on their skin. Its evaporation can cool the body enough to cause a chill.

The thermochemical equation given above indicates that when *one mole* of liquid water (at 25 °C and 1 bar) evaporates to form *one mole* of water vapor (at 25 °C and 1 bar), 44.0 kJ of energy must be transferred from the surroundings to the system to maintain the temperature at 25 °C. The size of the enthalpy change depends on how much process (in this case evaporation) takes place. The more water that evaporates, the more the surroundings are cooled. If 2 mol of $H_2O(\ell)$ is converted to 2 mol of $H_2O(g)$, 88 kJ of energy is transferred, and if 0.5 mol of $H_2O(\ell)$ is converted to 0.5 mol of $H_2O(g)$, only 22 kJ is required. The numerical value of $\Delta H°$ corresponds to the reaction as written, with the coefficients indicating moles of each reactant and moles of each product. For the equation

Usually the surroundings contain far more matter than the system and hence have a much greater heat capacity. Consequently, the temperature of the surroundings often does not change significantly, even though energy transfer has occurred. For evaporation of water at 25 °C the surroundings would not drop much below 25 °C.

$$2\ H_2O(\ell) \longrightarrow 2\ H_2O(g) \qquad \Delta H° = +88.0 \text{ kJ}$$

the process is evaporating 2 mol of $H_2O(\ell)$ to form 2 mol of $H_2O(g)$, both at 25 °C and 1 bar. For this process the enthalpy change is twice as great as for the case where there is a coefficient of 1 on each side of the equation.

Now consider water vapor condensing to form liquid. If 44.0 kJ of energy is required to do the work of separating the water molecules in 1 mol of the liquid as it vaporizes, the same quantity of energy will be released when the molecules move closer together as the vapor condenses to form liquid.

The idea here is similar to the example given earlier of water falling over a waterfall. The decrease in potential energy of the water when it falls from the top to the bottom of the waterfall is exactly equal to the increase in potential energy that would be required to take the same quantity of water from the bottom of the fall to the top. The signs are opposite because in one case potential energy is transferred *from* the water and in the other it is transferred *to* the water.

$$H_2O(g) \longrightarrow H_2O(\ell) \qquad \Delta H° = -44.0 \text{ kJ}$$

This thermochemical equation indicates that 44.0 kJ of energy is transferred to the surroundings *from* the system when 1 mol of water vapor condenses to liquid at 25 °C and 1 bar.

 Exercise 6.8 Thermochemical Equations

Why is it essential to specify the state (s, ℓ, or g) of each reactant and each product in a thermochemical equation?

Problem-Solving Example 6.7 Changes of State and $\Delta H°$

Calculate the energy transferred to the surroundings when water vapor in the air condenses at 25 °C to give rain in a thunderstorm. Suppose that one inch of rain falls over one square mile of ground, so that 6.6×10^{10} mL has fallen. (Assume $d_{H_2O(\ell)} = 1.0$ g/mL.)

Answer 1.6×10^{11} kJ

Explanation The thermochemical equation for condensation of 1 mol of water at 25 °C is

$$H_2O(g) \longrightarrow H_2O(\ell) \qquad \Delta H° = -44.0 \text{ kJ}$$

The standard enthalpy change tells how much heat transfer is required when 1 mol of water condenses at constant pressure, so we first calculate how many moles of water condensed.

$$\text{Amount of water condensed} = 6.6 \times 10^{10} \text{ g water} \times \frac{1 \text{ mol}}{18.0 \text{ g}} = 3.66 \times 10^9 \text{ mol water}$$

Next, calculate the quantity of energy transferred from the fact that 44.0 kJ is transferred per mole of water.

$$\text{Quantity of energy transferred} = 3.66 \times 10^9 \text{ mol water} \times \frac{44.0 \text{ kJ}}{1 \text{ mol}} = 1.6 \times 10^{11} \text{ kJ}$$

The negative sign of $\Delta H°$ in the thermochemical equation indicates transfer of the 1.6×10^{11} kJ *from the water* (system) to *the surroundings*.

✔ The quantity of water is about 10^{11} g. The energy transfer is 44 kJ for a mole (18 g) of water. Since 44 is about twice 18, this is about 2 kJ/g. Therefore, the energy transferred in kJ should be about twice the number of grams, or about 2×10^{11} kJ, and it is.

Problem-Solving Practice 6.7

The enthalpy change for sublimation of 1 mol of solid iodine at 25 °C and 1 bar is 62.4 kJ. (Sublimation means changing directly from solid to gas.)

$$I_2(s) \longrightarrow I_2(g) \qquad \Delta H° = +62.4 \text{ kJ}$$

(a) What quantity of energy must be transferred to vaporize 10.0 g of solid iodine?
(b) If 3.42 g of iodine vapor condenses to solid iodine, what quantity of energy is transferred?
(c) Is the process in part (b) exothermic or endothermic?

Agronomists and meteorologists measure quantities of water in units of acre-feet; an acre-foot is enough rainfall to cover an acre of land to a depth of one foot.

Since the explosion of 1000 tons of dynamite is equivalent to 4.2×10^9 kJ, the energy transferred by our hypothetical thunderstorm is about the same as that released when 38,000 tons of dynamite explodes! A great deal of energy can be stored in water vapor, which is one reason why storms can cause so much damage.

Like all examples in this chapter, this one assumes that the temperature of the system remains constant, so that all the energy transfer associated with the phase change goes to or from the surroundings.

Sublimation of iodine.

6.6 ENTHALPY CHANGES FOR CHEMICAL REACTIONS

Like phase changes, chemical reactions can be exothermic or endothermic, but reactions usually involve much larger energy transfers than do phase changes. Indeed, a temperature change is one piece of evidence that a chemical reaction has taken place. The large energy transfers that occur during chemical reactions are the result of breaking and making chemical bonds as reactants are converted into products. These energy transfers have important applications in living systems, in industrial processes, in heating or cooling your home, and in many other situations.

Hydrogen is an excellent fuel. It produces very little pollution when it burns in air, and its reaction with oxygen to form water is highly exothermic. It is used in the space shuttle, for example. The thermochemical equation for formation of one mole of water vapor from hydrogen and oxygen is

$$H_2(g) + \tfrac{1}{2} O_2(g) \longrightarrow H_2O(g) \qquad \Delta H° = -241.8 \text{ kJ} \quad [6.3]$$

Like all thermochemical equations, this one has four important characteristics:

• The *sign of* $\Delta H°$ indicates the direction of energy transfer.

 CD-ROM Screen 6.15: Enthalpy Changes for Chemical Reactions

Combustion of hydrogen gas. The combination reaction of hydrogen and oxygen produces water vapor in a highly exothermic process.

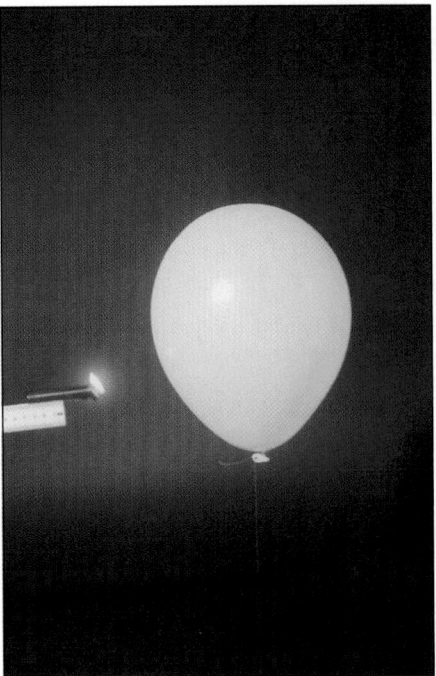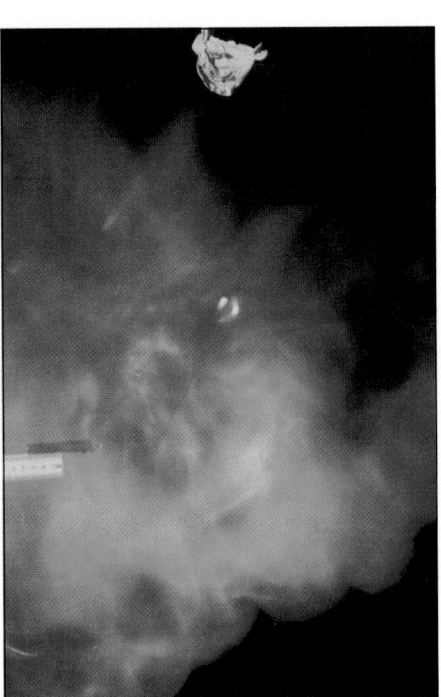

- The magnitude of $\Delta H°$ depends on the *states of matter* of the reactants and products.
- The *balanced equation represents moles* of reactants and of products.
- The *quantity of energy transferred is proportional to the quantity of reaction* that occurs.

The relationship of reaction heat transfer and enthalpy change:

Reactant $\longrightarrow$ product with transfer of thermal energy *from* system to surroundings. ΔH is negative; reaction is *exo*thermic.

Reactant $\longrightarrow$ product with transfer of thermal energy *into* system from surroundings. ΔH is positive; reaction is *endo*thermic.

Sign of $\Delta H°$. The thermochemical equation 6.3 tells us that this process is exothermic, because $\Delta H°$ is negative. Formation of 1 mol of water vapor transfers 241.8 kJ of energy from the reacting chemicals to the surroundings. If 1 mol of water vapor is decomposed to hydrogen and oxygen (the reverse process), the magnitude of $\Delta H°$ is the same, but the sign is opposite, indicating transfer of energy from the surroundings to the system:

$$H_2O(g) \longrightarrow H_2(g) + \tfrac{1}{2} O_2(g) \qquad \Delta H° = +241.8 \text{ kJ} \quad [6.4]$$

The reverse of an exothermic process is endothermic. The magnitude of the energy transfer is the same, but the direction of transfer is opposite.

States of Matter. If liquid water is involved instead of water vapor, the magnitude of $\Delta H°$ is different than that in Equation 6.3:

$$H_2(g) + \tfrac{1}{2} O_2(g) \longrightarrow H_2O(\ell) \qquad \Delta H° = -285.8 \text{ kJ} \quad [6.5]$$

Our discussion of phase changes ($\Longleftarrow$ *p. 226, Section 6.4*) showed that there is an enthalpy change when a substance changes state. Vaporizing 1 mol of $H_2O(\ell)$ requires 44.0 kJ. Formation of 1 mol of $H_2O(\ell)$ from $H_2(g)$ and $O_2(g)$ is 285.8 kJ − 241.8 kJ = 44.0 kJ more exothermic than is formation of 1 mol of $H_2O(g)$. Figure 6.14 shows the relationships among these quantities. The enthalpy of the reactants ($H_2(g)$ and $\tfrac{1}{2} O_2(g)$) is greater than that of the product ($H_2O(g)$). Because the system has less enthalpy after the reaction than before, the law of conservation of energy requires that 241.8 kJ must be transferred *to* the surroundings as the reaction takes place. $H_2O(\ell)$ has even less enthalpy than $H_2O(g)$, and so when $H_2O(\ell)$ is formed, even more energy, 285.8 kJ, must be transferred to the surroundings.

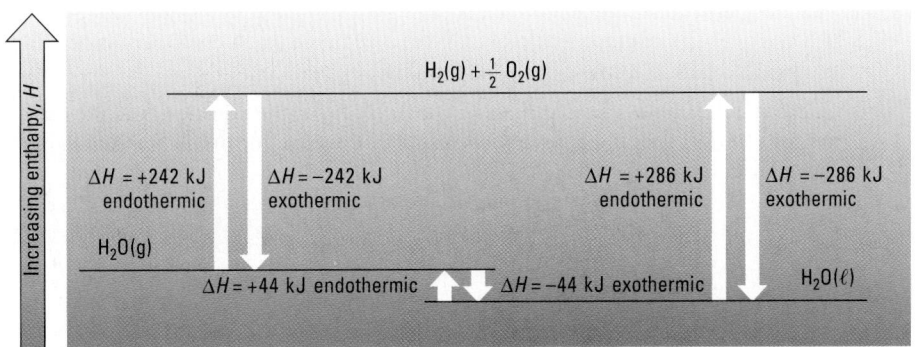

Figure 6.14   Enthalpy diagram.
Water vapor [1 mol of $H_2O(g)$] liquid water [1 mol of $H_2O(\ell)$] and a stoichiometric mixture of hydrogen and oxygen gases [1 mol of $H_2(g)$ and $\frac{1}{2}$ mol of $O_2(g)$] each have a different enthalpy value. The figure shows how these are related, with highest enthalpy at the top.

Balanced Equation Represents Moles.   In order to write an equation for the formation of 1 mol of H_2O it is necessary to use a fractional coefficient for O_2. This is acceptable in a thermochemical equation, because the coefficients mean moles, not molecules, and half a mole of O_2 is a perfectly reasonable quantity.

Quantity of Energy Is Proportional to Quantity of Reaction.   This property of thermochemical equations involves stoichiometry. The more reaction there is, the more energy is transferred. Because the balanced equation represents moles, we can calculate how much heat transfer occurs from the number of moles of a reactant that is consumed or the number of moles of a product that is formed. If the thermochemical equation for combination of hydrogen and oxygen is written without the fractional coefficient, so that 2 mol of $H_2O(g)$ is produced, then the energy transfer is twice as great, that is, $2(-241.8 \text{ kJ}) = -483.6 \text{ kJ}$.

The direct proportionality between quantity of reaction and quantity of heat transfer is in line with your everyday experience. Burning twice as much natural gas produces twice as much heating.

$$2\,H_2(g) + O_2(g) \longrightarrow 2\,H_2O(g) \qquad \Delta H° = -483.6 \text{ kJ} \quad [6.6]$$

Exercise 6.9   Enthalpy Change and Stoichiometry

Calculate the change in enthalpy if 0.5 mol of $H_2(g)$ reacts with an excess of oxygen to form water vapor at 25 °C.

Problem-Solving Example 6.8   Thermochemical Equations

Given the thermochemical equation

$$N_2(g) + 3\,H_2(g) \longrightarrow 2\,NH_3(g) \qquad \Delta H° = -92.22 \text{ kJ}$$

Write a thermochemical equation for

(a)   Formation of 1 mol of $NH_3(g)$
(b)   Decomposition of 4 mol of $NH_3(g)$
(c)   Combination of 1 mol of $H_2(g)$ with a stoichiometric quantity of nitrogen

Answer

(a) $\frac{1}{2} N_2(g) + \frac{3}{2} H_2(g) \rightarrow NH_3(g)$ .... $\Delta H° = -46.11 \text{ kJ}$
(b) $4\,NH_3(g) \rightarrow 2\,N_2(g) + 6\,H_2(g)$ .... $\Delta H° = +184.44 \text{ kJ}$
(c) $\frac{1}{3} N_2(g) + H_2(g) \rightarrow \frac{2}{3} NH_3(g)$ .... $\Delta H° = -30.74 \text{ kJ}$

Explanation

(a)   Producing 1 mol of $NH_3(g)$ requires that half the amount of each reactant and product be used and also halves the $\Delta H°$ value.
(b)   Decomposition of $NH_3(g)$ means that $NH_3(g)$ must be a reactant — changing the direction of the reaction changes the sign of $\Delta H°$; because 4 mol of $NH_3(g)$ is decomposed, each coefficient must be doubled, and the magnitude of $\Delta H°$ must also be doubled.

(c) If 1 mol of $H_2(g)$ reacts, only $\frac{1}{3}$ mol of $N_2(g)$ is required; each coefficient is one-third its original value, and $\Delta H°$ is also one third the original value.

✔ In each case examine the coefficients, the direction of the equation, and the sign of $\Delta H°$ to make certain that the appropriate quantity of reactant or product and the appropriate sign have been written.

Problem-Solving Practice 6.8

Given the thermochemical equation

$$Ca(s) + \tfrac{1}{2} O_2(g) \longrightarrow CaO(s) \qquad \Delta H° = -635.09 \text{ kJ}$$

Write the thermochemical equation for production of 4 mol of $O_2(g)$ by decomposition of solid calcium oxide.

In Section 4.4 we derived stoichiometric factors (mole ratios) from the coefficients in balanced chemical equations *(⇐ p. 135)*. Stoichiometric factors that relate quantity of energy transferred to quantity of reactant used up or quantity of product produced can be derived from a thermochemical equation. From Equation 6.6 these factors (and their reciprocals) can be derived:

$$2 H_2(g) + O_2(g) \longrightarrow 2 H_2O(g) \qquad \Delta H° = -483.6 \text{ kJ} \quad [6.6]$$

$$\frac{-483.6 \text{ kJ}}{2 \text{ mol } H_2 \text{ reacted}} \qquad \frac{-483.6 \text{ kJ}}{1 \text{ mol } O_2 \text{ reacted}} \qquad \frac{-483.6 \text{ kJ}}{2 \text{ mol } H_2O \text{ produced}}$$

The first factor says that 483.6 kJ of energy will transfer from system to surroundings whenever 2 mol of H_2 is consumed in Reaction 6.6. The reciprocal of the second factor says that if the reaction transfers 483.6 kJ to the surroundings, then 1 mol of O_2 must have been used up. We shall refer to stoichiometric factors that include thermochemical information as *thermostoichiometric factors.*

Exercise 6.10 Thermostoichiometric Factors from Thermochemical Equations

Write all of the thermostoichiometric factors (including their reciprocals) that can be derived from this equation:

$$N_2(g) + 3 H_2(g) \longrightarrow 2 NH_3(g) \qquad \Delta H° = -92.22 \text{ kJ}$$

Exercise 6.11 Hand Warmer

When the tightly sealed outer package is opened, the portable hand warmer shown in the margin transfers energy to its surroundings. In cold weather it can keep fingers or toes warm for several hours. Suggest a way that such a hand warmer could be designed. What chemicals might be used? Why is the tightly sealed package needed?

Portable hand warmer.

Enthalpy changes for reactions have many practical applications. For instance, when enthalpies of combustion are known, the quantity of energy transferred by the combustion of a given mass of fuel can be calculated. Suppose you are designing a heating system, and you want to know how much heating can be provided per pound (454 g) of propane, C_3H_8, burned in a furnace. The reaction that occurs is *exothermic* (which is not surprising, given that it is a combustion reaction).

$$C_3H_8(g) + 5 O_2(g) \longrightarrow 3 CO_2(g) + 4 H_2O(\ell) \qquad \Delta H° = -2220 \text{ kJ}$$

According to this thermochemical equation, 2220 kJ of energy transfers to the surroundings for every 1 mol of $C_3H_8(g)$ burned, for every 5 mol of $O_2(g)$ consumed,

CHEMISTRY You Can Do... Rusting and Heating

Chemical reactions can heat their surroundings, and a simple experiment demonstrates this very well. To do it you will need a steel wool pad (without soap), $\frac{1}{4}$ cup of vinegar, a cooking or outdoor thermometer, and a large jar with a lid. (The thermometer must fit inside the jar.)

Soak the steel wool pad in vinegar for several minutes. While doing so, place the thermometer in the jar, close the lid, and let it stand for several minutes. Read the temperature.

Squeeze the excess vinegar out of the steel wool pad, wrap the pad around the bulb of the thermometer, and place both in the jar. Close the lid. After about 5 min, read the temperature again. What has happened?

Repeat the experiment with another steel wool pad, but wash it with water instead of vinegar. Try a third pad that is not washed at all. Allow each pad to stand in air for a few hours or for a day and observe the pad carefully. Do you see any change in the metal? Suggest an explanation for your ob-

servations of temperature changes and appearance of the steel wool.

for every 3 mol of $CO_2(g)$ formed, and for every 4 mol of $H_2O(\ell)$ produced. We know that 454 g of $C_3H_8(g)$ has been burned, so we can calculate how many moles of propane that is.

$$\text{Amount of propane} = 454 \text{ g} \times \frac{1 \text{ mol } C_3H_8}{44.10 \text{ g}} = 10.29 \text{ mol } C_3H_8$$

Then multiply by the appropriate thermostoichiometric factor to find the total energy transferred.

$$\text{Energy transferred} = 10.29 \text{ mol } C_3H_8 \times \frac{-2220 \text{ kJ}}{1 \text{ mol } C_3H_8} = -22,900 \text{ kJ}$$

Burning a pound of fuel such as propane releases a substantial quantity of energy.

Problem-Solving Example 6.9 Calculating Energy Transferred

The reaction of iron with oxygen from the air provides the energy transferred by the hot pack described in Exercise 6.11. Assuming that the iron is converted to iron(III) oxide, how much heating can be provided by a hot pack that contains a tenth of a pound of iron? The thermochemical equation is

$$2 \text{ Fe}(s) + \tfrac{3}{2}O_2(g) \longrightarrow Fe_2O_3(s) \qquad \Delta H° = -824.2 \text{ kJ}$$

Answer −335 kJ

Explanation Begin by calculating how many moles of iron are present. A pound is 454 g, so

$$\text{Amount of iron} = 0.10 \text{ pound} \times \frac{454 \text{ g}}{1 \text{ pound}} \times \frac{1 \text{ mol Fe}}{55.84 \text{ g}} = 0.8130 \text{ mol Fe}$$

Then use a thermostoichiometric factor to calculate the energy transferred. The appropriate factor is 824.2 kJ transferred to the surroundings per 2 mol of Fe, so

$$\text{Energy transferred} = 0.8130 \text{ mol Fe} \times \frac{-824 \text{ kJ}}{2 \text{ mol Fe}} = -335 \text{ kJ}$$

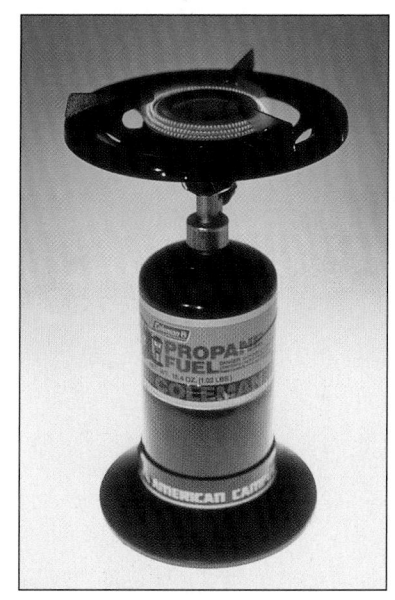

Propane burning. Propane is a major component of liquified petroleum (LP) gas, which is used for heating some houses.

Thus, 335 kJ is transferred by the reaction to heat your fingers. This is equal to 80.1 kcal.

✔ A tenth of a pound is about 45 g, which is a bit less than the molar mass of iron, so we are oxidizing less than a mole of iron. Two moles of iron gives about 800 kJ, so less than a mole should give less than 400 kJ, which makes 335 kJ a reasonable value. The sign should be negative because the hand warmer's (system's) enthalpy should go down when it transfers energy to your hand.

Problem-Solving Practice 6.9

How much thermal energy transfer is required to maintain constant temperature during decomposition of 12.6 g of liquid water to the elements hydrogen and oxygen? In what direction does the energy transfer?

$$H_2O(\ell) \longrightarrow H_2(g) + \tfrac{1}{2}O_2(g) \qquad\qquad \Delta H° = 285.8 \text{ kJ}$$

6.7 WHERE DOES THE ENERGY COME FROM?

CD-ROM Screen 9.10: Bond Energy and ΔH_{rxn}

During melting or boiling, nanoscale particles (atoms, molecules, or ions) that attract each other are separated, which increases their potential energy. This requires transfer of energy from the surroundings to enable the particles to overcome their mutual attractions. During a chemical reaction, chemical compounds are created or broken down; that is, reactant molecules are converted into product molecules. Atoms in molecules are held together by chemical bonds. When existing chemical bonds are broken and new chemical bonds are formed, atomic nuclei and electrons move farther apart or closer together, and their energy increases or decreases. These energy differences are usually much greater than for phase changes.

Consider the reaction of hydrogen gas with chlorine gas to form hydrogen chloride gas.

$$H_2(g) + Cl_2(g) \longrightarrow 2\,HCl(g) \qquad\qquad [6.7]$$

When this reaction occurs, the two hydrogen atoms in a H_2 molecule separate, as do the two chlorine atoms in a Cl_2 molecule. In the product the atoms are combined in a different way—as two HCl molecules. We can think of this change as involving two steps:

$$H_2(g) + Cl_2(g) \longrightarrow 2\,H(g) + 2\,Cl(g) \longrightarrow 2\,HCl(g) \qquad [6.8]$$

The first step is to break all bonds in the reactant H_2 and Cl_2 molecules. The second step is to form the bonds in the two product HCl molecules. The net effect of these two steps is the same as for Equation 6.7: One hydrogen molecule and one chlorine molecule change into two hydrogen chloride molecules.

The reaction of hydrogen with chlorine actually occurs by a complicated series of steps, but the details of how the atoms rearrange do not matter, because enthalpy is a state function and the initial and final states are the same. This means that we can concentrate on products and reactants and not worry about exactly what happens in between.

Another analogy for the enthalpy change for a reaction is the change in altitude when you climb a mountain. No matter which route you take to the summit (which atoms you separate or combine first), the difference in altitude between the summit and where you started to climb (the enthalpy difference between products and reactants) is the same.

Exercise 6.12   Reaction Pathways and Enthalpy Change

Suppose that the enthalpy change were different depending on the pathway a reaction took from reactants to products. For example, suppose that 190 kJ were released when a mole of hydrogen gas and a mole of chlorine gas combine to form two moles of hydrogen chloride (Equation 6.7), but suppose that only 185 kJ were released when the same reactant molecules are broken into atoms and the atoms are recombined to form hydrogen chloride (Equation 6.8). Would this violate the first law of thermodynamics? Explain why or why not.

Bond Enthalpies

Separating two atoms that are bonded together requires a transfer of energy into the system, because work must be done against the force holding the pair of atoms together. *The enthalpy change that occurs when two bonded atoms in a gas phase molecule are separated completely at constant pressure* is called the **bond enthalpy** (or the **bond energy**—the two terms are often used interchangeably.) The bond enthalpy is usually expressed per mole of bonds. For example, the bond enthalpy for a Cl_2 molecule is 243 kJ/mol, so we can write

$$Cl_2(g) \longrightarrow 2\,Cl(g) \qquad\qquad \Delta H° = 243\ \text{kJ}$$

Bond enthalpies are always positive, and they range in magnitude from about 150 kJ/mol to a little over 1000 kJ/mol. **Bond breaking is always endothermic,** because there is always a transfer of energy into the system (in this case, the mole of Cl_2 molecules) in order to separate pairs of bonded atoms. Conversely, when atoms come together to form a bond, energy will invariably be transferred to the surroundings, because the potential energy of the atoms is lower when they are bonded together. Conservation of energy requires that if the system's energy goes down, the energy of the surroundings must go up. Thus, **formation of bonds from separated atoms is always exothermic.** How these generalizations apply to the reaction of hydrogen with chlorine to form hydrogen chloride is shown in Figure 6.15.

Bond **breaking** is **endo**thermic.

Bond **making** is **exo**thermic.

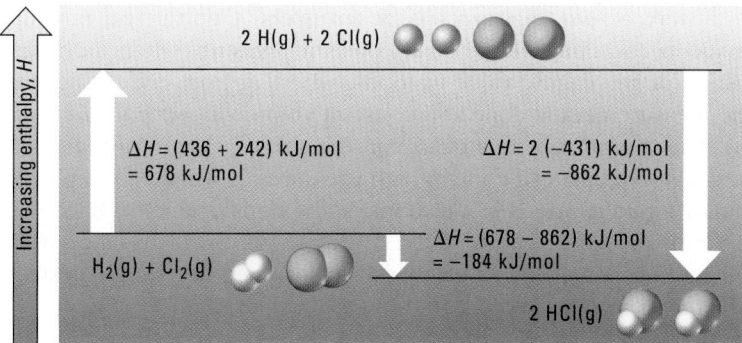

Figure 6.15   Stepwise energy changes in a reaction.   Breaking a mole of H_2 molecules into H atoms requires 436 kJ. Breaking a mole of Cl_2 molecules into Cl atoms requires 242 kJ. Putting 2 mol of H atoms together with 2 mol of Cl atoms to form 2 mol of HCl provides $2 \times (-431\ \text{kJ}) = -862\ \text{kJ}$, and so the reaction is exothermic. $\Delta H° = 436\ \text{kJ} + 242\ \text{kJ} - 862\ \text{kJ} = -184\ \text{kJ}$. The relatively weak Cl—Cl bond in the reactants accounts for the fact that this reaction is exothermic.

Bond enthalpies provide a way to see what makes a process exothermic or endothermic. If, as in Figure 6.15, the total energy transferred out of the system when new bonds are formed is greater than the total energy transferred in to break all of the bonds in the reactants, then the reaction is exothermic. In terms of bond enthalpies there are two ways for an exothermic reaction to happen:

• Weaker bonds are broken, stronger bonds are formed, and the number of bonds is the same.

• Bonds in reactants and products are of about the same strength, but more bonds are formed than are broken.

An endothermic reaction involves breaking stronger bonds than are formed, breaking more bonds than are formed, or both.

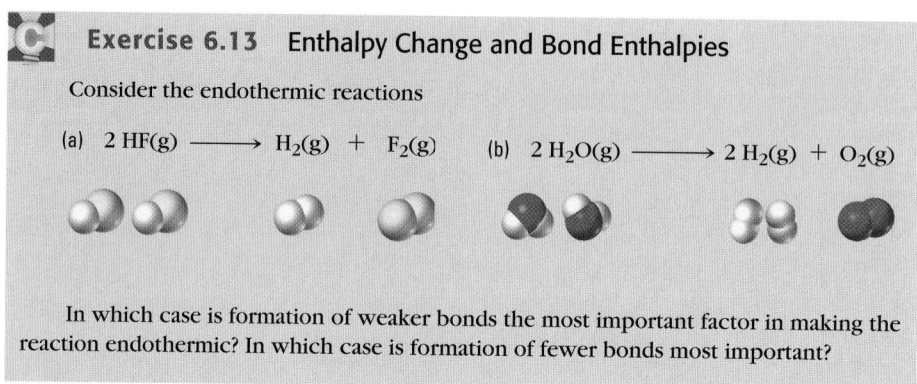

Exercise 6.13 Enthalpy Change and Bond Enthalpies

Consider the endothermic reactions

(a) $2 HF(g) \longrightarrow H_2(g) + F_2(g)$ (b) $2 H_2O(g) \longrightarrow 2 H_2(g) + O_2(g)$

In which case is formation of weaker bonds the most important factor in making the reaction endothermic? In which case is formation of fewer bonds most important?

6.8 MEASURING ENTHALPY CHANGES: CALORIMETRY

A thermochemical equation tells us how much energy is transferred as a chemical process occurs. Knowing this enables us to calculate the heat obtainable when a fuel is burned, as was done in the preceding section. Also, when reactions are carried out on a larger scale, say in a chemical plant that manufactures sulfuric acid, the surroundings must have enough cooling capacity to prevent an exothermic reaction from overheating, speeding up, and possibly damaging the plant. For these and many other reasons it is useful to know as many $\Delta H°$ values as possible.

For many reactions, direct experimental measurements can be made by using a **calorimeter,** *a device that measures heat transfers.* Calorimetric measurements can be made at constant volume or at constant pressure. Often, in finding heats of combustion or the caloric value of foods, where at least one of the reactants is a gas, the measurement is done at a constant volume in a *bomb calorimeter* (Figure 6.16). The "bomb" is a cylinder about the size of a large fruit juice can with heavy steel walls so that it can contain high pressures. A weighed sample of a combustible solid or liquid is placed in a dish inside the bomb, which is filled with pure O_2. The bomb is placed in a water-filled container with well-insulated walls, and the sample is ignited, usually by an electrical spark. When the sample burns, it warms the bomb and the water around it to the same temperature.

In this configuration, the oxygen and the compound represent the *system* and the bomb and the water around it are the *surroundings.* If we assume that the only energy transfer is from the reaction to the bomb and the water, the law of the conservation of energy requires that

$$q_{reaction} + q_{bomb} + q_{water} = 0$$

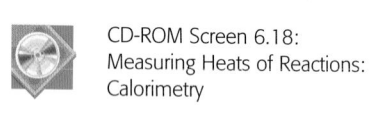

CD-ROM Screen 6.18:
Measuring Heats of Reactions:
Calorimetry

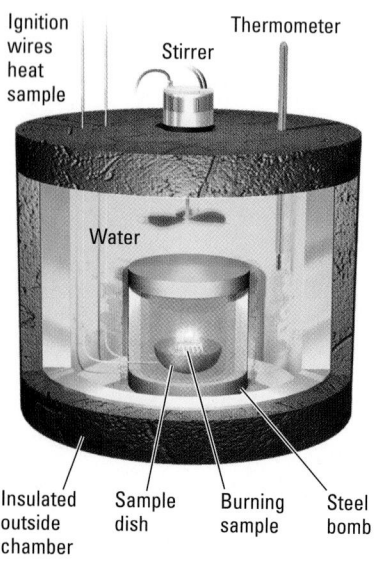

Ignition
wires
heat
sample

Stirrer

Thermometer

Water

Insulated   Sample   Burning   Steel
outside   dish   sample   bomb
chamber

Figure 6.16 **Combustion (bomb) calorimeter.**

That is, the net energy transfer must be zero. The temperature change of the water, which is the same as for the bomb, is measured. Then the energy transfer, $q_{bomb} + q_{water}$, can be calculated from the heat capacities of the bomb and the water. Rearranging the equation algebraically gives

$$-q_{reaction} = q_{bomb} + q_{water}$$

which says that the heat transfer out of the system ($-q_{reaction}$) equals the heat transfer into the bomb and water ($q_{bomb} + q_{water}$). Because the bomb is rigid, the heat transfer is measured at *constant volume* and is therefore equivalent to ΔE, the change in internal energy ($\Longleftarrow$ *p. 217*). Using $\Delta E = \Delta H + w_{expansion}$ and calculating the work that would be done under conditions of constant pressure allows us to calculate ΔH (and $\Delta H°$) from ΔE.

Problem-Solving Example 6.10   Measuring Energy Change with a Bomb Calorimeter

A 3.30-g sample of the sugar glucose, $C_6H_{12}O_6(s)$, is placed in a bomb calorimeter, ignited, and burned to form carbon dioxide and water. The temperature of the water changed from 22.4 °C to 34.1 °C. If the calorimeter contained 850. g of water and had a heat capacity of 847 J/°C, what is ΔE for combustion of 1 mol of glucose? (The heat capacity of the bomb is the quantity of energy transfer required to raise the bomb's temperature by 1 °C.)

Answer -2810 kJ

Explanation   Calculate the heat transfer to the calorimeter and the water from the temperature change and their heat capacities. (Look up the heat capacity of water in Table 6.1.) Use this to calculate the heat transfer from the reaction. Then use a proportion to find the heat transfer for 1 mol of glucose.

$$\Delta T = (34.1 - 22.4)\,°C = 11.7\,°C$$

$$q_{bomb} = \text{heat capacity of bomb} \times \Delta T = \frac{847\,J}{°C} \times 11.7\,°C = 9910\,J = 9.910\,kJ$$

$$q_{water} = c \times m \times \Delta T = \frac{4.184\,J}{g\,°C} \times 850.\,g \times 11.7\,°C = 41{,}610\,J = 41.61\,kJ$$

$$q_{reaction} = -(q_{bomb} + q_{water}) = -(9.910\,kJ + 41.61\,kJ) = -51.52\,kJ$$

This quantity of energy transfer corresponds to burning 3.3 g of glucose. To scale to 1 mol of glucose, first calculate how many moles of glucose were burned.

$$3.30\,g\,C_6H_{12}O_6 \times \frac{1\,mol}{180.16\,g} = 1.832 \times 10^{-2}\,mol\,C_6H_{12}O_6$$

Then set up the proportion

$$\frac{-51.52\,kJ}{1.832 \times 10^{-2}\,mol} = \frac{\Delta E}{1\,mol}; \quad \Delta E = 1\,mol \times \frac{-51.52\,kJ}{1.832 \times 10^{-2}\,mol} = -2.81 \times 10^3\,kJ$$

✔   The result is negative, which correctly reflects the fact that burning sugar is exothermic. A mole of glucose (180 g) is more than a third of a pound, and a third of a pound of sugar contains quite a bit of energy, so it is reasonable that the magnitude of the answer is in the thousands of kilojoules.

Problem-Solving Practice 6.10

When a 7.68-g sample of sulfur is burned in a bomb calorimeter with excess oxygen, the temperature increases from 20.34 °C to 36.75 °C. The calorimeter contains 815 g of water, and the bomb's heat capacity is 923 J/°C. Calculate ΔE for the reaction

$$S_8(s) + 8\,O_2(g) \longrightarrow 8\,SO_2(g)$$

Exercise 6.14 Comparing Enthalpy Change and Energy Change

Write a balanced equation for the combustion of glucose to form $CO_2(g)$ and $H_2O(\ell)$. Use what you learned in high school about the volume of a mole of any gas at a given temperature and pressure (or look in Section 10.5) to predict whether ΔH would differ significantly from ΔE for the reaction in Problem-Solving Example 6.10.

In a coffee cup calorimeter the masses of substances other than the solvent water are often so small that their heat capacities can be ignored; all of the energy of a reaction can be assumed to be transferred to the water.

When reactions take place in solution, it is much easier to use a calorimeter that is open to the atmosphere. An example, often encountered in introductory chemistry courses, is the *coffee cup calorimeter* shown in Figure 6.17. The nested coffee cups provide good thermal insulation; reactions can occur when solutions are poured together in the inner cup. Because a coffee cup calorimeter is a constant pressure device, the measured heat transfer is q_p, which equals ΔH.

(a)

(b)

Figure 6.17 Coffee cup calorimeter. (a) A simple constant pressure calorimeter can be made from two coffee cups that are good thermal insulators, a cork or other insulating lid, a temperature probe, and a stirrer. (b) Close-up of the nested cups that make up the calorimeter. A reaction carried out in an aqueous solution within the calorimeter will change the temperature of the solution. Because the thermal insulation is extremely good, essentially no energy transfer can occur to or from anything outside the calorimeter. Therefore, the heat capacity of the solution and its change in temperature can be used to calculate q_P and ΔH.

Problem-Solving Example 6.11 Measuring Enthalpy Change with a Coffee Cup Calorimeter

When 0.800 g of magnesium is added to 250. mL of 0.40 M HCl in a coffee cup calorimeter at 1 bar, the temperature of the solution increases from 23.4 °C to 37.9 °C. Determine the enthalpy change for the reaction and complete the thermochemical equation below. Assume that the heat capacities of the coffee cups, the temperature probe, and the stirrer are negligible and that the solution has the same density and the same specific heat capacity as water.

$$Mg(s) + 2\, HCl(aq) \longrightarrow H_2(g) + MgCl_2(aq) \qquad \Delta H = ?$$

Answer $\Delta H = -462$ kJ

Explanation Use the definition of specific heat capacity to calculate q, the heat transfer. The mass of solution is 250.8 g, the mass of the HCl(aq) plus the mass of the Mg.

$$q_{solution} = c \times m \times \Delta T$$

$$= (4.184\, J\, g^{-1}\, °C^{-1})(250.8\, g)(37.9 - 23.4)\, °C = 1.521 \times 10^4\, J = 15.21\, kJ$$

This quantity of heat transfer corresponds to consumption of all of the limiting reactant, but we do not know which is limiting. The quantities are 0.800 g of Mg and 250. mL of 0.40 M HCl, from which we can calculate the number of moles of Mg and the number of moles of HCl.

$$\text{Amount of Mg} = 0.800\, g \times \frac{1\, mol}{24.31\, g} = 3.291 \times 10^{-2}\, mol\, Mg$$

$$\text{Amount of HCl} = 250.\, mL \times \frac{0.40\, mol}{1000\, mL} = 1.00 \times 10^{-1}\, mol\, HCl$$

$$= 10.0 \times 10^{-2}\, mol\, HCl$$

From the balanced equation, 2 mol of HCl is required for each 1 mol of Mg. Since we actually have more than twice as many moles of HCl as Mg, Mg is the limiting reactant. Because the thermochemical equation involves 1 mol of Mg, the heat transfer must be scaled in proportion to this quantity of Mg. Because the temperature of the solution increased, the energy transfer was from the reaction to the surroundings. Therefore, the enthalpy change for the reaction is negative.

$$\frac{\Delta H}{1\, mol\, Mg} = \frac{-15.21\, kJ}{3.291 \times 10^{-2}\, mol\, Mg}$$

$$\Delta H = 1\, mol\, Mg \times \frac{-15.21\, kJ}{3.291 \times 10^{-2}\, mol\, Mg} = -462\, kJ$$

✔ The temperature of the surroundings increased, so the reaction is exothermic and $\Delta H°$ is negative. The temperature of about 250 g of solution went up about 15 °C, so the heat transfer was about $(250 \times 15 \times 4)$ J $= (250 \times 60)$ J $= 15,000$ J $= 15$ kJ. This corresponded to 0.1 mol or less of either reactant, so the heat transfer per mole must be at least 15 kJ/0.1 mol $= 150$ kJ, which it is.

Problem-Solving Practice 6.11

Suppose that 100. mL of 1.0 M HCl and 100. mL of 1.0 M NaOH, both at 20.4 °C, are mixed in a coffee cup calorimeter. If the thermochemical equation for the neutralization reaction is

$$H^+(aq) + OH^-(aq) \longrightarrow H_2O(\ell) \qquad \Delta H° = -58.7 \text{ kJ}$$

to what final temperature will the solution in the calorimeter be heated? Assume that the specific heat capacity and density of the solution are the same as for water and that the heat capacities of all other substances in contact with the reaction are negligible.

 ### Exercise 6.15   Calorimetry

In Problem-Solving Practice 6.11 you calculated $\Delta T = 7.0$ °C for mixing 100. mL of 1.0 M HCl and 100. mL of 1.0 M NaOH in a coffee cup calorimeter. What is ΔT for mixing

(a) 200. mL of 1.0 M HCl and 200. mL of 1.0 M NaOH?
(b) 100. mL of 0.50 M H_2SO_4 and 100. mL of 1.0 M NaOH?

6.9 HESS'S LAW

Calorimetry works well for some reactions, but for many others it is difficult. Besides, it would be very time-consuming to measure values for every conceivable reaction, and it would take a great deal of space to tabulate so many values. Fortunately, there is a better way. It is based on **Hess's law,** which states that, *if the equation for a reaction is the sum of the equations for two or more other reactions, then $\Delta H°$ for the first reaction must be the sum of $\Delta H°$ values of the other reactions.* Hess's law is another way of stating the law of conservation of energy. It works even if the overall reaction does not actually occur by way of the separate equations that are summed.

For example, in Figure 6.14 (⬅ *p. 223*) we noted that the formation of liquid water from its elements $H_2(g)$ and $O_2(g)$ could be thought of as two successive changes: (a) formation of water vapor from the elements and (b) condensation of water vapor to liquid water. As shown below, the equation for formation of liquid water can be obtained by adding algebraically the chemical equations for these two steps. Therefore, according to Hess's law, the $\Delta H°$ value can be found by adding the $\Delta H°$ values for the two steps.

(a) $H_2(g) + \frac{1}{2} O_2(g) \longrightarrow H_2O(g)$ $\Delta H°_1 = -241.8$ kJ

(b) $H_2O(g) \longrightarrow H_2O(\ell)$ $\Delta H°_2 = -44.0$ kJ

───

(a) + (b) $H_2(g) + \frac{1}{2} O_2(g) \longrightarrow H_2O(\ell)$ $\Delta H° = \Delta H°_1 + \Delta H°_2 = -285.8$ kJ

Here, 1 mol of $H_2O(g)$ is a product of the first reaction and a reactant in the second. Thus, $H_2O(g)$ can be canceled out. This is similar to adding two algebraic equations: If the same quantity or term appears on both sides of the equation, it cancels. The net result is an equation for the overall reaction and its associated enthalpy change. This overall enthalpy change applies even if the liquid water is formed directly from hydrogen and oxygen.

A useful approach to Hess's law is to analyze the equation whose $\Delta H°$ you are trying to calculate. Identify which reactants are desired in what quantities and also

 CD-ROM Screen 6.16: Hess's Law

Hess's law is based on a fact we mentioned earlier (⬅ *p. 229*). A system's enthalpy and internal energy will be the same no matter how the system is prepared.

Therefore, at 25 °C and 1 bar, the initial system, $H_2(g) + \frac{1}{2} O_2(g)$, has a particular enthalpy value. The final system, $H_2O(\ell)$, also has a characteristic (but different) enthalpy. Whether we get from initial system to final system by a single step or by the two-step process of equations (a) and (b), the enthalpy change will be the same.

Note that it takes 1 mol of $H_2O(g)$ to cancel 1 mol of $H_2O(g)$. If the coefficient of $H_2O(g)$ had been different on one side of the equation from the coefficent on the other side, $H_2O(g)$ could not have been completely canceled.

which products in what quantities. Then consider how the known thermochemical equations could be changed to give reactants and products in appropriate quantities. For example, suppose you want the thermochemical equation for the reaction

$$\tfrac{1}{2} CH_4(g) + O_2(g) \longrightarrow \tfrac{1}{2} CO_2(g) + H_2O(\ell) \qquad \Delta H° = ?$$

and you already know the equations

(a) $\quad CH_4(g) + 2\, O_2(g) \longrightarrow CO_2(g) + 2\, H_2O(g) \qquad \Delta H_a° = -560.52$ kJ

and

(b) $\quad H_2O(\ell) \longrightarrow H_2O(g) \qquad\qquad\qquad\qquad \Delta H_b° = 44.01$ kJ

The target equation has only $\tfrac{1}{2}$ mol of $CH_4(g)$ as a reactant; it also has $\tfrac{1}{2}$ mol of $CO_2(g)$ and 1 mol of $H_2O(\ell)$ as products. Equation (a) has the same reactants and products, but twice as many moles of each; also, water is in the gaseous state in equation (a). If we change each coefficient and the $\Delta H°$ value of equation (a) to one half their original values, we have the thermochemical equation

(a′) $\tfrac{1}{2} CH_4(g) + O_2(g) \longrightarrow \tfrac{1}{2} CO_2(g) + H_2O(g) \qquad \Delta H_{a'}° = -280.26$ kJ

which differs from the target equation only in the phase of water. Equation (b) has liquid water on the left and gaseous water on the right, but our target equation has liquid water on the right. If equation (b) is reversed (which changes the sign of $\Delta H°$), it becomes

(b′) $\quad H_2O(g) \longrightarrow H_2O(\ell) \qquad\qquad\qquad\qquad \Delta H_{b'}° = -44.01$ kJ

Summing the equations (a′) and (b′) gives the target equation, in which the $H_2O(g)$ has been canceled.

(a′ + b′) $\tfrac{1}{2} CH_4(g) + O_2(g) \longrightarrow \tfrac{1}{2} CO_2(g) + H_2O(\ell) \qquad \Delta H° = \Delta H_{a'}° + \Delta H_{b'}°$

$$\Delta H° = (-280.26 \text{ kJ}) + (-44.01 \text{ kJ}) = -324.27 \text{ kJ}$$

Polyethylene is a common plastic. Many products are packaged in polyethylene bottles.

Problem-Solving Example 6.12   Using Hess's Law

In designing a chemical plant for manufacturing the plastic polyethylene, you need to know the enthalpy change for the removal of H_2 from C_6H_6 (ethane) to give C_2H_4 (ethylene), a key step in the process.

$$C_2H_6(g) \longrightarrow C_2H_4(g) + H_2(g) \qquad \Delta H° = ?$$

From experiments you know the following thermochemical equations:

(a) $\quad 2\, C_2H_6(g) + 7\, O_2(g) \longrightarrow 4\, CO_2(g) + 6\, H_2O(\ell) \qquad \Delta H_a° = -3119.4$ kJ

(b) $\quad C_2H_4(g) + 3\, O_2(g) \longrightarrow 2\, CO_2(g) + 2\, H_2O(\ell) \qquad \Delta H_b° = -1410.9$ kJ

(c) $\quad 2\, H_2(g) + O_2(g) \longrightarrow 2\, H_2O(\ell) \qquad\qquad\qquad \Delta H_c° = -571.66$ kJ

Use this information to find the value of $\Delta H°$ for the formation of ethylene from ethane.

Answer $\quad \Delta H° = 137.0$ kJ

Explanation   Analyze reactions (a), (b), and (c). Equation (a) involves 2 mol of ethane on the reactant side, but only 1 mol is required in the desired equation. Equation (b) has C_2H_4 as a reactant, but C_2H_4 is a product in the desired equation. Equation (c) has 2 mol of H_2 as a reactant, but 1 mol of H_2 is a product in the desired equation.

First, since the desired equation has only 1 mol of ethane on the reactant side, we multiply equation (a) by $\tfrac{1}{2}$ to give an equation (a′) that also has 1 mol of ethane on the reactant side. Halving the equation also halves the enthalpy change.

(a′) = $\tfrac{1}{2}$ (a) $\quad C_2H_6(g) + \tfrac{7}{2} O_2(g) \longrightarrow 2\, CO_2(g) + 3\, H_2O(\ell) \qquad \Delta H_{a'}° = -1559.7$ kJ

Next, we reverse equation (b) so that C_2H_4 is on the product side, giving equation (b′).

This also reverses the sign of the enthalpy change.

(b') = −(b) $2 CO_2(g) + 2 H_2O(\ell) \longrightarrow C_2H_4(g) + 3 O_2(g)$

$$\Delta H_{b'}^\circ = -\Delta H_b^\circ = +1410.9 \text{ kJ}$$

To get 1 mol of $H_2(g)$ on the product side, we reverse equation (c) and multiply all coefficients by $\frac{1}{2}$. This changes the sign and halves the enthalpy change.

(c') = $-\frac{1}{2}$ (c) $H_2O(\ell) \longrightarrow H_2(g) + \frac{1}{2} O_2(g)$ $\Delta H_{c'}^\circ = -\frac{1}{2}\Delta H_c^\circ = +285.83 \text{ kJ}$

Now it is possible to add equations (a'), (b'), and (c') to give the desired equation.

(a') $C_2H_6(g) + \frac{7}{2}O_2(g) \longrightarrow 2 CO_2(g) + 3 H_2O(\ell)$ $\Delta H_{a'}^\circ = -1559.7 \text{ kJ}$

(b') $2 CO_2(g) + 2 H_2O(\ell) \longrightarrow C_2H_4(g) + 3 O_2(g)$ $\Delta H_{b'}^\circ = +1410.9 \text{ kJ}$

(c') $H_2O(\ell) \longrightarrow H_2(g) + \frac{1}{2} O_2(g)$ $\Delta H_{c'}^\circ = +285.83 \text{ kJ}$

Net equation: $C_2H_6(g) \longrightarrow C_2H_4(g) + H_2(g)$ $\Delta H_{net}^\circ = 137.0 \text{ kJ}$

When the equations are added, there is $\frac{7}{2}$ mol of $O_2(g)$ on the reactant side and $(3 + \frac{1}{2}) = \frac{7}{2}$ mol of $O_2(g)$ on the product side. There is 3 mol of $H_2O(\ell)$ on each side and 2 mol of $CO_2(g)$ on each side. Therefore, $O_2(g)$, $CO_2(g)$, and $H_2O(\ell)$ all cancel, and the equation for the conversion of ethane to ethylene and hydrogen remains.

✔ The overall process involves breaking a molecule apart into simpler molecules, which is likely to involve breaking bonds. Therefore it should be endothermic, and ΔH° should be positive.

Problem-Solving Practice 6.12

Given these two thermochemical equations,

$$C(s) + O_2(g) \longrightarrow CO_2(g) \qquad \Delta H^\circ = -393.5 \text{ kJ}$$

$$CO(g) + \frac{1}{2} O_2(g) \longrightarrow CO_2(g) \qquad \Delta H^\circ = -283.0 \text{ kJ}$$

calculate ΔH° for the reaction

$$C(s) + \frac{1}{2} O_2(g) \longrightarrow CO(g)$$

6.10 STANDARD MOLAR ENTHALPIES OF FORMATION

Hess's law makes it possible to tabulate ΔH° values for a relatively few reactions and, by suitable combinations of these few reactions, to calculate ΔH° values for a great many other reactions. To make such a tabulation we use standard enthalpies of formation. The **standard molar enthalpy of formation,** ΔH_f°, is the *standard enthalpy change for formation of one mole of a compound from its elements* in their standard states. The subscript f indicates *formation* of the compound. The **standard state** of an element or compound is the physical state in which it exists at 1 bar and the specified temperature. At 25 °C the standard state for hydrogen is $H_2(g)$ and for sodium chloride is NaCl(s). For an element that can exist in several different allotropic forms (⇐ *p. 26*) at 1 bar and 25 °C, the most stable form is usually selected as the standard. For example, graphite, not diamond or buckminsterfullerene, is the standard for carbon; $O_2(g)$, not $O_3(g)$, is the standard for oxygen.

Some examples of thermochemical equations involving standard molar enthalpies of formation are

$H_2(g) + \frac{1}{2} O_2(g) \longrightarrow H_2O(\ell)$ $\Delta H^\circ = \Delta H_f^\circ\{H_2O(\ell)\} = -285.8 \text{ kJ/mol}$

$2 C(graphite) + 2 H_2(g) \longrightarrow C_2H_4(g)$ $\Delta H^\circ = \Delta H_f^\circ\{C_2H_4(g)\} = 52.26 \text{ kJ/mol}$

$2 C(graphite) + 3 H_2(g) + \frac{1}{2} O_2(g) \longrightarrow C_2H_5OH(\ell)$

$$\Delta H^\circ = \Delta H_f^\circ\{C_2H_5OH(\ell)\} = -277.69 \text{ kJ/mol}$$

CD-ROM Screen 6.17:
Standard Enthalpy of Formation

The word "molar" means "per mole." Thus, the standard molar enthalpy of formation is the standard enthalpy of formation per mole of compound formed.

It is common to use the term "heat of formation" interchangeably with "enthalpy of formation." It is only the heat of reaction at constant pressure that is equivalent to the enthalpy change. If heat of reaction is measured under other conditions it may not equal the enthalpy change. For example, when measured at constant volume in a bomb calorimeter, heat of reaction corresponds to the change of internal energy, not enthalpy.

Burning charcoal. Charcoal is mainly carbon, and it burns to form mainly carbon dioxide gas. The energy transfer from a charcoal grill could be estimated from the mass of charcoal and the standard molar enthalpy of formation of $CO_2(g)$.

Notice that in each case *1 mol of a compound in its standard state is formed directly from appropriate amounts of elements in their standard states.*

Some examples of equations at 25 °C and 1 bar where $\Delta H°$ is *not* a standard molar enthalpy of formation (and the reason why it is not) are

$$MgO(s) + SO_3(g) \longrightarrow MgSO_4(s)$$

$$\Delta H° = -287.48 \text{ kJ} \quad \text{(reactants are not elements)}$$

and

$$P_4(s) + 6 Cl_2(g) \longrightarrow 4 PCl_3(\ell) \quad \Delta H° = -1278.8 \text{ kJ} \quad \text{(4 mol of product formed)}$$

Problem-Solving Example 6.13 Equations for Standard Molar Enthalpies of Formation

Given that the standard molar enthalpy of formation for $MgSO_4(s)$ is -1284.9 kJ/mol at 25 °C, rewrite each of the two equations immediately above so that they represent standard molar enthalpies of formation.

Answer

$$Mg(s) + \tfrac{1}{8} S_8(s) + 2O_2(g) \longrightarrow MgSO_4(s) \qquad \Delta H_f°\{MgSO_4(s)\} = -1284.9 \text{ kJ/mol}$$

and

$$\tfrac{1}{4} P_4(s) + \tfrac{3}{2} Cl_2(g) \longrightarrow PCl_3(\ell) \qquad \Delta H° = \Delta H_f°\{PCl_3(\ell)\} = -319.7 \text{ kJ/mol}$$

Explanation The left side of the first equation now contains the elements Mg, S_8, and O_2. Mg and S_8 are both solids, and O_2 is a gas at 25 °C and 1 bar. The right side of the second equation now involves only 1 mol of $PCl_3(\ell)$, and the left side has the appropriate amounts of the elements. All coefficients have been divided by four, so $\Delta H°$ is also divided by four to obtain

$$\Delta H_f°\{PCl_3(\ell)\} = -319.7 \text{ kJ/mol}$$

✔ Check each equation carefully to make certain the substance whose standard enthalpy of formation you want is on the right side and has a coefficient of one. For $PCl_3(\ell)$, $\Delta H_f°$ should be $\tfrac{1}{4}$ of about 1200 kJ, and it is.

Problem-Solving Practice 6.13

Write an appropriate thermochemical equation in each case. (You may need to use fractional coefficients.)

(a) The standard molar enthalpy of formation of $NH_3(g)$ at 25 °C is -46.11 kJ/mol.
(b) The standard molar enthalpy of formation of $CO(g)$ at 25 °C is -110.525 kJ/mol.

  Exercise 6.16 Standard Molar Enthalpies of Formation of Elements

Write the equation that corresponds to the standard molar enthalpy of formation of $N_2(g)$.

(a) What process, if any, takes place in this equation?
(b) What does this imply about the enthalpy change?

Table 6.2 and Appendix J list values of $\Delta H_f°$, obtained from the National Institute for Standards and Technology (NIST), for many compounds. Notice that there are no values listed in these tables for elements such as C(graphite) or $O_2(g)$. As you probably realized from Exercise 6.16, *standard enthalpies of formation for the elements in their standard states are zero,* because forming an element in its standard state from the same element in its standard state involves no chemical or physical change.

TABLE 6.2 Selected Standard Molar Enthalpies of Formation at 25 °C*

Formula	Name	Standard molar enthalpy of formation (kJ/mol)	Formula	Name	Standard molar enthalpy of formation (kJ/mol)
$Al_2O_3(s)$	Aluminum oxide	-1675.7	$HI(g)$	Hydrogen iodide	$+26.48$
$BaCO_3(s)$	Barium carbonate	-1216.3	$KF(s)$	Potassium fluoride	-567.27
$CaCO_3(s)$	Calcium carbonate	-1206.92	$KCl(s)$	Potassium chloride	-436.747
$CaO(s)$	Calcium oxide	-635.09	$KBr(s)$	Potassium bromide	-393.8
$CCl_4(\ell)$	Carbon tetrachloride	-135.44	$MgO(s)$	Magnesium oxide	-601.70
$CH_4(g)$	Methane	-74.81	$MgSO_4(s)$	Magnesium sulfate	-1284.9
$C_2H_5OH(\ell)$	Ethyl alcohol	-277.69	$Mg(OH)_2(s)$	Magnesium hydroxide	-924.54
$CO(g)$	Carbon monoxide	-110.525	$NaF(s)$	Sodium fluoride	-573.647
$CO_2(g)$	Carbon dioxide	-393.509	$NaCl(s)$	Sodium chloride	-411.153
$C_2H_2(g)$	Acetylene (ethyne)	$+226.73$	$NaBr(s)$	Sodium bromide	-361.062
$C_2H_4(g)$	Ethylene (ethene)	$+52.26$	$NaI(s)$	Sodium iodide	-287.78
$C_2H_6(g)$	Ethane	-84.68	$NH_3(g)$	Ammonia	-46.11
$C_3H_8(g)$	Propane	-103.8	$NO(g)$	Nitrogen monoxide	$+90.25$
$C_4H_{10}(g)$	Butane	-126.148	$NO_2(g)$	Nitrogen dioxide	$+33.18$
$C_6H_{12}O_6(s)$	α-D-Glucose	-1274.4	$PCl_3(\ell)$	Phosphorus trichloride	-319.7
$CuSO_4(s)$	Copper(II) sulfate	-771.36	$PCl_5(s)$	Phosphorus pentachloride	-443.5
$H_2O(g)$	Water vapor	-241.818	$SiO_2(s)$	Silicon dioxide (quartz)	-910.94
$H_2O(\ell)$	Liquid water	-285.830	$SnCl_2(s)$	Tin(II) chloride	-325.1
$HF(g)$	Hydrogen fluoride	-271.1	$SnCl_4(\ell)$	Tin(IV) chloride	-511.3
$HCl(g)$	Hydrogen chloride	-92.307	$SO_2(g)$	Sulfur dioxide	-296.830
$HBr(g)$	Hydrogen bromide	-36.40	$SO_3(g)$	Sulfur trioxide	-395.72

*From Wagman, D.D., Evans, W.H., Parker, V.B., Schumm, R.H., Halow, I., Bailey, S.M., Churney, K.L., and Nuttall, R.: The NBS Tables of Chemical Thermodynamic Properties. *Journal of Physical and Chemical Reference Data*, Vol. 11, Suppl. 2, 1982. (NBS, the National Bureau of Standards, is now NIST, the National Institute for Standards and Technology.)

Hess's law can be used to find the standard enthalpy change for any reaction if there is a set of reactions whose enthalpy changes are known and whose equations, when added together, will give the equation for the desired reaction. For example, suppose you are a chemical engineer and want to know how much heating is required to decompose limestone (calcium carbonate) to lime (calcium oxide) and carbon dioxide.

$$CaCO_3(s) \longrightarrow CaO(s) + CO_2(g) \qquad \Delta H° = ?$$

As a first approximation you can assume that all substances are in their standard states at 25 °C and look up the standard molar enthalpy of formation of each substance in a table such as Table 6.2 or Appendix J: (a) $\Delta H_f°\{CaCO_3(s)\} = -1206.9$ kJ/mol; (b) $\Delta H_f°\{CaO(s)\} = -635.1$ kJ/mol; (c) $\Delta H_f°\{CO_2(g)\} = -393.5$ kJ/mol. These standard molar enthalpies of formation correspond to the following equations:

(a) $\qquad Ca(s) + C(graphite) + \frac{3}{2}O_2(g) \longrightarrow CaCO_3(s) \qquad \Delta H_a° = -1206.9$ kJ

(b) $\qquad\qquad\qquad Ca(s) + \frac{1}{2}O_2(g) \longrightarrow CaO(s) \qquad \Delta H_b° = -635.1$ kJ

(c) $\qquad\qquad C(graphite) + O_2(g) \longrightarrow CO_2(g) \qquad \Delta H_c° = -393.5$ kJ

Now add the three equations in such a way that the resulting equation is the one given above for the decomposition of limestone. In equation (a), $CaCO_3(s)$ is a product, but it must appear in the desired equation as a reactant. Therefore, equa-

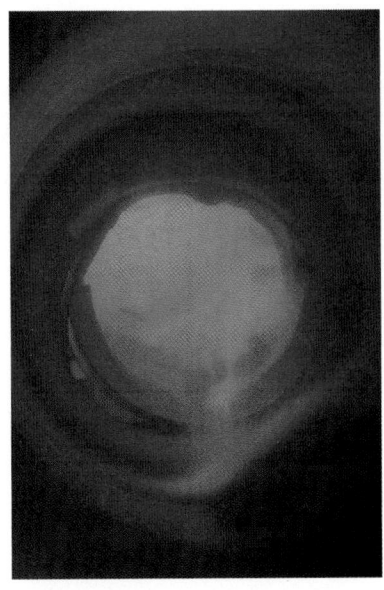

Lime production. At high temperature in a lime kiln, calcium carbonate (limestone, $CaCO_3$) decomposes to calcium oxide (lime, CaO) and carbon dioxide (CO_2).

tion (a) must be reversed, and the sign of ΔH_a° must also be reversed. On the other hand, CaO(s) and $CO_2(g)$ must appear as products in the desired equation, so equations (b) and (c) can be added with the same direction and sign of ΔH° as they have in the ΔH_f° equations:

(a') = $-$(a) $CaCO_3(s) \longrightarrow Ca(s) + C(graphite) + \frac{3}{2}O_2(g)$ $\Delta H_{a'}^\circ = +1206.9$ kJ

(b) $Ca(s) + \frac{1}{2}O_2(g) \longrightarrow CaO(s)$ $\Delta H_b^\circ = -635.1$ kJ

(c) $C(graphite) + O_2(g) \longrightarrow CO_2(g)$ $\Delta H_c^\circ = -393.5$ kJ

$CaCO_3(s) \longrightarrow CaO(s) + CO_2(g)$ $\Delta H^\circ = +178.3$ kJ

When the equations are added in this fashion, 1 mol each of C(graphite) and Ca(s) and $\frac{3}{2}$ mol of $O_2(g)$ appear on opposite sides and so are canceled out. Thus, the sum of these equations is the desired one for the decomposition of calcium carbonate, and the sum of the enthalpy changes of the three equations gives that for the desired equation.

Another very useful conclusion can be drawn from this example. The calculation can be written mathematically as

$$\Delta H^\circ = \Delta H_f^\circ\{CaO(s)\} + \Delta H_f^\circ\{CO_2(g)\} - \Delta H_f^\circ\{CaCO_3(s)\}$$

$$= (-635.1 \text{ kJ}) + (-393.5 \text{ kJ}) - (-1206.9 \text{ kJ}) = 178.3 \text{ kJ}$$

It involved adding the ΔH_f° values for the products of the reaction, CaO(s) and $CO_2(g)$, and subtracting the ΔH_f° value for the reactant, $CaCO_3(s)$. The mathematics of the problem can be summarized by the expression

$$\Delta H^\circ = \Sigma\{(\textbf{moles of product}) \times \Delta H_f^\circ \text{ (\textbf{product})}\}$$

$$- \Sigma\{(\textbf{moles of reactant}) \times \Delta H_f^\circ(\textbf{reactant})\} \quad [6.9]$$

This equation says that to get the standard enthalpy change of the reaction you should (1) multiply the standard molar enthalpy of formation of each product by the number of moles of that product and then sum over all products; (2) multiply the standard molar enthalpy of formation of each reactant by the number of moles of that reactant and sum over all reactants; and (3) subtract the sum for the reactants from the sum for the products. This is a useful shortcut to writing the equations for all appropriate formation reactions and applying Hess's law, as we did above.

Problem-Solving Example 6.14   Using Standard Molar Enthalpies of Formation

Benzene, C_6H_6, is a commercially important hydrocarbon that is present in gasoline, where it enhances the octane rating. Calculate its enthalpy of combustion per mole; that is, find the value of ΔH° for the reaction

$$C_6H_6(\ell) + \frac{15}{2}O_2(g) \longrightarrow 6\,CO_2(g) + 3\,H_2O(\ell)$$

For benzene, $\Delta H_f^\circ\{C_6H_6(\ell)\} = 49.0$ kJ/mol. Use Table 6.2 for any other values you may need.

Answer $\Delta H^\circ = -3169.5$ kJ

Explanation   To calculate ΔH° you need standard molar enthalpies of formation for all compounds (and possibly elements, if they are not in their standard states) involved in the reaction. Since $O_2(g)$ is in its standard state, it is not included. From Table 6.2,

$$C(graphite) + O_2(g) \longrightarrow CO_2(g) \qquad \Delta H_f^\circ = -393.509 \text{ kJ/mol}$$

$$H_2(g) + \frac{1}{2}O_2(g) \longrightarrow H_2O(\ell) \qquad \Delta H_f^\circ = -285.830 \text{ kJ/mol}$$

Using the equation given above,

$$\Delta H° = 6 \text{ mol} \times \Delta H_f°\{CO_2(g)\} + [3 \text{ mol} \times \Delta H_f°\{H_2O(\ell)\}] - \Delta H_f°\{C_6H_6(\ell)\}$$

$$= 6 \text{ mol} \times (-393.509 \text{ kJ/mol}) + [3 \text{ mol} \times (-285.830 \text{ kJ/mol})]$$

$$- 1 \text{ mol} \times (49.0 \text{ kJ/mol}) = -3169.5 \text{ kJ}$$

✔ As expected, the enthalpy change for combustion of a fuel is negative and large.

Problem-Solving Practice 6.14

Nitroglycerin is a powerful explosive because it decomposes exothermically and four different gases are formed.

$$2 \text{ C}_3\text{H}_5(\text{NO}_3)_3(\ell) \longrightarrow 3 \text{ N}_2(g) + \tfrac{1}{2} \text{O}_2(g) + 6 \text{ CO}_2(g) + 5 \text{ H}_2\text{O}(g)$$

For nitroglycerin, $\Delta H_f°\{C_3H_5(NO_3)_3(\ell)\} = -364$ kJ/mol. Using data from Table 6.2, calculate the energy transfer when 10.0 g of nitroglycerin explodes.

6.11 CHEMICAL FUELS FOR HOME AND INDUSTRY

A **chemical fuel** is any substance that will react exothermically with atmospheric oxygen and is available at reasonable cost and in reasonable quantity. It is desirable that when a fuel burns, the products create as little environmental damage as possible. As indicated in Figure 6.18, most of the fuels that supplied us with thermal energy in 1999 were fossil fuels: coal, petroleum, and natural gas; biomass fuels consisting of wood, peat, and other plant matter were a distant second among chemical fuels. A smaller but significant quantity of energy came from nuclear reactors and hydroelectric power plants. For specific applications, other fuels are sometimes chosen because of their special properties. For example, hydrogen is

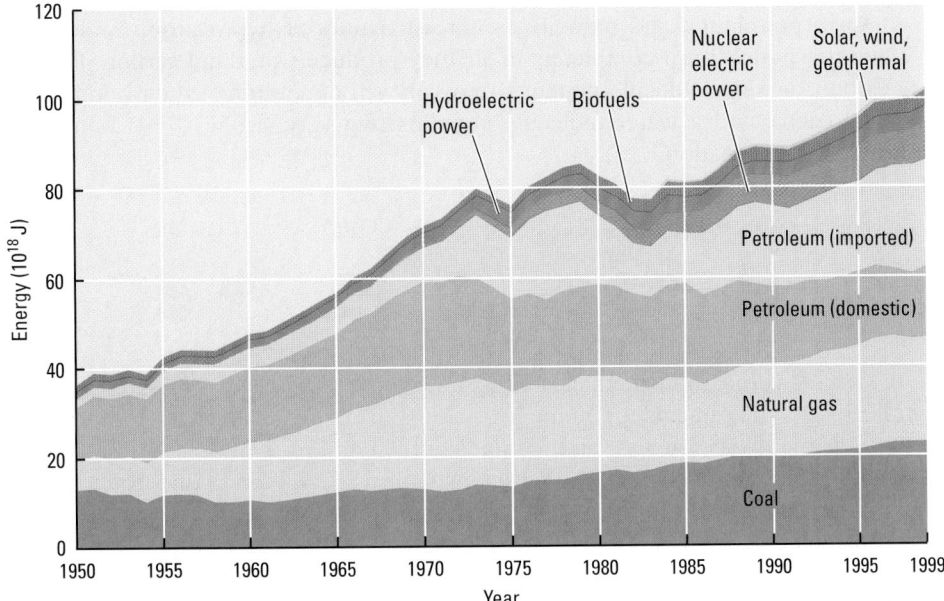

Figure 6.18 Use of energy resources in the United States. Use of energy resources in the United States is plotted from 1950 to 1999. (An energy resource is a naturally occurring fuel, such as petroleum, or a continuous supply, such as sunlight.) Halfway through the 20th century, coal and petroleum were almost equally important, with natural gas third. Today, petroleum and natural gas are used in greater quantities than coal, and more than half of the petroleum is imported. Nuclear electric power did not exist in 1949 but contributes significantly today, whereas hydroelectric electricity generation has grown slightly.

used as the fuel for the space shuttle, and hydrazine (N_2H_4) is used as a rocket fuel in some applications.

What nanoscale characteristics make a good fuel? If some or all of the bonds in the molecules of a fuel are weak, or if the bonds in the products of its combustion are strong, then the combustion reaction will be exothermic. An example of a molecule with a weak bond is hydrazine, $N_2H_4(g)$, which burns according to the equation

$$N_2H_4(g) \; + \; O_2(g) \longrightarrow N_2(g) \; + \; 2\,H_2O(g)$$

Its N—N bond enthalpy is only 160 kJ/mol, although its four N—H bonds are reasonably strong at 391 kJ/mol. The O_2 bond enthalpy is 498 kJ/mol. The reaction products are N_2, which has a very strong bond at 946 kJ/mol, and H_2O, which also has strong O—H bonds at 467 kJ/mol. In this case there are fewer bonds after the reaction than before, but the bonds are much stronger, and so the reaction is exothermic.

Exercise 6.17 Using Bond Enthalpies to Evaluate a Fuel

Based on the molecular structures and bond enthalpies given above for combustion of hydrazine:

(a) Calculate $\Delta H°$ when all the bonds in the reactant molecules are broken.
(b) Calculate $\Delta H°$ when all the bonds in the product molecules are formed.
(c) Hence, calculate $\Delta H°$ for the reaction and write the thermochemical equation.

Coal, petroleum, and natural gas consist mainly of hydrocarbon molecules. When these fuels burn completely in air they produce water and carbon dioxide. A carbon dioxide molecule contains two very strong carbon-oxygen bonds (803 kJ/mol each), and a water molecule contains two very strong O—H bonds. As shown by the equation,

$$CH_4(g) \; + \; 2\,O_2(g) \longrightarrow CO_2(g) \; + \; 2\,H_2O(g)$$

when the hydrocarbon methane (CH_4) burns, the number of bonds in reactant molecules is the same as the number of bonds in product molecules. Because the bonds formed are stronger than the bonds broken, the reaction is exothermic.

Another very good fuel is hydrogen. It burns in air to produce only water, which is a big advantage from an environmental point of view. Burning hydrocarbon fuels increases CO_2 levels in the atmosphere and therefore is partly responsible for global warming (see Section 12.4). The thermochemical equation for combustion of hydrogen corresponds to the formation of water from its elements, and so the standard enthalpy change is just $\Delta H_f°\{H_2O(g)\}$:

$$H_2(g) + \tfrac{1}{2}\,O_2(g) \longrightarrow H_2O(g) \qquad\qquad \Delta H_f° = -241.818 \text{ kJ}$$

On earth, little or no hydrogen is available naturally as the element; it is always combined in compounds. Therefore hydrogen fails to meet the criterion of availability in reasonable quantity mentioned at the beginning of this section. It is manufactured as a byproduct of petroleum refining at present, which makes it too

Recently scientists at the University of California, Berkeley, and the National Renewable Energy Laboratory have found that when sulfur is removed from the growing environment of *Chlamydomonas reinhardtii,* the algae generate hydrogen gas (*Plant Physiology,* Vol. 122, 2000; p. 127). Optimizing hydrogen production from algae might provide an inexpensive source of fuel.

expensive for most fuel applications. However, considerable research is aimed at finding ways to produce hydrogen either by electrolysis of water or chemically. For example, electricity supplied by solar cells could be used to electrolyze water and produce hydrogen, which could then be used as fuel. Or solar energy might be used directly to cause a series of chemical reactions in which hydrogen was produced from water. At present none of these ways of producing hydrogen is inexpensive enough to be competitive with fossil fuels, but as supplies of fossil fuels become exhausted, hydrogen may become much more important.

The fuel that was displaced by fossil fuels during the Industrial Revolution was wood. It is now referred to more generically as biomass, because plant matter other than wood can also be burned. Biomass is very important as a fuel in many less developed countries, and it is a renewable resource. Coal, petroleum, and natural gas will eventually be used up, but it is possible to continue growing plants to create biomass. Although biomass is a mixture of materials, it is primarily carbohydrate (cellulose in wood, for example) and can be represented by the empirical formula CH_2O. Combustion of biomass is highly exothermic:

$$CH_2O(s) + O_2(g) \longrightarrow CO_2(g) + H_2O(g) \qquad \Delta H° = -425 \text{ kJ}$$

Two important criteria for a fuel are the **fuel value,** which is the quantity of energy released when 1 g of fuel is burned to form carbon dioxide and water, and the **energy density,** which is the quantity of energy released per unit volume of fuel. For gaseous fuels the fuel value may be high, but the energy density will be low because the density of a gas is low. Fuels with low energy density take a large volume for storage and therefore are not convenient to use unless, as in the case of natural gas, they can be supplied on demand via pipes (mains). Often gaseous fuels, such as propane and butane, are condensed and stored as liquids under pressure. Fuels such as methane and hydrogen cannot be liquefied by pressure alone. If they are to be stored as liquids, the temperature must be kept low and the pressure high. This makes them less convenient than liquid fuels for use in cars and airplanes, for example.

Exercise 6.18 Comparing Fuels

Evaluate each of the following fuels on the basis of fuel value and energy density. Assume that water vapor is formed, use thermodynamic data from Table 6.2 or Appendix J, and use density data from the *CRC Handbook of Chemistry and Physics* (CRC Press, **http://www.crcpress.com/**). Which fuel provides the largest fuel value? Which provides the greatest energy density?

(a) Methane, $CH_4(g)$
(b) Octane, $C_8H_{18}(\ell)$
(c) Ethanol, $C_2H_5OH(\ell)$
(d) Hydrazine, $N_2H_4(\ell)$
(e) Hydrogen, $H_2(g)$
(f) Glucose, $C_6H_{12}O_6(s)$
(g) Biomass (assume that the fuel is entirely carbohydrate and has the same density as glucose)

 ## 6.12 FOODS: FUELS FOR OUR BODIES

Foods are similar to fuels, because the **caloric value** of fats and carbohydrates corresponds to the standard enthalpy of combustion per gram of the food. Foods consist mainly of carbohydrate, fat, and protein. Carbohydrates ($\Longleftarrow$ *p. 108*) have the general formula $C_x(H_2O)_y$. They are converted in the intestines to glucose, $C_6H_{12}O_6$, which is soluble in blood and thereby can be transported throughout the body. Glucose is metabolized in a complicated series of reactions that

This answers the question "where does the energy to make my muscles work come from?" that was raised in Chapter 1 (⇐ *p. 32*).

eventually produce $CO_2(g)$ and $H_2O(\ell)$, with release of energy. The net effect is the same as for combustion of glucose,

$$C_6H_{12}O_6(s) + 6\,O_2(g) \longrightarrow 6\,CO_2(g) + 6\,H_2O(\ell) \qquad \Delta H° = -2801.6 \text{ kJ}$$

Because enthalpy is a state function and the initial and final states are the same, it is appropriate and convenient to measure the caloric values for carbohydrates using a bomb calorimeter (⇐ *p. 238*). A sample of glucose or other carbohydrate is ignited inside the bomb, and the heat transfer to the calorimeter and surrounding water is measured. The average caloric value of carbohydrates in food is 4 Cal/g (17 kJ/g).

Exercise 6.19 Caloric Value of Carbohydrate

Use the thermochemical equation given above to verify that the caloric value of glucose corresponds to the average 4 Cal/g for carbohydrates.

Carbohydrates are metabolized quickly and are not stored in large amounts in the body. The energy released by carbohydrates is used to power muscles, to transmit nerve impulses, and to cause chemical reactions to occur that construct and repair tissues. Energy is also required to maintain body temperature. Energy from carbohydrates that is not needed for these purposes is stored in fats (⇐ *p. 110*), and fats also compose a significant portion of most people's diets. Like carbohydrates, fats are metabolized to $CO_2(g)$ and $H_2O(\ell)$. For example, tristearin is oxidized according to the equation

$$2\,C_{57}H_{110}O_6(s) + 163\,O_2(g) \longrightarrow 114\,CO_2(g) + 110\,H_2O(\ell) \quad \Delta H° = -75,520 \text{ kJ}$$

Again, bomb calorimetry simulates the metabolic process quite well. Fat molecules make excellent storehouses for energy. They are insoluble in water and therefore are not excreted easily, and they release 9 Cal/g (38 kJ/g) when metabolized—more than twice the energy from the same mass of carbohydrate or protein.

The third dietary component, protein, contains nitrogen in addition to carbon, hydrogen, and oxygen. The nitrogen is metabolized in the body to produce new proteins or urea, $(NH_2)_2CO(aq)$, which is excreted. The carbon, hydrogen, and oxygen are metabolized in pathways that release energy. On average, protein metabolism releases 4 Cal/g (17 kJ/g), the same caloric value as for carbohydrate.

A few substances may be part of your diet that are not protein, carbohydrate, or fat. For example, ethanol in alcoholic beverages contributes calories because ethanol can be oxidized, or metabolized, exothermically. Most food labels provide information about the caloric value of a typical serving and about the percentages of carbohydrate, fat, and protein in the food. An example is shown in Figure 6.19. The caloric value of the food can be estimated from the quantities of protein, carbohydrate, and fat, as long as no other component (such as alcohol) is present. Table 6.3 gives the content of fat, carbohydrate, and protein along with caloric values for some representative foods.

The release of energy upon combustion or metabolism of carbohydrates, fats, proteins, and other substances can be understood in terms of bond enthalpies. For example, fats consist almost entirely of long chains of carbon atoms to which hydrogen atoms are attached. Therefore, fats contain mostly C—H and C—C bonds. When fats burn or are metabolized, each carbon atom becomes bonded to oxygen in CO_2 and each hydrogen to oxygen in H_2O. The bond in each O_2 molecule that reacts with the fat must be broken, as must the C—H and C—C bonds in the fat. The respective bond enthalpies are 498 kJ/mol, 416 kJ/mol, and 356 kJ/mol. The strengths of the bonds formed in the products are 803 kJ/mol for carbon-oxygen bonds and 467 kJ/mol for hydrogen-oxygen bonds. Because the number of bonds

Figure 6.19 Nutrition facts from a food package.

TABLE 6.3	Composition and Caloric Values of Some Foods				
Food	**Approximate composition per 100. g**			**Caloric value**	
	Fat	**Carbohydrate**	**Protein**	**Cal/g**	**kJ/g**
All-purpose flour	0.0	73.3	13.3	3.33	13.95
Apple	0.5	13.0	0.4	0.59	2.47
Brownie with nuts	16.0	64.0	4.0	4.04	16.9
Cheese pizza	10.2	25.8	11.2	2.41	10.1
Egg	0.7	10.0	13.0	1.40	5.86
Egg noodle substitute	0.9	73.2	14.3	3.75	15.69
Grapes, white	0.6	17.5	0.6	0.7	2.9
Green beans	0.0	7.0	1.9	0.38	0.00
Hamburger	30.0	0.0	22.0	3.60	15.06
Microwave popcorn (popped)	7.1	11.4	2.9	1.00	4.18
Peanuts (unsalted)	50.0	21.4	28.6	5.71	23.91
Prunes (pitted)	0.0	65.0	2.5	2.75	11.51
Rice	1.0	77.6	8.2	3.47	14.52
Salad dressing (vinaigrette)	20.0	40.0	0.0	3.33	13.95
Tomato sauce	0.0	4.8	1.6	0.24	1.01
Wheat crackers	10.7	71.4	14.3	4.29	17.93

before and after reaction is nearly the same, and because the bonds formed are stronger, metabolism of fats transfers energy to the surroundings.

As we implied earlier, there is a balance between the quantity of food energy that is taken into our bodies and the quantity that is used for body functions. If food intake exceeds consumption, the body stores energy in fat molecules. If consumption exceeds intake, some fat is burned to provide the needed energy. The **basal metabolic rate (BMR)** is the minimum energy intake required to maintain a body that is awake and at rest, excluding the energy needed to digest, absorb, and metabolize the food, which is about 10% of the caloric intake. The BMR varies considerably depending on age, gender, and body mass. For a 70-kg (155-lb) human between 18 and 30 years old, the average BMR is 1750 Cal/day for a male and 1525 Cal/day for a female. The basal metabolic rate is approximately 1 Cal $kg^{-1}hr^{-1}$; that is, about 1 Calorie is expended per hour for each kilogram of body weight. Thus, the average 60-kg person has a daily BMR of

$$\frac{1 \text{ Cal}}{kg \times hr} \times 60 \text{ kg} \times \frac{24 \text{ hr}}{day} = 1440 \text{ Cal/day}$$

This value is multiplied by factors of up to 7 depending on the level of muscular activity. For example, walking or other light work requires 2.5 times the BMR. Heavy work, such as playing basketball or soccer, requires 7 times the BMR. Therefore, the appropriate food energy intake varies greatly from one individual to another.

Problem-Solving Example 6.15 Energy Value of Food

A 70-kg, 22-year old female eats 50. g of unsalted peanuts and then exercises by playing basketball for 30 min.

(a) What fraction of the food energy comes from fat, carbohydrate, and protein?
(b) Is the exercise sufficient to use up the energy provided by the food?

Answer

(a) 69% from fat, 13% from carbohydrate, and 18% from protein.

(b) No.

Explanation

(a) Data from Table 6.3 show that 100. g of unsalted peanuts contains 50.0 g of fat, 21.4 g of carbohydrate, and 28.6 g of protein. The fat provides 38 kJ/g and the carbohydrate and protein provide 17 kJ/g each. Therefore the energy provided is

$$\text{Energy from fat} = 50.\text{ g peanuts} \times \frac{50.0\text{ g fat}}{100.\text{ g peanuts}} \times \frac{38\text{ kJ}}{\text{g fat}} = 950\text{ kJ}$$

$$\text{Energy from carbohydrate} = 50.\text{ g peanuts} \times \frac{21.4\text{ g carbo}}{100.\text{ g peanuts}} \times \frac{17\text{ kJ}}{1\text{ g carbo}} = 182\text{ kJ}$$

$$\text{Energy from protein} = 50.\text{ g peanuts} \times \frac{28.6\text{ g protein}}{100.\text{ g peanuts}} \times \frac{17\text{ kJ}}{1\text{ g protein}} = 243\text{ kJ}$$

The total caloric intake is 1375 kJ, and the fractions are $(950/1375) \times 100\% = 69\%$ from fat, $(182/1375) \times 100\% = 13\%$ from carbohydrate, and $(243/1375) \times 100\% = 18\%$ from protein.

(b) Playing basketball requires seven times the BMR; that is, 7×1525 Cal/day.

$$\text{Energy required} = 7 \times \frac{1525\text{ Cal}}{\text{day}} \times \frac{1\text{ day}}{24\text{ hr}} \times \frac{1\text{ hr}}{60\text{ min}}$$

$$\times 30\text{ min} \times \frac{4.184\text{ kJ}}{1\text{ Cal}} = 930.\text{ kJ}$$

In addition, 10% of the caloric intake is required to digest, absorb, and metabolize the food. The quantity of energy required is thus

$$930\text{ kJ} + (0.10 \times 1375\text{ kJ}) = 1068\text{ kJ}$$

which is less than the caloric value of the 50. g of peanuts.

✔ As you might have expected, peanuts contain fat (oil), protein, and some carbohydrate, so the quantities of energy from these sources seem reasonable. However, it takes a lot of exercise to work off what we eat, so you may have been surprised by the answer to part (b). A little less than 2 oz of peanuts corresponds to 50. g, so eating peanuts without exercising can easily increase your weight.

Problem-Solving Practice 6.15

Whole milk contains 5.0% carbohydrate, 4.0% fat, and 3.3% protein by mass.

(a) Estimate the caloric value of an 8-oz (227-g) glass of milk.

(b) For how long would this caloric intake support a 70-kg male who was taking a leisurely walk?

SUMMARY PROBLEM

Sulfur dioxide, SO_2, is a major pollutant emitted by coal-fired electric power generating plants. A large power plant can produce 8.64×10^{13} J of electrical energy every day by burning 7000 tons of coal (1 ton = 9.08×10^5 g).

(a) Assume that the heating value of coal is approximately the same as for graphite. Calculate the quantity of energy transferred to the surroundings by the coal combustion reaction in a plant that burns 7000 tons of coal.

(b) What is the efficiency of this power plant in converting chemical energy to electrical energy; that is, what percentage of the thermal energy transfer shows up as electrical energy?

(c) When SO_2 is given off by a power plant, it can be trapped by reaction with MgO in the smokestack to form $MgSO_4$.

$$MgO(s) + SO_2(g) + \tfrac{1}{2}O_2(g) \longrightarrow MgSO_4(s)$$

If 140 tons of SO_2 is given off by coal-burning power plants each year, how much MgO would you have to supply to remove all of this SO_2? How much $MgSO_4$ would be produced?

(d) How much heat transfer does the reaction in part (c) add or take away from the heat effect of the coal combustion?

(e) Sulfuric acid comes from the oxidation of sulfur, first to SO_2 and then to SO_3. The SO_3 is then absorbed by water in 98% H_2SO_4 solution to make H_2SO_4.

$$S(s) + O_2(g) \longrightarrow SO_2(g)$$

$$SO_2(g) + \tfrac{1}{2}O_2(g) \longrightarrow SO_3(g)$$

$$SO_3(g) + H_2O \text{ (in 98\% } H_2SO_4) \longrightarrow H_2SO_4(\ell)$$

For which of these reactions can you calculate $\Delta H°$ using data in Table 6.2 or Appendix J? Do the calculation for each case where data are available.

IN CLOSING

Having studied this chapter, you should be able to . . .

- Understand the difference between kinetic energy and potential energy (Section 6.1).
- Be familiar with typical energy units and be able to convert from one unit to another (Section 6.1).
- Understand conservation of energy and energy transfer by heating and working (Section 6.2).
- Recognize and use thermodynamic terms: system, surroundings, heat, work, temperature, thermal equilibrium, exothermic, endothermic, and state function (Sections 6.2 and 6.4).
- Use specific heat capacity and the sign conventions for transfer of energy (Section 6.3).
- Distinguish between the change in internal energy and the change in enthalpy for a system (Section 6.4).
- Use thermochemical equations and derive thermostoichiometric factors from them (Sections 6.5 and 6.6).
- Use the fact that the standard enthalpy change for a reaction, $\Delta H°$, is proportional to the quantity of reactants consumed or products produced when the reaction occurs (Section 6.6).
- Understand the origin of the enthalpy change for a chemical reaction in terms of bond enthalpies (Section 6.7).
- Describe how calorimeters can measure the quantity of thermal energy transferred during a reaction (Section 6.8).
- Apply Hess's law to find the enthalpy change for a reaction (Sections 6.9 and 6.10).
- Use standard molar enthalpies of formation to calculate the thermal energy transfer when a reaction takes place (Section 6.10).
- Be able to define and give examples of some chemical fuels and to evaluate their abilities to provide heating (Section 6.11).
- Describe the main components of food and evaluate their contributions to caloric intake (Section 6.12).

KEY TERMS

basal metabolic rate (BMR) *(6.12)*

bond enthalpy (bond energy) *(6.7)*

caloric value *(6.12)*

calorimeter *(6.8)*

change of state *(6.4)*

chemical fuel *(6.11)*

conservation of energy, law of *(6.2)*

endothermic *(6.4)*

energy density *(6.11)*

enthalpy change *(6.4)*

exothermic *(6.4)*

first law of thermodynamics *(6.2)*

fuel value *(6.11)*

heat/heating *(6.2)*

heat capacity *(6.3)*

heat of fusion *(6.4)*

heat of vaporization *(6.4)*

Hess's law *(6.9)*

internal energy *(6.2)*

kinetic energy *(6.1)*

molar heat capacity *(6.3)*

phase change *(6.4)*

potential energy *(6.1)*

specific heat capacity *(6.3)*

standard enthalpy change *(6.4)*

standard molar enthalpy of formation *(6.10)*

standard state *(6.10)*

state function *(6.4)*

surroundings *(6.2)*

system *(6.2)*

thermal equilibrium *(6.2)*

thermochemical equation *(6.5)*

thermodynamics *(Introduction)*

work/working *(6.2)*

QUESTIONS FOR REVIEW AND THOUGHT

Conceptual Challenge Problems

CP-6.A (Section 6.2) Suppose a scientist discovered that energy was not conserved, but $1 \times 10^{-7}\%$ of the energy transferred from one system vanishes before it enters another system. How would this affect electric utilities, thermochemical experiments in scientific laboratories, and scientific thinking?

CP-6.B (Section 6.2) Suppose that someone were to tell your teacher during class that energy is not always conserved. This person states that he or she had previously learned that in the case of nuclear reactions, mass is converted into energy according to Einstein's equation $E = mc^2$. Hence, energy is continuously produced as mass is changed into energy. Your teacher quickly responds by giving the following assignment to the class, "Please write a paragraph or two to refute or clarify this student's thesis." What would you say?

CP-6.C (Section 6.3) The specific heat capacities at 25 °C for three metals with widely differing molar masses are 3.6 J g^{-1} $°C^{-1}$ for Li, 0.25 J g^{-1} $°C^{-1}$ for Ag, and 0.11 J g^{-1} $°C^{-1}$ for Th. Suppose that you have three samples, one of each metal and each containing the same number of atoms.

(a) Is the energy transfer required to increase the temperature of each sample by 1 °C significantly different from one sample to the next?

(b) What interpretation can you make about temperature based on the result you found in part (a)?

CP-6.D (Section 6.4) During one of your chemistry classes a student asks the professor, "Why does hot water freeze more quickly than cold water?"

(a) What do you expect the professor to say in answer to the student's question?

(b) In one experiment, two 100.-g samples of water were placed in identical containers on the same surface 1

decimeter apart in a room at −25 °C. One sample had an initial temperature of 78 °C, while the second was at 24 °C. The second sample took 151 min to freeze, and the first took 166 min (only 10% longer) to freeze. Clearly the cooler sample froze more quickly, but not nearly as quickly as one might have expected. How can this be so?

CP-6.F (Section 6.10) Assume that glass has the same properties as pure SiO_2. The thermal conductivity (rate at which heat transfer occurs through a substance) for aluminum is eight times that for SiO_2.

(a) Is it more efficient in time and energy to bake brownies in an aluminum pan or a glass pan?

(b) It is said that things cook more evenly in a glass pan than an aluminum pan; are there scientific data that indicate that this is reasonable?

CP-6.F (Section 6.10) On March 4, 1996, five railroad tank cars carrying liquid propane derailed in Weyauwega, Wisconsin, forcing evacuation of the town for more than a week. Residents who lived within the square mile centered on the accident were unable to return to their homes for over 2 weeks. Evaluate whether this evacuation was reasonable and necessary by considering the following questions.

(a) Estimate the volume of a railroad tank car. Obtain the density of liquid propane (C_3H_8) at or near its boiling point and calculate the mass of propane in the five tank cars. Obtain the data you need in order to calculate the energy transfer if all of that propane burned at once. (Assume that the reaction takes place at room temperature.)

(b) The enthalpy of decomposition of TNT ($C_7H_5N_3O_6$) to water, nitrogen, carbon monoxide, and carbon is −1066.1 kJ/mol. How many metric kilotons (1 metric ton = 1 Mg = 1×10^6 g) of TNT would provide energy transfer equivalent to that produced by combustion of propane in the five tank cars?

(c) Find the energy transfer (in kilotons of TNT) resulting from the nuclear fission bombs dropped on Hiroshima and Nagasaki, Japan, in 1945, and the energy transfer for modern fission weapons. What is the largest nuclear weapon thought to have been detonated to date? Compare the energy of the Hiroshima and Nagasaki bombs with the Weyauwega propane spill. What can you conclude about the wisdom of evacuating the town?

(d) Compare the energy that would have been released by burning the propane with the energy of a hurricane.

CP-6.G (Section 6.11) In some cities taxicabs run on lique-fied propane fuel instead of gasoline. This is done because it extends the lifetime of the vehicle and produces less pollu-tion. Given that it costs about $1500 to modify the engine of a taxicab to run on propane and that the cost of gasoline and liquid propane are $1.50 per gallon and $0.90 per gallon, re-spectively, make reasonable assumptions and figure out how many miles a taxi would have to go so that the decreased fuel cost would balance the added cost of modifying the taxi's motor. [For enthalpy calculations gasoline can be ap-proximated as octane, $C_8H_{18}(\ell)$.]

CP-6.H (Section 6.11) Suppose that you are an athlete and exercise a lot. Consequently you need a large caloric intake each day. Choose at least ten foods that you normally eat and evaluate each one to find which provides the most calo-ries per dollar.

Answers to questions in **bold** can be found in the back of the book.

Review Questions

1. Name two laws stated in this chapter and explain each in your own words.

2. For each of the following, define a system and its sur-roundings and give the direction of heat transfer:
   (a) Propane is burning in a Bunsen burner in the labora-tory.
   (b) After you have a swim, water droplets on your skin evaporate.
   (c) Water, originally at 25 °C, is placed in the freezing compartment of a refrigerator.
   (d) Two chemicals are mixed in a flask on a laboratory bench. A reaction occurs and heat is evolved.

3. What is the value of the standard enthalpy of formation for any element under standard conditions?

4. Criticize the following statements:
   (a) An enthalpy of formation refers to a reaction in which 1 mol of one or more reactants produces some quantity of product.
   (b) The standard enthalpy of formation of O_2 as a gas at 25 °C and a pressure of 1 atm is 15.0 kJ/mol.

5. Explain how a coffee cup calorimeter may be used to measure the enthalpy change of (a) a change in state and (b) a chemical reaction.

6. Describe how energy is changed from one form to an-other in the following processes:
   (a) At a July 4th celebration, a match is lit and ignites the fuse of a rocket firecracker, which fires off and explodes at an altitude of 1000 ft.
   (b) A gallon of gasoline is pumped from an underground storage tank into the fuel tank of your car, and you use it up by driving 25 mi.

7. Analyze transfer of energy from one form to another in each situation below.
   (a) In a space shuttle, hydrogen and oxygen combine to form water, boosting the shuttle into orbit above the earth.
   (b) You eat a package of Cheetos, go to class and listen to a lecture, walk back to your dorm, and climb the stairs to the fourth floor.

8. What is required in order for heat transfer of energy to occur from one sample of matter to another?

9. Name two exothermic processes and two endothermic processes that you encountered recently and that were not associated with your chemistry course.

10. Explain what is meant by (a) energy density of a fuel and (b) calorie value of a food. Why is each of these terms im-portant?

11. Explain in your own words why it is useful in thermody-namics to distinguish a system from its surroundings.

12. In a steelmaking plant, molten metal is poured from la-dles into furnaces. Considering the molten iron to be the system, what is the sign of q for the system shown in the photograph?

The Nature of Energy

13. (a) A 2-inch piece of two-layer chocolate cake with frost-ing provides 1670 kJ of energy. What is this in Cal?
   (b) If you were on a diet that calls for eating no more than 1200 Cal per day, how many joules would you con-sume per day?

14. Melting lead requires 5.91 cal/g. How many joules are re-quired to melt 1.00 lb (454 g) of lead?

15. Sulfur dioxide, SO_2, is found in wines and in polluted air. If a 32.1-g sample of sulfur is burned in the air to get 64.1 g of SO_2, 297 kJ of energy is released. Express this energy in (a) joules, (b) calories, and (c) kilocalories.

16. When an electrical appliance whose power usage is X watts is run for Y seconds, it uses $X \times Y$ J of energy. The energy unit used by electrical utilities in their monthly bills is the *kilowatt hour* (kWh) (1 kilowatt used for 1 hour). How many joules are there in a kilowatt hour? If electricity costs $.09 per kilowatt hour, how much does it cost per megajoule?

17. A 100-watt light bulb is left on for 14 h. How many joules of energy are used? With electricity at $.09 per kWh, how much does it cost to leave the light on for 14 h?

18. On a sunny day, solar energy reaches the earth at a rate of 4.0 J min^{-1} cm^{-2}. Suppose a house has a square flat roof of dimensions 12 m by 12 m. How much solar energy reaches this roof in 1 h? (*Note:* This is why roofs painted with light-reflecting paint are better than black unpainted roofs in warm climates: They reflect most of this energy rather than absorb it.)

Conservation of Energy

19. Suppose that you are studying the growth of a plant, and you want to apply thermodynamic ideas.
    (a) Make an appropriate choice of system and surroundings and describe it unambiguously.
    (b) Explain why you chose the system and surroundings you did.
    (c) Identify transfers of energy and material into and out of the system that would be important for you to monitor in your study.

20. Suppose that you are studying an ecosystem and want to apply thermodynamic ideas.
    (a) Make an appropriate choice of system and surroundings and describe it unambiguously.
    (b) Explain why you chose the system and surroundings you did.
    (c) Identify transfers of energy and material into and out of the system that would be important for you to monitor in your study.

21. Solid ammonium chloride is added to water in a beaker and dissolves. The beaker becomes cold to the touch.
    (a) Make an appropriate choice of system and surroundings and describe it unambiguously.
    (b) Explain why you chose the system and surroundings you did.
    (c) Identify transfers of energy and material into and out of the system that would be important for you to monitor in your study.
    (d) Is the process of dissolving $NH_4Cl(s)$ in water exothermic or endothermic?

22. A bar of Monel (an alloy of nickel, copper, iron, and manganese) is heated until it melts, poured into a mold, and solidifies.
    (a) Make an appropriate choice of system and surroundings and describe it unambiguously.
    (b) Explain why you chose the system and surroundings you did.

(c) Identify transfers of energy and material into and out of the system that would be important for you to monitor in your study.

23. If a system does 75.4 J of work on its surroundings and simultaneously there is 25.7 cal of heat transfer from the surroundings to the system, what is ΔE for the system?

24. A 20.0-g sample of water cools from 30 °C to 20.0 °C, which transfers 840 J to the surroundings. No work is done on the water. What is ΔE_{water}?

25. Make a diagram like the one in Figure 6.9 for a system in which 127.6 kJ of work is done on the surroundings and there is 843.2 kJ of heat transfer into the system. Use the diagram to help you determine ΔE_{system}.

26. Make a diagram like the one in Figure 6.9 for a system in which 876.3 J of work is done on the surroundings and there is 37.4 J of heat transfer into the system. Use the diagram to help you determine ΔE_{system}.

Heat Capacity

27. Which requires more energy: (a) warming 15.0 g of water from 25 to 37 °C or (b) warming 60.0 g of aluminum from 25 to 37 °C?

28. You hold a gram of copper in one hand and a gram of aluminum in the other. Each metal was originally at 0 °C. (Both metals are in the shape of a little ball that fits into your hand.) If they both take up heat at the same rate, which will warm to your body temperature first?

29. How much thermal energy is required to heat all the aluminum in a roll of aluminum foil (500. g) from room temperature (25 °C) to the temperature of a hot oven (250 °C)? Report your result in kilojoules.

30. How much thermal energy is required to heat all of the water in a swimming pool by 1 °C if the dimensions are 4 ft deep by 20 ft wide by 75 ft long? Report your result in megajoules.

31. Ethylene glycol, $(CH_2OH)_2$, is often used as an antifreeze in cars.
    (a) Which requires more thermal energy to warm from 25.0 °C to 100.0 °C, pure water or an equal mass of pure ethylene glycol?
    (b) If the cooling system in an automobile has a capacity of 5.00 quarts of liquid, compare the quantity of thermal energy the liquid in the system absorbs when its temperature is raised from 25.0 °C to 100.0 °C for water and ethylene glycol. (The densities of water and ethylene glycol are 1.00 g/cm^3 and 1.113 g/cm^3. 1 quart $= 0.946$ L. Report your results in joules.)

32. One way to cool a cup of coffee is to plunge an ice-cold piece of aluminum into it. Suppose a 20.0-g piece of aluminum is stored in the refrigerator at 32 °F (0.0 °C) and then put into a cup of coffee. The coffee's temperature drops from 90.0 °C to 75.0 °C. How much energy (in kilojoules) did the coffee transfer to the piece of aluminum?

33. A piece of iron (400. g) is heated in a flame and then plunged into a beaker containing 1.00 kg of water. The original temperature of the water was 20.0 °C, but it is 32.8 °C after the iron bar is put in and thermal equilibrium is reached. What was the original temperature of the hot iron bar?

34. A 192-g piece of copper was heated to 100.0 °C in a boiling water bath and then was put into a beaker containing 750. mL of water (density = 1.00 g/cm^3) at 4.0 °C. What is the final temperature of the copper and water after they come to thermal equilibrium?

35. A 12.3-g sample of iron requires heat transfer of 41.0 J to raise its temperature from 17.3 °C to 24.7 °C.
(a) Calculate the specific heat capacity of iron.
(b) Calculate the molar heat capacity of iron.

36. A diamond weighing 310. mg requires 2.38 J to raise its temperature from 23.4 °C to 38.7 °C.
(a) Calculate the specific heat capacity of diamond.
(b) Calculate the molar heat capacity of diamond.
(c) Is the specific heat capacity of diamond the same as the specific heat capacity of carbon in the form of graphite? Give one reason why you might expect them to be the same and one reason why you might expect them to be different.

37. An unknown metal requires 34.7 J to heat a 23.4-g sample of it from 17.3°C to 28.9 °C. Which of the metals in Table 6.1 is most likely to be the unknown?

38. An unknown metal requires 336.9 J to heat a 46.3-g sample of it from 24.3°C to 43.2°C. Which of the metals in Table 6.1 is most likely to be the unknown?

39. A 200.-g sample of Al is heated in a flame and then immersed in 500. mL of water in an insulated container. The initial temperature of the water was 22.0 °C, and after the Al and water had reached thermal equilibrium the temperature of both was 33.6 °C. What was the temperature of the Al just before it was plunged into the water? (Assume that the density of water is 0.98 g/mL and that there is no heat transfer other than between the water and aluminum.)

40. A 200.-g sample of copper is heated to 500. °C in a flame and then plunged into 1000. g of water in an insulated container. If the water's initial temperature was 23.4 °C, what is the temperature of water and Cu when thermal equilibrium has been reached? (Assume that the only energy transfer is between water and copper.)

41. A chemical reaction occurs, and 20.7 J is transferred from the chemical system to its surroundings. (Assume that no work is done.)
(a) What is the algebraic sign of $\Delta T_{surroundings}$?
(b) What is the algebraic sign of ΔE_{system}?

42. A physical process called a phase transition occurs in a sample of an alloy, and 437. kJ transfers from the surroundings to the alloy. (Assume that no work is done.)
(a) What is the algebraic sign of ΔT_{alloy}?
(b) What is the algebraic sign of ΔE_{alloy}?

Energy and Enthalpy

43. The thermal energy required to melt 1.00 g of ice at 0 °C is 333 J. If one ice cube has a mass of 62.0 g, and a tray contains 20 ice cubes, how much energy is required to melt a tray of ice cubes at 0 °C?

44. How much energy (in joules) would be required to raise the temperature of 1.00 lb of lead (1.00 lb = 454 g) from room temperature (25 °C) to its melting point, 327 °C, and then melt the lead completely at 327 °C? The specific heat capacity of lead is 0.159 J g^{-1} °C^{-1}, and its enthalpy of fusion is 24.7 J/g.

45. The hydrocarbon benzene, C_6H_6, boils at 80.1°C. How much energy is required to heat 1.00 kg of this liquid from 20.0°C to the boiling point and then change the liquid completely to a vapor at that temperature? The specific heat capacity of liquid C_6H_6 is 1.74 J g^{-1} °C^{-1}, and the enthalpy of vaporization is 395 J/g. Report your result in joules.

46. Calculate the quantity of heating required to convert the water in four ice cubes (60.1 g each) from H_2O(s) at 0 °C to H_2O(g) at 100 °C. The enthalpy of fusion of ice at 0 °C is 333 J/g and the enthalpy of vaporization of liquid water at 100 °C is 2260 J/g.

47. Mercury, with a freezing point of -39 °C, is the only metal that is liquid at room temperature. How much thermal energy must be transferred to its surroundings if 1.00 mL of the mercury is cooled from room temperature (23.0 °C) to -39 °C and then frozen to a solid? (The density of mercury is 13.6 g/cm^3. Its specific heat capacity is 0.138 J g^{-1} °C^{-1}, and its enthalpy of fusion is 11 J/g.)

48. On a cold day in winter, ice can sublime (go directly from solid to gas without melting). The heat of sublimation is approximately equal to the sum of the heat of fusion and the heat of vaporization (see Question 46). How much thermal energy in joules does it take to sublime 0.1 g of frost on a windowpane?

49. Draw a cooling graph for steam-to-water-to-ice.

50. Draw a heating graph for converting Dry Ice to carbon dioxide gas.

51. Based on the heating graph shown in the figure for a substance, X, at constant pressure,
(a) Which has the largest specific heat capacity, X(s), $X(\ell)$, or $X(g)$?
(b) Which is smaller, the heat of fusion or the heat of vaporization?
(c) What is the algebraic sign of the enthalpy of vaporization at the boiling point?

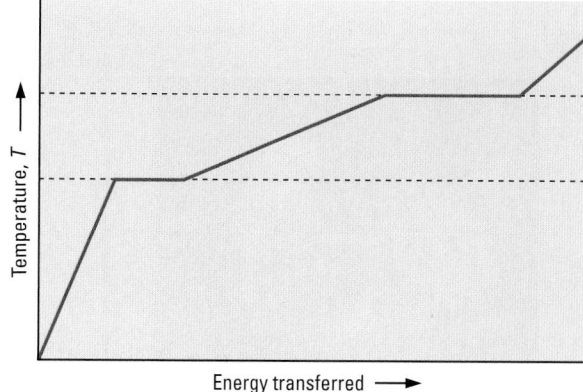

52. Based on the cooling graph shown in the figure on the next page for a substance, Y, at constant pressure,
(a) Which has the smallest specific heat capacity, Y(s), $Y(\ell)$, or $Y(g)$?
(b) Which is larger, the heat of fusion or the heat of vaporization?
(c) What is the algebraic sign of the enthalpy of fusion at the melting point?

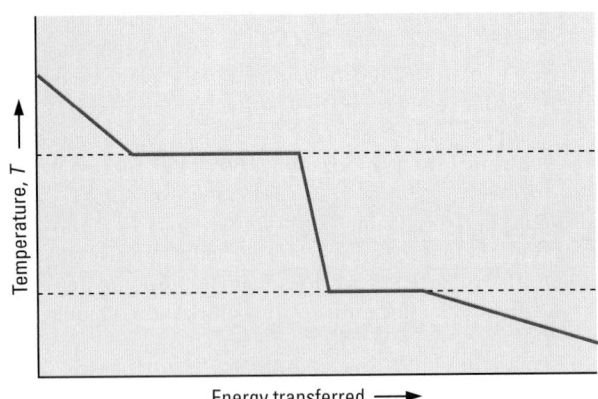

Thermochemical Equations

53. Energy is stored in the body in adenosine triphosphate, ATP, which is formed by the reaction between adenosine diphosphate, ADP, and dihydrogen phosphate ions.

$$ADP^{3-}(aq) + H_2PO_4^{2-}(aq) \longrightarrow ATP^{4-}(aq) + H_2O(\ell)$$
$$\Delta H° = 38 \text{ kJ}$$

Is the reaction endothermic or exothermic?

54. Calcium carbide, CaC_2, is manufactured by reducing lime with carbon at high temperature. (The carbide is used in turn to make acetylene, an industrially important organic chemical.)

$$CaO(s) + 3 C(s) \longrightarrow CaC_2(s) + CO(g) \quad \Delta H° = 464.8 \text{ kJ}$$

Is the reaction endothermic or exothermic?

55. A diamond can be considered a giant all-carbon supermolecule in which almost every carbon atom is bonded to four other carbons (⇐ *p. 27*). When a diamond cutter cleaves a diamond, carbon-carbon bonds must be broken. Is the cleavage of a diamond endothermic or exothermic? Explain.

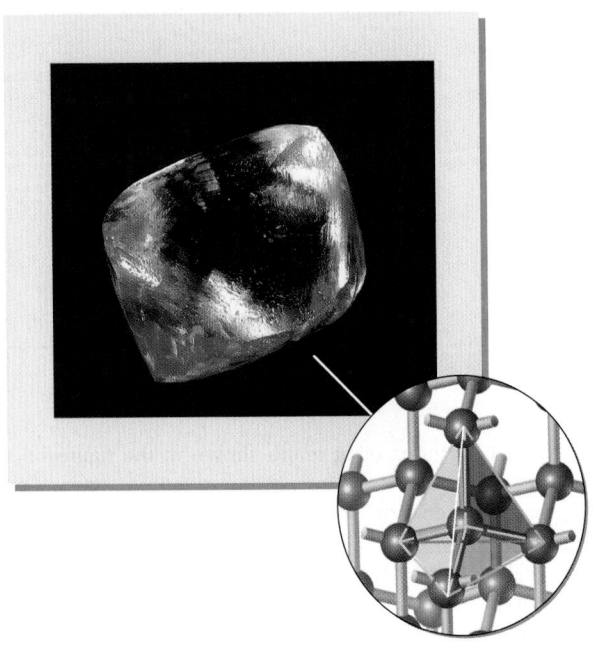

56. When table salt is dissolved in water, the temperature drops slightly. Write a chemical equation for this process, and indicate if it is exothermic or endothermic.

57. Write a word statement for the thermochemical equation

$$H_2O(s) \longrightarrow H_2O(\ell) \qquad \Delta H° = 6.0 \text{ kJ}$$

58. Write a word statement for the thermochemical equation

$$HI(\ell) \longrightarrow HI(s) \qquad \Delta H° = -2.87 \text{ kJ}$$

59. Given the thermochemical equation

$$H_2O(s) \longrightarrow H_2O(\ell) \qquad \Delta H° = 6.0 \text{ kJ}$$

what quantity of energy is transferred to the surroundings when
(a) 34.2 mol of liquid water freezes?
(b) 100.0 g of liquid water freezes?

60. Given the thermochemical equation

$$CaO(s) + 3 C(s) \longrightarrow CaC_2(s) + CO(g) \quad \Delta H° = 464.8 \text{kJ}$$

what quantity of energy is transferred when
(a) 34.8 mol of CO(g) is formed by this reaction?
(b) a metric ton (1000 kg) of CaC_2(s) is manufactured?
(c) 0.432 mol of carbon reacts with CaO(s)?

Enthalpy Changes for Chemical Reactions

61. Given the thermochemical equation for combustion of isooctane (a component of gasoline),

$$2 C_8H_{18}(\ell) + 25 O_2(g) \longrightarrow 16 CO_2(g) + 18 H_2O(\ell)$$
$$\Delta H° = -10,992 \text{ kJ}$$

write a thermochemical equation for
(a) production of 4.00 mol of CO_2(g).
(b) combustion of 100. mol of isooctane.
(c) combination of 1.00 mol of isooctane with a stoichiometric quantity of air.

62. Given the thermochemical equation for combustion of benzene,

$$2 C_6H_6(\ell) + 15 O_2(g) \longrightarrow 12 CO_2(g) + 6 H_2O(\ell)$$
$$\Delta H° = -6534.8 \text{ kJ}$$

write a thermochemical equation for
(a) combustion of 0.50 mol of benzene.
(b) consumption of 5 mol of O_2(g).
(c) production of 144 mol of CO_2(g).

63. Write all the thermostoichiometric factors that can be derived from the thermochemical equation

$$CaO(s) + 3 C(s) \longrightarrow CaC_2(s) + CO(g) \quad \Delta H° = 464.8 \text{kJ}$$

64. Write all the thermostoichiometric factors that can be derived from the thermochemical equation

$$2 CH_3OH(\ell) + 3 O_2(g) \longrightarrow 2 CO_2(g) + 4 H_2O(\ell)$$
$$\Delta H° = -1530 \text{ kJ}$$

65. Consider the decomposition of methanol into carbon monoxide and hydrogen (a possible method for producing combustible gaseous fuel).

$$CH_3OH(\ell) \longrightarrow CO(g) + 2 H_2(g) \qquad \Delta H° = 90.7 \text{ kJ}$$

(a) How much heat transfer into the system is required to decompose 100 kg of methanol?

(b) If 400. mol of CO(g) is produced, what quantity of energy must be transferred?

(c) Suppose that the reverse reaction were to occur, in which carbon monoxide and hydrogen combine to form methanol; if 43.0 g of CO(g) reacts with an excess of hydrogen, what is the heat transfer?

66. When $KClO_3(s)$, potassium chlorate, is heated it melts and decomposes to form oxygen gas. (Molten $KClO_3$ was shown reacting with a Cheetos chip earlier in this chapter [⟸ *p. 213*].) The thermochemical equation for decomposition of potassium chlorate is

$$2\ KClO_3(s) \longrightarrow 2\ KCl(s) + 3\ O_2(g) \qquad \Delta H° = -89.4\ kJ$$

Calculate q at constant pressure for
(a) formation of 97.8 g of KCl(s).
(b) production of 24.8 mol of $O_2(g)$.
(c) decomposition of 35.2 g of $KClO_3(s)$.

67. "Gasohol," a mixture of gasoline and ethanol, C_2H_5OH, is used as automobile fuel. The alcohol releases energy in a combustion reaction with O_2.

$$C_2H_5OH(\ell) + 3\ O_2(g) \longrightarrow 2\ CO_2(g) + 3\ H_2O(\ell)$$

If 0.115 g of alcohol evolves 3.62 kJ when burned at constant pressure, what is the molar enthalpy of combustion for ethanol?

68. White phosphorus, P_4, ignites in air to produce heat, light, and P_4O_{10}.

$$P_4(s) + 5\ O_2(g) \longrightarrow P_4O_{10}(s)$$

When 3.56 g of P_4 is burned, 85.8 kJ of thermal energy is evolved at constant pressure. What is the molar enthalpy of combustion of P_4?

69. A laboratory "volcano" can be made from ammonium dichromate. When ignited, the compound decomposes in a fiery display.

$$(NH_4)_2Cr_2O_7(s) \longrightarrow N_2(g) + 4\ H_2O(g) + Cr_2O_3(s)$$

If the decomposition produces 315 kJ per mole of ammonium dichromate at constant pressure, how much thermal energy would be produced by 28.4 g (1.00 oz) of the solid? (The substances involved in this reaction are poisons, so do not carry out the reaction.)

70. The thermite reaction, the reaction between aluminum and iron(III) oxide,

$$2\ Al(s) + Fe_2O_3(s) \longrightarrow Al_2O_3(s) + 2\ Fe(s)$$
$$\Delta H° = -851.5\ kJ$$

produces a tremendous amount of heat. If you begin with 10.0 g of Al and excess Fe_2O_3, how much energy is released at constant pressure?

71. When wood is burned we may assume that the reaction is the combustion of cellulose (empirical formula, CH_2O).

$$CH_2O(s) + O_2(g) \longrightarrow CO_2(g) + H_2O(g) \qquad \Delta H° = -425\ kJ$$

How much energy is released when a 10-lb wood log burns completely? (Assume the wood is 100% dry and burns via the reaction above.)

72. A plant takes CO_2 and H_2O from its surroundings and makes cellulose by the reverse of the reaction in the preceding problem. The energy provided for this process comes from the sun via photosynthesis. How much energy does it take for a plant to make 100 g of cellulose?

Where Does the Energy Come From?

Use the following bond enthalpy values to answer the questions below.

Bond	Bond enthalpy (kJ/mol)
H—F	566
H—Cl	431
H—Br	366
H—I	299
H—H	436
F—F	158
Cl—Cl	242
Br—Br	193
I—I	151

73. Which of the four molecules HF, HCl, HBr, and HI has the strongest chemical bond?
74. Which of the four molecules F_2, Cl_2, Br_2, and I_2 has the weakest chemical bond?
75. For the reactions of molecular hydrogen with fluorine and with chlorine:
(a) Calculate the enthalpy change for breaking all the bonds in the reactants.
(b) Calculate the enthalpy change for forming all the bonds in the products.
(c) From the results in parts (a) and (b), calculate the enthalpy change for the reaction.
(d) Which reaction is most exothermic?
76. For the reactions of molecular hydrogen with bromine and with iodine:
(a) Calculate the enthalpy change for breaking all the bonds in the reactants.
(b) Calculate the enthalpy change for forming all the bonds in the products.
(c) From the results in parts (a) and (b), calculate the enthalpy change for the reaction.
(d) Which reaction is most exothermic?

Measuring Enthalpy Changes: Calorimetry

77. Suppose you add a small ice cube to room temperature water in a coffee cup calorimeter. What is the final temperature when all of the ice is melted? Assume that you have 200. mL of water at 25 °C and that the ice cube weighs 15.0 g.
78. A coffee cup calorimeter can be used to investigate the "cold pack reaction," the process that occurs when solid ammonium nitrate dissolves in water:

$$NH_4NO_3(s) \longrightarrow NH_4^+(aq) + NO_3^-(aq)$$

25.0 g of solid NH_4NO_3 at 23.0 °C is added to 250. mL of H_2O at the same temperature, and after the solid is all dissolved the temperature is measured to be 15.6 °C. Calculate the enthalpy change for the cold pack reaction.

(*Hint:* Calculate the energy transferred per mole of NH_4NO_3.) Is the reaction endothermic or exothermic?

79. When a 13.0-g sample of NaOH(s) dissolves in 400.0 mL of water in a coffee cup calorimeter, the temperature of the water changes from 22.6 °C to 30.7 °C. Assuming that the specific heat capacity of the solution is the same as for water, calculate
    (a) The heat transfer from system to surroundings
    (b) $\Delta H°$ for the reaction

$$NaOH(s) \longrightarrow Na^+(aq) + OH^-(aq)$$

80. Suppose that you mix 200.0 mL of 0.200 M RbOH(aq) with 100 mL of 0.400 M HBr(aq) in a coffee cup calorimeter. If the temperature of each of the two solutions was 24.40 °C before mixing, and the temperature rises to 26.18 °C,
    (a) Calculate the heat transfer as a result of the reaction.
    (b) Write the thermochemical equation for the reaction.

81. How much thermal energy is evolved by a reaction in a bomb calorimeter (Figure 6.16) in which the temperature of the bomb and water increases from 19.50 °C to 22.83 °C? The bomb has a heat capacity of 650 J/°C; the calorimeter contains 320. g of water. Report your result in kJ.

82. Sulfur (2.56 g) was burned in a bomb calorimeter with excess $O_2(g)$. The temperature increased from 21.25 °C to 26.72 °C. The bomb had a heat capacity of 923 J °C^{-1} and the calorimeter contained 815 g of water. Calculate the heat evolved, per mole of SO_2 formed, in the course of the reaction

$$S(s) + O_2(g) \longrightarrow SO_2(g)$$

83. You can find the quantity of thermal energy evolved during combustion of carbon by carrying out the reaction in a combustion calorimeter. Suppose you burn 0.300 g of C(graphite) in an excess of $O_2(g)$ to give $CO_2(g)$.

$$C(graphite) + O_2(g) \longrightarrow CO_2(g)$$

The temperature of the calorimeter, which contains 775 g of water, increases from 25.00 °C to 27.38 °C. The heat capacity of the bomb is 893 J °C^{-1}. What quantity of thermal energy is evolved per mole of graphite?

84. Benzoic acid, $C_7H_6O_2$, occurs naturally in many berries. Suppose you burn 1.500 g of the compound in a combustion calorimeter and find that the temperature of the calorimeter increases from 22.50 °C to 31.69 °C. The calorimeter contains 775 g of water, and the bomb has a heat capacity of 893 J °C^{-1}. How much heat is evolved per mole of benzoic acid? What is the change in internal energy, ΔE, of the system?

Hess's Law

85. Calculate the standard enthalpy change, $\Delta H°$, for the formation of 1 mol of strontium carbonate (the material that gives the red color in fireworks) from its elements.

$$Sr(s) + C(graphite) + \tfrac{3}{2}O_2(g) \longrightarrow SrCO_3(s)$$

The information available is

$Sr(s) + \tfrac{1}{2}O_2(g) \longrightarrow SrO(s)$	$\Delta H° = -592$ kJ
$SrO(s) + CO_2(g) \longrightarrow SrCO_3(s)$	$\Delta H° = -234$ kJ
$C(graphite) + O_2(g) \longrightarrow CO_2(g)$	$\Delta H° = -394$ kJ

86. What is the standard enthalpy change for the reaction of lead(II) chloride with chlorine to give lead(IV) chloride?

$$PbCl_2(s) + Cl_2(g) \longrightarrow PbCl_4(\ell)$$

It is known that $PbCl_2(s)$ can be formed from the metal and $Cl_2(g)$,

$$Pb(s) + Cl_2(g) \longrightarrow PbCl_2(s) \qquad \Delta H° = -359.4 \text{ kJ}$$

and that $PbCl_4(\ell)$ can be formed directly from the elements.

$$Pb(s) + 2 Cl_2(g) \longrightarrow PbCl_4(\ell) \qquad \Delta H° = -329.3 \text{ kJ}$$

87. Using the following reactions, find the standard enthalpy change for the formation of 1 mol of PbO(s) from lead metal and oxygen gas.

$$PbO(s) + C(graphite) \longrightarrow Pb(s) + CO(g)$$
$$\Delta H° = 106.8 \text{ kJ}$$

$$2 C(graphite) + O_2(g) \longrightarrow 2 CO(g) \quad \Delta H° = -221.0 \text{ kJ}$$

If 250. g of lead reacts with oxygen to form lead(II) oxide, what quantity of thermal energy (in kJ) is absorbed or evolved?

88. Three reactions very important to the semiconductor industry are
    (a) the reduction of silicon dioxide to crude silicon,

$$SiO_2(s) + 2 C(s) \longrightarrow Si(s) + 2 CO(g) \quad \Delta H° = 689.9 \text{ kJ}$$

    (b) the formation of silicon tetrachloride from crude silicon,

$$Si(s) + 2 Cl_2(g) \longrightarrow SiCl_4(g) \qquad \Delta H° = -657.0 \text{ kJ}$$

    (c) and the reduction of silicon tetrachloride to pure silicon with magnesium,

$$SiCl_4(g) + 2 Mg(s) \longrightarrow 2 MgCl_2(s) + Si(s)$$
$$\Delta H° = -625.6 \text{ kJ}$$

What is the overall enthalpy change for changing 1.00 mol of sand (SiO_2) into very pure silicon?

Standard Molar Enthalpies of Formation

89. The standard molar enthalpy of formation of AgCl(s) is −127.1 kJ/mol. Write a balanced thermochemical equation for which the enthalpy of reaction is −127.1 kJ.

90. The standard molar enthalpy of formation of methanol, $CH_3OH(\ell)$, is −238.7 kJ/mol. Write a balanced thermochemical equation for which the enthalpy of reaction is −238.7 kJ.

91. For each compound below, write a balanced thermochemical equation depicting the formation of 1 mol of the compound. Standard molar enthalpies of formation are found in Appendix J.
    (a) $Al_2O_3(s)$ (b) $TiCl_4(\ell)$ (c) $NH_4NO_3(s)$

92. The standard molar enthalpy of formation of glucose, $C_6H_{12}O_6(s)$, is -1274.4 kJ/mol.
(a) Is the formation of glucose from its elements exothermic or endothermic?
(b) Write a balanced equation depicting the formation of glucose from its elements and for which the enthalpy of reaction is -1274.4 kJ.

93. An important reaction in the production of sulfuric acid is

$$SO_2(g) + \frac{1}{2} O_2(g) \longrightarrow SO_3(g)$$

It is also a key reaction in the formation of acid rain, beginning with the air pollutant SO_2. Using the data in Table 6.2, calculate the enthalpy change for the reaction.

94. In photosynthesis, the sun's energy brings about the combination of CO_2 and H_2O to form O_2 and a carbon-containing compound such as a sugar. In its simplest form, the reaction could be written

$$6 CO_2(g) + 6 H_2O(\ell) \longrightarrow 6 O_2(g) + C_6H_{12}O_6(s)$$

Using the enthalpies of formation in Table 6.2, (a) calculate the enthalpy of reaction and (b) decide whether the reaction is exothermic or endothermic.

95. The first step in the production of nitric acid from ammonia involves the oxidation of NH_3.

$$4 NH_3(g) + 5 O_2(g) \longrightarrow 4 NO(g) + 6 H_2O(g)$$

Use the information in Table 6.2 or Appendix J to find the enthalpy change for this reaction. Is the reaction exothermic or endothermic?

96. The Romans used CaO as mortar in stone structures. The CaO was mixed with water to give $Ca(OH)_2$, and this slowly reacted with CO_2 in the air to give limestone.

$$Ca(OH)_2(s) + CO_2(g) \longrightarrow CaCO_3(s) + H_2O(g)$$

Calculate the enthalpy change for this reaction.

97. A key reaction in the processing of uranium for use as fuel in nuclear power plants is the following:

$$UO_2(s) + 4 HF(g) \longrightarrow UF_4(s) + 2 H_2O(g)$$

Calculate the enthalpy change, $\Delta H°$, for the reaction using the data in Table 6.2, Appendix J, and the following: $\Delta H_f°$ for $UO_2(s) = -1085$ kJ/mol; $\Delta H_f°$ for $UF_4(s) = -1914$ kJ/mol.

98. Oxygen difluoride, OF_2, is a colorless, very poisonous gas that reacts rapidly and exothermically with water vapor to produce O_2 and HF.

$$OF_2(g) + H_2O(g) \longrightarrow 2 HF(g) + O_2(g) \qquad \Delta H° = -318 \text{ kJ}$$

Using this information and Table 6.2 or Appendix J, calculate the molar enthalpy of formation of $OF_2(g)$.

99. Iron can react with oxygen to give iron(III) oxide. If 5.58 g of Fe is heated in pure O_2 to give $Fe_2O_3(s)$, how much thermal energy is transferred out of this system (at constant pressure)?

100. The formation of aluminum oxide from its elements is highly exothermic. If 2.70 g of Al metal is burned in pure O_2 to give Al_2O_3, how much thermal energy is evolved in the process (at constant pressure)?

Chemical Fuels

101. If you want to convert 56.0 g of ice (at 0 °C) to water at 75.0 °C, how many grams of propane (C_3H_8) would you have to burn in order to supply the energy to melt the ice and then warm it to the final temperature (at 1 bar)?

102. Suppose you want to heat your house with natural gas (CH_4). Assume your house has 1800 ft^2 of floor area and that the ceilings are 8.0 ft from the floors. The air in the house has a molar heat capacity of 29.1 J mol^{-1} °C^{-1}. (The number of moles of air in the house can be found by assuming that the average molar mass of air is 28.9 g/mol and that the density of air at these temperatures is about 1.22 g/L.) How many grams of methane do you have to burn to heat the air from 15.0 °C to 22.0 °C?

103. Companies around the world are constantly searching for compounds that can be used as substitutes for gasoline in automobiles. Perhaps the most promising of these is methanol, CH_3OH, a compound that can be made relatively inexpensively from coal. The alcohol has a smaller energy per gram than gasoline, but with its higher octane rating it burns more efficiently than gasoline in internal combustion engines. (It also contributes smaller quantities of some air pollutants.) Compare the quantity of thermal energy produced per gram of CH_3OH and C_8H_{18} (octane), the latter being representative of the compounds in gasoline. The $\Delta H_f°$ for octane is -250.1 kJ/mol.

104. Hydrazine and 1,1-dimethylhydrazine both react spontaneously with O_2 and can be used as rocket fuels.

$$N_2H_4(\ell) + O_2(g) \longrightarrow N_2(g) + 2 H_2O(g)$$

hydrazine

$$N_2H_2(CH_3)_2(\ell) + 4 O_2(g) \longrightarrow$$
1,1-dimethylhydrazine $\qquad 2 CO_2(g) + 4 H_2O(g) + N_2(g)$

The molar enthalpy of formation of liquid hydrazine is 50.6 kJ/mol, and that of liquid dimethylhydrazine is 49.2 kJ/mol. By doing appropriate calculations, decide whether the reaction of hydrazine or dimethylhydrazine with oxygen gives more heat per gram (at constant pressure). (Other enthalpy of formation data can be obtained from Table 6.2.)

105. The four hydrocarbons of lowest molar mass are methane, ethane, propane, and butane. All are used extensively as fuels in our economy. Calculate the thermal energy transferred to the surroundings per gram (the fuel value) of each of these four fuels and rank them by this quantity.

106. For the four substances in the preceding question, calculate the energy density. Assume that substances that are gases at room temperature have been forced to condense to liquids by application of pressure and, if necessary, by lowering temperature. Obtain needed density information from the *CRC Handbook of Chemistry and Physics,* 80th ed. Boca Raton, FL: CRC Press, 2000 (http://www.crcpress.com/us/).

Food and Energy

107. M&M candies consist of 70.% carbohydrate, 21% fat, and 4.6% protein as well as other ingredients that do not have caloric value. What quantity of energy transfer would occur if 34.5 g of M&Ms were burned in a bomb calorimeter?

108. A particular kind of crackers contains 71.4% carbohydrate, 10.7% fat, and 14.3% protein. Calculate the caloric value in Cal/g and in kJ/g.

109. Suppose you eat a quarter-pound hamburger (no bread, cheese, or other items—just meat) and then take a walk. For how long will you need to walk to use up the caloric value of the hamburger?

110. If you eat a quarter-pound of cheese pizza and then go out and play soccer vigorously, how long will it take before you have used up the calories in the pizza you ate?

General Questions

111. The specific heat capacity of copper metal is 0.385 J g^{-1} °C^{-1}, whereas it is 0.128 J g^{-1} °C^{-1} for gold. Assume you place 100. g of each metal, originally at 25°C, in a boiling water bath at 100 °C. If each metal is heated at the same rate, which piece of metal will reach 100 °C first?

112. Calculate the molar heat capacity, in J mol^{-1} °C^{-1}, for the four metals in Table 6.1. What observation can you make about these values? Are they widely different or very similar? Using this information, can you calculate the heat capacity in the units J g^{-1} °C^{-1} for silver? (The correct value for silver is 0.23 J g^{-1} °C^{-1}.)

113. Suppose you add 100.0 g of water at 60.0 °C to 100.0 g of ice at 0.00 °C. Some of the ice melts and cools the warm water to 0.00 °C. When the ice/water mixture has come to a uniform temperature of 0.00 °C, how much ice has melted?

114. The combustion of diborane, B_2H_6, proceeds according to the equation

$$B_2H_6(g) + 3\ O_2(g) \longrightarrow B_2O_3(s) + 3\ H_2O(\ell)$$

and 2166 kJ is liberated per mole of $B_2H_6(g)$ (at constant pressure). Calculate the molar enthalpy of formation of $B_2H_6(g)$ using this information, the data in Table 6.2, and the fact that the standard molar enthalpy of formation for $B_2O_3(s)$ is -1273 kJ/mol.

115. In principle, copper could be used to generate valuable hydrogen gas from water.

$$Cu(s) + H_2O(g) \longrightarrow CuO(s) + H_2(g)$$

(a) Is the reaction exothermic or endothermic?
(b) If 2.00 g of copper metal reacts with excess water vapor at constant pressure, how much thermal energy transfer is involved (either into or out of the system) in the reaction?

116. P_4 ignites in air to give P_4O_{10} and a large thermal energy transfer to the surroundings. This is an important reaction, because the phosphorus oxide can then be treated with water to give phosphoric acid for use in making detergents, toothpaste, soft drinks, and other consumer products. About 500,000 tons of elemental phosphorus is made annually in the United States. If you oxidize just 1

ton of P_4 (9.08×10^5 g) to the oxide, how much thermal energy (in kJ) is evolved at constant pressure?

117. The enthalpy changes of the following two reactions are known:

$$2\ C(graphite) + 2\ H_2(g) \longrightarrow C_2H_4(g) \qquad \Delta H° = 52.3\ kJ$$

$$C_2H_4Cl_2(\ell) \longrightarrow Cl_2(g) + C_2H_4(g) \qquad \Delta H° = 217.5\ kJ$$

Calculate the molar enthalpy of formation of $C_2H_4Cl_2(\ell)$.

118. Given the following information and the data in Table 6.2, calculate the molar enthalpy of formation for liquid hydrazine, N_2H_4.

$$N_2H_4(\ell) + O_2(g) \longrightarrow N_2(g) + 2\ H_2O(g) \qquad \Delta H° = -534\ kJ$$

119. The combination of coke and steam produces a mixture called coal gas, which can be used as a fuel or as a starting material for other reactions. If we assume coke can be represented by graphite, the equation for the production of coal gas is

$$2\ C(s) + 2\ H_2O(g) \longrightarrow CH_4(g) + CO_2(g)$$

Determine the standard enthalpy change for this reaction from the following standard enthalpies of reaction:

$$C(s) + H_2O(g) \longrightarrow CO(g) + H_2(g) \qquad \Delta H° = 131.3\ kJ$$

$$CO(g) + H_2O(g) \longrightarrow CO_2(g) + H_2(g) \qquad \Delta H° = -41.2\ kJ$$

$$CH_4(g) + H_2O(g) \longrightarrow 3\ H_2(g) + CO(g) \qquad \Delta H° = 206.1\ kJ$$

120. Some years ago Texas City, Texas, was devastated by the explosion of a shipload of ammonium nitrate, a compound intended to be used as a fertilizer. When heated, ammonium nitrate can decompose exothermically to N_2O and water.

$$NH_4NO_3(s) \longrightarrow N_2O(g) + 2\ H_2O(g)$$

If the heat from this exothermic reaction is contained, higher temperatures are generated, at which point ammonium nitrate can decompose explosively to N_2, H_2O, and O_2.

$$2\ NH_4NO_3(s) \longrightarrow 2\ N_2(g) + 4\ H_2O(g) + O_2(g)$$

If oxidizable materials are present, fires can break out, as was the case at Texas City. Using the information in Appendix J, answer the following questions.
(a) How much thermal energy is evolved (at constant pressure and under standard conditions) by the first reaction?
(b) If 8.00 kg of ammonium nitrate explodes (the second reaction), how much thermal energy is evolved (at constant pressure and under standard conditions)?

121. Uranium-235 is used as a fuel in nuclear power plants. Since natural uranium contains only a small amount of this isotope, the uranium must be enriched in uranium-235 before it can be used. To do this, uranium(IV) oxide is first converted to a gaseous compound, UF_6, and the isotopes are separated by a gaseous diffusion technique. Some key reactions are

$$UO_2(s) + 4\ HF(g) \longrightarrow UF_4(s) + 2\ H_2O(g)$$

$$UF_4(s) + F_2(g) \longrightarrow UF_6(g)$$

How much thermal energy transfer (at constant pressure) would be involved in producing 225 tons of $UF_6(g)$ from

UO$_2$ (1 ton = 9.08×10^5 g)? Some necessary standard enthalpies of formation are

$$\Delta H_f^\circ\{UO_2(s)\} = -1085 \text{ kJ/mol}$$

$$\Delta H_f^\circ\{UF_4(s)\} = -1914 \text{ kJ/mol}$$

$$\Delta H_f^\circ\{UF_6(g)\} = -2147 \text{ kJ/mol}$$

122. One method of producing H$_2$ on a large scale is the following chemical cycle.

Step 1: SO$_2$(g) + 2 H$_2$O(g) + Br$_2$(g) $\rightarrow$ H$_2$SO$_4$(ℓ) + 2 HBr(g)

Step 2: H$_2$SO$_4$(ℓ) $\rightarrow$ H$_2$O(g) + SO$_2$(g) + $\frac{1}{2}$ O$_2$(g)

Step 3: 2 HBr(g) $\rightarrow$ H$_2$(g) + Br$_2$(g)

Using the table of standard enthalpies of formation in Appendix J, calculate ΔH° for each step. What is the equation for the overall process, and what is its enthalpy change? Is the overall process exothermic or endothermic?

123. One reaction involved in the conversion of iron ore to the metal is

$$\text{FeO(s) + CO(g)} \longrightarrow \text{Fe(s) + CO}_2\text{(g)}$$

Calculate the standard enthalpy change for this reaction from the following reactions of iron oxides with CO:

3 Fe$_2$O$_3$(s) + CO(g) $\longrightarrow$ 2 Fe$_3$O$_4$(s) + CO$_2$(g) $\Delta H^\circ = -47$ kJ

Fe$_2$O$_3$(s) + 3 CO(g) $\longrightarrow$ 2 Fe(s) + 3 CO$_2$(g) $\Delta H^\circ = -25$ kJ

Fe$_3$O$_4$(s) + CO(g) $\longrightarrow$ 3 FeO(s) + CO$_2$(g) $\Delta H^\circ = 19$ kJ

Applying Concepts

124. Based on your experience, when ice melts to liquid water is the process exothermic or endothermic? When liquid water freezes to ice at 0 °C, is this exothermic or endothermic? (Assume that the ice/water is the system in each case.)

125. You pick up a six-pack of soft drinks from the floor, but it slips from your hand and smashes onto your foot. Comment on the work and energy involved in this sequence. What forms of energy are involved and at what stages of the process?

126. Consider the following graph. Which substance has the highest specific heat capacity?

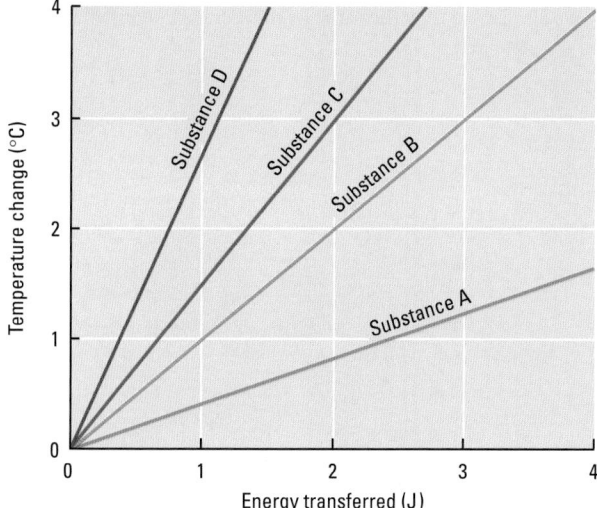

127. Based on the graph above, how much heat would need to be transferred to 10 g of substance B to raise its temperature from 35 °C to 38 °C?

128. The sketch below shows two identical beakers with different volumes of water at the same temperature.
Is the thermal energy content of Beaker 1 greater than, less than, or equal to that of Beaker 2? Explain your reasoning.

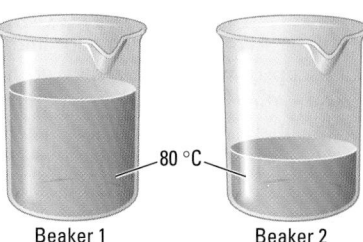

Beaker 1   Beaker 2

129. If the same quantity of thermal energy were transferred to each beaker in Question 128, would the temperature of Beaker 1 be greater than, less than, or equal to that of Beaker 2? Explain your reasoning.

130. Thermochemistry is often confusing because enthalpy change is referred to in different ways — for example, enthalpy of combustion, enthalpy of formation, enthalpy of reaction, and enthalpy of decomposition. What is similar about each of these situations? What is different?

131. In this chapter, the symbols ΔH_f° and ΔH° were used to denote a change in enthalpy. What is similar and what is different about the enthalpy changes they represent?

132. Consider the following equation

2 S(s) + 3 O$_2$(g) $\longrightarrow$ 2 SO$_3$(g) $\Delta H^\circ = -791$ kJ

and the standard molar enthalpy of formation for SO$_3$(g) listed in Table 6.2. Why are the enthalpy values different?

133. A student had five beakers, each containing 100. mL of 0.500 M NaOH (aq) and all at room temperature (20.0 °C). The student planned to add a carefully weighed quantity of solid ascorbic acid, C$_6$H$_8$O$_6$, to each beaker, stir until it dissolved, and measure the increase in temperature. After the fourth experiment, the student was interrupted and called away. The data table looked like this:

Experiment	Mass of ascorbic acid (g)	Final temperature (°C)
1	2.20	21.7
2	4.40	23.3
3	8.81	26.7
4	13.22	26.6
5	17.62	—

(a) Predict the temperature the student would have observed in Experiment 5. Explain why you predicted this temperature.

(b) For each experiment indicate which is the limiting reactant, sodium hydroxide or ascorbic acid.

(c) When ascorbic acid reacts with NaOH, how many hydrogen ions are involved? One, as in the case of HCl? Two, as in the case of H_2SO_4? Or three, as in the case of phosphoric acid, H_3PO_4? Explain clearly how you can tell, based on the student's calorimeter data.

134. In their home laboratory, two students do an experiment (a rather dangerous one — don't try it without proper safety precautions!) with drain cleaner (Drano, a solid) and toilet bowl cleaner (The Works, a liquid solution). The students measure 1 teaspoon (tsp) of Drano into each of four Styrofoam coffee cups and dissolve the solid in half a cup of water. Then they go have lunch. When they return, they measure the temperature of the solution in each of the four cups and find it to be 22.3 °C. Next they measure into separate small empty cups 1, 2, 3, and 4 tablespoons (Tbsp) of The Works. In each cup they add enough water to make the total volume 4 Tbsp. After a few minutes they measure the temperature of each cup and find it to be 22.3 °C. Finally the two students take each cup of The Works, pour it into a cup of Drano solution, and measure the temperature over a period of a few minutes. Their results are reported in the table below.

Experiment	Volume of The Works (Tbsp)	Highest temperature (°C)
1	1	28.0
2	2	33.6
3	3	39.3
4	4	39.4

Discuss these results and interpret them in terms of the thermochemistry and stoichiometry of the reaction. Is the reaction exothermic or endothermic? Why is more energy transferred in some cases than others? For each experiment, which reactant, Drano or The Works, is limiting? Why are the final temperatures nearly the same in Experiments 3 and 4? What can you conclude about the stoichiometric ratio between the two reactants?

135. A fully inflated Mylar party balloon (the kind that does not expand once it is fully inflated) is heated, and 310 J transfers to the gas in the balloon. Calculate ΔE and w for the gas.

136. A sample of Dry Ice, $CO_2(s)$, is placed in a flask that is then tightly stoppered. Because the Dry Ice is very cold, 350 J transfers from the surroundings to the $CO_2(s)$. Calculate ΔE and w for the contents of the flask.

 General Chemistry CD-ROM

CD6.1 Screen 6.5SB: A Closer Look: Heat . . .
(a) What is the difference between "heat" and "temperature"?
(b) How does the video illustrate heat transfer? How do you know that the heated copper bar is "hotter" than the water?

CD6.2 Screen 6.8: Heat Capacity of Pure Substances. The "tool" on Screen 6.8 allows you to do an experiment. Note that heat is added to each substance at the same rate, 50 J/s.
(a) What effect does heating the blocks for a longer time have on the final temperature?
(b) What is the effect of the mass of the blocks on the observed temperature change?
(c) Use this experiment to calculate the specific heat capacities of wood, copper, and glass. Do your answers agree with the "official" specific heat capacities for these substances? (For the values, click on the "Summary" button.)
(d) How does this "experiment" verify the relationship between heat transferred and the mass, specific heat, and temperature change of a substance? Explain fully.

CD6.3 Screen 6.10: Heat Transfer Between Substances.
(a) When a hotter object comes into contact with a cooler one, explain what happens in terms of molecular motions.
(b) What does it mean when we say that two objects have come to thermal equilibrium?

CD6.4 Screen 6.12: Energy Changes in Chemical Processes.
(a) A Gummi Bear has been placed in molten potassium chlorate (Screen 6.2). How might you define the changes in the system and the surroundings in this case?
(b) Is the observed chemical change exothermic or endothermic?

CD6.5 Screen 6.15: Enthalpy Changes for Chemical Reactions. The energies of the various changes in state or in the nature of the substances involved on this screen can be found by moving the cursor over the energy level diagram at the right.
(a) What is the enthalpy change when 1 mol of water evaporates at 25 °C?
(b) What is the enthalpy change for the condensation of water vapor at 25 °C to water at 25 °C?
(c) How does this compare with ΔH for the evaporation of water at 25 °C? Are they the same or different? If so, in what way?

7

As midnight arrived on December 31, 1999, fireworks displays burst forth around the world to celebrate the start of a new millennium. The vivid colors given off by the fireworks are due to transitions of electrons between energy levels in certain metal ions: strontium—crimson; copper—blue/green; calcium—red; barium—green. In this chapter we will consider the nature of such electron transitions, the energies associated with them, and their relationships to the electron configurations of atoms.

ELECTRON CONFIGURATIONS AND THE PERIODIC TABLE

7.1 Electromagnetic Radiation and Matter

7.2 Planck's Quantum Theory

7.3 The Bohr Model of the Hydrogen Atom

7.4 Beyond the Bohr Model: The Quantum Mechanical Model of the Atom

7.5 Quantum Numbers, Energy Levels, and Orbitals

7.6 Atom Electron Configurations

7.7 Ion Electron Configurations

7.8 Periodic Trends: Atomic Radii

7.9 Periodic Trends: Ionic Radii

7.10 Periodic Trends: Ionization Energies

7.11 Periodic Trends: Electron Affinities

The periodic table was created by Mendeleev to summarize experimental observations. He had no theory or model to explain why all alkaline earths combine with oxygen in a 1:1 atom ratio—they just do. In the early years of the twentieth century, however, it became evident that atoms contain electrons. As a result of these findings, explanations of periodic trends in physical and chemical properties began to be based on an understanding of the arrangement of electrons within atoms—on what we now call *electron configurations.* Studies of the interaction of light with atoms and molecules revealed that electrons in atoms are arranged roughly in concentric shells. Soon it was understood that the locations of electrons in these shells are determined by the energies of the electrons. Electrons in the outermost shell are called *valence electrons;* their number and location are the chief factors that determine chemical reactivity. In this chapter the relationship of the electron configurations of atoms to their properties will be described. Special emphasis is placed on the use of the periodic table to derive the electron configurations for atoms and ions.

7.1 ELECTROMAGNETIC RADIATION AND MATTER

CD-ROM Screen 7.3: Electromagnetic Radiation

Theories about the energy and arrangement of electrons in atoms are based on experimental studies of the interaction of matter with electromagnetic radiation, of which visible light is a familiar form. The human eye can distinguish the spectrum of colors that make up visible light. Interestingly, matter in some form is always associated with any color of light our eyes see. For example, the red glow of a neon sign comes from neon atoms excited by electricity, and fireworks displays are visible because of light from metal ions excited by the heat of explosive reactions. Have you ever wondered how these varied colors of light are produced?

When atoms gain energy from exposure to light, they are described as "excited." The added energy is absorbed by electrons, which then release it in the form of electromagnetic radiation, some of which is in the visible light region. Electromagnetic radiation and its applications are familiar to all of us; sunlight, automobile headlights, dental x-rays, microwave ovens, and radio waves that we use for communications are a few examples (Figure 7.1). These kinds of radiation seem very different, but they are actually very similar. All **electromagnetic radia-**

Figure 7.1 The electromagnetic spectrum. Visible light (enlarged section) is only a small part of the entire spectrum. The size of wavelengths in the electromagnetic spectrum is compared with common objects.

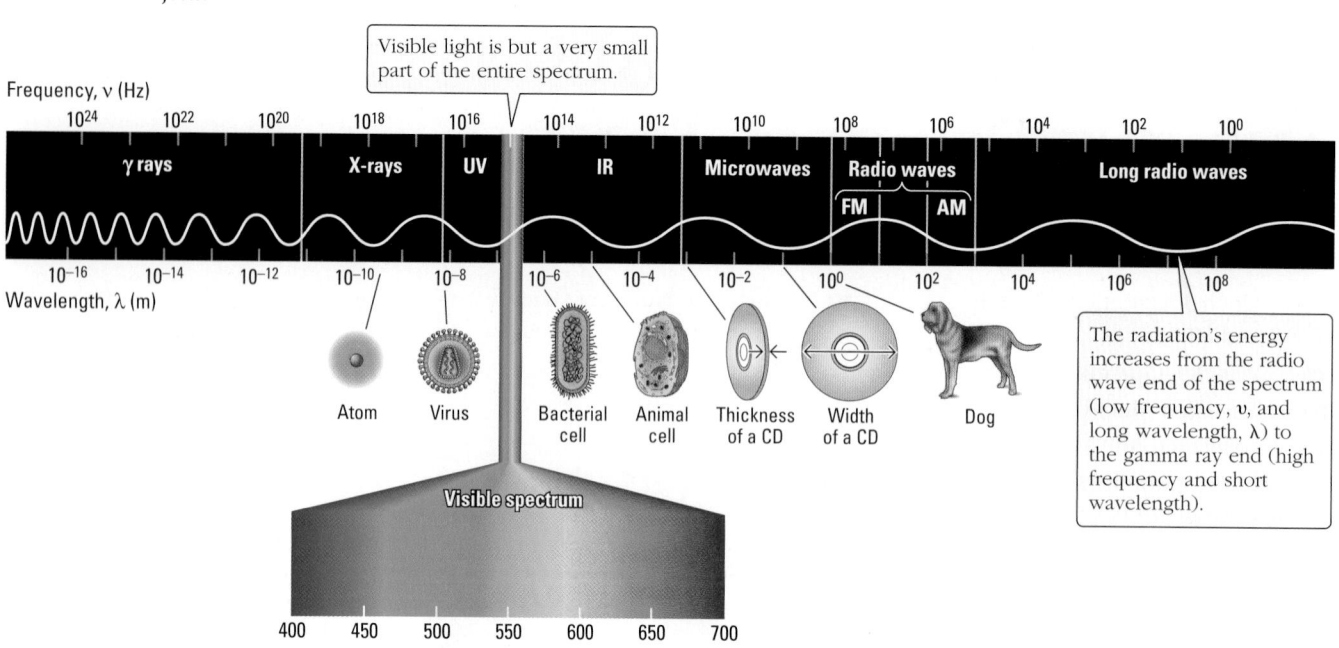

TABLE 7.1	Useful Wavelength Units for Different Regions of the Electromagnetic Spectrum	
Unit	**Length (m)**	**Radiation type**
Picometer, pm	10^{-12}	γ (gamma)
Angstrom, Å	10^{-10}	X-ray
Nanometer, nm	10^{-9}	Ultraviolet, visible
Micrometer, μm	10^{-6}	Infrared
Millimeter, mm	10^{-3}	Infrared
Centimeter, cm	10^{-2}	Microwave
Meter, m	1	TV, radio

tion consists of oscillating electric and magnetic fields that travel through space at the same rate (the "speed of light": 186,000 miles/second, or 2.998×10^8 m/s in a vacuum). Any of the various kinds of electromagnetic radiation can be described in terms of frequency (ν) and wavelength (λ). As illustrated in Figure 7.2, the **wavelength** is the distance between adjacent crests (or troughs) in a wave, and the **frequency** is the number of complete waves passing a point in a given amount of time — that is, cycles per second or simply per second, 1/s. Reciprocal units such as 1/s are often represented in the negative exponent form, s^{-1}, which means "per second." A frequency of 4.80×10^{14} s^{-1} means that 4.80×10^{14} waves pass a fixed point every second. The unit s^{-1} is given the name *hertz (Hz)*.

If light is thought of as electromagnetic waves, the intensity of radiation (brightness for visible light) is related to the amplitude (height of the wave crest) (Figure 7.2). The higher the amplitude, the more intense the radiation. For example, green light of a particular shade always has the same frequency and wavelength, but it can vary from bright (high amplitude) to dim (low amplitude).

The frequency of electromagnetic radiation is related to its wavelength by

$$\nu\lambda = c$$

where c is the speed of light, 2.998×10^8 m/s. A **spectrum** is the distribution of intensities of wavelengths or frequencies of electromagnetic radiation emitted or absorbed by an object. Figure 7.1 gives wavelength and frequency values for several regions of the electromagnetic spectrum.

As Figure 7.1 shows, visible light is only a small portion of the electromagnetic spectrum. Radiation with shorter wavelengths includes ultraviolet radiation (the type that leads to sunburn), x-rays, and gamma (γ) rays (emitted in the process of radioactive disintegration of some atoms). Infrared radiation, the type that is sensed as heat from a fire, has wavelengths longer than visible light. Longer still is the wavelength of the radiation in a microwave oven or of television and radio transmissions, such as that used for cellular telephones.

The speed of light through a substance (air, glass, or water, for example) depends on the chemical constitution of the substance and the wavelength of the light. This is the basis for using a glass prism to disperse light and is the explanation for rainbows.

The hertz unit (cycles/s, or s^{-1}) was named in honor of Heinrich Hertz (1857–1894), a German physicist.

Note that frequency and wavelength are *inversely* related; that is, as one decreases, the other increases. *Short*-wavelength radiation has a *high* frequency.

 CD-ROM Screen 7.4: The Electromagnetic Spectrum

Figure 7.2 Wavelength, frequency, and amplitude of water waves. The waves are moving from left to right toward the post. (a) The upper wave has a long wavelength (large λ) and low frequency (the number of times per second its peak hits the post). The lower wave has a shorter wavelength and a higher frequency (it hits the post more often per unit of time). (b) Variation in wavelength. (c) Variation in amplitude.

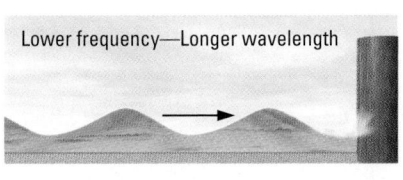
Lower frequency—Longer wavelength

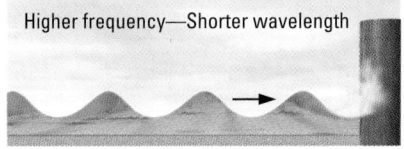
Higher frequency—Shorter wavelength

(a) Frequency (ν)

Longer wavelength

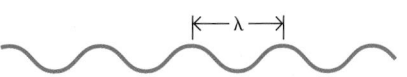
Shorter wavelength

(b) Wavelength (λ)

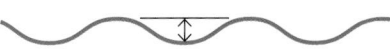
Lower amplitude

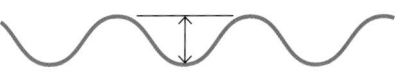
Higher amplitude

(c) Amplitude

Problem-Solving Example 7.1 Wavelength and Frequency

A new laser developed to use with digital video disc (DVD) players has a wavelength of 405 nm. What is the frequency of this light in hertz (Hz)?

Answer 7.40×10^{14} Hz

Explanation Rearranging the equation relating frequency, wavelength, and the speed of light allows the answer to be calculated in the proper units.

$$\nu = \frac{c}{\lambda} = \frac{2.998 \times 10^8 \text{ m/s}}{4.05 \times 10^{-7} \text{ m}} = 7.40 \times 10^{14} \text{ s}^{-1} \text{ or } 7.40 \times 10^{14} \text{ Hz}$$

Problem-Solving Practice 7.1

A processor allows laptop computers to operate at 650 MHz (1 MHz = 1 megahertz = 10^6 Hz = 10^6 s^{-1}). What is the wavelength (in meters) of the radiation that corresponds with the frequency used by this computer?

 ### Exercise 7.1 Frequency and Wavelength

A fellow chemistry student says that low-frequency radiation is short-wavelength radiation. You disagree. Explain why the other student is wrong.

 ### Exercise 7.2 Estimating Wavelengths

The size of a radio antenna is proportional to the wavelength of the radiation. Cellular phones have antennas often less than 0.076 m long, whereas submarine antennas are up to 2000 m long. Which is using higher frequency radio waves?

7.2 PLANCK'S QUANTUM THEORY

CD-ROM Screen 7.5: Planck's Equation

Figure 7.3 **The spectrum of radiation given off by a heated object.** At very high temperatures, the object becomes "white hot" as all wavelengths of visible light become equally intense.

Have you ever sat near an electric resistance heater as it warms up? Of course, you cannot see the metal atoms in the heater's wire, but as electric energy flows through the wire, the atoms gain energy and then emit it as radiation. First the wire emits a slight amount of heat that you can feel (infrared radiation). As the wire gets hotter, it begins to glow (emit visible light), emitting first red light, and then orange. If the wire gets very hot, it appears almost white. Figure 7.3 shows the spectrum of a typical heated object.

A metal heated to a high enough temperature gives off a red-orange glow.

(a)

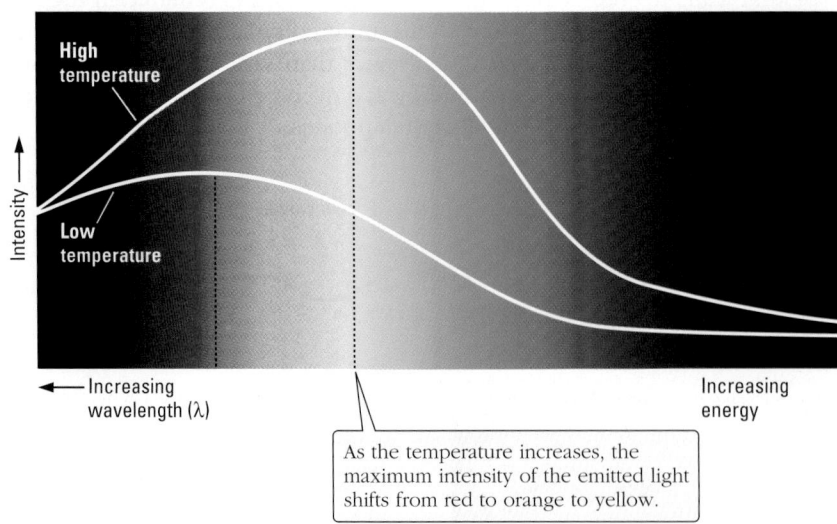

As the temperature increases, the maximum intensity of the emitted light shifts from red to orange to yellow.

(b)

At the close of the 19th century, scientists were trying to explain the nature of these emissions from hot objects. They assumed that vibrating atoms in a hot wire caused electromagnetic vibrations (light waves) to be emitted and that those light waves could have any frequency along a continuously varying range. The scientists were unable to predict the experimentally observed spectrum.

In 1900, Max Planck (1858–1947) offered an explanation for the spectrum of a heated body. His ideas contained the seeds of a revolution in scientific thought. He made what was at that time an incredible assumption: When an atom in a hot object emits radiation, it does so only in packets with a minimum amount of energy. That is, there must be a small packet of energy such that no smaller quantity can be emitted, just as an atom is the smallest packet of an element. Planck called this packet of energy a **quantum.** He further asserted that the energy of a quantum is proportional to the frequency of the radiation by the equation

$$E_{quantum} = h\nu_{radiation}$$

The proportionality constant h is called **Planck's constant** in his honor; it has a value of 6.626×10^{-34} J·s and relates the frequency of radiation to its energy *per particle* (*quantum*).

The frequency of orange light (quantum) is 4.80×10^{14} s^{-1}. The energy of *one* quantum of orange light is therefore

$$E = h\nu = (6.626 \times 10^{-34} \text{ J·s})(4.80 \times 10^{14} \text{ s}^{-1}) = 3.18 \times 10^{-19} \text{ J}$$

The theory based on Planck's work is called the **quantum theory.** By using his quantum theory, Planck was able to calculate results that agreed very well with the experimentally measured spectra of heated objects.

A quantum is the smallest possible unit of a distinct quantity—for example, the smallest possible unit of energy for electromagnetic radiation of a given frequency.

Max Planck won the 1918 Nobel Prize in Physics for his quantum theory.

Problem-Solving Example 7.2   Calculating Photon Energies

What is the energy associated with one quantum of laser light that has a frequency of 4.57×10^{14} s^{-1}?

Answer 3.03×10^{-19} J

Explanation   According to Planck's quantum theory, the energy and frequency of radiation are related by $E = h\nu$. Substituting the values for h and ν into the equation, followed by multiplying and canceling the units, gives the correct answer.

$$E = h\nu = (6.626 \times 10^{-34} \text{ J·s})(4.57 \times 10^{14} \text{ s}^{-1}) = 3.03 \times 10^{-19} \text{ J}$$

Problem-Solving Practice 7.2

Which has more energy,

(a)   One quantum of microwave radiation or one quantum of ultraviolet radiation?
(b)   One quantum of blue light or one quantum of green light?
(c)   Ten blue quanta of $\lambda = 460$ nm or 15 red quanta of $\lambda = 695$ nm?

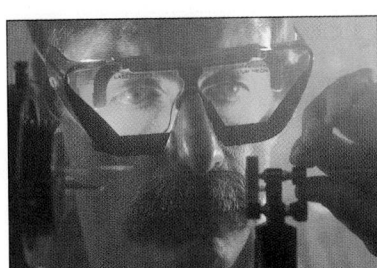

Red light from a helium-neon laser.

When a theory can accurately predict experimental results, the theory is usually regarded as useful. At first, however, Planck's quantum theory was not widely accepted because of its radical assumption that energy is quantized. The quantum theory of electromagnetic energy was firmly accepted after Planck's quanta were used by Albert Einstein to explain another phenomenon called the photoelectric effect.

There is an important relationship between the energy and wavelength of radiation. Since $E = h\nu$ and $\nu = \dfrac{c}{\lambda}$, then

$$E = \frac{hc}{\lambda}$$

where h is Planck's constant, c is the velocity of light, and λ is the wavelength of the radiation (in meters). Note that energy and wavelength are inversely proportional: ***The energy per quantum of radiation increases as the wavelength gets shorter.*** For example, red light and blue light have wavelengths of 656.3 nm and 434.1 nm, respectively. Therefore, because of its shorter wavelength, blue light has a higher energy than red light. We can calculate their different energies by applying the equation above.

$$E_{red} = \frac{(6.626 \times 10^{-34}\,\text{J}\cdot\text{s})(3.00 \times 10^{8}\,\text{m/s})}{(656.3\,\text{nm})(1\,\text{m}/10^{9}\,\text{nm})} = 3.03 \times 10^{-19}\,\text{J}$$

$$E_{blue} = \frac{(6.626 \times 10^{-34}\,\text{J}\cdot\text{s})(3.00 \times 10^{8}\,\text{m/s})}{(434.1\,\text{nm})(1\,\text{m}/10^{9}\,\text{nm})} = 4.58 \times 10^{-19}\,\text{J}$$

Exercise 7.3 Energy, Wavelength, and Radar

Police radar guns for traffic control emit radiation with an energy of 1.47×10^{-23} J per quantum, which is in the microwave region of the electromagnetic spectrum. Calculate the wavelength emitted by such a radar gun.

The Photoelectric Effect

In the early 1900s it was known that certain metals exhibit a **photoelectric effect**; that is, they emit electrons when illuminated by light of certain wavelengths (Figure 7.4a). For each metal there is a threshold wavelength below which no photoelectric effect is observed. For example, whereas cesium (Cs) emits electrons when illuminated by red light, some metals require yellow light, and others require ultraviolet light to emit electrons. The differences occur because each type of light has a different wavelength and thus a different energy per quantum. Red light, for example, has a lower energy (longer wavelength) than yellow light or ultraviolet light. Light with a wavelength (and energy) that is below the threshold needed to eject electrons from the metal will not cause an electric current to flow,

Figure 7.4 The photoelectric effect. (a) A photoelectric surface. (b) A photoelectric cell. Only photons with energies greater than the threshold energy will eject electrons. Photons with lower energy (left side of Figure 7.4c) do not have enough energy to remove electrons. Note that the *number* of electrons ejected (current) depends on the light's intensity, not its energy.

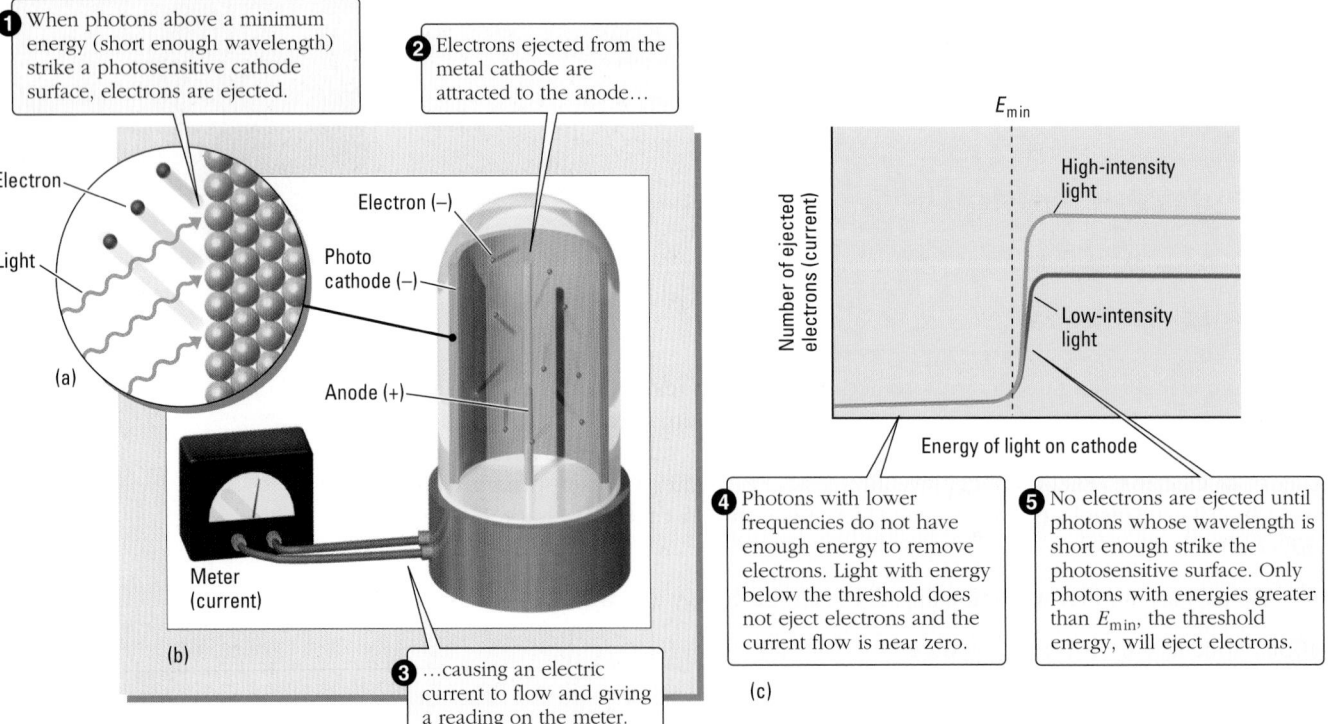

1 When photons above a minimum energy (short enough wavelength) strike a photosensitive cathode surface, electrons are ejected.

2 Electrons ejected from the metal cathode are attracted to the anode…

Electron

Light

Electron (−)

Photo cathode (−)

Anode (+)

Meter (current)

(a)

(b)

3 …causing an electric current to flow and giving a reading on the meter.

E_{min}

Number of ejected electrons (current)

High-intensity light

Low-intensity light

Energy of light on cathode

4 Photons with lower frequencies do not have enough energy to remove electrons. Light with energy below the threshold does not eject electrons and the current flow is near zero.

5 No electrons are ejected until photons whose wavelength is short enough strike the photosensitive surface. Only photons with energies greater than E_{min}, the threshold energy, will eject electrons.

(c)

no matter how bright the light. Figure 7.4b illustrates how the electrons ejected from the photosensitive cathode are attracted to the anode in the photoelectric cell, causing an electric current to flow, resulting in a display on the meter. Figure 7.4c shows how an electric current suddenly increases when light of a high enough energy per quantum shines on a photosensitive metal.

Einstein explained these observations by assuming that Planck's quanta were *massless* "particles" of light. He called them **photons** instead of quanta. That is, light could be described as a stream of photons that had particle-like properties as well as wave-like properties. To remove one electron from a photosensitive metal surface requires a certain minimum quantity of energy; we call it E_{min}. Since each photon has an energy given by $E = h\nu$, only photons whose E is greater than E_{min} will have enough energy to knock an electron loose. Thus, no electrons are ejected until photons with high enough frequencies (and short enough wavelengths) strike the photosensitive surface. Photons with lower frequencies (left side of Figure 7.4c) do not have enough energy to remove electrons. This means that if a metal requires photons of green light to eject electrons from its surface, then yellow light, red light, or light of any other lower frequency (longer wavelength) will not have sufficient energy to cause the photoelectric effect. This brilliant deduction about the quantized nature of light and how it relates to its interaction with matter won Einstein the Nobel Prize for physics in 1921.

The energy of photons is important for practical reasons. Photons of ultraviolet light can damage skin, while photons of visible light cannot. We use sunblocks containing molecules that selectively absorb ultraviolet photons to protect our skin. X-ray photons are even more energetic than ultraviolet photons and can disrupt molecules at the cellular level, causing genetic damage, among other effects. For this reason we try to limit our exposure to x-rays even more than we limit our exposure to ultraviolet light. Look again at Figure 7.4c. Notice how a higher intensity light source causes a higher photoelectric current. Higher intensities of ultraviolet light (or more time of exposure) can cause greater damage to the skin than lower intensities (or less exposure time). The same holds true for other high-energy forms of electromagnetic radiation such as x-rays.

Developments such as Einstein's explanation of the photoelectric effect led eventually to acceptance of what is referred to as the *dual nature* of light. Depending on the experimental circumstances, visible light and all other forms of electromagnetic radiation appear to have either "wave" or "particle" (photon) behavior. Although classical theory fails to explain the photoelectric effect, it does explain quite well the *refraction* of light by a prism or the diffraction of light by a diffraction grating, a device with a series of parallel, closely spaced small slits. When waves of light pass through such adjacent narrow slits, the waves are scattered so that the emerging light waves spread out, a phenomenon called *diffraction*. The semicircular waves emerging from the narrow slits can either amplify or cancel each other. Such behavior creates a diffraction pattern of dark and bright spots, as seen in Figure 7.5.

However, it is important to realize that this "dual nature" description arises because of our attempts to explain observations by inadequate models. Light is not changing back and forth from being a wave to being a particle, but has a single consistent nature that can be described by modern quantum theory. The dual-nature description arises when we try to explain our observations by using our familiarity with classical models for "wave" or "particle" behavior.

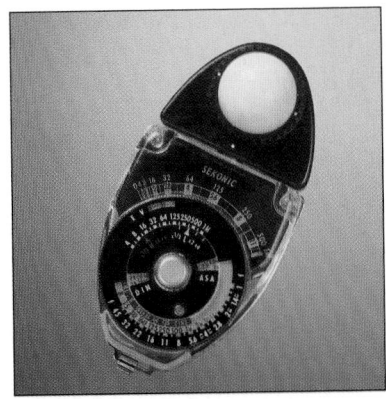

Use of photoelectric effect. The photoelectric effect is used in a light meter's sensors to determine exposure for photographic film.

A common misconception is that Einstein won the Nobel Prize for his theory of relativity.

In the quantum theory the intensity of radiation is proportional to the number of photons. The total energy is proportional to the number of photons times the energy per photon.

Prior to Einstein's explanation of the photoelectric effect, classical physics considered light as being only wave-like.

Refraction is the bending of light as it crosses the boundary from one medium to another, for example, from air to water. Diffraction is the bending of light around the edges of objects, such as slits in a diffraction grating.

7.3 THE BOHR MODEL OF THE HYDROGEN ATOM

Within just about a decade (1900–1911), three major discoveries were made about the atom and the nature of electromagnetic radiation. The first two, discussed in the previous section, were Max Planck's suggestion that radiation was

Figure 7.5 Electron diffraction pattern. Light waves passing through slits of a diffraction grating are scattered. When adjacent waves are in phase, they reinforce each other, producing a bright spot. When the waves are out of phase, they cancel each other and a dark spot occurs.

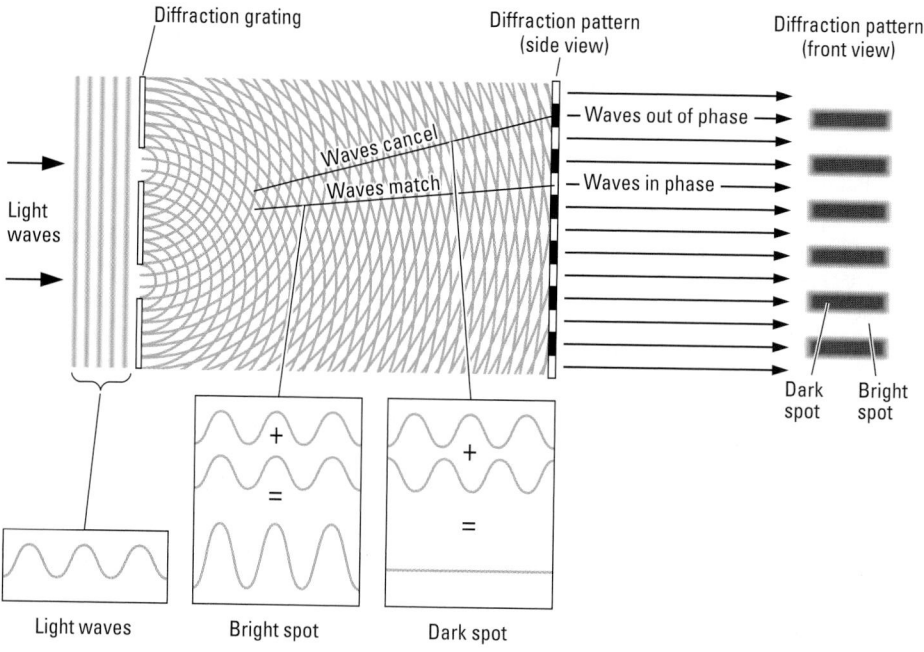

Figure 7.6 A continuous spectrum from white light. When white light is passed through slits to produce a narrow beam and then refracted in a glass prism, the various colors blend smoothly into one another.

quantized and, 5 years later, Albert Einstein's application of the quantum idea to explain the photoelectric effect. In 1911, Ernest Rutherford demonstrated experimentally that atoms contain a dense positive core (the nucleus) surrounded by electrons. Niels Bohr linked these three powerful ideas when, in 1913, he used quantum theory to explain the behavior of the electron in a hydrogen atom. In doing so, Bohr developed a mathematical model to explain how excited hydrogen atoms emit or absorb only certain wavelengths of light, a phenomenon that had been unexplained for nearly a century. To understand what Bohr did, we turn first to the visible spectrum.

The spectrum of white light, such as that from the sun or an incandescent light bulb, consists of a rainbow of colors, as shown in Figure 7.6. This rainbow spectrum, containing light of all wavelengths in the visible region, is called a **continuous spectrum.**

If a high voltage is applied to a gaseous element at low pressure, the atoms absorb energy and are "excited." The excited atoms then emit the extra energy as electromagnetic radiation. A neon advertising sign, in which the neon atoms emit

---

ESTIMATION Seeing Red

A photodetector is set up to receive only red light with a frequency of 4.51×10^{14} Hz. The detector records a signal of 6.455×10^{-15} J/hr. Approximately how many photons of red light are received per second by the detector, on average, from the source?

This particular red light has a frequency (ν) of 4.51×10^{14} Hz, or 4.51×10^{14} s^{-1}. The energy of a photon of the red light can be obtained from $E = h\nu$. To do this estimation, we can round off to order-of-magnitude values for Planck's constant (h) $\approx 10^{-33}$ J·s and the frequency $\approx 10^{15}$ s^{-1}, so that the energy of a photon of the red light is about $(10^{-33}$ J·s$)$ $(10^{15}$ s$^{-1}) = 10^{-18}$ J/photon.

The total energy per second (J/s) in the signal can be calculated from the energy per photon and the time. There are

about 10^4 seconds per hour (60 s/min $\times$ 60 min/hr = 3600 s), and the detector picks up about 10^{-14} J/hr (actually 6.455×10^{-15} J/hr). So the total energy per second is about $(10^{-14}$ J/hr$)$ $(1$ hr/10^4 s$) = 10^{-18}$ J/s. The number of red photons necessary to produce this amount of energy per second is approximately $(10^{-18}$ J/s$)$ $(1$ photon/10^{-18} J$) = 1$; the detector receives approximately one photon each second. This is the lower limit; the actual value is six photons. The variance comes in because we underestimated the energy per second (actually 1.79×10^{-18} J/s) and overestimated the energy per photon (actually 2.99×10^{-19} J, calculated from $E = h\nu$). Even so, the estimation is within one order of magnitude of the correct answer.

red-orange light, is an application of the light from excited neon atoms. When light from such a source passes through a prism onto a white surface, only a few colored lines are seen. This is called a **line emission spectrum,** which is characteristic of the element. The line spectra of the visible light emitted by excited atoms of hydrogen, mercury, and neon are shown in Figure 7.7.

Hydrogen has the simplest atomic emission spectrum, which was first studied in the 1880s. At that time, lines known as the Balmer series were discovered in the visible region of the spectrum and their wavelengths accurately measured (Figure 7.8, next page). Other series of lines in the ultraviolet region ($\lambda < 400$ nm) and in the infrared region ($\lambda > 700$ nm) were discovered subsequently.

Niels Bohr provided the first explanation of the line emission spectra of atoms by audaciously assuming that the single electron of a hydrogen atom moved in a circular orbit around its nucleus. He then related the energies of the electron to the radius of its orbit, but not by using the classic laws of physics. Such laws allowed the electron to have a wide range of energy. Instead, Bohr introduced quantum theory into his atomic model by invoking Planck's idea that energies are quantized. In the Bohr model, the electron could circle the nucleus in orbits of only certain radii, which correspond to specific energies. Thus, the energy of the electron is "quantized," and the electron is restricted to certain energy levels unless it gains or loses a certain amount of energy. Bohr referred to these energy levels as *orbits* and represented the energy difference between any two adjacent orbits as a specific quantity of energy. Each allowed orbit is assigned an integer, *n*, known as the **principal quantum number.**

In the Bohr model, the value of *n* for the possible orbits ranges by integers from 1 to infinity (∞). The energy of the electron and the size of its orbit increase as the value of *n* increases. The orbit of lowest energy, $n = 1$, is closest to the nucleus, and the electron of a hydrogen atom is normally in this energy level. Any atom with its electrons in their lowest energy levels is said to be in its **ground state.**

Because the positive nucleus attracts the negative electron, energy must be supplied to move the electron to a higher energy level, one farther from the nucleus. As the electron moves farther away from the nucleus, *n* increases, and the energy of the atom *increases*. The highest potential energy the electron can have is when sufficient energy has been added to separate the electron from the proton. Bohr designated this situation as zero energy and $n = \infty$. Consider this analogy to Bohr's arbitrarily selecting zero as the highest energy for the electron in a hydrogen atom. A book resting on a table is arbitrarily designated as having zero potential energy. As the book falls from the table to a chair to a stool and then to the

CD-ROM Screen 7.6: Atomic Line Spectra

CD-ROM Screen 7.7: Bohr's Model of the Hydrogen Atom

An electron must absorb energy to go from a lower energy state (lower *n* value) to a higher energy state (higher *n* value). Conversely, energy is emitted when the reverse process occurs— an electron going from a higher energy state to a lower energy state.

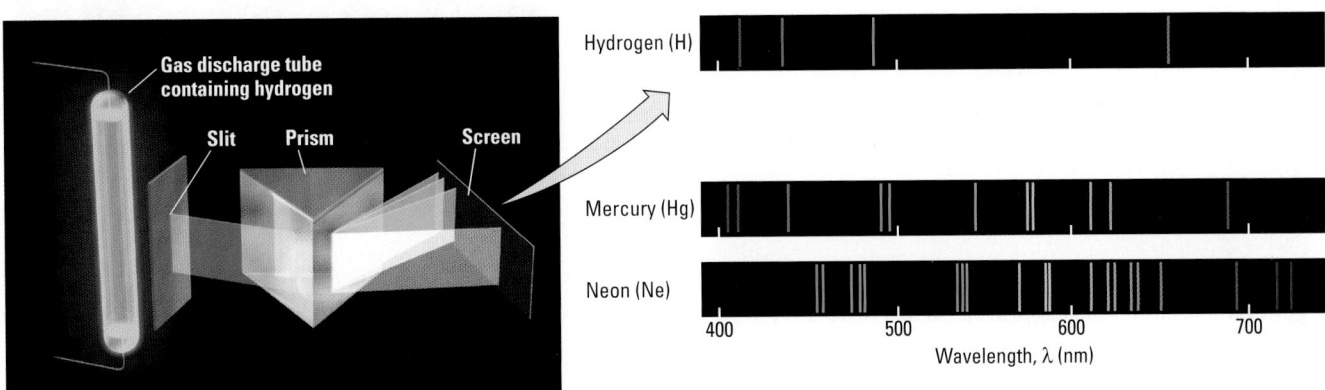

Figure 7.7 **Line emission spectra of hydrogen, mercury, and neon.** Excited gaseous elements produce characteristic spectra that can be used to identify the elements as well as to determine how much of an element is present in a sample.

CHEMISTRY
You Can Do...

Using a Compact Disc (CD) as a Diffraction Grating

Seeing the visible spectrum by using a prism to refract visible light is a familiar experiment (Figure 7.6). Less familiar is the use of a diffraction grating for the same purpose. A diffraction grating consists of many equally spaced parallel lines — thousands of lines per centimeter. A grating that transmits diffracted light is made by cutting thousands of grooves on a piece of glass or clear plastic; a grating that reflects diffracted light is made by cutting the grooves on a piece of metal or opaque plastic. You can get an idea of how diffraction gratings work by using a compact disc as a diffraction grating. Compare the spectra of light from two different sources — a mercury vapor lamp and a white incandescent light bulb. Stand about 20 to 60 m away from a mercury street lamp and hold the CD about waist high with the print side down. Tilt

the CD down until you see the reflected image of the street lamp. Close one eye, and view along the line from the light source to your body as you slowly tilt the CD up toward you. Use the same procedure with an incandescent light bulb as the light source.

1. What colors do you see with the mercury vapor lamp ?
2. Do you see the same colors with the incandescent light? Explain your results.

_____

Source:

Mebane, R.C., and Rybolt, T.R. Atomic Spectroscopy with a Compact Disc. *Journal of Chemical Education*, Vol. 69, 1992; p. 401.

floor, it loses energy so that when it is on the floor, its potential energy is negative with respect to when it was on the table. Such is also the case for an electron going from a higher energy level to a lower energy level. The electron must lose energy. Correspondingly, the electron's energy in all its allowed energy states within the atom must be less than zero; that is, the energy must be negative.

When the electron in a hydrogen atom gains a quantized amount of energy and moves to an orbit with $n > 1$, the atom is said to be in an **excited state** when

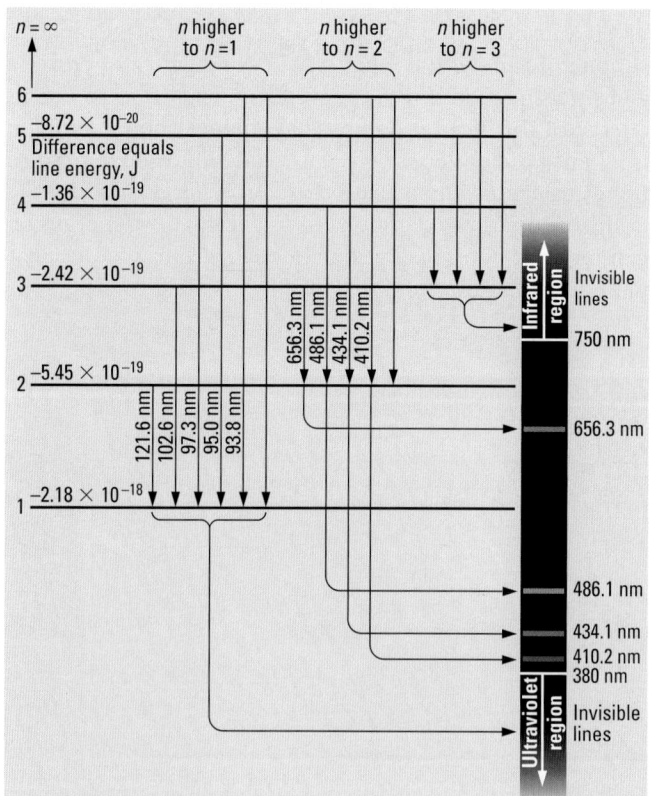

Figure 7.8 **Electron transitions in an excited H atom.** The lines in the ultraviolet region result from all transitions to the $n = 1$ level. Transitions from levels with values of $n > 2$ to the $n = 2$ level occur in the visible region. Lines in the infrared region result from the transitions from levels with values of $n > 3$ or 4 to the $n = 3$ or 4 levels (only the series for transitions to the $n = 3$ level is shown).

it has more energy than in its ground state. Any excited state of any atom is unstable. When the electron returns to the ground state from an excited state, the electron emits the energy gained when it moved from the ground state to an excited state. The emitted energy is given off as photons of light having an energy of hc/λ. Recall from Section 7.2 that h is Planck's constant, 6.626×10^{-34} J · s per particle; c is the velocity of light, 2.998×10^8 m/s; and λ is the wavelength. The energy of an emitted photon corresponds to the *difference* between two energy levels of the atom (Figure 7.8). Think of the Bohr orbit model as a set of stairs in which the higher stairs are closer together. Each step represents a quantized energy level; as you climb the stairs, you can stop at any step, *but not between steps.* Only whole steps are allowed. If you move down one step, you lose one quantum of energy. If you move down more than one step, you lose additional energy.

According to Bohr, the light forming the lines in the emission spectrum of hydrogen (Figure 7.7) comes from electrons in hydrogen atoms moving from higher energy orbits to lower energy orbits closer to the nucleus (Figure 7.8). Electrons returning from $n = 3, 4, 5, \ldots$ to $n = 2$ produce photons in the visible region of the spectrum (Balmer series). Photons in the ultraviolet region are the product of electrons dropping back to $n = 1$ from $n > 1$. Electron transitions from $n > 3$ back to $n = 3$ are responsible for the series of lines in the infrared region.

Electron transitions in the visible region of the spectrum are responsible for the impressive displays of color in fireworks.

Exercise 7.4 Many Spectral Lines, But Only One Kind of Atom

The hydrogen atom contains only one electron, but there are many lines in its line emission spectrum (Figure 7.7). How does the Bohr theory explain this?

Exercise 7.5 Electron Transitions

In which of these transitions in the hydrogen atom is energy emitted and in which is it absorbed?
(a) $n = 3 \rightarrow n = 2$ (b) $n = 1 \rightarrow n = 4$
(c) $n = 5 \rightarrow n = 3$ (d) $n = 6 \rightarrow n = 2$

Through his calculations, Bohr found that the allowed energies of the electron in a hydrogen atom are restricted by n, the principal quantum number, according to the equation

$$E = - \frac{2.179 \times 10^{-18}\,\text{J}}{n^2} \qquad (n = 1, 2, 3, \ldots)$$

where the constant, 2.179×10^{-18} J, called the Rydberg constant, can be calculated from theory. Bohr selected zero to be the energy of a completely separated electron and proton ($n = \infty$). This makes all energies for the electron in the atom negative, which is the origin of the negative sign in this equation.

The energy of a ground state hydrogen atom ($n = 1$) is

$$E = -2.179 \times 10^{-18}\,\text{J}\left(\frac{1}{n^2}\right) = -2.179 \times 10^{-18}\,\text{J}\left(\frac{1}{1^2}\right) = -2.179 \times 10^{-18}\,\text{J}$$

The difference in energy, ΔE, when the electron moves from its initial state, E_i, to a final state, E_f, can be calculated as $\Delta E = E_f - E_i$. When an electron moves from a higher energy state (greater n value) to a lower energy state (one with a lower n value), energy is emitted and ΔE is negative. Consequently, energy must be absorbed (ΔE is positive) when an electron moves from a lower energy state to

Division by infinity makes E equal to 0 when n = ∞.

Think of the atom as the system and everything else as the surroundings (⬅ **p. 215, Section 6.2**).

a higher one. This energy difference can be expressed in relation to the change in principal energy levels,

$$\Delta E = E_f - E_i = -2.179 \times 10^{-18} \, J \left(\frac{1}{n_f^2} - \frac{1}{n_i^2} \right)$$

where n_i and n_f are the initial and final energy states of the atom, respectively. This equation can be rewritten as

$$\Delta E = 2.179 \times 10^{-18} \, J \left(\frac{1}{n_i^2} - \frac{1}{n_f^2} \right)$$

In the Bohr model, only certain frequencies of light can be absorbed or emitted by an excited atom. The equation relating energy change and frequency, $\Delta E = h\nu$, and the previous equation can be rearranged into the following equation that relates the frequency of light absorbed or emitted to the n values of the initial (n_i) and final (n_f) states involved in the transition.

$$\nu = \frac{\Delta E}{h} = \frac{2.179 \times 10^{-18} \, J}{h} \left(\frac{1}{n_i^2} - \frac{1}{n_f^2} \right)$$

For example, the frequency of the $n = 2$ to $n = 3$ transition can be calculated using this equation.

$$\nu = \left(\frac{2.179 \times 10^{-18} \, J}{6.626 \times 10^{-34} \, J \cdot s} \right) \left(\frac{1}{2^2} - \frac{1}{3^2} \right)$$

$$= (3.289 \times 10^{15} \, s^{-1}) \left(\frac{1}{4} - \frac{1}{9} \right)$$

$$= (3.289 \times 10^{15} \, s^{-1})(0.250 - 0.111) = 4.567 \times 10^{14} \, s^{-1}$$

The positive sign indicates that energy was absorbed, which is as it should be because the electron moved from the $n = 2$ to the $n = 3$ level.

The wavelength (λ) of the light absorbed can be obtained from the frequency by using the relationship $\lambda = c/\nu$ where c is the velocity of light (2.998×10^8 m/s). For the $n = 2$ to $n = 3$ transition,

$$\lambda = \frac{2.998 \times 10^8 \, m/s}{4.567 \times 10^{14} \, s^{-1}} = 6.566 \times 10^{-7} \, m = 656.6 \, nm$$

There is exceptional agreement between the experimentally measured wavelengths and those calculated by the Bohr theory (Table 7.2). Thus, Niels Bohr had

Spectroscopy is the science of measuring spectra. Many kinds of spectroscopy have emerged since the first studies of simple line spectra. Some spectral measurements are done for quantitative analytical purposes and others are done to determine molecular structure.

TABLE 7.2	Agreement Between Bohr's Theory and the Lines of the Hydrogen Emission Spectrum*		
Changes in energy levels	**Wavelength predicted by Bohr's theory (nm)**	**Wavelength determined from laboratory measurement (nm)**	**Spectral region**
$2 \rightarrow 1$	121.6	121.7	Ultraviolet
$3 \rightarrow 1$	102.6	102.6	Ultraviolet
$4 \rightarrow 1$	97.28	97.32	Ultraviolet
$3 \rightarrow 2$	656.6	656.3	Visible red
$4 \rightarrow 2$	486.5	486.1	Visible blue-green
$5 \rightarrow 2$	434.3	434.1	Visible blue
$4 \rightarrow 3$	1876	1876	Infrared

* These lines are typical; other lines could be cited as well, with equally good agreement between theory and experiment. The unit of wavelength is the nanometer (nm), 10^{-9} m.

tied the unseen (the atom) to the seen (the observable lines of the hydrogen emission spectrum) — a fantastic achievement!

Problem-Solving Example 7.3   Electron Transitions

(a) Calculate the frequency and wavelength (nm) corresponding to the $n = 1$ to $n = 3$ transition in hydrogen.

(b) In what region of the spectrum does this occur?

Answer   (a) $2.924 \times 10^{15} \text{ s}^{-1}$; 102.6 nm   (b) ultraviolet region

Explanation   (a) We first calculate the frequency of the transition from which the wavelength can be determined. In this case, $n_i = 1$ and $n_f = 3$.

$$\nu = \frac{2.179 \times 10^{-18} \text{ J}}{h} \left(\frac{1}{n_i^2} - \frac{1}{n_f^2} \right)$$

$$= \left(\frac{2.179 \times 10^{-18} \text{ J}}{6.626 \times 10^{-34} \text{ J} \cdot \text{s}} \right) \left(\frac{1}{1^2} - \frac{1}{3^2} \right)$$

$$= \left(\frac{2.179 \times 10^{-18} \text{ J}}{6.626 \times 10^{-34} \text{ J} \cdot \text{s}} \right) \left(\frac{8}{9} \right) = (3.289 \times 10^{15} \text{ s}^{-1}) \left(\frac{8}{9} \right) = 2.924 \times 10^{15} \text{ s}^{-1}$$

Convert the frequency to wavelength.

$$\lambda = \frac{c}{\nu} = \frac{2.998 \times 10^8 \text{ m/s}}{2.924 \times 10^{15} \text{ s}^{-1}} = 1.026 \times 10^{-7} \text{ m} = 102.6 \text{ nm}$$

(b) Checking Figure 7.8 shows that this light is in the ultraviolet region of the spectrum (< 400 nm; $> 7.5 \times 10^{14} \text{ s}^{-1}$)

✔ Figure 7.8 indicates that electron transitions between any higher n level and the $n = 1$ level are in the ultraviolet region, corroborating the calculated answer.

Problem-Solving Practice 7.3

(a) Calculate the frequency and the wavelength of the line for the $n = 4$ to $n = 3$ transition.

(b) Is this wavelength longer or shorter than that of the $n = 5$ to $n = 3$ transition?

 Exercise 7.6   Conversions

Show that the value of the Rydberg constant per photon, 2.179×10^{-18} J, is equivalent to 1312 kJ/mol photons.

NIELS BOHR
(1885 – 1962)

Niels Bohr was born in Copenhagen, Denmark. While working with J.J. Thomson and Ernest Rutherford in England, Bohr began to develop the ideas that led to the publication of his explanation of atomic spectra. He received the Nobel Prize in physics in 1922 for this work.

As the director of the Institute of Theoretical Physics in Copenhagen, Bohr was a mentor to many young physicists, seven of whom later received Nobel Prizes for their studies in physics or chemistry, including Werner Heisenberg, Wolfgang Pauli, and Linus Pauling.

7.4 BEYOND THE BOHR MODEL: THE QUANTUM MECHANICAL MODEL OF THE ATOM

The Bohr atomic model was accepted almost immediately. Bohr's success with the hydrogen atom soon led, however, to attempts by him and others to extend the same model to more complex atoms. Before long it became apparent that line spectra for elements other than hydrogen had more lines than could be explained by the simple Bohr model. A totally different approach was needed to explain electron behavior in atoms or ions with more than one electron. The new approach again took a radical departure from classic physics.

In 1924, the young physicist Louis de Broglie posed the question: *If light can be viewed in terms of both wave and particle properties, why can't particles of matter, such as electrons, be treated the same way?* And so de Broglie proposed the revolutionary idea that electrons could have wave-like properties. The

 CD-ROM Screen 7.8: Wave Properties of the Electron

Chemistry in the News

KEEPING TIME BY PUMPING ATOMS

Are you a person who is chronically late, who can't show up on time? Maybe an atomic clock might help to keep you punctual. Atomic clocks that use the rhythmic pulsations of vibrating atoms have been around since 1949 and have been used since 1967 to define a second. At that time, one second was defined by an internationally accepted standard—how long it takes for an energized cesium atom to vibrate 9,192,631,770 times. Recently, scientists at the U.S. National Institute of Standards and Technology (NIST) and in France collaborated to develop a new atomic clock, a cesium "fountain" atomic clock, so accurate that it will not lose even one second in about 20 million years. Ultraprecise timing is essential to synchronize the time signals in telecommunications and for the extremely accurate navigation signals sent from the global positioning satellite system.

Previously, the atomic cesium clock used the normal frequencies of vibrating cesium atoms. The fountain clock takes that a step further. To do so, clusters of cooled cesium atoms in the clock are squeezed into spheres by six laser beams focused in opposite directions. This prevents the spheres from readily moving about, essentially holding them in place. Vertical lasers then squirt the cluster upward, much like a stream of water from a fountain to a maximum height (about 1 m). At this point, the lasers are turned off and the cluster falls downward through a microwave-filled chamber. The microwave radiation causes the cesium atoms in the cluster to vibrate, and another laser causes them to fluoresce, giving off light. Detectors measure the brightness of the fluorescing atoms, which is proportional to the number of atoms excited by the microwaves. The microwave frequency producing the

most intense fluorescence, which matches the natural vibration of cesium atoms, can be measured accurately. Dawn Meekhof, one of the builders of the new clock, commented on the development, saying "Think of each cesium atom as something like a little clock that ticks more than nine billion times a second. You count the ticks and then you can use the result to set the world's other clocks."

Source

Browne, M.W. This Clock Is Good for 20 Million Years. *The New York Times,* January 4, 2000; p. D3.

1. Why is such an ultraprecise timepiece as the cesium fountain atomic clock needed?
2. What is the importance of fluorescence to the workings of the cesium fountain atomic clock?

① Clusters of cooled cesium atoms in the clock are squeezed into spheres by six laser beams focused in opposite directions. This prevents the spheres from readily moving about, essentially holding them in place.

② Vertical lasers then squirt the cluster upward—much like a stream of water from a fountain—to a maximum height (about 1 m). At this point, the lasers are turned off and the cluster falls downward through a microwave-filled chamber.

③ The microwave radiation causes the cesium atoms in the cluster to vibrate...

④ ...and another laser causes them to fluoresce, giving off light.

⑤ Detectors measure the brightness of the fluorescing atoms. The brightest fluorescence matches the natural vibration of cesium atoms, 9,192,631,700 times in one second.

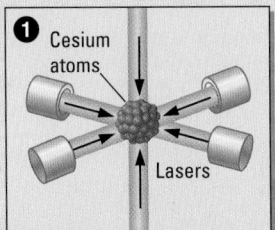

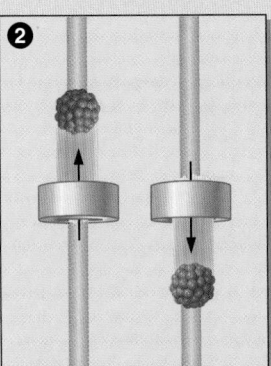

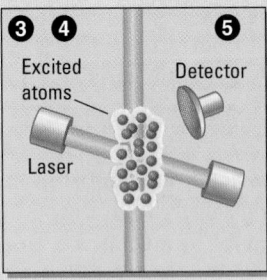

The cesium fountain atomic clock.

wavelength (λ) of the electron (or any other particle) would depend on its mass (m) and its velocity (v) according to the relationship

$$\lambda = \frac{h}{mv}$$

where h is Planck's constant. The product of $m \times v$ for any object (including electrons) is called its **momentum,** which is the mass of an object times its velocity. Note that particles of very low mass, such as the electron (mass = 9.11×10^{-28} g), have wavelengths that are measurable. Objects of ordinary size, such as a baseball (mass = 143 g), have wavelengths too short to be observed.

Momentum = mass × velocity

Problem-Solving Example 7.4   Baseballs and Electrons

Compare the de Broglie wavelength (nm) of an electron moving at a velocity of 5.0×10^6 m/s with that of a baseball traveling at 43.9 m/s (98 mi/hr). Masses: electron = 9.11×10^{-31} kg; baseball = 0.143 kg.

Answer   The wavelength of the electron is *much* longer than that of the baseball: electron = 0.15 nm; baseball = 1.1×10^{-25} nm.

Explanation   We can substitute the mass and velocity into the de Broglie wave equation to calculate the corresponding wavelength. Planck's constant, h, is 6.626×10^{-34} J·s, and 1 J = $\dfrac{1 \text{ kg} \cdot \text{m}^2}{\text{s}^2}$. So, $h = 6.626 \times 10^{-34}$ kg·m^2 s^{-1}.

For the electron:

$$\lambda = \frac{6.626 \times 10^{-34} \text{ kg} \cdot \text{m}^2 \text{ s}^{-1}}{(9.11 \times 10^{-31} \text{ kg})(5.0 \times 10^6 \text{ m/s})} = 1.5 \times 10^{-10} \text{ m} \times \frac{1 \text{ nm}}{10^{-9} \text{ m}} = 0.15 \text{ nm}$$

For the baseball:

$$\lambda = \frac{6.626 \times 10^{-34} \text{ kg} \cdot \text{m}^2 \text{ s}^{-1}}{(0.143 \text{ kg})(43.9 \text{ m/s})} = 1.1 \times 10^{-34} \text{ m} \times \frac{1 \text{ nm}}{10^{-9} \text{ m}} = 1.1 \times 10^{-25} \text{ nm}$$

The wavelength of the electron is in the x-ray region of the electromagnetic spectrum (Figure 7.1, p. 266). The wavelength of the baseball is far too short to observe.

Problem-Solving Practice 7.4

Calculate the de Broglie wavelength of a neutron moving at 10% the velocity of light. The mass of a neutron is 1.67×10^{-24} g.

A pitched baseball.   Even when the ball is pitched at 100 mi/hr, the wavelength of the baseball is too short to be observed.

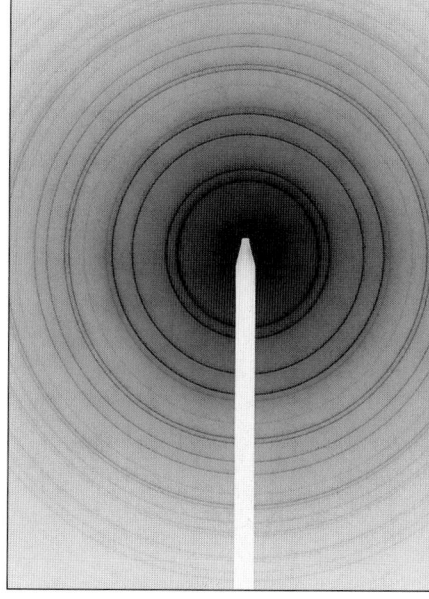

Many scientists found de Broglie's concept hard to accept, much less believe. But experimental proof was soon produced. In 1927, C. Davisson and L. H. Germer, working at the Bell Telephone Laboratories, found that a beam of electrons was diffracted by planes of atoms in a thin sheet of metal foil (Figure 7.9) in the same way that light waves are diffracted by a diffraction grating. Since diffraction is readily explained by the wave properties of light, it followed that electrons also can be described by the equations of waves under some circumstances.

A few years after de Broglie's hypothesis about the wave nature of the electron, Werner Heisenberg (1901–1976) proposed the **uncertainty principle,** which states that *it is impossible to simultaneously determine the exact position and the exact momentum of an electron.* This limitation is not a problem for a macroscopic object because the energy of photons used to locate such an object does not cause a measurable change in the position or momentum of that object. However, the very act of measurement would affect the position and momentum of the electron because of its very small size and mass. High-energy photons would be required to locate the small electron; when such photons collide with the electron, the momentum of the electron would be changed. If lower energy photons were used to avoid affecting the momentum, little information would be obtained

Figure 7.9   Electron diffraction pattern obtained for aluminum foil.

CD-ROM Screen 7.9:
Heisenberg's Uncertainty
Principle

A speeding race car. (a) Photo taken at high shutter speed. (b) Photo taken at low shutter speed.

(a)

(b)

about the location of the electron. Consider an analogy in photography. If you take a picture of a car race at a high shutter speed setting, you get a clear picture of the cars but you can't tell how fast they are going or even whether they are moving. With a slow shutter speed, you can tell from the blur of the car images something about the speed and direction, but you have less information about where the cars are.

The Heisenberg uncertainty principle illustrated another inadequacy in the Bohr model — its representation of the electron in the hydrogen atom in terms of well-defined orbits about the nucleus. In practical terms, the best we can do is to represent the *probability* of finding an electron of a given energy and momentum within a given space. This probability-based model of the atom is what chemists now use.

In 1926 Erwin Schrödinger (1877–1961) combined de Broglie's hypothesis with classic equations for wave motion. From these and other ideas he derived a new equation called the *wave equation* to describe the behavior of an electron in the hydrogen atom. Solutions to the wave equation, called *wave functions,* predict the allowed energy states of an electron and the probability of finding that electron in a given region of space.

Although each wave function is a complex mathematical equation, it is possible to represent the square of a wave function in graphic form — as a picture of the region in the atom where an electron with a given energy state is most likely to be found.

CD-ROM Screen 7.10: Schrödinger's Equation and Wave Functions

CD-ROM Screen 7.11: Shells, Subshells, and Orbitals

7.5 QUANTUM NUMBERS, ENERGY LEVELS, AND ORBITALS

The region in which an electron can be found within an atom is known as an **orbital.** One way to represent an orbital is to draw a surface within which there is a 90% probability that the electron will be found. That is, nine times out of ten an electron will be somewhere inside such a **boundary surface;** there is one chance in ten that the electron will be outside of it. A 100% probability isn't chosen because such a surface would have no definite boundary. Consider that a typical dart board has a finite size, and normally, players in a dart game hit the board more than 90% of the time. But if you wanted to be certain that any player, no matter how far away, would be able to hit the board on 100% of his or her throws, then the board would have to be considerably larger. By similar reasoning, a boundary surface in which there would be a 100% probability of finding a given kind of electron would have to be quite large.

A series of such three-dimensional boundary surface diagrams for the orbitals of a hydrogen atom is shown in Figure 7.10. Note that an *orbital* (quantum mechanical model) is not the same as an *orbit* (Bohr model). In the quantum mechanical model, the principal quantum number, n, is a measure of the most probable

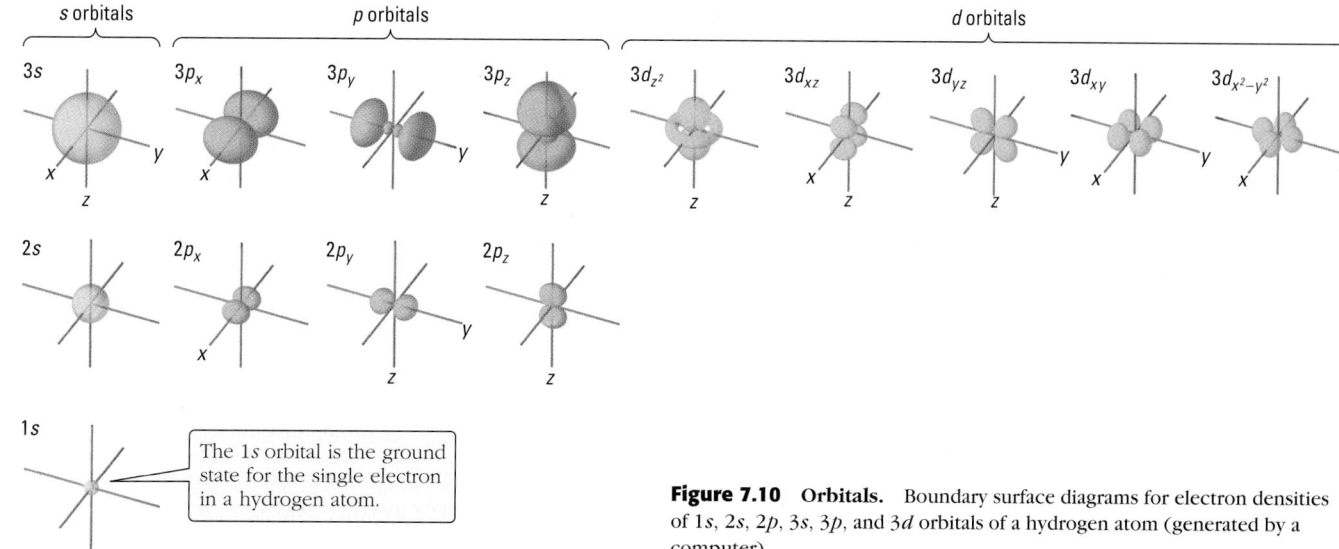

Figure 7.10 Orbitals. Boundary surface diagrams for electron densities of $1s$, $2s$, $2p$, $3s$, $3p$, and $3d$ orbitals of a hydrogen atom (generated by a computer).

distance of the electron from the nucleus, not the radius of a well-defined orbit. A collection of orbitals with the same principal quantum number value, n, is called an electron **shell.** When $n = 1$, there is only one kind of orbital possible; it is an s orbital. When $n = 2$, two kinds of orbitals are possible; s and p orbitals. When $n = 3$, three kinds of orbitals are possible: s, p, and d orbitals. When $n = 4$, four kinds of orbitals are possible: s, p, d, and f orbitals can exist.

As a result of his work, Schrödinger found that three quantum numbers—the principal quantum number n and two others, l and m_l—are needed to describe the orbitals in a hydrogen atom. These three quantum numbers describe the three-dimensional coordinates of an electron's motion. The need for a fourth quantum number, m_s, was identified in subsequent work by others. Thus, a set of four quantum numbers — n, l, m_l, and m_s — is used to denote the energy and the shape of the electron cloud for each electron in an atom. An electron cloud is a representation used to illustrate the probability of finding the electron in a particular region around the atom.

We can apply quantum numbers to electrons in any atom, not just hydrogen. The quantum numbers, their meanings, and the orbital notation used in Figure 7.10 are described in the following sections.

CD-ROM Screen 7.12:
Quantum Numbers and Orbitals

First Quantum Number, n: Principal Energy Levels

The first quantum number, n, the principal quantum number, is the most important one in determining the energy of an electron. The quantum number n has only integral values, starting with 1:

$$n = 1, 2, 3, 4, \ldots$$

The value of n corresponds to a **principal energy level** so that an electron is in the first principal energy level when $n = 1$, in the second principal level when $n = 2$, and so on. As n increases, the energy of the electron increases as well, and the electron, on average, is farther away from the nucleus and is not as tightly bound to it.

Second Quantum Number, l, and Subshells (s, p, d, f)

Each principal energy level, also known as a shell, has **subshells** within it. The $n = 1$ principal energy level has only one subshell; all other energy levels have more than one subshell. The subshells are designated by the second quantum number, *l* (sometimes refered to as the *azimuthal* quantum number). *The shape of the orbital is determined by the value of* *l*.

A subshell is one or more orbitals with the same *n* and *l* quantum numbers. Thus, the value of *l* is associated with an *n* value. The value of *l* is an integer that ranges from zero to a maximum of *n* − 1:

$$l = 0, 1, 2, 3, \ldots (n - 1)$$

According to this relationship, when *n* = 1, then *l* must be zero. This indicates that in the first principal energy level there is only one subshell, with an *l* value of 0. The second principal energy level has two subshells — one with an *l* value of 0 and a second with an *l* value of 1. Continuing in this manner,

$$n = 3; \quad l = 0, 1, \text{ or } 2 \qquad \text{(three subshells)}$$

$$n = 4; \quad l = 0, 1, 2, \text{ or } 3 \qquad \text{(four subshells)}$$

Rather than using *l* values, subshells are more commonly designated by letters: *s*, *p*, *d*, or *f*. The first four subshells are known as the *s* subshell, *p* subshell, *d* subshell, and *f* subshell.

The letters s, p, d, and f are derived historically from spectral lines called sharp, principal, diffuse, and fine.

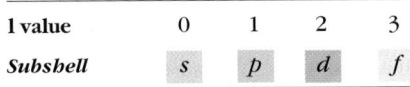

l value	0	1	2	3
Subshell	*s*	*p*	*d*	*f*

A number (the *n* value) and a letter (*s*, *p*, *d*, or *f*) are used to designate the principal energy level locations of specific subshells. For the first four principal energy levels, these designations are as follows.

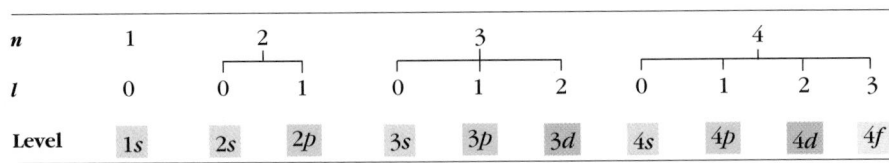

n	1	2		3			4			
l	0	0	1	0	1	2	0	1	2	3
Level	1*s*	2*s*	2*p*	3*s*	3*p*	3*d*	4*s*	4*p*	4*d*	4*f*

The energies of subshells within a given principal energy level always increase in the order *ns* < *np* < *nd* < *nf*. Consequently, a 3*p* subshell has a higher energy than a 3*s* subshell but less energy than a 3*d* subshell, which would have the highest energy in the *n* = 3 level.

Exercise 7.7 Subshell Designations

Give the subshell designation for an electron with the quantum numbers

(a) *n* = 5, *l* = 2   (b) *n* = 4, *l* = 3   (c) *n* = 6, *l* = 1

Exercise 7.8 Subshell Designations

Explain why 3*f* is an incorrect subshell designation. Likewise, why is *n* = 2, *l* = 2 incorrect for a subshell?

As mentioned earlier, a boundary surface for an orbital can be drawn within which there is a 90% probability that an electron can be found. Note in Figure 7.10 (p. 279) that as *n* increases, two things happen: (1) orbitals become larger and

(2) there is an increase in the number and types of orbitals within a given level. ***The number of orbital types within a principal energy level equals*** **n.** Thus, in the $n = 3$ level, there are three different types of orbitals — s, p, and d orbitals. *Atomic orbitals with the same* n *and* l *values are in the same subshell.*

Third Quantum Number, m_l: Atomic Orbitals

The magnetic quantum number, m_l, can have any integral value, including zero, between l and $-l$. Thus,

$$m_l = l, \ldots, +1, 0, -1, \ldots, -l$$

For an s sublevel, which has an l value of zero, m_l has only one value — zero. Therefore, an s subshell, regardless of its n value — $1s$, $2s$, and so on — contains only one orbital. For a p subshell, l equals 1, and so m_l can be $+1$, 0, or -1. This means that within each p subshell there are three *different* types of orbitals: one with $m_l = +1$, another with $m_l = 0$, and a third with $m_l = -1$. In general, for a subshell of quantum number l, there is a total of $2l + 1$ orbitals within that subshell. Orbitals within the same *subshell* have essentially the same energy.

Table 7.3 summarizes the relationships among n, l, and m_l. ***The total number of orbitals in a shell equals*** **n^2.** For example, an $n = 3$ shell has a total of nine orbitals: one $3s$ + three $3p$ + five $3d$.

The m_l value, in conjunction with the l value, is related to the shape and orientation of an orbital in space. The s orbitals are spherical; their size enlarges as n increases (Figure 7.10).

Beginning with the second principal energy level ($n = 2$), each energy level has three p orbitals. The three p orbitals in each of these levels are dumbell-shaped and oriented at right angles to each other, with maximum electron density directed along either the x-, y-, or z-axis, as shown in Figure 7.10. They are sometimes designated as p_x, p_y, and p_z orbitals. The five d orbitals in the $3d$ sublevel ($l = 2$) are also illustrated in Figure 7.10. We will discuss d orbitals further in Chapter 22 in connection with their role in the chemistry of transition metal ions.

In terms of an orbital, the quantum number

• n relates to its size.
• l relates to its shape.
• m_l relates to its orientation.

TABLE 7.3	Relationships Among n, l, and m_l for the First Four Principal Energy Levels				
n value	l value	Subshell designation	m_l values	No. of orbitals in subshell	Total number of orbitals in shell, n^2
1	0	$1s$	0	1	1
2	0	$2s$	0	1	
2	1	$2p$	1, 0, −1	3	4
3	0	$3s$	0	1	
3	1	$3p$	1, 0, −1	3	
3	2	$3d$	2, 1, 0, −1, −2	5	9
4	0	$4s$	0	1	
4	1	$4p$	1, 0, −1	3	
4	2	$4d$	2, 1, 0, −1, −2	5	
4	3	$4f$	3, 2, 1, 0, −1, −2, −3	7	16

In summary,

- *The principal energy level (shell) has a quantum number* $n = 1, 2, 3, \ldots$
- *Within each principal energy level there are subshells equal in number to* n *and designated as the* s, p, d, *and* f *subshells.*
- *Each subshell has these numbers of orbitals: one* s *orbital, three* p *orbitals, five* d *orbitals, and seven* f *orbitals.*
- *Within a principal energy level* n *there are* n^2 *orbitals.*

Problem-Solving Example 7.5   Quantum Numbers, Subshells, and Atomic Orbitals

Consider the $n = 4$ principal energy level.
(a) Without referring to Table 7.3, predict the number of subshells in this level.
(b) Identify each of the subshells and their l values.
(c) Use the $2l + 1$ rule to calculate how many orbitals each subshell has and identify the m_l value for each orbital.
(d) What is the total number of orbitals in the $n = 4$ level?

Answer   (a) Four subshells   (b) $4s$, $4p$, $4d$, and $4f$ (c) One $4s$ orbital, three $4p$ orbitals, five $4d$ orbitals, and seven $4f$ orbitals   (d) 16 orbitals

Explanation
(a) There are n subshells in the nth level. Thus, the $n = 4$ level contains four subshells.
(b) The subshells are $4s$, $4p$, $4d$, and $4f$. The number refers to the principal quantum number, n; the letter is associated with the l quantum number. The four sublevels correspond to the four possible l values: 0, 1, 2, and 3.
(c) There are a total of $2l + 1$ orbitals within a sublevel. Only one $4s$ orbital is possible ($l = 0$, so m_l must be zero). There are three $4p$ orbitals ($l = 1$) with m_l values of 1, 0, or -1. There are five $4d$ orbitals ($l = 2$) corresponding to the five allowed values for m_l: 2, 1, 0, -1, -2. There are seven $4f$ orbitals ($l = 3$), each with one of the seven permitted values of m_l: 3, 2, 1, 0, -1, -2, and -3.
(d) The total number of orbitals in a level is n^2. Therefore, the $n = 4$ level has a total of 16 orbitals.

Problem-Solving Practice 7.5

(a) Identify the subshell with $n = 6$ and $l = 2$.
(b) How many orbitals are in this subshell?
(c) What are the m_l values for these orbitals?

CD-ROM Screen 8.2: Electron Spin

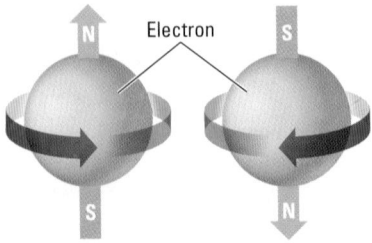

Figure 7.11   Spin directions. The electron can be pictured as though it were a charged sphere spinning about an axis through its center. The electron can have only two directions of spin; any other position is forbidden. Therefore, the spin of the electron is said to be quantized.

Fourth Quantum Number, m_s: Electron Spin

When spectroscopists more closely studied emission spectra of hydrogen and sodium atoms, they discovered that lines originally thought to be single were actually very closely spaced pairs of lines. In 1925 the Dutch physicists George Uhlenbeck and Samuel Goudsmit proposed that the line splitting could be explained by assuming that each electron in an atom can exist in one of two possible spin states. To visualize these states, consider an electron as a charged sphere rotating about an axis through its center (Figure 7.11). Such a spinning charge generates a magnetic field. This means that each electron acts like a tiny bar magnet with north and south magnetic poles. Only two directions of spin are possible in relation to the direction of an external magnetic field — clockwise or counterclockwise. Spins in opposite directions produce oppositely directed magnetic fields (Figure 7.11), which result in two slightly different energies. This slight difference in energy splits the spectral lines into closely spaced pairs.

Thus, to describe an electron in an atom completely, a fourth quantum number, m_s, called the *spin quantum number*, is needed in conjunction with the other three quantum numbers. The spin quantum number differs from the other three quantum numbers, which come from the solution to the wave equation for the hydrogen atom.

The spin quantum number can have just one of two values: $+\frac{1}{2}$ or $-\frac{1}{2}$. Electrons are said to have *parallel* spins if they have the *same* m_s quantum number (both $+\frac{1}{2}$ or both $-\frac{1}{2}$). Electrons are said to be *paired* when they are in the same orbital and have opposite spins—one has an m_s of $+\frac{1}{2}$ and the other has a $-\frac{1}{2}$ value.

To make quantum theory consistent with experiment, Wolfgang Pauli stated in 1925 what is now known as the **Pauli exclusion principle:** *no more than two electrons can be assigned to the same orbital in an atom, and these electrons must have opposite spins.* This is equivalent to saying that no two electrons in the same atom can have the same set of four quantum numbers n, l, m_l, and m_s. For example, two electrons can occupy a $3p$ orbital, but only if their spins are paired $(+\frac{1}{2}$ and $-\frac{1}{2})$. In such a case, the electrons would have the same first three quantum numbers, 3, 1, +1, but would differ in their m_s values.

CD-ROM Screen 8.4: The Pauli Exclusion Principle

Exercise 7.9   Quantum Numbers

Give the set of four quantum numbers for each electron in a $4s$ orbital in a sodium atom.

 ### Exercise 7.10   Orbitals and Quantum Numbers

Give sets of four quantum numbers for two electrons that are (a) in the same level and sublevel, but in different orbitals, (b) in the same level, but in different sublevels and in different orbitals.

Exercise 7.11   Quantum Number Comparisons

Two electrons in the same atom have these sets of quantum numbers: electron$_a$: 3, 1, 0, $+\frac{1}{2}$; electron$_b$: 3, 1, −1, $+\frac{1}{2}$. Show that these two electrons are not in the same orbital. Which subshell are these electrons in?

The restriction that only two electrons can occupy a single orbital has the effect of establishing the maximum number of electrons for each principal energy level, n, as summarized in Table 7.4. The $n = 1$ energy level has only one s orbital and therefore can accommodate only two electrons (if they are of opposite spin). The $n = 2$ energy level has one s and three p orbitals, each of which can accommodate a pair of paired electrons. Therefore, this level can accommodate a total of eight electrons (two in the $2s$ orbital and six in the three $2p$ orbitals). Since each orbital can hold a maximum of two paired electrons, the general rule is that *each principal energy level,* **n,** *can accommodate a maximum number of* $2n^2$ *electrons.*

Exercise 7.12   Maximum Number of Electrons

(a) What is the maximum number of electrons in the $n = 3$ level? Identify the orbital of each electron.

(b) What is the maximum number of electrons in the $n = 5$ level? Identify the orbital of each electron. (*Hint:* g orbitals follow f orbitals.)

 ### Exercise 7.13 g Orbitals

Using the same reasoning as was developed for s, p, d, and f orbitals, what should be the n value of the first shell that could contain g orbitals, and how many g orbitals would be in that shell?

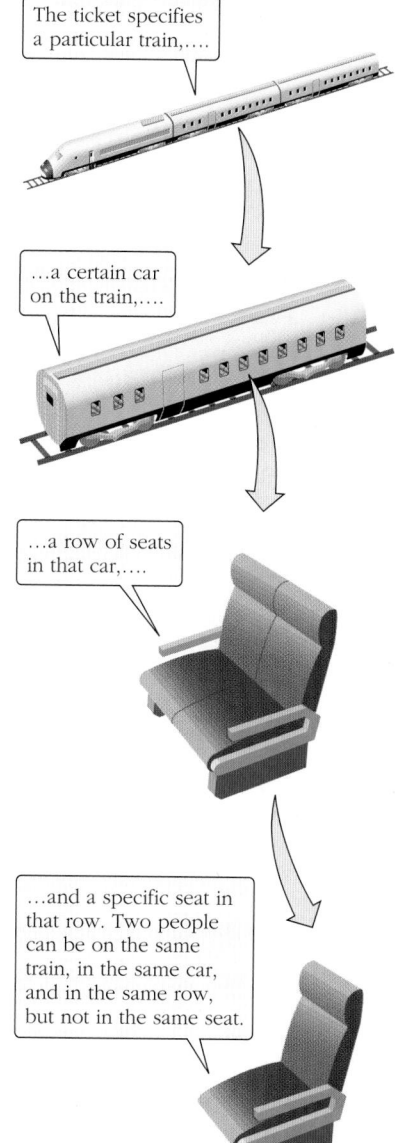

The ticket specifies a particular train,....

...a certain car on the train,....

...a row of seats in that car,....

...and a specific seat in that row. Two people can be on the same train, in the same car, and in the same row, but not in the same seat.

A reserved train ticket. A ticket for a reserved seat on a train is analogous to a set of four quantum numbers, n, l, m_l, and m_s. The ticket specifies a particular train, a certain car on the train, a row of seats in that car, and a specific seat in that row. The row is analogous to an atomic orbital, and the occupants of the row are like two electrons in the same orbital (same train, car, and row), but with opposite spins (different seats).

The results expressed in this table were predicted by the Schrödinger theory and have been confirmed by experiment.

TABLE 7.4	Number of Electrons Accommodated in Electron Shells and Subshells			
Electron shell (n)	Subshells available ($= n$)	Orbitals available	Number of electrons possible in subshell	Maximum electrons for nth shell ($= 2n^2$)
1	s	1	2	2
2	s	1	2	8
	p	3	6	
3	s	1	2	18
	p	3	6	
	d	5	10	
4	s	1	2	32
	p	3	6	
	d	5	10	
	f	7	14	
5	s	1	2	50
	p	3	6	
	d	5	10	
	f	7	14	
	g^*	9	18	
6	s	1	2	72
	p	3	6	
	d	5	10	
	f^*	7	14	
	g^*	9	18	
	h^*	11	22	
7	s	1	2	

* These orbitals are not used in the ground state of any known element.

7.6 ATOM ELECTRON CONFIGURATIONS

The complete description of the orbitals occupied by all the electrons in an atom or ion is called its **electron configuration.** Using quantum theory, it is possible to make some sense of the electron configurations of atoms of the elements. The periodic table will serve as a guide. As you will see, the similarities of elements in the same periodic table group are explained by their similar electron configurations.

Electron Configurations of Main Group Elements

The atomic numbers of the elements increase in numerical order throughout the periodic table. As a result, atoms of an element each contain one more electron (and proton) than atoms of the preceding element. How do we know which shell and orbital each new electron occupies? An important principle for answering this question is the following: *For an atom in its ground state, electrons are found in the energy shells, subshells, and orbitals that produce the lowest energy for the atom.* In other words, electrons fill orbitals starting with the 1s orbital and work upward in the subshell energy order starting with $n = 1$.

To better understand how this filling of orbitals works, consider the experimentally determined electron configurations of the first ten elements, which are

CD-ROM Screen 8.5: Atomic Subshell Energies

TABLE 7.5	Electron Configurations of the First Ten Elements						
	Electron configurations		**Orbital box diagrams**				
	Condensed	Expanded	$1s$	$2s$	$2p_x$	$2p_y$	$2p_z$
H	$1s^1$		↑				
He	$1s^2$		↑↓				
Li	$1s^2 2s^1$		↑↓	↑			
Be	$1s^2 2s^2$		↑↓	↑↓			
B	$1s^2 2s^2 2p^1$		↑↓	↑↓	↑		
C	$1s^2 2s^2 2p^2$	$1s^2 2s^2 2p_x^1 2p_y^1$	↑↓	↑↓	↑	↑	
N	$1s^2 2s^2 2p^3$	$1s^2 2s^2 2p_x^1 2p_y^1 2p_z^1$	↑↓	↑↓	↑	↑	↑
O	$1s^2 2s^2 2p^4$	$1s^2 2s^2 2p_x^2 2p_y^1 2p_z^1$	↑↓	↑↓	↑↓	↑	↑
F	$1s^2 2s^2 2p^5$	$1s^2 2s^2 2p_x^2 2p_y^2 2p_z^1$	↑↓	↑↓	↑↓	↑↓	↑
Ne	$1s^2 2s^2 2p^6$	$1s^2 2s^2 2p_x^2 2p_y^2 2p_z^2$	↑↓	↑↓	↑↓	↑↓	↑↓

CD-ROM Screen 8.7: Atomic Electron Configurations

written in three different ways in Table 7.5—condensed, expanded, and orbital box diagram. Since electrons assigned to the $n = 1$ shell are closest to the nucleus and therefore lowest in energy, electrons are assigned to it first (H and He). At the left in Table 7.5, the occupied orbitals and the number of electrons in each orbital are represented by the following notation:

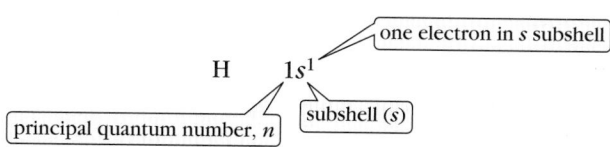

At the right in the table, each occupied orbital is represented by a box in which electrons are shown as arrows: ↑ for a single electron in an orbital and ↑↓ for paired electrons in an orbital.

In helium the two electrons are paired in the $1s$ orbital so that the lowest energy shell ($n = 1$) is filled. After the $n = 1$ shell is filled, electrons are assigned to the next lowest unoccupied energy level (Table 7.5), the $n = 2$ shell, beginning with lithium. This second shell can hold eight electrons, and its orbitals are occupied sequentially in the eight elements from lithium to neon. Notice in the periodic table inside the front cover that these are the eight elements of the second period.

As happens in each principal energy level (and each period), the first two electrons fill the s orbital. In the second period this occurs in Li ($1s^2 2s^1$—with the $2s$ orbital half-filled) and Be ($1s^2 2s^2$—with the $2s$ orbital completely filled). The next element is boron, B ($1s^2 2s^2 2p^1$), and the fifth electron goes into a $2p$ orbital. The three $2p$ orbitals are of equal energy, and it does not matter which $2p$ orbital is occupied first. Adding a second p electron for the next element, carbon ($1s^2 2s^2 2p^2$), presents a choice. Does the second $2p$ electron in the carbon atom pair with the existing electron in a p orbital, or does it occupy a p orbital by itself? It has been shown experimentally that both p electrons have the same spin. Hence, they must occupy different p orbitals, otherwise they would violate the Pauli exclusion principle. The expanded electron configurations in the middle of Table 7.5 show the locations of the p electrons individually in the boron, carbon, and nitrogen atoms' $2p$ orbitals (p_x, p_y, p_z).

Because electrons are negatively charged particles, electron configurations that produce maximum unpairing also minimize electron-electron repulsions, mak-

Figure 7.12 Partial orbital diagrams for Period 3 elements.

Atomic number/ element	Partial orbital diagram (3s and 3p sublevels only)		Electron configuration	Noble gas electron configuration
	3s	3p		
$_{11}$Na	↑	☐ ☐ ☐	$[1s^22s^22p^6]\,3s^1$	$[Ne]\,3s^1$
$_{12}$Mg	↑↓	☐ ☐ ☐	$[1s^22s^22p^6]\,3s^2$	$[Ne]\,3s^2$
$_{13}$Al	↑↓	↑ ☐ ☐	$[1s^22s^22p^6]\,3s^23p^1$	$[Ne]\,3s^23p^1$
$_{14}$Si	↑↓	↑ ↑ ☐	$[1s^22s^22p^6]\,3s^23p^2$	$[Ne]\,3s^23p^2$
$_{15}$P	↑↓	↑ ↑ ↑	$[1s^22s^22p^6]\,3s^23p^3$	$[Ne]\,3s^23p^3$
$_{16}$S	↑↓	↑↓ ↑ ↑	$[1s^22s^22p^6]\,3s^23p^4$	$[Ne]\,3s^23p^4$
$_{17}$Cl	↑↓	↑↓ ↑↓ ↑	$[1s^22s^22p^6]\,3s^23p^5$	$[Ne]\,3s^23p^5$
$_{18}$Ar	↑↓	↑↓ ↑↓ ↑↓	$[1s^22s^22p^6]\,3s^23p^6$	$[Ne]\,3s^23p^6$

ing the total energy of the set of electrons as low as possible. **Hund's rule** summarizes how subshells are filled: The most stable arrangement of electrons in the same subshell has the maximum number of unpaired electrons, all with the same spin. ***Electrons pair only after each orbital in a subshell is occupied by a single electron.*** The general result of Hund's rule is that in *p*, *d*, or *f* orbitals, each successive electron enters a different orbital of the subshell until the subshell is half-full, after which electrons pair in the orbitals one by one. We can see this in Table 7.5 for the elements that follow boron — carbon, nitrogen, oxygen, fluorine, and neon.

Suppose you need to know the electron configuration of a phosphorus atom. It can be done by using the periodic table, Table 7.5, and the general relationships among shells, subshells, and orbitals described in Section 7.5. Checking the periodic table inside the front cover, you find that phosphorus has atomic number 15 and therefore has 15 electrons. Phosphorus is in the third period. From Table 7.5 we see that the $n = 1$ and $n = 2$ shells are filled in the first two periods, giving the first ten electrons in phosphorus the configuration $1s^22s^22p^6$ (the same as neon). The next two electrons are assigned to the 3s orbital, giving $1s^22s^22p^63s^2$ so far. The final three electrons have to be assigned to the $3p$ orbitals, so the electron configuration for phosphorus is

$$P \qquad 1s^22s^22p^63s^23p^3$$

According to Hund's rule, the three electrons in the *p* orbitals of the phosphorus atom must be unpaired. To show this, you can write the *expanded* electron configuration

$$P \qquad 1s^22s^22p_x^22p_y^22p_z^23s^23p_x^13p_y^13p_z^1$$

or the orbital box diagram:

P	1s	2s	2p_x 2p_y 2p_z	3s	3p_x 3p_y 3p_z
	↑↓	↑↓	↑↓ ↑↓ ↑↓	↑↓	↑ ↑ ↑

Thus, all the electrons are paired except for the three electrons in $3p$ orbitals, and these occupy different orbitals and have parallel spins. The partial orbital box diagram of elements in Period 3 is given in Figure 7.12.

The relationship between periodic table position, shells, subshells, and electron configuration is summarized in Figure 7.13.

The arrangement of elements in the periodic table becomes more understandable once their electron configurations are understood. The electron configura-

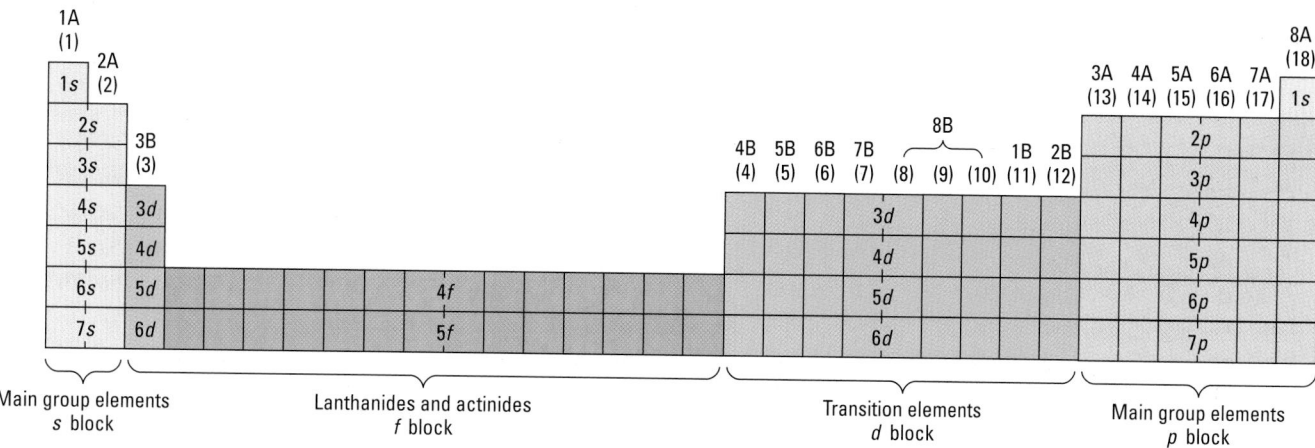

Figure 7.13 **Electron configuration and the periodic table.** Electron configurations can be generated by assigning electrons to the subshells starting at H and moving through this table in atomic number order until the desired element is reached. The electron configurations in atomic number order are given in Table 7.6.

tions of all the elements are given in Table 7.6. Notice that Table 7.6 uses an abbreviated representation of the configurations in which the symbol of the preceding noble gas represents filled subshells. This is called the **noble gas notation.** For Ca, atomic number 20 ($1s^22s^22p^63s^23p^64s^2$), the noble gas notation is $[Ar]4s^2$, where the symbol [Ar] represents the filled subshells in argon, $1s^22s^22p^63s^23p^6$.

At this point, you should be able to write electron configurations and orbital box diagrams for main group elements through Ca, atomic number 20, using the periodic table as a guide. Check your electron configurations against the ones given in Table 7.6.

Using the noble gas notation, the electron configuration of phosphorus is $[Ne]3s^23p^3$.

Problem-Solving Example 7.6 Electron Configurations

Using only the periodic table as a guide, give the complete electron configuration, orbital box diagram, and noble gas notation for silicon, Si.

	1s	2s	2p			3s	3p		
Answer $1s^22s^22p^63s^23p^2$ ↑↓ ↑↓ ↑↓ ↑↓ ↑↓ ↑↓ ↑ ↑ ↑ $[Ne]3s^23p^2$

Explanation The periodic table shows silicon in the third period with atomic number 14. Therefore, silicon has 14 electrons. The first ten are represented by $1s^22s^22p^6$, the electron configuration of neon, Ne. The last four electrons in silicon have the configuration $3s^23p^2$. The orbital box diagram follows from the total electron configuration.

Problem-Solving Practice 7.6

(a) Write the electron configuration of sulfur in the noble gas notation.
(b) Determine how many unpaired electrons a sulfur atom has by drawing the orbital box diagram.

Exercise 7.14 Highest Energy Electrons in Ground State Atoms

Give the electron configuration of electrons in the highest occupied energy level (highest n) in a ground state chlorine atom. Do the same for a sulfur atom.

Valence Electrons

The total number of electrons in each of its atoms is not what determines the chemical reactivity of an element. If such were the case, chemical reactivity would

 CD-ROM Screen 9.2: Valence Electrons

TABLE 7.6 Electron Configurations of Atoms in the Ground State

Z	Element	Configuration	Z	Element	Configuration	Z	Element	Configuration
1	H	$1s^1$	38	Sr	$[Kr]5s^2$	75	Re	$[Xe]4f^{14}5d^56s^2$
2	He	$1s^2$	39	Y	$[Kr]4d^15s^2$	76	Os	$[Xe]4f^{14}5d^66s^2$
3	Li	$[He]2s^1$	40	Zr	$[Kr]4d^25s^2$	77	Ir	$[Xe]4f^{14}5d^76s^2$
4	Be	$[He]2s^2$	41	Nb	$[Kr]4d^45s^1$	78	Pt	$[Xe]4f^{14}5d^96s^1$
5	B	$[He]2s^22p^1$	42	Mo	$[Kr]4d^55s^1$	79	Au	$[Xe]4f^{14}5d^{10}6s^1$
6	C	$[He]2s^22p^2$	43	Tc	$[Kr]4d^65s^2$	80	Hg	$[Xe]4f^{14}5d^{10}6s^2$
7	N	$[He]2s^22p^3$	44	Ru	$[Kr]4d^75s^1$	81	Tl	$[Xe]4f^{14}5d^{10}6s^26p^1$
8	O	$[He]2s^22p^4$	45	Rh	$[Kr]4d^85s^1$	82	Pb	$[Xe]4f^{14}5d^{10}6s^26p^2$
9	F	$[He]2s^22p^5$	46	Pd	$[Kr]4d^{10}$	83	Bi	$[Xe]4f^{14}5d^{10}6s^26p^3$
10	Ne	$[He]2s^22p^6$	47	Ag	$[Kr]4d^{10}5s^1$	84	Po	$[Xe]4f^{14}5d^{10}6s^26p^4$
11	Na	$[Ne]3s^1$	48	Cd	$[Kr]4d^{10}5s^2$	85	At	$[Xe]4f^{14}5d^{10}6s^26p^5$
12	Mg	$[Ne]3s^2$	49	In	$[Kr]4d^{10}5s^25p^1$	86	Rn	$[Xe]4f^{14}5d^{10}6s^26p^6$
13	Al	$[Ne]3s^23p^1$	50	Sn	$[Kr]4d^{10}5s^25p^2$	87	Fr	$[Rn]7s^1$
14	Si	$[Ne]3s^23p^2$	51	Sb	$[Kr]4d^{10}5s^25p^3$	88	Ra	$[Rn]7s^2$
15	P	$[Ne]3s^23p^3$	52	Te	$[Kr]4d^{10}5s^25p^4$	89	Ac	$[Rn]6d^17s^2$
16	S	$[Ne]3s^23p^4$	53	I	$[Kr]4d^{10}5s^25p^5$	90	Th	$[Rn]6d^27s^2$
17	Cl	$[Ne]3s^23p^5$	54	Xe	$[Kr]4d^{10}5s^25p^6$	91	Pa	$[Rn]5f^26d^17s^2$
18	Ar	$[Ne]3s^23p^6$	55	Cs	$[Xe]6s^1$	92	U	$[Rn]5f^36d^17s^2$
19	K	$[Ar]4s^1$	56	Ba	$[Xe]6s^2$	93	Np	$[Rn]5f^46d^17s^2$
20	Ca	$[Ar]4s^2$	57	La	$[Xe]5d^16s^2$	94	Pu	$[Rn]5f^67s^2$
21	Sc	$[Ar]3d^14s^2$	58	Ce	$[Xe]4f^15d^16s^2$	95	Am	$[Rn]5f^77s^2$
22	Ti	$[Ar]3d^24s^2$	59	Pr	$[Xe]4f^36s^2$	96	Cm	$[Rn]5f^76d^17s^2$
23	V	$[Ar]3d^34s^2$	60	Nd	$[Xe]4f^46s^2$	97	Bk	$[Rn]5f^97s^2$
24	Cr	$[Ar]3d^54s^1$	61	Pm	$[Xe]4f^56s^2$	98	Cf	$[Rn]5f^{10}7s^2$
25	Mn	$[Ar]3d^54s^2$	62	Sm	$[Xe]4f^66s^2$	99	Es	$[Rn]5f^{11}7s^2$
26	Fe	$[Ar]3d^64s^2$	63	Eu	$[Xe]4f^76s^2$	100	Fm	$[Rn]5f^{12}7s^2$
27	Co	$[Ar]3d^74s^2$	64	Gd	$[Xe]4f^75d^16s^2$	101	Md	$[Rn]5f^{13}7s^2$
28	Ni	$[Ar]3d^84s^2$	65	Tb	$[Xe]4f^96s^2$	102	No	$[Rn]5f^{14}7s^2$
29	Cu	$[Ar]3d^{10}4s^1$	66	Dy	$[Xe]4f^{10}6s^2$	103	Lr	$[Rn]5f^{14}6d^17s^2$
30	Zn	$[Ar]3d^{10}4s^2$	67	Ho	$[Xe]4f^{11}6s^2$	104	Rf	$[Rn]5f^{14}6d^27s^2$
31	Ga	$[Ar]3d^{10}4s^24p^1$	68	Er	$[Xe]4f^{12}6s^2$	105	Db	$[Rn]5f^{14}6d^37s^2$
32	Ge	$[Ar]3d^{10}4s^24p^2$	69	Tm	$[Xe]4f^{13}6s^2$	106	Sg	$[Rn]5f^{14}6d^47s^2$
33	As	$[Ar]3d^{10}4s^24p^3$	70	Yb	$[Xe]4f^{14}6s^2$	107	Bh	$[Rn]5f^{14}6d^57s^2$
34	Se	$[Ar]3d^{10}4s^24p^4$	71	Lu	$[Xe]4f^{14}5d^16s^2$	108	Hs	$[Rn]5f^{14}6d^67s^2$
35	Br	$[Ar]3d^{10}4s^24p^5$	72	Hf	$[Xe]4f^{14}5d^26s^2$	109	Mt	$[Rn]5f^{14}6d^77s^2$
36	Kr	$[Ar]3d^{10}4s^24p^6$	73	Ta	$[Xe]4f^{14}5d^36s^2$	110	—	$[Rn]5f^{14}6d^87s^2$
37	Rb	$[Kr]5s^1$	74	W	$[Xe]4f^{14}5d^46s^2$	111	—	$[Rn]5f^{14}6d^97s^2$
						112	—	$[Rn]5f^{14}6d^{10}7s^2$

increase sequentially with increasing atomic number. Rather, chemically similar behavior occurs among elements within a group in the periodic table, and differs among groups. As early as 1902, Gilbert N. Lewis (1875–1946) hit upon the idea that electrons in atoms might be arranged in shells, starting close to the nucleus and building outward. Lewis explained the similarity of chemical properties for elements in a given group by assuming that all elements in that group have the same number of electrons in their outer shell. These electrons are known as **valence electrons** for the main group elements. The electrons in the filled inner shells of these elements are the **core electrons.** For elements in the fourth and higher periods of Groups 3A to 7A, electrons in the d subshell are also core electrons. Using this notation the core electrons of As are represented by $[Ar]3d^{10}$. A few examples of core and valence electrons are the following:

Element	Core electrons	Total electron configuration	Valence electrons	Periodic group
Na	$1s^2 2s^2 2p^6$ ([Ne])	$[Ne]3s^1$	$3s^1$	1A
Si	$1s^2 2s^2 2p^6$ ([Ne])	$[Ne]3s^2 3p^2$	$3s^2 3p^2$	4A
As	$1s^2 2s^2 2p^6 3s^2 3p^6 3d^{10}$ ([Ar])	$[Ar]3d^{10}4s^2 4p^3$	$4s^2 4p^3$	5A

While teaching his students about atomic structure, G. N. Lewis used the element's symbol to represent the atomic nucleus together with the core electrons. He introduced the practice of representing the valence electrons as dots. The dots are placed around the symbol one at a time until they are used up or until all four sides are occupied; any remaining electron dots are paired with ones already there. The result is a **Lewis dot symbol.** Table 7.7 shows Lewis dot symbols for the atoms of the elements in Periods 2 and 3.

Main group elements in Groups 1A and 2A are known as *s*-**block elements,** and their valence electrons are s electrons (ns^1 for Group 1A, ns^2 for Group 2A). Elements in the main groups at the right in the periodic table, Groups 3A through 8A, are known as *p*-**block elements.** Their valence electrons include outermost s and p electrons. Notice in Table 7.7 how the Lewis dot symbols show that *in each A group the number of valence electrons is equal to the group number.*

 CD-ROM Screen 9.4: Lewis Electron Dot Structures

Problem-Solving Example 7.7   Valence Electrons

(a) Using the noble gas notation, write the electron configuration for iodine. Identify its core and valence electrons.

(b) Write the Lewis dot symbol for iodine.

Answer

(a) $[Kr]3d^{10}4s^2 4p^5$ Core electrons are $[Kr]3d^{10}$; $4s^2 4p^5$ are valence electrons. (b)

Explanation   Iodine is a Group 7A element, element 53. Its first 36 electrons are represented by [Kr]. These, along with the ten $3d$ electrons, make up the core electrons. Because it is in Group 7A, iodine has seven valence electrons, as given by $4s^2 4p^5$ and shown in the Lewis dot symbol.

Problem-Solving Practice 7.7

Use the noble gas notation to write electron configurations and Lewis dot symbols for Se and Te. What do these configurations illustrate about elements in the same main group?

TABLE 7.7   Lewis Dot Symbols for Atoms

1A ns^1	2A ns^2	3A $ns^2 np^1$	4A $ns^2 np^2$	5A $ns^2 np^3$	6A $ns^2 np^4$	7A $ns^2 np^5$	8A $ns^2 np^6$
Li	·Be·	·B·	·C·	·N·	:O·	:F·	:Ne:
Na·	·Mg·	·Al·	·Si·	·P·	:S·	:Cl·	:Ar:

CD-ROM Screen 8.7: Atomic Electron Configurations

Electron Configurations of Transition Elements

The **transition elements** are those in the B Groups in Periods 4 through 7 in the middle of the periodic table. These are metallic elements in which a d subshell is being filled as shown in Figure 7.13.

In each period in which they occur, the transition elements are immediately preceded by two s-block elements. As shown in Figure 7.13, once the $4s$ subshell is filled, the subshell with the next higher energy is $3d$ (not $4p$, as you might expect). *In general, (n − 1)d orbitals are filled after ns orbitals and before filling of np orbitals begins.*

The use of d orbitals begins with the first transition metal, scandium, which has the configuration $[Ar]3d^14s^2$. After scandium comes titanium with $[Ar]3d^24s^2$ and vanadium with $[Ar]3d^34s^2$. We would expect the configuration of the next element, chromium, to be $[Ar]3d^44s^2$, but that turns out to be incorrect. The correct configuration is $[Ar]3d^54s^1$, based on spectroscopic and magnetic measurements. This illustrates one of several anomalies in predicting electron configurations for transition and inner transition elements. When half-filled d subshells are possible, they are sometimes favored (an illustration of Hund's rule). As a result, elements that are close to the middle and the end of filling a d subshell have a stable configuration that leaves the s orbital half-filled and the d orbitals half-filled or filled. This kind of electron configuration is seen in copper, the next-to-last element in the first transition series, which has the electron configuration $[Ar]3d^{10}4s^1$ instead of the expected $[Ar]3d^94s^2$.

This also occurs with silver and gold, elements that are in the same group as copper.

The number of unpaired electrons for atoms of most transition elements can be predicted by placing valence electrons in orbital diagrams according to Hund's rule. For example, the electron configuration of Co ($Z = 27$) is $[Ar]3d^74s^2$, and the number of unpaired electrons is three, as seen from the following orbital box diagram.

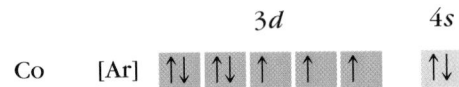

Exercise 7.15 Unpaired Electrons

Use orbital box diagrams to determine which chromium ground state configuration has the greater number of unpaired electrons: $[Ar]3d^44s^2$ or $[Ar]3d^54s^1$.

Elements with Valence Electrons in *f* and *p* orbitals

In the elements of the sixth and seventh periods, f subshell orbitals exist and can be filled (Figure 7.13). The elements (all metals) for which f subshells are filling are sometimes called the *inner transition* elements or, more usually, *lanthanides* (for lanthanum, the element just before those filling the $4f$ subshell) and *actinides* (for actinium, the element just before those filling the $5f$ subshell). The lanthanides start with lanthanum (La), which has the electron configuration $[Xe]5d^16s^2$. The next element, cerium (Ce), begins a separate row at the bottom of the periodic table, and it is with these elements that f orbitals are filled (Table 7.6). The electron configuration of Ce is $[Xe]4f^15d^16s^2$. Each of the lanthanide elements, from Ce to Lu, continues to add $4f$ electrons until the seven $4f$ orbitals are filled by 14 electrons in lutetium (Lu, $[Xe]4f^{14}5d^16s^2$). Note that both the $n = 5$ and $n = 6$ levels are partially filled before the $4f$ orbital starts to be occupied.

It is hard to overemphasize how useful the periodic table is as a guide to electron configurations. As another example, using Figure 7.13 as a guide we can find the configuration of Te, which is in Group 6A. Because tellurium is in Group 6A, by now you should immediately recognize that it has *six* valence electrons with a configuration of ns^2np^4. And, because Te is in the fifth period, $n = 5$. Thus, the complete electron configuration is given by starting with the electron configuration of krypton [$1s^22s^22p^63s^23p^63d^{10}4s^24p^6$], the noble gas at the end of the fourth period. Then add the filled $4d^{10}$ subshell and the six valence electrons of Te ($5s^25p^4$) to give $1s^22s^22p^63s^23p^63d^{10}4s^24p^64d^{10}5s^25p^4$, or [Kr]$4d^{10}5s^25p^4$. To predict the number of unpaired electrons, look at the outermost subshell's electron configuration, because the inner shells (represented by [Kr] in this case) are completely filled with paired electrons. For Te, [Kr]$4d^{10}5s^25p_x^25p_y^15p_z^1$ indicates two p orbitals with unpaired electrons for a total of two unpaired electrons.

7.7 ION ELECTRON CONFIGURATIONS

In studying ionic compounds (⬅ *p. 86*) you learned that atoms from Groups 1A through 3A form positive ions (cations) with charges equal to their group numbers — for example, Li^+, Mg^{2+}, and Al^{3+}. Nonmetals in Groups 5A through 7A that form ions do so by adding electrons to form negative ions (anions) with *charges equal to eight minus the A group number.* Examples of such anions are N^{3-}, O^{2-}, and F^-. Here's the explanation. When atoms from *s-* and *p*-block elements form ions, electrons are removed or added so that a noble gas configuration is achieved. Atoms from Groups 1A, 2A, and 3A *lose* 1, 2, or 3 valence electrons to form 1+, 2+, or 3+ ions, respectively, and have the electron configurations of the *preceding* noble gas. Atoms from Groups 7A, 6A, and some in 5A *gain* 1, 2, or 3 valence electrons to form 1−, 2−, or 3− ions, respectively, with the electron configurations of the *next* noble gas. **Metal atoms lose electrons to form cations with a charge equal to the group number; nonmetals gain electrons to form anions with a charge equal to the A group number minus eight.**

 CD-ROM Screen 8.8: Electron Configurations in Ions

This relationship holds for Groups 1A to 3A and 5A to 7A.

Problem-Solving Example 7.8 **Atoms and Their Ions**

Complete the following table.

Neutral atom	Neutral atom electron configuration	Ion	Ion electron configuration
Se	_____	_____	[Kr]
Ca	_____	Ca^{2+}	_____
Br	_____	Br^-	_____
_____	[Ar]$4s^1$	K^+	_____
_____	[Xe]$6s^2$	_____	[Xe]

Answer

Neutral atom	Neutral atom electron configuration	Ion	Ion electron configuration
Se	[Ar]$3d^{10}4s^24p^4$	Se^{2-}	[Kr]
Ca	[Ar]$4s^2$	Ca^{2+}	[Ar]
Br	[Ar]$3d^{10}4s^24p^5$	Br^-	[Kr]
K	[Ar]$4s^1$	K^+	[Ar]
Ba	[Xe]$6s^2$	Ba^{2+}	[Xe]

Explanation A neutral Se atom has the electron configuration $[Ar]3d^{10}4s^24p^4$; Se is in Group 6A, so it gains two electrons in the $4p$ sublevel to form Se^{2-} and achieve the noble gas configuration of krypton, Kr (36 electrons). Calcium is a Group 2A element and loses the two $4s$ electrons to acquire the electron configuration of argon (18 electrons), the preceding noble gas. Therefore, Ca is $[Ar]4s^2$ and Ca^{2+} is $[Ar]$. Bromine is in Group 7A and will gain one electron to form Br^-, which has the electron configuration of krypton, the next noble gas. An electron configuration of $[Ar]4s^1$ indicates 19 electrons in the neutral atom, which is a potassium atom, K. A neutral K atom loses the $4s$ electron to form a K^+ ion, which has an $[Ar]$ configuration. The $[Xe]6s^2$ configuration is for an element with 56 electrons, which is a barium atom, Ba. By losing the two $6s$ electrons, a neutral barium atom becomes a Ba^{2+} ion with the $[Xe]$ configuration.

Problem-Solving Practice 7.8

(a) What Period 3 anion with a 3− charge has the [Ar] electron configuration?
(b) What Period 4 2+ cation has the electron configuration of argon?

Noble
5A (15)	6A (16)	7A (17)	gas (18)	1A (1)	2A (2)	3A (3)
		H⁻	He	Li⁺	Be²⁺	
N³⁻	O²⁻	F⁻	Ne	Na⁺	Mg²⁺	Al³⁺
	S²⁻	Cl⁻	Ar	K⁺	Ca²⁺	Sc³⁺
	Se²⁻	Br⁻	Kr	Rb⁺	Sr²⁺	Y³⁺
	Te²⁻	I⁻	Xe	Cs⁺	Ba²⁺	La³⁺

Cations, anions, and atoms with ground state noble gas configurations.
Atoms and ions shown in the same color are isoelectronic, that is, they have the same electron configuration.

Ions with identical electron configurations are said to be **isoelectronic.** Table 7.8 lists some isoelectronic ions and the noble gas that has the same electron configuration. The table emphasizes that *metal ions are isoelectronic with the preceding noble gas, while nonmetal ions have the electron configuration of the next noble gas.*

Transition Metal Ions

The closeness in energy of the $4s$ and $3d$ subshells was mentioned earlier in connection with the electron configurations of transition metal atoms. The $5s$ and $4d$ subshells are also close to each other in energy, as are the $6s$, $4f$, and $5d$ subshells and the $7s$, $5f$, and $6d$ subshells. The ns and $(n-1)d$ subshells are so close in energy that once d electrons are added, the $(n-1)d$ subshell becomes slightly lower in energy than the ns subshell. As a result, *the ns electrons are at higher energy and are always removed before (n − 1)d electrons when transition metals form cations.*

For example, an Fe atom, $[Ar]3d^64s^2$, will lose its two $4s$ electrons to form an Fe^{2+} ion,

$$Fe \longrightarrow Fe^{2+} + 2\,e^-$$
$$[Ar]3d^64s^2 \qquad\quad [Ar]3d^6$$

and then Fe^{3+} is formed from Fe^{2+} by loss of a $3d$ electron.

$$Fe^{2+} \longrightarrow Fe^{3+} + e^-$$
$$[Ar]3d^6 \qquad\quad [Ar]3d^5$$

TABLE 7.8 Noble Gas Atoms and Their Isoelectronic Ions

(All $1s^2$)	He, Li^+, Be^{2+}, H^-
(All $1s^22s^22p^6$)	Ne, Na^+, Mg^{2+}, Al^{3+}, F^-, O^{2-}
(All $1s^22s^22p^63s^23p^6$)	Ar, K^+, Ca^{2+}, Ga^{3+}, Cl^-, S^{2-}
(All $1s^22s^22p^63s^23p^63d^{10}4s^24p^6$)	Kr, Rb^+, Sr^{2+}, Br^-, Se^{2-}
(All $1s^22s^22p^63s^23p^63d^{10}4s^24p^64d^{10}5s^25p^6$)	Xe, Cs^+, Ba^{2+}, I^-, Te^{2-}

There are five unpaired electrons in the Fe^{3+} ion, compared with only four unpaired electrons for Fe^{2+}, as shown by using orbital box diagrams for the d electrons:

Atoms or ions of inner transition elements can have as many as seven unpaired electrons in the f subshell, as occurs in Eu^{2+} and Gd^{3+} ions.

Problem-Solving Example 7.9   **Electron Configurations for Transition Elements and Ions**

(a) Write the electron configuration for the Co atom using the noble gas notation. Then draw the orbital box diagram for the electrons beyond the preceding noble gas configuration.

(b) Cobalt commonly exists as 2+ and 3+ ions. How does the orbital box diagram given in part (a) have to be changed to represent the outer electrons of Co^{2+} and Co^{3+}?

(c) How many unpaired electrons do Co, Co^{2+}, and Co^{3+} each have?

Answer

(a) $[Ar]3d^74s^2$; [Ar]   ↑↓ ↑↓ ↑ ↑ ↑   ↑↓ (3d 4s)

(b) For Co^{2+}, remove the two $4s$ electrons from Co to give

Co^{2+} [Ar] ↑↓ ↑↓ ↑ ↑ ↑ (3d)

For Co^{3+}, remove the two $4s$ electrons and one of the paired $3d$ electrons from Co to give

Co^{2+} [Ar] ↑↓ ↑ ↑ ↑ ↑ (3d)

(c) Co has three unpaired electrons; Co^{2+} has three unpaired electrons; Co^{3+} has four unpaired electrons

Explanation

(a) Co, with an atomic number of 27, is in the fourth period. It has nine more electrons than Ar, with two of the nine in the $4s$ subshell and seven in the $3d$ subshell, so its electron configuration is $[Ar]3d^74s^2$. For the orbital box diagram, all d orbitals get one electron before pairing occurs (Hund's rule).

(b) To form Co^{2+}, two electrons are removed from the $4s$ subshell. To form Co^{3+}, one of the paired electrons is removed from a $3d$ orbital.

(c) Looking at the orbital box diagrams, you should see that the numbers of unpaired electrons are three for Co, three for Co^{2+}, and four for Co^{3+}.

Problem-Solving Practice 7.9

(a) Write the electron configuration of a nickel atom, using the noble gas configuration.

(b) Draw an orbital box diagram for the outer electrons. How many unpaired electrons does nickel have?

(c) Draw an orbital box diagram for the Ni^{2+} ion. How many unpaired electrons does Ni^{2+} have?

Exercise 7.16 Cu^+ ion

Use the electron configuration of a neutral ground state copper atom to explain why copper readily forms the Cu^+ ion.

CD-ROM Screen 8.3: Spinning Electrons and Magnetism

Paramagnetism: The atoms or ions with magnetic moments are not aligned. If the substance is in a magnetic field, they align with and against the field. Magnetism is weak.

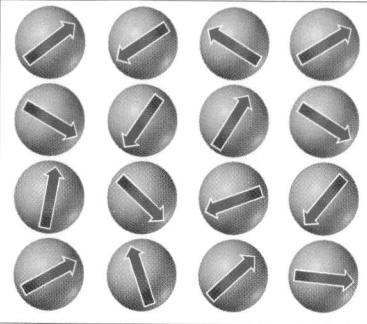

Ferromagnetism: The spins of unpaired electrons in clusters of atoms or ions are aligned in the same direction. In a magnetic field these domains all align and stay aligned when the field is removed.

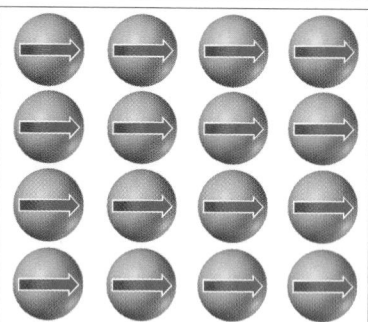

Figure 7.14 Types of magnetic behavior.

CD-ROM Screen 8.10: Atomic Size

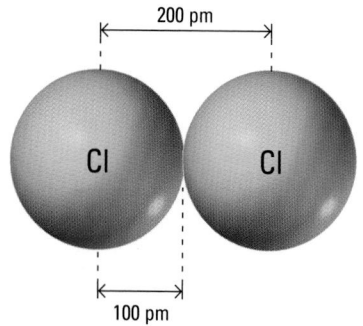

Atomic radius of chlorine. The atomic radius is taken to be one half of the internuclear distance in the Cl_2 molecule.

Paramagnetism and Unpaired Electrons

The magnetic properties of a spinning electron were described in Section 7.5. In atoms and ions that have filled shells, all the electrons are paired (opposite spins) and their magnetic fields effectively cancel each other. Such substances are called **diamagnetic.** Atoms or ions with unpaired electrons are attracted to a magnetic field; the more unpaired electrons, the greater is the attraction. Such substances are called **paramagnetic.** For example, the greater paramagnetism of chromium with six unpaired electrons rather than four is experimental evidence for the electron configuration being $[Ar]3d^54s^1$, rather than $[Ar]3d^44s^2$.

Ferromagnetic substances are permanent magnets; they retain their magnetism. The magnetic effect in ferromagnetic materials is much larger than for paramagnetic materials. Ferromagnetism occurs when the spins of unpaired electrons in a cluster of atoms (called a domain) in the solid are aligned in the same direction (Figure 7.14). Only the metals of the iron, cobalt, and nickel subgroups in the periodic table exhibit this property. They are also unique in that, once the domains are aligned in a magnetic field, the metal is permanently magnetized. In such a case, the magnetism can be eliminated only by heating or shaping the metal to rearrange the electron spin domains. Many alloys, such as alnico (an alloy of aluminum, nickel, and cobalt) exhibit greater ferromagnetism than do the pure metals themselves. Some metal oxides, such as CrO_2 and Fe_3O_4, are also ferromagnetic and are used in magnetic recording tape. Computer discs use ferromagnetic materials to store data in a binary code of zeroes and ones. Very small magnetic domains store bits of binary code as a magnetized region that represents a one and an unmagnetized region that represents a zero.

Exercise 7.17 Unpaired Electrons

The acetylacetonate ion, $acac^-$, which has no unpaired electrons, forms compounds with Fe^{2+} and with Fe^{3+} ions. Their formulas are $Fe(acac)_2$ and $Fe(acac)_3$, respectively. Which one will have the greater attraction to a magnetic field? Explain.

7.8 PERIODIC TRENDS: ATOMIC RADII

Using knowledge of electron configurations, we can now begin to answer fundamental questions of why atoms of different elements fit in as they do in the periodic table, as well as what trends in the table we observe among elements.

For atoms that form simple diatomic molecules, such as Cl_2, the **atomic radius** can be defined experimentally by finding the distance between the centers of the two atoms in the molecule. Assuming the atoms to be spherical, one half of this distance is a good estimate of the atom's radius. In the Cl_2 molecule, the atom-to-atom distance (the distance from the center of one atom to the center of the other) is 200 pm. Dividing by 2 shows that the Cl radius is 100 pm. Similarly, the C—C distance in diamond is 154 pm, and so the radius of the carbon atom is 77 pm. To test these estimates, we can add them together to estimate the distance between Cl and C in CCl_4. The estimated distance of 177 pm (100 pm + 77 pm) is in good agreement with the experimentally measured C—Cl distance of 176 pm.

This approach can then be extended to other atomic radii. The radii of O, C, and S atoms can be estimated by measuring the O—H, C—Cl, and H—S distances in H_2O, CCl_4, and H_2S and then subtracting the H and Cl radii found from H_2 and Cl_2. By this and other techniques, a reasonable set of atomic radii for main group elements has been assembled (Figure 7.15).

For the main group elements, atomic radii increase going down a group in the periodic table and decrease going across a period (Figure 7.15). These trends reflect three important effects:

- From the top to the bottom of a group in the periodic table, the atomic radii increase because electrons occupy orbitals that are successively farther from the nucleus as the value of n, the principal quantum number, increases.
- The atomic radius decreases from left to right across a period. The n value of the outermost orbitals stays the same, so that we might expect the radii of the occupied orbitals to remain approximately constant. However, in crossing a period, as each successive electron is added the nuclear charge also increases by the addition of one proton. The result is an increased attraction between the nucleus and electrons that is somewhat stronger than the increasing repulsion between electrons, causing atomic radii to decrease (Figure 7.15).
- There is a large increase in atomic radius going from any noble gas atom to the following Group 1A atom, where the outermost electron is assigned to the next higher energy level (the next larger shell). For example, compare the atomic radii of Ne (71 pm) and Na (186 pm) or Ar (98 pm) and K (227 pm).

The main group elements are those in the A groups in the periodic table (see inside front cover).

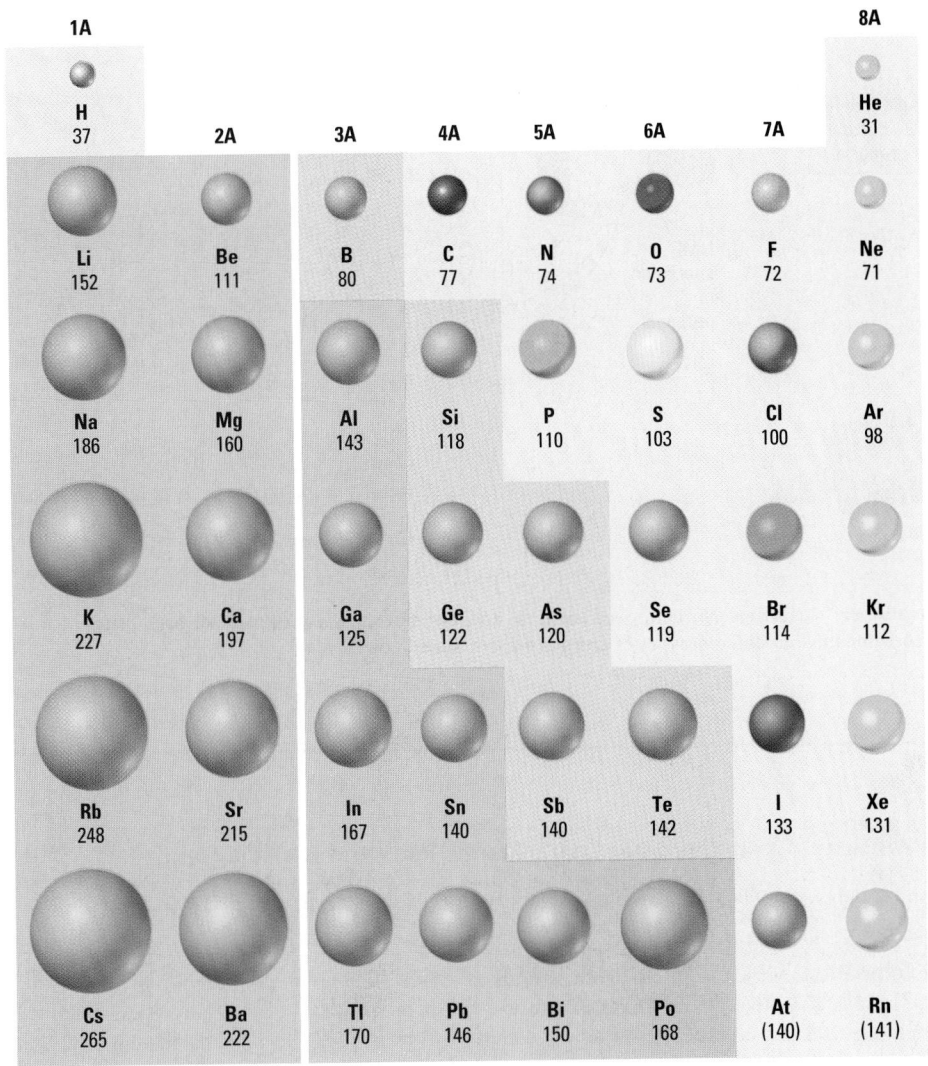

Figure 7.15 **Atomic radii of the main group elements** (in picometers; 1 pm = 10^{-12} m).

Tools of Chemistry — Nuclear Magnetic Resonance and Its Applications

In Section 7.5, the magnetic field created by the spinning electron was mentioned in connection with the discussion of the quantized spin of electrons. The nuclei of certain isotopes have the same property. For example, a ^1H nucleus (the hydrogen nucleus or proton) can spin in either of two directions. In the absence of a magnetic field, the two spin states have the same energy. When a strong external magnetic field is applied, those ^1H nuclei spinning in such a way as to be aligned with the external magnetic field are slightly lower in energy than those that are aligned against the magnetic field. At first it might appear that this energy difference, ΔE, between the two nuclear spin states would be the same for all of the hydrogen atoms in a molecule. This is not the case. Neighboring atoms of the hydrogen atoms in a molecule cause the hydrogens to see effective magnetic fields that are slightly different from the external magnetic field. However, the value of ΔE is small enough that radiation in the radiowave region of the electromagnetic spectrum (← Figure 7.1, p. 265) can change the direction of spin.

Picture the aligned hydrogen nuclei absorbing a particular radio frequency that changes their spin to the less stable direction. When the nuclei return to the more stable spin direction, the same radio frequency is emitted and can be measured with a radio receiver. This phenomenon is known as **nuclear magnetic resonance (NMR),** and it provides an extremely valuable tool for studying molecular structure.

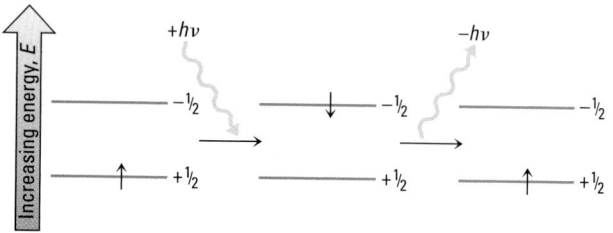

Absorption of radio frequency energy changes the direction of spin of a proton. Emission of the energy returns the proton to its original spin.

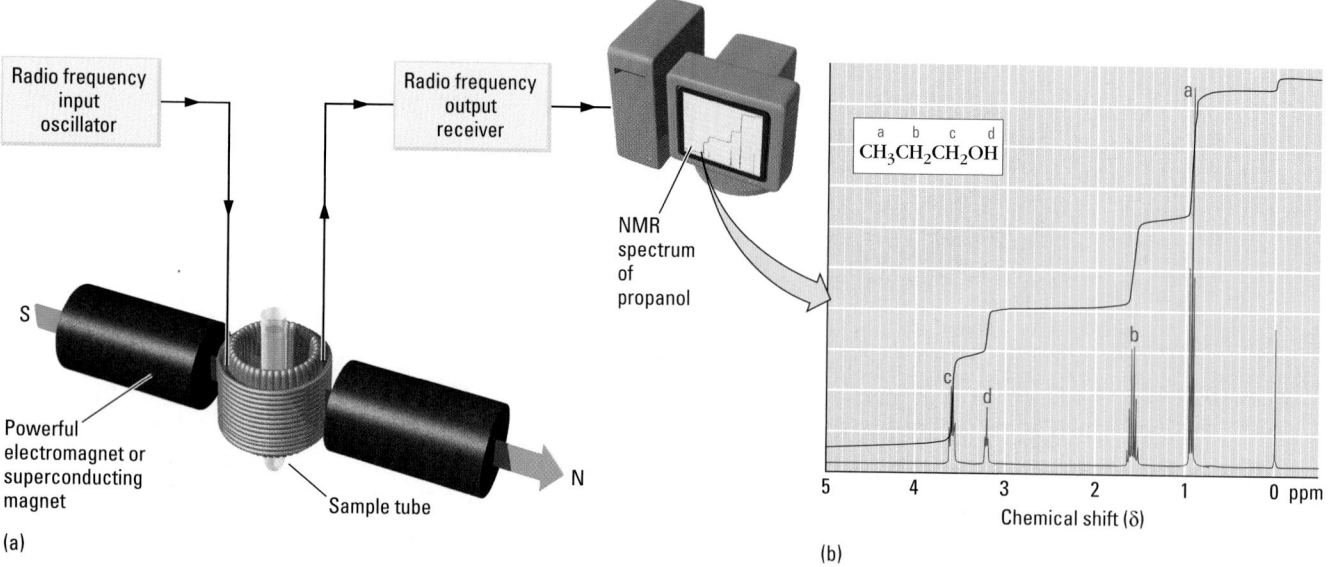

Schematic diagram of a modern NMR spectrometer. (a) A modern NMR spectrometer is a highly automated, computer-controlled instrument. (b) The spectrum pictured here is of propanol, showing the four different types of protons present in the molecule.

Problem-Solving Example 7.10 Atomic Radii and Periodic Trends

Using only a periodic table, list these atoms in order of decreasing size: Br, Cl, Ge, K, and Si.

Answer K > Ge > Br > Si > Cl

Explanation Based on the periodic trends in atomic size, we expect the radius of Cl to be smaller than that of Br, which is below it in Group 7A; likewise, Si should be smaller than Ge. Because size decreases across a period, we expect Cl to be smaller than Si. Of

NMR is used extensively by chemists because the radio frequency absorbed and then emitted depends on the chemical environment of the atoms with nuclear spin states in the sample. The study of hydrogen atoms with NMR (known as *proton* nuclear magnetic resonance, ^{1}H NMR) has quickly developed into an indispensable structural and analytical tool, particularly for organic chemists. Plots of the intensity of energy absorption versus the magnetic field strength are called *NMR spectra;* from them chemists deduce the kinds of atoms bonded to hydrogen as well as the number of hydrogen atoms present in a molecule.

Because hydrogen is the most common element in the body and H atoms give a strong NMR signal, ^{1}H is the most logical candidate for the application of NMR to medical imaging. The first use of NMR for this purpose was reported in 1973. The delay between the discovery of NMR and its application in medicine was related to the technical difficulties associated with getting a uniform magnetic field with a diameter large enough to enclose a patient. In addition, advances in computer technology for analysis of data and construction of an image from these data were needed.

NMR imaging in medicine is based on the time it takes for the protons (hydrogen nuclei) in the unstable high-energy nuclear spin position to "relax" or return to the low-energy nuclear spin position. These *relaxation times* are different for protons in fat (p. 110), muscle (see *Proteins,* Section 12.9), blood, and bone because of differences in the chemical environments. These differences in relaxation times are enhanced by the computer to produce a magnetic resonance image.

The use of NMR in medical imagining is now referred to as magnetic resonance imaging (MRI). The name "nuclear" was dropped because it frightened some people due to its association with weapons and radioactivity. In magnetic resonance imaging, the patient is placed in the opening of a large magnet. The magnetic field aligns the magnetic spin of the protons (as well as those of other magnetic nuclei). A radio frequency transmitter coil is placed in position near the region of the body to be examined. The radio frequency energy absorbed by the spinning hydrogen nuclei causes the aligned nuclei to flip to the less stable, high-energy spin direction. When the nuclei flip back to the more stable spin state, photons of radio frequency electromagnetic radiation are emitted. The intensity of the emitted signal is related to the density of hydrogen nuclei in the region being examined. The time it takes for the signal to be emitted (relaxation time) is related to the type of tissue. The emitted radio frequency signal is received by a radio receiver coil, which then sends it to a computer for mathematical construction of the image. MRI can readily image organs, detect small tumors or blockage in blood vessels, and assess damage to bones and vertebrae.

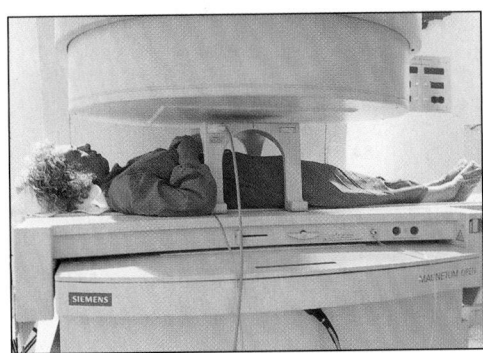

The MRI method. A patient undergoing an MRI scan. The external magnetic field aligns the nuclei. When the radio frequency coil is turned on, the resulting radio frequency flips the nuclei to the less stable spin state. When the nuclei flip back to the more stable spin state, radio frequency photons are generated, creating a signal that can be detected by a receiver.

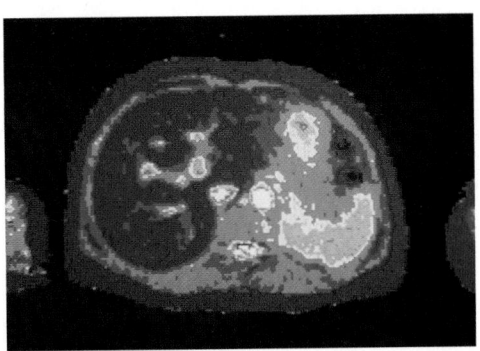

An MRI image. An MRI scan of the abdomen of a normal person. The liver is the large blue mass (*top and left*); the spleen is at the lower right (*yellow*). Stomach contents are at the top right (*red and yellow*). The purple outer layer is fat.

the Period 4 atoms, Ge is earlier in the period than Br and therefore larger than it but smaller than K, which starts the period. From these trends we can summarize the relative sizes of the radii, from largest to smallest, as K > Ge > Br > Si > Cl.

Problem-Solving Practice 7.10

Using just a periodic table, arrange these atoms in order of increasing atomic radius: B, Mg, K, Na.

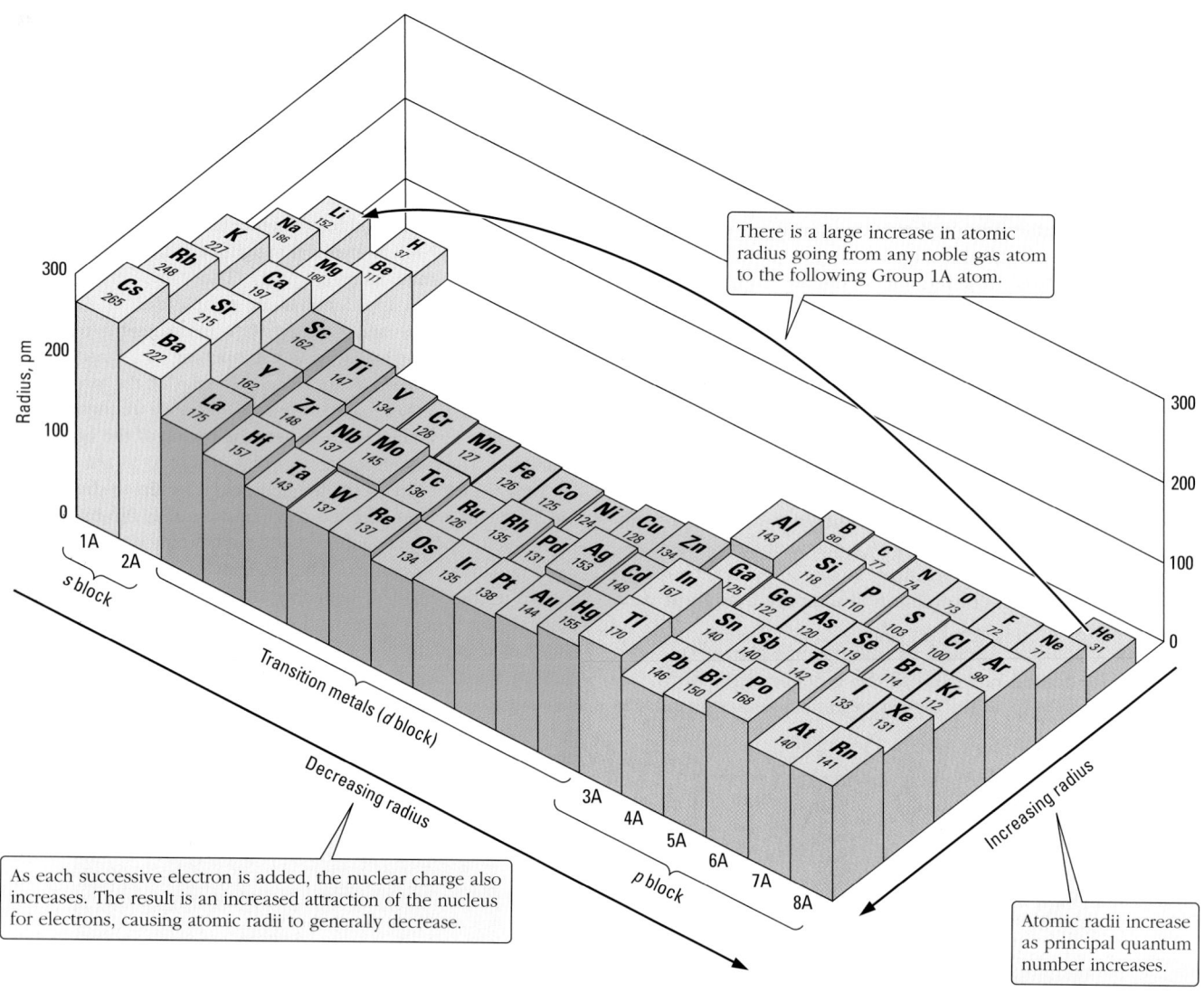

Figure 7.16 **Atomic radii of *s*-, *p*-, and *d*-block elements** (in picometers, pm; 1 pm = 10^{-12} m). Note the slight change in atomic radii for the transition elements (*d* block) across a period.

The periodic trend in the atomic radii of main group and transition metal atoms is illustrated in Figure 7.16. The sizes of transition metal atoms change very little across a period, especially beginning at Group 5 (V, Nb, or Ta), because the sizes are determined by the radius of an *ns* orbital (*n* = 4, 5, or 6) occupied by at least one electron. The variation in the number of electrons occurs instead in the (*n* − 1)*d* orbitals. As the number of electrons in these (*n* − 1)*d* orbitals increases, they increasingly repel the *ns* electrons, partly compensating for the increased nuclear charge across the periods. Consequently, the *ns* electrons experience only a slightly increasing nuclear attraction, and the radii remain nearly constant until the slight rise at the copper and zinc groups due to the continually increasing electron-to-electron repulsions as the *d* subshell is filled.

The similar radii of the transition metals and their ions have an important effect on their chemistry — they tend to be more alike in their properties than other elements in the same groups. The nearly identical radii of the fifth- and sixth-period transition elements lead to difficult problems in separating them from each other. The metals Ru, Os, Rh, Ir, Pd, and Pt are called the "platinum group metals" because they occur together in nature. Their radii and chemistry are so similar that their minerals are similar and are found in the same geologic zones.

7.9 PERIODIC TRENDS: IONIC RADII

Figure 7.17 shows clearly that the periodic trends in the ionic radii are the same as the trends in radii for neutral atoms: *positive or negative ions of elements in the same group increase in size down the group.* But pause and compare the two types of entries in Figure 7.17. When an electron is removed from an atom to form a cation, the size shrinks considerably; *the radius of a cation is always smaller than that of the atom from which it is derived.* The radius of Li is 152 pm, whereas that for Li⁺ is only 76 pm. This is understandable, because when an electron is removed from a lithium atom, the nuclear charge remains the same (3+), but there are fewer electrons repelling each other. Consequently, the positive nucleus can attract the two remaining electrons more strongly, causing the electrons to contract toward the nucleus. The decrease in ion size is especially great when the electron removed comes from a higher energy level than the new outer electron. This is the case for Li, for which the "old" outer electron was from a $2s$ orbital and the "new" outer electron is in a $1s$ orbital.

The shrinkage is also great when two or more electrons are removed — for example, Mg^{2+} and Al^{3+}.

Mg atom (radius = 160 pm) Mg^{2+} cation (radius = 66 pm)
$1s^2 2s^2 2p^6 3s^2$ $1s^2 2s^2 2p^6$

Al atom (radius 143 pm) Al^{3+} cation (radius 51 pm)
$1s^2 2s^2 2p^6 3s^2 3p^1$ $1s^2 2s^2 2p^6$

From Figure 7.17 you can see that *anions are always larger than the atoms from which they are derived.* Here the argument is the opposite of that used to explain the radii of positive ions: For anions, the nuclear charge is unchanged, but the added electron(s) introduces new repulsions and the electron

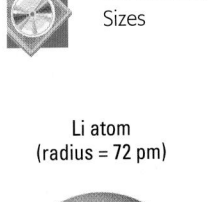

CD-ROM Screen 8.11: Ion Sizes

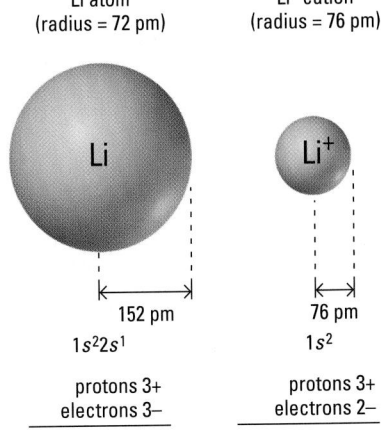

Li atom (radius = 72 pm)

Li⁺ cation (radius = 76 pm)

152 pm
$1s^2 2s^1$

protons 3+
electrons 3−
net charge 0

76 pm
$1s^2$

protons 3+
electrons 2−
net charge 1+

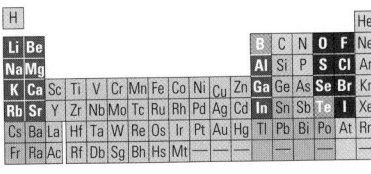

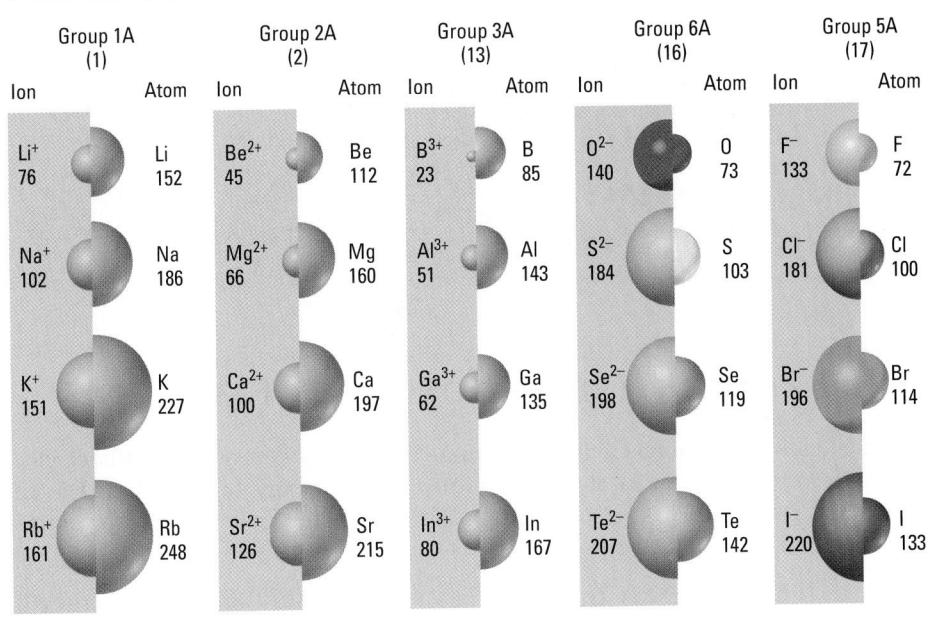

Figure 7.17 **Sizes of ions and their neutral atoms.** Radii are given in picometers (1 pm = 10^{-12} m).

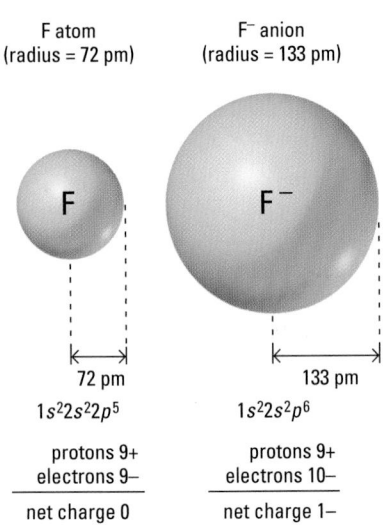

F atom
(radius = 72 pm)

F⁻ anion
(radius = 133 pm)

F

F⁻

72 pm

133 pm

$1s^22s^22p^5$

$1s^22s^2p^6$

protons 9+
electrons 9−

protons 9+
electrons 10−

net charge 0

net charge 1−

clouds swell. The F atom has nine protons and nine electrons. When it forms the F⁻ anion, the nuclear charge is still 9+, but there are now ten electrons in the anion. The F⁻ ion (133 pm) is much larger than the F atom (72 pm) because of increased electron-to-electron repulsions.

The oxide ion, O^{2-}, and fluoride ion, F^-, are *isoelectronic;* they both have the same electron configuration (the neon configuration). However, the oxide ion is larger than the fluoride ion because the O^{2-} ion has only eight protons available to attract ten electrons, whereas F^- has more protons (nine) to attract the same number of electrons. The effect of nuclear charge is evident when the sizes of isoelectronic ions across the periodic table are compared. Consider O^{2-}, F^-, Na^+, and Mg^{2+}.

Ion	O^{2-}	F^-	Na^+	Mg^{2+}
Ionic radius (pm)	140	133	102	66
Number of protons	8	9	11	12
Number of electrons	10	10	10	10

Each ion contains ten electrons. However, the O^{2-} ion has only eight protons in its nucleus to attract these electrons, while F^- has nine, Na^+ has eleven, and Mg^{2+} has twelve. As the electron-to-proton ratio decreases in an isoelectronic series of ions the electron-to-proton attraction increases and the ion size shrinks, as seen for O^{2-} to Mg^{2+}.

Problem-Solving Example 7.11 Trends in Ionic Sizes

For each of these pairs, choose the smaller atom or ion:

(a) An Au atom or an Au^{3+} ion
(b) A P atom or a P^{3-} ion
(c) An Fe^{2+} ion or an Fe^{3+} ion

Answer

(a) Au^{3+} ion (b) P atom (c) Fe^{3+} ion

Explanation

(a) A cation is smaller than its parent atom; therefore, Au^{3+} is smaller than an Au atom.
(b) Anions are larger than their parent atoms, so a P^{3-} ion is larger than a phosphorus atom.
(c) Both iron ions contain 26 protons, but Fe^{3+} has one fewer electron than does Fe^{2+}. Correspondingly, the radius of Fe^{3+} is smaller than that of Fe^{2+}.

Problem-Solving Practice 7.11

Which of these isoelectronic ions will be (a) the largest? (b) the smallest: Ba^{2+}, Cs^+, or La^{3+}?

7.10 PERIODIC TRENDS: IONIZATION ENERGIES

CD-ROM Screen 8.12:
Ionization Energy

An element's chemical reactivity is determined, in part, by how easily valence electrons are removed from its atoms. The **ionization energy** of an atom is the energy needed to remove an electron from that atom in the gas phase. For a gaseous sodium atom, the ionization process is

$$Na(g) \longrightarrow Na^+(g) + e^-$$

ΔE = ionization energy (IE)

Because energy is always required to remove an electron, the process is endothermic, and the sign of the ionization energy is always positive. The more difficult an electron is to remove, the greater its ionization energy. *For s- and p-block elements, first ionization energies (the energy needed to remove one electron from the neutral atom) generally decrease down a group and increase across a period* (Figures 7.18 and 7.19). The decrease down a group reflects the increasing radii of the atoms — it is easier to remove an electron from a larger atom. Thus, for example, the ionization energy of Rb (403 kJ/mol) is less than that of K (419 kJ/mol) or Na (496 kJ/mol). The increase in ionization energy across a period occurs because the radii decrease. As shown in Figure 7.19, the trend across a given period is not smooth, however, particularly in the second period. The single np electron of the Group 3A element is more easily removed than one of the two ns electrons in the preceding Group 2A element. Another deviation occurs for Group 6A elements (ns^2np^4), which have smaller ionization energies than the Group 5A elements that precede them. Beginning in Group 6A, two electrons are assigned to the same p orbital. Thus, greater electron repulsion is experienced by the fourth p electron, making it easier to remove.

Ionization energies increase much more gradually across the period for the transition and inner transition elements than for the main group elements (Figure 7.18). Since the atomic radii of transition elements change very little across a period, the ionization energy for the removal of an ns electron also shows small changes.

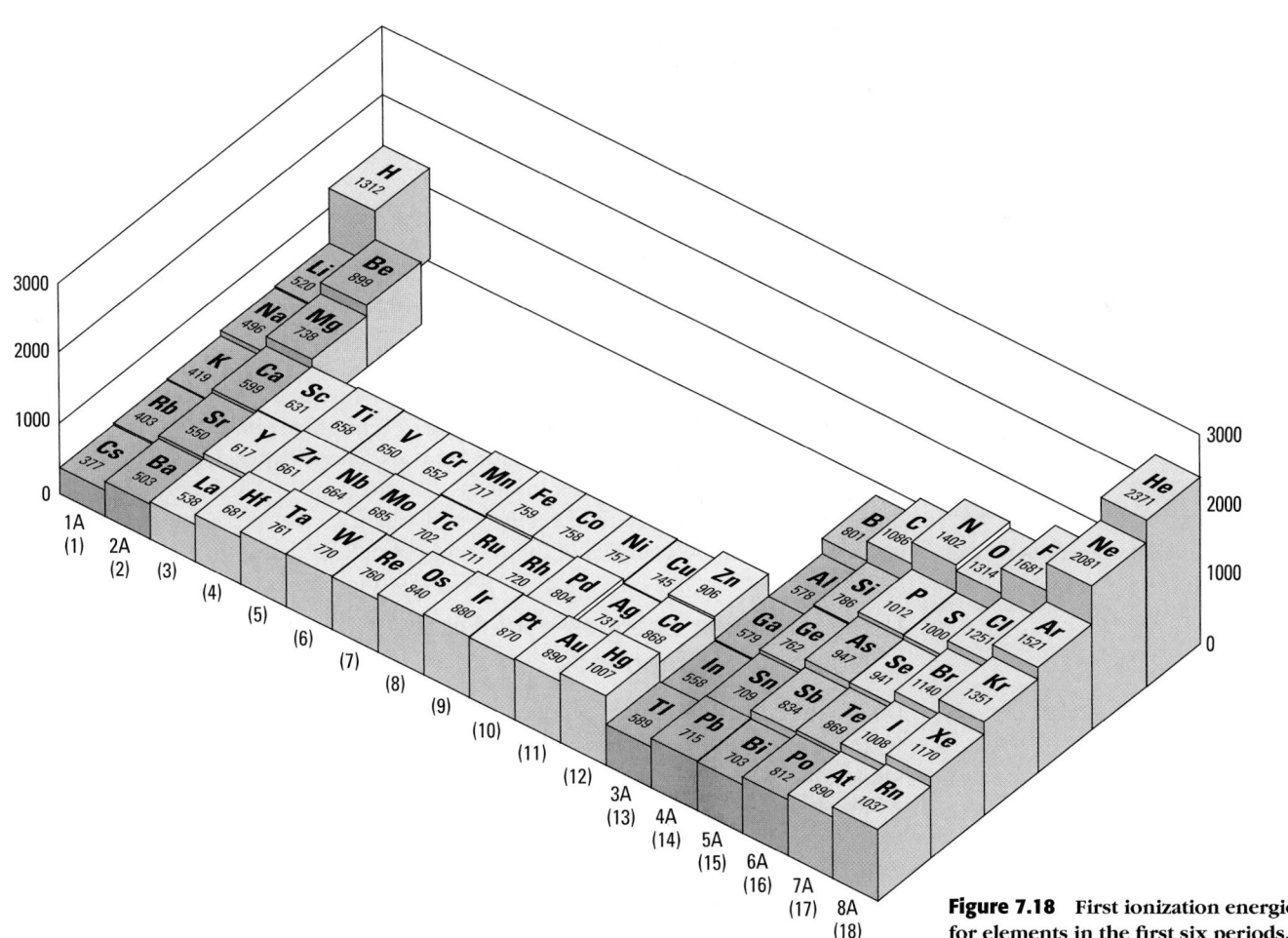

Figure 7.18 First ionization energies for elements in the first six periods.

Figure 7.19   First ionization energies for elements in the first six periods plotted against atomic number.

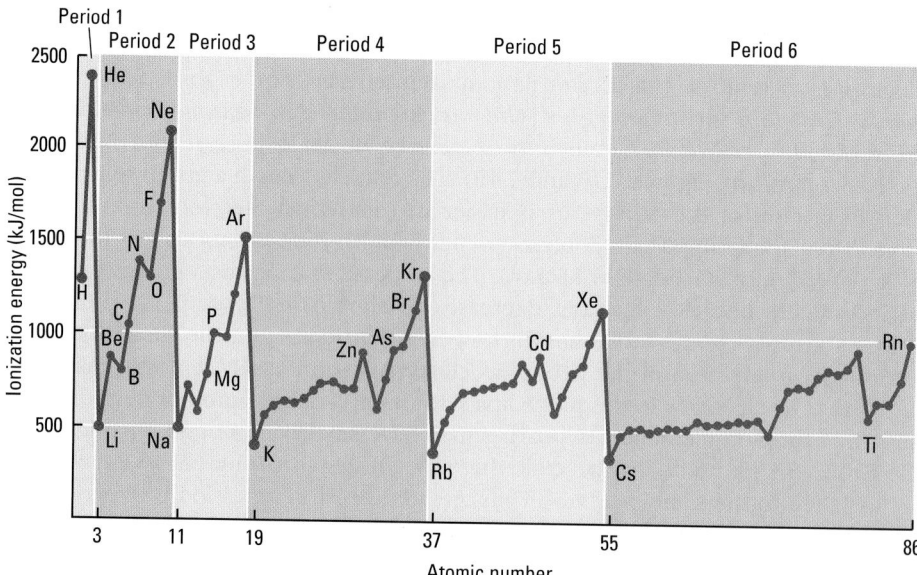

Every atom except hydrogen has a series of ionization energies, since more than one electron can always be removed. For example, the first three ionization energies of Mg in the gaseous state are

$$\underset{1s^2 2s^2 2p^6 3s^2}{\text{Mg(g)}} \longrightarrow \underset{1s^2 2s^2 2p^6 3s^1}{\text{Mg}^+(\text{g})} + \text{e}^- \qquad \text{IE}_1 = 738 \text{ kJ/mol}$$

$$\underset{1s^2 2s^2 2p^6 3s^1}{\text{Mg}^+(\text{g})} \longrightarrow \underset{1s^2 2s^2 2p^6}{\text{Mg}^{2+}(\text{g})} + \text{e}^- \qquad \text{IE}_2 = 1450 \text{ kJ/mol}$$

$$\underset{1s^2 2s^2 2p^6}{\text{Mg}^{2+}(\text{g})} \longrightarrow \underset{1s^2 2s^2 2p^5}{\text{Mg}^{3+}(\text{g})} + \text{e}^- \qquad \text{IE}_3 = 7734 \text{ kJ/mol}$$

Notice that removing each subsequent electron requires more energy, and the jump from the second (IE_2) to the third (IE_3) ionization energy of Mg is particularly great. The first electron removed from a magnesium atom comes from the $3s$ orbital. The second ionization energy corresponds to removing a $3s$ electron from a Mg^+ ion. As expected, the second ionization energy is higher than the first ionization energy because the electron is being removed from a positive ion, which strongly attracts the electron. The third ionization energy corresponds to removing a $2p$ electron from a filled p subshell in Mg^{2+}. The great difference between the second and third ionization energies for Mg is excellent experimental evidence for the existence of electron shells in atoms. Removal of the first core electron requires much more energy than removal of a valence electron. Table 7.9 gives the successive ionization energies of the second-period elements.

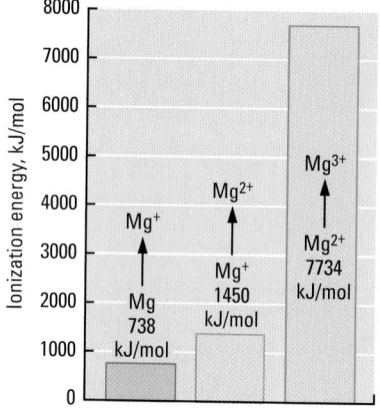

First three ionization energies for Mg.

Using Ionization Energies—The Discovery of Noble Gas Compounds

Earlier textbooks described these elements, all gases, as being inert—unreactive.

Note that Bartlett used the ionization energy for an O_2 molecule (1175 kJ/mol), not for an O atom (1314 kJ/mol).

Due to their high ionization energies, helium and the other members of Group 8A were once considered to be completely inert to normal chemical attack by oxidizing agents, acids, and other reactants. All that changed in 1962 when Neil Bartlett, while studying PtF_6, noticed, quite by accident, that it reacted with oxygen to form $[O_2^+][PtF_6^-]$. In this compound, an oxygen molecule has lost an electron to form the O_2^+ ion. Bartlett realized that the ionization energy of O_2 (1175 kJ/mol) is almost the same as the ionization energy for the noble gas xenon (1170 kJ/mol). He suspected that if an oxygen molecule could lose an electron to form O_2^+ when 1175 kJ/mol was added, then an Xe atom could also lose an electron. When he

Ionization Energy $\times 10^6$ J/mol	Li $1s^2 2s^1$	Be $1s^2 2s^2$	B $1s^2 2s^2 2p^1$	C $1s^2 2s^2 2p^2$	N $1s^2 2s^2 2p^3$	O $1s^2 2s^2 2p^4$	F $1s^2 2s^2 2p^5$	Ne $1s^2 2s^2 2p^6$
IE_1	0.52	0.90	0.80	1.09	1.40	1.31	1.68	2.08
IE_2	7.30	1.76	2.43	2.35	2.86	3.39	3.37	3.95
IE_3	11.81	14.85	3.66	4.62	4.58	5.30	6.05	6.12
IE_4		21.01	25.02	6.22	7.48	7.47	8.41	9.37
IE_5			32.82	37.83	9.44	10.98	11.02	12.18
IE_6				47.28	53.27	13.33	15.16	15.24
IE_7					64.37	71.33	17.87	20.00
IE_8			Core electrons			84.08	92.04	23.07
IE_9							106.43	115.38
IE_{10}								131.43

TABLE 7.9 Ionization Energies Required to Remove Successive Electrons from Second-Period Atoms

Note: In each of these elements the core electrons are those in the $1s$ orbital.

tried the experiment, PtF_6 quickly reacted with Xe gas to form a red crystalline solid, which was assigned the formula $[Xe^+][PtF_6^-]$.

As soon as word of his discovery was announced, chemists in other laboratories quickly discovered that xenon would even react directly with fluorine at room temperature in sunlight to form XeF_2.

$$Xe(g) + F_2(g) \longrightarrow XeF_2(s)$$

This discovery was particularly enlightening for chemists, because it made them realize that what they had so long believed to be so — that noble gases do not react to form compounds — was not true. It illustrated how careful use of knowledge of periodic properties of the elements can advance what we know about chemistry.

Later work indicated that the reaction at 25 °C gives a mixture of $[XeF^+][PtF_6^-]$ and PtF_5, which combines when heated to 60 °C to give $[XeF^+][Pt_2F_{11}^-]$.

Discovery by chance is sometimes attributed to serendipity, but Louis Pasteur observed that in science ". . . chance favors *only* the prepared mind."

Problem-Solving Example 7.12 Ionization Energies

Using only a periodic table, arrange these atoms in order of increasing first ionization energy: Al, Ar, Cl, Na, K, and Si.

Answer K < Na < Al < Si < Cl < Ar

Explanation Since ionization energy increases across a period, we expect the ionization energies for Ar, Al, Cl, Na, and Si, which are all in Period 3, to be in the order Na < Al < Si < Cl < Ar. Because K is below Na in Group 1A, the first ionization energy of K is less than that of Na. Therefore, the final order is K < Na < Al < Si < Cl < Ar.

Problem-Solving Practice 7.12

Use only a periodic table to arrange these elements in decreasing order of their first ionization energy: Na, F, N, and P.

7.11 PERIODIC TRENDS: ELECTRON AFFINITIES

The **electron affinity** of an element is the energy change when an electron is added to a gaseous atom to form a -1 ion. As the term implies, the electron affinity (EA) is a measure of the attraction an atom has for an additional electron. For example, the electron affinity of fluorine is -328 kJ/mol.

 CD-ROM Screen 8.13: Electron Affinity

$$F(g) + e^- \longrightarrow F^-(g) \qquad \Delta E = EA = -328 \text{ kJ/mol}$$
$$[He]\, 2s^2 2p^5 \qquad\quad [He]\, 2s^2 2p^6$$

TABLE 7.10	Electron Affinities (kJ/mol)						
1A (1)	2A (2)	3A (13)	4A (14)	5A (15)	6A (16)	7A (17)	8A (18)
H −73							He >0
Li −60	Be >0	B −27	C −122	N >0	O −141	F −328	Ne >0
Na −53	Mg >0	Al −43	Si −134	P −72	S −200	Cl −349	Ar >0
K −48	Ca −2	Ga −30	Ge −119	As −78	Se −195	Br −325	Kr >0
Rb −47	Sr −5	In −30	Sn −107	Sb −103	Te −190	I −295	Xe >0

This large negative value indicates that fluorine atoms readily accept an electron. As expected, fluorine and the rest of the halogens have large negative electron affinities because, by acquiring an electron, the halogen atoms achieve a stable octet of valence electrons (Table 7.10). Notice from Table 7.10 that electron affinities generally become more negative across a period toward the halogen, which is only one electron away from a noble gas configuration.

Some elements have a positive electron affinity, meaning that the negative ion is less stable than the neutral atom. Such electron affinity values are difficult to obtain experimentally, and the electron affinity is simply indicated as >0, as in the case of neon.

$$Ne + e^- \longrightarrow Ne^- \qquad\qquad EA > 0 \text{ kJ/mol}$$
$$[He]2s^2 2p^6 \qquad\quad [He]2s^2 2p^6 3s^1$$

The Ne⁻ ion would revert back to the neutral neon atom and an electron. From their electron configurations we can understand why the Ne⁻ ion would be less stable than the neutral neon atom; the anion would exceed the stable octet of valence electrons of Ne atoms.

There are two sign conventions used for electron affinity—the one used here and an alternative one. The latter uses the opposite sign, for example +349 kJ/mol for the electron affinity of chlorine.

SUMMARY PROBLEM

(a) Without looking back in the chapter, draw and label the first five energy levels of a hydrogen atom. Next, indicate the $2 \rightarrow 1$, the $3 \rightarrow 1$, the $5 \rightarrow 2$, and the $4 \rightarrow 3$ transitions in a hydrogen atom. Look in Table 7.2 to get the measured wavelengths and spectral regions for these transitions. Now calculate the frequencies (ν) for these transitions. Next, calculate the energies of the photons that are produced in these transitions.

(b) The photoelectric threshold is the minimum wavelength a photon must have to produce a photoelectric effect for a metal. These three metals exhibit photoelectric effects when photons of sufficient energies strike their surfaces.

Element	Photoelectric threshold (nm)
Lithium	540
Potassium	550
Cesium	660

Which photon energies calculated in part (a) would be sufficient to cause a photoelectric effect in lithium, in potassium, and in cesium?

(c) Vanadium, V, is a transition element named after Vanadis, the Scandinavian goddess of beauty, because of the range of beautiful colors among vanadium compounds. Vanadium forms a series of oxides in which vanadium ions are V^{2+} (in VO), V^{3+} (in V_2O_3), V^{4+} (in VO_2), and V^{5+} (in V_2O_5).

(i)   Write the full electron configuration for a vanadium atom.

(ii) Using the noble gas notation, write the electron configuration for a V^{2+} ion.

(iii) Write the orbital box diagram for a V^{3+} ion.

(iv) How many unpaired electrons are there in the V^{4+} ion?

(v)   Is the V^{5+} ion paramagnetic? Explain.

(d) An element is brittle with a steel-gray appearance. It is a relatively poor electrical conductor and forms a volatile molecular chloride and a molecular hydride that decomposes at room temperature. Is the element an alkaline earth metal, a transition metal, a metalloid, or a halogen? Explain your answer.

In Closing

Having studied this chapter, you should be able to . . .

- Use the relationships among frequency, wavelength, and the speed of light for electromagnetic radiation (Section 7.1).
- Explain the relationship between Planck's quantum theory and the photoelectric effect (Section 7.2).
- Use the Bohr model of the atom to interpret line emission spectra and the energy absorbed or emitted when electrons in atoms change energy levels (Section 7.3).
- Calculate the frequency, energy, or wavelength of an electron transition in a hydrogen atom and determine in what region of the electromagnetic spectrum the emission would occur (Section 7.3).
- Explain the use of the quantum mechanical model of the atom to represent the energy and probable location of electrons (Section 7.4).
- Apply quantum numbers (Section 7.5).
- Understand the spin properties of electrons and how they affect electron configurations and the magnetic properties of atoms (Section 7.5).
- Describe and explain the relationships among shells, subshells, and orbitals (Section 7.5).
- Use the periodic table to write the electron configurations of atoms and ions of main group and transition elements (Sections 7.6 and 7.7).
- Explain variations in valence electrons, electron configurations, ion formation, and paramagnetism of transition metals (Section 7.7).
- Explain how nuclear magnetic resonance works and how it is used in chemical analysis and medical diagnosis (*Tools of Chemistry*).
- Describe trends in atomic radii, based on electron configurations (Section 7.8).
- Describe trends in ionic radii and explain why ions differ in size from their atoms (Section 7.9).
- Use electron configurations to explain trends in the ionization energies of the elements (Section 7.10).
- Describe electron affinity (Section 7.11).

KEY TERMS

atomic radius *(7.8)*
boundary surface *(7.5)*
continuous spectrum *(7.3)*
core electrons *(7.6)*
diamagnetic *(7.7)*
electromagnetic radiation *(7.1)*
electron affinity *(7.11)*
electron configuration *(7.6)*
excited state *(7.3)*
ferromagnetic *(7.7)*
frequency *(7.1)*
ground state *(7.3)*
Hund's rule *(7.6)*
ionic radii *(7.9)*
ionization energy *(7.10)*

isoelectronic *(7.7)*
Lewis dot symbol *(7.6)*
line emission spectrum *(7.3)*
momentum *(7.4)*
noble gas notation *(7.6)*
nuclear magnetic resonance (NMR) *(p. 296)*
orbital *(7.5)*
paramagnetic *(7.7)*
Pauli exclusion principle *(7.5)*
p-block elements *(7.6)*
photoelectric effect *(7.2)*
photons *(7.2)*
Planck's constant *(7.2)*
principal energy level *(7.5)*

principal quantum number *(7.3)*
quantum *(7.2)*
quantum theory *(7.2)*
s-block elements *(7.6)*
shell *(7.5)*
spectrum *(7.1)*
subshell *(7.5)*
transition elements *(7.6)*
uncertainty principle *(7.4)*
valence electrons *(7.6)*
wavelength *(7.1)*

QUESTIONS FOR REVIEW AND THOUGHT

Conceptual Challenge Problems

CP-7.A (Section 7.2)   Planck stated in 1900 that the energy of a single photon of electromagnetic radiation was directly proportional to the frequency of the radiation ($E = h\nu$). The constant h is known as Planck's constant and has a value of 6.626×10^{-34} J·s. Soon after Planck's statement, Einstein proposed his famous equation ($E = mc^2$), which states that the total energy in any system is equal to its mass times the speed of light squared.

According to the de Broglie relation, what is the apparent mass of a photon emitted by an electron undergoing a change from the second to the first energy level in a hydrogen atom? How does the photon mass compare with the mass of the electron (9.109×10^{-31} kg)?

CP-7.B (Section 7.6)   When D. I. Mendeleev proposed a periodic law around 1870, he asserted that the properties of the elements are a periodic function of their atomic weights. Later, after H. G. J. Moseley measured the charge on the nuclei of atoms, the periodic law could be revised to state that the properties of the elements are a periodic function of their atomic numbers. What would be another way to define the periodic function that relates the properties of the elements?

CP-7.C (Section 7.3 and 7.10)   Figure 7.8 shows a diagram of the energy states that an electron can occupy in a hydrogen atom. Use this diagram to show that the first ionization energy for hydrogen, given in Figure 7.18, is correct.

Answers to questions in **bold** can be found in the back of the book.

Review Questions

1. How is the frequency of electromagnetic radiation related to its wavelength?
2. What is a photon? How are the energies of photons calculated?
3. Light is given off by a sodium- or mercury-containing streetlight when the atoms are excited in some way. The light you see arises for which of the following reasons?
   (a) Electrons moving from a given quantum level to one of higher n.
   (b) Electrons being removed from the atom, thereby creating a metal cation.
   (c) Electrons moving from a given quantum level to one of lower n.
   (d) Electrons whizzing about the nucleus in an absolute frenzy.
4. What is the Pauli exclusion principle?
5. What is Hund's rule? Give an example of the use of this rule.
6. Explain what it means when an electron occupies the $3p_x$ orbital.
7. How many electrons can be accommodated in the $n = 4$ shell?
8. Tell what happens to atomic size and ionization energy across a period and down a group.
9. Why is the radius of Na^+ so much smaller than the radius of Na? Why is the radius of Cl^- so much larger than the radius of Cl?
10. Write electron configurations to show the first two ion-

ization steps for potassium. Explain why the second ionization energy is much larger than the first.

11. Explain how the sizes of atoms change and why they change across a period of the periodic table.

12. (a) What is meant by the term "noble gas notation"?
    (b) Write an electron configuration using this notation.

13. Write the electron configurations for the valence electrons of the first three-period elements in Groups 1A through 8A.

Electromagnetic Radiation

14. When atoms absorb photons, which part (protons, electrons, neutrons) of the atom is affected?

15. Electromagnetic radiation is made up of two different "fields." What are they?

16. Electromagnetic radiation that is high in energy consists of waves that have _____ (long or short) wavelengths and _____ (high or low) frequencies. Give an example of radiation from the high-energy end of the electromagnetic spectrum.

17. Electromagnetic radiation that is low in energy consists of waves that have _____ (long or short) wavelengths and _____ (high or low) frequencies. Give an example of radiation from the low-energy end of the electromagnetic spectrum.

18. The regions of the electromagnetic spectrum are shown in Figure 7.1. Answer the following questions on the basis of this figure.
    (a) Which type of radiation involves less energy, radio waves or infrared light?
    (b) Which radiation has the higher frequency, radio waves or microwaves?

19. The colors of the visible spectrum and the wavelengths corresponding to the colors are given in Figure 7.1.
    (a) What colors of light involve less energy than yellow light?
    (b) Which color of visible light has photons of greater energy, green or violet?
    (c) Which color of light has the greater frequency, blue or green?

20. Assume that a microwave oven operates at a frequency of $1.00 \times 10^{11}\,s^{-1}$.
    (a) What is the wavelength of this radiation in meters?
    (b) What is the energy in joules per photon?
    (c) What is the energy per mole of photons?

21. The U.S. Navy has a system for communicating with submerged submarines. The system uses radio waves with a frequency of $76\,s^{-1}$.
    (a) What is the wavelength of this radiation in meters?
    (b) In miles? (1 mile = 1.61 km)

22. Place the following types of radiation in order of increasing energy per photon.
    (a) Green light from a mercury lamp
    (b) X-rays from a dental x-ray
    (c) Microwaves in a microwave oven
    (d) An FM music station at 89.1 MHz

23. Place the following types of radiation in order of increasing energy per photon.
    (a) Radio signals

(b) Radiation from a microwave oven
(c) Gamma rays from a nuclear reaction
(d) Red light from a neon sign
(e) Ultraviolet radiation from a sun lamp

24. If green light has a wavelength of 495 nm, what is its frequency?

25. What kind of radiation has a frequency of 5×10^{12} Hz? What is its wavelength?

26. What is the energy of one photon of blue light, which has a wavelength of 450 nm?

27. Which has more energy,
    (a) One photon of infrared radiation or one photon of microwave radiation?
    (b) One photon of yellow light or one photon of orange light?

28. Which kind of electromagnetic radiation can interact with molecules at the cellular level?

29. In the stratosphere, ultraviolet radiation with a frequency of $1.36 \times 10^{15}\,s^{-1}$ can break C—Cl bonds in chlorofluorocarbons (CFCs), which can lead to stratospheric ozone depletion. Calculate the wavelength and energy of this radiation.

30. Stratospheric ozone absorbs damaging UVC radiation from the sun, preventing the radiation from reaching the earth's surface. Calculate the frequency and energy of UVC radiation, which has a wavelength of 270 nm.

31. When someone uses a sunscreen, which kind of radiation will be blocked, and how does the sunscreen protect your skin from this type of radiation?

32. Calculate the energy of one photon of x-radiation having a wavelength of 2.36 nm, and compare that with the energy of one photon of orange light (3.18×10^{-19} J).

33. Green light of wavelength 516 nm is absorbed by an atomic gas. Calculate the energy difference between the two quantum states involved with this absorption.

34. Calculate the energy possessed by one mole of x-ray photons of wavelength 1.00×10^{-9} M.

Photoelectric Effect

35. Describe the role Einstein's explanation of the photoelectric effect played in the development of the quantum theory.

36. Light of very long wavelength strikes a photosensitive metallic surface and no electrons are ejected. Explain why increasing the intensity of this light on the metal still will not cause the photoelectric effect.

37. Photons of light with sufficient energy can eject electrons from a gold surface. To do so requires photons with a wavelength equal to or shorter than 257 nm. Will photons in the visible region of the spectrum dislodge electrons from a gold surface?

38. To eject electrons from the surface of potassium metal requires a minimum energy of 3.69×10^{-19} J. When 600.-nm photons shine on a potassium surface, will they cause the photoelectric effect?

39. A bright red light strikes a photosensitive surface and no electrons are ejected, even though dim blue light ejects electrons from the surface. Explain.

Atomic Spectra and the Bohr Atom

40. How does a line emission spectrum differ from sunlight?

41. Any atom with its electrons in their lowest energy levels is said to be in a _____ state.

42. Energy is emitted from an atom when an electron moves from the _____ state to the _____ state. The energy of the emitted radiation corresponds to the _____ between the two energy levels.

43. Which transition involves the emission of less energy in the H atom, an electron moving from $n = 4$ to $n = 3$ or an electron moving from $n = 3$ to $n = 1$? (See Figure 7.8.)

44. For which of these transitions in a hydrogen atom is energy absorbed? emitted?
  (a) $n = 1$ to $n = 3$ (b) $n = 5$ to $n = 1$
  (c) $n = 2$ to $n = 4$ (d) $n = 5$ to $n = 4$

45. For the transitions in Question 44:
  (a) Which ones involve the ground state?
  (b) Which one involves the greatest energy change?
  (c) Which one absorbs the most energy?

46. If energy is absorbed by a hydrogen atom in its ground state, the atom is excited to a higher energy state. For example, the excitation of an electron from the energy level with $n = 1$ to a level with $n = 4$ requires radiation with a wavelength of 97.3 nm. Which of these transitions would require radiation of a *wavelength longer* than this? (See Figure 7.8.)
  (a) $n = 2$ to $n = 4$ (b) $n = 1$ to $n = 3$
  (c) $n = 1$ to $n = 5$ (d) $n = 3$ to $n = 5$

47. (a) Calculate the wavelength, in nanometers, of the emission line that results from the $n = 2$ to $n = 1$ transition in hydrogen.
  (b) The emitted radiation is in what region of the electromagnetic spectrum?

48. In fireworks, barium atoms give a green-colored light at 555.4 nm due to an electron transition from a higher to a lower n value.
  (a) Calculate the energy difference for this transition.
  (b) Calculate n_i for this transition.

49. The Brackett series of emissions has $n_f = 4$.
  (a) Calculate the wavelength, in nanometers, of the $n = 7$ to $n = 4$ electron transition.
  (b) The emitted radiation is in what region of the electromagnetic spectrum?

50. The line spectrum of sodium has a vivid yellow-orange line at 589.6 nm. Calculate n_i for the transition that creates this line.

51. Calculate the energy and the wavelength of the electron transition from $n = 1$ to $n = 4$ in the hydrogen atom.

52. Calculate the energy and wavelength for the electron transition from $n = 2$ to $n = 5$ in the hydrogen atom.

de Broglie Wavelength

53. Calculate the de Broglie wavelength of a 4400-lb sport-utility vehicle moving at 75 miles per hour.

54. Calculate the de Broglie wavelength of an electron moving at 5% the speed of light.

Quantum Numbers

55. Assign a set of four quantum numbers for
  (a) *Each* electron in a nitrogen atom.
  (b) The valence electron in a sodium atom.
  (c) A 3*d* electron in a nickel atom.

56. Assign a set of four quantum numbers for
  (a) *Each* electron in a boron atom.
  (b) The 3*s* electrons in a magnesium atom.
  (c) A 3*d* electron in an iron atom.

57. One electron has the set of quantum numbers $n = 3$, $l = 1$, $m_l = -1$, and $m_s = +\frac{1}{2}$; another electron has the set $n = 3$, $l = 1$, $m_l = 1$, and $m_s = +\frac{1}{2}$.
  (a) Could the electrons be in the same atom? Explain.
  (b) Could they be in the same orbital? Explain.

58. Some of these sets of quantum numbers (n, l, m_l, m_s) could not occur. Explain why.
  (a) 2, 1, 2, $+\frac{1}{2}$ (b) 3, 2, 0, $-\frac{1}{2}$
  (c) 1, 0, 0, 1   (d) 3, 3, 2, $-\frac{1}{2}$
  (e) 2, 0, 0, $+\frac{1}{2}$

59. Give the n, l, and m_l values for
  (a) Each orbital in the 6*f* sublevel.
  (b) Each orbital in the $n = 5$ sublevel.

60. Assign a set of four quantum numbers for the circled electrons in these orbital diagrams.

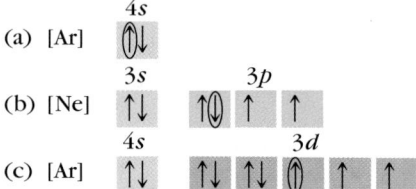

Quantum Mechanics

61. From memory, sketch the shape of the boundary surface for each of these atomic orbitals: (a) $2p_z$ (b) $4s$

62. How many subshells are there in the electron shell with the principal quantum number $n = 4$?

63. How many subshells are there in the electron shell with the principal quantum number $n = 5$?

64. Bohr pictured the electrons of the atom as being located in definite orbits about the nucleus, just as planets orbit the sun. Criticize this model in view of the quantum mechanical model.

65. Radiation is not the only thing that has both wave-like and particle-like characteristics. What component of matter can also be described in the same way?

66. How did the Heisenberg uncertainty principle illustrate the fundamental flaw in Bohr's model of the atom?

67. The three-dimensional boundary surfaces describing the energy and probability of finding an electron are called _____.

68. Which type of orbitals are found in the $n = 3$ shell? How many orbitals altogether are found in this shell?

Electron Configurations

69. Write electron configurations for Mg and Cl atoms.

70. Write electron configurations for Al and S atoms.

71. Write electron configurations for atoms of the following.
    (a) Strontium (Sr), named for a town in Scotland.
    (b) Tin (Sn), a metal used in the ancient world. Alloys of tin (solder, bronze, and pewter) are important.

72. Germanium had not been discovered when Mendeleev formulated his ideas of chemical periodicity. He predicted its existence, however, and it was found in 1886 by Winkler. Write the electron configuration of germanium.

73. Name an element of Group 3A. What does the group designation tell you about the electron configuration of the element?

74. Name an element of Group 6A. What does the group designation tell you about the electron configuration of the element?

75. (a) Which ions in the following list are likely to be formed a: K^{2+}, Cs^+, Al^{4+}, F^{2-}, Se^{2-}?
    (b) Which, if any, of these ions have a noble gas configuration?

76. These ground state orbital diagrams are incorrect. Explain why they are incorrect and how they should be corrected.

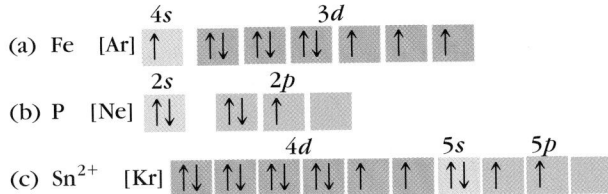

77. Write the orbital diagrams for
    (a) A nitrogen atom and a nitride, N^{3-}, ion.
    (b) The $3p$ electrons of a sulfur atom and a sulfide, S^{2-}, ion.

78. Write the orbital diagram for the $4s$ and $3d$ electrons in a
    (a) Vanadium atom.   (b) V^{2+} ion.   (c) V^{4+} ion.

79. Write the orbital diagram for the $4s$ and $3d$ electrons in a
    (a) Manganese atom.   (b) Mn^+ ion.   (c) Mn^{5+} ion.

80. How many elements are there in the fourth period of the periodic table? Explain why it is not possible for there to be another element in this period.

81. When transition metals form ions, electrons are lost first from which type of orbital? Why?

82. Give the electron configurations of Mn, Mn^{2+}, and Mn^{3+}. Use orbital box diagrams to determine the number of unpaired electrons for each species.

83. Write the electron configurations of chromium: Cr, Cr^{2+}, and Cr^{3+}. Use orbital box diagrams to determine the number of unpaired electrons for each species.

84. Write the electron configuration of vanadium (V). The name of the element was derived from Vanadis, a Scandinavian goddess.

85. Write electron configurations for these elements:
    (a) Zirconium (Zr). This metal is exceptionally resistant to corrosion and so has important industrial applications. Moon rocks show a surprisingly high zirconium content compared with rocks on earth.
    (b) Rhodium (Rh), used in jewelry and in industrial catalysts.

86. The lanthanides, or rare earths, are only "medium rare." All can be purchased for a reasonable price. Give electron configurations for atoms of these lanthanides.

(a) Europium (Eu), the most expensive of the rare earth elements; 1 g can be purchased for $50 to $100.
(b) Ytterbium (Yb). Less expensive than Eu, Yb costs only about $20 per gram. It was named for the village of Ytterby in Sweden, where a mineral source of the element was found.

Valence Electrons

87. Locate these elements in the periodic table, and draw a Lewis dot symbol that represents the number of valence electrons for an atom of each element.
    (a) F   (b) In   (c) Te   (d) Cs

88. Locate these elements in the periodic table, and draw a Lewis dot symbol that represents the number of valence electrons for an atom of each element.
    (a) Sr   (b) Br   (c) Ga   (d) Sb

89. Give the electron configurations of these ions, and indicate which ones are isoelectronic.
    (a) Na^+ (b) Al^{3+} (c) Cl^-

90. Give the electron configurations of these ions, and indicate which ones are isoelectronic.
    (a) Ca^{2+} (b) K^+ (c) O^{2-}

91. What is the electron configuration for (a) a bromine atom? (b) bromide ion?

92. (a) What is the electron configuration for an atom of tin?
    (b) What are the electron configurations for tin(II) and tin(IV) ions?

Paramagnetism and Unpaired Electrons

93. (a) In the first transition series (in row four of the periodic table), which elements would you predict to be *diamagnetic*?
    (b) Which element in this series has the greatest number of unpaired electrons?

94. What is ferromagnetism?

95. Which groups of elements in the periodic table are ferromagnetic?

96. How do the spins of unpaired electrons from paramagnetic and ferromagnetic materials differ in their behavior in a magnetic field?

Tools of Chemistry: NMR and MRI

97. What kind of electromagnetic radiation is used in nuclear magnetic resonance (NMR)?

98. In magnetic resonance imaging (MRI), the intensity of the emitted signal is related to the _____ of hydrogen nuclei and the relaxation time is related to the type of tissue being examined.

Periodic Trends

99. Arrange these elements in order of increasing size: Al, B, C, K, and Na. (Try doing it without looking at Figure 7.15 and then check yourself by looking up the necessary atomic radii.)

100. Arrange these elements in order of increasing size: Ca, Rb, P, Ge, and Sr. (Try doing it without looking at Figure 7.15 and then check yourself by looking up the necessary atomic radii.)

101. Select the atom or ion in each pair that has the larger radius.
   (a) Cl or Cl⁻   (b) Ca or Ca²⁺   (c) Al or N
   (d) Cl⁻ or K⁺   (e) In or Sn

102. Select the atom or ion in each pair that has the smaller radius.
   (a) Cs or Rb   (b) O²⁻ or O   (c) Br or As
   (d) Ba or Ba²⁺   (e) Cl⁻ or Ca²⁺

103. Write electron configurations to show the first two ionization steps for sodium. Explain why the second ionization energy is much larger than the first.

104. Arrange these atoms in order of increasing first ionization energy: F, Al, P, and Mg.

105. Arrange these atoms in order of increasing first ionization energy: Li, K, C, and N.

106. Which of these groups of elements is arranged correctly in order of increasing ionization energy?
   (a) C, Si, Li, Ne   (b) Ne, Si, C, Li
   (c) Li, Si,C, Ne   (d) Ne, C, Si, Li

107. Rank these ionization energies (IE) in order from the smallest value to the largest value. Briefly explain your answer.
   (a) First IE of Be   (b) First IE of Li
   (c) Second IE of Be   (d) Second IE of Na
   (e) First IE of K

108. Predict which of these elements would have the greatest difference between the first and second ionization energies: Si, Na, P, and Mg. Briefly explain your answer.

109. Compare the elements Li, K, C, and N.
   (a) Which has the largest atomic radius?
   (b) Place the elements in order of increasing ionization energy.

110. Compare the elements B, Al, C, and Si.
   (a) Which has the most metallic character?
   (b) Which has the largest atomic radius?
   (c) Place the three elements B, Al, and C in order of increasing first ionization energy.

111. Explain why the transition elements in a row are more alike in their properties than other elements in the same groups.

112. Select the atom or ion in each pair that has the larger radius. Explain your choice.
   (a) H⁺ or H⁻   (b) N or N³⁻   (c) F⁻ or Ne

113. Explain why nitrogen has a higher first ionization energy than does carbon, the preceding element in the periodic table.

General Questions

114. Arrange these colors of the visible region of the electromagnetic spectrum in order of increasing energy per photon: green, red, yellow, and violet.

115. A neutral atom has two electrons with $n = 1$, eight electrons with $n = 2$, eight electrons with $n = 3$, and one electron with $n = 4$. Assuming this element is in its ground state, supply its
   (a) Atomic number and name.
   (b) Total number of s electrons.
   (c) Total number of p electrons.
   (d) Total number of d electrons.

116. How many p orbital electron pairs are there in an atom of selenium (Se) in its ground state?

117. Give the symbol of all the ground state elements that have
   (a) No p electrons.
   (b) From two to four d electrons.
   (c) From two to four s electrons.

118. Give the symbol of the ground state element that
   (a) Is in Group 8 but has no p electrons.
   (b) Has a single electron in the $3d$ subshell.
   (c) Forms a 1+ ion with a $1s^2 2s^2 2p^6$ electron configuration.

119. Answer these questions about the elements X and Z, which have the electron configurations shown.

$$X = [Kr]4d^{10}5s^1 \qquad Z = [Ar]3d^{10}4s^2 4p^4$$

   (a) Is element X a metal or a nonmetal?
   (b) Which element has a larger atomic radius?
   (c) Which element would have the greater first ionization energy?

120. (a) Rank these in order of increasing atomic radius: O, S, and F.
   (b) Which has the largest first ionization energy, P, Si, S, or Se?

121. (a) Place these in order of increasing radius: Ne, O²⁻, N³⁻, and F⁻.
   (b) Place these in order of increasing first ionization energy: Cs, Sr, and Ba.

122. Name the element corresponding to each of these characteristics:
   (a) The element whose atoms have the electron configuration $1s^2 2s^2 2p^6 3s^2 3p^4$.
   (b) The element in the alkaline earth group that has the largest atomic radius.
   (c) The element in Group 5A whose atoms have the largest first ionization energy.
   (d) The element whose 2+ ion has the configuration $[Kr]4d^6$.
   (e) The element whose neutral atoms have the electron configuration $[Ar]3d^{10}4s^1$.

123. The ionization energies for the removal of the first electron from atoms of Si, P, S, and Cl are listed in the following table. Briefly rationalize this trend.

Element	First ionization energy (kJ/mol)
Si	780
P	1060
S	1005
Cl	1255

124. Answer these questions about the elements with the electron configurations shown.

$$X = [Ar]3d^84s^2 \qquad Z = [Ar]3d^{10}4s^24p^5$$

(a) An atom of which element is expected to have the larger first ionization energy?

(b) An atom of which element would be the smaller of the two?

125. Place these elements and ions in order of decreasing size: Ar, K^+, Cl^-, S^{2-}, and Ca^{2+}.

126. Which of these ions are unlikely, and why: Cs^+, In^{4+}, Fe^{6+}, Te^{2-}, Sn^{5+}, and I^-?

127. Rank these in order of increasing first ionization energy: Zn, Ca, Ca^{2+}, and Cl^-. Briefly explain your answer.

128. Classify these statements as being either true or false. If a statement is false, correct it to make it true.

(a) The wavelength of green light is longer than that of red light.

(b) Photons of green light have greater energy than those of red light.

(c) The frequency of green light is greater than that of red light.

(d) In the electromagnetic spectrum, frequency and wavelength of radiation are directly related.

129. Classify these statements as being either true or false. If a statement is false, correct it to make it true.

(a) A $3f$ orbital can hold a maximum of 14 electrons.

(b) The ground state electron configuration of a sulfur atom is $1s^22s^22p^63s^23p^4$.

(c) A ground state sulfur atom has four unpaired electrons.

(d) A Mg^{2+} ion has an argon electron configuration.

(e) An N^{3-} ion and a P^{3-} ion have the same ground state electron configuration.

130. Criticize the following statements.

(a) The energy of a photon is inversely related to its frequency.

(b) The energy of the hydrogen electron is inversely proportional to its principal quantum number n.

(c) Electrons start to enter the fourth energy level as soon as the third level is full.

(d) Light emitted by an $n = 4$ to $n = 2$ transition will have a longer frequency than that from an $n = 5$ to $n = 2$ transition.

131. A general chemistry student tells a chemistry classmate that when an electron goes from a $2d$ orbital to a $1s$ orbital, it emits more energy than that for a $2p$ to $1s$ transition. The other student is sceptical and says that such an energy change is not possible and explains why. What explanation was given?

132. Which of these types of radiation — infrared, visible, or ultraviolet — is required to ionize a hydrogen atom? Explain.

133. A certain minimum energy, E_{min}, is required to eject an electron from a photosensitive surface. Any energy absorbed beyond this minimum gives kinetic energy to the ejected electron. When 540-nm light falls on a cesium surface, an electron is ejected with a kinetic energy of 6.69×10^{-20} J. When the wavelength is 400 nm, the kinetic energy is 1.96×10^{-1} J.

(a) Calculate E_{min} for cesium, in joules.

(b) Calculate the longest wavelength, in nanometers, that will eject an electron from cesium.

134. Suppose a new element, extraterrestrium, tentatively given the symbol Et, has just been discovered. Its atomic number is 113.

(a) Write the electron configuration of the element.

(b) Name another element you would expect to find in the same group as Et.

(c) Give the formulas for the compounds of Et with O and Cl.

135. When sulfur dioxide reacts with chlorine, the products are thionyl chloride ($SOCl_2$) and dichlorine monoxide (Cl_2O).

$$SO_2(g) + 2\ Cl_2(g) \longrightarrow SOCl_2(g) + OCl_2(g)$$

(a) In what period of the periodic table is S located?

(b) Give the complete electron configuration of S. Do *not* use the noble gas notation.

(c) An atom of which element involved in this reaction (O, S, or Cl) should have the smallest first ionization energy? the smallest radius?

(d) If you want to make 675 g of $SOCl_2$, what mass in grams of Cl_2 is required?

(e) If you use 10.0 g of SO_2 and 20.0 g of Cl_2, what is the theoretical yield of $SOCl_2$?

Applying Concepts

136. Write the electron configuration for the product of the first ionization of the smallest halogen.

137. Write the electron configuration for the product of the second ionization of the third largest alkaline earth metal.

138. What compound will most likely form between chlorine and element X, if element X has the electronic configuration $1s^22s^22p^63s^1$?

139. Write the formula for the compound that most likely forms between potassium and element Z, if element Z has the electronic configuration $1s^22s^22p^63s^23p^4$.

140. Which of these electron configurations are for atoms in the ground state? in excited states? Which are impossible?

(a) $1s^22s^1$ (b) $1s^22s^22p^3$

(c) $[Ne]3s^23p^34s^1$ (d) $[Ne]3s^23p^64s^33d^2$

(e) $[Ne]3s^23p^64f^4$ (f) $1s^22s^22p^43s^2$

141. Which of these electron configurations are for atoms in the ground state? in excited states? Which are impossible?

(a) $1s^22s^2$ (b) $1s^22s^23s^1$

(c) $[Ne]3s^23p^84s^1$ (d) $[He]2s^22p^62d^2$

(e) $[Ar]4s^23d^3$ (f) $[Ne]3s^22p^54s^1$

142. Using the information in Table 7.6 and Figure 7.13, write the electron configuration for the undiscovered element with an atomic number of 164. Where would this element be located in the periodic table?

143. You are given the atomic radii of 110 pm, 118 pm, 120 pm, 122 pm, and 135 pm, but do not know to which element (As, Ga, Ge, P, and Si) these values correspond. Which must be the value of Ge?

144. These questions refer to the graph below.

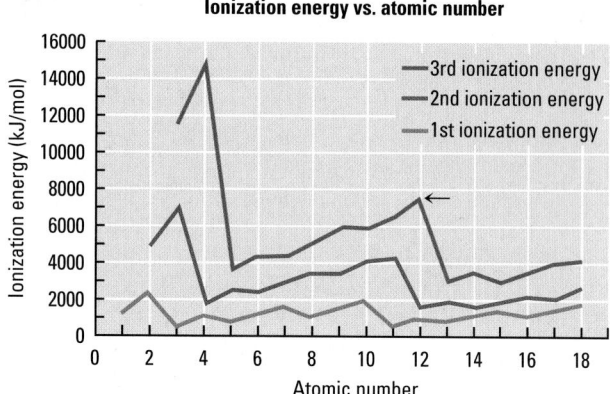

Ionization energy vs. atomic number

(a) Based on the graphic data, ionization energies _____ (decrease, increase) left to right and _____ (decrease, increase) top to bottom on the periodic table.

(b) Which element has the largest first ionization energy?

(c) A plot of the fourth ionization energy versus atomic number for elements 1 through 18 would have peaks at which atomic numbers?

(d) Why is there no third ionization energy for helium?

(e) What is the reason for the large second ionization energy for lithium?

(f) Find the arrow pointing to the third ionization energy curve. What is the equation for the process corresponding to this data point?

 General Chemistry CD-ROM

CD7.1 Screen 7.4: The Electromagnetic Spectrum. Use the spectrum "tool" to answer the following questions.

(a) As you move the slider from the blue region to the red region, what happens to the wavelength of the light? to its frequency?

(b) Place the slider somewhere in the blue region of the spectrum. What is the approximate wavelength of this radiation in nanometers? in meters?

(c) Which color of light in the visible spectrum has the longest wavelength? the shortest wavelength?

CD7.2 Screen 7.7: Bohr's Model of the Hydrogen Atom.

(a) How can we detect the movement of an electron from a higher energy level in an atom to a lower energy level?

(b) What does it mean when an electron is moved to an energy level with $n = $ infinity?

(c) What do we mean when we say that electrons occupy quantized energy levels?

CD7.3 Screen 7.11: Shells, Subshells, and Orbitals.

(a) An example of an orbital is $4p$. In what shell is the electron located? In what subshell? In what orbital within the subshell?

(b) What is the relation between the value of the quantum number n and the number of subshells in a given shell?

(c) What is the relation between the type of subshell and the number of orbitals in that subshell?

(d) Is there a subshell with four orbitals?

CD7.4 Screen 8.2: Electron Spin. Describe an experiment that would allow you to detect the presence of unpaired electrons in a molecule.

CD7.5 Screen 8.10: Atomic Size

(a) Why does the size of elements in a group increase on moving down the group?

(b) Why does the size of elements decrease on moving across a period?

CD7.6 Screen 8.12: Ionization Energy. Examine the ionization of magnesium to form its ions.

(a) As successive electrons are removed from magnesium, why does the ionization energy increase?

(b) Why is the ion depicted as smaller than the atom as an electron is removed?

(c) The difference between the first and second ionization energies of magnesium is about 700 kJ/mol. Why is the difference between the second and third ionization energies over 6000 kJ/mol? What implications does this have for the structure of atoms in general? What implications does this have for the chemistry of magnesium?

8

The human body consists of thousands of compounds, most of them made of atoms bonded by sharing electron pairs in covalent bonds. This chapter explains the nature of these bonds and describes some molecular compounds.

COVALENT BONDING

8.1   Covalent Bonding

8.2   Single Covalent Bonds and Lewis Structures

8.3   Single Covalent Bonds in Hydrocarbons

8.4   Multiple Covalent Bonds

8.5   Multiple Covalent Bonds in Hydrocarbons

8.6   Bond Properties: Bond Length and Bond Energy

8.7   Bond Properties: Bond Polarity and Electronegativity

8.8   Formal Charge

8.9   Lewis Structures and Resonance

8.10   Exceptions to the Octet Rule

8.11   Aromatic Compounds

Atoms of elements are only rarely found uncombined in nature. Only the noble gases consist of individual atoms. Most nonmetallic elements consist of molecules, and in a solid metallic element, each atom is closely surrounded by eight or twelve neighbors. What makes atoms stick to one another? Valence electrons form the glue, but how? In ionic compounds, the transfer of one or more valence electrons from an atom of a metal to an atom of a nonmetal produces ions whose opposite charges hold the ions in a crystal lattice. But many compounds do not conduct electricity when in the liquid state and correspondingly do not consist of ions. Examples are carbon monoxide (CO), water (H_2O), methane (CH_4), quartz (SiO_2), and the millions of organic compounds.

What holds atoms together in these molecular compounds? They are held together by bonds consisting of one or more pairs of electrons shared between the bonded atoms. The attraction of positively charged nuclei for electrons between them pulls the nuclei together. This simple idea can account for the bonding in all molecular compounds, allowing us to correlate their structures with their physical and chemical properties. This chapter describes the bonding found in molecules ranging from simple diatomic gases to hydrocarbons and more complex molecules.

8.1 COVALENT BONDING

 CD-ROM Screen 9.3: Chemical Bond Formation: Covalent Bonding

Hydrogen gas (H_2) and lithium hydride (LiH), with molar masses of 2 and 8 g/mol, respectively, have the two smallest molar masses of any known diatomic substances. Both H and Li have one valence electron. But H_2 and LiH are very different. Hydrogen, H_2, is a gas at room temperature, is practically insoluble in water, is an electric insulator, and burns easily in air. Lithium hydride, on the other hand, is a solid, reacts with water to form H_2, conducts electricity when in the molten state, and bursts into flame when exposed to moist air. Why is there such a difference in their properties?

H^-, He, and Li^+ are isoelectronic.

A lithium atom can lose its one valence electron (the 2s electron) relatively easily to form Li^+, which has the electron configuration of helium, the nearest noble gas. A hydrogen atom can add an electron to form a hydride ion, H^-, which also has the noble gas electron configuration of He. Hence, lithium hydride consists of Li^+ cations and H^- anions and has properties characteristic of an ionic compound—it is a crystalline solid at room temperature and has a high melting point. Hydrogen gas, on the other hand, has none of the properties of an ionic compound; it consists of H_2 molecules.

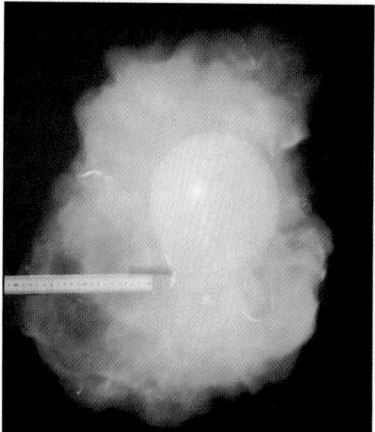

Hydrogen and lithium hydride. (a) A burning hydrogen-filled balloon; (b) A small pile of lithium hydride reacting with water on a watch glass.

(a)

(b)

316

G. N. Lewis assumed that each noble gas atom has a completely filled outermost shell, which he regarded as a stable configuration because of the lack of reactivity of noble gases. He suggested that valence electrons rearrange to give noble gas electron configurations when atoms join together chemically, such as in H_2 molecules. Lewis proposed that an attractive force called a **covalent bond** results when one or more pairs of electrons are shared between the bonded atoms. *The atoms in molecular compounds are connected by covalent bonds.* For example, the two H atoms in H_2 and the H and Cl atoms in HCl are held together by covalent bonds.

The shared electron pairs of covalent bonds occupy the same shell as the valence electrons of each atom. For main group elements, the electrons commonly contribute to a noble gas configuration on *each* atom. Lewis further proposed that by counting the valence electrons of an atom, it would be possible to predict how many bonds that atom can form. *The number of covalent bonds an atom can form is determined by the number of electrons that an atom must share to achieve a noble gas configuration.*

But why does sharing electrons provide an attractive force between bonded atoms? Consider the formation of the simplest stable molecule, H_2. If the two hydrogen atoms are widely separated, there is little if any interaction between them. When the two atoms get close enough, however, their $1s$ electron clouds overlap.

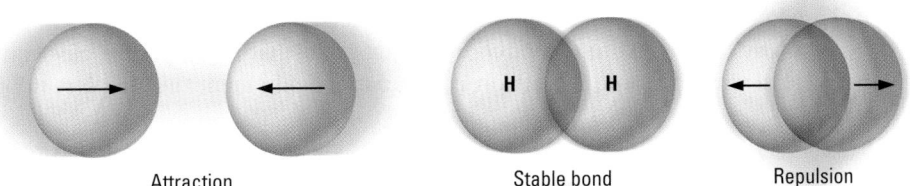

Attraction   Stable bond   Repulsion

This overlap allows the electron from each atom to be attracted by the other atom's nucleus, causing a net attraction between the two atoms. Experimental data and calculations indicate that an H_2 molecule has its lowest potential energy and is therefore most stable when the nuclei are 74 pm apart (Figure 8.1). At that dis-

GILBERT NEWTON LEWIS
(1875–1946)

In 1916, G.N. Lewis introduced the theory of the shared electron pair chemical bond in a paper published in the *Journal of the American Chemical Society*. The theory revolutionized chemistry, and it is in honor of this contribution that we refer to "electron dot" structures as Lewis structures. Of particular interest in this text is the extension of his theory of bonding to a generalized theory of acids and bases (Section 16.10).

Lowering potential energy is favorable to bond formation.

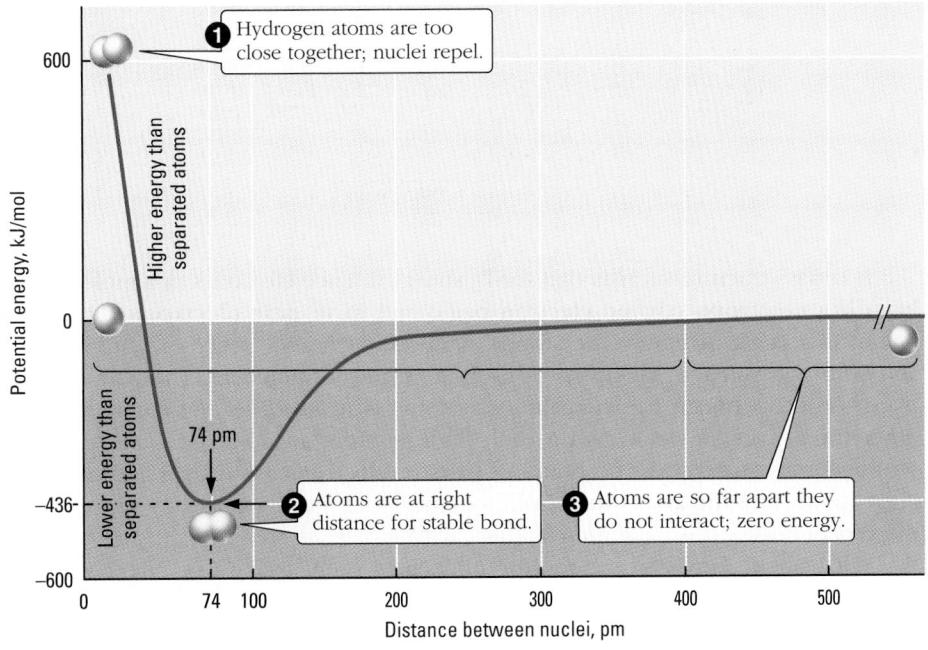

Figure 8.1 **H—H bond formation from isolated H atoms.** Energy is at a minimum at an internuclear distance of 74 pm, where there is a balance between electrostatic attractions and repulsions.

tance, the attractive and repulsive electrostatic forces are balanced. If the nuclei get closer than 74 pm, repulsion of each nucleus by the other begins to take over. When the H nuclei are 74 pm apart — *the bond length* — it takes 436 kJ of energy to break the hydrogen-to-hydrogen covalent bonds when a mole of gaseous H_2 molecules is converted into isolated H atoms. This energy is the *bond energy* of H_2.

8.2 SINGLE COVALENT BONDS AND LEWIS STRUCTURES

CD-ROM Screen 9.4: Lewis Electron Dot Structures

A **single covalent bond** is formed when two atoms share a pair of electrons. The simplest examples are the bonds in diatomic molecules such as H_2, F_2, and Cl_2, which, like other simple molecules, can be represented by Lewis structures. A **Lewis structure** for a molecule shows all valence electrons as dots or lines that represent covalent bonds.

Lewis structures are drawn by starting with the Lewis dot symbols for the atoms (⬅ *p. 289, Table 7.7*) and arranging the valence electrons until each atom in the molecule has a noble gas configuration. For example, the Lewis structure for H_2 shows two bonding electrons shared between two hydrogen atoms.

H:H   or   H—H

The shared electron pair of a *single* covalent bond is often represented by a line instead of a pair of dots. Note that *each* hydrogen atom shares the pair of electrons, thereby achieving the same 2-electron configuration as helium, the simplest noble gas.

Atoms with more than 2 valence electrons achieve a noble gas structure by sharing an octet of valence electrons. This is known as the **octet rule:** *To form bonds, main group elements gain, lose, or share electrons to achieve a stable electron configuration characterized by 8 valence electrons.* To obtain the Lewis structure for F_2, for example, we start with the Lewis dot symbol for a fluorine atom. Fluorine, in Group 7A, has seven valence electrons, so there are seven dots, one less than an octet. If each F atom in F_2 shares one valence electron with the other F atom to form a shared electron pair, a single covalent bond forms, and each fluorine atom achieves an octet. *Shared electrons are counted with each of the atoms in the bond.*

Achieving a noble gas configuration is also referred to as "obeying the octet rule" because all noble gases except helium have eight valence electrons.

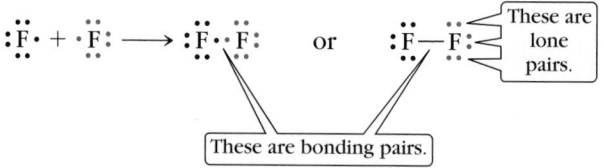

These are lone pairs.

These are bonding pairs.

The term "lone pairs" will be used in this text to refer to unshared electron pairs.

A Lewis structure such as that for F_2 shows valence electrons in a molecule as **bonding electrons** (shared electron pairs) and **lone pair electrons** (unshared pairs). In a Lewis structure the atomic symbols, such as F, represent the nucleus and core (nonvalence) electrons of each atom in the molecule. The pair of electrons shared between the two fluorine atoms is a *bonding electron pair*. The other three pairs of electrons on each fluorine atom are *lone pair electrons*. In writing Lewis structures, the bonding pairs of electrons are usually indicated by lines connecting the atoms they hold together; lone pairs are usually represented by pairs of dots.

Magnetic measurements (Section 7.7) support the concept that each electron in a pair of electrons (bonding or lone pair) has its spin opposite that of the other electron.

What about Lewis structures for molecules such as H_2O or NH_3? Oxygen (Group 6A) has six valence electrons and must share two electrons to satisfy the octet rule. This can be accomplished by forming covalent bonds with two hydrogen atoms.

$$H\cdot + H\cdot + \cdot\overset{..}{\underset{..}{O}}\cdot \quad \text{forms} \quad H:\overset{..}{\underset{..}{O}}:H \quad \text{or} \quad H-\overset{..}{\underset{..}{O}}-H$$

Nitrogen (Group 5A) in NH_3 must share three electrons to achieve a noble gas configuration, which can be done by forming covalent bonds with three hydrogen atoms.

$$H\cdot + H\cdot + H\cdot + \cdot\overset{..}{\underset{.}{N}}\cdot \quad \text{forms} \quad H:\overset{..}{N}:H \atop \ddot{H} \quad \text{or} \quad H-\overset{..}{\underset{|}{N}}-H \atop H$$

From the Lewis structures of F_2, H_2O, and NH_3, we can make an important generalization: ***The number of electrons that an atom of a main group element must share to achieve an octet equals 8 minus its A group number.*** Carbon, for example, which is in Group 4A, needs to share four electrons to reach an octet.

Main group elements are those in the groups labeled "A" in the periodic table inside the front cover of this book.

Group number	Number of valence electrons	Number of electrons shared to complete an octet (8 − group number)	Example
4A	4	4	C in CH_4
5A	5	3	N in NF_3 $:\overset{..}{\underset{..}{F}}-\overset{..}{\underset{\|}{N}}-\overset{..}{\underset{..}{F}}: \atop :\overset{..}{\underset{..}{F}}:$
6A	6	2	O in H_2O $H-\overset{..}{\underset{..}{O}}-H$
7A	7	1	F in HF $H-\overset{..}{\underset{..}{F}}:$

Many essential biochemical molecules contain carbon, hydrogen, oxygen, and nitrogen atoms. The structures of these molecules are dictated by the number of bonds C, O, and N need to complete their octets, as pointed out in the table. For example, glycerol, $C_3H_8O_3$,

$$\begin{array}{ccccc} & H & H & H & \\ & | & | & | & \\ H- & C & - C & - C & -H \\ & | & | & | & \\ & :\ddot{O}: & :\ddot{O}: & :\ddot{O}: & \\ & | & | & | & \\ & H & H & H & \end{array}$$

is vital to the formation and metabolism of fats. In glycerol, each carbon atom and each oxygen atom share eight electrons to complete an octet; hydrogen is satisfied by sharing two electrons. The same type of electron sharing occurs in deoxyribose, $C_5H_{10}O_4$, an essential component of DNA (deoxyribonucleic acid). Four of the carbon atoms are joined in a ring with an oxygen atom.

deoxyribose

Guidelines for Writing Lewis Structures

Guidelines have been developed for correctly writing Lewis structures, and we will illustrate using them to write the Lewis structure for PCl_3.

CD-ROM Screen 9.5: Drawing Lewis Structures

1. ***Count the total number of valence electrons in the molecule or ion.*** Use the A group number in the periodic table as a guide to indicate the number of valence electrons in each atom. The total of the A group numbers equals the total number of valence electrons of the atoms in a neutral molecule. For a negative ion, add electrons equal to the ion charge. For a positive ion, subtract the number of electrons equal to the charge. For example, add one electron for OH^-; subtract one electron for NH_4^+.

   Because PCl_3 is a neutral molecule, its number of valence electrons is 5 for P (it is in Group 5A) and 7 for *each* Cl (chlorine is in Group 7A): Total number of valence electrons = $5 + (3 \times 7) = 26$.

2. ***Use atomic symbols to draw a skeleton structure by joining the atoms with shared pairs of electrons (a single line).*** A skeleton structure indicates the attachment of terminal atoms to a central atom. The central atom is usually the one written first in the molecular formula and is the one that can form the most bonds, such as Si in $SiCl_4$ and P in PO_4^{3-}. Hydrogen, oxygen, and the halogens are often terminal atoms. In PCl_3, the central atom is phosphorus, so we draw a skeleton structure with P as the central atom and the terminal chlorine atoms arranged around it. The three bonding pairs account for 6 of the total of 26 valence electrons.

Although H is given first in the formulas of H_2O and H_2O_2, for example, it is not the central atom. H is never the central atom in a molecule or ion because hydrogen atoms form only one covalent bond.

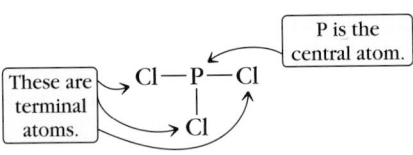

3. ***Place lone pairs of electrons around each atom (except H) to satisfy the octet rule, starting with the terminal atoms.*** Using lone pairs in this way on the Cl atoms accounts for 18 of the remaining 20 valence electrons, leaving 2 electrons for a lone pair on the P atom. When you check for octets, remember that shared electrons are counted as "belonging" to *each* of the atoms bonded by the shared pair; each P—Cl bond has two shared electrons that "count" for phosphorus and also "count" for chlorine.

   $$:\overset{..}{\underset{..}{Cl}}:\overset{..}{\underset{..}{P}}:\overset{..}{\underset{..}{Cl}}: \qquad or \qquad :\overset{..}{\underset{..}{Cl}}-\overset{..}{P}-\overset{..}{\underset{..}{Cl}}:$$
   $$:\overset{..}{\underset{..}{Cl}}: \qquad\qquad\qquad :\overset{}{\underset{..}{Cl}}:$$

   Counting dots and lines in the Lewis structure on the right shows 26 electrons, thus accounting for all valence electrons. It is the correct Lewis structure for PCl_3.

   The next two steps apply when Steps 1 through 3 result in a structure that does not use all the valence electrons or fails to give an octet of electrons to each atom that should have an octet.

4. ***Place any leftover electrons on the central atom, even if it will give the central atom more than an octet.*** If the central atom is from the third or higher period, it can accommodate more than an octet of electrons (Section 8.10.)

5. *If the number of electrons around the central atom is less than eight, change single bonds to the central atom to multiple bonds.* Some atoms can share more than one pair of electrons, resulting in a double covalent bond (two shared pairs) or a triple covalent bond (three shared pairs), known as *multiple bonds* (Section 8.4). Where multiple bonds are needed to complete an octet, use one or more lone pairs of electrons from the terminal atoms to form *double* (two shared pairs) or *triple* (three shared pairs) covalent bonds until the central atom and all terminal atoms have octets. This guideline does not apply to PCl_3, but will be illustrated in Section 8.4.

> Double bond: two shared pairs of electrons as in C=C; triple bond: three shared electron pairs, as in C≡N.

Problem-Solving Example 8.1   Lewis Structures

Write the Lewis structures of (a) dichlorine monoxide, Cl_2O, (b) hydrogen peroxide, H_2O_2, (c) borohydride ion, BH_4^-, (d) phosphonium ion, PH_4^+, and (e) PCl_5.

Answer

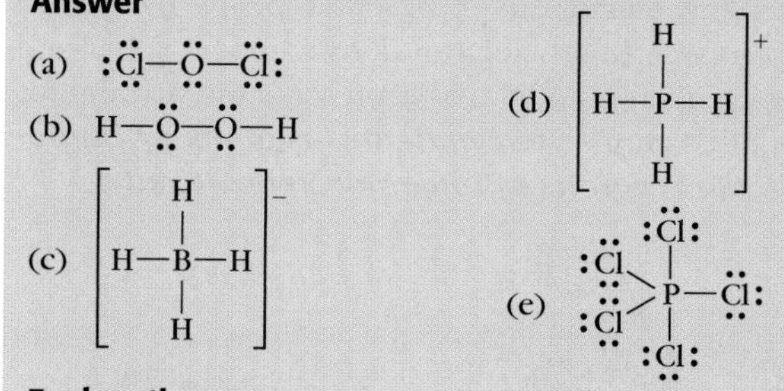

Explanation

(a) There are 20 valence electrons, including 14 from the two Cl atoms and 6 from the O atom (oxygen is in Group 6A). The central atom is O, and the skeleton structure is

$$Cl—O—Cl$$

Placing three lone pairs on each Cl and two lone pairs on the O atom uses all the electrons and satisfies the octet rule for all atoms.

(b) The atoms in H_2O_2 are H, which can never form more than one bond, and O, which can form two bonds, meaning that the two O atoms must be bonded to each other. The total number of valence electrons is 14, including 2 from the two H atoms and 12 from the two O atoms. The skeleton structure must be H—O—O—H, which uses 6 bonding electrons. Placing two lone pairs on each O atom satisfies the octet rule and uses all valence electrons.

(c) There are 8 valence electrons in the BH_4^- ion: 3 from B, 4 from 4 H, and 1 for the 1− charge. Each hydrogen forms one bond to boron, completing the octet around boron.

(d) Phosphorus can form more bonds than hydrogen, so phosphorus is the central atom. There are 8 valence electrons: 5 from P, 1 from each hydrogen, and 1 subtracted for the 1+ charge. The 8 electrons form four single P—H bonds.

(e) Because it is a Period 3 nonmetal, phosphorus can expand its octet to accommodate five chlorine atoms around the central phosphorus atom. The 40 valence electrons are distributed as five pairs in P—Cl bonds and 30 electrons as three lone pairs around each of the five chlorine atoms.

Problem-Solving Practice 8.1

Write the Lewis structures for (a) NF_3, (b) N_2H_4, and (c) SO_4^{2-}.

Although Lewis structures are useful for predicting the number of covalent bonds an atom will form, they do not give an accurate representation of where electrons are located in a molecule. Bonding electrons do not stay in fixed positions between nuclei, as Lewis's dots might imply. Instead, quantum mechanics tells us that there is a high probability of finding the bonding electrons between the nuclei. Also, Lewis structures are not meant to convey the shapes of molecules. The angle between the two O—H bonds in a water molecule is not 180°, as the Lewis structure on page 319 seems to imply. However, Lewis structures are used to predict geometries by a method based on the repulsions between valence shell electron pairs (Section 9.1).

8.3   SINGLE COVALENT BONDS IN HYDROCARBONS

In hydrocarbons, carbon's four valence electrons are shared with hydrogen atoms or other carbon atoms. In methane, CH_4, the simplest hydrocarbon, the four valence electrons are shared with electrons from four hydrogen atoms, forming four single covalent bonds. The bonding in methane is represented below by a Lewis structure and an *electron density model*. In an electron density model, a ball-and-stick model is surrounded by a space-filling model that represents the distribution of electron density around the bonded atoms. **Red indicates regions of higher electron density, and blue indicates regions of lower electron density.**

CD-ROM Screen 11.2: Carbon-Carbon Bonds: Chain Formation

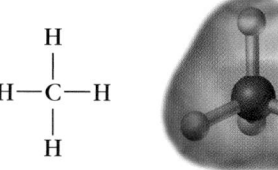

An electron density model of methane.

Carbon is unique among the elements because of the ability of its atoms to form strong bonds with one another as well as with atoms of hydrogen, oxygen, nitrogen, sulfur, and the halogens. The strength of the carbon-carbon bond permits long chains to form:

$$-\overset{|}{\underset{|}{C}}-\overset{|}{\underset{|}{C}}-\overset{|}{\underset{|}{C}}-\overset{|}{\underset{|}{C}}-\overset{|}{\underset{|}{C}}-\overset{|}{\underset{|}{C}}-\overset{|}{\underset{|}{C}}-\overset{|}{\underset{|}{C}}-\overset{|}{\underset{|}{C}}-\overset{|}{\underset{|}{C}}-\overset{|}{\underset{|}{C}}-\overset{|}{\underset{|}{C}}-$$

Review the discussion on alkanes in Section 3.4. See Table 3.4 for a list of selected alkanes.

Because each carbon atom can form four covalent bonds, such chains contain numerous sites to which other atoms (including more carbon atoms) can bond, leading to the great variety of carbon compounds.

Hydrocarbons contain only carbon and hydrogen atoms (⇐ *Section 3.3*). Alkanes, which contain only C—H and C—C single covalent bonds, are often referred to as **saturated hydrocarbons** because each carbon is bonded to a maximum number of hydrogen atoms (⇐ *Section 3.4, p. 83*).

CD-ROM Screen 11.3: Hydrocarbons

The carbon atoms in alkanes with four or more carbon atoms per molecule can be arranged in either a straight chain or a branched chain (⇐ *p. 83*).

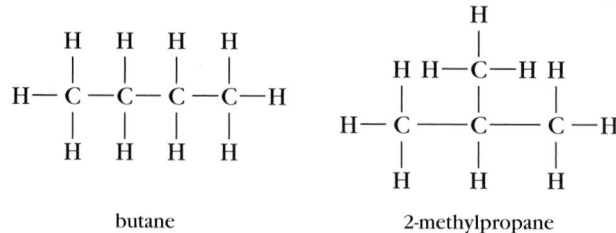

butane   2-methylpropane

Rules for naming organic compounds are given in Appendix E.

In addition to straight-chain and branched-chain alkanes, there are *cycloalkanes*, saturated hydrocarbon compounds consisting of carbon atoms joined in rings of —CH_2— units. Cycloalkanes are commonly represented by polygons in which each corner represents a carbon atom and two hydrogen atoms and each line represents a C—C bond. The C—H bonds usually are not shown, but are understood to be present. The simplest cycloalkane is cyclopropane; other common cycloalkanes include cyclobutane, cyclopentane, and cyclohexane.

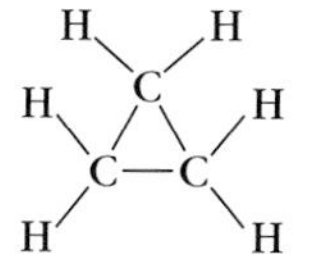

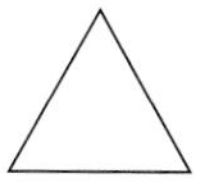

cyclopropane
C_3H_6

cyclobutane
C_4H_8

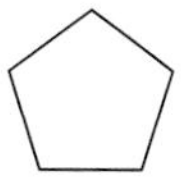

cyclopentane
C_5H_{10}

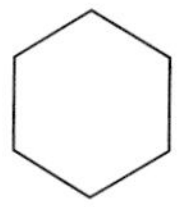

cyclohexane
C_6H_{12}

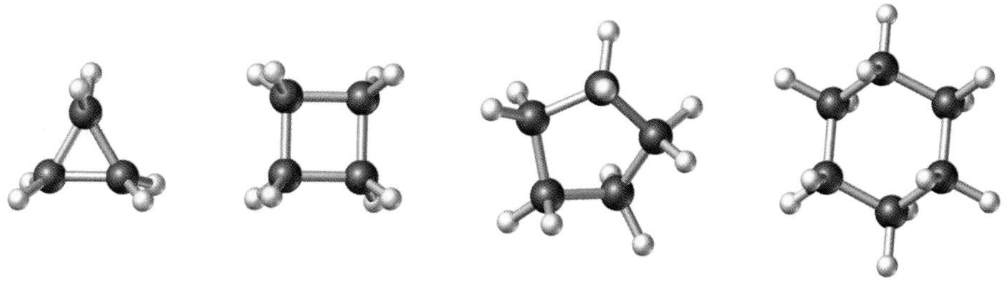

Exercise 8.1   Cyclic Hydrocarbons

Write the Lewis structure and the polygon that represents cyclooctane; write the molecular formula for this compound.

The great variety of organic compounds can also be accounted for by the fact that one or many carbon-hydrogen bonds in hydrocarbons can be replaced by bonds between carbon and other atoms. For example, the new bonds to carbon can connect to individual halogen atoms, thus creating entirely different compounds. Consider the new substances that result when a chlorine atom replaces a hydrogen atom in ethane, 2-methylbutane, and cyclopropane.

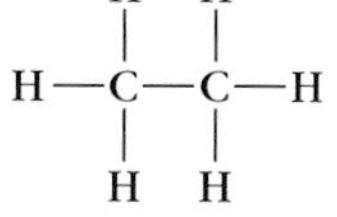

ethane

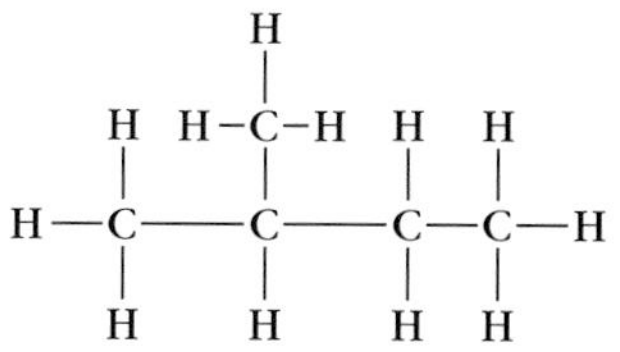

2-methylbutane

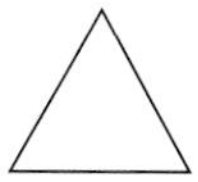

cyclopropane

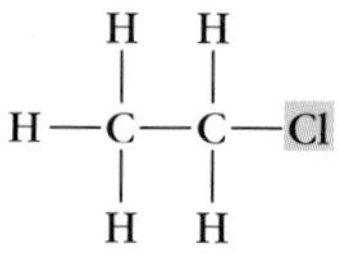

chloroethane

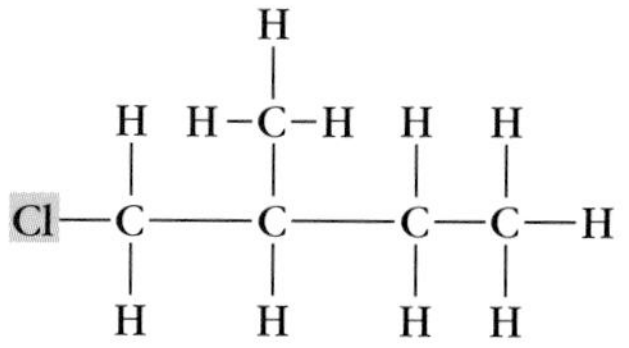

1-chloro-2-methylbutane

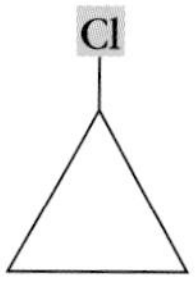

chlorocyclopropane

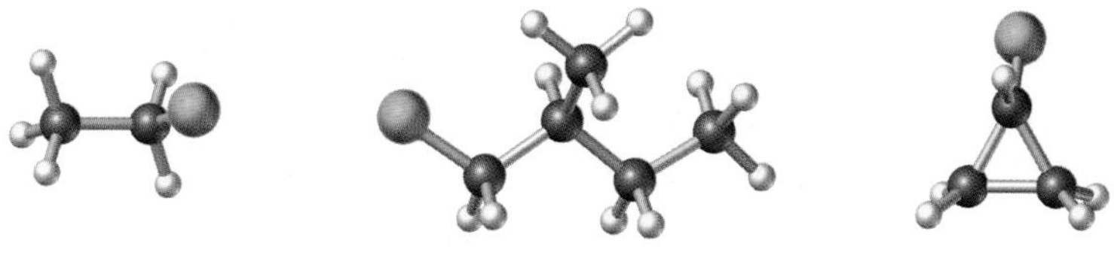

The —OH group in an alcohol is not an ion; it is different from the OH⁻ ion in a base.

Appendix E.2 includes a list of functional groups.

In another case, consider how an —OH functional group can replace one or more hydrogen atoms in an alkane. The —OH functional group is characteristic of alcohols. A molecule of ethanol can be thought of as a molecule of ethane in which a hydrogen atom has been replaced by an —OH group. Replacing two hydrogen atoms in ethane with —OH groups forms ethylene glycol. Glycerol is formed when three hydrogen atoms in propane are replaced by —OH groups.

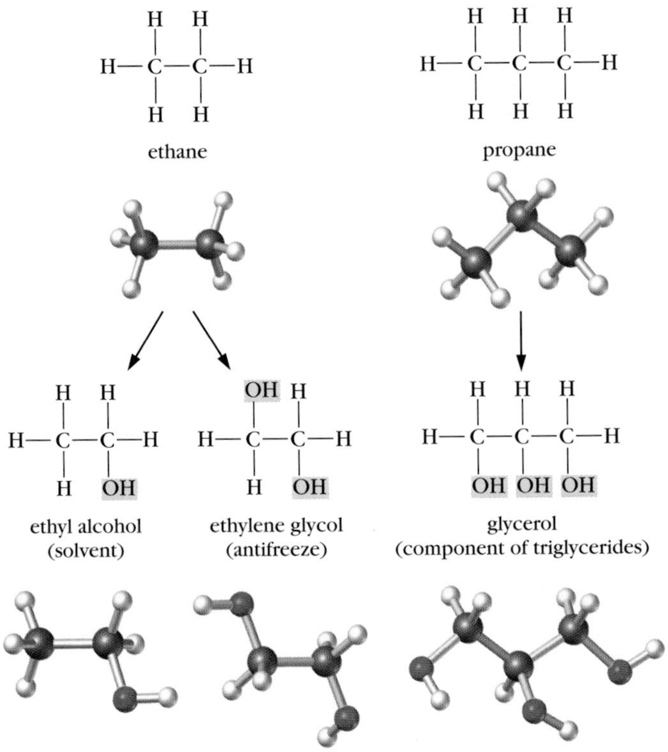

ethane

propane

ethyl alcohol
(solvent)

ethylene glycol
(antifreeze)

glycerol
(component of triglycerides)

Problem-Solving Example 8.2 Structural Formulas and Cl Substitution

Three hydrogens in propane can be replaced with Cl atoms to form different compounds. Draw the structural formulas for two such compounds.

Answer

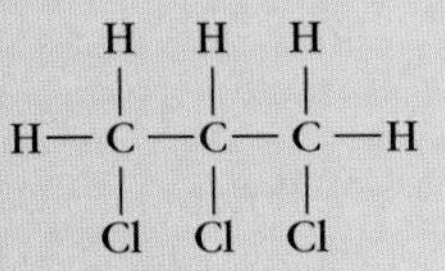

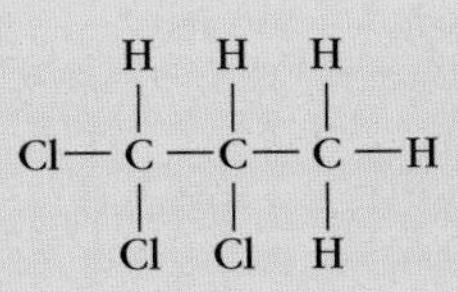

Explanation

One of the three Cl atoms can be placed on each of the carbons. Alternatively, two of the chlorine atoms can be bonded to one carbon atom, with the third on another carbon atom. Two possibilities are

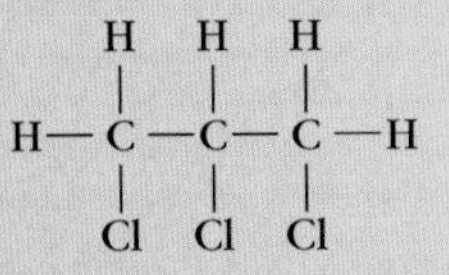

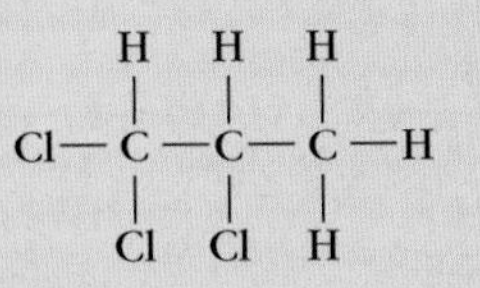

Both of these are known compounds.

Problem-Solving Practice 8.2

There are two other possible structures for three Cl atoms on a three-carbon chain. Draw these structures.

Chemistry in the News

KEEPING COOL, BUT SAFELY

Imagine a pet dog or cat sauntering into a driveway and licking a small puddle of a shiny, green liquid. The liquid's invitingly sweet taste appeals to the pet, which laps up more liquid. In a short time the pet is critically ill from drinking the green liquid.

The liquid is automobile antifreeze, which contains the toxic compound ethylene glycol (p. 324).

$$
\begin{array}{ccc}
 & H & H \\
 & | & | \\
H- & C- & C-H \\
 & | & | \\
 & OH & OH
\end{array}
$$

(Ethylene glycol is colorless; the green dye is added to distinguish antifreeze from other automobile fluids such as oil or transmission fluid.) Relatively small amounts of ethylene glycol can be fatal—only about 4 tablespoons can kill a large dog or adult human; fatal doses are even less for smaller pets (1 teaspoon for a cat) or children (2 tablespoons). More than 4000 people are poisoned by antifreeze annually in the United States, including several hundred children under the age of six.

Unless medical treatment is given promptly, ethylene glycol poisoning causes severe brain, heart, liver, and kidney damage because of toxic byproducts produced from ethylene glycol metabolism. Alcohol dehydrogenase, an enzyme

in cells, accelerates the conversion of ethylene glycol into several toxic organic acids, including oxalic acid. This acid combines with Ca^{2+} ions in blood to form insoluble calcium oxalate ($\Leftarrow$ *p. 324*). The calcium oxalate crystals block the flow of blood, causing brain damage. Kidneys are particularly susceptible to blockage by the precipitated calcium oxalate, which leads to kidney poisoning.

Recently a safe, effective antifreeze has been marketed that uses propylene glycol, rather than ethylene glycol.

$$
\begin{array}{cccc}
 & H & H & H \\
 & | & | & | \\
H- & C- & C- & C-OH \\
 & | & | & | \\
 & H & OH & H
\end{array}
$$

propylene glycol

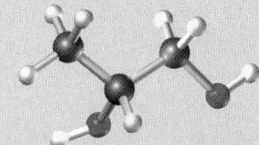

When mixed with an equal volume of water, propylene glycol prevents freezing down to $-35\ °C$, almost as good as an equal mixture of water and ethylene glycol ($-37\ °C$). Although a close molecular relative to ethylene glycol, propy-

Antifreeze. This antifreeze contains ethylene glycol.

lene glycol is not toxic to humans or animals. It is even used in human snack foods, cosmetics, candy, and in animal foods. The metabolism of propylene glycol is also catalyzed by alcohol dehydrogenase, but the breakdown products are nontoxic. This is evidence that sometimes just a slight change in molecular composition—in this case, just a difference of a CH_2 unit—can create dramatic differences in the chemical properties of two compounds.

Source:

Goldfarb, B. Antifreeze Antidote. *Chem Matters,* October 1996; pp. 4-6.

8.4 MULTIPLE COVALENT BONDS

A nonmetal atom with fewer than seven valence electrons can form covalent bonds in more than one way. The atom can share a single electron with another atom, which can also contribute a single electron; this process forms a shared electron pair—a *single* covalent bond. But the atom can also share two or three pairs of electrons with another atom, in which cases there will be two or three bonds, respectively, between the two atoms. When *two* shared pairs of electrons join the same pair of atoms, the bond is called a **double bond,** and when *three* shared pairs are involved, the bond is called a **triple bond.** Double and triple bonds are referred to as **multiple covalent bonds.**

In molecules where there are not enough electrons to complete all octets, one or more lone pairs of electrons from the terminal atoms can be shared with the central atom to form double or triple bonds, so that all atoms have octets of electrons (Guideline 5, ⇐ *p. 321*). Let's apply this guideline to the Lewis structure for formaldehyde, H_2CO. There are a total of 12 valence electrons (Guideline 1): 2 from two H atoms, 4 from the C atom (Group 4A), and 6 from the O atom (Group 6A). To complete noble gas configurations, H should form one bond, C four bonds, and O two bonds. Because C forms the most bonds, it is the central atom, and we can write a skeleton structure (Guideline 2).

$$H\!-\!\underset{\underset{O}{|}}{C}\!-\!H$$

You might have written the skeleton structure O—C—H—H, but remember that H forms only one bond. Another possible skeleton is H—O—C—H, but with this skeleton it is impossible to achieve an octet around carbon without having more than two bonds to oxygen.

Putting bonding pairs and lone pairs in the skeleton structure according to Guideline 3 yields a structure in which oxygen has an octet, but carbon does not.

$$H\!-\!\underset{\underset{:\ddot{O}:}{|}}{C}\!-\!H$$

We use one of the lone pairs on oxygen as a shared pair with carbon to change the C—O single bond to a double bond (Guideline 5).

$$H\!-\!\underset{:\ddot{O}\!\cdot\!)}{\overset{|}{C}}\!-\!H \qquad \text{to form} \qquad H\!-\!\underset{:\ddot{O}:}{\overset{|}{C}}\!-\!H \qquad \text{which is written} \qquad H\!-\!\underset{:\ddot{O}:}{\overset{\|}{C}}\!-\!H$$

This gives carbon and oxygen a share in an octet of electrons, and each hydrogen has a share of two electrons, accounting for all 12 valence electrons and verifying that this is the correct Lewis structure for formaldehyde.

The C=O combination, called the *carbonyl group,* is part of several functional groups that are very important in organic and biochemical molecules. The carbonyl-containing —CHO group that appears in formaldehyde is known as the *aldehyde functional group.*

Formaldehyde, a gas at room temperature, is the simplest compound with an aldehyde functional group.

Formaldehyde is produced in vast quantities (over a million tons annually in the United States), primarily to make hard plastics (Section 12.8) and in adhesives used to bind wood chips to produce particle board and plywood. An aqueous solution of formaldehyde (formalin) has been used as an antiseptic and embalming fluid, but its use for these purposes has declined because formaldehyde is a suspected carcinogen.

As another example of multiple bonds, let's write the Lewis structure for molecular nitrogen, N_2. There are a total of 10 valence electrons (5 from each N). If two nonbonding pairs of electrons (one pair from each N) become bonding pairs to give a triple bond, the octet rule is satisfied. This is the correct Lewis structure of N_2.

$$:\!\ddot{N}\!\overset{\cdot\,\downarrow\,\cdot}{\underset{\cdot\cdot}{}}\!\ddot{N}\!: \qquad \text{to form} \qquad :N\!\equiv\!N:$$

Exercise 8.2 Lewis Structures

Why is $:\!\ddot{N}\!-\!\ddot{N}\!:$ an incorrect Lewis structure for N_2?

A molecule can have more than one multiple bond, as in carbon dioxide, where carbon is the central atom. There are a total of 16 valence electrons in CO_2, and the skeleton structure uses 4 of them (2 shared pairs):

$$O\!-\!C\!-\!O$$

Adding lone pairs to give each O an octet of electrons uses up the remaining 12 electrons, but leaves C needing four more valence electrons to complete an octet.

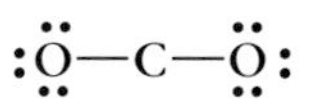

$$:\overset{..}{\underset{..}{O}}-C-\overset{..}{\underset{..}{O}}:$$

With no more valence electrons available, the only way that carbon can have four more valence electrons is to have a lone pair of electrons on each oxygen form a covalent bond to carbon. In this way the 16 valence electrons are accounted for, and each atom has an electron octet.

$$:\overset{..}{\underset{..}{O}}-C-\overset{..}{\underset{..}{O}}:\quad \text{forms}\quad :\overset{..}{O}=C=\overset{..}{O}:$$

Problem-Solving Example 8.3 Lewis Structures

Write Lewis structures for (a) carbon monoxide, CO, an air pollutant, (b) N_2O (laughing gas), a dental anesthetic, (c) hydrazoic acid, HN_3, an explosive compound; and (d) cyanide ion, CN^-, a poison.

Answer

(a) $:C\equiv O:$

(b) $:\overset{..}{N}=N=\overset{..}{O}:$

(c) $H-\overset{..}{N}=N=\overset{..}{N}:$

(d) $[:C\equiv N:]^-$

Explanation

(a) The molecule CO contains 10 valence electrons (4 from C and 6 from O). We start with a single C—O bond as a skeleton structure.

$$C-O$$

Putting lone pairs around the carbon and oxygen atoms satsfies the octet rule for one of the atoms, but not both. Therefore, lone pairs must become bonding pairs to make up this deficiency and achieve the correct Lewis structure.

$$:C-\overset{..}{\underset{..}{O}}:\quad \text{to form}\quad :C\equiv O:$$

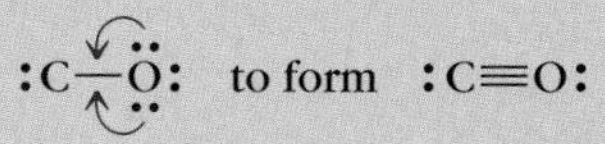

(b) In N_2O, nitrogen is the central atom (it can form more bonds than can oxygen) and there are 16 valence electrons (10 from two N and 6 from O). The skeleton structure is

$$N-N-O$$

Placing lone pairs on the terminal atoms uses the remaining valence electrons but leaves the central N atom without an octet.

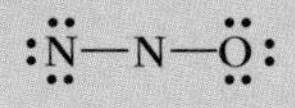

$$:\overset{..}{\underset{..}{N}}-N-\overset{..}{\underset{..}{O}}:$$

Converting two lone pairs to bonding pairs results in the correct Lewis structure.

$$:\overset{..}{N}=N=\overset{..}{O}:$$

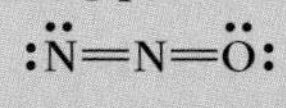

(c) There are 16 valence electrons in HN_3 (1 from H and 15 from three nitrogen atoms). Because hydrogen forms only one bond, a nitrogen atom must be the central atom. The skeleton structure with single bonds and lone pairs on the terminal nitrogen atoms uses all 16 valence electrons.

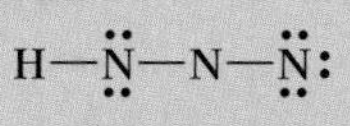

$$H-\overset{..}{\underset{..}{N}}-N-\overset{..}{\underset{..}{N}}:$$

However, the central nitrogen, with only four valence electrons, is four short of an octet. So a lone pair on each other nitrogen can be shifted to a bonding pair. Now each nitrogen atom has an octet, and hydrogen has a bonding pair.

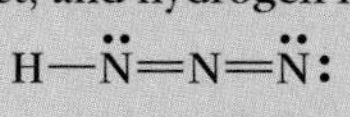

$$H-\overset{..}{N}=N=\overset{..}{N}:$$

(d) The cyanide ion has a total of 10 valence electrons (4 from C, 5 from N, plus 1 for the −1 charge). In this regard, it has the same number of valence electrons and the same Lewis structure as CO.

$$[:C\equiv N:]^-$$

Problem-Solving Practice 8.3

Write Lewis structures for the following: (a) nitrosyl ion, NO^+, (b) carbonyl chloride, $COCl_2$.

Exercise 8.3 Lewis Structures

Which of the following are correct Lewis structures and which are not? Explain what is wrong with the incorrect ones.

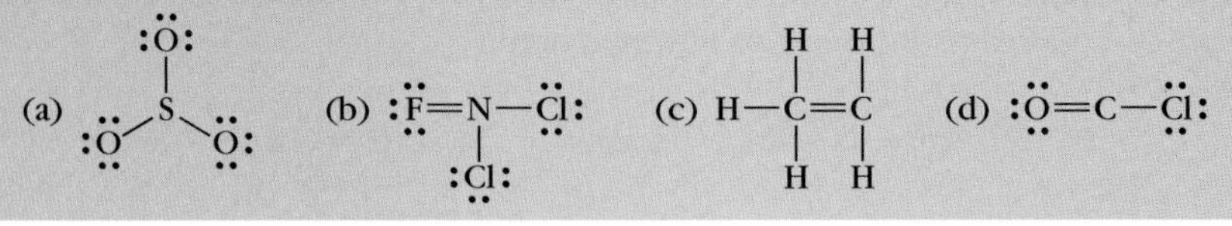

(a) :Ö:
     |
.. S ..
:Ö   Ö:
 ..   ..

(b) :F̈=N—C̈l:
         |
        :C̈l:

(c)
H   H
| |
H—C=C
| |
H   H

(d) :Ö=C—C̈l:

 8.5 MULTIPLE COVALENT BONDS IN HYDROCARBONS

 CD-ROM Screen 11.3: Hydrocarbons

Carbon atoms are connected by double bonds in some compounds and triple bonds in others. **Alkenes** *are hydrocarbons that have one or more carbon-carbon double bonds,* C=C. The general formula for alkenes with one double bond is C_nH_{2n} where n = 2, 3, 4, and so on. The first two members of the alkene series are ethene (CH$_2$=CH$_2$) and propene (CH$_3$CH=CH$_2$), commonly called ethylene and propylene, particularly when referring to the polymers polyethylene and polypropylene, which will be discussed in Section 12.8.

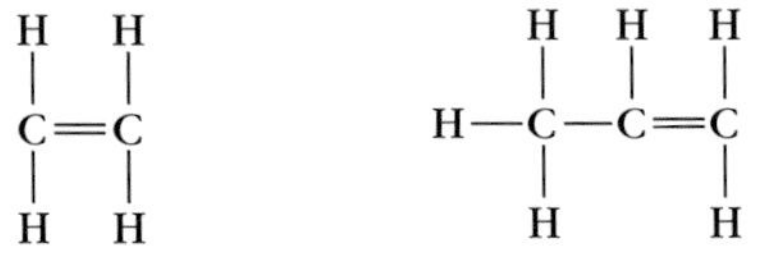

H   H
| |
C=C
| |
H   H

ethylene

H   H   H
| | |
H—C—C=C
| |
H   H

propylene

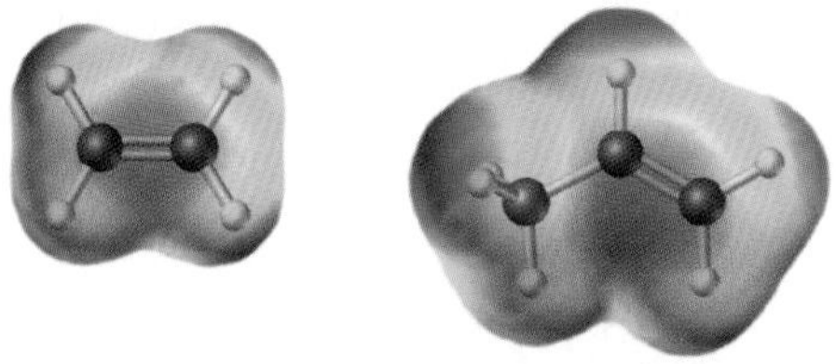

Alkenes are said to be **unsaturated hydrocarbons.** The carbon atoms connected by double bonds are the *unsaturated sites* — they contain fewer hydrogen atoms than the corresponding alkanes (ethylene, CH$_2$=CH$_2$; ethane, CH$_3$CH$_3$):

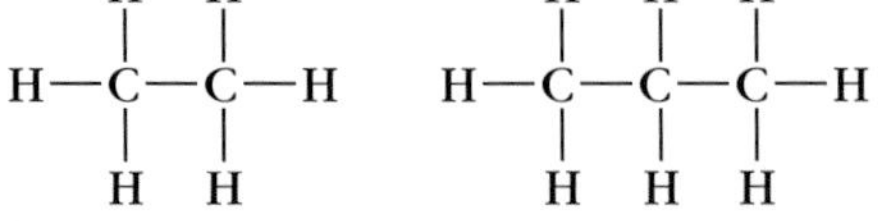

H   H
| |
H—C—C—H
| |
H   H

H   H   H
| | |
H—C—C—C—H
| | |
H   H   H

Alkenes are named by using the name of the corresponding alkane (⬅ *p. 83)* to indicate the number of carbons and the suffix *-ene* to indicate one or more double bonds. The first member, ethylene (ethene), is the most important raw material in the organic chemical industry, where it is used in making polyethylene, antifreeze (ethylene glycol), ethanol, and other chemicals.

 Exercise 8.4 Alkenes

(a) Write the molecular formula and structural formula of an alkene with five carbon atoms and one C=C double bond.
(b) How many different alkenes have five carbon atoms and one C=C double bond?

Hydrocarbons with one or more triple bonds, —C≡C—, per molecule are called **alkynes.** The general formula for alkynes with one triple bond is C_nH_{2n-2}, where $n = 2, 3, 4$, and so on. The simplest one is ethyne, commonly called acetylene (C_2H_2).

acetylene

An oxyacetylene torch cutting steel. An oxyacetylene torch cuts through the steel door of a Titan II missile silo, which is being retired as part of a disarmament treaty. A mixture of acetylene and oxygen burns with a flame hot enough (3000 °C) to cut steel.

Double Bonds and Isomerism

The C=C double bond creates an important difference between alkanes and alkenes—the degree of flexibility of the carbon-carbon bonds in the molecules. The C—C single bonds in alkanes allow the carbon atoms to rotate freely along the C—C bond axis (Figure 8.2). But in alkenes, the C=C double bond prevents such free rotation. This limitation is responsible for the *cis-trans* isomerism of alkenes.

Two or more compounds with the same molecular formula but different arrangements of atoms are called isomers (⬅ *p. 83*). *Cis-trans* isomerism occurs due to nonrotation around a C=C double bond. When two atoms or groups of atoms are attached to carbon atoms on the *same* side of the C=C bond, they are said to be *cis* to each other and the compound is the **cis isomer;** when they are on opposite sides, they are *trans* to each other and the compound is the **trans isomer,** as in the *cis* and *trans* isomers of ClHC=CHCl, 1,2-dichloroethene (the 1 and 2 indicate that the two chlorine atoms are attached to the first and second carbon atoms, respectively).

Cis-trans isomerism is also called geometric isomerism.

If free rotation could occur around a carbon-carbon double bond, these two molecules would be the same. Therefore, there would only be a single compound.

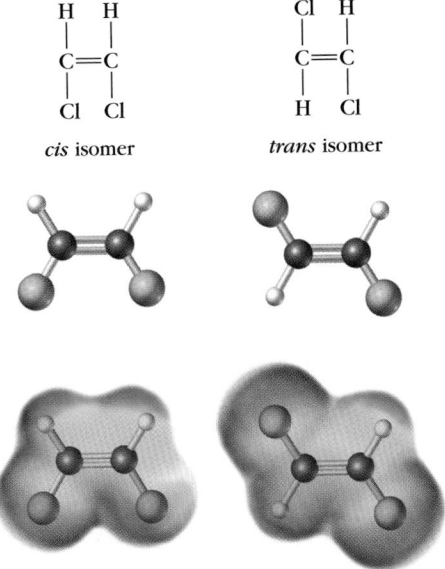

cis isomer   *trans* isomer

Experimental evidence confirms the existence of two compounds with the same set of bonds, but with a difference in the location in space of the two chlorine atoms. The *cis* isomer has two chlorine atoms on the *same* side of the double bond; the *trans* isomer has two chlorine atoms on *opposite* sides of the double

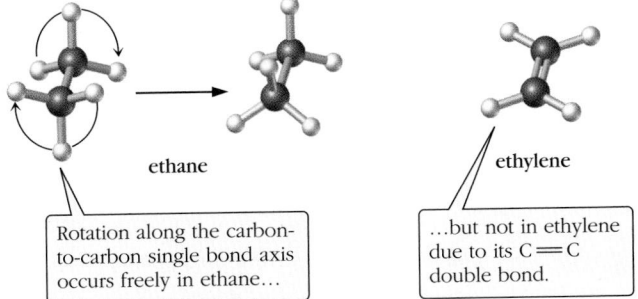

ethane

Rotation along the carbon-to-carbon single bond axis occurs freely in ethane...

ethylene

...but not in ethylene due to its C=C double bond.

Figure 8.2   Nonrotation around C=C. At room temperature rotation along the carbon-to-carbon axis occurs freely in ethane, but not in ethylene because of its double bond. This can be viewed by using the Chemscape CHIME program, a free plug-in available at **http://www.mdli.com.**

bond. Because of their different geometries, the two compounds have different physical properties, including melting point, boiling point, and density.

Physical property	*cis*-1,2-dichloroethene	*trans*-1,2-dichloroethene
Melting point	−80.5 °C	−50 °C
Boiling point (at 1 atm)	60.3 °C	47.5 °C
Density (at 20 °C)	1.284 g/mL	1.265 g/mL

 Exercise 8.5 *Cis-Trans* Isomerism and Biomolecules

Maleic acid and fumaric acid are biomolecules very important in metabolism that undergo different reactions because they are *cis-trans* isomers.

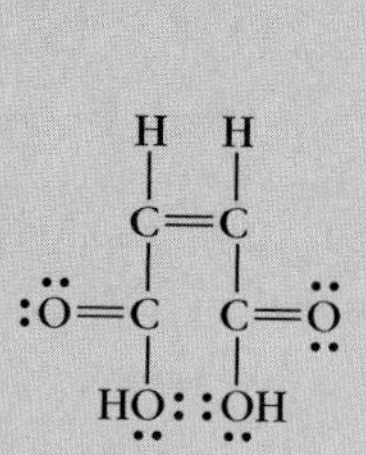

maleic
acid

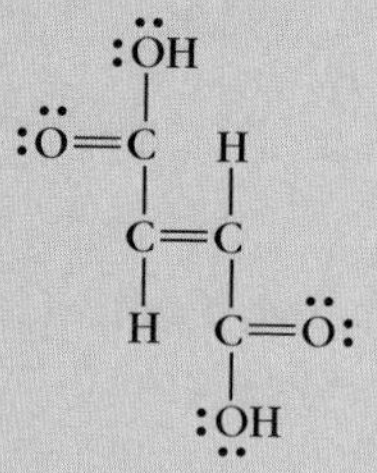

fumaric
acid

Identify the *cis* isomer and the *trans* isomer.

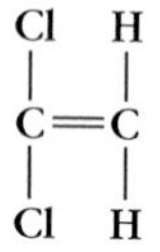

Cis-trans isomerism in alkenes is possible *only when each of the carbon atoms connected by the double bond has two different groups attached.* (For the sake of simplicity, the word "groups" refers to both atoms and groups of atoms.) For example, two chlorine atoms can also bond to the *first* carbon to give 1,1-dichloroethene, which does not have *cis* and *trans* isomers because each carbon atom is attached to two identical atoms (carbon 1 to two chlorines, carbon 2 to two hydrogens).

When there are four or more carbon atoms in an alkene, the possibility exists for *cis* and *trans* isomers even when only carbon and hydrogen atoms are present. For example, 2-butene has both *cis* and *trans* isomers. (The 2 indicates that the double bond is at the second carbon atom, with the straight carbon chain beginning with carbon 1.)

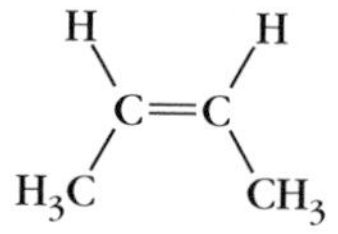

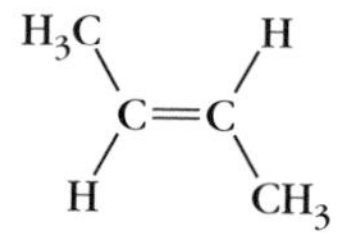

Cl   H
 | |
 C = C
 | |
Cl   H

1,1-dichloroethene

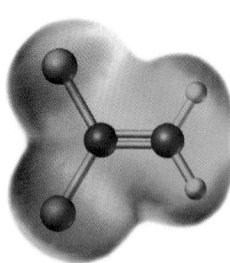

Physical property	*cis*-2-butene	*trans*-2-butene
Melting point	−138.9 °C	−105.5 °C
Boiling point (at 1 atm)	3.7 °C	0.88 °C

Problem-Solving Example 8.4 *Cis* and *Trans* Isomers

Which of these molecules can have *cis* and *trans* isomers? For those that do, write the structural formulas for the two isomers and label them *cis* and *trans*.

(a) $(CH_3)_2C=CCl_2$ (b) $CH_3ClC=CClCH_3$
(c) $CH_3BrC=CClCH_3$ (d) $(CH_3)_2C=CBrCl$

Answer (b) and (c)

(b)

H₃C   CH₃   Cl   CH₃
  \\   /   \\   /
   C=C   C=C
  /   \\   /   \\
 Cl   Cl   H₃C   Cl

      cis   *trans*

(c)

H₃C   CH₃   Br   CH₃
  \\   /   \\   /
   C=C   C=C
  /   \\   /   \\
 Br   Cl   H₃C   Cl

      cis   *trans*

Explanation Because both of the groups on each carbon are the same, (a) cannot have *cis* and *trans* isomers. In (b), the two —CH₃ groups and the two Cl atoms can both be on the same side of the C=C bond (the *cis* form) or on opposite sides (the *trans* form). The same holds true for the —CH₃ groups in (c). There are two CH₃ groups on the same carbon in (d), so *cis* and *trans* isomers are not possible.

Problem-Solving Practice 8.4

Which of these molecules can have *cis* and *trans* isomers? For those that do, write the structural formulas for the two isomers and label them *cis* and *trans*.

(a)
```
     H₃C   H
       |   |
H₃C — C = C — CH₃
```
2-methyl-2-butene

(b)
```
      H   H
      |   |
H₂C = C — C — CH₃
          |
          H
```
1-butene

(c)
```
      H   Cl  H
      |   |   |
Br — C — C = C — CH₃
      |
      H
```
1-bromo-2-chloro-2-butene

Cis-trans isomerism plays a vital role in vision. Visual pigments in our eyes contain rhodopsin, a combination of 11-*cis*-retinal and opsin, a protein. When rhodopsin absorbs radiation in the visible region of the spectrum, the 11-*cis*-retinal is converted to 11-*trans*-retinal. This change in structure causes opsin to separate from the 11-*trans*-retinal, causing a rapid cascade of other changes that results in sending an electrical impulse along the optic nerve to the brain. The brain interprets the signal and we see! An enzyme rapidly converts 11-*trans*-retinal back to the *cis* isomer, which combines with opsin to form rhodopsin, now ready to absorb light and "see" again. *A Chemistry You Can Do* in Chapter 13 (Kinetics and Vision) explores the *cis-trans* conversion of retinal.

Cis-Trans Isomerism in Fats and Oils

Fats are solids at room temperature; *oils* are obviously liquids at this temperature. Edible fats and oils (⬅ *Section 3.12*) are all triglycerides because they share the common structural feature of a three-carbon backbone from glycerol to which three long-chain fatty acids are bonded. Fatty acids generally have an even number of carbon atoms, ranging from 4 to 20. The three fatty acids in a triglyceride molecule can be the same or different.

Triglycerides are formed by the reaction of three moles of fatty acids with one mole of glycerol.

$$3 \text{ Fatty acids} + \text{glycerol} \longrightarrow \text{triglyceride} + 3 \text{ H}_2\text{O}$$

For example, the three fatty acids can be stearic acid, oleic acid, and linoleic acid.

Fatty acid 1

$$HOOC—CH_2CH_2CH_2CH_2CH_2CH_2CH_2CH_2CH_2CH_2CH_2CH_2CH_2CH_2CH_2CH_2CH_3$$

stearic acid, a saturated fatty acid

Fatty acid 2

$$HOOC—(CH_2)_7—CH=CH—(CH_2)_7—CH_3$$

oleic acid, a monounsaturated fatty acid

Fatty acid 3

$$HOOC—(CH_2)_7—CH=CH—CH_2—CH=CH—(CH_2)_4—CH_3$$

linoleic acid, a polyunsaturated fatty acid

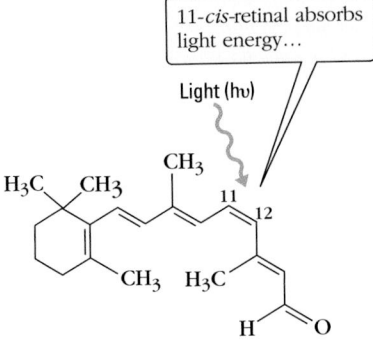

11-*cis*-retinal absorbs light energy…

Light (hυ)

11-*cis*-retinal

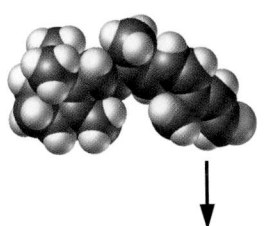

…converting it to 11-*trans*-retinal, which is not responsive to light. This small change initiates a cascade of events that sends a signal to the brain's visual centers and we see!

11-*trans*-retinal

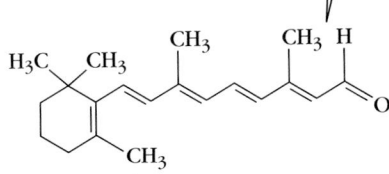

An enzyme quickly converts it back to the 11-*cis* form, ready to "see" again.

11-*cis*-retinal

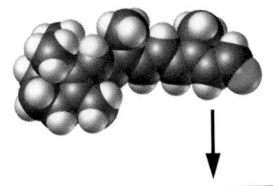

The resulting triglyceride is

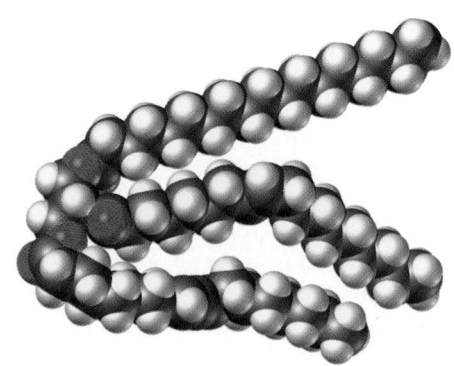

triglyceride

CD-ROM Screen 11.7:
Functional Groups (2); Fats and
Oils

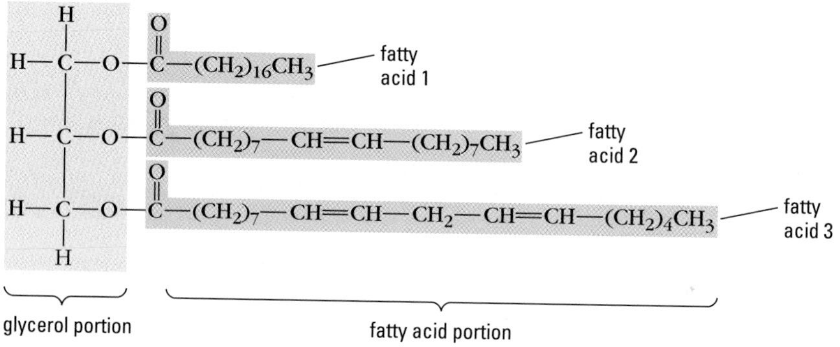

glycerol portion

fatty acid portion

fatty acid 1

fatty acid 2

fatty acid 3

Animal and vegetable fats and oils vary considerably because their fatty acids differ in the length of the hydrocarbon chain and the number of double bonds in the chain. **Saturated fats,** such as stearic acid, contain only C—C single bonds in their hydrocarbon chains along with C—H bonds; **unsaturated fats** contain one or more C=C double bonds in their hydrocarbon chains. Oleic acid, with one C=C bond per molecule, is classified as a *monounsaturated* fatty acid, whereas fatty acids such as linoleic acid, with two or more C=C double bonds per molecule, are termed *polyunsaturated.* Animal fats generally contain mainly saturated fatty acids, while most vegetable oils are high in unsaturated fatty acids. In general, the greater the percentage of unsaturated fatty acids in a triglyceride, the lower is its melting point. Thus, highly unsaturated fats are liquids at room temperature (oils); animal fats high in saturated fats are solids.

The melting points of fats reflect the shape of the molecules. The C—C single bonds in the hydrocarbon portion of saturated fats are all the same, and they allow the molecule to adopt a rather linear shape, which fits nicely into a solid packing arrangement. The C=C double bonds in unsaturated fats require a different shape. Most natural unsaturated fats are in the *cis* configuration at the C=C double bond, which puts a "kink" into the molecule, preventing unsaturated fat molecules from packing regularly into a solid. *Trans* fatty acid molecules can be more linear, similar to those of saturated fats.

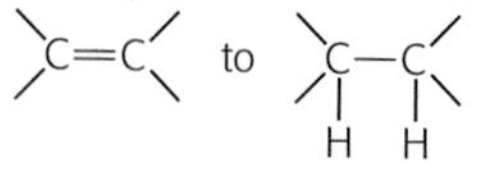

Vegetable oils can be converted to semisolids by *hydrogenation* — reaction of the oil with H_2 in the presence of a metal like palladium, which accelerates the reaction. The hydrogen reacts by addition to some of the C=C double bonds. The degree of saturation is limited by carefully controlling reaction conditions so that the product has the proper softness and spreadability. Margarine, shortening (Crisco), and many snack foods contain partially hydrogenated fats, noted on the label as "partially hydrogenated soybean oil" (or cottonseed or other vegetable oils).

Linoleic and linolenic acids are essential fatty acids; we must have them in our diet. The body cannot manufacture them.

Hydrogenation converts

Hydrogenation: H_2 added to C$=$C bond

A portion of an unsaturated vegetable oil

A portion of a hydrogenated vegetable oil

Diagram of a hydrogenation reaction.

PKB 1071

INGREDIENTS: Bleached Enriched Flour (Wheat Flour, Malted Barley Flour, Niacin, Reduced Iron, Thiamine Mononitrate [Vitamin B₁], Riboflavin [Vitamin B2], Folic Acid), Partially Hydrogenated Soybean Oil, Sugar, Salt, Dehydrated Cheddar Cheese (Pasteurized Cow's Milk, Cheese Cultures, Salt, Enzymes), Yeast Extract, Whey, Sodium Bicarbonate (Leavening), Natural Flavors, Citric Acid, Paprika, Sour Cream, Corn Syrup, Turmeric, Red Pepper, Monosodium Glutamate, Lactic Acid, Artificial Colors (FD & C Yellow #5 & #6) Reg. Penna. Dept. Agr.

STAUFFER BISCUIT COMPANY
YORK, PA 17405

Partially hydrogenated fats. Most snack foods contain partially hydrogenated fats.

8.6 BOND PROPERTIES: BOND LENGTH AND BOND ENERGY

Bond Length

The most important factor determining **bond length,** the distance between nuclei of two bonded atoms, is the sizes of the atoms themselves (⬅ *p. 294).* The bond length can be considered as the sum of the atomic radii of the two bonded atoms. When you compare bonds of the same type (single, double, or triple), the bond length will be greater for the larger atoms (Table 8.1). Thus, single bonds with carbon increase in length along the following series:

$$C-N < C-C < C-P$$

Increase in bond length ⟶

C—N   C—C   C—P

147 pm   154 pm   187 pm

Similarly, a C$=$O bond will be shorter than a C$=$S bond, and a C$\equiv$N bond will be shorter than a C$\equiv$C bond. Each of these trends can be predicted from the relative sizes shown in Figure 8.3 and confirmed by the average bond lengths given in Table 8.1.

The effect of bond type is evident when bonds between the same two atoms are compared. For example, structural data show that the bonds become shorter in the series C—O > C$=$O > C$\equiv$O. As the electron density between the atoms increases, the bond lengths decrease because the atoms are pulled together more strongly.

Bond	C—O	C=O	C≡O
Bond length (pm)	143 pm	122 pm	113 pm

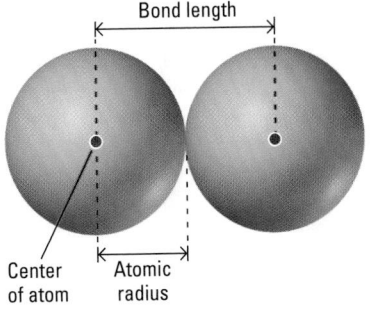

Bond length

Center of atom   Atomic radius

CD-ROM Screen 9.9: Bond Properties

Bond lengths are given in picometers (pm) in Table 8.1, but many scientists use nanometers (1 nm = 10^3 pm) or the older unit of angstroms (Å). 1 Å equals 100 pm. A C—C single bond is 0.154 nm, 1.54 Å, or 154 pm in length.

TABLE 8.1 Some Average Single and Multiple Bond Lengths (in picometers, pm)*

Single bonds

	I	Br	Cl	S	P	Si	F	O	N	C	H
H	161	142	127	132	138	145	92	94	98	110	74
C	210	191	176	181	187	194	141	143	147	154	
N	203	184	169	174	180	187	134	136	140		
O	199	180	165	170	176	183	130	132			
F	197	178	163	168	174	181	128				
Si	250	231	216	221	227	234					
P	243	224	209	214	220						
S	237	218	203	208							
Cl	232	213	200								
Br	247	228									
I	266										

Multiple bonds

$N{=}N$	120		$C{=}C$	134
$N{\equiv}N$	110		$C{\equiv}C$	121
$C{=}N$	127		$C{=}O$	122
$C{\equiv}N$	115		$C{\equiv}O$	113
$O{=}O$ (in O_2)	112		$N{\equiv}O$	108
$N{=}O$	115			

* 1 pm = 10^{-12} m.

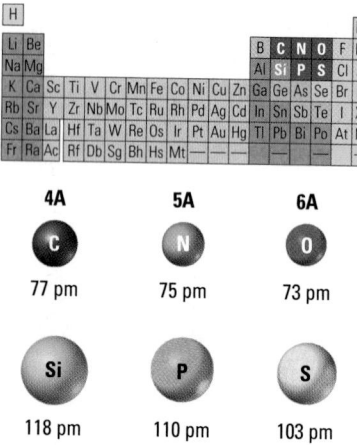

Figure 8.3 Relative atom sizes for second- and third-period elements in Groups 4A, 5A, and 6A.

The bond lengths in Table 8.1 are average values, because variations in neighboring parts of a molecule can affect the length of a particular bond. For example, the C—H bond has a length of 105.9 pm in acetylene, HC≡CH, but a length of 109.3 pm in methane, CH_4. Although there can be a variation of as much as 10% from the average values listed in Table 8.1, the average bond lengths are useful for estimating bond lengths and building models of molecules.

Problem-Solving Example 8.5   Bond Lengths

In each pair of bonds, predict which will be shorter.

(a) Si—O or P—O   (b) C=C or C—C   (c) C=C or C=O

Answer   The shorter bonds will be (a) P—O, (b) C=C, (c) C=O.

Explanation

(a) P—O is shorter than Si—O because a P atom is smaller than a Si atom.
(b) C=C is shorter than C—C because the more electrons that are shared by atoms, the more closely the atoms are pulled together.
(c) C=O is shorter than C=C because an O atom is smaller than a C atom.

Problem-Solving Practice 8.5

Explain the increasing order of bond lengths in these bond pairs.
(a) C—S is shorter than C—Si.
(b) C—Cl is shorter than C—Br.
(c) N≡O is shorter than N=O.

Chemistry in the News

TRANS FATTY ACIDS

Concerns have been raised by health officials about the high level of saturated fats and cholesterol in the typical American diet. The concerns focus on the likely connection between such fats and cholesterol with cardiovascular disease, particularly atherosclerosis, the narrowing of arteries that restricts blood flow. A diet high in saturated fats has been shown to increase total blood cholesterol and low-density lipoproteins (LDLs), compounds that transport cholesterol in blood serum. There is a strong positive correlation between heart disease and high levels of dietary saturated fatty acids with 12-, 14-, or 16-carbon atoms. The 14-carbon myristic acid is especially harmful, although no definitive mechanism has been established as to how the 12-carbon to 16-carbon saturated fatty acids cause an increase in blood cholesterol.

Unsaturated fats have long been considered to be far healthier than saturated ones. Such is the case, but apparently only if the unsaturated fatty acids are the *cis* isomers. The good news is that most natural fatty acids are in the *cis* form, not the *trans* form. The *cis* C=C double bonds cause the molecule to bend back on itself; the more *cis* double bonds, the greater the degree of bending. The shape of the *trans* isomer, on the other hand, is similar to that of saturated fats. Therein lies a possible problem. A recent clinical study reported that diets rich in *trans* fats are as likely as saturated fats to increase total cholesterol and LDLs, and also lower high-density lipoproteins (HDLs, the anticholesterol kind).

Trans fats show up in our diets mainly in the partially hydrogenated fats found in most processed foods such as snack foods, cookies, and crackers, as well as in margarine, spreads, and frying fats used by fast food establishments. During partial hydrogenation of vegetable oils, some *cis* double bonds are converted to the *trans* form. Stick mar-garine is more fully hydrogenated and harder than the softer, tub margarine. Correspondingly, stick margarine made from partially hydrogenated soybean and cottonseed oils contains 21% *trans* fats; there are only 11% *trans* fats in soft tub margarine, generally made from corn or canola oils. The American Heart Association recommends soft margarine rather than stick margarine as a butter substitute.

The effects of *trans* fats on cardiovascular health are controversial because, as this is written (2000), the jury is still out awaiting *the* definitive study that would prove dietary *trans* fats harmful. Recent clinical studies have come down on opposite sides of the issue. The methodology of these studies has been questioned, raising doubts about their validity. Some consumer advocates, not willing to wait for the defining clinical tests, have petitioned the Food and Drug Administration to require listing on food labels the mass of *trans* fat. Are *trans* fats the health menace some portray them to be? Stay tuned!

Source:

FDA Consumer Magazine, January-February 2000.

(a) stearic acid

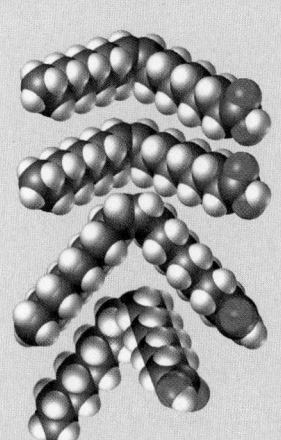

(b) *cis*-oleic acid

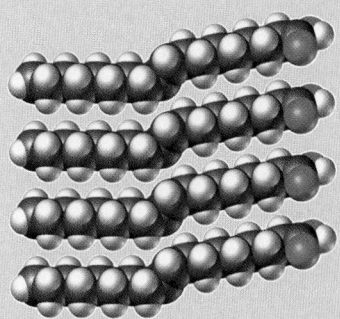

(c) *trans*-oleic acid

Space-filling models of (a) stearic acid, an 18-carbon saturated fatty acid, (b) *cis*-oleic acid, an 18-carbon unsaturated fatty acid, and (c) *trans*-oleic acid. The positions of substituents around the C=C double bond creates a difference between the shape of the *cis* form and that of the *trans* form, which more closely resembles a saturated fatty acid.

Bond Energy

The quantity of energy released when a mole of a particular bond is made equals that needed when 1 mol of that bond is broken. For example, the H—Cl bond energy is 431 kJ/mol, indicating that 431 kJ must be supplied to break 1 mol of H—Cl bonds ($\Delta H° = +431$ kJ); conversely, when 1 mol of H—Cl bonds is formed, 431 kJ is released ($\Delta H° = -431$ kJ).

The bond enthalpies in Table 8.2 are for gas phase reactions. If liquids or solids are involved there are additional energy transfers for the phase changes needed to convert the liquids or solids to the gas phase. We shall restrict our use of bond enthalpies to gas phase reactions for that reason.

In any chemical reaction, bonds are broken and new bonds are made. The energy required to break bonds (an **endothermic** reaction) and the energy released when bonds are formed (an **exothermic** reaction) contribute to the enthalpy change for the overall reaction (⬅ *Section 6.7*)

You've seen that as the number of bonding electrons between a pair of atoms increases, the bond length decreases. It is therefore reasonable to expect that multiple bonds are stronger than single bonds. *As the electron density (and number of bonds) between two atoms increases, the bond gets shorter and stronger.* For example, the bond energy of C=O in CO_2 is 803 kJ/mol and that of C≡O is 1073 kJ/mol. In fact, the C≡O triple bond in carbon monoxide is the strongest known covalent bond. Data on the strengths of bonds between atoms in gas phase molecules are summarized in Table 8.2.

The data in Table 8.2 can help us understand why an element such as nitrogen, which forms many compounds with oxygen, is unreactive enough to remain in the earth's atmosphere as N_2 molecules even though there is plenty of O_2 to react with. According to the table, the two N atoms are connected by a very strong bond (946 kJ/mol). Reactions in which N_2 combines with other elements are less likely to occur, because they require breaking a very strong N≡N bond. This allows us to inhale and exhale N_2 without its undergoing any chemical change. If this were not the case and N_2 reacted readily at body temperature (37 °C) to form oxides and other compounds, there would be severe consequences for us.

We can use data from Table 8.2 to estimate $\Delta H°$ for reactions such as the reaction of hydrogen gas with chlorine gas. Breaking the covalent bond of H_2 requires an input to the system of 436 kJ/mol; breaking the covalent bond in Cl_2 requires 242 kJ/mol. In both of these cases the sign of the enthalpy change is positive. Forming a covalent bond in HCl transfers 431 kJ/mol out of the system. Since 2

| TABLE 8.2 | Average Bond Energies (in kJ/mol)* |

Single bonds

	I	Br	Cl	S	P	Si	F	O	N	C	H
H	299	366	431	347	322	323	566	467	391	416	436
C	213	385	327	272	264	301	486	336	285	356	
N	—	—	193	—	~200	335	272	201	160		
O	201	—	205	—	~340	368	190	146			
F	—	—	255	326	490	582	158				
Si	234	310	391	226	—	226					
P	184	264	319	—	209						
S	—	213	255	226							
Cl	209	217	242								
Br	180	193									
I	151										

Multiple bonds

N=N	418	C=C		598
N≡N	946	C≡C		813
C=N	616	C=O (in CO_2, O=C=O)		803
C≡N	866	C=O (as in $H_2C=O$)		695
O=O (in O_2)	498	C≡O		1073

* Data from Cotton, F. A., Wilkinson, G., and Gaus, P. L. *Basic Inorganic Chemistry, 3rd ed.* New York: Wiley, 1995; p. 12.

mol of HCl bonds are formed, there will be 2 mol $\times$ 431 kJ/mol = 862 kJ transferred *out*, which makes the sign of this enthalpy change negative. If we represent bond enthalpy by the letter D, with a subscript to show which bond it refers to, the net transfer of energy is

$$\Delta H° = \{[(1 \text{ mol H—H}) \times D_{H-H}] + [(1 \text{ mol Cl—Cl}) \times D_{Cl-Cl}]\}$$

$$- [(2 \text{ mol H—Cl}) \times D_{H-Cl}]$$

$$= [(1 \text{ mol H—H})(436 \text{ kJ/mol})] + [(1 \text{ mol Cl—Cl})(242 \text{ kJ/mol})]$$

$$- (2 \text{ mol H—Cl})(431 \text{ kJ/mol})$$

$$= -184 \text{ kJ}$$

This differs only very slightly from the experimentally determined value of -184.614 kJ.

As illustrated in Figure 8.4, we can think of the process in the calculation in terms of breaking all the bonds in each reactant molecule and then forming all the bonds in the product molecules. Each bond enthalpy was multiplied by the number of moles of bonds that were broken or formed. For bonds in reactant molecules, we added the bond enthalpies because breaking bonds is endothermic. For products, we subtracted the bond enthalpies because bond formation is an exothermic process. This can be summarized in the following equation:

$$\Delta H° = \Sigma[(\text{moles of bonds}) \times D \text{ (bonds broken)}]$$

$$- \Sigma[(\text{moles of bonds}) \times D \text{ (bonds formed)}]$$

There are several important points about the bond enthalpies in Table 8.2:

- The enthalpies listed are often average bond enthalpies and may vary from molecule to molecule. For example, the enthalpy of a C—H bond is given as 413 kJ/mol, but C—H bond strengths are affected by other atoms and bonds in the same molecule. Depending on the molecule, the energy required to break a mole of C—H bonds may vary by 30 to 40 kJ/mol, and so the values in Table 8.2 can only be used to estimate an enthalpy change, not to calculate it exactly.

- The enthalpies in Table 8.2 are for breaking bonds in molecules in the gaseous state. If a reactant or product is in the liquid or solid state, the energy required

Bond enthalpy is represented by the letter D because it refers to dissociation of a bond.

CD-ROM Screen 9.10: Bond Energy and ΔH_{rxn}

The Σ (Greek capital letter *sigma*) represents summation. We add the bond enthalpies for all bonds broken, and we subtract the bond enthalpies for all bonds formed. This equation and the values in Table 8.2 allow us to estimate enthalpy changes for a wide variety of gas phase reactions.

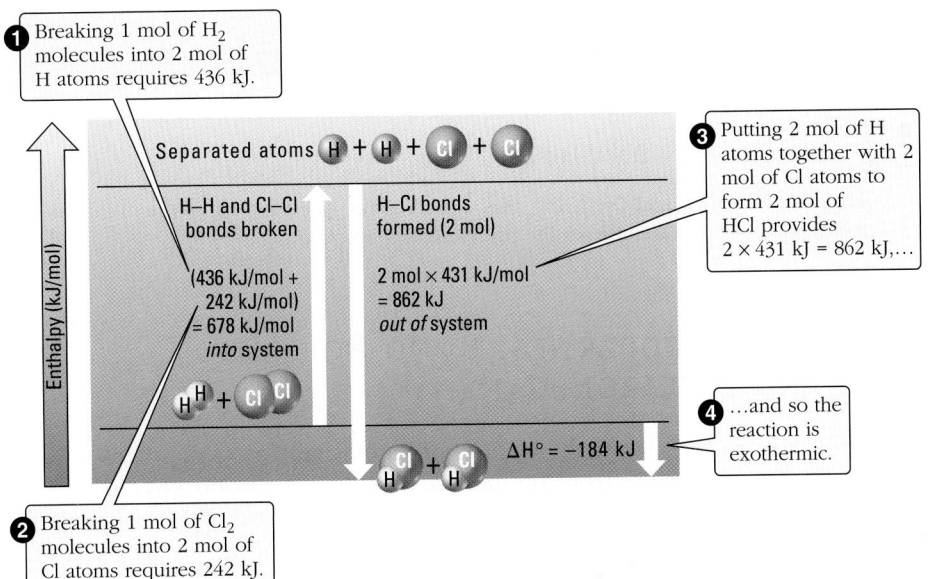

Figure 8.4 Stepwise energy changes in the reaction of hydrogen with chlorine. The enthalpy change for the reaction is -184 kJ; $\Delta H° = 436$ kJ + 242 kJ $-$ 862 kJ = -184 kJ.

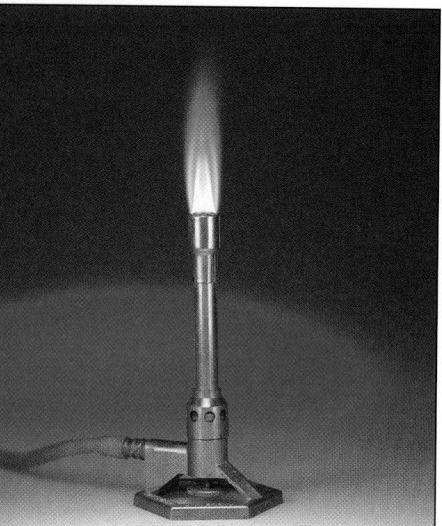

Burning methane in a Bunsen burner. Combustion of methane is highly exothermic, releasing energy to the surroundings for every mole of CH_4 that burns.

The measured value for the enthalpy of combustion of methane is −802 kJ; the average bond enthalpies used in Problem-Solving Example 8.6 give a good estimate.

to convert it to or from the gas phase will also contribute to the enthalpy change and should be accounted for.

• Multiple bonds, shown as double and triple lines between atoms, are listed at the bottom of Table 8.2. In some cases, different enthalpies are given for multiple bonds in specific molecules such as $O_2(g)$ or $CO_2(g)$.

Problem-Solving Example 8.6 Estimating $\Delta H°$ from Bond Enthalpies

Methane, CH_4, is the principal component of natural gas. Estimate $\Delta H°$ for burning a mole of methane according to the following equation:

$$CH_4(g) \;+\; 2\,O_2(g) \longrightarrow CO_2(g) \;+\; 2\,H_2O(g)$$

Answer $\Delta H° = -814\text{ kJ}$

Explanation The pictures of the molecules and their bonds show that there are four C—H bonds in a CH_4 molecule. In each O_2 molecule there is one O=O bond, in each H_2O molecule there are two O—H bonds, and in each CO_2 molecule there are two C=O bonds. The O_2 molecule and the CO_2 molecule are special cases; their bond enthalpies are listed at the bottom of Table 8.2.

$$\Delta H° = \Sigma[(\text{moles of bonds}) \times D\,(\text{bonds broken})]$$
$$- \Sigma[(\text{moles of bonds}) \times D\,(\text{bonds formed})]$$
$$= \{[(4\text{ mol C—H}) \times D_{C-H}] + [(2\text{ mol O=O}) \times D_{O=O}]\}$$
$$- \{[(2\text{ mol C=O}) \times D_{C=O}] + [(4\text{ mol O—H}) \times D_{O-H}]\}$$
$$= [(4\text{ mol C—H})(416\text{ kJ/mol}) + (2\text{ mol O=O})(498\text{ kJ/mol})]$$
$$- [(2\text{ mol C=O})(803\text{ kJ/mol}) + (4\text{ mol O—H})(467\text{ kJ/mol})]$$
$$= -814\text{ kJ}$$

 The reaction is a combustion and should be exothermic, so a negative $\Delta H°$ is expected. Also, there is the same number of bonds in reactants and products, but the product bonds are stronger, meaning that the potential energy of the products is lower than for reactants and $\Delta H°$ will be negative.

Problem-Solving Practice 8.6

Use values from Table 8.2 to estimate the enthalpy change when hydrogen and oxygen combine according to the following equation: $2\,H_2(g) + O_2(g) \rightarrow 2\,H_2O(g)$

C **Exercise 8.6** Bond Length

Arrange C=N, C≡N, and C—N in order of decreasing bond length. Is the order for decreasing bond energy the same or the reverse order? Explain.

8.7 BOND PROPERTIES: BOND POLARITY AND ELECTRONEGATIVITY

CD-ROM Screen 9.11: Bond Polarity and Electronegativity

In a molecule such as H_2 or F_2, where both atoms are the same, there is *equal* sharing of the bonding electron pair and the bond is a **nonpolar covalent bond.** When two different atoms are bonded, however, the sharing of the bonding electrons is generally *unequal* and results in a displacement of the bonding electrons toward one of the atoms. If the displacement is complete, electron transfer occurs

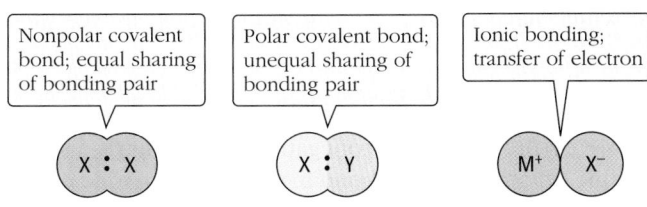

Figure 8.5 **Nonpolar covalent, polar covalent, and ionic bonding.**

and the bond is ionic. If the displacement is less than complete, the bonding electrons are shared *unequally,* and the bond is said to be a **polar covalent bond** (Figure 8.5). As you will see in Chapters 9 and 12, properties of molecules are dramatically affected by bond polarity.

Linus Pauling, in 1932, first proposed the concept of electronegativity based on an analysis of bond energies. **Electronegativity** represents the ability of an atom in a covalent bond to attract *shared* electrons to itself (Figure 8.6).

Pauling's electronegativity values (Figure 8.6) are relative numbers with an arbitrary value of 4.0 for fluorine, the most electronegative element. The nonmetal with the next highest electronegativity is oxygen, with a value of 3.5, followed by chlorine and nitrogen, each with the same value of 3.0. Elements with electronegativities of 2.5 or more are all nonmetals in the top right corner of the periodic table. By contrast, elements with electronegativities of 1.3 or less are all metals in the lower left of the periodic table. These elements are often referred to as the most electropositive elements; *they are the metals that invariably form ionic compounds.* Between these two extremes are most of the remaining metals (largely transition metals) with electronegativities between 1.4 and 1.9, the metalloids with electronegativities between 1.8 and 2.1, and some nonmetals with electronegativities between 2.1 and 2.4.

Electronegativities show a periodic trend (Figure 8.6). *In general, electronegativity increases diagonally upward and to the right in the periodic table.* That is, electronegativy *increases* across a period and *decreases* down a group. Because metals typically lose electrons, they are the least electronegative

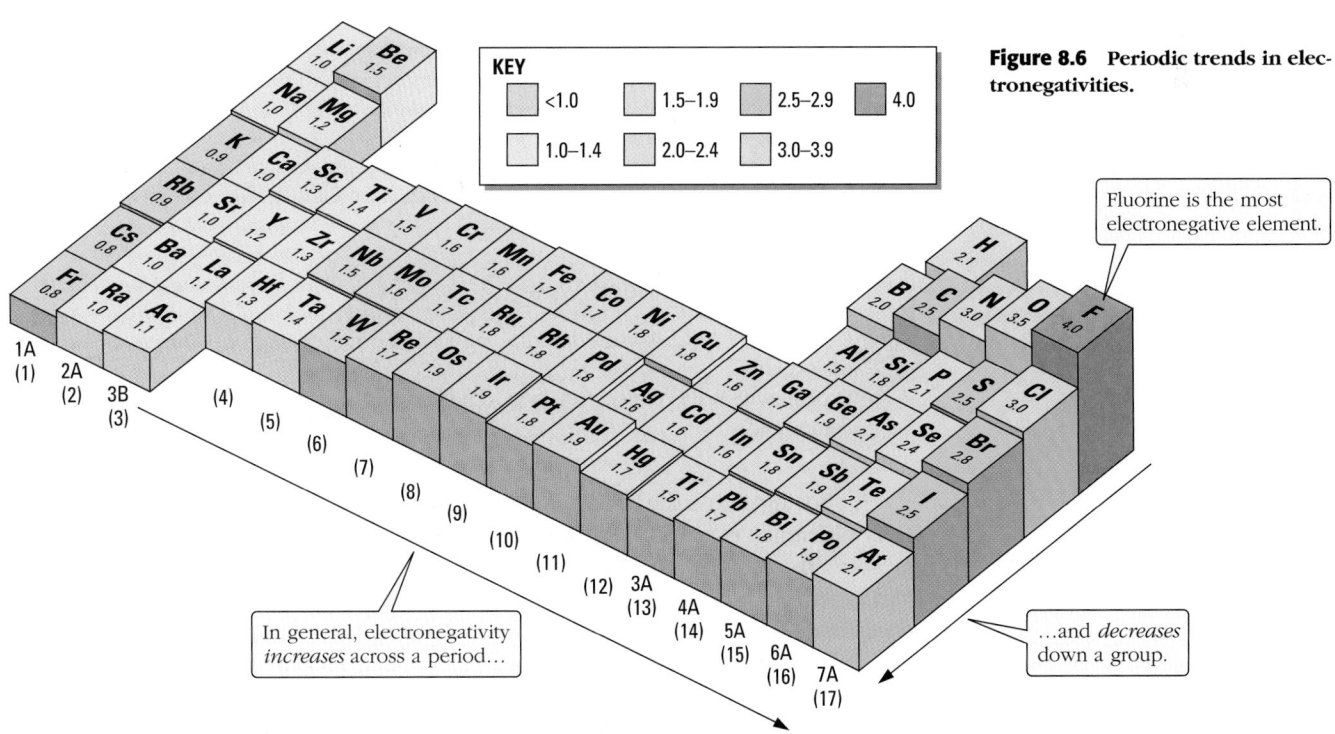

Figure 8.6 **Periodic trends in electronegativities.**

KEY
<1.0   1.5–1.9   2.5–2.9   4.0
1.0–1.4   2.0–2.4   3.0–3.9

Fluorine is the most electronegative element.

In general, electronegativity *increases* across a period…

…and *decreases* down a group.

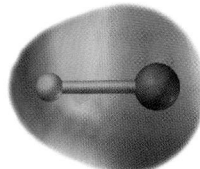

$$\overset{\delta+}{H}\!\!-\!\!\overset{\delta-}{Cl}$$

elements, and nonmetals, which have a tendency to gain electrons, are the most electronegative.

Electronegativity values are approximate and are primarily used to predict the polarity of covalent bonds. Bond polarity is indicated by writing $\delta+$ by the *less* electronegative atom and $\delta-$ by the *more* electronegative atom, where δ stands for partial charge. For example, the polar H—Cl bond in hydrogen chloride can be represented as shown to the left.

All bonds except those between identical atoms are polar to some extent, and the difference in electronegativity values is a qualitative measure of the degree of polarity. The change from nonpolar covalent bonds to slightly polar covalent bonds to very polar covalent bonds to ionic bonds can be regarded as a continuum (Figure 8.7). Examples are H_2 (nonpolar), HI (slightly polar), HF (very polar), and Na^+Cl^- (ionic). ***The greater the difference in electronegativity between two atoms, the more polar will be the bond between them.***

Problem-Solving Example 8.7   Bond Polarity

For each of these bond pairs, indicate the partial positive and negative atoms and tell which is the more polar bond.

(a) Cl—F and Cl—Br   (b) Si—Br and C—Br

Answer

(a) $\overset{\delta+}{Cl}\!\!-\!\!\overset{\delta-}{F}$ (more polar); $\overset{\delta-}{Cl}\!\!-\!\!\overset{\delta+}{Br}$

(b) $\overset{\delta+}{Si}\!\!-\!\!\overset{\delta-}{Br}$ (more polar); $\overset{\delta+}{C}\!\!-\!\!\overset{\delta-}{Br}$

Explanation

(a) The Cl—F bond is more polar because the difference in electronegativity is greater between Cl and F (1.0) than between Cl and Br (0.2). The F atom is the partial negative end in ClF, but because Cl is more electronegative than Br, the Cl atom is the partial negative end in ClBr.

(b) Si—Br is more polar than C—Br because Si is less electronegative than C, so the electronegativity difference is greater between Si and Br.

Problem-Solving Practice 8.7

For each of these pairs of bonds, decide which is the more polar. For each polar bond, indicate the partial positive and partial negative atoms.

(a) B—C and B—Cl   (b) N—H and O—H

Exercise 8.7   Bond Types

(a) Explain why NaCl is considered to be an ionic rather than a polar covalent compound.
(b) Explain why BrF is considered to be a polar covalent compound rather than an ionic compound.

In moving across the third period from sodium to argon, changes in electronegativity difference cause a significant shift in the properties of compounds of these elements with chlorine. The third period begins with the ionic compounds NaCl and $MgCl_2$; both are crystalline solids at room temperature. The electronegativity differences between the metal and chlorine are 2.0 and 1.8, respectively. Aluminum chloride is less ionic; with an aluminum-chlorine electronegativity difference of 1.5, its bonding is highly polar covalent. As electronegativity increases across the period, the remaining Period 3 chlorides—$SiCl_4$, PCl_3, and S_2Cl_2—are molecular compounds, with decreasing electronegativity differences between the nonmetal and chlorine

LINUS PAULING
(1901–1994)

The son of a druggist, Linus Pauling completed his Ph.D. in chemistry and then traveled to Europe, where he worked briefly with Erwin Schrödinger and Niels Bohr (see Chapter 7). For his bonding theories and his work with structural aspects of proteins, Pauling was awarded the Nobel Prize in Chemistry in 1954. Shortly after World War II, Pauling and his wife began a crusade to limit nuclear weapons, a crusade culminating in the limited test ban treaty of 1963. For this effort, Pauling was awarded the 1963 Nobel Peace Prize, the first time any person received two unshared Nobel Prizes.

Linus Pauling died in August 1994, leaving a remarkable legacy of breakthrough scientific research and a social consciousness to be remembered in guarding against the possible misapplications of technology.

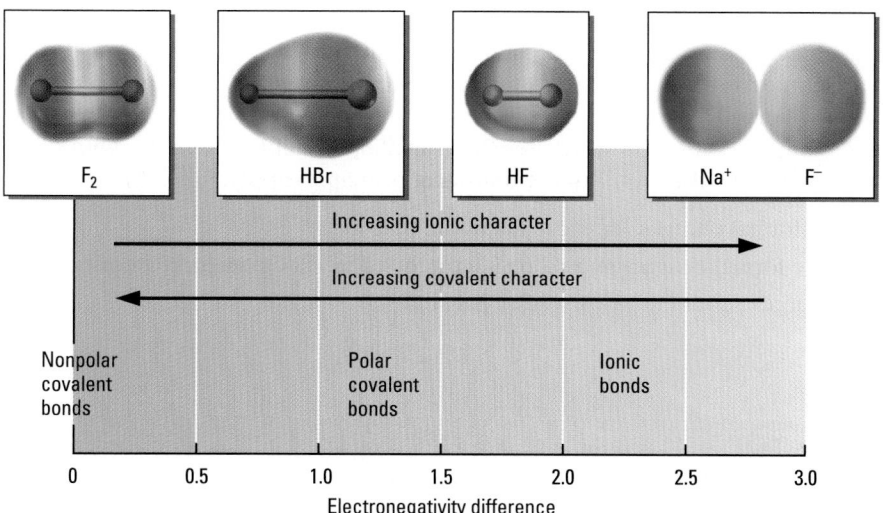

Figure 8.7 Bond character and electronegativity differences. This graph shows the relationship between electronegativity difference and the ionic character of a bond.

from Si to S. This results in a decrease in bond polarity from Si—Cl to P—Cl to S—Cl bonds, culminating in no electronegativity difference in Cl—Cl bonds.

	$SiCl_4$	PCl_3	S_2Cl_2	Cl_2
Electronegativity difference	1.2	0.9	0.5	0.0

Electronegativities: Si = 1.8; P = 2.1; S = 2.5; Cl = 3.0.

8.8 FORMAL CHARGE

Lewis structures depict how valence electrons are distributed in a molecule or ion. For some molecules or ions, more than one Lewis structure can be written, each of which obeys the octet rule. Which of the structures is more correct? How do you decide? Using formal charge is one way to do so. **Formal charge** is the charge a bonded atom would have ***if its bonding electrons were shared equally.*** In calculating formal charges, the following assignments are made:

CD-ROM Screen 9.13: Formal Charge

- *All of the lone pair electrons are assigned to the atom on which they are found.*
- *Half of the bonding electrons are assigned to each atom in the bond.*
- *The sum of the formal charges must equal the actual charge: zero for molecules and the ionic charge for an ion.*

Thus, in assigning formal charges to atoms in a Lewis structure,

Formal charge = *(number of valence electrons in an atom)* −

[*(number of lone pair electrons)* + ($\frac{1}{2}$ *number of bonding electrons)*]

Applying these rules, we can calculate the formal charges of the atoms in a cyanate ion, $[:N{\equiv}C{-}\overset{..}{\underset{..}{O}}:]^-$.

	N	**C**	**O**
Valence electrons	5	4	6
Lone pair electrons	2	0	6
$\frac{1}{2}$ shared electrons	3	4	1
Formal charge	0	0	−1

Note that the sum of the formal charges equals -1, the charge on the ion. The -1 formal charge is on oxygen, the most electronegative atom in the ion.

It is important to recognize that formal charges do *not* indicate actual charges on atoms. Formal charge is a useful way to determine the most likely structure from among several Lewis structures. In evaluating possible structures with different formal charge distributions, the following principles apply:

- Smaller formal charges are more favorable than larger ones.
- Negative formal charges should reside on the more electronegative atoms.
- Like charges should not be on adjacent atoms.

Problem-Solving Example 8.8   Formal Charges

(a) Two possible Lewis structures for N_2O are $:\ddot{O}=N=\ddot{N}:$ and $:\ddot{O}-N\equiv N:$. Determine the formal charges on each atom and which structure is preferred.

(b) Determine the formal charges on each atom of these two Lewis structures of chlorite ion, ClO_2^- . Which structure is preferred?

Structure 1: $\left[:\ddot{O}-\ddot{Cl}-\ddot{O}: \right]^-$

Structure 2: $\left[:\ddot{O}-\ddot{Cl}=\ddot{O}: \right]^-$

Answer

(a)
$$\begin{array}{ccc} 0 & +1 & -1 \\ :\ddot{O}=N=\ddot{N}: & & \end{array} \qquad \begin{array}{ccc} -1 & +1 & 0 \\ :\ddot{O}-N\equiv N: & & \text{(favored)} \end{array}$$

(b)
$$\begin{array}{ccc} -1 & +1 & -1 \\ \left[:\ddot{O}-\ddot{Cl}-\ddot{O}: \right]^- \end{array} \qquad \begin{array}{ccc} -1 & +1 & 0 \\ \left[:\ddot{O}-\ddot{Cl}=\ddot{O}: \right]^- \quad \text{(preferred)} \end{array}$$

Explanation

(a) In the first N_2O structure, $:\ddot{O}=N=\ddot{N}:$, the formal charges for the atoms in order in the structure are 0, +1, and -1.

	O	N	N
Valence electrons	6	5	5
Lone pair electrons	4	0	4
$\frac{1}{2}$ shared electrons	2	4	2
Formal charge	0	+1	-1

The sum of the formal charges is $0 + (+1) + (-1) = 0$. This is as it should be for a neutral molecule.

Applying the rules for assigning electrons to the second structure, $:\ddot{O}-N\equiv N:$, the formal charges are (in order) -1, +1, and 0.

	O	N	N
Valence electrons	6	5	5
Lone pair electrons	6	0	2
$\frac{1}{2}$ shared electrons	1	4	3
Formal charge	-1	+1	0

The sum of the formal charges in this structure is also zero. Formal charges are low in both structures, but the second structure is preferred because it has the negative charge on the more electronegative atom (O) rather than on the less electronegative N, as in the first structure.

(b)

	Structure 1			Structure 2		
	O	Cl	O	O	Cl	O
Valence electrons	6	7	6	6	7	6
Lone pair electrons	6	4	6	6	4	4
$\frac{1}{2}$ shared electrons	1	2	1	1	3	2
Formal charge	−1	+1	−1	−1	0	0

In both cases, the total formal charge is −1, the charge on the ion. Structure 2 is preferred because of the smaller formal charges.

The 10 electrons around Cl are an exception to the octet rule (Section 8.10).

Problem-Solving Practice 8.8

There is a third Lewis structure that can be written for N_2O, which also obeys the octet rule. Write this other Lewis structure and determine the formal charges on its atoms.

 Exercise 8.8   Formal Charge

Determine the formal charge of each atom in hydrazine, H_2NNH_2.

8.9 LEWIS STRUCTURES AND RESONANCE

Ozone, O_3, is an unstable, pale blue, diamagnetic gas with a pungent odor. Depending on its location, ozone is either beneficial or harmful. The ozone layer in the upper stratosphere protects the earth and its inhabitants from intense ultraviolet solar radiation, but ozone pollution in the lower atmosphere causes respiratory problems (Section 10.11).

As you have seen, the number of bonding electron pairs between two atoms is important in determining bond length and strength. The experimentally measured lengths for the two oxygen-oxygen bonds in ozone are the same, 128 pm, implying that both bonds contain the same number of bond pairs. However, using the guidelines for writing Lewis structures, you might come to a different conclusion. Two possible Lewis structures are

CD-ROM Screen 9.6: Resonance Structures

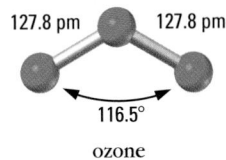

127.8 pm   127.8 pm

116.5°

ozone

$$:\ddot{O}{=}\ddot{O}{-}\ddot{O}: \quad \text{and} \quad :\ddot{O}{-}\ddot{O}{=}\ddot{O}:$$

Each structure shows a double bond on one side of the central O atom and a single bond on the other side. If either one were the actual structure of O_3, then one bond (O=O) should be shorter than the other (O—O), but this is not the case. That the oxygen-to-oxygen bond is neither an actual double or a single bond is supported by the fact that the 128 pm experimental bond length is longer than O=O (112 pm), but shorter than O—O (132 pm). Therefore, a single Lewis structure cannnot be written that is consistent with the the experimental data. When this situation arises, the concept of *resonance* is invoked to reconcile the experimental observation with two or more Lewis structures for the same molecule. Each of the Lewis structures, called **resonance structures,** is thought of as contributing to the true structure that cannot be written. The actual structure of O_3 is neither of the Lewis structures above, but a composite called a **resonance hybrid.** It is conventional to connect the resonance structures with a double-headed arrow, ↔ , to emphasize that the actual bonding is a composite of these structures.

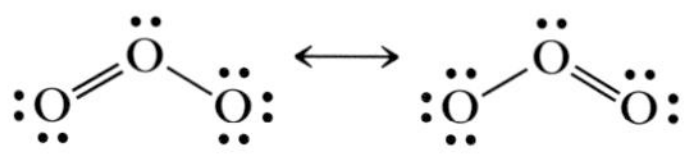

$$:O=\overset{..}{\underset{..}{O}}:O: \longleftrightarrow :O=\overset{..}{\underset{..}{O}}=O:$$

Resonance structures
of ozone

A resonance hybrid is often written as a composite picture in which a dotted line represents *delocalized* electrons, those spread evenly over the molecule but not associated as "double bonds" with any specific pair of bonded atoms. For ozone, such a structure is

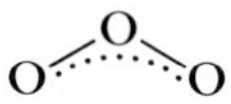

$$O \cdots \overset{O}{\cdots} O$$

The resonance concept is useful whenever there is a choice about which of two or three atoms contribute lone pairs to achieve an octet of electrons about a central atom by multiple bond formation.

When applying the concept of resonance, keep several important things in mind:

- *Lewis structures contributing to the resonance hybrid structure differ only in the assignment of electron pair positions, never atom positions.*
- *Contributing Lewis structures differ in the number of bond pairs between pairs of atoms.*
- *The resonance hybrid structure represents a single composite structure and not different structures that are continually changing back and forth.*

"Resonance" truly is an unfortunate term, because it implies that the molecule somehow "resonates," moving in some way to form different kinds of molecules, which is not true. There is only one kind of ozone molecule.

To illustrate the use of resonance, consider what happens in writing the Lewis structure of the carbonate ion, CO_3^{2-}, which has 24 valence electrons (4 from C, 18 from three O atoms, and 2 for the 2− charge). Writing the skeleton structure and putting in lone pairs so that each O has an octet uses 24 electrons but leaves carbon without an octet:

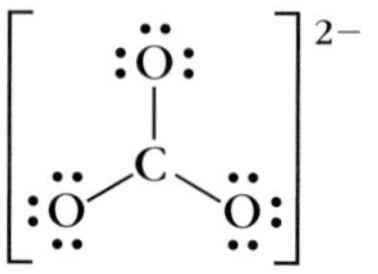

$$\left[\begin{array}{c} :\overset{..}{O}: \\ | \\ :\overset{..}{O}-\overset{}{C}-\overset{..}{O}: \\ \end{array}\right]^{2-}$$

Writing the Lewis structures of oxygen-containing anions often requires using resonance.

To give carbon an octet requires changing a single bond to a double bond, and this can be done in three equivalent ways:

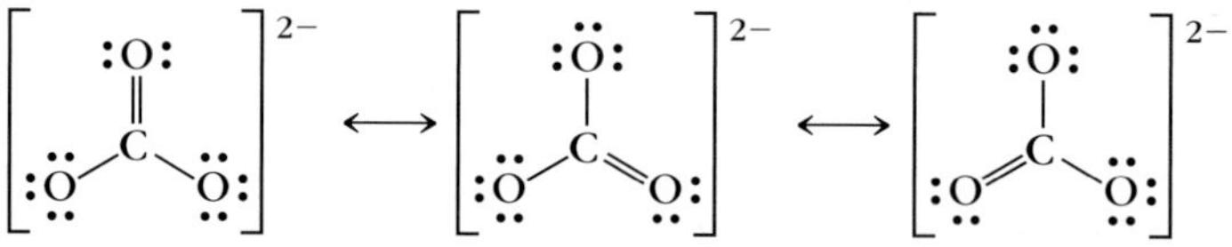

$$\left[\begin{array}{c} :O: \\ \| \\ :\overset{..}{O}-\overset{}{C}-\overset{..}{O}: \\ \end{array}\right]^{2-} \longleftrightarrow \left[\begin{array}{c} :\overset{..}{O}: \\ | \\ :\overset{..}{O}-\overset{}{C}=\overset{}{O}: \\ \end{array}\right]^{2-} \longleftrightarrow \left[\begin{array}{c} :\overset{..}{O}: \\ | \\ :\overset{}{O}=\overset{}{C}-\overset{..}{O}: \\ \end{array}\right]^{2-}$$

These three resonance structures contribute to the resonance hybrid, which is drawn with dotted lines representing the two delocalized electrons spread over the three C—O bonds.

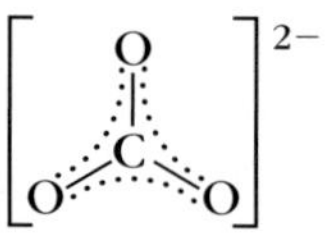

$$\left[\begin{array}{c} O \\ \vdots \\ O \cdots \overset{C}{\cdots} O \\ \end{array}\right]^{2-}$$

This representation is in agreement with experimental results: All three carbon-oxygen bond distances are 129 pm, intermediate between the C—O single bond (143 pm) and the C=O double bond (122 pm) distances.

The similar resonance structures of CO_3^{2-} contribute equally to the resonance hybrid for this ion. In some cases, resonance structures are significantly different and cannot make equal contributions to the hybrid. As demonstrated in Problem-Solving Example 8.9, formal charge is useful in determining which resonance structures are most plausible.

Problem-Solving Example 8.9 Writing Resonance Structures

The Lewis structure for cyanate ion was given previously as $\left[\, :N\equiv C-\ddot{\underset{..}{O}}: \right]^-$ (⇐ *p. 341*). This is just one of three possible resonance structures for cyanate ion.

(a) Draw the Lewis structures for the two other resonance structures.
(b) Use formal charges to determine which resonance structure is preferred.

Answer

(a) Structure 2: $\left[\, :\ddot{\underset{..}{N}}-C\equiv O: \right]^-$ Structure 3: $\left[\, :\ddot{N}=C=\ddot{\underset{..}{O}}: \right]^-$

(b) The initial structure, $:N\equiv C-\ddot{\underset{..}{O}}:$, is preferred.

Explanation

(a) The cyanate ion has 16 valence electrons: 5 from N, 4 from C, 6 from O, and 1 to account for the 1− charge. Both Structures 2 and 3 have 16 valence electrons and differ only in the placement of the multiple bonds, showing that they are correct resonance structures.

(b) The formal charges on the atoms are shown in the table.

	Structure 2			Structure 3		
	N	**C**	**O**	**N**	**C**	**O**
Valence electrons	5	4	6	5	4	6
Lone pair electrons	6	0	2	4	0	4
$\frac{1}{2}$ shared electrons	1	4	3	2	4	2
Formal charge	−2	0	+1	−1	0	0

The formal charges in Structures 2 and 3 add up to 1−, as they must for the 1− cyanate ion. Oxygen, the most electronegative atom, should bear the negative charge. Structure 2 would be least preferred because it has an atom with a high formal charge (−2 on N) and a positive charge on oxygen. The initial structure $\left[\, :N\equiv C-\ddot{\underset{..}{O}}: \right]^-$ is the one in which oxygen has the negative charge (⇐ *p. 341*). Thus, it is the preferred one.

Problem-Solving Practice 8.9

The nitrogen-oxygen bond lengths in NO_2^- are both 124 pm. Compare this with the bond distances given in Table 8.1 for N—O and N=O bond lengths. Account for any difference.

Exercise 8.9 Resonance Structures

Why is $\left[\, :\ddot{\underset{..}{N}}-O\equiv C: \right]^-$ not a resonance structure for cyanate ion?

8.10 EXCEPTIONS TO THE OCTET RULE

Many molecules and polyatomic ions have structures that are not consistent with the octet rule. Consideration of the electron configurations of their central atoms demonstrates why three kinds of exceptions occur: (1) molecules or ions with central atoms having fewer than eight electrons; (2) molecules or ions with an odd number of valence electrons; and (3) molecules or ions with central atoms having more than an octet of electrons.

Fewer than Eight Valence Electrons

CD-ROM Screen 9.7: Electron-Deficient Compounds: Exceptions to the Octet Rule

Boron trifluoride, BF_3, is a molecule with less than an octet of valence electrons around the central boron atom. Boron is a Group 3A element and has only three valence electrons; each fluorine contributes seven, for a total of 24 valence electrons. Although the Lewis structure has an octet around each fluorine atom, there are only six electrons around the B atom, an exception to the octet rule.

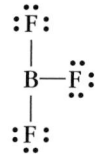

The type of bond in which both electrons are provided by the same atom is known as a coordinate covalent bond. This is discussed further in Section 22.6.

To achieve an octet around boron, BF_3 is very reactive, for example, readily combining with NH_3 to form a compound with the formula BF_3NH_3. The bonding between BF_3 and NH_3 can be explained by using the lone pair of electrons on N to form a covalent bond with B in BF_3. In this case, the nitrogen lone pair provides *both* of the shared electrons, resulting in an octet of electrons for both B and N.

$$
\begin{array}{ccc}
\text{H} & \text{F} & \text{H} \quad \text{F}\\
| & | & | \quad\;\; |\\
\text{H--N:} + \text{B--F} \longrightarrow & & \text{H--N--B--F}\\
| & | & | \quad\;\; |\\
\text{H} & \text{F} & \text{H} \quad \text{F}
\end{array}
$$

Odd Number of Valence Electrons

CD-ROM Screen 9.8: Free Radicals: Exceptions to the Octet Rule

Oxygen is more electronegative than nitrogen, so oxygen has the octet of electrons.

All the molecules we have discussed up to this point have contained only *pairs* of valence electrons. However, there are a few stable molecules that have an odd number of valence electrons. For example, NO has 11 valence electrons, and NO_2 has 17 valence electrons. The most plausible Lewis structures for these molecules are

$$:\dot{N}=\ddot{O} \qquad :\ddot{O}-\dot{N}=\ddot{O}:$$

Atoms and molecules that have an unpaired electron are known as **free radicals.** How do unpaired electrons affect reactivity? Simple free radicals such as atoms of H· and Cl· are very reactive and readily combine with other atoms to give molecules such as H_2, Cl_2, and HCl. Therefore, we would expect free radical molecules to be more reactive than molecules that have all paired electrons, and they are. A free radical either combines with another free radical to form a more stable molecule in which the electrons are paired, or it reacts with other molecules to produce new free radicals. These kinds of reactions are central to the formation of addition polymers (Section 12.8) and air pollutants (Section 10.10). For example, when gaseous NO and NO_2 are released in vehicle exhaust, the colorless

NO reacts with O_2 in the air to form brown NO_2. The NO_2 decomposes in the presence of sunlight to give NO and O, both of which are free radicals.

$$:\ddot{O}-\dot{N}=\ddot{O}: \xrightarrow{\text{sunlight}} :\dot{N}=\ddot{O} + \cdot\ddot{O}\cdot$$

The free O atom reacts with O_2 in the air to give ozone, O_3, an air pollutant that affects the respiratory system (Section 10.12). Free radicals also have a tendency to combine with themselves to form dimers, substances made from two smaller units. For example, when NO_2 gas is cooled it dimerizes to N_2O_4.

As expected, NO and NO_2 are paramagnetic (⬅ *p. 294*) because of the odd number of electrons. Experimental evidence indicates that O_2 is also paramagnetic (Figure 8.8) with two unpaired electrons and a double bond. The predicted Lewis structure for O_2 shows a double bond, but in that case all the electrons would be paired. It is impossible to write a conventional Lewis structure of O_2 that is in agreement with the experimental results.

More than Eight Valence Electrons

Exceptions to the octet rule are most common among molecules or ions with an "expanded octet," — that is, more than eight electrons in the valence shell around a central atom. For example, sulfur and phosphorus commonly form stable molecules and ions in which S or P are surrounded by more than an octet of valence electrons. Expanded octets are not found with elements in the first or second period. The octet rule is reliable for predicting stable molecules containing only H, C, N, O, or F.

Compounds with expanded octets occur only with elements in the third period and beyond. For example, the Period 3 elements P and S form the known compounds PF_5 and SF_4, but their Period 2 analogs, NF_5 and OF_4, do not exist. This is because elements in Period 2 have only $2s$ and $2p$ valence orbitals available for bonding electrons. The four orbitals in the second shell ($n = 2$) can only hold a maximum of eight electrons, thus limiting a second-period central atom to no more than an octet of valence electrons around it. Beginning with the third period ($n = 3$), five $3d$ orbitals are available in addition to the $3s$ and $3p$ orbitals. The $3d$ orbitals are of low enough energy to accommodate the extra electrons (⬅ *p. 292*). For phosphorus, its valence shell orbital diagram shows the vacant $3d$ orbitals available.

By using the $3d$ orbitals to accommodate additional electrons, phosphorus as a central atom can share five or six electron pairs to expand its octet as in PF_5 and PF_6^-.

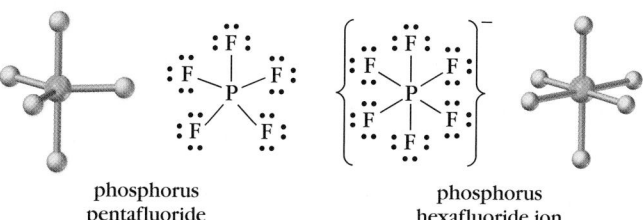

phosphorus   phosphorus
pentafluoride   hexafluoride ion

Table 8.3 illustrates molecules and ions of central atoms beyond the second period that have an expanded octet of valence electrons.

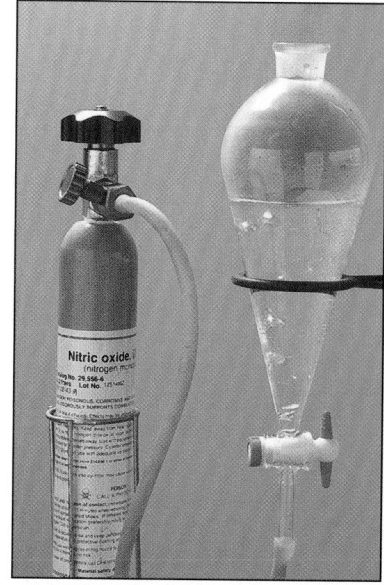

Nitrogen dioxide formation. Oxygen reacts with colorless NO to form reddish-brown NO_2.

NO_2 is a free radical; N_2O_4 is not because it has no unpaired valence electrons.

Figure 8.8 Paramagnetism of liquid oxygen. Liquid oxygen is suspended between the poles of a magnet because O_2 is paramagnetic. Paramagnetic substances are attracted into a magnetic field.

Third-period elements can also satisfy the octet rule by using just $3s$ and $3p$ orbitals, such as in PF_3.

Elements beyond the third period have nd orbitals to use for expanded octets. For example, bromine (Br) has the $4d$ and tellurium (Te) has the $5d$ valence orbitals available.

TABLE 8.3	Lewis Structures for Some Ions and Molecules with More than Eight Electrons Around the Central Atom*				
	Group 4A	**Group 5A**	**Group 6A**	**Group 7A**	**Group 8A**
Central atoms with five valence pairs	—	PF_5	SF_4	ClF_3	XeF_2
Bonding pairs	—	5	4	3	2
Lone pairs	—	0	1	2	3
Central atoms with six valence pairs	$SnCl_6^{2-}$	PF_6^-	SF_6	BrF_5	XeF_4
Bonding pairs	6	6	6	5	4
Lone pairs	0	0	0	1	2

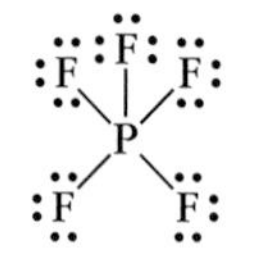

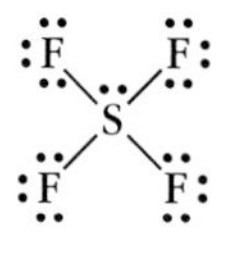

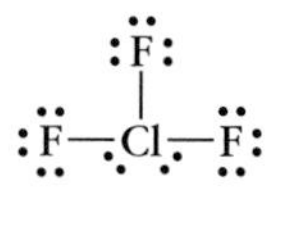

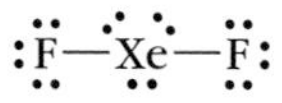

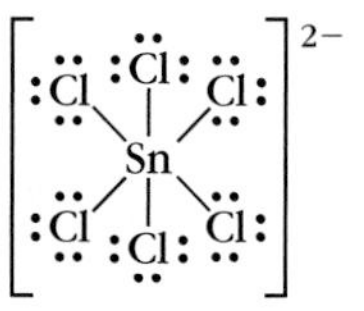

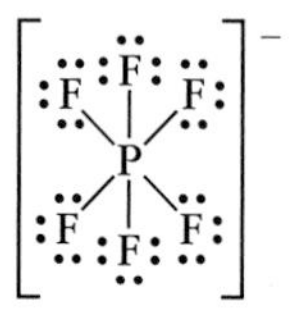

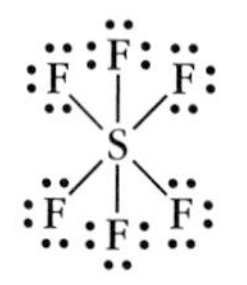

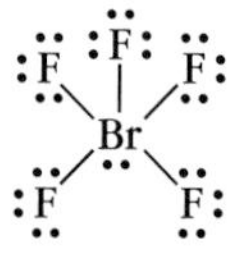

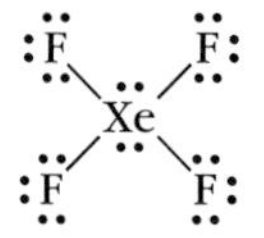

* In each case, the numbers of bond pairs and lone pairs about the central atom are given.

Problem-Solving Example 8.10   Exceptions to the Octet Rule

Write the Lewis structure for (a) tellurium tetrabromide, $TeBr_4$, (b) dichloroiodide ion, ICl_2^-, and (c) beryllium dichloride, $BeCl_2$.

Answer

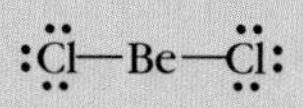

Explanation

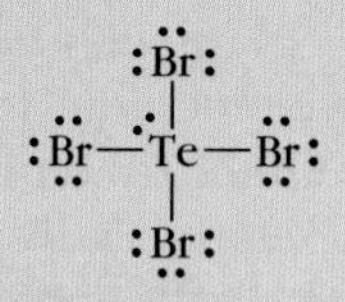

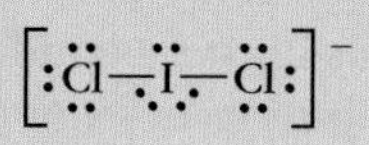

(a) $TeBr_4$ has 34 valence electrons, 8 of which are distributed among four Te—Br bonds. Of the remaining 26 lone-pair electrons, 24 complete octets for the Br atoms. The other 2 electrons form a lone pair on Te, which has a total of 10 electrons (five pairs: four shared, one unshared) around it, acceptable for a Period 5 element.

(b) There are a total of 22 valence electrons: 14 from two Cl atoms, 7 from I, and 1 for the 1− charge on the ion. Forming two I—Cl single bonds and then distributing six of the remaining nine electron pairs as lone pairs on Cl atoms to satisfy the octet rule uses a total of eight electron pairs. The remaining three electron pairs are placed on iodine, which can accommodate more than eight electrons because it is from the fifth period.

(c) $BeCl_2$ uses 4 of its 16 valence electrons to form two Be—Cl bonds. The remaining 12 electrons complete three lone pairs around each chlorine atom, giving them an octet and leaving Be with only two electron pairs. $BeCl_2$ does not have enough valence electrons for an octet and is an exception to the octet rule.

Problem-Solving Practice 8.10

Write the Lewis structure for each of the following molecules or ions. Indicate which central atoms break the octet rule, and why.

(a) BeF_2 (b) ClO_2 (c) PCl_5 (d) BH_2^+ (e) IF_7.

Chemistry in the News

BIOLOGICALLY REACTIVE OXYGEN AND ANTIOXIDANTS

The advertising headline for a nutritional supplement antioxidant touts its product: "Help Your Body Win the Battle Against Time." The advertisement goes on to say that "free radicals roam through your body—cell-destroying oxidizers that are like rust on metal—causing cell membranes to break down . . . . This product is the intelligent, fast way to provide antioxidants to protect your body." Is this advertising hype, or is there some basis to the idea that antioxidants combat the effects of free radicals? What are free radicals? Antioxidants?

Breathing is a natural act, something we do 12 to 16 times per minute. A very high percentage of the oxygen we breathe (> 90%) is used in a series of cellular oxidation-reduction reactions that make ATP (⬅ *Section 18.9*) and water, if everything goes properly. However, reactive oxygen species (ROS) may be formed instead. Three such highly reactive oxygen-containing species are hydrogen peroxide, H_2O_2, an effective oxidizing agent, plus two free radicals—superoxide ion, $O_2^- \cdot$, and hydroxyl, $OH \cdot$ radical. These free radicals pack an unpaired electron, which makes them very reactive. To find a mate for the unpaired electron, superoxide and hydroxyl radicals, and others, remove an electron from a covalent bond in another molecule, breaking the bond and leaving the molecule with an unpaired electron.

Superoxide radicals can be beneficial. A high concentration of them in white blood cells destroys invading bacteria and viruses. But superoxide radicals are also harmful because they can produce hydrogen peroxide through the action of the enzyme superoxide dismutase.

$$O_2^- \cdot + O_2^- \cdot + 2 H^+ \longrightarrow H_2O_2 + O_2$$

Hydrogen peroxide needs to be removed because it can form hydroxyl radicals, one of the most potent free radicals in the body.

$$H_2O_2 + O_2^- \cdot + H^+ \longrightarrow$$
$$H_2O + O_2 + OH \cdot$$

By removing electrons, free radicals can damage DNA (Section 9.6), causing mutations, degrade proteins, and break down polyunsaturated fatty acids (⬅ *p. 332*) in cell membranes. Such free radical attacks have been implicated in cancer, atherosclerosis (hardening of the arteries), and scores of other diseases.

Because harmful hydrogen peroxide and oxygen-containing free radicals are constantly created by cells, the body needs to destroy them before their concentrations become too high. We are protected, in part, by two very fast-acting enzymes—superoxide dismutase, which converts superoxide to hydrogen peroxide (see above), and catalase, which decomposes the hydrogen peroxide.

$$2 H_2O_2 \xrightarrow{\text{catalase}} 2 H_2O + O_2$$

Our bodies produce a number of other **antioxidants,** reducing agents (⬅ *p. 179*) that provide electrons to convert free radicals and other reactive oxygen species into less reactive substances. Vitamin E and vitamin C are two powerful antioxidants found in cells. Clinical studies have shown that diets high in fruits and vegetables rich in vitamin C are associated with a decrease in cancer and heart disease. Vitamin C is a particularly effective scavenger of hydroxyl radicals and hydrogen peroxide, converting them to water. Large-scale studies with vitamin E have shown its usefulness in lowering the risk of heart disease. Vitamin E is fat soluble and located near cell membranes, where it can protect against buildup of radicals from polyunsaturated fatty acid degradation.

Taking supplemental vitamins such as vitamins E and C as antioxidants is, however, controversial. A balanced diet provides the recommended daily amounts of the vitamins for most individuals. And so, taking vitamins to supplement a healthy diet may offer no distinct advantages. In spite of claims heralding the efficacy of vitamin supplements, continuing clinical studies seek to evaluate their possible benefits.

Source:

Adapted in part from Groff, J., and Gropper, S. *Advanced Nutrition and Human Metabolism,* 3rd ed. Minneapolis/St. Paul: Wadsworth Publishers, 2000.

1. What is an antioxidant?
2. What is the relationship between antioxidants and free radicals?

8.11 AROMATIC COMPOUNDS

Benzene, C_6H_6, is an important industrial chemical that is used, along with its derivatives, to manufacture plastics, detergents, pesticides, drugs, and other organic chemicals. It is the simplest member of a very large family of compounds known as **aromatic compounds,** which are compounds containing one or more benzene or benzene-like rings. The word "aromatic" is derived from "aroma," which describes the rather strong and often pleasant odors of these compounds.

CD-ROM Screen 11.3: Hydrocarbons

Resonance and the Structure of Benzene

To 19th-century chemists, the C_6H_6 molecular formula of benzene implied that it was an unsaturated compound because it lacked the ratio of carbon to hydrogen found in saturated noncyclic hydrocarbons (C_nH_{2n+2}). A six-membered ring structure with alternating double bonds uses all the available valence electrons and gives each carbon atom an octet of valence electrons.

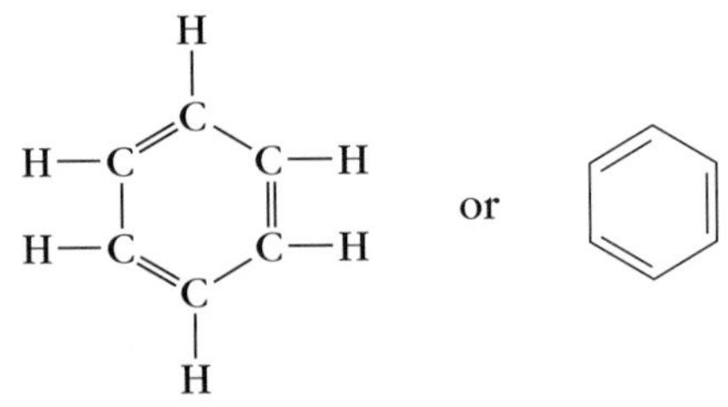

This structure implies that alternating C—C single bonds and C=C double bonds are present. But benzene incorporates bromine atoms by a *substitution* reaction, not an addition reaction as C=C bonds do under the same conditions. Rather, a bromine atom replaces (substitutes for) a hydrogen of benzene to produce bromobenzene, and the displaced hydrogen combines with the free bromine atom to form hydrogen bromide (HBr).

Addition reaction:

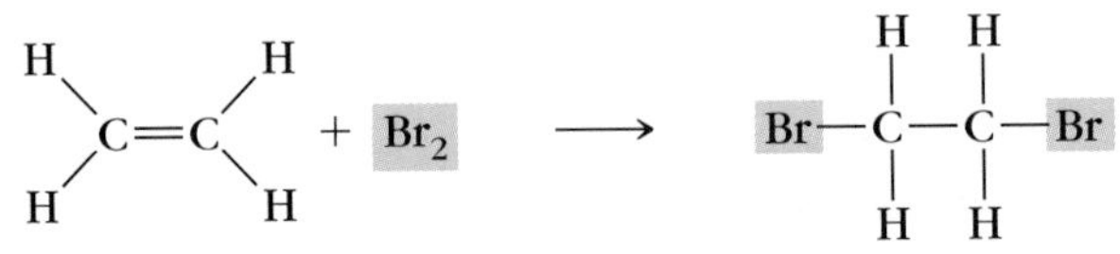

Substitution reaction:

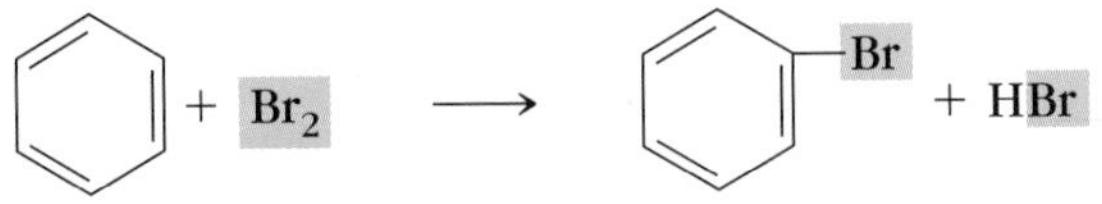

In 1872 Friedrich A. Kekulé proposed that benzene could be represented by a combination of two structures, which we now call resonance structures, indicated by the double-headed arrow.

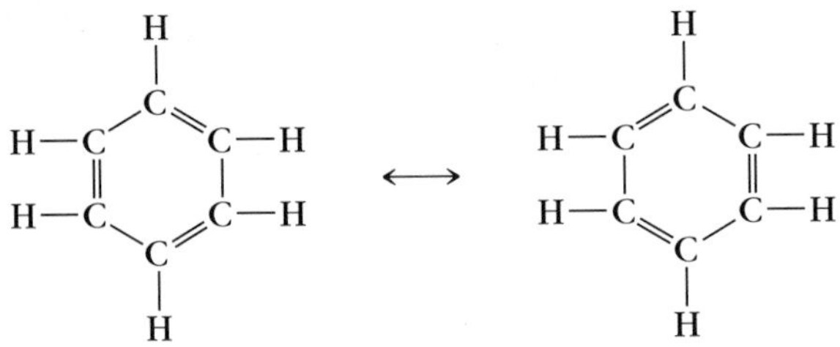

But neither of these alternating single- and double-bond resonance structures accurately represents benzene. Experimental structural data for benzene indicate a planar, symmetric molecule in which all carbon-carbon bonds are equivalent. Each carbon-carbon bond is 139 pm long, intermediate between the length of a C—C single bond (154 pm) and a C=C double bond (134 pm). Benzene is a resonance hybrid of these resonance structures — it is a molecule in which the six electrons of the suggested three double bonds are actually delocalized uniformly around the ring.

When hydrogen and carbon atoms are not shown, the benzene ring is written as a hexagon with a circle in the middle. Each corner in the hexagon represents one carbon atom and one hydrogen atom, and each line represents a single C—C bond. The circle represents the six delocalized electrons spread evenly over all of the carbon atoms.

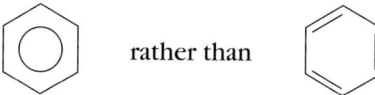

Whenever you see a formula with one or more carbon rings with central circles or alternating double and single bonds throughout, the compound is aromatic, like benzene. Benzaldehyde and toluene are examples of the many aromatic compounds with functional groups or alkyl groups bonded to the aromatic ring. Naphthalene is representative of a large group of aromatic compounds with more than one ring joined by common carbon-carbon bonds.

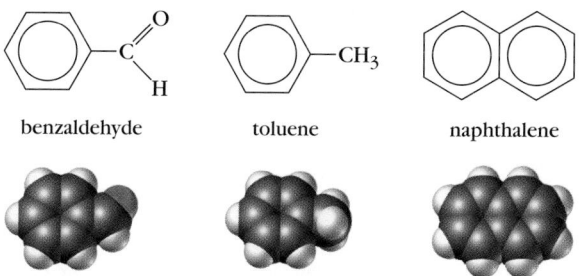

benzaldehyde   toluene   naphthalene

Benzaldehyde is used in synthetic almond and cherry food flavoring; toluene and benzene boost the octane rating of gasoline; and naphthalene is a moth repellant in one kind of moth balls.

Exercise 8.10   Aromatic Compounds

Write the Lewis structures for the resonance hybrids of toluene.

Constitutional Isomers of Aromatic Compounds

Because benzene is a planar molecule, constitutional isomers are possible when two or more groups are substituted for hydrogen atoms on the benzene ring. If two groups are substituted for two hydrogen atoms on the benzene ring, three constitutional isomers are possible. When the two groups are methyl groups, the compound is xylene. The prefixes *ortho-, meta-,* and *para-* are used to differentiate the three isomers of any disubstituted benzene.

Ortho—a prefix indicating that two substituents are on *adjacent* carbon atoms on a benzene ring. *Meta*—two substituents *separated by one carbon atom* on a benzene ring. *Para*—two substituents *separated by two carbon atoms* on a benzene ring.

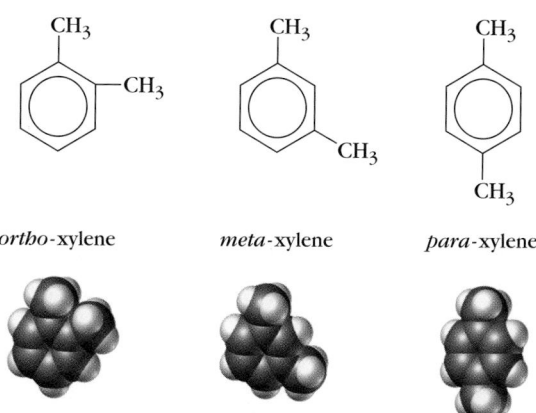

ortho-xylene   *meta*-xylene   *para*-xylene

These constitutional isomers differ in melting point, boiling point, density, and other physical properties.

Physical property	*ortho*-xylene	*meta*-xylene	*para*-xylene
Melting point, °C	−25.2	−47.8	13.2
Boiling point, °C	144.5	139.1	138.4
Density, g/mL	0.876	0.860	0.857

If more than two groups are attached to the benzene ring, numbers must be used to identify them and their positions, as for the three trichlorobenzenes:

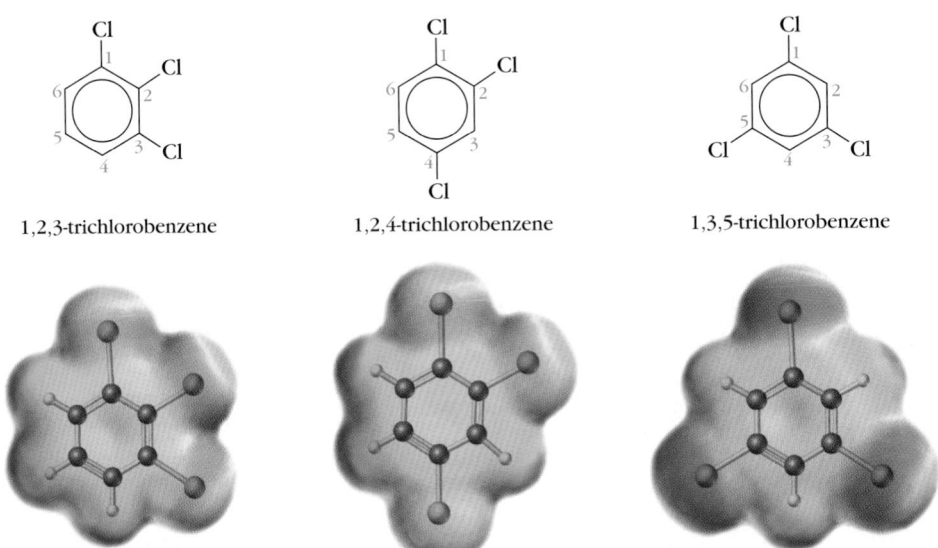

1,2,3-trichlorobenzene   1,2,4-trichlorobenzene   1,3,5-trichlorobenzene

There is no other way to attach three atoms of chlorine to a benzene ring, and only three trichlorobenzenes are known.

Exercise 8.11   Constitutional Isomers of Aromatic Compounds

Write the structural formula of 1,2,4-trimethylbenzene.

 Throughout this chapter, you have seen simple as well as complex structural formulas. The structural formulas of many biochemically active molecules are large and seemingly quite complicated. But, by closely examining the structural formulas, you can recognize the individual parts of the molecule. For example, by looking at the structural formulas of vitamins D and E that follow, you should recognize that the molecules are almost entirely assembled from structural parts that you now understand — rings and chains of carbon atoms, including an aromatic ring, C=C double bonds, and an alcohol group. There is also a six-membered ring with an oxygen atom instead of a carbon atom.

vitamin A

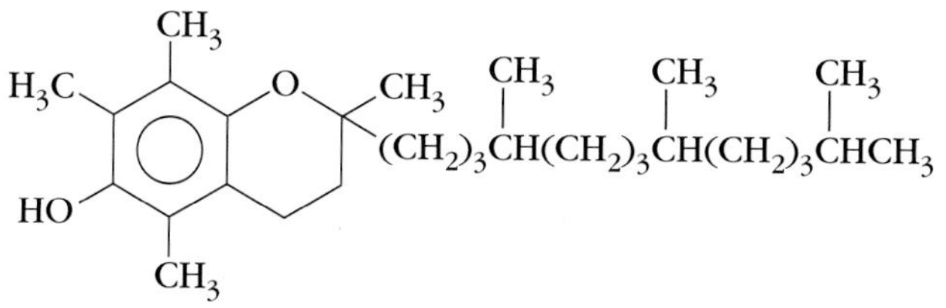

vitamin E

Exercise 8.12   Knowing Your Vitamins

1. Use the structural formula for vitamin A to answer.
   (a) How many carbon atoms and how many hydrogen atoms does it contain?
   (b) Locate the carbon atom that is bonded directly to four carbon atoms.
   (c) How many C=C double bonds are there in this molecule?
2. Consider the structural formula for vitamin E.
   (a) What is its molecular formula?
   (b) How many C=C double bonds does it have?
   (c) Which is likely to be the most polar bond in the molecule?

SUMMARY PROBLEM

Salicylic acid, $C_7H_6O_3$, is the starting compound from which aspirin (acetylsalicylic acid) is synthesized. The salicylic acid molecule consists of a benzene ring on which two adjacent hydrogen atoms have been replaced, one by an alcohol (!OH) functional group, the other by a carboxylic acid functional group.

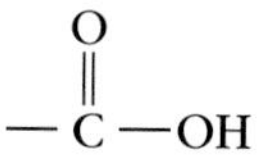

$$
\begin{array}{c}
O \\
\parallel \\
-C-OH
\end{array}
$$

carboxylic acid group

(a) Write the structural formula for salicylic acid.

(b) List the C—H, O—H, C—O, and C=O bonds in salicylic acid in order of increasing bond length and also increasing bond strength.

(c) Arrange these bonds in salicylic acid in order of increasing bond polarity: C—O, C—H, C—C, and O—H.

(d) The ionization of salicylic acid forms H^+ and salicylate ions. Write the structural formula for the salicylate ion.

(e) Salicylate ion is an example of a resonance hybrid. Write two Lewis structures that contribute to this resonance hybrid.

(g) Aspirin (acetylsalicylic acid) can be synthesized by reacting salicylic acid with acetic acid (CH_3COOH). Draw the Lewis structure of acetic acid and calculate the formal charges on each of its atoms.

IN CLOSING

Having studied this chapter, you should be able to . . .

- Recognize the different types of covalent bonding (Sections 8.1–8.3, 8.9).
- Use Lewis structures to represent covalent bonds in molecules and polyatomic ions (Sections 8.1–8.3, 8.8, 8.10).
- Describe multiple bonds in alkenes and alkynes (Section 8.5).
- Recognize molecules that can have *cis-trans* isomerism (Section 8.5).
- Predict bond lengths from periodic trends in atomic radii (Section 8.6).

- Relate bond energy to bond length (Section 8.6).
- Use bond enthalpies to calculate the enthalpy of a reaction (Section 8.6).
- Predict bond polarity from electronegativity trends (Section 8.7).
- Use formal charges to compare Lewis structures (8.8).
- Use resonance structures to model multiple bonding in molecules and polyatomic ions (Section 8.9).
- Explain why there are exceptions to the octet rule (Section 8.10).
- Describe bonding and constitutional isomerism in aromatic compounds (Section 8.11).

KEY TERMS

alkenes *(8.5)*	electronegativity *(8.7)*	resonance hybrid *(8.9)*
alkynes *(8.5)*	formal charge *(8.8)*	resonance structures
antioxidants (*Chemistry in the News* box)	free radicals *(8.10)*	*(8.9)*
	Lewis structure *(8.2)*	saturated fats *(8.5)*
aromatic compounds	lone pair electrons	saturated hydrocarbons
(8.11)	*(8.2)*	*(8.3)*
bond length *(8.6)*	multiple covalent	single covalent bond
bonding electrons *(8.2)*	bonds *(8.4)*	*(8.2)*
cis isomer *(8.5)*	nonpolar covalent	*trans* isomer *(8.5)*
cis-trans isomerism	bond *(8.7)*	triple bond *(8.4)*
(8.5)	octet rule *(8.2)*	unsaturated fats *(8.5)*
covalent bond *(8.1)*	polar covalent bond	unsaturated hydrocar-
double bond *(8.4)*	*(8.7)*	bons *(8.5)*

QUESTIONS FOR REVIEW AND THOUGHT

Conceptual Challenge Problems

CP-8.A (Section 8.2)   What deficiency is acknowledged by chemists when they write the formula for silicon dioxide as $(SiO_2)_n$ but write CO_2 for carbon dioxide?

CP-8.B (Section 8.7)   Without referring to the periodic table, write an argument to predict how the electronegativities of elements change based on the composition of their atoms.

CP-8.C (Section 8.7)   How would you rebut the statement, "There are no ionic bonds, only polar covalent bonds"?

Answers to questions in **bold** can be found in the back of the book.

Review Questions

1. Explain the difference between an ionic bond and a covalent bond.
2. What kind of bonding (ionic or covalent) would you predict for the products resulting from the following combinations of elements?
   (a) $Na + Br_2$ (b) $C + O_2$
   (c) $Ca + Cl_2$ (d) $N_2 + H_2$
3. What characteristics must atoms A and X have if they are able to form a covalent bond A—X with each other?
4. Boron compounds often do not obey the octet rule. Illustrate this with BCl_3. Show how the molecule can obey the octet rule by forming a coordinate covalent bond with NH_3.

5. Indicate the difference among alkanes, alkenes, and alkynes by giving the structural formula of a compound in each class that contains three carbon atoms.
6. Refer to Table 8.1 and answer the following questions:
   (a) Do any molecules with more than eight electrons have a second-period element as the central atom?
   (b) What is the maximum number of bond pairs and lone pairs that surround the central atom in any of these molecules?
7. While sulfur forms the compounds SF_4 and SF_6, no equivalent compounds of oxygen, OF_4 and OF_6, are known. Explain.
8. Which of the following molecules have an odd number of valence electrons: NO_2, SCl_2, NH_3, NO_3?
9. Write resonance structures for NO_2^-. Predict a value for the N—O bond length based on bond lengths given in Table 8.2, and explain your answer.
10. Consider the following structures for the formate ion,

HCO_2^-. Designate which two are resonance structures and which is equivalent to one of the resonance structures.

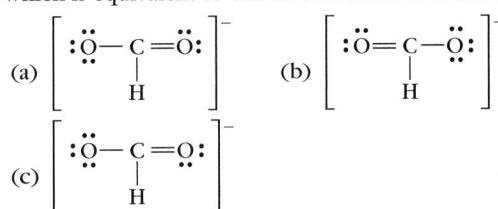

(a) $\begin{bmatrix} :\ddot{O} - C = \ddot{O}: \\ \quad\quad | \\ \quad\quad H \end{bmatrix}^-$ (b) $\begin{bmatrix} :\ddot{O} = C - \ddot{O}: \\ \quad\quad | \\ \quad\quad H \end{bmatrix}^-$

(c) $\begin{bmatrix} :\ddot{O} - C = O: \\ \quad\quad | \\ \quad\quad H \end{bmatrix}^-$

11. Consider a series of molecules in which the C atom is bonded to atoms of second-period elements: C—O, C—F, C—N, C—C, and C—B. Place these bonds in order of increasing bond length.

12. What are the trends in bond length and bond energy for a series of related bonds — for instance, single, double, and triple carbon-to-carbon bonds?

13. Why is *cis-trans* isomerism not possible for alkynes?

14. Define and give an example of a polar covalent bond. Give an example of a nonpolar covalent bond.

Lewis Structures

15. Write Lewis structures for the following molecules or ions.
(a) ClF_4^+ (b) ClO_3^- (c) $HOCl$ (d) SO_3^{2-}

16. Write Lewis structures for the following molecules or ions.
(a) ClF (b) H_2Se (c) BF_4^- (d) PO_4^{3-}

17. Write Lewis structures for these molecules.
(a) $CHClF_2$, one of the many chlorofluorocarbons that have been used in refrigeration
(b) Methyl alcohol, CH_3OH
(c) Methyl amine, CH_3NH_2

18. Write Lewis structures for these molecules or ions.
(a) CH_3Cl (b) SiO_4^{4-} (c) ClF_4^+ (d) C_2H_6

19. Write Lewis structures for these molecules.
(a) Formic acid, $HCOOH$, in which the atomic arrangement is

$$\begin{array}{c} \quad\quad O \\ \quad\quad \| \\ H - C - O - H \end{array}$$

(b) Acetonitrile, CH_3CN
(c) Vinyl chloride, CH_2CHCl, the molecule from which PVC plastics are made

20. Write Lewis structures for these molecules.
(a) Tetrafluoroethylene, C_2F_4, the molecule from which Teflon is made
(b) Acrylonitrile, CH_2CHCN, the molecule from which Orlon is made

21. Which of these are correct Lewis structures and which are incorrect? Explain what is wrong with the incorrect ones.

(a) $:N = \ddot{N}:$

N_2

(b) $\begin{array}{c} :\ddot{Cl} - \ddot{N} - \ddot{Cl}: \\ \quad\quad | \\ \quad\quad :\ddot{Cl}: \end{array}$

NCl_3

(c) $\begin{array}{c} \quad\quad :\ddot{O}: \\ :\ddot{O} - \ddot{Cl} - \ddot{O}: \end{array}$

ClO_3^-

(d) $\begin{array}{c} \quad H \quad\quad\quad H \\ \quad | \quad\quad\quad | \\ H - C - \ddot{O} - C - H \\ \quad | \quad\quad\quad | \\ \quad H \quad\quad\quad H \end{array}$

$(CH_3)_2O$

(e) $\begin{array}{c} \quad\quad H \\ \quad\quad | \\ H - \ddot{N} - H \\ \quad\quad | \\ \quad\quad H \end{array}$

NH_4^+

22. Which of these are correct Lewis structures and which are incorrect? Explain what is wrong with the incorrect ones.

(a) $F \; :\ddot{O}: \; F$

OF_2

(b) $:O \equiv O:$

O_2

(c) $\begin{array}{c} \quad\quad \ddot{O}: \\ \quad\quad \| \\ :\ddot{Cl} - C - \ddot{Cl}: \end{array}$

CCl_2O

(d) $\begin{array}{c} \quad\quad\quad H \\ H : \ddot{C} : H \; : \ddot{Cl} : \end{array}$

CH_3Cl

(e) $:\ddot{O} - N = \ddot{O}:$

NO_2^-

Bonding in Hydrocarbons

23. Write the structural formulas for all the branched-chain compounds with the molecular formula C_4H_{10}.

24. Write the structural formulas for all the branched-chain compounds with the molecular formula C_6H_{14}.

25. Write structural formulas for two straight-chain alkenes with the formula C_5H_{10}. Are these the only two structures that meet these specifications?

26. From their molecular formulas, classify each of these straight-chain hydrocarbons as an alkane, an alkene, or an alkyne.
(a) C_5H_8 (b) $C_{24}H_{50}$ (c) C_7H_{14}

27. From their molecular formulas, classify each of these straight-chain hydrocarbons as an alkane, an alkene, or an alkyne.
(a) $C_{21}H_{44}$ (b) C_4H_6 (c) C_8H_{16}

28. Write the *cis* and *trans* isomers of 2-pentene.

29. In each case, tell whether *cis* and *trans* isomers exist. If they do, write structural formulas for the two isomers and label each *cis* or *trans*.
(a) Br_2CH_2 (b) $CH_3CH_2CH=CHCH_2CH_3$
(c) $CH_3CH=CHCH_3$ (d) $CH_2=CHCH_2CH_3$

30. The structural formulas are given for these *cis* or *trans* alkenes.

(a) $\begin{array}{c} Cl \\ | \\ C = C - CH_3 \\ | \quad | \\ H \quad Cl \end{array}$ *trans*-1,2-dichloropropene

(b) $\begin{array}{c} \quad H \; H \; H \; H \; H \\ \quad | \; | \; | \; | \; | \\ H - C - C = C - C - C - H \\ \quad | \quad\quad\quad | \; | \\ \quad H \quad\quad\quad H \; H \end{array}$ *cis*-2-pentene

(c) $\begin{array}{c} \quad H \; H \; H \; H \; H \; H \\ \quad | \; | \; | \; | \; | \; | \\ H - C - C - C = C - C - C - H \\ \quad | \; | \quad\quad\quad | \; | \\ \quad H \; H \quad\quad\quad H \; H \end{array}$ *cis*-3-hexene

(d) $CH_3 - \underset{\underset{H}{|}}{\overset{\overset{H}{|}}{C}} = C - CH_2 - CH_2 - CH_3$ *trans*-2-hexene

Write the structural formula for
(a) *cis*-1,2-dichloropropene   (b) *trans*-2-pentene
(c) *trans*-3-hexene   (d) *cis*-2-hexene

31. Which of these molecules can have *cis* and *trans* isomers? For those that do, write the structural formulas of the two isomers and label them *cis* and *trans*. For those that cannot have these isomers, explain why.
   (a) $CH_3CH_2BrC{=}CBrCH_3$ (b) $(CH_3)_2C{=}C(CH_3)_2$
   (c) $CH_3CH_2IC{=}CICH_2CH_3$ (d) $CH_3ClC{=}CHCH_3$
   (e) $(CH_3)_2C{=}CHCH_3$

Bond Properties

32. For each pair of bonds, predict which will be the shorter.
   (a) B—Cl or Ga—Cl
   (b) C—O or Sn—O
   (c) P—S or P—O
   (d) The C=C or the C=O bond in acrolein,
   $H_2C{=}CH - \underset{\underset{H}{|}}{C} {=} O$

33. For each pair of bonds, predict which will be the shorter.
   (a) Si—N or P—O
   (b) Si—O or C—O
   (c) C—F or C—Br
   (d) The C=C or the C≡N bond in acrylonitrile, $H_2C{=}CH{-}C{\equiv}N$

34. Using only a periodic table (not a table of electronegativities), decide which of these is likely to be the strongest bond.
   (a) Si—F   (b) P—S   (c) P—O

35. Compare the nitrogen-nitrogen bonds in hydrazine, N_2H_4, and in "laughing gas," N_2O. In which molecule is the nitrogen-nitrogen bond shorter? In which should the nitrogen-nitrogen bond be stronger?

36. Consider the carbon-oxygen bonds in formaldehyde, H_2CO, and in carbon monoxide, CO. In which molecule is the C—O bond shorter?

37. Which bond will require more energy to break, the C—O bond in formaldehyde, H_2CO, or the CO bond in carbon monoxide, CO?

38. Compare the carbon-oxygen bond lengths in the formate ion, HCO_2^-, and in the carbonate ion, CO_3^{2-}. In which ion is the bond longer? Explain briefly.

39. Compare the nitrogen-oxygen bond lengths in NO_2^+ and in NO_3^-. In which ion are the bonds longer? Explain briefly.

Bond Energies and Enthalpy Changes

40. Estimate $\Delta H°$ for forming 2 mol of ammonia from molecular nitrogen and molecular hydrogen. Is this reaction exothermic or endothermic? (N_2 has a triple bond.)

41. Estimate $\Delta H°$ for the conversion of 1 mol of carbon monoxide to carbon dioxide by combination with molecular oxygen. Is this reaction exothermic or endothermic? (CO has a triple bond.)

42. Which of the four molecules HF, HCl, HBr, and HI has the strongest chemical bond? Using bond energies, estimate $\Delta H°$ for the reaction of molecular hydrogen with each of the gaseous molecular halogens: fluorine, chlorine, bromine, and iodine. Which is the most exothermic reaction?

Electronegativity and Bond Polarity

43. For each pair of bonds, indicate the more polar bond and use $\delta+$ and $\delta-$ to show the direction of polarity in each bond.
   (a) C—O and C—N
   (b) B—O and P—S
   (c) P—H and P—N
   (d) B—H and B—I

44. Given the bonds C—N, C—H, C—Br, and S—O,
   (a) Tell which atom in each is the more electronegative.
   (b) Which of these bonds is the most polar?

45. For each pair of bonds, identify the more polar one and use $\delta+$ and $\delta-$ to indicate the direction of polarity in each bond.
   (a) B—Cl and B—O   (b) O—F and O—Se
   (c) S—Cl and B—F(d) N—H and N—F

46. The molecule below is urea, a compound used in plastics and fertilizers.

$$\underset{H}{\overset{H}{\underset{|}{\overset{|}{N}}}} - \overset{\overset{\overset{O}{\|}}{}}{C} - \underset{H}{\overset{H}{\underset{|}{\overset{|}{N}}}}$$

urea

(a) Which bonds in the molecule are polar and which are nonpolar?
(b) Which is the most polar bond in the molecule? Which atom is the partial negative end of this bond?

47. The molecule below is acrolein, the starting material for certain plastics.

$H_2C{=}CH - \underset{\underset{H}{|}}{C} {=} O$

acrolein

(a) Which bonds in the molecule are polar and which are nonpolar?
(b) Which is the most polar bond in the molecule? Which atom is the partial negative end of this bond?

Formal Charge

48. What is the total formal charge of a molecule? of an ion?

49. What is the relationship between formal charge and electronegativity?

50. Write correct Lewis structures and assign a formal charge to each atom.
   (a) SO_3
   (b) C_2N_2 (atoms bonded in the order NCCN)
   (c) NO_2^-

51. Write correct Lewis structures and assign a formal charge to each atom.
   (a) OCS

(b) HNC (atoms bonded in that order)

(c) CH_3^-

52. Write correct Lewis structures and assign a formal charge to each atom.

(a) CH_3CHO

(b) N_3^-

(c) CH_3CN

53. Write correct Lewis structures and assign a formal charge to each atom.

(a) KrF_4

(b) ClO_3^-

(c) SO_2Cl_2

Resonance

54. The following have two or more resonance structures. Write all the resonance structures for each.

(a) Nitric acid

$$H-O-N\overset{\displaystyle O}{\underset{\displaystyle O}{}}$$

(b) Nitrate ion, NO_3^-

55. The following have two or more resonance structures. Write all the resonance structures for each molecule or ion.

(a) SO_3 (b) SCN^-

56. Several Lewis structures can be written for perbromate ion, BrO_4^-: the central Br with all single Br—O bonds, or with one, two, or three Br=O double bonds. Draw the Lewis structures of these possible resonance forms, and use formal charges to predict the most plausible one.

57. Use formal charges to predict which of the resonance forms is most plausible for

(a) SO_3 (b) HNO_3 (see Question 54)

58. Write the resonance structures for the following compound, adenine, which is the nitrogen-containing organic base portion of the nucleic acid ATP (adenosine triphosphate).

$$\begin{array}{c}
NH_2 \\
\end{array}$$

Exceptions to the Octet Rule

59. Write the Lewis structure for each of these molecules or ions.

(a) BrF_5 (b) IF_5 (c) IBr_2^-

60. Write the Lewis structure for each of these molecules or ions.

(a) BrF_3 (b) I_3^- (c) XeF_4

61. Which of these elements can form compounds with five or six pairs of valence electrons surrounding their atoms?

(a) C   (b) P   (c) O

(d) F   (e) Cl   (f) B

(g) Se   (h) Sn

Aromatic Compounds

62. Carbon-to-carbon double bonds (C=C) react by addition. Cite experimental evidence that benzene does not have C=C bonds.

63. All carbon-to-carbon bond lengths are identical in benzene. Does this argue for or against the presence of C=C bonds in benzene? Explain.

64. Three dibromobenzenes are known. Write the Lewis structure and name for each compound.

65. The structural formula of anthracene is

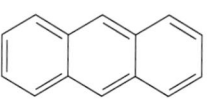

anthracene

What is its molecular formula?

66. Write the structural formula for 1,2-diiodobenzene (also known as *ortho*-diiodobenzene). Write the structural formulas for the *meta* and *para* isomers as well.

General Questions

67. Using just a periodic table (not a table of electronegativities), decide which of these is likely to be the most polar bond? Explain your answer.

(a) C—F   (b) S—F   (c) Si—F   (d) O—F

68. The C—Br bond length in CBr_4 is 191 pm; the Br—Br distance in Br_2 is 228 pm. Estimate the radius of a C atom in CBr_4. Use this value to estimate the C—C distance in ethane, H_3C—CH_3. How does your calculated bond length agree with the measured value of 154 pm? Are radii of atoms exactly the same in every molecule?

69. Is it a good generalization that elements that are close together in the periodic table form covalent bonds, whereas elements that are far apart form ionic bonds? Why or why not?

70. Write Lewis structures for

(a) CS_2 (b) NH_2OH

(c) S_2Cl_2 (contains S—S bond)   (d) NS_2^+

71. The molecule pictured below is acrylonitrile, the building block of the synthetic fiber Orlon.

$$H-\overset{\overset{\displaystyle H}{|}}{C}=\overset{\overset{\displaystyle H}{|}}{C}-\overset{\overset{\displaystyle H}{|}}{C}\equiv N\!:$$

(a) Which is the shorter carbon-carbon bond?

(b) Which is the stronger carbon-carbon bond?

(c) Which is the most polar bond and what is the partial negative end of the bond?

72. In nitryl chloride, NO_2Cl, there is no oxygen-oxygen bond. Write a Lewis structure for the molecule. Write any resonance structures for this molecule.

73. Write Lewis structures for

(a) SCl_2 (b) Cl_3^+

(c) $ClOClO_3$ (contains Cl—O—Cl bond)   (d) $SOCl_2$

74. Arrange the following bonds in order of increasing length (shortest first). List all the factors responsible for each placement: O—H, O—O, Cl—O, O=O, and O=C.

75. List the bonds in the previous problem in order of increasing bond *strength*.

76. Chlorine trifluoride, ClF_3, is one of the most reactive compounds known. Write the Lewis structure for ClF_3.

77. Judging from the number of carbon and hydrogen atoms in their formulas, which of these formulas represent alkanes? Which are likely aromatic? Which fall into neither category? (It may help to write structural formulas.)
    (a) C_8H_{10} (b) $C_{10}H_8$ (c) C_6H_{12}
    (d) C_6H_{14} (e) C_8H_{18} (f) C_6H_{10}

Applying Concepts

78. A student drew the following incorrect Lewis structure for ClO_3^-. What errors were made when determining the number of valence electrons?

79. The following Lewis structure for SF_5^+ is drawn incorrectly. What error was made when determining the number of valence electrons?

80. When asked to give an example of resonance structures, a student drew the following. Why is this example incorrect?

81. Why is this not an example of resonance structures?

82. How many bonds would you expect the elements in Groups 3A through 7A to form if they obeyed the octet rule?

83. In another universe, elements try to achieve a nonet (nine valence electrons) instead of an octet when forming chemical bonds. As a result, covalent bonds form when a trio of electrons are shared between two atoms. Draw Lewis structures for the compounds that would form between (a) hydrogen and oxygen and (b) hydrogen and fluorine.

84. Elemental phosphorus has the formula P_4. Propose a Lewis structure for this molecule. [*Hints:* (1) Each phosphorus atom is bonded to three other phosphorus atoms. (2) Visualize the structure three-dimensionally, not flat on a page.]

85. The elements As, Br, Cl, S, and Se have electronegativity values of 2.1, 2.4, 2.5, and 3.0, but not in that order. Using the periodic trend for electronegativity, assign the values to the elements. Which assignments are you certain about? Which are you not?

86. A substance is analyzed and found to contain 85.7% carbon and 14.3% hydrogen by weight. A gaseous sample of the substance is found to have a density of 1.87 g/L, and 1 mol of it occupies a volume of 22.4 L. What are two possible Lewis structures for molecules of the compounds? (*Hint:* First determine the empirical formula and molar mass of the substance.)

87. Which of these molecules is least likely to exist: NF_5, PF_5, SbF_5, or IF_5? Explain why.

88. When we estimate $\Delta H°$ from bond enthalpies we assume that all bonds of the same type (single, double, triple) between the same two atoms have the same energy, regardless of the molecule in which they occur. The purpose of this problem is to show you that this is only an approximation. You will need the following standard enthalpies of formation:

C(g)	$\Delta H° = 716.7$ kJ/mol
CH(g)	$\Delta H° = 596.3$ kJ/mol
CH_2(g)	$\Delta H° = 392.5$ kJ/mol
CH_3(g)	$\Delta H° = 146.0$ kJ/mol
H(g)	$\Delta H° = 218.0$ kJ/mol

(a) What is the average C—H bond energy in methane, CH_4?

(b) Using bond enthalpies, estimate $\Delta H°$ for the reaction
$$CH_4(g) \rightarrow C(g) + 2 H_2(g)$$

(c) By heating CH_4 in a flame it is possible to produce the reactive gaseous species CH_3, CH_2, CH, and even carbon atoms, C. Experiments give the following values of $\Delta H°$ for the reactions shown:

CH_3(g)	$\longrightarrow$ C(g) + H_2(g) + H(g)	$\Delta H° = 788.7$ kJ
CH_2(g)	$\longrightarrow$ C(g) + H_2(g)	$\Delta H° = 324.2$ kJ
CH(g)	$\longrightarrow$ C(g) + H(g)	$\Delta H° = 338.3$ kJ

For each of these reactions, draw a diagram similar to Figure 8.4. Then calculate the average C—H bond energy in CH_3, CH_2, and CH. Comment on any trends you see.

General Chemistry CD-ROM

CD8.1 Screen 9.3: Chemical Bond Formation: Covalent Bonding.
(a) What are the major coulombic interactions between two atoms?
(b) What is the difference between an ionic bond and a covalent bond?
(c) What is the difference in carbon-carbon bonding in ethane, ethene (ethylene), and acetylene?

CD8.2 Screen 9.4: Lewis Electron Dot Structures.
(a) How is the octet rule illustrated by the I atom in ICl?
(b) What is indicated by the pair of lines between the C atoms in C_2H_4?

CD8.3 Screen 9.9: Bond Properties.
(a) Examine the depiction of bond breaking on Screen 9.9. As two bonded atoms move apart, and the bond eventually breaks, what happens to the energy of the system?
(b) In the depiction of bond energy, the energy increases as the bond distance is less than 74 pm. Why does the energy increase?

CD8.4 Screen 9.11: Bond Polarity and Electronegativity.
(a) How is the electronegativity of an atom defined?
(b) Examine the table of electronegativity values. Is there a relation between the position of an element in the periodic table and its electronegativity?

CD8.5 Screen 9.7: Electron-Deficient Compounds.
(a) What is meant by an "electron-deficient" compound?
(b) Is a reaction between BF_3 and H_2O possible? Explain briefly. (*Hint:* In what way is water similar or dissimilar to ammonia in terms of their Lewis dot structures?)

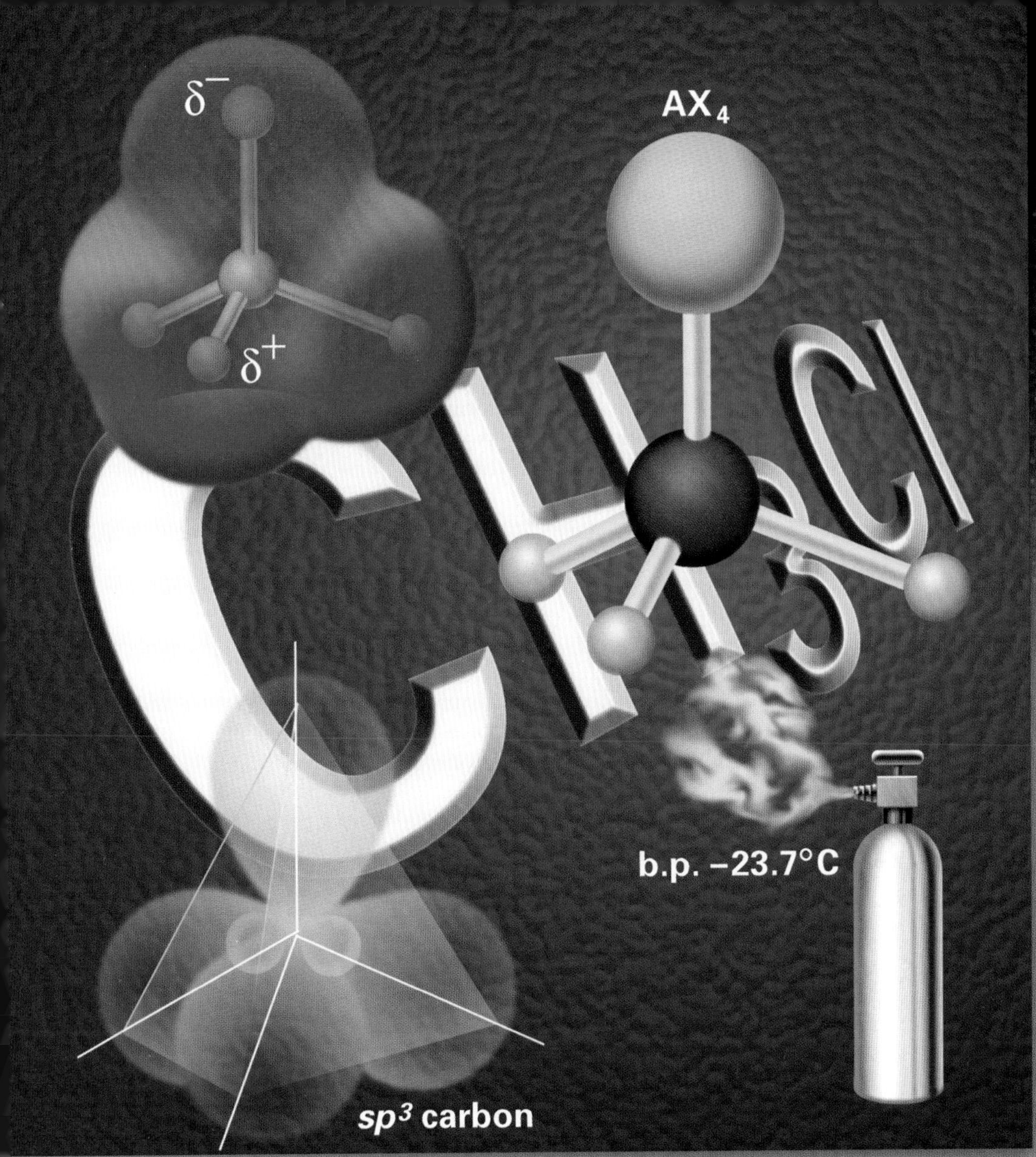

δ^-

AX_4

9

δ^+

CH₃Cl

b.p. −23.7°C

sp^3 **carbon**

Methyl chloride is a gas composed of tetrahedral polar molecules. The drawings here represent three ways of modeling compounds that are described in this chapter. From the top, clockwise, they are electron density modeling, valence shell electron pair repulsion modeling for geometric shape, and central atom hybridization of orbitals.

MOLECULAR STRUCTURES

9.1 Using Molecular Models

9.2 Predicting Molecular Shapes: VSEPR

9.3 Orbitals Consistent with Molecular Shapes: Hybridization

9.4 Molecular Polarity

9.5 Noncovalent Interactions and Forces Between Molecules

9.6 Biomolecules: DNA and the Importance of Molecular Structure

9.7 Chiral Molecules

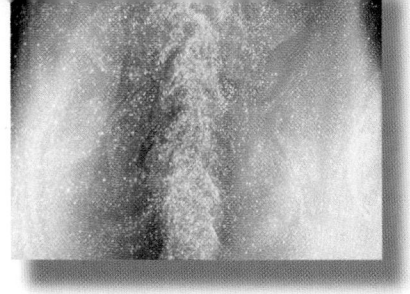

The composition, empirical formula, molecular formula, and Lewis structure of a substance provide important information. But, they are not sufficient to predict or explain the properties of most molecular compounds. The arrangement of the atoms and how they occupy three-dimensional space—the shape of a molecule—are also very important. Because of the arrangements of their atoms, molecules can have the same numbers of the same kinds of atoms and yet have different properties. For example, ethanol (in alcoholic beverages) and dimethyl ether (a refrigerant) have the same molecular formula, C_2H_6O. But the C, H, and O atoms are arranged so differently in these two compounds that their melting points differ by 27 °C and their boiling points by 103 °C.

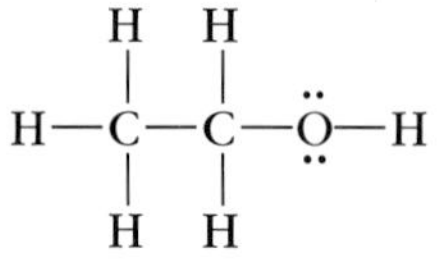

ethanol 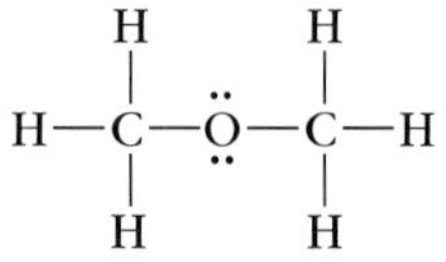   dimethyl ether

The ideas about molecular shape presented in this chapter are crucial to understanding the behavior of molecules in living organisms, the design of molecules that are effective drugs, and many other aspects of modern chemistry.

9.1 USING MOLECULAR MODELS

Molecules are three-dimensional aggregates of atoms and are too small to examine directly. Our ability to understand such three-dimensional structures is helped by the use of models. Probably the best example of the impact a model can have on the advancement of science is the double helix model of DNA used by James Watson and Francis Crick, which revolutionized the understanding of human heredity and genetic disease. We will discuss DNA (Section 9.6) after considering molecular shapes (Section 9.2) and noncovalent interactions (Section 9.5), both of which are essential to DNA function.

This book has used many computer-generated pictures of molecular models, like those for water shown below on the right. The computer programs that generate these pictures contain the most accurate experimentally derived data on atomic radii, bond lengths, and bond angles.

Before there were computers to generate molecular models, chemists relied on physical models assembled atom by atom, like those for water shown below on the left. The ball-and-stick model uses balls to represent atoms and short rods of wood or plastic to represent bonds. For example, the ball-and-stick model for water has a red ball representing oxygen, with holes at the correct angles connected by sticks to two white balls representing hydrogen atoms. In the space-filling physical model, the atomic models are scaled according to the experimental values for atom sizes, and the links between parts of the model are not visible when the

Ball-and-stick models are available in many campus bookstores. The models are easy to assemble and will help you to visualize the molecular geometries described in this chapter. Ball-and-stick models are relatively inexpensive compared with space-filling models.

Ball-and-stick

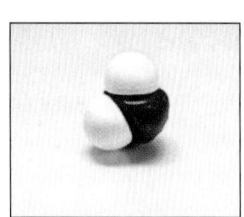

Space-filling

Physical molecular models

Ball-and-stick

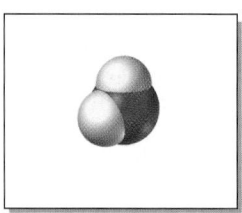

Space-filling

Computer-generated molecular models

360

model is assembled. This gives a better representation of the actual distances between atoms.

To convey a three-dimensional perspective for a molecule drawn on a flat surface, such as the page of a book, we can also make a perspective drawing that uses solid wedges (►) to represent bonds extending in front of the page, and dashed lines (----) to represent bonds behind the page. Bonds that lie in the plane of the page are indicated by a line (—), as illustrated in the following perspective drawing for the tetrahedral methane molecule:

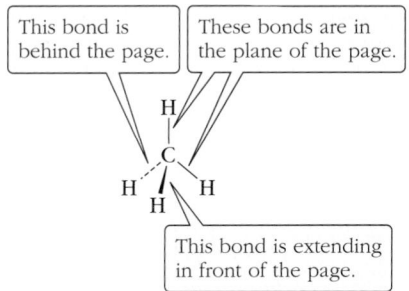

This bond is behind the page.

These bonds are in the plane of the page.

This bond is extending in front of the page.

Computers can draw and also rotate molecules so that they can be viewed from any angle, as illustrated in Figure 9.1 for ethane.

Advances in computer graphics have made it possible to draw scientifically accurate pictures of extremely complex molecules, as well as to study interactions between molecules. Figure 9.2 is a computer-generated model of hemoglobin, the remarkable and essential protein in the blood that takes up and releases oxygen. This vital protein is made up of two folded sets of pairs of long chains of atoms, indicated by the yellow and blue sets of ribbons. In a hemoglobin molecule, oxygen binds to Fe^{2+} ions at four sites shown in orange in Figure 9.2.

9.2 PREDICTING MOLECULAR SHAPES: VSEPR

A simple, reliable method for predicting the shapes of molecules and polyatomic ions is the **valence-shell electron-pair repulsion (VSEPR)** model. The VSEPR model is based on the idea that repulsions among pairs of bonding and lone pair electrons control the angles between bonds from a central atom to other atoms surrounding it. A central atom and its core electrons are represented by the atom's symbol.

How do repulsions among electron pairs result in different shapes? Imagine that the volume of a balloon represents the repulsive force of an electron pair that prevents other electron pairs from occupying the same space. When two or more balloons are tied together at a central point, the balloons assume the shapes shown in Figure 9.3. The central point represents the nucleus and the core electrons of a

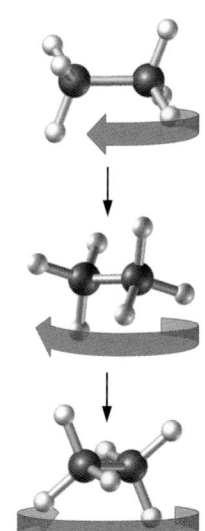

Figure 9.1 **An ethane molecule (C_2H_6) rotating.** The rotation of this molecule can be viewed using Chemscape CHIME, a free program available from MDL Information Systems, Inc., at **http://www.mdli.com.**

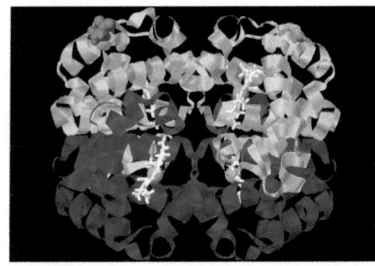

Figure 9.2 **Hemoglobin, a protein essential to human life.**

 CD-ROM Screen 9.14: Molecular Shape: VSEPR Theory

Figure 9.3 **Balloon models of the geometries predicted by the VSEPR theory.**

Linear

Triangular planar

Tetrahedral

Triangular bipyramidal

Octahedral

How are molecular structures determined? Many of the methods rely on the interaction of electromagnetic radiation with matter. Probing matter with electromagnetic radiation is called **spectroscopy,** and each area of the electromagnetic spectrum (⬅ *p. 265)* can be used as the basis for a particular spectroscopic method. Recall from Section 7.1 that electromagnetic radiation is emitted or absorbed in quantized packets of energy called photons and that the energy of the photon is represented by $E = h\nu$, where ν is the frequency of the light. Molecules may absorb several different electromagnetic radiation frequencies depending on the energy differences between their allowed energy levels. Each frequency absorbed must provide the exact package of energy needed to lift a molecule from one energy level to the next.

Infrared (IR) spectroscopy uses the interaction of infrared radiation with matter to study molecular structure. It is particularly useful for learning about molecules because the energy of the internal motions of molecules is similar to the energy of photons whose frequency is in the infrared region. Covalent bonds between atoms in a molecule can be considered to be like springs that can only bend or stretch in specified amounts.

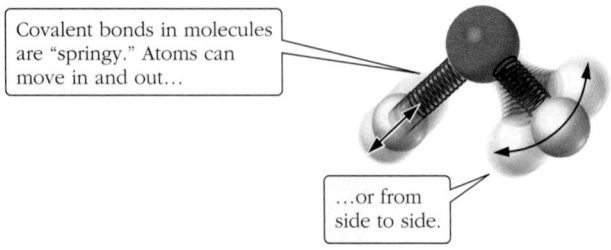

Covalent bonds in molecules are "springy." Atoms can move in and out…

…or from side to side.

Bending or stretching of the bonds of the water molecule occurs at specific frequencies corresponding to specific energy levels. The strength of the covalent bonds determines what frequency of infrared light is necessary for changing from one stretching or bending energy level to another. The molecule must be excited from one of these allowed energy states to another by an exact quantity of energy.

For example, a hydrogen chloride molecule vibrates at a specific energy. Photons with too low or too high an energy do not cause vibration at the next higher energy level, and the radiation passes through the molecule without being absorbed. Photons of the proper energy are absorbed and increase the molecule's vibrational energy.

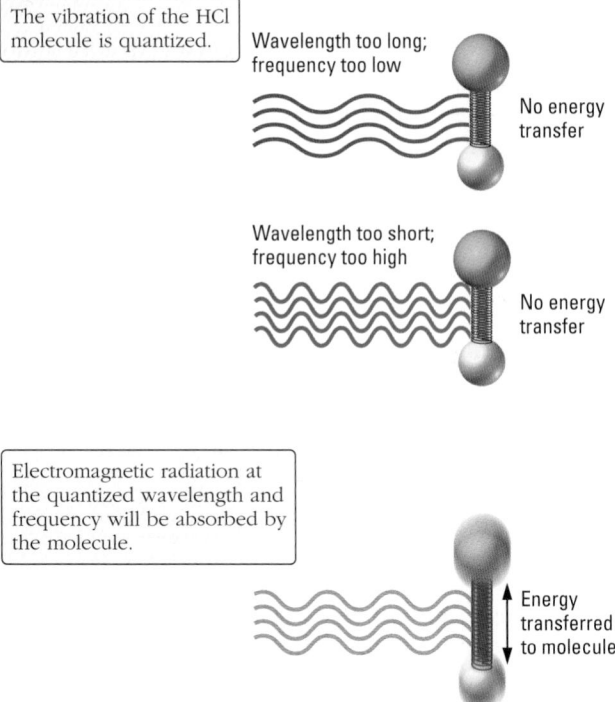

The vibration of the HCl molecule is quantized.

Wavelength too long; frequency too low

No energy transfer

Wavelength too short; frequency too high

No energy transfer

Electromagnetic radiation at the quantized wavelength and frequency will be absorbed by the molecule.

Energy transferred to molecule

Because the covalent bonds in molecules differ in strength and number, the molecular motions and the number of vibrational energy levels vary; hence, the infrared radiation that is absorbed by different molecules differs. As a result, infrared spectroscopy can be used to learn about the structures of molecules and even to analyze an unknown material by matching its infrared spectrum with that of a known compound. In fact, the infrared frequencies absorbed by a molecule are so characteristic of the bending and stretching of various bonds that the infrared spectrum of a molecule is regarded as its *fingerprint.* An example of the use of infrared spectroscopy in the identification of ethanol

$$\begin{array}{ccc} & \text{H} & \text{H} \\ & | & | \\ \text{H}-&\text{C}-\text{C}-&\ddot{\text{O}}-\text{H} \\ & | & | \\ & \text{H} & \text{H} \end{array}$$

is shown below where each of the absorption peaks is labeled with the type of bond action that causes it.

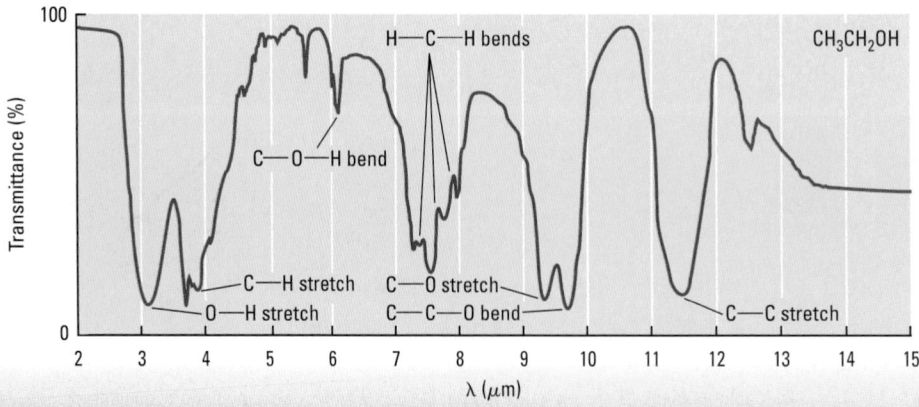

central atom. The arrangements of the balloons are those that minimize interactions among the balloons, and the electron pairs that the balloons represent.

Electron pairs are oriented to minimize electron-pair repulsions by having the electron pairs as far apart from each other as possible. Because electrons are constrained to be near the nucleus, the shapes resulting from minimizing the electron-pair repulsions are those predicted by VSEPR and illustrated in the balloon analogy.

Central Atoms with Only Bonding Pairs

The simplest application of VSEPR is to molecules having only shared pairs in single covalent bonds around a central atom that has no lone pairs. Figure 9.4 illustrates the geometries predicted by the VSEPR model for molecules of the types AX_2E_0 to AX_6E_0. These contain only single covalent bonds from X to A, the central atom, which has no lone pairs (E_0). Table 9.1 also summarizes this information. The **electron-pair geometry** is determined by the number of electron pairs around the central atom. The **molecular geometry** is the arrangement of the atoms in space. ***For molecules whose central atoms have no lone pairs, the electron-pair geometry and the molecular geometry are the same.***

The *linear* geometry for two bonding pairs and the *triangular planar* geometry for three bonding pairs contain a central atom that does not have an octet of electrons (⟸ *p. 346*). The central atom in a *tetrahedral* molecule obeys the octet rule with four bond pairs. The central atoms in *triangular bipyramidal* and *octahedral* molecules do not obey the octet rule because they have five and six bonding pairs, respectively. Hence, triangular bipyramidal and octahedral geometries would be expected only when the central atom is an element in Period 3 or higher (⟸ *p. 347*). The geometries illustrated in Figure 9.4 are by far the most common in molecules and polyatomic ions, and you should be thoroughly familiar with them.

Bond angles, as shown in Figure 9.4, are the angles between the bonds of two atoms that are bonded to the same third atom. In a methane molecule, for example, all the H—C—H bond angles are 109.5°.

CD-ROM Screen 9.15: Ideal Electron Repulsion Shapes

The term "molecular shape" is sometimes used rather than "molecular geometry."

The predicted bond angles given in the examples are in agreement with experimental values obtained from structural studies.

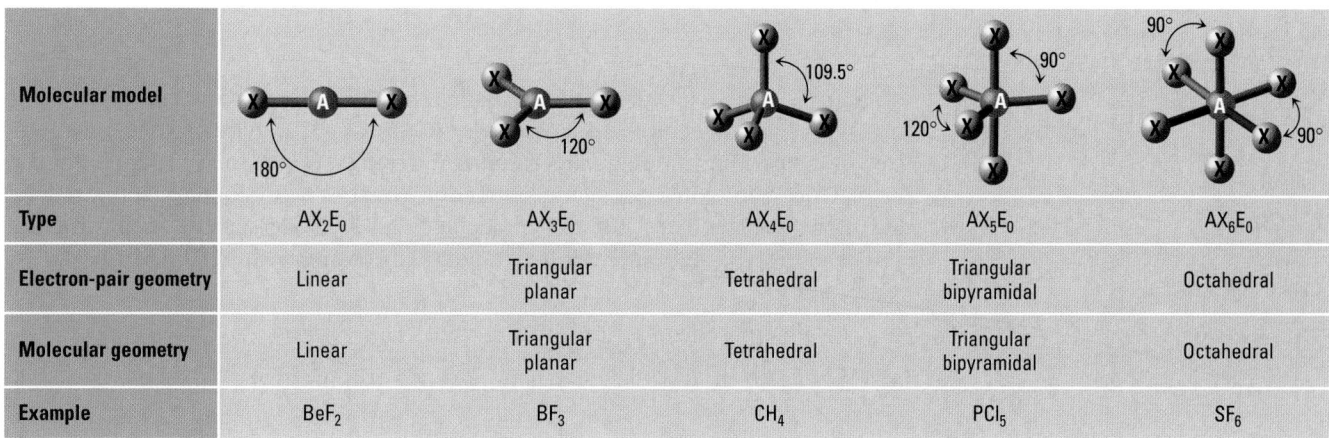

	AX₂E₀	AX₃E₀	AX₄E₀	AX₅E₀	AX₆E₀
Molecular model	180°	120°	109.5°	120° / 90°	90° / 90°
Type	AX_2E_0	AX_3E_0	AX_4E_0	AX_5E_0	AX_6E_0
Electron-pair geometry	Linear	Triangular planar	Tetrahedral	Triangular bipyramidal	Octahedral
Molecular geometry	Linear	Triangular planar	Tetrahedral	Triangular bipyramidal	Octahedral
Example	BeF_2	BF_3	CH_4	PCl_5	SF_6

Figure 9.4 Geometries predicted by the VSEPR model for molecules of the types AX_2E_0 through AX_6E_0 that contain only single covalent bonds and no lone pairs.

TABLE 9.1	Examples of Electron Pair Geometries and Molecular Geometries Predicted by VSEPR Model			
Type (X = atoms bonded to central atom A; E = lone pairs on central atom)	Number of lone pairs on central atom	Electron-pair geometry	Molecular geometry	Example
AX_2E_0	None	Linear	Linear	CO_2, $BeCl_2$
AX_2E_1	One	Triangular planar	Angular	$SnCl_2$
AX_2E_2	Two	Tetrahedral	Angular	H_2O, OCl_2
AX_2E_3	Three	Triangular bipyramidal	Linear	XeF_2
AX_3E_0	None	Triangular planar	Triangular planar	BCl_3, CO_3^{2-}
AX_3E_1	One	Tetrahedral	Triangular pyramidal	NCl_3
AX_3E_2	Two	Triangular bipyramidal	T-shaped	ClF_3
AX_4E_0	None	Tetrahedral	Tetrahedral	CH_4, $SiCl_4$
AX_4E_1	One	Triangular bipyramidal	Seesaw	SF_4
AX_4E_2	Two	Octahedral	Square planar	XeF_4
AX_5E_0	Five	Triangular bipyramidal	Triangular bipyramidal	PF_5
AX_5E_1	One	Octahedral	Square pyramidal	BrF_5
AX_6E_0	None	Octahedral	Octahedral	SF_6

CD-ROM Screen 9.16:
Determining Molecular Shape

Suppose you want to predict the shape of $SiCl_4$. First, draw the Lewis structure, with Si as the central atom. Because $SiCl_4$ is an AX_4E_0 type of molecule (Table 9.1), there are four bonding pairs forming four single bonds to Si and no lone pairs on Si. Thus, you would predict a tetrahedral electron-pair geometry and a tetrahedral molecular geometry for the molecule. This is in agreement with structural studies of $SiCl_4$, which indicate a tetrahedral molecule with all 109.5° Cl—Si—Cl bond angles.

Problem-Solving Example 9.1 Molecular Geometry

Use the VSEPR model to predict the electron-pair geometry, the molecular geometry, and bond angles of (a) $BeCl_2$ and (b) CF_4.

Answer

(a) Linear electron-pair and molecular geometries with 180° Cl—Be—Cl angles

(b) Tetrahedral electron-pair and molecular geometries with 109.5° F—C—F bond angles.

Explanation

(a) The Lewis structure is

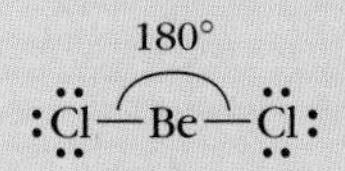

$$180°$$

$$:\ddot{Cl}—Be—\ddot{Cl}:$$

$BeCl_2$ is an AX_2E_0-type molecule (two bonding pairs, no lone pairs around the central Be atom; see Figure 9.4). The molecule has two Be—Cl single bonds around the central beryllium atom, arranged at 180° angles from each other. Therefore, $BeCl_2$ has a linear molecular geometry with all the atoms lying in the same plane.

(b) Being an AX_4E_0-type molecule, CF_4 will be a tetrahedral molecule with 109.5° F—C—F bond angles.

$$:\ddot{F}:$$
$$109.5° \diagdown | $$
$$:\ddot{F} \diagdown \overset{C}{\diagup} \ddot{F}:$$
$$:\ddot{F}:$$

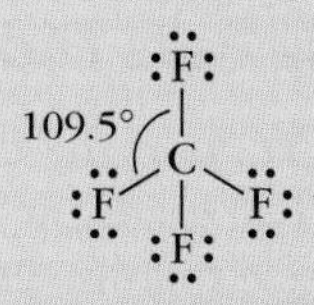

Problem-Solving Practice 9.1

Identify the electron-pair and molecular geometries and the bond angles for BF_3.

Multiple Bonds and Molecular Geometry

Although double bonds and triple bonds are shorter and stronger than single bonds *(⇐ p. 336),* they do not affect predictions of molecular shape. Why not? Electron pairs involved in a multiple bond are all shared between the same two nuclei and therefore occupy the same region. Because they must remain in that region, two electron pairs in a double bond or three in a triple bond are like a single balloon, rather than two or three balloons. Hence, *for the purpose of determining molecular geometry, the electron pairs in a multiple bond contribute to molecular geometry in the same manner as a single bond.* For example, compare BeF_2 (Figure 9.4), a linear molecule with the two Be—F single bonds 180° apart, with CO_2. In CO_2, the C=O double bonds act like the Be—F single bonds, and so the structure of CO_2 is also linear.

$$O=C=O$$

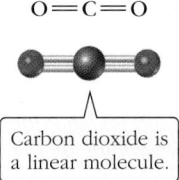

Carbon dioxide is a linear molecule.

When resonance structures are possible, the geometry can be predicted from any of the Lewis resonance structures or from the resonance hybrid structure. For example, the geometry of the CO_3^{2-} ion is predicted to be triangular planar because the carbon atom has three sets of bonds and no lone pairs. This can be predicted from either of the representations below.

$$\left[\begin{array}{c} :\ddot{O}: \\ | \\ :\ddot{O} = C \diagdown \ddot{O}: \end{array} \right]^{2-} \quad \text{or} \quad \left[\begin{array}{c} O \\ \vdots \\ O \cdots C \cdots O \end{array} \right]^{2-}$$

Exercise 9.1   Geometries

Based on the discussion so far, identify a characteristic that is common to all situations where electron-pair geometry and molecular geometry are the same for a molecule or a polyatomic ion.

Central Atoms with Bonding Pairs and Lone Pairs

How does the presence of lone pairs on the central atom affect the geometry of the molecule or polyatomic ion? The easiest way to visualize this situation is to return to the balloon model and notice that the electron pairs on the central atom do not all have to be bonding pairs. We can predict the electron-pair geometry, molecular geometry, and bond angles by applying the VSEPR model to the total number of valence electron pairs, that is, (bonding pairs as well as lone pairs) around the central atom. We will use NH_3 to illustrate guidelines for doing this.

1. ***Draw the Lewis structure.*** The Lewis structure for NH_3 is

$$H - \overset{..}{\underset{|}{N}} - H$$
$$H$$

2. ***Determine the number of bonds and the number of lone pairs around each atom.*** Any electron pairs in a multiple bond contribute to the molecular geometry in the same manner as those in a single bond. In the case of NH_3, there are three bonding pairs and one lone pair; thus, NH_3 is an AX_3E_1-type molecule (Table 9.1). The central N atom has no multiple bonds, only three H—N single covalent bonds.

3. ***Pick the appropriate electron-pair geometry around each central atom, and then choose the molecular shape that matches the total number of bonds and lone pairs.*** The *electron-pair geometry* around a central atom includes the spatial positions of all bond pairs and lone pairs. Because there are three bond pairs and one lone pair for a total of four pairs of electrons, we predict that the electron-pair geometry of NH_3 is tetrahedral. To represent this geometry, draw a tetrahedron with N as the central atom and the three bond pairs represented by lines, since they are single covalent bonds. The lone pair is drawn as a balloon shape to indicate its spatial position in the tetrahedron.

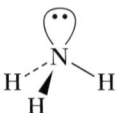

The positions of the lone pairs are *not* specified when describing the molecular geometry of molecules. The *molecular geometry* of ammonia is described as a triangular pyramid because the three hydrogen atoms form a triangular base with the nitrogen atom at the apex of the pyramid. (This can be seen by covering up the lone pair of electrons and looking at the molecular geometry — the location of the three H nuclei and the N nucleus.)

4. ***Predict the bond angles, remembering that lone pairs occupy more volume than do bonding pairs.*** Because the electron-pair geometry is tetrahedral, we would expect the H—N—H bond angles to be 109.5°. However, the experimentally determined bond angles in NH_3 are 107.5°. This is attributed to the bulkier lone pair forcing the bonding pairs closer together, reducing the bond angle from 109.5° to 107.5°.

Bonding pairs are concentrated in the bonding region between two atoms by the strong attractive forces of two positive nuclei and are, therefore, relatively compact, or "skinny." For a lone pair, there is only one nucleus attracting the electron pair. As a result, lone pairs are less compact. Using the balloon analogy, a lone pair is like a fatter balloon that takes up more room and squeezes the thinner balloons closer together. The relative strengths of electron-pair repulsions are

lone pair-lone pair > lone pair-bond pair > bond pair-bond pair

This predicts that lone pairs force bonding pairs closer together and decrease the angles between the bonding pairs. Recognizing this, we can predict that bond angles adjacent to lone pairs will be smaller than those predicted for perfect geometric shapes. The determination of electron-pair geometry and molecular geometry is summarized below:

> The success of the VSEPR model in predicting molecular shapes indicates that it is appropriate to account for the effects of lone pairs in this way.

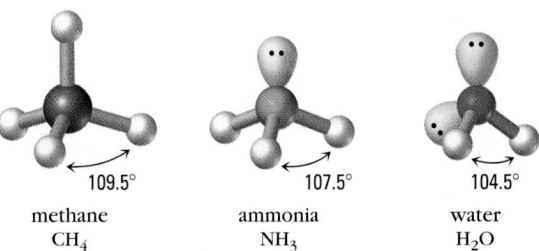

NH_3	H—N̈—H with H below	N̈ (tetrahedral)	N̈ molecular geometry	ball-and-stick model
Molecular formula	Lewis structure	Electron-pair geometry (tetrahedral)	Molecular geometry (triangular pyramid)	

The effect of lone pairs on bond angles as the number of lone pairs increases is shown for CH_4, NH_3, and H_2O.

109.5°	107.5°	104.5°
methane CH_4	ammonia NH_3	water H_2O

Methane, which has a tetrahedral shape, is the smallest member of the large family of saturated hydrocarbons called alkanes (⟵ *p. 80, Section 3.3)*. It is important to recognize that every carbon atom in an alkane has a tetrahedral environment. For example, notice that the carbon atoms in propane and in the much longer carbon chain of hexadecane do not lie in a straight line because of the tetrahedral geometry about each carbon atom. The tetrahedron is arguably the most important shape in chemistry because of its predominance in the chemistry of carbon and the carbon chains that form the backbone of many biomolecules.

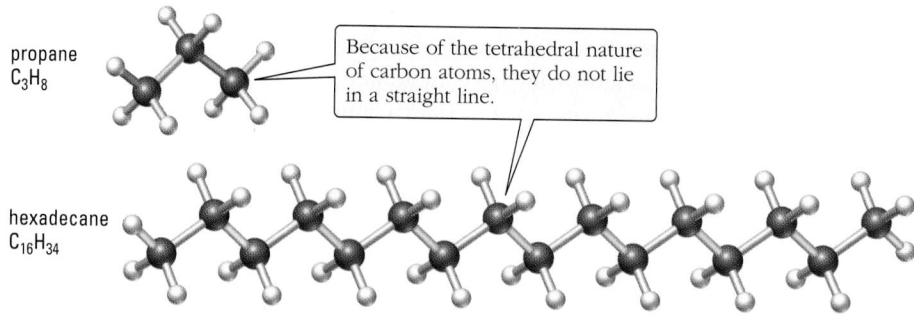

propane
C_3H_8

Because of the tetrahedral nature of carbon atoms, they do not lie in a straight line.

hexadecane
$C_{16}H_{34}$

Figure 9.5 gives additional examples of electron-pair and molecular geometries for molecules or ions with three and four electron pairs around the central atom. The experimentally determined bond angles are given for the examples. To check your understanding of the VSEPR model, try to explain the electron-pair geometry, the molecular geometry, and bond angles in each case.

Problem-Solving Example 9.2 Molecular Structure

Use the VSEPR model to predict the electron-pair geometry and the molecular geometry of (a) PH_4^+, (b) OCl_2, (c) SO_3, and (d) H_2CO.

Answer

Electron-pair geometry	Molecular geometry
(a) tetrahedral	tetrahedral
(b) tetrahedral	angular
(c) triangular planar	triangular planar
(d) triangular planar	triangular planar

Explanation   The Lewis structures are

(a) $\left[\begin{array}{c} H \\ | \\ H-P-H \\ | \\ H \end{array} \right]^+$ (b) $:\!\ddot{Cl}\!-\!\ddot{O}\!-\!\ddot{Cl}\!:$ (c) $:\!\ddot{O}\!-\!\underset{\underset{:\ddot{O}:}{\|}}{S}\!-\!\ddot{O}\!:$ (d) $\underset{H \quad\quad H}{\overset{:O:}{\underset{\|}{C}}}$

Sometimes the term "bent" is used rather than "angular" to describe this molecular geometry.

(a) The Lewis structure of PH_4^+ reveals that the central P atom has four electron pairs, all bonding pairs to terminal hydrogen atoms. Consequently, the ion is an AX_4E_0-type and has tetrahedral electron-pair and molecular geometries.

(b) The central oxygen is surrounded by two bonding pairs and two lone pairs (AX_2E_2 type). These four electron pairs give a tetrahedral electron-pair geometry. The molecular geometry is angular, not linear, because of lone pair–lone pair, lone pair–bonding pair, and bonding pair–bonding pair repulsions. The two lone pairs push the chlorine atoms closer together than the purely 109.5° tetrahedral angle expected for four bonding pair–bonding pair repulsions.

(c) The two bonding pairs in the S=O double bond are counted as one bond for determining the geometry of the molecule, giving three bonding regions around S (AX_3E_0 type). Therefore, the electron-pair and molecular geometries are triangular planar.

(d) For determining molecular structure, the double C=O bond is treated as one bond. There are no lone pairs on the central C atom. Therefore, the electron-pair geometry and molecular geometry are both triangular planar (AX_3E_0 type).

Problem-Solving Practice 9.2

Use Lewis structures and the VSEPR model to determine electron-pair and molecular geometries for (a) BrO_3^-, (b) SeF_2, and (c) NO_2^-.

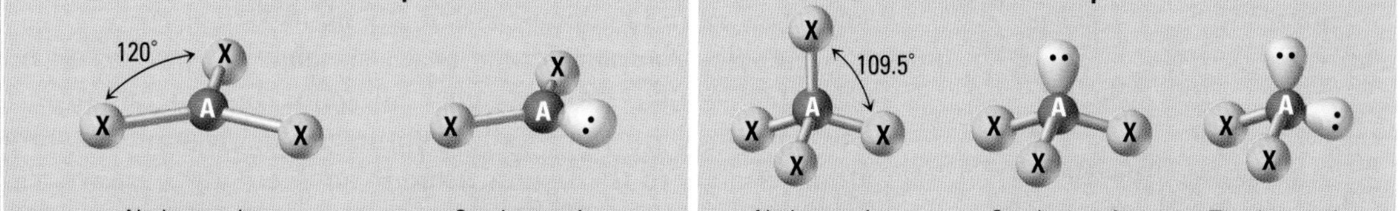

	Three electron pairs		Four electron pairs		
Molecular model	120°		109.5°		
	No lone pairs	One lone pair	No lone pairs	One lone pair	Two lone pairs
Type	AX_3E_0	AX_2E_1	AX_4E_0	AX_3E_1	AX_2E_2
Electron-pair geometry	Triangular planar	Triangular planar	Tetrahedral	Tetrahedral	Tetrahedral
Molecular geometry	Triangular planar	Angular	Tetrahedral	Triangular pyramidal	Angular
Example	BCl_3	$GeCl_2$	CCl_4	NCl_3	OF_2

Figure 9.5 Three and four electron pairs around a central atom. Examples are shown of electron-pair geometries and molecular shapes for molecules and polyatomic ions with three and four electron pairs around the central atom.

Expanded Octets: Central Atoms with Five or Six Electron Pairs

The situation becomes more complicated if a central atom has five or six electron pairs, some of which are lone pairs. Let's look first at the entries in Figure 9.6 for the case of five electron pairs. The three angles in the triangular plane are all 120°. The angles between any of the pairs in this plane and an upper or lower pair are only 90°. Thus, the *triangular bipyramidal* structure has two sets of positions that are not equivalent. Because the positions in the triangular plane lie in the equator of an imaginary sphere around the central atom, they are called **equatorial positions.** The north and south poles are called the **axial positions.** Each equatorial position is proximate to only two electron pairs at 90° angles, the axial ones, while an axial position is proximate to three electron pairs, the equatorial ones. This means that *any lone pairs, because they are fatter than bonding pairs, will occupy equatorial positions rather than axial positions.* For example, consider the ClF_3 molecule, which has three bonding pairs and two lone pairs, as shown in Figure 9.6. The two lone pairs in ClF_3 are equatorial; two bonding pairs are axial, with a third occupying an equatorial position, so the molecular geometry

Figure 9.6 Five and six electron pairs around a central atom. Molecules and polyatomic ions with five and six electron pairs around the central atom can have these electron-pair geometries and molecular shapes.

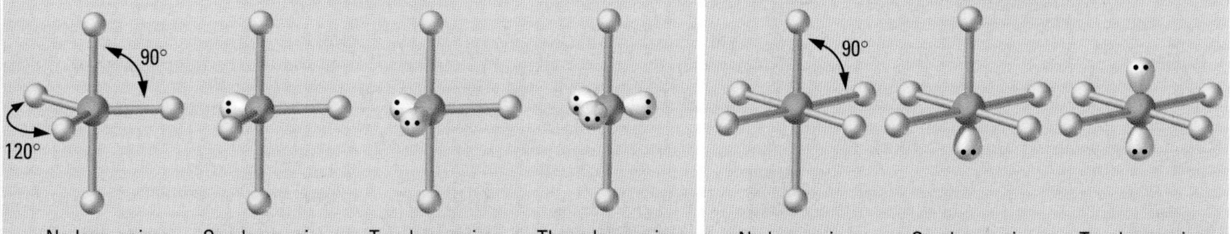

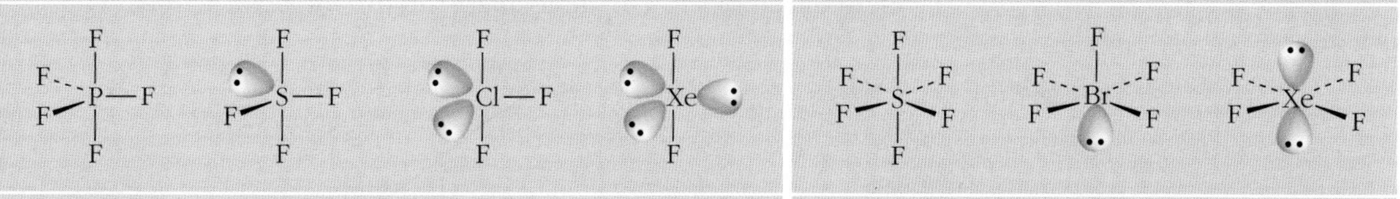

	Five electron pairs				Six electron pairs		
Molecular model	90° 120°				90°		
	No lone pairs	One lone pair	Two lone pairs	Three lone pairs	No lone pairs	One lone pair	Two lone pairs
Type	AX_5E_0	AX_4E_1	AX_3E_2	AX_2E_3	AX_6E_0	AX_5E_1	AX_4E_2
Example	F–P(F)(F)(F)F	F–S(F)(F)F	Cl–F(F)F	F–Xe(F)F	F–S(F)(F)(F)(F)F	F–Br(F)(F)(F)F	F–Xe(F)(F)F
Electron-pair geometry	Triangular bipyramidal	Triangular bipyramidal	Triangular bipyramidal	Triangular bipyramidal	Octahedral	Octahedral	Octahedral
Molecular geometry	Triangular bipyramidal	Seesaw	T shaped	Linear	Octahedral	Square pyramidal	Square planar

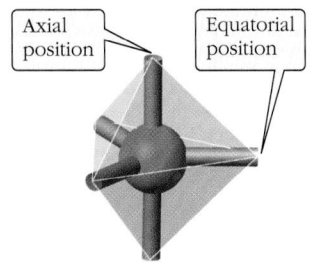

Triangular bipyramidal

In Figure 9.6, the lone pair in BrF_5 could have been located in any one of the six positions around Br. We arbitrarily chose the "down" position.

is T-shaped (Figure 9.6). (Given our viewpoint of axial positions lying on a vertical line, the T of the molecule is actually on its side ⊢.)

The electron-pair geometry for six electron pairs is an *octahedron* with each angle 90°. Unlike the triangular bipyramid, the octahedron has no distinct axial and equatorial positions; all positions are equivalent. As shown in Figure 9.6, six atoms bonded to the central atom in an octahedron can occupy three axes at right angles to one another (the *x*-, *y*-, and *z*-axes) and are *equidistant* from the central atom. Thus, if a molecule with octahedral electron-pair geometry has one lone pair, it makes no difference which apex it occupies. An example is BrF_5, whose molecular geometry is *square pyramidal* (Figure 9.6). There is a *square plane* containing the central atom and four of the atoms bonded to it, with the other bonded atom directly above the central atom and equidistant from the other four. If the molecule or ion has two lone pairs, as in ICl_4^-, each of the lone pairs needs as much room as possible. This is best achieved by placing the lone pairs at a 180° angle to each other above and below the square plane that contains the I atom and the four Cl atoms, so the molecular geometry of ICl_4^- is *square planar*.

An example of the power of the VSEPR model is the correct prediction of the shape of XeF_4. At one time the noble gases were not expected to form compounds because their atoms have a stable octet of valence electrons *(⇐ p. 302)*. The synthesis of XeF_4 was a surprise to chemists, but you can use the VSEPR model to predict the correct geometry. The molecule has 36 valence electrons (8 from Xe and 28 from four F atoms). There are 8 electrons in four bond pairs around Xe, and a total of 24 electrons in the lone pairs on the four F atoms. That leaves 4 electrons in two lone pairs on the Xe atom.

Because Xe is in Period 5, it can accommodate more than an octet of electrons (⇐ Section 8.8). The total of six electron pairs on Xe leads to a prediction of an octahedral *electron-pair* geometry (AX_4E_2 type). Where do you put the lone pairs? As explained above for the ICl_4^- ion, the lone pairs are placed at opposite corners of the octahedron to minimize repulsion by keeping them as far apart as possible. The result is a square planar *molecular geometry* for the XeF_4 molecule (cover the lone pairs in the XeF_4 drawing in Figure 9.6 to see this). This shape agrees with experimental structural results. A number of other xenon compounds have been prepared, and the VSEPR model has been useful in predicting their geometries as well.

 Exercise 9.2 Two Dissimilar Shapes

Triangular bipyramidal and square pyramidal molecular geometries both have a central atom with five electron pairs around it, but the molecular shapes are different. Explain how these two shapes differ.

Table 9.1 *(⇐ p. 364)* gives additional examples of molecules and ions whose shapes can be predicted by using the VSEPR model. The model also can be used to predict the geometry around atoms in molecules with more than one central atom. Consider, for example, lactic acid, a compound important in carbohydrate metabolism. The molecular structure of lactic acid is

$$
\begin{array}{ccccc}
 & H & H & :\!O\!: & \\
 & | & | & \| & \\
H\!-\!C\!-\!C\!-\!C\!-\!\ddot{O}\!-\!H \\
 & | & | & \\
 & H & :\!O\!: & \\
 & & | & \\
 & & H &
\end{array}
$$

Around C in the —CH_3 group are four bonding pairs and no lone pairs, so its molecular geometry is tetrahedral (109.5° bond angles). The middle carbon also has four bonding pairs associated with it and so also has tetrahedral geometry (109.5°

bond angles). The remaining carbon atom has four bonding pairs — two in single bonds and two in a C=O double bond — giving, for molecular geometry purposes, three bonding regions around it and a triangular planar geometry (120° bond angles). The single-bonded oxygens, in each case, have two bonding pairs and two lone pairs creating an angular molecular geometry around each oxygen atom (with a bond angle of less than 109.5°).

Lactic acid, a waste product of glucose metabolism, builds up rapidly in muscles during strenuous short-term exercise such as sprinting (100 to 800 m) and swimming (100 or 200 m). During such exercise, the accumulated lactic acid irritates the muscle tissue, causing muscle fatigue and soreness. But after the exercise stops, the lactic acid is converted to pyruvic acid, which is removed from the muscles, and the soreness and fatigue dissipate.

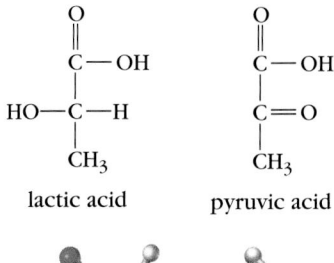

lactic acid   pyruvic acid

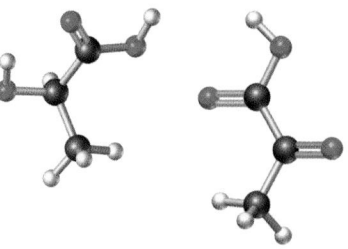

Problem-Solving Example 9.3   VSEPR and Molecular Shape

What are the electron-pair and molecular geometries of (a) SeF_6, (b) IF_3, and (c) $TeBr_4$?

Answer
(a) Octahedral electron-pair and molecular geometries
(b) Triangular bipyramidal electron-pair geometry and a T-shaped molecular geometry
(c) Triangular bipyramidal electron-pair geometry and seesaw shaped molecular geometry

Explanation   The Lewis structures are

(a) The 48 valence electrons are used for six single bonds from the central Se to the terminal fluorine atoms and three lone pairs around each fluorine. Because SeF_6 is an AX_6E_0-type of molecule, there are six bonding pairs around Se and no lone pairs, and the electron-pair geometry and the molecular geometry are octahedral.
(b) IF_3 is an AX_3E_2-type molecule. There are five electron pairs around I, giving a triangular bipyramidal electron-pair geometry. As noted in Figure 9.6, the two lone pairs will be in equatorial positions. One bonding pair is equatorial, while the remaining two bonding pairs are axial. The molecule is T-shaped.
(c) In $TeBr_4$, an AX_4E_1-type molecule, the central Te atom is surrounded by five electron pairs, giving it a triangular bipyramidal electron-pair geometry. The lone pair is in an equatorial position, creating a seesaw-shaped molecule (Figure 9.6).

Problem-Solving Practice 9.3

What is the electron-pair geometry and the molecular geometry of (a) ClF_2^- and (b) XeO_3?

Exercise 9.3   Electron Pairs and Molecular Shapes

Classify each of these species according to their AX_xE_y-type: (a) XeF_2 (b) NI_3 (c) I_3^-

9.3 ORBITALS CONSISTENT WITH MOLECULAR SHAPES: HYBRIDIZATION

Although Lewis structures are helpful in assigning molecular geometries, they indicate nothing about the orbitals occupied by the bonding and lone pair electrons. A theoretical model of covalent bonding, referred to as the **valence bond model,**

CD-ROM Screen 10.2: Models of Chemical Bonding

Chemistry in the News

A MOLECULE WITH A BANG

Nitroglycerine, dynamite, and TNT are widely used explosives, developed more than a century ago. Like all other explosives, they are chemically unstable compounds that decompose rapidly and violently into simpler, more stable products. Recently, Dr. Philip Eaton and Dr. Mao-Xi Gang at the University of Chicago have synthesized octanitrocubane (ONC), $C_8(NO_2)_8$, a com-

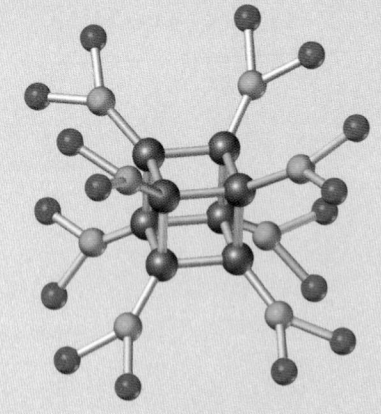

octanitrocubane
$C_8(NO_2)_8$

The structural formula of octanitrocubane.

pound that may be the most powerful non-nuclear explosive ever made.

An octanitrocubane molecule consists of a cube of carbon atoms, each of which is bonded to an NO_2 group. In that regard, ONC resembles TNT (trinitrotoluene), which also contains NO_2 groups bonded to carbon atoms, an unstable combination.

Central to ONC's explosive power is the 90° bond angle between the NO_2-bearing carbon atoms arranged in a cube. Carbon atoms in organic compounds generally have carbon-to-carbon bond angles that range from approximately 109.5° (tetrahedral) to 180° (linear). Forcing carbon atoms into arrangements with bond angles much less than 109° puts additonal strain on the carbon-to-carbon bonds, destabilizing them. In 1964, Dr. Eaton first synthesized cubane, C_8H_8, a molecule with a cube of eight carbon atoms, each bonded to a hydrogen atom. It was a breakthrough molecule, with any three adjacent carbon atoms having 90° bond angles among them. This sharp bond angle creates severe bond strain in cubane, a compound thought previously impossible to synthesize because of the required 90° bond angles.

Starting with cubane, Eaton and his colleagues learned how to sequentially replace each of its hydrogen atoms with NO_2 groups. This approach produced an intermediate compound, tetranitrocubane, a molecule with half the hydrogens of cubane replaced by NO_2 groups. Subsequently, heptanitrocubane was produced by replacing three of the remaining hydrogens of tetranitrocubane to create a compound just one NO_2 away from ONC. Heptanitocubane was then converted to octanitrocubane, which is thought to be up to 30% more powerful than the most powerful military explosive now used.

Source:

Browne, M.W. "Harnessing a Molecule's Explosive Powers." *The New York Times,* Tuesday, January 24, 2000; p. D3.

- When it explodes, ONC decomposes to N_2 and CO_2. Using molecular formulas, write the balanced chemical equation for this reaction.

- When TNT, $C_7H_5N_3O_6$, explodes, it also forms N_2, H_2O, and CO_2. Write the balanced equation for this decomposition using molecular formulas.

does so by describing a covalent bond as the result of an overlap of orbitals on the bonded atoms. For H_2, for example, the bond is a shared electron pair located in the overlapping *s* atomic orbitals.

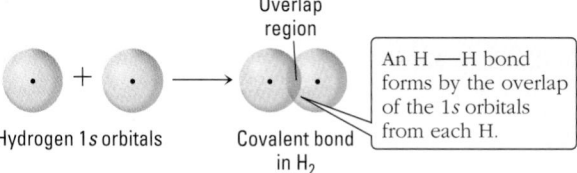

Hydrogen 1*s* orbitals   Covalent bond in H_2

An H —H bond forms by the overlap of the 1*s* orbitals from each H.

In hydrogen fluoride, HF, the overlap occurs between a 2*p* orbital with a single electron on fluorine and the single electron in the 1*s* orbital of a hydrogen atom.

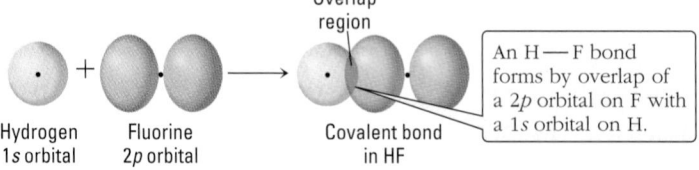

Hydrogen 1*s* orbital   Fluorine 2*p* orbital   Covalent bond in HF

An H — F bond forms by overlap of a 2*p* orbital on F with a 1*s* orbital on H.

The simple valence bond model of overlapping *s*, *p*, or *d* orbitals, however, must be modified to account for bonding molecules with central atoms, such as Be, B, and C. The electron configurations for Be, B, and C atoms are

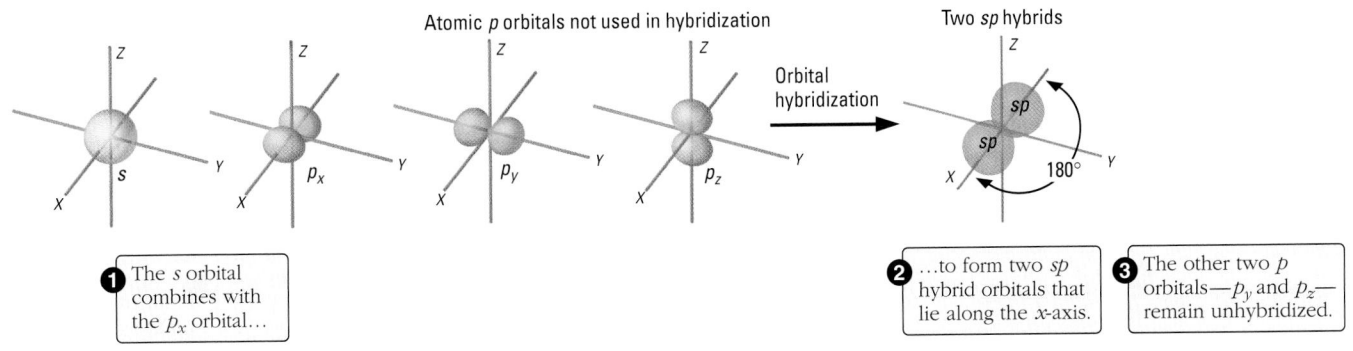

1 The s orbital combines with the p_x orbital...

2 ...to form two sp hybrid orbitals that lie along the x-axis.

3 The other two p orbitals—p_y and p_z—remain unhybridized.

Figure 9.7 **Formation of two sp hybrid orbitals.** An s orbital combines with a p orbital in the same atom, say, p_x, to form two hybrid orbitals that lie along the x-axis. The angle between the two sp orbitals is $180°$. The other two p orbitals remain unhybridized.

			1s	2s	2p$_X$	2p$_y$	2p$_z$
Be	$1s^2 2s^2$		↑↓	↑↓			
B	$1s^2 2s^2 2p^1$		↑↓	↑↓	↑		
C	$1s^2 2s^2 2p^2$	$1s^2 2s^2 2p_x^{\,1} 2p_y^{\,1}$	↑↓	↑↓	↑	↑	

The simple valence bond model would predict that Be, with an s^2 configuration like He, should form no compounds; B, with one unpaired electron ($2p^1$) should form only one bond; and C ($2p^2$) should form only two bonds with its two unpaired $2p$ electrons. But BeF_2, BF_3, and CF_4 exist, as well as other Be, B, and C compounds in which these atoms have two, three, and four bonds, respectively.

Hybrid Orbitals: sp, sp^2, sp^3

To account for molecules like CH_4, in which the actual molecular geometry is incompatible with simple overlap of s, p, and d orbitals, valence bond theory is modified to include a new kind of atomic orbital. Atomic orbitals of the proper energy and orientation in the same atom are **hybridized,** meaning that they are mixed to form **hybrid orbitals.** The resulting hybrid orbitals all have the same shape and energy. Consequently, they are better able to overlap with bonding orbitals on other atoms. ***The total number of hybrid orbitals formed is always equal to the number of atomic orbitals that are hybridized.***

The simplest hybrid orbitals are **sp hybrid orbitals** formed by the combination of one s orbital and one p orbital; these *two* atomic orbitals combine to form *two* sp hybrid orbitals (Figure 9.7).

One s atomic orbital + one p atomic orbital $\longrightarrow$ two sp hybrid orbitals

In BeF_2, for example, a $2s$ and a $2p$ orbital on Be are hybridized to form two sp hybrid orbitals that are $180°$ from each other.

 CD-ROM Screen 10.4: Hybrid Orbitals

Another theory, the *molecular orbital theory,* is also used to describe chemical bonding, as well as the magnetic properties of molecules. Molecular orbital theory is discussed in Appendix D.

Hybridization is described by a combination of mathematical functions that represent s, p, or d orbitals to give new functions that predict the energies and shapes of the hybrid orbitals.

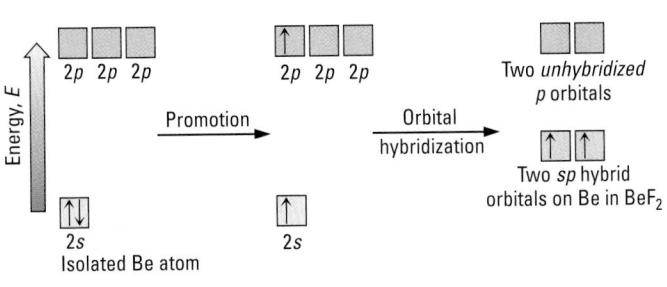

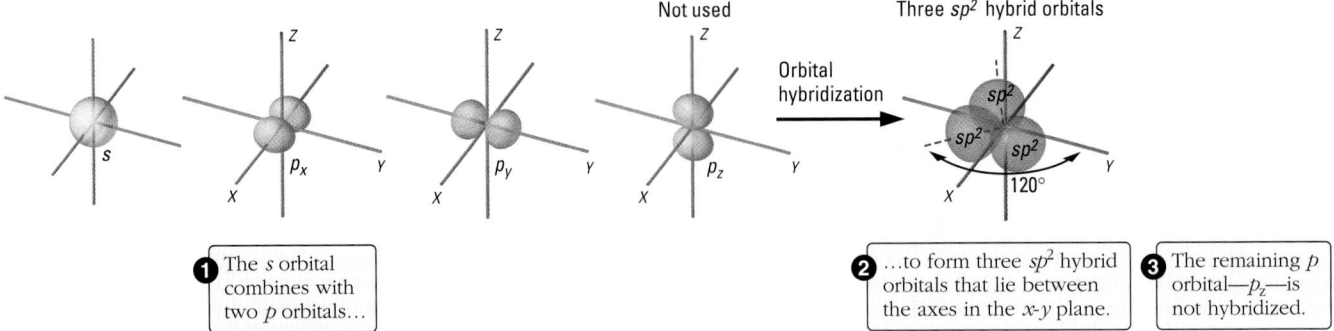

① The *s* orbital combines with two *p* orbitals…

② …to form three *sp²* hybrid orbitals that lie between the axes in the *x-y* plane.

③ The remaining *p* orbital—*pz*—is not hybridized.

Figure 9.8 Formation of three *sp²* hybrid orbitals. An *s* orbital combines with two *p* orbitals in the same atom, say p_x and p_y, to form three hybrid orbitals that lie in the *xy* plane. The angle between the three *sp²* orbitals is 120°. The other *p* orbital is unhybridized.

Each *sp* hybrid orbital has one electron that is shared with an electron from a fluorine atom to form two equivalent Be—F bonds. The remaining two 2*p* orbitals in Be are *unhybridized* and are 90° to each other and to the two hybrid orbitals.

In BF_3, three atomic orbitals on the central B atom—a 2*s* and two 2*p* orbitals—are hybridized to form *three **sp²** hybrid orbitals* (Figure 9.8). The superscript indicates the number of orbitals that have hybridized, one *s* and two *p* orbitals in this case.

One *s* atomic orbital + two *p* atomic orbitals ⟶ three *sp²* hybrid orbitals

The three *sp²* hybridized orbitals are 120° apart in a plane, each with an electron shared with a fluorine atom electron to form three equivalent B—F bonds. One of the boron 2*p* orbitals remains unhybridized.

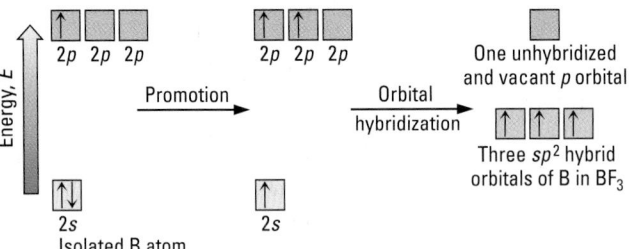

If an *s* and the three *p* orbitals on a central atom are hybridized, *four* hybrid orbitals, called **sp³ hybrid orbitals,** are formed (Figure 9.9).

One *s* atomic orbital + three *p* atomic orbitals ⟶ four *sp³* hybrid orbitals

The four *sp³* hybrid orbitals are equivalent and directed to the corners of a tetrahedron.

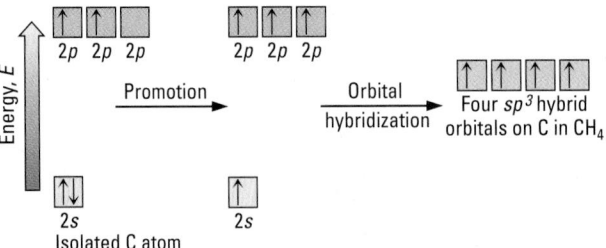

The *sp³* hybridization is consistent with the fact that carbon forms four tetrahedral bonds in CF_4 and in all other single-bonded carbon compounds. Overlap of the half-filled *sp³* hybrid orbitals on carbon with half-filled orbitals from four fluo-

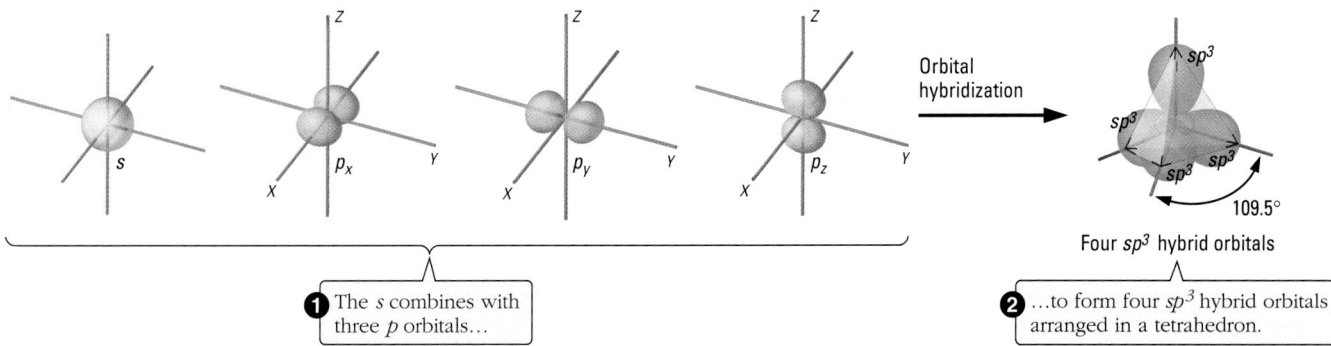

① The *s* combines with three *p* orbitals...

② ...to form four *sp³* hybrid orbitals arranged in a tetrahedron.

Four *sp³* hybrid orbitals

Figure 9.9 **Formation of four *sp³* hybrid orbitals.** An *s* orbital combines with three *p* orbitals to form four hybrid orbitals that are directed to the corners of a tetrahedron. The angle between the four *sp³* orbitals is 109.5°. The bonds of carbon atoms with four single bonds are described as formed by *sp³* hybrid orbitals.

rine atoms forms four equivalent C—F bonds. The central atoms of Periods 2 and 3 elements that obey the octet rule commonly have *sp³* hybridization.

Because the bond angles in NH_3 and H_2O are close to those in CF_4 and CH_4, this suggests that *in general, lone pairs as well as bonding electron pairs can occupy hybrid orbitals.* For example, according to the valence bond model, the four electron pairs surrounding the nitrogen and oxygen atoms in ammonia and water occupy *sp³* hybrid orbitals like those in CF_4 and CH_4. In NH_3, three shared pairs on nitrogen occupy three of these orbitals, with a lone pair filling the fourth. In H_2O, two of the *sp³* hybrid orbitals on oxygen contain bonding pairs, and two contain lone pairs (Figure 9.10).

It is important for you to recognize that because hybrid orbitals are oriented as far apart as possible to minimize repulsions, they are consistent with the molecular geometries predicted using VSEPR theory. Table 9.2 summarizes information about hybrid orbitals formed from *s* and *p* atomic orbitals.

Bonds in which there is head-to-head orbital overlap so that the electron density of the bond lies along the bonding axis are called **sigma bonds (σ bonds).** The single bonds in diatomic molecules such as H_2 and Cl_2 and also those in the molecules illustrated in Figure 9.10 and BeF_2, BF_3, and CF_4 are all sigma bonds.

CD-ROM Screen 10.5: Sigma Bonding; Screen 10.6: Determining Hybrid Orbitals

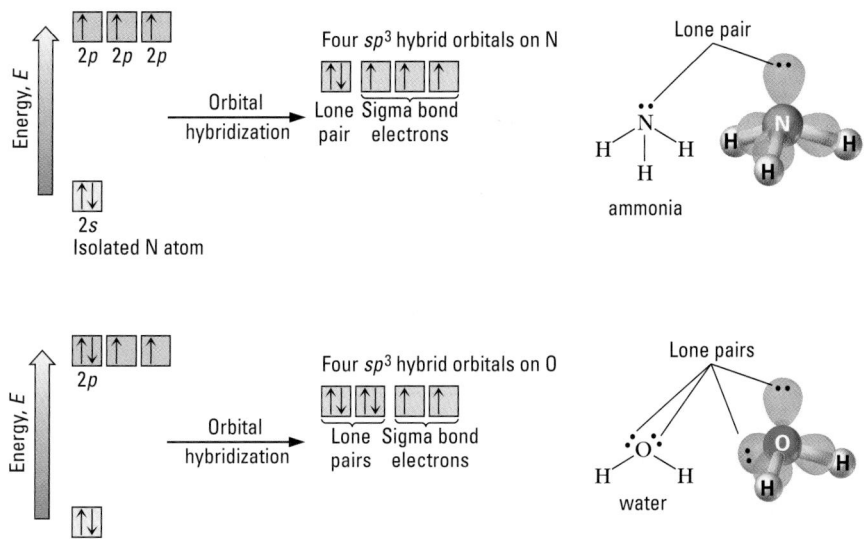

Figure 9.10 **Hybridization of nitrogen in ammonia and oxygen in water.** Nitrogen and oxygen are both *sp³* hybridized in these ammonia and water molecules.

TABLE 9.2	Hybrid Orbitals and Their Geometries		
	Linear	**Trigonal planar**	**Tetrahedral**
Atomic orbitals mixed	One *s* and one *p*	One *s* and two *p*	One *s* and three *p*
Hybrid orbitals formed	Two *sp*	Three *sp*^{2}	Four *sp*^{3}
Unhybridized orbitals remaining	Two *p*	One *p*	None

The N—H and O—H bonds in ammonia and water, respectively, are also sigma bonds.

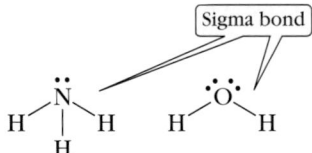

Hybrid atomic orbitals result in more bonds, stronger bonds, or both between atoms than do the unhybridized atomic orbitals from which they are formed.

Hybridization in Molecules with Multiple Bonds

CD-ROM Screen 10.7: Multiple Bonding: π Bonding

When using hybrid orbitals to account for molecular shapes, it must be recognized that in a multiple bond, the shared electron pairs do *not* all occupy hybrid orbitals. On a central atom with a multiple bond, the hybrid orbitals contain

- The electron pairs in single bonds to central atoms
- All lone pairs
- Only *one* of the shared electron pairs in a double or triple bond

It is also possible for *d* orbitals to overlap above and below a bond axis, forming a pi bond.

The electron pairs in *unhybridized* atomic orbitals are used to form the second bond of a double bond and the second and third ones in a triple bond. Such bonds, formed by the *sideways* (edgewise) overlap of *parallel* atomic *p* orbitals, are called **pi bonds (π bonds),** as shown in Figure 9.11. In contrast to sigma bonds, in which there is overlap along the bond axis, pi bonds result when *parallel p* orbitals (such as a p_y and another p_y) overlap above and below the bond axis.

The bonding in formaldehyde, H_2CO, exemplifies sigma and pi bonding. Formaldehyde has two C—H single bonds and one C=O double bond.

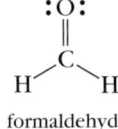

formaldehyde

The triangular planar electron-pair geometry suggests *sp*^{2} hybridization of the carbon atom to supply three *sp*^{2} orbitals for three sigma bonds. As seen in Figure

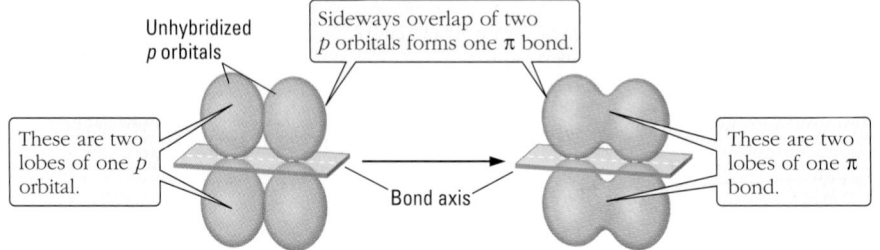

Figure 9.11 Pi bonding. Pi bonding occurs by sideways overlap of two adjacent *p* orbitals to make one pi (π) bond.

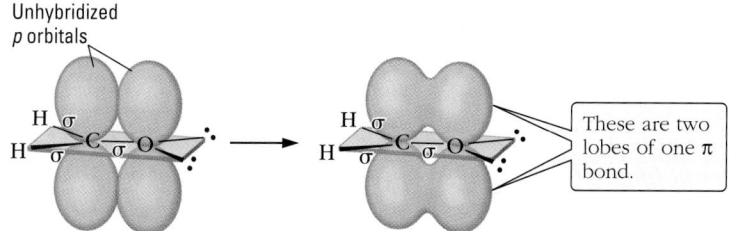

Unhybridized
p orbitals

These are two lobes of one π bond.

Figure 9.12 **Sigma and pi bonding in formaldehyde.** In formaldehyde, the C and O atoms are sp^2 hybridized. The C—O sigma bond forms from the end-to-end overlap of the hybrid orbitals. The sp^2 hybridization leaves a half-filled unhybridized p orbital on each C and O atom. These orbitals overlap to form the carbon-oxygen pi bond.

9.12, two of these sp^2 hybrid orbitals form two sigma bonds with half-filled $1s$ H orbitals; the third sp^2 hybrid orbital forms a sigma bond with a half-filled oxygen orbital. The *unhybridized p* orbital on carbon overlaps sideways with a p orbital on oxygen to form a pi bond, thus completing a double bond between carbon and oxygen.

Because pi bonds have less orbital overlap than sigma bonds, pi bonds are generally weaker than sigma bonds. Thus, a C=C double bond is stronger (and shorter) than a C—C single bond, but not twice as strong; correspondingly, a C≡C triple bond, although stronger (and shorter) than a C=C double bond, is not three times stronger than a C—C single bond. Also, a C=C double bond prevents rotation around the bond (under ordinary conditions). A consequence of this nonrotation is *cis-trans* isomerism (⇐ *p. 329*).

Bond energies are given in Table 8.2, p. 336.

Problem-Solving Example 9.4   Hybrid Orbitals, Sigma and Pi Bonding

Use hybridized orbitals and sigma and pi bonding to describe bonding in

(a) ethane (C_2H_6)   (b) ethylene (C_2H_4)   (c) acetylene (C_2H_2)

Answer

(a) On each carbon, sp^3 hybridized orbitals form sigma bonds to three H atoms and the other C atom.

(b) On each carbon, sp^2 hybridized orbitals form sigma bonds to two H atoms and to the other C atom; the unhybridized $2p$ orbitals form one pi bond between C atoms.

(c) The sp hybridized orbitals on each carbon atom form sigma bonds to one H atom and to the other C; the two unhybridized $2p$ orbitals on each carbon atom form two pi bonds between C atoms.

Explanation   The Lewis structures for the three compounds are

$$
\begin{array}{ccc}
\underset{\text{ethane}}{\overset{\displaystyle \text{H}\;\;\text{H}}{\text{H}-\overset{|}{\underset{|}{\text{C}}}-\overset{|}{\underset{|}{\text{C}}}-\text{H}}}
&
\underset{\text{ethylene}}{\overset{\text{H}}{}\text{C}=\text{C}\overset{\text{H}}{}}
&
\underset{\text{acetylene}}{\text{H}-\text{C}\equiv\text{C}-\text{H}}
\end{array}
$$

(a) Consistent with the tetrahedral bond angles for single-bonded carbon atoms, each carbon atom in ethane is sp^3 hybridized and all bonds are sigma bonds.

(b) The C=C double bond in ethylene, like the C=O double bond in formaldehyde, requires sp^2 hybridization of carbon. Two of the sp^2 hybrid orbitals on each carbon form sigma bonds with hydrogens; the third sp^2 hybrid orbital overlaps head-to-head with an sp^2 hybrid orbital on the other carbon, creating a C—C sigma bond. The double bond between the carbon atoms is completed by the sideways overlap of parallel unhybridized p orbitals from each carbon to form a pi bond. Thus, the C=C double bond (and all double bonds) consists of a sigma bond and a pi bond.

(c) In acetylene, sp hybridization of each carbon atom is indicated by the linear molecular geometry. One of the sp hybrid orbitals on each carbon forms a sigma bond between C and H; the other is used to join the carbon atoms by a sigma bond. The two

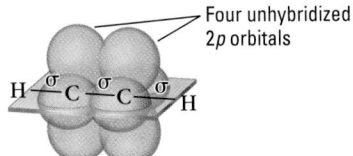

(a) Sigma bonds in acetylene

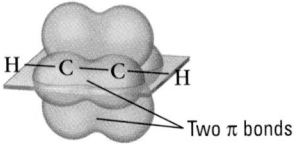

(b) Pi bonds in acetylene

Figure 9.13 Sigma and pi bonding in acetylene. (a) The carbon atoms in acetylene are sp hybridized. (b) Thus, each contains two unhybridized p orbitals, one from each carbon atom, which overlap sideways to form two pi bonds. There is also a sigma bond formed by the head-to-head overlap of two sp hybrid orbitals.

Hybridizing the appropriate five atomic orbitals yields five sp^3d hybrid orbitals; hybridization of the suitable six atomic orbitals yields six sp^3d^2 hybrid orbitals.

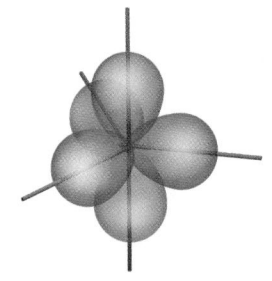

(a) Triangular bipyramidal sp^3d hybridization

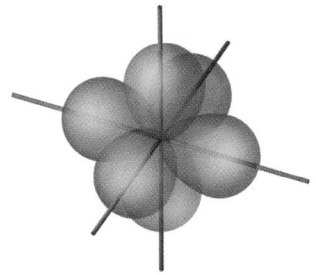

(b) Octahedral sp^3d^2 hybridization

Figure 9.14 Hybridization using d orbitals to form (a) sp^3d and (b) sp^3d^2 hybrid orbitals. The five sp^3d hybrid orbitals form from hybridizing a d orbital, an s orbital, and three p orbitals. Six sp^3d^2 hybrid orbitals come from hybridizing two d, an s, and three p orbitals.

unhybridized p orbitals on each carbon overlap sideways forming two pi bonds, completing the triple bond. Therefore, the triple bond consists of one sigma and two pi bonds in which the pi bonds are at right angles (90°) to each other (Figure 9.13).

Problem-Solving Practice 9.4

Using hybridization and sigma and pi bonding, explain the bonding in (a) HCN and (b) H₂CNH.

Exercise 9.4 Pi Bonding

Explain why pi bonding is not possible for an sp^3 hybridized carbon atom.

Hybridization in Expanded Octets

Central atoms from the third and subsequent periods can accommodate more than four electron pairs, that is, an expanded octet (⬅ *p. 347*). Bonding pairs and lone electron pairs are accommodated in these atoms by hybridizing d as well as s and p atomic orbitals. By including d atomic orbitals, two additional types of hybrid orbitals result, called sp^3d and sp^3d^2 **hybrid orbitals** (Figure 9.14).

One s atomic orbital + three p atomic orbitals + one d atomic orbital ⟶

five sp^3d hybrid orbitals

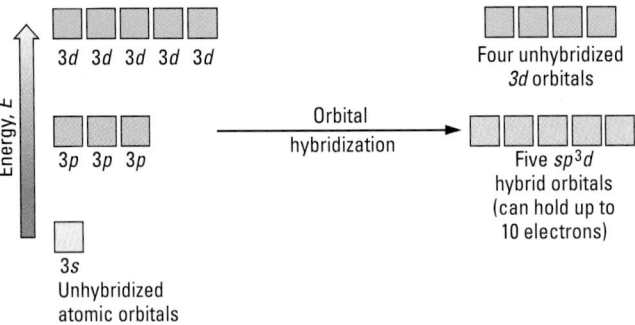

One s atomic orbital + three p atomic orbitals + two d atomic orbitals ⟶

six sp^3d^2 hybrid orbitals

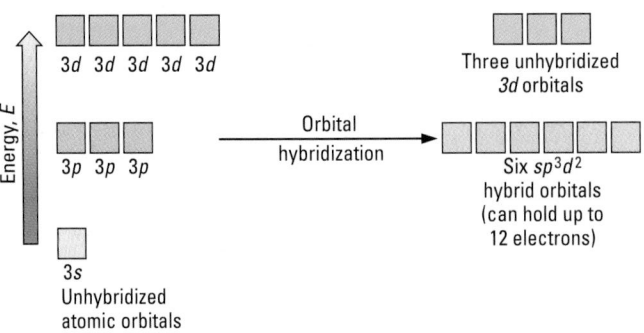

Hybridization of s, p, and d orbitals coincides with the molecular geometry expected from VSEPR theory (Figure 9.14). Table 9.3 summarizes information about hybrid orbitals formed from s, p, and d atomic orbitals.

TABLE 9.3	Hybrid Orbitals and Their Geometries	
	Trigonal bipyramidal	**Octahedral**
Atomic orbitals mixed	One s, three p, and one d	One s, three p, and two d
Hybrid orbitals formed	Five sp^3d	Six sp^3d^2
Unhybridized orbitals remaining	Four d	Three d

Problem-Solving Example 9.5 Hybridization and Expanded Octets

Describe the hybridization around the central sulfur atom and the bonding in SF_4 and SF_6.

Answer SF_4: sp^3d, four sigma bonds; SF_6: sp^3d^2 six sigma bonds

Explanation   We start with the Lewis structures of each compound.

$$SF_4 \qquad SF_6$$

In SF_4 there are five electron pairs around sulfur—four bonding pairs and one lone pair. This requires five hybrid orbitals, which can be produced by sp^3d hybridization. The four S—F sigma bonds are formed by head-to-head overlap of the half-filled sp^3d hybrid orbitals with half-filled fluorine orbitals; the lone pair is in the remaining hybrid orbital. SF_4 has triangular bipyramidal electron-pair geometry and a seesaw molecular geometry (Figure 9.6).

The six bonding pairs around sulfur in SF_6 are in six sp^3d^2 orbitals that form six sigma bonds with six fluorine atoms. The six bonding electron pairs are directed to the corners of an octahedron, so this molecule has octahedral electron-pair and molecular geometries (Figure 9.6).

Problem-Solving Practice 9.5

Describe the hybridization of the central atom and bonding in (a) PF_6^-, (b) IF_3, and (c) ICl_4^-.

9.4 MOLECULAR POLARITY

CD-ROM Screen 9.17: Molecular Polarity

A molecule that has polar bonds (⟸ Section 8.5) may or may not be polar. If the polar bonds are oriented so that their polarities cancel each other, the result is a **nonpolar molecule.** If electron density is concentrated at one end of the molecule, the result is a **polar molecule.** A polar molecule is a *permanent* dipole with a partial negative charge ($\delta-$) where the electron density is concentrated, and a partial positive ($\delta+$) charge at the opposite end, as illustrated in Figure 9.15.

Before examining the factors that determine whether a molecule is polar, let's look at the experimental measurement of the polarity of molecules. Polar molecules experience a force in an electric field that tends to align them with the field (Figure 9.16, on page 381). When an electric field is created by a pair of oppositely charged plates, the partial positive end of each polar molecule is attracted toward the negative plate, and the partial negative end is attracted toward the positive plate. The extent to which the molecules line up with the field depends on their **dipole moment** (μ), which is defined as the product of the magnitude of the par-

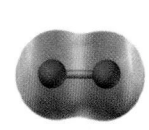

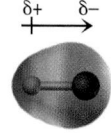

Nonpolar molecule, Cl_2 Polar molecule, HCl

Figure 9.15   Polar (HCl) and nonpolar (Cl₂) molecules.   The red color indicates higher electron density, the blue color lower electron density.

Tools of Chemistry Ultraviolet-Visible Spectroscopy

When a molecule absorbs radiation in the ultraviolet (UV) or visible region, the absorbed energy promotes an electron from a lower energy orbital in the ground state molecule to a higher energy orbital in the excited state molecule. The maximum UV-visible absorption occurs at a wavelength characteristic of the molecule's structure. It can be determined from a UV-visible spectrum, a plot of the absorbance, which is the intensity of radiation absorbed by the molecule, versus the wavelength at which the absorbance occurred.

Ultraviolet-visible spectral data can be used to account for several structural features, including (1) differentiation of *cis* from *trans* isomers, because a *trans* isomer generally absorbs UV-visible radiation at a longer wavelength than the *cis* isomer, (2) confirmation of the presence (or absence) of carbonyl

$\diagdown$C$=$O and aromatic groups in organic compounds, because

they absorb UV radiation at characteristic wavelengths, and (3)

specific transitions of electrons between *d* orbitals in transition metal compounds.

Ultraviolet-visible spectroscopy can also be used to study the molecular structures of colored compounds, those that absorb radiation in the visible region of the spectrum. Generally, organic compounds containing only sigma bonds are colorless; that is, they do not absorb light in the visible region. On the other hand, brightly colored pigments, such as beta-carotene, have an extended sequence of alternating single and double bonds, called a *conjugated* system. Beta-carotene, for example, contains 11 double bonds in its conjugated system. When visible light is absorbed, electron transitions occur among the pi electrons in the conjugated double bonds, resulting in the yellow-orange color in carrots (beta-carotene). Carrots look yellow-orange because electron transitions in their beta-carotene absorb in the 400- to 500-nm region, which is blue-green. We then see the remaining unabsorbed (reflected) portion of the visible spectrum, which is yellow-orange.

beta-carotene

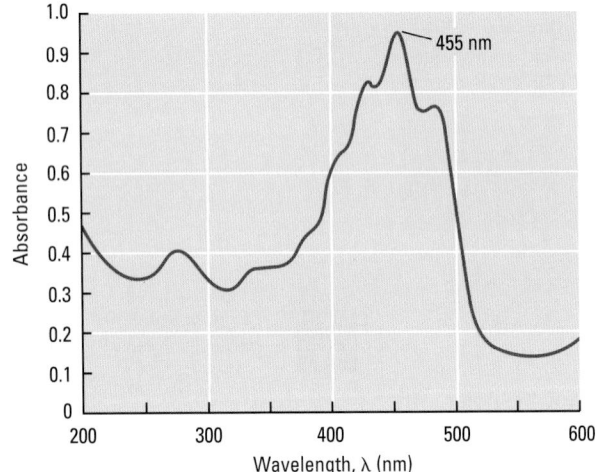

Some vegetables and fruits that are rich in beta-carotene.

The UV-visible spectrum of beta-carotene. Beta-carotene absorbs in the blue-green region of the visible spectrum indicated by a maximum absorbance at 455 nm. This peak is caused by transitions of pi electrons in the conjugated double bonds of the beta-carotene molecule.

tial charges ($\delta-$ and $\delta+$) times the distance of separation between them. The derived unit of the dipole moment is the coulomb-meter (C m); a more convenient derived unit is the debye (D), defined as 1 D = 3.34×10^{-30} C m. Some typical experimental values are listed in Table 9.4. Nonpolar molecules have a zero dipole moment ($\mu = 0$); *the dipole moments for polar molecules are always greater than zero and increase with greater molecular polarity.*

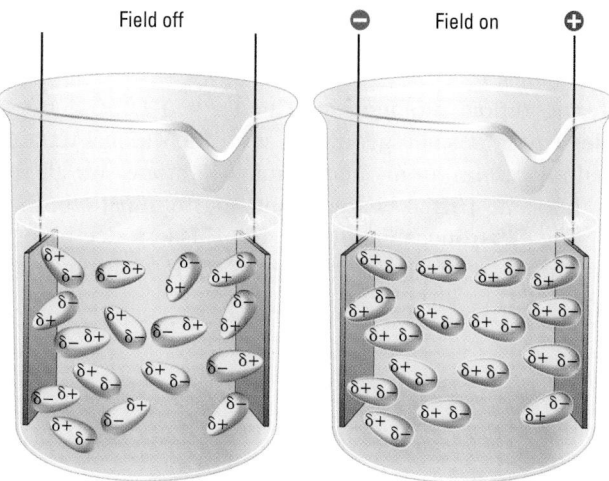

Field off   ⊖   Field on   ⊕

Figure 9.16 Polar molecules in an electric field. Polar molecules experience a force in an electric field that tends to align them so that oppositely charged ends of adjacent molecules are closer to each other.

To predict whether a molecule is polar, we need to consider whether the molecule has polar bonds and how those bonds are positioned relative to one another. A diatomic molecule is, of course, linear. If its atoms differ in electronegativity, then the bond and the molecule will be polar, with the partial negative charge at the more electronegative atom. In polyatomic molecules, the individual bonds can be polar, but because of molecular shape, there may be no net dipole ($\mu = 0$); if so, the *molecule* is nonpolar. In other nondiatomic molecules, the bond polarity and molecular shape combine to give a net dipole ($\mu > 0$) and a polar molecule. We can correlate the types of molecular geometry with dipole moment by applying a general rule to a molecule of the type AB_n (A is the central atom, B represents the terminal atom or groups of atoms, and n is the number of terminal atoms or groups). Such a molecule will *not* be polar if it meets *all* of the following conditions.

- All the terminal atoms or groups are the same, *and*
- All the terminal atoms or groups are symmetrically arranged around the central atom, A, in the molecular geometries given in Figure 9.4 (⇐ *p. 364*) or the AX_5E_0 and AX_6E_0 geometries given in Figure 9.6 (⇐ *p. 369*).

This means that molecules with the molecular geometries given in Figures 9.4 and 9.6 will never be polar if all their terminal atoms or groups are the same.

Alternatively, a molecule is *polar* if it meets *either* of the following conditions.

- One or more terminal atoms differs from the others, *or*
- The central atom has one or more lone pairs.

Consider, for example, carbon dioxide, CO_2, a linear triatomic molecule. Each C=O bond is polar because O is more electronegative than C, so O is the partial negative end of the bond dipole. The dipole moment contribution from each bond (the bond dipole) is represented by the symbol ↦, in which the plus sign indicates the partial positive charge and the arrow points to the partial negative end of the bond. We can use the arrows to help estimate whether a molecule is polar. In CO_2, the O atoms are at the same distance from the C atom, they both have the same $\delta-$ charge, and they are symmetrically on opposite sides of C. Therefore, their bond dipoles cancel each other, resulting in a molecule with a zero molecular dipole moment. Even though each C=O bond is polar, CO_2 is a nonpolar molecule due to its linear shape.

The situation is different for water, an angular triatomic molecule. Here, both O—H bonds are polar, with the H atoms having the same $\delta+$ charge. Note, however, that the two bond dipoles are not symmetrically arranged; they do not point

TABLE 9.4	Dipole Moments of Selected Molecules
Molecule	**Dipole moment, μ (D)**
H_2	0
HF	1.78
HCl	1.07
HBr	0.79
HI	0.38
ClF	0.88
BrF	1.29
BrCl	0.52
H_2O	1.85
H_2S	0.95
CO_2	0
NH_3	1.47
NF_3	0.23
NCl_3	0.39
CH_4	0
CH_3Cl	1.92
CH_2Cl_2	1.60
$CHCl_3$	1.04
CCl_4	0

Nonpolar molecules have a dipole moment of zero.

O=C=O   no net dipole
$\delta-$ $\delta+$ $\delta-$

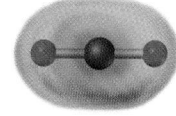

directly toward or away from each other, but augment each other to give a molecular dipole moment of 1.85 D (Table 9.4). Thus, water is a polar molecule (Figure 9.17).

Now consider the differing dipole moments of CF_4 ($\mu = 0$ D) and CF_3H ($\mu = 1.60$ D). Both molecules have the same geometry, with their atoms tetrahedrally arranged around a central carbon atom. The terminal F atoms are all the same in CF_4 and thus have the same partial charges. But, the terminal atoms in CF_3H are not the same; F is more electronegative than C, but H is less electronegative than C, so that the bond dipoles reinforce each other as CF_3H. Consequently, CF_4 is a nonpolar molecule and CF_3H is polar.

CF$_4$ is nonpolar   CHF$_3$ is polar

Figure 9.17 Water is a polar molecule.

Using a microwave oven to make popcorn or to heat dinner is an everyday application of the fact that water is a polar molecule. Most foods have a high water content. When water molecules in foods absorb microwave radiation, they rotate, turning so that the dipoles align with the crests and troughs of the oscillating microwave radiation. The radiation generator (magnetron) in the oven creates microwave radiation that oscillates at 2.45 GHz (2.45 gigahertz, $2.45 \times 10^9 \, s^{-1}$), very nearly the optimum frequency to rotate water molecules. So the leftover pizza warms up in a hurry for a late-night snack.

Problem-Solving Example 9.6   Molecular Polarity

Are dichloromethane (CH_2Cl_2) and boron trifluoride (BF_3) polar or nonpolar? If polar, indicate the direction of the net dipole.

Answer CH_2Cl_2 is polar, with the partial negative end at the Cl atoms. BF_3 is nonpolar.

Explanation   From the Lewis structure for CH_2Cl_2, you might be tempted to say that CH_2Cl_2 is nonpolar because the hydrogens appear across from each other, as do the chlorines. However, CH_2Cl_2 is a tetrahedral molecule in which neither hydrogen nor chlorine is directly across from each other, so no bond dipoles cancel.

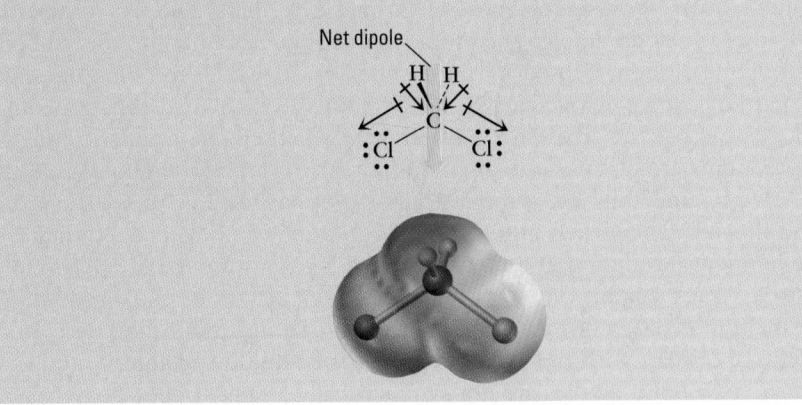

Because Cl is more electronegative than C, while H is less electronegative, negative charge is drawn away from H atoms toward Cl atoms. As a result, CH_2Cl_2 has a net dipole ($\mu = 1.60$ D), with the partial negative end at the Cl atoms and the partial positive end between the H atoms.

As predicted for a molecule with three electron pairs around the central atom, BF_3 is triangular planar.

Because F is more electronegative than B, the B—F bonds are polar, with F being the partial negative end. The molecule is nonpolar, though, because the three terminal F atoms are identical, they have the same partial charge, and they are arranged symmetrically around the central B atom. The bond dipole of each B—F is cancelled by the bond dipoles of the opposite two B—F bonds.

Problem-Solving Practice 9.6

For each of the following molecules, decide whether the molecule is polar, and if so, which side is partially positive and which is partially negative.

(a) $BFCl_2$ (b) NH_2Cl (c) SCl_2

Exercise 9.5   Dipole Moments

Explain the differences in the dipole moments of

(a) HI (0.38 D) and HBr (0.79 D)
(b) CH_3Cl (1.92 D), CH_3Br (1.81 D), and CH_3I (1.62 D)

9.5 NONCOVALENT INTERACTIONS AND FORCES BETWEEN MOLECULES

Molecules are attracted to one another; in a sense they are sticky. In some way, the stickiness of molecules is always the result of attraction between opposite charges, regardless of whether the charges are permanent or temporary. Polar molecules attract each other due to their permanent dipoles. Even nonpolar molecules have some attraction for each other, as you will see in this section.

Atoms within the same molecule are held together by covalent chemical bonds with strengths ranging from 150 to 1000 kJ/mol. Covalent bonds are strong forces. For example, it takes 1656 kJ to break 4 mol of C—H covalent bonds and separate the one C atom and four H atoms in all the molecules in 1 mol of methane molecules:

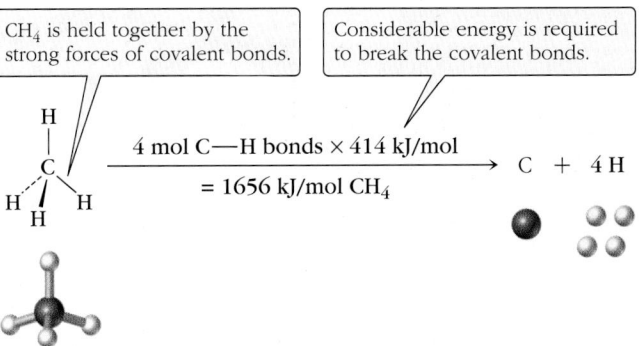

Weaker forces attract one molecule to another molecule. In contrast to the 1656 kJ/mol it takes to atomize methane, only 8.9 kJ is required to pull 1 mol of methane molecules that are close together in liquid methane away from each other to evaporate liquid methane to a gas.

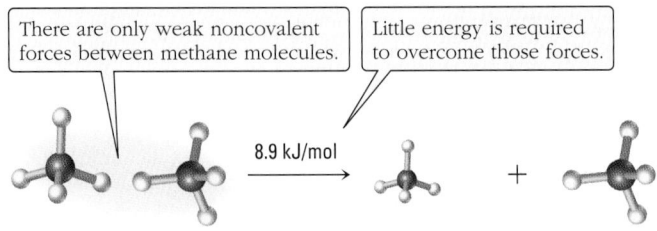

There are only weak noncovalent forces between methane molecules.

Little energy is required to overcome those forces.

8.9 kJ/mol

*Intra*molecular means *within* a single molecule; *inter*molecular indicates *between* two or more separate molecules.

CD-ROM Screen 13.3: Intermolecular Forces (1)

We refer collectively to all forces of attraction other than covalent, ionic, or metallic bonding as **noncovalent interactions** (*metallic bonding* is discussed in Section 11.8). When such forces act between molecules, they are referred to as **intermolecular forces.** Because intermolecular forces do not result from the sharing of electron pairs between atoms, they are weaker than covalent bonds. The strengths of noncovalent interactions between molecules (*inter*molecular forces) account for melting points, boiling points, and other properties of molecular substances. Noncovalent interactions between different parts of the same large molecule (*intra*molecular forces) maintain biologically important molecules in the exact shapes required to carry out their functions. For example, consider Figure 9.18, which shows the molecule chymotrypsin. The ribbon shows how chymotrypsin, a protein with a molecular weight of over 22,000, is folded into its biochemically active shape helped by noncovalent interactions between strategically placed atoms. Noncovalent interactions help maintain the shapes of thousands of proteins in our bodies.

The next few sections explore three types of noncovalent interactions: London forces, dipole-dipole attractions, and hydrogen bonding.

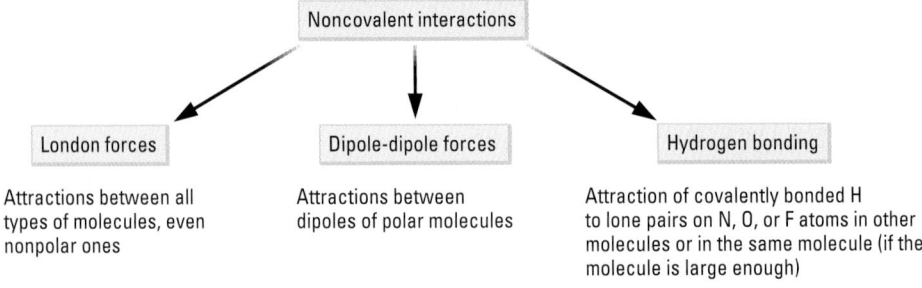

Noncovalent interactions

London forces

Attractions between all types of molecules, even nonpolar ones

Dipole-dipole forces

Attractions between dipoles of polar molecules

Hydrogen bonding

Attraction of covalently bonded H to lone pairs on N, O, or F atoms in other molecules or in the same molecule (if the molecule is large enough)

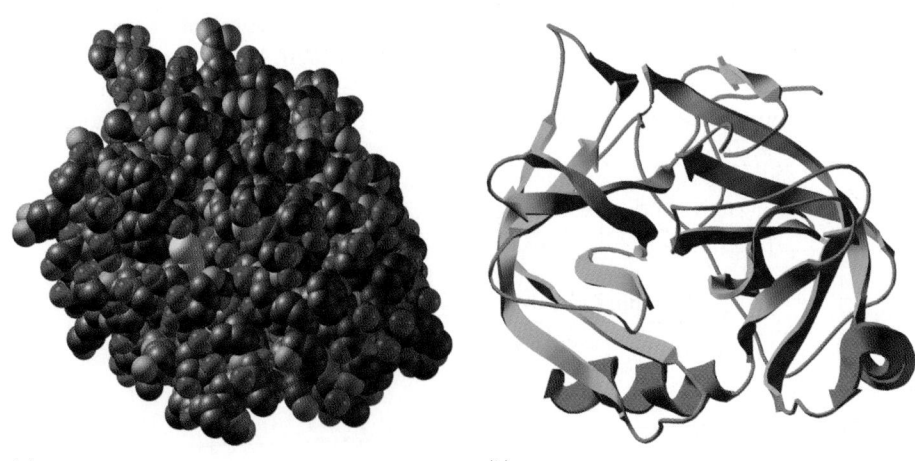

Figure 9.18 A space-filling model (a) and a ribbon model (b) of a chymotrypsin molecule. Noncovalent interactions help fold chymotrypsin into its biochemically active shape.

(a)   (b)

London Forces

London forces, also known as dispersion forces, occur in *all* molecular substances. They result from the attraction between the positive and negative ends of *induced* (nonpermanent) dipoles in adjacent molecules. An **induced dipole** is caused in one molecule when the electrons of a neighboring molecule are momentarily unequally distributed. The result is a *temporary* dipole in the first molecule. Figure 9.19 illustrates how one H_2 molecule with a momentary unevenness in its electron distribution can induce a dipole in a neighboring H_2 molecule. The process of inducing such a dipole between atoms or molecules is **polarization.** Due to polarization, even noble gas atoms, molecules of diatomic gases such as oxygen, nitrogen, and chlorine (which all must be nonpolar), and nonpolar hydrocarbon molecules such as CH_4 and C_2H_6 have these fleeting dipoles. Such London forces are the *only* noncovalent force among nonpolar molecules.

London forces range from approximately 0.05 to 40 kJ/mol. Their strength depends on how readily electrons in a molecule can be polarized, which depends on the number of electrons in a molecule and how tightly they are held by nuclear attraction. In general, the more electrons there are in a molecule, the more easily they can be polarized; *London forces increase with increased number of electrons in a molecule.* Thus, large molecules with many electrons, such as Br_2 and I_2, are relatively polarizable. In contrast, smaller molecules (F_2, N_2, O_2) are less polarizable because they have fewer electrons.

When we look at the boiling points of several groups of nonpolar molecules, the effect of the number of electrons becomes readily apparent (Table 9.5). (This effect also correlates with molar mass—the heavier an atom or molecule, the more electrons it has.) In a liquid, the molecules are relatively close to one another and are attracted to each other by their noncovalent intermolecular forces. For a liquid to boil, its molecules must have enough energy to overcome these attractive forces. Thus, *the boiling point of a liquid depends upon intermolecular forces.* If more energy is required to overcome the intermolecular attractions between molecules of liquid A than the intermolecular attractions between molecules of liquid B, then the boiling point of A will be higher than that of B. Conversely, lower intermolecular attractions result in lower boiling points. For example, the boiling point of Br_2 (59 °C) is higher than that of Cl_2 (-34 °C), indicating stronger London forces among Br_2 molecules than among Cl_2 molecules.

Interestingly, molecular shape can also play a role in London forces. Two of the isomers of pentane—straight-chain pentane and 2,2-dimethylpropane (both with the molecular formula C_5H_{12})—differ in boiling point by 27 °C. The linear shape of the *n*-pentane molecule allows close contact with adjacent molecules over its entire length, resulting in stronger London forces, while the more compact 2,2-dimethylpropane molecule does not allow as much close contact.

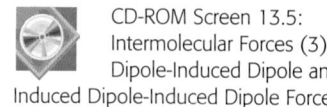

CD-ROM Screen 13.5: Intermolecular Forces (3): Dipole-Induced Dipole and Induced Dipole-Induced Dipole Forces

London forces are named to recognize the work of Fritz London, who extensively studied the origins and nature of such forces. London forces are also called *dispersion forces.*

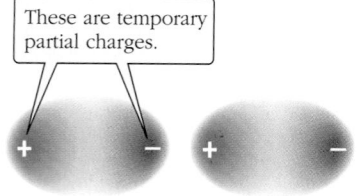

These are temporary partial charges.

Figure 9.19   Origin of London forces. Such attractive forces originate when electrons are momentarily distributed unevenly in the molecule on the right. It has a temporary positive charge that is close to the molecule on the left. The positive charge attracts electrons in the left-hand molecule, temporarily creating an induced dipole (an unbalanced electron distribution) in that molecule.

TABLE 9.5	Effect of Numbers of Electrons on Boiling Points of Nonpolar Molecular Substances							
Noble gases			**Halogens**			**Hydrocarbons**		
	No. e's	bp (°C)		No. e's	bp (°C)		No. e's	bp (°C)
He	2	-269	F_2	18	-188	CH_4	10	-161
Ne	10	-246	Cl_2	34	-34	C_2H_6	18	-88
Ar	18	-186	Br_2	70	59	C_3H_8	26	-42
Kr	36	-152	I_2	106	184	C_4H_{10}*	34	0

* Butane.

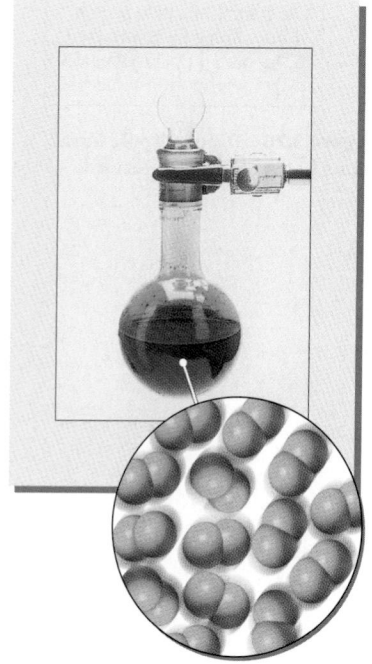

Vaporization of liquid bromine.

Structure and boiling point.
The boiling points of pentane and 2,2-dimethylpropane differ because of differences in their molecular structures even though the total number of electrons in each molecule is the same.

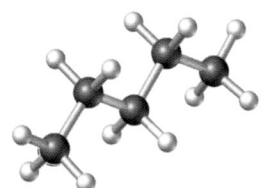 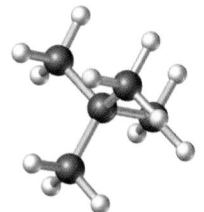

pentane, bp = 36.0 °C 2,2-dimethylpropane, bp = 9.5 °C

Dipole-Dipole Attractions

CD-ROM Screen 13.4: Intermolecular Forces (2): Ion-Dipole and Dipole-Dipole Forces

Polar molecules have *permanent* dipoles that create a **dipole-dipole attraction**, a noncovalent interaction between two *polar* molecules or two polar groups in the same large molecule. Molecules that are dipoles attract each other when the partial positive region of one is close to the partial negative region of another (Figure 9.20).

The boiling points of several nonpolar and polar substances with comparable numbers of electrons, and therefore comparable London forces, are given in Table 9.6. In general, the more polar its molecules, the higher the boiling point of a substance, provided the London forces are similar (compounds with similar numbers of electrons). The lower boiling points of nonpolar substances compared with those of polar substances in Table 9.6 reflect this. Dipole-dipole forces range from 5 to 25 kJ/mol, but London forces (0.05 to 40 kJ/mol) can be stronger. For example, the greater London forces in HI cause it to have a higher boiling point (−36 °C) than HCl (−85 °C), even though HCl is more polar. However, if their London forces are similar, a more polar substance will have stronger intermolecular attractions than a less polar one. An example of this is the difference in boiling points of polar ICl (97 °C) and nonpolar Br_2 (59 °C), even though the compounds have approximately the same London forces due to their similar number of electrons (Table 9.6).

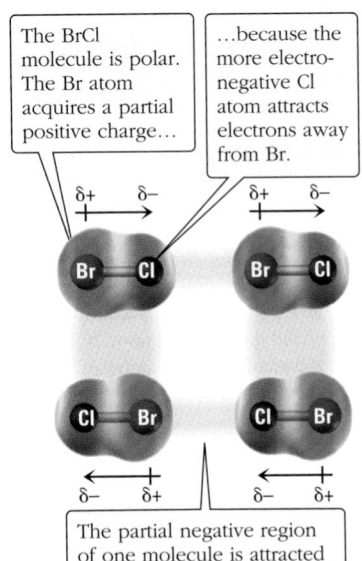

The BrCl molecule is polar. The Br atom acquires a partial positive charge...

...because the more electronegative Cl atom attracts electrons away from Br.

The partial negative region of one molecule is attracted to the partial positive region of the neighboring molecule.

Figure 9.20 Dipole-dipole attractions between BrCl molecules.

 Exercise 9.6 Dipole-Dipole Forces

Draw a sketch, like that in Figure 9.20, of four CO molecules to indicate dipole-dipole forces between the CO molecules.

Problem-Solving Example 9.7 Molecular Forces

What is the major type of force that must be overcome in these changes?

(a) The sublimation of solid iodine, I_2
(b) The melting of propane
(c) The decomposition of water into H_2 and O_2
(d) The evaporation of liquid PCl_3

Answer
(a) and (b) London forces
(c) Covalent bonds between O and H atoms
(d) Dipole-dipole forces

Explanation
(a) Nonpolar I_2 molecules in the solid are held to each other by London forces, which must be overcome for the I_2 molecules in the solid to escape from each other and become gaseous.

TABLE 9.6	Numbers of Electrons and Boiling Points of Nonpolar and Polar Substances					
Nonpolar molecules				**Polar molecules**		
	No. e's	bp (°C)			No. e's	bp (°C)
N_2	14	−196	CO		14	−192
SiH_4	18	−112	PH_3		18	−88
GeH_4	36	−90	AsH_3		36	−62
Br_2	70	59	ICl		70	97

(b) Solid propane melts when the thermal energy becomes great enough to overcome some of the noncovalent forces of attraction (London forces) among the nonpolar propane molecules (−187.7° C). Noncovalent forces still exist in the liquid propane.

(c) To decompose water into its component elements requires breaking the O—H covalent bonds in water molecules. Note that this change involves much greater energy (the bond energy) than that needed to melt or boil water, both of which involve overcoming noncovalent interactions.

(d) Dipole-dipole forces attract the polar PCl_3 molecules to each other in the lquid. These forces must be overcome so that PCl_3 molecules can get away from their neighbors at the surface of the liquid and enter the gaseous state. Because of the relatively large number of electrons in each PCl_3 molecule, London forces are also important in holding PCl_3 molecules close to each other and must be overcome.

Problem-Solving Practice 9.7

Explain the principal type of forces to overcome in order for

(a) Kr to melt   (b) Propane to release C and H_2

Hydrogen Bonds

A hydrogen bond is an especially significant type of dipole-dipole force. The **hydrogen bond** is the attraction between a partially positive hydrogen atom and a lone electron pair on a small, very electronegative atom (generally F, O, or N). Electron density within molecules shifts toward F, O, or N because of their high electronegativity, giving these atoms *partial negative charges*. As a result, a hydrogen atom bonded to the nitrogen, oxygen, or fluorine atom acquires a *partial positive charge*. In a hydrogen bond, a partially positive hydrogen atom bonded covalently to one of the electronegative atoms (X) is attracted electrostatically to the negative charge of a lone pair on the other atom (Z). Hydrogen bonds are typically shown as dotted lines (···) between the atoms in the following way:

CD-ROM Screen 13.6:
Hydrogen Bonding

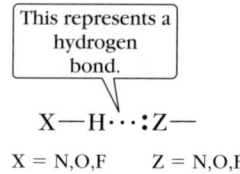

This represents a hydrogen bond.

X—H···:Z—

X = N,O,F   Z = N,O,F

The hydrogen bond forms a "bridge" between two highly electronegative atoms, which may be in different molecules or in the same large molecule. This type of bridge from hydrogen to nitrogen and oxygen plays an essential role in determining the folding of large protein molecules.

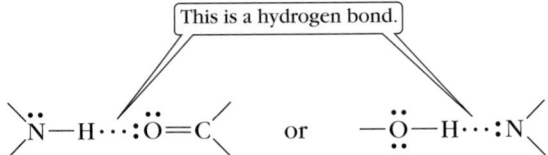

Hydrogen bonds can form between molecules or within a molecule. These are known as *inter*molecular and *intra*molecular hydrogen bonds, respectively.

Type of force	Magnitude of force (kJ/mol)
Covalent bond	150 – 1000
London force	0.05 – 40
Dipole-dipole	5 – 25
Hydrogen bond	10 – 40

Noncovalent interactions in ethanol and dimethyl ether. The molecules have the same number of electrons and so London forces are roughly the same. An ethanol molecule has an OH group, which means that there are both dipole forces and hydrogen bond forces attracting ethanol molecules to each other. A dimethyl ether molecule is polar, but there is no hydrogen bonding, and so the noncovalent intermolecular forces are weaker than in ethanol.

The greater the electronegativity of the atom connected to H, the greater is the partial positive charge on H and hence the stronger is the hydrogen bond.

The H atom is very small and its partial positive charge is concentrated in a very small volume, so it can come very close to the lone pair to form an especially strong dipole-dipole force through hydrogen bonding. Hydrogen-bond strengths range from 10 to 40 kJ/mol. However, a great many hydrogen bonds often occur in a sample of matter, and the overall effect can be very dramatic. An example of this effect can be seen in the melting and boiling points of ethanol. This chapter began by noting the very different melting and boiling points of ethanol and dimethyl ether, both of which have the same molecular formula, C_2H_6O, and thus the same number of electrons and roughly the same London forces.

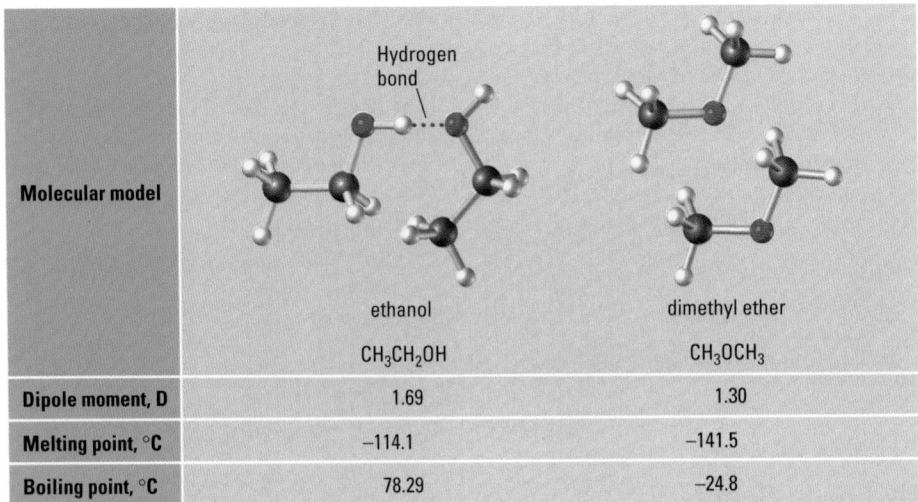

Molecular model	ethanol CH_3CH_2OH	dimethyl ether CH_3OCH_3
Dipole moment, D	1.69	1.30
Melting point, °C	−114.1	−141.5
Boiling point, °C	78.29	−24.8

Both molecules are polar (dipole moment > 0) and so there are dipole-dipole forces in each case. The differences in melting and boiling points arise because of hydrogen bonding in ethanol. The O—H bonds in ethanol make intermolecular hydrogen bonding possible, while this is not possible in dimethyl ether because it has no O—H bonds.

The hydrogen halides also illustrate the significant effects of hydrogen bonding (Figure 9.21). The boiling point of hydrogen fluoride (HF), the lightest hydrogen halide, is much higher than expected; this is attributed to hydrogen bonding, which does not occur significantly in the other hydrogen halides.

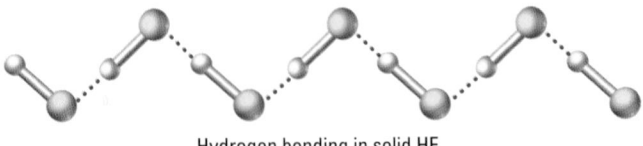

Hydrogen bonding in solid HF

CD-ROM Screen 13.7: The Weird Properties of Water: A Consequence of Hydrogen Bonding

Hydrogen bonding is especially strong among water molecules and is responsible for many of the unique properties of water (Section 11.4). Hydrogen compounds of oxygen's neighbors and family members in the periodic table are gases at room temperature: CH_4, NH_3, H_2S, H_2Se, H_2Te, PH_3, HF, and HCl. But H_2O is a liquid at room temperature, indicating a strong degree of intermolecular attraction.

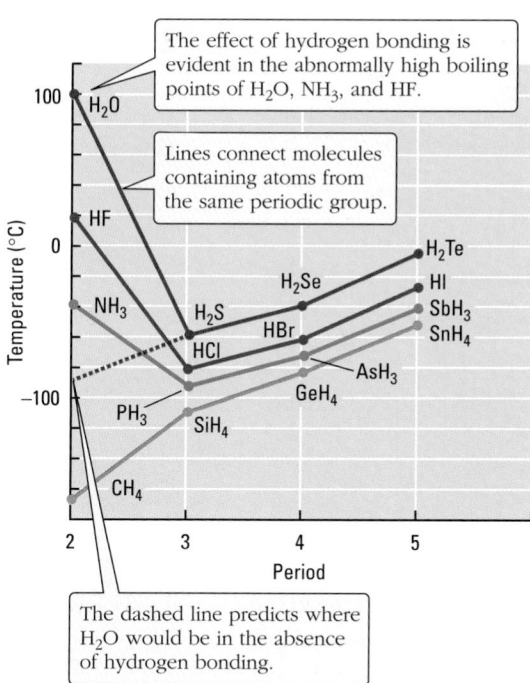

The effect of hydrogen bonding is evident in the abnormally high boiling points of H_2O, NH_3, and HF.

Lines connect molecules containing atoms from the same periodic group.

The dashed line predicts where H_2O would be in the absence of hydrogen bonding.

Figure 9.21 Boiling points of some simple hydrogen-containing binary compounds.

Figure 9.21 shows that the boiling point of H_2O is about 200 °C higher than would be predicted if hydrogen bonding were not present.

In liquid and solid water, where the molecules are close enough to interact, the hydrogen atom on one water molecule is attracted to the lone pair of electrons on the oxygen atom of an adjacent water molecule. Because each hydrogen atom can form a hydrogen bond to an oxygen atom in another water molecule, and because each oxygen atom has two lone pairs, every water molecule can participate in four hydrogen bonds to four other water molecules (Figure 9.22). The result is a tetrahedral cluster of water molecules around the central water molecule.

The structure of solid water is described in Section 11.4.

 Exercise 9.7 Strengths of Hydrogen Bonds

Of the three hydrogen bonds, F—H· · ·F—H, O—H· · ·N—C, and N—H· · ·O═C, which is the strongest? Explain why.

Problem-Solving Example 9.8 Molecular Forces

What are the types of forces, in addition to London forces, that are overcome in these changes? Using structural formulas, make a sketch representing the major type of force in each case.

(a) The evaporation of liquid methanol, CH_3OH
(b) The decomposition of hydrogen peroxide (H_2O_2) into water and oxygen
(c) The melting of urea, H_2NCONH_2
(d) The boiling of liquid HCl

Answer
(a) Hydrogen bonding:

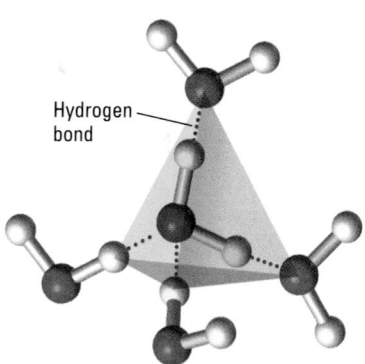

Figure 9.22 Hydrogen bonding between one water molecule and its neighbors. Each water molecule can participate in four hydrogen bonds — one for each hydrogen atom and two through the two lone pairs of oxygen. Since each hydrogen bond is shared between two water molecules, there are two hydrogen bonds per molecule.

(b) Covalent bonds between O—O and O—H:

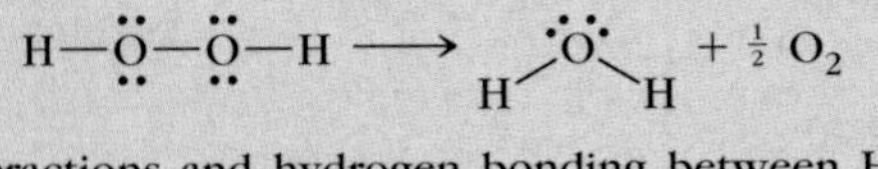

$$H-\overset{..}{\underset{..}{O}}-\overset{..}{\underset{..}{O}}-H \longrightarrow \overset{..}{\underset{H}{O}}{\overset{..}{\diagdown}}_{H} + \tfrac{1}{2}\,O_2$$

(c) Dipole-dipole interactions and hydrogen bonding between H and lone pair on O atoms:

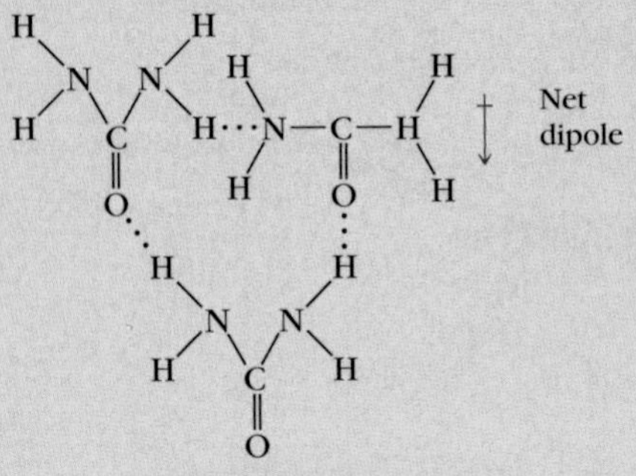

(d) Dipole-dipole interactions:

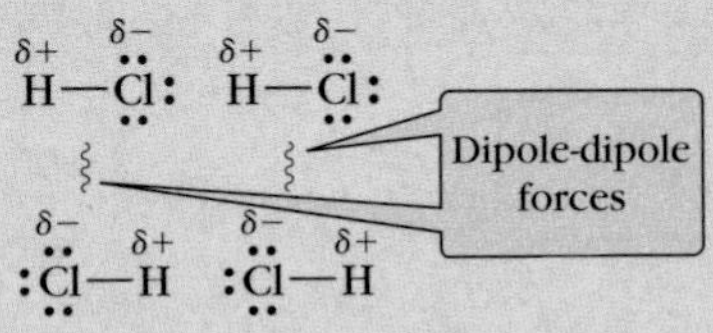

Explanation

(a) Methanol molecules have a hydrogen atom covalently bonded to a highly electronegative oxygen atom with a lone pair of electrons. Therefore, methanol molecules can hydrogen-bond to each other.

(b) This is not a case of overcoming an intermolecular force, but rather an *intra*molecular one, covalent bonding. The decomposition of hydrogen peroxide involves breaking the O—O covalent bond of the HOOH molecule.

(c) Urea is a polar molecule that also hydrogen-bonds to other urea molecules. Therefore, dipole-dipole interactions and hydrogen bonding are the principal forces among the molecules that must be overcome when urea melts at 132.7 °C.

(d) The linear HCl molecule is polar because the Cl atom is more electronegative than the H atom. Therefore, there are dipole-dipole forces to overcome between adjacent HCl molecules.

Problem-Solving Practice 9.8

Decide what types of intermolecular forces are involved in the attraction between (a) N_2 and N_2 (b) CO_2 and H_2O (c) CH_3OH and NH_3

Noncovalent Forces in Living Cells

Fatty acids were described in Section 8.5.

Noncovalent forces are important in biological structures, including cell membranes. The basic structure of a cell membrane known as a **lipid bilayer** is composed of two aligned layers of phospholipids. **Phospholipids** are glycerol derivatives having two long, nonpolar fatty acid chains and a polar phosphate derivative at the head of the molecule (Figure 9.23).

The polar end of phospholipid molecules is **hydrophilic,** that is, "water-loving," because of its attraction to water molecules. The polar end is oriented toward the aqueous environment of the cell and its surroundings. The long, nonpolar chains of the fatty acids are **hydrophobic,** or "water-hating." They point away from the aqueous environment and are held together by London forces. The result is separation of the interior of a cell from the surrounding fluids, a separation that allows control over which ions or molecules enter a cell and which chemical reactions can occur there.

Chemistry in the News

STICKING TO THE TASK AT HAND

Tropical lizards called Tokay geckos scamper up walls and cross ceilings upside down with reckless abandon.

It turns out that the structure of their feet, not friction or suction, allows them to perform such gravity-defying feats. This conclusion was drawn by Dr. Robert Full and his University of California-Berkeley colleagues from their research using a specially designed microelectrical mechanical sensor to measure lateral and perpendicular forces on a single seta, a tiny hair from a gecko's foot. Geckos use rows of thousands of setae on the bottom of their toe pads to crawl up and across improbable terrain. The tip of each seta is divided into hundreds of spatulae arranged in a leaf-like pattern, with each spatula ending in a cone-shaped structure. The setae are pointed toward the heel, but with each step, the gecko sets the setae backward to gain maximum adhesion to the surface. With just a slight change of the angle of the setae with the surface, the gecko's toe pad either sticks or lets go.

Since the toe pads stick to surfaces underwater or in a vacuum, Full and his colleagues ruled out friction or suction as the adhesive forces. The adhesive force values obtained from their studies are consistent with the hypothesis that individual seta function by London forces generated between the seta and the surfaces geckos climb. The Berkeley researchers have calculated that if all the setae of a gecko were stuck simultaneously to a surface, the force between the toe pads and the surface would be equivalent to about 150 pounds per square inch!

This news item answers the Chapter 1 question about how some animals walk on ceilings (⬅ *p. 32*).

Tokay gecko.

Sources:

Science, Vol. 288, 9 June 2000; pp. 1717–1718.
Nature, Vol. 405, 8 June 2000; pp. 631 and 681–684.

- Explain how London forces could be created between the setae and a surface.
- On what basis were friction and suction ruled out as the forces used by geckos?

In our bodies, noncovalent interactions play important roles in bringing metabolic products, drugs, hormones, and neurotransmitters (nervous system messengers) into contact with cellular receptors that in turn initiate vital chemical changes. The receptors are often proteins embedded in the cell membrane's lipid bilayer. Dipole-dipole forces, hydrogen bonding, and London forces help to align the target molecule with its particular site on the receptor surface.

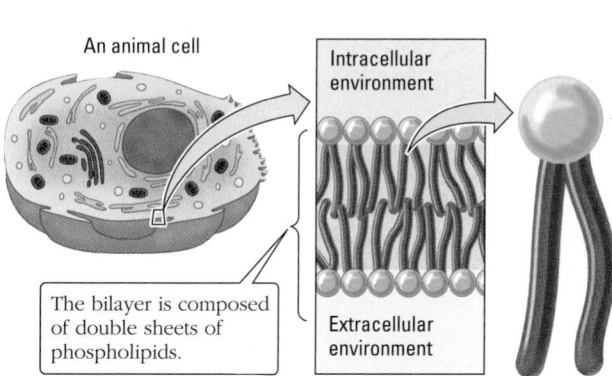

An animal cell

Intracellular environment

Extracellular environment

The bilayer is composed of double sheets of phospholipids.

The polar head of the phospholipid is attracted to water molecules. Fatty acid chains are in the boundary between outside and inside the cell.

The nonpolar fatty acid chains are attracted to each other and are oriented away from water molecules that are inside or outside the cell.

Figure 9.23 Lipid bilayer in cell membranes.

CHEMISTRY
You Can Do... Molecular Structure and Biological Activity

Noncovalent forces (London forces, hydrogen bonds, and dipole-dipole forces) are important in biochemical reactions for bringing together reactants in proper orientations. This is particularly the case for enzymes—biomolecules that help to speed up the rates of biochemical reactions *(Section 13.9)*. Noncovalent forces bind the reactant to an enzyme's *active site,* the particular region of the enzyme where its reactivity occurs.

Lactase, which acts on lactose, is such an enzyme. Lactose is a disaccharide *(⟵ Section 12.9)* also known as milk sugar because it is present in the milk of humans, cows, and other mammals. Lactose can make up to about 40% of an infant's diet during the first year. Lactase accelerates the decomposition of lactose to its component monosaccharides glucose and galactose. A portion of the lactase molecule has a region whose shape matches only that of lactose. Therefore, the enzyme acts selectively on lactose and not on sucrose or maltose, two other dietary disaccharides whose shapes do not match that required by lactase.

Infants have a more active form of the enzyme than do adults. Consequently, nearly 20% of adults in the United States and nearly 70% worldwide suffer some degree of lactose intolerance, the inability to rapidly digest lactose in milk and dairy products. Lactaid, an over-the-counter product that contains lactase, can be used by lactose-intolerant individuals to break down lactose in milk.

In this activity, you will use Lactaid to study the conversion of lactose in milk to glucose and galactose. Do this ex-

periment with several friends. You will need:

- One quart of 2% milk
- One package of Lactaid drops (not caplets), available at most pharmacies
- Two one-pint (16-oz) containers that can be closed
- Four to eight small glasses

Taste a drop of Lactaid and record its taste.

Pour two samples for yourself and two for each friend. For the samples pour half the milk into one container as a control and close the container. Pour the remainder of the milk into the second container, add 3 to 4 drops of Lactaid, close the container, and shake it to mix its contents. Mark the Lactaid containers to identify them. Refrigerate all containers for about 48 hours.

Take the containers from the refrigerator and pour samples of the Lactaid-treated and untreated milk into separate glasses. Each person should taste his or her two samples of milk and record any differences in their taste.

Source:

Richman, R. "Detection of Catalysis by Taste." *Journal of Chemical Education,* Vol. 75, 1998; p. 315.

- What difference in taste was there between the untreated and the Lactaid-treated samples?
- Explain the difference in taste to the others who have done this exercise.

Structural formulas of lactose, galactose, and glucose.

9.6 BIOMOLECULES: DNA AND THE IMPORTANCE OF MOLECULAR STRUCTURE

Nowhere do the shape of a molecule and noncovalent forces play a more intriguing and important role than in the structure and function of **DNA (deoxyribonucleic acid),** the molecule that stores the genetic code. Whether you are male or female, have blue or brown eyes, or have curly or straight hair depends on your genetic makeup. These and all your other physical traits are determined by the composition of the approximately 1.8 m of coiled DNA that makes up the 23 double-stranded chromosomes in the nucleus of each of your cells.

Nucleotides of DNA

DNA is a **polymer,** a molecule composed of many small, repeating units bonded together. Each repeating unit in DNA, called a **nucleotide,** has the three connected parts shown in Figure 9.24a—one sugar unit, one phosphate unit, and a cyclic nitrogen compound known as a *nitrogen base.* In DNA the sugar is always deoxyribose and the base is one of four bases—*adenine (A), thymine (T), guanine (G),* or *cytosine (C).* The bases are often referred to by their single-letter abbreviations.

A DNA segment with three nucleotides bonded together is shown in Figure 9.24b. The phosphate units join nucleotides into a polynucleotide chain that has a backbone of alternating deoxyribose and phosphate units in a long strand with the various bases extending out from the sugar-phosphate backbone. The order of the nucleotides (and thus the particular sequence of bases) along the DNA strand carries the genetic code from one generation to the next.

The deoxyribose units in the backbone chain are joined through phosphate diester linkages. Phosphoric acid is H_3PO_4, $O=P(OH)_3$.

phosphoric acid

$$HO-\!\!\!\begin{array}{c}HO\\|\\P\\|\\HO\end{array}\!\!\!=O$$

Replacing two H atoms with two CH_3 groups gives a phosphate diester, $O=P(OH)(OCH_3)_2$.

$$\begin{array}{c}H_3C-O\\HO\end{array}\!\!\!\!\begin{array}{c}\\P\\\end{array}\!\!\!=O \quad \begin{array}{c}\text{phosphate}\\\text{diester}\end{array}$$
$$H_3C-O$$

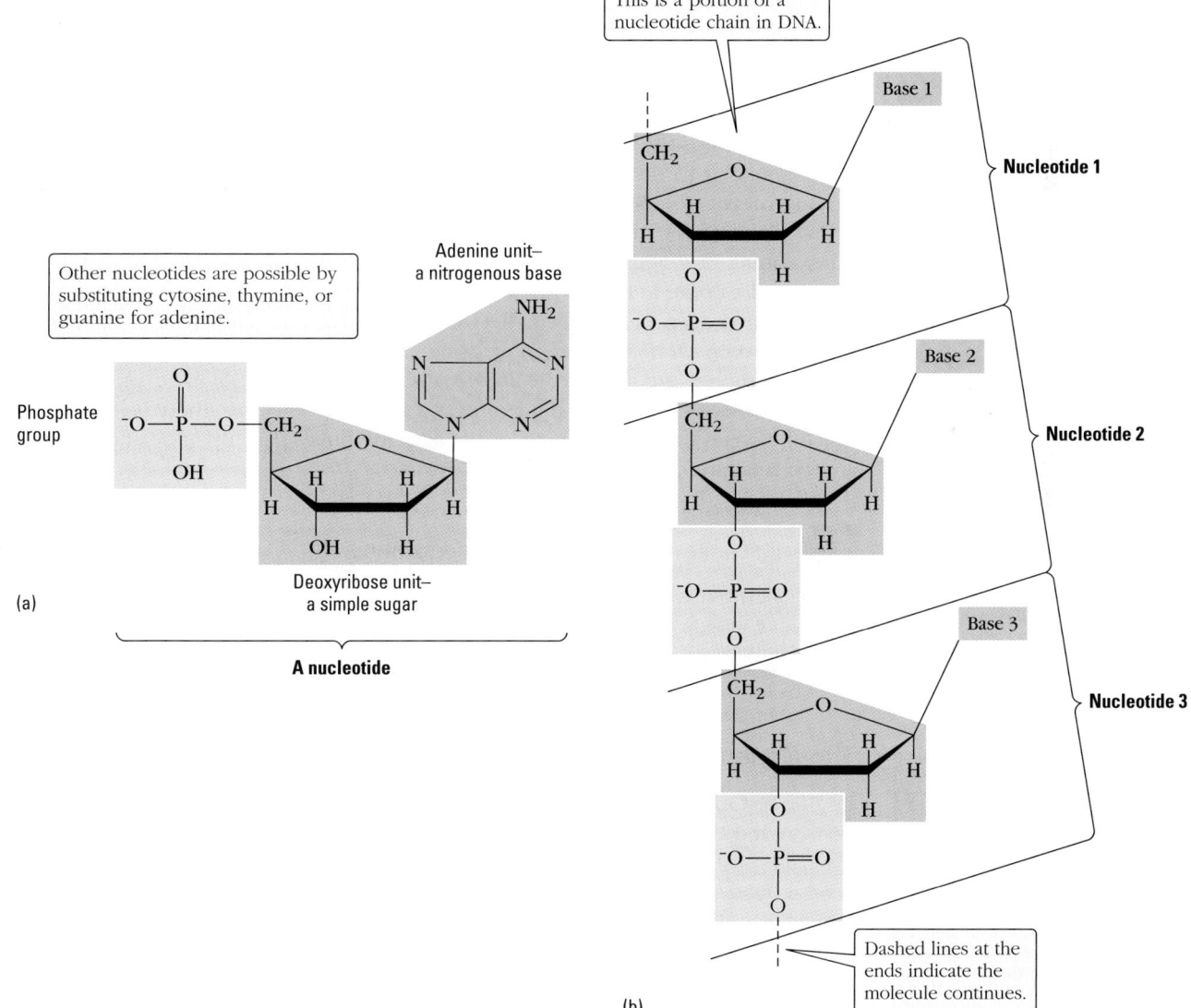

This is a portion of a nucleotide chain in DNA.

Base 1

Nucleotide 1

Other nucleotides are possible by substituting cytosine, thymine, or guanine for adenine.

Adenine unit– a nitrogenous base

Phosphate group

Deoxyribose unit– a simple sugar

(a)

A nucleotide

Base 2

Nucleotide 2

Base 3

Nucleotide 3

Dashed lines at the ends indicate the molecule continues.

(b)

Figure 9.24 Nucleotides and DNA. The light orange phosphate groups and dark orange deoxyribose sugar groups form the *backbone* of DNA. The genetic code is carried by the bases, shown in blue.

The Double Helix: The Watson-Crick Model

In the 1950s, Erwin Chargaff made the important discovery that the adenine/thymine and guanine/cytosine ratios are constant in any given organism, from a human genius to a bacterium. Based on this analysis, it was apparent that in any given organism,

- The base composition is the same in all cells of an organism and is characteristic of that organism.
- The amounts of adenine and thymine are equal, as are the amounts of guanine and cytosine.
- The amount of the bases adenine plus guanine equals that of the bases cytosine plus thymine.

This information implied that the bases occurred in pairs: adenine with thymine and guanine with cytosine.

Using x-ray crystallography data gathered by Rosalind Franklin and Maurice Wilkins on the relative positions of atoms in DNA, plus the concept of A-T and G-C base pairs, the American biologist James D. Watson and the British physicist Francis H. C. Crick proposed a double-helix structure for DNA in 1953.

In the three-dimensional Watson-Crick structure, two polynucleotide DNA strands wind around each other to form a double helix. Remarkable insight into how hydrogen bonding could stabilize DNA ultimately led Watson and Crick to propose the double helix structure. Hydrogen bonds form between specific base pairs lying opposite each other in the two polynucleotide strands. Adenine is hydrogen-bonded to thymine and guanine is hydrogen-bonded to cytosine to form **complementary base pairs**; that is, the bases of one strand match those of the other (Figure 9.25). The result is that the base pairs (A on one strand with T on the other strand, or C with G) are stacked one above the other in the interior of the double helix (*blue* in Figure 9.25).

Two hydrogen bonds occur between every adenine and thymine pair; three occur between each guanine and cytosine. These hydrogen bonds hold the double helix together. Note in Figure 9.26 the similar structures of thymine and cytosine, and of adenine and guanine. If adenine and guanine try to pair, there is insufficient space between the strands to accommodate the bulky pair; thymine and cytosine are too small to pair and to align properly.

The Double Helix by James Watson chronicles this discovery.

Francis H.C. Crick *(right)* and James Watson. Working in the Cavendish Laboratory at Cambridge, England, Watson and Crick built a scale model of the double-helical structure, based on x-ray data. Knowing distances and angles between atoms, they compared the task to working on a three-dimensional jigsaw puzzle. Watson, Crick, and Maurice Wilkins received the 1962 Noble Prize for their work relating to the structure of DNA.

 Exercise 9.8   Hydrogen Bonding and DNA

Only one hydrogen bond is possible between G and T or between C and A. Use the structural formulas for these compounds to indicate where such hydrogen bonding can occur in these base pairs.

The Genetic Code and DNA Replication

In the historic scientific journal article describing their revolutionary findings, Watson and Crick wrote, "It has not escaped our notice that the specific pairing we have postulated immediately suggests a possible copying mechanism for the genetic material." This bit of typical British understatement belies the enormous impact that deciphering DNA's structure had on establishing a molecular basis of heredity. It was now clear that the DNA sequence of base pairs in the nucleus of a cell represents a genetic code that controls the inherited characteristics of the next generation, as well as most of the life processes of an organism.

In humans, the double-stranded DNA forms 46 chromosomes (23 chromosome pairs). The **gene** that accounts for a particular hereditary trait is always lo-

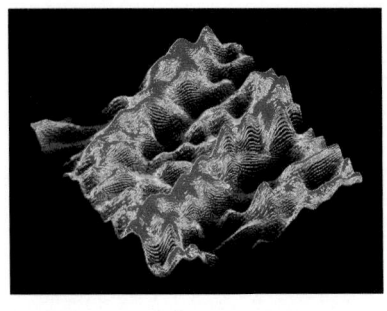

Scanning transmission microscope photo of double-stranded DNA.

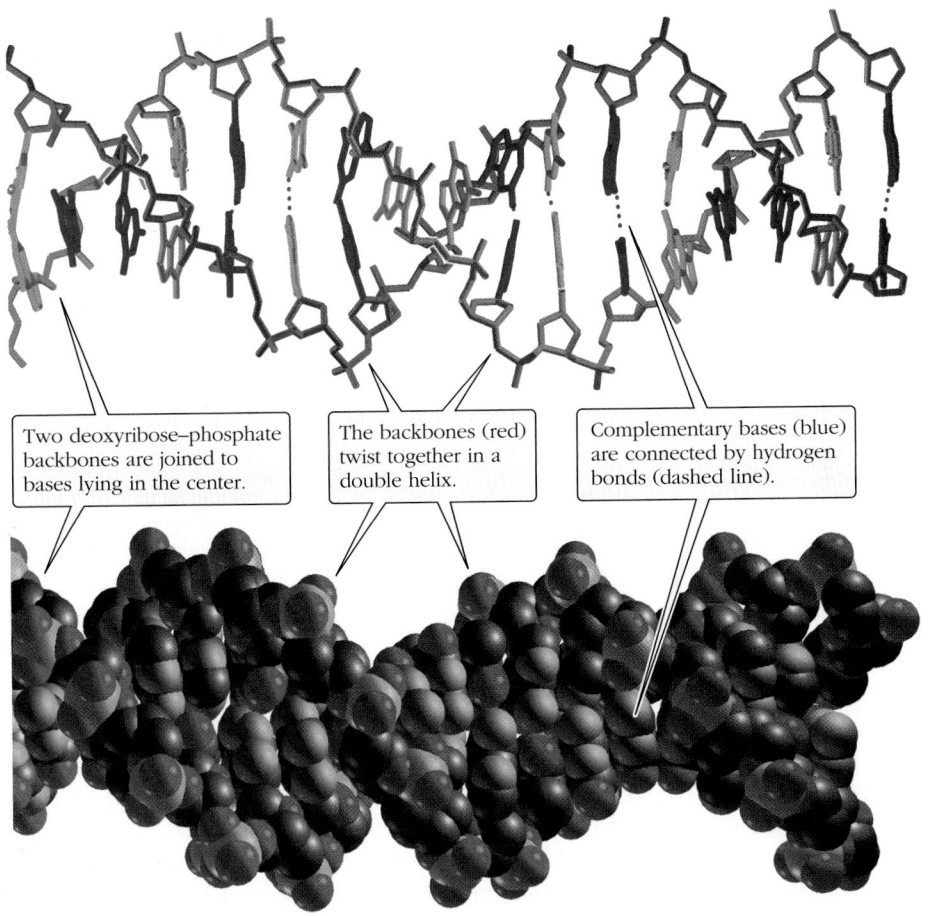

Two deoxyribose–phosphate backbones are joined to bases lying in the center.

The backbones (red) twist together in a double helix.

Complementary bases (blue) are connected by hydrogen bonds (dashed line).

Figure 9.25   Double-stranded DNA.
The illustration shows the deoxyribose-phosphate backbone of each strand as phosphate groups (orange and red) linking five-membered rings of deoxyribose joined to bases. The complementary bases *(shown in blue)* are connected by hydrogen bonds: two between adenine and thymine and three for guanine and cytosine.

ROSALIND FRANKLIN
(1920–1958)

Rosalind Franklin, daughter of a prominent London banking family, received her degrees from Cambridge University. Her most famous work was the first x-ray photographs of DNA, done in collaboration with Maurice Wilkins at King's College in London, showing two forms of DNA, one of them a helix. By interpreting these x-ray data, James Watson and Francis Crick derived their now-famous double-helix model of DNA for which they, along with Wilkins, received the 1962 Nobel Prize in Chemistry. Rosalind Franklin's untimely death of ovarian cancer at the age of 37 resulted in her not sharing that award. Nobel Prizes are not awarded posthumously.

cated in the same position on the same chromosome. Each gene is distinguished by a unique sequence of bases that codes for the synthesis of a single protein within the body. The protein then goes on to play its role in the growth and functioning of the individual. (Proteins are discussed in Section 12.9.)

Each organism (except viruses) begins life as a single cell with DNA consisting of a single strand from each parent. During regular cell division, both DNA strands are accurately copied, with a remarkably low incidence of error. The copying, called **replication**, takes place as the DNA helix unzips and new nucleotides are sequentially brought into the proper places on the new strand. Thus, each original DNA strand serves as a template from which the complementary strand is pro-

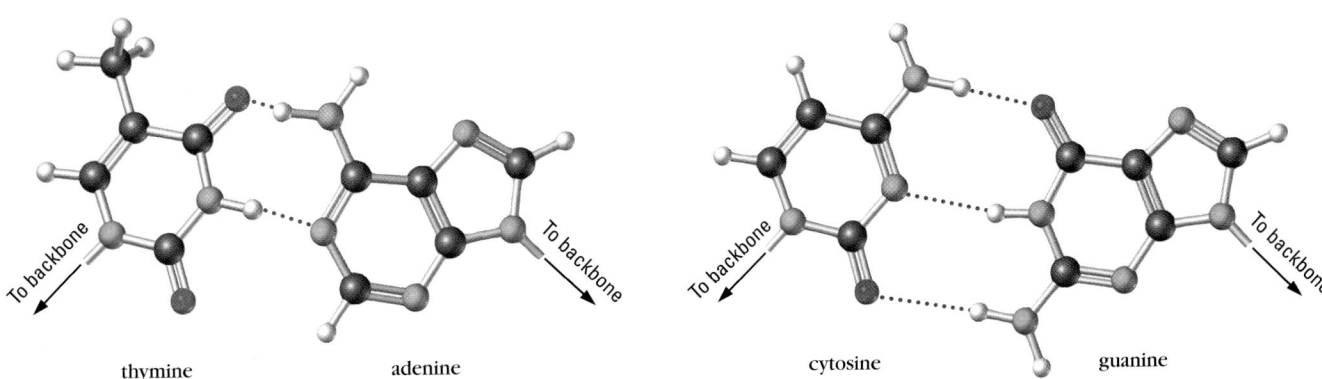

thymine   adenine   cytosine   guanine

Figure 9.26   Hydrogen bonding between T-A and C-G in DNA.

ESTIMATION   Base Pairs and DNA

The human genome DNA contains about three billion base pairs, which are an average distance of 0.34 nm apart in the DNA molecule. Only about 2% of this DNA consists of unique genes. The number of genes is estimated to be 3×10^4.

(a) What is the average number of base pairs per gene?
(b) Approximately how long (m) is a DNA molecule?

(a) To calculate the average number of base pairs/gene, we start with approximately 10^9 base pairs, of which there are 2 base pairs in genes per 100 base pairs. We get

$$\frac{10^9 \text{ base pairs}}{\text{DNA}} \times \frac{2 \text{ base pairs in genes}}{100 \text{ base pairs}}$$

$$= \frac{2 \times 10^7 \text{ base pairs in genes}}{\text{DNA}}$$

Thus, there are about 2×10^7 base pairs in genes/DNA, and the DNA contains 3×10^4 genes, so that there are roughly 1×10^2 base pairs/gene.

$$\frac{2 \times 10^7 \text{ base pairs in genes}}{\text{DNA}} \times \frac{1 \text{ DNA}}{3 \times 10^4 \text{ genes}}$$

$$\approx \frac{10^3 \text{ base pairs}}{\text{gene}}$$

(b) The distance between base pairs in DNA is 0.34 nm; 1 m = 10^9 nm and 1 m = 100 cm. Therefore,

$$\frac{10^9 \text{ base pairs}}{\text{DNA}} \times \frac{0.34 \text{ nm}}{\text{base pair}} \times \frac{1 \text{ m}}{10^9 \text{ nm}} \approx \frac{1 \text{ m}}{\text{DNA}}$$

This distance, 1 m, would be much too large to fit into a cell, and therefore the DNA molecule must be tightly coiled, which it is.

The total sequence of base pairs in a plant or animal cell is called its genome.

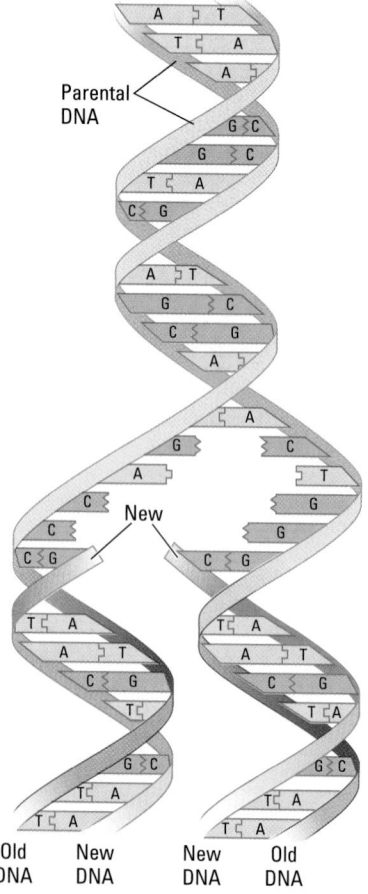

Figure 9.27 DNA replication. When the original DNA helix (orange) unwinds, each half is a template on which to assemble nucleotides from the cell environment to produce a complementary strand (green).

duced (Figure 9.27). The process is termed *semiconservative* because in each of the two new cells, each chromosome consists of one DNA strand from the parent cell and one newly made strand.

In June 2000 the Human Genome Project, a vast international consortium of scientists, and Celera Genomic Corporation jointly announced completion of the initial working draft of the human genome—the genetic blueprint of a human being. President Clinton hailed the achievement as "the most important, most wondrous map ever produced by humankind." The research succeeded in placing large segments of DNA in proper order for each of the 23 human chromosomes and determining the DNA sequences of those segments. At present, the fragments cover 97% of the human genome, of which 85% of the DNA sequence has been assembled. The ultimate goal of this work is to fully characterize each of the estimated 32,000 genes and their approximately three billion base pairs. Completion of this task is expected by 2003. Among the many expected benefits of this work—some already being realized—is a molecular-level understanding of the causes of and potential cures for inherited diseases.

Exercise 9.9   Replication and Base Pairing in DNA

How easily would the base pairs in DNA unpair during replication if they were linked by covalent bonds? How is it that the helix unzips so readily?

9.7 CHIRAL MOLECULES

You have seen one kind of isomerism that results from different arrangements of the same atoms in space—*cis-trans* isomerism (⇐ *p. 330*). There is another, somewhat more subtle, isomerism, one related to "handedness." Are you right-handed or left-handed? Regardless of the preference, we learn at a very early age that a right-handed glove doesn't fit the left hand, and vice versa. Our hands are mirror images of one another and are not superimposable (Figure 9.28).

An object that cannot be superimposed on its mirror image is called **chiral.** Objects that are superimposable on their mirror images are **achiral.** Stop and think about the extent to which chirality is a part of our everyday life. We've already discussed the chirality of hands (and feet). Helical seashells are chiral, and most spiral to the right like a right-handed screw. Many creeping vines show a chirality when they wind around a tree or post.

What is not as well known is that a large number of the molecules in plants and animals are chiral, and usually only one form (left-handed or right-handed) of the chiral molecule is found in nature. For example, all but one of the 20 amino acids from which proteins are made are chiral, and the left-handed amino acids predominate in nature! Most natural sugars are right-handed, including glucose and sucrose and deoxyribose, the sugar found in DNA.

A chiral molecule and its *nonsuperimposable* mirror image are called **enantiomers**; they are two different molecules, just as your left and right hands are different. Enantiomers are possible when a molecular structure is **asymmetric** (without symmetry). The simplest case of chirality is a tetrahedral carbon atom that is bonded to four *different* atoms or groups of atoms. Such a carbon atom is said to be a *chiral center,* and a molecule with just one chiral center is always a chiral molecule.

Consider lactic acid, which has its central C atom bonded to four different groups: $—CH_3$, $—OH$, $—H$, and $—COOH$ (Figure 9.29a). As a result of the tetrahedral geometry around the central carbon atom, it is possible to have two different arrangements of the four groups. If a lactic acid molecule is placed so that the C—H bond is vertical, as illustrated in Figure 9.29a, one possible arrangement of the remaining groups would be that in which $—OH$, $—CH_3$, and $—COOH$ are attached in a clockwise sequence (Isomer I). Alternatively, these groups can be attached in a counterclockwise sequence (Isomer II).

To see further that the arrangements are different, we place Isomer I in front of a mirror (Figure 9.29b). Now you see that Isomer II is the mirror image of Isomer I. What is important, however, is that these mirror image molecules *cannot be superimposed* on one another. *These two nonsuperimposable, mirror image chiral molecules are enantiomers.*

Lactic acid is one of the compounds found in nature in both enantiomeric forms under different circumstances. During the contraction of muscles, the body produces only one enantiomer of lactic acid. The other enantiomer is produced when milk sours.

In *A Midsummer Night's Dream,* Shakespeare wrote of the chirality of honeysuckle and woodbine plants. Queen Titania says to Bottom, ". . . Sleep thou, and I will wind thee in my arms . . . . So doth the woodbine the sweet honeysuckle gently entwist. . . ." Woodbine spirals clockwise and honeysuckle twists in a counterclockwise direction.

Figure 9.28 Nonsuperimposable mirror images. Your left hand is a nonsuperimposable mirror image of your right hand.

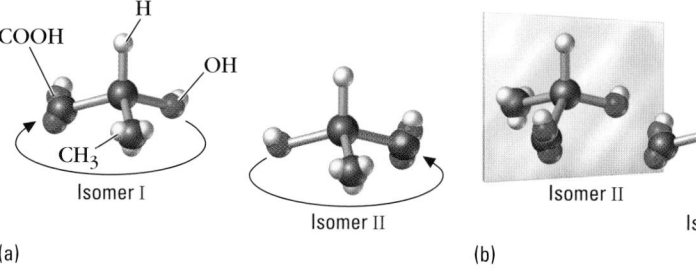

(a) (b)

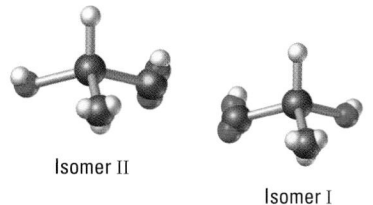

(c)

Figure 9.29 The enantiomers of lactic acid. (a) Isomer I: $—OH$, $—CH_3$, and $—COOH$ are attached in a clockwise manner. Isomer II: $—OH$, $—CH_3$, and $—COOH$ are attached in a counterclockwise manner. (b) Isomer I is placed in front of a mirror, and its mirror image is Isomer II. (c) The isomers are nonsuperimposable.

In *Through the Looking Glass* (the companion volume to *Alice in Wonderland*), Alice speculates to her cat that " . . . perhaps looking glass milk is not good to drink."

The "handedness" of enantiomers is sometimes represented by D for right-handed (D stands for "dextro" from the Latin *dexter,* meaning "right") and L for left-handed (L stands for "levo" from the Latin *laevus,* meaning "left"). In the case of lactic acid, the D form is found in souring milk and the L form is found in muscle tissue, where it accumulates during vigorous exercise and can cause cramps.

Nature's preference for one enantiomer of amino acids (the L form) has provoked much discussion and speculation among scientists since Louis Pasteur's discovery of molecular chirality in 1848. However, there is still no widely accepted explanation of this preference.

Large organic molecules may have a number of chiral carbon atoms within the same molecule. At each such carbon atom (a chiral center) there are two possible arrangements of the molecule. The total number of possible enantiomers, then, increases exponentially with the number of different chiral centers. With two different chiral carbon atoms there are 2^2, or four, possible enantiomers.

The widely used artificial sweetener aspartame (NutraSweet) has two chiral centers and thus four enantiomers. One enantiomer has a sweet taste, while another enantiomer is bitter, indicating that the receptor sites on our taste buds must be chiral, because they respond differently to the "handedness" of aspartame enantiomers! In another example, D-glucose is sweet and nutritious, while L-glucose is tasteless and cannot be metabolized by the body.

Problem-Solving Example 9.9   Handedness in Asparagine

Asparagine is a naturally occurring amino acid first isolated as a bitter-tasting white powder from asparagus juice. Later, a sweet-tasting second form of asparagine was isolated from a sprouting vetch plant. D-Asparagine is bitter; the L isomer is sweet, indicating that our taste receptors, like our odor receptors, must be chiral. The Lewis structure of asparagine is

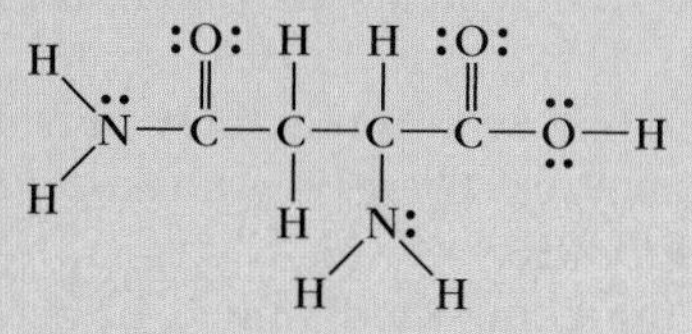

Identify which carbon atom in asparagine is the chiral center (asymmetric carbon).

Answer

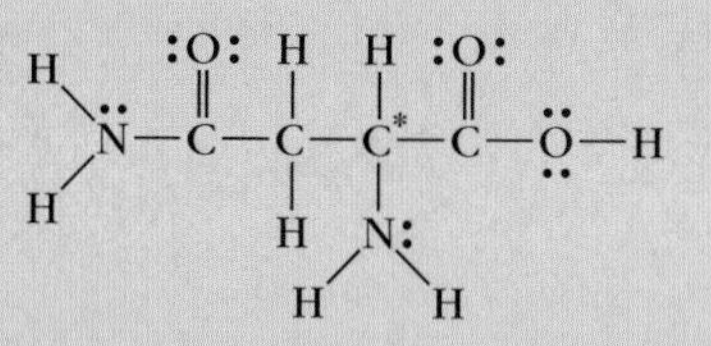

*Chiral center

Explanation   The carbon next to the —COOH group is the chiral center. It has four different atoms or groups of atoms attached to it: (1) a hydrogen atom, (2) a —COOH group, (3) an —NH$_2$ group, and (4) the rest of the molecule, starting with the —CH$_2$— group.

Problem-Solving Practice 9.9

Aspartame, a widely used sugar substitute, is a chiral molecule with the structural formula

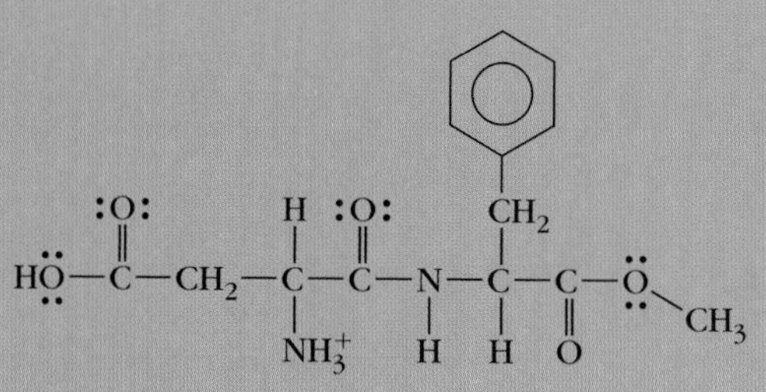

Identify the asymmetric (chiral) carbon atoms.

SUMMARY PROBLEM

Part 1

Write the Lewis structures and give the electron-pair geometry, the shape and bond angles, and the hybridization of the central atom of these polyatomic ions:

(a) BrF_2^+ (b) BrF_4^- (c) BrF_6^- (d) $IBrCl_3^-$

Part 2

Ketene, C_2H_2O, is a reactant for synthesizing cellulose acetate, which is used to make films, fibers, and fashionable clothing.

(a) Write the Lewis structure of ketene. Ketene does not contain an —OH bond.

(b) Identify the geometry around each carbon atom and all the bond angles in the molecule.

(c) Identify the hybridization of each carbon and oxygen atom.

(d) Is the molecule polar or nonpolar? Use appropriate data to support your answer.

Part 3

The structural formula for the open-chain form of glucose is

(a) Glucose dissolves readily in water. Use molecular structure principles to explain why glucose is so water-soluble.

(b) Identify the chiral centers in a glucose molecule. How many enantiomers of glucose are there?

IN CLOSING

Having studied this chapter, you should be able to . . .

- Recognize the various ways that the shapes of molecules are represented by models and on the printed page (Section 9.1).
- Predict shapes of molecules and polyatomic ions by using the VSEPR model (Section 9.2).
- Determine the orbital hybridization of a central atom and the associated molecular geometry (Section 9.3).
- Describe covalent bonding between two atoms in terms of sigma or pi bonds, or both (Section 9.3).
- Use molecular structure and electronegativities to predict the polarities of molecules (Section 9.4).
- Describe the different types of noncovalent interactions and use them to explain melting points and boiling points (Section 9.5).
- Identify the major components in the structure of DNA (Section 9.6).
- Define and describe the nature of chiral molecules and enantiomers (Section 9.7).
- Describe the basis of infrared spectroscopy *(Tools of Chemistry)* and UV-visible spectroscopy *(Tools of Chemistry)* and how they are used to determine molecular structures.

KEY TERMS

achiral *(9.7)*
asymmetric *(9.7)*
axial positions *(9.2)*
bond angle *(9.2)*
chiral *(9.7)*
complementary base
   pairs *(9.6)*
deoxyribonucleic acid
   (DNA) *(9.6)*
dipole-dipole attraction
   (9.5)
dipole moment *(9.4)*
electron-pair geometry
   (9.2)
enantiomers *(9.7)*
equatorial positions
   (9.2)
gene *(9.6)*
hybrid orbital *(9.3)*

hybridized *(9.3)*
hydrogen bond *(9.5)*
hydrophilic *(9.5)*
hydrophobic *(9.5)*
induced dipole *(9.5)*
intermolecular forces
   (9.5)
lipid bilayer *(9.5)*
London forces *(9.5)*
molecular geometry
   (9.2)
noncovalent interac-
   tions *(9.5)*
nonpolar molecule
   (9.4)
nucleotide *(9.6)*
phospholipid *(9.5)*
pi bond, π bond *(9.3)*
polar molecule *(9.4)*

polarization *(9.5)*
polymer *(9.6)*
replication *(9.6)*
sigma bond, σ bond
   (9.3)
sp hybrid orbital *(9.3)*
sp^2 hybrid orbital *(9.3)*
sp^3 hybrid orbital *(9.3)*
sp^3*d* hybrid orbital
   (9.3)
sp^3*d*^{2} hybrid orbital
   (9.3)
spectroscopy *(Tools of
   Chemistry, 9.1)*
valence bond model
   (9.3)
valence-shell electron-
   pair repulsion
   (VSEPR) model *(9.2)*

QUESTIONS FOR REVIEW AND THOUGHT

Conceptual Challenge Problems

CP-9.A   (Section 9.2)   What advantages does the VSEPR model of chemical bonding have compared with the Lewis dot formulas predicted by the octet rule?

CP-9.B   (Section 9.2)   The VSEPR model does not differentiate between single bonds and double bonds for predicting molecular shapes. What experimental evidence supports this?

CP-9.C   (Section 9.3)   What evidence could you present to show that two carbon atoms joined by a single sigma bond are able to rotate about an axis that coincides with the bond, but two carbon atoms bonded by a double bond cannot rotate about an axis along the double bond?

Answers to questions in **bold** can be found in the back of the book.

Review Questions

1. What is the VSEPR model? What is the physical basis of the model?

2. What is the difference between the electron-pair geometry and the molecular geometry of a molecule? Use the water molecule as an example in your discussion.

3. Designate the electron-pair geometry for each case from two to six electron pairs around a central atom.

4. What are the molecular geometries for each of the following?

   (a) H—$\overset{\cdot\cdot}{\underset{\cdot\cdot}{A}}$:   (b) H—$\overset{\cdot\cdot}{A}$—H

   (c) $\overset{\displaystyle H-\overset{\cdot\cdot}{A}-H}{\underset{\displaystyle H}{|}}$ (d) $\overset{\displaystyle H}{\underset{\displaystyle H}{\overset{|}{\underset{|}{H-A-H}}}}$

   Give the H—A—H bond angle for each of the last three.

5. If you have three electron pairs around a central atom, how

can you have a triangular planar molecule? An angular molecule? What bond angles are predicted in each case?

6. Draw a triangular bipyramid of electron pairs. Designate the axial and equatorial pairs. Are there axial and equatorial pairs in an octahedron?

7. Use VSEPR to explain why ethylene is a planar molecule.

8. How can a molecule with polar bonds be nonpolar? Give an example.

9. Give examples that illustrate the importance of the tetrahedral shape to a better understanding of chemistry.

10. Certain drug molecules are chiral. What implications does that have for how the drugs may function in the body?

11. Differentiate between a molecule being polar and being chiral.

12. Explain why the infrared spectrum of a molecule is referred to as its "fingerprint."

13. For infrared energy to be absorbed by a molecule, what frequency of motion must the molecule have?

14. One of the three isomers of dichlorobenzene, $C_6H_4Cl_2$, has a dipole moment of zero. Draw the structural formula of the isomer and explain your choice.

Molecular Shape

15. Use the different molecular modeling techniques (ball-and-stick, space-filling, two-dimensional pictures using wedges and dashed lines) to illustrate the following simple molecules:
    (a) NH_3 (b) H_2O (c) CO_2

16. All of the following molecules have central atoms with only bonding pairs of electrons. After drawing the Lewis structure, identify the molecular shape of each molecule.
    (a) BeH_2 (b) CH_2Cl_2 (c) BH_3
    (d) $SeCl_6$ (e) PF_5

17. Draw the Lewis structure for each of the following molecules or ions. Describe the electron-pair geometry and the molecular geometry.
    (a) NH_2Cl (b) OCl_2 (c) SCN^- (d) HOF

18. Determine the electron-pair geometry and molecular geometry for each of the following.
    (a) ClF_2^+ (b) $SnCl_3^-$ (c) PO_4^{3-} (d) CS_2

19. In each of the following molecules or ions, two oxygen atoms are attached to a central atom. Draw the Lewis structure for each one, and then describe the electron-pair geometry and the molecular geometry. Comment on similarities and differences in the series.
    (a) CO_2 (b) NO_2^- (c) SO_2
    (d) O_3 (e) ClO_2^-

20. In each of the following molecules or ions, three oxygen atoms are attached to a central atom. Draw the Lewis structure for each one, and then describe the electron-pair geometry and the molecular geometry. Comment on similarities and differences in the series.
    (a) BO_3^{3-} (b) CO_3^{2-} (c) SO_3^{2-} (d) ClO_3^-

21. The following are examples of molecules and ions that do not obey the octet rule. After drawing the Lewis structure, describe the electron-pair geometry and the molecular geometry for each.
    (a) ClF_2^- (b) ClF_3 (c) ClF_4^- (d) ClF_5

22. The following are examples of molecules and ions that do not obey the octet rule. After drawing the Lewis structure, describe the electron-pair geometry and the molecular geometry for each.
    (a) SiF_6^{2-} (b) SF_4 (c) PF_5 (d) XeF_4

23. Iodine forms three compounds with chlorine: ICl, ICl_3, and ICl_5. Draw the Lewis structures and determine the molecular shapes of these three molecules.

24. Give the approximate values for the indicated bond angles. The figures given are not the actual molecular shapes, but are used only to show the bonds being considered.
    (a) O—S—O angle in SO_2 (b) F—B—F angle in BF_3

(c) (d)

25. Give approximate values for the indicated bond angles. The figures given are not the actual molecular shapes, but are used only to show the bonds being considered.
    (a) Cl—S—Cl angle in SCl_2 (b) N—N—O angle in N_2O

(c) (d)

26. Give approximate values for the indicated bond angles.
    (a) F—Se—F angles in SeF_4
    (b) O—S—F angles in SOF_4 (The O atom is in an equatorial position.)
    (c) F—Br—F angles in BrF_5

27. Give approximate values for the indicated bond angles.
    (a) F—S—F angles in SF_6
    (b) F—Xe—F angle in XeF_2
    (c) F—Cl—F angle in ClF_2^-

28. Which would have the greater O—N—O bond angle, NO_2 or NO_2^-? Explain your answer.

29. Compare the F—Cl—F angles in ClF_2^+ and ClF_2^-. From Lewis structures, determine the approximate bond angle in each ion. Explain which ion has the greater angle and why.

Hybridization and Multiple Bonds

30. Describe the geometry and hybridization of chloroform, $CHCl_3$.

31. Describe the geometry and hybridization for each inner atom in ethylene glycol, $HOCH_2CH_2OH$, the main component in antifreeze.

32. Determine the Lewis structures, geometries, and hybridizations of GeF_4, SeF_4, and XeF_4.

33. Determine the geometry and hybridization of the three phosphorus-chlorine species: PCl_4^+, PCl_5, and PCl_6^-.

34. What designation is used for the hybrid orbitals formed by the following combinations of atomic orbitals?
    (a) One s and three p
    (b) One s, three p, and two d
    (c) One s and two p

35. How many hybrid orbitals are formed in each case in Question 34?

36. Describe the bond angles generally associated with each of the hybrid orbitals described in Question 34.

37. On a general chemistry exam, a student mistakenly drew up a set of sp^2d hybrid orbitals formed only from orbitals for which $n = 2$. What was the student's mistake?

38. The hybridization of the two carbon atoms differs in an acetic acid, CH_3COOH, molecule.
    (a) Designate the correct hybridization for each carbon atom in this molecule.
    (b) What is the approximate bond angle around each carbon?

39. The hybridization of the two nitrogen atoms differs in NH_4NO_3.
    (a) Designate the correct hybridization for each nitrogen atom.
    (b) What is the approximate bond angle around each nitrogen?

40. What are the hybridization and approximate bond angles for the N, C, and O atoms in alanine, an amino acid, whose Lewis structure is

41. Identify the type of hybridization, geometry, and approximate bond angle for each carbon atom in an alkane.

42. (a) Identify the type of hybridization and approximate bond angle for each carbon atom in CH_3CH_2CCH.
    (b) Which is the shortest carbon-to-carbon bond length in this molecule?
    (c) Which is the strongest carbon-to-carbon bond in this molecule?

43. Write the Lewis structure and designate which are sigma and pi bonds in each of these molecules.
    (a) HCN   (b) N_2H_2 (c) HN_3

44. Write the Lewis structure and designate which are sigma and pi bonds in each of these molecules.
    (a) OCS   (b) NH_2OH
    (c) CH_2CHCHO (d) $CH_3CH(OH)COOH$

45. Vanillin is the flavoring agent in vanilla extract and in vanilla ice cream. Its Lewis structure is

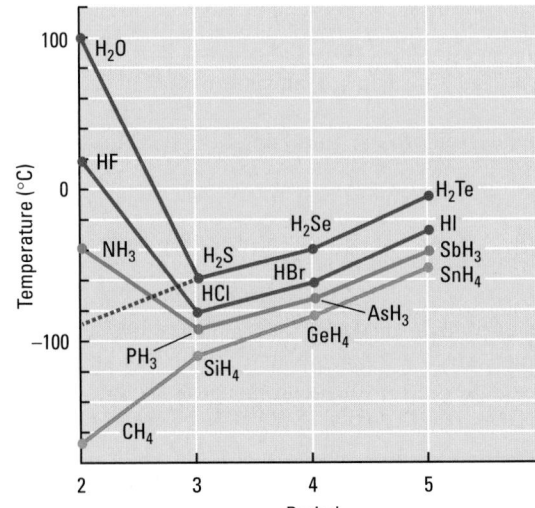

    (a) How many sigma bonds are in the molecule?
    (b) How many pi bonds are there?
    (c) What is the hybridization of the carbon atoms that are part of the ring in vanillin?
    (d) What is the hybridization of the carbon atoms not in the ring in vanillin?

46. Thymine, one of the nitrogen bases in DNA, has the Lewis structure

    (a) Locate the sigma bonds in the molecule. How many are there?
    (b) Locate the pi bonds in the molecule. How many are there?
    (c) What is the hybridization of the carbon atoms that are part of the ring?
    (d) What is the hybridization of the nitrogen atoms in the ring?
    (e) What is the hybridization of the carbon atom not in the ring?

Molecular Polarity

47. Consider these molecules: CH_4, NCl_3, BF_3, CS_2.
    (a) In which compound are the bonds most polar?

    (b) Which compounds in the list are not polar?

48. Consider these molecules: H_2O, NH_3, CO_2, ClF, CCl_4.
    (a) In which compound are the bonds most polar?
    (b) Which compounds in the list are not polar?
    (c) Which atom in ClF is more negatively charged?

49. Which of these molecules is (are) not polar? Which molecule has the most polar bonds?
    (a) CO   (b) PCl_3 (c) BCl_3
    (d) GeH_4 (e) CF_4

50. Which of these molecules is (are) polar? For each polar molecule, what is the direction of polarity; that is, which is the partial negative end and which is the partial positive end of the molecule?
    (a) CO_2 (b) HBF_2 (c) CH_3Cl (d) SO_3

51. Which of the following molecules have a dipole moment? For each of these polar molecules, indicate the direction of the dipole in the molecule.
    (a) XeF_2 (b) H_2S (c) CH_2Cl_2 (d) HCN

52. Explain the differences in the dipole moments of
    (a) BrF (1.29 D) and BrCl (0.52 D)
    (b) H_2O (1.86 D) and H_2S (0.95 D)

Noncovalent Interactions

53. Explain in terms of noncovalent interactions why water and ethanol are miscible, but water and cyclohexane are not.

54. Construct a table covering all the types of noncovalent interactions and comment about the distance dependence for each. (In general, the weaker the force, the closer together the molecules must be to feel the attractive force of nearby molecules.) You should also include an example of a substance that exhibits each type of noncovalent interaction in the table.

55. Explain the trends seen in the diagram below for the boiling points of some main group hydrogen compounds.
    (a) Group IV: CH_4, SiH_4, GeH_4, and SnH_4
    (b) Group V: NH_3, PH_3, AsH_3, and SbH_3
    (c) Group VI: H_2O, H_2S, H_2Se, and H_2Te
    (d) Group VII: HF, HCl, HBr, and HI

56. Explain why water "beads up" on a freshly waxed car but not on a dirty, unwaxed car.

57. Which of the following will form hydrogen bonds?
   (a) CH_2Br_2
   (b) $CH_3OCH_2CH_3$
   (c) H_2NCH_2COOH
   (d) H_2SO_3
   (e) CH_3CH_2OH

58. Arrange the noble gases in order of increasing boiling point. Explain your reasoning.

59. The structural formula for vitamin C is

Give a molecular level explanation why vitamin C is a water-soluble rather than a fat-soluble vitamin.

60. Which of the following would you expect to be most soluble in cyclohexane (C_6H_{12})? The least soluble? Explain your reasoning.
   (a) NaCl   (b) CH_3CH_2OH (c) C_3H_8

61. What is the major force that must be overcome to
   (a) Evaporate gasoline, a mixture of hydrocarbons.
   (b) Liquefy propane.
   (c) Decompose nitroglycerine, $CH_2NO_3CHNO_3CH_2NO_3$ into N_2, CO_2, and H_2O.
   (d) "Unzip" the DNA double helix during replication.

62. The structural formula of alpha-tocopherol, a form of vitamin E, is

Give a molecular level explanation why alpha-tocopherol dissolves in fat, but not in water.

Chirality in Organic Compounds

63. Circle the chiral centers, if any, in these molecules.

64. Circle the chiral centers, if any, in these molecules.

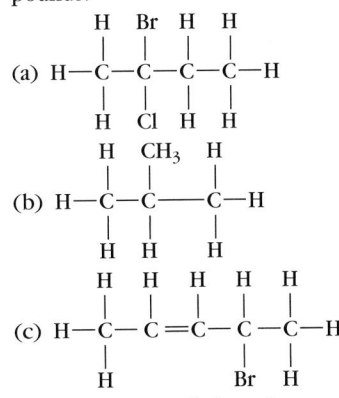

65. Circle the chiral centers, if any, in these compounds.

66. Circle the chiral centers, if any, in the following compounds.

67. How can you tell from its structural formula whether a compound can exist as a pair of enantiomers?

68. What conditions must be met for a molecule to be chiral?

69. Which of the following are not superimposable on its mirror image?
   (a) Nail   (b) Screw   (c) Shoe
   (d) Sock   (e) Golf club   (f) Football
   (g) Your ear   (h) Helix   (i) Baseball bat
   (j) Sweater

Molecular Structure Determination by Spectroscopy

70. Ultraviolet and infrared spectroscopies probe different aspects of molecules.
   (a) Ultraviolet radiation is energetic enough to cause transitions in the energies of which electrons in an atom?
   (b) What does infrared spectroscopy tell us about a given molecule?

71. The infrared spectrum of ethanol is given on page 363.
   (a) Which stretching requires the lowest amount of energy?
   (b) The highest amount of energy?

Biomolecules

72. Discuss the differences between the extent of hydrogen bonding for the pairs G-C and A-T in nucleic acids. If a strand of DNA has more G-C pairs than A-T pairs in the double helix, the melting point (unwinding point) increases. The melting point for strands with more A-T pairing will decrease in comparison. Explain.

73. One strand of DNA contains the base sequence T-C-G. Draw a structure of this section of DNA that shows the hydrogen bonding between the base pairs of this strand and its complementary strand.

General Questions

74. The formula for nitryl chloride is NO_2Cl. Draw the Lewis structure for the molecule, including all resonance structures. Describe the electron-pair and molecular geometries, and give values for all bond angles.

75. Vanillin is the flavoring agent in vanilla extract and in vanilla ice cream. Its structure is

(a) Give values for the three bond angles indicated.
(b) Indicate the most polar bond in the molecule.
(c) Circle the shortest carbon-oxygen bond.

76. Watson and Crick received the 1962 Nobel Prize in Chemistry for their simple but elegant model of the "heredity molecule," DNA. The key to their structure (the "double helix") was an understanding of the geometry and bonding capabilities of nitrogen-containing bases such as the thymine molecule below.
(a) Give approximate values for the indicated bond angles.
(b) Which are the most polar bonds in the molecule?

77. Cyanidin chloride is an anthocyanin found in strawberries, apples, and cranberrries.

Would you expect this compound to absorb in the visible region of the spectrum? Explain your answer.

78. Each of the following molecules has fluorine atoms attached to an atom from Groups 1A or 3A to 6A. Draw the Lewis structure for each one and then describe the electron-pair geometry and the molecular geometry. Comment on similarities and differences in the series.
(a) BF_3 (b) CF_4 (c) NF_3 (d) OF_2 (e) HF

79. The dipole moment of the HCl molecule is 3.43×10^{-30} C·m, and the bond length is 127.4 pm; the dipole moment of HF is 6.37×10^{-30} C·m, with bond length of 91.68 pm. Use the definition of dipole moment as a product of partial charge on each atom times the distance of separation (Section 9.4) to calculate the quantity of charge in coulombs that is separated by the bond length in each dipolar molecule. Use your result to show that fluorine is more electronegative than chlorine.

80. Sketch the geometry of a carbon-containing molecule or ion in which the angle between two atoms bonded to carbon is
(a) Exactly 109.5°.
(b) Slightly different from 109.5°.
(c) Exactly 120°.
(d) Exactly 180°.

81. In the gas phase, positive and negative ions form ion pairs that are like molecules. An example is KF, which is found to have a dipole moment of 28.7×10^{-30} C·m and a distance of separation between the two ions of 217.2 pm. Use this information and the definition of dipole moment to calculate the partial charge on each atom. Compare your result with the expected charge, which is the charge on an electron, -1.62×10^{-19} C. Based on your result, is KF really completely ionic?

82. The following compound is commonly called acetylacetone. As shown, it exists in two forms, one called the *enol* form and the other called the *keto* form.

enol form   keto form

While in the *enol* form, the molecule can lose H^+ from the —OH group to form an anion. One of the most interesting aspects of this anion (sometimes called the *acac* ion) is that one or more of them can react with a transition metal cation to give very stable, highly colored compounds.

(a) Using bond energies, calculate the enthalpy change for the *enol* to *keto* change. Is the reaction exothermic or endothermic?

(b) What are the electron-pair and "molecular" geometries around each C atom in the *keto* and *enol* forms? What changes (if any) occur when the *keto* form changes to the *enol* form?

(c) If you wanted to prepare 15.0 g of deep red $Cr(acac)_3$ using the following reaction,

$$CrCl_3 + 3\ H_3C—C(OH)=CH—C(O)—CH_3$$
$$+ 3\ NaOH \longrightarrow Cr(acac)_3 + 3\ H_2O + 3\ NaCl$$

how many grams of each of the reactants would you need?

83. How can a diatomic molecule be nonpolar? Polar?

84. How can a molecule have polar bonds but yet have a dipole moment of zero?

Applying Concepts

85. Complete this table.

Molecule or ion	Electron-pair geometry	Molecular geometry	Hybridization of the iodine atom
ICl_2^+			
I_3^-			
ICl_3			
ICl_4^-			
IO_4^-			
IF_4^+			
IF_5			
IF_6^+			

86. Complete this table.

Molecule or ion	Electron-pair geometry	Molecular geometry	Hybridization of the sulfur atom
SO_2			
SCl_2			
SO_3			
SO_3^{2-}			
SF_4			
SO_4^{2-}			
SF_5^+			
SF_6			

87. Name a Group 1A to 8A element that could be the central atom (X) in the following compounds.

(a) XH_3 with one lone pair of electrons
(b) XCl_3
(c) XF_5
(d) XCl_3 with two lone pairs of electrons

88. Name a Group 1A to 8A element that could be the central atom (X) in the following compounds.

(a) XCl_2
(b) XH_2 with two lone pairs of electrons
(c) XF_4 with one lone pair of electrons
(d) XF_4

89. How many water molecules could hydrogen-bond to an acetic acid molecule? Draw in the water molecules and use dotted lines to show the hydrogen bonds.

90. How many water molecules could hydrogen-bond to an ethylamine molecule? Draw in the water molecules and use dotted lines to show the hydrogen bonds.

91. The following are responses students wrote when asked to give an example of hydrogen bonding. Which are correct?

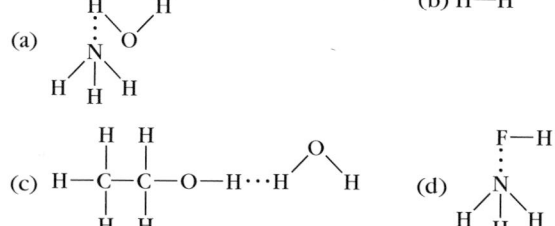

92. Which of these are examples of hydrogen bonding?

93. In another universe, elements try to achieve a nonet (nine valence electrons) instead of an octet when forming chemical bonds. As a result, covalent bonds form when a trio of electrons is shared between two atoms. Two compounds in this other universe are H_3O and H_2F. Draw their Lewis structures, then determine their electron-trio geometry and molecular geometry.

General Chemistry CD-ROM

CD9.1 Screen 10.2: Models of Chemical Bonding. This screen points out the main tenets of two theories of bonding. List aspects of the theories that are similar. List aspects of the theories that are dissimilar.

CD9.2 Screen 10.4: Hybrid Orbitals. Hybrid orbitals are an extension of our treatment of orbital shapes in Chapters 7 and 8. What observation leads us to propose this extension to the original wave mechanics theory?

CD9.3 Screen 10.6: Determining Hybrid Orbitals. Examine the Hybrid Orbitals tool on this screen. Use this tool to systematically combine atomic orbitals to form hybrid atomic orbitals.

(a) What is the relationship between the number of hybrid orbitals produced and the number of atomic orbitals used to create them?

(b) Do hybrid atomic orbitals form between different p orbitals without involving s orbitals?

(c) What is the relationship between the energy of hybrid atomic orbitals and the atomic orbitals from which they are formed?

CD9.4 Screen 10.7: Multiple Bonding. Ethylene is a flat molecule. Why do the two planar triangles centered on the C atoms align in the same plane even though twisting would lead to lower electron-electron repulsive forces?

CD9.5 Screen 13.3: Intermolecular Forces (1). The table of Intermolecular Force Strengths (IMFs) on this screen indicates that substances with strong intermolecular forces are solids, those with moderate IMFs are liquids, and those with weak IMFs are gases. Explain this in terms of the molecular scale view of how molecules act in each of these phases.

CD9.6 Screen 13.5: Intermolecular Forces (3).

(a) What does the term "polarizable" mean?

(b) Why are O_2 molecules attracted to molecules of water, despite the fact that O_2 is nonpolar and, as such, might be expected to be unaffected by coulombic forces?

10

Hot air balloons rise because the density of the gas within them is less than the density of the surrounding air. The density of the gas filling a balloon is decreased by heating, usually with a propane burner. How gases behave when they are heated, in addition to many other aspects of gas chemistry, is explained by the gas laws.

GASES AND THE ATMOSPHERE

10.1 Properties of Gases

10.2 The Atmosphere

10.3 Kinetic-Molecular Theory

10.4 The Behavior of Ideal Gases

10.5 The Ideal Gas Law

10.6 Quantities of Gases in Chemical Reactions

10.7 Gas Density and Molar Masses

10.8 Gas Mixtures and Partial Pressures

10.9 The Behavior of Real Gases

10.10 Chemical Reactions in the Atmosphere

10.11 Ozone and Ozone Depletion

10.12 Chemistry and Pollution in the Troposphere

Early chemists studied gases and their chemistry extensively. They carried out reactions that generated gases, bubbled the gases through water into glass containers, and transferred the gases to animal bladders for storage. They mixed gases together to see whether they would react and, if so, how much of each gas would be consumed or formed. They discovered that gases have a great many properties in common—much more so than liquids or solids. All gases are transparent, although some are colored. (F_2 is light yellow, Cl_2 is greenish yellow, Br_2 and NO_2 are reddish brown, and I_2 vapor is violet.) All gases are also mobile—they expand to fill all space available, and they mix together physically in any proportions. The volume of a gas sample can easily be altered by changing its temperature or pressure, or both. Thus, gas densities are quite variable, depending on the conditions. At first this might seem to make the study of gases complicated, but it really does not. The way that gas volume depends on temperature and pressure is the same for all gases, and so it is easy to formulate an expression that accounts for the dependence.

Earth is surrounded by a thin mixture of gases we call the atmosphere. A variety of chemical reactions take place in this mixture, many of them driven by energy from solar photons. Many of these reactions are beneficial to the earth's inhabitants, but some produce undesirable products. In this chapter, fundamental facts about gases are intermingled with information about atmospheric reactions. The relationships among pressure, temperature, and amount of a gas and its volume are expressed in the ideal gas law. Other quantitative treatments of gases in chemical reactions are introduced, the adjustments to the ideal gas law necessary for real gases are described, and the chapter concludes with discussions of chemistry in the atmosphere.

10.1 PROPERTIES OF GASES

Many molecular compounds are gases under everyday conditions of temperature and pressure. The air we breathe consists mostly of diatomic nitrogen and oxygen molecules. Other elements that exist as gases include H_2, Cl_2, F_2, and the noble gases. Some examples of small molecular compounds that are also gases are shown in Table 10.1. At sufficiently high temperatures, many other compounds that are normally liquids or solids become gases.

TABLE 10.1	Some Common Gases and Their Uses	
Gas	**Formula**	**Typical use**
Acetylene	C_2H_2	Welding torch fuel
Ammonia	NH_3	Fertilizer
Argon	Ar	Filling for light bulbs
Butane	C_4H_{10}	Fuel for space heating (LP gas)
Carbon dioxide	CO_2	Carbonation of beverages
Chlorine	Cl_2	Disinfectant, bleach
Ethylene	C_2H_4	Polymer manufacturing (polyethylene)
Helium	He	Lighter-than-air aircraft
Hydrogen	H_2	Hydrogenation of oils; possible fuel of the future
Methane	CH_4	Fuel (natural gas)
Nitrous oxide	N_2O	Anesthetic
Propane	C_3H_8	Fuel for space heating (LP gas)
Sulfur dioxide	SO_2	Preservative, disinfectant

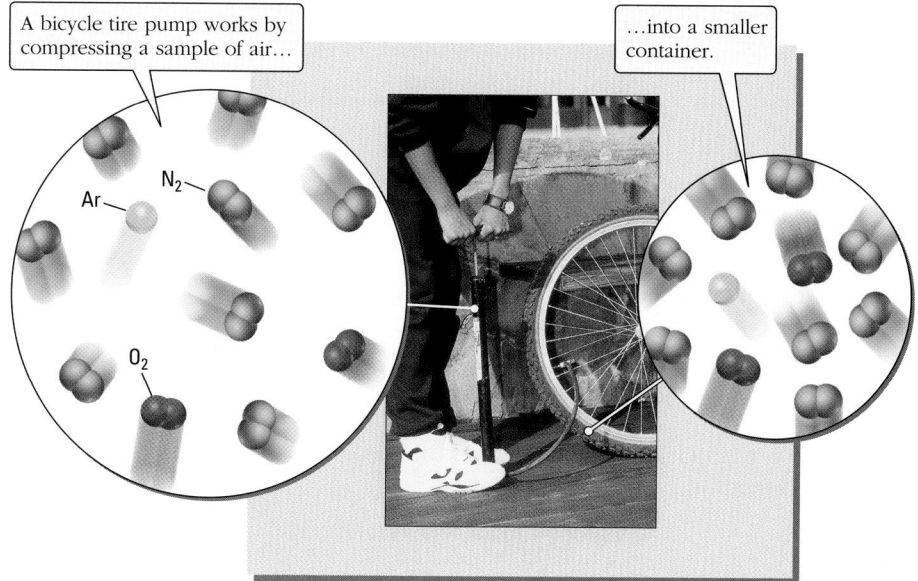

Figure 10.1 **Compression of gases.**

All gases, whatever their nature, have a number of properties in common:

- *Gases can be compressed.* We often pump compressed air into an automobile or bicycle tire. Compressed air occupies less volume than the noncompressed air (Figure 10.1).

- *Gases exert pressure on whatever surrounds them.* A gas sample inside a balloon or a storage cylinder exerts pressure on its surroundings. If the container cannot sustain the pressure, some of the gas sample will escape.

- *Gases expand into whatever volume is available.* The gaseous contents of an aerosol spray will continue to expand upon release from the can.

- *Gases mix completely with one another.* The atmospheric gases are an example. Once gases are mixed, they do not separate spontaneously.

- *Gases are described in terms of their temperature and pressure, the volume occupied, and the amount (numbers of molecules or moles) of gas present.* For example, a hot gas occupies a greater volume and exerts a greater pressure than does the same sample of gas when it is cold.

 CD-ROM Screen 12.2: Properties of Gases

The kinetic-molecular theory (⇐ *p. 19*) can be used to explain all these properties of gases on the nanoscale. You will see in Section 10.3 that using this theory makes gases fairly easy to understand and will enable you to predict accurately many of the properties of a sample of gas. Let's look at some gas properties in more detail.

Gases Exert Pressure

The firmness of a balloon filled with air indicates that the air inside exerts pressure. If too much gas is forced into a balloon, that pressure bursts the balloon, and the gas rapidly escapes. The balloon's firmness is caused by gas molecules striking its inner surface. Each collision of a gas molecule with the balloon's inner surface exerts a force on the surface. The force per unit area is the **pressure** of the gas. A gas exerts pressure on every surface it contacts, no matter what the direction of contact.

$$\text{Pressure} = \frac{\text{force}}{\text{area}}$$

TABLE 10.2 Pressure Units

SI unit: pascal (Pa)

$$1 \text{ Pa} = 1 \text{ kg m}^{-1} \text{ s}^{-2} = 1 \text{ N/m}^2$$

Other common units

$1 \text{ bar} = 10^5 \text{ Pa} = 100 \text{ kPa}$

$1 \text{ atm} = 1.01325 \times 10^5 \text{ Pa} = 101.325 \text{ kPa}$

$1 \text{ atm} = 760 \text{ torr} = 760 \text{ mm Hg}$ (this conversion is exact)*

$1 \text{ atm} = 14.7 \text{ lb/inch}^2 \text{ (psi)} = 1.01325 \text{ bar}$

*Exact conversion factors do not limit the number of significant figures in calculations.

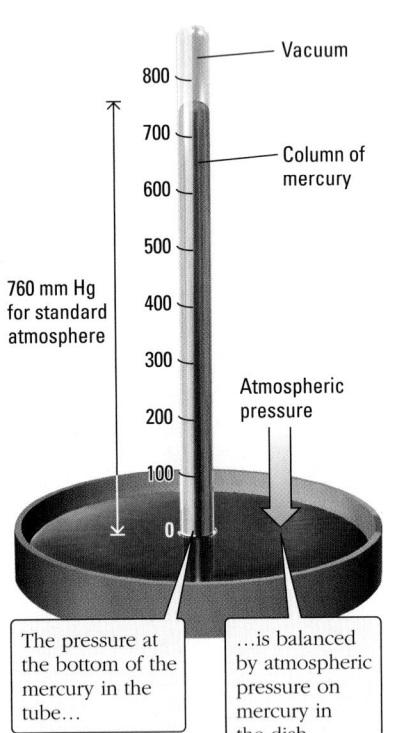

Figure 10.2 **A Torricellian barometer.**

$1 \text{kPa} = 10^3 \text{ Pa}$

A force can accelerate an object, and the force equals the mass of the object times its acceleration.

$$\text{Force} = \text{mass} \times \text{acceleration}$$

The SI units for mass and acceleration are kilograms (kg) and meters per second per second (m/s²), respectively, so force has the units kg m/s². A force of 1 kg m/s² is defined as a **newton (N)** in the SI system. A pressure of one newton per square meter (N/m²) is defined as a **pascal (Pa).** Table 10.2 provides units of pressure and conversion factors among units.

The earth's atmosphere exerts pressure on everything with which it comes into contact. Atmospheric pressure can be measured with a **barometer,** which can be made by filling a tube closed at one end with a liquid and then inverting the tube in a dish containing the same liquid. Figure 10.2 shows a mercury barometer—a glass tube filled with mercury, inverted, and placed in a container of mercury. At sea level the height of the mercury column is about 760 mm above the surface of the mercury in the dish. The pressure at the bottom of a column of mercury 760 mm tall is balanced by the pressure at the bottom of the column of air surrounding the dish—a column that extends to the top of the atmosphere. Pressure measured with a mercury barometer is usually reported in **millimeters of mercury (mm Hg),** a unit that is also called the **torr** after Evangelista Torricelli, who invented the mercury barometer in 1643. The **standard atmosphere (atm)** is defined as

$$1 \text{ Standard atmosphere} = 1 \text{ atm} = 760 \text{ mm Hg (exactly)} = 101.325 \text{ kPa}$$

The pressure of the atmosphere at sea level is about 101,300 Pa (101.3 kPa). A related unit, a **bar,** equal to 100,000 Pa, is sometimes used for atmospheric pressure, especially in weather reports. For a gaseous substance, the standard thermodynamic properties (⇐ *p. 243*) are given for a gas pressure of 1 bar.

Problem-Solving Example 10.1 Converting Pressure Units

Convert a pressure reading of 736 mm Hg to units of (a) atm, (b) torr, (c) kPa, (d) bar, and (e) psi.

Answer
(a) 0.968 atm   (b) 736 torr   (c) 98.1 kPa   (d) 0.981 bar   (e) 14.2 psi

Explanation Using the unit equivalences given in Table 10.2, we can write

(a) atm: $736 \text{ mm Hg} \times \left(\dfrac{1 \text{ atm}}{760. \text{ mm Hg}} \right) = 0.968 \text{ atm}$

(b) torr: $736 \text{ mm Hg} \times \left(\dfrac{760 \text{ torr}}{760. \text{ mm Hg}} \right) = 736 \text{ torr}$

(c) kPa: $736 \text{ mm Hg} \times \left(\dfrac{101.3 \text{ kPa}}{760. \text{ mm Hg}} \right) = 98.1 \text{ kPa}$

(d) bar: $736 \text{ mm Hg} \times \left(\dfrac{1.013 \text{ bar}}{760. \text{ mm Hg}} \right) = 0.981 \text{ bar}$

(e) psi: $736 \text{ mm Hg} \times \left(\dfrac{14.7 \text{ psi}}{760. \text{ mm Hg}} \right) = 14.2 \text{ psi}$

Problem-Solving Practice 10.1

A TV weather person says the barometric pressure is "29.5 inches of mercury." What is this pressure in (a) atm, (b) mm Hg, (c) bar, and (d) kPa?

Any liquid can be used in a barometer, but the height of the column depends on the density of the liquid. A water barometer would be about 34 ft high, far too tall to be practical. Many water wells (especially in regions of the world without readily available electricity) bring up water from rock strata less than 33 ft below the surface. A simple hand pump can reduce the pressure at the top of the well casing and thus enable the atmospheric pressure that is acting on the water at the bottom of the well to force the water upward. Before the invention of submersible electric pumps, well diggers knew from experience that if water was found deeper than 33 ft below the surface, a hole large enough for a bucket would have to be dug. The bucket would then have to be lowered and filled with water because atmospheric pressure alone could not push the water higher than 33 ft.

An old well pump. Old-fashioned water pumps like this one depend on atmospheric pressure to lift water from the well. They can only lift water from a depth of about 33 feet at sea level. At higher elevations this depth is lessened because the atmospheric pressure is lower.

Problem-Solving Example 10.2   Pressure Units

Show that 760.0 mm Hg equals 101.3 kPa. The density of mercury is 13.596 g/cm³. (The acceleration of gravity is 9.807 m/s².)

Answer See explanation.

Explanation Use a column of mercury 760.0 mm high with a cross-sectional area of 1.000 mm². Since the volume of the cylinder is the area of the base (1.000 mm²) times the height (760.0 mm), the volume of the mercury is 760.0 mm³. The density of mercury can be used to calculate the mass of the mercury.

$$\text{Mass of Hg} = (760.0 \text{ mm}^3) \left(\frac{1 \text{ cm}}{10 \text{ mm}} \right)^3 \left(\frac{13.596 \text{ g}}{1 \text{ cm}^3} \right) = 10.33 \text{ g} = 0.01033 \text{ kg}$$

The downward force on the mercury due to gravity is

$$\text{Force} = 0.01033 \text{ kg} \times 9.807 \text{ m/s}^2 = 0.1013 \text{ N}$$

So the pressure is

$$\text{Pressure} = \left(\frac{0.1013 \text{ N}}{1 \text{ mm}^2} \right) \left(\frac{1000 \text{ mm}}{1 \text{ m}} \right)^2$$

$$= 101.3 \times 10^3 \text{ N/m}^2 = 101.3 \times 10^3 \text{ Pa} = 101.3 \text{ kPa}$$

Problem-Solving Practice 10.2

Convert a pressure of 647 mm Hg into its corresponding value in units of bars, kilopascals (kPa), and atmospheres (atm).

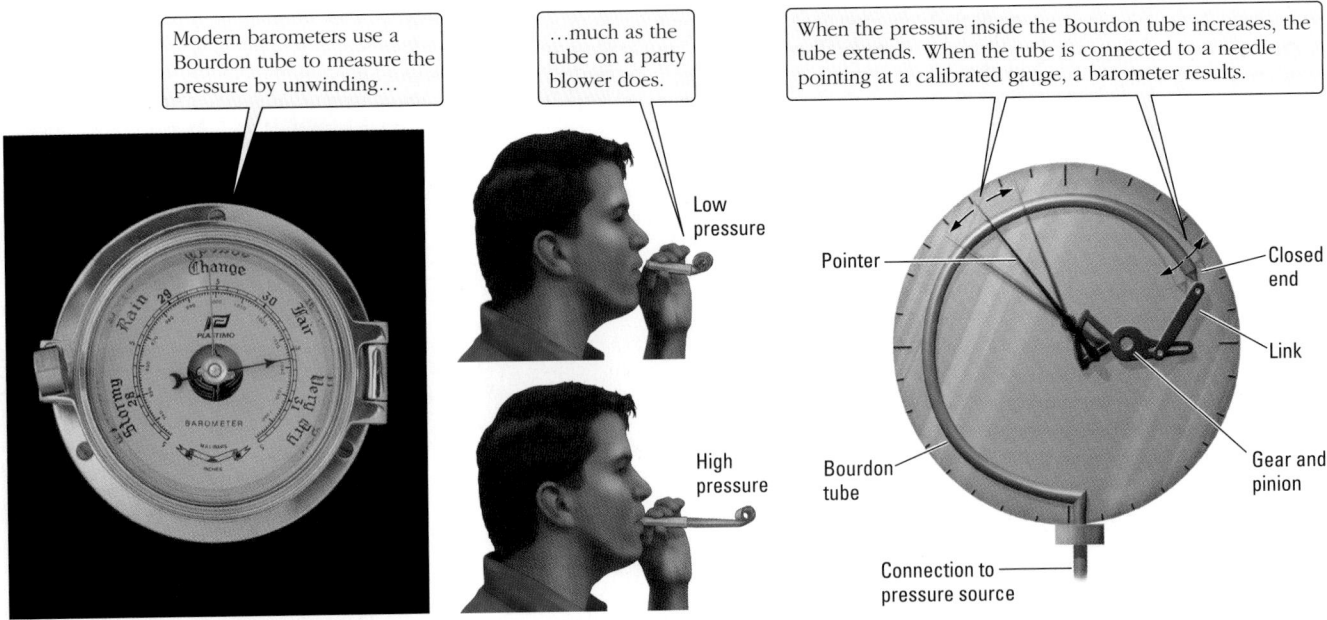

Barometers and the Bourdon gauge.

10.2 THE ATMOSPHERE

A metric ton is 1000 kg.

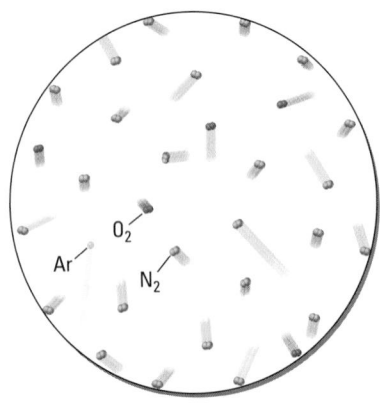

Nanoscale snapshot of air. This instantaneous snapshot of a sample of air at the nanoscale shows nitrogen, oxygen, and argon in approximately the correct proportions.

To convert percent to ppm, multiply by 10,000. Divide by 10,000 to convert ppm to percent.

Earth is enveloped by a few vertical miles of chemicals that compose the atmosphere—the gaseous medium in which we exist, a perfect place to begin the study of gases. Everything on the surface of the earth experiences the atmosphere's pressure. The atmosphere's total mass is approximately 5.3×10^{15} metric tons, a huge figure, but still only about one millionth of the earth's total mass. Close to the earth's surface the atmosphere is mostly nitrogen and life-sustaining oxygen, but a fraction of a percent of other chemicals can make a difference in the quality of life. Extra water in the atmosphere can support a rain forest; a little less water produces a balanced rainfall, and practically no water results in a desert. Air pollutants, present in such small quantities that they are outnumbered by oxygen and nitrogen molecules by 100,000 to 1 or more (Section 10.12), can lower our quality of life and generally be harmful to people with respiratory disorders.

The two major chemicals in our atmosphere are nitrogen, a rather unreactive gas, and oxygen, a highly reactive one. In dry air at sea level, nitrogen is the most abundant atmospheric gas, followed by oxygen, and then 13 other gases, each at less than 1% by volume (Table 10.3). For every 100 volume units of air, 21 units are oxygen. When it is pure, oxygen supports combustion at an explosive rate (Figure 10.3), but when diluted with nitrogen at moderate temperatures, oxygen's oxidizing capability is tamed somewhat. Except fo helium, the atmosphere is our only source for the noble gases—argon, neon, krypton, and xenon.

Due to gravity, molecules making up the atmosphere are most concentrated near the earth's surface. In addition to the percentages by volume used in Table 10.3, *parts per million (ppm)* and *parts per billion (ppb)* by volume are also used in describing the concentrations of components of the atmosphere. Since volume is proportional to the number of molecules, ppm and ppb units also give a ratio of molecules of one kind to another kind. For example, "10 ppm SO_2" means that for every 1 million air molecules, 10 of them are SO_2 molecules. This may not sound like much until you consider that in just 1 cm³ of air there are about 2.7×10^{13} million molecules. If this air contains 10 ppm SO_2, then there are 2.7×10^8 SO_2 molecules. That's a lot of SO_2 molecules, and we're only talking about 1 cm³ of air.

| | | TABLE 10.3 The Composition of Dry Air at Sea Level | | |

Gas	Percentage by volume	Gas	Percentage by volume
Nitrogen	78.084	Krypton	0.0001
Oxygen	20.948	Carbon monoxide[†]	0.00001
Argon	0.934	Xenon	0.000008
Carbon dioxide*	0.033	Ozone[†]	0.000002
Neon	0.00182	Ammonia	0.000001
Hydrogen	0.0010	Nitrogen dioxide[†]	0.0000001
Helium	0.00052	Sulfur dioxide[†]	0.00000002
Methane*	0.0002		

*The greenhouse gases carbon dioxide and methane are discussed in Section 12.4 as they relate to fuels and the burning of fuels for energy production.

[†]Trace gases of environmental importance.

Figure 10.3 Combustion in pure oxygen. When a glowing splint is inserted into the mouth of a beaker containing pure oxygen, the high oxygen concentration causes the splint to burn at an explosive rate.

Exercise 10.1   Calculating the Mass of Gases

Calculate the mass of the SO_2 molecules found in 1 cm^3 of air that contains 10 ppm SO_2.

The Troposphere and the Stratosphere

The earth's atmosphere can be roughly divided into layers, as shown in Figure 10.4. From the surface to about 10 km altitude, in the region named the **troposphere,** the temperature of the atmosphere decreases with increasing altitude. In

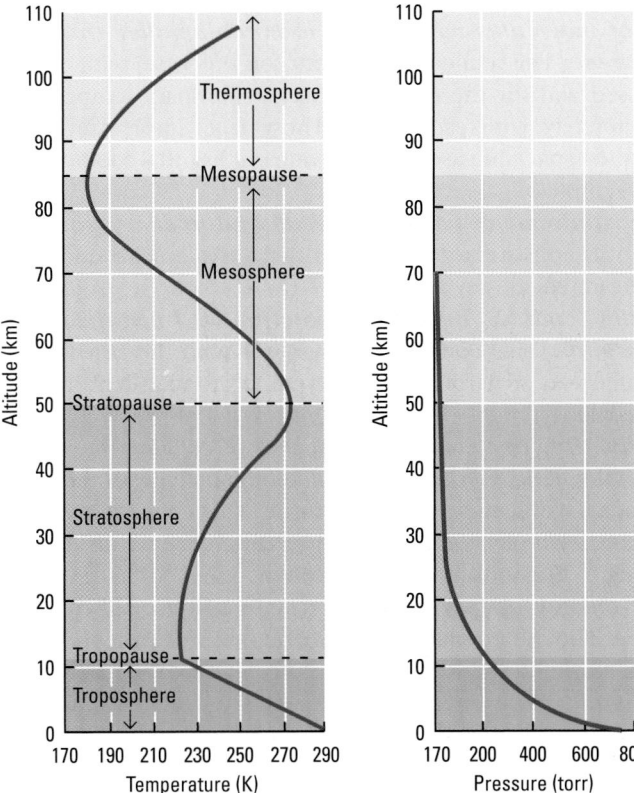

Figure 10.4 Pressure and temperature changes in our earth's atmosphere.

The troposphere was named by the British meteorologist Sir Napier Shaw from the Greek word *tropos,* meaning "turning."

The stratosphere was named by the French meteorologist Leon Phillipe Treisserenc de Bort, who believed this region consisted of orderly layers with no turbulence or mixing. *Stratum* is a Latin word meaning "layer."

CD-ROM Screen 12.9: The Kinetic-Molecular Theory of Gases

In the kinetic-molecular theory the word "molecule" is taken to include atoms of the monoatomic noble gases He, Ne, Ar, Kr, and Xe.

Gas pressure is caused by gas molecules bombarding a container's walls.

Figure 10.5 Kinetic-molecular theory and pressure.

this region the most violent mixing of air and the biggest variations in moisture content and temperature occur. Winds, clouds, storms, and precipitation are the result, the phenomena we know as weather. The troposphere is where we live. A commercial jet airplane flying at about 10 km (~33,000 ft) is still in the troposphere, although near its upper limits. The composition of the troposphere is roughly that of dry air near sea level (see Table 10.3), but the concentration of water vapor varies considerably; on average, it is about 10 ppm.

Just above the troposphere, from about 12 to 50 km above the earth's surface, is the **stratosphere.** If you take a ride in a Concorde supersonic aircraft, you will cruise in the stratosphere at about 20 km. The pressures in the stratosphere are extremely low, and there is little mixing between the stratosphere and the troposphere. The lower limit of the stratosphere varies from night to day over the globe, and at the polar regions it may be as low as 8 to 9 km above the earth's surface. About 75% of the mass of the atmosphere is in the troposphere, and 99.9% of the atmosphere's mass is below 50 km, in the troposphere and stratosphere.

10.3 KINETIC-MOLECULAR THEORY

To explain why gases behave as they do, we first look at the nanoscale behavior of gas molecules. The fact that all gases behave in very similar ways can be interpreted by means of the *kinetic-molecular theory,* a theory that applies to the properties of liquids and solids as well as gases. According to the kinetic-molecular theory, a gas consists of tiny molecules in constant rapid random motion. The pressure a gas exerts on the walls of its container results from the continual bombardment of the walls by rapidly moving gas molecules (Figure 10.5).

Four fundamental concepts form the foundation of the kinetic-molecular theory, and a fifth is closely associated with it. Each is consistent with the results of experimental studies of gases on the macroscale.

1. *A gas is composed of molecules whose size is much smaller than the distances between them.* This concept accounts for the ease with which gases can be compressed and for the fact that gases at ordinary temperature and pressure mix completely with each other. These facts imply that there must be much unoccupied space in gases that provides substantial room for additional molecules in a sample of gas.
2. *Gas molecules move randomly at various speeds and in every possible direction.* This concept is consistent with the fact that gases quickly and completely fill any container in which they are placed.
3. *Except when molecules collide, forces of attraction and repulsion between them are negligible.* This concept is consistent with the fact that all gases behave in the same way, regardless of the types of noncovalent interactions among their molecules.
4. *When collisions occur, they are elastic.* The speeds of colliding molecules may change, but the total kinetic energy of two colliding molecules is the same after a collision as before. That is, the collision is elastic. This concept is consistent with the fact that a gas sample at constant temperature never "runs down," with all molecules falling to the bottom of the container.
5. *The average kinetic energy of gas molecules is proportional to the absolute temperature.* Though not part of the kinetic molecular theory, this useful concept is consistent with the fact that gas molecules escape through a tiny hole faster as the temperature increases, and with the fact that rates of chemical reactions are faster at higher temperatures.

Like any moving object, a gas molecule has kinetic energy. An object's kinetic energy, E_k, depends on its mass, m, and its speed, v, according to the equation

ESTIMATION   Thickness of Earth's Atmosphere

The photograph in the margin scales the earth's diameter to about 5 cm. We have seen that 99.9% of the atmosphere is contained within 50 km of the earth's surface. To draw a circle properly relating the size of the earth with the thickness of the atmosphere, how thick should the line be drawn?

The diameter of the earth is about 7926 miles, which is 12,750 km. Set up the ratio

$$\frac{5 \text{ cm}}{12{,}750 \text{ km}} = \frac{x}{50 \text{ km}}$$

which gives $x = 0.020$ cm $= 0.2$ mm.

The circle representing the earth should be 5 cm in diameter with a line width of only 0.2 mm. The atmosphere of earth is really quite thin. The proportions are about the same as the skin on an apple in relation to the apple.

$$E_k = \tfrac{1}{2}(\text{mass})(\text{speed})^2 = \tfrac{1}{2}mv^2$$

All the molecules in a gas are moving, but they do not all move at the same speed, and so they do not all have the same kinetic energy. At any given time a few molecules are moving very quickly, most are moving at close to the average speed, and a few others may be in the process of colliding with a surface, in which case their speed is momentarily zero. The speed of any individual molecule changes as it collides with and exchanges energy with other molecules.

The relative number of molecules that have a given speed can be measured experimentally. Figure 10.6 is a graph of the number of molecules plotted versus their speed. The higher a point on the curve, the greater the number of molecules going at that speed. Notice in the plot that some molecules are moving quickly (have high kinetic energy) and some are moving slowly (have low kinetic energy). The maximum in the distribution curve is the most probable speed. For oxygen gas at 25°C, for example, the maximum in the curve occurs at a speed of 400 m/s (1000 mph), and most of the molecules' speeds are in the range from 200 m/s to 700 m/s. Notice that the curves are not symmetric. A consequence of this is that the average speed (shown in Figure 10.6 for O_2) is a little faster than the most probable speed.

Also notice in Figure 10.6 that as temperature increases, the most probable speed goes up, and the number of molecules traveling very quickly increases. The

Planet earth.

Graphs of molecular speeds (or energies) versus numbers of molecules are called Boltzmann distribution curves. They are named after Ludwig Boltzmann (1844–1906), an Austrian mathematician who helped develop the kinetic-molecular theory of gases.

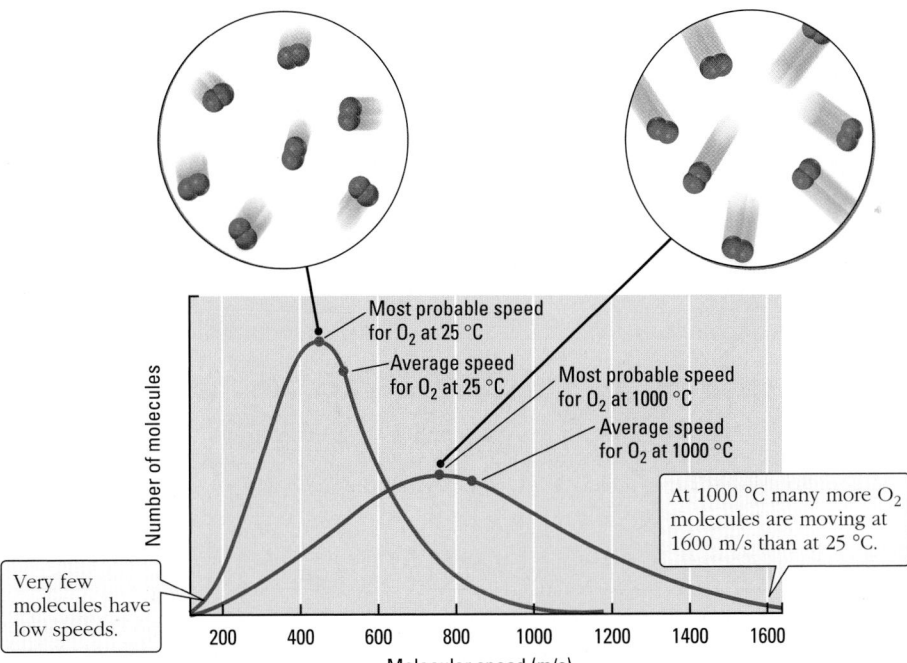

Most probable speed for O_2 at 25 °C

Average speed for O_2 at 25 °C

Most probable speed for O_2 at 1000 °C

Average speed for O_2 at 1000 °C

At 1000 °C many more O_2 molecules are moving at 1600 m/s than at 25 °C.

Very few molecules have low speeds.

Number of molecules

200   400   600   800   1000   1200   1400   1600

Molecular speed (m/s)

Figure 10.6   Distribution of molecular speeds. A plot of the relative number of gas molecules with a given speed versus that speed (in meters per second).

Figure 10.7 The effect of molar mass on the distribution of molecular speeds at a given temperature.

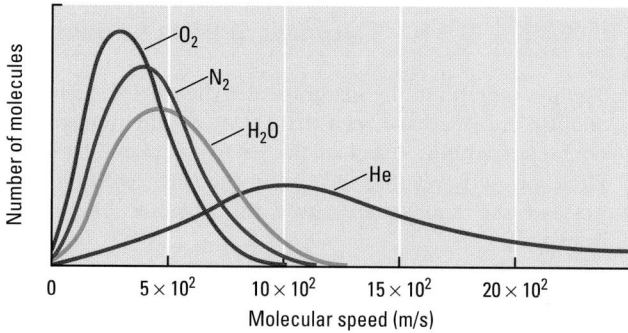

areas under the two curves representing the two gas samples at different temperatures are the same because the total number of molecules is the same in both samples.

Since $E_k = \frac{1}{2}mv^2$ and the average kinetic energy of the molecules of any gas is the same at a given temperature, the larger m is, the smaller the average v must be. That is, heavier molecules have slower average speed than lighter ones. Figure 10.7 illustrates this relationship. The peak in the curve for the heaviest molecule, O_2, occurs at a much lower speed than for the lightest, He. The average speed for each type of molecule is a little faster than its most probable speed because both curves are asymmetric. You can also see from the graph that average speeds range from a few hundred to a few thousand meters per second.

Problem-Solving Example 10.3   Kinetic-Molecular Theory

Consider a sample of $N_2(g)$ at 25 °C. Answer each question and use the kinetic-molecular theory to explain your answer.

(a) How would raising the temperature to 30 °C affect the pressure of the gas?
(b) How would adding more N_2 molecules to the sample, with fixed temperature and volume, affect the pressure? Explain your answers on the molecular level using the kinetic-molecular theory.

Answer

(a) The pressure would increase.   (b) The pressure would increase.

Explanation

(a) Raising the temperature causes the N_2 molecules to move at higher average speed and higher average energy. They would therefore hit the walls of the container harder and more often, causing a higher pressure.
(b) Molecules in the new sample of N_2, with more molecules than originally, but at the same temperature, would still have the same average speed. However, there would be more molecules per unit time hitting the walls, and the pressure would be greater.

Problem-Solving Practice 10.3

How does kinetic-molecular theory explain the change in pressure when a sample of gas has its volume decreased while the temperature remains constant?

 Exercise 10.2   Seeing Through Gases

Explain why gases are transparent to light. (*Hint:* You may need to refer to Chapter 7 in addition to using some of the concepts in this section.)

 Exercise 10.3   The Kinetic-Molecular Theory

Use the kinetic-molecular theory to explain why the pressure goes down when gas molecules are removed from a sample of gas in a fixed-volume container at constant temperature.

Exercise 10.4   Molecular Kinetic Energies

Arrange the following gaseous substances in order of increasing average kinetic energy of their molecules at 25 °C: Cl_2, H_2, NH_3, SF_6.

 Exercise 10.5   Molecular Kinetic Energies

Using Figure 10.6 as a source of information, first draw a plot of number of molecules versus molecular speed for a sample of helium at 25 °C. Now assume that an equal number of molecules of argon, also at 25 °C, are added to the helium. What would the distribution curve for the mixture of gases look like?

 Exercise 10.6   Molecular Motion

Suppose you have two helium-filled balloons of about equal size, put one of them in the freezer compartment of your refrigerator, and leave the other one out in your room. After a few hours you take the balloon from the freezer and compare it to the one left out in the room. Based on the kinetic-molecular theory, what differences would you expect to see (a) immediately after taking the balloon from the freezer and (b) after the cold balloon warms to room temperature?

10.4 THE BEHAVIOR OF IDEAL GASES

The kinetic-molecular theory explains why gases behave as they do on the nanoscale. On the macroscale, gases have been studied for hundreds of years, and the properties that all gases display have been summarized into *gas laws,* which are named for their discoverers. Using the variables pressure, volume, temperature, and amount (number of moles), we can write equations that explain how gases behave. A gas that behaves exactly as described by these equations is called an **ideal gas.** At room temperature and atmospheric pressure, most gases behave nearly ideally. However, at pressures much higher than 1 atm or temperatures just above the boiling point, gases deviate from ideal behavior (Section 10.9). Each of the gas laws described in this section can be explained by the kinetic-molecular theory.

The Pressure-Volume Relationship: Boyle's Law

Boyle's law states that the volume (V) *of an ideal gas varies inversely with the applied pressure* (P) *when temperature* (T) *and amount* (n, *moles) are constant.*

$$V \propto \frac{1}{P} \quad \text{(unchanging } T \text{ and } n\text{)}$$

$$V = \text{constant} \times \frac{1}{P} \qquad PV = \text{constant}$$

The value of the constant depends on the temperature and the amount of the gas. The relationship between V and P is shown graphically in Figure 10.8a. Plotting V versus $1/P$ yields a linear relationship shown in Figure 10.8b. For a gas sample under two sets of pressure and volume conditions, Boyle's law can be written as $P_1V_1 = P_2V_2$.

In terms of the kinetic-molecular theory, a decrease in volume of a gas increases its pressure because there is less room for the gas molecules to move around before they collide with the walls of the container. Thus, there are more frequent collisions with the walls. These collisions produce pressure on the con-

CD-ROM Screen 12.3: Gas Laws

The symbol $\propto$ means "proportional to."

Boyle's law explains how expanding and contracting your chest cavity leads to inhalation and exhalation of air.

Figure 10.8 Graphical illustration of Boyle's law. (a) Volume (V) versus pressure (P) (b) V versus $1/P$. These curves represent the inverse proportionality between volume and pressure. As pressure increases, volume decreases.

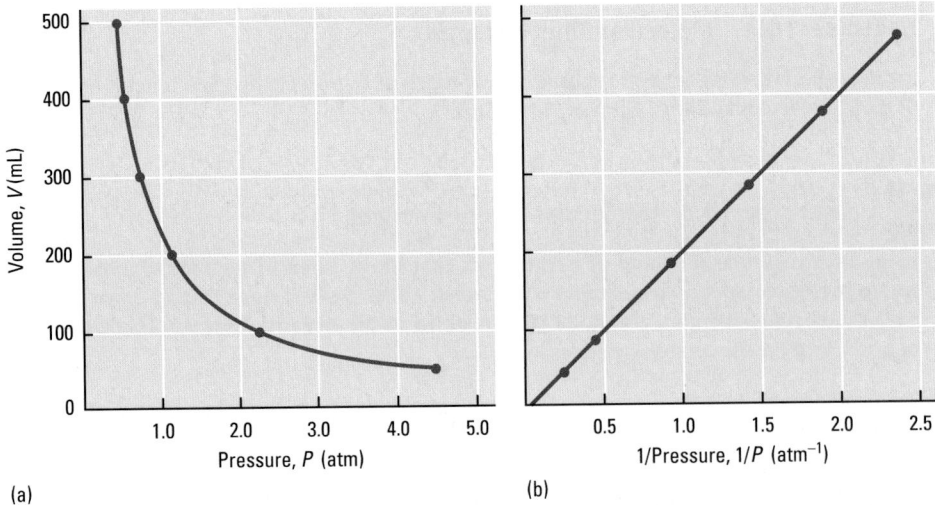

(a)

(b)

tainer walls, and more collisions mean a higher pressure. When you pump up a tire with a bicycle pump, the gas in the pump is squeezed into a smaller volume by application of pressure. This property is called *compressibility*. In contrast to gases, liquids and solids are only slightly compressible.

Robert Boyle studied the compressibility of gases in 1661 by pouring mercury into an inverted J-shaped tube containing a sample of trapped gas. Each time he added more mercury, the volume of the trapped gas decreased (Figure 10.9). The mercury additions increased the pressure on the gas and changed the gas volume in a predictable fashion.

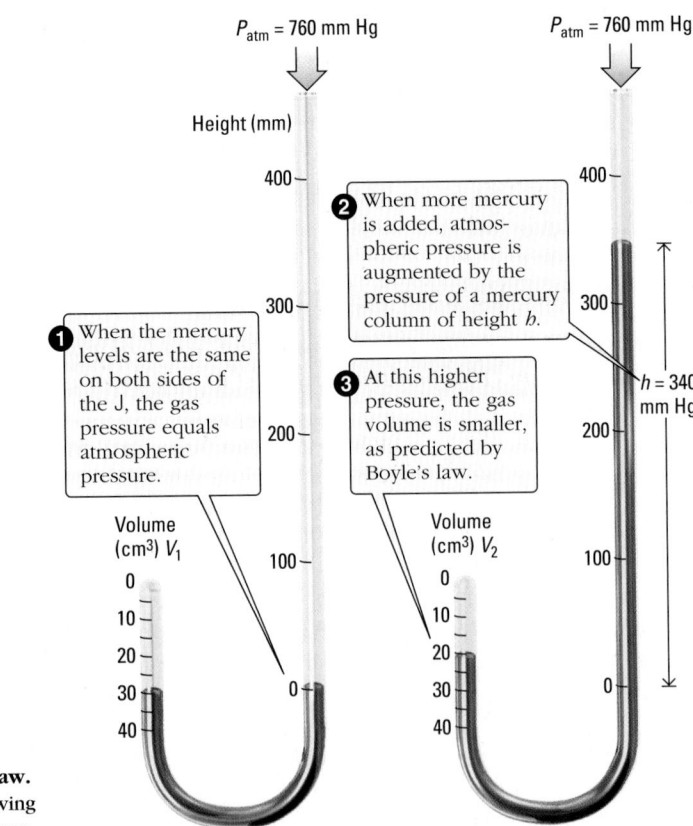

Figure 10.9 Boyle's law. Boyle's experiment showing the compressibility of gases.

Exercise 10.7   Visualizing Boyle's Law

Many cars have gas-filled shock absorbers to give the car and its occupants a smooth ride. If a four-passenger car is loaded with four NFL linemen, describe the gas inside the shock absorbers compared with when the car has no passengers.

The Temperature-Volume Relationship: Charles's Law

Charles's law *states that the volume* (V) *of an ideal gas varies directly with absolute temperature* (T) *when pressure* (P) *and amount* (n, *moles) are constant.*

CD-ROM Screen 12.3: Gas Laws

$$V \propto T \qquad \text{(unchanging } P \text{ and } n\text{)}$$

$$V = \text{constant} \times T \qquad \frac{V}{T} = \text{constant}$$

The value of the constant depends on pressure and the amount of the gas. If the volume, V_1, and temperature, T_1, of a sample of gas are known, then the volume, V_2, at some other temperature, T_2, at the same pressure is given by

$$\frac{V_1}{T_1} = \frac{V_2}{T_2} \qquad (P \text{ and } n \text{ constant})$$

In terms of the kinetic-molecular theory, higher temperature means faster molecular motion and a higher average kinetic energy. The more rapidly moving molecules therefore strike the walls of a container more often, and each collision exerts greater force. If the volume were held constant, this would result in a higher pressure. But for pressure to remain constant, the volume of the container must expand.

When using the gas law relationships, temperature must be expressed in terms of the **absolute temperature scale.** The zero on this scale is the lowest possible temperature. The unit of the absolute temperature scale is the kelvin, and its symbol is K (with no degree sign). The relationship between the absolute scale and the Celsius scale is shown in Figure 10.10. The kelvin is the same size as a degree

The absolute temperature scale is also called the **Kelvin temperature scale.** The kelvin is the SI unit of temperature.

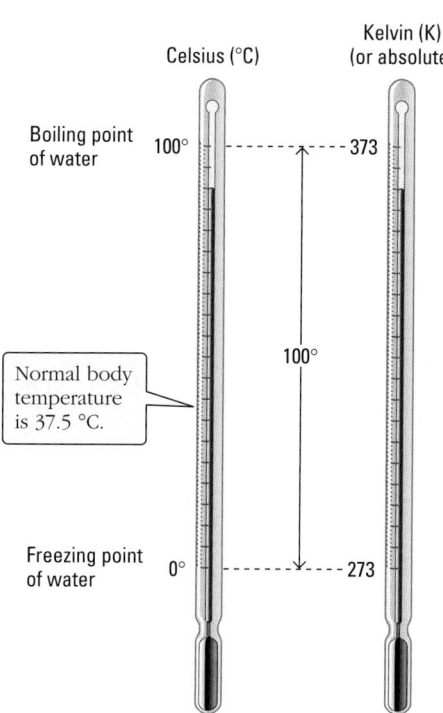

Figure 10.10 Temperature scales. The zero on the Celsius temperature scale is different from the zero on the Kelvin (absolute) scale. Zero on the absolute scale is the lowest possible temperature (0 K or −273.15 °C). The symbol for the absolute temperature unit is just K—there is no degree sign. The sizes of the kelvin and the degree Celsius are the same, so that calculated changes in temperature (Δ*T*) are the same on both scales.

Figure 10.11   Charles's law. The volumes of two different samples of gases decrease with decreasing temperature (at constant pressure). These graphs (as would those of all gases) intersect the temperature axis at about −273 °C.

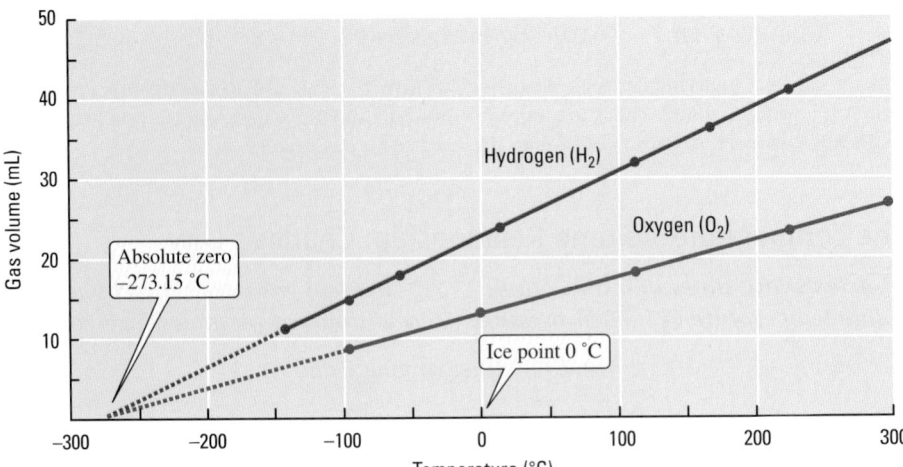

JACQUES ALEXANDRE CESAR CHARLES
(1746–1823)

The French scientist Charles was most famous in his lifetime for his experiments in ballooning. The first such flights were made by the Montgolfier brothers in June 1783, using a balloon of linen and paper and filled with hot air. In August 1783, however, Jacques Charles filled a silk balloon with hydrogen. Inflating the bag to its final diameter took several days and required nearly 500 lb of acid and 1000 lb of iron to generate the hydrogen gas. A huge crowd watched the ascent on August 27, 1783. The balloon stayed aloft for almost 45 min and traveled about 15 miles. However, when it landed in a village, the people there were so terrified they tore it to shreds.

The volume of any gas would appear to be zero at −273.15 °C. However, at sufficiently high pressure, all gases liquefy before reaching this temperature.

Celsius and so when ΔT is calculated by subtracting one temperature from another, the result is the same on both scales, even though the numbers involved are different. The lowest possible temperature, known as *absolute zero,* is −273.15 °C. Temperature on the Celsius scale T(°C) and on the absolute scale T(K) are related as follows.

$$T(K) = T(°C) + 273.15$$

Thus, 25.00 °C (a typical room temperature) is the same as 298.15 K.

In 1787, Jacques Charles discovered that the volume of a fixed quantity of a gas at constant pressure increases with increasing temperature. Figure 10.11 shows how the volumes of two different samples of the same number of moles of gas change with the temperature (pressure remains constant). When the plots of volume versus temperature for different gases are extended toward lower temperatures, they all reach zero volume at a common temperature, −273.15 °C.

Temperatures are expressed on the absolute scale, using kelvins, for gas law relationships. For example, suppose you want to calculate the new volume when 450.0 mL of a gas is cooled from 60.0 °C to 20.0 °C, at constant pressure. First, convert the temperatures to kelvins by adding 273.15 to the Celsius values: 60.0 °C becomes 333.2 K and 20.0 °C becomes 293.2 K. Using Charles's law for the two sets of conditions at constant pressure,

$$V_2 = \frac{V_1 T_2}{T_1} = \frac{450.0 \text{ mL} \times 293.2 \text{ K}}{333.2 \text{ K}} = 396.0 \text{ mL}$$

 Exercise 10.8   Visualizing Charles's Law

Consider a collection of gas molecules at some temperature, T_1. Now increase the temperature to T_2. Use the ideas of the kinetic-molecular theory and explain why the volume would have to be larger if the pressure remains constant. What would have to be done to maintain the volume constant if the temperature increased?

The Amount-Volume Relationship: Avogadro's Law

Avogadro's law states that the volume (V) of an ideal gas varies directly with amount (n) when temperature (T) and pressure (P) are constant.

$$V \propto n \qquad \text{(unchanging } T \text{ and } P\text{)}$$

$$V = \text{constant} \times n \qquad \frac{V}{n} = \text{constant}$$

 CD-ROM Screen 12.3: Gas Laws

The value of the constant depends on the temperature and the pressure. Avogadro's law means, for example, that at constant temperature and pressure, if the number of moles of gas doubles, the volume doubles. It also means that at the same temperature and pressure, the volumes of two different amounts of gases are related as follows:

$$\frac{V_1}{n_1} = \frac{V_2}{n_2} \qquad (T \text{ and } P \text{ constant})$$

In terms of the kinetic-molecular theory, increasing the number of gas molecules at a constant temperature means that the added molecules have the same average kinetic energy as the molecules to which they were added. As a result, the number of collisions with the container walls increases in proportion to a number of molecules. This would increase the pressure if the volume were held constant. To maintain constant pressure, the volume must increase.

In 1809 the French scientist Joseph Gay-Lussac (1778–1850) conducted experiments in which he measured the volumes of gases reacting with one another to form gaseous products. He found that, at constant temperature and pressure, the volumes of reacting gases were always in the *ratios of small whole numbers.* This is known as the **law of combining volumes.** For example, the reaction of 2 L of H_2 with 1 L of O_2 produces 2 L of water vapor. Similarly, the reaction of 4 L of H_2 with 2 L of O_2 produces 4 L of water vapor.

	2 H_2(g)	**+**	**O_2(g)**	**→**	**2 H_2O(g)**
Experiment 1:	2 L		1 L		2 L
Experiment 2:	4 L		2 L		4 L

In 1811 Amadeo Avogadro suggested that Gay-Lussac's observations actually showed that equal volumes of all gases under the same temperature and pressure conditions contain the same number of molecules. Viewed on a molecular scale, if a tiny volume contained only one molecule of a gas, then twice that volume would contain two molecules, and so on (Figure 10.12).

John Dalton (who devised the atomic theory) strongly opposed Avogadro's ideas and never did accept them. It took about 50 years—long after Avogadro and Dalton had died—for Avogadro's explanation of Gay-Lussac's experiments to be generally accepted.

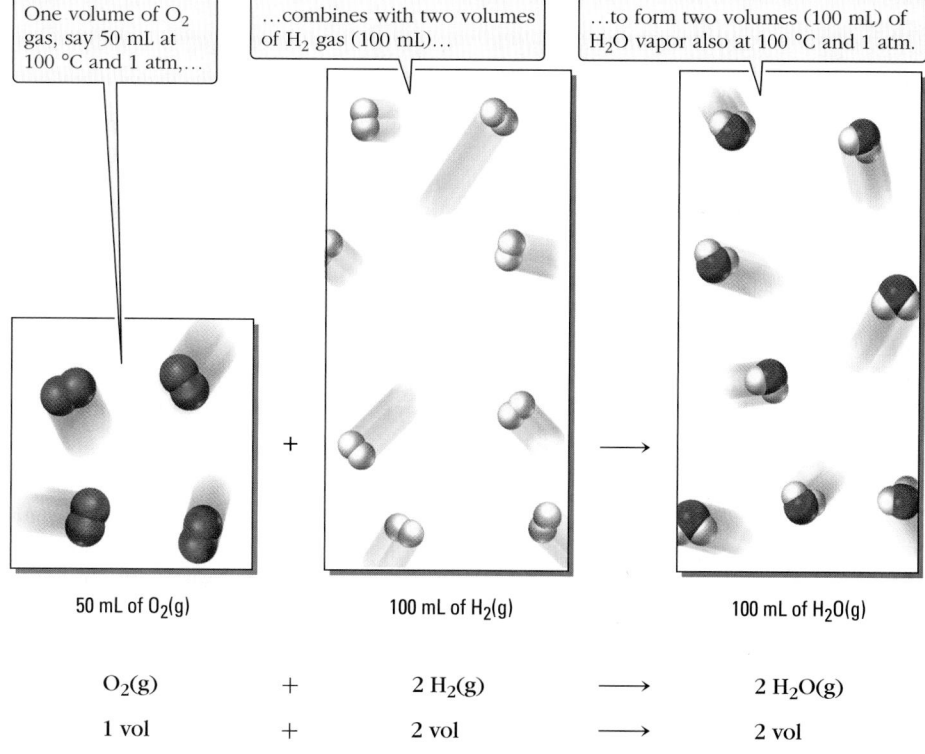

One volume of O_2 gas, say 50 mL at 100 °C and 1 atm,...

...combines with two volumes of H_2 gas (100 mL)...

...to form two volumes (100 mL) of H_2O vapor also at 100 °C and 1 atm.

50 mL of O_2(g)   +   100 mL of H_2(g)   →   100 mL of H_2O(g)

O_2(g)   +   2 H_2(g)   ⟶   2 H_2O(g)

1 vol   +   2 vol   ⟶   2 vol

Figure 10.12 Law of combining volumes. When gases at the same temperature and pressure combine with one another, their volumes are in the ratio of small whole numbers.

Problem-Solving Example 10.4 Using Avogadro's Law and the Law of Combining Volumes

Carbon monoxide burns in oxygen to form carbon dioxide.

$$2\ CO(g) + O_2(g) \longrightarrow 2\ CO_2(g)$$

If the volume occupied by the CO is 400. mL at 40 °C and 1 atm, what volume of O_2, at the same temperature and pressure, will be required in the reaction?

Answer   200. mL of O_2

Explanation   We want to find the volume of O_2, and the given information is the volume of the CO at the same temperature and pressure. Therefore, the problem can be solved by using the reaction coefficients of 1 for O_2 and 2 for CO as the volume ratio.

$$\text{Volume of } O_2 = 400.\ \text{mL CO} \times \left(\frac{1\ \text{mL } O_2}{2\ \text{mL CO}}\right) = 200.\ \text{mL } O_2$$

Problem-Solving Practice 10.4

Nitrogen monoxide (NO) combines with oxygen to form nitrogen dioxide.

$$2\ NO(g) + O_2(g) \longrightarrow 2\ NO_2(g)$$

If 1.0 L of oxygen gas at 30.25 °C and 0.975 atm is used, what volume of NO gas at the same temperature and pressure will be converted to NO_2?

 Exercise 10.9   Gas Burner Design

The gas burner in a stove or furnace admits enough air so that methane gas can react completely with oxygen according to the equation

$$CH_4(g) + 2\ O_2(g) \rightarrow CO_2(g) + 2\ H_2O(g)$$

Air is one-fifth oxygen by volume. Both air and methane gas are supplied to the flame by passing through separate small tubes. Compared with the tube for the methane gas, how much bigger does the cross section of the tube for the air need to be? Assume that both gases are at the same T and P.

Exercise 10.10   Filling Balloons

One hundred balloons of equal volume are filled with a total of 26.8 g of helium at 23 °C and 748 mm Hg. The total volume of these balloons is 168 L. Next, you are given 150 more balloons of the same size and 41.8 g of He gas. The temperature and pressure remain the same. Determine by calculation whether you will be able to fill all the balloons with the He you have available.

10.5 THE IDEAL GAS LAW

CD-ROM Screen 12.4: The Ideal Gas Law

The three gas laws focus on the effects of changes in P, T, or n on gas volume:

- Boyle's law and pressure $(V \propto 1/P)$
- Charles's law and temperature $(V \propto T)$
- Avogadro's law and amount (mol) $(V \propto n)$

The three gas laws can be combined to give the ideal gas law, which summarizes the relationships among them.

$$V \propto \frac{nT}{P} \quad \text{or} \quad PV \propto nT$$

To make this proportionality into an equation, a proportionality constant, *R*, named the **ideal gas constant,** is used. The equation becomes

$$V = R\frac{nT}{P}$$

and on rearranging, gives the equation called the **ideal gas law.**

$$PV = nRT$$

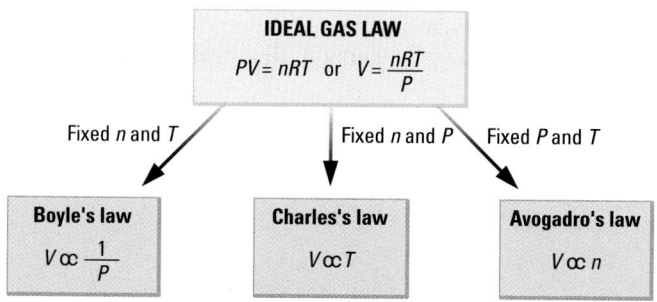

The ideal gas law correctly predicts the amount, pressure, volume, and temperature for samples of most gases at pressures of a few atmospheres or less and at temperatures well above their boiling points. The constant *R* can be calculated from the experimental fact that at 0 °C and 1 atm the volume of 1 mol of gas is 22.414 L. This temperature and pressure are called **standard temperature and pressure (STP),** and the volume is called the **standard molar volume.** Solving the ideal gas law for *R*, and substituting, we have

$$R = \frac{PV}{nT} = \frac{(22.414\ \text{L})(1\ \text{atm})}{(1\ \text{mol})(273.15\ \text{K})} = 0.082057\ \text{L atm mol}^{-1}\,\text{K}^{-1}$$

which is usually rounded to 0.0821 L atm mol^{-1} K^{-1}. The ideal gas constant has different numerical values in different units, as shown in Table 10.4.

The ideal gas law can be used to calculate *P*, *V*, *n*, or *T* whenever three of the four variables are known, provided that the conditions of temperature and pressure are not extreme.

TABLE 10.4	Values of *R*, in Different Units
$R = 0.08206\ \dfrac{\text{L atm}}{\text{mol K}}$	
$R = 62.36\ \dfrac{\text{torr L}}{\text{mol K}}$	
$R = 8.314\ \dfrac{\text{kPa dm}^3}{\text{mol K}}$	
$R = 8.314\ \dfrac{\text{J}}{\text{mol K}}$	

Problem-Solving Example 10.5   Using the Ideal Gas Law

What volume will 0.20 g of oxygen occupy at 1.0 atm and 20. °C?

Answer   0.15 L or 150 mL

Explanation   Use the ideal gas law, substitute known quantities and solve for the volume, *V*. When using $PV = nRT$ it is convenient to have all the variables in the same units as the gas constant, $R = 0.0821$ L atm mol^{-1} K^{-1}. Temperature should be in kelvins, pressure in atmospheres, volume in liters, and the amount of gas in moles.
   Begin by converting the mass of oxygen to moles.

$$(0.20\ \text{g O}_2)\left(\frac{1\ \text{mol O}_2}{32.00\ \text{g O}_2}\right) = 0.00625\ \text{mol O}_2$$

Next, convert the temperature to kelvins.

$$(20. + 273.15)\ \text{K} = 293.\ \text{K}$$

Now solve for *V* in the ideal gas law equation. The result will be in liters.

$$V = \frac{nRT}{P} = \frac{(0.00625\ \text{mol})(0.0821\ \text{L atm mol}^{-1}\,\text{K}^{-1})(293.\ \text{K})}{1.0\ \text{atm}} = 0.15\ \text{L} = 150\ \text{mL}$$

Notice how the units cancel. Note also that the answer is given with two significant digits because the mass of oxygen and the pressure had only two significant digits.

✔ The temperature and pressure are near STP. The mass of O_2 is a small fraction of a mole. So the volume should be a small fraction of 22.4 L, and it is.

Problem-Solving Practice 10.5

What volume will 2.64 mol of N_2 occupy at 0.640 atm pressure and 31 °C?

For many calculations involving a gas sample under two sets of conditions, it is convenient to use the ideal gas law in the following manner. For two sets of conditions (n_1, P_1, V_1, and T_1; n_2, P_2, V_2, and T_2), the ideal gas law can be written as

$$R = \frac{P_1V_1}{n_1T_1} \quad \text{and} \quad R = \frac{P_2V_2}{n_2T_2}$$

Since in both sets of conditions the quotient is equal to R, we can set the two quotients equal to each other.

$$\frac{P_1V_1}{n_1T_1} = \frac{P_2V_2}{n_2T_2}$$

When the amount of gas, n, is constant, so that $n_1 = n_2$, this equation simplifies to what is known as the **combined gas law:**

$$\frac{P_1V_1}{T_1} = \frac{P_2V_2}{T_2}$$

Problem-Solving Example 10.6   Pressure and Volume

Suppose you have a sample of a gas that occupies 100. mL at a pressure of exactly 3 atm. What volume will this gas occupy if the pressure is decreased to exactly 1 atm at the same temperature and if it obeys the ideal gas law?

Answer   300. mL

Explanation   The temperature (T) and the amount (n) are constant. The given conditions are $P_1 = 3$ atm, $P_2 = 1$ atm, $V_1 = 100.$ mL, and V_2 is unknown. The value of V_2 can be found by cancelling T in the combined gas law and rearranging $P_1V_1 = P_2V_2$ to solve for V_2.

$$V_2 = \frac{P_1V_1}{P_2} = \frac{3 \text{ atm} \times 100. \text{ mL}}{1 \text{ atm}} = 300. \text{ mL}$$

✔ The inverse relationship between P and V tells us that V_2 will be larger than V_1 because P_2 is smaller than P_1. The answer is reasonable.

Problem-Solving Practice 10.6

At a pressure of exactly 1 atm and some temperature, a gas sample occupies 400. mL. What will be the volume of the gas at the same temperature if the pressure is decreased to 0.750 atm?

Problem-Solving Example 10.7   Gas Thermometers

Because absolute temperature is proportional to volume at constant pressure and amount, a gas sample at constant pressure can serve as a thermometer. A certain gas sample occupies 100. mL at 25 °C. If the pressure is held constant and the temperature is changed, the sample occupies 175. mL. What is the final temperature in kelvins and in Celsius degrees?

Answer   521 K, or 248 °C

Explanation Since the volume of the gas increases, you should expect the final temperature to be higher. Begin by converting the starting temperature to kelvins.

$$T_1 = (25 + 273)K = 298 \text{ K}$$

The other variables are $V_1 = 100.$ mL and $V_2 = 175$ mL.

Solving the Charles's law relationship for the final temperature, T_2, gives

$$T_2 = \frac{T_1 V_2}{V_1} = \frac{(298 \text{ K})(175. \text{ mL})}{100. \text{ mL}} = 521 \text{ K, or } 248 \text{ °C}$$

✔ The final volume is larger, so the final temperature should be greater, and it is.

Problem-Solving Practice 10.7

If a gas sample occupies 236 mL at 31 °C, what volume will it occupy at 89 °C? The pressure remains constant.

Problem-Solving Example 10.8   The Combined Gas Law

Helium-filled balloons are used to carry scientific instruments high into the atmosphere. Suppose that such a balloon is launched on a summer day when the temperature at ground level is 22.5 °C and the barometer reads 754. mm Hg. If the balloon's volume is 1.00×10^6 L at launch, what will it be at a height of 37 km, where the pressure is 76.0 mm Hg and the temperature is 240. K?

Answer 8.05×10^6 L

Explanation Assume that no gas escapes from the balloon. Then only T, P, and V change. Using subscript 1 to indicate initial conditions (at launch) and subscript 2 to indicate final conditions (high in the atmosphere) gives

Initial: $P_1 = 754.$ mm Hg $V_1 = 1.00 \times 10^6$ L $T_1 = (22.5 + 273.15)$ K = 295.6 K

Final: $P_2 = 76.0$ mm Hg $T_2 = 240.$ K

Solving the combined gas law for V_2 gives

$$V_2 = \frac{P_1 V_1 T_2}{P_2 T_1} = \frac{(754. \text{ mm Hg})(1.00 \times 10^6 \text{ L})(240. \text{ K})}{(76.0 \text{ mm Hg})(295.6 \text{ K})} = 8.05 \times 10^6 \text{ L}$$

Thus, the final volume is about eight times larger. The volume has increased because the pressure has dropped. For this reason, weather balloons are never fully inflated at launch. A great deal of room has to be left so that the helium can expand at high altitudes.

✔ The pressure is dropping by a factor of about 10, and T is changing only 20%, so the volume should increase by a factor of about 10, and it does.

Problem-Solving Practice 10.8

A small sample of a gas is prepared in the laboratory and found to occupy 21 mL at a pressure of 710. mm Hg and a temperature of 22.3 °C. The next morning the temperature has changed to 26.5 °C, and the pressure is found to be 740. mm Hg. No gas has escaped from the container.

(a) What volume does the sample of gas now occupy?
(b) Assume that the pressure does not change. What volume does the gas occupy at the new temperature?

It is important to remember that Boyle's, Charles's, and Avogadro's laws do not depend on the identity of the gas being studied. These laws reflect properties of all gases and therefore must depend on the behavior of any gaseous atoms or molecules, regardless of their identities.

In summary, for problems involving gases, you have your choice of two useful equations:

- When three of the variables P, V, n, and T are given, and the value of the fourth variable is needed, use the ideal gas law, $PV = nRT$.

- When one set of conditions is given for a single gas sample and one of the variables under a new set of conditions is needed, then use the combined gas law and cancel any of the three variables that do not change.

$$\frac{P_1V_1}{T_1} = \frac{P_2V_2}{T_2} \qquad (n \text{ constant})$$

Exercise 10.11 Predicting Gas Behavior

Name three ways the volume occupied by a sample of gas can be decreased.

10.6 QUANTITIES OF GASES IN CHEMICAL REACTIONS

CD-ROM Screen 12.7: Gas Laws and Chemical Reactions

The law of combining volumes and the ideal gas law make it possible to use volumes as well as masses or molar amounts in calculations based on reaction stoichiometry ($\Longleftarrow$ *p. 124*). Consider the following balanced equation for the combination reaction of solid carbon and gaseous oxygen to give gaseous carbon dioxide.

$$C(s) + O_2(g) \longrightarrow CO_2(g)$$

Let's look at some of the questions that might be asked about this reaction, in which there is one gaseous reactant and one gaseous product. As with any problem to be solved, you have to recognize what information is known and what is needed. Then you may have to decide which relationship provides the connection between the two.

1. *How many liters of CO_2 are produced from 0.5 L of O_2 and excess carbon?* The known and needed information here are both volumes of gases, and the coefficients of the balanced chemical equation represent volumes of gases, so the law of combining volumes leads to the answer.
   Answer: 0.5 L of CO_2, because the reaction shows one mole of CO_2 formed for every mole of C that reacts — a 1:1 ratio.

   $$0.5 \text{ L } O_2 \times \frac{1 \text{ vol } CO_2}{1 \text{ vol } O_2} = 0.5 \text{ L } CO_2$$

2. *How many moles of O_2 are required to completely react with 4.00 g of C?* Here the known information is the mass of the solid reactant and the needed information is the moles of gaseous product. The answer is provided from the stoichiometric relationships between reactants and products from the balanced reaction, without using any of the gas laws.
   Answer: 0.333 mol of O_2, because 4.00 g of C is 0.333 mol of C, and 1 mol of O_2 is required for every 1 mol of C — a 1:1 ratio.

   $$4.00 \text{ g C} \times \frac{1 \text{ mol C}}{12.011 \text{ g C}} \times \frac{1 \text{ mol } O_2}{1 \text{ mol C}} = 0.333 \text{ mol } O_2$$

3. *What volume (in liters) of O_2 at STP is required to react with 12.011 g of C?* Because the needed information is the volume of a gaseous reactant, the law of

combining volumes can lead to the answer, but first the number of moles of carbon must be found.

Answer: 22.4 L of O_2, because 12.011 g of C is 1 mol of C which requires 1 mol of O_2, and the standard molar volume of any gas at STP is 22.4 L.

$$12.011 \text{ g C} \times \frac{1 \text{ mol C}}{12.011 \text{ g C}} \times \frac{1 \text{ mol O}_2}{1 \text{ mol C}} \times \frac{22.4 \text{ L O}_2}{1 \text{ mol O}_2} = 22.4 \text{ L O}_2$$

4. *What volume (in liters) of O_2 at 747 mm Hg and 21 °C is required to react with 12.011 g of C?* Realizing that 12.011 g of C is 1 mol of C and that 1 mol of O_2 will be required, you then use the ideal gas law to calculate the volume occupied by 1 mol of O_2 at the temperature and pressure given.

Answer: Convert the temperature to kelvins to get 294. K, then convert the pressure given to atmospheres to get 0.983 atm. Finally, use the ideal gas law to calculate the volume of oxygen required.

$$V_{O_2} = \frac{nRT}{P} = \frac{(1 \text{ mol})(0.0821 \text{ L atm mol}^{-1} \text{ K}^{-1})(294. \text{ K})}{0.983 \text{ atm}} = 24.5 \text{ L}$$

Problem-Solving Example 10.9 Gases and Stoichiometry

When a commercial drain cleaner that contains sodium hydroxide and small pieces of aluminum is poured into a clogged drain, the reaction that occurs is

$$2 \text{ Al(s)} + 2 \text{ NaOH(aq)} + 6 \text{ H}_2\text{O}(\ell) \longrightarrow 2 \text{ NaAl(OH)}_4\text{(aq)} + 3 \text{ H}_2\text{(g)}$$

The heat generated by the reaction helps the sodium hydroxide break up the grease, and the hydrogen gas being generated stirs up the mixture and speeds up the unclogging of the drain. If 6.5 g of Al and an excess of NaOH are used, what volume of gaseous H_2 measured at 742 mm Hg and 22.0 °C is produced?

Answer 9.0 L

Explanation The first step in solving this problem is to find the number of moles of H_2 generated, using the coefficients from the balanced equation. Once the number of moles of H_2 is known, the volume is obtained from the ideal gas law, $PV = nRT$.

$$6.5 \text{ g Al} \times \left(\frac{1.0 \text{ mol Al}}{27.0 \text{ g Al}} \right) \left(\frac{3 \text{ mol H}_2}{2 \text{ mol Al}} \right) = 0.362 \text{ mol H}_2$$

Next, solve $PV = nRT$ for V and substitute values for T, P, and n. When substituting into the rearranged gas equation, make sure the units of P, V, and T are compatible with the units of R. This means P must be converted to atmospheres.

$$P = (742 \text{ mm Hg}) \left(\frac{1 \text{ atm}}{760 \text{ mm Hg}} \right) = 0.976 \text{ atm}$$

$$V = \frac{nRT}{P} = \frac{(0.362 \text{ mol})(0.0821 \text{ L atm mol}^{-1} \text{ K}^{-1})(295.2 \text{ K})}{0.976 \text{ atm}} = 9.0 \text{ L}$$

✔ We have about $\frac{1}{4}$ mol of Al and therefore about $\frac{3}{8}$ mol of H_2 should be produced. Three-eighths of 22.4 L is about 8.4 L, which is comparable to our more accurate answer.

Problem-Solving Practice 10.9

Ammonium nitrate, NH_4NO_3, is an explosive that can undergo the following decomposition.

$$2 \text{ NH}_4\text{NO}_3\text{(s)} \longrightarrow 4 \text{ H}_2\text{O(g)} + \text{O}_2\text{(g)} + 2 \text{ N}_2\text{(g)}$$

If 10.0 g of NH_4NO_3 explodes, how many liters of gas are generated at 25 °C and 1 atm?

A drain cleaner containing sodium hydroxide and pieces of aluminum. When water is added, the hydroxide ion attacks the aluminum, and hydrogen gas is produced.

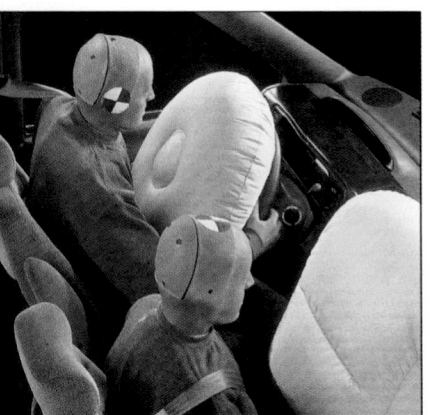

Air bags deploying on crash test dummies. The force of these bags deploying has caused the deaths of small children when they were placed in the front seat. Small children should always be placed in the rear seats of automobiles equipped with passenger-side air bags.

Problem-Solving Example 10.10   Stoichiometry with Gases—Air Bags

Automobile air bags are filled with N_2 from the decomposition of sodium azide.

$$2\ NaN_3(s) \longrightarrow 2\ Na(s) + 3\ N_2(g)$$

How many grams of sodium azide are needed to generate 60.0 L of N_2 at 1 atm and 26 °C?

Answer   106. g of NaN_3

Explanation   To calculate the grams of sodium azide required, we need the number of moles of N_2.

$$n = \frac{PV}{RT} = \frac{(1\ \text{atm})(60.0\ \text{L})}{(0.0821\ \text{L atm mol}^{-1}\ \text{K}^{-1})(299.15\ \text{K})} = 2.44\ \text{mol N}_2$$

Now we use the balanced equation to calculate the moles of sodium azide.

$$2.44\ \text{mol N}_2 \times \frac{2\ \text{mol NaN}_3}{3\ \text{mol N}_2} = 1.63\ \text{mol NaN}_3$$

We were asked for the mass of sodium azide.

$$1.63\ \text{mol NaN}_3 \times \frac{65.01\ \text{g NaN}_3}{1\ \text{mol NaN}_3} = 106.\ \text{g NaN}_3$$

(We should note that actual air bags use additional reactants because the reaction just described would produce elemental sodium, which is much too hazardous to remain in a consumer product. However, the actual reactions have nearly the same stoichiometric ratio of NaN_3 to N_2 as the simpler reaction used in this example.)

✔   Dividing the 106 g of NaN_3 by its molar mass of 65 g/mol gives an answer of about 1.6 mol of NaN_3. We get $\frac{3}{2}$ as many moles of N_2, or about 2.4 mol of N_2, which is about 54 L of N_2. This is close to the required amount of N_2.

Problem-Solving Practice 10.10

Lithium hydroxide is used in spacecraft to absorb the CO_2 exhaled by astronauts.

$$2\ LiOH(s) + CO_2(g) \longrightarrow Li_2CO_3(s) + H_2O(\ell)$$

What volume of CO_2 at 22 °C and 1 atm is absorbed per gram of LiOH?

Problem-Solving Example 10.11   Gas Law Stoichiometry and Temperature

Octane, C_8H_{18}, is one of the hydrocarbons in gasoline. In an automobile engine, octane burns to produce CO_2 and H_2O.

(a) What volume of oxygen, entering the engine at 0.950 atm and 20. °C, is required to burn 1.00 g of octane?

(b) Assuming the pressure is still 0.950 atm, what volume would be required on a cold winter day, when the temperature is −20. °C? The molar mass of octane is 114.2 g/mol.

Answer

(a) 2.77 L   (b) 2.39 L

Explanation

(a) First write the balanced equation for the combustion reaction.

$$2\ C_8H_{18}(\ell) + 25\ O_2(g) \longrightarrow 16\ CO_2(g) + 18\ H_2O(\ell)$$

Next, calculate the number of moles of O_2 required.

$$1.00\ \text{g octane} \left(\frac{1\ \text{mol octane}}{114.2\ \text{g octane}} \right) \left(\frac{25\ \text{mol O}_2}{2\ \text{mol octane}} \right) = 0.1095\ \text{mol O}_2$$

Then, using the ideal gas law, calculate the volume of oxygen.

$$V_{O_2} = \frac{nRT}{P} = \frac{(0.1095 \text{ mol } O_2)(0.0821 \text{ L atm mol}^{-1} \text{ K}^{-1})(293 \text{ K})}{0.950 \text{ atm}} = 2.77 \text{ L}$$

(b) At $-20.\,°C$ or 253 K, the same type of calculation gives

$$V_{O_2} = 2.39 \text{ L}$$

Because the volume of oxygen required (and hence of air) is significantly less in cold weather, many older gasoline engines have a choke, a valve that limits the air supply until the engine warms up. Current automobiles use electronics to control air flow.

✔ We have 1 g of octane, which is about 0.01 mol of octane. We require $\frac{25}{2}$ as much O_2, which is 0.125 mol of O_2 or about 2.8 L of O_2. This is close to our more accurate answer.

Problem-Solving Practice 10.11

If you carried out the Chemistry You Can Do stoichiometry experiment in Chapter 4 (⬅ *p. 25*), you observed the reaction of vinegar (acetic acid solution) with baking soda (sodium hydrogen carbonate) to generate carbon dioxide gas and inflate a balloon. The net ionic equation for an acid (⬅ *p. 165*), reacting with hydrogen carbonate ion is

$$H^+(aq) + HCO_3^-(aq) \longrightarrow CO_2(g) + H_2O(\ell)$$

If there is plenty of vinegar, what mass of $NaHCO_3$ is required to inflate a balloon to a diameter of 20. cm at room temperature, 20. °C? Assume that the balloon is a sphere. (The formula for the volume of a sphere is $V = \frac{4}{3}\pi r^3$.) Because the rubber is stretched, the pressure inside the balloon is twice the normal atmospheric pressure.

10.7 GAS DENSITY AND MOLAR MASSES

The ideal gas law can be used to relate gas density to molar mass. Density is mass per unit volume (⬅ *p. 7*). The densities of gases are extremely variable because the volume of a gas sample (but not its mass) varies with temperature and pressure. However, once T and P are specified, the density of a gas can be calculated from the ideal gas law. Additionally, because equal volumes of gas at the same T and P contain equal numbers of molecules, the densities of different gases are directly proportional to their molar masses. As a result, experimental gas densities can be used to determine molar masses.

To derive the relationship between gas density and molar mass, start with the ideal gas law, $PV = nRT$.

For any compound, the number of moles (n) equals its mass (m) divided by its molar mass (M), so we will substitute m/M for n in the ideal gas equation:

$$PV = \frac{m}{M} RT$$

Density (d) is defined as mass divided by volume (m/V). We can rearrange the equation so that m/V is on one side:

$$d = \frac{m}{V} = \frac{PM}{RT}$$

Thus, the density of a gas is directly proportional to its molar mass.

Let's use this equation to compute the approximate density of air at STP. The molar mass of air is estimated by the approximate weighted average of the molar masses of N_2 and O_2.

CD-ROM Screen 12.5: Gas Density

$$\text{Molar mass of air} = (0.80)(\text{molar mass of } N_2) + (0.20)(\text{molar mass of } O_2)$$

$$= (0.80)(28.01) + (0.20)(31.999) = 28.8 \text{ g/mol}$$

$$d = \frac{P}{RT} M = \frac{1 \text{ atm}}{(0.0821 \text{ L atm mol}^{-1} \text{ K}^{-1})(273.15 \text{ K})} (28.8 \text{ g/mol}) = 1.28 \text{ g/L}$$

Air has a density of approximately 1.28 g/L at STP.

Consider the densities of the three pure gases He, O_2, and SF_6. If we take 1 mol of each of these gases at 25 °C and 0.750 atm, we can see from the table below that density (at the same conditions) increases with molar mass.

Gas	Molar mass (g/mol)	Density (25 °C)
He	4.003	0.123 g/L
O_2	31.999	0.981 g/L
SF_6	146.06	4.48 g/L

Exercise 10.12   Calculating Gas Densities

Calculate the densities of Cl_2 and of SO_2 at 25 °C and 0.750 atm. Then calculate the density of Cl_2 at 35 °C and 0.750 atm and the density of SO_2 at 25 °C and 2.60 atm.

 ### Exercise 10.13   Comparing Densities

Express the gas density of He (0.123 g/L) in grams per milliliter (g/mL) and compare that value with the density of metallic lithium (0.53 g/mL). The mass of a Li atom is only a little more than that of a He atom. What do these densities tell you about how closely Li atoms are packed compared with He atoms when each element is in its standard state? To which of the concepts of the kinetic-molecular theory does this comparison apply?

 ### Exercise 10.14   Densities of Gas Mixtures

Assume a mixture of equal volumes of the two gases nitrogen and oxygen. Would the density of this gas mixture be higher, lower, or the same as the density of air at the same temperature and pressure? Explain your answer.

CD-ROM Screen 12.6: Using Gas Laws: Determining Molar Mass

Problem-Solving Example 10.12   Using the Ideal Gas Law to Calculate Molar Mass

A 0.100-g sample of a gaseous compound occupies 0.0470 L at 298 K and 755 mm Hg. The gas may be CH_2F_2, and evidence for this is sought by determining the molar mass of the gas. Use the information given to calculate the molar mass of the gas, and compare the calculated molar mass with the molar mass determined from the molecular formula.

Answer   Molar mass from gas laws = 52.4 g/mol. Molar mass from atomic weights = 52.02 g/mol.

Explanation   This problem is typical of the laboratory measurement of the molar mass of an unknown gas. Begin by organizing the data.

$$V = 0.0470 \text{ L} \qquad P = 755 \text{ mm Hg} \qquad T = 298 \text{ K} \qquad n = ?$$

Next, convert the pressure to atmospheres.

$$755 \text{ mm Hg} \left(\frac{1 \text{ atm}}{760 \text{ mm Hg}} \right) = 0.993 \text{ atm}$$

Then use the ideal gas law equation to find n:

$$\frac{(0.993 \text{ atm})(0.0470 \text{ L})}{(0.0821 \text{ L atm mol}^{-1} \text{ K}^{-1})(298 \text{ K})} = 0.00191 \text{ mol gas}$$

The mass of the sample is 0.100 g, so the molar mass is

$$\frac{0.100 \text{ g}}{0.00191 \text{ mol}} = 52.4 \text{ g/mol}$$

Finally, adding the atomic weights for the formula CH_2F_2 gives a molar mass of 52.02 g/mol, which is in close agreement with that determined using the gas laws.

✔ 0.05 L of CH_2F_2 is about 0.05/22.4 = 0.002 mol of CH_2F_2. If 0.1 g is 0.002 mol, then the molar mass is about 0.1/0.002 = 50, which is close to our more accurate answer.

Problem-Solving Practice 10.12

A flask contains 1.00 L of a gas at 0.850 atm and 20. °C. The mass of the gas is 1.13 g. What is the molar mass of the gas? What is its identity?

ESTIMATION Helium Balloon Buoyancy

How large a helium balloon is needed to lift a weather-instruments package weighing 500 lb? The buoyancy of He is the difference in density between air and He at the given temperature and pressure.

First, we must estimate the density of helium. Let's assume a temperature of 21 °C, and 1 atm pressure. Then 1 L of He contains

$$n = \frac{(1 \text{ atm})(1 \text{ L})}{(0.0821 \text{ L atm mol}^{-1}\text{K}^{-1})[(273 + 21)\text{K}]} = 0.0414 \text{ mol He}$$

To get helium's density, we need to know the mass of He in 1 L, so

$$0.0414 \text{ mol He} \times 4.00 \text{ g/mol} = 0.166 \text{ g He}$$

The density of He at 21 °C and 1 atm is 0.166 g/L. At these same conditions of temperature and pressure, the density of air (which we computed on p. 450) is approximately 1.20 g/L. The mass difference between 1 L of He and 1 L of air is 1.20 − 0.166 = 1.03 g. This is the buoyant (lifting) force of He per liter.

Now we can tackle the second part of the problem.

$$500 \text{ lb} \times 454 \text{ g/lb} = 2.27 \times 10^5 \text{ g}$$

Since He can lift 1.03 g/L, the total weight can be lifted by

$$\frac{2.27 \times 10^5 \text{ g}}{1.03 \text{ g/L}} = 2.20 \times 10^5 \text{ L He}$$

Let's change this to cubic meters.

$$2.20 \times 10^5 \text{ L He} \times 10^{-3} \text{ m}^3/\text{L} = 220 \text{ m}^3 \text{ He}$$

The volume and diameter of a spherical balloon are related by

$$V = \tfrac{4}{3}\pi r^3 = \tfrac{1}{6}\pi d^3$$

A weather balloon filled with helium. As it ascends into the troposphere, does the volume increase or decrease?

Rearranging this equation allows us to calculate the diameter of the balloon.

$$d = \left(\frac{6V}{\pi}\right)^{1/3} = \left(\frac{6 \times 220 \text{ m}^3}{3.14}\right)^{1/3} = 420^{1/3} \text{ m} = 7.49 \text{ m}$$

The diameter of a spherical He balloon that can lift 500 lb is approximately 7.5 m (approximately 25 feet).

10.8 GAS MIXTURES AND PARTIAL PRESSURES

CD-ROM Screen 12.8: Gas Mixtures and Partial Pressures

Our atmosphere is a mixture of nitrogen, oxygen, argon, carbon dioxide, water vapor, and small amounts of several other gases *(← p. 413, Table 10.3).* What we call atmospheric pressure is the sum of the pressures exerted by all these individual gases. The same is true of every gas mixture. Consider the mixture of nitrogen and oxygen illustrated in Figure 10.13. The pressure exerted by the mixture is equal to the sum of the pressures that the nitrogen alone and the oxygen alone would exert in the same volume at the same temperature and pressure. The pressure of one gas in a mixture of gases is called the **partial pressure** of that gas.

John Dalton was the first to observe that *the total pressure exerted by a mixture of gases is the sum of the partial pressures of the individual gases in the mixture*. This statement, known as **Dalton's law of partial pressures**, is a consequence of the fact that gas molecules behave independently of one another. As a demonstration of Dalton's law, consider the main components of our atmosphere, for which the total number of moles is approximately

$$n_{total} = n_{N_2} + n_{O_2} + n_{Ar}$$

If we replace n in the ideal gas law with n_{total}, the summation of the individual numbers of moles of gases, the equation becomes

$$P_{total}V = n_{total}RT$$

$$P_{total} = \frac{n_{total}RT}{V} = \frac{(n_{N_2} + n_{O_2} + n_{Ar})RT}{V}$$

Expanding the right side of this equation and rearranging,

$$P_{total} = \frac{n_{N_2}RT}{V} + \frac{n_{O_2}RT}{V} + \frac{n_{Ar}RT}{V} = P_{N_2} + P_{O_2} + P_{Ar}$$

The quantities P_{N_2}, P_{O_2}, and P_{Ar} are the partial pressures of the three major components of the atmosphere. Dalton's law means that the pressure exerted by the atmosphere is the sum of the pressures due to nitrogen, oxygen, argon, and the other much less abundant components.

We can write a ratio of the partial pressure of one of the components, A, of a gas mixture over the total pressure,

$$\frac{P_A}{P_{total}} = \frac{n_A(RT/V)}{n_{total}(RT/V)}$$

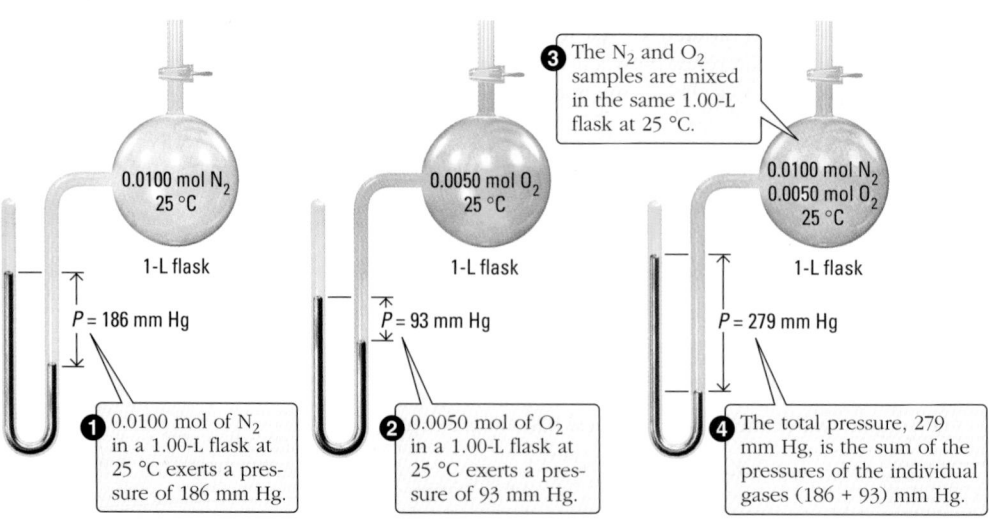

Figure 10.13 Dalton's law.

On canceling terms on the right-hand side of this equation, we get

$$\frac{P_A}{P_{total}} = \frac{n_A}{n_{total}}$$

The ratio of the pressures is the same as the ratio of the moles of gas A to the total number of moles. This ratio (n_A/n_{total}) is called the **mole fraction** of A and is given the symbol X_A. Hence, rearranging the equation gives

$$P_A = X_A P_{total}$$

In the gas mixture in Figure 10.13, the mole fractions of nitrogen and oxygen are

$$X_{N_2} = \frac{0.010 \text{ mol } N_2}{0.010 \text{ mol } N_2 + 0.0050 \text{ mol } O_2} = 0.67$$

$$X_{O_2} = \frac{0.0050 \text{ mol } O_2}{0.010 \text{ mol } N_2 + 0.0050 \text{ mol } O_2} = 0.33$$

Because these two gases are the only components of the mixture, the sum of the two mole fractions must equal 1:

$$X_{N_2} + X_{O_2} = 0.67 + 0.33 = 1.00$$

An interesting application of partial pressures is the composition of the breathing atmosphere in deep-sea–diving vessels. If normal air at 1 atm pressure, with an oxygen mole fraction of 0.21, is compressed to 2 atm, the partial pressure of oxygen becomes about 0.4 atm. Such high oxygen partial pressures are toxic, so a diluting gas must be added to lower the oxygen partial pressure to near-normal values. Nitrogen gas might seem the logical choice because it is the diluting gas in the atmosphere. The problem is that nitrogen is fairly soluble in the blood and at high concentrations causes *nitrogen narcosis*, a condition similar to alcohol intoxication.

Helium is much less soluble in the blood and is therefore a good substitute for nitrogen in a deep-sea–diving atmosphere. However, using helium leads to interesting side effects. Because He atoms, on average, move faster than the heavier nitrogen atoms at the same temperature (Figure 10.7), they strike a diver's skin more often than would the nitrogen atoms and are therefore more efficient at carrying away heat energy. This causes divers to complain of feeling chilled while breathing a helium/oxygen mixture.

An undersea explorer. Some deep-sea–diving vessels like this one use an atmosphere of oxygen and helium for their occupants.

Problem-Solving Example 10.13 Calculating Partial Pressures

Halothane ($F_3C—CHBrCl$) is a commonly used surgical anesthetic delivered by inhalation. What is the partial pressure of each gas if 15.0 g of halothane gas is mixed with 22.6 g of oxygen gas and the total pressure is 862 mm Hg?

Answer $P_{halothane} = 83.8$ mm Hg $P_{O_2} = 778$ mm Hg

Explanation First calculate the moles of each gas, then calculate the mole fractions.

$$15.0 \text{ g} \times \left(\frac{1 \text{ mol halothane}}{197.4 \text{ g}}\right) = 0.0760 \text{ mol halothane}$$

$$22.6 \text{ g} \times \left(\frac{1 \text{ mol } O_2}{32.00 \text{ g}}\right) = 0.706 \text{ mol } O_2$$

$$\frac{0.0706 \text{ mol halothane}}{0.782 \text{ total moles}} = 0.0972$$

Because the sum of the two mole fractions must equal 1.000, the mole fraction of O_2 is 0.903.

$$X_{halothane} + X_{O_2} = 1.000 = 0.0972 + X_{O_2}$$

$$X_{O_2} = 0.903$$

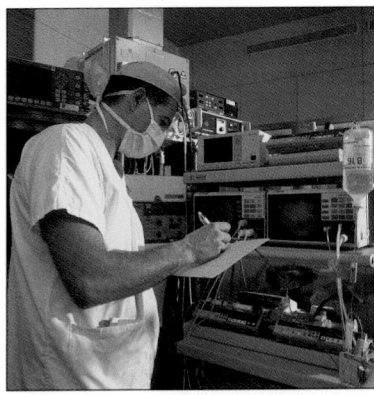

A gas-mixing manifold for anesthesia. Such equipment is used by an anesthesiologist to prepare a gas mixture to keep a patient unconscious during an operation. By proper mixing, the anesthetic gas can be added slowly to the breathing mixture. Near the end of the operation, the anesthetic gas can be replaced by air of normal composition or by pure oxygen.

Finally, calculate the partial pressure of each gas.

$$P_{halothane} = 0.0972 \times P_{total} = 0.0972 \times (862 \text{ mm Hg}) = 83.8 \text{ mm Hg}$$
$$P_{O_2} = 0.903 \times P_{total} = 0.903 \times (862 \text{ mm Hg}) = 778 \text{ mm Hg}$$

 We have a 10:1 mole ratio of O_2 to halothane, so the mole fraction should be 10:1 and the ratio of partial pressures roughly 10:1, and they are.

Problem-Solving Practice 10.13

A mixture of 7.0 g of N_2 and 6.0 g of H_2 is confined in a 5.0-L reaction vessel at 500. °C. Assume that no reaction occurs and calculate the total pressure. Then calculate the mole fraction and partial pressure of each gas.

Exercise 10.15 Pondering Partial Pressures

What happens to the partial pressure of each gas in a mixture when the volume is decreased by (a) lowering the temperature or (b) increasing the total pressure?

Exercise 10.16 Partial Pressures

A 355-mL flask contains 0.146 g of neon gas, Ne, and an unknown amount of argon gas, Ar, at 35 °C and a total pressure of 626 mm Hg. How many grams of Ar are in the flask?

Problem-Solving Example 10.14 Partial Pressure

The reaction between nitric oxide, NO, and oxygen gas to form NO_2 is environmentally important.

$$2 \text{ NO}(g) + O_2(g) \longrightarrow 2 \text{ NO}_2(g)$$

Suppose that the two reactant gases are kept in separate containers, as shown below. Then suppose the valve is opened and the reaction proceeds to completion. Assume a constant temperature of 25 °C.

(a) What gases remain at the end of the reaction, and how much of each of them is present?
(b) What are their partial pressures, and what is the total pressure in the system?

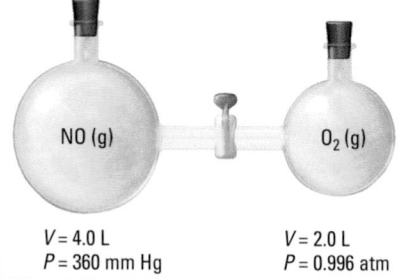

Answer
(a) No NO remains, 0.0427 mol O_2, and 0.0774 mol NO_2.
(b) $P_{NO_2} = 0.316$ atm; $P_{O_2} = 0.174$ atm; $P_{total} = 0.490$ atm.

Explanation (a) First, calculate the number of moles of each gas.

$$\frac{(360/760) \text{ atm } (4.0 \text{ L})}{(0.0821 \text{ L atm mol}^{-1}\text{K}^{-1})(298 \text{ K})} = 0.0774 \text{ mol NO}$$

$$\frac{(0.996 \text{ atm})(2.0 \text{ L})}{(0.0821 \text{ L atm mol}^{-1} \text{ K}^{-1})(298 \text{ K})} = 0.0814 \text{ mol O}_2$$

So NO is the limiting reactant, and it will be completely consumed by the reaction. The amount of oxygen consumed will be

$$0.0774 \text{ mol NO} \times \frac{1 \text{ mol O}_2}{2 \text{ mol NO}} = 0.0387 \text{ mol O}_2$$

Subtracting the amount used from the original amount gives $0.0814 - 0.0387 = 0.0427$ mol O_2 left when the reaction is complete. The amount of NO_2 formed is

$$0.0774 \text{ mol NO} \times \frac{2 \text{ mol NO}_2}{2 \text{ mol NO}} = 0.0774 \text{ mol NO}_2 \text{ formed}$$

At the end, we have 0.0427 mol of O_2, 0.0774 mol of NO_2, and no NO gas.
(b) The pressures are

$$\frac{(0.0774 \text{ mol})(0.0821 \text{ L atm mol}^{-1} \text{ K}^{-1})(298 \text{ K})}{6.0 \text{ L}} = 0.316 \text{ atm NO}_2$$

$$\frac{(0.0427 \text{ mol})(0.0821 \text{ L atm mol}^{-1} \text{ K}^{-1})(298 \text{ K})}{6.0 \text{ L}} = 0.174 \text{ atm O}_2$$

The total pressure of the system is the sum of the partial pressures.

$$P_{\text{total}} = 0.316 \text{ atm} + 0.174 \text{ atm} = 0.490 \text{ atm}$$

Problem-Solving Practice 10.14

Consider the case with the same reactants in the same vessels as above, but with the NO pressure being 1.00 atm and the O_2 pressure being 0.400 atm. What would the final pressure in the system be after the reaction is complete?

Collecting Gases Over Water

While studying chemical reactions that produce a gas as a product, it is often necessary to determine the number of moles of the product gas. One convenient way to do this, for gases that are insoluble in water, involves collecting the gas over water (Figure 10.14) The gas bubbles through the water and is collected by displacing the water in an inverted vessel that was filled with water to start. The levels of the water inside and outside the collection vessel are made equal. This ensures that the total pressure inside the vessel is equal to the barometric pressure. The volume of gas collected is then determined. Because some of the water evaporates,

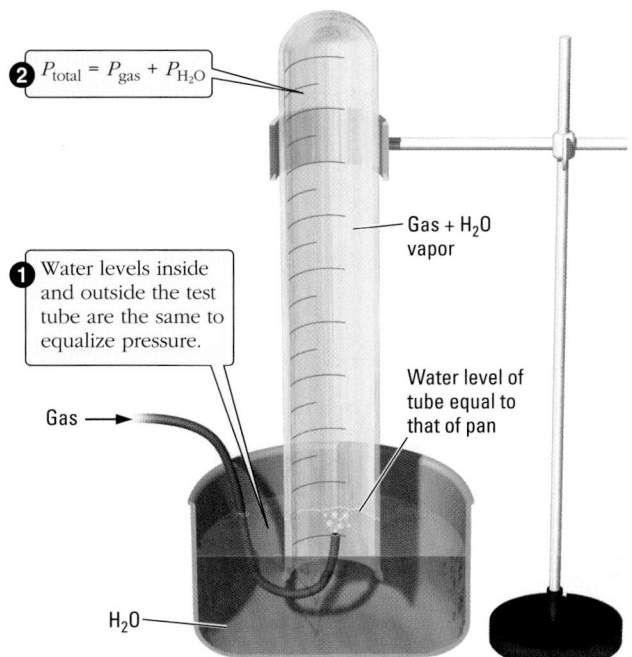

❷ $P_{\text{total}} = P_{\text{gas}} + P_{\text{H}_2\text{O}}$

Gas + H₂O vapor

❶ Water levels inside and outside the test tube are the same to equalize pressure.

Water level of tube equal to that of pan

Gas →

H₂O

Figure 10.14 Collecting a gas over water.

TABLE 10.5	Vapor Pressure of Water at Different Temperatures	
T(°C)	**P_{water} (mm Hg)**
0	4.6
20	17.5
21	18.6
22	19.8
23	21.1
24	22.4
25	23.8
100	760.0

forming water vapor, the total pressure of the mixture of gases equals the partial pressure of the gas of interest plus the partial pressure of the water. The partial pressure of water in the mixture is the vapor pressure of water, and it depends on the temperature of the liquid water (Table 10.5).

To calculate the amount of gaseous product, we first subtract the vapor pressure of the water from the total gas pressure (the barometric pressure), which yields the partial pressure of the gaseous product. Substituting this partial pressure and the known volume and temperature into the ideal gas law allows the calculation of the number of moles of the gaseous product.

$$H—C\equiv C—H$$

acetylene

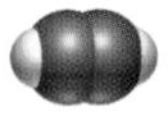

Problem-Solving Example 10.15 Collecting a Gas over Water

Before electric lamps were available, the following reaction between calcium carbide, CaC_2, and water was used to produce acetylene, C_2H_2, to burn a bright flame in miner's lamps.

$$CaC_2(s) + 2\,H_2O(\ell) \longrightarrow C_2H_2(g) + Ca(OH)_2(aq)$$

The reaction produced 385 mL of gas when the pressure had been adjusted to match barometric pressure (750 mm Hg). The water and gas temperature was 23 °C. How many milligrams of acetylene were generated? The vapor pressure of water at 23 °C is 21 mm Hg.

Answer 396 mg acetylene

Explanation We need to find the number of moles of acetylene, so we need the partial pressure of the acetylene. The total pressure is 750 mm Hg, and the vapor pressure of water is 21 mm Hg, so the partial pressure of the acetylene is

$$P_{acetylene} = P_{total} - P_{water} = 750 \text{ mm Hg} - 21 \text{ mm Hg} = 729 \text{ mm Hg}$$

Then the number of moles and the mass of acetylene are

$$\frac{(729/760 \text{ atm})(0.385 \text{ L})}{(0.0821 \text{ L atm mol}^{-1}\text{K}^{-1})(273 + 23)\text{K}} = 0.0152 \text{ mol acetylene}$$

$$0.0152 \text{ mol} \times 26.04 \text{ g/mol} = 0.396 \text{ g} = 396 \text{ mg acetylene}$$

✔ The quantities entered into the ideal gas equation can be approximated as

$$n = \frac{(1)(0.4)}{(0.0821)(300)} = 0.016 \text{ mol, which, when converted to milligrams, is close to our}$$

more accurate answer.

Problem-Solving Practice 10.15

Zinc metal reacts with HCl to produce hydrogen gas, H_2.

$$2\,HCl(aq) + Zn(s) \longrightarrow ZnCl_2(aq) + H_2(g)$$

which can be collected over water. If you collected 260 mL of H_2 at 23 °C, and the total pressure was 740 mm Hg, how many mg of H_2 were collected?

10.9 THE BEHAVIOR OF REAL GASES

The ideal gas law provides accurate predictions for the pressures, volumes, temperatures, and amounts of gases for pressures of a few atmospheres or less and temperatures well above a substance's boiling point. At STP (0 °C and 1 atm), most gases deviate only slightly from ideal behavior. However, at much higher pressures or much lower temperatures, the ideal gas law does not work nearly as well. We can illustrate this breakdown by plotting the quantity PV/RT for one mole of a gas as a function of pressure (Figure 10.15). For a gas that follows ideal gas law behavior, the quantity PV/RT must always equal 1. But we see that for real gases, the quantity deviates from 1, first by dipping to lower values and then rising to higher values as pressure increases.

To see what causes these deviations, we must revisit two of the fundamental concepts of kinetic-molecular theory (KMT) *(⟸ p. 414)*. KMT assumes that the molecules of a gas occupy no volume themselves and that gas molecules do not attract one another.

At standard temperature and pressure (STP), the volume occupied by a single molecule is very small relative to its share of the total gas volume. Recall that there are 6.02×10^{23} molecules in a mole and that 1 mol of a gas occupies about 22.4 L (22.4×10^{-3} m³) at STP *(⟸ p. 427)*. The volume, V, that each molecule has to move around in is given by

$$V = \frac{22.4 \times 10^{-3}\,\text{m}^3}{6.02 \times 10^{23}\,\text{molecules}} = 3.72 \times 10^{-26}\,\text{m}^3/\text{molecule}$$

If this volume is assumed to be a sphere, then the radius, r, of the sphere is about 2000 pm. The radius of the smallest gas molecule, the helium atom, is 31 pm *(⟸ p. 295)*, so a helium atom has a space to move around in that is similar to the room a pea has inside a basketball. Now suppose the pressure is increased significantly, to 1000 atm. The volume available to each molecule is now a sphere only about 200 pm in radius, which means the situation is now like that of a pea inside a sphere a bit larger than a ping-pong ball. The volume occupied by the gas molecules themselves relative to the volume of the sphere is no longer negligible. This violates the first concept of the kinetic-molecular theory. The kinetic-molecular theory and the ideal gas law deal with the volume available for the molecules to move around in, not the volume of the molecules themselves. However, the measured volume of the gas must include both. Therefore, at very high pressures, the measured volume will be larger than predicted by the ideal gas law, and the value of PV/RT is greater than 1.

The volume of a sphere is given by $\frac{4}{3}\pi r^3$. Solving for r gives

$$r = \sqrt[3]{\frac{3V}{4\pi}} = \sqrt[3]{\frac{3(3.72 \times 10^{-26}\,\text{m}^3)}{4(3.14)}}$$

$$= 2.11 \times 10^{-9}\,\text{m} = 2110\,\text{pm}$$

A pea in a basketball corresponds roughly to the relative volume that gas molecules have to move about in without striking another gas molecule at STP.

When the pressure is increased to 1000 atm, the relative volumes are like a pea in a ping-pong ball.

A cut-away view of a pea inside a basketball and inside a ping-pong ball.

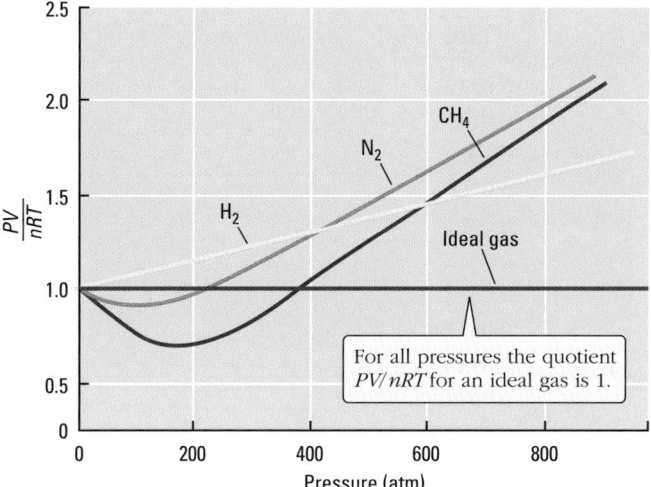

Figure 10.15 The nonideal behavior of real gases compared with that of an ideal gas.

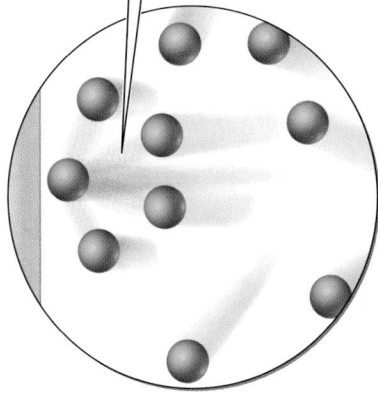

A gas molecule strikes the walls of a container with less force due to the attractive forces between it and its neighbors.

Figure 10.16 Nonideal gas behavior.

Attractions between molecules (⇐ *p. 383*). cause PV/RT to drop below the ideal value of 1 at low temperatures and medium pressures. Consider a gas molecule that is about to hit the wall of the container, as shown in Figure 10.16. The kinetic-molecular theory assumes that the other molecules exert no forces on the molecule, but in fact such forces do exist (⇐ *p. 383*). Their influence increases as higher pressures push gas molecules closer to each other. This means that when a molecule is about to hit the wall, the greater the pressure, the more the other molecules pull it *away* from the wall. This causes the molecule to hit the wall with less impact. The collision is softer than if there were no attraction among the molecules. Since all collisions with the walls are softer, the internal pressure is less than that predicted by the ideal gas law, and PV/RT is less than 1. As the external pressure increases, the gas volume decreases, the molecules are squeezed closer together, and the attractions among the molecules get stronger, which makes this deviation from ideal behavior larger.

An equation that predicts the behavior of real gases quantitatively by taking these two effects into account is the **van der Waals equation**.

$$\left(P + \frac{n^2 a}{V^2}\right)(V - nb) = nRT$$

Correction for molecular attraction (adjusts *P* up)

Correction for volume of molecules (adjusts *V* down)

The volume is decreased by the factor nb, which accounts for the volume occupied by the gas molecules themselves. The van der Waals constant b has units of L/mol. The pressure is increased by the factor n^2a/V^2, which accounts for the attractive forces between the gas molecules. The van der Waals constant a has units of L^2 atm/mol^2. The a correction has this form because the attractive forces being accounted for must occur between pairs of molecules, which means that the correction increases as the square of the number of molecules per unit volume $(n/V)^2$. The value of a shows how strongly the gas molecules attract each other. The constants a and b are different for each gas (Table 10.6) and must be experimentally determined. In general, the values for a and b increase as the gas mole-

TABLE 10.6	Van der Waals Constants for Some Common Gases	
Gas	$a \left(\dfrac{L^2 \, atm}{mol^2} \right)$	$b \left(\dfrac{L}{mol} \right)$
He	0.034	0.0237
Ne	0.211	0.0171
Ar	1.35	0.0322
H_2	0.244	0.0266
N_2	1.39	0.0391
O_2	1.36	0.0318
Cl_2	6.49	0.0562
CO_2	3.59	0.0427
CH_4	2.25	0.0428
NH_3	4.17	0.0371
H_2O	5.46	0.0305

cule's mass increases and as the molecule's complexity increases. Larger molecules occupy more volume, but they also tend to have larger intermolecular interactions.

Note that for an ideal gas, where the kinetic-molecular theory of gases holds, the attractive forces between molecules are negligible, and therefore $\left(P + \dfrac{n^2a}{V^2} \right) \approx P$. The volume of the molecules themselves is negligible compared with the container volume, so $(V - nb) \approx V$. Therefore, for ordinary conditions of P and T, the van der Waals equation reduces to the ideal gas law.

Problem-Solving Example 10.16 Using the van der Waals Equation

What is the pressure of 1.00 mol of CH_4 in 2.00 L at 273 K when calculated using the ideal gas law and the van der Waals equation? How much difference is there between the two calculated pressures? For CH_4, the values for the van der Waals constants are $a = 2.25\ L^2\ atm/mol^2$ and $b = 0.0428\ L/mol$.

Answer Ideal gas law pressure = 11.2 atm. van der Waals equation pressure = 10.9 atm. The difference is a 2.7% lower pressure from the van der Waals equation as compared with the ideal gas law.

Explanation First, we use the ideal gas law equation to find the pressure.

$$P = \frac{nRT}{V} = \frac{(1.00\ atm)(0.0821\ L\ atm\ mol^{-1}\ K^{-1})(273\ K)}{2.00\ L} = 11.2\ atm$$

Now we compute the pressure using the van der Waals equation.

$$P = \left(\frac{nRT}{V - nb} \right) - \left(\frac{n^2a}{V^2} \right)$$

$$P = \left(\frac{(1.00\ mol)(0.0821\ L\ atm\ mol^{-1}\ K^{-1})(273\ K)}{(2.00\ L) - (1.00\ mol)(0.0428\ L/mol)} \right)$$

$$- \left(\frac{(1.00\ mol)^2(2.25\ L^2\ atm/mol^2)}{(2.00\ L)^2} \right) = 10.9\ atm$$

The first term, containing the constant b, equals 11.5 atm, and the second term, containing the constant a, equals 0.56 atm. Thus, the term with a that accounts for intermolecular forces is more important. The overall percentage difference in pressure is

$$\text{Percentage difference in pressure} = \frac{11.2\ atm - 10.9\ atm}{11.2\ atm} \times 100\% = 2.7\%$$

Problem-Solving Practice 10.16

What is the percentage difference between the pressures predicted by the two equations if the volume is increased by a factor of ten to 20.0 L?

Exercise 10.17   Errors Caused by Deviations from Ideal Gas Behavior

In Problem-Solving Example 10.12, the molar mass calculated using the ideal gas law was slightly larger than that calculated using atomic weights. This means that the number of moles, n, must have been too small. Look at the Explanation for Problem-Solving Example 10.12 and explain why the number of moles is less than it should be.

10.10 CHEMICAL REACTIONS IN THE ATMOSPHERE

The molecules in the atmosphere are continually moving and colliding with one another, as described by the kinetic-molecular theory. At ordinary temperatures, most collisions fail to produce chemical reactions. However, the atmosphere is bathed in a constant flux of light photons during the daylight hours. The absorption of light energy by these molecules in the atmosphere can cause reactions, called **photochemical reactions**, that would not otherwise occur at normal atmospheric temperatures. Such reactions play an important role in determining the composition of the atmosphere itself and the fate of many chemical species that contribute to air pollution.

Nitrogen dioxide (NO_2) is one of the most photochemically active species in the atmosphere. The NO_2 molecule is an example of a *free radical* because it contains an unpaired electron ($\Leftarrow$ *p. 346*) represented by a · next to its formulas. When an NO_2 molecule absorbs a photon of light, $h\nu$, the molecule is raised to a higher energy level; it becomes an **electronically excited molecule**, designated by an asterisk (*).

$$\cdot NO_2(g) \xrightarrow{h\nu} \cdot NO_2^*$$

The excited molecule, $\cdot NO_2^*$, may quickly re-emit a photon of light, or the energy may break an N—O bond to form a nitrogen monoxide (NO) molecule and an oxygen atom (O).

$$\cdot NO_2^*(g) \longrightarrow \cdot NO(g) + \cdot O \cdot (g)$$

Both NO and O are free radicals, because they have one or more unpaired electrons.

Another mechanism of formation of radicals is **photodissociation**, in which a molecule absorbs an ultraviolet photon and produces two free radicals as products. Molecular oxygen can photodissociate to form two oxygen atoms

$$O_2(g) \xrightarrow{h\nu} \cdot O \cdot (g) + \cdot O \cdot (g)$$

Some free radicals, such as the oxygen atom, react with another atom or molecule almost immediately. Others, such as the NO_2 molecule, are not quite so reactive and are stable enough to exist for a longer time. Most radicals are highly reactive and short-lived. A radical usually reacts in either of two ways:

1. It combines with another radical. Each of the radicals contributes an electron to the formation of a bond, as when two NO_2 molecules combine to form N_2O_4 (dinitrogen tetraoxide).

$$\cdot NO_2(g) + \cdot NO_2(g) \longrightarrow N_2O_4(g)$$

2. It reacts with a molecule to form one or more new radicals or a new molecule, as illustrated by three important atmospheric reactions of the oxygen atom. The first is the formation of ozone, O_3, from O_2 and O.

$$\cdot O \cdot (g) + O_2(g) \longrightarrow O_3(g)$$

The second is the formation of two hydroxyl radicals (HO·) by the reaction with H_2O,

$$\cdot O \cdot (g) + H_2O(g) \longrightarrow 2 \cdot OH(g)$$

and the third is the formation of a new molecule by reacting with SO_2 to form SO_3, a precursor to acid rain (Section 17.3).

$$\cdot O \cdot (g) + SO_2(g) \longrightarrow SO_3(g)$$

Recall that Lewis dot structures for some simple molecules like NO and NO_2 have unpaired electrons ($\Leftarrow$ *p. 346*).

The oxygen atom is unusual because it contains two unpaired electrons. It has an electron configuration of $[He]2s^2 2p_x^2 2p_y^1 2p_z^1$.

Exercise 10.18   Reactions Involving Free Radicals

Predict the reaction products for

(a) The photodissociation of water, $H_2O \xrightarrow{h\nu}$
(b) A methane molecule reacting with a hydroxyl radical, $CH_4 + \cdot HO \longrightarrow$
(c) A hydrogen atom reacting with oxygen, $\cdot H + O_2 \longrightarrow$

10.11 OZONE AND OZONE DEPLETION

Ozone, O_3, is an important compound that is currently very much in the news because of its role in the troposphere. Ozone is an allotrope of oxygen and is a hybrid of two resonance structures. It has a bent (angular) geometry, with a bond angle of 116.8°. Ozone's ΔH_f° of 142.3 kJ/mol shows that it is less stable than O_2 (which is assigned an enthalpy of formation of zero).

$$\tfrac{3}{2} O_2(g) \longrightarrow O_3(g) \qquad \Delta H_f^\circ = 142.3 \text{ kJ/mol}$$

Ozone is highly reactive and does not survive long after formation. In the troposphere it is a pollutant. It irritates mucous membranes in the nose when it is inhaled. In the stratosphere, however, it is essential for life on earth.

Ozone is formed in the stratosphere by a natural photodissociation process in which ultraviolet solar radiation with $\lambda < 242$ nm splits an O_2 molecule to produce oxygen atoms. The oxygen atoms then combine with O_2 to form ozone.

Ozone resonance structures.

The bond energy of O_2 is 495 kJ/mol, so a calculation using $E = \nu c/\lambda$ shows that the energy in a photon of $\lambda < 242$ nm is needed to break the O_2 bond.

Formation of O_3

$$O_2(g) \xrightarrow{h\nu} \cdot O \cdot (g) + \cdot O \cdot (g)$$
$$\cdot O \cdot (g) + O_2(g) \longrightarrow O_3(g)$$

The region of maximum ozone concentration, known as the stratospheric **ozone layer**, is at an altitude of 25 to 30 km, where the concentration can reach 10 ppm. The ozone concentration in the stratosphere under natural conditions is maintained at a constant level by a reaction that destroys ozone at the same rate it is produced.

Natural destruction of O_3

$$O_3(g) + \cdot O \cdot (g) \longrightarrow 2 O_2(g)$$

Because ozone molecules and oxygen atoms are at very low concentrations, this reaction is relatively slow. However, it consumes ozone fast enough to maintain a balance with the ozone-forming reaction so that under normal circumstances the concentration of ozone in the stratosphere remains constant. Every day, 3×10^8 tons of stratospheric ozone are formed and destroyed by these two competing natural processes.

The importance of stratospheric ozone is that it prevents damaging ultraviolet radiation from reaching the earth's surface by the following photodissociation reaction.

Protection of earth's surface

$$O_3(g) \xrightarrow{h\nu} \cdot O \cdot (g) + O_2(g)$$

Wavelengths in the range of 200 to 310 nm are absorbed during this decomposition reaction. The oxygen atoms from the photodissociation of O_3 react with O_2 to regenerate O_3, and so this reaction results in no net O_3 loss.

Absorption of ultraviolet radiation in the stratosphere is essential for living things on earth. Stratospheric ozone prevents 95 to 99% of the sun's ultraviolet radiation from reaching the earth's surface. Photons in this 200- to 310-nm range have enough energy to cause skin cancer in humans and damage to living plants. For every 1% decrease in the stratospheric ozone, an additional 2% of this damaging radiation reaches the earth's surface. Stratospheric ozone depletion therefore has the potential for drastically damaging our environment.

This answers the question posed in Chapter 1 (⬅ **p. 32**): "Why is ozone depletion bad? I thought too much ozone was bad for your lungs."

Chlorofluorocarbons (CFCs)

Chlorofluorocarbons (CFCs), which are small molecules with central carbons bonded to halogens, play a major role in ozone depletion; $CFCl_3$ (CFC-11) and CF_2Cl_2 (CFC-12) are specific examples. They are nonflammable, relatively inert, volatile yet readily liquefied, and nontoxic (unless, of course, you breathe them instead of air)—an extremely useful set of properties. They have been used as

$CFCl_3$ (CFC-11) CF_2Cl_2 (CFC-12)

F. SHERWOOD ROWLAND
(1927–)

Rowland was born in Ohio and entered Ohio Wesleyan University at age 12. He received his Ph.D. from the University of Chicago in 1952. He held faculty appointments at several institutions and joined the newly formed University of California at Irvine in 1964 as Chair of the Chemistry Department. In 1974 he and Mario Molina recognized that CFCs could deplete the ozone layer in the atmosphere. Their work was received skeptically at first but was confirmed by others in later experiments, notably in 1985 by satellite data showing an "ozone hole" over Antarctica. Rowland, Molina, and Paul Crutzen shared the 1995 Nobel Prize in Chemistry ". . . for their work in atmospheric chemistry, particularly concerning the formation and decomposition of ozone."

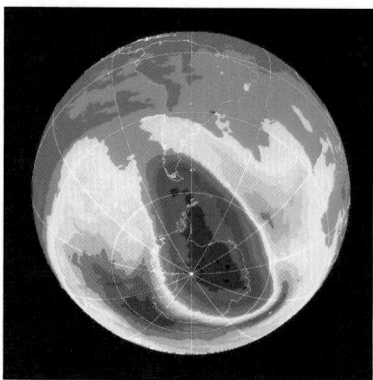

Figure 10.17 The Antarctic ozone hole.

Spring in Antarctica is fall in the Northern Hemisphere—seasons are reversed in the Southern Hemisphere.

coolants for refrigeration, in foam plastics manufacture, as aerosol propellants, and as industrial solvents. The properties that make them useful are also the reason for their destructive effect on the stratospheric ozone. Once gaseous CFCs are released into the atmosphere, they persist for a very long time in the troposphere because of their chemical nonreactivity. On average, a CF_2Cl_2 molecule survives roughly 100 years in the troposphere. Eventually, CFCs released into the troposphere mix and rise to the stratosphere, where they are decomposed by high-energy solar radiation. The decomposition products of CFCs participate in reaction mechanisms in the stratosphere that result in lowering of the concentration of ozone there.

Destruction of the stratospheric ozone layer by CFCs begins when a photon of high enough energy breaks a carbon-chlorine bond in a CFC molecule. This produces a chlorine atom, as shown here using CFC-12 as an example.

$$CF_2Cl_2(g) \xrightarrow{h\nu} CF_2Cl\cdot(g) + Cl\cdot(g)$$

The chlorine atom, a free radical, then participates in what is called a *chain reaction mechanism.* It first combines with an ozone molecule, producing a chlorine monoxide (ClO·) radical and an oxygen molecule.

Step 1: $Cl\cdot(g) + O_3(g) \rightarrow ClO\cdot(g) + O_2(g)$

Thus, an ozone molecule has been destroyed. If this were the only reaction that particular CFC molecule caused, there would be little danger to the ozone layer. However, once Step 1 has occurred twice, the two ClO· radicals produced can react further.

Step 2: $ClO\cdot + ClO\cdot \rightarrow ClOOCl$

Step 3: $ClOOCl \xrightarrow{h\nu} \cdot ClOO + \cdot Cl$

Step 4: $\cdot ClOO \xrightarrow{h\nu} \cdot Cl + O_2$

The net reaction obtained by adding twice Step 1 to Steps 2, 3, and 4 and canceling species that appear both as reactants and products is the conversion of two ozone molecules to three oxygen molecules.

$$2\,O_3(g) \longrightarrow 3\,O_2(g)$$

That is, this chain reaction increases the rate at which stratospheric ozone is destroyed, but it does not affect the rate at which ozone is formed. The reason this is called a chain reaction mechanism is that the reaction steps can repeat over and over. The two chlorine atoms that react when Step 1 occurs twice are regenerated in Steps 3 and 4, so there is no net change in concentration of Cl. It has been estimated that a single chlorine atom can destroy as many as 100,000 molecules of O_3 before it is inactivated or returned to the troposphere (probably as HCl). This sequence of ozone-destroying reactions upsets the balance in the stratosphere because ozone is being destroyed faster than it is being produced. Thus, the concentration of stratospheric ozone decreases.

The most prominent manifestation of stratospheric ozone depletion is the Antarctic **ozone hole.** In 1985 the British Antarctic Survey reported a startlingly large depletion of ozone in September and October, at the end of the Antarctic winter. This ozone hole has reappeared at this time each year (Figure 10.17). Subsequent measurements by several research teams showed that loss of ozone correlated with high concentrations of ClOOCl, supporting the theory that a chain mechanism involving chlorine atoms was responsible. However, the huge depletion of ozone in the Antarctic could not be explained solely by the reaction steps above. In the dark Antarctic winter, a vortex of intensely cold air containing ice crystals, unique to the region, builds up. On the surfaces of these crystals additional reactions produce hydrogen chloride and chlorine nitrate ($ClONO_2$). These can react with each other to form chlorine molecules.

$$HCl(g) + ClONO_2(g) \longrightarrow Cl_2(g) + HNO_3(g)$$

When sunlight returns in the spring, the Cl_2 molecules are readily photodissociated into chlorine atoms, which can then become involved in the ozone destruction reactions.

$$Cl_2(g) \xrightarrow{h\nu} 2\,Cl\cdot(g)$$

The first direct confirmation of the relationship between stratospheric ozone depletion and increased ultraviolet intensity on the earth was reported in the September 10, 1999 issue of *Science* by scientists from the New Zealand National Institute of Water and Atmospheric Research. They found that ". . . over the past 10 years, peak levels of skin-frying and DNA-damaging ultraviolet (UV) rays have been increasing in New Zealand, just as the concentrations of protective stratospheric ozone have decreased." Figure 10.18 shows a plot of declining ozone concentration and increasing UV radiation over the past 20 years, as reported by the New Zealand scientists.

Dealing with the Problem

Stratospheric ozone depletion is a global problem, and it has led to global treaties designed to solve the problem. In an effort to reduce the harm done by CFCs to the stratospheric ozone layer, the Montreal Protocol on Substances That Deplete the Ozone Layer was signed in 1989; it called for reductions in production and use of several of the long-lived CFCs. Later treaties called for complete phaseouts of all chemicals that can harm the ozone layer. Within the United States, in 1992 the U.S. Environmental Protection Agency (EPA) banned retail-store sales of small containers (less than 20 pounds) of CFCs for motor vehicle air-conditioning systems and prohibited service stations from emitting CFCs, requiring them to begin a recycling program for these compounds. These efforts are bearing fruit. The May 31, 1996, issue of *Science* contained a paper reporting that tropospheric chlorine from CFCs peaked near the beginning of 1994 and has been declining since. Nevertheless, it will be some time until the stratospheric chlorine level from CFCs, and its effect on ozone concentrations, begins to decline. The ozone hole in 2000 was the largest ever recorded.

Manufacturers began seeking alternatives to CFCs almost immediately after it became clear that they would eventually be phased out. One class of possible CFC substitutes includes molecules called HCFCs that have some C—H bonds as well as the C—Cl and C—F bonds of CFCs. An example of such a molecule is CH_2Cl_2. Because they contain C—H bonds, HCFC molecules are more reactive in the

SUSAN SOLOMON
(1956–)

Susan Solomon, a young NOAA (National Oceanic and Atmospheric Administration) scientist, was the first to propose a good explanation for the Antarctic ozone hole. Her chemist's intuition told her that the ice crystals that form during the Antarctic winter could provide a surface on which chemical reactions of CFC decomposition products could take place. In 1986, NASA chose Solomon (then 30 years old) to lead a team to Antarctica. Experiments during that visit showed that her cloud theory was correct and provided the first solid proof of a connection between CFCs and stratospheric ozone depletion. Currently, Susan Solomon is one of the youngest members of the National Academy of Sciences, and she was named a National Medal of Science winner in March 2000. She has said that her winters as a young girl in Chicago prepared her for her visits to Antarctica.

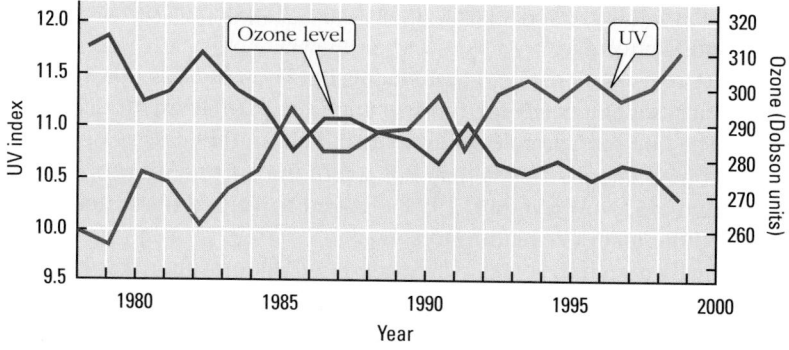

Figure 10.18   The ozone level is decreasing, and the UV intensity is increasing.   These measurements were made in Lauder, New Zealand. The red line indicates ultraviolet (UV) radiation at noon calculated from measured ozone levels, shown by the blue line. Dobson units, 1 DU = 2.69×10^{16} molecules/cm². The UV index is a standardized way to report UV radiation to the public. (Source: McKenzie, R, B. Connor, and G. Bodeker. "Increased Summertime UV Radiation in New Zealand in Response to Ozone Loss." *Science*, Vol. 285, September 10, 1999; p. 1709.)

troposphere than CFC molecules. Therefore, fewer HCFCs survive long enough to reach the stratosphere, where they could lead to ozone depletion. Even better are hydrofluorocarbons, HFCs. These contain only carbon, fluorine, and hydrogen, so cannot contribute chlorine atoms to catalyze ozone destruction. In refrigerators, for example, $C_2H_2F_4$ can be substituted for CFC-12. Older car air conditioners used CFC-12, but automakers switched the air-conditioning systems in their new cars to $C_2H_2F_4$ (HFC-134a) by 1995. In the United States, the phaseout of CFCs for auto air-conditioners, and the cost associated with switching older cars to new coolants, has led to a substantial amount of smuggling of the older CFCs into the United States.

10.12   CHEMISTRY AND POLLUTION IN THE TROPOSPHERE

An **air pollutant** is any substance that degrades air quality. Nature pollutes the air on a massive scale with volcanic ash, sulfur oxides, mercury vapor, hydrogen chloride, and hydrogen sulfide from volcanoes; and with reactive, odorous organic compounds from coniferous plants such as pine trees. Decaying vegetation, ruminant animals, and even termites add large quantities of methane gas to the atmosphere, and decaying animal carcasses and other protein materials add dinitrogen monoxide (N_2O). But automobiles, electric power plants, smelting and other metallurgic processes, and petroleum refining also add significant quantities of unwanted chemicals to the atmosphere, especially in heavily populated areas. Atmospheric pollutants, whatever their origin, cause burning eyes, coughing, decreased lung capacity, harm to vegetation, and even the destruction of ancient monuments. Millions of tons of soot, dust, smoke particles, and chemicals are discharged directly into the atmosphere every year. Such pollutants, which enter the environment directly from their sources, are called **primary pollutants.**

Primary Pollutants

Particle Pollutants

Pollutant particles range in size from fly ash particles, which are big enough to see, to individual molecules, ions, or atoms. Many pollutants are incorporated into water droplets and form **aerosols,** which are colloids (Sectopm 15.9) consisting of liquid droplets or finely divided solids dispersed in a gas. Fogs and smoke are common examples of aerosols. Larger solid particles in the atmosphere are called **particulates.** The solids in an aerosol or particulate may be metal oxides, soil particles, sea salt, fly ash from electrical generating plants and incinerators, elemental carbon, or even small metal particles. An aerosol particle can range in diameter from about 1 nm to about 10,000 nm (10 μm) and can contain 10^{12} atoms, ions, or small molecules. Particles in the 2000-nm range are largely responsible for the deterioration of visibility often observed in highly populated urban centers such as Houston and Mexico City.

In 1997 the EPA revised its national ambient air quality standards (NAAQS) for particulate matter and set standards for the first time for particles with diameters less than 2.5 μm. The fine-particle rule set an annual limit on the mass per unit volume of particulates with diameters less than 2.5 μm in the air to 65 μg/m^3, with a daily limit of 15 μg/m^3. However, in May 1999 a federal appeals court struck down these standards, and the matter remains unresolved.

Aerosols are small enough to remain suspended in the atmosphere for long periods. Such small particles are easily breathable and can cause lung disease. Because of their relatively large surface area, aerosol particles have great capacities

Chemistry in the News

MEXICO CITY AIR POLLUTION

To say Mexico City has an air pollution problem is an understatement. The 400-square-mile Mexico City Metropolitan Area (MCMA) is home for over 15 million people (20% of the country's population), more than 50% of its industry, and 60% of its automobiles. If you can imagine a city with three times the population density of Philadelphia, the car traffic of New York City, the truck traffic of a large distribution hub such as Chicago, and the petroleum industry of a city like Houston, you have a picture of Mexico City. All of this activity takes place in a high-altitude location surrounded by mountains that rise as much as 4000 ft above the city. If only the air in the city could be swept away periodically by the prevailing winds, things would not be nearly so bad. In effect, the city's air is isolated from the prevailing wind patterns. The high altitude of Mexico City makes the air about 25% thinner than at sea level, which means solar energy is much more intense than in cities at or near sea level. In addition, the sun is at a high angle most of the year. Since solar energy is responsible for the formation of tropospheric ozone, it is not surprising that the city has an ozone problem year-round, not just in the summer months as for many cities in the United States.

Tens of millions of commuter trips are made daily in the city, and NO_x emissions from the exhausts of internal combustion engines are sizable. Since NO_2 readily photodissociates to produce oxygen atoms, which in turn react with oxygen molecules to produce ozone, it is easy to see why ozone is such a problem. Besides being a health problem in its own right, ozone photodissociates to produce oxygen atoms that react with water to form hydroxyl radicals, and ozone itself can react with hydrocarbon molecules to form reactive aldehydes and ketones. The authorities in Mexico City must periodically take drastic action when ozone concentration reaches dangerous levels. On October 16, 1999, emergency action included ordering hundreds of thousands of cars to be parked and ordering factories to work at 30% of normal capacity.

The hydrocarbons in Mexico City's air come not only from the large number of automobiles that release unburned hydrocarbons in their exhausts, but from the vast petrochemical industry that has grown up in the city and the large number of propane heaters used to cook food and heat homes. In spite of all these problems, progress is being made. Automobile usage is being curtailed by limiting the driving of cars with a licensing system that allows cleaner cars to be driven every day but limits the driving of cars that produce more pollutants. In addition, the petrochemical industry is limiting its emissions of hydrocarbons. Furthermore, cleaner-burning fuels are becoming available that will produce fewer unburned hydrocarbons, and the public is being educated to the air-pollution potential of carelessly allowing unburned heating fuel to escape into the air. Perhaps someday the city will become entirely dependent on public transportation and electric cars.

Air pollution in Mexico City.

---

to *adsorb* and concentrate chemicals on their surfaces. Liquid aerosols or particles covered with a thin coating of water may also *absorb* air pollutants, thereby concentrating them and providing a medium in which reactions may occur.

To *adsorb* is to attract firmly to a surface. To *absorb* is to draw into the bulk of a solid or liquid.

Sulfur Dioxide

Sulfur dioxide (SO_2), the pollutant that is a major contributor to industrial smog and acid rain, is produced when sulfur or sulfur-containing compounds are burned in air. Most of the coal burned in the United States contains sulfur in the form of

CHEMISTRY
You Can Do...
Particle Size and Visibility

A common feature of all aerosols is that they decrease visibility. This can be observed in a city or along a busy highway, for example. Fog is an extreme example of disruption of visibility by aerosol particles. Here is a way to simulate the effect of air pollutants on visibility. You will need a flashlight, a little milk, a transparent container (if possible, with flat, parallel sides) full of water, and something with which to stir the water.

Turn off the room lights and shine the beam of the flashlight through the container perpendicular to the flat sides. What do you observe? Can you see the beam? Now add a couple of drops of milk to the container and stir. Can you see the

flashlight beam now? How does the beam's appearance change as the angle of viewing changes? What color is it? What color is the light that passes through the milky water? Keep adding milk dropwise, stirring and observing until the beam of the flashlight is no longer visible from the far side of the container.

1. Which wavelengths of light are scattered more by the particles? Which are scattered less?
2. Based on your observations of the milky water, devise an explanation of the fact that at midday on a sunny day the sun appears to be white or yellow, while at sunset it appears orange or red.

Oil-burning electrical generation plants produce quantities of SO_2 comparable to coal-burning facilities, since fuel oils can also contain up to 4% sulfur. The sulfur in oil is in the form of mercapto compounds, organic compounds in which sulfur atoms are bound to carbon and hydrogen atoms (the —SH functional group), such as CH_3CH_2SH. A mixture of low-molar-mass mercapto compounds is used as an odorant in natural gas so that gas leaks can be detected by smell.

the mineral pyrite (FeS_2). The weight percent of sulfur in this coal ranges from 1 to 4%. The pyrite is oxidized as the coal is burned.

$$4 \, FeS_2(s) + 11 \, O_2(g) \longrightarrow 2 \, Fe_2O_3(s) + 8 \, SO_2(g)$$

Sulfur is also part of the organic compounds making up the coal itself.

Exercise 10.19 Calculating SO_2 Emissions from Burning Coal

Large quantities of coal are burned in the United States to generate electricity. A 1000-MW coal-fired generating plant will burn 3.06×10^6 kg of coal per hour. For coal that contains 4% sulfur by weight, calculate the mass of SO_2 released (a) per hour and (b) per year.

Once in the atmosphere, SO_2 can be oxidized by reactions with O_2 or ozone, for example, to form SO_3. The SO_3 has a strong affinity for water and will dissolve in aqueous aerosol droplets to form sulfuric acid, which contributes to acid rain (Section 17.3).

$$SO_3(g) + H_2O(\ell) \longrightarrow H_2SO_4(aq)$$

Sulfur dioxide can be physiologically harmful to both plants and animals, although most healthy adults can tolerate fairly high levels of SO_2 without apparent lasting ill effects. Table 10.7 summarizes these effects. Individuals with chronic respiratory difficulties such as bronchitis or asthma tend to be much more sensitive to SO_2, accounting for many of the deaths during episodes of industrial smog. To reduce the hazards, many facilities have installed equipment that removes SO_2 from emitted gases.

Exercise 10.20 Calculations Involving Air Pollutants

The air on a smoggy day contains 5 ppm SO_2. What volume percent is that?

Nitrogen Oxides (NO_x)

Most of the nitrogen oxides found in the atmosphere originate from nitrogen monoxide, NO, formed whenever nitrogen and oxygen — always present in the at-

TABLE 10.7	Physiological and Corrosive Effects of SO_2 and National Ambient Air Quality Standards for SO_2	
SO_2 Exposure (ppm)	Duration	Effect
0.03–0.12	Annual average	Corrosion, especially in moist, temperate climates
0.3	8 hours	Vegetation damage (bleached spots, suppression of growth, leaf drop, and low yield)
0.47	<1 hour	Odor threshold (50% of subjects detect); varies with individuals
0.2	Daily average	Respiratory symptoms when community exposure exceeds 0.2 ppm more than 3% of the time

NAAQS for SO_2

An annual arithmetic mean of 0.03 ppm; a 24-hour level of 0.14 ppm; a 3-hour level of 0.50 ppm. The first two standards are primary (health-related standards); the third is a secondary (welfare-related) standard.

mosphere — are raised to high temperatures, as in an internal combustion engine. This nitrogen monoxide reacts rapidly with atmospheric oxygen to produce NO_2.

Vast quantities of nitrogen oxides are formed each year throughout the world, resulting in a global atmospheric concentration of NO_2 of a few parts per billion or less. In the United States, most oxides of nitrogen are produced during fossil-fuel combustion, including that of gasoline in automobile engines, with significantly less coming from natural sources such as the action of lightning. Elsewhere in the world, large amounts are also produced by the burning of trees and other biomass (Table 10.8). Most NO_2, either from human activities or from natural causes, eventually washes out of the atmosphere in precipitation. This is one way green plants obtain the nitrogen necessary for growth. In the troposphere, however, especially around urban centers, excessive NO_2 causes problems.

In laboratory studies, nitrogen dioxide in concentrations of 25 to 250 ppm inhibits plant growth and causes defoliation. Breathing NO_2 at a concentration of 3 ppm for 1 hour causes bronchial constriction in humans, and short exposures at high levels (150 to 220 ppm) causes changes in the lungs that can be fatal.

TABLE 10.8	Estimated Emissions of NO_x	
	Emissions (millions of tons)	
Source	United States	Global
Fossil-fuel combustion	66	231
Biomass burning	1.1	132
Lightning	3.3	88
Microbial activity in soil	3.3	88
Input from the stratosphere	0.3	5.5
TOTAL (uncertainty in estimates)	74.0 ± 1	545 ± 275*

* The large uncertainty for global emissions is due to incomplete data for much of the world.

During the day, as you saw in Section 10.10, nitrogen dioxide photodissociates to form nitrogen monoxide and free oxygen atoms. The oxygen atoms can then react to form ozone and regenerate NO_2. At night, reactions that lead to the formation of dinitrogen pentoxide, N_2O_5, predominate. The N_2O_5 can react with water to form nitric acid, HNO_3, a contributor to acid rain (Section 17.3). Normally nitrogen dioxide has a lifetime of about three days in the atmosphere.

> ### Exercise 10.21   Air Pollutant Stoichiometry
>
> If 400 metric tons of N_2 are converted to NO and then to HNO_3, what mass (in metric tons) of HNO_3 is produced?

Hydrocarbons

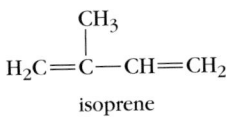

isoprene

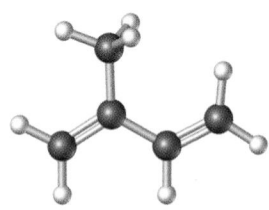

Hydrocarbons enter the atmosphere from both natural sources and human activities. Isoprene and α-pinene are hydrocarbons produced in large quantities by both coniferous and deciduous trees. Methane gas is produced by such diverse sources as ruminant animals, termites, ants, and decay-causing bacteria acting on dead plants and animals. Human activities such as using industrial solvents, refining and distribution of petroleum, and incomplete burning of gasoline and diesel fuel account for a large amount of hydrocarbons in the atmosphere. As a result, hydrocarbons ranging in size from ethane (C_2H_6) to 6- or 7-carbon compounds are abundant in urban air.

Hydroxyl radicals (⇐ *p. 440)* can oxidize hydrocarbons (R—H)

$$R{-}H(g) + \cdot OH(g) \longrightarrow \cdot R(g) + H_2O(g)$$

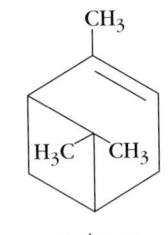

α-pinene

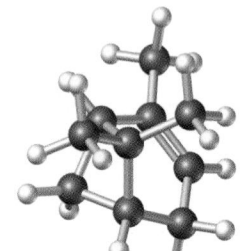

to produce hydrocarbon radicals $\cdot R(g)$. In a complex series of reactions involving O_2, NO_2, and oxygen-containing organic compounds, ozone is produced in the troposphere.

Ozone: A Secondary Pollutant

Ozone has a pungent odor that can be detected at concentrations as low as 0.02 ppm. We often smell the ozone produced by sparking electric appliances and photocopiers or after a thunderstorm, when lightning-caused ozone washes out with the rainfall. In the troposphere (the air we breathe), ozone is harmful because it is a component of photochemical smog and because it can damage human health and decompose materials such as plastics and rubber.

The most significant chemical reaction producing ozone in the atmosphere is the combination of molecular oxygen and atomic oxygen. In the troposphere the major source of oxygen atoms is the photodissociation of NO_2 (⇐ *p. 440).*

Ozone photodissociates to give an oxygen atom and an oxygen molecule whenever it is struck by photons in the near-ultraviolet range (240 to 310 nm), at any altitude.

$$O_3(g) \xrightarrow{h\nu} O_2(g) + \cdot O \cdot (g)$$

In the troposphere, oxygen atoms react with water to produce hydroxyl radicals ($\cdot OH$).

$$\cdot O \cdot (g) + H_2O(g) \longrightarrow 2 \cdot OH(g)$$

In the daytime, when they are produced in large numbers, hydroxyl radicals can react with nitrogen dioxide to produce nitric acid.

$$2 \cdot NO_2(g) + 2 \cdot OH(g) \longrightarrow 2\, HNO_3(g)$$

This three-step process is the primary means of removing NO_2 from the atmosphere. Summing the three steps gives the net reaction

$$O_3 + 2 \cdot NO_2 + H_2O \longrightarrow 2\,HNO_3 + O_2$$

Ozone is a **secondary air pollutant,** one that is formed by reaction of a primary pollutant. Ozone is a very difficult pollutant to control because its formation depends on ever-present sunlight, NO_2 and hydrocarbons, which almost every automobile emits to some degree. In cities with high ozone concentrations, the cause is always related to emissions of nitrogen oxides from automobiles, buses, and trucks. Most major urban areas operate vehicle inspection centers for passenger automobiles in an effort to control emissions of nitrogen oxides as well as those of carbon monoxide and unburned hydrocarbons.

As difficult as they are to attain, established ozone standards may not be low enough for good health. Exposure to concentrations of ozone at or near 0.12 ppm has been shown to lower the volume of air a person breathes out per second; studies of children who were exposed to ozone concentrations slightly below the old EPA standard (0.12 ppm) showed a significant decrease in exhaled air volume. In 1997 the EPA lowered the ozone standard to 0.08 ppm for an eight-hour average exposure. No matter what the official standard, ozone concentrations in many urban areas represent health hazards to children at play, joggers, others doing outdoor labor or exercise, and older persons who may have diminished respiratory capabilities.

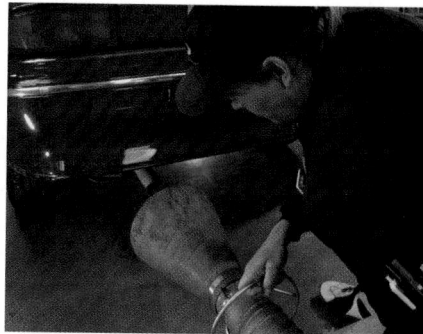

Automobile emissions testing. Such testing is mandated in many communities that have failed to meet EPA's ozone standards. Cars that fail the emissions standards for hydrocarbons and carbon monoxide are required to be repaired. Often the local government will not reissue operating licenses until satisfactory emissions levels are achieved.

Exercise 10.22   Photodissociation Reactions

Write the equations for the two photodissociation reactions that produce atomic oxygen. Then write the equation for a reaction in which ozone is formed.

Urban Air Pollution—Industrial Smog

The poisonous mixture of smoke (particulate matter), fog (an aerosol), air, and other chemicals was first called **smog** in 1911 by Dr. Harold de Voeux in his report on a London air pollution disaster that caused the deaths of 1150 people. The smog de Voeux identified is the *chemically reducing type* that is derived largely from the combustion of coal and oil and contains sulfur dioxide (a strong reducing agent) mixed with soot, fly ash, smoke, and partially oxidized organic compounds. This is an industrial smog, and it is common in many cities in the world where heavy industry and power plants are found. Thanks to the U.S. Clean Air Act, industrial smog is becoming less common in the United States as more pollution controls are installed. It is, unfortunately, still a major problem in some cities of the world.

Urban Air Pollution—Photochemical Smog

The other major kind of smog is the *chemically oxidizing type,* which contains strong oxidizing agents such as ozone and oxides of nitrogen, **NO_x**. It is known as **photochemical smog** because light—in this instance sunlight—is important in initiating the reactions that cause it.

Photochemical smog is typical in cities where sunshine is abundant and internal combustion engines exhaust large quantities of pollutants to the atmosphere. Los Angeles and Mexico City are examples (Figure 10.19). This type of smog is practically free of sulfur dioxide but contains substantial amounts of nitrogen oxides, ozone, the products of hydrocarbon oxidation, organic peroxides, and hydrocarbons of varying complexity. The concentrations of these substances vary during the day, building up in the morning hours and dropping off at night (Figure 10.20).

Figure 10.19   Photochemical smog over an urban area.

The common oxides of nitrogen found in air, NO and NO_2, are collectively called NO_x.

Figure 10.20   The average concentrations of pollutants NO, NO$_2$, and O$_3$ on a smoggy day in Los Angeles, California.   The NO concentration builds up during the morning rush hour. Later in the day the concentrations of NO$_2$ and O$_3$, which are secondary pollutants, build up.

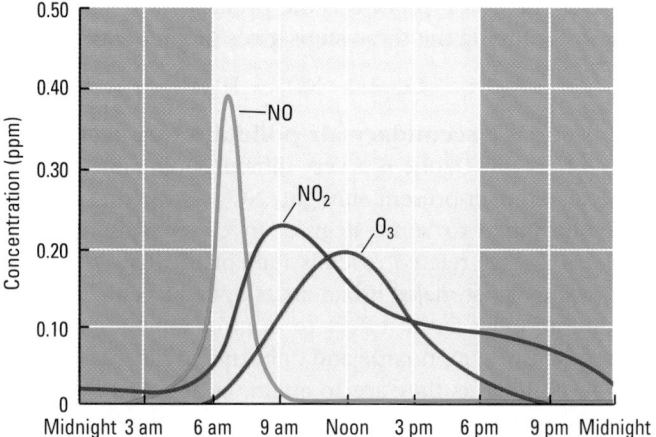

A city's atmosphere is an enormous mixing bowl of inter-related chemical reactions. Identifying the exact chemical reactions that produce photochemical smog was a challenge, but we now know that photochemical reactions are essential to the smog-making process and that aerosols serve to keep the primary pollutants together long enough to form secondary pollutants. Photons in the ultraviolet region of the spectrum are primarily responsible for the formation of photochemical smog.

The reaction scheme by which primary pollutants are converted into the secondary pollutants found in photochemical smog (Figure 10.21) is thought to begin with the photodissociation *(⇐ p. 440)* of nitrogen dioxide (Reaction 1). The very reactive atomic oxygen next reacts with molecular oxygen to form ozone

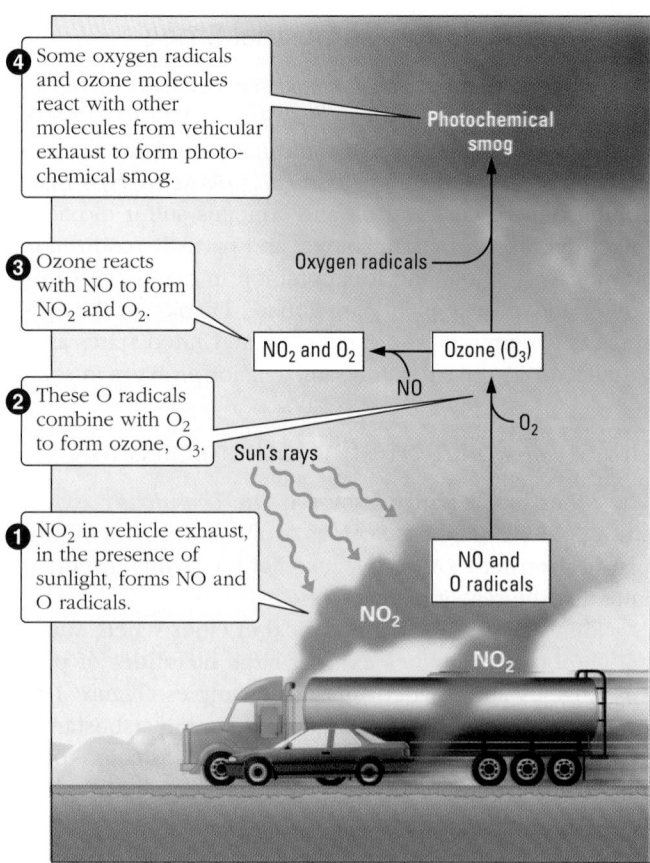

Figure 10.21   Photochemical smog formation.

(O$_3$) (Reaction 2), which is then consumed by reacting with nitrogen monoxide to form the original reactant — nitrogen dioxide (Reaction 3).

$$\cdot NO_2(g) \xrightarrow{\ h\nu\ } \cdot NO(g) + \cdot O \cdot (g) \tag{1}$$

$$\cdot O \cdot (g) + O_2(g) \longrightarrow O_3(g) \tag{2}$$

$$O_3(g) + \cdot NO(g) \longrightarrow \cdot NO_2(g) + O_2(g) \tag{3}$$

The net effect of these three reactions is absorption of energy without any net chemical change. If this were all that happened there would be no problem. Unfortunately, some of the reactive ozone molecules and oxygen atoms go on to react with other species in the atmosphere, and this leads to photochemical smog formation.

The air in an urban environment contains unburned hydrocarbons released from car exhausts and spillage at filling stations. Atomic oxygen produced in Reaction 1 above reacts with the more reactive of these hydrocarbons — unsaturated compounds and aromatics — to form other free radicals. These radicals, in turn, react to form yet other radicals and secondary pollutants such as aldehydes (e.g., formaldehyde, HCHO). In addition, reactions of the hydroxyl radical, $\cdot NO$, O_2, and other species can produce peroxyacetylnitrate (PAN).

PAN and related compounds are powerful eye irritants, helping to account for the discomfort we feel during episodes of photochemical smog. PAN formation stabilizes NO$_2$ so that it can be carried over great distances by prevailing winds. Eventually PAN decomposes and releases NO$_2$. In this way, urban pollution in the form of NO$_2$ may be carried to outlying areas, where it may do additional damage to vegetation, human tissue, and fabrics.

Exercise 10.23   Smog Ingredients

Write the formulas and give sources for three ingredients of photochemical smog.

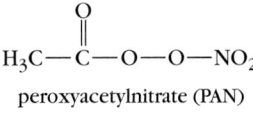

peroxyacetylnitrate (PAN)

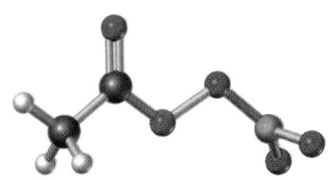

Where Do We Stand on Air Quality Now?

Average air quality across the United States is improving. A recent EPA report* on air quality in the United States says, "During the past 10 years (1988 through 1997), air quality has continued to improve." The report presented the data contained in the table to the right, which shows notable improvements in air quality, including a 38% decrease in CO concentrations, a 39% decrease in SO$_2$ concentrations, and a 67% decrease in Pb concentrations. Improvements were also seen in the other principal pollutants. The concentrations are based on actual measurements of pollutant concentrations in the ambient air at thousands of monitoring sites throughout the United States.

Pollutant	Percent decrease in concentrations (1988–1997)
CO	38
Pb	67
NO$_2$	1
O$_3$	19* 16†
SO$_2$	39

*Pre-existing NAAQS.
†Revised NAAQS.

SUMMARY PROBLEM

The air that enters an automobile engine contains the oxygen that reacts with the hydrocarbon fuel vapors to provide the energy to move the vehicle. Prior to the combustion process the fuel-air mixture is compressed, and then it is ignited with a spark. Most of the fuel is completely burned to CO$_2$ and H$_2$O, and some of the

* *Latest Findings on National Air Quality; 1997: Status and Trends.* Washington, DC: U.S. Environmental Protection Agency, December 1998.

nitrogen in air is converted to nitrogen monoxide (NO). For simplicity, assume the fuel has a formula of C_8H_{18} and a density of 0.760 g/mL.

(a) What are the partial pressures of N_2 and O_2 in the air before it goes into the engine if the atmospheric pressure is 734 mm Hg?

(b) If no fuel were added to the air and it was compressed to seven times atmospheric pressure (approximately the compression ratio in a modern engine), what would the partial pressures of N_2 and O_2 become?

(c) Assume the volume of each cylinder in the engine is 485 mL. If 0.050 mL of the fuel is added to the air in each cylinder just prior to compression, and the fuel is completely vaporized, what would be the partial pressure of the mixture in a cylinder due to the fuel molecules?

(d) How much oxygen would be required to burn the fuel completely to CO_2 and H_2O?

(e) If 10.% of the nitrogen in the combustion process is converted to NO, calculate how many grams of NO are produced in a single combustion reaction.

(f) Write a reaction that converts NO to a more photoreactive compound.

(g) Calculate the mass of this more photoreactive compound that would be formed from the NO calculated in part (f).

(h) What additional information would you need to calculate the NO emissions for an entire city for a year?

IN CLOSING

Having studied this chapter, you should be able to . . .

- Explain the properties of gases (Section 10.1).
- Describe the components of the atmosphere (Section 10.2).
- State the fundamental concepts of the kinetic-molecular theory and use them to explain gas behavior (Section 10.3).
- Solve problems using the appropriate gas laws (Sections 10.4 and 10.5).
- Calculate the quantities of gaseous reactants and products involved in chemical reactions (Section 10.6).
- Apply the ideal gas law to finding gas densities and molar masses (Section 10.7).
- Perform calculations using partial pressures of gases in mixtures (Section 10.8).
- Describe the differences between real and ideal gases (Section 10.9).
- Describe the main chemical reactions occurring in the atmosphere (Section 10.10).
- Explain the main features of stratospheric ozone depletion and the role of CFCs in it (Section 10.11).
- Explain the main chemicals found in and the reactions producing industrial pollution and urban pollution (Section 10.12).

KEY TERMS

absolute temperature scale *(10.4)*	**Boyle's law** *(10.4)*	**electronically excited molecule** *(10.10)*
aerosols *(10.12)*	**Charles's law** *(10.4)*	**ideal gas** *(10.4)*
air pollutant *(10.12)*	**chlorofluorocarbons (CFCs)** *(10.11)*	**ideal gas constant** *(10.5)*
Avogadro's law *(10.4)*	**combined gas law** *(10.5)*	**ideal gas law** *(10.5)*
bar *(10.1)*	**Dalton's law of partial pressures** *(10.8)*	**Kelvin temperature scale** *(10.4)*
barometer *(10.1)*		

law of combining vol-
umes *(10.4)*
millimeters of mercury
(mm Hg) *(10.1)*
mole fraction *(10.8)*
newton (N) *(10.1)*
NO$_x$ *(10.12)*
ozone hole *(10.11)*
ozone layer *(10.11)*
partial pressure *(10.8)*
particulates *(10.12)*
pascal (Pa) *(10.1)*

photochemical reac-
tions *(10.10)*
photochemical smog
(10.12)
photodissociation
(10.10)
pressure *(10.1)*
primary pollutant
(10.12)
secondary pollutant
(10.12)
smog *(10.12)*

standard atmosphere
(atm) *(10.1)*
standard molar volume
(10.5)
standard temperature
and pressure (STP)
(10.5)
stratosphere *(10.2)*
torr *(10.1)*
troposphere *(10.2)*
van der Waals equation
(10.9)

QUESTIONS FOR REVIEW AND THOUGHT

Conceptual Challenge Problems

CP-10.A (Section 10.2) How would you quickly estimate to see if the mass of the earth's atmosphere is 5.3×10^{15} metric tons?

CP-10.B (Section 10.4) Under what conditions would you expect to observe that the pressure of a confined gas at constant temperature and volume is *not* constant?

CP-10.C (Section 10.5) Suppose that the gas constant, R, were to be defined as 1.000 L atm mol^{-1} deg^{-1} where the "deg" referred to a newly defined Basic temperature scale. What would be the melting and boiling temperature of water in degrees Basic (°B)?

Answers to questions in **bold** can be found in the back of the book.

Review Questions

1. Name the three gas laws and explain how they interrelate P, V, and T. Explain the relationships in words and with equations.
2. What are the conditions represented by STP?
3. What is the volume occupied by 1 mol of an ideal gas at STP?
4. What is the definition of pressure?
5. State Avogadro's law. Explain why two volumes of hydrogen react with one volume of oxygen to form two volumes of steam.
6. State Dalton's law of partial pressures. If the air we breathe is 78% N$_2$ and 22% O$_2$ on a mole basis, what is the mole fraction of O$_2$? What is the partial pressure of O$_2$ if the total pressure is 720 mm Hg?
7. Explain Boyle's law on the basis of the kinetic-molecular theory.
8. Explain why gases at low temperature and high pressure do not obey the ideal gas equation.
9. Describe the difference between primary and secondary air pollutants and give two examples of each.

Properties of Gases

10. Gas pressures can be expressed in units of mm Hg, atm, torr, and kPa. Do the following conversions.
    (a) 720 mm Hg to atm   (b) 1.25 atm to mm Hg
    (c) 542 mm Hg to torr   (d) 740 mm Hg to kPa
    (e) 700 kPa to atm

11. Convert the following pressure measurements.
    (a) 120 mm Hg to atm   (b) 2 atm to mm Hg
    (c) 100 kPa to mm Hg   (d) 200 kPa to atm
    (e) 36 kPa to atm   (f) 600 kPa to mm Hg
12. Mercury has a density of 13.96 g/cm^3. A barometer is constructed using an oil with a density of 0.75 g/cm^3. If the atmospheric pressure is 1.0 atm, what will be the height in meters of the oil column in the barometer?
13. A vacuum pump is connected to the top of an upright tube whose lower end is immersed in a pool of mercury. How high will the mercury rise in the tube when the pump is turned on?
14. Why can't a hand-driven pump on a water well pull underground water from depths more than 33 ft? Would it help to have a motor-driven vacuum pump?
15. A scuba diver taking photos of a coral reef 60 ft below the ocean surface breathes out a stream of bubbles. What is the total gas pressure of these bubbles at the moment they are released? What is the gas pressure in the bubbles when they reach the surface of the ocean?

The Atmosphere

16. Explain the major roles played by nitrogen in the atmosphere. Do the same for oxygen.
17. Beginning nearest the earth, name the two most important layers or regions of the atmosphere. Describe, in general, the kinds of chemical reactions that occur in each layer.
18. Convert all of the "percentage by volume" figures in Table 10.1 into (a) parts per million and (b) parts per billion. Which atmospheric gases are present at concentrations of less than 1 ppb? Between 1 ppb and 1 ppm? Greater than 1 ppm?

19. The mass of the earth's atmosphere is 5.3×10^{15} metric tons. The atmospheric abundance of helium is 0.7 ppm when expressed as a fraction by *weight* instead of by *volume*, as in Table 10.1. How many metric tons of helium are there in the atmosphere? How many moles of helium is this?

20. Sulfur is about 2.5% of the mass of coal, and when coal is burned the sulfur is all converted to SO_2. In 1980, 3.1×10^9 metric tons of coal was burned worldwide. How many tons of SO_2 were added to the atmosphere? How many tons of SO_2 are currently in the atmosphere? (*Note:* The weight fraction of SO_2 in air is 0.4 ppb.)

Kinetic-Molecular Theory

21. List the five basic concepts of the kinetic-molecular theory. Which assumption is incorrect at very high pressures? Which one is incorrect at low temperatures? Which assumption is probably most nearly correct?

22. You are given two flasks of equal volume. Flask A contains H_2 at 0 °C and 1 atm pressure. Flask B contains CO_2 gas at 0 °C and 2 atm pressure. Compare these two samples with respect to each of the following.
    (a) Average kinetic energy per molecule
    (b) Average molecular velocity
    (c) Number of molecules

23. Place the following gases in order of increasing average molecular speed at 25 °C. Kr, CH_4, N_2, CH_2Cl_2

24. The reaction of SO_2 with Cl_2 to give dichlorine oxide is

$$SO_2(g) + 2\,Cl_2(g) \longrightarrow SOCl_2(g) + Cl_2O(g)$$

Place these molecules them in order of increasing average molecular speed.

25. From Table 10.1 list all the gases in the atmosphere in order of *decreasing* average molecular speed. Concern has been expressed about one of these gases escaping into outer space because a fraction of its molecules have velocities large enough to break free from earth's gravitational field. Which gas is it?

Gas Behavior and the Ideal Gas Law

26. What amount (number of moles) of CO is found in 1.0 L of air at STP that contains 950 ppm CO?

27. A sample of a gas has a pressure of 100. mm Hg in a 125-mL flask. If this gas sample is transferred to another flask with a volume of 200. mL, what will be the new pressure? Assume that the temperature remains constant.

28. A sample of a gas is placed in a 256-mL flask, where it exerts a pressure of 75.0 mm Hg. What is the pressure of this gas if it is transferred to a 125-mL flask? (The temperature stays constant.)

29. A sample of gas has a pressure of 62 mm Hg in a 100-mL flask. This sample of gas is transferred to another flask, where its pressure is 29 mm Hg. What is the volume of the new flask? (The temperature does not change.)

30. Some butane, the fuel used in backyard grills, is placed in a 3.50-L container at 25 °C; its pressure is 735 mm Hg. If you transfer the gas to a 15.0-L container, also at 25 °C, what is the pressure of the gas in the larger container?

31. A sample of gas at 30 °C has a pressure of 2 atm in a one-L container. What pressure will it exert in a 4-L container? The temperature does not change.

32. Suppose you have a sample of CO_2 in a gas-tight syringe with a movable piston. The gas volume is 25.0 mL at a room temperature of 20 °C. What is the final volume of the gas if you hold the syringe in your hand to raise the gas temperature to 37 °C?

33. A balloon is inflated with helium to a volume of 4.5 L at 23 °C. If you take the balloon outside on a cold day (−10 °C), what will be the new volume of the balloon?

34. A sample of gas has a volume of 2.50 L at 670 mm Hg pressure and a temperature of 80 °C. If the pressure remains constant but the temperature is decreased, the gas occupies 1.25 L. What is this new temperature, in degrees Celsius?

35. A sample of 9.0 L of CO_2 at 20 °C and 1 atm pressure is cooled so that it occupies 1.0 L at some new temperature. The pressure remains constant. What is the new temperature, in kelvins?

36. A bicycle tire is inflated to a pressure of 3.74 atm at 15 °C. If the tire is heated to 35 °C, what is the pressure in the tire? Assume the tire volume doesn't change.

37. An automobile tire is inflated to a pressure of 3.05 atm on a rather warm day when the temperature is 40 °C. The car is then driven to the mountains and parked overnight. The morning temperature is −5 °C. What will be the pressure of the gas in the tire? Assume the volume of the tire doesn't change.

38. A sample of gas occupies 754 mL at 22 °C and a pressure of 165 mm Hg. What is its volume if the temperature is raised to 42 °C and the pressure is raised to 265 mm Hg? Note that the number of moles does not change.

39. A balloon is filled with helium to a volume of 1.05×10^3 L on the ground, where the pressure is 745 mm Hg and the temperature is 20 °C. When the balloon ascends to a height of 2 miles, where the pressure is only 600 mm Hg and the temperature is −33 °C, what is the volume of the helium in the balloon?

40. What is the pressure exerted by 1.55 g of Xe gas at 20. °C in a 560-mL flask?

41. A 1.00-g sample of water is allowed to vaporize completely inside a 10.0-L container. What is the pressure of the water vapor at a temperature of 150. °C?

42. Which of the following gas samples contains the largest number of molecules and which contains the smallest?
    (a) 1.0 L of H_2 at STP
    (b) 1.0 L of N_2 at STP
    (c) 1.0 L of H_2 at 27 °C and 760 mm Hg
    (d) 1.0 L of CO_2 at 0 °C and 800 mm Hg

43. Ozone molecules attack rubber and cause cracks to appear. If enough cracks occur in a rubber tire, for example, it will be weakened, and it may burst from the pressure of the air inside. As little as 0.02 ppm O_3 will cause cracks to appear in rubber in about 1 hour. Assume that a 1.0-cm^3 sample of air containing 0.020 ppm O_3 is brought in contact with a sample of rubber that is 1.0 cm^2 in area. Calculate the number of O_3 molecules that are available to collide with the rubber surface. The temperature of the air sample is 25 °C and the pressure is 0.95 atm.

Quantities of Gases in Chemical Reactions

44. The yeast in rising bread dough converts sugar (sucrose, $C_{12}H_{22}O_{11}$) into carbon dioxide. A popular recipe for two loaves of French bread requires 1 package of yeast and $\frac{1}{4}$ teaspoon (about 2.4 g) of sugar. What volume of CO_2 at STP is produced by the complete conversion of this amount of sucrose into CO_2 by the yeast? Compare this volume with the typical volume of two loaves of French bread.

45. Water can be made by combining gaseous O_2 and H_2. If you begin with 1.5 L of $H_2(g)$ at 360 mm Hg and 23 °C, what volume in liters of $O_2(g)$ will you need for complete reaction if the O_2 gas is also measured at 360 mm Hg and 23 °C?

46. Gaseous silane, SiH_4, ignites spontaneously in air according to the equation

$$SiH_4(g) + 2\ O_2(g) \longrightarrow SiO_2(s) + 2\ H_2O(g)$$

If 5.2 L of SiH_4 is treated with O_2, what volume in liters of O_2 is required for complete reaction? What volume of H_2O vapor is produced? Assume all gases are measured at the same temperature and pressure.

47. Hydrogen can be made in the "water gas reaction."

$$C(s) + H_2O(g) \longrightarrow H_2(g) + CO(g)$$

If you begin with 250 L of gaseous water at 120 °C and 2.0 atm pressure and excess $C(s)$, what mass in grams of H_2 can be prepared?

48. If boron hydride, B_4H_{10}, is treated with pure oxygen, it burns to give B_2O_3 and H_2O.

$$2\ B_4H_{10}(s) + 11\ O_2(g) \longrightarrow 4\ B_2O_3(s) + 10\ H_2O(g)$$

If a 0.050-g sample of the boron hydride burns completely in O_2, what will be the pressure of the gaseous water in a 4.25-L flask at 30. °C?

49. If 1.0×10^3 g of uranium metal is converted to gaseous UF_6, what pressure of UF_6 would be observed at 62 °C in a chamber that has a volume of 3.0×10^2 L?

50. Metal carbonates decompose to the metal oxide and CO_2 on heating according to this general equation.

$$M_x(CO_3)_y(s) \longrightarrow M_xO_y(s) + y\ CO_2(g)$$

You heat 0.158 g of a white, solid carbonate of a Group 2A metal and find that the evolved CO_2 has a pressure of 69.8 mm Hg in a 285-mL flask at 25 °C. What is the molar mass of the metal carbonate?

51. Nickel carbonyl, $Ni(CO)_4$, can be made by the room-temperature reaction of finely divided nickel metal with gaseous CO. This is the basis for purifying nickel on an industrial scale. If you have CO in a 1.50-L flask at a pressure of 418 mm Hg at 25.0 °C, what is the maximum mass in grams of $Ni(CO)_4$ that can be made?

52. Assume that a car burns pure octane, C_8H_{18} ($d = 0.692$ g/cm³).
(a) Write the balanced equation for burning octane in air, forming CO_2 and H_2O.
(b) If the car has a fuel efficiency of 32 miles per gallon of octane, what volume of CO_2 at 25 °C and 1.0 atm is generated when the car goes on a 10.-mile trip?

53. Follow the directions in the previous question, but use methanol, CH_3OH ($d = 0.791$ g/cm³) as the fuel. Assume the fuel efficiency is 20. miles per gallon.

Gas Density and Molar Mass

54. A sample of gaseous SiH_4 weighing 4.25 g is placed in a 580-mL container. The resulting pressure is 1.2 atm. What is the temperature in °C?

55. To find the volume of a flask, it is first evacuated so it contains no gas at all. Next, 4.4 g of CO_2 is introduced into the flask. On warming to 27 °C, the gas exerts a pressure of 730 mm Hg. What is the volume of the flask in milliliters?

56. What mass of helium in grams is required to fill a 5.0-L balloon to a pressure of 1.1 atm at 25 °C?

57. A hydrocarbon with the general formula C_xH_y is 92.26% carbon. Experiment shows that 0.293 g of the hydrocarbon fills a 185-mL flask at 23 °C with a pressure of 374 mm Hg. What is the molecular formula for this compound?

58. Forty miles above the earth's surface the temperature is -23 °C, and the pressure is only 0.20 mm Hg. What is the density of air (molar mass = 29.0 g/mol) at this altitude?

59. A newly discovered gas has a density of 2.39 g/L at 23.0 °C and 715 mm Hg. What is the molar mass of the gas?

Partial Pressures of Gases

60. A sample of the atmosphere at a total pressure of 740. mm Hg is analyzed to give the following partial pressures: $P(N_2) = 575$ mm Hg; $P(Ar) = 6.9$ mm Hg; $P(CO_2) = 0.2$ mm Hg; $P(H_2O) = 4.0$ mm Hg.
(a) What is the partial pressure of O_2?
(b) What is the mole fraction of each gas?
(c) What is the composition of each component of this sample in percentage by volume? Compare your results with those of Table 10.1.

61. Gaseous CO exerts a pressure of 45.6 mm Hg in a 56.0-L tank at 22.0 °C. If this gas is released into a room with a volume of 2.70×10^4 L, what is the partial pressure of CO (in mm Hg) in the room at 22 °C?

62. The density of air at 20.0 km above the earth's surface is 92 g/m³. The pressure is 42 mm Hg and the temperature is -63 °C. Assuming the atmosphere contains only O_2 and N_2, calculate
(a) the average molar mass of the atmosphere.
(b) the mole fraction of each gas.

63. Benzene has acute health effects. For example, it causes mucous membrane irritation at a concentration of 100 ppm and fatal narcosis at 20,000 ppm (pressure ratios). Calculate the partial pressures in atmospheres at STP corresponding to these concentrations.

64. On a humid, rainy summer day the partial pressure of water vapor in the atmosphere can be as high as 25 mm Hg. What is the mole fraction of water vapor in the atmosphere under these conditions? Compare your results with the composition of dry air given in Table 10.1.

65. The mean fraction *by weight* of water vapor and cloud water in the earth's atmosphere is about 0.0025. Assume that the atmosphere contains two components: "air," with a molar mass of 29.2 g/mol, and water vapor. What is the *mean* mole fraction of water vapor in the earth's atmosphere? What is the *mean* partial pressure of water vapor?

Why is this so much smaller than the typical partial pressure of water vapor at the earth's surface on a rainy summer day?

66. Acetylene can be made by allowing calcium carbide to react with water.

$$CaC_2(s) + 2 H_2O(\ell) \longrightarrow C_2H_2(g) + Ca(OH)_2(aq)$$

Assume that you place 2.65 g of CaC_2 in excess water and collect the acetylene over water. The volume of the acetylene and water vapor is 795 mL at 25.0 °C and a barometric pressure of 735.2 mm Hg. Calculate the percent yield of acetylene. The vapor pressure of water at 25 °C is 23.8 mm Hg.

67. Potassium chlorate, $KClO_3$, can be decomposed by heating.

$$2 KClO_3(s) \longrightarrow 2 KCl(s) + 3 O_2(g)$$

If 465 mL of gas was collected over water at a total pressure of 750 mm Hg and a temperature of 23 °C, how many grams of O_2 were collected?

The Behavior of Real Gases

68. From the density of liquid water and its molar mass, calculate the volume that 1 mol of liquid water occupies. If water were an ideal gas at STP, what volume would a mole of water vapor occupy? Can we achieve the STP conditions for water vapor? Why?

69. At high temperatures and low pressures, gases behave ideally, but as the pressure is increased the product PV becomes greater than the product nRT. Give a molecular-level explanation of this fact.

70. At low temperatures and very low pressures, gases behave ideally, but as the pressure is increased the product PV becomes less than the product nRT. Give a molecular-level explanation of this fact.

71. The densities of liquid noble gases and their normal boiling points are given below.

Gas	Normal boiling point (K)	Liquid density (g/cm^3)
He	4.2	0.125
Ne	27.1	1.20
Ar	87.3	1.40
Kr	120.	2.42
Xe	165	2.95

Calculate the volume occupied by 1 mol of each of these liquids. Comment on any trend that you see. What is the volume occupied by 1 mol of each of these substances as an ideal gas at STP? On the basis of these calculations, which gas would you expect to show the largest deviations from ideality at room temperature?

72. Use the van der Waals constants in Table 10.6 to predict whether N_2 or CO_2 will behave more like an ideal gas at high pressures.

73. Calculate the pressure of 7.0 mol of CO_2 in a 2.00-L vessel at 50 °C using
(a) The ideal gas equation.
(b) The van der Waals equation.

Chemical Reactions in the Atmosphere

74. What is a free radical? Give an example of a chemical reaction that occurs in the atmosphere that produces a free radical.

75. What product is formed when two methyl radicals react with each other?

76. For the following forms of nitrogen most prevalent in the environment, write chemical formulas and give the oxidation number of the nitrogen.
(a) Ammonia   (b) Ammonium ion
(c) Dinitrogen monoxide   (d) Gaseous nitrogen
(e) Nitric acid   (f) Nitrous acid
(g) Nitrogen dioxide

77. Complete the following equations (all reactants and products are gases).
(a) $\cdot O \cdot + O_2 \longrightarrow$
(b) $O_3 \overset{h\nu}{\longrightarrow}$
(c) $\cdot O \cdot + SO_2 \longrightarrow$
(d) $H_2O \overset{h\nu}{\longrightarrow}$
(e) $\cdot NO_2 + \cdot NO_2 \longrightarrow$

78. Complete the following equations (all reactants and products are gases).
(a) $\cdot O \cdot + H_2O \longrightarrow$ (b) $\cdot NO_2 + OH \longrightarrow$
(c) $RH + OH \longrightarrow$ (d) $CO + OH \longrightarrow$

79. Describe and write an equation for a chemical reaction that
(a) Creates a free radical.
(b) Creates a hydroxyl radical.

Ozone and Ozone Depletion

80. Write the products for the following reactions that take place in the stratosphere.
(a) $CF_3Cl \overset{h\nu}{\longrightarrow}$
(b) $\cdot Cl + \cdot O \cdot \longrightarrow$
(c) $ClO \cdot + \cdot O \cdot \longrightarrow$

81. The molecule CF_4 is not implicated as having ozone-depletion potential. Can you explain why?

82. The molecule CH_3F has much less ozone-depletion potential than the corresponding molecule CH_3Cl. Explain why.

83. Write the chemical formulas of all possible CFCs with one carbon atom.

84. Are CFCs toxic? In considering this question, consider which compounds were used for refrigeration before CFCs were invented. Look up the toxicity of these compounds on the web.

85. Can CFCs catalyze the destruction of ozone in the stratosphere at night? Explain.

86. Can ozone form in the stratosphere at night? Explain why or why not.

Chemistry and Pollution in the Troposphere

87. Define air pollution in terms of the kinds of pollutants, their sources, and the ways they are harmful.

88. Explain how particulates can contribute to air pollution.

89. What is adsorption? What is absorption?

90. Assume that limestone ($CaCO_3$) is used to remove 90% of the sulfur from 4 metric tons of coal containing 2% S. The product is $CaSO_4$.

$$CaCO_3(s) + SO_3(g) \longrightarrow CaSO_4(s) + CO_2(g)$$

Calculate the mass of limestone required. Express your answer in metric tons.

91. If coal, on average, contains 2% S, how many metric tons of coal were burned to produce this much SO_2? A 1000-MW power plant burns about 700 metric tons of coal per hour. Calculate the number of hours the quantity of coal will burn in one of these power plants.

92. What mass of gasoline must be burned according to the reaction

$$C_8H_{18}(\ell) + O_2(g) \longrightarrow 8\,CO(g) + 9\,H_2O(g)$$

to cause a garage with dimensions of 7 m $\times$ 3 m $\times$ 3 m to contain a CO concentration of 1000. ppm? (Assume STP conditions.)

Urban Air Pollution

93. Give an example of a photochemical reaction. Do all photons of light have sufficient energy to cause photochemical reactions? Explain with examples.

94. What two reactions account for the production of O free radicals in the troposphere?

95. The reducing nature of industrial (London) smog is due to what oxide? The burning of which two fuels produces this oxide? Write an equation showing this oxide being further oxidized.

96. Photochemical smog contains quantities of which two oxidizing gases? What is the energy source for photochemical smog?

97. What atmospheric reaction favors the formation of nitrogen monoxide, NO? Explain how the formation of NO in a combustion chamber is similar to the formation of NH_3 in a reactor designed to manufacture ammonia.

98. Give an example of atmospheric ozone that is beneficial and an example of ozone that is harmful. Explain how ozone is beneficial and how it is harmful.

99. The air pollutant sulfur dioxide, SO_2, is known to increase mortality in people exposed to it for 24 hr at a concentration of 0.175 ppm.
   (a) What is the partial pressure of SO_2 when its concentration is 0.175 ppm?
   (b) What is the mole fraction of SO_2 at the same concentration?
   (c) Assuming the air is at STP, what mass in micrograms of SO_2 is present in 1 m^3?

General Questions

100. HCl can be made by the direct reaction of H_2 and Cl_2 in the presence of light. Assume that 3.0 g of H_2 and 140 g of Cl_2 are mixed in a 10-L flask at 28 °C. Before the reaction:
   (a) What are the partial pressures of the two reactants?
   (b) What is the total pressure due to the gases in the flask?

   After the reaction:
   (c) What is the total pressure in the flask?
   (d) What reactant remains in the flask? How many moles of it remain?

   (e) What are the partial pressures of the gases in the flask?
   (f) What will be the pressure inside the flask if the temperature is increased to 40 °C?

101. One of the major sources of SO_2 in the atmosphere is from the oxidation of H_2S, produced by the decay of organic matter. The reaction in which H_2S molecules are oxidized to SO_2 involves O_3. Write an equation showing that one molecule of each reactant combines to form two product molecules, one of them being SO_2. Then calculate the annual production in tons of H_2SO_4, assuming all of this SO_2 is converted to sulfuric acid.

Applying Concepts

102. Consider a sample of N_2 gas under conditions in which it obeys the ideal gas law exactly. Which of the following statements are true?
   (a) A sample of Ne(g) under the same conditions must obey the ideal gas law exactly.
   (b) The speed at which one particular N_2 molecule is moving changes from time to time.
   (c) Some N_2 molecules are moving more slowly than some of the molecules in a sample of $O_2(g)$ under the same conditions.
   (d) Some N_2 molecules are moving more slowly than some of the molecules in a sample of Ne(g) under the same conditions.
   (e) When two N_2 molecules collide, it is possible that both may be moving faster after the collision than they were before.

103. Which graph below would best represent the distribution of molecular speeds for the gases acetylene (C_2H_2) and N_2? Both gases are in the same flask with a total pressure of 750 mm Hg. The partial pressure of N_2 is 500 mm Hg.

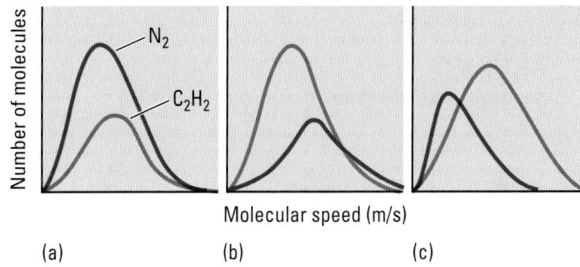

104. Draw a graph representing the distribution of molecular speeds for the gases ethane (C_2H_6) and F_2 when both are in the same flask with a total pressure of 720 mm Hg and a partial pressure of 540 mm Hg for F_2.

105. In this chapter Boyle's, Charles's, and Avogadro's laws were presented as word statements and mathematical relationships. Express each of these laws graphically.

106. The drawing on the next page represents a gas collected in a syringe (the needle end was sealed after collecting) at room temperature and pressure. Redraw the syringe and gas to show what it would look like under the following conditions. Assume that the plunger can move freely but no gas can escape.
   (a) The temperature of the gas is decreased by one half.
   (b) The pressure of the gas is decreased to one half its initial value.

(c) The temperature of the gas is tripled and the pressure is doubled.

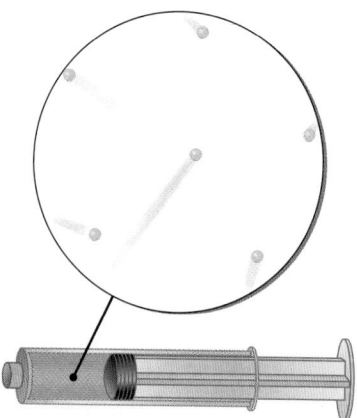

107. A gas phase reaction takes place in a syringe at a constant temperature and pressure. If the initial volume is 40 cm^3 and the final volume is 60 cm^3, which of the following general reactions took place? Explain your reasoning.
(a) $A(g) + B(g) \rightarrow AB(g)$
(b) $2 A(g) + B(g) \rightarrow A_2B(g)$
(c) $2 AB_2(g) \rightarrow A_2(g) + 2 B_2(g)$
(d) $2 AB(g) \rightarrow A_2(g) + B_2(g)$
(e) $2 A_2(g) + 4 B(g) \rightarrow 4 AB(g)$

108. The gas molecules in the box below undergo a reaction at constant temperature and pressure.

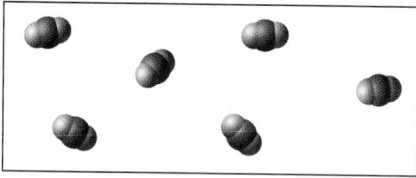

If the initial volume is 1.8 L and the final volume is 0.9 L, which of the boxes below could be the products of the reaction? Explain your reasoning.

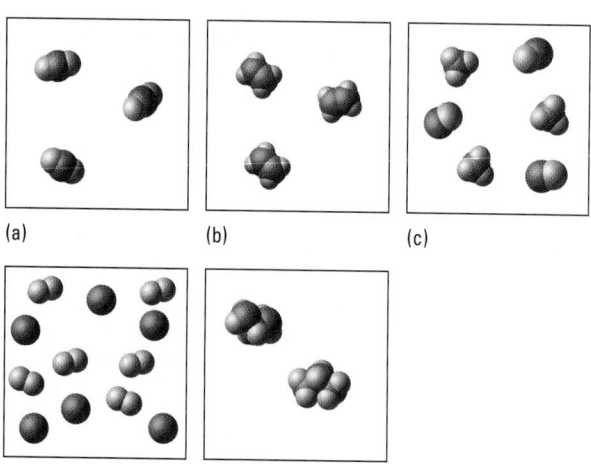

(a)   (b)   (c)

(d)   (e)

109. A substance is analyzed and found to contain 85.7% carbon and 14.3% hydrogen by mass. A gaseous sample of the substance is found to have a density of 1.87 g/L at STP.
(a) What is the molar mass of the compound?
(b) What are the empirical and molecular formulas of the compound?
(c) What are two possible Lewis structures for molecules of the compound?

110. A compound consists of 37.5% C, 3.15% H, and 59.3% F by mass. When 0.298 g of the compound is heated to 50 °C in an evacuated 125-mL flask, the pressure is observed to be 750 mm Hg. The compound has three isomers.
(a) What is the molar mass of the compound?
(b) What are the empirical and molecular formulas of the compound?
(c) Draw the Lewis structure for each isomer of the compound.

111. One very cold winter day you and a friend purchase a helium-filled balloon. As you leave the store and walk down the street, your friend notices the balloon is not as full as it was a moment ago in the store. He says the balloon is defective and he is taking it back. Do you agree with him? Explain why you do or do not agree.

 General Chemistry CD-ROM

CD10.1 Screen 12.2: Properties of Gases. List the four principal measurements we use to describe a gas along with the units with which we describe them.

CD10.2 Screen 12.9: The Kinetic-Molecular Theory of Gases.
(a) What are all the pictures on this screen intended to make you think about? What does this have to do with the kinetic theory of gases?
(b) Do the tenets of the kinetic-molecular theory include the shape, size, or chemical properties of gas molecules?

CD10.3 Screen 12.3: The Gas Laws. In the animation of Charles's law, it is stated that if we extrapolate to a temperature of absolute zero, a gas has, "in principle," no volume. Why can this experiment not be performed?

CD10.4 Screen 12.4: The Ideal Gas Law. The ideal gas law is a combination of the three laws described on Screen 12.3. It can be used to describe other relationships as well. Make up your own law describing the relationship between temperature and pressure. Is the pressure of a gas directly or inversely related to the gas temperature?

CD10.5 Screen 12.7: Gas Laws and Chemical Reactions.
(a) Examine the calculation on this screen. What is assumed about the temperature of the sample during the reaction?
(b) If the temperature increases during the reaction, would more or less sodium azide be required to inflate the bag to the required pressure? Explain.

Ice cubes floating in water. The hydrogen-bonding of water molecules in ice makes the solid less dense than liquid water. For a solid to float in a liquid of the same material is unusual, but this property of water is crucial for life on earth.

11

LIQUIDS, SOLIDS, AND MATERIALS

11.1   The Liquid State

11.2   Vapor Pressure

11.3   Phase Changes: Solids, Liquids, and Gases

11.4   Water: An Important Liquid with Unusual Properties

11.5   Types of Solids

11.6   Crystalline Solids

11.7   Materials Science

11.8   Metals, Semiconductors, and Insulators

11.9   Silicon and the Chip

11.10 Network Solids

11.11 Cement, Ceramics, and Glass

The properties of liquids and solids, like those of gases, can be understood on the molecular or nanoscale level. The noncovalent intermolecular forces between the atoms, molecules, or ions (⟵ *p. 383*) determine the liquid or solid state properties of substances made up of these particles.

In the condensed states of matter, atoms, molecules, or ions are close enough to have strong interactions with each other. The strength of these interactions accounts for the properties of liquids and solids. For example, the strong repulsions when the atoms, molecules, or ions in a liquid or solid are forced very close together account for the fact that solids and liquids can be compressed very little — as opposed to gases. Liquids are like gases in that their atoms, molecules, or ions (in the case of molten ionic compounds) can move freely. Liquids are therefore able to flow and fill a container to a given level. Solids, on the other hand, are composed of particles in relatively fixed positions. The atoms, molecules, or ions are not able to move any appreciable distances, but rather vibrate about in those positions, accounting for the fact that solids have definite shapes.

When the particles making up a solid are given enough energy to overcome the attractive forces holding them in place, the solid melts. In molecular liquids and solids, molecules with sufficient energy can also leave the surface and enter the gaseous state. In the reverse process, as energy is removed, gases condense into liquids or solids, and liquids solidify. All these properties of solids and liquids have great practical importance.

11.1 THE LIQUID STATE

CD-ROM Screen 13.2: Phases of Matter: The Kinetic-Molecular Theory

At low enough temperatures, gases condense to liquids. Condensation occurs when most molecules no longer have enough kinetic energy to overcome their intermolecular attractions (⟵ *p. 383*). Most liquids are substances whose condensation temperatures are above room temperature such as water, alcohol, and gasoline. There are, however, some liquids, such as molten salts and liquid polymers, that have no corresponding vapor state, so condensation of their vapors has no meaning.

In a liquid, the molecules are closer together than those in a gas. However, the molecules remain mobile enough that the liquid flows. At the nanoscale level, liquids have a regular structure only in very small regions; most of the molecules continue to move about randomly. Because they are difficult to compress and their molecules are moving in all directions, confined liquids can transmit applied pressure equally in all directions. This property is used in the hydraulic fluids that operate mechanical devices such as automotive brakes and airplane wing flaps and rudders.

Solid, liquid, and gas phases of water.

The resistance of a liquid to flow is called **viscosity**. Water flows smoothly and quickly (low viscosity), while motor oil or honey flow more slowly (higher viscosity; Figure 11.1). The viscosity of a liquid is related to its intermolecular forces, which determine how easily the molecules can move past each other, and also to the degree to which the molecules can become entangled. Viscosity decreases as temperature increases because the molecules have more kinetic energy at higher temperature and the attractive intermolecular forces are more easily overcome. Examples of common materials whose viscosity decreases as their temperature rises are motor oil, cooking oil, and honey.

Unlike gases, liquids have *surface properties*. Molecules beneath the surface of the liquid are completely surrounded by other liquid molecules and experience intermolecular attractions in all directions. By contrast, molecules at the surface are attracted only by molecules below or beside them (Figure 11.2). This unevenness of attractive forces at the liquid surface causes the surface to contract. The energy required to expand a liquid surface is called the **surface tension**, and it is higher for liquids that have strong intermolecular attractions. For example, water's surface tension is high compared with those of most other liquids (Table 11.1) because extensive hydrogen bonding holds water molecules together (⬅ *p. 389*). The surface tension of water, 7.29×10^{-2} J/m^2, is the quantity of energy (J) required to increase the surface area of water 1 m^2 at 20 °C. The very high surface tension of mercury (six times that of water) is due to the strong metallic bonding (Section 11.8) that holds Hg atoms together in the liquid.

Even though their densities are greater than that of water, water bugs can walk on the surface of water and small metal objects can "float." Surface tension prevents the objects from breaking through the surface and sinking. Surface tension also accounts for the nearly spherical shape of small water droplets falling as rain as well as the rounded shape of water droplets that bead up on waxed surfaces. A sphere has less surface area per unit volume than any other shape, so a spherical droplet has fewer surface H$_2$O molecules than any other shape would have; the surface tension is thus minimized.

CD-ROM Screen 13.11: Surface Tension/Capillary Action/Viscosity

Figure 11.1 Viscosity. Honey is viscous, so it builds up rather than spreading out, as less viscous water would.

The teardrop shape of raindrops you see on TV or in illustrations of rain bear little resemblance to actual raindrops, which are spherical. Air pressure begins to distort the sphere into an egg shape for 2- to 3-mm-radius raindrops. Larger raindrops assume other shapes, but they never look like teardrops.

Surface tension of water supports a water bug. This answers the question posed in Chapter 1 (⬅ *p. 32*) "Why can some insects walk on water?"

Exercise 11.1 Explaining Differences in Surface Tension

At 20 °C, chloroform (CHCl$_3$) has a surface tension of 2.68×10^{-2} J/m^2, while bromoform (CHBr$_3$) has a higher surface tension of 4.11×10^{-2} J/m^2. Explain this observation in terms of intermolecular forces.

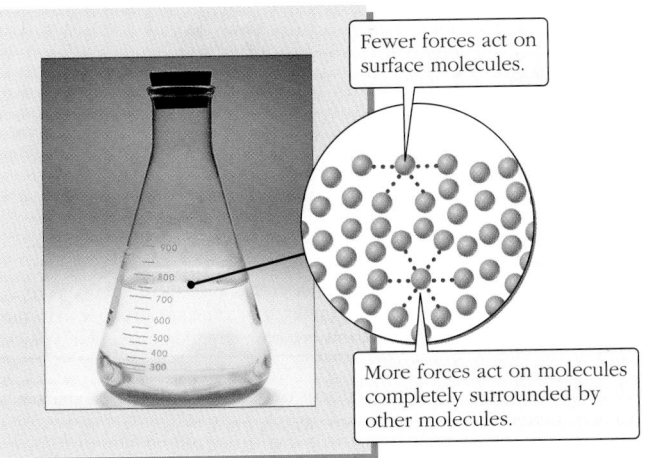

Fewer forces act on surface molecules.

More forces act on molecules completely surrounded by other molecules.

Figure 11.2 Surface tension. Surface tension is the energy required to increase the surface area of a liquid. It arises from the difference between the forces acting on a molecule within the liquid and those acting on a molecule at the surface of the liquid.

Surface tension answers the question posed in Chapter 1 (⬅ *p. 32*) "Why does rain fall as drops instead of cubes or cylinders?"

TABLE 11.1	Surface Tensions of Some Liquids	
Substance	**Formula**	**Surface tension (J/m² at 20 °C)**
Benzene	C_6H_6	2.85×10^{-2}
Chloroform	$CHCl_3$	2.68×10^{-2}
Ethanol	CH_3CH_2OH	2.23×10^{-2}
Octane	C_8H_{18}	2.16×10^{-2}
Mercury	Hg	46×10^{-2}
Water	H_2O	7.29×10^{-2}

A class of chemicals called surfactants (soaps and detergents are examples) can dissolve in water and dramatically lower its surface tension. When this happens, water becomes "wetter" and does a better job of cleaning. You can see the effect of a surfactant by comparing how water alone beads on a surface such as the hood of a car and how a soap solution fails to bead on the same surface. Surfactants are discussed in Section 15.10.

Exercise 11.2 Predicting Surface Tension

Predict which substance listed in Table 11.1 has a surface tension most similar to that of

(a) Glycerol, which has three —OH groups.
(b) Decane, $C_{10}H_{22}$.

When a glass tube with a small diameter is put into water, the water rises in the tube due to **capillary action**. The glass walls of the tube are largely silicon dioxide (SiO_2), so the water molecules form hydrogen bonds to the oxygen atoms in the glass wall. This attractive force is stronger than the attractive forces among water molecules (also hydrogen bonding). The result is that water creeps up the wall of the tube. Simultaneously, the surface tension of the water tries to keep the water's surface area small. The combination of the forces raises the water level in the tube, and a concave surface called a **meniscus** develops on the water (Figure 11.3). The water rises in the tube until the force of the water-to-wall hydrogen bonds is balanced by the pull of gravity. Capillary action is crucial to plant life because it helps water with its dissolved nutrients to move upward through plant tissues against the force of gravity.

Mercury behaves oppositely from water. Mercury in a small-diameter tube has a dome-shaped (convex) surface because the attractive forces between mercury atoms are stronger than the attraction between the mercury and the glass wall.

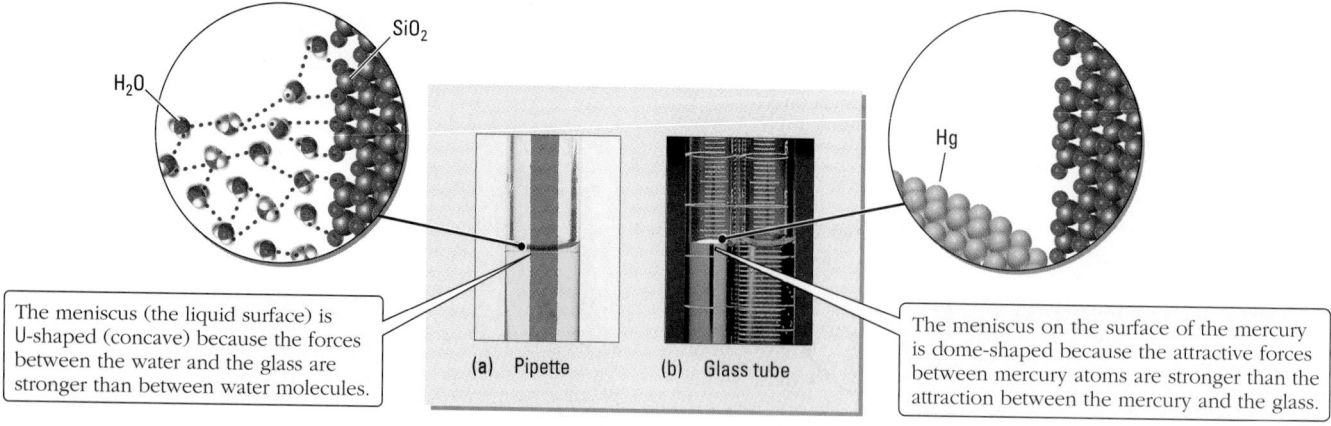

The meniscus (the liquid surface) is U-shaped (concave) because the forces between the water and the glass are stronger than between water molecules.

(a) Pipette (b) Glass tube

The meniscus on the surface of the mercury is dome-shaped because the attractive forces between mercury atoms are stronger than the attraction between the mercury and the glass.

Figure 11.3 Noncovalent forces at surfaces. (a) The meniscus of water in a pipette is concave. (b) The meniscus of mercury in a glass tube is dome-shaped (convex).

11.2 VAPOR PRESSURE

The tendency of a liquid to vaporize is called its **volatility**. Volatility increases with temperature. Everyday experiences such as heating water or soup on a stove, or evaporation of rain from hot pavement, demonstrate this phenomenom. Conversely, at lower temperatures the volatility of a liquid is lower.

The change of volatility with temperature change can be explained by taking the nanoscale view. The molecules in a liquid have varying kinetic energies and speeds (Figure 11.4). At any time, some fraction of the molecules has sufficient energy to escape from the liquid into the gas phase. When the temperature of the liquid is raised, a larger number of molecules exceed the energy threshold and evaporation proceeds more rapidly. Thus, a bowl of hot water evaporates more quickly than a bowl of warm water.

A liquid in an open container will eventually evaporate completely because air currents and diffusion take away most of the gas-phase molecules before they can re-enter the liquid phase. In a closed container, however, no molecules can escape. If a liquid is injected into a closed container that contains a perfect vacuum, so that no other substance is present, the rate of vaporization (number of molecules vaporizing per unit time) will at first far exceed the rate of condensation. The pressure of gas above the liquid will increase as the number of gas-phase molecules increases. Eventually the system will attain a state of *dynamic equilibrium*, in which molecules are entering and leaving the liquid state at equal rates. Once equilibrium is achieved it will appear that no further vaporization is occurring. At this point the pressure of the gas will no longer increase; this pressure is known as the **equilibrium vapor pressure** (or just the **vapor pressure**) of the liquid. As shown in Figure 11.5, the vapor pressure of a liquid increases with increasing temperature.

 CD-ROM Screen 13.9: Vapor Pressure

If a liquid is placed in an open container and heated, a temperature eventually is reached at which the vapor pressure of the liquid is equal to the atmospheric pressure. Below this temperature, only molecules at the surface of the liquid can go into the gas phase. But when the vapor pressure equals the atmospheric pressure, the liquid begins vaporizing throughout. Bubbles of vapor form and immediately rise to the surface due to their lower density. The liquid is said to be **boiling** (Figure 11.6). The temperature at which the equilibrium vapor pressure equals the atmospheric pressure is the **boiling point** of the liquid. If the atmospheric pressure is 1 atm, the temperature is designated the **normal boiling point**. The points in Figure 11.5 where the vapor pressure curves cross the 760 mm Hg line are the normal boiling points of the three liquids.

 CD-ROM Screen 13.10: Boiling Point

The lower the atmospheric pressure, the lower the vapor pressure at which boiling can occur. It takes longer to hard-boil an egg high in the mountains, where

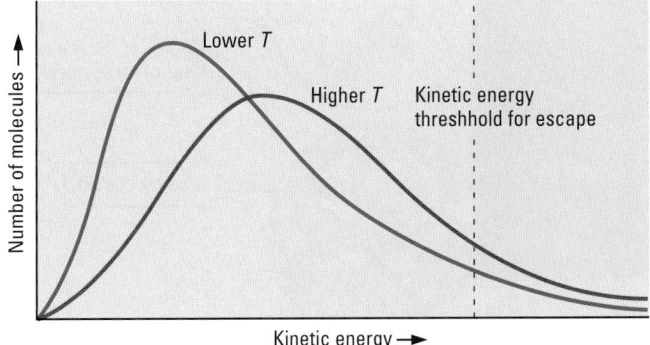

Figure 11.4 Kinetic energy and evaporation. At any given temperature, a fraction of the molecules at the surface of a liquid have sufficient energy to escape to the vapor phase. When the temperature of the liquid is raised, a larger fraction has energy greater than the kinetic energy threshold, so the vapor pressure increases.

Figure 11.5 **Vapor pressure curves for diethyl ether, $C_2H_5OC_2H_5$, ethyl alcohol, C_2H_5OH, and water.** Each curve represents the conditions of T and P where the two phases (pure liquid and its vapor) are in equilibrium. Each compound is a liquid in the temperature and pressure region to the left of its curve and a vapor for all temperatures and pressures to the right of the curve.

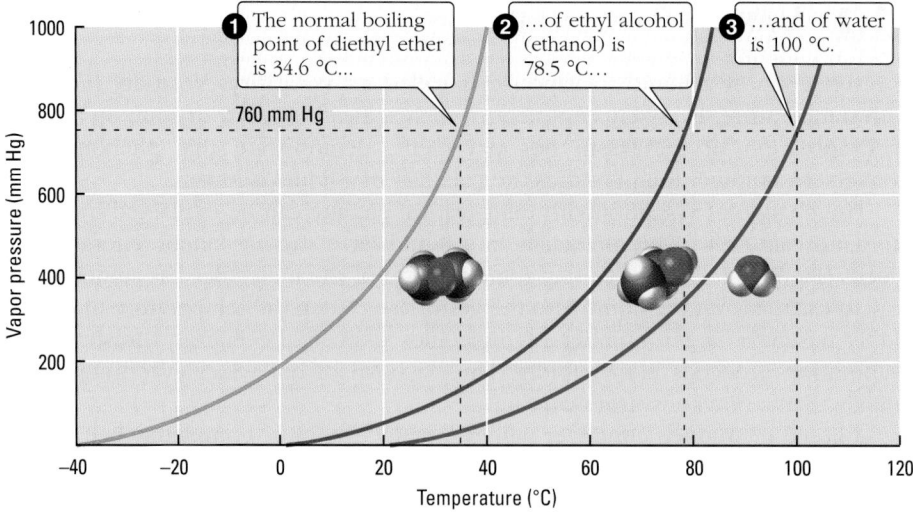

To shorten cooking times, one can use a pressure cooker. This is a sealed pot (with a relief valve for safety) that allows water vapor to build up pressures slightly greater than the external atmospheric pressure. At the higher pressure, the boiling point of water is higher, and foods cook faster.

the atmospheric pressure is lower, than it does at sea level, because the water at the higher elevation boils at a lower temperature. In Salt Lake City, Utah, where the average barometric pressure is about 650 mm Hg, water boils at about 95 °C. (Refer to Figure 11.5.)

 Exercise 11.3 Estimating Boiling Points

Use Figure 11.5 to estimate the boiling points of the liquids: (a) ethyl alcohol at 400 mm Hg; (b) diethyl ether at 200 mm Hg; (c) water at 400 mm Hg.

 Exercise 11.4 Explaining Bubbles

One of your classmates believes that the bubbles in a boiling liquid are air bubbles. Explain to him what is wrong with that idea and explain what the bubbles actually are. Suggest an experiment to show that the bubbles are not air.

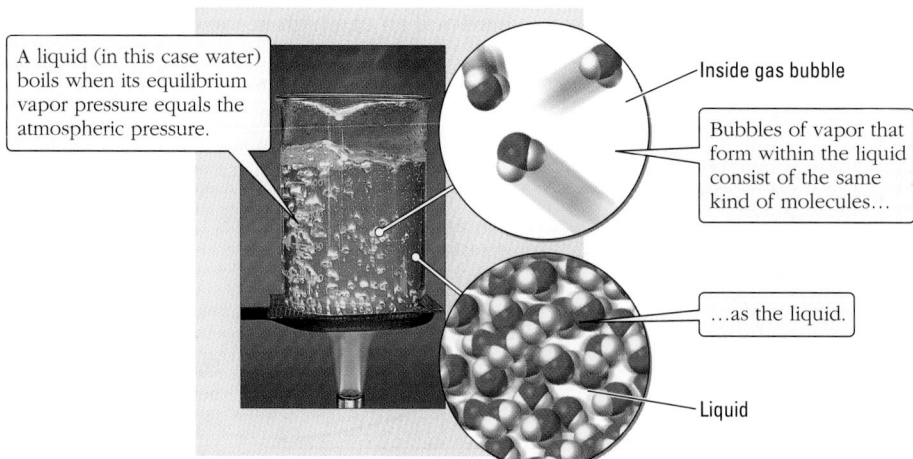

Figure 11.6 Boiling liquid.

11.3 PHASE CHANGES: SOLIDS, LIQUIDS, AND GASES

The three states of matter—solid, liquid, gas—can change from one to another through processes that are commonly seen in everyday life. Water evaporates from the street after a rain shower. Ice cubes melt in your cold drink as water vapor in the air condenses to water droplets on the outside wall of the glass. Each of the three states of matter can be converted to the other two states through six important processes, discussed in the following pages (Figure 11.7).

Vaporization and Condensation

Like molecules in a gas, the molecules in a liquid are in constant motion and have a range of kinetic energies like that shown in Figure 10.6 *(← p. 415)*. Some molecules have more kinetic energy than the potential energy of intermolecular attractive forces among the liquid molecules. If such a molecule is at the surface of the liquid and moving in the right direction, it will leave the liquid phase and enter the gaseous phase (Figure 11.8). This process is **vaporization** or **evaporation**.

As the high-energy molecules leave the liquid, they take some energy with them. Therefore, the process of vaporization is *endothermic*; the enthalpy change is called the heat of vaporization *(← p. 223)*. The higher the temperature, the faster the vaporization process because a larger fraction of the molecules has enough energy to vaporize.

 Exercise 11.5   Evaporative Cooling

In some countries where electric refrigeration is not readily available, drinking water is chilled by placing it in porous clay water pots. Water slowly passes through the clay, and when it reaches the outer surface, it evaporates. Explain how this cools the water inside.

A molecule in the gas phase will eventually transfer some of its kinetic energy by colliding with slower gaseous molecules and solid objects. If it should happen to come in contact with the liquid's surface again, it can reenter the liquid phase in a process called **condensation**. The overall effect of molecules reentering the liquid phase is the release of heat, making condensation an exothermic process. The heat evolved by condensation is equal to the heat absorbed by vaporization.

$$\text{Liquid} \underset{\text{heat of condensation}}{\overset{\text{heat of vaporization}}{\rightleftharpoons}} \text{Gas} \qquad \Delta H_{\text{vaporization}} = -\Delta H_{\text{condensation}}$$

 CD-ROM Screen 6.11: Heat Associated with Phase Changes

This answers the question posed in Chapter 1 *(← p. 32)* "Why do droplets of water form on the outside of a cold soft drink can?"

 CD-ROM Screen 13.8: Enthalpy of Vaporization

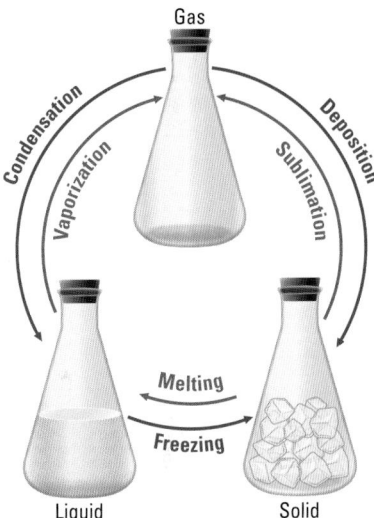

Figure 11.7   The three phases of matter and their six interconversions. The red arrows indicate endothermic conversions—processes that require an input of heat. The blue arrows indicate exothermic conversions—processes that require cooling.

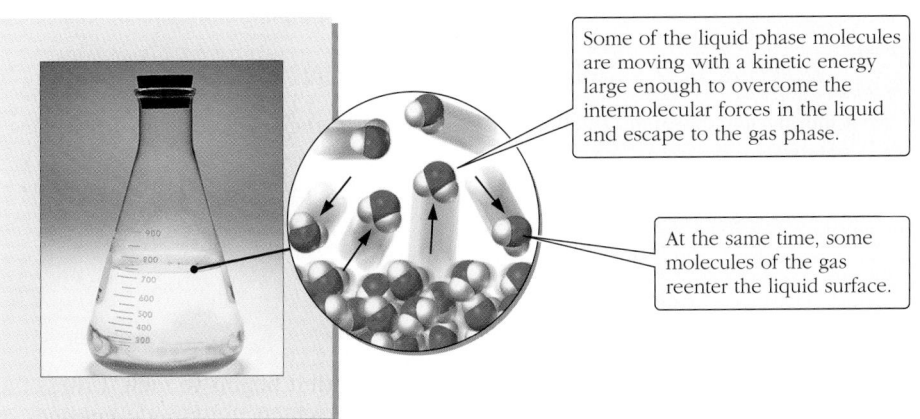

Some of the liquid phase molecules are moving with a kinetic energy large enough to overcome the intermolecular forces in the liquid and escape to the gas phase.

At the same time, some molecules of the gas reenter the liquid surface.

Figure 11.8   Molecules in the liquid and gas phases. All the molecules in both phases are moving, although the distances traveled in the liquid before collision with another molecule are much smaller.

The thermal energy absorbed during evaporation is sometimes called the latent heat of vaporization. It is somewhat dependent on the temperature: at 100 °C, ΔH°_{vap} (H_2O) = 40.7 kJ/mol, but at 25 °C the value is 44.0 kJ/mol.

For example, the quantity of heat required to completely vaporize 1 mol of liquid water once it has reached the boiling point of 100 °C (the *molar heat of vaporization*) and the quantity of heat released when 1 mol of water vapor at 100 °C condenses to liquid water at 100 °C (the *molar heat of condensation*) are (at 1 bar)

$$H_2O(\ell) \longrightarrow H_2O(g) \qquad\qquad \Delta H^\circ = \Delta H_{vap} = +40.7 \text{ kJ/mol}$$

$$H_2O(g) \longrightarrow H_2O(\ell) \qquad\qquad \Delta H^\circ = \Delta H_{cond} = -40.7 \text{ kJ/mol}$$

Table 11.2 illustrates the influence of noncovalent forces on heats of vaporization and boiling points. In the series of nonpolar molecules and noble gases, the increasing noncovalent forces (⇐ *p. 383*) with increasing molecular size and numbers of electrons are shown by the increasing ΔH°_{vap} and boiling point values. Comparison of HF and H_2O with CH_4, which have the same number of electrons, shows the effect of hydrogen bonding.

Problem-Solving Example 11.1 Enthalpy of Vaporization

You put 1.00 L of water (about 4 cups) in a pan at 100 °C , and it evaporates. How much thermal energy must have been transferred (at 1 bar) to the water for all of it to vaporize? (The density of liquid water at 100 °C is 0.958 g/mL.)

Answer 2.17×10^3 kJ

Explanation There are three pieces of information you need to solve this problem:
(a) ΔH°_{vap} for water is 40.7 kJ/mol at 100 °C (Table 11.2).
(b) The density of water is needed because ΔH°_{vap} has units of kJ/mol, so you must first find the mass of water and then the number of moles.
(c) The molar mass of water is 18.02 g/mol.

From the density of water, a volume of 1.00 L (or 1.00×10^3 cm^3) is equivalent to 958. g, and this mass in turn is equivalent to 53.2 mol of water. Therefore,

Heat required for vaporization = 53.2 mol H_2O × 40.7 kJ/mol = 2170 kJ

This enthalpy change of 2170 kJ is equivalent to about one quarter of the energy in the daily food intake of an average person in the United States.

Problem-Solving Practice 11.1

A rainstorm deposits 2.5×10^{10} kg of rain. Using the heat of vaporization of water at 25 °C of 44.0 kJ/mol, calculate the quantity of thermal energy, in joules, transferred when this much rain forms. Is this process exothermic or endothermic?

Exercise 11.6 Estimating ΔH°_{vap}

Using the data from Table 11.2, estimate the ΔH°_{vap} for Kr. Do the same for NO_2. (*Hint:* Use the periodic table or choose a substance that has a similar number of electrons.)

Exercise 11.7 Understanding Boiling Points

(a) Chlorine and bromine are both diatomic. Explain the difference in their boiling points.
(b) Methane and ammonia have the same number of electrons. Explain the difference in their boiling points.

Melting and Freezing

When a solid is heated, its temperature increases until it begins to melt. Unless the solid decomposes first, a temperature is reached at which the kinetic energies of

TABLE 11.2	Molar Enthalpies of Vaporization and Boiling Points for Some Common Substances		
Substance	Number of electrons	ΔH°_{vap} (kJ/mol)*	Boiling point (°C)†
Polar molecules			
SO_2	32	26.8	−10.0
HF	10	25.2	19.7
HCl	18	17.5	−84.8
NH_3	10	25.1	−33.4
H_2O	10	40.7	100.0
HBr	36	19.3	−66.5
HI	54	21.2	−35.1
Noble gases			
He	2	0.08	−269.0
Ne	10	1.8	−246.0
Ar	18	6.5	−185.9
Xe	54	12.6	−107.1
Nonpolar molecules			
H_2	2	0.90	−252.8
O_2	16	6.8	−183.0
F_2	18	6.54	−188.1
Cl_2	34	20.39	−34.6
Br_2	70	29.54	59.6
CH_4 (methane)	10	8.9	−161.5
CH_3—CH_3 (ethane)	18	15.7	−88.6
CH_3—CH_2—CH_3 (propane)	26	19.0	−42.1
CH_3—CH_2—CH_2—CH_3 (butane)	34	24.3	−0.5

*ΔH°_{vap} is given at the normal boiling point of the liquid.

†Boiling point is the temperature at which vapor pressure = 760 mm Hg.

the molecules or ions are sufficiently high that they push past one another out of their fixed positions. The solid's structure collapses, and the solid melts (a process also known as fusion) (Figure 11.9). This temperature is the melting point of the solid. Melting requires transfer of energy, the *heat of fusion*, from the surroundings into the system and so is always endothermic. The *molar heat of fusion* (⇐ p. 222) is the thermal energy required to melt 1 mol of a pure solid. Solids with high heats of fusion usually melt at high temperatures, and solids with low heats of fusion usually melt at low temperatures. The reverse of melting is *solidification* or *freezing* or **crystallization**, which is always an exothermic process. The *molar heat of crystallization* has the same magnitude as the molar heat of fusion, but the opposite sign.

$$\text{Solid} \xrightleftharpoons[\text{heat of crystallization}]{\text{heat of fusion}} \text{Liquid} \qquad \Delta H_{\text{fusion}} = -\Delta H_{\text{crystallization}}$$

Table 11.3 lists melting points and heats of fusion for examples of three classes of compounds: (a) nonpolar molecular solids, (b) polar molecular solids, some capable of hydrogen bonding, and (c) ionic solids. Solids composed of low-

Some solids do not have measurable melting points, and some liquids do not have measurable boiling points because increasing temperature causes them to decompose chemically before they melt or boil. Making peanut brittle provides an example: Melted sucrose chars before it boils, but it produces a great-tasting candy.

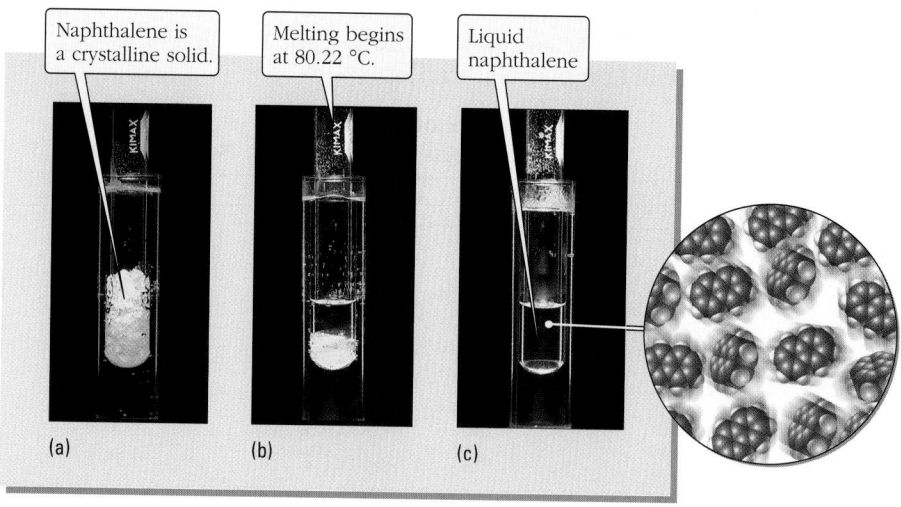

Figure 11.9 The melting of naphthalene, $C_{10}H_8$, at 80.22 °C.

molecular-weight nonpolar molecules have the lowest melting temperatures, because their intermolecular attractions are weakest. These molecules are held together by London forces only (⟵ *p. 385)*, and they form solids with melting points so low that we seldom encounter them in the solid state. Melting points and heats of fusion of nonpolar molecular solids increase with increasing number of electrons (which corresponds to increasing molar mass) as the London forces be-

TABLE 11.3	Melting Points and Enthalpies of Fusion of Some Solids		
Solid	Melting point (°C)	Enthalpy of fusion (kJ/mol)	Type of intermolecular forces
Molecular solids: Nonpolar molecules			
O_2	−248	0.445	These molecules have only
F_2	−220	1.020	London forces (which
Cl_2	−103	6.406	increase with the number
Br_2	−7.2	10.794	of electrons).
Molecular solids: Polar molecules			
HCl	−114	1.990	All of these molecules have
HBr	−87	2.406	London forces enhanced by
HI	−51	2.870	dipole-dipole forces. H_2O
H_2O	0	6.020	also has significant
H_2S	−86	2.395	hydrogen bonding.
Ionic solids			
NaCl	800	30.21	All ionic solids have strong
NaBr	747	25.69	attractions between
NaI	662	21.95	oppositely charged ions.

come stronger. The ionic compounds in Table 11.3 have the highest melting points and heats of fusion because of the very strong ionic bonding that holds the ions together. The polar molecular solids have intermediate melting points.

Problem-Solving Example 11.2   Heat of Fusion

The molar heat of fusion of NaI is 21.95 kJ/mol at its melting point. How much thermal energy will be absorbed when 41.65 g of NaI melts?

Answer 6.100 kJ

Explanation First it is necessary to determine how many moles of NaI are present in 41.65 g of NaI.

$$41.65 \text{ g NaI} \times \frac{1 \text{ mol NaI}}{149.894 \text{ g NaI}} = 0.2779 \text{ mol NaI}$$

Now the molar heat of fusion can be used to calculate the thermal energy required to melt this many moles of NaI.

$$0.2779 \text{ mol NaI} \times \frac{21.95 \text{ kJ}}{\text{mol NaI}} = 6.100 \text{ kJ}$$

✔ We have about $\frac{1}{4}$ mole of NaI, so the energy required should be about $\frac{1}{4}$ of the molar heat of fusion, and it is.

Problem-Solving Practice 11.2

Calculate the thermal energy transfer required to melt 100.0 g of NaCl at its melting point.

Exercise 11.8   Heat Liberated upon Crystallization

Which would liberate more thermal energy, the crystallization of 2 mol of liquid bromine or the crystallization of 1 mol of liquid water?

Sublimation and Deposition

Atoms or molecules can escape directly from a solid to the gas phase, a process known as **sublimation.** The enthalpy change is the *heat of sublimation*. The reverse process, in which a gas is converted directly to a solid, is called **deposition**. The enthalpy change for this exothermic process (the heat of deposition) has the same magnitude as the heat of sublimation, but the opposite sign.

$$\text{Solid} \underset{\text{heat of deposition}}{\overset{\text{heat of sublimation}}{\rightleftharpoons}} \text{Gas} \qquad \Delta H_{\text{sublimation}} = -\Delta H_{\text{deposition}}$$

A common substance that sublimes at normal atmospheric pressure is solid carbon dioxide, which has a vapor pressure of 1 atm at -78 °C (Figure 11.10). Having such a high vapor pressure below its melting point causes solid carbon dioxide to sublime rather than melt. Because of the high vapor pressure of the solid, carbon dioxide in the liquid state can exist only at pressures much higher than 1 atm.

In the reverse of sublimation, atoms or molecules in the gas phase can be made to deposit on the surface of a solid. Deposition is used to form thin coatings of metal atoms on surfaces. Audio CD and CD-ROM discs have shiny metallic surfaces of deposited aluminum or gold atoms. A metal filament is heated in a vacuum to a temperature at which metal atoms begin to sublime rapidly off the surface.

Figure 11.10 Dry Ice. In this photo the cold vapors of CO_2 are causing moisture, which is seen as wispy white clouds, to condense. Being more dense than air at room temperature, the CO_2 vapors glide slowly toward the table top or the floor.

Solid CO_2 is commonly known by the trade name Dry Ice.

Figure 11.11 Ice subliming.

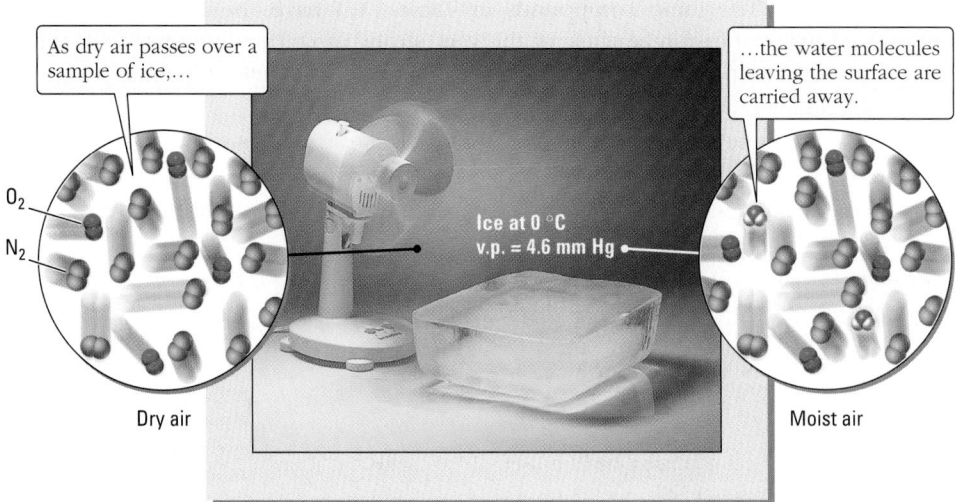

As dry air passes over a sample of ice,…

O_2

N_2

Ice at 0 °C
v.p. = 4.6 mm Hg

…the water molecules leaving the surface are carried away.

Dry air

Moist air

The plastic compact disc is cooler than the filament, so the metal atoms in the gas phase quickly deposit on the cool surface. The purpose of the metal coating is to provide a reflective surface for the laser beam that reads the pits and lands (unpitted areas) containing the digital audio or data information (⇐ *p. 272*).

Have you ever noticed that snow outdoors and ice cubes in a frost-free refrigerator slowly disappear even if the temperature never gets above freezing? The heat of sublimation of ice is 51 kJ/mol, and its vapor pressure at 0 °C is 4.60 mm Hg. Therefore, ice sublimes readily in dry air when the partial pressure of water vapor is below 4.6 mm Hg (Figure 11.11). Given enough air passing over it, a sample of ice will sublime completely, leaving no trace behind. In a frost-free refrigerator, a current of dry air periodically blows across any ice formed in the freezer compartment, taking away water vapor (and hence the ice) without warming the freezer enough to thaw the food.

Figure 11.7 (⇐ *p. 465)* summarizes the six transformations that have been discussed. Overcoming the attractive forces between atoms, molecules, or ions requires energy, so the transformations that involve separating (vaporization, sublimation, melting) are endothermic. Conversely, the transformations that involve bringing particles closer together give off energy, so these transformations are exothermic (condensation, deposition, freezing).

 Exercise 11.9 Frost-free Refrigeration

Sometimes, because of humidity conditions, a frost-free refrigerator doesn't work as efficiently as it should. Explain why.

 Exercise 11.10 Purification by Sublimation

Sublimation is an excellent means of purification for compounds that will readily sublime. Explain how purification by sublimation works at the nanoscale.

Heating Curve

Heating a solid or liquid will increase its temperature. The amount of temperature increase is governed by the quantity of heat added (in joules), the mass of the substance (in grams), and its specific heat capacity ($J g^{-1} °C^{-1}$) (⇐ *p. 217)*. The temperature will rise until a phase-transition temperature is reached, for example,

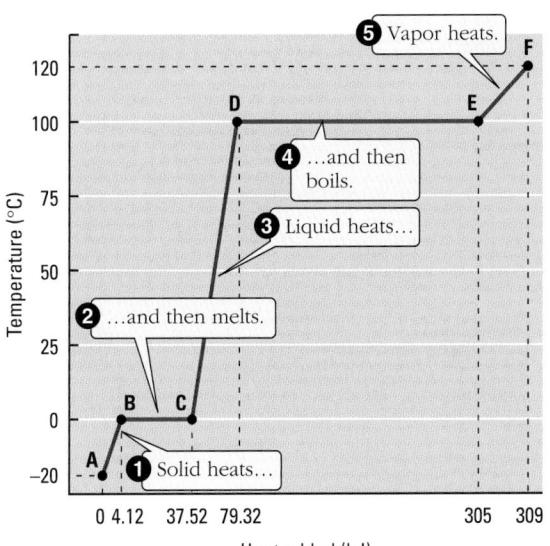

Figure 11.12 **Heating curve for water.**

100 °C for water. At this point, the temperature will no longer rise. Instead, the energy will go into changing the phase of the substance. Once all of the substance has changed phase, then the addition of still more heat will again cause the temperature to rise. A plot of the temperature of a substance versus the heat added is called a **heating curve** (Figure 11.12).

For example, consider heating 100. g of water from − 20 °C to 120 °C, as is illustrated in Figure 11.12. The constants used to construct the heating curve are

Specific heat capacity of liquid water: $4.184 \, \mathrm{J \, g^{-1} \, {}^{\circ}C^{-1}}$

Specific heat capacity of solid water: $2.06 \, \mathrm{J \, g^{-1} \, {}^{\circ}C^{-1}}$

Specific heat capacity of water vapor: $1.84 \, \mathrm{J \, g^{-1} \, {}^{\circ}C^{-1}}$

$\Delta H^{\circ}_{\text{fusion}} = 6.020 \, \mathrm{kJ/mol}$

$\Delta H^{\circ}_{\text{vaporization}} = 40.7 \, \mathrm{kJ/mol}$

The heating curve has five distinct parts. Three parts relate to heating the water in its three states (red lines). Two parts relate to phase changes (blue lines).

1. *Heat the ice to 0 °C (A → B).* $\Delta H^{\circ} = (100. \, \mathrm{g})(2.06 \, \mathrm{J \, g^{-1} \, {}^{\circ}C^{-1}})(20 \, {}^{\circ}C)$ $= 4120 \, \mathrm{J} = 4.12 \, \mathrm{kJ}$

2. *Melt the ice (B → C).* $\Delta H^{\circ} = 6.020 \, \mathrm{kJ/mol} \times \dfrac{100. \, \mathrm{g}}{18.015 \, \mathrm{g/mol}} = 33.4 \, \mathrm{kJ}$

3. *Heat the water to 100 °C (C → D).* $\Delta H^{\circ} = (100. \, \mathrm{g})(4.184 \, \mathrm{J \, g^{-1} \, {}^{\circ}C^{-1}})(100 \, {}^{\circ}C)$ $= 41800 \, \mathrm{J} = 41.8 \, \mathrm{kJ}$

4. *Evaporate the water (D → E).* $\Delta H^{\circ} = 40.7 \, \mathrm{kJ/mol} \times \dfrac{100. \, \mathrm{g}}{18.015 \, \mathrm{g/mol}} = 226. \, \mathrm{kJ}$

5. *Heat the water vapor to 120 °C (E → F).* $\Delta H^{\circ} = (100. \, \mathrm{g})(1.84 \, \mathrm{J \, g^{-1} \, {}^{\circ}C^{-1}})$ $(20 \, {}^{\circ}C) = 3680 \, \mathrm{J} = 3.68 \, \mathrm{kJ}$

The total energy required to complete the transformation is 309. kJ, the sum of the five steps. Notice that the largest portion of the energy (73%) goes into vaporizing water at 100 °C to steam at 100 °C.

Phase Diagrams

All three phases and the six interconversions among them shown in Figure 11.7 (⇐ *p. 465*) can be represented in a **phase diagram** (Figure 11.13). The temperature-pressure values at which each phase exists are also shown on the phase diagram.

CD-ROM Screen 13.17: Phase Diagrams

Figure 11.13 Generic phase diagram. The phase diagram shows the pressure-temperature regions in which the substance is solid, liquid, or vapor (gas). It also shows the melting point curve, the vapor pressure curve, and the six interconversions among the three phases.

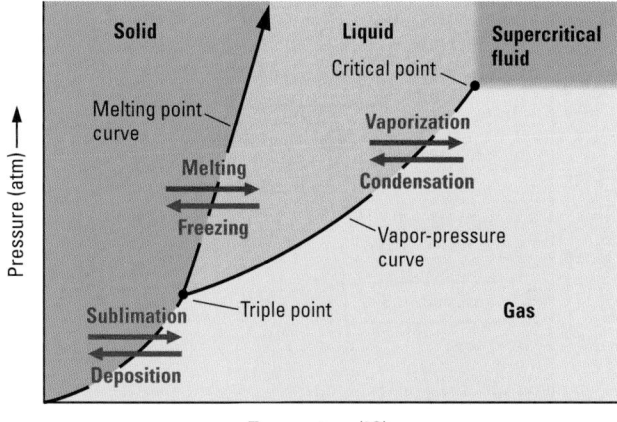

The phase diagram for water is shown in Figure 11.14. Note the three shaded regions. Each of the three phases — solid, liquid, and gas — is represented. The temperature scale has been exaggerated to better illustrate some of the features of water's phase diagram. Each line in the diagram represents the conditions of temperature and pressure at which equilibrium exists between the two phases shown on either side of the line. The temperatures and pressures along the line *AD* represent conditions at which liquid water and gaseous water are in equilibrium. This line, the *vapor pressure curve*, is the same one shown in Figure 11.5. The line *AC*, the *melting point curve*, represents the solid/liquid (ice/water) equilibrium, and the line *BA* represents the solid/gaseous (ice/water vapor) equilibrium. Note the point on the phase diagram at which all three phases are in equilibrium with one another. This temperature-pressure combination is called the **triple point**, which for water occurs at $P = 4.58$ mm Hg and $T = 0.01$ °C. Every pure substance that exists in all three phases has a characteristic phase diagram.

Problem-Solving Example 11.3 Phase Diagrams

Using Figure 11.14, the phase diagram for water, answer these questions. Starting at the triple point,
(a) What phase exists when the pressure is held constant and the temperature is increased to 0.5 °C?

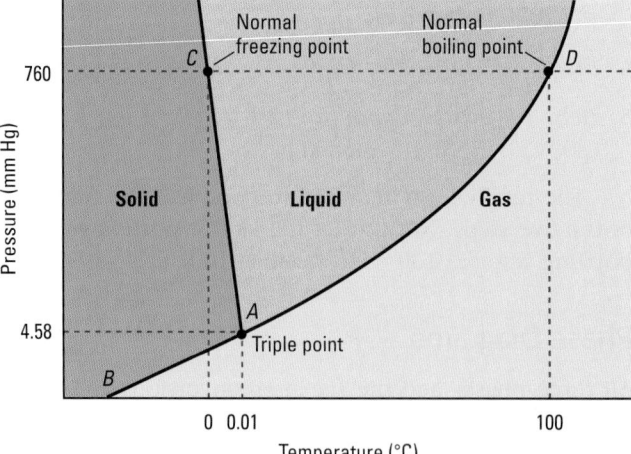

Figure 11.14 The phase diagram for water. The temperature and pressure scales are nonlinear for emphasis.

(b) What phase exists when the temperature is held constant and the pressure is increased to 20 mm Hg?

(c) What phase exists when the pressure is held constant and the temperature is decreased to 0 °C?

Answer (a) Vapor (b) Liquid (c) Solid

Explanation

(a) Increasing the temperature and holding the pressure constant means moving slightly to the right from the triple point. This is the gas region of the phase diagram.

(b) Increasing the pressure and holding the temperature constant means moving slightly up from the triple point. This is in the liquid region of the diagram.

(c) Decreasing the temperature and holding the pressure constant means moving to the left of the triple point. This is the solid region of the phase diagram.

Similar reasoning applies when changing from conditions of temperature and pressure along any line in the phase diagram.

Problem-Solving Practice 11.3

On the phase diagram of water (Figure 11.14), the point that corresponds to a temperature of 11 °C and a pressure of 9.8 mm Hg lies on the vapor pressure curve; that is, liquid water and water vapor are in equilibrium. What form of water exists when the pressure remains the same and the temperature increases by several degrees?

You are most likely familiar with the common situation in which ice and liquid water are in equilibrium because it is so useful for keeping things cool in an ice chest. If you fill an ice chest with ice, some of the ice melts to produce water. The ice-water mixture then remains at approximately 0 °C until all the ice has melted. (Thermal energy continually is transferred into the ice chest, no matter how well it is insulated.) This equilibrium between solid water and liquid water is illustrated in the phase diagram where pressure is 760 mm Hg and the temperature is 0 °C.

On a very cold day, ice can have a temperature well below 0 °C. Look along the temperature axis of the phase diagram for water and notice that below 1 atm and 0 °C the only equilibrium possible is between ice and water vapor. As a result, if the partial pressure of water in the air is low enough, ice will sublime on a cold day. Sublimation, which is endothermic, takes place more readily when solar energy warms the ice. The sublimation of ice allows snowfall to gradually disappear even though the temperature does not climb above freezing.

The phase diagram for water is unusual in that the line AC slopes in the opposite direction from that seen for almost every other substance. The right-to-left, or negative, slope of the solid-liquid equilibrium line is a consequence of the lower density of ice compared to that of liquid water (which is discussed in the next section). Thus, when ice and water are in equilibrium, one way to melt the ice is to apply greater pressure. This is evident from Figure 11.14. If you start at the normal freezing point (0 °C, 1 atm) and increase the pressure, you will move into the area of the diagram that corresponds to liquid water.

Ice skating was long thought to provide a practical example of this property of water. The melting caused by one's body weight applied to a very small area of a skate blade was thought to cause formation of sufficient water to make the ice slippery. However, more recent investigations show that the water molecules on the surface of ice are actually in a mobile state, resembling their mobility in liquid water, due to "surface melting." (See *Chemistry in the News, Melting Below Zero*, p. 476.) Ice is made much more slippery when a thin coating of liquid water is present. The thin coat of liquid water acts as a lubricant to help the skates move easily over the ice (Figure 11.15). Pressure plays very little role in making the ice slippery for skating. Although increasing the pressure on ice does cause it to melt, the effect is too small to contribute significantly in the context of ice skating.

Figure 11.15 Ice skating.

Figure 11.16 The phase diagram for carbon dioxide.

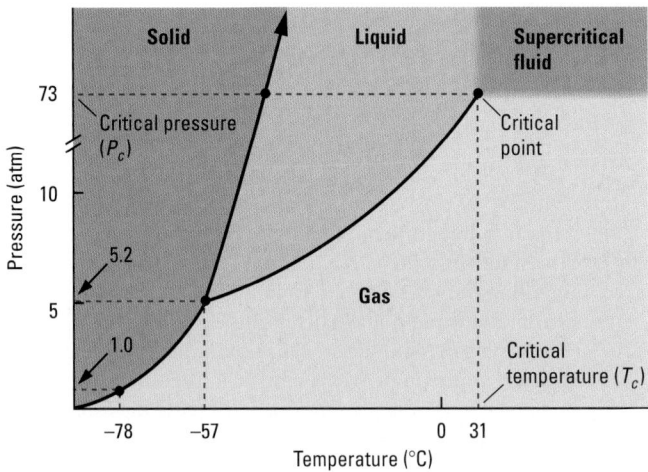

The phase diagram for CO_2 (Figure 11.16) is similar to that of water, except for one very important difference — the solid/liquid equilibrium line has a positive slope as it does for most solids. Solid CO_2 is more dense than liquid CO_2. Notice also that the pressure and temperature axes in the phase diagram for CO_2 are much different from those in the diagram for water. For pressures from about 1 to 5 atm, the only equilibrium that can exist is between solid and gaseous CO_2. This means solid CO_2 sublimes when heated. Liquid CO_2 can be produced only at pressures above 5 atm. The temperature range for liquid CO_2 is narrower than that for water. Tanker trucks marked "liquid carbonic" carry liquid CO_2, a convenient source of CO_2 gas for making carbonated beverages.

Exercise 11.11   Using Phase Diagrams

Find the pressure and temperature values for the triple point of CO_2. What phase exists when the temperature is held constant and the pressure is slightly increased over that of the triple point?

Critical Temperature and Pressure

On the phase diagram for CO_2, notice an upper point where the liquid/gas equilibrium line terminates (Figure 11.16). For CO_2 the conditions are 73 atm and 31 °C. For water the termination of the solid/liquid curve occurs at a pressure of 217.7 atm and a temperature of 374.0 °C (Figure 11.14). These conditions are called the **critical temperature** (T_c) and **critical pressure** (P_c). At any temperature *above* T_c, the molecules of a substance have sufficient kinetic energy to overcome any attractive forces, and no amount of pressure can cause them to act like a liquid again. Above T_c the substance becomes a **supercritical fluid**, which has a density characteristic of a liquid but the flow properties of a gas, thereby enabling it to diffuse through many substances easily.

Supercritical CO_2 is present in a fire extinguisher under certain conditions. As the phase diagram shows, at any temperature below 31 °C, the pressurized CO_2 (usually at about 73 atm) is a liquid and can be heard sloshing around in the container. On a hot day, however, the CO_2 becomes a supercritical fluid and cannot be heard sloshing. In fact, under these conditions the only way to know whether CO_2 is in the extinguisher — without discharging it — is to weigh it and compare its weight with the weight of the empty container, which is usually on a tag attached to the fire extinguisher.

The term "fluid" describes a substance that will flow. It is used for both liquids and gases.

Supercritical fluids are excellent solvents. Because it is nonpolar and a good solvent for nonpolar substances, supercritical CO_2 is used to extract caffeine from coffee beans. The use of supercritical CO_2 for dry cleaning is described in *Chemistry in the News: Green Chemistry: Being Critical About Grease* in Chapter 12. Supercritical water is used to extract toxic components from hazardous industrial wastes.

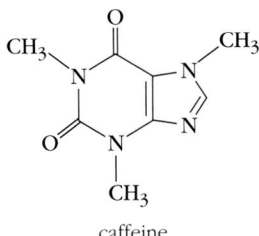

 Exercise 11.12   The Behavior of CO_2

If liquid CO_2 is slowly released to the atmosphere from a cylinder, what state will the CO_2 be in? If the liquid is suddenly released, as in the discharge of a CO_2 fire extinguisher, why is solid CO_2 seen? Can you explain this on the basis of the phase diagram alone, or do you need to consider other factors?

11.4 WATER: AN IMPORTANT LIQUID WITH UNUSUAL PROPERTIES

Earth is sometimes called the blue planet because the large quantities of water on its surface make it look blue from outer space. Three quarters of the globe is covered by oceans, with vast ice sheets at the poles. Large quantities of water are present in soils and rocks on the surface. Water is essential to almost every form of life, has played a key role in human history, and is a significant factor in weather and climate. The reason for this lies in water's unique properties (Table 11.4). Ice floats on water, although when most substances freeze, the solid sinks in its liquid. More thermal energy must be transferred to melt ice, heat water, and vaporize water than for the same amount of almost any other substance. Water has the largest thermal conductivity and the highest surface tension of any molecular substance in

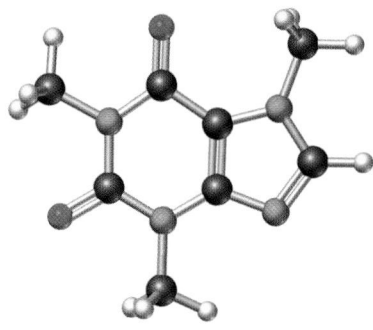

 CD-ROM Screen 13.7: The Weird Properties of Water

CHEMISTRY
You Can Do...   Melting Ice with Pressure

Do this experiment in a sink. Obtain a piece of thin, strong, single-strand wire about 50 cm long, two weights of about 1 kg each, and a piece of ice about 25 cm by 3 cm by 3 cm. The ice can be either a cylinder or a bar and can be made by pouring some water into a mold made of several thicknesses of aluminum foil and then placing it into a freezer. A piece of plastic pipe, not tightly sealed, will also work fine as a mold. You can use metal weights or make weights by filling two 1-qt milk jugs with water. (One quart is about 1 L, which is 1 kg of water.)

Fasten one weight to each end of the wire. Support each end of the bar of ice so that, without breaking it, you can hang the wire over it with one weight on each side. Observe the bar of ice and the wire every minute or so and record what happens. Suggest an explanation for what you have observed. The effect of pressure on the melting of ice and the heat of fusion of ice are good ideas to consider when thinking about your explanation.

1. Explain how the ice froze once again after the wire had passed through it.
2. What would be the effect of using a thinner piece of wire? Of using heavier weights?

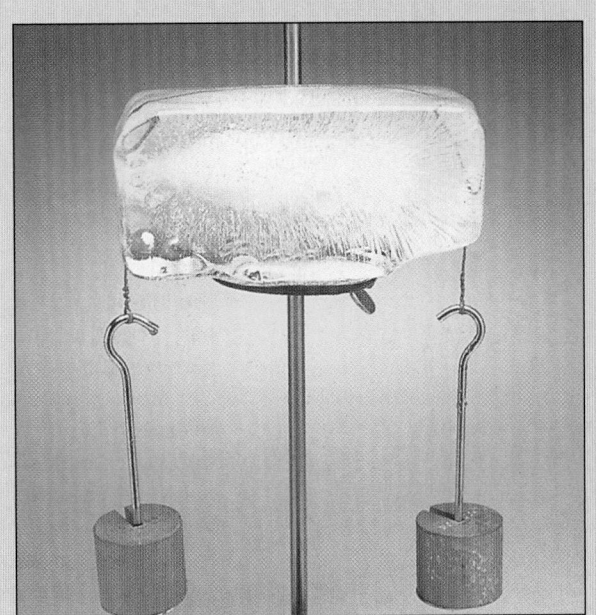

The apparatus used to melt ice with pressure.

Chemistry in the News

MELTING BELOW ZERO

Details on how the freezing of water and the thawing of ice proceed at the molecular level are being revealed by new research. The key lies in a very thin layer of water, only a few molecules thick, caused by a process termed "surface melting." This thin layer is very similar to the ice beneath it, but it has the mobility of the liquid state. Effects due to its presence include how ice skates slip over ice, how boulders heave from the ground during freeze-thaw cycles (frost heave), and the electrification of thunderclouds.

Surface melting, even at tens of degrees Celsius below the usual melting point, is an intrinsic property of the ice surface. It's due to the fact that the surface water molecules have interactions with fewer neighbors than do those molecules deeper in the solid. The ice lattice becomes less and less ordered closer to the surface because the surface molecules have the fewest hydrogen bonds holding them in place. They vibrate more than the molecules deeper in the solid. At temperatures well below the melting point, the surface molecules begin to take on the characteristics of a liquid. Surface melting was first described by Michael Faraday in 1842, and scientists are even now refining their understanding of this phenomenon.

The presence of the surface liquid phase permits you to develop a snowball. The liquid on the surface of ice particles refreezes and joins them together into a snowball. Truly dry snow, like beach sand, does not permit the formation of a snowball.

While the ice-water system is best known in the everyday world, this surface melting phenomenon occurs for virtually all solids.

Source:

Wettlaufer, J. S., and J. G. Dash. Melting Below Zero. *Scientific American*, February 2000, pp. 50–53.

At a sufficiently high temperature, well below their normal melting point, the water molecules become liquid-like.

Nearer the surface, the lattice is distorted, and the molecules are freer to move.

Deep in the ice crystal, water molecules are locked in a rigid lattice as each is H-bonded with its four nearest neighbors.

The surface of ice is slick due to a liquid-like film of water.

The six-sided symmetry of snowflakes corresponds to the symmetry of these hexagonal rings.

the liquid state. In other words, this most common of substances in our daily lives has properties that are highly unusual and that are crucial for the welfare of our planet and our species.

Most of water's unusual properties can be attributed to its unique capacity for hydrogen bonding. Referring back to Figure 9.22 (⇐ *p. 389*), you will recall that one water molecule can participate in four hydrogen bonds to other water molecules. When liquid water freezes, a three-dimensional network of water molecules forms to accommodate the maximum hydrogen bonding. In the crystal lattice of ice, the oxygen atoms lie at the corners of puckered, six-sided rings. Each of the six sides of a ring consists of one O—H covalent bond and one hydrogen bond between the H atom and an O atom of a different water molecule. In order to accommodate the linear hydrogen bond geometry, considerable open space is left within the rings, forming empty channels that run through the entire crystal lattice.

TABLE 11.4	Unusual Properties of Water	
Property	**Comparison with other substances**	**Importance in physical and biological environment**
Specific heat capacity ($4.18 \, J \, g^{-1} \, K^{-1}$)	Highest of all liquids and solids except NH_3	Moderates temperature in the environment and in organisms; climate affected by movement of water (e.g., Gulf Stream)
Heat of fusion (333 J/g)	Highest of all molecular solids except NH_3	Freezing water releases large quantity of thermal energy; used to save crops from freezing by spraying them with liquid water
Heat of vaporization (2250 J/g)	Highest of all molecular substances	Condensation of water vapor in clouds releases large quantities of thermal energy, fueling storms
Surface tension ($7.3 \times 10^{-2} \, J/m^2$)	Highest of all molecular liquids	Contributes to capillary action in plants; causes formation of spherical droplets; supports insects on water surfaces
Thermal conductivity ($0.6 \, J \, s^{-1} \, m^{-1} \, K^{-1}$)	Highest of all molecular liquids	Provides for transfer of thermal energy within organisms; rapidly cools organisms immersed in cold water, causing hypothermia

When ice melts to form liquid water, approximately 15% of the hydrogen bonds are broken and the rigid ice lattice collapses. Although large clusters of water molecules remain hydrogen-bonded, some molecules in the liquid phase are free to move into the empty spaces that were present in the ice lattice. Therefore there are more molecules, and hence more mass, in a given volume of liquid than in the same volume of solid. This makes the density of liquid water greater than for ice at the melting point. The density of ice at 0 °C is 0.917 g/mL, and that of liquid water at 0 °C is 0.998 g/mL. The density difference is not large, but it is enough so that ice floats on the surface of the liquid. This explains why about 90% of an iceberg is submerged and about 10% is above water.

As the liquid water is warmed further, more hydrogen bonds break, and more empty space is filled by water molecules. The density continues to increase until a temperature of 4 °C is reached. As the temperature rises beyond 4 °C , increased

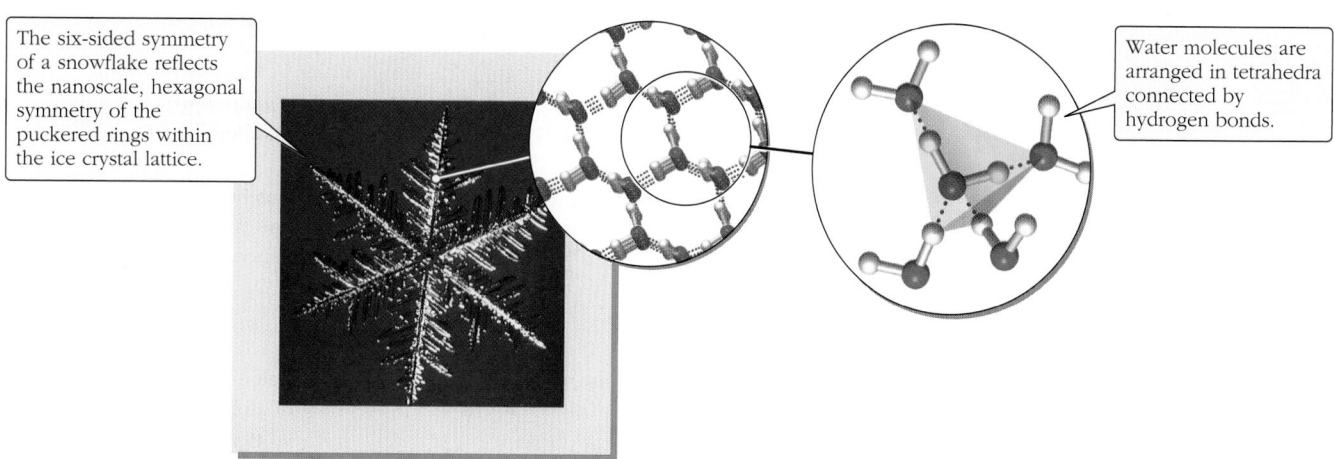

The six-sided symmetry of a snowflake reflects the nanoscale, hexagonal symmetry of the puckered rings within the ice crystal lattice.

Water molecules are arranged in tetrahedra connected by hydrogen bonds.

Open-cage structure of ice.

molecular motion causes the molecules to push each other aside more vigorously, the empty space between molecules increases, and the normal behavior of decreasing density with increasing temperature is observed. At 4 °C liquid water has a density of 1.0000 g/mL, and the density decreases by about 0.0001 g/mL for every 1 °C temperature rise above 4 °C.

Because of this unusual variation of density with temperature, when water in a lake is cooled by the air down to 4 °C, the higher density causes the cold water to sink to the bottom. Water cooled below 4 °C is less dense and stays on the surface, where it can be cooled even further. Consequently, the water at the bottom of the lake remains at 4 °C, while that on the surface freezes. Ice on the surface insulates the remaining liquid water from the cold air, and, unless the lake is quite shallow, not all of the water freezes. This allows fish and other organisms to survive without being frozen solid in winter. When water at 4 °C sinks to the bottom of a lake in the fall, it carries with it dissolved oxygen. Nutrients from the bottom are brought to the surface by the water it displaces. This is called "turnover" of the lake. The same thing happens in the spring, when the ice melts and water on the surface warms to 4 °C. Spring and fall turnovers are essential to maintain nutrient and oxygen levels required by fish and other lake-dwelling organisms.

When water is heated, the increased molecular motion breaks more and more hydrogen bonds. The strength of these intermolecular forces requires that considerable energy be transferred to raise the temperature of 1 g of water by 1 °C. That is, water's specific heat capacity is quite large. Many hydrogen bonds are broken when liquid water vaporizes, because the molecules are completely separated. This gives rise to water's very large heat of vaporization, which is a factor in humans' ability to regulate their body temperature by evaporation of sweat. As we have already described, hydrogen bonds are broken when ice melts, and this requires a large enthalpy of fusion. Its larger-than-normal enthalpy changes upon vaporization and freezing, together with its large specific heat capacity, allow water

TABLE 11.5	Structures and Properties of Various Types of Solid Substances	
Type	**Examples**	**Structural units**
Ionic (⇐ p. 86)	$NaCl$, K_2SO_4, $CaCl_2$, $(NH_4)_3PO_4$	Positive and negative ions (some polyatomic); no discrete molecules
Metallic (Section 11.8)	Iron, silver, copper, other metals and alloys	Metal atoms (or positive metal ions surrounded by an electron sea)
Molecular (⇐ p. 75)	H_2, O_2, I_2, H_2O, CO_2, CH_4, CH_3OH, CH_3COOH	Molecules with covalent bonds
Network (Section 11.10)	Graphite, diamond, quartz, feldspars, mica	Atoms held in an infinite one-, two-, or three-dimensional network
Amorphous (glassy)	Glass, polyethylene, nylon	Covalently bonded networks of atoms or collections of large molecules with no long-range regularity in their arrangement

to moderate climate and influence weather by a much larger factor than other liquids could. In the vicinity of a large body of water, summer temperatures do not get as high and winter temperatures do not get as low (at least until the water freezes over) as they do far away from water (⇐ *p. 217*). Seattle is farther north than Minneapolis, but Seattle has a much more moderate climate because it borders the Pacific Ocean.

11.5 TYPES OF SOLIDS

The relationship of nanoscale structure to macroscale properties is a central theme of this text. Nowhere is the influence of the nanoscale arrangement of atoms, molecules, or ions on properties more evident than in the study of solids. Practicing chemists, in collaboration with physicists, engineers, and other scientists, explore such relationships as they work to create new and useful materials. Table 11.5 summarizes the physical properties of the major types of solid substances. By classifying a substance as one of these types of solid, you will be able to form a reasonably good idea of what general physical properties to expect, even for a substance that you have never encountered before.

Unlike liquids, solids are rigid — they cannot transmit pressure in all directions. Solids have varying degrees of hardness that depend on the kinds of atoms in the solid and the types of forces that hold the atoms, molecules, or ions of the solid together. For example, talc (soapstone, Figure 11.17a), which is used as a lubricant and in talcum powder, is one of the softest solids known. Diamond is one of the hardest. At the atomic level, talc consists of layered sheets containing silicon, magnesium, and oxygen atoms. Attractive forces between these sheets are very weak, and so one sheet of talc can slide along another and be removed easily from the rest. In diamond (Figure 11.17b), each carbon atom is strongly bonded to

Forces holding units together	Typical properties
Ionic bonding; attractions among charges on positive and negative ions	Hard; brittle; high melting point; poor electrical conductivity as solid, good as liquid; often water-soluble
Metallic bonding; electrostatic attraction among metal ions and electrons	Malleability; ductility; good electrical conductivity in solid and liquid; good heat conductivity; wide range of hardness and melting points
London forces, dipole-dipole forces, hydrogen bonds	Low to moderate melting points and boiling points; soft; poor electrical conductivity in solid and liquid
Covalent bonds; directional electron-pair bonds	Wide range of hardnesses and melting points (three-dimensional bonding > two-dimensional bonding > one-dimensional bonding); poor electrical conductivity, with some exceptions
Covalent; directional electron-pair bonds	Noncrystalline; wide temperature range for melting; poor electrical conductivity, with some exceptions

(a)

(b)

Figure 11.17 **(a) Talc.** Although they do not appear to be soft, these talc crystals are so soft that they can be crushed between one's fingers. **(b) Diamond.** One of the hardest substances known.

CD-ROM Screen 13.12: Crystalline and Amorphous Solids

four neighbors; each of those neighbors is strongly bonded to four other carbon atoms, and so on throughout the solid (a network solid, Section 11.10). Because of the number and strength of the bonds holding each carbon atom to its neighbors, diamond is so hard that it can scratch or cut almost any other solid. For this reason diamonds are used in cutting tools and abrasives, which are more important commercially than diamonds used as gemstones.

Although all solids consist of atoms, molecules, or ions in relatively immobile positions, some solids exhibit a greater degree of regularity than others. In **crystalline solids**, the ordered arrangement of the individual particles is reflected in the planar faces and sharp angles of the crystals. Salt crystals, minerals, gemstones, and ice are examples of crystalline solids. **Amorphous solids**, on the other hand, are somewhat like liquids in that they exhibit very little long-range order, and yet they are like solids in being hard and having definite shapes. Ordinary glass is an amorphous solid, as are organic polymers such as polyethylene and polystyrene. The remainder of this chapter will be devoted to explaining the nanoscale structures that give rise to the properties summarized in Table 11.5.

Problem-Solving Example 11.4 Types of Solids

What type of solids are these substances?
(a) Solid $Ca(NO_3)_2$ has a high melting point (~560 °C) and has low electrical conductivity that becomes much higher when the solid melts.
(b) Solid naphthalene, $C_{10}H_8$, the primary constituent of mothballs, has a low melting point (80.22 °C), is soft, and is a poor electrical conductor both as a solid or a liquid.

Answer (a) Ionic solid; (b) molecular solid.

Explanation
(a) $Ca(NO_3)_2$ is made up of ions: Ca^{2+} and NO_3^-. Furthermore, the solid has properties that correspond to those of ionic solids. Ionic solids are poor electrical conductors as solids but good conductors as liquids.
(b) Naphthalene contains covalently bonded atoms, and its properties correspond to those of molecular solids. Molecular solids are poor electrical conductors as solids or liquids.

Problem-Solving Practice 11.4

What types of solids are these substances?

(a) The hydrocarbon decane, $C_{10}H_{22}$, has a melting point of -31 °C and is a poor electrical conductor.

(b) Solid $MgCl_2$ has a melting point of 708 °C and conducts electricity only when melted.

 Exercise 11.13   Lead into Gold?

Imagine samples of gold and lead, each with smoothly polished surfaces. These two surfaces are placed in contact with one another and held in place, under pressure, for about one year. After that time the two surfaces are analyzed. The gold surface is tested for the presence of lead, and the lead surface is tested for gold. Predict what the outcome of these two tests will be and explain what has happened.

11.6 CRYSTALLINE SOLIDS

The beautiful regularity of ice crystals, crystalline salts, and gemstones suggests that they must have some *internal* regularity. Toward the end of the 18th century, scientists found that shapes of crystals can be used to identify minerals. The angles at which crystal faces meet are characteristic of a crystal's composition, but do not depend on its size and shape. The shape of each crystalline solid reflects the shape of its **crystal lattice**—the orderly, repeating arrangement of ions, molecules, or atoms that shows the position of each individual particle. In such a lattice each ion, molecule, or atom is surrounded by neighbors in exactly the same arrangement (⇐ *p. 95*). Each crystal is built up from a three-dimensional repetition of the same pattern, which gives the crystal long-range order throughout.

Unit Cells

A convenient way to describe and classify the repeating pattern of atoms, molecules, or ions in a crystal is to define a small segment of a crystal lattice as a **unit cell**—the smallest part of the lattice that, when repeated along the directions defined by its edges, reproduces the entire crystal structure. To help understand the idea of the unit cell, look at the simple two-dimensional array of circles shown in Figure 11.18. The same size circle is repeated over and over, but a circle is not a good unit cell because it gives no indication of its relationship to all the other circles. A better choice is to recognize that the centers of four adjacent circles lie at the corners of a square and to draw four lines connecting those centers. A square unit cell results; four of them are drawn in Figure 11.18. As you look at the unit cell drawn in darker blue, notice that each of four circles contributes one quarter of itself to the unit cell, so a net of one circle is located within the unit cell. When this unit cell is repeated by moving the square parallel to its edges (that is, when unit cells are placed next to and above and below the first one), the two-dimensional lattice results. Notice that the corners of a unit cell are equivalent to each other and that collectively they define the crystal lattice.

The three-dimensional unit cells from which all known crystal lattices can be constructed fall into only seven categories. These seven types of unit cells have edges of different relative lengths that meet at different angles. Only cubic unit cells composed of atoms or monatomic ions are described here. These are quite common in nature and also are simpler and easier to visualize than the more complicated unit cells. The principles illustrated, however, apply to all unit cells and all

The unit cell is a square.

Each circle at the corner of the square contributes one quarter of its area to the unit cell.

There is a net of one circle per unit cell.

Figure 11.18 Unit cell for a two-dimensional solid made from flat, circular objects, such as coins.

Figure 11.19 The three different types of cubic unit cells. The top row shows the lattice points of the three cells superimposed on space-filling spheres centered on the lattice points. The bottom row shows the three cubic unit cells with only those atoms included that belong to the unit cell.

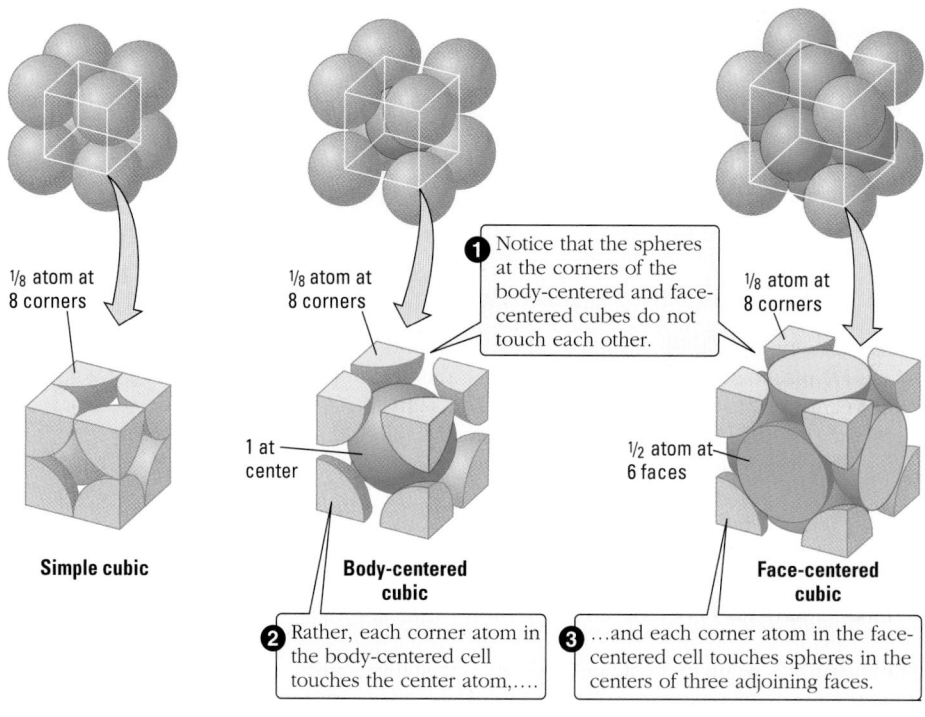

¹⁄₈ atom at 8 corners

¹⁄₈ atom at 8 corners

❶ Notice that the spheres at the corners of the body-centered and face-centered cubes do not touch each other.

¹⁄₈ atom at 8 corners

1 at center

½ atom at 6 faces

Simple cubic

Body-centered cubic

Face-centered cubic

❷ Rather, each corner atom in the body-centered cell touches the center atom,....

❸ ...and each corner atom in the face-centered cell touches spheres in the centers of three adjoining faces.

crystal structures, including those composed of polyatomic ions and complicated molecules.

Cubic unit cells have equal-length edges that meet at 90° angles. There are three types: *simple cubic* (sc), *body-centered cubic* (bcc), and *face-centered cubic* (fcc) (Figure 11.19). Both metals and ionic compounds crystallize in cubic unit cells. In metals, all three types of cubic unit cells have identical atoms centered at each corner of the cube.

When cubes pack into three-dimensional space, an atom at a corner is shared among eight cubes (Figure 11.19); this means that only one eighth of each corner atom is actually within the unit cell. Since a cube has eight corners and since one eighth of the atom at each corner belongs to the unit cell, the net result is $8 \times \frac{1}{8} =$ *1 atom within the simple cubic unit cell*. In the bcc unit cell an additional atom is at the center of the cube that lies entirely within the unit cell. This, combined with the net of one atom from the corners, gives a total of *two atoms per body-centered cubic unit cell*. In the face-centered cubic unit cell there are six atoms or ions that lie in the centers of the faces of the cube. One half of each of these belongs to the unit cell (Figure 11.19). In this case there is a net result of $6 \times \frac{1}{2} = 3$ atoms within the unit cell, in addition to the net of one atom contributed by the corners, for a total of *four atoms per face-centered cubic unit cell*. The number of atoms per unit cell helps to determine the density of a solid.

Simple cubic: 1 atom/unit cell
bcc: 2 atoms/unit cell
fcc: 4 atoms/unit cell

Exercise 11.14 Counting Atoms in Unit Cells of Metals

Crystalline polonium has a simple cubic unit cell, lithium has a body-centered cubic unit cell, and calcium has a face-centered cubic unit cell. How many Po atoms belong to one unit cell? How many Li atoms belong to one unit cell? How many Ca atoms belong to one unit cell? Draw each unit cell. Indicate on your drawing what fraction of each atom lies within the unit cell.

Closest Packing of Spheres

In crystalline solids, the atoms, molecules, or ions are arranged as close together as possible so that their interactions are maximized, which results in a stable structure. This arrangement is most easily illustrated for metals, in which the individual particles are identical atoms that can be represented as spheres. In the arrangement known as **closest packing**, there is one layer of equal-sized spheres, shown in Figure 11.20 (1), in which each sphere is surrounded by six neighbors. A second layer can be put on top of the first layer so the spheres nestle into the depressions of the first layer (Figure 11.20 (2)). Then a third layer can be added. However, there are two ways to add the third layer, each of which yields a different structure, as shown in Figure 11.20 (3 and 4). If the third layer is directly above the first layer, an *ababab* arrangement results, and the structure is known as a **hexagonal close-packed** structure. In this *ababab* structure, the spheres of the third layer are directly above the spheres of the first layer. On the other hand, if the third layer is *not* directly above the first layer, an *abcabc* arrangement results, and the structure is known as a **cubic close-packed** structure. The unit cell for a cubic close packed structure is face-centered cubic (fcc). In the *abcabc* structure, the spheres of the third layer are directly above the holes of the first layer.

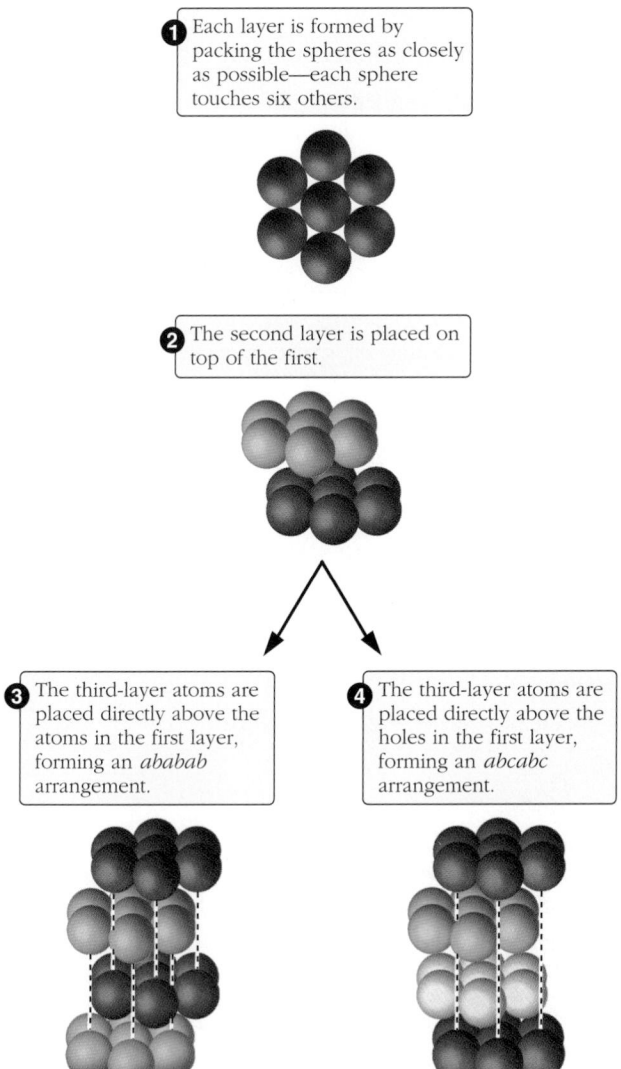

1 Each layer is formed by packing the spheres as closely as possible—each sphere touches six others.

2 The second layer is placed on top of the first.

3 The third-layer atoms are placed directly above the atoms in the first layer, forming an *ababab* arrangement.

4 The third-layer atoms are placed directly above the holes in the first layer, forming an *abcabc* arrangement.

Figure 11.20 Closest packing of spheres. Hexagonal close-packed structure (*left*) and cubic close-packed structrure (*right*).

Recently, a rigorous mathematical proof was developed that showed the impossibility of packing spheres to get a packing fraction greater than 74%.

The stacked fruits in markets are placed in closest packed arrangements for efficiency.

In each of these two close-packed structures, each sphere has 12 nearest neighbor spheres that are equidistant. In each of these arrangements, 74% of the total volume of the structure is occupied by spheres, and 26% is empty. Compare these percentages with those for the body-centered cubic structure, which is 68% occupied, and for the simple cubic structure, which is only 52% occupied.

Unit Cells and Density

What has been described about unit cells so far allows us to check whether a proposed unit cell is reasonable. Because a unit cell can be replicated to give the entire crystal lattice, the unit cell should have the same density as the crystal. Platinum has a density of 21.45 g/cm³ at 20 °C. A platinum atom has a radius of 139 pm. Are these data consistent with an experimental observation of an fcc unit cell for platinum? Starting with the mass of one Pt atom and a knowledge of the unit cell type, we can calculate the mass of one unit cell. From this mass and the density we can calculate the volume of a unit cell and then the length of its edge. From the length of the edge we can calculate the radius of one Pt atom. If the calculated radius is correct, this shows that the density and the unit cell type are consistent.

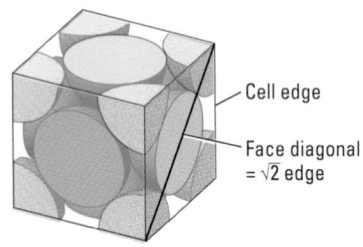

A face-centered cubic unit crystal.

Problem-Solving Example 11.5   Unit Cell Dimension, Type, and Density

Platinum crystals are fcc, with a density of 21.45 g/cm³. Platinum has an atomic weight of 195.08 amu, and platinum atoms have a 139-pm radius. What is the length of a unit cell edge? Is this consistent with Pt being fcc?

Answer   392 pm. Yes, it is consistent with fcc.

Explanation   We need to know the mass of one Pt atom, so

$$\frac{195.08 \text{ g Pt}}{1 \text{ mol Pt}} \times \frac{1 \text{ mol Pt}}{6.022 \times 10^{23} \text{ atoms Pt}} = 3.239 \times 10^{-22} \text{ g per Pt atom}$$

In an fcc unit cell there are four atoms per unit cell, so the mass of Pt atoms per unit cell is

$$4 \times 3.239 \times 10^{-22} \text{ g} = 1.296 \times 10^{-21} \text{ g Pt per unit cell}$$

We next calculate the volume of the unit cell.

$$\frac{1.296 \times 10^{-21} \text{ g}}{21.45 \text{ g/cm}^3} = 6.042 \times 10^{-23} \text{ cm}^3$$

The volume of a cube is its edge length cubed, $V = \ell^3$, so

$$\ell = \sqrt[3]{V} = \sqrt[3]{6.042 \times 10^{-23} \text{ cm}^3} = 3.924 \times 10^{-8} \text{ cm, or } 392 \text{ pm}$$

For an fcc unit cell, the relationship between the length of one side of the unit cell and its diagonal is (see figure in margin)

$$(\text{Diagonal distance})^2 = \text{edge}^2 + \text{edge}^2 = 2 \text{ edge}^2$$

We have the length of an edge, so

$$\text{Diagonal distance} = \sqrt{2} \times \text{edge} = 554 \text{ pm}$$

The radius of a Pt atom is four times the length of the diagonal (see figure in margin) so the Pt radius is 554 pm/4 = 139 pm. The values computed here are in agreement with the fcc nature of the Pt crystal.

✔   The density, atomic weight, unit cell edge length and diagonal length, and the Pt atom radius are all in agreement.

Problem-Solving Practice 11.5

Gold crystals have a bcc structure. The radius of a gold atom is 144 pm, and the atomic weight of gold is 196.97 amu. Calculate the density of gold.

CHEMISTRY
You Can Do...
Closest Packing of Spheres

In a metal the crystal lattice is occupied by identical atoms, which can be pictured as identical spheres. Obtain a bag of marbles, some expanded polystyrene foam spheres, some ping-pong balls, or some other set of reasonably small, identical spheres. Use them to construct models of each situation described here.

Consider how identical spheres can be arranged in two dimensions. Pack your spheres into a layer so that each sphere touches four other spheres; this should resemble the arrangement in part a of the figure. (You will find this somewhat hard to do; it may be necessary to use four pieces of wood to enclose the layer and keep the spheres in line. Making a square array is hard because the spheres are rather inefficiently packed, and they readily adopt a more efficient packing in which each sphere touches six other spheres.)

To make a three-dimensional lattice requires stacking layers of spheres. Start with the layer of spheres in a square array and stack another identical layer directly on top of the first; then stack another on that one, and so on. (This is really hard and will require vertical walls to hold the spheres as you stack the layers. If you are not able to do it, use sugar cubes, dice, or some other kind of small cubes instead of spheres.) This three-dimensional lattice will consist of *simple cubic* unit cells; in such a lattice 52.4% of the available space is occupied by the spheres.

Next, make the more efficiently packed hexagonal layer, in which each sphere is surrounded by six others (part b of the figure). Notice that in this layer, as in the square one, there are little holes left between the spheres. This time, however, the holes are smaller than they were in the square array; in fact, they are as small as possible, and for this reason the layer you have made is called a closest packed layer. Now add a second layer, and let the spheres nestle into the holes in the first layer. (Trying to prevent this from happening was probably the hardest task when you made your simple cubic array.) Now add a third layer. If you look carefully, you will see that there are two ways to put down the third layer. One of these places spheres directly above *holes* in the first layer. This is called *cubic close packing,* and the unit cell is face-centered cubic (fcc). The other arrangement, also found for many metals, places third-layer spheres directly above *spheres* in the first layer. This is *hexagonal close packing,* and its unit cell is *not* cubic. Cubic and hexagonal close packing are the most efficient ways known for filling space with spheres; 74% of the available space is occupied. The atoms in most metals are arranged in cubic close packing, in hexagonal close packing, or in lattices composed of body-centered cubic unit cells. Efficient packing of the atoms or ions in the crystal lattice allows stronger bonding, which gives greater stability to the crystal.

1. In either the cubic close-packed or the hexagonal close-packed arrangement, each atom has six neighbors in the same plane. How many closest neighbors does each atom have, considering all three dimensions?
2. How many closest neighbors does each atom have in the simple cubic structure?

(a)   (b)

One layer of marbles in (a) simple cubic arrangement and (b) hexagonal close-packed arrangement.

Ionic Crystal Structures

The crystal structures of many ionic compounds can be described as simple cubic or face-centered cubic lattices of spherical negative ions, with positive ions occupying spaces (called holes) among the negative ions. The number and locations of the occupied holes are the keys to understanding the relation between the lattice structure and the formula of an ionic compound. The simplest example is an ionic compound in which the hole in a simple cubic unit cell (Figure 11.19) is occupied. The ionic compound cesium chloride, $CsCl$, has such a structure. In it, each simple cube of Cl^- ions has a Cs^+ ion at its center (Figure 11.21). The spaces occupied by

CD-ROM Screen 13.13: Ionic Solids

Remember that negative ions are usually larger than positive ions. Therefore, building an ionic crystal is a lot like placing marbles (positive ions) in the spaces among larger ping-pong balls (negative ions).

It is not appropriate to describe the CsCl structure as bcc. The names "simple cubic," "body-centered cubic," and "face-centered cubic" refer to the arrangement of only one type of ion in the lattice, not both.

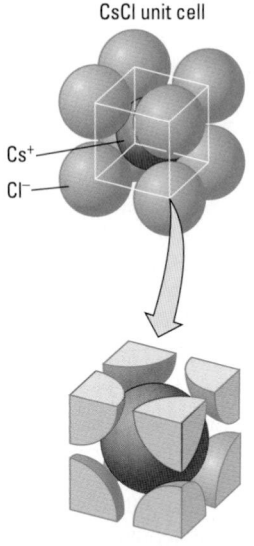

CsCl unit cell

Cs⁺

Cl⁻

Figure 11.21 Unit cell of the cesium chloride (CsCl) crystal lattice.

Since the edge of a cube in a cubic lattice is surrounded by four cubes, one fourth of a spherical ion at the midpoint of an edge is within any one of the cubes.

the Cs^+ ions are called cubic holes, and each Cs^+ has eight nearest-neighbor Cl^- ions.

The structure of sodium chloride, NaCl, is one of the most common ionic crystal lattices (Figure 11.22). It consists of an fcc lattice of the larger Cl^- ions, in which Na^+ ions occupy what are called *octahedral* holes—octahedral because each Na^+ ion is surrounded by six Cl^- ions at the corners of an octahedron. Likewise, each Cl^- ion is surrounded by six Na^+ ions. Figure 11.22 also shows a space-filling model of the NaCl lattice, in which each ion is drawn to scale based on its ionic radius.

If you look carefully at Figures 11.22 and 11.23, it is possible to determine the number of Na^+ and Cl^- ions in the NaCl unit cell. There is one eighth of a Cl^- ion at each corner of the unit cell and one half of a Cl^- ion in the middle of each face. The total number of Cl^- ions within the unit cell is

$$\tfrac{1}{8}\,Cl^- \text{ per corner} \times 8 \text{ corners} = 1\ Cl^-$$

$$\tfrac{1}{2}\,Cl^- \text{ per face} \times 6 \text{ faces} = 3\ Cl^-$$

Total of 4 Cl^- in a unit cell

There is one fourth of a Na^+ at the midpoint of each edge and a whole Na^+ in the center of the unit cell. For Na^+ ions, the total is

$$\tfrac{1}{4}\,Na^+ \text{ per edge} \times 12 \text{ edges} = 3\ Na^+$$

$$1\ Na^+ \text{ per center} \times 1 \text{ center} = 1\ Na^+$$

Total of 4 Na^+ in a unit cell

Thus, the unit cell contains four Na^+ and four Cl^- ions. This result agrees with the formula of NaCl for sodium chloride.

Exercise 11.15 Formulas and Unit Cells

Cesium chloride has a cubic unit cell, as seen in Figure 11.21. Show that the formula for the salt must be CsCl.

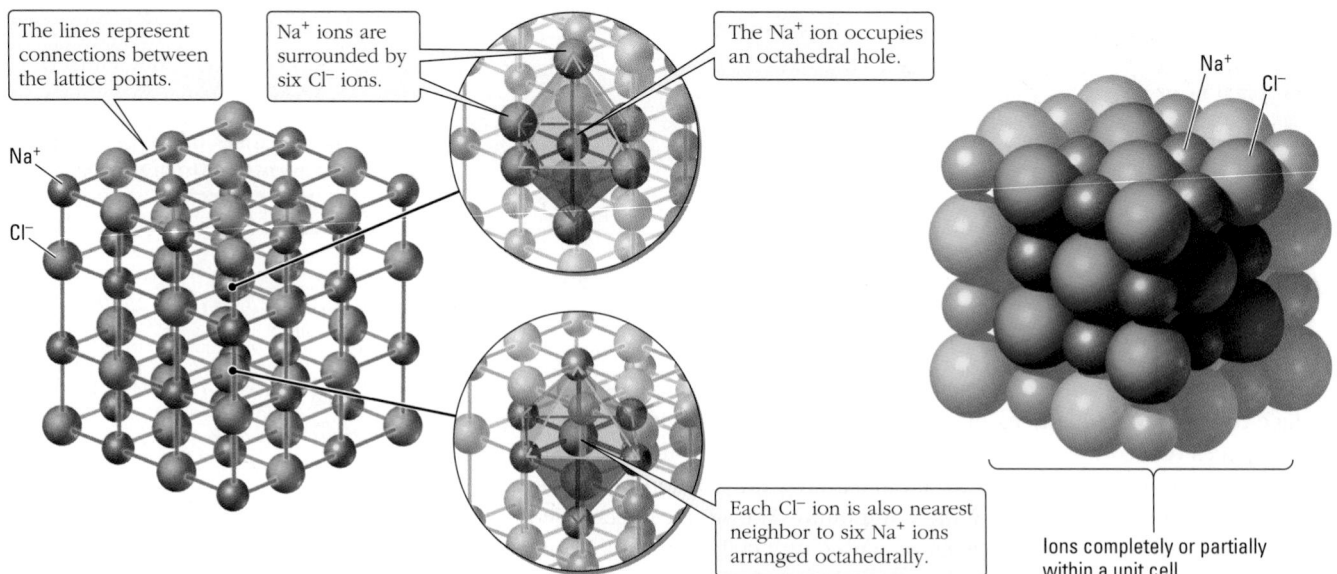

The lines represent connections between the lattice points.

Na⁺ ions are surrounded by six Cl⁻ ions.

The Na⁺ ion occupies an octahedral hole.

Na⁺

Cl⁻

Na⁺

Cl⁻

Each Cl⁻ ion is also nearest neighbor to six Na⁺ ions arranged octahedrally.

Ions completely or partially within a unit cell

Figure 11.22 The NaCl crystal lattice.

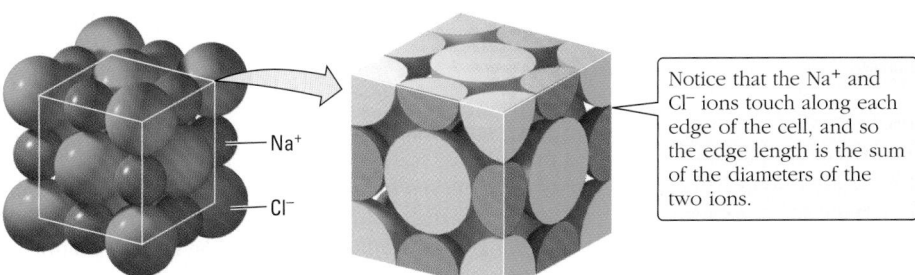

Figure 11.23   **A NaCl unit cell.**

Notice that the Na^+ and Cl^- ions touch along each edge of the cell, and so the edge length is the sum of the diameters of the two ions.

Problem-Solving Example 11.6   Calculating the Volume and Density of a Unit Cell

The unit cell of NaCl is as shown in Figure 11.23, and the ionic radii for Na^+ and Cl^- are 116 pm and 167 pm, respectively. Calculate the density of NaCl.

Answer   The calculated density is 2.14 g/cm³.

Explanation   The Na^+ and Cl^- ions are touching along the edge of the unit cell. Thus, the edge length is equal to two Cl^- radii plus two Na^+ radii.

$$Edge = 167 \text{ pm} + (2 \times 116 \text{ pm}) + 167 \text{ pm} = 566 \text{ pm}$$

The volume of the cubic unit cell is the cube of the edge length.

$$\text{Volume of unit cell} = (edge)^3 = (566 \text{ pm})^3 = 1.81 \times 10^8 \text{ pm}^3$$

Converting this to cm³,

$$1.81 \times 10^8 \text{ pm}^3 \times \left(\frac{10^{-10} \text{ cm}}{\text{pm}}\right)^3 = 1.81 \times 10^{-22} \text{ cm}^3$$

Next we can calculate the mass of a unit cell and divide it by the volume to get the density. With four NaCl formula units per unit cell,

$$4 \text{ NaCl formula units} \times \frac{58.44 \text{ g}}{\text{mol NaCl}} \times \frac{1 \text{ mol}}{6.022 \times 10^{23} \text{ formula units}} = 3.88 \times 10^{-22} \text{ g}$$

This means the density of a NaCl unit cell is

$$\frac{3.88 \times 10^{-22} \text{ g}}{1.81 \times 10^{-22} \text{ cm}^3} = 2.14 \text{ g/cm}^3$$

The experimental density is 2.164 g/cm³, which is in reasonable agreement with the calculated value. It is good to remember that all experiments have uncertainties associated with them. The density of NaCl calculated from unit cell dimensions could easily have given a value closer to the experimental density if the tabulated radii for the Na^+ and Cl^- ions were slightly smaller. There is, of course, a slight uncertainty in the published radii of all ions.

Problem-Solving Practice 11.6

KCl has the same crystal structure as NaCl. Calculate the volume of the unit cell for KCl, given that the ionic radii are K^+ = 152 pm and Cl^- = 167 pm. Now compute the density of KCl. Which has the larger unit cell, NaCl or KCl?

The general relationships among the unit cell type, ion or atom size, and solid density for most metallic or ionic solids are as follows:

Mass of 1 formula unit (e.g., NaCl) × number of formula units per unit cell (4)
= mass of unit cell
Mass of unit cell ÷ unit cell volume
(= edge³ for NaCl) = density

DOROTHY CROWFOOT HODGKIN
(1910–1994)

Dorothy Crowfoot Hodgkin was born in Egypt, where her father was a supervisor of schools and ancient monuments. Schooled in England and graduated from Oxford University in 1931, she was fascinated by the study of crystals and went to Cambridge University, where she eventually joined the faculty.

Her first major achievement in crystallography was the determination of the structure of penicillin in 1945. In 1964 she received the Nobel Prize in Chemistry for determining the molecular structure of vitamin B_{12}. Much of her work required pioneering and painstaking experimental technique that, as one biographer put it, "transformed crystallography from a black art into an indispensable scientific tool."

11.7   MATERIALS SCIENCE

Humans have long adapted or altered naturally occurring materials for their use in clothing, structures, and devices. The tanning of leather and firing of clay pots predate recorded history. Operations such as these alter the nanoscale structure of the material to improve its resulting properties. The names of the periods of early civilization reflect the defining materials of the time — Stone Age, Bronze Age, Iron

Tools of Chemistry — X-ray Crystallography

Scientists use x-ray crystallography to probe crystal structures. When x-rays strike crystals, the x-rays are scattered in different directions and with different intensities due to interference effects. **X-ray crystallography** is used to determine the absolute atomic arrangement of the atoms in solid samples. It is one of the most powerful tools of solid-state chemistry and is routinely applied to solids as varied as crystalline minerals, proteins, and semiconductors.

X-rays have wavelengths on the order of 1 nm, the same scale as atomic dimensions. When an x-ray beam strikes a crystal, the x-rays are scattered. When the interference is *constructive*, the waves combine to produce x-rays that can be observed indirectly because they expose a spot on photographic film. When the interference is *destructive*, the waves partly or entirely cancel each other, so little or no x-ray intensity is recorded.

Interference effects among the scattered x-rays from the different atoms cause the intensity of the scattered radiation to show maxima and minima in various directions, creating a diffraction pattern. The particular diffraction pattern obtained depends on the x-ray wavelength and the distances between the planes of atoms in the crystal. The Bragg equation explains these relationships, which are shown in the figure on the next page.

$$n\lambda = 2d \sin \theta$$

λ is the x-ray wavelength, d is the spacing between atomic layers, θ is the angle of scattering, and n is an integer (usually considered as one). Waves a and b reinforce each other when the Bragg equation is satisfied, that is, when the additional distance traveled by wave b to reach the detector is an integer multiple of the x-ray wavelength. Thus, the diffraction pattern can be related back to the atomic arrangement of the atoms in the crystal.

X-rays were discovered by W. C. Röntgen in 1895, and soon after that, Max von Laue and his co-workers correctly deduced that x-rays were light of very short wavelength and that the distances between layers of atoms in a crystal would be about the same as the wavelengths of x-rays. Röntgen and von Laue generated a diffraction pattern from a crystal of blue vitriol, copper(II) sulfate pentahydrate, and recorded it on a photographic plate. William and Lawrence Bragg carried forward the pioneering work in x-ray crystallography. Later, it was shown that the intensity of the scattered x-ray beam depends on the electron density of the atoms in the crystal. Therefore, a hydrogen atom is least effective in causing x-rays to scatter, while heavy atoms such as lead or mercury are quite effective.

Today, crystallographers use computer-controlled instruments, highly sensitive solid-state detectors, and other advanced instrumentation to measure the angles of reflection and the intensities of x-ray beams diffracted by crystals. Sophisticated soft-

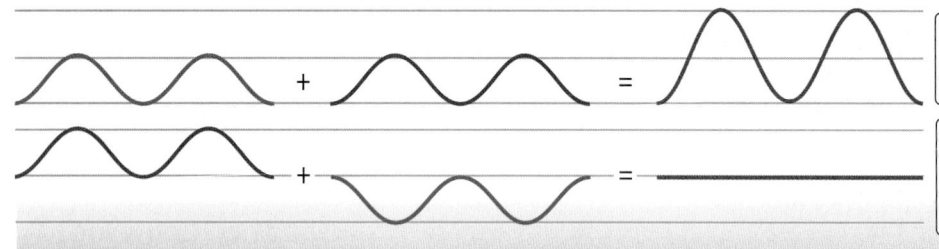

Constructive interference occurs when two in-phase waves combine to produce a wave of greater amplitude.

Destructive interference results from the combination of two waves of equal magnitude that are exactly out of phase. The result is zero amplitude.

Age. Decorative ceramic tiles and mosaics in ancient Greece and Rome were signature cultural symbols of these societies. This same close relationship between the nature of societies and their materials continues to hold in modern societies. The critical role played by materials in the modern world is a cornerstone of life today. Very recently, the *scientific* understanding of the relationships among the processing, structure, and properties of materials has been developed and exploited, and the evolution of materials has accelerated. Furthermore, new phenomena have been discovered that permit the creation of new materials. For example, superconducting ceramic oxides that conduct electricity without resistance at critical temperatures greater than the boiling point of liquid nitrogen may pave the way to astounding advances in information technology and computing in the coming decades (Section 11.9).

Just what are *materials* anyway? By materials we mean matter used to construct the devices and macroscopic structures of our highly developed technology. Application areas include transportation, communication and information technology, infrastructure, consumer goods, health care, and biomaterials. The phenomena that can be important in materials cover the entire range of characteristics of matter: electronic, optical, magnetic, superconducting, strength, and flexibility.

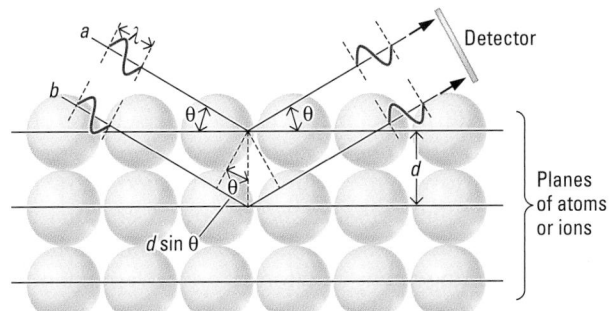

X-ray diffraction from a solid. The relationships among the variables in the Bragg equation are shown. For wave *b* to reach the detector in phase with wave *a*, *b* must travel farther by $2d \sin \theta$ and this distance must be an integral multiple of the x-ray wavelength.

ware allows the atomic arrangements in crystalline materials to be determined quickly enough that x-ray crystallography is used routinely.

In one of the most famous uses of x-ray crystal data, James Watson and Francis Crick in 1953 applied x-ray data obtained by Maurice Wilkins and Rosalind Franklin to discover the double-helix structure of DNA, the biological macromolecule that carries genetic information (⇐ *p. 392),* Watson, Crick, and Wilkins (Franklin was deceased) were awarded the Nobel Prize in Medicine and Physiology in 1962 for this discovery.

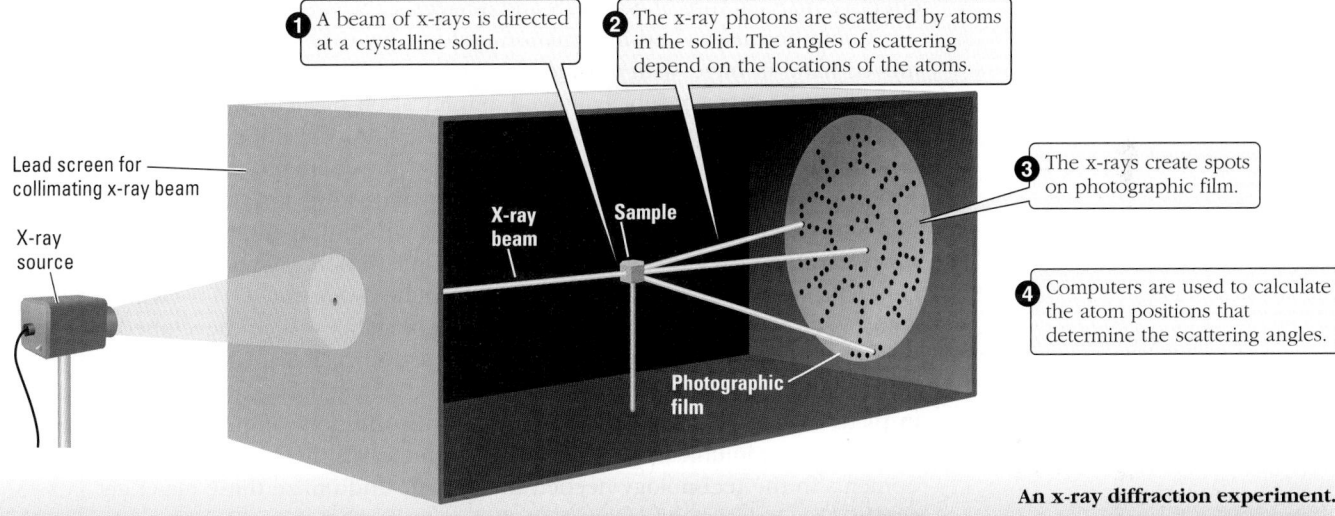

❶ A beam of x-rays is directed at a crystalline solid.

❷ The x-ray photons are scattered by atoms in the solid. The angles of scattering depend on the locations of the atoms.

❸ The x-rays create spots on photographic film.

❹ Computers are used to calculate the atom positions that determine the scattering angles.

Lead screen for collimating x-ray beam

X-ray source

X-ray beam

Sample

Photographic film

An x-ray diffraction experiment.

The three major classes of materials are metals, polymers, and ceramics. Under ceramics we include semiconductors, superconductors, and glasses, but no classification scheme is perfect — there are always exceptions. Some properties group well into the four major categories, as shown in the following list.

- **Metals**   Opaque with shiny surface unless covered by an oxidized coating, as in rusted iron. Usually crystalline. Generally ductile as with copper. Strong and stiff as with steel. Most are not magnetic, except Fe, Ni, and Co or in alloys. Usually conduct heat and electricity well, such as Cu. Form alloys such as stainless steel, bronze, brass.

- **Polymers**   Both natural (wool, cotton, rubber) and synthetic (polyethylene, styrofoam). Most are plastic and can be bent, but some are brittle. Thermoplastics become more flexible when heated; thermoset polymers become hard and brittle when heated. Usually thermal and electrical insulators. Not magnetic.

- **Ceramics**   Nonmetallic materials, often based on clays. Porcelain, bricks, glass, and cement are examples. Usually brittle and cannot be bent. Poor thermal conductors, can withstand high temperatures, and can function as thermal insulators. Usually poor electrical conductors and good electrical insulators. Some ceramics are magnetic. Usually crystalline. Many are transparent (glass). Semiconductors are ceramics with special electrical properties. Glasses are a special class of ceramics that are brittle yet often strong and are amorphous.

• **Composites** Materials combining components from the other three categories with properties that blend the best of both components.

What is **materials science**? By materials science we mean the science of the relationships between the structure and the chemical and physical *properties* of materials. How strong is the material? How dense? How easily formed? How does it react to acid, stress, magnetic fields, high pressure, heat? Does it conduct electricity? *Structure* refers to the organization of the material on scales ranging from the nanoscale (electron structure, arrangement of atoms, crystal arrangement) through the microscale and to the macroscale device.

The major breakthrough of the 20th century with respect to materials was that scientists can now *design* a material for many applications. That is, scientific understanding at the nanoscale permits synthesis and fabrication of new materials with precise properties desired for a particular application. When this capability is coupled with inventiveness and technological know-how, the result is cost-effective manufacture of materials and devices.

Communications and information systems provide examples of advances that moved hand in hand with advances in materials during the 20th century. The development of vacuum tubes allowed amplification of electrical signals and paved the way for modern electronics. Transistors were developed in the 1940s, followed in the 1960s by integrated circuits that put entire circuits on a silicon chip. A single silicon chip can contain millions of transistors and other electronic components. Our communications infrastructure and computational capabilities all depend on these materials. At the same time, lasers (which provide intense, coherent, monochromatic light) and optical fibers (Section 11.11) (which provide a means of carrying optical signals over long distances) were developed, and they have revolutionized communications and many other fields. Long distance communications are now almost universally carried by fiber optic cables. Thus, advances in materials have been one enabling factor in the development of instantaneous worldwide communication. Along with the materials themselves have come developments in the technology needed to fabricate and utilize these materials. For example, the industry of integrated chip manufacture has opened new areas of materials science, engineering, and manufacturing.

The 20th-century transportation revolution was also enabled by advances in materials. Airframes require light yet strong materials, and the advances from aluminum and its alloys to titanium to composite materials (graphite- and boron fiber-

Modern materials in action. Modern materials are employed in many ways to improve our lives, for example, in hip replacements, jet engine parts, and skis.

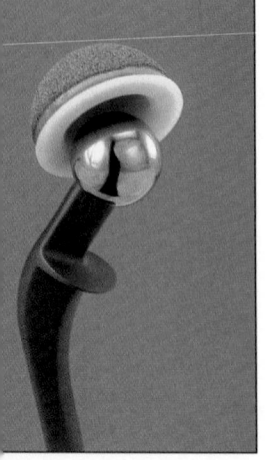

reinforced polymers, aluminum or titanium reinforced with fibers, and graphite-reinforced epoxy) have advanced the modern aircraft industry. Similar materials are used in high-performance mountain bike frames. At the same time, developments of materials that can withstand extreme temperatures and stresses in rocket and jet engine turbine blades have made modern aircraft and rockets possible. Automobiles have an ever-growing number of parts made from aluminum alloys, engineered plastics, and ceramics as well.

11.8 METALS, SEMICONDUCTORS, AND INSULATORS

All metals are solids at room temperature, except mercury (m.p. -38.8 °C). All metals exhibit common properties that we call metallic.

- **High electrical conductivity** Metal wires are used to carry electricity from power plants to homes and offices because electrons in metals are highly mobile.
- **High thermal conductivity** We learn early in life not to touch any part of a hot metal pot because it will transfer heat rapidly and painfully.
- **Ductility and malleability** Most metals are easily drawn into wire (ductility) or hammered into thin sheets (malleability); some metals (gold, for example) are more easily formed into shapes than others.
- **Luster** Polished metal surfaces reflect light. Most metals have a silvery-white color because they reflect all wavelengths equally well.
- **Insolubility in water and other common solvents** No metal dissolves in water, but a few (mainly from Groups 1A and 2A) react with water to form hydrogen gas and solutions of metal hydroxides.

As you will see, these properties are explained by the kinds of bonding that hold atoms together in metals.

The enthalpies of fusion and melting points of metals vary greatly. Low melting points correlate with low enthalpies of fusion, which implies weaker attractive forces holding the metal atoms together. Mercury, a liquid at room temperature, has an enthalpy of fusion of only 2.3 kJ/mol. The alkali metals and gallium also have very low enthalpies of fusion and notably low melting points (Table 11.6). Compare these values with those in Table 11.3 for nonmetals (⇐ *p. 468).*

TABLE 11.6	Enthalpies of Fusion and Melting Points of Some Metals	
Metal	$\Delta H^{\circ}_{\text{fusion}}$ **(kJ/mol)**	**Melting point (°C)**
Hg	2.3	-38.8
Li	3.0	180.5
Na	2.59	97.9
Ga	7.5	29.78*
Al	10.7	660.4
U	12.6	1132.1
Ti	20.9	1660.1
W	35.2	3410.1

*This means that gallium metal will melt in the palm of your hand from the warmth of your body (37 °C). It happens that gallium is a liquid over the largest range of temperature of any metal. Its boiling point is approximately 2250 °C.

Figure 11.24 Relative enthalpies of fusion for the metals in the periodic table. See Table 11.6 for some numerical values of enthalpies of fusion.

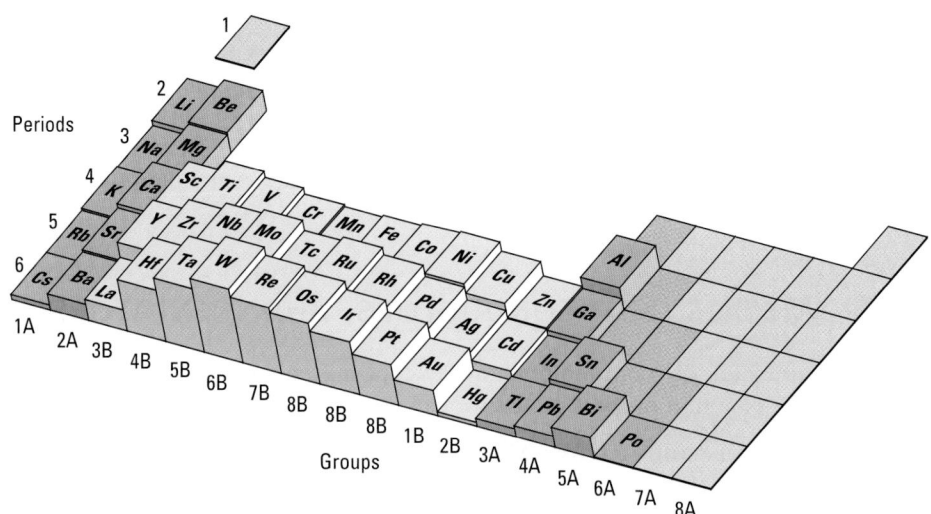

Figure 11.24 shows the relative sizes of enthalpies of fusion of the metals related to their positions in the periodic table. The transition metals, especially those of the third transition series, have very high melting points and extraordinarily high enthalpies of fusion. Tungsten (W) has the highest melting point (3410 °C) of all the metals, and among all the elements it is second only to carbon as graphite, which has a melting point of 3550 °C . Pure tungsten is used in incandescent light bulbs as the filament, the wire that glows white-hot. No other material has been found to be better since the invention of light bulbs in 1908 by Thomas Edison and his co-workers.

Problem-Solving Example 11.7   Calculating Heats of Fusion

Use data from Table 11.6 to calculate the thermal energy transfer required to melt 5.0 cm³ of titanium. The density of titanium is 4.5 g/cm³.

Answer   9.8 kJ

Explanation   First, determine how many moles of Ti are present in 5.0 cm³ of Ti.

$$5.0 \text{ cm}^3 \text{ Ti} \times \frac{4.5 \text{ g Ti}}{1 \text{ cm}^3 \text{ Ti}} \times \frac{1 \text{ mol Ti}}{47.88 \text{ g Ti}} = 0.470 \text{ mol Ti}$$

Now, calculate the quantity of heat required from the $\Delta H^\circ_{\text{fusion}}$ value in Table 11.6.

$$0.470 \text{ mol Ti} \times \frac{20.9 \text{ kJ}}{1 \text{ mol Ti}} = 9.8 \text{ kJ}$$

✔   We have about one-half mole of Ti, so the heat required should be about half of the molar heat of fusion, 20.9/2 or about 10 kJ, and this compares well to our more accurate answer.

Problem-Solving Practice 11.7

Use data from Table 11.6 to calculate the heat required to melt 1.45 g of aluminum.

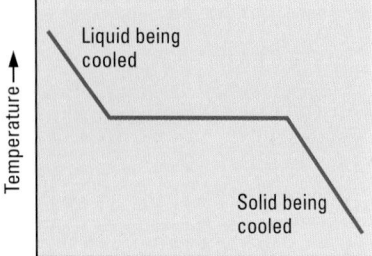

Exercise 11.16   Cooling a Liquid Metal Until It Solidifies

When a liquid metal is cooled to the temperature at which it solidifies, and the solid is then cooled to an even lower temperature, the "cooling curve," a plot of temperature against time, looks like the chart to the left. Account for the shape of this curve. Would all substances exhibit similar curves?

Exercise 11.17 Heats of Fusion and Electronic Configuration

Look at Table 7.6 and compare the electron configurations shown there with the heats of fusion for the metals shown in Table 11.6. Is there any correlation between these configurations and this property? Does strength of attraction among metal atoms correlate with number of valence electrons? Explain.

Electrons in Metals, Semiconductors, and Insulators

The properties of solids can be explained by the type of bonding that holds their constituent particles together. In molecular solids, molecules are held together by noncovalent forces; in ionic solids, ions are held together by electrostatic attractions between positive and negative ions. Metals behave as though metal cations exist in a "sea" of mobile electrons — the valence electrons of all the metal atoms. **Metallic bonding** is the nondirectional attraction between positive metal ions and the surrounding sea of negative charge. Each metal ion has a large number of near neighbors. The valence electrons are spread throughout the metal lattice, holding the positive metal ions together. When an electric field is applied to a metal, these valence electrons move toward the positively charged end, and the metal conducts electricity.

Because of the uniform charge distribution provided by the mobile valence electrons in a metal lattice, the positions of the positive ions can be changed without destroying the attractions among them. Thus, most metals can be bent and drawn into wire. Conversely, when we try to deform an ionic solid, the crystal usually shatters because the balance of positive ions surrounded by negative ions, and vice versa, is disrupted (⇐ *p. 96*).

To visualize how bonding electrons behave in a metal, first consider the arrangement of electrons in an individual atom far enough away from any neighbor so that there is no bonding. In such an atom the electrons occupy orbitals that have definite energy levels (⇐ *p. 285*). In a large number of separated, identical atoms, all of the energy levels are identical. If the atoms are brought closer together, they begin to influence one another. The identical energy levels shift up or down and become bands of energy levels characteristic of the large collections of metal atoms (Figure 11.25). An **energy band** is a large group of orbitals whose energies are closely spaced and whose average energy is the same as the energy of the corresponding orbital in an individual atom. In some cases, energy bands for different types of electrons (*s*, *p*, *d*, and so on) overlap; in other cases there is a gap between different energy bands.

Within each band, electrons fill the lowest energy orbitals much as electrons fill orbitals in atoms. The number of electrons in a given energy band depends on the number of metal atoms in the crystal. In considering conductivity and other metallic properties, it is usually necessary to consider only valence electrons, since other electrons all occupy completely filled bands in which two electrons occupy every orbital. In these, no electron can move from one orbital to another, because there is no empty spot for it.

A band containing the valence electrons (the **valence band**) that is partially filled, requires a little added energy to excite a valence electron to a slightly higher energy orbital. Such a small increment of energy can be provided by applying an electric field, for example. The presence of low-energy, empty orbitals that electrons can move into allows the electrons to be mobile and to conduct an electric current (Figure 11.26).

Another band containing higher energy orbitals (the **conduction band**) exists at an energy above the valence band. In a metal, the valence band and the conduction bands overlap, so electrons can move from the valence band to the conduction band freely. Thus, electrons in a metal are delocalized — they can move about the metal freely, which explains metals' electrical conductivity. Such a metal with

CD-ROM Screen 10.14:
Metallic Bonding: Band Theory

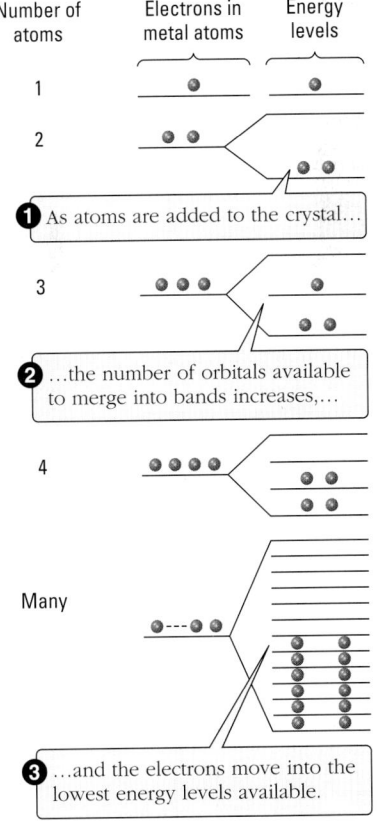

Figure 11.25 Formation of bands of electron orbitals in a metal crystal.

CD-ROM Screen 10.15:
Conductors and Insulators

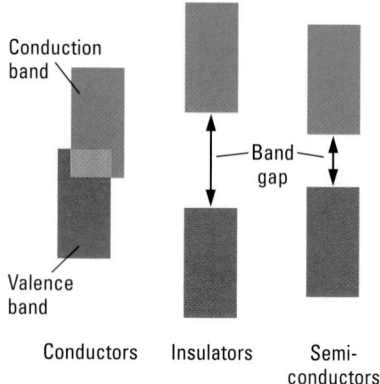

Conduction band

Valence band

Band gap

Conductors   Insulators   Semi-conductors

Figure 11.26 **Differences in the energy bands of available orbitals in conductors, insulators, and semiconductors.**

Recall from Section 7.4 that electrons have wave-like properties as well as particle-like properties.

an overlapping valence and conduction band is a **conductor**. The electrical conductivity of a metal decreases as the temperature rises because, as the metal atoms vibrate more vigorously with increasing temperature, they diminish the electrons' ability to move freely about within the overlapping bands.

With the conduction band so close in energy to the valence band, the electrons can absorb a wide range of wavelengths in the visible region of the spectrum. As the excited electrons fall back to their lower energy states, they emit their extra energy as visible light, producing the luster characteristic of metals.

The energy band theory also explains why some solids are **insulators**, which do not conduct electricity. In an insulator the valence band and conduction band do not overlap. Rather, there is a large band gap between them (Figure 11.26). There are very few electrons that have enough energy to jump across the large gap from a filled lower energy band to an empty higher energy band, and so no current flows through an insulator when an external electric field is applied.

In a **semiconductor** there is a very *narrow* energy gap between the valence band and the conduction band (Figure 11.26). At quite low temperatures, electrons remain in the filled lower energy valence band, and semiconductors are not good conductors. At higher temperatures, or when an electric field is applied, some electrons have enough energy to jump across the band gap into the conduction band. This allows an electric current to flow. This property of semiconductors—to switch from insulator to conductor with the application of an external electric field—is the basis for the operation of transistors, the cornerstone of modern electronics.

Superconductors

One interesting property of metals is that their electrical conductivity *decreases* with increasing temperature. In Figure 11.27b a metal in a burner flame shows higher resistance (lower conductivity) than the same metal at room temperature (Figure 11.27a). Lower conductivity of metals at higher temperatures can be explained by considering valence electrons in the lattice of metal ions as waves. As an electron wave moves through the metal crystal under the influence of an electrical voltage, it encounters lattice positions where the metal ions are close enough together to scatter it. This scattering is analogous to the scattering of x-rays caused by atoms in crystals. The scattered electron wave moves off in another direction,

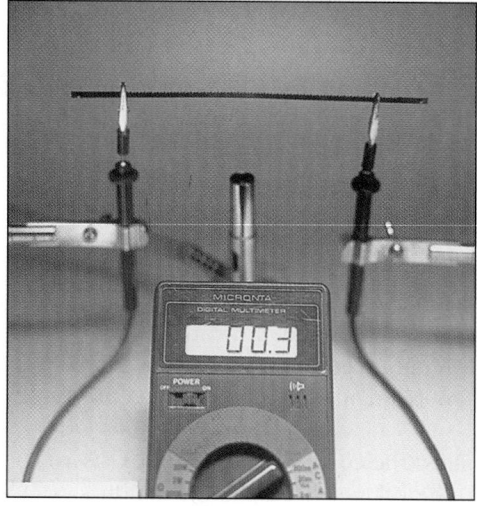

(a)

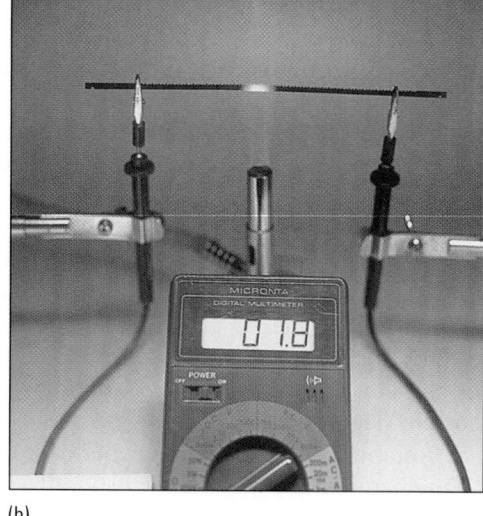

(b)

Figure 11.27 **Resistance, temperature, and electrical conductivity.** (a) A piece of metal at room temperature exhibits a small resistance (high electrical conductivity). (b) While being heated, this same piece of metal exhibits a higher resistance value, indicating a lower conductivity.

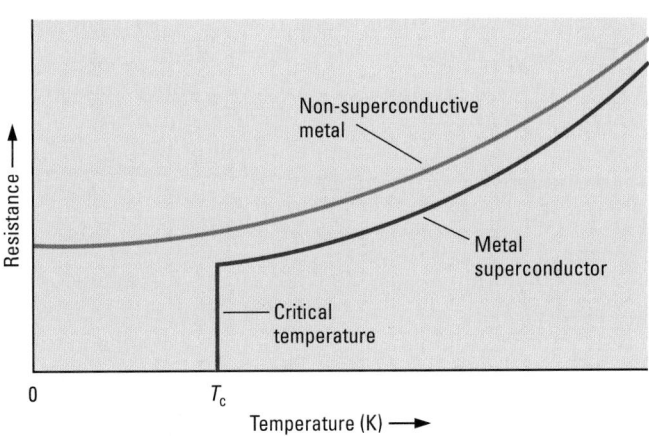

Figure 11.28 The resistance of a superconducting metal and a nonsuperconducting metal as a function of temperature.

only to be scattered again when it encounters some other occupied lattice position. All this scattering *lowers* the conductivity of the metal. At higher temperatures, the metal ions vibrate more, and the distances between lattice positions change more from their average values. This causes more scattering of electron waves as they move through the crystal, because there are now more possibilities of unfavorable lattice spacings. Hence there is lower electrical conductivity.

From this picture of electrical conductivity, it might be expected that the conductivity of a metal crystal at absolute zero (0 K or -273.15 °C) might be very large. In fact, the conductivity of a pure metal crystal approaches infinity as absolute zero is approached. But in some metals, a more interesting phenomenon occurs. At a critical temperature (T_c) that is low but finite, the conductivity abruptly increases to infinity, which means that the resistance drops to zero (Figure 11.28). The metal becomes a **superconductor** of electricity. A superconductor offers no resistance whatever to electric current. Once a current has been started in a superconducting circuit, it continues to flow indefinitely. No clear theory explaining superconductivity has emerged, but it appears that the scattering of electron waves by vibrating atoms is replaced by some cooperative action that allows the electrons to move through the crystal unhindered.

If a material could be made superconducting at a high enough temperature, it would find uses in the transmission of electrical energy, in high-efficiency motors, and in computers and other devices. Table 11.7 lists some metals that have super-

The critical temperature below which a superconductor has no electrical resistance is quite different from the critical point where the difference between liquid and vapor disappears.

TABLE 11.7	Superconducting Transition Temperatures of Some Metals*
Metal	**Superconducting transition temperature (K)**
Aluminum	1.175
Gallium	1.10
Tin	3.72
Mercury	4.15
Lanthanum	4.9
Lead	7.2

*Not all metals have superconducting properties. Those that can become superconductors at atmospheric pressure are Al, Ti, Zn, Ga, Zr, Mo, Tc, Ru, Cd, In, Sn, La, Hf, Ta, W, Re, Os, Ir, Hg, Tl, Pb, Th, Pa, and U.

Superconductivity was discovered in 1911 by the Dutch physicist Kamerlingh Onnes, who won a Nobel Prize for his work.

Müller and Bednorz received the 1987 Nobel Prize in Physics for their discovery of the superconducting properties of $LaBa_2Cu_3O_x$.

Figure 11.29 Levitation of a magnet above a superconductor. A magnet moving toward a superconductor induces a current in the superconductor, which induces a magnetic field. This magnetic field and that of the magnet oppose one another. The magnet remains at that height above the superconductor surface where the repulsive magnetic field strength is balanced by the force of gravity. The magnet remains in place as long as the current continues in the superconductor.

conducting transition temperatures. While some useful devices can be fabricated from these metals, the very low temperatures at which they become superconductors make them impractical for most applications. Shortly after superconductivity of metals was discovered, alloys were prepared that had higher transition temperatures. Niobium alloys were the best, but still required cooling to below 23 K (−250 °C) to exhibit superconductivity. Maintenance of such a low temperature requires liquid helium, which is expensive.

This picture abruptly changed in January 1986, when Alex Müller and Georg Bednorz, IBM scientists in Switzerland, discovered that a barium-lanthanum-copper oxide became superconducting at 35 K (−238 °C). This type of mixed metal oxide is a ceramic with the same structure as the mineral perovskite ($CaTiO_3$), and therefore would be expected to have insulating properties. The announcement of the superconducting properties of $LaBa_2Cu_3O_x$ rocked the world of science and provoked a flurry of activity to prepare related compounds in the hope that even higher transition temperatures could be found. Within four months a material ($YBa_2Cu_3O_7$) that became superconducting at 90 K (−183 °C), was announced. This was a major breakthrough because 90 K exceeds the boiling point of liquid nitrogen (b.p. = 77 K or −196 °C), which is easily obtainable and relatively cheap. The current record critical temperature of 133 K (−140 °C) at atmospheric pressure is for the ceramic $HgBa_2Ca_2Cu_3O_8$.

Why the great excitement over the potential of superconductivity? Superconducting materials will allow the building of more powerful electromagnets such as those used in nuclear particle accelerators and in magnetic resonance imaging (MRI) machines for medical diagnosis (⇐ *p. 296*). Such electromagnets will allow higher energies to be maintained for longer periods of time (and at lower cost) in the case of the particle accelerators, and allow better imaging of problem areas in a patient's body. Currently, the main barrier to wider application is the cost of cooling the magnets used in these devices. Many scientists say that the discovery of high-temperature superconducting materials is more important than the discovery of the transistor because of its potential effect on electrical and electronic technology. For example, the use of superconducting materials for transmission of electric power could save as much as 30% of the energy now lost because of the resistance of the wire. Superchips for computers could be up to 1000 times faster than existing silicon chips. Since a superconductor repels magnetic materials (Figure 11.29), trains can be levitated above a track and move with no friction other than air resistance to slow them down. A test rail line in Japan is now being tested, and on April 14, 1999, the test train attained a speed of 343 mph or 552 km/h.

Although the discovery of superconductivity is significant, translating the research into practical applications such as those described here may prove difficult. After all, these superconductors have many of the properties expected of ceramics — they are brittle and fragile. The technology is just beginning to be developed, but recently there has been some progress toward making ribbons and wire filaments from superconducting materials.

11.9 SILICON AND THE CHIP

Silicon is known as an "intrinsic" semiconductor because the element itself is a semiconductor. Silicon of about 98% purity can be obtained by heating silica (purified sand) and coke (an impure form of carbon) at 3000 °C in an electric arc furnace.

$$SiO_2(s) + 2 C(s) \xrightarrow{\text{heat}} Si(s) + 2 CO(g)$$

Silicon of this purity can be alloyed with aluminum and magnesium to increase the hardness and durability of the metals and is used for making silicone polymers. For

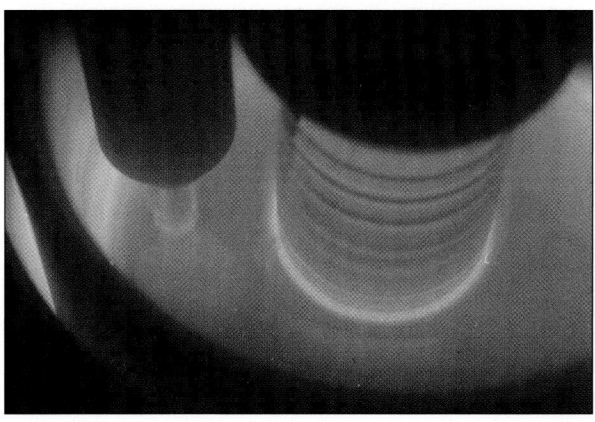 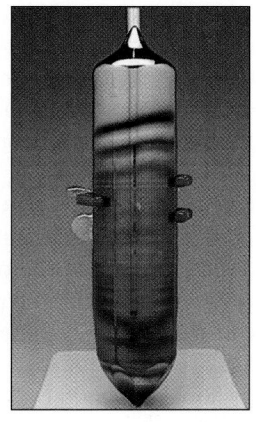 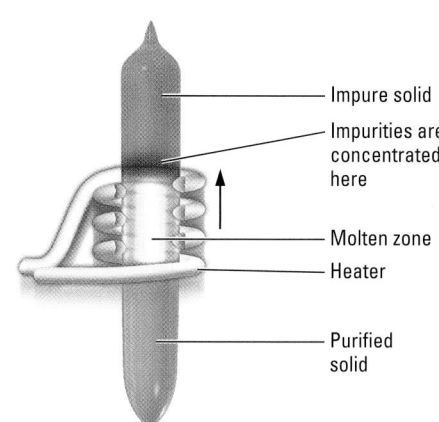

Figure 11.30 The zone refining process.

use in electronic devices, however, a much higher degree of purification is needed. High-purity silicon can be prepared by reducing $SiCl_4$ with magnesium.

$$SiCl_4(\ell) + 2\,Mg(s) \longrightarrow Si(s) + 2\,MgCl_2(s)$$

Magnesium chloride, which is water soluble, is washed from the silicon. The final purification of the silicon takes place by a melting process called **zone refining** (Figure 11.30), which produces silicon containing less than one part per billion of impurities such as boron, aluminum, and arsenic. Zone refining takes advantage of the fact that impurities are often more soluble in the liquid phase than in the solid phase. As a hot molten zone is moved through a sample being purified, the impurities move along in the liquefied portion of the sample. As the heated zone cools, the sample that resolidifies is purer than it was. Multiple passes of the hot molten zone are usually necessary to achieve the degree of purity necessary to fabricate semiconducting devices.

Like all semiconductors, high-purity silicon fails to conduct an electric current until a certain electrical voltage is applied, but at higher voltages it conducts moderately well. Silicon's semiconducting properties can be improved dramatically by a process known as doping. **Doping** is the addition of a tiny amount of some other element (a *dopant*) to the silicon.

For example, suppose a small number of boron atoms (or atoms of some other Group 3A element) replace silicon atoms (Group 4A) in solid silicon. Boron has only three valence electrons, whereas silicon has four. This leaves a deficiency of one electron around the B atom, creating what is called a *positive hole* for every B atom added. Hence silicon doped in this manner is referred to as positive-type or *p*-**type** silicon (Figure 11.31). In *p*-type semiconductors, the holes are the charge carriers.

 CD-ROM Screen 10.16: Semiconductors

A positive hole in an energy band of a semiconductor is a place where there is one less electron than normal and hence extra positive charge.

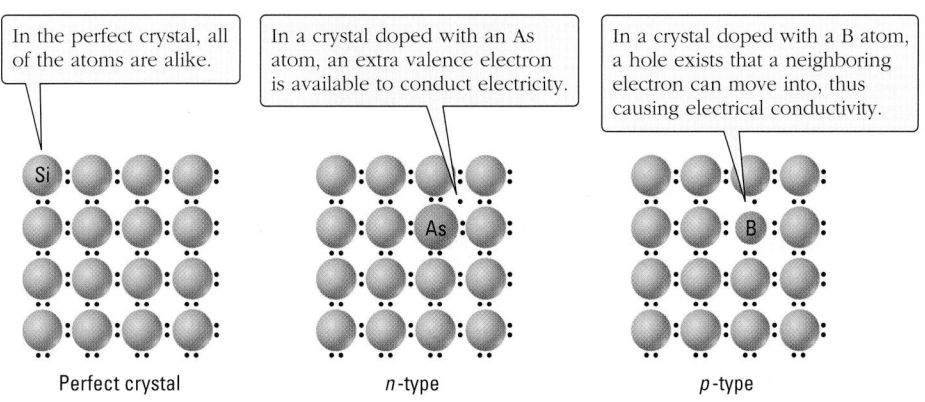

In the perfect crystal, all of the atoms are alike.

In a crystal doped with an As atom, an extra valence electron is available to conduct electricity.

In a crystal doped with a B atom, a hole exists that a neighboring electron can move into, thus causing electrical conductivity.

Perfect crystal   *n*-type   *p*-type

Figure 11.31 Schematic drawing of semiconductor crystals derived from silicon.

On the other hand, when a few atoms of a Group 5A element such as arsenic are added to silicon, only four of the five valence electrons of As are used for bonding with four Si atoms, leaving one electron relatively free to move. This type of doped silicon is referred to as negative-type or **n-type** silicon because it has extra (negative) valence electrons. In *n*-type semiconductors, the electrons are the charge carriers.

When *p*-type and *n*-type semiconductors are brought together, a **p-n junction** results. Such a junction can act as a rectifier; that is, it allows current to flow in one direction but not the other. When the two materials are joined, some of the excess electrons in the *n*-type material migrate across the junction and some of the holes in the *p*-type material migrate in the opposite direction. The result is the buildup of a negative charge on the *p*-type region and a positive charge on the *n*-type region. This charge is called the *junction potential*; it prevents the further migration of electrons or holes. However, if an external potential is applied to the *p-n* junction, one of two effects can result. If a negative charge is connected to the *p*-type side and a positive charge to the *n*-type side, the electrons and holes migrate away from the junction, and no current flows through the junction (Figure 11.32 *bottom*), a situation known as reverse bias. On the other hand, if a negative charge is connected to the *n*-type side and a positive charge to the *p*-type side, then both electrons and holes flow across the junction, and current flows through the device (Figure 11.32 *top*). This situation is known as forward bias. Thus, the *p-n* junction acts to pass current one way but not the other.

These *p-n* junctions can be joined together into larger composite structures to create transistors and integrated circuits. The simplest of these devices uses layers of *n-p-n-* or *p-n-p*-type silicon. Germanium, a Group 4A element just below silicon in the periodic table, is also used in place of silicon. The most revolutionary application of silicon's semiconductor properties has been the design of integrated electrical circuits (ICs) on tiny chips of silicon scarcely larger than a millimeter in diameter. In a single IC there are often thousands and even millions of transistors, as well as the circuits to carry the electrical signals. These devices, in the form of computer memories and central processing units (CPUs), permeate our whole society. They are in calculators, cameras, watches, toys, coin changers, cardiac pacemakers, and many other products. Truly, silicon is both the world we walk on (sand and silicate rock) and at the same time our constant companion in communications and electronic controls (the chip).

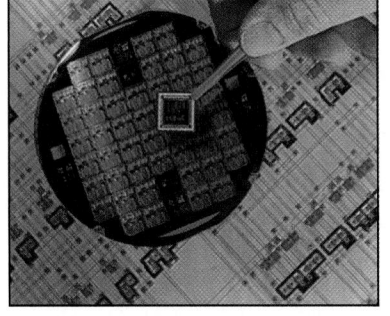

A microprocessor chip. As the chips become smaller and smaller, the purity of the silicon becomes more important, since impurities can prevent a circuit from working properly.

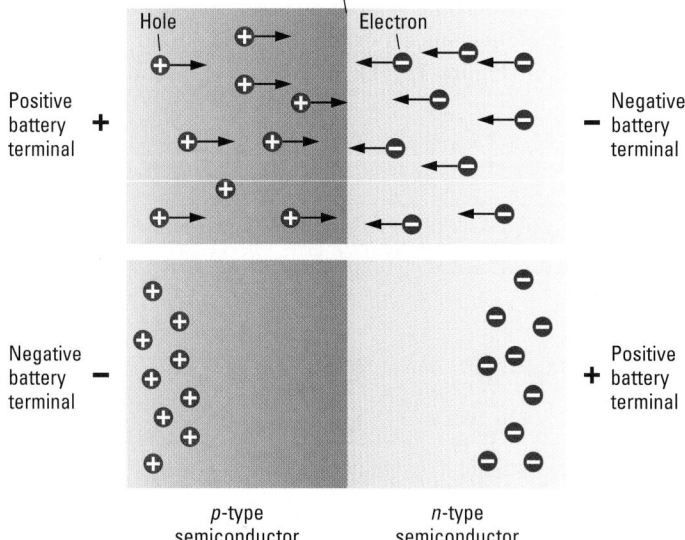

Figure 11.32 A *p-n* junction.   Placing a *p*-type semiconductor next to an *n*-type semiconductor creates a *p-n* junction. The charge carriers in the *p*-type region are holes, and the charge carriers in the *n*-type region are electrons. *Top:* The negative battery terminal is connected to the *n*-type region, and the positive battery terminal is connected to the *p*-type region. The holes and electrons move as indicated by the arrows. Electric current flows through the junction and the overall device. *Bottom:* If the battery terminals are reversed, the holes and electrons are attracted to the edges of the regions, there is charge depletion near the junction, and no current flows through the junction or the overall device.

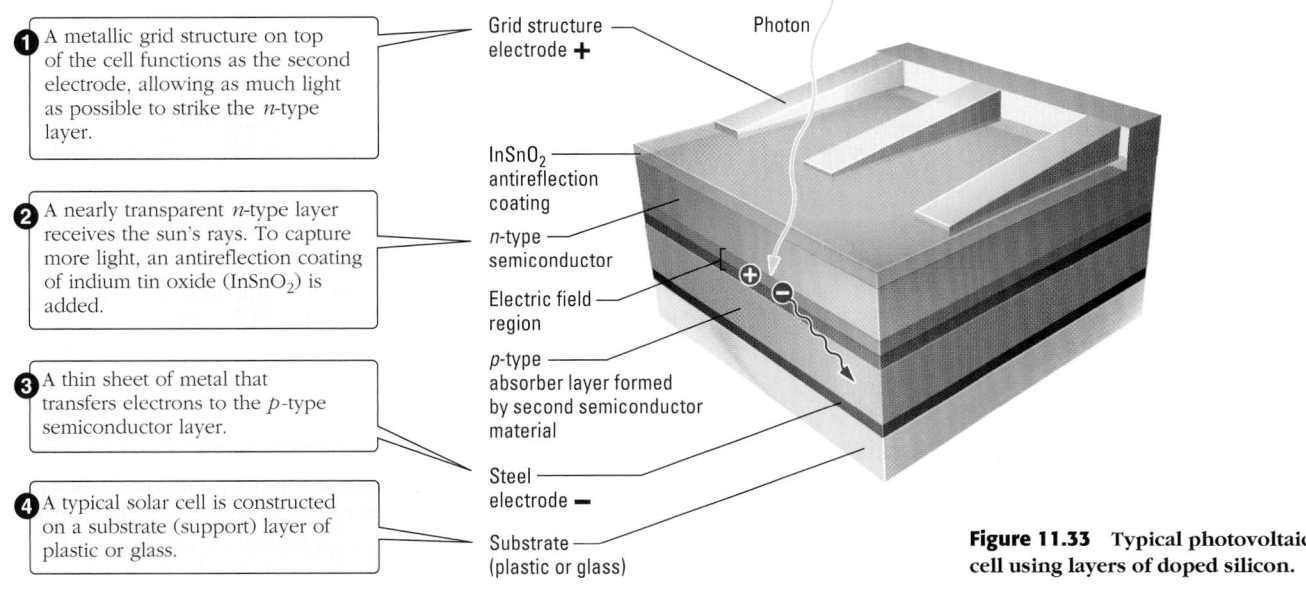

1 A metallic grid structure on top of the cell functions as the second electrode, allowing as much light as possible to strike the *n*-type layer.

2 A nearly transparent *n*-type layer receives the sun's rays. To capture more light, an antireflection coating of indium tin oxide ($InSnO_2$) is added.

3 A thin sheet of metal that transfers electrons to the *p*-type semiconductor layer.

4 A typical solar cell is constructed on a substrate (support) layer of plastic or glass.

Grid structure electrode **+**

Photon

$InSnO_2$ antireflection coating

n-type semiconductor

Electric field region

p-type absorber layer formed by second semiconductor material

Steel electrode **−**

Substrate (plastic or glass)

Figure 11.33 **Typical photovoltaic cell using layers of doped silicon.**

Doped silicon is also the basis of the **solar cell**. When a layer of *n*-type doped silicon is next to a *p*-type layer, there is a strong tendency for the extra electrons in the *n*-type layer to pair with the unpaired electrons in the holes of the *p*-type layer. If the two layers are connected by an external electrical circuit, light that strikes the silicon provides enough energy to cause an electrical current to flow. When a photon is absorbed, it excites an electron to a higher-energy orbital, allowing the electron to leave the *n*-type silicon layer and flow through the external circuit to the *p*-type layer. As the *p*-type layer becomes more negative (because of added electrons), the extra electrons are repelled *internally* back into the *n*-type layer (which has become positive because of loss of electrons via the circuit). This process can continue as long as the silicon layers are exposed to sunlight and the circuit remains closed. A typical solar cell is shown in Figure 11.33.

Solar cells are on the threshold of being the next great technological breakthrough, perhaps comparable to the computer chip. Although experimental solar-powered automobiles are now available and many novel applications of solar cells already exist, the real breakthrough will be the general use of banks of solar cells at utility power plants to produce huge quantities of electricity. One plant already

Solar power can provide electricity in remote areas.
These Brazilian houses and this water pump are powered by solar-generated electricity.

operating in California uses banks of solar cells to produce 20 megawatts (MW) of power—enough to supply the daily electricity needs of a city the size of Tampa, Florida.

11.10 NETWORK SOLIDS

CD-ROM Screen 13.15: Network Solids

A number of solids are composed of nonmetal atoms connected by a network of covalent bonds. Such **network solids** really consist of one huge molecule in which all the atoms are connected to all the others via a network of bonds. Separate small molecules do not exist in a network solid.

Graphite and Diamond

Graphite and diamond are allotropes of carbon *(⇐ p. 26)*. Graphite's name comes from the Greek *graphein* meaning "to write" because one of its earliest uses was for writing on parchment. Artists today still draw with charcoal, an impure form of graphite, and we write with graphite pencil leads. Graphite is an example of a *planar network solid* (Figure 11.34). Each carbon atom is covalently bonded to three other carbon atoms. The planes consist of six-membered rings of carbon atoms (like those in benzene) *(⇐ p. 349)*. Each hexagon shares all six of its sides with other hexagons around it, forming a *two-dimensional network*. Some of the bonding electrons are able to move freely around this network, and so graphite is a conductor of electricity. There are strong covalent bonds between carbon atoms in the same plane, but attractions between the planes are caused by London forces and hence are weaker. Because of this, the planes can easily slip across one another, which makes graphite an excellent solid lubricant for uses such as in locks, where greases and oils are undesirable.

The distance between graphite planes is more than twice the distance between the nearest carbon atoms within a plane.

Diamonds are also built of six-membered carbon rings, with each carbon atom bonded to four others by single covalent bonds. This forms a *three-dimensional network* (Figure 11.35). Because of the tetrahedral arrangement of bonds around each carbon atom, the six-membered rings in the diamond structure are puckered. In graphite the layers are much farther apart than normal C—C bond distances. As a result, diamond (3.51 g/cm^3) is denser than graphite (2.22 g/cm^3). Also, because its valence electrons are localized between carbon atoms, diamond does not conduct electricity. Diamond is one of the hardest materials and also one of the best conductors of heat known. It is also transparent to visible, infrared, and ultraviolet radiation.

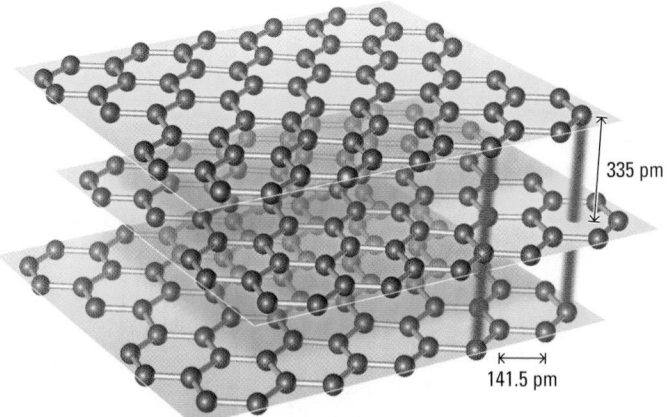

335 pm

141.5 pm

Figure 11.34 The structure of graphite. Three of the many layers of six-membered carbon rings are shown. These layers can slide past one another relatively easily, making graphite a good lubricant. In addition, some of the carbon valence electrons in the layers are delocalized, allowing graphite to be a good conductor of electricity parallel to the larger planes.

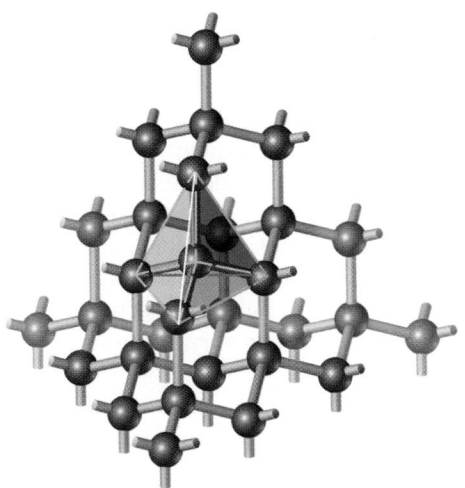

Figure 11.35 The structure of diamond. Each carbon atom is covalently bonded to four other carbon atoms in a tetrahedral arrangement.

 Exercise 11.18 How Would You Make Diamonds?

Given the properties of graphite and diamond, predict what conditions of temperature and pressure might be used if you wanted to convert inexpensive graphite into valuable diamonds within a reasonable period of time.

In the 1950s, scientists at General Electric in Schenectady, New York, first produced man-made diamonds. Their technique, still in use today, was to heat graphite to a temperature of 1500 °C in the presence of a metal catalyst, such as nickel or iron, and under a pressure of 50,000 to 65,000 atm. Under these conditions, the carbon dissolves in the metal and recrystallizes in its higher-density form, slowly becoming diamond. There is a $500 million worldwide market for diamonds made this way, many of which are used for abrasives and diamond-coated cutting tools for drilling. While gem-quality diamonds have been made from graphite, they are as yet too expensive compared with naturally occurring diamonds.

The most important network solids are the *silicates*. Many bonding patterns exist among the silicates, but extended arrays of covalently bonded silicon and oxygen atoms are common, such as quartz, SiO_2. Silicates will be discussed in depth in Chapter 21.

11.11 CEMENT, CERAMICS, AND GLASS

Cement, ceramics, and glass are examples of amorphous solids; that is, they lack crystalline structures with easily defined unit cells. They are all extremely important because of their useful properties. We build things with all three of these kinds of solids. They all are created at high temperatures, so making them from starting materials is highly energy-intensive.

Cement

Cement consists of microscopic particles containing compounds of calcium, iron, aluminum, silicon, and oxygen in varying proportions. In the presence of water cement reacts to form hydrated particles with large surface areas, which subsequently undergo recrystallization and reaction to bond to themselves as well as to the surfaces of bricks, stone, or other silicate materials.

Figure 11.36 Pouring concrete. After the concrete has been formed and exposed surfaces smoothed, it is allowed to set, a process hastened by moisture, in which strong bonds are formed between many of the tiny particles originally present in the dry concrete.

Compressive strength refers to a solid's resistance to shattering under compression.

Tensile strength refers to a substance's resistance to stretching.

Cement is made by roasting a powdered mixture of calcium carbonate (limestone or chalk), silica (sand), aluminosilicate mineral (kaolin, clay, or shale), and iron oxide at a temperature of up to 870 °C in a rotating kiln. As the materials pass through the kiln, they lose water and carbon dioxide and ultimately form a "clinker," in which the materials are partially fused. A small amount of calcium sulfate is added, and the cooled clinker is then ground to a very fine powder. A typical composition of cement is: 60% to 67% CaO, 17% to 25% SiO_2, 3% to 8% Al_2O_3, up to 6% Fe_2O_3, and small amounts of magnesium oxide, magnesium sulfate, and oxides of potassium and sodium. Cement is usually mixed with other substances. *Mortar* is a mixture of cement, sand, water, and lime. **Concrete** is a mixture of cement, sand, and aggregate (crushed stone or pebbles) in proportions that vary according to the application and the strength required.

In cement, the oxides are not isolated into ionic crystals. Rather, the entire nanoscale structure is a complex network of ions, each satisfying its charge requirements with ions of opposite charge. Many different reactions occur during the setting of cement. Various constituents react with water and with carbon dioxide in the air. The surface of each cement particle has many sites that attract water molecules, so when water is added to the dry solid, a stable suspension of the particles results. The initial reaction of cement with water is the hydrolysis of the calcium silicates, which forms a gel that sticks to itself and to the other particles (sand, crushed stone, or gravel). The gel has a very large surface area and ultimately is responsible for the great strength of concrete once it has set. The setting process also involves the formation of small, densely interlocked crystals after the initial solidification of the wet mass (Figure 11.36). This interlocking crystal formation continues for a long time after the initial setting, and it increases the compressive strength of the cement. For this reason, freshly poured concrete is kept moist for several days.

More than 800 million tons of cement are manufactured each year, most of which is used to make concrete. Concrete, like many other materials containing Si—O bonds, is highly noncompressible but lacks tensile strength. When concrete is to be used where it will be subject to tension, as in a bridge or building, it must be reinforced with steel.

Ceramics

Ceramics have been made since well before the dawn of recorded history. Much art of great historical significance is made of ceramic materials. **Ceramics** form an extremely large and diverse class of substances and are generally fashioned from clay or other natural earths at room temperature and then permanently hardened by heat in a baking ("firing") process that binds the particles together. *Silicate ceramics* include objects made from clays (aluminosilicates), such as pottery, bricks, and table china. The techniques developed with natural clay have been applied to a wide range of other inorganic materials in recent years.

China clay, or kaolin, is primarily kaolinite that is practically free of iron, which otherwise imparts a red color to the clay. As a result, china clay is almost colorless and particularly valuable in making fine pottery. The first pieces of fine Oriental clayware, which was named "china," arrived in Europe in the Middle Ages. European potters envied and admired the obviously superior product and struggled to discover how to duplicate it. Clays mixed with water form a moldable paste consisting of tiny silicate sheets that can easily slide past one another. Mixed to just the right consistency, this paste can be molded into almost any form. Then it is heated, the water is driven off, and new Si—O—Si bonds form so that the mass becomes permanently rigid. If silica (SiO_2) and feldspar are added in the right proportions, the mixture will not crack after heating. (Cracking occurs as a result of shrinkage when new bonds form between the silicate sheets.)

Oxide ceramics are produced from metal oxides such as alumina (Al_2O_3) and magnesia (MgO) by heating the powdered solids under pressure, causing the particles to bind to one another to form a rigid solid. Because it has a high electrical resistivity, alumina is used in spark plug insulators. High-density alumina has very high mechanical strength, and it is also used in armor plating and in high-speed cutting tools for machining metals. Magnesia is an insulator with a high melting point (2800 °C), so it is often used as insulation in electric heaters and electric stoves.

A third class of ceramics includes the *nonoxide ceramics* such as silicon nitride (Si_3N_4), silicon carbide (SiC), and boron nitride (BN). When the powdered solids are strongly heated under pressure, all of these compounds form ceramics that are hard and strong, but brittle. Boron nitride has the same average number of electrons per atom as does elemental carbon and exists in the graphite structure or the diamond structure, making it comparable in hardness with diamond and more resistant to oxidation. This is why boron nitride cups and tubes are used to contain molten metals that are being evaporated.

Silicon carbide has the trade name Carborundum and can be regarded as the diamond structure with half of the C atoms replaced by Si atoms. Widely used as an abrasive material, SiC is receiving more attention recently as a ceramic material, particularly for high-temperature engines.

A multibillion dollar market for advanced ceramics, industrial need for new materials, and accelerating fundamental research are causing an explosion of new ceramic materials. These include new forms of the ceramics described earlier, as well as *ceramic composites*, mixtures of ceramic materials with fibers, sometimes composed of various polymers, that improve the strength of the composite ceramics. Adding fibers to ceramic composites makes them less susceptible to brittleness and sudden fracture. The one severely limiting problem in using ceramics is their brittleness. Ceramics deform very little before they fail catastrophically, the failure resulting from a weak point in the bonding within the ceramic matrix. However, such weak points are not consistent from sample to sample, so that the failure is not very predictable. Since stress failure of ceramic composites is due to nanoscale irregularities resulting from impurities or disorder in the atomic arrangements, much attention is now being given to using purer starting materials and controlling the processing steps.

Glasses

Glasses, which are amorphous solids, occur naturally or can be prepared synthetically. The manufacture of glass goes back to at least 5000 BC, when Phoenician sailors used blocks of sodium carbonate, Na_2CO_3, and sand, SiO_2, to insulate fires from the wooden planks of their ships. The metal carbonate and sand melted in the heat of the fire and formed a material that resembled *obsidian*, a natural glassy material that has been valued since antiquity.

One of the more common glasses today is soda-lime glass, which is clear and colorless if the purity of the ingredients is carefully controlled. If, for example, too much iron oxide is present, the glass will have a green color. Of course, color may also be a desirable property. Adding certain metal oxides to the basic ingredients of a glass produces many beautiful colors (see Table 11.8).

The main glass-forming oxides are SiO_2, B_2O_3, GeO_2, and P_4O_{10}, all of which contain elements close to one another in the periodic table. Several other metal oxides, including Al_2O_3 and Na_2O, are also important in forming commercial glasses. The simplest glass is probably amorphous silica, SiO_2 (known as *vitreous silica*), which is built up of corner-sharing SiO_4 tetrahedra linked into a three-dimensional network that lacks symmetry or long-range order. Vitreous silica can be prepared by melting and quickly cooling either quartz or cristobalite.

TABLE 11.8	Substances Used to Color Glass
Substance	**Color**
Copper(I) oxide	Red, green, blue
Tin(IV) oxide	Opaque
Calcium fluoride	Milky white
Manganese(IV) oxide	Violet
Cobalt(II) oxide	Blue
Finely divided gold	Red, purple, blue
Uranium compounds	Yellow, green
Iron(II) compounds	Green
Iron(III) compounds	Yellow

"Lead glass," as the name implies, contains lead as PbO and is highly prized for its massive feel, acoustic properties, and high index of refraction.

If another oxide is added to SiO_2, the melting point of the SiO_2 mixture is lowered considerably from 1800 °C for quartz to about 800 °C if about 25 mole percent of Na_2O is added. The resulting melt cools to form a glass that is somewhat water-soluble and definitely soluble in strongly basic solutions. It is also prone to convert back to a crystalline solid. If other metal oxides such as CaO, MgO, or Al_2O_3 are added, the mixture still melts at a fairly low temperature, but it becomes resistant to chemical attack. Common glass like that used for windows, bottles, and lamps contains these metal oxides in addition to SiO_2.

It is important that glass be *annealed* properly during the manufacturing process. Annealing means cooling the glass slowly as it passes from a viscous liquid state to a solid at room temperature. If a glass is cooled too quickly, bonding forces become uneven because small regions of crystallinity develop. Poorly annealed glass may crack or shatter when subjected to mechanical shocks or sudden temperature changes. High-quality glass, such as that used in optics, must be annealed very carefully. The 200-in. mirror for the telescope at Mt. Palomar, California, was annealed from 500 °C to 300 °C over a period of nine months.

A type of *glass ceramic* with unusual and very valuable properties has been commercialized. Ordinary glass breaks because once a crack starts, there is nothing to stop the crack from spreading. It was discovered that if glass is treated by heating until many tiny crystals have developed throughout the sample, the resulting material, when cooled, is much more resistant to breaking than normal glass. In nanoscale terms, the randomness of the glass structure has been partially replaced by the order in a crystalline silicate. The process must be controlled carefully to obtain the desired properties. Materials produced in this way are generally opaque and are used for cooking utensils and kitchenware, as in products marketed under the name Pyroceram. The initial manufacturing process is similar to that for other glass objects, but once the materials have been formed into their final shapes, they are heat-treated to develop their special properties.

Colored glass. Adding small amounts of metal oxides to colorless glass creates various colors.

Mole percent is just another way of expressing concentration. If 0.75 mol of SiO_2 and 0.25 mol of Na_2O are mixed, there is 1.00 mol of matter present. The mole percent of SiO_2 is 75%, and the mole percent of Na_2O is 25%.

Optical Fibers

Our technological culture's dependence on instantaneous communication relies on the properties of many materials, but none more so than optical fibers. They have rapidly replaced copper wire for the transmission of data such as telephone conversations as well as Internet digital data. Optimization of the glass properties of optical fibers, and the manufacturing techniques developed to make optical fibers, are major modern success stories of materials science.

An **optical fiber** has three parts: core glass, which carries the light; cladding glass, which surrounds the core; and a polymer on the outside for physical protection and strength (Figure 11.37). The cladding glass has a slightly lower refractive index than the core glass, so there is total internal reflection of light travelling through the fiber as long as the light hits the surface of the core at a sufficiently small angle. The exact angle necessary for complete internal reflection is a function of the ratio of the refractive indices of the core glass and the cladding glass. If the core glass has a refractive index of 1.6 and the cladding glass has a refractive index of 1.5, then the angle between the light ray and the surface must be less than 15 degrees. A good optical fiber must have two properties: (1) It must contain the light; that is, it must produce total internal reflection and have no defects on the surface that would let light out of the fiber. (2) It must pass light long distances without significant loss of intensity — without attenuation. To pass light with no attenuation, the core glass must be pure, in fact, ultrapure. Until 1965, when light was put into the best available optical fiber 1000 m long, no light came out, because is was all absorbed. In the decades that followed, sophisticated manufacturing methods were developed that allowed the generation of core glass appropriate for optical fiber. Pure silica, SiO_2, is evaporated into a rod that can be

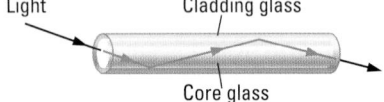

Light   Cladding glass

Core glass

Figure 11.37 Total internal reflection in an optical fiber.

heated and pulled (like taffy) to make a tiny optical fiber. The attenuation of the light signals was decreased by as much as 100 orders of magnitude by the materials scientists working on this problem. At present, optical fibers carrying signals over long distances need to have optical amplifiers only every 50 km or so to keep the signals from becoming too weak. Using such methodology, very long fiberoptic cables are now in routine use.

The two main advantages of optical fiber for transmission of voice and electronic information are the capability of the fibers to carry many more streams of data than copper wire can and the decreased interference when fibers rather than wires are used. Optical fibers are made from cheap, abundant materials.

A bundle of optical fibers. Optical fibers carry information in the form of light waves.

SUMMARY PROBLEM

Consider the phase diagram for xenon shown below. Answer these questions.

(a) In what phase is xenon found at room temperature and a pressure of 1.0 atm?

(b) If the pressure exerted on a sample of xenon is 0.75 atm, and the temperature is −114 °C, in what phase does xenon exist?

(c) If the vapor pressure of a sample of liquid xenon is 375 mm Hg, what is the temperature of the liquid phase?

(d) What is the vapor pressure of solid xenon at −122 °C?

(e) Which is the denser phase, solid or liquid? Explain.

(f) The critical temperature and pressure for xenon are 16.6 °C and 58 atm, respectively. Modify the phase diagram for xenon to reflect these data.

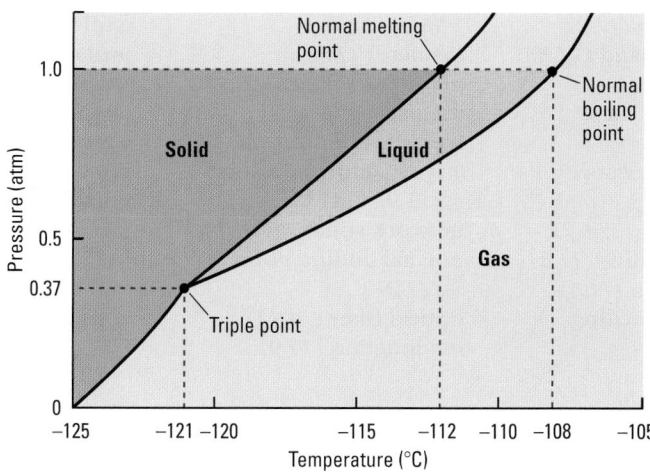

IN CLOSING

Having studied this chapter, you should be able to . . .

- Explain the properties of surface tension, capillary action, vapor pressure, and boiling point, and describe how these properties are influenced by intermolecular forces (Sections 11.1 and 11.2).
- Calculate the energy associated with vaporization and fusion (Section 11.3)
- Describe the changes of phase that occur between solids, liquids, and gases (Section 11.3).
- Use phase diagrams to predict what happens when temperatures and pressures are changed for a sample of matter (Section 11.3).

- Understand critical temperature and critical pressure (Section 11.3).
- Explain the unusual properties of water (Section 11.4).
- Do calculations based on knowledge of simple unit cells and the dimensions of atoms and ions that occupy positions in those unit cells (Section 11.5).
- Differentiate among the major types of solids (Section 11.6).
- Explain the basis of materials science (Section 11.7).
- Explain metallic bonding and how it results in the properties of metals and semiconductors (Section 11.8).
- Describe the phenomenon of superconductivity (Section 11.8).
- Explain the bonding in network solids and how it results in their properties (Section 11.10).
- Explain how the lack of regular structure in amorphous solids affects their properties (Section 11.11).

KEY TERMS

amorphous solids *(11.5)*
boiling *(11.2)*
boiling point *(11.2)*
capillary action *(11.1)*
cement *(11.11)*
ceramics *(11.11)*
closest packing *(11.6)*
composites *(11.7)*
concrete *(11.11)*
condensation *(11.3)*
conduction band *(11.8)*
conductor *(11.8)*
critical pressure (P_c) *(11.3)*
critical temperature (T_c) *(11.3)*
crystal lattice *(11.6)*
crystalline solids *(11.5)*
crystallization *(11.3)*
cubic close packing *(11.6)*

cubic unit cell *(11.6)*
deposition *(11.3)*
doping *(11.9)*
energy band *(11.8)*
equilibrium vapor pressure *(11.2)*
evaporation *(11.3)*
glass *(11.11)*
heating curve *(11.3)*
hexagonal close packing *(11.6)*
insulator *(11.8)*
materials science *(11.7)*
meniscus *(11.1)*
metallic bonding *(11.8)*
n-type semiconductor *(11.9)*
network solids *(11.10)*
normal boiling point *(11.2)*
optical fiber *(11.11)*
p-n junction *(11.9)*

p-type semiconductor *(11.9)*
phase diagram *(11.3)*
semiconductor *(11.8)*
solar cell *(11.9)*
sublimation *(11.3)*
superconductor *(11.8)*
supercritical fluid *(11.3)*
surface tension *(11.1)*
triple point *(11.3)*
unit cell *(11.6)*
valence band *(11.8)*
vapor pressure *(11.2)*
vaporization *(11.3)*
viscosity *(11.1)*
volatility *(11.2)*
x-ray crystallography *(Tools of Chemistry, p. 488)*
zone refining *(11.9)*

QUESTIONS FOR REVIEW AND THOUGHT

Conceptual Challenge Problems

CP-11.A (Section 11.1) In Section 11.1 you read that the enthalpy of vaporization of water "is somewhat dependent on the temperature." At 100 °C its value is 40.7 kJ/mol, but at 25 °C it is 44.0 kJ/mol, a difference of 3.3 kJ/mol. List three enthalpy changes whose sum would equal this difference. Remember, the sum of the changes for a cyclic process must be zero because the system is returned to its initial state.

CP-11.B (Section 11.3) For what reasons would you propose that two of the substances listed in Table 11.2 be con-

sidered better refrigerants for use in household refrigerators than the others listed there?

CP-11.C (Section 11.3) A table of enthalpies of sublimation is not given in Section 11.3, but the enthalpy of sublimation of ice at 0 °C is given as 51 kJ/mol. How was this value obtained? Tables 11.2 and 11.3 list the enthalpies of vaporization and fusion, respectively, for several substances. Determine from data in these tables the ΔH_{sub} for ice. Using the same method, estimate the enthalpies of sublimation of HBr and HI at their melting points.

Answers to questions in **bold** can be found in the back of the book.

Review Questions

1. Name three properties of solids that are different from those of liquids. Explain the differences for each.
2. List the concepts of the kinetic-molecular theory that apply to liquids.
3. What causes surface tension in liquids? Name a substance that has a very high surface tension. What kinds of intermolecular forces account for the high value?
4. Explain how the equilibrium vapor pressure of a liquid might be measured.
5. Define boiling point and normal boiling point.
6. What is the heat of crystallization of a substance, and how is it related to the substance's heat of fusion?
7. What is sublimation?
8. What is the unit cell of a crystal?
9. Assuming the same substance could form crystals with its atoms or ions in either simple cubic packing or hexagonal closest packing, which form would have the higher density?
10. How does conductivity vary with temperature for (a) a conductor, (b) a nonconductor, (c) a semiconductor, and (d) a superconductor? In your answer, begin at high temperatures and come down to low temperatures.

The Liquid State

11. Predict what compound in Table 11.1 has a surface tension most similar to that of the following liquids:
    (a) Ethylene glycol ($HOCH_2 - CH_2OH$)
    (b) Hexane (C_6H_{14})
    (c) Gallium metal at 40 °C
12. The surface tension of a liquid decreases with increasing temperature. Using the idea of intermolecular attractions, explain why this is so.
13. Explain on the molecular scale the process of condensation and vaporization.
14. How would you convert a sample of liquid to vapor without changing the temperature?
15. What is the heat of vaporization of a liquid? How is it related to the heat of condensation of that liquid? Using the idea of intermolecular attractions, explain why the process of vaporization is endothermic.
16. After exercising on a hot summer day and working up a sweat, you often become cool when you stop. What is the molecular-level explanation of this phenomenon?
17. The substances Ne, HF, H_2O, NH_3, and CH_4 all have the same number of electrons. In a thought experiment, you can make HF from Ne by removing a single proton from the nucleus and having the electrons follow the new arrangement of nuclei so as to make a new chemical bond. You can do the same for each of the other compounds. (Of course, none of these thought experiments can actually be done because of the enormous energies it takes to remove protons from nuclei.) For all of these substances, make a plot of (a) the boiling point in kelvin versus the number of hydrogen atoms and (b) the molar heat of va-

porization versus the number of hydrogen atoms. Explain any trend that you see in terms of intermolecular forces.
18. How much heat is required to vaporize 1.0 metric ton of ammonia? (1 metric ton = 10^3 kg.) The ΔH_{vap} for ammonia is 25.1 kJ/mol.
19. The chlorofluorocarbon CCl_3F has a heat of vaporization of 24.8 kJ/mol. To vaporize 1.00 kg of the compound, how much heat is required?
20. The molar heat of vaporization of methanol is 38.0 kJ/mol at 25 °C. How much heat is required to convert 250. mL of the alcohol from liquid to vapor? The density of CH_3OH is 0.787 g/mL at 25 °C.
21. Some camping stoves contain liquid butane (C_4H_{10}). They work only when the outside temperature is warm enough to allow the butane to have a reasonable vapor pressure (and so are not very good for camping in temperatures below about 0 °C). Assume the heat of vaporization of butane is 24.3 kJ/mol. If the camp stove fuel tank contains 190. g of liquid C_4H_{10}, how much heat is required to vaporize all of the butane?
22. Mercury is a highly toxic metal. Although it is a liquid at room temperature, it has a high vapor pressure and a low heat of vaporization (294 J/g). What quantity of heat is required to vaporize 0.500 mL of mercury at 357 °C, its normal boiling point? The density of Hg(ℓ) is 13.6 g/mL. Compare this heat with the amount needed to vaporize 0.500 mL of water. See Table 11.2 for the molar heat of vaporization of H_2O.
23. Rationalize the observation that 1-propanol ($CH_3CH_2CH_2OH$) has a boiling point of 97.2 °C, whereas a compound with the same empirical formula, ethyl methyl ether ($CH_3CH_2OCH_3$), boils at 7.4 °C.
24. Briefly explain the variations in the following boiling points. In your discussion be sure to mention the types of intermolecular forces involved.

Compound	Boiling point (°C)
NH_3	−33.4
PH_3	−87.5
AsH_3	−62.4
SbH_3	−18.4

Vapor Pressure

25. Give a molecular-level explanation of why the vapor pressure of a liquid increases with temperature.
26. Methanol, CH_3OH, has a normal boiling point of 64.7 °C and a vapor pressure of 100 mm Hg at 21.2 °C. Formaldehyde, $H_2C{=}O$, has a normal boiling point of −19.5 °C and a vapor pressure of 100 mm Hg at −57.3 °C. Explain why these two compounds have different boiling points and require different temperatures to achieve the same vapor pressure.
27. The lowest sea level barometric pressure ever recorded was 25.90 inches of mercury, recorded in a typhoon in the South Pacific. Suppose you were in this typhoon and, to calm yourself, boiled water to make yourself a cup of tea.

At what temperature would the water boil? Remember that 1 atmosphere is 760 mm (29.92 inches) of Hg, and use Figure 11.5.

28. The highest mountain in the Western hemisphere is Mt. Aconcagua, in the central Andes of Argentina (22,834 ft). If atmospheric pressure decreases at a rate of 3.5 millibar every 100 ft, estimate the atmospheric pressure at the top of Mt. Aconcagua, and then estimate from Figure 11.5 the temperature at which water would boil at the top of the mountain.

Phase Changes: Solids, Liquids, and Gases

29. What does a low heat of fusion for a solid tell you about the solid (its bonding or type)?

30. What does a high melting point and a high heat of fusion tell you about a solid (its bonding or type)?

31. Which would you expect to have the higher heat of fusion, N_2 or I_2? Explain your choice.

32. The heat of fusion for H_2O is about 2.5 times larger than the heat of fusion for H_2S. What does this say about the relative strengths of the forces between the molecules in their respective solids? Explain.

33. Benzene is an organic liquid that freezes at 5.5 °C to beautiful, feather-like crystals. How much heat is evolved when 15.5 g of benzene freezes at 5.5 °C? The heat of fusion of benzene is 127 J/g. If the 15.5-g sample is remelted, again at 5.5 °C, what quantity of heat is required to convert it to a liquid?

34. What is the total quantity of heat required to change 0.50 mol of ice at − 5 °C to 0.50 mol of steam at 100 °C?

35. The ions of NaF and MgO all have the same number of electrons, and the internuclear distances are about the same (235 pm and 212 pm). Why then are the melting points of NaF and MgO so different (992 °C and 2642 °C, respectively)?

36. For the pair of compounds LiF and CsI, tell which compound is expected to have the higher melting point, and briefly explain.

37. Which of these substances has the highest melting point? The lowest melting point? Explain your choice briefly.
(a) LiBr   (b) CaO
(c) CO   (d) CH_3OH

38. Which of these substances has the highest melting point? The lowest melting point? Explain your choice briefly.
(a) SiC   (b) I_2
(c) Rb   (d) $CH_3CH_2CH_2CH_3$

39. Why is solid CO_2 called Dry Ice?

40. During thunderstorms, very large hailstones can fall from the sky. To preserve some of these stones, place them in the freezer compartment of our frost-free refrigerator. Our friend, who is a chemistry student, tells us to use a non–frost-free model. Why?

41. From memory, sketch the phase diagram of water. Label all the regions as to the physical state of water. Draw either horizontal (constant pressure) or vertical (constant temperature) paths (i.e., lines with arrows indicating a direction) for the following changes of state:

(a) Sublimation   (b) Condensation to a liquid
(c) Melting   (d) Vaporization
(e) Crystallization

42. At the critical point for carbon dioxide, the substance is very far from being an ideal gas. Prove this statement by calculating the density of an ideal gas in g/cm^3 at the conditions of the critical point and comparing it with the experimental value. Compute the experimental value from the fact that a mole of CO_2 at its critical point occupies $94 cm^3$.

Types of Solids

43. Classify each of the following solids as ionic, metallic, molecular, network, or amorphous.
(a) KF   (b) I_2 (c) SiO_2 (d) BN

44. Classify each of the following solids as ionic, metallic, molecular, network, or amorphous.
(a) Tetraphosphorus decaoxide
(b) Brass
(c) Graphite
(d) Ammonium phosphate

45. On the basis of the description given, classify each of these solids as molecular, metallic, ionic, network, or amorphous, and explain your reasoning.
(a) A brittle, yellow solid that melts at 113 °C; neither its solid nor its liquid conducts electricity.
(b) A soft, silvery solid that melts at 40 °C; both solid and liquid conduct electricity.
(c) A hard, colorless, crystalline solid that melts at 1713 °C; neither solid nor liquid conducts electricity.
(d) A soft, slippery solid that melts at 63 °C; neither solid nor liquid conducts electricity.

46. On the basis of the description given, classify each of these solids as molecular, metallic, ionic, network, or amorphous, and explain your reasoning.
(a) A soft, slippery solid that has no definite melting point but decomposes at temperatures above 250 °C; the solid does not conduct electricity.
(b) Violet crystals that melt at 114 °C and whose vapor irritates the nose; neither solid nor liquid conducts electricity.
(c) Hard, colorless crystals that melt at 2800 °C; the liquid conducts electricity, but the solid does not.
(d) A hard solid that melts at 3410 °C ; both solid and liquid conduct electricity.

47. Describe how each of the following would behave as they were deformed by a hammer strike. Explain why they behave as they do.
(a) A metal, such as gold.
(b) A nonmetal, such as sulfur.
(c) An ionic compound, such as NaCl.

48. What type of solid exhibits each of these sets of properties?
(a) Melts below 100 °C and is insoluble in water.
(b) Conducts electricity only when melted.
(c) Insoluble in water and conducts electricity.
(d) Noncrystalline and melts over a wide temperature range.

Crystalline Solids

49. Each diagram below represents an array of like atoms that would extend indefinitely in two dimensions. Draw a two-dimensional unit cell for each array. How many atoms are there in each unit cell?

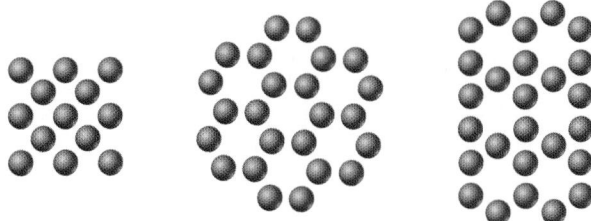

50. Name and draw the three cubic unit cells. Describe their similarities and differences.

51. Explain how the volume of a simple cubic unit cell is related to the radius of the atoms in the cell.

52. Solid xenon forms crystals with a face-centered unit cell that has an edge of 620 pm. Calculate the atomic radius of xenon.

53. Gold (atomic radius = 144 pm) crystallizes in a face-centered unit cell. What is the length of a side of the cell?

54. Consider the CsCl unit cell shown in Figure 11.21. How many Cs^+ ions are there per unit cell? How many Cl^- ions?

55. Using the NaCl structure shown in Figure 11.22, how many unit cells share each of the Na^+ ions in the front face of the unit cell? How many unit cells share each of the Cl^- ions in this face?

56. The ionic radii of Cs^+ and Cl^- are 169 and 181 pm, respectively. What is the length of the body diagonal in the CsCl unit cell? What is the length of the side of this unit cell? (See Figure 11.21.)

57. Thallium chloride, TlCl, crystallizes in either a simple cubic lattice or a face-centered cubic lattice of Cl^- ions with Tl^+ ions in the holes. If the density of the solid is 7.00 g/cm³ and the edge of the unit cell is 3.85×10^{-8} cm, what is the unit cell geometry?

58. Could $CaCl_2$ possibly have the NaCl structure? Explain your answer briefly.

59. A simple cubic unit cell is formed so that the spherical atoms or ions just touch one another along the edge. Prove mathematically that the percentage of empty space within the unit cell is 52.4%. (The volume of a sphere is $\frac{4}{3}\pi r^3$, where r is the radius of the sphere.)

60. Metallic lithium has a body-centered cubic structure, and its unit cell is 351 pm along an edge. Lithium iodide has the same crystal lattice structure as sodium chloride. The cubic unit cell is 600 pm along an edge.
    (a) Assume that the metal atoms in lithium touch along the body diagonal of the cubic unit cell, and estimate the radius of a lithium atom.
    (b) Assume that in lithium iodide the I^- ions touch along the face diagonal of the cubic unit cell and that the Li^+ and I^- ions touch along the edge of the cube; calculate the radius of an I^- ion and of an Li^+ ion.
    (c) Compare your results in parts (a) and (b) for the radius of a lithium atom and a lithium ion. Are your results rea-

sonable? If not, how could you account for the unexpected result? Could any of the assumptions that were made be in error?

Tools of Chemistry: X-Ray Crystallography

61. The surface of a CD-ROM disc contains narrowly separated lines that diffract light into its component colors. This means that the lines are spaced at distances approximately the same as the wavelength of the light. Taking the middle of the visible spectrum to be green light with a wavelength of 550 nm, calculate how many aluminum atoms (radius, 143 pm) touching its neighbors would make a straight line 550 nm long. Using this result, explain why an optical microscope using visible radiation will never be able to detect an individual aluminum atom (or, indeed, any other atom).

62. For a clear diffraction pattern to be seen from a regularly spaced lattice, the radiation falling on the lattice must have a wavelength less than the lattice spacing. From the unit cell size of the NaCl crystal, estimate the maximum wavelength of the radiation that would be diffracted by this crystal. Calculate the frequency of the radiation and the energy associated with (a) one photon and (b) one mole of photons of the radiation. In what region of the spectrum is this radiation?

Metals, Semiconductors, and Insulators

63. What is the principal difference between the orbitals that electrons occupy in individual, isolated atoms and the orbitals they occupy in solids?

64. In terms of band theory, what is the difference between a conductor and an insulator? Between a conductor and a semiconductor?

65. Name three properties of metals, and explain them by using a theory of metallic bonding.

66. Which substance has the greatest electrical conductivity? The smallest electrical conductivity? Explain your choice briefly.
    (a) Si   (b) Ge   (c) Ag   (d) P_4

67. Which substance has the greatest electrical conductivity? The smallest electrical conductivity? Explain your choices briefly.
    (a) $RbCl(\ell)$ (b) NaBr(s)
    (c) Rb   (d) Diamond

68. Define the term "superconductor." Give the chemical formulas of two kinds of superconductors and their associated transition temperatures.

69. What is the main technological or economic barrier to the widespread use of superconductors?

Silicon and the Chip

70. What are the two main chemical reactions involved in the production of electronic-grade silicon? Identify the element being reduced and being oxidized.

71. Extremely high-purity silicon is required to manufacture semiconductors such as the memory chips found in calculators and computers. If a silicon wafer is 99.99999999% pure, approximately how many silicon atoms per gram have been replaced by impurity atoms of some other element?

72. What is the process of doping, as applied to semiconductors? Why are Group IIIA and Group VA elements used to dope silicon?

73. Explain the difference between *n*-type semiconductors and *p*-type semiconductors.

Network Solids

74. Explain why diamond is denser than graphite.

75. Determine, by looking up data in a reference such as the *Handbook of Chemistry and Physics,* whether the examples of network solids given in the text are soluble in water or other common solvents. Explain your answer in terms of the chemical bonding in network solids.

76. Explain why diamond is an electrical insulator and graphite is an electrical conductor.

Cement, Ceramics, and Glass

77. Define the term "amorphous."

78. What makes a glass different from a solid such as NaCl? Under what conditions could NaCl become glass-like?

79. A typical cement contains, by weight, 65% CaO, 20% SiO_2, 5% Al_2O_3, 6% Fe_2O_3, and 4% MgO. Determine the mass percent of each element present. Then determine an empirical formula of the material from the percent composition, setting the coefficient of the least abundant element to 1.00. (Your result will contain fractional coefficients for all other elements).

80. Give two examples of (a) oxide ceramics and (b) nonoxide ceramics.

General Questions

81. The chlorofluorocarbon CCl_2F_2 has been used in air conditioners as the heat transfer fluid. Its normal boiling point is -30 °C, and the heat of vaporization is 165 J g^{-1}. The gas and the liquid have specific heat capacities of 0.61 J g^{-1} K^{-1} and 0.97 J g^{-1} K^{-1}, respectively. How much heat is evolved when 10.0 g of CCl_2F_2 is cooled from 40 to -40 °C?

82. Liquid ammonia, $NH_3(\ell)$, was used as a refrigerant fluid before the discovery of the chlorofluorocarbons and is still widely used today. Its normal boiling point is -33.4 °C, and the heat of vaporization is 23.5 kJ/mol. The gas and liquid have specific heat capacities of 2.2 J g^{-1} K^{-1} and 4.7 J g^{-1} K^{-1}, respectively. How much heat must be supplied to 10.0 kg of liquid ammonia to raise its temperature from -50.0 °C to -33.4 °C, and then to 0.0 °C?

83. Potassium chloride and rubidium chloride both have the sodium chloride structure. X-ray diffraction experiments indicate that their cubic unit cell dimensions are 629 pm and 658 pm, respectively.

(i) One mole of KCl and one mole of RbCl are ground together in a mortar and pestle to a very fine powder, and the x-ray diffraction pattern of the pulverized solid is measured. Two patterns are observed, each corresponding to a cubic unit cell—one with an edge length of 629 pm and one with an edge length of 658 pm. Call this Sample 1.

(ii) One mole of KCl and one mole of RbCl are heated until the entire mixture is molten and then cooled to room temperature. In this case there is a single x-ray diffraction pattern that indicates a cubic unit cell with an edge length of roughly 640 pm. Call this Sample 2.

(a) Suppose that Samples 1 and 2 were analyzed for their chloride content. What fraction of each sample is chloride? Could the samples be distinguished by means of chemical analysis?

(b) Interpret the two x-ray diffraction results in terms of the structures of the crystal lattices of Samples 1 and 2.

(c) What chemical formula should you write for Sample 1? For Sample 2?

(d) Suppose that you dissolved 1.00 g of Sample 1 in 100 mL of water in a beaker and did the same with 1.00 g of Sample 2. Which sample would conduct electricity better, or would both be the same? What ions would be present in each solution at what concentrations?

84. Sulfur dioxide, SO_2, is found in polluted air.

(a) What type of forces are responsible for binding SO_2 molecules to one another in the solid or liquid phase?

(b) Using the information below, place the compounds listed in order of increasing intermolecular attractions.

Compound	Normal boiling point (°C)
SO_2	-10
NH_3	-33.4
CH_4	-161.5
H_2O	100

85. Copper is an important metal in our economy, most of it being mined in the form of the mineral chalcopyrite, $CuFeS_2$.

(a) To obtain one metric ton (1000 kilograms) of copper metal, how many metric tons of chalcopyrite would you have to mine?

(b) If the sulfur in chalcopyrite is converted to SO_2, how many metric tons of the gas would you get from one metric ton of chalcopyrite?

(c) Copper crystallizes as a face-centered cube. Knowing that the density of copper is 8.95 g/cm^3, calculate the radius of the copper atom.

Applying Concepts

86. Refer to Figure 11.5 when answering these questions.

(a) What is the equilibrium vapor pressure for ethyl alcohol at room temperature?

(b) At what temperature does diethyl ether have an equilibrium vapor pressure of 400 mm Hg?

(c) If a pot of water were boiling at a temperature of 95 °C, what would be the atmospheric pressure?

(d) At 200 mm Hg and 60 °C, which of the three substances are gases?

(e) If you put a couple of drops of each substance on your hand, which would evaporate, and which would remain as a liquid?

(f) Which of the three substances has the greatest intermolecular attractions?

87. The normal boiling point of SO_2 is 263.1 K and that of NH_3 is 239.7 K. At -40 °C. Would you predict that ammonia has a vapor pressure greater than, less than, or equal to that of sulfur dioxide? Explain.

88. Butane is a gas at room temperature; however, if you look closely at a butane lighter you see it contains liquid butane. How is this possible?

89. While camping with a friend in the Rocky Mountains, you decide to cook macaroni for dinner. Your friend says the macaroni will cook faster in the Rockies because the lower atmospheric pressure will cause the water to boil at a lower temperature. Do you agree with your friend? Explain your reasoning.

90. Examine the nanoscale diagrams and the phase diagram below. Match each particulate diagram (1 through 8) to its corresponding point (A through H) on the phase diagram.

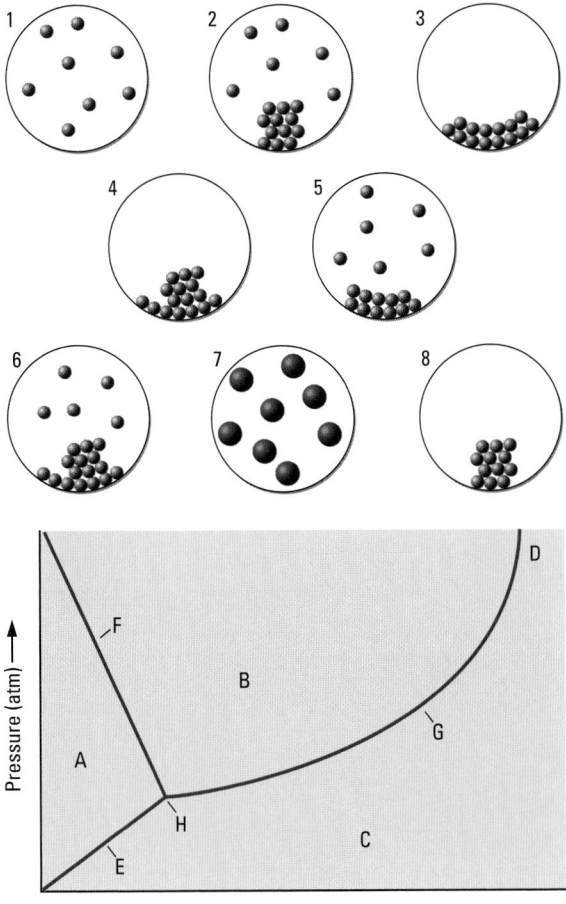

91. Consider the phase diagram below. Draw corresponding heating curves for T_1 to T_2 at Pressures P_1 and P_2. Label each phase and phase change on your heating curves.

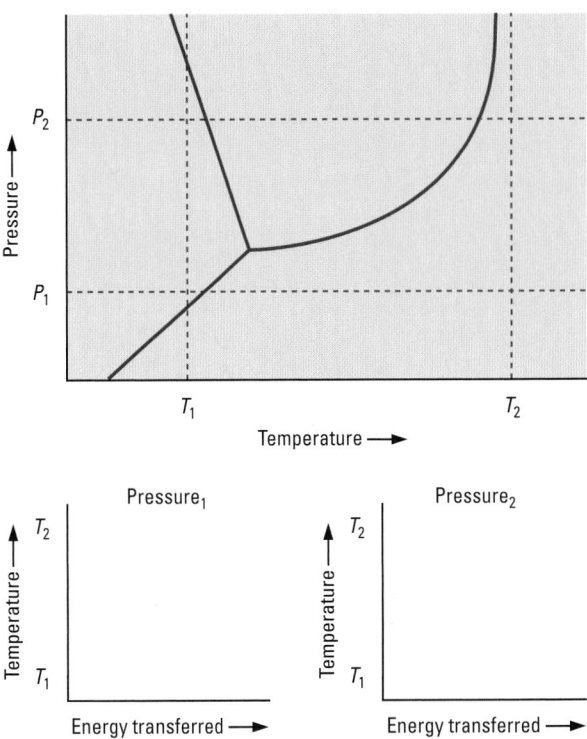

92. Consider three fish aquariums of equal volume. One is filled with tennis balls, another with golf balls, and the third with marbles. If a closest packing arrangement is used in each aquarium, which one has the most occupied space? Which one has the least occupied space? (Disregard the difference in filling space at the walls, bottom, and top of the aquarium.)

General Chemistry CD-ROM

CD11.1 Screen 13.2: Phases of Matter. List the similarities and differences between the four states of matter.

CD11.2 Screen 13.1: Surface Tension/Capillary Action/Viscosity. Think about toweling off after your daily shower. Explain this process. Indicate which forces are stronger, those between the towel and the water, or between your skin and the water.

CD11.3 Screen 13.9: Vapor Pressure. What relationship do you expect to exist between vapor pressure and enthalpy of vaporization for a series of compounds?

CD11.4 Screen 13.10: Boiling Point. Based on their boiling points, which liquid, water or ethanol, do expect to have the larger enthalpy of vaporization?

CD11.5 Screen 13.17: Phase Diagrams. Consider moving from left to right across the phase diagram beginning at a very low temperature and at a pressure of about 0.5 atm. Explain each step in terms of what occurs on the molecular scale and whether each step involves an endothermic or exothermic process.

CD11.6 Screen 13.12: Crystalline and Amorphous Solids.

(a) What is the difference between crystalline and amorphous solids?

(b) Watch the simple cubic unit cell animation on this screen. How many unit cells make up the rotating structure?

CD11.7 Screen 10.14: Metallic Bonding: Band Theory. Why is band theory not used to explain the bonding in molecules such as gaseous ammonia, NH_3, or nitrogen, N_2?

12

Oil wells and a pumping derrick *(foreground)* in Huntington Beach, California. Because of our huge dependence on oil as a fuel and as a raw material, the search for it and bringing it to the surface for refining is ongoing.

FUELS, ORGANIC CHEMICALS, AND POLYMERS

12.1 Petroleum

12.2 Natural Gas and Coal

12.3 Energy Conversions

12.4 Atmospheric Carbon Dioxide, the Greenhouse Effect, and Global Warming

12.5 Organic Chemicals

12.6 Alcohols and Their Oxidation Products

12.7 Carboxylic Acids and Esters

12.8 Synthetic Organic Polymers

12.9 Biopolymers: Proteins and Polysaccharides

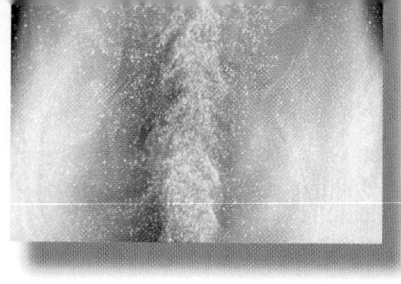

The combustion of fossil fuels — coal, natural gas, and petroleum — provides nearly 85% of all the energy used in the world. When these substances burn, the carbon they contain is released into the atmosphere as CO_2. Photosynthesis converts CO_2 back into other carbon-containing compounds. Many of the carbon compounds produced by photosynthesis are directly or indirectly very useful as energy sources for humans and animals. But they are not as convenient to use in power plants or automobiles as are fossil fuels, which are burned in prodigious quantities daily for these purposes. As fossil fuel becomes scarcer and if more stringent conservation measures are taken, conventional petroleum reserves could last for more than a century. But because of increasing scarcity, fuel costs will rise.

Fossil fuels are also the major source of hydrocarbons and their derivatives. Only about 3% of the petroleum refined today is the source of most of the organic chemicals used to make consumer products such as plastics, pharmaceuticals, synthetic rubber, synthetic fibers, and hundreds of other products we rely on. For this reason, the organic chemical industry is often referred to as the petrochemical industry. In this chapter we will discuss a few of the major classes of organic compounds and some of their reactions, especially those used to furnish energy and to make polymers, synthetic as well as natural ones.

The great Russian chemist Dimitri Mendeleev recognized the importance of petroleum as a source from which to make valuable carbon compounds and not merely to be used as a fuel. On visiting the oil fields of Pennsylvania and Azerbaijan, he supposedly remarked that burning petroleum as a fuel "would be akin to firing up a kitchen stove with bank notes."

12.1 PETROLEUM

Petroleum is a complex mixture of alkanes, cycloalkanes, alkenes, and aromatic hydrocarbons formed from the remains of plants and animals from millions of years ago. Thousands of compounds, almost all of them hydrocarbons, are present in petroleum. Its composition and color vary with the location in which it is found. Pennsylvania crude oils are primarily straight-chain hydrocarbons, whereas California crude oil contains a larger portion of aromatic hydrocarbons.

The early uses for petroleum were mainly for lubrication and for kerosene burned in lamps. The development of automobiles with internal combustion engines created the need for liquid fuels that would burn efficiently in these engines. To meet these needs, it is necessary to refine *crude oil,* the form of petroleum that is pumped from the ground.

The different classes of hydrocarbons (compounds of hydrogen and carbon) were introduced earlier (⇐ *p. 80).* Alkanes have C—C bonds; alkenes have one or more C=C bonds; aromatics include benzene-like rings.

Petroleum Refining

Distillation can separate substances that have different boiling points from liquid mixtures and solutions. For example, simple distillation can separate a mixture of the liquids cyclohexane (b.p. 80.7 °C) and toluene (b.p. 110.6 °C). The mixture is heated to slightly above 80.7 °C, at which point cyclohexane vaporizes and leaves the mixture. The cyclohexane vapor is condensed as a pure liquid and collected in a separate container (Figure 12.1).

Petroleum is a much more complex mixture. Because petroleum contains thousands of different hydrocarbons, their separation as individual pure compounds is neither economically feasible nor necessary. Instead, petroleum is separated by *fractional distillation* into **petroleum fractions,** which are mixtures of hydrocarbons that have boiling points in the same range. Such separation is possible because as the number of carbon atoms increases, the boiling point increases. Larger hydrocarbons have greater noncovalent forces and higher boiling points than smaller ones (⇐ *p. 385).*

The crude oil is first heated to about 400 °C to produce a hot vapor mixture that enters the bottom of the fractionating tower (Figure 12.2). The temperature decreases the farther up the tower the vapor goes, and different components condense at various points in the tower. The volatile lower boiling petroleum fractions remain in the vapor state longer than the less volatile higher boiling fractions.

Petroleum fractions are mixtures of hundreds of hydrocarbons with boiling points within a certain range.

The difference between simple distillation and fractional distillation is the degree of separation achieved.

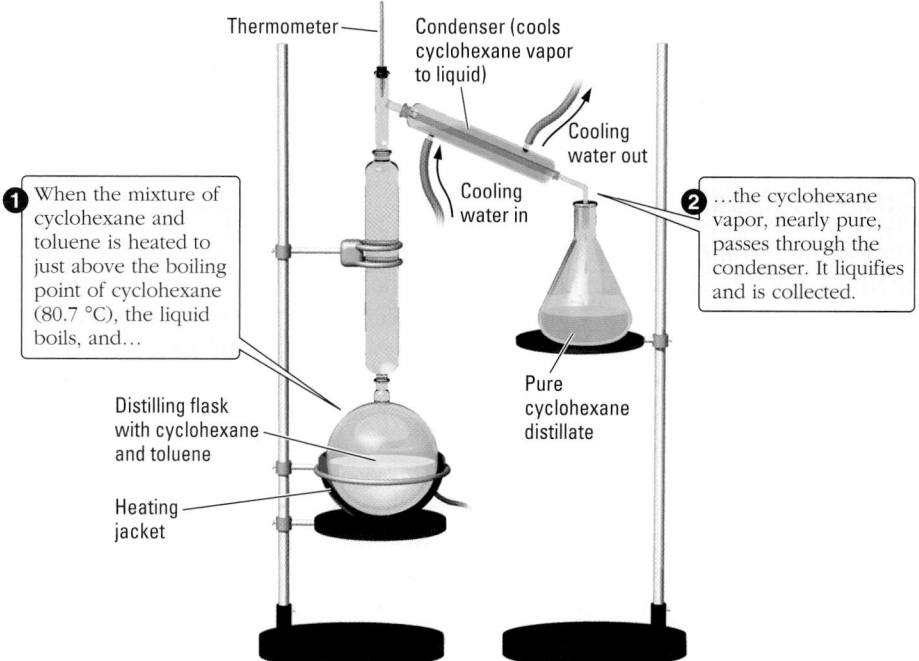

Figure 12.1 Distillation apparatus.

A petroleum fractional distillation tower.

The lightest hydrocarbons do not condense and are drawn off the top of the tower as gases. The heaviest hydrocarbons do not vaporize even at 400 °C and are collected at the bottom of the tower as liquids and dissolved solids.

The properties of various fractions differ, and so do their uses, as shown in Figure 12.2. About 83% (35 gallons) of the total refined barrel of crude oil is simply burned for transportation and heating. The remainder is used for nonfuel purposes,

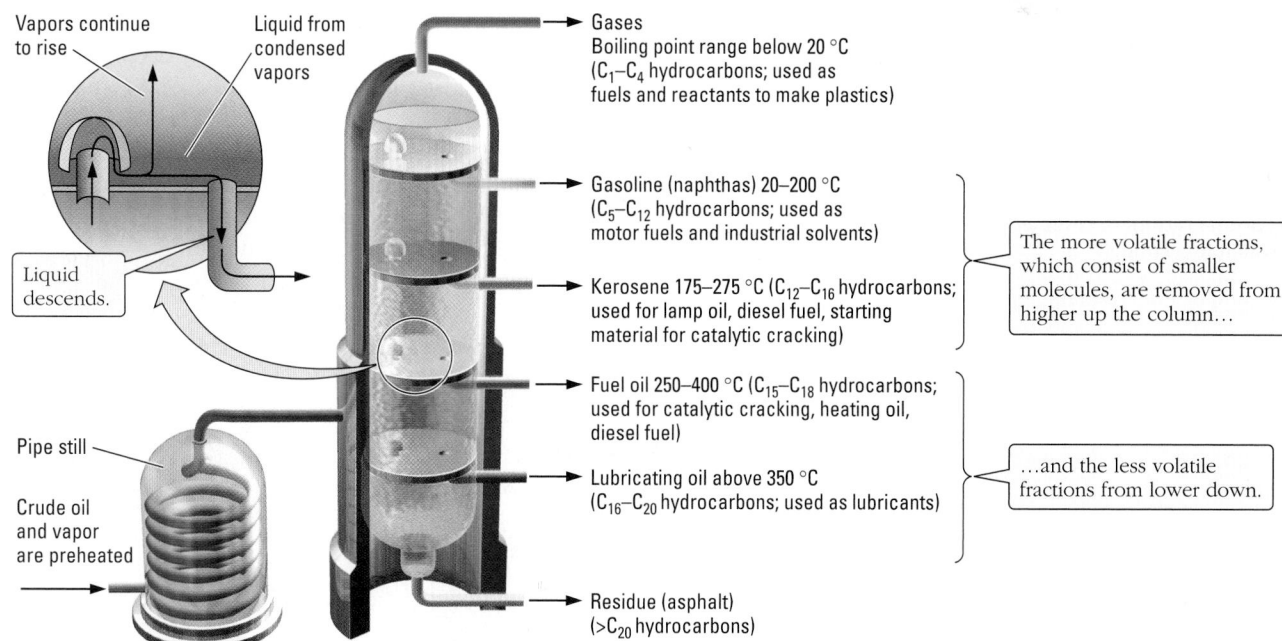

Figure 12.2 **Petroleum fractional distillation.** Crude oil is first heated to 400 °C in the pipe still. The vapors then enter the fractionation tower. As they rise, the vapors cool and condense so that different fractions can be removed at different heights in the tower.

Many years ago, petroleum was shipped in barrels. Today it seldom is seen in barrels, but rather is shipped in pipelines and ocean-going tankers. Nevertheless, the barrel remains as the common unit of volume measure for petroleum. A petroleum barrel contains 42 U.S. gallons.

Diesel engines do not have spark plugs, but use autoignition to ignite the fuel. Thus, diesel fuel is composed mostly of larger, straight-chain hydrocarbons.

Since the method for determining octane numbers was established, fuels that are superior to 2,2,4-trimethylpentane have been developed and thus have octane numbers greater than 100.

Typical octane ratings of commercially available gasoline.

including the very important petroleum feedstocks needed to make the synthetic fabrics, plastics, pharmaceuticals, and other synthetic organic chemicals we depend on.

Octane Number

The **octane number** of a gasoline is a measure of its ability to burn smoothly in an internal combustion engine. Smooth burning depends on the *autoignition temperature*, the temperature at which a liquid will ignite and burn without a source of ignition. The autoignition temperature of a hydrocarbon is related to its molecular composition. Large straight-chain hydrocarbon molecules have much lower autoignition temperatures than do branched-chain and smaller molecules. Because gasoline consists mainly of smaller molecules (C_5 to C_{12}), it has a fairly high autoignition temperature. Thus, to efficiently burn gasoline in an engine requires a source of ignition—a spark from the spark plug. In older cars, engine temperatures get sufficently high (nearly 1000 °C) to cause autoignition to occur before the spark ignites the fuel. When autoignition occurs, the engine "pings" or "knocks," which greatly decreases engine performance.

The octane-number rating of a gasoline is determined by comparing the knocking characteristics when the gasoline burns in a one-cylinder test engine with those obtained for mixtures of heptane and 2,2,4-trimethylpentane (often called isooctane). Heptane knocks considerably and is arbitrarily assigned an octane number of 0, whereas 2,2,4-trimethylpentane burns smoothly and is assigned an octane number of 100. Thus, a gasoline with the same knocking characteristics as a mixture of 13% heptane and 87% 2,2,4-trimethylpentane is assigned an octane number of 87. This corresponds to the octane number of regular unleaded gasoline currently available in the United States. Other, higher grades of gasoline available at gas stations have octane numbers of 89 (regular plus) and 92 (premium).

Straight-chain alkanes are less thermally stable and burn less smoothly than branched-chain alkanes. For example, the "straight-run" gasoline fraction obtained directly from the fractional distillation of petroleum is a poor motor fuel. It needs additional refining because it consists of primarily straight-chain hydrocarbons that autoignite too readily. The octane number of a gasoline can be increased either by increasing the percentage of branched-chain and aromatic hydrocarbon components or by adding octane enhancers. Table 12.1 lists octane numbers for some hydrocarbons and octane enhancers.

In addition to fractional distillation, petroleum refining also includes converting the components of various fractions into more economically important products through catalytic cracking and catalytic reforming. Among these products are compounds that can be added to gasoline to increase its octane rating.

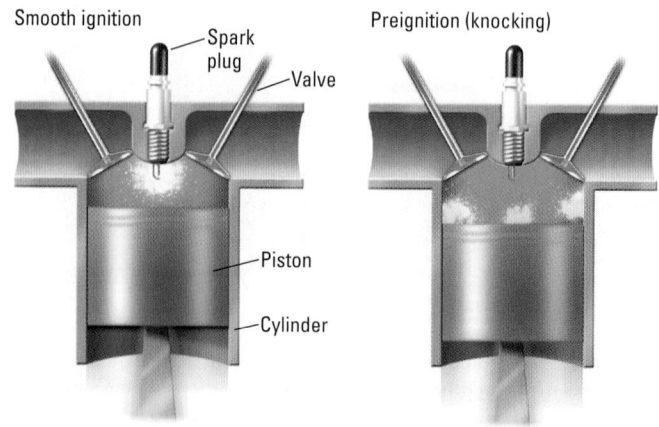

Smooth ignition and autoignition at the spark plugs. Smooth ignition occurs without preignition.

TABLE 12.1	Octane Numbers of Some Hydrocarbons and Gasoline Additives	
Name	**Class of compound**	**Octane number**
Octane	Alkane	− 20
Heptane	Alkane	0
Hexane	Alkane	25
Pentane	Alkane	62
1-Pentene	Alkene	91
2,2,4-Trimethylpentane (isooctane)	Alkane	100
Benzene	Aromatic hydrocarbon	106
Methanol	Alcohol	107
Ethanol	Alcohol	108
Tertiary-butyl alcohol	Alcohol	113
Toluene	Aromatic hydrocarbon	118

Catalytic Cracking

During petroleum refining, the percentage of each fraction collected is adjusted to match the market demand. For example, there is more demand for gasoline than for kerosene and diesel fuel. Demands also vary seasonally. In winter, the need for home heating oil is high; in summer, when more people take vacations, demand for gasoline is higher. In summer, refiners use chemical reactions to convert some of the larger, kerosene-fraction molecules (C_{12} to C_{16}) into smaller molecules in the gasoline range (C_5 to C_{12}) in a process called "cracking." **Catalytic cracking** uses a catalyst, high temperatures, and pressure to break long-chain hydrocarbons into shorter chain hydrocarbons that include both alkanes and alkenes, many in the gasoline range. A **catalyst** is a substance that increases the rate of a chemical reaction without being consumed as a reactant would be. (The role of catalysts in chemical reactions is further discussed in Section 13.8.)

$$C_{16}H_{34} \xrightarrow{\text{pressure and heat}} C_8H_{16} \quad + \quad C_8H_{18}$$

An alkane   An alkene   An alkane

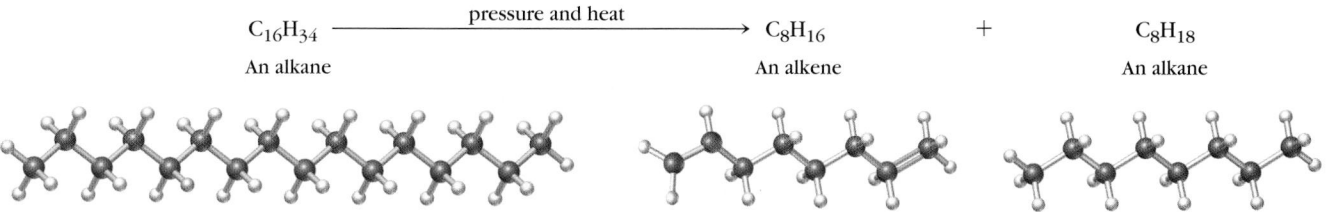

Since alkenes have higher octane numbers than alkanes, the catalytic cracking process also increases the octane number of the mixture. Catalytic cracking is beneficial in another way. Unlike alkanes, alkenes have C=C bonds, which makes them much more reactive than alkanes. Thus, alkenes such as ethylene and propylene can be used as starting materials to make other organic compounds, many of which are the raw materials for making plastics (Section 12.8).

Catalytic Reforming

After the gasoline fraction is separated from crude oil by fractional distillation, its octane number is only 50. This can be increased by converting straight-chain hydrocarbons to branched-chain hydrocarbons and aromatics in a process called

catalytic reforming. In the presence of certain catalysts, such as finely divided platinum on a support of Al_2O_3, straight-chain hydrocarbons with low octane numbers can be reformed into their branched-chain isomers, which have higher octane numbers.

$$CH_3CH_2CH_2CH_2CH_3 \xrightarrow{\text{catalyst}} \overset{\overset{\displaystyle CH_3}{|}}{CH_3CHCH_2CH_3}$$

pentane
62 octane
(C_5H_{12})

2-methylbutane
94 octane
(C_5H_{12})

The additional cost of higher octane gasoline is due to the extra processing required to make higher octane compounds.

By using different catalysts and petroleum mixtures, catalytic reforming is also used to produce aromatic hydrocarbons, which have high octane numbers. For example, catalytic reforming converts a high percentage of straight-run gasoline, kerosene, and light oil fractions into a mixture of aromatic hydrocarbons including benzene, toluene, and xylenes. This process can be represented by the equation for converting hexane into benzene.

$$CH_3CH_2CH_2CH_2CH_2CH_3 \xrightarrow{\text{catalyst}} C_6H_6 \quad + \quad 4\,H_2$$

hexane
25 octane
(C_6H_{14})

benzene
106 octane
(C_6H_6)

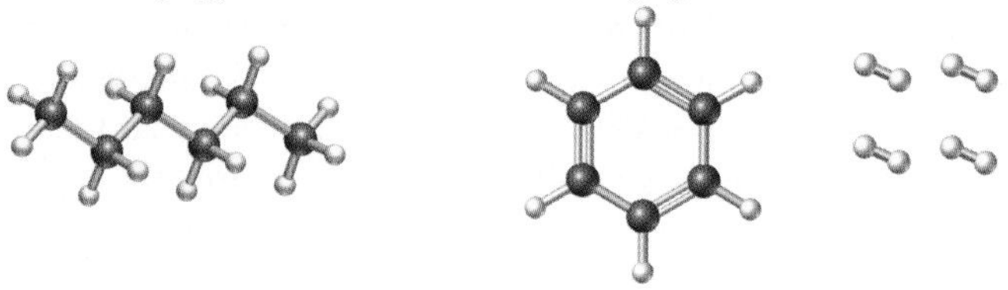

It is no exaggeration to state that our economy rests on the hydrocarbons present in petroleum or derived from it. In addition to being the source of our fuels for transportation, petroleum also provides raw materials to make an impressive array of consumer products, as shown in the figure at the top of the next page.

Octane Enhancers

The octane number of a given blend of gasoline is increased by adding octane enhancers, also known as "anti-knock" agents. In the United States, prior to 1975, the most widely used anti-knock agent was tetraethyllead, $(C_2H_5)_4Pb$. From 1925 until 1975, both regular and premium grades of gasoline contained an average of 3 g of $(C_2H_5)_4Pb$ per gallon.

The exhaust emissions of internal combustion engines contain carbon monoxide, nitrogen oxides, and unburned hydrocarbons, all of which contribute to air pollution (⬅ *Section 10.12*). As urban air pollution worsened, Congress passed the Clean Air Act of 1970, which required that 1975-model-year cars emit no more than 10% of the carbon monoxide and hydrocarbons emitted by 1970 models. The solution to lowering these emissions was a platinum-based catalytic converter, which accelerates the conversion of carbon monoxide to carbon dioxide and the more complete burning of hydrocarbons. The only problem was that the catalyic

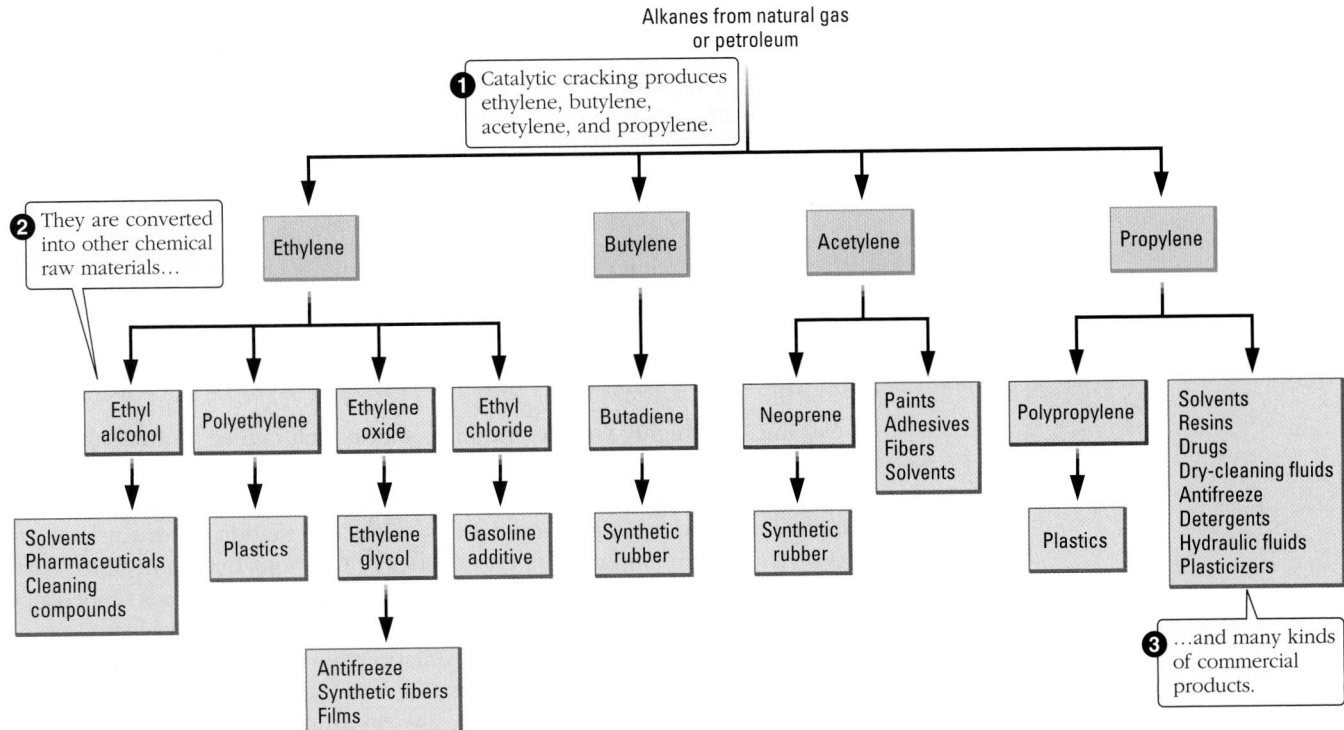

Some chemical raw materials and commercial products made from petroleum or natural gas.

converter required lead-free gasolines, since lead deactivates the platinum catalyst by coating its surface. As a result, automobiles manufactured since 1975 have been required to use lead-free gasoline to protect the catalytic converter. Taking lead out of gasoline has resulted in a dramatic decrease in lead as an air pollutant in the United States (Figure 12.3). With the phaseout of lead in gasoline, there has been a 98% decrease in annual airborne lead emissions in this country, from 220,000 tons of lead in 1970 to approximately 5000 tons in 1998.

Because tetraethyllead can no longer be used in the United States and a few other countries, other octane enhancers are now added to gasoline. These include toluene, 2-methyl-2-propanol (also called *tertiary*-butyl alcohol), methanol, and ethanol.

As little as two tanks of leaded gasoline can destroy the activity of a catalytic converter.

Regrettably, most countries of the world still allow the use of leaded gasoline, putting millions of children at risk for lead poisoning from automobile exhaust fumes.

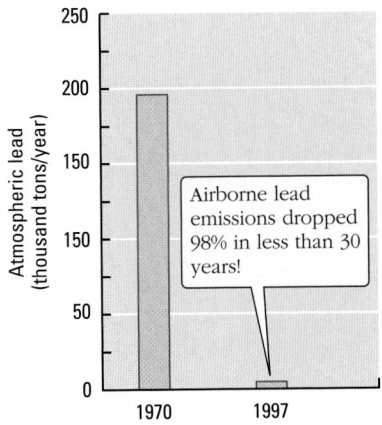

Figure 12.3 Reduction in atmospheric lead.

Exercise 12.1   Rearranging Hydrocarbons

Heptane (C_7H_{16}) can be catalytically reformed to make toluene ($C_6H_5CH_3$), another seven-carbon molecule.

(a) How many hydrogen molecules are produced for every toluene molecule derived from heptane?
(b) Write a balanced chemical equation for this reaction.
(c) Why would it be profitable to convert heptane into toluene?

Oxygenated and Reformulated Gasolines

The 1990 amendments to the Clean Air Act of 1970 require cities with excessive carbon monoxide pollution to use oxygenated gasolines during the winter. *Oxygenated gasolines* are blends of gasoline with organic compounds that contain oxygen, such as, methanol, ethanol, and *tertiary*-butyl alcohol. The 1990 regulations require oxygenated gasolines to contain 2.7% oxygen by weight. Tests conducted by the Environmental Protection Agency (EPA) indicate that in cold weather oxygenated gasolines burn more completely than nonoxygenated gasoline, thus potentially reducing carbon monoxide emissions in urban areas by up to 17%. The use of oxygenated gasoline is currently required in about 40 cities in the United States with excessive wintertime carbon monoxide emissions.

Exercise 12.2   Percent Oxygen in Ethanol

Calculate the percent oxygen in ethanol, CH_3CH_2OH. Is it less than 3%? How can ethanol help to meet the 2.7% oxygen requirement for oxygenated gasolines?

All gasolines are highly volatile and have vapors that can be ignited, allowing you to start your car even in the coldest of weather. However, this volatility means that some hydrocarbons get into the atmosphere as a result of accidental spills and evaporation during normal filling operations at the gas station. Hydrocarbons in the atmosphere play an important role in a series of reactions that contribute to urban air pollution (⬅ *p. 448*), including tropospheric ozone formation, especially in heavy-traffic metropolitan areas. *Reformulated gasolines* are oxygenated gasolines that contain a lower percentage of aromatic hydrocarbons and have a lower volatility than ordinary gasoline. Nine cities with the most serious ozone pollution were required by the 1990 regulations to use reformulated gasolines starting in 1995, and an additional 87 cities that are not meeting the ozone air quality standards can choose to use them.

Engineering improvements such as emission control systems have succeeded in decreasing the emissions of ozone-forming nitrogen oxides and hydrocarbons from automobiles per mile traveled. Today's cars typically emit 70% less nitrogen oxides and 80 to 90% less hydrocarbons over their lifetimes compared with automobiles produced 30 to 40 years ago. But tropospheric ozone levels continue to remain high for two reasons: the number of vehicles and their miles traveled, and the degraded emission controls of older vehicles. Since 1970 the number of vehicles in the United States has doubled, as has the number of miles they travel (Figure 12.4). The Environmental Protection Agency attributes a major portion (approximately 25%) of ozone forming hydrocarbons to a relatively small percentage of "super-dirty" vehicles with faulty emission control systems (Figure 12.5).

Oxygenated gasoline is produced by adding oxygen-containing organic compounds to refined gasoline. Reformulated gasoline requires changes in the refining process to alter the percentages of various hydrocarbons, particularly alkenes and aromatics.

The metropolitan areas with the most serious ozone pollution are Baltimore, Chicago, Hartford, Houston, Los Angeles, Milwaukee, New York, Philadelphia, Sacramento, and Ventura County (CA).

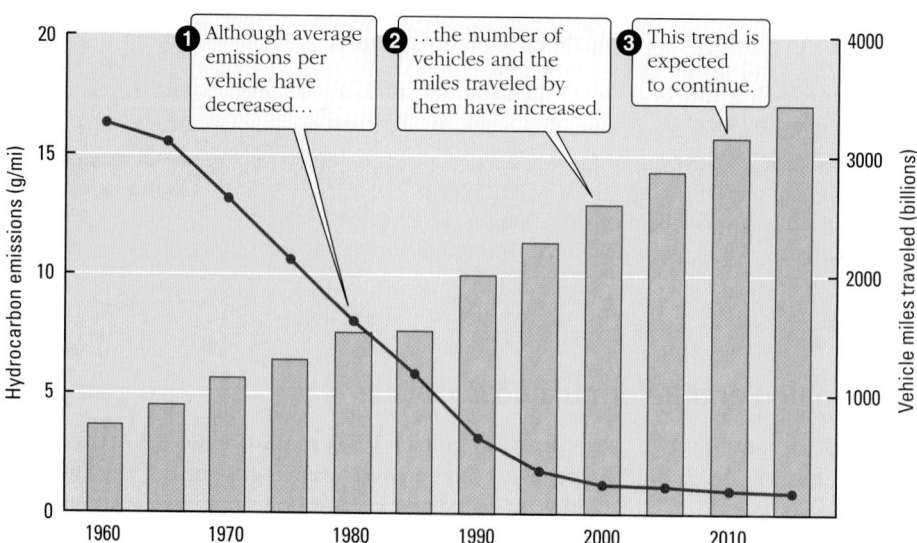

Figure 12.4 **Average per-vehicle emissions (grams of hydrocarbons/mile) and mileage increases since 1960.**

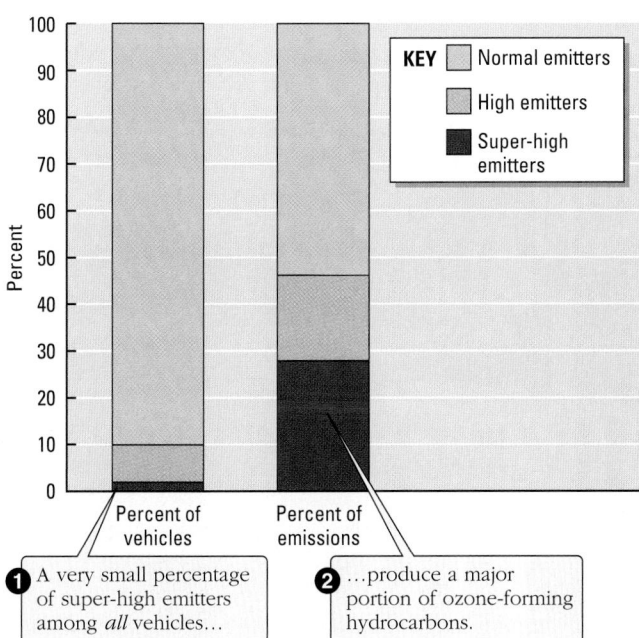

Figure 12.5 **Vehicle emissions.** Some badly tuned cars with faulty emission controls are "super-high-emitting" vehicles. Source: **http://www.epa.gov.**

Exercise 12.3 Carbon Monoxide from Ethanol and from Toluene

(a) The combustion of C_2H_5OH can produce carbon monoxide and water. Use the balanced chemical equation as the basis to calculate the mass in grams of CO produced per gram of ethanol burned.

(b) Repeat the calculation using toluene, C_7H_8, instead of ethanol.

(c) What conclusions can you draw from your answers about using ethanol to reduce carbon monoxide emissions?

12.2 NATURAL GAS AND COAL

Natural Gas

Natural gas is a mixture of low-molar-mass hydrocarbons and other gases trapped with petroleum in the earth's crust. It can be recovered from oil wells or from gas wells to which the gases have migrated through the surrounding rock. The natural gas found in North America is a mixture of C_1 to C_4 alkanes [methane, CH_4 (60 to 90%), ethane C_2H_6 (5 to 9%), propane, C_3H_8 (3 to 18%), and butane, C_4H_{10} (1 to 2%)] with small and varying amounts of other gases, such as CO_2, N_2, H_2S, and the noble gases, mainly helium. In Europe and Japan, natural gas is essentially only methane. The widespread use of natural gas as a fuel did not occur until a vast underground network of pipelines was built to distribute it from natural gas wells to consumers. Before that, most of it was simply burned at the wellheads.

For nearly three decades, the U.S. production of natural gas has exceeded that of U.S.-produced oil (Figure 12.6). Of all the natural gas produced, nearly half is used in industrial applications, 22% for home heating and cooling, 14% in commercial applications, and about 16% as a fuel to generate electricity. About half of the homes in the United States are heated by natural gas.

Natural gas is now also used as a vehicle fuel. Worldwide, there are nearly one million vehicles powered by natural gas, with 30,000 in the United States. The U.S. Postal Service is the largest user of natural gas vehicles, with a fleet of more than 3000 such vehicles. California and several other states are encouraging the use of

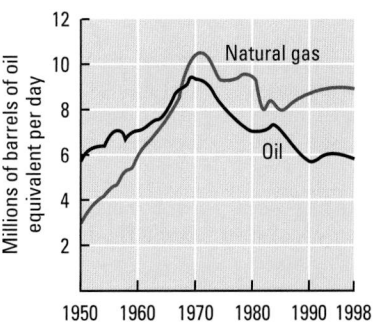

Figure 12.6 **Natural gas and oil production in the continental United States, 1950–1998.**

A natural gas–powered bus. Natural gas powers this bus, which can refuel at this natural gas pump.

Nonmethane hydrocarbons

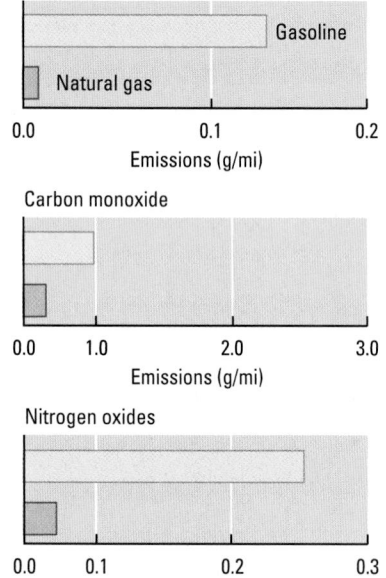

Figure 12.7 Comparisons of tailpipe emissions from natural gas– and gasoline-powered vehicles.

Over the past century, more than 100,000 miners have been killed in United States' mines, making coal mining one of the most dangerous occupations.

The mass percent of carbon in coal varies from about 40% to over 80%, depending on the type of coal.

The usefulness of coal tar to make other materials is summed up in this bit of doggerel from a 1905 organic chemistry textbook:

> You can make anything
> From a salve to a star
> If you only know how
> From black coal tar*

 * Cornish, *Organic Chemistry,* 1905.

natural gas vehicles to help meet new air quality regulations. It has been claimed that vehicles powered by natural gas emit sigificantly less carbon monoxide, hydrocarbons, and nitrogen oxides per mile compared to gasoline-powered cars and trucks (Figure 12.7). The main disadvantages of natural gas vehicles include the need for a pressurized gas tank and the lack of service stations that sell the gas in liquefied form.

Exercise 12.4   Natural Gas as a CO Source

It has been asserted that burning natural gas produces 30% less carbon monoxide than the burning of gasoline. Assume that natural gas is only methane and that gasoline is only octane, C_8H_{18}. Based on the (incorrect) assumption that all of the carbon in the fuel is converted to CO, do calculations that either confirm or refute the assertion.

Coal

About 90% of our annual coal production is burned to produce electricity. The use of coal as the fuel for commercial power plants is on the rise. But the burning of coal for home heating has declined because it is a relatively dirty fuel and bulky to handle. Burning coal is a major cause of air pollution because of its sulfur content. Known world coal reserves are far greater than petroleum reserves.

Coal consists of a complex and irregular array of partially hydrogenated six-membered carbon rings and other structures, some of which contain oxygen, nitrogen, and sulfur atoms. Like petroleum, coal supplies raw materials to the chemical industry. Most of the useful compounds obtained from coal are aromatic hydrocarbons. Heating coal at high temperatures in the absence of air, a process called *pyrolysis,* produces a mixture of coke (mostly carbon), coal tar, and coal gas. Fractional distillation of coal tar produces aromatic hydrocarbons such as benzene, toluene, and xylene. These are the starting materials for a large variety of important commercial products such as paints, solvents, pharmaceuticals, plastics, dyes, and synthetic detergents.

12.3 ENERGY CONVERSIONS

By now it should be apparent that our energy comes from a variety of sources. When you drive your car, the gasoline that moves it may have come from the United States or a foreign country. Some people are already driving cars powered by natural gas or electricity. The natural gas may be imported and the electricity may have come from burning coal, natural gas, fuel oil, or even from a nuclear reactor (Section 20.6).

The convenience of use and the amount of energy derived from different fuels vary depending on their chemical composition and properties. Two important measures of a fuel are its *fuel value* (quantity of energy released per gram of fuel burned) and its *energy density* (quantity of energy released per unit volume of fuel burned) (⬅ *p. 247*). Exercise 12.5 gives you the opportunity to investigate the relationship between these two criteria.

All forms of energy are, in principle, interchangeable. The work a given amount of energy can do is the same, no matter what the energy source. The energy contents of natural gas, fuel oil, and coal in terms of their equivalents in various units are compared in Table 12.2. As a consumer, it doesn't matter to you whether you travel using energy from a fossil fuel in the form of gasoline or whether you use electricity created by burning coal, although the energy is measured differently. A trip might require 1 gal of gasoline for the regular automobile. An electric automobile would require about 39 kilowatt-hours (kWh) of electricity to travel the same distance.

All fuel-burning engines, including automobile engines and electrical power plants, are less than 100% efficient; there is always wasted thermal energy. For example, the generation and distribution of electricity is only about 33% efficient overall. This means that for every 1000 J of energy produced by burning a fossil fuel like coal, only 330 J of electrical energy reaches the consumer.

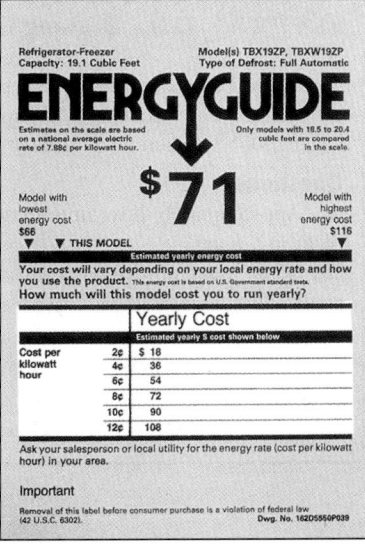

Energy efficiency comparison guide. Cost per kilowatt hour of a refrigerator-freezer, fully self-defrosting.

1 Btu (British thermal unit) is the quantity of energy required to raised the temperature of 1 lb of water from 63 °F to 64 °F; 1 Btu = 252 cal = 1.05×10^3 J.

One watt (W) of power is 1 J/s, so a 100-W electric lamp uses energy at the rate of 100 J every second. Using data from Table 12.2 we find that one cubic foot of natural gas supplies sufficient energy (1.055×10^6 J) to light a 100-W bulb for about 3 hr; a barrel of oil has enough energy (5.9×10^9 J) to light 1000 100-W light bulbs for almost 11 hr.

 Exercise 12.5   Comparing Fuels

Evaluate each of the fuels listed below on the basis of fuel energy and fuel density. Use data from Table 6.2 or Appendix J. Which fuel provides the largest fuel value? Which provides the greatest energy density? Assume that when the fuels burn, carbon is converted to gaseous CO_2, hydrogen to water vapor, and nitrogen to N_2 gas. The densities are given with the substances.

(a) Methane, $CH_4(g)$ (0.656 g/L)
(b) Octane, $C_8H_{18}(\ell)$ (0.699 g/mL)
(c) Ethanol, $C_2H_5OH(\ell)$ (0.785 g/mL)
(d) Hydrazine, $N_2H_4(\ell)$ (1.004 g/mL)
(e) Hydrogen, $H_2(g)$ (0.082 g/L)
(f) Glucose, $C_6H_{12}O_6(s)$ (1.56 g/mL)

TABLE 12.2	A Chart of Energy Units*				
Cubic feet of natural gas	**Barrels of oil**	**Tons of bituminous coal**	**Kilowatt hours of electricity**	**Joules**	**Btu**
1	0.00018	0.00004	0.293	1.055×10^6	1.00×10^3
1000	0.18	0.04	293	1.055×10^9	1.00×10^6
5556	1	0.22	1628	5.9×10^9	5.6×10^6
25,000	4.50	1	7326	26.4×10^9	25.0×10^7
1×10^6	180	40	293,000	1.055×10^{12}	1.00×10^9
3.41×10^6	614	137	1×10^6	3.6×10^{12}	3.4×10^9
1×10^9	180,000	40,000	293×10^6	1.055×10^{15}	1.00×10^{12}
1×10^{12}	180×10^6	40×10^6	293×10^9	1.055×10^{18}	1.00×10^{15}

*Based on normal fuel-heating values. 10^6 = 1 million; 10^9 = 1 billion; 10^{12} = 1 trillion; 10^{15} = 1 quadrillion (quad).

ESTIMATION Burning Coal

A coal-fired electric power plant burns about 1.5 million tons of coal a year. Coal has an approximate composition of $C_{135}H_{96}O_9NS$. When coal burns, it releases about 30 kJ/g.

Questions
(a) Approximately how much energy (kJ) is released by this plant in a year?
(b) What mass of CO_2 is released per year by burning this coal?

Answers
(a) There are 2.2 lb/kg and 2000 lb/ton, so 1.5×10^6 tons is approximately 1×10^9 kg, or 1×10^{12} g. Burning the coal at 30 kJ/g yields about 3×10^{13} kJ per year.

(b) There are 135 mol of carbon (1620 g of C) in 1 mol of coal (1906 g coal/mol). Thus, the ratio of grams of carbon to grams of coal is roughly 0.9. In 1×10^{12} g of coal there is approximately $0.9 \times 1 \times 10^{12}$ g of carbon, which is about 1×10^{12} g of carbon. When burned, each gram of carbon produces about 4 g of CO_2 because there are 44 g of CO_2 per 12 g of C. Thus, in 1 year the plant generates

$$\left(\frac{4 \text{ g } CO_2}{1 \text{ g C}}\right)(1 \times 10^{12} \text{ g of C}) = 4 \times 10^{12} \text{ g } CO_2.$$

In relation to Part (a), this means that 1 g of CO_2 is released from the generation of 8 kJ of energy from burning coal.

Problem-Solving Example 12.1   Energy Conversions

A large ocean-going tanker holds 1.5 million barrels (bbl) of crude oil.

(a) What is the energy equivalent of this oil in joules?
(b) How many tons of coal is this oil equivalent to?

Answer
(a) 8.9×10^{15} J
(b) 3.3×10^5 t coal

Explanation
(a) Table 12.2 gives the thermal energy equivalent to 1 barrel of oil as 5.9×10^9 J. Use this as a conversion factor to calculate the first answer.

$$1.5 \times 10^6 \text{ bbl} \times \frac{5.9 \times 10^9 \text{ J}}{1 \text{ bbl}} = 8.9 \times 10^{15} \text{ J}$$

(b) Look in Table 12.2 and find the conversion between barrels of oil and tons of coal. One bbl of oil is equivalent to 0.22 t of coal. Use this conversion factor to calculate the answer.

$$1.5 \times 10^6 \text{ bbl} \times \frac{0.22 \text{ t coal}}{1 \text{ bbl}} = 3.3 \times 10^5 \text{ t coal}$$

✔ It takes only 0.22 t of coal (approximately $\frac{1}{4}$ ton) to furnish the energy equivalent of a barrel of oil. Therefore, the coal equivalent of 1.5×10^6 barrels of oil should be about one fourth that number of barrels of oil, or about $(0.25)(1.5 \times 10^6) = 4 \times 10^5$, which is close to the calculated value.

Problem-Solving Practice 12.1

How much energy, in joules, can be obtained by burning 4.2×10^9 t of coal? This is equivalent to how many cubic feet of natural gas?

Exercise 12.6   Energy from Burning Oil

A large ocean-going oil tanker holds 1.5×10^6 barrels of crude oil. How many joules of energy can be obtained by burning this oil? How many kilowatt-hours of electricity would be delivered to the consumer by burning this much oil, assuming a 33% overall efficiency?

 ### Exercise 12.7   The Energy Value of CO_2

Explain why CO_2 has no fuel energy value.

12.4 ATMOSPHERIC CARBON DIOXIDE, THE GREENHOUSE EFFECT, AND GLOBAL WARMING

It took millions of years to form fossil fuels from organisms that had obtained their carbon from atmospheric CO_2. By burning carbon-based fuels over just a relatively short time, geologically speaking, we have been returning to the atmosphere carbon atoms that had been trapped for eons. This sudden increase in the release of CO_2 has upset the balance between natural production of CO_2 by combustion and animal respiration and its consumption by photosynthesis and other processes.

Exercise 12.8   Sources of CO_2

List as many natural sources of CO_2 as you can. List as many sources of CO_2 from human activities as you can.

The Greenhouse Effect

What is the problem with increasing atmospheric CO_2? Some of the electromagnetic radiation from the sun is reflected by the atmosphere back into space, and some is absorbed by the atmosphere. The remainder reaches the earth, warming its surface and oceans. The warmed surfaces (average temperature about 15 °C) then reradiate this energy into the troposphere as infrared radiation (⇐ *p. 265*). Carbon dioxide, water vapor, methane, and ozone all absorb radiation in various portions of the infrared region. By absorbing this reradiated energy they warm the atmosphere, creating what is called the **greenhouse effect** (Figure 12.8). Thus, all four are "greenhouse gases." Such gases constitute an absorbing blanket that reduces loss of heat by radiation back into space. Thanks to the greenhouse effect, the earth's average temperature is a comfortable 15 °C (59 °F).

There is such a vast reservoir of water in the oceans that human activity has a negligible influence on the concentration of water vapor in the atmosphere. In addition, methane is produced by natural processes in such large quantities that human contributions are negligible. Ozone is present in such small concentrations that its contribution to the greenhouse effect is small. So of these four greenhouse gases, most attention is focused on CO_2.

Carbon dioxide is a greenhouse gas because it absorbs infrared radiation, causing C=O bonds in the molecule to stretch and bend, much like springs attached

The greenhouse effect derives its name by analogy with a botanical greenhouse. However, the warming in a botanical greenhouse is much more dependent on the glass reducing convection than on blocking infrared radiation from leaving through the glass.

Figure 12.8   The greenhouse effect. Greenhouse gases form an effective barrier that prevents some heat from escaping the earth's surface. Without the greenhouse effect, earth's average temperature would be much lower.

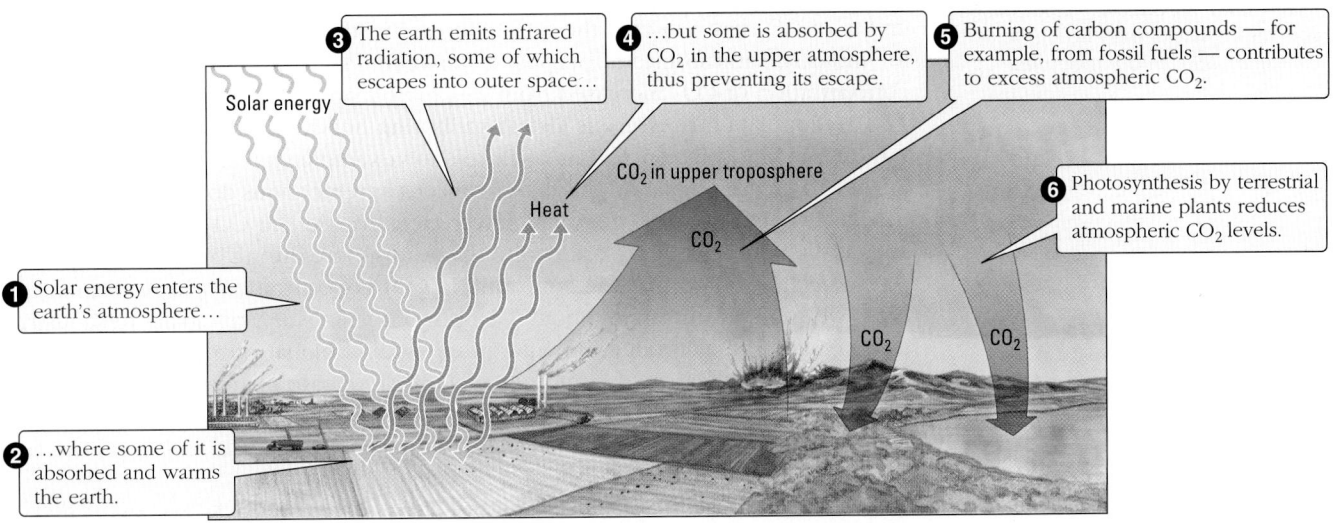

3 The earth emits infrared radiation, some of which escapes into outer space...

4 ...but some is absorbed by CO_2 in the upper atmosphere, thus preventing its escape.

5 Burning of carbon compounds — for example, from fossil fuels — contributes to excess atmospheric CO_2.

Solar energy

CO_2 in upper troposphere

Heat

CO_2

6 Photosynthesis by terrestrial and marine plants reduces atmospheric CO_2 levels.

1 Solar energy enters the earth's atmosphere...

CO_2 CO_2

2 ...where some of it is absorbed and warms the earth.

Figure 12.9 **Stretching and bending modes in the carbon dioxide gas infrared spectrum.** A *Tools of Chemistry* feature describes infrared spectroscopy *(⬅ Chapter 9, p. 362).*

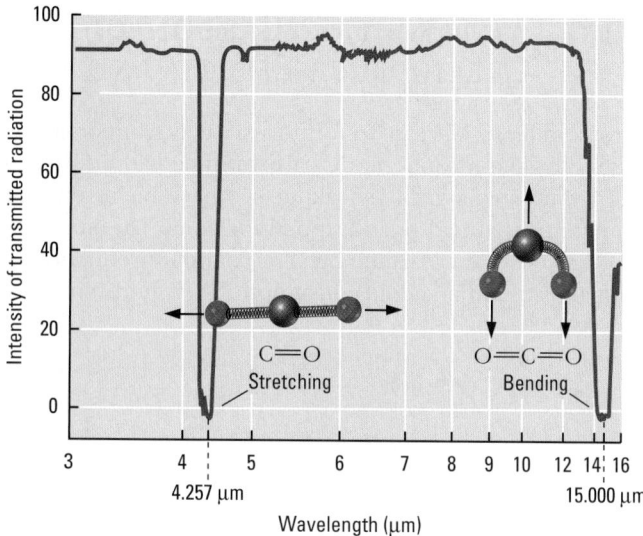

$1 \ \mu m = 1 \times 10^{-6} \ m = 1 \times 10^{3} \ nm$

Recall from Section 7.1 that energy and wavelength are inversely related.

Worldwide, nearly one third of all atmospheric CO_2 is released as a byproduct from fossil fuel–burning electric power plants.

Parts per million (ppm) is a convenient way to express low concentrations. One ppm means one part of something in one million things. A CO_2 concentration of 360 ppm means that for every million molecules of air, 360 are CO_2 molecules.

to balls (Figure 12.9). Stretching the C═O bonds requires more energy (lower wavelength) than does bending them. Thus, C═O stretching occurs when infrared radiation with a wavelength of 4.257 μm is absorbed, whereas the C═O bending vibrations occur at lower energy (longer wavelength) when the molecule absorbs 15.000-μm infrared radiation.

Counting all forms of fossil fuel combustion worldwide, about 6.2 billion metric tons of carbon as CO_2 are added to the atmosphere each year. About 45% is removed from the atmosphere by natural processes—some by plants during photosynthesis and the rest by dissolving in rainwater and the oceans to form bicarbonates and carbonates.

$$CO_2(g) + 2 \ H_2O(\ell) \longrightarrow H_3O^+(aq) + HCO_3^-(aq)$$

$$HCO_3^-(aq) + H_2O(\ell) \longrightarrow H_3O^+(aq) + CO_3^{2-}(aq)$$

The other 55% of the carbon dioxide from fossil fuel combustion remains in the atmosphere, increasing the global CO_2 concentration.

Without human influences, the flow of carbon dioxide among the air, plants, animals, and the oceans would be roughly balanced. In 1750, during the preindustrial era, CO_2 concentration in the atmosphere was 277 parts per million (ppm). During the next 130 years, as the Industrial Revolution took hold, the concentration increased to 291 ppm, a 5% increase. Since 1900, however, the rate of increase of CO_2 has reflected the rapid increase in the use of fossil fuel combustion for industrial and domestic purposes. Between 1900 and 1999, the atmospheric concentration of CO_2 increased from 296 to 369 ppm, an increase of 24% (Figure 12.10). Population pressure is also contributing heavily to increased CO_2 concentrations. In the Amazon region of Brazil, for example, forests are being cut and burned to create cropland. This activity places a tremendous double burden on the natural CO_2 cycle, since there are fewer trees to use the CO_2 in photosynthesis and, at the same time, CO_2 is added to the atmosphere during burning.

Expectations are that the atmospheric CO_2 concentration will continue to increase at the current rate of about 1.5 ppm per year, due principally to industrialization, agricultural production, and an expanding global population. As a result, according to sophisticated computer models of climate change, the CO_2 concentration is projected to reach 550 ppm, double its preindustrial value, between 2030 and 2050. The increase may perhaps not be so great if fuel costs increase substantially as fossil fuel supplies get tighter.

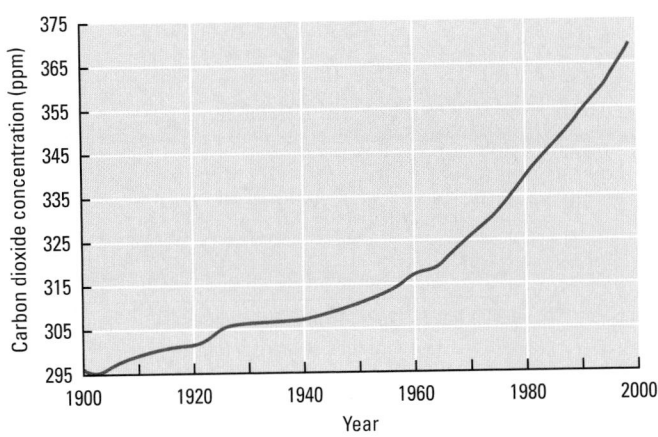

Figure 12.10 **Atmospheric carbon dioxide concentration.**
Atmospheric carbon dioxide concentrations have been rising steadily.
Source: **http://cdiac.esd.gov/ftp/maunaloa-co2/maunaloa.co2.**

Exercise 12.9   CO₂ Changes

Using data from Figure 12.10, calculate the percent increase in atmospheric CO_2 from

(a) 1900 to 1950   (b) 1950 to 1999   (c) 1990 to 1999

Which period showed the greatest percent increase in CO_2?

 ### Exercise 12.10   Annual CO₂ Changes

During a year, atmospheric CO_2 concentration fluctuates, building up to a high value, then dropping to a low value before building up again. Explain what causes this fluctuation and when during the year the high and the low occur.

To see how easily everyday activities affect the quantity of CO_2 being put into the atmosphere, consider a round-trip flight from New York to Los Angeles. Each passenger pays for about 200 gal of jet fuel, which weighs 1400 lb. When burned, each pound of jet fuel produces about 3.14 lb of carbon dioxide. So 4400 lb, or 2 metric tons, of carbon dioxide are produced per passenger during that trip.

 ### Exercise 12.11   Carbon Dioxide and Air Travel

Use an Internet search engine to find out the annual airline passenger miles for a recent year and calculate the tons of CO_2 produced by those flights. Compare this to the quantity of CO_2 produced for the same number of miles traveled in automobiles. (*Hint:* Assume a miles-per-gallon value and an average number of passengers per vehicle. Then use the same numbers for gasoline as were used in the previous paragraph.)

Global Warming

Global warming is the temperature increase due to the greenhouse effect amplified by increasing concentrations of CO_2 and other greenhouse gases. Figure 12.11 shows the CO_2 content of the atmosphere as determined by analysis of air pockets trapped in ice core samples dating back as far as 160,000 years. Notice the striking parallel relationship between average global temperature and atmospheric carbon dioxide change. From these measurements, there appears to be a direct correlation between the atmospheric carbon dioxide concentration and global temperatures in the same period (which were determined by other means): *As the CO_2 level increased, average global temperatures increased, and vice versa.*

Nearly a century ago (1896), Svante Arrhenius, the noted Swedish chemist, estimated the effect that increasing atmospheric carbon dioxide could have on

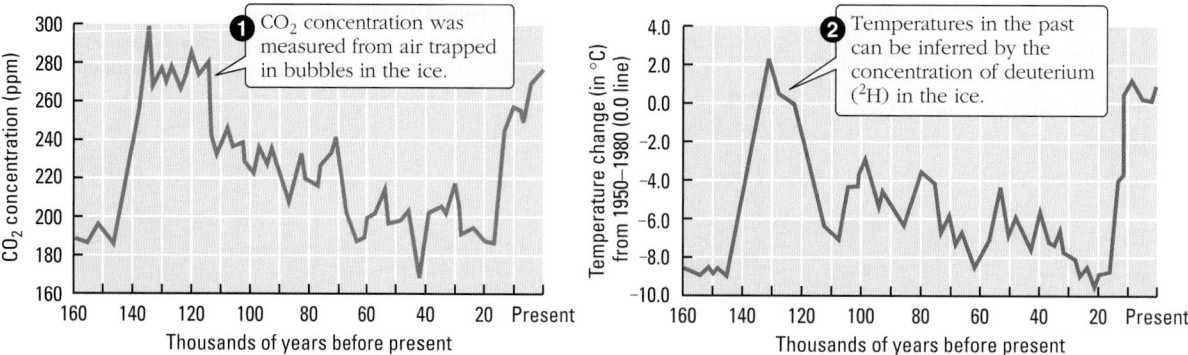

Figure 12.11 Atmospheric CO_2 concentrations and average global temperatures during the past 160,000 years, as derived from ice core data.
Source: *Chemical & Engineering News,* March 13, 1989, p. 36.

global temperature. He calculated that doubling the carbon dioxide concentration would increase the temperature by 5 to 6 °C. Disturbed by the prospects of his prediction, he described the possible scenario in picturesque terms: "We are evaporating our coal mines into the air."

 Exercise 12.12   Evaporating Coal Mines

Explain Arrhenius's quote in terms of global warming.

Many computer models predict that increasing atmospheric CO_2 to 600 ppm will increase average global temperature, and most climate experts who do global warming research agree. What is uncertain is the extent of the temperature increase, with estimates varying from 1.5 to 4.5 °C (or 2.7 to 8.1 °F). In its latest report, the prestigious United Nations' Intergovernmental Panel on Climate Change estimates a 1.0 to 3.5 °C (or 2 to 6 °F) increase by the year 2100. Warming by as little as 1.5 °C would produce the warmest climate seen on earth in the past 6000 years; an increase of 4.5 °C would produce world temperatures higher than any since the Mesozoic era, the time of the dinosaurs. Recent research suggests that the 1998 average global temperature was the highest in the past 1000 years. In particular, there has been nearly a 0.7 °C increase from the average since the late 1800s, a period of rapid industrial development.

> A 1.5 °C rise in global temperature is a significant change, requiring a very large input of energy.

Many scientists are concerned that rising temperatures will cause more of the polar ice caps to melt, raising the sea level and flooding coastal cities, and that atmospheric currents will change and produce significant changes in weather and agricultural productivity. Since the 1940s, average summertime temperatures in Antarctica have increased 2.5 °C to just above 0 °C. Ice shelves there lost nearly 13% of their total area during 1998 alone.

Computer models used to predict future global temperature changes have become more sophisticated and accurate during the past decade. When factors in addition to greenhouse gases, such as the presence of aerosols (tiny, suspended particles) and changes in the sun's irradiance are taken into account, the computer models predict temperature changes close to what has been observed since 1860 (Figure 12.12). Such work adds credibility to the proposed temperature increases of 1.5 to 4.5 °C due to global warming.

Global warming could become a major worldwide problem. Concerns about it prompted nearly 10,000 delegates from 159 countries to convene at the Kyoto Conference in December 1997. Goals were set there to reduce atmospheric greenhouse gases from 1990 levels. Under these targets, to be met between 2008 and 2012, industrialized nations would decrease their greenhouse gas emissions overall

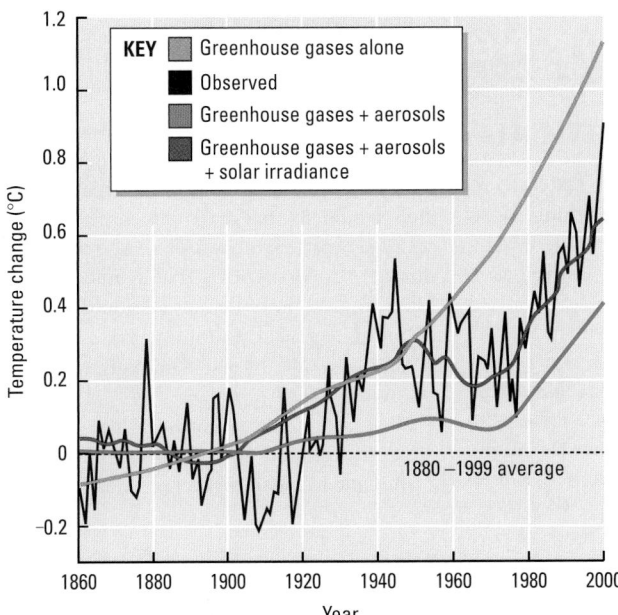

Figure 12.12 **Computer modeling of average global temperature change in relation to an 1880–1999 average.** When greenhouse gases, aerosols, and changes in solar irradiance are used as inputs into general circulation models, predicted temperatures are very close to those observed. In the past, only greenhouse gases were used. Doing so predicts temperatures higher than those observed. Source: Wigley, T.M.L. "The Science of Climate Change," Pew Center on Global Climate Change.

by about 5%. The United States would be expected to lower emissions to 7% below 1990 levels, the European Union nations by 8%, and Japan by 7%. A contentious issue is that no binding emission goals were set for developing countries. The Kyoto agreement does not take effect until 90 days after it is ratified by at least 55 nations that account for 55% of the total 1990 CO_2 emissions, which includes the United States. There is little likelihood that the U.S. Senate will ratify the treaty without a condition requiring key developing countries to limit their CO_2 emissions as well.

The most obvious measure to reduce global warming is to control CO_2 emissions throughout the world. Given our dependence on fossil fuels, however, getting and keeping these emissions under control will be very difficult. One direct way to decrease the quantity of atmospheric CO_2 is to plant more trees and replace those in deforested areas. A more dramatic method is deep storage of CO_2. Rather than release CO_2 to the atmosphere, it would be pumped into underground formations such as depleted oil or gas wells, unminable coal beds, or natural salt domes. Alternatively, it could be pumped more than 3000 m below the ocean surface, where it will liquefy into a carbon dioxide lake.

The projected extent of global warming remains a thorny issue. The uncertainty in the temperature changes caused by global warming are related to three fundamental, but difficult, questions: (1) To what levels will CO_2 and other greenhouse gases rise during the next decades? (2) How inherently responsive is the earth's climate to the warming created by the greenhouse effect? (3) Is the current global warming being caused by human activities, or is it part of a natural cycle of global temperature changes? These questions have yet to be answered unambiguously.

 ## 12.5 ORGANIC CHEMICALS

Organic chemicals were once obtained only from plants, animals, and fossil fuels. Prior to 1828, it was widely believed that chemical compounds found in living matter could not be made without living matter — a "vital force" was thought to be necessary for the synthesis. In 1828 a young German chemist, Friedrich Wöhler, destroyed the vital force myth when he prepared the organic compound urea, a

Suboceanic sequestration of CO_2. A Sleipner offshore natural gas production platform off the coast of Norway, where deep CO_2 sequestration is taking place.

Chemistry in the News

GREEN CHEMISTRY: BEING CRITICAL ABOUT GREASE

Because they remove grease and oil stains, chlorinated solvents such as tetrachlorethylene, $Cl_2C=Cl_2$, known as "perc," are used to dry clean soiled clothes. Perc, however, is considered to be a potential cancer-causing agent (carcinogen) and to have serious environmental effects.

$$
\begin{array}{ccc}
Cl & & Cl \\
 & C=C & \\
Cl & & Cl \\
\end{array}
$$

tetrachloroethylene

Given the massive amount of perc that is used annually by nearly 30,000 dry cleaners, it is beneficial to have an alternative that is noncarcinogenic and more environmentally friendly. One major alternative to perc is supercritical carbon dioxide (scCO$_2$), which is carbon dioxide at pressures and temperatures above its critical point (31.1 °C, 73 atm) (⇐ *p. 474*). An environmentally benign ("green") solvent, scCO$_2$ has replaced halogenated solvents to decaffeinate coffee because of the concerns that residues of these solvents remained in the coffee as a possible carcinogen.

A supercritical liquid combines the properties of a gas with those of a liquid—for example, the ability to expand to fill its container, but with a density and flow like those of a liquid. After extraction of the target solute from a mixture, the scCO$_2$ is separated from it as a gas simply by lowering the pressure be-low the critical point. To use scCO$_2$ to remove grease stains from clothing, however, requires a new kind of detergent, one that dissolves in scCO$_2$, as well as in grease and oil. Dr. Joseph DeSimone at North Carolina State University developed such a detergent based on a polymer having some groups that dissolve in grease and others that are soluble in scCO$_2$. For this work Dr. DeSimone received the 1997 Presidential Green Chemistry Challenge Award.

He has developed a chain of dry cleaning stores called Hangars that use a patented scCO$_2$ process that avoids the high heat of the conventional drying cycle, which is harsh to clothing. The scCO$_2$ is made from waste CO$_2$ from industrial processes, CO$_2$ that would have been vented into the atmosphere contributing to global warming.

• Why is scCO$_2$ by itself unlikely to dissolve grease?
• What structural features of the polymer developed by DeSimone make it effective in cleaning grease stains?

Source:

ChemMatters, April 2000, pp.14–15.

A scCO$_2$ cleaning machine (*ChemMatters*, April 2000, p. 15).

major product in urine, by heating an aqueous solution of ammonium cyanate, a compound obtained from mineral sources.

$$
[NH_4]^+[NCO]^-(aq) \xrightarrow{\text{heat}} \underset{\text{urea}}{\overset{\displaystyle O}{\underset{H_2N}{\overset{\|}{C}}}} NH_2
$$

ammonium cyanate   urea

The notion of a mysterious vital force declined as other chemists began to prepare more and more organic chemicals without the aid of a living system. About 85% of all known compounds are organic compounds. It is natural to ask, Why are there so many organic compounds, almost all of which contain carbon and hydrogen atoms (and commonly other kinds of atoms as well)? As discussed in earlier sections on bonding and isomerism *(⇐ p. 83, p. 329)*, two reasons are (1) the ability of up to thousands of carbon atoms to be linked to each other in a single molecule by stable C—C bonds, and (2) the occurrence of structural isomers. A third reason — the variety of functional groups that bond to carbon atoms — also was introduced earlier *(⇐ p. 81)* and is further illustrated in this chapter.

A few living organisms are still direct sources of useful hydrocarbons. Rubber trees produce latex that contains the familiar hydrocarbon, rubber, and some plants produce an oil that burns almost as well in a diesel engine as diesel fuel. As important as these examples are, the development of synthetic organic chemistry has led to cheaper methods of making copies of naturally occurring substances and to many substances that have no counterpart in nature.

Catalytic cracking is very important for the production of small, unsaturated hydrocarbons such as ethylene (CH_2=CH_2), propylene (CH_3CH=CH_2), butylene (CH_3CH_2CH=CH_2), and acetylene (HC≡CH) *(⇐ p. 519)*. These reactive molecules are starting materials in the organic chemical industry for the production of a substantial variety of substances we use daily *(⇐ p. 519)*. Aromatic compounds such as benzene, toluene, and xylene derived from coal tar and catalytic reforming are also important compounds from which a vast array of commercial products are produced. All organic molecules can be viewed as derived from hydrocarbons, and many of them are prepared in this manner. Among some of the most useful of these compounds are alcohols, acids, esters, and the natural and synthetic polymers that can be made from them.

A table of functional groups and further information on how organic compounds are named are given in Appendix E.

12.6 ALCOHOLS AND THEIR OXIDATION PRODUCTS

Alcohols, both natural and synthetic, contain one or more —OH functional groups bonded to carbon atoms and are a major class of organic compounds *(⇐ p. 81)*. Some examples of commercially important alcohols and their uses are listed in Table 12.3.

CD-ROM Screen 11.5: Functional Groups

Methanol and Ethanol

Methanol, CH_3OH, the simplest of all alcohols, has just one carbon atom. Because methanol is so useful in making other products, more than 8 billion pounds of it are produced annually in the United States by the reaction of carbon monoxide with hydrogen in the presence of a catalyst at 300 °C.

$$CO(g) + 2\,H_2(g) \xrightarrow{\text{catalyst, 300 °C}} CH_3OH(g)$$

Methanol is sometimes called *wood alcohol* for the old method of producing it by heating a hardwood such as beech, hickory, maple, or birch in the absence of air. Methanol is highly toxic. Drinking as little as 30 mL can cause death, and smaller amounts (10 to 15 mL) cause blindness.

About 50% of methanol is used in the production of formaldehyde (HCHO), which is used to make plastics, embalming fluid, germicides, and fungicides; 30% is used in the production of other chemicals; and the remaining 20% is used for jet fuels, antifreeze solvent mixtures, and as a gasoline additive.

Because methanol can be made from coal, it will likely increase in importance as petroleum and natural gas become too expensive as sources of both energy and

TABLE 12.3 Some Important Alcohols

Condensed formula	b.p. (°C)	Systematic name	Common name	Use
CH_3OH	65.0	Methanol	Methyl alcohol	Fuel, gasoline additive, making formaldehyde
CH_3CH_2OH	78.5	Ethanol	Ethyl alcohol	Beverages, gasoline additive, solvent
$CH_3CH_2CH_2OH$	97.4	1-Propanol	Propyl alcohol	Industrial solvent
$CH_3\overset{\displaystyle \mid}{\underset{\displaystyle OH}{C}}HCH_3$	82.4	2-Propanol	Isopropyl alcohol	Rubbing alcohol
$\underset{\displaystyle OH\ OH}{CH_2CH_2}$	198	1,2-Ethanediol	Ethylene glycol	Antifreeze
$\underset{\displaystyle OH\quad OH\quad OH}{CH_2-CH-CH_2}$	290	1,2,3-Propanetriol	Glycerol (glycerin)	Moisturizer in foods

chemicals. Since methanol is relatively cheap, its potential as a fuel and as a starting material for the synthesis of other chemicals is receiving more attention. Methanol is being considered as a replacement for gasoline in addition to its use in oxygenated gasoline. As a motor fuel, methanol burns more cleanly than gasoline, and levels of troublesome pollutants are reduced. Also, burning methanol emits no unburned hydrocarbons, which contribute significantly to air pollution (⇐ *p. 448)*.

The technology for methanol-powered vehicles has existed for many years, particularly for racing cars that burn methanol because of its high octane rating (107). However, methanol has only about half the energy content of the same volume of gasoline. Therefore, twice as much methanol must be burned to give the same distance per tankful as gasoline. This is partially compensated for by the fact that methanol costs about half as much to produce as gasoline. Another disadvantage is the tendency for methanol to corrode ordinary steel. Therefore, it is necessary to use stainless steel or a methanol-resistant coating for the fuel system. Until sufficient numbers of methanol-powered vehicles are on the road, cars equipped to run on both methanol and gasoline will be necessary because of the lack of service stations selling methanol.

Ethanol, also called *ethyl alcohol* or *grain alcohol,* is the "alcohol" of alcoholic beverages. For millennia it has been prepared for this purpose by fermentation of carbohydrates (starch, sugars) from a wide variety of plant sources. For example, glucose is converted into ethanol and carbon dioxide by the action of yeast in the absence of oxygen.

Racing cars burn methanol.

$$C_6H_{12}O_6 \xrightarrow{\text{Yeast}} 2\ C_2H_5OH + 2\ CO_2$$

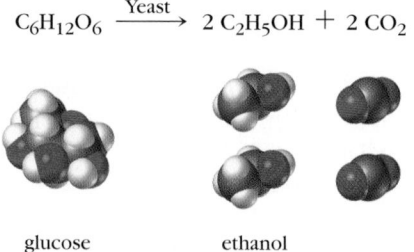

glucose   ethanol

Ninety-five percent ethanol is the maximum ethanol concentration that can be obtained by distillation of alcohol/water mixtures because ethanol and water form a mixture that boils without changing composition.

Fermentation eventually stops when the alcohol concentration reaches a level sufficient to inhibit the yeast cells. The "proof" of an alcoholic beverage is twice the volume percent of ethanol; 80-proof vodka, for example, contains 40% ethanol by volume. Although ethanol is not as toxic as methanol, one pint of pure ethanol, rapidly ingested, will kill most people. Ethanol is a depressant; the effects of various blood levels of alcohol are shown in Table 12.4. Rapid consumption of two 1-oz "shots" of 90-proof whiskey, two 12-oz beers, or two 4-oz glasses of wine can cause one's blood alcohol level to reach 0.05%.

In the United States, the federal tax on alcoholic beverages is about $20 per gallon. Since the cost of producing ethanol is only about $1 per gallon, ethanol intended for industrial use must be *denatured* to avoid the beverage tax and to prevent people from drinking the alcohol. Ethanol is denatured by adding to it small amounts of a toxic substance, such as methanol or gasoline, that cannot be removed easily by chemical or physical means.

Like methanol, ethanol is receiving increased attention as an alternative fuel. At present, most of it is used in a blend of 90% gasoline and 10% ethanol (known as *gasohol* when introduced in the 1970s). Ethanol is also used as an oxygenated fuel to add to gasoline. The fermentation of corn, an abundant (but not limitless) source of carbohydrate, is used to produce ethanol for fuel use.

TABLE 12.4	Blood Alcohol Levels and Their Effects
% by Volume	**Effect**
0.05 - 0.15	Lack of coordination
0.10	Commonly defined point for "driving while intoxicated"
0.15 - 0.20	Intoxication
0.30 - 0.40	Unconsciousness
0.50	Possible death

Problem-Solving Example 12.2 The Heat of Combustion of Ethanol Compared to that of Octane

Calculate the heat of combustion of ethanol in kilojoules per mole (kJ/mol) and compare its value with that of octane. Then, using the densities of the liquids, calculate the thermal energy liberated on burning a liter of each liquid fuel. The densities of octane and ethanol are 0.703 g/mL and 0.789 g/mL, respectively. The heat of formation of octane(g) is -208.0 kJ/mol.

Answer Ethanol: $\Delta H_{comb} = -1277.37$ kJ/mol; -2.19×10^4 kJ/L

Octane: $\Delta H_{comb} = -5116.4$ kJ/mol; -3.15×10^4 kJ/L

Explanation All internal combustion engines burn fuel vapors. The balanced equation for the combustion of ethanol vapor is

$$C_2H_5OH(g) + 3\ O_2(g) \longrightarrow 2\ CO_2(g) + 3\ H_2O(g)$$

Using the standard molar heats of formation from Appendix J and Hess's law ($\Longleftarrow$ **p. 241**) for both the products and reactants, we find that the heat of combustion, ΔH_{comb}, is

$$\Delta H_{comb} = 2[\Delta H_f^\circ\ CO_2(g)] + 3[\Delta H_f^\circ\ H_2O(g)] - 1[\Delta H_f^\circ\ C_2H_5OH(g)]$$

$$= 2(-393.509\ \text{kJ/mol}) + 3(-241.818\ \text{kJ/mol}) - (-235.10\ \text{kJ/mol})$$

$$= -1277.37\ \text{kJ/mol}$$

The balanced combustion reaction for octane vapor is

$$C_8H_{18}(g) + \tfrac{25}{2}\ O_2(g) \longrightarrow 8\ CO_2(g) + 9\ H_2O(g)$$

Using Hess's law, the heat of combustion of octane is

$$\Delta H_{comb} = 8[\Delta H_f^\circ\ CO_2(g)] + 9[\Delta H_f^\circ\ H_2O(g)] - 1[\Delta H_f^\circ\ C_8H_{18}(g)]$$

$$= 8(-393.509\ \text{kJ/mol}) + 9(-241.818\ \text{kJ/mol}) - (-208.0\ \text{kJ/mol})$$

$$= -5116.4\ \text{kJ/mol}$$

The molar masses of ethanol and octane are 46.069 and 114.23 g/mol, respectively. The thermal energy liberated per liter for each is calculated as follows:

Ethanol: $-1277.37\ \text{kJ/mol} \times \dfrac{1\ \text{mol}}{46.069\ \text{g}} \times \dfrac{0.789\ \text{g}}{\text{mL}} \times \dfrac{1000\ \text{mL}}{\text{L}} = -2.19 \times 10^4\ \text{kJ/L}$

Octane: $-5116.4\ \text{kJ/mol} \times \dfrac{1\ \text{mol}}{114.23\ \text{g}} \times \dfrac{0.703\ \text{g}}{\text{mL}} \times \dfrac{1000\ \text{mL}}{\text{L}} = -3.15 \times 10^4\ \text{kJ/L}$

✔ There are about 17 mol of ethanol in a liter of ethanol (789 g ethanol/L ethanol) (1 mol ethanol/46 g ethanol) = 17 mol and about 6 mol of octane in 1 L of octane

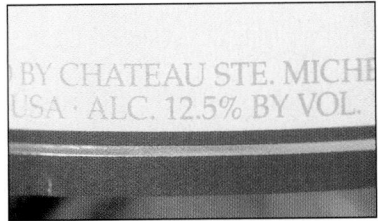

Alcohol concentration in wine.

Using ethanol as an automotive fuel. The ethanol is mixed with gasoline.

(703 g octane/L octane) (1 mol octane/114 g octane). But nearly four times as much energy is liberated by burning a mole of octane. Applying this factor on a liter-to-liter basis:

$$\frac{5000 \text{ kJ/mol octane}}{1200 \text{ kJ/mol ethanol}} \times \frac{6 \text{ mol octane/L}}{17 \text{ mol ethanol/L}} = 1.5$$

The 1.5 indicates that about 50% more energy is released from 1 L of octane than 1 L of ethanol. This is close to the 44% difference in the kJ/L calculated for the two fuels, $(3.15 \times 10^4 / 2.19 \times 10^4) \times 100\%$.

Problem-Solving Practice 12.2

Calculate the heat of combustion of methanol in kJ/mol and kJ/L using standard molar heats of formation from Appendix J. The density of methanol is 0.791 g/mL.

Hydrogen Bonding in Alcohols

Water molecules hydrogen-bond to each other *(⇐ p. 389)*, creating the unusual physical properties of water (Section 11.4). The physical properties of liquid alcohols are a direct result of the effects of hydrogen bonding among alcohol molecules. Alcohols can be considered as compounds in which alkyl groups such as CH_3— and CH_3CH_2— are substituted for one of the hydrogens of water. The change from the hydrogen in water to the alkyl group in an alcohol reduces the influence of hydrogen bonding in the progression from water (HOH) to methanol (CH_3OH) to ethanol (CH_3CH_2OH) and higher alcohols. As a result, the boiling points of methanol (32 g/mol; 65 °C) and ethanol (46 g/mol; 78.5 °C) are lower than that of water (18 g/mol; 100 °C) because methanol and ethanol have only one —OH hydrogen atom available for hydrogen bonding; water has two. The higher boiling point of ethylene glycol (198 °C) can be attributed to the presence of two —OH groups per molecule. Glycerol, with three —OH groups, has an even higher boiling point (290 °C). Hydrogen bonding among ethanol molecules explains why it is a liquid, while propane, which has a similar number of electrons but no hydrogen bonding, is a gas at the same temperature. The alcohols listed in Table 12.3 are very water-soluble because of hydrogen bonding between water molecules and the —OH alcohol group.

 Exercise 12.13 Water Solubility of Alcohols

What would you expect concerning the water solubility of an alcohol containing ten carbon atoms and one —OH group? Recall that hydrocarbons such as octane are not water-soluble.

 Exercise 12.14 One Alcohol in Another

Methanol dissolves in glycerol. Use Lewis structures to illustrate the hydrogen bonding between methanol and glycerol molecules.

Oxidation of Alcohols

In naming alcohols, the longest chain of carbon atoms is numbered so that the carbon with the —OH attached has the lowest possible number. A table of functional groups and further information on how organic compounds are named is given in Appendix E.

Alcohols are classified according to the number of carbon atoms *directly* bonded to the —C—OH carbon. If there is one carbon atom bonded directly, the compound is a *primary* alcohol. If there are two, it is a *secondary* alcohol; with three it is a *tertiary* alcohol, as illustrated below. The use of R, R′, and R″ to represent alkyl groups indicates that the alkyl groups can be different.

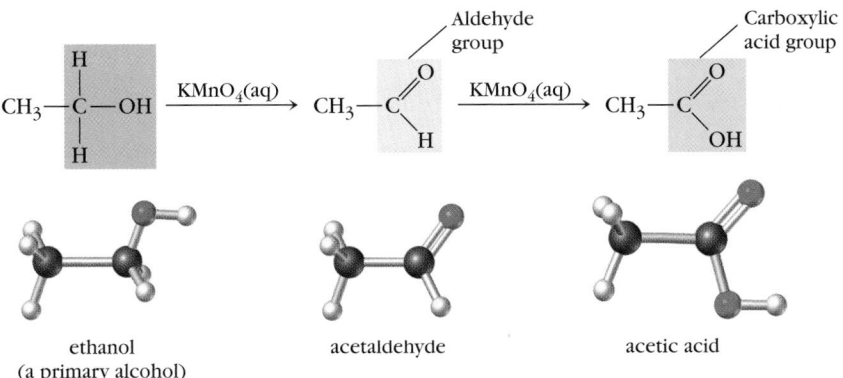

1-butanol,
a primary alcohol

2-butanol,
a secondary alcohol

2-methyl-2-propanol,
a tertiary alcohol

Stepwise oxidation of primary alcohols produces compounds called aldehydes, which are then oxidized to carboxylic acids.

$$\text{Primary alcohol} \longrightarrow \text{aldehyde} \longrightarrow \text{carboxylic acid}$$

Oxidation of organic compounds is generally either the loss of two H atoms or the gain of an oxygen. For example, the stepwise oxidation of ethanol with aqueous potassium permanganate as the oxidizing agent first produces acetaldehyde, a member of the **aldehyde** functional group class, which contains a —CHO group.

Oxidation of organic compounds is usually the removal of two hydrogens or the addition of one oxygen; reduction is usually the addition of two hydrogens or the removal of one oxygen.

ethanol
(a primary alcohol)

acetaldehyde

acetic acid

The acetaldehyde is then oxidized to acetic acid, a member of the **carboxylic acid** functional group class, which contains the —COOH group (also represented as —CO_2H). When ethanol is ingested, enzymes in the liver produce the same products. Acetaldehyde, the intermediate product, contributes to the toxic effects of alcoholism. In the presence of oxygen in air, ethanol in wine is oxidized naturally to acetic acid, converting the wine from a beverage into something more suitable for a salad dressing.

Vinegar is a 5% solution of acetic acid in water.

 Exercise 12.15   Looking at the Oxidation of Primary Alcohols

Look carefully at the formulas for ethanol and acetaldehyde. Explain how acetaldehyde is the oxidation product of ethanol. Do the same for the oxidation of acetaldehyde to acetic acid. How do your explanations differ for the different products?

Aldehydes contain the —CH (or —CHO) functional group. *Carboxylic acids* contain

the —C—OH (or —COOH, or —CO_2H) functional group.

Ketones contain the carbonyl functional group $\overset{O}{\underset{|}{\overset{||}{-C-}}}$, bonded to carbon atoms.

Alcohols, aldehydes, carboxylic acids, and ketones are important biologically. Oxidation of *secondary* alcohols produces **ketones,** which contain the C=O group.

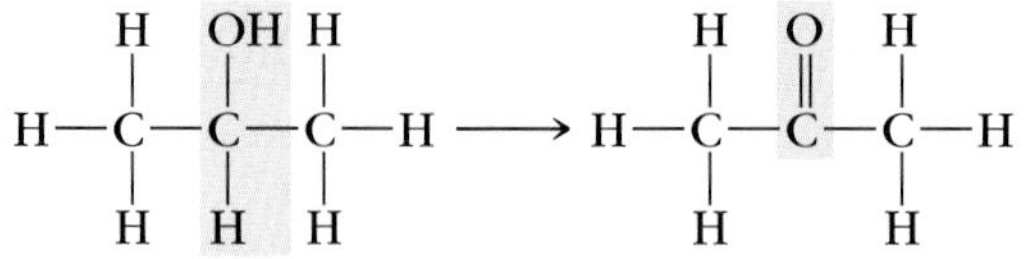

Secondary alcohol $\longrightarrow$ ketone

As a further example, this type of reaction is important during strenuous exercise, when the lungs and circulatory system are unable to deliver sufficient oxygen to the muscles. Under these conditions, lactic acid builds up in muscle tissue, causing soreness and exhaustion. After the exercising is finished, enzymes help to oxidize the secondary alcohol group in lactic acid to a ketone group, thereby converting the lactic acid to pyruvic acid. Pyruvic acid is an important biological compound that is further metabolized to provide energy.

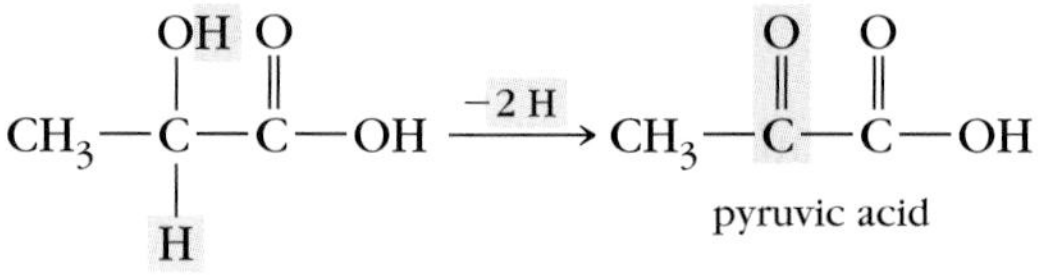

lactic acid $\xrightarrow{-2\,H}$ pyruvic acid

Tertiary alcohols, having no hydrogen atoms directly bonded to the —OH-bearing carbon, are not oxidized easily.

Functional group class	Functional group	Example
Aldehyde	$\overset{O}{\underset{}{\overset{||}{-CH}}}$	Formaldehyde $\quad$ $H-\overset{O}{\overset{||}{CH}}$
Carboxylic acid	$\overset{O}{\underset{}{\overset{||}{-C-OH}}}$	Acetic acid $\quad$ $CH_3-\overset{O}{\overset{||}{C}}-OH$
Ketone	$\overset{O}{\underset{}{\overset{||}{-C-}}}$	Acetone $\quad$ $CH_3-\overset{O}{\overset{||}{C}}-CH_3$

Problem-Solving Example 12.3   Oxidation of Alcohols

Write the condensed structural formulas of the alcohols that can be oxidized to make the following compounds:

(a) $CH_3CH_2CH_2\overset{O}{\overset{||}{C}}-H$ $\qquad$ (b) $CH_3CH_2\overset{O}{\overset{||}{C}}-CH_3$ $\qquad$ (c) $CH_3CH_2CH_2CH_2\overset{O}{\overset{||}{C}}-OH$

Answer

(a) $CH_3CH_2CH_2CH_2OH$ $\qquad$ (b) $CH_3CH_2\overset{OH}{\overset{|}{C}}HCH_3$ $\qquad$ (c) $CH_3CH_2CH_2CH_2CH_2OH$

Explanation
(a) The oxidation of the four-carbon primary alcohol, 1-butanol, will produce this aldehyde.
(b) To produce this ketone, choose the secondary alcohol, 2-butanol, which has two carbons to the left and one carbon to the right of the C—OH group.
(c) The oxidation of the five-carbon primary alcohol, 1-pentanol, will produce this carboxylic acid.

Problem-Solving Practice 12.3

Draw the structural formulas of the expected oxidation products of

(a) $CH_3CH_2CH_2OH$ $\qquad$ (b) $CH_3\overset{}{\underset{OH}{\overset{|}{C}H}}CH_2CH_3$

Exercise 12.16   Aldehydes and Combustion Products

Write the equation for the formation of an aldehyde by the oxidation of methanol. Critics of the use of methanol as a fuel or fuel oxygenate have cited the formation of this aldehyde as a major health threat. Use the Internet to find some of the toxic properties of the aldehyde.

 Exercise 12.17   Working Backward

When oxidized, a certain alcohol yields an aldehyde containing three carbon atoms. Further oxidation forms a three-carbon acid with the acid functional group located on the number one carbon of the chain. What are the structural formula and name of the alcohol?

 ## Large Molecules Containing Alcohol Groups

Many natural organic compounds are cyclic, based on hydrocarbons that consist of aromatic rings (⬅ *p. 349)* and cycloalkane (⬅ *p. 323)* or cycloalkene rings sharing some atoms in common (fused together). Steroids, all of which have the four-ring structure shown below, are an example of these fused-ring structures.

Cycloalkenes are ring compounds that contain at least one C=C double bond.

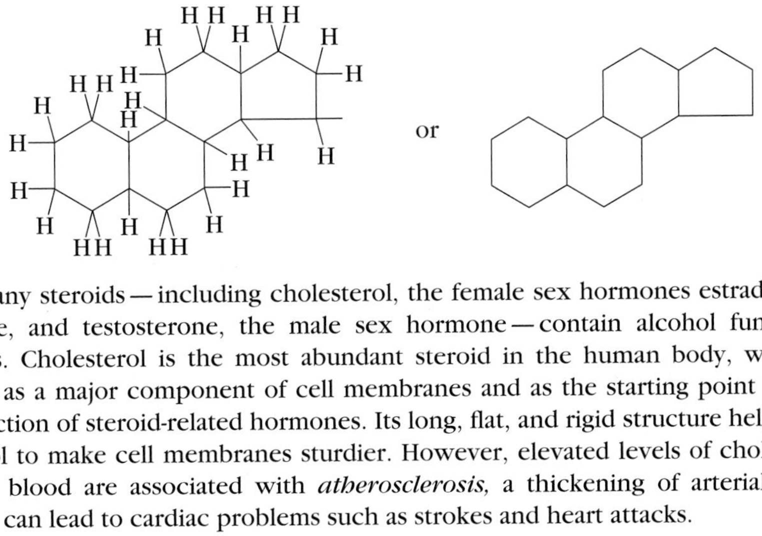

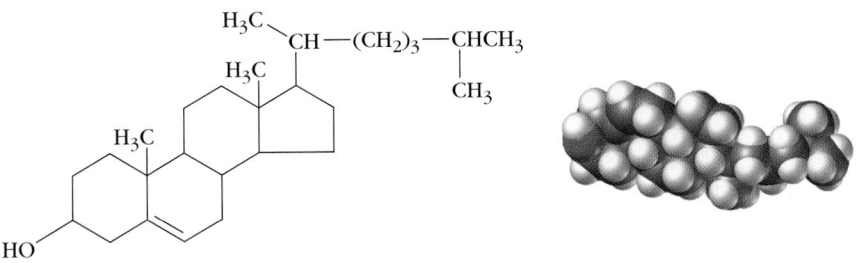

Many steroids—including cholesterol, the female sex hormones estradiol and estrone, and testosterone, the male sex hormone—contain alcohol functional groups. Cholesterol is the most abundant steroid in the human body, where it serves as a major component of cell membranes and as the starting point for the production of steroid-related hormones. Its long, flat, and rigid structure helps cholesterol to make cell membranes sturdier. However, elevated levels of cholesterol in the blood are associated with *atherosclerosis,* a thickening of arterial walls, which can lead to cardiac problems such as strokes and heart attacks.

cholesterol

Estradiol and estrone are female sex hormones responsible for the development of secondary sexual characteristics during puberty; testosterone is the male counterpart. From their structural formulas you can see that steroids are large molecules made up predominantly of carbon and hydrogen. Thus, steroids are not soluble in water.

estradiol
(female hormone,
an estrogen)

estrone
(female hormone,
an estrogen)

testosterone
(male hormone)

PERCY LAVON JULIAN
(1899–1975)

Percy Julian received a doctorate in chemistry in Vienna in 1931, spent 18 years as a research director in the chemical industry, and then directed his own research institute. He was granted over 100 patents and was the first to synthesize hydrocortisone (a steroid) and to isolate from soybean oil the compounds used to make the first synthetic sex hormone (progesterone). These amazing accomplishments came from a person who had to leave his home in Montgomery, Alabama, after eighth grade for further studies—no more public education was available there for a black man. He enrolled as a "subfreshman" at DePauw University in Indiana. On his first day, a white student welcomed him with a handshake. Julian later related his reaction: "In the shake of a hand my life was changed. I soon learned to smile and act like I believed they all liked me, whether they wanted to or not."

CD-ROM Screen 11.5:
Functional Groups

Exercise 12.18 A Closer Look

Look carefully at the structures for estradiol and estrone.

(a) What does the *-diol* suffix indicate in estradiol?
(b) What kind of alcohol group is on the five-membered ring in estradiol?
(c) What process converts that alcohol group to the functional group present in that same ring in estrone?
(d) Write a description of the differences in the molecular structures between the male and female sex hormones, testosterone and estradiol, respectively.

Exercise 12.19 Take Me Out To The Ball Game

Androstenedione, used by the home run record holder Mark McGwire and others, has been touted as a muscle-building nutritional supplement that can be bought without a prescription. The molecular structure of androstenedione is shown.

androstenedione

Compare it with that of testosterone. In the body, what simple chemical process converts androstenedione to testosterone?

12.7 CARBOXYLIC ACIDS AND ESTERS

Carboxylic Acids

Carboxylic acids contain the $-\overset{\overset{\displaystyle O}{\|}}{C}-OH$ (—COOH or —CO_2H) functional group and are prepared by the oxidation of aldehydes or the complete oxidation of primary alcohols (⇐ *p. 534*). All carboxylic acids react with bases to form salts; for example, sodium lactate is formed by the reaction of lactic acid and sodium hydroxide.

lactic acid   sodium lactate

Carboxylic acid molecules are polar and form hydrogen bonds with each other. This hydrogen bonding results in relatively high boiling points for the acids, even higher than those of alcohols of comparable molecular size. For example, formic acid (46 g/mol) has a boiling point of 101 °C, while ethanol (46 g/mol) has a boiling point of only 78.5 °C.

Exercise 12.20   Hydrogen Bonding in Formic Acid

Use the structural formula of formic acid given in Table 12.5 to illustrate hydrogen bonding in formic acid.

A large number of carboxylic acids are found in nature and have been known for many years. As a result, some of the familiar carboxylic acids are almost always referred to by their common names (Table 12.5).

Three carboxylic acids that when produced in large quantities have two acid groups and are known as *dicarboxylic acids.*

$$\text{HO}-\overset{\overset{\text{O}}{\|}}{\text{C}}-(\text{CH}_2)_4-\overset{\overset{\text{O}}{\|}}{\text{C}}-\text{OH}$$

adipic acid

$$\text{HO}-\overset{\overset{\text{O}}{\|}}{\text{C}}-\bigcirc-\overset{\overset{\text{O}}{\|}}{\text{C}}-\text{OH}$$

terephthalic acid

phthalic acid

These three acids are commercially important because they are used to make synthetic polymers (Section 12.8).

Acids in foods such as citrus fruits, strawberries, and vinegar cause them to have a sour taste.

Carboxylic acids in foods.   Lemon juice, vitamin C, and Coca-Cola (citric acid), apple (malic acid), and vinegar (acetic acid).

TABLE 12.5	Some Naturally Occurring Carboxylic Acids		
Structure	**Common name**	**b.p. (°C)**	**Natural source**
$\text{H}-\overset{\overset{\text{O}}{\|}}{\text{C}}-\text{OH}$	Formic acid	101	Ants
$\text{CH}_3-\overset{\overset{\text{O}}{\|}}{\text{C}}-\text{OH}$	Acetic acid	118	Fermented fruit
$\text{CH}_3\text{CH}_2-\overset{\overset{\text{O}}{\|}}{\text{C}}-\text{OH}$	Propionic acid	141	Dairy products
$\bigcirc-\overset{\overset{\text{O}}{\|}}{\text{C}}-\text{OH}$	Benzoic acid	250	Berries
		m.p. (°C)	
$\text{HOOC}-\text{CH}_2-\overset{\overset{\text{OH}}{\|}}{\underset{\underset{\text{COOH}}{\|}}{\text{C}}}-\text{CH}_2-\text{COOH}$	Citric acid	153	Citrus fruits
$\text{HOOC}-\text{CH}_2-\overset{}{\underset{\underset{\text{OH}}{\|}}{\text{CH}}}-\text{COOH}$	Malic acid	131	Apples
$\text{HOOC}-\overset{}{\underset{\underset{\text{OH}}{\|}}{\text{CH}}}-\overset{}{\underset{\underset{\text{OH}}{\|}}{\text{CH}}}-\text{COOH}$	Tartaric acid	168 - 170	Grape juice, wine

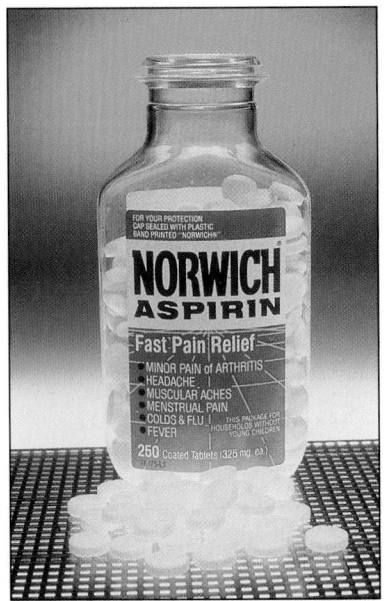

Aspirin (acetylsalicylic acid).

The nature and development of aspirin were discussed in Chapter 1.

The systematic name for aspirin (acetylsalicylic acid) is 2-(acetyloxy)-benzoic acid.

CD-ROM Screen 11.7: Functional Groups (2): Fats and Oils

The smell of crayons is due to the stearic acid in them.

Esters

Carboxylic acids react with alcohols in the presence of strong acids (such as sulfuric acid) to produce **esters**, which contain the

$$
\begin{array}{c} \text{O} \\ \| \\ -\text{C}-\text{O}-\text{R} \end{array}
$$

functional group. In an ester, the —OH of the carboxylic acid is replaced by the —OR group from the alcohol. The general equation for ester formation (esterification) is

$$
\text{R}-\text{OH} + \text{H}-\text{O}-\overset{\overset{\text{O}}{\|}}{\text{C}}-\text{R}' \xrightarrow{\text{H}^+} \text{R}-\text{O}-\overset{\overset{\text{O}}{\|}}{\text{C}}-\text{R}' + \text{H}_2\text{O}
$$

alcohol + organic acid → ester + water

Esterification is an example of a condensation reaction. In a **condensation reaction**, two molecules combine to form a larger molecule (in esterification it is the ester) while simultaneously splitting out a small molecule (such as water). For example, just over a century ago, Felix Hoffman, a chemist at the Bayer Chemical Company in Germany, synthesized aspirin, the world's most common pain reliever (⇐ *p. 3*), by esterifying an alcohol group on salicylic acid with acetic acid in strong acid solution to form acetylsalicylic acid. Aspirin (acetylsalicylic acid) is made by esterification of salicylic acid with acetic acid in strong acid solution.

salicylic acid · acetic acid · aspirin (acetylsalicyclic acid) · water

Note that the ester group in aspirin is formed from the reaction of the alcohol group originally on the salicylic acid with the —COOH group of acetic acid. The carboxylic acid originally in salicylic acid remains intact in aspirin. The ester group makes aspirin less irritating to the stomach lining than salicylic acid.

Esterification has vital importance biologically in the formation of triglycerides (⇐ *p. 331*), which are triesters formed by the reaction of long-chain fatty acids with glycerol, a trialcohol. Triglycerides are commonly known as fats (solids at room temperature) or oils (liquids at room temperature) and account for nearly 95% of dietary fat. The three fatty acids in a triglyceride can be alike (a simple triglyceride) or different (a mixed triglyceride); each of the fatty acids reacts with an —OH group of glycerol. For example, the reaction of three stearic acid molecules with glycerol produces the triglyceride glycerol tristearate, a fat.

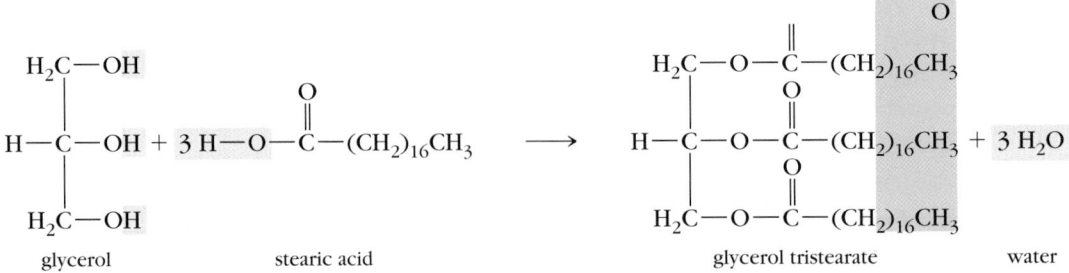

glycerol · stearic acid · glycerol tristearate · water

Notice that three water molecules are formed for each molecule of triglyceride, one for each ester linkage formed.

Fatty acids such as oleic acid, which contain only one —CH═CH— double bond, are referred to as **monounsaturated acids;** fatty acids such as linoleic acid with multiple —CH═CH— double bonds are **polyunsaturated acids.** Diets consisting of moderate amounts of fats and oils containing mono- and polyunsaturated fatty acids are considered better for good health than diets containing only saturated fats. In spite of this fact, there is a demand for solid or semisolid fats because of their texture and spreadability. The figure below indicates the relative proportions of saturated and unsaturated fats in some familiar cooking oils.

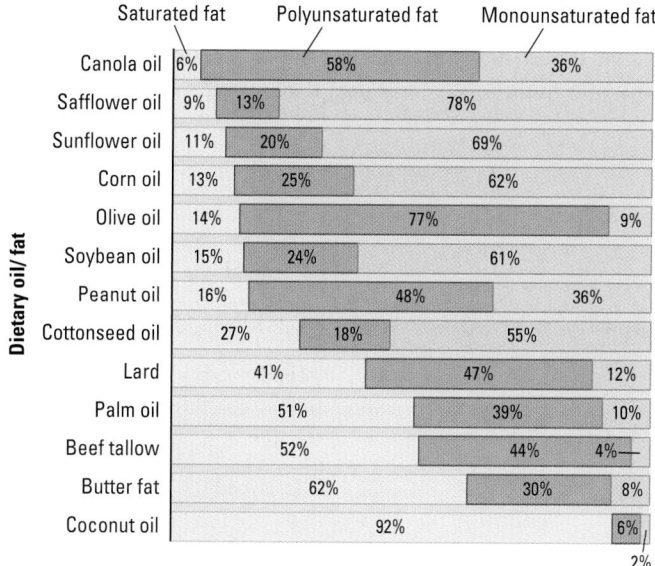

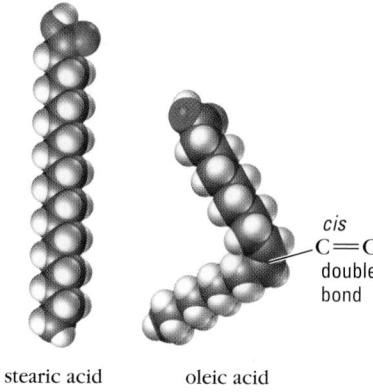

cis
C═C
double
bond

stearic acid   oleic acid

The American Heart Association recommends that no more than 30% of dietary calories should come from fat. Fats account for almost 40% of the calories in the average American diet.

A *cis* double
bond in a fatty acid

A *trans* double
bond in a fatty acid

Animal fats often have undesirable tastes, so vegetable oils, which have little or no taste, are converted to semisolid fats such as margarine and shortenings for use in home cooking and a variety of commercial food products. This is done by adding hydrogen to some of the —CH═CH— double bonds found in the oil. The process, known as **partial hydrogenation,** is the reaction of H_2 with the liquid triglyceride in the presence of a catalyst. Hydrogenation increases the degree of saturation in the oil as some, but not all, of the —CH═CH— bonds are converted into —CH$_2$—CH$_2$— bonds. An example is the partial hydrogenation of linoleic acid in peanut oil. In the process, however, some of the *cis* —CH═CH— double bonds in the original triglyceride are converted to *trans* —CH═CH— double bonds, causing concern about the effect of dietary *trans* fatty acids on health (⇐ *p. 332*).

Esters are not very reactive; their most important reaction is hydrolysis. In a **hydrolysis** reaction, bonds are broken by their reaction with water, and the H— and —OH of water add to the atoms that were in the bond broken by hydrolysis. Like other esters, triglycerides are hydrolyzed by strong acid or aqueous base. Triglycerides we eat are hydrolyzed during digestion in the same way by enzyme-catalyzed reactions. With strong acid hydrolysis, the three ester linkages in the triglyceride are broken; the products are glycerol and three moles of fatty acids, which is just the reverse of the ester formation reaction. For example, acid hydrolysis of one mole of glycerol tristearate produces one mole of glycerol and three moles of stearic acid.

Foods containing partially hydrogenated fats.

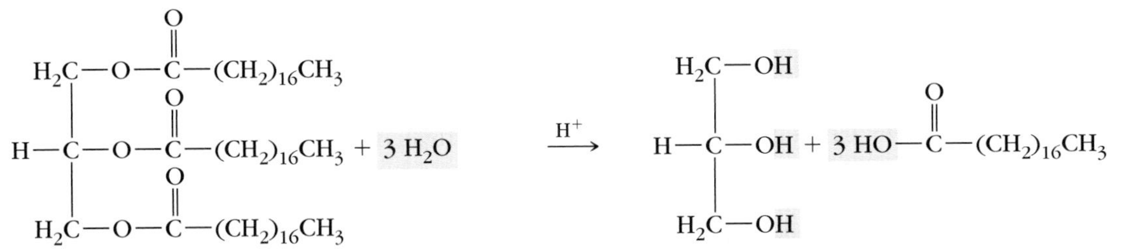

$$H_2C—O—\overset{\overset{\displaystyle O}{\|}}{C}—(CH_2)_{16}CH_3$$
$$H—C—O—\overset{\overset{\displaystyle O}{\|}}{C}—(CH_2)_{16}CH_3 + 3\ H_2O \overset{H^+}{\longrightarrow} H—C—OH + 3\ HO—\overset{\overset{\displaystyle O}{\|}}{C}—(CH_2)_{16}CH_3$$
$$H_2C—O—\overset{\overset{\displaystyle O}{\|}}{C}—(CH_2)_{16}CH_3$$

glycerol tristearate   glycerol   stearic acid

During hydrolysis of a triglyceride with an aqueous base such as NaOH, glycerol is formed and the fatty acids react with the base, converting them to their salts. In aqueous base, the hydrolysis process is called **saponification,** and the salts formed are called soaps. In biological systems, enzymes assist with saponification reactions that digest fats and oils. The saponification of one mole of glycerol tristearate forms one mole of glycerol and three moles of sodium stearate, a soap.

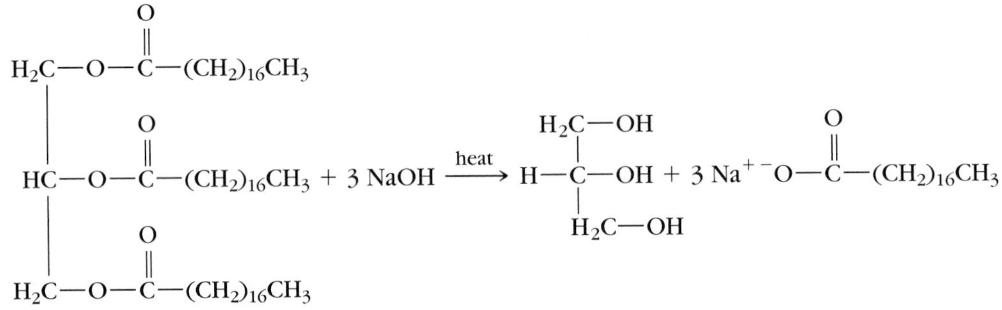

$$H_2C—O—\overset{\overset{\displaystyle O}{\|}}{C}—(CH_2)_{16}CH_3$$
$$HC—O—\overset{\overset{\displaystyle O}{\|}}{C}—(CH_2)_{16}CH_3 + 3\ NaOH \overset{heat}{\longrightarrow} H—C—OH + 3\ Na^{+\ -}O—\overset{\overset{\displaystyle O}{\|}}{C}—(CH_2)_{16}CH_3$$
$$H_2C—O—\overset{\overset{\displaystyle O}{\|}}{C}—(CH_2)_{16}CH_3$$

glycerol tristearate, a fat   NaOH   glycerol   sodium stearate, a soap

Note that the saponification reaction requires three moles of base per mole of triglyceride.

Exercise 12.21   Triglyceride Structures

(a) Draw the structural formula for the triglyceride formed when glycerol reacts with oleic acid. Circle the ester linkages in this triester molecule.
(b) Using structural formulas, write the equation for the hydrolysis in aqueous NaOH of the triglyceride formed in part (a).

 ## 12.8   SYNTHETIC ORGANIC POLYMERS

CD-ROM Screen 11.9:
Synthetic Organic Polymers (1):
Addition Polymerization

Nature makes many different polymers, including cellulose and starch in plants and proteins in both plants and animals.

As important as simple organic molecules are, there is another very important class of very large molecules made from these simpler molecules. **Polymers** (*poly,* many; *mer,* part) are large molecules composed of smaller repeating units, usually arranged in a chain-like structure. Polymers occur in nature and are synthesized by chemists as well. It is virtually impossible for us to get through a day without using a dozen or more synthetic organic polymers. Today, synthetic polymers are used to make our clothes, package our foods, and build our computers, phones, and cars. You, a friend, or a family member may be alive because of a medical application of synthetic polymers. Synthetic polymers are so important that approximately 80% of the organic chemical industry is devoted to their production.

There are two broad categories of the synthetic polymers commonly known as "plastics." One type, called **thermoplastics,** softens and flows when heated; when it is cooled, it hardens again. Common plastic materials that undergo such reversible changes when heated and cooled are polyethylene (milk jugs), polystyrene (inexpensive sunglasses and toys), and polycarbonates (CD audio discs). The other general type are **thermosetting plastics.** When first heated, a thermosetting plastic flows like a thermoplastic. But when heated further, a thermosetting plastic forms a rigid structure that will not remelt. Bowling balls, football helmets, and some kitchen countertops are examples of thermosetting plastics.

Some of the most useful synthetic polymers have resulted from copying natural polymers. Synthetic rubber, used in almost every automobile tire, is a copy of the molecule found in natural rubber. However, there are also many useful synthetic polymers that have no natural analogs, such as polystyrene, nylon, Teflon, and Dacron.

Both synthetic and natural polymers are made by chemically joining many small molecules, called **monomers,** into giant polymer molecules known as **macromolecules,** which have molar masses ranging from thousands to millions. In nature, polymerization reactions usually are controlled by enzymes, and in animal cells the reactions take place rapidly at body temperature. Making synthetic polymers often requires high temperatures and pressures and lengthy reaction times. Both synthetic and natural polymers can be classified as **addition polymers,** made by adding monomer units directly together, or **condensation polymers,** formed by monomer units combining so that a small molecule, usually water, is formed and released as the polymer forms from the monomer units.

Common household items made from plastic.

A macromolecule is a molecule with a very high molar mass.

Addition Polymers

In addition polymerization, the monomer units are added directly to each other, hooking together like boxcars on a train. The monomers for making addition polymers generally contain one or more C=C double bonds. The simplest monomer of this group is ethylene, CH_2=CH_2, which polymerizes to form *polyethylene.* When heated to 100 to 250 °C at a pressure of 1000 to 3000 atm in the presence of a catalyst, ethylene forms polyethylene chains with molar masses of up to a million.

The polymerization of ethylene usually begins with breaking one of the bonds in the carbon-carbon double bond, so that each carbon atom has an unpaired electron. Such an electron structure makes the molecule highly reactive. This step, the *initiation* of the polymerization, can be accomplished with chemicals such as organic peroxides (R—O—O—R) that are unstable and easily break apart into free radicals, ·OR, each with an unpaired electron *(⇐ p. 346).* The free radicals react readily with molecules containing carbon-carbon double bonds to produce new free radicals.

An organic peroxide, RO—OR, produces two free radicals, RO·, each with an unpaired electron.

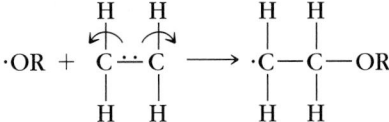

The polyethylene chain begins to grow as the unpaired electron bonds to a double-bond electron in another ethylene molecule.

This leaves an unpaired electron to bond with yet another ethylene molecule, and the process continues to form a huge polymer molecule:

$$n \ CH_2{=}CH_2 \longrightarrow {\left(\!\! \begin{array}{c} H \quad H \\ | \quad\; | \\ C{-}C \\ | \quad\; | \\ H \quad H \end{array} \!\!\right)}_{n}$$

Parentheses enclose the repeating unit of the polymer.

n ranges from 1000 to 50,000.

polyethylene

The C=C double bonds in the ethylene monomer have been changed to C—C single bonds in the polyethylene chain. Eventually, production of the polymer chain stops.

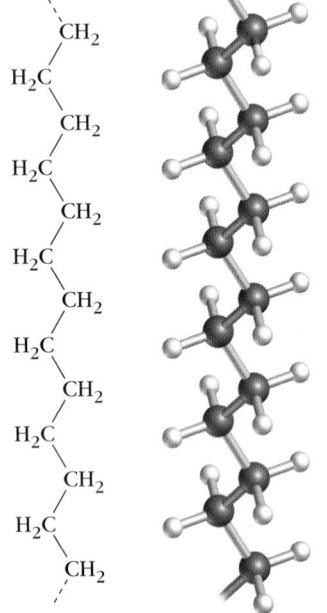

A portion of a polyethylene molecule

> ### Exercise 12.22 What Is at the Ends of the Polymer Chains?
>
> What do you think is attached at the ends of the polymer chains when all of the ethylene monomer molecules have been polymerized to form polyethylene?

Changing pressures and catalytic conditions produces polyethylenes of different molecular structures and hence different physical properties. For example, chromium oxide as a catalyst yields almost exclusively the linear polyethylene shown in the margin—a polymer with no branches on the carbon chain. The zigzag structure represents the shape of the chain more closely than does a linear drawing, because of the tetrahedral arrangement of bonds around each carbon in the saturated polyethylene chain. When ethylene is heated to 230 °C at a pressure of 200 atm without the chromium oxide catalyst, free radicals attack the chain at random positions, causing irregular branching.

Polyethylene is the world's most widely used polymer (Figure 12.13). Long linear chains of polyethylene can pack closely together to give a material with high density (0.97 g/mL) and high molar mass, referred to as high-density polyethylene (HDPE). This material is hard, tough, and semirigid; it is used in plastic milk jugs. Branched chains of polyethylene cannot pack as closely together as the linear chains in HDPE, so the resulting material has a lower density (0.92 g/mL) and is called low-density polyethylene (LDPE). This material is soft and flexible because of the weaker intermolecular forces. It is used to make sandwich bags, for example. If the linear chains of polyethylene are treated in a way that causes short chains of —CH₂— groups to connect adjacent chains, the result is cross-linked polyethylene (CLPE), a very tough material, used for synthetic ice rinks and soft drink bottle caps (Figure 12.14).

Many different kinds of addition polymers are made from monomers in which one or more of the hydrogen atoms in ethylene have been replaced with either halogen atoms or a variety of organic groups, represented by X in the reaction below.

Branched
(low density)

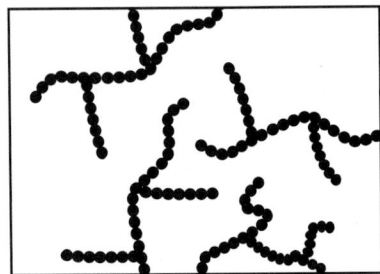

Linear
(high density)

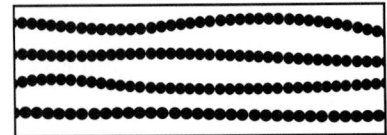

Linear and branched polyethylene.

$$n \ \begin{array}{c} H \qquad H \\ {\diagdown}C{=}C{\diagup} \\ H \qquad X \end{array} \longrightarrow {\left(\!\! \begin{array}{c} H \; H \; H \; H \; H \; H \\ |\;\; |\;\; |\;\; |\;\; |\;\; | \\ C{-}C{-}C{-}C{-}C{-}C \\ |\;\; |\;\; |\;\; |\;\; |\;\; | \\ H \; X \; H \; X \; H \; X \end{array} \!\!\right)}_{n}$$

The growing chain

X can be an atom, such as Cl in vinyl chloride, or a group of atoms, such as —CH₃ in propylene or —CN in acrylonitrile. Table 12.6 **(p. 546)** gives information on

Figure 12.13 Polyethylene. (a) Production of polyethylene film. (b) and (c) The wide range of properties of different structural types of polyethylene leads to a wide variety of applications.

some of these monomers and their addition polymers. For example, the monomer for making polystyrene is styrene. In polystyrene, *n* is typically about 5700.

Styrene → The growing chain

Polystyrene is a clear, hard, colorless, solid thermoplastic that can be molded easily at 250 °C. Nearly seven billion pounds of polystyrene are used annually in the United States alone to make food containers, toys, electrical parts, and many other items. The variation in properties shown by polystyrene products is typical of synthetic polymers. For example, a clear polystyrene drinking glass that is brittle and breaks into sharp pieces somewhat like glass is quite different from an expanded polystyrene coffee cup that is soft and pliable (Figure 12.15). A major use of polystyrene is in the production of Styrofoam by "expansion molding." In this process, polystyrene beads are placed in a mold and heated with steam or hot air. The tiny beads contain 4 to 7% by weight of a low-boiling liquid such as pentane.

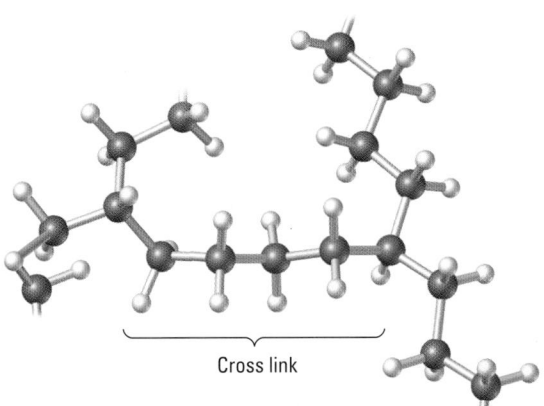

Cross link

Figure 12.14 Model of cross-linked polyethylene.

TABLE 12.6	Ethylene Derivatives that Undergo Addition Polymerization		

Formula	Monomer common name	Polymer name (trade names)	Uses
$\underset{H}{\overset{H}{>}}C=C\overset{H}{\underset{H}{<}}$	Ethylene	Polyethylene (Polythene)	Squeeze bottles, bags, films, toys and molded objects, electrical insulation
$\underset{H}{\overset{H}{>}}C=C\overset{H}{\underset{CH_3}{<}}$	Propylene	Polypropylene (Vectra, Herculon)	Bottles, films, indoor-outdoor carpets
$\underset{H}{\overset{H}{>}}C=C\overset{H}{\underset{Cl}{<}}$	Vinyl chloride	Poly(vinyl chloride) (PVC)	Floor tile, raincoats, pipe
$\underset{H}{\overset{H}{>}}C=C\overset{H}{\underset{CN}{<}}$	Acrylonitrile	Polyacrylonitrile (Orlon, Acrilan)	Rugs, fabrics
$\underset{H}{\overset{H}{>}}C=C\overset{H}{\underset{\text{⬡}}{<}}$	Styrene	Polystyrene (Styrene, Styrofoam, Styron)	Food and drink coolers, building material insulation
$\underset{H}{\overset{H}{>}}C=C\overset{H}{\underset{O-C-CH_3}{<}}$ $\overset{\|}{O}$	Vinyl acetate	Poly(vinyl acetate) (PVA)	Latex paint, adhesives, textile coatings
$\underset{H}{\overset{H}{>}}C=C\overset{CH_3}{\underset{C-O-CH_3}{<}}$ $\overset{\|}{O}$	Methyl methacrylate	Poly(methyl methacrylate) (Plexiglas, Lucite)	High-quality transparent objects, latex paints, contact lenses
$\underset{F}{\overset{F}{>}}C=C\overset{F}{\underset{F}{<}}$	Tetrafluoroethylene	Polytetrafluoroethylene (Teflon)	Gaskets, insulation, bearings, pan coatings

Figure 12.15 Polystyrene. Expanded polystyrene coffee cup (*left*) is soft. Clear polystyrene cup (*right*) is brittle.

The steam causes the low-boiling liquid to vaporize and expand the beads. As the foamed particles expand, they are molded into the shape of the mold cavity. Styrofoam is used for meat trays, coffee cups, and many kinds of packing material.

The numerous variations in chain length, branching, and cross linking make it possible to produce a variety of properties for each type of addition polymer. Chemists and chemical engineers can fine-tune the properties of the polymer to match the desired properties by appropriate selection of monomer and reaction conditions, thus accounting for the widespread and growing use of synthetic polymers.

Problem-Solving Example 12.4   Identify the Monomer

Use Table 12.6 to identify the monomer used to make each of these addition polymers, a portion of whose molecule is shown.

(a) PVC (b) Acrilan 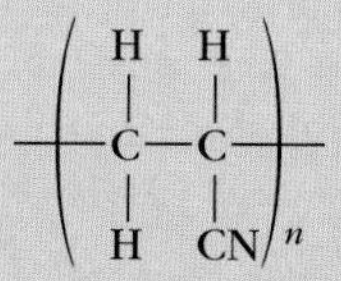 (c) Polypropylene

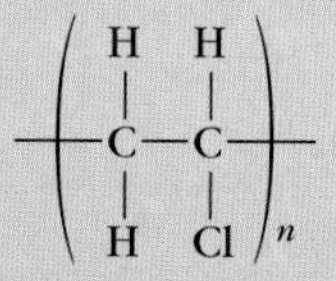

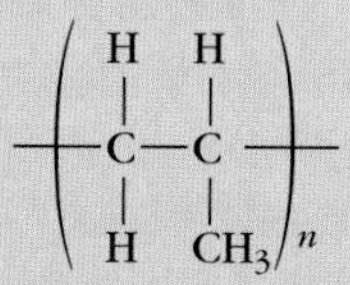

$$\left(\begin{matrix} H & H \\ | & | \\ -C - C - \\ | & | \\ H & Cl \end{matrix}\right)_n \qquad \left(\begin{matrix} H & H \\ | & | \\ -C - C - \\ | & | \\ H & CN \end{matrix}\right)_n \qquad \left(\begin{matrix} H & H \\ | & | \\ -C - C - \\ | & | \\ H & CH_3 \end{matrix}\right)_n$$

Answer

(a) Vinyl chloride, $CH_2 \!=\! CHCl$ (b) Acrylonitrile, $CH_2 \!=\! CHCN$
(c) Propylene, $CH_2 \!=\! CHCH_3$

Explanation   Each polymer has a repeating unit, which is derived from its monomer. Each of the monomers has a $C \!=\! C$ double bond, but they differ in the nonhydrogen atom or group attached to the carbon atom.

(a) $CH_2 \!=\! CHCl$ (b) $CH_2 \!=\! CHCN$ (c) $CH_2 \!=\! CHCH_3$

Problem-Solving Practice 12.4

Draw the structural formula of the repeating unit for each of these addition polymers:

(a) Polypropylene   (b) Poly(vinyl acetate)   (c) Poly(vinyl alcohol)

Natural and Synthetic Rubbers

Natural rubber is a hydrocarbon whose monomer unit has the empirical formula C_5H_8. When rubber is decomposed in the absence of oxygen, the monomer 2-methyl-1,3-butadiene (isoprene) is obtained.

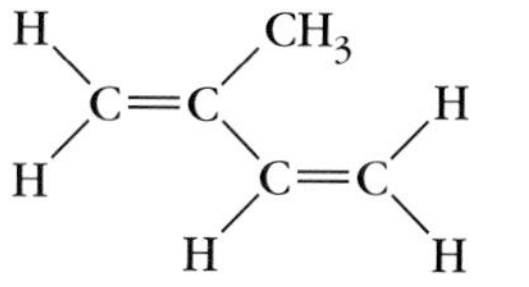

$$\begin{matrix} H & & CH_3 \\ & C \!=\! C & & H \\ H & & C \!=\! C \\ & & H & H \end{matrix}$$

2-methyl-1,3-butadiene (isoprene)

Natural rubber occurs as *latex* (an emulsion of rubber particles in water) that oozes from rubber trees when they are cut. Precipitation of the rubber particles yields a gummy mass that is not only elastic and water-repellent but also very sticky, especially when warm. In 1839, after five years' work on natural rubber, Charles Goodyear (1800–1860) discovered that heating gum rubber with sulfur produces a material that is no longer sticky but is still elastic, water-repellent, and resilient.

Vulcanized rubber, as the type of rubber Goodyear discovered is now known, contains short chains of sulfur atoms that bond together the polymer chains of the

CHEMISTRY
You Can Do...   Making "Gluep"

White school glue, such as Elmer's glue, contains poly(vinyl acetate) and other ingredients. A "gluep" similar to Silly Putty can be made by mixing $\frac{1}{2}$ cup of glue with $\frac{1}{2}$ cup of water and then adding $\frac{1}{2}$ cup of liquid starch and stirring the mixture. After stirring it, work the mixture in your hands until it has a putty consistency. Roll it into a ball and let it sit on a flat surface. Shape a piece into a ball and drop it on a hard surface.

The "gluep" can be stored in a sealed plastic bag for sev-eral weeks. Although "gluep" does not readily stick to clothes, it leaves a water mark on wooden furniture, so be careful where you set it. Mold will form on the "gluep" after a few weeks, but adding a few drops of Lysol to it will retard mold formation.

1. What did you observe about "gluep" when it was left sitting on a flat surface?
2. Did the ball of "gluep" bounce when it was dropped?

natural rubber and reduce its unsaturation. The sulfur chains help to align the polymer chains, so the material does not undergo a permanent change when stretched, but springs back to its original shape and size when the stress is removed. Substances that behave this way are called *elastomers*.

(a) Before stretching

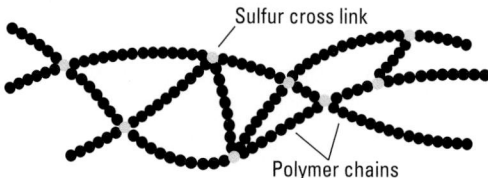

Sulfur cross link

Polymer chains

(b) After stretching

The behavior of natural rubber (polyisoprene) is due to the specific molecular geometry within the polymer chain. We can write the formula for polyisoprene with the —CH$_2$CH$_2$— groups on opposite sides of the double bond (the *trans* arrangement)

poly-*trans*-isoprene (the —CH$_2$—CH$_2$— groups are *trans*)

or with the —CH$_2$CH$_2$— groups on the same side of the double bond in a *cis* arrangement *(p. 330)*:

poly-*cis*-isoprene (the —CH$_2$—CH$_2$— groups are *cis*)

Natural rubber is poly-*cis*-isoprene. However, the *trans* material also occurs in nature in the leaves and bark of the sapotacea tree and is known as *gutta-percha*. It is brittle and hard and is used for golf ball covers and electrical insulation.

In 1955, chemists at the Goodyear and Firestone companies almost simultaneously discovered how to prepare synthetic poly-*cis*-isoprene. This material is structurally identical to natural rubber. Today, synthetic poly-*cis*-isoprene can be manufactured cheaply and is used when natural rubber is in short supply.

Many commercially important addition polymers are **copolymers,** polymers obtained by polymerizing a mixture of two or more different monomers. A copolymer of styrene with butadiene is the most important synthetic rubber produced in the United States. More than 1.5 million tons of styrene-butadiene rubber (SBR) are produced each year in the United States for making tires. A 3 : 1 mole ratio of butadiene to styrene is used to make SBR.

Saran Wrap is an example of a copolymer of vinyl chloride with 1,1-dichloroethylene.

$$3n \quad \overset{\text{H}}{\underset{\text{H}}{\diagdown}}\text{C}=\text{C}\diagup \overset{}{\underset{\text{H}}{}}\text{C}=\text{C}\overset{\text{H}}{\underset{\text{H}}{}} \quad + \quad n \quad \overset{\text{H}}{\underset{\text{H}}{}}\text{C}=\text{C}\overset{\text{H}}{\underset{}{}} \xrightarrow[\text{polymerization}]{\text{addition}}$$

1,3-butadiene   styrene

$$\left(\!-CH_2-\underset{H}{\underset{|}{C}}=\underset{H}{\underset{|}{C}}-CH_2-CH_2-\underset{H}{\underset{|}{C}}=\underset{H}{\underset{|}{C}}-CH_2-CH_2-\underset{H}{\underset{|}{C}}-\underset{H}{\underset{|}{C}}-CH_2-\underset{H}{\underset{|}{C}}=\underset{H}{\underset{|}{C}}-CH_2\!-\right)_n$$

styrene-butadiene rubber (SBR)

Another important copolymer is made by polymerizing mixtures of acrylonitrile, butadiene, and styrene (also called ABS) to produce a sturdy material used in car bumpers and computer cases.

Ethylene can also be copolymerized with vinyl acetate

$$CH_3-\overset{\overset{\displaystyle O}{\|}}{C}-O-CH=CH_2$$

vinyl acetate

to form poly(ethylene vinylacetate) (EVA), used for athletic shoe innersoles because of its resilience and durability.

$$\left(\!-CH_2-CH_2-\underset{\underset{\underset{CH_3}{|}}{\underset{C=O}{|}}}{\underset{|}{CH}}-CH_2-\underset{\underset{\underset{CH_3}{|}}{\underset{C=O}{|}}}{\underset{|}{CH}}-CH_2\!-\right)_n$$

poly(ethylene vinyl acetate), EVA

EVA innersole.

Condensation Polymers

The reactions of alcohols with carboxylic acids to give esters (⬅ *p. 540*) are examples of condensation reactions. Condensation polymers are formed by condensation polymerization reactions between monomers. Unlike addition polymerization, condensation polymerization reactions do not depend on the presence of a double bond in the reacting molecules. Rather, condensation polymerization reactions generally require two different functional groups present in two different monomers. Each of the two monomers has a functional group at each end of the monomer. As the functional groups at each end of one monomer react with the groups of the other monomer, long-chain condensation polymers are produced.

For example, molecules with two carboxylic acid groups (—COOH), such as terephthalic acid, and other molecules with two alcohol groups, such as ethylene glycol, can react with each other at both ends to form ester linkages. Water is the other product of this condensation polymerization reaction, produced during the formation of each ester linkage (shaded part of the structure) in the polymer.

CD-ROM Screen 11.10:
Synthetic Organic Polymers (2): Condensation Polymerization

$$2\ HO-\overset{\overset{\displaystyle O}{\|}}{C}-\bigcirc\!\!\!-\overset{\overset{\displaystyle O}{\|}}{C}-OH\ +\ 2\ HO-CH_2-CH_2-OH\ \longrightarrow$$

terephthalic acid   ethylene glycol

$$HO-\overset{\overset{\displaystyle O}{\|}}{C}-\bigcirc\!\!\!-\overset{\overset{\displaystyle O}{\|}}{C}-O-CH_2-CH_2-O-\overset{\overset{\displaystyle O}{\|}}{C}-\bigcirc\!\!\!-\overset{\overset{\displaystyle O}{\|}}{C}-O-CH_2-CH_2-OH\ +\ 3\ H_2O$$

terephthalic acid-ethylene glycol ester

Because this ester has a —COOH group on one end and an —OH group on the other end, the —COOH group can react with an —OH group of another ethylene glycol molecule. Similarly, the remaining alcohol group on the ester can react with another terephthalic acid molecule. This process continues, forming long chains of poly(ethylene terephthalate), a **polyester** commonly known as PET.

$$\left(\!\!-CH_2-CH_2-O-\overset{\displaystyle O}{\overset{\|}{C}}-\bigcirc-\overset{\displaystyle O}{\overset{\|}{C}}-O-\!\!\right)_n$$

the repeating unit poly(ethylene terephthalate), PET

Each year more than a billion pounds of PET are produced in the United States for making beverage bottles, apparel, tire cord, film for photography and magnetic recording, food packaging, coatings for microwave and conventional ovens, and home furnishings. A variety of trade names are associated with the various applications. PET textile fibers are marketed under such names as Dacron and Terylene. Films of the same polyester, when magnetically coated, are used to make audio and TV tapes. This film, Mylar, has unusual strength and can be rolled into sheets one-thirtieth the thickness of a human hair. The inert, nontoxic, noninflammatory, and non–blood-clotting characteristics of PET polymers make Dacron tubing an excellent substitute for human blood vessels in heart bypass operations. Dacron sheets are also used as a skin substitute for burn victims.

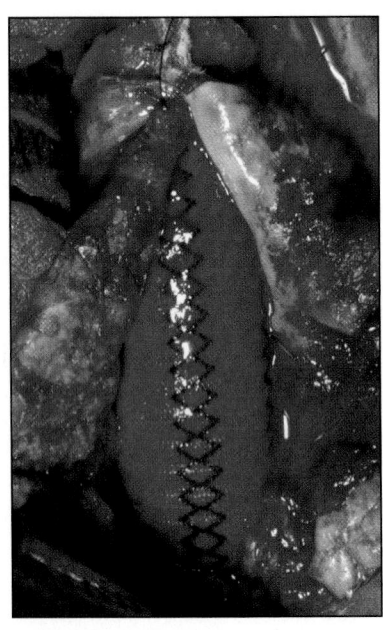

Medical uses of Dacron. A Dacron patch is used to close an atrial septal defect in a heart patient.

Problem-Solving Example 12.5 Condensation Polymerization

Poly(ethylene naphthalate) (PEN) is used for bar code labels. This condensation polymer is made by the reaction between naphthalic acid and ethylene glycol,

naphthalic acid   ethylene glycol

(a) Write the structural formula of the molecule formed after two ethylene glycol molecules have polymerized with two naphthalic acid molecules.

(b) Write the structural formula of the repeating unit of the polymer.

Answer

(a)

$$HO-\overset{O}{\overset{\|}{C}}-\bigcirc\!\bigcirc-\overset{O}{\overset{\|}{C}}-O-CH_2-CH_2-O-\overset{O}{\overset{\|}{C}}-\bigcirc\!\bigcirc-\overset{O}{\overset{\|}{C}}-O-CH_2-CH_2-OH$$

(b)

$$\left(\!\!-O-\overset{O}{\overset{\|}{C}}-\bigcirc\!\bigcirc-\overset{O}{\overset{\|}{C}}-O-CH_2-CH_2-O-\!\!\right)_n$$

Explanation

(a) The formation of PEN is similar to that of PET *(p. 549)*; both polymers are polyesters. Ethylene glycol is one of the monomers in both of them. The other monomer in PET is terephthalic acid; it is naphthalic acid in PEN. The formation of PEN occurs when the carboxylic acid groups of naphthalic acid react with the alcohol groups of ethylene glycol to form ester linkages.

HO—C(=O)— ... —O—CH₂—CH₂—OH

carboxylic acid group   ester linkage   ester linkage   ester linkage   alcohol group

Notice in this molecule that there are unreacted carboxylic acid and alcohol groups at each end. These react with other ethylene glycol and naphthalic acid groups, respectively. The growing polymer chain continues reacting at each end until it becomes very long.

(b) The repeating unit is the smallest unit that contains both monomers. In this case, the repeating unit is one formed by combining one naphthalic acid molecule with one molecule of ethylene glycol.

Problem-Solving Practice 12.5

A condensation polymer can be made from glycolic acid itself. A portion of the polymer is given below.

$$-O-CH_2-\overset{O}{\overset{\|}{C}}-O-CH_2-\overset{O}{\overset{\|}{C}}-O-CH_2-\overset{O}{\overset{\|}{C}}-O-CH_2-$$

Write the structural formula of glycolic acid.

Polyamides

Another useful and important type of condensation reaction is that between a carboxylic acid and a primary **amine,** which is an organic compound containing an —NH_2 functional group. Amines can be considered derivatives of ammonia (NH_3), and most of them are weak bases. An amine reacts with a carboxylic acid at high temperature to split out a water molecule and form an **amide:**

$$R-\overset{O}{\overset{\|}{C}}-OH + H-\overset{}{\underset{H}{N}}-R' \xrightarrow{\text{heat}} R-\overset{O}{\overset{\|}{C}}-\overset{}{\underset{H}{N}}-R' + H_2O$$

carboxylic acid   amine   amide   water

Dr. Wallace Carothers of the DuPont Company discovered a useful and important type of condensation polymerization reaction that occurs when diamines (compounds containing two —NH_2 groups) react with dicarboxylic acids (compounds containing two —COOH groups) to form polymers called **polyamides** or nylons. In February 1935 his research yielded a product known as nylon-66 (Figure 12.16), prepared from adipic acid (a dicarboxylic acid) and hexamethylenediamine (a diamine).

$$HO-\overset{O}{\overset{\|}{C}}-(CH_2)_4-\overset{O}{\overset{\|}{C}}-OH \quad + \quad H-\overset{}{\underset{H}{N}}-(CH_2)_6-NH_2 \longrightarrow$$

adipic acid   hexamethylenediamine

$$HO-\overset{O}{\overset{\|}{C}}-(CH_2)_4-\overset{O}{\overset{\|}{C}}-\overset{}{\underset{H}{N}}-(CH_2)_6-NH_2 \quad + \quad H_2O$$

CD-ROM Screen 11.8: Functional Groups (3): Amino Acids and Proteins

Amines are classified as primary, secondary, or tertiary according to how many of the H atoms in the —NH_2 group are replaced by alkyl groups: RNH_2 is primary; R_2NH is secondary; and R_3N is tertiary.

Amides contain the $-\overset{O}{\overset{\|}{C}}-\overset{H}{\overset{|}{N}}-$ functional group.

The name of nylon-66 is based on the number of carbon atoms in the diamine and diacid, respectively, that are used to make the polymer. Since hexamethylenediamine and adipic acid each have six carbon atoms, the product is called nylon-66.

Figure 12.16   Making nylon-66.

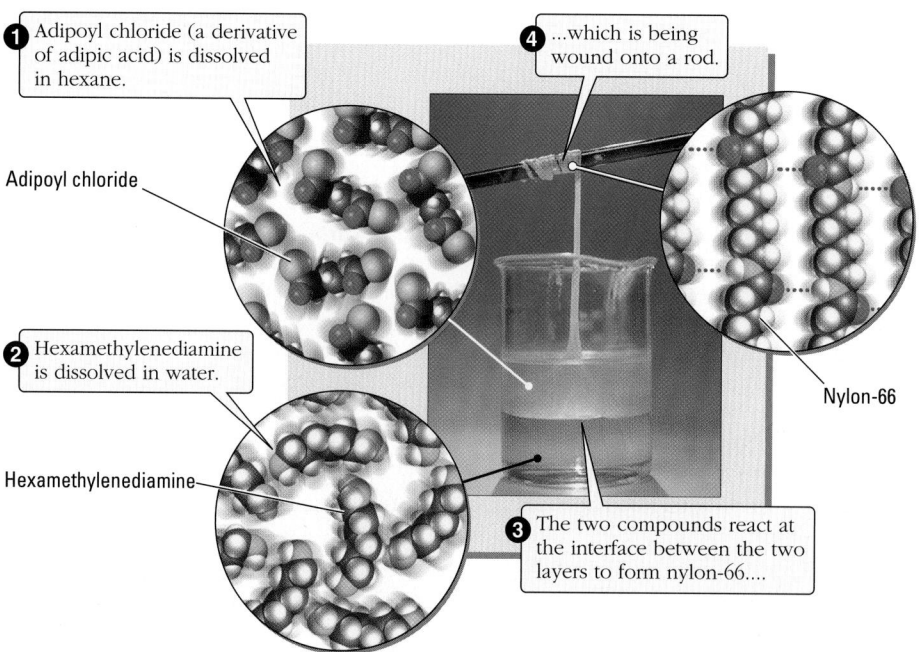

The reactions continue, extending the polymer chain, which consists of alternating adipic acid and hexamethylenediamine units. The overall equation is

$$n\ \text{HO}-\overset{\overset{\text{O}}{\|}}{\text{C}}-(\text{CH}_2)_4-\overset{\overset{\text{O}}{\|}}{\text{C}}-\text{OH} + n\ \text{H}_2\text{N}-(\text{CH}_2)_6-\text{NH}_2 \longrightarrow$$

$$-\overset{\overset{\text{O}}{\|}}{\text{C}}-(\text{CH}_2)_4-\overset{\overset{\text{O}}{\|}}{\text{C}}\Bigg(\!\!-\underset{\underset{\text{H}}{|}}{\text{N}}-(\text{CH}_2)_6-\underset{\underset{\text{H}}{|}}{\text{N}}-\overset{\overset{\text{O}}{\|}}{\text{C}}-(\text{CH}_2)_4-\overset{\overset{\text{O}}{\|}}{\text{C}}\!\!\Bigg)_{\!\!n}\underset{\underset{\text{H}}{|}}{\text{N}}-(\text{CH}_2)_6-\underset{\underset{\text{H}}{|}}{\text{N}}-\quad +\quad n\ \text{H}_2\text{O}$$

nylon

The $-\overset{\overset{\text{O}}{\|}}{\text{C}}-\underset{\underset{\text{H}}{|}}{\text{N}}-$ group between the monomers is called an **amide linkage;** hence, the polymers are known as polyamides.

Figure 12.17 illustrates another facet of the structure of nylon—hydrogen bonding—which explains why nylon makes such good fibers. To have good tensile strength, the chains of atoms in a polymer should be able to attract one another, but not so strongly that the plastic cannot initially be extended to form the fibers. Linking the chains together with covalent bonds would be too strong. Hydrogen bonds, with a strength about one tenth that of an ordinary covalent bond (⇐ *p. 388)*, join the chains in the desired manner.

Problem-Solving Example 12.6   Kevlar, A Condensation Polyamide

Kevlar, used to make bulletproof vests, canoes, and baseball batting gloves, is made from *p*-phenylenediamine and terephthalic acid.

p-phenylenediamine   terephthalic acid

(a) Using structural formulas, write the chemical equation for the formation of a segment of a Kevlar molecule containing three amide linkages.

Hydrogen bonds to adjacent nylon-66 molecules.

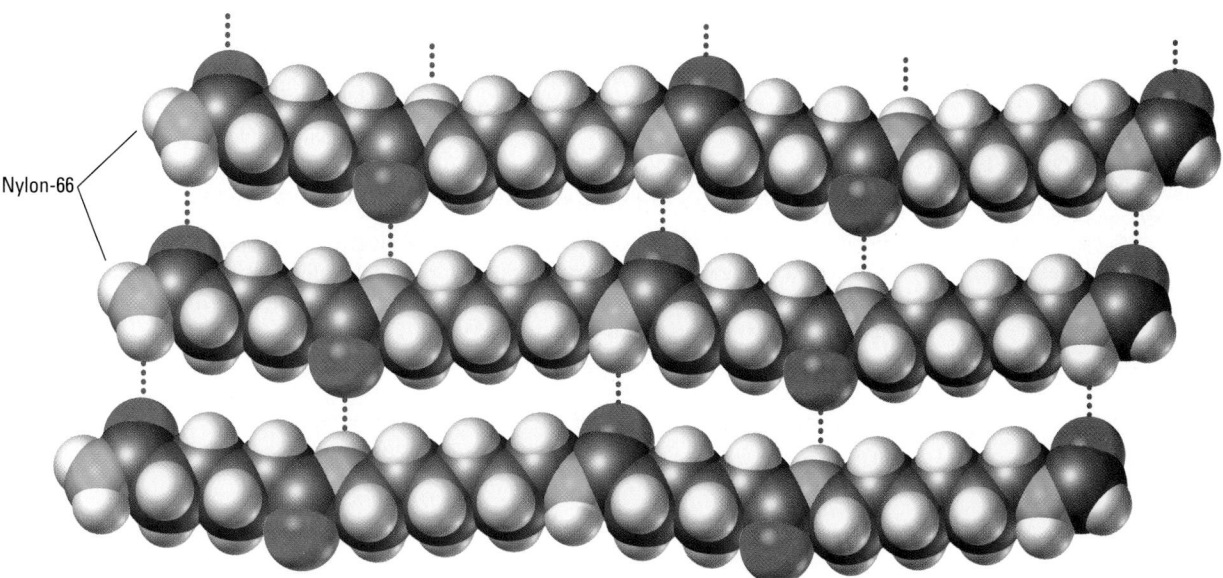

Nylon-66

Figure 12.17 Hydrogen bonding in nylon-66.

(b) Identify the repeating unit for Kevlar.

Answer

(a)

$$2\ HO-\overset{O}{\underset{}{C}}-\bigcirc-\overset{O}{\underset{}{C}}-OH + 2\ H_2N-\bigcirc-NH_2\ \longrightarrow$$

$$HO-\overset{O}{\underset{}{C}}-\bigcirc-\overset{O}{\underset{}{C}}-\overset{}{\underset{H}{N}}-\bigcirc-\overset{}{\underset{H}{N}}-\overset{O}{\underset{}{C}}-\bigcirc-\overset{O}{\underset{}{C}}-\overset{}{\underset{H}{N}}-\bigcirc-NH_2 + 3\ H_2O$$

(b)

$$\left(\overset{O}{\underset{}{C}}-\bigcirc-\overset{O}{\underset{}{C}}-\overset{}{\underset{H}{N}}-\bigcirc-\overset{}{\underset{H}{N}}\right)_n$$

Explanation The amide linkage joining Kevlar units together is formed by the condensation reaction between the amine functional group of *p*-phenylenediamine and the carboxylic acid group of terephthalic acid.

(a)

$$HO-\overset{O}{\underset{}{C}}-\bigcirc-\overset{O}{\underset{}{C}}-OH + H_2N-\bigcirc-NH_2\ \longrightarrow$$

$$HO-\overset{O}{\underset{}{C}}-\bigcirc-\overset{O}{\underset{}{C}}-\overset{}{\underset{H}{N}}-\bigcirc-NH_2 + H_2O$$

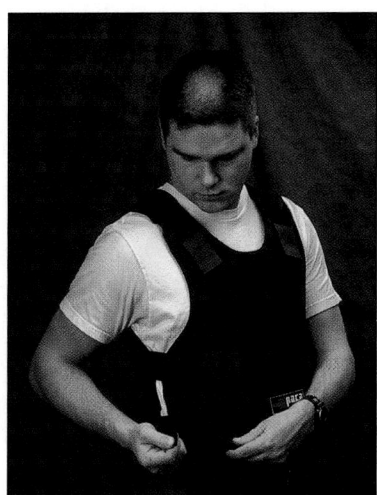

A Kevlar vest.

STEPHANIE LOUISE KWOLEK

(1923–)

Stephanie Kwolek received a Bachelor of Science degree from Carnegie Tech (now Carnegie Mellon University) in 1946. Although wanting to study medicine, she couldn't afford it and decided to take a temporary job at DuPont. She liked her work so well that she stayed for 40 years, retiring in 1986. Kwolek is best known for the development of Kevlar fiber, which is five times stronger than steel. She has received many awards, including the Perkin Medal in 1997, considered one of the most prestigious awards a chemist can receive in the United States.

The reaction continues between additional amine and acid groups to form three amide linkages as shown in (a).

(b) The repeating unit in Kevlar contains an amide linkage and the remaining portions of the reacting acid and amine molecules.

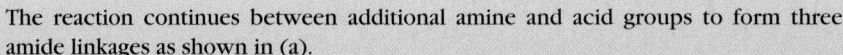

Problem-Solving Practice 12.6

How many moles of water are formed from the condensation reaction between 6 mol of *p*-phenylenediamine and 10 mol of terephthalic acid?

Exercise 12.23 Nylon from 3-Aminopropionic Acid

Polyamides can also be formed from a single monomer that contains both an amine and a carboxylic acid group. For example, the compound 3-aminopropionic acid can polymerize to form a nylon. Write the general formula for this polymer. Write a formula for the other product that is formed.

$$H_2N-CH_2-CH_2-\overset{\overset{\textstyle O}{\|}}{C}-OH$$
3-aminopropionic acid

Exercise 12.24 Functional Groups

We have discussed several functional groups. Shown below is the structural formula of aspartame, an artificial sweetener.

aspartame

Identify each of the numbered functional groups.

These jackets are made from recycled PET soda bottles.

Recycling Plastics

Polyethylene terephthalate (PET), widely used for soft drink bottles, and high-density polyethylene (HDPE) are the most commonly recycled plastics. Major end uses for recycled PET include fiberfill for ski jackets and sleeping bags, carpet fibers, and tennis balls. It takes just five recycled 2-L PET soft drink bottles to make a T-shirt or the insulation for a ski jacket. High-density polyethylene (HDPE) is the second most widely recycled plastic; principal sources are milk, juice, and water jugs. In the United States, more than 200 million recycled HDPE milk and water jugs have been converted into a fiber to make Tyvek, which is used for sportswear, insulating wrap for new buildings, and very durable shipping envelopes.

TABLE 12.7	Plastic Container Codes
Code	**Material**
PETE	Polyethylene terephthalate (PET)*
HDPE	High-density polyethylene
V	Poly(vinyl chloride) (PVC)*
LDPE	Low-density polyethylene
PP	Polypropylene
PS	Polystyrene
OTHER	All other resins and layered multimaterial

*Bottle codes are different from standard industrial identification to avoid confusion with registered trademarks.

Codes are stamped on plastic containers to help consumers identify and sort their recyclable plastics (Table 12.7). There has been a dramatic increase in the recycling of plastics in recent years. Still further increases require a sufficient demand for products made partially or completely from recycled materials. The use of some recycled (postconsumer) plastics has been mandated by law in some places. Since 1995 in California, for example, all HDPE packaging must contain 25% recycled material. A real challenge lies ahead to finding economically viable methods to recycle the plastics from the growing, massive numbers of obsolete personal computers, CD players, and cellular phones.

A Tyvek-wrapped house under construction.

12.9 BIOPOLYMERS: PROTEINS AND POLYSACCHARIDES

Biopolymers—naturally occurring polymers—are an integral part of living things. Many advances in creating and understanding synthetic polymers came from studying biopolymers.

Cellulose and starch, made by plants, resemble a synthetic polymer in that the monomer molecules—glucose—are all alike. On the other hand, proteins, which are made by both plants and animals, are very different from synthetic polymers because they include many different monomers. Also, the occurrences of the different monomers along the protein polymer chain are anything but regular. As a result, proteins are extremely complex copolymers.

Through recycling, over a half-billion tons of PET did not end up in landfills in 1997.

Amino Acids to Proteins

The monomer units in proteins, **amino acids,** each contain a carboxylic acid group and an amine group. The 20 amino acids from which proteins are made have the general formula

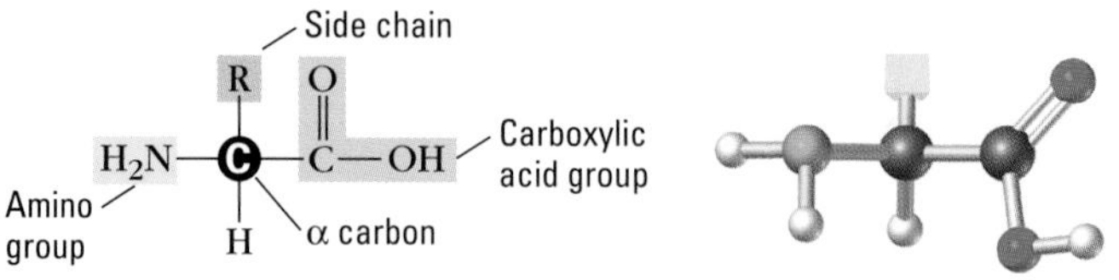

and are described as α-amino acids because the amine (—NH_2) group is attached to the **alpha carbon,** the first carbon next to the —COOH group. Each amino acid has a different R group, called a side chain (Table 12.8). Glycine, the simplest amino acid, has just hydrogen as its R group. Note from the table that some amino acid R groups contain only carbon and hydrogen, while others contain carboxylic acid, amine, or other functional groups. The amino acids are grouped in Table 12.8 according to whether the R group is nonpolar, polar, acidic, or basic. Each amino acid has a three-letter abbreviation for its name.

Like nylon, proteins are polyamides. The amide bond in a protein is formed by the condensation polymerization reaction between the amine group of one amino acid and the carboxylic acid group of another. In proteins, the amide linkage is called a **peptide linkage.** Relatively small amino acid polymers (up to about 50 amino acids) are known as **polypeptides.** Proteins are polypeptides containing hundreds to thousands of amino acids bonded together.

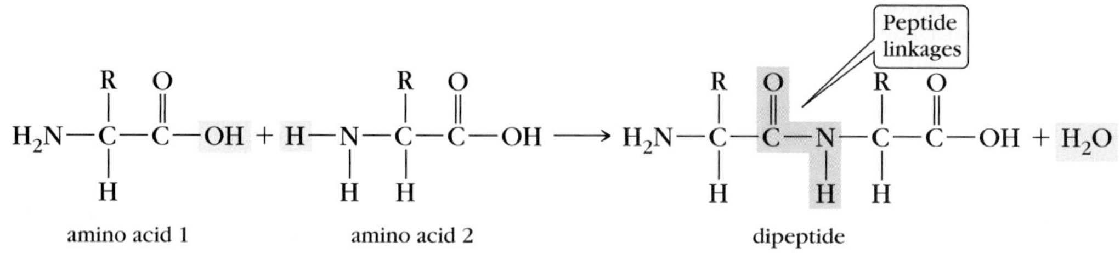

Peptide linkages are also referred to as peptide bonds.

Any two amino acids can react to form two different dipeptides, depending on which amine and acid groups react from each amino acid. For example, glycine and alanine can react in either of the following two ways.

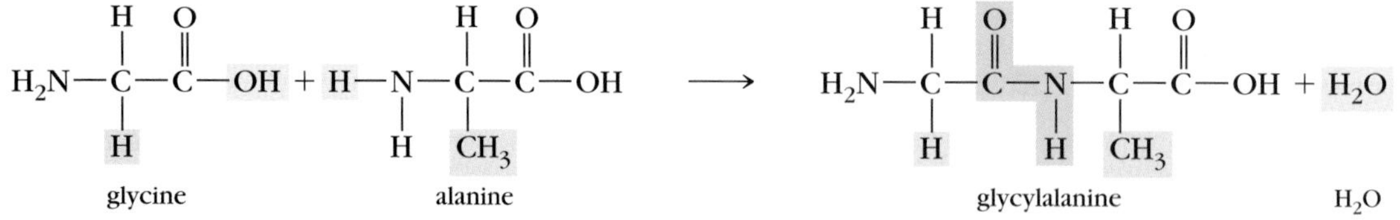

or

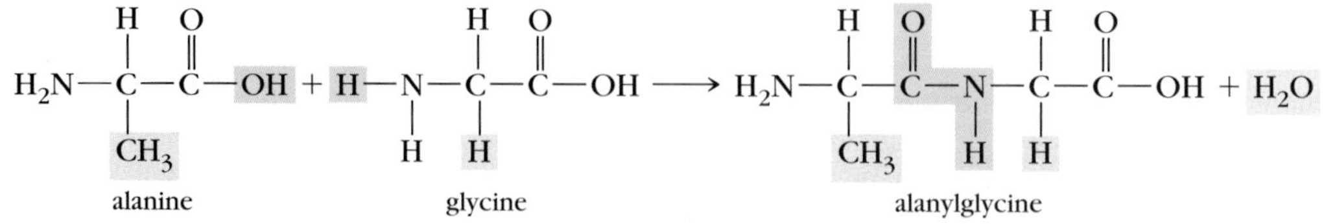

Either of these dipeptides can react with other amino acids at both ends. The extensive chains of amino acid units in proteins are built up by such condensation polymerization reactions.

| TABLE 12.8 | Common L-Amino Acids Found in Proteins† |

Nonpolar R groups

Amino Acid	Abbreviation	Structure	Amino acid	Abbreviation	Structure

Glycine	Gly	H—CH—COOH
		      │
		     NH₂
			*Isoleucine	Ile	CH₃—CH₂—CH—CH—COOH
					             │
					           CH₂

Alanine	Ala	CH₃—CH—COOH
		      │
		     NH₂
			Proline	Pro	(ring structure with N—H)

*Valine	Val	CH₃—CH—CH—COOH
		     │ │
		    CH₃ NH₂
			*Phenyl-alanine	Phe	(benzene ring)—CH₂—CH—COOH
					          │
					        NH₂

*Leucine	Leu	CH₃—CH—CH₂—CH—COOH
		     │ │
		    CH₃ NH₂
			*Methionine	Met	CH₃—S—CH₂CH₂—CH—COOH
					               │
					            NH₂

			*Tryptophan	Trp	(indole ring)—CH₂—CH—COOH
					           │
					         NH₂

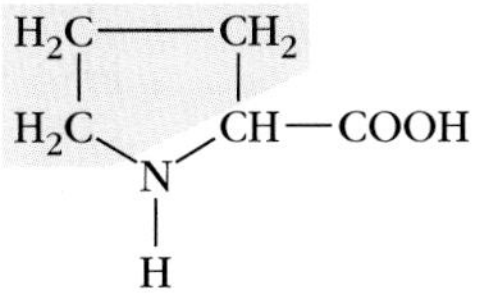

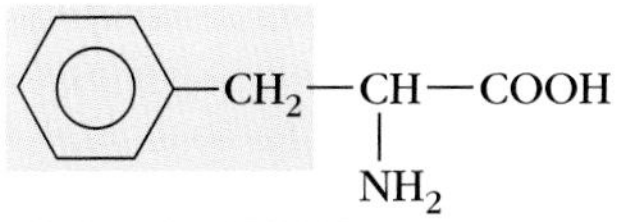

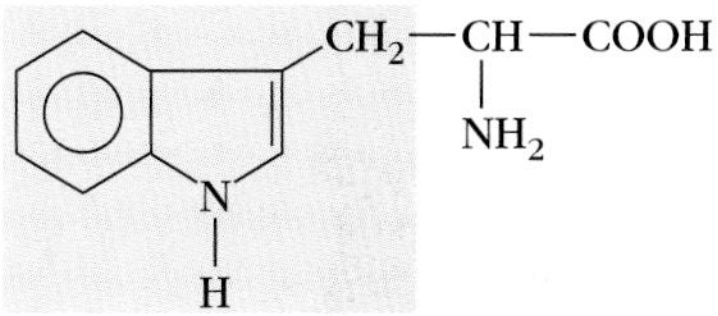

Polar but neutral R groups

Amino Acid	Abbreviation	Structure	Amino acid	Abbreviation	Structure

Serine	Ser	HO—CH₂—CH—COOH
		         │
		       NH₂
			Asparagine	Asn	H₂N—C—CH₂—CH—COOH
					      ‖ │
					      O NH₂

*Threonine	Thr	CH₃—CH—CH—COOH
		     │ │
		    OH NH₂
			Glutamine	Gln	H₂N—C—CH₂CH₂—CH—COOH
					      ‖ │
					      O NH₂

Cysteine	Cys
	│
	NH₂
		HS—CH₂—CH—COOH
		        │
		      NH₂
			Tyrosine	Tyr	HO—(benzene ring)—CH₂—CH—COOH
					                │
					             NH₂

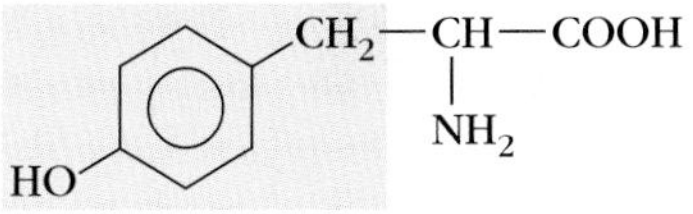

Acidic R groups | | | **Basic R groups** | | |

Amino Acid	Abbreviation	Structure	Amino acid	Abbreviation	Structure

Glutamic acid	Glu	HO—C—CH₂CH₂—CH—COOH
		     ‖ │
		    O NH₂
			*Lysine	Lys	H₂N—CH₂CH₂CH₂CH₂—CH—COOH
					                    │
					                 NH₂

Aspartic acid	Asp	HO—C—CH₂—CH—COOH
		     ‖ │
		    O NH₂
			‡Arginine	Arg	H₂N—C—NH—CH₂CH₂CH₂—CH—COOH
					     ‖ │
					    NH NH₂

			Histidine	His	(imidazole ring)—CH₂—CH—COOH
					               │
					           NH₂

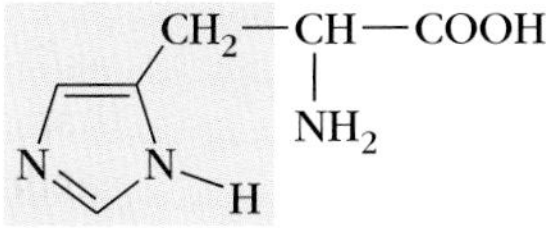

*Essential amino acids that must be part of the human diet. The other amino acids can be synthesized by the body.

†The R group in each amino acid is highlighted.

‡Growing children also require arginine in their diet.

Exercise 12.25 Hydrogen Bonding Between Amino Acids in Proteins

Pick two amino acids from Table 12.8 whose R groups could hydrogen-bond with one another if they were close together in a protein chain or in two adjacent protein chains. Then pick two whose R groups would not hydrogen-bond under similar circumstances.

As the the number of amino acids in the chain increases and the polypeptide chain lengthens, the number of variations quickly increases to a degree of complexity not found in synthetic polymers. As we have seen, two different dipeptides can form from two different amino acids. Six *tri*peptides are possible if three different amino acids (for example, phenylalanine, Phe; alanine, Ala; and serine, Ser) are linked in combinations that contain all three amino acids, if each is used only once. They are

Phe-Ala-Ser   Ser-Ala-Phe   Ala-Ser-Phe

Phe-Ser-Ala   Ser-Phe-Ala   Ala-Phe-Ser

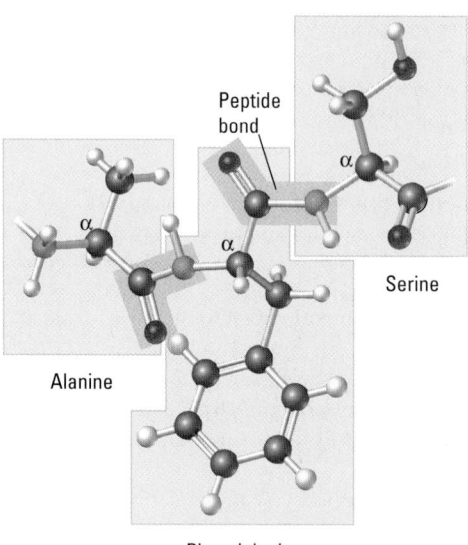

The Ala-Phe-Ser tripeptide

If *n* different amino acids are present, the number of arrangements is *n*! (*n* factorial). For four different amino acids, the number of different tetrapeptides is 4!, or $4 \times 3 \times 2 \times 1 = 24$. With five different amino acids, the number of different arrangements is 5!, or 120. If all 20 different naturally occurring amino acids were bonded in one polypeptide, the sequences would make $20! = 2.43 \times 10^{18}$ (2.43 quintillion) unique 20-monomer molecules. *Because proteins can also include more than one molecule of a given amino acid, the number of possible combinations is astronomical.* It is truly remarkable that of the many different proteins that could be made from a set of amino acids, a living cell makes only the relatively small number it needs.

Problem-Solving Example 12.7 Peptides

Draw the structural formula of the tripeptide represented by Ala-Ser-Gly. Explain why this is a different compound from that with the amino acids joined in the order Gly-Ala-Ser.

Answer

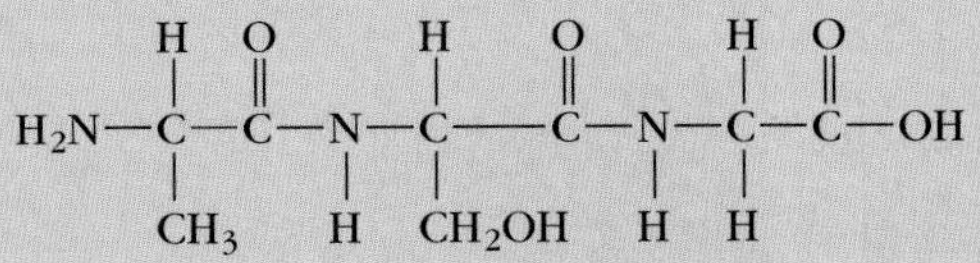

Ala-Ser-Gly

The structure Gly-Ala-Ser differs because the free —NH_2 group is on the glycine part of the molecule and the free —COOH group is on the serine part of the molecule.

Explanation The amino acid sequence in the abbreviated name shows that alanine should be written at the left with a free H_2N— group, glycine should be written at the right with a free —COOH group, and both should be connected to serine by peptide bonds. Writing the structure of Gly-Ala-Ser shows how the two tripeptides are different.

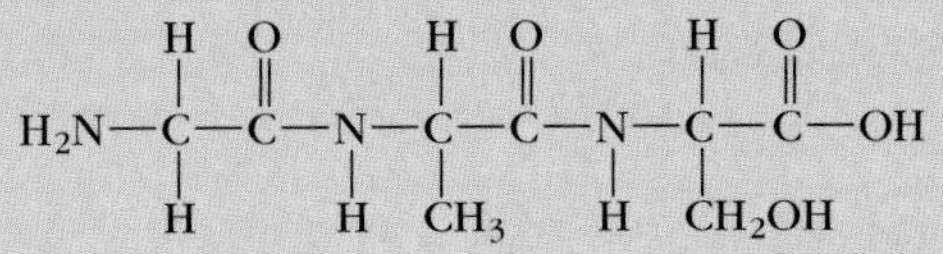

Gly-Ala-Ser

Problem-Solving Practice 12.7

Draw the structural formula of the tetrapeptide Cys-Phe-Ser-Ala.

Exercise 12.26 Peptide Sequences

Draw the structural formula of the tetrapeptide Ala-Ser-Phe-Cys.

The shape of a protein, and consequently its function, are determined by the sequence of amino acids along the chain, known as the **primary structure** of the protein. All proteins, no matter how large, have a peptide backbone of amino acid units covalently bonded to each other through peptide linkages.

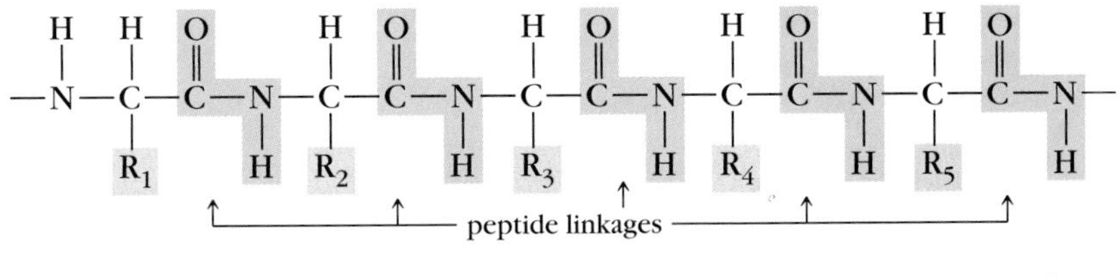

peptide linkages

The distinctions that determine the shape and function of a protein lie in the sequence of amino acids and the order of the R groups along the backbone. Even one out of place amino acid can create dramatic changes in the shape of a protein, which can lead to serious medical conditions. For example, sickle cell anemia results from an alteration in the primary structure of hemoglobin, the molecule that carries oxygen in red blood cells. Glutamic acid, the seventh amino acid in a 146-amino acid chain in hemoglobin, is replaced by valine. Replacement of the carboxylic acid side chain of glutamic acid with the nonpolar side chain of valine (Table 12.8) disturbs the intramolecular attractions in hemoglobin and changes its shape. As a result, hemoglobin molecules gather into fibrous chains that distort the red blood cells into a sickle shape.

Protein chains are not long and floppy, as you might imagine. Regular patterns of **secondary structure** are created by hydrogen bonding. There are two major types of secondary protein structure: (1) the alpha (α)-helix, which occurs within a protein molecule, and (2) the beta (β)-pleated sheet, which occurs between adjacent protein chains. In the alpha-helix, hydrogen bonding occurs between the

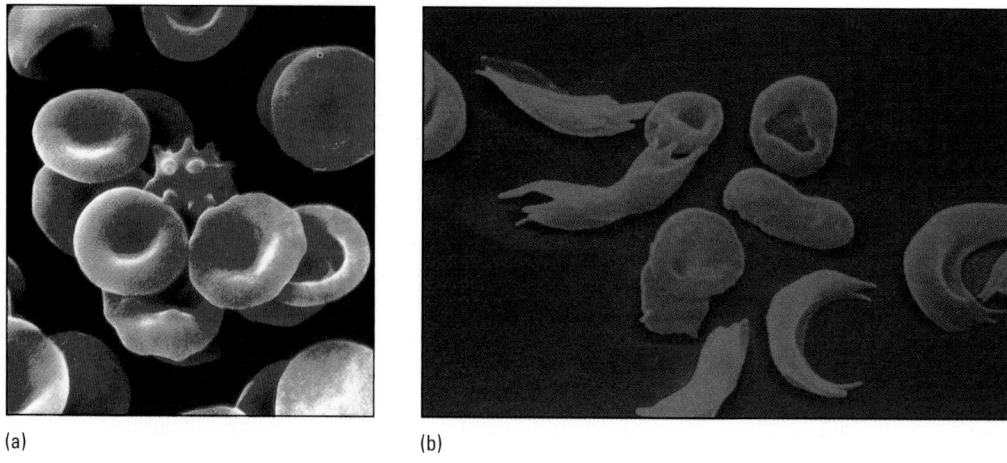

(a)   (b)

Red blood cells. (a) Normal cells. (b) Sickled cells.

N—H hydrogen of an amine group in a peptide bond with a lone electron pair on a C=O oxygen of a peptide bond four amino acid units further down the peptide backbone (Figure 12.18). In effect, the hydrogen in the hydrogen bond is "looking over its shoulder," causing the protein to curl into a helix, much like the coiling of a telephone cord. The R groups on the amino acid units point to the outside of the helix. In contrast, hydrogen bonding in the beta-pleated sheet occurs between the peptide backbones of neighboring protein chains, creating a zigzag pattern resembling a pleated sheet with the R groups above and below the sheet (Figure 12.18).

Wool is primarily the protein keratin, which has an alpha-helix secondary structure. Fibroin, the principal protein in silk, is largely beta-pleated sheets.

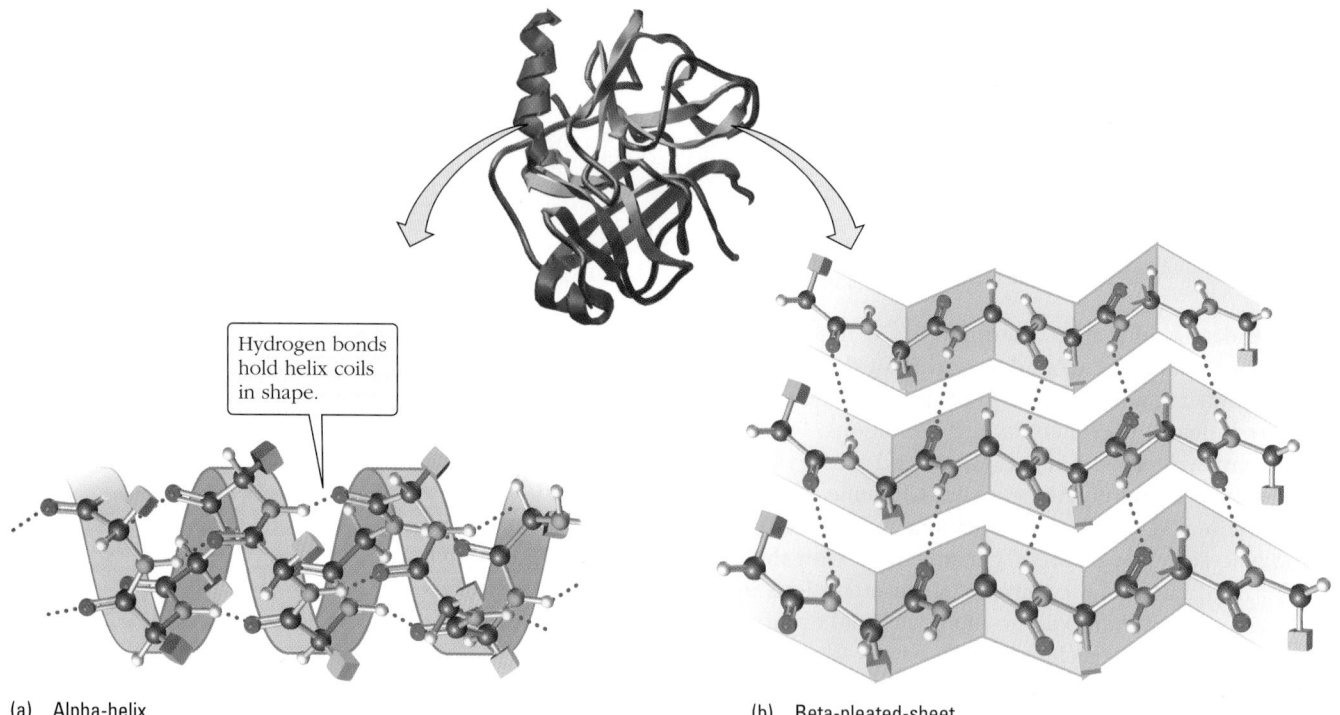

Hydrogen bonds hold helix coils in shape.

(a) Alpha-helix   (b) Beta-pleated-sheet

Figure 12.18 Alpha-helix (a) and beta-pleated (b) sheet structures of proteins. The enzyme molecule has regions of both alpha-helix and beta-pleated sheet protein structures.

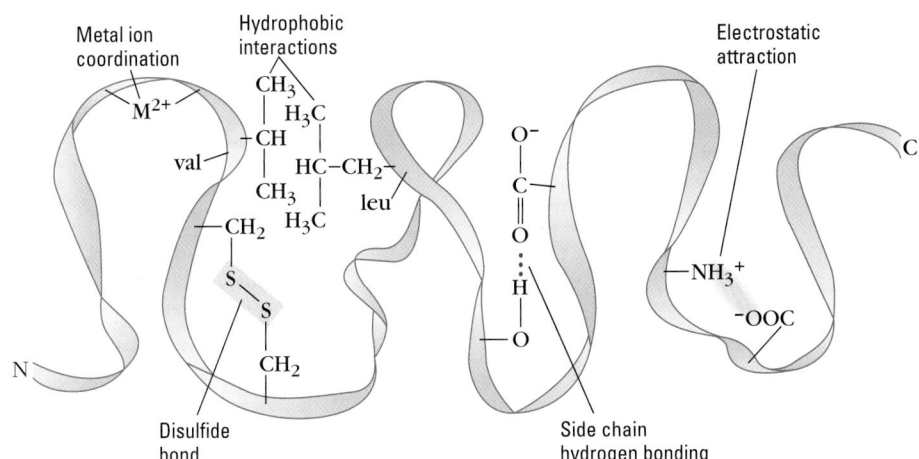

Figure 12.19 Noncovalent forces and bonds that stabilize protein tertiary structure. The tertiary structure of a protein is stabilized by several different forces.

The overall three-dimensional arrangement that accounts for all the twists and turns and folding of a protein is called its **tertiary structure.** The twists and turns of the tertiary structure result in a molecule of maximum stability. Tertiary structure is determined by interactions among the side chains strategically placed along the chain. These interactions, illustrated in Figure 12.19, include noncovalent forces of attraction and covalent bonding.

Noncovalent attractions	Covalent bonds
Hydrogen bonding between side-chain groups	Disulfide bonds (—S—S—)
Hydrophobic (water-hating) interactions between side-chain hydrocarbon groups	Coordinate covalent bonding between metal ions and electron pairs of side-chain N and O atoms
Electrostatic attractions between —NH$_3^+$ and —COO$^-$ side-chain groups	

The hydrogen bonding, hydrophobic interactions, and electrostatic attractions bring the groups closer together. Metal ions, such as Fe^{2+} in hemoglobin, are incorporated into a protein by the donation of lone pair electrons from oxygen or nitrogen atoms in side-chain groups to form coordinate covalent bonds to the metal ions. By loss of H, the —SH groups in proximate cysteine units form disulfide (—S—S—) covalent bonds, which cross-link regions of the peptide backbone. The number and proximity of disulfide bonds help to limit the flexibility of a protein.

Proteins can be divided into two broad categories — fibrous proteins and globular proteins — that reflect differences in their tertiary structures. *Fibrous proteins,* such as hair, muscle fibers, and fingernails, are rod-like, with the coils or sheets of protein aligned into parallel bundles, making for tough, water-insoluble materials. In contrast, *globular proteins* are highly folded, with hydrophilic (water-loving) side chains on the outside, making these proteins water-soluble. Hemoglobin and chymotrypsin (Figure 12.20, p. 562) are globular proteins, as are most enzymes.

Because of the complexity and the variety of properties provided by the different R groups and associated molecules or ions, proteins are able to perform widely diverse functions in the body. Consider some of them.

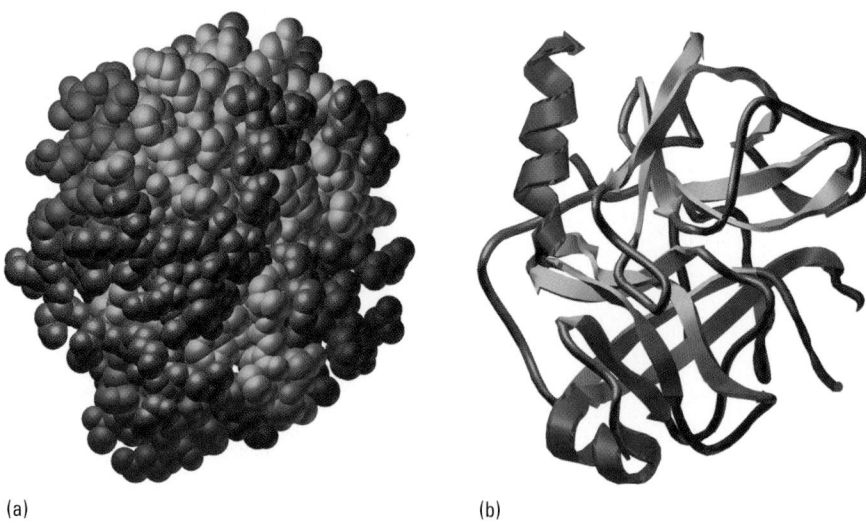

(a)   (b)

Figure 12.20   Protein folding.   The protein chymotrypsin is shown (a) in a space-filling model that illustrates how close together the atoms are in the molecule and (b) in a ribbon structure that illustrates only the polypeptide backbone. Alpha-helix regions are shown in blue, beta-pleated sheet regions are shown in green, and randomly coiled regions are shown in copper.

Class of protein	Example	Description of function
Hormones	Insulin, growth hormone	Regulate vital processes according to needs
Muscle tissue	Myosin	Does mechanical work
Transport proteins	Albumin	Carry fatty acids and other hydrophobic molecules through the bloodstream
Clotting proteins	Fibrin	Form blood clots
Enzymes	Chymotrypsin	Catalyze biochemical reactions
Messsengers	Endorphins	Transmit nerve impulses
Immune system proteins	Immunoglobulins	Fight off disease

Monosaccharides to Polysaccharides

Nature makes an abundance of compounds with the general formula $C_x(H_2O)_y$. These compounds are variously known as sugars, carbohydrates, and mono-, di-, or polysaccharides (from the Latin *saccharum*, "sugar" — because they taste sweet). The simplest carbohydrates, such as glucose, are monosaccharides, also called simple sugars, because they contain only one type of saccharide molecule. *Disaccharides* consist of two monosaccharide units joined together, such as glucose and fructose linked to form sucrose (table sugar). The monomer units of disaccharides and polysaccharides are joined by a C—O—C arrangement called a **glycosidic linkage** between carbons 1 and 4 of adjacent monosaccharide units. Polysaccharides contain many monosaccharide monomers joined together via glycosidic linkages into a very large polymer. The glycosidic linkages in disaccharides and polysaccharides are formed by a condensation reaction like the condensation reactions by which synthetic polymers are formed. A water molecule is released during the formation of each glycosidic bond from the reaction between the —OH groups of the monosaccharide monomers.

1,6-Glycosidic linkages are also possible.

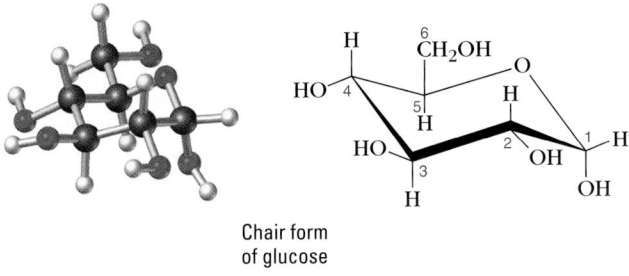

glucose   *glucose*   *maltose*   *water*

Exercise 12.27   Sucrose Solubility

Explain why table sugar (sucrose) is soluble in water.

Starches and cellulose are the most abundant natural polysaccharides. D-Glucose is the monomer in each of these polymers, which can contain as many as 5000 glucose units. Each D-glucose unit exists in a ring structure. To illustrate as closely as possible the shape of the glucose molecules in these polymers, the six-membered rings in monosaccharides are often written as follows.

Chair form
of glucose

Polysaccharides: Starches and Glycogen

Plant starch is stored in protein-covered granules until glucose is needed for synthesis of new molecules or for energy production. If these granules are ruptured by heat, they yield a starch, *amylose,* which is soluble in hot water, and an insoluble starch, *amylopectin.* Natural starches contain about 75% amylopectin and 25% amylose, both of which are polymers of glucose units joined by glycosidic linkages. Structurally, amylose is a condensation polymer with an average of about 200 glucose monomers per molecule arranged in a straight chain. A representative portion of the structure of amylose is shown in Figure 12.21. A typical amylopectin molecule has about 1000 glucose monomers arranged into branched chains analogous to the branched-chain synthetic polymers discussed earlier. Just as these polymers have different properties from their straight-chain counterparts, amylopectin is different from amylose. The main difference is in their water solubilities. A family of enzymes called amylases helps to break down starches sequentially into a mixture of small branched-chain polysaccharides called dextrins and ultimately into glucose. Dextrins are used as food additives and in mucilage, paste, and finishes for paper and fabrics.

Amylose turns blue-black when tested with iodine solution, whereas amylopectin turns red.

Figure 12.21 The structure of amylose. From 60 to 300 glucose units are bonded together to form amylose. The ring structures are drawn in what is called the "chair" form of glucose, which is a more accurate three-dimensional representation of the molecular shape than those given by "flat" two-dimensional drawings. The 1,4-glycosidic linkages between glucose units are all in the same orientation.

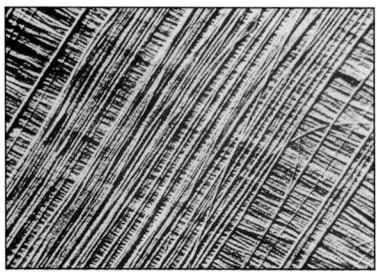

Figure 12.22 The structure of cellulose. About 280 glucose units are bonded together to form a chain structure in cellulose. The —OH groups at carbons 1 and 4 are in the *trans* position, so the glycosidic linkages between glucose units alternate in their orientation, unlike those in amylose.

Animals store energy as fats rather than carbohydrates because fats have more energy per gram.

Paper and cotton are essentially cellulose.

Parallel strands of cellulose in a plant fiber.

In animals, *glycogen* serves the same storage function as starch does in plants. Glycogen, the reserve carbohydrate in the body, is stored in the liver and muscle tissues and provides glucose for "instant" energy until the process of fat metabolism can take over and serve as the energy source ($\Leftarrow$ *p. 248*). The chains of D-glucose units in glycogen are more highly branched than those in amylopectin.

Cellulose, A Polysaccharide

Cellulose is the most abundant organic compound on earth, found as the woody part of trees and the supporting material in plants and leaves. Cotton is the purest natural form of cellulose. Like amylose, cellulose is composed of D-glucose units. The difference between the structures of cellulose and amylose lies in the orientation of the glycosidic linkages between the glucose units. In cellulose, the —OH groups at carbons 1 and 4 are in the *trans* position, so the glycosidic linkages between glucose units alternate in direction; thus, every other glucose unit is turned over (Figure 12.22). In amylose, the —OH groups at carbons 1 and 4 are in the *cis* position, so all the glycosidic linkages are in the same direction. This subtle structural difference allows humans to digest starch, but not cellulose; we lack the enzyme necessary to break the *trans* glycosidic linkages in cellulose. However, termites, a few species of cockroaches, and ruminant mammals such as cows, sheep, goats, and camels do have the proper internal chemistry for this purpose. Because cellulose is so abundant, it would be advantageous if humans could use it, as well as starch, for food.

Exercises 12.28   Digestion of Cellulose

Explain why humans cannot digest cellulose. Consult a reference on the Internet and explain why ruminant animals can digest cellulose.

 ### Exercise 12.29   What if Humans Could Digest Cellulose?

Think of some of the implications if humans could digest cellulose. What would be some desirable consequences? What would be some undesirable ones?

SUMMARY PROBLEM

Part 1

This chapter has described many kinds of carbon-containing compounds from a variety of sources. Several classes of carbon compounds have molecules with a carbon-to-oxygen bond, which gives the compound certain chemical properties. Use the information in the chapter to complete the sequence of chemical changes for the compound

$$CH_3(CH_2)_8CH_2OH$$

(a) Write the structural formulas of the initial and final oxidation products of this compound.

(b) Use structural formulas to write the equation for the reaction of the initial reactant in Question (a) with the final oxidation product in Question (a). Into what class of compounds does the product of the reaction in Question (b) fall?

(c) The final product in Question (b) reacts with aqueous sodium hydroxide. Write the equation for this reaction. What types of compounds are formed by this reaction? What would the products be if the reaction were done under acidic conditions instead of with aqueous base?

(d) The final oxidation product in Question (a) reacts with dimethylamine, $(CH_3)_2NH$. Write the structural formula of the product of this reaction. What class of compound is formed?

(e) Can the two reactants in Question (d) form a polymer? Explain.

Part 2

The citric acid cycle is a series of steps involving the degradation and reformation of oxaloacetic acid. Two steps in the cycle involve a reaction studied in this chapter. One step converts isocitric acid into oxalosuccinic acid.

In the last step of the cycle, L-malic acid is converted into oxaloacetic acid.

(a) Identify the general type of reaction in each step.
(b) What do the two steps have in common?

Part 3

(a) Select any four amino acids from Table 12.8. Write the structural formulas of four different tetrapeptides that could be formed from these four amino acids.

IN CLOSING

Having studied this chapter, you should be able to . . .

- Describe petroleum refining and methods used to improve the gasoline fraction (Section 12.1).
- Identify the major components of natural gas (Section 12.2).
- Identify processes used to obtain organic chemicals from coal, and name some of their products (Section 12.2).
- Convert equivalent energy units (Section 12.3).
- Relate atmospheric CO_2 concentration to the greenhouse effect and to global warming (Section 12.4).
- Identify major organic chemicals of industrial and economic importance (Section 12.5).
- Name and draw the structures of three functional groups produced by the oxidation of alcohols (Section 12.6).

- Name and give examples of the uses of some important alcohols (Section 12.6).
- List some properties of carboxylic acids, and write equations for the formation of esters from carboxylic acids and alcohols (Section 12.7).
- Explain the formation of polymers by addition or condensation polymerization; give examples of synthetic polymers formed by each type of reaction (Section 12.8).
- Draw the structures of the repeating units in some common types of synthetic polymers and identify the monomers that form them (Section 12.8).
- Identify or write the structures of the functional groups in alcohols, aldehydes, ketones, carboxylic acids, esters, and amines (Section 12.6–12.8).
- Identify the types of plastics most successfully being recycled (Section 12.8).
- Illustrate the basics of protein structures and how peptide linkages hold amino acids together in proteins (Section 12.9).
- Differentiate among the primary, secondary, and tertiary structures of proteins (Section 12.9).
- Identify polysaccharides, their sources, the different ways they are linked, and the different uses resulting from these linkages (Section 12.9).

KEY TERMS

addition polymer *(12.8)*
aldehyde *(12.6)*
alpha carbon *(12.9)*
amide *(12.8)*
amide linkage *(12.8)*
amine *(12.8)*
amino acid *(12.9)*
carboxylic acid *(12.7)*
catalytic cracking *(12.1)*
catalytic reforming
    (12.1)
catalyst *(12.1)*
condensation polymer
    (12.8)
condensation reaction
    (12.7)
copolymer *(12.8)*
ester *(12.7)*

global warming *(12.4)*
glycosidic linkage
    (12.9)
greenhouse effect
    (12.4)
hydrolysis *(12.7)*
ketone *(12.6)*
macromolecule *(12.8)*
monomer *(12.8)*
monosaturated acids
    (12.7)
octane number *(12.1)*
partial hydrogenation
    (12.7)
peptide linkage *(12.9)*
petroleum fractions
    (12.1)
polyamides *(12.8)*

polyester *(12.8)*
polymer *(12.8)*
polypeptide *(12.9)*
polyunsaturated acids
    (12.7)
primary structure
    (12.9)
saponification *(12.7)*
secondary structure
    (12.9)
tertiary structure *(12.9)*
thermoplastics *(12.8)*
thermosetting plastics
    (12.8)

QUESTIONS FOR REVIEW AND THOUGHT

Conceptual Challenge Problems

CP-12.A (Section 12.1) How are the boiling points of hydrocarbons during the distillation of petroleum related to their molecular size?

CP-12.B (Section 12.6) Even though there are millions of organic compounds and each compound may have 10, 100, or even thousands of atoms bonded together to make one molecule, the reactions of organic compounds can be stud-

ied and even predicted for compounds yet to be discovered. What characteristic of organic compounds allows their reactions to be studied and predicted?

CP-12.C (Section 12.7) What is the advantage of animals storing chemical potential energy in their bodies as triesters of glycerol and long-chain fatty acids, known as fats, instead of as carbohydrates?

Answers to questions in **bold** can be found in the back of the book.

Review Questions

1. Why is the organic chemical industry referred to as the *petrochemical industry?*
2. What products are produced by the petrochemical industry?
3. What is the difference between *catalytic cracking* and *catalytic reforming?*
4. Explain how the octane number of a gasoline is determined.
5. Why is coal receiving increased attention as a source of organic compounds?
6. What is synthesis gas? How can it be used to produce petrochemicals?
7. Methanol is one of the top 50 petrochemicals produced in the United States. What factors are likely to lead to an increased demand for methanol in the next decade?
8. What is the difference between *oxygenated gasoline* and *reformulated gasoline?* Why are they being produced?
9. Explain why world use of natural gas is predicted to double between 1990 and 2010.
10. Table 12.1 lists several compounds with octane numbers above 100 and one compound with an octane number below zero. Explain why such values are possible.
11. Explain why methanol has a lower boiling point (65.0 °C) than water (boiling point, 100.0 °C).
12. Explain why ethylene glycol has a higher boiling point than ethanol.
13. Outline the steps necessary to obtain 89-octane gasoline, starting with a barrel of crude oil.
14. What is the major difference between crude oil and coal as a source of hydrocarbons?
15. Describe the structural formula of a representative compound for each of these classes of organic compounds: alcohols, aldehydes, ketones, carboxylic acids, esters, and amines.
16. Explain why esters have lower boiling points than carboxylic acids of the same molecular weight.
17. What structural feature must a molecule have in order to undergo addition polymerization?
18. What feature do all condensation polymerization reactions have in common?
19. Give examples of (a) a synthetic addition polymer, (b) a synthetic condensation polymer, and (c) a natural addition polymer.
20. How does *cis-trans* isomerism affect the properties of rubber?
21. Discuss which two plastics are currently being recycled the most, and give examples of some products being made from these recycled plastics.
22. What is the difference between the formation of an addition polymer and a condensation polymer?

Fuel Sources and Products

23. What are petroleum fractions? What process is used to produce them?

24. (a) What is the boiling point range for the petroleum fraction containing the hydrocarbons that will provide fuel for your car?
    (b) Would you expect the octane rating from the "straight-run" gasoline obtained by fractional distillation of petroleum to be greater than 87 octane? Explain your answer.
    (c) Would you use this fraction to fuel your car? Why or why not?
25. (a) Draw the Lewis structure for the hydrocarbon that is assigned an octane rating of 0.
    (b) Draw the Lewis structure for the hydrocarbon that is assigned an octane rating of 100.
    (c) What is the boiling point for each of these hydrocarbons?
26. Explain what is meant by this statement: "All gasolines are highly volatile."
27. What would be the advantage of removing the higher octane components such as aromatics and alkenes from oxygenated gasolines?
28. Write the structural formula of ethanol, a common gasoline additive.
29. What are the components in natural gas?
30. What is the difference between the greenhouse effect and global warming? How are they related?
31. What are the four "greenhouse gases," and why are they called that?
32. Carbon dioxide is known to be the major contributor to the greenhouse effect. List some of its sources in our atmosphere and some of the processes that remove it. Currently, which predominates, CO_2 sources or removal processes?
33. Name a favorable effect of the global increase of CO_2 in the atmosphere.

Alcohols

34. Give an example of (a) a primary alcohol, (b) a secondary alcohol, and (c) a tertiary alcohol. Draw Lewis structures for each example.
35. Classify each of these alcohols as primary, secondary, or tertiary.
    (a) $CH_3CH_2CH_2CH_2OH$
    (b) $CH_3CHCH_2CH_3$ with OH
    (c) $CH_3CHCH_2CH_2OH$ with OH and CH_3
    (d) $CH_3CCH_2CH_3$ with CH_3 and OH
    (e) CH_3CCH_3 with CH_3 and OH
36. Write the condensed structural formula for each of these.
    (a) 2-methyl-2-pentanol
    (b) 2,3-dimethyl-1-butanol
    (c) 4-methyl-2-pentanol
    (d) 2-methyl-3-pentanol
    (e) *Tertiary*-butyl alcohol
    (f) Isopropyl alcohol

37. Explain what *oxidation* of organic compounds usually involves. What is meant by *reduction* of organic compounds?

38. Draw the structures of the first two oxidation products of each of these alcohols.
    (a) CH_3CH_2OH (b) $CH_3CH_2CH_2CH_2OH$

39. Draw the structures of the oxidation products of each of these alcohols.
    (a) 2-Butanol   (b) 4-Methyl-2-pentanol

40. Write the condensed structural formula of the alcohols that can be oxidized to make these compounds.

    (a) $CH_3CH{-}CH_2{-}\overset{\overset{\displaystyle O}{\|}}{C}{-}H$
            $\underset{CH_3}{|}$

    (b) $CH_3{-}CH_2{-}\overset{\overset{\displaystyle O}{\|}}{C}{-}CH_2{-}CH_3$

    (c) $CH_3{-}CH_2{-}CH{-}\overset{\overset{\displaystyle O}{\|}}{C}{-}OH$
                    $\underset{CH_3}{|}$

41. What is the percentage of ethanol in 90-proof vodka?

42. Explain the common names *wood alcohol* for methanol and *grain alcohol* for ethanol.

43. What is denatured alcohol? Why is it made?

44. Many biological molecules, including steroids and carbohydrates, contain many —OH groups. What need might biological systems have for this particular functional group?

Carboxylic Acids and Esters

45. Explain why the boiling points for carboxylic acids are higher than for alcohols with comparable molar masses.

46. Write the structural formula of the ester that can be formed from
    (a) $CH_3COOH + CH_3CH_2OH$
    (b) $CH_3CH_2COOH + CH_3CH_2CH_2OH$
    (c) $CH_3CH_2COOH + CH_3OH$

47. Write the structural formula for the esters that can be produced by these reactions.
    (a) Formic acid + methanol
    (b) Butyric acid + ethanol
    (c) Acetic acid + 1-butanol
    (d) Propionic acid + 2-propanol

48. Write the condensed formula of the alcohol and acid that will react to form each of these esters.

    (a) $CH_3CH_2\overset{\overset{\displaystyle O}{\|}}{C}{-}OCH_3$

    (b) $H\overset{\overset{\displaystyle O}{\|}}{C}{-}OCH_2CH_3$

    (c) $CH_3\overset{\overset{\displaystyle O}{\|}}{C}{-}OCH_2CH_3$

49. Explain why carboxylic acids are more soluble in water than are esters with the same molar mass.

Organic Polymers

50. What are some examples of *thermoplastics*? What are the properties of thermoplastics when heated and cooled?

51. What are some examples of *thermosetting plastics*? What are the properties of thermosetting plastics when heated and cooled?

52. Draw the structure of the repeating unit in a polymer in which the monomer is
    (a) 1-butene   (b) 1,1-dichloroethylene
    (c) Vinyl acetate

53. What is the principal structural difference between low-density and high-density polyethylene? Is polyethylene an addition or a condensation polymer?

54. Methyl methacrylate has the structural formula shown in Table 12.6. When polymerized it is very transparent, and it is sold in the United States under the trade names Lucite and Plexiglas. Draw the repeating unit for the poly(methyl methacrylate) chain.

55. What monomers are used to prepare these polymers?
    (a) $-CH_2CH_2CH_2CH_2CH_2CH_2CH_2CH_2CH_2-$

    (b) $\underset{\underset{CH_3}{|}}{-CH}CH_2\underset{\underset{CH_3}{|}}{CH}CH_2\underset{\underset{CH_3}{|}}{CH}CH_2-$

    (c) $-CH_2{-}\overset{\overset{H}{|}}{\underset{\underset{\bigcirc}{|}}{C}}{-}CH_2{-}\overset{\overset{H}{|}}{\underset{\underset{\bigcirc}{|}}{C}}{-}CH_2{-}\overset{\overset{H}{|}}{\underset{\underset{\bigcirc}{|}}{C}}{-}CH_2{-}\overset{\overset{H}{|}}{\underset{\underset{\bigcirc}{|}}{C}}{-}$

    (d) $-CH_2{-}\overset{\overset{CH_3}{|}}{\underset{\underset{Cl}{|}}{C}}{-}CH_2{-}\overset{\overset{CH_3}{|}}{\underset{\underset{Cl}{|}}{C}}{-}CH_2{-}\overset{\overset{CH_3}{|}}{\underset{\underset{Cl}{|}}{C}}{-}$

    (e) $-CH_2CH\underset{\underset{OC_2H_5}{\overset{\overset{C=O}{|}}{|}}}{}CH_2CH\underset{\underset{OC_2H_5}{\overset{\overset{C=O}{|}}{|}}}{}CH_2CH\underset{\underset{OC_2H_5}{\overset{\overset{C=O}{|}}{|}}}{}CH_2CH\underset{\underset{OC_2H_5}{\overset{\overset{C=O}{|}}{|}}}{}-$

56. What is the monomer in *natural rubber*? Which isomer is present in natural rubber, *cis* or *trans*?

57. What are the two monomers used to make SBR?

58. The formation of polyesters involves which two functional groups?

59. Name one important polyester polymer and its uses.

60. Polyamides are made by condensing which functional groups? Name the most common example of this class of polymers.

61. How are amide linkages and peptide linkages similar? How are they different?

62. State one major difference between proteins and polyamides.

63. Draw structures of monomers that could form each of these condensation polymers.

    (a) $-\overset{\underset{\|}{O}}{C}{-}(CH_2)_8{-}\overset{\underset{\|}{O}}{C}{-}NH(CH_2)_6NH-$

    (b) $-\overset{\underset{\|}{O}}{C}{-}\bigcirc{-}\overset{\underset{\|}{O}}{C}{-}OCH_2{-}\bigcirc{-}CH_2O-$

64. Orlon has this polymeric chain structure:

$$-CH_2-CH-CH_2-CH-CH_2-CH-$$
$$\qquad\quad |\qquad\qquad |\qquad\qquad |$$
$$\qquad\quad CN\qquad\quad CN\qquad\quad CN$$

    What is the monomer from which this structure can be made?

65. How many ethylene units are in a polyethylene molecule that has a molecular weight of approximately 42,000?

66. A polymer is formed by 4-hydroxybenzoic acid. Write the structural formula of three units of the polymer.

$$HO-\underset{\text{4-hydroxybenzoic acid}}{\bigcirc}-\overset{\overset{O}{\|}}{C}-OH$$

67. Write the structural formulas for the repeating units of
    (a) Natural rubber (b) Neoprene (c) Polybutadiene

68. What are some of the serious drawbacks to burning plastics? Make a list of harmful substances produced when plastics are incinerated.

Proteins and Polysaccharides

69. Which biological molecules have monomer units that are all alike, as in synthetic polymers?

70. Which biological molecules have monomer units that are not all alike, as in synthetic copolymers?

71. Identify and name all the functional groups in the tripeptide below.

72. Draw the structural formula of alanylglycylphenylalanine.

73. Draw the structural formula of leucylmethionylalanylserine.

74. Explain the difference between (a) monosaccharides and disaccharides, (b) disaccharides and polysaccharides.

75. What is the chief function of glycogen in animal tissue?

76. What polysaccharides yield only D-glucose upon complete hydrolysis?

77. (a) How do amylose and amylopectin differ?
    (b) How are they similar?
    (c) Are amylose and glycogen similar?

78. (a) Explain why humans can use glycogen but not cellulose for energy.
    (b) Why can cows digest cellulose?

General Questions

79. Compounds A and B both have the molecular formula C_2H_6O. The boiling points of compounds A and B are 78.5 °C and -23.7 °C, respectively. Use the table of func-

tional groups in Appendix E and write the structural formulas and names of the two compounds.

80. Explain why ethanol, CH_3CH_2OH, is soluble in water in all proportions, but decanol, $CH_3(CH_2)_9OH$, is almost insoluble in water.

81. Nitrile rubber (Buna N) is a copolymer of two parts 1,3-butadiene to one part acrylonitrile. Draw the repeating unit of this polymer.

82. How are rubber molecules modified by vulcanization?

83. Write the condensed structural formula for 3-ethyl-5-methyl-3-hexanol. Is this a primary, secondary, or tertiary alcohol?

84. Using structural formulas, write a reaction for the hydrolysis of a triglyceride that contains fatty acid chains, each consisting of 16 total carbon atoms.

85. Is the plastic wrap used in covering food a thermoplastic or thermosetting plastic? Explain.

86. Assume that a car burns pure octane,

$$C_8H_{18}\ (d = 0.692\ \text{g/cm}^3)$$

    (a) Write the balanced equation for burning octane in air, forming CO_2 and H_2O.
    (b) If the car has a fuel efficiency of 32 miles per gallon of octane, what volume of CO_2 at 25 °C and 1.0 atm is generated when the car goes on a 10.0-mile trip? (The volume of 1 mol of CO_2(g) at 25 °C and 1 atm is 24.5 L.)

87. Perform the same calculations as in Question 86, but use methanol, CH_3OH ($d = 0.791$ g/cm^3) as the fuel. Assume the fuel efficiency is 20.0 miles per gallon.

88. Show structurally why glycogen forms granules when stored in the liver but cellulose is found in cell walls as sheets.

89. Polytetrafluoroethylene (Teflon) is made by first treating HF with chloroform, then cracking the resultant difluorochloromethane.

$$CHCl_3 + 2\ HF \longrightarrow CHClF_2 + 2\ HCl$$
$$2\ CHClF_2 + heat \longrightarrow F_2C{=}CF_2 + 2\ HCl$$
$$F_2C{=}CF_2 + peroxide\ catalyst \longrightarrow Teflon$$

    If you wish to make 1.0 kg of Teflon, what mass of chloroform and HF must you use to make the starting material, $CHClF_2$? (Although it is not realistic, assume that each reaction step proceeds to a 100% yield.)

Applying Concepts

90. Hydrogen bonds can form between propanoic acid molecules and between 1-butanol molecules. Draw all the propanoic acid molecules that can hydrogen-bond to the one shown below. Draw all the 1-butanol molecules that can hydrogen-bond to the one shown below. Use dotted lines to represent the hydrogen bonds.

$$\underset{\text{propanoic acid}}{CH_3CH_2\overset{\overset{O}{\|}}{C}-OH}\qquad\underset{\text{1-butanol}}{CH_3CH_2CH_2CH_2-OH}$$

Based on your drawings, which should have the higher boiling point, propanoic acid or 1-butanol? Explain your reasoning.

91. Both propanoic acid and ethyl methanoate form hydrogen bonds with water. Draw all the water molecules that can hydrogen-bond to these molecules. Use dotted lines to represent the hydrogen bonds.

$$CH_3CH_2\overset{\displaystyle O}{\overset{\displaystyle \|}{C}}-OH \qquad H-\overset{\displaystyle O}{\overset{\displaystyle \|}{C}}-O-CH_2CH_3$$

propanoic acid   ethyl methanoate

Based on your drawings, which should be more soluble in water, propanoic acid or ethyl methanoate? Explain your reasoning.

92. What monomer formed this polymer?

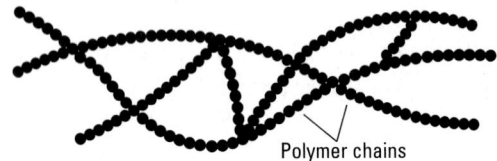

93. The illustrations below represent two different samples of polyethylene, each with the same number of monomer units. Based on the concept of density and not structure, which one is high-density polyethylene and which is low-density? Write a brief explanation.

(a) Before stretching

Polymer chains

(b) After stretching

94. The backbone of a DNA molecule is a polymer of alternating sugar (deoxyribose) and phosphoric acid units held together by a phosphate ester bond. Draw a segment of the polymer consisting of at least two sugar and two phosphate units. Circle the phosphate ester bonds.

phosphoric acid   deoxyribose

95. Draw the structure of a molecule that could undergo a condensation reaction with itself to form a polyester. Draw a segment of the polymer consisting of at least five monomer units.

96. It has been asserted that the photosynthesis of the trees in a forest the size of Australia would be needed to compensate for the additional CO_2 put into the atmosphere each year. Identify the data that would be required in order to check whether this assertion were valid.

97. A large oil refinery runs 400,000 barrels of crude oil a day through fractional distillation. Assume that all the crude oil was simply burned to produce energy. Use data from Table 12.2 to answer the following.
(a) How many 100-W light bulbs could be lighted by this amount of energy?
(b) For how long?
(c) The average toaster uses about 39 kWh per year. How many toasters would this amount of energy operate for a year?

General Chemistry CD-ROM

CD12.1 Screen 11.7: Fats and Oils.
(a) What is the primary structural difference between fats and oils?
(b) What types of functional groups do each contain?
(c) What class of hydrocarbon fragments does each contain?
(d) What structural feature of oil molecules prevents them from coiling up upon themselves as fat molecules do?

CD12.2 Screen 11.9: Addition Polymerization.
(a) What is the primary structural feature of the molecules used to form addition polymers?
(b) What controls the length of the polymer chains formed?

CD12.3 Screen 11.10: Condensation Polymerization.
(a) What is the primary structural feature necessary for a molecule to be useful in a condensation polymerization reaction?
(b) Why are these polymers called "condensation" polymers?

CD12.4 Screen 11.8: Amino Acids and Proteins.
(a) What two functional groups do all amino acids contain?
(b) The peptide shown in the animation is composed of three amino acids. Could this "tripeptide" further react with one or more additional amino acid molecules? Explain.

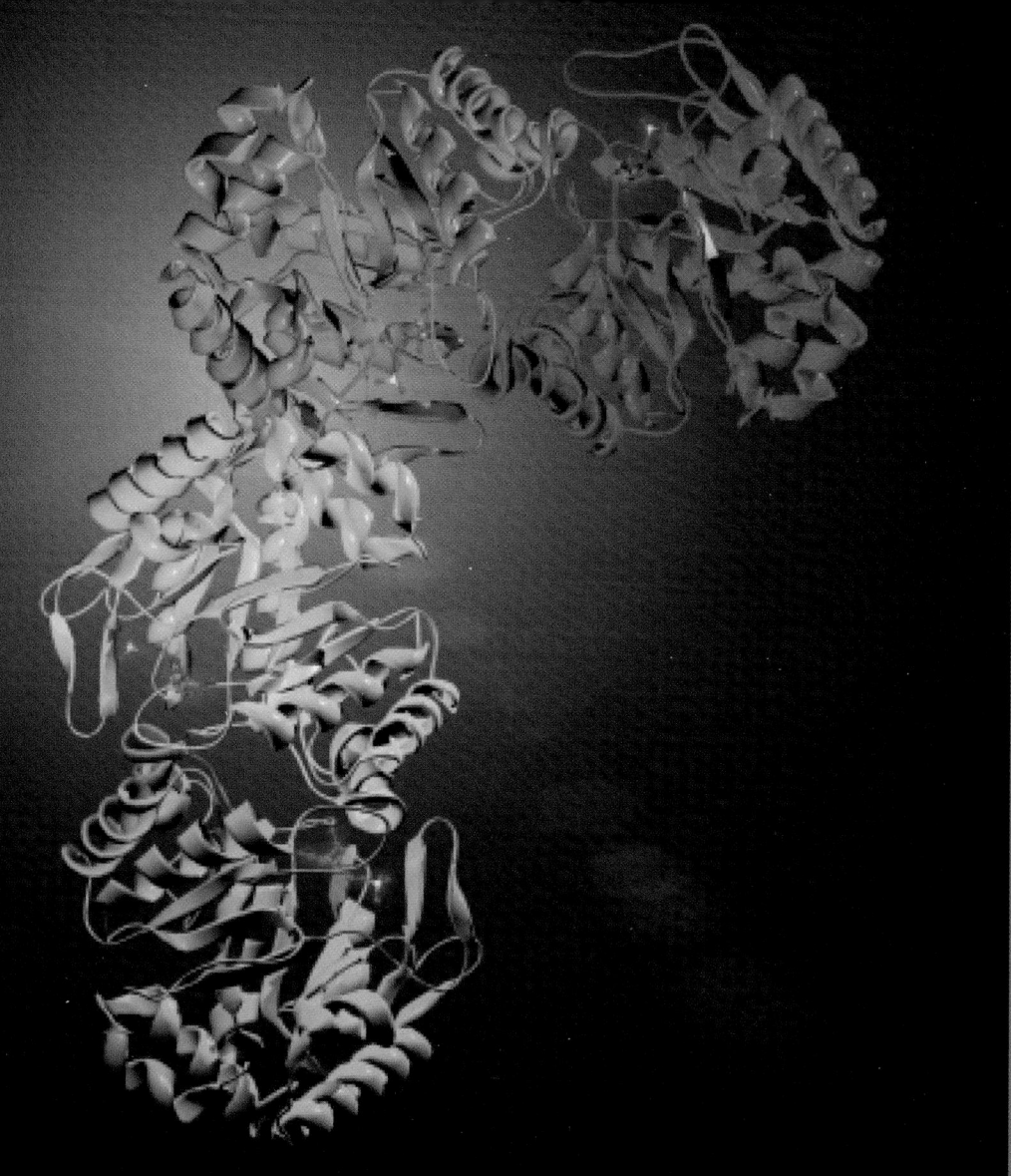

13

Enzymes are catalysts. With a catalyst a chemical reaction occurs many times faster than without it. OMP decarboxlyase, whose structure is shown here, is the most proficient enzyme known. It accelerates the reaction it catalyzes by a factor of 10^{17}. With the enzyme present the reaction occurs in a few seconds, but in the absence of the enzyme it would take millions of years. Enzymes enable our bodies to turn reactions on and off, thereby carrying out important functions such as maintaining proper concentrations of many substances and transmitting nerve impulses.

CHEMICAL KINETICS: RATES OF REACTIONS

13.1 Reaction Rate

13.2 Effect of Concentration on Reaction Rate

13.3 Rate Law and Order of Reaction

13.4 A Nanoscale View: Elementary Reactions

13.5 Temperature and Reaction Rate: The Arrhenius Equation

13.6 Rate Laws For Elementary Reactions

13.7 Reaction Mechanisms

13.8 Catalysts and Reaction Rate

13.9 Enzymes: Biological Catalysts

13.10 Catalysis in Industry

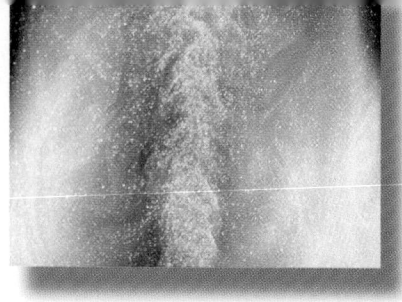

Turn on the valve of a Bunsen burner in your laboratory, bring up a lighted match, and a rapid combustion reaction begins with a whoosh:

$$CH_4(g) + 2\,O_2(g) \rightarrow CO_2(g) + 2\,H_2O(g) \qquad \Delta H° = -802.34\ kJ$$

But what would happen if you didn't put a lighted match in the methane-air stream? Nothing obvious. At room temperature the reaction of methane with oxygen is so slow that the two potential reactants can be mixed in a closed flask and stored unreacted for centuries. These facts about combustion of methane lie within the realm of **chemical kinetics** — *the study of the speeds of reactions and the nanoscale pathways or rearrangements by which atoms and molecules are transformed from reactants to products.*

Chemical kinetics is extremely important, because knowing about kinetics enables us to control reactions. For example, knowing that a fuel like methane burns very rapidly at high temperatures, but at low temperatures reacts extremely slowly with oxygen, allows us to control combustion. We can initiate it with a lighted match or a spark, but as long as there is nothing to speed up the reaction, we can handle the fuel safely. In pharmaceutical chemistry, an important problem is devising drugs that remain in their active form long enough to get to the site in the body where they are intended to act. Consequently, it is important to know whether a drug will react with other substances in the body and how long it will take to do so. The fact that you see motion on a TV screen depends on the speeds of reactions that take place in the retinas of your eyes. In environmental chemistry, there was more than a decade of controversy over whether stratospheric ozone is being depleted by chlorofluorocarbons. Much of this hinged on verifying the sequence of reactions by which stratospheric ozone is produced and consumed, and on accurate measurements of the rates of those reactions. Their careful studies of such reactions led to a Nobel Prize for Sherwood Rowland, Mario Molina, and Paul Crutzen (⇐ *p. 442*).

This chapter is about the factors that affect the speeds of reactions, the nanoscale basis for understanding them, and their importance in modern society, from industrial plants to the cells of our bodies.

13.1 REACTION RATE

CD-ROM Screen 15.3: Control of Reaction Rates: Surface Area

In order for a chemical reaction to occur, reactant molecules must come together so that atoms can be exchanged or rearranged. Atoms and molecules are mobile in the gas phase or in solution, and so reactions are often carried out in a mixture of gases or among solutes in a solution. For a **homogeneous reaction,** one in which reactants and products are all in the same phase (gas or solution, for example), four factors affect the speed of a reaction.

- The *properties* of reactants and products — in particular, their molecular structure and bonding.
- The *concentrations* of the reactants and sometimes the products.
- The *temperature* at which the reaction occurs.
- The presence of a *catalyst* (⇐ *p. 517*) and, if one is present, its concentration.

A catalyst (described in Section 13.8) speeds up a reaction but undergoes no net chemical change. The catalytic converter in an automobile speeds up reactions that remove pollutants from the exhaust gases.

Many important reactions, including the ones in catalytic converters that remove air pollutants from automobile exhaust, are **heterogeneous reactions.** They take place at a surface — at an interface between two different phases (solid and gas, for example). Their speeds depend on the factors listed above but also on the area and nature of the surface at which they occur. For example, very finely divided flour, cornstarch, or lycopodium powder can burn explosively, whereas a pile of powder with much less surface exposed to oxygen in the air is difficult to ignite (Figure 13.1). The much more rapid reaction when greater surface is exposed has been responsible for explosions in grain elevators and other situations where finely divided, combustible solids are exposed to air and a spark or flame.

(a)   (b)

Figure 13.1 Combustion of lycopodium powder. (a) The very finely divided spores of this common moss burn slowly when they are in a pile so that only a small surface area is exposed to air. (b) When the exposed surface is increased by spraying the powder through the air into a flame, combustion is rapid — even explosive.

The speed of any process is expressed as its **rate,** which is *the change in some measurable quantity per unit of time*. A car's rate of travel, for example, is found by measuring the change in its position, Δx, during a given time interval, Δt. Suppose you are driving on an interstate highway. If you pass mile marker 43 at 2:00 PM and mile marker 173 at 4:00 PM, $\Delta x = (173 - 43)$ mi = 130 mi and $\Delta t =$ 2.00 h. You are traveling at an average rate of $\Delta x/\Delta t = 65$ mi/h. For a chemical process, the **reaction rate** is defined as *the change in concentration of a reactant or product per unit time*. (Time can be measured in seconds, hours, days, or whatever unit is most convenient for the speed of the reaction.) As an example of measurements made in chemical kinetics, consider Figure 13.2, which shows a colored dye reacting to form a colorless product. The color disappears over time, and the intensity of color can be used to determine the concentration of the dye.

As a practical example, consider what happens to the cancer chemotherapy agent cisplatin, $Pt(NH_3)_2Cl_2$, in the presence of water.

Recall that the Greek letter Δ (delta) means that a change in some quantity has been measured ($\Leftarrow$ **p. 217**). As usual, Δ means to subtract the initial value of the quantity from the final value.

Change in concentration is used (rather than change in total amount of reactant) because this makes the rate independent of the volume of the reaction mixture.

 CD-ROM Screen 15.2: Rates of Chemical Reactions

$$H_2O(\ell) + Cl\!-\!\overset{\overset{\displaystyle NH_3}{|}}{\underset{\underset{\displaystyle Cl}{|}}{Pt}}\!-\!NH_3(aq) \longrightarrow H_2O\!-\!\overset{\overset{\displaystyle NH_3}{|}}{\underset{\underset{\displaystyle Cl}{|}}{Pt}}\!-\!NH_3{}^+(aq) + Cl^-(aq) \qquad [13.1]$$

(a)   (b)   (c)

Figure 13.2 Disappearance of a dye. Blue food dye in aqueous solution is reacting with bleach, which converts it into a colorless product. Over time, the intensity of color of the solution decreases and eventually the color disappears. The rate of the reaction could be determined by simultaneously measuring both the intensity of color and the time and then repeating the measurements many times during the course of the reaction. From the intensity of color the concentration of dye could be calculated, and so concentration could be determined as a function of time.

TABLE 13.1	Concentration-Time Data for Reaction of Cisplatin with Water at 25 °C	
Time, t (min)	Concentration [cisplatin] (mol/L)	Average rate (mol L^{-1} min^{-1})
0.0	0.01000	12.7×10^{-6}
200.0	0.00747	9.46×10^{-6}
400.0	0.00558	7.06×10^{-6}
600.0	0.00416	5.27×10^{-6}
800.0	0.00311	3.94×10^{-6}
1000.0	0.00232	2.94×10^{-6}
1200.0	0.00173	1.92×10^{-6}
1600.0	0.00097	1.08×10^{-6}
2000.0	0.00054	

It is obviously important for us to know something about the speed with which Cl^- is replaced by water in cisplatin. If the reaction is over in a few minutes, the drug will change into a different substance as soon as it is placed in the aqueous environment of the human body. The drug's perceived activity may well be due to the newly formed substance. In other cases, a drug (and its activity) can be destroyed by a chemical reaction that changes it into something else before it reaches the site where it carries out its function.

One of the Cl^- ions bound to the central Pt^{2+} ion is displaced by a water molecule. The rate at which this occurs is the change in concentration of cisplatin, Δ[cisplatin], divided by the elapsed time, Δt. For example, if the concentration of cisplatin is measured at some time t_1 to give [cisplatin]$_1$, and the measurement is repeated at time t_2 to give [cisplatin]$_2$, then the rate of reaction is

$$\text{Rate of change of [cisplatin]} = \frac{\text{change in concentration of cisplatin}}{\text{elapsed time}}$$

$$= \frac{\Delta[\text{cisplatin}]}{\Delta t} = \frac{[\text{cisplatin}]_2 - [\text{cisplatin}]_1}{t_2 - t_1}$$

The experimentally measured concentration of cisplatin as a function of time is shown in Table 13.1. Because the concentration of cisplatin decreases as time increases, [cisplatin]$_2$ is smaller than [cisplatin]$_1$. Therefore Δ[cisplatin]/Δt is negative. It is conventional to define reaction rate as a positive quantity. Therefore, for the cisplatin reaction the rate is

$$\text{Reaction rate} = -\frac{\Delta[\text{cisplatin}]}{\Delta t}$$

Table 13.1 also shows calculated values of reaction rates for each time interval. Because calculating the rate involves dividing a concentration difference by a time difference, the units of reaction rate are units of concentration divided by units of time, in this case mol/L divided by min, that is, mol L^{-1} min^{-1}.

Problem-Solving Example 13.1   Calculating Average Rates

Using the data in the first two columns of Table 13.1, calculate Δ[cisplatin], Δt, and the average rate of reaction for each time interval given. Use the numbers given in the third column to check your results. (A good way to do this is to use a computer spreadsheet program.)

Answer   See the third column in Table 13.1.

Explanation   The time interval from 1600 min to 2000 min provides an example of the calculation.

$$\Delta[\text{cisplatin}] = 0.00054 \text{ mol/L} - 0.00097 \text{ mol/L} = -0.00043 \text{ mol/L}$$

$$\Delta t = 2000 \text{ min} - 1600 \text{ min} = 400 \text{ min}$$

$$\text{Rate} = -\frac{\Delta[\text{cisplatin}]}{\Delta t} = -\frac{(-0.00043 \text{ mol/L})}{400 \text{ min}} = 1.08 \times 10^{-6} \text{ mol } L^{-1} \text{ min}^{-1}$$

Do the other calculations in a similar way.

✔ The rates of reaction should be positive numbers and should have units of concentration divided by time (such as mol L^{-1} min^{-1}). Both of these conditions are met.

Problem-Solving Practice 13.1

For the reaction of cisplatin with water, the rate of reaction is 4.54×10^{-6} mol L^{-1} min^{-1} when the concentration of cisplatin is 0.00311 mol/L.

(a) Estimate how long it will take for the concentration of cisplatin to drop from 0.00325 mol/L to 0.00279 mol/L.
(b) Could you use the same method to make an accurate estimate of how long it would take for the concentration of cisplatin to drop from 0.00325 mol/L to 0.00032 mol/L? Explain why or why not.

Exercise 13.1 Rates of Reaction

(a) From data in Table 13.1, calculate the rate of reaction for each time interval: (i) from 800 to 1200 min, (ii) from 400 to 1600 min, (iii) from 0 to 2000 min.
(b) Use all of the data in the first two columns of the table to draw a graph with time on the horizontal (x) axis and concentration on the vertical (y) axis. Draw a smooth curve through the data. Draw on the graph lines that correspond to $\Delta[\text{cisplatin}]/\Delta t$ for each interval.
(c) Why is the rate not the same for each time interval in part (a), even though the average time for each interval is 1000 min? Explain the reason to a friend who is taking this course.

Reaction Rates and Stoichiometry

From the stoichiometry of the cisplatin reaction (Reaction 13.1) it is clear that for every mole of cisplatin that reacts, a mole of chloride ion is formed, because the coefficients of $Pt(NH_3)_2Cl_2(aq)$ and $Cl^-(aq)$ are the same. This means that the rate of appearance of $Cl^-(aq)$ equals the rate of disappearance of $Pt(NH_3)_2Cl_2(aq)$. However, in many reactions the coefficients are not all the same. For example, in the reaction

$$2\,N_2O_5(g) \longrightarrow 4\,NO_2(g) + O_2(g) \qquad [13.2]$$

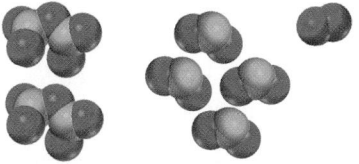

for every mole of O_2 formed, two moles of N_2O_5 react. This means that the rate of disappearance of N_2O_5 is twice as great as the rate of appearance of O_2. If we multiply $-\dfrac{\Delta[N_2O_5]}{\Delta t}$ by $\frac{1}{2}$ in the definition of the rate, then the rate is the same whether expressed in terms of $[O_2]$ or $[N_2O_5]$. Notice that $\frac{1}{2}$ is the reciprocal of the stoichiometric coefficient of N_2O_5 in Equation 13.2. For the general reaction equation,

$$a\,A + b\,B \longrightarrow c\,C + d\,D$$

where A, B, C, and D represent formulas of substances and a, b, c, and d are coefficients, the reaction rate can be defined uniformly in terms of each substance as

$$\text{Rate} = -\frac{1}{a}\frac{\Delta[A]}{\Delta t} = -\frac{1}{b}\frac{\Delta[B]}{\Delta t} = \frac{1}{c}\frac{\Delta[C]}{\Delta t} = \frac{1}{d}\frac{\Delta[D]}{\Delta t} \qquad [13.3]$$

That is, *the rate of change in concentration of any of the reactants or products is multiplied by the reciprocal of the stoichiometric coefficient to find the rate of reaction.* Because the concentrations of reactants decrease with time, their rates of change are given a negative sign.

Problem-Solving Example 13.2   Rates and Stoichiometry

Equation 13.2 shows the decomposition of $N_2O_5(g)$.

(a) Define the rate of reaction in terms of the rate of change in concentration of each reactant and product.
(b) If the rate of appearance of $O_2(g)$ is 0.023 mol L^{-1} s^{-1}, what is the rate of appearance of $NO_2(g)$?

Answer

(a) $\text{Rate} = -\dfrac{1}{2}\dfrac{\Delta[N_2O_5]}{\Delta t} = \dfrac{1}{4}\dfrac{\Delta[NO_2]}{\Delta t} = \dfrac{\Delta[O_2]}{\Delta t}$

(b) 0.092 mol L^{-1} s^{-1}

Explanation

(a) Define the rate of reaction as change in concentration divided by time elapsed (Δt). Multiply the rate for each substance by the reciprocal of the stoichiometric coefficient, and use a negative sign in front of the rate for each reactant.

(b) From the coefficients, 4 mol of NO_2 forms for every 1 mol of O_2. Therefore, the rate of formation of NO_2 is four times the rate of formation of O_2, which equals the rate of reaction. Algebraically,

$$\frac{1}{4}\frac{\Delta[NO_2]}{\Delta t} = \frac{1}{1}\frac{\Delta[O_2]}{\Delta t}$$

and

$$\frac{\Delta[NO_2]}{\Delta t} = 4 \times \frac{\Delta[O_2]}{\Delta t} = 4 \times 0.023 \text{ mol } L^{-1}s^{-1} = 0.092 \text{ mol } L^{-1}s^{-1}$$

Problem-Solving Practice 13.2

For the reaction

$$H_2(g) + I_2(g) \longrightarrow 2\,HI(g)$$

(a) Express the rate of formation of HI in terms of the rate of disappearance of H_2.
(b) If the rate of disappearance of I_2 is 0.0037 mol L^{-1} s^{-1}, what is the rate of formation of HI?

Average Rate and Instantaneous Rate

A reaction rate calculated from a change in concentration divided by a change in time is called the **average reaction rate** over the time interval from which it was calculated. For example, the average reaction rate at 25 °C for the cisplatin reaction over the interval from 0 to 200 min is 12.7×10^{-6} mol L^{-1} min^{-1}. The data in Table 13.1 indicate that as the concentration of cisplatin decreases, the average rate also decreases. Most reactions are like this: The average rate gets smaller as the concentration of one or more reactants decreases. Your results in Exercise 13.1 should have been different for each range of time over which you calculated. Because the average reaction rate changes over time, the rate you calculate depends on when, and for what range of time, you calculate. If you want to know the rate that corresponds to a particular concentration of cisplatin (and therefore to a particular time after the reaction began), the average rate is not appropriate, because it depends on the size of the time interval.

The **instantaneous reaction rate** is *the rate at a particular time after a reaction has begun.* To obtain it, the rate must be calculated over a very small interval around the time or concentration for which the rate is desired. For example, to

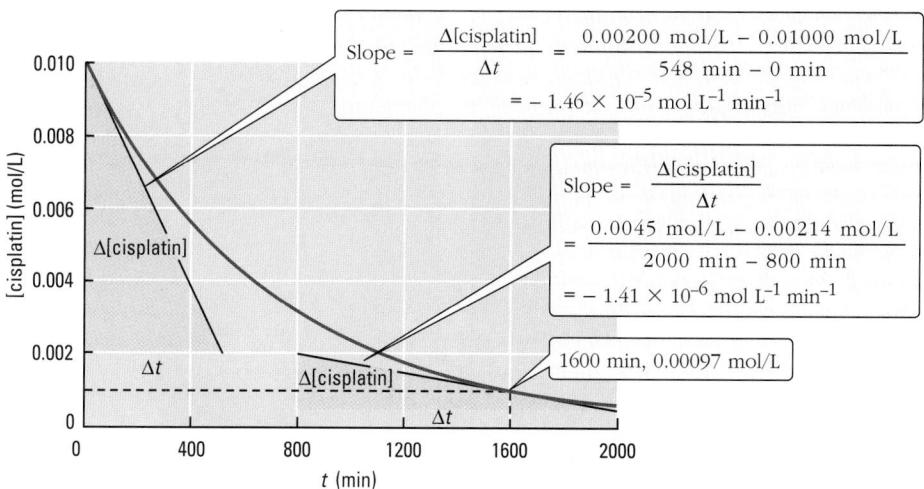

Figure 13.3 **Instantaneous reaction rates.** The experimentally measured concentration of $Pt(NH_3)_2Cl_2$ is plotted as a function of time during the reaction in which Cl^- is replaced by H_2O in aqueous solution. The slopes at time zero and at 1600 min are indicated on the graph. From these slopes the instantaneous rates 1.46×10^{-5} mol L^{-1} min^{-1} and 1.41×10^{-6} mol L^{-1} min^{-1} can be obtained.

calculate the rate at which cisplatin is disappearing when its concentration is 0.00097 mol/L, you would need to calculate $\Delta[\text{cisplatin}]/\Delta t$ at exactly 1600 min from the start of the reaction. A good way to do this is shown in Figure 13.3. *The instantaneous rate is the slope of a line tangent to the concentration-time curve at the point corresponding to the specified concentration and time.* For a particular concentration of the same reactant at the same temperature and the same concentrations of other species, the instantaneous rate will not vary as the average rate does.

If you are familiar with calculus, then you may recognize that in the limit of very small time intervals, $\Delta[A]/\Delta t$, where A represents a substance, becomes the same as the derivative of concentration with respect to time. That is,

$$\lim_{\Delta t \to 0} \frac{\Delta[A]}{\Delta t} = \frac{d[A]}{dt}$$

This also means that the rate of reaction at any time can be found from the *slope* (at that time) of the tangent to a curve of concentration versus time, such as the curve in Figure 13.3. How to determine the slope and intercept of a graph is discussed in Appendix A.8.

Exercise 13.2   Instantaneous Rates

Instantaneous rates for the reaction of water with cisplatin can be determined from the slope of the curve in Figure 13.3 at various concentrations. They are

(1) At 0.0080 mol/L, rate = 11.8×10^{-6} mol L^{-1} min^{-1}
(2) At 0.0060 mol/L, rate = 8.85×10^{-6} mol L^{-1} min^{-1}
(3) At 0.0040 mol/L, rate = 5.90×10^{-6} mol L^{-1} min^{-1}
(4) At 0.0030 mol/L, rate = 4.42×10^{-6} mol L^{-1} min^{-1}
(5) At 0.0020 mol/L, rate = 2.95×10^{-6} mol L^{-1} min^{-1}
(a) What is the relationship between the rates in (1) and (3)? between (2) and (4)? between (3) and (5)?
(b) What is the relationship between the concentrations in each of these cases?
(c) Is the rate of the reaction proportional to the concentration of cisplatin?

Exercise 13.3   Graphing Concentrations Versus Time

Consider the decomposition of $N_2O_5(g)$, Reaction 13.2. Assume that the initial concentration of $N_2O_5(g)$ is 0.02 mol/L and that none of the products are present. Make a graph that shows concentrations of $N_2O_5(g)$, $NO_2(g)$, and $O_2(g)$ as a function of time, all on the same set of axes and roughly to scale.

$$2\ N_2O_5(g) \longrightarrow 4\ NO_2(g) + O_2(g) \qquad [13.2]$$

13.2   EFFECT OF CONCENTRATION ON REACTION RATE

The rates of most reactions change when reactant concentrations change, just as we found for the cisplatin reaction. Figure 13.4 shows another example. The oxidation of hydrogen peroxide by potassium permanganate in aqueous solution

$$2\ MnO_4^-(aq) + 5\ H_2O_2(aq) + 6\ H_3O^+(aq) \longrightarrow 2\ Mn^{2+}(aq) + 5\ O_2(g) + 14\ H_2O(\ell)$$

Figure 13.4 Reaction of aqueous potassium permanganate with aqueous hydrogen peroxide. The rate of the reaction of potassium permanganate with hydrogen peroxide depends on the concentration of the permanganate. With dilute $KMnO_4$ the reaction is slow (a), but it is more rapid in more concentrated $KMnO_4$ (b). (In both cases the temperature is the same, so the difference in rate must be due to concentration of permanganate.)

(a)

(b)

is visibly more rapid when the concentration of permanganate is higher. One goal of chemical kinetics is to find out whether a reaction speeds up when the concentration of a reactant is increased and, if so, by how much.

The Rate Law

CD-ROM Screen 15.4:
Control of Reaction Rates:
Concentration Dependence

How the concentration of a reactant affects the rate can be determined by performing several experiments in which the concentration of that reactant is varied systematically (and temperature is held constant). Or a single experiment can be done in which concentration is determined continuously as a function of time. The latter approach gave the data for cisplatin shown in Table 13.1 and Figure 13.3, which you analyzed in Exercise 13.2. You should have discovered that if the concentration of cisplatin is halved, the reaction rate is also halved. If the concentration of cisplatin is doubled, then the reaction rate is doubled. This leads to the expression

The symbol ∝ means "proportional to."

$$\text{Rate} \propto [\text{cisplatin}]$$

This says that the rate is directly proportional to the concentration of one of the reactants, cisplatin.

This proportionality can be changed to a mathematical equation by including a proportionality constant, k. *A mathematical equation that summarizes the relationship between reactant concentration and reaction rate* is called a **rate law** (or *rate equation*). For the cisplatin reaction the rate law is

$$\text{Rate} = k \times [\text{cisplatin}]$$

The proportionality constant, k, is called the **rate constant.** The rate constant is independent of concentration, but it usually becomes larger the higher the temperature. The rate constant applies only to the specific reaction being studied, not to any other, and it applies at a specific temperature. Thus the chemical equation and the temperature for the reaction should be given along with the rate constant. In this case we write

$$H_2O(\ell) + \underset{\underset{Cl}{|}}{\overset{\overset{NH_3}{|}}{Cl-Pt-NH_3}} (aq) \longrightarrow \underset{\underset{Cl}{|}}{\overset{\overset{NH_3}{|}}{H_2O-Pt-NH_3^+}} (aq) + Cl^- (aq)$$

$$k = 1.46 \times 10^{-3} \text{ min}^{-1} \text{ (at 25 °C)}$$

Determining Rate Laws from Initial Rates

CD-ROM Screen 15.5:
Determination of Rate
Equations: Method of
Initial Rates

The relation between rate and concentration (the rate law) must be determined experimentally. One way to do this was illustrated in Exercise 13.3, but it is hard to determine rates from tangents to a curve such as that in Figure 13.3. Another way is to measure initial rates. The **initial rate** of a reaction is *the instantaneous rate determined at the very beginning of the reaction.* A good approximation to the initial rate is to calculate $-\Delta[\text{reactant}]/\Delta t$ after no more than 2% of the limiting reactant has been consumed.

An advantage of measuring initial rates is that as a reaction proceeds, more and more products are formed. In some cases products can alter the rate; comparing initial rates with rates when products are present can reveal such a complication.

Many reactions can be started by mixing two solutions or two different gas samples. Usually the concentrations of the reactants are known before they are mixed, and so the initial rate corresponds to this particular set of reactant concentrations. Several experiments can then be done in which initial concentrations are varied, and the change in the reaction rate can be correlated with changes in the concentration of each reactant. As an example, consider the reaction of a base with methyl acetate, an ester ($\Leftarrow$ *p. 538)* that is an industrial solvent. The reaction produces acetate ion and methanol.

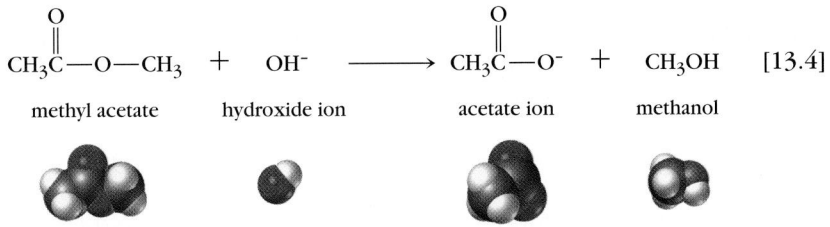

To control for the effect of temperature on rate, several experiments were done at the same temperature:

Experiment	Initial concentration (mol/L)		Initial rate (mol L^{-1} s^{-1})
	[CH$_3$COOCH$_3$]	[OH$^-$]	
1	0.040 M	0.040 M	0.00022
	no change	× 2	× 2
2	0.040 M	0.080 M	0.00045
	× 2	no change	× 2
3	0.080 M	0.080 M	0.00090

To determine the rate law, compare two experiments in which only a single initial concentration changed. In Experiments 1 and 2 the [CH$_3$COOCH$_3$] remained constant and the [OH$^-$] doubled. The rate also doubled, which means that the rate is *directly proportional* to the [OH$^-$]. In Experiments 2 and 3 the [OH$^-$] remained the same, the [CH$_3$COOCH$_3$] doubled, and the rate doubled, indicating that the rate is also proportional to the [CH$_3$COOCH$_3$]. Therefore, the rate is proportional to the *product* of the two concentrations, and the rate law is

$$\text{Rate} = k\,[\text{CH}_3\text{COOCH}_3]\,[\text{OH}^-]$$

This equation also tells us that doubling both concentrations at the same time would cause the rate to go up by a factor of 4, which it does from Experiment 1 to Experiment 3.

Exercise 13.4 Rates and Concentrations

Use the rate law for the reaction of methyl acetate with base to predict the effect on the rate of reaction if the concentration of methyl acetate is doubled and the concentration of hydroxide ions is halved.

Another way to approach this problem involves proportions. As before, choose two experiments in which one concentration did not change. Then calculate the ratio of the other concentrations and the ratio of rates. For the methyl acetate reaction, using Experiments 1 and 2 where the $[CH_3COOCH_3]$ was constant,

$$\frac{[OH^-]_2}{[OH^-]_1} = \frac{0.080\ M}{0.040\ M} = 2.0 \quad \text{and} \quad \frac{rate_2}{rate_1} = \frac{0.00045\ mol\ L^{-1}s^{-1}}{0.00022\ mol\ L^{-1}s^{-1}} = 2.0$$

it is clear that both the concentrations and the rates change in the same proportion. This same method could be applied to analyze results of Experiments 2 and 3, where the $[OH^-]$ was constant.

Once the rate law is known, a value for k, the rate constant, can be found by substituting rate and initial concentration data for any one experiment into the rate law. For example, a value of k for the methyl acetate–hydroxide ion reaction could be obtained from data for the first experiment,

$$0.00022\ mol\ L^{-1}\ s^{-1} = k(0.040\ mol/L)(0.040\ mol/L)$$

$$k = \frac{(0.00022\ mol\ L^{-1}s^{-1})}{(0.040\ mol/L)(0.040\ mol/L)} = 0.14\ L\ mol^{-1}\ s^{-1}$$

A better value for k could be obtained by using all available experimental data, that is, by calculating a k for each experiment and then averaging the k values to obtain an overall result.

Problem-Solving Example 13.3 Rate Law from Initial Rates

Initial rates for the reaction of nitrogen monoxide and oxygen

$$2\ NO(g) + O_2(g) \longrightarrow 2\ NO_2(g)$$

were measured at 25 °C starting with various concentrations of NO and O_2. These data were collected.

Experiment	Initial concentrations (mol/L)		Initial rate (mol L^{-1} s^{-1})
	[NO]	[O$_2$]	
1	0.020	0.010	0.028
2	0.020	0.020	0.057
3	0.020	0.040	0.114
4	0.040	0.020	0.227
5	0.010	0.020	0.014

Notice that for this reaction, where the rate is proportional to one concentration and to the square of another, the rate constant equals the rate (units of mol L^{-1} s^{-1}) divided by three concentration terms multiplied together (units of mol^3 L^{-3}). Thus, the rate constant has units of $\frac{mol\ L^{-1}\ s^{-1}}{mol^3\ L^{-3}} = mol^{-2}\ L^2\ s^{-1}$.

(a) What is the rate law?
(b) What is the value of the rate constant k?

Answer (a) Rate = $k\ [O_2]\ [NO]^2$ (b) $k = 7.1 \times 10^3\ L^2\ mol^{-2}\ s^{-1}$

Explanation

(a) Analyze data from experiments in which one concentration remains the same. In Experiments 1, 2, and 3 the concentration of NO is constant, while the O_2 concentration increases from 0.010 to 0.020 to 0.040 mol/L. Each time the $[O_2]$ is doubled, the initial rate also doubles. For example, when $[O_2]$ is doubled from 0.020 to 0.040 mol/L, in Experiments 2 and 3, the initial rate doubles from 0.057 to 0.114 mol L^{-1} s^{-1}. The initial rate is directly proportional to $[O_2]$.

In Experiments 2, 4, and 5 the $[O_2]$ is constant, while the $[NO]$ varies. From Experiments 2 and 4, the $[NO]$ is doubled, but the initial rate increases by a factor of 4, or 2^2.

$$\frac{\text{Experiment 4 rate}}{\text{Experiment 2 rate}} = \frac{0.227 \text{ mol } L^{-1} s^{-1}}{0.057 \text{ mol } L^{-1} s^{-1}} = \frac{4}{1} = \frac{2^2}{1}$$

This same result is found from Experiments 2 and 5. This means that the initial rate is proportional to the *square* of $[NO]$. Therefore, the rate law is

$$\text{Rate} = k [O_2] [NO]^2$$

(b) Once the rate law is known, the rate constant k can be calculated. For Experiment 1, for example,

$$0.028 \text{ mol } L^{-1} s^{-1} = k(0.010 \text{ mol/L})(0.020 \text{ mol/L})^2$$

$$k = \frac{0.028 \text{ mol } L^{-1}s^{-1}}{(0.010 \text{ mol/L})(0.020 \text{ mol/L})^2} = 7.0 \times 10^3 \text{ } L^2 \text{ mol}^{-2} s^{-1}$$

For Experiments 2, 3, 4, and 5 the rate constants are 7.1×10^3, 7.1×10^3, 7.1×10^3, and 7.0×10^3 L^2 mol^{-2} s^{-1}. These values average to give 7.1×10^3 L^2 mol^{-2} s^{-1}, which can be used to calculate the rate for any set of NO and O_2 concentrations at 25 °C.

✔ The five calculated k values are nearly equal. If the rate law were incorrect, or if an error were made in one or more calculations, some k values would be quite different from the others.

Problem-Solving Practice 13.3

At 25 °C, these data were collected for the cisplatin reaction

$$Pt(NH_3)_2Cl_2 + H_2O \longrightarrow [Pt(NH_3)_2(H_2O)Cl]^+ + Cl^-$$

Experiment	Initial concentrations (mol/L)		Initial rate (mol L^{-1} s^{-1})
	[cisplatin]	[H₂O]	
1	0.021	55.5	3.07×10^{-5}
2	0.040	55.5	5.84×10^{-5}
3	0.079	55.5	1.15×10^{-4}

(a) Is it possible to determine the rate law from the data given? Why or why not?
(b) Assume that the rate does not depend on the concentration of water. What is the rate law?
(c) Calculate the rate constant. Report your results in mol L^{-1} s^{-1}.
(d) Calculate the initial rate of reaction when the concentration of $Pt(NH_3)_2Cl_2$ is 0.100 M.
(e) Calculate the rate when the concentration has dropped to half this value.

In this reaction water is the solvent as well as a reactant. Because it is present in such a large excess, its concentration does not change significantly as the reaction takes place.

13.3 RATE LAW AND ORDER OF REACTION

For many (but not all) homogeneous reactions, the rate law has the general form

$$\text{Rate} = k [A]^m [B]^n \text{ . . .}$$

where concentrations of substances, $[A]$, $[B]$, . . . are raised to powers, m, n, ...

The substances A, B, . . . might be reactants, products, or catalysts. The exponents m, n, . . . are usually positive whole numbers but might be negative numbers or fractions. These exponents define the **order of the reaction** with respect to each reactant. If n is 1, for example, the reaction is first-order with respect to B; if m is 2, then the reaction is second-order with respect to A. The sum of m and n (plus the exponents on any other concentration terms) gives the **overall reaction order**. (The reaction in Problem-Solving Example 13.3 is first-order in O_2, second-order in NO, and third-order overall.) A very important point to remember is that *the rate law and reaction orders must be determined experimentally; they cannot be predicted from stoichiometric coefficients in the balanced overall chemical equation*.

Problem-Solving Example 13.4 Reaction Order and Rate Law

For each reaction and experimentally determined rate law listed below, determine the order with respect to each reactant and the overall order.

(a) $2\,H_2O_2(aq) \rightarrow 2\,H_2O(\ell) + O_2(g)$ Rate $= k\,[H_2O_2]$

(b) $14\,H_3O^+(aq) + 2\,HCrO_4^-(aq) + 6\,I^-(aq) \rightarrow 2\,Cr^{3+}(aq) + 3\,I_2(aq) + 22\,H_2O(\ell)$

$$\text{Rate} = k\,[HCrO_4^-]\,[I^-]^2\,[H_3O^+]^2$$

(c) *cis*-2-butene(g) $\rightarrow$ *trans*-2-butene(g) (with a catalytic concentration of I_2 present)

$$\text{Rate} = k\,[\textit{cis}\text{-2-butene}]\,[I_2]^{1/2}$$

Answer

(a) First-order in H_2O_2, first-order overall

(b) First-order in $HCrO_4^-$, second-order in I^-, second-order in H_3O^+, fifth-order overall

(c) First-order in *cis*-2-butene, $\frac{1}{2}$-order in I_2, 1.5-order overall.

Explanation

(a) The rate law has a single term that is raised to the first power, and so the reaction is first-order in H_2O_2, first-order overall.

(b) The rate law contains three terms. Since the $HCrO_4^-$ term is raised to the first power, the reaction is first-order in $HCrO_4^-$. The other two terms are squared, and so the reaction is second-order in I^-, and second-order in H_3O^+. The exponents sum to five, so the reaction is fifth-order overall.

(c) In this case the rate of reaction depends on the concentration of the reactant and also on the square root ($\frac{1}{2}$ power) of the concentration of a catalyst, I_2. The reaction is therefore first-order in *cis*-2-butene, $\frac{1}{2}$-order in I_2, and 1.5-order overall.

Problem-Solving Practice 13.4

The rate law for the reduction of NO to N_2 with hydrogen is

$$2\,NO(g) + 2\,H_2(g) \rightarrow N_2(g) + 2\,H_2O(g) \qquad \text{Rate} = k\,[NO]^2\,[H_2]$$

(a) What is the order of the reaction with respect to the NO? With respect to H_2?

(b) Suppose that you triple the concentration of NO and simultaneously decrease the concentration of H_2 by a factor of 8. Will the reaction be faster or slower under the new conditions? How much faster or slower? (Assume that the temperature is the same in both sets of conditions.)

For none of the reactions in this Problem-Solving Example could the rate law be derived correctly from the stoichiometric equation. For example, H_2O_2 has a coefficient of 2 in Reaction (a), but the rate law involves $[H_2O_2]$ to the first power.

The Integrated Rate Law

Another approach to experimental determination of the rate law and rate constant for a reaction uses calculus to derive what is called the integrated rate law. As an example of the integrated rate law method, suppose that we have a reaction in which a single substance A reacts to form products.

$$A \longrightarrow \text{products}$$

First-Order Reaction. If the rate law is first-order, then

$$\text{Rate} = -\frac{\Delta[A]}{\Delta t} = k[A]$$

This can be transformed, using calculus, to what is called the integrated first-order rate law,

$$\ln[A]_t = -kt + \ln[A]_0 \qquad [13.5]$$

where ln represents the natural logarithm function. (Logarithms are discussed in Appendix A.6.)

Equation 13.5 has the same form as the general equation for a straight line, $y = mx + b$, in which m is the slope and b is the y-intercept.

$$\underbrace{\ln[A]_t}_{\substack{y\text{-axis}\\\text{variable}}} = \underbrace{-k}_{\text{slope}}\underbrace{t}_{\substack{x\text{-axis}\\\text{variable}}} + \underbrace{\ln[A]_0}_{y\text{-intercept}} \qquad [13.5]$$

$$y \quad = \quad mx + \quad b$$

If the reaction is actually first-order, then a graph of ln[A] on the vertical (y) axis versus t on the horizontal (x) axis should be a straight line. A linear graph, such as the one in Figure 13.5a is evidence that the reaction is first-order.

Second-Order Reaction. For the same reaction

$$A \longrightarrow \text{products}$$

suppose that the rate depends on the square of the concentration of the reactant; that is, suppose the rate law is second-order.

$$\text{Rate} = k[A]^2$$

The integrated rate law derived using calculus is

$$\frac{1}{[A]_t} = kt + \frac{1}{[A]_0}$$

This equation is also of the form $y = mx + b$. If a reaction is second-order, a graph of $1/[A]_t$ versus t will be linear with slope $= k$ and y-intercept $= 1/[A]_0$. Such a straight-line graph is evidence that a reaction is second-order. An example graph is shown in Figure 13.5b.

Zeroth-Order Reaction. There are a few reactions for which the rate does not depend on the concentration of a reactant at all. These are called zeroth-order reactions, because the rate law can be written as a rate constant k times a concentration to the zeroth power. (Remember that anything raised to the zeroth power equals 1.)

$$\text{Rate} = -\frac{\Delta[A]}{\Delta t} = k[A]^0 = k$$

This rate law says that the rate is always the same no matter what the concentration of reactant. The rate is equal to the rate constant k. For a zeroth-order reaction you can derive the integrated form without calculus. Simply use algebra to rearrange the rate law just given.

$$\Delta[A] = -k\Delta t$$

$$[A]_t - [A]_0 = -k(t_t - t_0) = -kt$$

$$[A]_t = -kt + [A]_0$$

It is not necessary that you know calculus to use the results that constitute the integrated rate law method. If you do know calculus, however, you will be able to derive the results for yourself.

Using calculus,

$$-\frac{d[A]}{dt} = k[A] \quad \text{and} \quad \frac{d[A]}{[A]} = -kdt$$

$$\int_0^t \frac{d[A]}{[A]} = -k\int_0^t dt$$

$$\ln[A]_t - \ln[A]_0 = -k(t_t - t_0)$$

If the reaction starts at time t_0, then $t_t - t_0 = t$, the elapsed time, and

$$\ln[A]_t - \ln[A]_0 = -kt \quad \text{or}$$
$$\ln[A]_t = -kt + \ln[A]_0.$$

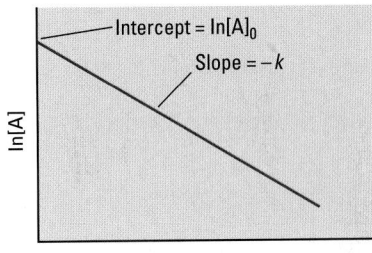

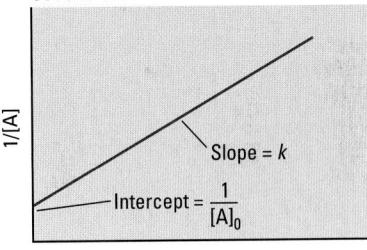

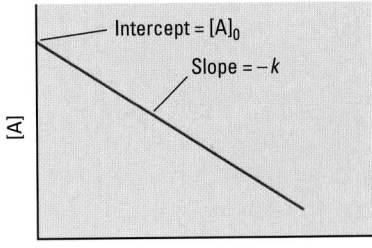

Figure 13.5 First-order, second-order, and zeroth-order plots. (a) If a reaction is first-order in reactant A, plotting ln[A]$_t$ versus t gives a straight line. (b) If a reaction is second-order in reactant A, plotting $1/[A]_t$ versus t gives a straight line. (c) If a reaction is zeroth-order in reactant A, plotting [A]$_t$ versus t gives a straight line.

TABLE 13.2 Integrated Rate Laws

Order	Rate equals	Integrated rate law*	Straight-line plot	Slope of plot	Units of k
0	$k[A]^0 = k$	$[A]_t = -kt + [A]_0$	$[A]_t$ vs t	$-k$	conc time^{-1}
1	$k[A]$	$\ln[A]_t = -kt + \ln[A]_0$	$\ln[A]_t$ vs t	$-k$	time^{-1}
2	$k[A]^2$	$\dfrac{1}{[A]_t} = kt + \dfrac{1}{[A]_0}$	$\dfrac{1}{[A]_t}$ vs t	k	conc^{-1} time^{-1}

*In the table, $[A]_0$ indicates the initial concentration of substance A, that is, the concentration of A at $t = 0$, the time when the reaction was started.

Again, the equation is of the form $y = mx + b$. If a reaction is zeroth order, graphing $[A]_t$ versus t gives a straight line with slope $= -k$ and y-intercept $= [A]_0$, as shown in Figure 13.5c.

To summarize these three situations, a rate law that involves powers of the reactant concentration can be written as

$$\text{Rate} = -\frac{\Delta[A]}{\Delta t} = k[A]^m$$

where m is the order of the reaction as we defined it before. The integrated rate law depends on the value of m, and the results for values of m from 0 up to 2 are given in Table 13.2.

CD-ROM Screen 15.7: Determination of Rate Equations: Graphic Methods

To determine the order of reaction, then, we collect concentration-time data and make the three plots listed in Table 13.2 and shown in Figure 13.5. Only one (or perhaps none) of the plots will be a straight line. If one is straight, then it indicates the order, and the rate constant can be calculated from its slope. The units of the rate constant are those of the slope of the line. They depend on the order of the reaction as indicated in the last column of Table 13.2.

Problem-Solving Example 13.5 Order and Rate Constant from Integrated Rate Law

These data were obtained for decomposition of aqueous hydrogen peroxide at 20 °C.

$$2\,H_2O_2(aq) \longrightarrow 2\,H_2O(\ell) + O_2(g)$$

Time (min)	[H₂O₂] (mol/L)	Time (min)	[H₂O₂] (mol/L)
0	0.0200	1000	0.0069
200	0.0160	1200	0.0054
400	0.0131	1600	0.0037
600	0.0106	2000	0.0024
800	0.0086		

Use the integrated rate law method to obtain the order of the reaction and the rate constant.

Answer The reaction is first-order in H_2O_2; $k = 1.06 \times 10^{-3}\,\text{min}^{-1}$.

Explanation The graphs for zeroth-, first-, and second-order plots are shown in Figure 13.6. The zeroth-order and second-order plots are curved, while the first-order plot is a straight line. Thus the reaction must be first-order. From the first-order plot a slope can be calculated using the points marked on the graph as open circles.

$$\text{Slope} = \frac{\{-5.82 - (-4.02)\}}{\{1800 - 100\}\,\text{min}} = -1.06 \times 10^{-3}\,\text{min}^{-1}$$

It is important to use two points on the straight line through the experimental data (circles in Figure 13.6b), not two of the data points themselves, when you calculate the slope. Making a graph is similar to averaging because the straight line and its slope are based on all nine points in the data table, not just two of them.

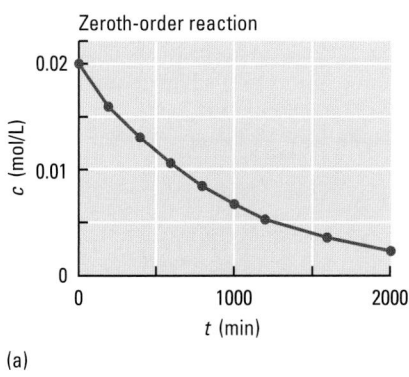

Zeroth-order reaction

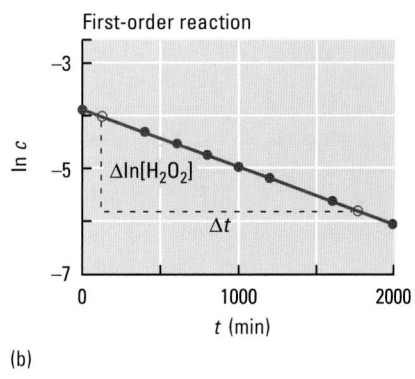

First-order reaction

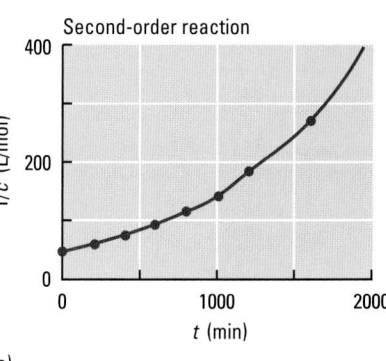

Second-order reaction

(a) (b) (c)

The slope of -1.06×10^{-3} min^{-1} is the negative of the rate constant (see Table 13.2), which means that $k = 1.06 \times 10^{-3}$ min^{-1}.

✔ The units are min^{-1} (per minute), which corresponds to the reciprocal time units indicated in Table 13.2 for a first-order rate constant.

Figure 13.6 **Integrated rate law plots for H$_2$O$_2$ reaction.** (a) Zeroth-order, (b) first-order, and (c) second-order plots for the decomposition reaction of hydrogen peroxide in aqueous solution at 20 °C.

Problem-Solving Practice 13.5

Use the concentration-time data in Table 13.1 for the reaction of cisplatin with water (← *p. 574*) to determine the order of the reaction with respect to cisplatin and the rate constant. (Assume that [H$_2$O] is constant during the reaction.)

Calculating Concentration or Time from Rate Law

Once the rate law has been determined experimentally, it provides a way to calculate the concentration of a reactant or product at any time after the reaction has begun. All that is needed is the integrated rate law (from Table 13.2), the value of the rate constant, and the initial concentration of reactant or product. These are related by the equations given in Table 13.2.

Problem-Solving Example 13.6 **Calculating Concentrations**

The first-order rate constant is 1.87×10^{-3} min^{-1} at 37 °C (body temperature) for reaction of cisplatin with water (Equation 13.1). Suppose that the concentration of cisplatin in the bloodstream of a cancer patient is 4.73×10^{-4} mol/L. What will the concentration be 24 hours later?

Answer 3.20×10^{-5} mol/L

Explanation Assume that the rate of reaction in the blood is the same as in water. Let [cisplatin] represent the concentration of cisplatin at any time and [cisplatin]$_0$ the initial concentration. The reaction is first-order, so from Table 13.2,

$$\ln[\text{cisplatin}] = -kt + \ln[\text{cisplatin}]_0$$

Rearrange the equation algebraically to

$$\ln[\text{cisplatin}] - \ln[\text{cisplatin}]_0 = -kt$$

$$\ln\left\{\frac{[\text{cisplatin}]}{[\text{cisplatin}]_0}\right\} = -kt$$

$$\frac{[\text{cisplatin}]}{[\text{cisplatin}]_0} = \text{anti } \ln(-kt) = e^{-kt}$$

$$[\text{cisplatin}] = [\text{cisplatin}]_0 e^{-kt}$$

$$= (4.73 \times 10^{-4} \text{ mol/L})e^{-(1.87 \times 10^{-3} \text{ min}^{-1} \times 24 \text{ h})}$$

$$= (4.73 \times 10^{-4} \text{ mol/L})e^{-\left(\frac{1.87 \times 10^{-3}}{\text{min}} \times 24 \text{ h} \times \frac{60 \text{ min}}{1 \text{ h}}\right)}$$

$$= (4.73 \times 10^{-4} \text{ mol/L})(6.77 \times 10^{-2}) = 3.20 \times 10^{-5} \text{ mol/L}$$

Mathematical operations involving logarithms and exponentials (antilogarithms) are discussed in Appendix A.6.

Because the solution to this example involves a ratio of concentrations in which the units divide out, the same approach can be taken in problems that involve the number of moles, the mass, or the number of atoms or molecules at two different times.

✔ Based on Figure 13.3, at 25 °C the concentration of cisplatin drops to about $\frac{1}{10}$ of its initial value in 24 h (1440 min). In this calculation the temperature is higher and the reaction should be faster, so the concentration after 24 h should be less than $\frac{1}{10}$ the initial value, and it is.

Problem-Solving Practice 13.6

The first-order rate constant for decomposition of an insecticide in the environment is 3.43×10^{-2} d^{-1}. How long does it take for the concentration of insecticide to drop to $\frac{1}{10}$ of its initial value?

Half-Life

CD-ROM Screen 15.8: Half-Life: First-Order Reactions

The **half-life** of a reaction, $t_{1/2}$, is *the time required for the concentration of a reactant A to fall to one half of its initial value.* That is, $[A]_{t_{1/2}} = \frac{1}{2}[A]_0$. For a first-order reaction the half-life is the same, no matter what the initial concentration is, but for other reaction orders this is not true, and the half-life concept is less useful.

The half-life is related to the first-order rate constant. To see how, rearrange Equation 13.5 algebraically:

$$\ln[A]_t = -kt + \ln[A]_0 \qquad [13.5]$$

$$\ln[A]_t - \ln[A]_0 = -kt$$

$$\ln[A]_{t_{1/2}} - \ln[A]_0 = -kt_{1/2} \qquad [13.6]$$

Because $[A]_{t_{1/2}} = \frac{1}{2}[A]_0$, Equation 13.5 can be rewritten as

$$\ln([A]_0/2) - \ln[A]_0 = -kt_{1/2}$$

and so

$$-kt_{1/2} = \ln[A]_0 - \ln(2) - \ln[A]_0 = -\ln 2$$

$$t_{1/2} = \frac{-\ln 2}{-k} = \frac{0.693}{k} \qquad [13.7]$$

This means that measuring the half-life of a first-order reaction determines the rate constant, and vice versa. Radioactive decay (Section 20.4) is a first-order process, and half-life is typically used to report the rates of decay of radioactive nuclei.

Problem-Solving Example 13.7 Half-Life and Rate Constant

In Problem-Solving Example 13.5, the rate constant for decomposition of aqueous hydrogen peroxide at 20 °C was found to be 1.06×10^{-3} min^{-1}. The reaction gave a linear first-order plot. What is the half-life in seconds for this reaction?

Answer 3.92×10^4 s

Explanation The reaction is first-order, so Equation 13.7 can be used.

$$t_{1/2} = \frac{0.693}{k} = \frac{0.693}{1.06 \times 10^{-3} \text{min}^{-1}} = 6.538 \times 10^2 \text{min}$$

Because the result is requested in seconds, a unit conversion must be made.

$$6.538 \times 10^2 \text{min} = 6.538 \times 10^2 \text{min} \times \frac{60 \text{ s}}{1 \text{ min}} = 3.92 \times 10^4 \text{s}$$

✔ The rate constant is about 0.001 min^{-1}. If the concentration of hydrogen peroxide were 1.0 mol L^{-1}, then the reaction rate would be 0.001 min^{-1} × 1.0 mol L^{-1} = 0.001 mol L^{-1} min^{-1} = 1 mmol L^{-1} min^{-1}. This means that 1 mmol of hydrogen peroxide would react every minute. For the concentration to drop to half of 1.0 mol L^{-1}, the change in concentration would be 0.500 mol L^{-1}, or 500 mmol L^{-1}. Therefore, it would take at least 500 min for the concentration to drop to half its initial value, and the half-life should be at least 500 min. (Because the rate decreases as concentration decreases, it will take longer than 500 min.) A half-life of 654 min is reasonable.

Problem-Solving Practice 13.7

From Figure 13.3 determine the time required for the concentration of cisplatin to fall to one-half the initial value. Verify that the same period is required for the concentration to fall from one-half to one-fourth the initial value. From this half-life, calculate the rate constant.

In Exercise 13.2 your results should have shown that the time required for the concentration of cisplatin to fall from 0.00800 mol/L to 0.00400 mol/L was the same as the time required to fall from 0.00400 mol/L to 0.00200 mol/L.

13.4 A NANOSCALE VIEW: ELEMENTARY REACTIONS

Macroscale experimental observations reveal that reactant concentrations, temperature, and catalysts can affect reaction rates. But how can we interpret such observations in terms of nanoscale models? We will use the *kinetic-molecular theory of matter,* which was first introduced in Section 1.7 (⇐ *p. 19)* and developed further in Section 10.3 (⇐ *p. 414)*, together with the ideas about molecular structure developed in Chapters 8 and 9. These concepts provide a good basis for understanding how atoms and molecules move and chemical bonds are made or broken during the very short time it takes for reactant molecules to be converted into product molecules.

According to kinetic-molecular theory, molecules are in constant motion. In a gas or liquid they bump into one another, while in a solid they vibrate about specific locations. Molecules also rotate, flex, or vibrate around or along the bonds that hold the atoms together. These motions produce the transformations of molecules that occur during chemical reactions. It turns out that **there are only two important types of molecular transformations**. They are described as unimolecular or bimolecular. In a **unimolecular reaction** *the structure of a single particle (atom, molecule, or ion) rearranges to produce a different particle or particles.* A unimolecular reaction might involve breaking a bond and forming two new molecules, or it might involve rearrangement of one isomeric structure into another. In a **bimolecular reaction** *two particles (atoms, molecules, or ions) collide and rearrange into products.* In a bimolecular reaction new bonds may be formed between the reactant particles, and existing bonds may be broken. Sometimes the two particles combine to form a new, larger one. Sometimes two new molecules are formed from the original two.

 CD-ROM Screen 15.12: Reaction Mechanisms

ESTIMATION   Pesticide Decay

There are usually several different ways that a pesticide can decompose in an ecosystem. It is difficult to define an accurate rate of decomposition and even more difficult to define a rate law. Often it is assumed that decomposition is first-order, and a rough half-life is reported. Organochlorine pesticides such as DDT, lindane, and dieldrin may have half-lives as long as 10 years in the environment. The maximum contaminant level (MCL) for lindane is 0.2 ppb (parts per billion). Suppose that an ecosystem has been contaminated with lindane at a concentration of 200 ppb. How long would you have to wait before it would be safe to enter the ecosystem without protection from the pesticide?

The level of contamination is 200 ppb/0.2 ppb = 1000 times the MCL. Presumably it would be safe to wait until the level had dropped to 1/1000 of its initial value. You can estimate how long this would be by using powers of 2, because the number of half-lives required is n, where $(1/2)^n = 1/1000$. That is, as soon as 2^n exceeds 1000, you have waited long enough. Computer scientists, who deal with binary arithmetic, can easily tell you that $2^{10} = 1024$, so $n = 10$. You can verify this using your calculator's y^x key. Or you could simply raise 2 to a power until a value greater than 1000 was calculated. If you did not have a calculator, you could multiply $2 \times 2 \times 2 \ldots$ in your head until you had enough factors to multiply out to a number bigger than 1000. Ten half-lives means 10×10 years, or 100 years, so after 100 years the ecosystem would be free of significant lindane contamination.

All chemical reactions can be understood in terms of simple reactions such as those just described. Very complicated reactions can be built up from combinations of unimolecular and bimolecular reactions, just as complicated compounds can be built from chemical elements. For example, hundreds of such reactions are needed to understand how smog is produced in a city such as Los Angeles or to understand why chlorofluorocarbons can deplete stratospheric ozone *(⇐ p. 441)*. Like the chemical elements, the simplest nanoscale reactions are building blocks, and so they are referred to as **elementary reactions.** *The equation for an elementary reaction shows exactly which molecules, atoms, or ions take part in the elementary reaction*. The next two sections describe the two important types of elementary reactions in more detail.

Exercise 13.5 **Unimolecular and Bimolecular Reactions**

For each of the nanoscale molecular diagrams below, write a balanced equation using chemical formulas. Which of the reactions are unimolecular reactions? Which are bimolecular?

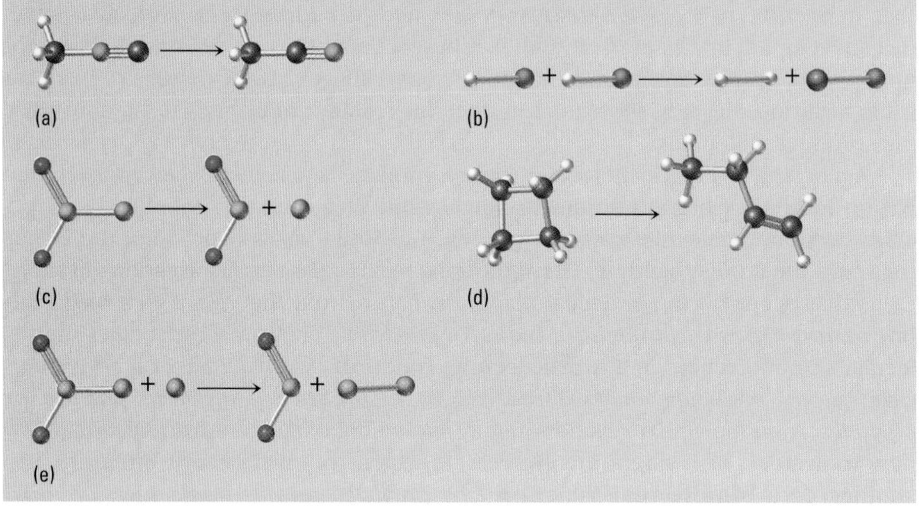

(a)

(b)

(c)

(d)

(e)

Unimolecular Reactions

An example of a unimolecular reaction is the conversion of *cis*-2-butene to *trans*-2-butene.

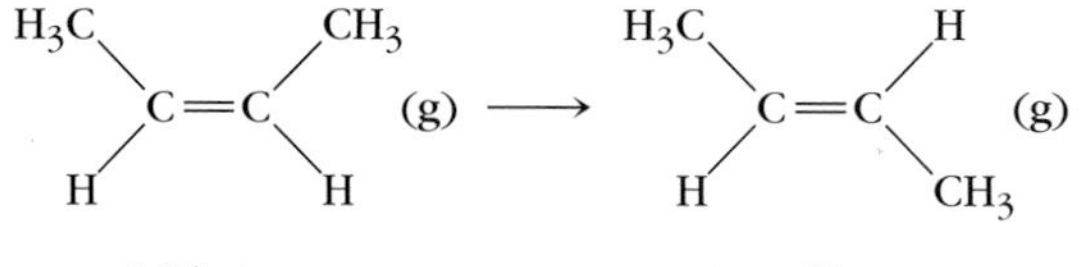

cis-2-butene   *trans*-2-butene

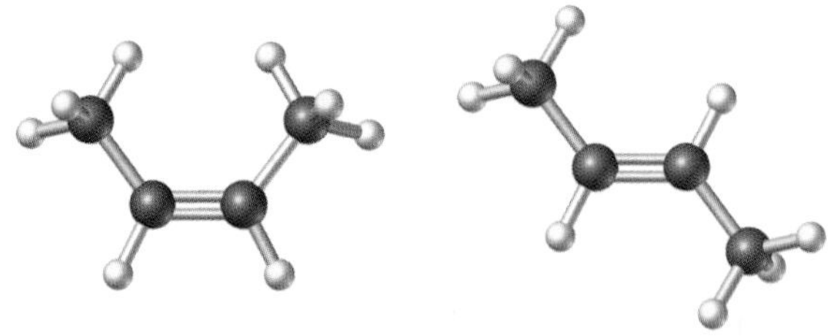

This equation says that the reaction is transformation of a *cis*-2-butene molecule into a *trans*-2-butene molecule. The difference between these two molecules, which are *cis-trans* stereoisomers *(⇐ p. 329)*, is the orientation of the methyl groups. They are on the same side of the double bond in the *cis* structure and on opposite sides in the *trans* structure. If we could grab one end of the molecule and twist it 180° around the axis of the double bond, we would get the other molecule. Thus, it is a reasonable hypothesis that the molecular pathway by which *cis*-2-butene changes to *trans*-2-butene involves twisting the molecule around the double bond. The angle of twist around the double bond axis is a measure of the progress of the reaction on the nanoscale. The greater the angle, the less the molecule is like *cis*-2-butene and the more it is like *trans*-2-butene, until an angle of 180° is reached and it has become *trans*-2-butene.

Such a twist requires that the reactant molecule have sufficient energy. Chemical bonds are like springs. They can be stretched, twisted, and bent, but this raises their potential energy. Consequently, some kinetic energy must be converted to potential energy when one end of the *cis*-2-butene molecule twists relative to the other, just as it would if a spring were twisted. At room temperature most of the molecules do not have enough energy to twist far enough to change *cis*-2-butene into *trans*-2-butene. Therefore, *cis*-2-butene can be kept in a sealed flask at room temperature for a long time without any appreciable quantity of *trans*-2-butene being formed. However, as the temperature is raised, more and more molecules have sufficient energy to react, and the reaction gets faster and faster.

Figure 13.7 shows a plot of potential energy versus the angle of twist in *cis*- and *trans*-2-butene. The potential energy is 435×10^{-21} J higher when one end of a *cis*-2-butene molecule is twisted by 90° from the initial flat molecule. This is similar to the increased potential energy that an object like a car has at the top of a hill compared with its energy at the bottom. Just as a car cannot reach the top of a hill unless it has enough energy, a molecule cannot reach the top of the "hill" for a reaction unless it has enough energy. Notice that the top of the hill can be approached from either side, and from the top a twisted molecule can go downhill

 CD-ROM Screen 15.9: Microscopic View of Reactions (1); Screen 15.10: Microscopic View of Reactions (2)

Figure 13.7 is similar to Figure 6.15 (⬅ *p. 237*), which showed the energy change when bonds were broken and formed as H_2 and Cl_2 changed into HCl.

Since molecules are very small, the energy required to twist one *cis*-2-butene molecule is also very small. However, if we wanted to twist a mole of molecules all at once, it would take a lot of energy. The energy required to reach the top of the "hill" is often reported per mole of molecules, that is, as (435×10^{-21} J/molecule) × (6.022×10^{23} molecules/mol) or 262 kJ/mol.

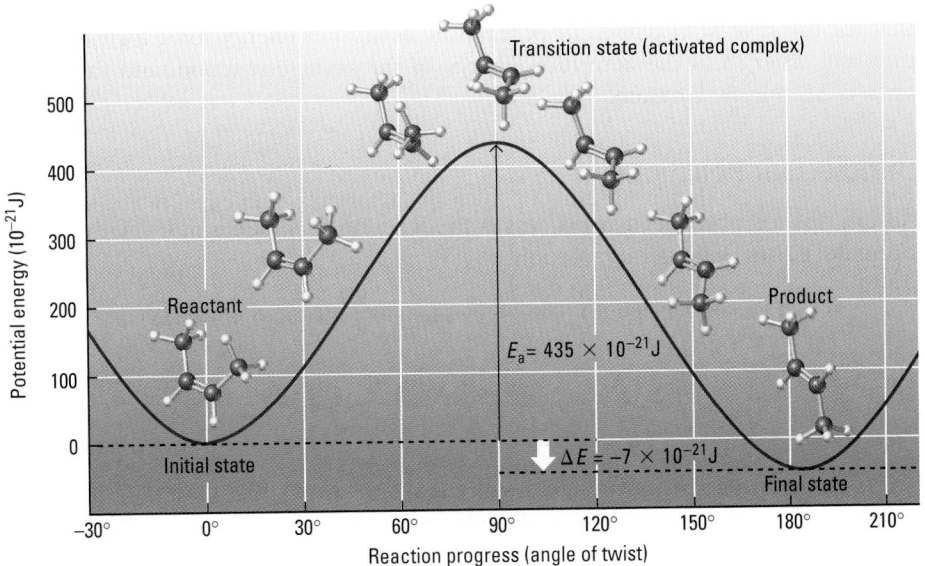

Figure 13.7 Energy diagram for the conversion of *cis*-2-butene to *trans*-2-butene. The double bond between the two central C atoms is stiff and resists twisting. However, if the molecule has enough energy, one end can twist with respect to the other end. When the angle between the ends of the molecule is 90°, the potential energy has risen by 435×10^{-21} J. A molecule of *cis*-2-butene must have at least this quantity of energy before it can twist past 90° to 180°, which converts it to *trans*-2-butene. The 90° twisted structure at the top of the diagram is called the transition state or the activated complex. The progress of the reaction (the change in the structure of this single molecule) is measured by the angle of twist.

energetically to either the *cis* or the *trans* form. *The structure at the top of an energy diagram like this one* is called the **transition state** or **activated complex**. In this case it is a molecule that has been twisted so that the methyl groups are at a 90° angle.

Exercise 13.6   Transition State

Methyl isonitrile reacts to form acetonitrile in a single-step elementary reaction.

$$CH_3NC(g) \longrightarrow CH_3CN(g)$$

During the reaction the nitrogen atom and one of the carbon atoms exchange places, but the rest of the molecule is unchanged. Suggest a structure for the transition state for this reaction. Draw this structure as a Lewis diagram and as a ball-and-stick molecular model.

The generality that higher activation energy results in slower reaction applies best if the reactions are similar. For example, it applies to a group of reactions that all involve twisting around a double bond. It also applies to reactions that involve collisions of one molecule with each of a group of similar molecules. It would be less applicable if we were comparing one reaction that involved collision of two molecules with another that involved twisting around a bond.

Every chemical reaction has an energy barrier that must be surmounted as reactant molecules change into product molecules. The heights of such barriers vary greatly — from almost zero to hundreds of kilojoules per mole. *At a given temperature, the higher the energy barrier the slower the reaction.* The minimum energy required to surmount the barrier is called the **activation energy, E_a**, for the reaction. For the *cis*-2-butene $\rightarrow$ *trans*-2-butene reaction the activation energy is 435×10^{-21} J/molecule, or 262 kJ/mol.

There is another interesting relationship shown in Figure 13.7 that connects kinetics and thermodynamics. The energy of the product, one molecule of *trans*-2-butene, is 7×10^{-21} J *lower* than that of reactant, one molecule of *cis*-2-butene. This means that the *cis* $\rightarrow$ *trans* reaction is *exothermic* by 7×10^{-21} J/molecule, which translates to 4 kJ/mol. Also, *cis*-2-butene is higher in energy by 7×10^{-21} J/molecule, and so the reverse reaction requires that 4 kJ/mol be absorbed from the surroundings; it is *endothermic*. The energy hill that has to be climbed when the reverse reaction occurs is $(435 + 7) \times 10^{-21}$ J or 442×10^{-21} J high (266 kJ/mol). Thus, the activation energy for the forward reaction is 4 kJ/mol less than that for the reverse reaction. In general the activation energy for a forward reaction will differ from the activation energy of the reverse reaction, and the difference is $\Delta E°$ for the reaction.

The actual relation is $\Delta E° = E_a$(forward) $- E_a$(reverse). Since $\Delta E°$ differs from $\Delta H°$ only when there is a change in volume of the reaction system (under constant pressure), the difference in activation energies is often equated with the enthalpy change (⟵ *p. 228)*.

Bimolecular Reactions

An example of a bimolecular process is the reaction of iodide ion, I^-, with methyl bromide, CH_3Br, in aqueous solution.

$$I^- (aq) + CH_3Br (aq) \longrightarrow ICH_3 (aq) + Br^- (aq)$$

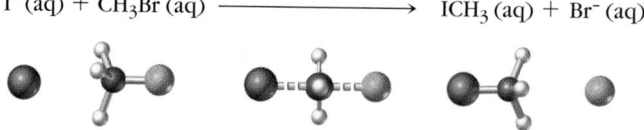

Here the equation for the elementary reaction shows that an iodide ion must collide with a methyl bromide molecule for the reaction to occur. The carbon-bromine bond does not break until after the iodine-carbon bond has begun to form. This makes sense, because just breaking a carbon-bromine bond would require a large increase in potential energy. Partially forming a carbon-iodine bond while the other bond is breaking lowers the potential energy. This helps keep the activation energy hill low. Figure 13.8 on p. 592 shows the energy-versus-reaction progress diagram for this reaction.

The methyl bromide molecule has a tetrahedral shape that is distorted because the Br atom is much larger than the H atoms. Numerous experiments suggest that

CHEMISTRY
You Can Do...

Kinetics and Vision

Vision is not an instantaneous process. It takes a little while after a bright light goes out before you stop seeing its image. The flash of a camera blinds you for a short time even though it is on only for an instant. Some sources of light flash on and off very rapidly, but you do not notice because your eyes continue to perceive their images while they are off. However, if you can focus such a source on different parts of your retina at different times, you can see whether it is flashing. Here's how.

Find a small mirror that you can hold easily in your hand. Now use the mirror to reflect the image of an incandescent light bulb onto your eye. You should be far enough away from the light so that its image is small. Now move the mirror quickly back and forth so that the image of the light bulb moves quickly across your eyeball. Does the light smear or do you see individual dots? Try the same experiment with the screen of a TV set. (Get really far away from it so the image is small.) Do you see separate images or just a smear of light? If you see individual images, it means that the light is flashing. Each time it flashes on, the moving mirror has caused it to hit a different part of your retina, and you see a separate image.

Repeat this experiment with as many different light sources as you can, and classify them as flashing or continuous. Try street lights of various kinds, car headlights, neon signs, fluorescent lights, and anything else you can think of, but don't try a very bright light, such as a laser pointer, that might damage your eyes. Record your observations.

Rotation around a double bond, as in the interconversion of *cis-* and *trans*-2-butene, occurs in the reactions that allow you to see (⬅ *p. 331*). A yellow-orange compound called β-carotene, the natural coloring agent in carrots,

breaks down in your body to produce vitamin A. This compound is converted in the liver to a compound called 11-*cis*-retinal. In the retina of your eye 11-*cis*-retinal combines with the protein opsin to form a light-sensitive substance called rhodopsin. When light strikes the retina, enough energy is transferred to a rhodopsin molecule to allow rotation around a carbon-carbon double bond, transforming rhodopsin into metarhodopsin II, a molecule whose shape is quite different, as you can see from the structural formulas below. This change in molecular shape causes a nerve impulse to be sent to your brain, and you see the light.

Eventually the metarhodopsin II reacts chemically to produce a different form of retinal, which is then converted back to vitamin A, and the cycle of chemical changes can begin again. However, decomposition of metarhodopsin II is not as rapid as its formation, and an image formed on the retina persists for a tenth of a second or so. This persistence of vision allows you to perceive videos as continuously moving images, even though they actually consist of separate pictures, each painted on a screen for a thirtieth of a second.

1. For each light source that you observed, what would happen if you focused a camera on the light source, opened the shutter, and moved the camera quickly while the shutter was open?
2. What would happen if you moved the camera more slowly or very quickly?

Note: This *Chemistry You Can Do* relates to the question "What are the molecules in my eyes doing when I watch a movie?" that was posed in Chapter 1 (⬅ *p. 32*).

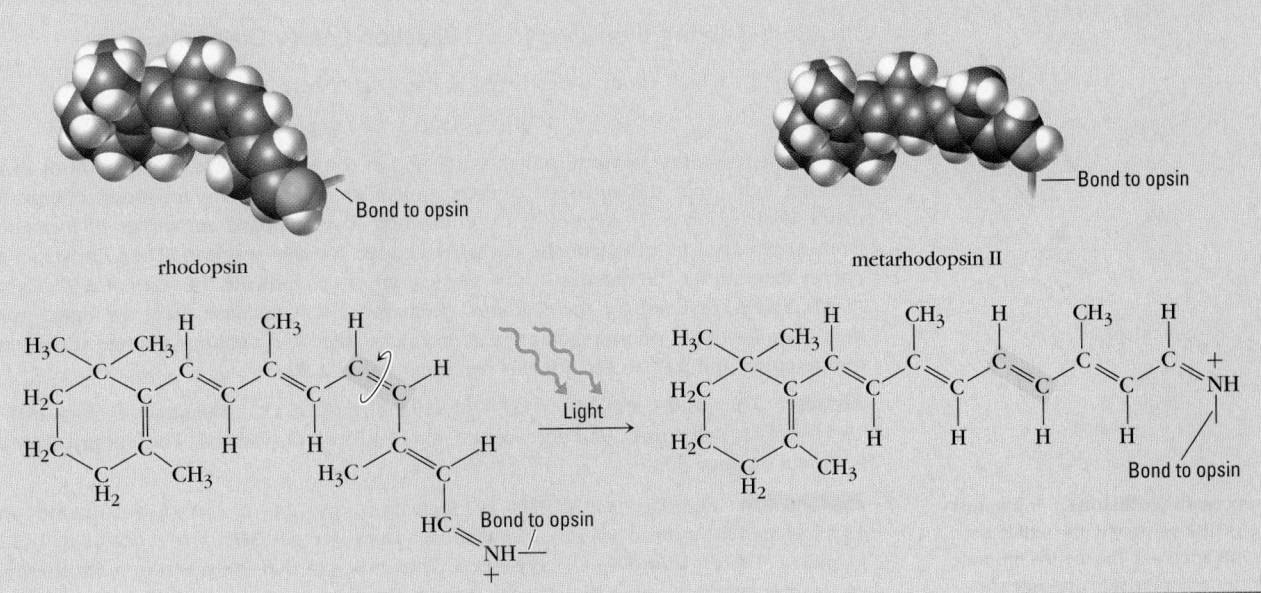

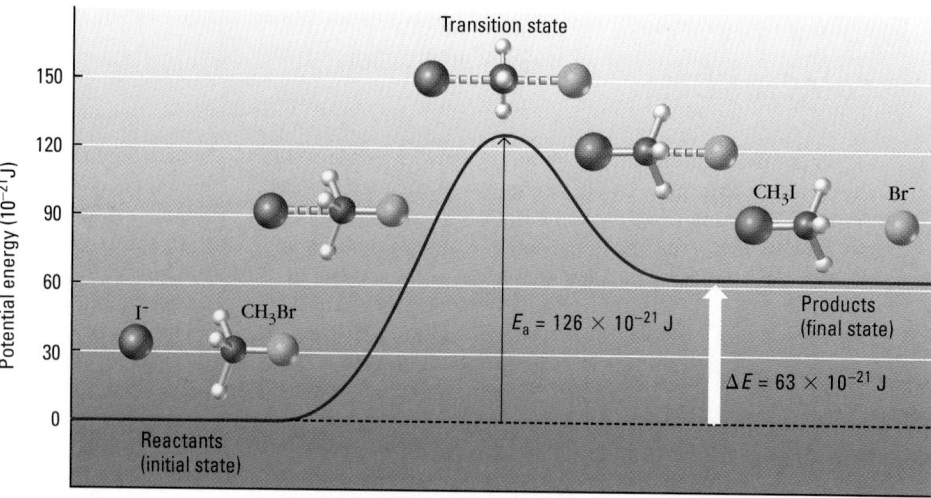

Figure 13.8 **Energy diagram for iodide–methyl bromide reaction.** During collision of an iodide ion with a methyl bromide molecule, a new iodine-carbon bond forms at the same time that the carbon-bromine bond is breaking. Forming the new bond lowers the potential energy, which otherwise would be raised a lot by breaking the carbon-bromine bond. This results in a lower activation energy and a faster reaction than would otherwise occur. In this case reaction progress is measured in terms of stretching of the carbon-bromine bond and formation of the iodine-carbon bond.

The word "steric" comes from the same root as the prefix "stereo-," which means three-dimensional.

the reaction occurs most rapidly in solution when the I^- ion approaches the methyl bromide from the side of the tetrahedron opposite the bromine atom. That is, approach to only one of the four sides of CH_3Br can be effective, which limits reaction to only one fourth of all the collisions at the most. This factor of one fourth is called a **steric factor** because it depends on the three-dimensional shapes of the reacting molecules. For molecules much more complicated than methyl iodide, such geometry restraints mean that only a very small fraction of the total collisions can lead to reaction. No wonder some chemical reactions are so slow.

Unsuccessful collisions

CH_3Br

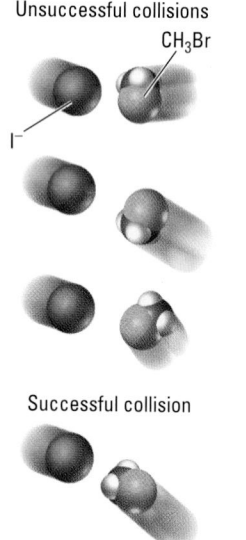

I^-

Successful collision

Unsuccessful collisions. In the first three collisions shown, the iodide ion does not approach the methyl bromide molecule from the side opposite the bromine atom. None of these collisions is as likely to result in a reaction as is the collision shown at the bottom.

Problem-Solving Example 13.8   Reaction Energy Diagrams

A reaction by which ozone is destroyed in the stratosphere (⇐ *p. 441*) is

$$O_3(g) + O(g) \rightarrow 2\,O_2(g)$$

(O represents atomic oxygen, which is formed in the stratosphere when photons of ultraviolet light from the sun split oxygen molecules in two.) The activation energy for ozone destruction is 19 kJ/mol of O_3 consumed. Use standard enthalpies of formation from Appendix J to calculate the enthalpy change for this reaction. Then construct an energy diagram for the reaction. Draw vertical arrows to indicate the sizes of $\Delta H°$, E_a(forward), and E_a(reverse) for the reaction. (Remember that because there are equal numbers of moles of gas phase reactants and products, there is no volume change at constant temperature, and $\Delta E° = \Delta H°$ for this reaction (⇐ *p. 228*).

Answer The values are $\Delta H° = \Delta E° = -392$ kJ/mol of O_3 consumed, E_a(forward) = 19 kJ/mol O_3 consumed, and E_a(reverse) = 411 kJ/mol O_3 formed. The energy diagram is shown on page 593.

Explanation Standard enthalpies of formation are 249.2 kJ/mol for ozone and 142.7 kJ/mol for atomic oxygen. Using these values, we get $\Delta H° = 0 - (249.2 + 142.7)$ kJ/mol $= -391.9$ kJ/mol of O_3 consumed. This indicates that the reaction is exothermic, and so the products must be lower in energy than the reactants by 391.9 kJ/mol. Since E_a(forward) $= 19$ kJ/mol, the transition state must be this much higher in energy than the reactants. Thus, the first two arrows on the left side of the diagram can be drawn.

Then the third arrow (from products to the transition state) can be drawn. It indicates that

$$E_a \text{ (reverse)} = -\Delta H° + E_a(\text{forward}) = (391.9 + 19) \text{ kJ/mol} = 411 \text{ kJ/mol}$$

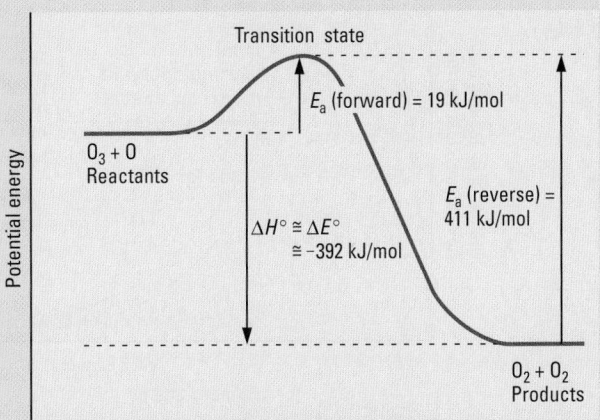

Problem-Solving Practice 13.8

For the hypothetical reaction A $\longrightarrow$ B, the activation energy is 24 kJ/mol. For the reverse reaction, B $\longrightarrow$ A, the activation energy is 36 kJ/mol. Draw a diagram similar to Figure 13.7 for this reaction. Is the reaction A $\longrightarrow$ B exothermic or endothermic?

 Exercise 13.7   **Successful and Unsuccessful Collisions**

The reaction

$$2 \text{ NOCl(g)} \longrightarrow 2 \text{ NO(g)} + \text{Cl}_2\text{(g)}$$

occurs in a single bimolecular step. Draw at least four possible ways that two NOCl molecules could collide, and rank them in order of greatest likelihood that a collision will be successful in producing products.

13.5 TEMPERATURE AND REACTION RATE: THE ARRHENIUS EQUATION

The most common way to speed up a reaction is to increase the temperature. A mixture of methane and air can be ignited by a lighted match, which raises the temperature of the mixture of reactants. This increases the reaction rate, the thermal energy evolved maintains the high temperature, and the reaction continues at a rapid rate. Reactions that speed up when the temperature is raised must slow down when the temperature is lowered. Foods are stored in refrigerators or freezers because the reactions that result in spoilage are slower at the lower temperature.

Reaction rates increase with temperature because at a higher temperature a greater fraction of reactant molecules has enough energy to surmount the activation energy barrier. Consider again the conversion of *cis*- to *trans*-2-butene (Figure 13.7, ($\Longleftarrow$ *p. 589*). You learned in Section 10.3 ($\Longleftarrow$ *p. 414*) that gas phase molecules are constantly in motion and have a wide distribution of speeds and energies. At room temperature relatively few *cis*-2-butene molecules are sufficiently energetic to surmount the energy barrier. However, as the temperature goes up, the

AHMED H. ZEWAIL
(1946–)

Ahmed H. Zewail of the California Institute of Technology received the 1999 Nobel Prize in Chemistry "for his studies of the transition states of chemical reactions using femtosecond spectroscopy." Zewail, who holds joint Egyptian and U.S. citizenship, pioneered the use of extremely short laser pulses—on the order of femtoseconds (10^{-15} s)—to study chemical kinetics. His technique has been called the world's fastest camera, and his research has enhanced understanding of many reactions, among them those involving rhodopsin and vision.

 CD-ROM Screen 15.11: Control of Reaction Rates: Temperature Dependence

As a rough rule of thumb, the reaction rate increases by a factor of 2 to 4 for each 10-K rise in temperature.

Figure 13.9 **Energy distribution curves.** The vertical axis gives the number of molecules that have the energy shown on the horizontal axis. Assume that a molecule reacts if it has more energy than the activation energy. The number of reactive molecules is given by the area under each curve to the right of the activation energy (262 kJ/mol). The two curves show that at 75 °C many more molecules have energies of 262 kJ/mol or higher than at 25 °C. (This graph is not to scale. Actually, 262 kJ/mol would be much farther to the right on the graph.)

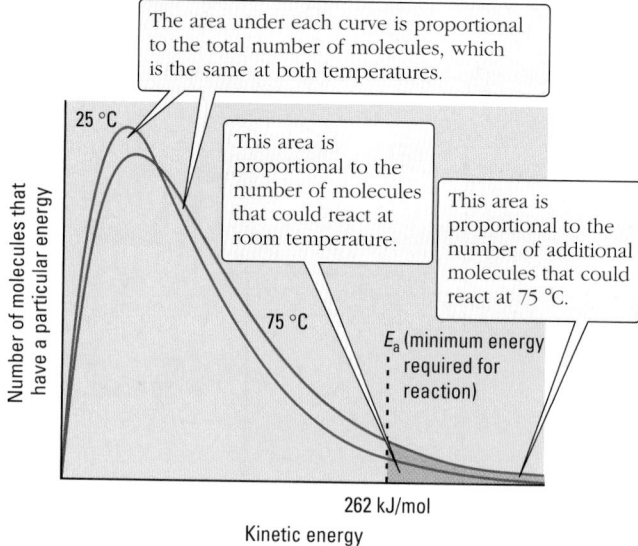

number of molecules that have enough energy goes up rapidly, and so the reaction rate increases rapidly.

The number of *cis*-2-butene molecules that have a given energy is shown by the curves in Figure 13.9. One curve is for 25 °C and the other for 75 °C. The higher a point on either curve, the greater the number of molecules that have the energy corresponding to that point. The areas under the two curves are the same and represent the total number of molecules. With a 50-°C rise in temperature, the number of molecules whose energy exceeds the activation energy is much higher, and so is the reaction rate.

A reaction is faster at a higher temperature because its rate constant is larger. That is, *a rate constant is constant only for a given reaction at a given temperature*. For example, for the reaction of iodide ion with methyl bromide, the data shown in Table 13.3 are found for the rate constant at different temperatures.

When the data from Table 13.3 are graphed (Figure 13.10), it is obvious that the rate constant increases very rapidly as temperature increases. A graph that is shaped like this is called an exponential curve. It can be represented by the equation

$$k = Ae^{-E_a/RT}$$

where A is the **frequency factor**, e is the base of the natural logarithm system (2.718 . . .), E_a is the activation energy, R is the gas law constant and has the value 8.314 J mol^{-1} K^{-1}, and T is the absolute (Kelvin) temperature ($\Leftarrow$ **p. 419**). This equation is called the **Arrhenius equation** after its discoverer, Svante Arrhenius, a Swedish chemist.

The Arrhenius equation can be interpreted as follows. The frequency factor, A, depends on how often molecules collide when all concentrations are 1 mol/L and

Exponential curves also represent the growth of populations (such as human population) over time and are important in many other scientific fields. Logarithms and exponentials are discussed in Appendix A.6.

R is the constant found in the ideal gas law, $PV = nRT$ ($\Leftarrow$ **p. 422**). Here it is expressed in units of J mol^{-1} K^{-1} instead of L atm mol^{-1} K^{-1}, which is the reason that the numerical value is not 0.0821 L atm mol^{-1} K^{-1}.

TABLE 13.3	Temperature Dependence of Rate Constant for Iodide Plus Methyl Bromide Reaction				
T (K)	k (L mol^{-1} s^{-1})	T (K)	k (L mol^{-1} s^{-1})	T (K)	k (L mol^{-1} s^{-1})
273	4.18×10^{-5}	310	2.31×10^{-3}	350	6.80×10^{-2}
280	9.68×10^{-5}	320	5.82×10^{-3}	360	1.41×10^{-1}
290	3.00×10^{-4}	330	1.39×10^{-2}	370	2.81×10^{-1}
300	8.60×10^{-4}	340	3.14×10^{-2}		

on whether the molecules are properly oriented when they collide. For example, in the case of the iodide + methyl bromide reaction, A includes the steric factor of $\frac{1}{4}$ that resulted because only one of the four sides of the CH_3Br molecule was appropriate for iodide to approach. The rest of the equation, $e^{-E_a/RT}$, gives the fraction of all the reactant molecules that have sufficient energy to surmount the activation energy barrier.

Determining Activation Energy

The activation energy and frequency factor can be obtained from experimental measurements of rate constants as a function of temperature (such as those in Table 13.3). When a large number of experimental data pairs are given, a graph is usually a good way of obtaining information from the data. This is easier to do if the graph is linear. The activation energy equation can be modified so that it gives a linear graph by taking natural logarithms of both sides.

$$k = Ae^{-E_a/RT}$$

$$\ln(k) = \ln(A) + \ln(e^{-E_a/RT}) = \ln(A) + (-E_a/RT)$$

Rearranging this equation gives the equation of a straight line.

$$\ln(k) = -\left(\frac{E_a}{R} \times \frac{1}{T}\right) + \ln(A)$$

$$y \quad = \quad (m \times x) \quad + \quad b$$

That is, if we graph $\ln(k)$ on the vertical (y) axis and $1/T$ on the horizontal (x) axis, the result should be a straight line whose slope is $-E_a/R$ and whose y-intercept is $\ln(A)$. For the data in Table 13.3, such a graph is shown in Figure 13.11. It is linear, its slope is -9.18×10^3 K, and the y-intercept is 23.53. Since the slope = $-E_a/R$, the activation energy can be calculated as

$$E_a = -\text{(slope)} \times R = -(-9.18 \times 10^3 \text{ K})\left(\frac{8.314 \text{ J}}{\text{K mol}}\right)\left(\frac{1 \text{ kJ}}{1000 \text{ J}}\right) = 76.3 \text{ kJ/mol}$$

The vertical axis plots $\ln(k)$, and k has units of L mol^{-1} s^{-1}. At the y-intercept, $\ln(A)$ equals $\ln(k)$, and therefore A must have the same units as k. Since $\ln(A) = 23.53$, the frequency factor, A, is $e^{23.53}$ L mol^{-1} s^{-1} = 1.66×10^{10} L mol^{-1} s^{-1}.

The Arrhenius equation can be used to calculate the rate constant at any temperature. For example, the rate constant for reaction of iodide ion with methyl bromide can be calculated at 50 °C by substituting the Kelvin temperature, the frequency factor, the activation energy, and the constant R into the equation.

$$k = Ae^{-E_a/RT} = (1.66 \times 10^{10} \text{ L mol}^{-1} \text{ s}^{-1})e^{(-76,300 \text{ J/mol})/(8.314 \text{ J K}^{-1} \text{ mol}^{-1})(273.15+50)\text{K}}$$

$$= (1.66 \times 10^{10} \text{ L mol}^{-1} \text{ s}^{-1})e^{-28.4} = 7.70 \times 10^{-3} \text{ L mol}^{-1} \text{ s}^{-1}$$

As a means of calculating rate constants, the Arrhenius equation works best within the range of temperatures over which the activation energy and frequency factor were determined. (For the reaction of iodide with methyl bromide, that range was 273 to 370 K.)

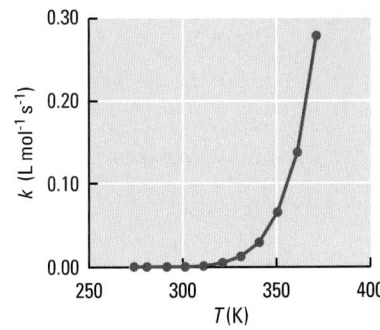

Figure 13.10 **Effect of temperature on rate constant.** The rate constant for the reaction of iodide ion with methyl bromide in aqueous solution is plotted as a function of temperature. The rate constant increases very rapidly with temperature, and the shape of the curve is characteristic of exponential increase.

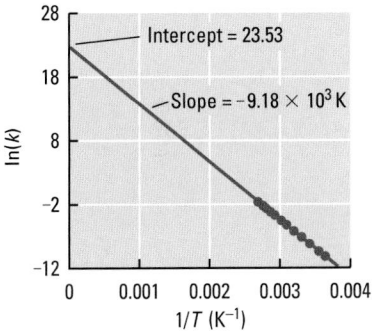

Figure 13.11 **Determining activation energy graphically.** A graph of $\ln(k)$ versus $1/T$ gives a straight line for the iodide + methyl bromide reaction. The activation energy can be obtained from the slope of the line, and the frequency factor from its y-intercept. Because the line through the experimental data has to be extrapolated a long way to reach the y-intercept, determining A accurately requires measuring k over a wide range of temperatures.

Problem-Solving Example 13.9 Temperature Dependence of Rate Constant

The rate constant for reaction of iodide ion with methyl bromide is 7.70×10^{-3} L mol^{-1} s^{-1} at 50 °C and 4.25×10^{-5} L mol^{-1} s^{-1} at 0 °C. Calculate the frequency factor and activation energy.

Answer $A = 1.66 \times 10^{10}$ L mol^{-1} s^{-1}; $E_a = 7.63 \times 10^4$ J mol^{-1}

Explanation This problem could be solved by making a graph of ln(k) versus $1/T$, but with only two data pairs, a graph is not the best way. Instead, write two equations, one for each data pair.

$$k_1 = Ae^{-E_a/RT_1}$$
$$k_2 = Ae^{-E_a/RT_2}$$

Now divide the first equation by the second to eliminate A.

$$\frac{k_1}{k_2} = \frac{Ae^{-E_a/RT_1}}{Ae^{-E_a/RT_2}} = e^{-E_a/RT_1} \times e^{+E_a/RT_2}$$

$$\frac{k_1}{k_2} = e^{\frac{E_a}{R}\left(\frac{1}{T_2}-\frac{1}{T_1}\right)}$$

Next take the natural logarithm of both sides.

$$\ln\left(\frac{k_1}{k_2}\right) = \frac{E_a}{R}\left(\frac{1}{T_2}-\frac{1}{T_1}\right)$$

This equation can be solved for E_a.

$$E_a = \frac{R\ln\left(\frac{k_1}{k_2}\right)}{\frac{1}{T_2}-\frac{1}{T_1}} = \frac{(8.314\,\text{J mol}^{-1}\,\text{K}^{-1})\ln\left(\frac{7.70\times10^{-3}}{4.25\times10^{-5}}\right)}{\frac{1}{273.15\,\text{K}}-\frac{1}{323.15\,\text{K}}}$$

$$= \frac{43.23\,\text{J mol}^{-1}\,\text{K}^{-1}}{5.66\times10^{-4}\,\text{K}^{-1}} = 7.63\times10^4\,\text{J mol}^{-1}$$

Finally, using the calculated value of E_a, solve one of the rate constant expressions for A.

$$k_1 = Ae^{-E_a/RT_1}$$
$$A = \frac{k_1}{e^{-E_a/RT_1}} = k_1 e^{E_a/RT_1}$$

$$= (7.70\times10^{-3}\,\text{L mol}^{-1}\,\text{s}^{-1})e^{\frac{7.63\times10^4\,\text{J mol}^{-1}}{(8.314\,\text{J mol}^{-1}\,\text{K}^{-1})(323.15\,\text{K})}}$$

$$= (7.70\times10^{-3}\,\text{L mol}^{-1}\,\text{s}^{-1})(2.156\times10^{12}) = 1.66\times10^{10}\,\text{L mol}^{-1}\,\text{s}^{-1}$$

Problem-Solving Practice 13.9

Calculate the rate constant for the reaction of iodide ion with methyl bromide at a temperature of 75 °C.

Recall that the collision frequency in the frequency factor is for 1 M concentrations. This equation summarizes the effects of both temperature and concentration on rate of a reaction. The temperature effect depends primarily on the large increase in the number of sufficiently energetic collisions as the temperature increases, and this shows up as larger values of k at higher temperatures. The effect of concentration is clearly indicated by the concentration terms in the rate law. If the rate law is known for a reaction, and if both the A and E_a values are known, then the rate can be calculated over a wide range of conditions.

If we substitute the Arrhenius equation for the rate constant into the rate law for the iodide + methyl bromide reaction, we have

$$\text{Rate} = k \times [\text{I}^-] \times [\text{CH}_3\text{Br}]$$

$$\text{Rate} = A \times e^{-E_a/RT} \times [\text{I}^-] \times [\text{CH}_3\text{Br}] \qquad [13.8]$$

Collision frequency × steric factor | Fraction of sufficiently energetic molecules | Concentrations of colliding molecules

Exercise 13.8 Activation Energy and Experimental Data

The frequency factor is 6.31×10^8 L mol^{-1} s^{-1} and the activation energy is 10 kJ/mol for the gas phase reaction

$$\text{NO}(g) + \text{O}_3(g) \rightarrow \text{NO}_2(g) + \text{O}_2(g)$$

which is important in the chemistry of stratospheric ozone depletion.

(a) Calculate the rate constant for this reaction at 370 K.
(b) Assuming that this is an elementary reaction, calculate the rate of the reaction at 370 K if [NO] = 0.0010 M and [O$_3$] = 0.00050 M.

13.6 RATE LAWS FOR ELEMENTARY REACTIONS

An elementary reaction is a one-step process whose equation describes which nanoscale particles break apart, rearrange their positions, or collide to make a reaction occur. Therefore it is possible to figure out what the rate law and reaction order are for an elementary reaction, without doing an experiment. By contrast, when an equation represents a reaction that we do not understand at the nanoscale, rate laws and reaction orders must be determined experimentally (⬅ *p. 579)*. Then the macroscale rate law can be used to help develop a theory about how a particular reaction takes place at the nanoscale.

Rate Law for a Unimolecular Reaction

In Section 13.4 we used the isomerization of *cis*-2-butene as an example of a reaction in which a single reactant molecule was converted to products — a unimolecular reaction (⬅ *p. 587)*.

$$cis\text{-2-butene} \longrightarrow trans\text{-2-butene}$$

Suppose a flask contains 0.005 mol/L of *cis*-2-butene vapor at room temperature. The molecules have a wide range of energies, but only a few of them have enough energy at this temperature to get over the activation energy barrier. Thus, during a given period only a few molecules twist sufficiently to become *trans*-2-butene. Now suppose that we double the concentration of *cis*-2-butene in the flask to 0.010 mol/L, keeping the temperature the same. The fraction of molecules with enough energy to cross over the barrier remains the same. However, since there are now twice as many molecules, there must be twice as many crossing the barrier in any given time. Therefore, the rate of the *cis* → *trans* reaction is twice as great. That is, the reaction rate is proportional to the concentration of *cis*-2-butene, and the rate law must be

$$\text{Rate} = k[cis\text{-2-butene}]$$

In the general case of any unimolecular elementary reaction,

$$\text{A} \longrightarrow \text{products} \qquad \text{the rate law is} \qquad \text{Rate} = k[\text{A}]$$

For any unimolecular reaction the nanoscale mechanism predicts that a first-order rate law will be observed in a macroscale laboratory experiment.

Rate Law for a Bimolecular Reaction

A good example of a reaction in which two molecules collide (a bimolecular reaction, (⬅ *p. 587)* is the gas phase reaction of nitrogen monoxide and ozone that is involved in stratospheric ozone depletion and was mentioned in Exercise 13.8.

$$\text{NO(g)} + \text{O}_3(\text{g}) \longrightarrow \text{NO}_2(\text{g}) + \text{O}_2(\text{g}) \qquad [13.9]$$

Here the equation shows that the elementary reaction involves the collision of one NO molecule and one O_3 molecule. Since the molecules must collide to exchange atoms, the rate depends on the number of collisions per unit time.

Figure 13.12a represents one NO molecule (the green ball) and many O_3 molecules (the purple balls) in a tiny region within a flask where Reaction 13.9 is taking place. In a given time, the NO molecule collides with five O_3 molecules. If the concentration of NO molecules is doubled to two NO molecules in the same portion

CD-ROM Screen 15.13: Reaction Mechanisms and Rate Equations

Suppose that the fraction of molecules that have enough energy to react is 0.1%, or 0.001. If there are 10,000 molecules in a given volume, then 0.001 × 10,000 gives only 10 that have enough energy to react. If there are twice as many molecules in the same volume — that is, 20,000 molecules — then 0.001 × 20,000 gives 20 with enough energy to react, and the number reacting per unit volume (the rate) is twice as great.

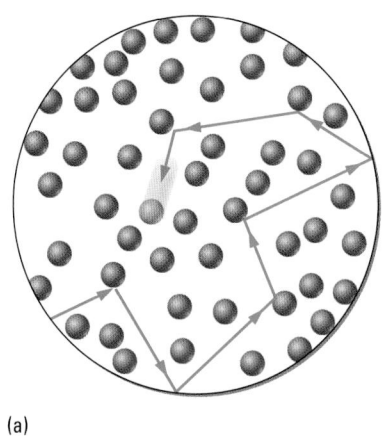

(a)

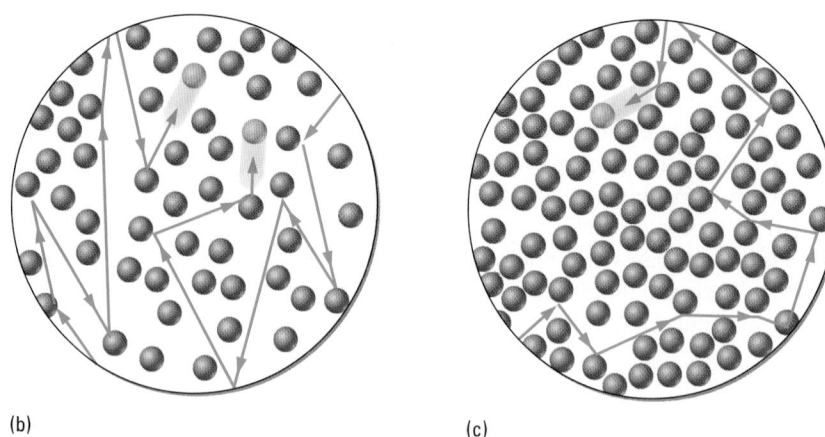

(b)   (c)

Figure 13.12 **Effect of concentration on frequency of bimolecular collisions.** (a) A single green molecule moves among 50 purple molecules and collides with five of them per second. (b) Two green molecules now move among 50 purple molecules, and there are 10 green-purple collisions per second. (c) If the number of purple molecules is doubled to 100, the frequency of green-purple collisions is also doubled, to 10 per second. The number of collisions is proportional to *both* the concentration of green molecules *and* the concentration of purple molecules. (In a real sample both the green and the purple molecules would be moving, but in this diagram the motion of the purple molecules is not shown.)

of the flask (Figure 13.12b), *each* NO molecule collides with five different O_3 molecules. Doubling the concentration of NO has doubled the number of collisions. This also doubles the rate, because the rate is proportional to the number of collisions (Equation 13.8 ⬅ *p. 596*). The number of collisions also doubles when the O_3 concentration is doubled (Figure 13.12c). Thus, the rate law for this reaction must be

$$\text{Rate} = k[\text{NO}][\text{O}_3]$$

This description of the NO + O_3 reaction applies in general to bimolecular elementary reactions, even if the two molecules that must collide are of the same kind. That is, for the elementary reaction

$$\text{A} + \text{B} \longrightarrow \text{products} \qquad \text{the rate law is} \qquad \text{Rate} = k[\text{A}][\text{B}]$$

and for the elementary reaction

$$\text{A} + \text{A} \longrightarrow \text{products} \qquad \text{the rate law is} \qquad \text{Rate} = k[\text{A}]^2$$

For the NO + O_3 reaction the experimentally determined rate law is the same as the one we just derived by assuming the reaction occurs in one step. The equation for the reaction is

$$\text{NO(g)} + \text{O}_3\text{(g)} \longrightarrow \text{NO}_2\text{(g)} + \text{O}_2\text{(g)}$$

and the experimental rate law is $\qquad \text{Rate} = k[\text{NO}][\text{O}_3]$

This experimental observation suggests, but does not prove, that the reaction does take place in a single step. There is other evidence that also suggests that this reaction is bimolecular.

In contrast, for the decomposition of hydrogen peroxide, which was mentioned in Problem-Solving Example 13.4 (⬅ *p. 582)*, the equation is

$$2\,\text{H}_2\text{O}_2\,\text{(aq)} \longrightarrow 2\,\text{H}_2\text{O}\,\text{(ℓ)} + \text{O}_2\,\text{(g)}$$

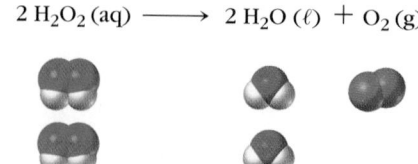

and the experimental rate law is Rate = $k[\text{H}_2\text{O}_2]$. This rate law proves that this reaction *cannot* occur in a single step that involves collision of two H_2O_2 molecules. A single-step bimolecular reaction would have a second-order rate law, but the observed rate law is first-order. This means that more than a single elementary step is needed when hydrogen peroxide decomposes.

Exercise 13.9   Rate Law and Elementary Reactions

For the reaction

$$NO_2(g) + CO(g) \rightarrow NO(g) + CO_2(g)$$

the experimentally determined rate law is

$$Rate = k[NO_2]^2$$

Does this reaction occur in a single step? Explain why or why not.

13.7   REACTION MECHANISMS

Most chemical reactions do not take place in a single step. Instead they involve a sequence of unimolecular or bimolecular elementary reactions. For each elementary reaction in the sequence we can write an equation. A set of such equations is called a **reaction mechanism** (or just a *mechanism*). For example, iodide ion can be oxidized by hydrogen peroxide in acidic solution to form iodine and water according to this overall equation:

$$2 I^-(aq) + H_2O_2(aq) + 2 H_3O^+(aq) \rightarrow I_2(aq) + 4 H_2O(\ell) \qquad [13.10]$$

When the acid concentration is between 10^{-3} M and 10^{-5} M, experiments show that the rate law is

$$Rate = k[I^-][H_2O_2]$$

The reaction is first-order in the concentrations of I^- and H_2O_2, and second-order overall.

Looking at the balanced equation for the oxidation of iodide ion by hydrogen peroxide, you might think that two iodide ions, one hydrogen peroxide molecule, and two hydronium ions would all have to come together at once. However, the rate law corresponds with a bimolecular collision of I^- and H_2O_2. It is highly unlikely that at the same time five ions or molecules would all be at the same place, be properly oriented, and have enough energy to react. Instead, those who have studied this reaction propose that first one H_2O_2 (HOOH) molecule and one I^- ion come together:

Step 1:

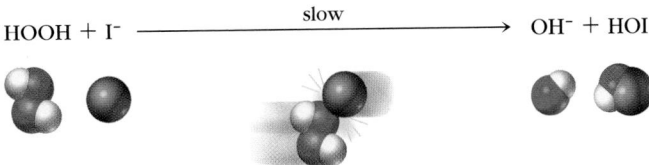

HOOH + I⁻ ————slow————→ OH⁻ + HOI

This step forms hypoiodous acid, HOI, and hydroxide ion, both known substances. The HOI then reacts with another I^- to form the product I_2:

Step 2:

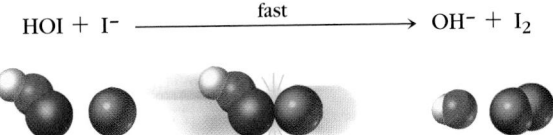

HOI + I⁻ ————fast————→ OH⁻ + I₂

In each of Steps 1 and 2, a hydroxide ion was produced. Since the solution is acidic, these OH^- ions react immediately with H_3O^+ ions to form water.

Step 3:

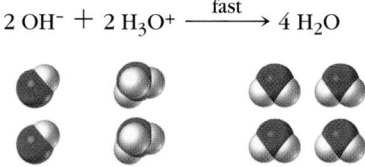

$$2\ OH^- + 2\ H_3O^+ \xrightarrow{\text{fast}} 4\ H_2O$$

Each of the three steps in this mechanism is an elementary reaction. Each has its own activation energy, E_a, and its own rate constant, k. When the three steps are summed (by putting all the reactants on the left, putting all the products on the right, and eliminating formulas that appear as both reactants and products), the overall stoichiometric equation (Equation 13.10) is obtained. ***Any valid mechanism must consist of a series of unimolecular or bimolecular elementary reaction steps that sum to the overall reaction.***

Step 1: $\qquad\qquad HOOH + I^- \xrightarrow{\text{slow}} HOI + OH^-$

Step 2: $\qquad\qquad HOI + I^- \xrightarrow{\text{fast}} I_2 + OH^-$

Step 3: $\qquad\qquad \underline{2\ OH^- + 2\ H_3O^+ \xrightarrow{\text{fast}} 4\ H_2O}$

Overall: $\qquad 2\ I^- + HOOH + 2\ H_3O^+ \longrightarrow I_2 + 4\ H_2O$ [13.10]

Step 1 of the mechanism is slow, while Steps 2 and 3 are fast. Step 1 is called the **rate-limiting step**; because it is the slowest in the sequence, it limits the rate at which I_2 and H_2O can be produced. Steps 2 and 3 are rapid and therefore not rate-limiting. As soon as some HOI and OH^- are produced by Step 1, they are transformed into I_2 and H_2O by Steps 2 and 3. ***The rate of the overall reaction is limited by, and equal to, the rate of the slowest step in the mechanism.***

Step 1 is a bimolecular elementary reaction. Therefore its rate must be first-order in HOOH and first-order in I^-. The mechanism predicts that the rate law should be

$$\text{Reaction rate} = k[\text{HOOH}][\text{I}^-]$$

which agrees with the experimentally observed rate law. ***A valid mechanism must correctly predict the experimentally observed rate law.***

> ### Exercise 13.10 Rate Law for an Elementary Reaction
>
> What is the rate law for Step 2 of the mechanism for the reaction of hydrogen peroxide and iodide ion?

The species HOI and OH^- are produced in Step 1 and used up in Step 2 or 3. In a mechanism, atoms, molecules, or ions that are produced in one step and used up in a later step (or later steps) are called **reaction intermediates** (or just *intermediates*). Very small concentrations of HOI and OH^- are produced while the reaction is going on. Once the HOOH, the I^-, or both are used up, the intermediates HOI and OH^- are consumed by Steps 2 and 3 and disappear. HOI and OH^- are crucial to the reaction mechanism, but neither of them appears in the overall stoichiometric equation. If an experimenter is proficient enough to demonstrate that a particular intermediate was present, this provides additional evidence that a mechanism involving that intermediate is the correct one.

An analogy to the rate-limiting or rate-determining step is that no matter how quickly you shop in the supermarket, it seems that the time it takes to get out of the store depends on the rate at which you move through the checkout line.

If significant concentrations of an intermediate build up while a reaction is occurring, then reactants may disappear faster than products are formed (because buildup of the intermediate stores some of the used reactant before it is converted to final product). In such a case, the definition of reaction rate given in Equation 13.3 will not be correct. One way to detect formation of an intermediate is to notice that products are not formed as fast as reactants are used up.

In summary, a valid reaction mechanism should

- Consist of a series of unimolecular or bimolecular elementary reactions.
- Consist of reaction steps that sum to the overall reaction equation.
- Correctly predict the experimentally observed rate law.

Mechanisms with a Fast Initial Step

An experimental rate law should include only concentrations that can be measured experimentally. Usually this means concentrations of reactants (and perhaps products), not the concentrations of intermediates, which are very small and therefore hard to measure. The first step in the mechanism in the previous section was rate-limiting and involved two of the reactants. Therefore it was easy to relate the overall rate to the concentrations of reactants. But what happens if the rate-limiting step is the second or a subsequent step?

Consider the reaction

$$2 \, NO(g) + Br_2(g) \longrightarrow 2 \, NOBr(g) \qquad \text{Rate (experimental)} = k[NO]^2[Br_2]$$

for which the currently accepted mechanism is

Step 1: $\qquad\qquad NO(g) + Br_2(g) \rightleftharpoons NOBr_2(g) \qquad$ fast

Step 2: $\qquad\qquad NOBr_2(g) + NO(g) \longrightarrow 2 \, NOBr(g) \qquad$ slow

Because the second step is slow, $NOBr_2$ can often separate and reform $NO + Br_2$ before it reacts to form products. We say that Step 1 is reversible, which is indicated by the double arrow ($\rightleftharpoons$). Once $NOBr_2$ forms, it can react in either of two ways: back to the reactants, $NO + Br_2$, or forward (in Step 2) to form the product, NOBr.

The overall rate of the reaction is the rate of the rate-limiting step (Step 2), namely,

$$\text{Rate} = k[NOBr_2][NO]$$

However, this is not a valid rate law to compare with experiment, because an experimental rate law should include only concentrations that can be measured experimentally. (Usually, the concentration of an intermediate, such as $NOBr_2$, cannot be measured.) Therefore, we need a way to relate the concentration of $NOBr_2$ to the concentrations of the reactants, NO and Br_2. This can be done by recognizing that in the mechanism there are three reactions and three rate constants:

Step $+1$ (forward), rate constant k_1

Step -1 (backward), rate constant k_{-1}

Step 2, rate constant k_2

The concentration of $NOBr_2$ depends on all three reactions. It is increased by Step $+1$ and decreased by Step -1 and Step 2. Initially, the concentration of $NOBr_2$ builds up because Step $+1$ produces $NOBr_2$. However, when $[NOBr_2]$ gets big enough, the rates of Step -1 and Step 2 get bigger, and eventually the $[NOBr_2]$ reaches what is called a steady state—it neither increases nor decreases until the reaction is nearly over.

In the steady state, the rate of the reaction by which $NOBr_2$ is formed must equal the sum of the two rates by which it is destroyed; that is,

$$\text{Rate of Step } +1 = \text{rate of Step } -1 + \text{rate of Step 2}$$

$$k_1[NO][Br_2] = k_{-1}[NOBr_2] + k_2[NOBr_2] \qquad [13.11]$$

Equation 13.11 can usually be simplified because Step 2 is much slower than either Step 1 or Step -1. This means that $k_2[NOBr_2] \ll k_{-1}[NOBr_2]$, and $k_2[NOBr_2]$ can be neglected in the calculation.

$$k_1[NO][Br_2] \cong k_{-1}[NOBr_2] \qquad [13.12]$$

Equation 13.12 can be rearranged algebraically to show that $[NOBr_2]$ is proportional to the concentration of each reactant.

$$[NOBr_2] = \frac{k_1}{k_{-1}}[NO][Br_2]$$

This allows the rate to be expressed in terms of the concentrations of reactants.

$$\text{Rate} = k_2[NOBr_2][NO] = k_2\left(\frac{k_1}{k_{-1}}[NO][Br_2]\right)[NO]$$

$$= \left(\frac{k_1 k_2}{k_{-1}}\right)[NO]^2[Br_2] = k'[NO]^2[Br_2] \qquad [13.13]$$

Equation 13.13 shows that the rate constant k' is actually a quotient of rate constants for three elementary reactions, but the rate is proportional to the concentration of Br_2 and to the square of the concentration of NO. That is, the mechanism predicts that the reaction is second-order in NO and first-order in Br_2, which agrees with the experimental rate law. For mechanisms in which the rate-limiting step is the second or a subsequent step, a mathematical relationship such as Equation 13.13 can usually be found that relates an overall rate constant, k', to the concentrations of the reactants and the rate constants for the steps up to and including the rate-limiting step.

Kinetics and Mechanism

Studying the kinetics of a chemical reaction involves collecting data on the concentrations of reactants as a function of time. From such data the rate law for the reaction and a rate constant can usually be obtained. The reaction can also be studied at several different temperatures to determine its activation energy. This allows us to predict how fast the macroscale reaction will go under a variety of experimental conditions, but it does not provide definitive information about the nanoscale mechanism by which the reaction takes place. A reaction mechanism is an educated guess—a hypothesis—about the way the reaction occurs. If the mechanism predicts correctly the overall stoichiometry of the reaction and the experimentally determined rate law, then it is a reasonable hypothesis. However, it is impossible to prove for certain that a mechanism is correct. Sometimes several mechanisms can agree with the same set of experiments. This is what makes kinetic studies one of the most interesting and rewarding areas of chemistry, but it also can provoke disputes among scientists who favor different possible mechanisms.

Suppose that
$k_{-1}[NOBr_2] = 0.20$ mol L^{-1}
and $k_2[NOBr_2] = 0.00020$ mol L^{-1}, which is much smaller. Then
$k_{-1}[NOBr_2] - k_2[NOBr_2] = (0.20 - 0.00020)$ mol $L^{-1} = 0.20$ mol L^{-1}. That is, the rate for Step 2 is negligible and can be ignored in the calculation.

Problem-Solving Example 13.10   Rate Law and Reaction Mechanism

The gas phase reaction between nitrogen monoxide and oxygen,

$$2 \, NO(g) + O_2(g) \rightarrow 2 \, NO_2(g)$$

is found experimentally to obey the rate law

$$Rate = k[NO]^2[O_2]$$

Decide which of the following mechanisms is (are) compatible with this rate law.

(a) $NO + NO \rightleftharpoons N_2O_2$ fast

$N_2O_2 + O_2 \rightarrow 2 \, NO_2$ slow

(b) $NO + NO \rightarrow NO_2 + N$ slow

$N + O_2 \rightarrow NO_2$ fast

(c) $NO + O \rightleftharpoons NO_2$ fast

$NO_2 + NO \rightarrow N_2O_3$ fast

$N_2O_3 + O \rightarrow 2 \, NO_2$ slow

(d) $NO + O_2 \rightleftharpoons NO_3$ fast

$NO_3 + NO \rightarrow 2 \, NO_2$ slow

(e) $NO + O_2 \rightarrow NO_2 + O$ slow

$NO + O \rightarrow NO_2$ fast

(f) $2 \, NO + O_2 \rightarrow 2 \, NO_2$

Answer   Mechanisms (a) and (d) are compatible with the rate law and stoichiometry.

Explanation   Examine each mechanism to see whether it (1) consists only of unimolecular and bimolecular steps, (2) agrees with the overall stoichiometry, and (3) predicts the experimental rate law. Eliminate those that do not. The remaining mechanism(s) may be correct.

Mechanism (f) involves collision of three molecules: two NO and one O_2. It can be eliminated because it does not consist of unimolecular or bimolecular steps. All other mechanisms consist of bimolecular steps.

Mechanism (c) does not have O_2 as a reactant in the overall stoichiometry, and so it can be eliminated. All other mechanisms predict the observed overall stoichiometry.

In mechanism (a) the first step is fast and reversible. Applying the idea that the rates are approximately equal for the forward and reverse reactions in that first step gives

$$k_1[NO]^2 = k_{-1}[N_2O_2] \quad \text{and} \quad [N_2O_2] = \frac{k_1}{k_{-1}} \, [NO]^2$$

Since the overall rate equals the rate of the rate-limiting step,

$$Rate = k_2[N_2O_2][O_2] = k_2 \, \frac{k_1}{k_{-1}} \, [NO]^2[O_2] = k'[NO]^2[O_2]$$

Consequently, mechanism (a) could be the actual mechanism.

The slow first step in mechanism (b) implies an overall rate = $k \, [NO]^2$, which eliminates it from consideration.

Continuing this kind of reasoning, mechanism (d) is seen to be a possibility, but mechanism (e) can be eliminated from further consideration. Because there are still two possible mechanisms, (a) and (d), additional experiments would need to be done to try to distinguish between them.

Problem-Solving Practice 13.10

The Raschig reaction produces the industrially important reducing agent hydrazine, N_2H_4, from ammonia, NH_3, and hypochlorite ion, OCl^-, in basic aqueous solution. A proposed mechanism is

Step 1: $NH_3(aq) + OCl^-(aq) \xrightarrow{\text{slow}} NH_2Cl(aq) + OH^-(aq)$

Step 2: $NH_2Cl(aq) + NH_3(aq) \xrightarrow{\text{fast}} N_2H_5^+(aq) + Cl^-(aq)$

Step 3: $N_2H_5^+(aq) + OH^-(aq) \xrightarrow{\text{fast}} N_2H_4(aq) + H_2O(\ell)$

(a) What is the overall stoichiometric equation?

(b) Which step is rate-limiting?

(c) What reaction intermediates are involved?

(d) What rate law is predicted by this mechanism?

Exercise 13.11 Rate Law and Mechanism

Consider the reaction mechanism

$$ICl(g) + H_2(g) \rightleftharpoons HI(g) + HCl(g) \qquad \text{fast}$$

$$HI(g) + ICl(g) \longrightarrow HCl(g) + I_2(g) \qquad \text{slow}$$

(a) What is the overall reaction equation?
(b) Derive the rate law predicted by this mechanism.
(c) Does the rate law depend on the concentration of one of the products of the reaction?
(d) Would the rate constant determined from the initial rate of this reaction equal the rate constant determined at a time when 80% of the reactants had been consumed? Explain why or why not.

13.8 CATALYSTS AND REACTION RATE

CD-ROM Screen 15.14: Catalysis and Reaction Rate

Raising the temperature increases a reaction rate because it increases the fraction of molecules that are energetic enough to surmount the activation energy barrier. Increasing reactant concentrations can also increase the rate because it increases the number of molecules per unit volume. A third way to increase reaction rates is to add a catalyst (⇐ *p. 517*).

For example, an aqueous solution of hydrogen peroxide can decompose to water and oxygen.

$$2\,H_2O_2(aq) \longrightarrow O_2(g) + 2\,H_2O(\ell)$$

If the peroxide is stored in a cool, dark place in a clean plastic container, it remains unreacted for months. The rate of the decomposition reaction is exceedingly slow. However, in the presence of a manganese salt, an iodide-containing salt, or a biological catalyst called an enzyme, the reaction can occur with explosive speed (Figure 13.13a). In fact, an insect called a bombardier beetle uses a very similar reaction as its defense mechanism (Figure 13.13b). By combining the organic compound hydroquinone with the peroxide in the presence of an enzyme, it produces a small, but sufficient, quantity of superheated steam and an irritating chemical to spray its enemies.

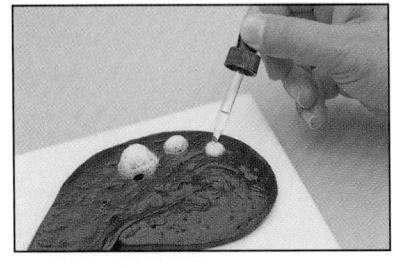

(a)

(b)

Figure 13.13 Enzyme catalysis of decomposition of hydrogen peroxide. (a) A 30% aqueous solution of H_2O_2 is dropped onto a piece of liver. The liquid foams as H_2O_2 rapidly decomposes to O_2 and H_2O. Liver contains an enzyme that catalyzes the decomposition of H_2O_2. (b) A bombardier beetle uses the enzyme-catalyzed decomposition of hydrogen peroxide as a defense mechanism. The rapid exothermic reaction lets the insect eject steam and other irritating chemicals with explosive force.

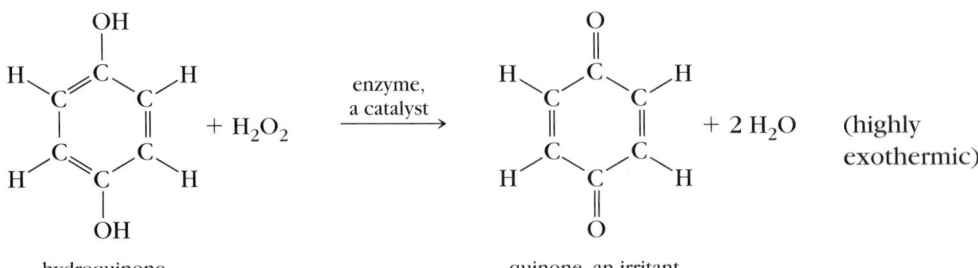

How does a catalyst or an enzyme help a reaction to go faster? It does so by participating in the reaction mechanism. That is, the mechanism for a catalyzed reaction is different from the mechanism of the same reaction without the catalyst. The rate-limiting step in the catalyzed mechanism has a lower activation energy and therefore is faster than the slow step for the uncatalyzed reaction. To see how this works, let us again consider conversion of *cis-* to *trans-*2-butene in the gas phase.

cis-2-butene   *trans*-2-butene   Rate $= k\,[\textit{cis}\text{-2-butene}]$

If a trace of gaseous molecular iodine, I_2, is added to a sample of *cis*-2-butene, the iodine accelerates the change to *trans*-2-butene. The iodine is neither consumed nor produced in the overall reaction, and so it does not appear in the overall balanced equation. However, because the reaction rate depends on the concentration of I_2, there is a term involving concentration of I_2 in the rate law for the catalyzed reaction.

$$\text{Rate} = k[\textit{cis-}2\text{-butene}][I_2]^{1/2}$$

The exponent of 1/2 for the concentration of I_2 in the rate law indicates the square root of the concentration. A square root dependence usually means that only half a molecule, in this case a single I atom, is involved in the mechanism.

The rate of the conversion of *cis*- to *trans*-2-butene changes because the presence of I_2 somehow changes the reaction mechanism. The best hypothesis is that iodine molecules first dissociate to form iodine atoms.

Step 1: I_2 dissociation

$$\tfrac{1}{2}[\, I_2(g) \longrightarrow 2\,I(g)\,]$$

(This equation is multiplied by $\tfrac{1}{2}$ because only one of the two I atoms from the I_2 molecule is needed in subsequent steps of the mechanism.) An iodine atom then attaches to the *cis*-2-butene molecule, breaking half of the double bond between the two central carbon atoms and allowing the ends of the molecule to twist freely relative to each other.

Step 2: Attachment of I atom to *cis*-2-butene

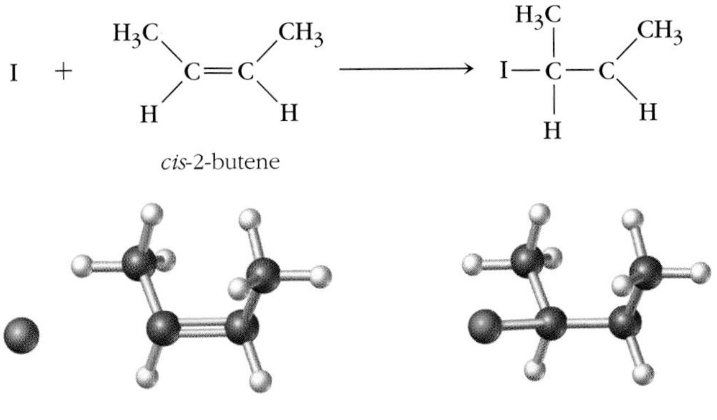

cis-2-butene

Step 3: Rotation around the C—C bond

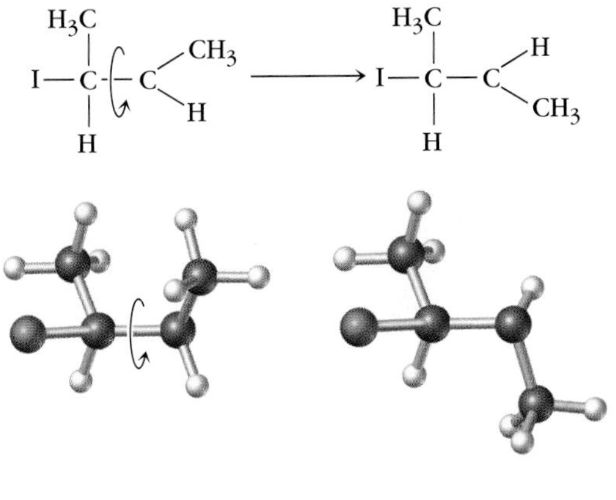

Step 4: Loss of an I atom and reformation of the carbon-carbon double bond

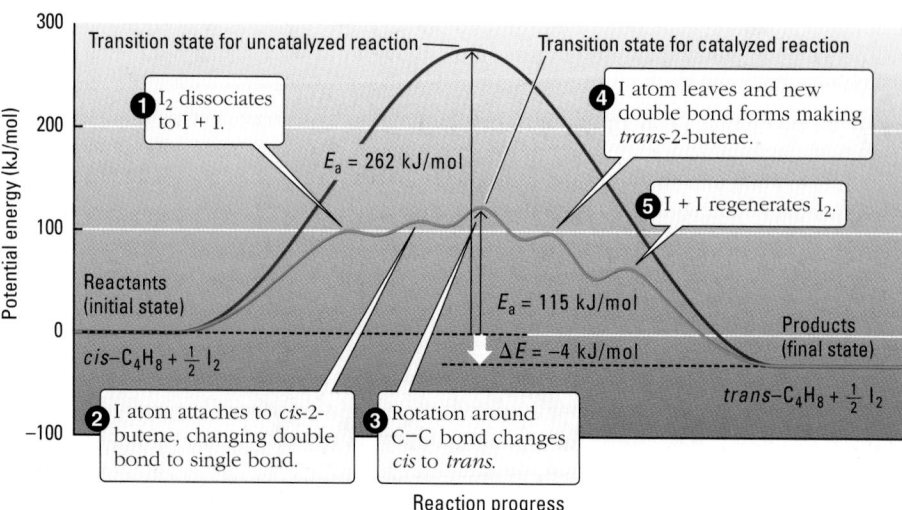

trans-2-butene

After the new double bond forms to give *trans*-2-butene and the iodine atom falls away, two iodine atoms come together to regenerate molecular iodine.

Step 5: I_2 regeneration

$$\tfrac{1}{2}[\, 2\, I(g) \longrightarrow I_2(g)\,]$$

$$\tfrac{1}{2}[\,\text{●●} \quad \text{●—●}\,]$$

There are five important points concerning this mechanism.

- The I_2 dissociates to atoms and then reforms. To an "outside" observer the concentration of I_2 is unchanged; I_2 is not involved in the balanced stoichiometric equation even though it has appeared in the mechanism. *This is generally true of catalysts.*

- Figure 13.14 shows that the activation energy barrier is significantly lower for the catalyzed reaction (because the mechanism changed). Consequently the reaction rate is much faster. Dropping the activation energy from 262 kJ/mol for the uncatalyzed reaction to 115 kJ/mol for the catalyzed process makes the catalyzed reaction 10^{15} times faster at a temperature of 500 K.

- The catalyzed mechanism has five reaction steps, and its energy-versus-reaction progress diagram (Figure 13.14) has five energy barriers (five humps appear in the curve).

Figure 13.14 Energy diagrams for catalyzed and uncatalyzed reactions. A catalyst accelerates a reaction by altering the mechanism so that the activation energy is reduced. With a smaller barrier to overcome, more reactant molecules have enough energy to cross the barrier, and reaction occurs more readily. (The steps involved are described in the text.) Notice that the shape of the barrier has changed because the mechanism has changed. This changes the activation energy, but not ΔE for the reaction.

[Figure: Potential energy (kJ/mol) versus Reaction progress]

Transition state for uncatalyzed reaction — Transition state for catalyzed reaction

❶ I_2 dissociates to I + I.

❹ I atom leaves and new double bond forms making *trans*-2-butene.

E_a = 262 kJ/mol

❺ I + I regenerates I_2.

Reactants (initial state)

cis–C_4H_8 + $\tfrac{1}{2}$ I_2

E_a = 115 kJ/mol

ΔE = –4 kJ/mol

Products (final state)

$trans$–C_4H_8 + $\tfrac{1}{2}$ I_2

❷ I atom attaches to *cis*-2-butene, changing double bond to single bond.

❸ Rotation around C–C bond changes *cis* to *trans*.

Reaction progress

- The catalyst I$_2$ and the reactant *cis*-2-butene are both in the gas phase during the reaction. When a catalyst is present in the same phase as the reacting substance or substances, it is called a **homogeneous catalyst**.

- Although the mechanism is different, the initial and final energies for the catalyzed reaction are the same as for the uncatalyzed reaction. This means that ΔE and ΔH are the same as for the uncatalyzed reaction.

Exercise 13.12   Catalysis

The oxidation of thallium(I) ion by cerium(IV) ion in aqueous solution has the equation

$$2\,Ce^{4+}(aq) + Tl^{+}(aq) \rightarrow 2\,Ce^{3+}(aq) + Tl^{3+}(aq)$$

The accepted mechanism for this reaction is

Step 1: $Ce^{4+}(aq) + Mn^{2+}(aq) \rightarrow Ce^{3+}(aq) + Mn^{3+}(aq)$

Step 2: $Ce^{4+}(aq) + Mn^{3+}(aq) \rightarrow Ce^{3+}(aq) + Mn^{4+}(aq)$

Step 3: $Mn^{4+}(aq) + Tl^{+}(aq) \rightarrow Mn^{2+}(aq) + Tl^{3+}(aq)$

(a) Verify that this mechanism predicts the overall reaction above.
(b) Identify all intermediates in this mechanism.
(c) Identify the catalyst in this mechanism.
(d) Suppose that the first step in this mechanism is rate-limiting. What would the rate law be?
(e) Suppose that the second step in this mechanism is rate-limiting. What would the rate law be?

 ## 13.9 ENZYMES: BIOLOGICAL CATALYSTS

Your body is a chemical factory that can manufacture a broad range of compounds that are needed so that you can move, breathe, digest food, see, hear, smell, and even think. But did you ever consider how the reactions that make those compounds are controlled? And how they can all occur reasonably quickly at the relatively low body temperature of 37 °C? Oxidation of glucose powers all the systems of your body, but you would not want it to take place at the temperature it does when cellulose (a polymer of glucose ⇐ *p. 564*) in wood burns in a fireplace. The chemical reactions of your body are catalyzed by enzymes. An **enzyme** is *a highly efficient catalyst for one or more chemical reactions in a living system.* The presence or absence of appropriate enzymes turns these reactions on or off by speeding them up or slowing them down. This allows your body to maintain nearly constant temperature and nearly constant concentrations of a variety of molecules and ions, an absolute necessity if you are to continue functioning.

Enzymes are almost always proteins — polymers of amino acids *(⇐ p. 556)*. Usually they are globular proteins, consisting of one or more long chains of amino acids folded into a nearly spherical shape. The shape of a globular protein is determined largely by noncovalent interactions *(⇐ p. 383)* among the amino-acid components (hydrogen bonds, attractions of opposite ionic charges, dipole-dipole and ion-dipole forces) plus a few weak covalent bonds. Also, nonpolar (hydrophobic) amino acid side groups congregate in the middle of the molecule to avoid the surrounding aqueous environment.

Enzymes are among the most effective catalysts known. They can increase reaction rates by factors of 10^9 to 10^{17}. Essentially every collision of the enzyme carbonic anhydrase with a carbonic acid molecule results in decomposition, and the enzyme can decompose about 36 million H$_2$CO$_3$ molecules every minute.

The weak covalent bonds are disulfide bonds. They occur between sulfur atoms in side chains of the amino acid cysteine. A cysteine side chain at one point in the protein can become bonded to a cysteine side chain much farther along the protein backbone *(⇐ p. 561)*.

$$H_2CO_3(aq) \xrightarrow{\text{carbonic anhydrase}} CO_2(g) + H_2O(\ell)$$

Most enzymes are highly specific catalysts. Some act on only one or two of the hundreds of different substances found in living cells. For example, carbonic anhydrase catalyzes only the decomposition of carbonic acid. Other enzymes can speed up several reactions, but usually these are all of the same type.

Some enzymes can act as catalysts entirely on their own. Others require one or more inorganic or organic molecules or ions called **cofactors** to be present before their catalytic activity becomes fully available. For example, many enzymes require nicotinamide adenine dinucleotide ion, NAD^+ (niacinamide ion). Molecules or ions that are cofactors are often derived from small quantities of minerals and vitamins in our diets. If the cofactor needed for an enzyme to catalyze a reaction is not available because of dietary deficiency, that reaction cannot occur when it is needed, and a bodily function will be impaired.

Enzyme Activity and Specificity

A molecule whose reaction is catalyzed by an enzyme is referred to as a **substrate.** In some cases there may be more than one substrate, as when an enzyme catalyzes transfer of a group from one molecule to another. Enzyme catalysis is so extremely effective and specific because the structure of the enzyme is finely tuned to minimize the activation energy barrier. Usually one part of the enzyme molecule, called the **active site**, interacts with the substrate via the same kinds of noncovalent attractions that hold the enzyme in its globular structure. The nanoscale structure of an enzyme's active site is specifically suited to attract and bind a substrate molecule and to help the substrate react.

When a substrate binds to an enzyme, both molecular structures can change. Each structure adjusts to fit closely with the other, and the structures become complementary. The change in shape of either the enzyme, the substrate, or both molecules when they bind is called **induced fit.** Enzymes catalyze reactions of only a few molecules because the structures of most molecules are not close enough to the structure of the active site for an induced fit to occur. The induced fit of a substrate to an enzyme also can lower the activation energy for a reaction.

CHEMISTRY *You Can Do...* Enzymes: Biological Catalysts

Raw potatoes contain an enzyme called *catalase*, which converts hydrogen peroxide to water and oxygen. You can demonstrate this by the following experiment:

Purchase a small bottle of hydrogen peroxide at a pharmacy or find one in your medicine chest. The peroxide is usually sold as a 3% solution in water. Pour about 50 mL of the peroxide solution into a clear glass or plastic cup. Add a small slice of a fresh potato to the cup. (Since potato is less dense than water, the potato will float.)

Almost immediately you will see bubbles of oxygen gas on the potato slice. (To make the bubbles more obvious you can add some dishwashing soap.)

1. Does the rate of evolution of oxygen change with time? If so, how does it change?
2. If you cool the hydrogen peroxide solution in a refrigerator and then do the experiment, is there a perceptible change in the initial rate of O_2 evolution?
3. Is there a difference between the time at which O_2 evolution begins for warm and for cold hydrogen peroxide?
4. What happens if you heat the slice of potato on a stove or in an oven before adding it to the peroxide solution?

For example, it may distort the substrate and stretch a bond that will be broken in the desired reaction. A schematic example of how this can work is shown in Figure 13.15. To see how this works in a specific case, consider the enzyme lysozyme, whose structure is shown in Figure 13.16 as a space-filling model with substrate in the active site. Lysozyme catalyzes hydrolysis reactions of polysaccharides (⬅ *p. 563*) found in bacterial cell walls. The reaction involved is

A hydrolysis reaction is the opposite of a condensation reaction. Most biopolymers are formed by condensation. Breaking them into their building block molecules requires hydrolysis, and many important enzymes catalyze hydrolysis reactions.

The section of polysaccharide shown in Figure 13.16 fits nicely into the cleft along the surface of the lysozyme, but many other long-chain molecules, such as polypeptides, might fit there as well. Shape is important, but so are noncovalent attractions and their positioning so that the substrate can make the most effective use of them. The enlarged portion of Figure 13.16 shows many hydrogen bonds between enzyme and substrate. It should be clear that the specificity of the enzyme depends not only on the shape of the active site, but also on the positions of hydrogen-bonding groups and groups that participate in other noncovalent interactions so that they can adjust to complementary sites on the substrate.

In Section 12.7 (⬅ *p. 541*) hydrolysis was described as a reaction in which a water molecule and some other molecule react, with both molecules splitting in two. The H from the water ends up with one part of the substrate molecule, and the OH ends up with the other part.

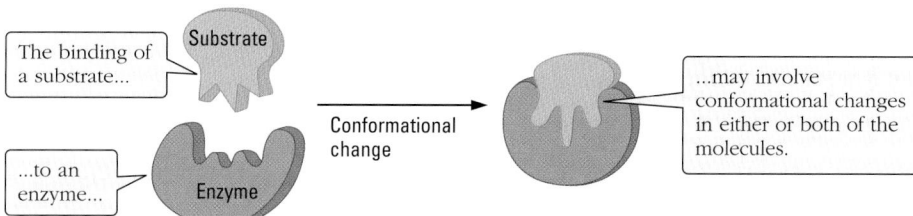

Figure 13.15 Induced fit of substrate to enzyme. Binding of a substrate to an enzyme may involve changing the shape of either or both molecules, thereby inducing them to fit together. In some cases a substrate molecule may be stretched or strained, helping bonds to break and reaction to occur.

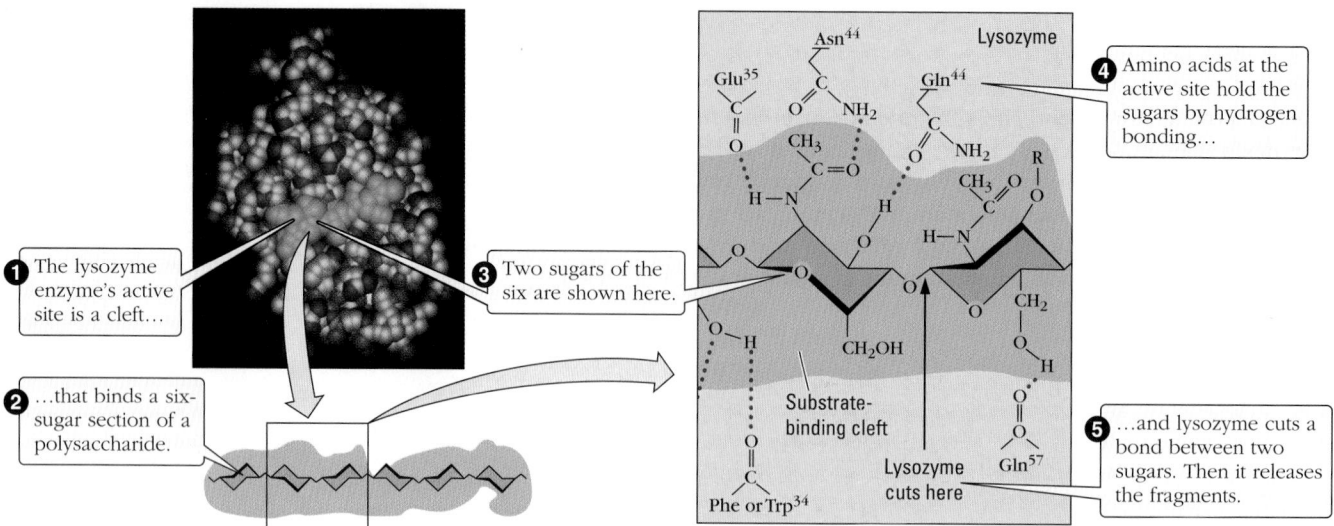

Figure 13.16 Lysozyme with substrate in the active site. The structure of lysozyme is shown at the left, with the atoms drawn as spheres occupying the space enclosed by their covalent radii. The active site is a cleft in the surface of the lysozyme that stretches horizontally across the middle of the enzyme. The active site is occupied by a portion of a polysaccharide molecule, the substrate (*green atoms*). (The part of the polysaccharide not bound to the active site has been omitted so that you can see the enzyme better.) The diagram at the right shows noncovalent interactions (*red dotted lines*) that hold the substrate to the enzyme. The bond that will be broken when the substrate is hydrolyzed is marked by an arrow.

To summarize, enzymes are extremely effective as catalysts for several reasons:

- Enzymes bring substrates into close proximity and hold them there while a reaction occurs.
- Enzymes hold substrates in the shape that is most effective for reaction.
- Enzymes can act as acids and bases during reaction, donating or accepting hydrogen ions from the substrate quickly and easily.
- The potential energy of a bond distorted by the induced fit of the substrate to the enzyme is already partway up the activation energy hill that must be surmounted for reaction to occur.
- Enzymes sometimes contain metal ions that are needed to help catalyze oxidation-reduction reactions.

Enzyme Kinetics

An enzyme changes the mechanism of a reaction, as does any catalyst. The first step in the mechanism for any enzyme-catalyzed reaction is binding of the substrate and enzyme. This is referred to as formation of an **enzyme-substrate complex.** Representing enzyme by E, substrate by S, and products by P, we can write a single-step uncatalyzed mechanism and a two-step enzyme-catalyzed mechanism as follows.

Uncatalyzed mechanism:$\qquad\qquad$ S $\longrightarrow$ P

Enzyme-catalyzed mechanism:

Step 1 (fast): S + E $\rightleftharpoons$ ES$\qquad$ (formation of enzyme-substrate complex)

Step 2 (slow):$\qquad$ ES $\longrightarrow$ P + E$\quad$ (formation of products and regeneration of enzyme)

That the enzyme is a catalyst is evident from the fact that it is a reactant in the first step and is regenerated in the second. This mechanism applies to nearly all enzyme-catalyzed reactions. Because the second step is slow, the enzyme-substrate

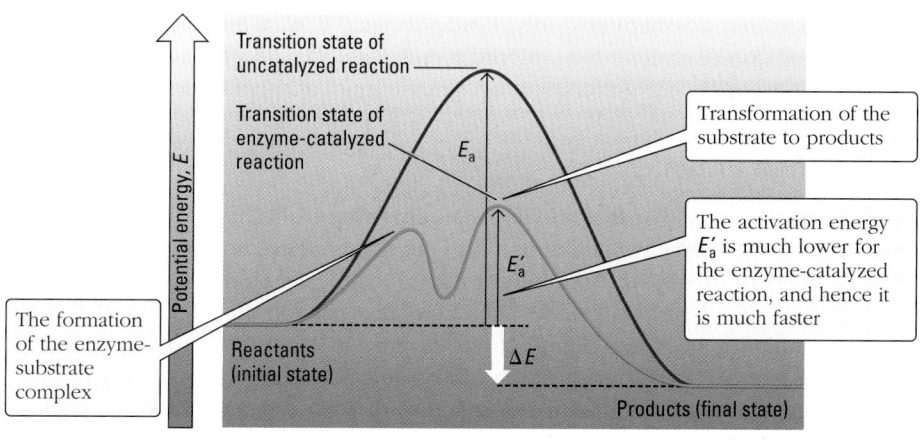

Figure 13.17 Energy diagram for enzyme-catalyzed reaction. (a) Energy profile for a typical reaction in a living system with no enzyme present. (b) Energy profile, drawn to the same scale, for the same reaction with enzyme catalysis.

complex can often separate and reform S + E before it reacts to form products. This is indicated by the double arrow in the first step. This mechanism for enzyme catalysis is similar to the mechanisms of reactions with a rapid, reversible first step that were discussed in Section 13.7 *(⇐ p. 601)*, and can be analyzed mathematically in the same way as those mechanisms were.

Because of the noncovalent interactions between enzyme and substrate, the activation energy is significantly lower for the enzyme-catalyzed reaction than it would be for the uncatalyzed process. This is shown in Figure 13.17. Even at temperatures only a little above room temperature, significant numbers of molecules have enough energy to surmount this lower barrier. Thus, enzyme-catalyzed reactions can occur reasonably quickly at body temperature.

Special Features of Enzyme Catalysis

Enzyme-catalyzed reactions obey the same principles of chemical kinetics that we discussed earlier in this chapter. However, there are some special features of both the enzyme itself and the mechanism of enzyme catalysis that you should be aware of. First, because of the form of the mechanism, either the enzyme or the substrate may be the limiting reactant in the first step. If the substrate is limiting, increasing the concentration of substrate produces more enzyme-substrate complex and makes the reaction go faster. This is the expected behavior: Increasing the concentration of a reactant should increase the rate in proportion. However, if the enzyme becomes the limiting reactant, it can become completely converted to enzyme-substrate complex, leaving no enzyme available for additional substrate. If this happens, further increase in concentration of substrate will not increase the rate of reaction. This means that there is a *maximum rate* (those who study enzyme kinetics call this the maximum velocity) for an enzyme-catalyzed reaction. The behavior of rate with increasing substrate concentration is shown in Figure 13.18.

Enzyme-catalyzed reactions also behave unusually with respect to temperature. The rate does increase with increasing temperature, but if the temperature gets high enough, there is a sudden decrease in rate, as shown in Figure 13.19. This happens because there is increased molecular and atomic motion as the temperature increases, and that motion can disrupt the structures of enzymes and other proteins. This change in protein structure is called **denaturation.** It occurs, for example, when an egg is boiled or fried. When a protein is denatured it loses its coiled globular structure, and its molecular structure becomes more linear. Once an enzyme's structure has changed, the active site is no longer available, enzyme catalysis is seriously impaired, and the reaction rate falls to its uncatalyzed

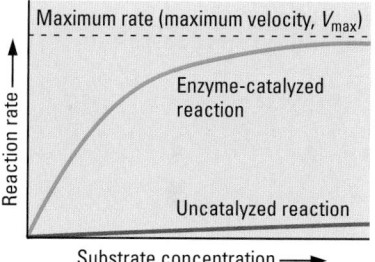

Figure 13.18 Maximum velocity for an enzyme-catalyzed reaction. Because there is only a limited quantity of enzyme available, increasing substrate concentration beyond the point at which the enzyme becomes the limiting reactant does not increase the rate further. There is a maximum rate (maximum velocity) for any enzyme-catalyzed reaction.

The enzymes in certain bacteria, such as those that inhabit hot springs as in Yellowstone National Park, are different from the enzymes in humans. They can withstand much higher temperatures without denaturing. Denaturation of enzymes answers the question "what happens when I hard boil an egg?" that was posed in Chapter 1 *(⇐ p. 32)*.

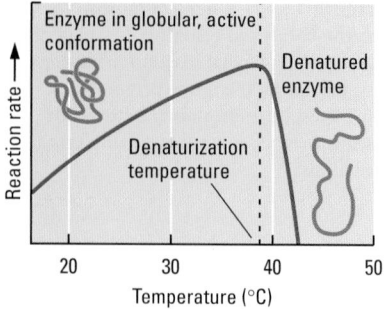

Figure 13.19 Enzyme activity destroyed by high temperature. At a temperature somewhat above normal body temperature there is sufficient molecular motion to overcome the noncovalent interactions that maintain protein structure. This disrupts the structure of an enzyme and thereby destroys its catalytic activity. The process by which the enzyme structure becomes disrupted is called denaturation.

value. Notice that this happens only a little above 37 °C, which is body temperature for humans. Enzymes have evolved to produce maximum rates at body temperature, and slightly higher temperatures cause most of them to denature.

Inhibition of Enzymes

There is another way that the activity of an enzyme can be destroyed. Some molecules or ions can fit an enzyme's active site, but remain there unreacted. Such a molecule is called an **inhibitor**. An inhibitor bound to an enzyme decreases its effective concentration and thereby decreases the rate of the reaction the enzyme catalyzes. If sufficient inhibitor becomes bound to an enzyme, the enzyme provides little catalytic effect because the concentration of available active sites becomes very small. An example of enzyme inhibition is the action of sulfa drugs on bacteria. Bacteria use *para*-aminobenzoic acid and an enzyme called dihydropteroate synthetase to synthesize folic acid, which is essential to their metabolism. Sulfa drugs bind to this enzyme, inhibit synthesis of folic acid, and destroy bacterial populations.

Problem-Solving Example 13.11   Enzyme Inhibition

The label of a container of methanol (methyl alcohol) invariably indicates that its contents are poisonous and should not be taken internally. This is because methanol, CH_3OH, which is not very toxic, is metabolized by the enzyme methanol oxidase to formaldehyde, $H_2C=O$, which is very toxic. Methanol poisoning is sometimes treated by giving the patient ethanol, CH_3CH_2OH, which inhibits the enzyme. Identify similarities and differences in the structures of methanol and ethanol that could account for ethanol's acting as an inhibitor.

Answer Both molecules are alcohols and can hydrogen-bond. Methanol has three hydrogens on the carbon next to the OH group. Ethanol has only two hydrogens on the carbon adjacent to the OH, and it has one more carbon and two more hydrogens.

Explanation Because its shape is similar to that of methanol and because it can form hydrogen bonds of similar strength, ethanol binds to the active sites of some of the methanol oxidase catalyst molecules. Because of the extra carbon atom or the difference in number of hydrogens adjacent to the OH group, the catalyst is unable to oxidize the ethanol, and ethanol remains bound. The smaller concentration of catalyst molecules decreases the rate of production of formaldehyde in the body, and the harmful effect is less.

Problem-Solving Practice 13.11

Bacteria need to use *p*-aminobenzoic acid to help synthesize folic acid in order to survive. Sulfa drugs interfere with this process. The structures of *p*-aminobenzoic acid and folic acid are

p-aminobenzoic acid

folic acid

pterin
(2-amino-4-oxopteridine)

p-aminobenzoic
acid

glutamates

Which of the following structures is most likely a sulfa drug? Explain your choice.

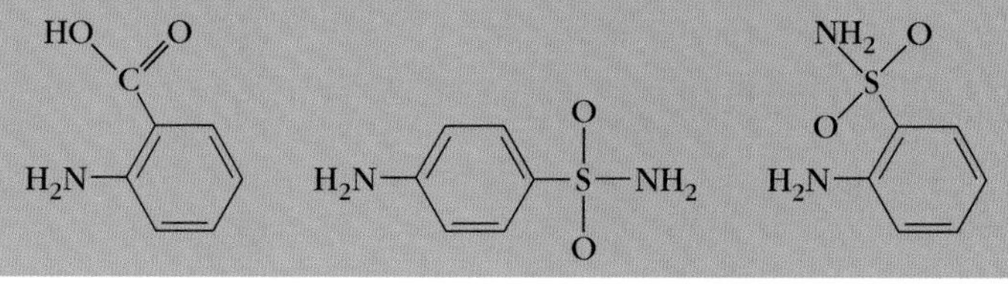

13.10 CATALYSIS IN INDUSTRY

An expert in the field of industrial chemistry has said that every year more than a trillion dollars' worth of goods is manufactured with the aid of manmade catalysts. Without them, fertilizers, pharmaceuticals, fuels, synthetic fibers, solvents, and detergents would be in short supply. Indeed, 90% of all manufactured items use catalysts at some stage of production. The major areas of catalyst use are in petroleum refining, industrial production of chemicals, and environmental controls. In this section we provide just a few examples of the many important industrial reactions that depend on catalysis.

Many industrial reactions use **heterogeneous catalysts**. These are catalysts that are present in a different phase from that of the reactants being catalyzed. Usually the catalyst is a solid and the reactants are in the gaseous or liquid phase. Heterogeneous catalysts are used in industry because they are more easily separated from the products and leftover reactants than are homogeneous catalysts. Catalysts for chemical processing are generally metal-based and often contain precious metals such as platinum and palladium. In the United States more than $600 million worth of such catalysts are used annually by the chemical processing industry, almost half of them in the preparation of polymers.

Manufacture of Acetic Acid

Acetic acid, CH_3COOH, has a place in the organic chemicals industry comparable to that of sulfuric acid in the inorganic chemicals industry; more than 4.7 billion pounds of acetic acid were made in the United States in 1995. Acetic acid is used widely in industry to make plastics and synthetic fibers, as a fungicide, and as the starting material for preparing many dietary supplements. One way of synthesizing the acid is an excellent example of homogeneous catalysis: rhodium(III) iodide is used to speed up the combination of carbon monoxide and methyl alcohol, both inexpensive chemicals, to form acetic acid.

$$CH_3OH \; + \; CO \xrightarrow{\text{RhI}_3 \text{ catalyst}} \; CH_3C\overset{\displaystyle O}{\overset{\|}{}}-OH$$

methyl alcohol   carbon monoxide   acetic acid

The role of the rhodium(III) iodide catalyst in this reaction is to bring the reactants together and allow them to rearrange to the products. Carbon monoxide and the methyl group from the alcohol become attached to the rhodium atom, which helps transfer the methyl group to the CO. After this rearrangement, the intermediate reacts with solvent water to form acetic acid.

Controlling Automobile Emissions

The largest growth in catalyst use is in *emissions control* for both automobiles and power plants. This market uses very large quantities of platinum group metals: platinum, palladium, rhodium, and iridium. In 1994, 52,800 kg of platinum—32%

CD-ROM Screen 15.15: Catalytic Converters

Chemistry in the News

PROTEASE INHIBITORS AND AIDS

A new class of drugs for treatment of AIDS has become available and widely publicized. These are called protease inhibitors, and they have slowed the spread of this disease considerably. They do this by inhibiting growth of HIV, human immunodeficiency virus.

As the name implies, protease inhibitors inhibit an enzyme, HIV-1 protease. This enzyme is essential for maturation of the virus because it catalyzes a reaction in which a long polypeptide chain is cut into shorter pieces. The cuts occur at specific locations along the chain, and the smaller pieces created by HIV-1 protease are proteins that are essential to the survival of HIV. Like plastic trash bags that have to be separated from a long roll before they

become useful, these proteins must be cut from the long polypeptide before they can carry out their functions in HIV. Several different proteins are produced this way by HIV-1 protease. Therefore, if this enzyme could be inhibited, reproduction of the virus would be interfered with in several different ways.

The action of HIV-1 protease, and its importance to HIV, is typical of how reactions are controlled in living organisms. It is much quicker to cut a long polypeptide chain into shorter, active protein molecules than it is to synthesize lots of protein molecules on short notice. Therefore, other enzymes work in advance together with the virus's DNA to synthesize the long polypeptide, called a pre-protein. HIV-1 protease then

chops the pre-protein into appropriate pieces whenever they are needed by HIV. If a lot of a particular protein is needed, it can be formed quickly by a few cuts in the pre-protein, instead of having to put together a large number of amino acids in the proper sequence.

HIV-1 protease actually consists of two polypeptide chains held together as a dimer by noncovalent attractive forces. The picture of HIV-1 protease shown here represents the polypeptide strands of the two monomers using ribbons and tubes instead of showing the individual atoms. This is done because showing the very large number of atoms in the enzyme would obscure your view of its overall structure. From the picture you can see that there is an open space in

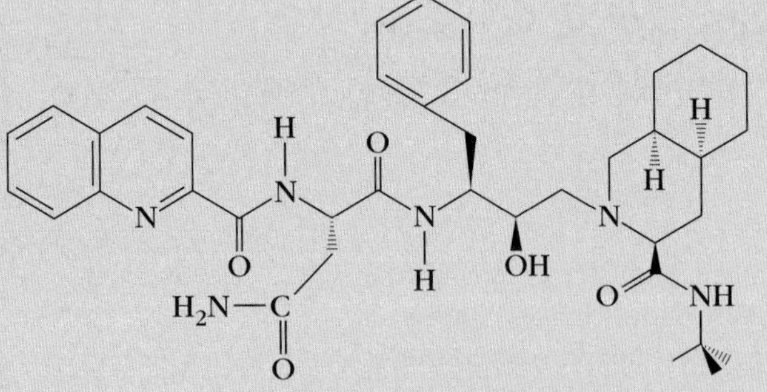

Invirase (saquinavir)

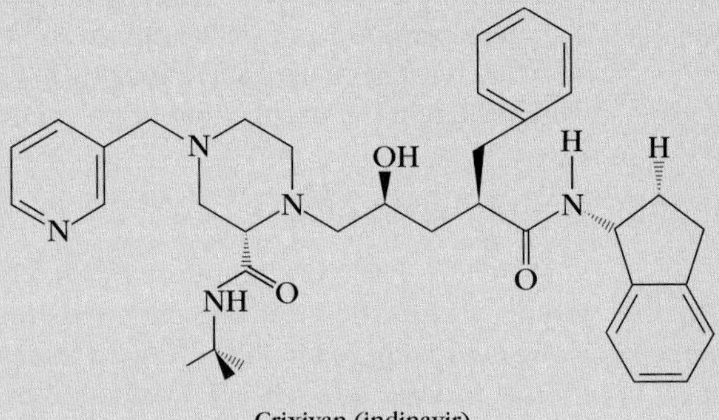

Crixivan (indinavir)

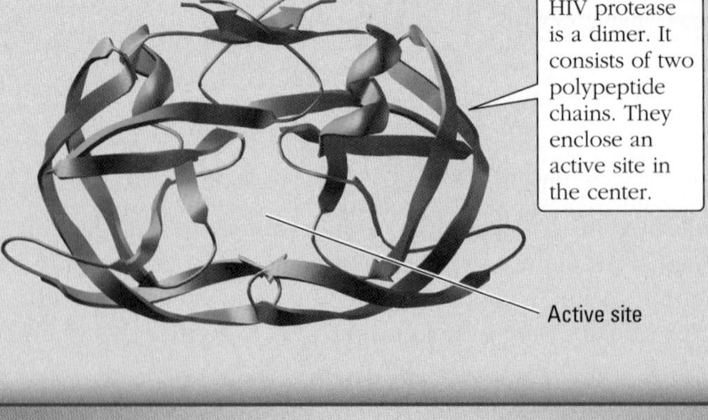

Structure of the HIV-1 protease dimer. HIV-1 protease consists of two similar parts that are held together by noncovalent attractions.

HIV protease is a dimer. It consists of two polypeptide chains. They enclose an active site in the center.

Active site

the middle—between the two halves of the enzyme. This is the active site. The enzyme works by having the two monomers come together to form an active site around the long pre-protein. This happens at a specific place along the pre-protein chain, and the active site cuts the polypeptide at that point by helping to break a peptide bond. Then the two monomers and the two pieces of polypeptide separate. The HIV-1 protease monomers can later cut another piece from the same or another polypeptide.

AIDS drugs that are protease inhibitors consist of molecules that can occupy the active site of HIV-1 protease, but their structures differ enough from the preprotein structure that the pro-

tease cannot cut them. They remain in the active site, as shown in the second figure, holding the dimer together and preventing HIV-1 protease from cutting any more preprotein molecules.

Several protease inhibitor molecules are now available for treatment of AIDS patients. The structures of four of them are shown here. These compounds are the result of much research to determine the structure and mechanism of action of HIV-1 protease, to design and synthesize molecules that block the active site, and to test these new drugs. Analyze the structures of the HIV-1 protease inhibitors shown here and note their similarities and differences. Try to identify hydrogen-bonding sites and sites for other noncovalent attractions that would

allow them to bind to the active site of an enzyme that cuts a protein (polypeptide) chain. Also note differences that might account for the protease's inability to cut the inhibitor molecules.

Source:

Based on information from *Science*, June 28, 1996; *FDA Consumer*, July–August 1999, and several Internet sites:
http://aids.org/immunet/atn.nsf/ homepage
http://www.aidsinfonyc.org/ network/trials/hiv.html
http://pharminfo.com/pubs/msb/ saquinavir.html
http://pharminfo.com/pubs/msb/ ritona.html

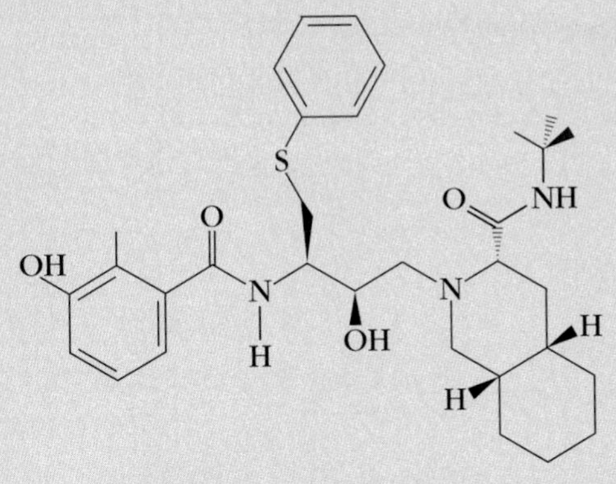

Viracept (nelfinavir mesylate)

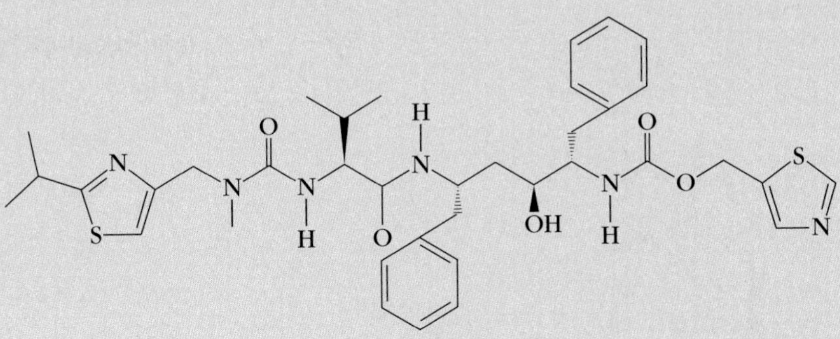

Norvir (ritonavir)

An inhibitor, bound in the active site, prevents HIV protease from acting as a catalyst.

HIV-1 protease dimer with inhibitor. An inhibitor molecule, drawn as a space-filling structure, occupies the active site of HIV-1 protease.

more than in 1993 — was sold in the United States for automotive uses. More than 7000 kg of palladium and rhodium was sold for this same purpose. All three metals are also used in chemical processing as catalysts, and the petroleum industry uses platinum and rhodium to catalyze refining processes.

The purpose of the catalysts in the exhaust system of an automobile is to ensure that the combustion of carbon monoxide and hydrocarbons is complete (Figure 13.20)

$$2\ CO(g) + O_2(g) \xrightarrow{\text{Pt-NiO catalyst}} 2\ CO_2(g)$$

$$2\ C_8H_{18}(g) + 25\ O_2(g) \xrightarrow{\text{Pt-NiO catalyst}} 16\ CO_2(g) + 18\ H_2O(g)$$

2,2,4-trimethylpentane,
a component of gasoline

and to convert nitrogen oxides to molecules less harmful to the environment. At the high temperature of combustion, some N_2 from air reacts with O_2 to give NO, a serious air pollutant. Thermodynamics informs us that nitrogen monoxide is unstable and should revert to N_2 and O_2. But remember that thermodynamics says nothing about rate. Unfortunately, the rate of reversion of NO to N_2 and O_2 is slow. Fortunately, catalysts have been developed that greatly speed this reaction.

$$2\ NO(g) \xrightarrow{\text{catalyst}} N_2(g) + O_2(g)$$

The role of the heterogeneous catalyst in the preceding reactions is probably to weaken the bonds of the reactants and to assist in product formation. For example, Figure 13.21 shows how NO molecules can dissociate into N and O atoms on the surface of a platinum metal catalyst.

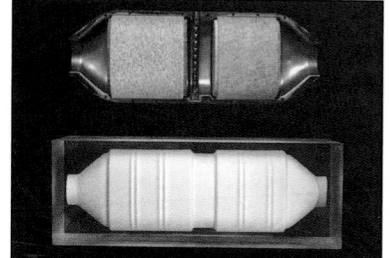

Figure 13.20 Automobile catalytic converter. Catalytic converters are standard equipment on the exhaust systems of all new automobiles. This one contains two catalysts: One converts nitrogen monoxide to nitrogen and the other converts carbon monoxide and hydrocarbons to carbon dioxide and water.

An NO molecule approaches the platinum surface...

...forms a bond with a platinum atom in the surface...

...and dissociates into an N atom and an O atom, each bonded to a platinum atom.

The N and O atoms migrate across the surface until they get close enough to another like atom to form N_2 or O_2...

...and the product molecules leave the surface.

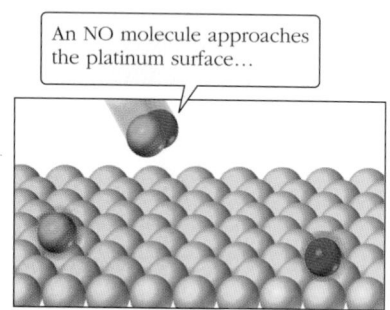
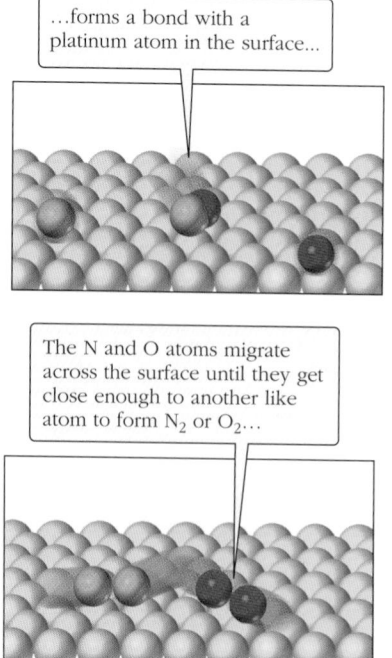
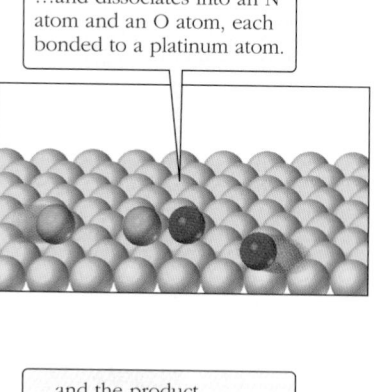
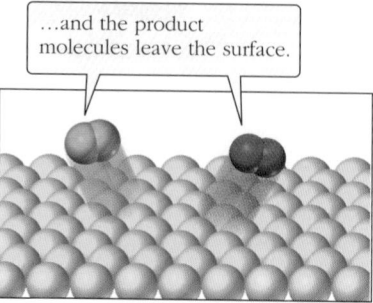

Figure 13.21 Catalytic conversion of NO to N_2 and O_2. A platinum surface can speed conversion of NO to N_2 and O_2 by helping to dissociate NO into N atoms and O atoms, which then travel across the surface and combine to form N_2 and O_2 molecules. The entire process of interaction with the surface and dissociation takes about 1.7×10^{-12} s.

Converting Methane to Liquid Fuel

Your home may be heated with natural gas, which consists largely of methane, CH_4. Although widely used, it is also widely wasted because much of it is found in geographical areas far removed from where fuels are consumed, and because transporting the flammable gas is expensive and dangerous. One solution to making methane useful is to convert it, where it is found, to a more readily transportable substance such as liquid methanol, CH_3OH. The methanol can then be used directly as a fuel, added to gasoline [as is currently done in some areas of the United States ($\Leftarrow$ *p. 532*)], or used to make other chemicals.

It has been known for some time that methane can be converted to carbon monoxide and hydrogen,

$$CH_4(g) + \tfrac{1}{2}O_2(g) \longrightarrow CO(g) + 2\,H_2(g)$$

and this mixture of gases can readily be turned into methanol in another step.

$$CO(g) + 2\,H_2(g) \longrightarrow CH_3OH(\ell)$$

Chemical engineers at the University of Minnesota have found that methane can in fact be converted to CO and H_2 under very mild conditions of temperature. They simply found the right catalyst. The photograph in Figure 13.22 shows what happens when a room temperature mixture of methane and oxygen flows through a heated, sponge-like ceramic disk coated with platinum or rhodium. Rather than oxidizing the methane all the way to water and carbon dioxide, the process produces a hot mixture of CO and H_2, which can be converted in good yield to methanol. It is also possible to produce other partially oxidized hydrocarbons by a similar catalytic process.

For more information on this discovery see *Science*, March 15, 1996, pp. 1560–1562, and *Science*, January 15, 1993, pp. 340–346.

Exercise 13.13   Catalysis

Which of the following statements is (are) true? If any are false, change the wording to make them true.

(a) The concentration of a homogeneous catalyst may appear in the rate law.
(b) A catalyst is always consumed in the overall reaction.
(c) A catalyst must always be in the same phase as the reactants.

Figure 13.22 Methane flowing through a catalyst.

SUMMARY PROBLEM

An excellent way to make pure nickel metal for use in specialized steel alloys is to decompose $Ni(CO)_4$ by heating it in a vacuum to slightly above room temperature.

$$Ni(CO)_4(g) \rightarrow Ni(s) + 4\,CO(g)$$

The reaction is proposed to occur in four steps, the first of which is

$$Ni(CO)_4(g) \rightarrow Ni(CO)_3(g) + CO(g)$$

Kinetic studies of this first-order decomposition reaction have been carried out between 47.3 °C and 66.0 °C to give the results in the table in the margin.*

(a) What is the activation energy for this reaction?

(b) $Ni(CO)_4$ is formed by the reaction of nickel metal with carbon monoxide. If you have 2.05 g of CO and you combine it with 0.125 g of nickel metal, what is the maximum quantity of $Ni(CO)_4$ (in grams) that can be formed?

Temperature (°C)	Rate constant (s^{-1})
47.3	0.233
50.9	0.354
55.0	0.606
60.0	1.022
66.0	1.873

*See Day, J. P., Basolo, F., and Pearson, R. G. *Journal of the American Chemical Society*, Vol. 90, 1968; p. 6933.

The replacement of CO by another molecule in $Ni(CO)_4$ (in the nonaqueous solvents toluene and hexane) was also studied to understand the general principles that govern the chemistry of such compounds.*

$$Ni(CO)_4 + P(CH_3)_3 \rightarrow Ni(CO)_3P(CH_3)_3 + CO$$

A detailed study of the kinetics of the reaction led to the mechanism

$$Ni(CO)_4 \longrightarrow Ni(CO)_3 + CO \qquad \text{Slow}$$

$$Ni(CO)_3 + P(CH_3)_3 \longrightarrow Ni(CO)_3P(CH_3)_3 \qquad \text{Fast}$$

(c) Which step in the mechanism is unimolecular? Which is bimolecular?

(d) Add the steps of the mechanism to show that the result is the balanced equation for the observed reaction.

(e) Is there an intermediate in this reaction? If so, what is it?

(f) It was found that doubling the concentration of $Ni(CO)_4$ led to an increase in reaction rate by a factor of 2. Doubling the concentration of $P(CH_3)_3$ had no effect on the reaction rate. Based on this information, write the rate equation for the reaction.

(g) Does the experimental rate equation support the proposed mechanism? Why or why not?

IN CLOSING

Having studied this chapter, you should be able to . . .

- Define reaction rate and calculate average rates (Section 13.1).
- Describe the effect reactant concentrations have on reaction rate, and determine rate laws and rate constants from initial rates (Section 13.2).
- Determine reaction orders from a rate law, and use the integrated rate law method to obtain orders and rate constants (Section 13.3).
- Calculate concentration from time, time to reach a certain concentration, and half-life for a first-order reaction (Section 13.3).
- Define and give examples of unimolecular and bimolecular elementary reactions (Section 13.4).
- Show by using an energy profile what happens as two reactant molecules interact to form product molecules (Section 13.4).
- Define activation energy and frequency factor, and use them to calculate rate constants and rates under different conditions of temperature and concentration (Section 13.5).
- Derive rate laws for unimolecular and bimolecular elementary reactions (Section 13.6).
- Define reaction mechanism and identify rate-limiting steps and intermediates (Section 13.7).
- Given several reaction mechanisms, decide which is (are) in agreement with experimentally determined stoichiometry and rate law (Section 13.7).

*See Day, J. P., Basolo, F., and Pearson, R. G. *Journal of the American Chemical Society*, Vol. 90, 1968; p. 6927.

- Explain how a catalyst can speed up a reaction; draw energy profiles for catalyzed and uncatalyzed reaction mechanisms (Section 13.8).
- Define the terms enzyme, substrate, and inhibitor, and identify similarities and differences between enzyme-catalyzed reactions and uncatalyzed reactions (Section 13.9).
- Describe several important industrial processes that depend on catalysts (Section 13.10).

KEY TERMS

activated complex *(13.4)*

activation energy (E_a) *(13.4)*

active site *(13.9)*

Arrhenius equation *(13.5)*

average reaction rate *(13.1)*

bimolecular reaction *(13.4)*

chemical kinetics *(Introduction)*

cofactor *(13.9)*

denaturation *(13.9)*

elementary reaction *(13.4)*

enzyme *(13.9)*

enzyme-substrate complex *(13.9)*

frequency factor *(13.5)*

half-life *(13.3)*

heterogeneous catalyst *(13.10)*

heterogeneous reaction *(13.1)*

homogeneous catalyst *(13.8)*

homogeneous reaction *(13.1)*

induced fit *(13.9)*

inhibitor *(13.9)*

initial rate *(13.2)*

instantaneous reaction rate *(13.1)*

intermediate *(13.7)*

order of reaction *(13.3)*

overall reaction order *(13.3)*

rate *(13.1)*

rate constant *(13.2)*

rate law *(13.2)*

rate-limiting step *(13.7)*

reaction intermediate *(13.7)*

reaction mechanism *(13.7)*

reaction rate *(13.1)*

steric factor *(13.4)*

substrate *(13.9)*

transition state *(13.4)*

unimolecular reaction *(13.4)*

QUESTIONS FOR REVIEW AND THOUGHT

Conceptual Challenge Problems

CP-13.A (Section 13.5)   A rule of thumb is that for a typical reaction, if concentrations are unchanged, a 10-K rise in temperature increases the reaction rate by 2 to 4 times. Use an average increase of 3 times to answer the questions below.

(a) What is the approximate activation energy of a "typical" chemical reaction at 298 K?

(b) If a catalyst increases a chemical reaction's rate by providing a mechanism that has a lower activation energy, then what change do you expect a 10-K increase in temperature to make in the rate of a reaction whose uncatalyzed activation energy of 75 kJ/mol has been lowered to one-half this value (at 298 K) by addition of a catalyst?

CP-13.B (Section 13.7)   A sentence in an introductory chemistry textbook reads, "Dioxygen reacts with itself to form trioxygen, ozone, according to the equation, 3 $O_2 \rightarrow$ 2 O_3." As a student of chemistry, what would you write to criticize this sentence?

CP-13.C (Section 13.7)   A classmate consults you about a problem concerning the reaction of nitrogen monoxide and dioxygen in the gas phase. She has been told that the reaction is second-order in nitrogen monoxide and first-order in dioxygen; hence, the rate law may be written as rate = $k[NO]^2[O_2]$. She has been asked to propose a mechanism for this reaction. She proposes that the mechanism is this single equation,

$$NO + NO + O_2 \longrightarrow NO_2 + NO_2$$

She asks your opinion about whether this is correct. What is your answer?

CP-13.D (Section 13.8)   Polypropylene (⇐ *p. 546)*, Table 12.6) is an important type of plastic, and about 3 million tons of it are produced every year in the United States. Properties of polypropylene can be changed by the way it is made. For example, melting points between 130 °C and 160 °C can be obtained by using an appropriate catalyst to polymerize propylene (propene, C_3H_6). Two important types of polypropylene are isotactic, in which the methyl groups are all on the same side of the polymer chain, and syndiotactic, in which the methyl groups alternate between one side and the other of the chain.

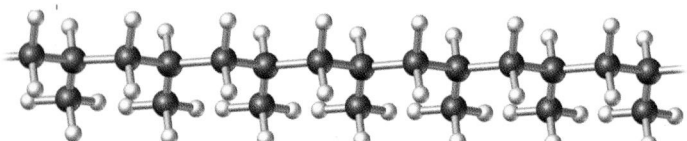

isotactic polypropylene

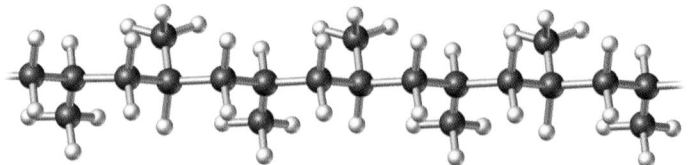

syndiotactic polypropylene

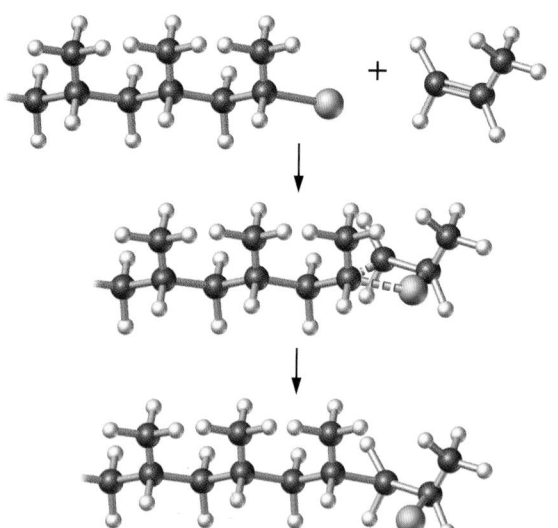

Suppose that you are part of a team designing a new catalyst to polymerize propylene. Your catalyst is going to have a zirconium atom at the center of a structure consisting of carbon and hydrogen atoms. The zirconium atom will hold the growing polypropylene chain by bonding to one end of it. The metal atom will also attract a propylene molecule and bond to it before transferring the growing polypropylene chain to the other end of the new propylene molecule. The process is shown here.

What would be a reasonable shape for the rest of the catalyst molecule surrounding the metal atom so that isotactic polypropylene would be produced? It may help to build molecular models to see how each new propylene molecule needs to be added to the growing polymer chain in order to get all the methyl groups on the same side.

Answers to questions in **bold** can be found in the back of the book.

Review Questions

1. Which of these is appropriate for determining the rate law for a chemical reaction?
   (a) Theoretical calculations based on balanced equations
   (b) Measuring the rate of the reaction as a function of the concentrations of the reacting species
   (c) Measuring the rate of the reaction as a function of temperature
2. Name at least three factors that affect the rate of a chemical reaction.
3. Using the rate law, rate $= k[A]^2[B]$, define the order of the reaction with respect to A and B and the overall reaction order.
4. Draw a reaction energy diagram for an exothermic process. Mark the positions of reactants, products, and activated complex. Indicate the activation energies of the forward and reverse processes and explain how ΔE for the reaction can be calculated from the diagram.
5. Draw a reaction energy diagram for an endothermic process. Mark the positions of reactants, products, and activated complex. Indicate the activation energies of the forward and reverse processes and explain how ΔE for the reaction can be calculated from the diagram.
6. Indicate whether each of these statements is true or false. Change the wording of each false statement to make it true.
   (a) It is possible to change the rate constant for a reaction by changing the temperature.
   (b) The reaction rate remains constant as a first-order reaction proceeds at a constant temperature.

(c) The rate constant for a reaction is independent of reactant concentrations.
   (d) As a second-order reaction proceeds at a constant temperature, the rate constant changes.
7. Consider the class of substances known as catalysts.
   (a) Define "catalyst."
   (b) What effect does a catalyst have on the energy barrier for a reaction?
   (c) What special characteristics do enzymes have that distinguish them from other catalysts?
8. Define the terms "enzyme," "substrate," and "inhibitor," and give an example of each kind of molecule.
9. Explain the difference between a homogeneous and a heterogeneous catalyst. Give an example of each.
10. Define the terms "unimolecular elementary reaction" and "bimolecular elementary reaction," and give an example of each.
11. Using an equation, define the terms "activation energy" and "frequency factor."
12. Why is reaction kinetics important to our understanding of depletion of stratospheric ozone (*p. 441*).

Reaction Rate

13. Consider the dissolving of sugar as a simple process in which kinetics is important. Suppose that you dissolve an equal mass of each kind of sugar listed. Which dissolves the fastest? Which dissolves the slowest? Explain why in terms of rates of heterogeneous reactions. (If you are not sure which is fastest or slowest, try them all out.)
    (a) Rock candy sugar (large sugar crystals)
    (b) Sugar cubes
    (c) Granular sugar
    (d) Powdered sugar

14. A cube of aluminum 1.0 cm on each edge is placed into 9 M NaOH(aq), and the rate at which H_2 gas is given off is measured.

(a) By what factor will this reaction rate change if the aluminum cube is cut exactly in half and the two halves are placed in the solution? Assume that the reaction rate is proportional to the surface area, and that all of the surface of the aluminum is in contact with the NaOH(aq).

(b) If you had to speed up this reaction as much as you could without raising the temperature, what would you do to the aluminum?

15. Experimental data in the table are for the hypothetical reaction

$$A \longrightarrow 2 B$$

Time (s)	[A] (mol/L)
0.00	1.000
10.0	0.833
20.0	0.714
30.0	0.625
40.0	0.555

(a) Make a graph of concentration as a function of time, draw a smooth curve through the points, and calculate the rate of change of [A] for each 10-s interval from 0 to 40 s. Why might the rate of change decrease from one time interval to the next?

(b) How is the rate of change of [B] related to the rate of change of [A] in the same time interval?

(c) Calculate the rate of change of [B] for the time interval from 10 to 20 s.

16. A compound called phenyl acetate reacts with water according to the equation

$$CH_3C{-}O{-}C_6H_5 + H_2O \longrightarrow$$
phenyl acetate

$$CH_3C{-}O{-}H + C_6H_5{-}O{-}H$$
acetic acid phenol

These data were collected at 5 °C.

Time (min)	[Phenyl acetate] (mol/L)
0	0.55
0.25	0.42
0.50	0.31
0.75	0.23
1.00	0.17
1.25	0.12
1.50	0.082

(a) Make a graph of concentration as a function of time, describe the shape of the curve, and compare it with Figure 13.3.

(b) Calculate the rate of change of the concentration of phenylacetate during the period from 0.20 to 0.40 min, and then during the period from 1.2 to 1.4 min. Compare the values and tell why one is smaller than the other.

(c) What is the rate of change of the phenol concentration during the period from 1.00 to 1.25 min?

17. Using data given in the table for the reaction

$$N_2O_5 \longrightarrow 2 NO_2 + \tfrac{1}{2} O_2$$

calculate the average rate of reaction during each of the following intervals:

(a) 0 to 0.5 h (b) 0.5 to 1.0 h
(c) 1.0 to 2.0 h (d) 2.0 to 3.0 h
(e) 3.0 to 4.0 h (f) 4.0 to 5.0 h

Time (h)	[N$_2$O$_5$] (mol/L)	Time (h)	[N$_2$O$_5$] (mol/L)
0	0.849	2.00	0.472
0.50	0.733	3.00	0.352
1.00	0.633	4.00	0.262

18. Using all your calculated rates from Question 17,

(a) Show that the reaction obeys the rate law

$$Rate = -\frac{\Delta[N_2O_5]}{\Delta t} = k[N_2O_5]$$

(b) Evaluate the rate constant k as an average of the values obtained for the six intervals.

19. Using data from Question 17, calculate the average rate over the interval 0 to 5.0 h. Compare your result with the average rates over the intervals 1.0 to 4.0 h and 2.0 to 3.0 h, all of which have the same midpoint (2.5 h from the start).

20. Using the rate law and the rate constant you calculated in Question 18, calculate the reaction rate exactly 2.5 h from the start. Do your results from this and Problem 19 agree with the statement in the text that the smaller the time interval, the more accurate the average rate?

21. For the reaction

$$2 NO_2(g) \longrightarrow 2 NO(g) + O_2(g)$$

make qualitatively correct plots of the concentrations of $NO_2(g)$, $NO(g)$, and $O_2(g)$ versus time. Draw all three graphs on the same axes, assume that you start with $NO_2(g)$ at a concentration of 1.0 mol/L, and assume that the reaction is first-order. Explain how you would determine, from these plots,

(a) The initial rate of the reaction, and

(b) The final rate (i.e., the rate as time approaches infinity)

22. For the reaction

$$O_3(g) + O(g) \longrightarrow 2 O_2(g)$$

make qualitatively correct plots of the concentrations of $O_3(g)$, $O(g)$, and $O_2(g)$ versus time. Draw all three graphs on the same axes, assume that you start with $O_3(g)$ and $O(g)$, each at a concentration of 1.0 μmol/L, and assume that the reaction is second-order. Explain how you would determine, from these plots,

(a) The initial rate of the reaction

(b) The final rate (i.e., the rate as time approaches infinity)

Effect of Concentration on Reaction Rates

23. If a reaction has the experimental rate law rate $= k[A]^2$, explain what happens to the rate when

(a) The concentration of A is tripled

(b) The concentration of A is halved

24. A reaction has the experimental rate law rate $= k[A]^2[B]$. If the concentration of A is doubled and the concentration of B is halved, what happens to the reaction rate?

25. The reaction of $CO(g) + NO_2(g)$ is second-order in NO_2 and zeroth-order in CO at temperatures less than 500 K.

(a) Write the rate law for the reaction.

(b) How will the reaction rate change if the NO_2 concentration is halved?

(c) How will the reaction rate change if the concentration of CO is doubled?

26. Nitrosyl bromide, NOBr, is formed from NO and Br_2.

$$2 NO(g) + Br_2(g) \longrightarrow 2 NOBr(g)$$

Experiment shows that the reaction is first-order in Br_2 and second-order in NO.

(a) Write the rate law for the reaction.

(b) If the concentration of Br_2 is tripled, how will the reaction rate change?

(c) What happens to the reaction rate when the concentration of NO is doubled?

27. For the reaction of $Pt(NH_3)_2Cl_2$ with water (Section 13.1),

$$Pt(NH_3)_2Cl_2 + H_2O \longrightarrow Pt(NH_3)_2(H_2O)Cl^+ + Cl^-$$

the rate law was given as rate $= k[Pt(NH_3)_2Cl_2]$ with $k = 0.090 \, h^{-1}$. Calculate the initial rate of reaction when the concentration of $Pt(NH_3)_2Cl_2$ is

(a) 0.010 M

(b) 0.020 M

(c) 0.040 M

(d) How does the rate of disappearance of $Pt(NH_3)_2Cl_2$ change with its initial concentration?

(e) How is this related to the rate law?

(f) How does the initial concentration of $Pt(NH_3)_2Cl_2$ affect the rate of appearance of Cl^- in the solution?

28. Methyl acetate, CH_3COOCH_3, reacts with base to break one of the C—O bonds.

$$CH_3\overset{O}{\overset{\|}{C}}-O-CH_3(aq) + OH^-(aq) \longrightarrow$$

$$CH_3\overset{O}{\overset{\|}{C}}-O^-(aq) + HO-CH_3(aq)$$

The rate law is rate $= k[CH_3COOCH_3][OH^-]$ where $k = 0.14 \, L \, mol^{-1} \, s^{-1}$ at 25 °C.

(a) What is the initial rate at which the methyl acetate disappears when both reactants, CH_3COOCH_3 and OH^-, have a concentration of 0.025 M?

(b) How rapidly (i.e., at what rate) does the methyl alcohol, CH_3OH, initially appear in the solution?

29. A study of the hypothetical reaction $2 A + B \rightarrow C + D$ gave these results:

Experiment	Initial concentration (mol/L)		Initial rate (mol L^{-1} s^{-1})
	[A]	[B]	
1	0.10	0.050	6.0×10^{-3}
2	0.20	0.050	1.2×10^{-2}
3	0.30	0.050	1.8×10^{-2}
4	0.20	0.150	1.1×10^{-1}

(a) What is the rate law for this reaction?

(b) Calculate the rate constant k and express it in appropriate units.

30. Measurements of the initial rate of reaction between triphenylmethyl hexachloroantimonate (**I**) and bis-(9-ethyl-3-carbazolyl)methane (**II**) in 1,2-dichloroethane at 40 °C yielded these data:

Initial concentration $\times 10^5$ (mole/L)		Initial rate $\times 10^9$ (mole L^{-1} s^{-1})
[I]	[II]	
1.65	10.6	1.50
14.9	10.6	17.7
14.9	7.10	11.2
14.9	3.52	6.30
14.9	1.76	3.10
4.97	10.6	4.52
2.48	10.6	2.70

(a) Derive the rate law for this reaction.

(b) Calculate the rate constant k and express it in appropriate units.

31. Measurements of the initial rate of hydrolysis of benzenesulfonyl chloride in aqueous solution at 15 °C in the presence of fluoride ion yielded the results in the table for a fixed concentration of benzenesulfonyl chloride of 2×10^{-4} M. The reaction rate is known to be proportional to the concentration of benzenesulfonyl chloride.

$[F^-] \times 10^2$ (mol/L)	Initial rate $\times 10^7$ (mole L^{-1} s^{-1})
0	2.4
0.5	5.4
1.0	7.9
2.0	13.9
3.0	20.2
4.0	25.2
5.0	32.0

Note that some reaction must be occurring in the absence of any fluoride ion, because at zero concentration of fluoride the rate is not zero. This residual rate should be subtracted from each observed rate to give the rate of the reaction being studied.

(a) Derive the complete rate law for the reaction.

(b) Calculate the rate constant k and express it in appropriate units.

32. The hypothetical reaction

$$2 A + 2 B \longrightarrow C + 3 D$$

was studied by measuring the initial rate of appearance of C. These data were obtained:

[A] (mol/L)	[B] (mol/L)	Initial rate (mol L^{-1} s^{-1})
6.0×10^{-3}	1.0×10^{-3}	0.012
6.0×10^{-3}	2.0×10^{-3}	0.024
2.0×10^{-3}	1.5×10^{-3}	0.0020
4.0×10^{-3}	1.5×10^{-3}	0.0080

(a) What is the order of the reaction with respect to substance A?

(b) What is the order with respect to B?

(c) What is the overall order?

(d) What is the rate law?

(e) Calculate the rate constant.

(f) If at a given instant A is disappearing at a rate of 0.034 mol L^{-1} s^{-1}, what is the rate of appearance of C? What is the rate of appearance of D?

33. For the reaction

$$2 NO(g) + 2 H_2(g) \longrightarrow N_2(g) + 2 H_2O(g)$$

these data were obtained at 1100 K:

[NO] (mol/L)	[H$_2$] (mol/L)	Initial rate (mol L^{-1} s^{-1})
5.00×10^{-3}	2.50×10^{-3}	3.0×10^{-3}
15.0×10^{-3}	2.50×10^{-3}	9.0×10^{-3}
15.0×10^{-3}	10.0×10^{-3}	3.6×10^{-3}

(a) What is the order with respect to NO? with respect to H$_2$?

(b) What is the overall order?

(c) Write the rate law.

(d) Calculate the rate constant.

(e) Calculate the initial rate of reaction when [NO] = [H$_2$] = 8.0×10^{-3} mol L^{-1}.

34. For the reaction of NO and O$_2$ at 660 K,

$$2 NO(g) + O_2(g) \longrightarrow 2 NO_2(g)$$

Concentration (mol/L)		Rate of disappearance of NO (mol L^{-1} s^{-1})
[NO]	[O$_2$]	
0.010	0.010	2.5×10^{-5}
0.020	0.010	1.0×10^{-4}
0.010	0.020	5.0×10^{-5}

(a) Determine the order of the reaction for each reactant.

(b) Write the rate equation for the reaction.

(c) Calculate the rate constant.

(d) Calculate the rate when [NO] = 0.025 mol/L and [O$_2$] = 0.050 mol/L.

(e) If O$_2$ disappears at a rate of 1.0×10^{-4} mol L^{-1} s^{-1}, what is the rate at which NO disappears? What is the rate at which NO$_2$ is forming?

35. Nitryl fluoride is an explosive compound that can be made by oxidizing nitrogen dioxide with fluorine:

$$2 NO_2(g) + F_2(g) \longrightarrow 2 NO_2F(g)$$

Several kinetics experiments involving formation of nitryl fluoride are summarized in the following table:

Experiment	Initial concentrations (mol/L)			Initial rate (mol L^{-1}s^{-1})
	[NO$_2$]	[F$_2$]	[NO$_2$F]	
1	0.0010	0.0050	0.0020	2.0×10^{-4}
2	0.0020	0.0050	0.0020	4.0×10^{-4}
3	0.0020	0.0020	0.0020	1.6×10^{-4}
4	0.0020	0.0020	0.0010	1.6×10^{-4}

(a) Write the rate law for the reaction.

(b) What is the order of the reaction with respect to each reactant and each product?

(c) Calculate the rate constant k and express it in appropriate units.

Rate Law and Order of Reaction

36. For each of these rate laws, state the reaction order with respect to the hypothetical substances A and B, and give the overall order.

(a) Rate = k[A][B]^3 (b) Rate = k[A][B]

(c) Rate = k[A]   (d) Rate = k[A]^3[B]

37. For each of the rate laws below, what is the order of the reaction with respect to the hypothetical substances X, Y, and Z? What is the overall order?

(a) Rate = k[X][Y][Z]   (b) Rate = k[X]^2[Y]$^{1/2}$[Z]

(c) Rate = k[X]$^{1.5}$[Y]$^{-1}$ (d) Rate = k[X]/[Y]^2

38. A reaction A + B → products is found to be second-order in B. Which rate equation cannot be correct?

(a) Rate = k[A][B]   (b) Rate = k[A][B]^2

(c) Rate = k[B]^2

39. For the reaction of phenyl acetate with water the concentration as a function of time was given in Question 16. Assume that the concentration of water does not change during the reaction. Analyze the data from Question 16 to determine

(a) The rate law

(b) The order of the reaction with respect to phenyl acetate

(c) The rate constant

(d) The rate of reaction when the concentration of phenyl acetate is 0.10 mol/L (assuming that the concentration of water is the same as in the experiments in the table in Question 16)

40. When phenacyl bromide and pyridine are both dissolved in methanol, they react to form phenacylpyridinium bromide.

$$C_6H_5-\overset{\overset{\displaystyle O}{\|}}{C}-CH_2Br + C_5H_5N \longrightarrow$$

$$C_6H_5-\overset{\overset{\displaystyle O}{\|}}{C}-CH_2NC_5H_5^+ + Br^-$$

When equal concentrations of reactants were mixed in methanol at 35 °C, these data were obtained.

Time (min)	[Reactant] (mol/L)	Time (min)	[Reactant] (mol/L)
0	0.0385	500	0.0208
100	0.0330	600	0.0191
200	0.0288	700	0.0176
300	0.0255	800	0.0163
400	0.0220	1000	0.0143

(a) What is the rate law for this reaction?
(b) What is the overall order?
(c) What is the rate constant?
(d) What is the rate of this reaction when the concentration of each reactant is 0.030 mol/L?

41. The compound p-methoxybenzonitrile N-oxide, $CH_3OC_6H_4CNO$, reacts with itself to form a dimer — a molecule that consists of two p-methoxybenzonitrile N-oxide units connected together $(CH_3OC_6H_4CNO)_2$. The reaction can be represented as

$$A + A \longrightarrow B \quad \text{or} \quad 2\,A \longrightarrow B$$

where A represents p-methoxybenzonitrile N-oxide and B represents the dimer $(CH_3OC_6H_4CNO)_2$. For the reaction in carbon tetrachloride at 40 °C with an initial concentration of 0.011 M, these data were obtained:

Time (min)	Percent reaction	Time (min)	Percent reaction
0	0	942	60.9
60	9.1	1080	64.7
120	16.7	1212	66.6
215	26.5	1358	68.5
325	32.7	1518	70.3
565	47.3		

(a) Determine the rate law for the reaction.
(b) What is the rate constant?
(c) What is the order of the reaction with respect to A?

42. The transfer of an oxygen atom from NO_2 to CO has been studied at 540 K:

$$CO(g) + NO_2(g) \longrightarrow CO_2(g) + NO(g)$$

These data were collected:

Initial rate (mol L^{-1} h^{-1})	Initial concentration (mol/L)	
	[CO]	[NO$_2$]
5.1×10^{-4}	0.35×10^{-4}	3.4×10^{-8}
5.1×10^{-4}	0.70×10^{-4}	1.7×10^{-8}
5.1×10^{-4}	0.18×10^{-4}	6.8×10^{-8}
1.0×10^{-3}	0.35×10^{-4}	6.8×10^{-8}
1.5×10^{-3}	0.35×10^{-4}	10.2×10^{-8}

Use the data in the table to
(a) Write the rate law.
(b) Determine the reaction order with respect to each reactant.
(c) Calculate the rate constant, and express it in appropriate units.

43. The bromination of acetone is catalyzed by acid.

$$CH_3COCH_3(aq) + Br_2(aq) + H_2O(\ell) \xrightarrow{\text{acid catalyst}}$$
$$CH_3COCH_2Br\,(aq) + H_3O^+(aq) + Br^-(aq)$$

The rate of disappearance of bromine was measured for several different initial concentrations of acetone, bromine, and hydronium ion.

Initial concentration (mol/L)			Initial rate of change of [Br$_2$] (mol L^{-1} s^{-1})
[CH$_3$COCH$_3$]	[Br$_2$]	[H$_3$O$^+$]	
0.30	0.05	0.05	5.7×10^{-5}
0.30	0.10	0.05	5.7×10^{-5}
0.30	0.05	0.10	12.0×10^{-5}
0.40	0.05	0.20	31.0×10^{-5}
0.40	0.05	0.05	7.6×10^{-5}

(a) Deduce the rate law for the reaction and give the order with respect to each reactant.
(b) What is the numerical value of k, the rate constant?
(c) If [H$_3$O$^+$] is maintained at 0.050 M, whereas both [CH$_3$COCH$_3$] and [Br$_2$] are 0.10 M, what is the rate of the reaction?

44. One of the major eye irritants in smog is formaldehyde, CH_2O, formed by reaction of ozone with ethylene.

$$C_2H_4(g) + O_3(g) \longrightarrow 2\,CH_2O(g) + \tfrac{1}{2}\,O_2(g)$$

These data were collected:

Initial concentration (mol/L)		Initial rate of formation of CH_2O (mol L^{-1} s^{-1})
$[O_3]$	$[C_2H_4]$	
0.50×10^{-7}	1.0×10^{-8}	1.0×10^{-12}
1.5×10^{-7}	1.0×10^{-8}	3.0×10^{-12}
1.0×10^{-7}	2.0×10^{-8}	4.0×10^{-12}

(a) Determine the rate law for the reaction using the data in the table.

(b) What is the reaction order with respect to O_3? What is the order with respect to C_2H_4?

(c) Calculate the rate constant, k.

(d) What is the rate of reaction when $[C_2H_4]$ and $[O_3]$ are both 2.0×10^{-7} M?

45. This problem requires working with the equations of Table 13.2. Using an initial concentration $[A]_0$ of 1.0 mol/L and a rate constant k with a numerical value of 1.0 in appropriate units, make plots of $[A]$ versus time over the time interval 0 to 5 s for each type of integrated rate law. Compare your results with Figure 13.5.

46. Studies of radioactive decay of nuclei show that the *decay rate* of a radioactive sample is proportional to the *amount* of the radioactive species present. Once half the radioactivity has disappeared, the radioactive decay *rate* is only half of its original value. Is radioactive decay a zeroth-order, first-order, or second-order process?

47. If the initial concentration of the reactant in a first-order reaction is 0.64 mol/L and the half-life is 30. s,

(a) Calculate the concentration of the reactant 60 s after initiation of the reaction.

(b) How long would it take for the concentration of the reactant to drop to one-eighth its initial value?

(c) How long would it take for the concentration of the reactant to drop to 0.040 mol L^{-1}?

48. If the initial concentration of the reactant in a first-order reaction is 0.50 mol/L and the half-life is 400 s,

(a) Calculate the concentration of the reactant 1600 s after initiation of the reaction.

(b) How long would it take for the concentration of the reactant to drop to one-sixteenth its initial value?

(c) How long would it take for the concentration of the reactant to drop to 0.062 mol L^{-1}?

49. Let x represent the number of half-lives that have elapsed during the course of a first-order reaction. That is, the elapsed time t is x times the half-life $t_{1/2}$: $t = xt_{1/2}$.

(a) Show that at time t the concentration $[A]_t$ of reactant is related to the initial concentration $[A]_0$ by

$$[A]_t = [A]_0(\tfrac{1}{2})^x$$

(b) Use your result in part (a) to show that

$$\log\left(\frac{[A]_t}{[A]_0}\right) = \frac{t}{t_{1/2}}(-\log 2) = \frac{t}{t_{1/2}}(-0.301).$$

(c) Use your result in part (b) to show that for any first-order reaction a plot of $\log [A]_t$ versus t will be a straight line and that half-life can be obtained from the slope of the line.

50. Given these data and your result from Question 49, determine the half-life and the initial concentration of the reactant for the reaction

$$trans\text{-}CHClCHCl(aq) \longrightarrow cis\text{-}CHClCHCl(g)$$

[*trans*-CHClCHCl] (mol/L)	Time (s)
9.23×10^{-4}	30
8.51×10^{-4}	60
7.86×10^{-4}	90
7.25×10^{-4}	120
6.19×10^{-4}	180
3.82×10^{-4}	360

51. The compound SO_2Cl_2 decomposes in a first-order reaction

$$SO_2Cl_2(g) \longrightarrow SO_2(g) + Cl_2(g)$$

that has a half-life of 1.47×10^4 s at 600 K. If you begin with 1.6×10^{-3} mol of pure SO_2Cl_2 in a 2.0-L flask, at what time will the amount of SO_2Cl_2 be 1.2×10^{-4} mol?

52. The decomposition of N_2O_5 is first-order with a rate constant of 2.5×10^{-4} s^{-1}.

(a) What is the half-life for decomposition of N_2O_5?

(b) How long does it take for the concentration of N_2O_5 to drop to $\frac{1}{32}$ of its original value?

(c) How long does it take for the concentration of N_2O_5 to drop from 3.4×10^{-3} mol/L to 2.3×10^{-5} mol/L?

53. The rate constant for decomposition of azomethane at 425 °C is 0.68 s^{-1}.

$$CH_3N{=}NCH_3(g) \longrightarrow N_2(g) + C_2H_6(g)$$

(a) Based on the units of the rate constant, is the reaction zeroth-, first-, or second order?

(b) If 2.0 g of azomethane is placed in a 2.0-L flask and heated to 425 °C, what mass remains after 5.0 s?

(c) How long does it take for the mass of azomethane to drop to 0.24 g?

(d) What mass of nitrogen would be found in the flask after 0.5 s of reaction?

A Nanoscale View: Elementary Reactions

54. Using a molecular model kit, build models of *cis*-2-butene, *trans*-2-butene, and the transition state, or activated complex. How much force do you need to apply to the models to change the reactant into the product by passing through the activated complex? The answer will of course depend on the kind of model kit that you use.

55. Which of these reactions are unimolecular and elementary, which are bimolecular and elementary, and which are not elementary?

(a) $CH_4(g) + 2 O_2(g) \longrightarrow CO_2(g) + 2 H_2O(g)$

(b) $O_3(g) + O(g) \longrightarrow 2 O_2(g)$

(c) $Mg(s) + 2 H_2O(\ell) \longrightarrow H_2(g) + Mg(OH)_2(s)$

(d) $O_3(g) \longrightarrow O_2(g) + O(g)$

56. Which of these reactions are unimolecular and elementary, which are bimolecular and elementary, and which are not elementary?
    (a) $HCl(g) + H_2O(g) \longrightarrow H_3O^+(g) + Cl^-(g)$
    (b) $I^-(g) + CH_3Cl(g) \longrightarrow ICH_3(g) + Cl^-(g)$
    (c) $C_2H_6(g) \longrightarrow C_2H_4(g) + H_2(g)$
    (d) $N_2(g) + 3 H_2(g) \longrightarrow 2 NH_3(g)$
    (e) $O_2(g) + O(g) \longrightarrow O_3(g)$

57. Assume that each gas phase reaction occurs via a single bimolecular step. For which reaction would you expect the steric factor to be more important? Why?

    $$Cl + O_3 \rightarrow ClO + O_2 \quad \text{or} \quad NO + O_3 \rightarrow NO_2 + O_2$$

58. Assume that each gas phase reaction occurs via a single bimolecular step. For which reaction would you expect the steric factor to be more important? Why?

    $$H_2C{=}CH_2 + H_2 \longrightarrow H_3C{-}CH_3 \quad \text{or}$$
    $$(CH_3)_2C{=}CH_2 + HBr \longrightarrow (CH_3)_2CBr{-}CH_3$$

Temperature and Reaction Rate

59. The rate of decay of a radioactive solid is independent of the temperature of that solid—at least for temperatures easily obtained in the laboratory. What does this observation imply about the activation energy for this process?

60. From Problem-Solving Example 13.8 (⬅ *p. 591*), where the energy profile of the ozone plus atomic oxygen reaction was derived, obtain the activation energy. Then determine the ratio of the reaction rate for this reaction at 50. °C to the reaction rate at room temperature (25 °C). Assume that the initial concentrations are the same at both temperatures.

61. Suppose a reaction rate constant has been measured at two different temperatures, T_1 and T_2, and its value is k_1 and k_2, respectively.
    (a) Write down the Arrhenius equation at each temperature.
    (b) By combining these two equations, derive an expression for the ratio of the two rate constants, k_1/k_2. Use this formula to solve the next two problems.

62. Suppose a chemical reaction rate constant has an activation energy of 76 kJ/mol, as in the example in Figure 13.11. By what factor is the rate of the reaction at 50. °C increased over its rate at 25 °C?

63. A chemical reaction has an activation energy of 30. kJ/mol. If you had to slow down this reaction a thousandfold by cooling it from room temperature (25 °C), what would the temperature be?

64. These data were obtained for the rate constant for reaction of an unknown compound with water:

T (°C)	k (s^{-1})	T (°C)	k (s^{-1})
25	7.95×10^{-8}	56.2	1.04×10^{-5}
30.	2.37×10^{-7}	78.2	1.45×10^{-4}

    (a) Calculate the activation energy and frequency factor for this reaction.

    (b) Estimate the rate constant of the reaction at a temperature of 100.0 °C.

65. *p*-Methylphenyl acetate reacts with imidazole to produce *p*-methylphenol and acetyl imidazole. The rate constants for this second-order reaction at a series of temperatures are given in the table.

T (°C)	k (L mol^{-1} s^{-1})
10	2.34×10^{-2}
18	3.25×10^{-2}
25	4.5×10^{-2}
35	5.83×10^{-2}
42	7.5×10^{-2}
60	1.52×10^{-1}

    (a) Calculate the activation energy and frequency factor for this reaction.
    (b) Estimate the rate constant for this reaction at a temperature of 100.0 °C.

66. For the reaction of iodine atoms with hydrogen molecules in the gas phase, these rate constants were obtained experimentally.

    $$2 I(g) + H_2(g) \longrightarrow 2 HI(g)$$

T (K)	$10^{-5} k$ (L^2 mol^{-2} s^{-1})
417.9	1.12
480.7	2.60
520.1	3.96
633.2	9.38
666.8	11.50
710.3	16.10
737.9	18.54

    (a) Calculate the activation energy and frequency factor for this reaction.
    (b) Estimate the rate constant of the reaction at 400.0 K.

67. Make an Arrhenius plot and calculate the activation energy for the gas phase reaction

    $$2 NOCl(g) \longrightarrow 2 NO(g) + Cl_2(g)$$

T (K)	Rate constant (L mol^{-1} s^{-1})
400.	6.95×10^{-4}
450.	1.98×10^{-2}
500.	2.92×10^{-1}
550.	2.60
600.	16.3

68. The activation energy E_a is 139.7 kJ mol^{-1} for the gas phase reaction

    $$HI + CH_3I \longrightarrow CH_4 + I_2$$

Calculate the fraction of the molecules whose collisions would be energetic enough to react at
(a) 100. °C   (b) 200. °C   (c) 500. °C   (d) 1000. °C

69. The activation energy E_a is 10. kJ mol^{-1} for the gas phase reaction

$$NO + O_3 \longrightarrow NO_2 + O_2$$

Calculate the fraction of the molecules whose collisions would be energetic enough to react at
(a) 400. °C   (b) 600. °C   (c) 800. °C   (d) 1000. °C

70. For the gas phase reaction

$$CH_3CH_2I(g) \longrightarrow CH_2CH_2(g) + HI(g)$$

the activation energy E_a is 221 kJ/mol and the frequency factor A is 1.2×10^{14} s^{-1}. If the concentration of CH_3CH_2I is 0.012 mol/L, what is the rate of reaction at
(a) 400. °C?   (b) 800. °C?

71. For the gas phase reaction

$$cis\text{-CHClCHCl}(g) \longrightarrow trans\text{-CHClCHCl}(g)$$

the activation energy E_a is 234 kJ/mol and the frequency factor A is 6.3×10^{12} s^{-1}. If the concentration of cis-CHClCHCl is 0.0043 mol/L, what is the rate of reaction at
(a) 400. °C?   (b) 800. °C?

72. For the reaction

$$N_2O_5(g) \longrightarrow 2\ NO_2(g) + \tfrac{1}{2}O_2(g)$$

the rate constant k at 25 °C is 3.46×10^{-5} s^{-1} and at 55 °C it is 1.5×10^{-3} s^{-1}. Calculate the activation energy, E_a.

73. Cyclopropane isomerizes to propene when heated. Rate constants for the reaction

$$cyclopropane \longrightarrow propene$$

are 1.10×10^{-4} s^{-1} at 470. °C and 1.02×10^{-3} s^{-1} at 510. °C. What is the activation energy, E_a, for this reaction?

74. When heated, cyclobutane, C_4H_8, decomposes to ethylene, C_2H_4.

$$C_4H_8(g) \longrightarrow 2\ C_2H_4(g)$$

The activation energy, E_a, for this reaction is 262 kJ/mol. If the rate constant $k = 0.032$ s^{-1} at 800. K, what is the value of k at 900. K?

Rate Laws for Elementary Reactions

75. For the hypothetic reaction A + B → C + D, the activation energy is 32 kJ/mol. For the reverse reaction (C + D → A + B), the activation energy is 58 kJ/mol. Is the reaction A + B → C + D exothermic or endothermic?

76. Use the diagram to answer these questions.

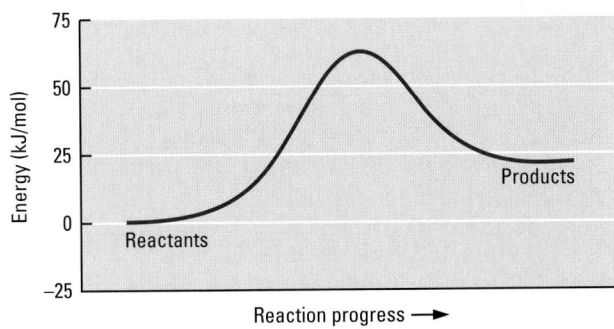

(a) Is the reaction exothermic or endothermic?
(b) What is the approximate value of ΔE for the forward reaction?
(c) What is the activation energy in each direction?
(d) A catalyst is found that lowers the activation energy of the reaction by about 10 kJ/mol. How will this catalyst affect the rate of the reverse reaction?

77. Draw an energy–reaction progress diagram (similar to the one in Question 76) for each of the reactions whose activation energy and enthalpy change are given below.
(a) $\Delta H° = -145$ kJ mol^{-1}; $E_a = 75$ kJ mol^{-1}
(b) $\Delta H° = -70$ kJ mol^{-1}; $E_a = 65$ kJ mol^{-1}
(c) $\Delta H° = 70$ kJ mol^{-1}; $E_a = 85$ kJ mol^{-1}

78. Draw an energy–reaction progress diagram (similar to the one in Question 76) for each of the reactions whose activation energy and enthalpy change are given below.
(a) $\Delta H° = 105$ kJ mol^{-1}; $E_a = 175$ kJ mol^{-1}
(b) $\Delta H° = -43$ kJ mol^{-1}; $E_a = 95$ kJ mol^{-1}
(c) $\Delta H° = 15$ kJ mol^{-1}; $E_a = 55$ kJ mol^{-1}

79. Which of the reactions in Problem 77 would be expected to (a) occur fastest?   (b) occur slowest?
(Assume equal temperatures, equal concentrations, equal frequency factors, and the same rate law for all reactions.)

80. Which of the reactions in Problem 78 would be expected to (a) occur fastest?   (b) occur slowest?
(Assume equal temperatures, equal concentrations, equal frequency factors, and the same rate law for all reactions.)

81. For which of the reactions in Problem 77 would the *reverse* reaction (a) be fastest?   (b) be slowest?
(Assume equal temperatures, equal concentrations, equal frequency factors, and the same rate law for all reactions.)

82. For which of the reactions in Problem 78 would the *reverse* reaction (a) be fastest?   (b) be slowest?
(Assume equal temperatures, equal concentrations, equal frequency factors, and the same rate law for all reactions.)

83. Assuming that each reaction is elementary, predict the rate law.
(a) $NO(g) + NO_3(g) \rightarrow 2\ NO_2(g)$
(b) $O(g) + O_3(g) \rightarrow 2\ O_2(g)$
(c) $(CH_3)_3CBr(aq) \rightarrow (CH_3)_3C^+(aq) + Br^-(aq)$
(d) $2\ HI(g) \rightarrow H_2(g) + I_2(g)$

84. Assuming that each reaction is elementary, predict the rate law.
(a) $Cl(g) + ICl(g) \rightarrow I(g) + Cl_2(g)$
(b) $Cl(g) + H_2(g) \rightarrow HCl(g) + H(g)$
(c) $2\ NO_2(g) \rightarrow N_2O_4(g)$
(d) $Cyclopropane(g) \rightarrow propene(g)$

Reaction Mechanisms

85. Experiments show that the reaction of nitrogen dioxide with fluorine

Overall reaction: $2\ NO_2(g) + F_2(g) \rightarrow 2\ FNO_2(g)$

has the rate law

$$\text{Initial reaction rate} = k[NO_2][F_2]$$

and the reaction is thought to occur in two steps.

Step 1: $NO_2(g) + F_2(g) \rightarrow FNO_2(g) + F(g)$
Step 2: $NO_2(g) + F(g) \rightarrow FNO_2(g)$

(a) Show that the sum of this sequence of reactions gives the balanced equation for the overall reaction.
(b) Which step is rate-determining?

86. Nitrogen oxide is reduced by hydrogen to give water and nitrogen

$$2\,H_2(g) + 2\,NO(g) \rightarrow N_2(g) + 2\,H_2O(g)$$

and one possible mechanism for this reaction is a sequence of three elementary steps.

Step 1 (fast): $2\,NO \rightleftharpoons N_2O_2$
Step 2 (slow): $N_2O_2(g) + H_2(g) \rightarrow N_2O(g) + H_2O(g)$
Step 3 (fast): $N_2O(g) + H_2(g) \rightarrow N_2(g) + H_2O(g)$

(a) Show that the sum of these steps gives the net reaction.
(b) What is the rate law for this reaction?

87. For the reaction

$$2\,NO(g) + Cl_2(g) \longrightarrow 2\,NOCl(g)$$

the currently accepted mechanism is

$$NO + Cl_2 \rightleftharpoons NOCl_2 \qquad \text{fast}$$
$$NOCl_2 + NO \longrightarrow 2\,NOCl \qquad \text{slow}$$

(a) What is the rate law for this mechanism? (Be sure to express it in terms of concentrations of reactants or products of the overall reaction, not in terms of intermediates.)
(b) Suggest another mechanism that agrees with the same rate law.
(c) Suggest another mechanism that does not agree with the same rate law.

88. For the reaction

$$2\,N_2O_5(g) + 4NO_2(g) \longrightarrow O_2(g)$$

the currently accepted mechanism is

$$N_2O_5 \rightleftharpoons NO_2 + NO_3 \qquad \text{fast}$$
$$NO_2 + NO_3 \longrightarrow NO_2 + O_2 + NO \qquad \text{slow}$$
$$NO + NO_3 \longrightarrow 2\,NO_2 \qquad \text{fast}$$

What is the rate law for this reaction?

89. For the reaction mechanism

$$CH_3\overset{\displaystyle O}{\overset{\|}{C}}-O-CH_3 + H_3O^+ \rightleftharpoons CH_3\overset{\displaystyle OH}{\overset{|}{\underset{+}{C}}}-O-CH_3 + H_2O \quad \text{fast}$$

$$CH_3\overset{\displaystyle OH}{\overset{|}{\underset{+}{C}}}-O-CH_3 + 2\,H_2O \longrightarrow CH_3\overset{\displaystyle O}{\overset{\|}{C}}-OH + HOCH_3$$
$$HOCH_3 \quad \text{slow}$$

(a) Write the equation for the overall reaction.
(b) Write the rate law for the reaction.
(c) Is there a catalyst involved in this reaction? If so, what is it.
(d) Identify all intermediates in the reaction.

90. For the reaction

$$CH_3-\overset{\displaystyle CH_3}{\overset{|}{\underset{|}{C}}}-Br + OH^- \longrightarrow CH_3-\overset{\displaystyle CH_3}{\overset{|}{\underset{|}{C}}}-OH + Br^-$$
$$\qquad\quad CH_3 \qquad\qquad\qquad CH_3$$

the rate law is

$$\text{Rate} = k[(CH_3)_3CBr]$$

Identify each mechanism that is compatible with the rate law.

(a) $(CH_3)_3CBr \rightarrow (CH_3)_3C^+ + Br^-$ slow
   $(CH_3)_3C^+ + OH^- \rightarrow (CH_3)_3COH$ fast
(b) $(CH_3)_3CBr + OH^- \rightarrow (CH_3)_3COH + Br^-$
(c) $(CH_3)_3CBr + OH^- \rightarrow (CH_3)_2(CH_2)CBr^- + H_2O$ fast
   $(CH_3)_2(CH_2)CBr^- \rightarrow (CH_3)_2(CH_2)C + Br^-$ slow
   $(CH_3)_2(CH_2)C + H_2O \rightarrow (CH_3)_3COH$ fast

91. Which of these mechanisms are compatible with the rate law (more than one may be chosen)

$$\text{Rate} = k[Cl_2]^{3/2}[CO]$$

(a) $\frac{1}{2}\,Cl_2 \rightleftharpoons Cl$ fast
   $Cl + Cl_2 \rightarrow Cl_3$ fast
   $Cl_3 + CO \rightarrow COCl_2 + Cl$ slow
   $Cl \rightleftharpoons \frac{1}{2}\,Cl_2$ fast
(b) $Cl_2 + CO \rightarrow CCl_2 + O$ slow
   $O + Cl_2 \rightarrow Cl_2O$ fast
   $Cl_2O + CCl_2 \rightarrow COCl_2 + Cl_2$ fast
(c) $\frac{1}{2}\,Cl_2 \rightleftharpoons Cl$ fast
   $Cl + CO \rightleftharpoons COCl$ fast
   $COCl + Cl_2 \rightarrow COCl_2 + Cl$ slow
   $Cl \rightleftharpoons \frac{1}{2}\,Cl_2$ fast
(d) $Cl_2 + CO \rightleftharpoons COCl + Cl$ fast
   $COCl + Cl_2 \rightarrow COCl_2 + Cl$ slow
   $Cl + Cl \rightarrow Cl_2$ fast

Catalysts and Reaction Rate

92. Which of these statements is (are) true?
(a) The concentration of a homogeneous catalyst may appear in the rate law.
(b) A catalyst is always consumed in the reaction.
(c) A catalyst must always be in the same phase as the reactants.
(d) A catalyst can change the course of a reaction and allow different products to be produced.

93. Hydrogenation reactions—processes in which H_2 is added to a molecule—are usually catalyzed. An excellent catalyst is a very finely divided metal suspended in the reaction solvent. Tell why finely divided rhodium, for example, is a much more efficient catalyst than a small block of the metal.

94. Which of these reactions appear to involve a catalyst? In those cases where a catalyst is present, tell whether it is homogeneous or heterogeneous.
(a) $CH_3CO_2CH_3(aq) + H_2O(\ell) \rightarrow$
                        $CH_3COOH(aq) + CH_3OH(aq)$
   Rate = $k[CH_3CO_2CH_3][H_3O^+]$
(b) $H_2(g) + I_2(g) \rightarrow 2\,HI(g)$
   Rate = $k[H_2][I_2]$
(c) $2\,H_2(g) + O_2(g) \rightarrow 2\,H_2O(g)$
   Rate = $k[H_2][I_2](\text{area of Pt surface})$
(d) $H_2(g) + CO(g) \rightarrow H_2CO(g)$
   Rate = $k[H_2]^{1/2}[CO]$

95. In acid solution, methyl formate forms methyl alcohol and formic acid.

$$HCO_2CH_3(aq) + H_2O(\ell) \longrightarrow HCOOH(aq) + CH_3OH(aq)$$

methyl formate   formic acid   methyl alcohol

The rate law is as follows: rate = $k[HCO_2CH_3][H_3O^+]$. Why does H_3O^+ appear in the rate law but not in the overall equation for the reaction?

96. Suppose a rate constant for an uncatalyzed reaction is described by the Arrhenius equation with an activation energy E_a. The introduction of a catalyst lowers the activation energy to the value E_a' and thus increases the rate constant to the value k'. By first writing down the Arrhenius equation for both k and k', derive an expression for the ratio k'/k in terms of E_a' and E_a; assume that the frequency factor A and the temperature T are constant. Use this equation to solve the next two problems.

97. Suppose a catalyst was found for the reaction in Exercise 13.8 that reduced the activation energy to zero. By what factor would this reaction rate be increased at 370. K?

98. In the discussion of Figure 13.15, this statement is made: "Dropping the activation energy from 262 kJ/mol for the uncatalyzed reaction to 115 kJ/mole for the catalyzed process makes the catalyzed reaction 10^{15} times faster." Prove this statement.

Enzymes: Biological Catalysts

99. In Section 13.9 there are quite a few chemical and biochemical terms that may be new to you. Write a one- to two-sentence definition for each of these terms, using other references besides this text if necessary:

enzyme	dimer
cofactor	hydrolysis
polypeptides	HIV protease
monomer	enzyme-substrate complex
polysaccharides	induced fit
lysozyme	globular proteins
substrate	denaturation
active site	maximum velocity
proteins	inhibition

100. When enzymes are present at very low concentration, their effect on reaction rate can be described by first-order kinetics. By what factor does the rate of an enzyme-catalyzed reaction change when the enzyme concentration is changed from 1.5×10^{-7} to 4.5×10^{-6} M?

101. When substrates are present at relatively high concentration and are catalyzed by enzymes, the effect on reaction rate of changing substrate concentration can be described by zero-order kinetics. By what factor does the rate of an enzyme-catalyzed reaction change when the substrate concentration is changed from 1.5×10^{-3} to 4.5×10^{-2} M?

102. The reaction

is catalyzed by the enzyme succinate dehydrogenase. When malonate ions or oxalate ions are added to the reaction mixture, the rate decreases significantly. Try to account for this observation in terms of the description of enzyme catalysis given in the text. The structures of malonate and oxalate ions are

malonate ion   oxalate ion

Catalysis in Industry

103. In the first paragraph of Section 13.10 of the text, an expert in the field of industrial chemistry is quoted. Explain the expert's statement, in view of your understanding of the nature of catalysts. Why are catalysts so important?

104. Why are homogeneous catalysts harder to separate from products and leftover reactants than are heterogeneous reactants?

105. Find all examples of reactions described in this chapter that are catalyzed by metals. Are these metals main group metals or transition metals? What type of chemical reactions are they, acid-base or oxidation-reduction? Can you reach some conclusions about metal-catalyzed chemical reactions from these examples?

General Questions

106. Nitrogen monoxide can be reduced with hydrogen.

$$2 H_2(g) + 2 NO(g) \longrightarrow 2 H_2O(g) + N_2(g)$$

Experiment shows that when the concentration of H_2 is halved, the reaction rate is halved. Furthermore, raising the concentration of NO by a factor of 3 raises the rate by a factor of 9. Write the rate equation for this reaction.

107. One reaction that may occur in air polluted with nitrogen monoxide is

$$2 NO(g) + O_2(g) \longrightarrow 2 NO_2(g)$$

Using the data in the table, answer the questions that follow.

	Initial concentration (mol/L)		Initial rate of formation of NO_2
Experiment	**[NO]**	**[O$_2$]**	**(mol L^{-1} s^{-1})**
1	0.0010	0.0010	7.0×10^{-6}
2	0.0010	0.0020	1.4×10^{-5}
3	0.0010	0.0030	2.1×10^{-5}
4	0.0020	0.0030	8.4×10^{-5}
5	0.0030	0.0030	1.9×10^{-4}

(a) What is the order of reaction with respect to each reactant?

(b) Write the rate law for the reaction.

(c) Calculate the rate of formation of NO_2 when [NO] = [O$_2$] = 0.005 mol/L.

108. The deep blue compound $CrO(O_2)_2$ can be made from the chromate ion by using hydrogen peroxide in an acidic solution.

$$HCrO_4^-(aq) + 2\ H_2O_2(aq) + H_3O^+(aq) \longrightarrow$$
$$CrO(O_2)_2(aq) + 4\ H_2O(\ell)$$

The kinetics of this reaction have been studied, and the rate equation is found to be

Rate of disappearance of $HCrO_4^-$ =
$$k[HCrO_4^-][H_2O_2][H_3O^+]$$

One of the mechanisms suggested for the reaction is

$$HCrO_4^- + H_3O^+ \rightleftharpoons H_2CrO_4 + H_2O$$

$$H_2CrO_4 + H_2O_2 \longrightarrow H_2CrO(O_2)_2 + H_2O$$

$$H_2CrO(O_2)_2 + H_2O_2 \longrightarrow CrO(O_2)_2 + 2\ H_2O$$

(a) Give the order of the reaction with respect to each reactant.
(b) Show that the steps of the mechanism agree with the overall equation for the reaction.
(c) Which step in the mechanism is rate-limiting?

109. Refer to Figure 13.3 and explain why the rate of change of the concentration of $Pt(NH_3)_2Cl_2$ decreases with time but the concentration of Cl^- increases with time.

110. How does a chemical reaction mechanism differ from other types of mechanisms, for example, the gear-changing mechanism of a bicycle, or the mechanism of an elevator? How is it similar to these mechanisms?

111. Why are catalysts important economically to the chemical industry?

112. Why is chemical kinetics important in understanding the environmental problem posed by chlorofluorocarbons in the ozone layer *(⇐ p. 441)*?

113. A reaction between molecules A and B (A + B → products) is found to be first-order in A. Which rate equation cannot be correct?
(a) Rate = $k[A][B]$ (b) Rate = $k[A][B]^2$
(c) Rate = $k[B]^2$

Applying Concepts

114. The graph shows the change in concentration as a function of time for the reaction

$$2\ H_2O_2(g) \longrightarrow 2\ H_2O(g) + O_2(g)$$

What do each of the curves A, B, and C represent?

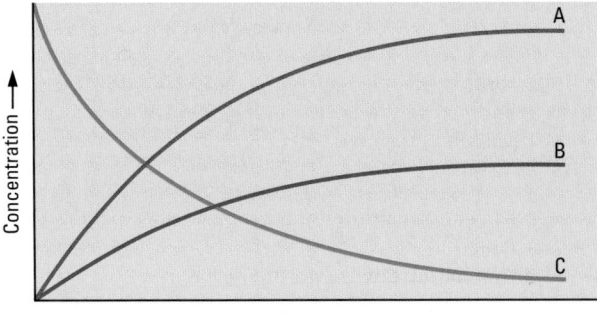

115. Draw a graph similar to the one in Question 114 for the reaction

$$2\ N_2O_5(g) \longrightarrow 4\ NO_2(g) + O_2(g)$$

116. The picture below is a "snapshot" of the reactants at time = 0 for the reaction

$$H_2(g) + I_2(g) \longrightarrow 2\ HI(g)$$

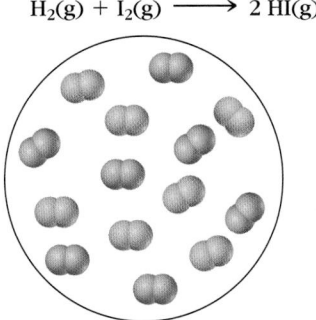

Suppose the reaction is carried out at two different temperatures and that another snapshot is taken after a constant time has elapsed.

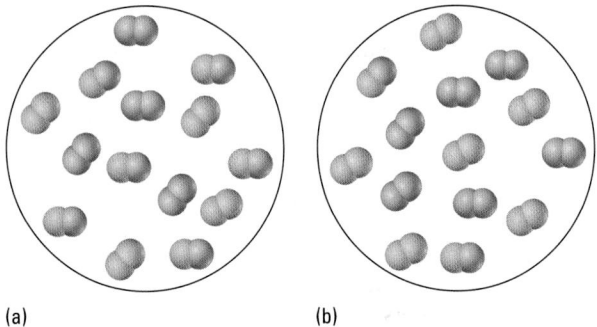

(a) (b)

Which of these two snapshots corresponds to the higher temperature reaction condition?

117. Consider the previous question again, only this time a catalyst is used instead of a higher temperature. Which of the two "snapshots" corresponds to the presence of a catalyst?

118. Initial rates for the reaction A + B + C → D + E were measured with various concentrations of A, B, and C as represented in the pictures below. Based on these data, what is the rate law?

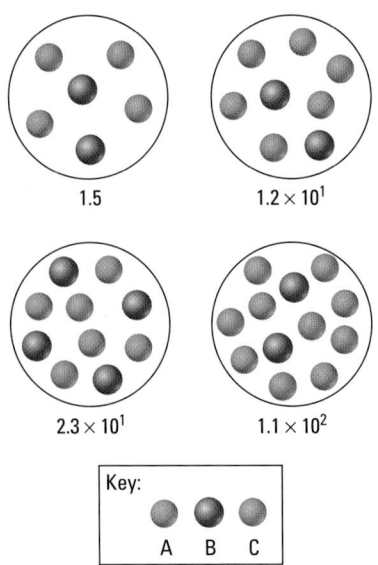

1.5 1.2×10^1

2.3×10^1 1.1×10^2

Key:
A B C

119. Platinum metal is used as a catalyst in the decomposition of NO(g) into N_2(g) and O_2(g). A graph of the rate of the reaction as a function of NO concentration looks like this (see below). Explain why the rate stops increasing and levels out.

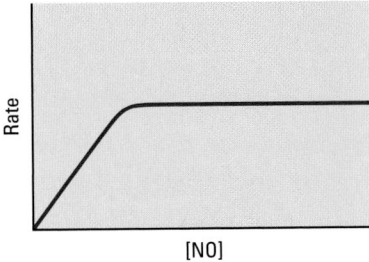

General Chemistry CD-ROM

CD13.1 Screen 15.2: Rates of Chemical Reactions.
(a) What is the difference between an instantaneous rate and an average rate?
(b) Observe the graph of food dye versus time on this screen. (Click on the "tool" icon on this screen.) The plot shows the concentration of dye as the reaction progresses. What does the steepness of the plot at any particular time tell you about the rate of the reaction at that time?
(c) As the reaction progresses, the concentration of dye decreases as it is consumed. What happens to the reaction rate as this occurs? What is the relationship between reaction rate and dye concentration?

CD13.2 Screen 15.5: Determination of Rate Equation: Method of Initial Rates. This screen describes how to determine experimentally a rate law using the method of initial rates.
(a) Why must the rate of reaction be measured at the very beginning of the process for this method to be valid?

(b) The first experiment shows that the initial rate of NH_4NCO degradation is 2.2×10^{-4} mol/L^{-1} s^{-1} when [NH_4NCO] = 0.14 mol/L. Using the rate law determined on this screen, predict the rate if [NH_4NCO] = 0.18 mol/L.

CD13.3 Screen 15.8: Half-Life: First-Order Reactions. Examine the graph and table of concentrations on the second portion of this screen.
(a) What will the concentration of H_2O_2 be after 3270 min? After 3924 min?
(b) What fraction of the original concentration of H_2O_2 remains after 3270 min? After 3924 min?

CD13.4 Screen 15.10: Microscopic View of Reactions (2). Examine the two animations on the Reaction Coordinate Diagrams sidebar to this screen. What is the difference between the way in which the two reactions occur?

CD13.5 Screen 15.11: Control of Reaction Rates: Temperature Dependence.
(a) What is the general effect of an increase in temperature on reaction rates?
(b) Describe the different portions of the Arrhenius equation and how they relate to the rate of a reaction.

CD13.6 Screen 15.13: Reaction Mechanisms and Rate Equations.
(a) What is the difference between an overall mechanism and an elementary step?
(b) What is the relationship between the stoichiometric coefficients of the reactants in an elementary step and the rate law for that step?
(c) What is the relationship between reactant concentrations and the rate law for that reaction?
(d) What is the rate law for Step 2 of Mechanism 2?

CD13.7 Screen 15.14: Catalysis and Reaction Rate.
(a) Examine the mechanism for the iodide ion–catalyzed decomposition of H_2O_2. Explain how the mechanism shows that I^- is a catalyst.
(b) How does the reaction coordinate diagram show that the catalyzed reaction is expected to be faster than the uncatalyzed reaction?

14

Ammonia fertilizer provides tremendous benefits to society by improving crop yields and enhancing the supply of food. But ammonia does not occur in large, concentrated deposits in nature. It has to be synthesized in chemical manufacturing plants. How to do this was discovered less than a century ago, at a time when shortages of naturally occurring fertilizers were expected to be a major factor that would limit the human population on earth. The principles of chemical equilibrium and chemical kinetics formed the basis for the discovery of how to make ammonia economically.

CHEMICAL EQUILIBRIUM

14.1 Characteristics of Chemical Equilibrium

14.2 The Equilibrium Constant

14.3 Determining Equilibrium Constants

14.4 The Meaning of the Equilibrium Constant

14.5 Using Equilibrium Constants

14.6 Shifting a Chemical Equilibrium: Le Chatelier's Principle

14.7 Equilibrium at the Nanoscale

14.8 Controlling Chemical Reactions: The Haber-Bosch Process

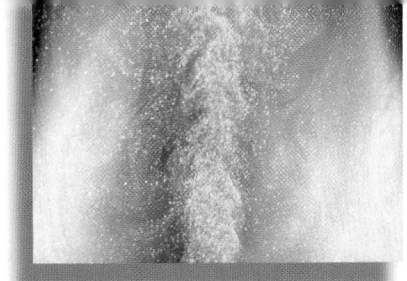

Chemical reactions that involve only pure solids or pure liquids are simpler than those that occur in the gas phase or in a solution. Either no reaction occurs between the solid and liquid reactants, or a reaction occurs in which at least one reactant is completely converted into products. (If one or more of the reactants is present in excess, those reactants will be left over, but the limiting reactant (⇐ *p. 140)* will be completely reacted away.) This happens because the concentration of a pure solid or a pure liquid does not change during a reaction, provided the temperature remains constant. If the initial concentrations of reactants are large enough to cause the reaction to occur, those same concentrations will be present throughout the reaction, and it will not stop until the limiting reactant is used up.

When a reaction occurs in the gas phase or in a solution, concentrations of reactants decrease as the reaction takes place (⇐ *p. 572)*. Eventually the concentrations decrease to the point at which conversion of reactants to products is no longer favored. Then the concentrations of reactants and products stop changing, but none of the concentrations has become zero. At least a tiny bit (and often a lot) of each reactant and each product is present in the reaction mixture. Because the concentrations have stopped changing, it is often relatively easy to measure them, thus providing quantitative information about *how much* product can be obtained from the reaction. It is also possible to predict how changes in temperature, pressure, and concentrations will affect the quantity of product produced. This kind of information, combined with what you learned in Chapter 13 about factors that affect the rates of chemical reactions, enables us to predict which reactions are useful for manufacturing a broad range of substances that enhance our quality of life.

As an example of the importance of such information, consider ammonia. The United States produces nearly 40 billion pounds of liquefied ammonia per year, mostly for use as fertilizer to provide nitrogen needed to support growth of a broad range of crops. Therefore, ammonia is a very important factor in providing people with food. Ammonia is synthesized directly from nitrogen and hydrogen by what is called the *Haber-Bosch process* (Section 14.8).

$$N_2(g) + 3 H_2(g) \rightleftharpoons 2 NH_3(g) \qquad \Delta H° = -92.2 \text{ kJ}$$

It is important to the chemists and chemical engineers who operate manufacturing plants that the maximum quantity of ammonia be obtained with the minimum input of reactants and the minimum consumption of energy resources. The German chemist Fritz Haber won the 1918 Nobel Prize for research that showed how to determine the best conditions for carrying out this reaction.

In this chapter you will learn the same principles that Haber used. With them you will be able to make both qualitative and quantitative predictions about how much product will be formed under a given set of reaction conditions.

14.1 CHARACTERISTICS OF CHEMICAL EQUILIBRIUM

When the concentrations of reactants stop decreasing and concentrations of products stop increasing, we say that a reaction has reached equilibrium. In a **chemical equilibrium,** *there are finite concentrations of reactants and products, and these concentrations remain constant.* An equilibrium reaction always results in smaller amounts of products than the theoretical yield predicts (⇐ *p. 145)*, and sometimes hardly any products are produced at all. The concentrations of reactants and products at equilibrium provide a quantitative way of determining how successful a reaction has been. *When products predominate over reactants,* the reaction is **product-favored.** *When most of the equilibrium mixture is reactants,* the reaction is **reactant-favored.**

> ### Exercise 14.1 Concentrations of Pure Solids and Liquids
>
> The introduction to this chapter states that at a given temperature the concentration of a pure solid or liquid does not depend on the quantity of substance present. Verify this by calculating the concentration (in mol/L) of these solids and liquids at 20 °C. Obtain densities from Table 1.1 *(⇐ p. 9).*
>
> (a) Iron (b) Ethanol (c) Water (d) Magnesium

Equilibrium Is Dynamic

CD-ROM Screen 16.2: The Principle of Microscopic Reversibility
Screen 16.3: The Equilibrium State

The "equi" in the word "equilibrium" means "equal." It refers to equal rates of forward and reverse reactions, not to equal quantities or concentrations of the substances involved. The "librium" part of the word comes from "libra," meaning balance. Chemical equilibrium is an equal balance between two reaction rates.

When equilibrium is reached and concentrations remain constant, it appears that a chemical reaction has stopped, but this is only true of the net, macroscopic reaction. On the nanoscale both forward and reverse reactions continue, but *the rate of the forward reaction exactly equals the rate of the reverse reaction.* To emphasize that *chemical equilibrium involves a balance between opposite reactions,* it is usually referred to as a **dynamic equilibrium,** and an equilibrium reaction is usually written with a double arrow (⇌) between reactants and products.

Weak acids and bases *(⇐ p. 168)* ionize only partially in water, providing a good example of chemical equilibrium.

$$CH_3COOH(aq) + H_2O(\ell) \rightleftharpoons CH_3COO^-(aq) + H_3O^+(aq)$$
$$\text{acetic acid} \qquad \text{water} \qquad \text{acetate ion} \qquad \text{hydronium ion}$$

At room temperature more than 90% of the acetic acid remains in molecular form after equilibrium has been reached, and the equilibrium concentrations of acetate ions and hydrogen ions are each less than one tenth the concentration of acetic acid molecules. Nevertheless, both the forward and reverse reactions continue. This can be demonstrated by adding a tiny quantity of sodium acetate in which radioactive carbon-14 has been substituted into the CH_3COO^- ion to give $^{14}CH_3COO^-$. Almost immediately the radioactivity can be found in acetic acid molecules as well. This would not happen if the reaction had come to a halt, but it can happen if the reverse reaction,

$$^{14}CH_3COO^-(aq) + H_3O^+(aq) \longrightarrow {}^{14}CH_3COOH(aq) + H_2O(\ell)$$

is still taking place. To a macroscopic observer nothing seems to be happening because the reverse reaction and the forward reaction are occurring at equal rates and there is no net change in each concentration.

Equilibrium Is Independent of Direction of Approach

A set of double arrows, ⇌, in an equation indicates a dynamic equilibrium in which forward and reverse reactions are occurring at equal rates; it also indicates that the reaction should be thought of in terms of the concepts of chemical equilibrium.

Another important characteristic of chemical equilibrium is that, *for a specific reaction at a specific temperature, the equilibrium state will be the same, no matter what the direction of approach to equilibrium.* As an example, consider again the synthesis of ammonia from N_2 and H_2.

$$N_2(g) + 3 H_2(g) \rightleftharpoons 2 NH_3(g)$$

Suppose that you introduce 1.0 mol of $N_2(g)$ and 3.0 mol of $H_2(g)$ into an evacuated (empty) 1.00-L container at 472 °C. Some (but not all) of the N_2 reacts with the H_2 to form NH_3. After equilibrium is established, you would find that the concentration of H_2 has fallen from its initial value of 3.0 mol/L to an equilibrium value of 0.89 mol/L. You would also find equilibrium concentrations of 0.30 mol/L for N_2 and 1.41 mol/L for NH_3.

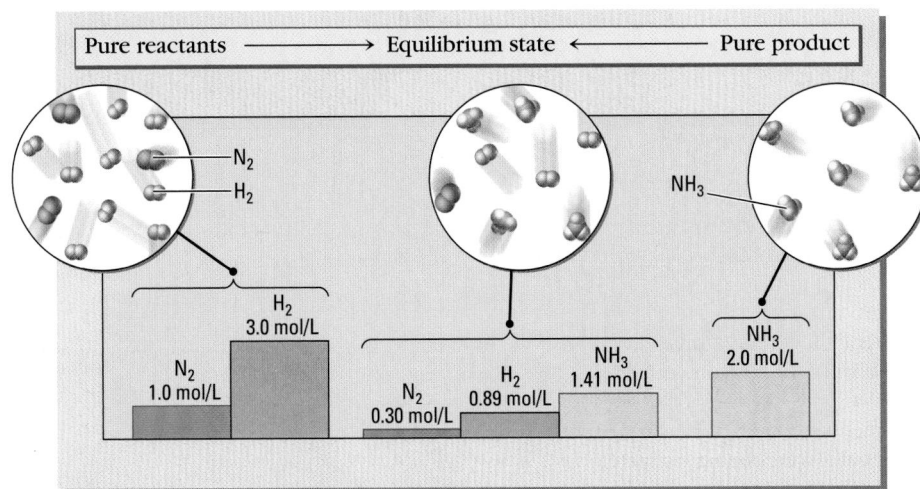

Figure 14.1   Reactants, products, and equilibrium. In the ammonia synthesis reaction, as in any equilibrium, it is possible to start with reactants or products and achieve the same equilibrium state. Here 1.0 mol of N_2 reacts with 3.0 mol of H_2 in a 1.0-L container to give the same equilibrium concentrations as when 2.0 mol of NH_3 is introduced into the same container at the same temperature (472 °C).

Now consider a second experiment at the same temperature in which you introduce 2.0 mol of $NH_3(g)$ into an empty 1.00-L container at 472 °C. Notice that 2.0 mol of NH_3 consists of 2.0 mol of N atoms and 6.0 mol of H atoms — the same number of N atoms and H atoms contained in the 1.0 mol of N_2 and 3.0 mol of H_2 used in the first experiment. Because there is nothing but NH_3 to start with, the reverse reaction occurs, producing some N_2 and H_2. Measuring the concentrations at equilibrium reveals that the concentration of NH_3 dropped from the initial 2.0 mol/L to 1.41 mol/L, and the concentrations of N_2 and H_2 built up to 0.30 mol/L and 0.89 mol/L. These equilibrium concentrations are the same as were achieved in the first experiment. Thus, whether you start with reactants or products, the same equilibrium state is achieved — as long as the number of atoms of each type, the volume of the container, and the temperature remain the same. This is shown schematically in Figure 14.1.

Catalysts Do Not Affect Equilibrium Concentrations

Another important characteristic of chemical equilibrium is that *if a catalyst is present, the same equilibrium state will be achieved, but more quickly.* A catalyst speeds up the forward reaction, but it also speeds up the reverse reaction. The overall effect is to produce exactly the same concentrations at equilibrium, whether or not a catalyst is in the reaction mixture. This means that a catalyst can be used to speed up production of products in an industrial process, but it will not result in greater equilibrium concentrations of products. (Of course the catalyst also will not reduce the concentration of product that is present when the system reaches equilibrium.)

Exercise 14.2   Recognizing an Equilibrium State

A mixture of hydrogen gas and oxygen gas is maintained at 25 °C for 1 year. On the first day of each month the mixture is sampled and the concentrations of hydrogen and oxygen measured. In every case they are found to be 0.50 mol/L H_2 and 0.50 mol/L O_2; that is, the concentrations are found not to change over a long period. Is this mixture at equilibrium? If you think not, how could you do an experiment to prove it?

EQUILIBRIUM AND ALZHEIMER'S DISEASE

Approximately two million Americans are afflicted by Alzheimer's disease, and more than 100,000 of them die each year. There is no definitive treatment or cure for this debilitating disease, which causes loss of memory and reasoning, disorientation, and eventually death. Alzheimer's affects between 20 and 30% of people over 80 and is the fourth leading cause of death in the United States.

One approach to fighting this disease is based on recognition that it is accompanied by formation of a tangle of toxic proteins, called plaques, in the brain. (Proteins are discussed in Section 12.9, ⬅ *p. 556*). A major component of the plaques is a protein called beta-amyloid. Its molecules attract each other strongly and at equilibrium form long clumps called fibrils. In this fibril form it is toxic and kills brain cells. The formation of fibrils can be represented approximately by an equilibrium equation:

$$n \text{ beta-amyloid} \rightleftharpoons (\text{beta-amyloid})_n$$

$$\begin{array}{cc} (n \text{ individual} & (\text{fibril}) \\ \text{protein molecules}) \end{array}$$

This is a product-favored reaction, so much of the beta-amyloid protein is converted into fibrils. Two scientists at the University of Wisconsin-Madison, Laura Kiessling and Regina Murphy, reasoned that if this product-favored reaction could be prevented from happening, the fibrils would not form and the protein would not be toxic.

Kiessling and Murphy realized that the protein molecules stick together to form fibrils because of noncovalent interactions (⬅ *p. 383*). Some of the protein's amino acid side chains are nonpolar. These are also called *hydrophobic*, meaning that they avoid water. The parts of the protein consisting mainly of nonpolar side chains tend to clump together, avoiding the aqueous solutions within cells. Other side chains are polar and/or capable of forming hydrogen bonds. These are called *hydrophilic*, meaning that they are attracted to water. Kiessling and Murphy reasoned that the hydrophobic segments of one beta-amyloid molecule were being attracted to hydrophobic segments of another beta-amyloid molecule, causing the formation of fibrils to be product-favored. If they could somehow avoid these attractions, perhaps the fibril-forming reaction would not occur.

They decided to try to trick the protein molecules into binding to something else, rather than to each other. Their plan was to create a new molecule that mimicked the hydrophobic parts of beta-amyloid. They expected that this new molecule would bind to the protein, preventing protein molecules from binding to each other. Their plan was to set up a second equilibrium between mimic molecules (MM) and beta-amyloid.

$$MM + \text{beta-amyloid} \rightleftharpoons$$
$$MM-(\text{beta-amyloid})$$

This new equilibrium, as predicted, is even more product-favored than the first one, and so it occurs preferentially. The beta-amyloid no longer forms fibrils. Instead it forms small bundles that turn out not to be toxic in cells.

According to John Cross, chief scientist for the American Health Assistance Foundation, this is an exciting development that "has the possibility of interfering with the progression of Alzheimer's. If there is something that can prevent nerve cell death, perhaps we can stop the disease." There is still a long way to go, though, before this basic research can be turned into an effective drug for treatment of Alzheimer's. Such a drug must be able to avoid the body's immune system, pass into the brain, and be very specific so that it prevents fibril formation without affecting the brain adversely in other ways. Kiessling and Murphy have devised a modular strategy for generating compounds that inhibit beta-amyloid toxicity. Eventually a pharmaceutical company may be able to manufacture and sell such a compound as a treatment for Alzheimer's disease. If that happens, this will become an even bigger story of chemistry in the news.

Source:

Wisconsin State Journal, January 12, 1997.
Doctor's Guide, **http://www.pslgroup. com/dg/23a86.htm**
Biochemistry, Vol. 38, 1999, pp. 3570–3578.

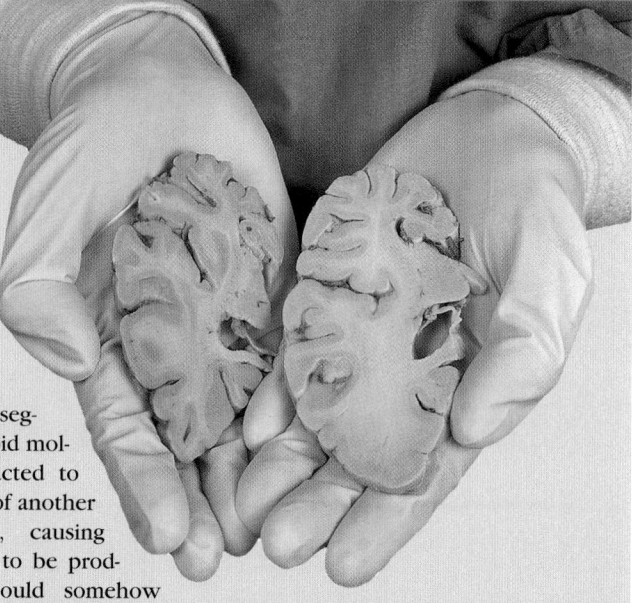

Diseased and normal brain tissue.

Beta-amyloid fibrils in brain tissue.

Regina Murphy and Laura Kiessling.

14.2 THE EQUILIBRIUM CONSTANT

CD-ROM Screen 16.4: The Equilibrium Constant

Consider again the isomerization reaction of *cis*-2-butene to *trans*-2-butene, whose rate we discussed previously (⇐ *p. 573*).

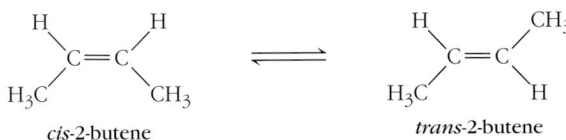

cis-2-butene   *trans*-2-butene

At 600 K the reaction reaches an equilibrium in which the concentration of *trans*-2-butene is 1.65 times the concentration of *cis*-2-butene. This is a single-step, elementary process (⇐ *p. 588*). Both the forward and reverse reactions in this system involve only a single molecule and therefore are unimolecular and first-order (⇐ *p. 581*). Thus, the rate equations for forward and reverse reactions can be derived from the reaction equations.

$$\text{Rate}_{\text{forward}} = k_{\text{forward}}\,[\textit{cis}\text{-2-butene}] \qquad \text{Rate}_{\text{reverse}} = k_{\text{reverse}}\,[\textit{trans}\text{-2-butene}]$$

Suppose that we start with 0.100 mol of *cis*-2-butene in a 5-L closed flask at 500 K. The *cis*-2-butene begins to react at a rate given by the forward rate equation above. Initially there is no *trans*-2-butene present, and so the initial rate of the reverse reaction is zero. As the forward reaction proceeds, the concentration of *cis*-2-butene decreases, and so the forward rate decreases. Also, as soon as some *trans*-2-butene has formed, the reverse reaction can begin, and as the concentration of *trans*-2-butene builds up, the reverse rate gets faster and faster. Eventually the forward rate slows down (and the reverse rate speeds up) to the point where the two rates are equal. At this time equilibrium has been achieved, and in the macroscopic system no further change in concentrations will be observed (Figure 14.2).

On the nanoscale, when equilibrium has been achieved, both reactions are still occurring, but the forward and reverse rates are equal. Therefore we can equate the two rates to give

$$\text{Rate}_{\text{forward}} = \text{Rate}_{\text{reverse}}$$

and, by substituting from the two previous rate equations,

$$k_{\text{forward}}\,[\textit{cis}\text{-2-butene}] = k_{\text{reverse}}\,[\textit{trans}\text{-2-butene}]$$

Figure 14.2 Approach to equilibrium. The graph shows the concentrations of *cis*-2-butene and *trans*-2-butene as the *cis* compound reacts to form the *trans* compound. The nanoscale diagrams above the graph show snapshots of the composition of a tiny portion of the reaction mixture. The same molecule has been circled in each diagram.

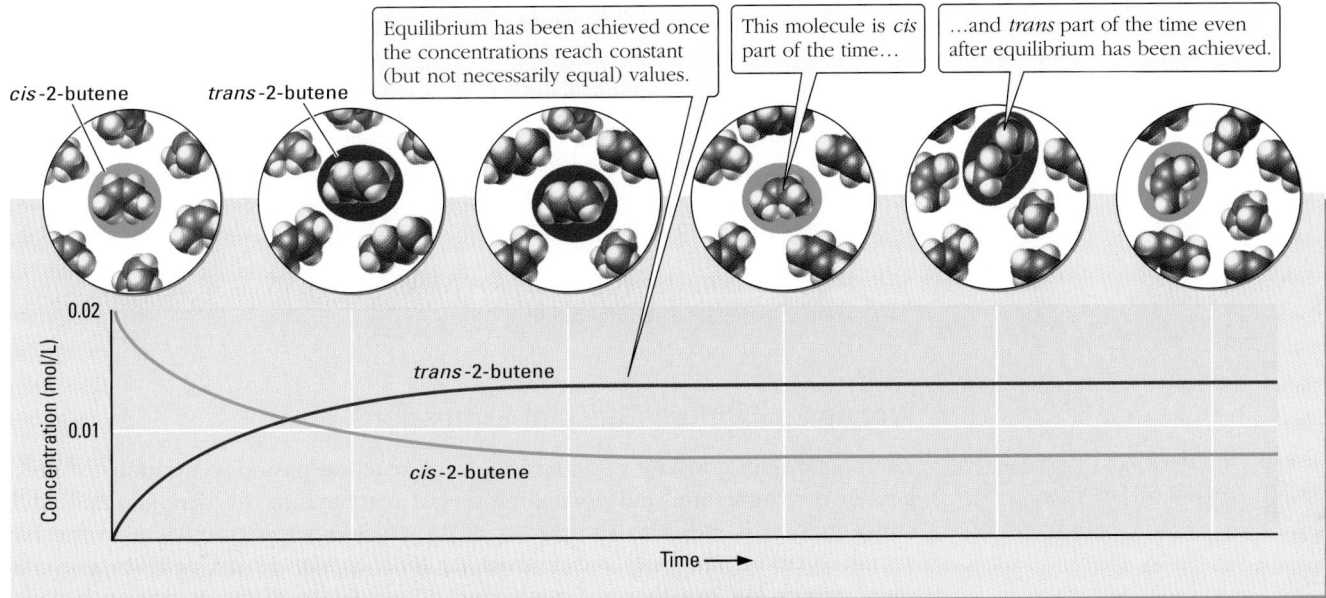

In this and subsequent equations square brackets distinguish unchanging equilibrium concentrations from the changing concentrations of reactants and products before equilibrium is achieved. The equation can be rearranged so that both rate constants are on one side and both equilibrium concentrations on the other.

$$\frac{k_{\text{forward}}}{k_{\text{reverse}}} = \frac{[\textit{trans-}2\text{-butene}]}{[\textit{cis-}2\text{-butene}]}$$

This shows that the ratio of equilibrium concentrations is equal to a ratio of rate constants. Because a ratio of two constants is also a constant, the ratio of equilibrium concentrations must also be constant. We call this ratio K_c, where the capital letter "K" is used to distinguish it from the rate constants, k_{forward} and k_{reverse}, and the subscript "c" indicates that it is a ratio of equilibrium *concentrations*.

$$K_c = \frac{k_{\text{forward}}}{k_{\text{reverse}}} = \frac{[\textit{trans-}2\text{-butene}]}{[\textit{cis-}2\text{-butene}]} = 1.65 \qquad \text{(at 500 K)}$$

Because the values of the rate constants vary with temperature (⬅ *p. 593*), the value of K_c also varies with temperature. For the butene isomerization reaction, K_c is 1.47 at 600 K and 1.36 at 700 K.

A quotient of equilibrium concentrations of reactant and product substances that has a constant value for a given reaction at a given temperature is called an **equilibrium constant** and is given the symbol K_c. In the next section we show how to derive a mathematical expression for the equilibrium constant directly from the chemical equation for any equilibrium process. The mathematical expression is called an **equilibrium constant expression.**

Equilibrium constants can be used to answer three important questions about a reaction.

- When equilibrium has been achieved, do products predominate over reactants?
- Given initial concentrations of reactants and products, in which direction will the reaction go?
- What concentrations of reactants and products are present at equilibrium?

If a reaction moves quickly to equilibrium, you can use equilibrium constants to determine the composition soon after reactants are mixed. Equilibrium constants are less valuable for slow reactions. Until equilibrium has been reached, only kinetics is capable of predicting the composition of a reaction mixture.

> In Chapter 13 we noted that the rate law for an overall chemical reaction cannot be derived from the balanced equation, but must be determined experimentally by kinetic studies. This is not true of the equilibrium constant expression. When a reaction takes place by a mechanism that consists of a sequence of steps, the equilibrium constant expression can be obtained by multiplying together the rate constants for the forward reactions in all steps and then dividing by the rate constants for the reverse reactions in all steps. This gives the same equilibrium constant expression that can be obtained from the coefficients of the balanced overall equilibrium equation.

Exercise 14.3   **Properties of Equilibrium**

After a mixture of *cis*-2-butene and *trans*-2-butene has reached equilibrium at 600 K, where $K_c = 1.47$, half of the *cis*-2-butene is suddenly removed. Answer these questions:

(a) Is the new mixture at equilibrium? Explain why or why not.

(b) In the new mixture, which rate is faster, *cis* → *trans* or the reverse? Or are both rates the same?

(c) In an equilibrium mixture, which concentration is larger, *cis*-2-butene or *trans*-2-butene?

(d) If the concentration of *cis*-2-butene at equilibrium is 0.10 mol/L, what will be the concentration of *trans*-2-butene?

Writing Equilibrium Constant Expressions

CD-ROM Screen 16.6: Writing Equilibrium Expressions

The equilibrium constant expression for any reaction has concentrations of products in the numerator and concentrations of reactants in the denominator. Each concentration is raised to the power of its stoichiometric coefficient in the balanced equation. *The only concentrations that appear in an equilibrium constant expression are those of gases and of solutes in dilute solutions,* because these are the only concentrations that can change as a reaction occurs. Concentrations of pure solids, pure liquids, and solvents in dilute solutions **do not** appear in equilibrium constant expressions.

To illustrate how this works, consider the general equilibrium reaction

$$a\text{A} + b\text{B} \rightleftharpoons c\text{C} + d\text{D} \qquad [14.1]$$

By convention we write the equilibrium constant expression for this reaction as

$$K_c = \frac{[\text{C}]^c[\text{D}]^d}{[\text{A}]^a[\text{B}]^b} \qquad [14.2]$$

Equilibrium constant

Product concentrations raised to powers of coefficients

Reactant concentrations raised to powers of coefficients

Let us apply these ideas to the combination of nitrogen and oxygen gases to form nitrogen monoxide.

$$\text{N}_2(g) + \text{O}_2(g) \rightleftharpoons 2\,\text{NO}(g)$$

> The reaction of nitrogen with oxygen occurs in automobile engines and other high-temperature combustion processes where air is present.

Because all substances in the reaction are gaseous, all concentrations appear in the equilibrium constant expression. The concentration of the product NO(g) is in the numerator and is squared because of the coefficient 2. The concentrations of the reactants $\text{N}_2(g)$ and $\text{O}_2(g)$ are in the denominator.

$$\text{Equilibrium constant} = K_c = \frac{[\text{NO}]^2}{[\text{N}_2][\text{O}_2]}$$

> Using square brackets for gas concentrations is less common than for aqueous solutions, but it still means mol/L.

Equilibria Involving Pure Liquids and Solids

As another example, consider the combustion of solid yellow sulfur, which produces sulfur dioxide gas.

$$\tfrac{1}{8}\text{S}_8(s) + \text{O}_2(g) \rightleftharpoons \text{SO}_2(g)$$

> This reaction occurs whenever a material that contains sulfur burns in air, and it is responsible for a good deal of sulfur dioxide air pollution. It is also the first reaction in a sequence by which sulfur is converted to sulfuric acid, the number-one industrial chemical in the world.

Placing product concentration in the numerator, reactant concentrations in the denominator, and stoichiometric coefficients as exponents, gives

$$K_c' = \frac{[\text{SO}_2(g)]}{[\text{O}_2(g)][\text{S}_8(s)]^{1/8}}$$

Product

Reactants

K_c' is inappropriate because concentration of a solid is included

Because sulfur is a solid the number of molecules per unit volume is fixed by the density of sulfur at any given temperature (see Exercise 14.1, (⇐ *p. 634*). Therefore, the sulfur concentration is not changed either by reaction or by addition or removal of solid sulfur. It is an experimental fact that, as long as there is some solid sulfur present, the equilibrium concentrations of O_2 and SO_2 are not affected by changes in the amount of sulfur. Therefore, the equilibrium constant expression should be written as

$$K_c = \frac{[\text{SO}_2(g)]}{[\text{O}_2(g)]} \qquad \text{which at 25 °C is equal to } 4.2 \times 10^{52}.$$

Equilibria in Dilute Solutions

Consider an aqueous solution of the weak base ammonia, which contains a small concentration of hydroxide ions because ammonia reacts with water.

$$\text{NH}_3(aq) + \text{H}_2\text{O}(\ell) \rightleftharpoons \text{NH}_4^+(aq) + \text{OH}^-(aq)$$

> It is very important to remember that concentrations of pure solids, pure liquids, and solvents for dilute solutions do not appear in the equilibrium constant expression.

Any molecule designated (ℓ) in the equilibrium reaction does not appear in the equilibrium constant expression.

If the concentration of ammonia molecules (and consequently of ammonium ions and hydroxide ions) is small, the number of water molecules per unit volume remains essentially the same as in pure water. Because the molar concentration of water is effectively constant for reactions involving dilute solutions, the concentration of water need not be included in the equilibrium constant expression. Thus, we write

$$K_c = \frac{[NH_4^+][OH^-]}{[NH_3]}$$

and the concentration of water is not included in the denominator. At 25 °C, $K_c = 1.8 \times 10^{-5}$.

In Exercise 14.1 you calculated the concentration of water in pure water as 55.4 mol/L. For equilibria in dilute aqueous solutions the K_c we write is actually the molar concentration of water (55.5 M) times a K_c' that includes the concentration of water in its denominator. Because the concentration of water changes very little as the concentrations of solutes change, the K_c also remains constant.

Notice that the equilibrium constant for the ammonia ionization reaction has no units. There are two concentrations in the numerator and only one in the denominator, which ought to give units of mol/L. However, if we always express concentrations in mol/L, the units of the equilibrium constant can be figured out from the equilibrium constant expression. Therefore it is customary to omit the units, and we shall follow that custom here.

Problem-Solving Example 14.1   Writing Equilibrium Constant Expressions

Write an equilibrium constant expression for each chemical equation.

(a) $4 NH_3(g) + 7 O_2(g) \rightleftharpoons 4 NO_2(g) + 6 H_2O(g)$
(b) $AgCl(s) \rightleftharpoons Ag^+(aq) + Cl^-(aq)$
(c) $Cu(NH_3)_4^{2+}(aq) \rightleftharpoons Cu^{2+}(aq) + 4 NH_3(aq)$
(d) $CaCO_3(s) \rightleftharpoons CaO(s) + O_2(g)$
(e) $H_2O(\ell) \rightleftharpoons H_2O(g)$

Answer

(a) $K_c = \dfrac{[NO_2]^4[H_2O]^6}{[NH_3]^4[O_2]^7}$ (b) $K_c = [Ag^+][Cl^-]$

(c) $K_c = \dfrac{[Cu^{2+}][NH_3]^4}{[Cu(NH_3)_4^{2+}]}$ (d) $K_c = [O_2]$ (e) $K_c = [H_2O(g)]$

Explanation Concentrations of products go in the numerator of the fraction, and concentrations of reactants go in the denominator. Each concentration is raised to the power of the stoichiometric coefficient of the species. In part (b), one species, AgCl(s), is a solid and does not appear in the expression. In part (d), two species, $CaCO_3(s)$ and CaO(s), are solids and do not appear. In part (e), one species, $H_2O(\ell)$, is a pure liquid and does not appear.

Problem-Solving Practice 14.1

Write an equilibrium constant expression for each equation.

(a) $H_2(g) + \frac{1}{8} S_8(s) \rightleftharpoons H_2S(g)$
(b) $HCl(g) + LiH(s) \rightleftharpoons H_2(g) + LiCl(s)$
(c) $CH_4(g) + H_2O(g) \rightleftharpoons CO(g) + 3 H_2(g)$
(d) $CN^-(aq) + H_2O(\ell) \rightleftharpoons HCN(aq) + OH^-(aq)$

Equilibrium Constant Expressions for Related Reactions

CD-ROM Screen 16.7: Manipulating Equilibrium Expressions

Consider the equilibrium among nitrogen, hydrogen, and ammonia.

$$N_2(g) + 3 H_2(g) \rightleftharpoons 2 NH_3(g) \qquad K_{c_1} = 3.5 \times 10^8 \quad \text{(at 25 °C)}$$

We could also write the equation so that 1 mol of NH_3 is produced.

$$\tfrac{1}{2} N_2(g) + \tfrac{3}{2} H_2(g) \rightleftharpoons NH_3(g) \qquad\qquad K_{c_2} = ?$$

Coefficients half as big

Is the value of the equilibrium constant, K_{c_2}, for the second equation the same as the value of the equilibrium constant, K_{c_1}, for the first equation? To see the relation between K_{c_1} and K_{c_2}, write the equilibrium constant expression for each balanced equation.

Concentrations raised to powers half as big

$$K_{c_1} = \frac{[NH_3]^2}{[N_2][H_2]^3} = 3.5 \times 10^8 \quad \text{and} \quad K_{c_2} = \frac{[NH_3]}{[N_2]^{1/2}[H_2]^{3/2}} = ?$$

This makes it clear that K_{c_1} is the square of K_{c_2}; that is, $K_{c_1} = (K_{c_2})^2$. Therefore, the answer to our question is

$$K_{c_2} = (K_{c_1})^{1/2} = (3.5 \times 10^8)^{1/2} = 1.9 \times 10^4$$

Whenever the stoichiometric coefficients of a balanced equation are multiplied by some factor, the equilibrium constant for the new equation (K_{c_2} in this case) is the old equilibrium constant (K_{c_1}) raised to the power of the multiplication factor.
What is the value of K_{c_3}, the equilibrium constant for the decomposition of ammonia to the elements, which is the reverse of the first equation?

$$2\,NH_3(g) \rightleftharpoons N_2(g) + 3\,H_2(g) \qquad K_{c_3} = ? = \frac{[N_2][H_2]^3}{[NH_3]^2}$$

Concentration of NH_3 is in denominator

It is clear that K_{c_3} is the reciprocal of K_{c_1}. That is, $K_{c_3} = 1/K_{c_1} = 1/(3.5 \times 10^8) = 2.9 \times 10^{-9}$. ***The equilibrium constant for a reaction and that for its reverse are the reciprocals of one another.*** This means that if a reaction has a very large equilibrium constant, the reverse reaction will have a very small one. That is, if a reaction is strongly product-favored, its reverse is strongly reactant-favored. In the case of production of ammonia from the elements at room temperature, the forward reaction has a large equilibrium constant (3.5×10^8). As expected, the reverse reaction, decomposition of ammonia to its elements, has a small equilibrium constant (2.9×10^{-9}).

Exercise 14.4 **Manipulating Equilibrium Constants**

The conversion of 1.5 mol of oxygen to 1.0 mol of ozone has a very small value of K_c.

$$\tfrac{3}{2}O_2(g) \rightleftharpoons O_3(g) \qquad\qquad K_c = 2.5 \times 10^{-29}$$

(a) What is the value of K_c if the equation is written using whole-number coefficients?

$$3\,O_2(g) \rightleftharpoons 2\,O_3(g)$$

(b) What is the value of K_c for the conversion of 2 mol of ozone to 3 mol of oxygen?

$$2\,O_3(g) \rightleftharpoons 3\,O_2(g)$$

Equilibrium Constant for a Reaction That Combines Two or More Other Reactions

If two chemical equations can be combined to give a third, the equilibrium constant for the combined reaction can be obtained from the equilibrium constants for the two original reactions. For example, air pollution is produced when nitrogen monoxide forms from nitrogen and oxygen and then combines with additional oxygen to form nitrogen dioxide.

(1) $\quad N_2(g) + O_2(g) \longrightarrow 2\,NO(g) \qquad\qquad K_{c_1} = \dfrac{[NO]^2}{[N_2][O_2]}$

(2) $2 NO(g) + O_2(g) \longrightarrow 2 NO_2(g)$
$$K_{c_2} = \frac{[NO_2]^2}{[NO]^2[O_2]}$$

The sum of these two equations is

Sum of Equations 1 and 2

(3) $N_2(g) + 2 O_2(g) \longrightarrow 2 NO_2(g)$

Product of equilibrium constants K_{c_1} and K_{c_2}

$$K_{c_3} = \frac{[NO_2]^2}{[N_2][O_2]^2} = \frac{[NO]^2}{[N_2][O_2]} \times \frac{[NO_2]^2}{[NO]^2[O_2]} = K_{c_1} \times K_{c_2}$$

That is, *if two chemical equations can be summed to give a third, the equilibrium constant for the overall equation equals the product of the two equilibrium constants for the equations that were summed.* This is a powerful tool for obtaining equilibrium constants for reactions without having to measure them experimentally for each individual reaction.

Problem-Solving Example 14.2   Manipulating Equilibrium Constants

Given these equilibrium reactions and constants,

(1) $\frac{3}{2} O_2(g) \rightleftharpoons O_3(g)$ $K_{c_1} = 2.5 \times 10^{-29}$

(2) $2 NO(g) + O_2(g) \rightleftharpoons 2 NO_2(g)$ $K_{c_2} = 2.25 \times 10^{12}$

calculate the equilibrium constant for a reaction that is important in the formation of photochemical smog air pollution.

$NO(g) + O_3(g) \rightleftharpoons NO_2(g) + O_2(g)$ $K_{c_3} = \dfrac{[NO_2][O_2]}{[NO][O_3]}$

Answer 6.0×10^{34}

Explanation   The target equation has O_3 on the left, so Equation 1 needs to be reversed. This means that we need to take the reciprocal of the equilibrium constant. The target equation has NO on the left and NO_2 on the right. Equation 2 need not be reversed, but each coefficient must be multiplied by $\frac{1}{2}$, which means taking the square root of K_{c_2}. Once this has been done, the two new equations sum to the target equation, and we can multiply their equilibrium constants.

(1') $O_3(g) \rightleftharpoons \frac{3}{2} O_2(g)$ $K'_{c_1} = \dfrac{1}{2.5 \times 10^{-29}} = 4.0 \times 10^{28}$

(2') $NO(g) + \frac{1}{2} O_2(g) \rightleftharpoons NO_2(g)$ $K'_{c_2} = \sqrt{2.25 \times 10^{12}} = 1.5 \times 10^6$

$NO(g) + O_3(g) \rightleftharpoons NO_2(g) + O_2(g)$ $K_{c_3} = K'_{c_1} \times K'_{c_2} = 6.0 \times 10^{34}$

✔   For Equation 1 the equilibrium constant was quite small, so for the reverse reaction it should be quite large, which it is. Also check the order of magnitude of the answer by checking the powers of 10. In the square root the power of 10 should be halved, which it is. And the sum of the powers of 10 for the two equilibrium constants that were multiplied should be close to the power of 10 in the answer, which it is.

Problem-Solving Practice 14.2

When carbon dioxide dissolves in water it produces carbonic acid, $H_2CO_3(aq)$, which can ionize in two steps.

$H_2CO_3(aq) + H_2O(aq) \rightleftharpoons HCO_3^-(aq) + H_3O^+(aq)$ $K_{c_1} = 4.2 \times 10^{-7}$

$HCO_3^-(aq) + H_2O(aq) \rightleftharpoons CO_3^{2-}(aq) + H_3O^+(aq)$ $K_{c_2} = 4.8 \times 10^{-11}$

Calculate the equilibrium constant for the reaction

$H_2CO_3(aq) + 2 H_2O(aq) \rightleftharpoons CO_3^{2-}(aq) + 2 H_3O^+(aq)$

Equilibrium Constants in Terms of Pressure

In a constant-volume system, when the concentration of a gas changes, the partial pressure ($\Leftarrow$ *p. 432*) of the gas also changes. This follows from the ideal gas equation

$$PV = nRT$$

Solving for the partial pressure P_A of a gaseous substance, A,

$$P_A = \frac{n_A}{V}\,RT = [A]RT \qquad [14.3]$$

| [A] = n_A/V, the number of moles of A per unit volume |

Equation 14.3 allows us to express the equilibrium constant for the general reaction in Equation 14.1 ($\Leftarrow$ *p. 639*) in a form similar to Equation 14.2, but in terms of partial pressures as

$$K_P = \frac{P_C^c \times P_D^d}{P_A^a \times P_B^b} \qquad [14.4]$$

Product pressures raised to powers of coefficients

Reactant pressures raised to powers of coefficients

The subscript on K_P indicates that the equilibrium constant has been expressed in terms of partial pressures. For some equilibria $K_c = K_P$, but for many it does not. Therefore it is useful to be able to relate one type of equilibrium constant to the other. This can be done by combining Equations 14.2, 14.3, and 14.4 to give

$\Delta n = c + d - a - b$ is the number of moles of gaseous products minus the number of moles of gaseous reactants

$$K_P = \frac{P_C^c \times P_D^d}{P_A^a \times P_B^b} = \frac{\{[C]RT\}^c\{[D]RT\}^d}{\{[A]RT\}^a\{[B]RT\}^b} = \frac{[C]^c[D]^d}{[A]^a[B]^b}(RT)^{c+d-a-b} = K_c(RT)^{\Delta n} \quad [14.5]$$

Because K_c and K_P for the same gas phase reaction are related, either can be used to calculate the composition of an equilibrium mixture. Most examples in this chapter involve K_c, but the same rules apply to solving problems with K_P.

As an example of this relation, consider the equilibrium

$$2\,NOCl(g) \rightleftharpoons 2\,NO(g) + Cl_2(g) \qquad K_c = 4.02 \times 10^{-2}\ \text{mol/L at 298 K}$$

For this reaction $\Delta n = 2 + 1 - 2 = 3 - 2 = 1$, because there are three moles of gas phase product molecules and only two moles of gas phase reactants. Therefore, for this reaction

$$K_P = K_c \times (RT)^1$$
$$= (4.02 \times 10^{-2}\ \text{mol/L}) \times (0.0821\ \text{L atm K}^{-1}\ \text{mol}^{-1}) \times (298\ \text{K}) = 0.984\ \text{atm}$$

Exercise 14.5 Relating K_c and K_P

For each of these reactions, calculate K_P from K_c.

(a) $N_2(g) + 3\,H_2(g) \rightleftharpoons 2\,NH_3(g)$ $K_c = 3.5 \times 10^8$ at 25 °C
(b) $2\,H_2(g) + O_2(g) \rightleftharpoons 2\,H_2O(g)$ $K_c = 3.2 \times 10^{81}$ at 25 °C
(c) $N_2(g) + O_2(g) \rightleftharpoons 2\,NO(g)$ $K_c = 1.7 \times 10^{-3}$ at 2300 K
(d) $2\,NO_2(g) \rightleftharpoons N_2O_4(g)$ $K_c = 1.7 \times 10^2$ at 25 °C

14.3 DETERMINING EQUILIBRIUM CONSTANTS

To determine an equilibrium constant it is necessary to know all of the concentrations that appear in the equilibrium constant expression. This is most commonly done by allowing a system to reach equilibrium and then measuring the

 CD-ROM Screen 16.8: Determining an Equilibrium Constant

A gas or a solution is colored if it absorbs visible light. The more light that is absorbed, the greater the concentration of the colored substance. An instrument known as a spectrophotometer can measure absorbance of light, thereby determining the concentration.

(a)

(b)

Figure 14.3 Formation of NO₂ from N₂O₄. Flasks containing an equilibrium mixture of dinitrogen tetraoxide (colorless) and nitrogen dioxide (red-brown) are shown (a) surrounded by ice and (b) immersed in water at 80 °C. Notice the much darker color at the higher temperature, indicating that the equilibrium has shifted some dinitrogen tetraoxide has reacted to form nitrogen dioxide. The intensity of color can be used to measure the concentration of nitrogen dioxide.

equilibrium concentration of one or more of the reactants or products. Algebra and stoichiometry are then used to obtain K_c.

Reaction Tables, Stoichiometry, and Equilibrium Concentrations

There is a systematic approach to calculations involving equilibrium constants. Make a table that shows initial conditions, changes that take place when a reaction occurs, and final (equilibrium) conditions. As an example, consider the colorless gas dinitrogen tetraoxide, $N_2O_4(g)$. When heated it dissociates to form red-brown $NO_2(g)$ according to the equation

$$N_2O_4(g) \rightleftharpoons 2\,NO_2(g)$$

Suppose that 2.00 mol of $N_2O_4(g)$ is placed into an empty 5.00-L flask and heated to 407 K. In a few seconds a dark red-brown color appears, indicating that much of the colorless gas has been transformed into NO_2 (Figure 14.3). The intensity of color indicates that the concentration of NO_2 at equilibrium is 0.525 mol/L. To use this information to find the equilibrium constant, follow these steps:

1. ***Write the balanced equation for the equilibrium reaction. From it derive the equilibrium constant expression.*** The balanced equation and equilibrium constant expression are

$$N_2O_4(g) \rightleftharpoons 2\,NO_2(g) \qquad K_c = \frac{[NO_2]^2}{[N_2O_4]}$$

2. ***Set up a table containing initial concentration, change in concentration, and equilibrium concentration for each substance included in the equilibrium constant expression. Enter all known information into this reaction table.*** In this case the number of moles of $N_2O_4(g)$ and the volume of the flask were given, so we first calculate the initial concentration of $N_2O_4(g)$ as

$$(\text{conc. } N_2O_4) = \frac{2.00\ \text{mol}}{5.00\ \text{L}} = 0.40\ \text{mol/L}.$$ Because the flask contained no NO_2, the initial concentration of NO_2 is zero. The equilibrium concentration of NO_2 was measured as 0.525 mol/L. The reaction table looks like this:

	N₂O₄(g) ⇌	2 NO₂(g)
Initial concentration (mol/L)	0.40	0
Change as reaction occurs (mol/L)	_____	_____
Equilibrium concentration (mol/L)	_____	0.525

3. ***Use x to represent the change in concentration of one substance. Use the stoichiometric coefficients in the balanced equilibrium equation to calculate the other changes in terms of x.*** When the reaction proceeds from left to right, the concentrations of reactants decrease. Therefore the change in concentration of a reactant is negative. The concentrations of products increase, so change in concentration of a product is positive. Usually it is best to begin with the reactant or product column that contains the most information. In this case that is the NO_2 column, where both initial and equilibrium concentrations are known. Therefore, we let x represent the unknown change in concentration of NO_2.

	N₂O₄(g) ⇌	2 NO₂(g)
Initial concentration (mol/L)	0.40	0
Change as reaction occurs (mol/L)	_____	x
Equilibrium concentration (mol/L)	_____	0.525

Next we use the mole ratio from the balanced equation to find the change in concentration of N_2O_4 in terms of x.

$$\Delta(\text{conc. } N_2O_4) = \frac{x \text{ mol } NO_2 \text{ formed}}{L} \times \frac{1 \text{ mol } N_2O_4 \text{ reacted}}{2 \text{ mol } NO_2 \text{ formed}}$$

$$= \tfrac{1}{2}x \text{ mol } N_2O_4 \text{ reacted/L}$$

The sign of the change in concentration of N_2O_4 is *negative,* because the concentration of N_2O_4 *decreases.* The table becomes

	$N_2O_4(g)$ $\rightleftharpoons$	$2 NO_2(g)$
Initial concentration (mol/L)	0.40	0
Change as reaction occurs (mol/L)	$-\tfrac{1}{2}x$	x
Equilibrium concentration (mol/L)	_____	0.525

For every 1 mol of N_2O_4 that decomposes, 2 mol of NO_2 forms. Therefore there is a $\tfrac{1}{2}$: 1 mol ratio of $N_2O_4 : NO_2$.

4. *From initial concentrations and the changes in concentrations, calculate the equilibrium concentrations in terms of* **x** *and enter them in the table.* The concentration of N_2O_4 at equilibrium is the sum of the initial 0.40 mol/L of N_2O_4 and the change due to reaction $-\tfrac{1}{2}x$ mol/L; that is, $[N_2O_4] = (0.40 - \tfrac{1}{2}x)$ mol/L. Similarly, the equilibrium concentration of NO_2 (which is already known to be 0.525 mol/L) is $0 + x$, and the table becomes

	$N_2O_4(g)$ $\rightleftharpoons$	$2 NO_2(g)$
Initial concentration (mol/L)	0.40	0
Change as reaction occurs (mol/L)	$-\tfrac{1}{2}x$	x
Equilibrium concentration (mol/L)	$0.40 - \tfrac{1}{2}x$	$0.525 = 0 + x$

Tables like this one are often called ICE tables from the initial letters of the labels on the rows.

5. *Use the simplest possible equation to solve for* **x**. *Then use* **x** *to calculate the unknown you were asked to find.* (Usually this is K_c or an unknown concentration.) In this case the simplest equation to solve for x is the last entry in the table, $0.525 = 0 + x$, and it is easy to see that $x = 0.525$. Calculate $[N_2O_4] = (0.40 - \tfrac{1}{2}x)$ mol/L $= (0.40 - \tfrac{1}{2} \times 0.525)$ mol/L $= 0.138$ mol/L. The problem stated that $[NO_2] = 0.525$ mol/L, so K_c is given by

$$K_c = \frac{[NO_2]^2}{[N_2O_4]} = \frac{(0.525 \text{ mol/L})^2}{(0.138 \text{ mol/L})} = 2.00 \qquad \text{(at 407 K)}$$

6. *Check your answer to make certain it is reasonable.* In this case the concentration of product is larger than the concentration of reactant. Since the products are in the numerator of the equilibrium constant expression, we expect a value greater than 1, and this is what we calculated.

Problem-Solving Example 14.3   Determining an Equilibrium Constant

Consider the gas phase reaction

$$H_2(g) + I_2(g) \rightleftharpoons 2 HI(g)$$

Suppose that a flask containing H_2 and I_2 has been heated to 425 °C and the initial concentrations of H_2 and I_2 were each 0.0175 mol/L. With time, the concentrations of H_2 and I_2 decline and the concentration of HI increases, and at equilibrium [HI] = 0.0276 mol/L. Use this experimental information to calculate the equilibrium constant.

Answer $K_c = 56$

Explanation   Use the information given and follow the six steps. In the table below, each entry has been color-coded to match the numbered steps.

1. Write the balanced equation and equilibrium constant expression.

$$H_2(g) + I_2(g) \rightleftharpoons 2\,HI(g) \qquad\qquad K_c = \frac{[HI]^2}{[H_2][I_2]}$$

2. Construct a reaction table (see below) and enter known information.

3. Represent changes in concentration in terms of x.

The best choice is to enter x in the third column, because both initial and equilibrium concentrations of HI are known, and this provides a simple way to calculate x. Next, derive the rest of the blanks in terms of x. If the concentration of HI increases by a given quantity, the mole ratios say that the concentrations of H_2 and I_2 must decrease only half as much:

$$\frac{x\ \text{mol HI produced}}{L} \times \frac{1\ \text{mol}\ H_2\ \text{consumed}}{2\ \text{mol HI produced}} = \frac{1}{2}x\ \text{mol/L}\ H_2\ \text{consumed}$$

Because the coefficients of H_2 and I_2 are equal, each of their concentrations decreases by $\frac{1}{2}x$ mol/L. The entries in the table are negative, because the concentrations decrease.

4. Calculate equilibrium concentrations and enter them in the table.

Color-coded entries in the table in this example correspond with color-coded steps in the explanation section.

	$H_2(g)$	+	$I_2(g)$	$\rightleftharpoons$	$2\,HI(g)$
Initial concentration (mol/L)	0.0175		0.0175		0
Change as reaction occurs (mol/L)	$-\frac{1}{2}x$		$-\frac{1}{2}x$		x
Equilibrium concentration (mol/L)	$0.0175 - \frac{1}{2}x$		$0.0175 - \frac{1}{2}x$		$0.0276 = 0 + x$

5. Solve the simplest equation for x. The last row and column in the table contains $0.0276 = 0 + x$, which gives $x = 0.0276$. Substitute this value into each equation in the last row of the table to get the equilibrium concentrations, and substitute them into the equilibrium constant expression:

$$K_c = \frac{[HI]^2}{[H_2][I_2]} = \frac{(0.0276)^2}{(0.0175 - [\frac{1}{2} \times 0.0276])(0.0175 - [\frac{1}{2} \times 0.0276])}$$

$$= \frac{(0.0276)^2}{(0.0037)(0.0037)} = 56 \qquad\qquad \text{(at 424 °C)}$$

✔ The equilibrium constant is larger than 1, so there should be more products than reactants when equilibrium is reached. The calculated equilibrium concentration of the product (0.0276 mol/L) is larger than those of the reactants (0.0037 mol/L each).

Problem-Solving Practice 14.3

Saying that 2.96% of the acetic acid molecules have ionized means that at equilibrium the concentration of acetate ions is 2.96/100 = 0.0296 times the initial concentration of acetic acid molecules.

Measuring the conductivity of an aqueous solution in which 0.0200 mol of CH_3COOH has been dissolved in 1.00 L of solution shows that 2.96% of the acetic acid molecules have ionized to CH_3COO^- ions and H_3O^+ ions. Calculate the equilibrium constant for ionization of acetic acid and compare your result with the value given in Table 14.1.

Experimentally determined equilibrium constants for a few reactions are given in Table 14.1. These reactions occur to widely differing extents, as shown by the wide range of values of K_c.

14.4   THE MEANING OF THE EQUILIBRIUM CONSTANT

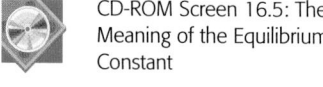

CD-ROM Screen 16.5: The Meaning of the Equilibrium Constant

The value of the equilibrium constant tells how far a reaction has proceeded by the time equilibrium has been achieved. In addition, it can be used to calculate how much product will be present at equilibrium. There are three important cases to consider.

TABLE 14.1	Selected Equilibrium Constants at 25 °C		
Reaction		K_c	K_P
Nonmetal reactions			
$\frac{1}{8} S_8(s) + O_2(g) \rightleftharpoons SO_2(g)$		4.2×10^{52}	4.2×10^{52}
$2 H_2(g) + O_2(g) \rightleftharpoons 2 H_2O(g)$		3.3×10^{81}	1.3×10^{80}
$N_2(g) + 3 H_2(g) \rightleftharpoons 2 NH_3(g)$		3.5×10^8	5.8×10^5
$N_2(g) + O_2(g) \rightleftharpoons 2 NO(g)$		4.5×10^{-31}	4.5×10^{-31}
		1.7×10^{-3} (at 2300 K)	
$H_2(g) + I_2(g) \rightleftharpoons 2 HI(g)$		2.5×10^1	2.5×10^1
$2 NO_2(g) \rightleftharpoons N_2O_4(g)$		1.7×10^2	7.0
$CH_4(g) + H_2O(g) \rightleftharpoons CO(g) + 3 H_2(g)$		9.4×10^{-1}	1.6×10^{-3}
cis-2-butene $\rightleftharpoons$ *trans*-2-butene		3.2	3.2
Weak acids and bases			
Formic acid			
$\quad HCOOH(aq) + H_2O(\ell) \rightleftharpoons H_3O^+(aq) + HCOO^-(aq)$		1.8×10^{-4}	—
Acetic acid			
$\quad CH_3COOH(aq) + H_2O(\ell) \rightleftharpoons H_3O^+(aq) + CH_3COO^-(aq)$		1.8×10^{-5}	—
Carbonic acid			
$\quad H_2CO_3(aq) + H_2O(\ell) \rightleftharpoons H_3O^+(aq) + HCO_3^-(aq)$		4.2×10^{-7}	—
Ammonia (weak base)			
$\quad NH_3(aq) + H_2O(\ell) \rightleftharpoons NH_4^+(aq) + OH^-(aq)$		1.8×10^{-5}	—
Very slightly soluble solids			
$CaCO_3(s) \rightleftharpoons Ca^{2+}(aq) + CO_3^{2-}(aq)$		3.8×10^{-9}	—
$AgCl(s) \rightleftharpoons Ag^+(aq) + Cl^-(aq)$		1.8×10^{-10}	—
$AgI(s) \rightleftharpoons Ag^+(aq) + I^-(aq)$		1.5×10^{-16}	—

Case 1. $K_c \gg 1$: Reaction is strongly product-favored; equilibrium concentrations of products are much greater than equilibrium concentrations of reactants.

A large value of K_c means that reactants have been converted almost entirely to products when equilibrium has been achieved. That is, the products are strongly favored over the reactants. An example is the reaction of NO(g) with O_3(g), which is one way that ozone is destroyed in the stratosphere ($\Leftarrow$ *p. 441*).

$$NO(g) + O_3(g) \rightleftharpoons NO_2(g) + O_2(g)$$

$$K_c = \frac{[NO_2][O_2]}{[NO][O_3]} = 6 \times 10^{34} \quad \text{(at 25 °C)}$$

The very large value of K_c tells us that if 1 mol each of NO and O_3 are mixed in a flask and allowed to come to equilibrium, $[NO_2][O_2] \gg [NO][O_3]$. Virtually none of the reactants will remain, and essentially only NO_2 and O_2 will be found in the flask. For practical purposes, this reaction goes to completion, and it would not be necessary to use the equilibrium constant to calculate the quantities of products that would be obtained. The simpler methods developed in Chapter 4 would work just fine ($\Leftarrow$ *p. 136*).

Case 2. $K_c \ll 1$: Reaction is strongly reactant-favored; equilibrium concentrations of reactants are greater than equilibrium concentrations of products.

The symbol $\gg$ means "much greater than."

Supersonic aircraft. NO emitted from the engines of a supersonic aircraft like this one can reduce the concentration of ozone in the stratosphere. NO reaches the stratosphere by several pathways, including emissions from aircraft that fly in the lower part of the stratosphere. The effect of NO on stratospheric ozone is much less than that of chlorofluorocarbons ($\Leftarrow$ *p. 441*).

The symbol $\ll$ means "much less than."

Conversely, *an extremely small K_c means that when equilibrium has been achieved, very little of the reactants have been transformed into products.* The reactants are favored over the products at equilibrium.

$$3\,O_2(g) \rightleftharpoons 2\,O_3(g) \qquad K_c = \frac{[O_3]^2}{[O_2]^3} = 6.25 \times 10^{-58} \quad \text{(at 25 °C)}$$

This means that $[O_3]^2 \ll [O_2]^3$ and if O_2 is placed in a flask, *very* little O_3 will be found when equilibrium is achieved. The concentration of O_2 would remain essentially unchanged. In the terminology of Chapter 5, we would write "N.R." and say that no reaction occurs.

Case 3. $K_c \cong 1$: *Equilibrium mixture contains significant concentrations of reactants and products; calculations are needed to determine equilibrium concentrations.*

If K_c is neither extremely large nor extremely small, the equilibrium constant must be used to calculate how far a reaction proceeds toward products. In contrast with the two reactions just described, dissociation of dinitrogen tetraoxide has neither a very large nor a very small equilibrium constant. At 391 K the value is 1.00, which means that significant concentrations of both N_2O_4 and NO_2 are present at equilibrium.

$$K_c = 1.00 = \frac{[NO_2]^2}{[N_2O_4]}, \text{ and so } [NO_2]^2 = [N_2O_4] \quad \text{(at 391 K)} \qquad [14.6]$$

What range of equilibrium constants comprises this middle ground depends on how small a concentration is significant and on the form of the equilibrium constant. If at equilibrium the concentrations of N_2O_4 and NO_2 are both 1.0 mol/L, then the ratio of $[NO_2]^2/[N_2O_4]$ does equal the K_c value of 1.0 at 391 K. But what if the concentrations were much smaller? Would they still be equal? You can verify this by using Equation 14.6. If the equilibrium concentration of NO_2 is 0.01, then the concentration of N_2O_4 must be $(0.01)^2$, which equals 0.0001. Thus, even though $K_c = 1.00$, the concentration of one substance can be much bigger than the concentration of the other. This happens because there is a squared term in the numerator of the equilibrium constant expression and a term to a different power (the first power) in the denominator. Whenever the total of the exponents in the numerator differs from the total in the denominator, it becomes very difficult to say whether the concentrations of products will exceed those of reactants.

By contrast, if the total of the exponents is the same, it is true that if $K_c > 1$, products predominate over reactants and if $K_c < 1$, reactants predominate over products. Examples in which this is true are

$$\textit{cis}\text{-2-butene} \rightleftharpoons \textit{trans}\text{-2-butene} \qquad K_c = \frac{[\textit{trans}]}{[\textit{cis}]} = 3.2 \quad \text{(at 25 °C)}$$

and

$$H_2(g) + I_2(g) \rightleftharpoons 2\,HI(g) \qquad K_c = \frac{[HI]^2}{[H_2][I_2]} = 25 \quad \text{(at 25 °C)}$$

Figure 14.4 diagrams this relationship for the isomerization of *cis*-2-butene.

You might wonder whether reactant-favored systems in which small quantities of products form are important. Many are. Examples are the acids and bases listed in Table 14.1. For acetic acid, the main ingredient in vinegar, the reaction is

$$CH_3COOH(aq) + H_2O(\ell) \rightleftharpoons H_3O^+(aq) + CH_3COO^-(aq)$$

$$K_c = \frac{[H_3O^+][CH_3COO^-]}{[CH_3COOH]} = 1.8 \times 10^{-5} \quad \text{(at 25 °C)}$$

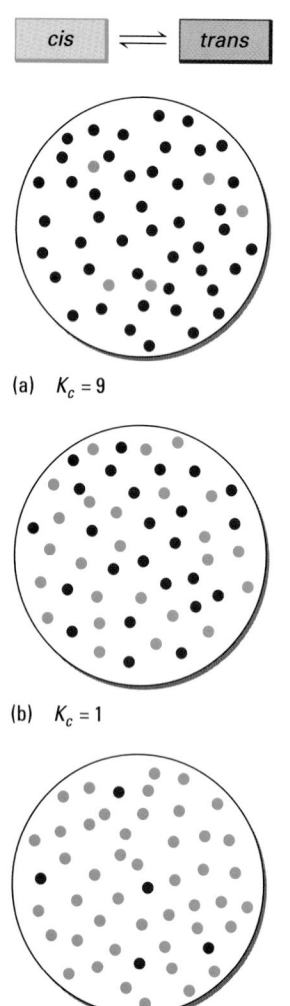

(a) $K_c = 9$

(b) $K_c = 1$

(c) $K_c = 1/9$

Figure 14.4 Equilibrium constants and concentrations of reactants and products. In the system *cis*-2-butene $\rightleftharpoons$ *trans*-2-butene, the equilibrium constant decreases as temperature increases. (a) At a low temperature, when $K_c = 9$, the ratio [*trans*]/[*cis*] is 9/1 or 45/5. (b) At 1000 K, when $K_c = 1$, the ratio is 1/1 or 25/25. (c) At a much higher temperature, when $K_c = 0.111 = 1/9$, the ratio is 1/9 or 5/45.

The value of K_c for acetic acid is small, and at equilibrium the concentrations of products (acetate ions and hydrogen ions) are small relative to the concentration of reactant (acetic acid molecules). Acetic acid is a weak acid. Nevertheless, vinegar is sour because a small percentage of the acetic acid molecules react with water to produce $H_3O^+(aq)$.

If the form of the equilibrium constant is the same for two or more different reactions, then the degree to which each of those reactions is product-favored can be compared quantitatively. For example, in Table 14.1 the equilibrium constant expressions for formic acid, acetic acid, and carbonic acid all have the same form,

$$K_c = \frac{[H_3O^+][anion^-]}{[acid]}$$

where anion$^-$ is $HCOO^-$, CH_3COO^-, or HCO_3^-, and acid is $HCOOH$, CH_3COOH, or H_2CO_3. Therefore, we can say that formic acid is stronger than acetic acid (has a larger K_c value) and carbonic acid is weakest of all.

If a reaction has a large tendency to occur in one direction, then the reverse reaction has little tendency to occur. This means that the equilibrium constant for the reverse of a strongly product-favored reaction will be extremely small. Table 14.1 shows that combustion of hydrogen to form water vapor has an enormous equilibrium constant (3.2×10^{81}). The reaction is strongly product-favored. The reverse reaction, decomposition of water to its elements,

$$2\,H_2O(g) \rightleftharpoons 2\,H_2(g) + O_2(g) \qquad K_c = \frac{[H_2]^2[O_2]}{[H_2O]^2} = 3.1 \times 10^{-82} \quad (\text{at } 25\,^\circ C)$$

is strongly reactant-favored, as indicated by the *very* small value of K_c.

 Problem-Solving Example 14.4 **Using Equilibrium Constants**

Use equilibrium constants (Table 14.1) to predict which of the reactions below will be product-favored at 25 °C. Then place all of the reactions in order from most reactant-favored to most product-favored.

(a) $H_2CO_3(aq) + H_2O(\ell) \rightleftharpoons H_3O^+(aq) + HCO_3^-(aq)$
(b) $N_2O_4(g) \rightleftharpoons 2\,NO_2(g)$
(c) $NH_3(aq) + H_2O(\ell) \rightleftharpoons NH_4^+(aq) + OH^-(aq)$

Answer All reactions are reactant-favored. The order from most reactant-favored to least reactant-favored is (a), (c), (b).

Explanation First check whether the reactions all have equilibrium constant expressions of the same form. If they do, then the smaller the equilibrium constant, the less product-favored (more reactant-favored) the reaction is. The equilibrium constant expressions are all of the form

$$K_c = \frac{[\text{product 1}][\text{product 2}]}{[\text{reactant}]}$$

because $H_2O(\ell)$ does not appear in the expressions for (a) and (c). The equilibrium constants for reactions (a) and (c) are 4.2×10^{-7} and 1.8×10^{-5} respectively. The equilibrium constant for reaction (b) is not given in Table 14.1, but K_c for the reverse reaction is given as 1.7×10^2. Because the reaction is reversed it is necessary to take the reciprocal, which gives an equilibrium constant for reaction (b) of 5.8×10^{-3}. Therefore the most reactant-favored reaction (smallest K_c) is (a), the next smallest K_c is for reaction (c), and the largest K_c is for reaction (b).

 Problem-Solving Practice 14.4

Suppose that solid AgCl and AgI are placed in 1.0 L of water in separate beakers.

(a) In which beaker would the silver ion concentration, $[Ag^+]$, be larger?
(b) Does the volume of water in which each compound dissolves affect the equilibrium concentration?

Exercise 14.6 Manipulating Equilibrium Constants

The equilibrium constant is 1.8×10^{-5} for reaction of ammonia with water.

$$NH_3(aq) + H_2O(\ell) \rightleftharpoons NH_4^+(aq) + OH^-(aq)$$

(a) Is the equilibrium constant large or small for the reverse reaction, the reaction of ammonium ion with hydroxide ion to give ammonia and water?

(b) What is the value of K_c for the reaction of ammonium ions with hydroxide ions?

(c) What does the value of this equilibrium constant tell you about the extent to which reaction can occur between ammonium ions and hydroxide ions?

(d) Predict what would happen if you added a 1.0 M solution of ammonium chloride to a 1.0 M solution of sodium hydroxide. What observations might allow you to test your prediction in the laboratory?

14.5 USING EQUILIBRIUM CONSTANTS

CD-ROM Screen 16.9: Systems at Equilibrium

Because equilibrium constants have numeric values, they can be used to predict quantitatively in which direction a reaction will proceed and how far it will go.

Predicting the Direction of a Reaction

Recall that the metric prefix "m" means $\frac{1}{1000}$. Therefore, 1 mmol = 1×10^{-3} mol.

Suppose that you have a mixture of 50 mmol of $NO_2(g)$ and 100 mmol of $N_2O_4(g)$ at 25 °C in a container with a volume of 10. L. Is the system at equilibrium? If not, in which direction will it react to achieve equilibrium? A useful approach to such questions is to use the **reaction quotient**, Q, which *has the same mathematical form as the equilibrium constant expression but is a ratio of actual concentrations in the mixture,* instead of equilibrium concentrations. For the reaction

$$2\,NO_2(g) \rightleftharpoons N_2O_4(g) \qquad K_c = \frac{[N_2O_4]}{[NO_2]^2} = 1.7 \times 10^2 \text{ at } 25\,°C$$

$$Q = \frac{(\text{conc. } N_2O_4)}{(\text{conc. } NO_2)^2} = \frac{(100 \times 10^{-3} \text{ mol}/10\text{ L})}{(50 \times 10^{-3} \text{ mol}/10\text{ L})^2} = \frac{1.00 \times 10^{-2}}{(5.0 \times 10^{-3})^2} = 4.0 \times 10^2$$

In the expression for Q, we have used (conc. N_2O_4) to represent the *actual* concentration of N_2O_4 at a given time. We use $[N_2O_4]$ to represent the equilibrium concentration of N_2O_4. When the reaction is at equilibrium (conc. N_2O_4) = $[N_2O_4]$.

If Q is equal to K_c, then the reaction is at equilibrium with the actual concentrations that were used to calculate Q. *If Q is less than K_c, then the concentrations of products are not as big as they would be at equilibrium.* The reaction will proceed from left to right to increase the concentrations until they reach their equilibrium values. *If Q is greater than K_c, then the product concentrations are higher than they would be at equilibrium,* and the reaction will proceed from right to left. This is shown schematically in Figure 14.5. In the case of the NO_2/N_2O_4 mixture above, Q is greater than K_c, and so some N_2O_4 will react to form NO_2 to establish equilibrium.

Problem-Solving Example 14.5 Predicting Direction of Reaction

Consider the equilibrium

$$N_2(g) + 3\,H_2(g) \rightleftharpoons 2\,NH_3(g) \qquad K_c = 0.105 \quad \text{at } 472\,°C$$

If 1.0 mol of N_2, 2.0 mol of H_2, and 3.0 mol of NH_3 are mixed in a 10.0-L container at 472 °C, will the concentration of N_2 be greater or less than 0.10 mol/L when equilibrium is reached?

Answer Greater

Explanation Calculate the initial concentration of each gas and thus evaluate Q. Then compare Q with K_c.

$$(\text{conc. N}_2) = \frac{1.0 \text{ mol}}{10.0 \text{ L}} = 0.10 \text{ mol/L} \qquad (\text{conc. H}_2) = \frac{2.0 \text{ mol}}{10.0 \text{ L}} = 0.20 \text{ mol/L}$$

$$(\text{conc. NH}_3) = \frac{3.0 \text{ mol}}{10.0 \text{ L}} = 0.30 \text{ mol/L}$$

Because the units are defined by the equilibrium constant expression, leave them out.

$$Q = \frac{(\text{conc. NH}_3)^2}{(\text{conc. N}_2)(\text{conc. H}_2)^3} = \frac{(0.30)^2}{(0.10)(0.20)^3} = 110, \text{ which is much greater than } 0.105,$$

the value of K_c. Because $Q > K_c$, the reverse reaction—decomposition of NH_3 to form N_2 and H_2—will occur until the equilibrium concentrations are reached. The initial concentration of N_2 was 0.10 mol/L, and when more N_2 forms, its concentration will increase.

✔ The equilibrium constant is less than 1 (about 0.1), so it is reasonable to expect that the concentrations of products will be less than the concentrations of reactants when equilibrium is reached. Because the initial concentration of NH_3 was bigger than the concentrations of N_2 and H_2, it is reasonable to expect that the reverse reaction will occur, raising the concentration of N_2 above its initial value of 0.10 mol/L.

Problem-Solving Practice 14.5

For the equilibrium

$$2 \text{ SO}_2(g) + \text{O}_2(g) \rightleftharpoons 2 \text{ SO}_3(g) \qquad\qquad K_c = 245 \quad \text{at } 1000 \text{ K}$$

the equilibrium concentrations are $[\text{SO}_2] = 0.102$, $[\text{O}_2] = 0.0132$, and $[\text{SO}_3] = 0.184$. Suppose that the concentration of SO_2 is suddenly doubled. Calculate Q and use it to show that the forward reaction would take place in order to reach a new equilibrium.

Exercise 14.7   Reaction Quotient and Pressure Equilibrium Constant

Is it possible to apply the idea of the reaction quotient to gas phase reactions where the equilibrium constant is given in terms of pressure (as in Equation 14.4, ⬅ *p. 543*)? Define Q for such a reaction and give an appropriate set of rules by which you can predict which direction a gas phase reaction will go to achieve equilibrium.

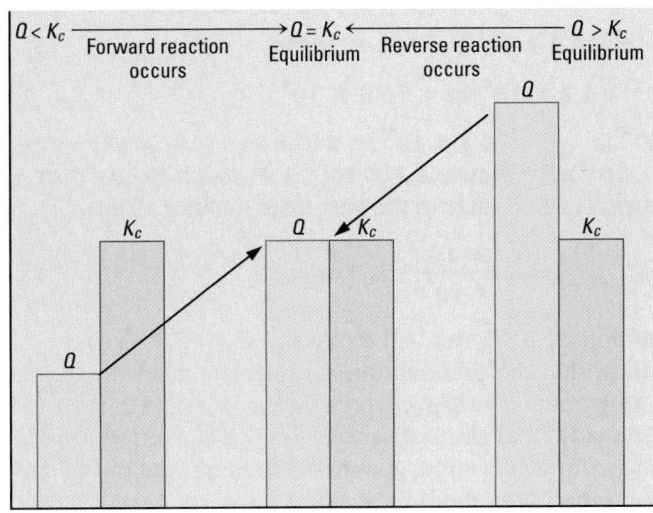

Figure 14.5 Predicting direction of a reaction. The relative sizes of the reaction quotient, Q, and the equilibrium constant, K_c, determine in which direction a mixture of substances will react in order to achieve equilibrium. The rule is

$Q < K_c$, reaction goes in forward direction ($\longrightarrow$)

$Q = K_c$, reaction at equilibrium

$Q > K_c$, reaction goes in backward direction ($\longleftarrow$)

The reaction will proceed until Q is equal to K_c.

Calculating Equilibrium Concentrations

CD-ROM Screen 16.10: Estimating Equilibrium Concentrations

Equilibrium constants from Table 14.1 can be used to calculate how much product is formed and how much of the reactants remain once a system has reached equilibrium. To verify our earlier statement that if an equilibrium constant is very large, essentially all of the reactants are converted to products *(⇐ p. 647)*, consider the reaction

$$\tfrac{1}{8} S_8(s) + O_2(g) \rightleftharpoons SO_2(g) \qquad K_c = 4.2 \times 10^{52} \quad \text{(at 25 °C)}$$

Suppose we place 4.0 mol/L of O_2 and a large excess of sulfur in an empty flask and allow the system to reach equilibrium. We can calculate the quantity of O_2 left and the quantity of SO_2 formed at equilibrium by summarizing information in a table. (Because S_8 is a solid and does not appear in the equilibrium constant expression, we do not need any entries under S_8 in the table.)

	$\tfrac{1}{8} S_8(s)$ +	$O_2(g)$	⇌	$SO_2(g)$
Initial concentration (mol/L)		4.0		0.0
Change in concentration on reaction (mol/L)		$-x$		$+x$
Equilibrium concentration (mol/L)		$4.0 - x$		$0.0 + x$

We know the concentrations of reactant and product before the reaction, but we do not know how many moles per liter of O_2 are consumed during the reaction, and so we designate this as x mol/L. (There is a minus sign in the table because O_2 is consumed.) Since the mole ratio is (1 mol SO_2)/(1 mol O_2), we know that x mol/L of SO_2 is formed when x mol/L of O_2 is consumed. To calculate the concentration of O_2 we take what was present initially (4.0 mol/L) minus what was consumed in the reaction (x mol/L). The equilibrium concentration of SO_2 must be the initial concentration (0.0 mol/L) plus what was formed by the reaction (x mol/L). Putting these values into the equilibrium constant expression, we have

$$K_c = \frac{[SO_2]}{[O_2]} = \frac{x}{4.0 - x} = 4.2 \times 10^{52}$$

Solving algebraically for x (and following the usual rules for significant figures), we find

$$x = (4.2 \times 10^{52})(4.0 - x)$$
$$x = (16.8 \times 10^{52}) - (4.2 \times 10^{52})x$$
$$x + (4.2 \times 10^{52})x = 16.8 \times 10^{52}$$

Notice that $x + (4.2 \times 10^{52})x = (1 + 4.2 \times 10^{52})x$, which to a very good approximation is equal to $(4.2 \times 10^{52})x$. (Because 4.2×10^{52} is so much bigger than 1, adding 1 to it makes no appreciable change in the very large number.) Thus,

$$x = \frac{16.8 \times 10^{52}}{4.2 \times 10^{52}} = 4.0$$

The fact that samples of sulfur at 25 °C can be exposed to oxygen in the air for long periods without being converted to sulfur dioxide shows the importance of chemical kinetics. This reaction is very slow at room temperature, and so there is only a faint odor of sulfur dioxide in the vicinity of solid sulfur.

The equilibrium concentration of SO_2 is $x = 4.0$ mol/L and that of O_2 is $(4.0 - x)$ mol/L, or 0 mol/L. That is, within the precision of our calculation, all the O_2 has been converted to SO_2. As we stated earlier, a very large K_c value (4.2×10^{52} in this case) implies that essentially all of the reactants have been converted to products. The reaction is strongly product-favored, goes to completion, and the calculation could have been done using the methods in Section 4.4 *(⇐ p. 134)*.

Problem-Solving Example 14.6 Calculating Equilibrium Concentrations

Consider the reduction of carbon dioxide by hydrogen to give water vapor and carbon monoxide at 420 °C.

$$H_2(g) + CO_2(g) \rightleftharpoons H_2O(g) + CO(g) \qquad K_c = 0.10 \quad \text{(at 420 °C)}$$

Suppose you place enough H_2 and CO_2 in a flask so that their concentrations are both 0.050 mol/L. You heat the mixture to 420 °C and wait for equilibrium to be achieved. What are the concentrations of reactants and products at equilibrium?

Answer $[H_2] = [CO_2] = 0.038$ mol/L; $[H_2O] = [CO] = 0.012$ mol/L.

Explanation   Follow the procedure on p. 644. The equation is written in the statement of the problem. In this equation all of the mole ratios are 1:1. This tells us that equal numbers of moles of H_2 and CO_2 are consumed as the reaction proceeds to equilibrium. Since both substances are in the same flask, this means that equal numbers of moles per liter (equal concentrations) of reactants must also be consumed, and we designate each of these as $- x$. The mole ratios also tell us that if the concentration of H_2 decreases by x mol/L, the concentration of H_2O (and the concentration of CO) must increase by the same quantity, x mol/L. Because their initial concentrations were the same and their mole ratio is 1:1, the equilibrium concentrations of H_2 and CO_2 must be the same. Each is equal to the initial concentration plus the change ($-x$) as the reactants are consumed.

	H_2	+	CO_2	$\rightleftharpoons$	H_2O	+	CO
Initial concentration (mol/L)	0.050		0.050		0.000		0.000
Change as reaction occurs (mol/L)	$-x$		$-x$		$+x$		$+x$
Equilibrium concentration (mol/L)	$0.050 - x$		$0.050 - x$		x		x

Substituting these values into the expression for K_c, we have

$$K_c = 0.10 = \frac{[H_2O][CO]}{[H_2][CO_2]} = \frac{(x)(x)}{(0.050 - x)(0.050 - x)} = \frac{x^2}{(0.050 - x)^2}$$

Solving this equation for x is not as hard as it might seem at first glance. Because the right-hand side is a perfect square, we can take the square root of both sides, giving

$$\sqrt{K_c} = \sqrt{0.10} = 0.316 = \frac{x}{(0.050 - x)}$$

and then solve for x.

$$x = (0.316)(0.050 - x)$$
$$x = 0.0158 - 0.316x$$
$$1.316x = 0.0158$$
$$x = 0.012$$

Thus, the concentrations of the products are both 0.012 mol/L, while the concentrations of the reactants that remain are both $(0.050 - x)$ mol/L = 0.038 mol/L.

✔   To check the calculation, substitute these concentrations into the equilibrium constant expression and compare the calculated K_c with the value given in the statement of the problem.

$$K_c = \frac{[H_2O][CO]}{[H_2O][CO_2]} = \frac{(0.012)^2}{(0.038)^2} = 0.099 \cong 0.10 \qquad \text{(indicates that the calculation is probably correct)}$$

This result demonstrates quantitatively that when $K_c < 1$ a reaction is reactant-favored, because at equilibrium the product concentrations are smaller than the reactant concentrations.

Problem-Solving Practice 14.6

The equilibrium constant for dissolving the insoluble substance gold(I) iodide, AuI(s), in aqueous solution is 1.6×10^{-23} at 25 °C. Write the equilibrium constant expression and calculate the concentration of $Au^+(aq)$ and $I^-(aq)$ ions in a solution in which 0.345 g of AuI(s) is in equilibrium with the aqueous ions.

Problem-Solving Example 14.7 Calculating Equilibrium Concentrations

When colorless hydrogen iodide gas is heated to 745 K, a beautiful purple color appears. This shows that some iodine gas has been formed, which means that the compound has been decomposed partially to its elements.

$$2\,HI(g) \rightleftharpoons H_2(g) + I_2(g) \qquad\qquad K_c = 0.0200 \quad \text{(at 745 K)}$$

Suppose that a mixture of 1.00 mol of HI(g) and 1.00 mol of H_2(g) is sealed into a 10.0-L flask and heated to 745 K. What will be the concentrations of all three substances when equilibrium has been achieved?

Answer [HI] = 0.096 M, [I_2] = 0.0018 M, [H_2] = 0.102 M

Explanation Follow the usual steps *(⬅ p. 644)*. The balanced equation was given, and the equilibrium constant expression follows the table (below). Because amounts are given instead of concentrations, divide each number of moles by the volume of the flask to get (conc. HI) = (conc. H_2) = 1.00 mol/10.0L = 0.100 mol/L. Because no I_2 is present at the beginning, $Q = 0$. This is much less than K_c, even though K_c is small. Therefore the reaction must go from left to right, so it makes sense to let x mol/L be the concentration of I_2 when equilibrium is reached. Since the coefficients of H_2 and I_2 are the same, if x mol/L of I_2 is produced, then x mol/L of H_2 must also be produced. Therefore the change in concentration of both H_2 and I_2 is $+x$ mol/L. Because the coefficient of HI is twice the coefficient of I_2, twice as many moles of HI must disappear as moles of I_2 formed; the change in concentration of HI is $-2x$ mol/L. The reaction table is then

	2 HI(g) $\rightleftharpoons$	**H_2(g)** +	**I_2(g)**
Initial concentration (mol/L)	0.100	0.100	0.000
Change as reaction occurs (mol/L)	$-2x$	$+x$	$+x$
Equilibrium concentration (mol/L)	$(0.100 - 2x)$	$(0.100 + x)$	x

Now write the equilibrium constant expression in terms of the equilibrium concentrations calculated in the third row of the table.

$$K_c = \frac{[H_2][I_2]}{[HI]^2} = \frac{(0.100 + x)x}{(0.100 - 2x)^2} = 0.0200$$

Since the ratio of terms involving x is not a perfect square, do not take a square root as was done in Problem-Solving Example 14.6. Multiply out the numerator and denominator to obtain

$$\frac{0.100x + x^2}{0.0100 - 0.400x + 4x^2} = 0.0200$$

Multiply both sides by the denominator and then multiply out the terms. This gives

$$0.100x + x^2 = 0.0200 \times (0.0100 - 0.400x + 4x^2) = 0.000200 - 0.00800x + 0.0800x^2$$

Collecting terms in x^2 and x we have

$$0.9200x^2 + 0.10800x - 0.000200 = 0$$

This is a quadratic equation of the form $ax^2 + bx + c = 0$, where $a = 0.9200$, $b = 0.10800$, and $c = -0.000200$. It can be solved using the quadratic formula (Appendix A.7).

$$x = \frac{-b \pm \sqrt{b^2 - 4ac}}{2a} = \frac{-0.10800 \pm \sqrt{0.10800^2 - 4 \times 0.9200 \times (-0.000200)}}{2 \times 0.9200}$$

$$= \frac{-0.10800 \pm 0.11135}{1.840}$$

This results in $x = 1.83 \times 10^{-3}$ or $x = -0.119$. The latter can be eliminated because it would result in a negative concentration of I_2 at equilibrium, which is clearly impossible. Using the first root and the equilibrium concentration row of the table gives

$$[HI] = (0.100 - 2x) \, mol/L = 0.100 - (2 \times 1.83 \times 10^{-3}) = 0.0963 \, mol/L$$

$$[H_2] = (0.100 + x) \, mol/L = 0.100 + (1.83 \times 10^{-3}) = 0.102 \, mol/L$$

$$[I_2] = x \, mol/L = 0.00183 \, mol/L$$

In solving this problem we might have chosen to let $-x$ be the change in concentration of HI, in which case the changes in concentrations of H_2 and I_2 would each have been $+\frac{1}{2}x$.

✔ Solving this problem involved a lot of algebra, and so it would be easy to make a mistake. Therefore it is very important to check the result of any such calculation. This is easily done by substituting the equilibrium concentrations into the equilibrium constant expression and verifying that the correct value of K_c is found.

$$K_c = \frac{[H_2][I_2]}{[HI]^2} = \frac{(0.102)(0.00183)}{(0.0963)^2} = 0.0201$$

which is acceptable agreement because it differs by 1 in the last significant figure.

Problem-Solving Practice 14.7

Obtain the equilibrium constant for dissociation of dinitrogen tetraoxide to form nitrogen dioxide from Table 14.1. If 1.00 mol of N_2O_4 and 0.500 mol of NO_2 are initially placed in a container with a volume of 4.00 L, calculate the concentrations of $N_2O_4(g)$ and $NO_2(g)$ present when equilibrium is achieved at 25 °C.

14.6 SHIFTING A CHEMICAL EQUILIBRIUM: LE CHATELIER'S PRINCIPLE

Suppose you are an environmental engineer, biologist, or geologist and that you have just measured the concentration of hydronium ion, H_3O^+, in a lake. You know that the H_3O^+ ions are involved in many different equilibrium reactions in the lake. How can you predict the influence of changing conditions? For example, what happens if there is a large increase in acid rainfall that has a hydronium ion concentration different from that of the lake? Or what happens if lime (calcium oxide), a strong base, is added to the lake? These questions and many others like them can be answered qualitatively by applying a useful guideline known as **Le Chatelier's principle:** *If a system is at equilibrium and the conditions are changed so that it is no longer at equilibrium, the system will react to reach a new equilibrium in a way that partially counteracts the change.* To adjust to a change, a system reacts in either the forward or the reverse direction until a new equilibrium state is achieved. Le Chatelier's principle applies to changes in the concentrations of reactants or products that appear in the equilibrium constant expression, the pressure or volume of a gas phase equilibrium, and the temperature. Changing conditions, thereby changing the equilibrium concentrations of reactants and products, is called **shifting an equilibrium.** If the reaction occurs in the forward direction, we say that the equilibrium reaction shifts to the right. If the system reacts in the reverse direction, the reaction has shifted to the left.

CD-ROM Screen 16.11: Disturbing a Chemical Equilibrium: Le Chatelier's Principle

Henri Le Chatelier (1850–1936) was a French chemist who, as a result of his studies of the chemistry of cement, developed his ideas about how altering conditions affects an equilibrium system.

Changing Concentrations of Reactants or Product

If the concentration of a reactant or product that appears in the equilibrium constant expression is changed, a system can no longer be at equilibrium because Q

CD-ROM Screen 16.13: Disturbing a Chemical Equilibrium: Addition or Removal of a Reagant

must have a different value from K_c. If the concentration of a reactant is increased, the reaction will shift in the forward direction, thereby consuming some of the extra reactant. If the concentration of a reactant is decreased, the system will react in the reverse direction. If the concentration of a product is increased, the system reacts in the reverse direction to decrease that concentration somewhat. If the concentration of a product is decreased, the forward reaction will occur to partially counteract the change.

To see why this happens, consider the simple equilibrium discussed in Section 14.2 (⬅ *p. 637*).

$$\textit{cis-2-butene} \rightleftharpoons \textit{trans-2-butene} \qquad K_c = \frac{[\textit{trans}]}{[\textit{cis}]} = 1.5 \quad \text{(at 600 K)}$$

Notice that as long as the temperature remains the same, the value of the equilibrium constant also remains the same. Adding or removing a reactant or product does not change the equilibrium constant value, but if the concentration of the substance added or removed appears in the equilibrium constant expression, the value of Q changes. Because Q no longer equals K_c, the system is no longer at equilibrium and must react to achieve a new equilibrium.

Suppose that 2 mmol of *cis*-2-butene is placed into a 1.0-L container at 1000 K. The forward reaction will be faster than the reverse reaction until the concentration of *trans*-2-butene builds up to 1.5 times the concentration of *cis*-2-butene. Then equilibrium is achieved. Now suppose that half the *cis*-2-butene is instantaneously removed from the container (see Exercise 14.3, ⬅ *p. 643*). Because the concentration of *cis*-2-butene suddenly drops to half its former value, the forward reaction rate will also drop to half its former value. The reverse rate will not be affected, because the concentration of *trans*-2-butene has not changed. This means that the reverse reaction is twice as fast as the forward reaction, and *cis*-2-butene molecules are being formed twice as fast as they are being used up. Therefore the concentration of *cis*-2-butene will increase and the concentration of *trans*-2-butene will decrease until the forward and reverse rates are again equal. This is shown by the graph in Figure 14.6.

The effect of changing concentration has many important consequences. For example, when the concentration of a substance needed by your body falls slightly, several enzyme-catalyzed chemical equilibria shift so as to increase the concentration of the essential substance. In industrial processes, reaction products are often continuously removed. This shifts one or more equilibria to produce more products and thereby maximize the yield of the reaction.

In nature, slight changes in conditions are responsible for effects such as formation of limestone stalactites and stalagmites in caves (Figure 14.7) and the crust of limestone that slowly develops in a tea kettle if you boil hard water in it. Both of

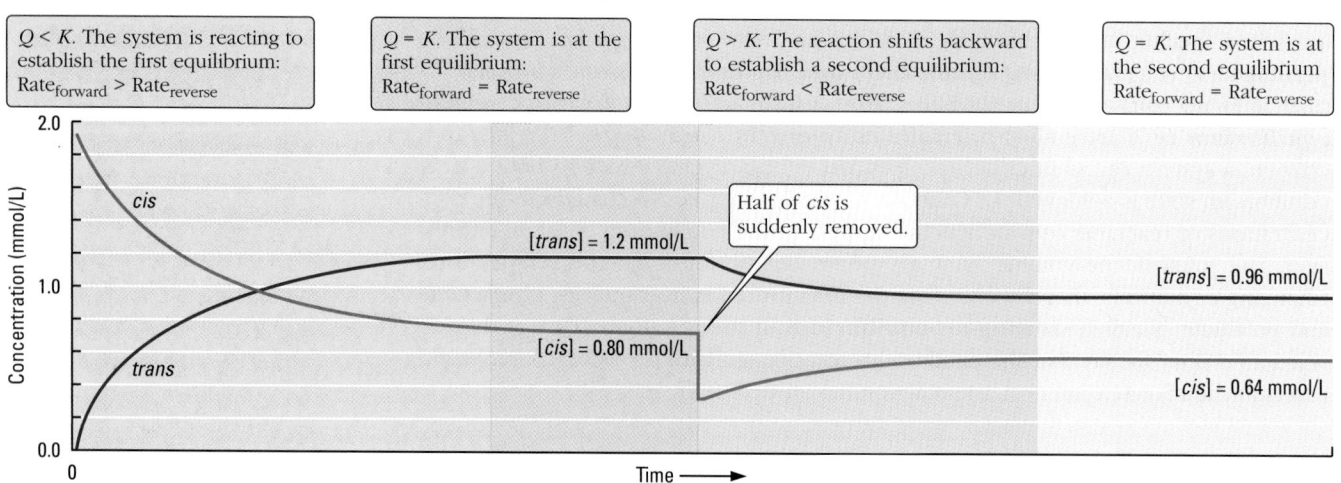

Figure 14.6. Approach to new equilibrium after a change in conditions.

these examples involve calcium carbonate. Limestone, a form of calcium carbonate, $CaCO_3$, is present in underground deposits, a leftover of ancient oceans from which it precipitated long ago. Limestone reacts with an aqueous solution of CO_2 and dissolves.

$$CaCO_3(s) + CO_2(aq) + H_2O(\ell) \rightleftharpoons Ca^{2+}(aq) + 2\ HCO_3^-(aq)$$

If groundwater that is saturated with CO_2 encounters a bed of limestone below the surface of the earth, the forward reaction can occur until equilibrium is reached, and the water subsequently contains significant concentrations of aqueous Ca^{2+} and HCO_3^- ions in addition to dissolved CO_2. These ions, often accompanied by $Mg^{2+}(aq)$, constitute hard water (Section 15.11).

As with all equilibria, there is a reverse reaction occurring as well as the forward reaction. This reverse reaction can be demonstrated by mixing aqueous solutions of $CaCl_2$ and $NaHCO_3$ (salts containing the Ca^{2+} and HCO_3^- ions) in an open beaker (Figure 14.8). You will eventually see bubbles of CO_2 gas and a precipitate of solid $CaCO_3$. Because the beaker is open to the air, any gaseous CO_2 that escapes the solution is swept away. This reduces the concentration of $CO_2(aq)$ and causes the equilibrium to shift to produce more CO_2. Eventually all of the dissolved Ca^{2+} and HCO_3^- ions disappear from the solution, having been converted to gaseous CO_2, solid $CaCO_3$, and water.

Suppose that water containing dissolved CO_2, Ca^{2+}, and HCO_3^- contacts the air in a cave (or hard water contacts air in your tea kettle). Carbon dioxide bubbles out of the solution, the concentration of CO_2 decreases on the reactant side, and the equilibrium shifts toward the reactants. There is a net reverse reaction that forms $CO_2(aq)$, compensating partially for the reduced concentration of $CO_2(aq)$. Some of the calcium ions and hydrogen carbonate ions combine, and some $CaCO_3(s)$ precipitates as a beautiful formation in the cave (or as scale in your kettle).

To summarize the effect of changing concentrations, for any reactant or product that appears in the equilibrium constant expression,

- Adding reactant will shift the equilibrium in the forward direction (to the right).

Hard water contains dissolved CO_2 and metal ions such as Ca^{2+} and Mg^{2+}.

Figure 14.7. Stalactites in a limestone cave. Stalactites hang from the ceilings of caves. Stalagmites grow from the floors of caves up toward the stalactites. Both consist of limestone, $CaCO_3$. The formations shown here resulted when stalactites and stalagmites met. The process that produces these lovely formations is an excellent example of chemical equilibrium.

❶ These two salt solutions provide Ca^{2+} and HCO_3^- ions, which are products in the net ionic equation for the reaction of calcium carbonate with the carbon dioxide.

❷ Increasing the concentrations of the products by mixing the two solutions shifts the equilibrium toward the reactants.

❸ The CO_2 produced by the reverse reaction bubbles out of the solution into the air, thereby reducing a reactant concentration in solution and shifting the equilibrium to the left.

❹ Therefore the reverse reaction continues until almost all of the Ca^{2+} and HCO_3^- ions have reacted to form CO_2, $CaCO_3$, and water.

Figure 14.8 **Reaction of $CaCl_2(aq)$ with $NaHCO_3(aq)$.**

- Removing reactant will shift the equilibrium in the reverse direction (to the left).
- Adding product will shift the equilibrium in the reverse direction (to the left).
- Removing product will shift the equilibrium in the forward direction (to the right).

Exercise 14.8   Effect of Adding a Substance

Solid phosphorus pentachloride decomposes when heated to form gaseous phosphorus trichloride and gaseous chlorine. Write the equation for the equilibrium that is set up when solid phosphorus pentachloride is introduced into a container, the container is evacuated and sealed, and the solid is heated. Once the system has reached equilibrium at a given temperature, what will be the effect on the equilibrium of the following?

(a) Adding chlorine to the container
(b) Adding phosphorus trichloride to the container
(c) Adding a small quantity of phosphorus pentachloride to the container

---

CHEMISTRY You Can Do...   Growing Crystals by Shifting Equilibria

Crystals that are grown slowly can become quite large and have interesting and beautiful shapes. This crystal-growing experiment takes days or weeks, but the results can be quite striking. Obtain the following materials from a grocery or hardware store:

- About $\frac{1}{8}$ cup (30 mL) of one or more of the following salts: alum ($KAl(SO_4)_2 \cdot 12H_2O$), blue vitriol or blue stone ($Cu(SO_4)_2 \cdot 5H_2O$), Epsom salt ($MgSO_4 \cdot 7H_2O$), or table salt ($NaCl$)
- For each salt you want to crystallize, two 4-oz or 6-oz clear plastic cups (or similar clear containers) and something to cover the cup (plastic wrap will work fine)
- Water, something to stir with, a plastic-coated paper clip, a pencil, string, and (optional) petroleum jelly (Vaseline)

Epsom salt is a laxative and copper salts are harmful if taken internally, so be careful with them, but all of these salts are safe to pour down the drain when you are finished.

Prepare a saturated solution of the salt by placing it in a cup, filling the cup $\frac{1}{3}$ to $\frac{1}{2}$ full with warm tap water, and stirring every 5 min or so for half an hour. Some of the solid salt should remain undissolved in the bottom of the cup. If all of the solid dissolves, add more solid and stir some more. While you wait for the salt to dissolve, tie one end of the string to the paper clip and the other to the pencil. Arrange the string so that you can place the pencil across the top of the second cup, suspending the paper clip so that it just sits flat on the bottom of the cup. If you want to grow a really big crystal, coat most of the string with petroleum jelly, leaving only a

small section uncoated about half an inch from the bottom of the cup. Once you are sure that the salt solution is saturated, decant it into the second cup, leaving all of the solid behind. Place the assemblage of paper clip, string, and pencil on the second cup, and cover everything with plastic wrap.

Put the cup somewhere where it will not be disturbed and its temperature will remain fairly constant. (Don't put it in a window where the sun would hit it, or on top of a radiator, for example.) Check it at least once a day. Some of the water will evaporate, but only very slowly because the cup is covered and water vapor cannot easily escape. Observe what happens over a period of several weeks or even months.

1. What do you see on the string?
2. Explain your observations in terms of an equilibrium between the ions of the salt you dissolved and the solid salt.
3. What shifts the equilibrium to achieve the results you observed?
4. Why is it important that the water evaporate slowly?

Changing Volume or Pressure in Gaseous Equilibria

One way to change the pressure of a gaseous equilibrium mixture is to keep the volume constant and add or remove one or more of the substances whose concentrations appear in the equilibrium constant expression. This is the situation we have just discussed. We consider here other ways of changing pressure or volume.

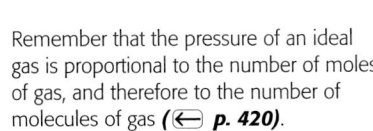

CD-ROM Screen 16.14: Disturbing a Chemical Equilibrium: Volume Changes

The pressures of all substances in a gaseous equilibrium can be changed by changing the volume of the container. Consider the effect of tripling the pressure on the equilibrium

$$N_2O_4(g) \rightleftharpoons 2\ NO_2(g) \qquad\qquad K_c = \frac{[NO_2]^2}{[N_2O_4]}$$

by reducing the volume of the container to one third of its original value (at constant temperature). This is shown in Figure 14.9. The effect of decreasing the volume is to increase the pressures of N_2O_4 and NO_2 to three times their equilibrium values. Decreasing the volume also increases the concentrations of N_2O_4 and NO_2 to three times their equilibrium values. Because $[NO_2]$ is squared in the equilibrium constant expression but $[N_2O_4]$ is not, tripling both concentrations increases the numerator of Q by $3^2 = 9$ but only increases the denominator by 3.

$$Q = \frac{(\text{conc. } NO_2)^2}{(\text{conc. } N_2O_4)} = \frac{(3 \times [NO_2])^2}{(3 \times [N_2O_4])} = \frac{9}{3} \times \frac{[NO_2]^2}{[N_2O_4]} = 3 \times K_c$$

Because Q is larger than K_c under the new conditions, the reaction should produce more reactant; that is, the equilibrium should shift to the left.

The same prediction is made using Le Chatelier's principle: The reaction should shift to partially compensate for the increase in pressure. This means decreasing the pressure, which can happen if the total number of gas phase molecules decreases. In the case of the N_2O_4/NO_2 equilibrium, the reverse reaction should occur, because one N_2O_4 molecule is produced for every two NO_2 molecules that react. A shift to the left reduces the number of gas phase molecules and hence the pressure.

Remember that the pressure of an ideal gas is proportional to the number of moles of gas, and therefore to the number of molecules of gas (◁ *p. 420*).

However, consider the situation with respect to another equilibrium we have already mentioned.

$$2\ HI(g) \rightleftharpoons H_2(g) + I_2(g)$$

Suppose that the pressure of this system were tripled by reducing its volume to one third of the original volume. What would happen to the equilibrium? In this case, all the concentrations triple, but because there are equal numbers of moles

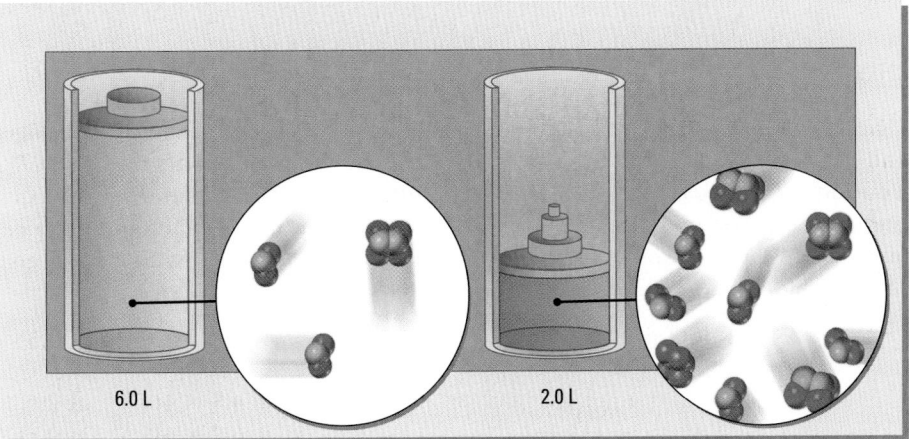

6.0 L   2.0 L

Figure 14.9 Shifting an equilibrium by changing pressure and volume. If the volume of an equilibrium mixture of NO_2 and N_2O_4 is decreased from 6.0 to 2.0 L, the equilibrium shifts toward the smaller number of molecules in the gas phase (the N_2O_4 side) to partially compensate for the increased pressure. When the new equilibrium is achieved, the concentrations of both NO_2 and N_2O_4 have increased, but the NO_2 concentration is less than three times as great, while the N_2O_4 concentration is more than three times as great. The total pressure goes from 1.00 to 2.62 atm (rather than to 3.00 atm, which was the new pressure before the equilibrium shifted).

of gaseous substances on both sides of the equation, Q still has the same numeric value as K_c. That is, the system is still at equilibrium, and no shift occurs. Thus, *changing pressure by changing the volume shifts an equilibrium only if the sum of the coefficients for gas phase reactants is different from the sum of the coefficients for gas phase products.*

Exercise 14.9 Changing Volume Does Not Always Shift an Equilibrium

In Problem-Solving Example 14.6 you found that for the reaction

$$2 HI(g) \rightleftharpoons H_2(g) + I_2(g) \qquad K_c = 0.0200 \quad (\text{at } 745 \text{ K})$$

the equilibrium concentrations were [HI] = 0.0963 mol/L, [H$_2$] = 0.102 mol/L, and [I$_2$] = 0.00183 mol/L. Use algebra to show that if each of these concentrations is tripled by reducing the volume of this equilibrium system to one third its initial value, the system is still at equilibrium, and therefore the pressure change causes no shift in the equilibrium.

Finally consider what happens if the pressure of the N_2O_4/NO_2 equilibrium system is increased by adding an inert gas such as nitrogen while retaining exactly the same volume. The total pressure of the system would increase, but since neither the amounts of N_2O_4 and NO_2 nor the volume would change, the concentrations of N_2O_4 and NO_2 would remain the same. Q still equals K_c, the system is still at equilibrium, and no shift occurs (or needs to). Thus, changing the pressure of an equilibrium system must change the concentrations of the substances in the equilibrium constant expression if a shift in the equilibrium is to occur.

Exercise 14.10 Effect of Changing Volume

Verify the statement in the text that, for the N_2O_4/NO_2 equilibrium system, decreasing the volume to one third of its original value increases the equilibrium concentration of N_2O_4 by more than a factor of 3 while it increases the equilibrium concentration of NO_2 by less than a factor of 3. Start with the equilibrium conditions you calculated in Problem-Solving Practice 14.7. Then decrease the volume of the system from 4.00 L to 1.33 L, and calculate the new concentrations of N_2O_4 and NO_2 assuming the shift in the equilibrium has not yet taken place. Then set up the usual table of initial concentrations, change, and equilibrium concentrations, and calculate the concentrations at the new equilibrium.

Changing Temperature

CD-ROM Screen 16.12: Disturbing a Chemical Equilibrium: Temperature Changes

When temperature changes, the values of most equilibrium constants also change, and systems will react to achieve new equilibria consistent with new values of K_c. You can make a qualitative prediction about the effect of temperature on an equilibrium if you know whether the reaction is exothermic or endothermic. As an example, consider the endothermic, gas phase reaction of N_2 with O_2 to give nitrogen monoxide, NO.

$$N_2(g) + O_2(g) \rightleftharpoons 2 NO(g) \qquad \Delta H° = 180.5 \text{ kJ}$$

$$K_c = \frac{[NO]^2}{[N_2][O_2]} \qquad \begin{array}{l} K_c = 4.5 \times 10^{-31} \quad \text{at } 298 \text{ K} \\ K_c = 6.7 \times 10^{-10} \quad \text{at } 900 \text{ K} \\ K_c = 1.7 \times 10^{-3} \quad \text{at } 2300 \text{ K} \end{array}$$

In this case the equilibrium constant increases very significantly as the temperature increases. At 298 K the equilibrium constant is so small that essentially no reaction occurs. Suppose that a room temperature equilibrium mixture were suddenly heated to 2300 K. What would happen? The equilibrium should shift to partially compensate for the temperature increase. This happens if the reaction shifts in the endothermic direction, since that would involve transfer of energy into the reaction system, cooling the surroundings. For the $N_2 + O_2$ reaction the forward process is endothermic ($\Delta H° > 0$), and so N_2 and O_2 should react to produce more NO. At the new equilibrium, the concentration of NO should be higher and the concentrations of N_2 and O_2 lower. Since this makes the numerator in the K_c expression bigger and the denominator smaller, K_c should be larger at the higher temperature. This corresponds with the experimental result.

The effect of temperature on the reaction of N_2 with O_2 has important consequences. This reaction produces NO in earth's atmosphere when lightning suddenly raises the temperature of the air. Because the reverse reaction is slow at room temperature, and because after the lightning bolt is over, the air rapidly cools back to normal temperatures, much of the NO that is produced does not react back to N_2 and O_2 as it would at equilibrium. This provides one natural mechanism by which nitrogen in the air can be converted into a form that can be used by plants. (Converting nitrogen into a useful form is called nitrogen fixation; see Section 21.3.) Humans have tried to use the same kind of process to produce NO and from it HNO_3 for use in manufacturing fertilizer. At the end of the 19th century, a chemical plant at Niagara Falls, New York (where there was plentiful electric power) operated an electric arc process for fixing nitrogen for several years. This electric arc plant was important because it was the first attempt to deal with the limitations on plant growth caused by lack of sufficient nitrogen in soils that had been heavily farmed. One hundred years ago scientists and many in the general public were worried that earth's farmland could not grow enough food to support a growing population. Consequently, strenuous efforts were made to adjust the conditions of the $N_2 + O_2$ reaction so that significant yields of NO could be obtained. The very high temperature of a lightning bolt or an electric arc was one way to do this.

The effect on K_c of raising the temperature of the $N_2 + O_2$ reaction leads us to a general conclusion. *For an endothermic reaction, an increase in temperature always means an increase in* $\mathbf{K_c}$; *the reaction will become more product-favored at higher temperatures.* When equilibrium is achieved at the higher temperature, the concentration of products is greater and that of the reactants is smaller. Likewise, and as illustrated by Problem-Solving Example 14.8, the opposite is true for an exothermic reaction. *For an exothermic reaction, an increase in temperature always means a decrease in* $\mathbf{K_c}$; *the reaction will become less product-favored at higher temperatures.*

Another consequence of the shift toward products in the $N_2 + O_2$ reaction at high temperatures is that automobile engines emit small concentrations of NO. The NO is rapidly oxidized to brown NO_2 in the air above cities, and the NO_2 in turn produces many further reactions that create air pollution problems. These are discussed in Section 10.10 (⇐ *p. 439)*. One of the functions of catalytic converters in automobiles is to speed up reduction of these nitrogen oxides back to elemental nitrogen.

Problem-Solving Example 14.8   Le Chatelier's Principle

Consider an equilibrium mixture of nitrogen, hydrogen, and ammonia in which the reaction is

$$N_2(g) + 3\,H_2(g) \rightleftharpoons 2\,NH_3(g) \qquad \Delta H° = -92.2 \text{ kJ at 25 °C}$$

For each of the changes listed below, tell whether the value of K_c increases or decreases, and tell whether more NH_3 or less NH_3 is present at the new equilibrium established after the change.

(a)   More H_2 is added (at a constant temperature of 25 °C and constant volume).

(b)   The temperature is increased.

(c)   The volume of the container is doubled (at constant temperature).

Answer

(a) K_c stays the same; more NH_3 is present.   (b) K_c decreases; less NH_3 is present.
(c) K_c stays the same; less NH_3 is present.

Explanation

(a) Since the temperature does not change, the value of K_c does not change. Adding a reactant to the equilibrium mixture will shift the equilibrium toward the product, producing more NH_3. This can be seen in another way by considering the reaction quotient.

$$Q = \frac{(\text{conc. } NH_3)^2}{(\text{conc. } N_2)(\text{conc. } H_2)^3}$$

When more H_2 is added, the denominator gets larger. This makes Q smaller than K_c, which predicts that the reaction should produce more product; that is, some of the added H_2 reacts with N_2 to make more NH_3. Notice also that the concentration of N_2 decreases, because some reacts with the H_2, but the concentration of H_2 in the new equilibrium is still greater than it was before.

(b) The reaction is exothermic. Increasing the temperature shifts the equilibrium in the endothermic direction, that is, to the left (toward the reactants.) This leads to a decrease in the NH_3 concentration, an increase in the concentrations of H_2 and N_2, and a decrease in the value of K_c.

(c) Since the temperature is constant, the value of K_c must be constant. Doubling the volume should cause the reaction to shift toward a greater number of moles of gaseous substance, that is, toward the left. Doubling the volume would normally halve the pressure, but the shift of the equilibrium partially compensates for this, and the final equilibrium will be at a pressure somewhat more than half the pressure of the initial equilibrium.

Problem-Solving Practice 14.8

Consider the equilibrium between N_2O_4 and NO_2 in a closed system.

$$N_2O_4(g) \rightleftharpoons 2\ NO_2(g)$$

Draw Lewis structures for the molecules involved in this equilibrium. Based on the bonding in the molecules, predict whether the reaction is exothermic or endothermic; hence, predict whether the concentration of N_2O_4 is larger in an equilibrium system at 25 °C or at 100 °C. Verify your prediction by looking at Figure 14.3.

 Exercise 14.11   Summarizing LeChatelier's Principle

Construct a table to summarize your understanding of LeChatelier's principle. Consider these changes in conditions:

(a) Addition of a reactant
(b) Removal of a reactant
(c) Addition of a product
(d) Removal of a product
(e) Increasing pressure by decreasing volume
(f) Decreasing pressure by increasing volume
(g) Increasing temperature
(h) Decreasing temperature

For each of these changes in conditions indicate (1) how the reaction system changes in order to achieve a new equilibrium, (2) which direction the equilibrium reaction shifts, and (3) whether the value of K_c changes and, if so, in what direction. For some of these changes there are qualifications. For example, increasing pressure by decreasing volume does not always shift an equilibrium. List as many of these qualifications as you can.

ESTIMATION   **Generating Gaseous Fuel**

The reaction of coke (mainly carbon) with steam is called the water-gas reaction. It was used for many years to generate gaseous fuel from coal. The thermochemical equation is

$$C(s) + H_2O(g) \rightleftharpoons CO(g) + H_2(g) \qquad \Delta H° = 131.293 \text{ kJ}$$

The equilibrium constant K_P has the value 9.5×10^{-17} at 298 K, 1.9×10^{-7} at 500 K, 1.35 at 1000 K, 2.6×10^2 at 1500 K, and 3.6×10^3 at 2000 K. Suppose that you have an equilibrium mixture in which the partial pressure of steam is 1.00 atm. Estimate the temperature at which the partial pressures of CO and H_2 would also equal 1 atm. (The reaction would need to be carried out at temperatures roughly this high to produce appreciable quantities of products.)

A good way to make this estimation is to plot the data and see which temperature corresponds to $K_P = 1$. This has been done in Graph A. It is clear that the range of values for

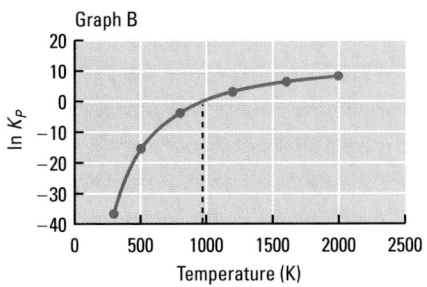

Graph B

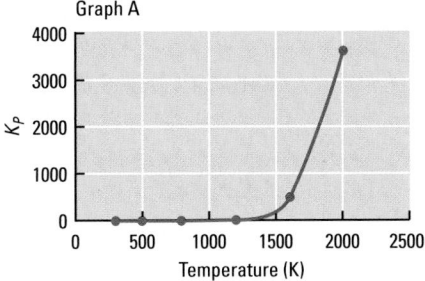

Graph A

the equilibrium constant is so wide that an accurate temperature cannot be read corresponding to $K_P = 1$. An estimate would be easier to make if the graph were a straight line. Because the values of K_P rise very rapidly, it is possible that taking the logarithm of each K_P value would generate a linear graph. This has been done in Graph B, where the natural logarithm of the equilibrium constant, $\ln(K_P)$, has been plotted on the vertical axis. Since $\ln(K_P) = 0$ when $K_P = 1$, we are looking for the temperature at which the graph crosses zero on

the vertical axis. Although the graph is not linear, the appropriate temperature can be read from the graph and is slightly less than 1000 K.

Experimenting with different functions of K_P on the vertical axis and different functions of T on the horizontal axis reveals that a linear graph is obtained when $\ln(K_P)$ is plotted against $1/T$ as in Graph C. From this graph the equation of the

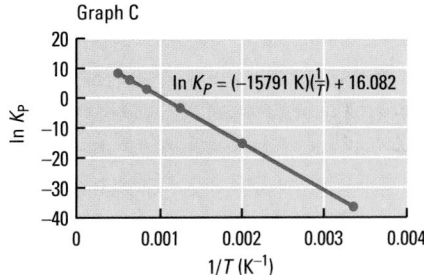

Graph C

$$\ln K_P = (-15791 \text{ K})\left(\frac{1}{T}\right) + 16.082$$

straight line is found to be

$$\ln(K_P) = (-1.57 \times 10^4 \text{ K})\left(\frac{1}{T}\right) + 16.1$$

Substituting $K_P = 1$, that is, $\ln(K_P) = 0$, we find that

$$T = \frac{1.57 \times 10^4 \text{ K}}{16.1} = 990 \text{ K}$$

14.7 EQUILIBRIUM AT THE NANOSCALE

In Section 14.2 ($\Longleftarrow$ *p. 637*), we used the isomerization of *cis*-2-butene to show that both forward and reverse reactions occur simultaneously in an equilibrium system.

$$cis\text{-2-butene} \rightleftharpoons trans\text{-2-butene} \qquad K_c = \frac{[trans]}{[cis]} = 2.0 \quad \text{(at 415 K)}$$

Because the equilibrium constant is 2.0, there are twice as many *trans* molecules as *cis* molecules at 415 K. In other words, two thirds of the molecules have the *trans* structure and one third have the *cis* structure. Because the molecules are continually reacting in both forward and reverse directions, another way to think about this is that each molecule is *trans* two thirds of the time and *cis* one third of the time.

Based on its structure, a 2-butene molecule ought to be just as likely to be *cis* as *trans,* so why is the *trans* isomer favored at 415 K? In Figure 13.7 (⬅ *p. 589*) we noted that 1 mol of *trans* isomer is 4 kJ/mol lower in energy than 1 mol of *cis,* and this is what makes the difference.

$$\text{\textit{cis}-2-butene} \rightleftharpoons \text{\textit{trans}-2-butene} \qquad \Delta H° = -4 \text{ kJ}$$

Consider the rate constants for the forward and reverse reactions in the isomerization of 2-butene. Based on Figure 13.7, $E_a(\text{forward}) = 262$ kJ/mol and $E_a(\text{reverse}) = 266$ kJ/mol. This means that the rate constant for the reverse reaction will be smaller than for the forward reaction. At equilibrium the forward and reverse rates are equal, so

| Larger k value and smaller concentration | | Smaller k value and larger concentration |

$$k_{\text{forward}} \times [\textit{cis}] = k_{\text{reverse}} \times [\textit{trans}]$$

If k_{reverse} is smaller than k_{forward}, then the concentration of *trans* must be bigger than the concentration of *cis,* or the rates would not be equal. Because *cis*-2-butene is 4 kJ/mol higher in energy than *trans*-2-butene, it occurs only half as often as *trans*-2-butene at 415 K. We can generalize that ***in an equilibrium system, molecules that are higher in energy occur less often.***

There is a second factor that affects how big an equilibrium constant is. It involves probability and can be illustrated by the dissociation of dinitrogen tetraoxide.

$$\text{N}_2\text{O}_4(g) \rightleftharpoons 2 \text{ NO}_2(g) \qquad \Delta H° = 57.2 \text{ kJ}$$

The two product molecules can be arranged in more different ways than can the single reactant molecule. For example, consider a container divided into two compartments (see Figure 14.10). The number of arrangements that consist of 2 NO_2 is higher than the number of arrangements that consist of N_2O_4. Therefore the probability is higher that the equilibrium mixture will consist of 2 NO_2. When atoms become spread out or dispersed over two molecules instead of a single one, or when matter expands into a larger volume, we say that the system has become more dispersed or more disordered. When there are many molecules (such as Avogadro's number of molecules) this probability effect can become extremely large. It is a major influence on the position of an equilibrium. ***If there are more different arrangements of product molecules than of reactant molecules, probability favors the products in an equilibrium system.***

Despite the fact that the probability factor favors NO_2 in the dissociation of N_2O_4, when data from Table 14.1 are used to calculate K_c at 25 °C (298 K) the result is $1/(1.7 \times 10^2) = 5.9 \times 10^{-3}$. This is because the dissociation of N_2O_4 is endothermic ($\Delta H° = 57.2$ kJ). The enthalpy of 2 mol of NO_2 is 57.2 kJ higher than the enthalpy of 1 mol of N_2O_4. From our first generalization that molecules with more energy occur less often, we expect that 2 NO_2 is less likely than N_2O_4 because of the energy effect. Therefore the equilibrium constant should be smaller

Figure 14.10 Probability and chemical equilibrium. Two NO_2 molecules (a) can be arranged in four different ways in a two-compartment container, but one N_2O_4 molecule (b) has only two arrangements. The greater the number of possible arrangements, the higher the probability that an equilibrium system will have that composition. Therefore, 2 NO_2 is favored over N_2O_4 on the basis of probability.

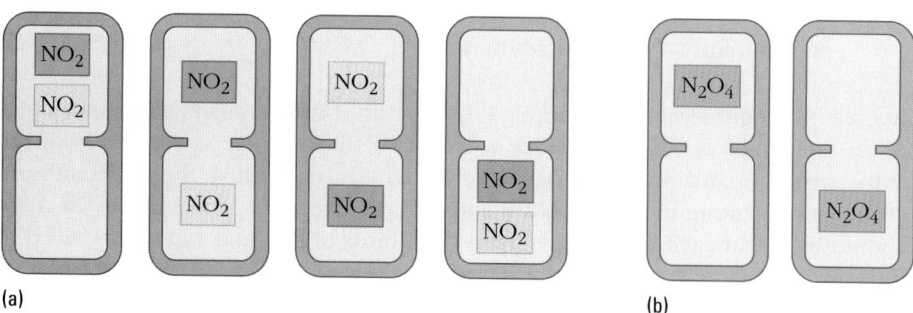

(a)   (b)

than the probability argument predicts. *Both the energy effect and the probability effect must be taken into account to predict an equilibrium constant value.*

These ideas can also help us to understand the temperature dependence of the equilibrium constant. K_c values for the dissociation of N_2O_4 are given in the table in the margin. As the temperature rises, K_c becomes much larger. At high temperatures the molecules have lots of energy, and the energy difference between reactants and products is smaller relative to the average energy per molecule. *The higher the temperature is, the less important the energy effect becomes, and the more the probability effect determines the position of equilibrium.*

We have shown that if the number of gas phase molecules increases when a reaction occurs, then the products will be favored by probability. This is not the only way for the products of a reaction to have greater probability than the reactants, but it is one of the most important. This probability factor is called the **entropy,** and *it is a measure of the disorder, dispersal, or probability of a system.* Entropy is described in more detail in Chapter 18, and there the probability factor will be made quantitative. This will enable us to use enthalpy changes and entropy changes for a reaction to calculate its equilibrium constant at a variety of temperatures.

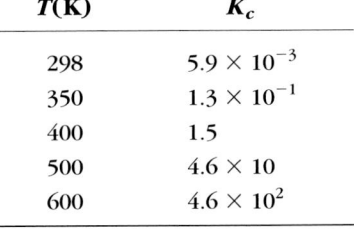

$T(K)$	K_c
298	5.9×10^{-3}
350	1.3×10^{-1}
400	1.5
500	4.6×10
600	4.6×10^2

Problem-Solving Example 14.9 Energy and Entropy Effects on Equilibria

For the equilibrium

$$CH_4(g) + H_2O(g) \rightleftharpoons CO(g) + 3\,H_2(g)$$

(a) Estimate whether the entropy increases, decreases, or remains the same when products form.
(b) Does the entropy effect favor reactants or products?
(c) Use data from Appendix J to calculate $\Delta H°$.
(d) Does the energy effect favor reactants or products?
(e) Is the reaction likely to be product-favored at high temperatures? Why or why not?

Answer
(a) Entropy increases.   (b) Entropy favors products.
(c) $\Delta H° = 206.103$ kJ.   (d) Reactants.   (e) Yes.

Explanation
(a) Because there are 4 mol of gas phase products and only 2 mol of gas phase reactants, the products will be more dispersed and have higher entropy.
(b) Higher entropy means higher probability, the products have higher entropy, and the products are favored.
(c) $\Delta H° = \Sigma\{(\text{moles of product})\,\Delta H_f°(\text{product})\} - \Sigma\{(\text{moles of reactant})\,\Delta H_f° (\text{reactant})\}$
   $= \{(1\text{ mol})(-110.525\text{ kJ/mol})\} -$
                $\{(1\text{ mol})(-74.81\text{ kJ/mol}) + (1\text{ mol})(-241.818\text{ kJ/mol})\}$
   $= -110.525\text{ kJ} - (-316.628\text{ kJ})$
   $= 206.103\text{ kJ}$
(d) Because the reactants are lower in energy, they are favored by the energy effect.
(e) The reaction is product-favored at high temperatures because the energy effect favoring reactants becomes less important, and there is a relatively large entropy effect (four product molecules for every two reactant molecules in the gas phase).

✔ In part (c) it is reasonable that the reaction is endothermic, because six bonds are broken and only four are formed when reactant molecules are changed into product molecules.

Problem-Solving Practice 14.9
For the ammonia synthesis reaction

$$N_2(g) + 3\,H_2(g) \rightleftharpoons 2\,NH_3(g)$$

(a) Does the entropy effect favor products?
(b) Does the energy effect favor products?
(c) Will there be a greater concentration of $NH_3(g)$ at high temperature or at low temperature? Explain.

14.8 CONTROLLING CHEMICAL REACTIONS: THE HABER-BOSCH PROCESS

The principles that allow us to control a reaction are based on our understanding of both equilibrium systems and the rates of chemical reactions. Some generalizations about equilibrium systems are

- *A product-favored reaction has an equilibrium constant larger than 1.*
- *If a reaction is exothermic, this favors the products.*
- *If there is an increase in entropy when a reaction occurs, this favors the products.*
- *Product-favored reactions at low temperatures are usually exothermic.*
- *Product-favored reactions at high temperatures are usually ones in which the entropy increases.*

With these general rules about equilibria we can often predict whether a reaction is capable of yielding products. But it is also important that those products be produced rapidly. Recall these useful generalizations about reaction rates from Chapter 13.

- *Reactions in the gas phase or in solution, where molecules of one reactant are completely mixed with molecules of another, occur more rapidly than do reactions between pure liquids or solids that do not dissolve in one another (⇐ p. 572).*
- *Reactions occur more rapidly at high temperatures than at low temperatures (⇐ p. 593).*
- *Reactions are faster when the reactant concentrations are high than when they are low (⇐ p. 577).*
- *Reactions between a solid and a gas, or a solid and something dissolved in solution, are usually much faster when the solid particles are as small as possible (⇐ p. 572).*
- *Reactions are faster in the presence of a catalyst. Often the right catalyst makes the difference between success and failure in industrial chemistry (⇐ p. 604).*

One of the best examples of the application of the principles of chemical reactivity is the chemical reaction we now use for the synthesis of ammonia from its elements. Even though the earth is bathed in an atmosphere that is about 80% N_2 gas, nitrogen cannot be used by most plants until it has been fixed—converted into biologically useful forms. Although nitrogen fixation is done naturally by organisms such as cyanobacteria and some field crops such as alfalfa and soybeans, most plants cannot fix N_2. They must obtain nitrogen from cyanobacteria or some other organism. Proper fertilization is especially important for recently developed varieties of wheat, corn, and rice that have resulted in much improved food production, particularly in developing countries.

Direct combination of nitrogen and oxygen was used at the beginning of the 20th century to provide fertilizer, but this process was not very efficient. A much better way of manufacturing ammonia was devised by Fritz Haber and Carl Bosch, who chose the *direct synthesis of ammonia from its elements* as the basis for an industrial process.

$$N_2(g) + 3 H_2(g) \rightleftharpoons 2 NH_3(g)$$

For gases, higher concentration corresponds to higher partial pressure.

Because small particles react more rapidly, coal is ground to a powder before it is burned for generating electricity.

The current interest in biotechnology is largely driven by the fact that naturally occurring enzymes are among the most effective catalysts known (⇐ *p. 607*).

Exercise 14.12 Ammonia Synthesis

For the ammonia synthesis reaction, predict
(a) Whether the reaction is exothermic or endothermic.
(b) Whether the reaction is favored by higher probability of the products (entropy).
(c) Whether the reaction produces more products at low or high temperatures.
(d) What would happen if you tried to increase the rate of the reaction by increasing the temperature.

FRITZ HABER
(1868–1934)

At first glance this reaction might seem to be a poor choice. Hydrogen is available naturally only in combined form—for example, in water or hydrocarbons—meaning that hydrogen must be extracted from these compounds at considerable expense in energy resources and money. As you discovered in Problem-Solving Practice 14.9, the ammonia synthesis reaction becomes less and less product-favored at higher temperatures. But higher temperatures are needed in order for ammonia to be produced fast enough for the process to be efficient. Nonetheless, the **Haber-Bosch process** (shown schematically in Figure 14.11) has been so well developed that ammonia is very inexpensive (about $150 per ton). For this reason it is widely used as a fertilizer and so is often among the "top five" chemicals produced in the United States. Annual U.S. production of NH_3 by the Haber-Bosch process is nearly 40 billion pounds.

Both the thermodynamics and the kinetics of the direct synthesis of ammonia have been carefully studied and fine-tuned by industry so that the maximum yield of product is obtained in a reasonable time and at a reasonable cost of both money and energy resources.

• The reaction is strongly exothermic, and there is a decrease in entropy when it takes place. Therefore it is predicted to be product-favored at low temperatures, but reactant-favored at high temperatures. (You should have verified this in Problem-Solving Practice 14.9, and it is predicted by Le Chatelier's principle for an exothermic process.)

The industrial chemical process by which ammonia is manufactured was developed by Fritz Haber, a chemist, and Carl Bosch, an engineer. Haber's studies in the early 1900s revealed that direct ammonia synthesis should be possible. In 1914 the engineering problems were solved by Bosch. Haber's contract with the manufacturer of ammonia called for him to receive 1 pfennig per kilogram of ammonia, and he soon became not only famous but rich. In 1918 he was awarded a Nobel Prize for the ammonia synthesis, but the choice was criticized because of his role in developing the use of poison gases for Germany during World War I.

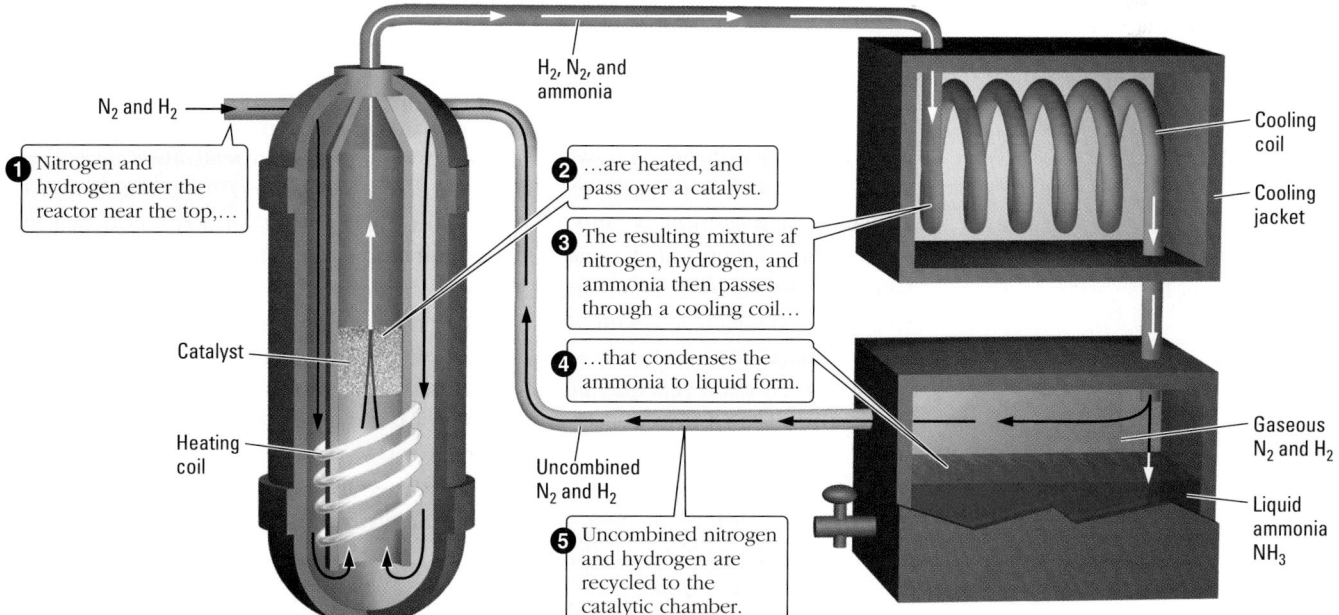

Figure 14.11 **Haber-Bosch process for synthesis of ammonia (schematic).**

- The reaction is quite slow at room temperature, and so the temperature must be raised to increase the rate. Although the rate increases with increasing temperature, the equilibrium constant declines. Thus the faster the reaction, the smaller the yield.

- To increase the equilibrium concentration of NH_3, the reaction is carried out at high pressure (200 atm). This does not change the value of K_c or K_P, but an increase in pressure can be compensated for by converting N_2 and H_2 to NH_3; 2 mol of $NH_3(g)$ exerts less pressure than a total of 4 mol of gaseous reactants $[N_2(g) + 3 H_2(g)]$ in the same-sized container.

- Ammonia is continually liquefied and removed from the reaction vessel, which reduces the concentration of the product of the reaction and shifts the equilibrium toward the right.

- The temperature cannot be raised too much in an attempt to increase the rate, but a rate increase can be achieved with a catalyst. An effective catalyst for the Haber-Bosch process is Fe_3O_4 mixed with KOH, SiO_2, and Al_2O_3. Since the catalyst is not effective below about $400\,°C$, the optimum temperature, considering all the factors controlling the reaction, is about 450 °C.

Making predictions about chemical reactivity is part of the challenge, the adventure, and the art of chemistry. Many chemists enjoy trying to make useful new materials, and this usually means choosing to make them by reactions that we believe will be product-favored and reasonably rapid. Such predictions are based on the ideas outlined in this chapter and Chapter 13.

SUMMARY PROBLEM

One approach to cleaner burning fuels and more efficient automobiles is to extract hydrogen from gasoline and other liquid fossil fuels. The extracted hydrogen could be combined with oxygen in fuel cells (Section 19.9) like those currently used in spacecraft to generate electricity. The electricity could be used in electric motors to power automobiles and also to provide air conditioning and other amenities expected by auto buyers. Because electric motors are far more efficient than current automobile engines, such a car might get 80 miles per gallon of fuel. By answering the following questions, you can explore how the ideas of chemical equilibrium and chemical kinetics can be applied to motive power for automobiles.

1. The hydrogen extracted from hydrocarbon fuels must be free from soot (solid carbon) and carbon monoxide, which would interfere with the operation of a fuel cell. Consider possible reactions by which hydrogen could be obtained from a hydrocarbon such as octane (C_8H_{18}). Write an equation for a reaction that you think would not be suitable and write an equation for one that you think would be suitable. Explain your choice in each case.

2. Use data from Appendix J to calculate the change in enthalpy for each of the two reactions you wrote in Question 1. Predict whether entropy increases or decreases when each reaction occurs. Is either of them ruled out because it is not product-favored? If not, continue. If either of the reactions is not product-favored, think about whether conditions could be altered to make it product-favored.

3. The reaction by which hydrogen is obtained for use in synthesizing ammonia (Haber-Bosch process, ⇐ *p. 666*) involves treating methane (from natural gas) with steam. The first step in this process is

$$CH_4(g) + H_2O(g) \rightleftharpoons CO(g) + 3 H_2(g)$$

(a) Write the equilibrium constant expression for this reaction.

(b) What is the relation between K_c and K_P for this reaction?

(c) Calculate the enthalpy change for this reaction.

(d) Based on the equation itself, predict the sign of the entropy change for this reaction.

(e) Is the reaction product-favored at high temperatures but not at lower temperatures? Or the other way around? Explain.

4. To remove carbon monoxide from the hydrogen destined for the Haber-Bosch process, the following reaction is used:

$$CO(g) + H_2O(g) \rightleftharpoons CO_2(g) + H_2(g)$$

(a) Use bond enthalpies to estimate the enthalpy change for this reaction.

(b) At 450 °C K_P for this reaction is 6.48. Calculate K_c.

(c) Suppose that 0.1 mol of CO and 0.1 mol of H_2O were introduced into an empty 10.0-L flask at 450 °C. Determine the concentration of $H_2(g)$ in the flask once equilibrium has been achieved.

(d) What is the concentration of CO(g) remaining in the flask in part (c)? Is this low enough that we can say that the hydrogen is free of carbon monoxide?

5. If you were in charge of designing a system for generating hydrogen gas for use in the Haber-Bosch process, how might you get pure hydrogen? Assume that the process is based on the two reactions given in Questions 3 and 4. Suggest a chemical reagent that could be used to react with $CO_2(g)$ and thereby remove it from the hydrogen generated as a product in each of the two reactions. Would this same reagent work if you needed to remove $SO_2(g)$?

6. To get the highest purity $H_2(g)$ and the maximum yield from the hydrogen-generating process, how would you adjust the concentrations of the reactants and products? What reactant concentration(s) would you increase or decrease? How would you adjust product concentrations?

7. In the hypothetical fuel cell system for an electric automobile, hydrocarbon fuel is vaporized and partially oxidized in a limited quantity of air. In a second step the products of the first reaction are treated with steam over copper oxide and zinc oxide catalysts. In a final purification step, more air is introduced and a platinum catalyst helps convert carbon monoxide to carbon dioxide. Write a balanced chemical equation for each of these three steps. Assume that the hydrocarbon fuel is octane.

8. What are the advantages of generating hydrogen gas from hydrocarbon fuel in an automobile rather than storing hydrogen in a fuel tank? What are the disadvantages of storing hydrogen in a car? What advantages does combining hydrogen with oxygen in a fuel cell have, as opposed to burning a hydrocarbon fuel in an internal combustion engine?

IN CLOSING

Having studied this chapter, you should be able to . . .

- Recognize a system at equilibrium and describe the properties of equilibrium systems (Section 14.1).
- Describe the dynamic nature of equilibrium and the changes in concentrations of reactants and products that occur as a system approaches equilibrium (Sections 14.1 and 14.2).
- Write equilibrium constant expressions, given balanced chemical equations (Section 14.2).
- Obtain equilibrium constant expressions for related reactions from the expression for one or more known reactions (Section 14.2).

- Calculate K_P from K_c or K_c from K_P for the same equilibrium (Section 14.2).
- Calculate a value of K_c for an equilibrium system, given information about initial concentrations and equilibrium concentrations (Section 14.3).
- Make qualitative predictions about the extent of reaction based upon equilibrium constant values; that is, be able to predict whether a reaction is product-favored or reactant-favored based on the size of the equilibrium constant (Section 14.4).
- Calculate concentrations of reactants and products in an equilibrium system if K_c and initial concentrations are known (Section 14.5).
- Use the reaction quotient Q to predict in which direction a reaction will go to reach equilibrium (Section 14.5).
- Show by using Le Chatelier's principle how changes in concentrations, pressure or volume, and temperature shift chemical equilibria (Section 14.6).
- Use the change in enthalpy and the change in entropy qualitatively to predict whether products are favored over reactants (Section 14.7).
- List the factors affecting chemical reactivity, and apply them to predicting optimal conditions for producing products (Section 14.8).

KEY TERMS

chemical equilibrium
(14.1)

dynamic equilibrium
(14.1)

entropy *(14.7)*

equilibrium concentra-
tion *(14.3)*

equilibrium constant
(14.2)

equilibrium constant
expression *(14.2)*

Haber-Bosch process
(14.8)

Le Chatelier's principle
(14.6)

product-favored *(14.1)*

reactant-favored *(14.1)*

reaction quotient *(14.5)*

shifting an equilibrium
(14.6)

QUESTIONS FOR REVIEW AND THOUGHT

Conceptual Challenge Problems

Conceptual Challenge Problems CP-14.A, CP-14.B, CP-14.D, CP-14.E, and CP-14.F are related to the information in this paragraph. Aqueous iron(III) ions, Fe^{3+}(aq), are nearly colorless. If their concentration is 0.001 M or below, a person cannot detect their color. Thiocyanate ions, SCN^-(aq), are colorless also, but monothiocyanatoiron(III) ions, $Fe(SCN)^{2+}$(aq), can be detected at very low concentrations because of their color. These ions are light amber in very dilute solutions, but as their concentration increases, the color intensifies and appears blood-red in concentrated solutions. Suppose you prepared a stock solution by mixing equal volumes of 1.0×10^{-3} M solutions of both iron(III) nitrate and potassium thiocyanate solutions. The equilibrium reaction is

$$Fe^{3+}(aq) + SCN^-(aq) \rightleftharpoons Fe(SCN)^{2+}(aq)$$

colorless   colorless   amber

CP-14.A (Section 14.1) Describe how you would use 5-mL samples of the stock solution and additional solutions of 0.01 M Fe^{3+}(aq) and 0.01 M SCN^-(aq) to show experimentally that the reaction between Fe^{3+}(aq) and SCN^-(aq) does not go to completion but instead reaches an equilib-

rium state in which appreciable quantities of reactants and product are present. (Refer to the first paragraph above for further information.)

CP-14.B (Section 14.1) Suppose that a person added 1 drop of 0.01 M Fe^{3+}(aq) to a 5-mL sample of the stock solution, followed by 10 drops of 0.01 M SCN^-(aq). This person treated a second 5-mL sample of the stock solution by first adding 10 drops of 0.01 M SCN^-(aq), followed by 1 drop of Fe^{3+}(aq). How would the color intensity of these two solutions compare after the same quantities of the same solutions were added in reverse order? (Refer to the first paragraph above for further information.)

CP-14.C (Section 14.4) Consider the equilibrium reaction between dioxygen and trioxygen (ozone). What is the minimum volume of air (21% dioxygen by volume) at 1.00 atm and 25 °C that you would predict to have at least one molecule of trioxygen, if the only source of trioxygen were its formation from dioxygen and if the atmospheric system were at equilibrium?

$$3\,O_2(g) \rightleftharpoons 2\,O_3(g) \qquad K_c = 6.3 \times 10^{-58} \quad \text{(at 25 °C)}$$

(The volume of 1 mol of air at 1 atm and 25 °C is 24.45 L.)

CP-14.D (Section 14.6) Predict what will happen if you add a small crystal of sodium acetate to a 5-mL sample of the stock solution (described in the first paragraph) so that some acetatoiron(III) ion, a coordination complex, is formed.

CP-14.E (Section 14.6) Predict what will happen if you begin to add a 0.01 M solution of Fe^{3+}(aq) drop by drop to a 5-mL sample of the stock solution (described in the first paragraph) until the total volume becomes 10 mL. (A 0.01 M solution of Fe^{3+} ions is pale yellow.) Predict what will happen if you do the same experiment adding 0.01 M SCN^-(aq) to the stock solution. Predict what will happen if you mix 0.01 M Fe^{3+}(aq) with 0.01 M SCN^-(aq). Explain why the results of these three experiments would be similar or different.

CP-14.F (Section 14.6) Predict what will happen if you put a 5-mL sample of the stock solution (described in the first paragraph) in a hot water bath. Predict what will happen if it is placed in an ice bath.

Answers to Questions in **bold** can be found in the back of the book.

Review Questions

1. Define the terms *chemical equilibrium* and *dynamic equilibrium.*

2. If an equilibrium is product-favored, is its equilibrium constant large or small with respect to 1? Explain.

3. List three characteristics that you would need to verify in order to determine that a chemical system is at equilibrium.

4. The decomposition of ammonium dichromate,

$$(NH_4)_2 Cr_2O_7(s),$$

yields nitrogen gas, water vapor, and solid chromium(III) oxide. The reaction is endothermic. In a closed container this process reaches an equilibrium state. Write a balanced equation for the equilibrium reaction. How is the equilibrium affected if

(a) More ammonium dichromate is added to the equilibrium system,

(b) More water vapor is added, and

(c) More chromium(III) oxide is added?

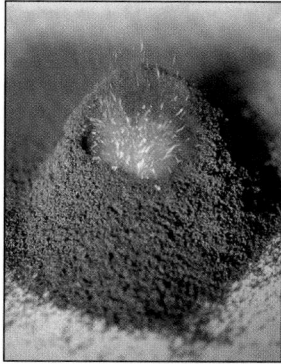

Decomposition of $(NH_4)_2Cr_2O_7$.

5. For the equilibrium reaction in Question 4, write the expression for the equilibrium constant.

(a) How would this equilibrium constant change if the total pressure on the system were doubled?

(b) How would the equilibrium constant change if the temperature were increased?

6. Indicate whether each statement below is true or false. If a statement is false, rewrite it to produce a closely related statement that is true.

(a) For a given reaction, the magnitude of the equilibrium constant is independent of temperature.

(b) If there is an increase in entropy and a decrease in enthalpy when reactants in their standard states are converted to products in their standard states, the equilibrium constant for the reaction will be negative.

(c) The equilibrium constant for the reverse of a reaction is the reciprocal of the equilibrium constant for the reaction itself.

(d) For the reaction

$$H_2O_2(\ell) \rightleftharpoons H_2O(\ell) + \tfrac{1}{2} O_2(g)$$

the equilibrium constant is one half the magnitude of the equilibrium constant for the reaction

$$2 H_2O_2(\ell) \rightleftharpoons 2 H_2O(\ell) + O_2(g)$$

7. Think of an experiment you could do to demonstrate that the equilibrium

$$2 NO_2(g) \rightleftharpoons N_2O_4(g)$$

is a dynamic process in which the forward and reverse reactions continue to occur after equilibrium has been achieved. Describe how such an experiment might be carried out.

8. Discuss the statement, "No true chemical equilibrium can exist unless reactant molecules are constantly changing into product molecules and vice versa."

Characteristics of Chemical Equilibrium

9. Suppose you drop a large piece of ice into a well-insulated thermos with some water in it, and it comes to equilibrium with part of the ice melted.

(a) What is the temperature of the equilibrium system?

(b) Is this a static or a dynamic equilibrium? Explain.

10. The atmosphere consists of about 80% N_2 and 20% O_2, yet there are many oxides of nitrogen that are stable and can be isolated in the laboratory.

(a) Is the atmosphere at chemical equilibrium with respect to forming NO?

(b) If not, why doesn't NO form? If so, how is it that NO can be made and kept in the laboratory for long periods?

The Equilibrium Constant

11. Consider the gas phase reaction of $N_2 + O_2$ to give 2 NO and the reverse reaction of 2 NO to give $N_2 + O_2$, discussed in Section 14.2. An equilibrium mixture of NO, N_2, and O_2 at 5000. K that contains equal concentrations of N_2 and O_2 has a concentration of NO about half as great. Make qualitatively correct plots of the concentration of reactants and products versus time for these two processes, showing the initial state and the final dynamic equilibrium state. Assume a temperature of 5000 K. Don't do any calculations — just sketch how you think the plots will look.

12. After 0.1 mol of pure *cis*-2-butene is allowed to come to equilibrium with *trans*-2-butene in a closed flask at 25 °C, another 0.1 mol of *cis*-2-butene is suddenly added to the flask.

    (a) Is the new mixture at equilibrium? Explain why or why not.

    (b) In the new mixture, immediately after addition of the *cis*-2-butene, which rate is faster, *cis* → *trans* or the reverse? Or are both rates the same?

    (c) After the second 0.1 mol of *cis*-2-butene has been added and the system is at equilibrium, if the concentration of *trans*-2-butene is 0.01 mol/L, what is the concentration of *cis*-2-butene?

13. Write the equilibrium constant expression for each reaction:

    (a) $2 H_2O_2(g) \rightleftharpoons 2 H_2O(g) + O_2(g)$
    (b) $PCl_3(g) + Cl_2(g) \rightleftharpoons PCl_5(g)$
    (c) $SiO_2(s) + 3 C(s) \rightleftharpoons SiC(s) + 2 CO(g)$
    (d) $H_2(g) + \frac{1}{8} S_8(s) \rightleftharpoons H_2S(g)$

14. Write the equilibrium constant expression for each reaction:

    (a) $3 O_2 \rightleftharpoons 2 O_3(g)$
    (b) $SiH_4(g) + 2 O_2(g) \rightleftharpoons SiO_2(s) + 2 H_2O(g)$
    (c) $MgO(s) + SO_2(g) + \frac{1}{2} O_2(g) \rightleftharpoons MgSO_4(s)$
    (d) $2 PbS(s) + 3 O_2(g) \rightleftharpoons 2 PbO(s) + 2 SO_2(g)$

15. Write the equilibrium constant expression for each reaction:

    (a) $TlCl_3(s) \rightleftharpoons TlCl(s) + Cl_2(g)$
    (b) $CuCl_4^{2-}(aq) \rightleftharpoons Cu^{2+}(aq) + 4 Cl^-(aq)$
    (c) $CO(g) + H_2O(g) \rightleftharpoons CO_2(g) + H_2(g)$
    (d) $4 H_3O^+(aq) + 2 Cl^-(aq) + MnO_2(s) \rightleftharpoons$
    $Mn^{2+}(aq) + 6 H_2O(\ell) + Cl_2(g)$

16. Write the equilibrium constant expression for each reaction:

    (a) The oxidation of ammonia with ClF_3 in a rocket motor.

    $$NH_3(g) + ClF_3(g) \rightleftharpoons 3 HF(g) + \frac{1}{2} N_2(g) + \frac{1}{2} Cl_2(g)$$

    (b) The simultaneous oxidation and reduction of a chlorite ion.

    $$3 ClO_2^-(aq) \rightleftharpoons 2 ClO_3^-(aq) + Cl^-(aq)$$

    (c) $IO_3^-(aq) + 6 OH^-(aq) + Cl_2(g) \rightleftharpoons$
    $IO_6^{5-}(aq) + 2 Cl^-(aq) + 3 H_2O(\ell)$

17. Write the equilibrium constant expression for each of these heterogeneous systems.

    (a) $CaSO_4 \cdot 5 H_2O(s) \rightleftharpoons CaSO_4 \cdot 3 H_2O(s) + 2 H_2O(g)$
    (b) $SiF_4(g) + 2 H_2O(g) \rightleftharpoons SiO_2(s) + 4 HF(g)$
    (c) $LaCl_3(s) + H_2O(g) \rightleftharpoons LaClO(s) + 2 HCl(g)$

18. Write the equilibrium constant expression for each of these heterogeneous systems.

    (a) $N_2O_4(g) + O_3(g) \rightleftharpoons N_2O_5(s) + O_2(g)$
    (b) $C(s) + 2 N_2O(g) \rightleftharpoons CO_2(g) + 2 N_2(g)$
    (c) $H_2O(\ell) \rightleftharpoons H_2O(g)$

19. In Section 14.2 the equilibrium constant for the reaction

    $$\frac{1}{8} S_8(s) + O_2(g) \rightleftharpoons SO_2(g)$$

    is given as 4.2×10^{52}. If this reaction is so product-favored, why can large piles of yellow sulfur exist in our environment (as they do in Louisiana and Texas)?

20. Consider these two equilibria involving $SO_2(g)$ and their corresponding equilibrium constants.

    $$SO_2(g) + \frac{1}{2} O_2(g) \rightleftharpoons SO_3(g) \qquad K_{c_1}$$
    $$2 SO_3(g) \rightleftharpoons 2 SO_2(g) + O_2(g) \qquad K_{c_2}$$

    Which of these expressions correctly relates K_{c_1} to K_{c_2}?

    (a) $K_{c_2} = K_{c_1}^2$ (b) $K_{c_2}^2 = K_{c_1}$ (c) $K_{c_2} = 1/K_{c_1}$
    (d) $K_{c_2} = K_{c_1}$ (e) $K_{c_2} = 1/K_{c_1}^2$

21. The reaction of hydrazine (N_2H_4) with chlorine trifluoride (ClF_3) has been used in experimental rocket motors.

    $$N_2H_4(g) + \frac{4}{3} ClF_3(g) \rightleftharpoons 4 HF(g) + N_2(g) + \frac{2}{3} Cl_2(g)$$

    How is the equilibrium constant, K_P, for this reaction related to K_P' for the reaction written in the following way?

    $$3 N_2H_4(g) + 4 ClF_3(g) \rightleftharpoons 12 HF(g) + 3 N_2(g) + 2 Cl_2(g)$$

    (a) $K_P = K_P'$ (b) $K_P = 1/K_P'$ (c) $K_P^3 = K_P'$
    (d) $K_P = (K_P')^3$ (e) $3 K_P = K_P'$

22. Hydrogen can react with elemental sulfur to give the smelly, toxic gas H_2S according to the reaction

    $$H_2(g) + \frac{1}{8} S_8(s) \rightleftharpoons H_2S(g)$$

    If the equilibrium constant K_c for this reaction is 7.6×10^5 at 25 °C, determine the value of the equilibrium constant for the reaction written as

    $$8 H_2(g) + S_8(s) \rightleftharpoons 8 H_2S(g)$$

23. At 450 °C, the equilibrium constant K_c for the Haber-Bosch synthesis of ammonia is 0.16 for the reaction written as

    $$3 H_2(g) + N_2(g) \rightleftharpoons 2 NH_3(g)$$

    Calculate the value of K_c for the same reaction written as

    $$\frac{3}{2} H_2(g) + \frac{1}{2} N_2(g) \rightleftharpoons NH_3(g)$$

24. For each reaction in Question 13, write the equilibrium constant expression for K_P.

25. For each reaction in Question 14, write the equilibrium constant expression for K_P.

26. The vapor pressure of water at 80. °C is 0.467 atm. Find the value of K_c for the reaction

    $$H_2O(\ell) \rightleftharpoons H_2O(g)$$

    at this temperature.

27. The value of K_c for the reaction

    $$N_2(g) + 3 H_2(g) \rightleftharpoons 2 NH_3(g)$$

    is 2.00 at 400 °C. Find the value of K_P for this reaction at this temperature using atmospheres as units.

Determining Equilibrium Constants

28. Isomer A is in equilibrium with isomer B, as in the reaction

    $$A(g) \rightleftharpoons B(g)$$

    Three experiments are done, each at a different temperature, and equilibrium concentrations are measured. For each experiment, calculate the equilibrium constant, K_c.

    (a) [A] = 0.74 mol/L, [B] = 0.74 mol/L
    (b) [A] = 2.0 mol/L, [B] = 2.0 mol/L
    (c) [A] = 0.01 mol/L, [B] = 0.01 mol/L

29. Two molecules of A react to form one molecule of B, as in the reaction

$$2 A(g) \rightleftharpoons B(g)$$

Three experiments are done, each at a different temperature, and equilibrium concentrations are measured. For each experiment, calculate the equilibrium constant, K_c.
(a) [A] = 0.74 mol/L, [B] = 0.74 mol/L
(b) [A] = 2.0 mol/L, [B] = 2.0 mol/L
(c) [A] = 0.01 mol/L, [B] = 0.01 mol/L
By comparing the results of Questions 28 and 29, what can you conclude about the statement, "If the concentrations of reactants and products are equal, then the equilibrium constant is always 1.0."

30. Consider the equilibrium

$$2 A(aq) \rightleftharpoons B(aq)$$

At equilibrium, [A] = 0.056 M and [B] = 0.21 M. Calculate the equilibrium constant for the reaction as written.

31. The following reaction was examined at 250 °C:

$$PCl_5(g) \rightleftharpoons PCl_3(g) + Cl_2(g)$$

At equilibrium, $[PCl_5] = 4.2 \times 10^{-5}$ M, $[PCl_3] = 1.3 \times 10^{-2}$ M, and $[Cl_2] = 3.9 \times 10^{-3}$ M. Calculate the equilibrium constant K_c for the reaction.

32. At high temperature, hydrogen and carbon dioxide react to give water and carbon monoxide.

$$H_2(g) + CO_2(g) \rightleftharpoons H_2O(g) + CO(g)$$

Laboratory measurements at 986 °C show that there is 0.11 mol each of CO and water vapor and 0.087 mol each of H_2 and CO_2 at equilibrium in a 1.0-L container. Calculate the equilibrium constant K_P for the reaction at 986 °C.

33. Carbon dioxide reacts with carbon to give carbon monoxide according to the equation

$$C(s) + CO_2(g) \rightleftharpoons 2 CO(g)$$

At 700. °C, a 2.0-L flask is found to contain at equilibrium 0.10 mol of CO, 0.20 mol of CO_2, and 0.40 mol of C. Calculate the equilibrium constant K_P for this reaction at the specified temperature.

34. Assume you place 0.010 mol of $N_2O_4(g)$ in a 2.0-L flask at 50. °C. After the system reaches equilibrium, $[N_2O_4] = 0.00090$ M. What is the value of K_c for this reaction?

$$N_2O_4(g) \rightleftharpoons 2 NO_2(g)$$

35. Nitrosyl chloride, NOCl, decomposes to NO and Cl_2 at high temperatures.

$$2 NOCl(g) \rightleftharpoons 2 NO(g) + Cl_2(g)$$

Suppose you place 2.00 mol of NOCl in a 1.00-L flask and raise the temperature to 462 °C. When equilibrium has been established, 0.66 mol of NO is present. Calculate the equilibrium constant K_c for the decomposition reaction from these data.

36. An equilibrium mixture contains 3.00 mol of CO, 2.00 mol of Cl_2, and 9.00 mol of $COCl_2$ in a 50. L reaction flask at 800. K. Calculate the value of the equilibrium constant K_c for the reaction

$$CO(g) + Cl_2(g) \rightleftharpoons COCl_2(g)$$

at this temperature.

37. At 667 K, HI is found to be 11.4% dissociated into its elements.

$$2 HI(g) \rightleftharpoons H_2(g) + I_2(g)$$

If 1.00 mol of HI is placed in a 1.00-L container and heated to 667 K, calculate (a) the equilibrium concentration of all three substances and (b) the value of K_c for this equilibrium at this temperature.

38. A sample of nitrosyl bromide is heated to 100. °C in a 10.00-L container in order to decompose it partially according to the equation

$$2 NOBr(g) \rightleftharpoons 2 NO(g) + Br_2(g)$$

The container is found to contain 6.44 g of NOBr, 3.15 g of NO, and 8.38 g of Br_2 at equilibrium.
(a) Find the value of K_c at 100. °C.
(b) Find the total pressure exerted by the mixture of gases.
(c) Calculate K_P for this reaction at 100. °C.

39. Exactly 5.0 mol of ammonia, NH_3, was placed in a 2.0-L flask that was then heated to 473 K. When equilibrium was established, 0.2 mol of nitrogen had been formed according to the decomposition reaction

$$2 NH_3(g) \rightleftharpoons N_2(g) + 3 H_2(g)$$

(a) Calculate the value of the equilibrium constant K_c for this reaction at 473 K.
(b) Calculate the total pressure exerted by the mixture of gases inside the 2.0-L flask at this temperature.
(c) Calculate K_P for this reaction at 473 K.

The Meaning of the Equilibrium Constant

40. Using the data of Table 14.1, predict which of the reactions below will be product-favored at 25 °C. Then place all the reactions in order from most reactant-favored to most product-favored.
(a) $2 NH_3(g) \rightleftharpoons N_2(g) + 3 H_2(g)$
(b) $NH_4^+(aq) + OH^-(aq) \rightleftharpoons NH_3(aq) + H_2O(\ell)$
(c) $2 NO(g) \rightleftharpoons N_2(g) + O_2(g)$

41. Using the data of Table 14.1, predict which of the reactions below will be product-favored at 25 °C. Then place all the reactions in order from most reactant-favored to most product-favored.
(a) $2 NO_2(g) \rightleftharpoons N_2O_4(g)$
(b) $H_2CO_3(aq) \rightleftharpoons HCO_3^-(aq) + H^+(aq)$
(c) $AgI(s) \rightleftharpoons Ag^+(aq) + I^-(aq)$

42. The equilibrium constants for dissolving silver sulfate and silver sulfide in water are 1.7×10^{-5} and 6×10^{-30}, respectively.
(a) Write the balanced reaction equation and the associated equilibrium constant expression for each process.
(b) Which compound is more soluble?
(c) Which is less soluble?

43. The equilibrium constants for dissolving calcium carbonate, silver nitrate, and silver chloride in water are 3.8×10^{-9}, 2.0×10^2, and 1.8×10^{-10}, respectively.
(a) Write the balanced reaction equation and the associated equilibrium constant expression for each process.
(b) Which compound is most soluble?
(c) Which is least soluble?

Using Equilibrium Constants

44. The hydrocarbon C_4H_{10} can exist in two forms, butane and 2-methylpropane. The value of K_c for the interconversion of the two forms is 2.5 at 25 °C.

$$CH_3—CH_2—CH_2—CH_3 \rightleftharpoons H—\underset{\underset{CH_3}{|}}{\overset{\overset{CH_3}{|}}{C}}—CH_3$$

butane   2-methylpropane

(a) Suppose that the initial concentrations of both butane and 2-methylpropane are 0.100 mol/L. Make up a table of initial concentration, change in concentration, and equilibrium concentration for this reaction.

(b) Write the equilibrium constant expression in terms of x, the change in the concentration of butane, then solve for x.

(c) If you place 0.017 mol of butane in a 0.50-L flask at 25 °C, what will be the equilibrium concentrations of the two isomers?

45. A mixture of butane and 2-methylpropane at 25 °C has [butane] = 0.025 mol/L and [2-methylpropane] = 0.035 mol/L. Is this mixture at equilibrium? If the *equilibrium* concentration of butane is 0.025 mol/L, what must [2-methylpropane] be at equilibrium? (See reaction in Question 44.)

46. Cyclohexane, C_6H_{12}, a hydrocarbon, can isomerize or change into methylcyclopentane, a compound with the same formula but with a different molecular structure.

$$C_6H_{12}(g) \rightleftharpoons C_5H_9CH_3(g)$$

cyclohexane methylcyclopentane

The equilibrium constant K_c has been estimated to be 0.12 at 25 °C. If you had originally placed 3.79 g of cyclohexane in a 2.80-L flask, how much cyclohexane (in grams) is present when equilibrium is established?

47. At room temperature, the equilibrium constant K_c for the reaction

$$2\ NO(g) \rightleftharpoons N_2(g) + O_2(g)$$

is 1.4×10^{30}.

(a) Is this reaction product-favored or reactant-favored?

(b) In the atmosphere at room temperature the concentration of N_2 is 0.33 mol/L, and the concentration of O_2 is about 25% of that value. Calculate the equilibrium concentration of NO in the atmosphere produced by the reaction of N_2 and O_2.

(c) Now revisit your answer to Question 10.

48. Hydrogen gas and iodine gas react via the equation

$$H_2(g) + I_2(g) \rightleftharpoons 2\ HI(g) \qquad K_c = 76 \quad \text{(at 600. K)}$$

If 0.05 mol of HI is placed in a 1.0-L flask at 600. K, what are the equilibrium concentrations of HI, I_2, and H_2?

49. The equilibrium constant K_c for the reaction

$$H_2(g) + I_2(g) \rightleftharpoons 2\ HI(g)$$

has the value 50.0 at 745 K.

(a) When 1.00 mol of I_2 and 3.00 mol of H_2 are allowed to come to equilibrium at 745 K in a flask of volume 10.00 L, what amount (in moles) of HI will be produced?

(b) What amount of HI is produced in a 5.00-L flask?

(c) What total amount of HI is present at equilibrium if an additional 3.00 mol of H_2 is added to the 10.00-L flask?

50. The equilibrium constant K_c for the *cis-trans* isomerization of gaseous 2-butene has the value 1.50 at 580. K.

$$\underset{\underset{H_3C}{\diagup}}{\overset{\overset{H}{\diagdown}}{C}}=\underset{\underset{CH_3}{\diagdown}}{\overset{\overset{H}{\diagup}}{C}} \rightleftharpoons \underset{\underset{H_3C}{\diagup}}{\overset{\overset{H}{\diagdown}}{C}}=\underset{\underset{H}{\diagdown}}{\overset{\overset{CH_3}{\diagup}}{C}}$$

(a) Is the reaction product-favored at 580. K?

(b) Calculate the amount (in moles) of *trans* isomer produced when 1 mol of *cis*-2-butene is heated to 580. K in the presence of a catalyst in a flask of volume 1.00 L.

(c) What would be the answer if the flask had a volume of 10.0 L?

51. The equilibrium constant K_c for the reaction

$$CO(g) + H_2O(g) \rightleftharpoons CO_2(g) + H_2(g)$$

has the value 4.00 at 500 K. If a mixture of 1.00 mol of CO and 1.00 mol of H_2O is allowed to come to equilibrium in a flask of volume 1.00 L at 500 K,

(a) Calculate the final concentrations of all four species: CO, H_2O, CO_2, and H_2.

(b) What would be the equilibrium concentrations if an additional 1.00 mol each of CO and H_2O were added to the flask?

52. At 503 K the equilibrium constant K_c for the dissociation of N_2O_4,

$$N_2O_4(g) \rightleftharpoons 2\ NO_2(g)$$

has the value 40.0.

(a) Calculate the fraction of N_2O_4 left undissociated when 1.00 mol of this gas is heated to 503 K in a 10.0-L container.

(b) If the volume is now reduced to 2.0 L, what will be the new fraction of N_2O_4 that is undissociated?

53. The equilibrium constant K_c for the reaction

$$N_2(g) + 3\ H_2(g) \rightleftharpoons 2\ NH_3(g)$$

has the value 5.97×10^{-2} at 500 °C.

(a) If 1.00 mol of N_2 gas and 1.00 mol of H_2 gas are heated to 500. °C in a flask of volume 10.00 L together with a catalyst, calculate the percentage of N_2 converted to NH_3. (*Hint*: Assume that only a very small fraction of the reactants is converted to products. Obtain an approximate answer and use this to obtain a more accurate result.)

54. What percentage of N_2 would be converted to NH_3 in the previous problem if the volume of the flask were 5.00 L?

55. The equilibrium constant K_c has a value of 3.30 at 760 K for the decomposition of phosphorus pentachloride,

$$PCl_5(g) \rightleftharpoons PCl_3(g) + Cl_2(g)$$

(a) Calculate the equilibrium concentrations of all three species arising from the decomposition of 0.75 mol of PCl_5 in a 5.00-L vessel.

(b) Calculate the equilibrium concentrations of all three species resulting from an initial mixture of 0.75 mol of PCl_5 and 0.75 mol of PCl_3 in a 5.00-L vessel.

56. A 1.00-mol sample of CO_2 is heated to 1000. K with excess solid graphite in a container of volume 40.0 L. At 1000. K, K_c is 2.11×10^{-2} for the reaction

$$C(\text{graphite}) + CO_2(g) \rightleftharpoons 2\ CO(g)$$

(a) What is the composition of the equilibrium mixture at 1000. K?

(b) If the volume of the flask is changed and a new equilibrium established in which the amount of CO_2 is equal to the amount of CO, what is the new volume of the flask?

Shifting a Chemical Equilibrium: Le Chatelier's Principle

57. Solid barium sulfate is in equilibrium with barium ions and sulfate ions in solution.

$$BaSO_4(s) \rightleftharpoons Ba^{2+}(aq) + SO_4^{2-}(aq)$$

What will happen to the barium ion concentration if more solid $BaSO_4$ is added to the flask? Explain your choice.

(a) It will increase.

(b) It will decrease.

(c) It will not change.

(d) It is not possible to tell from the information provided.

58. Consider the following equilibrium, established in a 2.0-L flask at 25 °C.

$$N_2O_4(g) \rightleftharpoons 2 NO_2(g) \qquad \Delta H° = +57.2 \text{ kJ}$$

What will happen to the concentration of N_2O_4 if the temperature is increased? Explain your choice.

(a) It will increase.

(b) It will decrease.

(c) It will not change.

(d) It is not possible to tell from the information provided.

59. Hydrogen, bromine, and HBr in the gas phase are in equilibrium in a container of fixed volume.

$$H_2(g) + Br_2(g) \rightleftharpoons 2 HBr(g) \qquad \Delta H° = -103.7 \text{ kJ}$$

How will each of the following changes affect the indicated quantities? Write "increase," "decrease," or "no change."

Change	$[Br_2]$	$[HBr]$	K_c	K_P
Some H_2 is added to the container.	___	___	___	___
The temperature of the gases in the container is increased.	___	___	___	___
The pressure of HBr is increased.	___	___	___	___

60. The equilibrium constant K_c for the following reaction is 0.16 at 25 °C, and the standard enthalpy change is 16.1 kJ.

$$2 NOBr(g) \rightleftharpoons 2 NO(g) + Br_2(\ell)$$

Predict the effect of each of the following changes on the position of the equilibrium; that is, state which way the equilibrium will shift (left, right, or no change) when each of the following changes is made.

(a) Adding more Br_2. (b) Removing some NOBr.

(c) Decreasing the temperature.

61. The formation of hydrogen sulfide from the elements is exothermic.

$$H_2(g) + \tfrac{1}{8} S_8(s) \rightleftharpoons H_2S(g) \qquad \Delta H° = -20.6 \text{ kJ}$$

Predict the effect of each of the following changes on the position of the equilibrium; that is, state which way the equilibrium will shift (left, right, or no change) when each of the following changes is made.

(a) Adding more sulfur. (b) Adding more H_2.

(c) Raising the temperature.

62. The oxidation of NO to NO_2,

$$2 NO(g) + O_2(g) \rightleftharpoons 2 NO_2(g)$$

is exothermic. Predict the effect of each of the following changes on the position of the equilibrium; that is, state which way the equilibrium will shift (left, right, or no change) when each of the following changes is made.

(a) Adding more O_2. (b) Adding more NO_2.

(c) Decreasing the temperature.

63. Consider the equilibrium.

$$PbCl_2(s) \rightleftharpoons Pb^{2+}(aq) + 2 Cl^-(aq)$$

(a) Will the equilibrium concentration of aqueous lead(II) ion increase, decrease, or remain the same if some solid NaCl is added to the flask?

(b) Make a graph like the one in Figure 14.6 to illustrate what happens to each of the concentrations after the NaCl is added.

64. Phosphorus pentachloride is in equilibrium with phosphorus trichloride and chlorine in a flask.

$$PCl_5(s) \rightleftharpoons PCl_3(g) + Cl_2(g)$$

What will happen to the concentration of Cl_2 if additional $PCl_5(s)$ is added to the flask?

(a) It will increase.

(b) It will decrease.

(c) It will not change.

(d) It is impossible to tell from the information provided.

65. Consider the transformation of butane into 2-methylpropane (see Question 44). The system is originally at equilibrium at 25 °C in a 1.0-L flask with [butane] = 0.010 M and [2-methylpropane] = 0.025 M. Suppose that 0.0050 mol of 2-methylpropane is suddenly added to the flask, and the system shifts to a new equilibrium.

(a) What is the new equilibrium concentration of each gas?

(b) Make a graph like the one in Figure 14.6 to show how the concentrations of the isomers change when the 2-methylpropane is added.

66. Predict whether the equilibria listed below will be shifted to the left or the right when the following changes occur: (i) the temperature is increased; (ii) the pressure is decreased; (iii) more of the substance indicated by color is added.

(a) $C(s) + H_2O(g) \rightleftharpoons CO(g) + H_2(g)$

$\Delta H° \text{ (298 K)} = +131.3 \text{ kJ mol}^{-1}$

(b) $3 Fe(s) + 4 H_2O(g) \rightleftharpoons Fe_3O_4(s) + 4 H_2(g)$

$\Delta H° \text{ (298 K)} = -149.9 \text{ kJ mol}^{-1}$

(c) $C(s) + CO_2(g) \rightleftharpoons 2 CO(g)$

$\Delta H° \text{ (298 K)} = +172.5 \text{ kJ mol}^{-1}$

(d) $N_2O_4(g) \rightleftharpoons 2\ NO_2\ (g)$

$$\Delta H° \text{ (298 K)} = +54.8 \text{ kJ mol}^{-1}$$

67. Predict whether the equilibria listed below will be shifted to the left or the right when the following changes are made: (i) the temperature is increased; (ii) the pressure is decreased; (iii) more of the substance indicated by color is added.

(a) $N_2(g) + O_2\ (g) \rightleftharpoons 2\ NO(g)$

$$\Delta H° \text{ (298 K)} = +180.0 \text{ kJ mol}^{-1}$$

(b) $CH_4(g) + 2\ O_2(g) \rightleftharpoons CO_2\ (g) + 2\ H_2O(g)$

$$\Delta H° \text{ (298 K)} = -802.3 \text{ kJ mol}^{-1}$$

(c) $CaCO_3(s) \rightleftharpoons CaO(s) + CO_2\ (g)$

$$\Delta H° \text{ (298 K)} = +177.9 \text{ kJ mol}^{-1}$$

Equilibrium at the Nanoscale

68. For each of these reactions at 25 °C, indicate whether the entropy effect, the energy effect, both, or neither favors the reaction.

(a) $N_2(g) + 3\ F_2(g) \rightleftharpoons 2\ NF_3(g)$ $\Delta H° = -249 \text{ kJ}$
(b) $N_2F_4(g) \rightleftharpoons 2\ NF_2(g)$ $\Delta H° = 93.3 \text{ kJ}$
(c) $N_2(g) + 3\ Cl_2(g) \rightleftharpoons 2\ NCl_3(g)$ $\Delta H° = 460 \text{ kJ}$

69. For each of these processes at 25 °C, indicate whether the entropy effect, the energy effect, both, or neither favors the process.

(a) $C_3H_8(g) + 5\ O_2(g) \rightleftharpoons 3\ CO_2(g) + 4\ H_2O(g)$

$$\Delta H° = -2045 \text{ kJ}$$

(b) $Br_2(g) \rightleftharpoons Br_2(\ell)$ $\Delta H° = -31 \text{ kJ}$
(c) $2\ Ag(s) + 3\ N_2(g) \rightleftharpoons 2\ AgN_3(g)$ $\Delta H° = 621 \text{ kJ}$

70. For each of these chemical reactions, predict whether the equilibrium constant at 25 °C is greater than 1, less than 1, or state that insufficient information is available. Also indicate whether each reaction is product-favored or reactant-favored.

(a) $2\ NO(g) + O_2(g) \rightleftharpoons 2\ NO_2(g)$ $\Delta H° = -115 \text{ kJ}$
(b) $2\ O_3(g) \rightleftharpoons 3\ O_2(g)$ $\Delta H° = -285 \text{ kJ}$
(c) $N_2(g) + 3\ Cl_2(g) \rightleftharpoons 2\ NCl_3(g)$ $\Delta H° = 460 \text{ kJ}$

71. For each of these chemical reactions, predict whether the equilibrium constant at 25 °C is greater than 1, less than 1, or state that insufficient information is available. Also indicate whether each reaction is product-favored or reactant-favored.

(a) $2\ NaCl(s) \rightleftharpoons 2\ Na(s) + Cl_2(g)$ $\Delta H° = 411 \text{ kJ}$
(b) $2\ CO(g) + O_2(g) \rightleftharpoons 2\ CO_2(g)$ $\Delta H° = -566 \text{ kJ}$
(c) $3\ CO_2(g) + 4\ H_2O(g) \rightleftharpoons C_3H_8(g) + 5\ O_2(g)$

$$\Delta H° = 2045 \text{ kJ}$$

Controlling Chemical Reactions: The Haber-Bosch Process

72. Using the signs of the enthalpy change and the entropy change for the Haber-Bosch process, explain why choosing the temperature at which to run this reaction is very important.

73. Although ammonia is made in enormous quantities by the Haber-Bosch process, sulfuric acid is made in even greater quantities. The *contact process* used to make sulfuric acid

can be simplified and represented by the following three reactions.

$$S(s) + O_2(g) \rightleftharpoons SO_2(g)$$

$$2\ SO_2(g) + O_2(g) \rightleftharpoons 2\ SO_3(g)$$

$$SO_3(g) + H_2O(\ell) \rightleftharpoons H_2SO_4(\ell)$$

(a) Use data from Appendix J to calculate $\Delta H°$ for each reaction.
(b) Which reactions are exothermic? Which are endothermic?
(c) In which of the reactions does the entropy increase? In which does it decrease? In which does it stay about the same?
(d) For which reaction(s) do low temperatures favor formation of products?

74. Lime, CaO(s), is produced by heating limestone, $CaCO_3(s)$, to cause a decomposition reaction.

(a) Write a balanced equation for the reaction.
(b) Predict the sign of the enthalpy change for the reaction.
(c) From the data in Appendix J, calculate $\Delta H°$ for this reaction at 25 °C to verify or contradict your prediction in part (b).
(d) Predict the sign of the entropy change for this reaction.
(e) Is the reaction favored by entropy, energy, both, or neither?
(f) Explain in terms of Le Chatelier's principle why limestone must be heated to make lime.

General Questions

75. Write equilibrium constant expressions, in terms of reactant and product concentrations, for each of these reactions.

$H_2O(\ell) \rightleftharpoons H^+(aq) + OH^-(aq)$ $K_c = 1.0 \times 10^{-14}$

$CH_3COOH(aq) \rightleftharpoons CH_3COO^-(aq) + H^+(aq)$

$$K_c = 1.8 \times 10^{-5}$$

$N_2(g) + 3\ H_2(g) \rightleftharpoons 2\ NH_3(g)$ $K_c = 3.5 \times 10^8$

$2\ O_3(g) \rightleftharpoons 3\ O_2(g)$ $K_c = 7 \times 10^{56}$

$2\ NO_2(g) \rightleftharpoons N_2O_4(g)$ $K_c = 1.7 \times 10^2$

$HCOO^-(aq) + H^+(aq) \rightleftharpoons HCOOH(aq)$ $K_c = 5.6 \times 10^3$

$Ag^+(aq) + I^-(aq) \rightleftharpoons AgI(s)$ $K_c = 6.7 \times 10^{15}$

Assume that all gases and solutes have initial concentrations of 1.0 mol/L. Then let the *first* reagent in each reaction change its concentration by $-x$.

(a) Using the reaction table approach, write equilibrium constant expressions in terms of the unknown variable x for each reaction.
(b) Which of these expressions yield quadratic equations?
(c) How would you go about solving the others for x?

76. Many common nonmetallic elements exist as diatomic molecules at room temperature ($\Leftarrow$ *p. 26*). When these elements are heated to 1500 K, the molecules break apart into atoms. A general equation for this type of reaction is

$$E_2(g) \rightleftharpoons 2\ E(g)$$

where E stands for each element. Equilibrium constants for dissociation of these molecules at 1500 K are

Species	K_c	Species	K_c
Br_2	8.9×10^{-2}	H_2	3.1×10^{-10}
Cl_2	3.4×10^{-3}	N_2	1×10^{-27}
F_2	7.4	O_2	1.6×10^{-11}
I_2	1.5		

(a) If 1.00 mol of each diatomic molecule is placed in a separate 1.0-L container and heated to 1500 K, what is the equilibrium concentration of the atomic form of each element at 1500 K?
(b) From these results, predict which of the diatomic elements has the lowest bond dissociation energy, and compare your results with thermochemical calculations and with Lewis structures.

77. The chemistry of compounds composed of a transition metal and carbon monoxide has been an interesting area of research for the past 50 years. $Ni(CO)_4$ is formed by the reaction of nickel metal with carbon monoxide.
(a) If you have 2.05 g of CO, and you combine it with 0.125 g of nickel metal, how many grams of $Ni(CO)_4$ can be formed?
(b) An excellent way to make pure nickel metal is to decompose $Ni(CO)_4$ in a vacuum at a temperature slightly higher than room temperature. If the molar enthalpy of formation of $Ni(CO)_4$ gas is -602.9 kJ/mol, what is the enthalpy change for this decomposition reaction?

$$Ni(CO)_4(g) \longrightarrow Ni(s) + 4\ CO(g)$$

In an experiment at 100 °C it is determined that with 0.01 mol of $Ni(CO)_4(g)$ initially present in a 1.0-L flask, only 0.00001 mol remains after decomposition.
(c) What is the equilibrium concentration of CO in the flask?
(d) What is the value of the equilibrium constant, K_c, for this reaction at 100 °C?
(e) Predict whether there is an increase or decrease in entropy when this reaction occurs.
(f) Calculate the equilibrium constant K_P for this reaction at 100 °C.

78. A small sample of *cis*-dichloroethene in which one carbon atom is the radioactive isotope ^{14}C is added to an equilibrium mixture of the *cis* and *trans* isomers at a certain temperature. Eventually, 40% of the radioactive molecules are found to be in the *trans* configuration at any given time.
(a) What is the value of K_c for the *cis* $\rightleftharpoons$ *trans* equilibrium?
(b) What would have occurred if a small sample of radioactive *trans* isomer had been added instead of the *cis* isomer?

79. In a 0.002 M solution of acetic acid any acetate species spends 9% of its time as an acetate ion, CH_3COO^-, and the remaining 91% of its time as acetic acid, CH_3COOH. What is the value of K_c for the reaction

$$CH_3COOH + H_2O \rightleftharpoons CH_3COO^- + H_3O^+$$

80. A sample of pure SO_3 weighing 0.8312 g was placed in a flask of volume 1.00 L and heated to 1100 K in order to decompose it partially.

$$2\ SO_3(g) \rightleftharpoons 2\ SO_2(g) + O_2(g)$$

If a total pressure of 1.295 atm was developed, find the value of K_c for this reaction at this temperature.

81. The following amounts of HI, H_2, and I_2 are introduced into a 10.00-L reaction flask and heated to 745 K.

	n HI (mol)	*n* H_2 (mol)	*n* I_2 (mol)
Case a	1.0	0.10	0.10
Case b	10.	1.0	1.0
Case c	10.	10.	1.0
Case d	5.62	0.381	1.75

The equilibrium constant for the reaction

$$2\ HI(g) \rightleftharpoons H_2(g) + I_2(g)$$

has the value 0.0200 at 745 K. In which cases will the concentration of HI increase as equilibrium is attained, and in which cases will the concentration of HI decrease?

Applying Concepts

82. Suppose that you have heated a mixture of *cis*- and *trans*-2-pentene to 600 K, and after 1 hr you find that the composition is 40% *cis*. After 4 hr the composition is found to be 42% *cis*, and after 8 hr it is 42% *cis*. Next you heat the mixture to 800 K and find that the composition changes to 45% *cis*. When the mixture is cooled to 600 K and allowed to stand for 8 hr, the composition is found to be 42% *cis*. Is this system at equilibrium at 600 K? Or would more experiments be needed before you could conclude that it was at equilibrium? If so, what experiments would you do?

83. When pressure is applied to ice at 0 °C, the ice melts. Explain how this is an example of Le Chatelier's principle.

84. For the reaction

$$\textit{cis-2-butene} \rightleftharpoons \textit{trans-2-butene}$$

K_c is 1.65 at 500 K, 1.47 at 600 K, and 1.36 at 700 K. Predict whether the conversion from the *cis* to the *trans* isomer of 2-butene is exothermic or endothermic.

Use this table to answer Questions 85 and 86 (p. 678).

Equilibrium Constants K_c for Some *Cis-Trans* Interconversions

Temperature (K)	R is F	R is Cl	R is CH_3
500	0.420	0.608	1.65
600	0.491	0.678	1.47
700	0.549	0.732	1.36

85. Based on the data in the table and the reaction equations below,

(a) Which reaction is most exothermic?

(b) Which reaction is most endothermic?

   (i) *cis*-(R = F) $\rightleftharpoons$ *trans*-(R = F)

   (ii) *trans*-(R = Cl) $\rightleftharpoons$ *cis*-(R = Cl)

   (iii) *cis*-(R = CH$_3$) $\rightleftharpoons$ *trans*-(R = CH$_3$)

86. Based on the data in the table and the reaction equations below,

(a) Which reaction is most product-favored at 500 K?

(b) Which reaction is most reactant-favored at 700 K?

   (i) *cis*-(R = F) $\rightleftharpoons$ *trans*-(R = F)

   (ii) *trans*-(R = Cl) $\rightleftharpoons$ *cis*-(R = Cl)

   (iii) *cis*-(R = CH$_3$) $\rightleftharpoons$ *trans*-(R = CH$_3$)

87. Figure 14.3 shows the equilibrium mixture of N$_2$O$_4$ and NO$_2$ at two different temperatures. Imagine that you can shrink yourself down to the size of the molecules in the two tubes and observe their behavior for a short period of time. Write a brief description of what you observe in each of the tubes.

88. Imagine yourself to be the size of ions and molecules inside a beaker containing the following equilibrium mixture with a K_c greater than 1.

Co(H$_2$O)$_6^{2+}$(aq) + 4 Cl$^-$(aq) $\rightleftharpoons$

   pink

                   CoCl$_4^{2-}$(aq) + 6 H$_2$O(ℓ)

                   blue

Write a brief description of what you observe around you before and after additional water is added to the mixture.

89. For the equilibrium

Co(H$_2$O)$_6^{2+}$(aq) + 4 Cl$^-$(aq) $\rightleftharpoons$

   pink

                   CoCl$_4^{2-}$(aq) + 6 H$_2$O(ℓ)

                   blue

K_c is somewhat greater than 1. If water is added to a blue solution of CoCl$_4^{2-}$(aq), the color changes from blue to pink.

(a) Does water appear in the equilibrium constant expression for this reaction?

(b) How can adding water shift the equilibrium to the left?

(c) Is this shift in the equilibrium in accord with Le Chatelier's principle? Why or why not? (*Hint*: Consider what happens to the concentrations of substances that appear in the equilibrium constant expression. It may help to calculate Q before and after water is added to the equilibrium mixture.)

90. Which of the diagrams represents equilibrium mixtures for the reaction

$$A_2(g) + B_2(g) \rightleftharpoons 2\ AB(g)$$

at a temperature where $10^2 > K_c > 0.1$?

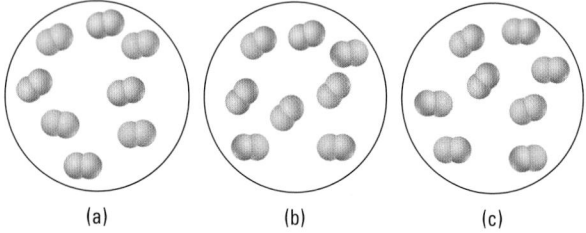

     (a)   (b)   (c)

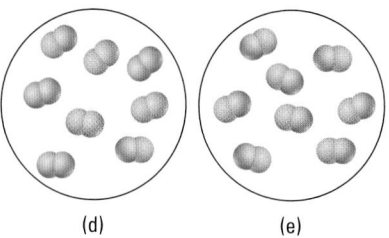

     (d)   (e)

91. Draw a nanoscale (particulate) level diagram for an equilibrium mixture of

$$2\ H_2O_2(g) \rightleftharpoons 2\ H_2O(g) + O_2(g)$$

92. Which diagram in Question 90 best represents an equilibrium mixture with an equilibrium constant of

(a) 0.44   (b) 4.0   (c) 36

93. The diagram below represents an equilibrium mixture for the reaction

$$N_2(g) + O_2(g) \rightleftharpoons 2\ NO(g)$$

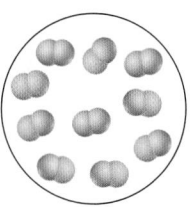

What is the equilibrium constant?

94. A sample of benzoic acid, a solid carboxylic acid, is in equilibrium with an aqueous solution of benzoic acid. A tiny quantity of D$_2$O, water containing the isotope ^{2}H, deuterium, is added to the solution. The solution is allowed to stand at constant temperature for several hours, after which some of the solid benzoic acid is removed and analyzed. The benzoic acid is found to contain a tiny quantity of deuterium, D, and the formula of the deuterium-containing molecules is C$_6$H$_5$COOD. Explain how this can happen.

95. In a second experiment with benzoic acid (see Question 94), a tiny quantity of water that contains the isotope ^{18}O is added to a saturated solution of benzoic acid in water. When some of the solid benzoic acid is analyzed, no ^{18}O is found in the benzoic acid. Compare this situation with the experiment involving deuterium, and explain how the results of the two experiments can differ as they do.

96. Samples of N$_2$O$_4$ can be prepared in which both nitrogen atoms are the heavier isotope ^{15}N. Designating this isotope as N*, we can write the formula of the molecules in such a sample as O$_2$N*—N*O$_2$ and the formula of typical N$_2$O$_4$ as O$_2$N—NO$_2$. When a tiny quantity of O$_2$N*—N*O$_2$ is introduced into an equilibrium mixture of N$_2$O$_4$ and NO$_2$, the ^{15}N immediately becomes distributed among both N$_2$O$_4$ and NO$_2$ molecules, and in the N$_2$O$_4$ it is invariably in the form O$_2$N*—NO$_2$. Explain how this observation supports the idea that equilibrium is dynamic.

97. Using the symbolism of Question 96, we can write an equilibrium

$$O_2N^*{-}N^*O_2 + O_2N{-}NO_2 \rightleftharpoons 2\ O_2N^*{-}NO_2$$

Assuming that all isotopes of nitrogen behave the same in N$_2$O$_4$ and NO$_2$ molecules, use the ideas of probability developed in Section 14.7 to figure out the value of the equilibrium constant for this process.

 General Chemistry CD-ROM

CD14.1 Screen 16.4: The Equilibrium Constant.

(a) The $Fe(SCN)^{2+}$ experiment seen on Screen 16.3 is used again on this screen. Examine the table of initial concentrations and the beakers of solutions formed by mixing these solutions. The quantity of iron used in all the experiments is the same, but additional SCN^- is used going from left to right. What effect does using more SCN^- have on the quantity of product formed?

(b) Realizing that in each case the SCN^- is the limiting reactant, calculate the percent yield for each of the four reactions.

CD14.2 Screen 16.5: The Meaning of the Equilibrium Constant.

(a) Is $PbI_2(s)$ expected to dissolve to an appreciable extent? How is this reflected in the value of the reaction's equilibrium constant?

(b) Would you expect the reaction $2\,NO_2(g) \rightleftharpoons 2\,NO(g) + O_2(g)$ to have a large or small equilibrium constant? Explain briefly.

CD14.3 Screen 16.11: Disturbing a Chemical Equilibrium: Le Chatelier's Principle.

(a) Examine the water tank animation on this screen. Describe how addition of water to the left-hand tank illustrates Le Chatelier's principle.

(b) Of the three potential changes to an equilibrium system described on the screen, which does the tank demonstration illustrate?

CD14.4 Screen 16.12: Disturbing an Equilibrium: Temperature Changes. Watch the photographs shown on this screen. Describe the difference between the two states shown, both in terms of temperature and in terms of the concentrations of the species in the flask.

CD14.5 Screen 16.13: Disturbing an Equilibrium: Addition or Removal of a Reagent.

(a) On the previous screen, changes in temperature resulted in a change in the equilibrium constant, resulting in a shift in the equilibrium. Does the addition or removal of a reagent described on this screen result in a change in the equilibrium constant?

(b) What is the difference between the reaction quotient, Q, and the equilibrium constant, K?

15

Loosening the cap on a bottle of carbonated beverage reduces the pressure of CO_2 gas inside. At a lower pressure, CO_2 is less soluble in water, and bubbles of CO_2 form within the liquid where they expand and rise. When they contact your tongue, they make the beverage taste fizzy. Knowing how pressure and temperature affect the solubility of CO_2 in water is useful for keeping the beverage from tasting flat.

THE CHEMISTRY OF SOLUTES AND SOLUTIONS

15.1 Solubility and Noncovalent Forces

15.2 Enthalpy, Entropy, and Dissolving Solutes

15.3 Solubility and Equilibrium

15.4 Temperature and Solubility

15.5 Pressure and Dissolving Gases in Liquids: Henry's Law

15.6 Solution Concentration: Keeping Track of Units

15.7 Vapor Pressures, Boiling Points, and Freezing Points of Solutions

15.8 Osmotic Pressure of Solutions

15.9 Colloids

15.10 Surfactants

15.11 Water: Natural, Clean, and Otherwise

W e all encounter many solutions, such as soft drinks, juices, coffee, and gasoline, every day. A *solution* is a homogeneous mixture of two or more substances (⟵ *Section 1.4*). The component present in greatest amount is the *solvent;* the other components are *solutes* (⟵ *Section 5.6*) In sweetened iced tea, for example, water is the solvent and sugar and soluble extracts of tea are the solutes.

Although solids dissolved in liquids or mixtures of liquids are the most common types of solutions, other kinds are possible as well, encompassing the three physical states of matter (Table 15.1). In general, chemistry focuses on liquid solutions, and in particular those in which water is the solvent (aqueous solutions).

In this chapter we will explore in some detail the macroscale to nanoscale connections that relate solutes, solvents, and their solutions to answer questions such as, Why does a particular solvent readily dissolve one kind of solute, but not another? Water, for example, dissolves NaCl but does not dissolve gasoline. In what ways can the concentration of a dissolved solute be expressed? Is thermal energy released or absorbed when a solute dissolves? How do factors such as temperature or pressure changes affect the solubility of a solute in a given solvent? Why, for example, does a cold, carbonated beverage become "flat" when it is opened and warmed to room temperature? To answer these questions, we will apply what we know from Chapter 9 about noncovalent forces — London forces, dipole-dipole forces, and hydrogen bonding — in terms of the interactions among solute and solvent molecules (⟵ *Section 9.5, p. 383)*. We will also utilize the thermodynamic and equilibrium principles studied in Chapter 6 and 14. These principles are important in understanding the effect that solutes have on the vapor pressures, melting points, and boiling points of solvents. In addition, the nature of unwanted solutes in natural and polluted water is discussed. Much of this chapter is devoted to aqueous solutions because water is the most important solvent on our planet.

15.1 SOLUBILITY AND NONCOVALENT FORCES

Whether or not water is the solvent, it is the interplay between solute and solvent particles that determines whether a solute will dissolve in a particular solvent and how much is dissolved.

Solute-Solvent Interactions

An old adage says that "oil and water don't mix." Chemists use a similar saying about solubility: "Like dissolves like," where "like" refers to solutes and solvents held togther by similar types of noncovalent forces. There are noncovalent forces between solute particles, as well as between solvent particles. Dissolving a solute in a solvent is favored when the stronger noncovalent forces are

 CD-ROM Screen 14.3: The Solution Process

TABLE 15.1	Types of Solutions
Type of solution	**Example**
Gas in gas	Air — a mixture principally of N_2 and O_2 but containing other gases as well
Gas in liquid	Carbonated beverages (CO_2 in water)
Gas in solid	Hydrogen in palladium metal
Liquid in liquid	Motor oil—a mixture of liquid hydrocarbons; coffee with cream; tea with lemon juice
Solid in liquid	The oceans (dissolved Na^+, Cl^-, and other ions)
Solid in solid	Bronze (copper and tin); pewter (tin, antimony, and lead)

between the solute and the solvent relative to the solute-solute and solvent-solvent forces.

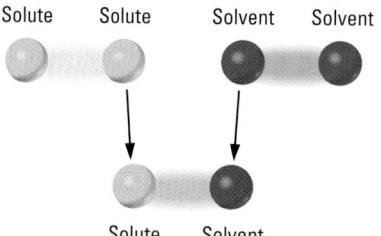

Consider, for example, the dissolving of the hydrocarbon octane, C_8H_{18}, in carbon tetrachloride, CCl_4 (Figure 15.1a). Both are nonpolar liquids that dissolve in each other to produce a colorless, clear solution because of the similar London forces present in each compound.

Liquids that dissolve in each other in any proportion are said to be **miscible.** In contrast, gasoline, a mixture of nonpolar hydrocarbons, does not dissolve in water, a polar substance. The nonpolar hydrocarbons cannot hydrogen-bond to water molecules, but rather stay attracted to each other through London forces (solute-solute attractions); the water molecules remain hydrogen-bonded to each other (solvent-solvent attractions). Liquids such as gasoline and water, with such different noncovalent attractions that they do not dissolve in each other, are described as **immiscible** (Figure 15.1b).

The solubilities of alcohols in water further illustrates the "like dissolves like" principle and the role of noncovalent forces. Simple, low-molar-mass alcohols dissolve in water due to hydrogen bonding between water molecules and the —OH group of the alcohol. The noncovalent forces in the solution (hydrogen bonding) are the same as those in pure water or pure ethanol. The hydrogen bonding between ethanol and water is illustrated in Figure 15.2.

Alcohol molecules contain a polar portion, the —OH group, and a nonpolar portion, the hydrocarbon part. The polar region is *hydrophilic* ("water loving"); any polar part of a molecule will be hydrophilic because of its attraction to polar water molecules. The nonpolar hydrocarbon region is *hydrophobic* ("water hating")(⇐, *p. 390)*. As the hydrophobic hydrocarbon chain attached to the —OH of the alcohol increases in length, the alcohol becomes less and less like water and more and more like a hydrocarbon. In a low-molar-mass alcohol such as methanol or ethanol (Figure 15.2), hydrogen bonding of the —OH group with water molecules is stronger than the London forces between the nonpolar parts

CD-ROM Screen 13.6: Hydrogen Bonding

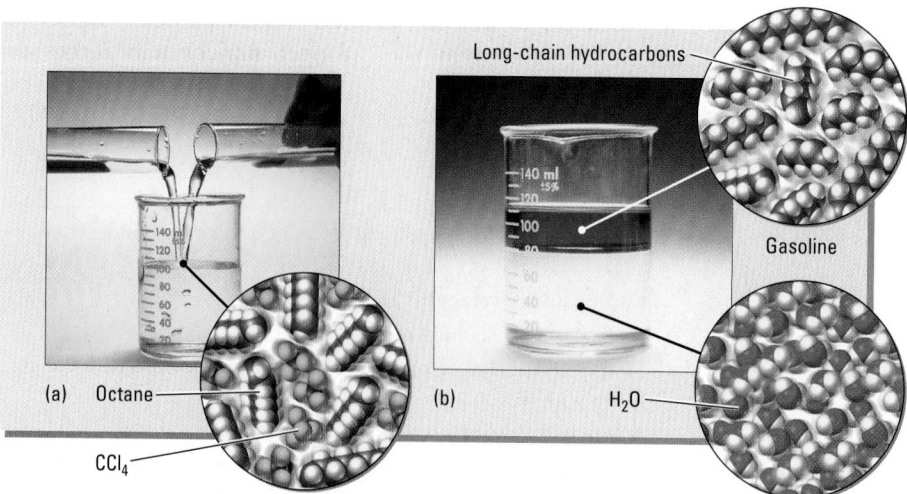

Figure 15.1 Miscible and immiscible liquids. When carbon tetrachloride and octane, both colorless, clear liquids, are mixed (a), each dissolves completely in the other, and there is no sign of an interface or boundary between them (b). When gasoline and water are put together, the mixture remains as two distinct layers.

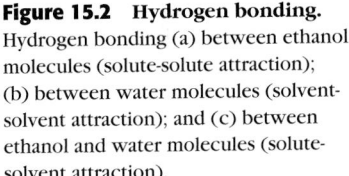

(a)

(b)

(c)

Figure 15.2 **Hydrogen bonding.** Hydrogen bonding (a) between ethanol molecules (solute-solute attraction); (b) between water molecules (solvent-solvent attraction); and (c) between ethanol and water molecules (solute-solvent attraction).

of the alcohol molecules. As the hydrocarbon portion of the alcohol gets larger in higher molar mass alcohols, the London forces between the nonpolar portion of the alcohol molecules become stronger. Consequently, a point is reached at which the London forces between the hydrocarbon portion of alcohol molecules (solute-solute attraction) become sufficiently large that the water solubility of the alcohol becomes small, such as that of 1-heptanol compared with ethanol (Table 15.2). The electron density models in Table 15.2 (p. 684) illustrate that the —OH group of the alcohol becomes a relatively smaller portion of the molecule from methanol to 1-heptanol compared with the hydrocarbon portion. Thus, methanol and ethanol, with just one and two carbon atoms, respectively, are infinitely soluble in water, whereas alcohols with more than six carbon atoms per molecue are virtually insoluble in water.

We can summarize the principle of "like dissolves like" as follows:

- ***Substances with similar noncovalent forces are likely to be soluble in each other.***
- ***Solutes do not readily dissolve in solvents whose noncovalent forces are quite different from their own.***
- ***Stronger solute-solvent attractions favor solubility. Stronger solute-solute or solvent-solvent attractions reduce solubility.***

 Exercise 15.1 **Predicting Solubility**

How could the data in Table 15.2 be used to predict the solubility of 1-octanol or 1-decanol?

 Exercise 15.2 **Predicting Water Solubility**

You have a sample of 1-octanol and a sample of methanol. Which is more water-soluble? Which is more soluble in gasoline? Explain your choices in terms of "like dissolves like."

Alaskan oil spill. This oil spill in Prince William Sound, Alaska, is a large-scale example of the nonsolubility of crude oil, a nonpolar material, in water, a polar substance. The oil floats on top of the water, but does not dissolve in it.

TABLE 15.2	Solubilities of Some Alcohols in Water		
Name	**Formula**	**Solubility in water (g/100 g H_2O at 20 °C)**	**Models**
Methanol	CH_3OH	Miscible	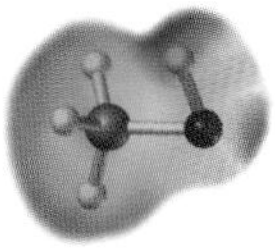
Ethanol	CH_3CH_2OH	Miscible	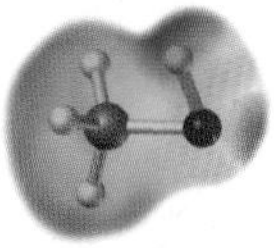
1-Propanol	$CH_3(CH_2)_2OH$	Miscible	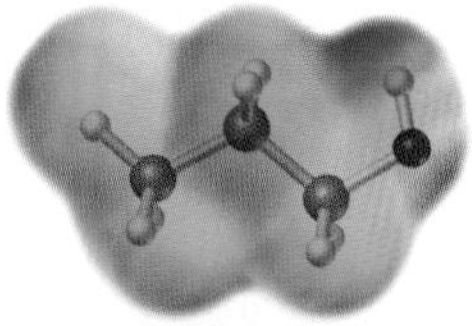
1-Butanol	$CH_3(CH_2)_3OH$	7.9	
1-Pentanol	$CH_3(CH_2)_4OH$	2.7	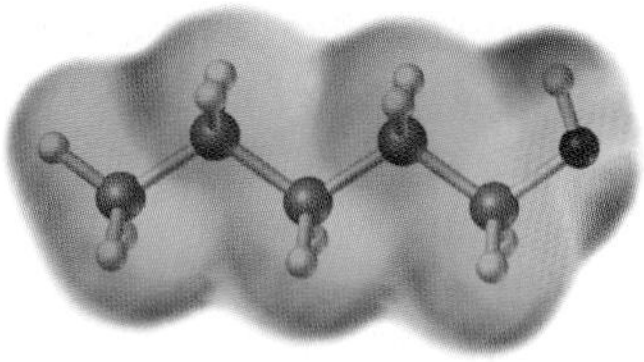
1-Hexanol	$CH_3(CH_2)_5OH$	0.6	

Problem-Solving Example 15.1   Predicting Solubilities

Using the principle of "like dissolves like," predict whether

(a) Ethylene glycol, $HOCH_2CH_2OH$, dissolves in gasoline
(b) Molecular iodine dissolves in carbon tetrachloride
(c) Motor oil dissolves in carbon tetrachloride.

Explain your predictions.

Answer
(a) No   (b) Yes   (c) Yes

Explanation
(a) Ethylene glycol molecules are polar and are attracted to each other by dipole-dipole attraction and hydrogen bonding. They will not dissolve in gasoline, a nonpolar substance.

(b) Molecular iodine, I_2, is nonpolar, as is CCl_4. Therefore, the iodine dissolves readily in carbon tetrachloride.
(c) Motor oil contains a mixture of nonpolar hydrocarbons that will dissolve in carbon tetrachloride, a nonpolar solvent.

Problem-Solving Practice 15.1

Explain why gasoline and motor oil are miscible, such as in the mixtures used in two-cycle lawn mower engines.

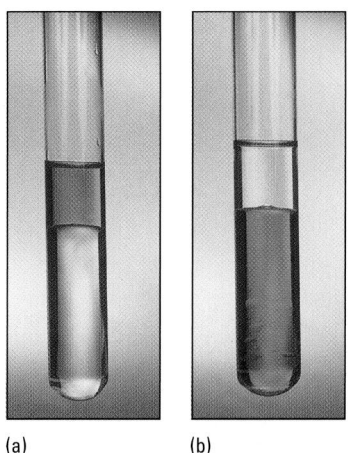

(a) (b)

Water, carbon tetrachloride (CCl_4), and iodine. (a) Water (polar molecules) and CCl_4 (nonpolar molecules) are not miscible, so that the less-dense water layer lies on top of the more dense CCl_4 layer. A small amount of iodine dissolves in the water to give a brown solution *(top)*. (b) The nonpolar I_2 molecules are more soluble in nonpolar CCl_4 and dissolve preferentially in CCl_4 to give a purple solution after the mixture is shaken.

Solids such as quartz (SiO_2) that are held together by an extensive network of covalent bonds are generally insoluble in polar and nonpolar solvents. In quartz, the silicon and oxygen atoms form SiO_4 tetrahedra linked through covalent bonds to shared oxygen atoms. These strong covalent bonds are not broken by weaker attractions to solvent molecules. Thus, quartz (and sand derived from it) is insoluble in water or any other solvent at room temperature. Sandy beaches may erode by wave action, but they do not dissolve in ocean water.

The solubility of a substance in water or in the nonpolar fats or oily substances in our bodies — the triglycerides (⬅ *p. 540*) —plays an important role in our body chemistry. Vitamins, for example, are either water-soluble or fat-soluble. Vitamins A, D, E, and K are known as fat-soluble vitamins because they dissolve in triglycerides. All the other vitamins are water-soluble. The major significance of this difference is the danger of overdosing on fat-soluble vitamins because they are stored in fatty tissues and may accumulate to harmful levels. By contrast, overdosing on water-soluble vitamins is not common because these vitamins are not stored in the body and any excess is excreted in urine.

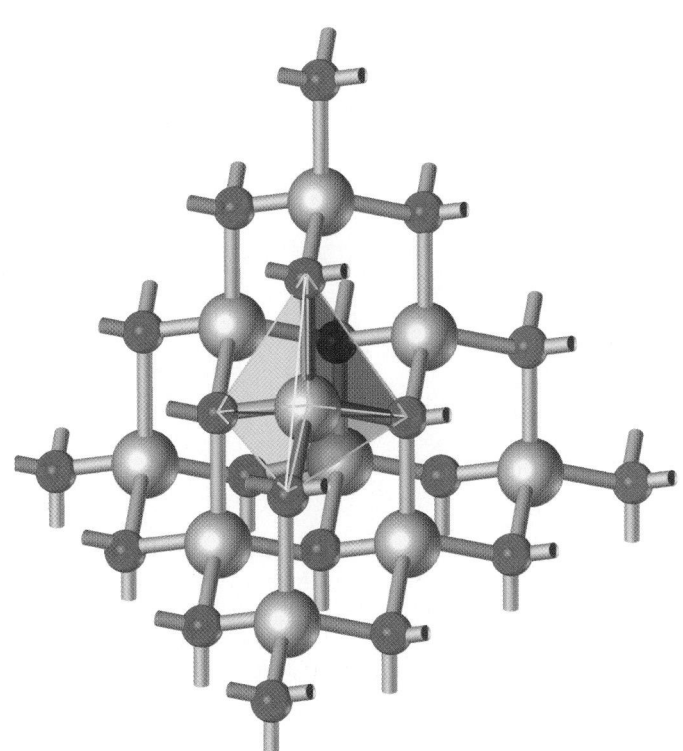

A portion of a quartz (SiO_2) structure. The structure is based on SiO_4 tetrahedra linked through shared oxygen atoms as shown by the shaded SiO_4 tetrahedron.

Problem-Solving Example 15.2 Solubility and Noncovalent Forces

Use the structural formulas of vitamin A and niacin (nicotinic acid) to determine which is more soluble in water and which is more soluble in fat.

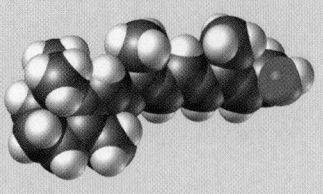

vitamin A

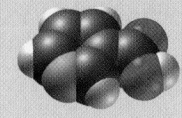

niacin
(nicotinic acid)

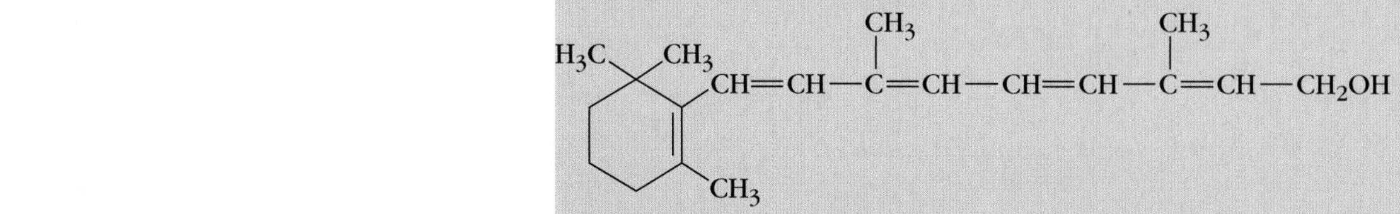

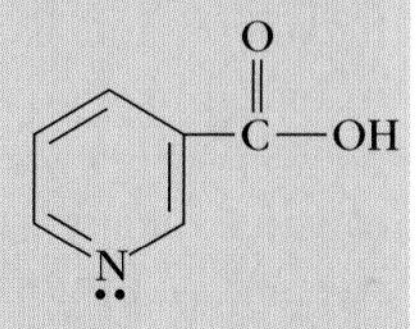

Answer Niacin is water-soluble; vitamin A is fat-soluble.

Explanation Niacin is water-soluble because of hydrogen bonding with water molecules.

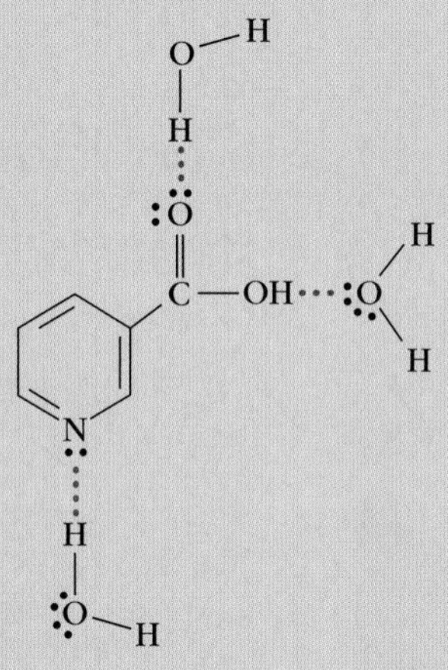

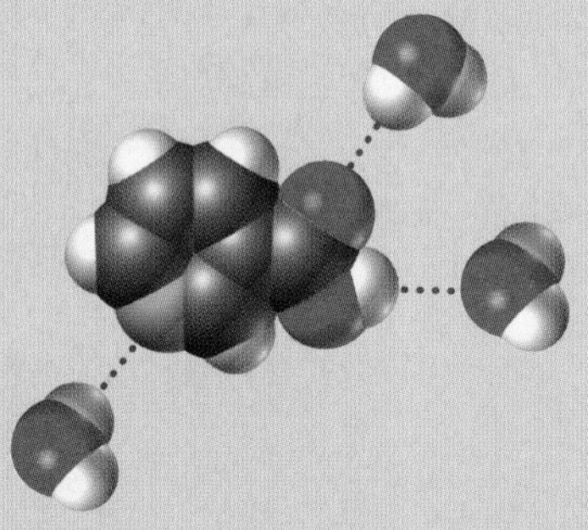

niacin
(nicotinic acid)

Niacin dissolves because the solute-solvent attraction between niacin and water is stronger than hydrogen bonding between water molecules or the dipole-dipole attractions between niacin molecules.

Vitamin A has an —OH group that can hydrogen-bond with water. However, as is the case with other long-chain alcohols, the hydrocarbon portion of the molecule is large enough to make the molecule hydrophobic. London forces between the hydrocarbon portions overcome the hydrogen bonding, and vitamin A is insoluble in water. On the other hand, the extended hydrocarbon portion of the molecule is soluble in the long-chain hydrocarbon portion of fats.

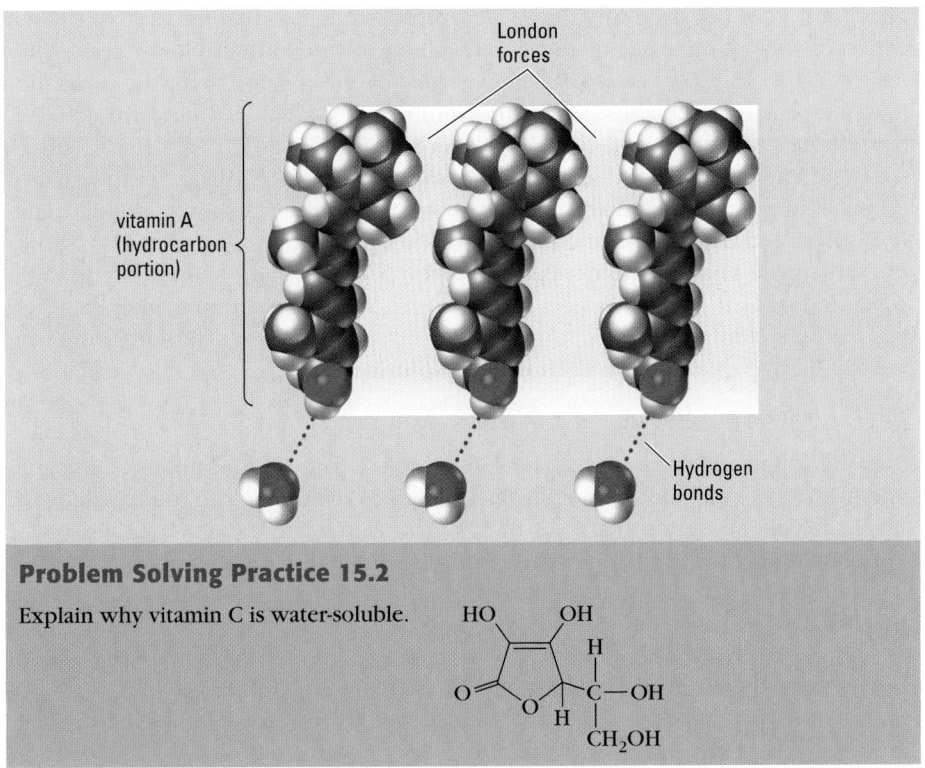

London forces

vitamin A (hydrocarbon portion)

Hydrogen bonds

Problem Solving Practice 15.2

Explain why vitamin C is water-soluble.

15.2 ENTHALPY, ENTROPY, AND DISSOLVING SOLUTES

When a solution forms, atoms, molecules, or ions of one kind mix with atoms, molecules, or ions of a different kind. Noncovalent forces attracting solvent molecules together and noncovalent forces attracting solute molecules together must be overcome (*steps a and b* in Figure 15.3) To separate solvent molecules (*step a*),

Figure 15.3 The solution-making process.

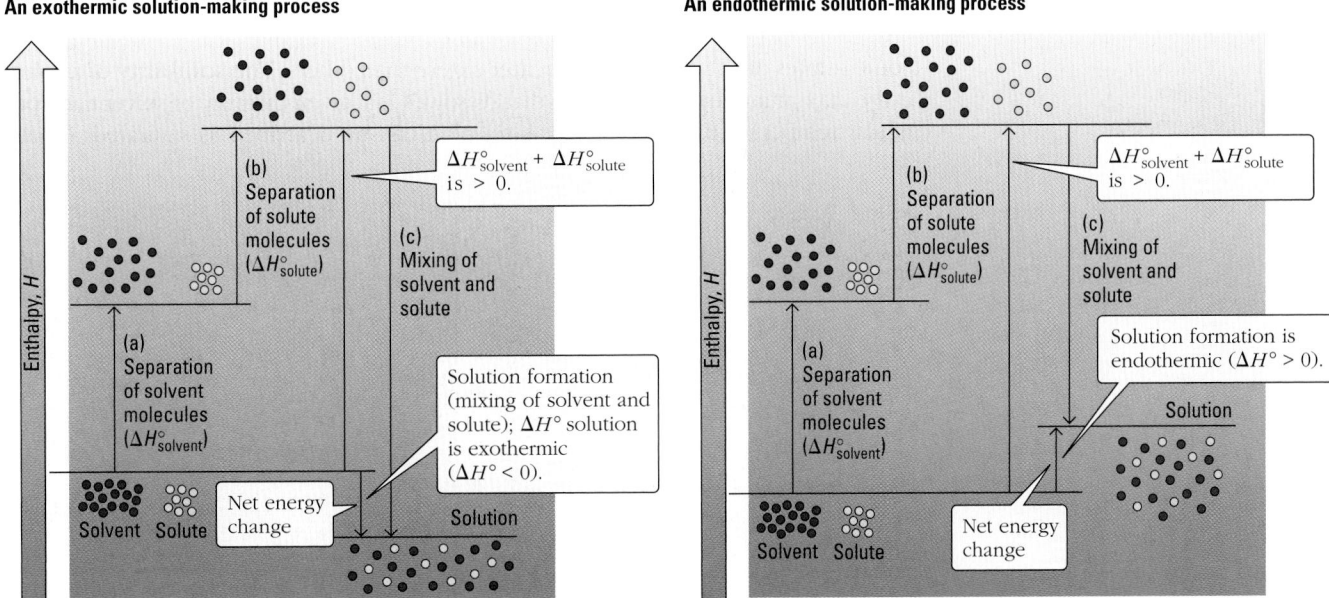

An exothermic solution-making process

(b) Separation of solute molecules ($\Delta H°_{solute}$)

(a) Separation of solvent molecules ($\Delta H°_{solvent}$)

(c) Mixing of solvent and solute

$\Delta H°_{solvent} + \Delta H°_{solute}$ is > 0.

Solution formation (mixing of solvent and solute); $\Delta H°$ solution is exothermic ($\Delta H° < 0$).

Net energy change

Solvent   Solute   Solution

Enthalpy, H

An endothermic solution-making process

(b) Separation of solute molecules ($\Delta H°_{solute}$)

(a) Separation of solvent molecules ($\Delta H°_{solvent}$)

(c) Mixing of solvent and solute

$\Delta H°_{solvent} + \Delta H°_{solute}$ is > 0.

Solution formation is endothermic ($\Delta H° > 0$).

Solution

Net energy change

Solvent   Solute

Enthalpy, H

CD-ROM Screen 14.3: The Solution Process

The enthalpy of solution is also known as the heat of solution.

attractive forces between them must be overcome, and so the enthalpy of the collection of solvent molecules increases, resulting in an endothermic process with a positive ΔH ($\Delta H > 0$). Likewise, this must also occur in step [b] for the separation of solute ions or molecules. When the solute and solvent particles mix (*step c*), they attract each other and there is a decrease in enthalpy; ΔH is negative ($\Delta H < 0$), and energy is released in this exothermic step. If the enthalpy of the final solution is lower than the initial solute and solvent enthalpy, as shown in Figure 15.3 (*left*), there is a net release of energy, and the solution-making process is *exothermic*. When the final solution enthalpy is greater than the enthalpies of the initial solute and solvent, the process is *endothermic*, and the surroundings transfer energy to the system (Figure 15.3, *right*). The net energy change, either exothermic or endothermic, is called the **enthalpy of solution** ΔH_{soln}.

$$\text{Enthalpy of solution} = \Delta H_{soln} = \Delta H_{step\ a} + \Delta H_{step\ b} + \Delta H_{step\ c}$$

After the solute and solvent mix, the solvent-solute forces (indicated by the magnitude of $\Delta H_{step\ c}$) may not be not strong enough to overcome the solute-solute and solvent-solvent attractions ($\Delta H_{step\ a} + \Delta H_{step\ b}$). Under these conditions, dissolving is not favored by the enthalpy effect (⟸ *Section 14.8*).

Most systems have a tendency to become more disordered. When solutes and solvents mix to form solutions, the increased disorder that occurs in the solution is a result of that natural tendency. As described in Section 14.7, *entropy* (symbolized by S) is a measure of a system's disorder; an increase in entropy ($\Delta S > 0$) indicates a process that creates more disorder, that is, greater randomness. *Processes in which the entropy of the system increases tend to be product-favored.* Therefore, in many cases of solution making there is a large entropy increase as solvent and solute molecules mix. In cases where the enthalpy of solution is rather small, entropy is the driving force behind solution formation, such as when octane and carbon tetrachloride are mixed. In other cases, even though the enthalpy of solution is positive and significantly large ($\Delta H_{soln} > 0$), the entropy increase is large enough to cause the solution to form. This is true for some ionic solutes such as ammonium nitrate, NH_4NO_3 (Section 15.3). Because gases are always much more disordered than liquids, there is a significant decrease in entropy ($\Delta S < 0$) as gas solute molecules are brought closer together between solvent molecules.

15.3 SOLUBILITY AND EQUILIBRIUM

CD-ROM Screen 14.2: Solubility

Some solutes dissolve to a much greater extent than others. The **solubility** of a solute is the maximum quantity of solute that dissolves in a given quantity of solvent at a particular temperature (Figure 15.4). A solution can be described as saturated, unsatu-

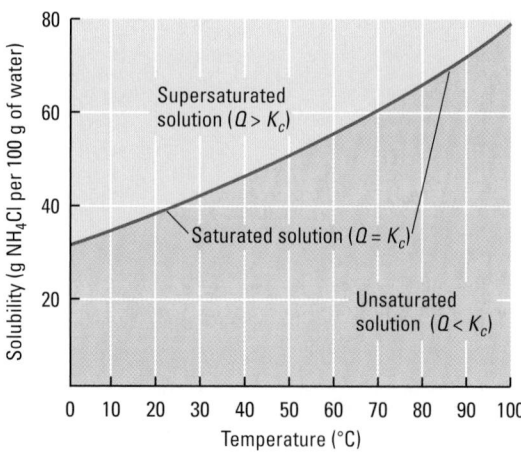

Figure 15.4 Types of solutions. The solubility of solid ammonium chloride and three types of solutions — saturated (the curve), unsaturated (below the curve), and supersaturated (above the curve).

rated, or supersaturated depending on the quantity of solute that is dissolved in it. A **saturated solution,** as its name implies, is one whose solute concentration equals its solubility (all points *along* the curve in Figure 15.4). When a saturated solution forms, there is a *dynamic equilibrium* between undissolved and dissolved solute.

$$\text{Solute} + \text{solvent} \rightleftharpoons \text{solution}$$

Some solute molecules or ions are going into solution, while others are separating from solvent molecules and entering the pure solute phase. Both processes are going on simultaneously, at identical rates. If the solute is a solid, we observe the presence of a solid associated with a saturated solution. A saturated solution is in equilibrium with its solute and solvent, and in this case $Q = K_c$ (⇐ *p. 650*).

An **unsaturated solution** is one in which the solute concentration is less than its solubility; that is, an unsaturated solution can accommodate additional solute at a given temperature without increasing the quantity of solvent (area *under* the curve in Figure 15.4). Thus, for an unsaturated solution, $Q < K_c$, and more solute dissolves. If solute continues to be added to an unsaturated solution at a given temperature, a point is reached where the solution becomes saturated, at which point $Q = K_c$.

For some solutes, there is a third case. It is possible to prepare solutions that contain *more* than the equilibrium concentration of solute at a given temperature; a solution like this is **supersaturated** (the area *above* the curve in Figure 15.4). For example, a supersaturated solution of ammonium chloride can be made by first making a saturated solution at 90 °C. This is done by adding 72 g of solid NH_4Cl to 100 g of water and heating the solution to 90 °C. At that point the solution is saturated. To make the supersaturated solution, the temperature of the saturated solution is lowered very slowly to 25 °C. If this is done carefully, none of the NH_4Cl will crystallize out of solution, and the resulting solution will be supersaturated, holding more dissolved NH_4Cl than a saturated solution at that temperature; 72 g of the solute is dissolved at 25 °C, whereas the solution should hold a maximum of 40 g at that temperature. In equilibrium terms, $Q > K_c$, and excess solute should crystallize out of solution, but it is very slow to do so. Some solutions, such as honey, can remain supersaturated for days or months. To form a crystal, several ions or molecules must be arranged very near the appropriate crystal lattice positions, and it can take a long time for such an alignment to occur by chance. However, precipitation of a solid from a supersaturated solution occurs rapidly if a tiny crystal of the solute is added to the solution (Figure 15.5). The lattice of the added crystal provides a template onto which more ions or molecules

Unsaturated solution: $Q < K_c$; saturated solution: $Q = K_c$; supersaturated solution: $Q > K_c$

Fudge is made using a supersaturated sugar solution. In smooth fudge, the sugar remains uncrystallized; poor-quality fudge has a gritty texture because it contains crystallized excess sugar.

(a)

(b)

(c)

(d)

(e)

Figure 15.5   A supersaturated solution. Sodium acetate ($NaCH_3COO$) easily forms supersaturated solutions in water. (a) The solution on the left looks ordinary, but it is supersaturated, holding more dissolved sodium acetate than a saturated solution at that temperature. The supersaturated solution was prepared by dissolving a quantity of sodium acetate in water at a much higher temperature and cooling the solution very slowly. (b) Upon adding a tiny seed crystal of sodium acetate, some of the excess dissolved sodium acetate immediately begins to crystallize. (c through e) Very soon, numerous sodium acetate crystals can be seen. If the solution remains uncovered for a long time, all of the water will evaporate, leaving behind the nonvolatile sodium acetate.

can be added. Sometimes other actions, such as stirring a supersaturated solution or scratching the inner walls of its container, will cause solute to precipitate rapidly.

Exercise 15.3 Crystallizing Out of Solution

Using Figure 15.4, how many grams of excess NH_4Cl would crystallize out of the supersaturated NH_4Cl solution at 25 °C?

 ### Exercise 15.4 Solubility

Refer to Figure 15.4 to determine whether each of these NH_4Cl solutions is unsaturated, saturated, or supersaturated:

(a) 30 g of NH_4Cl at 70 °C   (b) 60 g of solute at 60 °C   (c) 50 g of NH_4Cl at 50 °C

Dissolving Ionic Solids in Liquids

 CD-ROM Screen 14.4: Energetics of Solution Formation

Electron density of a water molecule. The red area has high electron density and partial negative charge; the blue area has low electron density and partial positive charge.

Sodium chloride is an ionic compound. Its crystal lattice consists of Na^+ and Cl^- ions in a cubic array (⬅ *p. 486*). Strong attractions between oppositely charged ions hold the ions tightly in the lattice. The enthalpy change when 1 mol of Na^+ and Cl^- ions is completely separated from a crystal lattice is referred to as the **lattice energy** of the ionic compound. The large lattice energy of NaCl (788 kJ/mol) accounts for sodium chloride's high melting point (800 °C). It is possible for solvent molecules to attract Na^+ away from Cl^- ions in a crystal, but they have to be the right kind of solvent molecules. Trying to dissolve NaCl (or any other ionic compound) with carbon tetrachloride or hexane (both nonpolar solvents) is a futile exercise because nonpolar molecules have very little attraction for ions. On the other hand, water dissolves NaCl.

Water is a good solvent for an ionic compound because water molecules are small and highly polar (⬅ *p. 382*). As shown in Figure 15.6, the partially negative

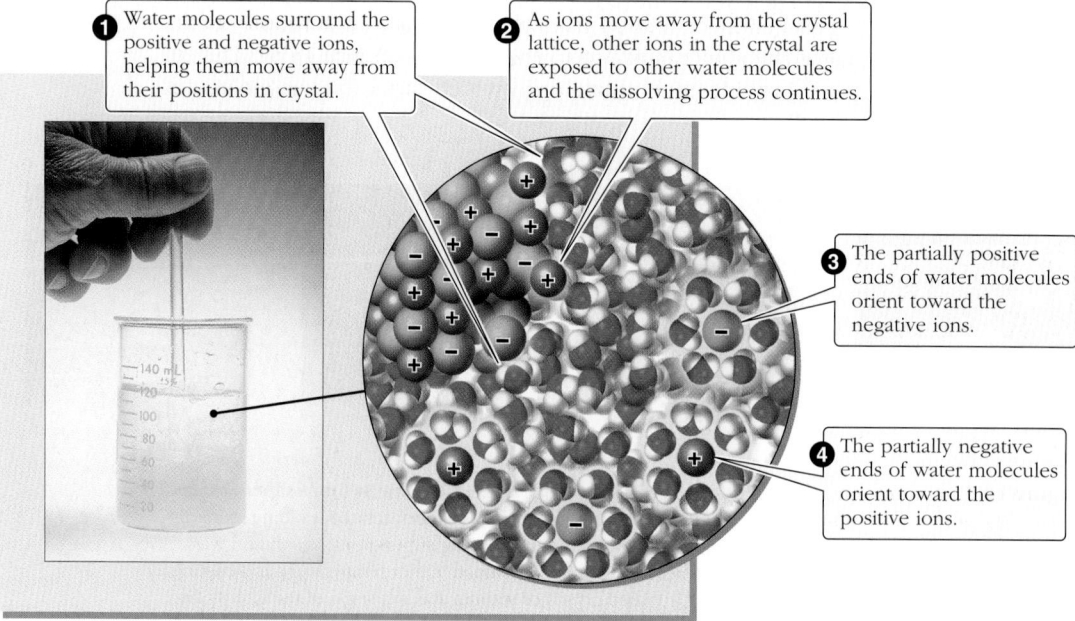

1 Water molecules surround the positive and negative ions, helping them move away from their positions in crystal.

2 As ions move away from the crystal lattice, other ions in the crystal are exposed to other water molecules and the dissolving process continues.

3 The partially positive ends of water molecules orient toward the negative ions.

4 The partially negative ends of water molecules orient toward the positive ions.

Figure 15.6 Water dissolving an ionic solid. Water molecules surround the positive (gray) and negative (green) ions, helping them move away from their positions in the crystal.

Small endothermic ΔH_solution

Na⁺(g) + Cl⁻(g)

Lattice
energy
(positive)   ΔH_hydration

Na⁺(aq) + Cl⁻(aq)

NaCl(s)   ΔH_solution

Enthalpy, H

Large exothermic ΔH_solution

Na⁺(g) + OH⁻(g)

Lattice
energy

NaOH(s)   ΔH_hydration

ΔH_solution

Na⁺(aq) + OH⁻(aq)

Enthalpy, H

Large endothermic ΔH_solution

NH₄⁺(g) + NO₃⁻(g)

Lattice
energy   ΔH_hydration

NH₄⁺(aq) + NO₃⁻(aq)

NH₄NO₃(s)   ΔH_solution

Enthalpy, H

Figure 15.7 **Heats of solution of three different ionic compounds dissolving in equal volumes of water.** (*Left*) For NaCl, the lattice energy that must be overcome is larger than the energy released on hydration of the ions. This causes the ΔH_{soln} of NaCl to be positive (endothermic). (*Center*) The lattice energy for NaOH is much smaller than the energy released when the ions become hydrated. Consequently, the ΔH_{soln} of NaOH is negative (exothermic). (*Right*) For NH_4NO_3, the heat of hydration of the ions is much smaller than the lattice energy, so the resulting ΔH_{soln} has a large positive value (highly endothermic).

oxygen atoms of water molecules are attracted to positive ions and help pull them away from the crystal lattice, while the partially positive hydrogen atoms of other water molecules are attracted to the negative ions in the lattice and help pull them away from the lattice. This process, in which water molecules surround positive and negative ions, is called **hydration.** Energy known as the *enthalpy of hydration* is released when these new attractions form between the ions and the water molecules mix and become close to one another. Energy is always required to separate the ions to overcome their attraction, and energy is always released when ions become hydrated because bonds are being formed. Whether dissolving a particular ionic compound is exothermic (ΔH_{soln} is negative) or endothermic (ΔH_{soln} is positive) depends on the relative sizes of the lattice energy of the ionic compound and the hydration enthalpies of its positive and negative ions. The relationship between the enthalpy of solution, the lattice energy of the ionic compound, and the enthalpies of hydration of the ions is

$$\Delta H_{soln} = \text{lattice energy} + \Delta H_{hydration}(\text{cations}) + \Delta H_{hydration}(\text{anions})$$

Figure 15.7 shows how lattice energy and the enthalpies of hydration combine to give the enthalpy of solution.

Practical applications of endothermic and exothermic dissolution include cold packs containing NH_4NO_3 used to treat athletic injuries and hot packs containing $CaCl_2$ used to warm foods (p. 692).

The solubility rules for ionic compounds in water (⬅ *p. 164*) remind us that not all ionic compounds are highly water-soluble in spite of the strong attractions between water molecules and ions. For some ionic compounds, the lattice energy is so large that water molecules cannot effectively pull ions away from the lattice. As a result, such compounds have large positive enthalpies of solution and usually are only slightly soluble.

Entropy and the Dissolving of Ionic Compounds in Water

The disorder introduced when a crystal lattice breaks down and the disorder introduced by the mixing of ions with solvent molecules both favor the dissolving process ($\Delta S > 0$). This entropy increase is counteracted by the ordering of solvent molecules around the ions ($\Delta S < 0$) (Figure 15.8). For 1+ and 1− charged ions, the overall entropy change is positive, and dissolving is favored. For some salts that

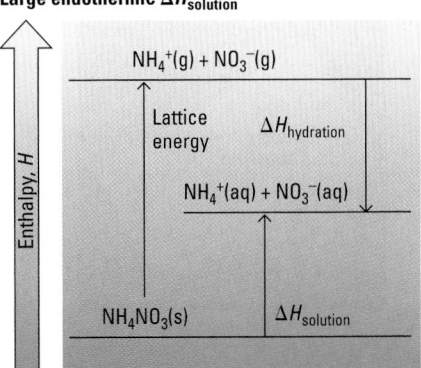

Enthalpies of Hydration of Selected Ions (kJ/mol)

Cations		Anions	
H^+	−1130	F^-	−483
Li	−558	Cl^-	−340
Na^+	−444	Br^-	−309
Mg^{2+}	−2003		
Ca^{2+}	−1557		
Al^{3+}	−2537		

Energy is always released when particles that attract each other get closer together; energy is always required to separate such particles.

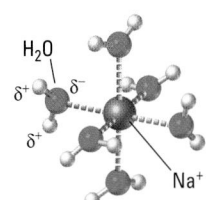

Figure 15.8 **Hydration of a sodium ion.** The arrangement of water molecules around this Na^+ ion is highly ordered.

Commercial cold and hot packs.
(*Top*) When this cold pack is used, ammonium nitrate in the inner container is brought into contact with water in the outer container when the desired cooling is needed by breaking the seal of the inner container. (*Bottom*) This hot pack releases heat when the inner pouch containing either $CaCl_2$ or $MgSO_4$ is punctured and either of these compounds dissolves in the water in the outer container.

 CD-ROM Screen 14.6: Factors Affecting Solubility (2): Temperature and Le Chatelier's Principle

contain $2+$ or $3+$ ions, the charges on the ions are so large and the ions so small that water molecules are aligned in a highly organized manner around the ions. When a large number of water molecules are locked into place by this strong hydration, the entropy of solution may be negative, which does not favor solubility. Calcium oxide (only 0.131 g of CaO dissolves in 100 mL of water at 10 °C) and aluminum oxide Al_2O_3 (insoluble) exemplify this effect.

15.4 TEMPERATURE AND SOLUBILITY

Solubility of Gases To understand how temperature affects gas solubility, we can apply Le Chatelier's principle to the dissolution equilibrium between a pure solute gas and a saturated solution of it.

$$\text{Gas + solvent} \rightleftharpoons \text{saturated solution} \qquad \text{Usually, } \Delta H_{soln} < 0 \text{ (exothermic)}$$

When a gas dissolves to form a saturated liquid solution, the process is almost always exothermic. Gas molecules that were relatively far apart are brought much closer in the solution to other molecules that attract them, lowering the potential energy and releasing some energy to the surroundings.

If the temperature of a solution of a gas in a liquid increases, the equilibrium shifts in the direction that partially counteracts the temperature rise. That is, the equilibrium shifts to the left in the preceding equation. Thus, a dissolved gas becomes less soluble with increasing temperature. Conversely, cooling a solution of a gas that is at equilibrium with undissolved gas will cause the equilibrium to shift to the right in the direction that liberates heat, and so more gas dissolves. This is illustrated in Figure 15.9 with data for the solubility of oxygen in water.

Cooler water in contact with the atmosphere contains more dissolved oxygen at equilibrium than water at a higher temperature. For this reason fish seek out cooler (usually deeper) waters in the summer. Fish have an easier time obtaining oxygen when its concentration in the water is higher. The decrease in gas solubility as temperature increases makes *thermal pollution* a problem for aquatic life in rivers and streams. Natural heating of water by sunlight and by warmer air can usually be accommodated. But excess heat from extended periods of very hot weather or from sources such as industrial facilities and electrical power plants can reduce the concentration of dissolved oxygen to the point where some species of fish die.

A fish kill caused by a lack of dissolved oxygen.

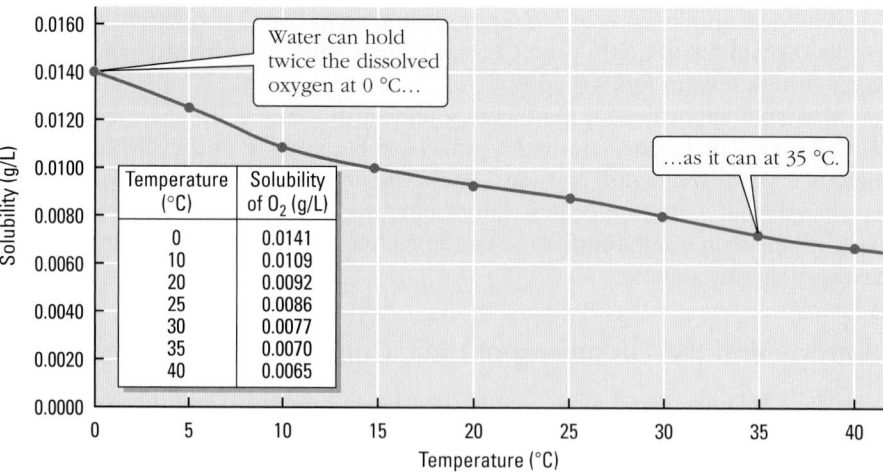

Figure 15.9 Solubility of oxygen in water at various temperatures. The solubility of oxygen, like that of other gases, decreases with increasing temperature.

Exercise 15.5 Carbonated Beverages

Explain on a molecular basis why a carbonated beverage goes "flat" once it is opened and warms to room temperature.

 Exercise 15.6 Warming Water

Explain why water that has been used to cool a nuclear power plant—and thus at a relatively high temperature—must be cooled before it is put back into the lake or river from which it came.

Solubility of Solids Common experience tells us that more sugar (sucrose) can dissolve in hot coffee or hot tea than in cooler coffee or tea. This is an example of the fact that the aqueous solubility of most solid solutes, including ionic compounds, increases with increasing temperature (Figure 15.10). There are exceptions, such as Li_2SO_4. Although predictions of solubility based on enthalpies of solution usually work, there are notable exceptions.

 Exercise 15.7 Temperature and Solubility

If a substance has a positive heat of solution, which would likely cause more of it to dissolve, hot solvent or cold solvent? Explain.

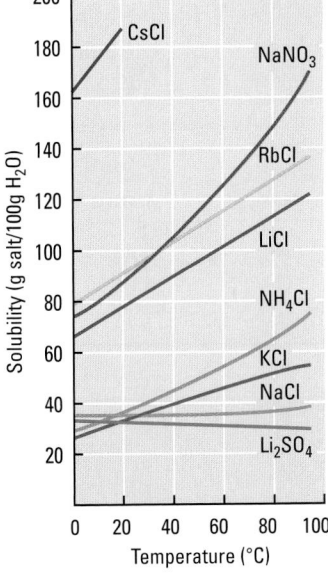

Figure 15.10 Solubility of ionic compounds and temperature. The solubility of ionic compounds in water depends on the temperature.

Although generally useful, Le Chatelier's principle does not always correctly predict how the solubility of ionic solutes changes with temperature.

15.5 PRESSURE AND DISSOLVING GASES IN LIQUIDS: HENRY'S LAW

Although pressure does not measurably affect the solubilities of solids or liquids in liquid solvents, *the solubility of any gas in a liquid increases as the pressure of the gas increases* (Figure 15.11). A dynamic equilibrium is established when a gas is in contact with a liquid—the rate at which gas molecules enter the liquid phase equals the rate at which gas molecules escape from the liquid. If the pressure is increased, gas molecules strike the surface of the liquid more often, increasing the rate of dissolution of the gas. A new equilibrium is established when the rate of escape increases to match the rate of dissolution. The rate of escape is first order (⬅ *p. 583*) in concentration of solute, so a higher rate of gas escape requires a higher concentration of solute gas molecules, that is, a higher gas solubility.

The relationship between gas pressure and solubility is known as **Henry's law:**

$$S_g = k_H P_g$$

where S_g is the solubility of the gas in the liquid, P_g is the pressure of the gas above the solution (or the partial pressure of the gas if the solution is in contact with a mixture of gases). The value of the constant, k_H, known as the Henry's law constant, depends on the identities of both the solute and the solvent and on the temperature (Table 15.3). It has units of moles per liter per millimeter of mercury (mm Hg).

Figure 15.12 illustrates how gas solubility depends on pressure. The behavior of a carbonated drink when the cap is opened is an everyday illustration of the solubility of gases in liquids under pressure. The drink fizzes when opened because the partial pressure of CO_2 over the solution drops, the solubility of the gas decreases, and dissolved gas escapes from the solution.

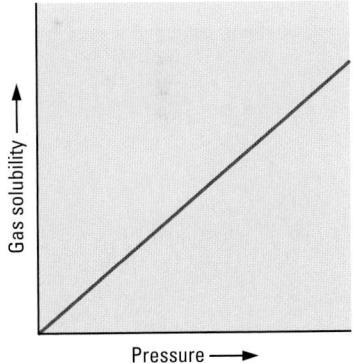

Figure 15.11 Pressure dependence of the solubility of a gas in a liquid. All gases that do not react with the solvent behave this way.

 CD-ROM Screen 14.5: Factors Affecting Solubility (1): Henry's Law and Gas Pressure

The partial pressure of a gas in a mixture of gases is the pressure that a pure sample of the gas would exert if it occupied the same volume as the mixture. Partial pressure is proportional to the mole fraction of the gas (⬅ *p. 433*).

TABLE 15.3	Henry's Law Constants (25 °C)
Gas	k_H $\left(\dfrac{mol/L}{mm\ Hg}\right)$
N_2	8.42×10^{-7}
O_2	1.66×10^{-6}
CO_2	4.45×10^{-5}

Henry's law. The greater the partial pressure of CO_2 over the soft drink in the bottle, the greater the concentration of dissolved CO_2. When the bottle is opened, the partial pressure of CO_2 drops and CO_2 bubbles out of the solution.

Gas solubility and scuba diving. When using scuba gear, a diver must be knowledgeable about the solubility of blood gases. Because N_2 and O_2 are soluble in blood, breathing pressurized air from a scuba tank means a high N_2 concentration in the blood. Ascending too rapidly causes dissolved N_2 to be released, forming bubbles in the blood, causing a painful condition known as "the bends." If the released N_2 bubbles block capillaries to the brain, the bends can be fatal.

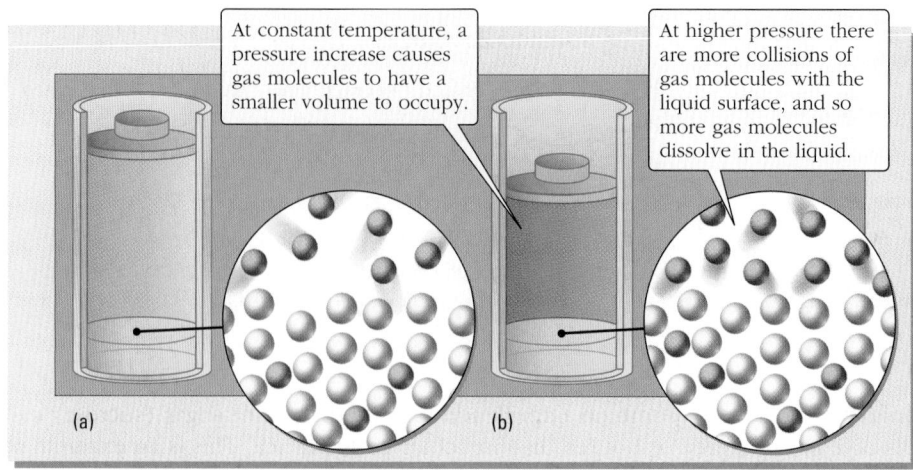

At constant temperature, a pressure increase causes gas molecules to have a smaller volume to occupy.

At higher pressure there are more collisions of gas molecules with the liquid surface, and so more gas molecules dissolve in the liquid.

(a)　　　　　(b)

Figure 15.12 Henry's law.

Problem-Solving Example 15.3 Using Henry's Law

The Henry's law constant for oxygen in water at 25 °C is 1.66×10^{-6} mol L^{-1} mm Hg^{-1}. Suppose that a trout stream is in equilibrium with air at normal atmospheric pressure. What is the concentration of O_2 in this stream? Express the result in grams per liter (g/L). The mole percent of oxygen in air is 21%.

Answer 0.0086 g/L or 8.6 mg/L

Explanation The solubility of oxygen can be calculated from Henry's law, but first we must calculate the partial pressure of oxygen in air, which is 21 mol percent oxygen. This means that the mole fraction of O_2 is 0.21. If the total pressure is 1.0 atm,

$$\text{Pressure of } O_2 = (1.0\ \text{atm})\left(\frac{760\ \text{mm Hg}}{1\ \text{atm}}\right)(0.21) = 160.\ \text{mm Hg}$$

Using Henry's law and the value of k_H given above,

$$S_g = k_H P_g = \left(1.66 \times 10^{-6}\ \frac{\text{mol/L}}{\text{mm Hg}}\right)(160.\ \text{mm Hg}) = 2.66 \times 10^{-4}\ \text{mol/L}$$

This concentration can be converted from molarity (mol/L) to the desired units by using molar mass.

$$(2.66 \times 10^{-4}\ \text{mol } O_2/L)\left(\frac{32.00\ \text{g } O_2}{\text{mol } O_2}\right) = 0.0085\ \text{g/L or 8.5 mg/L}$$

Although this oxygen concentration is very low, it is sufficient to provide the oxygen required by aquatic life.

Problem-Solving Practice 15.3

The Henry's law constant for N_2 in water at 25 °C is 8.4×10^{-7} mol L^{-1} mm Hg^{-1}. What is the solubility of N_2 in mol/L if its partial pressure is 1520 mm Hg? What is the solubility when the N_2 partial pressure is 20. mm Hg?

15.6 SOLUTION CONCENTRATION: KEEPING TRACK OF UNITS

The terms unsaturated, saturated, and supersaturated describe a solution with respect to the quantity of solute in a given quantity of solvent at a certain temperature. Sometimes the terms *concentrated* or *dilute* are also used to describe a solution. Although useful, these are all broad descriptors of solution concentration,

Chemistry in the News

BUBBLING AWAY: DELICATE AND STOUT

In the movie *Gigi,* Maurice Chevalier sings of "the night they invented champagne," that fizzy, celebratory beverage. As bubbles rise in a glass of champagne, their fizz and pop capture our attention. Recently, researchers in France and Australia have investigated more closely the bubbles in champagne and in beer (the body of the liquid, not the foamy "head" on beer). In both drinks the bubbles are carbon dioxide, produced as a byproduct of the natural process that produces alcohol via fermentation of sugars in grapes (wine, champagne) or grains (beer).

$$C_6H_{12}O_6 \longrightarrow 2\ CH_3CH_2OH + 2\ CO_2$$
glucose   ethanol

The properties of the bubbles are affected by the amount of dissolved gas, the alcohol content of the drink, and the concentration of large molecules, including proteins, present in the drink. Champagne has three times the CO_2 concentration and twice the alcohol content of beer. Beer has as much as 30 times more large molecules than in champagne.

Bubbles in beverages generally start to grow from what are called *nucleation sites* — places where the gas comes out of solution — such as on dust particles or on

 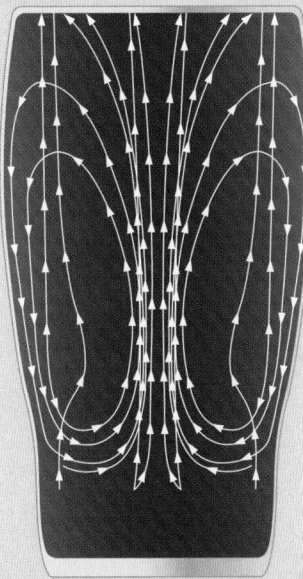

(a) Glass of Guiness Stout. (b) Computer simulation of bubbles traveling in Guinness stout.

tiny imperfections on the inside walls of a glass. When a bubble gets sufficiently buoyant, it breaks away from the nucleation site. As CO_2 diffuses into the bubble, it gets larger, its buoyancy increases, and it moves more quickly to the surface. Gerard Liger-Belair and his colleagues at the University of Reims in France have used high-speed cameras to measure the growth and ascent rate of champagne bubbles, whose radii grow at the rate of 120 to 240 μm per second as they ascend from the sides of a champagne flute.

Whereas bubbles in champagne always rise, those in the Irish beer Guinness stout rise and fall within the body of the beer, seemingly defying the laws of nature. For years, pub regulars argued whether the bubbles in stout actually were falling as well as rising. Was it perhaps an optical illusion or an alcohol-induced error of the observer? The higher protein concentration in beer causes more "drag" on the bubbles, slowing their rate of ascent much more than in champagne. At the higher protein concentration, the bubbles in beer act as rigid spheres because they collect a protein "skin" as they rise. The lower protein content in champagne does not

create bubbles that are as rigid. Champagne bubbles would have to travel more than 10 cm (about 4 in) to pick up enough protein to become as rigid as beer bubbles. This is longer than a typical champagne flute. That explains differences for rising bubbles. But do the bubbles in stout actually descend? Using computer modeling software, Australian researchers Clive Fletcher and colleagues demonstrated that, unlike those in champagne, bubbles in Guinness stout ascend *and* descend. The critical issue is the rigidity, size, and buoyancy of the bubbles. Beer bubbles with a radius of about 30 μm sink near the sides of the glass while 500-μm ones only rise. The barrel-chested shape of the typical Guinness glass also seems to be a factor.

Richard Zare, a Stanford University chemistry professor who also has published articles about beer bubbles, points out that "Once you begin to learn about the nature of beer bubbles, you will never again look at a glass of beer in quite the same way."

Source:

Weiss, P. "The Physics of Fizz." *Science News,* Vol. 157, May 6, 2000, p. 300.

Rising bubbles in champagne.

but other ways are needed to specify more precisely the concentration of a solute in a solution. Several concentration units are used to do so, including mass fraction, weight percent, molarity, and molality. Often it is useful to be able to quantitatively describe quantities of solute across a wide range of concentrations, from rather large to very small. For example, a variety of units are used to express the concentrations of dilute solutions containing unwanted—even potentially harmful—ionic solutes, such as lead, mercury, selenium, and nitrate, and organic compounds in drinking water. These units are also useful in discussing solute concentrations in other kinds of solutions, such as blood or urine.

Molality is discussed in Section 15.7.

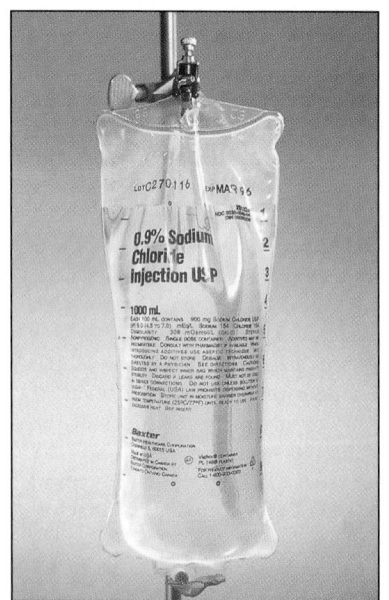

CD-ROM Screen 14.2: Solubility

Mass Fraction and Weight Percent

The **mass fraction** of a solute is the fraction of the total mass of the solution that a solute contributes, that is, the mass of a single solute divided by the total mass of all the solutes and the solvent. Mass fraction is commonly expressed as a percentage and called **weight percent,** which is the mass fraction multiplied by 100%. This is the same as the number of grams of solute per 100 g of solution. For example, the mass fraction of sucrose in a solution consisting of 25.0 g sucrose, 10.0 g fructose, and 300. g of water is

$$\text{Mass fraction of sucrose} = \frac{\text{mass of sucrose}}{\text{mass of sucrose} + \text{mass of fructose} + \text{mass of water}}$$

$$= \frac{25.0 \text{ g}}{25.0 \text{ g} + 10.0 \text{ g} + 300. \text{ g}} = 0.0746$$

Weight percent can be thought of as parts per hundred.

The weight percent of sucrose in the solution is $(0.0746) \times 100\% = 7.46\%$

Sterile saline solution. Saline solutions like this one are routinely given to patients who have lost body fluids.

The density of liquid water is essentially 1 g/mL, 10^3 g/L, or 1 kg/L.

Problem-Solving Example 15.4 Mass Fraction and Weight Percent

Sterile saline solutions containing NaCl in water are often used in medicine. What is the weight percent of NaCl in a solution made by dissolving 4.6 g of NaCl in 500. g of pure water?

Answer 0.91%

Explanation Using the definitions of mass fraction and weight percent, we have

$$\text{Mass fraction of NaCl} = \frac{4.6 \text{ g NaCl}}{4.6 \text{ g NaCl} + 500. \text{ g H}_2\text{O}} = 0.0091$$

Weight percent of NaCl = mass fraction of NaCl × 100% = 0.0091 × 100% = 0.91%

✔ The mass fraction is about 5 g of NaCl in about 500 g of solution, or about 0.01, which agrees with the accurate result.

Problem-Solving Practice 15.4

What is the weight percent of glucose in a solution containing 21.5 g of glucose ($C_6H_{12}O_6$) in 750. g of pure water?

 ### Exercise 15.8 Mass Fraction and Weight Percent

Ringer's solution is used in physiology experiments. One liter of the solution contains 6.5 g of NaCl, 0.20 g of $NaHCO_3$, 0.10 g of $CaCl_2$, and 0.10 g of KCl. Assume that one liter of water has been used to prepare the solution and that the density of water is 1.00 g/mL.

(a) Calculate the mass fraction and weight percent of $NaHCO_3$ in the solution.
(b) Which solute has the lowest weight percent in the solution?

Parts per Million, Billion, and Trillion

Solutes in very dilute solutions have very low mass fractions. Consequently, the mass fraction in such solutions is often expressed in **parts per million** (abbreviated ppm). One part per million is equivalent to one gram of solute per one million grams of solution, or proportionally, one milligram of solute per thousand grams of solution (1 mg/kg). Thus, a commercial bottled water with a calcium ion concentration of 66 ppm contains 66 mg of Ca^{2+} per thousand grams of water, essentially one liter. For even smaller mass fractions, **parts per billion** (1 ppb = one microgram, μg, of solute per thousand grams of solution) and **parts per trillion** (1 ppt = one nanogram, ng, of solute in one thousand grams of solution) are often used. As the names imply, a mass fraction converts to parts per billion by multiplying by 10^9 ppb and to parts per trillion by multiplying by 10^{12} ppt.

1 μg (microgram) = 10^{-6} g;
1 ng (nanogram) = 10^{-9} g.

To express the mass fraction as parts per million, multiply it by 10^6 ppm. This makes a very small number bigger and easier to handle.

Problem-Solving Example 15.5   Ppm, Ppb, and Mass Fraction

A sample of water is found to contain 0.010 ppm lead (Pb^{2+}).

(a) What is the mass of lead ion per liter of this solution? (Assume the density of the water solution is 1.0 g/mL.)
(b) What is the lead concentration in ppb?

Answer   (a) 10. μg/L   (b) 10. ppb

Explanation   (a) Since the solution is almost entirely water, its density will be the same as that of water, so 1 L of solution has a mass of 1.0×10^3 g. The mass of lead in this 1-L water sample is calculated from its mass fraction, expressed as the ratio of grams of lead per 10^6 g of solution.

$$\left(\frac{0.010 \text{ g Pb}^{2+}}{1 \times 10^6 \text{ g solution}} \right) \left(\frac{1.0 \times 10^3 \text{ g solution}}{\text{L solution}} \right) = 1.0 \times 10^{-5} \text{ g Pb}^{2+}/\text{L solution}$$

(b) This answer can be converted to ppb by first converting it to micrograms of Pb^{2+} per liter.

$$\left(\frac{1.0 \times 10^{-5} \text{ g Pb}^{2+}}{1 \text{ L}} \right) \left(\frac{1 \, \mu g}{1 \times 10^{-6} \text{ g}} \right) = 10. \, \mu g \, Pb^{2+}/L$$

Thus, a mass fraction of 1.0×10^{-2} g Pb^{2+} per 10^6 g of solution (0.010 ppm) corresponds to a concentration of 10. μg Pb^{2+}/L, which is the equivalent of 10. ppb.

1 ppm is equivalent to one penny in $10,000; 1 ppb is one penny in $10,000,000.

Problem-Solving Practice 15.5

Drinking water often contains small concentrations of selenium (Se). If a sample of water contains 30 ppb Se, how many micrograms of Se are present in 100. mL of this water?

 ### Exercise 15.9   Lead in Drinking Water

One drinking-water sample has a lead concentration of 20 ppb; another has a concentration of 0.003 ppm.

(a) Which sample has the higher lead concentration?
(b) The current EPA acceptable limit for lead in drinking water is 0.015 mg/L. Compare each of the water samples' lead concentration with the acceptable limit.

At the height of the Roman Empire, worldwide lead production was about 80,000 tons per year. Today it is about 3 million tons annually. Lead was first used for water pipes in ancient Rome. The Latin name for lead, *plumbum*, gave us the name "plumber."

Exercise 15.10   Bottled Water as a Magnesium Source

A 500-mL bottle of Evian bottled water provides 12 mg of magnesium. The recommended daily allowance of magnesium for adult women is 280 mg/day. How many 1-L bottles of Evian would a woman have to drink to obtain her total daily allowance of magnesium solely in this way?

Exercise 15.11   Striking It Rich in the Oceans?

The concentration of gold in sea water is about 1×10^{-3} ppm. The earth's oceans contain 3.5×10^{20} gal of sea water. Approximately how many pounds of gold are in the oceans? 1 gal = 3.785 L; 1 lb = 454 g.

Molarity

As defined in Section 5.6, the *molarity* of a solution is

$$\text{Molarity} = M = \frac{\text{moles of solute}}{\text{liters of solution}}$$

Multiplying the volume of a solution by its molarity yields the number of moles of solute in that volume of solution. For example, the number of moles of KNO_3 in 250. mL of 0.0200 M KNO_3 is

$$0.250 \text{ L} \times \left(\frac{0.0200 \text{ mol } KNO_3}{1 \text{ L}} \right) = 5.00 \times 10^{-3} \text{ mol } KNO_3$$

from which the number of grams can be determined.

$$5.00 \times 10^{-3} \text{ mol } KNO_3 \times \frac{101.1 \text{ g } KNO_3}{1 \text{ mol } KNO_3} = 0.506 \text{ g } KNO_3$$

Thus, to make up 250. mL of 0.0200 M KNO_3 solution, you would weigh 0.506 g of KNO_3, put it into a suitable container, and add to it sufficient water to bring the volume of the solution to 250. mL *(⇐ p. 191)*.

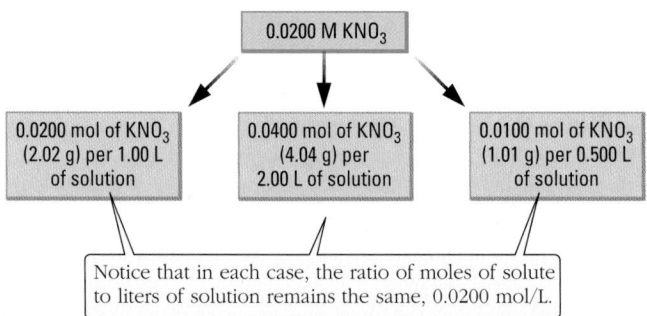

A 0.0200 M KNO_3 solution contains:

0.0200 M KNO_3

0.0200 mol of KNO_3 (2.02 g) per 1.00 L of solution

0.0400 mol of KNO_3 (4.04 g) per 2.00 L of solution

0.0100 mol of KNO_3 (1.01 g) per 0.500 L of solution

Notice that in each case, the ratio of moles of solute to liters of solution remains the same, 0.0200 mol/L.

Table 15.4 summarizes the types of concentration units we have used so far, comparing their units, advantages, and disadvantages.

Problem-Solving Example 15.6   Molarity, Mole Fraction, and Ppm

Chloride ion, at 0.550 M, is the most abundant anion in sea water. Express this concentration as (a) mass fraction and (b) ppm. Assume that the density of sea water is 1.03 g/mL.

Answer   (a) 0.019   (b) 19,000 ppm

Explanation

(a) Molarity can be used to find the grams of chloride ion per liter, which can then be translated into the mass fraction.

$$\frac{0.550 \text{ mol } Cl^-}{1 \text{ L}} \times \frac{35.5 \text{ g } Cl^-}{1 \text{ mol } Cl^-} = 19.53 \text{ g } Cl^-/\text{L}$$

The density of sea water is 1.03 g/mL, which is equivalent to 1.03×10^3 g/L. The mass fraction of chloride in 1.03×10^3 g of sea water is

$$\frac{19.53 \text{ g Cl}^-}{1.03 \times 10^3 \text{ g solution}} = 0.0190$$

(b) The parts per million (ppm) of Cl^- is determined by multiplying the mass fraction by 10^6 ppm.

$$0.0190 \times 10^6 \text{ ppm} = 19,000 \text{ ppm}$$

Problem-Solving Practice 15.6

Sea water contains 10,600 ppm Na^+, making sodium the most abundant cation in the oceans. Calculate the mass fraction of sodium ion in sea water and its molarity.

Problem-Solving Example 15.7   Weight Percent and Molarity

Concentrated nitric acid is a 70.0% solution of nitric acid, HNO_3, in water. The density of the solution is 1.41 g/mL at 25 °C. What is the molarity of nitric acid in this solution?

Answer   15.7 M

Explanation   To calculate the molarity, the number of moles of HNO_3 and the volume of the solution in liters must be determined. The 70.0% nitric acid solution means that 100. g of solution contains 70.0 g of HNO_3. Therefore,

$$\text{Moles of } HNO_3 = \frac{70.0 \text{ g } HNO_3}{100. \text{ g solution}} \times \frac{1 \text{ mol } HNO_3}{63.01 \text{ g } HNO_3}$$

$$= 1.11 \times 10^{-2} \text{ mol } HNO_3 \text{ g/solution}$$

The volume of the solution can be calculated using its density.

$$\frac{1.00 \text{ mL solution}}{1.41. \text{ g solution}} \times \frac{1 \text{ L solution}}{10^3 \text{ mL solution}} = 7.09 \times 10^{-4} \frac{\text{L solution}}{\text{g solution}}$$

TABLE 15.4   Comparison of Concentration Terms

Concentration term	Units of concentration	Advantages	Disadvantages
Mass fraction	None	Independent of temperature; used in special applications	Density must be known to convert mass fraction to molarity
Weight percent	Percent	Independent of temperature; useful in wide range of applications	Density must be known to convert weight percent to molarity
Parts per million, parts per billion, parts per trillion	ppm, ppb, ppt	Temperature independent; widely used in environmental applications	Must be determined by very exacting analytical methods
Molarity	$\dfrac{\text{Moles solute}}{\text{liter solution}}$	Volume measurements are easy and the results readily used in stoichiometric calculations	Temperature dependent; density must be known to determine solvent mass

With these two values, the molarity can be calculated.

$$\text{Molarity} = \frac{1.11 \times 10^{-2} \text{ mol HNO}_3}{7.09 \times 10^{-4} \text{ L}} = 15.7 \text{ M}$$

Problem-Solving Practice 15.7

The density of a commercial 30.0% hydrogen peroxide (H_2O_2) solution is 1.11 g/mL at 25 °C. Calculate the molarity of hydrogen peroxide in this solution.

15.7 Vapor Pressures, Boiling Points, and Freezing Points of Solutions

Up to this point, we have discussed solutions in terms of the nature of the solute and the nature of the solvent. There are some properties of solutions that do not depend on the nature of the solute or solvent, but rather depend only on the *number* of dissolved solute particles — ions or molecules — per unit volume.

In liquid solutions, solute molecules or ions disrupt solvent-solvent noncovalent attractions, causing changes in solvent properties that depend on these attractions. For example, when solute is added to a solvent, the freezing point of the resulting solution is lower and its boiling point is higher than that of the pure solvent. How much the properties of the solution differ from those of the pure solvent depends only on the concentration of the solute particles. **Colligative properties** of solutions are those that *depend only on the concentration of solute particles* (ions or molecules) in the solution, regardless of what kinds of particles are present. We will consider four colligative properties: vapor pressure lowering, boiling point elevation, freezing point depression, and osmotic pressure. These are all quite common and important in the world around us.

 CD-ROM Screen 14.7: Colligative Properties (1): Vapor Pressure and Raoult's Law

Osmotic pressure is discussed in Section 15.8.

Vapor Pressure Lowering

In a closed container there is a dynamic equilibrium between a pure liquid and its vapor — the rate at which molecules escape the liquid phase equals the rate at which vapor phase molecules return to the liquid. This equilibrium gives rise to a vapor pressure that is dependent upon the temperature. But the vapor pressure of a pure liquid differs from that of the liquid when a solute has been dissolved in it. Compare a small portion of the liquid/vapor boundary for pure water with that for sea water (mainly an aqueous sodium chloride solution) as shown at the molecular scale in Figure 15.13. For an aqueous solution such as sea water, in which sodium ions and chloride ions (and many other kinds of ions and molecules) are present, the vapor pressure is lower than for a sample of pure water.

The vapor pressure of any pure solvent will be lowered by the addition of a nonvolatile solute to the solvent. How much a dissolved solute lowers the vapor pressure of a solvent depends on the solute concentration and is expressed by **Raoult's law,**

Raoult's law works best with dilute solutions. Deviations from Raoult's law occur when solute-solvent forces are either much weaker or much stronger than the solvent-solvent and solute-solute forces.

$$P_1 = X_1 P_1^0$$

where P_1 is the vapor pressure of the solvent over the *solution*, P_1^0 is the vapor pressure of the *pure* solvent at the same temperature, and X_1 is the mole fraction of *solvent* in the solution. For example, suppose you want to know the vapor pressure over a sucrose solution at 25 °C in which the mole fraction of water is 0.986. The vapor pressure of pure water at 25 °C is 23.76 mm Hg. From these data, the vapor pressure of water over the solution, 23.42 mm Hg, can be calculated using Raoult's law.

Mole fraction of A, X_A,

$$= \frac{\text{moles of } A}{\text{total number of moles}}$$

$$= \frac{\text{moles } A}{\text{moles } A + \text{moles } B + \cdots}$$

$$P_{\text{water}} = (X_{\text{water}})(P_{\text{water}}^0) = (0.986)(23.76 \text{ mm Hg}) = 23.42 \text{ mm Hg}$$

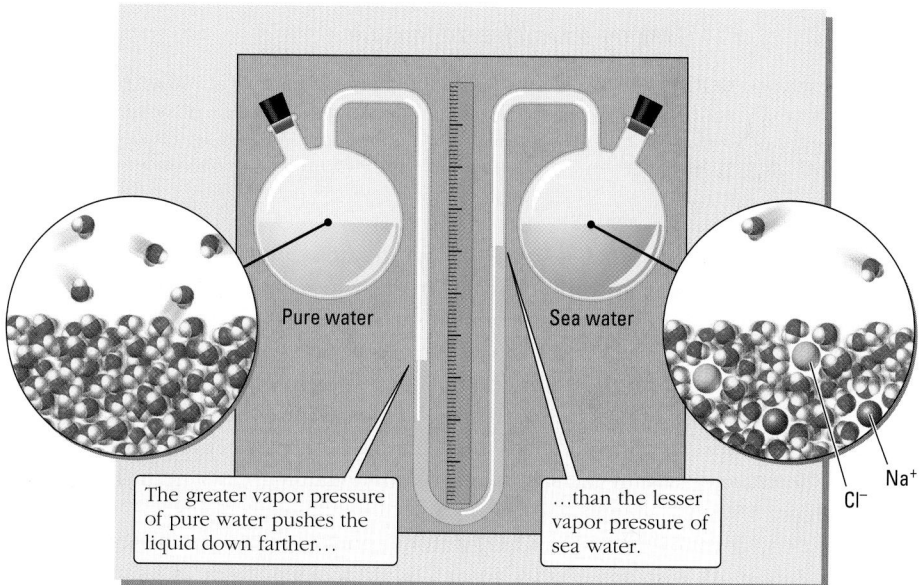

Figure 15.13 **The vapor pressure of pure water and that of sea water.** Sea water is an aqueous solution of NaCl and many other salts. The vapor pressure over an aqueous solution is not as great as that over pure water at the same temperature.

The greater vapor pressure of pure water pushes the liquid down farther...

...than the lesser vapor pressure of sea water.

Na^+

Cl^-

Pure water

Sea water

Therefore, the vapor pressure of water over the solution is only 98.6% that of pure water. The vapor pressure has been lowered by $(23.76 - 23.42)$ mm Hg = 0.34 mm Hg.

Raoult's law can also be applied to solutions in which the solvent and the solute are both volatile so that an appreciable amount of each can be in the vapor above the solution. We will not consider such cases.

Problem-Solving Example 15.8   Raoult's Law

Calculate the vapor pressure of water over a solution containing 50.0 g of sucrose, $C_{12}H_{22}O_{11}$, and 100.0 g of water at 45 °C. The vapor pressure of pure water at this temperature is 71.88 mm Hg.

Answer   70.0 mm Hg

Explanation   We first calculate the mole fraction of water and then use Raoult's law to determine the vapor pressure of water over the solution. For the mole fraction, the grams of sucrose and water must be converted to moles.

$$50.0 \text{ g sucrose} \times \left(\frac{1 \text{ mol sucrose}}{342 \text{ g sucrose}} \right) = 0.146 \text{ mol sucrose}$$

$$100.0 \text{ g water} \times \left(\frac{1 \text{ mol water}}{18.0 \text{ g water}} \right) = 5.56 \text{ mol water}$$

$$\frac{5.56}{5.56 + 0.146} = \frac{5.56}{5.706} = 0.974$$

Applying Raoult's law:

$$P_{water} = (X_{water})(P^0_{water}) = (0.974)(71.88 \text{ mm Hg}) = 70.0 \text{ mm Hg}.$$

The dissolved sucrose has lowered the vapor pressure of the pure water by nearly 2 mm Hg.

✔ Because there are nearly 40 times the number of moles of water than there are of sucrose, the mole fraction of water should be much greater than the mole fraction of sucrose, and it is—0.974 compared with $1 - 0.974 = 0.026$. Therefore, the vapor pressure of water over the solution should be about 97% that of pure water ($70/72 \times 100\%$), which it is. The answer is reasonable.

Problem-Solving Practice 15.8

The vapor pressure of an aqueous solution of urea, CH_4N_2O, is 291.2 mm Hg. The vapor pressure of water at that temperature is 355.1 mm Hg. Calculate the mole fraction of each component.

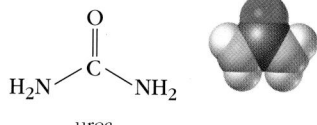

urea

ethylene glycol

HOCH₂CH₂OH

> ### Exercise 15.12   Making an Antifreeze Mixture
>
> Ethylene glycol, HOCH₂CH₂OH, is used as an antifreeze. What is the vapor pressure of water above a solution of 100.0 mL of ethylene glycol and 100.0 mL of water at 90 °C? Densities (g/mL): ethylene glycol, 1.15; water, 1.00. The vapor pressure of pure water at that temperature is 525.8 mm Hg.

The role of entropy (that is, of disorder) in vapor pressure lowering is illustrated by comparing the entropy change for the vaporization of pure water with that for the vaporization of a corresponding quantity of water from a sodium chloride solution. Because NaCl has a very low vapor pressure, very few sodium or chloride ions escape from the solution, and the vapor in equilibrium with the salt water consists almost entirely of water molecules. Thus, the entropy of a given amount of the vapor is approximately the same in both cases (pure water and salt water). The entropy within the salt solution, however, is greater than that in pure water (Figure 15.14).

Now consider what happens when water vaporizes from pure water and from the salt solution. As with any change from a liquid to a gas, there is a significant increase in entropy of the water vaporizing from either source (pure water or salt solution) (⇐ *p. 665*). But the entropy of vaporization is not the same in each case. As illustrated in Figure 15.14, the entropy of vaporization is larger for the water vapor from pure water than from the salt solution because the entropy of pure water was already smaller than that of the salt solution. The main point is that a bigger entropy increase corresponds to a more product-favored process. The result is that the vaporization from pure water creates a higher pressure of water vapor (that is, more water molecules per unit volume) in equilibrium with pure water than does the vaporization of water from a salt solution.

Boiling Point Elevation

CD-ROM Screen 14.8: Colligative Properties (2): Boiling Point and Freezing Point

As a result of vapor pressure lowering, the vapor pressure of an aqueous solution of a nonvolatile solute at 100 °C is less than 760 mm Hg (1 atm). For it to boil, the solution therefore must be heated *above* 100 °C. The **boiling point elevation,** ΔT_b, is the difference between the normal boiling point of water and the higher boiling point of an aqueous solution of a nonvolatile nonelectrolyte solute (Figure 15.15).

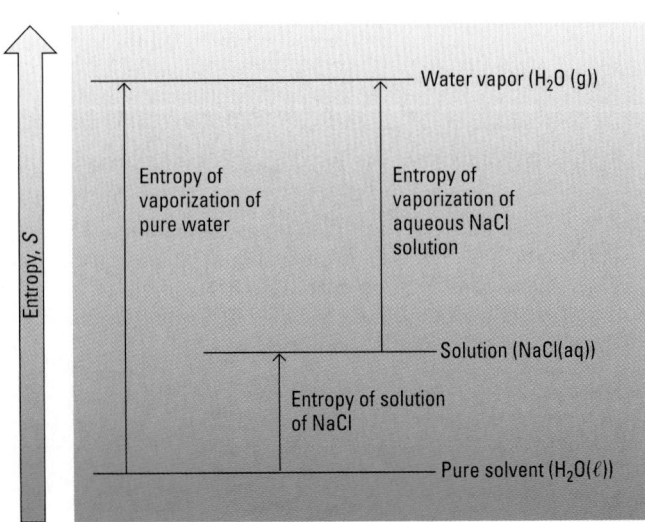

Figure 15.14 Vapor pressure lowering and entropies of solution and vaporization.   The entropy of vaporization of pure water is greater than the entropy of solution.

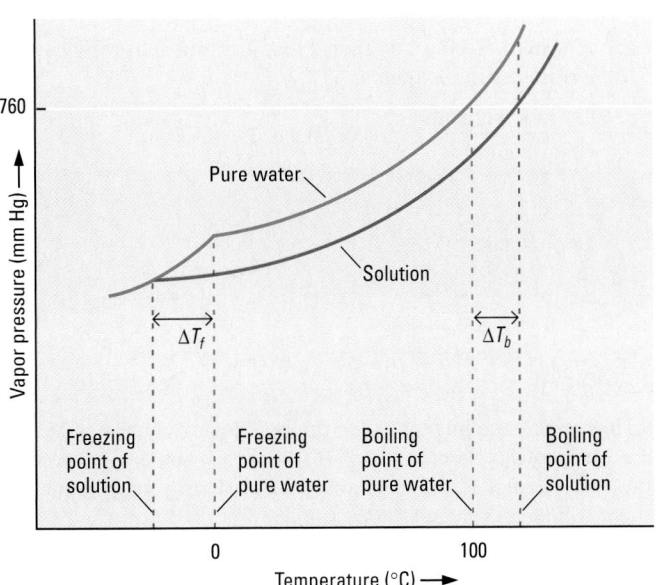

Figure 15.15 **Boiling point elevation (ΔT_b) and freezing point lowering (ΔT_f) for aqueous solutions.** Addition of solute to a pure solvent raises its boiling point and lowers its freezing point.

The increase in boiling point is proportional to the concentration of the solute. The solute concentration is expressed as **molality** (abbreviated *m*), defined as the moles of solute per kilogram of *solvent* (not solution).

$$\text{Molality of solute A} = m_A = \frac{\text{moles of solute A}}{\text{kilograms of solvent}}$$

For example, the molality of a solution made by dissolving 115.0 g of ethylene glycol, $HOCH_2CH_2OH$, in 500. mL of water is calculated by first determining the number of moles of solute, ethylene glycol, and the number of kilograms of solvent, water.

$$115.0 \text{ g} \times \left(\frac{1 \text{ mol ethylene glycol}}{62.07 \text{ g ethylene glycol}} \right) = 1.853 \text{ mol ethylene glycol}$$

The density of water is 1.00 g/mL, so 500. mL of water weighs 500 g, which is 0.500 kg. Substituting these values into the molality equation gives

$$\text{Molality} = \frac{1.853 \text{ mol ethylene glycol}}{0.500 \text{ kg water}} = 3.71 \text{ mol/kg}$$

Problem-Solving Example 15.9   Molality

(a) What is the molality of a solution prepared by dissolving 0.413 g of methanol (CH_3OH) in 1.50×10^3 g of water?
(b) What is the molarity of the solution?

Answer   (a) 0.00860 mol/kg   (b) 0.00860 mol/L

Explanation
(a) First, calculate the number of moles of the solute, methanol. Its molar mass is 32.042 g/mol.

$$0.413 \text{ g} \times \left(\frac{1 \text{ mol methanol}}{32.042 \text{ g methanol}} \right) = 0.0129 \text{ mol methanol}$$

The mass of solvent is 1.50 kg, so by the definition of molality,

$$\frac{0.0129 \text{ mol methanol}}{1.50 \text{ kg water}} = 0.00860 \text{ mol/kg}$$

Molarity and molality. The photo shows a 0.10 molal solution (0.10 mol/kg) of potassium chromate *(flask at right)* and a 0.10 molar solution (0.10 mol/L) of potassium chromate *(flask at left)*. Each solution contains 0.10 mol (19.4 g) of yellow K_2CrO_4, shown in the dish at the front. The 0.10 molar (0.10 mol/L) solution on the left was made by placing the solid in the flask and adding enough water to make 1.0 L of solution. The 0.10 molal (0.10 mol/kg) solution on the right was made by placing the solid in the flask and adding 1000 g (1 kg) of water. Adding 1 kg of water produces a solution that has a volume greater than 1 L.

Molality and molarity are *not* the same, although the difference becomes negligibly small for dilute solutions, those less than 0.01 mol/L.

(b) Assuming that the density of water is 1.00 g/mL, then 1500. g of water has a volume of 1500. mL, or 1.50 L. The molarity of the solution is

$$\frac{0.0129 \text{ mol methanol}}{1.50 \text{ L solution}} = 0.00860 \text{ mol/L}$$

Problem-Solving Practice 15.9

Calculate the molality of a solution made by dissolving 6.58 g of NaCl in 250. mL of water.

 Exercise 15.13 Molality

The "proof" of an alcoholic beverage is defined as twice the percent by volume of alcohol in the beverage. What is the molality of ethanol, C_2H_5OH, in a quart of a 90-proof alcoholic beverage? Assume that ethanol is the only solute. The density of ethanol is 0.79 g/mL; the density of the alcoholic beverage is 0.861 g/mL; 1 L = 1.057 qt.

 Exercise 15.14 Molality and Molarity

(a) What information is required to calculate the molality of a solution?
(b) What information is needed to calculate the molarity of a solution if the solution's composition is given in weight percent?

The increase in boiling point due to the presence of a solute can be calculated from this relationship.

$$\Delta T_b = T_b \text{ (solution)} - T_b \text{ (solvent)} = K_b m_{\text{solute}}$$

The units of the molal boiling point elevation constant are abbreviated °C kg mol⁻¹.

where ΔT_b = boiling point of solution − boiling point of pure solvent. The value of K_b, the *molal boiling point elevation constant* of the *solvent*, depends only on the solvent. For example, K_b for water is 0.52 °C kg mol⁻¹; that of benzene is 2.53 °C kg mol⁻¹.

Molality, rather than molarity, is used in boiling-point elevation determinations because the molality of a solution does not change with temperature changes, but the molarity of a solution does. Molality is based on the masses of solute *and* solvent, which are unaffected by temperature changes. The volume of a solution, as used in molarity, expands or contracts when the solution is heated or cooled.

Exercise 15.15 Calculating the Boiling Point of a Solution

The boiling point elevation constant for benzene is 2.53 °C kg mol⁻¹. The boiling point of pure benzene is 80.10 °C. If a solute's concentration in benzene is 0.10 mol/kg, what will be the boiling point of the solution?

Freezing Point Lowering

 CD-ROM Screen 14.8: Colligative Properties (2): Boiling Point and Freezing Point

A pure liquid begins to freeze when the temperature is lowered to the substance's freezing point and the first few molecules cluster together into a crystal lattice to form a tiny quantity of solid. As long as both solid and liquid phases are present and the temperature is at the freezing point, there is a dynamic equilibrium as the rate of crystallization equals the rate of melting. When a *solution* freezes, a few molecules of solvent cluster together to form pure solid *solvent* (Figure 15.16), and a dynamic equilibrium is set up between solution and solid solvent.

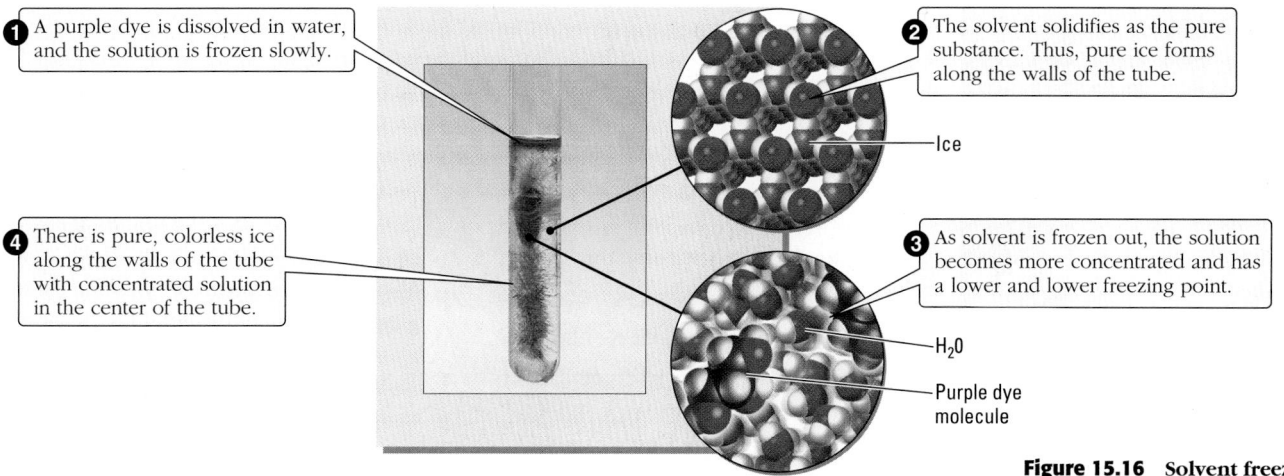

① A purple dye is dissolved in water, and the solution is frozen slowly.

② The solvent solidifies as the pure substance. Thus, pure ice forms along the walls of the tube.

④ There is pure, colorless ice along the walls of the tube with concentrated solution in the center of the tube.

③ As solvent is frozen out, the solution becomes more concentrated and has a lower and lower freezing point.

Ice

H₂O

Purple dye molecule

Figure 15.16   Solvent freezing.

In the case of a freezing solution, the molecules or ions in the liquid in contact with the frozen solvent are not all solvent molecules, causing a slower rate at which particles move from solution to solid than in the pure liquid. To achieve dynamic equilibrium, a correspondingly slower rate of escape of molecules from the solid crystal lattice must occur. According to the kinetic-molecular theory, this slower rate occurs at a lower temperature, and so the freezing point of the solution is lower than that of the pure liquid solvent (Figure 15.15, p. 703).

The **freezing point lowering**, ΔT_f, is proportional to the concentration of the solute in the same way as the boiling point elevation.

$$\Delta T_f = K_f m_{\text{solute}}$$

As with K_b, the proportionality constant K_f depends only on the solvent and not the type of solute. For water, the freezing point constant is $1.86 \,^\circ\text{C kg mol}^{-1}$; by comparison, that of benzene is $5.10 \,^\circ\text{C kg mol}^{-1}$ and that of cyclohexane is $20.2 \,^\circ\text{C kg mol}^{-1}$.

Using ethylene glycol, $HOCH_2CH_2OH$, a relatively nonvolatile alcohol, in automobile cooling systems is a practical application of boiling point elevation and freezing point lowering. Ethylene glycol raises the boiling temperature of the coolant mixture of ethylene glycol and water to a level that prevents "boil over" in hot weather. Ethylene glycol also lowers the freezing point of the coolant, thereby keeping the solution from freezing in the winter.

A practical application of freezing point lowering is adding salt (NaCl) to ice when making homemade ice cream. Lowering the freezing temperature of the ice-salt water mixture freezes the ice cream more quickly.

Problem-Solving Example 15.10   Boiling Point Elevation and Freezing Point Lowering

Calculate the boiling and freezing points of an aqueous solution containing 39.5 g of ethylene glycol ($HOCH_2CH_2OH$) dissolved in 750 mL of water. Assume the density of water to be 1.00 g/mL

Answer   Boiling point = 100.44 °C; freezing point = −1.58 °C.

Explanation   In order to use the equations for freezing point and boiling point changes, the molality of the solution must be determined. The molar mass of ethylene glycol is 62.07 g/mol, and the number of moles of ethylene glycol is

$$39.5 \text{ g} \times \frac{1 \text{ mol}}{62.07 \text{ g}} = 0.636 \text{ mol}$$

The mass of solvent is

$$750 \text{ mL} \times \frac{1.00 \text{ g}}{\text{mL}} \times \frac{1 \text{ kg}}{10^3 \text{ g}} = 0.750 \text{ kg}$$

Lowering the freezing point and raising the boiling point of a solution.
Adding an ethylene glycol – based antifreeze to an automobile's cooling system lowers the freezing point and raises the boiling point of the solution.

Ethylene glycol is toxic and should not be allowed to get into drinking water supplies.

The molality of the solution is

$$\frac{0.636 \text{ mol}}{0.750 \text{ kg}} = 0.848 \text{ mol/kg}$$

The boiling point elevation is

$$\Delta T_b = (0.52 \text{ °C kg mol}^{-1})(0.848 \text{ mol/kg}) = 0.44 \text{ °C}$$

Therefore, the solution boils at 100.44 °C.
The freezing point lowering is by 1.58 °C.

$$\Delta T_f = (1.86 \text{ °C kg mol}^{-1})(0.848 \text{ mol/kg}) = 1.58 \text{ °C}$$

and so the solution freezes at − 1.58 °C.

✔ There is a bit over half a mole of ethylene glycol in $\frac{3}{4}$ of a kilogram of solvent, so the molality should be less than 1 mol/kg, which it is. Because the concentration is less than 1 mol/kg, the boiling point should be raised less than 0.52 °C, and the freezing point should be lowered less than 1.86 °C, which they are. The answers are reasonable.

Problem-Solving Practice 15.10

A water tank contains 6.50 kg of water. Will the addition of 1.20 kg of ethylene glycol be sufficient to prevent the solution from freezing if the temperature drops to −25 °C?

Exercise 15.16   Protection Against Freezing

Suppose that you are closing a cabin in the north woods for the winter and you do not want the pipes to freeze. You know that the temperature might get as low as −30. °C, and you want to protect about 4.0 L of water in a toilet tank from freezing. What volume of ethylene glycol, density = 1.113 g/mL, molar mass = 62.1 g/mol, should be added to the 4.0 L of water?

Problem-Solving Example 15.11   Molar Mass From Freezing Point Lowering

The freezing point of p-dichlorobenzene is 53.1 °C, and its K_f is 7.10 °C kg mol^{-1}. A solution of 1.52 g of the drug sulfanilamide in 10.0 g of p-dichlorobenzene freezes at 46.7 °C. What is the molar mass of sulfanilamide?

Answer   169 g/mol

Explanation   To calculate the molar mass, we need to determine the change in freezing point and the molality of the solution. The change in the freezing point, ΔT_f, is 53.1 °C − 46.7 °C = 6.4 °C. Rearranging the freezing point lowering equation, we can solve for the molality of the solution.

$$\text{Molality} = \frac{\Delta T_f}{K_f} = \frac{6.4 \text{ °C}}{7.10 \text{ °C kg mol}^{-1}} = 0.90 \text{ mol/kg}$$

From the molality we can calculate how many moles of sulfanilamide were in solution.

$$\frac{0.90 \text{ mol sulfanilamide}}{1.00 \text{ kg } p\text{-dichlorobenzene}} \times 0.010 \text{ kg } p\text{-dichlorobenzene} = 9.0 \times 10^{-3} \text{ mol}$$

This number of moles corresponds with the 1.52 g that was dissolved, so the molar mass is

$$\frac{1.52 \text{ g}}{9.0 \times 10^{-3} \text{ mol}} = 1.7 \times 10^2 \text{ g/mol}$$

✔ About 1.5 g of sulfanilamide in 0.0100 kg of solvent forms a 0.90 mol/kg solution. Therefore, 150 g of solute in 1 kg of solvent would also be a 0.90 m solution, which is approximately a 1 m solution containing about 1 mol of sulfanilamide in 1 kg of solvent.

Thus, the molar mass of sulfanilamide is a bit more than 150 g/mol, and the answer of 170 g/mol is reasonable.

Problem-Solving Practice 15.11

To determine its molar mass, 1.50 g of a newly discovered compound is dissolved in 75.0 g of cyclohexane. Pure cyclohexane has a freezing point of 6.50 °C and a K_f of 20.2 °C kg mol^{-1}. The solution freezes at 2.70 °C. Calculate the molar mass of the new compound.

Colligative Properties of Electrolytes

Experimentally, the vapor pressures of 1 M aqueous solutions of sucrose, NaCl, and $CaCl_2$ are all less than that of water at the same temperature, which is to be expected because solutes lower the vapor pressure of the pure solvent. However,

$$\text{vp pure water} > \text{vp 1 M sucrose} > \text{vp 1 M NaCl} > \text{vp 1 M CaCl}_2$$

Because colligative properties of dilute solutions are proportional to the concentration of solute *particles,* this vapor pressure order is not surprising. Electrolytes such as NaCl and $CaCl_2$ contribute more particles per mole than do nonelectrolytes such as sucrose or ethanol. Whereas 1 mol of sucrose contributes 1 mol of particles to solution, 1 mol of NaCl contributes 2 mol of particles (1 mol of Na^+ and 1 mol of Cl^-), and 1 mol of $CaCl_2$ produces 3 mol of particles (1 mol of Ca^{2+} and 2 mol of Cl^-). Therefore, electrolytes have a greater effect on boiling point than nonelectrolytes do.

For solutions of electrolytes, the boiling point elevation equation can be written as

$$\Delta T_b = K_b m_{\text{solute}} i_{\text{solute}}$$

and the freezing point lowering equation becomes

$$\Delta T_f = K_f m_{\text{solute}} i_{\text{solute}}$$

The i_{solute} factor gives the number of particles per formula unit of solute. It is called the van't Hoff factor, named after Jacobus Henricus van't Hoff (1847–1930), who won the very first Nobel Prize in Chemistry (1901) for his work on the colligative properties of solutions. The value of i is 1 for nonelectrolytes because these molecular solutes, such as ethanol, sucrose, benzene, and carbon tetrachloride, do *not* dissociate in solution. For soluble ionic solutes (strong electrolytes), i equals the number of ions per formula unit of the ionic compound. In extremely dilute solutions $i_{\text{solute}} = 2$ for NaCl and $i_{\text{solute}} = 3$ for calcium chloride.

Another practical application of freezing point lowering can be seen in areas where winters produce lots of frozen precipitation. To remove snow and particularly ice, roads and walkways are often salted. Although sodium chloride is usually used, calcium chloride is particularly good for this purpose because it has three ions per formula unit and dissolves exothermically. Not only is the freezing point of water lowered, but the heat of solution helps melt the ice.

JACOBUS HENRICUS VAN'T HOFF
(1847–1930)

Jacobus van't Hoff was one of the founders of physical chemistry, the branch of chemistry that applies the laws of physics to understand chemical phenomena. While still a graduate student, van't Hoff proposed an explanation of optical isomerism based on the tetrahedral nature of the carbon atom ($\Leftarrow$ *Section 9.7*). He conducted seminal experimental studies in chemical kinetics, chemical equilibrium, osmotic pressure, and chemical affinity. Van't Hoff received the first Nobel Prize in Chemistry (1901) for his founding discoveries in physical chemistry.

The actual i_{solute} value must be determined experimentally. The theoretical i value assumes that the ions act independently in solution, which is achieved only in extremely dilute solutions where the interaction between cations and anions is minimal. In more concentrated solutions, cations and anions interact and $i_{\text{actual}} < i_{\text{theor}}$. For example, in 0.50 M MgSO$_4$, $i_{\text{expt}} = 1.07$; in 0.005 M MgSO$_4$ it is 1.72.

This answers the question raised in Chapter 1 ($\Leftarrow$ *p. 32)*: "Why does salt help to clear snow and ice from roads?"

A buildup of dissolved NaCl or CaCl$_2$ along roads is environmentally hazardous because the excessive Cl$^-$ concentration is harmful to roadside plants.

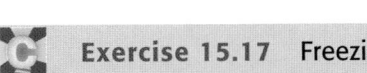 **Exercise 15.17   Freezing Point Lowering**

The freezing point of a 2.0-*m* solution of $CaCl_2$ is measured as -4.78 °C. Calculate the i factor and use it to approximate the degree of dissociation of $CaCl_2$ in this solution.

15.8 OSMOTIC PRESSURE OF SOLUTIONS

CD-ROM Screen 14.9:
Colligative Properties (3):
Osmosis

A *membrane* is a thin layer of material that allows molecules or ions to pass through it. A **semipermeable membrane** allows only certain kinds of molecules or ions to pass through while excluding others (Figure 15.17). Examples of semipermeable membranes are animal bladders, cell membranes in plants and animals, and cellophane, a polymer derived from cellulose. When two solutions containing the same solvent are separated by a membrane permeable only to solvent molecules, osmosis will occur. **Osmosis** is *the movement of a solvent through a semipermeable membrane from a region of lower solute concentration (higher solvent concentration) to a region of higher solute concentration (lower solvent concentration).* The **osmotic pressure** of a solution is *the pressure that must be applied to the solution to stop osmosis from a sample of pure solvent.*

Consider the osmosis example shown in Figure 15.18. A 5% aqueous sugar solution is placed in a bag attached to a glass tube. The bag is made of a semipermeable membrane that allows water but not sugar molecules to pass through it. When the bag is submerged in pure water, water flows into the bag by osmosis and raises the liquid level in the tube. When the bag is first submerged, there are more collisions of solvent molecules per unit area of the membrane on the pure solvent side than there are on the solution side (where there are fewer solvent molecules per unit volume). Hence, water moves through the membrane from the beaker where water is in greater concentration into the solution in the bag, where the water concentration is lower. As this continues, the number of water molecules in the solution increases, the number of collisions of water molecules within the solution increases, and water rises in the tube as pressure builds up in the bag. A dynamic equilibrium is achieved when the pressure in the bag equals the osmotic pressure, at which point the rate of passing water molecules is the same in both directions. The height of the water column then remains unchanged, and its height is a measure of Π, the osmotic pressure.

Osmotic pressure, like vapor pressure lowering, boiling point elevation, and freezing point lowering, results from the unequal rates at which solvent molecules pass through an interface or boundary. In the case of evaporation and boiling, it is the solution/vapor interface; for freezing, it is the solution/solid interface. The

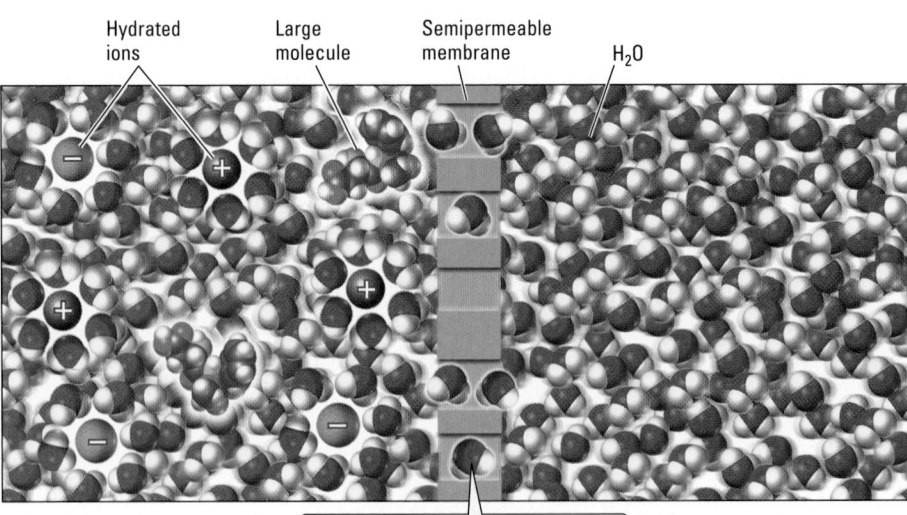

Hydrated ions   Large molecule   Semipermeable membrane H_2O

Water molecules pass through the membrane, but hydrated ions and large molecules do not.

Figure 15.17 Osmotic flow of a solvent through a semipermeable membrane to a solution. The semipermeable membrane is shown acting as a sieve. Many membranes operate in different ways, but the ultimate effect is the same.

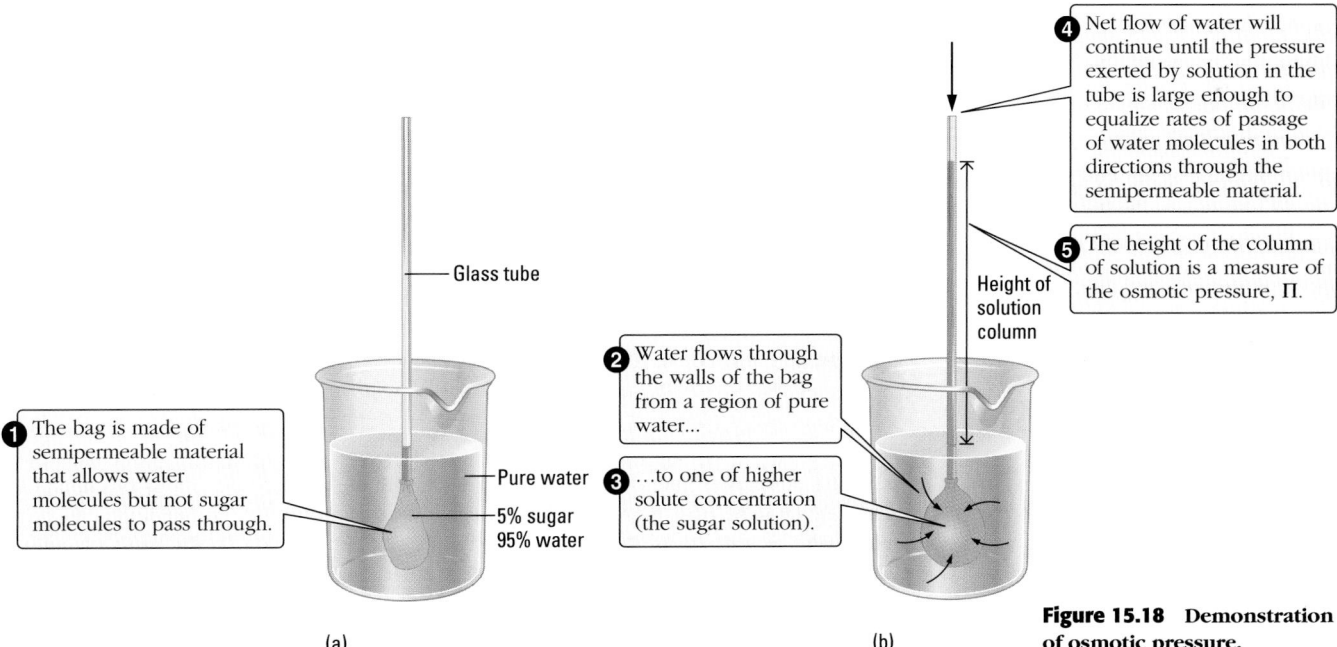

Figure 15.18 Demonstration of osmotic pressure.

semipermeable membrane is the interface for osmosis. All colligative properties can be understood in terms of differences in entropy between a pure solvent and a solution. This is perhaps most easily seen in the case of osmosis. When solvent and solute molecules mix, there is usually an increase in entropy. If pure solvent is added to a solution, a higher entropy state will be achieved as solvent and solute molecules diffuse among one another to form a more dilute solution. Unless there are strong noncovalent forces between the solute and solvent, there will be a negligible enthalpy change, and so the increase in entropy makes mixing of solvent and solution a product-favored process. A semipermeable membrane prevents solute molecules from passing into pure solvent, so the only way mixing can occur (and entropy can increase) is for solvent to flow into the solution, and it does.

The more concentrated the solution, the more product-favored the mixing is and the greater is the pressure required to prevent it. Osmotic pressure (Π) is proportional to the molarity of the solution, c,

$$\Pi = cRTi$$

where R is the gas constant, T is the absolute temperature (in kelvins), and i is the number of particles per formula unit of solute.

Even though the solution concentration is small, osmotic pressure can be quite large. For example, the osmotic pressure of a 0.020 M solution of a molecular solute ($i = 1$) at 25 °C is

$$\Pi = cRTi = \left(\frac{0.020 \text{ mol}}{L}\right)(0.0821 \text{ L atm mol}^{-1} \text{ K}^{-1})(298 \text{ K})(1) = 0.49 \text{ atm}$$

This pressure would support a water column more than 15 ft high. One way to determine osmotic pressure is to measure the height of a column of solution in a tube, as shown in Figure 15.18. Heights of a few centimeters can be measured accurately, and so quite small concentrations can be determined by osmotic-pressure experiments. If the mass of solute dissolved in a measured volume of solution is known, it is possible to calculate the molar mass of the solute by using the definition of molar concentration, $c = n/V =$ amount (mol)/volume (L). Osmotic pressure is especially useful in studying large molecules whose molar mass is difficult to determine by other means.

The osmotic pressure equation is similar to the ideal gas law equation, $PV = nRT$, which can be rearranged to $P = (n/V)RT = cRT$, where n/V is the molar concentration of the gas.

Freezing point lowering and boiling point elevation measurements can also be used to find the molar mass in the same manner as shown in Problem-Solving Example 15.12 for osmotic pressure measurements.

Problem-Solving Example 15.12   Molar Mass from Osmotic Pressure

The osmotic pressure of a solution of 5.0 g of horse hemoglobin in 1.0 L of water is 1.8×10^{-3} atm at 25 °C. What is the molar mass of the hemoglobin, a protein?

Answer   The molar mass is 6.8×10^4 g/mol.

Explanation   First, we use the osmotic pressure equation to calculate the molarity of the hemoglobin solution. Hemoglobin is a nonelectrolyte, so the i factor is 1.

$$c = \frac{\Pi}{RTi} = \frac{1.8 \times 10^{-3} \text{ atm}}{0.0821 \text{ L atm mol}^{-1} \text{ K}^{-1}(298 \text{ K})(1)} = 7.36 \times 10^{-5} \text{ mol/L}$$

Because the volume is 1.0 L, there must be 7.36×10^{-5} mol of hemoglobin present in the 5.0-g sample, and the molar mass is

$$\frac{5.0 \text{ g}}{7.36 \times 10^{-5} \text{ mol}} = 6.8 \times 10^4 \text{ g/mol}$$

✔ The molarity of the solution is very low, so the molar mass of hemoglobin must be relatively large to create an osmotic pressure of 1.8×10^{-3} atm. In Section 12.9, proteins were described as long chains of amino acid units, so a molar mass of about 70,000 g/mol is not unreasonable for a protein such as horse hemoglobin.

Problem-Solving Practice 15.12

The osmotic pressure at 25 °C is 1.79 atm for a solution prepared by dissolving 2.50 g of sucrose, empirical formula $C_{12}H_{22}O_{11}$, in enough water to give a solution volume of 100 mL. Use the osmotic pressure equation to show that the empirical formula for sucrose is the same as its molecular formula.

Blood and other fluids inside living cells contain many different solutes, and the osmotic pressures of these solutions play an important role in the distribution and balance of solutes within the body. Dehydrated patients are often given water and nutrients intravenously. However, pure water cannot simply be dripped into a patient's veins. The water would flow into the red blood cells by osmosis, causing them to burst (Figure 15.19c). A solution that causes this condition is called a **hypotonic** solution. To prevent cells from bursting, an **isotonic** (or iso-osmotic) intravenous solution must be used. Such a solution has the same total concentration of solutes and therefore the same osmotic pressure as the patient's blood (Figure 15.19a). A solution of 0.9% sodium chloride is isotonic with fluids inside cells in the body.

In a *hyper*tonic solution, the concentration of solutes outside the cell is greater than inside. There is a net flow of water out of the cell, causing the cell to dehydrate, shrink, and perhaps die.

In an *iso*tonic solution, the *net* movement of water in and out of the cell is zero because the concentration of solutes inside and outside the cell is the same.

In a *hypo*tonic solution, the concentration of solutes outside the cell is less than inside. There is a net flow of water into the cell, causing the cell to swell and perhaps to burst.

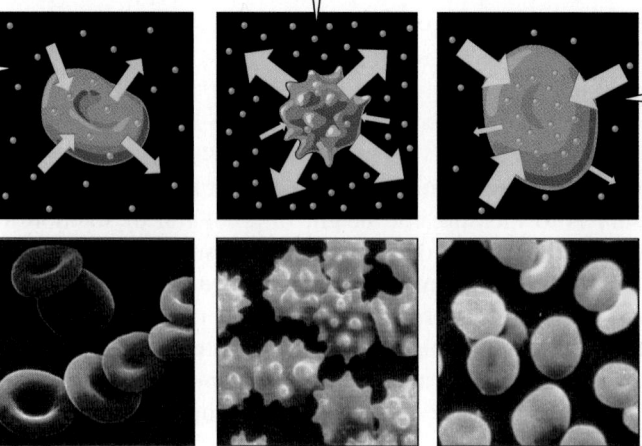

Figure 15.19   Osmosis and the living cell.

(a) Isotonic solution   (b) Hypertonic solution   (c) Hypotonic solution

Figure 15.20 **Osmosis in vegetables.** When a cucumber is soaked in a concentrated salt solution, water flows from the plant cells into the salt solution by osmosis, converting the cucumber into a pickle. A cucumber soaked in a concentrated salt solution *(right)* has lost much water and shrivels into a pickle. A cucumber soaked in pure water *(left)* is affected very little.

If an intravenous solution more concentrated than the solution inside a red blood cell were added to blood, the cell would lose water and shrivel up. A solution that causes this condition is a **hypertonic** solution (Figure 15.19b). Cell-shriveling by osmosis happens when vegetables or meats are cured in *brine,* a concentrated solution of NaCl. If you put a fresh cucumber into brine, water will flow out of its cells and into the brine, leaving behind a shriveled vegetable (Figure 15.20). With proper spices added to the brine, a cucumber will become a tasty pickle.

Reverse Osmosis

Reverse osmosis occurs when pressure greater than the osmotic pressure is applied and solvent flows through a semipermeable membrane from a concentrated solution to a dilute solution. In effect, the semipermeable membrane serves as a filter with very tiny pores through which only the solvent can pass. Reverse osmosis can be used to remove particles as small as molecules or ions to obtain highly purified water. Sea water contains a high concentration of dissolved salts; its osmotic pressure is 24.8 atm. If a pressure greater than 24.8 atm is applied to a chamber containing sea water, water molecules can be forced to flow from sea water through a semipermeable membrane to a region containing purer water (Figure 15.21). Pressures up to 100 atm are used to provide reasonable rates of sea-water purification. Sea water, which contains upwards of 35,000 ppm of dissolved salts (Table 15.5), can be purified by reverse osmosis to between 400 and 500 ppm of solutes, which is well within the World Health Organization's limits for drinking water. Large reverse osmosis plants in places like the Persian Gulf countries and Florida can purify more than 100 million gallons of water per day. Nearly 50% of Saudi Arabia's fresh water is provided by the world's largest desalination plant at Jubail. Reverse osmosis purifies nearly 15 million gallons of brackish underground water daily for the city of Cape Coral, Florida. That facility is one of 109 in the state of Florida.

Small reverse osmosis units are used to make the ultrapure water for some "spotless" car washes.

Brackish water is a mixture of salt water and fresh water.

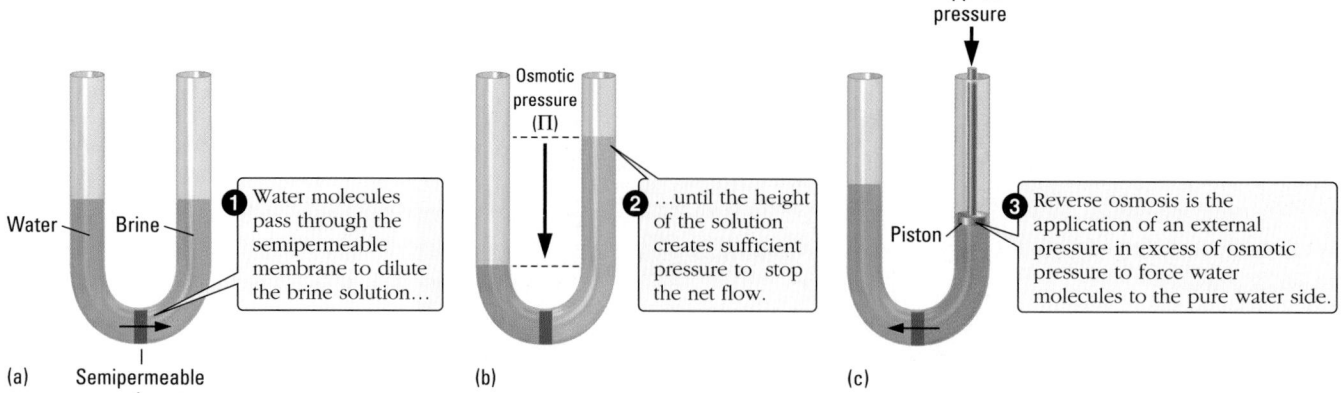

Figure 15.21 Normal and reverse osmosis. Normal osmosis is represented by (a) and (b). Reverse osmosis is represented in (c).

TABLE 15.5	Ions Present in Sea Water at 100 ppm or More	
	Mass fraction	
Ion	**g/kg**	**ppm**
Cl^-	19.35	19,350
Na^+	10.76	10,760
SO_4^{2-}	2.710	2710
Mg^{2+}	1.290	1290
Ca^{2+}	0.410	410
K^+	0.400	400
HCO_3^-, CO_3^{2-}	0.106	106
TOTAL	35.026	35,026

15.9 COLLOIDS

CD-ROM Screen 14.10: Colloids

Around 1860, Thomas Graham found that substances such as starch, gelatin, glue, and egg albumin diffuse only very slowly in water compared with sugar or salt. In addition, the former substances differ significantly from the latter ones in their ability to diffuse through a thin membrane; sugar or salt diffuse through many types of membranes, but glue, starch, albumin, and gelatin will not. Graham crystallized sugar and salt from their solutions, but he could not crystallize starch, glue, egg albumin, or gelatin. Therefore, Graham coined the word "colloid" (from the Greek meaning "glue") to describe a class of substances distinctly different from sugar and salt and similar materials.

Colloids are now understood to be mixtures in which relatively large particles, the **dispersed phase,** are distributed uniformly throughout a solvent-like medium called the **continuous phase,** or the dispersing medium. Like true solutions, colloids are found in the gas, liquid, and solid states. Although both true solutions and colloids appear homogeneous to the naked eye, at the microscopic level colloids are not homogeneous.

In colloids, the dispersed-phase particles might be as large as 10 to nearly 1000 times the size of a single small molecule. Colloidal particles are larger than those found in true solutions and smaller than the ones in suspensions.

	Smaller particles $\longrightarrow$ Larger particles		
	True solution	**Colloidal dispersion**	**Suspension**
Particles	Ions and molecules	Colloids	Large-sized particles
Particle size	0.2 – 2.0 nm	2 – 2000 nm	>2000 nm
Properties	• Don't settle out on standing	• Don't settle out on standing	• Settle out on standing
	• Not filterable	• Not filterable	• Filterable
Example	Sea water	Fog	River silt

Colloidal particles can be so large — about 2000 nm in diameter — that they scatter visible light passing through the continuous medium, a phenomenon known as the **Tyndall effect**. Figure 15.22 shows a beam of light passing through three glass bottles. The bottles on the left and right contain a colloidal mixture of a gelatin in water; the center bottle holds a solution of NaCl. The light scattering is clearly seen in the colloidal suspensions; particles in the NaCl solution are too small to scatter the light, and the Tyndall effect is not seen. Colloidal particles of dust and smoke in the air of a room can easily be observed in a beam of sunlight because they scatter the light; you have probably seen such a well-defined sunbeam many times. A common colloid, *fog,* consists of water droplets (the dispersed phase) in air (the continuous phase, and itself a solution). The lack of visibility in fog is due to the Tyndall effect.

Figure 15.22   Light scattering by a colloidal dispersion. A narrow beam of light passes through a colloidal mixture *(left),* then through a salt solution, and finally through another colloidal mixture. The Tyndall effect is seen in the colloidal mixtures, but not in the salt solution. This illustrates the light-scattering ability of colloid-sized particles.

Blue sky results from light scattering by molecule-sized particles in air. Because blue light is more strongly scattered than is red light, the sky appears to be a diffuse blue. This answers the question from Chapter 1 (⬅ *p. 32*), "Why is the sky blue?"

Types of Colloids

Colloids are classified according to the state of the dispersed phase (solid, liquid, or gas) and the state of the continuous phase. Table 15.6 lists several types of colloids and some examples of each. Liquid-liquid colloids form only in the presence of an emulsifier — a third substance that coats and stabilizes the particles of the dispersed phase. Such colloidal dispersions are called **emulsions.** In mayonnaise, for example, egg yolk contains a protein that stabilizes the tiny drops of oil that are dispersed in the aqueous continuous phase. As you can see from Table 15.6, colloids are very common in everyday life.

Colloids with water as the continuous phase are either hydrophilic or hydrophobic. In a hydrophilic colloid there is *a strong attraction between the dispersed phase and the continuous (aqueous) phase.* Hydrophilic colloids are formed when the molecules of the dispersed phase have multiple sites that interact with water through hydrogen bonding and dipole-dipole attraction. Proteins in aqueous media are hydrophilic colloids.

In a hydrophobic colloid there is a *lack of attraction between the dispersed phase and the continuous phase.* Although you might assume that such colloids would tend to separate quickly, hydrophobic colloids can be quite stable once they are formed. A colloidal solution (sol) of gold particles prepared in 1857 is still preserved in the British Museum. In hydrophobic colloids the surfaces of the colloidal particles apparently become electrically charged by some process that is not completely understood. Oppositely charged ions in solution are then attracted to the surfaces, forming a second layer. Because all the colloidal particles have the same kind of charge, they repel other particles of the same kind and are prevented from coming together to form larger particles.

A stable hydrophobic colloid coagulates when ions come into contact with the dispersed phase. Milk is a colloidal suspension of hydrophobic particles (principally butterfat). When milk ferments, lactose (milk sugar) is converted to lactic acid, which forms hydronium ions and lactate ions. The protective charge layer on

The Tyndall effect. Shafts of light are visible coming through the trees in the forest along the Oregon coast.

Figure 15.23 Silt formation in a river delta from colloidal soil particles.
Silt forms at a river delta as collidal particles in the river water meet the salt water as the river enters an ocean or a salt-water bay. The higher salt concentration causes the colloidal particles to coagulate.

TABLE 15.6	Types of Colloids		
Continuous phase	**Dispersed phase**	**Type**	**Examples**
Gas	Liquid	Aerosol	Fog, clouds, aerosol sprays
Gas	Solid	Aerosol	Smoke, airborne viruses, automobile exhaust
Liquid	Gas	Foam	Shaving cream, whipped cream
Liquid	Liquid	Emulsion	Mayonnaise, milk, face cream
Liquid	Solid	Solution	Gold in water, milk of magnesia, mud
Solid	Gas	Foam	Foam rubber, sponge, pumice
Solid	Liquid	Gel	Jelly, cheese, butter
Solid	Solid	Solid solution	Milk glass, many alloys such as steel, some colored gemstones

the surfaces of the colloidal particles is overcome and the milk coagulates—the milk solids separate in to clumps called "curds." Coagulated milk is used to make buttermilk and various kinds of cheese. Soil particles carried in rivers are hydrophobic soils. When river water containing large amounts of colloidally suspended soil particles meets sea water with a high ion concentration, the colloidal particles coagulate to form silt. The deltas of the Mississippi and Nile Rivers have been and continue to be formed in this way (Figure 15.23).

15.10 SURFACTANTS

Molecules that have a hydrophobic part and a hydrophilic part are called **surfactants** (surface-active agents) because they tend to act at the surface of a substance that is in contact with the solution that contains the surfactant. Soap, the classic surfactant, dates back to the Sumerians in 2500 BC. The ancient Greek physician Galen stated that soap removed dirt from the body as well as serving as a treatment for wounds. Chemically, soaps are salts of fatty acids and have always been made by the reaction of a fat with an alkali, a process known as *saponification (⟸ p. 542)*.

CD-ROM Screen 14.11: Surfactants

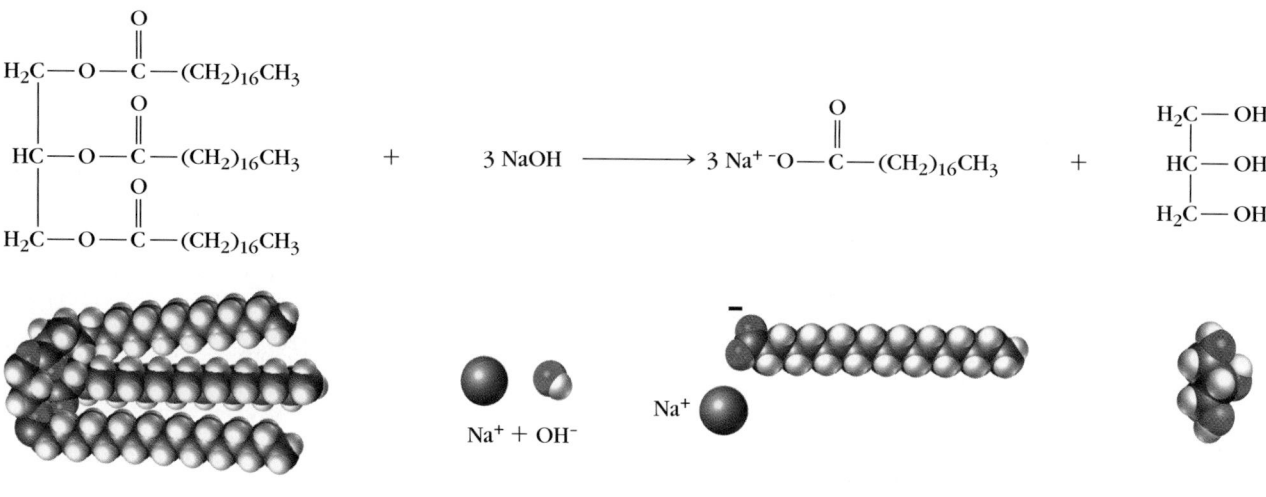

tristearin (glyceryl tristearate)   sodium hydroxide   sodium stearate (a soap)   glycerol

CHEMISTRY
You Can Do...
Curdled Colloids

Regular (whole) milk contains about 4% fat. Skim milk contains considerably less. In addition, milk contains protein. Both the fat and the proteins are in the form of colloids.

Add about 2 tablespoons of vinegar or lemon juice to about 100 mL of whole milk (do the same with skim milk and 1% or 2% butterfat milk, if you have it), stir, and watch what happens. Let it stand overnight at room temperature. Observe and record the results.

An additional experiment you can try is adding salt to similar samples of milk and recording your observations.

When you finish, discard this milk down the drain. You should *not* drink it because it has been unrefrigerated and might contain harmful bacteria.

1. Write an explanation for what you observed.
2. Does the salt have the same effect on milk as the acid does?

Sodium stearate is a typical soap. The long-chain hydrocarbon part of the molecule is hydrophobic, while the polar carboxylate group is hydrophilic.

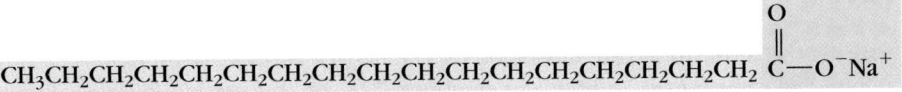

Hydrophobic end .... Hydrophilic end

sodium stearate

Hand soaps are pure soaps to which dyes and perfumes are added.

Bile salts are important surfactants in the body that help to break up ingested fats. The hydrophobic portion of bile salts is incorporated into the surface of the ingested fat molecules, where it acts as an emulsifying agent, keeping fat molecules dispersed from each other. The hydrophilic portion of the bile salt keeps the emulsified fats in solution.

In addition to soaps, which are made from naturally occurring fats and oils, many synthetic surfactants called **detergents** are made from refined petroleum or coal products. Detergents have a long hydrocarbon chain that is hydrophobic and a polar end that is hydrophilic, somewhat like those of soaps. One common detergent is sodium lauryl sulfate, which is used in many shampoos.

Water, oil, and a surfactant together form an emulsion. The surfactant acts as the emulsifying agent, such as in the case of bile salts emulsifying dietary fats.

$$CH_3CH_2CH_2CH_2CH_2CH_2CH_2CH_2CH_2CH_2CH_2CH_2OSO_3^- Na^+$$

sodium lauryl sulfate

 Exercise 15.18 **Estimating Osmotic Pressure**

Which solution would have the higher osmotic pressure, a 0.02 M sucrose solution or a 0.02 M typical soap solution? Explain your answer.

In water solutions, surfactants tend to aggregate to form hollow, colloid-sized spherical particles called *micelles* (Figure 15.24) that can transport various materials within them. The hydrophobic ends of the surfactant molecules point inward to the center of the micelle, while their hydrophilic heads point outward so that they interact with water molecules. Ordinary soap is a surfactant and, in water, forms micelles. Soaps cleanse because oil and grease (which are also hydrophobic) on clothing or skin link with the hydrophobic centers of the soap micelles and are rinsed away (Figure 15.24). When ordinary soaps are used with "hard water" containing Ca^{2+}, Mg^{2+}, and Fe^{2+} ions, the soaps react with these ions to form insoluble salts, producing an unsightly scum around the bathtub or on clothes. Detergents do not form these insoluble salts, even with hard water.

How does soap help clothes get clean (⬅ *p. 32*)? This is how.

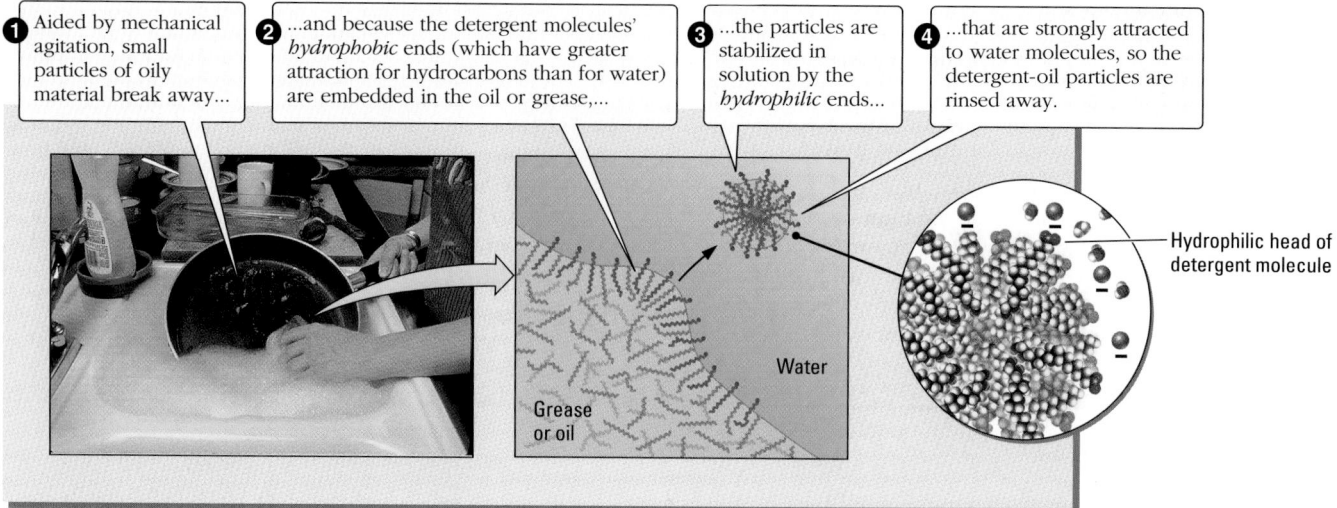

Figure 15.24 The cleansing action of soaps and detergents.

"... Water, water everywhere, nor any drop to drink." *The Rime of the Ancient Mariner* by Samuel Taylor Coleridge alludes to the fact that the dissolved salt in sea water renders it useless for drinking.

15.11 WATER: NATURAL, CLEAN, AND OTHERWISE

Water is the most abundant substance on the earth's surface. More than 97% of it is found in the oceans, which cover about 72% of the earth's surface (Figure 15.25). Glaciers, ice caps, and snow pack account for 2.16% of water, and *surface water*—lakes, rivers, streams, and reservoirs—makes up only a very small portion (0.0197%). *Groundwater*, water that is held in large underground natural *aquifers*, makes up 0.61% of water.

From Figure 15.25 note that fresh water makes up less than 3% of the earth's surface water, with most of it tied up in glaciers. Indeed, the amount of fresh water available to satisfy demands (we can't live without it) is relatively limited. In the United States, a bit more than half (57%) of fresh water is used in industry and about a third (34%) in agriculture to irrigate crops; only 9% is available for domestic and municipal purposes.

Even when a sufficient quantity of water is available, the quality of the water needs to be assured before it is safe to drink. Water must first be treated to make it potable.

Municipal Drinking Water Purification

Fresh water, even from natural sources, contains dissolved materials that must be removed or decreased to make water fit for domestic use, or for agricultural and

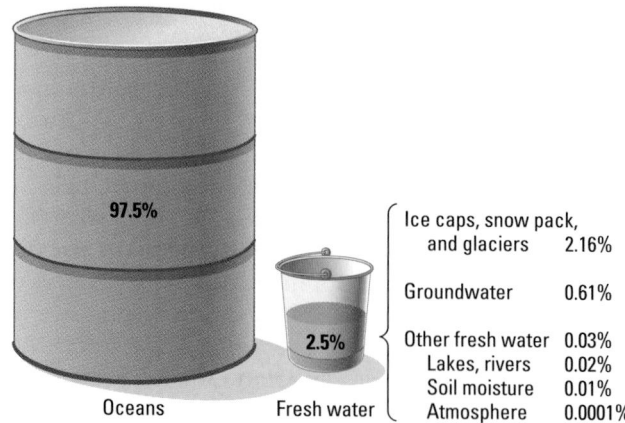

Figure 15.25 The earth's water supply. Of the 2.5% that is fresh water, less than 1% is available as groundwater or surface water.

industrial purposes. The Safe Water Drinking Act of 1974 established required standards of purity and safety for public water supplies. The EPA sets limits for contaminants that may be present in drinking water and requires continual monitoring of municipal water supplies.

Municipal water purification takes place in a series of steps (Figure 15.26). After a coarse filter (a screen) removes large objects such as tires, tree limbs, bottles, and so on, the water goes into a settling tank, where small clay and dirt particles settle out. To speed up the sedimentation, $Al_2(SO_4)_3$ (alum) and $Ca(OH)_2$ (slaked lime) are added. These compounds react to form a sticky gelatinous precipitate of aluminum hydroxide.

$$Al_2(SO_4)_3(aq) + Ca(OH)_2(aq) \longrightarrow 2\,Al(OH)_3(s) + 3\,CaSO_4(aq)$$

The $Al(OH)_3$ gel collects suspended clay and dirt particles as it sinks slowly down the settling tank. Particles not settled out are removed by passing the water from the settling tank through a sand filter.

In the next step, the water is aerated — sprayed into the air to oxidize organic substances dissolved in the water. To this point, nothing has been done to remove bacteria that might be harmful. The bacteria are killed in the final step by chemically treating the water, and *chlorination* is the most common method used in the United States. Chlorination is done by adding chlorine gas, sodium hypochlorite, $NaOCl$, or calcium hypochlorite, $Ca(OCl)_2$. In all three cases, the antibacterial agent generated is $HOCl$, hypochlorous acid. In the case of Cl_2, the $HOCl$ forms by the reaction of chlorine with water.

$$Cl_2(g) + H_2O(\ell) \longrightarrow HOCl(aq) + H^+(aq) + Cl^-(aq)$$

The extent of chlorination is adjusted so that between 0.075 and 0.600 ppm of $HOCl$ remain in solution as the water leaves the treatment plant, which is sufficient to ensure that bacterial contamination does not occur before the water reaches the user.

Chlorination was first used for drinking water supplies in the early 1900s, with a resulting drop in the number of deaths in the United States caused by typhoid and other water-borne diseases from 35/100,000 population in 1900 to 3/100,000 popu-

Calcium hydroxide (slaked lime) forms when lime (CaO) reacts with water.

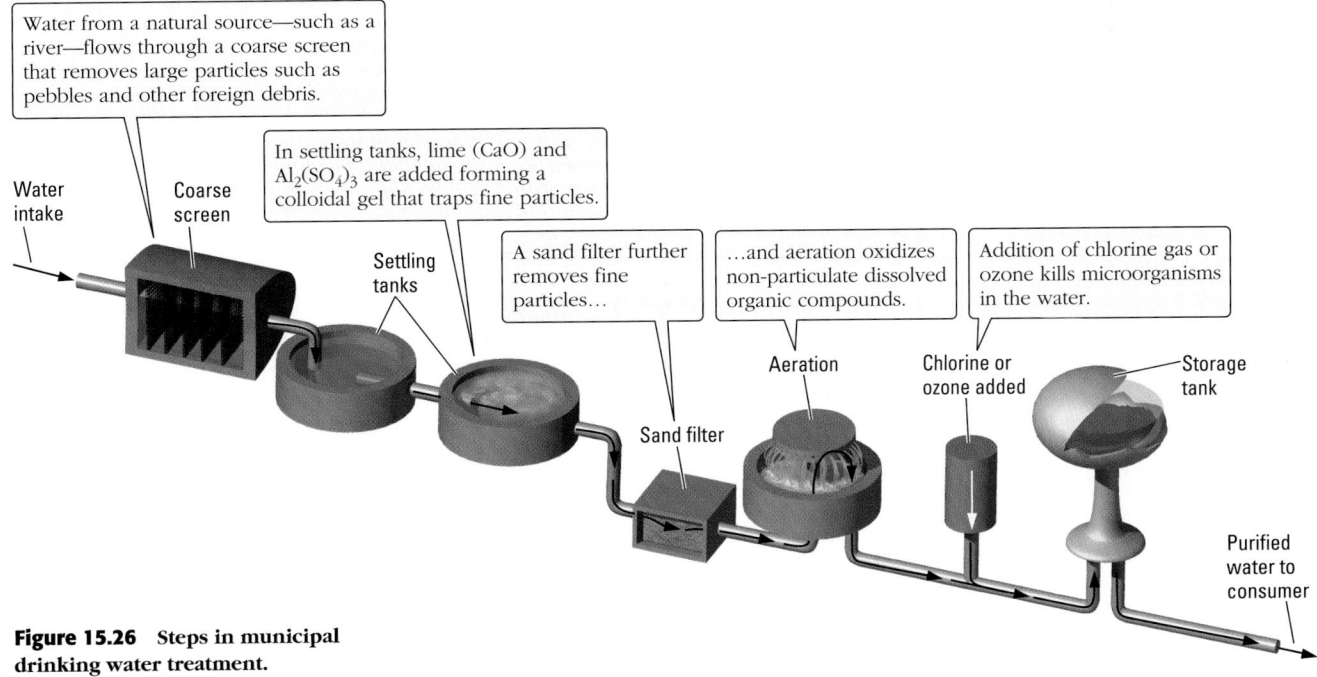

Water from a natural source—such as a river—flows through a coarse screen that removes large particles such as pebbles and other foreign debris.

In settling tanks, lime (CaO) and $Al_2(SO_4)_3$ are added forming a colloidal gel that traps fine particles.

A sand filter further removes fine particles…

…and aeration oxidizes non-particulate dissolved organic compounds.

Addition of chlorine gas or ozone kills microorganisms in the water.

Water intake

Coarse screen

Settling tanks

Sand filter

Aeration

Chlorine or ozone added

Storage tank

Purified water to consumer

Figure 15.26 Steps in municipal drinking water treatment.

lation in 1930. Chlorination is the principal means of preventing water-borne diseases spread by bacteria, including cholera, typhoid, paratyphoid, and dysentery.

In spite of chlorination, most city water supplies are not bacteria-free. But only rarely do these surviving bacteria cause disease. In the United States the most common water-borne bacterial disease is giardiasis, a gastrointestinal disorder. Most often this disease is caused by bacteria in surface water that has leaked into drinking water supplies, but on occasion it can be traced to city water systems.

Chlorination is not without its own small risk because of byproducts it forms. Even the best-designed purification systems allow some organic compounds to pass through, which then become chlorinated. In particular, humic acids, breakdown products of plant materials always present in surface water, react with residual HOCl. The reaction forms a class of compounds known as trihalomethanes (THMs), such as chloroform, $CHCl_3$. Most drinking water meets the current maximum contaminant level standard of 80 ppb THMs. Chloroform is the trihalomethane of chief concern because it is suspected of causing liver cancer. Information is still being evaluated about the seriousness of this potential threat, especially given the fact that the THMs level in drinking water is normally less than 1 ppm.

> The 80 ppb standard for THMs was established in 1998 by the U.S. Environmental Protection Agency. The previous standard had been 100 ppb. The national average for THMs in drinking water is 51 ppb.

Gaseous ozone, O_3, is used in many European cities to disinfect municipal water supplies. Ozone is an even more effective bactericide than chlorine, so less of it is needed to purify the water. The disadvantages of ozone are that it must be generated on site, and ozone does not remain in the water as long as chlorine, raising the issue of recontamination.

Disinfection of water by using ultraviolet radiation is another method becoming more popular. It is fast, economical for small installations such as rural homes, and leaves no residual byproducts. UV disinfection, like ozonation, however, does not protect against bacterial contamination after water leaves the treatment site unless the appropriate amount of chlorine is added.

Exercise 15.19 Drinking Water

Selenium poisoning can occur in individuals ingesting more than 400 μg of selenium per day. Calculate the mass of selenium ingested daily by an individual drinking 3 qt of water containing the maximum containment level of selenium, 0.050 ppm.

Exercise 15.20 Lead in Drinking Water

If the Pb^{2+} concentration in tap water is 0.025 ppm, how many liters of this water contain 100.0 μg of Pb^{2+}?

Hard Water: Natural Impurities

> Degrees of hardness are:
> Soft water: < 65 mg of metal ion/gal
> Slightly hard: 65–228 mg/gal
> Moderately hard: 228–455 mg/gal
> Hard: 455–682 mg/gal
> Very hard: > 682 mg/gal

A relatively high concentration of Ca^{2+}, Mg^{2+}, Fe^{3+}, or Mn^{2+} imparts "hardness" to water. Water hardness is objectionable because it (1) causes precipitates (called scale) to form in boilers and hot-water pipes, (2) causes soaps to form insoluble curds, and (3) may impart a disagreeable taste to the water.

Water hardness is produced when surface water containing carbon dioxide trickles through limestone or dolomite, releasing calcium or magnesium ions (⬅ p. 657) as their soluble bicarbonates.

$$CaCO_3(s) + CO_2(g) + H_2O(\ell) \longrightarrow Ca^{2+}(aq) + 2\,HCO_3^-(aq)$$
limestone

$$CaCO_3 \cdot MgCO_3(s) + 2\,CO_2(g) + 2\,H_2O(\ell) \longrightarrow$$
dolomite

$$Ca^{2+}(aq) + Mg^{2+}(aq) + 4\,HCO_3^-(aq)$$

Such hard water can be softened by removing these ions by two principal methods: (1) The lime-soda process and (2) ion exchange.

In the lime-soda process, hydrated lime, $Ca(OH)_2$, and soda, Na_2CO_3, are added to the water. Several reactions take place, which can be summarized as follows.

In hard	
water	**Added**

$$HCO_3^-(aq) + OH^-(aq) \longrightarrow CO_3^{2-}(aq) + H_2O(\ell)$$

$$Ca^{2+}(aq) + Na_2CO_3(aq) \longrightarrow CaCO_3(s) + 2\,Na^+(aq)$$

$$Mg^{2+}(aq) + 2\,OH^-(aq) \longrightarrow Mg(OH)_2(s)$$

The lime-soda process works because calcium carbonate, $CaCO_3$, is much less soluble than calcium bicarbonate, $Ca(HCO_3)_2$, and magnesium hydroxide, $Mg(OH)_2$, is much less soluble than magnesium bicarbonate, $Mg(HCO_3)_2$. The overall result of the lime-soda process is to precipitate almost all the calcium and magnesium ions and to leave sodium ions (non–hard water ions) as replacements.

Iron present as Fe^{2+} and manganese present as Mn^{2+} can be removed from water by aeration to produce higher oxidation states. If the water is neutral or slightly alkaline (either naturally or from the addition of lime), insoluble compounds $Fe(OH)_3$ and $MnO_2(H_2O)_x$ form and precipitate from solution.

Ion exchange is another way to remove ions causing water hardness. A cation-exchange resin containing either H^+ or Na^+ ions uses them to replace ions that cause hard water. Home water treatment ion-exchange units usually replace hardness ions with Na^+ ions. The exchange resin is a polymer containing numerous negatively charged $-SO_3^{2-}$ functional groups that have Na^+ ions attached. When smaller, more positive ions like Mg^{2+} and Ca^{2+} in hard water flow over the resin, they displace Na^+ ions from the resin. The process happens many times, and the result is water that contains Na^+ ions in place of the Ca^{2+} and Mg^{2+} ions, which are bound to the resin. Two $-SO_3H$ groups are required for every 2+ ion removed from solution; two Na^+ ions are released for every 2+ ion removed.

$$2\,(Polymer-SO_3^-)Na^+ + Ca^{2+} \longrightarrow (Polymer-SO_3^-)_2Ca^{2+} + 2\,Na^+(aq)$$

ion-exchange resin from hard water

Because Na^+ ions do not harden water, the resulting water is called "soft" water.

When all the sodium ions have been replaced, the ion-exchange resin becomes saturated with hard-water ions. The resin must be regenerated by treating it with a highly concentrated NaCl solution. This reverses the process given in the previous equation and the hard water ions released from the resin are rinsed down the drain.

The Impact of Household Wastes on Water Quality

Today, hazardous industrial wastes must either be put into secure landfills, incinerated, or treated in some way to render them nonhazardous.

We often don't think about what we as consumers of industrial products throw away, or how our household wastes can affect groundwater, lakes, rivers, and coastlines. Table 15.7 lists some common household products and the kinds of chemicals they contain. The bulk of this waste still goes into municipal landfills. Secure landfills usually have plastic linings to prevent leaching, are built on a thick layer of impervious clay, and have carefully spaced monitor wells for detecting any leaks.

If a landfill leaks, we are putting potentially hazardous substances into the groundwater. To reduce such occurrences, an increasing number of municipalities now have household hazardous waste cleanup days, when special trucks pick up paint, oil, batteries, and other hazardous household products, or citizens bring such household wastes to a central location for collection and proper disposal. Such activities greatly reduce the quantity of household hazardous waste that would otherwise end up in a landfill.

For persons on low-sodium diets, lime-soda–treated water may provide too high a daily dose of Na^+.

The same equation applies if Mg^{2+} is substituted for Ca^{2+}.

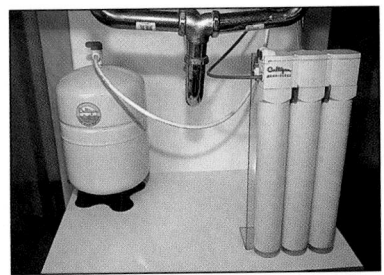

A home water softener. Hard water is softened by substituting sodium ions for calcium and magnesium ions.

The average household contains from 3 to 10 gallons of materials that are hazardous to human health or to the natural environment. If not stored carefully or disposed of properly, these materials can poison our water.

TABLE 15.7	Some Common Household Hazardous Wastes and Their Disposal	
Type of product	**Harmful ingredients**	**Disposal**[*]
Aerosol cans (empty)	Solvents, propellants	Trash
Antifreeze	Organic solvents, metals	Special; recycle
Auto battery	Sulfuric acid, lead	Special; recycle
Bathroom cleaners	Acids or strong bases	Drain
Batteries	Heavy metals such as Hg	Special
Bug sprays	Pesticides, organic solvents	Special
Drain cleaners	Strong bases	Drain
Furniture polish	Organic solvents	Special
Gasoline	Organic solvents	Special; recycle
Insecticides	Pesticides, solvents	Special
Nail polish	Solvents	Trash
Nail polish remover	Organic solvents	Special
Oven cleaner	Strong bases	Drain
Medicine (expired)	Organic compounds	Drain
Mercury batteries	Mercury	Special
Moth balls	Chlorinated organic compound	Special
Motor oil	Organic compounds, metals	Special; recycle
Paint (latex)	Organic polymers	Drain
Paint (oil-based)	Organic solvents	Special
Shoe polish	Waxes, solvents	Trash

[*]Most effective ways of disposal. *Special:* Professional disposal as a hazardous waste; save for community-wide collection day. *Drain:* Disposal down the kitchen or bathroom drain with plenty of water. *Recycle:* Take materials to area recycling program or dealer. *Trash:* Treat as normal trash for landfill. Unfortunately, in most households, the items marked special are disposed of as normal trash, which can result in groundwater pollution.

Source: Household Hazardous Waste Chart, Water Environment Federation.

Another approach to hazardous waste reduction is to consider the pollution potential of products when deciding what to buy. Alkaline batteries, for example, have less toxic ingredients than mercury batteries (Section 19.9), and often work just as well. When you change the oil in your automobile, buy it from a merchant who will take back and dispose the used oil properly. It is estimated that in the United States nearly 300,000 gallons of used oil are poured into the ground each year by people doing their own oil changes. This amount is equivalent to almost one third of the oil spilled into Prince William Sound, Alaska by the oil tanker *Exxon Valdez.*

SUMMARY PROBLEM

You are asked to prepare three mixtures at 25 °C. Mixture I: 25.0 g of CCl_4 and 100. mL of water; Mixture II: 15.0 g of $CaCl_2$ in 125 mL of water; Mixture III: 21 g of ethylene glycol ($HOCH_2CH_2OH$) in 150. mL of water. Answer these questions about these mixtures. If one of the solutes fails to dissolve in water, some of the questions will not be applicable.

(a) What is the weight percent of the mixture?

(b) What is the mass fraction of the mixture?

(c) Is a solution formed? (If a solution is formed, answer the remaining questions. You may assume a density of the solution of 1.0 g/mL.)

(d) Name the dissolved species in solution and draw a diagram representing how the solvent (water) molecules are interacting with these species.

(e) Express the concentration of the solution in ppm.

(f) Express the concentration of the solution in molality.

(g) Calculate the vapor pressure of water in equilibrium with the solution.

(h) Calculate the boiling point of the solution.

(i) Calculate the freezing point of the solution.

(j) Calculate the osmotic pressure of the solution.

IN CLOSING

Having studied this chapter, you should be able to . . .

- Describe how liquids, solids, and gases dissolve in a solvent (Section 15.1).
- Predict solubility based on properties of solute and solvent (Section 15.1).
- Interpret the dissolving of solutes in terms of enthalpy and entropy changes (Section 15.2).
- Differentiate among unsaturated, saturated, and supersaturated solutions (Section 15.3).
- Describe how ionic compounds dissolve in water (Section 15.3).
- Predict how temperature affects the solubility of ionic compounds (Section 15.4).
- Predict the effects of temperature (Section 15.4) and pressure on the solubility of gases in liquids (Section 15.5).
- Describe the compositions of solutions in terms of weight percent, mass fraction, parts per million, parts per billion, parts per trillion and molarity (Section 15.6).
- Interpret vapor pressure lowering in terms of Raoult's Law (15.7).
- Use molality to calculate the colligative properties: freezing point lowering, boiling point elevation, and osmotic pressure (Section 15.7).
- Differentiate the colligative properties of nonelectrolytes and electrolytes (15.7).
- Explain the phenomena of osmosis and reverse osmosis and calculate osmotic pressure (Section 15.8).
- Describe the various kinds of colloids and their properties (Section 15.9).
- Explain how surfactants work (Section 15.10).
- Discuss the earth's water supply and the sources of fresh water (Secton 15.11).
- Discuss how municipal drinking water is purified (Section 15.11).
- Describe what causes hard water and how it can be softened (Section 15.11).
- Explain how household wastes can contaminate groundwater (Section 15.11).

KEY TERMS

boiling point elevation
(15.7)
colligative properties
(15.7)
colloid *(15.9)*
continuous phase
(15.9)
detergents *(15.10)*

dispersed phase *(15.9)*
emulsion *(15.9)*
enthalpy of solution
(15.2)
freezing point lowering
(15.7)
Henry's law *(15.5)*
hydration *(15.3)*

hypertonic *(15.8)*
hypotonic *(15.8)*
immiscible *(15.1)*
isotonic *(15.8)*
lattice energy *(15.3)*
mass fraction *(15.6)*
miscible *(15.1)*
molality *(15.7)*

osmosis *(15.8)*
osmotic pressure *(15.8)*
parts per billion *(15.6)*
parts per million *(15.6)*
parts per trillion *(15.6)*
Raoult's law *(15.7)*
reverse osmosis *(15.8)*

saturated solution *(15.3)*
semipermeable membrane *(15.8)*
solubility *(15.3)*
supersaturated solution *(15.3)*

surfactants *(15.10)*
Tyndall effect *(15.9)*
unsaturated solution *(15.3)*
weight percent *(15.6)*

QUESTIONS FOR REVIEW AND THOUGHT

Conceptual Challenge Problems

CP-15.A (Section 15.6) Concentrations expressed in units of parts per million and parts per billion often have no meaning for people until they relate these small and large numbers to their own experiences.

(a) What time in seconds is 1 ppm of a year?

(b) What time in seconds is 1 ppb of a 70-year lifetime ?

CP-15.B (Section 15.4) Bodies of water with an abundance of nutrients that support a blooming growth of plants are said to be eutrophoric. In general, fish do not thrive for long in eutrophoric waters because there is little oxygen for fish.

Suppose someone asked you why this was true, given the fact that growing plants produce oxygen as a product of photosynthesis. How would you respond to this person's inquiry?

CP-15.C (Section 15.7) Suppose that you want to produce the lowest temperature possible by using ice, sodium chloride, and water to chill homemade ice cream made in a 1.5-L metal cylinder surrounded by a coolant held in a wooden bucket. You have all the ice, salt, and water you want. How would you plan to do this?

Answers to questions in **bold** can be found in the back of the book.

Review Questions

1. Which of these general types of substances would you expect to dissolve readily in water?
   (a) Alcohols   (b) Hydrocarbons
   (c) Metals   (d) Nonpolar molecules
   (e) Polar molecules (f) Salts

2. Explain on a molecular basis why the components of blended motor oils remain dissolved and do not separate.

3. Describe the differences among solutions that are unsaturated, saturated, and supersaturated in terms of amount of solute.

4. Describe the differences among unsaturated, saturated, and supersaturated solutions in terms of Q and K_c.

5. State Henry's law. Name three factors that govern the solubility of a gas in a liquid.

6. In general, how does the water solubility of most ionic compounds change as the temperature is increased?

7. How does the solubility of gases in liquids change with increased temperature? Explain why.

8. Which is the highest solute concentration: 50 ppm, 500 ppb, or 0.05% by weight?

9. Estimate your concentration on campus in parts per million and parts per thousand.

10. Define molality. How does it differ from molarity?

11. Explain the difference between the mass fraction and the mole fraction of solute in a solution.

12. Explain why the vapor pressure of a solvent is lowered by the presence of a nonvolatile solute.

13. Why is a higher temperature required for boiling a solution containing a nonvolatile solute than for boiling the pure solvent?

14. Which would have the lowest freezing point?
    (a) A 1.0 m NaCl solution   (b) A 1.0 m CaCl$_2$ solution
    (c) a 1.0 m methanol solution
    Explain your choice.

15. Write the osmotic pressure equation and explain all terms.

16. Explain the difference between (a) a hypotonic and an isotonic solution; (b) an isotonic and a hypertonic solution.

17. Explain how reverse osmosis works.

18. How do colloids differ from suspensions?

19. Explain why the Tyndall effect is not observed with solutions.

20. How can the presence of a strong electrolyte cause a hydrophobic colloid to coagulate?

21. Sketch an illustration of a soap molecule. Based on its structure, why is it considered a surfactant?

22. Surfactant molecules have what common structural features?

How Substances Dissolve

23. Explain why some liquids are miscible in each other while other liquids are immiscible. Using only three liquids, give an example of a miscible pair and an immiscible pair.

24. Why would the same solid readily dissolve in one liquid and be almost insoluble in another liquid? Give an example of such behavior.

25. Knowing that the solubility of oxalic acid at 25 °C is 1 g per 7 g of water, how would you prepare 1 L of a saturated oxalic acid solution?

26. A saturated solution of NH_4Cl was prepared by adding solid NH_4Cl to water until no more solid NH_4Cl would dissolve. The resulting mixture felt very cold and had a layer of undissolved NH_4Cl on the bottom. When the mixture reached room temperature, no solid NH_4Cl was present. Explain what happened. Was the solution still saturated?

27. The lattice energy of $CaCl_2$ is -2258 kJ/mol, and its enthalpy of hydration is $+2175$ kJ/mol. Is the process of dissolving $CaCl_2$ in water endothermic or exothermic?

28. Simple acids such as formic acid, $HCOOH$, and acetic acid, CH_3COOH, are very soluble in water; however, fatty acids such as stearic acid, $CH_3(CH_2)_{15}COOH$, and palmitic acid, $CH_3(CH_2)_{14}COOH$, are water-insoluble. Based on what you know about the solubility of alcohols, explain the solubility of these organic acids.

29. If a solution of a certain salt in water is saturated at some temperature and a few crystals of the salt are added to the solution, what do you expect will happen? What happens if the same quantity of the same salt crystals is added to an unsaturated solution of the salt? What would you expect to happen if the temperature of this second salt solution is slowly lowered?

30. Describe what happens when an ionic solid dissolves in water. Sketch an illustration that includes at least three positive ions, three negative ions, and a dozen or so water molecules in the vicinity of the ions.

31. The partial pressure of O_2 in your lungs varies from 25 mm Hg to 40 mm Hg. What concentration of O_2 (in grams per liter) can dissolve in water at 37 °C when the O_2 partial pressure is 40. mm Hg? The Henry's law constant for O_2 at 37 °C is 1.5×10^{-6} mol L^{-1} mm Hg^{-1}.

32. The Henry's law constant for nitrogen in blood serum is approximately 8×10^{-7} mol L^{-1} mm Hg^{-1}. What is the N_2 concentration in a diver's blood at a depth where the total pressure is 2.5 atm? The air the diver is breathing is 78% N_2 by volume.

Concentration Units

33. Convert 2.5 ppm to weight percent.

34. Convert 73.2 ppm to weight percent.

35. Mathematically show how 1 ppm is equivalent to 1 mg/1 kg.

36. Mathematically show how 1 ppb is equivalent to 1 μg/1 kg.

37. What mass (in grams) of sucrose is in 1.0 kg of a 0.25% sucrose solution?

38. How many grams of ethanol are in 750 mL of a 12% ethanol solution? (Assume its density is the same as for water.)

39. A sample of lead-based paint is found to contain 60.5 ppm lead. The density of the paint is 8.0 lb/gal. What mass of lead (in grams) would be present in 50. gal of this paint?

40. A paint contains 200. ppm lead. Approximately what mass of lead (in grams) will be in 1.0 cm^2 of this paint (density = 8.0 lb/gal) when 1 gal is uniformly applied to 500. ft^2 of a wall?

41. Hydrochloric acid is sold as a concentrated aqueous solution. The concentration of commercial HCl is 11.7 M and its density is 1.18 g/cm^3. Calculate the weight percent of HCl in the solution.

42. Concentrated sulfuric acid has a density of 1.84 g/cm^3 and is 18 M. What is the weight percent of H_2SO_4 in the solution?

43. You need a 0.050 m aqueous solution of methanol (CH_3OH). What mass of methanol would you need to dissolve in 500. g of water to make this solution?

44. You want to prepare a 1.0 m solution of ethylene glycol, $C_2H_4(OH)_2$, in water. What mass of ethylene glycol do you need to mix with 950. g of water?

45. A 23.2% by weight aqueous solution of sucrose has a density of 1.127 g/mL. Calculate the molarity of sucrose in this solution.

46. An aqueous beverage has a lead concentration of 25 ppb. Express this lead concentration as molarity.

Colligative Properties

47. What is the boiling point of a solution containing 0.200 mol of a nonvolatile nonelectrolyte solute in 100. g of benzene? The normal boiling point of benzene is 80.10 °C, and $K_b = 2.53$ °C kg mol^{-1}.

48. What is the boiling point of a solution composed of 15.0 g of urea, $(NH_2)_2CO$, in 0.500 kg of water?

49. Place the following aqueous solutions in order of increasing boiling point:
    (a) 0.10 mol/kg KCl (b) 0.10 mol/kg glucose
    (c) 0.080 mol/kg $MgCl_2$

50. List the following aqueous solutions in order of decreasing freezing point:
    (a) 0.10 m methanol   (b) 0.10 mol/kg KCl
    (c) 0.080 mol/kg $BaCl_2$ (d) 0.040 mol/kg Na_2SO_4
    (Assume that all of the salts dissociate completely into their ions in solution.)

51. Calculate the boiling point at 760 mm Hg and the freezing point of these solutions:
    (a) 20.0 g of citric acid, $C_6H_8O_7$, in 100.0 g of water
    (b) 3.00 g of CH_3I in 20.0 g of benzene. K_b benzene = 2.53 °C kg mol^{-1}; K_f benzene = 5.10 °C kg mol^{-1}

52. Calculate the freezing and boiling points (at 760 mm Hg) of a solution of 4.00 g of urea, $CO(NH_2)_2$, dissolved in 75.0 g of water.

53. Calculate the number of grams of urea that must be added to 150. g of water to give a solution whose vapor pressure is 2.5 mm Hg less than that of pure water at 40 °C (vp H_2O at 40 °C = 55.34 mm Hg).

54. At 60 °C the vapor pressure of pure water is 149.44 mm Hg and that above an aqueous sucrose ($C_{12}H_{22}O_{11}$) solution is 119.55 mm Hg. Calculate the mole fraction of water and the number of grams of sucrose in the solution if the mass of water is 150. g.

55. At 760 mm Hg, a solution of 5.52 g of glycerol in 40.0 g of water has a boiling point of 100.777 °C. Calculate the empirical molar mass of glycerol.

56. The boiling point of benzene is increased by 0.65 °C when 5.0 g of an unknown organic compound (a nonelectrolyte) is dissolved in 100. g of benzene. Calculate the approximate weight of the organic compound. K_b benzene = 2.53 °C kg mol^{-1}.

57. You add 0.255 g of an orange crystalline compound with an empirical formula of $C_{10}H_8Fe$ to 11.12 g of benzene.

The boiling point of the solution is 80.26 °C. The normal boiling point of benzene is 80.10 °C and its K_b = 2.53 °C kg mol^{-1}. What are the molar mass and molecular formula of the compound?

58. Anthracene, a hydrocarbon obtained from coal, has an empirical formula of C_7H_5. To find its molecular formula you dissolve 0.500 g of anthracene in 30.0 g of benzene. The boiling point of the solution is 80.34 °C. The normal boiling point of benzene is 80.10 °C, and K_b = 2.53 °C kg mol^{-1}. What are the molar mass and molecular formula of anthracene?

59. If you use only water and pure ethylene glycol, $HOCH_2CH_2OH$, in your car's cooling system, what mass (in grams) of the glycol must you add to each quart of water to give freezing protection down to −31.0 °C?

60. Some ethylene glycol, $HOCH_2CH_2OH$, was added to your car's cooling system along with 5.0 kg of water.
(a) If the freezing point of the solution is −15.0 °C, what mass (in grams) of the glycol must have been added?
(b) What is the boiling point of the coolant mixture?

61. Calculate the concentration of nonelectrolyte solute particles in human blood if the osmotic pressure is 7.53 atm at 37 °C, the temperature of the body.

62. The blood of cold-blooded animals and fish is isotonic with sea water. If sea water freezes at −2.3 °C, what is the osmotic pressure of the blood of these animals at 20.0 °C? (Assume the density is that of pure water.)

63. An osmotic pressure of 5.15 atm is developed by a solution containing 4.80 g of dioxane (a nonelectrolyte) dissolved in 250. mL of water at 15.0 °C. The empirical formula of dioxane is C_2H_4O. Use the osmotic pressure data to show that the empirical formula and the molecular formula of dioxane are not the same.

64. The molar mass of a polymer was determined by measuring the osmotic pressure, 7.6 mm Hg, of a solution containing 5.0 g of the polymer dissolved in 1.0 L of benzene. What is the molar mass of the polymer? Assume a temperature of 298.15 K.

Water: Purification and Solutions

65. When lead is present in drinking water, in what form does it exist?

66. Dietitians recommend drinking six 8-oz glasses of water each day. If your drinking water contains the maximum contamination level for arsenic, 0.050 ppm, how much arsenic would you consume in a week following this recommendation?

67. The maximum contamination level (MCL) for chlordane is 0.002 ppm. A sample of well water contained 5 ppb chlordane. Is the sample within the MCL for chlordane?

68. How do the lime-soda and ion-exchange processes differ in treating hard water?

69. In a home, 200. gallons of hard water containing 500 mg Ca^{2+}/gal passed through the Na$^+$-based ion-exchange water softener. If the ion-exchange resin operates at 100% efficiency, what mass of Na$^+$ ions is displaced from the resin?

70. Explain how hard water produces "ring around the bathtub."

71. During municipal drinking water treatment, water is sprayed into the air. Why is this done?

72. Describe the benefits and risks of chlorinating municipal drinking water.

73. Discuss the risks and benefits of using ozone to treat municipal drinking water.

74. Why is it necessary to bubble air through an aquarium?

General Questions

75. What is the difference between solubility and miscibility?

76. If 5 g of solvent, 0.2 g of solute A, and 0.3 g of solute B are mixed to form a solution, what is the weight percent concentration of A?

77. A chemistry classmate tells you that a supersaturated solution is also saturated. Is the student correct? What would you tell the student about her/his statement?

78. In *The Rime of the Ancient Mariner* the poet Samuel Taylor Coleridge wrote, ". . . Water, water everywhere/ And all the boards did shrink. . ." Explain this in terms of osmosis.

79. A 10.0 M aqueous solution of NaOH has a density of 1.33 g/cm^3 at 20 °C. Calculate the weight percent of NaOH in the solution.

80. Concentrated aqueous ammonia is 14.8 M and has a density of 0.90 g/cm^3. Calculate the weight percent of NH_3 in the solution.

81. Propylene glycol is used in "environmentally friendly" antifreezes. Assume you dissolve 45.0 g of propylene glycol, $C_3H_6(OH)_2$, in 0.500 L of water. Calculate the molality and weight percent of propylene glycol in the solution.

82. Dimethylglyoxime (DMG) reacts with nickel(II) ion in aqueous solution to form a bright red compound. However, DMG is insoluble in water. In order to get it into aqueous solution where it can encounter Ni^{2+} ions, it must first be dissolved in a suitable solvent such as ethanol. Suppose you dissolve 45.0 g of DMG ($C_4H_8N_2O_2$) in 500. mL of ethanol (C_2H_5OH, density = 0.7893 g/mL). What are the molality and weight percent of DMG in this solution?

83. Arrange the following aqueous solutions in order of increasing boiling point:
(a) 0.20 mol/kg ethylene glycol
(b) 0.12 mol/kg K_2SO_4
(c) 0.10 mol/kg $BaCl_2$
(d) 0.12 mol/kg KBr

84. Arrange the following aqueous solutions in order of decreasing freezing point
(a) 0.20 mol/kg ethylene glycol
(b) 0.12 mol/kg Na_2SO_4
(c) 0.10 mol/kg NaBr
(d) 0.12 mol/kg KI

85. The solubility of NaCl in water at 100 °C is 39.1 g/100. g of water. Calculate the boiling point of a saturated solution of NaCl.

86. The organic salt $[(C_4H_9)_4N][ClO_4]$ consists of the ions $(C_4H_9)_4N^+$ and ClO_4^-. The salt dissolves in chloroform. What mass (in grams) of the salt must have been dissolved if the boiling point of a solution of the salt in 25.0 g of chloroform is 63.20 °C? The normal boiling point of chloroform is 61.70 °C and K_b = 3.63 °C kg mol^{-1}. Assume that the salt dissociates completely into its ions in solution.

87. A solution, prepared by dissolving 9.41 g of $NaHSO_3$ in 1.00 kg of water, freezes at $-0.33\ °C$. From these data, decide which of the following equations is the correct expression for the ionization of the salt.
(a) $NaHSO_3(aq) \rightarrow Na^+(aq) + HSO_3^-(aq)$
(b) $NaHSO_3(aq) \rightarrow Na^+(aq) + H^+(aq) + SO_3^{2-}(aq)$

88. In chemical research, newly synthesized compounds are often sent to commercial laboratories for analysis that determines the weight percent of C and H by burning the compound and collecting the evolved CO_2 and H_2O. The molar mass is determined by measuring the osmotic pressure of a solution of the compound. Calculate the empirical and molecular formulas of a compound, C_xH_yCr, given the following information:
(a) The compound contains 73.94% C and 8.27% H; the remainder is chromium.
(b) At 25 °C, the osmotic pressure of 5.00 mg of the unknown dissolved in 100. mL of chloroform solution is 3.17 mm Hg.

89. The Ca^{2+} ions in hard water are often precipitated as $CaCO_3$ by adding soda ash, Na_2CO_3. If the calcium ion concentration in hard water is 0.010 mol/L, and if the Na_2CO_3 is added until the carbonate ion concentration is 0.050 mol/L, what percentage of the calcium ion has been removed from the water? (You may neglect the interaction of carbonate ions with water.)

90. Aluminum chloride reacts with phosphoric acid to give aluminum phosphate, $AlPO_4$, which is used industrially as the basis of adhesives, binders, and cements.
(a) Write a balanced equation for the reaction of aluminum chloride and phosphoric acid.
(b) If you begin with 152. g of aluminum chloride and 3.00 L of 0.750 M phosphoric acid, what mass of $AlPO_4$ can be isolated?
(c) If you place 25.0 g of $AlPO_4$ in enough pure water to have a volume of exactly 1 L, what are the concentrations of Al^{3+} and PO_4^{3-} at equilibrium? (Assume complete dissociation and no hydrolysis)

Applying Concepts

91. Using these symbols,

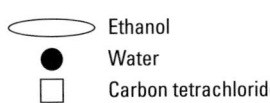

    Ethanol
    Water
    Carbon tetrachloride

draw nanoscale diagrams for the contents of a beaker containing
(a) Water and ethanol.
(b) Water and carbon tetrachloride.

92. Using these symbols,

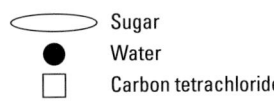

    Sugar
    Water
    Carbon tetrachloride

draw nanoscale diagrams for the contents of a beaker containing
(a) Water and sugar.
(b) Carbon tetrachloride and sugar.

93. Refer to Figure 15.10 to determine whether these situations would result in an unsaturated, saturated, or supersaturated solution.
(a) 40 g of NH_4Cl is added to 100 g of H_2O at 80 °C.
(b) 100 g of LiCl is dissolved in 100 g of H_2O at 30 °C.
(c) 120 g of $NaNO_3$ is added to 100 g of H_2O at 40 °C.
(d) 50 g of Li_2SO_4 is dissolved in 200 g of H_2O at 50 °C.

94. Refer to Figure 15.10 to determine whether these situations would result in an unsaturated, saturated, or supersaturated solution.
(a) 120 g of RbCl is added to 100 g of H_2O at 50 °C.
(b) 30 g of KCl is dissolved in 100 g of H_2O at 70 °C.
(c) 20 g of NaCl is dissolved in 50 g of H_2O at 60 °C.
(d) 150 g of CsCl is added to 100 g of H_2O at 10 °C.

95. Complete this table.

Compound	Mass of compound	Mass of water	Mass fraction	Weight percent	ppm
Table salt	52 g	175 g	____	____	____
Glucose	15 g	____	____	____	7×10^4
Methane	____	100. g	____	0.0025%	____

96. Complete this table.

Compound	Mass of compound	Mass of water	Mass fraction of solute	Weight percent of solute	ppm of solute
Lye	____	125 g	0.375	____	____
Glycerol	33 g	200. g	____	____	____
Acetylene	0.0015 g	____	____	0.0009%	____

97. If KI is added to a saturated solution of SrI_2, will the amount of solid SrI_2 present decrease, increase, or remain unchanged? What about the concentration of Sr^{2+} ion in solution?

98. What happens on the molecular level when a liquid freezes? What effect does a nonvolatile solute have on this process? Comment on the purity of water obtained by melting an iceberg.

99. Criticize these statements:
(a) A saturated solution is always a concentrated one.
(b) A 0.10 mol/kg sucrose solution and a 0.10 mol/kg KCl solution have the same osmotic pressure.

100. In your own words, explain why
(a) Sea water has a lower freezing point than fresh water.
(b) Salt is added to the ice in an ice cream maker to freeze the ice cream faster.

General Chemistry CD-ROM

CD15.1 Screen 14.3: The Solution Process. Examine the problem screen associated with this screen. What do you think would be found if the solubility of hexane (C_6H_{14}) was examined in these same two solvents, water and carbon tetrachloride? In which would hexane be more soluble?

CD15.2 Screen 14.4: Energetics of Solution Formation. Examine the *Oil and Water* sidebar on this screen. What prevents water and oil from mixing intimately: strong attractions between oil molecules, strong attractions between water molecules, or weak attractions between oil and water molecules?

CD15.3 Screen 14.7: Colligative Properties (1).

(a) Why is the vapor pressure of a liquid lowered upon dissolution of a solute?

(b) Why, on the molecular scale, is the vapor pressure of a solution proportional to the mole fraction of the solute?

(c) What substance would have the greatest influence on the vapor pressure of water when added to 1000 g of the solvent: 10.0 g of sucrose ($C_{12}H_{22}O_{11}$), 10.0 g of ethylene glycol [$C_2H_4(OH)_2$], or 10.0 g of AgCl?

CD15.4 Screen 14.9: Colligative Properties (3).

(a) This screen explains the observations first seen on the Chemical Puzzler screen. Osmotic pressure is found to be responsible for the changes in the egg's size. What part of the egg acts as a semipermeable membrane?

(b) If the egg were put in concentrated salt water, what would happen to its size?

CD15.5 Screen 14.11: Surfactants. Explain how surfactants act to help oil and water form a solution.

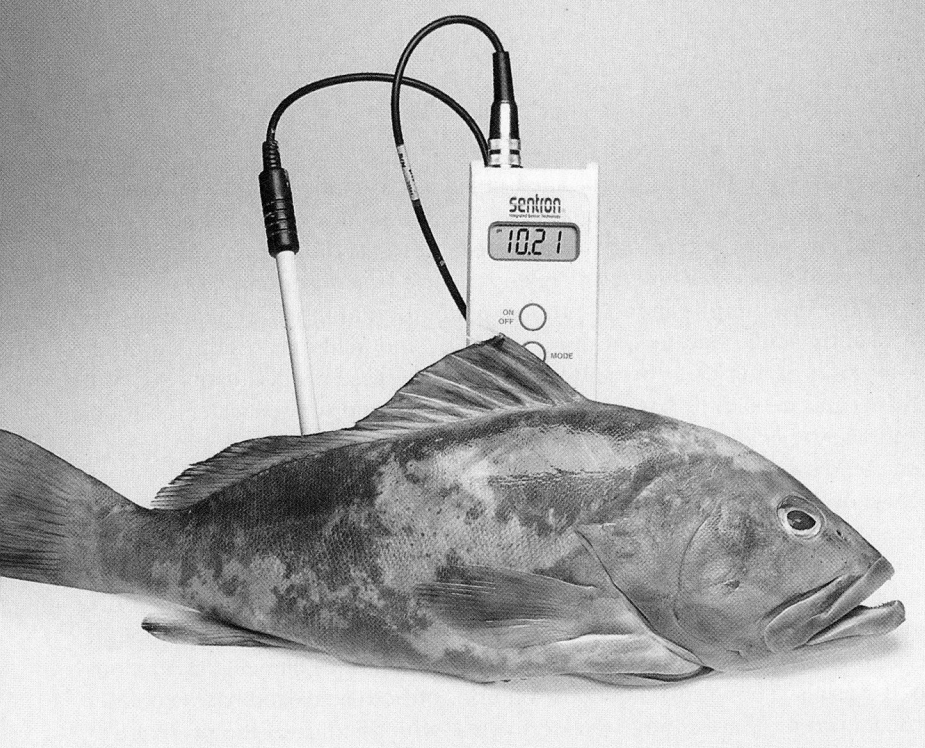

Measurement of the acidity or alkalinity of substances is important in commerce, medicine, and studies of the environment. Such measurements can be expressed in terms of pH. For example, the pH of this fish's meat is 10.21, as indicated by the pH meter. The measurement and uses of pH are discussed in this chapter.

ACIDS AND BASES

16.1 The Brønsted-Lowry Concept of Acids and Bases

16.2 Carboxylic Acids and Amines

16.3 The Autoionization of Water

16.4 The pH Scale

16.5 Ionization Constants of Acids and Bases

16.6 Problem Solving Using K_a and K_b

16.7 Molecular Structure and Acid Strength

16.8 Acid-Base Reactions of Salts

16.9 Practical Acid-Base Chemistry

16.10 Lewis Acids and Bases

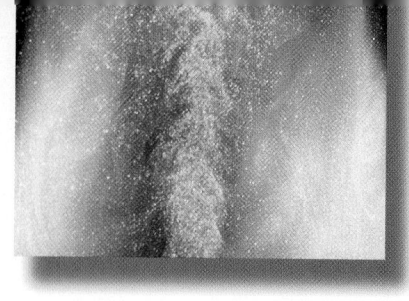

It is difficult to overstate the importance of acids and bases. Aqueous solutions, which abound in our environment and in all living organisms, are almost always acidic or basic to some degree. Photosynthesis and respiration, the two most important biological processes on earth, depend on acid-base reactions. Carbon dioxide (CO_2) is the most important acid-producing compound in nature. Rainwater is generally slightly acidic because of dissolved CO_2, and acid rain results from further acidification of rainwater by acids formed by the gaseous pollutants SO_2 and NO_2. The oceans are slightly basic, as are many ground and surface waters. Natural waters can also be acidic; the more acidic the water, the more easily metals such as lead can be dissolved from pipes or soldered joints.

Because of their importance, the properties of acids and bases have been studied extensively. In 1677 Antoine Lavoisier proposed that oxygen made an acid acidic; he even derived the name *oxygen* from Greek words meaning "acid former." But in 1808 it was discovered that the gaseous compound HCl, which dissolves in water to give hydrochloric acid, contains only hydrogen and chlorine. It later became clear that hydrogen, not oxygen, is common to all acids in aqueous solution. It was also shown that aqueous solutions of both acids and bases conduct an electrical current; that is, they are electrolytes, which indicates the presence of ions. In 1887 the Swedish chemist Svante Arrhenius proposed that acids *ionize in aqueous solution to produce hydrogen ions (protons) and anions;* bases *ionize to produce hydroxide ions and cations.* We now rely on a more general acid-base concept: *hydronium ions, H_3O^+, are responsible for the properties of acidic aqueous solutions, and hydroxide ions, OH^-, are responsible for the properties of basic aqueous solutions.*

Because of their small size and high charge density, hydrogen ions (protons) are always associated with water molecules in aqueous solution and are usually represented as H_3O^+, the hydronium ion.

CD-ROM Screen 17.2: Brønsted Acids and Bases

16.1 THE BRØNSTED-LOWRY CONCEPT OF ACIDS AND BASES

A major problem with the Arrhenius acid-base concept is that certain substances, such as ammonia, NH_3, produce basic solutions and react with acids yet contain no hydroxide ions. In 1923 J. N. Brønsted in Denmark and T. M. Lowry in England independently proposed a new way of defining acids and bases in aqueous solutions:

- **Brønsted-Lowry acids** are hydrogen ion donors.
- **Brønsted-Lowry bases** are hydrogen ion acceptors.

According to the Brønsted-Lowry concept, an acid can donate a hydrogen ion, H^+ (a proton), to another substance, while a base can accept an H^+ ion from another substance. In **acid-base reactions,** acids donate H^+ ions, and bases accept them. To accept an H^+ and serve as a Brønsted-Lowry base, a molecule or ion must have an *unshared pair of electrons.* For example, ammonia, NH_3, is a base because it accepts an H^+ from water (an acid) to form an ammonium ion, NH_4^+. Water, having lost an H^+, is converted into a hydroxide ion, OH^-.

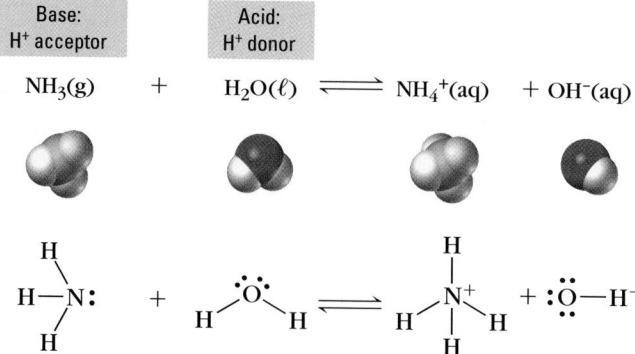

Ammonia establishes an equilibrium with water, ammonium ions, and hydroxide ions and is therefore a weak base (⬅ *p. 172*). An aqueous solution of ammonia is sometimes called an ammonium hydroxide solution, but this is not a good name because the solution contains far more un-ionized ammonia molecules than ammonium and hydroxide ions.

In aqueous solutions, H^+ ions from acids such as nitric acid, HNO_3, react with water to form hydronium ions, H_3O^+.

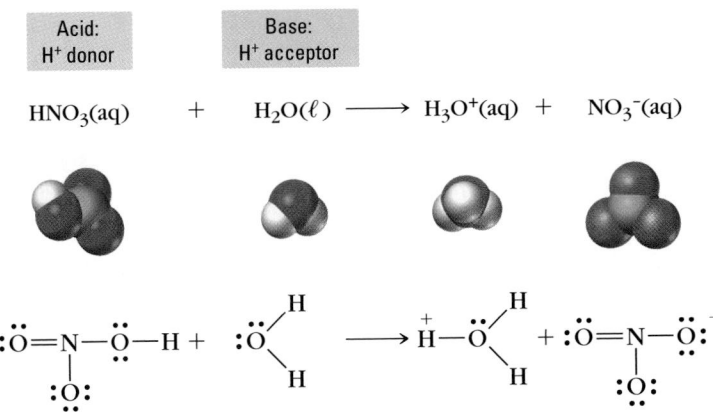

Acid: H^+ donor		Base: H^+ acceptor			
$HNO_3(aq)$	$+$	$H_2O(\ell)$	$\longrightarrow$	$H_3O^+(aq)$ $+$	$NO_3^-(aq)$

In reacting with an acid, water acts as a Brønsted-Lowry base by using an unshared electron pair to accept the H^+. Because nitric acid is a strong acid, it is 100% ionized in aqueous solution, and the above equation is written with a single arrow.

In contrast, a weak acid does *not* ionize completely and therefore is a weak electrolyte that establishes an equilibrium with water. Hydrogen fluoride, for example, is a weak acid, and its ionization in water is written as

$$HF(aq) \quad + H_2O(\ell) \rightleftharpoons H_3O^+(aq) + F^-(aq)$$

hydrogen fluoride
weak acid

An ionic substance that dissolves in water is a strong electrolyte (⬅ *p. 170*). A strong acid ionizes completely and is a strong electrolyte.

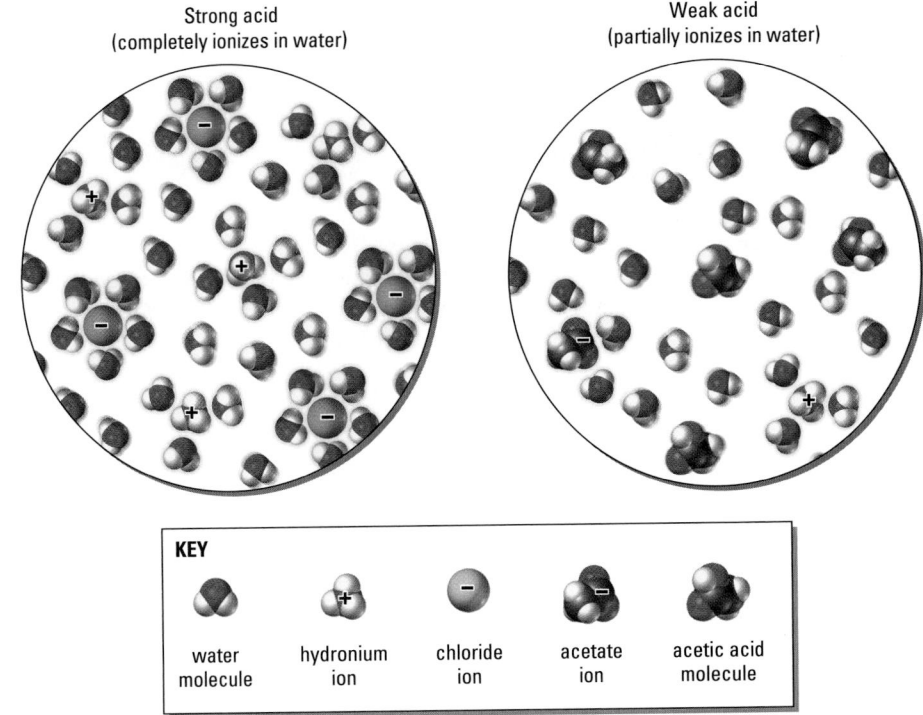

Strong acid
(completely ionizes in water)

Weak acid
(partially ionizes in water)

KEY

water molecule	hydronium ion	chloride ion	acetate ion	acetic acid molecule

Ionization of acids in water. A strong acid such as hydrochloric acid (HCl) is completely ionized in water; a weak acid such as acetic acid (CH_3COOH) is only partially ionized in water.

The double arrow indicates an equilibrium between the reactants and the products. Because ionization of HF is much less than 100%, this means that at equilibrium most HF molecules are un-ionized, and there are relatively few hydronium and fluoride ions in solution. Thus, the ionization of *weak* acids is a reactant-favored process.

 Exercise 16.1   Brønsted-Lowry Acids and Bases

Identify each molecule or ion as a Brønsted-Lowry acid or base.

(a) HBr   (b) Br⁻   (c) HNO_2 (d) CH_3NH_2

 Exercise 16.2   Using Le Chatelier's Principle

Use Le Chatelier's principle to explain why a larger percentage of NH_3 will be ionized in a very dilute solution than in a less dilute solution.

Water's Role as Acid or Base

In aqueous solution, all Brønsted-Lowry acids and bases react with water molecules. As we have seen, a water molecule *accepts an H⁺* from an acid such as nitric acid, while a water molecule *donates an H⁺* to a base like an ammonia molecule. According to the Brønsted-Lowry definitions, water serves as a base (an H⁺ acceptor) when an acid is present and as an acid (an H⁺ donor) when a base is present. Therefore, water displays both acid and base properties—*it can donate or accept H⁺* ions, depending on the circumstances. The general reactions of water with acids (HA) and molecular bases (B) are

> A substance that can donate or accept H⁺ is said to be amphiprotic.

Water acting as a base

$$HA + H_2O \rightleftharpoons H_3O^+ + A^-$$
$$\text{acid} \qquad \text{base}$$

Water acting as an acid

> If the base is an anion, B⁻, it accepts H⁺ to give BH.

$$B + H_2O \rightleftharpoons BH^+ + OH^-$$
$$\text{base} \qquad \text{acid}$$

Exercise 16.3   Acids and Bases

Complete these equations. (*Hint:* CH_3NH_2 and $(CH_3)_2NH$ are bases.)

(a) $HCN + H_2O \rightarrow$ (b) $HBr + H_2O \rightarrow$
(c) $CH_3NH_2 + H_2O \rightarrow$ (d) $(CH_3)_2NH + H_2O \rightarrow$

Conjugate Acid-Base Pairs

> CD-ROM Screen 17.2:
> Brønsted Acids and Bases

Whenever an acid donates H⁺ to a base, a new acid and a new base are formed. This can be understood by looking at the reaction between acetic acid, CH_3COOH, and water. The products of the reaction are a new acid, H_3O^+, and a new base, CH_3COO^-.

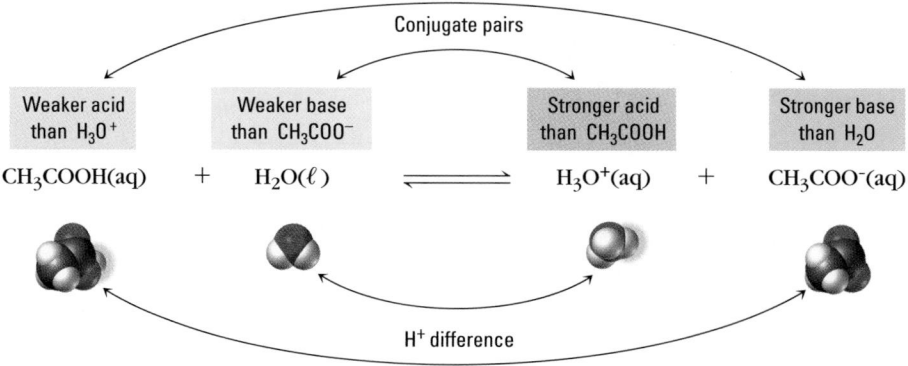

The structures of CH_3COOH and CH_3COO^- differ from one another by only a single H^+, just as the structures of H_2O and H_3O^+ do.

A pair of molecules or ions *related to one another by the loss or gain of a single H^+* is called a **conjugate acid-base pair.** Every Brønsted-Lowry acid has its conjugate base, and every Brønsted-Lowry base has its conjugate acid.

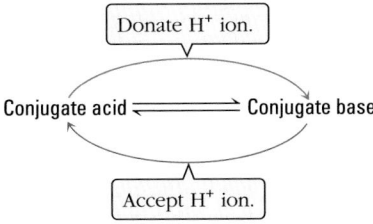

Removing an H^+ ion from the acid and making the charge of the remaining portion of the acid one unit more negative gives the conjugate base. For example, the conjugate base of the acid HF is the F^- ion; HF and F^- are a conjugate acid-base *pair.*

$$\text{Donate } H^+ \longrightarrow$$

$$HF(aq) + H_2O \rightleftharpoons H_3O^+(aq) + F^-(aq)$$

ACID   base   acid   BASE

$$\longleftarrow \text{ Accepts } H^+$$

In the forward reaction, HF is a Brønsted-Lowry acid. It donates an H^+ to water, which, by accepting the H^+, acts as a Brønsted-Lowry base. In the reverse reaction, fluoride ion is the base, accepting an H^+ from H_3O^+, the acid. As noted in the equation, there are two conjugate acid-base pairs: (1) HF and F^- and (2) H_2O and H_3O^+. One member of a conjugate acid-base pair is always a reactant and the other is always a product; they are never both products or both reactants.

The conjugate acid can be derived from the formula of its conjugate base by adding an H^+ ion to the conjugate base and making the charge of the base one unit more positive. So the conjugate acid of the Cl^- ion is HCl; the conjugate acid of NH_3 is NH_4^+.

$$\text{Accept } H^+ \longrightarrow$$

$$NH_3(aq) + H_2O(\ell) \rightleftharpoons NH_4^+(aq) + OH^-(aq)$$

BASE   acid   ACID   base

$$\longleftarrow \text{ Donate } H^+$$

Problem-Solving Example 16.1   Conjugate Acid-Base Pairs

Complete the following table.

Acid	Its conjugate base	Base	Its conjugate acid
$H_2PO_4^-$	_____	_____	HPO_4^{2-}
_____	H^-	NH_2^-	_____
HSO_3^-	_____	ClO_4^-	_____
HF	_____	_____	HBr

Answer

Acid	Its conjugate base	Base	Its conjugate acid
$H_2PO_4^-$	HPO_4^{2-}	PO_4^{3-}	HPO_4^{2-}
H_2	H^-	NH_2^-	NH_3
HSO_3^-	SO_3^{2-}	ClO_4^-	$HClO_4$
HF	F^-	Br^-	HBr

Explanation In each of these cases, apply the relationship

$$\text{Conjugate acid} \quad \underset{\longleftarrow \text{Accept } H^+}{\overset{\text{Donate } H^+ \longrightarrow}{\rightleftharpoons}} \quad \text{Conjugate base}$$

The conjugate acid can be identified by adding H^+ to the conjugate base; the conjugate base forms by loss of H^+ from the conjugate acid. For example, because F^- has no H^+ to donate, it must be a base, the conjugate base of HF, its conjugate acid. HPO_4^{2-} is the conjugate base of its conjugate acid, $H_2PO_4^-$. The other conjugate acid-base pairs can be worked out similarly.

Problem-Solving Practice 16.1

Identify the conjugate acid or base of these acids or bases.

(a) HClO   (b) O^{2-} (c) HCOOH
(d) OH^- (e) IO_3^- (f) PH_3

Exercise 16.4 HSO_4^- as a Base

(a) Write the equation for HSO_4^- ion acting as a base in water.
(b) Identify the conjugate acid-base pairs.

Relative Strengths of Acids and Bases

Strong acids are better H^+ ion donors than weak acids. Correspondingly, strong bases are better H^+ ion acceptors than weak bases. Thus, *stronger acids have weaker conjugate bases and weaker acids have stronger conjugate bases.* For comparison, consider the two acids HCl, a strong acid, and HF, a weak acid.

$$HCl(aq) + H_2O(\ell) \longrightarrow H_3O^+(aq) + Cl^-(aq)$$

$$HF(aq) + H_2O(\ell) \rightleftharpoons H_3O^+(aq) + F^-(aq)$$

The ionization of HCl is virtually 100%; essentially the reverse reaction does not occur. The Cl^- ion exhibits virtually no tendency to accept H^+ from H_3O^+. On the other hand, the reverse of the ionization of HF is significant; F^- ion readily accepts H^+ from H_3O^+, and hydrofluoric acid is a weak acid that is mainly un-ionized.

By measuring the extent to which various acids donate H^+ ions to water, chemists have developed an extensive tabulation of the relative strengths of acids and their conjugate bases. An abbreviated table is given in Figure 16.1.

The strongest acids are at the top left of Figure 16.1, with the weakest bases at the top right. The weakest acids are at the bottom left along with the strongest bases at the bottom right. From Figure 16.1 we can draw two important generalizations regarding conjugate acid-base pairs:

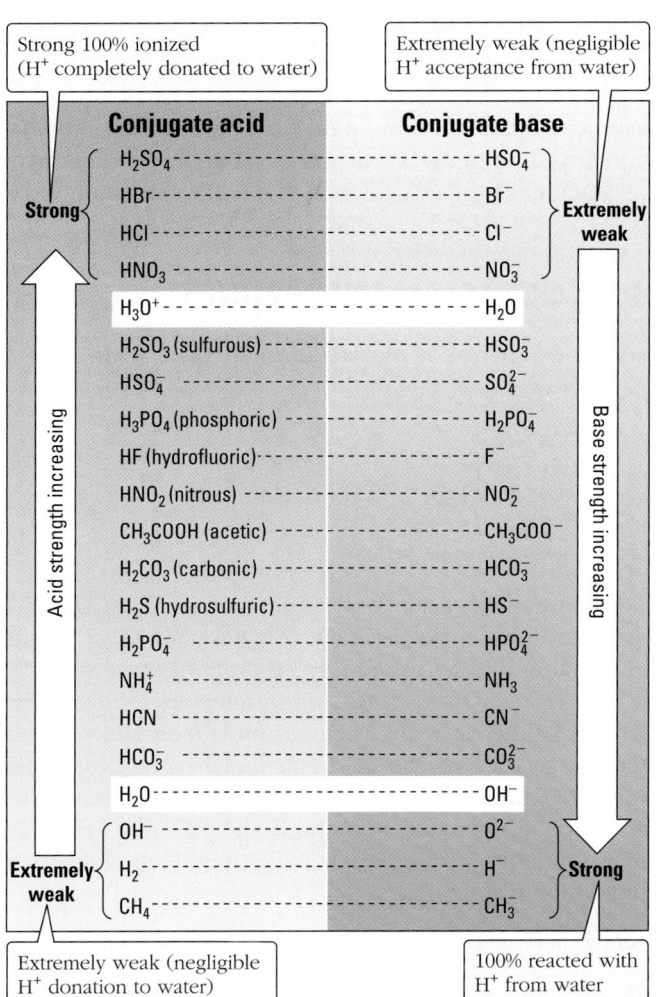

Figure 16.1 Relative strengths of conjugate acids and bases.

Strong 100% ionized (H^+ completely donated to water)

Extremely weak (negligible H^+ acceptance from water)

Conjugate acid		Conjugate base
H_2SO_4	--------	HSO_4^-
HBr	--------	Br^-
HCl	--------	Cl^-
HNO_3	--------	NO_3^-
H_3O^+	--------	H_2O
H_2SO_3 (sulfurous)	--------	HSO_3^-
HSO_4^-	--------	SO_4^{2-}
H_3PO_4 (phosphoric)	--------	$H_2PO_4^-$
HF (hydrofluoric)	--------	F^-
HNO_2 (nitrous)	--------	NO_2^-
CH_3COOH (acetic)	--------	CH_3COO^-
H_2CO_3 (carbonic)	--------	HCO_3^-
H_2S (hydrosulfuric)	--------	HS^-
$H_2PO_4^-$	--------	HPO_4^{2-}
NH_4^+	--------	NH_3
HCN	--------	CN^-
HCO_3^-	--------	CO_3^{2-}
H_2O	--------	OH^-
OH^-	--------	O^{2-}
H_2	--------	H^-
CH_4	--------	CH_3^-

Strong — Extremely weak

Acid strength increasing — Base strength increasing

Extremely weak — Strong

Extremely weak (negligible H^+ donation to water)

100% reacted with H^+ from water

- *As acid strength decreases, base strength increases; the weaker the acid, the stronger its conjugate base.*
- *As base strength decreases, acid strength increases; the weaker the base, the stronger its conjugate acid.*

Knowing the relative acid and base strengths of the reactants, we can predict the direction of an acid-base reaction. ***The stronger acid and the stronger base will always react to form a weaker conjugate base and a weaker conjugate acid.*** Strong Brønsted-Lowry bases such as hydride ion, H^-, sulfide ion, S^{2-}, oxide ion, O^{2-}, amide ion, NH_2^-, and hydroxide ion, OH^-, readily accept H^+ ions, while weaker Brønsted-Lowry bases do so less readily. For example, the reaction of calcium hydride, CaH_2, with water is highly exothermic because of the extremely strong basic properties of the hydride ion, H^-, which avidly accepts H^+ from water to produce H_2, its weak conjugate acid. In this reaction, hydride ion is a stronger base than OH^-, and water is a stronger acid than H_2, so the forward reaction is favored.

$$H^-(aq) + H_2O(\ell) \longrightarrow H_2(g) + OH^-(aq)$$

There are a great many weak bases that are anions, such as CN^- (cyanide), F^- (fluoride), and CH_3COO^- (acetate).

We can apply information from Figure 16.1 to consider whether the forward or the reverse reaction is favored in the equilibrium

$$HSO_4^-(aq) + CO_3^{2-}(aq) \rightleftharpoons HCO_3^-(aq) + SO_4^{2-}(aq)$$

From Figure 16.1 we note that HSO_4^- is a stronger acid than HCO_3^-, and CO_3^{2-} is a stronger base than SO_4^{2-}. Since acid-base reactions favor going from the stronger to

The strongly basic properties of hydride ion. The reaction of calcium hydride with water is highly exothermic due to the strongly basic hydride ion.

the weaker member of each conjugate acid-base pair, the forward reaction is favored and H^+ will be transferred from HSO_4^- to CO_3^{2-}. Therefore, addition of HSO_4^- ions to a carbonate-containing aqueous solution favors the formation of bicarbonate and sulfate ions.

 ## 16.2 CARBOXYLIC ACIDS AND AMINES

Many weak acids are organic acids, such as acetic, lactic, and pyruvic acids.

$$CH_3-\overset{\overset{\displaystyle O}{\|}}{C}-OH \qquad CH_3\overset{\overset{\displaystyle OH}{|}}{C}-\overset{\overset{\displaystyle O}{\|}}{C}-OH \qquad CH_3\overset{\overset{\displaystyle O}{\|}}{C}-\overset{\overset{\displaystyle O}{\|}}{C}-OH$$

acetic acid   lactic acid   pyruvic acid

> Carboxylic acid group

These acids all contain the carboxylic acid (—COOH) functional group (⟸ *p. 538*). Although carboxylic acid molecules generally contain many other hydrogen atoms, only the hydrogen atom bound to the oxygen atom of the carboxylic acid group is sufficiently positive to donate an H^+ ion in aqueous solution. The oxygen atom of the —OH part of the carboxylic acid group is more electronegative than its hydrogen, and the second oxygen of the carboxylic acid group also pulls electron density away from the hydrogen atom. Together, these make the —OH group even more polar and its hydrogen atom more acidic. The C—H bonds in organic acids are relatively nonpolar and strong, making these hydrogen atoms *not* acidic. This can be seen with butanoic acid.

$$H-\overset{\overset{\displaystyle H}{|}}{\underset{\underset{\displaystyle H}{|}}{C}}-\overset{\overset{\displaystyle H}{|}}{\underset{\underset{\displaystyle H}{|}}{C}}-\overset{\overset{\displaystyle H}{|}}{\underset{\underset{\displaystyle H}{|}}{C}}-\overset{\overset{\displaystyle O}{\|}}{C}-O-H$$

$\underbrace{}$ ↑
nonacidic hydrogens   acidic hydrogen

butanoic acid

Anions formed by loss of an H^+ from a —COOH group, such as acetate ion, CH_3COO^-, from acetic acid, CH_3COOH, are stabilized by resonance (⟸ *p. 344*).

$$CH_3-C\overset{\displaystyle \ddot{O}}{\underset{\displaystyle \ddot{O}H}{}} \qquad\qquad CH_3-C\overset{\displaystyle O}{\underset{\displaystyle O^-}{}} \longleftrightarrow CH_3-C\overset{\displaystyle \ddot{O}:^-}{\underset{\displaystyle \ddot{O}}{}}$$

← electron-attracting oxygen atom

← acidic hydrogen

acetic acid   acetate ion

Amines are compounds that, like ammonia, have a nitrogen atom with three of its valence electrons in covalent bonds and an unshared electron pair on the nitrogen atom. The lone pair of electrons can accept an H^+, and so, like ammonia, amines react as weak bases with water, accepting an H^+ from water.

$$R-NH_2 + H_2O(\ell) \rightleftharpoons R-NH_3^+(aq) + OH^-(aq)$$

Amines can have one, two, or three groups covalently bonded to the nitrogen atom, for example as in methylamine, dimethylamine, and trimethylamine.

$$CH_3-\ddot{N}H_2 \qquad CH_3-\ddot{N}H-CH_3 \qquad CH_3-\overset{\overset{\displaystyle CH_3}{|}}{\underset{\underset{\displaystyle \cdot\cdot}{}}{N}}-CH_3$$

methylamine   dimethylamine   trimethylamine

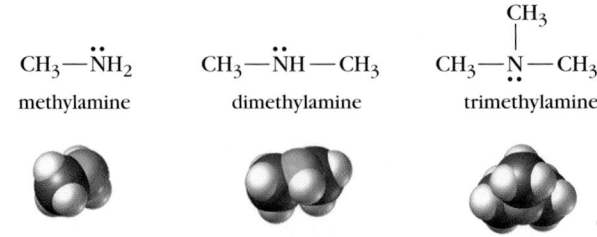

Many biochemically active compounds, including natural biochemicals and drugs, are amines, such as epinephrine (adrenaline, a natural hormone) and Novocaine (a local anesthetic).

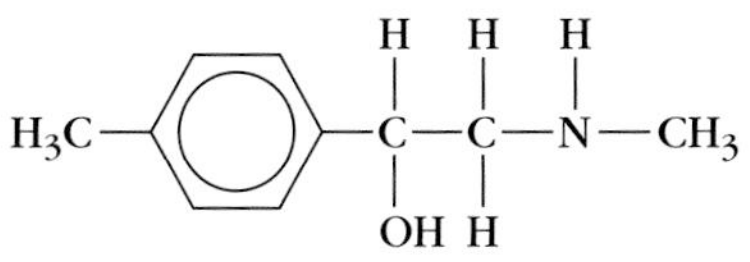

epinephrine

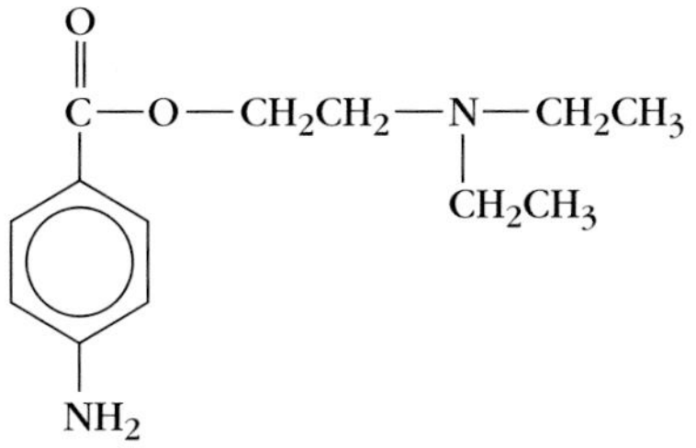

Novocaine

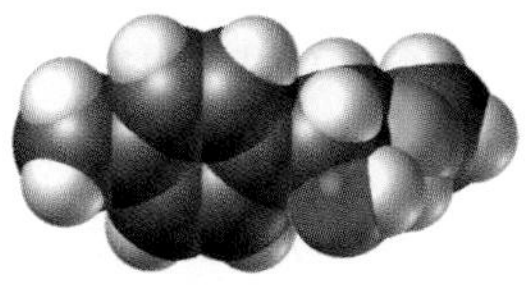

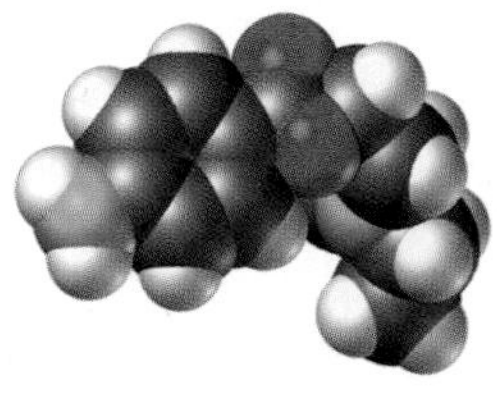

Exercise 16.5 Conjugate Acid-Base Strength

Use Figure 16.1 to predict whether the forward or reverse reaction is favored for the equilibrium

$$CH_3COOH(aq) + SO_4^{2-}(aq) \rightleftharpoons HSO_4^-(aq) + CH_3COO^-(aq)$$

Exercise 16.6 Pyridine, An Analog of Ammonia

Write the equation for the reaction of pyridine with (a) water, (b) hydrochloric acid.

pyridine

Exercise 16.7 Classifying Compounds

Identify each of these as a carboxylic acid, an amine, or neither.

(a) $CH_3-CH_2-CH(NH_2)-CH_3$

(b) $CH_3-CH_2-N(H)-C(=O)-CH_3$

(c)

(d)

(e) $CH_3CH_2-N(H)-CH_2C(=O)-H$

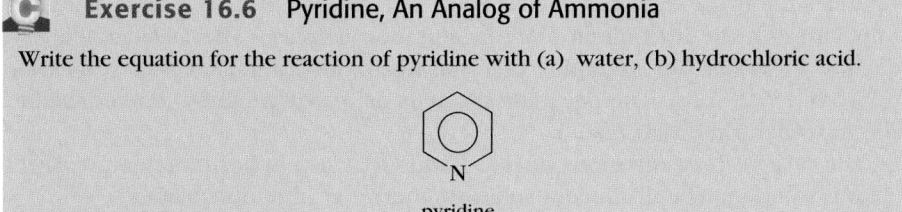

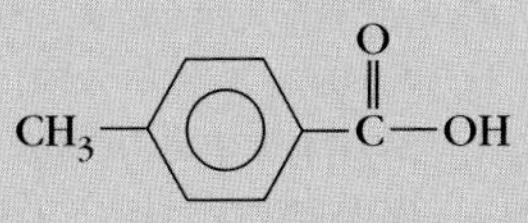

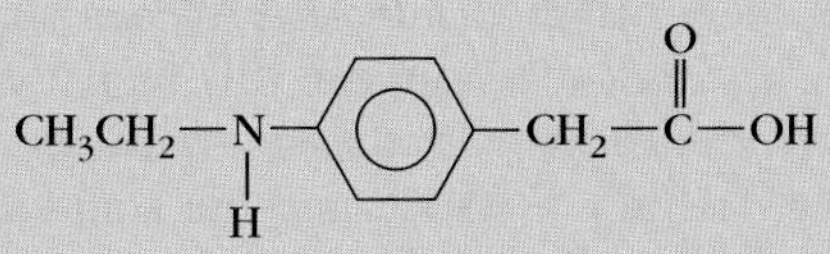

16.3 THE AUTOIONIZATION OF WATER

CD-ROM Screen 17.3: The Acid-Base Properties of Water

Even carefully purified water conducts a very tiny electrical current. This indicates that pure water contains a very small concentration of ions, which are formed when water molecules react to produce hydronium ions and hydroxide ions in a process called **autoionization.**

$$H_2O(\ell) + H_2O(\ell) \rightleftharpoons H_3O^+(aq) + OH^-(aq)$$

BASE   acid   ACID   base

In this reaction, one water molecule serves as an H^+ acceptor (base) while the other is an H^+ donor (acid). The equilibrium between the water molecules and the hydronium and hydroxide ions is very reactant-favored. Therefore, the concentrations of the ions in pure water are *very* low. Nevertheless, autoionization of water is very important to understanding how acids and bases function in aqueous solutions. As in the case of any equilibrium reaction, an equilibrium constant expression can be written for autoionization of water.

Like all equilibrium constant expressions, that for K_w includes concentrations of solutes but not the concentration of the solvent, which in this case is water.

$$2\,H_2O(\ell) \rightleftharpoons H_3O^+(aq) + OH^-(aq) \qquad K_w = [H_3O^+][OH^-]$$

This equilibrium constant K_w is known as the **ionization constant for water.** From electrical conductivity measurements of pure water, we know that $[H_3O^+] = [OH^-] = 1.0 \times 10^{-7}$ M at 25 °C. Hence

$$K_w = [H_3O^+][OH^-] = (1.0 \times 10^{-7})(1.0 \times 10^{-7}) = 1.0 \times 10^{-14} \qquad \text{(at 25 °C)}$$

The equation $K_w = [H_3O^+][OH^-]$ applies to pure water and any aqueous solution. Like other equilibrium constants, the value of K_w is temperature-dependent (Table 16.1.)

In aqueous solutions, the $[H_3O^+]$ and $[OH^-]$ concentrations are inversely related; as one increases, the other must decrease. Their product must always equal 1.0×10^{-14} at 25 °C.

According to the K_w expression, the product of the hydronium ion concentration times the hydroxide ion concentration will always remain the same at a given temperature. If the hydronium ion concentration increases (because an acid was added to the water, for example), then the hydroxide ion concentration must decrease, and vice versa. The equation also tells us that if we know one concentration, the other can be calculated.

The relative concentrations of H_3O^+ and OH^- also indicate the nature of the aqueous solution. For all aqueous solutions there are three possibilities.

Neutral solution: $[H_3O^+] = [OH^-]$	both equal to 1.0×10^{-7} M	
Acidic solution: $[H_3O^+] > 1.0 \times 10^{-7}$ M	$[OH^-] < 1.0 \times 10^{-7}$ M	
Basic solution: $[H_3O^+] < 1.0 \times 10^{-7}$ M	$[OH^-] > 1.0 \times 10^{-7}$ M	

TABLE 16.1 Temperature Dependence of K_w for Water

T (°C)	K_w
10	0.29×10^{-14}
15	0.45×10^{-14}
20	0.68×10^{-14}
25	1.01×10^{-14}
30	1.47×10^{-14}
50	5.48×10^{-14}

When the concentrations of $[H_3O^+]$ and $[OH^-]$ are equal, a solution is said to be **neutral.** If either an acid or a base is added to a neutral solution, the autoionization equilibrium between H_3O^+ and OH^- will be disturbed. Recall that according to Le Chatelier's principle ($\Leftarrow$ *p. 655*), an equilibrium shifts in such a way as to offset the effect of any disturbance. When an acid is added, the concentration of H_3O^+ ions increases. To oppose this increase, added H_3O^+ ions react with OH^- ions in water to form H_2O, thereby reducing the $[OH^-]$. When equilibrium is re-established, $[H_3O^+] > [OH^-]$ and the solution is **acidic;** however, $[H_3O^+][OH^-]$ is still equal to 1.0×10^{-14} at 25 °C. Similarly, if a base is added to water, the added OH^- ions react with H_3O^+ ions in water to form H_2O, thereby decreasing the $[H_3O^+]$. When equilibrium is re-established, $[H_3O^+] < [OH^-]$ and the solution is **basic;** the mathematical product $[H_3O^+][OH^-]$ still equals 1.0×10^{-14}.

The term alkaline is also used to describe basic solutions.

Problem-Solving Example 16.2 $[H_3O^+]$ and $[OH^-]$ Concentrations

Calculate the hydroxide ion concentration at 25 °C in 6.0 M nitric acid and the hydronium ion concentration in 6.0 M NaOH, a strong base.

Answer $[OH^-]$ of 6.0 M nitric acid $= 1.7 \times 10^{-15}$ M; $[H_3O^+]$ of 6.0 M NaOH $= 1.7 \times 10^{-15}$ M

Explanation   Nitric acid is a strong acid (100% ionized), so a 6.0-M nitric acid solution has a $[H_3O^+]$ of 6.0 M. Therefore, its $[OH^-]$ can be calculated.

$$[H_3O^+][OH^-] = (6.0)[OH^-] = 1.0 \times 10^{-14}$$

$$[OH^-] = \frac{1.0 \times 10^{-14}}{6.0} = 1.7 \times 10^{-15} \text{ M}$$

Being a strong base, 6.0 M sodium hydroxide has a $[OH^-]$ of 6.0 M.

$$[H_3O^+][OH^-] = [H_3O^+](6.0) = 1.0 \times 10^{-14}$$

$$[H_3O^+] = \frac{1.0 \times 10^{-14}}{6.0} = 1.7 \times 10^{-15} \text{ M}$$

✔   Note that at the high hydronium ion concentration of 6.0 M nitric acid, the hydroxide ion concentration is very, very low, which is to be expected of a highly acidic solution. In contrast, the hydronium ion concentration is exceedingly low in 6.0 M NaOH, a highly basic solution.

Problem-Solving Practice 16.2

Which will be more acidic, a solution whose H_3O^+ concentration is 2.0×10^{-5} M or one that has an OH^- concentration of 5.0×10^{-9} M?

16.4 THE pH SCALE

The $[H_3O^+]$ and $[OH^-]$ in an aqueous solution vary widely depending on the acid or base present and its concentration. In general, the $[H_3O^+]$ in aqueous solutions can range from about 10 mol/L down to about 10^{-15} mol/L. The $[OH^-]$ can also vary over the same range in aqueous solution.

Because these concentrations can be so small, they have very large negative exponents. It is more convenient to express these concentrations in terms of logarithms. The **pH** of a solution is defined as *the negative of the base 10 logarithm (log) of the hydronium ion concentration.*

$$\textbf{pH} = -\textbf{log}(\textbf{H}_3\textbf{O}^+)$$

The *negative* logarithm of the small concentration values is used since it gives a positive pH value. Thus, the pH of pure water at 25 °C is given by

$$\text{pH} = -\log[1.0 \times 10^{-7}] = -(-7.00) = 7.00$$

In terms of pH, for solutions at 25 °C we can write

Neutral solution	pH = 7.00
Acidic solution	pH < 7.00
Basic (alkaline) solution	pH > 7.00

Figure 16.2 shows the pH values along with the corresponding H_3O^+ and OH^- concentrations of some common solutions. Notice that $-\log(1 \times 10^{-x}) = x$.

Lemon juice: $[H_3O^+] = 1 \times 10^{-2}$ M;   pH $= -\log(1 \times 10^{-2}) = 2$

Black coffee: $[H_3O^+] = 1 \times 10^{-5}$ M;   pH $= -\log(1 \times 10^{-5}) = 5$

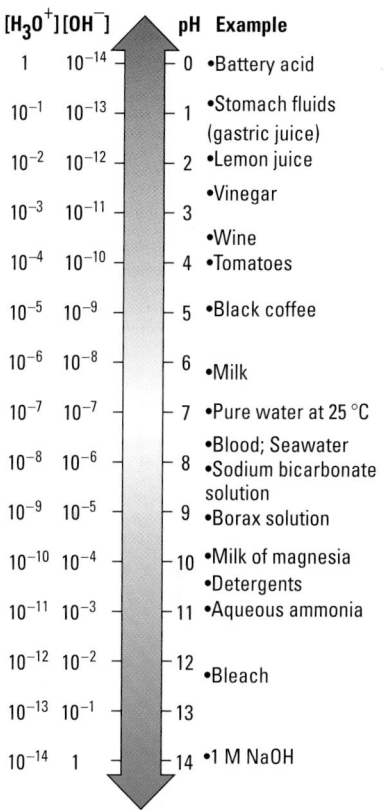

Figure 16.2   The pH of aqueous solutions.   The relationship of pH to the concentrations of H_3O^+ and OH^- (in moles/liter at 25 °C) is shown. The pH values of some common substances are also included in the diagram.

 CD-ROM Screen 17.4: The pH Scale

Keep in mind that a change of 1 pH unit represents a 10-fold change in H_3O^+ concentration, 2 pH units represent a 100-fold change, and so on. Thus, according to Figure 16.2, the $[H_3O^+]$ in lemon juice (pH = 2) is more than 100 times greater than that in tomato juice (pH < 5).

For solutions in which $[H_3O^+]$ or $[OH^-]$ has a value other than an exact power of 10 (1, 1×10^{-1}, 1×10^{-2}, . . .) a calculator is convenient for finding the pH (see Appendix A.6). For example, the pH is 2.30 for a solution that contains 0.0045 mol of the strong acid HNO_3 per liter.

$$pH = -\log(4.50 \times 10^{-3}) = 2.30$$

Problem-Solving Example 16.3   Calculating pH from $[H_3O^+]$

Calculate the pH of an aqueous HCl solution that has a volume of 500. mL and contains 1.25 g of HCl.

Answer   1.164 (*Note:* This pH has been calculated to three significant figures. In actual measurements, pH values are seldom obtainable to this degree of accuracy.)

Explanation   Hydrochloric acid is a strong acid, so every mole of HCl that dissolves produces a mole of H_3O^+ and a mole of Cl^-. First, determine the number of moles of HCl.

$$1.25 \text{ g HCl} \times \frac{1 \text{ mol HCl}}{36.461 \text{ g HCl}} = 0.03428 \text{ mol HCl}$$

Next, calculate the H_3O^+ concentration.

$$[H_3O^+] = \frac{0.03428 \text{ mol HCl}}{0.500 \text{ L}} = 0.0686 \text{ M}$$

Then express this concentration as pH.

$$pH = -\log(6.86 \times 10^{-2}) = 1.164$$

✔ If the $[H_3O^+]$ were 0.10 M, the pH would be 1.00; the pH would be 2.00 for an H_3O^+ concentration of 0.010 M. Therefore, a solution with an H_3O^+ concentration of 0.0686 M, which is between these two values, should have a pH between 1.00 and 2.00, which it does.

Solving Practice 16.3

Calculate the pH of a 0.040 M NaOH solution.

The digits to the left of the decimal point in a pH represent a power of 10. Only the digits to the right of the decimal point are significant. In Problem-Solving Example 16.3 where pH = $-\log(6.86 \times 10^{-2})$ = $-\log(6.86) + (-\log 10^{-2})$ = $-0.836 + 2.000 = 1.164$, there are three significant figures in 0.164 the result because there are three significant figures in -0.836.

Notice that, as in the case of equilibrium constants, the concentration units of mol/L are ignored when the logarithm is taken. It is not possible to take the logarithm of a unit.

The calculation done in Problem-Solving Example 16.3 can be reversed; the hydronium ion concentration of a solution can be calculated from its pH value.

Problem-Solving Example 16.4   Calculating $[H_3O^+]$ from pH

The measured pH of a sample of sea water is 8.30.

(a) What is the H_3O^+ concentration?   (b) Is the sample acidic or basic?

Answer   (a) 5.0×10^{-9} M   (b) basic

Explanation
(a) Substituting into the definition of pH,

$$-\log[H_3O^+] = 8.30, \quad \text{so} \quad \log[H_3O^+] = -8.30$$

By the rules of logarithms, $10^{\log(x)} = x$, so we can write $10^{\log [H_3O^+]} = 10^{-pH} = [H_3O^+]$. Finding $[H_3O^+]$ therefore requires finding the antilogarithm of -8.30 (Appendix A.6).

$$[H_3O^+] = 10^{-8.30} = 5.0 \times 10^{-9} \text{ M}$$

(b) Because the pH is greater than 7.0, the sample is basic.

Problem-Solving Practice 16.4

A human blood sample has a pH of 7.40.

(a) What is its H_3O^+ concentration?   (b) What is its OH^- concentration?
(c) Is the sample acidic, neutral, or basic?

ARNOLD BECKMAN
(1900–)

Arnold Beckman revolutionized pH measurement when he invented the first electronic pH meter in 1934. At the time Beckman, a professor at the California Institute of Technology, developed the instrument in response to a request from the California Fruit Growers' Association for a quicker, more accurate way to measure the acidity of lemon juice. He went on to found the highly successful Beckman Instrument Company, a firm that invented the first widely used infrared and ultraviolet spectrophotometers and other laboratory instruments. Arnold Beckman and his wife Mabel have donated millions of dollars to advance chemical research and education nationwide.

Exercise 16.8   pH of Solutions of Different Acids

Would the pH of a 0.1 M solution of the strong acid HNO_3 be the same as the pH of a 0.1 M solution of the strong acid HCl? Explain.

Exercise 16.9   Super acidic

Recently, a pH sensor has been developed that operates under extremely acidic conditions such as those found in environments like Iron Mountain, California (see *Chemistry in the News,* p. 740). At this abandoned mine site, the groundwater has a pH of −3.6 (not a typo; it is a minus!!). Calculate the H^+ concentration (molarity) of this groundwater.

The OH^- concentration can also be expressed in exponential terms as pOH.

$$pOH = -\log[OH^-]$$

The $[OH^-]$ of pure water at 25 °C is 1.0×10^{-7} M, and therefore its pOH is

$$-\log(1 \times 10^{-7}) = -(-7.00) = 7.00$$

Because the values of $[H_3O^+]$ and $[OH^-]$ are related by the K_w expression, for all aqueous solutions at 25 °C, we can write

$$K_w = [H_3O^+][OH^-] = 1.0 \times 10^{-14}$$

This equation can be rewritten as

$$-\log K_w = -\log[H_3O^+] + (-\log[OH^-]) = -\log(1.0 \times 10^{-14})$$

or

$$pK_w = pH + pOH = 14.00$$

The relation between pH and pOH can be used to find one value when the other is known. A 0.0010 M solution of the strong base NaOH, for example, has an OH^- concentration of 0.0010 M and a pOH given by

$$pOH = -\log(1.0 \times 10^{-3}) = 3.00$$

and therefore

$$pH = 14.00 - pOH = 14.00 - 3.00 = 11.00$$

If you know the pH, then the pOH is just 14.00 − pH; knowing the pOH, the pH is 14.00 − pOH.

Exercise 16.10   pOH and pH

Which solution is more basic, one that has a pH of 5.5 or one with a pOH of 8.5? What is the H_3O^+ concentration in each solution?

Measuring pH

The pH of a solution is readily measured using a pH meter (Figure 16.3). The device consists of a pair of electrodes that detect the H^+ concentration of the test solution, convert it into an electrical signal, and display it directly as the pH value.

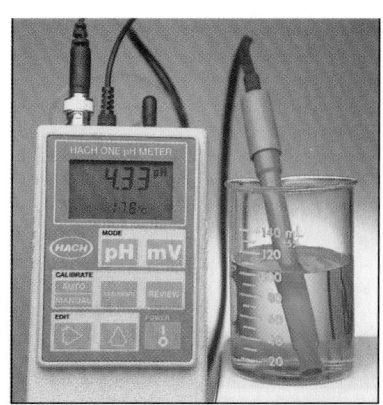

Figure 16.3 A pH meter. A pH meter can quickly and accurately determine the pH of a sample. The functioning of a pH meter is described in Section 19.7.

Chemistry in the News

HOW LOW CAN YOU GO? ULTRAACIDIC WATER

The pH range of zero to 14 is normally sufficient to account for acidic or basic aqueous solutions in the laboratory or in natural environments. That range is set arbitrarily with a symmetry such that a neutral pH of 7.0 is exactly in the middle of the scale. Solutions with a negative pH are theoretically possible but generally beyond the detection range of most pH sensors. Recently, a calibrated pH sensor has been developed that functions in ultraacidic conditions, those with a pH below zero. Such a sensor is useful to determine the pH of groundwater such as that at Iron Mountain, California, an abandoned mine site. Groundwater at the site has a pH of -3.6. The very high hydrogen-ion concentration is created by the reaction between groundwater and iron pyrite (FeS_2) to produce sulfuric acid. This is a two-step process: The first step dissolves the iron pyrite; the second step oxidizes Fe^{2+} to Fe^{3+}.

$$2\,FeS_2(s) + 2\,H_2O(\ell) + 7\,O_2(g) \longrightarrow$$
$$4\,H^+(aq) + 4\,SO_4^{2-}(aq) + 2\,Fe^{2+}(aq)$$

$$4\,Fe^{2+}(aq) + O_2(g) + 4\,H^+(aq) \longrightarrow$$
$$4\,Fe^{3+}(aq) + 2\,H_2O(\ell)$$

The overall reaction is

$$4\,FeS_2(s) + 2\,H_2O(\ell) + 15\,O_2(g) \longrightarrow$$
$$4\,H^+ + 8\,SO_4^{2-} + 4\,Fe^{3+}$$

The relatively high temperature of the mine (at least 47 °C) evaporates water, decreasing the volume of solution, which further concentrates the acid, raises the hydronium ion concentration, and decreases the pH.

Source(s):

Nordstrom, D.K. *Environmental Science and Technology,* Vol. 34, 2000; pp. 254–257.

Science, Vol. 287, February 11, 2000; p. 933.

Ultraacidic water at Iron Mountain, California.

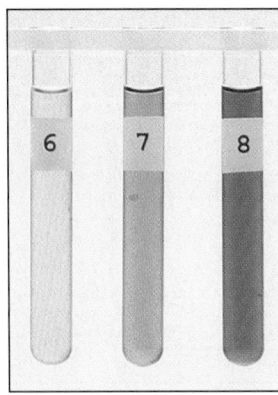

Figure 16.4 Bromthymol blue indicator. Below a pH of 6 the indicator is yellow. At pH 7 it is pale green, and at pH 8 and above, the color is blue.

The meter is initially calibrated using standard solutions of known pH. The pH of body fluids, soil, environmental and industrial samples, and other substances can be measured easily and accurately with a pH meter.

A much older and less precise (but convenient) method to determine the pH of a sample is the use of acid-base indicators, substances that change color within a narrow pH range, generally 1 to 2 pH units. Loss or gain of an H^+ ion changes the indicator's molecular structure so that it absorbs light in different regions of the visible spectrum. The indicator is one color at a lower pH (its "acid" form), and it is a different color at a higher pH (its "base" form). Consider the indicator bromthymol blue (Figure 16.4). Below pH 6 it is yellow (its acid form); at pH 8 and above it is blue (its base form). Between pH 6 and 7, the indicator changes from pure yellow to a yellow-green color. At pH 7, it is a mixture of 50% yellow and 50% blue, so it appears green. As the pH changes from 7 to 8, the color becomes pure blue. Thus, the pH of a sample that turns bromthymol blue to green has a pH of about 7. If the indicator color is blue, the pH of the sample is at least 8, and could be much higher (see Figure 16.4).

Strips of paper impregnated with acid-base indicators are also used to test the pH of many substances. The color of the paper after it has been dampened by the solution to be tested is compared with a set of colors at known pHs.

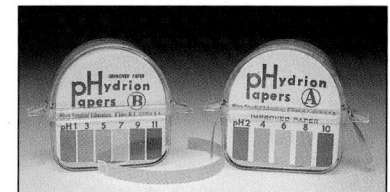

Indicator paper strips. Strips of paper impregnated with indicator are used to find an approximate pH.

16.5 IONIZATION CONSTANTS OF ACIDS AND BASES

Earlier you learned that the greater the equilibrium constant for a reaction, the more product-favored that reaction is. In an acid-base reaction, the stronger the re-actant acid and the base, the more product-favored the reaction is ($\Leftarrow$ *p. 733*). Consequently, the magnitude of equilibrium constants can give us an idea about the relative strengths of weak acids and bases. For example, the larger the equilibrium constant for its ionization, the stronger the acid.

CD-ROM Screen 17.5: Strong Acids and Bases.

CD-ROM Screen 17.6: Weak Acids and Bases.

Acid Ionization Constants

An ionization equation for the transfer to water of H^+ from any acid represented by the general formula HA is

$$\underset{\text{conjugate acid}}{HA(aq)} + H_2O(\ell) \rightleftharpoons H_3O^+(aq) + \underset{\text{conjugate base}}{A^-(aq)}$$

The corresponding **acid ionization constant expression** is

$$K_a = \frac{[H_3O^+][\text{conjugate base}]}{[\text{conjugate acid}]} = \frac{[H_3O^+][A^-]}{[HA]}$$

In the acid ionization constant expression, the *equilibrium* concentrations of conjugate base and hydronium ion appear in the numerator; the *equilibrium* concentration of *un-ionized* conjugate acid appears in the denominator. As with other equilibrium constant expressions, pure solids and liquids, such as water, are not included.

The equilibrium constant K_a is called the **acid ionization constant.** The stronger the acid, the larger the acid ionization constant. As more acid ionizes, the [HA] denominator term in the acid ionization constant expression gets smaller as the numerator terms increase. In contrast with strong acids, weak acids such as acetic acid ionize to a much smaller extent, establishing equilibria in which significant concentrations of un-ionized weak acid molecules are still present in the solution. The ionization of a weak acid is reactant-favored, and all weak acids have K_a values less than 1. They are weak electrolytes. For strong acids such as hydrochloric acid, the equilibrium is so product-favored that the acid ionization constant value is much larger than 1.

Acid ionization constants are also called acid dissociation constants.

The common strong acids are hydrochloric (HCl), nitric (HNO_3), sulfuric (H_2SO_4), perchloric ($HClO_4$), hydrobromic (HBr), and hydroiodic (HI). The ionization reaction of a strong acid is also an equilibrium, but double arrows are not used because the equilibrium lies so far to the right.

Problem-Solving Example 16.5 Acid Ionization Constant Expressions

Write the ionization equation and the ionization constant expression for these weak acids.

(a) HF   (b) HBrO   (c) $H_2PO_4^-$

Answer

(a) $HF(aq) + H_2O(\ell) \rightleftharpoons H_3O^+(aq) + F^-(aq)$ $K_a = \dfrac{[H_3O^+][F^-]}{[HF]}$

(b) $HBrO(aq) + H_2O(\ell) \rightleftharpoons H_3O^+(aq) + BrO^-(aq)$ $K_a = \dfrac{[H_3O^+][BrO^-]}{[HBrO]}$

(c) $H_2PO_4^-(aq) + H_2O(\ell) \rightleftharpoons H_3O^+(aq) + HPO_4^{2-}(aq)$ $\qquad K_a = \dfrac{[H_3O^+][HPO_4^{2-}]}{[H_2PO_4^-]}$

Explanation In each case, the ionization equation represents the transfer of an H^+ ion from an acid to water, creating a hydronium ion and the conjugate base of the acid.

Problem-Solving Practice 16.5

Write the ionization equation and ionization constant expression for each of these acids:

(a) Hydrazoic acid, HN_3 (b) Formic acid, $HCOOH$ (c) Chlorous acid, $HClO_2$

Weak acids are only slightly ionized, which can be shown by measuring the pH of their aqueous solutions. The pH of a 0.10 M acetic acid solution is 2.88. This means that the concentration of H_3O^+ is only 1.3×10^{-3} M. Compare this value with the 0.10 M concentration of H_3O^+ ions in a 0.10 M solution of HCl, a strong acid. In a 0.10-M acetic acid solution, only 1.3% of the initial concentration of acetic acid is ionized:

$$\underset{\substack{\text{Stronger acid} \\ \text{than } CH_3COOH}}{} \qquad \underset{\substack{\text{Stronger base} \\ \text{than } H_2O}}{}$$

$$CH_3COOH(aq) + H_2O(\ell) \rightleftharpoons H_3O^+(aq) + CH_3COO^-(aq)$$

$$\% \text{ ionization} = \frac{[H_3O^+] \text{ at equilibrium}}{\text{initial acid conc.}} \times 100\% = \frac{1.3 \times 10^{-3}}{1.0 \times 10^{-1}} \times 100\% = 1.3\%$$

Therefore, almost 99% of the acetic acid remains in the un-ionized molecular form, CH_3COOH. This is why weak acids (and bases) are weak electrolytes.

In an acetic acid solution, or a solution of any weak acid, there are two different bases competing for H^+ ions that can be donated from two different acids. In the equation above, the two bases are water and acetate ion; the two acids are acetic acid and hydronium ions. Since the K_a is much less than 1, the equilibrium favors the reactants. The acetate ion must be a stronger H^+ acceptor than the water molecule. Another way of looking at the same reaction is that the hydronium ion must be a stronger H^+ donor than the acetic acid molecule. Both of these statements are true. Recall from Section 16.1 that acid-base reactions favor going from the stronger to the weaker member of each conjugate acid-base pair. Thus, the acetic acid equilibrium is reactant-favored; a significant concentration of un-ionized acetic acid molecules is present.

Base Ionization Constants

A general equation analogous to that for the donation of H^+ to water by acids can be written for the acceptance of a H^+ *from* water by a molecular base, B, to form its conjugate acid, BH^+.

$$\underset{\substack{\text{conjugate} \\ \text{base}}}{B(aq)} + H_2O(\ell) \rightleftharpoons \underset{\substack{\text{conjugate} \\ \text{acid}}}{BH^+(aq)} + OH^-(aq)$$

If the base B were NH_3, then BH^+ would be NH_4^+. The corresponding equilibrium constant expression is

$$K_b = \frac{[\text{conjugate acid}][OH^-]}{[\text{conjugate base}]} = \frac{[BH^+][OH^-]}{[BH]}$$

The equilibrium constant K_b is called the **base ionization constant,** a term that can be misleading. Notice from the chemical equation that the base does not ionize. Rather, K_b and its equilibrium constant expression refer to *the reaction in which a base forms its conjugate acid by removing an H^+ ion from water.*

When the base is an anion, A^- (such as the anion of a weak acid), the general equation is

$$A^-(aq) + H_2O(\ell) \rightleftharpoons HA(aq) + OH^-(aq)$$

conjugate   conjugate
base   acid

If the base A^- were CH_3COO^-, then HA would be CH_3COOH. The corresponding **base ionization constant expression** is

$$K_b = \frac{[\text{conjugate acid}][OH^-]}{[\text{conjugate base}]} = \frac{[HA][OH^-]}{[A^-]}$$

The K_b value indicates the extent to which the base reacts with water to produce OH^- ions. The larger the base ionization constant, K_b, the stronger the base, the more product-favored the H^+ transfer reaction, and the greater the OH^- concentration produced. For a strong base the ionization constant is greater than 1. For a weak base the ionization constant is less than 1, sometimes considerably less than 1, because at equilibrium there is a significant concentration of unreacted weak conjugate base and a much smaller concentration of its conjugate acid and OH^- ions.

Problem-Solving Example 16.6   Base Ionization

Write the ionization equation and the K_b expression for these weak bases:

(a) CH_3NH_2 (b) Phosphine, PH_3 (c) NO_2^-

Answer

(a) $CH_3NH_2(aq) + H_2O(\ell) \rightleftharpoons CH_3NH_3^+(aq) + OH^-(aq)$ $K_b = \dfrac{[CH_3NH_3^+][OH^-]}{[CH_3NH_2]}$

(b) $PH_3(aq) + H_2O(\ell) \rightleftharpoons PH_4^+(aq) + OH^-(aq)$ $K_b = \dfrac{[PH_4^+][OH^-]}{[PH_3]}$

(c) $NO_2^-(aq) + H_2O(\ell) \rightleftharpoons HNO_2(aq) + OH^-(aq)$ $K_b = \dfrac{[HNO_2][OH^-]}{[NO_2^-]}$

Explanation For each of the bases the general reaction is the same — the removal of an H^+ from water by the base to form its corresponding conjugate acid. In (a) and (b), an H^+ ion is added to a base that is a neutral molecule to form the positively charged conjugate acid $CH_3NH_3^+$ in (a) and PH_4^+ in (b). In part (c), the H^+ ion is added to a negatively charged ion, NO_2^-. The resulting conjugate acid, HNO_2, has no charge.

Problem-Solving Practice 16.6

Write the chemical equation and the K_b expression for these bases.

(a) CN^- (b) $C_6H_5NH_2$ (c) HS^-

Hydrated Metal Ions as Acids

Some hydrated metal ions, especially those of the transition metals, are also weak acids. When a salt containing a metal ion dissolves in water, the metal ion becomes hydrated, often by having six water molecules around it, $[M(H_2O)_6]^{n+}$, where M represents a metal ion whose charge is $n+$. There are M—O—H bonds in the hydrated ion. Metal ions other than those in Groups 1A and 2A have great enough charges and small enough sizes to attract the shared electron pair in the M—O bond to themselves. This weakens the O—H bond, making the hydrogens in the M—O—H bonds more acidic than they would be in a water molecule that is not

Lone pairs of electrons on water molecules form coordinate covalent bonds with the metal ion ($\Leftarrow$ *p. 768*).

Look carefully at the Fe-containing reactant and product in the equation. There are six water molecules in the reactant ion and only five in the product ion. The other water molecule has lost an H^+ to become an OH^- ion. As a result, the net charge on the product ion is one less than on the reactant ion.

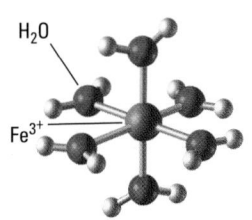

Water molecules bonded to an Fe^{3+} ion.

bonded to a metal ion. Thus, the $[M(H_2O)_6]^{n+}$ ion can donate H^+, the solution becomes acidic, and the positive charge of the remaining hydrated metal ion has been decreased by one.

The ionization reaction and acid ionization constant expression for a hydrated metal ion such as Fe^{3+} can be written as

$$[Fe(H_2O)_6]^{3+}(aq) + H_2O(\ell) \rightleftharpoons [Fe(H_2O)_5(OH)]^{2+}(aq) + H_3O^+(aq)$$

$$K_a = \frac{[Fe(H_2O)_5(OH)^{2+}][H_3O^+]}{[Fe(H_2O)_6^{3+}]} = 6.3 \times 10^{-3}$$

This K_a value shows that a solution of $FeCl_3$ will have about the same pH as a solution of phosphoric acid ($K_a = 7.5 \times 10^{-3}$) of equal concentration. Many metal ions form weakly acidic aqueous solutions, and this property is important in the chemistry of such ions in the environment. For example, Al^{3+} ions in soils react with water to produce an acidic environment that can be detrimental to tree growth.

Exercise 16.11 The pH of a Solution of $Ni(NO_3)_2$

When anhydrous nickel(II) nitrate dissolves in water, the Ni^{2+} ions become hydrated, forming $[Ni(H_2O)_6]^{2+}$ ions. What is the pH of a solution that is 0.15 M in nickel nitrate? K_a for $[Ni(H_2O)_6]^{2+}$ is 2.5×10^{-11}.

Values of Acid and Base Ionization Constants

Table 16.2 summarizes the ionization constants for a number of acids and their conjugate bases. The ionization constants for strong acids (those above H_3O^+ in Table 16.2) and strong bases (those below OH^- in Table 16.2) are too large to be measured easily. Fortunately, since their ionization reactions are virtually complete, these K_a and K_b values are hardly ever needed. For weak acids, K_a values show relative strengths quantitatively; for weak bases, K_b values do the same.

Consider acetic acid and boric acid. Since boric acid is below acetic acid in Table 16.2, boric acid must be a weaker acid than acetic acid; the K_a values tell us how much weaker. The K_a for boric acid is 7.3×10^{-10}; that for acetic acid is 1.8×10^{-5}, which shows that boric acid is somewhat more than 10^4 times weaker than acetic acid. In fact, boric acid is such a weak acid that a dilute solution of it can be used safely as an eyewash. Don't try that with acetic acid!

Exercise 16.12 Acid Strengths

The K_a of lactic acid is 1.5×10^{-4}; that of pyruvic acid is 3.2×10^{-3}.

(a) Which of these acids is the stronger acid?
(b) Which acid's ionization reaction is more reactant-favored?

sulfuric acid

oxalic acid

phosphoric acid

K_a Values for Polyprotic Acids

So far we have concentrated on **monoprotic acids** such as hydrogen fluoride (HF), hydrogen chloride (HCl), and nitric acid (HNO_3)—acids that can donate a single H^+ per molecule.

Some acids, called **polyprotic acids,** donate more than one H^+ per molecule. These include sulfuric acid (H_2SO_4), carbonic acid (H_2CO_3), phosphoric acid (H_3PO_4), oxalic acid ($H_2C_2O_4$ or HOOCCOOH), and other organic acids with two or more carboxylic acid (—COOH) groups (Table 16.3, on p. 746).

| TABLE 16.2 | Ionization Constants for Some Acids and Their Conjugate Bases at 25 °C |

Acid name	Acid	$K_a = \dfrac{[H_3O^+]\begin{bmatrix} \text{conj} \\ \text{base} \end{bmatrix}}{[\text{conj acid}]}$	Base name	Base	$K_b = \dfrac{\begin{bmatrix} \text{conj} \\ \text{base} \end{bmatrix}[OH^-]}{[\text{conj base}]}$
Perchloric acid	$HClO_4$	Large	Perchlorate ion	ClO_4^-	Very small
Sulfuric acid	H_2SO_4	Large	Hydrogen sulfate ion	HSO_4^-	Very small
Hydrochloric acid	HCl	Large	Chloride ion	Cl^-	Very small
Nitric acid	HNO_3	≈ 20	Nitrate ion	NO_3^-	$\approx 5 \times 10^{-16}$
Hydronium ion	H_3O^+	1.0	Water	H_2O	1.0×10^{-14}
Sulfurous acid	H_2SO_3	1.2×10^{-2}	Hydrogen sulfite ion	HSO_3^-	8.3×10^{-13}
Hydrogen sulfate ion	HSO_4^-	1.2×10^{-2}	Sulfate ion	SO_4^{2-}	8.3×10^{-13}
Phosphoric acid	H_3PO_4	7.5×10^{-3}	Dihydrogen phosphate ion	$H_2PO_4^-$	1.3×10^{-12}
Hexaaquairon(III) ion	$Fe(H_2O)_6^{3+}$	6.3×10^{-3}	Pentaaquahydroxoiron(III) ion	$Fe(H_2O)_5OH^{2+}$	1.6×10^{-12}
Hydrofluoric acid	HF	7.2×10^{-4}	Fluoride ion	F^-	1.4×10^{-11}
Nitrous acid	HNO_2	4.5×10^{-4}	Nitrite ion	NO_2^-	2.2×10^{-11}
Formic acid	HCOOH	1.8×10^{-4}	Formate ion	$HCOO^-$	5.6×10^{-11}
Benzoic acid	C_6H_5COOH	6.3×10^{-5}	Benzoate ion	$C_6H_5COO^-$	1.6×10^{-10}
Acetic acid	CH_3COOH	1.8×10^{-5}	Acetate ion	CH_3COO^-	5.6×10^{-10}
Propanoic acid	CH_3CH_2COOH	1.4×10^{-5}	Propanoate ion	$CH_3CH_2COO^-$	7.1×10^{-10}
Hexaaquaaluminum ion	$Al(H_2O)_6^{3+}$	7.9×10^{-6}	Pentaaquahydroxoaluminum ion	$Al(H_2O)_5OH^{2+}$	1.3×10^{-9}
Carbonic acid	H_2CO_3	4.2×10^{-7}	Hydrogen carbonate ion	HCO_3^-	2.4×10^{-8}
Hexaaquacopper(II) ion	$Cu(H_2O)_6^{2+}$	1.6×10^{-7}	Pentaaquahydroxocopper(II) ion	$Cu(H_2O)_5OH^+$	6.25×10^{-8}
Hydrogen sulfide	H_2S	1×10^{-7}	Hydrogen sulfide ion	HS^-	1×10^{-7}
Dihydrogen phosphate ion	$H_2PO_4^-$	6.2×10^{-8}	Hydrogen phosphate ion	HPO_4^{2-}	1.6×10^{-7}
Hydrogen sulfite ion	HSO_3^-	6.2×10^{-8}	Sulfite ion	SO_3^{2-}	1.6×10^{-7}
Hypochlorous acid	HClO	3.5×10^{-8}	Hypochlorite ion	ClO^-	2.9×10^{-7}
Hexaaqualead(II) ion	$Pb(H_2O)_6^{2+}$	1.5×10^{-8}	Pentaaquahydroxolead(II) ion	$Pb(H_2O)_5OH^+$	6.7×10^{-7}
Hexaaquacobalt(II) ion	$Co(H_2O)_6^{2+}$	1.3×10^{-9}	Pentaaquahydroxocobalt(II) ion	$Co(H_2O)_5OH^+$	7.7×10^{-6}
Boric acid	$B(OH)_3(H_2O)$	7.3×10^{-10}	Tetrahydroxoborate ion	$B(OH)_4^-$	1.4×10^{-5}
Ammonium ion	NH_4^+	5.6×10^{-10}	Ammonia	NH_3	1.8×10^{-5}
Hydrocyanic acid	HCN	4.0×10^{-10}	Cyanide ion	CN^-	2.5×10^{-5}
Hexaaquairon(II) ion	$Fe(H_2O)_6^{2+}$	3.2×10^{-10}	Pentaaquahydroxoiron(II) ion	$Fe(H_2O)_5OH^+$	3.1×10^{-5}
Hydrogen carbonate ion	HCO_3^-	4.8×10^{-11}	Carbonate ion	CO_3^{2-}	2.1×10^{-4}
Hexaaquanickel(II) ion	$Ni(H_2O)_6^{2+}$	2.5×10^{-11}	Pentaaquahydroxonickel(II) ion	$Ni(H_2O)_5OH^+$	4.0×10^{-4}
Hydrogen phosphate	HPO_4^{2-}	3.6×10^{-13}	Phosphate ion	PO_4^{3-}	2.8×10^{-2}
Water	H_2O	1.0×10^{-14}	Hydroxide ion	OH^-	1.0
Hydrogen sulfide ion	HS^-	1×10^{-19}	Sulfide ion	S^{2-}	1×10^5
Ethanol	C_2H_5OH	Very small	Ethoxide ion	$C_2H_5O^-$	Large
Ammonia	NH_3	Very small	Amide ion	NH_2^-	Large
Hydrogen	H_2	Very small	Hydride ion	H^-	Large
Methane	CH_4	Very small	Methide ion	CH_3^-	Large

Increasing Acid Strength

Increasing Base Strength

TABLE 16.3 Polyprotic Acids	
Acid form	**Conjugate base form**
H_2S (hydrosulfuric acid)	HS^- (hydrogen sulfide or bisulfide ion)
H_3PO_4 (phosphoric acid)	$H_2PO_4^-$ (dihydrogen phosphate ion)
$H_2PO_4^-$ (dihydrogen phosphate ion)	HPO_4^{2-} (hydrogen phosphate ion)
H_2CO_3 (carbonic acid)	HCO_3^- (hydrogen carbonate or bicarbonate ion)
$H_2C_2O_4$ (oxalic acid)	$HC_2O_4^-$ (hydrogen oxalate ion)
$C_3H_5(COOH)_3$ (citric acid)	$C_3H_5(COOH)_2COO^-$ (monocitrate ion)

In aqueous solution, a polyprotic acid donates its H^+ ions to water molecules in a stepwise manner. In the first step for sulfuric acid, hydrogen sulfate ion, HSO_4^-, is formed. Sulfuric acid is a strong acid, and so this first ionization is complete.

$$H_2SO_4(aq) + H_2O(\ell) \longrightarrow H_3O^+(aq) + HSO_4^-(aq)$$
ACID   base   acid   BASE

Hydrogen sulfate ion is the conjugate base of sulfuric acid.

In the next step, hydrogen sulfate ion donates an H^+ ion to another water molecule. Hydrogen sulfate ion is a weak acid and in this case, as with other weak acids, an equilibrium is established.

$$HSO_4^-(aq) + H_2O(\ell) \rightleftharpoons H_3O^+(aq) + SO_4^{2-}(aq)$$
ACID   base   acid   BASE

Many chemical reactions occur in steps that can be represented by individual equations. Sometimes only the overall equation is written.

There are polyprotic bases, those that can accept more than one H^+ per molecule of base. We will not discuss polyprotic bases here.

Exercise 16.13   Explaining Acid Strengths

Look at the charge on the hydrogen sulfate ion. What does this have to do with the fact that this ion is a weaker acid than H_2SO_4?

The weak acid H_3PO_4, for example, has three H^+ ions per molecule to donate and hence three ionization reactions.

First ionization
$$H_3PO_4(aq) + H_2O(\ell) \rightleftharpoons H_3O^+(aq) + H_2PO_4^-(aq) \quad K_a = 7.5 \times 10^{-3}$$

Second ionization
$$H_2PO_4^-(aq) + H_2O(\ell) \rightleftharpoons H_3O^+(aq) + HPO_4^{2-}(aq) \quad K_a = 6.2 \times 10^{-8}$$

Third ionization
$$HPO_4^{2-}(aq) + H_2O(\ell) \rightleftharpoons H_3O^+(aq) + PO_4^{3-}(aq) \quad K_a = 3.6 \times 10^{-13}$$

The successive K_a values for the ionization of a polyprotic acid decrease by a factor of 10^4 to 10^5, indicating that each ionization step occurs to a lesser extent than the one before it. The $H_2PO_4^-$ ion is a much weaker acid ($K_a = 6.2 \times 10^{-8}$) than phosphoric acid ($K_a = 7.5 \times 10^{-3}$), and the HPO_4^{2-} ($K_a = 3.6 \times 10^{-13}$) ion is an even weaker acid than $H_2PO_4^-$. The K_a values indicate that it is more difficult to remove H^+ from a negatively charged $H_2PO_4^-$ ion than from a neutral H_3PO_4 molecule and even more difficult to remove H^+ from a doubly negative HPO_4^{2-} ion.

Exercise 16.14   Polyprotic Acids

Write equations for the stepwise ionization in aqueous solution of (a) oxalic acid and (b) citric acid. (Formulas for these acids are given in Table 16.3.)

16.6 PROBLEM SOLVING USING K_a and K_b

Calculations with K_a or K_b follow the same patterns as those of other equilibrium calculations illustrated earlier (Section 14.5). Similar important relationships apply in these calculations.

- *Starting with only reactants, equilibrium can be achieved only if some amount of the reactants is converted to products; that is, products are formed at the expense of reactants.*
- *The chemical equilibrium equation for the ionization of the acid or base is the basis for the acid ionization or base ionization constant expression.*
- *The concentrations in the acid ionization or base ionization expression, expressed as molarity (mol/L), are those at equilibrium.*
- *The magnitude of the K_a or K_b value indicates how far the forward reaction occurs at equilibrium (K_a: H^+ donation **to** water by an acid; K_b: H^+ gain by a base **from** water).*

There are several experimental methods for determining acid or base ionization constants. The simplest is based on measuring the pH of an acid solution of known concentration. If both the acid concentration and the pH are known, the K_a for the acid can be calculated.

 Problem-Solving Example 16.7 K_a from pH

The pH of a 0.10 M solution of propanoic acid, CH_3CH_2COOH, a weak organic acid, is measured at equilibrium and found to be 2.93 at 25 °C. What is the K_a of this acid?

 CD-ROM Screen 17.7:
Determining K_a and K_b Values

Answer 1.5×10^{-5}

Explanation   The acid ionizes according to the balanced equation

$$CH_3CH_2COOH(aq) + H_2O(\ell) \rightleftharpoons H_3O^+(aq) + CH_3CH_2COO^-(aq)$$

The pH gives us the equilibrium concentration of H_3O^+. Using the definition of pH,

$$[H_3O^+] = 10^{-pH} = 10^{-2.93} = 0.0012 \text{ M}$$

The equilibrium concentrations of the other species are represented by using a reaction table.

	$CH_3CH_2COOH + H_2O \rightleftharpoons$	H_3O^+	$+ CH_3CH_2COO^-$
Initial concentration (mol/L)	0.10	1.0×10^{-7} (from water)*	0
Change as reaction occurs (mol/L)	-0.0012	$+0.0012$	$+0.0012$
Equilibrium concentration (mol/L)	$0.10 - 0.0012$	0.0012	0.0012

* This concentration can be ignored because it is so small.

From the measured pH we have calculated $[H_3O^+]$ to be 1.2×10^{-3} mol/L, which is also the $CH_3CH_2COO^-$ concentration at equilibrium because the ions are formed in equal amounts as propanoic acid ionizes. The concentration of the un-ionized acid at equilibrium is $0.10 - 0.0012 = 0.0988$ M. Using these values, we can now calculate K_a for propanoic acid.

$$K_a = \frac{[H_3O^+][CH_3CH_2COO^-]}{[CH_3CH_2COOH]} = \frac{[x][x]}{0.10 - x} = \frac{[0.0012][0.0012]}{0.0988} = 1.5 \times 10^{-5}$$

✔ This K_a is small, indicating that propanoic acid is a weak acid, as reflected by the fact that a 0.10 M propanoic acid solution has an $[H_3O^+]$ of just 0.0012 M. Thus, the answer makes sense; propanoic acid is only slightly ionized. Propanoic acid is similar to acetic acid in strength, as expected from its similar structure, and a 0.10 M solution has a pH nearly the same as that of 0.10 M acetic acid (pH = 2.87).

Problem-Solving Practice 16.7

Lactic acid is a monoprotic acid that occurs naturally in sour milk and also forms by metabolism in the human body. A 0.10 M aqueous solution of lactic acid, $CH_3CH(OH)COOH$, has a pH of 2.43. What is the value of K_a for lactic acid? Is lactic acid stronger or weaker than propanoic acid?

CD-ROM Screen 17.8: Estimating the pH of Weak Acid Solutions

Acid-base ionization constants such as those in Table 16.2 can be used to calculate the pH of a solution of a weak acid or a weak base from its concentration.

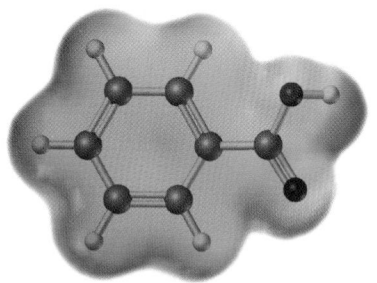

benzoic acid

Problem-Solving Example 16.8   pH from K_a

(a) What is the pH of a 0.050-M solution of benzoic acid, C_6H_5COOH ($K_a = 6.3 \times 10^{-5}$ at 25 °C)?

(b) What percent of the acid has ionized in this solution?

Answer

(a) pH = 2.74   (b) 3.6% ionized

Explanation

(a) First write the equilibrium equation and equilibrium constant expression.

$$C_6H_5COOH(aq) + H_2O(\ell) \rightleftharpoons H_3O^+(aq) + C_6H_5COO^-(aq)$$

$$K_a = \frac{[H_3O^+][C_6H_5COO^-]}{[C_6H_5COOH]}$$

Next, define equilibrium concentrations and organize the known information in the usual table. In this case, x is the H_3O^+ concentration at equilibrium. At equilibrium, the benzoate ion ($C_6H_5COO^-$) concentration is also equal to x because the reaction produces H_3O^+ and $C_6H_5COO^-$ in equal amounts.

	C_6H_5COOH + H_2O $\rightleftharpoons$	H_3O^+	+ $C_6H_5COO^-$
Initial concentration (mol/L)	0.050	1.0×10^{-7} (from water)*	0
Change as reaction occurs (mol/L)	$-x$	$+x$	$+x$
Concentration at equilibrium (mol/L)	$0.050 - x$	x	x

Since all equilibrium concentrations are defined in terms of the single unknown, x, the equilibrium constant expression can be rewritten as

$$K_a = \frac{[H_3O^+][C_6H_5COO^-]}{[C_6H_5COOH]} = \frac{[x][x]}{0.050 - x} = 6.3 \times 10^{-5}$$

Because K_a is very small, the reaction must be reactant-favored. This means not very much product will form, and the concentrations of H_3O^+ and $C_6H_5COO^-$ will be very small when equilibrium is reached. Therefore, x must be quite small compared with 0.050. When x is subtracted from 0.050, the result will still be almost exactly 0.050, and so we can approximate $0.050 - x$ as 0.050 to get

$$\frac{x^2}{0.050} \approx 6.3 \times 10^{-5}$$

Solving for x gives

$$x = \sqrt{(6.3 \times 10^{-5})(0.050)} = \sqrt{3.2 \times 10^{-6}} = 1.8 \times 10^{-3} = [H_3O^+]$$

$$pH = -\log[H_3O^+] = -\log(1.8 \times 10^{-3}) = 2.74$$

The solution is acidic.

(b) The ionization of the acid is the major source of H_3O^+ ions (the concentration from water is insignificant). Therefore, the percent ionization is calculated by comparing the H_3O^+ concentration at equilibrium with the initial concentration of the acid.

$$\% \text{ ionization} = \frac{[H_3O^+]}{(C_6H_5COOH)_{\text{initial}}} \times 100\% = \frac{1.8 \times 10^{-3}}{0.050} \times 100\% = 3.6\%$$

 Its K_a value of 6.3×10^{-5} indicates that benzoic acid is a weak acid, similar to acetic acid ($K_a = 1.8 \times 10^{-5}$) in strength and should be only slightly ionized. That 0.050-M benzoic acid is only 3.6% ionized and has a $[H_3O^+]$ of 1.8×10^{-3} M and a pH of 2.74 is reasonable.

Problem-Solving Practice 16.8

Boric acid is a weak acid often used as an eyewash. K_a for boric acid is 7.3×10^{-10}. Find the pH of a 0.10 M solution of boric acid.

An analogous calculation can be done to find the pH of a solution of a weak base, such as methylamine, CH_3NH_2.

 CD-ROM Screen 17.9: Estimating the pH of Weak Base Solutions

Problem-Solving Example 16.9   pH of a Weak Base From K_b

Calculate the OH^- concentration and the pH of a 0.025-M methylamine solution. $K_b = 4.2 \times 10^{-4}$.

Answer $[OH^-] = 3.0 \times 10^{-3}$ M; pH = 11.48

Explanation   In such cases we

- Use the K_b expression and value to calculate $[OH^-]$.
- Calculate pOH from the OH^- concentration.
- Derive pH from the relation pH + pOH = 14.

Methylamine reacts with water according to the equation

$$CH_3NH_2(aq) + H_2O(\ell) \rightleftharpoons CH_3NH_3^+(aq) + OH^-(aq)$$

$$K_b = \frac{[CH_3NH_3^+][OH^-]}{[CH_3NH_2]} = 4.2 \times 10^{-4}$$

We can set up a table like the one in Problem-Solving Example 16.8, letting x be the concentration of OH^- and of methylammonium ion, $CH_3NH_3^+$, at equilibrium, because the forward reaction produces them in equal amounts. The equilibrium concentration of *unreacted* methylamine will be its initial concentration, 0.025 mol/L, minus x, the amount per liter that has reacted.

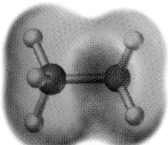

methylamine

	CH_3NH_2 + $H_2O(\ell)$ $\rightleftharpoons$	$CH_3NH_3^+$ +	OH^-
Initial concentration (mol/L)	0.025	0	$1.0 \times 10^{-7*}$
Change as reaction occurs (mol/L)	$-x$	$+x$	$+x$
Equilibrium concentration (mol/L)	$(0.025 - x)$	x	x

*The low concentration can be ignored, as it was in the K_a calculations.

Substitution into the base ionization constant expression gives

$$K_b = \frac{[CH_3NH_3^+][OH^-]}{[CH_3NH_2]} = \frac{x^2}{0.025 - x} = 4.2 \times 10^{-4}$$

In this case the OH^- concentration, x, must be found by using the quadratic formula because methylamine reacts with water sufficiently so that x is not negligible compared with 0.025. Generally in K_a and K_b calculations, if $\dfrac{x}{\text{initial concentration}} \times 100\% > 5\%$, the x term cannot be dropped from the denominator in the equilibrium constant expression and the quadratic equation is used.

Multiplying out the terms gives Equation A.

$$x^2 = (0.025 - x)(4.2 \times 10^{-4}) = 1.05 \times 10^{-5} - (4.2 \times 10^{-4}x) \qquad [A]$$

Rearranging Equation A into the quadratic form $ax^2 + bx + c = 0$ then gives Equation B.

$$x^2 + (4.2 \times 10^{-4}x) - (1.05 \times 10^{-5}) = 0 \qquad [B]$$

Solving for x using the quadratic formula,

$$x = \frac{-(4.2 \times 10^{-4}) \pm \sqrt{(4.2 \times 10^{-4})^2 - (4 \times 1)(-1.05 \times 10^{-5})}}{2(1)}$$

$$= \frac{-(4.2 \times 10^{-4}) \pm \sqrt{4.2 \times 10^{-5}}}{2}$$

$$= \frac{-(4.2 \times 10^{-4}) \pm (6.5 \times 10^{-3})}{2}$$

$$= \frac{6.1 \times 10^{-3}}{2}$$

Therefore, $x = 3.0 \times 10^{-3}$ M = $[OH^-]$. (The negative root in the solution of the quadratic equation is disregarded because concentration cannot be negative; you can't have less than nothing.)

Note that $\dfrac{3.0 \times 10^{-3}}{0.025} \times 100\% = 12\%$, which is greater than 5%, so the quadratic equation was necessary in this case; the approximation of

$$\frac{x^2}{0.025 - x} \approx \frac{x^2}{0.025}$$

would not have given the correct answer.

The pOH can be calculated from the OH^- concentration.

$$pOH = -\log(3.0 \times 10^{-3}) = 2.52$$

$$pH = 14.00 - pOH = 14.00 - 2.52 = 11.48$$

Therefore, methylamine reacts sufficiently with water to generate a fairly basic solution.

✔ Both the K_b and the initial concentration of methylamine are small. Thus, the pH should be less than that of a 0.025 M solution of a strong base like NaOH, which would be 12.40.

$$[OH^-] = 0.025 \text{ M}; pOH = -\log(0.025) = 1.60; pH = 12.40$$

A pH of 11.48 for 0.025 M methylamine is reasonable.

Problem-Solving Practice 16.9

Calculate the pH of a 0.0050 M solution of dimethylamine, $(CH_3)_2NH$, whose K_b is 5.9×10^{-4}.

Relationship Between K_a and K_b Values

The right-hand side of Table 16.2 gives K_b values for the conjugate base of each acid. Try an experiment with these data: Multiply a few of the K_a values by their

K_b values for the conjugate bases. What do you find? Within a very small error you ought to find that $K_a \times K_b = 1.0 \times 10^{-14}$. This value is the same as K_w, the autoionization constant for water. To see why, multiply the equilibrium constant expressions for K_a and K_b.

$$K_a \times K_b = \left(\frac{[H_3O^+][A^-]}{[HA]} \right) \left(\frac{[HA][OH^-]}{[A^-]} \right)$$

Canceling like terms in the numerator and denominator of this expression gives

$$K_a \times K_b = \left(\frac{[H_3O^+][\cancel{A^-}]}{\cancel{[HA]}} \right) \left(\frac{\cancel{[HA]}[OH^-]}{\cancel{[A^-]}} \right) = [H_3O^+][OH^-] = K_w$$

This relation shows that if you know K_a for an acid, you can find K_b for its conjugate base by using K_w. Furthermore, the larger the K_a, the smaller the K_b, and vice versa (because they always have to give the same product when multiplied, K_w). For example, K_a for HCN is 4.0×10^{-10}. The value of K_b for the conjugate base, CN^-, is

$$K_b \text{ (for } CN^-) = \frac{K_w}{K_a \text{ (for HCN)}} = \frac{1.0 \times 10^{-14}}{4.0 \times 10^{-10}} = 2.5 \times 10^{-5}$$

HCN has a relatively small K_a and lies fairly far down in Table 16.2, which means it is a relatively weak acid. However, CN^- is a fairly strong weak base; its K_b of 2.5×10^{-5} is nearly the same as the K_b for ammonia (1.8×10^{-5}), making CN^- a slightly stronger base than ammonia. In general, if $K_a > K_b$, the acid is stronger than its conjugate base. Alternatively, if $K_b > K_a$, the conjugate base is stronger than its conjugate acid. For example, hypochlorite ion, ClO^- ($K_b = 2.9 \times 10^{-7}$) is a stronger base than hypochlorous acid is an acid ($K_a = 3.5 \times 10^{-8}$).

> **Exercise 16.15** K_b from K_a
>
> Phenol, or carbolic acid, C_6H_5OH, is a weak acid, $K_a = 1.3 \times 10^{-10}$. Calculate K_b for the phenolate ion, $C_6H_5O^-$. Which base in Table 16.2 is closest in strength to the phenolate ion? How did you make your choice?

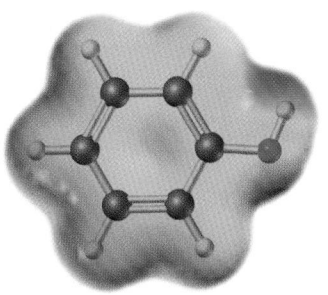

phenol

16.7 MOLECULAR STRUCTURE AND ACID STRENGTH

If all acids donate H^+ ions, why are some acids strong while others are weak? Why is there such a broad range of K_a values? To answer these questions we turn to the relationship of an acid's strength to its molecular structure. In doing so, we will consider a wide range of acids, from simple binary ones like HBr, to more complex ones containing oxygen, carbon, and other elements as well.

Factors Affecting Acid Strength

All acids have their acidic hydrogen bonded to some other atom, call it A, which can be bonded to other atoms as well. The H—A bond must be broken for the acid to transfer its hydrogen as an H^+ to water, and that will occur only if the H—A bond is polar.

$$\overset{\delta^+ \quad \delta^-}{\underset{\text{H—A}}{\longleftrightarrow}}$$

A nonpolar H—A bond, such as H—C in methane, CH_4, makes the hydrogens nonacidic, and methane is not an H^+ donor to water.

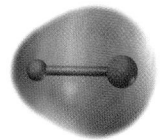

HBr

The simplest case of an acid is a *binary acid,* such as HBr, one that contains just hydrogen and one other element. In this case, A is bromine.

$$\overset{\delta^+ \quad \delta^-}{\underset{\text{H}-\text{Br}}{\xrightarrow{\hspace{1cm}}}}$$

The H—Br bond is polar and HBr is a strong acid ($K_a \approx 10^8$). The H—A bond energy is the most important determinant when comparing acid strengths for binary acids in which A is in the same group, for example HF, HCl, HBr, and HI. Note that *as H—A bond energies decrease down a group, the H—A bond weakens, and binary acid strengths increase.*

For a series of binary acids for which A is in the same period, H—A bond energies do not vary considerably. The H—A bond polarity is the principal factor affecting acid stengths for binary acids in the same row of the periodic table: As the electronegativity of A increases across a period, the H—A bond polarity also gets larger due to greater electronegativity differences. Correspondingly, the hydrogen becomes more acidic, as seen for Period 3 nonmetals:

- The H—Si bond is relatively nonpolar and SiH_4 is nonacidic.
- The H—P bond is only slightly polar and the unshared electron pair on phosphorus makes PH_3 accept H^+ ions rather than donate them.
- The H—S bond is slightly polar and H_2S is a weak acid.
- The H—Cl bond is very polar and HCl is a strong acid.

Strengths of Oxoacids

Acids in which the acidic hydrogen is bonded directly to oxygen in an H—O— bond are called **oxoacids.** They have at least one hydrogen bonded to an oxygen and have the general formula

$$\text{H}-\text{O}-\text{Z}\!\!\diagup\!\!\diagdown$$

The three strong acids nitric acid, perchloric acid, and sulfuric acid are oxoacids.

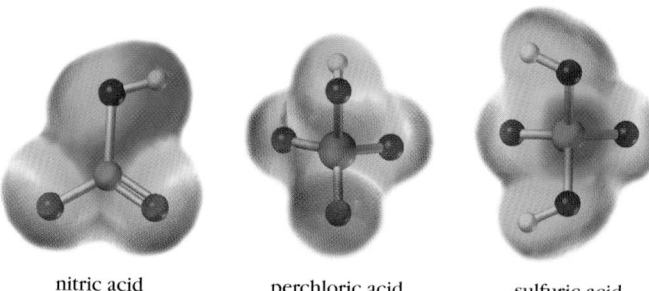

nitric acid   perchloric acid   sulfuric acid

The nature of Z and other atoms that may be attached to it are important in determining the strength of the H—O bond and thus the strength of an oxoacid. In general, *acid strength decreases with the decreasing electronegativity of Z.* This is reflected in the differences among the K_a values of HClO, HBrO, and HIO, as the electronegativity of the halogen decreases from chlorine (3.0) to bromine (2.8) to iodine (2.5).

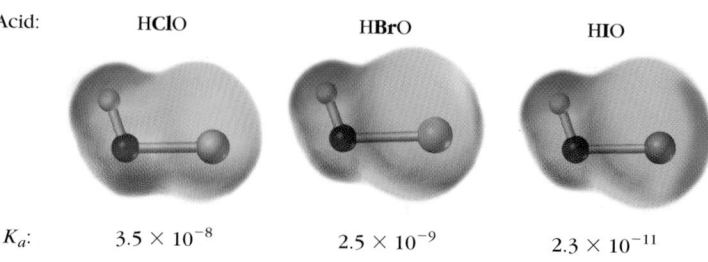

Acid:	HClO	HBrO	HIO
K_a:	3.5×10^{-8}	2.5×10^{-9}	2.3×10^{-11}

The number of oxygen atoms attached to Z also significantly affects the strength of the H—O bond and oxoacid strength: ***The acid strength increases as the number of oxygen atoms attached to Z increases.*** The terminal oxygen atoms (those not in an H—O bond) are sufficiently electronegative, along with Z, to withdraw electron density from the H—O bond. This weakens the bond, promoting the transfer of an H^+ ion to water. The more terminal oxygen atoms, the greater the electron density shift and the greater the acid strength. A particularly striking example of this trend is seen with the oxoacids of chlorine from the weakest, hypochlorous acid, HClO, to the strongest, perchloric acid, $HClO_4$.

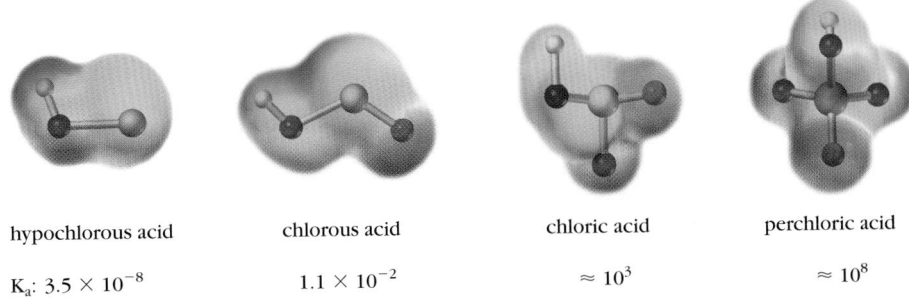

hypochlorous acid	chlorous acid	chloric acid	perchloric acid
K_a: 3.5×10^{-8}	1.1×10^{-2}	$\approx 10^3$	$\approx 10^8$

To be a strong acid, an inorganic oxoacid has to have at least two more oxygen atoms than hydrogen atoms in the molecule. Thus, sulfuric acid is a strong acid. Although an inorganic oxoacid, phosphoric acid, H_3PO_4, is a weak acid because it has only four oxygen atoms for three hydrogen atoms.

Strengths of Carboxylic Acids

All carboxylic acids contain the carboxylic acid group, —COOH, although all such acids do not have the same acid strength.

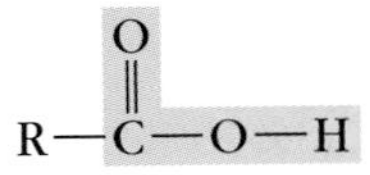

$$
\begin{array}{c}
\quad\quad O \\
\quad\quad \| \\
R—C—O—H
\end{array}
$$

The differences in carboxylic acid strength are due to differences in the R group attached to the —COOH group. When R is simply a hydrocarbon group, there is little effect on acid strength; K_a values are similar, as seen by comparing acetic acid, CH_3COOH, and hexanoic acid, $CH_3(CH_2)_4COOH$.

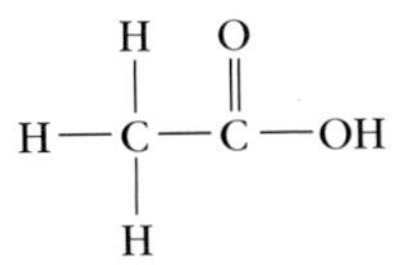

acetic acid
$K_a = 1.8 \times 10^{-5}$

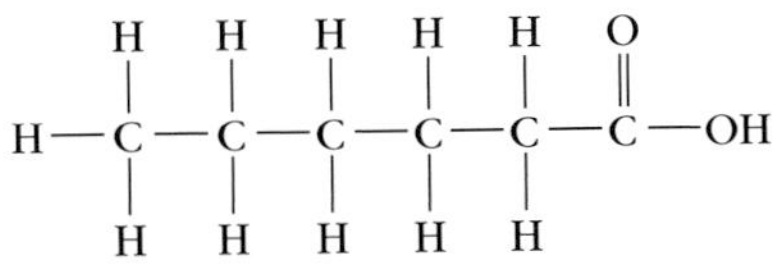

hexanoic acid
$K_a = 1.4 \times 10^{-5}$

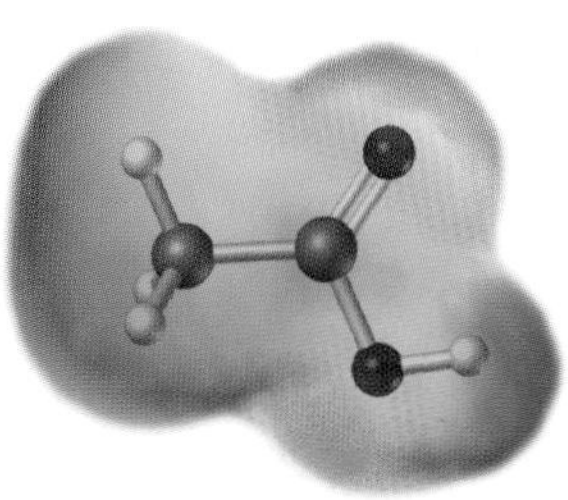

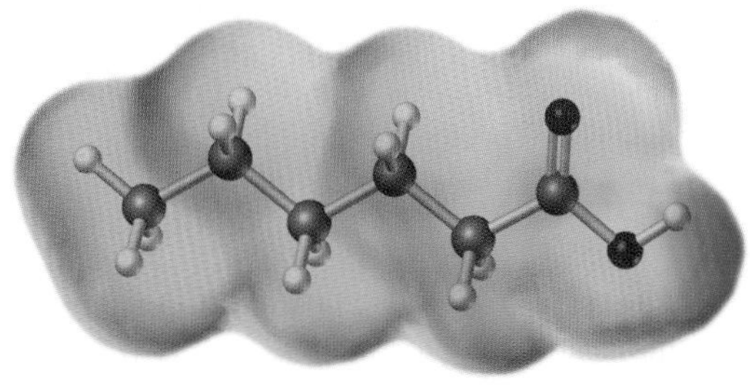

Acid strength, however, is affected by the addition of a highly electronegative atom to the R group. The electronegative atom causes electron density to shift from the O—H bond, weakening it and thus increasing the acid's strength. For example, replacing a hydrogen in the CH_3 group of acetic acid with chlorine, a more electronegative atom, forms chloroacetic acid, $ClCH_2COOH$. The K_a for this acid is 1.4×10^{-3}, nearly 100 times larger than that of acetic acid. By replacing all three hydrogens with chlorines, acetic acid is converted to trichloroacetic acid, Cl_3CCOOH, an acid that is about 10,000 times stronger than acetic acid and 100 times stronger than chloroacetic acid.

acetic acid
$K_a = 1.8 \times 10^{-5}$

chloroacetic acid
$K_a = 1.4 \times 10^{-3}$

trichloroacetic acid
$K_a = 3 \times 10^{-1}$

 Exercise 16.16 Molecular Structure and Acid Strength

Which has the larger K_a

(a) Fluorobenzoic acid, C_6H_4FCOOH, or benzoic acid, C_6H_5COOH?
(b) Chloroacetic acid or bromoacetic acid, $BrCH_2COOH$?

Explain your answer.

 Exercise 16.17 Molecular Structure and Acid Strength

Consider the acids acetic acid, CH_3COOH, and oxalic acid, $HOOC—COOH$. Which is the stronger acid? (Consider only the first ionization.) Explain your answer.

 ## Amino Acids and Zwitterions

Amino acids are the monomers from which proteins are assembled *(⬅ p. 556)*. The general formula for amino acids is

Unlike acetic acid (CH_3COOH) and ethylamine ($CH_3CH_2NH_2$), which are liquids at 25 °C, amino acids are crystalline solids that generally melt above 200 °C. What makes amino acids more like salts than simple organic compounds? The general formula given for amino acids does not lead to the prediction of such properties.

But consider the functional groups in an amino acid. All amino acids contain at least one acidic carboxylic acid group *and* one basic amine group. An intermolecular Brønsted-Lowry acid-base reaction occurs in amino acids in which H^+ is transferred from the carboxylic acid group to the amine group. In the case of alanine the change is

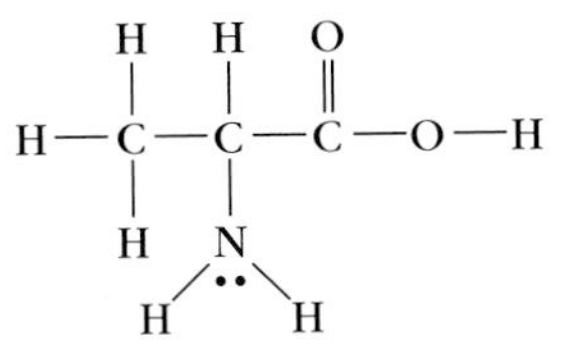

alanine, molecular form 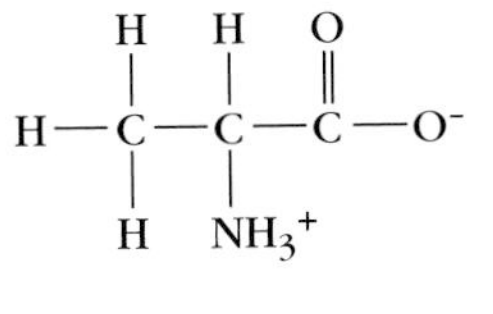   alanine, zwitterion form

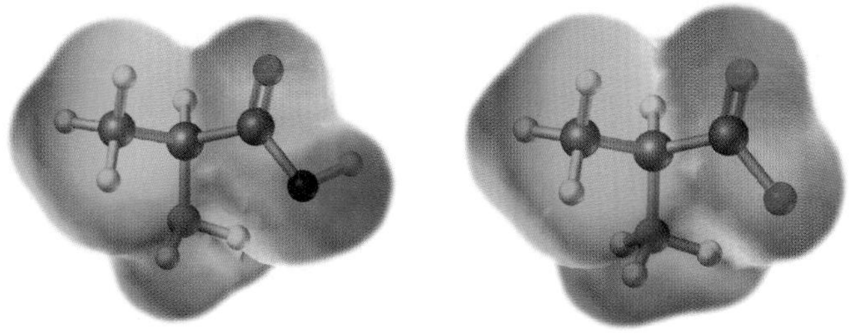

This resulting dipolar structure, called a **zwitterion,** creates a salt-like substance. The structure shown for the alanine zwitterion has no net charge, although regions of positive and negative charge exist in the molecule.

The term "zwitterion" is derived from the German word *zwitter,* meaning "a hybrid."

Alanine and other amino acids in solution can undergo additional acid-base reactions depending on the pH of the solution. Under acidic conditions, the H^+ concentration is large enough to add an H^+ ion to the negative carboxylate group ($-COO^-$), resulting in the formation of a $-COOH$ group. Under these conditions, the amino acid has a net positive charge. In basic solution, the concentration of OH^- ions is sufficient to remove the H^+ attached to the nitrogen of the zwitterion, resulting in a net negative charge on alanine.

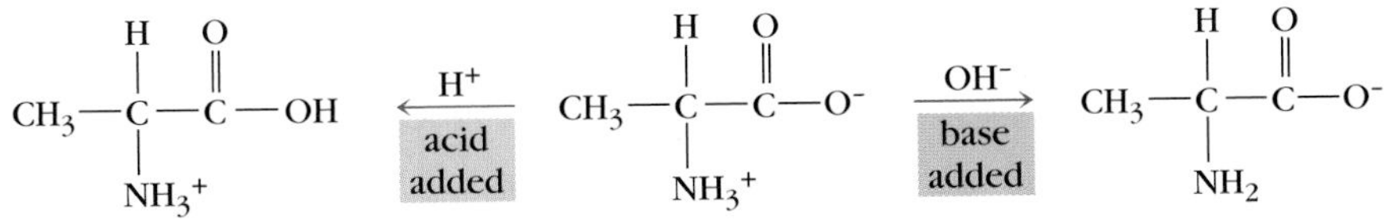

Positively   alanine zwitterion   Negatively
charged   no net charge   charged

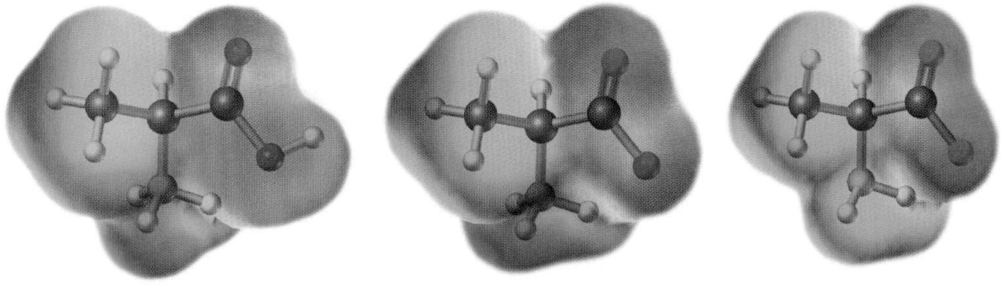

The relative amounts of the three alanine forms, or the analogous forms of every amino acid, depend on the pH of the body fluids in which they are present.

Exercise 16.18 Glycine and Its Zwitterion

Glycine is the simplest amino acid.

$$
\underset{\displaystyle \underset{\text{NH}_2}{|}}{\overset{\displaystyle \overset{\text{H}}{|}}{\text{H}-\text{C}}}-\overset{\displaystyle \overset{\text{O}}{\|}}{\text{C}}-\text{OH}
$$

(a) Write the structural formula for its zwitterion form.
(b) Write the structural formula for glycine in solution at a pH of 2 and at a pH of 10.

Exercise 16.19 H⁺ Transfers

The following tripeptide, made from three amino acids, is shown in its molecular form.

Write the structural formula for the tripeptide with all the acidic or basic groups in their charged form.

16.8 ACID-BASE REACTIONS OF SALTS

CD-ROM Screen 17.10: Acid-Base Properties of Salts

An exchange reaction between an acid and a base produces a salt plus water (⇐ p. 173). Recall that a *salt* is any ionic compound that can be formed by the reaction of an acid with a base; a salt's positive ion comes from the base and its negative ion comes from the acid. In the case of a metal hydroxide as a base, the salt-forming general reaction is

$$
\underset{\text{acid}}{\text{HX(aq)}} + \underset{\text{base}}{\text{MOH(aq)}} \longrightarrow \underset{\text{salt}}{\text{MX(aq)}} + \text{HOH}(\ell)
$$

Now that you know more about the Brønsted-Lowry acid-base concept and the strengths of acids and bases, it is useful to consider acid-base reactions and salt formation in more detail.

Salts of Strong Bases and Strong Acids

CD-ROM Screen 18.3: Acid-Base Reactions (Strong Acids + Strong Bases)

The strong acid HCl reacts with the strong base NaOH to form the salt NaCl. If the amounts of HCl and NaOH are in the correct stoichiometric ratio (1 mol of HCl per 1 mol of NaOH), this reaction occurs with the complete neutralization of the acidic properties of HCl and the basic properties of NaOH. The reaction can be described first by an overall equation, then by a complete ionic equation, and finally by a net ionic equation (⇐ p. 175). Each of these equations contains useful information.

$$
\text{HCl(aq)} + \text{NaOH(aq)} \longrightarrow \text{NaCl(aq)} + \text{H}_2\text{O}(\ell)
$$

$$
\text{H}_3\text{O}^+(\text{aq}) + \text{Cl}^-(\text{aq}) + \text{Na}^+(\text{aq}) + \text{OH}^-(\text{aq}) \longrightarrow \text{Na}^+(\text{aq}) + \text{Cl}^-(\text{aq}) + 2\,\text{H}_2\text{O}(\ell)
$$

$$
\underset{\text{ACID}}{\text{H}_3\text{O}^+(\text{aq})} + \underset{\text{base}}{\text{OH}^-(\text{aq})} \longrightarrow \underset{\text{BASE}}{\text{H}_2\text{O}(\ell)} + \underset{\text{acid}}{\text{H}_2\text{O}(\ell)}
$$

The overall equation shows the substances that were dissolved or that could be recovered at the end of the reaction. The complete ionic equation indicates all of the

TABLE 16.4	Some Salts Formed by Neutralization of Strong Acids with Strong Bases		
	Base		
Acid	**NaOH**	**KOH**	**Ba(OH)$_2$**
HCl	NaCl	KCl	BaCl$_2$
HNO$_3$	NaNO$_3$	KNO$_3$	Ba(NO$_3$)$_2$
H$_2$SO$_4$	Na$_2$SO$_4$	K$_2$SO$_4$	BaSO$_4$
HClO$_4$	NaClO$_4$	KClO$_4$	Ba(ClO$_4$)$_2$

ions that are present before and after reaction. The net ionic equation emphasizes that a Brønsted-Lowry acid (H$_3$O$^+$) is reacting with a Brønsted-Lowry base (OH$^-$); the spectator ions, Na$^+$ and Cl$^-$, are omitted. This reaction goes to completion because H$_3$O$^+$ is a strong acid, OH$^-$ is a strong base, and water is a very weak acid and a very weak base.

The resulting solution contains only sodium ions and chloride ions, with a few more water molecules than before. Its properties are the same as if it had been prepared by simply dissolving some NaCl(s) in water. It has a neutral pH because it contains no significant acids or bases. The Cl$^-$ ion is the conjugate base of a strong acid and hence is such a weak base that it does not react with water. The Na$^+$ ion also does not react as either an acid or a base with water. Examples of some other salts of this type are given in Table 16.4. These salts all form neutral solutions.

Salts of Strong Bases and Weak Acids

Suppose, for example, that 0.010 mol of NaOH is added to 0.010 mol of the weak acid acetic acid in 1 L of solution. The three equations are

CD-ROM Screen 18.5: Acid-Base Reactions (Weak Acids + Strong Bases)

$$CH_3COOH(aq) \quad + \quad NaOH(aq) \longrightarrow \quad NaCH_3COO(aq) + H_2O(\ell)$$

$$CH_3COOH(aq) + Na^+(aq) + OH^-(aq) \longrightarrow Na^+(aq) + CH_3COO^-(aq) + H_2O(\ell)$$

$$\underset{\text{weak acid}}{CH_3COOH(aq)} \quad + \quad \underset{\text{strong base}}{OH^-(aq)} \longrightarrow \quad \underset{\text{base}}{CH_3COO^-(aq)} + \underset{\text{acid}}{H_2O(\ell)}$$

In this case, acetate ion, a weak base, has been formed by the reaction. Therefore, the solution is slightly basic (pH > 7), even though exactly the stoichiometric amount of acetic acid was added to the sodium hydroxide. The reaction that makes the solution basic is the reaction of acetate ion as a weak Brønsted-Lowry base with water.

$$CH_3COO^-(aq) + H_2O(\ell) \rightleftharpoons CH_3COOH(aq) + OH^-(aq)$$

This is a **hydrolysis** reaction, one in which a water molecule is split — in this case, into an H$^+$ ion and a hydroxide ion. An H$^+$ ion is donated to the acetate ion to form acetic acid. The extent of hydrolysis is determined by the value of K_b for acetate ion.

All of the weak bases in Table 16.2, except for the very weak bases above water as a base, undergo hydrolysis reactions in aqueous solution. The larger their K_b values, the more basic the solutions they produce. The pH of a solution of a salt of a strong base and a weak acid can be estimated from K_b, as shown in Problem-Solving Example 16.10.

The term "hydrolysis" is derived from *hydro*, meaning "water," and *lysis*, meaning "to break apart." The hydrolysis reaction of a molecular compound results in the addition of H— and —OH to the molecules produced by breaking a covalent bond.

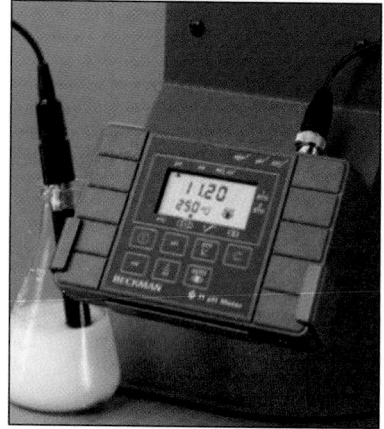

Hydrolysis of salts of strong bases and weak acids in aqueous solution. The pH meter readings indicate that aqueous solutions of sodium acetate ($NaCH_3COO$, *top*) and sodium cyanide (NaCN, *bottom*) are basic. The NaCN solution is more basic because cyanide ion, CN^-, is a stronger base than is acetate ion, CH_3COO^-; that is, the K_b of $CN^- > K_b$ of CH_3COO^-. An inert, insoluble solid has been added to each flash to enhance the visibility of liquid in the flash.

Problem-Solving Example 16.10   pH of a Salt Solution

Sodium hypochlorite, NaClO, is used as a source of chlorine in some laundry bleaches, swimming pool disinfectants, and water treatment plants. Calculate the pH of a 0.010 M solution of NaClO ($K_b = 2.9 \times 10^{-7}$).

Answer   pH = 9.73

Explanation   Sodium hypochlorite consists of sodium ions and hypochlorite ions; it is the salt of a strong base and a weak acid. The Na^+ ion does not react with water, but hypochlorite ion, ClO^-, is the conjugate base of a weak acid (HClO) and reacts with water to produce a basic solution.

$$ClO^-(aq) + H_2O(\ell) \rightleftharpoons HClO(aq) + OH^-(aq)$$

$$K_b = 2.9 \times 10^{-7} = \frac{[\text{conjugate acid}][OH^-]}{[\text{conjugate base}]} = \frac{[HClO][OH^-]}{[ClO^-]}$$

The concentrations of hypochlorite ion, hypochlorous acid, and hydroxide ion initially and at equilibrium are summarized in the following table. We let x be equal to the equilibrium concentration of OH^- as well as that of ClO^- because they are formed in equal amounts.

	ClO^-	$+ H_2O \rightleftharpoons$	$HClO +$	OH^-
Initial concentration (mol/L)	0.010		0	1.0×10^{-7} (from water)
Change as reaction occurs (mol/L)	$-x$		$+x$	$+x$
Equilibrium concentration (mol/L)	$0.010 - x$		x	x

Hypochlorite ion has a very small K_b (2.9×10^{-7}) and thus is a very weak base, so it is safe to assume that x will be negligibly small compared with 0.010, and $0.010 - x \approx 0.010$.

$$K_b = 2.9 \times 10^{-7} \approx \frac{x^2}{0.010}$$

Solving for x gives $x = 5.4 \times 10^{-5}$. Since $0.010 - 5.4 \times 10^{-5} = 0.010$ (using the significant figures rules), our assumption that x is negligible compared with 0.010 is justified. Therefore, at equilibrium

$$[OH^-] = [HClO] = 5.4 \times 10^{-5} \text{ mol/L and } [ClO^-] = 0.010 \text{ mol/L}$$

Finally, the pH of the solution is found as follows:

$$K_w = [H_3O^+][OH^-] = [H_3O^+](5.4 \times 10^{-5}) = 1.0 \times 10^{-14}$$

$$[H_3O^+] = \frac{1.0 \times 10^{-14}}{5.4 \times 10^{-5}} = 1.9 \times 10^{-10}$$

$$pH = -\log(1.9 \times 10^{-10}) = 9.72$$

As expected, the solution is basic.

✔   The reaction of hypochlorite ion with water produces hydroxide ions in addition to those from the dissociation of water. The excess hydroxide ions cause the solution to become basic, as indicated by the pH greater than 7. This is expected because the salt is formed from a strong base and a weak acid.

Problem-Solving Practice 16.10

Sodium carbonate is an environmentally safe paint stripper. It is water-soluble, and carbonate ion is a strong enough base to loosen paint so it can be scraped off. What is the pH of a 1.0 M solution of Na_2CO_3?

Exercise 16.20   pH of Soap Solutions

Ordinary soaps are often sodium salts of fatty acids, which are weak organic acids. Would you expect the pH of a soap solution to be <7 or >7? Explain your answer.

Salts of Weak Bases and Strong Acids

When a weak base reacts with a strong acid, the resulting salt solution is acidic. The conjugate acid of the weak base determines the pH of the solution. For example, suppose equal volumes of 0.10 M NH_3 and 0.10 M HCl are mixed. The reaction, shown in overall, complete ionic, and net ionic forms, is

CD-ROM Screen 18.4: Acid-Base Reactions (Strong Acids + Weak Bases)

$$NH_3(aq) \quad + \quad HCl(aq) \longrightarrow NH_4Cl(aq)$$

$$NH_3(aq) + H_3O^+(aq) + Cl^-(aq) \longrightarrow NH_4^+(aq) + Cl^-(aq) + H_2O(\ell)$$

$$\underset{\text{weak base}}{NH_3(aq)} \quad + \quad \underset{\text{strong acid}}{H_3O^+(aq)} \longrightarrow \underset{\text{acid}}{NH_4^+(aq)} \quad + \quad \underset{\text{base}}{H_2O(\ell)}$$

As soon as it is formed, the weak acid NH_4^+ reacts with water and establishes an equilibrium. The resulting solution is slightly acidic, because of the reaction

$$NH_4^+(aq) + H_2O(\ell) \rightleftharpoons NH_3(aq) + H_3O^+(aq)$$

Many drugs, such as Novocain, are high molecular weight amines that are weak bases. Such amines are not soluble in water, which limits the ways they can be administered and also means that they are not soluble in body fluids such as blood plasma and cerebrospinal fluid. By reaction with hydrochloric acid, the amines are converted to soluble hydrochlorides that can be administered by injection or dissolved in liquid oral medications. The resulting salts have the general formula BH^+Cl^-, where B represents the basic amine. This formula is like that of ammonium chloride, $NH_4^+Cl^-$. Two examples are

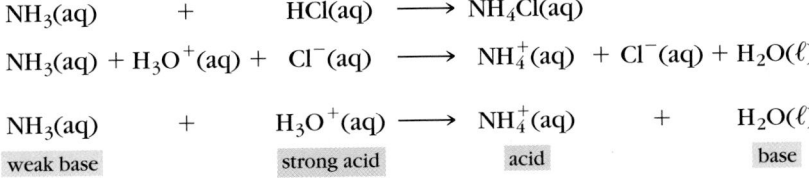

phenylephrine hydrochloride, a decongestant

diphenhydramine hydrochloride (Benadryl), an antihistamine

The amine hydrochloride salt of a drug is much more water-soluble than the amine form of the drug. For example, only 0.5 g of Novocain dissolves in 100 g of water, whereas 100 g of Novocain hydrochloride dissolves in the same amount of water.

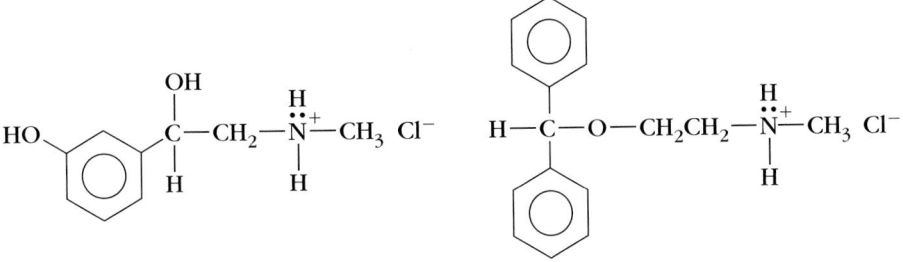

Novocain hydrochloride

Exercise 16.21 Forming A Drug Hydrochloride

Using structural formulas, write the equation for the formation of pseudoephedrine hydrochloride from pseudoephedrine, a decongestant.

pseudoephedrine

Salts of Weak Bases and Weak Acids

CD-ROM Screen 18.6: Acid-Base Reactions (Weak Acids + Weak Bases)

What is the pH of a solution of a salt containing an acidic cation and a basic anion, such as NH_4F or $Ni(CH_3COO)_2$? There are two possible reactions that can determine the pH of the solution: formation of H_3O^+ by H^+ transfer from the cation, and formation of OH^- by hydrolysis of the anion. In the case of NH_4F,

$$NH_4^+(aq) + H_2O(\ell) \rightleftharpoons H_3O^+(aq) + NH_3(aq) \qquad K_a(NH_4^+) = 5.6 \times 10^{-10}$$

$$F^-(aq) + H_2O(\ell) \rightleftharpoons HF(aq) + OH^-(aq) \qquad K_b(F^-) = 1.4 \times 10^{-11}$$

Since $K_a(NH_4^+) > K_b(F^-)$, the reaction of ammonium ions with water to produce hydronium ions is the more favorable reaction. Therefore, the resulting solution is slightly acidic. For $Ni(CH_3COO)_2$ the possible reactions are

$$Ni(H_2O)_6^{2+}(aq) + H_2O(\ell) \rightleftharpoons Ni(H_2O)_5(OH)^+(aq) + H_3O^+(aq)$$

$$K_a(Ni(H_2O)_6^{2+}) = 2.5 \times 10^{-11}$$

$$CH_3COO^-(aq) + H_2O(\ell) \rightleftharpoons CH_3COOH(aq) + OH^-(aq)$$

$$K_b(CH_3COO^-) = 5.6 \times 10^{-10}$$

Since $K_b(CH_3COO^-) > K_a(Ni(H_2O)_6^{2+})$, the hydrolysis of CH_3COO^- ions is more favorable, and the resulting solution is slightly basic.

Exercise 16.22 Hydrolysis of a Salt of a Weak Acid and a Weak Base

Name a salt of a weak acid and a weak base where $K_a = K_b$. What should the pH of a solution of this salt be?

The following generalizations can be made about acid-base reactions in aqueous solution and the pH of the resulting salt solutions.

• Solution of strong acid + solution of strong base $\rightarrow$ salt solution with pH = 7
• Solution of strong acid + solution of weak base $\rightarrow$
    salt solution with pH < 7 (acidic)
• Solution of weak acid + solution of strong base $\rightarrow$
    salt solution with pH > 7 (basic)
• Solution of weak acid + solution of weak base $\rightarrow$ salt solution with pH determined by relative strengths of conjugate base and conjugate acid formed

Table 16.5 summarizes the acid-base behavior of many different ions in aqueous solution.

Acidic pH of an aqueous copper(II) sulfate solution. The blue solution of this copper salt is acidic due to hydrolysis of the Cu^{2+} ion.

TABLE 16.5	Acid-Base Properties of Typical Ions in Aqueous Solution					
	Neutral		**Basic**			**Acidic**
Anions	Cl^-	NO_3^-	CH_3COO^-	CN^-	SO_4^{2-}	HSO_4^-
	Br^-	ClO_4^-	$HCOO^-$	PO_4^{3-}	HPO_4^{2-}	$H_2PO_4^-$
	I^-		CO_3^{2-}	HCO_3^-	SO_3^{2-}	HSO_3^-
			S^{2-}	HS^-	ClO^-	
			F^-	NO_2^-		
Cations	Li^+	Mg^{2+}	None			Al^{3+}
	Na^+	Ca^{2+}				NH_4^+
	K^+	Ba^{2+}				Transition metal ions

16.9 PRACTICAL ACID-BASE CHEMISTRY

In addition to their uses in industry, various acids and bases find many applications around the home. Antacids are used to neutralize stomach acidity; gardeners use acid salts such as sodium hydrogen sulfate ($NaHSO_4$) to help acidify soil and bases such as lime (CaO) to make soil more basic. In the kitchen, baking soda and baking powders are used to make biscuit dough and cake batter rise. Mild acids and bases are used to clean everything from dishes and clothes to vehicles and the family dog.

Neutralizing Stomach Acidity

Human stomach fluids have a pH of approximately 1. This very acidic pH is caused by HCl, which is secreted by thousands of cells in the stomach lining that specialize in transporting $H_3O^+(aq)$ and $Cl^-(aq)$ from the blood. The main purpose of this acid is to suppress the growth of bacteria and to aid in the digestion of certain foods. The hydrochloric acid does not harm the stomach because its inner lining is replaced at the rate of about half a million cells per minute. However, when too much food is eaten and the stomach is stretched, or when the stomach is irritated by very spicy food, some of its acidic contents can flow back into the esophagus (gastroesophageal reflux), producing a burning sensation called *heartburn.*

An antacid is a base that is used to neutralize excess stomach acid. The recommended dose is the amount of the base required to neutralize *some,* but not all, of the stomach acid. Several antacids and their acid-base reactions are shown in Table 16.6. People who need to restrict the quantity of sodium (Na^+) in their diets should avoid sodium containing antacids such as sodium bicarbonate.

Chloride ions secreted by the stomach lining come mostly from the salty foods we eat and the salt we add to our foods.

 Exercise 16.23 Strong Antacids?

Explain why strong bases such as NaOH or KOH are never used as antacids.

Problem-Solving Example 16.11 Neutralizing Stomach Acid

How many moles (and what mass) of HCl could be neutralized by 0.750 g of the antacid $CaCO_3$?

Answer 1.50×10^{-2} mol of HCl; 0546 g of HCl

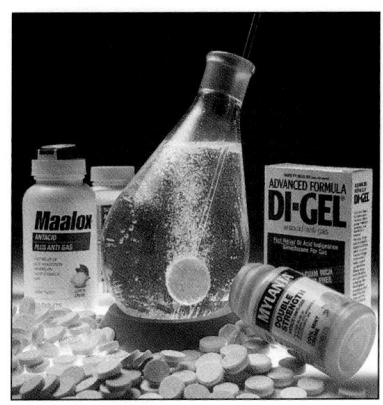

Commercial antacid products. The Alka-Seltzer tablet in the flask contains sodium hydrogen carbonate, which neutralizes excess stomach acid and relieves heartburn.

TABLE 16.6 The Acid-Base Chemistry of Some Antacids

Compound	Reaction in stomach	Examples of commercial products
Milk of magnesia: $Mg(OH)_2$ in water	$Mg(OH)_2(s) + 2 H_3O^+(aq) \longrightarrow Mg^{2+}(aq) + 4 H_2O(\ell)$	Phillips' Milk of Magnesia
Calcium carbonate: $CaCO_3$	$CaCO_3(s) + 2 H_3O^+(aq) \longrightarrow Ca^{2+}(aq) + 3 H_2O(\ell) + CO_2(g)$	Tums, Di-Gel
Sodium bicarbonate: $NaHCO_3$	$NaHCO_3(s) + H_3O^+(aq) \longrightarrow Na^+(aq) + H_2O(\ell) + CO_2(g)$	Baking soda, Alka-Seltzer
Aluminum hydroxide: $Al(OH)_3$	$Al(OH)_3(s) + 3 H_3O^+(aq) \longrightarrow Al^{3+}(aq) + 6 H_2O(\ell)$	Amphojel
Dihydroxyaluminum sodium carbonate: $NaAl(OH)_2CO_3$	$NaAl(OH)_2CO_3(s) + 4 H_3O^+(aq) \longrightarrow Na^+(aq) + Al^{3+}(aq) + 7 H_2O(\ell) + CO_2(g)$	Rolaids

Explanation The balanced equation for this reaction is given in Table 16.6. From the equation we see that 1 mol of the antacid reacts with 2 mol of HCl, molar mass = 36.46 g.

$$0.750 \text{ g} \times \frac{1 \text{ mol } CaCO_3}{100.1 \text{ g } CaCO_3} = 7.49 \times 10^{-3} \text{ mol } CaCO_3$$

Using the stoichiometric mole ratio,

$$7.49 \times 10^{-3} \text{ mol } CaCO_3 \times \frac{2 \text{ mol HCl}}{1 \text{ mol } CaCO_3} \times \frac{36.46 \text{ g HCl}}{1 \text{ mol HCl}} = 0.546 \text{ g HCl}$$

✔ Two moles of HCl are required to react with each mole of $CaCO_3$, so to neutralize 0.00749 mol of $CaCO_3$ requires 0.0150 mol of HCl, which is just over 0.5 g (0.546 g) of HCl.

Problem-Solving Practice 16.11

Using the reactions in Table 16.6, determine which antacid, on a per gram basis, can neutralize the most stomach acid (assume 1.0 M HCl).

ESTIMATION **Using an Antacid**

Estimate how many Rolaids tablets (Table 16.6) it would take to neutralize the acidity in one glass (250 mL) of a regular cola drink. Assume the pH of the cola is 3.0. One Rolaids tablet contains 334 mg of $NaAl(OH)_2CO_3$.

With a pH of 3.0, the cola has 1×10^{-3} mol of acid per liter of cola, so 0.250 L of cola has one-fourth that much acid, or about 3×10^{-4} mol of acid. To neutralize this amount of acid requires 3×10^{-4} mol of base (1 mol of base for every 1 mol of acid). There are two bases in Rolaids — hydroxide ions and carbonate ions. Each mole of hydroxide neutralizes 1 mol of acid, and each mole of carbonate neutralizes 2 mol of acid.

$$H^+(aq) + OH^-(aq) \longrightarrow H_2O(\ell)$$

$$2 H^+(aq) + CO_3^{2-}(aq) \longrightarrow H_2O(\ell) + CO_2(g)$$

Because each mole of $NaAl(OH)_2CO_3$ contains 2 mol of OH^- ions and 1 mol of CO_3^{2-} ions, 1 mol of $NaAl(OH)_2CO_3$ neutral-

izes 4 mol of acid. The molar mass of $NaAl(OH)_2CO_3$ is 144 g/mol, so one Rolaids tablet contains about 0.002 mol of the antacid.

$$\frac{0.344 \text{ g antacid}}{1 \text{ antacid tablet}} \times \frac{1 \text{ mol antacid}}{144 \text{ g antacid}}$$

$$= 0.00239 \text{ mol antacid/tablet}$$

This tablet can neutralize four times that many moles of acid, or about 0.008 mol of acid. To neutralize the 3×10^{-4} mol of acid in the cola requires about 0.04 tablet.

$$3 \times 10^{-4} \text{ mol acid} \times \frac{1 \text{ tablet}}{0.008 \text{ mol acid}} \approx 0.04 \text{ tablet}$$

It would take only a small portion of a tablet to do the job.

Acid-Base Chemistry in the Kitchen

Vinegar is an approximately 4 to 5% aqueous acetic acid solution present in almost all salad dressings. Lemon juice, handy for flavoring tea and cooked fish and for making salad dressings, contains citric acid, as do all citrus fruits. One of the most useful substances in the kitchen is carbon dioxide gas, produced by chemical reactions as it is needed. Pockets of the gas are generated in bread dough and cake batter. The expanding gas makes the resulting biscuits, breads, and cakes rise, producing lighter and more palatable baked goods.

Various methods are used to generate CO_2. One is the addition of yeast, which causes bread dough to rise by catalyzing the fermentation of carbohydrates to produce ethyl alcohol and carbon dioxide.

$$C_6H_{12}O_6 \xrightarrow{\text{yeast}} 2\,CH_3CH_2OH + 2\,CO_2$$

Many commercial breads and homemade dinner rolls use this method to make these doughs rise.

Because CO_2 production by fermentation is slow, it is sometimes necessary to use another method, the reaction of a bicarbonate salt such as sodium bicarbonate, $NaHCO_3$ (also known as *baking soda*) with acid. But which acid should be used? A weak acid is needed; if a strong acid were used, complete neutralization of the acid would be required to make the food safe to eat. Although vinegar could be used, it would impart an undesirable taste. Long ago it was discovered that lactic acid ($CH_3CO(OH)COOH$), present in milk and formed in larger quantities when milk sours to form buttermilk, is a good source of acid for reacting with bicarbonate.

$$\underset{\underset{OH}{|}}{CH_3CHCOOH(aq)} + HCO_3^-(aq) \longrightarrow \underset{\underset{OH}{|}}{CH_3CHCOO^-(aq)} + H_2CO_3(aq)$$

lactic acid   lactate ion

$$\underset{\text{carbonic acid}}{H_2CO_3(aq)} \longrightarrow CO_2(g) + H_2O(\ell)$$

When buttermilk is not available, or when a different taste is desired, dihydrogen phosphate ion, $H_2PO_4^-$, is a convenient acid to react with the bicarbonate ion. Baking powders are a mixture of sodium or potassium dihydrogen phosphate and sodium bicarbonate. When dry, the two salts in baking powder do not react with one another. But when mixed with water in the dough, they react to produce CO_2, a reaction which occurs even more quickly in a heated oven.

$$H_2PO_4^-(aq) \rightleftharpoons H^+(aq) + HPO_4^{2-}(aq)$$

$$H^+(aq) + HCO_3^-(aq) \longrightarrow H_2O(\ell) + CO_2(g)$$

Net reaction: $H_2PO_4^-(aq) + HCO_3^-(aq) \longrightarrow H_2O(\ell) + CO_2(g) + HPO_4^{2-}(aq)$

Action of baking powder. Baking powder contains the weak acid dihydrogen phosphate ion and the weak base bicarbonate ion. When mixed together in water, the acid and base react to produce carbon dioxide gas. Bubbles of the gas are seen in the picture.

Exercise 16.24   Acids and Muffins

Consider the list of ingredients for buttermilk blueberry muffins in the margin. There are two sources of acid and two sources of bicarbonate, the reaction of which produces the CO_2 to make these muffins rise. What are the sources?

Household Cleaners

Most cleaning compounds such as dishwashing detergents, scouring powders, laundry detergents, and oven cleaners are basic. Synthetic detergents are derived

Buttermilk Blueberry Muffins
$2\frac{1}{2}$ C flour
$1\frac{1}{2}$ tsp baking powder
$\frac{1}{2}$ tsp baking soda
$\frac{1}{2}$ C sugar
$\frac{1}{4}$ tsp salt
2 eggs, beaten
1 C buttermilk
3 oz butter
$1\frac{1}{2}$ C blueberries (added for flavor)

from organic molecules designed to have even better cleaning action than soaps, but less reaction with the doubly positive ions (Mg^{2+} and Ca^{2+}) found in hard water (⇐ *p. 718*).

The molecular structure of a synthetic detergent molecule, like that of a soap, consists of a long oil-soluble (hydrophobic group), and a water-soluble hydrophilic group (⇐ *p. 390*).

A Typical Synthetic Detergent Molecule

$$CH_3CH_2CH_2CH_2CH_2CH_2CH_2CH_2CH_2CH_2CH_2CH_2CH_2CH_2 \!-\!\!\bigcirc\!\!-\! SO_3^-Na^+$$

$$\underbrace{\hphantom{CH_3CH_2CH_2CH_2CH_2CH_2CH_2CH_2CH_2CH_2CH_2CH_2CH_2CH_2}}_{\substack{\text{Oil-soluble part} \\ \text{(hydrophobic)}}} \qquad \underbrace{\hphantom{SO_3^-Na^+}}_{\substack{\text{Water-soluble part} \\ \text{(hydrophilic)}}}$$

Typical hydrophilic groups include negatively charged sulfate ($-OSO_3^{2-}$), sulfonate ($-SO_3^-$), and phosphate ($-OPO_3^{3-}$) groups. Compounds with these groups are called *anionic surfactants. Cationic surfactants* are almost all quaternary ammonium halides (four groups attached to the central nitrogen atom) with the general formula

$$R_1\!-\!\overset{\displaystyle R_2}{\underset{\displaystyle R_3}{\overset{\displaystyle |}{\underset{\displaystyle |}{N^+}}}}\!-\!R_4\ X^-$$

where one of the R groups is a long hydrocarbon chain and another frequently includes an $-OH$ group. The X^- in the formula represents a halide ion such as chloride (Cl^-) or bromide (Br^-).

Some detergents are *nonionic;* they have an uncharged hydrophilic polar group attached to a large organic group of low polarity, for example,

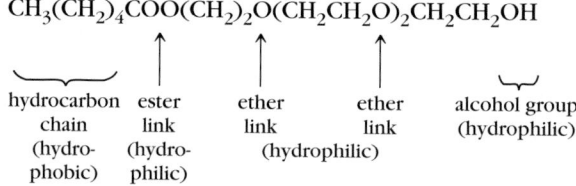

$$CH_3(CH_2)_4COO(CH_2)_2O(CH_2CH_2O)_2CH_2CH_2OH$$

hydrocarbon chain (hydro-phobic)	ester link (hydro-philic)	ether link	ether link	alcohol group (hydrophilic)
		(hydrophilic)		

The carbon chain in this molecule is oil-soluble and hydrophobic; the rest of the molecule is hydrophilic, the combination of properties needed for a molecule to be a detergent. Nonionic detergents have several advantages over ionic detergents. With no ionic groups, nonionics cannot form salts with calcium, magnesium, and iron ions and consequently are totally unaffected by hard water. For the same reason, nonionic detergents do not react with acids and may be used even in relatively strong acid solutions, which makes them useful in toilet bowl cleaners.

Many decades ago homemade lye soap was used to clean clothes as well as people's skins. This type of soap was made using either pure lye (NaOH) or sodium carbonate (Na_2CO_3, also called caustic soda) and potassium carbonate, K_2CO_3 (also called potash) from wood ashes. Most of the soap made this way contained considerable amounts of unreacted base. This was considered desirable because it helped raise the pH and break up the heavy soil particles common on fabrics in those days. In addition, the fabrics then were much more durable than fabrics today.

The closest thing to lye soap we see today is the typical dishwashing detergent. The first three ingredients in Table 16.7 contain anions that react with water to produce OH^-. Together, these three salts produce solutions inside the dishwasher that have a pH near 12.5, high enough to quickly break away animal and

TABLE 16.7	Formulation of a Dishwashing Detergent
Ingredient	**% by Weight**
Sodium carbonate, Na_2CO_3	37.5
Sodium tripolyphosphate, $Na_5P_3O_{10}$	30
Sodium metasilicate, Na_2SiO_3	30
Low-foam surfactant	0.5
Sodium dichloroisocyanurate (Cl_2 source)	1.5
Other ingredients such as colorants	0.5

vegetable oils from surfaces during the agitation cycle. The detergent helps to dissolve these oily particles and carry them away in the rinse water.

> ### Exercise 16.24   pH of a Basic Cleaning Solution
>
> Calculate the pH of a solution that is 5.2 M in sodium carbonate.

Corrosive Household Cleaners

Really tough cleaning jobs around the home require chemically aggressive cleaners. In those circumstances, either very acidic or highly basic cleaners are used to get rid of dirt, grease, or stains. Hydrochloric acid along with phosphoric acid and oxalic acid are used in toilet bowl cleaners to help get rid of stains. This combination of acids makes the pH of such cleaners very low, around 2. They should be handled cautiously because of their high acidity and reactivity. For example, labels on bottles of bleach warn against mixing bleach with other cleaners such as acidic toilet bowl cleaners. Bleach commonly contains sodium hypochlorite, $NaOCl$, and generally has a pH above 8; ClO^- and Cl^- are present in the bleach solution. In the presence of acidic toilet bowl cleaners, however, ClO^- and Cl^- are converted to toxic Cl_2 gas, which can erupt from the mixture. Bleach should also not be mixed with any cleaning agents containing ammonia. The chlorine in the bleach reacts with ammonia to produce fumes of chloramines such as NH_2Cl and $NHCl_2$, which can cause respiratory distress.

Drain and oven cleaners are at the other end of the pH spectrum, having pHs of 12 or higher. Deposits of hair, grease, and fats build up inside pipes and eventually clog the drain. When this occurs, the drain has to be dismantled and cleaned (a messy job), a long flexible piece of metal called a plumber's snake is used to physically grind through the material, or a drain cleaner is added to dissolve the clog.

Drain cleaners generally contain a strong base such as NaOH that reacts with fats and grease to form a soluble soap *(⇐ p. 714)*.

The pH of some household substances. The liquids in the flasks are club soda, vinegar, and a household cleaner. The colors of the acid-base indicators in the flasks show that vinegar is more acidic than club soda, and the cleaner is much more basic than the other two liquids.

Muriatic acid (hydrochloric acid) is used for cleaning bricks and concrete in new home construction or when remodeling is done. The strong acid should be handled with extreme caution.

$$CH_3(CH_2)_{16}\overset{\overset{\displaystyle O}{\|}}{C}-O-CH_2$$
$$CH_3(CH_2)_{16}\overset{\overset{\displaystyle O}{\|}}{C}-O-CH \; + \; 3\,NaOH \; \longrightarrow \; 3\,CH_3(CH_2)_{16}\overset{\overset{\displaystyle O}{\|}}{C}-O^-Na^+ \; + \; \begin{matrix} HO-CH_2 \\ HO-CH \\ HO-CH_2 \end{matrix}$$
$$CH_3(CH_2)_{16}\overset{\overset{\displaystyle O}{\|}}{C}-O-CH_2$$

tristearin (glyceryl tristearate)   sodium stearate (a soap)   glycerol

This reaction converts the grease or fat into soluble products (a sodium salt and glycerol) that are washed down the now-opened drain. The reaction is exothermic, and the released heat helps soften the grease or fats, which helps their removal. The strong base also decomposes and rinses away any hair trapped in the clog.

Grocery and hardware stores have an abundance of sodium hydroxide–based drain cleaners. Some are almost pure solid NaOH (pellets or flakes). Liquid drain cleaners are often 50% or more NaOH by weight in water. These solutions, being more dense than water, sink to the bottom of the drain trap to start working quickly. Although easy to use, the liquid cleaners become diluted when running water is put into the drain, reducing their efficiency. A particularly aggressive form of drain cleaner contains small bits of aluminum metal along with solid NaOH. In water, the aluminum reacts with the sodium hydroxide to produce hydrogen gas, whose bubbles help to unseat the clogged material. The net ionic equation for the reaction is

$$\text{Al(s)} + 2\,\text{OH}^-\text{(aq)} \longrightarrow \text{H}_2\text{(g)} + \text{AlO}_2^-\text{(aq)}$$

The hydrogen gas is flammable, so no flames or sparks should be present when this type of cleaner is used. If a pipe is weak and the clog very strong, sufficient gas pressure can build up to rupture the pipe (a way — but not a very desirable one — to unclog the drain).

Spray-on oven cleaners contain an NaOH solution mixed with a detergent and a propellant to apply the mixture to the soiled places. This mixture is sufficiently viscous to adhere to the oven surfaces long enough for the strongly basic solution to react with the baked-on food. If baked at a high temperature, the food has probably carbonized. If so, the cleaner will not be effective, and only scraping will remove the deposits.

You should always be cautious with household acids and bases. They are usually just as concentrated and harmful as industrial chemicals. All strongly acidic and basic solutions, both in the lab as well as in the home, can be hazardous. Acids, interestingly, are somewhat less dangerous than solutions of bases because the H_3O^+ ion tends to *denature (⇐ p. 611)* proteins in skin. The denatured proteins harden, forming a protective layer that prevents further attack by the acid, unless it is hot or highly concentrated. Basic solutions, on the other hand, tend to dissolve proteins slowly and so produce little, if any, pain, causing considerable harm before any problem is noticed. If you should get acids or bases on your skin, wash with water for at least 15 min. If acid or base splashes into your eyes, have someone call a physician while you begin gently washing the affected area with lots of water. The international warning placard that is required for shipments of acids and bases in quantities over 1000 lb illustrates schematically the personal dangers of these substances.

A corrosive chemical warning placard. This type of placard is required by the U.S. Department of Transportation on loads of 1001 lb or more of acids or bases transported by highway or rail. Such a placard makes fairly clear the reactions with human skin and metals.

16.10 LEWIS ACIDS AND BASES

CD-ROM Screen 17.11: Lewis Acids and Bases

In 1923 when Brønsted and Lowry independently proposed their acid-base concept, Gilbert N. Lewis also was developing a new concept of acids and bases. By the early 1930s Lewis had proposed definitions of acids and bases that are more general than those of Brønsted and Lowry because they are based on sharing of electron pairs rather than on H^+ ion transfers. A **Lewis acid** is *a substance that can accept a pair of electrons to form a new bond,* and a **Lewis base** is *a substance that can donate a pair of electrons to form a new bond.* This means that in the Lewis sense, an acid-base reaction occurs when there is a molecule (or ion) with a lone pair of electrons that can be donated (a Lewis base) and a molecule (or ion) that can accept an electron pair (a Lewis acid). In general, Lewis acids are

CHEMISTRY
You Can Do... Aspirin and Digestion

Aspirin is a potent drug capable of relieving pain, fever, and inflammation. Recent studies indicate it may also decrease blood clotting and heart disease. It is made from salicylic acid, which is naturally found in a variety of plants. The effect of pure salicylic acid on your stomach, however, makes it quite unpleasant as a pain remedy. Commercial aspirin is a derivative of salicylic acid, called acetylsalicylic acid (⟵ *p. 3*), which has all the benefits of salicylic acid with less discomfort.

Aspirin is still somewhat acidic, however, and can sometimes cause discomfort or worse in people susceptible to stomach irritation. As a result, there are several different forms of aspirin on the market today. The first and most common is plain aspirin. For people with stomach problems, there is also buffered aspirin, which includes a buffer in the aspirin tablet to lessen the effect of aspirin's acidity (Section 17.1). A more recent development is enteric-coated aspirin, which is plain aspirin in a tablet with a coating that prevents the aspirin from dissolving in stomach acid, but does allow it to dissolve in the small intestine, which is alkaline.

For this experiment, obtain at least three tablets of each kind of aspirin. Examples of regular aspirin are Bayer and Anacin. Buffered aspirin can be found as Bufferin or Bayer Plus. Enteric aspirin is most commonly available as Bayer

Enteric or Ecotrin. Fill three transparent cups or glasses with water. Drop one intact tablet of plain aspirin into the first glass, one tablet of buffered aspirin into the second glass, and one tablet of enteric aspirin into the third glass. Note the changes in the tablets at 1-min intervals until no further changes occur. Now repeat this experiment using vinegar instead of water to dissolve the tablets. Observe each of the tablets in the vinegar at 30-s intervals until no further change is seen. For the final experiment, fill the glasses with water and add 2 tsp of baking soda and stir until dissolved. Once you have made your baking soda solution, add one of each type of aspirin tablet and observe what happens.

1. How did each of the tablets react to each of the different solutions? The acidity of the vinegar should have mimicked the acidity of your stomach. The basic nature of the baking soda should have mimicked the environment of your intestines, which follow your stomach in the digestive system.
2. How do you suppose each type of aspirin works according to its ability to dissolve in your experiments?
3. Does this lead you to think about the kind of aspirin you take?

cations or neutral molecules with an available, empty orbital; Lewis bases are anions or neutral molecules with a lone pair of electrons.

$$A + :B \longrightarrow A-B$$

Acid
(electron pair acceptor)

Base
(electron pair donor)

Coordinate covalent bond

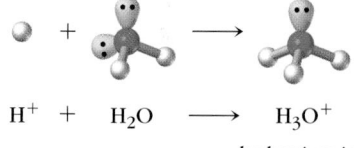

$$H^+ + H_2O \longrightarrow H_3O^+$$

hydronium ion

This type of bond, defined as a **coordinate covalent bond,** is present in many neutral molecular compounds and also in metal complex ions.

A simple example of a Lewis acid-base reaction is formation of a hydronium ion from H^+ and water. The H^+ ion has no electrons, while the water molecule has two unshared pairs of electrons on the oxygen atom. One of the electron pairs can be shared between H^+ and water, thus forming an —OH bond.

A Brønsted-Lowry base (H^+ ion acceptor) must also be a Lewis base by donating an electron pair to bond with the H^+.

Positive Metal Ions as Lewis Acids

All metal cations are potential Lewis acids. Not only do they attract electrons due to their positive charge, but all have at least one empty orbital. This empty orbital can accommodate an electron pair donated by a base and thereby form a two-electron chemical bond. Consequently, metal ions readily form coordination complexes (Section 22.6) and also are hydrated in aqueous solution (Section 16.5). When the metal ion becomes hydrated, one of the lone pairs on the oxygen atom in each of several water molecules forms a coordinate covalent bond to a metal ion; the ion is a Lewis acid, and water is a Lewis base.

 CD-ROM Screen 17.12: Cationic Lewis Acids

Figure 16.5 The amphoteric nature of Al(OH)₃. (a) Adding aqueous ammonia to a solution of Al^{3+} causes formation of a precipitate of $Al(OH)_3$. (b) Adding a strong base (NaOH) to the $Al(OH)_3$ dissolves the precipitate. Here the aluminum hydroxide acts as a Lewis acid toward the Lewis base OH^- and forms a soluble salt of the complex ion $[Al(OH)_4]^-$. (c) If we begin again with freshly precipitated $Al(OH)_3$, it dissolves as strong acid (HCl) is added. In this case $Al(OH)_3$ acts as a Brønsted-Lowry base and forms a soluble aluminum salt and water.

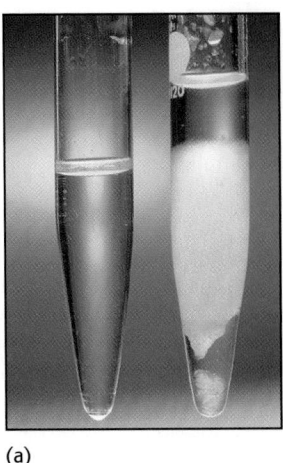
(a)

(b)

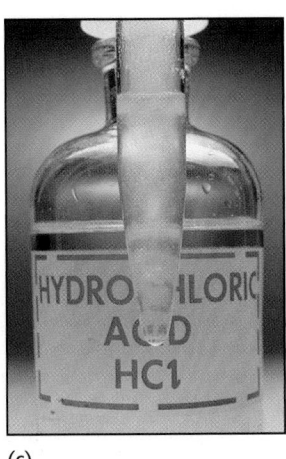
(c)

The hydroxide ion (OH^-) is an excellent Lewis base and so binds readily to metal cations to give metal hydroxides. An important feature of the chemistry of many metal hydroxides is that they are **amphoteric,** meaning that they can react as both a base and an acid. The amphoteric aluminum hydroxide, for example, behaves as a Lewis acid when it dissolves in a basic solution to form a complex ion containing one additional OH^- ion.

$$Al(OH)_3(s) + OH^-(aq) \rightleftharpoons [Al(OH)_4]^-(aq)$$

This reaction is shown in Figure 16.5. The same compound behaves as a Brønsted-Lowry base when it reacts with a Brønsted-Lowry acid (Table 16.8).

$$Al(OH)_3(s) + 3 H_3O^+(aq) \rightleftharpoons Al^{3+}(aq) + 6 H_2O(\ell)$$

Metal ions also form many complex ions with the Lewis base ammonia, $:NH_3$. For example, silver ion readily forms a water-soluble, colorless complex ion in liquid ammonia or in aqueous ammonia. Indeed, this complex is so stable that the very water-insoluble compound AgCl can be dissolved in aqueous ammonia.

$$AgCl(s) + 2 :NH_3(aq) \longrightarrow [H_3N:Ag:NH_3]^+(aq) + Cl^-(aq)$$

Complex ions are discussed in detail in Section 22.6.

Neutral Molecules as Lewis Acids

 CD-ROM Screen 17.13: Neutral Lewis Acids

Lewis's ideas about acids and bases account nicely for the fact that oxides of nonmetals behave as acids. Two important examples are carbon dioxide and sulfur dioxide, whose Lewis structures are

$$:\ddot{O}=C=\ddot{O}: \qquad :\ddot{O}=\ddot{S} \quad \longleftrightarrow \quad :\ddot{O}-\ddot{S}$$
$$\qquad\qquad\qquad :\ddot{O}: \qquad\qquad :O:$$

carbon dioxide   sulfur dioxide

In each case, there is a double bond; an "extra" pair of electrons is being shared between an oxygen atom and the central atom. Because oxygen is highly electronegative, electrons in these bonds are attracted away from the central atom,

TABLE 16.8	Some Common Amphoteric Metal Hydroxides	
Hydroxide	**Reaction as a base**	**Reaction as an acid**
$Al(OH)_3$	$Al(OH)_3(s) + 3 H_3O^+(aq) \longrightarrow Al^{3+}(aq) + 6 H_2O(\ell)$	$Al(OH)_3(s) + OH^-(aq) \longrightarrow [Al(OH)_4]^-(aq)$
$Zn(OH)_2$	$Zn(OH)_2(s) + 2 H_3O^+(aq) \longrightarrow Zn^{2+}(aq) + 4 H_2O(\ell)$	$Zn(OH)_2(s) + 2 OH^-(aq) \longrightarrow [Zn(OH)_4]^{2-}(aq)$
$Sn(OH)_4$	$Sn(OH)_4(s) + 4 H_3O^+(aq) \longrightarrow Sn^{4+}(aq) + 8 H_2O(\ell)$	$Sn(OH)_4(s) + 2 OH^-(aq) \longrightarrow [Sn(OH)_6]^{2-}(aq)$
$Cr(OH)_3$	$Cr(OH)_3(s) + 3 H_3O^+(aq) \longrightarrow Cr^{3+}(aq) + 6 H_2O(\ell)$	$Cr(OH)_3(s) + OH^-(aq) \longrightarrow [Cr(OH)_4]^-(aq)$

which becomes slightly positively charged. This makes the central atom a likely site to attract a pair of electrons. A Lewis base such as OH^- can bond to the carbon atom in CO_2 to give bicarbonate ion, HCO_3^-. This displaces one double-bond pair of electrons back onto an oxygen atom.

bicarbonate ion

Carbon dioxide from the air can react to form sodium carbonate around the mouth of a bottle of sodium hydroxide. Sulfur dioxide can react similarly with hydroxide ion.

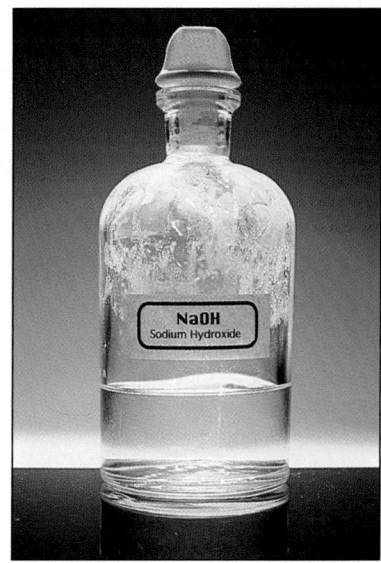

Carbon dioxide in air reacts with spilled base such as NaOH to form Na_2CO_3. If the mouth of a glass-stoppered bottle such as the one shown here is not routinely cleaned, the sodium carbonate formed can virtually cement the top of the bottle to the neck, making it difficult to open the bottle.

Exercise 16.26   Lewis Acids and Bases

Predict whether each of the following is a Lewis acid or a Lewis base. Drawing a Lewis structure for a molecule or ion is often helpful in making such a prediction.

(a) PH_3 (b) BCl_3 (c) H_2S (d) NO_2 (e) Ni^{2+} (f) CO

SUMMARY PROBLEM

Lactic acid $CH_3CH(OH)COOH$ is a weak, monoprotic acid with a melting point of 53 °C. It exists as two enantiomers (⬅ *Section 9.7*) that have slightly different K_a values. The L form has a K_a of 1.6×10^{-4} and the D form has a K_a of 1.5×10^{-4}. The D form is found in molasses, beer, wines, and souring milk. The L form is produced in muscle cells during anaerobic metabolism in which glucose molecules are broken down into lactic acid and molecules of adenosine triphosphate (ATP) are formed. When lactic acid builds up too rapidly in muscle tissue, severe pain results.

(a) Which form of lactic acid (D or L) is the stronger acid?

(b) What would be the measured pK_a of a 50:50 mixture of the two forms of lactic acid?

(c) A solution of D-lactic acid is prepared. Use HL as a general formula for lactic acid, and write the equation for the ionization of lactic acid in water.

(d) If 0.1 M solutions of these two acids (D and L) were prepared, what would be the pH of each solution?

(e) Before any lactic acid dissolves in the water, what reaction determines the pH?

(f) Calculate the pH of a solution made by dissolving 4.46 g of D-lactic acid in 500. mL of water.

(g) How many milliliters of a 0.115 M NaOH solution would be required to completely neutralize 4.46 g of pure lactic acid?

(h) What would be the pH of the solution made by the neutralization if the lactic acid were the D form? The L form? A 50:50 mixture of the two forms?

IN CLOSING

Having studied this chapter, you should be able to . . .

• Describe water's role in aqueous acid-base chemistry (Section 16.1).

• Identify the conjugate base of an acid, the conjugate acid of a base, and the relationship between conjugate acid and base strengths (Section 16.1).

- Recognize how amines act as bases and how carboxylic acids ionize in aqueous solution (Section 16.2).
- Use the autoionization of water and show how this equilibrium takes place in aqueous solutions of acids and bases (Section 16.3).
- Classify an aqueous solution as acidic, neutral, or basic based on its concentration of H_3O^+ or OH^- (Section 16.4).
- Calculate pH (or pOH) given $[H_3O^+]$, or $[OH^-]$ given pH (or pOH) (Section 16.4).
- Estimate acid and base strengths from K_a and K_b values (Sections 16.5).
- Describe the acidic behavior of hydrated metal ions (Section 16.5).
- Write the ionization steps of polyprotic acids (Section 16.5).
- Calculate pH from K_a or K_b values and solution concentration (Section 16.6).
- Describe the relationships between acid strength and molecular structure (16.7).
- Explain the nature of zwitterions (Section 16.7).
- Describe the hydrolysis of salts in aqueous solution (Section 16.8).
- Apply acid-base principles to the chemistry of antacids, kitchen chemistry, and household cleaners (Section 16.9).
- Recognize Lewis acids and bases and how they react (Section 16.10).

KEY TERMS

acid ionization con-
  stant *(16.5)*
acid ionization con-
  stant expression
  (16.5)
acid-base reaction
  (16.1)
acidic solution *(16.3)*
amines *(16.2)*
amphoteric *(16.10)*
autoionization *(16.3)*
base ionization con-
  stant *(16.5)*

base ionization con-
  stant expression
  (16.5)
basic solution *(16.3)*
Brønsted-Lowry acid
  (16.1)
Brønsted-Lowry base
  (16.1)
conjugate acid-base
  pair *(16.2)*
coordinate covalent
  bond *(16.10)*
hydrolysis *(16.8)*

ionization constant for
  water *(16.3)*
Lewis acid *(16.10)*
Lewis base *(16.10)*
monoprotic acids
  (16.5)
neutral solution *(16.3)*
oxoacids *(16.7)*
pH *(16.4)*
polyprotic acids *(16.5)*
zwitterion *(16.7)*

QUESTIONS FOR REVIEW AND THOUGHT

Conceptual Challenge Problems

CP-16.A (Section 16.4) Is it possible for an aqueous solution to have a pH of 0 or even less than 0? Explain your answer mathematically as well as practically based on what you know about acid solubilities.

CP-16.B (Section 16.4) What is the pH of water at 200 °C? Liquid water this hot would have to be under a pressure greater than 1.0 atm and might be found in a pressurized water reactor located in a nuclear power plant.

CP-16.C (Section 16.5) Develop a set of rules by which you could predict the pH for solutions of strong or weak acids and strong or weak bases *without using a calculator*. Your predictions need to be accurate to ±1 pH units. Assume that you know the concentration of the acid or base and that for the weak acids and bases you can look up the pK_a ($-\log K_a$) or K_a values. What rules would work to predict pH?

Answers to questions in **bold** can be found in the back of the book.

Review Questions

1. Define a Brønsted-Lowry acid and a Brønsted-Lowry base.
2. Explain in your own words what 100% ionization means.
3. Write the chemical equation for the autoionization of water. Write the equilibrium constant expression for this reaction. What is the value of the equilibrium constant at 25 °C? What is this constant called?
4. When OH^- is the base in a conjugate acid-base pair, the acid is _____; when OH^- is the acid, the base is _____.
5. Write balanced chemical equations that show phosphoric acid, H_3PO_4, ionizing stepwise as a polyprotic acid.
6. Write ionization equations for a weak acid and its conjugate base. Show that adding these two equations gives the autoionization equation for water.
7. Designate the acid and the base on the left side of the following equations, and designate the conjugate partner of each on the right side.
   (a) $HNO_3(aq) + H_2O(\ell) \rightarrow H_3O^+(aq) + NO_3^-(aq)$
   (b) $NH_4^+(aq) + CN^-(aq) \rightarrow NH_3(aq) + HCN(aq)$
8. Dissolving ammonium bromide in water gives an acidic solution. Write a balanced equation showing how that can occur.
9. Solution A has a pH of 8 and solution B a pH of 10. Which has the greater hydronium ion concentration? How many times greater is its concentration?
10. Contrast the main ideas of the Brønsted-Lowry and Lewis acid-base concepts. Name and write the formula for a substance that behaves as a Lewis acid but not as a Brønsted-Lowry acid.

The Brønsted-Lowry Concept of Acids and Bases

11. Write an equation to describe the proton transfer that occurs when each of these acids is added to water.
    (a) HBr (b) CF_3COOH (c) HSO_4^- (d) HNO_2
12. Write an equation to describe the proton transfer that occurs when each of these acids is added to water.
    (a) HCO_3^- (b) HCl (c) CH_3COOH (d) HCN
13. Write an equation to describe the proton transfer that occurs when each of these acids is added to water.
    (a) $HClO$ (b) CH_3CH_2COOH
    (c) $HSeO_3^-$ (d) HO_2^-
14. Write an equation to describe the proton transfer that occurs when each of these acids is added to water.
    (a) HIO (b) $CH_3(CH_2)_4COOH$
    (c) $HOOCCOOH$ (d) $CH_3NH_3^+$
15. Write an equation to describe the proton transfer that occurs when each of these bases is added to water.
    (a) H^- (b) HCO_3^- (c) NO_2^-
16. Write an equation to describe the proton transfer that occurs when each of these bases is added to water.
    (a) HSO_4^- (b) CH_3NH_2 (c) I^- (d) $H_2PO_4^-$
17. Write an equation to describe the proton transfer that occurs when each of these bases is added to water.
    (a) PO_4^{3-} (b) SO_3^{2-} (c) HPO_4^{2-}

18. Write an equation to describe the proton transfer that occurs when each of these bases is added to water.
    (a) AsO_4^{3-} (b) S^{2-} (c) N_3^-
19. Based on formulas alone, classify each of the following oxoacids as strong or weak.
    (a) H_3PO_4 (b) H_2SO_4 (c) $HClO$ (d) $HClO_4$
    (e) HNO_3 (f) H_2CO_3 (g) HNO_2
20. Based on formulas alone, which is the stronger acid?
    (a) H_2CO_3 or H_2SO_4 (b) HNO_3 or HNO_2
    (c) $HClO_4$ or H_2SO_4 (d) H_3PO_4 or $HClO_3$
    (e) H_2SO_3 or H_2SO_4
21. Write the formula and name for the conjugate partner for each acid or base.
    (a) CN^- (b) SO_4^{2-}
    (c) HS^- (d) S^{2-}
    (e) HSO_3^- (f) $HCOOH$ (formic acid)
22. Write the formula and name for the conjugate partner for each acid or base.
    (a) HI (b) NO_3^- (c) CO_3^{2-}
    (d) H_2CO_3 (e) HSO_4^- (f) SO_3^{2-}
23. Which are conjugate acid-base pairs?
    (a) H_2O and H_3O^+ (b) H_3O^+ and OH^-
    (c) NH_2^- and NH_4^+ (d) NH_3 and NH_4^+
    (e) O^{2-} and H_2O
24. Which are conjugate acid-base pairs?
    (a) NH_2^- and NH_4^+ (b) NH_3 and NH_2^-
    (c) H_3O^+ and H_2O (d) OH^- and O^{2-}
    (e) H_3O^+ and OH^-
25. Identify the acid and the base that are reactants in each equation; identify the conjugate base and conjugate acid on the product side of each equation.
    (a) $HI(aq) + H_2O(\ell) \rightarrow H_3O^+(aq) + I^-(aq)$
    (b) $OH^-(aq) + NH_4^+(aq) \rightarrow H_2O(\ell) + NH_3(aq)$
    (c) $NH_3(aq) + H_2CO_3(aq) \rightarrow NH_4^+(aq) + HCO_3^-(aq)$
26. Identify the acid and the base that are reactants in each equation; identify the conjugate base and conjugate acid on the product side of the equation.
    (a) $HS^-(aq) + H_2O(\ell) \rightarrow H_2S(aq) + OH^-(aq)$
    (b) $S^{2-}(aq) + NH_4^+(aq) \rightarrow NH_3(aq) + HS^-(aq)$
    (c) $HCO_3^-(aq) + HSO_4^-(aq) \rightarrow H_2CO_3(aq) + SO_4^{2-}(aq)$
    (d) $NH_3(aq) + NH_2^-(aq) \rightarrow NH_2^-(aq) + NH_3(aq)$
27. Identify the acid and the base that are reactants in each equation; identify the conjugate base and conjugate acid on the product side of the equation.
    (a) $H_2PO_4^-(aq) + HCO_3^-(aq) \rightarrow$
          $H_2CO_3(aq) + HPO_4^{2-}(aq)$
    (b) $NH_3(aq) + NH_2^-(aq) \rightarrow NH_2^-(aq) + NH_3(aq)$
    (c) $HSO_4^-(aq) + CO_3^{2-}(aq) \rightarrow SO_4^{2-}(aq) + HCO_3^-(aq)$
28. Identify the acid and the base that are reactants in each equation; identify the conjugate base and conjugate acid on the product side of the equation.
    (a) $CN^-(aq) + CH_3COOH(aq) \rightarrow$
          $CH_3COO^-(aq) + HCN(aq)$
    (b) $O^{2-}(aq) + H_2O(\ell) \rightarrow 2\,OH^-(aq)$
    (c) $HCO_2^-(aq) + H_2O(\ell) \rightarrow HCOOH(aq) + OH^-(aq)$
29. Write stepwise equations for protonation or deprotonation of each of these polyprotic acids and bases.
    (a) H_2SO_3 (b) S^{2-}
    (c) $NH_3CH_3COOH^+$ (glycinium ion, a diprotic acid)

30. Write stepwise equations for protonation or deprotonation of each of these polyprotic acids and bases.
    (a) CO_3^{2-} (b) H_3AsO_4
    (c) $NH_2CH_2COO^-$ (glycinate ion, a diprotic base)

pH Calculations

31. The pH of a popular soft drink is 3.30. What is its hydronium ion concentration? Is the drink acidic or basic?
32. Milk of magnesia, $Mg(OH)_2$, has a pH of 10.5. What is the hydronium ion concentration of the solution? Is this solution acidic or basic?
33. A cup of coffee has a pH of 4.3. Calculate the hydronium ion concentration in this coffee.
34. A solution of lactic acid has a pH of 2.44. What is its hydronium ion concentration?
35. What is the pH of a 0.0013 M solution of HNO_3? What is the pOH of this solution?
36. What is the pH of a solution that is 0.025 M in NaOH? What is the pOH of this solution?
37. A solution of benzyl amine, $C_7H_7NH_2$, has a hydroxide ion concentration of 2.4×10^{-3} M. What is the pH of the solution? What is its pOH?
38. The hydronium ion concentration of a cyanoacetic acid solution is 0.032 M. What is its pOH?
39. The pH of a $Ba(OH)_2$ solution is 10.66 at 25 °C. What is the hydroxide ion concentration of this solution? If the solution volume is 250. mL, how many grams of $Ba(OH)_2$ must have been used to make this solution?
40. A 1000.-mL solution of hydrochloric acid has a pH of 1.3. How many grams of HCl are dissolved in the solution?
41. Make the following interconversions. In each case tell whether the solution is acidic or basic.

	pH	$[H_3O^+]$ (M)	$[OH^-]$ (M)
(a)	1.00	_____	_____
(b)	10.5	_____	_____
(c)	_____	1.8×10^{-4}	_____
(d)	_____	_____	2.3×10^{-5}

42. Make the following interconversions. In each case tell whether the solution is acidic or basic.

	pH	$[H_3O^+]$ (M)	$[OH^-]$ (M)
(a)	_____	6.1×10^{-7}	_____
(b)	_____	_____	2.2×10^{-9}
(c)	4.67	_____	_____
(d)	_____	2.5×10^{-2}	_____
(e)	9.12	_____	_____

43. Figure 16.2 shows the pH of some common solutions. How many times more acidic or basic are the following compared with a neutral solution?
    (a) Milk   (b) Sea water
    (c) Blood   (d) Battery acid

44. Figure 16.2 shows the pH of some common solutions. How many times more acidic or basic are the following compared with a neutral solution?
    (a) Black coffee   (b) Household ammonia
    (c) Baking soda   (d) Vinegar

Acid-Base Strengths

45. Write ionization equations and ionization constant expressions for these acids and bases.
    (a) CH_3COOH (b) HCN   (c) SO_3^{2-}
    (d) PO_4^{3-} (e) NH_4^+ (f) H_2SO_4
46. Write ionization equations and ionization constant expressions for these acids and bases.
    (a) F^- (b) NH_3 (c) H_2CO_3
    (d) H_3PO_4 (e) CH_3COO^- (f) S^{2-}
47. Which solution will be more acidic?
    (a) 0.10 M H_2CO_3 or 0.10 M NH_4Cl
    (b) 0.10 M HF or 0.10 M $KHSO_4$
    (c) 0.1 M $NaHCO_3$ or 0.1 M Na_2HPO_4
    (d) 0.1 M H_2S or 0.1 M HCN
48. Which solution will be more basic?
    (a) 0.10 M NH_3 or 0.10 M NaF
    (b) 0.10 M K_2S or 0.10 M K_3PO_4
    (c) 0.10 M $NaNO_3$ or 0.10 M CH_3COONa
    (d) 0.10 M NH_3 or 0.10 M KCN
49. Without doing any calculations, assign each of the following 0.10 M aqueous solutions to one of these pH ranges: pH 2; pH between 2 and 6; pH between 6 and 8; pH between 8 and 12; pH 12.
    (a) HNO_2 (b) NH_4Cl
    (c) NaF   (d) $Mg(CH_3COO)_2$
    (e) BaO   (f) $KHSO_4$
    (g) $NaHCO_3$ (h) $BaCl_2$
50. Calculate the pH of each solution in Question 49 to verify your prediction.
51. A 0.015 M solution of cyanic acid has a pH of 2.67. What is the ionization constant, K_a, of the acid?
52. What is the K_a of butyric acid if a 0.025 M solution has a pH of 3.21?
53. What are the equilibrium concentrations of H_3O^+, acetate ion, and acetic acid in a 0.20 M aqueous solution of acetic acid (CH_3COOH)?
54. The ionization constant of a very weak acid HA is 4.0×10^{-9}. Calculate the equilibrium concentrations of H_3O^+, A^-, and HA in a 0.040 M solution of the acid.
55. The weak base methylamine, CH_3NH_2, has $K_b = 5.0 \times 10^{-4}$. It reacts with water according to the equation

    $$CH_3NH_2(aq) + H_2O(\ell) \rightleftharpoons CH_3NH_3^+(aq) + OH^-(aq)$$

    What is the pH of a 0.23 M methylamine solution?
56. Calculate the pH of a 0.12 M aqueous solution of the base aniline, $C_6H_5NH_2$ ($K_b = 4.2 \times 10^{-10}$).
57. Now you wish you had an aspirin. Aspirin is a weak acid with $K_a = 3.27 \times 10^{-4}$ for the reaction

    $$HC_9H_7O_4(aq) + H_2O(\ell) \rightleftharpoons H_3O^+(aq) + C_9H_7O_4^-(aq)$$

    Two aspirin tablets, each containing 0.325 g of aspirin (along with a nonreactive "binder" to hold the tablet together), are dissolved in 200.0 mL of water. What is the pH of this solution?

58. Lactic acid, $C_3H_6O_3$, occurs in sour milk as a result of the metabolism of certain bacteria. What is the pH of a solution of 56 mg of lactic acid in 250 mL of water? K_a for lactic acid is 1.4×10^{-4}.

Acid-Base Reactions

59. Complete each of these reactions by filling in the blanks. Predict whether each reaction is product-favored or reactant-favored, and explain your reasoning.
(a) _____(aq) + Br^-(aq) $\rightleftharpoons$ NH_3(aq) + HBr(aq)
(b) CH_3COOH(aq) + CN^-(aq) $\rightleftharpoons$
_____(aq) + HCN(aq)
(c) _____(aq) + $H_2O(\ell)$ $\rightleftharpoons$ NH_3(aq) + OH^-(aq)

60. Complete each of these reactions by filling in the blanks. Predict whether each reaction is product-favored or reactant-favored, and explain your reasoning.
(a) _____(aq) + HSO_4^-(aq) $\rightleftharpoons$ HCN(aq) + SO_4^{2-}(aq)
(b) H_2S(aq) + $H_2O(\ell)$ $\rightleftharpoons$ H_3O^+(aq) + _____(aq)
(c) H^-(aq) + $H_2O(\ell)$ $\rightleftharpoons$ OH^-(aq) + _____(aq)

61. Predict which of the following acid-base reactions are product-favored and which are reactant-favored. In each case write a balanced equation for any reaction that might occur, even if the reaction is reactant-favored. Consult Table 16.2 if necessary.
(a) $H_2O(\ell)$ + HNO_3(aq)   (b) H_3PO_4(aq) + $H_2O(\ell)$
(c) CN^-(aq) + HCl(aq)   (d) NH_4^+(aq) + F^-(aq)

62. Predict which of the following acid-base reactions are product-favored and which are reactant-favored. In each case write a balanced equation for any reaction that might occur, even if the reaction is reactant-favored. Consult Table 16.2 if necessary.
(a) NH_4^+(aq) + HPO_4^{2-}(aq)
(b) CH_3COOH(aq) + OH^-(aq)
(c) HSO_4^-(aq) + $H_2PO_4^-$(aq)
(d) CH_3COOH(aq) + F^-(aq)

63. For each salt, predict whether an aqueous solution will have a pH less than, equal to, or greater than 7. Explain your answer.
(a) $NaHSO_4$ (b) NH_4Br (c) $KClO_4$

64. For each salt, predict whether an aqueous solution will have a pH less than, equal to, or greater than 7. Explain your answer.
(a) $AlCl_3$ (b) Na_2S (c) $NaNO_3$

65. For each salt, predict whether an aqueous solution will have a pH less than, equal to, or greater than 7. Explain your answer.
(a) NaH_2PO_4 (b) NH_4NO_3 (c) $SrCl_2$

66. For each salt, predict whether an aqueous solution will have a pH less than, equal to, or greater than 7. Explain your answer.
(a) Na_2HPO_4 (b) $(NH_4)_2S$ (c) KCH_3COO

67. Explain why $BaCO_3$ is soluble in aqueous HCl, but $BaSO_4$, which is used to make the intestines visible in x-ray photographs, remains sufficiently insoluble in the HCl in a human stomach so that poisonous barium ions do not get into the bloodstream.

68. For which of the following substances would solubility be greater at pH = 2 than at pH = 7?
(a) $Cu(OH)_2$ (b) $CuSO_4$ (c) $CuCO_3$
(d) CuS   (e) $Cu_3(PO_4)_2$

Practical Acid-Base Chemistry

69. Double-acting baking powder contains two salts, sodium hydrogen carbonate and potassium dihydrogen phosphate, whose anions react in water to form CO_2 gas. Write a balanced chemical equation for the reaction. Which anion is the acid and which is the base?

70. Common soap is made by reacting sodium carbonate with stearic acid $(CH_3(CH_2)_{16}COOH)$. Write a balanced equation for the reaction.

71. If 1 g of each antacid in Table 16.6 reacted with equal volumes of stomach acid, which would neutralize the most stomach acid?

72. If 1 g each of vinegar, lemon juice, and lactic acid react with equal masses of baking soda, which will produce the most CO_2 gas?

73. Why do cleaning products containing sodium hydroxide feel slippery when you get them on your skin?

74. Why is it not a good idea to substitute dishwashing detergent for automobile-washing detergent?

Lewis Acids and Bases

75. Which of these is a Lewis acid? a Lewis base?
(a) NH_3 (b) $BeCl_2$ (c) BCl_3

76. Which of these is a Lewis acid? a Lewis base?
(a) O^{2-} (b) CO_2 (c) H^-

77. Which of these is a Lewis acid? a Lewis base?
(a) Al^{3+} (b) H_2O (c) SCN^-

78. Which of these is a Lewis acid? a Lewis base?
(a) Cr^{3+} (b) SO_3 (c) CH_3NH_2

79. Identify the Lewis acid and the Lewis base in each reaction.
(a) $H_2O(\ell)$ + SO_2(aq) $\rightarrow$ H_2SO_3(aq)
(b) H_3BO_3(aq) + OH^-(aq) $\rightarrow$ $B(OH)_4^-$(aq)
(c) Cu^{2+}(aq) + 4 NH_3(aq) $\rightarrow$ $[Cu(NH_3)_4]^{2+}$(aq)
(d) 2 Cl^-(aq) + $SnCl_2$(aq) $\rightarrow$ $SnCl_4^{2-}$(aq)

80. Identify the Lewis acid and the Lewis base in each reaction.
(a) I_2(s) + I^-(aq) $\rightarrow$ I_3^-(aq)
(b) SO_2(g) + BF_3(g) $\rightarrow$ O_2SBF_3(s)
(c) Au^+(aq) + 2 CN^-(aq) $\rightarrow$ $[Au(CN)_2]^-$(aq)
(d) CO_2(g) + $H_2O(\ell)$ $\rightarrow$ H_2CO_3(aq)

81. Trimethylamine, $(CH_3)_3N$:, interacts readily with diborane, B_2H_6. The diborane dissociates to two BH_3 fragments, each of which can react with trimethylamine to form a complex, $(CH_3)_3N:BH_3$. Write an equation for this reaction and interpret it in terms of Lewis's acid-base theory.

82. Draw a Lewis structure for ICl_3. Predict the shape of this molecule. Does it function as a Lewis acid or base when it reacts with chloride ion to form ICl_4^-? What is the structure of this ion?

General Questions

83. Classify each of the following as a strong acid, weak acid, strong base, weak base, amphiprotic substance, or neither acid nor base.
(a) HCl   (b) NH_4^+ (c) H_2O
(d) CH_3COO^- (e) CH_4 (f) CO_3^{2-}

84. Classify each of the following as a strong acid, weak acid, strong base, weak base, amphiprotic substance, or neither acid nor base.
    (a) CH_3COOH (b) Na_2O (c) H_2SO_4
    (d) NH_3 (e) $Ba(OH)_2$ (f) $H_2PO_4^-$

85. Several acids and their respective equilibrium constants are:

    $$HF(aq) + H_2O(\ell) \rightleftharpoons H_3O^+(aq) + F^-(aq)$$
    $$K_a = 7.2 \times 10^{-4}$$

    $$HS^-(aq) + H_2O(\ell) \rightleftharpoons H_3O^+(aq) + S^{2-}(aq)$$
    $$K_a = 8 \times 10^{-18}$$

    $$CH_3COOH(aq) + H_2O(\ell) \rightleftharpoons$$
    $$H_3O^+(aq) + CH_3COO^-(aq)$$
    $$K_a = 1.8 \times 10^{-5}$$

    (a) Which is the strongest acid? Which is the weakest acid?
    (b) Which acid has the weakest conjugate base?
    (c) Which acid has the strongest conjugate base?

86. State whether equal molar amounts of the following would have a pH equal to 7, less than 7, or greater than 7.
    (a) A weak base and a strong acid react.
    (b) A strong base and a strong acid react.
    (c) A strong base and a weak acid react.

87. Sulfurous acid, H_2SO_3, is a weak diprotic acid ($K_{a_1} = 1.2 \times 10^{-2}$, $K_{a_2} = 6.2 \times 10^{-8}$). What is the pH of a 0.45 M solution of H_2SO_3? (Assume that only the first ionization is important in determining pH.)

88. Ascorbic acid (vitamin C, $C_6H_8O_6$) is a diprotic acid ($K_{a_1} = 7.9 \times 10^{-5}$, $K_{a_2} = 1.6 \times 10^{-12}$). What is the pH of a solution that contains 5.0 mg of the acid per mL of water? (Assume that only the first ionization is important in determining pH.)

89. Does the pH of the solution increase, decrease, or stay the same when you
    (a) Add solid ammonium chloride to 100 mL of 0.10 M NH_3?
    (b) Add solid sodium acetate to 50.0 mL of 0.015 M acetic acid?
    (c) Add solid NaCl to 25.0 mL of 0.10 M NaOH?

90. Does the pH of the solution increase, decrease, or stay the same when you
    (a) Add solid sodium oxalate, $Na_2C_2O_4$, to 50.0 mL of 0.015 M oxalic acid?
    (b) Add solid ammonium chloride to 100 mL of 0.016 M HCl?
    (c) Add 20.0 g of NaCl to 1.0 L of 0.012 M sodium acetate, $NaCH_3COO$?

Applying Concepts

91. When a 0.1 M aqueous ammonia solution is tested with a conductivity apparatus (⇐ *p. 96, Fig. 3.5*), the bulb glows dimly. When a 0.1 M hydrochloric acid solution is tested, the bulb glows brightly. As water is added to each of the solutions would you expect the bulb to glow brighter, stop glowing, or stay the same? Explain your reasoning.

92. What is the pH of pure water at 10 °C, 25 °C, and 50 °C? Classify the water at each temperature as either acidic, neutral or basic.

93. If you evaporate the water in a sodium hydroxide solution, you will end up with solid sodium hydroxide. However, if you evaporate the water in an ammonium hydroxide solution, you will not end up with solid ammonium hydroxide. Explain why. What will remain after the water is evaporated?

94. For each aqueous solution, predict what ions and molecules will be present. Without doing any calculations, list the ions and molecules in order of decreasing concentration.
    (a) HCl   (b) $NaClO_4$ (c) HNO_2
    (d) NaClO   (e) NH_4Cl (f) NaOH

95. The diagrams below are nanoscale representations of different acids.

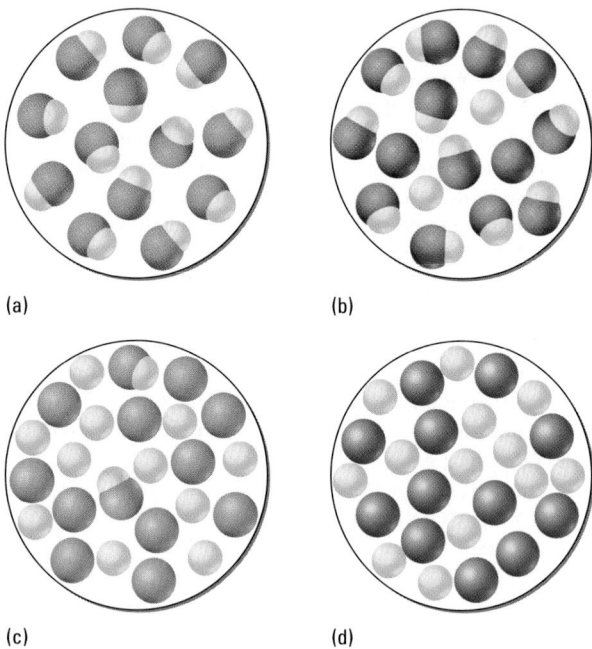

(a)   (b)

(c)   (d)

(a) Which diagram best represents hydrochloric acid? (The yellow circles are H^+ ions and the other circles are Cl^- ions.)
(b) Which diagram best represents acetic acid? (The yellow circles are H^+ ions and the other circles are CH_3COO^- ions.)

96. When asked to identify the conjugate acid-base pairs in the reaction

    $$HCO_3^-(aq) + HSO_4^-(aq) \rightleftharpoons H_2CO_3(aq) + SO_4^{2-}(aq)$$

    a student incorrectly wrote: "HCO_3^- is a base and HSO_4^- is its conjugate acid. H_2CO_3 is an acid and SO_4^- is its conjugate base." Write a brief explanation to the student telling why the answer is incorrect.

97. A person claimed that his stomach ruptured when he took a spoonful of baking soda in a glass of water to relieve heartburn after a full meal ($\frac{1}{2}$ tsp = 2.5 g $NaHCO_3$). Assume that the pH of stomach acid is 1 and that the stomach has a volume of 1 L when expanded fully. Body temperature is 37 °C. What volume of carbon dioxide gas was generated by the reaction of baking soda with stomach acids? Might his stomach have ruptured from this volume of CO_2?

98. Explain how the Arrhenius acid-base theory and the Brønsted-Lowry theory of acids and bases are explained by the Lewis acid-base theory.

99. Calculate the K_b for the conjugate base of gallic acid, found in tea. $K_a = 3.9 \times 10^{-5}$. Identify a base from Table 16.2 with a K_b close to that of the conjugate base of gallic acid.

100. It is determined that 0.1 M solutions of the sodium salts NaM, NaQ, and NaZ have pH values of 7.0, 8.0, and 9.0, respectively. Arrange the acids HM, HQ, and HZ in order of decreasing strength. Where possible, find the K_a values of these acids.

 ## General Chemistry CD-ROM

CD16.1 Screen 17.2: Brønsted Acids and Bases. What is the difference between a Brønsted-Lowry acid and a Brønsted-Lowry base? How are the two related?

CD16.2 Screen 17.3: The Acid-Base Properties of Water. Explain how water acts as both an acid and a base in its autoionization reaction.

CD16.3 Screen 17.4: The pH Scale.

(a) In other calculations involving species in solution, we use concentrations in units of moles per liter. Why do we bother with the seemingly more complicated concentration unit, pH?

(b) In which solution is the concentration of H_3O^+ larger, one with a pH of 3 or one with a pH of 5?

CD16.4 Screen 17.6: Weak Acids and Bases.

(a) What is the difference between a strong acid and a weak acid?

(b) The acids and their conjugate bases are listed in the same row in the table on this screen. How do the values of K_b for the bases change as the values of K_a for the acids decrease going down the table?

(c) K_w is the product of the ionization constant for a weak acid (K_a) and the constant for its conjugate base (K_b), that is, $K_a \times K_b = K_w$. Use the equilibrium expressions for K_a and K_b for HF and F^- to show that this is true.

CD16.5 Screen 17.11: Lewis Acids and Bases.

(a) What is the difference between a Lewis acid and a Brønsted-Lowry acid?

(b) On Screen 17.5 you learned that the hydride ion, H^-, is a strong base. Is it also a Lewis base? Use the electron dot structure of H^- to prove your point.

CD16.6 Screen 17.12: Cationic Lewis Acids.

(a) Explain how Cu^{2+} ion acts as a Lewis acid on this screen.

(b) Name another compound that could take the part of ammonia in this reaction.

(c) Examine the *Blood and CO* sidebar to this screen. Why is carbon monoxide a poison?

17

The striking, dark vertical stripes on the lower cliff face are due to metal oxide deposits. Deposits precipitate from run-off water as it flows over the cliff face, and the water evaporates. The stripes are sometimes called "desert varnish."

ADDITIONAL AQUEOUS EQUILIBRIA

17.1 Buffer Solutions

17.2 Acid-Base Titrations

17.3 Acid Rain

17.4 Solubility Equilibria and the Solubility Product Constant, K_{sp}

17.5 Factors Affecting Solubility

17.6 Precipitation: Will It Occur?

From the environment within a cell to the watery depths of the oceans, important interactions occur among solutes in aqueous solution. This chapter extends the concepts covered in Chapters 15 and 16 about aqueous solutions—conjugate acid-base behavior, the neutralization of an acid by a base, and the link between solubility and precipitation. We will now consider the quantitative aspects of dealing with (1) buffers, which are combinations of a weak acid and its conjugate base, (2) acid-base titrations, which are neutralization reactions of an acid with a base, and (3) equilibria associated with solutions of slightly soluble salts.

17.1 BUFFER SOLUTIONS

Adding a small amount of acid or base to pure water radically affects the pH. Consider what happens if 0.01 mol of HCl is added to 1 L of water. The pH changes from 7 to 2 because $[H_3O^+]$ changes from 10^{-7} M to 10^{-2} M. This pH change represents a *100,000-fold increase* in $[H_3O^+]$. Similarly, if 0.01 mol of NaOH is added to 1 L of pure water, the pH goes from 7 to 12, a *100,000-fold decrease* in $[H_3O^+]$ (Figure 17.1). Many aquatic organisms could not survive such dramatic pH changes; the organisms can survive only within a narrow pH range. For example, if acid rain lowers the pH of a lake or stream, fish such as trout may die.

Unlike aqueous NaOH or HCl, there are chemical systems that maintain a relatively constant pH because they contain a **buffer**—*a system that resists changes in pH when limited amounts of base or acid are added to it.* Solutions like these are referred to as **buffer solutions.** For example, a solution that contains 0.50 mol of acetic acid, CH_3COOH, and 0.50 mol of sodium acetate, $NaCH_3COO$, in 1.0 L of solution is a buffer with a pH of 4.74. When 0.010 mol of a strong acid is added to it, the pH changes by only 0.02 pH units to 4.72; adding 0.010 mol of strong base to this buffer changes the pH to 4.76. These are only slight pH changes. How do the sodium acetate–acetic acid buffer and other buffers offset such additions of acid or base without changing pH significantly?

CD-ROM Screen 18.8: Buffer Solutions

Buffer Action

To maintain a relatively constant pH, a buffer must contain *a weak acid that can react with any added base,* and at the same time it must contain *a weak base that can react with any added acid.* It is also necessary that the acid and base components of a buffer solution not react with each other. A conjugate acid-base pair, such as acetic acid and acetate ion (from sodium acetate) satisfies this need.

Figure 17.1   Addition of HCl or NaOH to water.   (a) The addition of 0.01 mol of NaOH to a liter of water causes a tremendous increase in pH from 7.00 to 12.00. (b) The pH of pure water is 7. (c) Addition of 0.01 mol of HCl to a liter of water drastically decreases the pH to 2.

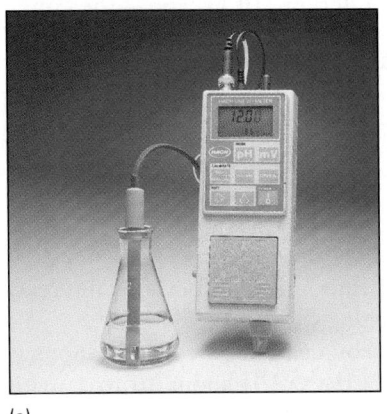

(a)

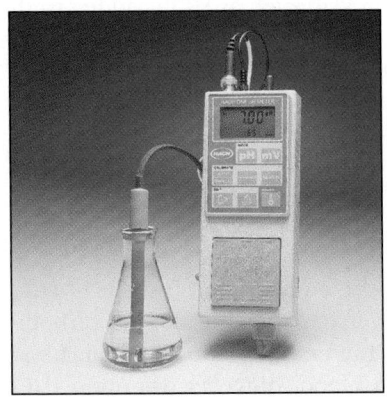

(b)

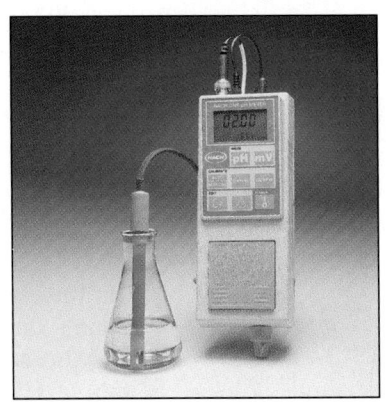

(c)

In a conjugate pair, if the acid and base react with each other they just produce conjugate base and conjugate acid—no observable change occurs. For example, acetic acid reacts with acetate ion to form acetate ion and acetic acid.

$$CH_3COOH(aq) + CH_3COO^-(aq) \rightleftharpoons CH_3COO^-(aq) + CH_3COOH(aq)$$

| CONJ ACID | conj base | CONJ BASE | conj acid |

Buffers usually consist of approximately equal quantities of a weak acid and its conjugate base, or a weak base and its conjugate acid.

To see how a buffer works, consider human blood. The blood of mammals is an aqueous solution that maintains a constant pH. The normal pH of human blood is 7.40 ± 0.05. If pH decreases below 7.35, a condition known as *acidosis* occurs; increasing the pH above 7.45 causes *alkalosis*. Both of these conditions can be life-threatening. Acidosis, for example, causes a decrease in oxygen transport by hemoglobin and also depresses the central nervous system, leading in extreme cases to coma and death by creating weak and irregular cardiac contractions—symptoms of heart failure. To prevent such problems your body must keep the pH of your blood nearly constant.

Carbon dioxide provides the most important blood buffer (but not the only one). In solution, CO_2 reacts with water to form H_2CO_3, which ionizes to produce H_3O^+ and HCO_3^- ions. The equilibria are

$$CO_2(aq) + H_2O(\ell) \rightleftharpoons H_2CO_3(aq)$$

$$H_2CO_3(aq) + H_2O(\ell) \rightleftharpoons H_3O^+(aq) + HCO_3^-(aq)$$

The normal concentrations of H_2CO_3 and HCO_3^- in blood are 0.0025 M and 0.025 M, a 1:10 ratio. Since H_2CO_3 is a weak acid and HCO_3^- is its conjugate weak base, they constitute a buffer. As long as the ratio of H_2CO_3 to HCO_3^- concentrations remains about 1 to 10, the pH of the blood remains near 7.4. (We'll calculate this pH in Problem-Solving Example 17.1.)

If a strong base such as NaOH is added to this buffer, carbonic acid in the buffer will react with the added OH^-. Since OH^- is the strongest base that can exist in water solution, this reaction is essentially complete.

$$H_2CO_3(aq) + OH^-(aq) \longrightarrow HCO_3^-(aq) + H_2O(\ell)$$

$$K = \frac{1}{K_b(HCO_3^-)} = \frac{1}{2.4 \times 10^{-8}} = 4.2 \times 10^7$$

Here the equilibrium constant is $1/K_b$ of hydrogen carbonate ion, because the reaction is the reverse of the reaction of hydrogen carbonate ion with OH^- ion.

If a strong acid such as HCl is added to this buffer, the HCO_3^- ion—the conjugate base in the buffer—will react with the hydronium ions from HCl. Since the H_3O^+ ion is such a strong acid, the reaction between HCO_3^- and H_3O^+ is essentially complete.

$$HCO_3^-(aq) + H_3O^+(aq) \longrightarrow H_2CO_3(aq) + H_2O(\ell)$$

$$K = \frac{1}{K_a(H_2CO_3)} = \frac{1}{4.2 \times 10^{-7}} = 2.4 \times 10^6$$

In this case, the equilibrium constant is $1/K_a$ of carbonic acid, because the reaction is the reverse of the ionization of carbonic acid.

In many buffers, the ratio of conjugate acid:conjugate base concentrations is about 1:1. In blood, however, the ratio $[HCO_3^-]/[H_2CO_3]$ is about 10 to 1, with good reason. There are more acidic byproducts of metabolism in the blood that must be neutralized than there are basic ones.

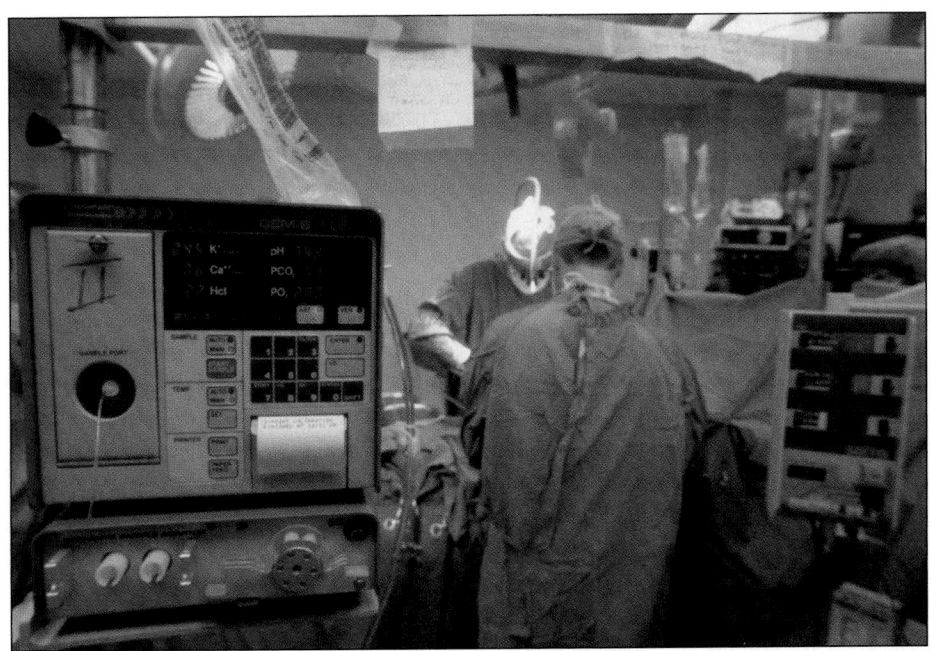

A blood gas analyzer. This instrument measures CO_2 levels in blood, blood pH, and oxygen level. These values are related and must be within a narrow range for good health.

 Exercise 17.1   Possible Buffers?

Could a solution of equimolar amounts of HCl and NaCl be a buffer? What about a solution of equimolar amounts of KOH and KCl? Explain each of your answers.

The pH of Buffer Solutions

The pH of a buffer solution can be calculated in two ways: (a) by using $[H_3O^+]$ in the K_a expression if the K_a and the concentrations of the conjugate acid and the conjugate base are known (Problem-Solving Example 17.1) and (b) by using the Henderson-Hasselbalch equation (Problem-Solving Example 17.2).

Problem-Solving Example 17.1   The pH of a Buffer from K_a

Calculate the pH of blood containing 0.0025 M carbonic acid and 0.025 M hydrogen carbonate ion. Assume that this is the only buffer system present in blood. The K_a of carbonic acid is 4.2×10^{-7}.

Answer   The pH is 7.38.

Explanation   To calculate the pH using the $[H_3O^+]$ in the K_a expression, we start with the ionization constant equation for carbonic acid and its K_a value.

$$H_2CO_3(aq) + H_2O(\ell) \rightleftharpoons H_3O^+(aq) + HCO_3^-(aq)$$

$$K_a = 4.2 \times 10^{-7} = \frac{[H_3O^+][HCO_3^-]}{[H_2CO_3]} = \frac{[H_3O^+](0.025)}{0.0025}$$

$$[H_3O^+] = \frac{(4.2 \times 10^{-7})(0.0025)}{0.025} = 4.2 \times 10^{-8}$$

$$pH = -\log(4.2 \times 10^{-8}) = 7.38$$

✔   The amount of conjugate base is ten times that of the conjugate acid, and so the $[H_3O^+]$ should be less than the K_a, which it is.

Problem-Solving Practice 17.1

Calculate the pH of blood containing 0.0020 M carbonic acid and 0.025 M hydrogen carbonate ion.

 CD-ROM Screen 18.9: pH of Buffer Solutions

The pH of a buffer containing known concentrations of conjugate base and conjugate acid or the ratio of conjugate base to conjugate acid concentrations needed to achieve a buffer of a given pH can also be conveniently calculated by using the *Henderson-Hasselbalch equation*. This equation is derived by writing the acid ionization constant expression for a weak acid, HA, and solving for $[H_3O^+]$; A^- is the conjugate base of HA.

$$HA(aq) + H_2O(\ell) \rightleftharpoons H_3O^+(aq) + A^-(aq)$$

$$K_a = \frac{[H_3O^+][\text{conj base}]}{[\text{conj acid}]} = \frac{[H_3O^+][A^-]}{[HA]}$$

$$[H_3O^+] = \frac{K_a[\text{conj acid}]}{[\text{conj base}]} = \frac{K_a[HA]}{[A^-]}$$

The next steps convert $[H_3O^+]$ to pH: Taking the base 10 logarithm of each side of this equation gives

$$\log[H_3O^+] = \log K_a + \log \frac{[\text{conj acid}]}{[\text{conj base}]} = \log K_a + \log \frac{[HA]}{[A^-]}$$

Multiplying both sides of the equation by -1 and using the relation $-\log(x) = \log(1/x)$, we get

Because $-\log x = \log\left(\frac{1}{x}\right)$,

$$-\log \frac{[\text{conj acid}]}{[\text{conj base}]} = \log \frac{[\text{conj base}]}{[\text{conj acid}]}$$

and $-\log \frac{[HA]}{[A^-]} = \log \frac{[A^-]}{[HA]}$.

$$-\log[H_3O^+] = -\log K_a + \log \frac{[\text{conj base}]}{[\text{conj acid}]} = -\log K_a + \log \frac{[A^-]}{[HA]}$$

Using the definition of pH and defining $-\log K_a$ as pK_a (analogous to the definition of pH), the equation becomes

$$pH = pK_a + \log \frac{[\text{conj base}]}{[\text{conj acid}]} = pK_a + \log \frac{[A^-]}{[HA]}$$ **Henderson-Hasselbalch Equation**

Problem-Solving Example 17.2 Calculating the pH of a Buffer Using the Henderson-Hasselbalch Equation

Using the Henderson-Hasselbalch equation, calculate the pH of blood containing 0.0025 M carbonic acid and 0.025 M hydrogen carbonate ion. Assume that this is the only buffer system present in blood. The K_a of carbonic acid is 4.2×10^{-7}.

Answer The pH is 7.38.

Explanation With 0.025 M and 0.0025 M concentrations of HCO_3^- and H_2CO_3, respectively, the Henderson-Hasselbach equation gives

$$pH = -\log K_a + \log\frac{[HCO_3^-]}{[H_2CO_3]} = -\log(4.2 \times 10^{-7}) + \log \frac{0.0250}{0.00250}$$

$$= 6.38 + 1 = 7.38$$

✔ The concentration of conjugate base is ten times that of the conjugate acid and so the pH should be greater than the pK_a by 1 pH unit, and it is. The answer is reasonable.

Problem-Solving Practice 17.2

Calculate the pH of blood containing 0.0025 M carbonic acid and 0.020 M hydrogen carbonate ion.

Exercise 17.2 A Buffer Solution

Calculate the pH of a buffer containing 0.050 mol/L pyruvic acid, $CH_3COCOOH$, and 0.060 mol/L of sodium pyruvate, $Na^+CH_3COCOO^-$. K_a of pyruvic acid = 3.2×10^{-3}.

Exercise 17.3 A Blood Buffer

Calculate the ratio of HPO_4^{2-} to $H_2PO_4^-$ in blood at a normal pH of 7.40.

From the Henderson-Hasselbalch equation note that when the concentrations of conjugate base and conjugate acid are equal

$$\frac{[\text{conj base}]}{[\text{conj acid}]} = 1 \qquad \log \frac{[\text{conj base}]}{[\text{conj acid}]} = \log(1) = 0$$

and so the pH = pK_a. Thus, **a buffer's pH equals the pK_a of its weak acid when the concentrations of the acid and its conjugate base in the buffer are equal.**

We can use this relationship in determining what buffer to use for a desired pH. A buffer for maintaining a desired pH can be chosen easily by examining pK_a values, which are often tabulated along with K_a values. Table 17.1 lists pK_a values of several common acids that could be used to prepare buffers over the pH range from 4 to 10.

To have comparable quantities of both acid and conjugate base in a buffer solution, the ratio of conjugate base to conjugate acid cannot get much smaller than 1:10 or much bigger than 10:1. **The pH range of a buffer is limited to about one pH unit above or below the pK_a of the conjugate acid.** In the carbonate/bicarbonate case, that would be a pH from 5.38 to 7.38 because $pK_a = -\log K_a = -\log 4.2 \times 10^{-7} = 6.38$. Other acid-base pairs can be used to prepare buffers with much different pH ranges, as determined by the K_a value of the acid in the buffer (Table 17.1).

Notice that as K_a decreases, the pK_a increases; therefore, the weaker the acid, the larger its pK_a.

Problem-Solving Example 17.3 Selecting an Acid-Base Pair for a Buffer Solution of Known pH

An experiment requires a buffer solution that has a pH of 9.56. You have available the solutions needed to make buffers with these acid-base pairs: H_2CO_3/HCO_3^- ; $H_2PO_4^-/HPO_4^{2-}$; HPO_4^{2-}/PO_4^{3-} ; and NH_4^+/NH_3. Which acid-base pair should you use to

CD-ROM Screen 18.10:
Preparing Buffer Solutions

TABLE 17.1 Buffer Systems That Are Useful at Various pH Values*

Desired pH	Weak acid	Weak base	K_a (Weak acid)	pK_a
4	Lactic acid ($CH_3CHOHCOOH$)	Lactate ion ($CH_3CHOHCOO^-$)	1.4×10^{-4}	3.85
5	Acetic acid (CH_3COOH)	Acetate ion (CH_3COO^-)	1.8×10^{-5}	4.74
6	Carbonic acid (H_2CO_3)	Hydrogen carbonate ion (HCO_3^-)	4.2×10^{-7}	6.38
7	Dihydrogen phosphate ion ($H_2PO_4^-$)	Hydrogen phosphate ion (HPO_4^{2-})	6.2×10^{-8}	7.21
8	Hypochlorous acid ($HClO$)	Hypochlorite ion (ClO^-)	3.5×10^{-8}	7.46
9	Ammonium ion (NH_4^+)	Ammonia (NH_3)	5.6×10^{-10}	9.25
10	Hydrogen carbonate ion (HCO_3^-)	Carbonate ion (CO_3^{2-})	4.8×10^{-11}	10.32

*Adapted from Masterton, W. L., and Hurley, C. N. *Chemistry—Principles and Reactions,* 4th ed. Philadelphia: Saunders College Publishing, 2001; p. 416.

make such a buffer solution, and what molar ratios of the compounds should you use? Use Table 17.1 for K_a and pK_a values.

Answer The NH_4^+/NH_3 acid-base pair with a molar ratio of 0.20 mol of NH_3 to 0.098 mol of NH_4^+ would work.

Explanation Because you need an acid-base pair that has a pK_a near the pH you are trying to achieve (9.56) you can evaluate the available acid-base pairs in terms of which pK_a is closest to that pH. Once that acid-base pair has been selected, you can use the Henderson-Hasselbalch equation to calculate the necessary conjugate acid and conjugate base concentrations to give a buffer with pH 9.56.

The acid with the pK_a closest to the target pH is NH_4^+. The pK_a's of the other acids are not close enough to 9.56.

Substituting this pK_a and the ammonium ion–ammonia conjugate acid-base pair concentration terms into the Henderson-Hasselbalch equation gives

$$pH = 9.56 = 9.25 + \log \frac{[NH_3]}{[NH_4^+]}$$

$$\log \frac{[NH_3]}{[NH_4^+]} = 9.56 - 9.25 = 0.31$$

The antilog of 0.31 is $10^{0.31} = 2.04$; thus, $[NH_3]/[NH_4^+] = 2.04$. This means that, roughly, the required concentration of NH_3 will have to be twice that of NH_4^+. The buffer could be made using NH_3 and a soluble ammonium salt, such as NH_4Cl. If the concentration of NH_3 is 0.20 mol/L, the concentration of the ammonium chloride needed is 0.098 mol/L; $0.20/0.098 = 2.04$.

For example, if the NH_3 concentration is 0.10 M, the NH_4Cl concentration required for a NH_3/NH_4^+ ratio of 2.04 is 0.049 M.

✔ The target pH is 9.56. We can see from the Henderson-Hasselbalch equation that with equal concentrations of conjugate acid and base, the pH would equal the pK_a, 9.25. Therefore, to reach the target pH of 9.56, the concentration of NH_3, the conjugate base, must be greater than that of NH_4^+, the conjugate acid.

Problem-Solving Practice 17.3

Use the data in Table 17.1 to select an acid-base conjugate pair you could use to make buffer solutions having each of these hydrogen ion concentrations.

(a) 3.2×10^{-4} M (b) 5.0×10^{-5} M (c) 7.0×10^{-8} M (d) 6.0×10^{-11} M

Exercise 17.4 Solution to a Buffer Solution

Calculate the ratio of sodium acetate and acetic acid needed to make a buffer of pH 4.68.

Exercise 17.5 Buffers and pH

Use data from Table 17.1 to calculate the pH of these buffers.

(a) H_2CO_3 (0.10 M)/HCO_3^- (0.25 M) (b) $H_2PO_4^-$ (0.10 M)/HPO_4^{2-} (0.25 M)

Buffer Capacity

When acid (H_3O^+) is added to a buffer, the acid reacts with the conjugate base of the buffer to form its conjugate acid:

Conjugate base in buffer + H_3O^+ added $\longrightarrow$

conjugate acid of the buffer base + water

If base (OH^-) is added to a buffer, it reacts with the conjugate acid of the buffer, which is converted into its conjugate base.

Conjugate acid in buffer + OH^- added $\longrightarrow$

conjugate base of the buffer acid + water

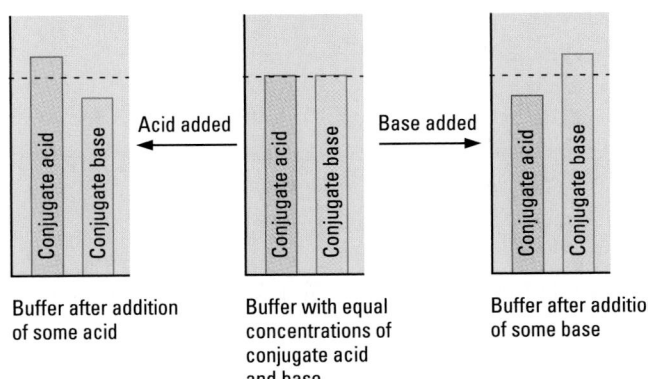

Figure 17.2 **The effects of adding acid or base to a buffer.** When H_3O^+ or OH^- ions are added to a buffer, the amounts of conjugate acid and conjugate base in the buffer change.

Acid added ←   Base added →

Buffer after addition of some acid

Buffer with equal concentrations of conjugate acid and base

Buffer after addition of some base

Figure 17.2 summarizes these relationships. For a buffer made from acetic acid and sodium acetate, the changes are

$$CH_3COO^-(aq) + H_3O^+(aq) \longrightarrow CH_3COOH(aq) + H_2O(\ell)$$

Conjugate base   added   Conjugate acid
of the buffer   of the buffer

$$CH_3COOH(aq) + OH^-(aq) \longrightarrow CH_3COO^-(aq) + H_2O(\ell)$$

Conjugate acid   added   Conjugate base
of the buffer   of the buffer

The *amounts* of conjugate acid and conjugate base of the buffer in solution determine the **buffer capacity**—the quantity of acid or base the buffer can accommodate without a significant pH change (more than 1 pH unit). When nearly all of the conjugate acid in a buffer has reacted with added base, adding a little more base can increase the pH significantly, because there is almost no conjugate acid left in the buffer to consume the base. Similarly, if enough acid is added to a buffer to react with all of the buffer's conjugate base and excess acid remains, the pH will decrease significantly. In either case, the buffer capacity has been exceeded. For example, 1 L of a buffer solution that is 0.25 M in CH_3COOH and 0.25 M in CH_3COO^- contains 0.25 mol of CH_3COOH and 0.25 mol of CH_3COO^-. This buffer can accommodate the addition of up to 0.25 mol of H_3O^+ or OH^-, at which point it has used up its buffer capacity. Thus, the initial buffer is not able to accommodate the addition of 0.30 mol of strong acid or 0.30 mol of strong base without a major change in pH. Such additions would use up all of the buffer's conjugate base or all of its conjugate acid, respectively, and exceed the buffer's capacity. The pH would drop or rise accordingly.

Problem-Solving Example 17.4   **Buffer Capacity**

A buffer is prepared using 0.25 mol of $H_2PO_4^-$ and 0.15 mol of HPO_4^{2-} in 500. mL of solution. Will the buffer capacity be exceeded if 6.2 g of KOH is added to it? What will be the pH of the new solution?

Answer   No, the buffer capacity will not be exceeded.

Explanation   The initial pH of the buffer can be calculated using the Henderson-Hasselbalch equation. From Table 17.1, the pK_a of $H_2PO_4^-$ is 7.21.

$$pH = 7.21 + \log\frac{[HPO_4^{2-}]}{[H_2PO_4^-]} = 7.21 + \log\frac{(0.15/0.500)}{(0.25/0.500)}$$

$$= 7.21 + \log(0.60) = 7.21 - 0.22 = 6.99$$

The 6.2 g of KOH is 0.11 mol of KOH, which contributes 0.11 mol of OH^- to be neutralized by the buffer.

$$6.2 \text{ g KOH} \left(\frac{1 \text{ mol KOH}}{56.1 \text{ g KOH}} \right) = 0.11 \text{ mol KOH}$$

The 0.11 mol of OH^- added is neutralized by reacting with 0.11 mol of $H_2PO_4^-$, the conjugate acid of the buffer, to form 0.11 mol of HPO_4^{2-}.

$$H_2PO_4^-(aq) + OH^- \longrightarrow HPO_4^{2-}(aq) + H_2O(\ell)$$

0.11 mol	0.11 mol	0.11 mol
from buffer	from KOH	formed

The reaction changes the amounts of HPO_4^{2-} and $H_2PO_4^-$ remaining in the buffer:

	Before reaction	**After reaction**
Moles HPO_4^{2-}	0.15	$0.15 + 0.11 = 0.26$
Moles $H_2PO_4^-$	0.25	$0.25 - 0.11 = 0.14$

Because there is still some $H_2PO_4^-$ remaining (0.14 mol), the buffer's capacity was not exceeded by adding 6.2 g of KOH. The pH after the KOH addition is

$$pH = 7.21 + \log \frac{[HPO_4^{2-}]}{[H_2PO_4^-]} = 7.21 + \log \frac{(0.26/0.500)}{(0.14/0.500)}$$

$$= 7.21 + \log(1.9) = 7.21 + 0.27 = 7.48$$

✔ Addition of base to a buffer should increase the pH to some extent depending on the amount of base added. The amount of KOH (0.11 mol) added was less than the amount of conjugate acid (0.25 mol) in the buffer. Therefore, the buffer's capacity was not exceeded, and the pH change should be less than 1 pH unit, which it is.

Problem-Solving Practice 17.4

Calculate the minimum number of grams of KOH that would have to be added to the initial buffer in Problem-Solving Example 17.4 to exceed its buffer capacity.

CD-ROM Screen 18.11: Adding Reagents to Buffer Solutions

The pH Change on Addition of an Acid or a Base to a Buffer

When acid or base is added to a buffer, its pH changes because of shifts in the amounts of conjugate acid and conjugate base in the buffer. The extent of the pH change depends on the amount of acid or base added and on the buffer's capacity. Problem-Solving Example 17.4 illustrated this for the addition of base to a buffer. The Henderson-Hasselbalch equation provides a very convenient way to calculate the change in pH of a buffer when base or acid is added to it.

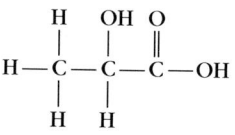

lactic acid

Problem-Solving Example 17.5   pH Changes in a Buffer

A buffer is prepared by adding 0.15 mol of lactic acid, $CH_3CHOHCOOH$, and 0.20 mol of sodium lactate, $Na^+CH_3CHOHCOO^-$, to sufficient water to make 1.00 L of buffer solution. The K_a of lactic acid is 1.4×10^{-4}.

(a) Calculate the pH of the buffer.
(b) Calculate the pH of the buffer after 0.050 mol of HCl has been added (neglect volume changes).
(c) Calculate the pH of the buffer after 0.10 mol of NaOH has been added (neglect volume changes).

Answer   (a) 3.97   (b) 3.72   (c) 4.63

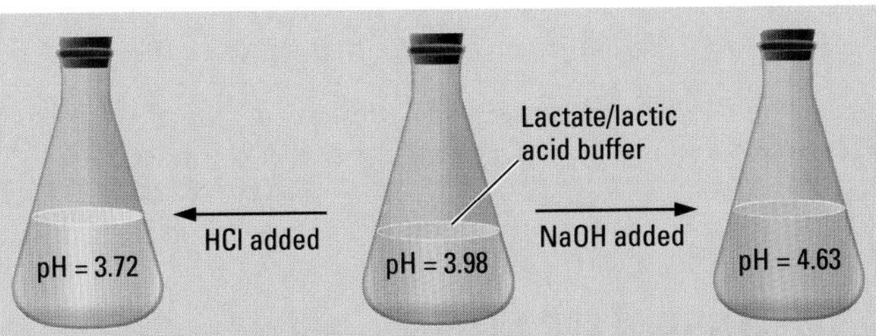

Comparison of pH of lactate/lactic acid buffer after addition of HCl or NaOH.

Explanation

(a) First find the pK_a

$$pK_a = -\log K_a = -\log(1.4 \times 10^{-4}) = 3.85$$

and then use the Henderson-Hasselbalch equation to calculate the pH.

$$pH = 3.85 + \log \frac{(0.20 \text{ mol/L lactate})}{(0.15 \text{ mol/L lactic acid})} = 3.85 + \log(1.33) = 3.85 + 0.13 = 3.97$$

(b) The 0.050 mol of HCl added reacts with 0.050 mol of lactate ion to form 0.050 mol of lactic acid and water.

$$CH_3CHOHCOO^-(aq) + H_3O^+(aq) \longrightarrow CH_3CHOHCOOH(aq) + H_2O(\ell)$$

| 0.050 mol | 0.050 mol | 0.050 mol |
| from buffer | added | formed in buffer |

This changes the original lactate/lactic acid ratio, so the pH changes because of the added HCl:

$$pH = 3.85 + \log \frac{[\text{lactate}]}{[\text{lactic acid}]} = 3.85 + \log \frac{(0.20 - 0.050)}{(0.15 + 0.050)} = 3.85 + \log \frac{0.15}{0.20}$$

$$= 3.85 + \log(0.75) = 3.85 + (-0.13) = 3.72$$

The pH drops from 3.97 to 3.72. The buffer's capacity for added acid has not yet been exceeded because 0.15 mol of lactate remains to react with additional acid that might be added.

(c) To offset the addition of 0.10 mol of NaOH requires 0.10 mol of lactic acid.

$$CH_3CHOHCOOH(aq) + OH^-(aq) \longrightarrow CH_3CHOHCOO^-(aq) + H_2O(\ell)$$

| 0.10 mol | 0.10 mol | 0.10 mol |
| from buffer | added | formed in buffer |

This changes the initial lactate/lactic acid ratio, and a pH change occurs.

$$pH = 3.85 + \log \frac{[\text{lactate}]}{[\text{lactic acid}]} = 3.85 + \log \frac{(0.20 + 0.10)}{(0.15 - 0.10)} = 3.85 + \log \frac{0.30}{0.05}$$

$$= 3.85 + \log(6) = 3.85 + 0.78 = 4.63$$

✔ The pH of the buffer is reasonable because the concentration of conjugate base is slightly greater than that of the conjugate acid; therefore, the ratio is greater than 1 and the log of the ratio is positive. We would expect that when acid is added to the buffer, if the pH of the buffer changes, it should decrease, and it does—from 3.97 to 3.72. The pH change that occurs when NaOH is added is greater than that for the addition of HCl because more moles of base than acid were added. But in neither case was the buffer capacity exceeded; the changes were less than 1 pH unit, so the answers are reasonable.

Problem-Solving Practice 17.5

For the lactate/lactic acid buffer given above, calculate the pH when the following amounts are added to it:

(a) 0.075 mol HCl   (b) 0.025 mol NaOH (neglect any volume changes)

Exercise 17.6 Which Reaction in this Blood Buffer?

If an abnormally high CO_2 concentration is present in blood, which phospate ion— $H_2PO_4^-$ or HPO_4^{2-} —will be used to counteract the presence of excess CO_2?

17.2 Acid-Base Titrations

CD-ROM Screen 18.12:
Titration Curves

For example, a standardized acid could be added from a buret to a known volume of base whose concentration is to be determined.

In Section 5.8 (← *p. 199*) acid-base titrations were described as a method by which the concentration of an acid or a base could be determined. An acid-base titration is carried out by slowly adding a measurable amount of an aqueous solution of a base or acid to a known volume of an aqueous acid or base whose concentration is to be determined. The apparatus generally used for this type of titration is shown in Figure 17.3. A *standard solution,* (← *p. 199*) one whose concentration is known accurately, is added from the buret, which allows the required volume of a solution to be measured accurately. The solution in the buret is known as the **titrant.** The *equivalence point* (← *p. 199*) is reached when the stoichiometric amount of titrant has been added, the amount that exactly neutralizes the acid or base being titrated.

Detection of the Equivalence Point

A method is needed in a titration to detect the equivalence point. This can be done by using a pH meter (Section 19.7), which monitors the pH of the solution as the titration proceeds. Alternatively, the color change of an acid-base indicator can be used to detect when sufficient titrant has been added (Figure 17.4). The **end point** of a titration occurs when the indicator changes color. The goal is to use an indicator that gives an end point close to the equivalence point.

Acid-base indicators were described in Section 16.4. Such indicators are typically weak organic acids (HIn) that differ in color from their conjugate bases (In$^-$).

$$HIn(aq) + H_2O(\ell) \rightleftharpoons H_3O^+(aq) + In^-(aq)$$

Color 1   Color 2

Removal of an H$^+$ from the indicator produces a different molecular structure that absorbs light in a different region of the visible spectrum. For methyl red the reaction is

Figure 17.3 An acid-base titration set-up for titrating an acid sample with NaOH as the titrant. A standard NaOH solution in the buret is slowly added to the acid solution to be titrated. An acid-base indicator is added to the solution before the titration begins; it will change color when the end point has been reached.

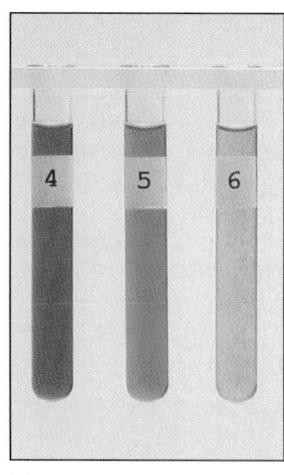
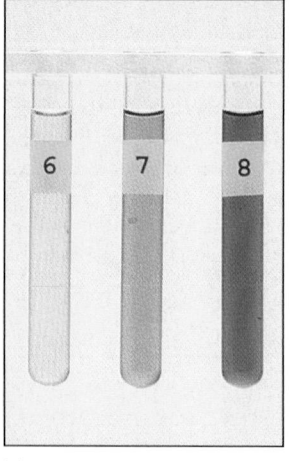
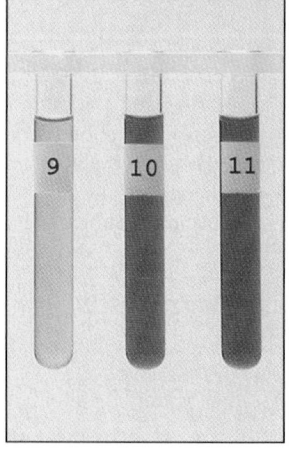

(a)   (b)   (c)

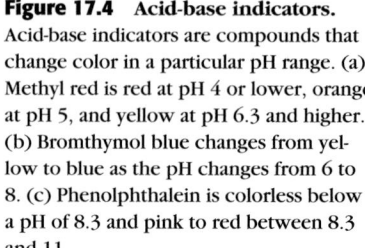

Figure 17.4 Acid-base indicators. Acid-base indicators are compounds that change color in a particular pH range. (a) Methyl red is red at pH 4 or lower, orange at pH 5, and yellow at pH 6.3 and higher. (b) Bromthymol blue changes from yellow to blue as the pH changes from 6 to 8. (c) Phenolphthalein is colorless below a pH of 8.3 and pink to red between 8.3 and 11.

The structure shows a molecular equation:

$$\text{(red)} \ (aq) + H_2O(l) \rightleftharpoons \text{(yellow)} \ (aq) + H_3O^+(aq)$$

red — yellow

The color observed for the indicator during an acid-base titration depends on the $\dfrac{[\text{HIn}]}{[\text{In}^-]}$ ratio, for which three cases apply:

- When $\dfrac{[\text{HIn}]}{[\text{In}^-]} \geq 10$ the indicator solution is the acid color (HIn).

- When $\dfrac{[\text{HIn}]}{[\text{In}^-]} \leq 0.1$ the indicator solution is the conjugate base color (In$^-$).

- When $\dfrac{[\text{HIn}]}{[\text{In}^-]} \approx 1$ the indicator solution color is intermediate between the acid and the conjugate base colors.

Bromthymol blue, for example, changes from yellow to blue as the pH changes from 6 to 8 (Figure 17.4).

We will consider in some detail three types of acid-base titrations: (1) a strong acid (HCl) titrated by a strong base (NaOH), (2) a weak acid (CH$_3$COOH) titrated by a strong base (NaOH), and (3) a weak base (NH$_3$) titrated by a strong acid (HCl). In all cases, we will examine the **titration curve** for the titration, a graph of pH as a function of the volume of titrant added. In each case, we will be interested particularly in the pH at four stages of the titration:

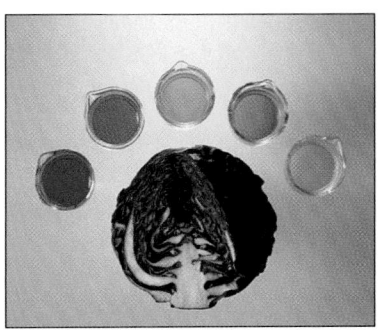

Red cabbage juice is a naturally occurring acid-base indicator. From left to right are solutions of pH 1, 4, 7, 10, and 13.

CHEMISTRY
You Can Do...   Making An Acid-Base Indicator

Red cabbage juice is a very useful natural indicator that changes color across the entire pH range. For this experiment you will need these items:

- A few red cabbage leaves
- A food blender or processor
- Coffee filter or food strainer
- 6–8 colorless glasses
- A selection of household chemicals — vinegar, soap, baking soda, ammonia, detergent, colorless carbonated beverage, milk, and so forth

Use the blender or food processor to make red cabbage juice. Tear the cabbage leaves into small pieces and put them into the blender. Add two cups of water, turn on the blender, and run the blender long enough so that the water becomes colored by the cabbage juice. Use the filter or strainer to filter the mixture from the blender. Save the juice and discard the solid.

First, test the pH of vinegar by adding two teaspoonfuls of red cabbage juice to one-third cup of vinegar; record the color of the solution. Save this sample to compare it with other test samples. Next, test the pH of liquid soap by adding a few squirts of the soap into a glass containing about one-fourth cup of warm water and add two teaspoonfuls of the red cabbage juice. Record the color of the solution and save the solution for comparison with other test samples. Test the pH of samples of several other household products by adding two teaspoonfuls of red cabbage juice to one-third cup of sample. Use fresh cabbage juice with each new sample. Record the color of each. After completing the experiment, rinse the samples down the drain, using plenty of water to flush them away.

1. Compare the colors of each sample with those in the figure on this page to determine the approximate pH of the samples.
2. Which sample had the lowest pH?
3. Which sample had the highest pH?
4. Which sample had the highest H$_3$O$^+$ concentration?

Figure 17.5 Curve for titration of 0.100 M HCl with 0.100 M NaOH. This strong acid reacts with this strong base to form a solution with a pH of 7.0 at the equivalence point. Acceptable indicators are methyl red, bromthymol blue, phenolphthalein, or any other indicator whose color changes within the pH range at the equivalence point.

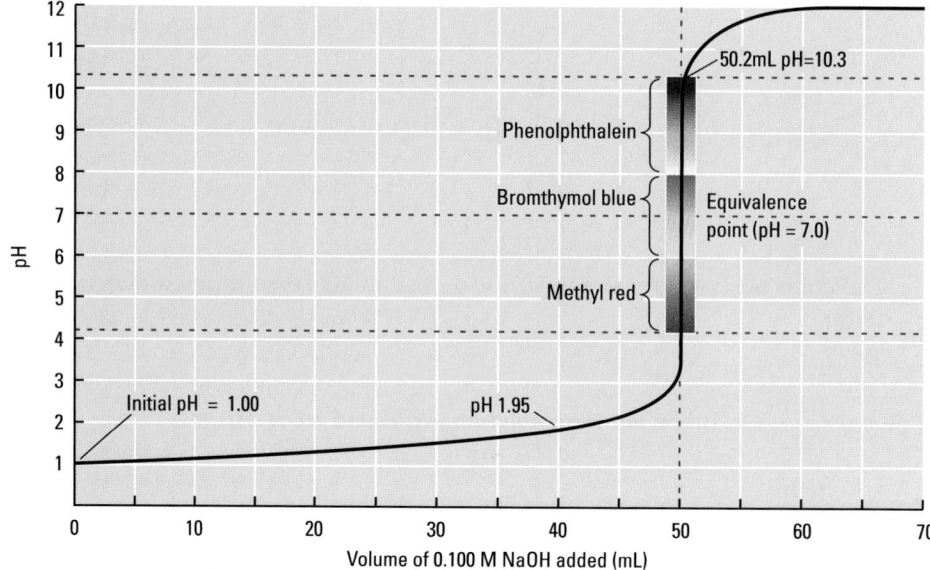

- Prior to the addition of titrant
- After addition of titrant, but prior to the equivalence point
- At the equivalence point
- After the equivalence point

Titration of a Strong Acid with a Strong Base

The titration curve for the titration of 50.0 mL of 0.100 M HCl with 0.100 M NaOH is given in Figure 17.5. Problem-Solving Example 17.6 illustrates calculation of the four points marked on the curve.

Problem-Solving Example 17.6   Titration of HCl with NaOH

A 0.100 M NaOH solution is used to titrate 50.0 mL of 0.100 M HCl. Calculate the pH of the solution at these four points:

(a) Before any titrant is added.
(b) After 40.0 mL of titrant has been added.
(c) After 50.0 mL of NaOH has been added (what indicator — methyl red, bromthymol blue, or phenolphthalein — can be used to detect the equivalence point?)
(d) After 50.2 mL of NaOH has been added.

Answers
(a) 1.00   (b) 1.95
(c) 7.00 methyl red, bromthymol blue, or phenolphthalein   (d) 10.3

Explanation
(a) Because HCl is a strong acid, the initial H_3O^+ concentration is 0.100 M and the pH is $-\log(0.100) = 1.00$. (We will round the pH values to two significant figures; see Appendix A.6 for treatment of significant figures when using logarithms.)
(b) The initial 50.0-mL solution of acid contains

$$(0.0500 \text{ L}) (0.100 \text{ mol/L}) = 5.00 \times 10^{-3} \text{ mol } H_3O^+$$

As NaOH is added, the number of moles of H_3O^+ decreases due to the reaction of added OH^- ions with H_3O^+ ions in the acid.

$$H_3O^+(aq) + OH^-(aq) \longrightarrow H_2O(\ell)$$

from HCl   from NaOH

In general, prior to the equivalence point in a strong acid–strong base titration, the $[H_3O^+]$ can be calculated for any volume of base added by the relation

$$[H_3O^+] = \frac{\text{original moles acid} - \text{total moles base added}}{\text{volume acid (L)} + \text{volume base added (L)}}$$

After 40.0 mL of 0.100 M NaOH is added to the original 50.0 mL of 0.100 M HCl, the $[H_3O^+]$ is

$$[H_3O^+] = \frac{(5.00 \times 10^{-3}) - (4.00 \times 10^{-3})}{0.0500\ L + 0.0400\ L} = 1.11 \times 10^{-2}\ M \qquad pH = 1.95$$

(c) At the equivalence point, 50.0 mL of 0.100 M NaOH has been added. This amounts to $(0.0500\ L)(0.100\ mol\ OH^-/L) = 5.00 \times 10^{-3}\ mol\ OH^-$, which exactly neutralizes the $5.00 \times 10^{-3}\ mol$ of H_3O^+ initially in the solution. No residual acid or excess NaOH is present. The NaCl produced is a neutral salt and so the pH is 7.00 at the equivalence point. Because the pH rises so rapidly near the equivalence point, methyl red, bromthymol blue, or phenolphthalein can be used as the indicator in a strong acid–strong base titration.

(d) Adding 50.2 mL of 0.100 M NaOH to the solution puts $5.02 \times 10^{-3}\ mol$ of OH^- into the solution: $(0.0502\ L)(0.100\ mol\ OH^-/L) = 5.02 \times 10^{-3}\ mol$ of OH^-. As seen in part (c), $5.00 \times 10^{-3}\ mol$ of OH^- neutralized all of the HCl in the initial sample. Therefore, the additional $0.02 \times 10^{-3}\ mol$ of OH^-, now in 100.2 mL of solution, is not neutralized; the pH of the solution is

$$\frac{0.02 \times 10^{-3}\ mol\ of\ OH^-}{0.1002\ L} = 2.0 \times 10^{-4}\ M$$

$$pOH = -\log(2.0 \times 10^{-4}) = 3.70 \qquad pH = 14.00 - pOH = 14.00 - 3.70 = 10.3$$

Notice that the addition of just 0.2 mL of excess NaOH dramatically raises the pH, as seen from the titration curve (Figure 17.5).

A volume of 0.2 mL is approximately 6 drops.

Problem-Solving Practice 17.6

For the HCl-NaOH titration described above, calculate the pH when these volumes of NaOH have been added:

(a) 10.0 mL   (b) 25.00 mL   (c) 45.0 mL   (d) 50.5 mL

Exercise 17.7   Titration Curve

Draw the titration curve for the titration of 50.0 mL of 0.100 M NaOH using 0.100 M HCl as the titrant.

Titration of a Weak Acid with a Strong Base

As noted in Section 16.8, the reaction of a weak acid, such as acetic acid, with a strong base, like NaOH, produces a salt — sodium acetate in this case — that has a basic anion. As a result, when a weak acid is titrated with a strong base, the pH of the solution at the equivalence point will be above 7 due to formation of the basic anion. The titration curve in Figure 17.6 for the titration of 50.0 mL of 0.100 M acetic acid with 0.100 M NaOH represents this type of titration, and Problem-Solving Example 17.7 illustrates the calculations associated with the titration curve.

Notice in Figure 17.6 that the initial pH of 0.100 M acetic acid is higher than that of the 0.100 M HCl in Figure 17.5. This is to be expected because acetic acid is a weaker acid than HCl. Acetic acid is only slightly ionized (K_a of acetic acid = 1.8×10^{-5}), and the pH of 0.100 M acetic acid is 2.88 ($\Leftarrow$ *p. 742*). Also notice from Figure 17.6 that the rapidly rising portion of the titration curve near the equivalence point is shorter than it is for the NaOH/HCl titration. The equivalence

Figure 17.6   Curve for titration of
**50.0 mL of 0.100 M acetic acid with
0.100 M NaOH.**

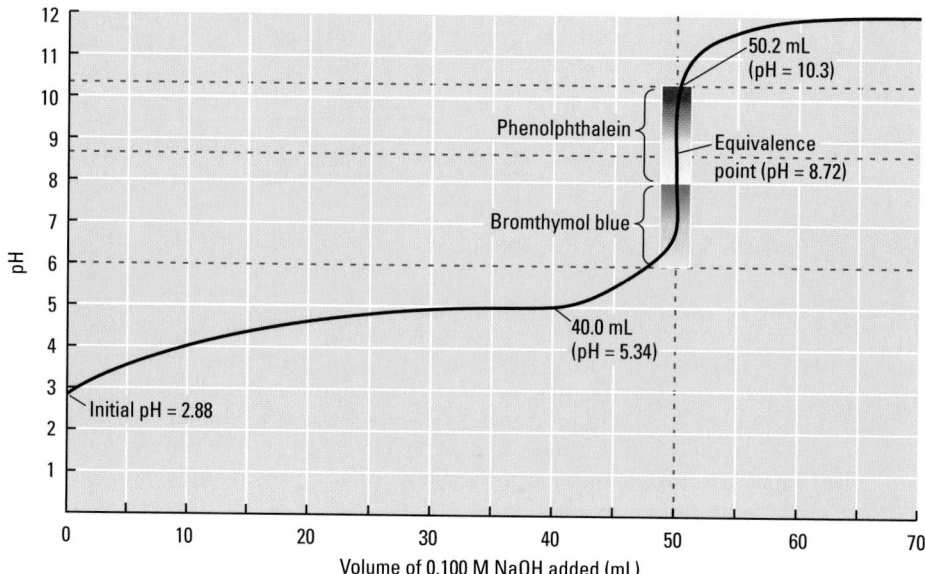

point is 8.72, making methyl red an unsuitable indicator because its color changes
before the equivalence point. Bromthymol blue or phenolphthalein can be used.

Problem-Solving Example 17.7   Titration of CH₃COOH with NaOH

A 0.100 M NaOH solution is used to titrate 50.0 mL of 0.100 M acetic acid. Calculate the
pH of the solution at these three points:

(a)   After 40.0 mL of titrant has been added
(b)   After 50.0 mL of NaOH has been added
(c)   After 50.2 mL of NaOH has been added

Answers
(a)   5.34   (b) 8.72   (c) 10.3

Explanation
(a)   The acetic acid sample contains $(0.0500 \text{ L})(0.100 \text{ mol acetic acid/L}) = 5.00 \times 10^{-3}$ mol acetic acid. Adding 40.0 mL of 0.100 M NaOH puts 4.00×10^{-3} mol of OH^- ions into the solution, which neutralizes 4.00×10^{-3} mol of the acid.

$$CH_3COOH(aq) + OH^-(aq) \longrightarrow CH_3COO^-(aq) + H_2O(\ell)$$

$$\begin{array}{ccc} 5.00 \times 10^{-3} \text{ mol} & 4.00 \times 10^{-3} \text{ mol} & 4.00 \times 10^{-3} \text{ mol} \\ \text{in acid soln} & \text{added} & \text{formed} \end{array}$$

Therefore, 4.00×10^{-3} mol of acetate ions is present along with 1.00×10^{-3} mol
of unneutralized acetic acid. The total volume of the solution is now 90.0 mL, and
the concentrations are

$$\text{Acetic acid} = \frac{1.00 \times 10^{-3} \text{ mol}}{0.0900 \text{ L}} = 0.0111 \text{ M}$$

$$\text{Acetate ion} = \frac{4.00 \times 10^{-3} \text{ mol}}{0.0900 \text{ L}} = 0.0444 \text{ M}$$

The pH can be calculated using the Henderson-Hasselbalch equation. The pK_a of
acetic acid $= -\log(1.8 \times 10^{-5}) = 4.74$.

$$pH = 4.74 + \log\frac{0.0444}{0.0111} = 4.74 + \log(4) = 4.74 + 0.602 = 5.34$$

(The pH can also be calculated using the K_a value and expression for acetic acid to
solve for $[H_3O^+]$ and then pH.)

(b) This is the equivalence point, so the stoichiometric amount of base has been added to exactly neutralize the acid in the sample: 5.00×10^{-3} mol of OH^- has been added to 5.0×10^{-3} mol of acetic acid initially present to produce 5.00×10^{-3} mol of acetate ion. The solution at the equivalence point is 0.0500 M sodium acetate.

$$\frac{5.00 \times 10^{-3} \text{ mol acetate}}{0.100 \text{ L solution}} = 0.0500 \text{ M}$$

The pH at the equivalence point is governed by the hydrolysis of acetate ion:

$$CH_3COO^-(aq) + H_2O(\ell) \longrightarrow CH_3COOH(aq) + OH^-(aq)$$

The K_b for acetate ion can be calculated from K_w and the K_a for acetic acid.

$$K_b = \frac{K_w}{K_a} = \frac{1.0 \times 10^{-14}}{1.8 \times 10^{-5}} = 5.6 \times 10^{-10}$$

We can use the K_b expression to calculate $[OH^-]$ and from it the pH. Let $x = [OH^-] = [CH_3COOH]$:

$$K_b = 5.6 \times 10^{-10} = \frac{[CH_3COOH][OH^-]}{[CH_3COO^-]} = \frac{x^2}{0.0500 - x}$$

Because K_b is small, we can approximate $0.0500 - x$ to be 0.0500. Solving for x,

$$K_b = 5.6 \times 10^{-10} = \frac{x^2}{0.0500} \qquad x = 5.3 \times 10^{-6} = [OH^-]$$

which converts to a pOH of 5.28 and a pH of 8.72. This is in marked contrast to the equivalence point of 7.00 for the NaOH/HCl titration, where neutral NaCl was the titration product.

(c) The pH beyond the equivalence point is controlled by the OH^- concentration from excess NaOH, which is greater than the OH^- contributed by the hydrolysis of acetate ion. Therefore, the pH beyond the equivalence point is like that for the NaOH/HCl titration with excess NaOH.

$$\text{Final } OH^- \text{ concentration} = \frac{0.02 \times 10^{-3} \text{ mol of } OH^-}{0.1002 \text{ L}} = 2.00 \times 10^{-4} \text{ M}$$

$$pOH = -\log(2.00 \times 10^{-4}) = 3.70 \qquad pH = 14.00 - pOH = 14.00 - 3.70 = 10.3$$

Problem-Solving Practice 17.7

Calculate the pH when these volumes of 0.100 M NaOH have been added when titrating 50.0 mL of 0.100 M acetic acid:

(a) 10.0 mL   (b) 25.00 mL   (c) 45.0 mL   (d) 51.0 mL

Exercise 17.8   Titration of Acetic Acid with NaOH

Use the K_a expression and value for acetic acid to calculate the pH after 30.0 mL of 0.100 M NaOH has been added to 50.0 mL of 0.100 M acetic acid.

 ## Exercise 17.9   Shape of the Titration Curve

Explain why the NaOH/acetic acid titration curve in Figure 17.6 has a relatively flat region between 20.0 and 30.0 mL of NaOH added.

As seen from Figures 17.5 and 17.6 and their associated Problem-Solving Examples, there are differences in the titration curves for a strong base with a strong acid or a weak acid of equal concentration. In particular, the differences are

Figure 17.7 **The effect of acid strength on the shape of the titration curve.** Each curve is for titration of 10 mL of a 1.0 M acid with 1.0 M NaOH.

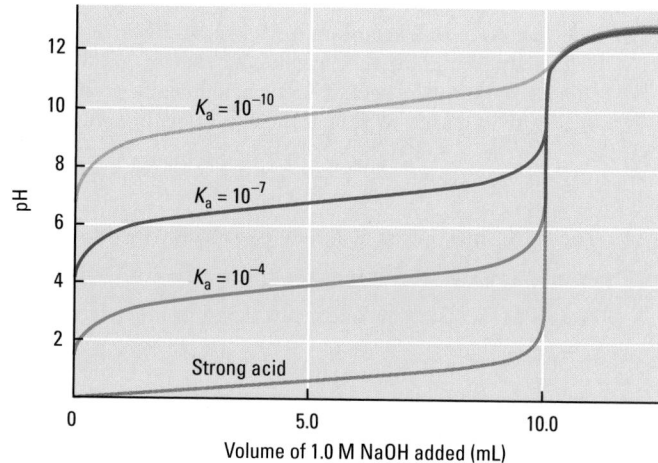

- Before the titration, the initial pH of the solution is higher for the weak acid.
- Very near the equivalence point, the length of the rapid rise of the curve is shorter for the weak acid.
- The pH at the equivalence point is higher for the weak acid titration.

These features are shown in Figure 17.7. Note especially that near the equivalence point, the weaker the acid, the higher the pH and the less abrupt the rise in pH.

Titration of a Weak Base with a Strong Acid

The titration of a weak base (NH_3) with a strong acid (HCl) has the titration curve shown in Figure 17.8. The reaction produces ammonium chloride, NH_4Cl. Notice that the starting pH is above 7.0, but less than it would be for 0.100 M NaOH (13.00) because NH_3 is a weak base. Also notice that the pH at the equivalence point, 5.28, is below 7.0 because ammonium chloride is an acidic salt. Beyond the equivalence point the pH continues to drop as excess acid is added. Although methyl red is a suitable indicator for this titration, phenolphthalein is not suitable because its color change occurs well before the equivalence point.

Figure 17.8 **The titration of a weak base, NH_3, by a strong acid, HCl.** This titration curve is for the titration of 50.0 mL of 0.100 M NH_3 by 0.100 M HCl. It is essentially the inverse of the curve for the titration of a weak acid by a strong base, Figure 17.6.

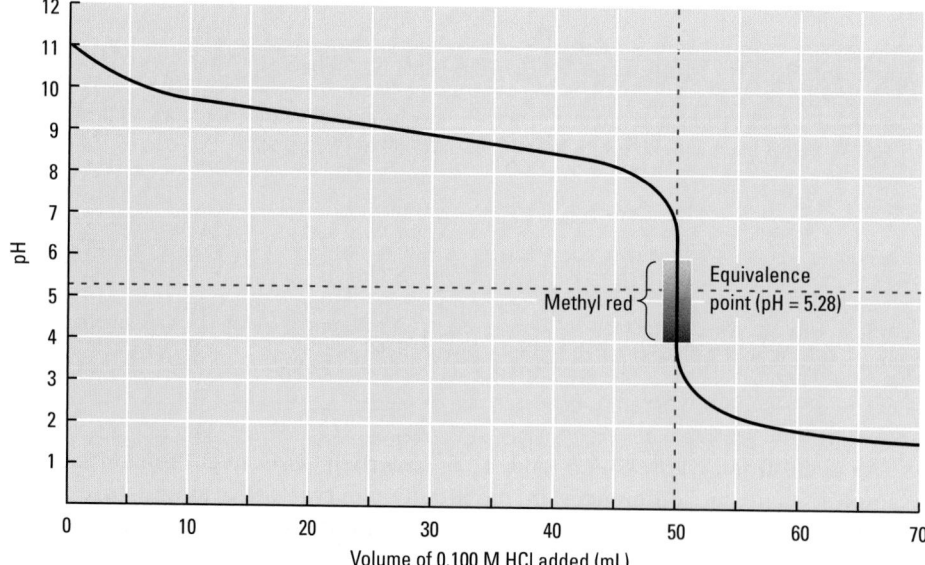

Titration of a Polyprotic Acid with Base

When titrated, polyprotic acids—those with more than one ionizable hydrogen—react stepwise with bases. If the K_a values of the ionizable forms of the acid are sufficiently different, the titration curve has an equivalence point for each of the hydrogens removed from the acid molecule by titration. For example, maleic acid, HOOC—CH=CH—COOH, is a diprotic acid with two ionizable hydrogens, one from each —COOH group. Its titration with NaOH occurs in two steps:

HOOC—CH=CH—COOH(aq) + OH$^-$(aq) $\rightleftharpoons$

$$\text{HOOC—CH=CH—COO}^-\text{(aq)} + \text{H}_2\text{O}(\ell)$$

HOOC—CH=CH—COO$^-$(aq) + OH$^-$(aq) $\rightleftharpoons$

$$^-\text{OOC—CH=CH—COO}^-\text{(aq)} + \text{H}_2\text{O}(\ell)$$

As shown in Figure 17.9, the two equivalence points are at pH = 4.1 and pH = 9.4.

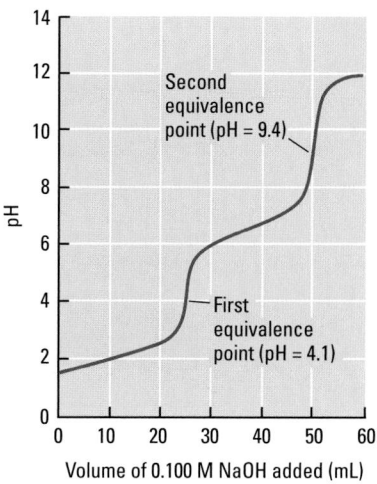

Figure 17.9 **Titration of a polyprotic acid.** The curve is for the titration of 25.00 mL of 0.100 M maleic acid with 0.100 M NaOH.

17.3 ACID RAIN

The term *acid rain* was first used in 1872 by Robert Angus Smith, an English chemist and climatologist. In his book *Air and Rain* he used the term to describe the acidic precipitation that fell on Manchester, England, at the start of the Industrial Revolution. Although neutral water has a pH of 7, rainwater becomes acidified naturally from dissolved carbon dioxide, a normal component of the atmosphere. The carbon dioxide reacts reversibly with water to form a solution of carbonic acid, a weak acid, which ionizes into hydronium and hydrogen carbonate ions.

$$2\,\text{H}_2\text{O}(\ell) + \text{CO}_2(g) \rightleftharpoons \text{H}_2\text{CO}_3 + \text{H}_2\text{O} \rightleftharpoons \text{H}_3\text{O}^+\text{(aq)} + \text{HCO}_3^-\text{(aq)}$$

The pH of water in equilibrium with CO_2 from the air is about 5.6, which is the pH of natural, unpolluted rainwater. Any precipitation with a pH below 5.6 is considered to be **acid rain.**

Nitrogen dioxide (NO_2) from industrial as well as natural sources reacts with water in the atmosphere to produce acids; NO_2 produces nitric acid (HNO_3) and nitrous acid (HNO_2).

$$2\,\text{NO}_2(g) + \text{H}_2\text{O}(\ell) \longrightarrow \text{HNO}_3\text{(aq)} + \text{HNO}_2\text{(aq)}$$

Atmospheric sulfur dioxide (SO_2), produced from burning fossil fuels, reacts with water to produce sulfurous acid (H_2SO_3) and, if oxygen is present, sulfuric acid (H_2SO_4).

$$\text{SO}_2(g) + \text{H}_2\text{O}(\ell) \longrightarrow \text{H}_2\text{SO}_3\text{(aq)}$$

$$2\,\text{SO}_2(g) + \text{O}_2(g) \longrightarrow 2\,\text{SO}_3(g)$$

$$\text{SO}_3(g) + \text{H}_2\text{O}(\ell) \longrightarrow \text{H}_2\text{SO}_4\text{(aq)}$$

The resulting acidic water droplets precipitate as rain or snow with a pH less than 5.6. Ice core samples taken in Greenland and dating back to 1900 contain sulfate (SO_4^{2-}) and nitrate (NO_3^-) ions. This indicates that acid rain has been commonplace, at least from 1900 onward.

Acid rain is a problem today due to the large amounts of these acidic oxides being put into the atmosphere by human activities every year (Figure 17.10). When such precipitation falls on areas without naturally occurring bases such as limestone and other carbonate minerals to offset the acidity, serious environmental damage can occur. The average annual pH of precipitation falling on

Although the term "acid rain" is commonly used, the more accurate term is "acid deposition," which takes into account acidic snow, sleet, rain, and fog.

Approximately 200 million metric tons of SO_2 and 150 million metric tons of nitrogen oxides are emitted annually into the atmosphere from human activities, largely the burning of fossil fuels.

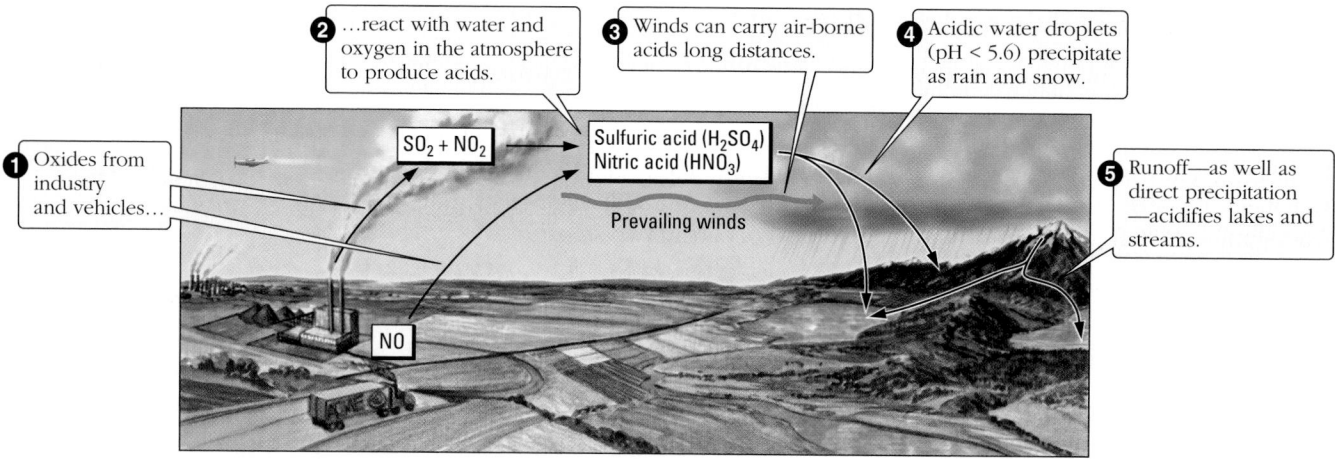

Figure 17.10 How acid deposition occurs.

much of the northeastern United States and northeastern Europe is between 4 and 4.5 (Figure 17.11). Rainfall in those areas have had rain with pH values as low as 1.5. To further complicate matters, acid rain is an international problem because precipitation carried by winds does not observe international borders. Canadian residents are offended by the fact that much of the acid precipitation falling on Canadian cities and forests results from acidic oxides produced in the United States.

17.4 SOLUBILITY EQUILIBRIA AND THE SOLUBILITY PRODUCT CONSTANT, K_{sp}

CD-ROM Screen 19.4:
Solubility Product Constant

Many ionic compounds are only modestly or slightly soluble, producing *saturated* solutions of 0.001 M or less, far less in some cases. Consider the case of a saturated aqueous solution of silver bromide, AgBr, a light-sensitive ionic compound used in photographic film. Sufficient AgBr is added to water so that some of it dissolves to form a saturated solution, in which undissolved AgBr is present. In solution, the

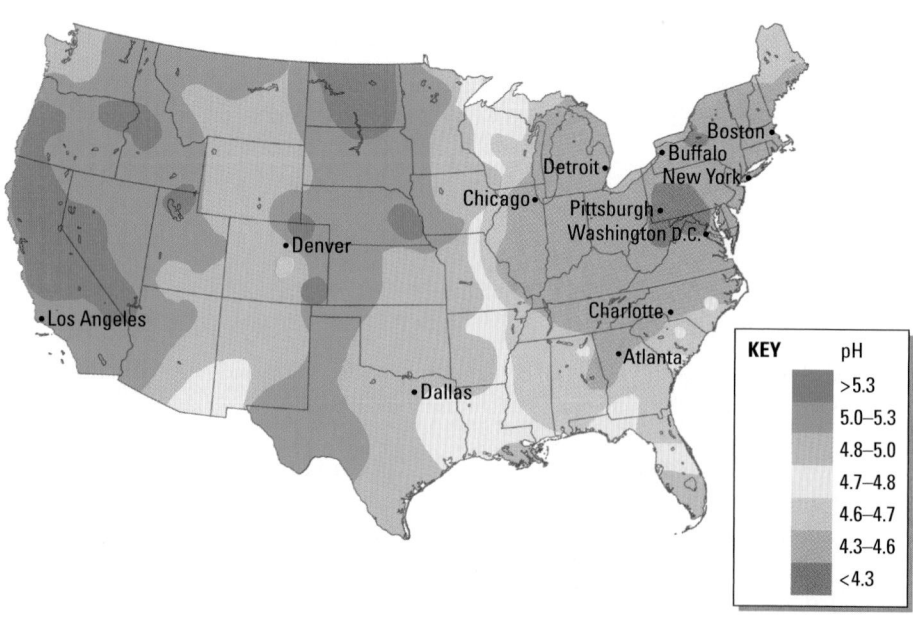

Figure 17.11 pH of precipitation within the United States.

dissolved AgBr forms aqueous Ag^+ and Br^- ions that are in equilibrium with the undissolved solid AgBr.

$$AgBr(s) \rightleftharpoons Ag^+(aq) + Br^-(aq)$$

This balanced chemical equation represents a solubility equilibrium. As with other equilibria *(⇐ p. 638)*, we can derive an equilibrium constant from the chemical equation. In this case, the equilibrium constant is called the **solubility product constant, K_{sp},** a value whose magnitude indicates the extent to which the solid solute dissolves to give ions in solution. To evaluate the equilibrium constant, we first must write a **solubility product constant expression.** For the chemical equation given above, the solubility product constant expression is

$$K_{sp} = [Ag^+][Br^-]$$

In general, the equilibrium expression for dissolving a slightly soluble salt with the general formula $A_x B_y$ is

$$A_x B_y(s) \rightleftharpoons x\, A^{n+}(aq) + y\, B^{m-}(aq)$$

This results in the general K_{sp} expression

$$K_{sp} = [A^{n+}]^x [B^{m-}]^y$$

Notice that

- The chemical equation related to the solubility product constant expression is written for a solid solute compound as a reactant and its aqueous ions as the products.
- The concentration of the pure solid solute reactant is omitted from the K_{sp} expression.
- The K_{sp} equals the product of the molar concentrations of the cation and the anion, each raised to the power given by the coefficient in the balanced chemical equation representing the solubility equilibrium.

> The solubility product constant is commonly just called the solubility product.

> A solubility product expression has the same general form that other equilibrium constant expressions have, except there is no denominator in the K_{sp} expression.

Problem-Solving Example 17.8 Writing K_{sp} Expressions

Write the K_{sp} expressions for the following slightly soluble salts:

(a) $BaCrO_4$ (b) $Mn(OH)_2$ (c) Ag_2SO_4

Answer

(a) $K_{sp} = [Ba^{2+}][CrO_4^{2-}]$ (b) $K_{sp} = [Mn^{2+}][OH^-]^2$ (c) $K_{sp} = [Ag^+]^2[SO_4^{2-}]$

Explanation

(a) The equilibrium reaction for the solubility of $BaCrO_4$ in water is

$$BaCrO_4(s) \rightleftharpoons Ba^{2+}(aq) + CrO_4^{2-}(aq)$$

Since the K_{sp} expression contains only the soluble ions, raised to the balancing coefficients in the equilibrium reaction, $K_{sp} = [Ba^{2+}][CrO_4^{2-}]$.

(b) The equilibrium reaction for the solubility of $Mn(OH)_2$ in water is

$$Mn(OH)_2(s) \rightleftharpoons Mn^{2+}(aq) + 2\, OH^-(aq)$$

In this example, for every Mn^{2+} ion there are two OH^- ions produced, so their concentration is raised to the power of 2 in the K_{sp} expression: $K_{sp} = [Mn^{2+}][OH^-]^2$.

(c) The equilibrium reaction for the solubility of Ag_2SO_4 in water is

$$Ag_2SO_4(s) \rightleftharpoons 2\, Ag^+(aq) + SO_4^{2-}(aq)$$

Since two Ag^+ ions are produced for every SO_4^{2-} ion, the K_{sp} expression is written as $K_{sp} = [Ag^+]^2[SO_4^{2-}]$.

Problem-Solving Practice 17.8

Write the K_{sp} expressions for the slightly soluble salts CuBr, HgI_2, and $SrSO_4$.

CD-ROM Screen 19.6: Estimating Salt Solubility

TABLE 17.2	K_{sp} Values for Some Slightly Soluble Salts
Compound	**K_{sp} at 25 °C**
AgBr	3.3×10^{-13}
AuBr	5.0×10^{-17}
AuBr$_3$	4.0×10^{-36}
CuBr	5.3×10^{-9}
Hg$_2$Br$_2$*	1.3×10^{-22}
PbBr$_2$	6.3×10^{-6}
AgCl	1.8×10^{-10}
AuCl	2.0×10^{-23}
AuCl$_3$	3.2×10^{-25}
CuCl	1.9×10^{-7}
Hg$_2$Cl$_2$*	1.1×10^{-18}
PbCl$_2$	1.7×10^{-5}
AgI	1.5×10^{-13}
AuI	1.6×10^{-13}
AuI$_3$	1.0×10^{-46}
CuI	5.1×10^{-12}
Hg$_2$I$_2$*	4.5×10^{-29}
HgI$_2$	4.0×10^{-29}
PbI$_2$	8.7×10^{-9}
Ag$_2$SO$_4$	1.7×10^{-5}
BaSO$_4$	1.1×10^{-10}
PbSO$_4$	1.8×10^{-8}
Hg$_2$SO$_4$*	6.8×10^{-7}
SrSO$_4$	2.8×10^{-7}

*These compounds contain the diatomic ion Hg$_2^{2+}$.

Solubility and K_{sp}

The solubility of a sparingly soluble solute and its solubility product constant, K_{sp}, are not the same thing, but they are related. The solubility is the amount of solute per unit volume of solution that dissolves to form a saturated solution. On the other hand, the solubility product constant is the equilibrium constant for the chemical equilibrium that exists between an ionic solute and its ions in a saturated solution. If the equilibrium concentrations of the ions are known, they can be used to calculate the K_{sp} value for the solute. For example, in a saturated AgCl solution at 10 °C, the molar concentrations of Ag$^+$ and Cl$^-$ each are experimentally determined to be 6.3×10^{-6} M. This means that K_{sp} at 10 °C is

$$K_{sp} = [\text{Ag}^+][\text{Cl}^-] = [6.3 \times 10^{-6}][6.3 \times 10^{-6}] = 4.0 \times 10^{-11}$$

The K_{sp} values for selected ionic compounds are listed in Table 17.2. A more extensive listing is in Appendix H.

Problem-Solving Example 17.9 Solubility and K_{sp}

The K_{sp} of BaSO$_4$ is 1.1×10^{-10} at 25 °C. Calculate the solubility of BaSO$_4$, expressing the result in moles per liter.

Answer 1.0×10^{-5} M

Explanation The solubility product constant expression for barium sulfate is derived from the chemical equation

$$\text{BaSO}_4(s) \rightleftharpoons \text{Ba}^{2+}(aq) + \text{SO}_4^{2-}(aq) \qquad K_{sp} = [\text{Ba}^{2+}][\text{SO}_4^{2-}] = 1.1 \times 10^{-10}$$

The ionization of solid BaSO$_4$ forms Ba^{2+} ions and SO$_4^{2-}$ ions in equal amounts. Therefore, if we let S equal the solubility of BaSO$_4$, then at equilibrium the concentration of Ba^{2+} and SO$_4^{2-}$ ions will each be S.

$$1.1 \times 10^{-10} = (S)(S) = S^2$$
$$S = \sqrt{1.1 \times 10^{-10}} = 1.0 \times 10^{-5} \text{ M}$$

Consequently, the aqueous solubility of BaSO$_4$ at 25 °C is 1.0×10^{-5} M.

Problem-Solving Practice 17.9

The K_{sp} of AgBr at 100 °C is 5×10^{-10}. Calculate the solubility of AgBr at that temperature in moles per liter.

A note of caution is in order here. Although it might seem perfectly straightforward to calculate the solubility of an ionic compound from its K_{sp}—that is, the calculation just completed in Problem-Solving Example 17.9—or to calculate the K_{sp} from the solubility, doing so will often lead to incorrect answers. This approach is too simplified and overlooks several complicating factors. One is that ionic solids such as PbCl$_2$ dissociate stepwise, so that PbCl$^+$ ions as well as Pb^{2+} and Cl$^-$ are present in PbCl$_2$ solution. Also, ion pairs such as PbCl$^+$Cl$^-$ can exist, reducing the concentrations of unassociated Pb^{2+} and Cl$^-$. The solubilities of some solutes, such as metal hydroxides, depend on the acidity or alkalinity of the solution. And solutes containing anions such as CO$_3^{2-}$ and PO$_4^{3-}$ that react with water are more soluble than predicted by their K_{sp} values. Solubilities calculated from K_{sp} values and K_{sp} values calculated from solubilities best agree with the experimentally measured solubilities of compounds with +1 and −1 charged ions and ions that do not react with water.

Problem-Solving Example 17.10   Solubility and K_{sp}

In a saturated CaF_2 solution at 25 °C, the calcium concentration is analyzed to be 9.1 mg/L. Use this value to calculate the K_{sp} of CaF_2 assuming that the solute dissociates completely into Ca^{2+} and F^- ions and that neither ion reacts with water.

Answer $K_{sp} = 4.9 \times 10^{-11}$

Explanation   The chemical equilibrium in the solution is

$$CaF_2(s) \rightleftharpoons Ca^{2+}(aq) + 2\,F^-(aq)$$

and the solubility product expression is $K_{sp} = [Ca^{2+}][F^-]^2$. To calculate the value of K_{sp}, the *molar* concentrations of Ca^{2+} and F^- must first be determined. The first step is to convert the calcium concentration from milligrams per liter to molarity, mol/L.

You can review molarity calculations in Section 5.7.

$$\left(\frac{9.1\text{ mg Ca}^{2+}}{1\text{ L}}\right)\left(\frac{1\text{ g Ca}^{2+}}{10^3\text{ mg Ca}^{2+}}\right)\left(\frac{1\text{ mol Ca}^{2+}}{40.0\text{ g Ca}^{2+}}\right) = 2.3 \times 10^{-4}\text{ M Ca}^{2+}$$

From the balanced chemical equation we see that when dissociation of the solute occurs, two moles of fluoride ions are produced for each mole of calcium ions. Therefore, the fluoride ion concentration is 4.6×10^{-4} M, twice that of calcium ion. The K_{sp} value can be calculated directly from these concentrations and the solubility product equilibrium constant expression.

$$K_{sp} = [Ca^{2+}][F^-]^2 = (2.3 \times 10^{-4})(4.6 \times 10^{-4})^2 = 4.9 \times 10^{-11}$$

Note that $[F^-]$ is *twice* that of $[Ca^{2+}]$; it is also *squared* in the K_{sp} expression.

✔   This value is in close agreement with the value listed in Appendix H.

Problem-Solving Practice 17.10

A saturated solution of silver oxalate, $Ag_2C_2O_4$, contains 6.9×10^{-5} M $C_2O_4^{2-}$ at 25 °C. Calculate the K_{sp} of silver oxalate at that temperature, assuming that the ions do not react with water.

   Exercise 17.10   Solubility and Le Chatelier's Principle

At 25 °C, 0.014 g of calcium carbonate dissolves in 100 mL of water. Two equilibria are present in this solution.

(a) $CaCO_3(s) \rightleftharpoons Ca^{2+}(aq) + CO_3^{2-}(aq)$
(b) $CO_3^{2-}(aq) + H_2O(\ell) \rightleftharpoons HCO_3^-(aq) + OH^-(aq)$

Suppose (b) occurs to an appreciable extent. Use Le Chatelier's principle (⬅ *p. 655, Section 14.6*) to predict how the extent of (b) will affect the solubility of $CaCO_3$.

17.5 Factors Affecting Solubility

The aqueous solubility of ionic compounds is affected by a number of factors, some of which have already been discussed—temperature (⬅ *p. 692, Section 15.4*), the formation of ion pairs and competing equilibria. In this section we will consider four other factors affecting the aqueous solubility of ionic compounds: the effect of acids and pH, the presence of common ions, the formation of complex ions, and amphoterism.

pH and Dissolving Slightly Soluble Salts Using Acids

Many salts are only slightly soluble in water (⬅ *p. 164*). An insoluble salt can be dissolved by acid if one or both of its ions are moderately basic. Consider calcium carbonate, $CaCO_3$, which is found in minerals such as limestone and marble. $CaCO_3$ is not very soluble in pure water.

CD-ROM Screen 19.11:
Solubility and pH

$$\text{(a) } CaCO_3(s) \rightleftharpoons Ca^{2+}(aq) + CO_3^{2-}(aq)$$

Since the solubility of calcium carbonate is so low, the equilibrium concentrations of Ca^{2+} and CO_3^{2-} must also be small. However, if acid is added, calcium carbonate will dissolve and CO_2 will be released from the solution. Adding acid adds hydronium ions, which react with carbonate and bicarbonate ions.

$$\text{(b)} \quad CO_3^{2-}(aq) + H_3O^+(aq) \longrightarrow HCO_3^-(aq) + H_2O(\ell)$$

$$\text{(c)} \quad HCO_3^-(aq) + H_3O^+(aq) \longrightarrow H_2CO_3(aq) + H_2O(\ell)$$

Reaction (b) is the reaction of a fairly strong base, CO_3^{2-}, with a strong acid, and so nearly all of the carbonate is converted to hydrogen carbonate ion. Reaction (c) produces a product, carbonic acid, that is unstable. It breaks down to CO_2 gas and water in a very product-favored reaction.

$$\text{(d)} \quad H_2CO_3(aq) \longrightarrow CO_2(g) + H_2O(\ell) \qquad\qquad K \approx 10^5$$

Reactions (a) through (d) are linked through carbonate, bicarbonate, and carbonic acid. As CO_2 gas escapes from the solution, the H_2CO_3 concentration decreases. This shifts Reaction (c) to the right, which decreases the concentration of HCO_3^-. This in turn shifts Reaction (b) to the right, decreasing the concentration of CO_3^{2-} to an even lower value. To oppose this decrease in carbonate ion concentration, Reaction (a) shifts to the right, and more $CaCO_3(s)$ dissolves. Because the acidity of the solution determines the positions of equilibria (b) and (c), small changes in pH can cause limestone and marble to dissolve or precipitate (⇐ *p. 657*). Acid rain (Section 17.3) can dissolve a marble statue as well as underground limestone deposits, which causes massive cave formations. Impressive stalactite and stalagmite formations in caves result from such changes. In addition, limestone has been precipitated as layers of sedimentary rock on the ocean floor where there has been a slight increase in the pH of sea water.

In general, ***insoluble salts containing anions that are Brønsted-Lowry bases dissolve in solutions of low pH.*** This rule covers carbonates, sulfides (which produce $H_2S(g)$), phosphates, and other anions listed as bases in Table 16.5 (⇐ *p. 761*). The principal exceptions to this rule are a few sulfides, such as HgS, CuS, and CdS, that have extremely low solubilities and therefore do not dissolve even when the pH is extremely low.

In contrast, an insoluble salt such as AgCl, which contains the conjugate base of a strong acid, is not soluble in strongly acidic solution, because Cl^- is a very weak base and so does not react with H_3O^+.

Stalactites and stalagmites. Calcium carbonate precipitates from solution in caves to form stalactites and stalagmites.

Solubility and the Common Ion Effect

CD-ROM Screen 19.8: The Common Ion Effect

It is often desirable to remove a particular ion from solution by forming a precipitate of one of its insoluble compounds. For example, barium ions readily absorb x-rays and so are quite effective in making the intestinal tract visible when x-ray photographs are taken. But barium ions are poisonous and must not be allowed to dissolve in body fluids. The insoluble compound barium sulfate can be used as an x-ray absorber, but both physician and patient want to be certain that no harmful amounts of barium ions will be in solution.

The solubility of $BaSO_4$ in water at 25 °C is 1.0×10^{-5} mol/L, which means that the concentration of Ba^{2+} ions is 1.0×10^{-5} M.

$$BaSO_4(s) \rightleftharpoons Ba^{2+}(aq) + SO_4^{2-}(aq)$$

The concentration of aqueous Ba^{2+} ions can be reduced by adding a soluble sulfate salt, such as Na_2SO_4. The solubility of $BaSO_4$ decreases because of the increased concentration of SO_4^{2-} ions. The sulfate ion from Na_2SO_4 is called a "common ion" because it is common to both substances dissolved in the solution — barium sulfate and sodium sulfate. The common ion displaces the equilibrium to the left by what is called the **common ion effect.** *The presence of a second solute that provides a common ion lowers the solubility of an ionic compound.*

The common ion effect can be interpreted by using Le Chatelier's principle *(⬅ p. 655, Section 14.6)*. Adding sodium sulfate to the barium sulfate solution causes a stress on the equilibrium, which shifts to offset the added sulfate ions. To use up some of them, the equilibrium shifts to the left, removing an equal concentration of Ba^{2+} ions from solution as well forming solid $BaSO_4$. The outcome is that less $BaSO_4$ is dissolved; the salt's solubility is lower in the presence of the common ion.

$$BaSO_4(s) \rightleftharpoons Ba^{2+}(aq)Na_2SO_4(aq) + SO_4^{2-}(aq)$$

Solubility decreases; additional solid forms

The common ion effect. The tube at the left contains a saturated solution of silver acetate, $AgCH_3CO_2$. When 1 M $AgNO_3$ is added to the tube, the equilibrium

$$AgCH_3CO_2(s) \rightleftharpoons Ag^+(aq) + CH_3CO_2^-(aq)$$

shifts to the left, as evidenced by the tube at the right, where more silver acetate has formed.

 Exercise 17.11   Common Ion Effect

Consider 0.0010 M solutions of these sparingly soluble solutes in equilibrum with their ions.

$$BaSO_4(s) \rightleftharpoons Ba^{2+}(aq) + SO_4^{2-}(aq)$$
$$AgI(s) \rightleftharpoons Ag^+(aq) + I^-(aq)$$
$$PbI_2(s) \rightleftharpoons Pb^{2+}(aq) + 2\,I^-(aq)$$

(a) Using Le Chatelier's principle, explain what would be likely to happen if a saturated solution of sodium iodide were added to the last two solutions.
(b) What would be likely to occur if a saturated solution of potassium sulfate were added to the $BaSO_4$ solution?

Problem-Solving Example 17.11   The Common Ion Effect

The solubility of AgCl in pure water is 1.3×10^{-5} mol/L at 25 °C. If you put some solid AgCl into a solution that is 0.55 M in NaCl, what mass of AgCl will dissolve per liter of this solution? The K_{sp} of AgCl is 1.8×10^{-10} at 25 °C.

Answer   About 4.7×10^{-8} g AgCl/L

Explanation   We predict that because of the common ion effect the solubility of AgCl, S, will be smaller than it would be in pure water. The solubility of AgCl, S, equals the silver ion concentration $[Ag^+]$ at equilibrium.

$$AgCl(s) \rightleftharpoons Ag^+(aq) + Cl^-(aq)$$

If the AgCl(s) were dissolved in *pure* water, the $[Ag^+]$ and the $[Cl^-]$ would both be equal because there is no other source of chloride ion. Thus, S would be equal to $[Ag^+] = [Cl^-]$. In water containing AgCl and NaCl, both are sources of the common ion Cl^-. Thus, the value of S is equal only to $[Ag^+]$ and not to $[Cl^-]$ because Cl^- comes from two sources.

The following table shows the concentrations of Ag^+ and Cl^- when equilibrium is attained in the presence of extra Cl^-.

	$AgCl(s) \rightleftharpoons Ag^+(aq) + Cl^-(aq)$	
Initial concentration (mol/L)	0	0.55
Change as reaction occurs (mol/L)	+S	+S
Equilibrium concentration (mol/L)	S	S + 0.55

The total chloride ion concentration at equilibrium is the amount from AgCl (equals S) *plus* what was already there (0.55 M) from the NaCl. Because NaCl is a soluble salt, far more Cl^- comes from NaCl than from the AgCl.

Using the equilibrium concentrations from the table gives

$$K_{sp} = 1.8 \times 10^{-10} = [Ag^+][Cl^-] = (S)(S + 0.55)$$

The easiest approach to solve an equation like this is to approximate that S is *very* small compared to 0.55; that is, the answer will be approximately the same if we assume that $(S + 0.55) \approx 0.55$. Such an assumption is very reasonable because we know that the solubility of AgCl equals only 1.3×10^{-5} mol/L *without* the Cl^- added from NaCl. When NaCl is added, it will further decrease the solubility of AgCl due to the presence of the common ion Cl^-. Therefore,

$$(S)(S + 0.55) \approx (S)(0.55) = K_{sp}$$

or

$$K_{sp} = (S)(0.55) = 1.8 \times 10^{-10}$$

Solving for S, we get

$$S = \frac{1.8 \times 10^{-10}}{0.55} = 3.3 \times 10^{-10} \text{ M} = [Ag^+]$$

Therefore, the $[Ag^+]$, which is the same as S, is approximately 3.3×10^{-10} mol/L.

Using the molar mass for AgCl, 143.4 g/mol, the solubility, S, of AgCl in 0.55 M NaCl is

$$S = \left(\frac{3.3 \times 10^{-10} \text{ mol AgCl}}{1 \text{ L}} \right) \left(\frac{143.3 \text{ g AgCl}}{1 \text{ mol AgCl}} \right) = 4.7 \times 10^{-8} \text{ g AgCl/L}$$

As predicted by Le Chatelier's principle, the solubility of AgCl in the presence of Cl^-, added from another source is clearly less (3.3×10^{-10} M) than in pure water (1.3×10^{-5} M).

✔ As a final step, let us check the approximation we made. To do this, we substitute the approximate value of S into the exact expression $K_{sp} = (S)(S + 0.55)$. Then, if the product $(S)(S + 0.55)$ is the same as the given value of K_{sp}, the approximation is valid.

$$K_{sp} = (S)(S + 0.55) = (3.3 \times 10^{-10})(3.3 \times 10^{-10} + 0.55) \approx 1.8 \times 10^{-10}$$

A more accurate solution to this problem can be obtained by solving for S using the quadratic equation described in Appendix A.7 and used in Problem-Solving Example 14.7 Sec. 14.5 (⬅ *p. 654*). When the quadratic equation is used, its answer, to two significant figures, is the same as our approximation.

> **Problem-Solving Practice 17.11**
>
> Calculate the solubility of $PbCl_2$ at 25 °C in a solution that is 0.50 M in NaCl.

Generally, you can neglect S in the denominator if doing so does not create an error of more than 5%

Complex Ion Formation

CD-ROM Screen 19.12: Complex Ion Formation and Solubility

As pointed out in Section 16.10 (⬅ *p. 766*), all metal cations are potential Lewis acids because they can accept an electron pair donated by a Lewis base. The combination of a metal ion and a Lewis base forms a **complex ion,** such as $Ag(NH_3)_2^+$ in which a silver ion (Lewis acid) is bonded to ammonia molecules (Lewis base). Complex ion formation can dissolve an insoluble metal salt, such as AgBr. The solubility of AgBr is very low in water, 0.135 mg/L, but AgBr dissolves readily in a sodium thiosulfate ($Na_2S_2O_3$) solution due to the formation of the $Ag(S_2O_3)_2^{3-}$ complex ion (Figure 17.12).

$$AgBr(s) + 2 S_2O_3^{2-}(aq) \rightleftharpoons Ag(S_2O_3)_2^{3-}(aq) + Br^-(aq)$$

The dissolving of AgBr in this way can be considered as the sum of two reactions—the solubility equilibrium of aqueous AgBr and the formation of the complex ion.

Figure 17.12 Sodium thiosulfate dissolves silver bromide. (a) Silver bromide (white solid) is insoluble in water. (b) When aqueous sodium thiosulfate is added, the AgBr dissolves.

$$AgBr(s) \rightleftharpoons Ag^+(aq) + Br^-(aq)$$

$$Ag^+(aq) + 2\,S_2O_3^{2-}(aq) \rightleftharpoons Ag(S_2O_3)_2^{3-}(aq)$$

Net reaction: $AgBr(s) + 2\,S_2O_3^{2-}(aq) \longrightarrow Ag(S_2O_3)_2^{3-}(aq) + Br^-(aq)$

This reaction is commercially important for removing unreacted AgBr from photographic film ($\Leftarrow$ *p. 197*).

The extent to which complex ion formation occurs can be evaluated from the magnitude of the equilibrium constant for the formation of the complex ion, K_f, called the **formation constant.** For example, the formation constant for $Ag(S_2O_3)_2^{3-}$ is 2×10^{13}.

$$Ag^+(aq) + 2\,S_2O_3^{2-}(aq) \rightleftharpoons Ag(S_2O_3)_2^{3-}(aq)$$

$$K_f = \frac{[Ag(S_2O_3)_2^{3-}]}{[Ag^+][S_2O_3^{2-}]^2} = 2 \times 10^{13}$$

Formation constants for some metal complex ions are given in Table 17.3. In general, metal salts that are insoluble in water are brought into solution by complex ion formation with Lewis bases such as $S_2O_3^{2-}$, NH_3, OH^-, and CN^-. The formation and structure of complex ions, which are very important in biochemistry and metallurgy, are considered in more detail in Chapter 22.

Problem-Solving Example 17.12   Solubility and Complex Ion Formation

The K_{sp} of AgBr is 3.3×10^{-13}. The K_f of $Ag(S_2O_3)_2^{3-}$ is 2×10^{13}. Use these data to show that dissolving AgBr by $Ag(S_2O_3)_2^{3-}$ complex ion formation is a product-favored process.

Answer   The net equilibrium constant is 7, which indicates that dissolving AgBr by complex ion formation is favored.

Explanation   The magnitude of an equilibrium constant indicates whether a reaction is product-favored ($\Leftarrow$ *p. 647, Section 14.4*). The net reaction for dissolving AgBr by $Ag(S_2O_3)_2^{3-}$ complex ion formation is the sum of the K_{sp} and K_f equations.

$$AgBr(s) \rightleftharpoons Ag^+(aq) + Br^-(aq) \qquad K_{sp} = 3.3 \times 10^{-13}$$

$$Ag^+(aq) + 2\,S_2O_3^{2-}(aq) \rightleftharpoons Ag(S_2O_3)_2^{3-}(aq) \quad K_f\ Ag(S_2O_3)_2^{3-} = 2 \times 10^{13}$$

Net reaction: $AgBr(s) + 2\,S_2O_3^{2-}(aq) \longrightarrow Ag(S_2O_3)_2^{3-}(aq) + Br^-(aq)$

Therefore, the equilibrium constant for the net reaction is the product of K_{sp} and K_f: $K_{net} = K_{sp} \times K_f = (3.3 \times 10^{-13})(2 \times 10^{13}) = 7$. Because K_{net} is greater than 1, the net reaction is product-favored, and AgBr is much more soluble in a $Na_2S_2O_3$ solution than it is in water.

TABLE 17.3	Formation Constants for Some Complex Ions in Aqueous Solution	
Formation equilibrium		K_f
$Ag^+ + 2\,Br^- \rightleftharpoons [AgBr_2]^-$		1.3×10^7
$Ag^+ + 2\,Cl^- \rightleftharpoons [AgCl_2]^-$		2.5×10^5
$Ag^+ + 2\,CN^- \rightleftharpoons [Ag(CN)_2]^-$		5.6×10^{18}
$Ag^+ + 2\,S_2O_3^{2-} \rightleftharpoons [Ag(S_2O_3)_2]^{3-}$		2.0×10^{13}
$Ag^+ + 2\,NH_3 \rightleftharpoons [Ag(NH_3)_2]^+$		1.6×10^7
$Al^{3+} + 6\,F^- \rightleftharpoons [AlF_6]^{3-}$		5.0×10^{23}
$Al^{3+} + 4\,OH^- \rightleftharpoons [Al(OH)_4]^-$		7.7×10^{33}
$Au^+ + 2\,CN^- \rightleftharpoons [Au(CN)_2]^-$		2.0×10^{38}
$Cd^{2+} + 4\,CN^- \rightleftharpoons [Cd(CN)_4]^{2-}$		1.3×10^{17}
$Cd^{2+} + 4\,Cl^- \rightleftharpoons [CdCl_4]^{2-}$		1.0×10^4
$Cd^{2+} + 4\,NH_3 \rightleftharpoons [Cd(NH_3)_4]^{2+}$		1.0×10^7
$Co^{2+} + 6\,NH_3 \rightleftharpoons [Co(NH_3)_6]^{2+}$		7.7×10^4
$Cu^+ + 2\,CN^- \rightleftharpoons [Cu(CN)_2]^-$		1.0×10^{16}
$Cu^+ + 2\,Cl^- \rightleftharpoons [CuCl_2]^-$		1.0×10^5
$Cu^{2+} + 4\,NH_3 \rightleftharpoons [Cu(NH_3)_4]^{2+}$		6.8×10^{12}
$Fe^{2+} + 6\,CN^- \rightleftharpoons [Fe(CN)_6]^{4-}$		7.7×10^{36}
$Hg^{2+} + 4\,Cl^- \rightleftharpoons [HgCl_4]^{2-}$		1.2×10^{15}
$Ni^{2+} + 4\,CN^- \rightleftharpoons [Ni(CN)_4]^{2-}$		1.0×10^{31}
$Ni^{2+} + 6\,NH_3 \rightleftharpoons [Ni(NH_3)_6]^{2+}$		5.6×10^8
$Zn^{2+} + 4\,OH^- \rightleftharpoons [Zn(OH)_4]^{2-}$		2.9×10^{15}
$Zn^{2+} + 4\,NH_3 \rightleftharpoons [Zn(NH_3)_4]^{2+}$		2.9×10^9

Problem-Solving Example 17.12

The K_{sp} of AgCl is 1.8×10^{-10}. The K_f of $Ag(CN)_2^-$ is 5.6×10^{18}. Use these data to show that AgCl will dissolve in aqueous NaCN.

Amphoterism

The majority of metal hydroxides are insoluble in water, but many dissolve in highly acidic or basic solutions. This is because these hydroxides are *amphoteric;* that is, they can react with both H_3O^+ ions and OH^- ions (⬅ *p. 768, Section 16.10*). Aluminum hydroxide, $Al(OH)_3$, is an example of an amphoteric hydroxide (Figure 17.13). When reacted with acid, $Al(OH)_3$ dissolves by acting as a base, donating OH^- ions.

$$Al(OH)_3(s) + 3\,H_3O^+(aq) \longrightarrow Al^{3+}(aq) + 6\,H_2O(\ell)$$

In highly basic solutions, $Al(OH)_3$ is dissolved through complex ion formation.

$$Al(OH)_3(s) + OH^-(aq) \longleftrightarrow Al(OH)_4^-(aq)$$

17.6 PRECIPITATION: WILL IT OCCUR?

CD-ROM Screen 19.7: Can a Precipitation Reaction Occur?

Earlier, when writing net ionic equations, you used the solubility rules to predict whether a precipitate forms when ions in two solutions are mixed (⬅ *p. 164*). Those rules apply to situations where the ions involved are at concentrations of

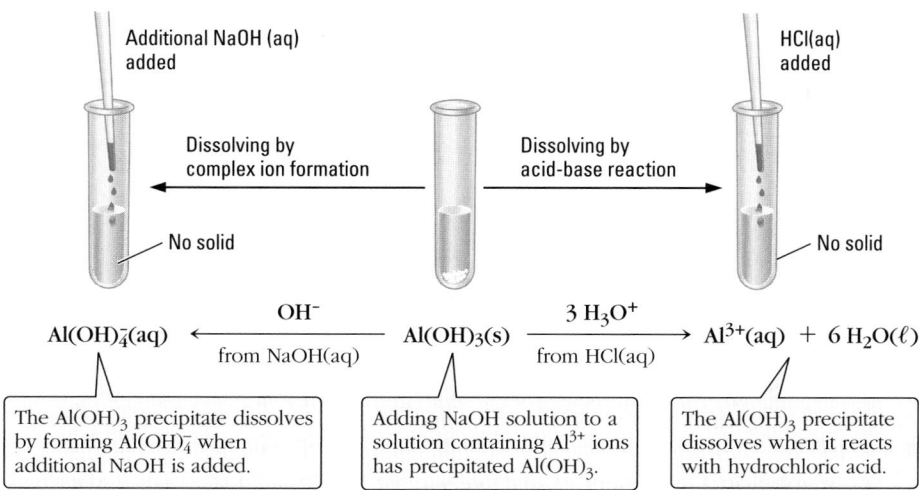

Figure 17.13 The amphoteric nature of Al(OH)$_3$.

$$Al(OH)_4^-(aq) \xleftarrow[\text{from NaOH(aq)}]{OH^-} Al(OH)_3(s) \xrightarrow[\text{from HCl(aq)}]{3\,H_3O^+} Al^{3+}(aq) + 6\,H_2O(\ell)$$

The Al(OH)$_3$ precipitate dissolves by forming Al(OH)$_4^-$ when additional NaOH is added.	Adding NaOH solution to a solution containing Al^{3+} ions has precipitated Al(OH)$_3$.	The Al(OH)$_3$ precipitate dissolves when it reacts with hydrochloric acid.

0.1 M or greater. In circumstances where the ion concentrations are considerably less than 0.1 M, precipitation may or may not occur when solutions of the ions are mixed. The result depends on the concentrations of the ions in the resulting solution and the K_{sp} value for any precipitate that might form.

For example, AgBr might precipitate when a water-soluble silver salt, such as AgNO$_3$, is added to an aqueous solution of a bromide salt, such as KBr. The net ionic equation for the reaction is

$$Ag^+(aq) + Br^-(aq) \rightleftharpoons AgBr(s)$$

To determine whether or not a precipitate will form, we compare a value called the **ion product,** Q, with the solubility product constant, K_{sp}. The Q expression has the same form as that for K_{sp}. For Q, however, the *original* concentrations are used, not those at equilibrium as in K_{sp}. For AgBr the two expressions are

$$Q = (\text{conc. of Ag}^+)(\text{conc. of Br}^-) \qquad K_{sp} = [Ag^+][Br^-]$$

When Q is compared with K_{sp}, three cases are possible (Figure 17.14).

Q is related to the reaction quotient introduced in Chapter 14.

The bracket notation, [ ], represents molarity at equilibrium.

1. **$Q < K_{sp}$ *The solution is unsaturated and no precipitate forms.*** In this case, the solution contains ions at a concentration lower than required for equilibrium with the solid. An equilibrium is not established between a solid solute and its ions because no solid solute is present; more solute can be added to the solution before precipitation occurs.
2. **$Q > K_{sp}$** The solution contains a higher concentration of ions than it can hold at equilibrium; that is, the solution is supersaturated. ***To reach equilibrium, a***

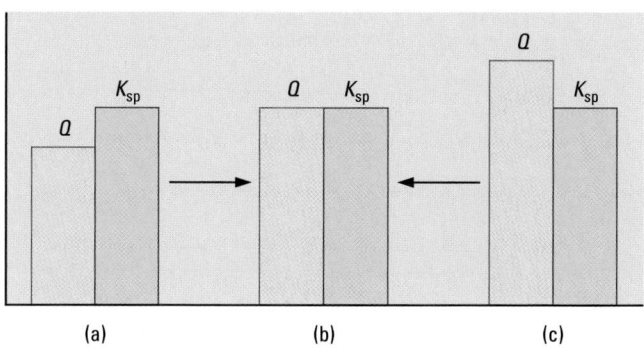

Figure 17.14 Predicting precipitation. (a) When $Q < K_{sp}$, the solution is unsaturated, and no precipitation occurs. (b) $Q = K_{sp}$; the solution is saturated and just at the point of precipitation. (c) $Q > K_{sp}$; the solution is supersaturated, and precipitation occurs until $Q = K_{sp}$.

precipitate forms, decreasing the concentration of ions until the ion product equals the K_{sp}.

3. $Q = K_{sp}$ *The solution is saturated with ions* and is at equilibrium and at the point of precipitation.

Consider the case of two solutions, each made by combining $Pb(NO_3)_2$ and Na_2SO_4 solutions. In each of the the two solutions, the products are $NaNO_3$ and $PbSO_4$. The solubility rules indicate that $NaNO_3$ is soluble and remains in solution as Na^+ and NO_3^- ions, whereas $PbSO_4$ is insoluble. Will a precipitate of $PbSO_4$ form in either or both solutions? We can determine this by using Q and K_{sp} for $PbSO_4$.

In one case (solution 1) when the solutions are mixed the initial concentrations of Pb^{2+} and SO_4^{2-} are each 1.0×10^{-4} M. In the other case (solution 2) these concentrations are each 2.0×10^{-4}, twice that of the first solution. K_{sp} $PbSO_4$ = 1.8×10^{-8}.

Q of solution 1 = (conc. of Pb^{2+})(conc. of SO_4^{2-}) = $(1.0 \times 10^{-4}$ M)$(1.0 \times 10^{-4}$ M) = 1.0×10^{-8}.

Q of solution 2 = (conc. of Pb^{2+})(conc. of SO_4^{2-}) = $(2.0 \times 10^{-4}$ M)$(2.0 \times 10^{-4}$ M) = 4.0×10^{-8}. Since Q of solution 1 is less than the K_{sp}, no precipitate will form. In contrast, Q of solution 2 exceeds the K_{sp}, and precipitation will occur.

Problem-Solving Example 17.13 Q, K_{sp}, and Precipitation

A chemistry student mixes 0.200 L of 4.5×10^{-3} M $AgNO_3$ with 0.100 L of 7.5×10^{-2} M $NaBrO_3$. The final volume is 0.300 L. Will a precipitate of $AgBrO_3$ form? The K_{sp} of $AgBrO_3 = 6.7 \times 10^{-5}$.

Answer Yes, $AgBrO_3$ precipitates.

Explanation For a precipitate to form, Q must be equal to or greater than the K_{sp}. The chemical equilibrium is

$$AgBrO_3(s) \rightleftharpoons Ag^+(aq) + BrO_3^-(aq)$$

The ion product expression is Q = (conc. of Ag^+)(conc. of BrO_3^-); the K_{sp} expression is $K_{sp} = [Ag^+][BrO_3^-]$. To determine whether a precipitate forms, substitute the original concentrations into the Q expression and compare the value with K_{sp}. In calculating the original concentrations, the total volume of 0.300 L must be taken into account. The number of moles of Ag^+ in 0.200 L of 4.5×10^{-3} M $AgNO_3$ is

$$(0.200 \text{ L}) \left(\frac{4.5 \times 10^{-3} \text{ mol Ag}^+}{1 \text{ L}} \right) = 9.0 \times 10^{-4} \text{ mol Ag}^+$$

The Ag^+ concentration in the 0.300-L mixture is

$$\frac{9.0 \times 10^{-4} \text{ mol Ag}^+}{0.300 \text{ L}} = 3.0 \times 10^{-3} \text{ M Ag}^+$$

Likewise for the moles of BrO_3^- and its concentration in the 0.300-L mixture:

$$(0.100 \text{ L}) \left(\frac{7.5 \times 10^{-2} \text{ mol BrO}_3^-}{1 \text{ L}} \right) = 7.5 \times 10^{-3} \text{ mol BrO}_3^-$$

$$\frac{7.5 \times 10^{-3} \text{ mol BrO}_3^-}{0.300 \text{ L}} = 2.5 \times 10^{-2} \text{ M BrO}_3^-$$

$$Q = (3.0 \times 10^{-3} \text{ M})(2.5 \times 10^{-2} \text{ M}) = 7.5 \times 10^{-5}$$

which is greater than the $K_{sp} = 6.7 \times 10^{-5}$. Precipitation occurs until the ion product equals K_{sp}.

Problem-Solving Practice 17.13

(a) Will $AgCl$ precipitate from a solution containing 1.0×10^{-5} M Ag^+ and 1.0×10^{-5} M Cl^-? K_{sp} $AgCl = 1.8 \times 10^{-10}$.

(b) An $AgCl$ precipitate forms from a solution that is 1.0×10^{-5} M Ag^+. What must be the minimum Cl^- concentration in this solution for precipitation to occur?

Kidney Stones—Common Ion Effect and Le Chatelier's Principle

Many ions circulate in our bloodstream, some combinations of which can precipitate to form kidney stones. Such stones can become large enough to be extremely painful and even life-threatening, requiring treatment by drugs, lasers, or surgical removal. Kidney stones generally consist of insoluble calcium and magnesium compounds such as calcium oxalate (CaC_2O_4), calcium phosphate $Ca_3(PO_4)_2$, magnesium ammonium phosphate ($MgNH_4PO_4$), or a mixture of these. For calcium oxalate kidney stones, the equilibrium $CaC_2O_4(s) \rightleftharpoons Ca^{2+}(aq) + C_2O_4^{2-}(aq)$ applies. High intake of foods rich in calcium or oxalate can cause a rise in the urinary concentration of either ion (or both) sufficient to shift the equilibrium to the left and $Q > K_{sp}$. The result is precipitation of calcium oxalate as a kidney stone. Thus, foods rich in Ca^{2+}, such as milk, ice cream, and cheese, or high in $C_2O_4^{2-}$, such as chocolate, spinach, celery, and black tea, can trigger the onset of a kidney stone through the common ion effect. Such foods are restricted in the diets of individuals prone to developing kidney stones. A high-sugar diet may also create kidney stones because excessive sugar promotes excretion of Ca^{2+} and Mg^{2+}, which increases the concentrations of these ions passing through the kidneys. This can cause kidney stone formation, such as through calcium phosphate precipitation:

$$3\,Ca^{2+}(aq) + 2\,PO_4^{3-}(aq) \longrightarrow Ca_3(PO_4)_2(s)$$

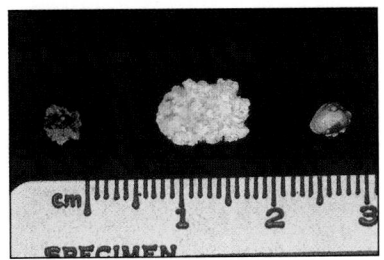

Kidney stones. These kidney stones were surgically removed from a patient.

Leaving chocolate out of a diet seems far more punishment than foregoing spinach.

Selective Precipitation of Ions

If their solubilities are sufficiently different, ionic compounds can be precipitated selectively from solution. The more soluble compound remains in solution as the less soluble one starts to precipitate. For example, silver chloride (AgCl) and silver chromate (Ag_2CrO_4) are only slightly soluble in water. Their solubilities differ enough, however, to precipitate one, leaving the other in solution.

 CD-ROM Screen 19.9: Using Solubility

Problem-Solving Example 17.14   Selective Precipitation

Consider a solution containing 0.020 M Cl^- and 0.010 M CrO_4^{2-} ions to which Ag^+ ions are added slowly. Which precipitate forms first—AgCl or Ag_2CrO_4? K_{sp} AgCl $= 1.8 \times 10^{-10}$; K_{sp} $Ag_2CrO_4 = 9 \times 10^{-12}$.

Answer   AgCl precipitates first.

Explanation   To answer this question we first find the minimum Ag^+ concentration required to precipitate each one, which is the molar concentration of ions that just barely exceeds the K_{sp}. To precipitate AgCl,

$$K_{sp}\ AgCl = [Ag^+][Cl^-] = 1.8 \times 10^{-10}$$

$$[Ag^+] = \frac{1.8 \times 10^{-10}}{[Cl^-]} = \frac{1.8 \times 10^{-10}}{2.0 \times 10^{-2}} = 9.0 \times 10^{-9}\ M$$

An Ag^+ concentration of slightly greater than 9.0×10^{-9} M will precipitate some AgCl from the solution. To precipitate Ag_2CrO_4,

$$K_{sp}\ Ag_2CrO_4 = [Ag^+]^2\,[CrO_4^-] = 9 \times 10^{-12}$$

$$[Ag^+]^2 = \frac{9 \times 10^{-12}}{[CrO_4^{2-}]} = \frac{9 \times 10^{-12}}{1.0 \times 10^{-2}} = 9 \times 10^{-10}\ M \qquad [Ag^+] = 3 \times 10^{-5}\ M$$

Silver chromate will precipitate when the chromate concentration slightly exceeds 3×10^{-5} M. Because a *much* smaller concentration of Ag^+ (9.0×10^{-9} M) is required to precipitate AgCl, it will precipitate before Ag_2CrO_4. In fact, the difference is so great that essentially all of the AgCl will precipitate before Ag_2CrO_4 precipitation begins.

Problem-Solving Practice 17.14

Hydrochoric acid is slowly added to a solution that is 0.10 M in Pb^{2+} and 0.01 M in Ag^+. Which precipitate occurs first, AgCl or $PbCl_2$?

SUMMARY PROBLEM

I. (a) Describe how to prepare a pH 3.70 buffer using formic acid (HCOOH) and sodium formate (Na^+HCOO^-).

(b) Calculate the pH of this buffer after the addition of 0.0050 mol of HCl.

(c) How many grams of NaOH could be added to the buffer before its buffer capacity is just exceeded?

II. (a) The K_a of nitrous acid, HNO_2, is 4.5×10^{-4}. In a titration, 50.0 mL of 1.00 M HNO_2 is titrated with 0.750 M NaOH. Calculate the pH of the solution:

(i) Before the titration begins

(ii) When sufficient NaOH has been added to neutralize half of the nitrous acid originally present

(iii) At the equivalence point

(iv) When 0.05 mL of NaOH less than that required to reach the equivalence point has been added

(v) When 0.05 mL of NaOH more than that required to reach the equivalence point has been added

(vi) Can bromthymol blue be used as the indicator for this titration?

(vii) Will methyl red be a satisfactory indicator here?

(b) Use these data to plot a graph of pH versus volume of titrant.

III. A 0.500-L solution contains 0.025 mol of Ag^+.

(a) Calculate the minimum mass of NaCl that must be added to precipitate AgCl from the solution.

(b) If excess Cl^- is added to the solution, the AgCl precipitate dissolves due to the formation of $AgCl_2^-$. K_f $AgCl_2^- = 2.5 \times 10^5$. Calculate the minimum amount of Cl^- that must be added to dissolve the precipitate.

IN CLOSING

Having studied this chapter, you should be able to . . .

- Explain how buffers maintain pH, how to calculate their pH, how they are prepared, and the importance of buffer capacity (Section 17.1).
- Use the Henderson-Hasselbalch equation to calculate the pH of a buffer and the pH change after acid or base has been added to the buffer (Section 17.1).
- Interpret acid-base titration curves and calculate the pH of the solution at various stages of the titration (17.2).
- Explain how acid rain is formed and its effects on the environment (Section 17.3).
- Relate a K_{sp} expression to its chemical equation (Section 17.4).
- Use the solubility of a slightly soluble solute to calculate its solubility product (Section 17.4).
- Describe the factors affecting the aqueous solubility of ionic compounds (17.5).
- Apply Le Chatelier's principle to the common ion effect (Section 17.5).
- Use the solubility product to calculate the solubility of a sparingly soluble solute in pure water and in the presence of a common ion (Section 17.5).
- Describe the effect of complex ion formation on the solubility of a sparingly soluble ionic compound (Section 17.5).
- Relate Q, the ion product, to K_{sp} to determine whether precipitation will occur (Section 17.6.).
- Predict which of two ionic solutes will precipitate first (Section 17.6).

KEY TERMS

acid rain *(17.3)*
buffer *(17.1)*
buffer capacity *(17.1)*
buffer solution *(17.1)*
common ion effect
   (17.5)
complex ion *(17.5)*

endpoint *(17.2)*
formation constant, K_f
   (17.5)
Henderson-Hasselbalch
   equation *(17.1)*
ion product, Q *(17.6)*

solubility product con-
   stant, K_{sp} *(17.4)*
solubility product con-
   stant expression
   (17.4)
titrant *(17.2)*
titration curve *(17.2)*

QUESTIONS FOR REVIEW AND THOUGHT

Conceptual Challenge Problem

CP-17.A (Section 17.2) Suppose you were asked on a laboratory test to outline a procedure to prepare a buffered solution of pH 8.0 using hydrocyanic acid, HCN. You realize that
a pH of 8.0 is basic, and you find that the K_a of hydrocyanic
acid is 4.0×10^{-10}. What is your response?

Answers to questions in **bold** can be found in the back of the
book.

Review Questions

1. What is meant by the term "buffer capacity"?
2. Which would form a buffer?
   (a) HCl and CH_3COOH
   (b) NaH_2PO_4 and Na_2HPO_4
   (c) H_2CO_3 and $NaHCO_3$
3. Which would form a buffer?
   (a) NaOH and NaCl
   (b) NaOH and NH_3
   (c) Na_3PO_4 and Na_2HPO_4
4. Briefly describe how a buffer solution can control the pH of
   a solution when strong acid is added and when strong base
   is added. Use NH_3/NH_4Cl as an example of a buffer and HCl
   and NaOH as the strong acid and strong base.
5. What is the difference between the endpoint and the equivalence point in an acid-base titration?
6. What is meant by an indicator range for an acid-base indicator?
7. What are the characteristics of a good acid-base indicator?
8. A strong acid is titrated with a strong base, such as KOH.
   Describe the changes in the composition of the solution as
   the titration proceeds: prior to the equivalence point, at the
   equivalence point, and beyond the equivalence point.
9. Repeat the description for Question 8, but use a weak acid
   rather than a strong one.
10. Use Le Chatelier's principle to explain why $PbCl_2$ is less sol-
    uble in 0.010 M $Pb(NO_3)_2$ than in pure water.
11. Describe what a complex ion is and give an example.
12. What is amphoterism?
13. Distinguish between the ion product and the solubility
    product constant expression of a sparingly soluble solute.
14. Describe how the solubility of a sparingly soluble metal hy-
    droxide can be changed.

Buffer Solutions

15. Many natural processes can be studied in the laboratory but
    only in an environment of controlled pH. Which of the fol-
    lowing combinations would be the best choice to buffer
    the pH at approximately 7?
    (a) H_3PO_4/NaH_2PO_4
    (b) NaH_2PO_4/Na_2HPO_4
    (c) Na_2HPO_4/Na_3PO_4
16. Which of the following combinations would be the best to
    buffer the pH at approximately 9?
    (a) $CH_3COOH/NaCH_3COO$
    (b) HCl/NaCl
    (c) NH_3/NH_4Cl
17. Without doing calculations, determine the pH of a
    buffer made from equimolar amounts of these acid-base
    pairs.
    (a) Nitrous acid and sodium nitrite
    (b) Ammonia and ammonium chloride
    (c) Formic acid and potassium formate
18. Without doing calculations, determine the pH of a
    buffer made from equimolar amounts of these acid-base
    pairs.
    (a) Phosphoric acid and sodium dihydrogen phosphate
    (b) Sodium hydrogen phosphate and sodium dihydrogen
    phosphate
    (c) Sodium phosphate and sodium hydrogen phosphate
19. Select from Table 17.1 an acid-base conjugate pair that
    would be suitable for preparing a buffer solution whose
    concentration of hydronium ions is
    (a) 4.5×10^{-3} M   (b) 5.2×10^{-8} M
    (c) 8.3×10^{-6} M   (d) 9.7×10^{-11} M
20. Select from Table 17.1 an acid-base conjugate pair that
    would be suitable for preparing a buffer solution with pH
    equal to
    (a) 3.45   (b) 5.48
    (c) 8.32   (d) 10.15

21. In order to buffer a solution at a pH of 4.57, what mass of sodium acetate, $NaCH_3COO$, should you add to 500. mL of a 0.150 M solution of acetic acid, CH_3COOH?

22. How many grams of ammonium chloride, NH_4Cl, would have to be added to 500. mL of 0.10 M NH_3 solution to have a pH of 9.00?

23. A buffer solution can be made from benzoic acid (C_6H_5COOH) and sodium benzoate (NaC_6H_5COO). How many grams of the acid would you have to mix with 14.4 g of the sodium salt in order to have a liter of a solution with a pH of 3.88?

24. If a buffer solution is prepared from 5.15 g of NH_4NO_3 and 0.10 L of 0.15 M NH_3, what is the pH of the solution?

25. You dissolve 0.425 g of NaOH in 2.00 L of a solution that originally had $[H_2PO_4^-] = [HPO_4^{2-}] = 0.132$ M. Calculate the resulting pH.

26. A buffer solution is prepared by adding 0.125 mol of ammonium chloride to 500. mL of 0.500 M aqueous ammonia. What is the pH of the buffer? If 0.0100 mol of HCl gas is bubbled into 500. mL of the buffer, what is the new pH of the solution?

27. If added to 1 L of 0.20 M acetic acid, CH_3COOH, which of these would form a buffer?
(a) 0.10 mol $NaCH_3COO$ (b) 0.10 mol NaOH
(c) 0.10 mol HCl   (d) 0.30 mol NaOH
Explain your answers.

28. If added to 1 L of 0.20 M NaOH, which of these would form a buffer?
(a) 0.10 mol acetic acid   (b) 0.30 mol acetic acid
(c) 0.20 mol HCl   (d) 0.10 mol $NaCH_3COO$
Explain your answers.

29. Calculate the pH change when 10.0 mL of 0.10 M NaOH is added to 90.0 mL of pure water, and compare the pH change with that when the same amount of NaOH solution is added to 90.0 mL of a buffer consisting of 1.0 M NH_3 and 1.0 M NH_4Cl. Assume that the volumes are additive. K_b $NH_3 = 1.8 \times 10^{-5}$.

30. Calculate the pH change when 1.0 mL of 1.0 M NaOH is added to 0.100 L of a solution of
(a) 0.10 M acetic acid and 0.10 M sodium acetate
(b) 0.010 M acetic acid and 0.010 M sodium acetate
(c) 0.0010 M acetic acid and 0.0010 M sodium acetate

31. Calculate the pH change when 1.0 mL of 1.0 M HCl is added to 0.100 L of a solution of
(a) 0.10 M acetic acid and 0.10 M sodium acetate
(b) 0.010 M acetic acid and 0.010 M sodium acetate
(c) 0.0010 M acetic acid and 0.0010 M sodium acetate

32. A buffer consists of 0.20 M propanoic acid ($K_a = 1.4 \times 10^{-5}$) and 0.30 M sodium propanoate.
(a) Calculate the pH of this buffer.
(b) Calculate the pH after the addition of 1.0 mL of 0.10 M HCl to 0.010 L of the buffer.
(c) Calculate the pH after the addition of 3.0 mL of 1.0 M HCl to 0.010 L of the buffer.

Titrations and Titration Curves

33. The titration curves for two acids with the same base are

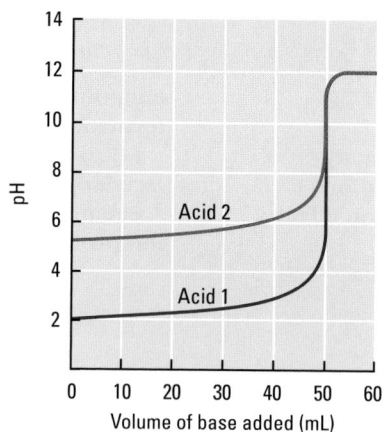

(a) Which is the curve for the weak acid? Explain your choice.
(b) Give the approximate pH at the equivalence point for the titration of each acid.
(c) Explain why the pH at the equivalence point differs for each acid.
(d) Explain why the starting pH values of the two acids differ.
(e) Which indicator could be used for the titration of Acid 1 and for the titration of Acid 2? Explain your choices.

34. Explain why it is that the weaker the acid being titrated, the more alkaline the pH is at the equivalence point.

35. Sketch the titration curve for the titration of 20.0 mL of a 0.100 M solution of a strong acid by a 0.100 M weak base; that is, the base is the titrant. In particular, note the pH of the solution:
(a) Prior to the titration
(b) When half the required volume of titrant has been added
(c) At the equivalence point
(d) 10 mL beyond the equivalence point.

36. Consider the acid-base indicators discussed in this chapter. Which of these indicators would be suitable for the titration of the following?
(a) NaOH with $HClO_4$ (b) Acetic acid with KOH
(c) NH_3 solution with HBr   (d) KOH with HNO_3
Explain your choices.

37. Which of the acid-base indicators discussed in this chapter would be suitable for the titration of
(a) HNO_3 with KOH   (b) KOH with acetic acid
(c) HCl with NH_3 (d) KOH with HNO_2
Explain your answers.

38. It required 22.6 mL of 0.0140 M $Ba(OH)_2$ solution to titrate a 25.0-mL sample of HCl to the endpoint. Calculate the molarity of the HCl solution.

39. It took 12.4 mL of 0.205 M H_2SO_4 solution to titrate 20.0 mL of a sodium hydroxide solution to the endpoint. Calculate the molarity of the original NaOH solution.

40. Vitamin C is a monoprotic acid. To analyze a vitamin C capsule weighing 0.505 g by titration took 24.4 mL of 0.110 M NaOH. Calculate the percentage of vitamin C, $C_6H_8O_6$, in the capsule. Assume that vitamin C is the only substance in the capsule that reacts with the titrant.

41. An acid-base titration was used to find the percentage of $NaHCO_3$ in 0.310 g of a powdered commercial product used to relieve upset stomachs. The titration required 14.3 mL of 0.101 M HCl to titrate the powder to the endpoint. Assume that the $NaHCO_3$ in the powder is the only substance that reacted with the titrant. Calculate the percentage of $NaHCO_3$ in the powder.

42. What volume of 0.150 M HCl is required to titrate to the endpoint each of these samples?
(a) 25.0 mL of 0.175 M KOH
(b) 15.0 mL of 6.00 M NH_3
(c) 15.0 mL of propylamine, $CH_3CH_2CH_2NH_2$, which has a density of 0.712 g/mL
(d) 40.0 mL of 0.0050 mL of $Ba(OH)_2$

43. What volume of 0.225 M HCl is required to titrate to the endpoint each of these samples?
(a) 20.0 mL of 0.315 M HBr
(b) 30.0 mL of 0.250 M $HClO_4$
(c) 6.00 g of concentrated acetic acid, CH_3COOH, which is 99.7% pure.

44. A 30.00-mL solution of 0.100 M benzoic acid, a monoprotic acid, is titrated with 0.100 M NaOH. The K_a of benzoic acid is 6.3×10^{-5}. Determine the pH after each of the following volumes of titrant has been added:
(a) 10.00 mL   (b) 30.00 mL   (c) 40.00 mL

45. The titration of 50.00 mL of 0.150 NaOH with 0.150 M HCl is carried out in a chemistry laboratory. Calculate the pH of the solution after these volumes of the titrant have been added:
(a) 0.00 mL   (b) 25.00 mL   (c) 49.9 mL
(d) 50.00 mL   (e) 50.1 mL   (f) 75.00 mL
Use the results of your calculations to plot a titration curve for this titration. On the curve indicate the position of the equivalence point.

46. The titration of 50.00 mL of 0.150 HCl with 0.150 M NaOH is carried out in a chemistry laboratory. Calculate the pH of the solution after these volumes of the titrant have been added:
(a) 0.00 mL   (b) 25.00 mL   (c) 49.9 mL
(d) 50.00 mL   (e) 50.1 mL   (f) 75.00 mL
Use the results of your calculations to plot a titration curve for this titration. On the curve indicate the position of the equivalence point.

Acid Rain

47. Explain why rain with a pH of 6.7 is not classified as acid rain.

48. Identify two oxides that are key producers of acid rain. Write chemical equations that illustrate how these oxides form acid rain.

49. Acid rain with a pH of 1.5 has been measured. Calculate the hydrogen ion concentration of this rain.

50. Write a chemical equation that shows how limestone neutralizes acid rain.

Solubility Product

51. Write a balanced chemical equation for the equilibrium occurring when each of these solutes is added to water, then write the K_{sp} expression for each solute.
(a) Lead(II) carbonate   (b) Nickel(II) hydroxide
(c) Strontium phosphate   (d) Mercury(I) sulfate

52. Write a balanced chemical equation for the equilibrium occurring when each of these solutes is added to water, then write the K_{sp} expression.
(a) Iron(II) carbonate   (b) Silver sulfate
(c) Calcium phosphate   (d) Mn(II) hydroxide

53. A saturated solution of silver arsenate, Ag_3AsO_4, contains 8.5×10^{-6} g of Ag_3AsO_4 per mL. Calculate the K_{sp} of silver arsenate.

54. At 20. °C, 2.03 g of $CaSO_4$ dissolves per liter of water. From these data calculate the K_{sp} of calcium sulfate at 20. °C.

55. The water solubility of strontium fluoride, SrF_2, is 0.011 g/100 mL. Calculate its solubility product constant.

56. The solubility of silver chromate, Ag_2CrO_4, in water is 2.7×10^{-3} g/100. mL. Estimate the K_{sp} of silver chromate.

57. Calculate the K_{sp} for HgI_2 given that its solubility in water is 4.0×10^{-29} M.

58. The solubility of $PbCl_2$ in water is 1.62×10^{-2} M. Calculate the K_{sp} for $PbCl_2$.

Common Ion Effect

59. What is the Cl^- concentration (in mol/L) in a solution that is 0.05 M in $AgNO_3$ and contains some undissolved AgCl?

60. What is the molarity of Zn^{2+} ion in a saturated solution of $ZnCO_3$ that contains 0.25 M Na_2CO_3?

61. Calculate the solubility of $ZnCO_3$ in
(a) Water
(b) 0.050 M $Zn(NO_3)_2$
(c) 0.050 M K_2CO_3. K_{sp} $ZnCO_3$ = 3×10^{-8}.

62. Calculate the solubility (mol/L) of $SrSO_4$ (K_{sp} = 3.1×10^{-7}) in 0.010 M Na_2SO_4.

63. Iron (II) hydroxide, $Fe(OH)_2$, has a solubility in water of 6.0×10^{-1} mg/L.
(a) Calculate the K_{sp} of iron(II) hydroxide.
(b) Calculate the hydroxide concentration needed to precipitate Fe^{2+} ions such that no more than 1.0 μg of Fe^{2+} per liter remains in the solution.

64. The solubility of $Mg(OH)_2$ in water is approximately 9 mg/L.
(a) Calculate the K_{sp} of magnesium hydroxide.
(b) Calculate the hydroxide concentration needed to precipitate Mg^{2+} ions such that no more than 5.0 μg of Mg^{2+} per liter remains in the solution.

Factors Affecting the Solubility of Sparingly Soluble Solutes

65. Calculate the maximum concentration of Mg^{2+} (molarity) that can exist in a solution of pH = 12.00.

66. What is the maximum concentration of Zn^{2+} in a solution of pH = 10.00?

67. Name the maximum concentration of Mn^{2+} in two solutions with pH as follows:
    (a) 7.81 (b) 11.15

68. Hydrochloric acid is added to dissolve 5.00 g of $Mg(OH)_2$ in a liter of water. To what value must the pH be adjusted to do so?

69. When a few drops of 1×10^{-5} M $AgNO_3$ are added to 0.01 M NaCl, a white precipitate immediately forms. When a few drops of 1×10^{-5} M $AgNO_3$ are added to 5 M NaCl, no precipitate forms. Explain these observations.

Complex Ion Formation

70. For these complex ions, write the chemical equation for the formation of the complex ion, and write its formation constant expression.
    (a) $Ag(CN)_2^-$ (b) $Cd(NH_3)_4^{2+}$

71. For these complex ions, write the chemical equation for the formation of the complex ion and write its formation constant expression.
    (a) $CoCl_6^{3-}$ (b) $Zn(OH)_4^{2-}$

72. Calculate how many moles of $Na_2S_2O_3$ must be added to dissolve 0.020 mol of AgBr in 1.0 L of water.

73. Gaseous ammonia is added to a 0.063 M solution of $AgNO_3$ until the aqueous ammonia concentration rises to 0.18 M. Calculate the concentrations of $Ag(NH_3)_2^+$ and Ag^+ in the solution.

74. Write chemical equations to illustrate the amphoteric behavior of
    (a) $Zn(OH)_2$ (b) $Sb(OH)_3$

75. Write chemical equations to illustrate the amphoteric behavior of
    (a) $Cr(OH)_3$ (b) $Sn(OH)_2$

General Questions

76. A buffer solution was prepared by adding 4.95 g of sodium acetate, $NaCH_3COO$, to 250. mL of 0.150 M acetic acid, CH_3COOH.
    (a) What ions and molecules are present in the solution? List them in order of decreasing concentration.
    (b) What is the pH of the buffer?
    (c) What is the pH of 100. mL of the buffer solution if you add 80. mg of NaOH? (Assume negligible change in volume.)
    (d) Write a net ionic equation for the reaction that occurs to change the pH.

77. How many grams of NH_4Cl must be added to 400 mL of a 0.93 M solution of NH_3 to prepare a pH = 9.00 buffer?

78. Calculate the relative concentrations of *o*-ethylbenzoic acid ($pK_a = 3.79$) and potassium *o*-ethylbenzoate that are needed to prepare a pH = 4.0 buffer.

79. Calculate the relative concentrations of aniline ($pK_a = 9.42$) and anilinium chloride that are required to prepare a buffer with a pH = 5.00.

80. A solution contains 7.50 g of KNO_2 per liter. How much HNO_2 must be added to prepare a buffer of pH = 4.00? (Assume there is no volume change.)

81. Which of these buffers has the greater resistance to change in pH?
    (a) Conjugate acid concentration = 0.100 M = conjugate base concentration
    (b) Conjugate acid concentration = 0.300 M = conjugate base concentration
    Explain your answer.

82. (a) What is the pH of a 0.15 M acetic acid solution?
    (b) If you add 83 g of sodium acetate to 1.50 L of the 0.15 M acetic acid solution, what is the new pH of the solution?

83. (a) Calculate the pH of a 0.050 M solution of HF.
    (b) What is the pH of the solution if you add 1.58 g of NaF to 250. mL of the 0.050 M solution?

84. When 40.00 mL of a weak monoprotic acid solution is titrated with 0.100 M NaOH, the equivalence point is reached when 35.00 mL of base has been added. After 20.00 mL of NaOH solution has been added, the titration mixture has a pH of 5.75. Calculate the ionization constant of the acid.

85. Pyridine, C_5H_5N ($K_b = 1.5 \times 10^{-9}$), is a base like ammonia. A 25.0-mL sample of 0.085 M pyridine is titrated with 0.102 M HCl. The equivalence point occurs at 21.1 mL. Calculate the pH when 5.5 mL of acid has been added.

86. What is the effect on the equilibrium if more solid AgCl is added to a solution saturated with Ag^+ and Cl^- ions?

87. At 20. °C, 2.03 g of $CaSO_4$ dissolve per liter of water. From these data calculate the K_{sp} of calcium sulfate at 20. °C.

88. The solubility of silver chromate, Ag_2CrO_4, in water is 2.7×10^{-3} g per 100. mL. Estimate the K_{sp} of silver chromate.

Applying Concepts

89. The average normal concentration of Ca^{2+} in urine is 5.33 g/L.
    (a) What concentration of oxalate is needed to precipitate calcium oxalate to initiate formation of a kidney stone? K_{sp} calcium oxalate = 2.3×10^{-9}.
    (b) What minimum phosphate concentration would it take to precipitate a calcium phosphate kidney stone? K_{sp} calcium phosphate = 1×10^{-25}.

90. Explain why even though an aqueous acetic acid solution contains acetic acid and acetate ions, it cannot be a buffer.

91. Vinegar must contain at least 4% acetic acid (0.67 M). A 5.00-mL sample of commercial vinegar required 33.5 mL of 0.100 M NaOH to reach the equivalence point. Does the vinegar meet the legal limit of 4% acetic acid?

92. An unknown acid is titrated with base, and the pH is 3.64 at the point where exactly half of the acid in the original sample has been neutralized. What is the value of the ionization constant of the acid?

93. When asked to prepare a carbonate buffer with a pH = 10, a lab technician wrote the following equation to determine the ratio of weak acid to conjugate base needed:

$$10 = 10.25 + \log \frac{[HCO_3^-]}{[H_2CO_3]}$$

What is wrong with this setup? If the technician prepared a solution containing equimolar concentrations of HCO_3^- and H_2CO_3, what would be the pH of the resulting buffer?

94. When you hold your breath, carbon dioxide gas is trapped in your body. Does this increase or decrease your blood pH? Does this lead to acidosis or alkalosis? Explain your answers.

95. Choose the words that make this statement true: During a recent television medical drama, a person went into cardiac arrest and stopped breathing. A doctor quickly injected sodium hydrogen carbonate into the heart. This would indicate that cardiac arrest leads to (acidosis or alkalosis) and that the sodium hydrogen carbonate helps to (increase or decrease) the pH. Explain your choices clearly.

96. You are given four different aqueous solutions and are told that they each contain $NaOH$, Na_2CO_3, $NaHCO_3$, or a mixture of these solutes. You do some experiments and gather the following data about the samples.

Sample A: Phenolphthalein is colorless in the solution.

Sample B: The sample was titrated with HCl until the pink color of phenolphthalein disappeared then methyl orange was added to the solution. The solution then became a pink color. Methyl orange changes color from pH 3.01 (red) to pH 4.4 (orange).

Sample C: Equal volumes of the sample were titrated with standardized acid. Using phenolphthalein as an indicator required 15.26 mL of standardized acid to change the phenolphthalein color; it required 17.90 mL for a color change using methyl orange as the indicator.

Sample D: Two equal volumes of the sample were titrated with standardized HCl. Using phenolphthalein as the indicator, it took 15.00 mL of acid to reach the endpoint; using methyl orange as the indicator required 30.00 mL of the HCl to achieve neutralization.

Identify the solute in each of the solutions.

 General Chemistry CD-ROM

CD17.1 Screen 18.8: Buffer Solutions. What are some practical uses of buffer solutions? How do the videos on this screen illustrate their uses?

CD17.2 Screen 18.9: pH of Buffer Solutions.
(a) Consider the Henderson-Hasselbalch equation shown when you press the *General Rule* button on this screen. Clearly, the pH of a buffer solution depends on the pK_a of the weak acid and on the relative concentrations of the conjugate acid and base. Which of these two is more important in controlling buffer pH?
(b) Examine the sidebar to this screen. Suggest a reason why the body does not use the acetic acid/acetate buffer system for holding blood pH near 7.4

CD17.3 Screen 19.4: Solubility Product Constant.
(a) Why does the K_{sp} expression shown on the screen for the dissolution of $PbCl_2$ not have a denominator, such as those of other equilibrium reactions we have studied?
(b) Explain the difference between the K_{sp} value of a salt and the salt's solubility.

CD17.4 Screen 19.6: Estimating Salt Solubility. This screen describes how the solubility of $BaSO_4$ is estimated using its K_{sp} value. The calculation assumes that the ions produced, Ba^{2+} and SO_4^{2-}, do not react further in solution. What would happen to the solubility of $BaSO_4$ if some of the sulfate ion reacted with acid in solution?

$$SO_4^{2-}(aq) + H_3O^+(aq) \longrightarrow HSO_4^-(aq) + OH^-(aq)$$

CD17.5 Screen 19.7: Can a Precipitation Reaction Occur?
(a) Three cases are given on this screen: $Q < K_{sp}$, $Q > K_{sp}$, and $Q = K_{sp}$. Which of these represents a system at equilibrium?
(b) Can a precipitation system be at equilibrium if there is no solid present? Explain briefly.

CD17.6 Screen 19.11: Solubility and pH.
(a) Explain how the example on this screen, the solubility of $Co(OH)_2(s)$, is an example of Le Chatelier's principle.
(b) Notice that the solubility of the compound increases and the pH decreases. What effects do you think a pH decrease would have on the solubility of $MgCO_3$, considering the fact that carbonate ion, CO_3^{2-}, is a relatively strong base?

18

When water drips onto a sample of metallic potassium, a violent exothermic reaction occurs. Sparks fly, and you see a purple flame characteristic of very hot potassium compounds. The water and potassium are transformed into potassium hydroxide and hydrogen. The reaction begins as soon as the water contacts the potassium. It continues until either the water or the potassium has been completely consumed.

THERMODYNAMICS: DIRECTIONALITY OF CHEMICAL REACTIONS

18.1 Reactant-Favored and Product-Favored Processes

18.2 Probability and Chemical Reactions

18.3 Measuring Dispersal or Disorder: Entropy

18.4 Calculating Entropy Changes

18.5 Entropy and the Second Law of Thermodynamics

18.6 Gibbs Free Energy

18.7 Gibbs Free Energy Changes and Equilibrium Constants

18.8 Gibbs Free Energy, Maximum Work, and Energy Resources

18.9 Gibbs Free Energy and Biological Systems

18.10 Conservation of Gibbs Free Energy

18.11 Thermodynamic and Kinetic Stability

M any chemical reactions behave as the reaction of potassium and water does. They begin when the reactants come into contact and continue until at least one reactant (the limiting reactant ⇐ *p. 140*) is completely used up. Other reactions, such as the rusting of iron at room temperature, happen much more slowly, but reactants are still converted completely to products. After many years and enough flaking of hydrated iron(III) oxide (rust) from its surface, a piece of iron exposed to air will rust away. Still other reactions are even slower at room temperature. Gasoline reacts so slowly with air at room temperature that it can be stored safely for long periods, although it may go bad after a very long time. However, if its temperature is raised by a spark or flame, gasoline vapor burns rapidly and is essentially all converted to CO_2 and H_2O.

By contrast with the reaction of potassium and water, there are many chemical reactions that do not occur by themselves. For example, table salt, NaCl, does not of its own accord decompose into sodium and chlorine. Neither does water change into hydrogen and oxygen all by itself. These reactions take place only if another reaction occurs simultaneously and transfers energy to them. It may surprise you to learn that a significant portion of our energy resources is used to cause desirable reactions to occur—reactions that transform inexpensive, readily available substances into new substances with more useful properties. Examples are the manufacture of rubber and the manufacture of medicines. It is important to differentiate between a reaction that is so slow that it *appears* not to occur (such as air oxidation of gasoline) and one that cannot take place of its own accord, such as decomposition of sodium chloride. The principles of chemical kinetics *(⇐ Chapter 13)* can be applied to find ways to speed up a slow reaction, but they are of no use in dealing with one that cannot occur by itself.

In Chapter 6 *(⇐ p. 231)* you learned that thermal energy is transferred when most reactions occur. You also learned how to predict whether a reaction is exothermic or endothermic and how to calculate what quantity of energy transfer takes place as a reaction occurs. In this chapter you will learn how thermodynamics helps us to predict what will happen when potential reactants are mixed. Will most or all of the reactants be converted to products, as in the case of potassium and water? Will some be converted? Or virtually none?

18.1 REACTANT-FAVORED AND PRODUCT-FAVORED PROCESSES

In Chapter 14 *(⇐ p. 633)* we introduced the idea that a chemical process can be described as reactant-favored or product-favored. When products predominate over reactants, we designate the reaction as a *product-favored process.* Examples are the reaction of potassium with water, rusting of iron, and combustion of gasoline. If a process is product-favored, most or all of the reactants will eventually be converted to products without continuous outside intervention, although "eventually" may mean a very, very long time.

There are other reactions that have virtually no tendency to occur by themselves. (Some examples are the reactions for which we wrote "N.R." for "no reaction" in Chapter 5 ⇐ *p. 163*.) For example, nitrogen and oxygen have coexisted in the earth's atmosphere for at least a billion years without appreciable concentrations of nitrogen oxides such as N_2O, NO, or NO_2 building up. Similarly, deposits of salt, NaCl(s), have existed on earth for millions of years without forming the elements Na(s) and $Cl_2(g)$. If, when equilibrium has been reached, reactants predominate over products, we categorize a chemical reaction as a *reactant-favored process.*

A reactant-favored process is always a transformation that is exactly the opposite of the transformation in a product-favored process. For example, the equation

$$2 \text{ Na(s)} + \text{Cl}_2(g) \longrightarrow 2 \text{ NaCl(s)}$$

Because it can oxidize slowly, gasoline is usually drained from lawn mowers or cars that are going to be stored for many months, or else a substance is added to prevent very slow oxidation.

 CD-ROM Screen 20.2: Reaction Spontaneity: Thermodynamics and Kinetics

The term "product-favored" designates reactions that many scientists refer to as "spontaneous"; many people will use the two terms interchangeably. We prefer "product-favored" because some reactions do begin spontaneously, but produce only tiny quantities of products when equilibrium is reached. "Product-favored" describes clearly a situation in which products predominate over reactants. Also, the normal usage of "spontaneous" implies a rapid change; if the rate of a product-favored reaction is very slow, the reaction does not appear spontaneous at all.

Although earth's atmosphere is 78% N_2 and 21% O_2, the concentration of N_2O, the most abundant oxide of nitrogen, is more than two million times smaller than the concentration of N_2.

describes a product-favored reaction, because sodium metal and chorine gas react readily to produce salt. However, if we had written the equation in the reverse direction

$$2\,NaCl(s) \longrightarrow 2\,Na(s) + Cl_2(g)$$

the system would be designated as reactant-favored. This equation represents decomposition of sodium chloride to form sodium and chlorine, a reaction that does not occur of its own accord. The designations "product-favored" and "reactant-favored" indicate the direction in which a chemical reaction will take place—either forward or backward based on a given equation.

Unless there is some continuous outside intervention, a reactant-favored process does not produce large quantities of products. What do we mean by continuous outside intervention? Usually it is some flow of energy. For example, if enough energy is provided to a sample of air to keep it at a very high temperature, small but significant quantities of NO can be formed from the N_2 and O_2. Such high temperatures are found in lightning bolts, electric power generating plants, and automobile engines. A large number of automobiles and power plants can produce enough NO and other nitrogen oxides to cause significant air pollution problems. Salt can be decomposed to its elements by continuously heating it to keep it molten and passing electricity through it to separate the ions and form the elements.

$$2\,NaCl(\ell) \xrightarrow{\text{electricity}} 2\,Na(\ell) + Cl_2(g)$$

In each case, a reactant-favored process can be forced to produce products if sufficient energy is continuously supplied. This is in contrast to the situation for a product-favored process such as combustion of gasoline, which requires only a brief spark to initiate the reaction. Once started, gasoline combustion continues of its own accord without a supply of energy from outside.

> ### Exercise 18.1 Reactant-Favored and Product-Favored Processes
>
> Write a chemical equation for each process described below, and classify each as reactant-favored or product-favored.
>
> (a) A puddle of water evaporates on a summer day.
> (b) Silicon dioxide (sand) decomposes to the elements silicon and oxygen.
> (c) Paper, which is mainly cellulose ($C_6H_{10}O_5$)$_n$, burns at a temperature of 451 °F.
> (d) A pinch of salt dissolves in water at room temperature.

18.2 PROBABILITY AND CHEMICAL REACTIONS

Most exothermic reactions are product-favored at room temperature. The reason for this is very similar to the reason that there is a one-way transfer of energy from a hotter to a colder sample (⇐ p. 216). When an exothermic reaction takes place, energy is transferred to the surroundings, as seen for the reaction of potassium with water in the photo at the beginning of this chapter. Chemical potential energy that has been stored in bonds between relatively few atoms and molecules (the reactants) spreads over many more atoms and molecules as the surroundings (as well as the products) are heated. Because there are a great many more atoms, molecules, and ions in the surroundings than in the reactants and products, it is always true that after an exothermic reaction, energy will be distributed more randomly—dispersed over a much larger number of particles—than it was before.

The air in the immediate vicinity of a lightning bolt can be heated enough to cause a small fraction of the nitrogen and oxygen to combine to form NO, but this takes place only while the lightning is present. A similar reaction can occur in the engine of an automobile, but again only a small fraction of the air is converted to nitrogen oxides, and only while the temperature is high.

CD-ROM Screen 20.3: Directionality of Reactions: Matter and Energy Dispersal

Remember that a chemical reaction system is usually defined as the collection of atoms that make up the reactants. These same atoms also make up the products, but there they are bonded in a different way. Everything else is designated as the surroundings (⇐ *p. 217*).

Dispersal of Energy

Dispersal of energy occurs because the probability is much higher that energy will be spread over many particles than that it will be concentrated in a few. To better understand energy dispersal and probability, consider the hypothetical case of a very small sample of matter consisting of two atoms, A and B, and suppose that this sample contains two units of energy, each designated by *. There are three ways the energy can be distributed over the two atoms: Atom A could have both units of energy, atom A and atom B could each have one, or atom B could have both. Designate these three situations as

A**   A*B*   B**

Now suppose that atoms A and B come into contact with two other atoms, C and D, that have no energy. There are ten possibilities for distributing the two units of energy over all four atoms.

A**   A*B*   A*C*   A*D*   B**   B*C*   B*D*   C**   C*D*   D**

Only three of these (A**, A*B*, and B**) have all the energy in atoms A and B, which was the initial situation. When all four atoms are in contact, there are seven chances out of ten that some energy will have transferred from A and B to C and D. Thus, there is a probability of $7/10 = 0.70$ that the energy will become spread out over more than just the two atoms A and B. When this dispersal of energy occurs, we say that the atoms A, B, C, and D have gone from a situation where the energy was ordered to one where it has become more disordered.

The low probability that a lot of energy will be associated with only a few particles makes a substance with a lot of chemical potential energy valuable. We call substances such as coal, oil, and natural gas "energy resources" and sometimes fight wars over them because of their concentrated energy.

Exercise 18.2   Probability of Energy Dispersal

Suppose that you have three units of energy to distribute over two atoms, A and B. Designate each possible arrangement. Now suppose that atoms A and B come into contact with three more atoms, C, D, and E. From the possible arrangements of energy over the five atoms, calculate the probability that all the energy will remain confined to atoms A and B.

The probability that energy will become disordered becomes overwhelming when large numbers of atoms or molecules are involved. For example, suppose that atoms A and B had been brought into contact with a mole of other atoms. There would still be only three arrangements in which all the energy was associated with atoms A and B, but there would be many, many more arrangements (more than 10^{47}) in which all the energy had been transferred to other atoms. In such a case it is essentially certain that energy will be transferred. ***If energy can be dispersed over a very much larger number of particles, it will be.***

Recall that for the same amount of substance or number of particles, higher thermal energy corresponds to higher temperature (⬅ *p. 217*). Therefore, a substance at a higher temperature has greater energy per particle on average. Dispersal of energy over a larger number of particles corresponds to transfer of energy from a substance at a higher temperature to another at a lower temperature.

Dispersal of Matter

Just as there is a tendency for concentrated energy to disperse, ***concentrated matter also tends to disperse, unless there is something to prevent it.*** For example, a characteristic property of gases is that they expand until they fill a container. Recall that molecules in the gas phase at room temperature are essentially independent of one another and that there are only weak forces attracting the molecules together (⬅ *p. 414*). Suppose a sample of bromine gas is confined within one flask that is connected through a tube with a barrier to a second flask of equal size from which all gas molecules have been removed (Figure 18.1 on page 816). What happens if the barrier is removed? The confined bromine expands to fill the vacuum in the second flask.

CD-ROM Screen 20.3: Directionality of Reactions: Matter and Energy Dispersal

The conclusions drawn about dispersal of bromine molecules in this paragraph apply in general to particles of a gas, whether they are atoms or molecules.

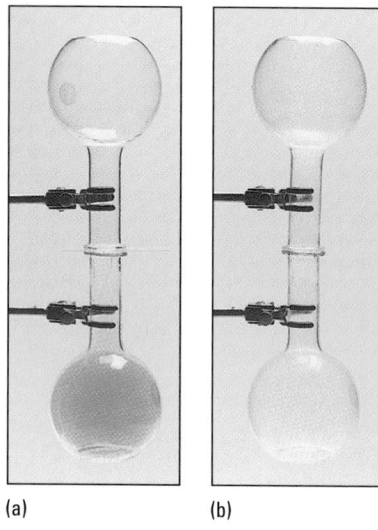

(a)   (b)

Figure 18.1 Expansion of a gas. It is well known that a gas will expand to fill any container. (a) Bromine vapor is confined in the lower flask. There is a vacuum in the upper flask. (b) When the barrier between the flasks is removed, the bromine molecules begin to rush into flask B, and eventually bromine is evenly distributed throughout both flasks.

Atoms or molecules of a material that is a solid at room temperature (like the glass flask) do not disperse, because there are strong attractive forces between them. Their tendency to disperse becomes more obvious if the temperature is raised so that they can vaporize.

The dispersal of matter also involves dispersal of energy. When matter spreads out, the number of ways of arranging the energy associated with that matter becomes larger.

Dispersal of the gaseous bromine can be analyzed in the same way as dispersal of energy. Suppose that a single bromine molecule occupies the two-flask system in Figure 18.2. The molecule moves around within one flask until it hits the opening to the other flask; then it occupies the second flask for a while before returning to the first. Because the volumes are equal and the bromine molecule's motion is random, it will spend half its time in each flask, on average. The probability is 1/2 that it can be found in flask A and 1/2 that it is in flask B. Now consider two bromine molecules within the same two-flask system. There are four equally probable arrangements, all shown in Figure 18.2. Only one of them has both molecules in flask A. The probability that both molecules are in flask A is thus 1/4 or $(1/2)^2$. By making a similar diagram you can verify that for three molecules there are eight arrangements. Only one of these eight corresponds to all three molecules in flask A, giving a probability of 1/8 or $(1/2)^3$ for this arrangement. In general (with no barrier between the flasks) the probability that all bromine molecules will be in flask A is $(1/2)^N$, where N is the number of atoms.

As the number of particles increases, the probability of their all being in the same flask (a more highly ordered state) becomes very, very small. For a mole of gas, the probability that all particles are in flask A is incredibly small, and so it becomes absolutely certain that some particles will move from flask A to flask B. *Expansion of a gas from one half of a container into the whole container is a product-favored process. The final arrangement with gas in both flasks is much more probable than the initial one, and so the process occurs of its own accord.* If we wanted to reverse the expansion by concentrating all the particles into flask A, a continuous outside influence such as a pump would be required — the pump could do work on the gas to force it into a highly improbable arrangement. The work done by the pump would be stored in the gas and could later be used for some other purpose.

To summarize, there are two ways that the final state of a system can be more probable than the initial one:

• having energy dispersed over a greater number of atoms, molecules, or ions
• having the atoms and molecules themselves more disordered

If both of these happen, then a reaction will definitely be product-favored, since the products and the distribution of energy will both be more probable. On the other hand, a process that spreads out neither matter nor energy will be reactant-favored — the initial substances will remain no matter how long we wait. If one of

... or ...   ... or ...   ... or ...

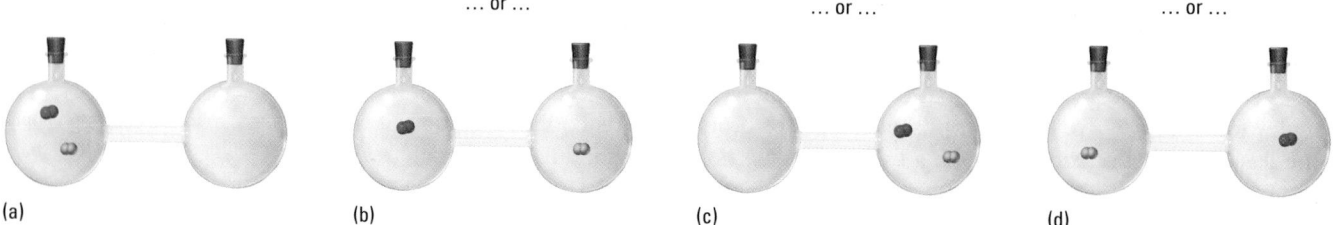

(a)   (b)   (c)   (d)

Figure 18.2 Gas expansion and probability. When two molecules occupy two flasks of equal volume, four different, equally probable arrangements are possible.

these happens but not the other, then quantitative information is needed to decide which effect is greater. The remainder of this chapter is devoted to developing that quantitative information.

CD-ROM Screen 20.4: Entropy: Matter Dispersal or Disorder

18.3 MEASURING DISPERSAL OR DISORDER: ENTROPY

The nanoscale dispersal or disorder in a sample of matter is called *entropy* and is symbolized by S *(⇐ p. 665)*. Entropy changes can be measured with a calorimeter *(⇐ p. 238)*, the same instrument used to measure the enthalpy change when a reaction occurs. For a process that takes place at constant temperature and pressure, the entropy change can be calculated by dividing the thermal energy transferred, q_{rev}, by the absolute temperature, T,

$$\Delta S = S_{final} - S_{initial} = \frac{q_{rev}}{T}$$

The subscript "rev" has been added to q to indicate that the equation applies only to processes that can be reversed by a very small change in conditions. An example of such a process is melting of ice at 0 °C and normal atmospheric pressure. If the temperature is just a tiny bit below 0 °C, so that energy is transferred from the water to its surroundings, the water will freeze. If the temperature is a tiny bit above 0 °C, the ice will melt. *Any process like this, for which a very small change in conditions can reverse its direction,* is called a **reversible process.**

The symbol Δ was defined in Section 6.2 *(⇐ p. 214)*. The quantity of energy transferred by heating, q, was defined in Section 6.3 *(⇐ p. 218)*.

The absolute temperature scale is also called the *Kelvin temperature scale* or the *thermodynamic temperature scale*. It was defined in Section 10.4 *(⇐ p. 417)*. The kelvin unit should be used in all thermodynamic calculations involving temperature.

LUDWIG BOLTZMANN
(1844–1906)

Ludwig Boltzmann was an Austrian mathematician who gave us the useful interpretation of entropy as probability. Engraved on his tombstone in Vienna is his equation

$$S = k \log W$$

It relates entropy, S, and thermodynamic probability, W—the number of different arrangements of nanoscale particles that correspond to a given macroscale system. The proportionality constant, k, is called Boltzmann's constant, and log stands for the natural logarithm (ln in modern symbolism).

Problem-Solving Example 18.1 Calculating Entropy Change

The standard molar enthalpy of fusion of water is 6.02 kJ/mol at 0 °C and 1 bar. Calculate $\Delta S°$ for the process

$$H_2O(s) \longrightarrow H_2O(\ell)$$

Express your answer in units of joules per kelvin.

Answer 22.0 J/K

Explanation Melting ice is reversible, so use the equation $\Delta S = q_{rev}/T$. The pressure is 1 bar, the standard-state pressure *(⇐ p. 243)*, so $\Delta S° = \Delta S$. Remember that for a constant-pressure process, thermal energy transfer, q_{rev}, is the same as the enthalpy change, $\Delta H°$ *(⇐ p. 228)*. Therefore, the entropy change can be calculated from the enthalpy of fusion and the temperature.

$$\Delta S° \text{ (melting ice)} = \frac{q_{rev}}{T} = \frac{\Delta H° \text{ (melting ice)}}{T} = \frac{\Delta H_{fusion}}{T}$$

$$= \frac{6.02 \text{ kJ}}{273.15 \text{ K}} = 2.20 \times 10^{-2} \text{ kJ/K} = 22.0 \text{ J/K}$$

✔ The result is positive, meaning that entropy increased when ice was converted to liquid water. Since larger entropy corresponds to greater disorder and since molecules in a liquid are arranged less regularly than in a solid *(⇐ p. 19)*, this is reasonable.

Problem-Solving Practice 18.1

The enthalpy of vaporization of benzene (C_6H_6) is 30.8 kJ/mol at the boiling point of 80.1 °C. Calculate the entropy change for 1.0 mol of benzene going from (a) liquid to vapor and (b) vapor to liquid at 80.1 °C and 1 bar. Express your answer in joules per kelvin.

Entropy changes are usually reported in units of joules per kelvin (J/K), whereas enthalpy changes are usually given in kilojoules (kJ). This means that you need to be careful about the units to avoid being wrong by a factor of 1000.

Exercise 18.3 The Importance of Absolute Temperature

Consider what would happen if the Celsius temperature scale were used when calculating entropy change by means of $\Delta S = q_{rev}/T$. Suppose, for example, that energy were transferred reversibly to $H_2O(s)$ at a temperature 10° below its melting point and we wanted to calculate the entropy change. Would the value calculated from the Celsius temperature agree with the fact that transfer of thermal energy to a sample always increases its entropy?

Absolute Entropy Values

In Chapter 6 we mentioned that there is no way to measure the total energy content of a sample of matter. Therefore, to summarize a large number of calorimetric measurements of enthalpy changes, we tabulated standard molar enthalpies of formation in Table 6.2 (← *p. 245*). The standard enthalpy of formation is the difference between the enthalpy of a substance in its standard state and the enthalpies of the elements that make up that substance, all in their standard states. The elements have enthalpy values, but we do not know what they are. Therefore, the standard enthalpies of formation of elements are arbitrarily set to zero.

For entropy the situation is simpler, because it is possible to define conditions for which it is logical to assume the entropy of a substance has its lowest possible value, namely zero. Then, measuring ΔS for a change from those zero-entropy conditions tells us the absolute entropy value for the substance under new conditions. This follows from the definition $\Delta S = S_{final} - S_{initial}$, because if $S_{initial} = 0$, then $\Delta S = S_{final}$, the absolute entropy of the substance. Because decreasing temperature corresponds to decreasing molecular motion, the minimum possible temperature can reasonably be expected to correspond to minimum motion and thus minimum disorder. Thus it is logical to assume that a perfect crystal of a substance at 0 K has an entropy value of zero. (In a perfect crystal every nanoscale particle is in exactly the right position in the crystal lattice (← *p. 481*), and there are no empty spaces or discontinuities.) By starting as close as possible to absolute zero, repeatedly introducing small quantities of energy, and calculating ΔS from the equation $\Delta S = q_{rev}/T$, an entropy change can be determined for each small increase in energy (and therefore temperature). These entropy changes can then be summed to give the total (or absolute) entropy of a substance at any desired temperature.

The results of such measurements for several substances at 298.15 K are given in Table 18.1. These are standard molar entropy values, and so they apply to 1 mol

Though it is impossible to cool anything all the way to absolute zero, it is possible to get very close. Temperatures of a few nanokelvins can be achieved in a Bose-Einstein condensate—the coldest thing known to science.

The idea that a perfect crystal of any substance at 0 K has minimum entropy is called the ***third law of thermodynamics.*** Even if absolute zero cannot be achieved, there are ways of estimating the disorder that is in a substance near 0 K. Thus, accurate entropy values can be obtained for many substances.

The process of introducing small quantities of energy, calculating an entropy increase, and then summing these small entropy increases is actually done by measuring the heat capacity of a substance as a function of temperature and then using integral calculus to calculate the integral of the function q_{rev}/T between the limits of 0 K and the desired temperature.

TABLE 18.1	Some Standard Molar Entropy Values at 298.15 K*				
Compound or element	**Entropy, $S°$ ($J\ K^{-1}\ mol^{-1}$)**	**Compound or element**	**Entropy, $S°$ ($J\ K^{-1}\ mol^{-1}$)**	**Compound or element**	**Entropy, $S°$ ($J\ K^{-1}\ mol^{-1}$)**
C(graphite)	5.740	$Br_2(\ell)$	152.231	Ca(s)	41.42
C(g)	158.096	$I_2(s)$	116.135	NaF(s)	51.5
$CH_4(g)$	186.264	Ar(g)	154.7	MgO(s)	26.94
$CH_3CH_3(g)$	229.60	$H_2(g)$	130.684	NaCl(s)	72.13
$CH_3CH_2CH_3(g)$	269.9	$N_2(g)$	191.61	KOH(s)	78.9
$CH_3OH(\ell)$	126.8	$O_2(g)$	205.138	$MgCO_3(s)$	65.7
CO(g)	197.674	$NH_3(g)$	192.45	$NH_4NO_3(s)$	151.08
$CO_2(g)$	213.74	HCl(g)	186.908	NaCl(aq)	115.5
$F_2(g)$	202.78	$H_2O(g)$	188.825	$NH_4NO_3(aq)$	259.8
$Cl_2(g)$	223.066	$H_2O(\ell)$	69.91	KOH(aq)	91.6

* Data from Wagman, D. D., Evans, W. H., Parker, V. B., Schumm, R. H., Halow, I., Bailey, S. M., Churney, K. L., and Nuttall, R. The NBS Tables of Chemical Thermodynamic Properties, *Journal of Physical and Chemical Reference Data*, Vol. 11, Suppl. 2, 1982.

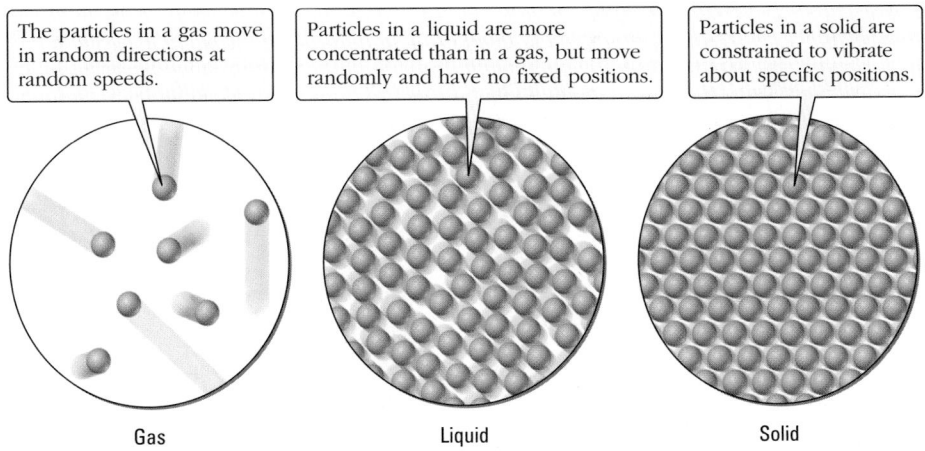

The particles in a gas move in random directions at random speeds.

Particles in a liquid are more concentrated than in a gas, but move randomly and have no fixed positions.

Particles in a solid are constrained to vibrate about specific positions.

Gas   Liquid   Solid

Figure 18.3 Entropies of solid, liquid, and gas phases.

of each substance at the standard pressure of 1 bar and the specified temperature of 25 °C. The units are joules per kelvin per mole ($J\ K^{-1}\ mol^{-1}$). Because there is a real zero on the entropy scale, the values in Table 18.1 are not measured relative to elements in their most stable form under standard-state conditions. Therefore, absolute entropies can be determined for elements as well as compounds.

Qualitative Guidelines for Entropy

Some useful guidelines can be drawn from the data given in Table 18.1.

- *Entropies of gases are usually much larger than those of liquids, which in turn are usually larger than those of solids.* In a solid the particles can vibrate only around their lattice positions. When a solid melts, its particles can move around more freely, and there is an increase in molar entropy. When a liquid vaporizes, the position restrictions due to forces between the particles nearly disappear, and there is another large entropy increase (Figure 18.3). For example, the entropies (in $J\ K^{-1}\ mol^{-1}$) of the halogens $I_2(s)$, $Br_2(\ell)$, and $Cl_2(g)$ are 116.1, 152.2, and 223.0, respectively. Similarly, the entropies of C(s, graphite) and C(g) are 5.7 and 158.1.

- *Entropies of more complex molecules are larger than those of simpler molecules, especially in a series of closely related compounds.* In a more complicated molecule there are more ways for the atoms to move about in three-dimensional space, and hence there is greater entropy. For example, the entropies (in $J\ K^{-1}\ mol^{-1}$) of methane (CH_4), ethane (CH_3CH_3), and propane ($CH_3CH_2CH_3$) are 186.26, 229.6, and 269.9, respectively. For atoms or molecules of similar molar mass, we have Ar, CO_2, and $CH_3CH_2CH_3$ with entropies of 154.7, 213.74, and 269.9 (Figure 18.4).

Name	Propane	Carbon dioxide	Argon
Molecular model	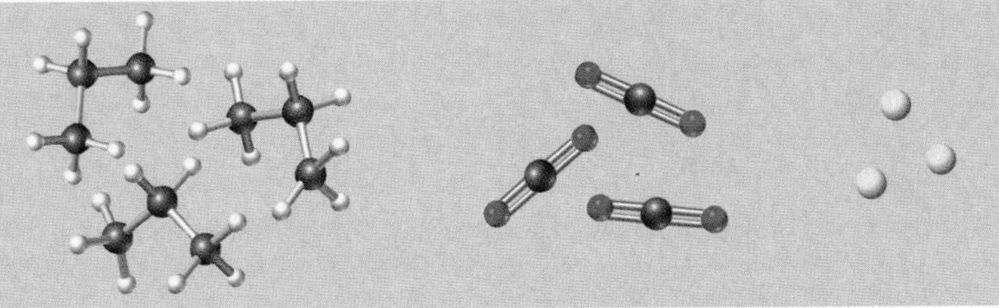		
Entropy ($J\ K^{-1}\ mol^{-1}$)	269.9	213.74	154.7
Molar mass (g/mol)	44.1	44.0	39.9

Figure 18.4 Entropy and molecular structure. Three groups of particles of similar molar mass are shown. In propane, $CH_3CH_2CH_3$, there are many different conformations (different ways to arrange the atoms relative to one another within a molecule), and there are many different bond stretching and bending vibrations possible (⬅ *p. 362).* In CO_2 there is a single conformation and many fewer vibrations. The individual Ar atoms can move about, but there are no conformations or vibrations possible.

Figure 18.5 Entropy and dissolving. There is usually an increase in entropy when a solid or liquid dissolves in a liquid solvent, because the solute particles become dispersed among the solvent particles.

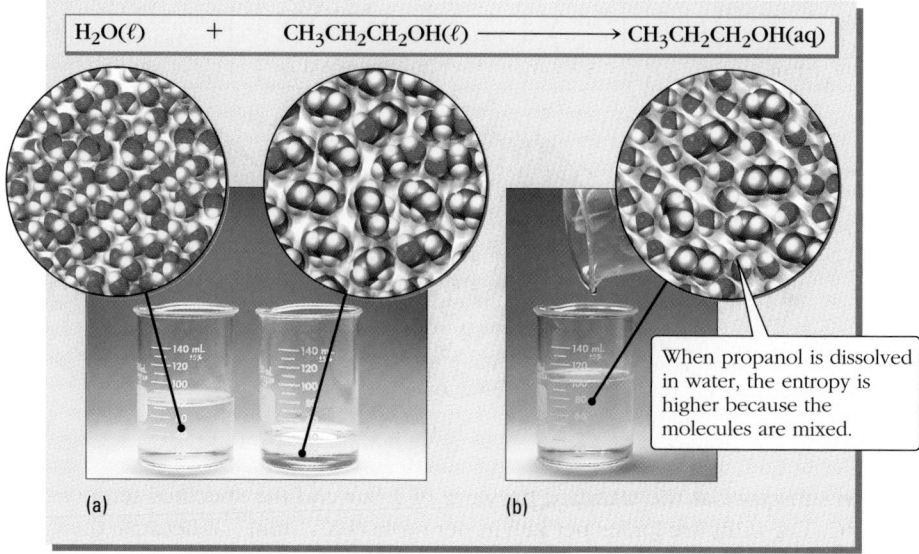

$H_2O(\ell)$ + $CH_3CH_2CH_2OH(\ell)$ ⟶ $CH_3CH_2CH_2OH(aq)$

(a)   (b)

When propanol is dissolved in water, the entropy is higher because the molecules are mixed.

When ionic solids that consist of very small ions (such as Li^+) or of ions that carry two or more units of charge (such as Mg^{2+} or Al^{3+}) dissolve in water, there is often a decrease in entropy. This happens because very small ions and highly charged ions have greater attraction for water dipoles. The water molecules surround such ions in a highly ordered structure. Although the ions disperse among the water molecules, the reduced disorder of the water molecules results in an overall decrease in entropy.

- *Entropies of ionic solids are larger the weaker the attractions among the ions are.* The weaker such forces are, the easier it is for ions to vibrate about their lattice positions and the greater the entropy is. The entropy of NaF(s) is $51.5\ J\ K^{-1}\ mol^{-1}$, and that of MgO(s) is $26.8\ J\ K^{-1}\ mol^{-1}$; Na^+ and F^-, with unit positive and negative charges, attract each other less than Mg^{2+} and O^{2-} (⟸ p. 95), and therefore NaF(s) has lower entropy. NaF(s) and NaCl(s) have entropies of 51.5 and $72.8\ J\ K^{-1}\ mol^{-1}$. Fluoride ions, F^-, are smaller than Cl^- ions, and attractions are smaller when the ions are farther apart.

- *Entropy usually increases when a pure liquid or solid dissolves in a solvent.* Matter usually becomes more dispersed or disordered when a substance dissolves and different kinds of molecules mix together (Figure 18.5). An example is $NH_4NO_3(s)$ and $NH_4NO_3(aq)$ with standard molar entropies of $151.08\ J\ K^{-1}\ mol^{-1}$ and $259.8\ J\ K^{-1}\ mol^{-1}$. Some ionic compounds dissolving in water are exceptions to this generalization because the ions are strongly hydrated.

- *Entropy decreases when a gas dissolves in a liquid.* Although gas molecules are dispersed among solvent molecules in solution, the very large entropy of the gas phase is lost when the widely separated gas particles become crowded together with solvent particles in the liquid solution (Figure 18.6).

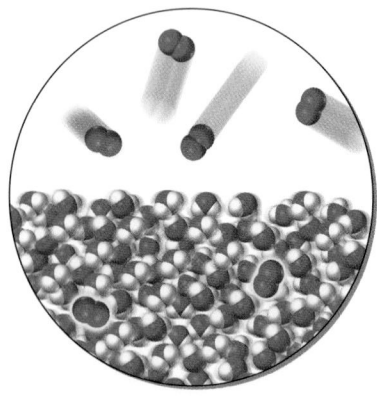

Figure 18.6 Entropy of solution of a gas. The very large entropy of the gas exceeds that of the solution. Even though particles are dispersed among each other in the liquid solution, the gas particles are much more widely spread out and have much higher entropy.

Problem-Solving Example 18.2 Relative Entropy Values

For each pair of substances below, predict which has greater entropy and give a reason for your choice. (Assume 1-mol samples at 25 °C and 1 bar.)

(a) $CH_3CH_2CH_2CH_3(g)$ or $CH_3CH_3(g)$ (b) $O_2(g)$ or $O_2(aq)$
(c) NaF(s) or KCl(s) (d) $CO_2(s)$ or $CO_2(g)$

Answer

(a) $CH_3CH_2CH_2CH_3(g)$ (b) $O_2(g)$ (c) KCl(s) (d) $CO_2(g)$

Explanation Use the rules given above.

(a) Larger, more complex molecules have greater entropy than similar smaller ones, so the entropy of $CH_3CH_2CH_2CH_3(g)$ is greater.

(b) The molecules of a gas are highly disordered, so when a gas dissolves in a liquid, the entropy decreases; therefore $O_2(g)$ has greater entropy.

(c) These are ionic solids, both with +1 and −1 ions; the attractive forces are greater the closer the ions are to each other, and the ions are smaller in NaF (see table of ionic radii, ⟵ *p. 301)*, so KCl has greater entropy.

(d) Entropy increases from solid to liquid to gas for the same substance, so $CO_2(g)$ has greater entropy.

Problem-Solving Practice 18.2

In each case, predict which of the two substances has greater entropy, assuming 1-mol samples at 25 °C and 1 bar. Then check your prediction by looking up each substance's absolute entropy in Table 18.1.

(a) C(g) or C(s, graphite)   (b) Ca(s) or Ar(g)   (c) KOH(s) or KOH(aq)

Predicting Entropy Changes

The general guidelines about entropy can be used to predict whether there will be an increase or decrease in entropy when reactants are converted to products. For both of the processes

$$H_2O(s) \longrightarrow H_2O(\ell) \quad \text{and} \quad H_2O(\ell) \longrightarrow H_2O(g)$$

an entropy increase is expected. Water molecules in the solid phase are more ordered than in the liquid, and water molecules in the liquid are much more ordered than in the gas. This is confirmed by the data in Table 18.1, where $S°(H_2O[g]) > S°(H_2O[\ell]) > S°(H_2O[s])$. For the decomposition of iron(III) oxide to its elements,

$$2 Fe_2O_3(s) \longrightarrow 4 Fe(s) + 3 O_2(g)$$

an increase in entropy is also predicted, because 3 mol of gaseous oxygen is present in the products and the reactant is a solid. This is confirmed by the experimental $\Delta S°$, which is 551.7 J/K. (Because gases have much higher entropy than solids or liquids, gaseous substances are most important in determining entropy changes.)

An example where a decrease in entropy can be predicted is

$$2 CO(g) + O_2(g) \longrightarrow 2 CO_2(g) \qquad \Delta S° = -173.0 \text{ J/K}$$

Here there is 3 mol of gaseous substance (2 mol CO and 1 mol O_2) at the beginning but only 2 mol of gaseous substance at the end of the reaction. Two moles of gas almost always contains less entropy than 3 mol of gas, and so $\Delta S°$ is negative (experimentally, $\Delta S° = -173.0$ J/K). Another example in which entropy decreases is the process

$$Ag^+(aq) + Cl^-(aq) \longrightarrow AgCl(s)$$

Here the reactant ions are free to move about among water molecules in aqueous solution, but those same ions are held in a crystal lattice in the solid, a situation with much greater constraint.

Predicting entropy changes for chemical processes is usually easier than predicting enthalpy changes. For gas-phase reactions the guideline is that having more bonds or stronger bonds (or both) in the products gives a negative $\Delta H°$ (⟵ *p. 238);* however, a table of bond enthalpies is usually needed to tell which bonds are stronger.

Exercise 18.4 Predicting Entropy Changes

For each process, tell whether entropy increases or decreases, and explain how you arrived at your prediction.

(a) $2 CO_2(g) \rightarrow 2 CO(g) + O_2(g)$

(b) NaCl(s) → NaCl(aq)

(c) $MgCO_3(s) \xrightarrow{\text{heat}} MgO(s) + CO_2(g)$

18.4 CALCULATING ENTROPY CHANGES

CD-ROM Screen 20.5: Calculating ΔS for a Chemical Reaction

The standard molar entropy values given in Table 18.1 can be used to calculate entropy changes for physical and chemical processes. Assume that each reactant and each product is at the standard pressure of 1 bar and at the temperature given (298.15 K). The number of moles of each substance is specified by its stoichiometric coefficient in the equation for the process. Multiply the entropy of each product substance by the number of moles of that product and add the entropies of all products. Calculate the total entropy of the reactants in the same way and subtract it from the total entropy of the products. This is summarized in the equation

Notice that the equation for calculating $\Delta S°$ has the same form as that for calculating $\Delta H°$ for a reaction (⟸ *p. 246*).

$$\Delta S° = \Sigma\{(\text{moles of product}) \times S°(\text{product})\} - \Sigma\{(\text{moles of reactant}) \times S°(\text{reactant})\}$$

It is important to note that this calculation gives the entropy change for the chemical reaction *system* only. It tells whether the atoms that make up the system are more dispersed or less dispersed after the reaction than before it. It does not take account of any entropy change in the surroundings.

Problem-Solving Example 18.3 Calculating an Entropy Change from Tabulated Values

The reaction

$$CO(g) + 2 H_2(g) \longrightarrow CH_3OH(\ell)$$

is being evaluated as a possible way to manufacture liquid methanol, $CH_3OH(\ell)$, for use in motor fuel. Calculate $\Delta S°$ for the reaction.

Answer $\Delta S° = -332.3 \, J \, K^{-1} \, mol^{-1}$

Explanation Use information in Table 18.1, subtracting the entropies of the reactants from the entropy of the product. Because all substances, including elements, have nonzero absolute entropy values, elements as well as compounds must be included.

$$\Delta S° = \Sigma\,\{(\text{moles of product}) \times S°(\text{product})\} - \Sigma\,\{(\text{moles of reactant}) \times S°(\text{reactant})\}$$

$$= (1 \text{ mol}) \times S°[(CH_3OH(\ell)] - \{(1 \text{ mol}) \times S°[CO(g)] + (2 \text{ mol}) \times S°[H_2(g)]\}$$

$$= (1 \text{ mol}) \times (126.8 \, J \, K^{-1} \, mol^{-1}) - \{(1 \text{ mol}) \times (197.7 \, J \, K^{-1} \, mol^{-1})$$

$$+ (2 \text{ mol}) \times (130.7 \, J \, K^{-1} \, mol^{-1})\}$$

$$= -332.3 \, J/K$$

✔ $\Delta S°$ is negative, which is reasonable because 3 mol of gaseous reactants is converted to 1 mol of a liquid-phase product. Even though the product molecule is more complicated, the fact that it is a liquid makes its entropy much smaller than that of 1 mol of gaseous carbon monoxide and 2 mol of gaseous hydrogen.

Problem-Solving Practice 18.3

Use absolute entropies from Table 18.1 to calculate the entropy change for each of the following processes, hence verifying the predictions made in Exercise 18.4.

(a) $2 CO_2(g) \rightarrow 2 CO(g) + O_2(g)$

(b) $NaCl(s) \rightarrow NaCl(aq)$

(c) $MgCO_3(s) \xrightarrow{\text{heat}} MgO(s) + CO_2(g)$

18.5 ENTROPY AND THE SECOND LAW OF THERMODYNAMICS

CD-ROM Screen 20.6: The Second Law of Thermodynamics

A great deal of experience with many chemical reactions and other processes in which energy is transferred is consistent with the conclusion that *whenever a product-favored chemical or physical process occurs, matter, energy, or both be-*

come more dispersed or disordered. This is summarized in the **second law of thermodynamics,** which states that the *total entropy of the universe (the system plus its surroundings) is continually increasing.* Evaluating whether this will happen during a proposed chemical reaction allows us to predict whether reactants will form appreciable quantities of products.

Predicting whether a reaction is product-favored can be done in three steps:

1. Calculate how much entropy is created or destroyed by dispersal or concentration of energy ($\Delta S_{surroundings}$).
2. Calculate how much entropy is created or destroyed by dispersal or concentration of matter (ΔS_{system}).
3. Add these two results to get $\Delta S_{universe} = \Delta S_{system} + \Delta S_{surroundings}$.

Let us apply these steps to the reaction

$$CO(g) + 2\,H_2(g) \longrightarrow CH_3OH(\ell)$$

for which we calculated the entropy change in Problem-Solving Example 18.3. If the reaction is product-favored, it would be a good way to produce methanol for use as automotive fuel. The reactants can be obtained from plentiful resources: coal and water. We base our prediction upon having as reactants 1 mol of $CO(g)$ and 2 mol of $H_2(g)$ and as product 1 mol of liquid methanol, with all substances at 1 bar and 298.15 K (25 °C). Then data from Table 6.4 *(⇐ p. 245)* and Table 18.1 will apply. If the entropy of the universe is predicted to be higher after the product has been produced, then the reaction is product-favored under these conditions and might be useful. If not, perhaps some other conditions could be used, or perhaps we should consider some other reaction altogether.

Step 1: *Calculate the reaction's dispersal or concentration of energy by calculating ΔH° and assuming that this quantity of thermal energy is transferred reversibly to or from the surroundings.* The entropy change for the surroundings can be calculated as

$$\Delta S^\circ_{surroundings} = \frac{q_{rev}}{T} = \frac{q_{surroundings}}{T} = \frac{\Delta H^\circ_{surroundings}}{T} = \frac{-\Delta H^\circ_{system}}{T} = \frac{-\Delta H^\circ}{T}$$

The minus sign in this equation comes from the fact that the direction of energy transfer for the surroundings is opposite from the direction of energy transfer for the system. For an exothermic reaction (negative ΔH°) there will be an increase in entropy of the surroundings, a fact that we have already mentioned. For the proposed methanol-producing reaction, ΔH° (calculated from Table 6.4) is -128.1 kJ, and so the entropy change is

$$\Delta S^\circ_{surroundings} = \frac{-(-128.1\ \text{kJ})}{298\ \text{K}} \times \frac{1000\ \text{J}}{\text{kJ}} = 430.\ \text{J/K}$$

Step 2: *Calculate the entropy change for dispersal of matter. This is just the entropy change for the system itself* (the atoms involved in the methanol-producing reaction). This entropy change can be evaluated from the absolute entropies of the products and reactants as described in the previous section and has already been calculated in Problem-Solving Example 18.3 to be

$$\Delta S^\circ_{system} = \Delta S^\circ = -332.3\ \text{J/K}$$

Step 3: *Calculate the total entropy change for system and surroundings,* $\Delta S^\circ_{universe}$. Because the universe includes both system and surroundings, $\Delta S^\circ_{universe}$ is the sum of the entropy change for the system and the entropy change for the surroundings. (We assume that nothing else but our

Steps 1 and 2 can be carried out by assuming that reactants under standard conditions of 1 bar and a specified temperature are converted to products under the same standard conditions.

We shall consider the effect of changing temperature later in this chapter. The effects of changing other conditions can often be predicted qualitatively using Le Chatelier's principle *(⇐ p. 655).*

Entropy diagram for combination of carbon monoxide and hydrogen to form methanol.

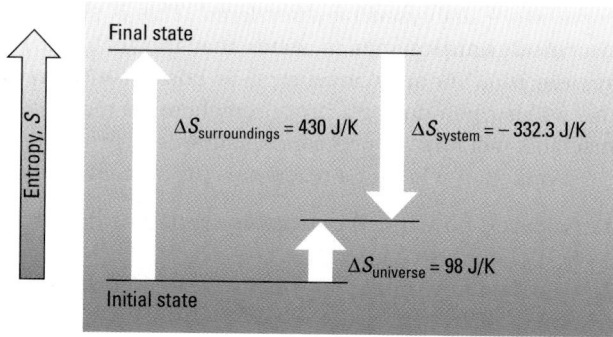

reaction happens, and so there are no other entropy changes.) This total entropy change is

$$\Delta S^\circ_{universe} = \Delta S^\circ_{surroundings} + \Delta S^\circ_{system} = \frac{-\Delta H^\circ_{system}}{T} + \Delta S^\circ_{system} \quad [18.1]$$

$$= (430. - 332.3) \text{ J/K} = 98 \text{ J/K}$$

Combination of carbon monoxide and hydrogen to form methanol is accompanied by an increase in entropy of the universe. The process is product-favored and might be useful for manufacturing methanol.

CHEMISTRY
You Can Do... Rubber Bands and Thermodynamics

Here are two experiments that you can analyze with the laws of thermodynamics. For the first experiment you need a wide rubber band and a stiff upper lip. (Your lip is a good detector of small temperature changes.) For the second experiment you will need the same rubber band, something heavy enough to stretch the rubber band, like a hammer or a coffee mug, and a hair dryer.

In this experiment the rubber band is the thermodynamic system; you and your lip are part of the surroundings. Hold the rubber band against your lip and quickly stretch it to its limit. What did you feel? Was the process of stretching the rubber band exothermic or endothermic? What was the

sign of q for this process? When you stretched the rubber band you had to work to do it. For the rubber band, what is the sign of w for the stretching process? Can you determine the sign of ΔE for the rubber band? If so, what is it? If not, why not? What kinds of experiments might you devise to make possible a determination of both the sign and the value of ΔE? Since you carried out this experiment at constant (atmospheric) pressure, what was the sign of ΔH?

Now for the second experiment. Use the rubber band to hang the hammer or coffee mug from a door knob. (You could do this on a larger scale with a bungee cord!) Once the suspended object has stopped bouncing up and down, turn on the hair dryer and heat the rubber band. What happens? Is this what you expected? Do most objects do this when heated?

Now think about what happened when you released the stretched rubber band in the first experiment.

1. What is the sign of the enthalpy change for unstretching the rubber band?
2. Was what happened when you released the rubber band a product-favored process?
3. Is that process favored by the enthalpy change?
4. What must be the sign of the entropy change for unstretching a rubber band?
5. Use your conclusions about the entropy and enthalpy changes for unstretching the rubber band to account for the observation you made in the second experiment.

Exercise 18.5 Effect of Temperature on Entropy Change

The reaction of carbon monoxide with hydrogen to form methanol is quite slow at room temperature. As a general rule, reactions go faster at higher temperatures. Suppose that you tried to speed up this reaction by increasing the temperature.

(a) Assuming that $\Delta H°$ does not change very much as the temperature changes, what effect would increasing the temperature have on $\Delta S°_{surroundings}$?

(b) Assuming that $\Delta S°$ for a reaction system does not change much as the temperature changes, what effect would increasing the temperature have on $\Delta S°_{universe}$?

Problem-Solving Example 18.4 Calculating $\Delta S°_{universe}$

Calculate $\Delta S°_{universe}$ for the combustion of methane in air, and from your result confirm that this reaction is product-favored. (Assume that reactants and products are at 298.15 K and use data from Appendix J.)

Answer 2745.4 J/K

Explanation Begin by writing a balanced equation for the reaction:

$$CH_4(g) + 2\,O_2(g) \longrightarrow CO_2(g) + 2\,H_2O(\ell)$$

Even though water vapor might be produced initially, write $H_2O(\ell)$, because at a temperature of 25 °C, water vapor would condense to the liquid form. Next calculate $\Delta H°$ and $\Delta S°$ using the equations involving products minus reactants.

$$\Delta H° = \sum \{(\text{moles of product}) \times \Delta H_f° \,(\text{product})\}$$

$$- \sum \{(\text{moles of reactant}) \times \Delta H_f° \,(\text{reactant})\}$$

$$= \{-393.509 \text{ kJ} + 2 \times (-285.830 \text{ kJ})\} - \{-74.81 \text{ kJ} + 2 \times 0 \text{ kJ}\} = -890.36 \text{ kJ}$$

$$\Delta S° = \sum \{(\text{moles of product}) \times S°(\text{product})\}$$

$$- \sum \{(\text{moles of reactant}) \times S°(\text{reactant})\}$$

$$= \{213.74 \text{ J/K} + 2 \times (69.91 \text{ J/K})\} - \{186.264 \text{ J/K} + 2 \times (205.138 \text{ J/K})\}$$

$$= -242.98 \text{ J/K}$$

Now use Equation 18.1 to calculate $\Delta S°_{universe}$

$$\Delta S°_{universe} = \frac{-\Delta H°}{T} + \Delta S° = \frac{-(-890.36 \text{ kJ/K})}{298.15 \text{ K}} + (-242.98 \text{ J/K})$$

$$= 2.9863 \text{ kJ/K} - 242.98 \text{ J/K}$$

$$= 2986.3 \text{ J/K} - 242.98 \text{ J/K} = 2743.3 \text{ J/K}$$

Because $\Delta S°_{universe}$ is positive, the reaction is product-favored.

Notice that for $-\Delta H°/T$ the units were kilojoules per kelvin (kJ/K), while for $\Delta S°$ the units were joules per kelvin (J/K). Therefore it was necessary to convert kilojoules to joules by multiplying the first term by 1000 J/kJ.

✔ A positive result is reasonable, because you know that methane burns in air, which means that the reaction is product-favored. Even though the entropy change for the system is unfavorable, the reaction is highly product-favored because it is strongly exothermic.

Problem-Solving Practice 18.4

Use Data from Appendix J to determine whether the synthesis of ammonia from nitrogen and hydrogen is product-favored at 298.15 K and 1 bar.

Exercise 18.6 Variation of $\Delta H°$ and $\Delta S°$ with Temperature

Suppose that the combustion of methane were carried out at 150 °C.

(a) How would this affect the chemical equation for the reaction?
(b) What effect would the change in the chemical equation have on $\Delta H°$ and $\Delta S°$ for the reaction?
(c) When is it definitely *not* safe to assume that $\Delta H°$ and $\Delta S°$ for a reaction will be almost the same over a broad range of temperatures?

Predictions of the sort we have just made by calculating $\Delta S°_{universe}$ can also be made qualitatively, without calculating, if we know whether a reaction is exothermic and if we can predict whether there is a dispersal of matter when the reaction takes place. *A reaction is certain to be product-favored if it is exothermic and the atoms of the reaction system are more disordered afterwards than before. Also, a reaction is certainly not product-favored if it is endothermic and there is a decrease in entropy for the system.* There are two other possible cases, as indicated in Table 18.2, but they are more difficult to predict without quantitative information.

As examples, consider the reactions of carbonates with acids *(⬅ p. 177)*. These reactions are product-favored because they are exothermic and produce highly disordered gases and solutions. Reaction of limestone with hydrochloric acid is typical.

$$CaCO_3(s) + 2\ HCl(aq) \longrightarrow CaCl_2(aq) + H_2O(\ell) + CO_2(g) \qquad \text{(exothermic)}$$

Similarly, combustion reactions of hydrocarbons such as butane, $CH_3CH_2CH_2CH_3$, are product-favored because they are exothermic and produce a larger number of gas-phase product molecules than there were gas-phase reactant molecules.

$$2\ CH_3CH_2CH_2CH_3(g) + 13\ O_2(g) \longrightarrow 8\ CO_2(g) + 10\ H_2O(g) \qquad \text{(exothermic)}$$

But what about a reaction such as the production of ethylene, $CH_2{=}CH_2$, from ethane, CH_3CH_3? Although entropy is predicted to increase (one gas-phase molecule forms two), the reaction is very endothermic.

$$CH_3CH_3(g) \longrightarrow H_2(g) + CH_2{=}CH_2(g) \qquad \Delta H° = +137\ kJ \quad \Delta S° = +121\ J/K$$

Enthalpy change predicts that this process is reactant-favored, while entropy change predicts the opposite. Which is more important? It depends on the temperature.

Calculating $\Delta S°_{surroundings}$ $(=\Delta H°/T)$ involves dividing the enthalpy change by the temperature. Because $\Delta H°$ stays pretty much the same at different temperatures, the bigger the temperature the smaller the absolute value of $\Delta S°_{surroundings}$ $(=\Delta H°/T)$. At room temperature $\Delta S°_{surroundings}$ is usually bigger in absolute value than $\Delta S°_{system}$, and so exothermic reactions are expected to be product-favored and endothermic reactions (like this one) are expected to be reactant-favored. The

A product-favored reaction.
Carbonates react rapidly with acid to produce carbon dioxide and water.

TABLE 18.2	Predicting Whether a Reaction is Product-Favored	
Sign of ΔH_{system}	**Sign of ΔS_{system}**	**Product-favored?**
Negative (exothermic)	Positive	Yes
Negative (exothermic)	Negative	Yes at low T; no at high T
Positive (endothermic)	Positive	No at low T; yes at high T
Positive (endothermic)	Negative	No

ethylene-producing reaction is reactant-favored at 25 °C, because $\Delta S^\circ_{universe} = -339$ J/K. To make a successful industrial process, chemical engineers have designed plants that carry out this reaction at about 1000 °C. At this higher temperature, $\Delta S^\circ_{surroundings}$ is smaller in magnitude than ΔS°_{system}. Thus $\Delta S^\circ_{universe} = 13$ J/K, and products are predicted to predominate over reactants.

At 25 °C:

$$\Delta S^\circ_{surroundings} = \frac{-137 \text{ kJ}}{298 \text{K}}$$
$$= -460 \text{ J/K}$$
$$\Delta S^\circ_{system} = 121 \text{ J/K}$$
$$\Delta S^\circ_{universe} = (-460 + 121) \text{ J/K}$$
$$= -339 \text{ J/K}$$

At 1000 °C:

$$\Delta S^\circ_{surroundings} = \frac{-137 \text{ kJ}}{(1000 + 273)\text{K}}$$
$$= -108 \text{ J/K}$$
$$\Delta S^\circ_{system} = 121 \text{ J/K}$$
$$\Delta S^\circ_{universe} = 13 \text{ J/K}$$

Exercise 18.7 Predicting the Direction of a Reaction

Complete the table and then classify each reaction into one of the four types in Table 18.2. From this, predict whether each reaction is product-favored or reactant-favored at room temperature.

Reaction	ΔH°, 298 K (kJ)	ΔS°, 298 K (J/K)
(a) $C_2H_4(g) + 3 O_2(g) \rightarrow$ $2 H_2O(\ell) + 2 CO_2(g)$	_____	_____
(b) $2 Fe_2O_3(s) + 3 C(graphite) \rightarrow$ $4 Fe(s) + 3 CO_2(g)$	_____	_____
(c) $C(graphite) + O_2(g) \rightarrow CO_2(g)$	_____	_____
(d) $2 Ag(s) + 3 N_2(g) \rightarrow 2 AgN_3(s)$	_____	_____

Exercise 18.8 Product- or Reactant-Favored?

(a) Is the combination reaction of hydrogen gas and chlorine gas to give hydrogen chloride gas (at 1 bar) predicted to be product-favored or reactant-favored at 298 K?

(b) What is the value for $\Delta S^\circ_{universe}$?

18.6 GIBBS FREE ENERGY

Calculations of the sort done in the previous section would be simpler if we did not have to separately evaluate the entropy change of the surroundings from T and a table of ΔH°_f values and the entropy change of the system from a table of S° values. To simplify such calculations, a new thermodynamic function was defined by J. Willard Gibbs (1838–1903). It is now called **Gibbs free energy** and given the symbol G. Gibbs defined his free energy so that $\Delta G_{system} = -T\Delta S_{universe}$. Notice that because of the minus sign, if the entropy of the universe increases, the Gibbs free energy of the system must decrease. That is, *a decrease in Gibbs free energy of a system is characteristic of a product-favored process.*

In the previous section, Equation 18.1 showed that the total entropy change accompanying a chemical reaction carried out at constant temperature and pressure is

$$\Delta S_{universe} = \Delta S_{surroundings} + \Delta S_{system} = \frac{-\Delta H_{system}}{T} + \Delta S_{system} \qquad [18.1]$$

Combining this algebraically with Gibbs's definition of free energy, we have

$$\Delta G_{system} = -T\Delta S_{universe} = -T\left[\frac{-\Delta H_{system}}{T} + \Delta S_{system}\right] = \Delta H_{system} - T\Delta S_{system}$$

or, under standard-state conditions,

$$\Delta G^\circ_{system} = \Delta H^\circ_{system} - T\Delta S^\circ_{system} \qquad [18.2]$$

CD-ROM Screen 20.7: Gibbs Free Energy: Putting ΔH and ΔS Together

In Chapter 14 we stated that if a reaction is exothermic or involves an increase in entropy of the system, this favors the products. The entropy effect becomes more important the higher the temperature is.

This equation summarizes the ideas about chemical equilibrium that were developed in Chapter 14 ($\Leftarrow$ *p. 665*). A negative value of $\Delta G°_{system}$ indicates that a reaction is product-favored, and the equation says that two conditions will make $\Delta G°_{system}$ more negative: (1) if the reaction is exothermic, $\Delta H°_{system}$ will be negative, thereby favoring the products, and (2) if the products have greater entropy than the reactants, then $\Delta S°_{system}$ will be positive, the $-T\Delta S°_{system}$ term will be negative, and this favors the products. Because $\Delta S°_{system}$ is multiplied by T, the entropy of the system is more important at higher temperatures.

 ### Exercise 18.9   Predicting Whether a Process is Product-Favored

Make a table similar to Table 18.2, but add a column for $\Delta G°$. Based on the value of $\Delta G°$, predict whether the reaction is product-favored. If there is insufficient information, indicate whether the products would be favored more at high temperatures or at low temperatures. Check your results against Table 18.2.

The Gibbs free energy change provides a way of predicting whether a reaction will be product-favored that depends only on the system—the chemical substances undergoing reaction. Therefore, we can tabulate values of the standard Gibbs free energy of formation, $\Delta G°_f$ for a variety of substances, and from them calculate

For $\Delta H°$, $\Delta S°$, and $\Delta G°$ you can assume that the values apply to the system; that is, $\Delta G° = \Delta G°_{system}$, unless a subscript is attached to indicate that a value is for the surroundings.

$$\Delta G° = \Sigma \ \{(\text{moles of product}) \times \Delta G°_f (\text{product})\} \qquad [18.3]$$
$$- \ \Sigma \ \{(\text{moles of reactant}) \times \Delta G°_f (\text{reactant})\}$$

for a great many reactions. The calculation is similar to using $\Delta H°_f$ values from Table 6.2 or Appendix J to calculate $\Delta H°$ for a reaction ($\Leftarrow$ *p. 246*). As was the case for $\Delta H°_f$ values, there are no $\Delta G°_f$ values for elements, because forming an element from itself constitutes no change at all. Appendix J contains a table that includes $\Delta G°_f$ values for many compounds.

It is important to realize that $\Delta G°$ varies significantly as the temperature changes (because of the $-T\Delta S°$ term). Therefore, values of $\Delta G°$ calculated from Equation 18.3 apply only to the temperature specified in the table of $\Delta G°_f$ values. Appendix J specifies a temperature of 25 °C.

Combustion of natural gas (methane).

Problem-Solving Example 18.5   Using Standard Free Energies of Formation

Calculate the standard Gibbs free energy change for combustion of methane using values of $\Delta G°_f$ from Appendix J.

Answer -800.9 kJ

Explanation   Write a balanced equation for the combustion reaction and look up $\Delta G°_f$ values in Appendix J.

$$CH_4(g) \ + \ 2 O_2(g) \ \longrightarrow \ 2 H_2O(g) \ + \ CO_2(g)$$

$\Delta G°_f$(kJ/mol): -50.7 0 -228.6 -394.4

(Notice that elements in their standard states have $\Delta G°_f = 0$, just as they have $\Delta H°_f = 0$.) Now calculate

$$\Delta G° = \Sigma \ \{(\text{moles of product}) \times \Delta G°_f (\text{product})\}$$
$$- \ \Sigma \ \{(\text{moles of reactant}) \times \Delta G°_f (\text{reactant})\}$$
$$= \ \{(2 \ \text{mol} \times \Delta G°_f [H_2O(g)]\} + \{1 \ \text{mol} \times \Delta G°_f [CO_2(g)]\}$$
$$- \ \{1 \ \text{mol} \times \Delta G°_f [CH_4(g)]\} - \{2 \ \text{mol} \times \Delta G°_f [O_2(g)]\}$$
$$= \ [2 \ \text{mol} \times (-228.6 \ \text{kJ/mol}\} + \{1 \ \text{mol} \times (-394.4 \ \text{kJ/mol})\}$$
$$- \ \{1 \ \text{mol} \times (250.7 \ \text{kJ/mol})\} - \{2 \ \text{mol} \times (0 \ \text{kJ/mol})\}$$
$$= -800.9 \ \text{kJ}$$

The Gibbs free energy change for this combustion reaction, $\Delta G°$, is a large negative number, clearly indicating that the reaction is product-favored under standard conditions.

Problem-Solving Practice 18.5

In the text we concluded that the reaction to produce methanol from CO and H_2 is product-favored.

$$CO(g) + 2 H_2(g) \longrightarrow CH_3OH(\ell)$$

(a) Verify this result by calculating $\Delta G°$ from $\Delta H°$ and $\Delta S°$ for the system. Use values of $\Delta H_f°$ and $S°$ from Appendix J.
(b) Compare your result in part (a) with the calculated value of $\Delta G°$ obtained from $\Delta G_f°$ values from Appendix J.
(c) Is the sign of $\Delta G°$ positive or negative? Is the reaction product-favored? At all temperatures?

The Effect of Temperature on Reaction Direction

Many reactions are product-favored at some temperatures and reactant-favored at others. This means that it might be possible to make the reaction produce products by increasing or decreasing the temperature. There is a simple, approximate way to estimate the temperature at which a reactant-favored process becomes product-favored.

CD-ROM Screen 20.8: Free Energy and Temperature

In Exercises 18.5 and 18.6 ($\Leftarrow$ **p. 825 and p. 826**) we developed the idea that $\Delta H°$ and $\Delta S°$ have nearly constant values over a broad range of temperatures, provided that each of the substances involved in a chemical reaction remains in the same state of matter (solid, liquid, or gas). Because $\Delta H°$ and $\Delta S°$ are nearly constant, the T on the right-hand side of the equation $\Delta G° = \Delta H° - T\Delta S°$ implies that $\Delta G°$ must vary with temperature. It also implies that if we know $\Delta H°$ and $\Delta S°$ at one temperature, we can estimate $\Delta G°$ over a range of temperatures. For example, suppose we are interested in whether the reaction

$$2 HgO(s) \longrightarrow 2 Hg(\ell) + O_2(g)$$

will produce products at a temperature of 350. °C and a pressure of 1 bar. To find out, calculate $\Delta H°$ and $\Delta S°$ at 298 K and then estimate $\Delta G°$, assuming that $\Delta H°$ and $\Delta S°$ have the same values at 350. °C (623 K) that they do at 298 K. The boiling point of mercury is 356 °C, and mercury(II) oxide does not melt until well above 500 °C, so the substances are all in the same states at 350. °C that they were at 0 °C.

$$\Delta S°(298.15 \text{ K}) = \{2 \text{ mol} \times S°[Hg(\ell)] + (1 \text{ mol}) \times S°[O_2(g)]\}$$
$$- \{2 \text{ mol} \times S°[HgO(s)]\}$$
$$= 2 \times 76.02 \text{ J/K} + 205.138 \text{ J/K} - 2 \times 70.29 \text{ J/K} = 216.60 \text{ J/K}$$

$$\Delta H°(298.15 \text{ K}) = \{2 \text{ mol} \times \Delta H_f°[Hg(\ell)] + 1 \text{ mol} \times \Delta H_f°[O_2(g)]\}$$
$$- \{2 \text{ mol} \times \Delta H_f°[HgO(s)]\}$$
$$= 0 \text{ kJ} + 0 \text{ kJ} - 2 \times (-90.83) \text{ kJ} = 181.66 \text{ kJ}$$

$$\Delta G°(623 \text{ K}) = \Delta H°(298.15 \text{ K}) - T \times \Delta S°(298.15 \text{ K})$$
$$= 181.66 \text{ kJ} - 623 \text{ K} \times 216.60 \text{ J/K}$$
$$= 181.66 \text{ kJ} - 134,942 \text{ J} = 181.66 \text{ kJ} - 134.94 \text{ kJ} = 47 \text{ kJ}$$

The reaction is not product-favored at 623 K, because $\Delta G°$ is positive. But $\Delta G°$ has a smaller positive value than at 298 K, where it has the value

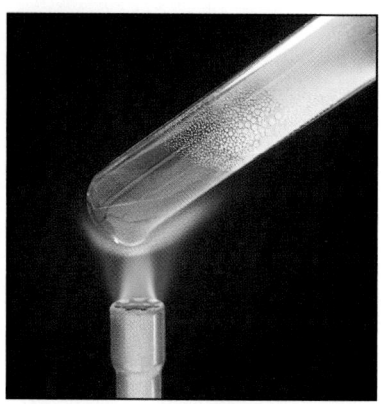

Figure 18.7   Decomposition of HgO(s).   When heated, red mercury(II) oxide decomposes to liquid mercury metal and oxygen gas.

Joseph Priestley's discovery of oxygen involved heating mercury(II) oxide to decompose it.

$\Delta G° = -(2 \text{ mol}) \times \Delta G_f°(\text{HgO[s]}) = +117.1 \text{ kJ}$. That is, the reaction is not as reactant-favored at 623 K as it was at room temperature. Because the calculated values of $\Delta H°$ and $\Delta S°$ are both positive, the reaction is expected to be product-favored at high temperatures but not at low temperatures. Heating to an even higher temperature than 623 K does decompose mercury(II) oxide (see Figure 18.7).

Exercise 18.10   High-Temperature Decomposition

Suppose that a sample of HgO(s) is heated above the boiling point of mercury (356 °C).

(a) Could you use the same method to estimate the Gibbs free energy change that was used in the preceding paragraph? Why or why not? (*Hint:* Write the equation for the process that would occur if the temperature were 400 °C.)

(b) At 400 °C, would you expect the reaction to be more or less product-favored than it was at 350 °C? Give two reasons for your choice.

For reactions such as decomposition of mercury(II) oxide that are reactant-favored at low temperatures and product-favored at high temperatures, it is possible to calculate the minimum temperature to which the system must be heated to make it product-favored. Below that temperature $\Delta G°$ is positive, and above that temperature $\Delta G°$ is negative. Therefore, $\Delta G°$ must equal zero at the desired temperature. Because $\Delta G° = \Delta H° - T\Delta S°$ (Equation 18.2), we can set $\Delta H° - T\Delta S° = 0$. Solving for T gives

$$T(\text{at which } \Delta G° \text{ changes sign}) = \frac{\Delta H°}{\Delta S°} \qquad [18.4]$$

Equation 18.4 also applies to reactions that are product-favored at low temperatures and reactant-favored at high temperatures. For such reactions Equation 18.4 gives the temperature *below* which the reaction is product-favored. Heating the system above this temperature will probably result in insufficient quantities of products being produced. As an example, the contributions of $\Delta H°$ and $-T\Delta S°$ to $\Delta G°$ for the decomposition of silver(I) oxide are shown graphically in Figure 18.8.

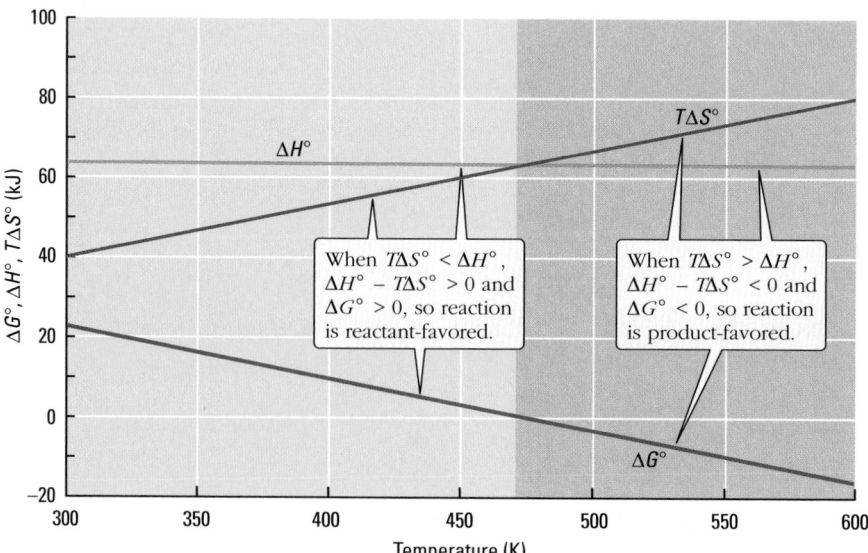

Figure 18.8   Effect of temperature on reaction spontaneity.   The two terms, $\Delta H°$ and $-T\Delta S°$, that contribute to the Gibbs free energy change are plotted as a function of temperature for the reaction of silver ore, $Ag_2O(s)$, to form silver metal and oxygen gas.

Problem-Solving Example 18.6 Effect of Temperature on
Gibbs Free Energy Change

Calculate the temperature to which silver ore, $Ag_2O(s)$, must be heated to decompose the ore to oxygen gas and silver metal.

Answer 468 K (741 °C)

Explanation Begin by writing the equation for the reaction, and use values from Appendix J to calculate $\Delta H°$ and $\Delta S°$ at 25 °C. Then use Equation 18.4 to calculate the desired temperature

$$2\,Ag_2O(s) \longrightarrow 4\,Ag(s) + O_2(g)$$

$\Delta H° = (4\text{ mol}) \times (0\text{ kJ/mol}) + (1\text{ mol}) \times (0\text{ kJ/mol})$

$\qquad\qquad\qquad\qquad - (2\text{ mol}) \times (-31.05\text{ kJ/mol}) = 62.10\text{ kJ}$

$\Delta S° = (4\text{ mol}) \times (42.55\text{ J K}^{-1}\text{ mol}^{-1}) + (1\text{ mol}) \times (205.138\text{ J K}^{-1}\text{ mol}^{-1})$

$\qquad\qquad\qquad\qquad - (2\text{ mol}) \times (121.3\text{ J K}^{-1}\text{ mol}^{-1}) = 132.7\text{ J/K}$

Because both $\Delta H°$ and $\Delta S°$ are positive, the reaction will become more product-favored as temperature increases.

$$T = \frac{62.1\text{ kJ}}{132.7\text{ J/K}} \times \frac{1000\text{ J}}{1\text{ kJ}} = \frac{62.1\text{ kJ}}{0.1327\text{ kJ/K}} = 467.9\text{ K} = 468\text{ K}$$

✔ The reaction produces a gas and a solid from a solid, so the entropy change should be positive. Silver is one of the few metals that can be found as the element in nature, so its oxide probably does not require an extremely high temperature to decompose, and 468 K (741 °C) is reasonable. Check that each reactant and product is in the same physical state at 741 °C as it was at room temperature. *The CRC Handbook* gives the melting point of silver as 962 °C and reports that silver(I) oxide decomposes before it melts, so the assumption of nearly constant $\Delta S°$ and $\Delta H°$ is valid.

Problem-Solving Practice 18.6

For the reaction

$$2\,CO(g) + O_2(g) \longrightarrow 2\,CO_2(g)$$

(a) Predict the temperature at which the reaction changes from being reactant-favored to being product-favored.

(b) If you wanted this reaction to produce $CO_2(g)$, what temperature conditions would you choose?

18.7 GIBBS FREE ENERGY CHANGES AND EQUILIBRIUM CONSTANTS

The difference between the standard Gibbs free energies of products and reactants determines whether a reaction is product-favored, but so far we have only considered pure reactants and pure products. It is also important to consider what happens to the Gibbs free energy *during a reaction* (when some, but not all, of the reactants have been converted to products), because that determines the position of equilibrium.

 CD-ROM Screen 20.9:
Thermodynamics and the
Equilibrium Constant

Variation of Gibbs Free Energy During a Reaction

One way to remove carbon dioxide from air is to pass the air over solid sodium hydroxide.

$$NaOH(s) + CO_2(g) \longrightarrow NaHCO_3(s) \qquad \Delta G° \text{ (at 25 °C)} = -77.2\text{ kJ} \qquad [18.5]$$

Figure 18.9 Gibbs free energy as a function of extent of reaction. For the reaction of NaOH(s) with CO_2(g), the Gibbs free energy decreases linearly as the reaction proceeds. A similar linear graph is obtained whenever all reactants and products are in their standard states throughout a reaction. The slope of such a graph is equal to $\Delta G°$. The Gibbs free energy of the reactants has been arbitrarily set to zero so that it is easier to see how big the differences in Gibbs free energy are.

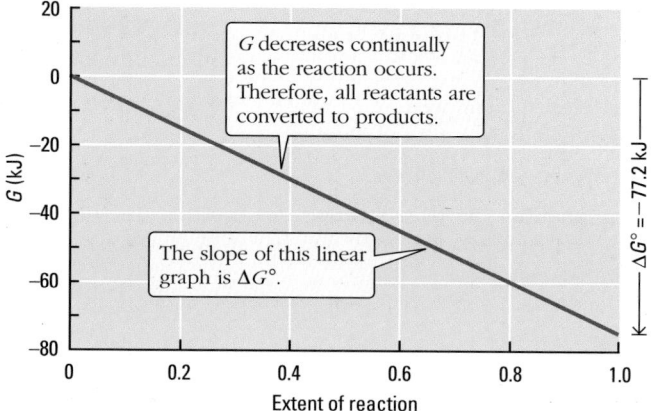

Recall that the standard-state pressure is 1 bar, which is nearly the same as 1 atm.

Recall that a thermochemical equation refers to the exact number of moles of reactants and products indicated by the coefficients.

If we start with a mol of A, then the maximum quantity of A that can be consumed is a mol of A. Therefore $z \le a$ and z/a is always a fraction. Its maximum value is 1.

Suppose that this reaction is carried out so that the pressure remains at 1 bar and the temperature remains at 25 °C throughout the change from NaOH and CO_2 to $NaHCO_3$. Then all three substances will remain in their standard states throughout the reaction. The concentration of CO_2, which is proportional to its pressure, remains constant, and the concentrations of the two solids also remain constant, as you showed in Exercise 14.1 (⇐ p. 634).

When the reaction is halfway over, half of the reactants have been converted to products, so 0.5 mol of NaOH(s) and 0.5 mol of CO_2(g) have been converted to 0.5 mol of $NaHCO_3$(s). The equation for what has happened so far is

$$0.5\ NaOH(s) + 0.5\ CO_2(g) \longrightarrow 0.5\ NaHCO_3(s)$$

$$\Delta G°\ (\text{at } 25\ °C) = (0.5)(-77.2\ kJ) = -38.6\ kJ$$

When the reaction is halfway over, we say that the extent of reaction is 0.5. The **extent of reaction,** which we will represent by x, *is the fraction of the reactants that has been converted to products.* For the general equation

$$a\,A + b\,B \longrightarrow c\,C + d\,D$$

if z mol of reactant A has been consumed, then the extent of reaction is z/a. If y mol of B has been consumed, the extent of reaction is y/b, and if w mol of D has been formed, the extent of reaction is w/d. For a reaction such as the combination of NaOH with CO_2, in which all substances remain in their standard states throughout the chemical change, an extent of reaction of x corresponds to a Gibbs free energy change of x times the Gibbs free energy change for the complete reaction. This is shown graphically in Figure 18.9.

Exercise 18.11 Gibbs Free Energy and Extent of Reaction

For Reaction 18.5,

(a) Calculate the Gibbs free energy change when the extent of reaction is 0.10, 0.40, and 0.80.
(b) Verify the statement in the text that an extent of reaction of x corresponds to a Gibbs free energy change of x times $\Delta G°$ for the complete reaction.
(c) Show that the slope of the line in Figure 18.9 is $\Delta G°$.

Reactions that Reach Equilibrium

When a reaction occurs in which two or more substances become mixed together, the situation is different. The reaction will reach an equilibrium state in which both reactants and products are present. Consider an equilibrium reaction we discussed in Chapter 14 (⬅ *p. 637*), the isomerization of *cis*-2-butene.

$$cis\text{-}2\text{-butene(g)} \rightleftharpoons trans\text{-}2\text{-butene(g)} \qquad \Delta G°(\text{at } 500 \text{ K}) = -2.08 \text{ kJ}$$

If we start with 1.0 mol of *cis*-2-butene at 500 K and 1 bar, then as soon as some *trans*-2-butene forms it will mix with the *cis*-2-butene. In the mixture of gases, neither gas is at its standard pressure of 1 bar, which means that neither gas is in its standard state. Because of this, the method you used in Exercise 18.11 to calculate ΔG as a function of extent of reaction no longer works. However, we can tell something about how G varies as the reaction proceeds. Mixing two ideal gases involves no change in enthalpy, but it does involve an increase in entropy because of the mixing. (This is in addition to any difference in entropy between products and reactants.) Since ΔS_{mixing} is positive and ΔH_{mixing} is zero, ΔG_{mixing} must be negative. Therefore, in Figure 18.10, there is *not* a straight line from reactants to products. Because there is a negative component (ΔG_{mixing}) in addition to x times $\Delta G°$, there is a curved line below where the straight line would have been.

The graph in Figure 18.10 is steeper at both ends than a straight line from $G_{reactants}$ to $G_{products}$ and passes through a minimum at $x = 0.62$. This value of x means that the fraction of *trans* molecules is 0.62 and the fraction of *cis* molecules is $1 - 0.62 = 0.38$. Since both *cis*- and *trans*-2-butene occupy the same container, their concentrations and partial pressures are proportional to the number of molecules. Therefore, we can substitute these fractions into the equilibrium constant expression.

$$K_P = K_c = \frac{P_{trans}}{P_{cis}} = \frac{[trans]}{[cis]} = \frac{0.62}{0.38} = 1.6$$

This is the same value that was reported in Chapter 14 (⬅ *p. 638*). That is, the equilibrium concentrations correspond to the minimum in the graph of G versus extent of reaction. This makes sense, because any set of concentrations that differs from $x = 0.62$ has a higher value of G and therefore could change to the equilibrium concentrations with a decrease in Gibbs free energy (which corresponds to an increase in entropy of the universe).

At the far left of Figure 18.10 the graph drops faster than a straight line from $G_{reactants}$ to $G_{products}$. The slope at any point on the curve differs from the slope of the straight line ($\Delta G°$) by a factor that depends on the reaction quotient, Q (⬅ *p. 650*).

$$\text{Slope} = \Delta G° + RT \ln Q \qquad\qquad [18.6]$$

Remember that if there is the same number of gas-phase molecules in the reactants as in the products, $K_c = K_P$.

When $x = 0.62$, the system is at equilibrium and $Q = K$. When x differs from 0.62, $Q \neq K$ and the system must react to reach equilibrium.

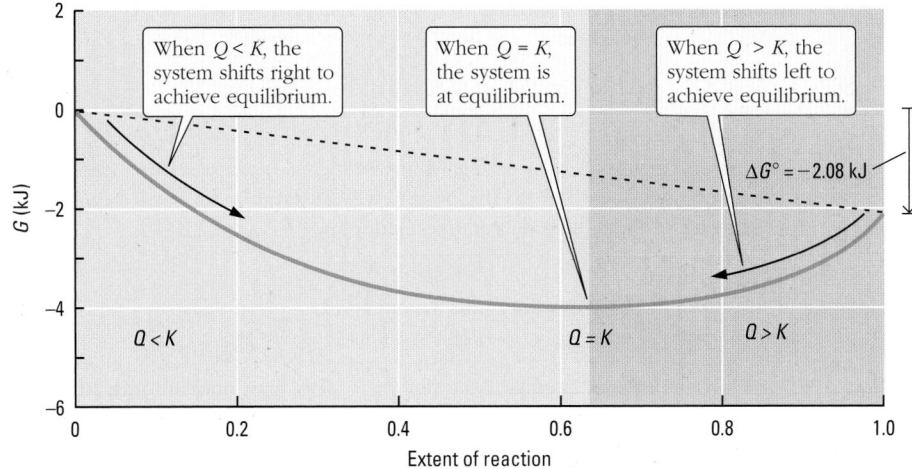

Figure 18.10 Gibbs free energy versus extent of reaction when there is mixing. For the isomerization of *cis*-2-butene, mixing of *cis*-2-butene with *trans*-2-butene increases entropy and decreases Gibbs free energy. This causes the graph of Gibbs free energy versus extent of reaction to curve below a straight line from reactants to products.

Relation Between $\Delta G°$ and $K°$ at 25 °C

$\Delta G°$	$K°$
200	9×10^{-36}
100	3×10^{-18}
10	2×10^{-2}
1	7×10^{-1}
0	1
−1	1.5
−10	6×10^{1}
−100	3×10^{17}
−200	1×10^{35}

$K°$	$\Delta G°$ ($-RT \ln K°$)	Product-favored?
<1	Positive	No
>1	Negative	Yes
=1	0	Neither

At equilibrium the slope is zero, and $Q = K°$, where the superscript indicates that $K°$ must be expressed in the same units as the standard state (pressure in bars for gases and concentration in mol/L for solutes in solutions). Substituting into Equation 18.6 gives

$$0 = \Delta G° + RT \ln K°$$

which rearranges to

$$\Delta G° = -RT \ln K° \qquad [18.7]$$

$K°$ is called the **standard equilibrium constant.** If the reaction occurs in solution, $K°$ has the same form (and the same value) as the concentration equilibrium constant. For gases, $K°$ relates pressures, not concentrations, and has the same value as K_P.

Regardless of the choice of standard state, Equation 18.7 indicates that the Gibbs free energy change for a reaction is the negative of a constant times the temperature times the natural logarithm of the equilibrium constant. If $K°$ is larger than 1, then $\ln K°$ is positive and $\Delta G°$ will be negative because of the minus sign. Both of these, a negative $\Delta G°$ and $K° > 1$, indicate that the reaction is product-favored under standard-state conditions. Conversely, if $K° < 1$, then $\ln K°$ is negative, and $\Delta G°$ must be positive, indicating a reactant-favored system.

$\Delta G°$ is the difference in Gibbs free energy between products in their standard states and reactants in their standard states. For ionization of formic acid in water,

$$\text{HCOOH(aq)} \rightleftharpoons \text{HCOO}^-\text{(aq)} + \text{H}^+\text{(aq)} \qquad \Delta G° = 21.4 \text{ kJ}$$
$$\text{1 mol/L} \qquad\qquad \text{1 mol/L} \qquad \text{1 mol/L}$$

the free energy change of 21.4 kJ is for converting 1 mol of HCOOH(aq) at a concentration of 1 mol/L into 1 mol of HCOO⁻(aq) and 1 mol of H⁺(aq), each at a concentration of 1 mol/L. Since $\Delta G°$ is positive, we predict that the process will be reactant-favored.

Problem-Solving Example 18.7 Gibbs Free Energy and Equilibrium Constant

In the preceding paragraph you learned that $\Delta G° = 21.4$ kJ at 25 °C for the reaction

$$\text{HCOOH(aq)} \rightleftharpoons \text{HCOO}^-\text{(aq)} + \text{H}^+\text{(aq)}$$

Use this information to calculate the equilibrium constant for ionization of formic acid in aqueous solution at 25 °C.

Answer $K_c = K° = 1.8 \times 10^{-4}$

Explanation The relation between Gibbs free energy and $K°$ was given above as

$$\Delta G° = -RT \ln K°$$

To obtain $K°$ from $\Delta G°$, we first divide both sides of the equation by $-RT$:

$$-\Delta G°/RT = \ln K°$$

Next we make use of the properties of logarithms (which are discussed in Appendix A.6). Since ln represents a logarithm to the base e, we can remove the logarithm function by using each side of the equation as an exponent of e.

$$e^{-\Delta G°/RT} = e^{\ln K°} = K°$$

Now we can substitute the known values into the equation.

$$K° = e^{-\Delta G°/RT} = e^{-(21.4 \text{ kJ/mol})(1000 \text{ J/kJ})/(8.314 \text{ J K}^{-1}\text{mol}^{-1})(298\text{K})} = 1.8 \times 10^{-4}$$

Thus, the positive value of $\Delta G°$ results in a value of $K°$ less than one and indeed indicates a reactant-favored system. Because this reaction occurs in aqueous solution, the standard states of reactants and products involve concentrations. Therefore $K° = K_c$.

Make certain to check units in calculations like this. Standard Gibbs free energy changes involve kilojoules, and the gas constant R involves joules, so a unit conversion is needed.

✔ Formic acid is a weak acid and is not expected to have a very large equilibrium constant. The value calculated appears to be reasonable.

Problem-Solving Practice 18.7

For each of the following reactions, evaluate $K°$ at 298 K from the standard free energy change. If necessary, obtain data from Appendix J to calculate $\Delta G°$. Check your results against the K_c and K_P values in Table 14.1 (⬅ *p. 647*). For which of these reactions is $K_c = K°$?

(a) $CaCO_3(s) \rightleftharpoons Ca^{2+}(aq) + CO_3^{2-}(aq)$
(b) $H_2CO_3(aq) \rightleftharpoons HCO_3^-(aq) + H^+(aq)$
(c) $2 NO_2(g) \rightleftharpoons N_2O_4(g)$

Remember also that $\Delta G°$ can be calculated from the equation

$$\Delta G° = \Delta H° - T\Delta S°$$

If we know or can estimate changes in enthalpy and entropy for a reaction, then we can calculate or estimate the Gibbs free energy change and hence the equilibrium constant. And because $\Delta H°$ and $\Delta S°$ have nearly constant values over a wide range of temperatures, we can estimate equilibrium constants at a variety of temperatures, not just at 25 °C.

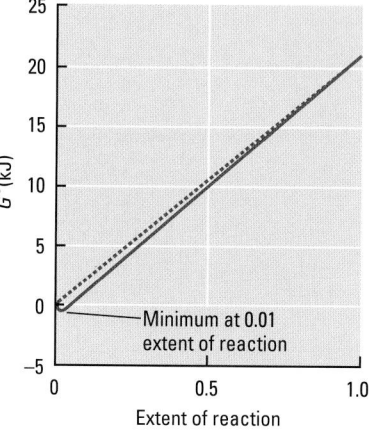

For the reaction in Problem-Solving Example 18.7, $\Delta G° = 21.4$ kJ, a positive value. The minimum in the curve is very close to zero extent of reaction; that is, the system is reactant-favored. The extent of reaction at equilibrium is 0.01.

Problem-Solving Example 18.8 Estimating K at Different Temperatures

Use data from Appendix J to obtain values of $\Delta H°$ and $\Delta S°$ for the reaction

$$N_2(g) + O_2(g) \rightleftharpoons 2 NO(g)$$

From these data estimate the value of $\Delta G°$ and hence the value of $K°$ at (a) 298 K (b) 1000. K   (c) 2300. K

Answer

(a) $\Delta G° = 173.12$ kJ; $K° = 4.51 \times 10^{-31}$
(b) $\Delta G° = 155.73$ kJ; $K° = 7.33 \times 10^{-9}$
(c) $\Delta G° = 123.53$ kJ; $K° = 1.565 \times 10^{-3}$

Explanation At each temperature, use the Gibbs equation to calculate $\Delta G° = \Delta H° - T\Delta S°$. Then calculate $K°$ as was done in Problem-Solving Example 18.6. Part (c) is done below to illustrate the calculations:

$$\Delta G° = \Delta H° - T\Delta S° = (180,500 \text{ J}) - (2300.\text{ K})(24.772 \text{ J/K}) = 123,530 \text{ J} = 123.53 \text{ kJ}$$

$$K° = e^{-\Delta G°/RT} = e^{-(123,530 \text{ J})/(8.314 \text{ J/K})(2300.\text{ K})} = 1.565 \times 10^{-3}$$

Problem-Solving Practice 18.8

For the ammonia synthesis reaction,

$$N_2(g) + 3 H_2(g) \rightleftharpoons 2 NH_3(g)$$

estimate the equilibrium constant at (a) 298. K   (b) 450. K   (c) 800. K

Gibbs Free Energy Changes Under Nonstandard Conditions

In previous sections we calculated $\Delta G°$ for reactions in which reactants in their standard states were converted to products in their standard states. However, substances usually are not at a pressure of 1 bar or at a concentration of 1 mol/L. How do we calculate a ΔG if the reactants and products are not at standard concentration or standard pressure? A simple correction can be made to $\Delta G°$ to account for the difference between actual pressures or concentrations and standard-state pressures or concentrations. The equation from which ΔG (for nonstandard conditions) can be calculated from $\Delta G°$ (for standard conditions) is

$$\Delta G = \Delta G° + RT \ln Q \qquad [18.8]$$

For standard conditions (concentration of 1 mol/L or pressure of 1 bar) $Q = 1$. Substituting this into Equation 18.8, we get $\ln Q = \ln(1) = 0$ and $\Delta G = \Delta G°$, the correct value for standard conditions. According to Equation 18.8, the bigger Q becomes, the more positive the correction factor $RT \ln Q$ becomes, and the more positive ΔG becomes also. This makes sense, because the larger Q is, the more the concentrations (or pressures) of products exceed those of reactants. According to LeChatelier's principle, increasing the concentrations of products (or decreasing concentrations of reactants) causes the reaction to shift in the reverse direction. A shift toward more reactants is expected, because the more positive ΔG is, the more reactant-favored a process is.

To determine the change in Gibbs free energy when reactants at nonstandard concentrations or pressures are converted to products at nonstandard concentrations or pressures, we first write an appropriate chemical equation, then calculate $\Delta G°$ and Q, and finally use Equation 18.8 to correct the value of $\Delta G°$ for the nonstandard conditions.

Problem-Solving Example 18.9 Gibbs Free Energy Change for Nonstandard Conditions

For the ammonia synthesis reaction at 25 °C, calculate the change in Gibbs free energy if 1 mol of $N_2(g)$ at 0.23 bar and 3 mol of $H_2(g)$ at 0.42 bar are converted to 2 mol of $NH_3(g)$ at 1.45 bar.

Answer $\Delta G = -20.96$ kJ/mol

Explanation First write a balanced equation. Then calculate $\Delta G°$ and Q. Finally, use Equation 18.8 to calculate ΔG.

$$N_2(g) + 3 H_2(g) \rightleftharpoons 2 NH_3(g)$$

$$\Delta G° = 2(\Delta G_f°(NH_3[g]) - \{(\Delta G_f°(N_2[g]) + 3(\Delta G_f°(H_2[g])\}$$

$$= 2(-16.45) \text{ kJ/mol} - (0 + 0) = -32.90 \text{ kJ/mol}$$

$$Q = \frac{P_{NH_3}^2}{P_{N_2}P_{H_2}^3} = \frac{(1.45)^2}{(0.23)(0.42)^3} = 123.4$$

$$\Delta G = \Delta G° + RT \ln Q = -32.90 \text{ kJ/mol} + (8.314 \text{ J mol}^{-1} \text{ K}^{-1})(298.15 \text{ K}) \times \ln(123.4)$$

$$= -32.90 \text{ kJ/mol} + 11936 \text{ J/mol}$$

$$= -32.90 \text{ kJ/mol} + 11.936 \text{ kJ/mol} = -20.96 \text{ kJ/mol}$$

✔ The nonstandard conditions involve a concentration of ammonia (the product) well above standard pressure and concentrations of nitrogen and hydrogen (the reactants) well below standard pressure. According to LeChatelier's principle, if an equilibrium is disturbed by increasing concentrations of products or decreasing concentrations of reactants (both of which apply here), then the equilibrium will shift toward the left, that is, in a reactant-favored direction. The value of ΔG (-20.96 kJ/mol) is negative, but it is less negative than the value of $\Delta G°$ (-32.90 kJ/mol). A less negative ΔG corresponds to a less product-favored (that is, more reactant-favored) process, which corresponds with the prediction from LeChatelier's principle.

Problem-Solving Practice 18.9

Calculate ΔG at 25 °C for a reaction in which $Ca^{2+}(aq)$ combines with $CO_3^{2-}(aq)$ to form a precipitate of $CaCO_3(s)$ if the concentrations of $Ca^{2+}(aq)$ and $CO_3^{2-}(aq)$ are 0.023 M and 0.13 M, respectively.

18.8 GIBBS FREE ENERGY, MAXIMUM WORK, AND ENERGY RESOURCES

An important interpretation of the Gibbs free energy is that **ΔG *represents the maximum useful work that can be done by a product-favored system on its surroundings. ΔG also represents the minimum work that must be done to force a reactant-favored process to occur.*** Consider the product-favored reaction of hydrogen with oxygen to form liquid water under standard conditions.

$$2 H_2(g) + O_2(g) \rightleftharpoons 2 H_2O(\ell) \qquad \Delta G° = -474.258 \text{ kJ}$$

This thermochemical equation tells us that for every 2 mol of $H_2O(\ell)$ produced, as much as 474.258 kJ of useful work could be done. The negative sign of $\Delta G°$ tells us that the work is done on the surroundings. (Because the system has less Gibbs free energy after the reaction than before it, the surroundings will have more energy.) Even if reactants and products are not at standard pressure or concentration, ΔG still equals $-w_{max}$, the maximum work the system can do on its surroundings.

$$\Delta G = w_{system} = -w_{max} \text{ (work done on the surroundings)} \qquad [18.9]$$

Now consider the decomposition of water to form hydrogen and oxygen, which is the reverse of the previous reaction.

$$2 H_2O(\ell) \rightleftharpoons 2 H_2(g) + O_2(g) \qquad \Delta G° = 474.258 \text{ kJ}$$

The positive value of $\Delta G°$ indicates that this process is reactant-favored. Because the Gibbs free energy of the products is greater than the Gibbs free energy of the reactant, a least 474.258 kJ must be supplied for every 2 mol of $H_2O(\ell)$ that decomposes. This 474.258 kJ is the minimum work that must be done to change liquid water into hydrogen gas and oxygen gas. One way to supply this work is to use a direct electric current to carry out electrolysis of the water. In general, a continuous supply of energy is required in order for a reactant-favored process, such as decomposition of liquid water, to continue.

It is important to remember that w_{max} is the maximum work the system can do on the surroundings. That is, w_{max} is defined in terms of the surroundings, not the system. Therefore the sign of w_{max} is opposite to that of ΔG.

Because transformations of energy from one form to another are not 100% efficient, we seldom observe anything close to the maximum quantity of useful work given by the value of $\Delta G°$.

Problem-Solving Example 18.10 Gibbs Free Energy Change and Maximum Work

Predict whether each reaction is product-favored or reactant-favored at 25 °C and 1 bar. For each product-favored reaction, calculate the maximum useful work the reaction could do. For each reactant-favored process, calculate the minimum work needed to force it to occur.

(a) $2 Al_2O_3(s) \rightarrow 4 Al(s) + 3 O_2(g)$
(b) $Cl_2(g) + Mg(s) \rightarrow MgCl_2(s)$

Answer
(a) Reactant-favored; at least 3164.6 kJ must be supplied.
(b) Product-favored; can do up to 591.79 kJ of useful work.

Explanation Use data from Appendix J to calculate $\Delta G°$ for each reaction. If $\Delta G°$ is negative, the process is product-favored, and the value of $\Delta G°$ gives the maximum work that can be done. If $\Delta G°$ is positive, the process is reactant-favored, and the value tells the minimum work that has to be done to force the reaction to occur.

(a) $\Delta G° = 0 + 0 - 2(-1582.3) \text{ kJ} = 3164.6 \text{ kJ}$
(b) $\Delta G° = -591.79 \text{ kJ} - 0 - 0 = -591.79 \text{ kJ}$

✔   Reaction (a) involves decomposing an oxide to a metal and oxygen. Because metals are good reducing agents and oxygen is a strong oxidizing agent, the reverse of this reaction is likely to be product-favored, which would make the reaction reactant-favored, which agrees with the calculation. Reaction (b) involves combining an alkaline earth element with a halogen, which should form a stable ionic compound. Therefore reaction (b) should be product-favored, which agrees with the calculation. In both cases the value of $\Delta G°$ is large, which also would be expected based on the arguments just given.

Problem-Solving Practice 18.10

Predict whether each reaction is reactant-favored or product-favored at 298 K and 1 bar, and calculate the minimum work that would have to be done to force it to occur, or the maximum work that could be done by the reaction.

(a) $2 CO_2(g) \longrightarrow 2 CO(g) + O_2(g)$
(b) $4 Fe(s) + 3 O_2(g) \longrightarrow 2 Fe_2O_3(s)$

Charging a dead battery.

Coal-fired electric power plant.

All plants and animals, and indeed earth's ecosystem as a whole, depend on coupling of chemical reactions for their very existence. We will consider this topic in more detail in the next section.

Coupling Reactant-Favored Processes with Product-Favored Processes

A dead car battery will not charge itself. The process that takes place when a battery is charged is reactant-favored. But a battery can be charged if it is connected to a charger that is in turn powered by electricity generated in a power plant that burns coal. Coal, which is mainly carbon, burns in air according to the equation

$$C(s) + O_2(g) \longrightarrow CO_2(g) \qquad \Delta G° = -394.4 \text{ kJ}$$

If enough coal is burned, the large negative Gibbs free energy change for its combustion more than offsets the positive Gibbs free energy change of the battery-charging process. There is an overall decrease in Gibbs free energy, even though the battery-charging part we are interested in has an increase. Once a battery has been charged, the reverse of the charging reaction (which is product-favored) can supply electricity to start a car's engine or play its radio. Some of the Gibbs free energy lost when the coal was burned has been stored in the car's battery for use later.

Charging a battery is an example of coupling a product-favored reaction with a reactant-favored process to cause the latter to take place. Both processes occur at the same time and in a way that allows the Gibbs free energy released by the product-favored reaction to be used by the reactant-favored reaction. Other examples are obtaining aluminum or iron from their ores; synthesizing large, complicated molecules from simple reactants to make medicines, plastics, and other useful materials; and maintaining a comfortable temperature in a house on a day when the outside temperature is below zero. All of these processes involve decreasing entropy (increasing order) in the region of our interest, but all can be made to occur provided that there is a larger increase in entropy at a power plant or somewhere else.

The Gibbs free energy change indicates a chemical reaction's capacity to drive a reactant-favored system so that it produces products. The word "free" in the name indicates not "zero cost," but rather "available." ***Gibbs free energy is available to do useful tasks that would not happen on their own.*** Another way of saying this is that Gibbs free energy is a measure of the *quality* of the energy contained in a chemical system. If it contains a lot of Gibbs free energy, a chemical system can do a lot of useful work for us; the energy is of high quality — potentially useful to humankind. When the system's reactants are transformed into products, that available free energy can do useful work, but only if the reaction is coupled to some other, reactant-favored process we want to carry out. If systems are not coupled, then the free energy released by a reaction will be wasted.

Exercise 18.12   Coupling Reactions

One way to produce iron metal is to reduce iron(III) oxide with aluminum. You can think of the reaction as occurring in two steps. The first is the loss of oxygen from iron(III) oxide,

(i) $Fe_2O_3(s) \longrightarrow 2\,Fe(s) + \frac{3}{2}\,O_2(g)$

and the second is the combination of aluminum with the oxygen.

(ii) $2\,Al(s) + \frac{3}{2}\,O_2(g) \longrightarrow Al_2O_3(s)$

(a) Calculate the enthalpy, entropy, and Gibbs free energy changes for each step. Decide whether each step is product- or reactant-favored. Comment on the signs of $\Delta H°$, $\Delta S°$, and $\Delta G°$ for each step.

(b) What is the overall net reaction that occurs when aluminum is combined with iron(III) oxide? What are the enthalpy, entropy, and Gibbs free energy changes for the overall reaction? Is it product- or reactant-favored? Comment on the signs of $\Delta H°$, $\Delta S°$, and $\Delta G°$ for the overall reaction.

(c) Discuss briefly the effect of coupling Reaction (i) with Reaction (ii), on our ability to obtain iron metal from iron(III) oxide by reacting it with aluminum.

(d) Suggest a reaction other than oxidation of aluminum that might be used to reduce iron(III) oxide to iron. Test your selection by calculating the Gibbs free energy change for the coupled system.

Thermite reaction.   Exercise 18.12 deals with the reaction of aluminum with iron(III) oxide. It is very product-favored and releases a large quantity of Gibbs free energy.

18.9   GIBBS FREE ENERGY AND BIOLOGICAL SYSTEMS

Have you ever thought about how unlikely it is that a human being can exist? Your body contains about 100 trillion (10^{14}) cells, all working together to make you what you are. Each of those cells contains trillions of molecules, and many of the molecules contain tens of thousands of atoms. Those molecules and cells are arranged in structures such as organs, bones, and skin that provide for all the functions of your body and that determine its overall shape and size. When you need to, you can synthesize molecules on very short notice. For example, it does not take long to generate the surge of adrenalin your body makes when you are scared. Your body is a very highly organized system, which means that its entropy must be very low. This in turn means that, thermodynamically speaking, you are very, very improbable.

Human Metabolism and Gibbs Free Energy

How can it be then, that you exist? How can all the molecules from which you are made be synthesized and organized into the organs and other tissues of your body? The answer lies in the coupling of reactions described in the preceding section. Since you are a very low-entropy system, you must be very high in Gibbs free energy. Your body extracts that Gibbs free energy from the food you eat. In the processes of metabolism, foods rich in Gibbs free energy are oxidized, and their oxidation is coupled to other reactions that store Gibbs free energy in specific molecules within your body. Later those molecules can release the Gibbs free energy to cause muscles to contract, nerve signals to be sent, important molecules to be synthesized, and other processes to occur. **Metabolism** refers to all of the chemical changes that occur as food nutrients are converted by an organism into Gibbs free energy and the complex chemical constituents of living cells. **Nutrients** are the chemical raw materials needed for survival of an organism.

As an example of metabolism, consider the single nutrient glucose, also known as dextrose or blood sugar (⇐ *p. 108)*. Glucose can be oxidized to carbon dioxide and water according to the equation

$$C_6H_{12}O_6(aq) + 6\,O_2(g) \longrightarrow 6\,CO_2(g) + 6\,H_2O(\ell) \qquad \Delta G°' = -2870 \text{ kJ}$$

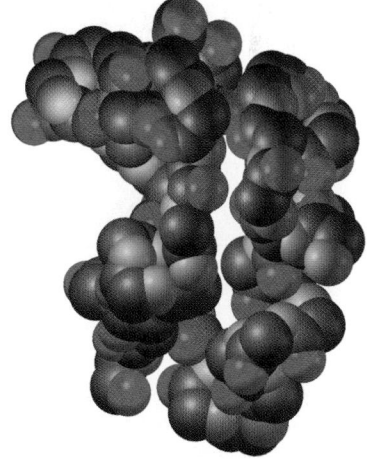

insulin

Insulin, a protein.   Like all protein molecules, the insulin molecule is highly ordered. It contains 51 amino acids connected in exactly the correct order and folded into exactly the molecular shape needed for its function in the metabolism of glucose. Hydrogen atoms are not shown.

Chemistry in the News

THE *HINDENBURG* DISASTER — HYDROGEN OR THERMITE?

The *Hindenburg* was the largest, most famous lighter-than-air, rigid airship (dirigible). It was 245 meters (804 feet) long, had a top speed of 135 km/h (84 mi/h), and was kept aloft by the buoyancy of 200,000 m³ of hydrogen gas contained in 16 gas cells made of specially treated cotton fabric. In 1936 the *Hindenburg* provided the first commercial air service across the Atlantic, carrying 1002 passengers on ten scheduled round trips between Germany and the United States.

On May 6, 1937, at the end of its first crossing of that year, the *Hindenburg* burst into flames while attempting to land at the U.S. Naval Air Station at Lakehurst, New Jersey. The fire engulfed the entire aircraft and destroyed it completely (see photograph). Thirty-six of the 97 people on board were killed. The cause of the fire has never been established beyond doubt, but the most common theory has been that vented or leaking hydrogen gas was ignited, probably by an electrical spark (there were thunderstorms in the area). The *Hindenburg* disaster ended the use of rigid airships in commercial air transportation and also served as a vivid ex-

ample of the extreme flammability of hydrogen. When hydrogen is suggested as a portable fuel for vehicles or for other uses, the *Hindenburg* disaster is often cited as an example of how unsafe such a fuel would be.

In a May 17, 2000, telecast on the Public Broadcasting System (PBS), Addison Bain, an advocate for the safe use of hydrogen, suggested an alternative theory. He interprets testimony from eyewitnesses who could see the *Hindenburg's* starboard side to indicate that the fire did not start near a hydrogen vent as originally supposed. (When the airship was ready to land, some hydrogen was released to decrease its buoyancy.) Instead, the fire appears to have started on the surface of the airship, well away from any of the vents. Only after the fire burned through the cotton of the gas cells, or heated the gas enough to burst them, did the hydrogen begin to burn. But why would the surface of the airship catch fire, and how could the fire spread so fast that the entire ship was consumed in 34 seconds?

In normal operation of the dirigible, it was important to maintain the hydro-

gen at roughly constant temperature. Warming the gas would build up pressure in the gas cells (⬅ *p. 419*) and might rupture them. To keep the gas from overheating, the *Hindenburg's* surface was coated with aluminized paint that was highly reflective. In addition to aluminum, the paint contained iron oxide. The surface consisted of a large number of separate panels that were fastened to the rigid frame by ropes. Each panel was supposed to be connected electrically to all of the others so that static charge that built up on any panel could be conducted to the ground through the mooring cables that were dropped as the ship approached its mooring mast.

Bain theorizes that one or more of the panels was not grounded and therefore electric charge built up on it. When the charge became large enough, a spark jumped from the first panel to another, igniting the aluminum and iron oxide mixture. That is, there may have been a thermite reaction on the surface of the *Hindenburg*. The high temperature produced by the large release of Gibbs free energy from this reaction would have

glucose

The prime symbol (′) on $\Delta G°'$ indicates that the value of the Gibbs free energy change is for pH = 7, the same concentration of $H^+(aq)$ ions as in a typical cell. When aqueous solutions are involved, $\Delta G°$ values in tables such as Appendix J refer to H_3O^+ concentrations of 1 mol/L, but such a high concentration of acid would destroy a typical cell. Consequently, biochemists have calculated a set of $\Delta G°'$ values that apply to solutions at pH = 7. These values are usually reported for a temperature of 37 °C, human body temperature.

Thus, a large quantity of Gibbs free energy can be released when glucose is oxidized. This reaction is strongly product-favored. This is an example of an **exergonic** reaction — *one that releases Gibbs free energy.* The same quantity of Gibbs free energy is available whether glucose is burned in air or reacts in your body. However, burning glucose would release all of the Gibbs free energy as thermal energy. This would not be appropriate in your body because it would raise the temperature rapidly, which in turn would kill cells. Instead, your body makes use of a large number of reactions that allow the Gibbs free energy to be released in small steps and stored in small quantities that can be used later.

By far the most important way that Gibbs free energy is stored in your body is through formation of adenosine triphosphate (ATP) from adenosine diphosphate (ADP).

$$\text{ADP}^{3-}(aq) + \text{H}_2\text{PO}_4^-(aq) \longrightarrow \text{ATP}^{4-}(aq) + \text{H}_2\text{O}(\ell) \qquad \Delta G°' = 30.5 \text{ kJ}$$

The structures of ADP and ATP are shown in Figure 18.11, on page 842; note that they are closely related. This is an example of an **endergonic** reaction — *one that uses up Gibbs free energy* and is therefore reactant-favored. In a typical human cell, this reaction takes place 32 times for each molecule of glucose that is oxidized. In bacterial cells, it takes place 38 times for each molecule of glucose oxi-

burned through the cotton gas cells, or heated the hydrogen enough to burst the cells, thereby expanding the conflagration. Movies of the *Hindenburg* show rapid advance of flame along the surface of the craft, which is consistent with Bain's theory.

The cause of the *Hindenburg* fire may never be known with certainty, but Bain's alternative theory calls into question whether the disaster is a compelling argument against proposals to use hydrogen as a fuel. Combustion of hydrogen produces much less pollution than combustion of hydrocarbon fuels, and hydrogen can be used in fuel cells (Section 19.10) to generate electricity. If the fear of accidental ignition of hydrogen is less well-founded than has been thought in the past, hydrogen might become more useful to society than it already is.

Source:

This news feature includes information from the PBS program Secrets of the Dead: The Hindenburg, *which aired May 17, 2000 (see* **http://www.pbs.org/wnet/secrets/html/e3-menu.html**)*, and from the* Encyclopedia Britannica (**http://www.britannica.com/**).

The *Hindenburg* disaster.

dized. That is, in a human cell the Gibbs free energy released by the exergonic glucose oxidation is used to force the endergonic, reactant-favored process of forming ATP from ADP to occur 32 times. The overall process is

$$C_6H_{12}O_6(aq) + 6\ O_2(g) + 32\ ADP^{3-}(aq) + 32\ H_2PO_4^-(aq) \longrightarrow$$
$$6\ CO_2(g) + 32\ ATP^{4-}(aq) + 38\ H_2O(\ell) \qquad \Delta G^{\circ\prime} = -1894\ kJ$$

Since it is exergonic, it must be product-favored, and therefore appreciable quantities of products can be obtained.

The words "exergonic" and "endergonic" have nearly the same prefixes as "exothermic" and "endothermic." In both cases "ex" means *out* and "end" means *into*. "Thermic" indicates that it is thermal energy that is released or taken up. "Ergonic" indicates that it is Gibbs free energy that is released or used up.

Exercise 18.13 Coupled Metabolic Reactions

Add the Gibbs free energy change for oxidation of glucose to the appropriate Gibbs free energy change for 32 conversions of ADP to ATP. Hence, verify that the Gibbs free energy change given above for the overall reaction is correct. What happens to the 1894 kJ of Gibbs free energy released by the overall reaction?

Because ATP is high in Gibbs free energy, it is said to be a high-energy molecule (or ion). Sometimes the bonds in ATP are called high-energy bonds, but this is a misnomer. Actually, the bonds have low bond energies and can break easily to form ADP and release Gibbs free energy.

The metabolic process by which the Gibbs free energy contained in nutrients is stored in ATP is far more complicated than the overall equation given above

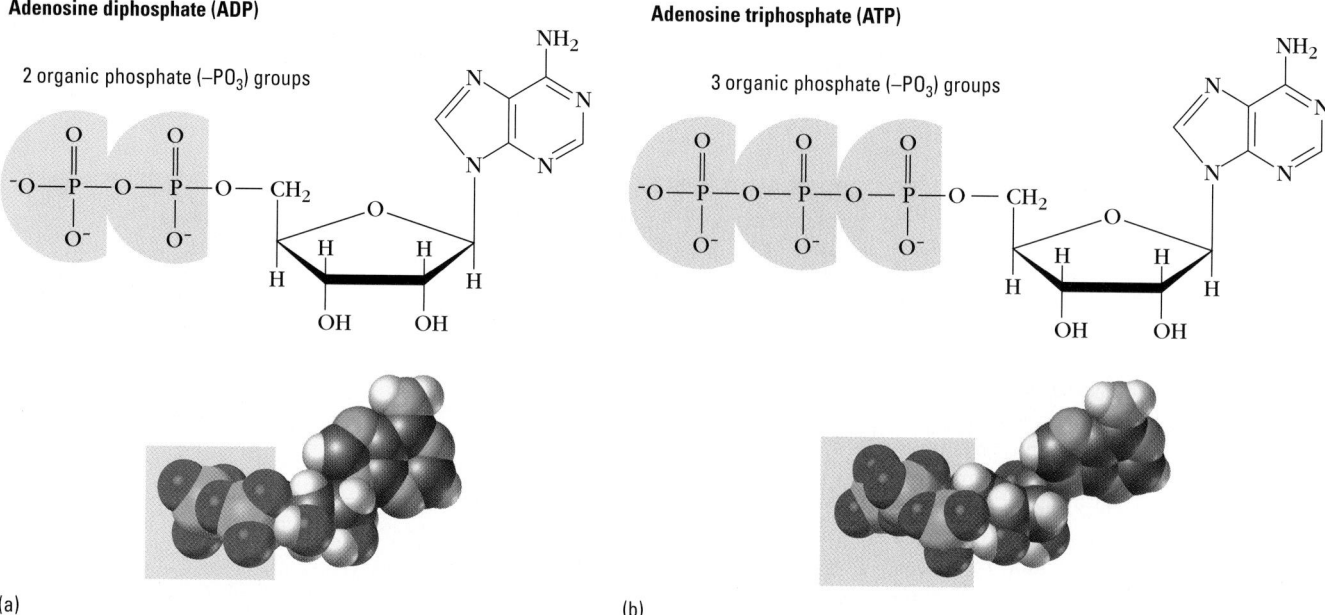

Adenosine diphosphate (ADP)

2 organic phosphate (–PO₃) groups

(a)

Adenosine triphosphate (ATP)

3 organic phosphate (–PO₃) groups

(b)

Figure 18.11 Biochemical storage of Gibbs free energy. Structures of (a) adenosine diphosphate (ADP) and (b) adenosine triphosphate (ATP). Notice that ATP has one more organic phosphate (—PO₃) group at the left end of the molecule, but otherwise the structures are identical. Notice also that ADP has three negatively charged oxygen atoms and ATP has four.

The citric acid cycle is also known as the Krebs cycle or the tricarboxylic acid (TCA) cycle.

Conversion of ATP to ADP is the source of the energy that causes muscles to contract. This answers the question "Where does the energy come from to make my muscles work?" that was posed on p. 32.

makes it seem. It can be divided into three stages that were first clearly identified by Hans Krebs. The first stage is digestion, which breaks apart large molecules, such as carbohydrates (polysaccharides), fats, or proteins, into smaller molecules, such as glucose, glycerol and fatty acids, or amino acids. These smaller molecules are more easily transferred into the blood by the digestive system. In the second stage, the smaller molecules are changed into a few simple units that play a central role in metabolism. The most important of these is the acetyl group in acetyl coenzyme A (acetyl CoA). The structure of acetyl CoA is shown in Figure 18.12. The third stage consists of oxidation of the acetyl group from acetyl CoA to form carbon dioxide and water. This takes place in an eight-step cycle of reactions called the citric acid cycle, which also produces substances that cause ADP to be transformed into ATP, a process called oxidative phosphorylation. The overall three-stage process is diagrammed in Figure 18.13.

Because conversion of ADP to ATP is endergonic, ATP contains stored Gibbs free energy. In your body, ATP generated from glucose or other nutrients is a convenient and readily available Gibbs free energy resource, just as electricity generated from coal or natural gas is a convenient and readily available Gibbs free energy resource in modern society. ATP can release Gibbs free energy in packets of 30.5 kJ for each ATP converted to ADP. This size is convenient for driving many biochemical processes in your body. For example, as part of the metabolism of glucose, it is necessary to attach a phosphate group to the glucose molecule.

The "6" in glucose 6-phosphate indicates that the phosphate group has been added to the oxygen atom attached to carbon number 6 of glucose.

glucose + $H_2PO_4^-$ ⟶ glucose 6-phosphate + H_2O + H^+

$$\Delta G^{\circ\prime} = 13.8 \text{ kJ}$$

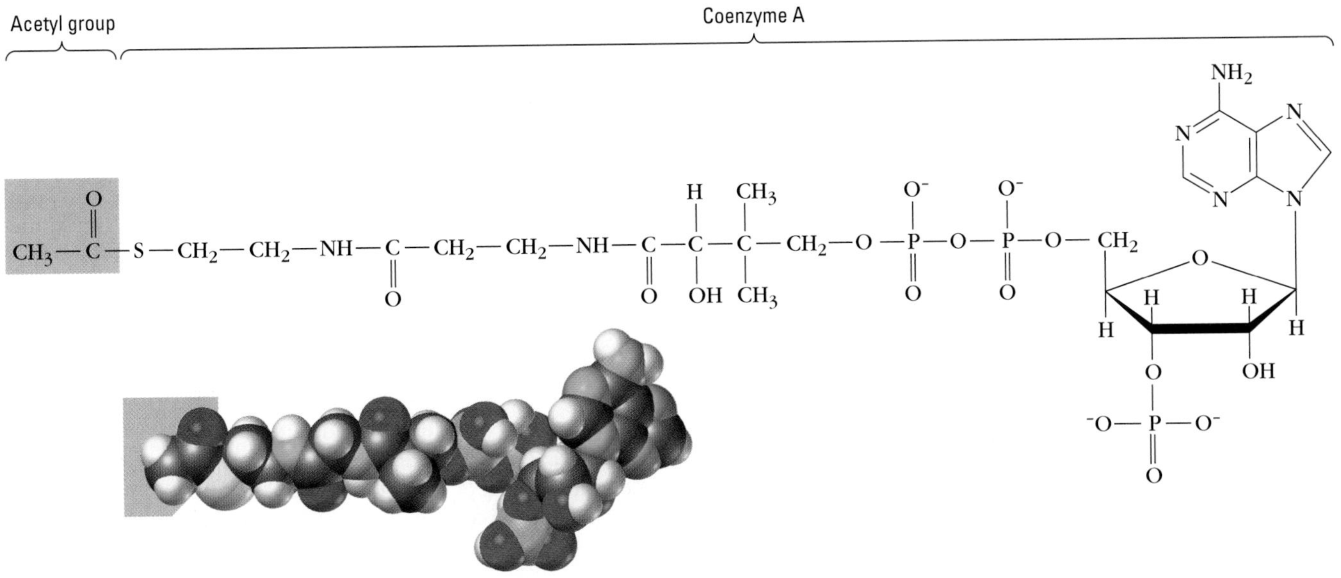

Acetyl group   Coenzyme A

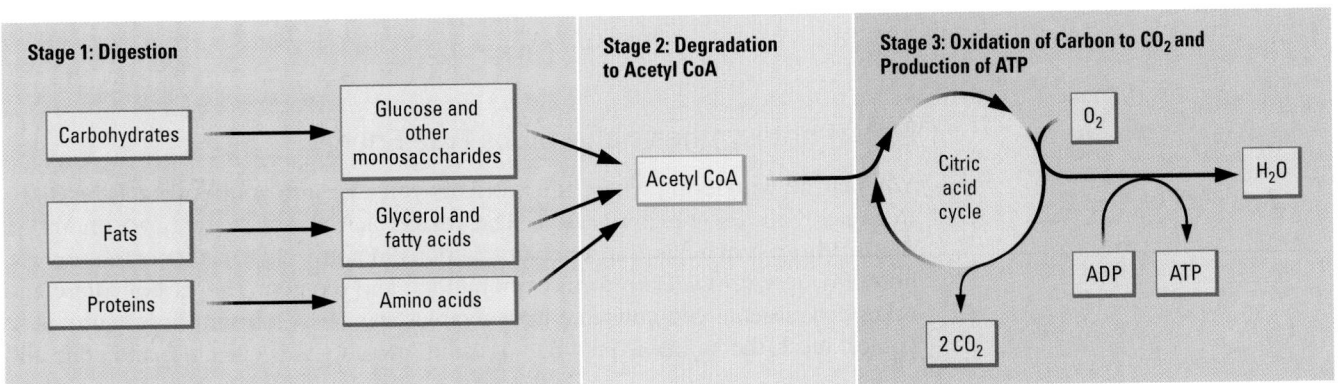

Figure 18.12   Structure of acetyl coenzyme A.   Chemical formula and space-filling model of acetyl coenzyme A. The acetyl group $\left(\begin{smallmatrix} & O \\ & \| \\ CH_3C- \end{smallmatrix}\right)$ at the far left end of the molecule can be formed from glucose that came originally from starch.

This reaction is endergonic by 13.8 kJ and therefore is not product-favored. It will not occur unless forced to do so.

The endergonic reaction can be forced to occur by coupling it to the transformation of ATP to ADP.

$$\text{ATP(aq)}^{4-} + \text{H}_2\text{O}(\ell) \longrightarrow \text{ADP}^{3-}(\text{aq}) + \text{H}_2\text{PO}_4^-(\text{aq}) \qquad \Delta G^{\circ\prime} = -30.5 \text{ kJ}$$

Notice that H_2PO_4^- produced by this reaction can be used to react with glucose in the first reaction, coupling the two directly. Also, water produced in the first reaction is used up in the second one. The overall process is

$$\text{Glucose} + \text{ATP}^{4-} \longrightarrow \text{glucose 6-phosphate}^{2-} + \text{ADP}^{3-} + \text{H}_3\text{O}^+$$

$$\Delta G^{\circ\prime} = (-30.5 + 13.8) \text{ kJ} = -16.7 \text{ kJ}$$

The negative value of $\Delta G^{\circ\prime}$ indicates that the overall process is exergonic and product-favored. Thus the ATP $\rightarrow$ ADP transformation can force glucose to condense with dihydrogen phosphate. The 16.7 kJ of Gibbs free energy released appears as thermal energy transferred from the system to its surroundings.

Figure 18.13   Gibbs free energy and nutrients.   Extraction of Gibbs free energy from nutrients is a three-stage process. In stage 1, digestion, large molecules are broken down into smaller ones. In stage 2, smaller molecules are converted to acetyl groups attached to coenzyme A. In stage 3, the citric acid cycle, the acetyl groups are oxidized to carbon dioxide and water.

Stage 1: Digestion

Carbohydrates → Glucose and other monosaccharides

Fats → Glycerol and fatty acids

Proteins → Amino acids

Stage 2: Degradation to Acetyl CoA

Acetyl CoA

Stage 3: Oxidation of Carbon to CO₂ and Production of ATP

Citric acid cycle

O_2

H_2O

ADP   ATP

$2 CO_2$

In biochemistry it is conventional to write these equations in a shorthand notation that indicates that they are coupled. The process just described would be represented as follows:

Glucose $\longrightarrow$ Glucose 6-phosphate

ATP   ADP

The curved line indicates that the transformation of ATP to ADP occurs simultaneously with the glucose reaction and that the two are coupled.

Problem-Solving Example 18.11   Biochemical Standard State

We noted earlier that many biochemical processes involve reactions that take place at a temperature of 37 °C and at pH = 7 in body fluids. Under these conditions the Gibbs free energy change is specified as $\Delta G^{\circ\prime}$, where the prime specifies that all substances are at their standard-state concentrations except for H_3O^+, which is at a biological concentration of 10^{-7} mol/L (pH = 7). What is ΔG° (1 mol/L H_3O^+) for the reaction

Glucose + ATP^{4-} $\longrightarrow$ glucose 6-phosphate^{2-} + ADP^{3-} + H_3O^+ $\Delta G^{\circ\prime} = -16.7$ kJ/mol

Answer $\Delta G^\circ = 24.8$ kJ/mol

Explanation   The $\Delta G^{\circ\prime}$ value differs from ΔG° because one of the concentrations (that of H_3O^+) has the nonstandard value of 10^{-7} mol/L. That is, $\Delta G^{\circ\prime}$ is ΔG for conditions such that every concentration is 1 mol/L except for the concentration of H_3O^+. Therefore, set $\Delta G = \Delta G^{\circ\prime}$, calculate Q, and use Equation 18.8 to calculate ΔG°.

$$\Delta G = \Delta G^{\circ\prime} = -16.7 \text{ kJ/mol} = \Delta G^\circ + RT \ln Q$$

$$\Delta G^\circ = -16.7 \text{ kJ/mol} - RT \ln Q$$

$$Q = \frac{(\text{conc. glucose 6-phosphate}^{2-})(\text{conc. ADP}^{3-})(\text{conc. H}_3\text{O}^+)}{(\text{conc. glucose})(\text{conc. ATP}^{4-})}$$

$$= \frac{(1)(1)(1 \times 10^{-7})}{(1)(1)} = 1 \times 10^{-7}$$

$$\Delta G^\circ = -16.7 \text{ kJ/mol} - RT \ln(1 \times 10^{-7})$$

$$= -16.7 \text{ kJ/mol} - (8.314 \text{ J mol}^{-1} \text{ K}^{-1+})\{(273 + 37) \text{ K}\}(-16.12)$$

$$= -16.7 \text{ kJ/mol} + 41541 \text{ J/mol} = -16.7 \text{ kJ/mol} + 41.5 \text{ kJ/mol} = 24.8 \text{ kJ/mol}$$

✔   Using LeChatelier's principle, we predict less driving force toward products for a system in which the concentration of H_3O^+, a product, is 1 mol/L than there would be for one in which the concentration of H_3O^+ is 1×10^{-7} mol/L. Less driving force means a more positive ΔG, and the value of ΔG° is indeed more positive than $\Delta G^{\circ\prime}$.

Problem-Solving Practice 18.11

Will ΔG° be larger than, smaller than, or the same size as $\Delta G^{\circ\prime}$ for the reaction

$$C_6H_{12}O_6(aq) + 6\,O_2(g) \longrightarrow 6\,CO_2(g) + 6\,H_2O(\ell) \qquad \Delta G^{\circ\prime} = -2870 \text{ kJ}$$

Explain why you chose the response you did.

 ## Photosynthesis and Gibbs Free Energy

You may be wondering where the nutrients you take into your body get the Gibbs free energy they so obviously have. The answer is from solar energy via photosynthesis. **Photosynthesis** is a series of reactions in a green plant that combines carbon dioxide with water to form carbohydrate and oxygen. The carbohydrate and other constituents you consume in vegetables are derived from photosynthesis. If you eat meat, the animal from which it came probably ate vegetables and grain and therefore derived its nutrients from plant photosynthesis. The overall reaction in photosynthesis is just the opposite of oxidation of glucose.

$$6\ CO_2 + 6\ H_2O \longrightarrow C_6H_{12}O_6 + 6\ O_2 \qquad \Delta G^{\circ\prime} = 2870\ kJ$$

It is endergonic and can occur only because of an influx of energy in the form of sunlight. That is, the energy in the sunlight causes this reactant-favored process to form appreciable quantities of products, and the sunlight's energy is stored as Gibbs free energy in the glucose and oxygen that are formed. This is diagrammed in Figure 18.14.

Organisms that can carry out photosynthesis are called *phototrophs* (literally, "light-feeders") because they can use sunlight to supply needed energy. Phototrophs include all green plants, all algae, and some groups of bacteria. The phototrophs capture light by means of photosynthetic pigment systems and store the light energy in chemical bonds in molecules such as glucose. Nearly all other organisms belong to the class of *chemotrophs* (literally, "chemical-feeders"), which must depend on the chemical bonds created by the phototrophs for their energy. All animals, fungi, and most bacteria are chemotrophs. A world composed only of chemotrophs would not last long because without the phototrophs, food supplies would disappear almost immediately. Without sunlight and its ability to drive a reactant-favored system to form products (carbohydrate and oxygen), organisms such as ourselves and indeed almost the entire biosphere of planet earth could not exist.

Both phototrophs and chemotrophs make use of the Gibbs free energy stored up in photosynthesis by using oxidation of glucose to drive a large number of conversions of ADP to ATP and then using the ATP to couple to desired endergonic reactions and force them to occur. Thus, ATP is the minute-to-minute energy currency of living cells. The Gibbs free energy released in these reactions either contributes to synthesis of molecules needed by the cell, causes some desirable process such as muscle contraction, or is dissipated as thermal energy. If more Gibbs free energy is taken in than the organism needs, then the excess Gibbs free energy can be stored long-term through the synthesis of fats, which have nearly twice as much Gibbs free energy as an equal mass of carbohydrate.

It is significant that when ATP reacts and causes other reactions to occur, the product ADP is very similar to the reactant and can easily be recycled to ATP. A reasonable estimate of the quantity of ATP converted to ADP during one day in the

A few organisms that are found near deep-ocean volcanic vents, the chemautotrophs, do not depend on phototrophs for their energy supply.

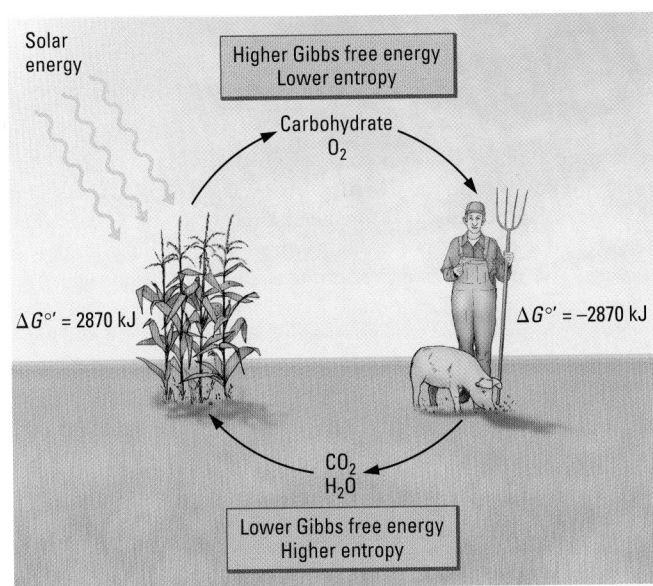

Figure 18.14 Solar energy storage by photosynthesis. The energy in the foods we eat is derived from solar energy via photosynthesis. Organisms that can photosynthesize combine carbon dioxide with water to form carbohydrates and oxygen, which have a much higher Gibbs free energy. That Gibbs free energy is released when the carbohydrates are oxidized in metabolic processes.

life of an average human is 117 mol. Since the molar mass of the sodium salt of ATP is 551 g/mol, we can calculate that

$$117 \text{ mol} \times \frac{551 \text{ g}}{\text{mol}} = 64{,}500 \text{ g ATP}$$

is converted to ADP every day. This is 64.5 kg, which is close to the 70-kg body weight of an average person. Obviously ATP is not a long-term storage molecule for Gibbs free energy. Instead it is recycled from ADP as needed and used almost immediately for some necessary process. The typical 70-kg human body contains only 50 g of ATP and ADP total. If we actually had to take in 64.5 kg of ATP per day to provide Gibbs free energy, it would be a very expensive habit. The price of ATP from a laboratory supplier is currently about $10 per gram, which would put the cost of supplying each of us with our daily energy currency at more than half a million dollars.

Problem-Solving Example 18.12   Coupling of Biological Reactions

ATP undergoes hydrolysis with release of Gibbs free energy according to the equation

(i) adenosine triphosphate + $H_2O(\ell) \longrightarrow$

$\qquad$ adenosine diphosphate + dihydrogen phosphate $\qquad \Delta G°'$ (i) = -30.5 kJ

Other organophosphates undergo similar hydrolysis reactions. For creatine phosphate and glycerol 3-phosphate the hydrolysis reactions are

(ii) creatine phosphate + $H_2O(\ell) \longrightarrow$ creatine + dihydrogen phosphate

$\qquad\qquad\qquad\qquad\qquad\qquad\qquad\qquad\qquad\qquad \Delta G°'$ (ii) = -43.1 kJ

(iii) glycerol 3-phosphate + $H_2O(\ell) \longrightarrow$ glycerol + dihydrogen phosphate

$\qquad\qquad\qquad\qquad\qquad\qquad\qquad\qquad\qquad\qquad \Delta G°'$ (iii) = -9.7 kJ

For each reaction below, predict whether the reaction is product-favored and, if it is, calculate the maximum work that could be done if the reaction took place as written.

(a) adenosine triphosphate + creatine $\rightarrow$
$\qquad$ creatine phosphate + adenosine diphosphate
(b) glycerol + adenosine triphosphate $\rightarrow$
$\qquad$ glycerol 3-phosphate + adenosine diphosphate

Answer

(a) Not product-favored
(b) Product-favored; $\Delta G°' = -20.8$ kJ, so up to 20.8 kJ of work could be done on the surroundings.

Explanation Use the same procedure as for Hess's law calculations *(⬅ p. 241)* to write overall reactions that couple two of the reactions for which $\Delta G°'$ values are known. Then sum the $\Delta G°'$ values to obtain $\Delta G°'$ for the desired reaction. If $\Delta G°'$ for the desired reaction is negative, the process is product-favored and the magnitude of $\Delta G°'$ gives the maximum work. Use part (a) as an example of the calculation. Because ATP is a reactant in the desired equation, use Reaction (i) as written.

(i) adenosine triphosphate + $H_2O(\ell) \longrightarrow$

$\qquad$ adenosine diphosphate + dihydrogen phosphate $\qquad \Delta G°'$ (i) = -30.5 kJ

Because creatine phosphate is a product in the desired reaction, reverse Reaction (ii) and change the sign of $\Delta G°'$.

$\qquad$ reverse of (ii) creatine + dihydrogen phosphate $\longrightarrow$ creatine phosphate + $H_2O(\ell)$

$\qquad\qquad\qquad\qquad\qquad\qquad\qquad\qquad\qquad \Delta G°'$ = $-\Delta G°'$ (ii) = $+43.1$ kJ

The overall reaction is

adenosine triphosphate + creatine $\longrightarrow$ adenosine diphosphate + creatine phosphate

$$\Delta G^{\circ\prime} = \Delta G^{\circ\prime}(i) - \Delta G^{\circ\prime}(ii) = -30.5 \text{ kJ} + 43.1 \text{ kJ} = 12.6 \text{ kJ}$$

Therefore, the process (a) is reactant-favored, and at least 12.6 kJ would have to be supplied to force it to occur.

Problem-Solving Practice 18.12

ATP, creatine phosphate, and glycerol 3-phosphate could be thought of as phosphate donors (just as Brønsted acids can be thought of as proton donors).

(a) Which of the three substances is the strongest phosphate donor?
(b) Which is the weakest?
(c) Explain your choices.

Exercise 18.14   Recycling of ATP

From the figures given above for the daily quantity of ATP converted to ADP by an average human and the quantity of ATP and ADP actually present in the body, calculate the number of times each ADP molecule must be recycled to ATP each day.

18.10   CONSERVATION OF GIBBS FREE ENERGY

When a ton of coal is burned its energy has not been used up. The law of conservation of energy ($\Leftarrow$ *p. 214*) summarizes many experiments whose results verify that energy cannot be destroyed. When coal is burned in a power plant, its chemical energy is changed to an equal quantity of energy in other forms. These are mainly electric energy, which can be very useful, and thermal energy in the gases going up the smokestack and in the immediate surroundings of the plant, which is much less useful. However, an *energy resource* has been used up: the coal's ability to store energy and release it to do useful work. When coal burns in air, some of the Gibbs free energy that was in the coal and the oxygen that combined with it has been used up. This is indicated by the negative value of ΔG° for combustion of coal. The same is true of any other product-favored reaction.

What we commonly refer to as **energy conservation** *is actually conservation of useful energy: Gibbs free energy.* Energy conservation does not mean conserving energy—nature takes care of that automatically. But nature does not automatically conserve Gibbs free energy. Substances with high Gibbs free energies are energy resources, and it is their *useful* energy that we must take pains to conserve. Once a product-favored reaction with a negative ΔG° has taken place, it cannot be reversed, thereby restoring the Gibbs free energy of its reactants, without coupling the reverse reaction with some other product-favored reaction. That is, once we have used an energy resource, it cannot be restored, except by using some other energy resource. Analysis of chemical systems in terms of Gibbs free energy can lead to important insights into how energy resources can be conserved effectively.

By comparing Gibbs free energy changes calculated using the equations in this chapter with the actual loss of Gibbs free energy in industrial processes, environmentalists and industrialists can suggest ways to minimize loss of Gibbs free energy. For example, there is a very large quantity of Gibbs free energy stored in aluminum metal and oxygen gas compared with aluminum ore, Al_2O_3. This can be seen from the thermochemical equation

$$2 \text{ Al}_2\text{O}_3(s) \longrightarrow 4 \text{ Al}(s) + 3 \text{ O}_2(g) \qquad \Delta G^{\circ} = 3164.6 \text{ kJ}$$

Diamond: a material resource. Energy resources are like other natural resources in that they contain high-quality, concentrated energy. An analogy is a material resource such as a diamond, which is pure carbon with the atoms bonded so that each is surrounded tetrahedrally by four others. A diamond is valuable because it consists of a single crystal with atoms arranged in a specific way. If you were to grind a diamond into dust and spread the same quantity of carbon over the area of a city block, the carbon would be nearly worthless, because it would require tremendous expense to collect the carbon and convert it back to diamond. Similarly, an energy resource is valuable not for the energy it contains, but because that energy is concentrated and available to do useful work.

which shows that the Gibbs free energy of 4 mol of Al(s) and 3 mol of O_2(g) is 3164.6 kJ higher than the Gibbs free energy of 2 mol of Al_2O_3(s). If 4 mol of Al(s) is oxidized to aluminum oxide, 3164.6 kJ of Gibbs free energy is lost — energy that was expended to manufacture the aluminum is wasted if the aluminum is oxidized. It is not surprising, then, that throughout the United States there are major programs for recycling aluminum. A similar statement can be made about almost every metal: Once reduced from their ores, metals are storehouses of Gibbs free energy that should be maintained in their reduced forms in order to avoid the expenditure of Gibbs free energy needed to separate them from chemical combination with oxygen.

Energy Conservation and Coupled Reactions

In the previous section we mentioned that in a typical human cell, oxidation of 1 mol of glucose to carbon dioxide and water can cause 32 conversions of adenosine diphosphate to adenosine triphosphate. In Exercise 18.12, you calculated the overall change in Gibbs free energy when 1 mol of glucose is metabolized and 32 mol of ADP is transformed into ATP.

$$C_6H_{12}O_6(aq) + 6\ O_2(g) \longrightarrow 6\ CO_2(g) + 6\ H_2O(\ell) \qquad \Delta G^{\circ\prime} = -2870\ kJ$$

$$\underline{32\ ADP^{3-}(aq) + 32\ H_2PO_4^-(aq) \longrightarrow 32\ ATP^{4-}(aq) + 32\ H_2O(\ell) \qquad \Delta G^{\circ\prime} = 976.0\ kJ}$$

$$C_6H_{12}O_6(aq) + 6\ O_2(g) + 32\ ADP^{3-}(aq) + 32\ H_2PO_4^-(aq) \longrightarrow 6\ CO_2(g) + 32\ ATP^{4-}(aq) + 38\ H_2O(\ell)$$

$$\Delta G^{\circ\prime} = -1894\ kJ$$

Some of the Gibbs free energy released when the glucose was oxidized did useful work by causing synthesis of ATP, the energy-storage medium of living cells. However, nearly two thirds of the Gibbs free energy did no useful work and ended up as thermal energy. That is, about two thirds of the original Gibbs free energy was lost, and one third was stored in ATP. The energy-storage process had an efficiency of about 33%, because only about 33% of the Gibbs free energy change actually did useful work.

The more efficient a biochemical process (or an industrial process) is, the less Gibbs free energy is lost and the more energy conservation takes place. Efficiency is defined as the useful work done per 100 units of energy input. Current energy-conversion systems have a broad range of efficiencies. For example, the generators in a large electric generating plant convert about 99% of the mechanical energy input to electric energy output, but based on the energy of the coal combustion reaction, the overall efficiency of such a plant is only about 40%. An incandescent electric light bulb converts only about 5% of the electric energy it receives into light energy, but a fluorescent light converts 20% — about four times as much. The higher the energy efficiency of the devices we use, the less Gibbs free energy is lost.

Like ATP in your body, many compounds can store Gibbs free energy. An example is ethylene. About 50 billion pounds of this gas are produced in the United States every year from the dehydrogenation of ethane in chemical plants like the one shown in Figure 18.15.

$$C_2H_6(g) \longrightarrow H_2(g) + C_2H_4(g) \qquad \Delta G^\circ = 100.97\ kJ$$

When a mole of hydrogen and a mole of ethylene are produced from a mole of ethane, at least 100.97 kJ of Gibbs free energy must be supplied from an external source. This Gibbs free energy becomes stored in the hydrogen and ethylene. Ethylene production is the largest single consumer of Gibbs free energy in the chemical industry, so there has been great interest in improving the process to save energy and money. Since 1960 there has been a 60% decline in the Gibbs free energy requirement per pound of ethylene produced. Even so, the energy re-

Figure 18.15 Polyethylene production. A chemical plant in Houston, Texas, that produces ethylene.

Much of the ethylene is transformed into polyethylene, a plastic used in many consumer items (⇐ *p. 543*).

Ethylene manufacturing is only 25% energy-efficient.

ESTIMATION Gibbs Free Energy and Automobile Travel

Given that $\Delta G° = -5295.74$ kJ per mole of octane burned, estimate the quantity of Gibbs free energy consumed when a typical car makes a 1000-mile round trip on interstate highways.

Assume that the typical car averages 20 miles per gallon and that combustion of gasoline can be approximated by combustion of octane. Because the trip is a round trip, the car ends up exactly where it started out, which means that it has done no useful work. Therefore all of the Gibbs free energy released by combustion of the fuel is lost. The combustion reaction is

$$C_8H_{18}(\ell) + \tfrac{25}{2} O_2(g) \longrightarrow 8 CO_2(g) + 9 H_2O(\ell)$$

$$\Delta G° = -5295.74 \text{ kJ}$$

Fuel economy of 20 miles per gallon means that five gallons of fuel will be used in 100 miles and 50 gallons in 1000 miles. One gallon is four quarts and a quart is about a liter, so the volume of octane is about $4 \times 50 = 200$ L. The density of gasoline is less than that of water, because gasoline floats on water. Assume that it is about 80% as big. Then the density is 0.8 g/mL or 800 g/L. The 200 L of fuel weighs about $200 \times 800 = 160,000$ g.

The molar mass of octane, C_8H_{18}, is about $8 \times 12 + 18 = 114$ g/mol. To make the arithmetic easier, round this to 100 g/mol. Then 160,000 g of octane corresponds to 1600 mol of octane. The Gibbs free energy released by combustion is about 5000 kJ for every mole of octane burned. Therefore $1600 \times 5000 = 8,000,000$ kJ. Thus, about 8 million kJ of useful energy is consumed for every 1000 miles a car is driven. Most of us drive ten times that far every year, and there are a lot of cars in the United States, so the energy resources consumed by automobile travel are huge.

---

sources used to make ethylene from ethane are four times the minimum required (100.97 kJ). This is largely due to inefficiencies in energy transfer from external sources to the reaction system.

It is important to recognize that completely eliminating consumption of Gibbs free energy is impossible. Whenever anything happens, whether a chemical reaction or a physical process, the final state must have less Gibbs free energy than was available initially. This is the same as saying that the entropy of the universe must have increased during the change. This is true of any system in which the initial substances are changed into something new—any product-favored system. Thus, there will always be losses of Gibbs free energy. The aim of energy conservation is to minimize, not eliminate, them. This can be done by maximizing the efficiency of coupling exergonic reactions to endergonic processes we want to force to occur. The ideas of thermodynamics help us figure out how to do that and are the most powerful tool we have for conserving energy while maintaining a high standard of living.

18.11 THERMODYNAMIC AND KINETIC STABILITY

Chemists often say that substances are "stable," but what exactly does this mean? Usually it means that the substance in question does not decompose or react with other substances that normally are in contact with it. Most chemists, for example, would say that the aluminum can that holds the soda you drink is stable. It will be around for quite a long time. The fact that aluminum cans thrown by the roadside do not decompose rapidly is one of the reasons you are encouraged to recycle them. Some aluminum cans have emerged almost unchanged from landfills after 40 or 50 years.

Strictly speaking, there are two kinds of stability. We discussed one of them in this chapter. A substance is *thermodynamically stable* if it does not undergo product-favored reactions. Such reactions release energy and increase disorder. Although we just said it was stable, the aluminum in a soda can is *thermodynamically* unstable, because its reaction with oxygen in air has a negative Gibbs free energy change.

$$4 Al(s) + 3 O_2(g) \longrightarrow 2 Al_2O_3(s) \qquad \Delta G° = -3164.6 \text{ kJ}$$

However, the aluminum is *kinetically stable.* Although it has the potential to undergo a product-favored oxidation reaction, it does this so slowly that it remains essentially unchanged for a long time. This happens because a thin coating of aluminum oxide forms on the surface and prevents oxygen from reaching the rest of the aluminum atoms below the surface. If we grind the aluminum into a fine powder and throw it into a flame, the powder will burn and the evolved heat will lead to an entropy increase in the little piece of the universe around the burning metal.

Another substance that is *thermodynamically unstable* but *kinetically stable* is diamond. If you look up the data in Appendix J, you will find that the conversion of diamond to graphite has a negative Gibbs free energy change. But diamonds don't change into graphite. Engagement rings contain diamonds precisely because the diamond (like the love it represents) is expected to last for a long time. It does so because there is a very high activation barrier for the change from the diamond structure to the graphite structure. When a chemist says something is stable, it usually means that it is kinetically stable — only an activation-energy barrier prevents it from reacting fast enough for us to see a change.

Problem-Solving Example 18.13 Thermodynamic and Kinetic Stability

Whenever air is heated to a very high temperature, the reaction between nitrogen and oxygen to form nitrogen monoxide occurs. It is an important source of nitrogen-containing air pollutants that can be formed in the cylinders of an automobile engine.

(a) Write a balanced equation with minimum whole-number coefficients for the equilibrium reaction of N_2 with O_2 to form NO.
(b) Is this reaction product-favored at room temperature?
(c) Estimate the temperature at which the standard equilibrium constant for this reaction equals one.
(d) If NO is formed at high temperature in an automobile engine, why does it not all change back to N_2 and O_2 when the mixture of gases enters the exhaust system and its temperature falls?
(e) How might the concentration of NO in automobile exhaust be reduced?

Answer

(a) $N_2(g) + O_2(g) \rightleftharpoons 2\,NO(g)$ (b) No   (c) 7301 K
(d) The reverse reaction is too slow.   (e) Use a suitable catalyst.

Explanation

(a) See answer.
(b) Calculate $\Delta G°$ at 25 °C using data from Appendix J.
$\Delta G° = 2\{\Delta G_f°[NO(g)]\} = 2(86.55\ kJ) = 173.10\ kJ$
(c) If $K° = 1$, then $\Delta G° = -RT \ln K° = -RT \ln(1) = 0$.
Because $\Delta G° = \Delta H° - T\Delta S° = 0$, $\Delta H° = T\Delta S°$ and $T = \Delta H°/\Delta S°$.
Using data from Appendix J gives $\Delta H° = 2\{\Delta H°[NO(g)]\} = 2(90.25\ kJ) = 180.50\ kJ$, and
$\Delta S° = 2\{S°[NO(g)]\} - \{S°[N_2(g)] + S°[O_2(g)]\}$
$= 2(210.76\ J/K) - 191.66\ J/K - 205.138\ J/K = 24.722\ J/K$.
Therefore $T = 180.50\ kJ/24.722\ J/K = 7301\ K$.
(d) When the mixture of gases, which contains some NO as well as N_2 and O_2, leaves the cylinder of the automobile engine and enters the exhaust system, it cools very rapidly to a temperature below 500 K. The reverse reaction should occur, according to thermodynamics, but it does not. The activation energy for the decomposition of NO is quite high, because NO contains a double bond, and it is very difficult to separate the two atoms (which must be done to form N_2 and O_2). Therefore, the reaction rate is greatly affected by temperature, and at low temperatures the reaction is very slow. This means that significant concentrations of NO exist in automobile exhaust.
(e) With a suitable catalyst, decomposition of NO to its elements can take place at appreciable rates even at relatively low temperatures. Catalytic converters are installed in the exhaust systems of cars partly to reduce the concentration of NO.

✔ It is reasonable that $\Delta G°$ for the reaction of N_2 with O_2 is positive, because N_2 and O_2 are the principal components of the atmosphere, and they do not react with each other. It is reasonable that $\Delta S°$ for the reaction is small and positive. The total number of gas-phase molecules does not change, but the product molecules have two different atoms, and both reactant molecules have two atoms that are the same, making the product molecules slightly more probable. It is reasonable that the reaction is endothermic, because the reactant molecules have a triple bond and a double bond, and the product molecules have two double bonds. The bonds broken are therefore expected to be stronger than the bonds formed.

Problem-Solving Practice 18.13

All of these substances are stable with respect to decomposition to their elements at 25 °C. Which are kinetically stable and which are thermodynamically stable?

(a) $MgO(s)$ (b) $N_2H_4(\ell)$ (c) $C_2H_6(g)$ (d) $N_2O(g)$

Finally, think about whether you yourself are stable (thermodynamically or kinetically). From a thermodynamic standpoint, most of the substances you are made of are unstable with respect to oxidation to carbon dioxide, water, and other substances. That is, based on Gibbs free energy changes, most of the substances that you are made of should undergo product-favored reactions that would completely destroy them. Your protein, fat, carbohydrate, and even DNA should spontaneously change into much smaller, simpler molecules with evolution of thermal energy. Fortunately for you, the reactions by which this would happen are very slow at room temperature and body temperature. Only when enzymes catalyze those reactions do they occur with reasonable speed. It is the combination of thermodynamic instability and kinetic stability that allows those enzymes to control the reactions in your body or in any living organism. Were it not for the kinetic stability of a wide variety of substances, everything would be quickly converted to a small number of very thermodynamically stable substances. Life and the environment as we know them would be impossible.

The roles of thermodynamics and kinetics in determining chemical reactivity can be summarized by saying that ***thermodynamics tells whether a reaction can produce predominantly products under standard conditions and, if it does, how much useful work can be accomplished by coupling the reaction to another process.*** If a reaction involves mixing of substances in the gas phase or in solution, ***thermodynamics tells the value of the standard equilibrium constant and allows quantitative prediction of how much product is formed.*** Thermodynamics also can be used to predict what will happen under nonstandard conditions. ***Chemical kinetics tells how fast a given reaction goes and indicates how we can control the rate of reaction.*** Together, thermodynamics and kinetics provide the intellectual foundation on which modern chemical industries are based and the principles upon which understanding of physiology and medicine depends.

SUMMARY PROBLEM

In a blast furnace for making iron from iron ore, large quantities of coke (which is mainly carbon) are dumped into the top of the furnace along with iron ore (which can be assumed to be Fe_2O_3) and limestone (which is used to help remove impurities from the iron). The overall process is

$$2\ Fe_2O_3(s) + 3\ C(s) \longrightarrow 4\ Fe(s) + 3\ CO_2(g)$$

Which can be thought of as a combination of several individual steps.

$$2\,Fe_2O_3(s) \longrightarrow 4\,FeO(s) + O_2(g)$$

$$2\,FeO(s) \longrightarrow 2\,Fe(s) + O_2(g)$$

$$2\,C(s) + O_2(g) \longrightarrow 2\,CO(g)$$

$$2\,CO(g) + O_2(g) \longrightarrow 2\,CO_2(g)$$

(a) Calculate the enthalpy change for each step, assuming a temperature of 25 °C. Which steps are exothermic and which are endothermic?

(b) Based on the equations, predict which of the individual steps would involve an increase and which a decrease in the entropy of the system.

(c) Based on your results in parts (a) and (b), what can you say about whether each step is reactant-favored or product-favored at room temperature? At a much higher temperature (>1000 K)?

(d) Calculate the entropy change and the Gibbs free energy change for each reaction step, assuming a temperature of 25 °C.

(e) Keeping in mind the equation $\Delta G° = \Delta H° - T\Delta S°$ and the fact that the enthalpy change and entropy change for a reaction do not vary much with temperature, what would be the slope of a graph of $\Delta G°$ versus T for each of the reactions? For which of the reactions does $\Delta G°$ become more negative the higher the temperature? For which does it become more positive? Does this agree with what you predicted in part (c)?

(f) For which of these reactions might the assumption of nearly constant $\Delta H°$ and $\Delta S°$ not be true as the temperature increases from 25 °C? For each reaction you choose, explain why the assumption might not be correct.

(g) Use your results from previous parts of this problem to estimate the Gibbs free energy change for each of these reactions at a temperature of 1500 K.

(h) Which of the two iron oxides is more easily reduced at 1500 K? Which of the reactions involving carbon compounds is more product-favored at 1500 K? What chemical reactions do you think are taking place in the hottest part of the blast furnace?

(i) In portions of the furnace where the temperature is about 800 K, would you predict the same reactions would be occurring as in the higher-temperature part of the furnace? Why or why not?

(j) Show that the individual steps can be combined to give the overall reaction. From the enthalpy, entropy, and Gibbs free energy changes already calculated, calculate these changes for the overall reaction.

(k) In a typical blast furnace every kilogram of iron produced requires 2.5 kg of iron ore, 1 kg of coke, and nearly 6 kg of air (to provide oxygen for oxidation of the coke to heat the furnace). How much Gibbs free energy would be destroyed if the coke were simply burned to form carbon dioxide? Given the quantity of iron produced in a typical furnace, how much Gibbs free energy is stored by coupling the oxidation of coke to the reduction of iron oxides? What percentage of the Gibbs free energy available from combustion of coke is wasted per kilogram of iron produced?

IN CLOSING

Having studied this chapter, you should be able to. . .

- Understand and be able to use the terms product-favored and reactant-favored (Section 18.1).

- Explain why there is a higher probability that both matter and energy will be dispersed than that they will be concentrated in a small number of nanoscale particles (Section 18.2).
- Calculate the entropy change for a process occurring at constant temperature (Section 18.3).
- Use qualitative rules to predict the sign of the entropy change for a process (Section 18.3).
- Calculate the entropy change for a chemical reaction, given a table of standard molar entropy values for elements and compounds (Section 18.4).
- Use entropy and enthalpy changes to predict whether a reaction is product-favored (Section 18.5).
- Describe the connection between enthalpy and entropy changes for a reaction and the Gibbs free energy change; use this relation to estimate quantitatively how temperature affects whether a reaction is product-favored (Section 18.6).
- Calculate the Gibbs free energy change for a reaction from values given in a table of standard molar free energies of formation (Section 18.6).
- Relate Gibbs free energy change and standard equilibrium constant for the same reaction and be able to calculate one from the other (Section 18.7).
- Describe how a reactant-favored system can be coupled to a product-favored system so that a desired reaction can be carried out (Section 18.8).
- Explain how biological systems make use of coupled reactions to maintain the high degree of order found in all living organisms; give examples of coupled reactions that are important in biochemistry (Section 18.9).
- Explain the relationship between Gibbs free energy and energy conservation (Sections 18.8 and 18.10).
- Distinguish between thermodynamic stability and kinetic stability and describe the effect of each on whether a reaction is useful in producing products (Section 18.11).

KEY TERMS

endergonic *(18.9)*
energy conservation *(18.10)*
exergonic *(18.9)*
extent of reaction *(18.7)*
Gibbs free energy *(18.6)*

metabolism *(18.9)*
nutrients *(18.9)*
photosynthesis *(18.9)*
reversible process *(18.3)*
second law of thermodynamics *(18.5)*

standard equilibrium constant *(18.7)*
third law of thermodynamics *(18.3)*

QUESTIONS FOR REVIEW AND THOUGHT

Conceptual Challenge Problems

CP-18.A (Section 18.2) Suppose that you are invited to play a game as either the "player" or the "house." A pair of dice is used to determine the winner. Each die is a cube having a different number, one through six, showing on each face. The player rolls two dice and sums the numbers showing on the top side of each die to determine the number rolled. Obviously, the number rolled has a minimum value of 2 (both dice showing a 1) and a maximum of 12 (both dice showing a 6). The player begins the game with his or her initial roll of the dice. If the player rolls a 7 or an 11, he or she wins on the first roll and the house loses. If the player does not roll a 7 or an 11 on the initial roll, then whatever number was rolled is called the point, and the player must roll again. For the player to win, he or she must roll the point again before either a 7 or an 11 is rolled. Should the player roll a 7 or an 11 before rolling the point a second time, the house wins. Which would you choose to be, player or house? Explain clearly in terms of the probabilities of rolling the dice why you chose the role you did.

CP-18.B (Section 18.2) Suppose a button is placed in the middle of a football field and a penny is flipped to decide which direction to move the button, up or down the field. Each time the penny comes up heads, the button is moved 10 cm toward your opponent's goal line; and each time it comes up tails, the button is moved 10 cm toward your goal line. Your friend concludes that after many flips of the penny the button is likely to remain within 10 cm of the middle of the field, because numerous flips of the penny will produce heads just as often as tails. You doubt this because you know that perfume molecules and particles diffuse away from their original source, even though, like the button, they are just as likely to be hit from one direction as from any other by the moving molecules around them. How would you explain the error of your friend's conclusion about the movement of the button?

CP-18.C (Section 18.3) When thermal energy is transferred to a substance at its standard melting point or boiling point, the substance melts or vaporizes, but its temperature does not change while it is doing so. It is clear then that temperature cannot be a measure of "how much energy is in a sample of matter" or the "intensity of energy in a sample of matter." In Qualitative Guidelines for Entropy (*Section 18.3* ⬅ *p. 819*) we noted that atoms and molecules are not stationary, but rather are in constant motion. When heated, their motion increases. If this is true, what can you infer that temperature measures about a sample of matter?

CP-18.D (Section 18.10) Suppose that you are a member of an environmental group and have been assigned to evaluate various ways of delivering milk to consumers with respect to Gibbs free energy conservation. Think of all the ways that milk could be delivered, the kinds of containers that could be used, the ways they could be transported, and whether the containers could be reused (refillable) or recycled. Define the problem in terms of the kinds of information you would need to collect, how you would analyze the information, and the criteria you would use to decide which systems are more efficient in use of Gibbs free energy. Do not try to collect the actual data you would use, but define the problem well enough so that someone could collect the necessary data based on your statement of the problem.

Answers to questions in **bold** can be found in the back of the book.

Review Questions

1. Define the terms "product-favored system" and "reactant-favored system." Give one example of each.
2. What are the two ways that a final chemical state of a system can be more probable than its initial state?
3. Define the term "entropy," and give an example of a sample of matter that has zero entropy. What are the units of entropy? How do they differ from the units of enthalpy?
4. State five useful qualitative rules for predicting entropy changes when chemical or physical changes occur.
5. State the second law of thermodynamics.
6. In terms of values of $\Delta H°$ and $\Delta S°$, under what conditions can you be sure that a reaction is product-favored? When can you be sure that it is not product-favored?
7. Define the Gibbs free energy change of a chemical reaction in terms of its enthalpy and entropy changes. Why is the Gibbs free energy change especially useful in predicting whether a reaction is product-favored?
8. Why are materials of high Gibbs free energy useful to society? Give two examples of such materials.
9. How are materials of high Gibbs free energy important to *you?* Give two examples of such materials.
10. Define the terms "endergonic" and "exergonic."
11. What is the citric acid cycle, and why is it important to organisms?
12. Define the following important biochemistry terms: metabolism, nutrients, ATP, ADP, oxidative phosphorylation, coupled reactions, phototrophs, chemotrophs, photosynthesis.

13. Describe two ways to get reactant-favored reactions to form products.
14. Describe the process by which sunlight is employed to convert high-entropy, low-Gibbs-free-energy substances into low-entropy, high-Gibbs-free-energy substances.
15. Name an important substance that stores Gibbs free energy in your body, and write the chemical reaction by which it is formed.

Reactant-Favored and Product-Favored Processes

16. For each process, write a chemical equation and classify the process as reactant-favored or product-favored.
 (a) Water decomposes to its elements, hydrogen and oxygen.
 (b) Gasoline spilled on the ground evaporates (use octane, C_8H_{18}, to represent gasoline).
 (c) Sugar dissolves in water at room temperature.
17. For each process, write a chemical equation and classify the process as reactant-favored or product-favored.
 (a) Carbon dioxide gas decomposes to its elements, carbon and oxygen.
 (b) The steel (mostly iron) body of an automobile rusts.
 (c) Gasoline reacts with oxygen to form carbon dioxide and water (use octane, C_8H_{18}, to represent gasoline).

Probability and Chemical Reactions

18. Suppose you flip a coin.
 (a) What is the probability that the coin will come up heads?

(b) What is the probability that it will come up tails?

(c) If you flip the coin 100 times, what is the most likely number of heads and tails you will see?

19. Consider two equal-sized flasks connected as in Figure 18.2.

(a) Suppose you put one molecule inside. What is the probability that the molecule will be in flask A? What is the probability that it will be in flask B?

(b) If you put 100 molecules into the two-flask system, what is the most likely arrangement of molecules? Which arrangement has the highest entropy?

20. Suppose you have four identical molecules labeled 1, 2, 3, and 4. Draw 16 simple two-flask diagrams as in Figure 18.2, and draw all possible arrangements of the four molecules in the two flasks. How many of these arrangements have two molecules in each flask? How many have no molecules in one flask? From these results, what is the most probable arrangement of molecules? Which arrangement has the highest entropy?

Measuring Dispersal or Disorder: Entropy

21. For each process, tell whether the entropy change of the system is positive or negative.

(a) Water vapor (the system) deposits as ice crystals on a cold windowpane.

(b) A can of carbonated beverage loses its fizz. (Consider the beverage but not the can as the system. What happens to the entropy of the dissolved gas?)

(c) A glassblower heats glass (the system) to its softening temperature.

22. For each process, tell whether the entropy change of the system is positive or negative.

(a) Water boils.

(b) A teaspoon of sugar dissolves in a cup of coffee. (The system consists of both sugar and coffee.)

(c) Calcium carbonate precipitates out of water in a cave to form stalactites and stalagmites. (Consider only the calcium carbonate to be the system.)

23. For each situation described in Question 16, tell whether the entropy of the system increases or decreases.

24. For each situation described in Question 17, tell whether the entropy of the system increases or decreases.

25. For each pair of items, tell which has the higher entropy, and explain why.

(a) Item 1, a sample of solid CO_2 at -78 °C, or item 2, CO_2 vapor at 0 °C.

(b) Item 1, solid sugar, or item 2, the same sugar dissolved in a cup of tea.

(c) Item 1, a 100-mL sample of pure water and a 100-mL sample of pure alcohol, or item 2, the same samples of water and alcohol after they had been poured together and stirred.

26. For each pair of items, tell which has the higher entropy, and explain why.

(a) Item 1, a sample of pure silicon (to be used in a computer chip), or item 2, a piece of silicon having the same mass but containing a trace of some other element, such as B or P.

(b) Item 1, an ice cube at 0 °C, or item 2, the same mass of liquid water at 0 °C.

(c) Item 1, a sample of pure I_2 at room temperature, or item 2, the same mass of iodine vapor at room temperature.

27. Comparing the formulas or states for each pair of substances, tell which you would expect to have the higher entropy per mole at the same temperature, and explain why.

(a) NaCl(s) or CaO(s)   (b) Cl_2(g) or P_4(g)

(c) NH_4NO_3(s) or NH_4NO_3(aq)

28. Comparing the formulas or states for each pair of substances, tell which you would expect to have the higher entropy per mole at the same temperature, and explain why.

(a) CH_3NH_2(g) or $(CH_3)_2NH$(g)   (b) Au(s) or Hg(ℓ)

(c) Kr(g) or C_6H_{14}(g)

29. From each pair of substances listed below, select the one having the larger standard molar entropy at 25 °C. Give reasons for your choice.

(a) Ga(s) or Ga(ℓ)   (b) AsH_3(g) or Kr(g)

(c) NaF(s) or MgO(s)

30. From each pair of substances listed below, select the one having the larger standard molar entropy at 25 °C. Give reasons for your choice.

(a) H_2O(g) or H_2S(g)   (b) CH_3OH(ℓ) or C_2H_5OH(ℓ)

(c) Butane or cyclobutane

31. Without doing a calculation, predict whether the entropy change will be positive or negative when each reaction occurs in the direction it is written.

(a) C_2H_4(g) + H_2(g) $\longrightarrow$ C_2H_6(g)

(b) CH_3OH(ℓ) + $\frac{3}{2}$ O_2(g) $\longrightarrow$ CO_2(g) + 2 H_2O(g)

(c) N_2(g) + 3 H_2(g) $\longrightarrow$ 2 NH_3(g)

(d) $CaCO_3$(s) $\longrightarrow$ CaO(s) + CO_2(g)

32. Without doing a calculation, predict whether the entropy change will be positive or negative when each reaction occurs in the direction it is written.

(a) CH_3OH(ℓ) $\longrightarrow$ CO(g) + 2 H_2(g)

(b) Br_2(ℓ) + H_2(g) $\longrightarrow$ 2 HBr(g)

(c) C_3H_8(g) $\longrightarrow$ C_2H_4(g) + CH_4(g)

(d) Ag^+(aq) + I^-(aq) $\longrightarrow$ AgI(s)

33. Without consulting a table of standard molar entropies, predict whether $\Delta S°_{system}$ will be positive or negative for each of these reactions.

(a) 2 CO(g) + O_2(g) $\longrightarrow$ 2 CO_2(g)

(b) 2 H_2(g) + O_2(g) $\longrightarrow$ 2 H_2O(ℓ)

(c) 2 O_3(g) $\longrightarrow$ 3 O_2(g)

34. Without consulting a table of standard molar entropies, predict whether $\Delta S°_{system}$ will be positive or negative for each of these reactions.

(a) 2 NH_3(g) $\longrightarrow$ N_2(g) + 3 H_2(g)

(b) 2 Na(s) + Cl_2(g) $\longrightarrow$ 2 NaCl(s)

(c) H_2(g) + I_2(s) $\longrightarrow$ 2 HI(g)

Calculating Entropy Changes

35. Calculate the entropy change, $\Delta S°$, for the vaporization of ethanol, C_2H_5OH, at the boiling point of 78.3 °C. The heat of vaporization of the alcohol is 39.3 kJ/mol.

$$C_2H_5OH(\ell) \longrightarrow C_2H_5OH(g) \qquad \Delta S° = ?$$

36. Diethyl ether, $(C_2H_5)_2O$, was once used as an anesthetic. What is the entropy change, $\Delta S°$, for the vaporization of ether if its heat of vaporization is 26.0 kJ/mol at the boiling point of 35.0 °C?

37. Calculate $\Delta S°$ for each substance when the quantity of thermal energy indicated is transferred reversibly to the system at the temperature specified. Assume that you have enough of each substance so that its temperature remains constant as the thermal energy is transferred.
    (a) $H_2(g)$, 0.775 kJ, 295 K   (b) $KCl(s)$, 500 kJ, 500. K
    (c) $N_2(g)$, 2.45 kJ, 1000. K

38. Calculate $\Delta S°$ for each of these substances when the quantity of thermal energy indicated is transferred reversibly to the system at the temperature specified. Assume that you have enough of each substance so that its temperature remains constant as the thermal energy is transferred.
    (a) $NaCl(s)$, 5.00 kJ, 500 K   (b) $N_2O(g)$, 0.30 kJ, 300 K

39. The standard molar entropy of methanol vapor, $CH_3OH(g)$, is 239.8 J K^{-1} mol^{-1}. Calculate
    (a) the entropy change for the vaporization of 1 mol of methanol (use data from Table 18.1 or Appendix J)
    (b) the enthalpy of vaporization of methanol, assuming that $\Delta S°$ doesn't depend on temperature and taking the boiling point of methanol to be 64.6 °C.

40. The standard molar entropy of iodine vapor, $I_2(g)$, is 260.7 J K^{-1} mol^{-1} and the standard enthalpy of formation is 62.4 kJ/mol. Calculate
    (a) the entropy change for vaporization of 1 mol of solid iodine (use data from Table 18.1 or Appendix J)
    (b) the enthalpy change for sublimation of iodine;
    (c) assuming that $\Delta S°$ does not change with temperature, estimate the temperature at which iodine would sublime (change directly from solid to gas).

41. Check your predictions in Question 31 by calculating the entropy change for each reaction. Standard molar entropies not in Table 18.1 can be found in Appendix J.

42. Check your predictions in Question 32 by calculating the entropy change for each reaction. Standard molar entropies not in Table 18.1 can be found in Appendix J.

43. Check your predictions in Question 33 by calculating the entropy change for each reaction. Standard molar entropies not in Table 18.1 can be found in Appendix J.

44. Check your predictions in Question 34 by calculating the entropy change for each reaction. Standard molar entropies not in Table 18.1 can be found in Appendix J.

45. When calculating $\Delta S°$ from $S°$ values, it is necessary to look up all substances, including elements in their standard state, such as $O_2(g)$, $H_2(g)$, and $N_2(g)$. When calculating $\Delta H°$ from $\Delta H_f°$ values, however, elements in their standard state can be ignored. Why is the situation different for $S°$ values?

46. Calculate the entropy change for formation of exactly 1 mol of each of these gaseous hydrocarbons under standard conditions from carbon (graphite) and hydrogen. What trend do you see in these values? Does $\Delta S°$ increase or decrease on adding H atoms?
    (a) acetylene, $C_2H_2(g)$
    (b) ethylene, $C_2H_4(g)$
    (c) ethane, $C_2H_6(g)$

Entropy and the Second Law of Thermodynamics

47. Calculate $\Delta S°_{system}$ at 25 °C for the reaction

$$C_2H_4(g) + H_2O(g) \longrightarrow C_2H_5OH(\ell)$$

Can you tell from the result of this calculation whether this reaction is product-favored? If you cannot tell, what additional information do you need? Obtain that information and decide whether the reaction is product-favored.

48. Calculate $\Delta S°_{system}$ at 25 °C for the reaction

$$C_6H_6(\ell) + 4 H_2(g) \longrightarrow C_6H_{14}(\ell)$$

Can you tell from the result of this calculation whether this reaction is product-favored? If you cannot tell, what additional information do you need? Obtain that information and decide whether the reaction is product-favored.

49. Is this reaction predicted to favor the products at low temperatures, at high temperatures, or both? Explain your answer briefly.

$$Mg(s) + \tfrac{1}{2} O_2(g) \longrightarrow MgO(s) \qquad \Delta H° = -601.70 \text{ kJ}$$

50. Is this reaction predicted to favor the products at low temperatures, at high temperatures, or both? Explain your answer briefly.

$$MgCO_3(s) \longrightarrow MgO(s) + CO_2(g) \qquad \Delta H° = 116.48 \text{ kJ}$$

51. Explain briefly why the exothermic combustion of propane is product-favored.

$$C_3H_8(g) + 5 O_2(g) \longrightarrow 3 CO_2(g) + 4 H_2O(g)$$

52. Explain briefly why the exothermic reaction of a metal carbonate with an acid is product-favored.

$$CuCO_3(s) + H_2SO_4(aq) \longrightarrow$$
$$CuSO_4(aq) + CO_2(g) + H_2O(\ell)$$

53. Sodium reacts violently with water according to the equation

$$Na(s) + H_2O(\ell) \longrightarrow NaOH(aq) + \tfrac{1}{2} H_2(g)$$

    (a) Predict the signs of $\Delta H°$ and $\Delta S°$ for the reaction.
    (b) Verify your predictions with calculations.

54. Hydrogen burns in air with considerable heat transfer to the surroundings. Consider the decomposition of water to gaseous hydrogen and oxygen. Without doing any calculations, and basing your prediction on the enthalpy change and the entropy change, is this reaction product-favored at 25 °C? Explain your answer briefly.

55. Calcium hydroxide, $Ca(OH_2(s)$, can be dehydrated to form lime, CaO, by heating. Without doing any calculations, and basing your prediction on the enthalpy change and the entropy change, is this reaction product-favored at 25 °C? Explain your answer briefly.

56. Octane is the product of adding hydrogen to 1-octene.

$$C_8H_{16}(g) + H_2(g) \longrightarrow C_8H_{18}(g)$$
$$\text{1-octene} \qquad\qquad \text{octane}$$

The enthalpies of formation are

$$\Delta H_f°(C_8H_{16}[g]) = -82.93 \text{ kJ/mol}$$
$$\Delta H_f°(C_8H_{18}[g]) = -208.45 \text{ kJ/mol}$$

Predict whether this reaction is product-favored or reactant-favored at 25 °C and explain your reasoning.

57. For each reaction, calculate $\Delta H°$ and $\Delta S°$ and predict whether the reaction is always product-favored, product-favored only at low temperatures, product-favored only at high temperatures, or never product-favored.
 (a) $Fe_2O_3(s) + 2\,Al(s) \rightarrow 2\,Fe(s) + Al_2O_3(s)$
 (b) $N_2(g) + 2\,O_2(g) \rightarrow 2\,NO_2(g)$

58. For each reaction, calculate $\Delta H°$ and $\Delta S°$ and predict whether the reaction is always product-favored, product-favored only at low temperatures, product-favored only at high temperatures, or never product-favored.
 (a) $C_6H_{12}O_6(s) + 6\,O_2(g) \rightarrow 6\,CO_2(g) + 6\,H_2O(\ell)$
 (b) $MgO(s) + C(graphite) \rightarrow Mg(s) + CO(g)$

Gibbs Free Energy

59. Determine whether the combustion of ethane, C_2H_6, is product-favored at 25 °C.

$$C_2H_6(g) + \tfrac{7}{2}\,O_2(g) \longrightarrow 2\,CO_2(g) + 3\,H_2O(\ell)$$

 (a) Calculate $\Delta S_{universe}$. Required values of $\Delta H_f°$ and $S°$ are in Appendix J.
 (b) Verify your result by calculating the value of $\Delta G°$ for the reaction.
 (c) Do your calculated answers in (a) and (b) agree with your preconceived idea of this reaction?

60. The reaction of magnesium with water can be used as a means for heating food.

$$Mg(s) + 2\,H_2O(\ell) \longrightarrow Mg(OH)_2(s) + H_2(g)$$

Determine whether this reaction is product-favored at 25 °C.
 (a) Calculate $\Delta S_{universe}$. See Appendix J for the needed data.
 (b) Verify your result by calculating $\Delta G°$ for the reaction.

61. Add a column for the Gibbs free energy to Table 18.2.
 (a) For the first and last lines in the table, tell whether ΔG is positive or negative.
 (b) When ΔH_{system} and ΔS_{system} are both negative, is ΔG positive or negative, or does it depend on temperature? If it is temperature-dependent, what is that dependence?

62. Use a mathematical equation to show how the statement leads to the conclusion cited: If a reaction is exothermic (negative ΔH) and if the entropy of the system increases (positive ΔS), then ΔG must be negative, and the reaction will be product-favored.

63. Use a mathematical equation to show how the statement leads to the conclusion cited: If ΔH and ΔS have the same sign, then the magnitude of T determines whether ΔG will be negative and whether the reaction will be product-favored.

64. Predict whether the reaction below is product-favored or reactant-favored by calculating $\Delta G°$ from the entropy and enthalpy changes for the reaction at 25 °C.

$$H_2(g) + CO_2(g) \longrightarrow H_2O(g) + CO(g)$$
$$\Delta H° = 41.17\text{ kJ} \qquad S° = 42.08\text{ J/K}$$

65. If this reaction were product-favored, it would be a good way to make pure silicon, crucial in the semiconductor industry, from sand (SiO_2).

$$SiO_2(s) + C(s) \longrightarrow Si(s) + CO_2(g)$$

Calculate $\Delta G°$ from data in Appendix J and decide whether it can be used to produce silicon at 25 °C.

66. From data in Appendix J, calculate $\Delta G°$ for the reactions of sand with hydrogen fluoride and hydrogen chloride. Explain why hydrogen fluoride attacks glass, whereas hydrogen chloride does not.

$$SiO_2(s) + 4\,HF(g) \longrightarrow SiF_4(g) + 2\,H_2O(g)$$
$$SiO_2(s) + 4\,HCl(g) \longrightarrow SiCl_4(g) + 2\,H_2O(g)$$

67. Use data from Appendix J to calculate $\Delta G°$ for each reaction at 25 °C. Which are product-favored?
 (a) $C_2H_2(g) + H_2(g) \rightarrow C_2H_4(g)$
 (b) $2\,SO_3(g) \rightarrow 2\,SO_2(g) + O_2(g)$
 (c) $4\,NH_3(g) + 5\,O_2(g) \rightarrow 4\,NO(g) + 6\,H_2O(g)$

68. Evaluate $\Delta H°$ for each reaction in Problem 67. Use your results to calculate standard molar entropies at 25.00 °C for
 (a) $C_2H_2(g)$ (b) $SO_3(g)$ (c) $NO(g)$

69. If a system falls within the second or third category in Table 18.2 ($\Leftarrow$ *p. 826*), then there must be a temperature at which it shifts from being reactant-favored to being product-favored. For each reaction, obtain data from Appendix J and calculate what that temperature is.
 (a) $CO(g) + 2\,H_2(g) \rightleftharpoons CH_3OH(\ell)$
 (b) $2\,Fe_2O_3(s) + 3\,C(graphite) \rightleftharpoons 4\,Fe(s) + 3\,CO_2(g)$

70. If a system falls within the second or third category in Table 18.2 ($\Leftarrow$ *p. 826*), then there must be a temperature at which it shifts from being reactant-favored to being product-favored. For each reaction, obtain data from Appendix J and calculate what that temperature is.
 (a) $2\,H_2O(g) \rightleftharpoons 2\,H_2(g) + O_2(g)$
 (b) $N_2(g) + 3\,H_2(g) \rightleftharpoons 2\,NH_3(g)$

71. Estimate $\Delta G°$ at 2000. K for each reaction in Question 67.

72. Many metal carbonates can be decomposed to the metal oxide and carbon dioxide by heating.

$$CaCO_3(s) \longrightarrow CaO(s) + CO_2(g)$$

 (a) What are the enthalpy, entropy, and Gibbs free energy changes for this reaction at 25.00 °C?
 (b) Is it product-favored or reactant-favored?
 (c) Based on the signs of $\Delta H°$, and $\Delta S°$, predict whether the reaction is product-favored at all temperatures.
 (d) Predict the lowest temperature at which appreciable quantities of products can be obtained.

73. Some metal oxides, such as silver oxide, can be decomposed to the metal and oxygen at relatively low temperatures.

$$Ag_2O(s) \longrightarrow 2\,Ag(s) + \tfrac{1}{2}\,O_2(g)$$

 (a) Is the decomposition of silver oxide product-favored at 25 °C?
 (b) If not, can it become so if the temperature is raised?
 (c) As the temperature increases, at what temperature does the reaction first become product-favored?

74. Use the thermochemical equation

$$CaC_2(s) + 2\,H_2O(\ell) \longrightarrow C_2H_2(g) + Ca(OH)_2(aq)$$
$$\Delta G° = -119.282\text{ kJ}$$

to calculate $\Delta G_f°$ for $Ca(OH)_2(aq)$.

75. Use the thermochemical equation

$$PCl_3(g) + Cl_2(g) \longrightarrow PCl_5(g) \qquad \Delta G° = -37.2\text{ kJ}$$

to calculate $\Delta G_f°$ for $PCl_5(g)$.

76. From data in Appendix J, estimate
    (a) the boiling point of bromine.
    (b) the boiling point of tin(IV) chloride.
77. From data in Appendix J, estimate
    (a) the boiling point of titanium(IV) chloride.
    (b) the boiling point of carbon disulfide, CS_2, which is a liquid at 25 °C and 1 bar.

Gibbs Free Energy Changes and Equilibrium Constants

78. Suppose that at a certain temperature T a chemical reaction is found to have a standard equilibrium constant $K°$ of 1.0. Indicate whether each statement is true or false.
    (a) The enthalpy change for the reaction, $\Delta H°$, is zero.
    (b) The entropy change for the reaction, $\Delta S°$, is zero.
    (c) The Gibbs free energy change for the reaction, $\Delta G°$, is zero.
    (d) $\Delta H°$ and $\Delta S°$ have the same sign.
    (e) $\Delta H°/T = \Delta S°$ at the temperature T.
79. Use data from Appendix J to obtain the equilibrium constant K_p for each reaction at 298.15 K.
    (a) $2\,HCl(g) \rightleftharpoons H_2(g) + Cl_2(g)$
    (b) $N_2(g) + O_2(g) \rightleftharpoons 2\,NO(g)$
80. Use data from Appendix J to obtain the equilibrium constant K_p for each of these reactions at 298 K.
    (a) $CH_4(g) + 2\,O_2(g) \rightleftharpoons CO_2(g) + 2\,H_2O(g)$
    (b) $2\,NO_2(g) \rightleftharpoons N_2O_4(g)$
81. Ethylene reacts with hydrogen to produce ethane.

    $$H_2C{=}CH_2(g) + H_2(g) \rightleftharpoons H_3C{-}CH_3(g)$$

    (a) Using the data in Appendix J, calculate $\Delta G°$ for the reaction at 25 °C. Is the reaction predicted to be product-favored under standard conditions?
    (b) Calculate K_p from $\Delta G°$. Comment on the connection between the sign of $\Delta G°$ and the magnitude of K_p.
82. Use the data in Appendix J to calculate $\Delta G°$ and K_p at 25 °C for the reaction

    $$2\,HBr(g) + Cl_2(g) \rightleftharpoons 2\,HCl(g) + Br_2(\ell)$$

    Comment on the connection between the sign of $\Delta G°$ and the magnitude of K_p.
83. For each chemical reaction, calculate the standard equilibrium constant at 298 K and at 1000. K from the thermodynamic data in Appendix J. Indicate whether each reaction is product-favored or reactant-favored at each temperature.
    (a) The conversion of nitric oxide to nitrogen dioxide in the atmosphere,

    $$2\,NO(g) + O_2(g) \rightleftharpoons 2\,NO_2(g)$$

    (b) The reaction of an alkali metal with a halogen to produce an alkali metal halide salt,

    $$2\,Na(s) + Cl_2(g) \rightleftharpoons 2\,NaCl(s)$$

84. For each chemical reaction, calculate the standard equilibrium constant at 298 K and at 1000 K from the thermodynamic data in Appendix J. Indicate whether each reaction is product-favored or reactant-favored at each temperature.
    (a) The oxidation of carbon monoxide to carbon dioxide,

    $$2\,CO(g) + O_2(g) \rightleftharpoons 2\,CO_2(g)$$

    (b) The first step in the production of electronic-grade silicon from sand,

    $$SiO_2(s) + 2\,C(s) \rightleftharpoons Si(s) + 2\,CO(g)$$

85. For each reaction, estimate $K°$ at the temperature indicated.
    (a) $2\,H_2(g) + O_2(g) \rightleftharpoons 2\,H_2O(g)$ at 800. K
    (b) $2\,SO_2(g) + O_2(g) \rightleftharpoons 2\,SO_3(g)$ at 500. K
    (c) $2\,HF(g) \rightleftharpoons H_2(g) + F_2(g)$ at 2000. K
86. For each reaction, an equilibrium constant at 298 K is given. Calculate $\Delta G°$ for each reaction.
    (a) $Br_2(\ell) + H_2(g) \rightleftharpoons 2\,HBr(g)$ $K_p = 4.4 \times 10^{18}$
    (b) $H_2O(\ell) \rightleftharpoons H_2O(g)$ $K_p = 3.17 \times 10^{-2}$
    (c) $N_2(g) + 3\,H_2(g) \rightleftharpoons 2\,NH_3(g)$ $K_c = 3.5 \times 10^8$
87. For each reaction, an equilibrium constant at 298 K is given. Calculate $\Delta G°$ for each reaction.
    (a) $\frac{1}{8}\,S_8(s) + O_2(g) \rightleftharpoons SO_2(g)$ $K_p = 4.2 \times 10^{52}$
    (b) $2\,H_2(g) + O_2(g) \rightleftharpoons 2\,H_2O(g)$ $K_c = 3.3 \times 10^{81}$
    (c) $CH_4(g) + H_2O(g) \rightleftharpoons CO(g) + 3\,H_2(g)$
                              $K_c = 9.4 \times 10^{-1}$
88. Nitric oxide and chlorine combine at 25 °C to produce nitrosyl chloride, NOCl.

    $$2\,NO(g) + Cl_2(g) \longrightarrow 2\,NOCl(g)$$

    (a) Calculate the equilibrium constant K_p for the reaction.
    (b) Is the reaction product-favored or reactant-favored?
    (c) Calculate the equilibrium constant K_c for the reaction.
89. Hydrogen for use in the Haber-Bosch process for ammonia synthesis is generated from natural gas by the reaction

    $$CH_4(g) + H_2O(g) \rightleftharpoons CO(g) + 3\,H_2(g)$$

    (a) Calculate $\Delta G°$ for this reaction at 25 °C.
    (b) Calculate K_p for the reaction at 25 °C.
    (c) Is the reaction product-favored under standard conditions? If not, at what temperature will it become so?
    (d) Estimate K_c for the reaction at 1000 K.

Gibbs Free Energy, Maximum Work, and Energy Resources

90. Which of these reactions are capable of being harnessed to do useful work at 298 K and 1 bar? Which require that work be done to make them occur?
    (a) $2\,C_6H_6(\ell) + 15\,O_2(g) \rightarrow 12\,CO_2(g) + 6\,H_2O(g)$
    (b) $2\,NF_3(g) \rightarrow N_2(g) + 3\,F_2(g)$
    (c) $TiO_2(s) \rightarrow Ti(s) + O_2(g)$
91. Which of these reactions are capable of being harnessed to do useful work at 298 K and 1 bar? Which require that work be done to make them occur?
    (a) $Al_2O_3(s) \rightarrow 2\,Al(s) + 3\,O_2(g)$
    (b) $2\,CO(g) + O_2(g) \rightarrow 2\,CO_2(g)$
    (c) $C_2H_6(g) \rightarrow C_2H_4(g) + H_2(g)$
92. For each of the reactions in Questions 90 and 91 that require work to be done, calculate the minimum mass of graphite that would have to be oxidized to $CO_2(g)$ to provide the necessary work.
93. It would be very useful if we could use the inexpensive carbon in coal to make more complex organic molecules such as gaseous or liquid fuels. The formation of methane from coal and water is reactant-favored and thus cannot occur unless there is some energy transfer from outside. This problem examines the feasibility of other reactions using coal and water.

(a) Write three balanced equations for the reactions of coal (carbon) and steam to make ethane gas, $C_2H_6(g)$, propane gas, $C_3H_8(g)$, and liquid methanol, $CH_3OH(\ell)$, with carbon dioxide as a byproduct.

(b) Using the data in Appendix J, calculate $\Delta H°$, $\Delta S°$, and $\Delta G°$ for each reaction, and then comment on whether any of them would be a feasible way to make the stated products.

94. Titanium is obtained from its ore, $TiO_2(s)$, by heating the ore in the presence of chlorine gas and coke (carbon) to produce gaseous titanium(IV) chloride and carbon monoxide.

(a) Write a balanced equation for this process.

(b) Calculate $\Delta H°$, $\Delta S°$, and $\Delta G°$ for the reaction.

(c) Is this reaction product-favored or reactant-favored at 25 °C?

(d) Does the reaction become more product-favored or more reactant-favored as the temperature increases?

95. To obtain a metal from its ore, the decomposition of the metal oxide to form the metal and oxygen is often coupled with oxidation of coke (carbon) to carbon monoxide. For each metals oxide listed, write a balanced equation for the decomposition of the oxide and for the overall reaction when the decomposition is coupled to oxidation of coke to carbon monoxide. Calculate the overall value of $\Delta G°$ for each coupled reaction at 25 °C. Which of the metals could be obtained from these ores at 25 °C by this method?

(a) CuO(s)   (b) $Ag_2O(s)$ (c) HgO(s)

(d) MgO(s)   (e) PbO(s)

96. From which of the metal oxides in Question 95 could the metal be obtained by coupling with oxidation of carbon to carbon monoxide at 800 °C?

97. From which of the metal oxides in Question 95 could the metal be obtained by coupling with oxidation of carbon to carbon monoxide at 1500 °C?

Gibbs Free Energy in Biological Systems

98. When you eat a candy bar, how does your body store the Gibbs free energy that is released during oxidation of the sugars (glucose and other carbohydrates) in the candy bar? What was the original source of the Gibbs free energy needed to synthesize the sugars before they went into the candy bar?

99. The molecular structure of one form of glucose, $C_6H_{12}O_6$, looks like this:

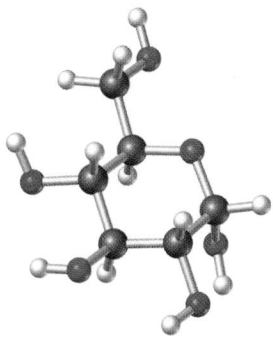

glucose

Glucose can be oxidized to carbon dioxide and water according to the equation

$$C_6H_{12}O_6(s) + 6\,O_2(g) \longrightarrow 6\,CO_2(g) + 6\,H_2O(\ell)$$

(a) Using the method described in Section 8.6 (⇐ p. 336) for estimating enthalpy changes from bond energies, estimate $\Delta H°$ for the oxidation of this form of glucose. Make a list of all bonds broken and all bonds formed in this process.

(b) Compare your result with the experimental value of -2816 kJ for combustion of a mole of glucose. Why might there be a difference between this value and the one you calculated in part (a)?

100. Another step in the metabolism of glucose, which occurs after the formation of glucose-6-phosphate, is the conversion of fructose-6-phosphate to fructose-1,6-bisphosphate ("bis" means *two*):

fructose-6-phosphate(aq) + $H_2PO_4^-$(aq) $\longrightarrow$

fructose-1,6-bisphosphate(aq) + H_3O^+(aq)

(a) This reaction has a Gibbs free energy change of $+16.7$ kJ per mole of fructose-6-phosphate. Is it endergonic or exergonic?

(b) Write the equation for the formation of 1 mol of ADP from ATP, for which $\Delta G° = -30.5$ kJ.

(c) Couple these two reactions to get an exergonic process; write its overall chemical equation, and calculate the Gibbs free energy change.

101. In muscle cells under the condition of vigorous exercise, glucose is converted to lactic acid ("lactate"), $CH_3CHOHCOOH$, by the chemical reaction

$C_6H_{12}O_6 \longrightarrow 2\,CH_3CHOHCOOH \qquad \Delta G° = -197$ kJ

(a) If all of the Gibbs free energy from this reaction were used to convert ADP to ATP, how many moles of ATP could be produced per mole of glucose?

(b) The actual reaction involves the production of 3 mol of ATP per mole of glucose. What is the $\Delta G°$ for this reaction?

(c) Is the overall reaction in part (b) reactant-favored or product-favored?

102. The biological oxidation of ethanol, C_2H_5OH, is also a source of Gibbs free energy.

(a) Does the oxidation of 1 g of ethanol give more or less energy than the oxidation of 1 g of glucose? (*Hint:* Write the balanced equation for the production of carbon dioxide and water from ethanol and oxygen, and use Appendix J.)

(b) Comment on potential problems of replacing glucose with ethanol in your diet.

Conservation of Gibbs Free Energy

103. What are the resources human society uses to supply Gibbs free energy? (*Hint:* Consider information you learned in Chapter 6.)

104. For one day, keep a log of all the activities you undertake that consume Gibbs free energy. Distinguish between Gibbs free energy provided by nutrient metabolism and that provided by other energy resources.

Thermodynamic and Kinetic Stability

105. Billions of pounds of acetic acid are made each year, much of it by the reaction of methanol with carbon monoxide.

$$CH_3OH(\ell) + CO(g) \rightarrow CH_3COOH(\ell)$$

(a) By calculating the standard Gibbs free energy change $\Delta G°$ for this reaction, show that it is product-favored.
(b) Determine the standard Gibbs free energy change $\Delta G°$ for the reaction of acetic acid with oxygen to form gaseous carbon dioxide and liquid water.
(c) Based on this result, is acetic acid thermodynamically stable?
(d) Is acetic acid kinetically stable?

106. Determine the standard Gibbs free energy change $\Delta G°$ for the reactions of methanol, of carbon monoxide, and of ethene, $C_2H_2(g)$, with oxygen to form gaseous carbon dioxide and liquid water. Use your calculations to decide which of these substances are kinetically stable and which are thermodynamically stable: $CH_3OH(\ell)$, $CO(g)$, $C_2H_2(g)$, $CO_2(g)$, $H_2O(\ell)$.

107. How can kinetically stable substances exist at all, if they are not thermodynamically stable?

108. This is a group project: Estimate or look up, to the nearest order of magnitude,
(a) the number of kg of CH_3OH made each year
(b) the number of kg of CO in the entire atmosphere
(c) the number of kg of CH_3COOH made each year
(d) the number of kg of H_2O on earth
(e) the the number of kg of CO_2 in the atmosphere
What do these facts tell you about the difference between kinetic stability and thermodynamic stability?

109. Actually, the carbon in CO_2 is thermodynamically unstable with respect to the carbon in calcium carbonate, or limestone. Verify this by determining the standard Gibbs free energy change for the reaction of lime, CaO(s), with $CO_2(g)$ to make $CaCO_3(s)$.

110. There are millions of organic compounds known, and new ones are being discovered or made at a rate of over 100,000 compounds per year. Organic compounds burn readily in air at high temperatures to form carbon dioxide and water. Several classes of organic compounds are listed, with a simple example of each. Write a balanced chemical equation for the combustion in O_2 of each of these compounds, and then use the data in Appendix J to show that each reaction is product-favored at room temperature.

Class of organics	Simple example
Aliphatic hydrocarbons	Methane, CH_4
Aromatic hydrocarbons	Benzene, C_6H_6
Alcohols	Methanol, CH_3OH

From these results, it is reasonable to hypothesize that *all* organic compounds are thermodynamically unstable in an oxygen atmosphere (i.e., their room-temperature reaction with O_2 to form CO_2 and H_2O is product-favored). If this hypothesis is true, how can organic compounds exist on earth?

General Questions

111. This problem will help you understand the dependence of the American economy on energy. Referring to the figure, calculate the energy (in joules) used by the agricultural, mining, and construction industries

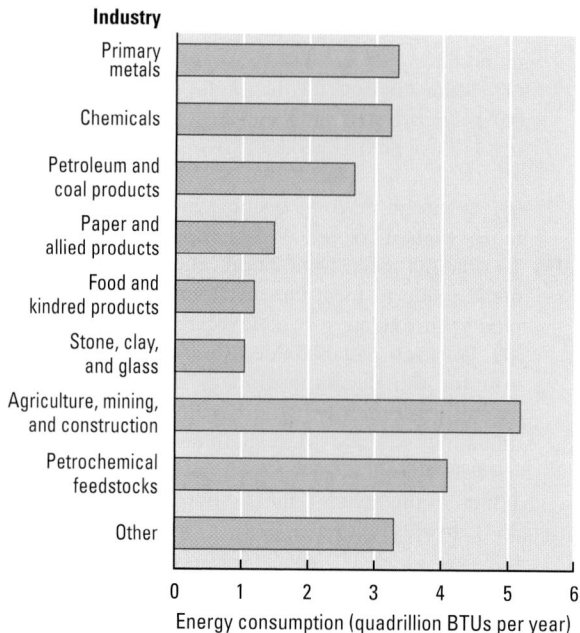

Energy consumption.

(a) in one year.
(b) in one day.
(c) in one second.
(d) Remembering that 1 watt is the expenditure of 1 joule every second, calculate the average power needs of these industries in watts.
(e) Assuming a U.S. population of 300 million people, calculate the power needed by the agricultural, mining, and construction industries *per person in the United States.*

112. Suppose you signed a contract to provide to the agricultural, mining, and construction industries the energy they use each year by eating glucose and giving them the resulting energy from its oxidation in your body.
(a) How much glucose would you have to eat each day to meet your contract? Assume that it is someone else's job to figure out how to get the energy stored in your ATP to the industries!
(b) An Olympic sprinter uses energy at the rate of 700 to 900 watts in a sprint. Compare this figure with the one you calculated in (a), and draw conclusions about the feasibility of keeping your contract.

113. You are exploring the marketing possibilities of a scheme by which every family in the United States produces enough water for its own needs by the combustion of hydrogen and oxygen. Would the release of Gibbs free energy from the combination of hydrogen and oxygen be sufficient to supply the family's energy needs? Do not try

to collect the actual data you would use, but define the problem well enough so that someone else could collect the necessary data and do the calculations that would be needed.

Reaction	Chemical equation	K_c	$\Delta H°$ (kJ)
1	$CH_3OH(g) + H_2(g) \rightleftharpoons$ $CH_4(g) + H_2O(g)$	3.6×10^{20}	−115.4
2	$Mg(OH)_2(s) \rightleftharpoons$ $MgO(s) + H_2O(g)$	1.24×10^{-5}	81.1
3	$2\, CH_4(g) \rightleftharpoons$ $C_2H_6(g) + H_2(g)$	9.5×10^{-13}	64.9
4	$2\, H_2(g) + CO(g) \rightleftharpoons$ $CH_3OH(g)$	3.76	−90.7
5	$H_2(g) + Br_2(g) \rightleftharpoons$ $2\, HBr(g)$	1.9×10^{24}	−103.7

114. The table above provides data at 25 °C for five reactions. For which (if any) of the reactions 1 through 5 is
(a) K_P greater than K_c?
(b) the reaction product-favored?
(c) there only a single concentration in the K_c expression?
(d) there an increase in concentrations of products when the temperature increases?
(e) there a change in the sign of $\Delta G°$ if water is liquid instead of gas?

115. The table above provides data at 25 °C for five reactions. For which (if any) of the reactions 1 through 5 is
(a) K_P less than K_c?
(b) there a decrease in concentration of products when the pressure increases?
(c) the value of $\Delta S°$ positive?
(d) the sign of $\Delta G°$ dependent on temperature?

116. Consider the gas-phase decomposition of sulfur trioxide to sulfur dioxide and oxygen.
(a) Calculate $\Delta G°$ for the reaction at 25 °C.
(b) Is the reaction product-favored under standard conditions at 25 °C?
(c) If the reaction is not product-favored at 25 °C, is there a temperature at which it will become so?
(d) Estimate K_P for the reaction at 1500 °C.
(e) Estimate K_c for the reaction at 1500 °C.

117. The Haber process for the synthesis of ammonia involves the reaction

$$N_2(g) + 3\, H_2(g) \rightleftharpoons 2\, NH_3(g)$$

Using data from Appendix J, estimate the amount (in moles) of $NH_3(g)$ that would be produced from 1 mol of $N_2(g)$ and 3 mol of $H_2(g)$ once equilibrium is reached at 450 °C and a total pressure of 1000 atm.

118. Quite often a graph of $\ln K°$ versus $1/T$ is a straight line. Use Equation 18.7 (⬅ *p. 834*) to show how $\Delta H°$ and $\Delta S°$ can be determined from such a graph. Does the fact that such a graph is straight tell you anything about the dependence of $\Delta H°$ and $\Delta S°$ on temperature?

119. Assuming that $\Delta H°$ and $\Delta S°$ do not vary with temperature, use Equation 18.7 (⬅ *p. 834*) to derive a formula relating $K_1°$ at temperature T_1 to $K_2°$ at temperature T_2.

120. Mercury is a poison, and its vapor is readily absorbed through the lungs. Therefore it is important that the partial pressure of mercury be kept as low as possible in any area where people could be exposed to it (such as a dentist's office). The relevant equilibrium reaction is

$$Hg(\ell) \rightleftharpoons Hg(g)$$

For $Hg(g)$, $\Delta H_f° = 61.4$ kJ/mol, $S° = 175.0$ J k^{-1} mol^{-1}, and $\Delta G_f° = 31.8$ kJ/mol. Use data from Appendix J and these values to evaluate the vapor pressure of mercury at different temperatures. (Remember that concentrations of pure liquids and solids do not appear in the equilibrium constant expression, and for gases $K°$ involves pressures in bars.)
(a) Calculate $\Delta G°$ for vaporization of mercury at 25 °C.
(b) Write the equilibrium constant expression for vaporization of mercury.
(c) Calculate $K°$ for this reaction at 25 °C.
(d) What is the vapor pressure of mercury at 25 °C?
(e) Estimate the temperature at which the vapor pressure of mercury reaches 10 mm Hg.

Applying Concepts

121. A friend of yours says that the boiling point of water is twice that of cyclopentane, which boils at 50 °C. Write a brief statement about the validity of this observation.

122. Using the second law of thermodynamics, explain why it is very difficult to unscramble an egg. Who was Humpty-Dumpty? Why did his moment of glory illustrate the second law of thermodynamics?

123. Appendix J lists standard molar entropies $S_m°$, not standard entropies of formation $\Delta S_f°$. Why is this possible for entropy but not for internal energy, enthalpy, or Gibbs free energy?

124. In the *Chemistry You Can Do* experiment with the rubber band, you found that the relaxation of the rubber is endothermic. The molecules that make up the rubber are long chains of carbon atoms, and when the band is stretched the chains are straightened out. When the rubber relaxes, the chains return to a tangled, curled state. What happens to the entropy of the rubber band when a stretched band relaxes?

125. In the *Chemistry You Can Do* experiment in Chapter 6 (⬅ *p. 235*) you explored the heat from the rusting of iron to form iron oxide. Look at the enthalpies of formation of other metal oxides in Table 6.3 or Appendix J and comment on your observations. Are oxidations of metals generally endothermic or exothermic? Are they usually reactant-favored or product-favored?

126. Using the reactions

$$2\, H_2(g) + O_2(g) \rightarrow 2\, H_2O(\ell)$$
$$2\, H_2(g) + O_2(g) \rightarrow 2\, H_2O(g)$$

as an example, explain why it is dangerous to assume for reactions involving solids or liquids that $\Delta S°$ and $\Delta H°$ do not change appreciably with increasing temperature.

127. Without consulting tables of ΔH_f°, S°, or ΔG_f° values, predict which of these reactions will be

(i) always product-favored;

(ii) product-favored at low temperatures, but not product-favored at high

(iii) not product-favored at low temperatures, but product-favored at high

(iv) never product-favored.

(a) $2 NO_2(g) \rightarrow N_2O_4(g)$

(b) $C_5H_{12}(g) + 8 O_2(g) \rightarrow 5 CO_2(g) + 6 H_2O(g)$

(c) $P_4(g) + 10 F_2(g) \rightarrow 4 PF_5(g)$

[*Hint:* Use the qualitative rules regarding bond enthalpies in Section 6.7 (⇐ *p. 238)* to predict the sign of ΔH°.]

128. Explain how the entropy of the universe is increased when an aluminum metal can is made from aluminum ore. The first step is to extract the ore, which is primarily a form of Al_2O_3, from the ground. After it is purified by freeing it from oxides of silicon and iron, aluminum oxide is changed to the metal by an input of electrical energy.

$$2 Al_2O_3(s) \xrightarrow{\text{electrical energy}} 4 Al(s) + 3 O_2(g)$$

129. Explain why the entropy of the system increases when solid NaCl dissolves in water.

130. Explain how biological systems make use of coupled reactions to maintain the high degree of order found in all living organisms.

131. Criticize the following statement: Provided it occurs at an appreciable rate, any chemical reaction for which $\Delta G < 0$ will proceed until all reactants have been converted to products.

132. Reword the statement in Question 131 so that it is always true.

 General Chemistry CD-ROM

CD18.1 Screen 20.2: Reaction Spontaneity: Thermodynamics and Kinetics.

(a) What is the relationship between the terms "spontaneous" and "product-favored"?

(b) What is the relationship between a reaction's spontaneity and its rate?

(c) If gaseous H_2 and O_2 are carefully mixed and left alone, they can remain intact for millions of years. Is this "stability" a function of thermodynamics or of kinetics?

(d) Can a reaction be "driven" only by thermodynamics or only by kinetics?

CD18.2 Screen 20.4: Entropy.

(a) Describe your concept of entropy.

(b) Which of the dispersal mechanisms described on Screen 20.3 involved changes in a system's entropy?

(c) Why does the entropy of a substance increase with temperature?

CD18.3 Screen 20.5: Calculating ΔS for a Chemical Reaction.

(a) Will reactions with products that are much more disordered than the reactants have a positive or a negative value of ΔS?

(b) Which do you expect to have a greater entropy, potassium chloride as a solid, or potassium chloride dissolved in water?

CD18.4 Screen 20.8: Free Energy and Temperature.

(a) Are reactions that occur only at high temperature but not at low temperature enthalpy-favored, entropy-favored, or both? Or, can't you tell?

(b) Are reactions that occur only at low temperature but not at high temperature enthalpy-favored, entropy-favored, or both? Or, can't you tell?

Advanced batteries supply the electrical energy that powers many of our electronic devices, such as cellular telephones, laptop computers, and electronic organizers.

ELECTROCHEMISTRY AND ITS APPLICATIONS

19.1 Redox Reactions

19.2 Using Half-Reactions to Understand Redox Reactions

19.3 Electrochemical Cells

19.4 Electrochemical Cells and Voltage

19.5 Using Standard Cell Potentials

19.6 $E°$ and Gibbs Free Energy

19.7 Effect of Concentration on Cell Potential

19.8 Neuron Cells

19.9 Common Batteries

19.10 Fuel Cells

19.11 Electrolysis — Forcing Reactant-Favored Reactions to Occur

19.12 Counting Electrons

19.13 Corrosion — Product-Favored Reactions

Many of our modern devices are powered by batteries—portable storage devices for electrochemical energy that is produced by product-favored redox reactions. Making these reactions do useful work is the goal of much chemical research. What chemistry goes on inside a battery? How does it produce electricity? Is this chemistry similar or different from the other kinds of reactions you have studied?

Oxidation-reduction (redox) reactions are an important class of chemical reactions ($\Leftarrow$ *p. 176, Section 5.3)*. Many redox reactions involve the transfer of electrons from one atom, molecule, or ion to another. **Electrochemistry** is the study of the relationship between electron flow and redox reactions. Applications of electrochemistry are numerous and important. In electrochemical cells (commonly called batteries), electrons from a product-favored redox reaction are generated and transferred as an electrical current through an external circuit. We rely on batteries to power many useful devices such as CD players, cellular telephones, calculators, flashlights, portable computers, heart pacemakers, and golf carts.

The voltage produced by an electrochemical cell depends on the nature of the reactants as oxidizing agents and reducing agents. A knowledge of the strengths of oxidizing and reducing agents helps in the design of better batteries. Product-favored electrochemical reactions also cause their share of problems. Corrosion of metals, for example, is product-favored. Preventing and keeping corrosion under control is very costly.

By contrast, electroplating and electrolysis are applications of redox reactions that are reactant-favored. In an electrolysis cell, an external source of energy causes an electrical current to force a reactant-favored process to occur. Electrolysis is important in the manufacture of many products, including chrome-plated objects and chlorine used to disinfect water supplies. Electrochemistry and its applications are the subject of this chapter.

CD-ROM Screen 21.2: Redox Reactions: Electron Transfer

19.1 REDOX REACTIONS

Redox reactions form a large class of chemical reactions in which the reactants may be atoms, ions, or molecules ($\Leftarrow$ *p. 176)*. How do you know when a reaction involves oxidation-reduction?

- *By identifying the presence of strong oxidizing or reducing agents as reactants* ($\Leftarrow$ *p. 185; Table 5.5)*.
- *By recognizing a change in oxidation number* ($\Leftarrow$ *p. 181)*. This means you have to determine the oxidation number of each element as it appears in a reactant or a product.
- *By the presence of an uncombined element as a reactant or product.* Producing a free element or incorporating one in a compound always results in a change in oxidation number.

An uncombined element is always assigned an oxidation number of 0.

To review the definitions of oxidation and reduction, consider the displacement reaction between magnesium (a relatively reactive metal, ($\Leftarrow$ *p. 185; Table 5.5)* and hydrochloric acid. The oxidation numbers of the elements are shown above their symbols

You may want to review the definitions of oxidation and reduction in Section 5.3 and the rules for assigning oxidation numbers in Section 5.4.

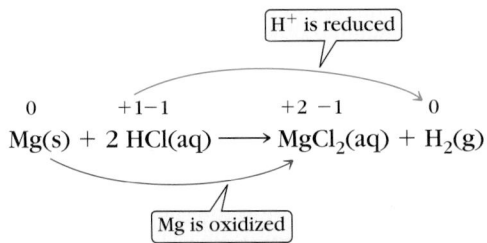

$$\overset{0}{\text{Mg(s)}} + 2\,\overset{+1\,-1}{\text{HCl(aq)}} \longrightarrow \overset{+2\,-1}{\text{MgCl}_2\text{(aq)}} + \overset{0}{\text{H}_2\text{(g)}}$$

H⁺ is reduced

Mg is oxidized

The presence of the uncombined elements Mg and H_2 indicates a redox reaction, as do the changes in oxidation number. Mg(s) is oxidized, indicated by an *increase* in its oxidation number (from 0 to +2). Hydrogen is reduced, as shown by a *decrease* in its oxidation number (from +1 to 0). Magnesium is the reducing agent, and it causes the hydrogen to be reduced. Hydrochloric acid is the oxidizing agent, and it causes the magnesium to be oxidized. Note that oxidation and reduction always occur together, with one reactant the oxidizing agent and another the reducing agent. *The oxidizing agent is reduced, and the reducing agent is oxidized.*

Redox reactions such as the one between Mg and HCl involve complete gain and loss of electrons by reactants, products, or both. It is this type of reaction that is utilized in electrochemistry. A flow of electrons through the reaction system and an external circuit is necessary for the reaction to proceed.

Oxidation: loss of electron(s) and increase in oxidation number

Reduction: gain of electron(s) and decrease in oxidation number

Problem-Solving Example 19.1 Identifying Oxidizing and Reducing Agents in Redox Reactions

Manganese dioxide, MnO_2 (used in many common batteries) reacts with hydrogen gas as follows:

$$2\,MnO_2(s) + H_2(g) \longrightarrow Mn_2O_3(s) + H_2O(\ell)$$

Is this a redox reaction? Give the oxidation numbers of all the atoms that change oxidation number. What gets reduced? What gets oxidized? What is the oxidizing agent? What is the reducing agent?

Answer Yes, this is a redox reaction. The oxidation number changes are Mn: +4 to +3; H: 0 to +1. Molecular hydrogen is oxidized to form water, and the Mn in MnO_2 is reduced. The oxidizing agent is MnO_2 and the reducing agent is H_2.

Explanation First, determine the oxidation number of each element on the reactant side of the equation. Oxygen in compounds is normally −2 ($\Longleftarrow$ *p. 181*). Since the sum of the oxidation states of all the atoms in a formula must equal the charge on the formula, this means that the Mn in MnO_2 is +4 and each oxygen is −2.

$$\underset{MnO_2}{+4 \quad -4 = 2 \times (-2)}$$

The oxidation state for hydrogen in H_2 is 0, as it is for all uncombined elements.

On the products side, the Mn in Mn_2O_3 has an oxidation number of +3.

$$\underset{Mn_2O_3}{2 \times (+3) = +6 \quad -6 = 3 \times (-2)}$$

Similarly, for H_2O,

$$\underset{H_2O}{2 \times (+1) = +2 \quad -2}$$

Thus, the elements that change oxidation number are

$$\overset{+4}{2\,MnO_2(s)} + \overset{0}{H_2(g)} \longrightarrow \overset{+3}{Mn_2O_3(s)} + \overset{+1}{H_2O(\ell)}$$

The oxidation state of the Mn atoms decreased, while the oxidation state of the H atoms increased. Thus, the MnO_2 has been reduced and the H_2 has been oxidized. Consequently, MnO_2 is the oxidizing agent and H_2 is the reducing agent. Oxygen is neither oxidized nor reduced; its oxidation number remains −2.

Problem-Solving Practice 19.1

Give the oxidation number for each atom and identify the oxidizing and reducing agents in the following balanced chemical equations.

(a) $2\,Fe(s) + 3\,Cl_2(g) \longrightarrow 2\,FeCl_3(s)$
(b) $2\,H_2(g) + O_2(g) \longrightarrow 2\,H_2O(\ell)$
(c) $Cu(s) + 2\,NO_3^-(aq) + 4\,H_3O^+(aq) \longrightarrow Cu^{2+}(aq) + 2\,NO_2(g) + 6\,H_2O(\ell)$
(d) $C(s) + O_2(g) \longrightarrow CO_2(g)$
(e) $6\,Fe^{2+}(aq) + Cr_2O_7^{2-}(aq) + 14\,H_3O^+(aq) \longrightarrow$
$$6\,Fe^{3+}(aq) + 2\,Cr^{3+}(aq) + 21\,H_2O(\ell)$$

CD-ROM Screen 21.3:
Balancing Equations for Redox
Reactions

19.2 USING HALF-REACTIONS TO UNDERSTAND REDOX REACTIONS

Now consider the redox reaction between zinc metal and copper(II) ions pictured in Figure 19.1, for which the net ionic equation is

$$Zn(s) + Cu^{2+}(aq) \longrightarrow Zn^{2+}(aq) + Cu(s)$$

Zinc metal is oxidized to Zn^{2+} ions, and Cu^{2+} ions are reduced to copper metal.

In order to see more clearly how electrons are transferred, this overall reaction can be thought of as the result of two simultaneous **half-reactions**: one half-reaction for the oxidation of Zn and one half-reaction for the reduction of Cu^{2+} ions. The oxidation half-reaction

$$Zn(s) \longrightarrow Zn^{2+}(aq) + 2\,e^-$$

Note how the sum of the charges on the left side of the reaction equals the sum of the charges on the right side, even for half-reactions.

shows that each atom of Zn loses two electrons when it is oxidized to a Zn^{2+} ion. These two electrons are accepted by a Cu^{2+} ion in the reduction half-reaction,

$$Cu^{2+}(aq) + 2\,e^- \longrightarrow Cu(s)$$

As Cu^{2+} ions are converted to Cu(s) in this half-reaction, the blue color of the solution becomes less intense and metallic copper forms on the surface of the zinc.

The net reaction is the sum of the oxidation and reduction half-reactions.

$$
\begin{array}{ll}
Zn(s) \longrightarrow Zn^{2+}(aq) + 2\,e^- & \text{(oxidation half-reaction)} \\
\underline{Cu^{2+}(aq) + 2\,e^- \longrightarrow Cu(s)} & \text{(reduction half-reaction)} \\
Zn(s) + Cu^{2+}(aq) \longrightarrow Zn^{2+}(aq) + Cu(s) & \text{(net reaction)}
\end{array}
$$

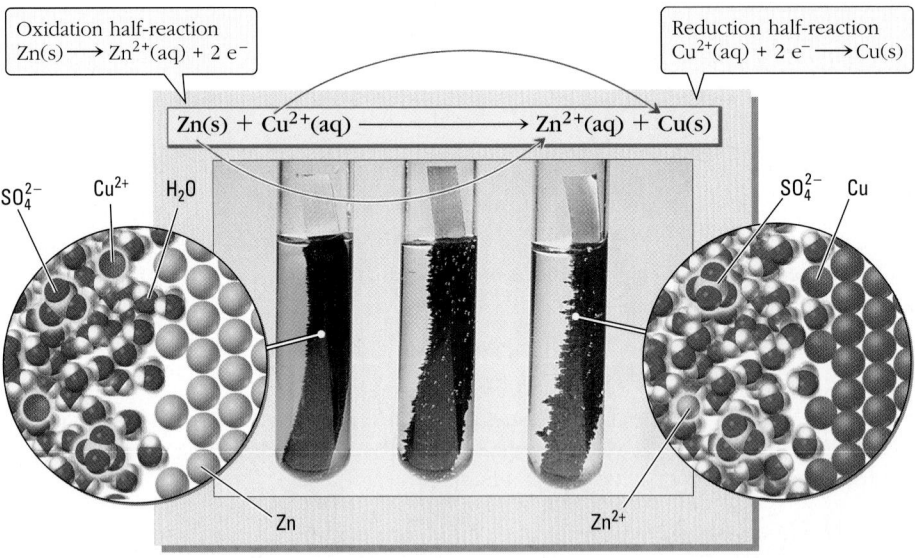

Figure 19.1 An oxidation-reduction reaction. A strip of zinc is placed in a solution of copper(II) sulfate (*left*). The zinc reacts with the copper(II) ions to produce copper metal (the brown-colored deposit on the zinc strip) and zinc ions in solution.

$$Zn(s) + Cu^{2+}(aq) \longrightarrow Zn^{2+}(aq) + Cu(s)$$

As copper metal accumulates on the zinc strip, the blue color due to the aqueous copper ions gradually fades (*middle and right*) as Cu^{2+} ions are reduced to metallic copper. The zinc ions in aqueous solution are colorless.

Notice that no electrons appear in the equation for the net reaction—the number of electrons produced by the oxidation half-reaction equals the number of electrons gained by the reduction half-reaction. This must always be true in a net reaction. Otherwise, electrons would be created or destroyed, violating the law of conservation of mass.

Consider another example, as shown in Figure 19.2. A piece of metallic copper screen is immersed in a solution of silver nitrate. As the reaction proceeds, the solution gradually turns blue, and fine, silvery, hair-like crystals form on the copper screen. Knowing that Cu^{2+} ions in aqueous solution appear blue, we can conclude that the copper metal is being oxidized to Cu^{2+}. Reduction must also be taking place, so it is reasonable to conclude that the silvery whiskers are the result of the reduction of Ag^+ ions to metallic silver. The two half-reactions are

$$Cu(s) \longrightarrow Cu^{2+}(aq) + 2\,e^- \qquad \text{(oxidation half-reaction)}$$

$$Ag^+(aq) + e^- \longrightarrow Ag(s) \qquad \text{(reduction half-reaction)}$$

In this case, two electrons are produced in the oxidation half-reaction, but only one is needed for the reduction half-reaction. *One* atom of copper provides enough electrons (two) to reduce *two* Ag^+ ions, and so the reduction half-reaction must occur twice every time the oxidation half-reaction occurs once. To indicate this, we multiply the reduction half-reaction by 2.

$$2\,Ag^+(aq) + 2\,e^- \longrightarrow 2\,Ag(s) \quad \text{(reduction half-reaction)} \times 2$$

Adding this half-reaction to the oxidation half-reaction gives the net equation

$$Cu(s) + 2\,Ag^+(aq) \longrightarrow Cu^{2+}(aq) + 2\,Ag(s)$$

The method shown here is a general one. A net equation can always be generated by writing oxidation and reduction half-reactions, using coefficients to adjust the half-reaction equations so that the number of electrons lost by the oxidation equals the number gained by the reduction, and then adding the two half-reactions to give the equation for the net reaction.

In all redox reactions, the number of electrons lost equals the number gained.

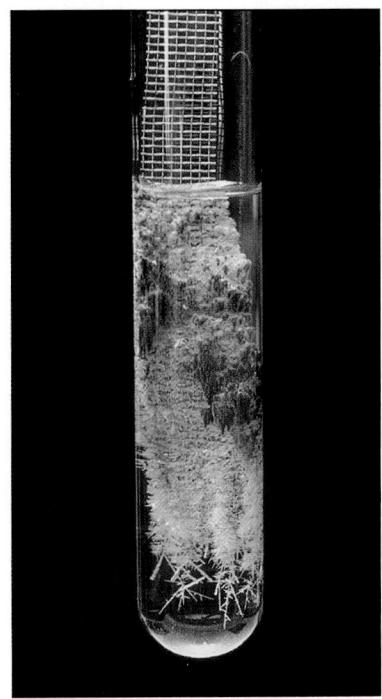

Figure 19.2 **Copper metal screen in a solution of $AgNO_3$.** The blue color intensifies as more copper is oxidized to aqueous Cu^{2+} ion.

Problem-Solving Example 19.2   Determining Half-Reactions from Net Redox Reactions

When a piece of aluminum is immersed in an aqueous solution containing a small quantity of dissolved bromine, aqueous $AlBr_3$ is formed. The reaction is

$$3\,Br_2(aq) + 2\,Al(s) \longrightarrow 2\,Al^{3+}(aq) + 6\,Br^-(aq)$$

This is a redox reaction. The oxidation number of Al changes from 0 to +3, and the oxidation number of Br changes from 0 to −1. Write the oxidation half-reaction and reduction half-reaction that can be combined to give this net reaction.

Answer
Oxidation half-reaction: $Al(s) \rightarrow Al^{3+}(aq) + 3\,e^-$
Reduction half-reaction: $Br_2(aq) + 2\,e^- \rightarrow 2\,Br^-(aq)$

Explanation   Aluminum is oxidized, as shown by the production of Al^{3+} ions.

$$Al(s) \longrightarrow Al^{3+}(aq) + 3\,e^-$$

The half-reaction must have 3 electrons on the right to balance the +3 charge on the aluminum ion and give a net 0 charge on the right to equal the 0 charge on the left.

Bromine is reduced, as shown by the conversion of elemental bromine to Br^- ions.

$$Br_2(aq) + 2\,e^- \longrightarrow 2\,Br^-(aq)$$

The half-reaction must have 2 electrons on the left to balance the 2 Br^- ions on the right.

Now notice that these two half-reactions contain different numbers of electrons. The half-reactions are multiplied by 2 and 3, respectively, so 6 e^- appear in each half-reaction.

$$2\,[Al(s) \longrightarrow Al^{3+}(aq) + 3\,e^-] \quad \text{gives} \quad 2\,Al(s) \longrightarrow 2\,Al^{3+}(aq) + 6\,e^-$$

$$3\,[Br_2(aq) + 2\,e^- \longrightarrow 2\,Br^-(aq)] \quad \text{gives} \quad 3\,Br_2(aq) + 6\,e^- \longrightarrow 6\,Br^-(aq)$$

$$\text{Net reaction: } 3\,Br_2(aq) + 2\,Al(s) \longrightarrow 2\,Al^{3+}(aq) + 6\,Br^-(aq)$$

The net reaction is the sum of these two half-reactions taken the proper number of times to make the number of electrons lost in the oxidation half-reaction equal to the number gained in the reduction half-reaction.

Problem-Solving Practice 19.2

Write an oxidation and a reduction half-reaction for the following net redox equations. Show that their sum is the net reaction.

(a) $Cd(s) + Cu^{2+}(aq) \rightarrow Cu(s) + Cd^{2+}(aq)$
(b) $Zn(s) + 2\,H^+(aq) \rightarrow Zn^{2+}(aq) + H_2(g)$
(c) $2\,Al(s) + 3\,Zn^{2+}(aq) \rightarrow 2\,Al^{3+}(aq) + 3\,Zn(s)$

CD-ROM Screen 21.3:
Balancing Equations for Redox
Reactions

Balancing Redox Equations by Using Half-Reactions

All of the equations in Problem-Solving Practice 19.2 are balanced. While these particular equations could be balanced by inspection, this is not always the case. Equations for redox reactions often involve ions, water, hydronium ions (H_3O^+), and hydroxide ions as reactants or products. It is difficult to tell by observing the unbalanced equation just how many H_2O, H_3O^+, and OH^- are involved, or whether they will be reactants, or products, or even present at all. But there is a way to figure this out.

Consider the reaction of permanganate ion with oxalic acid in acidic solution. The products are manganese(II) ion and carbon dioxide, so the unbalanced equation is

Oxalic acid, HOOC—COOH, is the simplest organic acid containing two carboxylic acid groups.

$$\text{(unbalanced equation)} \quad \underset{\substack{\text{oxalic} \\ \text{acid}}}{H_2C_2O_4(aq)} + \underset{\substack{\text{permanganate} \\ \text{ion}}}{MnO_4^-(aq)} \longrightarrow Mn^{2+}(aq) + CO_2(g)$$

If you try to balance this equation by trial and error, you will almost certainly have a hard time with hydrogen and oxygen. You have probably already noticed that no hydrogen-containing species appears on the product side of the unbalanced equation. Because the reaction is taking place in an aqueous acidic solution, water and hydronium ions may be involved. Generating the balanced equation for a reaction like this is best done by a series of steps. In each step you must use what you know about the oxidation and reduction half-reactions, as well as conservation of matter and conservation of electrical charge. The steps that will produce a balanced equation for a redox reaction that occurs in acidic solution are illustrated in Problem-Solving Example 19.3.

Problem-Solving Example 19.3   Balancing Redox Equations for
Reactions in Acidic Solution

Balance the equation for the oxidation of oxalic acid in acidic permanganate solution. The products of this reaction are CO_2 and Mn^{2+} ions.

Answer $5\,H_2C_2O_4(aq) + 6\,H_3O^+(aq) + 2\,MnO_4^-(aq) \longrightarrow$
$$10\,CO_2(g) + 2\,Mn^{2+}(aq) + 14\,H_2O(\ell)$$

Explanation   It is best to follow a series of steps to balance the equation for this reaction.

Step 1: Recognize whether the reaction is an oxidation-reduction process. If it is, then determine what is reduced and what is oxidized. This is a redox reaction because the oxidation number of Mn changes from $+7$ in MnO_4^- to $+2$ in Mn^{2+}, so MnO_4^- is reduced. The oxidation number of C changes from $+3$ in $H_2C_2O_4$ to $+4$ in CO_2, so $H_2C_2O_4$ is oxidized.

Step 2: Break the overall unbalanced equation into half-reactions.

$$H_2C_2O_4(aq) \longrightarrow CO_2(g) \qquad \text{(oxidation half-reaction)}$$

$$MnO_4^-(aq) \longrightarrow Mn^{2+}(aq) \qquad \text{(reduction half-reaction)}$$

Step 3: Balance the atoms in each half-reaction. First balance all atoms except for O and H; then balance O by adding H_2O and balance H by adding H^+. (Hydroxide ion, OH^-, cannot be used here because the reaction occurs in an acidic solution and the OH^- concentration is very low.)

Oxalic acid half-reaction: First, balance the carbon atoms in the half-reaction.

$$H_2C_2O_4(aq) \longrightarrow 2\,CO_2(g)$$

This step balances the O atoms as well (no H_2O needed here), so only H atoms remain to be balanced. Because the product side is deficient by two H atoms, we put $2\,H^+$ there.

$$H_2C_2O_4(aq) \longrightarrow 2\,CO_2(g) + 2\,H^+(aq) \qquad \text{(oxalic acid half-reaction)}$$

(Strictly speaking, we ought to use H_3O^+ instead of H^+, but this would result in adding eight more water molecules to each side of the equation, which is rather cumbersome; it is simpler to add just H^+ now, and add the water molecules at the end.)

Permanganate half-reaction: The Mn atoms are already balanced, but the oxygen atoms are not balanced until H_2O is added. Adding $4\,H_2O$ on the product side takes care of the needed oxygen atoms.

$$MnO_4^-(aq) \longrightarrow Mn^{2+}(aq) + 4\,H_2O(\ell)$$

But now there are eight H atoms on the right and none on the left. To balance hydrogen atoms, eight H^+ are placed on the left side of the half-reaction.

$$8\,H^+(aq) + MnO_4^-(aq) \longrightarrow$$
$$Mn^{2+}(aq) + 4\,H_2O(\ell) \qquad \text{(permanganate half-reaction)}$$

Step 4: Balance the half-reactions for charge. The net oxalic acid half-reaction has a net charge of 0 on the left side and $+2$ on the right. The reactants have lost two electrons. To show this, $2\,e^-$ should appear on the right side.

$$H_2C_2O_4(aq) \longrightarrow 2\,CO_2(g) + 2\,H^+(aq) + 2\,e^-$$

This confirms that $H_2C_2O_4$ is the reducing agent (it loses electrons and gets oxidized). The loss of two electrons is also in keeping with the increase in the oxidation number of each of two C atoms by 1, from $+3$ to $+4$. The $2\,e^-$ also balance charge on the product side of the equation.

The net MnO_4^- half-reaction has a charge of $+7$ on the left and $+2$ on the right. Therefore, to achieve a net $+2$ charge on each side, $5\,e^-$ must appear on the left. The gain of electrons shows that MnO_4^- is the oxidizing agent.

$$5\,e^- + 8\,H^+(aq) + MnO_4^-(aq) \longrightarrow Mn^{2+}(aq) + 4\,H_2O(\ell)$$

Step 5: Multiply the half-reactions by appropriate factors so that the oxidation half-reaction produces as many electrons as the reduction half-reaction accepts. In this case, one half-reaction involves two electrons, and the other half-reaction involves five electrons. To make each half-reaction balance takes ten electrons. The oxalic acid reaction should be multiplied by 5, and the MnO_4^- reaction by 2.

$$5\,[H_2C_2O_4(aq) \longrightarrow 2\,CO_2(g) + 2\,H^+(aq) + 2\,e^-]$$

$$2\,[5\,e^- + 8\,H^+(aq) + MnO_4^-(aq) \longrightarrow Mn^{2+}(aq) + 4\,H_2O(\ell)]$$

Step 6: Add the half-reactions to give the overall reaction and cancel equal amounts of reactants and products that appear on both sides of the reaction arrow.

$$5 \, H_2C_2O_4(aq) \longrightarrow 10 \, CO_2(g) + 10 \, H^+(aq) + 10 \, e^-$$

$$\underline{10 \, e^- + 16 \, H^+(aq) + 2 \, MnO_4^-(aq) \longrightarrow 2 \, Mn^{2+}(aq) + 8 \, H_2O(\ell)}$$

$$5 \, H_2C_2O_4(aq) + 16 \, H^+(aq) + 2 \, MnO_4^-(aq) \longrightarrow$$
$$10 \, CO_2(g) + 10 \, H^+(aq) + 2 \, Mn^{2+}(aq) + 8 \, H_2O(\ell)$$

Since 16 H$^+$ appear on the left and 10 H$^+$ appear on the right, 10 H$^+$ are canceled, leaving 6 H$^+$ on the left.

$$5 \, H_2C_2O_4(aq) + 6 \, H^+(aq) + 2 \, MnO_4^-(aq) \longrightarrow 10 \, CO_2(g) + 2 \, Mn^{2+}(aq) + 8 \, H_2O(\ell)$$

Step 7: *Check the final results to make sure both atoms and charge are balanced.*
Atom balance: Each side of the equation has 2 Mn, 28 O, 10 C, and 16 H atoms.
Charge balance: Each side has a net charge of +4. On the left side, $(6 \times 1+) + (2 \times 1-) = +4$. On the right side, $2(2+) = +4$.

Step 8: *Add enough water molecules to both sides of the equation to convert all H$^+$ to H$_3$O$^+$.* In this case, six water molecules are needed ($6 \, H_2O + 6 \, H^+ \longrightarrow 6 \, H_3O^+$).

$$5 \, H_2C_2O_4(aq) + 6 \, H_3O^+(aq) + 2 \, MnO_4^-(aq) \longrightarrow$$
$$10 \, CO_2(g) + 2 \, Mn^{2+}(aq) + 14 \, H_2O(\ell)$$

Problem-Solving Practice 19.3

Balance this equation for the reaction of Zn with $Cr_2O_7^{2-}$ in acidic aqueous solution.

$$Zn(s) + Cr_2O_7^{2-}(aq) \longrightarrow Cr^{3+}(aq) + Zn^{2+}(aq)$$

 Exercise 19.1   Electrons Lost = Electrons Gained

Why must the number of electrons lost equal the number gained? (*Hint:* Look at Section 1.8.)

Problem-Solving Example 19.4   Balancing Redox Equations for Reactions in Basic Solution

In a nickel-cadmium (nicad) battery, cadmium forms $Cd(OH)_2$ and Ni_2O_3 forms $Ni(OH)_2$ in an alkaline solution. Write the balanced equation for this reaction.

Answer $Cd(s) + Ni_2O_3(s) + 3 \, H_2O(\ell) \rightarrow Cd(OH)_2(s) + 2 \, Ni(OH)_2(s)$

Explanation

Step 1: *Recognize whether the reaction is an oxidation-reduction process. Then determine what is reduced and what is oxidized.* This is a redox reaction because the oxidation number of Cd changes from 0 in Cd to +2 in $Cd(OH)_2$, so Cd is oxidized. The oxidation number of Ni changes from +3 in Ni_2O_3 to +2 in $Ni(OH)_2$, so the Ni is reduced.

Step 2: *Break the overall unbalanced equation into half-reactions.*

$$Cd(s) \longrightarrow Cd(OH)_2(s) \qquad \text{(oxidation half-reaction)}$$

$$Ni_2O_3(s) \longrightarrow Ni(OH)_2(s) \qquad \text{(reduction half-reaction)}$$

Step 3: *Balance the atoms in each half-reaction.* First balance all atoms except the O and H atoms; do them last. Balance each half-reaction as if it were in acidic solution to start. Balance O by adding H_2O and balance H by adding H^+. We will revert to the OH^- characteristic of basic solutions later in the process.

In the Cd half-reaction, the Cd atoms are balanced. Adding two water molecules on the left balances the two O atoms on the right but leaves four H atoms on the left and only two on the right. Adding two H^+ ions on the right balances H atoms in this half-reaction.

$$2 \, H_2O(\ell) + Cd(s) \longrightarrow Cd(OH)_2(s) + 2 \, H^+(aq)$$

For the Ni_2O_3 half-reaction, a coefficient of 2 is needed for $Ni(OH)_2$ because there are two Ni atoms on the left. To balance the four O atoms and four H atoms now on the right with the three O atoms on the left requires one water molecule and two additional H^+ ions on the left.

$$2\,H^+(aq) + H_2O(\ell) + Ni_2O_3(s) \longrightarrow 2\,Ni(OH)_2(s)$$

Step 4: *Balance the half-reactions for charge.* The Cd half-reaction requires $2\,e^-$ as a product.

$$2\,H_2O(\ell) + Cd(s) \longrightarrow Cd(OH)_2(s) + 2\,H^+(aq) + 2\,e^- \qquad \text{(balanced)}$$

The Ni_2O_3 half-reaction requires $2\,e^-$ as a reactant.

$$2\,H^+(aq) + H_2O(\ell) + Ni_2O_3(s) + 2\,e^- \longrightarrow 2\,Ni(OH)_2(s) \qquad \text{(balanced)}$$

Step 5: *Multiply the half-reactions by appropriate factors so that the reducing agent produces as many electrons as the oxidizing agent accepts.* The Cd half-reaction produces two electrons and the Ni_2O_3 half-reaction accepts two, so the electrons are balanced.

Step 6: *Because H^+ does not exist at any appreciable concentration in a basic solution, remove it by adding an appropriate amount of OH^- to both sides of the equation.* In the Cd half-reaction, add two OH^- ions to get

$$2\,OH^-(aq) + 2\,H_2O(\ell) + Cd(s) \longrightarrow Cd(OH)_2(s) + 2\,H_2O(\ell) + 2\,e^-$$

On the product side, two OH^- ions plus two H^+ ions form two H_2O molecules. For the Ni_2O_3 half-reaction, add two OH^- ions to get

$$3\,H_2O(\ell) + Ni_2O_3(s) + 2\,e^- \longrightarrow 2\,Ni(OH)_2(s) + 2\,OH^-(aq)$$

Step 7: *Add the half-reactions to give the overall reaction, and cancel reactants and products that appear on both sides of the reaction arrow.*

$$2\,\cancel{OH^-(aq)} + 2\,\cancel{H_2O(\ell)} + Cd(s) \longrightarrow Cd(OH)_2(s) + 2\,\cancel{H_2O(\ell)} + 2\,e^-$$
$$\underline{3\,H_2O(\ell) + Ni_2O_3(s) + 2\,e^- \longrightarrow 2\,Ni(OH)_2(s) + 2\,\cancel{OH^-(aq)}}$$
$$Cd(s) + Ni_2O_3(s) + 3\,H_2O(\ell) \longrightarrow Cd(OH)_2(s) + 2\,Ni(OH)_2(s)$$

Step 8: *Check the final results to make sure both atoms and charge are balanced.* The equation is balanced. In the final equation, there are no charges on either side of the reaction arrow, and the numbers of atoms of each kind on each side of the reaction arrow are equal.

Problem-Solving Practice 19.4

In basic solution, aluminum metal forms $Al(OH)_4^-$ ion as it reduces NO_3^- ion to NH_3. Write the balanced equation for this reaction using the steps outlined in Problem-Solving Example 19.4.

19.3 ELECTROCHEMICAL CELLS

CD-ROM Screen 21.4: Electrochemical Cells

In a redox reaction, electrons are *transferred* from one kind of atom, molecule, or ion to another. It is easy to see by the color changes in the two redox reactions shown in Figures 19.1 and 19.2 that these reactions favor the formation of products—as soon as the reactants are mixed, changes take place. All product-favored reactions release Gibbs free energy *(⇐ p. 827)*, energy that can do useful work. An electrochemical cell makes this possible for redox reactions.

An **electrochemical cell** is an arrangement of an oxidizing agent and a reducing agent pair in such a way that they can react only if electrons flow through an outside conductor. Such electrochemical cells are also known as **voltaic cells** or **batteries**. Figure 19.3 diagrams how this can be done for the Zn/Cu^{2+} reaction that was shown in Figure 19.1. The two half-reactions are allowed to occur in separate beakers, each of which is called a **half-cell**. When Zn atoms are oxidized, the electrons that are given up by zinc pass through the wire and a lamp (this could also be a voltmeter or a small motor) to the copper metal. There the electrons are

On a flashlight battery, the anode is marked "−" because oxidation produces electrons that make the anode negative. Conversely, the cathode is marked "+" because reduction consumes electrons, leaving the metal electrode positive.

Figure 19.3   A simple electrochemical cell.   The cell consists of a copper electrode in a solution containing Cu^{2+} ions (*right*), a zinc electrode in a solution containing Zn^{2+} ions (*left*), and a salt bridge that allows ions to flow into and out of the two solutions. When the two metal electrodes are connected by a conducting circuit, electrons flow from the zinc electrode, where zinc is oxidized, to the copper electrode, where copper ions are reduced.

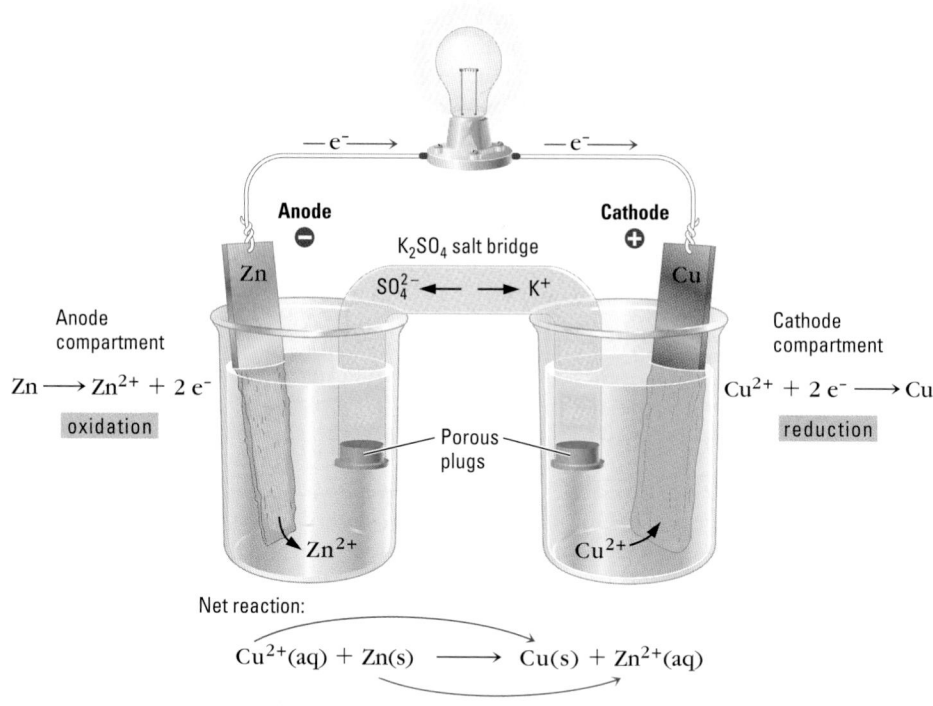

Net reaction:

$$Cu^{2+}(aq) + Zn(s) \longrightarrow Cu(s) + Zn^{2+}(aq)$$

Strictly speaking, many devices we call batteries consist of several voltaic cells connected together, but the term "battery" has taken on the same meaning as a "voltaic cell."

Electrochemical cells are sometimes referred to as *galvanic cells* in recognition of the work of Luigi Galvani, who discovered, before Volta, that a frog's leg would twitch when it was touched simultaneously in salt water by two dissimilar metals.

To identify the anode and cathode, remember that **o**xidation takes place at the **a**node (both words begin with vowels), and **r**eduction takes place at the **c**athode (both words begin with consonants).

Figure 19.4   A grapefruit battery.   A voltaic cell can be made by inserting zinc and copper electrodes into a grapefruit. A potential of 0.95 V is obtained. (The water and citric acid of the fruit allow for ion conduction between electrodes.) This cell is more complicated than the one in Figure 19.3. To learn how it works, see *Jour. Chem. Ed.*, 78, 516 (2000).

available to reduce Cu^{2+} ions from the solution. The zinc and copper strips are called electrodes. An **electrode** conducts electrical current (electrons) into or out of something — in this case, a solution. An electrode is most often a metal plate or wire, but it can also be a piece of graphite or another conductor. The electrode where oxidation occurs is named the **anode;** the electrode where reduction takes place is called the **cathode.**

The voltaic cell is named after the Italian scientist Alessandro Volta, who, in about 1800, constructed the first electrochemical cell — a stack of alternating zinc and silver disks separated by pieces of paper soaked in salt water (an electrolyte). Later, Volta showed that any two different metals and an electrolyte could be used to make a battery. Figure 19.4 shows a cell constructed by sticking a strip of zinc and a strip of copper into a grapefruit, for example.

In the cell diagrammed in Figure 19.3, electrons are transferred to the anode by the half-reaction

$$Zn(s) \longrightarrow Zn^{2+}(aq) + 2\,e^- \qquad \text{(anode reaction)}$$

They then flow from the anode through the filament in the bulb, causing it to glow, and eventually travel to the cathode, where they react with copper(II) ions in the half-reaction

$$Cu^{2+}(aq) + 2\,e^- \longrightarrow Cu(s) \qquad \text{(cathode reaction)}$$

If nothing else but electron flow took place, the concentration of Zn^{2+} ions in the anode compartment would increase as zinc metal is oxidized, building up positive charge in the solution. The concentration of Cu^{2+} ions in the cathode compartment would decrease as they are reduced to metallic copper, making that solution less positive. Because of this charge buildup, the flow of electrons would very quickly stop. In order for the cell to work, there has to be a way for the positive charge buildup in the anode compartment to be balanced by addition of negative ions or removal of positive ions, and vice versa for the cathode compartment.

The charge buildup can be avoided by using a salt bridge to connect the two compartments. A **salt bridge** is a solution of a salt (K_2SO_4 in Figure 19.3) arranged so that the bulk of the solution cannot flow into the cell solutions, but the ions (K^+ and

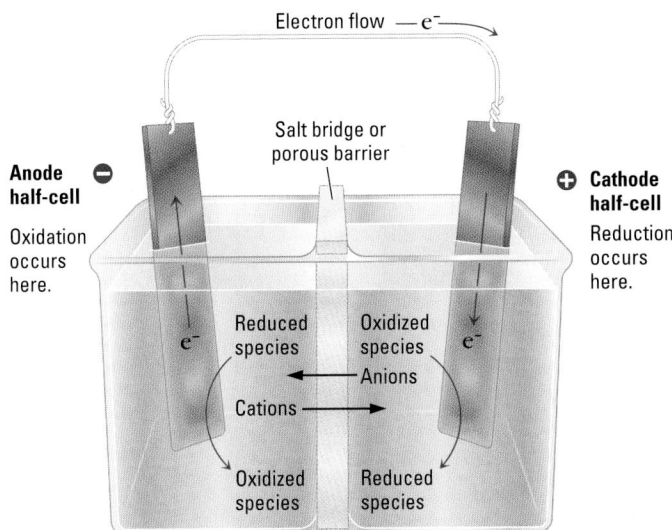

Electron flow — e⁻ →

Salt bridge or
porous barrier

Anode
half-cell ⊖

Oxidation
occurs
here.

⊕ **Cathode**
half-cell

Reduction
occurs
here.

Reduced
species

Oxidized
species

Anions

Cations →

Oxidized
species

Reduced
species

Figure 19.5 **Summary of the terminology used in voltaic cells.**
Oxidation occurs at the anode, and reduction occurs at the cathode.
Electrons move from the negative electrode (anode) to the positive
electrode (cathode) through the external wire. The circuit is completed by the movement of ions in solution. Anions move from the
cathode compartment to the anode compartment, and cations move
from the anode compartment to the cathode compartment. The compartments can be separated either by a salt bridge or a porous barrier.

SO_4^{2-}) can pass freely. As electrons flow through the wire from the zinc electrode to
the copper electrode, negative ions (SO_4^{2-}) move through the salt bridge toward the
solution in the anode compartment and positive ions (K^+) move in the opposite direction into the cathode compartment. In general, anions from the salt bridge always
flow into the anode cell, and cations from the salt bridge always flow into the cathode
cell. This flow of ions completes the electrical circuit, allowing current to flow. If the
salt bridge is removed from this battery, the flow of electrons will stop.

In commercial batteries, the salt bridge is
often a porous polymer membrane.

All voltaic cells and batteries operate in a similar fashion.

- The oxidation-reduction reaction must favor the formation of products.
- There must be an external circuit through which electrons flow.
- There must be a salt bridge, porous barrier, or some other means of allowing
  ions to flow between the electrode compartments.

The components of an electrochemical cell are summarized in Figure 19.5.

Problem-Solving Example 19.5 **Electrochemical Cells**

A simple voltaic cell is assembled with Ni(s) and Ni(NO₃)₂(aq) in one compartment and
Cd(s) and Cd(NO₃)₂(aq) in the other. An external wire connects the two electrodes, and
a salt bridge containing NaNO₃ connects the two solutions. The net reaction is

$$Ni^{2+}(aq) + Cd(s) \longrightarrow Ni(s) + Cd^{2+}(aq)$$

(a) What is the reaction at the anode?
(b) What is the reaction at the cathode?
(c) What are the directions of electron flow in the external wire and of the ion flow in
    the salt bridge?
(d) Complete the cell diagram by indicating the directions of electron flow and ion flow.

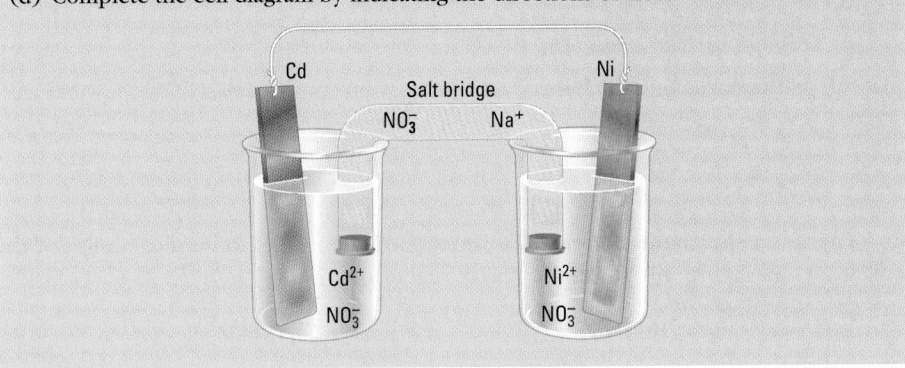

Cd

Salt bridge

NO_3^- Na^+

Ni

Cd²⁺

NO_3^-

Ni²⁺

NO_3^-

This voltaic cell is shown without a meter
in the external unit for simplicity.

Answer

(a) Anode: $Cd(s) \rightarrow Cd^{2+}(aq) + 2\,e^-$ (oxidation)

(b) Cathode: $Ni^{2+}(aq) + 2\,e^- \rightarrow Ni(s)$ (reduction)

(c) The completed voltaic cell is shown below.

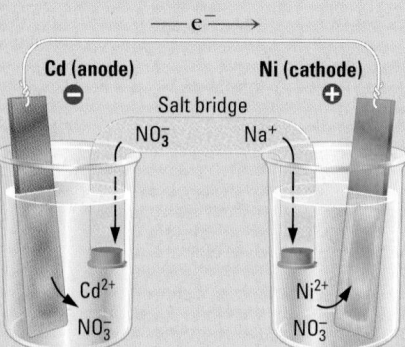

Explanation First, since the net reaction shows the oxidation of Cd to Cd^{2+} and the reduction of Ni^{2+} to Ni, let's decide at which electrode these reactions take place. The Cd^{2+} ions and Cd metal are in the left compartment in the figure, and by the net reaction this is where the oxidation of Cd occurs. The Ni^{2+} ions and Ni metal are in the right compartment, and by the net reaction this is where the reduction takes place.

The half-reactions are

$$Cd(s) \longrightarrow Cd^{2+}(aq) + 2\,e^- \quad \text{(left compartment — oxidation at the anode)}$$

$$Ni^{2+}(aq) + 2\,e^- \longrightarrow Ni(s) \quad \text{(right compartment — reduction at the cathode)}$$

Electrons flow *from* their source (the oxidation of cadmium at the Cd electrode — the anode) through the wire *to* the electrode where they are used (the Ni electrode — the cathode). Because Cd^{2+} ions are being formed in the anode compartment, the NO_3^- ions in the salt bridge must move into that compartment from the salt bridge. In the cathode compartment the Ni^{2+} ions are being reduced, so Na^+ ions move from the salt bridge into that compartment to take the place of Ni^{2+} ions. There is now a complete electrical circuit.

Problem-Solving Practice 19.5

A voltaic cell is assembled to use the following net reaction.

$$Ni(s) + 2\,Ag^+(aq) \longrightarrow Ni^{2+}(aq) + 2\,Ag(s)$$

(a) Write half-reactions for this cell, and indicate which is the oxidation reaction and which is the reduction reaction.

(b) Name the electrodes at which these reactions take place.

(c) What is the direction of flow of electrons in an external wire connected between the electrodes?

(d) If a salt bridge connecting the two electrode compartments contains KNO_3, what is the direction of flow of the K^+ ions and the NO_3^- ions?

 Exercise 19.2 Battery Design

Devise an internal on-off switch for a battery that would not be a part of the flow of electrons.

There is a shorthand notation for representing an electrochemical cell. For the cell shown in Figure 19.3 with the redox reaction

$$Zn(s) + Cu^{2+}(aq) \longrightarrow Zn^{2+}(aq) + Cu(s)$$

the representation is

$$Zn(s)\,|\,Zn^{2+}(aq)\,||\,Cu^{2+}(aq)\,|\,Cu(s)$$

The single vertical lines denote phase boundaries, and the double vertical lines denote the salt bridge. The anode half-cell is represented on the left, and the cathode half-cell is represented on the right. The electrodes are written on the extreme left (anode, Zn) and extreme right (cathode, Cu) of the notation. Within each half-cell the reactants are written first, followed by the products. The electron flow is from left to right.

19.4 ELECTROCHEMICAL CELLS AND VOLTAGE

CD-ROM Screen 21.5: Electrochemical Cells and Potentials

Because electrons flow from the anode to the cathode in an electrochemical cell, they can be thought to be "driven" or "pushed" by an **electromotive force** or **emf.** The emf is produced by the difference in electrical potential energy between the two electrodes. Just as water flows downhill in response to a difference in gravitational potential energy, so an electron moves from an electrode of higher electrical potential energy to another of lower potential energy. The moving water can do work, and so can moving electrons—for example, they could run a motor.

The quantity of electrical work done is proportional to the number of electrons that go from higher to lower potential energy and to the size of the potential energy difference.

Electrical work = charge × potential energy difference

or

Electrical work = number of electrons × potential energy difference

Electrical charge is measured in coulombs (C). The charge on a single electron is very small (1.6022×10^{-19} C), so it takes 6.24×10^{18} electrons to produce 1 coulomb of charge. A **coulomb (C)** is the quantity of charge that passes a fixed point in an electrical circuit when a current of 1 ampere flows for 1 second. The **ampere (A)** is the unit of electrical current, defined as the flow of one coulomb of charge per second.

Coulombs = amperes × seconds

This is similar to comparing the amount of work a few drops of water can do when falling 100 m with that when a few tons of water fall the same distance.

Exercise 19.3   Large Charges

Which has the larger charge, a coulomb of charge, or Avogadro's number of electrons?

Electrical potential energy difference is measured in volts. The **volt (V)** is defined such that one joule of work is performed when one coulomb of charge moves through a potential difference of one volt:

$$1\ volt = \frac{1\ joule}{1\ coulomb} \qquad or \qquad 1\ joule = 1\ volt \times 1\ coulomb$$

Therefore, the electromotive force of an electrochemical cell, commonly called its **cell voltage,** shows how much work a cell can produce for each coulomb of charge that the chemical reaction produces.

The voltage of an electrochemical cell depends on the substances that make up the cell—their pressures if they are gases or their concentrations if they are solutes in solution—and the temperature. The quantity of charge (coulombs) depends on how much of each substance reacts. Look at the 1.5-V batteries shown in

When a single electron moves through a potential of 1 V, the work done is one electron-volt, abbreviated eV.

Figure 19.6 **Dry cell 1.5-V batteries.** The larger batteries are capable of more work since they contain more oxidizing and reducing agents.

CD-ROM Screen 21.6: Standard Potentials

The convention of assigning voltages to half-reactions is similar to the convention of tabulating standard enthalpies of formation; in both cases a relatively small table of data can provide information about a large number of different reactions.

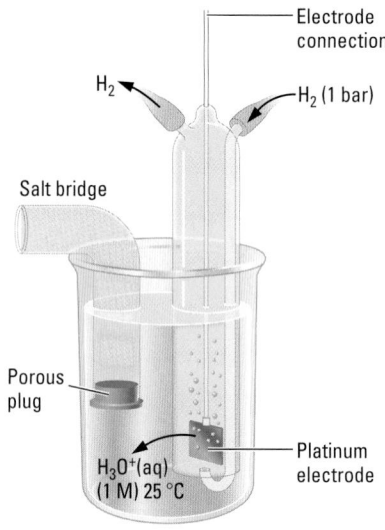

Figure 19.7 **The standard hydrogen electrode.** Hydrogen gas at 1 bar pressure bubbles over an inert platinum electrode that is immersed in a solution containing exactly 1 M H_3O^+ ions at 25 °C. The potential for this electrode is defined as exactly 0 V.

Quantities defined as exact, such as the voltage of the hydrogen electrode at standard conditions, do not limit the number of significant figures in the answer when they are used in calculations.

Figure 19.6. They have the same voltage because they have electrodes with the same potential difference between them. Yet a larger battery is capable of far more work than a smaller one, because it contains a larger quantity of reactants. In this section and the next, we consider how cell voltage depends on the materials from which a cell is made. In Section 19.5, we will return to the question of how much electrical work a cell can do.

Cell Voltage

A cell's voltage is readily measured by inserting a voltmeter into the circuit. Because the voltage depends on concentrations, **standard conditions** are defined for voltage measurements. These are the same as those used for $\Delta H°$ ($\Longleftarrow$ *p. 241*): All reactants and products must be present as pure solids, pure liquids, gases at 1 bar pressure, or solutes at 1 M concentrations. Voltages measured under these conditions are **standard voltages,** symbolized by $E°$. Unless specified otherwise, all values of $E°$ are for 25 °C (298 K). By definition, cell voltages for product-favored electrochemical reactions are *positive.* For example, the standard cell voltage for the Zn/Cu^{2+} cell, discussed earlier, is $+1.10$ V at 25 °C.

Since every redox reaction can be thought of as the sum of two half-reactions, it is convenient to assign a voltage to every possible half-reaction. Then the cell voltage for any reaction can be obtained by adding the voltages of the oxidation and reduction half-reactions.

$$E°_{cell} = E°_{ox} + E°_{red}$$

If $E°_{cell}$ is positive, the reaction is product-favored. If $E°_{cell}$ is negative, the reaction is reactant-favored. However, because only *differences* in potential energy can be measured, it is not possible to measure the voltage for a single half-reaction. Instead, one half-reaction is chosen as the standard, and then all others are compared to it. The half-reaction chosen as the standard is the one that occurs at the **standard hydrogen electrode,** in which hydrogen gas at a pressure of 1 bar is bubbled over a platinum electrode immersed in 1 M aqueous acid (Figure 19.7.)

$$2\,H_3O^+(aq,\,1\,M) + 2\,e^- \longrightarrow H_2(g,\,1\,bar) + 2\,H_2O(\ell)$$

A voltage of exactly 0 V is *assigned* to this half-cell. In a cell that combines another half-reaction with the standard hydrogen electrode, the overall cell voltage is the sum of the potential between the two electrodes. Because the potential of the hydrogen electrode is assigned to be 0, the overall cell voltage equals the voltage of the other electrode.

When the standard hydrogen electrode is paired with a half-cell that contains a better reducing agent than H_2, $H_3O^+(aq)$ is reduced to H_2.

H_3O^+ reduced: $2\,H_3O^+(aq) + 2\,e^- \longrightarrow H_2(g,\,1\,bar) + 2\,H_2O(\ell)$ $E°_{red} = 0.0$ V

When the standard hydrogen electrode is paired with a half-cell that contains a better oxidizing agent than H_3O^+, then H_2 is oxidized to H_3O^+.

H_2 oxidized: $H_2(g,\,1\,bar) + 2\,H_2O(\ell) \longrightarrow 2\,H_3O^+(aq,\,1\,M) + 2\,e^-$ $E°_{ox} = 0.0$ V

The reaction that occurs at this electrode is reversible. In either case, the standard hydrogen electrode, by definition, has a potential of 0 V.

Figure 19.8 diagrams a cell in which one compartment contains the standard hydrogen electrode and the other contains a zinc electrode immersed in a 1 M solution of Zn^{2+}. The voltmeter is connected between the two electrodes to measure the difference in electrical potential energy. For this cell the voltage is 0.76 V. After this cell operates for a time, the zinc electrode decreases in mass as zinc is oxidized to $Zn^{2+}(aq)$. Therefore, the Zn electrode must be the anode; that is, it is

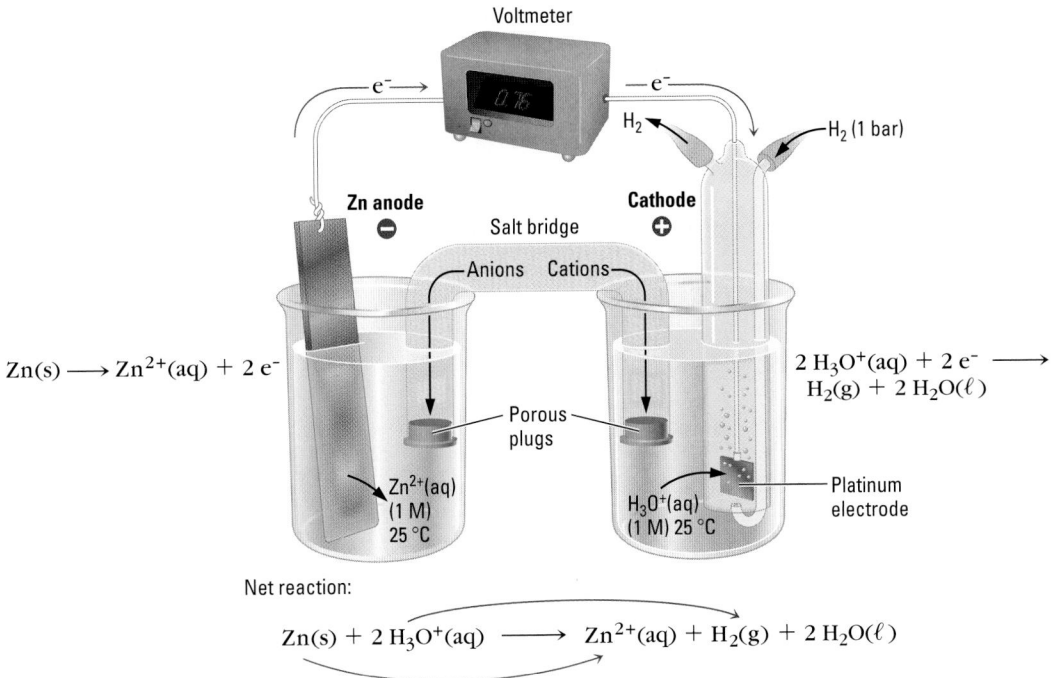

$$Zn(s) \longrightarrow Zn^{2+}(aq) + 2\,e^-$$

$$2\,H_3O^+(aq) + 2\,e^- \longrightarrow H_2(g) + 2\,H_2O(\ell)$$

Net reaction:

$$Zn(s) + 2\,H_3O^+(aq) \longrightarrow Zn^{2+}(aq) + H_2(g) + 2\,H_2O(\ell)$$

Figure 19.8 **An electrochemical cell using $Zn^{2+}/Zn(s)$ half-cell and a standard hydrogen electrode.** The zinc electrode is the anode and the standard hydrogen electrode is the cathode in this cell, which has a voltage of $+0.76$ V. Zinc is the reducing agent and is oxidized to Zn^{2+}; H_3O^+ is the oxidizing agent and is reduced to H_2. In the standard hydrogen electrode, reaction occurs only where the three phases — gas, solution, and solid electrode — are in contact. The platinum electrode does not undergo any chemical change, and in the cell pictured here the half-cell reaction is $2\,H_3O^+(aq) + 2e^- \rightarrow H_2(g) + 2\,H_2O(\ell)$. When the standard hydrogen electrode is the anode, the half-cell reaction is $H_2(g) + 2\,H_2O(\ell) \rightarrow 3\,H_3O^+(aq) + 2e^-$.

the electrode where oxidation is taking place. The hydrogen electrode must be the cathode, where reduction is taking place. The cell reaction is therefore the sum of the half-cell reactions.

$$Zn(s) \longrightarrow Zn^{2+}(aq, 1\ M) + 2\,e^- \qquad E^\circ_{ox} = ?\ V\ (anode)$$

$$2\,H_3O^+(aq, 1\ M) + 2\,e^- \longrightarrow H_2(g, 1\ bar) + 2\,H_2O(\ell) \qquad E^\circ_{red} = 0\ V\ (cathode)$$

$$Zn(s) + 2\,H_3O^+(aq, 1\ M) \longrightarrow Zn^{2+}(aq, 1\ M) + H_2(g, 1\ bar) + 2\,H_2O(\ell)$$

$$E^\circ_{cell} = +0.76\ V\ (cell\ reaction)$$

The voltmeter tells us that the potential at the Zn electrode is 0.76 V higher than at the hydrogen electrode. Because the half-cell potential for the hydrogen electrode is assigned to be 0 V, the half-cell potential for the oxidation of zinc must be $+0.76$ V.

$$Zn(s) \longrightarrow Zn^{2+}(aq, 1\ M) + 2\,e^- \qquad E^\circ_{ox} = +0.76\ V$$

 Exercise 19.4   **What Is Going on Inside the Electrochemical Cell?**

Devise an experiment that would show that Zn is being oxidized in the electrochemical cell shown in Figure 19.8.

The half-cell potentials of many different half-reactions can be measured by comparing them with the hydrogen electrode. For example, in a cell consisting of the Cu^{2+}/Cu half-reaction connected to a standard hydrogen electrode, the mass of

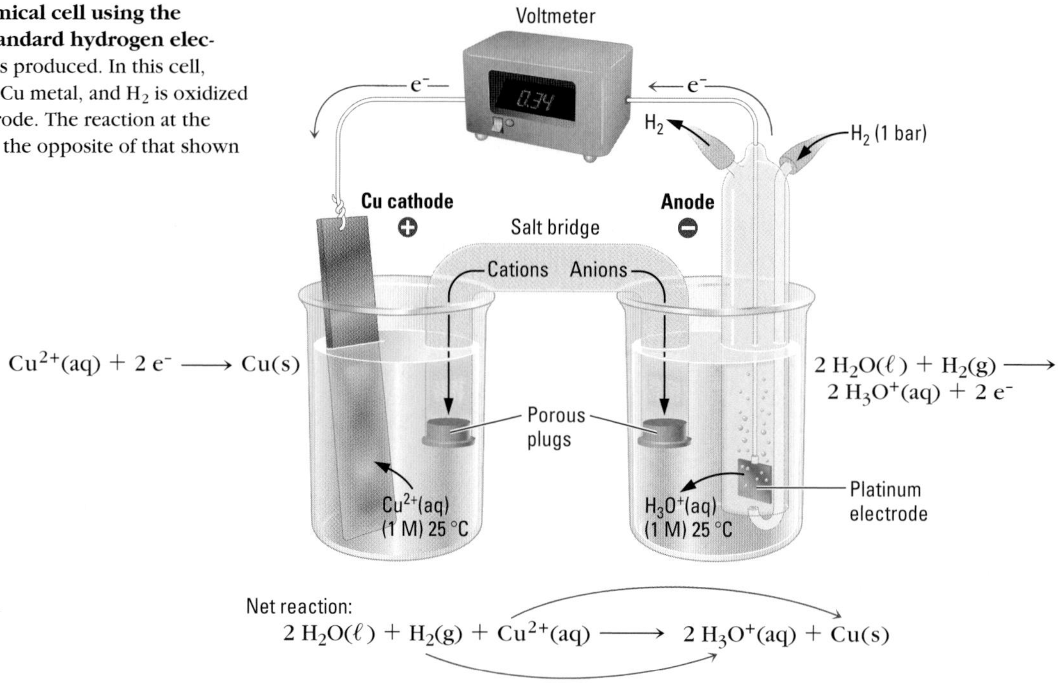

Figure 19.9 **An electrochemical cell using the Cu^{2+}/Cu half-cell and the standard hydrogen electrode.** A voltage of $+0.34$ V is produced. In this cell, Cu^{2+} ions are reduced to form Cu metal, and H_2 is oxidized at the standard hydrogen electrode. The reaction at the standard hydrogen electrode is the opposite of that shown in Figure 19.8.

Voltmeter

H_2

H_2 (1 bar)

Cu cathode
⊕

Salt bridge

Anode
⊖

Cations   Anions

$Cu^{2+}(aq) + 2\,e^- \longrightarrow Cu(s)$

$2\,H_2O(\ell) + H_2(g) \longrightarrow$
$2\,H_3O^+(aq) + 2\,e^-$

Porous
plugs

$Cu^{2+}(aq)$
(1 M) 25 °C

$H_3O^+(aq)$
(1 M) 25 °C

Platinum
electrode

Net reaction:
$$2\,H_2O(\ell) + H_2(g) + Cu^{2+}(aq) \longrightarrow 2\,H_3O^+(aq) + Cu(s)$$

the copper electrode increases and the voltmeter reads $+0.34$ V (Figure 19.9). This means that the reactions are

$$H_2(g, 1\text{ bar}) + 2\,H_2O(\ell) \longrightarrow 2\,H_3O^+(aq, 1\text{ M}) + 2\,e^-$$
$$E^\circ_{ox} = 0\text{ V (anode)}$$

$$Cu^{2+}(aq, 1\text{ M}) + 2\,e^- \longrightarrow Cu(s) \qquad E^\circ_{red} = ?\text{ V (cathode)}$$

$$H_2(g, 1\text{ bar}) + Cu^{2+}(aq, 1\text{ M}) + 2\,H_2O(\ell) \longrightarrow 2\,H_3O^+(aq, 1\text{ M}) + Cu(s)$$
$$E^\circ_{cell} = +0.34\text{ V}$$

The half-cell potential for $Cu^{2+}(aq, 1\text{ M}) + 2\,e^- \rightarrow Cu(s)$ must be $+0.34$ V. Note that in this cell the standard hydrogen electrode is the anode.

We can now return to the first electrochemical cell we looked at in which Zn reduces Cu^{2+} ions to Cu. Using the potentials for the half-reactions, we can write

$$Zn(s) \longrightarrow Zn^{2+}(aq, 1\text{ M}) + 2\,e^- \qquad E^\circ_{ox} = +0.76\text{ V (anode)}$$

$$Cu^{2+}(aq, 1\text{ M}) + 2\,e^- \longrightarrow Cu(s) \qquad E^\circ_{red} = +0.34\text{ V (cathode)}$$

$$Zn(s) + Cu^{2+}(aq, 1\text{ M}) \longrightarrow Zn^{2+}(aq, 1\text{ M}) + Cu(s) \qquad E^\circ_{cell} = +1.10\text{ V}$$

This is an important result because the sum of the potentials of the two half-reactions equals the measured potential for the cell reaction.

$$E^\circ_{cell} = E^\circ_{ox} + E^\circ_{red} = (+0.76\text{ V}) + (+0.34\text{ V}) = +1.10\text{ V}$$

Problem-Solving Example 19.6   Determining a Half-Cell Potential

The cell illustrated in the following drawing generates a potential of $E^\circ = 0.51$ V under standard conditions at 25 °C . The net cell reaction is

$$Zn(s) + Ni^{2+}(aq, 1\text{ M}) \longrightarrow Zn^{2+}(aq, 1\text{ M}) + Ni(s)$$

The half-cell potential for $Zn(s)/Zn^{2+}(aq, 1\text{ M})$ is 0.76 V.

(a) Determine which electrode is the anode and which is the cathode.

(b) Show the direction of electron flow outside the cell, and complete the cell diagram.
(c) Calculate the half-cell potential for $Ni^{2+}(aq) + 2\,e^- \rightarrow Ni(s)$.

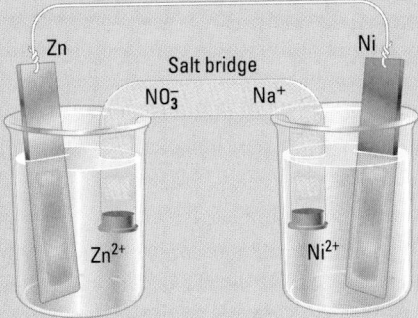

This voltaic cell is shown without a meter in the external circuit for simplicity.

Answer

(a) Zinc is the anode, nickel is the cathode.
(b) The cell diagram is as shown in the drawing below.
(c) -0.25 V

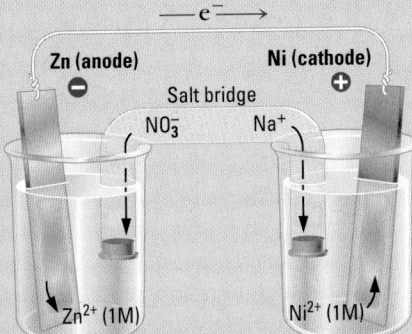

Explanation The electrode where oxidation occurs is the anode. Because Zn(s) is oxidized to $Zn^{2+}(aq)$, the Zn electrode is the anode. Nickel(II) ions are reduced at the Ni electrode, so it is the cathode.

Since the cell potential and the potential for the $Zn(s)/Zn^{2+}(aq, 1\,M)$ half-cell are known, the value of $E°$ for $Ni^{2+}(aq, 1\,M) + 2\,e^- \rightarrow Ni(s)$ can be calculated.

$$Zn(s) \longrightarrow Zn^{2+}(aq) + 2\,e^- \qquad E°_{ox} = +0.76\text{ V (anode)}$$
$$\underline{Ni^{2+}(aq) + 2\,e^- \longrightarrow Ni(s) \qquad\qquad E°_{red} = ?\text{ V (cathode)}}$$
$$Zn(s) + Ni^{2+}(aq) \longrightarrow Zn^{2+}(aq) + Ni(s) \qquad E°_{cell} = +0.51\text{ V}$$

Using $E°_{cell} = E°_{ox} + E°_{red}$, solve for $E°_{red}$.

$$E°_{red} = E°_{cell} - E°_{ox} = 0.51\text{ V} - 0.76\text{ V} = -0.25\text{ V}$$

At 25 °C, the value of $E°$ for the $Ni^{2+}(aq, 1\,M) + 2\,e^- \rightarrow Ni(s)$ half-reaction is -0.25 V.

Problem-Solving Practice 19.6

Given that the reaction of aqueous copper(II) ions with iron metal has an $E°_{cell}$ value of $+0.78$ V, what is the value of $E°$ for the half-cell $Fe(s) \rightarrow Fe^{2+}(aq) + 2\,e^-$?

$$Fe(s) + Cu^{2+}(aq, 1\,M) \longrightarrow Fe^{2+}(aq, 1\,M) + Cu(s) \qquad E°_{cell} = +0.78\text{ V}$$

19.5 USING STANDARD CELL POTENTIALS

The results of a great many measurements of cell potentials such as the ones just described are summarized as **standard reduction potentials** in Table 19.1. A much longer and more complete list of standard reduction potentials is given in

CD-ROM Screen 21.6: Standard Potentials

Silverware tarnishes when exposed to air because the silver reacts with the hydrogen sulfide gas in the air to form a thin coating of black silver sulfide, Ag_2S. You can chemically remove the tarnish from silverware and other silver utensils by using a solution of baking soda and some aluminum foil. The chemical cleaning of silver is an electrochemical process in which electrons move from aluminum atoms to silver ions in the tarnish, reducing the silver ions to silver atoms while aluminum atoms are oxidized to aluminum ions. Look up the position of Ag^+ ion with respect to aluminum metal in Table 19.1. The sodium bicarbonate provides a conductive ionic solution for the flow of electrons and also helps to remove the aluminum oxide coating from the surface of the aluminum foil.

Get a large pan. Put 1 to 2 L of water in the pan. Add 7 to 8 Tbsp of baking soda. Heat the solution, but do not boil it. Place some aluminum foil in the bottom of the pan, and put the tarnished silverware on the aluminum foil. Make sure the silverware is covered with water. Heat the water almost to boiling. After a few minutes remove the silverware and rinse it in running water.

This method of cleaning silverware is better than using polish, because polish removes the silver sulfide, including the silver it contains; instead, the process described here restores the silver from the tarnish to the surface. If you have aluminum pie pans or aluminum cooking pans, you can use them as both the container and the aluminum source. You may notice that devices for removing silver tarnish are sometimes advertised on television. These devices are actually little more than a piece of aluminum metal and some salt. Would you be willling to pay very much (plus shipping and handling) for this, after you have done this simple experiment?

1. In the reaction between silver and H_2S, what is being oxidized? What is being reduced?
2. What metal other than aluminum could be used for this reaction?

Appendix I. The values reported in the table are called standard reduction potentials because they are the potentials, reported as voltages, that would be measured for a cell in which a half-reaction *occurred as a reduction* when paired with the standard hydrogen electrode. If a half-reaction would occur as an oxidation when paired with the standard hydrogen electrode, the half-reaction voltage is given a negative sign. For example, we saw earlier that the oxidation half-reaction

$$Zn(s) \longrightarrow Zn^{2+}(aq) + 2\,e^-$$

has a half-cell potential of $+0.76$ V ($E^\circ_{ox} = +0.76$ V). But this reaction appears in Table 19.1 as the *reduction* half-reaction

$$Zn^{2+}(aq) + 2\,e^- \longrightarrow Zn(s) \qquad E^\circ_{red} = -0.76\,V$$

Its standard potential is equal in magnitude but *opposite* in sign to that for the oxidation reaction. It is always true that *if a half-reaction is written in the reverse direction, the sign of the corresponding* **E°** *must be changed.* That is, if any half-reaction given in Table 19.1 is written as an oxidation reaction, the sign of the E° in the table must be reversed. Here are some important points to notice about Table 19.1:

1. *Each half-reaction is written as a reduction.* This means that the species on the left-hand side of each half-reaction is in the higher oxidation state, and the species on the right-hand side is in the lower oxidation state.
2. *Each half-reaction listed in the table can occur in either direction.* A given substance can react at the anode or the cathode, depending on the conditions. We have already seen examples in which H_2 is oxidized to H_3O^+ and in which H_3O^+ is reduced to H_2 by different reactants.
3. *The more positive the value of the reduction potential,* E°_{red}, *the more easily the substance on the left side of a half-reaction can be reduced.* When a substance is easy to reduce, it is a strong oxidizing agent. (Remember that an oxidizing agent must be reduced when it oxidizes something else.) Thus, $F_2(g)$ is the best oxidizing agent in the table, and Li^+ is the poorest oxi-

TABLE 19.1	Standard Reduction Potentials in Aqueous Solution at 25 °C*

Reduction half-reaction		$E°$ (V)
$F_2(g) + 2\,e^-$	$\rightarrow 2\,F^-(aq)$	+2.87
$H_2O_2(aq) + 2\,H_3O^+(aq) + 2\,e^-$	$\rightarrow 4\,H_2O(\ell)$	+1.77
$PbO_2(s) + SO_4^{2-}(aq) + 4\,H_3O^+(aq) + 2\,e^-$	$\rightarrow PbSO_4(s) + 6\,H_2O(\ell)$	+1.685
$MnO_4^-(aq) + 8\,H_3O^+(aq) + 5\,e^-$	$\rightarrow Mn^{2+}(aq) + 12\,H_2O(\ell)$	+1.52
$Au^{3+}(aq) + 3\,e^-$	$\rightarrow Au(s)$	+1.50
$Cl_2(g) + 2\,e^-$	$\rightarrow 2\,Cl^-(aq)$	+1.360
$Cr_2O_7^{2-}(aq) + 14\,H_3O^+(aq) + 6\,e^-$	$\rightarrow 2\,Cr^{3+}(aq) + 21\,H_2O(\ell)$	+1.33
$O_2(g) + 4\,H_3O^+(aq) + 4\,e^-$	$\rightarrow 6\,H_2O(\ell)$	+1.229
$Br_2(\ell) + 2\,e^-$	$\rightarrow 2\,Br^-(aq)$	+1.08
$NO_3^-(aq) + 4\,H_3O^+ + 3\,e^-$	$\rightarrow NO(g) + 6\,H_2O(\ell)$	+0.96
$OCl^-(aq) + H_2O(\ell) + 2\,e^-$	$\rightarrow Cl^-(aq) + 2\,OH^-(aq)$	+0.89
$Hg^{2+}(aq) + 2\,e^-$	$\rightarrow Hg(\ell)$	+0.855
$Ag^+(aq) + e^-$	$\rightarrow Ag(s)$	+0.80
$Hg_2^{2+}(aq) + 2\,e^-$	$\rightarrow 2\,Hg(\ell)$	+0.789
$Fe^{3+}(aq) + e^-$	$\rightarrow Fe^{2+}(aq)$	+0.771
$I_2(s) + 2\,e^-$	$\rightarrow 2\,I^-(aq)$	+0.535
$O_2(g) + 2\,H_2O(\ell) + 4\,e^-$	$\rightarrow 4\,OH^-(aq)$	+0.40
$Cu^{2+}(aq) + 2\,e^-$	$\rightarrow Cu(s)$	+0.337
$Sn^{4+}(aq) + 2\,e^-$	$\rightarrow Sn^{2+}(aq)$	+0.15
$2\,H_3O^+(aq) + 2\,e^-$	$\rightarrow H_2(g) + 2\,H_2O(\ell)$	0.00
$Sn^{2+}(aq) + 2\,e^-$	$\rightarrow Sn(s)$	−0.14
$Ni^{2+}(aq) + 2\,e^-$	$\rightarrow Ni(s)$	−0.25
$PbSO_4(s) + 2\,e^-$	$\rightarrow Pb(s) + SO_4^{2-}(aq)$	−0.356
$Cd^{2+}(aq) + 2\,e^-$	$\rightarrow Cd(s)$	−0.40
$Fe^{2+}(aq) + 2\,e^-$	$\rightarrow Fe(s)$	−0.44
$Zn^{2+}(aq) + 2\,e^-$	$\rightarrow Zn(s)$	−0.763
$2\,H_2O(\ell) + 2\,e^-$	$\rightarrow H_2(g) + 2\,OH^-(aq)$	−0.8277
$Al^{3+}(aq) + 3\,e^-$	$\rightarrow Al(s)$	−1.66
$Mg^{2+}(aq) + 2\,e^-$	$\rightarrow Mg(s)$	−2.37
$Na^+(aq) + e^-$	$\rightarrow Na(s)$	−2.714
$K^+(aq) + e^-$	$\rightarrow K(s)$	−2.925
$Li^+(aq) + e^-$	$\rightarrow Li(s)$	−3.045

* In volts (V) versus the standard hydrogen electrode

dizing agent in the table. Other strong oxidizing agents are at the top left of the table:

$$H_2O_2(aq), \quad PbO_2(s), \quad Au^{3+}(aq), \quad Cl_2(g), \quad O_2(g)$$

4. *The more negative the value of the reduction potential, $E°_{red}$, the less likely the reaction will occur as a reduction, and the more likely the reverse reaction (an oxidation) will occur.* The farther down we go in the table, the better the reducing ability of the atom, ion, or molecule on the *right*. Thus, Li(s) is the strongest reducing agent in the table, and F$^-$ is the weakest reducing agent in the table. Other strong reducing agents are alkali and alkaline earth metals and hydrogen at the lower right of the table.

5. *Under standard conditions, any species on the left of a half-reaction will oxidize any species on the right that is farther down in the table.* For example, we can predict that $Fe^{3+}(aq)$ will oxidize $Al(s)$; $Br_2(\ell)$ will oxidize $Mg(s)$; and even $Na^+(aq)$ will oxidize $Li(s)$. The net reaction and the cell voltage are obtained by adding the half-reactions and their voltages; for example,

$$Br_2(\ell) + 2\,e^- \longrightarrow 2\,Br^-(aq) \qquad E^o_{red} = +1.08\ V$$

$$\underline{\qquad\qquad Mg(s) \longrightarrow Mg^{2+}(aq) + 2\,e^- \qquad E^o_{ox} = +2.37\ V}$$

$$Br_2(\ell) + Mg(s) \longrightarrow Mg^{2+}(aq) + 2\,Br^-(aq) \qquad E^o_{cell} = +3.45\ V$$

A positive cell potential denotes a product-favored reaction.

6. *Electrode potentials depend on the nature and concentration of reactants and products, but not on the quantity of each that reacts.* This means that changing the stoichiometric coefficients for a half-reaction does *not* change the value of E°. For example, the reduction of Fe^{3+} has an E° of $+0.771$ V whether the reaction is written as

$$Fe^{3+}(aq,\ 1\ M) + e^- \longrightarrow Fe^{2+}(aq,\ 1\ M) \qquad E^o_{red} = +0.771\ V$$

or as

$$2\,Fe^{3+}(aq,\ 1\ M) + 2\,e^- \longrightarrow 2\,Fe^{2+}(aq,\ 1\ M) \qquad E^o_{red} = +0.771\ V$$

This fact about half-cell potentials seems unusual at first. It arises because a half-cell voltage is energy per unit charge (1 volt = 1 joule/1 coulomb). When a half-reaction is multiplied by some number, both the energy and the charge are multiplied by that number. Thus the ratio of the energy to the charge (voltage) does not change.

Using the preceding guidelines and the table of standard reduction potentials, we will make some *predictions* about whether reactions will occur and then check our results by calculating E°.

Problem-Solving Example 19.7 Predicting Redox Reactions

(a) Will aluminum metal react with a 1 M tin(IV) solution? If so, what is E° for the reaction?

(b) What about a 1 M solution of Cd^{2+} oxidizing metallic Cu? Do you predict that this reaction will occur? Explain.

Answer

(a) Yes, the E^o_{cell} value is 1.81 V.

(b) This reaction does not occur; E^o_{cell} is negative.

Explanation

(a) First locate the reactants in Table 19.1. $Sn^{4+}(aq)$ is about halfway down the table on the left, while $Al(s)$ is fifth up from the bottom on the right. Since $Sn^{4+}(aq)$ is above $Al(s)$, we predict that it can oxidize aluminum, causing metallic Al atoms to form Al^{3+} ions. To be certain, we can add the half-cell reactions to give the balanced equation. Adding the half-cell potentials gives a positive E^o_{cell}, so this reaction is product-favored, as we predicted.

$$2\,[Al(s) \longrightarrow Al^{3+}(aq,\ 1\ M) + 3\,e^-] \qquad E^o_{ox} = +1.66\ V\ (anode)$$

$$\underline{3\,[Sn^{4+}(aq,\ 1\ M) + 2\,e^- \longrightarrow Sn^{2+}(aq,\ 1\ M)] \qquad E^o_{red} = +0.15\ V\ (cathode)}$$

$$2\,Al(s) + 3\,Sn^{4+}(aq,\ 1\ M) \longrightarrow 2\,Al^{3+}(aq,\ 1\ M) + 3\,Sn^{2+}(aq,\ 1\ M) \qquad E^o_{cell} = +1.81\ V$$

(b) The reaction between Cd^{2+} and Cu is evaluated the same way. $Cd^{2+}(aq)$ is about three quarters of the way down Table 19.1 on the left, but $Cu(s)$ is a little less than halfway down on the right. Therefore, $Cd^{2+}(aq)$ is not a strong enough oxidizing agent to oxidize Cu(s), and we predict that the reaction will not occur. When the half-cell potentials are added, we get

$$Cu(s) \longrightarrow Cu^{2+}(aq) + 2 e^- \qquad E^o_{ox} = -0.34 \text{ V (anode)}$$
$$Cd^{2+} + 2 e^- \longrightarrow Cd(s) \qquad E^o_{red} = -0.40 \text{ V (cathode)}$$
$$\overline{Cd^{2+}(aq) + Cu(s) \longrightarrow Cd(s) + Cu^{2+}(aq) \qquad E^o_{cell} = -0.74 \text{ V}}$$

The negative E^o_{cell} value shows that this process is reactant-favored and does not form appreciable quantities of products under standard conditions. Rather, cadmium metal will reduce Cu^{2+} (the reverse of the reaction given).

Problem-Solving Practice 19.7

Look at Table 19.1 and determine which two half-reactions would produce the largest value of E^o_{cell}. Write the two half-reactions and the overall cell reaction, and give the E^o for the reaction.

Exercise 19.5   Using E^o Values

Transporting chemicals is of great practical and economic importance. Suppose that you have a large volume of mercury(II) chloride, $HgCl_2$, solution that needs to be transported. A driver brings a tanker truck made of aluminum to the loading dock. Will it be okay to load the truck with your solution? Explain your answer fully.

Standard reduction potentials can be used to explain an annoying experience many of us have had, a pain in a tooth when a filling is touched with a stainless steel fork or a piece of aluminum foil (Figure 19.10). A common material for filling tooth cavities is dental amalgam — tin and silver dissolved in mercury to form solid solutions having compositions approximating Ag_2Hg_3, Ag_3Sn, and Sn_xHg (where x ranges from 7 to 9). All of these may undergo electrochemical reactions; for example,

Stainless steel is an alloy of iron.

$$3 Hg_2^{2+}(aq) + 4 Ag(s) + 6 e^- \longrightarrow 2 Ag_2Hg_3(s) \qquad E^o_{red} = +0.85 \text{ V}$$
$$Sn^{2+}(aq) + 3 Ag(s) + 2 e^- \longrightarrow Ag_3Sn(s) \qquad E^o_{red} = -0.05 \text{ V}$$

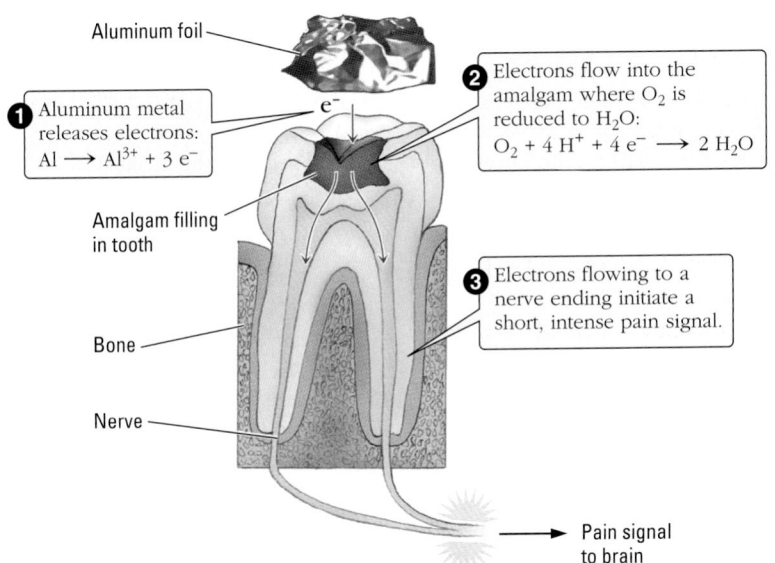

Figure 19.10   A dental voltaic cell. Pain usually results from accidentally touching an active metal, such as a piece of aluminum foil, to metallic filling materials commonly used in dentistry.

The $E°$ values in Table 19.1 indicate that both iron and aluminum have much more negative reduction potentials and therefore are much better reducing agents than any of the solid amalgam solutions. If a piece of iron or aluminum comes in contact with a dental filling, the saliva and gum tissue act as a salt bridge, and an electrochemical cell results. The iron or aluminum donates electrons, producing a tiny electrical current that results in pain.

Exercise 19.6 Predicting Redox Reactions Using $E°$ Values

Consider the following half-reactions:

Half-reaction	$E°$ (V)
$Cl_2(g) + 2\,e^- \rightarrow 2\,Cl^-(aq)$	+1.36
$I_2(s) + 2\,e^- \rightarrow 2\,I^-(aq)$	+0.535
$Pb^{2+}(aq) + 2\,e^- \rightarrow Pb(s)$	−0.126
$V^{2+}(aq) + 2\,e^- \rightarrow V(s)$	−1.18

(a) Which is the weakest oxidizing agent?
(b) Which is the strongest oxidizing agent?
(c) Which is the strongest reducing agent?
(d) Which is the weakest reducing agent?
(e) Will Pb(s) reduce V^{2+}(aq) to V(s)?
(f) Will I_2(g) oxidize Cl^-(aq) to Cl_2(g)?
(g) Name the molecules or ions in the above reactions that can be reduced by Pb(s).

Before we leave this discussion of Table 19.1, consider again the activity series of metals shown in Table 5.5 *(⬅ p. 185)*, which has a lot in common with Table 19.1. They contain many of the same elements, for example. Looking closely, however, you will notice that the most active metal, lithium, in Table 5.5 is at the very bottom right of Table 19.1. That is because Table 19.1 is arranged by *reduction potential,* and lithium has the lowest tendency to be reduced. Table 5.5, on the other hand, lists the metals in order of activity, that is, their tendency to be oxidized. Since oxidation is the opposite of reduction, it is reasonable that lithium is in opposite positions in the two tables.

 ### Exercise 19.7 Predicting $E°$ Values

The two elements on either side of hydrogen in Table 5.5 are not listed in Table 19.1. Indicate where they would be in Table 19.1 and, based on values from the table, estimate the reduction potentials for their positive ions being reduced to the metal atom.

19.6 $E°$ AND GIBBS FREE ENERGY

The sign of $E°_{cell}$ indicates whether a redox reaction is product-favored (positive $E°$) or reactant-favored (negative $E°$). You have learned another way to decide whether a reaction is product-favored: the change in standard Gibbs free energy, $\Delta G°$, must be negative *(⬅ p. 827)*. Since both $E°_{cell}$ and $\Delta G°$ tell something about whether a reaction will occur, it should be no surprise that there is a relationship between them.

The "free" in Gibbs free energy indicates that it is energy available to do work. The energy available for electrical work from a cell can be calculated by multiply-

CD-ROM Screen 21.11:
Coulometry: Counting Electrons

ing the quantity of electrical charge transferred times the cell voltage, *E°*. The quantity of charge is given by the number of moles of electrons transferred in the overall reaction, *n*, multiplied by the number of coulombs per mole of electrons.

Quantity of charge = moles of electrons × coulombs per mole of electrons

The charge on 1 mol of electrons can be calculated from the charge on one electron (⟸ *p. 44*) and Avogadro's number.

$$\text{Charge on 1 mol of } e^- = \left(\frac{1.60218 \times 10^{-19}\text{ C}}{\text{electron}}\right)\left(\frac{6.02214 \times 10^{23}\text{ electrons}}{\text{mol}}\right)$$

$$= 9.6485 \times 10^4 \text{ C/mol } e^-$$

The quantity 9.6485×10^4 C/mol of electrons (commonly rounded to 96,500 C/mol of electrons) is known as the **Faraday constant** (*F*) in honor of Michael Faraday, who first explored the quantitative aspects of electrochemistry.

The electrical work that can be done by a cell is equal to the Faraday constant (*F*) multiplied by the number of moles of electrons transferred (*n*) and by the cell voltage (E°_{cell}).

$$\text{Electrical work} = nFE^\circ_{\text{cell}}$$

Notice that, unlike the cell voltage, the electrical work a cell can do *does* depend on the quantity of reactants in the cell reaction. More reactants mean more moles of electrons transferred and hence more work. Equating the electrical work of a cell at standard conditions with ΔG°, we get

$$\Delta G^\circ = -nFE^\circ_{\text{cell}}$$

The negative sign on the right side of the equation accounts for the fact that ΔG° *is always negative for a product-favored process, but* E_{cell} *is always positive for a product-favored process.* Thus their signs must be opposite.

Using this equation we can calculate ΔG° for the Cu^{2+}/Zn cell. This represents the maximum work that the cell can do. The reaction is

$$Cu^{2+}(aq) + Zn(s) \longrightarrow Cu(s) + Zn^{2+}(aq) \qquad E^\circ = +1.10 \text{ V}$$

so 2 mol of electrons are transferred per mole of copper ions reduced. The Gibbs free energy change when this quantity of reactants is converted is

$$\Delta G^\circ = -(2 \text{ mol electrons transferred}) \times$$

$$\left(\frac{9.65 \times 10^4 \text{ C}}{\text{mol } e^-}\right)\left(\frac{1 \text{ J}}{1 \text{ V} \times 1 \text{ C}}\right)\left(\frac{1 \text{ kJ}}{10^3 \text{ J}}\right)(1.10 \text{ V}) = -212 \text{ kJ/mol}$$

MICHAEL FARADAY
(1791–1867)

Michael Faraday became fascinated by science as an apprentice to a London bookbinder when he was a boy. At 22 he was appointed as a laboratory assistant at the Royal Institution and became its director within 12 years. He became a skilled experimenter in chemistry and physics, making many important discoveries, the most important of which was electromagnetic induction, the basis of modern electromagnetic technology. Faraday built the first electric motor, generator, and transformer. A popular speaker and educator, he also performed chemical and electrochemical experiments, and he discovered benzene.

Problem-Solving Example 19.8   Determining E°_{cell} and ΔG°

Consider the redox reaction

$$Zn^{2+}(g) + H_2(g) + 2 H_2O(\ell) \longrightarrow Zn(s) + 2 H_3O^+(aq)$$

Use the standard reduction potentials in Table 19.1 to calculate E°_{cell} and ΔG° and to determine whether the reaction as written favors products.

Answer $E^\circ_{\text{cell}} = -0.763$ V. $\Delta G^\circ = 147$ kJ. The reaction as written does not favor products.

Explanation   The first step is to write the two half-reactions for the oxidation and reduction that occur and to look up the standard reduction potentials in Table 19.1.

Reduction: $Zn^{2+}(aq) + 2 e^- \longrightarrow Zn(s)$ $E^\circ_{\text{red}} = -0.763$ V

Oxidation: $H_2(g) + 2 H_2O(\ell) \longrightarrow 2 H_3O^+(aq) + 2 e^-$ $E^\circ_{\text{ox}} = 0$ V

The E°_{cell} for the reaction is the sum of the standard potentials for the half-reactions

$$E^\circ_{cell} = E^\circ_{red} + E^\circ_{ox} = -0.763 \text{ V} + 0 \text{ V} = -0.763 \text{ V}$$

Since the value of E°_{cell} is negative, the reaction is not product-favored in the direction written. The reverse reaction is product-favored; that is, zinc metal reacts with acid.

From E°_{cell} we can calculate ΔG°.

$$\Delta G^\circ = -nFE^\circ_{cell} = -(2 \text{ mol e}^-) \times$$

$$\left(\frac{9.65 \times 10^4 \text{ C}}{1 \text{ mol e}^-}\right)\left(\frac{1 \text{ J}}{1 \text{ V} \times 1 \text{ C}}\right)(-0.763 \text{ V}) = 1.47 \times 10^5 \text{ J} = 147 \text{ kJ}$$

The positive value for ΔG° also shows that the reaction as written is not product-favored.

✔ The oxidation is the reaction at the standard hydrogen electrode with a potential of zero, so the overall reaction is governed by the Zn reduction. The negative cell potential and the positive ΔG° are consistent with the reaction not being product-favored as written.

Problem-Solving Practice 19.8

Using standard reduction potentials, determine whether the following reaction is product-favored as written.

$$Hg^{2+}(aq) + 2\,I^-(aq) \longrightarrow Hg(\ell) + I_2(s)$$

We have seen that the standard free energy change is directly proportional to the E°_{cell} for an electrochemical cell at standard conditions

$$\Delta G^\circ = -nFE^\circ_{cell}$$

Recall from Chapter 18 (⬅ *p. 834*) that the standard free energy change is directly proportional to the equilibrium constant of a reaction.

$$\Delta G^\circ = -RT \ln K^\circ$$

Putting these two equations together yields

$$-nFE^\circ_{cell} = -RT \ln K^\circ$$

which, when solved for E°_{cell}, yields

$$E^\circ_{cell} = \frac{RT}{nF} \ln K^\circ$$

Thus, using electrochemistry, one can measure E°_{cell} and then use its value to determine directly the values of ΔG° and K°. The relationships linking ΔG°, E°_{cell}, and K° are summarized in Figure 19.11.

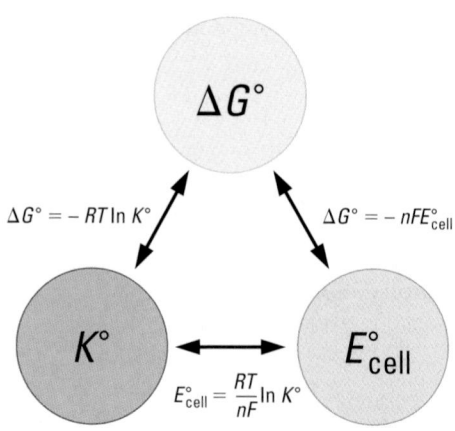

Figure 19.11 The relationships linking ΔG°, E°_{cell}, and K°. Given any one of the values, the other two can be calculated.

The E°_{cell} expression can be simplified by substituting numerical values for R (8.314 J mol^{-1} K^{-1}) and F (96,485 J V^{-1} mol^{-1}) and assuming the temperature to be 25 °C (298 K). For n moles of electrons we get

$$E^\circ_{cell} = \frac{RT}{nF} \ln K = \frac{0.0257 \text{ V}}{n} \ln K^\circ$$

Changing from natural logarithms to base 10 logarithms (multiplying by 2.303) yields

$$E^\circ_{cell} = \frac{0.0592 \text{ V}}{n} \log K^\circ \qquad \text{and} \qquad \log K^\circ = \frac{nE^\circ_{cell}}{0.0592 \text{ V}} \qquad \text{(at 25 °C)}$$

This equation holds for standard states of all reactants and products.

Problem-Solving Example 19.9   Equilibrium Constant for a Redox Reaction

Calculate the equilibrium constant for the reaction

$$Fe(s) + Cu^{2+}(aq) \rightleftharpoons Fe^{2+}(aq) + Cu(s)$$

using the standard reduction potentials listed in Table 19.1.

Answer $K = 2.2 \times 10^{26}$

Explanation   We first need to calculate E°_{cell}, and to do this we break the reaction into two half-reactions.

$$
\begin{array}{ll}
Fe(s) \longrightarrow Fe^{2+}(aq) + 2\,e^- & E^\circ_{ox} = 0.44 \text{ V} \\
Cu^{2+}(aq) + 2\,e^- \longrightarrow Cu(s) & E^\circ_{red} = 0.34 \text{ V} \\
\hline
Fe(s) + Cu^{2+}(aq) \longrightarrow Fe^{2+}(aq) + Cu(s) & E^\circ_{cell} = 0.78 \text{ V}
\end{array}
$$

Two moles of electrons are transferred.

$$\log K^\circ = \frac{nE^\circ_{cell} \text{ V}}{0.0592 \text{ V}} = \frac{2 \times 0.78 \text{ V}}{0.0592 \text{ V}} = 26.35 \quad \text{and} \quad K^\circ = 10^{26.35} = 2.2 \times 10^{26}$$

The large value of K° indicates that the reaction is strongly product-favored as written.

✔   The value for E°_{cell} is positive, which indicates a product-favored reaction, and the large value of K° indicates this as well.

Problem-Solving Practice 19.9

Using the standard reduction potentials listed in Table 19.1, calculate the equilibrium constant for the reaction

$$I_2(s) + Sn^{2+}(aq) \longrightarrow 2\,I^-(aq) + Sn^{4+}(aq)$$

19.7 EFFECT OF CONCENTRATION ON CELL POTENTIAL

The electrochemical cells discussed previously, all voltaic cells, are based on product-favored chemical reactions. When an electrochemical cell is used, reactants are consumed and products are generated, so the concentrations of the species change continuously. As the reactant concentrations decrease, the voltage produced by the cell drops and finally reaches zero when the reactants and products are at equilibrium.

We can relate the voltage of a voltaic cell to the concentrations of the reactants and products of its chemical reaction through the Nernst equation. To do so,

CD-ROM Screen 21.7:
Electrochemical Cells at
Nonstandard Conditions

we start with the relationship between free energy change and concentration ($\Leftarrow$ *p. 836*)

$$\Delta G = \Delta G° + RT \ln Q$$

where Q is the reaction quotient. Q has the same form as the equilibrium constant, but it refers to a reaction mixture at a given instant in time, that is, a reaction mixture not yet at equilibrium.

We know that $\Delta G = -nFE_{cell}$ and $\Delta G° = -nFE°_{cell}$, so

$$\Delta G = -nFE_{cell} = -nFE°_{cell} + RT \ln Q$$

Rearranging this for E_{cell} gives the relationship we seek:

$$E_{cell} = E°_{cell} - \frac{RT}{nF} \ln Q$$

This is the **Nernst equation.** We can change to base 10 logarithms (multiply by 2.303) to get

$$E_{cell} = E°_{cell} - \frac{2.303\ RT}{nF} \log Q$$

We can simplify further by substituting numerical values for R and F and by assuming 25 °C (298 K) to get

$$E_{cell} = E°_{cell} - \frac{0.0592\ V}{n} \log Q \qquad (T = 298\ K)$$

Notice that if all the concentrations in Q are equal to 1 (which is the standard state), then $Q = 1$ and $\log(1) = 0$, so the Nernst equation reduces to $E_{cell} = E°_{cell}$.

The Nernst equation can be used to calculate the voltage produced by an electrochemical cell under nonstandard conditions. It can also be used to calculate the concentration of a reactant or product in an electrochemical reaction from the measured value of the voltage produced.

Problem-Solving Example 19.10   Using the Nernst Equation

Consider the following electrochemical reaction that we discussed earlier:

$$Zn(s) + Ni^{2+}(aq) \longrightarrow Zn^{2+}(aq) + Ni(s)$$

The standard cell potential $E°_{cell} = 0.51$ V. Find the cell potential if the Ni^{2+} concentration is 5.0 M and the Zn^{2+} concentration is 0.050 M.

Answer   0.57 V

Explanation   We use the Nernst equation to solve the problem. Two moles of electrons are transferred from 1 mol of Zn to 1 mol of Ni^{2+}, giving $n = 2$. At 298 K,

$$E_{cell} = 0.51\ V - \frac{0.0592\ V}{2} \log \left(\frac{0.050}{5.0} \right)$$

$$= 0.51\ V - \frac{0.0592\ V}{2} (-2.00) = 0.57\ V$$

✔   The reactant concentration of Ni^{2+} is 5.0 M, larger than the standard state value of 1.0 M, and the product concentration of Zn^{2+} is 0.050 M, smaller than the standard state value. Each of these changes tends to make the voltage under these conditions larger than the standard cell potential ($E°_{cell} = 0.51$ V), and it is.

Problem-Solving Practice 19.10

What would the cell potential become if $[Zn^{2+}] = 3.0$ M and $[Ni^{2+}] = 0.010$ M?

Concentration Cells

The voltaic cells discussed to this point have different reactions proceeding at the anode and the cathode. However, since the voltage of a cell depends on the concentrations of the reactants, we can construct a cell that uses the same species in both the anode and cathode compartments but with different concentrations. A **concentration cell** is a cell in which the voltage is generated because of a difference in concentrations.

Let's look at a concentration cell constructed with two identical Cu/Cu^{2+} half-reactions occurring in separated compartments, as shown in Figure 19.12. If the same half-reactions occurred in the two compartments *at standard conditions* of 1 M concentrations, the standard potentials for the two would be the same but with opposite signs, $E^{\circ}_{red} = +0.337$ V and $E^{\circ}_{ox} = -0.337$ V, so the cell potential would be zero. However, in a concentration cell the half-reactions are the same but the concentrations in the two cells are different. Let's take as an example Cu^{2+} concentrations of 0.050 M in the anode half-cell and 0.50 M in the cathode half-cell. The two half-reactions are

$$Cu(s) \longrightarrow Cu^{2+}(aq, 0.050\ M) + 2\ e^{-} \qquad \text{(anode, oxidation)}$$

$$\underline{Cu^{2+}(aq, 0.50\ M) + 2\ e^{-} \longrightarrow Cu(s) \qquad\qquad\qquad \text{(cathode, reduction)}}$$

$$Cu^{2+}(aq, 0.50\ M) \longrightarrow Cu^{2+}(aq, 0.050\ M) \qquad\qquad \text{(net reaction)}$$

The cell potential is expressed by the Nernst equation:

$$E_{cell} = E^{\circ}_{cell} - \frac{0.0592\ V}{2} \log \frac{(conc\ Cu^{2+})_{dilute}}{(conc\ Cu^{2+})_{concentrated}}$$

$$= 0 - 0.0296 \times \log \frac{0.050\ M}{0.50\ M} = 0 - 0.0296 \times (-1.00) = 0.0296\ V$$

The cell potential is entirely determined by the ratio of concentrations of Cu^{2+} ions in the two half-reaction cells. The value of 0.0296 V is the potential of the cell when the concentrations of Cu^{2+} are at the initial conditions of 0.050 M and 0.50 M. As the cell operates, the concentration of Cu^{2+} in each half-cell changes — the concentration in the dilute cell increases and the concentration in the concentrated cell decreases. Eventually, the two concentrations become equal, and the cell potential becomes zero.

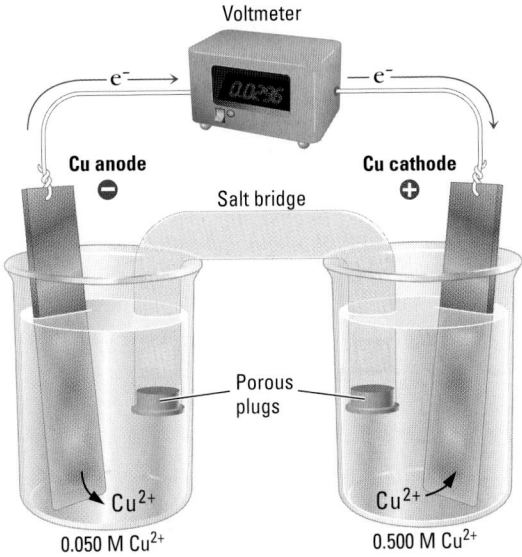

Figure 19.12 Concentration cell based on Cu/Cu^{2+} half-reactions. The cell has a positive net cell voltage and operates because the concentrations of Cu^{2+} ion are different in the two half-reaction compartments.

Measurement of pH

The concentration of H^+ in a solution can be measured using the principles of a concentration cell. Consider a concentration cell based on the H_2/H^+ half-reaction. The cathode compartment contains the standard hydrogen electrode, and the anode compartment contains the same type of electrode in contact with a solution of unknown H^+ concentration. The half-reactions and the overall reaction that occur are

$$H_2(g, 1 \text{ bar}) \longrightarrow 2\,H^+(aq, \text{unknown}) + 2\,e^- \qquad \text{(anode, oxidation)}$$

$$2\,H^+(aq, 1 \text{ M}) + 2\,e^- \longrightarrow H_2(g, 1 \text{ bar}) \qquad \text{(cathode, reduction)}$$

$$2\,H^+(aq, 1 \text{ M}) \longrightarrow 2\,H^+(aq, \text{unknown}) \qquad E_{cell} = ?$$

The *standard* potential of the cell would be zero, $E_{cell}^\circ = 0$. However, the two half-cells have different hydrogen ion concentrations, so E_{cell} is not zero.

To analyze the cell further we use the Nernst equation with $n = 2$.

$$E_{cell} = E_{cell}^\circ - \frac{0.0592 \text{ V}}{2} \log \frac{(H^+ \text{ conc.})^2_{\text{unknown}}}{(H^+ \text{ conc.})^2_{\text{standard}}}$$

$[H^+]_{\text{standard}} = 1 \text{ M}$ and $E_{cell}^\circ = 0$, so

$$E_{cell} = -\frac{0.0592 \text{ V}}{2} \log (H^+ \text{ conc.})^2_{\text{unknown}}$$

Since $\log x^2 = 2 \log x$,

$$E_{cell} = \frac{-0.0592 \text{ V}}{2} \times 2 \log (H^+ \text{ conc.})_{\text{unknown}}$$

$$= -0.0592 \text{ V} \times \log (H^+ \text{ conc.})_{\text{unknown}}$$

Because $-\log[H^+] = pH$, the final expression is

$$E_{cell} = 0.0592 \text{ V} \times pH \qquad \text{or} \qquad pH = \frac{E_{cell}}{0.0592 \text{ V}}$$

Therefore, by measuring E_{cell}, pH can be measured.

Problem-Solving Example 19.11   Measuring pH with a Concentration Cell

Consider a concentration cell consisting of two hydrogen electrodes, which can be used to measure pH. One of the cells is a standard hydrogen electrode, and the second is in contact with an aqueous solution having an unknown pH. If the unknown $[H^+]$ is less than 1.0 M (which is generally true), then reduction occurs at the standard hydrogen electrode (cathode), and oxidation occurs at the nonstandard hydrogen electrode (anode). If the measured potential of the cell is 0.366 V, what is the pH of the unknown solution?

Answer   pH = 6.18

Explanation   We write the two half-reactions to start.

$$2\,H^+(1 \text{ M}) + 2\,e^- \longrightarrow H_2(g, 1 \text{ bar}) \qquad \text{(reduction)}$$

$$H_2(g, 1 \text{ bar}) \longrightarrow 2\,H^+(\text{unknown M}) + 2\,e^- \qquad \text{(oxidation)}$$

$$2\,H^+(1 \text{ M}) \longrightarrow 2\,H^+(\text{unknown M}) \qquad E_{cell} = 0.366 \text{ V}$$

The Nernst equation is

$$E_{cell} = -\frac{0.0592 \text{ V}}{2} \log \frac{(\text{unknown } H^+)^2}{(H^+ \text{ 1 M})^2}$$

$$= -0.0296 \text{ V} \times \log (\text{unknown } H^+ \text{ conc.})^2 = 0.366 \text{ V}$$

Since $\log x^2 = 2 \log x$, we can simplify the expression to

$$E_{cell} = -0.0592 \text{ V} \times \log (\text{unknown H}^+ \text{ conc.}) = 0.366 \text{ V}$$

By definition pH $= -\log [\text{H}^+]$, so

$$0.0592 \text{ V} \times \text{pH} = 0.366 \text{ V}$$

and pH $= 6.18$.

✔ The general relationship between the variables is pH $= E_{cell}/0.0592$, so using approximate values gives $0.4/0.06 \approx 6.7$, which is close to our more exact answer.

Problem-Solving Practice 19.11

If the same type of concentration cell were used with a solution of pH 3.66, what E_{cell} would be measured?

The pH Meter

A concentration cell utilizing two hydrogen electrodes is not the best practical choice for routine pH measurement because it is bulky and difficult to maintain. Commercial pH meters are based on electrochemical principles similar to those described previously, but with more rugged and economical half-cells. A **pH meter** has two electrodes (Figure 19.13). One is a glass electrode using an Ag/AgCl half-cell dipped in an HCl solution of known concentration. At the tip of this indicator electrode is a very thin glass membrane that is sensitive to H^+ ion concentration. The other electrode is a reference electrode known as a *saturated calomel electrode*. It consists of a Pt wire dipped in a paste of Hg_2Cl_2 (calomel), liquid Hg, and saturated KCl solution. The glass electrode measures the H^+ ion concentration of the solution relative to its internal hydrogen ion concentration. The difference in voltage between the two electrodes is then converted electronically to the pH of the solution.

In commercial instruments, the indicator and reference electrodes are generally combined in what is known as a combination electrode.

The pH of aqueous solutions is an extremely important indicator of their chemistry. Medical applications of pH measurements abound, as do environmental applications such as measuring the pH of acid rain *(⇐ p. 793)*.

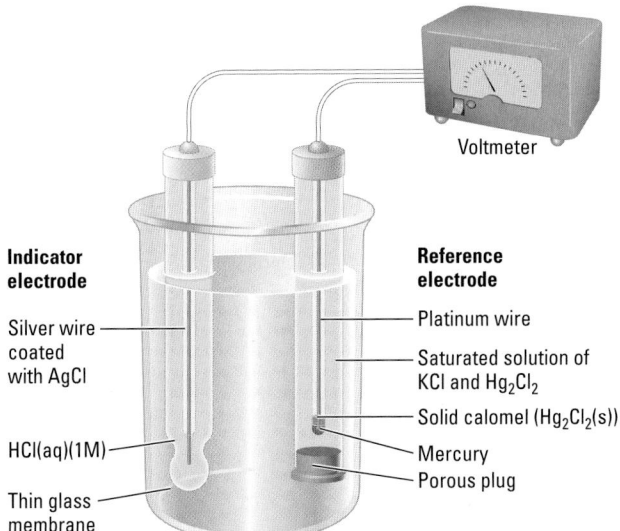

Voltmeter

Indicator electrode

Silver wire coated with AgCl

HCl(aq)(1M)

Thin glass membrane

Reference electrode

Platinum wire

Saturated solution of KCl and Hg_2Cl_2

Solid calomel (Hg_2Cl_2(s))

Mercury

Porous plug

Figure 19.13   The electrodes and reactions of the pH meter.   The pH meter has a glass electrode, which is a Ag/AgCl half-cell in a standard HCl solution and enclosed by a glass membrane. It is sensitive to the external $[\text{H}^+]$ in the solution relative to the $[\text{H}^+]$ in the internal standard HCl solution. The saturated calomel electrode is the reference electrode.

Figure 19.14 A mammalian neuron cell.

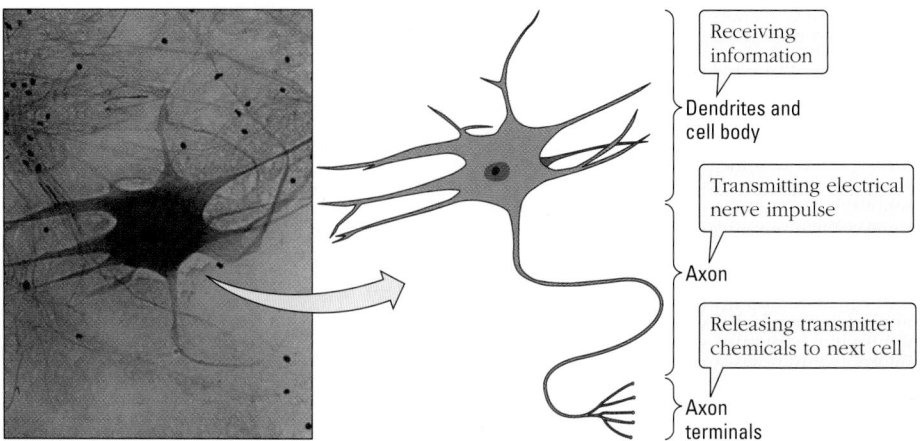

19.8 Neuron Cells

Animals' nervous systems are partly comprised of specialized cells called **neurons** whose actions can be understood using electrochemical principles (Figure 19.14). Communication between neurons occurs through electrical signals generated by millisecond alterations in voltage due to changes in ion concentration. These electrical signals depend on the cell membrane of the neuron functioning to separate different concentrations of ions inside and outside the cell at rest, as shown in Figure 19.15. These different resting ion concentrations are maintained by a number of processes that move ions across the cell membrane. The main ions of importance are Na^+, K^+, Cl^-, and Ca^{2+}. The intracellular versus extracellular concentrations of these four ions are markedly different in mammalian neurons.

Recall (⟵ *p. 390, Section 9.5*) that a cell membrane is composed of lipid molecules that separate the aqueous environment inside the cell from the aqueous environment outside the cell.

$1 \text{ mM} = 10^{-3}$ M and $1 \text{ } \mu\text{M} = 10^{-6}$ M
$1 \text{ mV} = 10^{-3}$ V

Ion	Intracellular conc. (mM)	Extracellular conc. (mM)	Potentials (mV)
Na^+	18	150	+56
K^+	135	3	−102
Cl^-	7	120	−76
Ca^{2+}	0.0001	1.2	+125

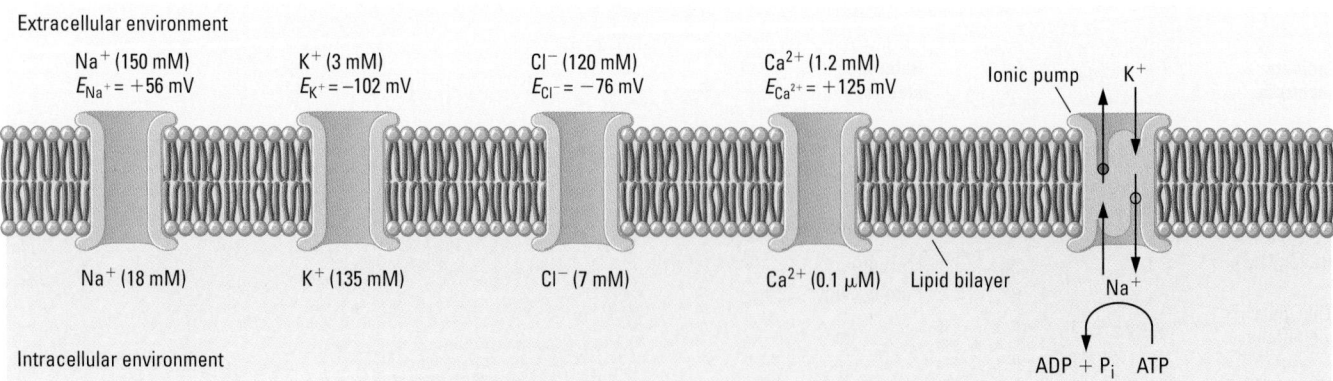

Figure 19.15 Ion concentrations inside and outside a mammalian neuron cell. Ion channels for Na^+, K^+, Cl^-, and Ca^{2+} are shown, as is the Na^+-K^+ ion pump. Concentrations are given in millimolar, except for intracellular Ca^{2+}.

These differences in concentration create potentials across the neuronal cell membrane. An associated equilibrium potential for each ion is given by the Nernst equation.

$$E_{\text{ion}} = \frac{2.303\,RT}{nF} \log \frac{(\text{ion})_{\text{outside}}}{(\text{ion})_{\text{inside}}} \qquad \text{(in V)}$$

E° has been omitted from the initial Nernst equation because its value is zero.

Substituting numerical values for the constants, and assuming $n = 1$ and a body temperature of 37 °C, we can simplify this expression to

$$E_{\text{ion}} = 61.5 \log \frac{(\text{ion})_{\text{outside}}}{(\text{ion})_{\text{inside}}} \qquad \text{(in mV)}$$

$1\ \text{mV} = 10^{-3}\ \text{V}$

This expression computes the potential outside the cell membrane relative to the potential inside the cell membrane for each individual ion. Applying the equation to the specific case of K^+, we have $[K^+]_{\text{outside}} = 3$ mM and $[K^+]_{\text{inside}} = 135$ mM, so

$$E_{K^+} = 61.5 \log \left(\frac{[K^+]_{\text{outside}}}{[K^+]_{\text{inside}}} \right)$$

$$= 61.5 \log \left(\frac{3}{135} \right) \text{mV} = 61.5\,(-1.65)\,\text{mV} = -102\,\text{mV}$$

The cell membrane has a potential 102 mV (0.102 V) more negative on the inside than the outside due to the much higher K^+ concentration inside the cell (Figure 19.16).

Just as for K^+, each important ion shown in Figure 19.15 has a potential that depends on the concentrations of that ion inside and outside the cell membrane. The values for Na^+, K^+, Cl^-, and Ca^{2+} shown in the figure can be used with the Nernst equation to calculate the contribution of each ionic species to the final resting potential of the neuron. The resting membrane potential for a cell — that is, the potential when no nerve impulse is being transmitted — depends on each of the individual ion potentials. The equilibrium potentials for each of the ions involved are averaged in proportion to the relative permeability of the cell membrane for each ion. The resting membrane potential is different for different types of neuronal cells and is in the range of −60 to −75 mV, with the outside of the cell being more positive than the inside.

How do the concentrations of ions inside and outside the cell membrane get to be different? Ions move across the cell membrane by several mechanisms, but

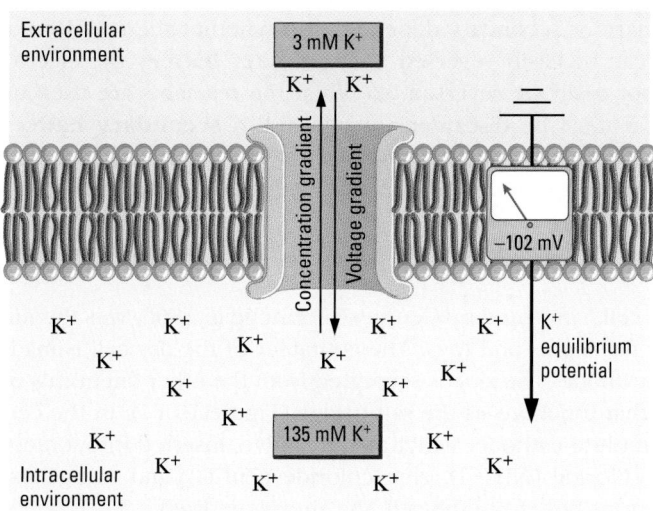

Figure 19.16 **The equilibrium potential for K^+ across a neuronal cell membrane.** The K^+ concentration is higher inside than outside. The Nernst equation explains the −102 mV equilibrium potential for K^+.

all the ions are undergoing continual movement. In general, ions will move by diffusion from a region of higher concentration to a region of lower concentration if they can, and some ions move across cell membranes in response to such a concentration gradient. In other words, such movement is product-favored. In addition, active *ion pumps* move Na^+ and K^+ *against* their concentration gradients; that is, they move Na^+ from inside to outside the cell membrane and they move K^+ from outside to inside the cell membrane. These active ion pumps require energy to perform this task, and this energy comes from the hydrolysis of ATP (⇐ *p. 843*). When a neuron is at rest, the passive movement of Na^+ and K^+ ions is exactly counterbalanced by the active movement of Na^+ and K^+ ions via the ion pumps. Thus, the concentrations of the ions remain constant, although passive and active transport are both functioning at all times.

When a neuron is stimulated, a voltage change occurs from the resting value in the -60 to -75 mV range toward more positive values, and a large flow of Na^+ ions moves into the cell. This causes the membrane potential to become more positive. This change in the membrane potential is then followed by a sustained flow of K^+ out of the cell, moving the potential back toward its resting value. These flows of ions generate *action potentials,* the brief electrical signals that result in signaling along neurons. The sequence leading to the generation of the action potential occurs during a period of about 1 msec (10^{-3} sec). Generation of action potentials by this mechanism is the basis for transmission of signals along nerve cells in animals and humans. During the process, the bulk concentrations of K^+ and Na^+ change little, either inside or outside the cell membrane. The number of ions moving across the cell membrane is much less than the number present. The action potential is an electrochemical event related to the change in ion concentrations inside and outside the cell and not to bulk concentrations of the ions.

Exercise 19.8 Neuron Equilibrium Potential

What would the membrane potential, E_{Na^+}, across a neuron cell membrane be if Na^+ were the only ion to be considered?

CD-ROM Screen 21.8: Batteries

19.9 COMMON BATTERIES

Voltaic cells include the convenient, portable sources of energy that we call *batteries*. Some batteries, such as the common flashlight battery, consist of a single cell, while others, such as automobile batteries, contain multiple cells. Batteries can be classified as primary or secondary depending on whether the reactions at the anode and cathode can be easily reversed. In a **primary battery** the electrochemical reactions cannot easily be reversed, so when the reactants are used up the battery is "dead" and must be discarded. In contrast, a **secondary battery** (sometimes called a storage battery or a rechargeable battery) uses an electrochemical reaction that can be reversed, so this type of battery can be recharged.

Primary Batteries

For a long time the "dry cell," invented by Georges Leclanché in 1866, was the major source of energy for flashlights and toys. The container of the dry cell is made of zinc, which acts as the anode. The zinc is separated from the other chemicals by a liner of porous paper that functions as the salt bridge (Figure 19.17). In the center of the dry cell is a graphite cathode, which is unreactive, inserted into a moist mixture of ammonium chloride (NH_4Cl), zinc chloride ($ZnCl_2$), and manganese dioxide (MnO_2). As electrons flow from the cell, the zinc is oxidized

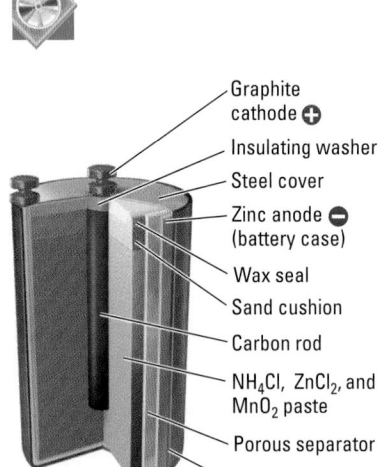

Graphite
cathode ⊕

Insulating washer

Steel cover

Zinc anode ⊖
(battery case)

Wax seal

Sand cushion

Carbon rod

NH_4Cl, $ZnCl_2$, and
MnO_2 paste

Porous separator

Wrapper

Figure 19.17 The Leclanché dry cell. It contains a zinc anode (the battery container), a graphite cathode, and an electrolyte consisting of a moist paste of NH_4Cl, $ZnCl_2$, and MnO_2.

$$Zn(s) \longrightarrow Zn^{2+}(aq) + 2\,e^- \qquad \text{(anode, oxidation)}$$

and the ammonium ions are reduced.

$$2\,NH_4^+(aq) + 2\,e^- \longrightarrow 2\,NH_3(g) + H_2(g) \qquad \text{(cathode, reduction)}$$

The ammonia reacts with zinc ions to form a zinc-ammonia complex ion (Section 17.5), which prevents a buildup of gaseous ammonia.

$$Zn^{2+}(aq) + 2\,NH_3(g) \longrightarrow [Zn(NH_3)_2]^{2+}(aq)$$

The hydrogen is oxidized by the MnO_2 in the cell, which prevents hydrogen accumulation.

$$H_2(g) + 2\,MnO_2(s) \longrightarrow Mn_2O_3(s) + H_2O(\ell)$$

The overall cell reaction, which produces 1.5 V, is

$$2\,MnO_2(s) + 2\,NH_4^+(aq) + Zn(s) \longrightarrow Mn_2O_3(s) + H_2O(\ell) + [Zn(NH_3)_2]^{2+}(aq)$$

A major disadvantage of the Leclanché cell is the occurrence of a slow reaction even when current is not being drawn, so stored cells run down and tend to have a short shelf life.

Some of the problems of the dry cell are overcome by the newer, more expensive alkaline battery. An *alkaline battery,* which produces 1.54 V, also uses the oxidation of zinc as the anode reaction, but under alkaline (pH > 7) conditions.

$$Zn(s) + 2\,OH^-(aq) \longrightarrow ZnO(aq) + H_2O(\ell) + 2\,e^- \qquad \text{(anode, oxidation)}$$

The electrons that pass through the external circuit are consumed by reduction of manganese dioxide at the cathode.

$$MnO_2(s) + H_2O(\ell) + e^- \longrightarrow MnO(OH)(s) + OH^-(aq) \qquad \text{(cathode, reduction)}$$

In the *mercury battery* (Figure 19.18) the oxidation of zinc is again the anode reaction. The cathode reaction, however, is the reduction of mercury(II) oxide.

$$HgO(s) + H_2O(\ell) + 2\,e^- \longrightarrow Hg(\ell) + 2\,OH^-(aq)$$

The voltage of this battery is about 1.35 V. Mercury batteries are used in calculators, watches, hearing aids, cameras, and other devices in which small size is an advantage. However, mercury and its compounds are poisonous, and careless disposal of mercury batteries can lead to environmental problems such as contamination of groundwater or even mercury vapor in the atmosphere.

Mercury batteries are hermetically sealed (to prevent leakage of mercury) and should never be heated. Heating increases the pressure of vapor within the battery, ultimately causing the battery to explode.

Secondary Batteries

Secondary batteries are rechargeable because, as they discharge, the oxidation products remain at the anode and the reduction products remain at the cathode. As a result, if the direction of electron flow is reversed, the anode and cathode reactions are reversed and the reactants are regenerated. Under favorable conditions, secondary batteries may be discharged and recharged hundreds or thousands of times. Examples of secondary batteries include automobile batteries and nicad batteries.

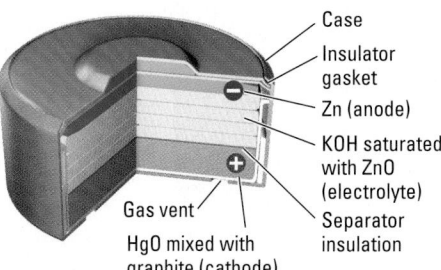

Case
Insulator gasket
Zn (anode)
KOH saturated with ZnO (electrolyte)
Separator insulation
Gas vent
HgO mixed with graphite (cathode)

Figure 19.18 The mercury battery. The reducing agent is zinc and the oxidizing agent is mercury(II) oxide.

WILSON GREATBATCH
(1919–)

In the early 1960s, Wilson Great-batch had an idea of how a battery might be used to help an ailing heart keep pumping. His story, told in his own words, is fascinating:

I quit all my jobs, and with two thousand dollars I went out in the barn in the back of my house and built 50 pacemakers in two years. I started making the rounds of all the doctors in Buffalo who were working in this field, and I got consistently negative results. The answer I got was, well, these people all die in a year, you can't do much for them. . . . When I first approached Dr. Shardack with the idea of the pacemaker, he alone thought that it really had a future. He said, "You know — if you can do that — you can save a thousand lives a year."

After the first ten years, we were still only getting one or two years out of pacemakers . . . and the failure mechanism was always the battery. The human body is a very hostile environment. . . . You're trying to run things in a warm salt water environment. . . . So we started looking around for new power sources. And we finally wound up with this lithium battery. It really revolutionized the pacemaker business. The doctors have told me that the introduction of the lithium battery was more significant than the invention of the pacemaker in the first place.

Source:

The World of Chemistry video, Program 15, The Annenberg/CPB Collection.

The lead-acid battery was first described to the French Academy of Sciences in 1860 by Gaston Planté.

Lead-Acid Storage Battery

The familiar automobile battery, the lead-acid storage battery, is a secondary battery consisting of six cells, each containing porous lead electrodes and lead(IV) oxide electrodes immersed in aqueous sulfuric acid (Figure 19.19). As this battery produces an electric current, metallic lead is oxidized to lead sulfate at the anode, and lead(IV) oxide is reduced to lead sulfate at the cathode.

$$Pb(s) + HSO_4^-(aq) + H_2O(\ell) \longrightarrow PbSO_4(s) + H_3O^+(aq) + 2\,e^-$$

$$E° = -0.356\text{ V (anode reaction)}$$

$$PbO_2(s) + 3\,H_3O^+(aq) + HSO_4^-(aq) + 2\,e^- \longrightarrow PbSO_4(s) + 5\,H_2O(\ell)$$

$$E° = 1.685\text{ V (cathode reaction)}$$

$$Pb(s) + PbO_2(s) + 2\,H_3O^+(aq) + 2\,HSO_4^-(aq) \longrightarrow 2\,PbSO_4(s) + 4\,H_2O(\ell)$$

$$E°_{cell} = +2.041\text{ V}$$

The combined voltage from the six cells connected in series in a typical automobile battery gives a voltage of 12 V.

During discharge, sulfuric acid is consumed in both the anode and the cathode reactions, causing the concentration of the sulfuric acid electrolyte to decrease. Before the introduction of modern sealed automotive batteries, the measured density of this battery acid was routinely used to indicate the state of charge of the battery. Since sulfuric acid has a density greater than water, the density of the battery acid decreases as the battery discharges. Consequently, the lower the density, the lower the charge of the battery. It is now harder or impossible to measure the density of acid in modern sealed batteries.

To understand why the lead storage battery is rechargeable, consider that the lead sulfate formed at both electrodes is an insoluble compound that mostly *stays on the electrode surface.* This keeps it available for the reverse reaction. To recharge a secondary battery, a source of direct electrical current is supplied so that electrons are forced to flow in the direction opposite from when the battery was discharging. This causes the overall battery reaction to be reversed and regenerates the reactants that originally produced the battery's voltage and current. For the lead-acid storage battery, the overall reaction is

Discharging battery produces electricity

$$Pb(s) + PbO_2(s) + 2\,HSO_4^-(aq) + 2\,H_3O^+(aq) \rightleftharpoons 2\,PbSO_4(s) + 4\,H_2O(\ell)$$

Charging battery requires electricity

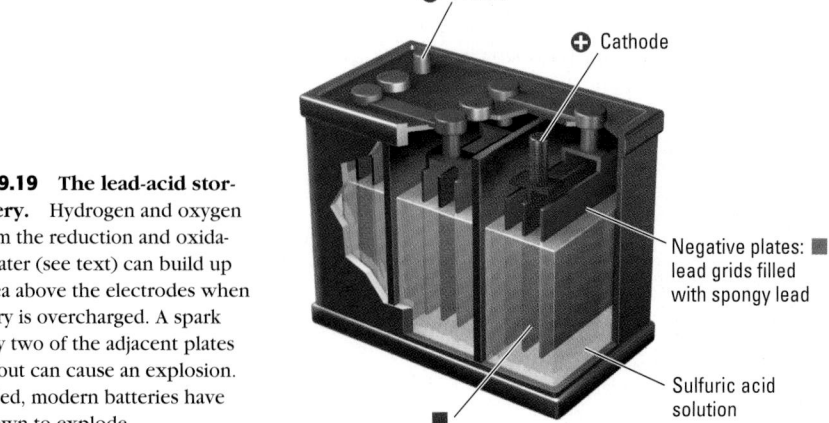

Figure 19.19 The lead-acid storage battery. Hydrogen and oxygen gases from the reduction and oxidation of water (see text) can build up in the area above the electrodes when the battery is overcharged. A spark caused by two of the adjacent plates shorting out can cause an explosion. Even sealed, modern batteries have been known to explode.

Normal charging of an automobile lead-acid storage battery occurs during driving. In addition to reversing the overall battery reaction, charging reduces a little water at the cathode and oxidizes a little water at the anode,

Reduction of water: $\quad 4\,H_2O(\ell) + 4\,e^- \longrightarrow 2\,H_2(g) + 4\,OH^-(aq)$

Oxidation of water: $\quad 6\,H_2O(\ell) \longrightarrow O_2(g) + 4\,H_3O^+(aq) + 4\,e^-$

These reactions produce a hydrogen-oxygen mixture inside the battery, which, if accidentally ignited, can explode. Therefore, no sparks or open flames should be brought near a lead-acid storage battery, even the sealed kind.

The lead-acid storage battery is relatively inexpensive, reliable, and simple, and it has an adequate lifetime. High weight is its major fault. A typical automobile battery contains about 15 to 20 kg of lead, which is required to provide the large number of electrons needed to crank an automobile engine, especially on a cold morning. (Recall that the number of electrons that a battery can move from the anode to the cathode is proportional to the amount of reactants involved.) Another problem with lead batteries is that mining and manufacturing the lead and disposing of the batteries can contaminate air and groundwater. Auto batteries should be recycled by companies equipped with the proper safeguards.

Nickel-Cadmium Battery

Nickel-cadmium ("nicad") secondary batteries are lightweight, can be quite small, and produce a constant voltage until completely discharged — making them useful in cordless appliances, video camcorders, portable radios, and other applications (Figure 19.20).

Nicad batteries can be recharged because the reaction products are insoluble hydroxides that remain at the electrode surfaces. The anode reaction during the discharge cycle is the oxidation of cadmium, and the cathode reaction is the reduction of nickel oxyhydroxide $NiO(OH)$.

Figure 19.20 Nickel-cadmium (nicad) batteries.

$$Cd(s) + 2\,OH^-(aq) \longrightarrow Cd(OH)_2(s) + 2\,e^-$$

$$E^\circ = 0.809\ V\ \text{(anode reaction)}$$

$$2\,[NiO(OH)(s) + H_2O(\ell) + e^- \longrightarrow Ni(OH)_2(s) + OH^-(aq)]$$

$$E^\circ = 0.490\ V\ \text{(cathode reaction)}$$

$$Cd(s) + 2\,NiO(OH)s + 2\,H_2O(\ell) \longrightarrow Cd(OH)_2(s) + 2\,Ni(OH)_2(s)$$

$$E^\circ_{cell} = 1.299\ V$$

Like mercury batteries, nicad batteries should be disposed of properly because of the toxicity of cadmium and its compounds.

Lithium Battery

Lithium batteries (Figure 19.21) benefit from the low density and strong reducing strength of lithium metal (Table 19.1). The anode in a lithium battery is made of lithium metal that has been mixed with a conducting carbon polymer. The polymer has tiny spaces in its structure that can hold the lithium atoms and lithium ions formed by the oxidation reaction.

$$Li(s)\ \text{(in polymer)} \longrightarrow Li^+\text{(in polymer)} + e^- \qquad \text{(anode reaction)}$$

The cathode also contains lithium ions, but in the lattice of a metal oxide such as CoO_2. This oxide lattice, like the carbon-polymer electrode, has holes in it that can accommodate Li^+ ions. The reduction reaction is

$$Li^+\text{(in }CoO_2) + e^- + CoO_2 \longrightarrow LiCoO_2 \qquad \text{(cathode reaction)}$$

The overall reaction is therefore

$$Li(s) + CoO_2(s) \longrightarrow LiCoO_2(s) \qquad E_{cell} = 3.4\ V$$

Figure 19.21 The lithium battery. The lithium battery finds many uses in which a high energy density and light weight are desired.

Lithium battery in a cell phone.

Lithium batteries have a large voltage (3.4 V per cell) and very high energy output for their weight (1 mol of Li, 6.94 g/mol, can produce 1 mol e⁻). They can be recharged many hundreds of times. Because of these desirable characteristics, lithium batteries are used in cellular telephones, laptop computers, and cameras.

Exercise 19.9 Recharging a Nicad Battery

Write the electrode reactions that take place when a nicad battery is recharged; identify the anode and cathode reactions.

 ### Exercise 19.10 Emergency Batteries

You are stranded on an island and need to communicate your location for help. You have a battery-powered radio transmitter, but the lead batteries are discharged. There is a swimming pool nearby and you find a tank of chlorine gas and some plastic tubing that can withstand being oxidized by chlorine. Devise a battery that might be used to power the radio using these things.

19.10 FUEL CELLS

A **fuel cell** is an electrochemical cell that converts the chemical energy of fuels directly into electricity. It functions somewhat like a battery, but in contrast to a battery, its reactants are continually supplied from an external reservoir.

The best-known fuel cell is the *alkaline fuel cell* shown in Figure 19.22, which is used in the space shuttles. The net cell reaction is simply the oxidation of hydrogen to give water. A stream of pure H_2 gas is pumped into the anode compartment of the cell, and pure O_2 gas is directed onto the cathode. The cell contains concentrated KOH, so the electrochemical half-reactions are

$$2\,H_2(g) + 4\,OH^-(aq) \longrightarrow 4\,H_2O(\ell) + 4\,e^- \qquad \text{(anode, oxidation)}$$

$$\underline{O_2(g) + 2\,H_2O(\ell) + 4\,e^- \longrightarrow 4\,OH^-(aq) \qquad \text{(cathode, reduction)}}$$

$$2\,H_2(g) + O_2(g) \longrightarrow 2\,H_2O(\ell) \qquad E_{cell} = 0.9\text{ V at 70 to }140\text{ °C}$$

Figure 19.22 An H_2/O_2 fuel cell.
The anode chamber oxidizes H_2. The cathode chamber reduces O_2. The water produced is often purified for drinking purposes.

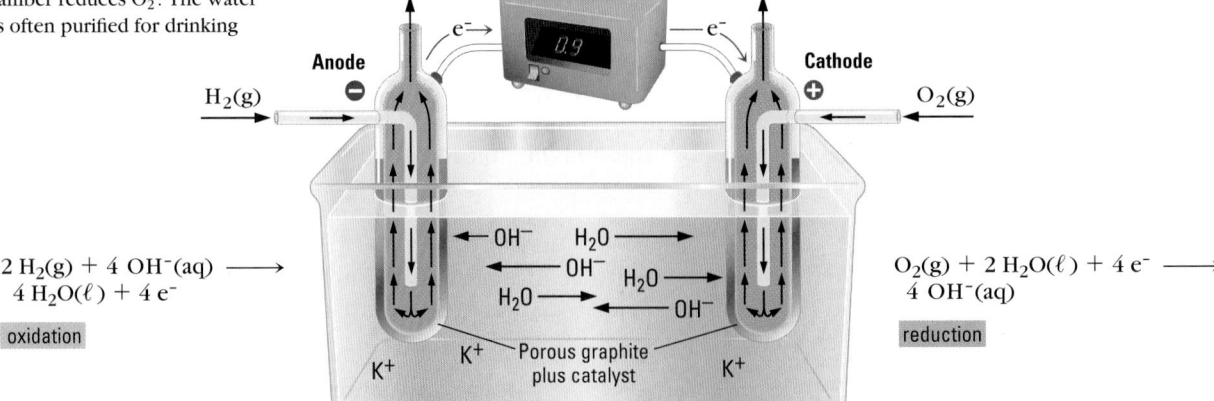

The electrons lost by the hydrogen molecules at the anode flow out of the fuel cell, through a circuit, and then back into the cell at the cathode, where oxygen is reduced. This electron flow powers the electrical needs of the spacecraft or whatever else is connected to the fuel cell. The water produced in the fuel cell can be purified for drinking purposes.

Because of their light weight and their high efficiency compared with batteries, fuel cells have proved valuable in the space program. Beginning with Gemini 5, through the Apollo program, and now in the Space Shuttle program, alkaline fuel cells have logged more than 10,000 hours of operation in space. The fuel cells used aboard the Space Shuttle deliver the same power that batteries weighing ten times as much would provide. On a typical seven-day mission, the Shuttle fuel cells consume 1500 lb of hydrogen and generate 190 gal of potable water (suitable for drinking).

Other types of fuel cells that have been developed use air as the oxidizer and hydrogen or carbon monoxide as the fuel. Considerable research is currently aimed at developing fuel cells capable of direct air oxidation of cheap gaseous fuels such as natural gas. Fuel cells using H_2 as their fuel produce only water, and they are therefore environmentally benign. However, the H_2 must be manufactured—gasoline, natural gas, methanol, biomass, or water are starting sources—and the method used determines the overall pollution. If renewable energy (solar, wind, or hydroelectric) is used to produce the H_2, then a truly pollution-free energy system results. Such an approach is not economically competitive at the present time, but this area is undergoing rapid change as new technologies are explored and developed. The field of fuel cell research is competitive, with advocates of different technologies pushing their types of cells. Some forecasts now predict the manufacture of hundred of thousands of fuel-cell-based automobiles within the coming five years. Presently, it is far from clear exactly which types of fuel cells will emerge as the best choices in the future.

The movie *Apollo 13* deals with an in-flight explosion of the fuel cell storage tanks.

19.11 ELECTROLYSIS—FORCING REACTANT-FAVORED REACTIONS TO OCCUR

Electrons forced into an electrochemical system from a source of electrical current such as a battery can cause reactant-favored redox systems to produce products. This process, called **electrolysis,** provides a way to carry out reactant-favored reactions that will not take place by themselves. Electrolytic processes are even more important in our economy than the redox reactions that power batteries. Such electrolytic processes are used in the production and purification of many metals, including copper and aluminum (Sections 22.3 and 21.4), and in electroplating processes that produce a thin coating of metal on many different kinds of items.

Like voltaic cells, electrolysis cells have electrodes in contact with a conducting medium and an external circuit. As in a voltaic cell, the electrode where reduction takes place is called the cathode, and the electrode where oxidation takes place is called the anode. The electrodes in electrolysis cells are often inert, and their function is to furnish a path for electrons to enter and leave the cell. In contrast to voltaic cells, however, the external circuit connected to an electrolysis cell must contain a direct current *source* of electrons. A battery can be used as a source of electrical current when an electrolysis is carried out on a small scale. The battery forces electrons into one of the electrodes (which becomes negative) and removes electrons from the other electrode (which becomes positive). There is often no need for a physical separation of the two electrode reactions, so there is usually no salt bridge. The conducting medium in contact with the electrodes is often the same for both electrodes, and it can be a molten salt or an aqueous solution.

CD-ROM Screen 21.10: Electrolysis: Chemical Change from Electrical Energy

Lysis means "splitting," so electrolysis means "splitting with electricity." Electrolysis reactions are chemical reactions caused by the flow of electricity.

Chemistry in the News

BATTERIES FOR ELECTRIC CARS

Electric automobiles are now being driven in the United States, especially in California. General Motors was producing and leasing EV1 automobiles for several years. They were available in two models: (1) a longer range (75–130 miles), higher cost model with nickel-metal hydride batteries from GM Ovonics, and (2) a lower cost, shorter range model (55–95 miles) with advanced lead-acid storage batteries. The EV1 vehicles were built in blocks of about 500 cars each in 1996–1999. Toyota is also manufacturing the RAV4 electric vehicle.

These cars represent many technologic breakthroughs. Their biggest drawback, and the one that has perhaps the greatest impact on their acceptance, is the lack of availability of lightweight storage batteries that can be recharged quickly and can store enough energy to power a car for 100 miles or more without recharging. Such a power system is necessary to make electric-powered vehicles attractive to consumers.

Several batteries that are expected to have these characteristics are incorporated into the latest electric cars. One such battery was developed by Energy Conversion Devices of Troy, Michigan, under a grant from the U.S. Advanced Battery Consortium, which was set up by the U.S. Department of Energy, the three large U.S. automakers, and the Electric Power Research Institute. Called the Ovonic battery, it has an anode made of nickel alloyed with several other metals—vanadium, titanium, zirconium, and chromium—instead of the usual pure metal electrode. The half-cell reactions for this battery are

$$MH(s) + OH^-(aq) \underset{\text{charge}}{\overset{\text{discharge}}{\rightleftharpoons}} M(s) + H_2O(\ell) + e^-$$

$$NiOOH(s) + H_2O(\ell) + e^- \underset{\text{charge}}{\overset{\text{discharge}}{\rightleftharpoons}} Ni(OH)_2(s) + OH^-(aq)$$

Net equation:

$$MH(s) + NiOOH(s) \underset{\text{charge}}{\overset{\text{discharge}}{\rightleftharpoons}} M(s) + Ni(OH)_2(s)$$

M is an intermetallic alloy that is capable of forming a metal hydride phase. MH(s) represents a metal hydride compound similar to those used to store hydrogen in other applications. The amount of hydrogen that can be absorbed into this electrode determines the number of electrons that the battery can deliver as it discharges and hence the energy storage capacity of the battery. The metal hydride electrode has the advantage of being metallic and hence electrically conducting. In many other batteries metal oxides, which do not conduct electricity, are formed on the electrode surface and decrease the number of times the battery can be recharged.

Each component of the metal alloy anode plays a role in the battery's performance. Vanadium, titanium, and zirconium are in the alloy because they readily absorb hydrogen. For an effective metal hydride battery, the strength of

Electric vehicles participating in the American Tour de Sol, May 2000.

Electrolysis can decompose molten sodium chloride. A pair of electrodes dip into pure sodium chloride that has been heated above its melting temperature (Figure 19.23). In the liquid, the Na^+ and Cl^- ions are free to move. The Na^+ ions are attracted to the negative electrode, and the Cl^- ions are attracted to the positive electrode. Reduction of Na^+ ions to Na atoms occurs at the negative electrode (the cathode). Oxidation of Cl^- ions occurs at the positive electrode (the anode).

$$2\,Na^+ + 2\,e^- \longrightarrow 2\,Na(\ell) \qquad \text{(cathode, reduction)}$$

$$2\,Cl^- \longrightarrow Cl_2(g) + 2\,e^- \qquad \text{(anode, oxidation)}$$

$$\overline{2\,Na^+ + 2\,Cl^- \longrightarrow 2\,Na(\ell) + Cl_2(g)} \qquad \text{(net cell reaction)}$$

The electrolysis of molten salts is an energy-intensive reaction because energy is needed to melt the salt and to cause the anode and cathode reactions to take place.

the bonding between the metal atoms and hydrogen atoms must be just right—from 25 to 50 kJ/mol. If it is too low, recharging will release hydrogen as $H_2(g)$, instead of incorporating hydrogen atoms into holes in the metal crystal lattice. If hydrogen-to-metal bonding is too strong, the electrode metal will be oxidized instead of the hydrogen atoms, and the battery will not be able to discharge properly. Alloying the other metals with nickel allows the strength of the bond to hydrogen to be carefully adjusted for maximum efficiency. Chromium limits corrosion of vanadium in the alloy, and both zirconium and chromium affect the structure of the alloy, leading to a high surface area that promotes rapid cell reactions and hence high power output.

The metal hydride electrode consists of amorphous metal. Its atoms are in an irregular, disordered structure instead of the orderly, crystalline arrangement of most metals. This disorder increases hydrogen storage capacity and speeds up electrode reactions. According to Energy Conversion Devices spokesperson Stanford R. Ovshinsky, the use of an amorphous material is "a fundamentally different approach" to battery design. The company claims that the battery can be charged in as little as 15 min and can undergo more than 1000 charge-discharge cycles, which translates to a lifetime of 10 years and more than 100,000 miles of travel in an automobile.

Electric cars (and hybrid cars that use both batteries and internal combustion engines) are growing in use, especially in California, where there are currently about 2300 electric vehicles on the road. The use of these cars is driven by the State of California Air Resources Board (ARB), which has mandated a California zero-emission vehicle (ZEV) requirement whereby at least 10% of the cars and light trucks for sale in California by 2003 must be ZEVs. This mandate was upheld at an ARB meeting in September 2000, where "The California ARB . . . held fast to its mandate requiring automakers to market thousands of ZEVs in the state starting in 2003." A service in California has an Internet site that lists alternative fuel vehicle (AFV) refueling sites. Currently, it lists 395 electric inductive and 297 electric conductive sites where electric vehicles can be recharged, and the number is growing. Inductive sites involve magnetic coupling of electrical systems, whereas conductive sites involve metal-to-metal plugs.

In the American Tour de Sol held in May 2000, an EV1 finished second in the production category. This race, from New York City to Washington DC, rated cars on fuel economy and "global warming potential." The EV1 had an equivalent fuel ecomony rating of 68 mpg and had a superior rating for reducing global warming.

Tour de Sol electric cars in action.

Source:

Electric Vehicle Association of America, **http://www.evaa.org.**

What happens if we pass electricity through an *aqueous solution* of a salt, such as potassium iodide, KI? To predict the outcome of the electrolysis we must first decide what is in the solution that can be oxidized and reduced. For KI(aq), the solution contains K^+ ions, I^- ions, and H_2O molecules. In K^+, potassium is already in its highest common oxidation state, so it cannot be involved in an oxidation reaction. However, both the I^- ion and the H_2O could be oxidized. The possible anode half-reaction oxidations are

$$2\,I^-(aq) \longrightarrow I_2(s) + 2\,e^- \qquad\qquad E^\circ_{ox} = -0.535\text{ V}$$

$$6\,H_2O(\ell) \longrightarrow O_2(g) + 4\,H_3O^+(aq) + 4\,e^- \qquad E^\circ_{ox} = -1.229\text{ V}$$

Whenever two or more electrochemical reactions are possible at a single electrode, the one with the more positive E° will usually occur under standard state conditions. Judging by the values of E°_{ox} (which are from Table 19.1), the iodide ion will be oxidized more readily than will water.

When a redox reaction written as a reduction has its direction reversed, then E°_{red} becomes E°_{ox}, and the sign is switched as well.

Figure 19.23 Electrolysis of molten sodium chloride.

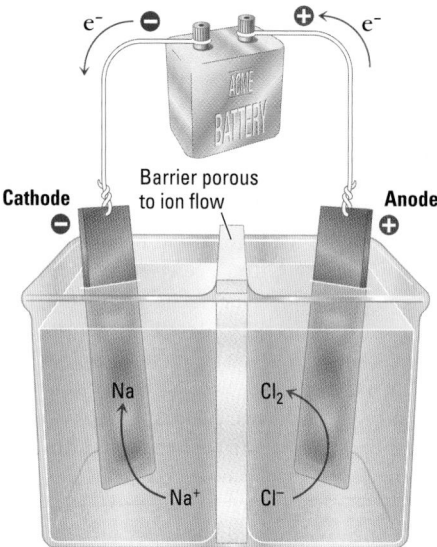

Since I^- is already in a reduced form, there are only two species that can be reduced at the cathode: K^+ ions and water molecules. The reduction reactions and their potentials are

$$K^+(aq) + e^- \longrightarrow K(s) \qquad\qquad E°_{red} = -2.925 \text{ V}$$

$$2\,H_2O(\ell) + 2\,e^- \longrightarrow H_2(g) + 2\,OH^-(aq) \qquad\qquad E°_{red} = -0.8277 \text{ V}$$

In this case we predict that $H_2O(\ell)$ will be reduced because it has the less-negative $E°$ value. The overall reaction is

$$2\,I^-(aq) + 2\,H_2O(\ell) \longrightarrow I_2(s) + H_2(g) + 2\,OH^-(aq) \qquad E°_{cell} = -1.363 \text{ V}$$

An experiment in which electrons are passed through aqueous KI (Figure 19.24) shows that this prediction is correct. At the anode, on the right, the I^- ion is oxidized to I_2, which produces a yellow-brown color in the solution. At the cathode, water is reduced and hydroxide ions are formed, as shown by the pink color of the phenolphthalein indicator that has been added to the solution. A close look at Figure 19.24 reveals that hydrogen gas is also being produced at the surface of the inert platinum electrode.

When an electrolysis is carried out by passing electrical current through an aqueous solution, the electrode reactions most likely to take place are those that require the least voltage, that is, the half-reactions that combine to give the least negative overall cell voltage. This means that in aqueous solution the following conditions apply.

1. *A metal ion or other species can be reduced if it has a reduction potential more positive than -0.8 V, the potential for reduction of water.* Table 19.1 shows that most metal ions are in this category. If a species has a reduction potential more negative than -0.8 V, then water will be reduced to $H_2(g)$ preferentially. Metal ions in this latter category include Na^+, K^+, Mg^{2+}, and Al^{3+}. Producing these metals from their ions requires electrolysis of a molten salt in which no water is present.

2. *A species can be oxidized in aqueous solution if it has an oxidation potential more positive than -1.2 V, the potential for oxidation of water to $O_2(g)$.* Most species on the right-hand side of the half-equations in Table 19.1 are in this category. If a species has an oxidation potential more negative than -1.2 V (that is, if its half-equation is above the water-oxygen

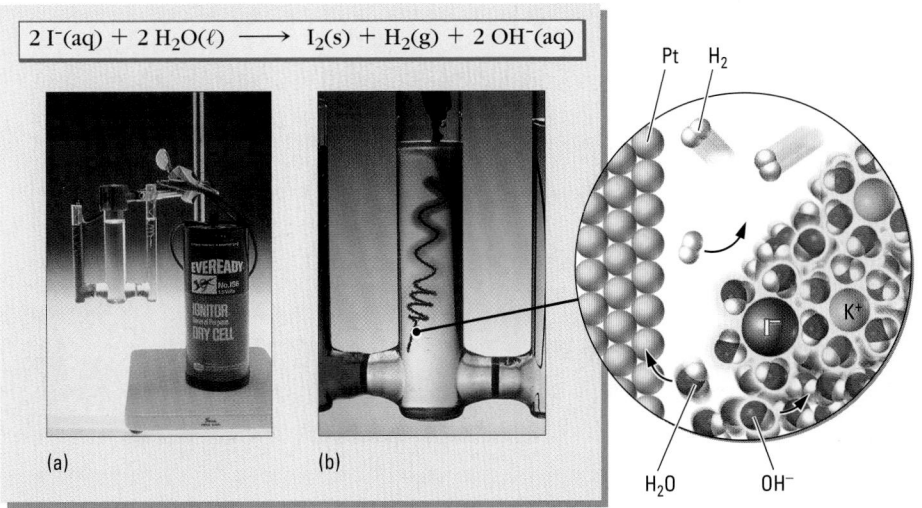

$$2 \, I^-(aq) + 2 \, H_2O(\ell) \longrightarrow I_2(s) + H_2(g) + 2 \, OH^-(aq)$$

(a) (b)

Pt H₂

H₂O OH⁻

Figure 19.24 The electrolysis of aqueous potassium iodide. (a) Aqueous KI is contained in all three compartments of the cell, and both electrodes are platinum. At the positive electrode, or anode (*right*), the I⁻ ion is oxidized to iodine, which gives the solution a yellow-brown color.

$$2 \, I^-(aq) \longrightarrow I_2(aq) + 2 \, e^-$$

At the negative electrode, or cathode (*left*), water is reduced, and the presence of OH⁻ ion is indicated by the pink color of the acid-base indicator, phenolphthalein.

$$2 \, H_2O(\ell) + 2 \, e^- \longrightarrow H_2(g) + 2 \, OH^-(aq)$$

(b) A close-up of the cathode of a different cell running the same reaction. Bubbles of H₂ and evidence of OH⁻ generation at the electrode are evident.

half-equations in Table 19.1), water will be oxidized preferentially. Thus, for example, F⁻(aq) cannot be oxidized electrolytically to F₂(g), because water will be oxidized to O₂(g) instead.

The voltage that must be applied to an electrolysis cell is always greater than what is calculated from standard reduction potentials. An *overvoltage* is required, which is an additional voltage that is needed to overcome limitations in the electron transfer rate at the interface between electrode and solution. Redox reactions that involve the formation of O₂ or H₂ are especially prone to have large overvoltages. Since overvoltages cannot be predicted accurately, the only certain way to determine which half-reaction will occur when two possible reactions have similar standard reduction potentials is to perform the experiment.

Problem-Solving Example 19.12 Electrolysis of Aqueous NaOH

Predict the result of passing a direct electrical current through an aqueous solution of NaOH.

Answer The net cell reaction is $2 \, H_2O(\ell) \longrightarrow 2 \, H_2(g) + O_2(g)$. Hydrogen is produced at the cathode and oxygen is produced at the anode.

Explanation First, list all the species in the solution. In this case they are Na⁺, OH⁻, and H₂O. Next, use Table 19.1 to decide which of these species can be oxidized and which can be reduced, and note the potential of each possible reaction.

Reductions:

$$Na^+(aq) + e^- \longrightarrow Na(s) \qquad\qquad E^\circ_{red} = -2.71 \text{ V}$$

$$2 \, H_2O(\ell) + 2 \, e^- \longrightarrow H_2(g) + 2 \, OH^-(aq) \qquad E^\circ_{red} = -0.83 \text{ V}$$

Oxidations:

$$4 \, OH^-(aq) \longrightarrow O_2(g) + 2 \, H_2O(\ell) + 4 \, e^- \qquad E°_{ox} = -0.40 \, V$$

$$6 \, H_2O(\ell) \longrightarrow O_2(g) + 4 \, H_3O^+(aq) + 4 \, e^- \qquad E°_{ox} = -1.229 \, V$$

It is evident that water will be reduced to H_2 at the cathode and that OH^- will be oxidized at the anode, because these are the reactions with the least negative $E°$ values. The net cell reaction is

$$2 \, H_2O(\ell) \longrightarrow 2 \, H_2(g) + O_2(g)$$

and the potential under standard conditions is $(-0.83 \, V) + (-0.40 \, V) = -1.23 \, V$.

Problem-Solving Practice 19.12

Predict the results of passing a direct electrical current through (a) molten NaBr, (b) aqueous NaBr, and (c) aqueous $SnCl_2$.

  Exercise 19.11 Making F_2 Electrolytically

In 1886, Moissan was the first to prepare F_2 by the electrolysis of F^- ions. He electrolyzed the salt KF dissolved in pure HF. No water was present, so only F^- ions were available at the anode. What was produced at the cathode? Write the half-equations for the oxidation and the reduction reactions, then write the net cell reaction.

An extremely important industrial reaction is the chlor-alkali process that involves the electrolysis of aqueous NaCl (brine). The process is the major commercial source of chlorine gas and sodium hydroxide; it is described in detail in Section 21.4.

CD-ROM Screen 21.11:
Coulometry: Counting Electrons

19.12 COUNTING ELECTRONS

When an electric current is passed through an aqueous solution of the soluble salt $AgNO_3$, metallic silver is produced at the cathode. One mole of electrons is required for every mole of Ag^+ reduced.

$$Ag^+(aq) + e^- \longrightarrow Ag(s)$$

If a Cu(II) salt were in aqueous solution, 2 mol of electrons would be required to produce 1 mol of metallic copper from copper(II) ions.

$$Cu^{2+}(aq) + 2 \, e^- \longrightarrow Cu(s)$$

Each of these balanced half-reactions is like all other balanced chemical equations. They illustrate the fact that both matter and charge are conserved in chemical reactions. This means that if you could measure the number of moles of electrons flowing through an electrolysis cell, you would know the number of moles of silver or copper produced. Conversely, if you knew the amount of silver or copper produced, you could calculate the number of moles of electrons that had passed through the circuit.

The number of moles of electrons transferred during a redox reaction is usually determined by measuring the current flowing in the external electrical circuit during a given time. The product of the current (measured in amperes, A) and the time interval (in seconds, s) equals the electric charge (coulombs, C) of electricity that has flowed through the circuit.

Large electric currents, like those needed to run a hair dryer or refrigerator, are measured in amperes. Smaller currents, in the milliampere (mA) range, are more commonly used in laboratory electrolysis experiments. $1 \, mA = 10^{-3} \, A$.

$$\text{Charge} = \text{current} \times \text{time}$$

$$1 \, \text{coulomb} = 1 \, \text{ampere} \times 1 \, \text{second}$$

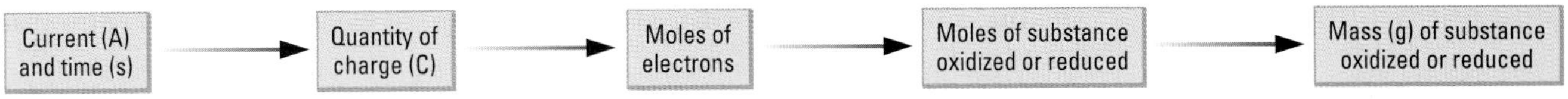

| Current (A) and time (s) | → | Quantity of charge (C) | → | Moles of electrons | → | Moles of substance oxidized or reduced | → | Mass (g) of substance oxidized or reduced |

Figure 19.25 Calculation steps for electrolysis. These steps relate the quantity of electrical charge used in electrolysis to the amounts of substances oxidized or reduced.

The Faraday constant (96,500 C/mol of electrons, ⬅ *p. 885*) can then be used to find the number of moles of electrons from a known number of coulombs of charge. This information is of practical significance in chemical analysis and synthesis.

Figure 19.25 shows the relationship between quantity of charge used and the quantities of substances that are oxidized or reduced during electrolysis.

Problem-Solving Example 19.13   Using the Faraday Constant

What mass of nickel will be deposited at the cathode of an electrolysis cell if a current of 20. mA passes through an aqueous solution containing Ni^{2+} ions for 1.0 hr (3600 s)?

Answer   0.022 g Ni

Explanation   The reaction at the cathode is

$$Ni^{2+}(aq) + 2\,e^- \longrightarrow Ni(s)$$

The charge that passes through the cell is

$$\text{Charge} = 20. \times 10^{-3}\,A \times 3600\,s = 72.\,C$$

Use the Faraday constant, the coefficients of the balanced cathode half-reaction, and the molar mass of nickel as conversion factors to find the mass of nickel deposited:

$$72.\,C \times \left(\frac{1\text{ mol }e^-}{9.65 \times 10^4\,C}\right)\left(\frac{1\text{ mol Ni}}{2\text{ mol }e^-}\right)\left(\frac{58.7\text{ g Ni}}{1\text{ mol Ni}}\right) = 0.022\text{ g Ni}$$

Problem-Solving Practice 19.13

In the commercial production of sodium by electrolysis, the cell operates at 7.0 V and a current of 25×10^3 A. What mass of sodium can be produced in 1 hr?

   Exercise 19.12   How many Faradays?

Which would require more Faradays of electricity?

(a) Making 1 mol of Al from Al^{3+}
(b) Making 2 mol of Na from Na^+
(c) Making 2 mol of Cu from Cu^{2+}

Electrolytic Production of Hydrogen

Hydrogen holds great promise as a fuel in our economy because it is a gas and can be transported through pipelines; it burns without producing pollutants; and it could be used in fuel cells to generate electricity on demand. Hydrogen can be produced by the electrolysis of dilute sulfuric acid. Both water and sulfuric acid are in plentiful supply. The major problem with producing hydrogen in quantities large enough to meet a nation's energy demands is the source of electricity.

The minimum voltage required for this reaction is 1.24 V. Let's consider how much electrical energy would be required to produce 1.00 kg of gaseous H_2 (about 11,200 L at STP) and at what cost. We will first calculate the required charge in coulombs by using Faraday's constant, and then use the definition 1 joule = 1 volt × 1 coulomb to get energy units.

The reduction half-reaction shows that 2 mol of electrons is required to produce 1 mol (2.02 g) of $H_2(g)$.

$$2 H_3O^+(aq) + 2 e^- \longrightarrow H_2(g) + 2 H_2O(\ell)$$

The amount (number of moles) of electrons required to produce 1.00 kg of H_2 is found as follows:

$$1.00 \text{ kg } H_2 \times \left(\frac{1 \times 10^3 \text{ g}}{\text{kg}} \right) \left(\frac{1 \text{ mol } H_2}{2.016 \text{ g } H_2} \right) \left(\frac{2 \text{ mol } e^-}{1 \text{ mol } H_2} \right) = 9.92 \times 10^2 \text{ mol } e^-$$

Now we can calculate the charge by using Faraday's constant.

$$9.92 \times 10^2 \text{ mol } e^- \times \left(\frac{9.65 \times 10^4 \text{ C}}{1 \text{ mol } e^-} \right) = 9.57 \times 10^7 \text{ C}$$

The energy (in joules) can be calculated from the charge and the cell voltage.

$$\text{Energy} = \text{charge} \times \text{voltage} = (9.57 \times 10^7 \text{ C})(1.24 \text{ V}) = 1.19 \times 10^8 \text{ J}$$

The kilowatt-hour (kWh) is a unit of energy.

We convert joules to kilowatt-hours (kWh), which is the unit we see when we pay the electric bill. The conversion factor is 1 kWh = 3.60×10^6 J.

$$1.19 \times 10^8 \text{ J} \times \left(\frac{1 \text{ kWh}}{3.60 \times 10^6 \text{ J}} \right) = 33.1 \text{ kWh}$$

At a rate of 10 cents per kilowatt-hour, the production of 1.00 kg of hydrogen costs $3.31.

Exercise 19.13   Calculations Based on Electrolysis

In the production of aluminum metal, Al^{3+} is reduced to Al metal with currents of about 50,000 A and a low voltage of about 4.0 V. How much energy (in kilowatt-hours) is required to produce 2000. tons of aluminum metal?

Exercise 19.14   How Many Joules?

Think of a battery you just purchased at the store as an energy source containing some number of joules. Name the two pieces of information you need to calculate the number of joules of energy in this battery. Which one is obviously available as you read the label on the battery? Devise a means of determining the other information needed.

Electroplating

If a metal or other electrical conductor is made the cathode in an electrolysis cell, it can be plated with another metal to decorate it or protect it against corrosion. To plate an object with copper, we have only to render the surface conducting and make the object the cathode in a cell containing a solution of a soluble copper salt. The object will become coated with copper, and the copper coating will grow thicker as the electrolysis continues and electrons reduce more Cu^{2+} ions to Cu atoms. If the plated object is a metal, it will conduct electricity by itself. If the object is a nonmetal, its surface can be lightly dusted with graphite powder to make it conducting.

Precious metals such as gold are often plated onto cheaper metals such as copper to make jewelry. If the current and duration of the plating reaction are known, it is possible to calculate the mass of gold that will be reduced onto the cathode surface. For example, suppose the object to be plated is immersed in a solution of $AuCl_3$ and is made a cathode by connecting it to the negative pole of a battery. The

---

ESTIMATION   **The Amount of Aluminum in a Soda Can**

How much does it cost to generate the mass of aluminum in one soda can? The aluminum in these cans is produced by reducing Al^{3+} to $Al(s)$. The reaction is run commercially at 50,000 A at a voltage of 4 V. A soda can weighs about 14 g, 1 kWh of electricity costs about 10 cents, and 1 kWh = 3.60×10^6 J.

The charge needed to generate 14 g of Al is

$$(14 \text{ g Al})\left(\frac{1 \text{ mol Al}}{26.98 \text{ g Al}}\right)\left(\frac{3 \text{ mol e}^-}{1 \text{ mol Al}}\right)\left(\frac{96,500 \text{ C}}{1 \text{ mol e}^-}\right)$$

$$= 1.5 \times 10^5 \text{ C}$$

The energy used is

$$(1.5 \times 10^5 \text{ C})(4 \text{ V})\left(\frac{1 \text{ J}}{1 \text{ C} \times 1 \text{ V}}\right)\left(\frac{1 \text{ kWh}}{3.60 \times 10^6 \text{ J}}\right)$$

$$= 0.17 \text{ kWh}$$

The cost of 0.17 kWh at 10 cents/kWh is 1.7 cents. So the mass of Al in one soda can could be generated by electrolysis for less than 2 cents.

It is estimated that recycling of aluminum for cans requires overall less than 1% of the cost of manufacturing new aluminum.

---

circuit is completed by immersing an inert anode in the solution, and gold is reduced at the cathode for 60. min at a current of 0.25 A.

$$Au^{3+}(aq) + 3 \text{ e}^- \longrightarrow Au(s)$$

The mass of gold that is reduced is calculated by

$$0.25 \text{ A} \times 60. \text{ min} \times \left(\frac{60 \text{ s}}{1 \text{ min}}\right)\left(\frac{1 \text{ C}}{1 \text{ A s}}\right)\left(\frac{1 \text{ mol e}^-}{9.65 \times 10^4 \text{ C}}\right)$$

$$\times \left(\frac{1 \text{ mol Au}}{3 \text{ mol e}^-}\right)\left(\frac{197 \text{ g Au}}{1 \text{ mol Au}}\right) = 0.61 \text{ g Au}$$

That's about $7.83 worth of gold, assuming that gold is selling for $400 per ounce.

Exercise 19.15   **Electroplating Silver**

Calculate the mass of silver that could be plated from solution with a current of 0.50 A for 20. min. The cathode reaction is $Ag^+(aq) + \text{e}^- \longrightarrow Ag(s)$.

Oscar is gold-plated. Most gold jewelry is made by plating a thin coating of gold onto a base metal.

CD-ROM Screen 21.9: Corrosion: Redox Reactions in the Environment

19.13 CORROSION—PRODUCT-FAVORED REACTIONS

Corrosion is the oxidation of a metal that is exposed to the environment. Visible corrosion on the steel supports of a bridge, for example, indicates possible structural failure. Corrosion reactions are invariably product-favored. This means that $E°$ for the reaction is positive and $\Delta G° < 0$. Corrosion of iron, for example, takes place quite readily and is difficult to prevent. It results in the red-brown substance we call rust, which is hydrated iron(III) oxide ($Fe_2O_3 \cdot xH_2O$, where x varies from 2 to 4). The rust that forms when iron corrodes does not adhere to the surface of the metal, so it can easily fall away and expose more metal surface to corrosion (Figure 19.26). The corrosion of aluminum, a metal that is even more reactive than iron, is also very product-favored. The aluminum oxide that forms as a result of corrosion adheres tightly as a thin coating on the surface of the metal and actually forms a protective coating that prevents further corrosion.

For corrosion of a metal (M) to occur, there must be an anodic area where the oxidation of the metal can occur. The general reaction is

Anode reaction: $$M(s) \longrightarrow M^{n+} + n \text{ e}^-$$

There must also be a cathodic area where electrons are consumed. Frequently, the cathode reactions are reductions of oxygen or water.

Corrosion is so commonplace that about 25% of the annual steel production in the United States is destined for replacement of material lost to corrosion.

Figure 19.26 Rusting. The formation of rust destroys the structural integrity of objects made of iron and steel. Given time, this chain will completely rust away.

Cathode reactions:

$$O_2(g) + 2\,H_2O(\ell) + 4\,e^- \longrightarrow 4\,OH^-(aq)$$

$$2\,H_2O(\ell) + 2\,e^- \longrightarrow 2\,OH^-(aq) + H_2(g)$$

Anodic areas can occur at cracks in the oxide coating that protects the surfaces of many metals, and they may also occur around impurities. Cathodic areas occur at the metal oxide coating, at less reactive metallic impurity sites, or around other metal compounds trapped at the surface, such as sulfides or carbides.

The other requirements for corrosion are an electrical connection between the anode and cathode and an electrolyte with which both anode and cathode are in contact. Both requirements are easily fulfilled—the metal itself is the conductor, and ions dissolved in moisture from the environment provide the electrolyte.

In the corrosion of iron, the anodic reaction is the oxidation of metallic iron (Figure 19.27). If both water and O_2 gas are present, the cathode reaction is the reduction of oxygen, giving the net reaction

$$2\,[Fe(s) \longrightarrow Fe^{2+}(aq) + 2\,e^-] \quad \text{(anode reaction)}$$

$$\underline{O_2(g) + 2\,H_2O(\ell) + 4\,e^- \longrightarrow 4\,OH^-(aq)} \quad \text{(cathode reaction)}$$

$$2\,Fe(s) + O_2(g) + 2\,H_2O(\ell) \longrightarrow 2\,Fe(OH)_2(s)$$

iron(II) hydroxide

In the presence of an ample supply of oxygen and water, as in the open air or in flowing water, the iron(II) hydroxide is oxidized to the red-brown iron(III) oxide (Figure 19.26).

$$4\,Fe(OH)_2(s) + O_2(g) \longrightarrow 2\,Fe_2O_3 \cdot 2H_2O(s)$$

Red-brown

This hydrated iron oxide is the familiar rust you see on iron and steel objects and the substance that colors the water red in some mountain streams or in your home water supply at times. It is easily removed by mechanical shaking, rubbing, or even the action of rain or freeze-thaw cycles, thus exposing more iron at the surface and allowing the objects to eventually deteriorate completely.

Other substances in air and water can hasten corrosion. Metal salts, such as the chlorides of sodium and calcium from sea air or from salt spread on roadways in the winter, function as salt bridges between anodic and cathodic regions, thus speeding up corrosion reactions.

 Exercise 19.16   Do All Metals Corrode?

Do all metals corrode as readily as iron and aluminum? Name three metals that you would expect to corrode about as readily as iron and aluminum, and name three metals that do not corrode readily. Name a use for each of the three noncorroding metals. Explain why metals fall into these two groups.

Figure 19.27 Corroding iron nails. Two nails were placed in an agar gel, which also contained the indicator phenolphthalein and $[Fe(CN)_6]^{3-}$. The nails began to corrode and produced Fe^{2+} ions at the tip and where the nail is bent. (These are points of stress and corrode more quickly.) These points are the anode, as indicated by the formation of the blue-colored compound called Prussian blue ($Fe_3[Fe(CN)_6]_2$). The remainder of the nail is the cathode, since oxygen is reduced in water to give OH^-. The presence of OH^- ions causes the phenolphthalein to turn pink.

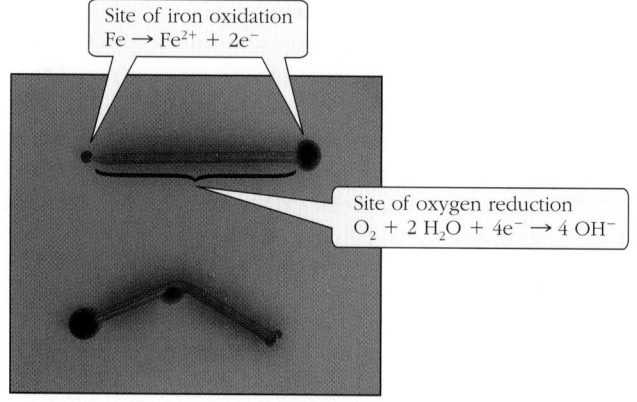

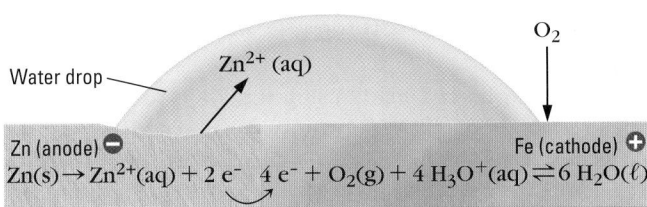

Figure 19.28 Cathodic protection of an iron-containing object. The iron is coated with a film of zinc, a metal more easily oxidized than iron. Therefore, the zinc acts as an anode and forces iron to become the cathode, thereby preventing the corrosion of the iron.

Corrosion Protection

How can you stop a metal object from corroding? The general approaches are to (a) inhibit the anodic process, (b) inhibit the cathodic process, or (c) do both. The most common method is **anodic inhibition,** which directly limits or prevents the oxidation half-reaction by painting the metal surface, coating it with grease or oil, or allowing a thin film of metal oxide to form. More recently developed methods of anodic protection are illustrated by the following reaction, in which the surface is treated with a solution of sodium chromate.

$$2\,Fe(s) + 2\,Na_2CrO_4(aq) + 2\,H_2O(\ell) \longrightarrow Fe_2O_3(s) + Cr_2O_3(s) + 4\,NaOH(aq)$$

The surface iron is oxidized by the chromate salt to give Fe(III) and Cr(III) oxides. These together form a coating that is impervious to O_2 and water, and further atmospheric oxidation is inhibited.

Cathodic protection is accomplished by forcing the metal to become the cathode instead of the anode. Usually, this is achieved by attaching another, more readily oxidized metal to the metal being protected. The best example of this is galvanized iron — iron that has been coated with a thin film of zinc (Figure 19.28). $E°_{ox}$ for zinc oxidation is considerably more positive than $E°_{ox}$ for iron oxidation. Therefore, the zinc metal film is oxidized before any of the iron and the zinc coating forms what is called a *sacrificial anode.* In addition, when the zinc is corroded, $Zn(OH)_2$ forms an insoluble film on the surface (K_{sp} of $Zn(OH)_2 = 4.5 \times 10^{-17}$) that further slows corrosion.

Galvanized objects. A thin coating of zinc helps prevent the oxidation of iron.

To get $E°_{ox}$ values, look at Table 19.1, but remember to change the sign of $E°_{red}$, since the reactions in the table are written as reductions.

Exercise 19.17   Corrosion Rates

Rank the following environments for their relative rates of corrosion of iron. Place the fastest first. Explain your answers.

(a) Moist clay   (b) Sand by the seashore
(c) The surface of the moon   (d) Desert sand in Arizona

SUMMARY PROBLEM

Many kinds of secondary batteries are known. Arguably the most useful is the lead-acid storage battery. If it were not for the density of lead, this battery would find far greater acceptance. Many electric vehicles currently use lead-acid storage batteries as their source of power, but automotive engineers look longingly at other batteries because of their lower weights. When you look at Table 19.1 and consider the chemical properties of all of the other oxidizing and reducing agents shown there, you might be tempted to try and create a hybrid battery that would combine some of the desirable features of, say, a PbO_2 cathode and some other kind of anode rather than the anode made of Pb found in the lead-acid storage battery. In that way, at least the high reduction potential of the half-reaction involving PbO_2 might still be used.

(a) What would be the $E°$ value of a cell made using the PbO_2 reduction reaction and magnesium metal as the reducing agent? Write the two half-reactions and the net cell reaction. Would this cell be the basis for a secondary battery? Explain your answer.

(b) What would be the $E°$ value of a cell made using the PbO_2 reduction reaction and nickel metal as the reducing agent? Write the two half-reactions and the net cell reaction. Would this cell have a voltage greater than or less than that of a single cell of a lead-acid storage battery? Would this cell be the basis of a secondary battery? Explain. What could you do to the chemistry in the anode compartment to make it a secondary battery?

(c) If your Ni/PbO_2 hybrid battery were a success and it was manufactured for use in electric automobiles, how many amperes could it produce, assuming that 500.0 g of Ni reacted in exactly 30 min? How much PbO_2 would be reduced during this same amount of time?

(d) Of course, batteries must be recharged, so how much time would be required to recharge your Ni/PbO_2 battery to its original state (the 500.0 g of Ni being converted back to its original form) if a current of 25.5 A is passed through the battery?

(e) Just as you are getting ready to cash in on the success of your new battery, someone announces that it has some serious environmental problems. What could these be? Explain.

IN CLOSING

Having studied this chapter, you should be able to . . .

- Identify the oxidizing and reducing agents in a redox reaction (Section 19.1).
- Write equations for the oxidation and reduction half-reactions, and use them to balance the net equation (Section 19.2).
- Identify and describe the functions of the parts of an electrochemical cell; describe the direction of electron flow outside the cell and the direction of the ion flow inside the cell (Section 19.3).
- Describe how standard reduction potentials are defined and use them to predict whether a reaction will be product-favored as written (Sections 19.4 and 19.5).
- Calculate $\Delta G°$ from the value of $E°$ for a redox reaction (Section 19.6).
- Explain how product-favored electrochemical reactions can be used to do useful work, and list the requirements for using such reactions in rechargeable batteries (Section 19.6).
- Explain how the Nernst equation relates concentrations of redox reactants to E_{cell} (Section 19.7).
- Use the Nernst equation to calculate the potentials of cells that are not at standard conditions (Section 19.7).
- Explain the source of the equilibrium potential across the membrane of a neuron cell (Section 19.8).
- Describe the chemistry of the dry cell, the mercury battery, and the lead-acid storage battery (Section 19.9).
- Describe how a fuel cell works, and indicate how it is different from a battery (Section 19.10).
- Use standard reduction potentials to predict the products of electrolysis of an aqueous salt solution (Section 19.11).
- Calculate the quantity of product formed at an electrode during an electrolysis reaction, given the current passing through the cell and the time during which the current flows (Section 19.12).
- Explain how electroplating works (Section 19.12).
- Describe what corrosion is and how it can be prevented by cathodic protection (Section 19.13).

KEY TERMS

ampere *(19.4)*

anode *(19.3)*

anodic inhibition
(19.13)

battery *(19.3)*

cathode *(19.3)*

cathodic protection
(19.13)

cell voltage *(19.4)*

concentration cell
(19.7)

corrosion *(19.13)*

coulomb *(19.4)*

electrochemical cell
(19.3)

electrochemistry
(Introduction)

electrode *(19.3)*

electrolysis *(19.11)*

electromotive force
(emf) *(19.4)*

emf *(19.4)*

Faraday constant *(19.6)*

fuel cell *(19.10)*

half-cell *(19.3)*

half-reaction *(19.1)*

Nernst equation *(19.7)*

neurons *(19.8)*

pH meter *(19.7)*

primary battery *(19.9)*

salt bridge *(19.3)*

secondary battery
(19.9)

standard conditions
(19.4)

standard hydrogen
electrode *(19.4)*

standard reduction po-
tential *(19.5)*

standard voltages (E^o)
(19.4)

volt (V) *(19.4)*

voltaic cell *(19.3)*

QUESTIONS FOR REVIEW AND THOUGHT

Conceptual Challenge Problems

CP-19.A (Section 19.6) Automobiles run on internal combustion engines, in which the energy used to run the vehicle is obtained from the combustion of gasoline. The main component of gasoline is octane (C_8H_{18}). An automobile manufacturer has recently announced a chemical method for generating hydrogen gas from gasoline and proposes to develop a car in which an H_2/O_2 fuel cell powers an electric propulsion motor, thus eliminating the internal combustion engine with its problems (for example, the generation of unwanted byproducts that pollute the air). The hydrogen for the fuel cell would be directly generated from gasoline on board the vehicle. There are two steps in this hydrogen generation process:

  (i) Partial oxidation of octane by oxygen to carbon monoxide and hydrogen.

  (ii) Combination of carbon monoxide with additional gaseous water to form carbon dioxide and more hydrogen (the water-gas shift reaction).

(a) Write the chemical equation for the complete combustion of 1 mol of octane.

(b) Write balanced chemical equations for the two-step hydrogen generation process. How many moles of H_2 are produced per mole of octane? (Remember that water is a reactant in the two-step process.)

(c) By combining these equations, show that the net *overall* reaction is the same as in the combustion of octane.

(d) By assuming that the entire Gibbs free energy change of the H_2/O_2 fuel cell reaction is available for use by the electric propulsion motor, calculate the energy produced by a fuel cell when it consumes all of the hydrogen produced from 1 mol of octane. Compare this energy with the Gibbs free energy change for the combustion of 1 mol of octane. (*Note:* The Gibbs free energy of formation, ΔG_f°, for $C_8H_{18}(\ell)$ is 6.14 kJ/mol.)

CP-19.B (Section 19.4) People obtain energy by oxidizing food. Glucose is a typical foodstuff. It is a carbohydrate that is oxidized to water and carbon dioxide.

$$C_6H_{12}O_6(aq) + 6\,O_2(g) \longrightarrow 6\,CO_2(g) + 6\,H_2O(\ell)$$

The heat of combustion of glucose is 2.80×10^3 kJ/mol, which means that as glucose is oxidized, its electrons lose 2.80×10^3 kJ/mol as they give up potential energy in a complicated series of chemical steps.

(a) Assume that a person requires 2400 food Calories per day and that they are obtained from the oxidation of glucose. How much O_2 must a person breathe each day to react with this much glucose?

(b) Each mol of O_2 requires 4 mol of electrons, regardless of whether the O atoms become part of CO_2 or H_2O. What would be the average electrical current (C/s) in a human body using the above amount of energy per day?

(c) Use the answer from part (b) and calculate the electrical potential this current flows through in a day to produce the 2400 food Calories. (1 Calorie = 4.18 kJ.)

CP-19.C (Section 19.11) A piece of chromium metal is attached to a battery and dipped into 50 mL of 0.3 M KOH solution in a 250-mL beaker. A stainless steel electrode is connected to the other electrode of the battery and immersed in the same solution. A steady current of 0.50 A is maintained for exactly 2 hr. Several samples of a gas formed at the stainless steel electrode during the electrolysis are captured, and all are found to ignite in air. After the electrolysis, the chromium electrode is weighed and found to have decreased in weight by 0.321 g. The mass of stainless steel electrode does not change.

After electrolysis, the KOH solution is neutralized with nitric acid to a pH of just below 7, then is heated and reacted with 0.151 M lead(II) nitrate solution. As the lead(II) nitrate solution is added, a yellow precipitate quickly forms

from the hot solution. The formation of precipitate stops after 40.4 mL of the lead(II) nitrate solution has been added. The yellow solid is then filtered, dried, and weighed. Its mass is 1.97 g.
(a) How much electrical charge passes through the cell?
(b) How many moles of Cr react?
(c) What is the oxidation state of the Cr after reacting?

(d) Assuming that the yellow compound that precipitates from the solution during the titration contains both Pb and Cr, what do you conclude to be the ratio of the numbers of atoms of Pb and Cr?
(e) If the yellow compound contains an element other than Pb and Cr, what is it and how much is in the compound? What is the formula for the yellow compound?

Answers to questions in **bold** can be found in the back of the book.

Review Questions

1. Describe the principal parts of an electrochemical cell by drawing a hypothetical cell, indicating the cathode, the anode, the direction of electron flow outside the cell, and the direction of ion flow within the cell.
2. Explain how product-favored electrochemical reactions can be used to do useful work.
3. Explain how reactant-favored electrochemical reactions can be induced to proceed and make products.
4. Explain how electroplating works.
5. Tell whether each of the following statements is true or false. If false, rewrite it to make it a correct statement.
   (a) Oxidation always occurs at the anode of an electrochemical cell.
   (b) The anode of a battery is the site of reduction and is negative.
   (c) Standard conditions for electrochemical cells are a concentration of 1.0 M for dissolved species and a pressure of 1 bar for gases.
   (d) The potential of a cell does not change with temperature.
   (e) All product-favored oxidation-reduction reactions have a standard cell voltage $E°_{cell}$ with a negative sign.

Redox Reactions

6. In each of the following reactions, tell which substance is oxidized and which is reduced. Tell which is the oxidizing agent and which is the reducing agent. Assign oxidation numbers to all species.
   (a) $2 Al(s) + 3 Cl_2(g) \rightarrow 2 AlCl_3(s)$
   (b) $8 H_3O^+(aq) + MnO_4^-(aq) + 5 Fe^{2+}(aq) \rightarrow$
   $5 Fe^{3+}(aq) + Mn^{2+}(aq) + 12 H_2O(\ell)$
   (c) $FeS(s) + 3 NO_3^-(aq) + 4 H_3O^+(aq) \rightarrow$
   $3 NO(g) + SO_4^{2-}(aq) + Fe^{3+}(aq) + 6 H_2O(\ell)$
7. In each of the following reactions, tell which substance is oxidized and which is reduced. Tell which is the oxidizing agent and which is the reducing agent. Assign oxidation numbers to all species.
   (a) $Fe(s) + Br_2(\ell) \rightarrow FeBr_2(s)$
   (b) $8 HI(aq) + H_2SO_4(aq) \rightarrow$
   $H_2S(aq) + 4 I_2(s) + 4 H_2O(\ell)$
   (c) $H_2O_2(aq) + 2 Fe^{2+}(aq) + 2 H_3O^+(aq) \rightarrow$
   $2 Fe^{3+}(aq) + 4 H_2O(\ell)$
8. Choose four elements: a regular metal, a transition metal, a nonmetal, and a metalloid. Using the index to this text, find a chemical reaction in which each element occurs as a reactant. Assign oxidation numbers to all elements on the reactant and product sides, and identify the oxidizing agent and the reducing agent.
9. Answer Question 8 again, only this time find a chemical reaction in which each element is produced.

Using Half-Reactions to Understand Redox Reactions

10. Write half-reactions for the following:
    (a) Oxidation of zinc to Zn^{2+} ion
    (b) Reduction of H_3O^+ ion to hydrogen gas
    (c) Reduction of Sn^{4+} ion to Sn^{2+} ion
    (d) Reduction of chlorine to Cl^- ion
    (e) Oxidation of sulfur dioxide to sulfate ion in acidic solution
11. Write half-reactions for the following:
    (a) Reduction of MnO_4^- ion to Mn^{2+} ion in acid solution
    (b) Reduction of $Cr_2O_7^{2-}$ ion to Cr^{3+} ion in acid solution
    (c) Oxidation of hydrogen gas to H_3O^+ ion
    (d) Reduction of hydrogen peroxide to water in acidic solution
    (e) Oxidation of nitric oxide to nitrogen monoxide in acidic solution
12. For each reaction in Question 6, write balanced half-reactions.
13. For each reaction in Question 7, write balanced half-reactions.
14. Balance the following redox reactions, and identify the oxidizing agent and the reducing agent.
    (a) $CO(g) + O_3(g) \rightarrow CO_2(g)$
    (b) $H_2(g) + Cl_2(g) \rightarrow HCl(g)$
    (c) $H_2O_2(aq) + Ti^{2+}(aq) \rightarrow H_2O(\ell) + Ti^{4+}(aq)$ in acidic solution
    (d) $Cl^-(aq) + MnO_4^-(aq) \rightarrow Cl_2(g) + MnO_2(s)$ in acidic solution
    (e) $FeS_2(s) + O_2(g) \rightarrow Fe_2O_3(s) + SO_2(g)$
    (f) $O_3(g) + NO(g) \rightarrow O_2(g) + NO_2(g)$
    (g) $Zn(Hg)(amalgam) + HgO(s) \rightarrow ZnO(s) + Hg(\ell)$ in basic solution (This is the reaction in the mercury battery.)
15. Balance the following redox reactions, and identify the oxidizing agent and the reducing agent.
    (a) $FeO(s) + O_3(g) \rightarrow Fe_2O_3(s)$
    (b) $P_4(s) + Br_2(\ell) \rightarrow PBr_5(\ell)$
    (c) $H_2O_2(aq) + Co^{2+}(aq) \rightarrow H_2O(\ell) + Co^{3+}(aq)$ in acidic solution
    (d) $Cl^-(aq) + Cr_2O_7^{2-}(aq) \rightarrow Cl_2(g) + Cr^{3+}(aq)$ in acidic solution
    (e) $CuFeS_2(s) + O_2(g) \rightarrow Cu_2S(s) + FeO(s) + SO_2(g)$
    (f) $H_2CO(g) + O_2(g) \rightarrow CO_2(g) + H_2O(\ell)$

(g) $C_3H_8(g) + O_2(g) \rightarrow CO_2(g) + H_2O(\ell)$ in acidic solution (This is the reaction occurring in a propane fuel cell.)

Electrochemical Cells

16. For the redox reaction $Cu^{2+}(aq) + Zn(s) \rightarrow Cu(s) + Zn^{2+}(aq)$, why can't you generate electrical current by placing a piece of copper metal and a piece of zinc metal in a solution containing $CuCl_2(aq)$ and $ZnCl_2(aq)$?

17. Explain the function of a salt bridge in an electrochemical cell.

18. Are standard half-cell reactions always written as oxidation reactions or reduction reactions?

19. Tell whether the following statement is true or false. If false, rewrite it to make it a correct statement: The value of an electrode potential changes when the half-reaction is multiplied by a factor. That is, $E°$ for $Li^+ + e^- \rightarrow Li$ is different from that for $2 Li^+ + 2 e^- \rightarrow 2 Li$.

20. A voltaic cell is assembled with $Pb(s)$ and $Pb(NO_3)_2(aq)$ in one compartment and $Zn(s)$ and $ZnCl_2(aq)$ in the other. An external wire connects the two electrodes, and a salt bridge containing KNO_3 connects the two solutions.
(a) In the product-favored reaction, zinc metal is oxidized to Zn^{2+}. Write a balanced net ionic equation for this reaction.
(b) Which half-reaction occurs at each electrode? Which is the anode and which is the cathode?
(c) Draw a diagram of the cell, indicating the direction of electron flow outside the cell and of ion flow within the cell.

21. A voltaic cell is assembled with $Sn(s)$ and $Sn(NO_3)_2(aq)$ in one compartment and $Ag(s)$ and $AgCl(aq)$ in the other. An external wire connects the two electrodes, and a salt bridge containing KNO_3 connects the two solutions.
(a) In the product-favored reaction, Ag^+ is reduced to silver metal. Write a balanced net ionic equation for this reaction.
(b) Which half-reaction occurs at each electrode? Which is the anode and which is the cathode?
(c) Draw a diagram of the cell, indicating the direction of electron flow outside the cell and of ion flow within the cell.

Electrochemical Cells and Voltage

22. You light a 25-W light bulb with the current from a 12-V lead-acid storage battery. After 1.0 hr of operation, how much energy has the light bulb utilized? How many coulombs of charge have been withdrawn from the battery? Assume 100% efficiency. (A watt is the transfer of 1 J of energy in 1 s.)

23. Draw a diagram of a standard hydrogen electrode and describe how it works.

24. Copper can reduce silver ion to metallic silver, a reaction that could in principle be used in a battery.

$$Cu(s) + 2 Ag^+(aq) \longrightarrow Cu^{2+}(aq) + 2 Ag(s)$$

(a) Write equations for the half-reactions involved.
(b) Which half-reaction is an oxidation and which is a reduction? Which half-reaction occurs in the anode compartment and which in the cathode compartment?

25. Chlorine gas can oxidize zinc metal in a reaction that has been suggested as the basis of a battery. Write the half-reactions involved. Label which is the oxidation and which is the reduction reaction.

Using Standard Cell Potentials

26. What is the strongest oxidizing agent in Table 19.1? What is the strongest reducing agent? What is the weakest oxidizing agent? What is the weakest reducing agent?

27. Using the reduction potentials in Table 19.1, place the following elements in order of increasing ability to function as reducing agents:
(a) Cl_2 (b) Fe   (c) Ag   (d) Na   (e) H_2

28. Using the reduction potentials in Table 19.1, place the following elements in order of increasing ability to function as oxidizing agents:
(a) O_2 (b) H_2O_2 (c) $PbSO_4$ (d) H_2O

29. One of the most energetic redox reactions is that between F_2 gas and lithium metal.
(a) Write the half-reactions involved. Label which is the oxidation and which is the reduction reaction.
(b) According to data from Table 19.1, what is $E°$ for this reaction?

30. Calculate the value of $E°$ for each of the following reactions. Decide whether each is product-favored.
(a) $I_2(s) + Mg(s) \rightarrow Mg^{2+}(aq) + 2 I^-(aq)$
(b) $Ag(s) + Fe^{3+}(aq) \rightarrow Ag^+(aq) + Fe^{2+}(aq)$
(c) $Sn^{2+}(aq) + 2 Ag^+(aq) \rightarrow Sn^{4+}(aq) + 2 Ag(s)$
(d) $2 Zn(s) + O_2(g) + 2 H_2O(\ell) \rightarrow$
$ 2 Zn^{2+}(aq) + 4 OH^-(aq)$

31. Consider the following half-reactions:

Half-reaction	$E°$ (V)
$Au^{3+}(aq) + 3 e^- \rightarrow Au(s)$	1.50
$Pt^{2+}(aq) + 2 e^- \rightarrow Pt(s)$	1.2
$Co^{2+}(aq) + 2 e^- \rightarrow Co(s)$	-0.28
$Mn^{2+}(aq) + 2 e^- \rightarrow Mn(s)$	-1.18

(a) Which is the weakest oxidizing agent?
(b) Which is the strongest oxidizing agent?
(c) Which is the strongest reducing agent?
(d) Which is the weakest reducing agent?
(e) Will $Co^{2+}(aq)$ reduce $Pt^{2+}(aq)$ to $Pt(s)$?
(f) Will $Pt(s)$ reduce $Co^{2+}(aq)$ to $Co(s)$?
(g) Which ions can be reduced by $Co(s)$?

32. Consider the following half-reactions:

Half-reaction	$E°$ (V)
$Ce^{4+}(aq) + e^- \rightarrow Ce^{3+}(aq)$	1.61
$Ag^+(aq) + e^- \rightarrow Ag(s)$	0.80
$Hg_2^{2+}(aq) + 2 e^- \rightarrow 2 Hg(\ell)$	0.79
$Sn^{2+}(aq) + 2 e^- \rightarrow Sn(s)$	-0.14
$Ni^{2+}(aq) + 2 e^- \rightarrow Ni(s)$	-0.25
$Al^{3+}(aq) + 3 e^- \rightarrow Al(s)$	-1.66

(a) Which is the weakest oxidizing agent?

(b) Which is the strongest oxidizing agent?

(c) Which is the strongest reducing agent?

(d) Which is the weakest reducing agent?

(e) Will $Sn(s)$ reduce $Ag^+(aq)$ to $Ag(s)$?

(f) Will $Hg(\ell)$ reduce $Sn^{2+}(aq)$ to $Sn(s)$?

(g) Name the ions that can be reduced by $Sn(s)$.

(h) What metals can be oxidized by $Ag^+(aq)$?

33. In principle, a battery could be made from aluminum metal and chlorine gas.

(a) Write a balanced equation for the reaction that would occur in a battery using $Al^{3+}(aq)/Al(s)$ and $Cl_2(g)/Cl^-(aq)$ half-reactions.

(b) Tell which half-reaction occurs at the anode and which at the cathode. What are the polarities of these electrodes?

(c) Calculate the standard potential, $E°$, for the battery.

E° and Gibbs Free Energy

34. Choose the correct answers: In a product-favored chemical reaction, the standard cell potential, $E°$, is (greater/less) than zero, and the Gibbs free energy change, $\Delta G°$, is (greater/less) than zero.

35. For each of the reactions in Question 30, compute the Gibbs free energy change, $\Delta G°$.

36. Hydrazine, N_2H_4, can be used as the reducing agent in a fuel cell.

$$N_2H_4(aq) + O_2(g) \longrightarrow N_2(g) + 2 H_2O(\ell)$$

(a) If $\Delta G°$ for the reaction is -598 kJ, calculate the value of $E°$ expected for the reaction.

(b) Suppose the reaction is written with all coefficients doubled. Determine $\Delta G°$ and $E°$ for this new reaction.

37. The standard cell potential for the oxidation of Mg by Br_2 is 3.45 V.

$$Br_2(\ell) + Mg(s) \longrightarrow Mg^{2+}(aq) + 2 Br^-(aq)$$

(a) Calculate $\Delta G°$ for this reaction.

(b) Suppose the reaction is written with all coefficients doubled. Determine $\Delta G°$ and $E°$ for this new reaction.

38. The standard cell potential, $E°$, for the reaction of $Zn(s)$ and $Cl_2(g)$ is 2.12 V. Write the chemical equation for the reaction of 1 mol of zinc. What is the standard free energy change, $\Delta G°$, for this reaction?

39. What is the equilibrium constant and $\Delta G°$ for the reaction between $Cd(s)$ and $Cu^{2+}(aq)$?

40. What is the equilibrium constant and $\Delta G°$ for the reaction between $I_2(s)$ and $Br^-(aq)$?

41. What is the equilibrium constant and $\Delta G°$ for the reaction between $Ag(s)$ and $Zn^{2+}(Aq)$?

42. What is the equilibrium constant and $\Delta G°$ for the reaction between $Cl_2(g)$ and $Br^-(aq)$?

Effect of Concentration on Cell Potential

43. Consider the voltaic cell

$$Zn(s) + Cd^{2+}(aq) \longrightarrow Zn^{2+}(aq) + Cd(s)$$

operating at 298 K.

(a) What is the $E°_{cell}$ for this cell?

(b) If $E_{cell} = 0.390$ and $[Cd^{2+}] = 2.00$ M, what is $[Zn^{2+}]$?

(c) If $[Cd^{2+}] = 0.068$ M and $[Zn^{2+}] = 1.00$ M, what is E_{cell}?

44. Consider the voltaic cell

$$2 Ag^+(aq) + Cd(s) \longrightarrow 2 Ag(s) + Cd^{2+}(aq)$$

operating at 298 K.

(a) What is the $E°_{cell}$ for this cell?

(b) If $[Cd^{2+}] = 2.0$ M and $[Ag^+] = 0.25$ M, what is E_{cell}?

(c) If $E_{cell} = 1.25$ V and $[Cd^{2+}] = 0.100$ M, what is $[Ag^+]$?

45. For the reaction

$$H_2(g) + Sn^{4+}(aq) \longrightarrow 2 H^+(aq) + Sn^{2+}(aq)$$

operating at 298 K,

(a) What is the $E°_{cell}$ for this cell?

(b) What is the E_{cell} for $P_{H_2} = 1.0$ bar, $[Sn^{2+}] = 6.0 \times 10^{-4}$, $[Sn^{4+}] = 5.0 \times 10^{-4}$, and pH = 3.60?

46. What is the cell potential of a concentration cell that contains two hydrogen electrodes if the cathode contacts a solution with pH = 7.8 and the anode contacts a solution with $[H^+] = 0.05$ M?

Common Batteries

47. What are the advantages and disadvantages of lead-acid storage batteries?

48. Nicad batteries are rechargeable and are commonly used in cordless appliances. Although such batteries actually function under basic conditions, imagine an electrochemical cell using the following setup.

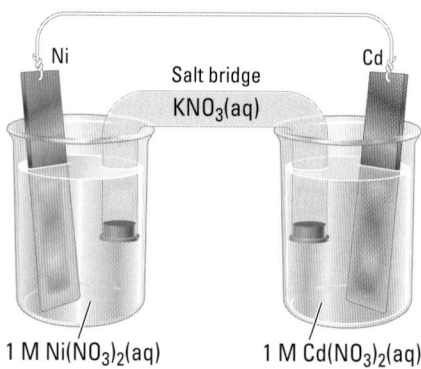

(a) Write a balanced net ionic equation depicting the reaction occurring in the cell.

(b) What is oxidized? What is reduced? What is the reducing agent and what is the oxidizing agent?

(c) Which is the anode and which is the cathode?

(d) What is $E°$ for the cell?

(e) What is the direction of electron flow in the external wire?

(f) If the salt bridge contains KNO_3, toward which compartment will the NO_3^- ions migrate?

49. Consider the nicad cell in the previous question.

(a) If the concentration of Cd^{2+} is reduced to 0.010 M, and $[Ni^{2+}] = 1.0$ M, will the cell emf be smaller or larger than when the concentration of $Cd^{2+}(aq)$ was 1.0 M? Explain your answer in terms of Le Chatelier's principle.

(b) Begin with 1.0 L of each of the solutions, both initially 1.0 M in dissolved species. Each electrode weighs 50.0 g at the start. If 0.050 A is drawn from the battery, how long can it last?

Fuel Cells

50. How does a fuel cell differ from a battery?

51. Describe the principal parts of an H_2/O_2 fuel cell. What is the reaction at the cathode? At the anode? What is the product of the fuel cell reaction?

52. Hydrazine, N_2H_4, has been proposed as the fuel in a fuel cell in which oxygen is the oxidizing agent. The reactions are

$$N_2H_4(aq) + 4\,OH^-(aq) \longrightarrow N_2(g) + 4\,H_2O(\ell) + 4\,e^-$$

$$O_2(g) + 2\,H_2O(\ell) + 4\,e^- \longrightarrow 4\,OH^-(aq)$$

(a) Which reaction occurs at the anode and which at the cathode?

(b) What is the net cell reaction?

(c) If the cell is to produce 0.50 A of current for 50.0 hr, what mass in grams of hydrazine must be present?

(d) What mass in grams of O_2 must be available to react with the mass of N_2H_4 determined in part (c)?

Electrolysis: Reactant-Favored Reactions

53. Write chemical equations for the electrolysis of molten salts of three different alkali halides to produce the corresponding halogens and alkali metals.

54. From Table 19.1 write down all of the aqueous metal ions that can be reduced by electrolysis to the corresponding metal.

55. From Table 19.1 write down all of the species that can be oxidized by electrolysis, and determine the products.

56. What are the products of the electrolysis of a 1 M aqueous solution of NaBr? What species are present in the solution? What is formed at the cathode? What is formed at the anode?

57. For each of the following solutions, tell what reactions take place at the anode and at the cathode during electrolysis.

(a) $NiBr_2(aq)$ (b) $NaI(aq)$ (c) $CdCl_2(aq)$

(d) $CuI_2(aq)$ (e) $MgF_2(aq)$ (f) $HNO_3(aq)$

Counting Electrons

58. A current of 0.015 A is passed through a solution of $AgNO_3$ for 155 min. What mass of silver is deposited at the cathode?

59. Current is passed through a solution containing $Ag^+(aq)$. How much silver was in the solution if all the silver was removed as Ag metal by electrolysis for 14.5 min at a current of 1.0 mA?

60. A current of 2.50 A is passed through a solution of $Cu(NO_3)_2$ for 2.00 hr. What mass of copper is deposited at the cathode?

61. A current of 0.0125 A is passed through a solution of $CuCl_2$ for 2.00 hr. What mass of copper is deposited at the cathode and what volume of Cl_2 gas (in mL at STP) is produced at the anode?

62. The major reduction half-reaction occurring in the cell in which Al_2O_3 and aluminum salts are electrolyzed is $Al^{3+}(aq) + 3\,e^- \longrightarrow Al(s)$. If the cell operates at 5.0 V and 1×10^5 A, what mass (in grams) of aluminum metal can be produced in 8.0 hr?

63. The vanadium(II) ion can be produced by electrolysis of a vanadium(III) salt in solution. How long must you carry out an electrolysis if you wish to convert completely 0.125 L of 0.015 M $V^{3+}(aq)$ to $V^{2+}(aq)$ using a current of 0.268 A?

64. The reactions occurring in a lead-storage battery are given in Section 19.9. A typical battery might be rated at 50 ampere-hours (A-hr). This means that it has the capacity to deliver 50. A for 1.0 hr or 1.0 A for 50. hr. If it does deliver 1.0 A for 50. hr, what mass of lead would be consumed to accomplish this?

65. It has been demonstrated that an effective battery can be built using the reaction between Al metal and O_2 from the air. If the Al anode of this battery consists of a 3-oz piece of aluminum (84 g), for how many hours can the battery produce 1.0 A of electricity?

66. A dry cell is used to supply a current of 250 mA for 20 min. What mass of Zn is consumed?

67. If the same current as in Question 66 were supplied by a mercury battery, what mass of Hg would be produced at the cathode?

68. Assuming that the anode reaction for the lithium battery is

$$Li(s) \longrightarrow Li^+(aq) + e^-$$

and the anode reaction for the lead-acid storage battery is

$$Pb(s) + HSO_4^-(aq) + H_2O(\ell) \longrightarrow$$
$$PbSO_4(s) + 2\,e^- + H_3O^+(aq)$$

compare the masses of metals consumed when each of these batteries supplies a current of 1.0 A for 10. min.

69. A hydrogen-oxygen fuel cell operates on the simple reaction

$$2\,H_2(g) + O_2(g) \longrightarrow 2\,H_2O(\ell)$$

If the cell is designed to produce 1.5 A of current, how long can it operate if there is an excess of oxygen and only sufficient hydrogen to fill a 1.0-L tank at 200. bar pressure at 25 °C?

70. Fluorine, F_2, is made by the electrolysis of anhydrous HF.

$$2\,HF(\ell) \longrightarrow H_2(g) + F_2(g)$$

Typical electrolysis cells operate at 4000 to 6000 A and 8 to 12 V. A large-scale plant can produce about 9 metric tons of F_2 gas per day.

(a) What mass in grams of HF is consumed?

(b) Using the conversion factor of 3.60×10^6 J/kWh, how much energy in kilowatt-hours is consumed by a cell operating at 6.0×10^3 A at 12 V for 24 hr?

Corrosion — Product-Favored Reactions

71. Explain how rust is formed from iron materials by corrosion.

72. What common metal does not corrode readily under normal conditions?

73. Why does coating a steel object with chromium stop corrosion of the iron?

74. Explain how galvanizing iron stops corrosion of the underlying iron.

General Questions

75. A 12-V automobile battery consists of six cells of the type described in Section 19.9. The cells are connected in series so that the same current flows through all of them. Calculate the theoretical minimum electrical potential difference needed to recharge an automobile battery. (Assume standard state concentrations.) How does this compare with the maximum voltage that could be delivered by the battery? Assuming that the lead plates in an automobile battery each weigh 2.5 kg and that there is sufficient PbO_2 available, what is the maximum possible work that could be obtained from the battery?

76. Three electrolytic cells are connected in series, so that the same current flows through all of them for 20. min. In cell A, 0.0234 g of Ag plates out from a solution of $AgNO_3(aq)$; cell B contains $Cu(NO_3)_2(aq)$; cell C contains $Al(NO_3)_3(aq)$. What mass of Cu will plate out in cell B? What mass of Al will plate out in cell C?

77. Fluorinated organic compounds are important commercially, as they are used as herbicides, flame retardants, and fire-extinguishing agents, among other things. A reaction such as

$$CH_3SO_2F + 3\,HF \longrightarrow CF_3SO_2F + 3\,H_2$$

is actually carried out electrochemically in liquid HF as the solvent.
(a) Draw the structural formula for CH_3SO_2F. (S is the "central" atom with the O atoms, F atom, and CH_3 group bonded to it.) What is the geometry around the S atom? What are the O—S—O and O—S—F bond angles?
(b) If you electrolyze 150 g of CH_3SO_2F, how many grams of HF are required and how many grams of each product can be isolated?
(c) Is H_2 produced at the anode or the cathode of the electrolysis cell?
(d) A typical electrolysis cell operates at 8.0 V and a low current, such as 250 A. How many kilowatt-hours of energy does one such cell consume in 24 hr?

Applying Concepts

78. Four metals, A, B, C, and D, exhibit the following properties:
(a) Only A and C react with 1.0 M HCl to give H_2 gas.
(b) When C is added to solutions of ions of the other metals, metallic A, B, and D are formed.
(c) Metal D reduces B^{n+} ions to give metallic B and D^{n+} ions.
On the basis of this information, arrange the four metals in order of increasing ability to act as reducing agents.

79. The table below lists the cell potentials for the ten possible electrochemical cells assembled from the elements A, B, C, D, and E and their respective ions in solutions.

Using the data in the table, establish a standard reduction potential table similar to Table 19.1. Assign a reduction potential of 0.00 V to the element that falls in the middle of the series.

	A(s) in A^{n+}(aq)	B(s) in B^{n+}(aq)
E(s) in E^{n+}(aq)	+0.21 V	+0.68 V
D(s) in D^{n+}(aq)	+0.35 V	+1.24 V
C(s) in C^{n+}(aq)	+0.58 V	+0.31 V
B(s) in B^{n+}(aq)	+0.89 V	—

	C(s) in C^{n+}(aq)	D(s) in D^{n+}(aq)
E(s) in E^{n+}(aq)	+0.37 V	+0.56 V
D(s) in D^{n+}(aq)	+0.93 V	—
C(s) in C^{n+}(aq)	—	—
B(s) in B^{n+}(aq)	—	—

80. When the electrochemical cell shown below runs for several hours, the green solution gets lighter and the yellow solution gets darker.

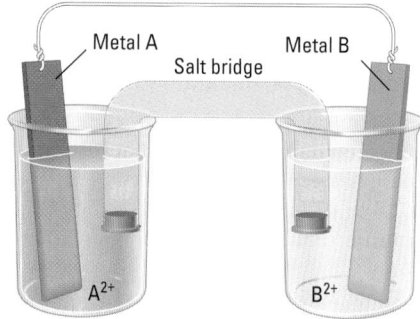

(a) What is oxidized, and what is reduced?
(b) What is the oxidizing agent, and what is the reducing agent?
(c) What is the anode, and what is the cathode?
(d) Write equations for the half-reactions.
(e) Which metal gains mass?
(f) What is the direction of the electron transfer through the external wire?
(g) If the salt bridge contains $KNO_3(aq)$, into which solution will the K^+ ions migrate?

81. An electrolytic cell is set up with Cd(s) in $Cd(NO_3)_2(aq)$ and Zn(s) in $Zn(NO_3)_2(aq)$. Initially both electrodes weigh 5.00 g. After running the cell for several hours the electrode in the left compartment weighs 4.75 g.
(a) Which electrode is in the left compartment?
(b) Does the mass of the electrode in the right compartment increase, decrease, or stay the same? If the mass changes, what is the new mass?
(c) Does the mass of the solution in the right compartment increase, decrease, or stay the same?
(d) Does the volume of the electrode in the right compartment increase, decrease, or stay the same? If the volume changes, what is the new volume? (Density of Cd is 8.65 g/cm^3)

General Chemistry CD-ROM

CD19.1 Screen 21.2: Redox Reactions: Electron Transfer.

(a) What is the difference between an oxidizing agent and a reducing agent?

(b) What is the difference between a direct redox reaction and an indirect redox reaction?

CD19.2 Screen 21.3: Balancing Equations for Redox Reactions.

(a) When first learning about balancing equations for reactions, you learned that the number of atoms of each element in the products and reactants must be equivalent. What are some additional factors that must be taken into account when balancing equations for redox reactions?

(b) What are half-reactions? What two aspects of these equations must be balanced?

CD19.3 Screen 21.4: Electrochemical Cells.

(a) What electron transfer reaction occurs at the anode of an electrochemical cell?

(b) What electron transfer reaction occurs at the cathode of an electrochemical cell?

(c) Why can an indirect redox reaction not occur if the wire connecting the two compartments is not attached to the electrodes?

CD19.4 Screen 21.7: Electrochemical Cells at Non-standard Conditions. Two factors are important in determining the potential of a cell: the nature of the reactants and the concentration of those reactants. Which of these plays the more important role? Explain your answer in terms of the Nernst equation.

CD19.5 Screen 21.8: Batteries.

(a) In what practical way do the batteries shown differ from the "two-beaker" cells shown thus far?

(b) What is the difference between a primary battery and a secondary battery?

CD19.6 Screen 21.9: Corrosion.

(a) When iron rusts, what is the reducing agent? What is the oxidizing agent?

(b) What is responsible for the fact that aluminum does not "rust": thermodynamics or kinetics?

20

A fuel element being removed at the High Flux Isotope Reactor at Oak Ridge National Laboratory. The fuel element and the reactor head are submerged in water in this design. The blue glow is due to Cerenkov radiation, emitted by energetic charged particles traveling through the water. This 100-megawatt reactor is used to conduct research on synthetic heavy elements, and its principal product is the isotope ^{252}Cf.

NUCLEAR CHEMISTRY

20.1 The Nature of Radioactivity

20.2 Nuclear Reactions

20.3 Stability of Atomic Nuclei

20.4 Rates of Disintegration Reactions

20.5 Artificial Transmutations

20.6 Nuclear Fission

20.7 Nuclear Fusion

20.8 Nuclear Radiation: Effects and Units

20.9 Applications of Radioactivity

Nuclear chemistry, a subject that bridges chemistry and physics, has a significant impact on our society. Radioactive isotopes are now widely used in medicine; PET (positron emission tomography) scans depend on radioactivity. Your home may be protected with a smoke detector that uses a radioactive element, and research in all fields of science employs radioactive elements and their compounds. The national security of the United States since World War II has depended on nuclear weapons, and more than 30 nations depend on nuclear reactors as a source of electricity. No matter what your reason for taking a college course in chemistry — to prepare for a career in one of the sciences or simply to gain knowledge as a concerned citizen — you should know about nuclear chemistry. Therefore, this chapter considers changes in atomic nuclei and their effects, the fissioning and fusion of nuclei and the energy that can be derived from such changes, the units used to measure radioactivity, and the beneficial uses of radioactive isotopes.

On August 2, 1939, with the world on the brink of World War II, Albert Einstein sent a letter to President Franklin D. Roosevelt. In this letter, which profoundly changed the course of history, Einstein called attention to work being done on the physics of the atomic nucleus. He said he and others believed this work suggested the possibility that "uranium may be turned into a new and important source of energy . . . and [that it was] conceivable . . . that extremely powerful bombs of a new type may thus be constructed. . . . " Einstein's letter was the impetus for the Manhattan Project, which led to the detonation of the first atomic bomb at 5:30 AM on July 16, 1945, in the desert of New Mexico.

Since World War II, more powerful nuclear weapons have been developed and stockpiled by a number of nations. With the end of the Cold War, fears of a nuclear holocaust are fading, and recent nuclear disarmament treaties have been signed between the United States and the former Soviet Union for removing the plutonium-239 and other nuclear fuel from existing nuclear warheads. But those fears have been replaced to some extent by the concern that a great many other nations have developed or acquired nuclear weapons. For many years, the respected magazine *The Bulletin of Atomic Scientists* has used the symbol of a clock with its hands near the fateful midnight hour (representing nuclear annihilation) to illustrate the danger faced by the world from atomic weapons. Even with the end of the Cold War, the hands have moved back only a little from midnight.

The Bulletin of the Atomic Scientists can be found on-line at **http://www.bullatomsci.org/**.

20.1 THE NATURE OF RADIOACTIVITY

Many minerals, called phosphors, glow for some time after being stimulated by exposure to sunlight or ultraviolet light. (You may have a phosphor on the hands of a wristwatch or on a "black light" poster.) In 1896, French physicist Antoine Henri Becquerel was studying this phenomenon, called *phosphorescence,* when he accidentally made an important and totally unexpected observation that led to him to the discovery of radioactivity. While waiting for a sunny day, Becquerel stored a photographic plate wrapped in black paper along with a uranium salt (a material known to phosphoresce) in a dark drawer. To his amazement, the image of the uranium salt appeared on the plate that had been in the drawer, unexposed to sunlight. Becquerel realized that he had observed penetrating radiation from matter that had not been stimulated by light.

Becquerel performed many more experiments and found that pure uranium metal produced the same emissions as uranium salts did, but even more strongly. This would be expected if the radiation were the property of the metal and not dependent on its form of chemical combination. But no pure metal had been observed to phosphoresce. Becquerel was mystified. Where did all of this energy come from? The radiation had nothing to do with phosphorescence. Becquerel gave up his work for several years, but it was taken up by Marie Curie and her husband Pierre, who later named the phenomenon *radioactivity.*

TABLE 20.1		Characteristics of α, β, and γ Emissions		
Name	**Symbol**	**Charge**	**Mass (g/particle)**	**Penetrating power***
Alpha	$^4_2\text{He}^{2+}$, $^4_2\alpha$	+2	6.65×10^{-24}	0.03 mm
Beta	$^0_{-1}e$, $^0_{-1}\beta$	−1	9.11×10^{-28}	2 mm
Gamma	$^0_0\gamma$, γ	0	0	10 cm

* Distance at which half the radiation has been stopped by water.

Encouraged by the Curies, Becquerel returned to the study of radiation. He found that the radiation from uranium was affected by magnetic fields and consisted of two kinds of particles, which we now know to be alpha and beta particles.

For more on experiments done by Becquerel and the Curies, see Walton, H. F., *Journal of Chemical Education*, Vol. 69, 1992; p. 10.

One of Marie Curie's first findings was to confirm Becquerel's observation that uranium metal itself was radioactive and that the degree to which a uranium-containing sample was radioactive depended on the percentage of uranium present. When she tested pitchblende, a common ore containing uranium and other metals (such as lead, bismuth, and copper) she was surprised to find that it was even more radioactive than pure uranium. There was only one explanation: Pitchblende contained an element (or elements) more radioactive than uranium. Eventually, the Curies discovered the next element after bismuth in the periodic table, which they named *polonium* after Marie's homeland of Poland. They also isolated another new, highly radioactive element, *radium.*

In England at about the same time, Sir J. J. Thomson and his student Ernest Rutherford were studying the radiation from uranium and thorium (⬅ *p. 45*). Rutherford found that "There are present at least two distinct types of radiation—one that is readily absorbed, which will be termed for convenience alpha (α) radiation, and the other of a more penetrative character, which will be termed beta (β) radiation." **Alpha radiation,** he discovered, was composed of particles that, when passed through an electric field, were attracted to the negative side of the field (⬅ *p. 43, Figure 2.1*). Indeed, his later studies showed these **alpha (α) particles** to be helium nuclei, $^4_2\text{He}^{2+}$, which were ejected at high speeds from a radioactive element (Table 20.1). As might be expected for particles with the mass of a helium nucleus, they have limited penetrating power and can be stopped by skin, clothing, or several sheets of ordinary paper.

In the same experiment, Rutherford also found that **beta radiation** must be composed of negatively charged particles, since the beam of beta radiation was attracted to the electrically positive plate. Later work by Becquerel showed that these particles have an electric charge and mass equal to those of an electron. Thus, **beta (β) particles** are electrons ejected at high speeds from some radioactive nuclei. They are more penetrating than alpha particles (Table 20.1), and a $\frac{1}{8}$-inch-thick piece of aluminum is necessary to stop them. They penetrate 1 to 2 cm into living bone or tissue.

A third type of radiation was later discovered by P. Villard, a Frenchman, who named it **gamma (γ) radiation,** using the third letter in the Greek alphabet in keeping with Rutherford's scheme. Unlike alpha and beta particles, which are particulate in nature, gamma rays are a form of electromagnetic radiation like x-rays, and they not affected by an electric field. Gamma radiation is the most penetrating, since it can pass completely through the human body. Thick layers of lead or concrete are required to stop a beam of gamma rays completely.

20.2 NUCLEAR REACTIONS

Equations for Nuclear Reactions

Ernest Rutherford found that radium not only emits alpha particles but also produces the radioactive gas radon in the process. Such observations led Rutherford and Frederick Soddy, in 1902, to propose the revolutionary theory that *radioactiv-*

ity is the result of a natural change of a radioactive isotope of one element into an isotope of a different *element.* In such changes, called **nuclear reactions** or *transmutations,* an unstable nucleus (the *parent nucleus*) spontaneously emits radiation and is converted into a more stable nucleus of a different element (the *daughter product*). Thus, a nuclear reaction results in a change in atomic number and often a change in mass number as well. For example, the reaction of radium studied by Rutherford can be written as

$$^{226}_{88}\text{Ra} \longrightarrow {}^{4}_{2}\text{He} + {}^{222}_{86}\text{Rn}$$

In this representation the subscripts are the atomic numbers and the superscripts are the mass numbers.

In a chemical change, the atoms in molecules and ions are rearranged, but they are not created or destroyed; the number of atoms remains the same. Similarly, in nuclear reactions the total number of nuclear particles, or **nucleons** (protons and neutrons), remains the same. The essence of nuclear reactions, however, is that one nucleon can change into a different nucleon. A proton can change to a neutron or a neutron can change to a proton, but the total number of nucleons remains the same. Therefore, *the sum of the mass numbers of reacting nuclei must equal the sum of the mass numbers of the nuclei produced.* Furthermore, to maintain charge balance, *the sum of the atomic numbers of the products must equal the sum of the atomic numbers of the reactants.* These principles may be verified for the preceding nuclear equation.

	$^{226}_{88}\text{Ra}$		$^{4}_{2}\text{He}$	+	$^{222}_{86}\text{Rn}$
	radium-226		alpha particle		radon-222
Mass number:	226	$\longrightarrow$	4	+	222
Atomic number:	88	$\longrightarrow$	2	+	86

> Be sure to notice that when a radioactive atom decays, the emission of a charged particle leaves a charged atom. Thus, when radium-226 decays, it gives a helium-4 cation ($^{4}_{2}\text{He}^{2+}$) and a radon-222 anion (Rn^{2-}). By convention, the ion charges are not shown in balanced nuclear equations.

> Recall that atomic number is the number of protons in an atom's nucleus; mass number is the sum of protons and neutrons in a nucleus.

Alpha and Beta Particle Emission

One way a radioactive isotope can decay is to eject an alpha particle from the nucleus. This is illustrated by the conversion of uranium to thorium by the following reaction.

	$^{234}_{92}\text{U}$		$^{4}_{2}\text{He}$	+	$^{230}_{90}\text{Th}$
	uranium-234		alpha particle		thorium-230
Mass number:	234	$\longrightarrow$	4	+	230
Atomic number:	92	$\longrightarrow$	2	+	90

You will notice that in alpha emission *the atomic number decreases by two units and the mass number decreases by four units for each alpha particle emitted.*

Emission of a beta particle is another way for a radioactive isotope to decay. For example, loss of a beta particle by uranium-239 is represented by

	$^{239}_{92}\text{U}$		$^{0}_{-1}\text{e}$	+	$^{239}_{93}\text{Np}$
	uranium-239		beta particle		neptunium-239
Mass number:	239	$\longrightarrow$	0	+	239
Atomic number:	92	$\longrightarrow$	−1	+	93

How does a nucleus, composed only of protons and neutrons, increase its number of protons by ejecting an electron during beta emission? It is generally accepted that a series of reactions is involved, but the net process is

	$^{1}_{0}\text{n}$		$^{0}_{-1}\text{e}$	+	$^{1}_{1}\text{p}$
	neutron		electron		proton

where we use the symbol p for a proton and n for a neutron. In this process, a neutron is converted to a proton along with the release of a beta particle. *The ejection of a beta particle always means that a different element is formed. The new element has an atomic number one unit greater than that of the decaying nucleus.* The mass number does not change, however, because no proton or neutron has been emitted.

In many cases, the emission of an alpha or beta particle results in the formation of an isotope that is also unstable and therefore radioactive. The new radioactive isotope may therefore undergo a number of successive transformations until a stable, nonradioactive isotope is finally produced. Such a series of reactions is called a **radioactive series.** One such series begins with uranium-238 and ends with lead-206, as illustrated in Figure 20.1. The first step in the series is

$$^{238}_{92}U \longrightarrow {}^{4}_{2}He + {}^{234}_{90}Th$$

and the final step, the conversion of polonium-210 to lead-206, is

$$^{210}_{84}Po \longrightarrow {}^{4}_{2}He + {}^{206}_{82}Pb$$

A nucleus formed as a result of alpha or beta emission is often in an excited state and so emits a gamma ray.

Problem-Solving Example 20.1   Radioactive Series

The second, third, and fourth steps in the uranium-238 series in Figure 20.1 are emission of a β particle, then another β particle, and finally an α particle. Write equations to show the products of these steps.

Answer   Second step: $^{234}_{90}Th \rightarrow {}^{0}_{-1}e + {}^{234}_{91}Pa$

Third step: $^{234}_{91}Pa \rightarrow {}^{0}_{-1}e + {}^{234}_{92}U$

Fourth step: $^{234}_{92}U \rightarrow {}^{4}_{2}He + {}^{230}_{90}Th$

Explanation   The product of the first step, thorium-234, is our starting point. Figure 20.1 shows that the mass number remains the same during the second step but that the atomic number increases by 1 to 91, indicating emission of a β particle, a result shown by the balanced equation

$$^{234}_{90}Th \longrightarrow {}^{0}_{-1}e + {}^{234}_{91}Pa$$

thorium-234   protactinium-234

In the third step, Figure 20.1 shows that the mass number again stays constant, and the atomic number increases once again by 1 due to beta emission:

$$^{234}_{91}Pa \longrightarrow {}^{0}_{-1}e + {}^{234}_{92}U$$

protactinium-234   uranium-234

Finally, the fourth step is alpha-particle emission, so both the mass number and atomic number decline. This is again confirmed in Figure 20.1.

$$^{234}_{92}U \longrightarrow {}^{4}_{2}He + {}^{230}_{90}Th$$

uranium-234   thorium-230

Problem-Solving Practice 20.1

(a) Write an equation showing the emission of an alpha particle by an isotope of neptunium, $^{237}_{93}Np$, to produce an isotope of protactinium.

(b) Write an equation showing the emission of a beta particle by sulfur-35, $^{35}_{16}S$, to produce an isotope of chlorine.

Exercise 20.1   Radioactive Series

The actinium series begins with uranium-235, $^{235}_{92}U$, and ends with lead-207, $^{207}_{82}Pb$. The first five steps involve the successive emission of α, β, α, α, and β particles. Identify the radioactive isotope produced in each of the steps, beginning with uranium-235.

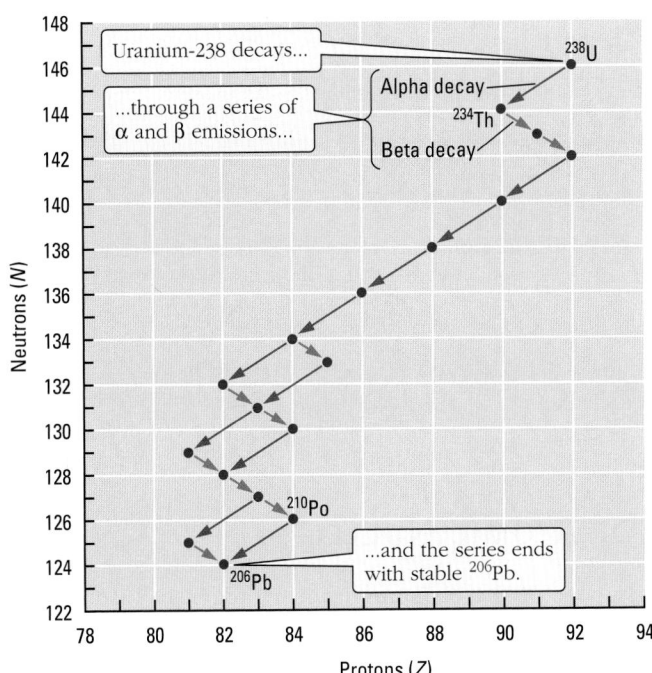

Figure 20.1 The ^{238}U decay series.

Other Types of Radioactive Decay

In addition to radioactive decay by emission of alpha, beta, or gamma radiation, other decay processes are known. Some nuclei decay, for example, by emission of a **positron,** $_{+1}^{0}e$ or β^{+}, which is effectively a positively charged electron. For example, positron emission by polonium-207 leads to the formation of bismuth-207.

$$^{207}_{84}\text{Po} \longrightarrow {}^{0}_{+1}e + {}^{207}_{83}\text{Bi}$$

polonium-207   positron   bismuth-207

Mass number:	207	$\longrightarrow$	0	+	207
Atomic number:	84	$\longrightarrow$	+1	+	83

Notice that this is the opposite of beta decay, because positron decay leads to a *decrease* in the atomic number.

The atomic number is also reduced by one when **electron capture** occurs. In this process an inner shell electron is captured by the nucleus.

$$^{7}_{4}\text{Be} + {}^{0}_{-1}e \longrightarrow {}^{7}_{3}\text{Li}$$

beryllium-7   electron   lithium-7

Mass number:	7	+	0	$\longrightarrow$	7
Atomic number:	4	+	-1	$\longrightarrow$	3

In the old nomenclature of atomic physics, the innermost shell ($n = 1$ principal quantum number) was called the K-shell, so the electron capture mechanism is sometimes called *K-capture.*

In summary, there are four common ways that a radioactive nucleus can decay, as summarized in the figure at right. In nuclear chemistry, the radioactive isotope that begins a process is called the "parent," and the product is called a "daughter" isotope.

The positron was discovered by Carl Anderson in 1932. It is sometimes called an "antielectron," one of a group of particles that have become known as "antimatter." Contact between an electron and a positron leads to mutual annihilation of both particles with production of two high-energy photons (gamma rays). This is the basis of positron emission tomography scanning to detect tumors (Section 20.9).

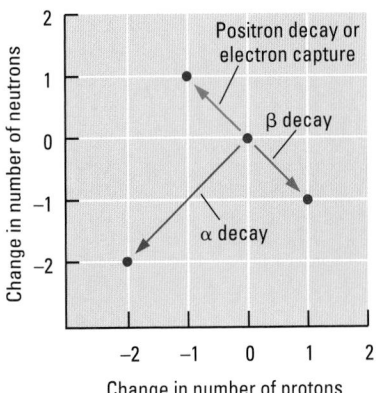

A memory aid for predicting the product of alpha, beta, or positron emission and electron capture.

Exercise 20.2   Nuclear Reactions

Balance each of the following nuclear reactions by filling in the missing symbol, mass number, and atomic number.

(a) $^{13}_{7}\text{N} \rightarrow {}^{13}_{6}\text{C} + ?$ (b) $^{41}_{20}\text{Ca} + {}^{0}_{-1}\text{e} \rightarrow ?$

(c) $^{90}_{38}\text{Sr} \rightarrow {}^{90}_{39}\text{Y} + ?$ (d) $^{11}_{6}\text{C} \rightarrow {}^{11}_{5}\text{B} + ?$

(e) $^{43}_{21}\text{Sc} \rightarrow ? + {}^{0}_{+1}\text{e}$

Exercise 20.3   Nuclear Reactions

Aluminum-26 can undergo either positron emission or electron capture. Write the balanced nuclear equation for each case.

20.3   STABILITY OF ATOMIC NUCLEI

The naturally occurring isotopes of the elements from hydrogen to bismuth are shown in Figure 20.2, where the radioactive isotopes are represented by orange dots and the stable isotopes are represented by purple and green dots. It is surprising that there are so few stable isotopes. Why are there not hundreds more?

In its simplest and most abundant form, hydrogen has only one nuclear particle, a single proton. In addition, the element has two other well-known isotopes: nonradioactive deuterium, with one proton and one neutron, $^{2}_{1}\text{H} = \text{D}$, and radioactive tritium, with one proton and two neutrons, $^{3}_{1}\text{H} = \text{T}$. Helium, the next element, has two protons and two neutrons in its most stable isotope. At the end of the actinide series is element 103, lawrencium, one isotope of which has a mass number of 257 and 154 neutrons. From hydrogen to lawrencium, except for $^{1}_{1}\text{H}$ and $^{3}_{2}\text{He}$, *the mass numbers of stable isotopes are always at least twice as large as the atomic number.* In other words, except for $^{1}_{1}\text{H}$ and $^{3}_{2}\text{He}$, every isotope of every element has a nucleus containing *at least* one neutron for every proton. Apparently the tremendous *repulsive* forces between the positively charged protons in the nucleus are moderated by the presence of neutrons which have no electrical charge. Figure 20.2 illustrates

1. For light elements up to Ca ($Z = 20$), the stable isotopes usually have equal numbers of protons and neutrons, or perhaps one more neutron than protons. Examples include $^{7}_{3}\text{Li}$, $^{12}_{6}\text{C}$, $^{16}_{8}\text{O}$, and $^{32}_{16}\text{S}$.

2. Beyond calcium the neutron/proton ratio becomes increasingly greater than 1. The band of stable isotopes deviates more and more from the line $N = Z$ (number of neutrons = number of protons). It is evident that more neutrons are needed for nuclear stability in the heavier elements. For example, whereas one stable isotope of Fe has 26 protons and 30 neutrons, one of the stable isotopes of platinum has 78 protons and 117 neutrons.

3. For elements beyond bismuth (83 protons and 126 neutrons), all isotopes are unstable and radioactive. Furthermore, the rate of disintegration becomes greater the heavier the nucleus. For example, half of a sample of $^{238}_{92}\text{U}$ disintegrates in 4.5 billion years, whereas half of a sample of $^{256}_{103}\text{Lr}$ decays in only 28 seconds.

4. A careful look at Figure 20.2 shows additional interesting features. First, elements of even atomic number have more stable isotopes than do those of odd atomic number. Second, stable isotopes generally have an even number of neutrons. For elements of odd atomic number, the most stable isotope has an even number of neutrons. In fact, of the nearly 300 stable isotopes represented in Figure 20.2, roughly 200 have an even number of neutrons *and* an even num-

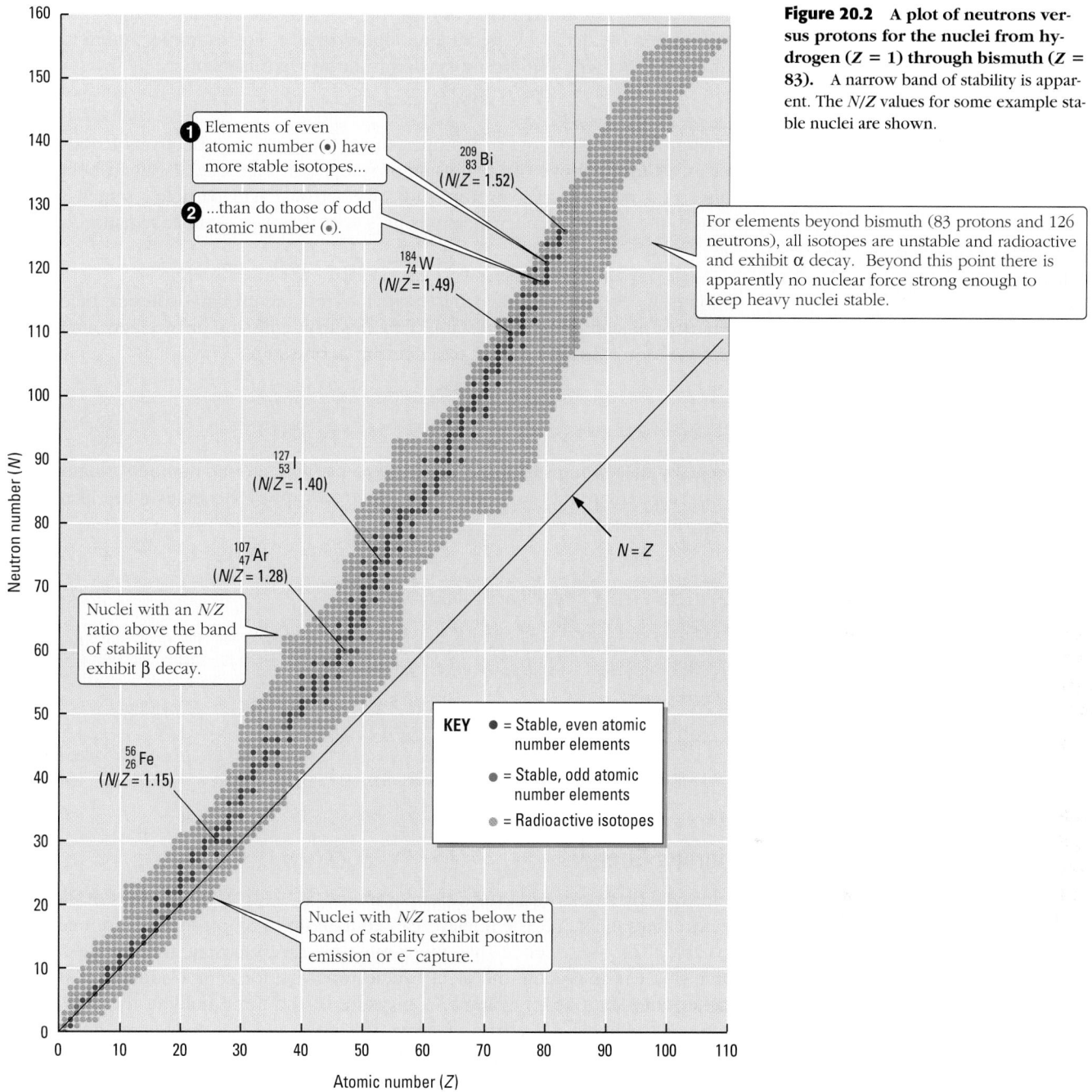

Figure 20.2 A plot of neutrons versus protons for the nuclei from hydrogen ($Z = 1$) through bismuth ($Z = 83$).   A narrow band of stability is apparent. The N/Z values for some example stable nuclei are shown.

❶ Elements of even atomic number (●) have more stable isotopes...

❷ ...than do those of odd atomic number (●).

$^{209}_{83}$Bi ($N/Z = 1.52$)

For elements beyond bismuth (83 protons and 126 neutrons), all isotopes are unstable and radioactive and exhibit α decay. Beyond this point there is apparently no nuclear force strong enough to keep heavy nuclei stable.

$^{184}_{74}$W ($N/Z = 1.49$)

$^{127}_{53}$I ($N/Z = 1.40$)

$N = Z$

$^{107}_{47}$Ar ($N/Z = 1.28$)

Nuclei with an N/Z ratio above the band of stability often exhibit β decay.

$^{56}_{26}$Fe ($N/Z = 1.15$)

KEY ● = Stable, even atomic number elements

● = Stable, odd atomic number elements

● = Radioactive isotopes

Nuclei with N/Z ratios below the band of stability exhibit positron emission or e⁻ capture.

Neutron number (N)

Atomic number (Z)

ber of protons. Only about 120 have an odd number of either protons or neutrons. Only four isotopes ($^2_1$H, $^6_3$Li, $^{10}_5$B, and $^{14}_7$N) have odd numbers of *both* protons and neutrons.

The Band of Stability and Type of Radioactive Decay

The narrow "band" of stable isotopes in Figure 20.2 (the purple and green dots) is sometimes called the *peninsula of stability* in a "sea of instability." Any isotope (the orange dots) not on this peninsula will decay in such a way that it can come ashore on the peninsula, and the chart can help us predict what type of decay will be observed.

The nuclei of all elements beyond Bi ($Z = 83$) are unstable — that is, radioactive — and most decay by ejecting an alpha particle. For example, americium, the radioactive element used in smoke alarms, decays in this manner.

$$^{243}_{95}\text{Am} \longrightarrow {}^{4}_{2}\text{He} + {}^{239}_{93}\text{Np}$$

Beta emission occurs in isotopes that have too many neutrons to be stable, that is, isotopes above the peninsula of stability in Figure 20.2. When beta decay converts a neutron to a proton and an electron (beta particle), which is then ejected, the atomic number increases by 1, and the mass number remains constant.

$$^{60}_{27}\text{Co} \longrightarrow {}^{0}_{-1}\text{e} + {}^{60}_{28}\text{Ni}$$

Conversely, lighter isotopes that have too few neutrons — isotopes below the peninsula of stability — attain stability by positron emission or by electron capture, because they convert a proton to a neutron in one step.

$$^{13}_{7}\text{N} \longrightarrow {}^{0}_{+1}\text{e} + {}^{13}_{6}\text{C}$$

$$^{41}_{20}\text{Ca} + {}^{0}_{-1}\text{e} \longrightarrow {}^{41}_{19}\text{K}$$

Decay by these routes is observed for elements with atomic numbers ranging from 4 to greater than 100; as Z increases, electron capture becomes more likely than positron emission.

Exercise 20.4   Nuclear Stability

For each of these unstable isotopes, write an equation for its probable mode of decay.

(a) Silicon-32
(b) Titanium-43
(c) Plutonium-239

Binding Energy

As proved by Ernest Rutherford's experiment *(⇐ p. 45)*, the nucleus of the atom is extremely small. Yet the nucleus can contain up to 83 protons before becoming unstable. This is evidence that there must be a very strong short-range binding force that can overcome the electrostatic repulsive force of a number of protons packed into such a tiny volume. A measure of the force holding the nucleus together is the nuclear **binding energy.** This energy (E_b) is defined as the negative of the energy change (ΔE) that would occur if a nucleus were formed directly from its component protons and neutrons. For example, if a mole of protons and a mole of neutrons directly formed a mole of deuterium nuclei, the energy change would be more than 200 million kJ, the equivalent of exploding 73 tons of TNT.

$$^{1}_{1}\text{H} + {}^{1}_{0}\text{n} \longrightarrow {}^{2}_{1}\text{H} \qquad \Delta E = -2.15 \times 10^{8} \text{ kJ}$$

$$\text{Binding energy} = -\Delta E = E_b = 2.15 \times 10^{8} \text{ kJ}$$

This nuclear synthesis reaction is highly exothermic (and so E_b is very positive), an indication of the strong attractive forces holding the nucleus together. The deuterium nucleus is more stable than an isolated proton and an isolated neutron.

To understand the enormous energy released during the formation of atomic nuclei, we turn to an experimental observation and a theory. The experimental observation is that the mass of a nucleus is always slightly less than the sum of the masses of its constituent protons and neutrons.

$$\text{\(_1^1\)H} + \text{\(_0^1\)n} \longrightarrow \text{\(_1^2\)H}$$

$$\text{1.007825 g/mol} \quad \text{1.008665 g/mol} \quad \text{2.01410 g/mol}$$

Change in mass = Δm = mass of product − sum of masses of reactants

$$= 2.01410 \text{ g/mol} - 2.016490 \text{ g/mol}$$

$$= -0.00239 \text{ g/mol}$$

The theory is that the "missing mass," Δm, has been converted to energy, and it is this energy that we described as the binding energy.

The relationship between mass and energy is contained in Albert Einstein's 1905 theory of special relativity, which holds that mass and energy are simply different manifestations of the same quantity. Einstein stated that the energy of a body is equivalent to its mass times the square of the speed of light, $E = mc^2$. So, to calculate the energy change in a process in which the mass has changed, the equation becomes

$$\Delta E = (\Delta m)c^2$$

We can calculate ΔE in joules if the change in mass is given in kilograms and the velocity of light is in meters per second (because 1 J = 1 kg m^2/s^2). For the formation of 1 mol of deuterium nuclei from 1 mol of protons and 1 mol of neutrons, we have

$$\Delta E = (-2.39 \times 10^{-6} \text{ kg}) (3.00 \times 10^8 \text{ m/s})^2$$

$$= -2.15 \times 10^{11} \text{ J} = -2.15 \times 10^8 \text{ kJ}$$

This is the value of ΔE given at the beginning of this section for the change in energy when a mole of protons and a mole of neutrons form a mole of deuterium nuclei.

A helium-4 nucleus is composed of two protons and two neutrons, and its binding energy, E_b, is very large, even larger than that for deuterium.

$$2\,_1^1\text{H} + 2\,_0^1\text{n} \longrightarrow \,_2^4\text{He} \qquad E_b = +2.73 \times 10^9 \text{ kJ/mol of helium nuclei}$$

To compare nuclear stabilities more directly, nuclear scientists generally calculate the **binding energy per nucleon.** For 1 mol of helium-4 atoms this is

$$E_b \text{ per mol nucleons} = \frac{2.73 \times 10^9 \text{ kJ}}{4 \text{ mol nucleons}} = 6.83 \times 10^8 \text{ kJ/mol nucleons}$$

The greater the binding energy per nucleon, the greater is the stability of the nucleus. Scientists have calculated the binding energies per nucleon of a great number of nuclei and have plotted them as a function of mass number (Figure 20.3). It is very interesting—and important—that the point of maximum stability occurs in the vicinity of iron-56, $_{26}^{56}\text{Fe}$. This means that *all elements are thermodynamically unstable with respect to iron.* That is, very heavy nuclei may split, or *fission,* to give more stable nuclei with atomic numbers nearer iron, and simultaneously release enormous quantities of energy (Section 20.6). In contrast, two very light nuclei may come together and undergo *fusion* exothermically to form heavier nuclei (Section 20.7). Finally, it is because of its high nuclear stability that *iron is the most abundant of the heavier elements in the universe.*

Exercise 20.5 Binding Energy

Calculate the binding energy, in kJ/mol, for the formation of lithium-6.

$$3\,_1^1\text{H} + 3\,_0^1\text{n} \longrightarrow \,_3^6\text{Li}$$

The necessary masses are $_1^1\text{H}$ = 1.00783 g/mol, $_0^1\text{n}$ = 1.00867 g/mol, and $_3^6\text{Li}$ = 6.015125 g/mol. Is the binding energy greater than or less than that for helium-4? Finally, compare the binding energy per nucleon of lithium-6 and helium-4. Which nucleus is more stable?

Figure 20.3 **Binding energy per nucleon.** The values plotted were derived by calculating the binding energy per nucleon in million electron volts (MeV) for the most abundant isotope of each element from hydrogen to uranium (1 MeV = 1.602×10^{-13} J). The nuclei at the top of the curve are most stable.

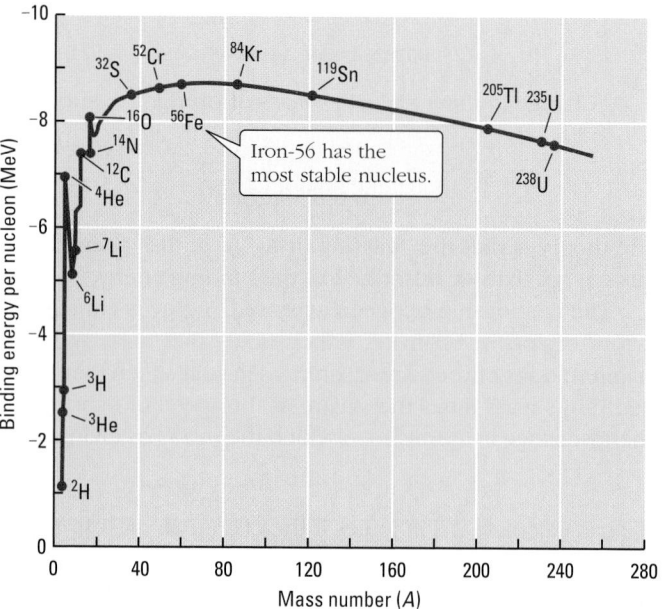

Exercise 20.6   **Binding Energy**

By interpreting the shape of the curve in Figure 20.3, determine which is more exothermic per gram—fission or fusion. Explain your answer.

20.4   RATES OF DISINTEGRATION REACTIONS

Cobalt-60 is used as a source of β particles and γ rays to treat malignancies in the human body. ^{60}Co is radioactive, but only half of a sample of cobalt-60 will change via β decay into nickel-60 in a little over five years. On the other hand, copper-64, which is used in the form of copper acetate to detect brain tumors, decays much more rapidly; half of the radioactive copper decays in slightly less than 13 hours. These two radioactive isotopes are clearly different in their rates of decay.

Half-Life

The relative instability of a radioactive isotope is expressed as its half-life, the time required for one half of a given quantity of the isotope to undergo decay. In terms of reaction kinetics (Section 13.3), radioactive decay is a first-order reaction. Therefore, the rate of decay is given by the following rate law:

$$\ln[A]_t = -kt + \ln[A]_0$$

where $[A]_0$ is the initial concentration of isotope A, $[A]_t$ is the concentration of A after time t has passed, and k is the first-order rate constant. Because radioactive decay is first order, the half-life ($t_{1/2}$) of an isotope is the same no matter what the initial concentration, and it is given by

$$t_{1/2} = \frac{0.693}{k}$$

TABLE 20.2	Half-Lives of Some Common Radioactive Isotopes	
Isotope	Decay process	Half-life
$^{238}_{92}U$	$^{238}_{92}U \rightarrow ^{234}_{90}Th + ^4_2He$	4.15×10^9 y
^3_1H (tritium)	$^3_1H \rightarrow ^3_2He + ^0_{-1}e$	12.3 y
$^{14}_6C$ (carbon-14)	$^{14}_6C \rightarrow ^{14}_7N + ^0_{-1}e$	5730 y
$^{131}_{53}I$	$^{131}_{53}I \rightarrow ^{131}_{54}Xe + ^0_{-1}e$	8.04 d
$^{123}_{53}I$	$^{123}_{53}I + ^0_{-1}e \rightarrow ^{123}_{52}Te$	13.2 h
$^{57}_{24}Cr$	$^{57}_{24}Cr \rightarrow ^{57}_{25}Mn + ^0_{-1}e$	21 s
$^{28}_{15}P$	$^{28}_{15}P \rightarrow ^{28}_{14}Si + ^0_{+1}e$	0.270 s
$^{90}_{38}Sr$	$^{90}_{38}Sr \rightarrow ^{90}_{39}Y + ^0_{-1}e$	28.8 y
$^{60}_{27}Co$	$^{60}_{27}Co \rightarrow ^{60}_{28}Ni + ^0_{-1}e$	5.26 y

As illustrated by Table 20.2, isotopes have widely varying half-lives; some take years, even millennia, for half of the sample to decay (^{238}U, ^{14}C), whereas others decay to half the original number of atoms in fractions of seconds (^{28}P). The half-life is given in terms of whatever time unit is most appropriate. It can be anything from years to seconds.

As an example of the concept of half-life, consider the decay of plutonium-239, the alpha-emitting isotope formed in nuclear reactors.

$$^{239}_{94}Pu \longrightarrow ^4_2He + ^{235}_{92}U$$

The half-life of plutonium-239 is 24,400 years. This half-life means that half of the quantity of $^{239}_{94}Pu$ present at any given time will disintegrate every 24,400 years. Thus, if we begin with 1.00 g of $^{239}_{94}Pu$, 0.500 g of the isotope will remain after 24,400 years. After 48,800 years (two half-lives), only half of the remainder, 0.250 g, will still be there. After 73,200 years (three half-lives), only half of the 0.250 g will still be present, or 0.125 g. The amounts of $^{239}_{94}Pu$ present at various times are illustrated in Figure 20.4. All radioactive isotopes follow this type of decay curve as they decay.

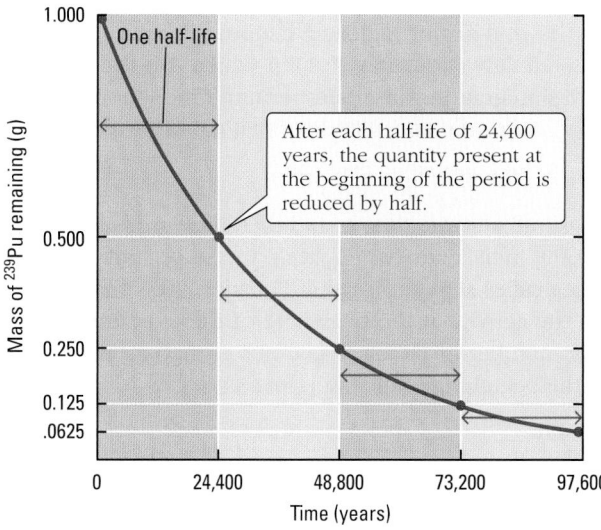

Figure 20.4   The decay of 1.00 g of plutonium-239.   After each half-life of 24,400 years, the quantity present at the beginning of the period is reduced by half.

Problem-Solving Example 20.2 Half-Life

Tritium ($_1^3$H), a radioactive isotope of hydrogen, has a half-life of 12.3 years.

$$_1^3\text{H} \longrightarrow {}_{-1}^0\text{e} + {}_2^3\text{He}$$

If you begin with 1.5 mg of the isotope, how many milligrams remain after 49.2 years?

Answer Only 0.094 mg of tritium remains.

Explanation First, we find the number of half-lives in the given time period of 49.2 years. Since the half-life is 12.3 years, the number of half-lives is

$$49.2 \text{ years} \times \frac{1 \text{ half-life}}{12.3 \text{ years}} = 4.00 \text{ half-lives}$$

This means that the initial quantity of 1.5 mg is reduced by $\frac{1}{2}$ four times.

$$1.5 \text{ mg} \times \tfrac{1}{2} \times \tfrac{1}{2} \times \tfrac{1}{2} \times \tfrac{1}{2} = 1.5 \text{ mg} \times \left(\tfrac{1}{2}\right)^4 = 0.094 \text{ mg}$$

After 49.2 years, only 0.094 mg of the original 1.5 mg will remain.

✔ The remaining tritium should be a small fraction of the starting amount, and it is.

Problem-Solving Practice 20.2

Strontium-90, $_{38}^{90}$Sr, is a radioisotope ($t_{1/2}$ = 29 years) produced in atomic bomb explosions. Its long half-life and tendency to concentrate in bone marrow by replacing calcium make it particularly dangerous to people and animals.

(a) The isotope decays with loss of a β particle. Write a balanced equation showing the other product of decay.
(b) A sample of the isotope emits 2000 β particles per minute. How many half-lives and how many years are necessary to reduce the emission to 125 β particles per minute?

Exercise 20.7 Half-Lives

The radioactivity of formerly highly radioactive isotopes is essentially negligible after ten half-lives. What percentage of the original radioisotope remains after this amount of time (ten half-lives)?

Rate of Radioactive Decay

To determine the half-life of a radioactive element, its *rate of decay*, the number of atoms that disintegrate in a given time — per second, per hour, or per year — must be measured.

Radioactive decay is a first-order process (⇐ *p. 583, Section 13.3*), with a rate that is directly proportional to the number of radioactive atoms present (N). This proportionality is expressed as a rate law (Equation 20.1) in which A is the **activity** of the sample — the number of disintegrations observed per unit time — and k is the first-order rate constant or *decay constant* characteristic of that radioisotope.

$$A = kN \qquad [20.1]$$

Let us say the activity is measured at some time t_0 and then measured again after a few minutes, hours, or days. If the initial activity is A_0 at t_0, then a second measurement at a later time t will give a smaller activity A. Using Equation 20.1, the ratio of the activity A at some time t to the activity at the beginning of the experiment (A_0) must be equal to the ratio of the number of radioactive atoms N that are present at time t to the number present at the beginning of the experiment (N_0).

$$\frac{A}{A_0} = \frac{kN}{kN_0} \qquad \text{or} \qquad \frac{A}{A_0} = \frac{N}{N_0}$$

Thus, either A/A_0 or N/N_0 expresses the fraction of radioactive atoms remaining in a sample after some time has elapsed.

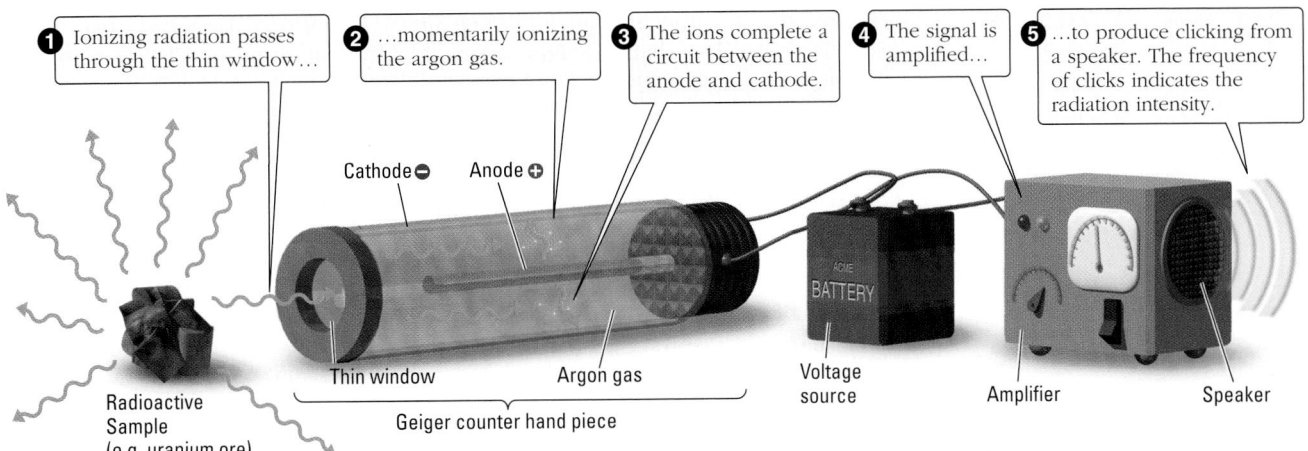

❶ Ionizing radiation passes through the thin window…

❷ …momentarily ionizing the argon gas.

❸ The ions complete a circuit between the anode and cathode.

❹ The signal is amplified…

❺ …to produce clicking from a speaker. The frequency of clicks indicates the radiation intensity.

Cathode⊖   Anode⊕

ACME BATTERY

Thin window   Argon gas   Voltage source   Amplifier   Speaker

Radioactive Sample (e.g. uranium ore)   Geiger counter hand piece

Figure 20.5 **A Geiger counter.**

The activity of a sample can be measured with a device such as a Geiger counter (Figure 20.5). It detects radioactive emissions as they ionize a gas to form free electrons and cations that can be attracted to a pair of electrodes. A metal tube is filled with low-pressure argon gas. The inside of the tube acts as the cathode. A thin wire running through the center of the tube is the anode. When radioactive emissions enter the tube through the thin window at the end, they collide with argon atoms, which produces free electrons and argon cations. As the free electrons accelerate toward the anode, they collide with other argon atoms to generate more free electrons. The free electrons all go to the anode, and they constitute a pulse of current. This current pulse is counted; the rate of pulses per unit time is the output of the Geiger counter.

The **curie** (Ci) is commonly used as a unit of activity. One curie represents a decay rate of 3.7×10^{10} disintegrations per second (dps), which is the decay rate of 1 g of radium. One millicurie (mCi) $= 10^{-3}$ Ci $= 3.7 \times 10^{7}$ dps. Another unit of radioactivity is the **becquerel** (Bq); 1 becquerel is equal to one nuclear disintegration per second (1 Bq = 1 dps).

The *curie* was named for Pierre Curie by his wife, Marie; the *becquerel* honors Henri Becquerel.

The change in activity of a radioactive sample over a period of time, or the fraction of radioactive atoms still present in a sample after some time has elapsed, can be calculated using the integrated rate equation for a first-order reaction (⬅ *p. 584, Table 13.2*)

 CD-ROM Screen 15.6: Concentration-Time Relationships

$$\ln A = -kt + \ln A_0$$

which can be rearranged to

$$\ln \frac{A}{A_0} = -kt \qquad [20.2]$$

where A/A_0 is the ratio of activities. Equation 20.2 can also be stated in terms of the fraction of radioactive atoms present in the sample after some time, t, has passed.

$$\ln \frac{N}{N_0} = -kt \qquad [20.3]$$

In words, Equation 20.3 says

Equation 20.3 can be derived from Equation 20.1 using calculus.

Natural logarithm $\left(\dfrac{\text{number of radioactive atoms at time } t}{\text{number of radioactive atoms at start of experiment}} \right)$

= natural logarithm (fraction of radioactive atoms remaining at time t)

= −(decay constant)(time)

As radioactive atoms decay, N becomes less than N_0.

CD-ROM Screen 15.8: Half-Life: First-Order Reactions

Notice the negative sign in Equation 20.3. The ratio N/N_0 is less than 1 because N is always less than N_0. This means that the logarithm of N/N_0 is negative, and so the other side of the equation has a compensating negative sign because k and t are always positive.

The half-life of an isotope is inversely proportional to the first-order rate constant k:

$$t_{1/2} = \frac{0.693}{k}$$

Thus, the half-life can be found by calculating k from Equation 20.3 using N and N_0 from laboratory measurements over the time period t.

Problem-Solving Example 20.3 Determination of Half-Life

A sample of radon initially undergoes 7.0×10^4 alpha particle disintegrations per second (dps). After 6.6 days, it undergoes only 2.1×10^4 alpha particle dps. What is the half-life of this isotope of radon?

Answer This isotope has a 3.8-day half-life.

Explanation Experiment has provided us with both A and A_0 ($A = 2.1 \times 10^4$ dps, $A_0 = 7.0 \times 10^4$ dps) and the time ($t = 6.6$ days). Therefore, we can find the value of k.

Since $\dfrac{N}{N_0} = \dfrac{A}{A_0}$, we can use Equation 20.3.

$$\ln\left(\frac{2.1 \times 10^4 \text{ dps}}{7.0 \times 10^4 \text{ dps}}\right) = \ln(0.30) = -k(6.6 \text{ d})$$

$$k = -\frac{\ln(0.30)}{6.6 \text{ d}} = -\left(\frac{-1.20}{6.6 \text{ d}}\right) = 0.18 \text{ d}^{-1}$$

and from k we can obtain $t_{1/2}$.

$$t_{1/2} = \frac{0.693}{k} = \frac{0.693}{0.18 \text{ d}^{-1}} = 3.8 \text{ d}$$

✔ The decay rate fell to about one third of its starting value in 6.6 days, so the half-life must be somewhat less than the 6.6-day period, and it is.

Problem-Solving Practice 20.3

The decay of iridium-192, a radioisotope used in cancer radiation therapy, has a rate constant of $9.3 \times 10^{-3} \text{ d}^{-1}$.

(a) What is the half-life of ^{192}Ir?
(b) What fraction of an ^{192}Ir sample remains after 100 days?

Problem-Solving Example 20.4 Time and Radioactivity

Some high-level radioactive waste with a half-life, $t_{1/2}$, of 200 years is stored in underground tanks. What time is required to reduce an activity of 6.50×10^{12} disintegrations per minute (dpm) to a fairly harmless activity of 3.00×10^{-3} dpm?

Answer 1.02×10^4 years.

Explanation The data give you the initial activity ($A_0 = 6.50 \times 10^{12}$ dpm) and the activity after some elapsed time ($A = 3.00 \times 10^{-3}$ dpm). In order to find the elapsed time t, you must first find k from the half-life.

$$k = \frac{0.693}{t_{1/2}} = \frac{0.693}{200 \text{ y}} = 0.00347 \text{ y}^{-1}$$

With k known, the time t can be calculated using Equation 20.2.

$$\ln\left(\frac{3.00 \times 10^{-3} \text{ dpm}}{6.50 \times 10^{12} \text{ dpm}}\right) = -(0.00347 \text{ y}^{-1})t \qquad 35.312 = -(0.00347 \text{ y}^{-1})t$$

$$t = \frac{-35.312}{-0.00347 \text{ y}^{-1}} = 1.02 \times 10^4 \text{ y}$$

Problem-Solving Practice 20.4

In 1921 the women of America honored Marie Curie by giving her a gift of 1.00 g of pure radium, which is now in Paris at the Curie Institute of France. The principal isotope, ^{226}Ra, has a half-life of 1.60×10^3 years. How many grams of radium still remain?

Carbon-14 Dating

In 1946 Willard Libby developed a technique for measuring the age of archaeological objects using radioactive carbon-14. Carbon is an important building block of all living systems, and so all organisms contain the three isotopes of carbon: ^{12}C, ^{13}C, and ^{14}C. The first two are stable (nonradioactive) and have been present for billions of years. Carbon-14, however, is radioactive and decays to nitrogen-14 by beta emission.

$$^{14}_{6}\text{C} \longrightarrow {}^{0}_{-1}\text{e} + {}^{14}_{7}\text{N}$$

Since the half-life of ^{14}C is known by experiment to be 5.73×10^3 years, the fraction of the radioactive isotope present (N) can be measured from the activity of a sample. If the fraction of ^{14}C originally in the sample (N_0) is known, then the age of the sample can be found from Equation 20.3.

This method of age determination clearly depends on knowing how much ^{14}C was originally in the sample. The answer to this question comes from work by physicist Serge Korff, who discovered in 1929 that ^{14}C is continually generated in the upper atmosphere. High-energy cosmic rays collide with gas molecules in the upper atmosphere and cause them to eject neutrons. These free neutrons collide with nitrogen atoms to produce carbon-14.

$$^{14}_{7}\text{N} + {}^{1}_{0}\text{n} \longrightarrow {}^{14}_{6}\text{C} + {}^{1}_{1}\text{H}$$

Throughout the *entire* atmosphere, only about 7.5 kg of ^{14}C is produced per year. However, this tiny amount of radioactive carbon is incorporated into CO_2 and other carbon compounds and then is distributed worldwide as part of the carbon cycle. The continual formation of ^{14}C, transfer of the isotope within the oceans, atmosphere, and biosphere, and decay of living matter keep the supply of ^{14}C constant.

Plants absorb carbon dioxide from the atmosphere, convert it into food via photosynthesis *(⇐ p. 844)*, and so incorporate the ^{14}C into living tissue, where radioactive ^{14}C atoms and nonradioactive ^{12}C atoms in CO_2 chemically react in the same way. It has been established that the beta activity of carbon-14 in *living* plants and in the air is constant at 15.3 disintegrations per minute per gram (d m^{-1} g^{-1}) of carbon. However, when the plant dies, carbon-14 disintegration continues *without the ^{14}C being replaced.* Consequently, the ^{14}C activity decreases with passage of time. The smaller the activity of carbon-14 in the plant, the longer the period of time between the death of the plant and the present. Assuming that ^{14}C activity was about the same hundreds or thousands of years ago as it is now, measurement of the ^{14}C beta activity of an artifact can be used to date an article containing carbon. Fluctuations in the ^{14}C activity for the past several thousand years have been measured by studying growth rings of long-lived trees, and the carbon-14 dates of objects can be corrected accordingly.

Willard Libby won the 1960 Nobel prize in chemistry for his discovery of radiocarbon dating.

Willard Libby and his apparatus for carbon-14 dating.

The time scale accessible to carbon-14 dating is determined by the half-life of ^{14}C. Therefore, this method for dating objects can be extended back approximately 50,000 years. This is almost nine half-lives, during which the number of disintegrations per minute per gram of carbon would fall by a factor of $(\frac{1}{2})^9 = 9.8 \times 10^{-4}$ from about 15.3 d m^{-1} g^{-1} to about 0.030 d m^{-1} g^{-1}, which is a disintegration rate so low that it is difficult to measure accurately.

Problem-Solving Example 20.5   Carbon-14 Dating

The so-called Dead Sea Scrolls, Hebrew manuscripts of the books of the Old Testament, were found in 1947. The activity of carbon-14 in the linen wrappings of the book of Isaiah is about 12 disintegrations per minute per gram (d m^{-1} g^{-1}). Calculate the approximate age of the linen.

Answer   The cloth is about 2000 years old.

Explanation   We will use Equation 20.3

$$\ln\frac{N}{N_0} = -kt$$

where N is proportional to the activity at the present time (12 d m^{-1} g^{-1}) and N_0 is proportional to the activity of carbon-14 in the living material (15.3 d m^{-1} g^{-1}). In order to calculate the time elapsed since the linen wrappings were made from a living plant, we first need k, the rate constant. From the text, you know that $t_{1/2}$ of ^{14}C is 5.73×10^3 years, so

$$k = \frac{0.693}{t_{1/2}} = \frac{0.693}{5.73 \times 10^3 \text{ y}} = 1.21 \times 10^{-4} \text{ y}^{-1}$$

Now everything is in place to calculate t.

$$\ln\left(\frac{12 \text{ d m}^{-1} \text{ g}^{-1}}{15.3 \text{ d}^{-1} \text{ g}^{-1}}\right) = -0.24 = -kt = -(1.21 \times 10^{-4} \text{ y}^{-1})t$$

$$t = \frac{-0.24}{-1.21 \times 10^{-4} \text{ y}^{-1}} = 2.0 \times 10^3 \text{ y}$$

Therefore, the linen is about 2000 years old.

✔   The disintegration rate has fallen a small percentage (about 20%) from the modern rate, so the linen must be a fraction of one half-life in age, and 2000 years is a fraction of 5730 years.

Problem-Solving Practice 20.5

Tritium, ^3H, ($t_{1/2} = 12.3$ y) is produced in the atmosphere and incorporated in much the same way as ^{14}C. Estimate the age of a sealed sample of Scotch whiskey that has a tritium content 0.60 times that of the water in the area where the whiskey was produced.

 Exercise 20.8   Radiochemical Dating

The radioactive decay of uranium-238 to lead-206 provides a method of radiochemically dating ancient rocks by using the ratio of lead-206 atoms to uranium-238 atoms in a sample. Using this method, a moon rock was found to have a ^{206}Pb-to-^{238}U ratio of 1.00 to 1.09, that is, 100 lead-206 atoms for every 109 uranium-238 atoms. No other lead isotopes were present in the rock, indicating that all of the lead-206 was produced by uranium-238 decay. Estimate the age of the moon rock. The half-life of uranium-238 is 4.51×10^9 years.

 Exercise 20.9   Radiochemical Dating

Ethanol, C_2H_5OH, is produced by the fermentation of grains or by the reaction of water with ethylene, which is made from petroleum. The alcohol content of wines can be increased fraudulently beyond the usual 12% from fermentation by adding ethanol produced from ethylene. How can carbon dating techniques be used to differentiate the ethanol sources in these wines?

20.5 ARTIFICIAL TRANSMUTATIONS

In the course of his experiments, Rutherford found in 1919 that alpha particles ionize atomic hydrogen, knocking off an electron from each atom. Using atomic nitrogen instead, he found that bombardment with alpha particles *also produced protons.* Quite correctly he concluded that the alpha particles had knocked a proton out of the nitrogen nucleus and that an isotope of another element had been produced. Nitrogen had undergone a *transmutation* to oxygen.

$$\ce{^4_2He + ^{14}_7N \longrightarrow ^{17}_8O + ^1_1H}$$

Rutherford had proposed that protons and neutrons are the fundamental building blocks of nuclei. Although Rutherford's search for the neutron was not successful, it was found by James Chadwick in 1932 as a product of the alpha-particle bombardment of beryllium.

$$\ce{^9_4Be + ^4_2He \longrightarrow ^{12}_6C + ^1_0n}$$

Changing one element into another by alpha-particle bombardment has its limitations. Before a positively charged bombarding particle (such as the alpha particle) can be captured by a positively charged nucleus, the bombarding particle must have sufficient kinetic energy to overcome the repulsive forces developed as the particle approaches the nucleus. But the neutron is electrically neutral, so Enrico Fermi (1934) reasoned that a nucleus would not oppose its entry. By this approach, practically all elements have since been transmuted, and a number of *transuranium elements* (elements beyond uranium) have been prepared. For example, plutonium-239 forms americium-241 by neutron bombardment.

$$\ce{^{239}_{94}Pu + ^1_0n \longrightarrow ^{240}_{94}Pu}$$

$$\ce{^{240}_{94}Pu + ^1_0n \longrightarrow ^{241}_{94}Pu}$$

$$\ce{^{241}_{94}Pu \longrightarrow ^{241}_{95}Am + ^{\ \ 0}_{-1}e}$$

Of the elements identified at present, only elements up to uranium exist in nature (except for Tc and Pm). The transuranium elements, those with atomic numbers greater than 92, are all synthetic. Up to element 101, mendelevium, all the elements can be made by bombarding the nucleus of a lighter element with small particles such as $\ce{^4_2He}$ or $\ce{^1_0n}$. Beyond element 101, though, special techniques using heavier particles are required and are still being developed. For example, element 111, synthesized in 1994, was made by bombarding bismuth-209 with nickel-64 nuclei (⇐ *p. 65, Chemistry in the News*).

$$\ce{^{64}_{28}Ni + ^{209}_{83}Bi \longrightarrow ^{272}_{111}E + ^1_0n}$$

Element 118, the latest to be discovered (in 1999 at the Lawrence Berkeley Laboratory), was made by firing krypton-86 atoms into lead-208 nuclei.

$$\ce{^{86}_{36}Kr + ^{208}_{82}Pb \longrightarrow ^{293}_{118}E + ^1_0n}$$

Elements beyond 109 have not yet been assigned names and symbols.

GLENN SEABORG
(1912–1999)

Seaborg was a pioneer in developing radioisotopes for medical use (Section 20.9). He was the first to produce iodine-131, used subsequently to treat his mother's abnormal thyroid condition. As a result of Seaborg's further research, it became possible to predict accurately the properties of many of the as yet undiscovered transuranium elements. In a remarkable 21-year span (1940–1961), Seaborg and his colleagues synthesized ten new transuranic elements (plutonium to lawrencium). He received the Nobel Prize in 1951 for his creation of new elements. In the 1990s he was honored by having element 106 named for him.

Striking the lead-208 target with 10^{18} krypton-86 atoms produced just one atom of element 118. Bombarding the target for three weeks generated three atoms of element 118.

Exercise 20.10   Nuclear Transmutation and Radioactive Decay

Element 118 (mass number 293) was identified by its radioactive decay products. Identify the element (and its mass number) formed after six sequential alpha emissions starting with element 118 (mass number 293).

DARLEANE C. HOFFMAN
(1926–)

Hoffman achieved recognition for her investigation of the chemical properties of the heaviest elements at Los Alamos National Laboratory, the University of California, Berkeley, and the Lawrence Berkeley Laboratory. She was awarded the National Medal of Science in 1997 and the American Chemical Society's highest honor, the Priestley Medal, in 2000. She became interested in nuclear chemistry as a student at Iowa State University and searched for new heavy elements and isotopes in nuclear test debris while working at Los Alamos National Laboratory. Later she moved to Berkeley to continue her research and was co-director of the team that announced the discovery of new, super-heavy elements 114, 116, and 118.

In 1938 Meitner, who was Jewish, had to flee her Austrian homeland to Sweden to escape being persecuted by the anti-Semitic policies of the Nazis.

 Exercise 20.11   Nuclear Transmutations

Balance the following equations for nuclear reactions, indicating the symbol, the mass number, and the atomic number of the remaining product.

(a) $^{13}_{6}C + ^{1}_{0}n \rightarrow ^{4}_{2}He + ?$ (b) $^{14}_{7}N + ^{4}_{2}He \rightarrow ^{1}_{0}n + ?$ (c) $^{253}_{99}Es + ^{4}_{2}He \rightarrow ^{1}_{0}n + ?$

Exercise 20.12   Element Synthesis

In 1999 an international team of scientists at the Lawrence Livermore National Laboratory reported that they had produced element 116. Bombardment of lead-208 produced a nucleus that emitted an alpha particle to produce element 116 (mass number = 290). Write a balanced nuclear equation for this process, and identify the bombarding nucleus.

20.6   NUCLEAR FISSION

In 1938 the nuclear chemists Otto Hahn and Fritz Strassman found barium in a sample of uranium that had been bombarded with neutrons. Further work by Lise Meitner, Otto Frisch, Niels Bohr, and Leo Szilard confirmed that a uranium-235 nucleus had captured a neutron and undergone **nuclear fission;** that is, the nucleus had split in two (Figure 20.6).

$$^{235}_{92}U + ^{1}_{0}n \longrightarrow ^{236}_{92}U \longrightarrow ^{141}_{56}Ba + ^{92}_{36}Kr + 3\,^{1}_{0}n \qquad \Delta E = -2 \times 10^{10}\ \text{kJ}$$

The fact that the fission reaction produces more neutrons than are required to begin the process is important. The preceding nuclear reaction shows that bombardment with a single neutron produces three neutrons. Each of these neutrons is capable of inducing three more fission reactions, which release nine neutrons to induce nine more fissions, from which 27 neutrons are obtained, and so on. Since the neutron-induced fission of uranium-235 is extremely rapid, this sequence of reactions can be an explosive chain reaction, as illustrated in Figure 20.7.

The fission of uranium-235 produces a variety of products. Thirty-four elements have been detected among the fission products, including those shown in the figure. If the amount of uranium-235 is small, so few neutrons are captured by ^{235}U nuclei that the chain reaction cannot be sustained. In an atomic bomb, two small pieces of uranium-235, neither capable of sustaining a chain reaction, are brought together to form one piece capable of supporting a chain reaction, and an atomic explosion results.

Rather than allow a fission reaction to run away explosively, engineers can control it by limiting the number and energy of the neutrons available so that energy can be derived safely and used as a heat source in a power plant (Figure

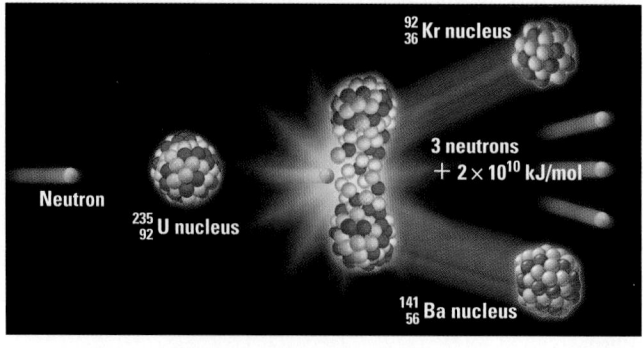

Figure 20.6 The fission of a $^{235}_{92}U$ nucleus from its bombardment with a neutron.

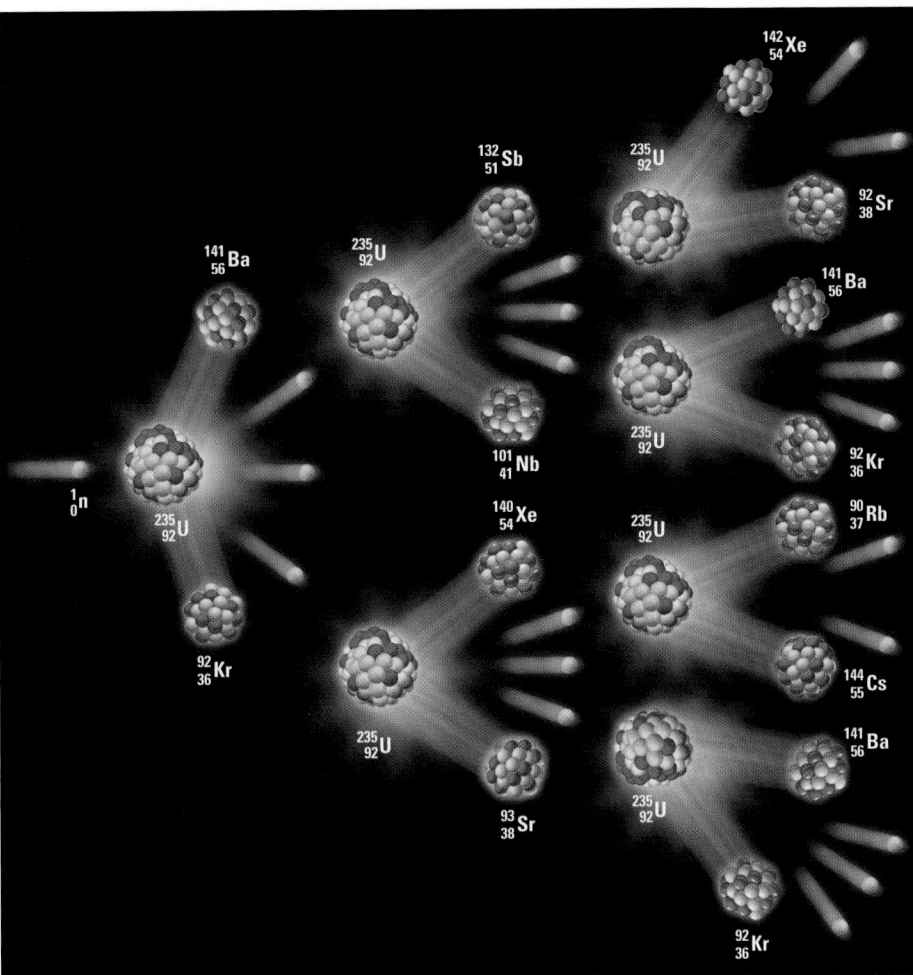

Figure 20.7 A self-propagating nuclear chain reaction initiated by capture of a neutron. The fission of uranium-235 produces a variety of products. Thirty-four elements have been detected among the fission products, including those shown in the figure. Each fission event produces two lighter nuclei plus two or three neutrons.

20.8). In a **nuclear reactor,** the rate of fission is controlled by inserting cadmium rods or other "neutron absorbers" into the reactor. The rods absorb the neutrons that cause fission reactions. The rate of the fission reaction can be increased or decreased by withdrawing or inserting the control rods.

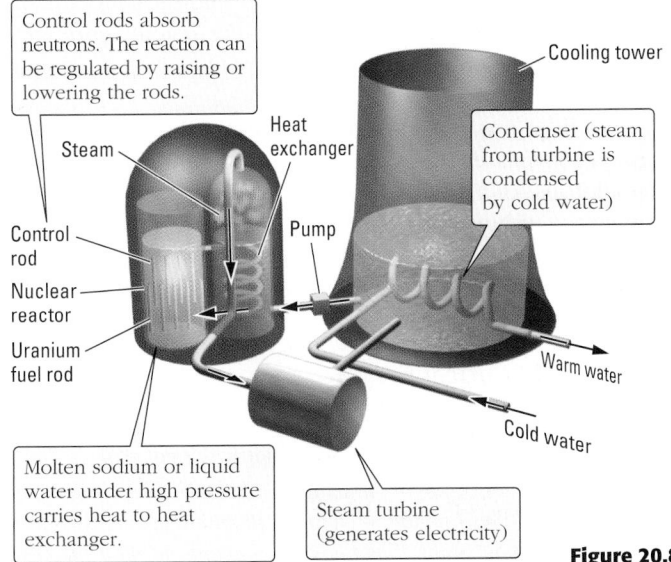

Control rods absorb neutrons. The reaction can be regulated by raising or lowering the rods.

Steam

Heat exchanger

Cooling tower

Condenser (steam from turbine is condensed by cold water)

Pump

Control rod

Nuclear reactor

Uranium fuel rod

Warm water

Cold water

Molten sodium or liquid water under high pressure carries heat to heat exchanger.

Steam turbine (generates electricity)

Figure 20.8 A nuclear power plant.

A nuclear power plant with four prominent cooling towers.

The tremendous amount of heat from the fissioning nuclei is transferred to the primary coolant, a substance with a very high heat capacity, usually water or liquid sodium. The primary coolant is at a pressure of more than 150 atm, so it does not boil, even though the temperature is higher than its normal boiling point ($\Longleftarrow$ *p. 464*). The hot primary coolant is pumped in a closed loop from the reaction vessel to the steam generators and back to the reaction vessel. By giving up its heat to water that runs the steam generators, the temperature of the primary coolant is lowered, enabling it to be heated again. This closed loop is what links the nuclear reactor and the rest of the power plant.

The primary coolant transfers heat to water in the steam generators, sometimes referred to as the secondary coolant. As the water in the steam generators is vaporized, the steam strikes the large turbine blades, causing them to spin. The turbine shaft is connected to a large metal rod in the generator, which is surrounded by a magnetic field. The rapid revolution of the metal core rod in a magnetic field produces electricity.

After striking the turbine blades, the steam must be condensed so the heating/cooling cycle can be repeated to create additional electricity. The steam is condensed by cooling water pumped from a neighboring river or lake to the secondary coolant loop. Enormous amounts of outside cooling water are needed to condense the vast quantity of steam produced by such power plants. For example, the nuclear power reactor at the Entergy Arkansas Unit 1 uses 750,000 gal of cooling water per minute. Having picked up heat from the secondary coolant, the cooling water must be cooled before being returned to its source. Such cooling is done in many nuclear power plants by passing the water through large concave cooling towers, which are often mistaken for the nuclear reactors themselves.

Not all nuclei can be made to fission on colliding with a neutron, but ^{235}U and ^{239}Pu are two fissionable isotopes. Natural uranium contains an average of only 0.72% of the fissionable ^{235}U isotope; more than 99% of the natural element is nonfissionable uranium-238. Since the percentage of natural ^{235}U is too small to sustain a chain reaction, uranium for nuclear power fuel must be enriched to about 3% uranium-235. To accomplish this, some of the ^{238}U isotope in a sample is effectively discarded, thereby raising the concentration of ^{235}U. If sufficient fissionable uranium-235 is present, it can capture enough neutrons to sustain the fission chain reaction. The minimum amount of fissionable material required for a self-sustaining chain reaction is termed the **critical mass.** Approximately one third of the fuel rods are replaced annually because fission byproducts absorb neutrons, reducing the efficiency of the fission reactions.

Because the amount of uranium-235 in the fuel rods of a nuclear power plant is lower than the critical mass needed for an atomic bomb, the reactor core *cannot* undergo an uncontrolled chain reaction to convert the reactor into an atomic bomb.

Nuclear fission produces a huge amount of energy. For example, the fissioning of 1.0 kg (2.2 lb) of uranium-235 releases 9.0×10^{13} J of energy, the equivalent of exploding 33,000 tons (33 kilotons) of TNT. Each UO_2 fuel pellet used in a nuclear reactor has the energy equivalent to burning 136 gal of oil, 2.5 tons of wood, or 1 ton of coal.

Conventional (nonnuclear) power plants burn fossil fuel to generate the heat to produce steam to drive the turbine.

Cooling towers are also used by fossil fuel–burning power plants.

Nuclear fuel rods that have become depleted in U-235 are known as *spent* fuel.

Uranium-238 can fission, but only when bombarded by fast neutrons, not like those in nuclear reactors. Thus, we consider uranium-238 to be nonfissionable in the context of nuclear reactors.

A sample is considered to be of weapons-grade quality only if its uranium-235 content is over 90%. Even in reactors using weapons-grade quality, the U-235 is still too dispersed to produce uncontrolled fission.

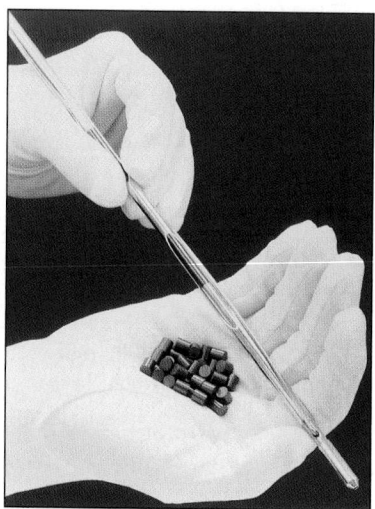

Uranium oxide pellets used in nuclear fuel rods.

Exercise 20.13 Energy of Nuclear Fission

The burning of 1.0 kg of high-grade coal produces 2.8×10^4 kJ of energy; the fissioning of 1.0 mol of uranium-235 generates 2.1×10^{10} kJ. How many metric tons of coal (1 metric ton = 10^3 kg) are needed to produce the same amount of energy as that released by the fission of 1.0 kg of uranium-235? (Assume that the processes have equal efficiency.)

There is, of course, substantial controversy surrounding the use of nuclear power plants, and not just in the United States. Nuclear power plant proponents argue that the health of our economy and our standard of living are dependent on inexpensive, reliable, and safe sources of energy. Just within the past few years the demand for electric power has once again begun to exceed the supply in the United States, so many believe nuclear power plants should be built to meet the demand. Nuclear power plants are capable of supplying these demands, and they can be the source of "clean" energy in that they do not pollute the atmosphere with ash, smoke, or oxides of sulfur, nitrogen, or carbon as coal-fired plants do. In addition, nuclear plants help to ensure that our supplies of fossil fuels will not be depleted as quickly in the near future, and they reduce our dependency on buying such fuels from other countries. There are currently 103 operating nuclear plants in the United States; 435 nuclear power plants worldwide in 30 nations produce about 17% of the world's electricity. The nuclear plants in the United States supply about 20% of our nation's electric energy (Figure 20.9). Around the world, 17 countries use nuclear power to generate at least 25% of their electricity. France uses 59 nuclear plants to generate three out of every four kilowatts of electricity (75%) in that country. In contrast, 73% of the electricity in Lithuania is produced by just two nuclear plants, indicating the differences in demand for electricity between the two countries.

Since the 1979 accident at the Three Mile Island nuclear power plant near Harrisburg, Pennsylvania, *no* construction of new nuclear power plants has begun in the United States. One problem associated with nuclear power plants is the highly radioactive fission products in the waste fuel. Since the 1950s, commercial nuclear reactors in the United States have produced about 30,000 tons of waste fuel; this amount is expected to reach 52,000 tons by 2005. Although some of the products are put to various uses (Section 20.9), many are not suitable as a fuel or for other purposes. Because these products are often highly radioactive and some have long half-lives (plutonium-239; $t_{1/2} = 24,400$ y), proper disposal of this high-level nuclear waste poses an enormous problem. Perhaps the most reasonable suggestion is that high-level radioactive wastes can be converted to a glassy material having a volume of about 2 m^3 per reactor per year. In 1996 a Department of Energy facility in Savannah River, South Carolina, began encapsulating radioactive waste in glass, a process called vitrification. A mixture of glass particles and radioactive waste is heated to 1200 °C. The molten mixture is poured into stainless steel cannisters, cooled, and stored. Eventually, such high-level nuclear wastes may be stored underground in geological formations, such as salt deposits, that are known to be stable for hundreds of millions of years. A site at Yucca Mountain, Nevada, is the designated national repository for high-level nuclear waste, such as that from spent nuclear reactor cores (see *Chemistry in the News*, p. 940).

In the United States, coal-fired plants generate more than 56% of electricity.

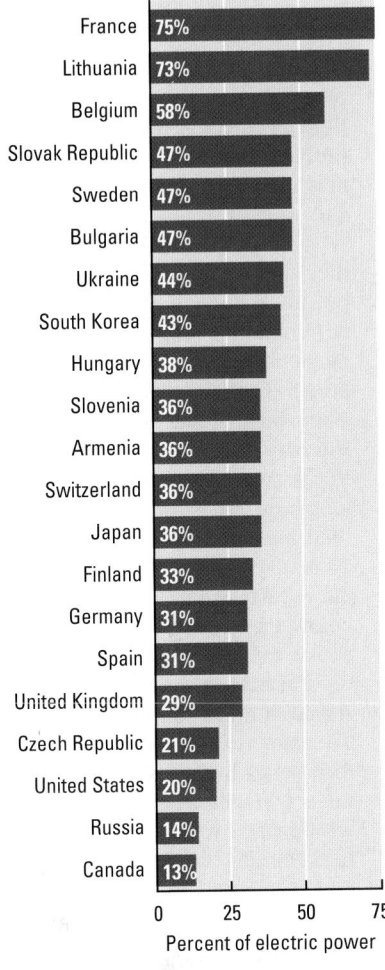

Figure 20.9 **The approximate share of electricity generated by nuclear power in various countries.** About 20% of the electricity in the United States is produced by nuclear power.

Exercise 20.14   Radioactive Decay of Fission Products

Unlike the 1979 incident at Three Mile Island, the accident at the Chernobyl nuclear plant in the former Soviet Union in 1986 released significant quantities of radioisotopes into the atmosphere. One of those radioisotopes was strontium-90 ($t_{1/2} = 29.1$ y). What fraction of strontium-90 released at that time remains?

 ### Exercise 20.15   Nuclear Waste

Cesium-137 ($t_{1/2} = 30.2$ y) is produced by ^{235}U fission. If ^{137}Cs is part of nuclear waste stored deep underground, how long will it take for the initial ^{137}Cs activity when it was first buried to drop

(a) By 60%?   (b) By 90%?

Nuclear waste is vitrified at the Savannah River Plant in South Carolina.

Chemistry in the News

WHERE TO PUT HIGH-LEVEL NUCLEAR WASTE FOR LONG-TERM STORAGE?

High-level radioactive waste disposal is a problem that has existed ever since the first nuclear power plants came online and nuclear weapons were developed. What to do with these wastes? In 1957, the National Academy of Sciences first proposed burying such wastes deep underground for long-term storage, and geologic formations suitable for such deep burial were identified in six states. Ultimately, a site at Yucca Mountain, Nevada, 100 miles northwest of Las Vegas, was designated as the national long-term repository, to be designed to hold high-level nuclear wastes safely for 10,000 years. The plan is to bury the nuclear waste in chambers about 1000 feet below the surface and at least 1000 feet above the water table. The wastes will be encased in ceramic or glass and stored in metal canisters for deep burial. To date, tunnels have been dug 1400 feet beneath Yucca Mountain to determine the adequacy of the site. The site would have to accommodate burial of beta-emitting wastes such as Sr-90 and Cs-137 for 300 to 500 years, which is about ten half-lives for such radioactive isotopes. The repository would also have to store safely radioisotopes with much longer half-lives, such as Pu-239 (24,400 y), but with lower radiation intensity. If completed, the repository would have a capacity to store 70,000 tons of spent nuclear fuel and 8000 tons of high-level military nuclear waste. At the rate of 20 shipments per day, it would take at least 20 to 25 years just to transport the accumulated waste to Yucca Mountain. Once buried, the nuclear waste could ostensibly be recovered and reprocessed to extract fissionable material from it, such as

The tunnel-boring machine used in the geological study of Yucca Mountain underground nuclear waste repository.

plutonium-239. There is currently a moratorium in the United States on the reprocessing of waste nuclear fuel.

To fulfill the requirements of the 1982 Nuclear Waste Policy Act, the Department of Energy expected to begin receiving nuclear waste at Yucca Mountain in 1998. However, due to political maneuvering and technologic delays, no nuclear waste is stored as yet at the site. The state of Nevada has passed legislation opposing the burial of nuclear waste within its borders. At present, an environmental impact statement and a

site characterization report are being prepared by the U.S. Department of Energy. If the approval processes proceed according to plan, waste disposal could begin as soon as 2010. It is apparent that political as well as technical problems are associated with the Yucca Mountain project, and the ultimate fate of the project cannot be predicted at this time.

Source:

www.epa.gov/radiation/yucca

20.7 NUCLEAR FUSION

Tremendous amounts of energy are generated when very light nuclei combine to form heavier nuclei. Such a reaction is called **nuclear fusion,** and one of the best examples is the fusion of hydrogen nuclei (protons) to give helium nuclei.

$$4\,^{1}_{1}H \longrightarrow\, ^{4}_{2}He + 2\,^{0}_{+1}e \qquad\qquad \Delta E = -2.5 \times 10^{9}\ \text{kJ}$$

The helium nucleus produced by this reaction is more stable than the reactant hydrogen nuclei, as shown in Figure 20.3. This reaction is the source of the energy from our sun and other stars, and it is the beginning of the synthesis of the elements in the universe (Section 21.1). Temperatures of 10^6 to 10^7 K, found in the core and radiative zone of the sun, are required to bring the positively charged ^1_1H nuclei together with enough kinetic energy to overcome nuclear repulsions and react.

Deuterium can also be fused to give helium-3,

$$^2_1\text{H} + {}^2_1\text{H} \longrightarrow {}^3_2\text{He} + {}^1_0\text{n} \qquad\qquad \Delta E = -3.2 \times 10^8 \text{ kJ}$$

which can undergo further fusion with a proton to give helium-4.

$$^1_1\text{H} + {}^3_2\text{He} \longrightarrow {}^4_2\text{He} + {}^0_{+1}\text{e} \qquad\qquad \Delta E = -1.9 \times 10^9 \text{ kJ}$$

Each of these reactions releases an enormous quantity of energy, so it has been the dream of nuclear physicists to harness them to provide energy for the people of the world.

At the very high temperatures that allow fusion reactions to occur rapidly, atoms do not exist as such. Instead, there is a **plasma,** which consists of unbound nuclei and electrons. In order to achieve the high temperatures required for the fusion reaction of the hydrogen bomb, a fission bomb (atomic bomb) is first set off. One type of hydrogen bomb depends on the production of tritium in the bomb. In this type, lithium-6 deuteride (LiD, a solid salt) is placed around an ordinary ^{235}U or ^{239}Pu fission bomb, and the fission reaction is set off in the usual way. A ^6Li nucleus absorbs one of the neutrons produced and splits into tritium and helium. The temperature

$$^1_0\text{n} \quad + \quad {}^6_3\text{Li} \quad \longrightarrow \quad {}^3_1\text{H} \quad + \quad {}^4_2\text{He}$$

reached by the fission of uranium or plutonium is high enough to bring about the fusion of tritium and deuterium with the release of 1.7×10^9 kJ per mole of ^3H. A 20-megaton hydrogen bomb usually contains about 300 lb of lithium deuteride, as well as a considerable amount of plutonium and uranium.

Development of nuclear fusion as a commercial energy source is inviting because hydrogen isotopes are available (from water), and fusion products are generally nonradioactive or have short half-lives, which eliminates problems associated with the disposal of high-level radioactive fission reactor products. Controlling a nuclear fusion reaction for peaceful commercial uses, however, has proven to be extraordinarily difficult. Three critical requirements must be met for controlled fusion. First, the temperature must be high enough for fusion to occur. The fusion of deuterium and tritium, for example, requires a temperature of 100 million °C or more. Second, the plasma must be confined long enough to release a net output of energy. Third, the energy must be recovered in some usable form.

Magnetic "bottles" (enclosures in space bounded by magnetic fields) have confined the plasma so that controlled fusion has been achieved. But the energy generated by it has been less than that required to produce and control the fusion reaction. Using more energy to produce less energy is not a commercially appealing investment. Thus, commercial fusion reactors are not likely in the near future without a dramatic breakthrough in fusion technology.

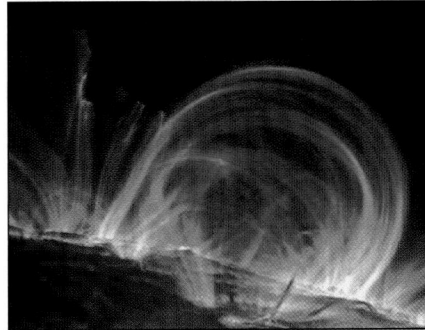

Nuclear fusion reactions power the sun, whose surface is shown here.

Containment is one of the biggest problems in developing controlled nuclear fusion.

Exercise 20.16 Nuclear Fusion

Complete the equations for these nuclear fusion reactions.

(a) $^7_3\text{Li} + {}^1_1\text{H} \rightarrow {}^1_0\text{n} + \underline{\quad}$ (b) $^2_1\text{H} + \underline{\quad} \rightarrow {}^4_2\text{He} + {}^1_1\text{H}$

20.8 NUCLEAR RADIATION: EFFECTS AND UNITS

The use of nuclear energy and radiation is a double-edged sword that carries risks and benefits. It can be used to harm (nuclear armaments) or to cure (radioisotopes in medicine).

Alpha, beta, and gamma radiation disrupt normal cell processes in living organisms by interacting with the atoms comprising their biomolecules, breaking covalent bonds, and producing energetic radicals and ions that can lead to further disruptive reactions. The potential for serious radiation damage to humans is well known. The biological effects of the atomic bombs exploded at Hiroshima and Nagasaki, Japan, at the close of World War II in 1945 have been well documented. However, controlled exposure to nuclear radiation can be beneficial in destroying malignant tissue, as in radiation therapy for treating some cancers.

Radiation Units

To quantify radiation and its effects, particularly on humans, several units have been developed. For example, the **roentgen (R)** is used to give the dosage of x-rays and γ rays. One roentgen corresponds to the *deposition* of 93.3×10^{-7} J per gram of tissue. The **rad** (radiation *a*bsorbed *d*ose) is similar to the roentgen, but it measures the amount of radiation *absorbed;* 1 rad represents a dose of 1.00×10^{-2} J absorbed per kilogram of material. The SI unit of absorbed radiation dose is the **gray** (Gy), which is equal to the absorption of 1 J per kilogram of material. Thus, 1 Gy = 100 rad.

The biological effects of radiation per rad or gray differ with the kind of radiation, which can be quantified more generally using the **rem** (standing for *r*oentgen *e*quivalent applied to *m*ammals).

Effective dose in rems = quality factor × dose in rads

The quality factor depends on the type of radiation and other factors. The quality factor is arbitrarily set as 1 for beta and gamma radiation; it is between 10 and 20 for alpha particles, depending on total dose, dose rate, and type of tissue. Since one rem is a large amount of radiation, the millirem (mrem) is commonly used (1 mrem = 10^{-3} rem). The SI unit of effective dose is the **sievert** (Sv), which is defined similarly to the rem, except that the absorbed dose is in grays, not rads. Consequently, 1 Sv = 100 rem.

Background Radiation

Humans are constantly exposed to natural and artificial **background radiation,** estimated to be about 360 mrem per year (Figure 20.10), well below 500 mrem,

All technologies carry risks as well as benefits. In the 1800s, railroads were new and the poet William Wordsworth wrote of their risks and benefits in terms of "Weighing the mischief with the promised gain. . . ."

The roentgen is named to honor Wilhelm Röntgen, the German physicist who discovered x-rays.

There are no observable physiologic effects from a single dose of radiation less than 25 rem (25×10^3 mrem).

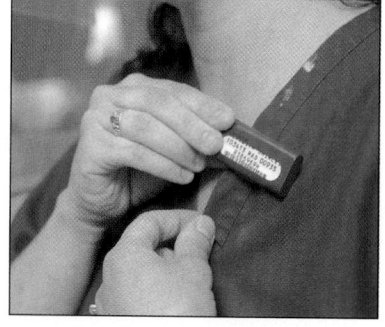

Individuals who work where there is potential danger from exposure to excessive nuclear radiation wear film badges to monitor their radiation dose.

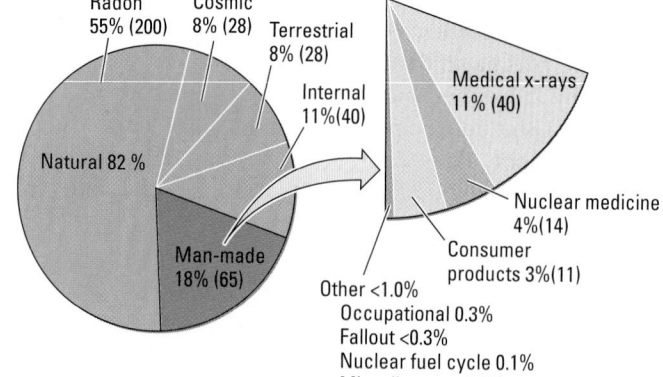

Figure 20.10 Sources of average background radiation exposure in the United States. The sources are expressed as percentages of the total, as well as in millirems per year, the values in parentheses. The background radiation from natural sources far exceeds that from artificial sources.

CHEMISTRY
You Can Do...

Counting Millirems: Your Radiation Exposure

The Committee on Biological Effects of Ionizing Radiation of the National Academy of Sciences issued a report in 1987 that contained a survey for an individual to evaluate his or her exposure to ionizing radiation. The table below is adapted from this report and updated. By adding up your exposure, you can compare your annual dose to the United States annual average of 360 mrem.

	Common sources of radiation	Your annual dose (mrem)
Where you live	**Location:** Cosmic radiation at sea level	27
	For your elevation (in feet), add this number of mrem	____
	Elevation *mrem* *Elevation* *mrem* *Elevation* *mrem* 1000 2 4000 15 7000 40 2000 5 5000 21 8000 53 3000 9 6000 29 9000 70	
	Ground: U.S. average	26
	Radon: U.S. average	200
	House construction: For stone, concrete, or masonry building, add 7; wood add 30	____
What you eat, drink and breathe	**Radioisotopes** in the body from	
	Food, air, water: U.S. average	40
	Weapons test fallout	4
How you live	**X-ray and radiopharmaceutical diagnosis**	
	Number of chest x-rays ____ × 10	____
	Number of lower gastrointestinal tract x-rays ____ × 500	____
	Number of radiopharmaceutical examinations ____ × 300	____
	(Average dose to total U.S. population = 53 mrem)	
	Jet plane travel: For each 2500 miles add 1 mrem	____
	TV viewing: Number of hours per day ____ × 0.15	____
How close you live to a nuclear plant	**At site boundary:** Average number of hours per day ____ × 0.2	____
	One mile away: Average number of hours per day ____ × 0.02	____
	Five miles away: Average number of hours per day ____ × 0.002	____
	Over 5 miles away: . . . . none	____
	Note: Maximum allowable dose determined by "as low as reasonably achievable" (ALARA) criteria established by the U.S. Nuclear Regulatory Commission. Experience shows that your actual dose is substantially less than these limits.	
	Your total annual dose in mrem:	____

Compare your annual dose with the U.S. annual average of 360 mrem.

* Based on the BEIR Report III. National Academy of Sciences, Committee on Biological Effects of Ionizing Radiation. The Effects on Populations of Exposure to Low Levels of Ionizing Radiation," Washington, DC: National Academy of Sciences, 1987.

the federal government's background radiation standard for the general public. Note that *most* background radiation, about 300 mrem per year (82%), comes from *natural* background radiation sources: cosmic radiation and radioactive elements and minerals found naturally in the earth, air, and materials around and within us. The remaining 18% comes from artificial sources.

Cosmic radiation, emitted by the sun and other stars, continually bombards the earth and accounts for about 8% of natural background radiation. The remainder comes from radioactive isotopes such as ^{40}K. Potassium is present to the extent of about 0.3 g/kg of soil and is essential to all living organisms. We all acquire some radioactive potassium from the foods we eat. For example, a hamburger contains 960 mg of ^{40}K, giving off 29 dps; a hot dog contains 200 mg of ^{40}K and gives off 6 dps; a serving of French fries has 650 mg of ^{40}K and gives off 20 dps. Other radioactive elements found in some abundance on the earth are thorium-232 and uranium-238. Approximately 8% of the natural background radiation arises from Th-232 and U-238 in rocks and soil. Thorium, for example, is found to the extent of 12 g/1000 kg of soil. Most natural background radiation comes from radon, a byproduct of radium decay, as discussed in the next subsection.

On average, roughly 15% of our annual exposure comes from medical procedures such as diagnostic x-rays and the use of radioactive compounds to trace the body's functions. Consumer products account for 3% of our total annual exposure. Contrary to popular belief, less than 1% comes from sources such as the radioactive products from testing nuclear explosives in the atmosphere, nuclear power plants and their wastes, nuclear weapons manufacture, and nuclear fuel processing.

Burning fossil fuels (coal and oil) releases into the atmosphere considerable amounts of naturally occurring radioactive isotopes originally in the fossil fuel. Thus, fossil fuel plants add significantly to the background radiation. Far more thorium and uranium are released annually into the atmosphere from fossil fuel-burning plants than from nuclear power plants.

Radon

Radon is a chemically inert gas in the same periodic table group as helium, neon, argon, and krypton. Radon-222 is produced in the decay series of uranium-238 ($\Leftarrow$ *p. 923, Figure 20.1*). Other isotopes of Rn are products of other decay series. Although chemically inert, radon is problematic because it is radioactive; Figure 20.10 shows that radon accounts for 55% of natural background radiation.

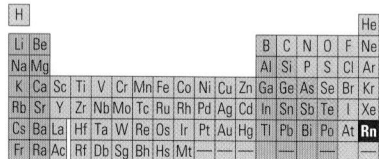

It should be kept in mind that radon occurs naturally in our environment. Because it comes from natural uranium deposits, the amount of radon depends on the nature of the geology of the rocks and soil in a given locality. Furthermore, since the gas is chemically inert and has a half-life of 3.82 days, it is not trapped by chemical processes in the soil or water. Thus, it is free to seep up from the ground and into underground mines or into homes through pores in concrete block walls, cracks in basement floors or walls, and around pipes. Radon-222 decays to give polonium-218, a radioactive, heavy metal element that is not a gas and is not chemically inert. If radon is inhaled, this decay will occur deep in the lungs, and ^{218}Po will be generated in the lungs.

$$^{222}_{86}\text{Rn} \longrightarrow {}^{4}_{2}\text{He} + {}^{218}_{84}\text{Po} \qquad\qquad t_{1/2} = 3.82 \text{ d}$$

$$^{218}_{84}\text{Po} \longrightarrow {}^{4}_{2}\text{He} + {}^{214}_{82}\text{Pb} \qquad\qquad t_{1/2} = 3.10 \text{ m}$$

Therefore, polonium-218 can lodge in lung tissues, where it undergoes α decay to give lead-214, itself a radioactive isotope. The range of an α particle is quite small, perhaps 0.7 mm (about the thickness of a sheet of paper). However, this is approximately the thickness of the epithelial cells of the lungs, so the radiation can damage these tissues and induce lung cancer.

Most homes in the United States are believed to have some level of radon gas. There is currently a great deal of controversy over the level of radon that is considered "safe." Estimates indicate that only about 6% of U.S. homes have radon levels above 4 picocuries per liter (pCi/L) of air, the "action level" standard set by the U.S. Environmental Protection Agency. There are some who believe 1.5 pCi/L is more likely the average level and that only about 2% of the homes will contain over 8 pCi/L. To test for the presence of radon, you can purchase testing kits of various kinds. If your home shows higher levels of radon gas than 4 pCi/L, you should probably have it tested further and perhaps take corrective actions such as sealing cracks around the foundation and in the basement. But keep in mind the relative risks involved ($\Leftarrow$ *p. 31*). A 1.5 pCi/L level of radon leads to a lung cancer risk about the same as the risk of your dying in an accident in your home.

1 picocurie, pCi = 10^{-12} Ci.

Exercise 20.17   Radon Levels

Calculate how long it will take for the activity of a radon-222 sample ($t_{1/2}$ = 3.82 d) initially at 8 pCi to drop

(a) To 4 pCi, the EPA action level.
(b) To 1.5 pCi, approximately the U.S. average.

A commercially available kit to test for radon gas in a residence.

20.9 APPLICATIONS OF RADIOACTIVITY

Food Irradiation

In some parts of the world, spoilage of stored food may claim up to 50% of the food crop. In our society, refrigeration, canning, and chemical additives lower this figure considerably. Still, there are problems with food spoilage, and food preservation costs are a substantial fraction of the final cost of food. Food and grains can also be preserved by gamma irradiation. Contrary to some popular opinion, such irradiation does *not* make foods radioactive, just as a dental x-ray does not make you radioactive.

Food irradiation with gamma rays from ^{60}Co or ^{137}Cs sources is allowed in 40 countries and is endorsed by the World Health Organization and the American Medical Association. Astronauts' food has been preserved by gamma irradiation. The United States and several other countries require that foods preserved by irradiation be labeled with the international symbol for irradiated food.

Bacteria, molds, and yeasts are killed or their growth retarded by irradiation. As a result, the shelf life of irradiated foods during refrigeration is prolonged in much the same way that heat pasteurization protects milk. In recent years, outbreaks of foodborne illnesses caused by new types of harmful bacteria or inappropriate food handling have heightened interest in the benefits of irradiation as a safety measure, especially for use with meat.

The U.S. Food and Drug Administration has approved irradiation of meat, poultry, and a variety of fresh fruits and vegetables and spices. Except for spices, however, foods treated with irradiation are not as yet widely available.

The FDA permits irradiation up to 300 kilorads for the pasteurization of poultry. Radiation levels in the 1- to 5-megarad (1 megarad = 10^6 rads) range sterilize, killing every living organism. Foods irradiated at these levels will keep indefinitely when sealed in plastic or aluminum foil packages. However, FDA approval is unlikely for irradiation sterilization of foods in the near future because of potential problems caused by as yet undiscovered, but possible, "unique radiolytic products." For example, irradiation sterilization might produce a substance that is capable of causing genetic damage. To prove or disprove the presence of these substances, animal feeding studies using irradiated foods are now being conducted in the United States.

In addition to preserving foods, gamma irradiation is used to sterilize bandages, contact lens solutions, and many cosmetics.

Radioactive Tracers

The chemical behavior of a radioisotope is essentially identical to that of the nonradioactive isotopes of the same element. Compounds containing radioactive atoms are formed and undergo chemical reactions in exactly the same way as compounds containing no radioactive atoms. Therefore, chemists can use radioactive isotopes as **tracers** in nonbiological and biological chemical reactions. To use a tracer, a chemist prepares a reactant compound in which one of the elements consists of both radioactive and stable (nonradioactive) isotopes, and introduces it into the reaction (or feeds it to an organism). After the reaction, the chemist

Strawberries preserved by gamma irradiation.

TABLE 20.3	Radioisotopes Used as Tracers	
Isotope	**Half-life**	**Use**
^{14}C	5730 y	CO_2 for photosynthesis research
^{3}H	12.33 y	Tag hydrocarbons
^{35}S	87.2 d	Tag pesticides, measure air flow
^{32}P	14.3 d	Measure phosphorus uptake by plants

Melvin Calvin used ^{14}C to monitor the uptake and release of $^{14}CO_2$ in order to determine the basic biochemical pathways of photosynthesis. This groundbreaking work earned him the 1961 Nobel Prize in Chemistry.

measures the radioactivity of the products (or determines which parts of the organism contain the radioisotope) by using a Geiger counter or other radiation detector. Several radioisotopes commonly used as tracers are listed in Table 20.3.

For example, plants take up phosphorus-containing compounds from the soil through their roots. The use of the radioactive phosphorus isotope ^{32}P, a beta emitter, provides a way to detect the uptake of phosphorus by a plant, and also to measure the speed of uptake under various conditions. Plant biologists can grow hybrid strains of plants that absorb phosphorus, an essential nutrient, quickly. They can test the new plants by measuring their uptake of the radioactive ^{32}P tracer. This type of research leads to faster maturing crops, better yields per acre, and more food or fiber at less expense.

Medical Imaging

Radioactive isotopes are used in **nuclear medicine** in two different ways: diagnosis and therapy. In the diagnosis of internal disorders such as tumors, physicians need information on the locations of abnormal tissue. This is done by imaging, a technique in which the radioisotope, either alone or combined with some other substance, accumulates at the site of the disorder. There, the radioisotope decays, emitting its characteristic radiation, which is detected. Modern medical diagnostic instruments not only determine where the radioisotope is located in the patient's body, but also construct an image of the volume within the body where the radioisotope is concentrated.

Four of the most common diagnostic radioisotopes are given in Table 20.4. Most are made in a particle accelerator in which heavy, charged nuclear particles are made to react with other atoms. Each of these radioisotopes produces gamma radiation, which in low doses is less harmful to the tissue than ionizing radiations such as beta or alpha particles because gamma rays pass through the tissue. By the use of special carrier compounds, these radioisotopes can be made to accumulate

TABLE 20.4	Diagnostic Radioisotopes		
Radioisotope	**Name**	**Half-life (hours)**	**Site for diagnosis**
$^{99m}Tc^{*}$	Technetium-99m	6.0	As $^{99m}TcO_4^-$ to the thyroid
^{201}Tl	Thallium-201	72.9	To the heart
^{123}I	Iodine-123	13.2	To the thyroid
^{67}Ga	Gallium-67	78.2	To various tumors and abscesses

* The technetium-99m isotope is the radioisotope most commonly used for diagnostic purposes. The m stands for "metastable."

in specific areas of the body. For example, the pyrophosphate ion, $P_4O_7^{4-}$, can bond to the technetium-99m radioisotope. Together they accumulate in the skeletal structure where abnormal bone metabolism is occurring (Figure 20.11). The technetium-99m radioisotope is metastable, as denoted by the letter m; this term means that the nucleus loses energy by disintegrating to a more stable version of the same isotope,

$$^{99m}\text{Tc} \longrightarrow {}^{99}\text{Tc} + \gamma$$

and the gamma rays are detected. Such investigations often pinpoint bone tumors.

Exercise 20.18 Rate of Radioactive Decay

Gallium citrate containing radioactive gallium-67 is used medically as a tumor-seeking agent. It has a half-life of 78.2 hours. How much time is needed for a gallium citrate sample to reach 10% of its original activity?

Exercise 20.19 Half-Life

Chromium-51 is a radioisotope ($t_{1/2} = 27.7$ days) used to evaluate the lifetime of red blood cells; the radioisotope iron-59 ($t_{1/2} = 44.5$ days) is used to assess bone marrow function. A hospital laboratory has 80 mg of iron-59 and 100 mg of chromium-51. After 90 days, which radioisotope is present in greater mass—chromium-51 or iron-59?

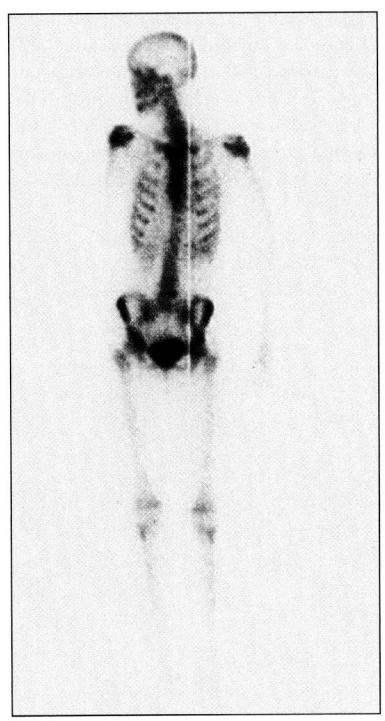

Figure 20.11 A whole-body scan. Phosphate with technetium-99m was injected into the blood and then absorbed by the bones and kidneys. This picture was taken three hours after injection.

Paradoxically, high-energy radiation can kill healthy cells, but it is also used therapeutically to kill malignant, cancerous cells—those exhibiting rapid, uncontrolled growth. Because they divide more rapidly than normal cells, malignant cells are killed because they are more susceptible to radiation damage. For external radiation therapy, a narrow beam of high-energy gamma radiation from a cobalt-60 or cesium-137 source is focused on the cancerous cells. Internal radiation therapy uses gamma-emitting salts of radioisotopes such as ^{192}Ir ($t_{1/2} = 73.8$ days). The radioactive salts are encapsulated in platinum or gold "seeds" or needles and surgically implanted into the body. Because the thyroid gland uses iodine, thyroid cancer can be treated internally by administering orally a sodium iodide solution containing a relatively high concentration of radioactive iodine-131.

Positron emission tomography (PET) is a form of nuclear imaging that uses positron emitters, such as carbon-11, fluorine-18, nitrogen-13, or oxygen-15. All these radioisotopes are neutron-deficient, have short half-lives, and therefore must be prepared in a cyclotron immediately before use. When these radioisotopes decay, a proton is converted into a neutron, a positron, and a neutrino; the neutrino is generally not shown in the equation.

$$^{1}_{1}\text{p} \longrightarrow {}^{1}_{0}\text{n} + {}^{0}_{+1}\text{e}$$

Since matter is essentially transparent to neutrinos, they escape undetected, but the positron, $^{0}_{+1}\text{e}$, travels on average less than a few millimeters before it encounters an electron, $^{0}_{-1}\text{e}$, and undergoes antimatter-matter annihilation.

$$^{0}_{+1}\text{e} + {}^{0}_{-1}\text{e} \longrightarrow 2\,\gamma$$

The annihilation event produces two gamma rays that move in opposite directions and are detected by detectors located 180° apart in the PET scanner. By detecting several million annihilation gamma rays within a circular field around the subject

The neutrino, first discovered experimentally in 1950, is a subatomic particle with zero electric charge and a mass less than that of an electron.

Figure 20.12 **PET (positron emission tomography) of an axial section through a normal human brain.** PET scans are obtained by injecting a tracer labeled with a short-lived radioisotope into the bloodstream. The isotope concentrates in brain tissue and emits positrons. The positrons react with electrons to create gamma rays that are recorded by a circular detector when the scan is performed. Here, radioactive methionine (an amino acid) has been used to show the level of activity of protein synthesis in the brain.

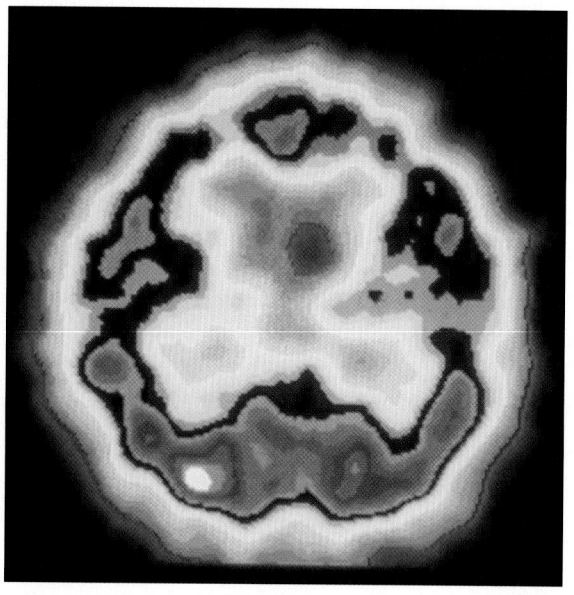

over approximately 10 min, the region of tissue containing the radioisotope can be imaged with computer signal-averaging techniques (Figure 20.12).

SUMMARY PROBLEM

(a) Seaborgium, element 106, is an α emitter. Write the nuclear equation for α emission by ^{269}Sg.

(b) Naturally occurring uranium contains more than 99% uranium-238 and less than 1% uranium-235. Therefore, sooner or later the supplies of ^{235}U will be depleted. Plutonium-239 is formed indirectly in nuclear reactors by the bombardment of uranium-238 with neutrons. The bombardment initially forms uranium-239 that is converted to plutonium-239 through two consecutive beta emissions. Write nuclear equations for the conversion of ^{238}U to ^{239}Pu.

(c) Neutrons from the fission of uranium-235 in so-called breeder reactors can bombard nonfissionable ^{238}U and convert it into fissionable ^{239}Pu, such as that used in the atomic bomb exploded over Nagasaki, Japan, in World War II. Plutonium-239 can fission to form, for example, strontium-90 and barium-147. Write a nuclear equation for this fission reaction.

(d) Hydrogen bombs that use fusion reactions were developed following World War II. One reaction used in a hydrogen bomb was

$$^2_1H + {^3_1}H \longrightarrow {^4_2}He + {^1_0}n$$

Calculate the energy released, in kilojoules per gram of reactants, for this fusion reaction. The necessary nuclear masses are $^2_1H = 2.01355$ g/mol; $^3_1H = 3.01550$ g/mol; $^4_2He = 4.00150$ g/mol; $^1_0n = 1.00867$ g/mol.

(e) The goal of recent nuclear arms treaties has been to dismantle the stockpiles of nuclear weapons built up by the United States and the former Soviet Union since World War II, including those containing plutonium-239 ($t_{1/2} = 2.44 \times 10^4$ y). How long will it take for the activity of plutonium-239 in a nuclear warhead to decrease (i) to 75% of its initial activity? (ii) to 10% of its initial activity?

(f) Deep underground burial has been proposed for long-term storage of the ^{239}Pu waste removed from nuclear weapons. Based on the answers to Summary

Problem (e), comment on factors that need to be considered for the storage and burial of such nuclear waste.

(g) Cobalt-60, used in cancer treatments, is a beta and gamma emitter ($t_{1/2}$ = 5.27 y). A cobalt-60 sample has an activity of 7.0×10^{11} dps.

(i) Write a nuclear equation for the conversion of cobalt-60 to nickel-59 by beta emission.

(ii) Calculate the activity of the cobalt-60 sample after 25.0 years.

IN CLOSING

Having studied this chapter, you should be able to . . .

- Characterize the three major types of radiation observed in radioactive decay: alpha (α), beta (β), and gamma (γ) (Section 20.1).
- Write a balanced equation for a nuclear reaction or transmutation (Section 20.2).
- Decide whether a particular radioactive isotope will decay by α, β, or positron emission or by electron capture (Sections 20.2 and 20.3).
- Calculate the binding energy for a particular isotope and understand what this energy means in terms of nuclear stability (Section 20.3).
- Use the equation $\ln\dfrac{N}{N_0} = -kt$ (Equation 20.2), which relates (through the decay constant k) the time period over which a sample is observed (t) to the number of radioactive atoms present at the beginning (N_0) and end (N) of the time period (Section 20.4).
- Calculate the half-life of a radioactive isotope ($t_{1/2}$) from the activity of a sample, or use the half-life to find the time required for an isotope to decay to a particular activity (Section 20.4).
- Describe nuclear chain reactions, nuclear fission, and nuclear fusion (Sections 20.6 and 20.7).
- Describe the basic functioning of a nuclear power reactor (Section 20.6).
- Describe some sources of background radiation and the units used to measure radiation (Section 20.8).
- Give examples of some uses of radioisotopes (Section 20.9).

KEY TERMS

activity *(20.4)*
alpha (α) particles *(20.1)*
alpha radiation *(20.1)*
background radiation *(20.8)*
becquerel (Bq) *(20.4)*
beta (β) particles *(20.1)*
beta radiation *(20.1)*
binding energy *(20.3)*
binding energy per nucleon *(20.3)*

critical mass *(20.6)*
curie (Ci) *(20.4)*
electron capture *(20.2)*
gamma (γ) radiation *(20.1)*
gray (Gy) *(20.8)*
nuclear fission *(20.6)*
nuclear fusion *(20.7)*
nuclear medicine *(20.9)*
nuclear reactions *(20.2)*
nuclear reactor *(20.6)*

nucleons *(20.2)*
plasma *(20.7)*
positron *(20.2)*
rad *(20.8)*
radioactive series *(20.2)*
rem *(20.8)*
roentgen (R) *(20.8)*
sievert *(20.8)*
tracers *(20.9)*

QUESTIONS FOR REVIEW AND THOUGHT

Conceptual Challenge Problems

CP-20.A (Section 20.4) The half-life for the alpha decay of uranium-238 to thorium-234 is 4.5×10^9 years, which happens to be the estimated age of the earth.
(a) How many atoms were decaying per second in a 1.0-g sample of uranium-238 that existed 1.0×10^6 years ago?
(b) How would you find the number of atoms now decaying per second in this sample?

CP-20.B (Section 20.4) If the earth is 4.5×10^9 years old and the amount of radioactivity in a sample becomes smaller with time, how is it possible for there to be any radioactive elements left on earth that have half-lives less than a few million years?

CP-20.C (Section 20.4) Using experiments based on a sample of living wood, a nuclear chemist estimates that the uncertainty of her measurements of the carbon-14 radioactivity in the sample is 1.0%. The half-life of carbon-14 is 5730 y.
(a) How long must a sample of wood be separated from a living tree before the chemist's radioactivity measurements on the sample support the time when it died?
(b) Suppose that the chemist's uncertainty in the radioactivity of carbon-14 continues to be 1.0% of the radioactivity

of living wood. How long must a sample of wood be dead before the chemist's measurements support the claim that the time since the wood was separated from the tree is not changing?

CP-20.D (Section 20.8) You have read that alpha radiation is the least penetrating type of radiation, followed by beta radiation. Gamma radiation penetrates matter well, and thick samples of matter are required to contain gamma radiation. Knowing these facts, what can you correctly deduce about the harmful effects of these three types of radiation on living tissue?

CP-20.E (Section 20.8) Death will likely occur within weeks to a 150-lb person who receives 500,000 mrem of radiation over a short time, an exposure that is 1000 times the federal government's standard for 1 year (500 mrem/y). A student realizes that 500,000 mrem is 500 rem, and that 500 rem has the effect of depositing 317 J of energy on the body of the 150-lb person. The student is puzzled. How can the deposition of only 317 J of energy from nuclear radiation, much less energy than that deposited by cooling a cup of coffee 1 °C within a person's body, have such a disastrous effect on the person?

Answers to questions in **bold** can be found in the back of the book.

Review Questions

1. Complete the tables.

	Symbol	Mass	Charge
α particle	___	___	___
β particle	___	___	___
γ radiation	___	___	___

	Ionizing power	Penetrating power
α particle	___	___
β particle	___	___
γ radiation	___	___

2. Compare nuclear and chemical reactions in terms of changes in reactants, type of products formed, and conservation of matter and energy.
3. What is meant by "the band of stability"?
4. What is the binding energy of a nucleus?
5. If the mass number of an isotope is much greater than twice the atomic number, what type of radioactive decay might you expect?

6. If the number of neutrons in an isotope is much less than the number of protons, what type of radioactive decay might you expect?
7. Define critical mass and chain reaction.
8. What is the difference between nuclear fission and fusion? Illustrate your answer with an example of each.
9. Use the Internet to locate the nuclear reactor power plant nearest to your college residence. Do you consider it a threat to your health and safety? If so, why? If not, why not?
10. Name at least two uses of radioactive isotopes (outside of their use in power reactors and weapons).

Nuclear Reactions

11. Fill in the mass number, atomic number, and symbol for the missing particle in each nuclear equation.
(a) $^{242}_{94}\text{Pu} \rightarrow {}^4_2\text{He} +$ _____
(b) _____ $\rightarrow {}^{32}_{16}\text{S} + {}^0_{-1}\text{e}$
(c) $^{252}_{98}\text{Cf} +$ _____ $\rightarrow 3\,{}^1_0\text{n} + {}^{259}_{103}\text{Lr}$
(d) $^{55}_{26}\text{Fe} +$ _____ $\rightarrow {}^{55}_{25}\text{Mn}$
(e) $^{15}_{8}\text{O} \rightarrow$ _____ $+ {}^0_{+1}\text{e}$
12. Fill in the mass number, atomic number, and symbol for the missing particle in each nuclear equation.
(a) _____ $\rightarrow {}^{22}_{10}\text{Ne} + {}^0_{+1}\text{e}$
(b) $^{122}_{53}\text{I} \rightarrow {}^{122}_{54}\text{Xe} +$ _____
(c) $^{210}_{84}\text{Po} \rightarrow$ _____ $+ {}^4_2\text{He}$
(d) $^{195}_{79}\text{Au} +$ _____ $\rightarrow {}^{195}_{78}\text{Pt}$
(e) $^{241}_{94}\text{Pu} + {}^{16}_{8}\text{O} \rightarrow 5\,{}^1_0\text{n} +$ _____
13. Write balanced nuclear equations for each word statement.
(a) Magnesium-28 undergoes β emission.

(b) When uranium-238 is bombarded with carbon-12, four neutrons are emitted and a new element forms.

(c) Hydrogen-2 and helium-3 react to form helium-4 and another particle.

(d) Argon-38 forms by positron emission.

(e) Platinum-175 forms osmium-171 by spontaneous radioactive decay.

14. Write balanced nuclear equations for each word statement.

(a) Einsteinium-253 combines with an alpha particle to form a neutron and a new element.

(b) Nitrogen-13 undergoes positron emission.

(c) Iridium-178 captures an electron to form a stable nucleus.

(d) A proton and boron-11 fuse together, forming three identical particles.

(e) Nobelium-252 and six neutrons form when carbon-12 collides with a transuranium isotope.

15. One radioactive series that begins with uranium-235 and ends with lead-207 undergoes the following sequence of emission reactions: α, β, α, β, α, α, α, α, β, β, α. Identify the radioisotope produced in each of the *first five steps*.

16. One radioactive series that begins with uranium-235 and ends with lead-207 undergoes the following sequence of emission reactions: α, β, α, β, α, α, α, α, β, β, α. Identify the radioisotope produced in each of the *last six steps*.

Nuclear Stability

17. Write a nuclear equation for the type of decay each of these unstable isotopes is most likely to undergo.

(a) Neon-19   (b) Thorium-230

(c) Bromine-82   (d) Polonium-212

18. Write a nuclear equation for the type of decay each of these unstable isotopes is most likely to undergo.

(a) Silver-114   (b) Sodium-21

(c) Radium-226   (d) Iron-59

19. Boron has two stable isotopes, ^{10}B (abundance = 19.78%) and ^{11}B (abundance = 80.22%). Calculate the binding energies per nucleon of these two nuclei and compare their stabilities.

$$5\,^1_1H + 5\,^1_0n \longrightarrow\ ^{10}_5B$$

$$5\,^1_1H + 6\,^1_0n \longrightarrow\ ^{11}_5B$$

The required masses (in g/mol) are $^1_1H = 1.00783$; $^1_0n = 1.00867$; $^{10}_5B = 10.01294$; and $^{11}_5B = 11.00931$.

20. Calculate the binding energy in kJ per mole of P for the formation of $^{30}_{15}P$ and $^{31}_{15}P$.

$$15\,^1_1H + 15\,^1_0n \longrightarrow\ ^{30}_{15}P$$

$$15\,^1_1H + 16\,^1_0n \longrightarrow\ ^{31}_{15}P$$

Which is the more stable isotope? The required masses (in g/mol) are $^1_1H = 1.00783$; $^1_0n = 1.00867$; $^{30}_{15}P = 29.97832$; and $^{31}_{15}P = 30.97376$.

Rates of Disintegration Reactions

21. Sodium-24 is a diagnostic radioisotope used to measure blood circulation time. How much of a 20-mg sample remains after 1 d and 6 h if sodium-24 has a $t_{1/2} = 15$ h?

22. Iron-59 in the form of iron(II) citrate is used in iron metabolism studies. Its half-life is 45.6 d. If you start with 0.56 mg of iron-59, how much would remain after 1 y?

23. Iodine-131 is used in the form of sodium iodide to treat cancer of the thyroid.

(a) The isotope decays by ejecting a β particle. Write a balanced equation to show this process.

(b) The isotope has a half-life of 8.05 d. If you begin with 25.0 mg of radioactive $Na^{131}I$, what mass remains after 32.2 d?

24. Phosphorus-32 is used in the form of $Na_2H^{32}PO_4$ in the treatment of chronic myeloid leukemia, among other things.

(a) The isotope decays by emitting a β particle. Write a balanced equation to show this process.

(b) The half-life of ^{32}P is 14.3 d. If you begin with 9.6 mg of radioactive $Na_2H^{32}PO_4$, what mass remains after 28.6 d?

25. What is the half-life of a radioisotope if it decays to 12.5% of its radioactivity in 12 y?

26. After 2 hr, tantalum-172 has $\frac{1}{16}$ of its initial radioactivity. How long is its half-life?

27. Radioisotopes of iodine are widely used in medicine. For example, iodine-131 ($t_{1/2} = 8.05$ days) is used to treat thyroid cancer. If you ingest a sample of $Na^{131}I$, how much time is required for the isotope to fall to 5.0% of its original activity?

28. The rare gas radon has been the focus of much attention recently because it may be found in homes. Radon-222 emits α particles and has a half-life of 3.82 d.

(a) Write a balanced equation to show this process.

(b) How long does it take for a sample of radon to decrease to 10.0% of its original activity?

29. A sample of wood from a Thracian chariot found in an excavation in Bulgaria has a ^{14}C activity of 11.2 disintegrations per minute per gram. Estimate the age of the chariot and the year it was made. ($t_{1/2}$ for ^{14}C is 5.73×10^3 y, and the activity of ^{14}C in living material is 15.3 disintegrations per minute per gram.)

30. A piece of charred bone found in the ruins of an American Indian village has a ^{14}C-to-^{12}C ratio of 0.72 times that found in living organisms. Calculate the age of the bone fragment. (See Question 29 for required data on carbon-14.)

Artificial Transmutations

31. There are two isotopes of americium, both with half-lives sufficiently long to allow the handling of massive quantities. Americium-241, for example, has a half-life of 248 y as an α emitter, and it is used in gauging the thickness of materials and in smoke detectors. The isotope is formed from ^{239}Pu by absorption of two neutrons followed by emission of a β particle. Write a balanced equation for this process.

32. Americium-240 is made by bombarding plutonium-239 atoms with α particles. In addition to ^{240}Am, the products are a proton and two neutrons. Write a balanced equation for this process.

33. To synthesize the heavier transuranium elements, one must bombard a lighter nucleus with a relatively large particle. If you know the products are californium-246 and four neutrons, with what particle would you bombard uranium-238 atoms?

34. Until 1999, the element with the highest known atomic number was 112. Still heavier elements have been made, including one with $Z = 114$ and $N = 184$. To this end, serious attempts have been made to force calcium-40 and curium-248 to merge. What would be the atomic number of the element formed?

Nuclear Fission and Fusion

35. Name the fundamental parts of a nuclear fission reactor and describe their functions.

36. Explain why no commercial fusion reactors are in operation today.

37. The average energy output of a good grade of coal is 2.6×10^7 kJ/ton. Fission of 1 mol of ^{235}U releases 2.1×10^{10} kJ. Find the number of tons of coal needed to produce the same energy as 1.0 lb of ^{235}U.

38. A concern in the nuclear power industry is that, if nuclear power becomes more widely used, there may be serious shortages in worldwide supplies of fissionable uranium. One solution is to build "breeder" reactors that manufacture more fuel than they consume. One such cycle works as follows:

   (i) A ^{238}U nucleus collides with a neutron to produce ^{239}U.

   (ii) ^{239}U decays by β emission ($t_{1/2} = 24$ min) to give an isotope of neptunium.

   (iii) This neptunium isotope decays by β emission to give a plutonium isotope.

   (iv) The plutonium isotope is fissionable. On collision of one of these plutonium isotopes with a neutron, fission occurs with energy, at least two neutrons, and other nuclei as products.

   Write an equation for each of the steps, and explain how this process can be used to breed more fuel than the reactor originally contained and still produce energy.

Effects of Nuclear Radiation

39. Two common units of radiation used in newspaper and news magazine articles are the rad and rem. What do each measure? Which would you use in an article describing the damage an atomic bomb would have on a human population? What relationship does the unit gray have with these units?

40. Which electrical power plant—fossil fuel or nuclear—exposes a community to more radiation? Explain why.

41. Explain how our own bodies are sources of radiation.

42. What is the source of radiation during jet plane travel?

Uses of Radioisotopes

43. Why are foods irradiated with gamma rays instead of alpha or beta particles?

44. X-rays and PET scans are two medical imaging techniques. How are they similar and different?

45. In order to measure the volume of the blood system of an animal, the following experiment was done. A 1.0-mL sample of an aqueous solution containing tritium with an activity of 2.0×10^6 disintegrations per second (dps) was injected into the bloodstream. After time was allowed for complete circulatory mixing, a 1.0-mL blood sample was withdrawn and found to have an activity of 1.5×10^4 dps. What was the volume of the circulatory system? (The half-life of tritium is 12.3 years, so this experiment assumes that only a negligible amount of tritium has decayed during the experiment.)

46. Radioactive isotopes are often used as "tracers" to follow an atom through a chemical reaction, and the following is an example. Acetic acid reacts with methanol, CH_3OH, by eliminating a molecule of H_2O to form methyl acetate, $CH_3CO_2CH_3$. Explain how you would use the radioactive isotope ^{18}O to show whether the oxygen atom in the water product comes from the —OH of the acid or the —OH of the alcohol.

$$CH_3COOH + HOCH_3 \longrightarrow CH_3COOCH_3 + H_2O$$

acetic acid   methanol   methyl acetate

General Questions

47. Complete the following nuclear equations.
   (a) $^{214}Bi \rightarrow$ _____ $+ ^{214}Po$
   (b) $4\,^1_1H \rightarrow$ _____ $+ 2$ positrons
   (c) $^{249}Es +$ neutron $\rightarrow 2$ neutrons $+$ _____ $+ ^{161}Gd$
   (d) $^{220}Rn \rightarrow$ _____ $+$ alpha particle
   (e) $^{68}Ge +$ electron $\rightarrow$ _____

48. Complete the following nuclear equations.
   (a) _____ $+$ neutron $\rightarrow 2$ neutrons $+ ^{137}Tc + ^{97}Zr$
   (b) $^{45}Ti \rightarrow$ _____ $+$ positron
   (c) _____ $\rightarrow$ beta particle $+ ^{59}Co$
   (d) $^{24}Mg +$ neutron $\rightarrow$ _____ $+$ proton
   (e) $^{131}Cs +$ _____ $\rightarrow ^{131}Xe$

49. Radioactive nitrogen-13 has a half-life of 10 min. After an hour, how much of this isotope remains in a sample that originally contained 96 mg?

50. The half-life of molybdenum-99 is 67.0 h. How much of a 1.000-mg sample of ^{99}Mo is left after 335 h? How many half-lives did it undergo?

51. The oldest known fossil cells form a biological cluster found in South Africa. The fossil has been dated by the reaction

$$^{87}Rb \longrightarrow ^{87}Sr + ^{\;\,0}_{-1}e \qquad\qquad t_{1/2} = 4.9 \times 10^{10}\,y$$

If the ratio of the present quantity of ^{87}Rb to the original quantity is 0.951, calculate the age of the fossil cells.

52. Cobalt-60 is a therapeutic radioisotope used in treating certain cancers. If a sample of cobalt-60 initially disintegrates at a rate of 4.3×10^6 dps and after 21.2 y the rate has dropped to 2.6×10^5 dps, what is its half-life?

53. Balance the following equations used for the synthesis of transuranium elements.
   (a) $^{238}_{92}U + ^{14}_{7}N \rightarrow$ _____ $+ 5\,^1_0n$
   (b) $^{238}_{92}U +$ _____ $\rightarrow ^{249}_{100}Fm + 5\,^1_0n$
   (c) $^{253}_{99}Es +$ _____ $\rightarrow ^{256}_{101}Md + ^1_0n$
   (d) $^{246}_{96}Cm +$ _____ $\rightarrow ^{254}_{102}No + 4\,^1_0n$
   (e) $^{252}_{98}Cf +$ _____ $\rightarrow ^{257}_{103}Lr + 5\,^1_0n$

54. On December 2, 1942, the first manmade self-sustaining nuclear fission chain reactor was operated by Enrico Fermi

and others under the University of Chicago stadium. In June 1972 natural fission reactors, which operated billions of years ago, were discovered in Oklo, Gabon. At present, natural uranium contains 0.72% ^{235}U. How many years ago did natural uranium contain 3.0% ^{235}U, sufficient to sustain a natural reactor? ($t_{1/2}$ for ^{235}U is 7.04×10^8 y.)

Applying Concepts

55. If a radioisotope is used for diagnosis (e.g., detecting cancer), it should decay by gamma radiation. However, if its use is therapeutic (e.g., treating cancer), it should decay by alpha or beta radiation. Explain why in terms of ionizing and penetrating power.

56. During the Three Mile Island incident, people in central Pennsylvania were concerned that strontium-90 (a beta emitter) released from the reactor could become a health threat (it did not). Where would this isotope collect in the body? What types of problems could it cause?

57. Classify the isotopes ^{17}Ne, ^{20}Ne, and ^{23}Ne as stable or unstable. What type of decay would you expect the unstable isotope(s) to have?

58. The following demonstration was carried out to illustrate the concept of a nuclear chain reaction. Explain the connections between the demo and the reaction.

Eighty mousetraps are arranged side by side in eight rows of ten traps each. Each trap is set with two rubber stoppers for bait. A small plastic mouse is tossed into the middle of the traps, setting off one trap, which in turn sets off two traps and so on until all the traps are sprung.

59. Most students have no trouble understanding that 1.5 g of a 24-g sample of a radioisotope would remain after 8 h if it had a $t_{1/2} = 2$ h. What they don't always understand is where the other 22.5 g disappeared to. How would you explain this disappearance to another student?

60. Nuclear chemistry is a topic that raises many debatable issues. Briefly discuss your views on the following.

(a) Twice a year the general public is allowed to visit the Trinity Site in Alamogordo, New Mexico, where the first atomic bomb was tested. If you had the opportunity to do so, would you visit the site? Explain your answer.

(b) Now that the Cold War has ended, should the United States stockpile nuclear weapons? Explain your answer.

(c) The FDA allows irradiated grapefruit to be exported from, but not sold in, the United States. Is it acceptable to sell to other nations food that is not approved for our domestic consumption? Explain your answer.

21

A supernova occurs as the outer layers of a neutron star explode from the surface of the star.

THE CHEMISTRY OF SELECTED MAIN GROUP ELEMENTS

21.1   Formation of the Elements

21.2   Terrestrial Elements

21.3   Some Elements Extracted by Physical Methods: Nitrogen, Oxygen, and Sulfur

21.4   Some Main Group Elements Extracted by Electrolysis: Sodium, Chlorine, Magnesium, and Aluminum

21.5   Some Main Group Elements Extracted by Chemical Oxidation-Reduction: Phosphorus, Bromine, and Iodine

Elements have been discussed throughout this book. Their names, symbols, physical properties, and chemical reactivity have been noted, demonstrating their enormous diversity. Some elements (such as sodium and fluorine) react violently, while others (the noble gases) are so quiescent that they enter into very few or no chemical combinations. In spite of this diversity, elements in each group of the periodic table have predictable chemical similarities based on their number of valence electrons.

A fundamental question that has not yet been discussed concerns the origin of the elements: How did they form? How, for example, did magnesium atoms acquire a different number of protons from atoms of calcium or helium? This chapter will answer those questions, as well as describe how selected elements are extracted from natural sources, the chemical principles associated with such processes, and the properties of those elements.

21.1 FORMATION OF THE ELEMENTS

Cosmologists — scientists who study the formation of the universe — use spectral evidence from various parts of the universe to develop theories about its origin. They believe that about 15 billion years ago all the matter in the universe was contained in a pinpoint-sized region that exploded at inconceivably high temperatures (estimated to be about 10^{32} K) in what is called the "Big Bang." This explosion produced an expanding universe that within one second cooled sufficiently (to 10^9 K) to form the fundamental subatomic particles — neutrons, protons, and electrons. Within two hours, the temperature had dropped to temperatures suitable for the formation of the initial light nuclei — ^{2}H, ^{3}He, and ^{4}He.

Nuclear Burning

The elements from hydrogen to iron are formed inside stars by **nuclear burning,** a sequence of nuclear fusion reactions not to be confused with chemical combustion. The fusion of protons (hydrogen-1 nuclei) to form helium-4 nuclei is called **hydrogen burning.**

$$4\,^1_1\text{H} \longrightarrow\ ^4_2\text{He} + 2\,^0_{+1}\text{e} + 2\,\gamma$$

After billions of years of hydrogen burning, the star contracts and the core becomes dense and hot enough for **helium burning** to occur, the fusion of helium-4 nuclei.

$$^4_2\text{He} +\ ^4_2\text{He} \longrightarrow\ ^8_4\text{Be}$$

Beryllium-8 is unstable, but by fusing with another helium-4 nucleus, beryllium-8 is converted to stable carbon-12.

$$^8_4\text{Be} +\ ^4_2\text{He} \longrightarrow\ ^{12}_6\text{C}$$

The low natural abundance of beryllium is evidence of the instability of beryllium-8.

When helium burning stops, the star contracts further with the result that the temperature becomes high enough for heavier nuclei to form by fusion. Starting with carbon-12, three successive fusions with helium-4 nuclei form oxygen-16, then neon-20, and then magnesium-24. The process continues up to the formation of calcium-40. Carbon and oxygen "burning" also occurs:

$$^{12}_6\text{C} +\ ^{12}_6\text{C} \longrightarrow\ ^{23}_{11}\text{Na} +\ ^1_1\text{H}$$

$$^{12}_6\text{C} +\ ^{16}_8\text{O} \longrightarrow\ ^{28}_{14}\text{Si}$$

Starting with silicon-28, fusion reactions build up heavier nuclei all the way to iron-56 and nickel-58, which are very stable nuclei with the highest binding energies per nucleon *(◂⃗ p. 927)*. Elements heavier than iron cannot be formed by such nuclear fusion reactions.

Exercise 21.1 Fusing Nuclei

Write balanced nuclear equations representing the formation of oxygen-16, neon-20, and magnesium-24 starting from carbon-12 and helium-4 nuclei.

Formation of Heavier Elements

The amount of time a star spends in the various stages of elemental syntheses in relation to the central temperature of the star is given in Figure 21.1. Following helium burning, the formation of elements heavier than iron occurs by neutron capture in massive stars in which the core collapses rapidly. Stable nuclei such as those of iron decompose into neutrons and protons, and the protons are converted into additional neutrons by capturing electrons. The result is the formation of a neutron star whose outer layers explode away as a supernova.

Heavier elements form during supernovas by one of two processes. In the *s process,* the slow capture of neutrons takes place over many years. Because this capture shifts the neutron/proton ratio, eventually the nuclei will become beta emitters. As noted in Section 20.3, beta emission causes an increase in the atomic number and so a new element is formed from the parent nucleus. Such is the case for $^{98}_{42}$Mo, for example, which is converted to technetium-99 by this two-step process.

$$^{98}_{42}\text{Mo} + {}^{1}_{0}\text{n} \longrightarrow {}^{99}_{42}\text{Mo}$$

$$^{99}_{42}\text{Mo} \longrightarrow {}^{99}_{43}\text{Tc} + {}^{0}_{-1}\text{e}$$

Isotopes with masses up to 209 can form by the *r process.*

Some elements, including the radioactive acitines, are produced in the very rapid *r process,* which occurs during the explosive stage of a star. Because of the speed at which a series of neutrons can be captured one by one, new elements can be produced from nuclei with very short half-lives, too short to react by the *s process.* A nucleus may capture many neutrons in an extremely short time in a series of reactions that produce a nucleus much heavier than the original one. Suppose, for example, that $^{130}_{48}$Cd is produced in the *r process.* This isotope is highly unstable because it has far too many neutrons; cadmium-116 is the most stable known isotope of cadmium. The cadmium-130 can undergo a rapid series of beta decays, increasing in atomic number until it reaches tellurium-130, the most abundant isotope of tellurium.

$$^{130}_{48}\text{Cd} \xrightarrow{\text{4 beta decays}} {}^{130}_{52}\text{Te}$$

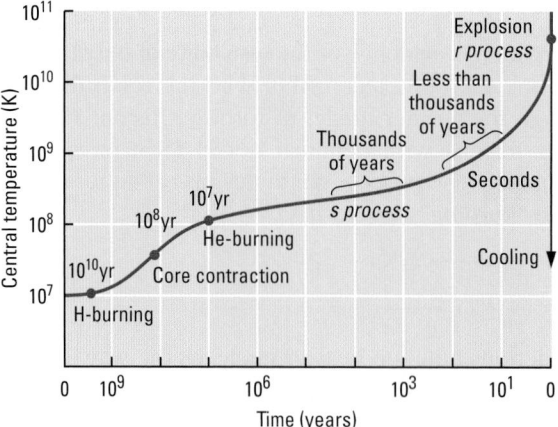

Figure 21.1 The time scale for various stages of elemental syntheses in stars.

21.2 TERRESTRIAL ELEMENTS

Beginning with this section and continuing for the remainder of the chapter, we take up how selected elements are obtained from their natural sources and what properties and uses those elements have. We obtain large quantities of nitrogen and oxygen from the atmosphere, as well as much smaller amounts of those noble gases for which the atmosphere is the sole source. Helium is found in the atmosphere and in underground deposits. Ocean water is treated to extract commercial quantities of magnesium, bromine, and sodium chloride. And the earth's crust is an indispensable source of most of the other elements. Figure 21.2 illustrates the relation between the earth's crust, mantle, and core. The crust, which extends only from the surface to a depth of about 35 km, is but a tiny fraction of the entire depth of the earth. If you think of the earth as an apple, the crust is akin to the thickness of the skin.

The average composition of the earth's crust is given in Figure 21.3. All the elements shown in the pie chart are in compounds; they do not exist "free" in the earth's crust. Note the preponderance of oxygen and silicon; these are the major components of silicate minerals, clays, and sand. Aluminum is the most abundant metal, followed by iron and several alkali and alkaline earth metals—calcium, sodium, potassium, and magnesium.

Most elements in the crust of the earth are in chemically combined forms, which can be simple molecular compounds such as water and ammonia, but more often are complex solids known as minerals. A **mineral** is commonly defined as a naturally occurring inorganic compound with a characteristic composition and crystal structure. The major chemical form in which each element occurs on the earth's surface is shown in Figure 21.4. In particular, note the predominance of oxygenated compounds, either binary ones such as MgO and TiO_2, or complex

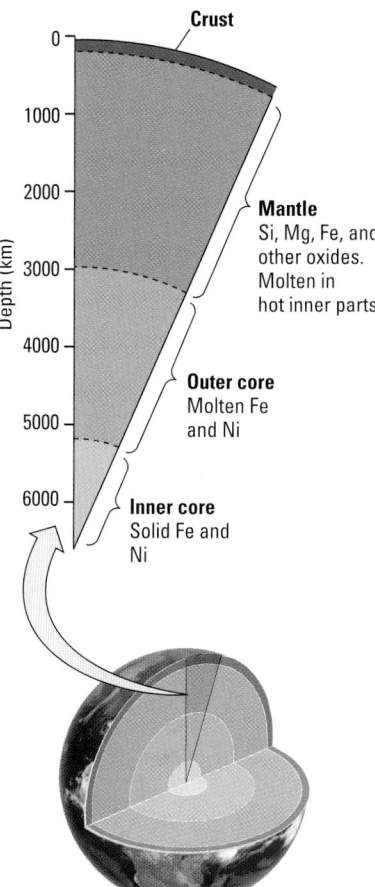

Figure 21.2 A cross section of the earth. Note the thinness of the crust compared with that of the mantle and the core.

Most of the iron on earth is in the core and the mantle, not the crust.

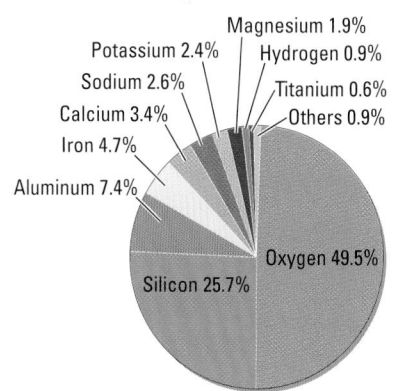

Figure 21.3 Elemental composition of the earth's crust.

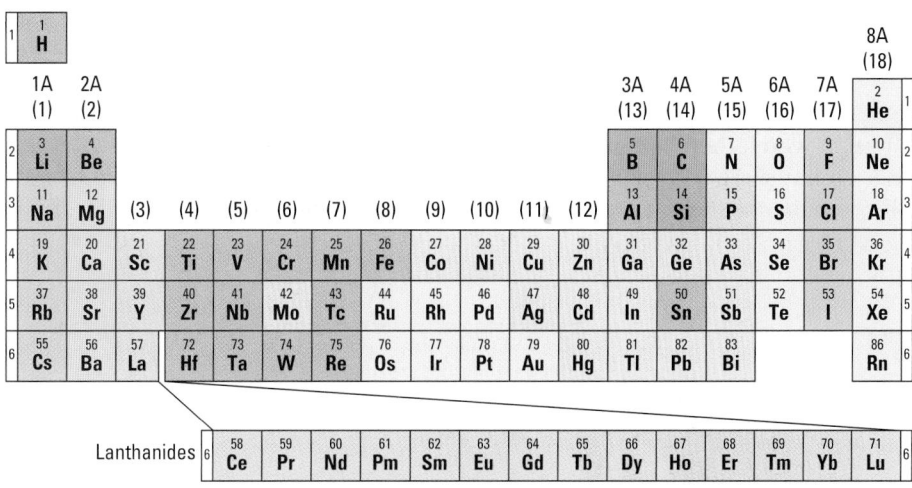

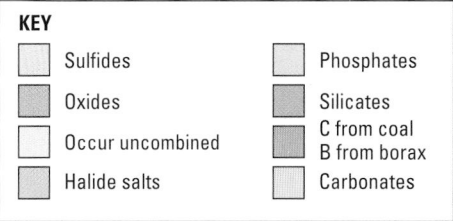

Figure 21.4 Main types of minerals in the earth's crust.

ones such as carbonates and silicates. The preponderance of such compounds is testimony to the abundance of oxygen and silicon in the earth's surface. The lanthanides and naturally occurring actinides also form oxide minerals. Many transition metals and heavier elements of Groups 3A (13) to 6A (16) are found as sulfides, such as ZnS and Sb_2S_3.

Silica and Silicates

Silica is pure SiO_2. Its most common form is α-quartz, which is a major component of many rocks such as granite and sandstone. α-Quartz also occurs as a pure rock crystal (Figure 21.5) and in several less pure forms. Silicon and oxygen, the two most abundant elements in the earth's crust, are combined in the crust as silicate minerals. Such minerals all contain SiO_4^{4-} ions in which four oxygen atoms are arranged tetrahedrally around a central Si atom. The SiO_4^{4-} ions are the fundamental building block for all silicate minerals. In silicate minerals these tetrahedra typically share one or more oxygens to form chains, sheets, and rings; quartz has a three-dimensional structure. Some silicate minerals also contain cations such as Mg^{2+} or Na^+, for example, as in olivine, Mg_2SiO_4, to balance the negative charges of the silicate ions.

In olivine, one out of every ten Mg^{2+} ions is replaced by Fe^{2+}.

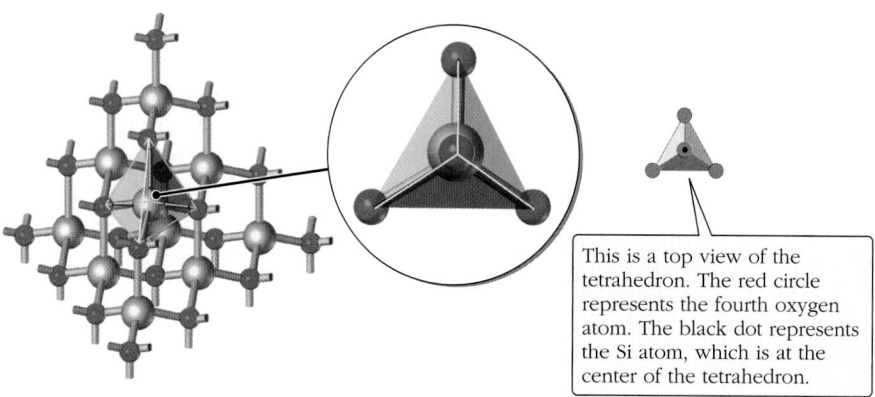

This is a top view of the tetrahedron. The red circle represents the fourth oxygen atom. The black dot represents the Si atom, which is at the center of the tetrahedron.

The simplest silicate minerals, such as olivine and willemite, Zn_2SiO_4, are those in which there are independent tetrahedra that contain discrete SiO_4^{4-} units that do not share oxygens. *Condensed silicates* are those formed by two or more SiO_4 tetrahedra sharing oxygen atoms. *Pyroxenes* are the simplest class of con-

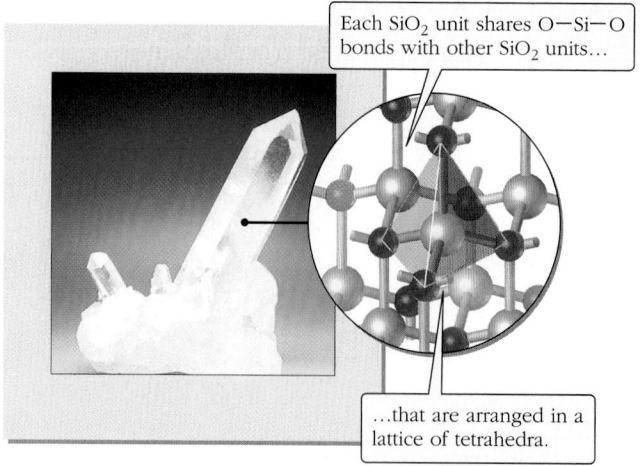

Each SiO_2 unit shares O—Si—O bonds with other SiO_2 units...

...that are arranged in a lattice of tetrahedra.

Figure 21.5   A pure quartz crystal (pure SiO_2).   The formula is derived from the fact that each oxygen atom of a SiO_4 tetrahedron is shared by two silicon atoms. Correspondingly, only half of the oxygens "belong" to a given Si, thus making the formula SiO_2, not SiO_4. Quartz crystals are used as oscillators in watches, radios, VCRs, and computers.

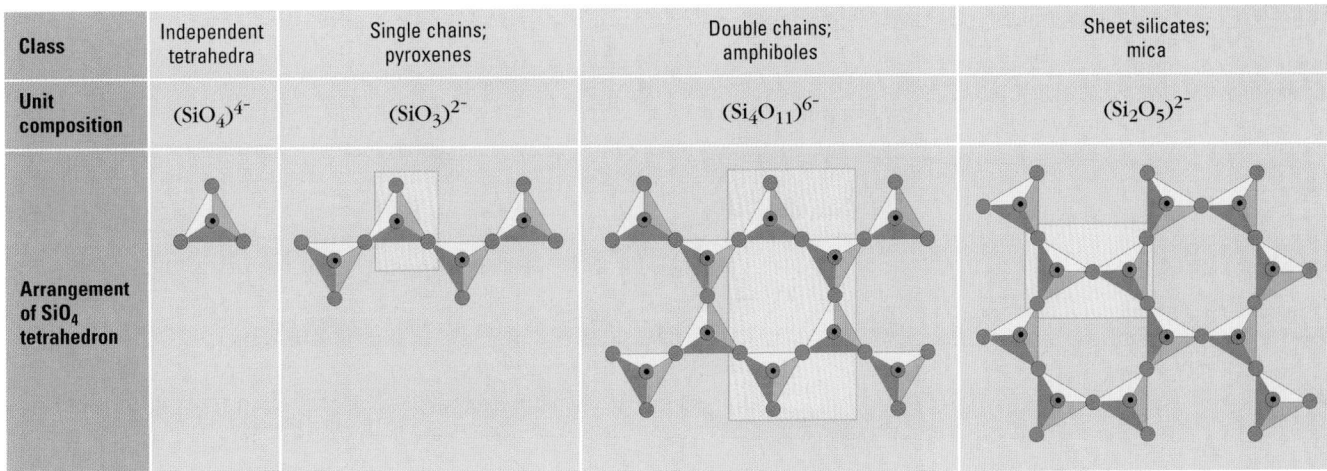

Class	Independent tetrahedra	Single chains; pyroxenes	Double chains; amphiboles	Sheet silicates; mica
Unit composition	$(SiO_4)^{4-}$	$(SiO_3)^{2-}$	$(Si_4O_{11})^{6-}$	$(Si_2O_5)^{2-}$
Arrangement of SiO$_4$ tetrahedron				

Figure 21.6 **Silicate structures.** These structures are all based on the tetrahedral SiO_4^{4-} unit. The repeating unit of each structure is shown with an orange background.

densed silicates and contain extended chains of linked SiO$_4$ tetrahedra sharing oxygen atoms. Their formulas appear to contain the metasilicate ion, SiO_3^{2-}, to give a typical formula such as Na$_2$SiO$_3$ (Figure 21.6). Pyroxenes are abundant in the ocean's floor and the earth's mantle. If two pyroxene chains are laid side by side, they can link together by sharing oxygen atoms in adjoining chains to form a type of silicate called an *amphibole,* with the typical formula $(Si_4O_{11})^{6-}$ (Figure 21.6).

One form of *asbestos* is an excellent example of an amphibole. What is called "asbestos" is not a single substance; the name asbestos applies broadly to a family of naturally occurring hydrated silicates that crystallize as fibers. Asbestos minerals are generally subdivided into two forms, serpentine and amphibole fibers. Approximately 5 million tons of the serpentine form of asbestos, chrysotile, are mined each year, chiefly in Canada and the former Soviet Union. Chrysotile is essentially the only form used commercially in the United States. Another form, the amphibole crocidolite, is mined in small quantities, mainly in South Africa. The two minerals differ greatly in composition, color, shape, solubility, and persistence in human tissue. This last property is important in determining the toxicity of asbestos. Crocidolite is blue, relatively insoluble, and persists in tissue. Its long, thin, straight fibers can penetrate narrow lung passages. In contrast, chrysotile is white, and it tends to be soluble and disappear in tissue. Chrysotile fibers are curly, so they ball up like yarn and are more easily rejected by the body. Long-term occupational exposure to certain asbestos minerals can lead to lung cancer. Although there is some disagreement in the medical and scientific communities, evidence strongly suggests that amphiboles such as crocidolite are much more potent cancer-causing agents than the serpentines such as chrysotile. Most asbestos in public buildings is the chrysotile type, and so initiatives to remove asbestos insulation may be misguided overreaction in many cases. Nevertheless, most asbestos-containing materials have been removed from the market, and strict standards now exist for the handling and use of asbestos.

When silicate chains continue to link in two dimensions, extended sheets of SiO$_4$ tetrahedral units result (Figure 21.6), with the formula $(Si_2O_5)^{2-}$. All of the atoms within each sheet are strongly covalent bonded, but each sheet is only weakly bonded to those above and below it. Various clay minerals and mica have this sheet-like silicate structure. Mica, for example, is used to prepare "metallic" looking paint on new automobiles.

Clays are essential components of soils that come from the weathering of igneous rocks. Since early in human history, clays have been used for pottery, bricks, tiles, and writing materials. Clays are actually *aluminosilicates,* in which some Si^{4+} ions are replaced by Al^{3+} ions, such as in feldspar, KAlSi$_3$O$_8$, a component of many

Because of their double-stranded chain structure, asbestos minerals are fibrous and can even be woven into a cloth-like material.

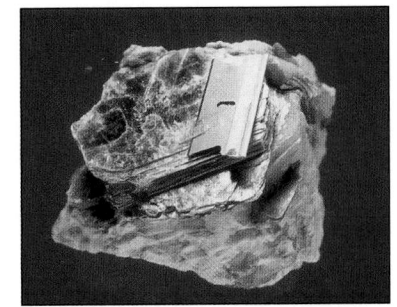

Sheets of mica can be peeled away from each other because the sheets are weakly bonded to each other.

Figure 21.7   A mudslide caused by shifting clay.   During heavy rains, clays become saturated with water, causing the aluminosilicate layers to shift, sliding over each other. This photo was taken along the Caribbean coast of Venezuela.

rocks. When Al^{3+} ions are replaced by other 3+ metal ions, the clay may become colored. For example, a red clay contains Fe^{3+} ions in place of some Al^{3+} ions.

Artists who work with clay first wet the clay and mold it into a shape. Water molecules strongly interact with the oxygen atoms as well as the metal ions near the surface of clay particles, and the silicate layers slide over one another, making the clay pliable. After the clay has been formed into the desired shape, the object is heated in an oven to remove the water. Bonds form between the exposed oxygen atoms and ions on the surfaces and adjacent particles, which causes the clay to harden. Too much water in the wet clay can make it unstable. This occurs not only in clay for pottery, but also on a much larger scale in nature. During very heavy rains, entire hillsides of clay can shift and slide downhill causing massive destruction of property (Figure 21.7).

   Exercise 21.2   Linking Tetrahedra

Explain how the silicate unit in pyroxenes has the general formula SiO_3^{2-}, not SiO_4^{4-}.

Methods for Obtaining Pure Elements

Except for those relatively few elements in nature that are available directly in their uncombined form, all other elements needed for practical applications must be extracted from their compounds. The types of extraction methods used are listed in Table 21.1. Metals exist in minerals as cations that must be reduced to the elemental form. Therefore, chemical and electrochemical oxidation-reduction reactions are needed in the production of metals.

In this chapter we will describe some elements that illustrate extraction from their naturally occurring forms by physical methods (Section 21.3: nitrogen, oxygen, and sulfur), by electrochemical redox reactions (Section 21.4: sodium, chlorine, magnesium, and aluminum), and by chemical redox reactions (Section 21.5: phosphorus, bromine, and iodine).

Ores are minerals that contain a sufficiently high concentration of an element to make extraction of it profitable. Not all elements are used to the same extent, so

TABLE 21.1	Methods for Extraction of Elements from Their Compounds
Extraction method	**Examples of elements extracted by this method**
Carbon reduction of oxide	Si, Fe, Sn
Oxidation with Cl_2	Br, I
Reaction of sulfide with O_2	Cu, Hg
Conversion with sulfide to oxide, then reduction with C	Zn, Pb
Halide reduction with sodium or other electropositive metal	K, Ti, Cr, Cs, U
Halide or oxide reduction with H_2	B, Ni, Mo, W
Electrolysis of solution or molten salt	H, Li, F, Na, Ca, Al, Cl

the amount of different elements taken from the earth's crust and the amount remaining vary with market demands. A metal is extracted from its ore in response to such demands. Current known reserves of some common elements such as aluminum and iron are sufficient to last hundreds of years at the current rate of use, whereas the known reserves of other widely used elements, such as copper, tin, and lead, are rather slim (Table 21.2). Notice from Table 21.2 that the United States does not have major reserves of several critical metals, for example, the chromium and manganese needed for making steel and other alloys. We must import and stockpile such metals.

21.3 SOME ELEMENTS EXTRACTED BY PHYSICAL METHODS: NITROGEN, OXYGEN, AND SULFUR

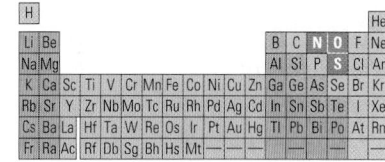

Some elements occur in nature in their elemental form, that is, not combined with any other element. Among these aloof elements are some metals—gold, silver, mercury, and copper—and sulfur, a nonmetal. Clearly, nitrogen and oxygen, the principal components of the atmosphere, are such elements, as are the noble gases

TABLE 21.2	Known Reserves of Selected Elements		
Element	**Reserves (10^9 kg)**	**Lifetime (yr)**	**Locations of major reserves**
Al	20,000	220	Australia, Brazil, Guinea
Fe	66,000	120	Australia, Canada, CIS*
Mn	800	100	CIS,* Gabon, S. Africa
Cr	400	100	CIS,* S. Africa, Zimbabwe
Cu	300	36	Chile, CIS,* USA, Zaïre
Zn	150	21	Australia, Canada, USA
Pb	71	20	Australia, Canada, CIS,* USA
Ni	47	55	Canada, CIS,* Cuba, New Caledonia
Sn	5	28	Brazil, China, Indonesia, Malaysia
U	2.8	58	Australia, CIS,* S. Africa, USA

*No individual breakdown is available for nations constituting the Commonwealth of Independent States (formerly the USSR).

in the atmosphere. Large quantities of nitrogen, oxygen, and to a lesser extent, argon are extracted from the atmosphere by the liquefaction of air.

Elements From the Atmosphere

The composition of the atmosphere is given in Table 21.3, which shows that nitrogen is by far the most abundant component, its concentration being nearly four times that of oxygen, the next most abundant. The gases of the atmosphere can be separated from each other by liquefying and fractionating air, as can the refining of petroleum fractions *(⟵ p. 514)*, except at much lower temperatures.

By lowering the temperature and raising the pressure it is possible to use the nonideal behavior of gases to liquefy air. Under these conditions, the attractive forces between molecules become sufficient for them to condense from vapor to liquid. Because of their different boiling points, the liquid components can then be separated from one another by distillation. Before pure oxygen and nitrogen can be obtained from air, water vapor and carbon dioxide must be removed. The air is then compressed to more than 100 times normal atmospheric pressure, cooled to room temperature, and allowed to expand into a chamber. This expansion produces a cooling effect (the *Joule-Thompson effect*) because energy is required to overcome intermolecular forces to move molecules farther apart. The expanding gas absorbs kinetic energy from the motion of its own molecules, which cools the gas. If this expansion is repeated and controlled properly, the expanding air cools to the point of liquefaction (Figure 21.8).

The temperature of the liquid air is usually well below the normal boiling points of nitrogen (-195.8 °C), oxygen (-183 °C), and argon (-189 °C). The liquid air is again allowed to vaporize partially, and since N_2 is more volatile and has a lower boiling point than O_2 or Ar, the N_2 evaporates and the remaining liquid becomes more concentrated in O_2 and Ar. This process, known as the *Linde process*, produces high-purity nitrogen (>99.5%) and oxygen with a purity of 99.5%. Further processing produces pure Ar and Ne (bp -246 °C).

Properties and Uses of Oxygen

Most of the oxygen produced by liquid air fractionation is used as an oxidizing agent and in steel making (Section 22.2), although some is used in rocket propulsion (to oxidize hydrogen) and in controlled oxidation reactions of other types. Liquid oxygen (LOX) can be shipped and stored at its boiling temperature of -183 °C under atmospheric pressure. Substances this cold are called **cryogens** (from the Greek *kryos*, meaning "icy cold"). Cryogens present special hazards since contact with them produces instantaneous frostbite and structural materials such as plastics, rubber gaskets, and some metals become brittle and fracture eas-

Properties of Oxygen

Atomic number	8
Density (g/L)*	1.43
Melting point (°C)	−219
Boiling point (°C)	−183
Atomic radius (pm)	73
Ionic radius, O^{2-} (pm)	140
Electronegativity	3.5

*At 25 °C.

TABLE 21.3	Composition of Clean, Dry Air at Sea Level		
Element	**Percent by volume**	**Element**	**Percent by volume**
N_2	78.09	He, Ne, Kr, Xe	0.002
O_2	20.04	CH_4,	0.00015*
Ar	0.93	H_2	0.00005
CO_2	0.03*	All others combined	<0.00004

*Variable.

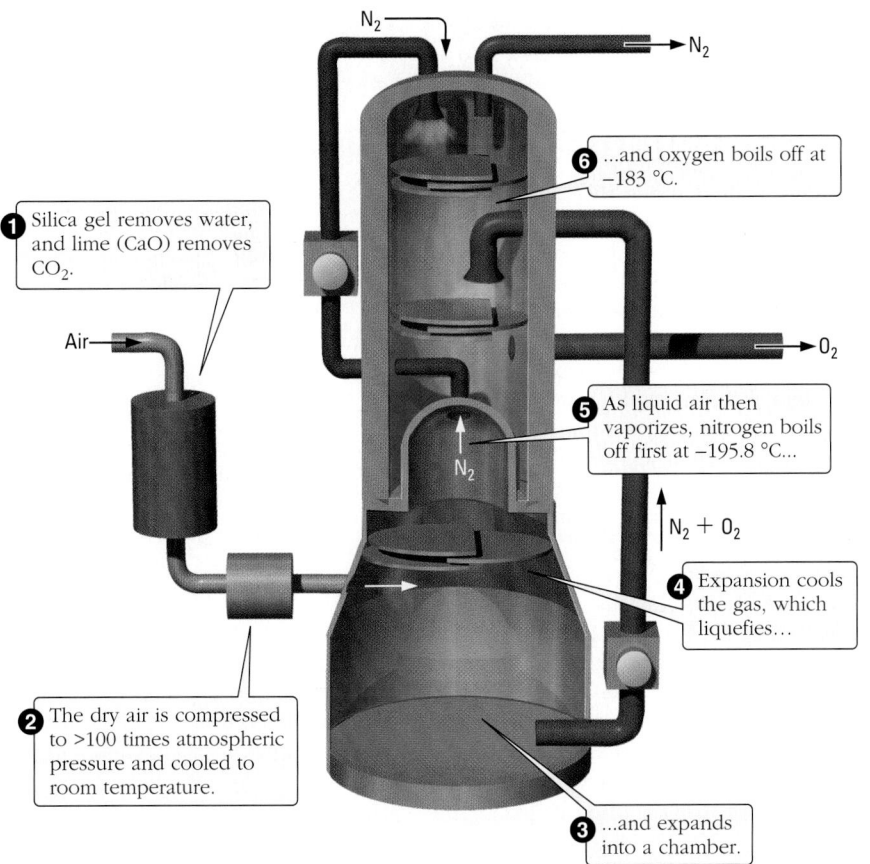

Figure 21.8 **Fractional distillation of air.** Air can be liquefied by low temperatures and high pressure. The components of the liquefied air are then separated by taking advantage of their distinctive boiling points.

❶ Silica gel removes water, and lime (CaO) removes CO_2.

❻ ...and oxygen boils off at −183 °C.

❺ As liquid air then vaporizes, nitrogen boils off first at −195.8 °C...

❷ The dry air is compressed to >100 times atmospheric pressure and cooled to room temperature.

❹ Expansion cools the gas, which liquefies...

❸ ...and expands into a chamber.

Air

N_2

O_2

$N_2 + O_2$

ily at these low temperatures. The high oxygen concentration in the presence of liquid oxygen can accelerate oxidation reactions to the point of explosion, in spite of the low temperature. For this reason, contact between liquid oxygen and substances that will ignite and burn in air must be prevented.

Special cryogenic containers holding liquid oxygen incorporate huge vacuum-walled bottles much like those used to carry hot soup or hot coffee. These containers can be seen outside hospitals and industrial complexes, on highways and railroads, and even aboard ocean-going vessels. In hospitals as well as in homes, supplemental oxygen is used to help patients who have difficulty breathing.

Exercise 21.3 Liquefied Gases

A cryogenic flask contains 5.0 L of liquid oxygen, which has a density of 1.4 g/mL. What volume will this oxygen occupy at STP if it is allowed to boil?

Most atmospheric oxygen comes from photosynthesis, in which green plants convert water and carbon dioxide into glucose and oxygen. The concentration of oxygen in the atmosphere has upper and lower limits that are essential for our safety. If the concentration goes above 25%, the rates of oxidation reactions would increase significantly, potentially endangering us by the increased rates of oxygen-requiring metabolic processes. With too little atmospheric oxygen, below 17%, we would suffocate.

Properties and Uses of Nitrogen

Liquid nitrogen is also a cryogen. It is used in medicine in cryosurgery, for example, to cool an area of skin prior to removal of a wart or other unwanted or

Properties of Nitrogen

Atomic number	7
Density (g/L)*	1.25
Melting point (°C)	−210
Boiling point (°C)	−196
Atomic radius (pm)	74
Ionic radius, N^{3-} (pm)	171
Electronegativity	3.0

*At 25 °C.

Chemistry in the News

HArF: GETTING TOGETHER WITH ARGON

Chemists in Finland have reported the synthesis of HArF, the first stable argon compound. The new compound, which exists only at a very chilly 7 to 27 K (−266 to −246 °C), was made by irradiating a solidified mixture of argon and hydrogen fluoride with very energetic, short-wavelength (127 to 160 nm) light. The irradiation breaks H—F bonds, making H and F atoms available for possible bonding with argon atoms. Using infrared spectroscopy to study the reaction, Leonid Khriachtchev and his colleagues observed three new bands in the infrared spectrum following the irradiation of the mixture. The chemists attribute the new bands to H—Ar, Ar—F stretching and H—Ar—F bending vibrations, indicative of chemical bonding between hydrogen, argon, and fluorine atoms (⬅ *p. 362, Tools of Chemistry*).

To test the hypothesis that an H—Ar bond formed, the researchers substituted deuterium for hydrogen and observed the shift in the D—Ar and D—Ar—F stretching frequencies expected when a heavier $^2_1$H atom is substituted for a lighter $^1_1$H one. Likewise, they used two different argon isotopes—argon-36 and argon-40—with hydrogen and with deuterium to study the nature of the H—Ar, D—Ar and Ar—F bonding. The calculated bond distances are 133 pm for H—Ar and 197 pm for Ar—F.

Khriachtchev and his researchers characterize the bonding in HArF as a combination of ionic and covalent bonds. If that is the case, it raises the possibility that HNeF and even HHeF might be synthesized. Recent theoretical calculations have predicted the exis-

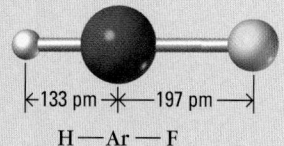

Bond distances in HArF.

tence of a stable HHeF molecule. No compounds of neon or helium are known.

Source:

Khriachtchev, L., *Nature,* "A Stable Argon Compound." Vol. 406, 2000; pp. 836–837, 874–876.

1. What experimental evidence is given for the formation of HArF?

2. Why were various argon and hydrogen isotopes used?

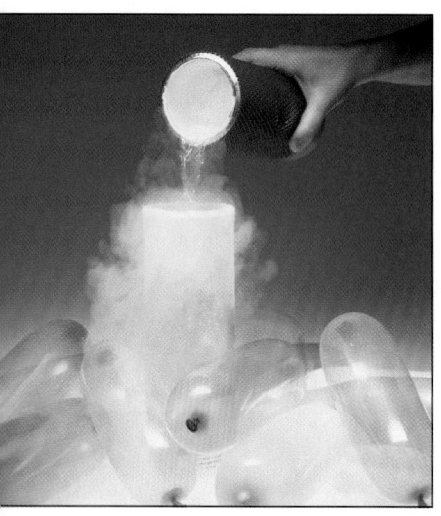

Liquid nitrogen.

pathogenic tissue. Since nitrogen is so chemically unreactive, it is used as an inert atmosphere for applications such as welding. Because of its low temperature and inertness, liquid nitrogen has found wide use in frozen-food preparation and preservation during transit. Containers with nitrogen atmospheres, such as railroad boxcars or truck vans, present potential safety hazards since they contain little (if any) oxygen to support life.

Nitrogen, phosphorus, and potassium are primary nutrients for plants. Although bathed in an atmosphere containing abundant nitrogen, most plants are unable to use the air directly as a supply of this vital element due to the energy required to break the N≡N triple bond, one of the strongest known. **Nitrogen fixation** is the process of changing atmospheric nitrogen into compounds that can be dissolved in water, absorbed through the plant roots, and assimilated by the plant (Figure 21.9). Nitrogen-fixing bacteria convert N_2 into NH_3. Ammonia is converted by other bacteria into nitrate, NO_3^-, which is used by plants. When an organism dies, bacteria reverse the process by converting nitrate to N_2 and organic nitrogen compounds to ammonia. Most plants thrive on soils rich in nitrates, but many plants that grow in swamps, where there is a lack of oxidized materials, can use reduced forms of nitrogen such as the ammonium ion. The nitrate ion is the most highly oxidized form of combined nitrogen, and the ammonium ion is the most reduced form of nitrogen.

 Exercise 21.4   Chemically Combined Nitrogen

Show that the nitrate ion is the most highly oxidized form of combined nitrogen and that the ammonium ion is the most reduced form of combined nitrogen.

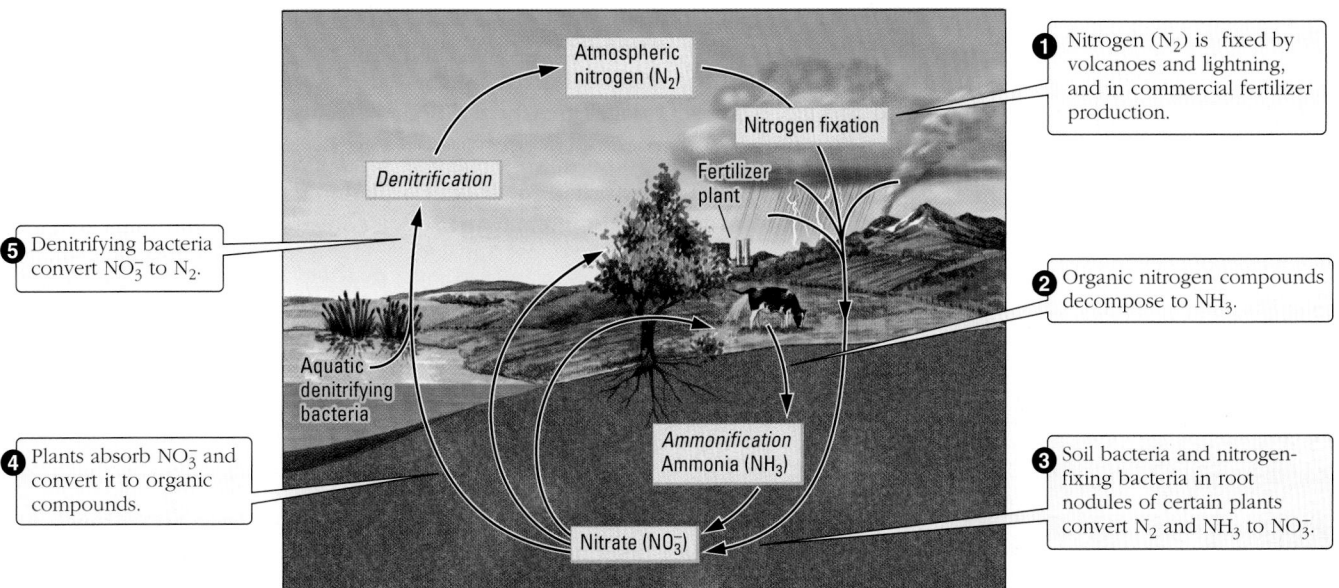

Figure 21.9 **Natural nitrogen chemical pathways: The nitrogen cycle.** Nitrogen-fixing bacteria convert N_2 into NH_3. Ammonia is converted by other bacteria into nitrate, NO_3^-, which is used by plants. When an organism dies, bacteria reverse the process converting nitrate to N_2 and nitrogen compounds to ammonia.

Nitrogen is fixed by natural processes on a massive scale in two ways. In the first method, nitrogen is oxidized under highly energetic conditions in the discharge of lightning or, to a lesser extent, in a fire. The initial reaction, which takes place in the atmosphere, is the reaction of nitrogen with oxygen to form nitrogen monoxide, NO, a colorless, reactive gas

$$N_2(g) + O_2(g) \longrightarrow 2 NO(g)$$

This reaction also occurs in automobile engines ($\Leftarrow$ **p. 447**).

Once formed, nitrogen monoxide is easily oxidized in air to nitrogen dioxide, NO_2, which dissolves in water to form nitrous acid, HNO_2, and nitric acid, HNO_3.

$$H_2O(\ell) + 2 NO_2(g) \longrightarrow HNO_2(aq) + HNO_3(aq)$$

nitrous acid   nitric acid

These acids are readily soluble in rain, clouds, or ground moisture, and thus increase nitrogen concentration in soil. They also contribute to the formation of acid rain ($\Leftarrow$ **p. 793, Section 17.3**).

In the second natural method of nitrogen fixation, bacteria that live on the roots of plants called *legumes,* such as clover, beans, and peas, convert atmospheric nitrogen into ammonia. This complex series of reactions depends on enzyme catalysis. Under ideal conditions, legume fixation can add more than 100 lb of nitrogen per acre of soil in one growing season.

The main industrial use of nitrogen at present is in the Haber-Bosch process ($\Leftarrow$ **p. 666**), in which it is combined with hydrogen to form ammonia.

$$N_2(g) + 3 H_2(g) \longrightarrow 2 NH_3(g)$$

Millions of tons of ammonia are produced annually by this method, which synthetically fixes nitrogen. Pure gaseous ammonia is condensed and the liquid anhydrous ammonia is applied directly to fields as a fertilizer. Ammonia is also reacted with nitric acid to produce ammonium nitrate, the major solid fertilizer in the world.

About 15% of the ammonia made by the Haber process is converted to nitric acid through a process developed by a German chemist, Wilhelm Ostwald. This three-step process is carried out at pressures of 1 to 10 atm.

Step 1: The ammonia is burned in air over a platinum-rhodium catalyst at about 1000 °C, achieving a greater than 95% conversion of ammonia to nitric oxide, NO.

$$4\ NH_3(g) + 5\ O_2(g) \xrightarrow[\text{catalyst}]{1000\ °C} 4\ NO(g) + 6\ H_2O(g)$$

Step 2: More air is added to the gaseous mixture, which lowers the temperature and causes the reaction

$$2\ NO(g) + O_2(g) \rightleftharpoons 2\ NO_2(g)$$

Step 3: The nitrogen dioxide produced in the second step is passed through water to produce nitric acid.

$$3\ NO_2(g) + H_2O(\ell) \longrightarrow 2\ HNO_3(aq) + NO(g)$$

The resulting aqueous solution is about 60% nitric acid by mass. The anhydrous acid is produced by adding sulfuric acid and boiling the mixture to distill nearly pure nitric acid from the mixture.

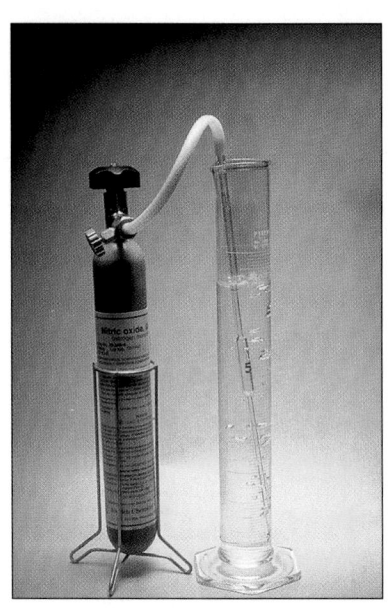

Conversion of colorless NO to red-brown NO_2.

Problem-Solving Example 21.1   Producing Nitric Acid

Consider the second step in the Ostwald process.

$$2\ NO(g) + O_2(g) \rightleftharpoons 2\ NO_2(g) \qquad\qquad \Delta H = -113.0\ kJ$$

What would happen to the yield of NO_2 at equilibrium if

(a) The pressure were increased?   (b) The temperature were increased?
(c) A catalyst were used?

Answer

(a) The yield should increase.   (b) The yield should decrease.   (c) No effect on yield.

Explanation   Le Chatelier's principle applies in each case.

(a) Increasing the pressure increases the concentration of reactants, so the yield would increase because the pressure change favors the formation of the lesser number of moles of gas (3 mol of gaseous reactants, 2 mol of gaseous products).
(b) An increase in temperature favors the endothermic reverse reaction, decreasing the yield.
(c) A catalyst will speed up the reaction, but like all catalysts, it has no effect on yield.

Problem-Solving Practice 21.1

Use Le Chatelier's principle to explain why in the manufacturing of nitric acid

(a) Lowering the temperature after step one favors NO_2 formation.
(b) Coupling step two with step three of the Ostwald process favors the formation of nitric acid.

A Lifesaving Use of N_2: Automobile Air Bags

Expanding air bags.

Azides are ionic compounds of nitrogen containing azide ions, N_3^-, which decompose rapidly to liberate N_2. In addition to being used as explosives, azides are also used in automobile air bags. An uninflated air bag has a small cylinder containing a carefully formulated mixture of the solids sodium azide, NaN_3, potassium nitrate, and silicon dioxide. When a car decelerates rapidly, as in a collision, a sensor sends an electrical signal to the mixture, igniting and decomposing the sodium azide and releasing nitrogen gas, which inflates the air bag.

$$2\,NaN_3(s) \longrightarrow 2\,Na(s) + 3\,N_2(g)$$

Because the residual sodium could react vigorously with any water, it must be removed. This is done by reacting it with potassium nitrate in a reaction that produces additional nitrogen.

$$10\,Na(s) + 2\,KNO_3(s) \longrightarrow K_2O(s) + 5\,Na_2O(s) + N_2(g)$$

The heat released by these reactions melts the solid products and the silicon dioxide (sand), fusing them into an unreactive glass.

$$K_2O(s) + Na_2O(s) + SiO_2(s) \xrightarrow{\text{heat}} glass$$

 Exercise 21.5 Expanding Air Bags

Calculate the volume of N_2 released in an air bag at STP when 150 g of sodium azide decomposes.

Sulfur

Sulfur, the element known biblically as brimstone, is a lemon-yellow solid. Very pure sulfur has been obtained from large deposits in salt domes along the coast of the Gulf of Mexico in the United States and Mexico, and in underground deposits in Poland. Such deposits of sulfur are believed to have been formed by bacterial reduction of sulfur in gypsum, which is hydrated calcium sulfate, a naturally occurring mineral. Millions of tons of sulfur have been recovered from such deposits by the **Frasch process,** developed in the 1890s by Herman Frasch, a petroleum engineer (Figure 21.10). Large quantities of sulfur are now produced by extracting it from petroleum and natural gas, thereby avoiding the formation of sulfur dioxide, an atmospheric pollutant formed when petroleum burns. High-sulfur natural gas from Alberta, Canada, an especially large source of recovered sulfur, has now displaced the Frasch process as the chief source of sulfur (Figure 21.11).

Properties and Uses of Sulfur

Sulfur exists in two common allotropic forms—rhombic (mp 115 °C) and monoclinic (mp 119 °C), both consisting of S_8 rings in the solid. When sulfur is heated above 150 °C, the S_8 rings break open, forming chains that become entangled, in-

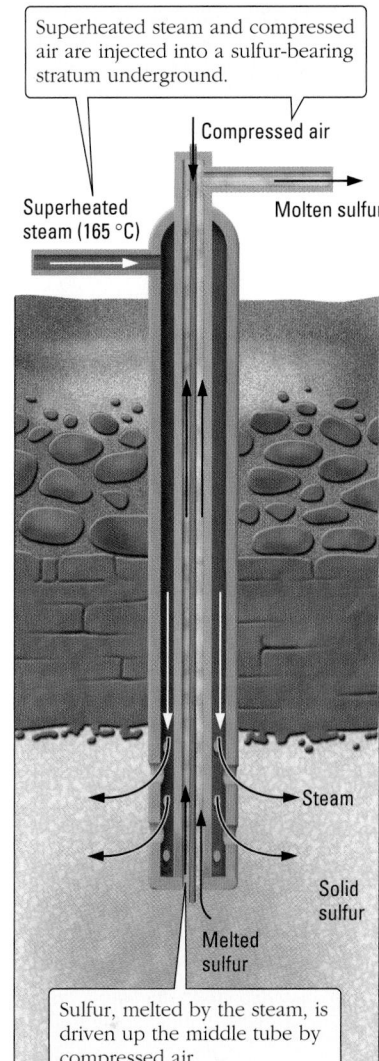

Superheated steam and compressed air are injected into a sulfur-bearing stratum underground.

Compressed air

Superheated steam (165 °C)

Molten sulfur

Steam

Solid sulfur

Melted sulfur

Sulfur, melted by the steam, is driven up the middle tube by compressed air.

Figure 21.10 The Frasch process for mining sulfur. The molten sulfur froths up out of the inner pipe. Most sulfur is used to manufacture sulfuric acid.

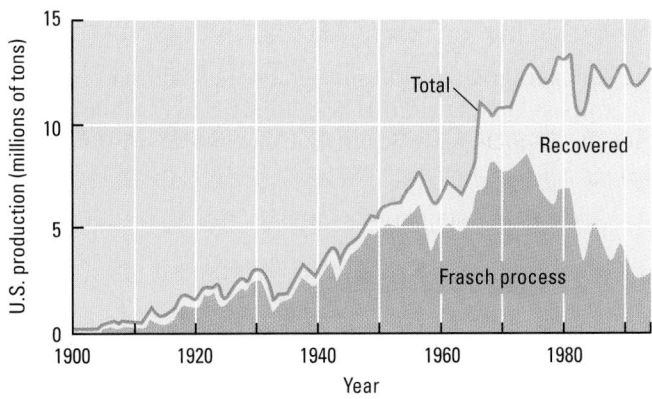

Figure 21.11 Changes in annual sulfur production.

Pure sulfur. Huge blocks of recently mined sulfur await shipment.

Figure 21.12 **Sulfur allotropes.** (a) At room temperature, sulfur is a bright yellow solid. At the nanoscale level it consists of rings of eight sulfur atoms. (b) When melted, the rings break open to form long chains.

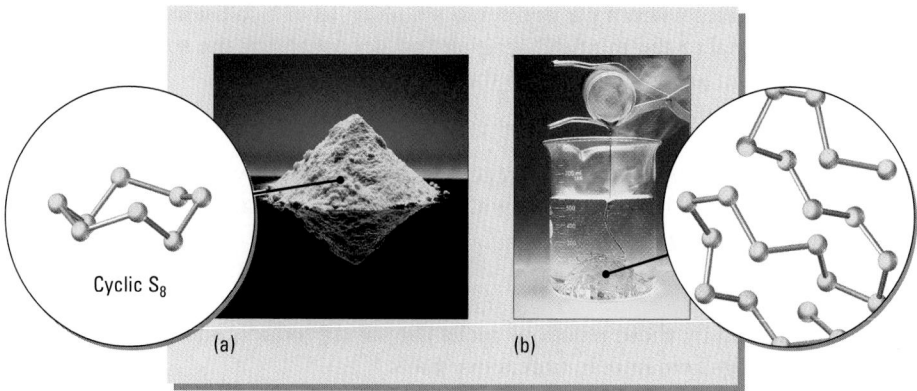

Cyclic S_8

(a)   (b)

Properties of Sulfur

Atomic number	16
Density (g/L)*	2.07
Melting point (°C)	115
Boiling point (°C)	445
Atomic radius (pm)	103
Ionic radius, S^{2-} (pm)	184
Electronegativity	2.4

*At 25 °C.

creasing the viscosity of the molten sulfur. Upon continued heating, the color of sulfur changes from yellow to a dark red because of unpaired electrons at the ends of the chains. If heated to 210 °C and poured into cold water, the sulfur forms an uncrystallized polymer called "plastic sulfur," which reverts back to the common crystalline forms at 25 °C (Figure 21.12).

Sulfur is a critical element in the body, necessary for the formation of methionine, an essential amino acid (⟵ *p. 557*). In proteins and enzymes, sulfur forms —S—S— disulfide linkages among chains of amino acids, which help to create their essential molecular shapes. Sulfur is also used to cross-link polymer chains in the vulcanization of rubber. The sulfur helps to align the polymer chains, which makes the rubber more elastic and prevents it from becoming sticky in warm weather.

Sulfuric Acid Production

Most sulfur is used to produce sulfuric acid, the workhorse industrial chemical used in steel production, in automobile batteries, in the petroleum industry, and in the manufacture of fertilizers, plastics, drugs, dyes, and many other products. Since sulfuric acid costs less to make than any other acid, it is the first to be considered when an acid is needed in an industrial process.

Sulfur is converted to sulfuric acid in four steps, collectively called the *contact process*. In the first step the sulfur is burned in air to give mostly sulfur dioxide.

$$S_8(s) + 8\,O_2(g) \longrightarrow 8\,SO_2(g)$$

The SO_2 is then converted to SO_3 over a heated catalyst, such as platinum metal or vanadium pentaoxide.

$$2\,SO_2(g) + O_2(g) \xrightarrow{\text{catalyst}} 2\,SO_3(g)$$

The next step converts the sulfur trioxide to sulfuric acid by the addition of water. The best way to do this is to pass the SO_3 into H_2SO_4 to form pyrosulfuric acid, $H_2S_2O_7$, and then to dilute the $H_2S_2O_7$ with water. The net reaction is 1 mol of H_2SO_4 for every 1 mol of SO_3.

$$SO_3(g) + H_2SO_4(\ell) \longrightarrow H_2S_2O_7(\ell)$$

$$H_2S_2O_7(\ell) + H_2O(\ell) \longrightarrow 2\,H_2SO_4(aq)$$

Net reaction: $SO_3(g) + H_2O(\ell) \longrightarrow H_2SO_4(aq)$

Sulfur dioxide for the contact process can also be obtained as a byproduct from copper or lead smelting. Unless this sulfur dioxide is recovered, it pollutes the atmosphere.

Problem-Solving Example 21.2   Sulfur and Sulfuric Acid

In 1994, 1.3×10^{10} kg of sulfur was produced in the United States. If all of this had been converted to sulfuric acid, how many kilograms of sulfuric acid would it have produced?

Answer 4.0×10^{10} kg H_2SO_4

Explanation   The equations for the formation of sulfuric acid provide the mole-to-mole relationships for the conversion of S_8 to H_2SO_4. The net reaction is the formation of one mole of H_2SO_4 per mole of SO_3. Each mole of SO_3 requires 1 mol of SO_2, and each mole of S_8 forms 8 mol of SO_2

$$1.3 \times 10^{10} \text{ kg } S_8 \left(\frac{10^3 \text{ g } S_8}{1 \text{ kg } S_8} \right) \left(\frac{1 \text{ mol } S_8}{256.5 \text{ g } S_8} \right) \left(\frac{8 \text{ mol } SO_2}{1 \text{ mol } S_8} \right) \left(\frac{1 \text{ mol } SO_3}{1 \text{ mol } SO_2} \right)$$
$$= 4.05 \times 10^{11} \text{ mol } SO_3$$

which can be converted to kilograms of sulfuric acid.

$$4.05 \times 10^{11} \text{ mol } SO_3 \left(\frac{1 \text{ mol } H_2SO_4}{1 \text{ mol } SO_3} \right) \left(\frac{98.08 \text{ g } H_2SO_4}{1 \text{ mol } H_2SO_4} \right) \left(\frac{1 \text{ kg } H_2SO_4}{10^3 \text{ g } H_2SO_4} \right)$$
$$= 4.0 \times 10^{10} \text{ kg } H_2SO_4$$

✔   Sulfuric acid is approximately 33% sulfur by mass ($[32 \text{ g } S/98 \text{ g } H_2SO_4] \times 100\%$). Therefore, 33 kg of sulfur would be able to form about 100 kg of sulfuric acid, and 1×10^{10} kg of S forms about 3×10^{10} kg of sulfuric acid, which is close to the calculated value of 4.0×10^{10} kg H_2SO_4.

Problem-Solving Practice 21.2

Calculate the number of kilograms of SO_3 produced by 1.3×10^{10} kg of S_8.

21.4 SOME MAIN GROUP ELEMENTS EXTRACTED BY ELECTROLYSIS: SODIUM, CHLORINE, MAGNESIUM, AND ALUMINUM

Chapter 19 described how electrolysis is used to make reactant-favored chemical reactions occur (⇐ *p. 899*). Electrolysis is applied commercially on a vast scale to extract several elements such as sodium, magnesium, aluminum, and chlorine from their natural sources. These are reactive elements that do not exist naturally in elemental form. Consequently, the metals must be obtained by reduction from their compounds, and chlorine must be oxidized from Cl^- to Cl_2.

Sodium

The industrial production of sodium described in this subsection and that of sodium hydroxide and chlorine (described in the next subsection) are linked by their naturally occurring raw material — sodium chloride — and by the use of electrolysis to produce them.

Sodium metal was discovered by Humphrey Davy in 1807 by the electrolysis of molten NaOH. The half-reactions are

$$4 \, OH^-(\text{in the melt}) \longrightarrow O_2(g) + 2 \, H_2O(g) + 4 \, e^- \qquad \text{(anode, oxidation)}$$

$$4 \, [Na^+(\text{in the melt}) + e^- \longrightarrow Na(\text{in the melt})] \qquad \text{(cathode, reduction)}$$

$$4 \, Na^+(\text{in the melt}) + 4 \, OH^-(\text{in the melt}) \longrightarrow 4 \, Na(\text{in the melt}) + O_2(g) + 2 \, H_2O(g) \qquad \text{(net cell reaction)}$$

By the early 1900s, commercial uses for sodium increased so that a large-scale production method was needed. In 1921 the Downs process was developed to meet this demand. In a Downs cell, molten NaCl is electrolyzed at 7 to 8 V and

Properties of Sodium

Atomic number	11
Density (g/L)*	0.97
Melting point (°C)	98
Boiling point (°C)	883
Atomic radius (pm)	186
Ionic radius, Na$^+$ (pm)	102
Electronegativity	1.0

*At 25 °C.

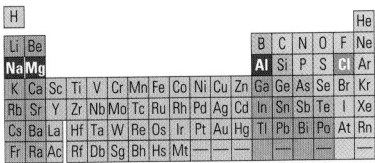

Figure 21.13 **The Downs cell for the electrolysis of molten NaCl.**

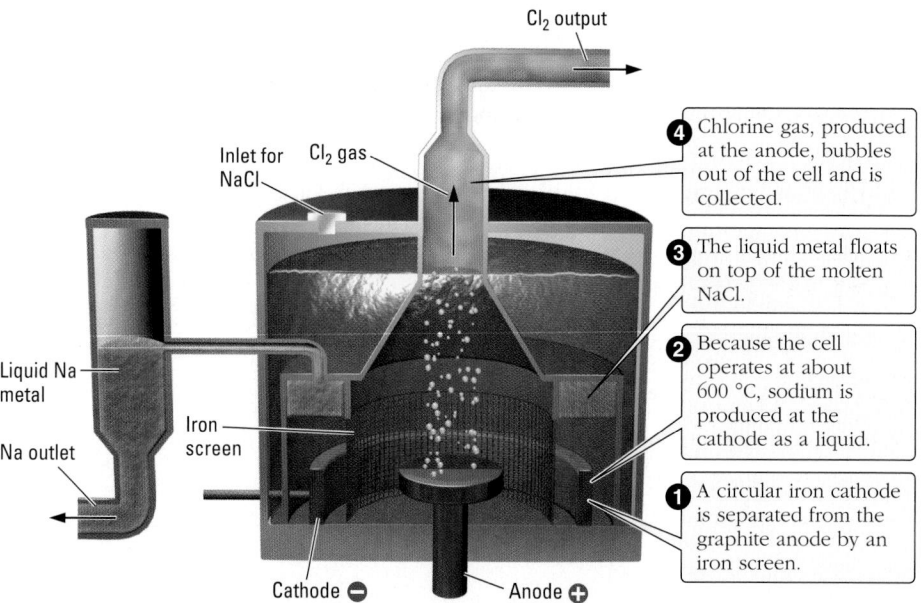

Cl₂ output

Inlet for NaCl   Cl₂ gas

❹ Chlorine gas, produced at the anode, bubbles out of the cell and is collected.

❸ The liquid metal floats on top of the molten NaCl.

❷ Because the cell operates at about 600 °C, sodium is produced at the cathode as a liquid.

❶ A circular iron cathode is separated from the graphite anode by an iron screen.

Liquid Na metal

Na outlet

Iron screen

Cathode ⊖   Anode ⊕

25,000 to 40,000 A (Figure 21.13). The cell is filled with a 1:3 mixture of NaCl and CaCl$_2$. Pure NaCl is not used because of its high melting point (800 °C). Mixing the two salts lowers the melting point of the mixture to approximately 600 °C.

In the Downs cell, sodium is produced at a cathode made of copper or iron that surrounds a cylindric graphite anode. Directly over the cathode is an inverted trough through which the molten sodium flows (sodium melts at 97.8 °C); liquid sodium is less dense than the molten mixture and therefore floats on top of it. Gaseous chlorine, the other product, passes through an inverted cone of nickel metal extending through the molten salt mixture and is collected, cooled, and liquefied.

$$Cl^-(\ell) \longrightarrow \tfrac{1}{2} Cl_2(g) + e^- \qquad \text{(anode, oxidation)}$$

$$Na^+(\text{in the melt}) + e^- \longrightarrow Na(\text{in the melt}) \qquad \text{(cathode, reduction)}$$

$$Na^+(\text{in the melt}) + Cl^-(\text{in the melt}) \longrightarrow Na(\text{in the melt}) + \tfrac{1}{2} Cl_2(g) \qquad \text{(net cell reaction)}$$

Exercise 21.6   The Downs Cell

How many tons of sodium can be produced in one day by a Downs cell operating at 2.0×10^4 A? How many tons of Cl$_2$ are produced in this same time?

Manufacturing facilities in the United States have the capacity to produce about 76,000 tons of sodium per year. Much of the manufacturing is located near Niagara Falls, New York because of the relatively low-cost electricity available from hydroelectric plants. A major use for sodium has been in the production of tetraethyllead [Pb(C$_2$H$_5$)$_4$], the octane enhancer in leaded gasoline. Although leaded gasoline is still used in some other countries, tetraethyllead is no longer used as a gasoline additive in the United States, causing a decline in sodium production. Because sodium is a strong reducing agent, it is used to obtain metals from metal halides. In particular, titanium, an element essential in aircraft production, can be prepared from its chloride by reduction with sodium.

A hydroelectric plant on the Niagara River near Niagara Falls.

$$TiCl_4(s) + 4\,Na(s) \longrightarrow Ti(s) + 4\,NaCl(s)$$

Liquid sodium has high thermal conductivity and an anomalously high heat capacity. Metallic sodium has a low melting point and can be liquefied easily. These properties make liquid sodium an excellent heat exchange liquid in nuclear reactors (⟵ *Section 20.6*).

Problem-Solving Example 21.3 Titanium Production

Assume that the annual production of sodium in the United States is 76,000 tons. If half of this amount were used to produce titanium from $TiCl_4$, how many tons of titanium could be produced?

Answer 2.0×10^4 tons

Explanation Half of the sodium produced would be 38,000 tons. From the balanced equation, we see that 4 mol of sodium are needed to produce 1 mol of titanium. Using this mole ratio we can calculate the number of moles of titanium, from which we can then find the mass of titanium.

$$3.8 \times 10^4 \text{ tons Na} \left(\frac{2000 \text{ lb Na}}{1 \text{ ton Na}} \right) \left(\frac{454 \text{ g Na}}{1 \text{ lb Na}} \right) \left(\frac{1 \text{ mol Na}}{23.0 \text{ g Na}} \right) = 1.5 \times 10^9 \text{ mol Na}$$

$$1.5 \times 10^9 \text{ mol Na} \left(\frac{1 \text{ mol Ti}}{4 \text{ mol Na}} \right) = 3.8 \times 10^8 \text{ mol Ti}$$

and

$$3.8 \times 10^8 \text{ mol Ti} \left(\frac{47.9 \text{ g Ti}}{1 \text{ mol Ti}} \right) \left(\frac{1 \text{ lb Ti}}{454 \text{ g Ti}} \right) \left(\frac{1 \text{ ton Ti}}{2000 \text{ lb Ti}} \right) = 2.0 \times 10^4 \text{ tons Ti}$$

✔ It takes about 100 g of sodium to produce about 50 g of titanium, or approximately 100 tons of sodium per 50 tons of titanium, that is, about half the mass of titanium per mass of sodium. Therefore, 38,000 tons of sodium would produce about 19,000 tons of Ti, which is close to the calculated answer.

Problem-Solving Practice 21.3

Under the same conditions as in Problem-Solving Example 21.3, how many tons of sodium chloride are produced?

Metallic sodium.

Chlorine and Sodium Hydroxide

Chlorine is produced by the electrolysis of aqueous sodium chloride in the **chlor-alkali process;** the alkali produced in the same process is sodium hydroxide. More than 24 billion pounds of sodium hydroxide are produced annually in the United States alone, along with a similar quantity of chlorine. These large amounts testify to the usefulness of these two products: The oxidizing and bleaching ability of chlorine is utilized in many industrial and everyday applications, and it is a raw material in the manufacture of chlorine-containing chemicals. Sodium hydroxide is the base of choice in many industrial chemistry applications because it is cheap. It is also used widely to produce soaps and detergents as well as other compounds.

The chlor-alkali process electrolyzes brine (saturated aqueous NaCl), as illustrated in Figure 21.14. Chloride ions are oxidized at the anode, and water is reduced at the cathode. The anode and cathode compartments are separated by a special polymeric membrane that allows only cations to pass through it. The brine solution is added to the anode compartment, and sodium ions pass through the membrane into the cathode compartment. The half-reactions are

$$2\,Cl^-(aq) \longrightarrow Cl_2(g) + 2\,e^- \qquad \text{(anode, oxidation)}$$

$$\underline{2\,H_2O(\ell) + 2\,e^- \longrightarrow 2\,OH^-(aq) + H_2(g) \qquad \text{(cathode, reduction)}}$$

$$2\,Cl^-(aq) + 2\,H_2O(\ell) \longrightarrow Cl_2(g) + 2\,OH^-(aq) + H_2(g) \qquad \text{(net cell reaction)}$$

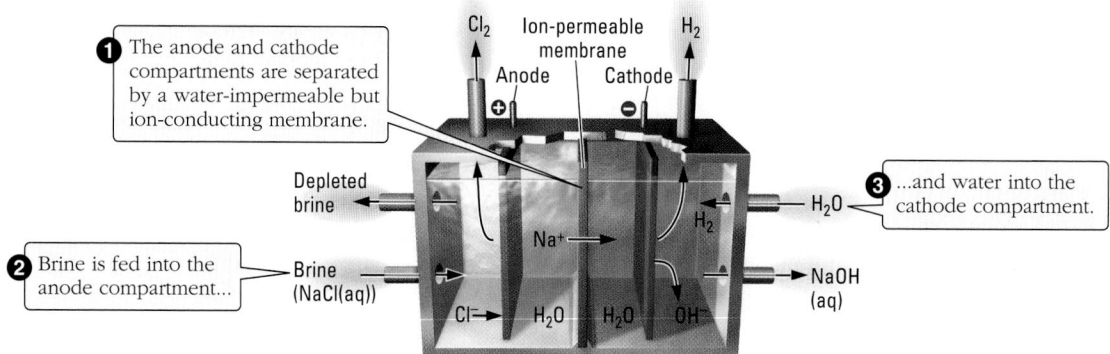

Figure 21.14 A membrane cell used in the chlor-alkali process.

Properties of Chlorine

Atomic number	17
Density (g/L)*	3.21
Melting point (°C)	−101
Boiling point (°C)	−34
Atomic radius (pm)	99
Ionic radius, Cl⁻ (pm)	181
Electronegativity	3.0

*At 25 °C.

The anode is activated titanium, and the cathode is stainless steel or nickel. The membrane is not permeable to water and acts as a salt bridge. So, as chloride ions are reduced in the anode compartment, sodium ions must migrate from there to the cathode compartment to maintain charge balance. The resulting NaOH solution in the cathode compartment is about 21 to 30% NaOH by weight.

The membrane cell was developed to replace the mercury cell that had been used in the chlor-alkali process. A major problem with mercury cells is the environmental damage caused by mercury spills. In the past when mercury cells were cleaned, mercury was routinely allowed to run into neighboring bodies of water.

Chlorine is a toxic, pale greenish-yellow gas with an irritating odor. It is the most important halogen used in industry. Chlorine is used to purify water (⇐ *p. 717*) to bleach paper and textiles, to manufacture herbicides, insecticides, and other chlorinated organic compounds, to produce polyvinyl chloride (PVC ⇐ *p. 546*) and to extract titanium metal from its ores. A strong oxidizing agent, chlorine is also used to oxidize Br^- and I^- to Br_2 and I_2 (Section 21.5).

Exercise 21.7 NaOH Production

A chlor-alkali membrane cell operates at 2.0×10^4 A for 100. hours. How many tons of NaOH are produced in this time?

Magnesium from Sea Water

With a concentration of 1.35 mg of Mg^{2+} per liter, the oceans provide a nearly limitless supply of magnesium, containing approximately 6 million tons of it per cubic

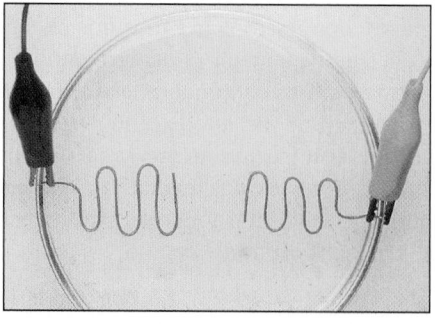

The microscale electrolysis of aqueous sodium chloride to produce chlorine gas.

Chlorine gas.

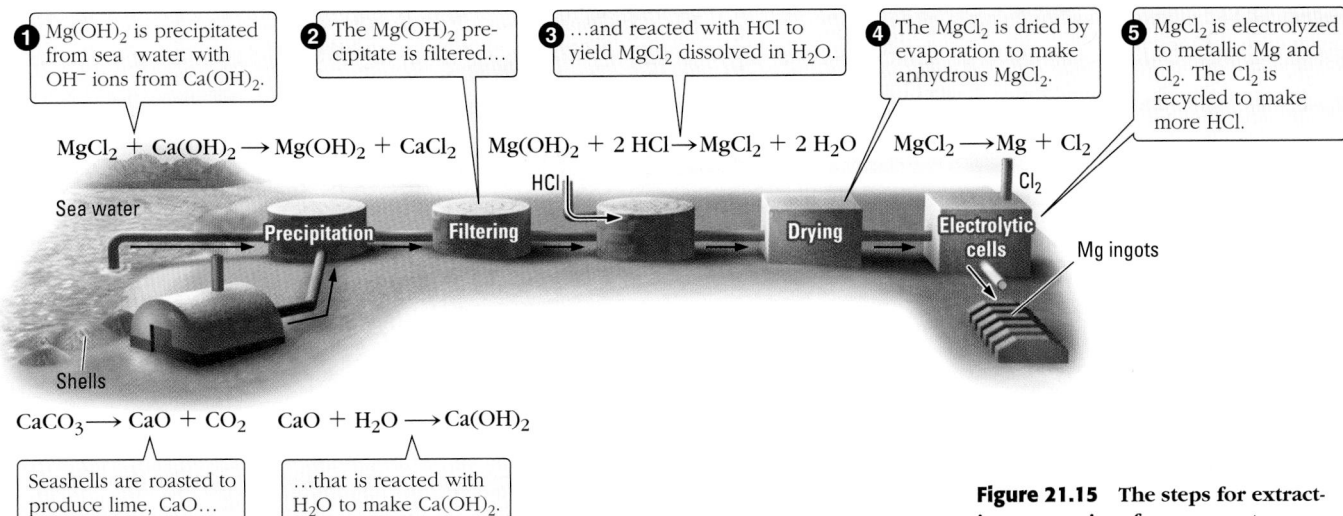

1 $Mg(OH)_2$ is precipitated from sea water with OH^- ions from $Ca(OH)_2$.

2 The $Mg(OH)_2$ precipitate is filtered...

3 ...and reacted with HCl to yield $MgCl_2$ dissolved in H_2O.

4 The $MgCl_2$ is dried by evaporation to make anhydrous $MgCl_2$.

5 $MgCl_2$ is electrolyzed to metallic Mg and Cl_2. The Cl_2 is recycled to make more HCl.

$$MgCl_2 + Ca(OH)_2 \longrightarrow Mg(OH)_2 + CaCl_2 \quad\quad Mg(OH)_2 + 2\,HCl \longrightarrow MgCl_2 + 2\,H_2O \quad\quad MgCl_2 \longrightarrow Mg + Cl_2$$

HCl

Sea water

Precipitation   Filtering   Drying   Electrolytic cells

Cl_2

Mg ingots

Shells

$$CaCO_3 \longrightarrow CaO + CO_2 \quad\quad CaO + H_2O \longrightarrow Ca(OH)_2$$

Seashells are roasted to produce lime, CaO...

...that is reacted with H_2O to make $Ca(OH)_2$.

Figure 21.15 The steps for extracting magnesium from sea water.

mile. As with other reactive metals, the conversion of the metal ion to the metal is not a product-favored reaction and so electrolysis is required.

The *Dow process* is used to extract magnesium metal from sea water. It begins with the precipitation of Mg^{2+} as its insoluble hydroxide ($K_{sp} = 1.5 \times 10^{-11}$). Hydroxide ions come from an inexpensive base, $Ca(OH)_2$, produced by roasting seashells to form calcium oxide, which then reacts with water to produce calcium hydroxide (Figure 21.15).

$$CaCO_3(s) \xrightarrow{\text{heat}} CaO(s) + CO_2(g)$$
$$\text{seashells} \quad\quad\quad \text{lime}$$

$$CaO(s) + H_2O(\ell) \longrightarrow Ca(OH)_2(aq)$$

$$Mg^{2+}(aq) + Ca(OH)_2(aq) \longrightarrow Mg(OH)_2(s) + Ca^{2+}(aq)$$

The magnesium hydroxide is filtered and neutralized by hydrochloric acid, another inexpensive chemical, to produce magnesium chloride.

$$Mg(OH)_2(s) + 2\,HCl \longrightarrow MgCl_2(aq) + 2\,H_2O(\ell)$$

The dried, anydrous magnesium chloride is then melted and electrolyzed in a steel pot, which serves as the cathode (Figure 21.16). The electrode reactions are

$$2\,Cl^-(\text{in the melt}) \longrightarrow Cl_2(g) + 2\,e^- \quad\quad (\text{anode, oxidation})$$

$$Mg^{2+}(\text{in the melt}) + 2\,e^- \longrightarrow Mg(\text{in the melt}) \,\,(\text{cathode, reduction})$$

$$Mg^{2+}(\text{in the melt}) + 2\,Cl^-(\text{in the melt}) \longrightarrow Mg(\text{in the melt}) + Cl_2(g)$$
$$(\text{net cell reaction})$$

The molten magnesium is less dense than the molten $MgCl_2$ and floats at the surface, where it can be removed. Chlorine produced at the anode is converted to HCl by mixing Cl_2 with methane from natural gas and burning the mixture.

$$4\,Cl_2(g) + 2\,CH_4(g) + O_2(g) \longrightarrow 2\,CO(g) + 8\,HCl(g)$$

The HCl is recycled to neutralize $Mg(OH)_2$, which forms $MgCl_2$

Magnesium metal itself has limited use, being found in flashbulbs, fireworks, and flares because the metal burns with a brilliant white light. Its most important use is in making alloys, principally with aluminum. Magnesium is the least dense structural material; lightweight, strong magnesium alloys are used to make aircraft wheels, truck bodies, and ladders, among other things.

Properties of Magnesium

Atomic number	12
Density (g/L)*	1.74
Melting point (°C)	650
Boiling point (°C)	1090
Atomic radius (pm)	160
Ionic radius, Mg^{2+} (pm)	72
Electronegativity	1.2

*At 25 °C.

Magnesium burning.

Figure 21.16 Electrolysis of molten magnesium chloride.

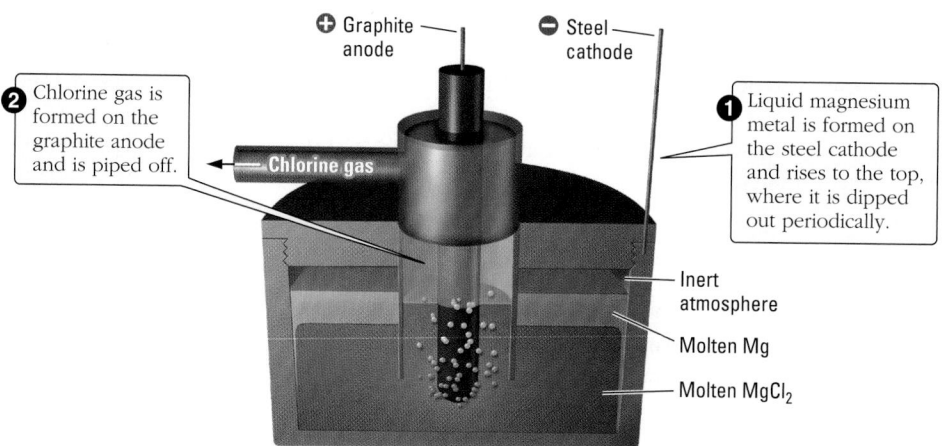

② Chlorine gas is formed on the graphite anode and is piped off.

⊕ Graphite anode

⊖ Steel cathode

① Liquid magnesium metal is formed on the steel cathode and rises to the top, where it is dipped out periodically.

Chlorine gas

Inert atmosphere

Molten Mg

Molten MgCl₂

Aluminum and copper with nitric acid. The aluminum oxide coating protects the underlying aluminum metal from attack by nitric acid (*left*). Copper without an oxide coating reacts vigorously (*right*).

Properties of Aluminum

Atomic number	13
Density (g/L)*	2.70
Melting point (°C)	661
Boiling point (°C)	2520
Atomic radius (pm)	143
Ionic radius, Al³⁺ (pm)	54
Electronegativity	1.5

*At 25 °C.

Remarkably, these two men, linked through their common discovery, also shared the same birth year (1863) and died the same year (1914).

Exercise 21.8 Lighting Things Up

When magnesium metal burns in air, magnesium nitride and magnesium oxide are produced.

(a) Write the formula for magnesium nitride.
(b) Write a balanced chemical equation for the formation of magnesium nitride from the elements.

Aluminum Production

Aluminum is an economically important, useful metal because of its low density (2.70 g/cm³), high strength when alloyed, and formability. It can be fashioned into wire, food wrapping sheets, step ladders, aircraft and automotive parts, and many other useful items. Aluminum metal is corrosion resistant because of a transparent, chemically inactive film of aluminum oxide that clings avidly to the metal's surface.

$$2\,Al(s) + 3\,O_2(g) \longrightarrow 2\,Al_2O_3(s)$$

Aluminum is the most abundant metal in the earth's surface (7.4%), but it is present there as Al^{3+} ions, from which the metal must be obtained by reduction. Aluminum was first isolated in metallic form in 1825 by an expensive and potentially dangerous method—using metallic sodium or potassium to reduce Al^{3+} ions in aluminum chloride, $AlCl_3$. Because of this difficulty, metallic aluminum was very expensive and considered to be a precious metal, like gold or platinum. An early use was in jewelry, including the Danish crown. In 1884, a 2.8-kg aluminum cap, produced by sodium reduction, topped the Washington monument as ornamentation and the tip of a lightning rod system. At that time, the aluminum cap cost considerably more than the same mass of silver. In the 1855 Exposition in Paris, some of the first aluminum metal pieces produced were displayed along with the French crown jewels. Napoleon II saw the advantages of using aluminum for military purposes because of its low density, and he commissioned studies on improving its production. Near the town of Les Baux, France, was a ready source of the aluminum-containing ore bauxite (Al_2O_3 combined with oxides of Si, Fe, and other elements); but how to extract aluminum from it readily? In 1886 Paul Héroult, a Frenchman, conceived of how to do so by an electrochemical process that is still used today. In a curious coincidence, Charles Martin Hall, an American, independently came up with the identical process two months earlier. Hence, the commercial method is known as the *Hall-Héroult process*. Just five years after the

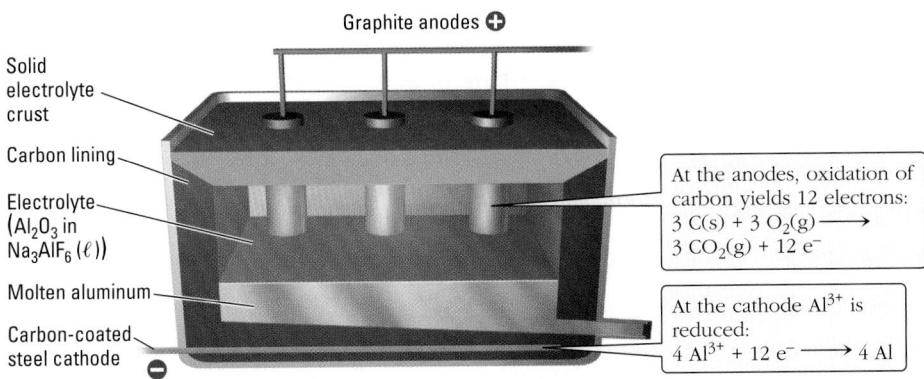

Figure 21.17 A Hall-Héroult process electrolytic cell. Molten aluminum is drawn off from the bottom of the cell into molds.

process was first used to produce aluminum commercially, the price of the metal plummeted from $12 per pound, a substantial sum at that time, to 70 cents per pound. What was once a jewelry metal now became commonplace.

In the Hall-Héroult process, metallic aluminum is obtained by electrolysis of Al_2O_3 in molten cryolite, Na_3AlF_6, in which a considerable amount of aluminum oxide dissolves. The cryolite allows the electrolysis to be carried out at a lower temperature (1000 °C) than would be required for molten Al_2O_3 (mp 2030 °C). The aluminum oxide–cryolite mixture is electrolyzed in a cell using carbon anodes and a carbon cell lining that serves as the cathode on which aluminum metal deposits (Figure 21.17). The half-reactions for extracting aluminum are

$$3\ C(s) + 6\ O_2(g) \longrightarrow 3\ CO_2(g) + 12\ e^- \qquad \text{(anode, oxidation)}$$

$$\underline{4\ Al^{3+}(\text{in the melt}) + 12\ e^- \longrightarrow 4\ Al(\text{in the melt}) \qquad \text{(cathode, reduction)}}$$

$$4\ Al^{3+}(\text{in the melt}) + 3\ C(s) + 6\ O_2(g) \longrightarrow 4\ Al(\text{in the melt})) + 3\ CO_2(g) \qquad \text{(net cell reaction)}$$

As the cell operates, molten aluminum sinks to the bottom of the cell, from which it is removed from time to time. The cells operate at a very low voltage of 4.0 to 5.5 V, but at a very high current of 50,000 to 150,000 A.

Aluminum production uses extremely large quantities of electricity, so aluminum production plants are located near hydroelectric power sources, such as those in the Pacific Northwest, because electricity from hydroelectric plants is generally less expensive than that from fossil fuel power plants. Production of each kilogram of aluminum requires about 13 to 16 kWh of electric energy, excluding that required to heat the molten mixture. Because of the high energy cost to extract aluminum metal from its ore, there is much interest in recycling aluminum beverage containers and other aluminum objects. It takes far less energy to process recycled aluminum than to produce the metal from bauxite. Putting this into perspective, you could run your television set for three hours on the energy saved by recycling just one aluminum can!

Problem-Solving Example 21.4 Aluminum Production

If electricity costs $0.080 per kilowatt hour (kWh), how much does the electricity cost to produce 1.00 ton of aluminum in a Hall-Héroult cell operating at 5.00 V? 1 kWh = 3.60×10^6 J; 1 V = 1 J/C.

Answer $1100

Explanation First, calculate the moles of aluminum produced.

$$1.00\ \text{ton Al} \left(\frac{2000\ \text{lb Al}}{1\ \text{ton Al}} \right) \left(\frac{454\ \text{g Al}}{1\ \text{lb Al}} \right) \left(\frac{1\ \text{mol Al}}{26.98\ \text{g Al}} \right) = 3.37 \times 10^4\ \text{mol Al}$$

CHARLES MARTIN HALL

(1863–1914)

While a student at Oberlin College (OH), Charles Martin Hall became intrigued with trying to separate aluminum from its ores cheaply. When just 22 years old, using batteries and a blacksmith's forge, Hall succeeded in reducing Al_2O_3 dissolved in cryolite to metallic aluminum. To take advantage of his discovery, he formed the Aluminum Corporation of America (ALCOA), an enterprise that made Hall a multimillionaire.

Properties of Phosphorus

Atomic number	15
Density (g/L)*	1.82
Melting point (°C)	44
Boiling point (°C)	280
Atomic radius (pm)	110
Ionic radius, p⁻ (pm)	212
Electronegativity	2.1

*At 25 °C.

Because the reduction reaction is $Al^{3+} + 3\,e^- \longrightarrow Al$, 1 mol of aluminum requires 3 mol of electrons. Therefore,

$$\text{Total charge} = 3.37 \times 10^4 \text{ mol Al} \times \frac{3 \text{ mol e}^-}{1 \text{ mol Al}} \times \frac{9.65 \times 10^4 \text{ C}}{1 \text{ mol e}^-} = 9.74 \times 10^9 \text{ C}$$

The number of kilowatt hours is

$$9.74 \times 10^9 \text{ C} \times \frac{5.00 \text{ J}}{1 \text{ C}} \times \frac{1 \text{ kWh}}{3.60 \times 10^6 \text{ J}} = 1.35 \times 10^4 \text{ kWh}$$

$$\text{Cost} = 1.35 \times 10^4 \text{ kWh} \times \frac{\$0.080}{1 \text{ kWh}} = \$1.1 \times 10^3, \text{ or } \$1100.$$

Problem-Solving Practice 21.4

How long would it take a Hall-Héroult cell operating at 1.00×10^5 A to produce 1.00 ton of aluminum metal?

21.5 SOME MAIN GROUP ELEMENTS EXTRACTED BY CHEMICAL OXIDATION-REDUCTION: PHOSPHORUS, BROMINE, AND IODINE

This section describes the extraction of phosphorus, bromine, and iodine, all of which are produced by redox reactions.

Phosphorus

Elemental phosphorus is extracted from phosphate rock by heating the rock with sand (SiO_2) and coke in an electric furnace (Figure 21.18). At 1400 to 1500 °C, the following reaction produces phosphorus, which evaporates from the mixture, leaving behind insoluble calcium silicate.

$$2\,Ca_3(PO_4)_2(\ell) + 10\,C(s) + 6\,SiO_2 \longrightarrow P_4(g) + 10\,CO(g) + 6\,CaSiO_3(\ell)$$

The mixture of phosphorus vapor and carbon monoxide gas is passed through water, where the phosphorus condenses and the CO bubbles out.

Exercise 21.9 Phosphorus Extraction

The extraction of phosphorus from phosphate rock involves oxidation and reduction. Identify what is oxidized and what is reduced.

About 90% of the elemental phosphorus produced is oxidized in air to P_4O_{10}, which reacts with water to produce phosphoric acid, H_3PO_4.

$$P_4(s) + 5\,O_2(g) \longrightarrow P_4O_{10}(s)$$

$$P_4O_{10}(s) + 6\,H_2O(\ell) \longrightarrow 4\,H_3PO_4(aq)$$

Some phosphoric acid is used in soft drinks, baking powder, and detergents.

Curiously, the principal use of phosphate rock is to make fertilizers directly rather than to produce the element (Figure 21.19). Phosphate rock is reacted with sulfuric acid and converted into a soluble fertilizer. The mixture of hydrated calcium dihydrogen phosphate and sulfate is called "superphosphate."

$$\underset{\text{phosphate rock}}{Ca_3(PO_4)_2(s)} + 2\,H_2SO_4(aq) + 5\,H_2O(\ell) \longrightarrow$$

$$\underset{\text{superphosphate}}{Ca(H_2PO_4)_2(aq)\cdot H_2O(s) + 2\,CaSO_4\cdot 2\,H_2O(s)}$$

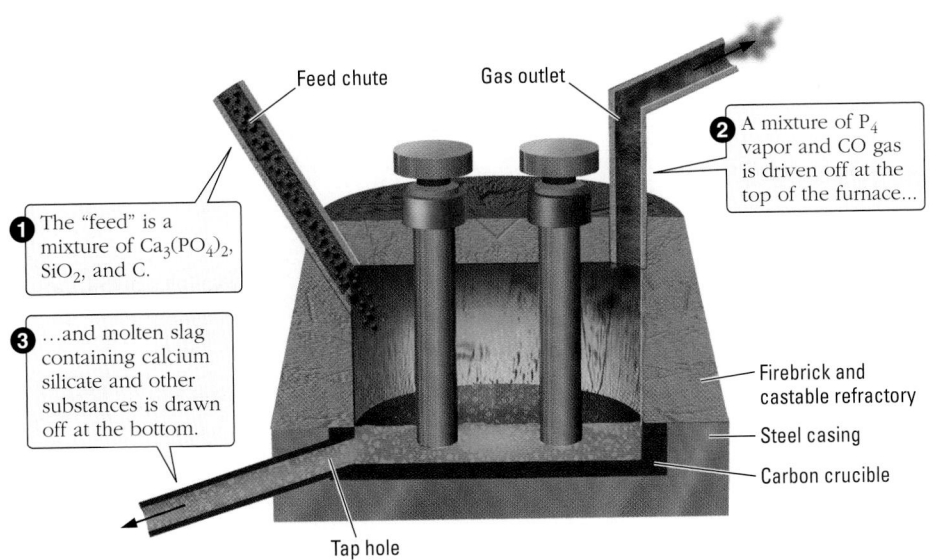

Figure 21.18 The production of phosphorus in an electric furnace.

① The "feed" is a mixture of $Ca_3(PO_4)_2$, SiO_2, and C.

② A mixture of P_4 vapor and CO gas is driven off at the top of the furnace...

③ ...and molten slag containing calcium silicate and other substances is drawn off at the bottom.

Feed chute

Gas outlet

Firebrick and castable refractory

Steel casing

Carbon crucible

Tap hole

Fertilizer is also made from phosphoric acid by neutralizing the acid with ammonia to form ammonium hydrogen phosphate, $(NH_4)_2HPO_4$.

A typical soft drink contains phosphoric acid.

Exercise 21.10 Phosphorus in Phosphate Rock

Calculate the mass percent of phosphorus present in another form of phosphate rock, hydroxyapatite, $Ca_5(PO_4)_3OH$.

Phosphorus has two main allotropes, *white* phosphorus and *red* phosphorus, which have very different properties. White phosphorus is highly reactive, igniting spontaneously in air at room temperature. For this reason, white phosphorus is stored under water. The waxy, nonpolar, solid white phosphorus is soft and easily cut, reflecting the fact that it consists of P_4 tetrahedra held together by weak noncovalent forces. Because it is nonpolar, white phosphorus is not soluble in water, but dissolves readily in nonpolar liquids such as carbon disulfide, CS_2, and hexane, C_6H_{14}. Unlike white phosphorus, red phosphorus does not oxidize in air at room temperature and is nontoxic (p. 978).

Although white phosphorus is toxic, phosphorus is an essential dietary mineral because of the many ways it is used by the body. Phosphorus is part of phosphate groups that link alternately with deoxyribose units to form the backbone of the DNA double helix (⟵ *p. 393*). Phosphate anhydride linkages, which contain

In a very limited oxygen supply, white phosphorus glows with a greenish light, the source of the term "phosphorescence."

A third, less common allotrope called black phosphorus is produced by heating white phosphorus at high pressure.

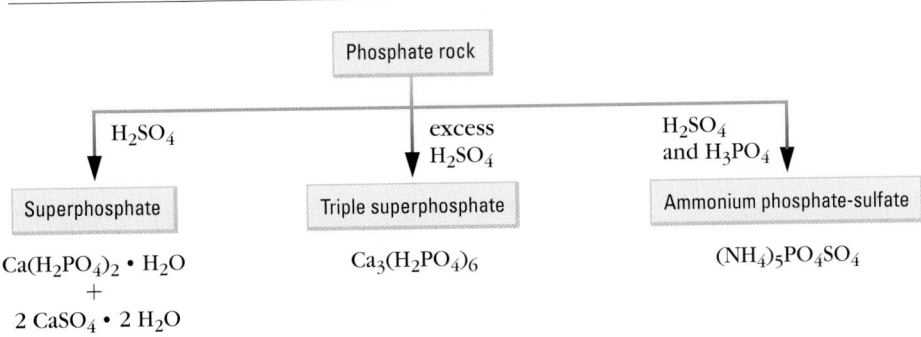

The discovery of phosphorus by Herman Brand.

$$\underset{O^-}{\overset{O}{-O-\overset{\|}{P}-O-\overset{\|}{\underset{O^-}{P}}-O-}}$$

Phosphate rock

H_2SO_4

excess H_2SO_4

H_2SO_4 and H_3PO_4

Superphosphate

Triple superphosphate

Ammonium phosphate-sulfate

$Ca(H_2PO_4)_2 \cdot H_2O$
+
$2\ CaSO_4 \cdot 2\ H_2O$

$Ca_3(H_2PO_4)_6$

$(NH_4)_5PO_4SO_4$

Figure 21.19 Fertilizers produced from phosphate rock.

ATP in cells (electron microscope photo).

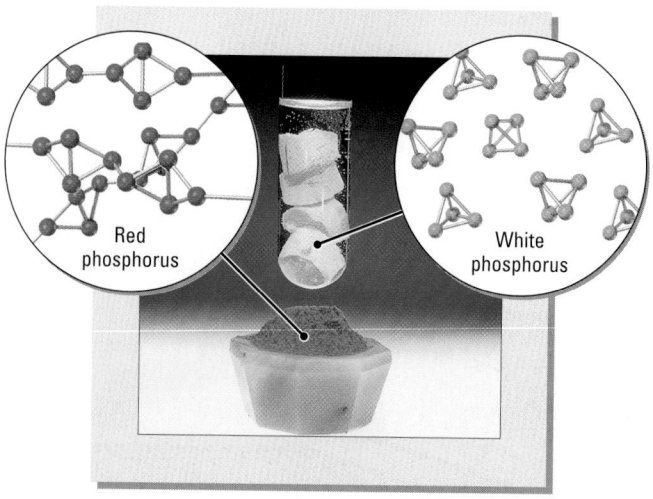

Phosphorus allotropes. Because white phosphorus (*top*) reacts with air at room temperature, it must be stored under water. Red phosphorus (*bottom*) does not react with air at room temperature.

Properties of Bromine

Atomic number	35
Density (g/L)*	3.12
Melting point (°C)	−7
Boiling point (°C)	59
Atomic radius (pm)	114
Ionic radius, Br⁻ (pm)	196
Electronegativity	2.7

*At 25 °C.

Properties of Iodine

Atomic number	53
Density (g/L)*	4.92
Melting point (°C)	114
Boiling point (°C)	184
Atomic radius (pm)	133
Ionic radius, I⁻ (pm)	220
Electronegativity	2.2

*At 25 °C.

bonds, are responsible for the way in which cellular energy is stored in ATP ($\Longleftarrow$ *p. 843)*.

Tooth enamel and bone contain the mineral hydroxyapatite, $Ca_5(PO_4)_3OH$. Water fluoridation reduces tooth decay because fluoride ions from the fluoridated water substitute for OH^- ions in tooth enamel to form fluoroapatite, $Ca_5(PO_4)_3F$, which is more resistant to decay than hydroxyapatite.

Bromine and Iodine

Bromine and iodine are halogens with similar but different properties. Like the other halogens, they are too reactive to be found free in nature. Consequently, Br_2 and I_2 must be extracted by the oxidation of their anions.

Bromine and iodine are extracted from sea water or *brines* (underground natural salt water deposits) by treating the solution with chlorine gas, which oxidizes Br^- to Br_2 and I^- to I_2. This is a case of a more reactive halogen, chlorine, displacing a less reactive one (bromine or iodine) from solution.

$$Cl_2(g) + 2\,Br^-(aq) \longrightarrow Br_2(\ell) + 2\,Cl^- \qquad E° = +0.293\ V$$

$$Cl_2(g) + 2\,I^-(aq) \longrightarrow I_2(aq) + 2\,Cl^- \qquad E° = +0.823\ V$$

Bromine and iodine.

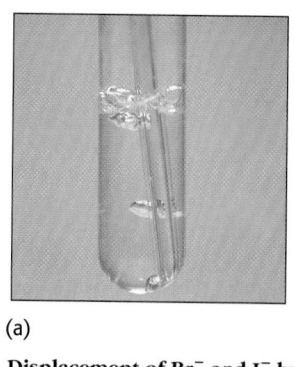
(a)

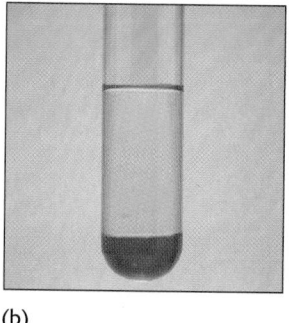
(b)

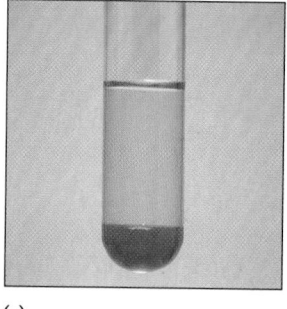
(c)

Displacement of Br⁻ and I⁻ by Cl₂. (a) Chlorine gas is bubbled into a colorless NaBr or NaI solution. (b) Br^- is oxidized by Cl_2 to give Br_2. (c) I^- is oxidized by Cl_2 to give I_2. Carbon tetrachloride, a dense liquid, is added and extracts the Br_2 and I_2 from the upper aqueous layer into the bottom CCl_4 layer, concentrating the Br_2 (*orange*) and I_2 (*purple*).

Another source of iodine is iodate ions, IO_3^-, in Chilean ore deposits, which are converted to I_2 in a two-step process using hydrogen sulfite ions.

Step 1: $2\,IO_3^-(aq) + 6\,HSO_3^-(aq) \longrightarrow 2\,I^-(aq) + 3\,SO_4^{2-}(aq) + 3\,H_2SO_4(aq)$

Step 2: $5\,I^-(aq) + IO_3^-(aq) + 3\,H_2SO_4(aq) \longrightarrow$
$$3\,I_2(aq) + 3\,SO_4^{2-}(aq) + 3\,H_2O(\ell)$$

Problem-Solving Example 21.5 Oxidation-Reduction Reactions

Identify the oxidizing and reducing agents in Step 1 of the extraction of iodine from IO_3^--bearing ores.

Answer Oxidizing agent: IO_3^-; reducing agent: HSO_3^-

Explanation In the reaction, the oxidation number of iodine changes from +5 in IO_3^- to −1 in I^-. Recall from Chapter 19 that in reduction there is a decrease in oxidation number due to a gain of electrons. The reduction requires a reducing agent, which in this case is hydrogen sulfite ion, HSO_3^- which donates the electrons. The +4 oxidation number of sulfur in HSO_3^- is increased to +6 in SO_4^{2-} This is oxidation, an increase in oxidation number, indicating a loss of electrons. Thus, the reducing agent is oxidized (+4 sulfur to +6 sulfur), while simultaneously iodine in the oxidizing agent, IO_3^-, is reduced from oxidation number +5 to oxidation number −1.

Problem-Solving Practice 21.5

Identify the oxidizing and reducing agents in Step 2 of the extraction of I_2 from Chilean ores.

 ### Exercise 21.11 Bromine Conversion

Use the terms oxidation, reduction, oxidizing agent, and reducing agent to explain the extraction of bromine from brines.

Exercise 21.12 Iodine Production

What does the $E°$ of +0.823 V indicate about the extraction of iodine by chlorine?

HERBERT H. DOW
(1866–1930)

Herbert H. Dow was the first to produce bromine by the electrolysis of brine (1891). In the 1920s, the demand for bromine rose sharply in order to make ethylene bromide, which was starting to be used in the higher octane gasoline required by high-performance automobile engines. Dow realized that ethylene bromide demand would be so great that brine sources could not supply enough bromine. And so he told the head of General Motors that to meet the demand, " . . . we'll have to go to sea and extract bromine from ocean water."* Herbert Dow died four years before achieving this goal, which was accomplished by his son Willard.

*Brandt, E. N. *Chemical Heritage*, Vol. 18, Number 3, Fall 2000; p. 39.

Bromine is the only liquid nonmetal at room temperature. It is used to prepare methyl bromide, CH_3Br, an efficient fire extinguisher and pesticide. Bromine is also used to make light-sensitive silver bromide for photographic films *(⇐ p. 197)*.

Iodine is the only common halogen that is a solid at room temperature. It is a dark gray, metallic-looking solid that sublimes to a violet-colored vapor. Iodine was discovered by burning dried seaweed, which concentrates iodide ions. Iodine is an essential dietary mineral for humans because the ions are necessary for the production of thyroxine, a growth-controlling hormone produced by the thyroid gland.

Sublimation of iodine.

Insufficient dietary iodine causes enlargement of the thyroid gland, a condition known as *goiter*. Sodium iodide (0.01%) is added to table salt (iodized salt) to prevent goiter.

SUMMARY PROBLEM

(a) In 1999, the United States used about 7.3×10^9 barrels of crude oil, enough for 28 barrels per person (1 barrel = 42 gallons; the density of crude oil is 1.3 g/mL). Assume that the crude oil is 3% sulfur by mass and that all of the sulfur was removed from the crude oil before it was used. How many liters of SO_2 at 25 °C and 1 atm from just your share of crude oil in 1999 would have been prevented from entering the atmosphere?

(b) Consider the conversion of $SO_2(g)$ to $SO_3(g)$.

$$SO_2(g) + \frac{1}{2} O_2(g) \longrightarrow SO_3(g)$$

(i) Use data from Appendix J to calculate ΔH for this reaction.

(ii) The reaction reaches equilibrium. Explain what effect each of these would have on the amount of sulfur trioxide formed.

    1. The pressure is increased.

    2. The temperature is decreased.

    3. A catalyst is used.

    4. Sulfur dioxide is added.

    5. Sulfur trioxide is removed as it forms.

(iii) At 1000 K, the equilibrium constant for this reaction is 1.7×10^1. Initially there were 0.250 mol of SO_2, 0.210 mol of O_2, and no SO_3 in a 10.0-L reaction chamber. Calculate the equilibrium concentrations of all species.

(c) Using data from Appendix J, calculate $\Delta G°$ for the conversion of sulfur dioxide gas to sulfur trioxide gas. Then calculate the value for K_c at 800 °C and at 1000 °C.

IN CLOSING

Having studied this chapter, you should be able to . . .

- Give a general explanation of how elements form in stars (Section 21.1).
- Know the principal elements in the earth's crust (Section 21.2).
- Describe the general structure of silicates (21.2).
- Identify the general methods by which elements are extracted from the earth's crust (Section 21.2).
- Identify the major components of the atmosphere (Section 21.3).
- Explain how elements are obtained by the liquefaction of air (Section 21.3).
- Describe the Frasch process for obtaining sulfur (Section 21.3).
- Explain how sulfuric acid is produced (Section 21.3).
- Describe how electrolysis is used to obtain sodium, chlorine, magnesium, and aluminum (Section 21.4).
- Explain how chemical redox reactions are used to extract bromine, iodine, and phosphorus from compounds (Section 21.5).
- Apply chemical principles to the processes for extracting and purifying elements (Sections 21.3–21.5).

KEY TERMS

chlor-alkali process *(21.4)*

cryogen *(21.3)*

Frasch process *(21.3)*

helium burning *(21.1)*

hydrogen burning *(21.1)*

mineral *(21.2)*

nitrogen fixation *(21.3)*

nuclear burning *(21.1)*

ores *(21.2)*

QUESTIONS FOR REVIEW AND THOUGHT

Answers to questions in **bold** can be found in the back of the book.

Review Questions

1. What is meant by hydrogen burning and helium burning in relation to the formation of elements?
2. Identify the most abundant nonmetallic element in the earth's crust. Identify the most abundant metallic element in the earth's crust.
3. Describe the difference between an ore and a mineral.
4. Give a simple explanation for the abundance of clay minerals in the earth's crust.
5. Differentiate among pyroxenes, amphiboles, and silica.
6. Explain how the silicate unit in amphiboles has the general formula $Si_4O_{11}^{6-}$.
7. Explain how the silicate unit in mica and other sheet silicates has the general formula $Si_2O_5^{2-}$.
8. Identify two major differences between white phosphorus and red phosphorus.
9. Identify
   (a) Two elements obtained from the atmosphere.
   (b) Two elements obtained from the sea.
   (c) Two elements obtained from the earth's crust.
10. Write balanced equations for the recovery of magnesium from sea water. Begin with the precipitation of magnesium hydroxide by addition of calcium hydroxide to sea water.
11. Why are nitrogen and oxygen important industrial chemicals?
12. Describe how nature fixes nitrogen. Why is nitrogen fixation necessary?
13. Briefly explain why different products are obtained from the electrolysis of molten NaCl and the electrolysis of aqueous NaCl.
14. Identify two uses of phosphate rock.
15. Describe the structural changes that occur in sulfur as it goes from room temperature to 210 °C.
16. Identify the substance or substances produced by each of these commercial processes:
    (a) Hall-Héroult (b) Contact
    (c) Ostwald (d) Dow
17. Identify the substance or substances produced by each of these commercial processes:
    (a) Frasch (b) Chlor-alkali (c) Downs cell
18. Why is phosphate rock not applied directly as a phosphorus fertilizer?

Electrolytic Methods

19. (a) Write the balanced chemical equation for the electrolysis of aqueous NaCl.
    (b) In 1995, 1.2×10^{10} kg of NaOH and 1.1×10^{10} kg of chlorine were produced in the United States. Does the ratio of these masses agree with the ratio of masses from the balanced chemical equation? If not, what does that suggest about the ways that NaOH and Cl_2 are produced?
20. To produce magnesium metal, 1000. kg of molten $MgCl_2$ is electrolyzed.

(a) At which electrode is magnesium produced?
(b) What is produced at the other electrode?
(c) How many faradays of electricity are used in the process?
(d) An industrial process uses 8.4 kWh per pound of Mg. How much energy is required per mole of magnesium?
21. A Downs cell operates at 7.0 V and 4.0×10^4 A.
    (a) How much Na(s) and Cl_2(g) can be produced in 24 hours by such a cell?
    (b) Assuming 100% efficiency, what is the energy consumption (kWh) of this cell?
22. How much energy (kWh) is required to prepare a ton of sodium in a typical Downs cell operating at 25,000 A and 7.0 V?
23. What mass of aluminum can be produced when 6.0×10^4 A is passed through a series of 100 Hall-Héroult electrolytic cells operating at an 85% efficiency for 24 hours?
24. What mass of aluminum can be produced from the electrolysis of molten $AlCl_3$ in an electrolytic cell operating at 100. A for 2.00 hr?

General Questions

25. Complete this table.

Formula	Name	Oxidation state of nitrogen
_____	Nitrogen	_____
NH_3	_____	_____
_____	Nitrous acid	_____
_____	Nitrogen dioxide	_____
NH_4^+	_____	_____
_____	Ammonium nitrate	_____

26. Complete this table.

Formula	Name	Oxidation state of phosphorus
_____	Phosphorus	_____
$(NH_4)_2HPO_4$	_____	_____
_____	Phosphoric acid	_____
_____	Pentaphosphorus decaoxide	_____
$Ca_3(PO_4)_2$	_____	_____
_____	Calcium dihydrogen phosphate	_____

27. Molten NaCl is electrolyzed in a Downs cell operating at 1.00×10^4 A for 24 hr.
    (a) How much sodium is produced?

(b) What volume of Cl_2 in liters is collected from the outlet tube at 20 °C and 15 atm?

28. Bauxite, the principal source of aluminum oxide, contains 55% Al_2O_3. How much bauxite is required to produce the 5.0×10^6 tons of Al produced annually by electrolysis?

29. Write a plausible Lewis structure for azide ion, N_3^-.

30. Write a plausible Lewis structure for P_4O_{10}.

31. Write the chemical equation for the
    (a) Combustion of white phosphorus.
    (b) Reaction of the combustion product with water.

32. Calculate the temperature at which the conversion of white phosphorus to red phosphorus occurs. $\Delta H° = -17.6$ kJ/mol; $\Delta S° = -18.3$ J/K.

33. There are two common oxides of sulfur. Name these oxides, and write chemical equations for the reaction of each of them with water. Identify the products.

34. What raw materials are used to produce sulfuric acid? Write chemical equations to represent the steps in the contact process to produce sulfuric acid.

35. Write Lewis structures for all the resonance forms of sulfuric acid.

36. Write the Lewis structures for all the resonance forms of nitric acid.

37. Iodine trichloride, ICl_3, is an interhalogen compound.
    (a) Write the Lewis structure of ICl_3.
    (b) Does the central atom have more than an octet of valence electrons?
    (c) Using VSEPR theory, predict the molecular shape of ICl_3.

38. At some temperature, a gaseous mixture in a 1.00-L vessel originally contained 1.00 mol of SO_2 and 5.00 mol of O_2. When equilibrium was reached, 77.8% of the SO_2 had been converted to SO_3. Calculate the equilibrium constant (K_c) for this reaction at this temperature.

Applying Concepts

39. Assume that the radius of the earth is 6400 km, the crust is 50 km thick, the density of the crust is 3.5 g/cm³, and 25.7% of the crust is silicon by mass. What is the total mass of silicon in the crust of the earth?

40. The K_{sp} of $Ca(OH)_2$ is 7.9×10^{-6}; that for $Mg(OH)_2$ is 1.5×10^{-11}. Calculate the equilibrium constant for the reaction

$$Ca(OH)_2(s) + Mg^{2+}(aq) \longrightarrow Ca^{2+}(aq) + Mg(OH)_2(s)$$

and use it to explain why this reaction can be used commercially to extract magnesium from sea water.

41. Commercial concentrated nitric acid contains 69.5 mass percent HNO_3 and has a density of 1.42 g/mL.
    (a) Calculate the molarity of this solution.
    (b) What volume of the concentrated acid must be used to prepare 10.0 L of 6.00 M HNO_3?

42. The compound nitrosyl azide, N_4O, is a covalent compound with an NNNNO atomic arrangement. Write a plausible Lewis structure for this compound.

43. Hydrazoic acid, HN_3, is very explosive in its pure state but can be studied in aqueous solution. The acid is prepared by the reaction of hydrazine with nitrous acid.

$$N_2H_4(\ell) + HNO_2(aq) \longrightarrow HN_3(aq) + 2\,H_2O(\ell)$$

(a) Determine the oxidation states of nitrogen in the compounds in this reaction.
(b) What is the oxidizing agent in this reaction?
(c) The K_a of hydrazoic acid is 2.4×10^{-5} at 25 °C. Calculate the pH of a 0.010 M solution of HN_3.

44. (a) Write two plausible resonance structures for hydrazoic acid, HN_3.
    (b) Use bond energy data (Table 8.2) to calculate the ΔH_f for each resonance form; $\Delta H_f = 218.0$ kJ/mol for H(g) and 472.7 kJ/mol for N(g).

45. Given the reaction

$$Cl_2(g) + H_2O(\ell) \rightleftharpoons H^+(aq) + Cl^-(aq) + HOCl(aq)$$

(a) Identify the oxidizing agent and the reducing agent.
(b) Write the equilibrium constant expression for the reaction.
(c) Calculate the concentration of HOCl in equilibrium with $Cl_2(g)$ at 1.0 atm. $K = 2.7 \times 10^{-5}$.

46. Dinitrogen trioxide, N_2O_3, is a blue liquid formed by the reaction of NO_2 and NO.
    (a) Write a balanced chemical equation for the formation of N_2O_3.
    (b) Write the Lewis structure of N_2O_3 and any plausible resonance forms.
    (c) Predict the O—N—O and the N—N—O bond angles.

47. (a) Write the resonance forms of SO_3.
    (b) Predict the molecular shape of SO_3 and the O—S—O bond angle.

48. The density of sulfur vapor at 700 °C and 1.00 atm is 0.8012 g/L. What is the molecular formula of sulfur in the vapor?

49. Iodine can be produced by the oxidation of iodide ion by permanganate ion.

$$MnO_4^- + 2\,I^-(aq) + 4\,H^+(aq) \longrightarrow$$
$$I_2(s) + Mn^{2+}(aq) + 2\,H_2O(\ell)$$

Excess HI is added to 0.200 g of MnO_4^-. Assuming 100% yield, how many grams of iodine are produced?

50. In a Downs cell, molten NaCl is electrolyzed to sodium metal and chlorine gas.

$$2\,NaCl(\ell) \longrightarrow 2\,Na(\ell) + Cl_2(g)$$

The $\Delta H°$ and $\Delta S°$ for the reaction are +820 kJ and +180 J/K, respectively.
(a) Calculate $\Delta G°$ at 600 °C, the electrolysis temperature.
(b) Calculate the voltage required for the electrolysis.

51. A 425-gal tank of water contains 175 g of NaI. Calculate the number of liters of chlorine gas at 758 mm Hg and 25 °C required to convert all the iodide to iodine.

The exterior walls of the Guggenheim Museum in Bilbao, Spain, are a beautiful and extraordinarily architectural application of pure titanuim, a transition element.

CHEMISTRY OF SELECTED TRANSITION ELEMENTS AND COORDINATION COMPOUNDS

22.1 Properties of the Transition (*d*-block) Elements

22.2 Iron and Steel: The Use of Pyrometallurgy

22.3 Copper: A Coinage Metal

22.4 Silver and Gold: The Other Coinage Metals

22.5 Chromium

22.6 Coordinate Covalent Bonds: Complex Ions and Coordination Compounds

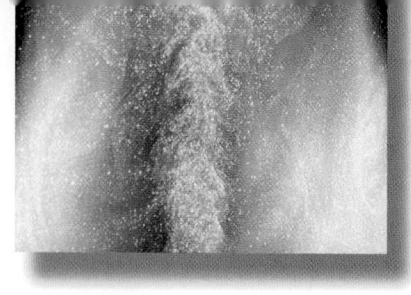

It is hard to overstate how important metals have been to the development of civilizations. Transitions from the Stone Age (400,000 to 7000 BC) to the Bronze Age (began 4000 to 3500 BC) to the Iron Age (1800s and beyond) and to the Computer Age (late 20th century) have been marked by the ability to extract metals from their ores and to process the metals into tools and objects useful in industry, warfare, and homes. In this chapter we consider the transition metals, the *d* block of elements in the periodic table. Some of these elements and their compounds are of major economic importance. The precious metals gold, silver, and platinum are transition elements used in coinage and jewelry. Others, such as iron and its alloy, steel, are valuable for their structural uses. We will first consider the transition metals in overview and then look more closely at a few of them, including iron, the most economically important. The chapter closes with coverage of coordination compounds, in which ions or molecules surround transition metal ions or atoms. Such compounds run the gamut from being responsible for the vivid colors of famous oil paintings to their role in significant biomolecules such as hemoglobin, vitamin B_{12}, and critical enzymes.

22.1 PROPERTIES OF THE TRANSITION (*d*-BLOCK) ELEMENTS

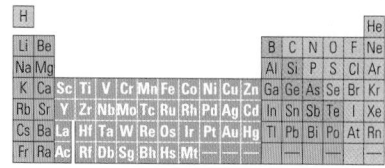

The four series of *d*-block elements called the *transition elements* are in the center of the periodic table in Periods 4 through 7. As the name indicates, these elements lie between the very active metals of the *s* blocks and the less reactive metals of the *p*-block elements. The transition elements share the properties in this section and are discussed in this chapter. (The transition elements in Period 7 beginning with rutherfordium, element 104, are all radioactive. They have been made synthetically in *very* small amounts, just several atoms in some cases, and therefore, less is currently known about their properties.)

These generalities apply to the transition elements:

- All are metals that conduct electricity well, but to varying degrees.
- Most are ductile (able to be drawn into a wire) and malleable (able to be hammered into thin sheets).
- Except for gold and copper, they are silvery-white or bluish.
- They generally have higher melting and boiling points than the main group elements; tungsten has the highest melting point of any metal (3410 °C). Mercury is an exception, being the only liquid metal at room temperature.
- They generally have high densities; osmium (22.61 g/cm³) and iridium (22.65 g/cm³) are the most dense metals, even more dense than gold (19.3 g/cm³).
- Many form brightly colored compounds.
- Some are paramagnetic; a few are ferromagnetic (⇐ *p. 296*).
- They form complex ions (Section 22.6).
- Most have multiple oxidation states; the scandium (+3) and zinc (+2) groups are exceptions.

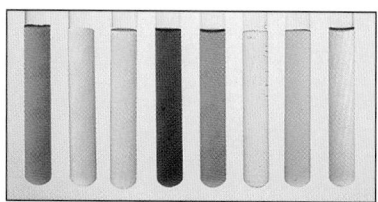

Many transition metal ions form colored aqueous solutions. *(Left to right)*: $FeCl_3$, $CuSO_4$, $MnCl_2$, $CoCl_2$, $Cr(NO_3)_3$, $FeSO_4$, $NiSO_4$, and $K_2Cr_2O_7$.

The concept of oxidation states (numbers) is described in Sections 5.4 and 19.1.

The transition elements at the ends of each series show considerable differences in their chemical behavior. At the left side of the series, members of the scandium group of elements (Sc, Y, La) are reactive metals like the alkaline earth metals, their predecessors in each period. On the other hand, the zinc family members (Zn, Cd, Hg) are not like other transition elements in that the zinc family members have filled $(n - 1)d$ and ns sublevels. In fact, the elements of the zinc family are sometimes not classified as transition elements. Such differences in chemical behavior can be attributed to the number and distribution of *d* orbital electrons.

TABLE 22.1   Outermost Electron Configurations of *d*-Block Elements

Deviations (marked in color) occur when a different configuration is more stable.

Configuration	$(n-1)d\,ns^2$	$(n-1)d^2\,ns^2$	$(n-1)d^3\,ns^2$	$(n-1)d^4\,ns^2$	$(n-1)d^5\,ns^2$	$(n-1)d^6\,ns^2$	$(n-1)d^7\,ns^2$	$(n-1)d^8\,ns^2$	$(n-1)d^9\,ns^2$	$(n-1)d^{10}\,ns^2$
First series:	$_{21}$**Sc** $3d^1\,4s^2$	$_{22}$**Ti** $3d^2\,4s^2$	$_{23}$**V** $3d^3\,4s^2$	$_{24}$**Cr** $3d^5\,4s^1$	$_{25}$**Mn** $3d^5\,4s^2$	$_{26}$**Fe** $3d^6\,4s^2$	$_{27}$**Co** $3d^7\,4s^2$	$_{28}$**Ni** $3d^8\,4s^2$	$_{29}$**Cu** $3d^{10}\,4s^1$	$_{30}$**Zn** $3d^{10}\,s^2$
Second series:	$_{39}$**Y** $4d^1\,5s^2$	$_{40}$**Zr** $4d^2\,5s^2$	$_{41}$**Nb** $4d^4\,5s^1$	$_{42}$**Mo** $4d^5\,5s^1$	$_{43}$**Tc** $4d^5\,5s^2$	$_{44}$**Ru** $4d^7\,5s^1$	$_{45}$**Rh** $4d^8\,5s^1$	$_{46}$**Pd** $4d^{10}$	$_{47}$**Ag** $4d^{10}\,5s^1$	$_{48}$**Cd** $4d^{10}\,s^2$
Third series:	$_{57}$**La** $5d^1\,6s^2$	$_{72}$**Hf** $4f^{14}5d^2\,6s^2$	$_{73}$**Ta** $4f^{14}5d^3\,6s^2$	$_{74}$**W** $4f^{14}5d^4\,6s^2$	$_{75}$**Re** $4f^{14}5d^5\,6s^2$	$_{76}$**Os** $4f^{14}5d^6\,6s^2$	$_{77}$**Ir** $4f^{14}5d^7\,6s^2$	$_{78}$**Pt** $4f^{14}5d^9\,6s^1$	$_{79}$**Au** $4f^{14}5d^{10}\,6s^1$	$_{80}$**Hg** $4f^{14}5d^{10}\,6s^2$

4f-elements intervene

Electron Configurations

All transition elements have the electron configuration

$$[\text{noble gas}](n-1)d^x ns^y$$

where n is the period number (4 through 7), x is the number of d electrons (1 through 10), and y is the number of s electrons (1 or 2, except in palladium), as summarized in Table 22.1. The number of d electrons increases from left to right across a transition metal series. Elements in the zinc group (Zn, Cd, Hg) at the end of the series have filled d^{10} sublevels. In the preceding group the elements Cu, Ag, and Au also have filled d^{10} sublevels, along with half-filled ns^1 sublevels, as described below.

The progressive filling of the d orbitals from Sc to Zn in the first series is not uniform, as seen in Table 22.2. The first three elements of the series—Sc, Ti, and

TABLE 22.2   Orbital Occupancy of the First Transition Series Elements

Element	Partial orbital diagram	Unpaired electrons
	3d *4s* *4p*	
Sc	↑ _ _ _ _ ↑↓ _ _ _	1
Ti	↑ ↑ _ _ _ ↑↓ _ _ _	2
V	↑ ↑ ↑ _ _ ↑↓ _ _ _	3
Cr	↑ ↑ ↑ ↑ ↑ ↑ _ _ _	6
Mn	↑ ↑ ↑ ↑ ↑ ↑↓ _ _ _	5
Fe	↑↓ ↑ ↑ ↑ ↑ ↑↓ _ _ _	4
Co	↑↓ ↑↓ ↑ ↑ ↑ ↑↓ _ _ _	3
Ni	↑↓ ↑↓ ↑↓ ↑ ↑ ↑↓ _ _ _	2
Cu	↑↓ ↑↓ ↑↓ ↑↓ ↑↓ ↑ _ _ _	1
Zn	↑↓ ↑↓ ↑↓ ↑↓ ↑↓ ↑↓ _ _ _	0

V — have $[Ar]3d^14s^2$, $[Ar]3d^24s^2$, and $[Ar]3d^34s^2$ electron configurations, respectively. The filling sequence changes at chromium, which has a ground state electron configuration of $[Ar]3d^54s^1$ with *two* half-filled sublevels, which is a lower energy state than $[Ar]3d^44s^2$. The sequence reverts back to normal from manganese through nickel, with the pairing of $3d$ electrons. But it changes again with copper, which fills up the $3d$ sublevel before the $4s$. This gives copper a filled d^{10} sublevel and a half-filled s^1 sublevel (Table 22.2), which is more stable than an $[Ar]3d^94s^2$ configuration. The first series ends with zinc, which has filled $4s$ *and* $3d$ sublevels, $[Ar]3d^{10}4s^2$.

When transition metal atoms lose electrons to form ions, the* ns *electrons are lost before the* (n − 1)d *electrons. Thus, both $4s$ electrons are lost when Fe^{2+} and Fe^{3+} form. These ions differ in their number of $3d$ electrons, not $4s$ electrons.

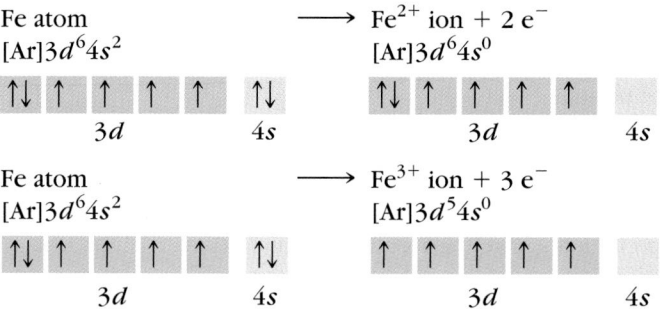

Magnetic measurements confirm the electron configurations of the first row and other transition metal ions (Figure 22.1). The *magnetic moment* is a value calculated from the measured paramagnetism of a sample and is indicative of the number of unpaired electrons in the sample (⇐ *p. 296*). As seen from Figure 22.2, the greater the number of unpaired electrons, the greater the magnetic moment of the substance. Notice from Figure 22.2 the confirming experimental evidence that the $4s$ electrons are removed first to give Fe^{2+} and Fe^{3+} four and five unpaired electrons, respectively. If the $3d$ electrons were removed first, these ions would have four (Fe^{2+}) and three (Fe^{3+}) unpaired electrons.

If a substance is diamagnetic (has no unpaired electrons), its apparent mass is unaffected (or slightly reduced) when the magnetic field is "on."

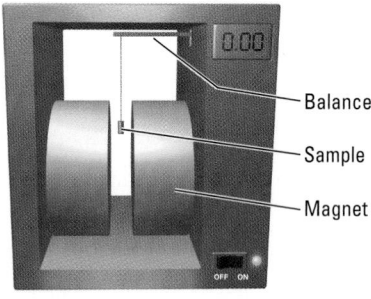

(a)

If a substance is paramagnetic (has unpaired electrons), its apparent mass increases when the field is "on" because the balance arm feels an additional force as the sample is attracted by the magnetic field.

(b)

Figure 22.1 Measurement of magnetic behavior.

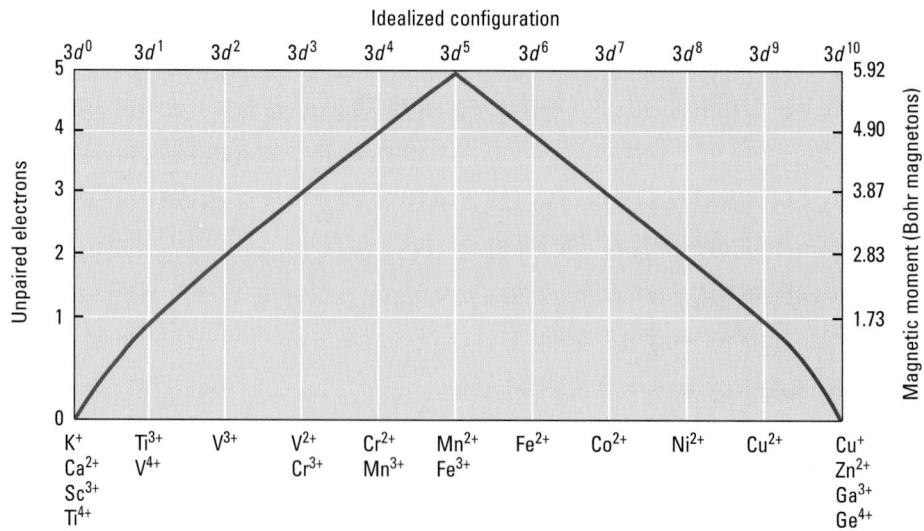

Figure 22.2 The number of unpaired electrons in the first-row transition metal ions and their magnetic moments.

Problem-Solving Example 22.1   Transition Metal Ion Electron Configurations

Use orbital box diagrams to explain the number of unpaired electrons shown in Figure 22.2 for

(a) Ti^{3+} (b) Cr^{2+} (c) Cu^{2+}

Answer

(a) $[Ar]3d^14s^0$

$\boxed{\uparrow}\,\boxed{}\,\boxed{}\,\boxed{}\,\boxed{}\qquad\boxed{}$
$\qquad\qquad 3d\qquad\qquad\qquad 4s$

(b) $[Ar]3d^44s^0$

$\boxed{\uparrow}\,\boxed{\uparrow}\,\boxed{\uparrow}\,\boxed{\uparrow}\,\boxed{}\qquad\boxed{}$
$\qquad\qquad 3d\qquad\qquad\qquad 4s$

(c) $[Ar]3d^94s^0$

$\boxed{\uparrow\downarrow}\,\boxed{\uparrow\downarrow}\,\boxed{\uparrow\downarrow}\,\boxed{\uparrow\downarrow}\,\boxed{\uparrow}\qquad\boxed{}$
$\qquad\qquad 3d\qquad\qquad 4s$

Explanation   When an ion is formed from a Period 4 transition metal, $4s$ electrons are removed before $3d$ electrons, giving rise to the number of unpaired electrons in the ion.

Ti atom $\longrightarrow$ Ti^{3+} ion $+$ $3\,e^-$

$[Ar]3d^24s^2$ $[Ar]3d^14s^0$

$\boxed{\uparrow}\,\boxed{\uparrow}\,\boxed{}\,\boxed{}\,\boxed{}\quad\boxed{\uparrow\downarrow}\qquad\boxed{\uparrow}\,\boxed{}\,\boxed{}\,\boxed{}\,\boxed{}\quad\boxed{}$
$\qquad 3d\qquad\qquad 4s\qquad\qquad\quad 3d\qquad\qquad\qquad 4s$

Cr atom $\longrightarrow$ Cr^{2+} ion $+$ $2\,e^-$

$[Ar]3d^54s^1$ $[Ar]3d^44s^0$

$\boxed{\uparrow}\,\boxed{\uparrow}\,\boxed{\uparrow}\,\boxed{\uparrow}\,\boxed{\uparrow}\quad\boxed{\uparrow}\qquad\boxed{\uparrow}\,\boxed{\uparrow}\,\boxed{\uparrow}\,\boxed{\uparrow}\,\boxed{}\quad\boxed{}$
$\qquad 3d\qquad\qquad 4s\qquad\qquad\quad 3d\qquad\qquad\qquad 4s$

Cu atom $\longrightarrow$ Cu^{2+} ion $+$ $2\,e^-$

$[Ar]3d^{10}4s^1$ $[Ar]3d^94s^0$

$\boxed{\uparrow\downarrow}\,\boxed{\uparrow\downarrow}\,\boxed{\uparrow\downarrow}\,\boxed{\uparrow\downarrow}\,\boxed{\uparrow\downarrow}\quad\boxed{\uparrow}\qquad\boxed{\uparrow\downarrow}\,\boxed{\uparrow\downarrow}\,\boxed{\uparrow\downarrow}\,\boxed{\uparrow\downarrow}\,\boxed{\uparrow}\quad\boxed{}$
$\qquad 3d\qquad\qquad 4s\qquad\qquad\quad 3d\qquad\qquad\quad 4s$

Problem-Solving Practice 22.1

Use orbital box diagrams to explain the number of unpaired electrons shown in Figure 22.2 for

(a) Mn^{2+} (b) Cr^{3+}

Atomic Radii

Transition metals have less variation in atomic radii across a period than do main group elements. Across a row of transition elements, the atomic radii decrease steadily and then increase slightly (Figure 22.3). The decrease occurs because the added *d* electrons only partially shield valence electrons from the increasing nuclear charge. As a result, the *effective* nuclear charge felt by the valence electrons increases and they are more strongly attracted toward the nucleus, causing a decrease in the size of the atoms. Toward the end of each transition metal series, the radii increase slightly due to several factors, including electron-electron repulsion as electrons are paired in *d* orbitals.

The radii of the second transition series elements (Period 5) are, as expected, greater than those of the first-row transition metals. What is unexpected is what occurs with the atomic radii in going from the second row to the third row (Period 6). Instead of the third row radii being larger, they are nearly the same as

Figure 22.3 Radii of transition metals (in picometers).

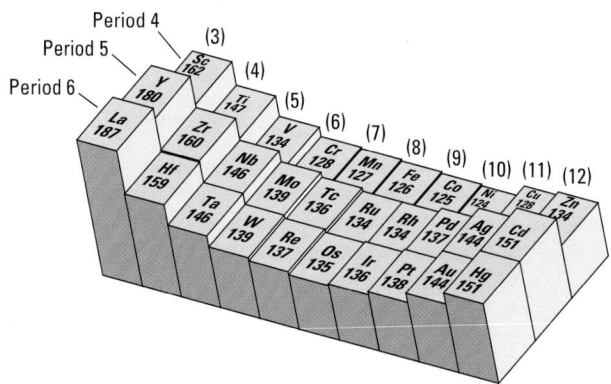

those of the second row. This is a consequence of the lanthanide series of elements (La, element 57, to Lu, element 71) intervening between barium, the Period 6 alkaline earth element, and hafnium, the next transition element of Period 6. In the lanthanide series elements, the effective nuclear charge builds up, causing a decrease in their atomic radii because all the additional electrons go into the $4f$ orbitals, which do not effectively screen valence electrons from the increasing nuclear charge. The increased nuclear charge pulls the valence electrons closer to the nucleus, decreasing the atomic radii. The decrease in size, known as the **lanthanide contraction,** just offsets the expected size increase going from the second to the third row of transition elements. Consequently, the second- and third-row transition elements are of similar size, which causes them to have similar chemical properties. The transition elements of the second and third row occur together in ores, and because of their chemical similarities, they are very difficult to separate from each other.

Oxidation States

Except for the scandium ($+3$) and zinc ($+2$) group elements, all other transition metals have multiple oxidation states. The oxidation states of the first transition series elements are listed in Figure 22.4. Manganese, for example, has three common oxidation states: $+2$ in Mn^{2+}, $+4$ in MnO_2, and $+7$ in MnO_4^-; and iron has two: $+2$ in FeO and $+3$ in Fe_2O_3. Less common oxidation states are noted as well.

Transition metals that form $+2$ ions do so, in general, by losing two ns electrons before losing any $(n-1)d$ electrons. Higher oxidation states involve losing $(n-1)d$ electrons as well. The maximum oxidation state for the first five elements in the first series—Sc through Mn—equals the sum of the $(n-1)d$ plus ns electrons. Thus, the maximum oxidation state of chromium is $+6$, which is found in

Figure 22.4 Oxidation states of transition series elements.

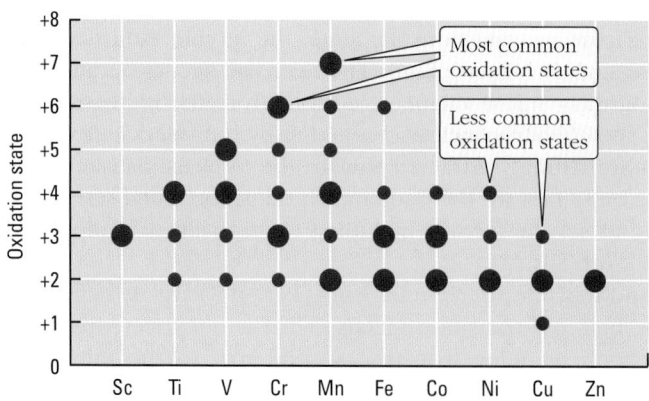

CrO_4^{2-} and $Cr_2O_7^{2-}$, and that for manganese is $+7$, which is found in MnO_4^-. These high oxidation states make $Cr_2O_7^{2-}$ (in acidic solution) and MnO_4^- strong oxidizing agents. In general, compounds in which the transition metal has a low oxidation state tend to be ionic, whereas compounds with transition metals in high oxidation states are relatively covalent. Thus, $MnCl_2$ (m.p. 650 °C) is an ionic solid containing Mn^{2+} and Cl^- ions. On the other hand, MnO_4^- is a polyatomic ion containing covalent Mn—O bonds.

Compounds of transition elements with partially filled d orbitals can accept or donate electrons, a property that makes them effective catalysts. In iron compounds, for example, iron can be present as Fe^{2+} (reduced form) or Fe^{3+} (oxidized form). The iron ions act as an electron shuttle, losing or gaining electrons between the oxidized and reduced forms when catalyzing electron transfer reactions, such as the production of ammonia by the Haber process (⇐ *Section 14.8)*.

Exercise 22.1   Oxidation States of Transition Metals

Determine the oxidation state of the transition metal in these compounds:

(a) V_2O_5 (b) $K_2Cr_2O_7$ (c) MnO_2 (d) OsO_4

Exercise 22.2   Oxidation State

Identify the oxidation state of iron in $KFe[Fe(CN)_6]$. Explain your answer.

22.2 IRON AND STEEL: THE USE OF PYROMETALLURGY

Iron is the most abundant transition metal and the second most abundant metallic element in the earth's crust (4.7%). Pure iron is a silvery-white, rather soft metal. The great commercial importance of iron comes with the addition of small amounts of carbon or other alloying elements to it to form steel.

In air $Fe^{2+}(aq)$ is oxidized to $Fe^{3+}(aq)$.

$$4\,Fe^{2+}(aq) + O_2(g) + 4\,H^+(aq) \longrightarrow 4\,Fe^{3+}(aq) + 2\,H_2O(\ell)$$

Aqueous Fe^{3+} reacts with water to form a hydrated oxide known as *rust*.

$$2\,Fe^{3+}(aq) + 4\,H_2O(\ell) \longrightarrow Fe_2O_3\cdot H_2O(s) + 6\,H^+(aq)$$

Iron reacts with nonoxidizing acids such as HCl and acetic acid to form the pale green $Fe(H_2O)_6^{2+}$ ion.

$$Fe(s) + 2\,H^+(aq) \longrightarrow Fe^{2+}(aq) + H_2(g) \qquad E° = +0.44\ V$$

When reacted with oxidizing acids such as dilute nitric acid, the metal is oxidized directly to Fe^{3+}.

$$Fe(s) + 4\,H^+(aq) + NO_3^-(aq) \longrightarrow Fe^{3+}(aq) + NO(g) + 2\,H_2O \qquad E° = +1.00\ V$$

The principal iron ores are hematite, Fe_2O_3, and magnetite, Fe_3O_4, which are found in large deposits in Minnesota, Russia, France, England, and Australia. Iron production involves steps to concentrate and purify the ores. Iron ions in the oxide ores are reduced to the metal by using carbon in the form of coke as the reducing agent at high temperatures in a blast furnace (Figure 22.5). **Pyrometallurgy** is the extraction of a metal from its ore using chemical reactions carried out at high temperatures.

A mixture of iron ore, coke, and limestone ($CaCO_3$) is fed into the top of the blast furnace, and a blast of heated air or oxygen is forced into the bottom of the

Iron also occurs in nature as the sulfide, FeS_2. This mineral is not used as an ore because in steel making it is difficult to remove all the sulfur, which makes the steel brittle.

Figure 22.5 Diagram of a blast furnace. Iron ore is reduced to iron in a blast furnace.

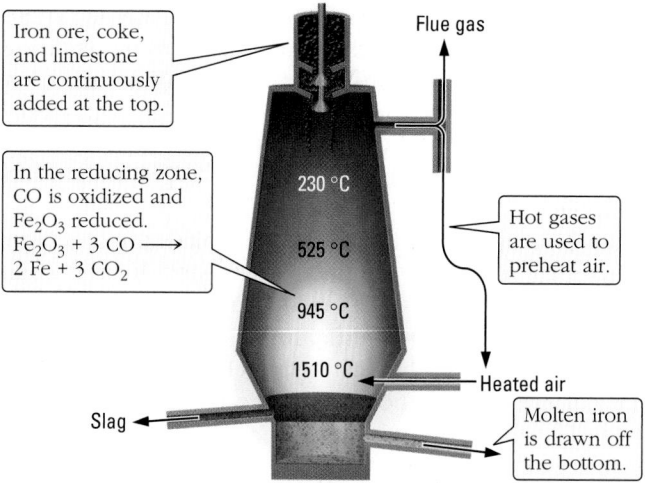

Iron ore, coke, and limestone are continuously added at the top.

In the reducing zone, CO is oxidized and Fe_2O_3 reduced.
$Fe_2O_3 + 3\ CO \longrightarrow 2\ Fe + 3\ CO_2$

230 °C

525 °C

945 °C

1510 °C

Flue gas

Hot gases are used to preheat air.

Heated air

Molten iron is drawn off the bottom.

Slag

furnace. The coke reacts exothermically with the heated air, producing a high temperature that speeds up the iron-forming reactions, thus making the process economical. The iron ore is reduced to metallic iron by the reactions

$$2\ C(s) + O_2(g) \longrightarrow 2\ CO(g) \quad \text{exothermic}$$

$$Fe_2O_3(s) + 3\ CO(g) \longrightarrow 2\ Fe(s) + 3\ CO_2(g) \quad \text{exothermic}$$

Limestone is added to remove silica-containing impurities in the ore.

$$CaCO_3(s) \longrightarrow CaO(s) + CO_2(g) \quad \text{endothermic}$$

$$CaO(s) + SiO_2(s) \longrightarrow CaSiO_3(\ell) \quad \text{endothermic}$$

Calcium silicate and other metal silicates form *slag,* which is a liquid at the temperature of the blast furnace. The result is the formation of two layers at the bottom of the furnace. The lower, more dense liquid is molten iron that contains a substantial amount of dissolved carbon and smaller amounts of other impurities. The upper liquid layer is the slag. Periodically, the blast furnace is tapped from the bottom to draw off the molten iron. The liquid slag is drawn off at a port higher in the furnace.

The iron withdrawn from a blast furnace is *pig iron,* which is a brittle material due to impurities of up to 4.5% carbon, 1.7% manganese, 0.3% phosphorus, 0.04%

ESTIMATION Up and Down the East Coast

Railroad rails are made of steel, which for this problem can be considered to be pure iron. The iron for the rails is made from an iron ore. Each rail weighs an average of 125 lb/yd. Estimate how much of an iron ore containing approximately 2% Fe_3O_4 would be needed to produce enough steel to build a railroad line from Boston to Washington DC, a distance of approximately 500 miles.

In the 500 miles of rails (two rails per track), there are approximately 2×10^8 lb of iron.

$$\left(\frac{5 \times 10^2\ \text{mi}}{1\ \text{track}}\right)\left(\frac{5 \times 10^3\ \text{ft}}{1\ \text{mi}}\right)\left(\frac{1\ \text{yd}}{3\ \text{ft}}\right) \times$$

$$\left(\frac{125\ \text{lb Fe}}{1\ \text{yd}}\right)\left(\frac{2\ \text{rails}}{1\ \text{track}}\right) \approx 2 \times 10^8\ \text{lb Fe}$$

Each 100 lb of iron ore contains 2 lb of Fe_3O_4 in which there is 168 lb of Fe per 232 lb of Fe_3O_4. Thus, it would require about 1×10^{10} lb (5×10^6 tons) of the iron ore to make the rails.

$$2 \times 10^8\ \text{lb Fe} \left(\frac{232\ \text{lb Fe}_3\text{O}_4}{168\ \text{lb Fe}}\right)\left(\frac{100\ \text{lb ore}}{2\ \text{lb Fe}_3\text{O}_4}\right)$$

$$\approx 1 \times 10^{10}\ \text{lb ore}$$

sulfur, and 1% silicon. The principal embrittling material is cementite, an iron carbide formed at the temperatures of the blast furnace.

$$3\ Fe(s) + C(s) \longrightarrow Fe_3C(s)$$

Molten pig iron that is poured into molds of a desired shape is called *cast iron*. It can be used directly to make molded automobile engine blocks, brake drums, transmission housings, and the like. Cast iron, however, like pig iron, contains too much carbon and other impurities for most structural uses. To make **steel**, a much stronger material, from cast iron or pig iron, the phosphorus, sulfur, and silicon impurities must be removed and the carbon content reduced to about 1.3%.

Steel

Many iron alloys are known collectively as *steels*, each with its own particular structural properties. One of the most common is carbon steel, an iron alloy containing about 0.5 to 1.3% carbon. To convert pig iron to steel, the excess carbon is oxidized away using oxygen. Thus, whereas extracting iron from an ore is a reduction process, steel making is an oxidation. One of several techniques used to make steel is the *basic oxygen process* (Figure 22.6), in which pure oxygen is blown through a ceramic tube that is pushed below the surface of molten, impure iron. At the high temperatures of the melt, the dissolved carbon reacts rapidly with the oxygen to form gaseous carbon monoxide and carbon dioxide, which are vented. The scale of the basic oxygen process operation is impressive. About 200 tons of molten pig iron, 100 tons of scrap iron, and 20 tons of limestone are loaded into the furnace at a time. The steel is produced within an hour using such a process.

The composition of steel is varied by adding silicon, chromium, manganese, molybdenum, nickel, or other metals to give the steel specific physical, chemical, and mechanical properties. Table 22.3 lists the composition and uses of some common steel alloys. Magnetic alloys can be permanent magnets, such as those in audio speakers, or temporary magnets such as those in electric motors, generators, and transformers. Alnico is the general name for a series of popular permanent magnets containing Al, Ni, Co, Fe, and sometimes Cu and Ti. Alnico V, for example, contains 51% Fe, as well as four other elements — 14% Ni, 24% Co, 8% Al, and 3% Cu.

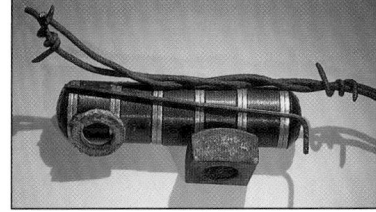

An Alnico magnet picking up iron or steel objects.

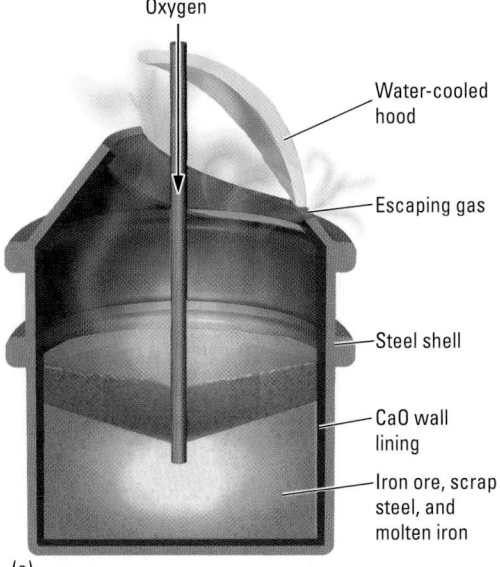

(a)

Oxygen

Water-cooled hood

Escaping gas

Steel shell

CaO wall lining

Iron ore, scrap steel, and molten iron

(b)

Figure 22.6 **The basic oxygen process for making steel.** (a) Most of the steel manufactured today is produced by the basic oxygen process. (b) Molten steel being poured from a basic oxygen furnace.

TABLE 22.3 Some Steels and Their Uses

Name	Composition	Properties	Uses
Carbon steel	1.3% C, 98.7% Fe	Hard	Sheet steel, tools
Manganese steel	10–18% Mn, 90–82% Fe, 0.5% C	Hard, resistant to wear	Railroad rails, safes, armor plate
Stainless steel	14–18% Cr, 7–9% Ni, 79–73% Fe, 0.2% C	Resistant to corrosion	Cutlery, instruments
Nickel steel	2–4% Ni, 98–96% Fe, 0.5% C	Hard, elastic, resistant to corrosion	Drive shafts, gears, cables
Invar steel	36% Ni, 64% Fe, 0.5% C	Low coefficient of expansion	Meter scales, measuring tapes
Silicon steel	1–5% Si, 99–95% Fe, 0.5% C	Hard, strong, highly magnetic	Magnets
Duriron	12–15% Si, 88–85% Fe, 0.85% C	Resistant to corrosion, acids	Pipes
High-speed steel	14–20% W, 86–80% Fe, 0.5% C	Retains temper at high speeds	High-speed cutting tools

The artist Michelangelo wrote about using fire to transform substances:
"It is with fire that blacksmiths iron subdue
Unto fair form, the image of their
thought . . . "
Sonnet 59

Blacksmiths and other steel fabricators have long known that the properties of steel are also affected by the processing temperature, cooling rates, and hammering, rolling, and extrusion. If hot steel is cooled quickly by immersing it in water or oil, the carbon in the steel will remain as cementite, Fe_3C, resulting in hard, but brittle steel. Because the formation of cementite is endothermic, slower cooling favors crystals of carbon (graphite) rather than cementite to form in the steel. Such steel is more ductile. By further rapid cooling, followed by controlled reheating, a process called *tempering,* the cementite-to-graphite ratio is adjusted and the properties of the resultant steel varied further.

Problem-Solving Example 22.2 Iron Production

(a) How much carbon monoxide is needed to form 1.00×10^3 kg of iron from hematite, Fe_2O_3, and from magnetite, Fe_3O_4?
(b) How much carbon in the form of coke must be used to prepare the total amount of CO needed for the two reductions in part (a)?

Answer
(a) 7.52×10^5 g CO for hematite and 6.68×10^5 g CO for magnetite
(b) 6.09×10^5 g C

Explanation
(a) The equation for the reduction of hematite to iron is

$$Fe_2O_3(s) + 3\ CO(g) \longrightarrow 2\ Fe(s) + 3\ CO_2(g)$$

Therefore, the mass of CO required to form 1.00×10^3 kg of Fe from hematite is

$$1.00 \times 10^3 \text{ kg Fe} \left(\frac{10^3 \text{ g Fe}}{1 \text{ kg Fe}} \right) \left(\frac{1 \text{ mol Fe}}{55.85 \text{ g Fe}} \right) \left(\frac{3 \text{ mol CO}}{2 \text{ mol Fe}} \right) \left(\frac{28.00 \text{ g CO}}{1 \text{ mol CO}} \right)$$

$$= 7.52 \times 10^5 \text{ g CO}$$

The CO required to reduce magnetite can be calculated the same way based on the equation

$$Fe_3O_4(s) + 4\,CO(g) \longrightarrow 3\,Fe(s) + 4\,CO_2(g)$$

$$1.00 \times 10^3\,\text{kg Fe}\left(\frac{10^3\,\text{g Fe}}{1\,\text{kg Fe}}\right)\left(\frac{1\,\text{mol Fe}}{55.85\,\text{g Fe}}\right)\left(\frac{4\,\text{mol CO}}{3\,\text{mol Fe}}\right)\left(\frac{28.00\,\text{g CO}}{1\,\text{mol CO}}\right)$$
$$= 6.68 \times 10^5\,\text{g CO}$$

(b) The carbon monoxide is prepared by the reaction

$$2\,C(s) + O_2(g) \longrightarrow 2\,CO(g)$$

From part (a), the total mass of CO required is $(7.52 \times 10^5\,\text{g}) + (6.68 \times 10^5\,\text{g}) = 1.42 \times 10^6\,\text{g}$. The required amount of carbon is

$$1.42 \times 10^6\,\text{g CO}\left(\frac{1\,\text{mol CO}}{28.00\,\text{g CO}}\right)\left(\frac{2\,\text{mol C}}{2\,\text{mol CO}}\right)\left(\frac{12.01\,\text{g C}}{1\,\text{mol C}}\right) = 6.09 \times 10^5\,\text{g C}$$

✔ (a) Hematite: 10^3 kg of iron is equivalent to about

$$10^6\,\text{g Fe} \times \frac{1\,\text{mol Fe}}{56\,\text{g Fe}} = 1.8 \times 10^4\,\text{mol of Fe}$$

which requires 1.5 times that number of moles of CO, about 2.7×10^4 mol of CO. This is equivalent to approximately 8×10^5 g of CO.
Magnetite: Using the same approach, we estimate that magnetite requires 7×10^5 mol of CO. The estimated values for CO needed for each ore are close to the calculated values, so the answers are reasonable.

(b) The sum of the estimated masses of CO is 1.5×10^6 g of CO. This is approximately 6×10^5 g of C, close to the actual calulated answer, which is reasonable.

Problem-Solving Practice 22.2

Calculate the total mass of iron ore needed to produce 1.00×10^3 kg of iron by the process described in Problem-Solving Example 22.2.

Exercise 22.3 High-Speed Steel

Ultrahigh-speed steel is used in some saw blades. Such a saw blade, weighing 500. g, contains 0.6% C, 4.0% Cr, 18% W, 1.0% Mo, 1.5% V, and 6.0% Co along with iron.

(a) Calculate the mass of W and of Co in the saw blade.
(b) Which alloying metal is present in the greatest mole per cent, that is $\dfrac{\text{moles alloying metal}}{\text{moles all alloying metals}} \times 100\%$?

22.3 COPPER: A COINAGE METAL

Copper is sometimes found in metallic form in nature and evidence suggests that such naturally occuring copper was known and used during the Stone Age. As early as 10,000 years ago, the metal was hammered into useful items such as coins, jewelry, tools, and weapons. Nearly six millennia later, the Bronze Age was ushered in when humans learned how to alloy copper with tin to make bronze.

The Metallurgy of Copper

Native copper, that which is found "free" in nature, is not available in sufficient supply to meet the demands for the metal, and so chemical methods have been developed to extract copper from its ores. The principal ores are chalcocite, Cu_2S, and chalcopyrite, $CuFeS_2$, which occur along with iron sulfides, FeS_2 and FeS.

Native copper.

An open-pit copper mine near Bagdad, Arizona.

Modern methods to extract copper from its ores begin with crushing the ore and separating it from rocks. The ore is then heated to temperatures high enough to drive off the sulfur as sulfur dioxide, a process called *roasting*.

$$3\ FeS_2(s) + 8\ O_2(g) \xrightarrow{\text{heat}} Fe_3O_4(s) + 6\ SO_2(g)$$

$$2\ CuFeS_2(s) + O_2(s) \xrightarrow{\text{heat}} Cu_2S(s) + 2\ FeS(s) + SO_2(g)$$

Copper is separated from the iron by melting the Cu_2S and Fe_3O_4 mixture and combining it with oxygen and SiO_2 to form a liquid iron silicate slag and molten Cu_2S. The slag is less dense than the molten copper(I) sulfide and floats on it, where it can be drawn off periodically.

The conversion of copper(I) sulfide to metallic copper takes place in a process similar to the basic oxygen process for steel making. After the iron silicate slag is removed, air is blown through the molten Cu_2S, converting it to Cu_2O, which reacts with the remaining Cu_2S to form copper metal.

$$2\ Cu_2S(\ell) + 3\ O_2(g) \longrightarrow 2\ Cu_2O(\ell) + 2\ SO_2(g)$$

$$2\ Cu_2O(\ell) + Cu_2S(\ell) \longrightarrow 6\ Cu(\ell) + SO_2(g)$$

Copper Hill, Tennessee, a wasteland as a result of SO_2 released by copper smelting.

The resulting copper, called *blister copper,* which is about 96 to 99.5% copper, can be further purified by electrolysis.

The electrorefining of copper is carried out in large electrolytic cells in which anodes of blister copper bars alternate with very thin sheets of pure copper, which are the cathodes (Figure 22.7). A mixture of copper(II) sulfate and dilute sulfuric acid is the electrolyte. Impure copper is oxidized at the anode to Cu^{2+}, and Cu^{2+} in the electrolyte is reduced to pure metallic copper at the cathode.

Anode: $Cu(s,\ \text{impure blister copper}) \longrightarrow Cu^{2+}(aq) + 2\ e^-$

Cathode: $Cu^{2+}(aq) + 2\ e^- \longrightarrow Cu(s,\ \text{pure})$

By controlling the voltage, only copper and those impurities (zinc, lead, iron) in the blister copper anode that have a high enough oxidation potential are oxidized and dissolved in the electrolyte. Any less electropositive metal impurities, such as metallic gold and silver, in the anode are essentially unaffected and drop from the anode as it is consumed during electrolysis, forming an *anode sludge.* The voltage is regulated so that only copper, the least electropositive of the metals, is plated out onto the pure copper cathode. Electrorefined copper is greater than 99.9% pure, a purity required for copper used in electrical applications. Generally, enough gold, silver, and platinum are recovered from the anode sludge to pay for the cost of copper electrorefining.

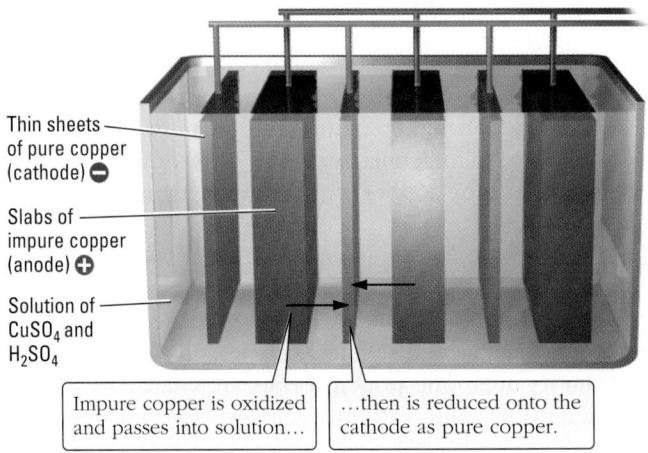

Thin sheets of pure copper (cathode) ⊖

Slabs of impure copper (anode) ⊕

Solution of $CuSO_4$ and H_2SO_4

Impure copper is oxidized and passes into solution...

...then is reduced onto the cathode as pure copper.

Figure 22.7 **Electrolytic cell for refining copper.**

Problem-Solving Example 22.3 Copper Electrorefining

Copper is electrorefined by removing impurities such as lead, silver, gold, and zinc. Based on these standard reduction potentials, explain how copper can be separated by electrolysis from these impurities.

A copper electrorefining facility, showing the refined copperplate out on cathodes.

$$E°_{red}, V$$

$$Ag^+(aq) + e^- \longrightarrow Ag(s) \quad +0.799$$

$$Au^+(aq) + e^- \longrightarrow Au(s) \quad +1.691$$

$$Cu^{2+}(aq) + 2\,e^- \longrightarrow Cu(s) \quad +0.337$$

$$Pb^{2+}(aq) + 2\,e^- \longrightarrow Pb(s) \quad -0.126$$

$$Zn^{2+}(aq) + 2\,e^- \longrightarrow Zn(s) \quad -0.763$$

Answer At the voltage that oxidizes the blister copper in the anode (-0.337 V), zinc and lead also are oxidized to Zn^{2+} and Pb^{2+} and dissolve, but silver and gold are not oxidized and drop off to form the anode sludge.

Explanation The voltage must be high enough to oxidize copper from the blister copper anode: $Cu(s) \rightarrow Cu^{2+}(aq) + 2\,e^-$. Reduction occurs at the cathode. Of the ions in solution, Cu^{2+} has the most positive standard reduction potential and is more readily reduced than Zn^{2+} or Pb^{2+}. Therefore, by carefully controlling the voltage, Cu^{2+} is exclusively reduced and only copper is plated out at the cathode, leaving Zn^{2+} or Pb^{2+} in solution.

Problem-Solving Practice 22.3

Explain how zinc and lead could be separated from each other without plating out copper in the electrolytic cell in Problem-Solving Example 22.3.

The amount of pure copper deposited on a cathode by electrorefining can be calculated. For example, a copper electrorefining cell operates at 200. amperes (A) for 24 hours a day for a year. How many kilograms of pure copper are produced? This amount can be calculated by determining the number of coulombs of charge used and recognizing that two moles of electrons are needed for each mole of Cu^{2+} reduced to copper metal. From Section 19.12 we know that

$$1 \text{ ampere} = 1 \text{ coulomb/second (C/s)} \quad \text{and} \quad 1 \text{ mol e}^- = 9.65 \times 10^4 \text{ C}$$

$$365 \text{ days} \times 200.\text{ A} \times \left(\frac{1 \text{ C/s}}{1 \text{ A}}\right)\left(\frac{24 \text{ h}}{1 \text{ day}}\right)\left(\frac{3600 \text{ s}}{1 \text{ h}}\right) = 6.31 \times 10^9 \text{ C}$$

$$6.31 \times 10^9 \text{ C}\left(\frac{1 \text{ mol e}^-}{9.65 \times 10^4 \text{ C}}\right)\left(\frac{1 \text{ mol Cu}}{2 \text{ mol e}^-}\right) = 3.27 \times 10^4 \text{ mol Cu}$$

$$3.27 \times 10^4 \text{ mol Cu}\left(\frac{63.55 \text{ g Cu}}{1 \text{ mol Cu}}\right)\left(\frac{1 \text{ kg Cu}}{10^3 \text{ g Cu}}\right) = 2.08 \times 10^3 \text{ kg Cu}$$

Exercise 22.4 Refining copper

What mass of pure copper is deposited during electrorefining in a cell operating at 250. A for 12.0 h?

Bronze and Brass

About 3800 BC the discovery was made in the Middle East that bronze formed when tin combined with copper, thus starting the Bronze Age. The discovery was

An ancient bronze piece of art.

Patina on the Statue of Liberty. The copper hydroxycarbonate coating on the statue is responsible for the green color.

The Susan B. Anthony dollar coin is copper coated with a 87.5% copper/12.5% nickel alloy.

Pre-1982 (left) and post-1982 (right) pennies. A copper-clad penny (right) with some of the copper cladding removed to expose the silvery-appearing zinc core.

likely accidental, brought on by the fact that copper ores and tin ores are often found together. In a stroke of good fortune, during the reduction of copper ore in the charcoal of a wood fire some tin ore was present, and the heated mixture of the resulting copper and tin metals formed bronze. The advantage of bronze is that it is sufficiently hard to keep a cutting edge, something copper cannot do. Bronze usually contains 7 to 10% tin, but bronzes with up to 20% tin are known. Because of its properties, all early civilizations that produced bronze used it to create weapons as well as exquisite works of art.

On prologed exposure to moist air, copper or bronze forms a green outer coating (a patina) of copper hydroxycarbonate, $Cu_2(OH)_2CO_3$, often seen on statuary, such as the Statue of Liberty, and on copper-clad roofs.

$$2\ Cu(s) + O_2(g) + CO_2(g) + H_2O(g) \longrightarrow Cu_2(OH)_2CO_3(s)$$

Brass is an alloy made of varying proportions of copper and zinc (20 to 45% Zn), which becomes harder as the percentage of zinc increases. Because it is easy to forge, cast, and stamp, brass is widely used for pipes, valves, and fittings.

Copper is used in all U.S. coins. By 1982, the price of copper rose to where it was costing the U.S. Treasury Department more than 1 cent to make a penny. Since then, to conserve copper and reduce costs, the penny has been made of 97.5% zinc and 2.5% copper, with the zinc core sandwiched between two thin layers of copper. The silver-colored coins actually contain no silver. The nickel is made of a copper (75%) and nickel (25%) alloy of uniform composition throughout the coin. The other silver-colored coins are a "sandwich" made of a pure copper core covered with a thin layer of a copper and nickel alloy (91.67% copper). The new Sacagawea dollar coin, which looks like gold, contains no gold (see *Chemistry in the News*).

Chemistry in the News

THE SACAGAWEA DOLLAR COIN: IN SEARCH OF GOLD

Sacagawea, the young Shoshone woman who was essential to the success of Lewis and Clark's exploration of the vast territory of the Lousiana purchase (1804–1806), is commemorated by a new dollar coin. One side of the coin shows a profile of Sacagawea and her son, who was born on the Lewis and Clark expedition; an eagle is shown on the reverse side.

The coin is masterful metallurgy. To be acceptable, the coin must be able to be used in vending machines, which is not a simple matter. The machines identify a coin by its weight, size, and electromagnetic properties. Typically, vending machines test a coin's electrical conductivity by passing an alternating current through it and detecting the induced magnetic field. The electrical conductivity of the coin depends on the kinds of metals used in the coin and on its construction.

The Sacagawea dollar was designated to be a golden color to distinguish it easily from silver-colored coins, yet to contain no gold so that it could be made cheaply. But none of the 25 different gold-colored test alloys manufactured by Olin Brass worked in vending machines; the conductivity of the coins was too high. That's when Olin Brass metallurgists hit upon the idea that because some manganese alloys have low conductivities, adding manganese might work. But when manganese was added, the coins looked pink. This problem was solved by adding enough manganese and copper, and a bit of nickel for tarnish resistance, so that the resulting gold-colored coin had the correct conductivity. The coin, which has an expected 30-year "lifetime," is fabricated with a pure copper core around which are wrapped two layers of manganese brass, making the overall composition 88.5% Cu, 6.0% Zn, 3.5% Mn, and 2.0% Ni. In spite of its complex composition, the Sacagawea dollar coin costs only 12 cents to make, thus generating a robust 88-cent profit on each dollar coin sold by the U.S. Treasury. Once again, Sacagewea has proven to be a pathfinder.

- Is the percent of copper in the Sacagewea dollar greater than that in the U.S. nickel?

- What metals are used as an alloy in the Sacagawea dollar that are not present in other U.S. coins?

The Sacagawea dollar coin.

Source:

Science News, Vol. 157, April 2000; pp. 226–227.

Metallic copper is not attacked by most acids, although it does react with nitric acid. The nitrate ion, NO_3^-, acts as the oxidizing agent. In dilute nitric acid it is reduced to NO; concentrated nitric acid yields NO_2.

$$3\,Cu(s) + 2\,NO_3^-(aq) + 8\,H^+(aq) \longrightarrow 3\,Cu^{2+}(aq) + 2\,NO(g) + 4\,H_2O(\ell)$$

$$Cu(s) + 2\,NO_3^-(aq) + 4\,H^+(aq) \longrightarrow Cu^{2+}(aq) + 2\,NO_2(g) + 2\,H_2O(\ell)$$

Copper(II) sulfate pentahydrate, $CuSO_4 \cdot 5H_2O$, is the most widely used copper compound. Commonly called blue vitriol, this compound is used to kill algae and fungi. In the solid it contains the hydrated copper(II) complex ion,

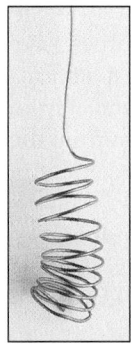
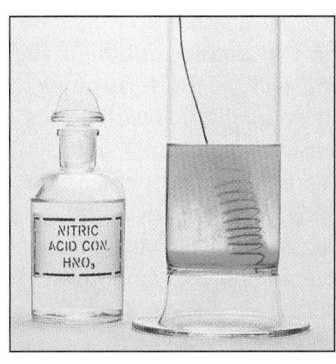
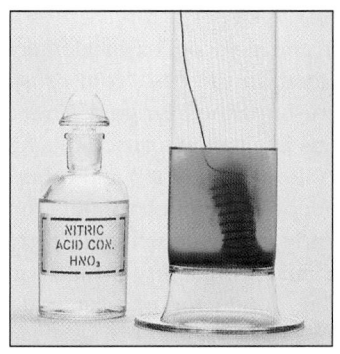

The reaction of metallic copper with concentrated nitric acid produces brown fumes of NO_2. The solution takes on the green color of the other reaction product, copper nitrate.

Hydrated copper(II) sulfate (a) and partially anhydrous copper sulfate (b). Hydrated copper(II) sulfate is converted to anhydrous copper(II) sulfate by heating, which drives off the waters of hydration.

(a)

(b)

$[Cu(H_2O)_4]^{2+}$. The fifth water molecule is bound to the sulfate ion through hydrogen bonding. The water of hydration can be removed by gentle heating or by putting the hydrate into a desiccator. Aqueous solutions containing copper(II) ions are blue due to the presence of the $[Cu(H_2O)_6]^{2+}$ ion.

Copper also exists in a +1 oxidation state, Cu^+, although aqueous copper chemistry involves principally Cu^{2+} because Cu^+ is unstable in aqueous solutions and disproportionates into Cu and Cu^{2+}.

> In a disproportionation reaction, the same substance is oxidized as well as reduced.

$$Cu^+(aq) + e^- \longrightarrow Cu(s) \qquad E° = +0.52 \text{ V}$$
$$\underline{Cu^+(aq) \longrightarrow Cu^{2+}(aq) + e^- \qquad E° = -0.15 \text{ V}}$$
$$2\,Cu^+(aq) \longrightarrow Cu(s) + Cu^{2+}(aq) \qquad E° = +0.37 \text{ V}$$

 In basic solution, however, Cu^{2+} is reduced to copper(I) oxide, Cu_2O, a reaction that serves as a traditional test for glucose in urine. When heated, a basic solution containing blue aqueous Cu^{2+} and a reducing sugar such as glucose react to form a brick-red precipitate of Cu_2O. The reducing sugar is represented here with its aldehyde functional group, RCHO.

$$2\,Cu^{2+}(aq) + RCHO(aq) + 5\,OH^-(aq) \longrightarrow Cu_2O(s) + RCOO^-(aq) + 3\,H_2O(\ell)$$
blue   reducing sugar   brick red

At elevated temperatures copper reacts with oxygen. Below 1000 °C it forms black copper(II) oxide, CuO. Above 1000 °C copper reacts with oxygen to form red copper(I) oxide, Cu_2O, which is found in the mineral cuprite.

22.4 SILVER AND GOLD: THE OTHER COINAGE METALS

Samples of copper(I) oxide, Cu_2O, and copper(II) oxide, CuO.

Silver and gold occur in elemental form in nature, although such sources have dwindled. In the past, gold prospectors, such as the forty-niners in the California gold rush, panned for gold. They simply swirled gold-bearing rock and gravel from streams in a pan. Because of its high density (19.3 g/cm³), gold settles out from the sand (about 2.5 g/cm³) and other rocky impurities. At present, the gold content in such deposits is much too low for panning to be effective.

Early civilizations highly prized silver and gold for their luster, corrosion resistance, and workability in making jewery, art objects, and coins. Silver is the best metallic conductor of heat and electricity. Long synonymous with wealth and power, gold was thought to be a part of the sun, thus its Latin name *aurum,* meaning bright dawn, from which its chemical symbol is derived. Gold is the most malleable metal, so malleable that it can be rolled thinly enough to be transparent. Very thin gold leaf is used to cover domes of churches and capitol buildings.

The lack of reactivity of gold and silver is explained by their oxidation potentials, which are well below that of hydrogen (0.00 V).

$$Au(s) \longrightarrow Au^+(aq) + e^- \qquad E° = -1.68 \text{ V}$$

$$Ag(s) \longrightarrow Ag^+(aq) + e^- \qquad E° = -0.7994 \text{ V}$$

Consequently, neither metal reacts with nonoxidizing acids such as hydrochloric acid, nor readily with oxygen at normal temperatures. When heated in air, gold does not react with oxygen, even at high temperatures. Silver, however, slowly forms silver(I) oxide, Ag_2O, which is unstable and decomposes back to the elements when heated strongly. Silver does react with oxidizing acids such as nitric acid (HNO_3). It takes aqua regia, a 3:1 mixture of concentrated HCl and HNO_3, to dissolve gold.

$$3 Ag(s) + 4 H^+(aq) + NO_3^-(aq) \longrightarrow 3 Ag^+(aq) + NO(g) + 2 H_2O(\ell)$$

$$Au(s) + 6 H^+(aq) + 3 NO_3^-(aq) + 4 Cl^-(aq) \longrightarrow$$

$$AuCl_4^-(aq) + 3 NO_2(g) + 3 H_2O(\ell)$$

The oxidation of metallic gold to Au^{3+} is highly unfavorable ($E° = -1.40$ V), but the reaction in aqua regia occurs because as Au^{3+} ions form, they are tied up in the complex ion, $AuCl_4^-$. (Complex ions such as $AuCl_4^-$ were discussed in Section 17.5.)

Gold can be hammered into ultrathin sheets for decorative purposes.

The ability of aqua regia (Latin, *royal water*) to dissolve gold has been known since the 1300s.

Exercise 22.5   Putting Silver and Gold into Solution

(a) Identify the oxidizing and reducing agents in the reaction of nitric acid with silver.

(b) Do the same for the reaction of gold with aqua regia.

Gold and silver are both relatively soft metals whose hardness is increased by alloying with other metals. Sterling silver, for example, is 92.5% silver and 7.5% copper. The proportion of gold in its alloys is expressed in carats. Pure gold (24 carats) is too soft to be used in jewelry. It is alloyed with copper or other metals to make the 18-carat and 14-carat gold for jewelry, which are 75% (18/24 × 100) and 58% (14/24 × 100) gold, respectively.

As noted earlier ($\Leftarrow$ *p. 994*), silver and gold are byproducts of the electrorefining of copper. They are also obtained from ores. Silver is obtained from its principal ore argentite, Ag_2S, by cyanide extraction. The ore is ground and put into a 0.5% solution of NaCN through which air is blown.

$$Ag_2S(s) + 4 CN^-(aq) \longrightarrow 2 Ag(CN)_2^-(aq) + S^{2-}(aq)$$

Powdered zinc, a good reducing agent, is added to the solution to convert Ag^+ to metallic silver.

$$Zn(s) + 2 Ag(CN)_2^-(aq) \longrightarrow 2 Ag(s) + Zn(CN)_4^{2-}(aq)$$

The silver can be further purified by electrolytic refining.

Like silver, gold is also obtained from ores by cyanide extraction followed by reduction with zinc.

$$Au(s) + 8 CN^-(aq) + O_2(g) + 2 H_2O(\ell) \longrightarrow 4 Au(CN)_2^-(aq) + 4 OH^-(aq)$$

$$Zn(s) + 2 Au(CN)_2^-(aq) \longrightarrow 2 Au(s) + Zn(CN)_4^{2-}(aq)$$

The cyanide extraction of silver and gold depends on the formation of the $Ag(CN)_2^-$ and $Au(CN)_2^-$ complex ions. Because of the toxicity of CN^-, waste cyanide solutions must be disposed of properly. In places where such has not been the case and the solutions were simply dumped near the processing site, serious environmental damage has occurred.

22.5 CHROMIUM

Chromium is characteristic of the middle transition elements that exhibit multiple oxidation states *(⬅ p. 988)*. In compounds, chromium has oxidation states from +2 to +6 because of its five $3d$ electrons and one $4s$ electron, although +2 and +3 are the most common oxidation states. The +6 oxidation state is found in CrO_4^{2-} and $Cr_2O_7^{2-}$ ions; the +4 and +5 oxidation states are uncommon.

Chromium is a hard, brittle metal that is extremely corrosion resistant due to a chromium oxide surface layer that passivates and protects the metal from further oxidation. In this regard, chromium resembles aluminum *(⬅ p. 974)*.

Chromium is used in stainless steel and is plated on truck bumpers to give them a bright surface. Its oxides are used in magnetic recording tapes (CrO_2), in abrasives, and as a glass pigment (Cr_2O_3).

The chief chromium ore is chromite, $FeCr_2O_4$, which is treated with lime (CaO), oxygen, and sodium carbonate to form sodium chromate. Water is added to remove the soluble sodium chromate produced.

$$4\,FeCr_2O_4(s) + 8\,CaO(s) + 8\,Na_2CO_3(s) + 7\,O_2(g) \longrightarrow$$
$$8\,Na_2CrO_4(s) + 2\,Fe_2O_3(s) + 8\,CaCO_3(s)$$

Sodium chromate is used widely in chrome plating and as an intermediate in the formation of many chromium compounds, one of which is chromium(III) oxide, from which chromium metal is obtained by reduction with aluminum metal.

$$2\,CrO_4^{2-}(aq) + 3\,SO_2(g) + H_2O(\ell) \longrightarrow Cr_2O_3(s) + 3\,SO_4^{2-} + 2\,H^+(aq)$$

$$Cr_2O_3(s) + 2\,Al(s) \longrightarrow 2\,Cr(s) + Al_2O_3(s)$$

Exercise 22.6   Production of Chromium Metal

Use data from Appendix J to calculate the enthalpy change and the Gibbs free energy change for the reduction of chromium(III) oxide by aluminum.

In aqueous solution, chromate ions, CrO_4^{2-}, and dichromate ions, $Cr_2O_7^{2-}$, exist in a highly pH-dependent equilibrium.

$$H^+(aq) + CrO_4^{2-}(aq) \rightleftharpoons HCrO_4^-(aq)$$
$$\text{yellow}$$

$$2\,HCrO_4^{2-}(aq) \rightleftharpoons Cr_2O_7^{2-}(aq) + H_2O(\ell)$$
$$\text{orange}$$

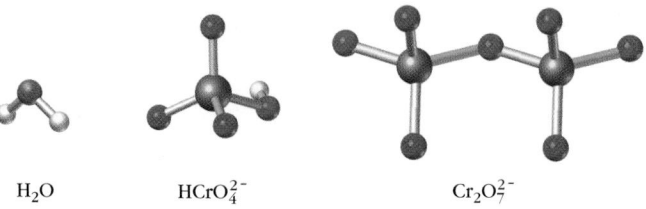

H_2O $HCrO_4^{2-}$ $Cr_2O_7^{2-}$

Figure 22.8 Chromium ions in solution. The two flasks at the left contain solutions of chromium +3 ions: $Cr(NO_3)_3$ (violet) and $CrCl_3$ (green). The flasks at the right show the colors of the two chromium anions with oxidation state +6: yellow chromate ion (CrO_4^{2-}) and orange dichromate ion ($Cr_2O_7^{2-}$).

Notice from the second equilibrium that dichromate is formed by a condensation reaction in which two $HCrO_4^-$ units are joined by splitting out water.

The net equilibrium is

$$2\,H^+(aq) + 2\,CrO_4^{2-}(aq) \rightleftharpoons Cr_2O_7^{2-}(aq) + H_2O(\ell) \qquad K = 4 \times 10^{14}$$

From the very large value of K, it should be obvious that in acid, chromate is converted to dichromate; chromate is stable in basic or neutral solution (Figure 22.8).

 Exercise 22.7 Applying Le Chatelier's Principle

Use Le Chatelier's principle to explain how the addition of acid or base shifts the equilibrium to favor chromate or dichromate.

In acidic solution the dichromate ion is a powerful oxidizing agent.

$$Cr_2O_7^{2-}(aq) + 14\,H^+(aq) + 6\,e^- \longrightarrow 2\,Cr^{3+}(aq) + 7\,H_2O(\ell) \qquad E^\circ_{red} = +1.33\text{ V}$$

It is sufficiently strong to oxidize aldehydes to carboxylic acids for example, acetaldehyde to acetic acid.

$$Cr_2O_7^{2-}(aq) + 3\,CH_3CHO(aq) + 8\,H^+(aq) \longrightarrow$$
$$2\,Cr^{3+}(aq) + 3\,CH_3COOH(aq) + 4\,H_2O(\ell)$$

The oxidizing strength of dichromate decreases as the pH increases, as shown in Problem-Solving Example 22.4.

Problem-Solving Example 22.4 Dichromate and pH

Use the Nernst equation *(⟸ p. 888)* to calculate the E_{red} for the reduction of dichromate ion at a pH of 4.00 and 25 °C with the concentrations of dichromate and Cr^{3+} both 1.0 M. Use the redox equation just discussed.

Answer $E_{red} = +0.78$ V

Explanation The Nernst equation expresses how a standard voltage changes with changing conditions, in this case a change in pH from 0.00 (H^+ concentration = 1.00 M at standard conditions). At a pH of 4.00, the hydrogen ion concentration is 1.00×10^{-4} M. To calculate the reduction potential of dichromate at a 1.00×10^{-4} M hydrogen ion concentration, substitute this and the other concentrations into the Nernst equation based on Equation 22.1.

$$E_{red} = E^{\circ}_{red} - \left(\frac{0.0591}{n}\right) \log\left(\frac{[Cr^{3+}]^2}{[Cr_2O_7^{2-}][H^+]^{14}}\right)$$

where n is the number of moles of electrons transferred and $E^{\circ}_{red} = +1.33$ V. Substituting the concentration values,

$$E_{red} = +1.33\ V - \left(\frac{0.0591}{6}\right) \log\left(\frac{1}{[1 \times 10^{-4}]^{14}}\right) = +1.33\ V - 0.00985 \log\left(\frac{1}{10^{-56}}\right)$$

$$= +1.33\ V - 0.00985 \log (10^{-56}) = +1.33\ V - (0.00985)(56)$$

$$= +1.33\ V - 0.5516 = 0.78\ V$$

Therefore, the reduction potential drops by 0.55 V, from 1.33 to 0.78 V, as the pH increases from 0 to 4.00, and the oxidation strength of dichromate drops likewise.

Problem-Solving 22.4

At what pH does the E_{red} equal 1.00 V for the reduction of dichromate in acid solution, assuming Cr^{3+} concentration = dichromate concentration = 1.0 M?

22.6 COORDINATE COVALENT BONDS: COMPLEX IONS AND COORDINATION COMPOUNDS

In Section 8.10, the formation of BF_3NH_3 was described as occurring by the sharing of a lone pair from NH_3 with BF_3. This type of covalent bond, in which one atom contributes *both* electrons for the shared pair, is called a **coordinate covalent bond.** Atoms with lone pairs of electrons, such as nitrogen, phosphorus, and sulfur, can use those lone pairs to form coordinate covalent bonds. For example, the formation of the ammonium ion from ammonia results from formation of a coordinate covalent bond between H^+ and the lone pair of electrons of nitrogen in NH_3.

$$
\begin{array}{ccc}
\underset{\underset{\displaystyle H}{|}}{\overset{\overset{\displaystyle H}{|}}{H-N:}} + H^+ & \longrightarrow &
\left[\underset{\underset{\displaystyle H}{|}}{\overset{\overset{\displaystyle H}{|}}{H-N:H}}\right]^+ \quad \text{or} \quad
\left[\underset{\underset{\displaystyle H}{|}}{\overset{\overset{\displaystyle H}{|}}{H-N-H}}\right]^+
\end{array}
$$

Once the coordinate covalent bond is formed, it is impossible to distinguish which of the equivalent N—H bonds it is.

Metals and Coordination Compounds

Much of the chemistry of *d*-block transition metals is related to their ability to form coordinate covalent bonds with molecules or ions that have lone pair electrons. Transition metals have incompletely filled *d* orbitals that can accept the lone pairs.

You have seen that metal ions in aqueous solution are surrounded by water molecules, for example, the Ni^{2+} ion surrounded by six water molecules. This type of ion, in which several molecules or ions are connected to a central metal ion or atom by coordinate covalent bonds, is known as a *complex ion*. The molecules or ions bonded to the central metal ion are called **ligands,** from the Latin verb *ligare,* "to bind." Each ligand (a water molecule in this example) has one or more atoms with lone pairs that can form coordinate covalent bonds. To write the formula of a complex ion, the ligand formulas are placed in parentheses following the metal ion. The entire complex ion formula is enclosed by brackets. For the nickel complex ion with water ligands this gives $[Ni(H_2O)_6]^{2+}$.

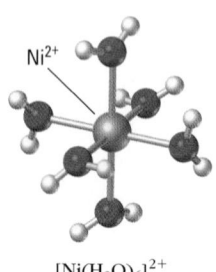

Ni^{2+}

$[Ni(H_2O)_6]^{2+}$

The charge of a complex ion is determined by the charge of the metal ion and the charge of its ligands. In $[Ni(H_2O)_6]^{2+}$ the water ligands have no net charge, so the charge of the complex ion is that of the Ni^{2+} ion. In the complex ion formed by Ni^{2+} with four chloride ions, $[NiCl_4]^{2-}$, the net $2-$ charge of this complex ion results from the $4-$ charge of four chloride ions and the $2+$ charge of the nickel ion.

Compounds in which complex ions are combined with oppositely charged ions *(counter ions)* to form neutral compounds are known as **coordination compounds.** Such compounds are generally brightly colored as solids or in solution (Figure 22.12, p. 1008). The complex ion part of a coordination compound's formula is enclosed in brackets; counter ions are outside the brackets, as in the formula of the compound of chloride ions with the $[Ni(H_2O)_6]^{2+}$ complex ion, $[Ni(H_2O)_6]Cl_2$. The two Cl^- ions compensate for the $2+$ charge of the complex ion. $[Ni(H_2O)_6]Cl_2$ is an ionic compound analogous to $CaCl_2$, which also contains a $2+$ cation and two Cl^- ions. Occasionally, no compensating ions are needed outside the brackets for a coordination compound. For example, the anticancer drug $[Pt(NH_3)_2Cl_2]$ (cisplatin) is a coordination compound containing NH_3 and Cl^- ligands coordinated to a central Pt^{2+} ion. The two Cl^- ions compensate for the charge of the Pt^{2+} ion and the result is a neutral coordination compound rather than a complex ion.

> Brackets here do not mean concentration.

> When the word "coordinated" is used in chemistry, such as in "the chloride ions in $[NiCl_4]^{2-}$ are coordinated to the nickel ion," it means that coordinate covalent bonds have been formed.

Problem-Solving Example 22.5 Coordination Compounds

For the coordination compound $K_3[Fe(CN)_6]$, identify

(a) The central metal ion.
(b) The ligands.
(c) The formula and charge of the complex ion and the central metal ion.

Answer
(a) Iron; (b) six cyanide ions, CN^-; (c) $[Fe(CN)_6]^{3-}$, Fe^{3+}

Explanation
(a) The iron ion is the central metal ion, as shown by its placement inside the brackets.
(b) Cyanide ions, CN^-, are coordinated to the central iron ion.
(c) The charge on three potassium ions is $(3 \times 1+) = 3+$. Therefore, the compensating charge of the complex ion must be $3-$, arising from the $6-$ charge of six cyanide ions $(6 \times 1- = 6-)$, combined with the $3+$ charge of the central iron(III) ion: $(6-) + (3+) = 3-$.

Problem-Solving Practice 22.5

For the coordination compound $[Cu(NH_3)_4]SO_4$, identify

(a) The counter ion
(b) The central metal ion.
(c) The ligands
(d) The formula and charge of the complex ion.

Exercise 22.8 Coordination Compounds

In a complex ion a central Cr^{3+} ion is bonded to two ammonia molecules, two water molecules, and two hydroxide ions. Give the formula and the net charge of this complex ion.

Naming Complex Ions and Coordination Compounds

As with other compounds, coordination compounds early on were known by common names, for example "roseo" salt and Zeise's salt. Since those times, a systematic nomenclature has been developed for complex ions and coordination

TABLE 22.4	Names and Formulas of Some Common Ligands		
Neutral ligand	**Ligand name**	**Anionic ligand**	**Ligand name**
NH_3	Ammine	Br^-	Bromo
CO	Carbonyl	CO_3^{2-}	Carbanato
$H_2NCH_2CH_2NH_2$	Ethylenediamine, en	Cl^-	Chloro
H_2O	Aqua	CN^-	Cyano
		F^-	Fluoro
		OH^-	Hydroxo
		$C_2O_4^{2-}$	Oxalato
		NCS^-	Isothiocyanato
		SCN^-	Thiocyanato

compounds. The systematic nomenclature indicates the central metal ion and its oxidation state, as well as the number and kinds of ligands. Table 22.4 lists the names and formulas of some common ligands. Although there are extensive rules for such nomenclature, we will consider only some basic aspects of the system by interpreting the names of a neutral coordination compound and two other coordination compounds, one containing a complex cation, the other a complex anion.

Consider the coordination compound $Co(NH_3)_3(OH)_3$, which is named triamminetrihydroxocobalt(III). From Table 22.4, we see that the name and formula indicate that three ammonia molecules and three hydroxide ions are bonded to a central cobalt ion. The three hydroxide ions carry a total 3− charge; ammonia molecules have no net charge, and thus cobalt must be Co^{3+} because the compound has no net charge. In naming any coordination compound or complex ion, the ligands are named in alphabetical order—in this case ammine for ammonia precedes hydroxo for hydroxide (for anions -*ide* is changed to *o*). The name and oxidation state (in parentheses) of the metal ion are given last. Greek prefixes *di, tri, tetra,* and so on are used to denote the number of times each of these ligands is used. Such prefixes are ignored when determining the alphabetical order of the ligands.

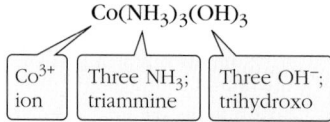

triamminetrihydroxocobalt(III)

Next, consider $[Fe(H_2O)_2(NH_3)_4]Cl_3$, a coordination compound that consists of a complex *cation*, $[Fe(H_2O)_2(NH_3)_4]^{3+}$, and three chloride ions as counter ions. In such cases the complex cation is always named first, followed by the name of the anionic counter ions. The compound's name is tetraamminediaquairon(III) chloride. From Table 22.4, we see that the ligands are ammine (NH_3, four of them) and aqua (H_2O, two of them). For complex cations, the metal ion and its oxidation state follow the names of the ligands.

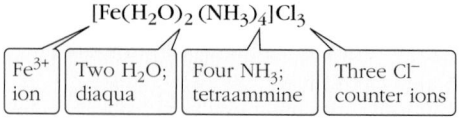

diaquatetraammineiron(III) chloride

The compound $K_2[PtCl_4]$ contains a complex *anion*, $[PtCl_4]^{2-}$, and two K^+ ions as counter ions and is named potassium tetrachloroplatinate(II). As with any ionic compound, the cation is named first, followed by the anion name. For complex *anions* the central metal ion's name ends in *-ate* followed by its oxidation state in parentheses.

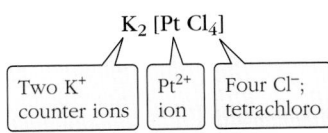

$$K_2 [Pt\ Cl_4]$$

Two K^+ counter ions | Pt^{2+} ion | Four Cl^-; tetrachloro

potassium tetrachloroplatinate(II)

Problem-Solving Example 22.6   Formulas and Names of Coordination Compounds

(a) Write the formula of diamminetriaquahydroxochromium(III) nitrate.
(b) Name $K[Cr(NH_3)_2(C_2O_4)_2]$.

Answer
(a) $[Cr(NH_3)_2(H_2O)_3(OH)](NO_3)_2$
(b) Potassium diamminedioxalatochromate(III)

Explanation   Use the names and formulas of the ligands in Table 22.4. Compound (a) contains a complex cation, and compound (b) contains a complex anion.

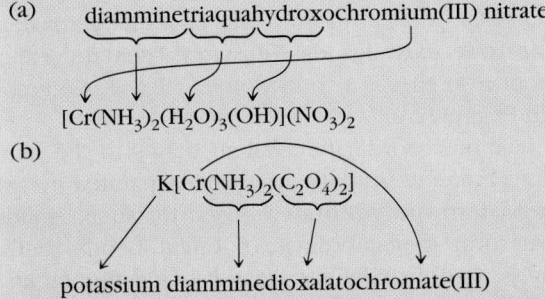

(a)   diamminetriaquahydroxochromium(III) nitrate

$$[Cr(NH_3)_2(H_2O)_3(OH)](NO_3)_2$$

(b)

$$K[Cr(NH_3)_2(C_2O_4)_2]$$

potassium diamminedioxalatochromate(III)

Problem-Solving Practice 22.6

(a) Name this coordination compound: $[Ag(NH_3)_2]NO_3$.
(b) Write the formula of pentaaquaisothiocyanatoiron(III) chloride

---

 Exercise 22.9   Coordination Compounds

$CaCl_2$ and $[Ni(H_2O)_6]Cl_2$ have the same formula type, MCl_2. Give the formula of a simple ionic compound (noncoordination) that has a formula analogous to $K_2[NiCl_4]$.

Types of Ligands and Coordination Number

The number of coordinate covalent bonds between the ligands and the central metal ion is the **coordination number** of the metal ion, usually 2, 4, or 6.

Coordination number	Examples
2	$[Ag(NH_3)_2]^+$, $[AuCl_2]^-$
4	$[NiCl_4]^{2-}$, $[Pt(NH_3)_4]^{2+}$
6	$[Fe(H_2O)_6]^{2+}$, $[Co(NH_3)_6]^{3+}$

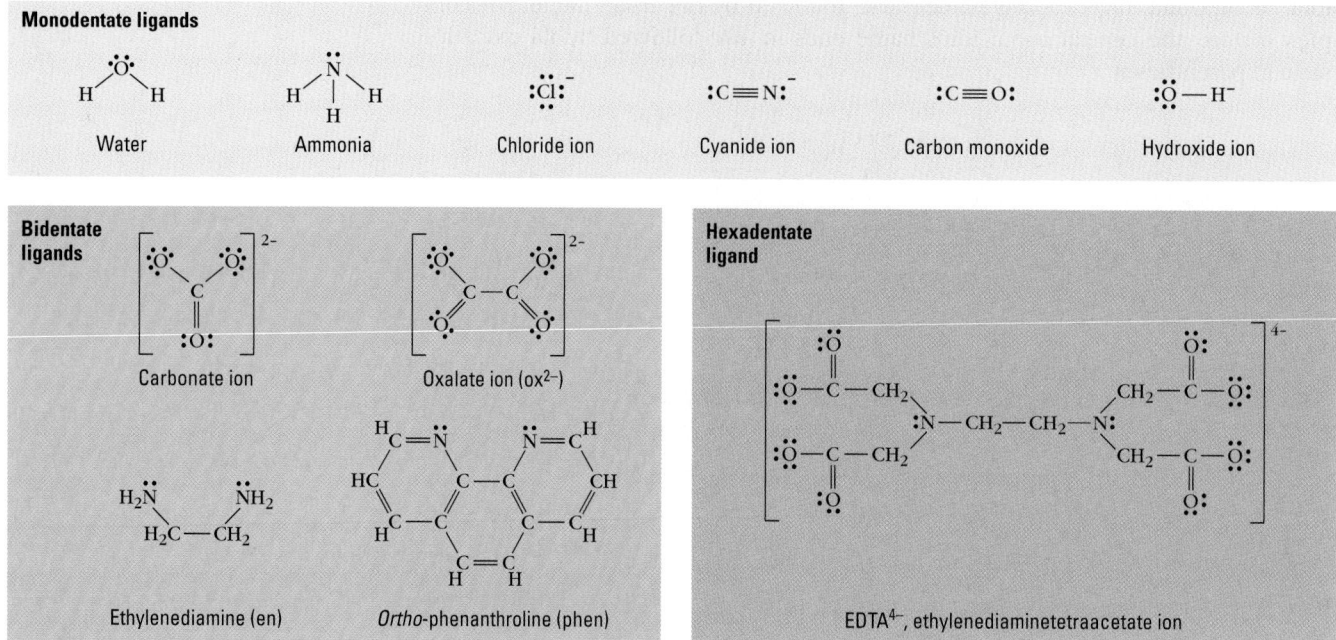

Figure 22.9 **Monodentate ligands; bidentate and hexadentate chelating ligands.** Ligands with two (bidentate) or more lone pairs to share with a central metal ion are chelating ligands.

Ligands such as H_2O, NH_3, and Cl^- that form only one coordinate covalent bond to the metal are termed **monodentate** ligands. The word "dentate" stems from the Latin word *dentis,* for tooth, so NH_3 is a "one-toothed" ligand. Common monodentate ligands are shown in Figure 22.9.

Some ligands can form two or more coordinate covalent bonds to the same metal ion because they have two or more atoms with lone pairs separated by several intervening atoms. The general term **polydentate** is used for such ligands. **Bidentate** ligands are those that form *two* coordinate covalent bonds to the central metal ion. A good example is the bidentate ligand 1,2-diaminoethane, $H_2NCH_2CH_2NH_2$, commonly called ethylenediamine and abbreviated *en*. When lone pairs of electrons from both nitrogen atoms in en coordinate to a metal ion, a stable five-membered ring is formed (Figure 22.10). Notice that Co^{3+} has a coordination number of 6 in this complex ion.

The word "chelating," derived from the Greek *chele,* "claw," describes the pincer-like way in which a ligand can grab a metal ion. Some common **chelating ligands,** those that are polydentate ligands and can share two or more electron pairs with the central metal ion, are also shown in Figure 22.9.

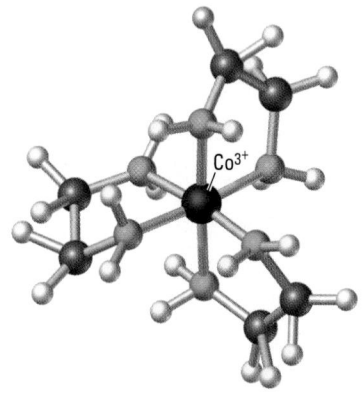

Figure 22.10 **The $[Co(en)_3]^{3+}$ complex ion.** Cobalt ion (Co^{3+}) forms a coordination complex ion with three ethylenediamine ligands.

Problem-Solving Example 22.7 Chelating Agents

Two ethylenediamine ligands and two chloride ions form a complex ion with Co^{3+}.

(a) Write the formula for this complex ion.
(b) What is the coordination number of the Co^{3+} ion?
(c) Write the formula of the coordination compound formed by Cl^- counter ions and the Co^{3+} complex ion.

Answer
(a) $[Co(en)_2Cl_2]^+$
(b) 6
(c) $[Co(en)_2Cl_2]Cl$

Explanation
(a) Two en molecules and two chloride ions are bonded to the central cobalt ion, so the formula of the complex ion is $[Co(en)_2Cl_2]^+$. Ethylenediamine is a neutral lig-

and, each chloride ion is 1−, and cobalt has a 3+ charge. The charge on the complex ion is $2(0) + 2(1-) + (3+) = 1+$.

(b) The coordination number is 6 because there are six coordinate covalent bonds to the central Co^{3+} ion—two from each bidentate ethylenediamine and one from each monodentate chloride ion.

(c) The 1+ charge of the complex ion requires one chloride ion as a counter ion: $[Co(en)_2Cl_2]Cl$.

Note that Cl_2 in the complex ion's formula represents two chloride ions, not a diatomic chlorine molecule.

Problem-Solving Practice 22.7

The dimethylglyoximate anion (abbreviated DMG),

$$CH_3C-CCH_3$$
$$\underset{HO-\overset{\cdot\cdot}{N}}{\parallel}\quad\underset{\overset{\cdot\cdot}{N}-O^-}{\parallel}$$

is a bidentate ligand used to test for the presence of nickel because it reacts with Ni^{2+} to form a beautiful red solid in which the Ni^{2+} has a coordination number of 4. DMG coordinates to Ni^{2+} by the lone pairs on the nitrogen atoms.

(a) How many DMG ions are needed to satisfy a coordination number of 4 on the central Ni^{2+} ion?

(b) What is the net charge after coordination occurs?

(c) How many atoms are in the rings formed by DMG and Ni^{2+}?

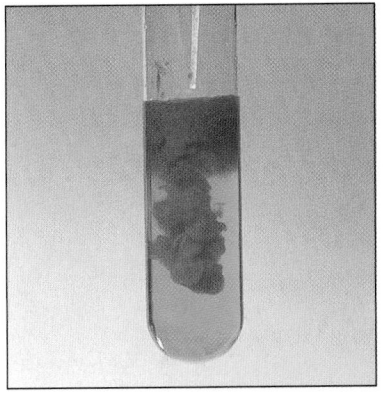

The nickel-dimethylgloxime complex.

 ### Exercise 22.10 Chelating and Complex Ions

Oxalate ion forms a complex ion with Mn^{2+} by coordinating at the oxygen lone pairs (see Figure 22.9).

(a) How many oxalate ions are needed to satisfy a coordination number of 6 on the central Mn^{2+} ion?

(b) What is the charge on this complex ion?

(c) How many atoms are in the rings formed between the ligand and the central metal ion?

CHEMISTRY
You Can Do...   A Penny For Your Thoughts

You will need the following items to do the experiment:

- Two glasses or plastic cups that will each hold about 50 mL of liquid
- About 30 to 40 mL of household vinegar
- About 30 to 40 mL of household ammonia
- A copper penny

Place the penny in one cup and add 30 to 40 mL of vinegar to clean the surface of the penny. Let the penny remain in the vinegar until the surface of the penny is cleaner (reddish-coppery) than it was before (darker copper color). Pour off the vinegar and wash the penny thoroughly in running water.

Then place the penny in the other cup and add 30 to 40 mL of household ammonia. Observe the color of the solution over several hours.

1. What did you observe happening to the penny in the ammonia solution?
2. What did you observe happening to the ammonia solution?
3. Interpret what you observed happening to the solution on the nanoscale level, citing observations to support your conclusions.
4. What is necessary to form a complex ion?
5. Are all of these kinds of reactants present in the solution in this experiment? If so, identify them.
6. How do the terms "ligand," "central metal ion," and "coordination complex" apply to your experiment?
7. Try to write a formula for a complex ion that might form in this experiment.

Some household products that contain EDTA. Check the label on your shampoo container. It will likely list disodium EDTA as an ingredient. The EDTA in this case has a 2− charge because two of the four organic acid groups have each lost an H^+, but $EDTA^{2-}$ still coordinates to metal ions in the shampoo.

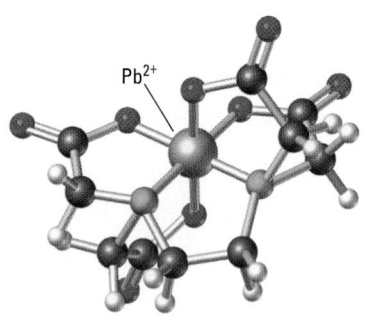

Figure 22.11 A Pb^{2+}-EDTA complex ion. The structure of the chelate formed when the $EDTA^{4-}$ anion forms a complex with Pb^{2+}.

For metals that display a coordination number of 6, an especially effective ligand is the **hexadentate** ethylenediaminetetraacetate ion (abbreviated EDTA, Figure 22.9) that encapsulates and firmly binds metal ions. It has six lone pair donor atoms (two O atoms and four N atoms) that can coordinate to a single metal ion, so $EDTA^{4-}$ is an excellent chelating ligand. It is often added to commercial salad dressing to remove traces of metal ions from solution, because these metal ions could otherwise accelerate the oxidation of oils in the product and make it rancid.

Another use of $EDTA^{4-}$ is in bathroom cleansers, where it removes hard water deposits of insoluble $CaCO_3$ and $MgCO_3$ by chelating Ca^{2+} or Mg^{2+} ions, allowing them to be rinsed away. EDTA is also used in the treatment of lead and mercury poisoning because it has the ability to chelate these metals and aid in their removal from the body (Figure 22.11).

Coordination compounds of d-block transition metals are often colored, and the colors of the complexes of a given transition metal ion depend on both the metal ion and the ligand (Figure 22.12). Many transition metal coordination compounds are used as pigments in paints and dyes. For example, Prussian blue, $Fe_4[Fe(CN)_6]_3$, a deep-blue compound known for hundreds of years, is the "bluing agent" in engineering blueprints.

Exercise 22.11 Complex Ions

Prussian blue contains two iron ions. What is the charge of the iron in

(a) The complex ion $[Fe(CN)_6]^{4-}$?

(b) The iron ion not in the complex ion?

Geometry of Coordination Compounds and Complex Ions

The geometry of a complex ion or coordination compound is dictated by the arrangement of the electron donor atoms of the ligands attached to the central metal ion. Although other geometries are possible, we will discuss only the four most common ones, those associated with coordination numbers of 2, 4, and 6. To simplify matters, we will consider only monodentate ligands, L, bonded to a central metal ion, M^{n+}.

(a)

(b)

Figure 22.12 Color of transition metal compounds. (a) Concentrated aqueous solutions of the nitrate salts containing hydrated transition metal ions of (*left to right*) Fe^{3+}, Co^{2+}, Ni^{2+}, Cu^{2+}, and Zn^{2+}. (b) The colors of the complexes of a given transition metal ion depend on the ligand(s). All of the complexes pictured here contain the Ni^{2+} ion. The green solid is $[Ni(H_2O)_6](NO_3)_2$; the purple solid is $[Ni(NH_3)_6]Cl_2$; the red solid is $Ni(dimethylglyoximate)_2$.

Coordination Number = 2, ML_2^{n+}

All such complex ions have a *linear* geometry with the two ligands on opposite sides of the central metal ion to give an L—M—L bond angle of 180°, such as that in $[Ag(NH_3)_2]^+$. Other examples are $[CuCl_2]^-$ and $[Au(CN)_2]^-$, the complex ion used to extract gold.

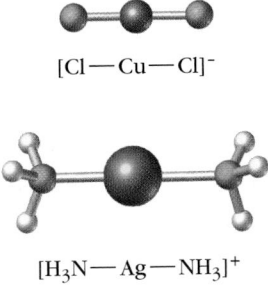

$$[Cl—Cu—Cl]^-$$

$$[H_3N—Ag—NH_3]^+$$

Coordination Number = 4: ML_4^{n+}

Four-coordinate complex ions have either tetrahedral or square planar geometries. In the *tetrahedral* case, the four monodentate ligands are at the corners of a tetrahedron, such as in $[Zn(NH_3)_4]^{2+}$. In *square planar* geometry, the ligands lie in a plane at the corners of a square as in $[Ni(CN)_4]^{2-}$ and $[Pt(NH_3)_4]^{2+}$ ions.

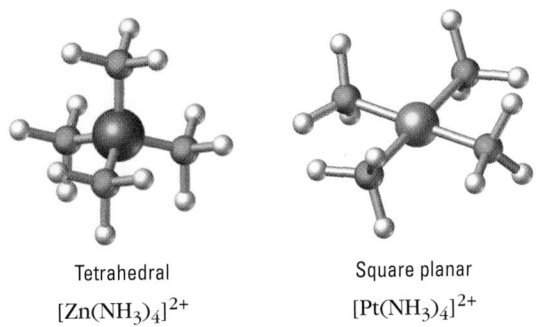

Tetrahedral
$[Zn(NH_3)_4]^{2+}$

Square planar
$[Pt(NH_3)_4]^{2+}$

Coordination Number = 6, ML_6^{n+}

Octahedral geometry is characteristic of this coordination number. The six ligands are at the corners of an octahedron with the central metal ion at its center. Octahedral geometry can be regarded as derived from a square planar geometry by adding two ligands, one above and one below the square plane. Two common octahedral complex ions are $[Co(NH_3)_6]^{3+}$ and $[Fe(CN)_6]^{3-}$, in which the six ligands are equidistant from the central metal ion and all six ligand sites are equivalent.

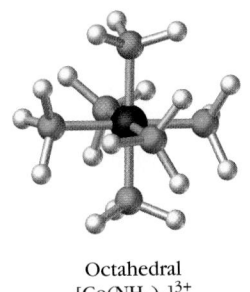

Octahedral
$[Co(NH_3)_6]^{3+}$

Isomerism in Coordination Compounds and Complex Ions

Various types of isomerism have been discussed previously with regard to organic compounds. *Constitutional isomerism* occurs with molecules that have the same molecular formula but differ in the arrangement of their atoms, such as occurs

ALFRED WERNER
(1866–1919)

In 1893, while still a young professor, Alfred Werner published a revolutionary paper about transition metal compounds. He asserted that transition metal ions could exhibit a secondary valence as well as a primary one, such as in $CoCl_3 \cdot 6NH_3$ (now written as $[Co(NH_3)_6]Cl_3$). The primary valence was represented by the ionic bonds between Co^{3+} and the chloride ions; the secondary valence was represented by the coordinate covalent bonds between the metal ion and six NH_3 molecules, what we now called the coordination sphere around the central metal ion. Werner also made the inspired proposal that the ammonia molecules were octahedrally coordinated around the Co^{3+} ion, thereby laying the foundation for understanding the geometry of complex ions. For his groundbreaking work, Werner received the 1913 Nobel Prize in Chemistry.

with butane and 2-methylpropane *(⬅ p. 83)*. *Stereoisomerism* is a second general category of isomerism in which the isomers have the same bonds, but the atoms are arranged differently in space. One type of stereoisomerism is *geometric isomerism,* such as that found in *cis-* and *trans-*1,2-dichloroethene *(⬅ p. 329)*. The other type of stereoisomerism is *optical isomerism,* which occurs when mirror images are nonsuperimposable *(⬅ p. 397)*. Constitutional, geometric, and optical isomers also occur with coordination complex ions and coordination compounds.

Linkage Isomerism, a Type of Constitutional Isomerism

Linkage isomerism occurs when a ligand can bond to the central metal using either of two different electron-donating atoms. Thiocyanato (SCN)⁻ and isothiocyanato (NCS)⁻ are examples of such ligands with coordination to a metal ion by sulfur in the first case and by nitrogen in the second, as illustrated for Co^{3+} in the margin.

Geometric Isomerism

Geometric isomerism does not exist in tetrahedral complex ions because all the corners of a tetrahedron are equivalent. But geometric isomerism occurs with square planar complex ions and compounds of the type Ma_2b_2 or Ma_2bc, where M is the central metal ion and a, b, and c are different ligands. The square planar coordination compound $[Pt(NH_3)_2Cl_2]$, an Ma_2b_2 type, occurs in two geometric forms. The *cis-*$[Pt(NH_3)_2Cl_2]$ isomer has the chloride ligands as close as possible. In *trans-*$[Pt(NH_3)_2Cl_2]$ the chloride ions are as far apart as possible, diagonally across the plane of the molecule from each other.

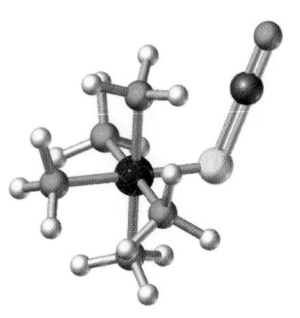

pentaamminine-
thiocyanatocobalt(III) ion

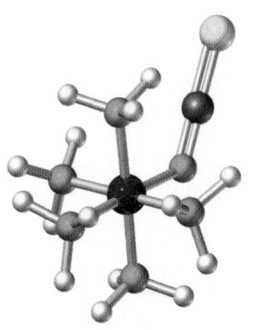

pentaammine-
isothiocyanatocobalt(III) ion

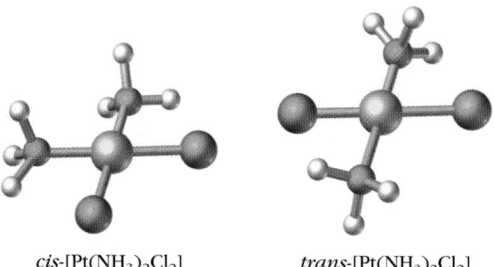

cis-$[Pt(NH_3)_2Cl_2]$ *trans-*$[Pt(NH_3)_2Cl_2]$

These two isomers differ in water solubility, color, melting point, and chemical reactivity. The *cis* isomer is used in cancer chemotherapy, whereas the *trans* form is not effective against cancer.

Cis-trans isomerism is also possible in octahedral complex ions and compounds, as illustrated with $[Co(NH_3)_4Cl_2]^+$. In this complex ion the *cis* isomer has the chloride ions adjacent to each other; the *trans* isomer has them opposite each other. The differences in properties are striking, particularly the color. The *cis* isomer is violet, whereas the *trans* form is green.

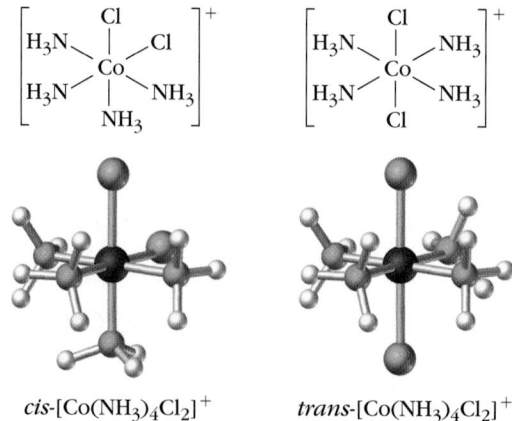

cis-$[Co(NH_3)_4Cl_2]^+$ *trans-*$[Co(NH_3)_4Cl_2]^+$

Problem-Solving Example 22.8   Geometric Isomerism

How many geometric isomers are there for $[Co(en)_2Cl_2]^+$?

Answer   Only two geometric isomers are possible, *cis* and *trans*.

trans   *cis*

Explanation   Start by putting the two Cl^- ions in *trans* positions, that is, one at the "top" of the octahedron and the other at the "bottom." The two ethylenediamine ligands (en), represented here as N⁀N, occupy the other four sites around the cobalt ion. This is the *trans* isomer. The *cis* isomer has the Cl^- ions in adjacent *(cis)* positions.

Problem-Solving Practice 22.8

How many geometric isomers are there for the square planar compound $[Pt(NH_3)_2ClBr]$?

 ### Exercise 22.12   Geometric Isomerism

How many isomers are possible for $[Co(NH_3)_3Cl_3]$? Write the structural formulas of the isomers.

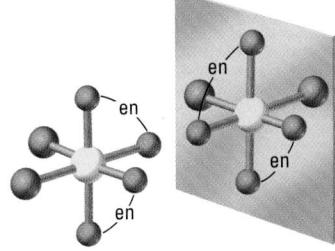

$[Cr(en)_2Cl_2]^+$

Figure 22.13   Optical isomerism in $[Cr(en)_2Cl_2]^+$. The ion on the left cannot be superimposed on its mirror image *(right)*.

Optical Isomerism

Optical isomers are mirror images that are not superimposible. Such nonsuperimposable mirror images are known as *enantiomers* (⇐ *p. 397, Section 9.7)*. An example of a complex ion that has optical isomerism is $[Cr(en)_2Cl_2]^+$. There are two enantiomers, as shown in Figure 22.13. No matter how they are twisted and turned, the two enantiomers are nonsuperimposable.

Optical isomerism is not possible for square planar complexes based on the geometry around the metal ion; the mirror images are superimposable. Although optical isomers of tetrahedral complex ions with four different ligands are theoretically possible, no such stable complexes are known.

 ## Coordination Compounds and Life

Bioinorganic chemistry, the study that applies chemical principles to inorganic ions and compounds in biological systems, is a rapidly growing field, centered mainly around coordination compounds. This is because living systems depend for their existence on many coordination compounds in which metal ions are chelated to the nitrogen atoms in proteins and especially in enzymes. Copper-containing proteins, for example, give the blood of crabs, lobsters, and snails its blue color, as well as transport oxygen.

In humans, molecular oxygen (O_2) is carried by hemoglobin, a very large protein (molecular weight of about 68,000 amu) in red blood cells. Hemoglobin is blue but becomes red when oxygenated. This is why arterial blood is bright red (high O_2 concentration) and blood in veins is bluish (low O_2 concentration).

A hemoglobin molecule carries four O_2 molecules, each of which forms a coordinate covalent bond to an Fe^{2+} ion. The Fe^{2+} ion is at the center of heme, a nonprotein part of the hemoglobin molecule that consists of four linked nitrogen-containing rings (Figure 22.14). Bound in this way, molecular oxygen is carried to the cells, where it is released as needed by breaking the $Fe—O_2$ bond.

The blue blood of horseshoe crabs is used to test for bacterial contamination of drugs.

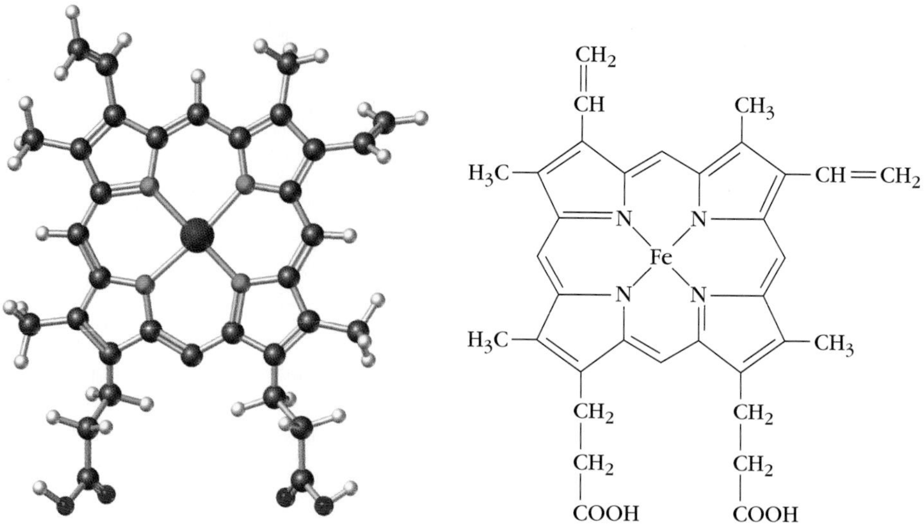

Figure 22.14 Heme, the carrier of Fe²⁺ in hemoglobin. Fe²⁺ is coordinated to four nitrogen atoms in heme.

It is curious (and fortunate) that N≡N does not behave chemically like C≡O.

Other substances that can donate an electron pair can also bond to the Fe²⁺ in heme. Carbon monoxide is such a ligand and forms an exceptionally strong Fe²⁺—CO bond, nearly 200 times stronger than the O₂—Fe²⁺ bond. Therefore, when a person breathes in CO, it displaces O₂ from hemoglobin and prevents red blood cells from carrying oxygen. The initial effect is drowsiness. But if CO inhalation continues, cells deprived of oxygen can no longer function and the person suffocates to death.

Structures similar to the oxygen-carrying unit in hemoglobin are also found in other biologically important compounds, including such diverse ones as myoglobin and vitamin B-12. Myoglobin, like hemoglobin, contains Fe²⁺ and carries and stores molecular oxygen, principally in muscles. At the center of a vitamin B-12 molecule is a Co³⁺ ion bonded to the same type of group as in hemoglobin. Vitamin B-12 is the only known dietary use of cobalt, but it makes cobalt an essential mineral (⇐ *p.105*).

The dietary necessity of zinc for humans has only become established since the 1980s. Zinc, in the form of Zn²⁺ ions, is essential to the functioning of several hundred enzymes, including those that catalyze the breaking of P—O—P bonds in adenosine triphosphate (ATP), an important energy-releasing compound in cells (⇐ *p. 840*).

Copper ranks third among biologically important transition metal ions in humans, trailing only iron and zinc. Although we generally excrete any dietary excess of copper, a genetic defect causes Wilson's disease, a condition in which Cu²⁺ accumulates in the liver and brain. Fortunately, Wilson's disease can be treated by administering chelating agents that coordinate excess Cu²⁺ ions, allowing them to be excreted harmlessly.

SUMMARY PROBLEM

I. Solid iron(II) sulfide, FeS, is roasted to form either $Fe_2O_3(s)$ or $Fe_3O_4(s)$.

(a) Write the balanced equations for the roasting of FeS(s) to $Fe_2O_3(s)$ and to $Fe_3O_4(s)$.

(b) Use thermochemical data from Appendix J to calculate the enthalpy change for these reactions at 25 °C, given FeS(s): $\Delta H_f^\circ = -101.671$ kJ/mol.

(c) Use thermochemical data from Appendix J to calulate the Gibbs free energy change for these reactions at 25 °C, given FeS(s): $S° = 82.81$ J mol^{-1} K^{-1}. Which reaction is more favored at this temperature?

(d) Calculate the Gibbs free energy change for the conversion of FeS to Fe$_2$O$_3$ and to Fe$_3$O$_4$ at 600 °C. Which reaction is favored at this temperature?

II. Iron(III) forms a purple coordination compound [Fe(salicylate)$_3$] with salicylate ions.

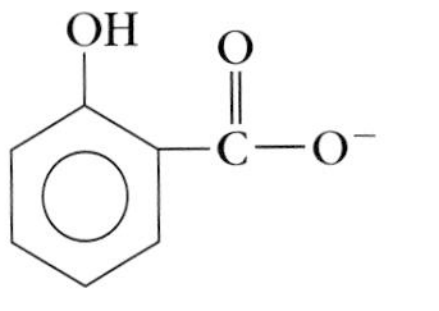

salicylate ion

The salicylate anion is a bidentate ligand that furnishes lone pairs from the C—O$^-$ and the OH oxygens.

(a) Write a structural formula for [Fe(salicylate)$_3$].

(b) Give the coordination number of Fe^{3+} in this compound.

(c) Account for the fact that there is no net charge on [Fe(salicylate)$_3$].

(d) Why are no counter ions needed for this compound?

IN CLOSING

Having studied this chapter, you should be able to . . .

- Recognize the general properties of transition metals (Section 22.1).
- Write electron configurations and orbital box diagrams for transition metals and their ions (Section 22.1).
- Explain how most transition metals have multiple oxidation states (Section 22.1).
- Explain trends in sizes of transition metal atomic radii (Section 22.1).
- Describe how iron ore is processed into iron and then into steel (Section 22.2).
- Discuss how copper is extracted from its ores and purified (Section 22.3).
- Discuss the chemistry of gold and silver (Section 22.4).
- Explain the coodinate covalent bonding of ligands in coordination compounds and complexes (Section 22.6).
- Interpret the names and formulas of coordination complex ions and compounds (Section 22.6).
- Discuss isomerism in coordination compounds and complex ions (Section 22.6).
- Give examples of coordination compounds and their uses (Section 22.6).

KEY TERMS

bidentate *(22.6)*
chelating ligands *(22.6)*
coordinate covalent bond *(22.6)*
coordination compound *(22.6)*

coordination number *(22.6)*
hexadentate *(22.6)*
lanthanide contraction *(22.1)*
ligands *(22.6)*

monodentate *(22.6)*
polydentate *(22.6)*
pyrometallurgy *(22.2)*
steel *(22.2)*

Questions for Review and Thought

Answers to questions in **bold** can be found in the back of the book.

Review Questions

1. What is the primary reducing agent in the production of iron from its ores? Write a balanced chemical equation for this reduction process.
2. Why is lime necessary in the blast furnace reduction of iron ore?
3. What is the difference between pig iron and cast iron?
4. Explain the purpose of each of these materials in the blast-furnace conversion of iron ore to iron.
   (a) Air   (b) Limestone   (c) Coke
5. Identify what is produced by each of these processes or operations.
   (a) Blast furnace   (b) Basic oxygen process
   (c) Roasting
6. Identify a common use for Cr, Cu, Fe, Au, and Ag.
7. Name three transition metals that are found "free" in nature.
8. What is the lanthanide contraction? Why does it occur?
9. In general, how do the atomic radii change across the first transition series (Period 4)?
10. What is the distinguishing chemical feature of a ligand?
11. Distinguish between
    (a) A monodentate and a bidentate ligand
    (b) A *cis* and a *trans* isomer
    (c) A coordination compound and a coordination complex ion
    (d) A geometric isomer and an optical isomer
12. Define the following words or phrases and give an example for each.
    (a) Coordination compound   (b) Complex ion
    (c) Ligand   (d) Chelate
    (e) Bidentate ligand

Transition Metals

13. Write electron configurations for the 2+ ions of
    (a) Iron   (b) Copper   (c) Chromium
14. Write electron configurations for the common oxidation states of
    (a) Silver   (b) Gold
15. Which first transition metal series ions are isoelectronic with
    (a) Zn^{2+}?   (b) Mn^{2+}?   (c) Cr^{3+}?   (d) Fe^{3+}?
16. Which two oxidation states of chromium are paramagnetic?
17. Arrange these substances in decreasing strengths as oxidizing agents: Mn^{2+}, MnO_4^-, MnO_2. Explain the trend.
18. Arrange these substances in decreasing strengths as oxidizing agents: $Cr_2O_7^{2-}$ (in acid), Cr^{2+}, Cr^{3+}. Explain the trend.
19. Write a balanced equation to represent
    (a) The roasting of Cu_2S to copper metal.
    (b) The reduction of Fe_2O_3 with aluminum.
20. Write a balanced equation to represent

(a) The reduction of Fe_2O_3 with carbon monoxide in a blast furnace.
(b) The production of hydrogen gas when hydrochloric acid reacts with an iron nail.

21. Balance this redox reaction.
    $$Cu(s) + NO_3^-(aq) \longrightarrow$$
    $$Cu^{2+}(aq) + NO_2(g) \quad \text{(acidic solution)}$$

22. Balance this redox reaction.
    $$Fe(s) + NO_3^-(aq) \longrightarrow$$
    $$Fe^{3+}(aq) + NO_2(g) \quad \text{(acidic solution)}$$

Coordination Compounds

23. In a complex ion a central ruthenium atom (Ru(III)) is bonded to six ammonia molecules.
    (a) Give the formula and net charge for this complex ion.
    (b) Balance the net charge on this complex ion with chloride ions. How many chloride ions are needed?
    (c) Write the formula for the complex, including the chloride ions that are not part of the complex ion.
24. In a complex ion, a central Cr^{3+} ion is bonded to two ammonia molecules, three water molecules, and a hydroxide ion.
    (a) Give the formula and charge of the complex ion.
    (b) What kind of counter ion would be needed?
25. Consider the complex ion $[Cr(NH_3)_2(H_2O)_2Br_2]^+$.
    (a) Identify the ligands and their charges (if any).
    (b) What is the charge on the central metal ion?
    (c) What is the formula of the sulfate salt of this cation?
26. Consider the complex ion $[Co(C_2O_4)_2Cl_2]^{3-}$.
    (a) Identify the ligands and their charges (if any).
    (b) What is the charge on the central metal ion?
    (c) What would be the formula and charge of the complex ion if the $C_2O_4^{2-}$ ions were replaced by NH_3 molecules?
27. Determine the charge of the central metal ion in each case.
    (a) $[Zn(H_2)_3(OH)]^+$ (b) $[Pt(NH_3)_3Cl_3]^-$
    (c) $[Cr(CN)_6]^{3-}$
28. For coordination compounds $Na_3[IrCl_6]$ and $[Mo(CO)_4Br_2]$, identify in each case
    (a) The ligands.
    (b) The central metal ion and its charge.
    (c) The formula and charge of the complex ion.
    (d) Ions not in the complex ion.
29. Give the coordination number of the central metal ion in
    (a) $[Cu(en)_2(NH_3)_2]^{2+}$ (b) $[Fe(en)(ox)Cl_2]^-$
30. Give the coordination number of the central metal ion in
    (a) $[Pt(en)_2]^{2+}$ (b) $[Cu(ox)_2]^{2+}$
31. Write a structural formula for the coordination compound $[Cr(en)(NH_3)_2I_2]$, and give the coordination number for the central Cr^{2+} ion.
32. Give the formula of each of these coordination compounds formed with Pt^{2+}.
    (a) Two ammonia molecules and two bromide ions.
    (b) One ethylenediamine molecule and two nitrite ions, NO_2^-.

(c) One chloride ion, one bromide ion, and two ammonia molecules.

33. Give the charge on the central metal ion in each of these.
    (a) $[VCl_6]^{4-}$ (b) $[Sc(H_2O)_3Cl_3]$
    (c) $[Mn(NO)(CN)_5]^{3-}$ (d) $[Cu(en)_2(NH_3)_2]^{2+}$

34. Identify the coordination number of the metal ion in these coordination complexes.
    (a) $[FeCl_4]^-$ (b) $[PtBr_4]^{2-}$
    (c) $[Mn(en)_3]^{2+}$ (d) $[Cr(NH_3)_5H_2O]^{3+}$

35. Using structural formulas, show how the carbonate ion can be either a monodentate or bidentate ligand to a transition metal cation.

36. Classify these ligands as monodentate, bidentate, and so on.
    (a) $(CH_3)_3P$
    (b)

    (c) $H_2N—(CH_2)_2—NH—(CH_2)_2—NH_2$
    (d) H_2O

37. Which of these would be expected to be effective chelating agents?
    (a) CH_3CH_2OH
    (b) $H_2N—(CH_2)_3—NH_2$
    (c)

    (d) PH_3

38. Give an analogous (noncoordination) simple ionic compound to $[Rh(en)_3]Cl_3$.

Naming Complex Ions and Coordination Compounds

39. Write the formula for
    (a) Potassium diaquadioxalatocobaltate(III)
    (b) Diamminetriaquahydroxochromium(II) nitrate
    (c) Ammonium tetrachlorocuprate(II)
    (d) Tetrachloroethylenediaminecobaltate(III)
    (e) Triaquatrifluorocobalt(III)

40. Interpret these formulas.
    (a) $[MnCl_4]^{2-}$ (b) $K_3[Fe(C_2O_4)_3]$
    (c) $[Pt(NH_3)_2(CN)_2]$ (d) $[Fe(H_2O)_5(OH)]^{2+}$
    (e) $[Mn(en)_2Cl_2]$

Geometry of Coordination Complexes

41. Sketch the geometry of
    (a) *cis*-$[Cu(H_2O)_2Br_4]^{2-}$ (b) *trans*-$[Ni(NH_3)_2(en)_2]^{2+}$

42. Sketch the geometry of
    (a) *cis*-$[Ni(H_2O)_2Cl_2]$ (b) *trans*-$[Cr(H_2O)_4Cl_2]^+$

43. The ligand 1,2-diaminocyclohexane

is abbreviated "dech." Sketch the geometry of *cis*-$[Pd(H_2O)_2(dech)]^{2+}$

44. The acetylacetonate ion (acac$^-$)

forms a complex with Fe^{3+}. Sketch the geometry of $Fe(acac)_3$.

45. Which of these octahedral coordination complexes can exhibit geometric isomerism?
    (a) $[Fe(H_2O)_2Cl_2Br_2]^{2-}$ (b) $[Fe(H_2O)_2Cl_3Br]^-$

46. Which of these octahedral coordination complexes can exhibit geometric isomerism?
    (a) $[Cr(H_2O)_3Cl_3]$ (b) $[Cr(H_2O)_4Cl_2]^+$

47. Draw the possible geometric isomers of
    (a) $[Ni(NH_3)_4Cl_2]$
    (b) $[Pt(NH_3)_2(NCS)Br]$ (The S in NCS is bonded to Pt^{2+}.)
    (c) $[Co(en)Cl_4]^-$

48. In which of these is geometric isomerism possible?
    (a) $[Co(H_2O)_4Cl_2]^+$ (b) $[Pt(NH_3)Cl_3]^-$
    (c) $[Co(H_2O)_3Cl_3]$ (d) $[Co(en)_2(NH_3)Br]^{2+}$

General Questions

49. Give the electron configuration of
    (a) Ti^{3+} (b) V^{2+} (c) Ni^{3+} (d) Cu^+

50. Give the electron configuration of
    (a) Cr^{2+} (b) Zn^{2+} (c) Co^{2+} (d) Mn^{4+}

51. Write an orbital box diagram and determine the number of unpaired electrons for each species in Question 49.

52. Write an orbital box diagram and determine the number of unpaired electrons for each species in Question 50.

53. Assuming 100% recovery of the metal, which would yield the greater number of grams of copper?
    (a) An ore containing 3.60 mass % azurite, $Cu(OH)_2 \cdot 2CuCO_3$, or
    ((b) An ore containing 4.95 mass % chalcopyrite, $CuFeS_2$.

54. What mass of copper could be electroplated from a $CuSO_4$ solution using an electric current of 2.50 A for 5.00 h? Assume 100% efficiency.

55. Copper is obtained directly by roasting covelite, CuS.
    (a) Write a balanced equation for this process.
    (b) Assume that the roasting is 90.0% efficient. How many tons of SO_2 would be released into the air by roasting 500. tons of covelite?

56. What mass of SO_2 is produced when 1.0 ton of chalcocite, Cu_2S, is roasted to Cu_2O?

57. What is the coordination number of the central metal ion in
    (a) $[Ni(NH_3)_2Br_2]$ (b) $[Fe(CN)_6]^{3-}$
    (c) $[Ti(H_2O)Cl_5]^{2-}$ (d) $[Mn(C_2O_4)_3]^{4-}$

58. What is the coordination number of the central metal ion in
    (a) $[Ni(en)Cl_2]$ (b) $[Mo(CO)_4Br_2]$
    (c) $[Cd(CN)_4]^{2-}$ (d) $[Co(CN)_5(OH)]^{3-}$

59. Draw sketches for as many octahedral complexes as you can using only ethylenediamine and/or Cl$^-$ as ligands.

60. Draw sketches for as many octahedral complexes as you can for the formula $Co(NH_3)_4Cl_2Br$.

61. In your own words explain why
    (a) $H_2N-(CH_2)_3-NH_2$ is a bidentate ligand.
    (b) AgCl dissolves in NH_3.
    (c) There are no geometric isomers of tetrahedral complexes.

62. Determine whether each statement is true or false. If it is false, correct the statement.
    (a) The coordination number of the Fe^{3+} ion in $[Fe(H_2O)_4(C_2O_4)]^+$ is five.
    (b) Cu^+ has two unpaired electrons.
    (c) The net charge of a coordination complex of Cr^{3+} with two NH_3, one en, and two H_2O, is 2+.

63. Determine whether each statement is true or false. If it is false, correct the statement.
    (a) In $[Pt(NH_3)_4Cl_4]$, platinum has a +4 charge and a coordination number of six.
    (b) In general, Cu^{2+} is more stable than Cu^+ in aqueous solutions.

64. The metal ion in $[Pt(NH_3)_2(C_2O_4)]$ is surrounded by a square planar array of coordinating atoms.
    (a) Give the oxidation number of the central metal ion.
    (b) Draw the structural formula of this coordination compound.

65. Chromium(III) forms three different compounds with water and chloride ions, all of which have the same composition: 19.51% Cr, 39.92% Cl, and 40.57% water. One of the compounds is violet and dissolves in water to give a complex ion with a 3+ charge plus three chloride ions. All three chloride ions precipitate immediately as AgCl when $AgNO_3$ is added to the solution. Draw the structural formula of this complex ion.

Applying Concepts

66. Iron nails are put into Fe^{2+} aqueous solutions to reduce any Fe^{3+} that forms back to Fe^{2+}. Write a balanced chemical equation for this preventative reaction.

67. Use VSEPR theory to predict the shape and bond angles around chromium in
    (a) Chromate ions.   (b) Dichromate ions.

68. The structure of cyclam is given below.

Cyclam can act as a ligand. How many coordinate covalent bonds can one cyclam molecule form with a central metal ion?

69. The compound 1,10-phenanthroline is a chelating agent used in analytical chemistry. Its isomer 4,7-phenanthroline is not. Use these structural formulas to explain this difference.

1,10-phenanthroline   4,7-phenanthroline

70. Consider

$$M \xrightarrow{+0.20\ V} M^{2+} \xrightarrow{+0.50\ V} M^+ \xrightarrow{-0.20\ V} M$$

Show by calculations which, if any, of these species should disproportionate.

71. An electrochemical cell is constructed by immersing a strip of chromium into a 1.0 M solution of Cr^{3+} and a strip of gold into a 1.0 M solution of Au^{3+}. The half-cells are connected by a salt bridge, and a wire completes the circuit.
    (a) Write the balanced chemical equation for the reaction that is product-favored.
    (b) Calculate the cell potential.
    (c) Draw a sketch of the cell and indicate the anode, cathode, and direction of electron flow.

72. Repeat the directions for Question 71 using a cell constructed of a strip of nickel immersed in a 1.0 M Ni^{2+} solution and a strip of silver dipping into a 1.0 M Ag^+ solution.

73. Calculate $\Delta G°$ for the reduction of Fe_2O_3 with CO gas at 25 °C and at 1000 °C. What application does this have to the conversion of iron ore to iron in a blast furnace?

74. To determine the percent iron in an ore, a 1.500-g sample of the ore containing Fe^{2+} is titrated to the equivalence point with 18.6 mL of 0.05012 M $KMnO_4$. The products of the titration are Fe^{3+} and Mn^{2+}. What is the percent of iron in the ore?

75. Consider the reaction

$$2\ Cu^+(aq) \longrightarrow Cu^{2+}(aq) + Cu(s)$$

for which the $E° = +0.37$ V. Use the Nernst equation to calculate
    (a) E when the Cu^{2+} concentration is equal to the Cu^+ concentration $= 1 \times 10^{-4}$ M.
    (b) The concentration of Cu^+ when the Cu^{2+} concentration $= 1.0$ M and $E = 0.00$ V.

76. Consider the reaction

$$2\ Ag^+(aq) \longrightarrow Ag(s) + Ag^{2+}(aq)$$

for which the $E° = -1.18$ V. Use the Nernst equation to calculate
    (a) E when the Ag^+ concentration $= 1 \times 10^{-4}$ M, which is five times the concentration of Ag^{2+}.
    (b) The concentration of Ag^{2+} when the Ag^+ concentration $= 1.0$ M and $E = 0.00$ V.

77. Use the Nernst equation to calculate E_{red} for $Cr_2O_7^{2-}$ in 6.0 M H^+ when the concentration of $Cr_2O_7^{2-}$ = concentration of $Cr^{3+} = 0.10$ M.

78. If 1.00 mol of each of these compounds is dissolved in separate samples of water sufficient to dissolve the compound, how many moles of ions are present in each solution?
    (a) $[Pt(en)Cl_2]$ (b) $Na[Cr(en)_2(SO_4)_2]$
    (c) $K_3[Au(CN)_4]$ (d) $[Ni(H_2O)_2(NH_3)_4]Cl_2$

79. For each of the compounds in Question 78, state which it would most likely resemble in colligative properties and conductivity: $CO(NH_2)_2$ (urea), KCl, K_2SO_4, or K_3PO_4.

80. Analysis of a coordination compound gives these results: 22.0% Co, 31.4% N, 6.78% H, and 39.8% Cl. One mol of the compound dissociates in water to form 4 mol of ions.
    (a) What is the formula of the compound?
    (b) Write the equation for its dissociation in water.

81. A chemist synthesizes two coordination compounds. One compound decomposes at 210 °C, the other at 240 °C. When analyzed, the compounds give the same mass percent data: 52.6% Pt, 7.6% N, 1.63% H, and 38.2% Cl. Both compounds contain a $+4$ central metal ion.

    (a) What is the simplest formula of the compounds?
    (b) Draw structural formulas for the complexes present.

82. A coordination compound has the simplest formula $PtN_2H_6Cl_2$ with a molar mass of about 600 g/mol. It contains a complex cation and a complex anion. Draw its structural formula.

83. The glycinate ion (gly) is $H_2NCH_2CO_2^-$. It can act as a ligand coordinating through the nitrogen and one of the oxygens. Using $\stackrel{\frown}{N \quad O}$ to represent glycinate ion, draw structural formulas for four stereoisomers of $[Co(gly)_3]$.

APPENDICES

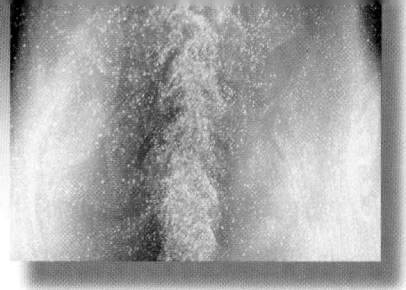

A PROBLEM SOLVING AND MATHEMATICAL OPERATIONS *A.2*

B UNITS, EQUIVALENCES, AND CONVERSION FACTORS *A.17*

C PHYSICAL CONSTANTS *A.22*

D MOLECULAR ORBITALS *A.23*

E NAMING SIMPLE ORGANIC COMPOUNDS *A.28*

F IONIZATION CONSTANTS FOR WEAK ACIDS AT 25 °C *A.34*

G IONIZATION CONSTANTS FOR WEAK BASES AT 25 °C *A.35*

H SOLUBILITY PRODUCT CONSTANTS FOR SOME INORGANIC COMPOUNDS AT 25 °C *A.36*

I STANDARD REDUCTION POTENTIALS IN AQUEOUS SOLUTION AT 25 °C *A.38*

J SELECTED THERMODYNAMIC VALUES *A.41*

ANSWERS TO PROBLEM-SOLVING PRACTICE PROBLEMS *A.49*

ANSWERS TO EXERCISES *A.69*

ANSWERS TO SELECTED QUESTIONS FOR REVIEW AND THOUGHT *A.93*

FIGURE CREDITS *C.1*

GLOSSARY *G.1*

INDEX *I.1*

APPENDIX A

Problem Solving and Mathematical Operations

Contents

A.1 General Problem-Solving Strategies *A.2*

A.2 Numbers, Units, and Quantities *A.3*

A.3 Significant Figures *A.6*

A.4 Electronic Calculators *A.10*

A.5 Exponential or Scientific Notation *A.10*

A.6 Logarithms *A.13*

A.7 Quadratic Equations *A.15*

In this book we have provided many illustrations of problem solving and many problems for practice. Some are numerical problems that must be solved by mathematical calculations. Others are conceptual problems that must be solved by applying an understanding of the principles of chemistry. Often, it is necessary to use chemical concepts to relate what we know about matter at the nanoscale to the properties of matter at the macroscale. The problems throughout this book are representative of the kinds of problems chemists and other scientists must regularly solve to pursue their goals, although our problems are often not as difficult as those encountered in the real world.

Problem solving is not a simple skill that can be mastered in a few hours of study or practice. Because there are many different kinds of problems and many different kinds of people who are problem solvers, there are no hard and fast rules that are guaranteed to lead you to solutions. The general guidelines presented in this Appendix are, however, helpful in getting you started on any kind of problem and in checking to see if your answers are correct. The problem-solving skills you develop in a chemistry course such as this can later be applied to difficult and important problems that may arise in your profession, your personal life, or the society in which you live.

In getting a clear picture of a problem and asking appropriate questions regarding the problem, you need to keep in mind all the principles of chemistry and other subjects that you think may apply. In many real-life problems there is not enough information available for you to arrive at an unambiguous solution; in such cases, try to look up or estimate what is needed and then go ahead, noting assumptions you have made. Often the hardest part is deciding which principle or idea is most likely to help solve the problem and what information is needed. To some degree this can be a matter of luck or chance. Nevertheless, in the words of Louis Pasteur, "In the field of observation chance only favors those minds which have been prepared." The more practice you have had, and the more principles and facts you can keep in mind, the more likely you are to be able to solve the problems that you face.

A.1 GENERAL PROBLEM-SOLVING STRATEGIES

1. **Define the problem.** Carefully review the information contained in the problem. What is the problem asking you to find? What key principles are involved? What known information is necessary to solving the problem and what information is there only to place the question in context? Organize the information to see what is necessary and to see the relationships among the known data. Try writing the information down in a table. If it is numerical information, be sure to include proper units. Can you picture the situation under consideration? Try sketching it and including any relevant dimensions in the sketch.

2. **Develop a plan.** Have you solved a problem of this type before? If you recognize the new problem as similar to ones you know how to solve, you can use the same method that worked before. Try reasoning backward from the units of what is being sought. What data are needed to find an answer in those units?

Can the problem be broken down into smaller pieces, each of which can be solved separately to produce information that can be assembled to solve the entire problem? When a problem can be divided into simpler problems, it often helps to write down a plan that lists the simpler problems and the order in which they must be put together to arrive at an overall solution. Many major problems in chemical research have to be solved in this way. In problems in this book we have mostly provided the needed numerical data, but in the laboratory, the first aspect of solving a problem is often devising experiments to gather the data or searching databases to find needed information.

   If you are still unsure about what to do, do something anyway. It may not be the right thing to do, but as you work on it, the way to solve the problem may become apparent, or you may see what is wrong with the approach you have started with, thereby making clearer what a good plan would be.

3. **Execute the plan.** Carefully write down each step of a mathematical problem, being sure to properly keep track of the units. Do the units cancel to give you the answer in the desired units? Don't skip steps. Don't do any but the simplest steps in your head. Once you've written down the steps for a mathematical problem, check what you've written — is it all correct? Students often say they got a problem wrong because they "made a stupid mistake." Teachers — and textbook authors — make mistakes, too. It is usually because they don't take the time to write down the steps of a problem clearly and correctly. In solving a mathematical problem, remember to apply the principles of dimensional analysis and significant figures. Dimensional analysis is introduced in Sections 1.3 *(⇐ p. 6)* and 2.3 *(⇐ p. 46)*; it is reviewed below. A review of significant figures is in Appendix A.3.

4. **Check the answer to see whether it is reasonable.** As a final check of your solution to any problem, ask yourself whether the answer is reasonable: Are the units of a numerical answer correct? Is a numerical answer of about the right size? Don't just copy a result from your calculator without thinking about whether it makes sense.

   Suppose you have been asked to convert 100. yards to a distance in meters. Using dimensional analysis and some well-known factors for converting from the English system to the metric system, you could write

$$100. \text{ yd} \times \frac{3 \text{ ft}}{1 \text{ yd}} \times \frac{12 \text{ in.}}{1 \text{ ft}} \times \frac{2.54 \text{ cm}}{1 \text{ in.}} \times \frac{1 \text{ m}}{100 \text{ cm}} = 91.4 \text{ m}$$

To check that a distance of 91.4 m is about right, recall that a yard is a little shorter than a meter. Therefore 100 yd should be a little less than 100 m. If you had mistakenly divided instead of multiplied by 3 ft/yd in the first step, your final answer would have been a little more than 10 m. This is equivalent to only about 30 ft, and you probably know a 100-yd football field is longer than that.

A.2   NUMBERS, UNITS, AND QUANTITIES

Many scientific problems require you to use mathematics to calculate a result or draw a conclusion. Therefore, knowledge of mathematics and its application to problem solving is important. However, one aspect of scientific calculations is often absent from pure mathematical work: Science deals with *measurements* in which an unknown quantity is compared with a standard or unit of measure. For example, using a balance to determine the mass of an object involves comparing the object's mass with standard masses, usually in multiples or fractions of one gram; the result is reported as some number of grams, say 4.357 g. *Both the number and the unit are important.* If the result had been 123.5 g, this would clearly be different, but a result of 4.357 oz (ounces) would also be different, because the unit "ounce" is different from the unit "gram." A *result that describes the quantitative measurement of a property,* such as 4.357 g, is called a *quantity* (or physical quantity), and it consists of a number and a unit. Chemical problem solving requires calculating with quantities. Notice that whether a quantity is large or small depends on the units as well as the number; the two quantities 123.5 g and 4.357 oz represent the *same* mass.

A quantity is always treated as though the number and the units are multiplied together; that is, 4.357 g can be handled mathematically as $4.357 \times g$. Keeping in mind this simple rule, you will see that calculations involving quantities follow the normal rules of algebra and arithmetic: $5\ g + 7\ g = (5 + 7) \times g = 12\ g$; or $6\ g \div 2\ g = (6\ g)/(2\ g) = 3$. (Notice that in the second calculation the unit g appears in numerator and denominator and cancels out, leaving a pure number, 3.) Treating units as algebraic entities has the advantage that *if a calculation is set up correctly, the units will cancel or multiply together so that the final result has appropriate units.* For example, if you measured the size of a sheet of paper and found it to be 8.5 in. by 11 in., the area A of the sheet could be calculated as area = length × width = 11 in. × 8.5 in. = 94 in.^{2}, or 94 square inches. If a calculation is set up incorrectly, the units of the result will be inappropriate. Using units to check whether a calculation has been properly set up is called *dimensional analysis* (⇐ *p. 8*).

This idea of using algebra on units as well as numbers is useful in all kinds of situations. For example, suppose you are having a party for some friends who like pizza. A large pizza consists of 12 slices and costs \$10.75. You expect to need 36 slices of pizza and want to know how much you will have to spend. A strategy for solving the problem is first to figure out how many pizzas you need and then to figure the cost in dollars. This solution could be diagrammed as

$$\text{Slices} \xrightarrow[\text{step 1}]{\text{slices per pizza}} \text{pizzas} \xrightarrow[\text{step 2}]{\text{dollars per pizza}} \text{dollars}$$

Step 1: Find the number of pizzas required by dividing the number of slices per pizza into the number of slices, thus converting "units" of slices to "units" of pizzas:

$$\text{Number of pizzas} = 36\ \text{slices} \left(\frac{1\ \text{pizza}}{12\ \text{slices}} \right) = 3\ \text{pizzas}$$

Notice that if you had multiplied the number of slices times the number of slices per pizza, the result would have been labeled pizzas × slices^2, which does not make sense. In other words, the labels indicate whether multiplication or division is appropriate.

Strictly speaking, slices and pizzas are not units in the same sense that a gram is a unit; however, labeling things this way will often help you keep in mind what a number refers to—pizzas, slices, or dollars in this case.

Step 2: Find the total cost by multiplying the cost per pizza by the number of pizzas needed, thus converting "units" of pizzas to "units" of dollars:

$$\text{Total price} = 3\ \text{pizzas} \left(\frac{\$10.75}{1\ \text{pizza}} \right) = \$32.25$$

Notice that in each step you have multiplied by a factor that allowed the initial units to cancel algebraically, giving the answer in the desired units. A factor such as (1 pizza/12 slices) or (\$10.75/pizza) is referred to as a *proportionality factor* (⇐ *p. 9*). This name indicates that it comes from a proportion. For instance, in the pizza problem you could set up the proportion

$$\frac{x\ \text{pizzas}}{36\ \text{slices}} = \frac{1\ \text{pizza}}{12\ \text{slices}} \quad \text{or} \quad x\ \text{pizzas} = 36\ \text{slices} \left(\frac{1\ \text{pizza}}{12\ \text{slices}} \right) = 3\ \text{pizzas}$$

A proportionality factor such as (1 pizza/12 slices) is also called a *conversion factor,* which indicates that it converts one kind of unit or label to another; in this case the label "slices' is converted to the label "pizzas."

Many everyday scientific problems involve proportionality. For example, the bigger the volume of a solid or liquid substance, the bigger its mass. When the volume is zero, the mass is zero also. These facts indicate that mass, m, is directly proportional to volume, V, or, symbolically,

$$m \propto V$$

where the symbol $\propto$ means "is proportional to." Whenever a proportion is expressed this way, it can also be expressed as an equality by using a proportionality constant, for example,

$$m = d \times V$$

In this case the proportionality constant, d, is called the density of the substance. This equation embodies the definition of density as mass per unit volume, since it can be rearranged algebraically to

$$d = \frac{m}{V}$$

As with any algebraic equation involving three variables, it is possible to calculate any one of the three quantities m, V, or d, provided the other two are known. If density is wanted, simply use the definition of mass per unit volume; if mass or volume is to be calculated, the density can be used as a proportionality factor.

Suppose that you are going to buy a ton of gravel and want to know how big a bin you will need to store it. You know the mass of gravel and want to find the volume of the bin; this implies that density will be useful. If the gravel is primarily limestone, you can assume that its density is about the same as for limestone and look it up. Limestone has the chemical formula $CaCO_3$, and its density is 2.7 kg/L. However, these mass units are different from the units for mass of gravel, namely tons. Therefore you need to recall or look up the mass of 1 ton (exactly 2000 pounds, lb) and the fact that there are 2.20 lb per kg. This provides enough information to calculate the volume needed. Here is a "roadmap" plan for the calculation:

Mass of gravel in tons $\xrightarrow[\text{step 1}]{\text{change units}}$ mass of gravel in kilograms

$\xrightarrow[\text{step 2}]{\text{density}}$ volume of bin

Step 1: Figure out how many kilograms of gravel are in a ton.

$$m_{\text{gravel}} = 1 \text{ ton} = 2000 \text{ lb} = 2000 \text{ lb} \left(\frac{1 \text{ kg}}{2.20 \text{ lb}} \right) = 909 \text{ kg}$$

The fact that there are 2.20 lb per kg implies two proportionality factors: (2.20 lb/1 kg) and (1 kg/2.20 lb). The latter was used because it results in appropriate cancellation of units.

Step 2: Use the density to calculate the volume of 909 kg of gravel.

$$V_{\text{gravel}} = \frac{m_{\text{gravel}}}{d_{\text{gravel}}} = \frac{909 \text{ kg}}{2.7 \text{ kg/L}} = 909 \text{ kg} \left(\frac{1 \text{ L}}{2.7 \text{ kg}} \right) = 340 \text{ L}$$

In this step we used the definition of density, solved algebraically for volume, substituted the two known quantities into the equation, and calculated the result. However, it is quicker simply to remember that mass and volume are related by a proportionality factor called density and to use the units of the quantities to decide whether to multiply or divide by that factor. In this case we divided mass by density because the units kilograms canceled, leaving a result in liters, which is a unit of volume.

Also, it is quicker and more accurate to solve a problem like this by a single setup. Then all the calculations can be done at once, and no intermediate results need to be written down. The "roadmap" plan given above can serve as a guide to the single-setup solution, which looks like this:

$$V_{\text{gravel}} = 1 \text{ ton} \left(\frac{2000 \text{ lb}}{1 \text{ ton}} \right) \left(\frac{1 \text{ kg}}{2.20 \text{ lb}} \right) \left(\frac{1 \text{ L}}{2.7 \text{ kg}} \right) = 340 \text{ L}$$

To calculate the result, then, you would enter 2000 on your calculator, divide by 2.20, and divide by 2.7. A setup like this makes it easy to see what to multiply and divide by, and the calculation goes more quickly when it can be entered into a calculator all at once.

The liter is not the most convenient volume unit for this problem, however, because it does not relate well to what we want to find out — how big a bin to make. A liter is about the same volume as a quart, but whether you are familiar with liters or quarts or both, 300 of them is not easy to visualize. Let's convert liters to something we can understand better. A liter is a volume equal to a cube one tenth of a meter (1 dm) on a side; that is, a liter is 1 dm^3. Consequently,

$$340 \text{ L} = 340 \text{ L} \left(\frac{1 \text{ dm}^3}{1 \text{ L}} \right) \left(\frac{1 \text{ m}}{10 \text{ dm}} \right)^3 = 340 \text{ dm}^3 \left(\frac{1 \text{ m}^3}{1000 \text{ dm}^3} \right) = 0.34 \text{ m}^3$$

Thus, the bin would need to have a volume of about one third of a cubic meter; that is, it could be a meter wide, a meter long, and about a third of a meter high and it would hold the ton of gravel.

One more thing should be noted about this example. We don't need to know the volume of the bin very precisely, because being off a bit will make very little difference; it might mean getting a little too much wood to build the bin, or not making the bin quite big enough and having a little gravel spill out, but this isn't a big deal. In other cases, such as calculating the quantity of fuel needed to get a space shuttle into orbit, being off by a few percent could be a life-or-death matter. Because it is important to know how precise data are and to be able to evaluate how important precision is, scientific results usually indicate precision. The simplest way to do this is by means of significant figures.

A.3 SIGNIFICANT FIGURES

The **precision** of a measurement indicates how well several determinations of the same quantity agree. Precision is illustrated by the results of throwing darts at a bull's-eye (Figure A.1). In part (a) the darts are scattered all over the board; the dart thrower was apparently not very skillful (or threw the darts from a long distance away from the board), and the precision of their placement on the board is low. In part (b) the darts are all clustered together, indicating much better reproducibility on the part of the thrower, that is, greater precision. In addition, every

Figure A.1 Precision and accuracy. (a) poor precision, (b) good precision and good accuracy, (c) good precision and poor accuracy.

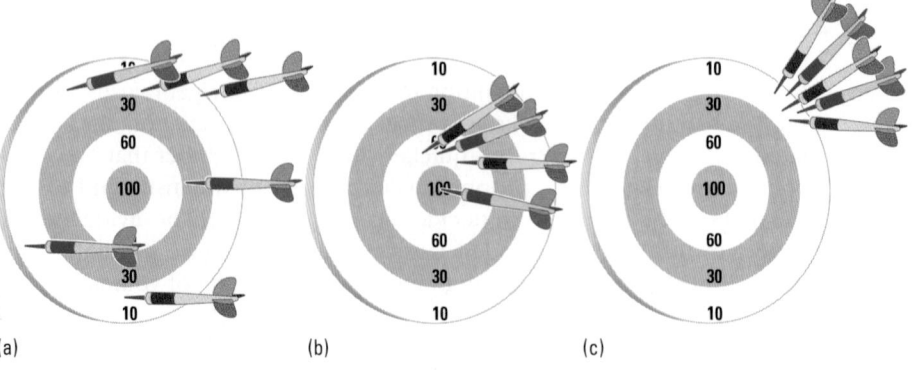

(a)   (b)   (c)

dart has come very close to the bull's-eye; this is described by saying that the thrower has been quite **accurate**—the average of all throws is very close to the accepted position, namely, the bull's-eye. Figure A.1c illustrates that it is possible to be precise without being accurate—the dart thrower has consistently missed the bull's-eye, although all darts are clustered very precisely around the wrong point on the board. This third case is like an experiment with some flaw (either in its design or in a measuring device) that causes all results to differ from the correct value by the same amount.

In the laboratory we attempt to set up experiments so that the greatest possible accuracy can be obtained. As a further check on accuracy, results are usually compared among different laboratories so that any flaw in experimental design or measurement can be detected. For each individual experiment, several measurements are usually made and their precision determined. Usually, better precision is taken as an indication of better experimental work, and it is necessary to know precision in order to compare results among different experimenters. If two different experimenters both had results like those in Figure A.1a, their average values could differ quite a lot before they would say that their results did not agree within experimental error.

In most experiments several different kinds of measurements must be made, and some can be done more precisely than others. It is common sense that *a calculated result can be no more precise than the least precise piece of information that went into the calculation.* This is where the rules for significant figures come in. In the preceding example the quantity of gravel was described as "a ton." Usually gravel is measured by weighing a truck empty, putting some gravel in the truck, weighing the truck again, and subtracting the weight of the truck from the weight of the truck plus gravel. The quantity of gravel is not adjusted if there is a bit too much or a bit too little, because this would be a lot of trouble. You might end up with as much as 2200 pounds or as little as 1800 pounds, even though you asked for a ton. In terms of significant figures this would be expressed as 2.0×10^3 lb.

The quantity 2.0×10^3 lb is said to have two significant figures; it designates a quantity in which the 2 is taken to be exactly right but the 0 is not known precisely. (In this case, the number could be as large as 2.2 or as small as 1.8, and so the 0 obviously is not exactly right.) In general, in a number that represents a scientific measurement, the last digit on the right is taken to be inexact, but all digits farther to the left are assumed to be exact. When you do calculations using such numbers, you must follow some simple rules so that the results will reflect the precision of all the measurements that go into the calculations. Here are the rules:

Rule 1: To determine the number of significant figures in a measurement, read the number from left to right and count all digits, starting with the first digit that is *not* zero.

Example	Number of significant figures
1.23 g	3
0.00123 g	3; the zeros to the left of the 1 simply locate the decimal point. The number of significant figures is more obvious if you write numbers in scientific notation; thus, $0.00123 = 1.23 \times 10^{-3}$.
2.0 g and 0.020 g	2; both have two significant digits. When a number is greater than 1, *all zeros to the right of the decimal point are significant.* For a number less than 1, only zeros to the right of the first significant digit are significant.

For a number written in scientific notation, all digits are significant.

Example	Number of significant figures
100 g	1; in numbers that do not contain a decimal point, "trailing" zeros may or may not be significant. To eliminate possible confusion, the practice followed in this book is to include a decimal point if the zeros are significant. Thus, 100. has three significant digits, while 100 has only one. Alternatively, we write in scientific notation 1.00×10^2 (three significant digits) or 1×10^2 (one significant digit). For a number written in scientific notation, all digits are significant.
100 cm/m	Infinite number of significant figures, because this is a defined quantity. There are *exactly* 100 centimeters in one meter.
$\pi = 3.1415926\ \ldots$	The value of π is known to a greater number of significant figures than any data you will ever use in a calculation.

The number π is now known to 1,011,196,691 digits. It is doubtful that you will need this much precision in this course—or ever.

Rule 2: When adding or subtracting, the number of decimal places in the answer should be equal to the number of decimal places in the number with the *fewest* places. Suppose you add three numbers:

0.12	2 significant figures	2 decimal places
1.6	2 significant figures	1 decimal place
+ 10.976	5 significant figures	3 decimal places
12.696		

This sum should be reported as 12.7, a number with one decimal place, because 1.6 has only one decimal place.

Rule 3: In multiplication or division, the number of significant figures in the answer should be the same as that in the quantity with the fewest significant figures.

$$\frac{0.1208}{0.0236} = 0.512,\ \text{or, in scientific notation,}\quad 5.12 \times 10^{-1}$$

Since 0.0236 has only three significant figures, while 0.01208 has four, the answer is limited to three significant figures.

Rule 4: When a number is rounded (the number of digits is reduced), the last digit retained is increased by 1 only if the following digit is 5 or greater. If there are no following digits or if all following digits are zeros, then increase the last digit by 1 if it is *odd* or leave the last digit unchanged if it is *even*. Thus, both 18.35 and 18.45 are rounded to 18.4.

Full number	Number rounded to three significant figures
12.696	12.7
16.249	16.2
18.35	18.4
18.351	18.4

One last word regarding significant figures and calculations. In working problems on a pocket calculator, you should do the calculation using all the digits allowed by the calculator and round only at the end of the problem. Rounding in the middle can introduce small errors. If your answers do not quite agree with those in the Appendices of this book, this may be the source of the disagreement.

Now let us consider a problem that is of practical importance and that makes use of all the rules. Suppose you discover that young children are eating chips of paint that flake off a wall in an old house. The paint contains 200. ppm lead (200. mg of Pb per kg of paint). Suppose that a child eats five such chips. How much lead has the child gotten from the paint?

As stated, this problem does not include enough information for a solution to be obtained; however, some reasonable assumptions can be made, and they can lead to experiments that could be used to obtain the necessary information. The statement does not say how big the paint chips are. Let's assume that they are 1.0 cm by 1.0 cm so that the area is 1.0 cm^2. Then eating five chips means eating 5.0 cm^2 of paint. (This assumption could be improved by measuring similar chips from the same place.) Since the concentration of lead is reported in units of mass of lead per mass of paint, we need to know the mass of 5.0 cm^2 of paint. This could be determined by measuring the areas of several paint chips and determining the mass of each. Suppose that the results of such measurements were

Mass of chip (mg)	Area of chip (cm^2)	Mass per unit area (mg/cm^2)
29.6	2.34	12.65
21.9	1.73	12.66
23.6	1.86	12.60

$$\text{Average mass per unit area} = \frac{(12.65 + 12.66 + 12.69)\ \text{mg/cm}^2}{3}$$

$$= 12.67\ \text{mg/cm}^2 = 12.7\ \text{mg/cm}^2$$

The average has been rounded to three significant figures because each experimental number has three significant figures. (Notice that more than three significant figures were kept in the intermediate calculations so as not to lose precision.) Now we can use this information to calculate how much lead the child has consumed.

$$m_{\text{paint}} = 5.0\ \text{cm}^2\ \text{paint} \left(\frac{12.67\ \text{mg paint}}{1\ \text{cm}^2\ \text{paint}} \right) \left(\frac{1\ \text{g}}{1000\ \text{mg}} \right) \left(\frac{1\ \text{kg}}{1000\ \text{g}} \right)$$

$$= 6.335 \times 10^{-5}\ \text{kg paint}$$

$$m_{\text{Pb}} = 6.335 \times 10^{-5}\ \text{kg paint} \left(\frac{200.\ \text{mg Pb}}{1\ \text{kg paint}} \right) = 1.267 \times 10^{-2}\ \text{mg Pb}$$

$$= 1.3 \times 10^{-2}\ \text{mg Pb} = 0.013\ \text{mg Pb}$$

Notice that the final result was rounded to two significant figures because there were only two significant figures in the initial area of the paint chip. This is quite adequate precision, however, for you to determine whether this quantity of lead is likely to be harmful to the child.

The methods of problem solving presented here have been developed over time and represent a good way of keeping track of the precision of results, the units in which those results were obtained, and the correctness of calculations. These methods are not the only way that such goals can be achieved, but they do work well. We recommend that you include units in all calculations and check that they cancel appropriately. It is also important not to overstate the precision of results by keeping too many significant figures. By solving many problems, you

The ppm unit stands for "parts per million." If a substance is present with a concentration of 1 ppm, there is 1 gram of the substance in 1 million grams of sample.

should be able to develop your problem-solving skills so that they become second nature and you can do them without thinking about the mechanics. This will allow you to devote all your thought to the logic of a problem solution.

A.4 ELECTRONIC CALCULATORS

The directions for calculator use in this section are given for calculators using "algebraic" logic. Such calculators are the most common type used by students in introductory courses. For calculators using RPN logic (such as those made by Hewlett-Packard), the procedure will differ slightly.

The advent of inexpensive electronic calculators has made calculations in introductory chemistry much more straightforward. You are well advised to purchase a calculator that has the capability of performing calculations in scientific notation, has both base 10 and natural logarithms, and is capable of raising any number to any power and of finding any root of any number. In the discussion below, we shall point out in general how these functions of your calculator can be used. You should practice using your calculator to carry out arithmetic operations and make certain that you are able to use all of the calculator's functions correctly.

Although electronic calculators have greatly simplified calculations, they have also forced us to focus again on significant figures. A calculator easily handles eight or more significant figures, but real laboratory data are rarely known to this accuracy. Therefore, you are urged to review Appendix A.3 on significant figures, precision, and rounding numbers.

The mathematical skills required to read and study this textbook successfully involve algebra, some geometry, scientific notation, logarithms, and solving quadratic equations. The next three sections review each of the last three topics.

A.5 EXPONENTIAL OR SCIENTIFIC NOTATION

In exponential or scientific notation, a number is expressed as a product of two numbers: $N \times 10^n$. The first number, N, is called the digit term and is a number between 1 and 10. The second number, 10^n, the exponential term, is some integer power of 10. For example, 1234 would be written in scientific notation as 1.234×10^3 or 1.234 multiplied by 10 three times.

$$1234 = 1.234 \times 10^1 \times 10^1 \times 10^1 = 1.234 \times 10^3$$

Conversely, a number less than 1, such as 0.01234, would be written as 1.234×10^{-2}. This notation tells us that 1.234 should be divided twice by 10 in order to obtain 0.01234.

$$0.01234 = \frac{1.234}{10^1 \times 10^1} = 1.234 \times 10^{-1} \times 10^{-1} = 1.234 \times 10^{-2}$$

Some other examples of scientific notation are

$$10000 = 1 \times 10^4 \qquad 12345 = 1.2345 \times 10^4$$
$$1000 = 1 \times 10^3 \qquad 1234 = 1.234 \times 10^3$$
$$100 = 1 \times 10^2 \qquad 123 = 1.23 \times 10^2$$
$$10 = 1 \times 10^1 \qquad 12 = 1.2 \times 10^1$$
$$1 = 1 \times 10^0 \qquad \text{(any number to the zeroth power} = 1)$$
$$1/10 = 1 \times 10^{-1} \qquad 0.12 = 1.2 \times 10^{-1}$$
$$1/100 = 1 \times 10^{-2} \qquad 0.012 = 1.2 \times 10^{-2}$$
$$1/1000 = 1 \times 10^{-3} \qquad 0.0012 = 1.2 \times 10^{-3}$$
$$1/10000 = 1 \times 10^{-4} \qquad 0.00012 = 1.2 \times 10^{-4}$$

When converting a number to scientific notation, notice that the exponent n is positive if the number is greater than 1 and negative if the number is less than 1. The value of n is the number of places by which the decimal was shifted to obtain the number in scientific notation.

$$1\underset{\curvearrowleft}{\,2\,3\,4\,5\,} . = 1.2345 \times 10^4$$

Decimal shifted 4 places to the left. Therefore, n is positive and equal to 4.

$$0.0\underset{\curvearrowright}{\,0\,1\,2} = 1.2 \times 10^{-3}$$

Decimal shifted 3 places to the right. Therefore, n is negative and equal to 3.

If you wish to convert a number in scientific notation to the usual form, the procedure above is simply reversed.

$$6\underset{\curvearrowright}{\,.\,2\,7\,}3 \times 10^2 = 627.3$$

Decimal point shifted 2 places to the right, since n is positive and equal to 2.

$$\underset{\curvearrowleft}{0\,0\,}6.273 \times 10^{-3} = 0.006273$$

Decimal point shifted 3 places to the left, since n is negative and equal to 3.

There are two final points to be made concerning scientific notation. First, if you are used to working on a computer you may be in the habit of writing a number such as 1.23×10^3 as 1.23E3, or 6.45×10^{-5} as 6.45E − 5. Second, some electronic calculators allow you to convert numbers readily to scientific notation. If you have such a calculator, you can change a number shown in the usual form to scientific notation by pressing an appropriate key or keys.

Usually you will handle numbers in scientific notation with a calculator, but if you need to work without a calculator, these rules are important to follow.

1. Adding and Subtracting

When adding or subtracting numbers in scientific notation without using a calculator, first convert the numbers to the same powers of ten. Then add or subtract the digit terms as appropriate.

$$(1.234 \times 10^{-3}) + (5.623 \times 10^{-2}) = (0.1234 \times 10^{-2}) + (5.623 \times 10^{-2})$$
$$= 5.746 \times 10^{-2}$$
$$(6.52 \times 10^2) - (1.56 \times 10^3) = (6.52 \times 10^2) - (15.6 \times 10^2)$$
$$= -9.1 \times 10^2$$

Note that in this calculation, there are only two significant figures in the result, although each of the original numbers had three. Subtracting two numbers that are nearly the same can reduce the number of significant figures appreciably.

2. Multiplying

The digit terms are multiplied in the usual manner, and the exponents are added algebraically. The result is expressed with a digit term with only one nonzero digit to the left of the decimal.

$$(1.23 \times 10^3)(7.60 \times 10^2) = (1.23)(7.60) \times 10^{3+2}$$
$$= 9.35 \times 10^5$$
$$(6.02 \times 10^{23})(2.32 \times 10^{-2}) = (6.02)(2.32) \times 10^{23-2}$$
$$= 13.966 \times 10^{21} = 14.0 \times 10^{21}$$
$$= 1.40 \times 10^{22} \text{ (rounded to 3 significant figures)}$$

3. Dividing

The digit terms are divided in the usual manner, and the exponents are subtracted algebraically. The quotient is written with one nonzero digit to the left of the decimal in the digit term.

$$\frac{7.60 \times 10^3}{1.23 \times 10^2} = \frac{7.60}{1.23} \times 10^{3-2} = 6.18 \times 10^1$$

$$\frac{6.02 \times 10^{23}}{9.10 \times 10^{-2}} = \frac{6.02}{9.10} \times 10^{(23)-(-2)} = 0.662 \times 10^{25} = 6.62 \times 10^{24}$$

4. Raising Numbers in Scientific Notation to Powers

When raising a number in scientific notation to a power, treat the digit term in the usual manner. The exponent is then multiplied by the number indicating the power.

$$(1.25 \times 10^3)^2 = (1.25)^2 \times 10^{3 \times 2}$$
$$= 1.5625 \times 10^6 = 1.56 \times 10^6$$
$$(5.6 \times 10^{-10})^3 = (5.6)^3 \times 10^{(-10) \times 3}$$
$$= 175.6 \times 10^{-30} = 1.8 \times 10^{-28}$$

Electronic calculators usually have two methods of raising a number to a power. To square a number, enter the number and then press the "x^2" key. To raise a number to any power, use the "y^x" key. For example, to raise 1.42×10^2 to the 4th power,

(a) Enter 1.42×10^2.
(b) Press "y^x".
(c) Enter 4 (this should appear on the display).
(d) Press " = " and 4.0658 . . . $\times 10^8$ will appear on the display. (The number of digits will depend upon the calculator.)

As a final step, express the number in the correct number of significant figures (4.07×10^8 in this case).

5. Taking Roots of Numbers in Scientific Notation

Unless you use an electronic calculator, the number must first be put into a form in which the exponential is exactly divisible by the root. The root of the digit term is found in the usual way, and the exponent is divided by the desired root.

$$\sqrt{3.6 \times 10^7} = \sqrt{36 \times 10^6} = \sqrt{36} \times \sqrt{10^6} = 6.0 \times 10^3$$
$$\sqrt[3]{2.1 \times 10^{-7}} = \sqrt[3]{210 \times 10^{-9}} = \sqrt[3]{210} \times \sqrt[3]{10^{-9}} = 5.9 \times 10^{-3}$$

To take a square root on an electronic calculator, enter the number and then press the "$\sqrt{x}$" key. To find a higher root of a number, such as the fourth root of 5.6×10^{-10},

On some calculators steps (a) and (c) may be interchanged.

(a) Enter the number, 5.6×10^{-10} in this case.
(b) Press the "$\sqrt[x]{y}$" key. (On most calculators, the sequence you actually use is to press "2ndF" and then "$\sqrt[x]{y}$." Alternatively, you press "INV" and then "y^x.")
(c) Enter the desired root, 4 in this case.
(d) Press "=". The answer here is 4.8646×10^{-3} or 4.9×10^{-3}.

A general procedure for finding any root is to use the "y^x" key. For a square root, x is 0.5 (or $\frac{1}{2}$), whereas it is 0.33 (or $\frac{1}{3}$) for a cube root, 0.25 (or $\frac{1}{4}$) for a fourth root, and so on.

A.6 LOGARITHMS

There are two types of logarithms used in this text: common logarithms (abbreviated log) whose base is 10, and natural logarithms (abbreviated ln) whose base is e ($= 2.71828$).

$$\log x = n \qquad\qquad \text{where } x = 10^n$$

$$\ln x = m \qquad\qquad \text{where } x = e^m$$

Logarithms to the base 10 are needed when dealing with pH.

Most equations in chemistry and physics were developed in natural or base e logarithms, and this practice is followed in this text. The relation between log and ln is

$$\ln x = 2.303 \log x$$

Aside from the different bases of the two logarithms, they are used in the same manner. What follows is largely a description of the use of common logarithms.

A common logarithm is the power to which you must raise 10 to obtain the number. For example, the log of 100 is 2, since you must raise 10 to the second power to obtain 100. Other examples are

$$\log 1000 = \log (10^3) = 3$$

$$\log 10 = \log (10^1) = 1$$

$$\log 1 = \log (0^0) = 0$$

$$\log 1/10 = \log (10^{-1}) = -1$$

$$\log 1/10000 = \log (10^{-4}) = -4$$

To obtain the common logarithm of a number other than a simple power of 10, you must resort to a log table or an electronic calculator. For example,

$$\log 2.10 = 0.3222, \text{ which means that } 10^{0.3222} = 2.10$$

$$\log 5.16 = 0.7126, \text{ which means that } 10^{0.7126} = 5.16$$

$$\log 3.125 = 0.49485, \text{ which means that } 10^{0.49485} = 3.125$$

To check this on your calculator, enter the number and then press the "log" key. When using a log table, the logs of the first two numbers above can be read directly from the table. The log of the third number (3.125), however, must be interpolated. That is, 3.125 is midway between 3.12 and 3.13, so the log is midway between 0.4942 and 0.4955.

To obtain the natural logarithm, ln, of the numbers above, use a calculator having this function. Enter each number and press "ln."

$$\ln 2.10 = 0.7419, \text{ which means that } e^{0.7419} = 2.10$$

$$\ln 5.16 = 1.6409, \text{ which means that } e^{1.6409} = 5.16$$

To find the common logarithm of a number greater than 10 or less than 1 with a log table, first express the number in scientific notation. Then find the log of each part of the number and add the logs. For example,

Nomenclature of Logarithms: The number to the left of the decimal in a logarithm is called the *characteristic,* and the number to the right of the decimal is the *mantissa.*

$$\log 241 = \log (2.41 \times 10^2) = \log 2.41 + \log 10^2$$

$$= 0.382 + 2 = 2.382$$

$$\log 0.00573 = \log (5.73 \times 10^{-3}) = \log 5.73 + \log 10^{-3}$$

$$= 0.758 + (-3) = -2.242$$

Significant Figures and Logarithms. Notice that the mantissa (digits to the right of the decimal point) has as many significant figures as the number whose log was found. (So that you could more clearly see the result obtained with a calculator or a table, this rule was not strictly followed until the last two examples.)

Obtaining Antilogarithms. If you are given the logarithm of a number and need to find the number from it, you need to obtain the "antilogarithm" or "antilog" of the number. There are two common procedures used by electronic calculators to do this:

Procedure A

(a) Enter the log or ln (a number).

(b) Press 2ndF.

(c) Press 10^x or e^x.

Procedure B

(a) Enter the log or ln (a number).

(b) Press INV.

(c) Press log or ln x.

Test one or the other of these procedures with the following examples:

Example 1 Find the number whose log is 5.234.

Recall that log $x = n$ where $x = 10^n$. In this case n = 5.234. Enter that number in your calculator and find the value of 10^n, the antilog. In this case,

$$10^{5.234} = 10^{0.234} \times 10^5 = 1.71 \times 10^5$$

Notice that the characteristic (5) sets the decimal point; it is the power of 10 in the exponential form. The mantissa (0.234) gives the value of the number x. Thus, if you use a log table to find x, you need only look up 0.234 in the table and see that it corresponds to 1.71.

Example 2 Find the number whose log is –3.456.

$$10^{-3.456} = 10^{0.544} \times 10^{-4} = 3.50 \times 10^{-4}$$

Notice here that -3.456 must be expressed as the sum of -4 and $+0.544$.

Mathematical Operations Using Logarithms. Because logarithms are exponents, operations involving them follow the same rules as the use of exponents. Thus, multiplying two numbers can be done by adding logarithms.

$$\log xy = \log x + \log y$$

For example, we multiply 563 by 125 by adding their logarithms and finding the antilogarithm of the result.

$$\log 563 = 2.751$$

$$\log 125 = 2.097$$

$$\log (563 \times 125) = 2.751 + 2.097 = 4.848$$

$$563 \times 125 = 10^{4.848} = 10^4 \times 10^{0.848} = 7.05 \times 10^4$$

One number (x) can be divided by another (y) by subtraction of their logarithms.

$$\log \frac{x}{y} = \log x - \log y$$

For example, to divide 125 by 742,

$$\log 125 = 2.097$$

$$\log 742 = 2.870$$

$$\log (125/742) = 2.097 - 2.870 = -0.773$$

$$125/742 = 10^{-0.773} = 10^{0.227} \times 10^{-1} = 1.69 \times 10^{-1}$$

Similarly, powers and roots of numbers can be found using logarithms.

$$\log x^y = y(\log x)$$

$$\log \sqrt[y]{x} = \log x^{1/y} = \frac{1}{y} \log x$$

As an example, find the fourth power of 5.23. We first find the log of 5.23 and then multiply it by 4. The result, 2.874, is the log of the answer. Therefore, we find the antilog of 2.874.

$$(5.23)^4 = ?$$

$$\log (5.23)^4 = 4 \log 5.23 = 4(0.719) = 2.874$$

$$(5.23)^4 = 10^{2.874} = 748$$

As another example, find the fifth root of 1.89×10^{-9}.

$$\sqrt[5]{1.89 \times 10^{-9}} = (1.89 \times 10^9)^{1/5} = ?$$

$$\log (1.89 \times 10^{-9})^{1/5} = \frac{1}{5} \log (1.89 \times 10^{-9}) = \frac{1}{5}(-8.724) = -1.745$$

The answer is the antilog of –1.745.

$$(1.89 \times 10^{-9})^{1/5} = 10^{-1.745} = 1.80 \times 10^{-2}$$

A.7 QUADRATIC EQUATIONS

Algebraic equations of the form $ax^2 + bx + c = 0$ are called **quadratic equations.** The coefficients a, b, and c may be either positive or negative. The two roots of the equation may be found using the *quadratic formula.*

$$x = \frac{-b \pm \sqrt{b^2 - 4ac}}{2a}$$

As an example, solve the equation $5x^2 - 3x - 2 = 0$. Here $a = 5$, $b = -3$, and $c = -2$. Therefore,

$$x = \frac{3 \pm \sqrt{(-3)^2 - 4(5)(-2)}}{2(5)}$$

$$= \frac{3 \pm \sqrt{9 - (-40)}}{10} = \frac{3 \pm \sqrt{49}}{10} = \frac{3 \pm 7}{10}$$

$$x = 1 \text{ and } -0.4$$

How do you know which of the two roots is the correct answer? You have to decide in each case which root has physical significance. However, it is usually true in this course that negative values are not significant.

When you have solved a quadratic expression, you should always check your values by substitution into the original equation. In the example above, we find that $5(1)^2 - 3(1) - 2 = 0$ and that $5(-0.4)^2 - 3(-0.4) - 2 = 0$.

The most likely place you will encounter quadratic equations is in the chapters on chemical equilibria, particularly in Chapters 14, 16, and 17. Here you may be faced with solving an equation such as

$$1.8 \times 10^{-4} = \frac{x^2}{0.0010 - x}$$

This equation can certainly be solved by using the quadratic equation (to give $x = 3.4 \times 10^{-4}$). However, you may find the method of successive approximations to be especially convenient. Here you begin by making a reasonable approximation of x. This approximate value is substituted into the original equation, and this is solved to give what is hoped to be a more correct value of x. This process is repeated until the answer converges on a particular value of x, that is, until the value of x derived from two successive approximations is the same.

Step 1: First assume that x is so small that $(0.0010 - x) \approx 0.0010$. This means that

$$x^2 = 1.8 \times 10^{-4}(0.0010)$$

$$x = 4.2 \times 10^{-4} \text{ (to 2 significant figures)}$$

Step 2: Substitute the value of x from Step 1 into the denominator (but not the numerator) of the original equation and again solve for x.

$$x^2 = (1.8 \times 10^{-4})(0.0010 - 0.00042)$$

$$x = 3.2 \times 10^{-4}$$

Step 3: Repeat Step 2 using the value of x found in that step.

$$x = \sqrt{1.8 \times 10^{-4}(0.0010 - 0.00032)} = 3.5 \times 10^{-4}$$

Step 4: Continue by repeating the calculation, using the value of x found in the previous step.

Step 5. $x = \sqrt{1.8 \times 10^{-4}(0.0010 - 0.00034)} = 3.4 \times 10^{-4}$

Here we find that iterations after the fourth step give the same value for x, indicating that we have arrived at a valid answer (and the same one obtained from the quadratic formula).

Some final thoughts on using the method of successive approximations: First, there are cases where the method does not work. Successive steps may give answers that are random or that diverge from the correct value. For quadratic equations of the form $K = x^2/(C - x)$, the method of approximations will work only as long as $K < 4C$ (assuming one begins with $x = 0$ as the first guess; that is, $K \approx x^2/C$). This will always be true for weak acids and bases.

Second, values of K in the equation $K = x^2/(C - x)$ are usually known only to two significant figures. Therefore, we are justified in carrying out successive steps until two answers are the same to two significant figures.

Finally, we highly recommend this method of solving quadratic equations. If your calculator has a memory function, successive approximations can be carried out easily and rapidly.

Units, Equivalences, and Conversion Factors

B.1 UNITS OF THE INTERNATIONAL SYSTEM (SI)

Contents
B.1 Units of the International System (SI) *A.17*
B.2 Conversion of Units for Physical Quantities *A.19*

The metric system was begun by the French National Assembly in 1790 and has undergone many modifications. The International System of Units, or *Systeme International* (SI), which represents an extension of the metric system, was adopted by the 11th General Conference on Weights and Measures in 1960. It is constructed from seven base units, each of which represents a particular physical quantity (Table B.1). More information about the SI is available at **http://physics.nist.gov/cuu/Units/index.html.**

The first five units listed in Table B.1 are particularly useful in chemistry. They are defined as follows:

1. The *meter* is the length of the path traveled by light in a vacuum during a time interval of 1/299,792,458 of a second.
2. The *kilogram* represents the mass of a platinum-iridium block kept at the International Bureau of Weights and Measures in Sevres, France.
3. The *second* is the duration of 9,192,631,770 periods of a certain line in the microwave spectrum of cesium-133.
4. The *kelvin* is 1/273.16 of the temperature interval between absolute zero and the triple point of water (the temperature at which liquid water, ice, and water vapor coexist).
5. The *mole* is the amount of substance that contains as many elementary entities (atoms, molecules, ions, or other particles) as there are atoms in exactly 0.012 kg of carbon-12 (12 g of ^{12}C atoms).

Decimal fractions and multiples of metric and SI units are designated by using the **prefixes** listed in Table B.2. The prefix *kilo-*, for example, means that a unit is multiplied by 10^3,

$$1 \text{ kilogram} = 1 \times 10^3 \text{ grams} = 1000 \text{ grams}$$

and the prefix *centi-* means that the unit is multiplied by the factor 10^{-2}:

$$1 \text{ centigram} = 1 \times 10^{-2} \text{ gram} = 0.01 \text{ gram}$$

The prefixes are added to give units of a magnitude appropriate to what is being measured. The distance from New York to London (5.6×10^3 km = 5600 km) is much easier to comprehend measured in kilometers than in meters (5.6 ×

TABLE B.1 SI Fundamental Units

Physical quantity	Name of unit	Symbol
Length	Meter	m
Mass	Kilogram	kg
Time	Second	s
Temperature	Kelvin	K
Amount of substance	Mole	mol
Electric current	Ampere	A
Luminous intensity	Candela	cd

TABLE B.2	Prefixes for Metric and SI Units*				
Factor	**Prefix**	**Symbol**	**Factor**	**Prefix**	**Symbol**
10^{12}	tera-	T	10^{-1}	*deci-*	d
10^{9}	giga-	G	10^{-2}	*centi-*	c
10^{6}	mega-	M	10^{-3}	*milli-*	m
10^{3}	*kilo-*	k	10^{-6}	micro-	μ
10^{3}	hecto-	h	10^{-9}	*nano-*	n
10^{1}	deka-	da	10^{-12}	*pico-*	p
			10^{-15}	femto-	f
			10^{-18}	atto-	a

*The most common prefixes are shown in italics.

10^6 m = 5,600,000 m). The following is a list of units for measuring very small and very large distances:

picometer (pm)	0.000000000001 meter
nanometer (nm)	0.000000001 meter
micrometer (μm)	0.000001 meter
millimeter (mm)	0.001 meter
centimeter (cm)	0.01 meter
decimeter (dm)	0.1 meter
meter (m)	1 meter
dekameter (dam)	10 meters
hectometer (hm)	100 meters
kilometer (km)	1000 meters
megameter (Mm)	1,000,000 meters

TABLE B.3	Derived SI Units			
Physical quantity	**Name of unit**	**Symbol**	**Definition**	**Symbol**
Area	Square meter	m^2	—	
Volume	Cubic meter	m^3	—	
Density	Kilogram per cubic meter	kg/m^3	—	
Force	Newton	N	(kilogram)(meter)/(second)^2	$kg\ m/s^2$
Pressure	Pascal	Pa	newton/square meter	N/m^2
Energy	Joule	J	(kilogram)(meter)^2/(second)^2	$kg\ m^2/s^2$
Electric charge	Coulomb	C	(ampere)(second)	A s
Electric potential difference	Volt	V	joule/(ampere)(second)	J/(A s)

In the International System of Units, all physical quantities are represented by appropriate combinations of the base units listed in Table B.1. The result is a derived unit for each kind of measured quantity. The most common derived units are listed in Table B.3. It is easy to see that the derived unit for area is length $\times$ length = meter $\times$ meter = square meter, m^2, or that the derived unit for volume is length $\times$ length $\times$ length = meter $\times$ meter $\times$ meter = cubic meter, m^3. The more complex derivations are arrived at by the same kind of combination of units. Units such as the one for force (the *newton*) have been given simple names that represent the units by which they are defined.

B.2   CONVERSION OF UNITS FOR PHYSICAL QUANTITIES

The result of a measurement is a physical quantity, which consists of a number and a unit. Algebraically, a physical quantity can be treated as if the number is multiplied by the unit. To convert a physical quantity from one unit of measure to another requires a conversion factor (proportionality factor) based on equivalences between units of measure such as those given in Table B.4. (See Appendix A.2 for more about physical quantities and proportionality factors.) Each equivalence provides two conversion factors that are the reciprocals of each other. For example, the equivalence between a quart and a liter, 1 quart = 0.9463 liter, gives

$$\frac{1 \text{ quart}}{0.9463 \text{ liter}} \qquad \text{There is 1 quart per 0.9463 liter.}$$

$$\frac{0.9463 \text{ liter}}{1 \text{ quart}} \qquad \text{There is 0.9463 liter per 1 quart.}$$

The method of canceling units described in Appendix A.2 provides the basis for choosing which conversion factor is needed: It is always the one that allows the unit being converted to be canceled and leaves the new unit uncanceled.

To convert 2 quarts to liters:

$$2 \text{ quarts } \times \frac{0.9463 \text{ liter}}{1 \text{ quart}} = 1.893 \text{ liters}$$

To convert 2 liters to quarts:

$$2 \text{ liters } \times \frac{1 \text{ quart}}{0.9463 \text{ liter}} = 2.113 \text{ quarts}$$

Because of the definitions of Celsius degrees and Fahrenheit degrees, conversions between these temperature scales are a bit more complicated. Both units are based on the properties of water. The Celsius unit is defined by assigning 0 °C as the freezing point of pure water and 100 °C as its boiling point, when the pressure is exactly 1 atm. The size of the Fahrenheit degree is equally arbitrary. Fahrenheit defined 0 °F as the freezing point of a solution in which he had dissolved the maximum quantity of ammonium chloride (because this was the lowest temperature he could reproduce reliably), and he intended 100 °F to be the normal human body temperature (but this turned out to be 98.6 °F). Today, the reference points are set at exactly 32 °F and 212 °F (the freezing and boiling points of pure water, at 1 atm). The number of units between these two Fahrenheit temperatures is 180 °F. Thus, the Celsius degree is almost twice as large as the Fahrenheit degree; it takes only 5 Celsius degrees to cover the same temperature range as 9 Fahrenheit degrees.

To be entirely correct, we must specify that water boils at 100 °C and freezes at 0 °C only when the pressure of the surrounding atmosphere is 1 atm.

$$\frac{100 \text{ °C}}{180 \text{ °F}} = \frac{5 \text{ °C}}{9 \text{ °F}}$$

Mass and weight

1 pound = 453.59 grams = 0.45359 kilogram

1 kilogram = 1000 grams = 2.205 pounds

1 gram = 10 decigrams = 100 centigrams = 1000 milligrams

1 gram = 6.022×10^{23} atomic mass units

1 atomic mass unit = 1.6605×10^{-24} gram

1 short ton = 2000 pounds = 907.2 kilograms

1 long ton = 2240 pounds

1 metric tonne = 1000 kilograms = 2205 pounds

Length

1 inch = 2.54 centimeters (exactly)

1 mile = 5280 feet = 1.609 kilometers

1 yard = 36 inches = 0.9144 meter

1 meter = 100 centimeters = 39.37 inches = 3.281 feet = 1.094 yards

1 kilometer = 1000 meters = 1094 yards = 0.6215 mile

1 Angstrom = 1.0×10^{-8} centimeter = 0.10 nanometer = 100 picometers
$= 1.0 \times 10^{-10}$ meter = 3.937×10^{-9} inch

Volume

1 quart = 0.9463 liter

1 liter = 1.0567 quarts

1 liter = 1 cubic decimeter = 1000 cubic centimeters = 0.001 cubic meter

1 milliliter = 1 cubic centimeter = 0.001 liter = 1.056×10^{-3} quart

1 cubic foot = 28.316 liters = 29.924 quarts = 7.481 gallons

Force and pressure

1 atmosphere = 760 millimeters of mercury = 1.03×10^5 pascals
= 14.70 pounds per square inch

1 bar = 10^5 pascals = 0.98692 atmosphere

1 torr = 1 millimeter of mercury

1 pascal = $1 \text{ kg m}^{-1} \text{ s}^{-2}$ = 1 N/m^2

Energy

1 joule = 1×10^7 ergs

1 thermochemical calorie* = 4.184 joules = 4.184×10^7 ergs
$= 4.129 \times 10^{-2}$ liter-atmospheres
$= 2.612 \times 10^{19}$ electron volts

1 erg = 1×10^{-7} joule = 2.3901×10^{-8} calorie

1 electron volt = 1.6022×10^{-19} joule = 1.6022×10^{-12} erg = 96.85 kJ/mol†

1 liter-atmosphere = 24.217 calories = 101.32 joules = 1.0132×10^9 ergs

1 British thermal unit = 1055.06 joules = 1.05506×10^{10} ergs = 252.2 calories

Temperature

0 K = −273.15 °C

If T_K is the numerical value of the temperature in kelvins, t_C the numerical value of the temperature in °C, and t_F the numerical value of the temperature in °F, then

$$T_K = t_C + 273.15$$
$$t_C = (\tfrac{5}{9})(t_F - 32)$$
$$t_F = (\tfrac{9}{5})t_C + 32$$

*The heating required to raise the temperature of one gram of water from 14.5 °C to 15.5 °C.

†Note that the other units in this line are per particle and must be multiplied by 6.022×10^{23} to be strictly comparable.

This relationship is the basis for converting a temperature on one scale to a temperature on the other. If t_C is the numerical value of the temperature in °C and t_F the numerical value of the temperature in °F, then

$$t_C = (\tfrac{5}{9})(t_F - 32)$$

$$t_F = (\tfrac{9}{5})t_C + 32$$

For example, to show that your normal body temperature of 98.6 °F corresponds to 37.0 °C, use the first equation.

$$t_C = (\tfrac{5}{9})(t_F - 32) = (\tfrac{5}{9})(98.6 - 32) = (\tfrac{5}{9})(66.6) = 37.0$$

Thus, body temperature in °C = 37.0 °C.

Laboratory work is almost always done using Celsius units, and we rarely need to make conversions to and from Fahrenheit. It is best to try to calibrate your senses to Celsius units; to help you do this, it is useful to know that water freezes at 0 °C, a comfortable room temperature is about 22 °C, your body temperature is 37 °C, and the hottest water you could stand to leave your hand in for some time is about 60 °C.

Physical Constants*

Quantity	Symbol	Traditional units	SI units
Acceleration of gravity	g_n	980.665 cm/s^2	$9.806\ 65 \text{ m/s}^2$
Atomic mass unit ($\frac{1}{12}$ the mass of ^{12}C atom)	amu or u	$1.660\ 538\ 73 \times 10^{-24}$ g	$1.660\ 538\ 73 \times 10^{-27}$ kg
Avogadro constant	N_A, L	$6.022\ 141\ 99 \times 10^{23}$ particles/mol	$6.022\ 141\ 99 \times 10^{23}$ particles/mol
Bohr radius	a_o	$0.529\ 177\ 208\ 3$ Å	$5.291\ 772\ 083 \times 10^{-11}$ m
Boltzmann constant	k	$1.380\ 650\ 3 \times 10^{-16}$ erg/K	$1.380\ 650\ 3 \times 10^{-23}$ J/K
Charge-to-mass ratio of electron	e/m	$-1.758\ 820\ 174 \times 10^8$ C/g	$-1.758\ 820\ 174 \times 10^{11}$ C/kg
Elementary charge (electron or proton charge)	e	$1.602\ 176\ 462 \times 10^{-19}$ C	$1.602\ 176\ 462 \times 10^{-19}$ C
Electron rest mass	m_e	$9.109\ 381\ 88 \times 10^{-28}$ g	$9.109\ 381\ 88 \times 10^{-31}$ kg
Faraday constant	F	$96\ 485.3415 \text{ C/mol e}^-$ $23.06 \text{ kcal V}^{-1} \text{ mol}^{-1}$	$96\ 485.3415 \text{ C/mol e}^-$ $96\ 485 \text{ J V}^{-1} \text{ mol}^{-1}$
Gas constant	R	$0.082\ 057 \text{ L atm mol}^{-1} \text{ K}^{-1}$ $1.987 \text{ cal mol}^{-1} \text{ K}^{-1}$	$8.314\ 472 \text{ dm}^3 \text{ Pa mol}^{-1} \text{ K}^{-1}$ $8.314\ 472 \text{ J mol}^{-1} \text{ K}^{-1}$
Molar volume (STP)	V_m	22.414 L/mol	$22.414 \times 10^{-3} \text{ m}^3\text{/mol}$ $22.414 \text{ dm}^3\text{/mol}$
Neutron rest mass	m_n	$1.674\ 927\ 16 \times 10^{-24}$ g $1.008\ 664\ 916$ amu	$1.674\ 927\ 16 \times 10^{-27}$ kg
Planck's constant	h	$6.626\ 068\ 76 \times 10^{-27}$ erg s	$6.626\ 068\ 76 \times 10^{-34}$ J s
Proton rest mass	m_p	$1.672\ 621\ 58 \times 10^{-24}$ g $1.007\ 276\ 467$ amu	$1.672\ 621\ 58 \times 10^{-27}$ kg
Rydberg constant	R_∞	$3.289\ 841\ 960\ 368 \times 10^{15} \text{ s}^{-1}$ $2.179\ 871\ 90 \times 10^{-11}$ erg	$1.097\ 373\ 156\ 854\ 9 \times 10^7 \text{ m}^{-1}$ $2.179\ 871\ 90 \times 10^{-18}$ J
Velocity of light (in a vacuum)	c, c_0	$2.997\ 924\ 58 \times 10^{10}$ cm/s $186\ 282$ mile/s	$2.997\ 924\ 58 \times 10^8$ m/s

*Data from the National Institute for Standards and Technology reference on constants, units, and uncertainty, **http://physics.nist.gov/cuu/Constants/index.html.**

Molecular Orbitals

In Chapter 8, we used electron-pair bonds to explain bonding in molecules, accounting qualitatively for the stability of the covalent bond in terms of the overlap of atomic orbitals. In Section 9.3 *(⇐ p. 371)* we introduced the *valence bond model* and hybridization to rationalize the molecular geometries predicted by electron-pair repulsion and VSEPR theory *(⇐ p. 361, Section 9.2)*. Where more than one Lewis structure could be drawn for the same arrangement of atomic nuclei, as in SO_2, the concept of resonance *(⇐ p. 343, Section 8.9)* was used to account for observed properties of molecules.

A major weakness of the theories presented in Chapters 8 and 9 is that they do not always correctly predict the magnetic properties of substances. An important example is O_2, which is paramagnetic. This means that O_2 molecules must have unpaired electrons. Diatomic oxygen has an even number of valence electrons (12), and the octet rule predicts that all of these electrons should be paired. According to valence bond theory, O_2 should be diamagnetic, but it is not.

Contents

D.1 Molecular Orbitals   *A.23*
D.2 Molecular Orbitals for Diatomic Molecules   *A.24*
D.3 Polyatomic Molecules: Delocalized π Electrons   *A.27*

D.1 MOLECULAR ORBITALS

This discrepancy between experiment and theory for O_2 (and many others) can be resolved by using an alternative model of covalent bonding, the molecular orbital (MO) approach. *Molecular orbital theory* treats bonding in terms of orbitals that can extend over an entire molecule. The orbitals are not confined to two atoms at a time. The MO approach involves three basic operations.

Step 1: Combine the valence atomic orbitals of all atoms in the molecule to give a new set of MOs characteristic of the molecule as a whole. *The number of MOs formed is equal to the number of atomic orbitals combined.* When two H atoms combine to form H_2, two *s* orbitals, one from each atom, yield two molecular orbitals. For O_2 there are one *s* orbital and three *p* orbitals in the valence shell of each of the two atoms. This gives eight atomic orbitals, which combine to give eight MOs.

Step 2: Arrange the MOs in order of increasing energy. The relative energies of MOs are usually deduced from experiments involving spectra and magnetic properties.

Step 3: Distribute the *valence electrons* of the molecule among the available MOs, filling the lowest energy MO first and continuing to build up the electron configuration of the molecule in the same way that electron configurations of atoms are built up *(⇐ p. 286, Section 7.6)*. As electrons are added,

(a) *Each MO can hold a maximum of two electrons.* In a filled MO the two electrons have opposed spins, in accordance with the Pauli principle *(⇐ p. 285, Section 7.5)*.

(b) *Electrons go into the lowest energy MO available.* A higher energy orbital starts to fill only when each orbital below it has its quota of two electrons.

(c) *Hund's rule is obeyed.* When two orbitals of equal energy are available to two electrons, one electron goes into each, giving two half-filled orbitals.

D.2   MOLECULAR ORBITALS FOR DIATOMIC MOLECULES

To illustrate molecular orbital theory, we apply it to the diatomic molecules of the elements in the first two periods of the periodic table.

Hydrogen and Helium (Combining Two 1s Orbitals)

When two hydrogen atoms (or two helium atoms) come close together, two 1s orbitals overlap and combine to give two MOs. One MO has an energy lower than that of the atomic orbitals from which it was formed; the other has higher energy (Figure D.1). A molecule that has two electrons in the lower energy MO is lower in energy (more stable) than the isolated atoms. That lowering of energy corresponds to the bond energy. For that reason the lower energy MO in Figure D.1(a) is called a *bonding molecular orbital*. If electrons are placed in the higher energy MO, the molecule's energy is higher than the energy of the isolated atoms. This is an unstable situation that is the opposite of bonding; the higher energy MO is called an *antibonding molecular orbital.*

The electron density in these MOs is shown in part (b) of Figure D.1. The bonding MO has higher electron density between the nuclei than the sum of the individual atomic orbitals. This attracts the nuclei together and forms the bond. In the antibonding orbital, the electron density between the nuclei is smaller than it would have been in the atoms alone. The positively charged nuclei have less electron "glue" to hold them together, they repel each other, and the molecule flies apart — the opposite of a bond.

The electron density in both MOs is symmetrical about the axis between the two nuclei. This means that both of these are sigma orbitals. In MO notation, the 1s bonding orbital is designated as σ_{1s}. The antibonding orbital is given the symbol σ_{1s}^*. An asterisk designates an antibonding orbital.

In the H_2 molecule, there are two 1s electrons. They fill the σ_{1s} orbital, giving a single bond. If an He_2 molecule could form, there would be four electrons — two from each atom. These would fill the bonding and antibonding orbitals. One bond and one antibond give a *bond order* (number of bonds) of zero in He_2. The bond order is calculated as

$$\text{Bond order} = \text{number of bonds} = \frac{n_B - n_A}{2}$$

where n_B is the number of electrons in bonding orbitals, n_A is the number of electrons in antibonding orbitals, and dividing by 2 accounts for the fact that two electrons are needed for a bond. In H_2, $n_B = 2$ and $n_A = 0$, so we have one bond. For He_2, $n_B = n_A = 2$, so the number of bonds is zero. The MO theory predicts that the He_2 molecule should not exist, and it does not.

Figure D.1   Molecular orbital formation.   Two MOs are formed by combining two 1s atomic orbitals. (a) Energy level diagram. (b) Orbital overlap diagram.

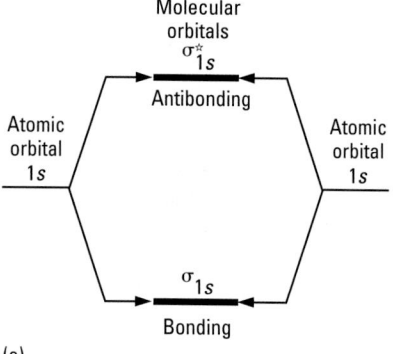

(a)

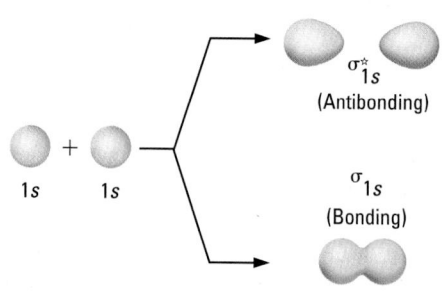

(b)

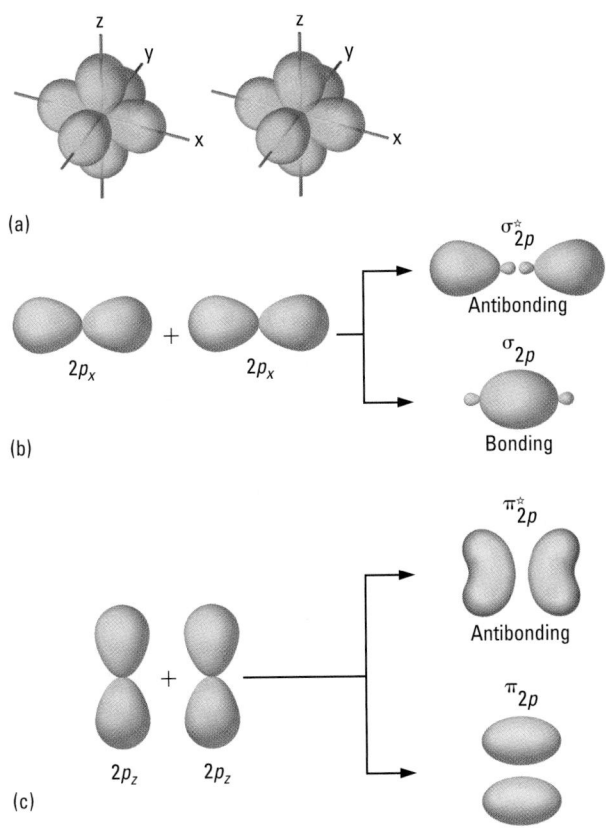

(a)

(b)

(c)

Figure D.2 **Forming molecular orbitals from** p **orbitals.** When p orbitals from two different atoms (a) overlap, there are two quite different possibilities. If they overlap head to head (b), two σ MOs are produced. If, on the other hand, they overlap side to side (c), two π MOs result. In this example, both p_z and p_y orbitals can overlap to form π MOs. Only p_z overlap is shown.

Second-Period Elements (Combining 2s and 2p Orbitals)

Three of the elements in the second period form familiar diatomic molecules: N_2, O_2, and F_2. Less common, but also known, are Li_2, B_2, and C_2, which have been observed as gases. The molecules Be_2 and Ne_2 are either highly unstable or nonexistent. To see what MO theory predicts about the stability of diatomic molecules from the second period, consider the valence atomic orbitals, $2s$ and $2p$.

Combining two $2s$ atomic orbitals, one from each atom, gives two MOs. These are very similar to the ones shown in Figure D.1. They are designated as σ_{2s} (sigma, bonding, 2s) and σ_{2s}^* (sigma, antibonding, 2s).

In an isolated atom, there are three $2p$ orbitals, oriented at right angles to each other. We call these atomic orbitals, p_x, p_y, and p_z (Figure D.2a). Assume that two atoms approach along the x axis. The two p_x atomic orbitals overlap head to head to form two orbitals that are symmetric around the line connecting the two nuclei (Figure D.2b). That is, they form a sigma bonding orbital, σ_{2p}, and a sigma antibonding orbital, σ_{2p}^*.

The situation is quite different when the p_z orbitals overlap. Because they are oriented parallel to one another, they overlap side to side (Figure D.2c). The two MOs formed in this case are pi orbitals; one is a bonding MO, π_{2p}, the other an anti-bonding MO, π_{2p}^*. Similarly, the p_y orbitals of the two atoms interact to form another pair of pi MOs, π_{2p} and π_{2p}^*, which are not shown in Figure D.2.

The relative energies of the MOs available for occupancy by the valence electrons of diatomic molecules formed from second-period atoms are shown in Figure D.3. To obtain the MO structure of the diatomic molecules of the elements in the second period, we fill the available MOs in order of increasing energy. The results are shown in Table D.1. Note that the MO theory correctly predicts the number of unpaired electrons in each molecule.

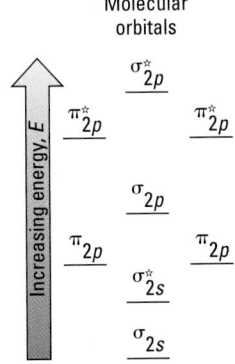

Molecular orbitals

Figure D.3 **Relative order of filling molecular orbitals.** The energies of MOs formed by combining $2s$ and $2p$ atomic orbitals increase from bottom to top of the diagram. This order of energies applies to Li_2 through N_2, but for O_2 and F_2 the energy of the σ_{2p} orbital drops below the energy of the two π_{2p} orbitals. This does not affect the filling order, because in O_2 and F_2 the σ_{2p} and π_{2p} orbitals are filled.

TABLE D.1	Predicted and Observed Properties of Diatomic Molecules of Second-Period Elements

Occupancy of orbitals

	σ_{2s}	σ_{2s}^*	π_{2p}	π_{2p}	σ_{2p}	π_{2p}^*	π_{2p}^*	σ_{2s}^*
Li_2	(↑↓)	( )	( )	( )	( )	( )	( )	( )
Be_2	(↑↓)	(↑↓)	( )	( )	( )	( )	( )	( )
B_2	(↑↓)	(↑↓)	(↑)	(↑)	( )	( )	( )	( )
C_2	(↑↓)	(↑↓)	(↑↓)	(↑↓)	( )	( )	( )	( )
N_2	(↑↓)	(↑↓)	(↑↓)	(↑↓)	(↑↓)	( )	( )	( )
O_2	(↑↓)	(↑↓)	(↑↓)	(↑↓)	(↑↓)	(↑)	(↑)	( )
F_2	(↑↓)	(↑↓)	(↑↓)	(↑↓)	(↑↓)	(↑↓)	(↑↓)	( )
Ne_2	(↑↓)	(↑↓)	(↑↓)	(↑↓)	(↑↓)	(↑↓)	(↑↓)	(↑↓)

	Predicted properties		**Observed properties**	
	Number of unpaired e⁻	Bond order	Number of unpaired e⁻	Bond energy (kJ/mol)
Li_2	0	1	0	105
Be_2	0	0	0	Unstable
B_2	2	1	2	289
C_2	0	2	0	628
N_2	0	3	0	946
O_2	2	2	2	498
F_2	0	1	0	158
Ne_2	0	0	0	Nonexistent

There is also a general correlation between the predicted bond order, $\frac{(n_B - n_A)}{2}$, and the experimental bond energy. The bond order of two for C_2 and O_2 implies a double bond, which is expected to be stronger than the single bonds in Li_2, B_2, and F_2 (bond order = 1). The bond order of three for N_2 (triple bond) implies a still stronger bond. A major triumph of MO theory is its ability to explain the properties of O_2. The bond order of two and the presence of two unpaired electrons in the two π_{2p}^* MOs explains how the molecule can have a double bond and at the same time be paramagnetic.

Problem-Solving Example D.1   Electron Structure of Peroxide Ion

Using MO theory, predict the bond order and number of unpaired electrons in the peroxide ion, O_2^{2-}.

Answer   Bond order = 1; there are no unpaired electrons.

Explanation   First, find the number of valence electrons. Then construct an orbital diagram, filling the available MOs (Figure D.3) in order of increasing energy.

Recall that oxygen is in Group VIA of the periodic table, so an oxygen atom has six valence electrons. For the peroxide ion, there are two oxygen atoms and two extra electrons (2− charge) The total number of valence electrons is 2(6) + 2 = 14. The orbital diagram is

	σ_{2s}	σ_{2s}^*	π_{2p}	π_{2p}	σ_{2p}	π_{2p}^*	π_{2p}^*	σ_{2p}^*
O_2^{2-}	(↑↓)	(↑↓)	(↑↓)	(↑↓)	(↑↓)	(↑↓)	(↑↓)	( )

The bond order is (8 − 6)/2 = 1. There are no unpaired electrons. These conclusions are in agreement with the Lewis structure (valence bond theory) for the peroxide ion: $(:\ddot{O}-\ddot{O}:)^{2-}$.

D.3 POLYATOMIC MOLECULES; DELOCALIZED π ELECTRONS

The bonding in molecules containing more than two atoms can also be described in terms of MOs. We will not attempt to do this; the energy level structure is considerably more complex than that for diatomic molecules. However, one point is worth mentioning. In polyatomic species, *the MOs can be spread over the entire molecule* rather than being localized between two atoms.

The nitrate ion is known from experiment to be triangular planar, with three N—O bonds whose lengths are equal. However, a single Lewis structure for nitrate ion does not predict three equal-length bonds.

Valence bond theory *(⇐ p. 343, Section 8.9)* invokes three contributing resonance structures to explain the fact that the three N—O bonds are identical. In each contributing structure there is a double bond from N to a different O atom. MO theory, on the other hand, considers that the skeleton of the nitrate ion involves three sigma bonds, but the fourth electron pair (the second pair in the double bond) is in a pi orbital that is delocalized; it is shared by all of the atoms in the ion. That is, this pair of electrons is shared equally among all three O atoms, making the bond order 1.33 between each N and O. According to MO theory, a similar interpretation applies with all of the resonance hybrids described in Chapter 8.

Another species in which delocalized pi orbitals play an important role is benzene, C_6H_6. There are 30 valence electrons in the molecule, 24 of which are required to form the sigma bond framework:

The remaining six electrons are located in three π orbitals, which according to MO theory extend over the entire molecule. Figure D.4 is one way of representing this structure; more commonly it is shown simply as

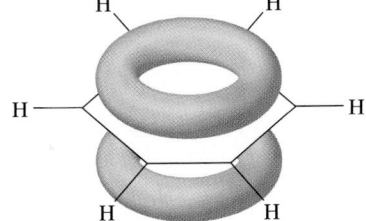

Figure D.4 Electron delocalization in benzene. Three of the 15 valence electron pairs in the benzene molecule are not localized between particular carbon atoms. Instead, they are spread out over two doughnut-shaped regions of electron density, one above the plane of the benzene ring and the other below it. These three electron pairs are in π molecular orbitals that add 0.5 to the bond order between each pair of carbon atoms.

APPENDIX E

Naming Simple Organic Compounds

Contents
E.1 Hydrocarbons   *A.28*
E.2 Functional Groups   *A.31*

The systematic nomenclature for organic compounds was proposed by the International Union of Pure and Applied Chemistry (IUPAC). The IUPAC set of rules provides different names for the more than 10 million known organic compounds, and allows names to be assigned to new compounds as they are synthesized. Many organic compounds also have *common* names. Usually the common name came first and is widely known. Many consumer products are labeled with the common name, and when only a few isomers are possible, the common name adequately identifies the product for the consumer. However, as illustrated in Section 3.4 (⬅ *p. 83*), a system of common names quickly fails when several structural isomers are possible.

E.1   HYDROCARBONS

The name of each member of the hydrocarbon classes has two parts. The first part, the prefix (*meth-, eth-, prop-, but-*, and so on), reflects the number of carbon atoms. When more than four carbons are present, the Greek or Latin number prefixes are used: *pent-, hex-, hept-, oct-, non-*, and *dec-*. The second part of the name, or the suffix, tells the class of hydrocarbon. Alkanes have carbon-carbon single bonds, alkenes have carbon-carbon double bonds, and alkynes have carbon-carbon triple bonds.

Unbranched Alkanes and Alkyl Groups

The names of the first 20 unbranched (straight-chain) alkanes are given in Table E.1.

Alkyl groups are named by dropping *-ane* from the parent alkane and adding *-yl* (see Table 3.5 for examples).

Branched-Chain Alkanes

The rules for naming branched-chain alkanes are as follows:

1. *Find the longest continuous chain of carbon atoms: this chain determines the*

TABLE E.1	Names of Unbranched Alkanes		
CH_4	Methane	$C_{11}H_{24}$	Undecane
C_2H_6	Ethane	$C_{12}H_{26}$	Dodecane
C_3H_8	Propane	$C_{13}H_{28}$	Tridecane
C_4H_{10}	Butane	$C_{14}H_{30}$	Tetradecane
C_5H_{12}	Pentane	$C_{15}H_{32}$	Pentadecane
C_6H_{14}	Hexane	$C_{16}H_{34}$	Hexadecane
C_7H_{16}	Heptane	$C_{17}H_{36}$	Heptadecane
C_8H_{18}	Octane	$C_{18}H_{38}$	Octadecane
C_9H_{20}	Nonane	$C_{19}H_{40}$	Nonadecane
$C_{10}H_{22}$	Decane	$C_{20}H_{42}$	Eicosane

parent name for the compound. For example, the following compound has two methyl groups attached to a *heptane* parent.

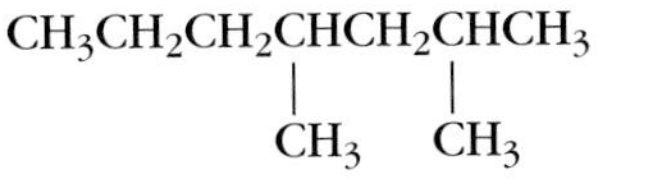

$$CH_3CH_2CH_2CHCH_2CHCH_3$$
$$||$$
$$CH_3CH_3$$

The longest continuous chain may not be obvious from the way the formula is written, especially for the straight-line format that is commonly used. For example, the longest continuous chain of carbon atoms in the following chain is *eight,* not *four* or *six.*

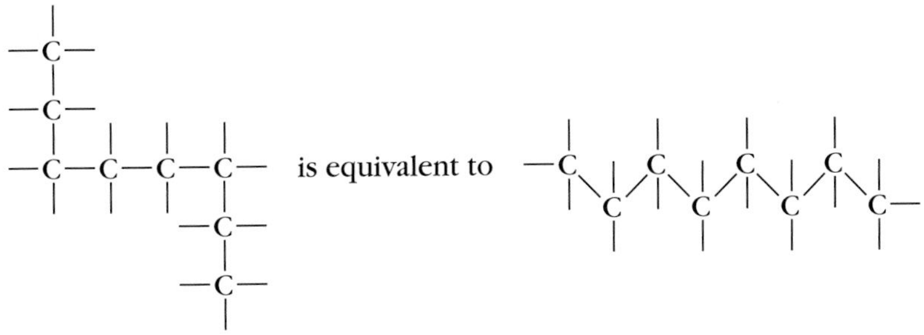

2. *Number the longest chain beginning with the end of the chain nearest the branching. Use these numbers to designate the location of the attached group. When two or more groups are attached to the parent, give each group a number corresponding to its location on the parent chain.* For example, the name of

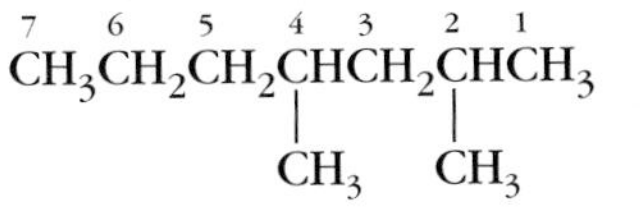

$$\overset{7}{C}H_3\overset{6}{C}H_2\overset{5}{C}H_2\overset{4}{C}HCH_2\overset{2}{C}H\overset{1}{C}H_3$$
$$\underset{CH_3}{|}\underset{CH_3}{|}$$

is 2,4-dimethylheptane. The name of the compound below is 3-methylheptane, not 5-methylheptane or 2-ethylhexane.

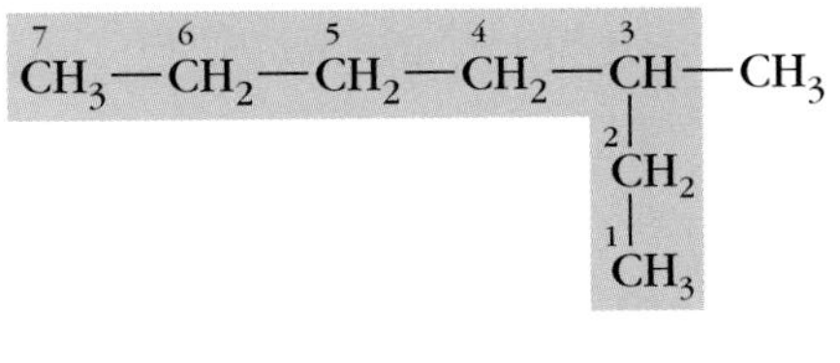

3. *When two or more substituents are identical, indicate this by the use of the prefixes* di-, tri-, tetra-, *and so on. Positional numbers of the substituents should have the smallest possible sum.*

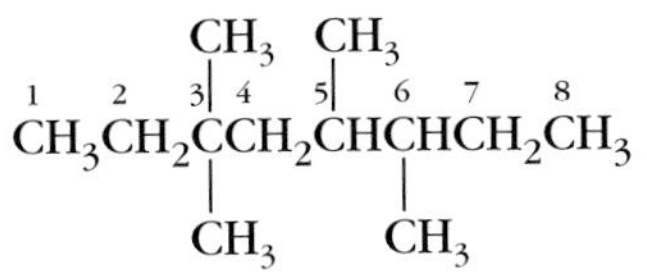

The correct name of this compound is 3,3,5,6-tetramethyloctane.

4. *If there are two or more different groups, the groups are listed alphabetically.*

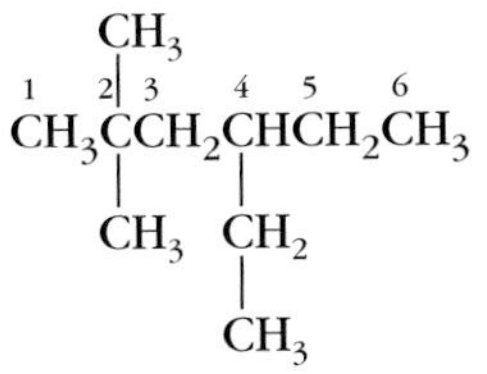

The correct name of this compound is 4-ethyl-2,2-dimethylhexane. Note that the prefix *di-* is ignored in determining alphabetical order.

Alkenes

Alkenes are named by using the prefix to indicate the number of carbon atoms and the suffix *-ene* to indicate one or more double bonds. The systematic names for the first two members of the alkene series are *ethene* and *propene*.

$$CH_2{=}CH_2 \qquad CH_3CH{=}CH_2$$

When groups, such as methyl or ethyl, are attached to carbon atoms in an alkene, the longest hydrocarbon chain is numbered from the end that will give the double bond the lowest number, and then numbers are assigned to the attached groups. For example, the name of

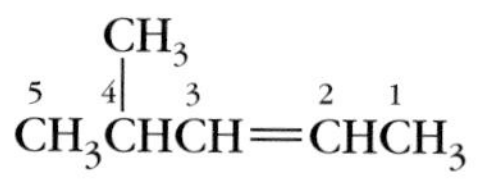

$$\overset{\displaystyle CH_3}{\underset{\overset{5}{CH_3}\ \overset{4}{C}H\overset{3}{C}H{=}\overset{2}{C}H\overset{1}{C}H_3}{|}}$$

is 4-methyl-2-pentene. See Section 8.5 for a discussion of *cis-trans* isomers of alkenes.

Alkynes

The naming of alkynes is similar to that of alkenes, with the lowest number possible being used to locate the triple bond. For example, the name of

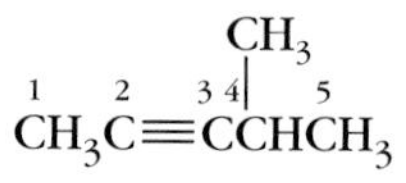

$$\overset{\displaystyle CH_3}{\underset{\overset{1}{CH_3}\overset{2}{C}{\equiv}\overset{3}{C}\overset{4}{C}H\overset{5}{C}H_3}{|}}$$

is 4-methyl-2-pentyne.

Benzene Derivatives

Monosubstituted benzene derivatives are named by using a prefix for the substituent. Some examples are

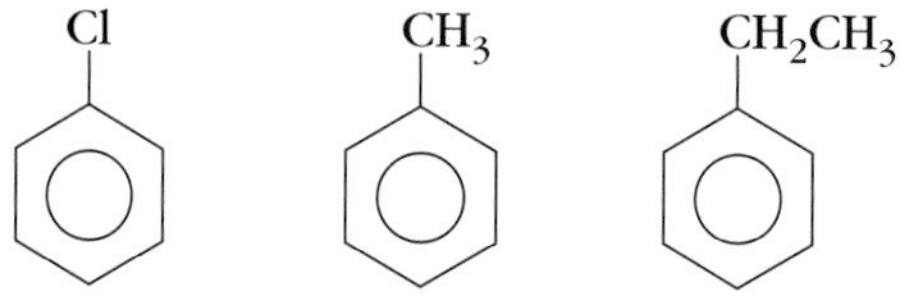

chlorobenzene   methylbenzene   ethylbenzene
(toluene)

Three isomers are possible when two groups are substituted for hydrogen atoms on the benzene ring. The relative positions of the substituents are indicated either by the prefixes *ortho-*, *meta-*, and *para-* (abbreviated *o-*, *m-*, *p-*) or by numbers. For example,

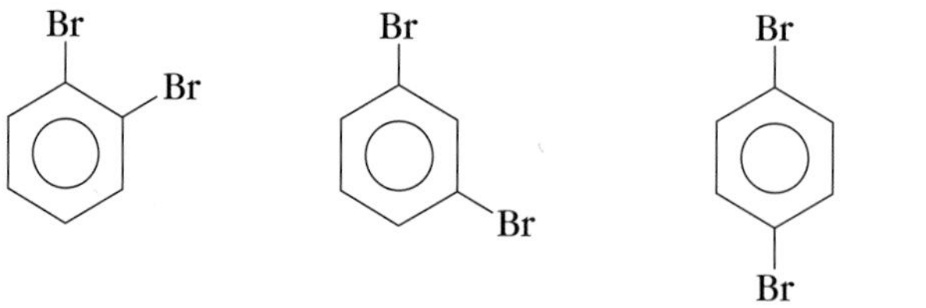

1,2-dibromobenzene   1,3-dibromobenzene   1,4-dibromobenzene
(*o*-dibromobenzene)   (*m*-dibromobenzene)   (*p*-dibromobenzene)

The dimethylbenzenes are called *xylenes*.

If more than two groups are attached to the benzene ring, numbers must be used to identify the positions. The benzene ring is numbered to give the lowest possible numbers to the substituents.

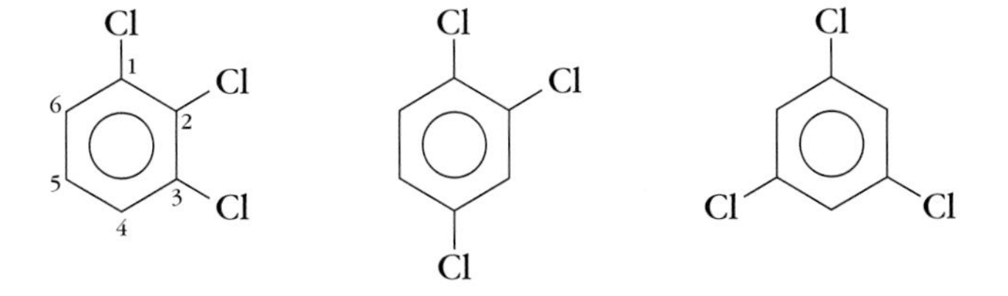

1,2,3-trichlorobenzene   1,2,4-trichlorobenzene   1,3,5-trichlorobenzene

E.2   FUNCTIONAL GROUPS

An atom or group of atoms that defines the structure of a specific class of organic compounds and determines their properties is called a *functional group* (⬅ *p. 76)*. The millions of organic compounds include classes of compounds that are obtained by replacing hydrogen atoms of hydrocarbons with functional groups (Sections 3.1, 12.5, 12.6, 12.7). The important functional groups are shown in Table E.2.

The "R" attached to the functional group represents the hydrocarbon framework with one hydrogen removed for each functional group added. The IUPAC system provides a systematic method for naming all members of a given class. For example, alcohols end in *-ol* (methan*ol*); aldehydes end in *-al* (methan*al*); carboxylic acids end in *-oic* (ethan*oic* acid); and ketones end in *-one* (propan*one*).

Alcohols

Isomers are also possible for molecules containing functional groups. For example, three different alcohols are obtained when a hydrogen atom in pentane is replaced by —OH, depending on which hydrogen atom is replaced. The rules for naming the "R" or hydrocarbon framework are the same as those for hydrocarbon compounds.

$$CH_3CH_2CH_2CH_2CH_2OH \qquad \text{1-pentanol}$$

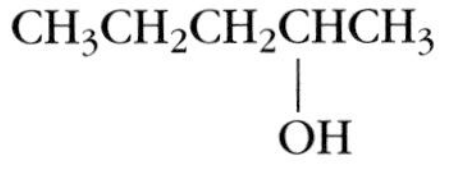

$$\underset{\underset{OH}{|}}{CH_3CH_2CH_2CHCH_3} \qquad \text{2-pentanol}$$

$$\underset{\underset{OH}{|}}{CH_3CH_2CHCH_2CH_3} \qquad \text{3-pentanol}$$

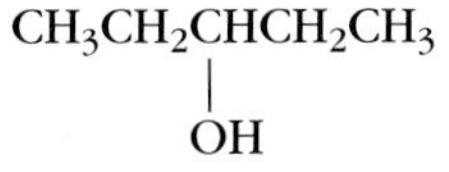

Compounds with one or more functional groups (Table E.2) and alkyl substituents are named so as to give the functional groups the lowest numbers. For example, the correct name of

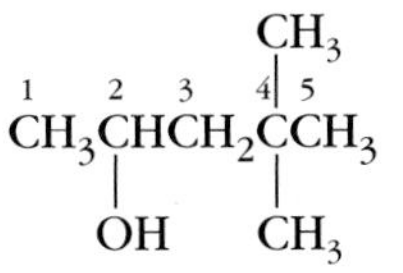

$$\overset{1}{C}H_3\overset{2}{C}H\overset{3}{C}H_2\overset{4}{\underset{\underset{CH_3}{|}}{C}}\overset{5}{C}H_3$$

(with CH₃ above C4 and OH below C2)

is 4,4-dimethyl-2-pentanol.

Aldehydes and Ketones

The systematic names of the first three aldehydes are methanal, ethanal, and propanal.

TABLE E.2   Classes of Organic Compounds Based on Functional Groups*

General formulas of class members	Class name	Typical compound	Compound name	Common use of sample compound
R—X	Halide	H—C(H)(Cl)—Cl	Dichloromethane (methylene chloride)	Solvent
R—OH	Alcohol	H—C(H)(H)—OH	Methanol (wood alcohol)	Solvent
R—C(=O)—H	Aldehyde	H—C(=O)—H	Methanal (formaldehyde)	Preservative
R—C(=O)—OH	Carboxylic acid	H—C(H)(H)—C(=O)—OH	Ethanoic acid (acetic acid)	Vinegar
R—C(=O)—R′	Ketone	H—C(H)(H)—C(=O)—C(H)(H)—H	Propanone (acetone)	Solvent
R—O—R′	Ether	C₂H₅—O—C₂H₅	Diethyl ether (ethyl ether)	Anesthetic
R—C(=O)—O—R′	Ester	CH₃—C(=O)—O—C₂H₅	Ethyl ethanoate (ethyl acetate)	Solvent in fingernail polish
R—N(H)(H)	Amine	H—C(H)(H)—N(H)(H)	Methylamine	Tanning hides (foul odor)
R—C(=O)—N(H)—R′	Amide	CH₃—C(=O)—N(H)(H)	Acetamide	Plasticizer

*R stands for an H or a hydrocarbon group such as —CH₃ or —C₂H₅. R′ could be a different group from R.

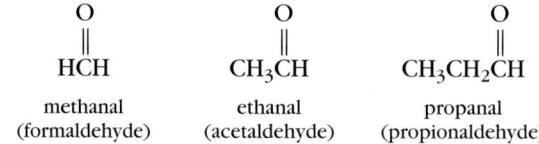

| $\underset{\text{methanal}}{\underset{\text{(formaldehyde)}}{\text{HCH}}}$ | $\underset{\text{ethanal}}{\underset{\text{(acetaldehyde)}}{\text{CH}_3\text{CH}}}$ | $\underset{\text{propanal}}{\underset{\text{(propionaldehyde)}}{\text{CH}_3\text{CH}_2\text{CH}}}$ |

For ketones, a number is used to designate the position of the carbonyl group, and the chain is numbered in a way that gives the carbonyl carbon the smallest number.

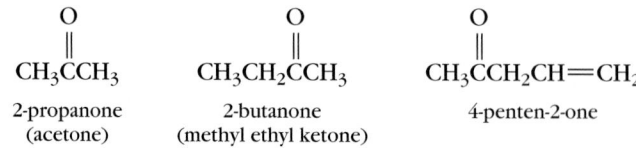

2-propanone (acetone)   2-butanone (methyl ethyl ketone)   4-penten-2-one

Carboxylic Acids

The systematic names of carboxylic acids are obtained by dropping the final *e* of the name of the corresponding alkane and adding *-oic acid.* For example, the name of

$$CH_3CH_2CH_2CH_2CH_2COOH$$

is hexanoic acid. Other examples are

$$
\overset{\displaystyle CH_3}{\underset{\underset{\text{2-methylbutanoic acid}}{\overset{4}{CH_3}\overset{3}{CH_2}\overset{2}{C}\overset{1}{H}COOH}}{|}}
\qquad
\underset{\text{2-butenoic acid}}{\overset{4}{CH_3}\overset{3}{CH}=\overset{2}{CH}\overset{1}{C}OOH}
$$

Esters

The systematic names of esters are derived from the names of the alcohol and the acid used to prepare the ester. The general formula for esters is

$$
R{-}\overset{\overset{\displaystyle O}{\|}}{C}{-}OR'
$$

As shown in Section 12.5, the $R{-}\overset{\overset{\displaystyle O}{\|}}{C}$ comes from the acid and the $R'O$ comes from the alcohol. The alcohol part is named first, followed by the name of the acid changed to end in *-ate.* For example,

$$
CH_3CH_2\overset{\overset{\displaystyle O}{\|}}{C}{-}OCH_3
$$

is named methyl propanoate and

$$
CH_3\overset{\overset{\displaystyle O}{\|}}{C}{-}OCH{=}CH_2
$$

is named ethenyl ethanoate.

Ionization Constants for Weak Acids at 25 °C

Acid	Formula and ionization equation	K_a
Acetic	$CH_3COOH \rightleftharpoons H^+ + CH_3COO^-$	1.8×10^{-5}
Arsenic	$H_3AsO_4 \rightleftharpoons H^+ + H_2AsO_4^-$	$K_1 = 2.5 \times 10^{-4}$
	$H_2AsO_4^- \rightleftharpoons H^+ + HAsO_4^{2-}$	$K_2 = 5.6 \times 10^{-8}$
	$HAsO_4^{2-} \rightleftharpoons H^+ + AsO_4^{3-}$	$K_3 = 3.0 \times 10^{-13}$
Arsenous	$H_3AsO_3 \rightleftharpoons H^+ + H_2AsO_3^-$	$K_1 = 6.0 \times 10^{-10}$
	$H_2AsO_3^- \rightleftharpoons H^+ + HAsO_3^{2-}$	$K_2 = 3.0 \times 10^{-14}$
Benzoic	$C_6H_5COOH \rightleftharpoons H^+ + C_6H_5COO^-$	6.3×10^{-5}
Boric	$H_3BO_3 \rightleftharpoons H^+ + H_2BO_3^-$	$K_1 = 7.3 \times 10^{-10}$
	$H_2BO_3^- \rightleftharpoons H^+ + HBO_3^{2-}$	$K_2 = 1.8 \times 10^{-13}$
	$HBO_3^{2-} \rightleftharpoons H^+ + BO_3^{3-}$	$K_3 = 1.6 \times 10^{-14}$
Carbonic	$H_2CO_3 \rightleftharpoons H^+ + HCO_3^-$	$K_1 = 4.2 \times 10^{-7}$
	$HCO_3^- \rightleftharpoons H^+ + CO_3^{2-}$	$K_2 = 4.8 \times 10^{-11}$
Citric	$H_3C_6H_5O_7 \rightleftharpoons H^+ + H_2C_6H_5O_7^-$	$K_1 = 7.4 \times 10^{-3}$
	$H_2C_6H_5O_7^- \rightleftharpoons H^+ + HC_6H_5O_7^{2-}$	$K_2 = 1.7 \times 10^{-5}$
	$HC_6H_5O_7^{2-} \rightleftharpoons H^+ + C_6H_5O_7^{3-}$	$K_3 = 4.0 \times 10^{-7}$
Cyanic	$HOCN \rightleftharpoons H^+ + OCN^-$	3.5×10^{-4}
Formic	$HCOOH \rightleftharpoons H^+ + HCOO^-$	18×10^{-4}
Hydrazoic	$HN_3 \rightleftharpoons H^+ + N_3^-$	1.9×10^{-5}
Hydrocyanic	$HCN \rightleftharpoons H^+ + CN^-$	4.0×10^{-10}
Hydrofluoric	$HF \rightleftharpoons H^+ + F^-$	7.2×10^{-4}
Hydrogen peroxide	$H_2O_2 \rightleftharpoons H^+ + HO_2^-$	2.4×10^{-12}
Hydrosulfuric	$H_2S \rightleftharpoons H^+ + HS^-$	$K_1 = 1.0 \times 10^{-7}$
	$HS^- \rightleftharpoons H^+ + S^{2-}$	1×10^{-19}
Hypobromous	$HOBr \rightleftharpoons H^+ + OBr^-$	2.5×10^{-9}
Hypochlorous	$HOCl \rightleftharpoons H^+ + OCl^-$	3.5×10^{-8}
Nitrous	$HNO_2 \rightleftharpoons H^+ + NO_2^-$	4.5×10^{-4}
Oxalic	$H_2C_2O_4 \rightleftharpoons H^+ + HC_2O_4^-$	$K_1 = 5.9 \times 10^{-2}$
	$HC_2O_4^- \rightleftharpoons H^+ + C_2O_4^{2-}$	$K_2 = 6.4 \times 10^{-5}$
Phenol	$HC_6H_5O \rightleftharpoons H^+ + C_6H_5O^-$	1.3×10^{-10}
Phosphoric	$H_3PO_4 \rightleftharpoons H^+ + H_2PO_4^-$	$K_1 = 7.5 \times 10^{-3}$
	$H_2PO_4^- \rightleftharpoons H^+ + HPO_4^{2-}$	$K_2 = 6.2 \times 10^{-8}$
	$HPO_4^{2-} \rightleftharpoons H^+ + PO_4^{3-}$	$K_3 = 3.6 \times 10^{-13}$
Phosphorous	$H_3PO_3 \rightleftharpoons H^+ + H_2PO_3^-$	$K_1 = 1.6 \times 10^{-2}$
	$H_2PO_3^- \rightleftharpoons H^+ + HPO_3^{2-}$	$K_2 = 7.0 \times 10^{-7}$
Selenic	$H_2SeO_4 \rightleftharpoons H^+ + HSeO_4^-$	$K_1 = \text{very large}$
	$HSeO_4^- \rightleftharpoons H^+ + SeO_4^{2-}$	$K_2 = 1.2 \times 10^{-2}$
Selenous	$H_2SeO_3 \rightleftharpoons H^+ + HSeO_3^-$	$K_1 = 2.7 \times 10^{-3}$
	$HSeO_3^- \rightleftharpoons H^+ + SeO_3^{2-}$	$K_2 = 2.5 \times 10^{-7}$
Sulfuric	$H_2SO_4 \rightleftharpoons H^+ + HSO_4^-$	$K_1 = \text{very large}$
	$HSO_4^- \rightleftharpoons H^+ + SO_4^{2-}$	$K_2 = 1.2 \times 10^{-2}$
Sulfurous	$H_2SO_3 \rightleftharpoons H^+ + HSO_3^-$	$K_1 = 1.7 \times 10^{-2}$
	$HSO_3^- \rightleftharpoons H^+ + SO_3^{2-}$	$K_2 = 6.4 \times 10^{-8}$
Tellurous	$H_2TeO_3 \rightleftharpoons H^+ + HTeO_3^-$	$K_1 = 2 \times 10^{-3}$
	$HTeO_3^- \rightleftharpoons H^+ + TeO_3^{2-}$	$K_2 = 1 \times 10^{-8}$

Ionization Constants for Weak Bases at 25 °C

Base	Formula and ionization equation	K_b
Ammonia	$NH_3 + H_2O \rightleftharpoons NH_4^+ + OH^-$	1.8×10^{-5}
Aniline	$C_6H_5NH_2 + H_2O \rightleftharpoons C_6H_5NH_3^+ + OH^-$	4.2×10^{-10}
Dimethylamine	$(CH_3)_2NH + H_2O \rightleftharpoons (CH_3)_2NH_2^+ + OH^-$	7.4×10^{-4}
Ethylenediamine	$(CH_2)_2(NH_2)_2 + H_2O \rightleftharpoons (CH_2)_2(NH_2)_2H^+ + OH-$	$K_1 = 8.5 \times 10^{-5}$
	$(CH_2)_2(NH_2)_2H^+ + H_2O \rightleftharpoons (CH_2)_2(NH_2)_2H_2^{2+} + OH^-$	$K_2 = 2.7 \times 10^{-8}$
Hydrazine	$N_2H_4 + H_2O \rightleftharpoons N_2H_5^+ + OH^-$	$K_1 = 8.5 \times 10^{-7}$
	$N_2H_5^+ + H_2O \rightleftharpoons N_2H_6^{2+} + OH^-$	$K_2 = 8.9 \times 10^{-16}$
Hydroxylamine	$NH_2OH + H_2O \rightleftharpoons NH_3OH^+ + OH^-$	6.6×10^{-9}
Methylamine	$CH_3NH_2 + H_2O \rightleftharpoons CH_3NH_3^+ + OH^-$	5.0×10^{-4}
Pyridine	$C_5H_5N + H_2O \rightleftharpoons C_5H_5NH^+ + OH^-$	1.5×10^{-9}
Trimethylamine	$(CH_3)_3N + H_2O \rightleftharpoons (CH_3)_3NH^+ + OH^-$	7.4×10^{-5}

APPENDIX H

Solubility Product Constants for Some Inorganic Compounds at 25 °C*

Substance	K_{sp}
Aluminum compounds	
$AlAsO_4$	1.6×10^{-16}
$Al(OH)_3$	1.9×10^{-33}
$AlPO_4$	1.3×10^{-20}
Barium compounds	
$Ba_3(AsO_4)_2$	1.1×10^{-13}
$BaCO_3$	8.1×10^{-9}
$BaC_2O_4 \cdot 2H_2O\dagger$	1.1×10^{-7}
$BaCrO_4$	2.0×10^{-10}
BaF_2	1.7×10^{-6}
$Ba(OH)_2 \cdot 8H_2O\dagger$	5.0×10^{-3}
$Ba_3(PO_4)_2$	1.3×10^{-29}
$BaSeO_4$	2.8×10^{-11}
$BaSO_3$	8.0×10^{-7}
$BaSO_4$	1.1×10^{-10}
Bismuth compounds	
$BiOCl$	7.0×10^{-9}
$BiO(OH)$	1.0×10^{-12}
$Bi(OH)_3$	3.2×10^{-40}
BiI_3	8.1×10^{-19}
$BiPO_4$	1.3×10^{-23}
Cadmium compounds	
$Cd_3(AsO_4)_2$	2.2×10^{-32}
$CdCO_3$	2.5×10^{-14}
$Cd(CN)_2$	1.0×10^{-8}
$Cd_2[Fe(CN)_6]$	3.2×10^{-17}
$Cd(OH)_2$	1.2×10^{-14}
Calcium compounds	
$Ca_3(AsO_4)_2$	6.8×10^{-19}
$CaCO_3$	3.8×10^{-9}
$CaCrO_4$	7.1×10^{-4}
$CaC_2O_4 \cdot H_2O\dagger$	2.3×10^{-9}
CaF_2	3.9×10^{-11}
$Ca(OH)_2$	7.9×10^{-6}
$CaHPO_4$	2.7×10^{-7}
$Ca(H_2PO_4)_2$	1.0×10^{-3}
$Ca_3(PO_4)_2$	1.0×10^{-25}
$CaSO_3 \cdot 2H_2O\dagger$	1.3×10^{-8}
$CaSO_4 \cdot 2H_2O\dagger$	2.4×10^{-5}
Chromium compounds	
$CrAsO_4$	7.8×10^{-21}

Substance	K_{sp}
$Cr(OH)_3$	6.7×10^{-31}
$CrPO_4$	2.4×10^{-23}
Cobalt compounds	
$Co_3(AsO_4)_2$	7.6×10^{-29}
$CoCO_3$	8.0×10^{-13}
$Co(OH)_2$	2.5×10^{-16}
$Co(OH)_3$	4.0×10^{-45}
Copper compounds	
$CuBr$	5.3×10^{-9}
$CuCl$	1.9×10^{-7}
$CuCN$	3.2×10^{-20}
$Cu_2O(Cu^+ + OH^-)\ddagger$	1.0×10^{-14}
CuI	5.1×10^{-12}
$CuSCN$	1.6×10^{-11}
$Cu_3(AsO_4)_2$	7.6×10^{-36}
$CuCO_3$	2.5×10^{-10}
$Cu_2[Fe(CN)_6]$	1.3×10^{-16}
$Cu(OH)_2$	1.6×10^{-19}
Gold compounds	
$AuBr$	5.0×10^{-17}
$AuCl$	2.0×10^{-13}
AuI	1.6×10^{-23}
$AuBr_3$	4.0×10^{-36}
$AuCl_3$	3.2×10^{-25}
$Au(OH)_3$	1×10^{-53}
AuI_3	1.0×10^{-46}
Iron compounds	
$FeCO_3$	3.5×10^{-11}
$Fe(OH)_2$	7.9×10^{-15}
FeS	4.9×10^{-18}
$Fe_4[Fe(CN)_6]_3$	3.0×10^{-41}
$Fe(OH)_3$	6.3×10^{-38}
Lead compounds	
$Pb_3(AsO_4)_2$	4.1×10^{-36}
$PbBr_2$	6.3×10^{-6}
$PbCO_3$	1.5×10^{-13}
$PbCl_2$	1.7×10^{-5}
$PbCrO_4$	1.8×10^{-14}
PbF_2	3.7×10^{-8}
$Pb(OH)_2$	2.8×10^{-16}
PbI_2	8.7×10^{-9}

Substance	K_{sp}
$Pb_3(PO_4)_2$	3.0×10^{-44}
$PbSeO_4$	1.5×10^{-7}
$PbSO_4$	1.8×10^{-8}
Magnesium compounds	
$Mg_3(AsO_4)_2$	2.1×10^{-20}
$MgCO_3 \cdot 3H_2O$†	4.0×10^{-5}
MgC_2O_4	8.6×10^{-5}
MgF_2	6.4×10^{-9}
$MgNH_4PO_4$	2.5×10^{-12}
Manganese compounds	
$Mn_3(AsO_4)_2$	1.9×10^{-11}
$MnCO_3$	1.8×10^{-11}
$Mn(OH)_2$	4.6×10^{-14}
$Mn(OH)_3$	$\sim 1 \times 10^{-36}$
Mercury compounds	
Hg_2Br_2	1.3×10^{-22}
Hg_2CO_3	8.9×10^{-17}
Hg_2Cl_2	1.1×10^{-18}
Hg_2CrO_4	5.0×10^{-9}
Hg_2I_2	4.5×10^{-29}
$Hg_2O \cdot H_2O(Hg_2^{2+} + 2OH^-)$†‡	1.6×10^{-23}
Hg_2SO_4	6.8×10^{-7}
$Hg(CN)_2$	3.0×10^{-23}
$Hg(OH)_2$	2.5×10^{-26}
HgI_2	4.0×10^{-29}
Nickel compounds	
$Ni_3(AsO_4)_2$	1.9×10^{-26}
$NiCO_3$	6.6×10^{-9}
$Ni(CN)_2$	3.0×10^{-23}
$Ni(OH)_2$	2.8×10^{-16}
Silver compounds	
Ag_3AsO_4	1.1×10^{-20}
$AgBr$	3.3×10^{-13}

Substance	K_{sp}
Ag_2CO_3	8.1×10^{-12}
$AgCl$	1.8×10^{-10}
Ag_2CrO_4	9.0×10^{-12}
$AgCN$	1.2×10^{-16}
$Ag_4[Fe(CN)_6]$	1.6×10^{-41}
$Ag_2O (Ag^+ + OH^-)$‡	2.0×10^{-8}
AgI	1.5×10^{-16}
Ag_3PO_4	1.3×10^{-20}
Ag_2SO_3	1.5×10^{-14}
Ag_2SO_4	1.7×10^{-5}
$AgSCN$	1.0×10^{-12}
Strontium compounds	
$Sr_3(AsO_4)_2$	1.3×10^{-18}
$SrCO_3$	9.4×10^{-10}
$SrC_2O_4 \cdot 2H_2O$†	5.6×10^{-8}
$SrCrO_4$	3.6×10^{-5}
$Sr(OH)_2 \cdot 8H_2O$†	3.2×10^{-4}
$Sr_3(PO_4)_2$	1.0×10^{-31}
$SrSO_3$	4.0×10^{-8}
$SrSO_4$	2.8×10^{-7}
Tin compounds	
$Sn(OH)_2$	2.0×10^{-26}
SnI_2	1.0×10^{-4}
$Sn(OH)_4$	1×10^{-57}
Zinc compounds	
$Zn_3(AsO_4)_2$	1.1×10^{-27}
$ZnCO_3$	1.5×10^{-11}
$Zn(CN)_2$	8.0×10^{-12}
$Zn_3[Fe(CN)_6]$	4.1×10^{-16}
$Zn(OH)_2$	4.5×10^{-17}
$Zn_3(PO_4)_2$	9.1×10^{-33}

*No metallic sulfides are listed in this table because sulfide ion is such a strong base that the usual solubility product equilibrium equation does not apply. See Myers, R.J. *Journal of Chemical Education*, Vol. 63, 1986; pp. 687–690.

†Since [H_2O] does not appear in equilibrium constants for equilibria in aqueous solution in general, it does *not* appear in the K_{sp} expressions for hydrated solids.

‡Very small amounts of oxides dissolve in water to give the ions indicated in parentheses. Solid hydroxides are unstable and decompose to oxides as rapidly as they are formed.

APPENDIX I

Standard Reduction Potentials in Aqueous Solution at 25 °C

Acidic solution	Standard reduction potential, $E°$ (volts)
$F_2(g) + 2e^- \longrightarrow 2F^-(aq)$	2.87
$Co^{3+}(aq) + e^- \longrightarrow Co^{2+}(aq)$	1.82
$Pb^{4+}(aq) + 2e^- \longrightarrow Pb^{2+}(aq)$	1.8
$H_2O_2(aq) + 2H^+(aq) + 2e^- \longrightarrow 2H_2O$	1.77
$NiO_2(s) + 4H^+(aq) + 2e^- \longrightarrow Ni^{2+}(aq) + 2H_2O$	1.7
$PbO_2(s) + SO_4^{2-}(aq) + 4H^+(aq) + 2e^- \longrightarrow PbSO_4(s) + 2H_2O$	1.685
$Au^+(aq) + e^- \longrightarrow Au(s)$	1.68
$2HClO(aq) + 2H^+(aq) + 2e^- \longrightarrow Cl_2(g) + 2H_2O$	1.63
$Ce^{4+}(aq) + e^- \longrightarrow Ce^{3+}(aq)$	1.61
$NaBiO_3(s) + 6H^+(aq) + 2e^- \longrightarrow Bi^{3+}(aq) + Na^+(aq) + 3H_2O$	~1.6
$MnO_4^-(aq) + 8H^+(aq) + 5e^- \longrightarrow Mn^{2+}(aq) + 4H_2O$	1.51
$Au^{3+}(aq) + 3e^- \longrightarrow Au(s)$	1.50
$ClO_3^-(aq) + 6H^+(aq) + 5e^- \longrightarrow \frac{1}{2}Cl_2(g) + 3H_2O$	1.47
$BrO_3^-(aq) + 6H^+(aq) + 6e^- \longrightarrow Br^-(aq) + 3H_2O$	1.44
$Cl_2(g) + 2e^- \longrightarrow 2Cl^-(aq)$	1.358
$Cr_2O_7^{2-}(aq) + 6e^- \longrightarrow 2Cr^{3+}(aq) + 7H_2O$	1.33
$N_2H_5^+(aq) + 3H^+(aq) + 2e^- \longrightarrow 2NH_4^+(aq)$	1.24
$MnO_2(s) + 4H^+(aq) + 2e^- \longrightarrow Mn^{2+}(aq) + 2H_2O$	1.23
$O_2(g) + 4H^+(aq) + 4e^- \longrightarrow 2H_2O$	1.229
$Pt^{2+}(aq) + 2e^- \longrightarrow Pt(s)$	1.2
$IO_3^-(aq) + 6H^+(aq) + 5e^- \longrightarrow \frac{1}{2}I_2(aq) + 3H_2O$	1.195
$ClO_4^-(aq) + 2H^+(aq) + 2e^- \longrightarrow ClO_3^-(aq) + H_2O$	1.19
$Br_2(\ell) + 2e^- \longrightarrow 2Br^-(aq)$	1.066
$AuCl_4^-(aq) + 3e^- \longrightarrow Au(s) + 4Cl^-(aq)$	1.00
$Pd^{2+}(aq) + 2e^- \longrightarrow Pd(s)$	0.987
$NO_3^-(aq) + 4H^+(aq) + 3e^- \longrightarrow NO(g) + 2H_2O$	0.96
$NO_3^-(aq) + 3H^+(aq) + 2e^- \longrightarrow HNO_2(aq) + H_2O$	0.94
$2Hg^{2+}(aq) + 2e^- \longrightarrow Hg_2^{2+}(aq)$	0.920
$Hg^{2+}(aq) + 2e^- \longrightarrow Hg(\ell)$	0.855
$Ag^+(aq) + e^- \longrightarrow Ag(s)$	0.7994
$Hg_2^{2+}(aq) + 2e^- \longrightarrow 2Hg(\ell)$	0.789
$Fe^{3+}(aq) + e^- \longrightarrow Fe^{2+}(aq)$	0.771
$SbCl_6^-(aq) + 2e^- \longrightarrow SbCl_4^-(aq) + 2Cl^-(aq)$	0.75
$[PtCl_4]^{2-}(aq) + 2e^- \longrightarrow Pt(s) + 4Cl^-(aq)$	0.73
$O_2(g) + 2H^+(aq) + 2e^- \longrightarrow H_2O_2(aq)$	0.682
$[PtCl_6]^{2-}(aq) + 2e^- \longrightarrow [PtCl_4]^{2-}(aq) + 2Cl^-(aq)$	0.68
$H_3AsO_4(aq) + 2H^+(aq) + 2e^- \longrightarrow H_3AsO_3(aq) + H_2O$	0.58
$I_2(s) + 2e^- \longrightarrow 2I^-(aq)$	0.535

Acidic solution	Standard reduction potential, $E°$ (volts)
$TeO_2(s) + 4\,H^+(aq) + 4\,e^- \longrightarrow Te(s) + 2\,H_2O$	0.529
$Cu^+(aq) + e^- \longrightarrow Cu(s)$	0.521
$[RhCl_6]^{3-}(aq) + 3\,e^- \longrightarrow Rh(s) + 6\,Cl^-(aq)$	0.44
$Cu^{2+}(aq) + 2\,e^- \longrightarrow Cu(s)$	0.337
$HgCl_2(s) + 2\,e^- \longrightarrow 2\,Hg(\ell) + 2\,Cl^-(aq)$	0.27
$AgCl(s) + e^- \longrightarrow Ag(s) + Cl^-(aq)$	0.222
$SO_4^{2-}(aq) + 4\,H^+(aq) + 2e^- \longrightarrow SO_2(g) + 2\,H_2O$	0.20
$SO_4^{2-}(aq) + 4\,H^+(aq) + 2e^- \longrightarrow H_2SO_3(aq) + H_2O$	0.17
$Cu^{2+}(aq) + e^- \longrightarrow Cu^+(aq)$	0.153
$Sn^{4+}(aq) + 2\,e^- \longrightarrow Sn^{2+}(aq)$	0.15
$S(s) + 2\,H^+(aq) + 2\,e^- \longrightarrow H_2S(aq)$	0.14
$AgBr(s) + e^- \longrightarrow Ag(s) + Br^-(aq)$	0.0713
$2\,H^+(aq) + 2\,e^- \longrightarrow H_2(g)$ (reference electrode)	0.0000
$N_2O(g) + 6\,H^+(aq) + H_2O + 4\,e^- \longrightarrow 2\,NH_3OH^+(aq)$	−0.05
$Pb^{2+}(aq) + 2\,e^- \longrightarrow Pb(s)$	−0.126
$Sn^{2+}(aq) + 2\,e^- \longrightarrow Sn(s)$	−0.14
$AgI(s) + e^- \longrightarrow Ag(s) + I^-(aq)$	−0.15
$[SnF_6]^{2-}(aq) + 4\,e^- \longrightarrow Sn(s) + 6\,F^-(aq)$	−0.25
$Ni^{2+}(aq) + 2\,e^- \longrightarrow Ni(s)$	−0.25
$Co^{2+}(aq) + 2\,e^- \longrightarrow Co(s)$	−0.28
$Tl^+(aq) + e^- \longrightarrow Tl(s)$	−0.34
$PbSO_4(s) + 2\,e^- \longrightarrow Pb(s) + SO_4^{2-}(aq)$	−0.356
$Se(s) + 2\,H^+(aq) + 2\,e^- \longrightarrow H_2Se(aq)$	−0.40
$Cd^{2+}(aq) + 2\,e^- \longrightarrow Cd(s)$	−0.403
$Cr^{3+}(aq) + e^- \longrightarrow Cr^{2+}(aq)$	−0.41
$Fe^{2+}(aq) + 2\,e^- \longrightarrow Fe(s)$	−0.44
$2\,CO_2(g) + 2\,H^+(aq) + 2\,e^- \longrightarrow (COOH)_2(aq)$	−0.49
$Ga^{3+}(aq) + 3\,e^- \longrightarrow Ga(s)$	−0.53
$HgS(s) + 2\,H^+(aq) + 2\,e^- \longrightarrow Hg(\ell) + H_2S(g)$	−0.72
$Cr^{3+}(aq) + 3\,e^- \longrightarrow Cr(s)$	−0.74
$Zn^{2+}(aq) + 2\,e^- \longrightarrow Zn(s)$	−0.763
$Cr^{2+}(aq) + 2\,e^- \longrightarrow Cr(s)$	−0.91
$Mn^{2+}(aq) + 2\,e^- \longrightarrow Mn(s)$	−1.18
$V^{2+}(aq) + 2\,e^- \longrightarrow V(s)$	−1.18
$Zr^{4+}(aq) + 4\,e^- \longrightarrow Zr(s)$	−1.53
$Al^{3+}(aq) + 3\,e^- \longrightarrow Al(s)$	−1.66
$H_2(g) + 2\,e^- \longrightarrow 2\,H^-(aq)$	−2.25
$Mg^{2+}(aq) + 2\,e^- \longrightarrow Mg(s)$	−2.37
$Na^+(aq) + e^- \longrightarrow Na(s)$	−2.714
$Ca^{2+}(aq) + 2\,e^- \longrightarrow Ca(s)$	−2.87
$Sr^{2+}(aq) + 2\,e^- \longrightarrow Sr(s)$	−2.89
$Ba^{2+}(aq) + 2\,e^- \longrightarrow Ba(s)$	−2.90
$Rb^+(aq) + e^- \longrightarrow Rb(s)$	−2.925
$K^+(aq) + e^- \longrightarrow K(s)$	−2.925
$Li^+(aq) + e^- \longrightarrow Li(s)$	−3.045

Basic solution	Standard reduction potential, E° (volts)
$ClO^-(aq) + H_2O + 2\,e^- \longrightarrow Cl^-(aq) + 2\,OH^-(aq)$	0.89
$OOH^-(aq) + H_2O + 2\,e^- \longrightarrow 3\,OH^-(aq)$	0.88
$2\,NH_2OH(aq) + 2\,e^- \longrightarrow N_2H_4(aq) + 2\,OH^-(aq)$	0.74
$ClO_3^-(aq) + 3\,H_2O + 6\,e^- \longrightarrow Cl^-(aq) + 6\,OH^-(aq)$	0.62
$MnO_4^-(aq) + 2\,H_2O + 3\,e^- \longrightarrow MnO_2(s) + 4\,OH^-(aq)$	0.588
$MnO_4^-(aq) + e^- \longrightarrow MnO_4^{2-}(aq)$	0.564
$NiO_2(s) + 2\,H_2O + 2\,e^- \longrightarrow Ni(OH)_2(s) + 2\,OH^-(aq)$	0.49
$Ag_2CrO_4(s) + 2\,e^- \longrightarrow 2\,Ag(s) + CrO_4^{2-}(aq)$	0.446
$O_2(g) + 2\,H_2O + 4\,e^- \longrightarrow 4\,OH^-(aq)$	0.40
$ClO_4^-(aq) + H_2O + 2\,e^- \longrightarrow ClO_3^-(aq) + 2\,OH^-(aq)$	0.36
$Ag_2O(s) + H_2O + 2\,e^- \longrightarrow 2\,Ag(s) + 2\,OH^-(aq)$	0.34
$2\,NO_2^-(aq) + 3\,H_2O + 4\,e^- \longrightarrow N_2O(g) + 6\,OH^-(aq)$	0.15
$N_2H_4(aq) + 2\,H_2O + 2\,e^- \longrightarrow 2\,NH_3(aq) + 2\,OH^-(aq)$	0.10
$[Co(NH_3)_6]^{3+}(aq) + e^- \longrightarrow [Co(NH_3)_6]^{2+}(aq)$	0.10
$HgO(s) + H_2O + 2\,e^- \longrightarrow Hg(\ell) + 2\,OH^-(aq)$	0.0984
$O_2(g) + H_2O + 2\,e^- \longrightarrow OOH^-(aq) + OH^-(aq)$	0.076
$NO_3^-(aq) + H_2O + 2\,e^- \longrightarrow NO_2^-(aq) + 2\,OH^-(aq)$	0.01
$MnO_2(s) + 2\,H_2O + 2\,e^- \longrightarrow Mn(OH)_2(s) + 2\,OH^-(aq)$	−0.05
$CrO_4^{2-}(aq) + 4\,H_2O + 3\,e^- \longrightarrow Cr(OH)_3(s) + 5\,OH^-(aq)$	−0.12
$Cu(OH)_2(s) + 2\,e^- \longrightarrow Cu(s) + 2\,OH^-(aq)$	−0.36
$Fe(OH)_3(s) + e^- \longrightarrow Fe(OH)_2(s) + OH^-(aq)$	−0.56
$2\,H_2O + 2\,e^- \longrightarrow H_2(g) + 2\,OH^-(aq)$	−0.8277
$2\,NO_3^-(aq) + 2\,H_2O + 2\,e^- \longrightarrow N_2O_4(g) + 4\,OH^-(aq)$	−0.85
$Fe(OH)_2(s) + 2\,e^- \longrightarrow Fe(s) + 2\,OH^-(aq)$	−0.877
$SO_4^{2-}(aq) + H_2O + 2\,e^- \longrightarrow SO_3^{2-}(aq) + 2\,OH^-(aq)$	−0.93
$N_2(g) + 4\,H_2O + 4\,e^- \longrightarrow N_2H_4(aq) + 4\,OH^-(aq)$	−1.15
$[Zn(OH)_4]^{2-}(aq) + 2\,e^- \longrightarrow Zn(s) + 4\,OH^-(aq)$	−1.22
$Zn(OH)_2(s) + 2\,e^- \longrightarrow Zn(s) + 2\,OH^-(aq)$	−1.245
$[Zn(CN)_4]^{2-}(aq) + 2\,e^- \longrightarrow Zn(s) + 4\,CN^-(aq)$	−1.26
$Cr(OH)_3(s) + 3\,e^- \longrightarrow Cr(s) + 3\,OH^-(aq)$	−1.30
$SiO_3^{2-}(aq) + 3\,H_2O + 4\,e^- \longrightarrow Si(s) + 6\,OH^-(aq)$	−1.70

Selected Thermodynamic Values*

Species	ΔH_f° (298.15 K) (kJ/mol)	S° (298.15 K) (J K^{-1} mol^{-1})	ΔG_f° (298.15 K) (kJ/mol)
Aluminum			
Al(s)	0	28.275	0
Al^{3+}(aq)	−531	321.7	−485
AlCl$_3$(s)	−704.2	110.67	−628.8
Al$_2$O$_3$(s, corundum)	−1675.7	50.92	−1582.3
Argon			
Ar(g)	0	154.843	0
Ar(aq)	−12.1	59.4	16.4
Barium			
BaCl$_2$(s)	−858.6	123.68	−810.4
BaO(s)	−553.5	70.42	−525.1
BaSO$_4$(s)	−1473.2	132.2	−1362.2
BaCO$_3$(s)	−1216.3	112.1	85.35
Beryllium			
Be(s)	0	9.5	0
Be(OH)$_2$(s)	−902.5	51.9	−815
Bromine			
Br(g)	111.884	175.022	82.396
Br$_2$(ℓ)	0	152.231	0
Br$_2$(g)	30.907	245.463	3.110
Br$_2$(aq)	−2.59	130.5	3.93
Br$^-$(aq)	−121.55	82.4	−103.96
BrCl(g)	14.64	240.10	−0.98
BrF$_3$(g)	−255.6	292.53	−229.43
HBr(g)	−36.40	198.695	−53.45
Calcium			
Ca(s)	0	41.42	0
Ca(g)	178.2	158.884	144.3
Ca^{2+}(g)	1925.9	—	—
Ca^{2+}(aq)	−542.83	53.1	−553.58
CaC$_2$(s)	−59.8	69.96	−64.9
CaCO$_3$(s, calcite)	−1206.92	92.9	−1128.79
CaCl$_2$(s)	−795.8	104.6	−748.1
CaF$_2$(s)	−1219.6	68.87	−1167.3
CaH$_2$(s)	−186.2	42	−147.2
CaO(s)	−635.09	39.75	−604.03
CaS(s)	−482.4	56.5	−477.4
Ca(OH)$_2$(s)	−986.09	83.39	−898.49
Ca(OH)$_2$(aq)	−1002.82	−74.5	−868.07
CaSO$_4$(s)	−1434.11	106.7	−1321.79

*Taken from Wagman, D. D., Evans, W. H., Parker, V. B., Schumm, R. H., Halow, I., Bailey, S. M., Churney, K. L., and Nuttall, R. The NBS Tables of Chemical Thermodynamic Properties. *Journal of Physical and Chemical Reference Data*, Vol. 11, Suppl. 2, 1982.

Species	ΔH_f° (298.15 K) (kJ/mol)	S° (298.15 K) (J K^{-1} mol^{-1})	ΔG_f° (298.15 K) (kJ/mol)
Carbon			
C(s, graphite)	0	5.74	0
C(s, diamond)	1.895	2.377	2.9
C(g)	716.682	158.096	671.257
CCl$_4$(ℓ)	−135.44	216.4	−65.21
CCl$_4$(g)	−102.9	309.85	−60.59
CHCl$_3$(ℓ)	−134.47	201.7	−73.66
CHCl$_3$(g)	−103.14	295.71	−70.34
CH$_4$(g, methane)	−74.81	186.264	−50.72
C$_2$H$_2$(g, ethyne)	226.73	200.94	209.2
C$_2$H$_4$(g, ethene)	52.26	219.56	68.15
C$_2$H$_6$(g, ethane)	−84.68	229.6	−32.82
C$_3$H$_8$(g, propane)	−103.8	269.9	−23.49
C$_4$H$_{10}$(g, butane)	−126.148	310.227	−16.985
C$_6$H$_6$(ℓ, benzene)	49.03	172.8	124.5
C$_6$H$_{14}$(ℓ, hexane)	−198.782	296.018	−4.035
C$_8$H$_{18}$(g, octane)	−208.447	466.835	16.718
C$_8$H$_{18}$(ℓ, octane)	−249.952	361.205	6.707
CH$_3$OH(ℓ, methanol)	−238.66	126.8	−166.27
CH$_3$OH(g, methanol)	−200.66	239.81	−161.96
CH$_3$OH(aq, methanol)	−245.931	133.1	−175.31
C$_2$H$_5$OH(ℓ, ethanol)	−277.69	160.7	−174.78
C$_2$H$_5$OH(g, ethanol)	−235.1	282.7	−168.49
C$_2$H$_5$OH(aq, ethanol)	−288.3	148.5	−181.64
C$_6$H$_{12}$O$_6$(s, glucose)	−1274.4	235.9	−917.2
CH$_3$COO$^-$(aq)	−486.01	86.6	−369.31
CH$_3$COOH(aq)	−485.76	178.7	−396.46
CH$_3$COOH(ℓ)	−484.5	159.8	−389.9
CO(g)	−110.525	197.674	−137.168
CO$_2$(g)	−393.509	213.74	−394.359
H$_2$CO$_3$(aq)	−699.65	187.4	−623.08
HCO$_3^-$(aq)	−691.99	91.2	−586.77
CO$_3^{2-}$(aq)	−677.14	−56.9	−527.81
HCOO$^-$(aq)	−425.55	92	−351.0
HCOOH(aq)	−425.43	163	−372.3
HCOOH(ℓ)	−424.72	128.95	−361.35
CS$_2$(g)	117.36	237.84	67.12
CS$_2$(ℓ)	89.70	151.34	65.27
COCl$_2$(g)	−218.8	283.53	−204.6
Cesium			
Cs(s)	0	85.23	0
Cs$^+$(g)	457.964	—	—
CsCl(s)	−443.04	101.17	−414.53
Chlorine			
Cl(g)	121.679	165.198	105.68
Cl$^-$(g)	−233.13	—	—
Cl$^-$(aq)	−167.159	56.5	−131.228
Cl$_2$(g)	0	223.066	0

Species	ΔH_f° (298.15 K) (kJ/mol)	S° (298.15 K) (J K^{-1} mol^{-1})	ΔG_f° (298.15 K) (kJ/mol)
$Cl_2(aq)$	−23.4	121	6.94
$HCl(g)$	−92.307	186.908	−95.299
$HCl(aq)$	−167.159	56.5	−131.228
$ClO_2(g)$	102.5	256.84	120.5
$Cl_2O(g)$	80.3	266.21	97.9
$ClO^-(aq)$	−107.1	42.	−36.8
$HClO(aq)$	−120.9	142.	−79.9
$ClF_3(g)$	−163.2	281.61	−123.0
Chromium			
$Cr(s)$	0	23.77	0
$Cr_2O_3(s)$	−1139.7	81.2	−1058.1
$CrCl_3(s)$	−556.5	123	−486.1
Copper			
$Cu(s)$	0	33.15	0
$CuO(s)$	−157.3	42.63	−129.7
$CuCl_2(s)$	−220.1	108.07	−175.7
$CuSO_4(s)$	−771.36	109.	−661.8
Fluorine			
$F_2(g)$	0	202.78	0
$F(g)$	78.99	158.754	61.91
$F^-(g)$	−255.39	—	—
$F^-(aq)$	−332.63	−13.8	−278.79
$HF(g)$	−271.1	173.779	−273.2
HF(aq, un-ionized)	−320.08	88.7	−296.82
HF(aq, ionized)	−332.63	−13.8	−278.79
Hydrogen†			
$H_2(g)$	0	130.684	0
$H_2(aq)$	−4.2	57.7	17.6
$HD(g)$	0.318	143.801	−1.464
$D_2(g)$	0	144.960	0
$H(g)$	217.965	114.713	203.247
$H^+(g)$	1536.202	—	—
$H^+(aq)$	0	0	0
$OH^-(aq)$	−229.994	−10.75	−157.244
$H_2O(\ell)$	−285.83	69.91	−237.129
$H_2O(g)$	−241.818	188.825	−228.572
$H_2O_2(\ell)$	−187.78	109.6	−120.35
$H_2O_2(aq)$	−191.17	143.9	−134.03
$HO_2^-(aq)$	−160.33	23.8	−67.3
$HDO(\ell)$	−289.888	79.29	−241.857
$D_2O(\ell)$	−294.600	75.94	−243.439
Iodine			
$I_2(s)$	0	116.135	0
$I_2(g)$	62.438	260.69	19.327
$I_2(aq)$	22.6	137.2	16.40
$I(g)$	106.838	180.791	70.25

Species	ΔH_f° (298.15 K) (kJ/mol)	S° (298.15 K) (J K^{-1} mol^{-1})	ΔG_f° (298.15 K) (kJ/mol)
I$^-$(g)	−197	—	—
I$^-$(aq)	−55.19	111.3	−51.57
I$_3^-$(aq)	−51.5	239.3	−51.4
HI(g)	26.48	206.594	1.70
HI(aq, ionized)	−55.19	111.3	−51.57
IF(g)	−95.65	236.17	−118.51
ICl(g)	17.78	247.551	−5.46
ICl$_3$(s)	−89.5	167.4	−22.29
ICl(ℓ)	−23.89	135.1	−13.58
IBr(g)	40.84	258.773	3.69
Iron			
Fe(s)	0	27.78	0
FeO(s)	−272.0	—	—
Fe$_2$O$_3$(s, hematite)	−824.2	87.4	−742.2
Fe$_3$O$_4$(s, magnetite)	−1118.4	146.4	−1015.4
FeCl$_2$(s)	−341.79	117.95	−302.3
FeCl$_3$(s)	−399.49	142.3	−344
FeS$_2$(s, pyrite)	−178.2	52.93	−166.9
Fe(CO)$_5$(ℓ)	−774	338.1	−705.3
Lead			
Pb(s)	0	64.81	0
PbCl$_2$(s)	−359.41	136	−314.1
PbO(s, yellow)	−217.32	68.7	−187.89
PbS(s)	−100.4	91.2	−98.7
Lithium			
Li(s)	0	29.12	0
Li$^+$(g)	685.783	—	—
LiOH(s)	−484.93	42.8	−438.95
LiOH(aq)	−508.48	2.8	−450.58
LiCl(s)	−408.701	59.33	−384.37
Magnesium			
Mg(s)	0	32.68	0
Mg^{2+}(aq)	−466.85	−138.1	−454.8
MgCl$_2$(g)	−400.4	—	—
MgCl$_2$(s)	−641.32	89.62	−591.79
MgCl$_2$(aq)	−801.15	−25.1	−717.1
MgO(s)	−601.70	26.94	−569.43
Mg(OH)$_2$(s)	−924.54	63.18	−833.51
MgS(s)	−346	50.33	−341.8
MgSO$_4$(s)	−1284.9	91.6	−1170.6
MgCO$_3$(s)	−1095.8	65.7	−1012.1
Mercury			
Hg(ℓ)	0	29.87	0
HgCl$_2$(s)	−224.3	146	−178.6
HgO(s, red)	−90.83	70.29	−58.539
HgS(s, red)	−58.2	82.4	−50.6

Species	ΔH_f° (298.15 K) (kJ/mol)	S° (298.15 K) (J K^{-1} mol^{-1})	ΔG_f° (298.15 K) (kJ/mol)
Nickel			
Ni(s)	0	29.87	0
NiO(s)	−239.7	37.99	−211.7
NiCl$_2$(s)	−305.332	97.65	−259.032
Nitrogen			
N$_2$(g)	0	191.61	0
N$_2$(aq)	−10.8	—	—
N(g)	472.704	153.298	455.563
NH$_3$(g)	−46.11	192.45	−16.45
NH$_3$(aq)	−80.29	111.3	−26.50
NH$_4^+$(aq)	−132.51	113.4	−79.31
N$_2$H$_4$(ℓ)	50.63	121.21	149.34
NH$_4$Cl(s)	−314.43	94.6	−202.87
NH$_4$Cl(aq)	−299.66	169.9	−210.52
NH$_4$NO$_3$(s)	−365.56	151.08	−183.87
NH$_4$NO$_3$(aq)	−339.87	259.8	−190.56
NO(g)	90.25	210.761	86.55
NO$_2$(g)	33.18	240.06	51.31
N$_2$O(g)	82.05	219.85	104.20
N$_2$O$_4$(g)	9.16	304.29	97.89
N$_2$O$_4$(ℓ)	−19.50	209.2	97.54
NOCl(g)	51.71	261.69	66.08
HNO$_3$(ℓ)	−174.10	155.60	−80.71
HNO$_3$(g)	−135.06	266.38	−74.72
HNO$_3$(aq)	−207.36	146.4	−111.25
NO$_3^-$(aq)	−205.0	146.4	−108.74
NF$_3$(g)	−124.7	260.73	−83.2
Oxygen†			
O$_2$(g)	0	205.138	0
O$_2$(aq)	−11.7	110.9	16.4
O(g)	249.170	161.055	231.731
O$_3$(g)	142.7	238.93	163.2
OH$^-$(aq)	−229.994	−10.75	−157.244
Phosphorus			
P$_4$(s, white)	0	164.36	0
P$_4$(s, red)	−70.4	91.2	−48.4
P(g)	314.64	163.193	278.25
PH$_3$(g)	5.4	310.23	13.4
PCl$_3$(g)	−287	311.78	−267.8
PCl$_3$(ℓ)	−319.7	217.1	−272.3
PCl$_5$(s)	−443.5	—	—
P$_4$O$_{10}$(s)	−2984	228.86	−2697.7
H$_3$PO$_4$(s)	−1279	110.5	−1119.1
Potassium			
K(s)	0	64.18	0
KF(s)	−567.27	66.57	−537.75

Species	ΔH_f° (298.15 K) (kJ/mol)	S° (298.15 K) (J K^{-1} mol^{-1})	ΔG_f° (298.15 K) (kJ/mol)
KCl(s)	−436.747	82.59	−409.14
KCl(aq)	−419.53	159.0	−414.49
KBr(s)	−393.798	95.90	−380.66
KI(s)	−327.900	106.32	−324.892
KClO$_3$(s)	−397.73	143.1	−296.25
KOH(s)	−424.764	78.9	−379.08
KOH(aq)	−482.37	91.6	−440.5
Silicon			
Si(s)	0	18.83	0
SiBr$_4$(ℓ)	−457.3	277.8	−443.8
SiC(s)	−65.3	16.61	−62.8
SiCl$_4$(g)	−657.01	330.73	−616.98
SiH$_4$(g)	34.3	204.62	56.9
SiF$_4$(g)	−1614.94	282.49	−1572.65
SiO$_2$(s, quartz)	−910.94	41.84	−856.64
Silver			
Ag(s)	0	42.55	0
Ag$^+$(aq)	105.579	72.68	77.107
Ag$_2$O(s)	−31.05	121.3	−11.2
AgCl(s)	−127.068	96.2	−109.789
AgI(s)	−61.84	115.5	−66.19
AgN$_3$(s)	620.60	99.22	591.0
AgNO$_3$(s)	−124.39	140.92	−33.41
AgNO$_3$(aq)	−101.8	219.2	−34.16
Sodium			
Na(s)	0	51.21	0
Na(g)	107.32	153.712	76.761
Na$^+$(g)	609.358	—	—
Na$^+$(aq)	−240.12	59.0	−261.905
NaF(s)	−573.647	51.46	−543.494
NaF(aq)	−572.75	45.2	−540.68
NaCl(s)	−411.153	72.13	−384.138
NaCl(g)	−176.65	229.81	−196.66
NaCl(aq)	−407.27	115.5	−393.133
NaBr(s)	−361.062	86.82	−348.983
NaBr(aq)	−361.665	141.4	−365.849
NaI(s)	−287.78	98.53	−286.06
NaI(aq)	−295.31	170.3	−313.47
NaOH(s)	−425.609	64.455	−379.484
NaOH(aq)	−470.114	48.1	−419.15
NaClO$_3$(s)	−365.774	123.4	−262.259
NaHCO$_3$(s)	−950.81	101.7	−851.0
Na$_2$CO$_3$(s)	−1130.68	134.98	−1044.44
Na$_2$SO$_4$(s)	−1387.08	149.58	−1270.16

Species	ΔH_f° (298.15 K) (kJ/mol)	S° (298.15 K) (J K^{-1} mol^{-1})	ΔG_f° (298.15 K) (kJ/mol)
Sulfur			
S(s, monoclinic)	0.33	—	—
S(s, rhombic)	0	31.80	0
S(g)	278.805	167.821	238.250
S^{2-}(aq)	33.1	−14.6	85.8
S$_2$Cl$_2$(g)	−18.4	331.5	−31.8
SF$_6$(g)	−1209.	291.82	−1105.3
SF$_4$(g)	−774.9	292.03	−731.3
H$_2$S(g)	−20.63	205.79	−33.56
H$_2$S(aq)	−39.7	121	−27.83
HSO$_4^-$(aq)	−887.34	131.8	−755.91
SO$_2$(g)	−296.830	248.22	−300.194
SO$_3$(g)	−395.72	256.76	−371.06
SOCl$_2$(g)	−212.5	309.77	−198.3
SO$_4^{2-}$(aq)	−909.27	20.1	−744.53
H$_2$SO$_4$(ℓ)	−813.989	156.904	−690.003
H$_2$SO$_4$(aq)	−909.27	20.1	−744.53
Tin			
Sn(s, white)	0	51.55	0
Sn(s, gray)	−2.09	44.14	0.13
SnCl$_2$(s)	−325.1	—	—
SnCl$_4$(ℓ)	−511.3	258.6	−440.1
SnCl$_4$(g)	−471.5	365.8	−432.2
SnO$_2$(s)	−580.7	52.3	−519.6
Titanium			
Ti(s)	0	30.63	0
TiCl$_4$(ℓ)	−804.2	252.34	−737.2
TiCl$_4$(g)	−763.2	354.9	−726.7
TiO$_2$(s)	−939.7	49.92	−884.5
Uranium			
U(s)	0	50.21	0
UO$_2$(s)	−1084.9	77.03	−1031.7
UO$_3$(s)	−1223.8	96.11	−1145.9
UF$_4$(s)	−1914.2	151.67	−1823.3
UF$_6$(g)	−2147.4	377.9	−2063.7
UF$_6$(s)	−2197.0	227.6	−2068.5
Zinc			
Zn(s)	0	41.63	0
ZnCl$_2$(s)	−415.05	111.46	−369.398
ZnO(s)	−348.28	43.64	−318.3
ZnS(s, sphalerite)	−205.98	57.7	−201.29

†Many hydrogen-containing and oxygen-containing compounds are listed only under other elements; for example, HNO$_3$ appears under nitrogen.

ANSWERS TO PROBLEM-SOLVING PRACTICE PROBLEMS

Chapter 1

1.1 (1) *Define the problem:* You know that you have one gallon of mercury and want to calculate the mass in pounds.
(2) *Develop a plan:* Density relates mass and volume and is the appropriate conversion factor, so look up the density of mercury. The density is in units of g/mL, so convert from gallons to mL before using the density. The mass will be calculated in g, so convert from g to pounds at the end. In each step, use the units to tell how to set up the conversion factor.
(3) *Execute the plan:*

$$1 \text{ gallon} \times \frac{4 \text{ quart}}{1 \text{ gallon}} = 4 \text{ quart}$$

$$4 \text{ quart} \times \frac{946.3 \text{ mL}}{1 \text{ quart}} = 3785.2 \text{ mL}$$

$$3785.2 \text{ mL} \times \frac{13.55 \text{ g}}{1 \text{ mL}} = 5.1289 \times 10^4 \text{ g}$$

$$5.1289 \times 10^4 \text{ g} \times \frac{1 \text{ pound}}{453.59 \text{ g}} = 113.1 \text{ pound}$$

It is quicker and more accurate to set up a solution like this as a single step.

$$1 \text{ gallon} \times \frac{4 \text{ quart}}{1 \text{ gallon}} \times \frac{946.3 \text{ mL}}{1 \text{ quart}}$$

$$\times \frac{13.55 \text{ g}}{1 \text{ mL}} \times \frac{1 \text{ pound}}{453.59 \text{ g}} = 113.1 \text{ pound}$$

(4) *Check your answer:* A pint of water weighs about a pound. There are 2 pints in a quart and 4 quarts in a gallon, so a gallon of water has $2 \times 4 = 8$ pints and weighs about 8 pounds. The density of mercury is more than 10 times the density of water, so the mercury should weigh more than 10 times as much. The answer is somewhat more than 80 pounds, which is reasonable.

1.2 You cannot conclude that the black substance is an element. The mass increased, suggesting that the black substance combined with something in the air (probably oxygen, maybe nitrogen) to form the red-orange substance. However, either an element or a compound could react with oxygen or with nitrogen; therefore, insufficient information is available to decide whether the black substance is an element.

1.3 Oxygen is O_2; ozone is O_3. Oxygen is a colorless, odorless gas; ozone is a pale blue gas with a pungent odor.

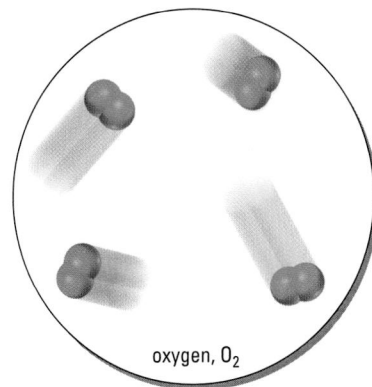

oxygen, O_2

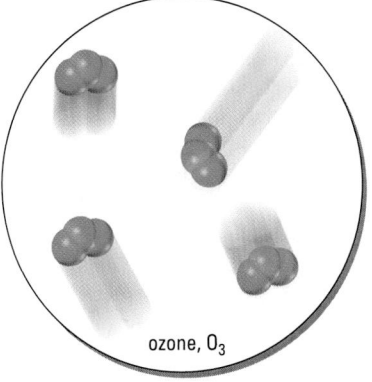

ozone, O_3

Chapter 2

2.1 (a) 10 gal = 40 qt. There are 1.0567 quarts per liter, so

$$40 \text{ qt} \times \frac{1 \text{ L}}{1.0567 \text{ qt}} = 37.9 \text{ L}$$

(b) $3 \text{ pt} \times \frac{1 \text{ qt}}{2 \text{ pt}} \times \frac{1 \text{ L}}{1.057 \text{ qt}} \times \frac{1000 \text{ ml}}{1 \text{ L}} = 1420 \text{ ml}$

(c) $\frac{1.420 \text{ L}}{5 \text{ L}} \times 100\% = 28\%$

2.2 (a) 1 lb = 453.59 g, so 5 lb = 2268. g

2.3 Work with the numerator first: 165 mg is 0.165 g. Work with the denominator next: 1 dL = 0.1 L. Therefore, the concentration is $\frac{0.165 \text{ g}}{0.1 \text{L}} = 1.65 \text{ g/L}$.

2.4 (a) A phosphorus atom (Z = 15) with 16 neutrons has A = 31.

A.49

(b) A neon-22 atom has A = 22 and Z = 10, so the number of electrons must be 10 and the number of neutrons must be A − Z = 22 − 10 = 12 neutrons.

(c) The periodic table shows us that the element with 82 protons is lead. The atomic weight of this isotope of lead is 82 + 125 = 207, so the correct symbol is $^{207}_{82}Pb$.

2.5 The magnesium isotope with 12 neutrons has 12 protons, so Z = 12 and the notation is $^{24}_{12}Mg$; the isotope with 13 neutrons has Z = 12 and $^{25}_{12}Mg$; the isotope with 14 neutrons has Z = 12 and $^{26}_{12}Mg$.

2.6 75 g wire × (fraction Ni) = g Ni, so 75 × 0.80 = 60 g Ni. For Cr we have 75 × 0.20 = 15 g Cr. Or we could have solved for the mass of Cr from 75 g wire − 60 g Ni = 15 g Cr.

2.7 (a) 1 mg Mo = 1×10^{-3} g Mo

$$1 \times 10^{-3} \text{ g Mo} \times \frac{1 \text{ mol Mo}}{95.94 \text{ g Mo}} = 1.04 \times 10^{-5} \text{ mol Mo}$$

(b) 5.00×10^{-3} mol Au $\times \dfrac{196.97 \text{ g Au}}{1 \text{ mol Au}} = 0.985$ g Au

Chapter 3

3.1 (a) $C_{10}H_{11}O_{13}N_5P_3$
   (b) $C_{18}H_{27}O_3N$
   (c) $C_2H_2O_4$

3.2 (a) carbon disulfide, (b) dinitrogen tetrafluoride, (c) phosphorus trichloride

3.3 Methane to methanol: −161.6 °C to 65 °C, a difference of 227 °C. Ethane to ethanol: −88.6 °C to 78 °C, a difference of 167 °C. Propane to propanol: −42.1 °C to 97 °C, a difference of 139 °C. As the carbon chains get longer, the difference in boiling points decreases.

3.4 (a) A Ca^{4+} charge is unlikely because calcium is in Group 2A, the elements of which lose two electrons to form 2+ ions.
   (b) Cr^{2+} is possible because chromium is a transition metal ion that forms 2+ and 3+ ions.
   (c) Strontium is a Group 2A metal and forms 2+ ions; thus, a Sr^- ion is highly unlikely.

3.5 (a) CH_4 is formed from two nonmetals and is molecular.
   (b) $CaBr_2$ is formed from a metal and a nonmetal, so it is ionic.
   (c) $MgCl_2$ is formed from a metal and a nonmetal, so it is ionic.
   (d) PCl_3 is formed from two nonmetals and is molecular.
   (e) KCl is formed from a metal and a nonmetal and is ionic.

3.6 (a) $In_2(SO_4)_3$ contains two In^{3+} and three SO_4^{2-} ions. There are 17 atoms in this formula unit.
   (b) $(NH_4)_3PO_4$ contains three ammonium ions, NH_4^+, and one phosphate ion, PO_4^{3-}, collectively containing 20 atoms.

3.7 (a) One Ca^{2+} ion and two fluoride (F^-) ions
   (b) One cobalt 2+ ion and two chloride (Cl^-) ions
   (c) Two potassium ions (K^+) and one monohydrogen phosphate (HPO_4^{2-}) ion.
   (d) CuBr is copper(I) bromide, and $CuBr_2$ is copper(II) bromide
   (e) K_2O and K_2SO_4; SrO and $SrSO_4$

3.8 (a) KNO_2 is potassium nitrite
   (b) $NaHSO_3$ is sodium hydrogen sulfite or sodium bisulfite
   (c) $Mn(OH)_2$ is manganese hydroxide
   (d) $Mn_2(SO_4)_3$ is manganese sulfate
   (e) Ba_3N_2 is barium nitride
   (f) LiH is lithium hydride

3.9 (a) KH_2PO_4 (b) CuOH
   (c) NaClO   (d) NH_4ClO_4
   (e) $CrCl_3$ (f) $FeSO_3$

3.10 (a) The molar mass of $K_2Cr_2O_7$ is 294.2 g/mol.

$$\frac{12.5 \text{ g}}{294.2 \text{ g/mol}} = 4.25 \times 10^{-2} \text{ mol}$$

(b) The molar mass of $KMnO_4$ is 158.0 g/mol.

$$\frac{12.5 \text{ g}}{158.0 \text{ g/mol}} = 7.91 \times 10^{-2} \text{ mol}$$

(c) The molar mass of $(NH_4)_2CO_3$ is 96.1 g/mol.

$$\frac{12.5 \text{ g}}{96.1 \text{ g/mol}} = 1.30 \times 10^{-1} \text{ mol}$$

3.11 (a) The molar mass of sucrose, $C_{12}H_{22}O_{11}$, is 342.3 g/mol.

$$5.0 \times 10^{-3} \text{ mol} \times \frac{342.3 \text{ g}}{1 \text{ mol}} = 1.7 \text{ g}$$

(b) The molar mass of the protein is 7100 g/mol.

$$5.0 \times 10^{-3} \text{ mol} \times \frac{7100 \text{ g}}{1 \text{ mol}} = 35.5 \text{ g}$$

3.12 (a) The molar mass of cholesterol is 386.7 g/mol.

$$\frac{10.0 \text{ g}}{386.7 \text{ g/mol}} = 2.59 \times 10^{-2} \text{ mol}$$

The molar mass of $Mn_2(SO_4)_3$ is 398.1 g/mol.

$$\frac{10.0 \text{ g}}{398.1 \text{ g/mol}} = 2.51 \times 10^{-2} \text{ mol}$$

(b) The molar mass of K_2HPO_4 is 174.2 g/mol.

$$0.25 \text{ mol} \times \frac{174.2 \text{ g}}{1 \text{ mol}} = 43.55 \text{ g}$$

The molar mass of caffeine is 194.2 g/mol.

$$0.25 \text{ mol} \times \frac{194.2 \text{ g}}{1 \text{ mol}} = 48.55 \text{ g}$$

3.13 The mass of Si in 1 mol of SiO_2 is 28.0855 g. The mass of O in 1 mol of SiO_2 is 31.9988 g.

$$\% \text{ Si in } SiO_2 = \frac{28.0855 \text{ g}}{60.08 \text{ g}} \times 100\% = 46.7\% \text{ Si}$$

$$\% \text{ O in } SiO_2 = \frac{31.9988 \text{ g}}{60.08 \text{ g}} \times 100\% = 53.3\% \text{ O}$$

3.14 Urea's molar mass is 60.0554 g.

$$\% \text{ N in urea} = \frac{\text{mass of N in 1 mol of urea}}{\text{mass of urea in 1 mol of urea}} \times 100\%$$

$$= \frac{28.0134 \text{ g N}}{60.0554 \text{ g urea}} \times 100\% = 46.6\%$$

So, to have 5 lb of N would require (5 lb)/0.466 = 10.7 lb urea.

3.15 A 100-g sample of the phosphorus oxide contains 43.64 g of P and 56.36 g of O.

$$43.64 \text{ g P} \times \frac{1 \text{ mol P}}{30.9738 \text{ g P}} = 1.41 \text{ mol P}$$

$$56.36 \text{ g O} \times \frac{1 \text{ mol O}}{15.9994 \text{ g O}} = 3.52 \text{ mol O}$$

The mole ratio is

$$\frac{3.52 \text{ mol O}}{1.41 \text{ mol P}} = \frac{2.50 \text{ mol O}}{1.00 \text{ mol P}}$$

There are 2.5 oxygen atoms for every phosphorus atom. Thus, the empirical formula is P_2O_5. The molar mass corresponding to this empirical formula is

$$\left(2 \text{ mol P} \times \frac{30.9738 \text{ g P}}{1 \text{ mol P}}\right) +$$

$$\left(5 \text{ mol O} \times \frac{15.994 \text{ g O}}{1 \text{ mol O}}\right) = 141.9 \text{ g/mol}$$

The known molar mass is 283.89 g/mol. The molar mass is twice as large as the empirical formula mass, so the molecular formula of the oxide is P_4O_{10}.

3.16 Find the number of moles of each element in 100.0 g of vitamin C.

$$40.9 \text{ g C} \times \frac{1 \text{ mol C}}{12.011 \text{ g C}} = 3.405 \text{ mol C}$$

$$4.58 \text{ g H} \times \frac{1 \text{ mol H}}{1.0079 \text{ g H}} = 4.544 \text{ mol H}$$

$$54.5 \text{ g O} \times \frac{1 \text{ mol O}}{15.9994 \text{ g O}} = 3.406 \text{ mol O}$$

Find the mole ratios.

$$\frac{4.544 \text{ mol H}}{3.406 \text{ mol O}} = \frac{1.334 \text{ mol H}}{1.000 \text{ mol O}}$$

The same ratio holds for H to C. Using whole numbers, we have $C_3H_4O_3$ for the empirical formula. The empirical formula weight is $(3 \times 12.011) + (4 \times 1.0079) + (3 \times 15.9994) = 88.06$ g. The molar mass, however, is 176.13 g/mol, so the molecular formula must be twice the empirical formula: $C_6H_8O_6$.

Chapter 4

4.1 (a) $2 \text{ NO (g)} + O_2(g) \rightarrow 2 \text{ NO}_2(g)$
(b) $4 \text{ Fe(s)} + 3 O_2(g) \rightarrow 2 \text{ Fe}_2O_3(s)$
(c) $2 \text{ NaN}_3(s) \rightarrow 2 \text{ Na(s)} + 3 N_2(g)$

4.2 (a) Decomposition reaction:
$2 \text{ Al(OH)}_3(s) \rightarrow \text{Al}_2O_3(s) + 3 H_2O(g)$
(b) Combination reaction:
$\text{Na}_2O(s) + H_2O(\ell) \rightarrow 2 \text{ NaOH(aq)}$
(c) Combination reaction: $S_8(s) + 24 F_2(g) \rightarrow 8 \text{ SF}_6(g)$
(d) Exchange reaction:
$3 \text{ NaOH(aq)} + H_3PO_4(aq) \rightarrow \text{Na}_3PO_4(aq) + 3 H_2O(\ell)$

(e) Single displacement:
$3 \text{ C(s)} + \text{Fe}_2O_3(s) \rightarrow 3 \text{ CO(g)} + 2 \text{ Fe}(\ell)$

4.3 (a) $\text{Xe(g)} + 2 F_2(g) \rightarrow \text{XeF}_4(g)$
(b) $\text{As}_2O_3(s) + 3 H_2(g) \rightarrow 2 \text{ As(s)} + 3 H_2O(\ell)$

4.4 (a) $C_2H_5OH + 3 O_2 \rightarrow 2 \text{ CO}_2 + 3 H_2O$
(b) $C_2H_5OH + 2 O_2 \rightarrow 2 \text{ CO} + 3 H_2O$

4.5 We have 0.55 mol of CH_4, and we need twice as many moles of O_2: $2 \times 0.55 = 1.10 \text{ mol } O_2$.

$$1.10 \text{ mol } O_2 \times \frac{31.99 \text{ g } O_2}{1 \text{ mol } O_2} = 35.2 \text{ g } O_2$$

4.6 (a) $0.300 \text{ mol cassiterite} \times \dfrac{1 \text{ mol Sn}}{1 \text{ mol cassiterite}}$
$$\times \frac{118.7 \text{ g Sn}}{1 \text{ mol Sn}} = 35.6 \text{ g Sn}$$

(b) $35.6 \text{ g Sn} \times \dfrac{1 \text{ mol Sn}}{118.7 \text{ g Sn}} \times \dfrac{2 \text{ mol C}}{1 \text{ mol Sn}}$
$$\times \frac{12.01 \text{ g C}}{1 \text{ mol C}} = 7.20 \text{ g C}$$

4.7 (a) $57. \text{ g C} \times \dfrac{1 \text{ mol C}}{12.01 \text{ g C}} \times \dfrac{1 \text{ mol } O_2}{2 \text{ mol C}}$
$$\times \frac{32.0 \text{ g } O_2}{1 \text{ mol } O_2} = 76. \text{ g } O_2$$

(b) $57. \text{ g C} \times \dfrac{1 \text{ mol C}}{12.01 \text{ g C}} \times \dfrac{2 \text{ mol CO}}{2 \text{ mol C}}$
$$\times \frac{28.0 \text{ g CO}}{1 \text{ mol CO}} = 1.3 \times 10^2 \text{ g CO}$$

4.8

$$6.46 \text{ g MgCl}_2 \times \frac{1 \text{ mol MgCl}_2}{95.2104 \text{ g MgCl}_2}$$
$$= 6.78 \times 10^{-2} \text{ mol MgCl}_2$$

The same number of moles of Mg as $MgCl_2$ are involved, so

$$6.78 \times 10^{-2} \text{ mol Mg} \times \frac{24.3050 \text{ g Mg}}{1 \text{ mol Mg}} = 1.65 \text{ g Mg}$$

$$\frac{1.65 \text{ g Mg}}{1.72\text{-g sample}} \times 100\% = 95.9\% \text{ Mg in sample}$$

4.9 Concentrate on the conversion of H_2 to NH_3 since H_2 is the limiting reagent.

$$1.7 \text{ mol } H_2 \times \frac{2 \text{ mol NH}_3}{3 \text{ mol } H_2} = 1.13 \text{ mol NH}_3$$

4.10 (a) $CS_2(\ell) + 3 O_2(g) \rightarrow CO_2(g) + 2 SO_2(g)$
(b) Determine the quantity of CO_2 produced by each reactant; the limiting reactant produces the lesser quantity.

$$3.5 \text{ g CS}_2 \times \frac{1 \text{ mol CS}_2}{76.0 \text{ g CS}_2} \times \frac{1 \text{ mol CO}_2}{1 \text{ mol CS}_2}$$
$$\times \frac{44.01 \text{ g CO}_2}{1 \text{ mol CO}_2} = 2.0 \text{ g CO}_2$$

$$17.5 \text{ g } O_2 \times \frac{1 \text{ mol } O_2}{31.998 \text{ g } O_2} \times \frac{1 \text{ mol CO}_2}{3 \text{ mol } O_2}$$
$$\times \frac{44.01 \text{ g CO}_2}{1 \text{ mol CO}_2} = 8.02 \text{ g CO}_2$$

Therefore, CS_2 is the limiting reagent.

(c) The yield of SO_2 must be calculated using the limiting reagent, CS_2.

$$3.5 \text{ g } CS_2 \times \frac{1 \text{ mol } CS_2}{76.0 \text{ g } CS_2} \times \frac{2 \text{ mol } SO_2}{1 \text{ mol } CS_2}$$

$$\times \frac{64.1 \text{ g } SO_2}{1 \text{ mol } SO_2} = 5.9 \text{ g } SO_2$$

4.11 Find the number of moles of each reactant.

$$100. \text{ g } SiCl_4 \times \frac{1 \text{ mol } SiCl_4}{169.90 \text{ g } SiCl_4} = 0.589 \text{ mol } SiCl_4$$

$$100. \text{ g } Mg \times \frac{1 \text{ mol } Mg}{24.3050 \text{ g } Mg} = 4.11 \text{ mol } Mg$$

Find the mass of Si produced, based on the mass available of each reactant.

$$0.589 \text{ mol } SiCl_4 \times \frac{1 \text{ mol } Si}{1 \text{ mol } SiCl_4} \times \frac{28.0855 \text{ g } Si}{1 \text{ mol } Si}$$

$$= 16.5 \text{ g } Si$$

$$4.11 \text{ mol } Mg \times \frac{1 \text{ mol } Si}{2 \text{ mol } Mg} \times \frac{28.0855 \text{ g } Si}{1 \text{ mol } Si} = 57.7 \text{ g } Si$$

Thus, $SiCl_4$ is the limiting reactant, and the mass of Si produced is 16.5 g.

4.12 To make 1.0 kg of CH_3OH with 85% yield will require using enough reactant to produce 1000/0.85, or 1180 g of CH_3OH.

$$1180 \text{ g } CH_3OH \times \frac{1 \text{ mol}}{32.042 \text{ g}} = 36.83 \text{ mol } CH_3OH$$

$$36.83 \text{ mol } CH_3OH \times \frac{2 \text{ mol } H_2}{1 \text{ mol } CH_3OH} = 73.65 \text{ mol } H_2$$

$$73.65 \text{ mol } H_2 \times \frac{2.0158 \text{ g } H_2}{1 \text{ mol } H_2} = 148.5 \text{ g } H_2$$

4.13 Calculate the mass of Cu_2S you should have produced and compare it with the amount actually produced.

$$2.50 \text{ g } Cu \times \frac{1 \text{ mol } Cu}{63.546 \text{ g } Cu} = 3.93 \times 10^{-2} \text{ mol } Cu$$

$$3.93 \times 10^{-2} \text{ mol } Cu \times \frac{8 \text{ mol } Cu_2S}{16 \text{ mol } Cu}$$

$$= 1.97 \times 10^{-2} \text{ mol } Cu_2S$$

$$1.97 \times 10^{-2} \text{ mol } Cu_2S \times \frac{159.16 \text{ g } Cu_2S}{1 \text{ mol } Cu_2S} = 3.14 \text{ g } Cu_2S$$

$$\frac{2.53 \text{ g}}{3.14 \text{ g}} \times 100\% = 80.6\% \text{ yield was obtained}$$

Your synthesis met the standard.

4.14 (a) $491 \text{ mg } CO_2 \times \dfrac{1 \text{ g } CO_2}{10^3 \text{ mg } CO_2} \times \dfrac{1 \text{ mol } CO_2}{44.01 \text{ g } CO_2}$

$$\times \frac{1 \text{ mol } C}{1 \text{ mol } CO_2} = 1.116 \times 10^{-2} \text{ mol } C$$

$$1.116 \times 10^{-2} \text{ mol } C \times \frac{12.01 \text{ g } C}{1 \text{ mol } C}$$

$$= 0.1340 \text{ g } C = 134.0 \text{ mg } C$$

$$100 \text{ mg } H_2O \times \frac{1 \text{ g } H_2O}{10^3 \text{ mg } H_2O} \times \frac{1 \text{ mol } H_2O}{18.02 \text{ g } H_2O}$$

$$\times \frac{2 \text{ mol } H}{1 \text{ mol } H_2O} = 1.110 \times 10^{-2} \text{ mol } H$$

$$1.110 \times 10^{-2} \text{ mol } H \times \frac{1.008 \text{ g } H}{1 \text{ mol } H}$$

$$= 1.119 \times 10^{-2} \text{ g } H = 11.2 \text{ mg } H$$

The mass of oxygen in the compound

= total mass − (mass C + mass H)

= 175 mg − (134.0 mg C + 11.2 mg H) = 29.8 mg O

The moles of oxygen are

$$29.8 \text{ mg } O \times \frac{1 \text{ g } O}{10^3 \text{ mg } O} \times \frac{1 \text{ mol } O}{16.00 \text{ g } O}$$

$$= 1.862 \times 10^{-3} \text{ mol } O$$

The empirical formula can be derived from the mole ratios of the elements:

$$\frac{1.116 \times 10^{-2} \text{ mol } C}{1.862 \times 10^{-3} \text{ mol } O} = 5.993 \text{ mol } C/\text{mol } O$$

$$\frac{1.110 \times 10^{-2} \text{ mol } H}{1.862 \times 10^{-3} \text{ mol } O} = 5.961 \text{ mol } H/\text{mol } O$$

$$\frac{1.862 \times 10^{-3} \text{ mol } O}{1.862 \times 10^{-3} \text{ mol } O} = 1.000 \text{ mol } O$$

The empirical formula of phenol is C_6H_6O.
(b) The molar mass is needed to determine the molecular formula.

4.15 $0.569 \text{ g } Sn \times \dfrac{1 \text{ mol } Sn}{118.7 \text{ g } Sn} = 4.794 \times 10^{-3} \text{ mol } Sn$

$$2.434 \text{ g } I_2 \times \frac{1 \text{ mol } I_2}{253.81 \text{ g } I_2} \times \frac{2 \text{ mol } I}{1 \text{ mol } I_2}$$

$$= 1.918 \times 10^{-2} \text{ mol } I$$

$$\frac{1.918 \times 10^{-2} \text{ mol } I}{4.794 \times 10^{-3} \text{ mol } Sn} = \frac{4.001 \text{ mol } I}{1.000 \text{ mol } Sn}$$

Therefore, the empirical formula is SnI_4.

Chapter 5

5.1 (a) NaF is soluble.
   (b) $Ca(CH_3CO_2)_2$ is soluble.
   (c) $SrCl_2$ is soluble.
   (d) MgO is not soluble.
   (e) $PbCl_2$ is not soluble.
   (f) HgS is not soluble.
5.2 (a) This exchange reaction forms insoluble nickel hydroxide and aqueous sodium chloride:
   $NiCl_2(aq) + 2 \text{ NaOH}(aq) \rightarrow Ni(OH)_2(s) + 2 \text{ NaCl}(aq)$
   (b) This is an exchange reaction that forms aqueous potassium bromide and a precipitate of calcium carbonate: $K_2CO_3(aq) + CaBr_2(aq) \rightarrow CaCO_3(s) + 2 \text{ KBr}(aq)$
5.3 (a) $2 \text{ Na}^+(aq) + 2 \text{ F}^-(aq) + Ca^{+2}(aq) + 2 \text{ CH}_3CO_2^-(aq)$
   $\rightarrow CaF_2(s) + 2 \text{ Na}^+(aq) + 2 \text{ CH}_3CO_2^-(aq)$

$$Ca^{+2}(aq) + 2 \text{ F}^-(aq) \longrightarrow CaF_2(s)$$

(b) $2 NH_4^+(aq) + S^{2-}(aq) + Fe^{2+}(aq) + 2 Cl^-(aq) \rightarrow$
$FeS(s) + 2 NH_4^+(aq) + 2 Cl^-(aq)$

$$Fe^{2+}(aq) + S^{2-}(aq) \longrightarrow FeS(s)$$

5.4 Any of the strong acids in Table 5.2 would also be strong electrolytes. Any of the weak acids or bases in Table 5.2 would be weak electrolytes. Any organic compound that yields no ions on dissolution would be a nonelectrolyte.

5.5 $H_3PO_4(aq) + 3 NaOH(aq) \rightarrow Na_3PO_4(aq) + 3 H_2O(\ell)$

5.6 (a) Sulfuric acid and magnesium hydroxide
(b) Carbonic acid and strontium hydroxide

5.7 $2 HCN(aq) + Ca(OH)_2(aq) \longrightarrow Ca(CN)_2(aq) + 2 H_2O(\ell)$

$$HCN(aq) + OH^-(aq) \longrightarrow CN^-(aq) + H_2O(\ell)$$

5.8 The oxidation numbers of Fe and Sb are 0 (Rule 1). The oxidation numbers in Sb_2S_3 are +3 for Sb^{3+} and −2 for S^{2-} (Rules 2 and 4); the oxidation numbers in FeS are +2 for Fe^{2+} and −2 for S^{2-} (Rules 2 and 4).

5.9 In the reaction $PbO(s) + CO(g) \rightarrow Pb(s) + CO_2(g)$, Pb^{2+} is reduced to Pb; Pb^{2+} is the oxidizing agent. C^{2+} is oxidized to C^{4+}; C^{2+} is the reducing agent.

5.10 Reactions (a) and (b) will occur. Aluminum is above copper and chromium in Table 5.5; therefore, aluminum will be oxidized and acts as the reducing agent in reactions (a) and (b). In reaction (a), Cu^{2+} is reduced, and Cu^{2+} is the oxidizing agent. Cr^{3+} is the oxidizing agent in reaction (b) and is reduced to Cr metal. Reactions (c) and (d) do not occur because Pt cannot reduce H^+, and Au cannot reduce Ag^+.

5.11 $36.0 \text{ g } Na_2SO_4 \times \dfrac{1 \text{ mol } Na_2SO_4}{142.1 \text{ g } Na_2SO_4} = 0.2533 \text{ mol } Na_2SO_4$

$$\text{Molarity} = \dfrac{0.2533 \text{ mol}}{0.750 \text{ L}} = 0.338 \text{ molar}$$

5.12

$$V(conc) = \dfrac{0.150 \text{ molar} \times 0.050 \text{ L}}{0.500 \text{ molar}} = 0.015 \text{ L}$$

5.13 (a) 1.00 L of 0.125 M Na_2CO_3 contains 0.125 mol of Na_2CO_3.

$$0.125 \text{ mol} \times \dfrac{105.99 \text{ g}}{1 \text{ mol}} = 13.2 \text{ g } Na_2CO_3$$

Prepare the solution by adding 13.2 g of Na_2CO_3 to a volumetric flask, dissolving it and mixing thoroughly, and adding sufficient water until the solution volume is 1.0 L.
(b) Use water to dilute a specific volume of the 0.125 M solution to 100 mL.

$$V(conc) = \dfrac{0.0500 \text{ M} \times 0.100 \text{ L}}{0.125 \text{ M}}$$

$$= 0.040 \text{ L} = 40 \text{ mL of } 0.125 \text{ M solution}$$

Therefore, put 40 mL of the more concentrated solution into a container and add water until the solution volume equals 100 mL.
(c) 500 L of 0.215 M $KMnO_4$ contains 1.70 g of $KMnO_4$.

$$0.500 \text{ L} \times \dfrac{0.0215 \text{ mol } KMnO_4}{1 \text{ L}} = 0.01075 \text{ mol } KMnO_4$$

$$0.01075 \text{ mol } KMnO_4 \times \dfrac{158.0 \text{ g } KMnO_4}{1 \text{ mol } KMnO_4} = 1.70 \text{ g } KMnO_4$$

Put 1.70 g of $KMnO_4$ into a container and add water until the solution volume is 500 mL.
(d) Dilute the more concentrated solution by adding sufficient water to 52.3 mL of 0.0215 M $KMnO_4$ until the solution volume is 250 mL.

$$V(conc) = \dfrac{0.00450 \text{ M} \times 0.250 \text{ L}}{0.0215 \text{ M}} = 0.0523 \text{ L} = 52.3 \text{ mL}$$

5.14 $1.2 \times 10^{10} \text{ kg NaOH} \times \dfrac{1 \text{ mol NaOH}}{0.040 \text{ kg NaOH}} \times \dfrac{2 \text{ mol NaCl}}{2 \text{ mol NaOH}}$

$\times \dfrac{58.5 \text{ g NaCl}}{1 \text{ mol NaCl}} \times \dfrac{1 \text{ L brine}}{360 \text{ g NaCl}} = 4.9 \times 10^{10} \text{ L}$

5.15 The net ionic equation is $AgBr(s) + 2 S_2O_3^{2-}(aq) \rightarrow$ $Ag(S_2O_3)_2^{3-}(aq) + Br^-(aq)$. Moles $Na_2S_2O_3$ = (0.0200 M) (0.125 L) = 0.0025 moles $Na_2S_2O_3$.

2 mol of $Na_2S_2O_3$ dissolves 1 mol of AgBr, so 0.0025 mol of $Na_2S_2O_3$ dissolves 0.00125 mol of AgBr. The molar mass of AgBr is 187.8 g/mol, so 0.00125 mol AgBr $\times$ 187.8 g/mol = 0.235 g AgBr or 235 mg AgBr.

5.16 $H_2SO_4(aq) + 2 NaOH(aq) \longrightarrow Na_2SO_4(aq) + 2 H_2O(\ell)$

moles NaOH = (0.0413 L)(0.100 M) = 0.00413 mol NaOH

moles $H_2SO_4 = \frac{1}{2} \times 0.00413 = 0.002065$ mol H_2SO_4

$$\text{Molarity} = \dfrac{0.002065 \text{ mol}}{0.020 \text{ L}} = 0.103 \text{ M } H_2SO_4$$

Chapter 6

6.1 (a) $160 \text{ Cal} \times \dfrac{1000 \text{ cal}}{\text{Cal}} \times \dfrac{4.184 \text{ J}}{\text{cal}} = 6.7 \times 10^5 \text{ J}$

(b) $75 \text{ W} = 75 \text{ J/s}$; $75 \text{ J/s} \times 3.0 \text{ hr} \times 60 \text{ min/hr} \times 60 \text{ s/min} = 8.1 \times 10^5 \text{ J}$

(c) $16 \text{ kJ} \times \dfrac{1 \text{ kcal}}{4.184 \text{ kJ}} = 3.8 \text{ kcal}$

6.2 $\Delta E = 2400 \text{ J} = q + w = -1.89 \text{ kJ} + w$

$w = 2400 \text{ J} + 1.89 \text{ kJ} = 2.4 \text{ kJ} + 1.89 \text{ kJ} = 4.3 \text{ kJ}$

6.3 $q = c \times m \times \Delta T = c \times m \times (T_{final} - T_{initial})$

$T_{final} = T_{initial} + \dfrac{q}{c \times m} = 5 \,°C + \dfrac{24{,}100 \text{ J}}{(0.902 \text{ J g}^{-1}°C^{-1})(250. \text{ g})}$

$= 5 \,°C + 106.8 \,°C = 112 \,°C$

6.4 $q = c \times m \times \Delta T = 4.184 \text{ J g}^{-1}°C^{-1} \times 250.0 \text{ g}$

$\times (65 - 37) \,°C = 29.2 \text{ kJ}$

$\Delta E = q + w = 29.2 \text{ kJ} + 0 = 29.2 \text{ kJ}$

6.5 $q_{water} = -q_{iron}$

$(4.184 \text{ J/g }°C)(1000 \text{ g})(32.8 \,°C - 20.0 \,°C)$

$= -(0.451 \text{ J/g }°C)(400 \text{ g})(32.8 \,°C - T_i)$

$T_i = (297 + 32.8) \,°C = 330 \,°C$

6.6 $1.00 \text{ g K(s)} \times \dfrac{1 \text{ mL}}{0.86 \text{ g}} = 1.16 \text{ mL}$;

$1.00 \text{ g K}(\ell) \times \dfrac{1 \text{ mL}}{0.82 \text{ g}} = 1.22 \text{ mL}$

The change in volume is $(1.22 - 1.16)$ mL $= 0.06$ mL.

$$w = 0.06 \text{ mL} \times \frac{0.10 \text{ J}}{1 \text{ mL}} = 6 \times 10^{-3} \text{ J}$$

$$\Delta H = \frac{14.6 \text{ cal}}{1 \text{ g}} \times 1.00 \text{ g} \times \frac{4.184 \text{ J}}{1 \text{ cal}} = 61.1 \text{ J}$$

$$\Delta E = \Delta H + w = 61.1 \text{ J} + (5 \times 10^{-3} \text{ J}) = 61.1 \text{ J}$$

6.7 (a) $10.0 \text{ g I}_2 \times \dfrac{1 \text{ mol I}_2}{253.8 \text{ g I}_2} \times \dfrac{62.4 \text{ kJ}}{1 \text{ mol I}_2} = 2.46 \text{ kJ}$

(b) $3.42 \text{ g I}_2 \times \dfrac{1 \text{ mol I}_2}{253.8 \text{ g I}_2} \times \dfrac{-62.4 \text{ kJ}}{1 \text{ mol I}_2}$
$$= -0.841 \text{ kJ} = -841 \text{ J}$$

(c) This process is the reverse of the one in part (a), so $\Delta H°$ is negative. This means that the process is exothermic. The quantity of energy transferred is 841 J.

6.8 $8 \text{ CaO(s)} \rightarrow 8 \text{ Ca(s)} + 4 \text{ O}_2\text{(g)}$
$\Delta H° = 8(+635.09 \text{ kJ}) = +5080.7 \text{ kJ}$

6.9 According to the thermochemical equation, the reaction is endothermic, so 285.8 kJ of energy is transferred into the system per mole of $H_2O(\ell)$ decomposed. Thus,

$$12.6 \text{ g H}_2\text{O} \times \frac{1 \text{ mol H}_2\text{O}}{18.02 \text{ g H}_2\text{O}} \times \frac{285.8 \text{ kJ}}{1 \text{ mol H}_2\text{O}} = 200. \text{ kJ}$$

6.10 $\Delta T = (36.75 - 20.34) \text{ °C} = 16.41 \text{ °C}$

$$q_{\text{calorimeter}} = \frac{923 \text{ J}}{1 \text{ °C}} \times 16.41 \text{ °C} = 15.15 \times 10^3 \text{ J} = 15.15 \text{ kJ}$$

$$q_{\text{water}} = 4.184 \text{ J g}^{-1}\text{°C}^{-1} \times 815 \text{ g} \times 16.41 \text{ °C} = 55.96 \text{ kJ}$$

$$q_{\text{reaction}} = -(q_{\text{calorimeter}} + q_{\text{water}}) = -(15.15 + 55.96) \text{ kJ}$$
$$= -71.11 \text{ kJ}$$

$$7.68 \text{ g S}_8 \times \frac{1 \text{ mol S}_8}{256.1 \text{ g S}_8} = 3.00 \times 10^{-2} \text{ mol S}_8$$

$$\frac{-71.11 \text{ kJ}}{3.00 \times 10^{-2} \text{ mol S}_8} = \frac{\Delta E}{1 \text{ mol S}_8}$$

$$\Delta E = -2.37 \times 10^3 \text{ kJ}$$

6.11 The total volume of the initial solutions is 200. mL, which corresponds to 200. g of solution. The quantities of reactants are 0.10 mol of H^+(aq) and 0.10 mol of OH^-(aq), so 0.10 mol of H_2O is formed.

$$0.10 \text{ mol H}_2\text{O} \times \frac{-58.7 \text{ kJ}}{1 \text{ mol H}_2\text{O}} = -5.87 \text{ kJ}$$

Since $\Delta H°$ is negative, energy is transferred to the water, and its temperature will rise.

$$\Delta T = \frac{q}{c \times m} = \frac{5.87 \times 10^3 \text{ J}}{(4.184 \text{ J g}^{-1}\text{°C}^{-1})(200. \text{ g})} = 7.0 \text{ °C}$$

The final temperature will be $(20.4 + 7.0) \text{ °C} = 27.4 \text{ °C}$.

6.12 $C\text{(s)} + O_2\text{(g)} \rightarrow CO_2\text{(g)}$ $\qquad \Delta H° = -393.5 \text{ kJ}$

$\dfrac{CO_2\text{(g)} \rightarrow CO\text{(g)} + \frac{1}{2} O_2\text{(g)} \qquad \Delta H° = 283.0 \text{ kJ}}{}$

$C\text{(s)} + \frac{1}{2} O_2\text{(g)} \rightarrow CO\text{(g)} \qquad \Delta H° = -110.5 \text{ kJ}$

6.13 (a) $\frac{1}{2}\text{N}_2\text{(g)} + \frac{3}{2}\text{H}_2\text{(g)} \rightarrow \text{NH}_3\text{(g)}$ $\quad \Delta H_f° = -46.11 \text{ kJ/mol}$
(b) $C\text{(graphite)} + \frac{1}{2} O_2\text{(g)} \rightarrow CO\text{(g)}$
$$\Delta H_f° = -110.525 \text{ kJ/mol}$$

6.14 For the reaction given,

$\Delta H° = \{6 \text{ mol CO}_2\text{(g)}\} \times \Delta H_f°\{CO_2\text{(g)}\}$
$\qquad + \{5 \text{ mol H}_2\text{O}(\ell)\} \times \Delta H_f°\{H_2O(\ell)\}$
$\qquad - (2 \text{ mol } \{C_3H_5(NO_3)_3(\ell)\} \times \Delta H_f°\{C_3H_5(NO_3)_3(\ell)\}$
$\qquad = \{6(-393.509) + 5(-285.830) - 2(-364)\} \text{ kJ}$
$\qquad = -3.06 \times 10^3 \text{ kJ}$

For 10.0 g of nitroglycerin (nitro),

$$q = 10.0 \text{ g} \times \frac{1 \text{ mol nitro}}{227.09 \text{ g}}$$

$$\times \frac{-3.06 \times 10^3 \text{ kJ}}{2 \text{ mol nitro}} = -67.4 \text{ kJ}$$

(The 2 mol of nitro in the last factor comes from the coefficient of 2 associated with nitroglycerin in the chemical equation.)

6.15 (a) $227 \text{ g milk} \times \dfrac{5.0 \text{ g carbohydrate}}{100 \text{ g milk}}$

$$\times \frac{17 \text{ kJ}}{1 \text{ g carbohydrate}} = 193 \text{ kJ}$$

$227 \text{ g milk} \times \dfrac{4.0 \text{ g fat}}{100 \text{ g milk}} \times \dfrac{38 \text{ kJ}}{1 \text{ g fat}} = 345 \text{ kJ}$

$227 \text{ g milk} \times \dfrac{3.3 \text{ g protein}}{100 \text{ g milk}} \times \dfrac{17 \text{ kJ}}{1 \text{ g protein}} = 127 \text{ kJ}$

Total caloric intake $= (193 + 345 + 127) \text{ kJ} = 665 \text{ kJ}$

Reduce by 10% to digest, absorb, and metabolize, giving $665 \text{ kJ} - 66.5 \text{ kJ} = 598 \text{ kJ}$.

Walking requires $2.5 \times \text{BMR} = 2.5 \times \dfrac{1750 \text{ Cal}}{1 \text{ day}}$

$$\times \frac{4.184 \text{ kJ}}{1 \text{ Cal}} \times \frac{1 \text{ day}}{24 \text{ h}} \times \frac{1 \text{ h}}{60 \text{ min}} = 12.7 \text{ kJ/min}$$

$$598 \text{ kJ} \times \frac{1 \text{ min}}{12.7 \text{ kJ}} = 47 \text{ min}$$

Chapter 7

7.1 $\lambda = \dfrac{c}{\nu} = \dfrac{2.998 \times 10^8 \text{ m/s}}{650 \times 10^6 \text{ s}^{-1}} = 4.61 \times 10^{-1} \text{ m}$

7.2 (a) One photon of ultraviolet radiation has more energy because ν is larger in the UV spectral region than in the microwave region.

(b) One photon of blue light has more energy because the blue portion of the visible spectrum has a higher frequency than the green portion of the visible spectrum.

(c) Ten blue photons of $\lambda = 460$ nm would have an energy of $10 \, hc/\lambda$.

$$6.626 \times 10^{-34} \text{ J·s}^{-1} \left(\frac{2.998 \times 10^8 \text{ m·s}^{-1}}{460 \times 10^{-9} \text{ m}} \right) \times 10$$

$$= 4.32 \times 10^{-18} \text{ J}$$

while 15 red photons of $\lambda = 695$ nm would have an energy of $15\,hc/\lambda$

$$6.626 \times 10^{-34}\,\text{J}\cdot\text{s}^{-1}\left(\frac{2.998 \times 10^{8}\,\text{m}\cdot\text{s}^{-1}}{695 \times 10^{-9}\,\text{m}}\right) \times 15$$

$$= 4.29 \times 10^{-18}\,\text{J}$$

So the 10 blue photons have slightly more energy.

7.3 (a) $$\nu = \frac{2.179 \times 10^{-18}\,\text{J}}{h}\left(\frac{1}{n_i^2} - \frac{1}{n_f^2}\right)$$

$$= \left(\frac{2.179 \times 10^{-18}\,\text{J}}{6.626 \times 10^{-34}\,\text{J}\cdot\text{s}}\right)\left(\frac{1}{4^2} - \frac{1}{3^2}\right)$$

$$= (3.289 \times 10^{15}\,\text{s}^{-1})\left(\frac{1}{16} - \frac{1}{9}\right)$$

$$= (3.289 \times 10^{15}\,\text{s}^{-1})(-4.86 \times 10^{-2})$$

$$= -1.599 \times 10^{14}\,\text{s}^{-1}$$

The negative sign indicates that energy is emitted.

(b) Longer than that of the $n = 5$ to $n = 3$ transition.

7.4 $$\lambda = \frac{h}{mv} = \frac{6.626 \times 10^{-34}\,\text{J}\cdot\text{s}}{1.67 \times 10^{-27}\,\text{kg} \times 2.998 \times 10^{7}\,\text{m/s}}$$

$$= 1.33 \times 10^{-14}\,\text{m}$$

7.5 (a) 6*d* (b) 10 (c) 2, 1, 0, −1, −2

7.6 (a) [Ne] $3s^2 3p^4$ (b) [Ne] 3*s*: [↑↓] 3*p*: [↑↓] [↑] [↑]

7.7 The electron configurations for :Se: and :Te: are $[Ar]4s^2 3d^{10} 4p^4$ and $[Kr]5s^2 4d^{10} 5p^4$, respectively. Elements in the same main group have similar electron configurations.

7.8 (a) P^{3-} (b) Ca^{2+}

7.9 (a) The electron configuration for the nickel atom is $[Ar]4s^2 3d^8$.

(b) The orbital box diagram for the Ni atom's outer electrons is 3*d*: [↑↓] [↑↓] [↑↓] [↑] [↑] 4*s*: [↑↓]. The Ni atom has two unpaired electrons.

(c) The orbital box diagram for the Ni^{2+} ion is 3*d*: [↑↓] [↑↓] [↑↓] [↑] [↑] (only the two 4*s* electrons are removed in forming the Ni^{2+} ion), so it also has two unpaired electrons.

7.10 B < Mg < Na < K

7.11 (a) Cs^+ (b) La^{3+}

7.12 F > N > P > Na

Chapter 8

8.1 (a) 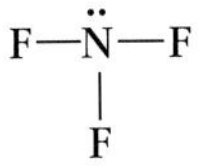 (b) 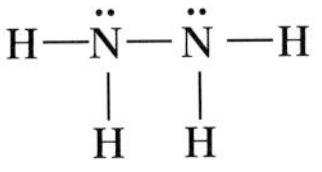

(c) 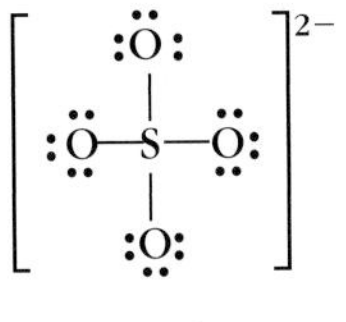

8.2 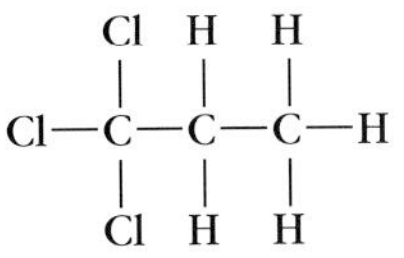

8.3 (a) $[N{\equiv}O:]^+$ (b) H—C≡N:

8.4 Only (c) can have geometric isomers. Molecules in (a) and (b) each have the same two groups on one of the double-bonded carbons.

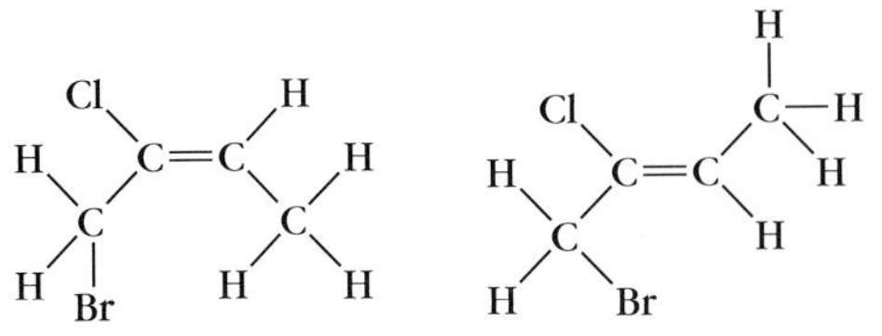

cis-1-bromo-2-chloro-2-butene *trans*-1-bromo-2-chloro-2-butene

8.5 (a) Si is a larger atom than S. (b) Br is a larger atom than Cl. (c) The greater electron density in the triple bond brings the N≡O atoms closer together than the smaller electron density in the N═O double bond does.

8.6 $\Delta H° = [(2\text{ mol H})(436\text{ kJ/mol}) + (1\text{ mol } O_2)(498\text{ kJ/mol})] - (4\text{ mol O—H})(467\text{ kJ/mol}) = (1370\text{ kJ}) - (1868\text{ kJ}) = -498\text{ kJ}$

8.7 (a) B—Cl is more polar; $\overset{\delta+}{B} — \overset{\delta-}{Cl}$; O—H is more polar; $\overset{\delta-}{O} — \overset{\delta+}{H}$.

8.8 The other Lewis structure is :O≡N—N̈:

	O	N	N
Valence electrons	6	5	5
Lone pair electrons	2	0	6
$\frac{1}{2}$ shared electrons	3	4	1
Formal charge	+1	+1	−2

8.9 The N-to-O bond length in NO_2^- is 124 pm. From Table 8.1, N—O is 136 pm; N═O is 115 pm. Thus, the nature of the bond in NO_2^- is between that of a N—O single bond and a N═O double bond.

8.10 (a) :F̤̈—Be—F̤̈: (b) :Ö—C̤̈l—Ö: (c) 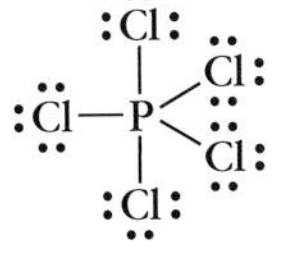 (d) $[H—B—H]^+$

(e) 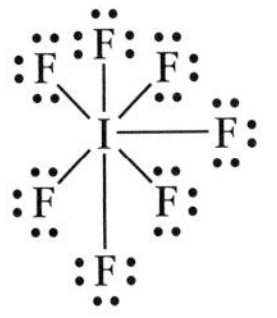

BeF_2—not an octet around the central Be atom; ClO_2—an odd number of electrons around Cl; PCl_5—more than four electron pairs around the central phosphorus atom; BH_2^+—only two electron pairs around the central B atom; IF_7—iodine has seven shared electron pairs.

Chapter 9

9.1 There are three B—F bonds in BF_3 and no lone pairs on boron. Therefore, the electron-pair geometry and the molecular geometry are the same, triangular planar, with 120° B—F bond angles.

9.2

Central atom (underlined)	Bond pairs	Lone pairs	Electron-pair geometry	Molecular shape
$\underline{Br}O_3^-$	3	1	Tetrahedral	Triangular pyramid
$\underline{Se}F_2$	2	2	Tetrahedral	Angular
$\underline{N}O_2^-$	3	1	Triangular planar	Angular

9.3 (a) ClF_2^-: triangular bipyramidal electron-pair geometry and linear molecular geometry

(b) XeO_3: tetrahedral electron-pair geometry and triangular pyramidal molecular geometry

9.4 (a) In HCN, the *sp* hybridized carbon atom is sigma bonded to H and to N, as well as having two pi bonds to N. The sigma and two pi bonds form the C≡N triple bond. The nitrogen is *sp* hybridized with a sigma and two pi bonds to carbon; a lone pair is in the nonbonding *sp* hybrid orbital on N.

(b) The double-bonded carbon and nitrogen are both sp^2 hybridized. The sp^2 hybrid orbitals on C form sigma bonds to H and to N; the unhybridized *p* orbital on C forms a pi bond with the unhybridized *p* orbital on N. The sp^2 hybrid orbitals on N form sigma bonds to carbon and to H; the N lone pair is in the nonbonding sp^2 hybrid orbital.

9.5 (a) The central P atom has six bond pairs, no lone pairs, and is sp^3d^2 hybridized.

(b) The central I atom has three bond pairs and two lone pairs; these five electron pairs are in sp^3d hybridized orbitals on I.

(c) In ICl_4^-, the central I has four bond pairs, two lone pairs; these six pairs are in sp^3d^2 orbitals.

9.6 (a) $BFCl_2$ is a triangular planar molecule with polar B—F and B—Cl bonds. The molecule is polar because the B—F bond is more polar than the B—Cl bonds, resulting in a net dipole.

(b) NH_2Cl is a triangular pyramidal molecule with polar N—H bonds. (N—Cl is a nonpolar bond; N and Cl have the same electronegativity.) It is a polar molecule because the N—H dipoles do not cancel and produce a net dipole.

(c) SCl_2 is an angular polar molecule. The polar S—Cl bond dipoles do not cancel each other because they are not symmetrically arranged due to the two lone pairs on S.

9.7 (a) London forces between Kr atoms must be overcome for krypton to melt.

(b) The C—H covalent bonds in propane must be broken to form C and H atoms; the H atoms covalently bond to form H_2.

9.8 (a) London forces occur between N_2 molecules.

(b) CO_2 is nonpolar, and London forces occur between it and polar water molecules.

(c) London forces occur between the two molecules, but the principal intermolecular forces are the hydrogen bonds between the H on NH_3 with the lone pairs on the OH oxygen, and the hydrogen bonds between the H on the oxygen in CH_3OH and lone pair on nitrogen in NH_3.

9.9 The chiral centers are identified by an asterisk in the structural formula. Each of those carbon atoms has four different groups or atoms attached to it.

Chapter 10

10.1 (a) Pressure in atm =
$$29.5 \text{ in Hg} \times \frac{1 \text{ atm}}{76.0 \text{ cm Hg}} \times \frac{2.54 \text{ cm}}{1 \text{ in}} = 0.986 \text{ atm}$$

(b) Pressure in mm Hg =
$$29.5 \text{ in Hg} \times \frac{25.4 \text{ mm}}{1 \text{ in}} = 749 \text{ mm Hg}$$

(c) Pressure in bar =
$$29.5 \text{ in Hg} \times \frac{1.013 \text{ bar}}{760 \text{ mm Hg}} \times \frac{25.4 \text{ mm}}{1 \text{ in}} = 0.999 \text{ bar}$$

(d) Pressure in kPa =
$$29.5 \text{ in Hg} \times \frac{101.3 \text{ kPa}}{760 \text{ mm Hg}} \times \frac{25.4 \text{ mm}}{1 \text{ in}} = 99.9 \text{ kPa}$$

10.2 Pressure in bar =
$$647 \text{ mm Hg} \times \frac{1.013 \text{ bar}}{760 \text{ mm Hg}} = 0.862 \text{ bar}$$

Pressure in kPa =
$$647 \text{ mm Hg} \times \frac{101.3 \text{ kPa}}{760 \text{ mm Hg}} = 86.2 \text{ kPa}$$

Pressure in atm =
$$647 \text{ mm Hg} \times \frac{1 \text{ atm}}{760 \text{ mm Hg}} = 0.851 \text{ atm}$$

10.3 The temperature remains constant, so the average energy of the gas molecules remains constant. If the volume is decreased, then the gas molecules must hit the walls more frequently, and the pressure is increased.

10.4 Volume of NO gas $= 1.0 \text{ L O}_2 \times \dfrac{2 \text{ L NO}}{1 \text{ L O}_2} = 2 \text{ L NO}$

10.5 $V = \dfrac{nRT}{P}$

$= \dfrac{(2.64 \text{ mol})(0.0821 \text{ L atm mol}^{-1} \text{ K}^{-1})(304 \text{ K})}{0.640 \text{ atm}} = 103 \text{ L}$

10.6 $V_2 = \dfrac{P_1 V_1}{P_2} = \dfrac{(1.00 \text{ atm})(400. \text{ mL})}{0.750 \text{ atm}} = 533 \text{ mL}$

10.7 $V_2 = \dfrac{V_1 T_2}{T_1} = \dfrac{(236 \text{ mL})(362 \text{ K})}{304 \text{ K}} = 281 \text{ mL}$

10.8 (a) $V_2 = \dfrac{P_1 V_1 T_2}{P_2 T_1}$

$= \dfrac{(710 \text{ mm Hg})(21 \text{ mL})(299.6 \text{ K})}{(740 \text{ mm Hg})(295.4 \text{ K})} = 20 \text{ mL}$

(b) $V_2 = \dfrac{(21 \text{ mL})(299.6 \text{ K})}{(295.4 \text{ K})} = 21 \text{ mL}$

10.9 $\dfrac{10.0 \text{ g NH}_4\text{NO}_3}{80.043 \text{ g/mol}} = 0.1249 \text{ mol NH}_4\text{NO}_3$

$0.1249 \text{ mol NH}_4\text{NO}_3 \times \dfrac{7 \text{ mol product gases}}{2 \text{ mol NH}_4\text{NO}_3}$
$= 0.437 \text{ mol produced}$

$V = \dfrac{(0.437 \text{ mol})(0.0821 \text{ L atm mol}^{-1} \text{ K}^{-1})(298 \text{ K})}{1 \text{ atm}} = 10.7 \text{ L}$

10.10 $\dfrac{1.0 \text{ g LiOH}}{23.94 \text{ g/mol}} = 0.0418 \text{ mol LiOH}$

$0.0418 \text{ mol LiOH} \times \dfrac{1 \text{ mol CO}_2}{2 \text{ mol LiOH}} = 0.0209 \text{ mol CO}_2$

$V = \dfrac{(0.0209 \text{ mol})(0.0821 \text{ L atm mol}^{-1} \text{ K}^{-1})(295 \text{ K})}{1 \text{ atm}}$
$= 0.51 \text{ L CO}_2$

10.11 $V = \frac{4}{3}\pi r^3 = \frac{4}{3}\pi (10. \text{ cm})^3 = 4190 \text{ cm}^3 = 4.19 \text{ L}$

Amount of CO_2 gas, $n = \dfrac{PV}{RT}$

$= \dfrac{(2.00 \text{ atm})(4.19 \text{ L})}{(0.0821 \text{ L atm mol}^{-1} \text{ K}^{-1})(293 \text{ K})} = 0.348 \text{ mol CO}_2$

Mass of $NaHCO_3 = 0.348 \text{ mol CO}_2 \times$

$\dfrac{1 \text{ mol NaHCO}_3}{1 \text{ mol CO}_2} \times \dfrac{84.00 \text{ g NaHCO}_3}{1 \text{ mol NaHCO}_3} = 29 \text{ g NaHCO}_3$

10.12 Amount of gas, $n = \dfrac{PV}{RT} =$

$\dfrac{(0.850 \text{ atm})(1.00 \text{ L})}{(0.0821 \text{ L atm mol}^{-1} \text{ K}^{-1})(293 \text{ K})} = 0.0353 \text{ mol}$

Molar mass $= \dfrac{1.13 \text{ g}}{0.0353 \text{ mol}} = 32.0 \text{ g/mol}$

The gas is probably oxygen.

10.13 Amount of $N_2 = 7.0 \text{ g N}_2 \times \dfrac{1 \text{ mol N}_2}{28.10 \text{ g N}_2} = 0.25 \text{ mol N}_2$

Amount of $H_2 = 6.0 \text{ g H}_2 \times \dfrac{1 \text{ mol H}_2}{2.02 \text{ g H}_2} = 3.0 \text{ mol H}_2$

Total number of moles $= 3.0 + 0.25 = 3.25 \text{ mol}$

$X_{N_2} = \dfrac{0.25 \text{ mol}}{3.23 \text{ mol}} = 0.078 \qquad X_{H_2} = \dfrac{3.0 \text{ mol}}{3.25 \text{ mol}} = 0.94$

$P_{N_2} =$

$\dfrac{(0.25 \text{ mol})(0.0821 \text{ L atm mol}^{-1} \text{ K}^{-1})(773 \text{ K})}{5.0 \text{ L}} = 3.2 \text{ atm}$

$P_{H_2} =$

$\dfrac{(3.0 \text{ mol})(0.0821 \text{ L atm mol}^{-1} \text{ K}^{-1})(773 \text{ K})}{5.0 \text{ L}} = 38 \text{ atm}$

10.14 For NO:

$n = \dfrac{(1.0 \text{ atm})(4.0 \text{ L})}{(0.0821 \text{ L atm mol}^{-1} \text{ K}^{-1})(298 \text{ K})} = 0.163 \text{ mol NO}$

For O_2:

$n = \dfrac{(0.40 \text{ atm})(2.0 \text{ L})}{(0.0821 \text{ L atm mol}^{-1} \text{ K}^{-1})(298 \text{ K})} = 0.0327 \text{ mol O}_2$

All the O_2 is used.

$0.0327 \text{ mol O}_2 \times \dfrac{2 \text{ mol NO}}{1 \text{ mol O}_2} = 0.0654 \text{ mol NO used}$

$0.163 \text{ mol} - 0.0654 \text{ mol} = 0.0976 \text{ mol NO remains}$

$0.0327 \text{ mol O}_2 \times \dfrac{2 \text{ mol NO}_2}{1 \text{ mol O}_2} = 0.0654 \text{ mol NO}_2 \text{ formed}$

$n_{\text{total}} = 0.0976 + 0.0654 = 0.163 \text{ mol of gas}$

$P_{\text{total}} = \dfrac{nRT}{V}$

$= \dfrac{(0.163 \text{ mol})(0.0821 \text{ L atm mol}^{-1} \text{ K}^{-1})(298 \text{ K})}{6.0 \text{ L}}$

$= 0.661 \text{ atm}$

10.15 $P_{HCl} = P_{\text{total}} - P_{\text{water}} = 740 \text{ mm Hg} - 21 \text{ mm Hg}$
$= 719 \text{ mm Hg}$

$n = \dfrac{PV}{RT} = \dfrac{(719/760 \text{ atm})(0.260 \text{ L})}{(0.0821 \text{ L atm mol}^{-1} \text{ K}^{-1})(296 \text{ K})}$
$= 0.0101 \text{ mol HCl}$

$0.0101 \text{ mol} \times 2.0158 \text{ g/mol} = 0.0204 \text{ g} = 20.4 \text{ mg HCl}$

10.16

$P = \dfrac{(1 \text{ mol})(0.0821 \text{ L atm mol}^{-1} \text{ K}^{-1})(273 \text{ K})}{20 \text{ L}} = 1.121 \text{ atm}$

$P = \left(\dfrac{nRT}{V - nb}\right) - \left(\dfrac{n^2 a}{V^2}\right)$

$= \dfrac{(1)(0.0821)(273)}{20.0 - (1)(0.0428)} - \dfrac{(1)^2 (2.25)}{20^2} = 1.117 \text{ atm}$

% difference $= \dfrac{1.121 - 1.117}{1.121} \times 100\% = 0.36\%$

Chapter 11

11.1 Heat $= 2.5 \times 10^{10} \text{ kg H}_2\text{O} \times \dfrac{10^3 \text{ g}}{1 \text{ kg}}$

$\times \dfrac{1 \text{ mol H}_2\text{O}}{18.02 \text{ g H}_2\text{O}} \times \dfrac{-44.0 \text{ kJ}}{\text{mol}} = -6.10 \times 10^{13} \text{ kJ}$

This process is exothermic as water vapor condenses, forming rain.

11.2 Heat required to melt NaCl $= 100.0 \text{ g NaCl}$

$\times \dfrac{1 \text{ mol NaCl}}{58.442 \text{ g NaCl}} \times \dfrac{30.21 \text{ kJ}}{\text{mol NaCl}} = 51.69 \text{ kJ}$

11.3 Only water vapor could exist under these conditions.

11.4 (a) Solid decane is a molecular solid.

(b) Solid $MgCl_2$ is composed of Mg^{2+} and Cl^- ions and is an ionic solid.

11.5 There are 2 atoms per bcc unit cell. The diagonal of the bcc unit cell is 4 times the radius of the atoms in the unit cell, so, solving for the edge,

$$\text{edge} = \frac{4 \times 144 \text{ pm}}{\sqrt{3}} = 332 \text{ pm}$$

$$\text{density} = \frac{\text{mass}}{\text{volume}}$$

$$= \frac{(2 \text{ Au atoms})(196.67 \text{ g Au}/6.022 \times 10^{23} \text{ Au atoms})}{[(332 \text{ pm})(1 \text{ m}/10^{12} \text{ pm})(10^2 \text{ cm/m})]^3}$$

$$= 17.9 \text{ g/cm}^3$$

11.6 The edge of the KCl unit cell would be 2×152 pm $+$ 2×167 pm $= 638$ pm
The unit cell of KCl is larger than that of NaCl.

Volume of the unit cell $= (638 \text{ pm})^3 =$

$$2.60 \times 10^{18} \text{ pm}^3 \times \left(\frac{10^{-10} \text{ cm}}{\text{pm}}\right)^3 = 2.60 \times 10^{-22} \text{ cm}^3$$

11.7 Energy transfer required $= 1.45 \text{ g Al} \times \dfrac{1 \text{ mol Al}}{26.98 \text{ g Al}} \times$
$\dfrac{10.7 \text{ kJ}}{\text{mol}} = 0.575 \text{ kJ}$

Chapter 12

12.1 Energy $= 4.2 \times 10^9 \text{ t coal} \times \dfrac{26.4 \times 10^9 \text{ J}}{1 \text{ t coal}}$

$$= 1.1 \times 10^{20} \text{ J}$$

$\text{ft}^3 \text{ natural gas} = 1.1 \times 10^{20} \text{ J} \times \dfrac{1 \text{ ft}^3 \text{ natural gas}}{1.055 \times 10^6 \text{ J}} =$

$$1.0 \times 10^{14} \text{ ft}^3 \text{ natural gas}$$

12.2 The balanced combustion reaction for methanol vapor is

$$CH_3OH(g) + \tfrac{3}{2} O_2(g) \longrightarrow CO_2(g) + 2 H_2O(g)$$

Using Hess's law, we see that the heat of combustion of methanol vapor is
$\Delta H_{comb} = [\Delta H_f^\circ\, CO_2(g)] + 2[\Delta H_f^\circ\, H_2O(g)]$
$$- 1[\Delta H_f^\circ\, CH_3OH(g)]$$
$= -393.509 \text{ kJ/mol} + 2(-241.818 \text{ kJ/mol})$
$$- (-200.66 \text{ kJ/mol})$$
$= -676.48 \text{ kJ/mol}$
For methanol,

$$-676.48 \text{ kJ/mol} \left(\frac{1 \text{ mol}}{32.04 \text{ g}}\right)\left(\frac{0.791 \text{ g}}{1 \text{ mL}}\right)\left(\frac{1000 \text{ mL}}{\text{L}}\right) =$$

$$-1.67 \times 10^4 \text{ kJ/L}$$

Methanol yields less energy per liter than ethanol.

12.3 (a) The first oxidation product of $CH_3CH_2CH_2OH$ is the

$$\overset{O}{\overset{\|}{}}$$
aldehyde CH_3CH_2CH. The second oxidation product

$$\overset{O}{\overset{\|}{}}$$
of $CH_3CH_2CH_2OH$ is the acid $CH_3CH_2C{-}OH$

(b) The oxidation product of this secondary alcohol is

$$\overset{O}{\overset{\|}{}}$$
the ketone $CH_3{-}C{-}CH_2CH_3$.

12.4 (a)

(b)

(c)

12.5 $HO{-}CH_2{-}\overset{O}{\overset{\|}{C}}{-}OH$

12.6 Twelve moles of water

12.7

Chapter 13

13.1 (a)

$$\text{Rate} = \frac{-\Delta[\text{cisplatin}]}{\Delta t} = 4.54 \times 10^{-6} \text{ mol L}^{-1} \text{ min}^{-1}$$

$$\Delta t = \frac{-\Delta[\text{cisplatin}]}{4.54 \times 10^{-6} \text{ mol L}^{-1} \text{ min}^{-1}}$$

$$= \frac{(0.00325 - 0.00279) \text{ mol/L}}{4.54 \times 10^{-6} \text{ mol L}^{-1} \text{ min}^{-1}}$$

$$= 101 \text{ min}$$

(b) No. The rate of reaction depends on the concentration of cisplatin and therefore gets slower as the reaction progresses. Therefore, the method used in part (a) works only over a small range of concentrations.

13.2 (a) $\dfrac{1}{2}\dfrac{\Delta[\text{HI}]}{\Delta t} = -\dfrac{1}{1}\dfrac{\Delta[\text{H}_2]}{\Delta t}$

$$\frac{\Delta[\text{HI}]}{\Delta t} = -2\frac{\Delta[\text{H}_2]}{\Delta t}$$

(b) $\dfrac{\Delta[\text{HI}]}{\Delta t} = -2\dfrac{\Delta[\text{I}_2]}{\Delta t} = -2(-0.0037 \text{ mol L}^{-1}\text{s}^-)$

$$= 0.0074 \text{ mol L}^{-1} \text{ s}^{-1}$$

13.3 (a) The effect of $[H_2O]$ on the rate of reaction cannot be determined, because the $[H_2O]$ is the same in all three experiments.

(b) Rate $= k$[cisplatin]

(c) $k_1 = \dfrac{\text{rate}}{\text{[cisplatin]}} = \dfrac{3.07 \times 10^{-5}\,\text{mol L}^{-1}\,\text{min}^{-1}}{0.021\,\text{mol/L}}$

$= 1.46 \times 10^{-3}\,\text{min}^{-1}$

$k_2 = 1.46 \times 10^{-3}\,\text{min}^{-1}$

$k_3 = 1.46 \times 10^{-3}\,\text{min}^{-1}$

$k = \dfrac{k_1 + k_2 + k_3}{3} = 1.5 \times 10^{-3}\,\text{min}^{-1}$

(d) Rate $= k$[cisplatin] $= (1.5 \times 10^{-3}\,\text{min}^{-1})$
$(0.100\,\text{mol/L}) = 1.5 \times 10^{-4}\,\text{mol L}^{-1}\,\text{min}^{-1}$

(e) Rate $= (1.5 \times 10^{-3}\,\text{min}^{-1})(0.050\,\text{mol/L}) = 7.5 \times 10^{-5}\,\text{mol L}^{-1}\,\text{min}^{-1}$

13.4 (a) Second-order in NO; first-order in H_2.

(b) Let $[NO]_1$ be the original concentration of NO and $[H_2]_1$ be the original concentration of H_2. Then $[NO]_2 = 3[NO]_1$ and $[H_2]_2 = \frac{1}{8}[H_2]_1$

$\text{Rate}_1 = k[NO]_1^2[H_2]_1$

$\text{Rate}_2 = k[NO]_2^2[H_2]_2 = k(3[NO]_1)^2(\frac{1}{8}[H_2]_1)$

$= k \times 9 \times \frac{1}{8}[NO]_1^2[H_2]_1$

$= 1.125\,k[NO]_1^2[H_2]_1 = 1.125\,\text{Rate}_1$

The reaction is 12.5% faster.

13.5 Make three plots of the data.

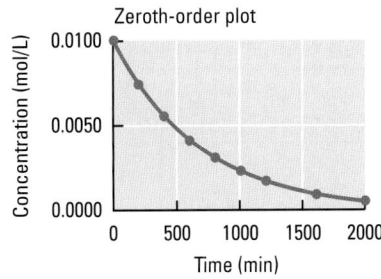

Zeroth-order plot

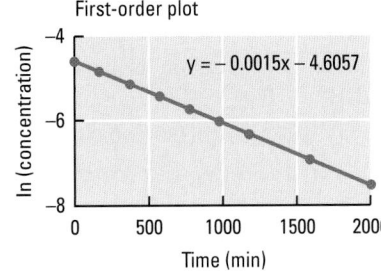

First-order plot

$y = -0.0015x - 4.6057$

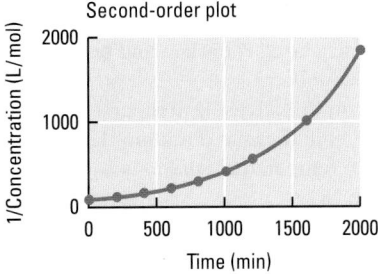

Second-order plot

The first-order plot is a straight line and the others are curved, so the reaction is first order. The slope of the first-order plot is $-0.0015\,\text{min}^{-1}$, so $k = -\text{slope} = 0.0015\,\text{min}^{-1}$.

13.6 Use the integrated first-order rate law from Table 13.2.

$$\ln[A]_t = -kt + \ln[A]_0$$

$$\ln\frac{[A]_t}{[A]_0} = -kt$$

$$t = -\frac{1}{k}\ln\frac{[A]_t}{[A]_0} = -\left(\frac{1}{3.43 \times 10^{-2}\,\text{d}^{-1}}\right)\ln\left(\frac{0.1}{1.0}\right)$$

$$= -(29.15\,\text{d})(-2.303) = 67.1\,\text{d}$$

13.7 In Figure 13.3 the [cisplatin] falls from 0.010 M to 0.0050 M in 470 min. The concentration falls from 0.0050 M to 0.0025 M between 470 and 936 min, or 467 min. Estimating from the graph yields only 2 significant figures, so the two times are equal within the precision of the graph. Therefore $t_{1/2} = 470$ min.

$$k = \frac{0.693}{t_{1/2}} = \frac{0.693}{470} = 1.5 \times 10^{-3}\,\text{min}^{-1}$$

13.8 Reaction is exothermic.

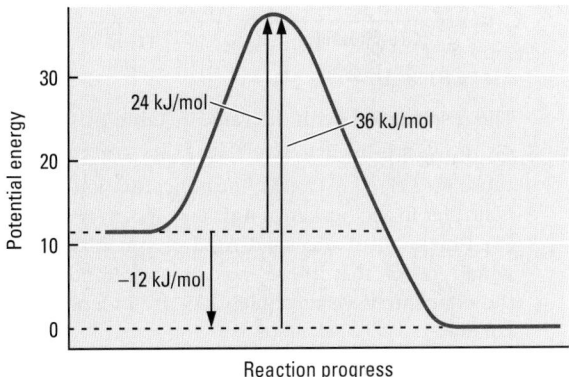

13.9 Obtain the value $E_a = 76.3$ kJ/mol from the discussion and analysis of the data in Figure 13.10 and the value $k = 4.18 \times 10^{-5}$ L mol^{-1} s^{-1} at 273 K from Table 13.3.

$$\frac{k_1}{k_2} = e^{\left[\frac{E_a}{R}\left(\frac{1}{T_2} - \frac{1}{T_1}\right)\right]} = e^{\left[\frac{76{,}300\,\text{J/mol}}{8.314\,\text{J K}^{-1}\text{mol}^{-1}}\left(\frac{1}{348\,\text{K}} - \frac{1}{273\,\text{K}}\right)\right]}$$

$$= e^{-7.245} = 7.137 \times 10^{-4}$$

$$k_2 = \frac{k_1}{7.138 \times 10^{-4}} = \frac{4.18 \times 10^{-5}\,\text{L mol}^{-1}\,\text{s}^{-1}}{7.138 \times 10^{-4}}$$

$$= 5.86 \times 10^{-2}\,\text{L mol}^{-1}\,\text{s}^{-1}$$

13.10 (a) $2\,NH_3(aq) + OCl^-(aq) \longrightarrow$

$N_2H_4(aq) + Cl^-(aq) + H_2O(\ell)$

(b) Step 1

(c) NH_2Cl, OH^-, $N_2H_5^+$

(d) Rate = rate of step 1 = $k[NH_3][OCl^-]$

13.11 Choose the structure that is most similar to the structure of p-aminobenzoic acid. An enzyme might be inhibited by this molecule, which could fit the active site but not be converted to a product similar to folic acid. Or the molecule might react in an enzyme-catalyzed process, producing a product whose biological function was different from folic acid. The best choice is

Chapter 14

14.1 (a) $K_c = \dfrac{[H_2S]}{[H_2]}$ (b) $K_c = \dfrac{[H_2]}{[HCl]}$

(c) $K_c = \dfrac{[CO][H_2]^3}{[CH_4][H_2O]}$ (d) $K_c = \dfrac{[HCN][OH^-]}{[CN^-]}$

14.2 $K_c = K_{c_1} \times K_{c_2} = (4.2 \times 10^{-7})(4.8 \times 10^{-11})$
$= 2.0 \times 10^{-17}$

14.3 $[CH_3COO^-] = \dfrac{2.96}{100} \times 0.0200 \text{ mol/L} = 5.92 \times 10^{-4}\,M$

$[H_3O^+] = 5.92 \times 10^{-4}\,M$

$[CH_3COOH] = \dfrac{100 - 2.96}{100} \times 0.0200 \text{ mol/L}$
$= 1.94 \times 10^{-2}\,M$

$K_c = \dfrac{[H_3O^+][CH_3COO^-]}{[CH_3COOH]} = \dfrac{(5.92 \times 10^{-4})(5.92 \times 10^{-4})}{1.94 \times 10^{-2}}$
$= 1.81 \times 10^{-5}$

The result agrees with the value in Table 14.1.

14.4 (a) $K_c(AgCl) = 1.8 \times 10^{-10}$; $K_c(AgI) = 1.5 \times 10^{-16}$. Because $K_c(AgI) < K_c(AgCl)$, the concentration of silver ions is smaller in the beaker of AgI.
(b) Unless all of the solid AgCl or AgI dissolves (which would mean that there was no equilibrium reaction), the concentrations at equilibrium are independent of the volume.

14.5 $Q = \dfrac{(\text{conc. } SO_3)^2}{(\text{conc. } SO_2)^2 (\text{conc. } O_2)}$
$= \dfrac{(0.184)^2}{(0.102 \times 2)^2(0.0132)} = 61.6$

Since $Q < K_c$ the forward reaction should occur.

14.6

	AuI(s) $\rightleftharpoons$ Au$^+$(aq) + I$^-$(aq)	
Initial concentration (mol/L)	0	0
Change as reaction occurs (mol/L)	$+x$	$+x$
Equilibrium concentration (mol/L)	x	x

$K_c = 1.6 \times 10^{-23} = [Au^+][I^-] = x^2$

$x = \sqrt{1.6 \times 10^{-23}} = 4.0 \times 10^{-12} = [Au^+] = [I^-]$

14.7 The reaction is the reverse of the one in Table 14.1, so

$K_c = \dfrac{1}{1.7 \times 10^2} = 5.9 \times 10^{-3}$

$Q = \dfrac{(\text{conc. } NO_2)^2}{(\text{conc. } N_2O_4)} = \dfrac{\left(\dfrac{0.500 \text{ mol}}{4.00 \text{ L}}\right)^2}{\left(\dfrac{1.00 \text{ mol}}{4.00 \text{ L}}\right)} = 6.25 \times 10^{-2}$

Because $Q > K_c$, the reaction should go in the reverse direction. Therefore, let x be the change in concentration of N_2O_4, giving the ICE table

	N$_2$O$_4$	$\rightleftharpoons$	**2 NO$_2$**
Initial concentration (mol/L)	$\dfrac{1.00}{4.00} = 0.250$		$\dfrac{0.500}{4.00} = 0.125$
Change as reaction occurs (mol/L)	x		$-2x$
Equilibrium concentration (mol/L)	$0.250 + x$		$0.125 - 2x$

$K_c = 5.9 \times 10^{-3} = \dfrac{(0.125 - 2x)^2}{0.250 + x}$
$= \dfrac{(1.56 \times 10^{-2}) - 0.500x + 4x^2}{0.250 + x}$

$(1.48 \times 10^{-3}) + (5.9 \times 10^{-3})x$
$= (1.56 \times 10^{-2}) - 0.500x + 4x^2$

$4x^2 - 0.5059x + (1.412 \times 10^{-2}) = 0$

$x = \dfrac{-(0.5059) \pm \sqrt{(-0.5059)^2 - 4 \times 4 \times 1.412 \times 10^{-2}}}{2 \times 4}$

$= \dfrac{0.5059 \pm \sqrt{3.001 \times 10^{-2}}}{8}$

$= \dfrac{0.5059 \pm 0.1732}{8}$

$x = 8.49 \times 10^{-2}$ or $x = 4.16 \times 10^{-2}$

If $x = 8.49 \times 10^{-2}$, then $[N_2O_4] = 0.250 + x = 0.335$ and $[NO_2] = 0.125 - 2x = 0.125 - (2 \times 7.86 \times 10^{-2}) = -0.0448$.

A negative concentration is impossible, so x must be 4.16×10^{-2}. Then

$[N_2O_4] = 0.250 + 0.0416 = 0.292$

$[NO_2] = 0.125 - (2 \times 0.0416) = 0.0418$

As predicted by Q, the reverse reaction has occurred, and the concentration of $[N_2O_4]$ has increased.

14.8

Because a bond is broken and because bond breaking is always endothermic ($\Leftarrow$ p. 234), the reaction must be endothermic. Increasing temperature shifts the equilibrium in the endothermic direction. Figure 14.3 shows that at a higher temperature there is a greater concentration of brown NO_2.

14.9 (a) There are more moles of gas phase reactants than products, so entropy favors the reactants.

(b) Data from Appendix J show that $\Delta H° = -46.11$ kJ. The reaction is exothermic, so the energy effect favors the products.

(c) As T increases the reaction shifts in the endothermic direction, which is toward reactants. The entropy effect also becomes more important at high T, and it favors reactants.

Chapter 15

15.1 Motor oil and gasoline are miscible because they are both nonpolar substances containing hydrocarbons, which are mutually soluble in each other.

15.2 The —OH groups attached to the ring and to the side chain of vitamin C hydrogen bond to water molecules. The oxygen atoms in the ring also form hydrogen bonds to water.

15.3 Solubility $= (8.4 \times 10^{-7}$ mol L^{-1} mm Hg$^{-1})(1520$ mm Hg$)$
$= 1.3 \times 10^{-3}$ mol/L
Solubility $= (8.4 \times 10^{-7}$ mol L^{-1} mm Hg$^{-1})(20$ mm Hg$)$
$= 1.7 \times 10^{-5}$ mol/L

15.4 Total mass is $750 + 21.5 = 771.5$ g.
Weight percent glucose $= \dfrac{21.5\ \text{g}}{771.5\ \text{g}} \times 100\% = 2.79\%$

15.5 $\left(\dfrac{30\ \text{g Se}}{10^9\ \text{g H}_2\text{O}}\right)\left(\dfrac{1\ \text{g H}_2\text{O}}{1\ \text{mL H}_2\text{O}}\right)\left(\dfrac{10^6\ \mu\text{g Se}}{1\ \text{g Se}}\right)$
$= 3.0 \times 10^{-2}\ \mu\text{g Se/mL H}_2\text{O}$

Se in 100 mL of water
$= \left(\dfrac{3.0 \times 10^{-2}\ \mu\text{g Se}}{1\ \text{mL H}_2\text{O}}\right)(100\ \text{mL H}_2\text{O}) = 3.0\ \mu\text{g Se}$

15.6 (a) Mass fraction $\times 10^6 = $ ppm; $\dfrac{\text{ppm}}{10^6} = $ mass fraction

$\dfrac{10,600\ \text{ppm}}{10^6} = 0.0106\ \text{Na}^+$

(b) Molarity $= \dfrac{\text{moles Na}^+}{\text{L solution}}$;

$10,600\ \text{ppm Na}^+ = \dfrac{10,600\ \text{mg}}{\text{kg solution}}$

$\dfrac{10,600\ \text{mg Na}^+}{\text{kg solution}} \times \dfrac{1.03\ \text{kg solution}}{1\ \text{L solution}} \times \dfrac{1\ \text{g Na}^+}{10^3\ \text{mg Na}^+}$
$\times \dfrac{1\ \text{mol Na}^+}{22.99\ \text{g Na}^+}$
$= \dfrac{1.09 \times 10^4\ \text{mol Na}^+}{2.299 \times 10^4\ \text{L solution}} = 0.474\ \text{M Na}^+$

15.7 Molarity $\text{H}_2\text{O}_2 = \dfrac{\text{moles H}_2\text{O}_2}{\text{L Solution}}$
$= \dfrac{30.0\ \text{g H}_2\text{O}_2}{100.\ \text{g solution}} \times \dfrac{1.11\ \text{g solution}}{1\ \text{mL solution}} \times \dfrac{10^3\text{mL solution}}{1\ \text{L solution}}$
$\times \dfrac{1\ \text{mol H}_2\text{O}_2}{34.0\ \text{g H}_2\text{O}_2} = \dfrac{3.33 \times 10^4\ \text{mol H}_2\text{O}_2}{3.40 \times 10^3\ \text{L}} = 9.79\ \text{M}$

15.8 $P_{\text{water}} = (X_{\text{water}})(P°_{\text{water}})$
291.2 mm Hg $= (X_{\text{water}}) (355.1$ mmHg$)$
$X_{\text{water}} = \dfrac{291.2\ \text{mm Hg}}{355.1\ \text{mm Hg}} = 0.8201$
$X_{\text{urea}} = 1.000 - X_{\text{water}} = 1.000 - 0.8201 = 0.1799$

15.9 Amount of NaCl $= 6.58$ g NaCl $\times \left(\dfrac{1\ \text{mol NaCl}}{58.44\ \text{g NaCl}}\right)$
$= 1.13 \times 10^{-1}$ mol NaCl
Molality $= \dfrac{\text{moles solute}}{\text{kg solvent}} = \left(\dfrac{1.13 \times 10^{-1}\ \text{mol NaCl}}{250.0\ \text{mL H}_2\text{O}}\right)$
$\times \left(\dfrac{1\ \text{mL H}_2\text{O}}{1\ \text{g H}_2\text{O}}\right)\left(\dfrac{1000\ \text{g H}_2\text{O}}{1\ \text{kg H}_2\text{O}}\right) = 0.452\ \text{mol/kg}$

15.10 Molality of solution $= \left(\dfrac{1.20\ \text{kg ethylene glycol}}{6.50\ \text{kg H}_2\text{O}}\right)$
$\left(\dfrac{1000\ \text{g}}{\text{kg}}\right)\left(\dfrac{1\ \text{mol ethylene glycol}}{62.068\ \text{g ethylene glycol}}\right) = 2.97\ \text{mol/kg}$

Next, calculate the freezing point depression of a 2.97 mol/kg solution.

$\Delta T_f = (1.86\ °\text{C kg mol}^{-1})(2.97\ \text{mol/kg}) = 5.52\ °\text{C}$

This solution will freeze at $-5.52\ °$C, so this quantity of ethylene glycol will not protect the 6.5 kg of water in the tank if the temperature drops to $-25\ °$C.

15.11 $\Delta T_f = 3.80\ °$C;
Molality $= \dfrac{3.80\ °\text{C}}{20.2\ °\text{C kg mol}^{-1}} = 0.188\ \text{mol/kg}$
$\dfrac{1.50\ \text{g}}{0.0750\ \text{kg}} \times \dfrac{1\ \text{kg}}{0.188\ \text{mol}} = 106\ \text{g/mol}$

15.12 Assume $i = 1$ because sucrose is a nonelectrolyte. Calculate c, from which the molar mass can be determined.

$c = \dfrac{\Pi}{RT} = \dfrac{1.79\ \text{atm}}{(0.0821\ \text{L atm mol}^{-1}\ \text{K}^{-1})(298\ \text{K})} = 0.0732\ \text{mol/L}.$
There are 2.50 g of sucrose in 100 mL.
$(0.0732\ \text{mol/L})(0.100\ \text{L}) = 0.00732$ mol sucrose.
Molar mass $= \dfrac{2.50\ \text{g sucrose}}{0.00732\ \text{mol sucrose}} = 342\ \text{g/mol},$ and so the empirical and molecular formulas of sucrose are the same.

Chapter 16

16.1 (a) ClO^- is the conjugate base of HClO.
(b) O^{2-} is the conjugate base of OH^-.
(c) The conjugate base of HCOOH is HCOO^-.
(d) The conjugate acid of OH^- is H_2O.
(e) The conjugate acid of IO_3^- is HIO_3.
(f) The conjugate acid of PH_3 is PH_4^+.

16.2 The 2.0×10^{-5} M H^+ solution is more acidic.

16.3 In a 0.040 M solution of NaOH, the $[\text{OH}^-]$ is 0.040 because the NaOH is 100% dissociated; pH $= 12.6$.

16.4 (a) $[\text{H}_3\text{O}^+] = 10^{-7.40} = 3.98 \times 10^{-8}$
(b) $[\text{OH}^-] = \dfrac{1.0 \times 10^{-14}}{[\text{H}_3\text{O}^+]} = 2.51 \times 10^{-7}\ \text{M}$
(c) A pH of 7.40 is slightly basic.

16.5 (a) $\text{HN}_3(aq) \rightleftharpoons \text{H}^+(aq) + \text{N}_3^-(aq)$ $K_a = \dfrac{[\text{H}^+][\text{N}_3^-]}{[\text{HN}_3]}$

(b) $\text{HCOOH}(aq) \rightleftharpoons \text{H}^+(aq) + \text{HCOO}^-(aq)$
$K_a = \dfrac{[\text{H}^+][\text{HCOO}^-]}{[\text{HCOOH}]}$

(c) $HClO_2(aq) \rightleftharpoons H^+(aq) + ClO_2^-(aq)$

$$K_a = \frac{[H^+][ClO_2^-]}{[HClO_2]}$$

16.6 (a) $CN^-(aq) + H_2O(\ell) \rightleftharpoons HCN(aq) + OH^-(aq)$

$$K_b = \frac{[HCN][OH^-]}{[CN^-]}$$

(b) $C_6H_5NH_2(aq) + H_2O(\ell) \rightleftharpoons$
$$C_6H_5NH_3^+(aq) + OH^-(aq)$$

$$K_b = \frac{[C_6H_5NH_3^+][OH^-]}{[C_6H_5NH_2]}$$

(c) $HS^-(aq) + H_2O(\ell) \rightleftharpoons H_2S(aq) + OH^-(aq)$

$$K_b = \frac{[H_2S][OH^-]}{[HS^-]}$$

16.7 Setting up a small table for lactic acid, HLa:

	$HLa + H_2O \rightleftharpoons H_3O^+ + La^-$		
Initial concentration (mol/L)	0.10	10^{-7}	0
Concentration change due to reaction (mol/L)	$-x$	$+x$	$+x$
Equilibrium concentration (mol/L)	$0.10 - x$	x	x

But $x = 10^{-2.43} = 3.7 \times 10^{-3}$ because $x = [H_3O^+]$. Substituting in the K_a expression,

$$K_a = \frac{[H_3O^+][La^-]}{[HLa]} = \frac{(3.7 \times 10^{-3})^2}{0.10 - (3.7 \times 10^{-3})}$$

$$= \frac{1.4 \times 10^{-5}}{0.1} = 1.4 \times 10^{-4}$$

Lactic acid is a stronger acid than propionic acid, with a K_a of 1.5×10^{-5}.

16.8 Using the same methods as shown in the example,

$$\frac{x^2}{0.10} = 7.3 \times 10^{-10}$$

Solving for x, which is $[H_3O^+]$, we get $x = \sqrt{7.3 \times 10^{-11}} = 8.54 \times 10^{-6} = [H_3O^+]$.
So the pH of this solution is $-\log(8.54 \times 10^{-6}) = 5.07$.

16.9 $(CH_3)_2NH(aq) + H_2O(\ell) \rightarrow (CH_3)_2NH_2^+(aq) + OH^-(aq)$

$$K_b = 5.9 \times 10^{-4} = \frac{[(CH_3)_2NH_2^+][OH^-]}{[(CH_3)_2NH]}$$

Let $x = [(CH_3)_2NH_2^+] = [OH^-]$.

$$5.9 \times 10^{-4} = \frac{x^2}{0.0050 - x}$$

Assume that $0.0050 - x \approx 0.0050$.
$x^2 = (5.9 \times 10^{-4})(0.0050) = 2.95 \times 10^{-6}$
$= 1.72 \times 10^{-3} = [OH^-]$

$$[H^+] = \frac{1.0 \times 10^{-14}}{1.72 \times 10^{-3}} = 5.81 \times 10^{-12} \text{ M}$$

pH $= -\log(5.81 \times 10^{-12}) = 11.23$

16.10 Using the same methods as those used in the example, letting $x = [OH^-]$ and $[HCO_3^-]$, and using the value of

2.1 × 10^{-4} for K_b for CO_3^{2-}, we get

$$\frac{x^2}{1.0} = 2.1 \times 10^{-4} \quad x = \sqrt{2.1 \times 1.0^{-4}} = 1.45 \times 10^{-2}$$

pOH $= 1.84$ and pH $= 12.16$

16.11 The formula weights and moles of acid per gram for the five antacids are as follows:

	Formula weight	Mol acid/gram
$Mg(OH)_2$	58.32	1 mol acid/29.16 g antacid
$CaCO_3$	100.10	1 mol acid/50.05 g antacid
$NaHCO_3$	84.00	1 mol acid/84.00 g antacid
$Al(OH)_3$	78.0034	1 mol acid/26.00 g antacid
$NaAl(OH)_2CO_3$	143.99	1 mol acid/36.00 g antacid

Of these antacids, $Al(OH)_3$ neutralizes the most stomach acid per gram.

Chapter 17

17.1 $K_a = \dfrac{[H^+][HCO_3^-]}{[H_2CO_3]} = \dfrac{H^+(0.025)}{(0.0020)} = [H^+] \times 12.5$
$= 4.2 \times 10^{-7}$

$$[H^+] = \frac{4.2 \times 10^{-7}}{12.5} = 3.4 \times 10^{-8}$$

pH $= -\log(3.4 \times 10^{-8}) = 7.47$

17.2 pH $= 6.38 + \log \dfrac{(0.020)}{(0.0025)} = 6.38 + \log(8.0)$
$= 6.38 + 0.90 = 7.28$

17.3 (a) Lactic acid-lactate (b) Acetic acid-acetate
(c) Hypochlorous acid-hypochlorite (d) CO_3^{2-}-HCO_3^-

17.4 The buffer capacity will be exceeded when just over 0.25 mol of KOH is added, which will have reacted with the 0.25 mol of $H_2PO_4^-$. 0.25 mol $OH^- = 0.25$ mol
KOH $\times \dfrac{56 \text{ g KOH}}{1 \text{ mol KOH}} = 14$ g. Thus, slightly more than 14 g of KOH will exceed the buffer capacity.

17.5 (a) 0.075 mol HCl converts 0.075 mol of lactate to lactic acid (0.075 mol).

pH $= 3.85 + \log \dfrac{(0.20 - 0.075)}{(0.15 + 0.075)} = 3.85 + \log \dfrac{(0.125)}{(0.225)}$

$= 3.85 + \log(0.556) = 3.85 + (-0.25) = 3.60$

(b) 0.025 mol NaOH converts 0.025 mol of lactic acid to 0.025 mol of lactate.

pH $= 3.85 + \log \dfrac{(0.20 + 0.025)}{(0.15 - 0.025)} = 3.85 + \log \dfrac{(0.225)}{(0.125)}$

$= 3.85 + \log(1.8) = 3.85 + (0.26) = 4.11$

17.6 (a) $[H_3O^+] = \dfrac{(5.00 \times 10^{-3}) - (1.00 \times 10^{-3})}{0.0500 + 0.0100}$

$= \dfrac{4.00 \times 10^{-3}}{0.0600} = 6.67 \times 10^{-2}$ M

pH $= 1.176 = 1.18$

(b) $[H_3O^+] = \dfrac{(5.00 \times 10^{-3}) - 0.00250}{0.0500 + 0.0250}$

$= \dfrac{2.50 \times 10^{-3}}{0.0750} = 3.33 \times 10^{-2}$ M

pH $= -\log(3.33 \times 10^{-2}) = 1.48$

(c) $[H_3O^+] = \dfrac{(5.00 \times 10^{-3}) - 0.00450}{0.0500 + 0.0450}$

$= \dfrac{5.00 \times 10^{-4}}{0.0950} = 5.26 \times 10^{-3}$

pH $= -\log(5.26 \times 10^{-3}) = 2.28$

(d) $[OH^-] = \dfrac{0.05 \times 10^{-3}\,\text{mol}}{0.0500\,\text{L} + 0.0505\,\text{L}}$

$= 5.0 \times 10^{-4}$ mol/L

pOH $= -\log(5.0 \times 10^{-4}) = 3.30$

pH $= 14.00 - 3.30 = 10.70$

17.7 (a) Adding 10.0 mL of 0.100 M NaOH is adding (0.100 mol/L)(0.0100 L) = 0.00100 mol OH$^-$, which neutralizes 0.00100 mol of acetic acid, converting it to 0.00100 mol of acetate ion.

pH = pH $+ \log \dfrac{[\text{acetate}]}{[\text{acetic acid}]}$

$= 4.74 + \log \dfrac{(0.00100/0.0600)}{(0.00400/0.0600)}$

$= 4.74 + \log(0.25) = 4.74 + (-0.602) = 4.14$

(b) pH $= 4.74 + \log \dfrac{(0.00250/0.0750)}{(0.00250/0.0750)}$

$= 4.74 + \log(1) = 4.74$

(c) pH $= 4.74 + \log \dfrac{(0.00450/0.0950)}{(0.00050/0.0950)}$

$= 4.74 + \log(9) = 4.74 + 0.95 = 5.69$

(d) $[OH^-] = \dfrac{0.10 \times 10^{-3}\,\text{mol}}{0.0500\,\text{L} + 0.0510\,\text{L}}$

$= 9.9 \times 10^{-4}$ mol/L

pOH $= -\log(9.9 \times 10^{-4})$

$= 3.00$

pH $= 14.00 - 3.00 = 11.00$

17.8 K_{sp} CuBr $= [Cu^+][Br^-]$
K_{sp} HgI$_2$ $= [Hg^{2+}][I^-]^2$
K_{sp} SrSO$_4$ $= [Sr^{2+}][SO_4^{2-}]$

17.9 AgBr(s) $\rightleftharpoons$ Ag$^+$(aq) + Br$^-$(aq)
$K_{sp} = [Ag^+][Br^-] = 5 \times 10^{-10} = x^2$; $x = $ solubility
$x = \sqrt{5 \times 10^{-10}} \cong 2 \times 10^{-5}$

17.10 Ag$_2$C$_2$O$_4$(s) $\rightleftharpoons$ 2 Ag$^+$(aq) + C$_2$O$_4^{2-}$(aq)
$K_{sp} = (1.38 \times 10^{-4})^2(6.9 \times 10^{-5}) = 1.3 \times 10^{-12}$

17.11

	PbCl$_2$ $\rightleftharpoons$	Pb^{2+}	+ 2 Cl$^-$
Initially (mol/L)		0	0.5
Change due to dissolving (mol/L)		$+S$	$0.5 + 2S$
At equilibrium (mol/L)		S	0.5 (ignore 2S because S will be small)

$K_{sp} = (S)(0.5) = 1.7 \times 10-5$

$S = \dfrac{1.7 \times 10^{-5}}{0.5} = 3.4 \times 10^{-5}$ mol/L

17.12   AgCl(s) $\rightleftharpoons$ Ag$^+$(aq) + Cl$^-$(aq)
$K_{sp} = 1.8 \times 10^{-10}$

AgCl(aq) + 2 CN$^-$(aq) $\rightleftharpoons$ Ag(CN)$_2^-$(aq)

$K_f = 5.6 \times 10^{18}$

AgCl(s) + 2 CN$^-$(aq) $\longrightarrow$ Ag(CN)$_2^-$(aq) + Cl$^-$(aq)

$K_{net} = (1.8 \times 10^{-10})(5.6 \times 10^{18}) = 1.0 \times 10^9$. K_{net} is large enough so that AgCl dissolves in aqueous NaCN.

17.13 (a) $Q = (1.0 \times 10^{-5})(1.0 \times 10^{-5}) = 1.0 \times 10^{-10} < K_{sp}$; no precipitation.

(b) For precipitation to occur, $Q \geq K_{sp}$; $Q = $ conc Ag$^+ \times$ conc Cl$^-$; $K_{sp} = [Ag^+][Cl^-]$; $1.8 \times 10^{-10} = [Ag^+][Cl^-]$;

$[Cl^-] = \dfrac{1.8 \times 10^{-10}}{1.0 \times 10^{-5}} = 1.8 \times 10^{-5}$ M, the minimum for AgCl precipitation.

17.14 AgCl will precipitate first. [Cl$^-$] needed to precipitate AgCl:

$[Cl^-] = \dfrac{1.8 \times 10^{-10}}{1.0 \times 10^{-2}} = 1.8 \times 10^{-8}$ M

[Cl$^-$] needed to precipitate PbCl$_2$:

$[Cl^-] = \sqrt{\dfrac{1.7 \times 10^{-5}}{1.0 \times 10^{-1}}} = 1.3 \times 10^{-2}$ M

Chapter 18

18.1 (a) $\Delta S^\circ = \dfrac{q_{rev}}{T} = \dfrac{\Delta H_{vap}}{T} = \dfrac{30.8\,\text{kJ}}{(80.1 + 273.15)\text{K}}$

$= \dfrac{30,800\,\text{J}}{353.25\,\text{K}} = 87.2$ J/K

(b) Because the process is opposite, ΔS° has the opposite sign and equals -87.2 J/K.

18.2 (a) C(g) has higher S°, 158.096 J K^{-1} mol^{-1}, versus 5.740 J K^{-1} mol^{-1} for C (graphite).

(b) Ar(g) has higher S°, 154.7 J K^{-1} mol^{-1}, versus 41.42 J K^{-1} mol^{-1} for Ca(s)

(c) KOH(aq) has higher S°, 91.6 J K^{-1} mol^{-1}, versus 78.9 J K^{-1} mol^{-1} for KOH(s).

18.3 (a) $\Delta S^\circ = 2$ mol CO(g) $\times S^\circ$ [CO(g)] + 1 mol O$_2$(g)
$\times S^\circ$ [O$_2$(g)] $-$ 2 mol CO$_2$(g) $\times S^\circ$[CO$_2$(g)]
$= \{2 \times (197.674) + (205.138) - 2 \times (213.74)\}$ J/K
$= 173.01$ J/K

(b) $\Delta S^\circ = 1$ mol NaCl(aq) $\times S^\circ$(NaCl(aq))
$- 1$ mol NaCl(s) $\times S^\circ$(NaCl(s))
$= (115.5 - 72.8)$ J/K $= 42.7$ J/K

(c) $\Delta S^\circ = 1$ mol MgO(s) $\times S^\circ$(MgO[s]) + 1 mol CO$_2$(g)
$\times S^\circ$ (CO$_2$[g]) $-$ 1 mol MgCO$_3$(s) $\times S^\circ$(MgCO$_3$[s])
$= (26.94 + 213.74 - 65.854)$ J/K
$= 174.83$ J/K

18.4 N$_2$(g) + 3 H$_2$(g) $\rightarrow$ 2 NH$_3$(g)

$\Delta H^\circ = 2$ mol NH$_3 \times \Delta H_f^\circ$ [NH$_3$(g)]

$= 2(-46.11)$ kJ $= -92.22$ kJ

$$\Delta S° = 2 \text{ mol NH}_3 \times S° \text{ (NH}_3[g]) - 1 \text{ mol N}_2(g)$$
$$\times S° \text{ (N}_2[g]) - 3 \text{ mol H}_2 \times S° \text{ (H}_2[g])$$
$$= 2(192.45) \text{ J/K} - (191.61) \text{ J/K} - 3(130.684) \text{ J/K}$$
$$= -198.76 \text{ J/K}$$

$$\Delta S°_{universe} = \frac{-\Delta H°}{T} + \Delta S° = \frac{92.2 \text{ kJ}}{298.15 \text{ K}} + (-198.76 \text{ J/K})$$

$$= \frac{92,200 \text{ J}}{298.15 \text{ K}} - 198.76 \text{ J/K} = 110.5 \text{ J/K}$$

The process is product-favored.

18.5 (a) $\Delta H° = \{(-238.66) - (-110.525)\}\text{kJ} = -128.14 \text{ kJ}$
$\Delta S° = \{(126.8) - 197.674 - 2 \times 130.684\}$ J/K
$= -332.2$ J/K
$\Delta G° = \Delta H° - T\Delta S°$
$= -128.14 \times 10^3 \text{ J} - 298.15 \text{ K} \times (-332.2 \text{ J/K})$
$= -29.09 \times 10^3 \text{ J} = -29.09 \text{ kJ}$
(b) $\Delta G° = [-166.27 - (-137.168)]\text{kJ} = -29.10 \text{ kJ}$.
The two results agree.
(c) $\Delta G°$ is negative. The reaction is product-favored at 298.15 K. Because $\Delta S°$ is negative, at very high temperatures the reaction will become reactant-favored.

18.6 (a) $T = \Delta H°/\Delta S° = (-565,968 \text{ J})/(-173.01 \text{ J/K}) = 3271 \text{ K}$
(b) The reaction is exothermic and therefore is product-favored at temperatures lower than 3271 K.

18.7 (a) $\Delta G° = \{-553.04 - 528.10 - (-1128.79)\}$ kJ
$= 47.65 \text{ kJ}$
$K° = e^{-\Delta G°/RT} = e^{-(47.65 \text{ kJ/mol})/(8.314 \text{ J K}^{-1} \text{ mol}^{-1})(298 \text{ K})}$
$= e^{-(47,650 \text{ J/mol})/(8.314 \text{ J K}^{-1} \text{ mol}^{-1})(298 \text{ K})}$
$= e^{-19.23} = 4.4 \times 10^{-9}$ (close to K_c)
(b) $K° = e^{-14.68} = 4.2 \times 10^{-7}$ (agrees with K_c)
(c) $K° = e^{-(-1.909)} = 6.75$ (agrees with K_P)
For reactions (a) and (b), $K_c = K°$

18.8 (a) At 298 K,
$\Delta G° = 2 \times (-16.45) \text{ kJ} = -32.9 \text{ kJ}$
$K° = e^{-(-32,900 \text{ J})/(8.314 \text{ J K}^{-1} \text{ mol}^{-1})(298 \text{ K})} = e^{13.28} = 5.8 \times 10^5$
(b) At 450. K,
$\Delta G° = \Delta H° - T\Delta S° = -92.22 \text{ kJ} - (450.)(-0.19876) \text{ kJ}$
$= -2.78 \text{ kJ}$
$K° = e^{-(-2780 \text{ J})/(8.314 \text{ J K}^{-1})(450. \text{ K})} = 2.10$
(c) At 800. K,
$\Delta G° = -92.22 \text{ kJ} - (800.)(-0.19876) \text{ kJ} = 66.79 \text{ kJ}$
$K° = e^{-(66,790 \text{ J})/(8.314 \text{ J K}^{-1} \text{ mol}^{-1})(800. \text{ K})} = 4.3 \times 10^{-5}$

18.9 $\Delta G = \Delta G° + RT \ln Q$
$\Delta G°$ was calculated in Problem-Solving Practice 18.7 to be 47.65 kJ for the reverse of this reaction. Therefore
$\Delta G° = -47.65 \text{ kJ}$
$\Delta G = -47.65 \text{ kJ/mol}$
$\quad + (8.314 \text{ J K}^{-1} \text{ mol}^{-1})(298 \text{ K}) \ln\left(\frac{1}{(0.023)(0.13)}\right)$
$= -47.65 \text{ kJ/mol} + 1.44 \times 10^4 \text{ J/mol}$
$= -47.65 \text{ kJ/mol} + 14.4 \text{ kJ/mol} = -33.2 \text{ kJ/mol}$

18.10 (a) $\Delta G° = 2(-137.168) \text{ kJ} - 2(-394.359) \text{ kJ} = 514.382 \text{ kJ}$.
The reaction is reactant-favored, and at least 514.382 kJ of work must be done to make it occur.
(b) $\Delta G° = 2(-742.2) \text{ kJ} = -1484.4 \text{ kJ}$. The reaction is product-favored and could do up to 1484.4 kJ of useful work.

18.11 $\Delta G°' = \Delta G°$ for this reaction because none of the reactants or products requires a standard state different from 1 bar or 1 mol/L.

18.12 (a) The strongest phosphate donor has the most negative $\Delta G°$ for its reaction with water to produce dihydrogen phosphate. The $\Delta G°'$ values are given in Problem-Solving Example 18.12. Creatine phosphate, at -43.1 kJ, has the most negative value and is the strongest phosphate donor.
(b) Glycerol 3-phosphate has the least negative $\Delta G°'$ at -9.7 kJ and therefore is the weakest phosphate donor.
(c) See parts (a) and (b) for explanation.

18.13 (a) $\Delta G°_f (\text{MgO}[s]) = -569.43 \text{ kJ}$, so formation of MgO(s) is product-favored and MgO(s) is thermodynamically stable.
(b) $\Delta G°_f (\text{N}_2\text{H}_4[\ell]) = 149.34 \text{ kJ}$; kinetically stable.
(c) $\Delta G°_f (\text{C}_2\text{H}_6[g]) = -32.82 \text{ kJ}$; thermodynamically stable.
(d) $\Delta G°_f (\text{N}_2\text{O}[g]) = 104.20 \text{ kJ}$; kinetically stable.

Chapter 19

19.1 Reducing agents are indicated by "red" and oxidizing agents are indicated by "ox." Oxidation numbers are shown above the symbols for the elements.

(a) $\overset{0}{2 \text{ Fe(s)}} + \overset{0}{3 \text{ Cl}_2(g)} \rightarrow \overset{+3 \; -1}{2 \text{ FeCl}_3(s)}$
$\quad\;$ red $\quad\quad$ ox

(b) $\overset{0}{2 \text{ H}_2(g)} + \overset{0}{\text{O}_2(g)} \rightarrow \overset{+1 \; -2}{2 \text{ H}_2\text{O}(\ell)}$
$\quad\;$ red $\quad\;$ ox

(c) $\overset{0}{\text{Cu(s)}} + \overset{+5 \; -2}{2 \text{ NO}_3^-(aq)} + \overset{+1 \; -2}{4 \text{ H}_3\text{O}^+(aq)} \rightarrow$
$\quad\;$ red $\quad\quad$ ox
$\quad\quad\quad\quad\quad \overset{+2}{\text{Cu}^{2+}(aq)} + \overset{+4 \; -2}{2 \text{ NO}_2(g)} + \overset{+1 \; -2}{6 \text{ H}_2\text{O}(\ell)}$

(d) $\overset{0}{\text{C(s)}} + \overset{0}{\text{O}_2(g)} \rightarrow \overset{+4 \; -2}{\text{CO}_2(g)}$
$\quad\;$ red $\quad\;$ ox

(e) $\overset{+2}{6 \text{ Fe}^{2+}(aq)} + \overset{+6 \; -2}{\text{Cr}_2\text{O}_7^{2-}(aq)} + \overset{+1 \; -2}{14 \text{ H}_3\text{O}^+(aq)} \rightarrow$
$\quad\;$ red $\quad\quad\quad$ ox
$\quad\quad\quad \overset{+3}{6 \text{ Fe}^{3+}(aq)} + \overset{+3}{2 \text{ Cr}^{3+}(aq)} + \overset{+1 \; -2}{21 \text{ H}_2\text{O}(\ell)}$

19.2

(a) Ox: $\quad\quad\quad\quad\quad\quad \text{Cd(s)} \rightarrow \text{Cd}^{2+}(aq) + 2 \text{ e}^-$
Red: $\quad \text{Cu}^{2+}(aq) + 2 \text{ e}^- \rightarrow \text{Cu(s)}$
Net: $\quad \text{Cd(s)} + \text{Cu}^{2+}(aq) \rightarrow \text{Cd}^{2+}(aq) + \text{Cu(s)}$

(b) Ox: $\quad\quad\quad\quad\quad\quad \text{Zn(s)} \rightarrow \text{Zn}^{2+}(aq) + 2 \text{ e}^-$
Red: $\quad 2 \text{ H}_3\text{O}^+(aq) + 2 \text{ e}^- \rightarrow \text{H}_2(g) + 2 \text{ H}_2\text{O}(\ell)$
Net: $\quad \text{Zn(s)} + 2 \text{ H}_3\text{O}^+(aq) \rightarrow$
$\quad\quad\quad\quad\quad\quad \text{Zn}^{2+}(aq) + \text{H}_2(g) + 2 \text{ H}_2\text{O}(\ell)$

(c) Ox: $\quad\quad\quad\quad\quad 2 \text{ Al(s)} \rightarrow 2 \text{ Al}^{3+}(aq) + 6 \text{ e}^-$
Red: $\quad 3 \text{ Zn}^{2+}(aq) + 6 \text{ e}^- \rightarrow 3 \text{ Zn(s)}$
Net: $\quad 2 \text{ Al(s)} + 3 \text{ Zn}^{2+}(aq) \rightarrow 2 \text{ Al}^{3+}(aq) + 3 \text{ Zn(s)}$

19.3 **Step 1.** This is an oxidation-reduction reaction. It is obvious that Zn is oxidized by its change in oxidation state.

Step 2. The half-reactions are

$$Zn(s) \longrightarrow Zn^{2+}(aq) \quad \text{(This is the oxidation reaction.)}$$

$$Cr_2O_7^{2-}(aq) \longrightarrow 2\,Cr^{3+}(aq) \quad \text{(This is the reduction reaction.)}$$

Step 3. Balance the atoms in the half-reactions. The atoms are balanced in the Zn half-reaction. We need to add water and H in the $Cr_2O_7^{2-}$ half-reaction. Fourteen H^+ ions are required on the right to combine with the seven O atoms.

$$Cr_2O_7^{2-}(aq) + 14\,H^+(aq) \longrightarrow 2\,Cr^{3+}(aq) + 7\,H_2O(\ell)$$

Step 4. Balance the half-reactions for charge. Write the Zn half-reaction as

$$Zn(s) \longrightarrow Zn^{2+}(aq) + 2\,e^-$$

and write the $Cr_2O_7^{2-}$ half-reaction as

$$Cr_2O_7^{2-}(aq) + 14\,H^+(aq) + 6\,e^- \longrightarrow$$
$$2\,Cr^{3+}(aq) + 7\,H_2O(\ell)$$

Step 5. Multiply the half-reactions by factors to make the number of electrons gained equal to the number lost.

$$3\,[Zn(s) \longrightarrow Zn^{2+}(aq) + 2\,e^-]$$
$$1\,[Cr_2O_7^{2-}(aq) + 14\,H^+(aq) + 6\,e^- \longrightarrow$$
$$2\,Cr^{3+}(aq) + 7\,H_2O(\ell)]$$

Step 6. Add the two half-reactions, canceling the electrons.

$$3\,Zn(s) \longrightarrow 3\,Zn^{2+}(aq) + 6\,e^-$$
$$Cr_2O_7^{2-}(aq) + 14\,H^+(aq) + 6\,e^- \longrightarrow$$
$$2\,Cr^{3+}(aq) + 7\,H_2O(\ell)$$

$$\overline{Cr_2O_7^{2-}(aq) + 3\,Zn(s) + 14\,H^+(aq) \longrightarrow}$$
$$2\,Cr^{3+}(aq) + 3\,Zn^{2+}(aq) + 7\,H_2O(\ell)$$

Step 7. Everything checks.

Step 8. Water was added in Step 3. The balanced equation is

$$Cr_2O_7^{2-}(aq) + 3\,Zn(s) + 14\,H^+(aq) \longrightarrow$$
$$2\,Cr^{3+}(aq) + 3\,Zn^{2+}(aq) + 7\,H_2O(\ell)$$

19.4 **Step 1.** This is an oxidation-reduction reaction. The wording of the question says Al reduces NO_3^- ion. Al is oxidized.

Step 2. The half-reactions are:

$$Al(s) \longrightarrow Al(OH)_4^-(aq) \quad \text{(This is the oxidation reaction.)}$$

$$NO_3^-(aq) \longrightarrow NH_3(aq) \quad \text{(This is the reduction reaction.)}$$

Step 3. Balance the atoms in the half-reactions. For the Al half-reaction, add four H^+ ions on the right and four water molecules on the left.

$$Al(s) + 4\,H_2O(\ell) \longrightarrow Al(OH)_4^- + 4\,H^+(aq)$$

For the NO_3^- half-reaction,

$$NO_3^-(aq) + 9\,H^+(aq) \longrightarrow NH_3(aq) + 3\,H_2O(\ell)$$

Step 4. Balance the half-reactions for charge. Put 3 e^- on the right in the Al half-reaction.

$$Al(s) + 4\,H_2O(\ell) \longrightarrow Al(OH)_4^- + 4\,H^+(aq) + 3\,e^-$$

and put 8 e^- on the left side of the NO_3^- half-reaction

$$NO_3^-(aq) + 9\,H^+(aq) + 8\,e^- \longrightarrow NH_3(aq) + 3\,H_2O(\ell)$$

Step 5. Multiply the half-reactions by factors to make the electrons gained equal to those lost.

$$8\,[Al(s) + 4\,H_2O(\ell) \longrightarrow Al(OH)_4^- + 4\,H^+(aq) + 3\,e^-]$$
$$3\,[NO_3^-(aq) + 9\,H^+(aq) + 8\,e^- \longrightarrow$$
$$NH_3(aq) + 3\,H_2O(\ell)]$$

Step 6. Remove $H^+(aq)$ ions by adding an appropriate amount of OH^-. For the Al half-reaction, add 32 OH^- ions to get

$$8\,Al(s) + 32\,OH^-(aq) + 32\,H_2O(\ell) \longrightarrow$$
$$8\,Al(OH)_4^- + 32\,H_2O + 24\,e^-$$

For the NO_3^- half-reactions, add 27 OH^- ions to get

$$3\,NO_3^-(aq) + 27\,H_2O + 24\,e^- \longrightarrow$$
$$3\,NH_3(aq) + 9\,H_2O(\ell) + 27\,OH^-(aq)$$

Step 7. Add both half-reactions and cancel the electrons.

$$8\,Al(s) + 32\,OH^-(aq) + 32\,H_2O(\ell) \longrightarrow$$
$$8\,Al(OH)_4^- + 32\,H_2O + 24\,e^-$$
$$3\,NO_3^-(aq) + 27\,H_2O + 24\,e^- \longrightarrow$$
$$3\,NH_3(aq) + 9\,H_2O(\ell) + 27\,OH^-(aq)$$

$$\overline{3\,NO_3^-(aq) + 8\,Al(s) + 59\,H_2O(\ell) + 32\,OH^-(aq) \longrightarrow}$$
$$8\,Al(OH)_4^-(aq) + 3\,NH_3(aq) + 27\,OH^-(aq) + 41\,H_2O(\ell)$$

Step 8. Make a final check. Since there are OH^- ions and water molecules on both sides of the equation, cancel them out. This gives the final balanced equation.

$$3\,NO_3^-(aq) + 8\,Al(s) + 18\,H_2O(\ell) + 5\,OH^-(aq) \longrightarrow$$
$$8\,Al(OH)_4^-(aq) + 3\,NH_3(aq)$$

(This is a fairly complicated equation to balance. If you got this one with a minimum of effort, your understanding of balancing redox equations is rather good. If you had to struggle with one or more of the steps, go back and repeat them.)

19.5 (a) $Ni(s) \rightarrow Ni^{2+}(aq) + 2\,e^-$ (This is the oxidation half-reaction.)

$2\,Ag^+(aq) + 2\,e^- \rightarrow 2\,Ag(s)$ (This is the reduction half-reaction.)

(b) The oxidation of Ni takes place at the anode and the reduction of Ag^+ takes place at the cathode.

(c) Electrons would flow through an external circuit from the anode (where Ni is oxidized) to the cathode (where Ag^+ ions are reduced).

(d) Nitrate ions would flow through the salt bridge to the anode compartment. Potassium ions would flow into the cathode compartment.

19.6 Oxidation half-reaction: $Fe(s) \rightarrow Fe^{2+}(aq, 1\,M) + 2\,e^-$

Reduction half-reaction: $Cu^{2+}(aq, 1\,M) + 2\,e^- \rightarrow Cu(s)$

$$E^\circ_{net} = +0.78\,V = E^\circ_{ox} + E^\circ_{red}$$

Since $E^\circ_{red} = +0.34\,V$, E°_{ox} must be $+0.44\,V$.

19.7

$$F_2(g) + 2\,e^- \longrightarrow 2\,F^-(aq) \qquad\qquad E^\circ_{red} = +2.87\text{ V}$$

$$\underline{2\,Li(s) \longrightarrow 2\,Li^+(aq) + 2\,e^- \qquad E^\circ_{ox} = -(-3.045\text{ V})}$$

$$2\,Li(s) + F_2(g) \longrightarrow 2\,Li^+(aq) + 2\,F^-(aq)$$

$$E^\circ_{cell} = +5.91\text{ V}$$

19.8 The two half-reactions are

$$Hg_2^{2+}(aq) + 2\,e^- \longrightarrow 2\,Hg(\ell) \qquad E^\circ_{red} = +0.789\text{ V}$$

$$2\,I^-(aq) \longrightarrow I_2(s) + 2\,e^- \qquad\qquad E^\circ_{ox} = -0.535\text{ V}$$

The E°_{cell} is the sum of the standard potentials for the half-reactions:

$$E^\circ_{cell} = E^\circ_{red} + E^\circ_{ox} = +0.789 + (-0.535) = +0.254\text{ V}$$

The reaction is product-favored as written.

19.9 We first need to calculate E°_{net}, and to do this we break the reaction into two half-reactions.

$$Ox:\ Sn^{2+}(aq) \longrightarrow Sn^{4+}(aq) + 2\,e^- \qquad E^\circ_{ox} = -0.13\text{ V}$$

$$\underline{Red:\ \ I_2(s) + 2\,e^- \longrightarrow 2\,I^-(aq) \qquad E^\circ_{red} = +0.53\text{ V}}$$

$$I_2(s) + Sn^{2+}(aq) \longrightarrow 2\,I^-(aq) + Sn^{4+}(aq)$$

$$E^\circ_{cell} = +0.40\text{ V}$$

E°_{net} is 0.40 V, and 2 mol of electrons are transferred.

$$\log K = \frac{nE^\circ\text{ V}}{0.0592\text{ V}} = \frac{2 \times 0.40\text{ V}}{0.0592\text{ V}}$$

$$= 13.51 \text{ and } K = 3 \times 10^{13}$$

The large value of K indicates that the reaction is strongly product-favored as written.

19.10

$$E_{cell} = 0.51\text{ V} - \left(\frac{0.0592\text{ V}}{2} \times \log\frac{3}{0.010}\right)$$

$$= 0.51\text{ V} - (0.0296\text{ V} \times \log(300))$$

$$= 0.51\text{ V} - 0.073\text{ V} = +0.44\text{ V}$$

19.11 If the pH = 3.66, then the $E_{cell} = 0.217$ V.

19.12 (a) The net cell reaction would be

$$2\,Na^+ + 2\,Br^- \longrightarrow 2\,Na + Br_2$$

Sodium ions would be reduced at the cathode and bromide ions would be oxidized at the anode.

(b) H_2 would be produced at the cathode for the same reasons given in Problem-Solving Example 19.8. That reaction is

$$2\,H_2O(\ell) + 2\,e^- \longrightarrow H_2(g) + 2\,OH^-(aq)$$

At the anode, two reactions are possible: the oxidation of water and the oxidation of Br^- ions.

$$6\,H_2O(\ell) \longrightarrow O_2(g) + 4\,H_3O^+(aq) + 4\,e^-$$

$$E^\circ_{ox} = -1.229\text{ V}$$

$$2\,Br^-(aq) \longrightarrow Br_2(\ell) + 2\,e^- \qquad E^\circ_{ox} = -1.08\text{ V}$$

The oxidation of bromide ion has the least negative oxidation potential, so that reaction will occur. The net cell reaction is

$$2\,H_2O(\ell) + 2\,Br^-(aq) \longrightarrow$$

$$Br_2(\ell) + H_2(g) + 2\,OH^-(aq)$$

(c) Sn metal will be formed at the cathode because its reduction potential (-0.14 V) is less negative than the potential for the reduction of water. O_2 will form at the anode because the E°_{ox} value for the oxidation of water is less negative than the E°_{ox} value for the oxidation of Cl^-. The net cell reaction is

$$2\,Sn^{2+}(aq) + 6\,H_2O(\ell) \longrightarrow$$

$$2\,Sn(s) + O_2(g) + 4\,H_3O^+(aq)$$

19.13 First, calculate the quantity of charge:

$$\text{Charge} = (25 \times 10^3\text{ A})(1\text{ hr})\left(\frac{60\text{ s}}{1\text{ min}}\right)\left(\frac{60\text{ min}}{1\text{ hr}}\right)$$

$$= 9.0 \times 10^7\text{ A}\cdot\text{s} = 9.0 \times 10^7\text{ C}$$

Then calculate the mass of Na:

Mass of Na $= (9.0 \times 10^7\text{ C}) \times$

$$\left(\frac{1\text{ mol }e^-}{96,500\text{ C}}\right)\left(\frac{1\text{ mol Na}}{1\text{ mol }e^-}\right)\left(\frac{22.99\text{ g Na}}{1\text{ mol Na}}\right)$$

$$= 2.1 \times 10^4\text{ g Na}$$

Chapter 20

20.1 (a) $^{237}_{93}Np \rightarrow ^{4}_{2}He + ^{233}_{91}Pa$

(b) $^{35}_{16}S \rightarrow ^{0}_{-1}e + ^{35}_{17}Cl$

20.2 (a) $^{90}_{38}Sr \rightarrow ^{0}_{-1}e + ^{90}_{39}Y$

(b) It takes four half-lives (4×29 y = 116 y) for the activity to decrease to 125 beta particles emitted per minute:

Number of half-lives	Change of activity	Total elapsed time (y)
1	2000 to 1000	29
2	1000 to 500	58
3	500 to 250	87
4	250 to 125	116

20.3 (a) $t_{1/2} = \dfrac{0.693}{9.3 \times 10^{-3}\text{ d}^{-1}} = 75$ d

(b) ln(fraction remaining) $= -k \times t$
$= -(9.3 \times 10^{-3}\text{ d}^{-1}) \times (100\text{ d}) = -0.930$
Fraction of iridium-192 remaining $= e^{-0.930} = 0.39$.
Therefore, 39% of the original iridium-192 remains.

20.4 $k = \dfrac{0.693}{1.60 \times 10^3\text{ y}} = 4.33 \times 10^{-4}\,y^{-1}$

As of 2001: ln(fraction remaining) $= -k \times t$
$= -(4.33 \times 10^{-4}\,y^{-1}) \times 80\text{ y} = -3.46 \times 10^{-2}$
Fraction of radium-226 remaining $= e^{-0.0346} = 0.966$.
Therefore, 96.6% of the original radium-226 remains;
$0.966 \times 1.00\text{ g} = 0.966$ g.

20.5 ln(0.60) $= -0.510 = -k \times t$

$$k = \frac{0.693}{t_{1/2}} = \frac{0.693}{12.3\text{ y}} = 0.0563\,y^{-1}$$

$$t = \frac{-0.510}{-0.0563\,y^{-1}} = 9.1\,y$$

Chapter 21

21.1 (a) The formation of NO_2 from NO is exothermic. Thus, lowering the temperature favors the forward reaction (NO_2 formation).

(b) By reacting with water, NO_2 is converted to HNO_3, thereby removing NO_2 from the reaction mixture.

21.2 $S_8(s) + 12\,O_2(g) \rightarrow 8\,SO_3(g)$

$$1.3 \times 10^{10}\ \text{kg }S_8 \times \frac{1\ \text{mol }S_8}{0.256\ \text{kg }S_8} \times \frac{8\ \text{mol }SO_3}{1\ \text{mol }S_8}$$

$$\times\ \frac{0.080\ \text{kg }SO_3}{1\ \text{mol }SO_3} = 3.3 \times 10^{10}\ \text{kg }SO_3$$

21.3 $1.5 \times 10^9\ \text{mol Na} \times \dfrac{4\ \text{mol NaCl}}{4\ \text{mol Na}} \times \dfrac{58.5\ \text{g NaCl}}{1\ \text{mol NaCl}}$

$$\times\ \frac{1\ \text{lb NaCl}}{454\ \text{g NaCl}} \times \frac{1\ \text{ton NaCl}}{2000\ \text{lb NaCl}} = 9.7 \times 10^4\ \text{tons NaCl}$$

21.4 mol of Al $= 1.00\ \text{ton Al} \times \dfrac{2000\ \text{lb Al}}{1\ \text{ton Al}} \times \dfrac{454\ \text{g Al}}{1\ \text{ton Al}}$

$$\times\ \frac{1\ \text{mol Al}}{26.98\ \text{g Al}} = 3.37 \times 10^4\ \text{mol Al}$$

$Al^{3+} + 3e^- \rightarrow Al$; therefore 3 mol of electrons is needed to produce 1 mol of Al.

Number of moles of electrons

$$= 3.37 \times 10^4\ \text{mol Al} \times \frac{3\ \text{mol }e^-}{1\ \text{mol Al}}$$

$$= 1.01 \times 10^5\ \text{mol }e^-$$

Charge $= 1.01 \times 10^5\ \text{mol }e^- \times \dfrac{9.65 \times 10^4\ C}{1\ \text{mol }e^-}$

$$= 9.75 \times 10^9\ C$$

Time $= 9.75 \times 10^9\ C \times \dfrac{1\ s}{1.00 \times 10^5\ C} = 9.75 \times 10^4\ s$

$$= 1.63 \times 10^3\ \text{min} = 27.0\ \text{h}$$

21.5 I^- is oxidized to I_2; IO_3^- is reduced to I_2. IO_3^- is the oxidizing agent and I^- is the reducing agent.

Chapter 22

22.1 (a) Mn [Ar] $3d^5 4s^2$

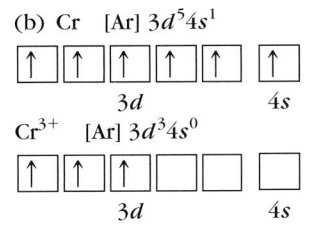

(b) Cr [Ar] $3d^5 4s^1$

Cr^{3+} [Ar] $3d^3 4s^0$

22.2 Assume that 50% of the iron comes from each ore.

Fe_2O_3: $5.00 \times 10^2\ \text{kg Fe} \times \dfrac{159.7\ \text{kg }Fe_2O_3}{111.7\ \text{kg Fe}}$

$$= 7.15 \times 10^2\ \text{kg }Fe_2O_3$$

Fe_3O_4: $5.00 \times 10^2\ \text{kg Fe} \times \dfrac{231.6\ \text{kg }Fe_3O_4}{167.6\ \text{kg Fe}}$

$$= 6.91 \times 10^2\ \text{kg }Fe_3O_4$$

Total $= 7.15 \times 10^2\ \text{kg} + 6.91 \times 10^2\ \text{kg} = 1.41 \times 10^3\ \text{kg}$

22.3 By controlling the voltage, the zinc could be removed and then the lead.

22.4 $E_{red} = 1.00\ \text{V} = +1.33\ \text{V} - 0.00985 \log\left(\dfrac{1}{[H^+]^{14}}\right)$

$$-0.33 = -0.00985 \log \frac{1}{[H^+]^{14}}$$

$$\log \frac{1}{[H^+]^{14}} = \frac{-0.33}{-0.00985} = 33.5$$

$$\frac{1}{[H^+]^{14}} = 10^{33.5} = 3.16 \times 10^{33}$$

$$1 = (3.16 \times 10^{33})[H^+]^{14}$$

$$[H^+]^{14} = \frac{1}{3.16 \times 10^{33}} = 3.16 \times 10^{-34}$$

$$14 \log[H^+] = \log 3.16 \times 10^{-34}$$

$$\log[H^+] = \frac{-33.5}{14} = -2.39$$

$$-\log[H^+] = pH = 2.39$$

22.5 (a) SO_4^{2-} (b) Cu^{2+}

(c) NH_3 (d) $[Cu(NH_3)_4]^{2+}$

22.6 (a) diamminesilver(I) nitrate

(b) $[Fe(H_2O)_5(NCS)]Cl_2$

22.7 (a) Two   (b) Zero   (c) Five

22.8 Two

ANSWERS TO EXERCISES

Chapter 1

1.1 (a) 37 °C is body temperature and therefore is higher than 85 °F.
(b) 20 °F is below the freezing point of water and therefore is lower than 0 °C.
(c) Since body temperature is 37 °C, the substance will be heated above 15 °C and will boil.

1.2 (a) Water is most dense because it is at the bottom of the tube; less dense substances can float on water.
(b) Kerosene is least dense (top layer).
(c) There will be no permanent change in the order of layers.
(d) The vegetable oil layer in the middle will be twice as big.

1.3 (a) Properties: blue (qualitative), melts at 99 °C (quantitative)
Change: melting
(b) Properties: white, cubic (both qualitative)
Change: none
(c) Properties: mass of 0.123 g, melts at 327 °C (both quantitative)
Change: melting
(d) Properties: colorless, vaporizes easily (both qualitative), boils at 78 °C, density of 0.789 g/mL (both quantitative).
Changes: vaporizing, boiling

1.4 Physical change: boiling water
Chemical changes: combustion of propane, cooking the egg

1.5 (a) Homogeneous mixture (solution)
(b) Heterogeneous mixture (contains cement, sand, and stone)
(c) Heterogeneous mixture of dirt and water
(d) Element; diamond is pure carbon.
(e) Element or heterogeneous mixture; pennies minted before 1982 are fairly pure copper; pennies minted since 1982 are zinc with a copper coating. Old pre-1982 pennies are less pure than newly minted ones, because the copper surface reacts slowly with air.
(f) Compound; contains sodium and chlorine

1.6 (a) Energy from the sun warms the ice and the water molecules vibrate more; eventually they break away from their fixed positions in the solid and liquid water forms. As the temperature of the liquid increases, some of the molecules have enough energy to become widely separated from the other molecules, forming water vapor (gas).
(b) Some of the water molecules in the clothes have enough speed and energy to escape from the liquid state and become water vapor; these molecules are carried away from the clothes by breezes or air currents. Eventually nearly every water molecule in the clothes vaporizes, and the clothes become dry.
(c) Water molecules from the air come into contact with the cold glass, and their speeds are decreased, allowing them to become liquid. As more and more molecules enter the liquid state, droplets form on the glass.
(d) Some water molecules escape from the liquid state, forming water vapor. As more and more molecules escape, the ratio of sugar molecules to water molecules becomes larger and larger, and eventually some sugar molecules start to stick together. As more and more sugar molecules stick to each other, a visible crystal forms. Eventually all of the water molecules escape, leaving sugar crystals behind.

1.7 (a) Tellurium, Te, earth (Latin *tellus* means earth); uranium, U, for Uranus; neptunium, Np, for Neptune; and plutonium, Pu, for Pluto. (Mercury, like the planet Mercury, is named for a Roman god.)
(b) Californium, Cf
(c) Curium, Cm, for Marie Curie; and meitnerium, Mt, for Lise Meitner
(d) Scandium, Sc, for Scandinavia; gallium, Ga, for France (Latin *Gallia* means France); germanium, Ge, for Germany; ruthenium, Ru, for Russia; europium, Eu, for Europe; polonium, Po, for Poland; francium, Fr, for France; americium, Am, for America; californium, Cf, for California
(e) H, He, C, N, O, F, Ne, P, S, Cl, Ar, Se, Br, Kr, I, Xe, At, Rn, element 118

1.8 (a) Elements that consist of diatomic molecules are H, N, O, F, Cl, Br, and I; At is radioactive and there is probably less than 50 mg of naturally occurring At on earth, but it does form diatomic molecules H, N, O, plus group 7A.
(b) Metalloids are B, Si, Ga, As, Sb, and Te.

1.9 Tin and lead are two different elements; allotropes are two different forms of the same element, so tin and lead are not allotropes.

1.10 Compare your list of risks with that of your friend.

Chapter 2

2.1 The movement of the comb though your hair removes some electrons, leaving slight charges on your hair and the comb. The charges must sum to zero; therefore, one must be slightly positive and one must be slightly negative, so they attract each other.

2.2 (a) A nucleus is about one ten-thousandth as large as an atom, so $100 \text{ m} \times (1 \times 10^{-4}) = 1 \times 10^{-2} \text{ m} = 1 \text{ cm}$. (b) Many everyday objects are about 1 cm in size — for example, a grape.

2.3 The statement is wrong because two atoms that are isotopes always have the same number of protons. It is the number of neutrons that varies from one isotope of an element to another.

2.4 Atomic weight of lithium = $(0.07500)(6.015121 \text{ amu}) + (0.9250)(7.016003 \text{ amu}) = 0.451134 \text{ amu} + 6.489802 \text{ amu} = 6.940936 \text{ amu}$, or 6.941 amu

2.5 Because the most abundant isotope is magnesium-24 (78.70%), the atomic weight of magnesium is closer to 24 than to 25 or 26, the mass numbers of the other magnesium isotopes, which make up approximately 21% of the remaining mass. The simple arithmetic average is $(24 + 25 + 26)/3 = 25$, which is larger than the atomic weight. In the arithmetic average, the relative abundance of each magnesium isotope is 33%, far less than the actual percent abundance of magnesium-24, and much more than the natural percent abundances of magnesium-25 and magnesium-26.

2.6 If there were two isotopes of gallium, and if their relative abundances were both 50%, then the atomic weight of gallium would have to be halfway between two integers. But the actual atomic weight is 69.72, which is far from being halfway between two integers.

2.7 Start by calculating the number of moles in 10.00 g of each element.

$$10.00 \text{ g Li} \times \frac{1 \text{ mol Li}}{6.941 \text{ g Li}} = 1.441 \text{ mol Li}$$

$$10.00 \text{ g Ir} \times \frac{1 \text{ mol Ir}}{192.22 \text{ g Ir}} = 0.05202 \text{ mol Ir}$$

Multiply the number of moles of each element by Avogadro's number.

$1.441 \text{ mol Li} \times 6.022 \times 10^{23}$ atoms/mol

$$= 8.678 \times 10^{23} \text{ atoms Li}$$

$0.05202 \text{ mol Ir} \times 6.022 \times 10^{23}$ atoms/mol =

$$3.133 \times 10^{22} \text{ atoms Ir}$$

Find the difference.

$(8.678 \times 10^{23}) - (0.03133 \times 10^{23})$

$$= 8.365 \times 10^{23} \text{ fewer atoms of Ir than Li}$$

2.8 1. (a) 13 metals: potassium (K), calcium (Ca), scandium (Sc), titanium (Ti), vanadium (V), chromium (Cr), manganese (Mn), iron (Fe), cobalt (Co), nickel (Ni), copper (Cu), zinc (Zn), and gallium (Ga)

(b) 3 nonmetals: selenium (Se), bromine (Br), and krypton (Kr)

(c) 2 metalloids: germanium (Ge) and arsenic (As)

2. (a) Groups 1A (except hydrogen), 2A, 1B, 2B, 3B, 4B, 5B, 6B, 7B, 8B

(b) Groups 7A and 8

(c) None

3. Period 6

2.9 The question should ask, Which of the following men does not have an *element* named after him? The correct answer to the question, after it is properly posed, is Isaac Newton.

Chapter 3

3.1 Propylene glycol structural formula:

Condensed formula: CH_3CHCH_2 (with OH, OH)

Molecular formula: $C_3H_8O_2$

3.2 (a) CS_2 (b) PCl_3 (c) SBr_2 (d) SeO_2 (e) OF_2 (f) XeO_3

3.3 (a) $C_{16}H_{34}$, $C_{28}H_{58}$ (b) $C_{14}H_{30}$, 14 carbon atoms and 30 hydrogen atoms

3.4 Propanol structural formula:

3.5 The molecular formula for propanol is C_3H_8O, which has the —OH group attached to the terminal carbon. If the —OH group is attached to the middle carbon, a different alcohol results. Isopropanol structural formula:

3.6 The structural and condensed formulas for the three constitutional isomers of five-carbon alkanes (pentanes) are

$CH_3CH_2CH_2CH_2CH_3$

$CH_3CH_2CHCH_3$

CH_3

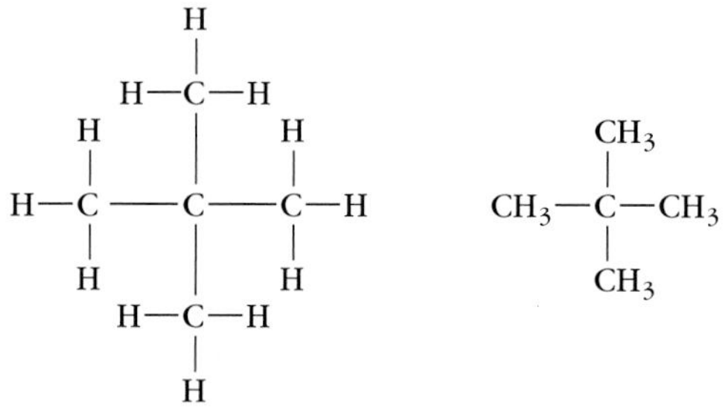

```
           H
           |
      H — C — H
  H        |        H
  |        |        |
H—C ——— C ——— C—H
  |        |        |
  H        |        H
      H — C — H
           |
           H
```

3.7 (a) 2-methylpentane

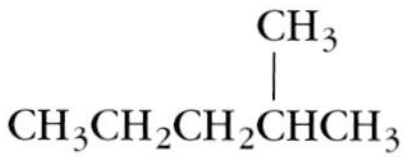

$$CH_3CH_2CH_2CHCH_3$$
with CH_3 branch

(b) 3-methylpentane

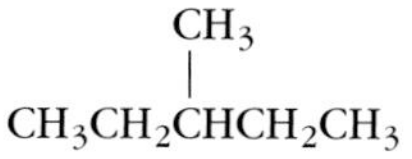

$$CH_3CH_2CHCH_2CH_3$$
with CH_3 branch

(c) 2,2-dimethylbutane

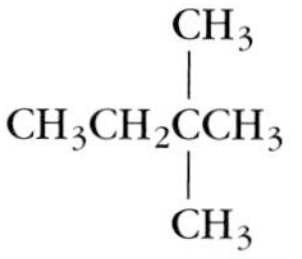

$$CH_3CH_2CCH_3$$
with CH_3 above and CH_3 below

(d) 2,3-dimethylbutane

$$CH_3CHCHCH_3$$
with H_3C and CH_3 branches

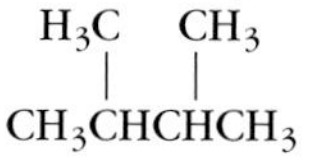

3.8 The compound is a solid at room temperature and is soluble in water, so it is likely to be an ionic compound.

3.9 (a) 174.16 g/mol   (b) 386.64 g/mol   (c) 398.06 g/mol
(d) 194.2 g/mol

3.10 The statement is true. Because both compounds have the same formula, they have the same molar mass. Thus, 100 g of each compound contains the same number of moles.

3.11 Epsom salt is $MgSO_4 \cdot 7H_2O$, which has a molar mass of 246 g/mol.

$$20 \text{ g} \times \frac{1 \text{ mol}}{246 \text{ g}} = 8.1 \times 10^{-2} \text{ mol Epsom salt}$$

3.12 (a) SF_6 molar mass is 146.06 g/mol; 1.000 mol of SF_6 contains 32.07 g of S and $18.9984 \times 6 = 113.99$ g of F. The mass percents are

$$\frac{32.07 \text{ g S}}{146.06 \text{ g } SF_6} \times 100\% = 21.96\% \text{ S}$$

$$100.0\% - 21.96\% = 78.04\% \text{ F}$$

(b) $C_{12}H_{22}O_{11}$ has a molar mass of 342.3 g/mol; 1.000 mol of $C_{12}H_{22}O_{11}$ contains

$$12.011 \times 12 = 144.13 \text{ g C}$$

$$1.0079 \times 22 = 22.174 \text{ g H}$$

$$15.9994 \times 11 = 175.99 \text{ g O}$$

The mass percents of the three elements are

$$\frac{144.13 \text{ g C}}{342.3 \text{ g}} \times 100\% = 42.117\% \text{ C}$$

$$\frac{22.174 \text{ g H}}{342.3 \text{ g}} \times 100\% = 6.478\% \text{ H}$$

$$\frac{175.99 \text{ g O}}{342.3 \text{ g}} \times 100\% = 51.41\% \text{ O}$$

(c) $Al_2(SO_4)_3$ molar mass is 342.15 g/mol; 1.000 mol of $Al_2(SO_4)_3$ contains

$$26.9815 \times 2 = 53.96 \text{ g Al}$$

$$32.066 \times 3 = 96.20 \text{ g S}$$

$$15.9994 \times 12 = 192.0 \text{ g O}$$

The mass percents of the three elements are

$$\frac{53.96 \text{ g Al}}{342.15 \text{ g}} \times 100\% = 15.77\% \text{ Al}$$

$$\frac{96.20 \text{ g S}}{342.15 \text{ g}} \times 100\% = 28.12\% \text{ S}$$

$$\frac{192.0 \text{ g O}}{342.15 \text{ g}} \times 100\% = 56.12\% \text{ O}$$

(d) $U(OTeF_5)_6$ molar mass is 1669.6 g/mol; 1.000 mol of $U(OTeF_5)_6$ contains 238.0289 g of U and

$$15.0004 \times 6 = 96.00 \text{ g O}$$

$$127.60 \times 6 = 765.6 \text{ g Te}$$

$$18.9984 \times 30 = 570.0 \text{ g F}$$

The mass percents of the four elements are

$$\frac{238.0289 \text{ g U}}{1669.6 \text{ g}} \times 100\% = 14.26\% \text{ U}$$

$$\frac{96.00 \text{ g O}}{1669.6 \text{ g}} \times 100\% = 5.750\% \text{ O}$$

$$\frac{765.6 \text{ g Te}}{1669.6 \text{ g}} \times 100\% = 45.86\% \text{ Te}$$

$$\frac{570.0 \text{ g F}}{1669.6 \text{ g}} \times 100\% = 34.41\% \text{ F}$$

3.13 (a) Carbon, nitrogen, oxygen, phosphorus, hydrogen, selenium; (b) calcium and strontium; (c) chloride and iodide; (d) iron, copper, zinc, vanadium (also chromium, manganese, cobalt, nickel, molybdenum, and cadmium)

3.14 Carbohydrates have the general formula $C_x(H_2O)_y$, so they have a hydrogen-to-oxygen ratio of 2:1. Compounds (a), (b), and (d) have this formula and could be carbohydrates. Compound (c) does not have this formula, since it contains N, so it is not a carbohydrate.

Chapter 4

4.1 One mole of methane reacts with two moles of oxygen to produce one mole of carbon dioxide and two moles of water.

4.2 (a) The total mass of reactants {4 Fe(s) + 3 O_2 (g)} must equal the total mass of products {2 Fe_2O_3(s)}, which is 2.50 g.
(b) The stoichiometric coefficients are 4, 3, and 2.
(c) 1.000×10^4 O atoms $\times \dfrac{1 \text{ } O_2 \text{ molecule}}{2 \text{ O atoms}} \times$

$$\frac{4 \text{ Fe atoms}}{3 \text{ } O_2 \text{ molecules}} = 6.667 \times 10^3 \text{ Fe atoms}$$

4.3 (a) Not balanced; the number of oxygen atoms do not match.

(b) Not balanced; the number of bromine atoms do not match.

(c) Balanced.

4.4 (a) To predict the product of a combination reaction between two elements, we need to know the ion that will be formed by each element when combined. (b) For calcium, Ca^{2+} ions are formed, and for fluorine, F^- ions are formed. (c) The product is CaF_2.

4.5 (a) Magnesium chloride, $MgCl_2$

(b) Magnesium oxide, MgO, and carbon dioxide, CO_2

4.6 $\dfrac{2 \text{ mol Al}}{3 \text{ mol Br}_2}$, $\dfrac{2 \text{ mol Al}}{1 \text{ mol Al}_2\text{Br}_6}$, $\dfrac{3 \text{ mol Br}_2}{1 \text{ mol Al}_2\text{Br}_6}$ and their reciprocals

4.7 $0.300 \text{ mol CH}_4 \times \dfrac{2 \text{ mol H}_2\text{O}}{1 \text{ mol CH}_4} = 0.600 \text{ mol H}_2\text{O}$

$0.600 \text{ mol H}_2\text{O} \times \dfrac{18.02 \text{ g H}_2\text{O}}{1 \text{ mol H}_2\text{O}} = 10.8 \text{ g H}_2\text{O}$

4.8 (a) $300. \text{ g urea} \times \dfrac{1 \text{ mol urea}}{60.06 \text{ g urea}} \times \dfrac{2 \text{ mol NH}_3}{1 \text{ mol urea}}$
$\times \dfrac{17.03 \text{ g NH}_3}{1 \text{ mol NH}_3} = 170. \text{ g urea}$

$100. \text{ g H}_2\text{O} \times \dfrac{1 \text{ mol H}_2\text{O}}{18.02 \text{ g H}_2\text{O}} \times \dfrac{2 \text{ mol NH}_3}{1 \text{ mol H}_2\text{O}}$
$\times \dfrac{17.03 \text{ g NH}_3}{1 \text{ mol NH}_3} = 189. \text{ g NH}_3$

Therefore, urea is the limiting reactant.

(b) $176. \text{ g NH}_3$

$300. \text{ g urea} \times \dfrac{1 \text{ mol urea}}{60.06 \text{ g urea}} \times \dfrac{1 \text{ mol H}_2\text{O}}{1 \text{ mol urea}}$
$\times \dfrac{44.01 \text{ g CO}_2}{1 \text{ mol CO}_2} = 220. \text{ g CO}_2$

(c) $300. \text{ g urea} \times \dfrac{1 \text{ mol urea}}{60.06 \text{ g urea}}$
$\times \dfrac{1 \text{ mol H}_2\text{O}}{1 \text{ mol urea}} \times \dfrac{18.02 \text{ g H}_2\text{O}}{1 \text{ mol H}_2\text{O}} = 90.0 \text{ g H}_2\text{O}$

$100. \text{ g} - 90.0 \text{ g} = 10.0 \text{ g H}_2\text{O}$ remains

4.9 (1) Impure reactants, (2) inaccurate weighing of reactants and products

4.10 Assuming that the nicotine is pure, weigh a sample of nicotine and burn the sample. Separately collect and weigh the carbon dioxide and water generated, and calculate the moles and grams of carbon and hydrogen collected. By mass difference, determine the mass of nitrogen in the original sample, then calculate the moles of nitrogen. Calculate the mole ratios of carbon, hydrogen, and nitrogen in nicotine to determine its empirical formula.

Chapter 5

5.1 It is possible for an exchange reaction to form two different precipitates—for example, the reaction between barium hydroxide and iron(II) sulfate: $Ba(OH)_2(aq) + FeSO_4(aq) \rightarrow BaSO_4(s) + Fe(OH)_2(s)$.

5.2 $H_3PO_4(aq) \rightleftharpoons H_2PO_4^-(aq) + H^+(aq)$

$H_2PO_4^-(aq) \rightleftharpoons HPO_4^{2-}(aq) + H^+(aq)$

$HPO_4^{2-}(aq) \rightleftharpoons PO_4^{3-}(aq) + H^+(aq)$

5.3 (a) Hydrogen ions and perchlorate ions:
$HClO_4(aq) \rightarrow H^+(aq) + ClO_4^-(aq)$
(b) $Ca(OH)_2(aq) \rightarrow Ca^{2+}(aq) + 2\ OH^-(aq)$

5.4 (a) $H^+(aq) + Cl^-(aq) + K^+(aq) + OH^-(aq) \rightarrow H_2O(\ell) + K^+(aq) + Cl^-(aq)$

$H^+(aq) + OH^-(aq) \rightarrow H_2O(\ell)$
(b) $2H^+(aq) + SO_4^{2-}(aq) + Ba^{2+}(aq) + 2OH^-(aq) \rightarrow 2\ H_2O(\ell) + BaSO_4(s)$

$H^+(aq) + OH^-(aq) \rightarrow H_2O(\ell)$
$Ba^{2+}(aq) + SO_4^{2-}(aq) \rightarrow BaSO_4(s)$
(c) $2\ H^+(aq) + 2\ CH_3CO_2^-(aq) + Ca^{2+}(aq) + 2\ OH^-(aq) \rightarrow Ca^{2+}(aq) + 2\ CH_3CO_2^-(aq) + 2\ H_2O(\ell)$
$H^+(aq) + OH^-(aq) \rightarrow H_2O(\ell)$

5.5 $Al^{3+}(aq) + 3\ OH^-(aq) + 3\ H^+(aq) + 3\ Cl^-(aq) \rightarrow 3\ H_2O(\ell) + Al^{3+}(aq) + 3\ Cl^-(aq)$
$H^+(aq) + OH^-(aq) \rightarrow H_2O(\ell)$

5.6 (a) The products are aqueous sodium sulfate, water, and carbon dioxide gas.

$2\ Na^+(aq) + CO_3^{2-}(aq) + 2\ H^+(aq) + SO_4^{2-}(aq) \longrightarrow 2\ Na^+(aq) + SO_4^{2-}(aq) + H_2O(\ell) + CO_2(g)$

$2\ H^+(aq) + CO_3^{2-}(aq) \longrightarrow H_2O(\ell) + CO_2(g)$

(b) The products are aqueous iron(II) chloride and hydrogen sulfide gas.

$FeS(s) + 2\ HCl(aq) \longrightarrow FeCl_2(aq) + H_2S(g)$

$2\ H^+(aq) + S^{2-}(aq) \longrightarrow H_2S(g)$

(c) Aqueous potassium chloride, water, and sulfur dioxide gas are produced.

$K_2SO_3(aq) + 2\ HCl(aq) \longrightarrow 2\ KCl(aq) + H_2O(\ell) + SO_2(g)$

$2\ H^+(aq) + SO_3^{2-}(aq) \longrightarrow H_2O(\ell) + SO_2(g)$

5.7 (a) Gas-forming reaction; the products are aqueous nickel sulfate, water, and carbon dioxide gas.

$NiCO_3(s) + H_2SO_4(aq) \longrightarrow NiSO_4(aq) + H_2O(\ell) + CO_2(g)$

$NiCO_3(s) + 2\ H^+(aq) \longrightarrow Ni^{2+}(aq) + H_2O(\ell) + CO_2(g)$

(b) Acid-base reaction; nitric acid reacts with strontium hydroxide, a base, to produce water and strontium nitrate, a salt.

$2\ HNO_3(aq) + Sr(OH)_2(s) \longrightarrow Sr(NO_3)_2(aq) + 2\ H_2O(\ell)$

$Sr(OH)_2(s) + 2\ H^+(aq) \longrightarrow Sr^+(aq) + 2\ H_2O(\ell)$

(c) Precipitation reaction; aqueous sodium chloride and insoluble barium oxalate are produced.

$BaCl_2(aq) + Na_2C_2O_4(aq) \longrightarrow BaC_2O_4(s) + 2\ NaCl(aq)$

$Ba^{2+}(aq) + C_2O_4^{2-}(aq) \longrightarrow BaC_2O_4(s)$

(d) Precipitation and gas-forming reaction; lead sulfate precipitates and carbon dioxide gas is released.

$PbCO_3(aq) + H_2SO_4(aq) \longrightarrow PbSO_4(s) + H_2O(\ell) + CO_2(g)$

$Pb^{2+}(aq) + SO_4^{2-}(aq) \longrightarrow PbSO_4(s)$

$2\ H^+(aq) + CO_3^{2-}(aq) \longrightarrow H_2O(\ell) + CO_2(g)$

5.8 In the reaction $2\ Ca(s) + O_2(g) \rightarrow 2\ CaO(s)$, Ca loses electrons, is oxidized, and is the reducing agent; O gains electrons, is reduced, and is the oxidizing agent.

5.9 $Cl_2(g) + Ca(s) \rightarrow CaCl_2(s)$. $Cl_2(g)$ is the oxidizing agent.

5.10 (a) This is not a redox reaction. Nitric acid is a strong oxidizing agent, but here it serves as an acid.

(b) In this redox reaction, chromium metal (Cr) is oxidized (loses electrons) to form Cr^{3+} ions in Cr_2O_3; oxygen (O_2) is reduced (gains electrons) to form oxide ions, O^{2-}. Oxygen is the oxidizing agent, and chromium is the reducing agent.

(c) This is an acid-base reaction, but not a redox reaction; there are no strong oxidizing or reducing agents present.

(d) Copper is oxidized and chlorine is reduced in this redox reaction, in which copper is the reducing agent and chlorine is the oxidizing agent. The equations are $Cu \rightarrow Cu^{2+} + 2e^-$ and $Cl_2 + 2e^- \rightarrow 2\,Cl^-$.

5.11 (a) Carbon in oxalate ion, $C_2O_4^{2-}$ (oxidation state $= +3$), is oxidized to oxidation state $+4$ in CO_2.

(b) Carbon is reduced from $+4$ in CCl_2F_2 to 0 in C(s).

5.12 (a) $CH_3CH_2OH(\ell) + 3\,O_2(g) \rightarrow 3\,H_2O(\ell) + 2\,CO_2(g)$
Redox

(b) $2\,Fe(s) + 6\,HNO_3(aq) \rightarrow 2\,Fe(NO_3)_3(aq) + 3\,H_2(g)$
Redox

(c) $AgNO_3(aq) + KBr(aq) \rightarrow AgBr(s) + KNO_3(aq)$ Not redox

5.13 Molar mass of cholesterol $= 386.7$ g/mol

$$240\ \text{mg} \times \frac{1\ \text{g}}{10^3\ \text{mg}} \times \frac{1\ \text{mol}}{386.7\ \text{g}} = 6.21 \times 10^{-4}\ \text{mol cholesterol}$$

$$\frac{6.21 \times 10^{-4}\ \text{mol}}{0.100\ \text{L}} = 6.21 \times 10^{-3}\ \text{M}$$

5.14 (a) $6.37\ \text{g Al(NO}_3)_3 \times \dfrac{1\ \text{mol Al(NO}_3)_3}{213.0\ \text{g Al(NO}_3)_3} =$

$0.0299\ \text{mol Al(NO}_3)_3;\ \dfrac{0.0299\ \text{mol}}{0.250\ \text{L}} = 0.120\ \text{M Al(NO}_3)_3$

(b) Molarity: $Al^{3+} = 0.120$; $NO_3^- = 3(0.120) = 0.360$

5.15 If the description of solution preparation always is worded in terms of adding enough solvent to make a specific volume of solution, then any possible expansion or contraction has no effect on the molarity of the solution. The denominator of the definition of molarity is liters of *solution.*

5.16 The moles of HCl in the concentrated solution is given by (6.0 mol/L)(0.100 L) = 0.6 mol HCl. The moles of HCl in the dilute solution are given by (1.20 moles/L)(0.500 L) = 0.6 mol HCl.

5.17 The molarity could be increased by evaporating some of the solvent.

5.18 $0.0193\ \text{L} \times \dfrac{0.200\ \text{mol AgNO}_3}{1\ \text{L}} \times \dfrac{1\ \text{mol Ag}^+}{1\ \text{mol AgNO}_3}$

$\times \dfrac{1\ \text{mol Cl}^-}{1\ \text{mol Ag}^+} \times \dfrac{1\ \text{mol NaCl}}{1\ \text{mol Cl}^-} = 3.86 \times 10^{-3}\text{mol NaCl}$

$\dfrac{3.86 \times 10^{-3}\ \text{mol NaCl}}{0.0250\ \text{L}} = 0.154\ \text{M NaCl}$

Chapter 6

6.1 You transfer some mechanical energy to the ball to accelerate it upward. The ball's potential energy increases the higher it gets, but its kinetic energy decreases by an equal quantity, and eventually it stops rising and begins to fall. As it falls, some of the ball's potential energy changes to kinetic energy, and the ball goes faster and faster until it

hits the floor. When the ball hits the floor, some of its kinetic energy is transferred to atoms, molecules, or ions that make up the floor, causing them to move faster. Eventually all of the ball's kinetic energy is transferred, and the ball stops moving. The nanoscale particles in the floor (and some in the air that the ball fell through) are moving faster on average, and the temperature of the floor (and the air) is slightly higher. The energy has spread out over a much larger number of particles.

6.2

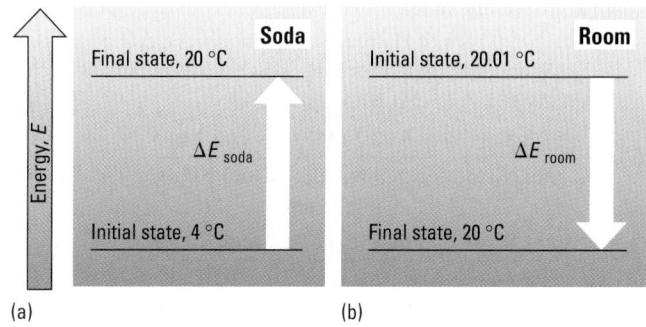

(a) (b)

6.3 From Table 6.1, $c_{\text{granite}} = 0.79$ J g^{-1} °C^{-1} and $c_{\text{H}_2\text{O}(\ell)} = 4.184$ J g^{-1} °C^{-1}. Since water has a larger specific heat capacity, the same quantity of heating will raise its temperature less. Therefore, the rock (granite) will be hotter.

6.4

Metal	Molar heat capacity (J mol^{-1} °C^{-1})	Metal	Molar heat capacity (J mol^{-1} °C^{-1})
Al	24.3	Cu	24.5
Fe	25.2	Au	25.2

The molar heat capacities of most metals are close to 25 J mol^{-1} °C^{-1}. This rule does not work for ethanol.

6.5 (a) Since the heat of vaporization is almost seven times larger than the heat of fusion, the temperature stays constant at 100 °C almost seven times longer than it stays constant at 0 °C. It stays constant at 0 °C for slightly less time than it takes to heat the water from 0 °C to 100 °C (see graph).

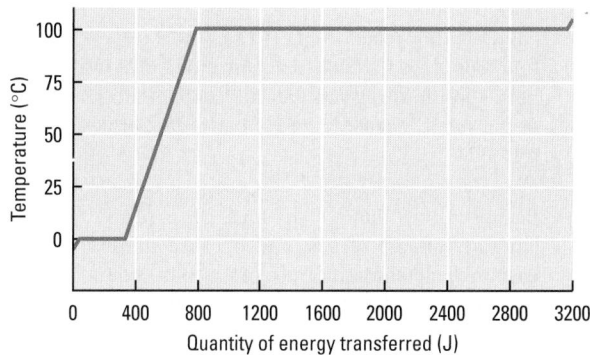

(b) The mass of water is half as great as in part (a), so each process takes half as long. A graph to the same scale as in part (a) begins at 105 °C and reaches −5 °C with half the quantity of energy transferred.

6.6 Heat of fusion: 237 g × 333 J/g = 78.9 kJ
Heating liquid: 237 g × 4.184 J g^{-1} °C^{-1} × 100. °C = 99.2 kJ
Heat of vaporization: 237 g × 2260 J/g = 536 kJ
Total heating = (78.9 + 99.2 + 536) kJ = 714 kJ

6.7 The direction of energy transfer is indicated by the sign of the enthalpy change. Transfer to the system corresponds to a positive enthalpy change.

6.8 Because of heats of fusion and heats of vaporization, the enthalpy change is different when a reactant or product is in a different state.

6.9 When 1.0 mol of H_2 reacts (Equation 6.3), $\Delta H = -241.8$ kJ. When half that much H_2 reacts, ΔH is half as great; that is, 0.5 × (−241.8) = −120.9 kJ.

6.10 $\dfrac{-92.2 \text{ kJ}}{1 \text{ mol } N_2(g)}$ $\dfrac{-92.2 \text{ kJ}}{3 \text{ mol } H_2(g)}$ $\dfrac{-92.2 \text{ kJ}}{2 \text{ mol } NH_3(g)}$

$\dfrac{1 \text{ mol } N_2(g)}{-92.2 \text{ kJ}}$ $\dfrac{3 \text{ mol } H_2(g)}{-92.2 \text{ kJ}}$ $\dfrac{2 \text{ mol } NH_3(g)}{-92.2 \text{ kJ}}$

6.11 The reaction used must be exothermic. Because it can be started by opening the package, it probably involves oxygen from the air, and the sealed package prevents it from occurring before it is needed. Many metals can be oxidized easily and exothermically. The reaction of iron with oxygen (as in the *Chemistry You Can Do* experiment) is a good candidate.

6.12 Yes, it would violate the first law of thermodynamics. According to the supposition, we could create energy by starting with 2 mol of HCl, breaking all the molecules apart, recombining the atoms to form 1 mol of H_2 and 1 mol of Cl_2, and then reacting the H_2 and Cl_2 to give 2 HCl.

2 HCl ⟶ atoms ⟶ H_2 + Cl_2 $\Delta H° = +185$ kJ

H_2 + Cl_2 ⟶ 2 HCl $\Delta H° = -190$ kJ

The net effect of these two processes is that there is still 2 mol of HCl, but 5 kJ of energy has been created. This is impossible according to the first law of thermodynamics.

6.13 (a) In the reaction 2HF → H_2 + F_2 there are two bonds in the two reactant molecules and two bonds in the two product molecules. Since the reaction is endothermic, the bonds in the reactant molecules must be stronger than in the products.
(b) For the reaction 2 H_2O → 2 H_2 + O_2, there are four bonds in the two reactant molecules but only three bonds in the three product molecules. The reaction is endothermic because more bonds are broken than are formed.

6.14 $C_6H_{12}O_6(s)$ + 6 $O_2(g)$ → 6 $CO_2(g)$ + 6 $H_2O(\ell)$
Because the volume of any ideal gas is proportional to the amount (moles) of gas, and because there are 6 mol of gaseous reactant and 6 mol of gaseous product, there will be very little change in volume. Almost no work will be done, and $\Delta H \cong \Delta E$.

6.15 (a) The volumes of acid and base are double those in Problem-Solving Practice 6.11, and the concentrations are the same. Twice as much heat transfer will take place, but there is twice as much water to heat, so ΔT will be the same.

(b) The volumes of acid and base are the same as in Problem-Solving Practice 6.11, but the concentration of H_2SO_4 is half as great as the concentration of HCl. However, the equation for the reaction is

$$\tfrac{1}{2} H_2SO_4(aq) + NaOH(aq) \longrightarrow \tfrac{1}{2} Na_2SO_4(aq) + H_2O(\ell)$$

which means that all of the NaOH can react with only half as much H_2SO_4. ΔT will therefore be the same.

6.16 $N_2(g) \rightarrow N_2(g)$
(a) The product is the same as the reactant, so there is no change — nothing happens.
(b) Since product and reactant are the same, $\Delta H = 0$.

6.17 (a) It takes 160 kJ to break 1 mol of N—N bonds, 4 × 391 kJ to break 4 mol of N—H bonds, and 498 kJ to break 1 mol of bonds in O_2.
(b) Forming 1 mol of bonds in N_2 releases 946 kJ, and forming 4 mol of O—H bonds releases 4 × 467 kJ.
(c) Therefore, ΔH = {160 + (4 × 391) + 498} kJ
          − {946 + (4 × 467)} kJ = −592 kJ

6.18 (a) $CH_4(g)$ + 3 $O_2(g)$ → $CO_2(g)$ + 2 $H_2O(g)$
$\Delta H°$ = {−393.509 + 2(−241.818) − (−74.81)} kJ
   = −802.34 kJ

$\dfrac{802.34 \text{ kJ}}{1 \text{ mol } CH_4} \times \dfrac{1 \text{ mol}}{16.0426 \text{ g}}$ = 50.013 kJ/g CH_4

(b) $C_8H_{18}(\ell)$ + $\tfrac{25}{2} O_2(g)$ → 8 $CO_2(g)$ + 9 $H_2O(g)$
$\Delta H°$ = {8(−393.509) + 9(−241.818) − (−249.952)} kJ
   = −5074.48 kJ

$\dfrac{5074.48 \text{ kJ}}{1 \text{ mol } C_8H_{18}} \times \dfrac{1 \text{ mol}}{114.23 \text{ g}}$ = 44.423 kJ/g C_8H_{18}

(c) $C_2H_5OH(\ell)$ + 3 $O_2(g)$ → 2 CO_2 + 3 $H_2O(g)$
$\Delta H°$ = {2(−393.509) + 3(−241.818) − (−277.69)} kJ
   = −1234.782 kJ

$\dfrac{1234.782 \text{ kJ}}{1 \text{ mol } C_2H_5OH(\ell)} \times \dfrac{1 \text{ mol}}{46.068 \text{ g}}$ = 26.8 kJ/g C_2H_5OH

(d) $N_2H_4(\ell)$ + $O_2(g)$ → $N_2(g)$ + 2 $H_2O(g)$
$\Delta H°$ = {2(−241.818) − 50.63} kJ = −534.26 kJ

$\dfrac{534.26 \text{ kJ}}{1 \text{ mol } N_2H_4} \times \dfrac{1 \text{ mol}}{32.045 \text{ g}}$ = 16.672 kJ/g N_2H_4

(e) $H_2(g)$ + $\tfrac{1}{2} O_2(g)$ → $H_2O(g)$ $\Delta H°$ = −241.818 kJ

$\dfrac{241.818 \text{ kJ}}{1 \text{ mol } H_2} \times \dfrac{1 \text{ mol}}{2.0158 \text{ g}}$ = 119.96 kJ/g H_2

(f) $C_6H_{12}O_6(s)$ + 6 $O_2(g)$ → 6 $CO_2(g)$ + 6 $H_2O(g)$
$\Delta H°$ = {6(−393.509) + 6(−241.818) − (−1274.4)} kJ
   = −2537.6 kJ

$\dfrac{2537.6 \text{ kJ}}{1 \text{ mol } C_6H_{12}O_6(s)} \times \dfrac{1 \text{ mol}}{180.16 \text{ g}}$ = 14.085 kJ/g $C_6H_{12}O_6$

(g) Biomass gives the same result as glucose.
Hydrogen has the greatest fuel value. Octane has the greatest energy density. Its fuel value is more than twice that of the other liquids and solids, and its density is far

greater than for $CH_4(g)$ and $H_2(g)$. Energy density values are

$$CH_4(g) \quad \frac{50.013 \text{ kJ}}{1 \text{ g}} \times \frac{6.9 \times 10^{-4} \text{ g}}{1 \text{ mL}} = 3.4 \times 10^{-2} \text{ kJ/mL}$$

$$C_8H_{18}(\ell) \quad \frac{44.423 \text{ kJ}}{1 \text{ g}} \times \frac{0.70 \text{ g}}{1 \text{ mL}} = 31 \text{ kJ/mL}$$

$$C_2H_5OH(\ell) \quad \frac{26.8 \text{ kJ}}{1 \text{ g}} \times \frac{0.80 \text{ g}}{1 \text{ mL}} = 21 \text{ kJ/mL}$$

$$N_2H_4(\ell) \quad \frac{16.672 \text{ kJ}}{1 \text{ g}} \times \frac{1.00 \text{ g}}{1 \text{ mL}} = 16.7 \text{ kJ/mL}$$

$$H_2(g) \quad \frac{119.96 \text{ kJ}}{1 \text{ g}} \times \frac{8.2 \times 10^{-5} \text{ g}}{1 \text{ mL}} = 9.8 \times 10^{-3} \text{ kJ/mL}$$

$$\text{Carbohydrate} \quad \frac{14.1 \text{ kJ}}{1 \text{ g}} \times \frac{1.56 \text{ g}}{1 \text{ mL}} = 22.0 \text{ kJ/mL}$$

6.19 According to the thermochemical equation, 2801.6 kJ is released per mole of glucose.

$$\frac{2801.6 \text{ kJ}}{1 \text{ mol}} \times \frac{1 \text{ mol}}{180.16 \text{ g}} \times \frac{1 \text{ kcal}}{4.184 \text{ kJ}} \times \frac{1 \text{ Cal}}{1 \text{ kcal}} = 3.717 \text{ Cal}$$

This rounds to 4 Cal.

Chapter 7

7.1 Wavelength and frequency are inversely related. Therefore, low-frequency radiation has long-wavelength radiation.

7.2 Cellular phones use higher frequency radio waves.

7.3 $E = \dfrac{hc}{\lambda}; \lambda = \dfrac{hc}{E}$

$$= \frac{(6.626 \times 10^{-34} \text{ J} \cdot \text{s})(2.998 \times 10^8 \text{ m/s})}{1.47 \times 10^{-23} \text{ J}} = 1.35 \times 10^{-3} \text{ m}$$

7.4 In a sample of excited hydrogen gas there are many atoms, and each can exist in one of the excited states possible for hydrogen. The observed spectral lines are a result of all the possible transitions of all these hydrogen atoms.

7.5 (a) Emitted   (b) Absorbed
(c) Emitted   (d) Emitted

7.6 $(2.179 \times 10^{-18}$ J/photon$)$
$\left(\dfrac{1 \text{ kJ}}{10^3 \text{ J}}\right)\left(\dfrac{6.022 \times 10^{23} \text{ photons}}{1 \text{ mol}}\right)$
$= 1312$ kJ/mol photons

7.7 (a) $5d$ (b) $4f$ (c) $6p$

7.8 The $n = 3$ level can have only three types of sublevels — s, p, and d. The $n = 2$ level can have only s and p sublevels, not d sublevels ($l = 2$).

7.9 $4, 0, 0, +\frac{1}{2}$

7.10 (a) $3, 0, 0, +\frac{1}{2}$ (b) $3, 1, 1, +\frac{1}{2}$

7.11 Electron a is in the $3p_y$ orbital. Electron b is in the $3p_z$ orbital.

7.12 (a) The maximum number of electrons in the $n = 3$ level is 18 (2 electrons per orbital). The orbitals would be designated $3s$, $3p_x$, $3p_y$, $3p_z$, $3d_{z^2}$, $3d_{xy}$, $3d_{yz}$, $3d_{xz}$, and $3d_{x^2-y^2}$.

(b) The maximum number of electrons in the $n = 4$ level is 32. The orbitals would be designated $4s$, $4p_x$, $4p_y$, $4p_z$, $4d_{z^2}$, $4d_{xy}$, $4d_{yz}$, $4d_{xz}$, $4d_{x^2-y^2}$, and the

seven $4f$ orbitals, which are not designated by name in the text.

7.13 The first shell that could contain g orbitals would be the $n = 5$ shell. There would be nine g orbitals.

7.14 For the chlorine atom, $n = 3$, and there are seven electrons in this highest energy level. The configuration is

$3s$ $3p$

[↑↓] [↑↓][↑↓][↑] . For the sulfur atom, the highest energy level is $n = 3$, and there is one less electron.

$3s$ $3p$

The configuration is [↑↓] [↑↓][↑][↑] .

7.15 The $[\text{Ar}]3d^4 4s^2$ configuration for chromium has four unpaired electrons, and the $[\text{Ar}]3d^5 4s^1$ configuration has six unpaired electrons.

7.16 The ground state Cu atom has a configuration $[\text{Ar}]4s^1 3d^{10}$. When it loses one electron, it becomes the Cu^+ ion with configuration $[\text{Ar}]3d^{10}$. There is an added stability for the completely filled set of $3d$ orbitals.

7.17 The Fe(acac)_2 contains an Fe^{2+} ion with a $3d$ electron configuration of [↑↓][↑][↑][↑][↑] . This configuration has four unpaired electrons. The compound Fe(acac)_3 contains an Fe^{3+} ion, with a $3d$ electron configuration of [↑][↑][↑][↑][↑] . This configuration has five unpaired electrons. The Fe(acac)_3, with more unpaired electrons per molecule, would be attracted more strongly into a magnetic field.

Chapter 8

8.1 C_8H_{16}

8.2 N_2 has only 10 valence electrons. The Lewis structure shown has 14 valence electrons.

8.3 None of the structures is correct. (a) is incorrect because sulfur does not have an octet of electrons (has only six); (b) is incorrect because, although it shows the correct number of valence electrons (26), there is a double bond between F and N rather than a single bond with a lone pair on N; (c) is incorrect because the left carbon has five bonds; (d) is incorrect because COCl should have 17 valence electrons, not 16 as shown.

8.4 (a) C_5H_{10}
(b) Two

8.5

maleic acid
(the *cis* isomer)

fumaric acid
(the *trans* isomer)

8.6 C—N > C=N > C≡N. The order of decreasing bond energy is the reverse order: C≡N > C=N > C—N.

8.7 (a) The electronegativity difference between sodium and chlorine is 2.0, sufficient to cause electron transfer from sodium to chlorine to form Na^+ and Cl^- ions. Molten NaCl conducts an electric current, indicating the presence of ions.

(b) There is an electronegativity difference of 1.2 in BrF, which is sufficient to form a polar covalent bond, but not great enough to cause electron transfer leading to ion formation.

8.8 The Lewis structure of hydrazine is

	H	**H**	**N**	**N**	**H**	**H**
Valence electrons	1	1	5	5	1	1
Lone pair electrons	0	0	2	2	0	0
$\frac{1}{2}$ shared electrons	1	1	3	3	1	1
Formal charge	0	0	0	0	0	0

8.9 Atoms cannot be rearranged to derive a resonance structure. There is no N-to-O bond in cyanate ion; therefore, such an arrangement cannot be a resonance structure of cyanate ion.

8.10

8.11

1,2,4-trimethylbenzene

8.12 1. (a) 20 carbons atoms and 30 hydrogen atoms   (b) The carbon atom at the top of the six-membered ring (c) Five C=C double bonds.
2. (a) $C_{29}H_{50}O_2$ (b) No C=C double bonds   (c) The H—O bond.

Chapter 9

9.1 When the central atom has no lone pairs.

9.2 The triangular bipyramidal shape has three of the five pairs situated in equatorial positions 120° apart and the remaining two pairs in axial positions. The square pyramidal shape has four of the atoms bonded to the central atom in a square plane, with the other bonded atom di-

rectly above the central atom and equidistant from the other four.

9.3 (a) AX_2E_3 (b) AX_3E_1 (c) AX_2E_3

9.4 Pi bonding is not possible for a carbon atom with sp^3 hybridization because it has no unhybridized $2p$ orbitals. All of its $2p$ orbitals have been hybridized.

9.5 (a) Bromine is more electronegative than iodine, and the H—Br bond is more polar than the H—I bond.

(b) Chlorine is more electronegative than the other two halogens; therefore, the C—Cl bond is more polar than the C—Br and C—I bonds.

9.6

9.7 The F—H · · · F—H hydrogen bond is the strongest because the electronegativity difference between H and F produces a more polar F—H bond than does the lesser electronegativity difference between O and H or N and H in the O—H or N—H bonds.

9.8

9.9 Replication would be very difficult because covalent bonds would have to be broken. DNA replicates easily because it is hydrogen bonds, not covalent bonds, that occur between the base pairs and hold the two DNA strands together.

Chapter 10

10.1 $(2.7 \times 10^8 \text{ molecules}) \times \left(\dfrac{64.06 \text{ g SO}_2}{6.02 \times 10^{23} \text{ molecules}} \right)$
$= 2.9 \times 10^{-14} \text{ g SO}_2$

10.2 First, gas molecules are far apart. This allows most light to pass through. Second, molecules are much smaller than the wavelengths of visible light. This means that the waves are not reflected or diffracted by the molecules.

10.3 As more gas molecules are added to a container of fixed volume, there will be more collisions of all of the gas molecules with the container walls. This causes the observed pressure to rise.

10.4 All have the same kinetic energy at the same temperature.

10.5 For a sample of helium, the plot would look like the curve marked He in Figure 10.7. When an equal number of argon molecules, which are heavier, are added to the helium, the distribution of molecular speeds would look like the sum of the curves marked He and O_2 in Figure 10.7, except that the curve for Ar would have its peak a little to the left of the O_2 curve.

10.6 (a) The balloon placed in the freezer will be smaller than the one kept at room temperature because its sample of helium is colder.

(b) Upon warming, the helium balloon that had been in the freezer will be either the same size as the balloon kept at room temperature or perhaps slightly larger because there is a greater chance that He atoms leaked out of the room temperature balloon during the time the other balloon was kept in the freezer. This would be caused by the faster moving He atoms in the room temperature balloon having more chances to escape from tiny openings in the balloon's walls.

10.7 The gas in the shock absorbers will be more highly compressed. The gas molecules will be closer together. The gas molecules will collide with the walls of the shock absorber more often, and the pressure exerted will be larger.

10.8 Increasing the temperature of a gas causes the gas molecules to move faster, on average. This means that each collision with the container walls involves greater force, because on average a molecule is moving faster and hits the wall harder. If the container remained the same (constant volume), there would also be more collisions with the container wall because faster moving molecules would hit the walls more often. Increasing the volume of the container, on the other hand, requires that the faster-moving molecules must travel a greater distance before they strike the container walls. Increasing the volume enough would just balance the greater numbers of harder collisions caused by increased temperature. To maintain a constant volume requires that the pressure increase to match the greater pressure due to more and harder collisions of gas molecules with the walls.

10.9 Two moles of O_2 gas are required for the combustion of one mole of methane gas. If air were pure O_2, the oxygen delivery tube would need to be twice as large as the delivery tube for methane. Since air is only one-fifth O_2, the air delivery tube would need to be 10 times larger than the methane delivery tube to ensure complete combustion.

10.10 Using the ratio of 100 balloons/26.8 g He, calculate the number of balloons 41.8 g of He can fill:

$$\text{Balloons} = (41.8 \text{ g He})\left(\frac{100 \text{ balloons}}{26.8 \text{ g He}}\right) = 155 \text{ balloons}$$

This much He will fill more balloons than needed.

10.11 1. Increase the pressure.
2. Decrease the temperature.
3. Remove some of the gas by reaction to form a non-gaseous product.

10.12

$$\text{Density of Cl}_2 \text{ at 25 °C and 0.750 atm} = \frac{PM}{RT}$$

$$= \frac{(0.750 \text{ atm})(70.905 \text{ g/mol})}{(0.0821 \text{ L atm mol}^{-1} \text{ K}^{-1})(298 \text{ K})} = 2.17 \text{ g/L}$$

$$\text{Density of SO}_2 \text{ at 25 °C and 0.750 atm} = \frac{PM}{RT}$$

$$= \frac{(0.750 \text{ atm})(64.06 \text{ g/mol})}{(0.0821 \text{ L atm mol}^{-1} \text{ K}^{-1})(298 \text{ K})} = 1.96 \text{ g/L}$$

$$\text{Density of Cl}_2 \text{ at 35 °C and 0.750 atm} = \frac{PM}{RT}$$

$$= \frac{(0.750 \text{ atm})(70.905 \text{ g/mol})}{(0.0821 \text{ L atm mol}^{-1} \text{ K}^{-1})(308 \text{ K})} = 2.10 \text{ g/L}$$

$$\text{Density of SO}_2 \text{ at 25 °C and 2.60 atm} = \frac{PM}{RT}$$

$$= \frac{(2.60 \text{ atm})(64.06 \text{ g/mol})}{(0.0821 \text{ L atm mol}^{-1} \text{ K}^{-1})(298 \text{ K})} = 6.81 \text{ g/L}$$

10.13 Density of He = 1.23×10^{-4} g/mL
Density of Li = 0.53 g/mL
Since the density of He is so much less than that of Li, the atoms in a sample of He must be much farther apart than the atoms in a sample of Li. This idea is in keeping with the general principle of the kinetic-molecular theory that the particles making up a gas are far from one another.

10.14 A 50-50 mixture of N_2 and O_2 would have less N_2 in it than does air. Since O_2 molecules have greater mass than N_2 molecules, this 50-50 mixture has greater density than air.

10.15 (a) If lowering the temperature causes the volume to decrease, by $PV = nRT$, the pressure can be assumed to be constant. The value of n is unchanged. Since both P and n remain unchanged, the partial pressures of the gases in the mixture remain unchanged.

(b) When the total pressure of a gas mixture increases, the partial pressure of each gas in the mixture increases because the partial pressure of each gas in the mixture is the product of the mole fraction for that gas and the total pressure.

10.16 We can calculate the total number of moles of gas in the flask from the given information.

$$n = \frac{PV}{RT} = \frac{\left(\frac{626}{760}\right) \text{atm} (0.355 \text{ L})}{(0.0821 \text{ L atm mol}^{-1} \text{ K}^{-1})(308 \text{ K})}$$
$$= 0.01156 \text{ mol gas}$$

The number of moles of Ne is

$$0.146 \text{ g Ne} \times \frac{1 \text{ mol Ne}}{20.18 \text{ g/mol}} = 0.007235 \text{ mol Ne}$$

We find the number of moles of Ar by subtraction.

$$0.01156 \text{ mol gas} - 0.007235 \text{ mol Ne} = 0.004325 \text{ mol Ar}$$

$$0.004325 \text{ mol Ar} \times 35.95 \text{ g/mol} = 0.155 \text{ g Ar}$$

10.17 The value of n depends directly on the measured pressure, P. Intermolecular attractions in a real gas would cause the measured P to be slightly smaller than for an ideal gas. The lower value of P would cause the calculated number of moles to be somewhat smaller. Using this slightly smaller value of n in the denominator would cause the calculated molar mass to be a little larger than it should be.

10.18 (a) $HO\cdot + H\cdot$

(b) $CH_3\cdot + HOH$

(c) $HO\cdot + \cdot O\cdot$

10.19

Mass of S burned per hour
$$= (3.06 \times 10^6 \text{ kg})(0.04) = 1 \times 10^5 \text{ kg}$$

Mass of SO_2 per hour
$$= (1 \times 10^5 \text{ kg})\left(\frac{64.06 \text{ kg } SO_2}{32.07 \text{ kg } S}\right) = 2 \times 10^5 \text{ kg}$$

Mass of SO_2 per year
$$= (2 \times 10^5 \text{ kg/hr})(8760 \text{ hr/yr}) = 2 \times 10^9 \text{ kg/yr}$$

10.20 Vol percent SO_2
$$= (5 \text{ parts } SO_2/10^6 \text{ parts air}) \times 100\% = 5 \times 10^{-4} \text{ vol\%}$$

10.21 1 metric ton $= 1000 \text{ kg} = 10^6 \text{ g} = 1 \text{ Mg}$

Mass of $HNO_3 = (400 \text{ Mg } N_2)\left(\dfrac{2 \text{ Mg } NO}{\text{Mg } N_2}\right)$
$$\left(\frac{1 \text{ Mg } NO_2}{1 \text{ Mg } NO}\right)\left(\frac{1 \text{ Mg } HNO_3}{1 \text{ Mg } NO_2}\right) = 800 \text{ Mg } HNO_3$$

10.22 $\cdot NO_2 \xrightarrow{h\nu} NO\cdot + \cdot O\cdot$

$O_3 \xrightarrow{h\nu} O_2 + \cdot O\cdot$

$\cdot O\cdot + O_2 \longrightarrow O_3$

10.23 Among the many possible molecules would be: NO, which comes from automobile combustion; NO_2, which comes from reactions of NO and O_2 in the atmosphere; and O_3, which comes from reactions of NO, NO_2, and O_2 in the atmosphere. Also, hydrocarbons, and oxidation products of hydrocarbon reactions with O_2 and other reactive species.

Chapter 11

11.1 The London forces are greater between bromoform molecules than between chloroform molecules because the bromoform molecules have more electrons. This stronger intermolecular attraction causes the $CHBr_3$ molecules to exhibit a greater surface tension. (The dipole in each molecule contributes less than the London forces to the intermolecular attractions.)

11.2 Water and glycerol would have similar surface tensions because of extensive hydrogen bonding. Octane and decane would have similar surface tensions because both are alkane hyydrocarbons.

11.3 (a) 62 °C   (b) 0 °C   (c) 80 °C

11.4 Bubbles form within a boiling liquid when the vapor pressure of the liquid equals the pressure of the surroundings of the liquid sample. The bubbles are actually filled with vapor of the boiling liquid. One way to prove this would be to trap some of these bubbles and allow them to condense. They would condense to form the liquid that had boiled.

11.5 The evaporating water carries with it thermal energy from the water inside the pot. In addition, a large quantity of thermal energy is required to cause the water to evaporate. Much of this thermal energy comes from the water inside the pot.

11.6 Estimated ΔH°_{vap} for Kr $\simeq$ 20 kJ/mol based on HBr, Cl_2, and C_4H_{10}, but based on Xe, the value is probably closer to 10 kJmol. Estimated ΔH°_{vap} for $NO_2 \approx$ 20 kJ/mol based on propane and HBr.

11.7 (a) Bromine molecules have more electrons than chlorine molecules. Therefore, bromine molecules are held together by stronger intermolecular attractions.

(b) Ammonia molecules are attracted to one another by hydrogen bonds. This causes ammonia to have a higher boiling point than that of methane, which has no hydrogen bonding.

11.8 Two moles of liquid bromine crystallizing liberates 21.59 kJ of heat. One mole of liquid water crystallizing liberates 6.02 KJ of heat.

11.9 High humidity conditions make the evaporation of water or the sublimation of ice less favorable. Under these conditions, the sublimation of ice required to make the frost-free refrigerator work is less favorable, so the defrost cycle is less effective.

11.10 The impurity molecules are less likely to be converted from the solid phase to the vapor phase. This causes them to be left behind as the molecules that sublime go into the gas phase and then condense at some other place. The molecules that condense are almost all of the same kind, so the sublimed sample is much purer than the original.

11.11 The triple point of CO_2, from Figure 11.16, is at 5.2 atm and -57 °C. Increasing the pressure from 5.2 atm while holding the temperature at -57 °C would result in solid CO_2 being formed.

11.12 (a) If liquid CO_2 is slowly released from a cylinder of CO_2, gaseous CO_2 is formed. The temperature remains constant (at room temperature) because there is time for energy to be transferred from the surroundings to separate the CO_2 molecules from their intermolecular attractions. This can be seen from the phase diagram as the phase changes from liquid to vapor as the pressure decreases.

(b) If the pressure is suddenly released, the attractive forces between a large number of CO_2 molecules must be overcome, which requires energy. This energy comes from the surroundings as well as from the CO_2 molecules themselves, causing the temperature of both the surroundings and the CO_2 molecules to decrease. (There are the other factors that must be considered.) On the phase diagram for CO_2, a decrease in both temperature and pressure moves into a region where only solid CO_2 exists.

11.13 It is predicted that a small concentration of gold will be found in the lead and that a small concentration of lead will be found in the gold. This will occur because of the movement of the metal atoms with time, as predicted by the kinetic molecular theory.

11.14 One Po atom belongs to its unit cell. Two Li atoms belong to its unit cell. Four Ca atoms belong to its unit cell.

Simple cubic

= 1 atom
Each of the 8 atoms
contributes 1/8 to unit cell

Body-centered cubic

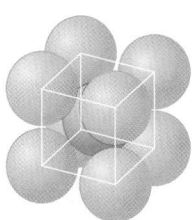

= 2 atoms
8/8 from corner atoms
+ 1 atom at center

Face-centered cubic

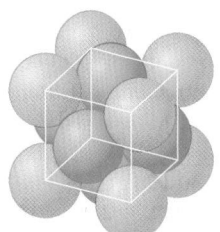

= 4 atoms
8/8 from corner atoms
+ 6/2 from each atom on
the 6 faces contributing 1/2 atom

11.15 Each Cs^+ ion at the center of the cube has eight Cl^- ions as its neighbors. One eighth of each Cl^- ion belongs to that Cs^+ ion. So the formula for this salt must be a 1:1 ratio of Cs^+ ions to Cl^- ions, or CsCl.

11.16 Cooling a liquid above its freezing point causes the temperature to decrease. When the liquid begins to solidify, energy is released as atoms, molecules, or ions move closer together to form in the solid crystal lattice. This causes the temperature to remain constant until all the molecules in the liquid have positioned themselves in the lattice. Further cooling then causes the temperature to decrease. The shape of this curve is common to all substances that can exist as liquids.

11.17 Increasing strength of metallic bonding is related to increasing numbers of valence electrons. In the transition metals, the presence of *d*-orbital electrons causes stronger metallic bonding. Beyond a half-filled set of *d*-orbitals, however, extra electrons have the effect of decreasing the strength of metallic bonding. See Figure 11.25.

11.18 Diamond is denser than graphite, so high pressure is required to convert the carbon atoms from the less dense form. Higher temperatures would allow the atoms to move more rapidly relative to one another and therefore give them a greater opportunity to rearrange.

Chapter 12

12.1 (a) $4 H_2$

(b) $C_7H_{16} \rightarrow C_7H_8 + 4 H_2$

(c) Toluene has a much higher octane number and can be used as an octane enhancer.

12.2 % oxygen = $\left(\dfrac{16.0}{46.0}\right) \times 100\% = 34.8\%$

Ethanol is more highly oxygenated than hydrocarbons in gasoline.

12.3 (a) $C_2H_6O + 2 O_2 \rightarrow 2 CO + 3 H_2O$

$$\frac{g\ CO}{g\ ethanol} = \frac{2(28.0\ g)}{46.0\ g} = 1.22\ g\ CO/g\ ethanol$$

(b) $C_7H_8 + \frac{11}{2} O_2 \rightarrow 7 CO + 4 H_2O$

$$\frac{g\ CO}{g\ toluene} = \frac{7(28.0\ g)}{92.0\ g} = 2.13\ g\ CO/g\ toluene$$

(c) Ethanol produces less CO per gram than does toluene.

12.4 $CH_4(g) + \frac{3}{2} O_2(g) \rightarrow CO(g) + 2 H_2O(g)$

Thus, 28.0 g of CO is produced per 16.0 g of CH_4;

$$\frac{28.0\ g\ CO}{16.0\ g\ CH_4} = 1.75\ g\ CO/g\ CH_4$$

$C_8H_{18}(\ell) + \frac{17}{2} O_2(g) \rightarrow 8 CO(g) + 9 H_2O(g)$

114.0 g of octane produces 224.0 g of CO;

$$\frac{224\ g\ CO}{114.0\ g\ octane} = 1.96\ g\ CO/g\ C_8H_{18}$$

$\dfrac{1.96}{1.75} = 1.12$, which indicates 12% more CO is produced per gram of octane than per gram of methane.

12.5 (a) $CH_4(g) + 2 O_2(g) \rightarrow CO_2(g) + 2 H_2O(g)$

$\Delta H° = [-393.509 + 2(-241.818) - (-74.81)]$ kJ
$\qquad\qquad = -802.34$ kJ

$$\frac{802.34\ kJ}{1\ mol\ CH_4} \times \frac{1\ mol}{16.0426\ g} = 50.013\ kJ/g\ CH_4$$

$50.013\ kJ/g\ CH_4 \times 0.656\ g/L\ CH_4 = 3.28 \times 10^{-2}\ kJ/L\ CH_4$

(b) $C_8H_{18}(\ell) + \frac{25}{2} O_2(g) \rightarrow 8 CO_2(g) + 9 H_2O(g)$

$\Delta H° = [8(-393.509) + 9(-241.818) - (-249.952)]$ kJ
$\qquad\qquad = -5074.48$ kJ

$$\frac{5074.48\ kJ}{1\ mol\ C_8H_{18}} \times \frac{1\ mol}{114.23\ g} = 44.423\ kJ/g\ C_8H_{18}$$

$44.423\ kJ/g\ C_2H_{18} \times 0.699\ g/mL$
$\qquad\qquad = 3.11 \times 10^4\ kJ/L\ C_8H_{18}$

(c) $C_2H_5OH(\ell) + 3 O_2(g) \rightarrow 2 CO_2(g) + 3 H_2O(g)$

$\Delta H° = [2(-393.509) + 3(-241.818)] - (-277.69)$
$\qquad\qquad = -1234.78$ kJ

$$\frac{1234.78\ kJ}{1\ mol\ C_2H_5OH} \times \frac{1\ mol}{46.062\ g} = 26.807\ kJ/g\ C_2H_5OH$$

$26.807\ kJ/g\ C_2H_5OH \times 0.785\ g/mL$
$\qquad\qquad = 2.10 \times 10^4\ kJ/L\ C_2H_5OH$

(d) $N_2H_4(\ell) + O_2(g) \rightarrow N_2(g) + 2 H_2O(g)$

$\Delta H° = [2(-241.818) - 50.63]$ kJ $= -534.26$ kJ

$$\frac{534.26\ kJ}{1\ mol\ N_2H_4} \times \frac{1\ mol}{32.045\ g} = 16.672\ kJ/g\ N_2H_4$$

$16.672\ kJ/g\ N_2H_4 \times 1004\ g/L = 1.67 \times 10^4\ kJ/L\ N_2H_4$

(e) $H_2(g) + \frac{1}{2} O_2(g) \rightarrow H_2O(g)$ $\qquad \Delta H° = -241.818$ kJ

$$\frac{241.818 \text{ kJ}}{1 \text{ mol } H_2} \times \frac{1 \text{ mol}}{2.0158 \text{ g}} = 119.96 \text{ kJ/g } H_2$$

119.96 kJ/g H_2 × 8.20 × 10^{-2} g/L = 9.84 kJ/L H_2

(f) $C_6H_{12}O_6(g) + 6 O_2(g) \rightarrow 6 CO_2(g) + 6 H_2O(g)$

$\Delta H^\circ = [6(-393.509) + 6(-241.818) - (-1274.4)]$ kJ
  $= -2537.56$ kJ

$$\frac{2537.56 \text{ kJ}}{\text{mol } C_6H_{12}O_6} \times \frac{1 \text{ mol}}{180.158 \text{ g}} = 14.085 \text{ kJ/g}$$

14.085 kJ/g × 15.60 g/L = 2.20 × 10^{-4} kJ/L

Hydrogen provides the most thermal energy per gram. Octane has the greatest energy density.

12.6 Thermal energy = 1.5 × 10^6 bbl oil × $\dfrac{5.9 \times 10^9 \text{ J}}{1 \text{ bbl oil}} =$

8.9×10^{15} J

Electricity delivered = 8.9 × 10^{15} J × $\dfrac{1 \times 10^6 \text{ kWh}}{2.6 \times 10^{12} \text{ J}}$ ×

$0.33 = 1.1 \times 10^9$ kWh

12.7 Carbon dioxide cannot be burned.

12.8 Natural sources: animal respiration, forest fires, decay of cellulose products, partial digestion of carbohydrates, volcanoes. Human sources: burning fossil fuels, burning agricultural wastes and refined cellulose products such as paper, decay of carbon compounds in landfills.

12.9 (a) 4.7%, (b) 19.0%, (c) 4.2%; 1950–1999 showed the greatest percent increase in CO_2.

12.10 The fluctuations occur due to the seasons. Photosynthesis, which uses CO_2, is greatest during the spring and summer, accounting for lower CO_2 levels.

12.11 For this calculation, we can use the figure of 450 × 10^9 passenger miles. Using the ratio of 2 × 10^3 kg CO_2/3000 passenger miles, we can calculate the CO_2 released.

Quantity of CO_2 = 450 × 10^9 passenger mi ×

$$\frac{2 \times 10^3 \text{ kg } CO_2}{3 \times 10^3 \text{ passenger mi}} = 3 \times 10^{11} \text{ kg } CO_2$$

If a typical automobile gets 20 mi/gal of fuel and about 1.5 passengers are transported for every mile the automobile travels, then an automobile gets 30 passenger miles per gallon. The number of gallons used would be

Volume of gasoline = 450 × 10^9 passenger miles ×

$$\frac{1 \text{ gal gasoline}}{30 \text{ passenger mi}} = 2 \times 10^{10} \text{ gal gasoline}$$

Assume the gasoline produces about the same mass of CO_2 per gallon as does jet fuel, or 2 × 10^3 kg CO_2/200 gallons.

Quantity of CO_2 from gasoline =

2 × 10^{10} gal gasoline × $\dfrac{2 \times 10^3 \text{ kg } CO_2}{200 \text{ gal gasoline}}$

$= 2 \times 10^{11}$ kg CO_2

So the numbers are about the same for these two modes of transportation.

12.12 He meant that burning coal converted its carbon into carbon dioxide in the air, thus increasing atmospheric CO_2 concentration.

12.13 Ten or so carbon atoms in an alcohol molecule will make it much less water-soluble than alcohols with fewer numbers of carbon atoms.

12.14 (see chemical structure at bottom of page)

12.15 The acetaldehyde molecule has two fewer hydrogen atoms compared with the ethanol molecule. Loss of hydrogen is oxidation. The acetaldehyde molecule is more oxidized than the ethanol molecule. Comparing the formulas for acetaldehyde and acetic acid, the hydrogen atoms are the same, but the acetic acid molecule has one additional oxygen atom. Gain of oxygen is oxidation. So the acetic acid molecule is more oxidized than the acetaldehyde molecule.

12.16

12.17 $CH_3CH_2CH_2OH$, propanol

12.18 (a) Estradiol contains two alcohol groups (−OH).
   (b) Secondary alcohol
   (c) Oxidation (removal of two hydrogens)
   (d) Estradiol: Two alcohol groups, aromatic ring, one −CH_3 group
   Testosterone: One alcohol group; one ketone group; C=C double bond in ring; two −CH_3 groups

12.19 Conversion of a ketone on the five-membered ring to a secondary alcohol by reduction (addition of 2 H atoms)

12.20

(Question 12.14)

12.21 (a)

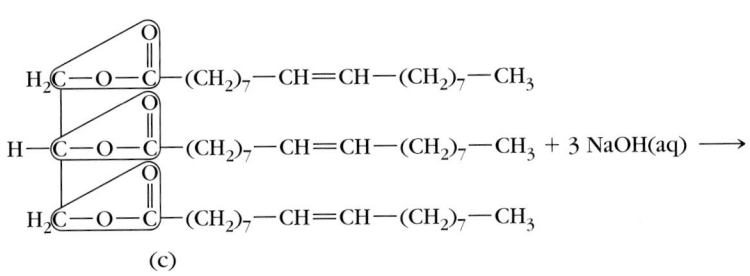

(b)

H₂C—O—C(=O)—(CH₂)₇—CH=CH—(CH₂)₇—CH₃
H—C—O—C(=O)—(CH₂)₇—CH=CH—(CH₂)₇—CH₃ + 3 NaOH(aq) ⟶
H₂C—O—C(=O)—(CH₂)₇—CH=CH—(CH₂)₇—CH₃

(c)

$$\begin{array}{l}CH_2OH\\HC-OH\\CH_2OH\end{array} + 3\,Na^+[CH_3-(CH_2)_7-CH=CH-(CH_2)_7-COO]^-$$

12.22 The ends of the chains are possibly occupied by the OR groups from the initiator molecules.

12.23

$$H_2N-CH_2-CH_2-\overset{O}{\overset{\|}{C}}\left(-\overset{H}{\overset{|}{N}}-CH_2-CH_2-\overset{O}{\overset{\|}{C}}\right)_n-\overset{H}{\overset{|}{N}}-CH_2-CH_2-\overset{O}{\overset{\|}{C}}-OH;\ H_2O$$

12.24 (1) Amine (2) carboxylic acid (3) amide (4) ester

12.25 Serine and glutamine could hydrogen-bond to one another if they were close in two adjacent protein chains because they have polar groups containing H atoms in their R groups. Glycine and valine would not because they have no additional polar groups in their R groups.

12.26

$$H_2N-\overset{H}{\underset{CH_3}{\overset{|}{C}}}-\overset{O}{\overset{\|}{C}}-N-\overset{H}{\underset{CH_2OH}{\overset{|}{C}}}-\overset{O}{\overset{\|}{C}}-N-\overset{H}{\underset{CH_2}{\overset{|}{C}}}-\overset{O}{\overset{\|}{C}}-N-\overset{H}{\underset{CH_2}{\overset{|}{C}}}-\overset{O}{\overset{\|}{C}}-OH$$

12.27 The OH groups in this molecule allow it to be extensively hydrogen-bonded with solvent water molecules.

12.28 Cellulose contains glucose molecules linked together by *trans* 1,4 linkages. Ruminant animals have large colonies of bacteria and protozoa that live in the forestomach and digest cellulose.

12.29 If humans could digest cellulose, then common plants that are easy to grow could become food. There might be less reliance upon cultivation of plants for food. In addition, the entire plant could be used for food rather than just certain parts eaten and the other parts wasted. On the other hand, in times of famine, there might not be enough cellulose to go around. Destroying trees and other plants for food might cause enlargements of desert regions and the disappearance of entire species of plants.

Chapter 13

13.1 (a)

i. Rate $= \dfrac{-\Delta[\text{cisplatin}]}{\Delta t} = \dfrac{-(0.00173-0.00311)\ \text{mol/L}}{(1200-800)\ \text{min}}$

$= \dfrac{0.00138\ \text{mol/L}}{400\ \text{min}} = 3.45\times10^{-6}\ \text{mol L}^{-1}\,\text{min}^{-1}$

ii. Rate $= \dfrac{-(0.00097-0.00558)\ \text{mol/L}}{(1600\text{-}400)\ \text{min}}$

$= 3.84\times10^{-6}\ \text{mol L}^{-1}\,\text{min}^{-1}$

iii. Rate $= \dfrac{-(0.00054-0.01000)\ \text{mol/L}}{(2000-0)\ \text{min}}$

$= 4.73\times10^{-6}\ \text{mol L}^{-1}\,\text{min}^{-1}$

(b)

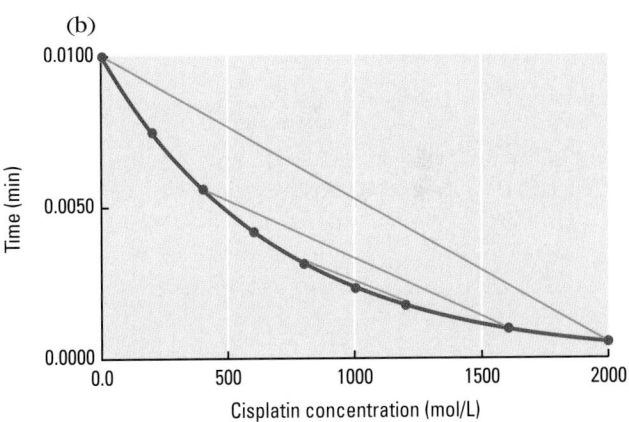

(c) The rate is faster when the concentration of cisplatin is larger. As the reaction takes place, the rate gets slower. There is a much larger change in concentration from 0 min to 1000 min than from 1000 min to 2000 min. Therefore, the Δ[cisplatin] is more than three times bigger for the time range from 400 to 1600 min than it is for the time range 800 to 1200 min, even though Δt is exactly three times larger.

13.2 (a) Rate (1) is twice rate (3); rate (2) is twice rate (4); rate (3) is twice rate (5).

(b) In each case the [cisplatin] is twice as great when the rate is twice as great.

(c) Yes, the rate doubles when [cisplatin] doubles.

13.3

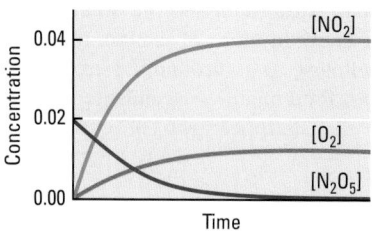

13.4 $\text{Rate}_1 = k[CH_3COOCH_3][OH^-]$
$\text{Rate}_2 = k(2[CH_3COOCH_3])(\frac{1}{2}[OH^-])$
$\qquad\qquad = k[CH_3COOCH_3][OH^-]$
The rate is unchanged.

13.5 (a) $CH_3NC \rightarrow CH_3CN$ unimolecular
(b) $2\,HI \rightarrow H_2 + I_2$ bimolecular
(c) $NO_2Cl \rightarrow NO_2 + Cl$ unimolecular
(d) $C_4H_8 \rightarrow C_4H_8$ unimolecular
(e) $NO_2Cl + Cl \rightarrow NO_2 + Cl_2$ bimolecular

13.6

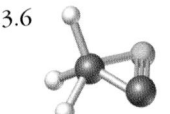

The Lewis structure has five pairs of electrons around one C atom, which would not be stable. However, this is a transition state, which is by definition unstable.

13.7 (1) :Ö=N̈   N̈=Ö:
          Cl::Cl
    (2) :Ö=N̈
          :Cl:
          :Cl:
    :O=N
    (3)   N=Ö::Ö=N
    :Cl   Cl:
    (4) :Cl
         N=Ö:
      :O=N
           Cl:

(1) and (2) are much more likely to result in a reaction than are (3) and (4).

13.8 (a) $\quad k = Ae^{-E_a/RT}$
$$= (6.31 \times 10^8 \text{ L mol}^{-1}\text{s}^{-1})e^{\frac{-10,000 \text{ J/mol}}{(8.314 \text{ J K}^{-1}\text{mol}^{-1})(370 \text{ K})}}$$
$\quad k = 2.4 \times 10^7 \text{ L mol}^{-1}\text{s}^{-1}$
(b) $\text{Rate} = k[NO][O_3]$
$\qquad = (2.4 \times 10^7 \text{ L mol}^{-1}\text{s}^{-1})(1.0 \times 10^{-3} \text{ mol/L})$
$\qquad\quad \times (5.0 \times 10^{-4} \text{ mol/L})$
$\qquad = 12 \text{ mol L}^{-1}\text{s}^{-1}$

13.9 The reaction does not occur in a single step. If it did, the rate law would be $\text{Rate} = k[NO_2][CO]$.

13.10 $\text{Rate} = k[HOI][I^-]$

13.11 (a) $2\,ICl(g) + H_2(g) \rightarrow 2\,HCl(g) + I_2(g)$
(b) $\text{Rate} = k_2[HI][ICl]$
However, HI is an intermediate. Assume that the concentration of HI reaches a steady state.

$\qquad$ Rate of step $+1$ = rate of step -1 + rate of step 2

Also assume that step 2 is much slower than step $+1$ and step -1. Then

$$k_1[ICl][H_2] = k_{-1}[HI][HCl]$$

$$[HI] = \frac{k_1[ICl][H_2]}{k_{-1}[HCl]}$$

$$\text{Rate} = k_2[ICl][HI] = k_2[ICl] \times \frac{k_1[ICl][H_2]}{k_{-1}[HCl]}$$

$$= \frac{k_1k_2}{k_{-1}}[ICl]^2[H_2][HCl]^{-1}$$

(c) The rate is inversely proportional to the concentration of the product, HCl.
(d) Because [HCl] increases as the reaction proceeds, the rate of reaction will decrease more quickly over time than it would if $[HCl]^{-1}$ were not in the rate law. However, the rate constant will not change.

13.12 (a) $Ce^{4+} + Mn^{2+} \rightarrow Ce^{3+} + Mn^{3+}$
$\quad\quad\quad Ce^{4+} + Mn^{3+} \rightarrow Ce^{3+} + Mn^{4+}$
$\quad\quad\quad \underline{Mn^{4+} + Tl^+ \rightarrow Mn^{2+} + Tl^{3+}}$
$\quad\quad\quad 2\,Ce^{4+} + Tl^+ \rightarrow 2\,Ce^{3+} + Tl^{3+}$
(b) Intermediates are Mn^{3+} and Mn^{4+}.
(c) The catalyst is Mn^{2+}.
(d) $\text{Rate} = k[Ce^{4+}][Mn^{2+}]$
(e) $\text{Rate} = k[Ce^{4+}]^2[Mn^{2+}][Ce^{3+}]^{-1}$

13.13 (a) The concentration of a homogeneous catalyst *must* appear in the rate law.
(b) A catalyst does not appear in the equation for an overall reaction.
(c) A *homogeneous* catalyst must always be in the same phase as the reactants.

Chapter 14

14.1 (a) $(\text{conc. Fe}) = \dfrac{7.86 \text{ g}}{\text{mL}} \times \dfrac{1000 \text{ mL}}{1 \text{ L}} \times \dfrac{1 \text{ mol}}{55.85 \text{ g}}$
$\qquad\qquad\qquad\quad = 141 \text{ mol/L}$
(b) $(\text{conc. ethanol}) = \dfrac{0.789 \text{ g}}{\text{mL}} \times \dfrac{1000 \text{ mL}}{1 \text{ L}} \times \dfrac{1 \text{ mol}}{46.07 \text{ g}}$
$\qquad\qquad\qquad\qquad\quad = 17.1 \text{ mol/L}$
(c) $(\text{conc. water}) = \dfrac{0.998 \text{ g}}{\text{mL}} \times \dfrac{1000 \text{ mL}}{1 \text{ L}} \times \dfrac{1 \text{ mol}}{18.02 \text{ g}}$
$\qquad\qquad\qquad\qquad = 55.4 \text{ mol/L}$
(d) $(\text{conc. Mg}) = \dfrac{1.74 \text{ g}}{\text{mL}} \times \dfrac{1000 \text{ mL}}{1 \text{ L}} \times \dfrac{1 \text{ mol}}{24.30 \text{ g}}$
$\qquad\qquad\qquad\quad = 71.6 \text{ mol/L}$

14.2 The mixture is not at equilibrium, but the reaction is so slow that there is no change in concentrations. You could show that the system was not at equilibrium by providing a catalyst or by raising the temperature to speed up the reaction.

14.3 (a) The new mixture is not at equilibrium because the quotient (conc. *trans*)/(conc. *cis*) no longer equals the equilibrium constant. Because (conc. *cis*) was halved, the quotient is twice K_c.

(b) The rate *trans* $\rightarrow$ *cis* remains the same as before, because (conc. *trans*) did not change. The rate *cis* $\rightarrow$ *trans* is only half as great, because (conc. *cis*) is half as great as at equilibrium.

(c) At 600 K, K_c is 1.47. Thus, $[trans] = 1.47[cis]$.

(d) 0.15 mol/L

14.4 (a) $K'_c = \dfrac{[O_3]^2}{[O_2]^3} = (K_c)^2 = (2.5 \times 10^{-29})^2$

$= 6.2 \times 10^{-58}$

(b) $K''_c = \dfrac{[O_2]^3}{[O_3]^2} = (K'_c)^{-1} = \dfrac{1}{6.2 \times 10^{-58}}$

$= 1.6 \times 10^{57}$

14.5 (a) $K_P = K_c \times (RT)^{\Delta n} = (3.5 \times 10^8 \text{ L}^2 \text{ mol}^{-2})$
$\{(0.082057 \text{ L atm K}^{-1} \text{ mol}^{-1})(298)\}^{-2}$
$= 5.8 \times 10^5 \text{ atm}^{-2}$

(b) $K_P = (3.2 \times 10^{81} \text{ L mol}^{-1})$
$\{(0.082057 \text{ L atm mol}^{-1} \text{ K}^{-1})(298 \text{ K})\}^{-1}$
$= 1.3 \times 10^{80} \text{ atm}^{-1}$

(c) $K_P = K_c = 1.7 \times 10^{-3}$

(d) $K_P = (1.7 \times 10^2 \text{ L mol}^{-1})$
$\{(0.082057 \text{ L atm mol}^{-1} \text{ K}^{-1})(298 \text{ K})\}^{-1}$
$= 7.0 \text{ atm}^{-1}$

14.6 (a) Because K_c for the forward reaction is small, K_c for the reverse reaction is large.

(b) $K'_c = \dfrac{1}{K_c} = \dfrac{1}{1.8 \times 10^{-5}} = 5.6 \times 10^4$

(c) Ammonium ions and hydroxide ions should react, using up nearly all of whichever is the limiting reactant.

(d) $NH_4^+(aq) + OH^-(aq) \rightleftharpoons NH_3(aq) + H_2O(\ell)$
$NH_3(aq) \rightleftharpoons NH_3(g)$

You might detect the odor of $NH_3(g)$ above the solution. A piece of moist red litmus paper above the solution would turn blue.

14.7 Q should have the same mathematical form as K_P, so for the general Equation 14.1,

$$Q_P = \dfrac{P_C^c \times P_D^d}{P_A^a \times P_B^b}$$

The rules for Q_P and K_P are analogous to those for Q and K_c:

If $Q_P > K_P$, then the reverse reaction occurs.
If $Q_P = K_P$, then the system is at equilibrium.
If $Q_P < K_P$, the forward reaction occurs.

14.8 $PCl_5(s) \rightleftharpoons PCl_3(g) + Cl_2(g)$

(a) Adding Cl_2 shifts the equilibrium to the left.

(b) Adding PCl_3 to the container shifts the equilibrium to the left.

(c) Because $PCl_5(s)$ does not appear in the K_c expression, adding some will not affect the equilibrium.

14.9 $Q = \dfrac{(\text{conc. } H_2)(\text{conc. } I_2)}{(\text{conc. } HI)^2} = \dfrac{(0.102 \times 3)(0.0183 \times 3)}{(0.0963 \times 3)^2}$

$= \dfrac{(0.102)(0.0183)(9)}{(0.0963)^2(9)} = K_c$

Since $Q = K_c$, the system is at equilibrium under the new conditions. No shift is needed and none occurs.

14.10 From Problem-Solving Practice 14.7, $[NO_2] = 0.0418$ mol/L and $[N_2O_4] = 0.292$ mol/L.

Decreasing the volume from 4.00 to 1.33 L increases the concentrations as

$$(\text{conc. } NO_2) = 0.0418 \times \dfrac{4.00}{1.33} = 0.1257 \text{ mol/L}$$

$$(\text{conc. } N_2O_4) = 0.292 \times \dfrac{4.00}{1.33} = 0.8782 \text{ mol/L}$$

$$Q = \dfrac{(\text{conc. } NO_2)^2}{(\text{conc. } N_2O_4)} = \dfrac{(0.1257)^2}{0.8782} = 1.80 \times 10^{-2}$$

which is greater than K_c. Therefore the equilibrium should shift to the left. Let x be the change in concentration of NO_2.

	N_2O_4	$\rightleftharpoons$	$2 NO_2$
Initial concentration (mol/L)	0.8872		0.1059
Change as reaction occurs (mol/L)	$-\frac{1}{2}x$		x
New equilibrium concentration (mol/L)	$0.8872 - \frac{1}{2}x$		$0.1059 + x$

$K_c = 5.9 \times 10^{-3} = \dfrac{(0.1059 + x)^2}{0.8872 - 0.500x}$

$= \dfrac{(1.121 \times 10^{-2}) + (2.118 \times 10^{-1})x + x^2}{0.8872 - 0.500x}$

$5.23 \times 10^{-3} - (2.95 \times 10^{-3})x$
$= (1.121 \times 10^{-2}) + (2.118 \times 10^{-1})x + x^2$

$x^2 + 0.2148x + (5.98 \times 10^{-3}) = 0$

$x = \dfrac{-0.2148 \pm \sqrt{4.612 \times 10^{-2} - (4 \times 1 \times 5.98 \times 10^{-3})}}{2}$

$x = \dfrac{(-2.413 \times 10^{-2}) \pm 0.1490}{2}$

$x = 0.0624$ or $x = -0.0852$

The first (positive) root is mathematically reasonable, but disagrees with the conclusion based on Q that the reaction shifts to the left.

The new equilibrium concentrations are

$[NO_2] = 0.1059 - 0.0852 = 0.0207$ mol/L

$[N_2O_4] = 0.8872 - \frac{1}{2}(-0.0852) = 0.930$ mol/L

Compared to the initial equilibrium, the concentrations have changed by

$NO_2: \dfrac{0.0207}{0.0352} = 0.588 \qquad N_2O_4: \dfrac{0.930}{0.295} = 3.15$

The concentration of N_2O_4 did increase by more than a factor of 3. The concentration of NO_2 actually decreased, which is clearly less than a factor of 3 increase.

	How reaction system changes	Equilibrium shifts	Change in K_c?
(a) Add reactant	Some reactants consumed	To right	No
(b) Remove reactant	More reactants formed	To left	No
(c) Add product	More reactants formed	To left	No
(d) Remove product	More products formed	To right	No
(e) Increase P by decreasing V	Total pressure decreases	Toward fewer gas molecules	No
(f) Decrease P by increasing V	Total pressure increases	Toward more gas molecules	No
(g) Increase T	Heat transfer into system	In endothermic direction	Yes
(h) Decrease T	Heat transfer out of system	In exothermic direction	Yes

(Question 14.6)

14.11 See table at top of page.

If a substance is added or removed, the equilibrium is affected only if the substance's concentration appears in the equilibrium constant expression or if its addition or removal changes concentrations that appear in the equilibrium constant expression.

Changing pressure by changing volume affects an equilibrium only for gas phase reactions in which there is a difference in the number of moles of reactants and products.

14.12 (a) The reaction is exothermic. $\Delta H° = -46.11$ kJ.

(b) The reaction is not favored by entropy.

(c) The reaction produces more products at low temperatures.

(d) If you increase T the reaction will go faster, but a smaller amount of products will be produced.

Chapter 15

15.1 The data in Table 15.2 indicate that the solubility of alcohols decrease as the hydrocarbon chain lengthens. Thus, 1-octanol is less soluble in water than 1-heptanol and 1-decanol should be even less soluble than 1-octanol.

15.2 Methanol is more water soluble than is octanol, but octanol is more soluble in gasoline. The octanol molecule is more hydrocarbon-like, so this explains its solubility in gasoline. The methanol molecule is more water-like and this explains its greater solubility in water.

15.3 32 g of NH_4Cl would crystallize from solution at 25 °C.

15.4 (a) Unsaturated; (b) supersaturated; (c) saturated

15.5 The solubility of CO_2 decreases with increasing temperature, and the beverage loses its carbonation, causing it to go "flat."

15.6 Putting back water that is too warm would decrease the solubility of oxygen in the lake or river, thereby decreasing the oxygen concentration. This could cause a fish kill if the oxygen concentration dropped sufficiently.

15.7 Hot solvent would cause more of the solute to dissolve because Le Chatelier's principle states that at higher temperature an equilibrium will shift in the endothermic direction.

15.8 (a) Mass fraction of $NaHCO_3$
$$= \frac{0.20}{1000 + 6.5 + 0.20 + 0.10 + 0.10} = 2.0 \times 10^{-4}$$
Wt. fraction $= 2.0 \times 10^{-4} \times 100\% = 2.0 \times 10^{-2}$

(b) KCl and $CaCl_2$ each have the lowest mass fraction, 0.015.

15.9 (a) The 20-ppb sample has the higher lead concentration. (The other sample is 3 ppb lead.)

(b) 0.015 mg/L is equivalent to 0.015 ppm, which is 15 ppb. The 20-ppb sample exceeds the EPA limit; the 3-ppb sample does not.

15.10 $\dfrac{280 \text{ mg}}{1 \text{ d}} \times \dfrac{0.500 \text{ L}}{12 \text{ mg}} = \dfrac{12 \text{ L}}{\text{d}} = 12$ bottles per day

15.11 3.5×10^{-20} gal $\times \dfrac{3.785 \text{ L}}{\text{gal}} \times \dfrac{1 \times 10^{-3} \text{ mg Au}}{1 \text{ L}} =$
1×10^{18} mg Au $= 1 \times 10^{15}$ g Au
1×10^{15} g Au $\times \dfrac{1 \text{ lb Au}}{454 \text{ g Au}} = 2 \times 10^{12}$ lb Au

15.12 100.0 ml $\times$ 1.15 g/mL $=$ 115. g ethylene glycol;
$115. \text{ g} \times \dfrac{1 \text{ mol}}{62.0 \text{ g}} = 1.85$ mol ethylene glycol
100.0 mL $\times$ 1.00 g/mL $=$ 100. g water $\equiv 5.56$ mol water
$$X_{\text{water}} = \frac{5.56}{1.85 + 5.56} = \frac{5.56}{7.41} = 0.750$$
$P_{\text{water}} = (X_{\text{water}})(P°_{\text{water}})$
$= (0.750)(525.8 \text{ mm Hg})$
$= 394 \text{ mm Hg}$

15.13 90 proof = 45 mL ethanol/100 mL beverage

$$\frac{45 \text{ mL ethanol}}{100 \text{ mL BVG}} \times \frac{1 \text{ mL BVG}}{0.861 \text{ g BVG}} \times \frac{0.79 \text{ g ethanol}}{1 \text{ mL ethanol}}$$

$$= \frac{0.414 \text{ g ethanol}}{\text{g BVG}}$$

Mass of BVG: $1 \text{ qt} \times \dfrac{1 \text{ L}}{1.057 \text{ qt}} \times \dfrac{861 \text{ g}}{1 \text{ L}} = 814.5 \text{ g BVG}.$

Moles of ethanol: $814.5 \text{ g BVG} \times \dfrac{0.414 \text{ g ethanol}}{\text{g BVG}}$

$$\times \frac{1 \text{ mol ethanol}}{46.0 \text{ g ethanol}} = 7.32 \text{ mol ethanol}$$

Mass of ethanol: $814.5 \text{ g BVG} \times \dfrac{0.414 \text{ g ethanol}}{\text{g BVG}}$

$$= 336.8 \text{ g ethanol}$$

Mass of solvent: $814.5 \text{ g BVG} - 336.8 \text{ g ethanol}$

$$= 477.7 \text{ g solvent}$$
$$= 0.4777 \text{ kg solvent}$$

$$m_{\text{ethanol}} = \frac{7.32 \text{ mol ethanol}}{0.4777 \text{ kg solvent}} = 15.3 \text{ mol/kg}$$

15.14 (a) Moles of solute and kilograms of solvent
(b) Molar mass of the solute and the density of the solution.

15.15 $\Delta T_b = (2.53 \text{ °C kg mol}^{-1})(0.10 \text{ mol/kg}) = 0.25 \text{ °C}$.
The boiling point of the solution is $80.10 \text{ °C} + 0.25 \text{ °C} = 80.35 \text{ °C}$.

15.16 First, calculate the required molality of the solution that would have a freezing point of -30 °C.

$$\Delta T_f = -30 \text{ °C} = (-1.86 \text{ °C} \cdot \text{kg mol}^{-1}) \times m$$

$$m = \frac{-30 \text{ °C}}{-1.86 \text{ °C kg mol}^{-1}} = 16.1 \text{ mol/kg}$$

To protect 4 kg of water from this freezing temperature, you would need 4×16.1 mol of ethylene glycol, or 65 mol.

$$65 \text{ mol} \left(\frac{62.1 \text{ g}}{\text{mol}}\right) \left(\frac{1 \text{ mL}}{1.113 \text{ g}}\right) \left(\frac{1 \text{ L}}{1000 \text{ mL}}\right) = 3.6 \text{ L}$$

15.17 $\Delta T_f = K_f \times m \times i$;
$4.78 \text{ °C} = (1.86 \text{ °C kg mol}^{-1})(2.0 \text{ mol/kg})(i)$

$$i = \frac{4.78 \text{ °C}}{(0.20 \text{ mol/kg})(1.86 \text{ °C kg mol}^{-1})} = 1.28$$

Degree of dissociation: If completely dissociated, 1 mol $CaCl_2$ should yield 3 mol of ions:

$$CaCl_2 \longrightarrow Ca^{2+} + 2 Cl^-$$

and i should be 3. The degree of dissociation in this solution is $\frac{1.28}{3} = 0.427 \approx 43\%$.

15.18 The 0.02 mol/kg solution of ordinary soap would contain more particles since a soap is a salt of a fatty acid while sucrose is a nonelectrolyte.

15.19 $\dfrac{3 \text{ qt}}{\text{d}} \times \dfrac{1 \text{ L}}{1.06 \text{ qt}} \times \dfrac{0.050 \text{ mg Se}}{\text{L}} = \dfrac{0.14 \text{ mg Se}}{\text{d}}$

15.20 0.025 ppm Pb means that for every liter (1 kg) of water, there is 0.025 mg, or 25 μg of Pb. Using this factor,

$$\text{Volume of water} = 100.0 \ \mu\text{g Pb} \left(\frac{1 \text{ L}}{25 \ \mu\text{g Pb}}\right) = 4.0 \text{ L}$$

Chapter 16

16.1 (a) Acid (b) Base (c) Acid (d) Base

16.2 More water molecules are available per NH_3 molecule in a very dilute solution of NH_3.

16.3 (a) $H_3O^+(aq) + CN^-(aq)$
(b) $H_3O^+(aq) + Br^-(aq)$
(c) $CH_3NH_3^+(aq) + OH^-(aq)$
(d) $(CH_3)_2NH_2^+(aq) + OH^-(aq)$

16.4 (a) $HSO_4^-(aq) + H_2O(\ell) \rightarrow H_2SO_4(aq) + OH^-(aq)$
(b) HSO_4^- is the conjugate base of H_2SO_4; water is the conjugate acid of OH^-.

16.5 The reverse reaction is favored because HSO_4^- is a stronger acid than CH_3COOH.

16.6 (a) $N(aq) + H_2O(\ell) \rightleftharpoons$ $NH^+(aq) + OH^-(aq)$

(b) $N(aq) + HCl(aq) \longrightarrow$ $NH^+(aq) + Cl^-(aq)$

16.7 (a) Amine (b) Neither (c) Acid (d) Amine and acid
(e) Amine

16.8 The pH values of 0.1 M solutions of these two strong acids would be essentially the same since they both are 100% ionized, resulting in $[H_3O^+]$ values that are the same.

16.9 $[H^+] = 10^{-pH} = 10^{-(-3.6)} = 10^{3.6} = 4 \times 10^3$ M

16.10 Because pH + pOH = 14.0, both solutions have a pOH of 8.5. The $[H_3O^+] = 10^{-pH} = 10^{-5.5} = 3.16 \times 10^{-6}$ M.

16.11 A reaction had the following conditions:

$$Ni(H_2O)_6^{2+}(aq) + H_2O(\ell) \rightleftharpoons$$

	$Ni(H_2O)_5(OH)^+(aq) + H_3O^+(aq)$		
Initial concentration (mol/L)	0.15	0	10^{-7}
Change in concentration on reaction (mol/L)	$-x$	$+x$	$+x$
Concentration at equilibrium (mol/L)	$0.15 - x$	x	x

Substituting these values in the equilibrium constant expression, and simplifying $0.15 - x$ to be 0.15 because the value of K_a is so small,

$$K_a = \frac{[Ni(H_2O)_5(OH)^+][H_3O^+]}{[Ni(H_2O)_6^{2+}]}$$

$$= \frac{(x)(x)}{(0.15 - x)} \approx \frac{x^2}{0.15} = 2.5 \times 10^{-11}$$

Solving for x, which is the $[H_3O^+]$,

$$x = \sqrt{(0.15)(2.5 \times 10^{-11})} = 1.9 \times 10^{-6}$$

So, the pH of this solution is $-\log (1.9 \times 10^{-6}) = 5.72$.

16.12 Pyruvic acid is the stronger acid, as indicated by its larger K_a value. Lactic acid's ionization reaction is more reactant-favored (less acid ionizes).

16.13 Being negatively charged, the HSO_4^- ion has a lower tendency to lose a positively charged proton because of the electrostatic attractions of opposite charges.

16.14 (a) Step 1: $HOOC-COOH(aq) + H_2O(\ell) \rightleftharpoons$
$$H_3O^+(aq) + HOOC-COO^-(aq)$$
Step 2: $HOOC-COO^-(aq) + H_2O(\ell) \rightleftharpoons$
$$H_3O^+(aq) + {}^-OOC-COO^-(aq)$$
(b) Step 1: $C_3H_5(COOH)_3(aq) + H_2O \rightleftharpoons$
$$H_3O^+(aq) + C_3H_5(COOH)_2COO^-(aq)$$
Step 2: $C_3H_5(COOH)_2COO^-(aq) + H_2O \rightleftharpoons$
$$H_3O^+(aq) + C_3H_5(COOH)(COO)_2^{2-}(aq)$$
Step 3: $C_3H_5(COOH)(COO)_2^{2-}(aq) + H_2O \rightleftharpoons$
$$H_3O^+(aq) + C_3H_5(COO)_3^{3-}(aq)$$

16.15 $K_b = \dfrac{1.0 \times 10^{-14}}{K_a} = \dfrac{1.0 \times 10^{-14}}{1.3 \times 10^{-10}} = 7.7 \times 10^{-5}$;
carbonate ion; by comparing K_b values.

16.16 (a) Fluorobenzoic acid (b) Chloroacetic acid.
In both cases, the more electronegative halogen atom increases electron withdrawal from the acidic hydrogen, thereby increasing its partial positive charge.

16.17 Oxalic acid is the stronger acid because it has a greater number of oxygens.

16.18 (a)

16.19

16.20 The pH of soaps is >7 due to the reaction with water of the conjugate base in the soap to form a basic solution.

16.21

16.22 Ammonium acetate. The pH of a solution of this salt will be 7.

16.23 Strong bases would cause damage to tissue.

16.24 Baking powder and baking soda.

16.25 Set up a small table for the hydrolysis reaction.

	$CO_3^{2-} + H_2O \rightleftharpoons HCO_3^- + OH^-$		
Initial concentration (mol/L)	5.2	0	10^{-7}
Concentration change due to reaction (mol/L)	$-x$	$+x$	$+x$
Concentration at equilibrium (mol/L)	$5.2 - x$	x	x

Using the K_b expression and substituting the values from the table,

$$K_b = \frac{K_w}{K_a(HCO_3^-)} = \frac{[HCO_3^-][OH^-]}{[CO_3^{2-}]} = \frac{x^2}{5.2 - x}$$
$$\approx \frac{x^2}{5.2} = 2.1 \times 10^{-4}$$

$x = [OH^-] = \sqrt{(5.2)(2.1 \times 10^{-4})} = 3.3 \times 10^{-2}$

$pOH = -\log(3.3 \times 10^{-2}) = 1.48$

$pH = 14.00 - 1.48 = 12.52$

16.26 (a) Lewis base   (b) Lewis acid
(c) Lewis acid and base   (d) Lewis acid
(e) Lewis acid   (f) Lewis base

Chapter 17

17.1 HCl and NaCl: no; has no significant H^+ acceptor (Cl^- is a very poor base).
KOH and KCl: no; has no H^+ donor.

17.2 $pH = 2.49 + \log \dfrac{0.060}{0.050} = 2.49 + \log(1.2)$
$= 2.49 + 0.079 = 2.57$

17.3 $7.40 = 7.21 + \log(\text{ratio}) = 7.21 + \log \dfrac{[HPO_4^{2-}]}{[H_2PO_4^-]}$
$\log \dfrac{[HPO_4^{2-}]}{[H_2PO_4^-]} = 7.40 - 7.21 = 0.19$
$\dfrac{[HPO_4^{2-}]}{[H_2PO_4^-]} = 10^{0.19} = 1.5$
Therefore, $[HPO_4^{2-}] = 1.5 \times [H_2PO_4^-]$.

17.4 $pH = pK_a + \log \dfrac{[\text{acetate}]}{[\text{acetic acid}]}$
$4.68 = 4.74 + \log \dfrac{[\text{acetate}]}{[\text{acetic acid}]}$
$\log \dfrac{[\text{acetate}]}{[\text{acetic acid}]} = 4.68 - 4.74 = -0.06$;
$\dfrac{[\text{acetate}]}{[\text{acetic acid}]} = 10^{-0.06} = 0.86$
Therefore, $[\text{acetate}] = 0.86 \times [\text{acetic acid}]$.

17.5 (a) $pH = 6.38 + \log \dfrac{[0.25]}{[0.10]} = 6.38 + 0.398 = 6.78$
(b) $pH = 7.21 + \log \dfrac{[HPO_4^{2-}]}{[H_2PO_4^-]}$
$= 7.21 + \log \dfrac{(0.25)}{(0.10)} = 7.21 + 0.398 = 7.61$

17.6 Since CO_2 reacts to form an acid, H_2CO_3, the phosphate ion that is the stronger base, HPO_4^{2-}, will be used to counteract its presence.

17.7

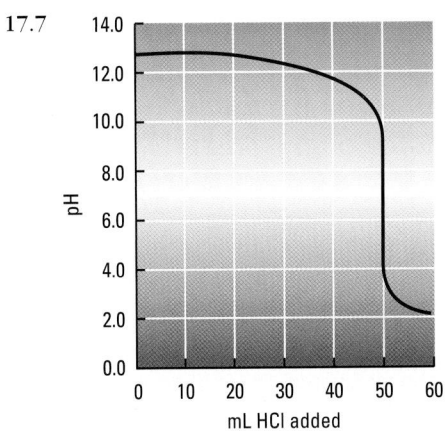

17.8 The addition of 30.0 mL of 0.100 M NaOH neutralizes 30.0 mL of 0.100 M acetic acid, forming 0.0030 mol of acetate ions, which is in 80.0 mL of solution. There is (0.0200 L)(0.100 M) = 0.00200 mol of acetic acid that is unreacted.

$$K_a = 1.8 \times 10^{-5} = \frac{[H^+][C_2H_3O_2^-]}{[HC_2H_3O_2]}$$

$$= \frac{[H^+] \times \left(\dfrac{0.00300 \text{ mol}}{0.0800 \text{ L}}\right)}{\left(\dfrac{0.00200 \text{ mol}}{0.0800 \text{ L}}\right)}$$

$$1.8 \times 10^{-5} = \frac{[H^+] \times (0.0375)}{(0.025)} = [H^+] \times 1.5$$

$$[H^+] = \frac{1.8 \times 10^{-5}}{1.5} = 1.2 \times 10^{-5}$$

$$pH = -\log(1.2 \times 10^{-5}) = 4.92$$

17.9 As NaOH is added, it reacts with acetic acid to form sodium acetate. After 20.0 mL of NaOH has been added, just less than half of the acetic acid has been converted to sodium acetate; when 30.0 mL of NaOH has been added, just over half of the acetic acid has been neutralized. Thus, after 20.0 mL and 30.0 mL of base have been added, the solution contains approximately equal amounts of acetic acid and acetate ion, its conjugate base, which acts as a buffer.

17.10 Because Reaction (b) occurs to an appreciable extent, CO_3^{2-} is used as it forms by Reaction (a), causing additional $CaCO_3(s)$ to dissolve.

17.11 (a) The excess iodide would create a stress on the equilibrium and shift it to the left; some AgI and some PbI_2 would precipitate from solution.
(b) The added SO_4^{2-} would cause the precipitation of $BaSO_4$.

Chapter 18

18.1 (a) $H_2O(\ell) \rightarrow H_2O(g)$ Product-favored
(b) $SiO_2(s) \rightarrow Si(s) + O_2(g)$ Reactant-favored
(c) $(C_6H_{10}O_5)_n + 6n\, O_2(g) \rightarrow 6n\, CO_2(g) + 5n\, H_2O(g)$
Product-favored
(d) $NaCl(s) \rightarrow NaCl(aq)$ Product-favored

18.2 A*** A**B* A*B** B***
If C, D, and E are added, there are many more arrangements in addition to these:

A*B*C* A*B*D* A*B*E* A*C*D* A*C*E*
A*D*E* B*C*D* B*C*E* B*D*E* C*D*E*
A**C* A**D* A**E* B**C* B**D*
B**E* C**A* C**B* C**D* C**E*
D**A* D**B* D**C* D**E* E**A*
E**B* E**C* E**D* C*** D***
E***

There are 35 possible arrangements, but only 4 of them have the energy confined to atoms A and B. The probability that all energy remains with A and B is thus 4/35 = 0.114, or a little more than 11%.

18.3 Using Celsius temperature and $\Delta S = q_{rev}/T$, if the temperature were $-10\,°C$, the value of ΔS would be negative, in disagreement with the fact that transfer of energy to a sample should increase molecular motion and hence entropy.

18.4 (a) The reactant is a gas. The products are also gases, but the number of molecules has increased, so entropy is greater for products. (Entropy increases.)
(b) The reactant is a solid. The product is a solution. Mixing sodium and chloride ions among water molecules results in greater entropy for the product. (Entropy increases.)
(c) The reactant is a solid. The products are a solid and a gas. The much larger entropy of the gas results in greater entropy for the products. (Entropy increases.)

18.5 (a) Because $\Delta S_{\text{surroundings}} = -\Delta H/T$ at a given temperature, the larger the value of T the smaller the value of $\Delta S_{\text{surroundings}}$.
(b) If ΔS_{system} does not change much with temperature, then S_{universe} must also get smaller. In this case, because ΔS_{system} is negative, $\Delta S_{\text{universe}}$ would become negative at a high enough temperature.

18.6 (a) The reaction would have gaseous water as a product.
(b) Both $\Delta H°$ and $\Delta S°$ would change. $\Delta H° = -802.36$ kJ and $\Delta S° = -5.15$ J/K.
(c) If any of the reactants or products change to a different phase (s, ℓ, or g) over the range of temperature, $\Delta H°$ and $\Delta S°$ will change at the temperature of the phase transition.

18.7

Reaction	$\Delta H°$, 298 K (kJ)	$\Delta S°$, 298 K (J/K)
(a)	-1410.94	-267.67
(b)	467.87	560.32
(c)	-393.509	2.862
(d)	620.60	-461.50

Reaction (a) is product-favored at low T (room temperature) and reactant-favored at high T.
Reaction (b) is reactant-favored at low T, product-favored at high T.
Reaction (c) is product-favored at all values of T.
Reaction (d) is reactant-favored at all values of T.

18.8 $\Delta S^\circ_{\text{system}} = 2 \text{ mol HCl(g)} \times S^\circ(\text{HCl}[g]) - 1 \text{ mol } H_2[g]$
$\times S^\circ(H_2[g]) - 1 \text{ mol } Cl_2[g] \times S^\circ(Cl_2[g])$
$= (2 \times 186.908 - 130.684 - 223.066) \text{ J/K}$
$= 20.066 \text{ J/K}$

$\Delta S^\circ_{\text{surroundings}} = -\Delta H^\circ/T = -[2 \text{ mol HCl(g)}$
$\times (-92.307 \text{ kJ/mol})]/298.15 \text{ K}$
$= 619.20 \text{ J/K}$

$\Delta S^\circ_{\text{universe}} = (619.20 + 20.066) \text{ J/K} = 639.27 \text{ J/K}$

18.9

Sign of ΔH°	Sign of ΔS°	Sign of ΔG°	Product-favored?
Negative (exothermic)	Positive	Negative	Yes
Negative (exothermic)	Negative	Depends on T	Yes at low T; no at high T
Positive (endothermic)	Positive	Depends on T	No at low T; yes at high T
Positive (endothermic)	Negative	Positive	No

18.10 (a) At 400 °C the equation is

$$2 \text{ HgO(s)} \longrightarrow 2 \text{ Hg(g)} + O_2(g)$$

Because Hg(g) is a product, instead of Hg(ℓ), both ΔH° and ΔS° will have significantly different values above 356 °C from their values below 356 °C. Therefore the method of estimating ΔG° would not work above 356 °C.
(b) At 400 °C the entropy change should be more positive, which would make the reaction more product-favored.

18.11 (a) If the extent of reaction is 0.10, then 0.10 mol of NaOH has reacted with 0.10 mol of CO_2 to produce 0.10 mol of $NaHCO_3$.

$\Delta G^\circ(0.10 \text{ extent}) = 0.10 \text{ mol} \times \Delta G^\circ_f(NaHCO_3[s])$
$- 0.10 \text{ mol} \times \Delta G^\circ_f(NaOH[s])$
$- 0.10 \text{ mol} \times \Delta G^\circ_f(CO_2[g])$
$= -0.10 \times 851.0 \text{ kJ} +$
$0.10 \times 379.484 \text{ kJ} +$
$0.10 \times 394.359 \text{ kJ}$
$= -7.72 \text{ kJ}$

Similarly,

$\Delta G^\circ (0.40 \text{ extent}) = -30.9 \text{ kJ} (= 0.40[-72.2 \text{ kJ}])$

$\Delta G^\circ (0.80 \text{ extent}) = -61.8 \text{ kJ} (= 0.80[-72.2 \text{ kJ}])$

(b) In each case $\Delta G^\circ(x \text{ extent}) = x\Delta G^\circ(\text{full extent})$, which verifies the statement.
(c) Since $\Delta G^\circ (x \text{ extent}) = x\Delta G^\circ (\text{full extent})$,

$$y = xm + b$$

where $b = 0$ and $m = \Delta G^\circ(\text{full extent}) = \text{slope}$.

18.12 (a) $\Delta S^\circ(i) = \{2 \times (27.78) + \frac{3}{2} \times (205.138) - 87.40]\} \text{ J/K}$
$= 275.86 \text{ kJ}$

$\Delta H^\circ(i) = -\Delta H^\circ_f(Fe_2O_3[s]) = 824.2 \text{ kJ}$

$\Delta G^\circ(i) = -\Delta G^\circ_f(Fe_2O_3[s]) = 742.2 \text{ kJ}$

$\Delta S^\circ(ii) = \{50.92 - 2 \times (28.3) - \frac{3}{2} \times (205.138)\} \text{ J/K}$
$= -313.4 \text{ kJ}$

$\Delta H^\circ(ii) = \Delta H^\circ_f(Al_2O_3[s]) = -1675.7 \text{ kJ}$

$\Delta G^\circ(ii) = -\Delta G^\circ_f(Al_2O_3[s]) = -1582.3 \text{ kJ}$

Step (i) is reactant-favored. Step (ii) is product-favored.
(b) Net reaction:
$Fe_2O_3(s) + 2 \text{ Al(s)} \longrightarrow 2 \text{ Fe(s)} + Al_2O_3(s)$

$\Delta S^\circ = 275.86 \text{ J/K} + (-313.4 \text{ J/K}) = -37.5 \text{ J/K}$

$\Delta H^\circ = 824.2 \text{ kJ} + (-1675.7 \text{ kJ}) = -851.5 \text{ kJ}$

$\Delta G^\circ = 742.2 \text{ kJ} + (-1582.3 \text{ kJ}) = -840.1 \text{ kJ}$

The net reaction has negative ΔG° and is therefore product-favored. For the *net* reaction, ΔS°, ΔH°, and ΔG° are all negative.
(c) If the two reactions are coupled, it is possible to obtain iron from iron(III) oxide even though that reaction is not product-favored by itself. The large negative ΔG° for formation of $Al_2O_3(s)$ makes the overall ΔG° negative for the coupled reactions.
(d) $Mg(s) + \frac{3}{2} O_2(g) \rightarrow MgO(s)$
$\Delta G^\circ = \Delta G^\circ_f(MgO[s]) = -569.43 \text{ kJ}$
Coupling the reactions, we have

$FeO_3(s) \longrightarrow 2 \text{ Fe(s)} + \frac{3}{2} O_2(g) \qquad \Delta G^\circ_1 = 742.2 \text{ kJ}$

$3 \times (Mg[s] + \frac{1}{2} O_2[g] \longrightarrow MgO[s])$
$\Delta G^\circ_2 = 3(-569.43) \text{ kJ} = -1708.29 \text{ kJ}$

$Fe_2O_3(s) + 3 \text{ Mg(s)} \longrightarrow 2 \text{ Fe(s)} + 3 \text{ MgO(s)}$
$\Delta G^\circ_3 = -966.1 \text{ kJ}$

18.13 $\Delta G^\circ = -2870 \text{ kJ} + 32 \times (30.5 \text{ kJ}) = -1894 \text{ kJ}$. The 1894 kJ of Gibbs free energy is transformed into thermal energy.

18.14 64,500 g ATP/50 g ATP = 1290 times each ADP must be recycled to ATP on average each day.

Chapter 19

19.1 This is an application of the law of conservation of matter. If the number of electrons gained were different from the number of electrons lost, some electrons must have been created or destroyed.

19.2 Removal of the salt bridge would effectively switch off the flow of electricity from the battery.

19.3 Avogadro's number of electrons is 96,500 coulombs of charge, so it is 96,500 times as large as one coulomb of charge.

19.4 The zinc anode could be weighed before the battery was put into use. After a period of time, the zinc anode could be dried and reweighed. A loss in weight would be interpreted as being caused by the loss of Zn atoms from the surface through oxidation.

19.5 No, because Hg^{2+} ions can oxidize Al metal to Al^{3+} ions. The net cell reaction is

$$2\ Al(s) + 3\ Hg^{2+}(aq) \longrightarrow 2\ Al^{3+}(aq) + 3\ Hg(\ell)$$

$$E_{cell} = +2.51\ V$$

19.6 For this table,
   (a) V^{2+} ion is the weakest oxidizing agent.
   (b) Cl_2 is the strongest oxidizing agent.
   (c) V is the strongest reducing agent.
   (d) Cl^- is the weakest reducing agent.
   (e) No, E_{net} for that reaction would be < 0.
   (f) No, E_{net} for that reaction would be < 0.
   (g) Pb can reduce I_2 and Cl_2.

19.7 In Table 19.1, Sb would be above H_2 and Pb would be below H_2. For Sb, the reduction potential would be between 0.00 and $+0.337$ V, and for Pb the value would be between 0.00 and -0.14 V.

19.8 For Na^+: $E_{ion} = 61.5 \log\left(\dfrac{150}{18}\right) = 61.5 \log(8.33)$

$$= 61.5 \times 0.921 = 56\ mV$$

19.9 During charging, the reactions at each electrode are reversed. At the electrode that is normally the anode, the charging reaction is

$$Cd(OH)_2(s) + 2\ e^- \longrightarrow Cd(s) + 2\ OH^-(aq)$$

This is reduction, so this electrode is now a cathode.
At the electrode that is normally the cathode, the charging reaction is

$$Ni(OH)_2 + OH^-(aq) \longrightarrow NiO(OH)(s) + H_2O(\ell) + e^-$$

This is oxidation, so this electrode is now an anode.

19.10 Remove the lead cathodes and as much sulfuric acid as you can from the discharged battery. Find some steel and construct a battery with Cl_2 gas flowing across a piece of steel. The two half-reactions would be

$Cl_2(g) + 2\ e^- \rightarrow 2\ Cl^-(aq)$ $\qquad +1.36\ V$
$Pb(s) + SO_4^{2-}(aq) \rightarrow PbSO_4(s) + 2\ e^-$ $\qquad +0.356\ V$
$E_{cell} = 1.36 + 0.356 = 1.71\ V$

19.11 Potassium metal was produced at the cathode.
Oxidation reaction: $2\ F^-(molten) \rightarrow F_2(g) + 2\ e^-$
Reduction reaction: $2(K^+[molten] + e^- \rightarrow K[\ell])$
Net cell reaction: $2\ K^+(molten) + 2\ F^-(molten) \rightarrow 2\ K(\ell) + F_2(g)$

19.12 (c) Making 2 mol of Cu from Cu^{2+} would require 4 Faradays of electricity. Two F are required for part (b), and 3 F are required for part (a).

19.13 First, calculate how many coulombs of electricity are required to make this much aluminum.

$$(2000.\ t\ Al)\left(\frac{2000\ lb\ Al}{1\ t\ Al}\right)\left(\frac{454.3\ g\ Al}{1\ lb\ Al}\right)\left(\frac{1\ mol\ Al}{26.982\ g\ Al}\right) \times$$

$$\left(\frac{3\ mol\ e^-}{1\ mol\ Al}\right)\left(\frac{96500\ C}{1\ mol\ e^-}\right) = 1.950 \times 10^{13}\ C$$

Next, using the product of charge and voltage, calculate how many joules are required; then convert to kilowatt-hours.

$$\text{Energy} = (1.950 \times 10^{13}\ C)(4.0\ V)\left(\frac{1\ J}{1\ C \times 1\ V}\right) \times$$

$$\left(\frac{1\ kWh}{3.60 \times 10^6\ J}\right) = 2.2 \times 10^7\ kWh$$

19.14 To calculate how much energy is stored in a battery, you need the voltage and the number of coulombs of charge the battery can provide. The voltage is generally given on the battery label. To determine the number of coulombs available, you would have to disassemble the battery and determine the masses of the chemicals at the cathode and anode.

19.15 $(0.50\ A)(20.\ min)\left(\dfrac{60\ s}{1\ min}\right)\left(\dfrac{1\ C}{1\ A\ s}\right)\left(\dfrac{1\ mol\ e^-}{96,500\ C}\right) \times$

$$\left(\frac{1\ mol\ Ag}{1\ mol\ e^-}\right)\left(\frac{107.9\ g\ Ag}{1\ mol\ Ag}\right) = 0.67\ g\ Ag$$

19.16 No. Not all metals. Three metals that would corrode about as readily as Fe and Al are Zn, Mg, and Cd. Three metals that do not corrode as readily as Fe and Al are Cu, Ag, and Au. These three metals are used in making coins and jewelry. Metals fall into these two broad groups because of their relative ease of oxidation compared with the oxidation of H_2. In Table 19.3, you can see this breakdown easily.

19.17 (b) $>$ (a) $>$ (d) $>$ (c)
Sand by the seashore, (b), would contain both moisture and salts, which would aid corrosion. Moist clay, (a), would contain water, but less dissolved salts. If an iron object were embedded within the clay, its impervious nature might prevent oxygen from getting to the iron, which would also lower the rate of corrosion. Desert sand in Arizona, (d), would be quite dry, and this low-moisture environment would not lead to a rapid rate of corrosion. On the moon, (c), there would be a lack of moisture and oxygen. This would lead to a very low rate of corrosion.

Chapter 20

20.1 $^{235}_{92}U \rightarrow {}^4_2He + {}^{231}_{90}Th$
$^{231}_{90}Th \rightarrow {}^0_{-1}e + {}^{231}_{91}Pa$
$^{231}_{91}Pa \rightarrow {}^4_2He + {}^{227}_{89}Ac$
$^{227}_{89}Ac \rightarrow {}^4_2He + {}^{223}_{87}Fr$
$^{223}_{87}Fr \rightarrow {}^0_{-1}e + {}^{223}_{88}Ra$

20.2 (a) $^{13}_7N \rightarrow {}^{13}_6C + {}^0_{+1}e$
   (b) $^{41}_{20}Ca + {}^0_{-1}e \rightarrow {}^{41}_{19}K$
   (c) $^{90}_{38}Sr \rightarrow {}^{90}_{39}Y + {}^0_{-1}e$
   (d) $^{11}_6C \rightarrow {}^{11}_5B + {}^0_{+1}e$
   (e) $^{43}_{21}Sc \rightarrow {}^{43}_{20}Ca + {}^0_{+1}e$

20.3 Positron emission: $^{26}_{13}Al \rightarrow {}^0_{+1}e + {}^{26}_{12}Mg$
   Electron capture: $^{26}_{13}Al + {}^0_{-1}e \rightarrow {}^{26}_{12}Mg$

20.4 (a) $^{32}_{14}Si \rightarrow {}^0_{-1}e + {}^{32}_{15}P$
   (b) $^{43}_{22}Ti \rightarrow {}^0_{+1}e + {}^{43}_{21}Sc$
   (c) $^{239}_{94}Pu \rightarrow {}^4_2He + {}^{235}_{92}U$

20.5 Mass difference = $\Delta m = -0.03438\ g/mol$

$$\Delta E = (-3.438 \times 10^{-5}\ kg/mol)(2.998 \times 10^8\ m/s)^2$$

$$= -3.090 \times 10^{12}\ J/mol$$

$$= -3.090 \times 10^9\ kJ/mol$$

E_b per nucleon = $5.150 \times 10^8\ kJ/nucleon$

E_b for ^{6}Li is smaller than E_b for ^{4}He; therefore, helium-4 is more stable than lithium-6.

20.6 From the graph it can be seen that the binding energy per nucleon increases more sharply for the fusion of lighter elements than it does for heavy elements undergoing fission. Therefore, fusion is more exothermic per gram than fission.

20.7 $(\frac{1}{2})^{10} = 9.8 \times 10^{-4}$; this is equivalent to 0.098% of the radioisotope remaining.

20.8 All the lead came from the decay of ^{238}U; therefore, at the time the rock was dated, $N = 100$ and $N_0 = 209$. The decay constant, k, can be determined:

$$k = \frac{0.693}{4.51 \times 10^9 \text{ y}} = 1.54 \times 10^{-10} \text{ y}^{-1}$$

The age of the rock (t) can be calculated using Equation 20.3:

$$\ln \frac{100}{209} = -(1.54 \times 10^{-10} \text{ y}^{-1}) \times t$$

$$t = 4.80 \times 10^9 \text{ y}$$

20.9 Ethylene is derived from petroleum, which was formed millennia ago. The half-life of ^{14}C is 5730 y, and thus much of ethylene's ^{14}C would have decayed and would be much less than that of the ^{14}C alcohol produced by fermentation.

20.10 We start with element 118 with mass 293. Six alpha particles emissions would decrease the atomic number by 12 and the isotopic mass by 24, which leaves $^{269}_{106}$Sg, or seaborgium-269.

20.11 (a) $^{13}_{6}$C + $^{1}_{0}$n → $^{4}_{2}$He + $^{10}_{4}$Be
(b) $^{14}_{7}$N + $^{4}_{2}$He → $^{1}_{0}$n + $^{17}_{9}$F
(c) $^{253}_{99}$Es + $^{4}_{2}$He → $^{1}_{0}$n + $^{256}_{101}$Md

20.12 $^{208}_{82}$Pb + $^{86}_{36}$Kr → $^{294}_{118}$E → $^{290}_{116}$E + α

20.13 Burning a metric ton of coal produces 2.8×10^7 kJ of energy.

$$\left(\frac{2.8 \times 10^4 \text{ kJ}}{1.0 \text{ kg}}\right)\left(\frac{10^3 \text{ kg}}{\text{metric ton}}\right) = 2.8 \times 10^7 \text{ kJ of energy}$$

The fission of 1.0 kg of ^{235}U produces

$$\frac{2.1 \times 10^{10} \text{ kJ}}{0.235 \text{ kg } ^{235}\text{U}} = 8.93 \times 10^{10} \text{ kJ}$$

It would require burning 3.2×10^3 metric tons of coal to equal the amount of energy from 1.0 kg of ^{235}U:

$$8.93 \times 10^{10} \text{ kJ from } ^{235}\text{U} \times \frac{1 \text{ metric ton coal}}{2.9 \times 10^7 \text{ kJ}}$$

$$= 3.2 \times 10^3 \text{ metric tons}$$

20.14 $k = \dfrac{0.693}{29.1 \text{ y}} = 2.38 \times 10^{-2} \text{ y}^{-1}$

$\ln(\text{fraction}) = -(2.38 \times 10^{-2} \text{ y}^{-1})(15 \text{ y, as of } 2001)$
$= -0.262$
$\text{fraction} = e^{-0.357} = 0.700 = 70.0\%$

20.15 $k = \dfrac{0.693}{30.2 \text{ y}} = 2.29 \times 10^{-2} \text{ y}^{-1}$

(a) 60% drop in activity; 40% activity remaining
$\ln(0.40) = -0.916 = -(2.29 \times 10^{-2} \text{ y}^{-1}) \times t$
$t = \dfrac{-0.916}{-2.29 \times 10^{-2} \text{ y}^{-1}} = 40 \text{ y}$

(b) 90% drop in activity, 10% remains
$\ln(0.10) = -2.30 = -(2.29 \times 10^{-2} \text{ y}^{-1}) \times t$
$t = \dfrac{-2.30}{2.29 \times 10^{-2} y^{-1}} = 100 \text{ y}$

20.16 (a) $^7_3$Li + $^1_1$H → $^1_0$n + $^7_4$Be
(b) $^2_1$H + $^3_2$He → $^4_2$He + $^1_1$H

20.17 $k = \dfrac{0.693}{3.82 \text{ d}} = 0.181 \text{ d}^{-1}$

(a) The drop from 8 to 4 pCi represents one half-life, 3.82 days.

(b) $\ln\left(\dfrac{1.5}{8}\right) = -1.67 = -(0.181 \text{ d}^{-1}) \times t$

$t = \dfrac{-1.67}{-0.181 \text{ d}^{-1}} = 9.25 \text{ d}$

20.18 $k = \dfrac{0.693}{78.2 \text{ h}} = 8.86 \times 10^{-3} \text{ h}^{-1}$

$\ln(0.10) = -2.30 = -(8.86 \times 10^{-3} \text{ h}^{-1}) \times t$
$t = 260 \text{ h}$

20.19 Iron-59 $k = \dfrac{0.693}{44.5 \text{ d}} = 1.557 \times 10^{-2} \text{ d}^{-1}$

Chromium-51 $k = \dfrac{0.693}{27.7 \text{ d}} = 0.0250 \text{ d}^{-1}$

Fractions remaining:
^{59}Fe: $\ln(\text{fraction}) = -(1.557 \times 10^{-2} \text{ d}^{-1}) \times 90 \text{ d}$
fraction $= e^{-1.40} = 0.246$; 80 mg $\times$ 0.246
$= 19.7 \text{ mg left}$
^{51}Cr: $\ln(\text{fraction}) = -(0.0250 \text{ d}^{-1}) \times 90 \text{ d}$; 10.5 mg left
Alternatively, consider the fact that 90 days is approximately two half-lives of ^{90}Fe. Therefore, approximately $\frac{3}{4}$ of it (about 60 mg) has decayed after 90 days, and about 20 mg remains. In that same time, ^{51}Cr has undergone more than three half-lives, so that less than $\frac{1}{8}$ remains (less than 12.5 mg).

Chapter 21

21.1 $^{12}_{6}$C + $^4_2$He → $^{16}_{8}$O
$^{16}_{8}$O + $^4_2$He → $^{20}_{10}$Ne
$^{20}_{10}$Ne + $^4_2$He → $^{24}_{12}$Mg

21.2 Two of the four oxygens in an SiO_4 unit are shared with other SiO_4 tetrahedra. Therefore, for each SiO_4 unit,

1 Si + two oxygen not shared + 2 oxygen shared

$$= SiO_{2+1} = SiO_3^{2-}$$

21.3 $5.0 \text{ L} \times \dfrac{1.4 \text{ g}}{\text{mL}} \times \dfrac{10^3 \text{ mL}}{1 \text{ L}} = 7.0 \times 10^3 \text{ g O}_2$

$7.0 \times 10^3 \text{ g} \times \dfrac{1 \text{ mol}}{32.0 \text{ g}} = 2.2 \times 10^2 \text{ mol}; PV = nRT$

$$V = \frac{(2.2 \times 10^2 \text{ mol})(0.0821 \text{ L·atm·mol}^{-1}\cdot\text{K}^{-1})(273 \text{ K})}{1 \text{ atm}}$$

$$= 4.9 \times 10^3 \text{ L}$$

21.4 NO_3^-; oxidation number of oxygen is -2; oxidation number of N is $+5$.
NH_4^+; oxidation number of hydrogen is $+1$; oxidation number of nitrogen is -3.

21.5 $150 \text{ g NaN}_3 \times \dfrac{1 \text{ mol NaN}_3}{65.0 \text{ g NaN}_3} \times \dfrac{3 \text{ mol N}_2}{2 \text{ mol NaN}_3}$

$= 3.46 \text{ mol } N_2$

$V = \dfrac{nRT}{P} = \dfrac{(3.46)(0.0821)(273)}{(1)} = 77.6 \text{ L}$

21.6 $2 \text{ NaCl}(\ell) \rightarrow 2 \text{ Na}(\ell) + \text{Cl}_2(g)$

Coulombs (C) = (A)(s) =

$2.0 \times 10^4 \text{ A} \times 24 \text{ hr} \times \dfrac{3600 \text{ s}}{\text{hr}} = 1.7 \times 10^9 \text{ C}$

$1.7 \times 10^9 \text{ C} \times \dfrac{1 \text{ mol e}^-}{9.65 \times 10^4 \text{ C}} = 1.8 \times 10^4 \text{ mol e}^-$

$1.8 \times 10^4 \text{ mol e}^- \times \dfrac{1 \text{ mol Na}}{1 \text{ mol e}^-}$

$\times \dfrac{23.0 \text{ g Na}}{1 \text{ mol Na}} \times \dfrac{1 \text{ lb}}{454 \text{ g}} \times \dfrac{1 \text{ ton}}{2000 \text{ lb}} = 0.46 \text{ tons Na}$

$1.8 \times 10^4 \text{ mol e}^- \times \dfrac{1 \text{ mol Cl}_2}{2 \text{ mol e}^-}$

$\times \dfrac{70.9 \text{ g Cl}_2}{1 \text{ mol Cl}_2} \times \dfrac{1 \text{ lb}}{454 \text{ g}} \times \dfrac{1 \text{ ton}}{2000 \text{ lb}} = 0.70 \text{ tons Cl}_2$

21.7 $\text{C} = 2.00 \times 10^4 \text{ A} \times 100. \text{ hr} \times \dfrac{3600 \text{ s}}{\text{hr}} = 7.20 \times 10^9 \text{ C}$

$7.20 \times 10^9 \text{ C} \times \dfrac{1 \text{ mol e}^-}{9.65 \times 10^4 \text{ C}}$

$\times \dfrac{1 \text{ mol NaOH}}{1 \text{ mol e}^-} \times \dfrac{40.00 \text{ g NaOH}}{1 \text{ mol NaOH}} \times \dfrac{1 \text{ lb}}{454 \text{ g}} \times \dfrac{1 \text{ ton}}{2000 \text{ lb}}$

$= 3.29 \text{ tons NaOH}$

21.8 (a) Mg_3N_2

(b) $3 \text{ Mg}(s) + \text{N}_2(g) \rightarrow \text{Mg}_3\text{N}_2(s)$

21.9 Phosphorus in $\text{Ca}_3(\text{PO}_4)_2$ has an oxidation state of $+5$; it is reduced to zero in P_4. Carbon is oxidized to CO (oxidation state changes from zero to $+2$).

21.10 $\dfrac{93 \text{ g}}{502 \text{ g}} \times 100\% = 19\% \text{ P}$

21.11 Br^- is oxidized to Br_2; Cl_2 is reduced to Cl^-.

21.12 The positive voltage indicates that oxidation of I^- by Cl_2 is product-favored; I^- is converted to I_2.

Chapter 22

22.1 (a) V^{5+}　　(b) Cr^{6+}　　(c) Mn^{4+}　　(d) Os^{8+}

22.2 $\text{KFe}^{2+}[\text{Fe}^{3+}(\text{CN})_6]^{3-}$

Potassium is K^+ and the iron counter ion is Fe^{2+}. Therefore, the complex ion must have a $3-$ charge. Each

CN ion is -1, so the central iron ion in the complex ion is Fe^{3+}.

22.3 (a) $500 \text{ g} \times \dfrac{18 \text{ g W}}{100 \text{ g}} = 90 \text{ g W}$

$500 \text{ g} \times \dfrac{6.0 \text{ g Co}}{100 \text{ g}} = 30 \text{ g Co}$

(b) Iron is present in the greatest mole percent.

$500 \text{ g} \times \dfrac{68.9 \text{ g Fe}}{100 \text{ g}} = 344 \text{ g Fe} = 6.2 \text{ mol}$

$\dfrac{6.2}{7.8} \times 100\% = 79.5 \% \text{ Fe}$

22.4 $\text{Cu}^{2+}(aq) + 2 \text{ e}^- \rightarrow \text{Cu}(s)$

$12.0 \text{ h} \left(\dfrac{3600 \text{ s}}{\text{h}}\right)\left(\dfrac{250 \text{ C}}{\text{s}}\right)\left(\dfrac{1 \text{ mol e}^-}{9.65 \times 10^4 \text{ C}}\right) \times$

$\left(\dfrac{1 \text{ mol Cu}}{2 \text{ mol e}^-}\right)\left(\dfrac{63.55 \text{ g Cu}}{1 \text{ mol Cu}}\right) = 3.56 \times 10^3 \text{ g Cu}$

22.5 (a) NO_3^- is the oxidizing agent; silver metal is the reducing agent.

(b) NO_3^- is the oxidizing agent; gold is the reducing agent.

22.6 $2 \text{ Al}(s) + \text{Cr}_2\text{O}_3(s) \rightarrow \text{Al}_2\text{O}_3(s) + 2 \text{ Cr}(s)$

$\Delta H_f^\circ = \Delta H_{f_{\text{Al}_2\text{O}_3(s)}}^\circ - \Delta H_{f_{\text{Cr}_2\text{O}_3(s)}}^\circ$

$= (-1675.7 \text{ kJ}) - (-1139.7 \text{ kJ})$

$= -536.0 \text{ kJ}$

$\Delta G^\circ = \Delta G_{f_{\text{Al}_2\text{O}_3(s)}}^\circ - \Delta G_{f_{\text{Cr}_2\text{O}_3(s)}}^\circ$

$= (-1582.3 \text{ kJ}) - (-1058.1 \text{ kJ})$

$= -524.2 \text{ kJ}$

22.7 The addition of acid converts CrO_4^{2-} to $\text{Cr}_2\text{O}_7^{2-}$, shifting the equilibrium to the right. Added base reacts with H^+ to form water, causing the equilibrium to shift to the left, converting $\text{Cr}_2\text{O}_7^{2-}$ to CrO_4^{2-}.

22.8 $[\text{Cr}(\text{H}_2\text{O})_2(\text{NH}_3)_2(\text{OH})_2]^+$

22.9 K_2SO_4

22.10 (a) Three　　(b) $4-$　　(c) Four

22.11 (a) Fe^{2+}　　(b) Fe^{3+}

22.12 Two isomers

isomer 1　　　　isomer 2

ANSWERS TO SELECTED QUESTIONS FOR REVIEW AND THOUGHT

Chapter 1

14. Some qualitative observations: The object is solid and roughly cubic in shape. It appears to have cubic subunits. It appears to be a silvery metal. Some quantitative observations: The side of the cube is about 3.5 inches or about 9.0 cm across.

16. (a) Quantitative   (b) Qualitative
    (c) Quantitative   (d) Qualitative

18. Liquid, because body temperature is greater than 29.8 °C

19. (a) 20 °C   (b) 100 °C   (c) 180 °F   (d) 20 °F

21. Charlotte is warmest; Montreal is coldest.

22. 8.94 g/mL

24. Al

26. 3.9×10^3 g

27. (a) Physical   (b) Chemical   (c) Chemical   (d) Physical

29. (a) Chemical   (b) Chemical   (c) Physical

31. (a) An outside source of energy is forcing a chemical reaction to occur.
    (b) A chemical reaction is releasing energy and causing work to be done.
    (c) A chemical reaction is releasing energy and causing work to be done.
    (d) An outside source of energy is forcing a chemical reaction to occur.

32. Heterogeneous; use a magnet.

34. (a) Homogeneous   (b) Heterogeneous
    (c) Heterogeneous   (d) Heterogeneous

37. (a) A compound decomposed.   (b) A compound decomposed.

38. (a) Elements combined to make a compound.
    (b) Elements combined to make a compound.

39. (a) Heterogeneous mixture   (b) Compound
    (c) Heterogeneous mixture   (d) Homogeneous mixture

41. (a) Heterogeneous mixture   (b) Compound
    (c) Element   (d) Homogeneous mixture

43. Macroscopic world; parallelepiped shape; atom crystal arrangement is parallelepiped shape.

45. Microscale world

47. Carbon dioxide molecules are crowded in the unopened can. When the can is opened, the molecules quickly escape through the hole.

50. Because atoms in the starting materials must all be accounted for in the substances produced, there would be no change in the mass.

52. All matter is composed of atoms, which are extremely tiny. All atoms of a given element have the same chemical properties. Compounds are formed by the chemical combination of two or more different kinds of atoms. A chemical reaction involves joining, separating, or rearranging atoms.

54. If two compounds contain the same elements and samples of those two compounds both contain the same mass of one element, then the ratio of the masses of the other elements will be small whole numbers.

56. Many responses are equally valid here. Common examples given here:
    (a) Fe, iron; Au, gold   (b) C, carbon; H, hydrogen
    (c) B, boron; Si, silicon   (d) Nitrogen, N_2; oxygen, O_2

58.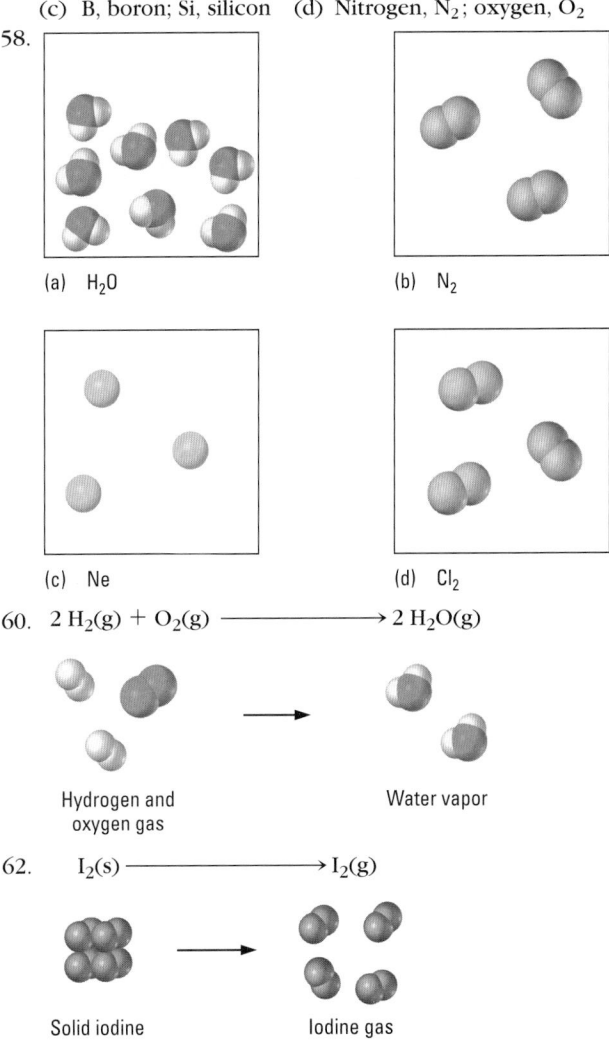

(a) H_2O (b) N_2

(c) Ne   (d) Cl_2

60. $2 H_2(g) + O_2(g) \longrightarrow 2 H_2O(g)$

Hydrogen and oxygen gas → Water vapor

62. $I_2(s) \longrightarrow I_2(g)$

Solid iodine → Iodine gas

64. The answers to these questions are very subjective. The answers given here may not reflect your own opinion.
    (a) The benefits outweigh the risks. Drug companies and their stockholders and the HIV-positive patients would benefit. The HIV-positive individuals are also at risk.
    (b) The benefits outweigh the risks. The manufacturer of Olestra and the general chip-eating public would be the main benefactors. People with "sensitive" digestive systems are at risk.

66. See Section 1.12.

68. (a) Mass is quantitative and related to a physical property. Colors are qualitative and related to physical properties. Reaction is qualitative and related to a chemical property.
(b) Mass is quantitative and related to a physical property. The fact that a chemical reaction occurs between substances is qualitative information and related to a chemical property.

70. In solid calcium, smaller radius atoms are more closely packed, making a smaller volume. In solid potassium, larger radius atoms are less closely packed, making a larger volume.

73. (a) Bromobenzene (b) Gold (c) Lead

75. (a) 2.7×10^2 mL ice (b) Bulging, cracking, deformed, or broken

76. Gold

78. (a) Water on top of bromobenzene (b) Benzene on top of water (c) Ethanol on top of benzene (d) Ethanol and water will dissolve together and sit on top of the bromobenzene and benzene, which also dissolve together

79. Liquid molecules are moving faster and they are farther apart, spreading the molecules out and making the volume larger, even though they have the same mass. The gas molecules are moving faster and they are farther apart, spreading the molecules out and making the volume larger, even though they have the same mass.

82. Look through magazines, local and national newspapers, and advertisements.

Chapter 2

9. 6×10^{13} cm

11. 614 cm, 242 in, 20.1 ft

13. 123 m

15. 4.10 L

18. 80.1% silver, 19.9% copper

20. 245 g sulfuric acid

24. 95 protons, 95 electrons, 146 neutrons

26. number

28. mass; electron mass is negligible.

30. The number of neutrons

32. (a) 56 (b) 243 (c) 184

34. (a) $^{15}_{7}N$ (b) $^{64}_{30}Zn$ (c) $^{129}_{54}Xe$

36. (a) Electrons = 6, protons = 6, neutrons = 7
(b) Electrons = 24, protons = 24, neutrons = 26
(c) Electrons = 83, protons = 83, neutrons = 122

40. $^{57}_{27}Co$, $^{58}_{27}Co$, $^{60}_{27}Co$

42. 24.31 amu/atom

44. 69.51% ^{63}Cu and 30.49% ^{65}Cu

46. Est. 40 amu/atom; 39.95 amu/atom

47. Pair (2), dozen (12), gross (144), hundred (100)

51. (a) 1.19×10^3 g Au (b) 11 g U
(c) 315 g Ne (d) 0.0886 g Pu

53. (a) 0.696 mol Na (b) 1.7×10^{-5} mol Pt
(c) 0.0497 mol P (d) 0.0117 mol As
(e) 7.49×10^{-3} mol Xe

57. 1.45 cm

59. 5.93×10^{21} Au atoms

61. 7.951×10^{-23} g Ti

62. In a group, elements share the same vertical column, whereas in a period, elements share the same horizontal row.

70. 18 elements; metals: potassium (K), calcium (Ca), scandium (Sc), titanium (Ti), vanadium (V), chromium (Cr), manganese (Mn), iron (Fe), cobalt (Co), nickel (Ni), copper (Cu), zinc (Zn), and gallium (Ga); metalloids: germanium (Ge) and arsenic (As); nonmetals: selenium (Se), bromine (Br), and krypton (Kr)

72. Two have 8 elements, two have 18 elements, and one period of the current periodic table has 32 elements (period 6). Period 7 will have 32 elements when all are isolated and characterized.

74. (a) zinc (Zn) (b) xenon (Xe) (c) lead (Pb) (d) sulfur (S) (e) sodium (Na) (f) xenon (Xe) (g) selenium (Se), nonmetal (h) antimony (Sb)

76. Lighter elements are more abundant than the heavier ones. Even-numbered elements are often more abundant than the odd-numbered elements on either side of them. (Both of these general trends have exceptions.)

79. 0.197 nm, 197 pm

81. 0.178 nm³, 1.78×10^{-22} cm³

83. (a) 272 mL (b) No

85. 89 tons of sodium fluoride per year

87. ^{39}K

89. (a) Ti, atomic number = 22, atomic weight = 47.88
(b) Titanium is in Period 4, Group 4B. The other Group 4B elements are zirconium (Zr), hafnium (Hf), and rutherfordium (Rf).
(c) Titanium is lightweight and strong, making it a good choice for something that needs to be sturdy and small.
(d) See a dictionary or handbook.

92. 0.038 mol C

94. $2150

96. 3.4 mol Cu, 2.0×10^{24} Cu atoms

98. (a) The same number (b) 1 mol of O_2 has more.
(c) The same number (d) The same number
(e) The same number (f) 159.8 g of Br_2 has more.
(g) The same number (h) 58.9 g of Co has more.
(i) 6.022×10^{23} calcium atoms has more.
(j) 1 g of Cl has more.

Chapter 3

12. Butanol, $C_4H_{10}O$, $CH_3CH_2CH_2CH_2OH$,

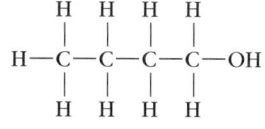

pentanol, $C_5H_{12}O$, $CH_3CH_2CH_2CH_2CH_2OH$,

$$
\begin{array}{ccccc}
H & H & H & H & H \\
| & | & | & | & | \\
H-C-&C-&C-&C-&C-OH \\
| & | & | & | & | \\
H & H & H & H & H
\end{array}
$$

14. Sucrose has more oxygen atoms and more atoms of all kinds.

17. (a) 1 calcium atom, 2 carbon atoms, and 4 oxygen atoms
(b) 8 carbon atoms and 8 hydrogen atoms
(c) 2 nitrogen atoms, 8 hydrogen atoms, 1 sulfur atom, and 4 oxygen atoms
(d) 1 platinum atom, 2 nitrogen atoms, 6 hydrogen atoms, and 2 chlorine atoms

(e) 4 potassium atoms, 1 iron atom, 6 carbon atoms, and 6 nitrogen atoms

21. (a) Lithium (Group 1A), Li^+
    (b) Strontium (Group 2A), Sr^{2+}
    (c) Aluminum (Group 3A), Al^{3+}
    (d) Calcium (Group 2A), Ca^{2+}
    (e) Zinc (Group 2B), Zn^{2+}

23. Ba^{2+} and Br^-

25. (a) Se^{2-} (b) F^- (c) Ni^{2+} (d) N^{3-}

27. $PbCl_2$ and $PbCl_4$

29. (b) and (d) are correct. (a) CaO (c) FeO or Fe_2O_3

31. (a) 1 calcium ion (Ca^{2+}) and 2 acetate ions ($CH_3CO_2^-$ or CH_3COO^-)
    (b) 1 cobalt (III) ion (Co^{3+}) and 3 sulfate ions (SO_4^{2-})
    (c) 1 aluminum ion (Al^{3+}) and 3 hydroxide ions (OH^-)
    (d) 2 ammonium ions (NH_4^+) and 1 carbonate ion (CO_3^{2-})

33. (a) $Ni(NO_3)_2$ (b) $NaHCO_3$ (c) $LiClO$ (d) $Mg(ClO_3)_2$
    (e) $CaSO_3$

35. (a) Not ionic (b) Not ionic (c) Ionic (d) Not ionic
    (e) Ionic

37. (a) $Ca(HCO_3)_2$ (b) $KMnO_4$ (c) $Mg(ClO_4)_2$
    (d) $(NH_4)_2HPO_4$

39. (a) Calcium acetate (b) Cobalt(III) sulfate
    (c) Aluminum hydroxide

41. MgO; it has higher ionic charges and smaller ion sizes.

43. An electrolyte is a compound that conducts electricity when dissolved in water. A dissolved strong electrolyte (such as NaCl) ionizes completely (making Na^+ and Cl^-). A dissolved weak electrolyte (such as acetic acid, CH_3COOH) dissolves in water but ionizes very little.

45. Molecular compounds are generally not ionic compounds and therefore would not ionize in water.

47. (a) K^+ and OH^- (b) K^+ and SO_4^{2-} (c) Na^+ and NO_3^-
    (d) NH_4^+ and Cl^-

49.

	CH_3OH	**Carbon**
No. of moles	1	1
No. of molecules or atoms	6.022×10^{23} molecules	6.022×10^{23} atoms
Molar mass	32.042 g/mol M	12.011 g/mol M

	Hydrogen	**Oxygen**
No. of moles	4	1
No. of molecules or atoms	2.409×10^{24} atoms	6.022×10^{23} atoms
Molar mass	4.0316 g/mol M	15.9994 g/mol M

51. (a) 159.692 g/mol (b) 67.806 g/mol
    (c) 44.0128 g/mol (d) 197.906 g/mol
    (e) 176.126 g/mol

52. (a) 122.221 g/mol (b) 227.133 g/mol
    (c) 300.05 g/mol (d) 90.189 g/mol
    (e) 324.422 g/mol

53. (a) 0.0312 mol (b) 0.0101 mol (c) 0.0125 mol
    (d) 0.00406 mol (e) 0.00599 mol

55. (a) 179.857 g/mol (b) 36.0 g (c) 0.0259 mol

57. (a) 0.00180 mol $C_9H_8O_4$; 0.02266 mol $NaHCO_3$; 0.005205 mol $C_6H_8O_7$
    (b) 1.08×10^{21} $C_9H_8O_4$ molecules

59. (a) 5.67 mol SO_3 (b) 3.41×10^{24} SO_3 molecules
    (c) 3.41×10^{24} S atoms (d) 1.02×10^{25} O atoms

60. (a) 0.250 mol CF_3CH_2F (b) 6.02×10^{23} F atoms

61. 1.7×10^{21} H_2O molecules

63. (a) 86.60% Pb and 13.40% S
    (b) 79.889% C and 20.111% H
    (c) 40.002% C, 6.735% H, and 53.285% O
    (d) 34.9979% C, 5.0368% H, and 59.9654% O

65. 58.0% M in MO

67. 2.20×10^5 g/mol

69. $x = 6$

71. Percentages given on food labels are usually related to percent daily value (% DV) based on a 2000-calorie diet. For example, a product with 19 grams of carbohydrates will have a label that says 6% DV carbohydrates because we should eat no more than 300 grams of carbohydrates on a 2000-calories-per-day diet.

73. Empirical formula shows the simplest whole-number ratio of the atoms, whereas molecular formula gives the actual number of atoms of each element in one formula unit. C_2H_6 is the molecular formula, whereas CH_3 is the empirical formula of ethane.

75. $C_4H_8N_2O_2$

77. B_5H_7

79. C_3H_4, C_9H_{12}

81. (a) C_5H_7N (b) $C_{10}H_{14}N_2$

83. $C_5H_{14}N_2$

85. $x = 7$

87. H, O, C, N, Ca, P, Cl, S, Na, K

89. (a) Ions (b) Metals are part of solid structural components, such as Ca^{2+} in bones. Metals also are part of large biomolecules that perform critical functions, such as Fe^{2+} in hemoglobin, which helps transport oxygen.

91. Se is toxic at high levels. Fe, Cu, and I_2 are also toxic at high levels.

93. (a) Glucose (b)

(c) Plants make glucose and store it as starch. Glucose is also used in plants to form cellulose for structural support. Animals rely on the metabolism of glucose as their major source of energy.

96. (a) $C_6H_5N_3O_6$ (b) $C_3H_7NO_3$

98. (b) $LiTe_2$ (d) MgF_2 (f) In_2S_3

100. (a) NaClO, ionic (b) $Al(ClO_4)_3$, ionic
     (c) $KMnO_4$, ionic (d) KH_2PO_4, ionic
     (e) ClF_3, not ionic (f) BBr_3, not ionic
     (g) $Ca(CH_3CO_2)_2$ or $Ca(CH_3COO)_2$, ionic
     (h) Na_2SO_3, ionic (i) S_2Cl_4, not ionic
     (j) PF_3, not ionic

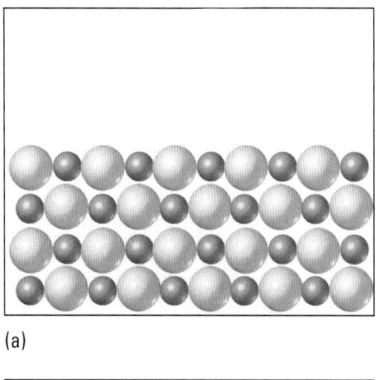

(a)

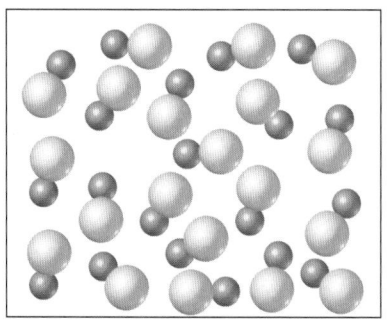

(b)

(c)

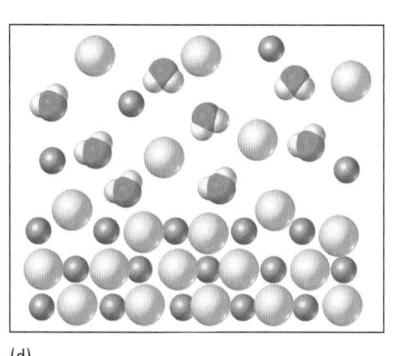

(d)

(Question 115)

102. 14.9 cm
104. (a) C_5H_4 (b) $C_{10}H_8$
106. (a) ICl_3 (b) I_2Cl_6
108. 7.35 kg Fe
110. 67.1 g Sb_2S_3
112. (a) 0.0130 mol Ni (b) NiF_2
113. (a) 7.60×10^{-4} mol U, U_2O_5, 3.79×10^{-4} mol U_2O_5
    (b) Five
115. See above
117. 2,4-dimethylpentane
119. Tl_2CO_3, Tl_2SO_4
121. (a) Perbromate, bromate, bromite, hypobromite
    (b) Selenate, selenite
123. Sample (a)

Chapter 4

14. (a) 1.00 g (b) 2 for Mg, 1 for O_2, and 2 for MgO
    (c) 50 atoms of Mg
16. Equation (b)
18. $2 Sb + 3 Cl_2 \longrightarrow 2 SbCl_3$, (c) represents products.
19. (a) $2 C(s) + O_2(g) \longrightarrow 2 CO(g)$
    (b) $2 Ni(s) + O_2(g) \longrightarrow 2 NiO(s)$
    (c) $4 Cr(s) + 3 O_2(g) \longrightarrow 2 Cr_2O_3(s)$
21. (a) $BeCO_3(s) \longrightarrow BeO(s) + CO_2(g)$
    (b) $NiCO_3(s) \longrightarrow NiO(s) + CO_2(g)$
    (c) $Al_2(CO_3)_3(s) \longrightarrow Al_2O_3(s) + 3 CO_2(g)$
23. (a) $2 C_4H_{10}(g) + 13 O_2(g) \longrightarrow 8 CO_2(g) + 10 H_2O(g)$
    (b) $C_6H_{12}O_6(s) + 6 O_2(g) \longrightarrow 6 CO_2(g) + 6 H_2O(g)$
    (c) $2 C_4H_8O(\ell) + 11 O_2(g) \longrightarrow 8 CO_2(g) + 8 H_2O(g)$
24. (a) $2 Mg(s) + O_2(g) \longrightarrow 2 MgO(s)$

(b) $2 Ca(s) + O_2(g) \longrightarrow 2 CaO(s)$
(c) $4 In(s) + 3 O_2(g) \longrightarrow 2 In_2O_3(s)$
26. (a) $2 K(s) + Cl_2(g) \longrightarrow 2 KCl(s)$
    (b) $Mg(s) + Br_2(\ell) \longrightarrow MgBr_2(s)$
    (c) $2 Al(s) + 3 F_2(g) \longrightarrow 2 AlF_3(s)$
29. (a) $2 Fe(s) + 3 Cl_2(g) \longrightarrow 2 FeCl_3(s)$
    (b) $SiO_2(s) + 2 C(s) \longrightarrow Si(s) + 2 CO(g)$
    (c) $3 Fe(s) + 4 H_2O(g) \longrightarrow Fe_3O_4(s) + 4 H_2(g)$
31. (a) $3 MgO(s) + 2 Fe(s) \longrightarrow Fe_2O_3(s) + 3 Mg(s)$
    (b) $2 H_3BO_3(s) \longrightarrow B_2O_3(s) + 3 H_2O(\ell)$
    (c) $2 NaNO_3(s) + H_2SO_4(aq) \longrightarrow$
    $$Na_2SO_4(aq) + 2 HNO_3(g)$$
33. (a) $CaNCN(s) + 3 H_2O(\ell) \longrightarrow CaCO_3(s) + 2 NH_3(g)$
    (b) $2 NaBH_4(s) + H_2SO_4(aq) \longrightarrow$
    $$B_2H_6(g) + 2 H_2(g) + Na_2SO_4(aq)$$
    (c) $8 H_2S(aq) + 8 Cl_2(aq) \longrightarrow S_8(s) + 16 HCl(aq)$
35. (a) $Mg + 2 HNO_3 \longrightarrow H_2 + Mg(NO_3)_2$
    (b) $2 Al + Fe_2O_3 \longrightarrow Al_2O_3 + 2 Fe$
    (c) $2 S + 3 O_2 \longrightarrow 2 SO_3$
    (d) $SO_3 + H_2O \longrightarrow H_2SO_4$
36. 50.0 mol HCl
38. 1.1 mol O_2, 35 g O_2, 1.0×10^2 g NO_2
40. 12.7 g Cl_2, 0.179 mol $FeCl_2$, 22.7 g $FeCl_2$ expected

42.

$(NH_4)_2PtCl_6$	Pt	HCl
12.35 g	5.428 g	5.410 g
0.02782 mol	0.02782 mol	0.1484 mol

44. (a) 0.148 mol H_2O (b) 5.90 g TiO_2, 10.8 g HCl
46. 793 g W
48. 3.5×10^2 g N_2, 4.5×10^2 g H_2O, 2.0×10^2 g O_2

52. (a) $NH_4NO_3 \longrightarrow N_2O + 2\,H_2O$
    (b) 5.50 g N_2O, 4.50 g H_2O
54. (a) $K_2PtCl_4 + 2\,NH_3 \longrightarrow Pt(NH_3)_2Cl_2 + 2\,KCl$
    (b) 3.46 g K_2PtCl_4, 0.284 g NH_3
55. (a) Cl_2 is limiting   (b) 5.08 g $AlCl_3$
    (c) 1.67 g Al unreacted
58. 0 mol CH_4, 15 mol H_2O, 62.0 mol CO_2, and 248 mol H_2
60. 1.40 kg Fe
62. 73.5%
64. 67.1%
66. 5.3 g SCl_2
68. CH
70. $C_3H_6O_2$
72. 21.6 g N_2
74. 86.3 g Al_2Br_6
76. 12.5 g $Pt(NH_3)_2Cl_2$
78. SiH_4
80. KOH is the limiting reactant in both cases.
82. Two butane molecules react with 13 diatomic oxygen molecules to produce eight carbon dioxide molecules and ten water molecules. Two mol of gaseous butane molecules react with 13 mol of gaseous diatomic oxygen molecules to produce 8 mol of gaseous carbon dioxide molecules and 10 mol of liquid water molecules.
84. A_3B
86. (a) $\dfrac{x}{y} = \dfrac{2}{1}$ (b) Carbon
88. Figure (4), (b) is true.
90. Reaction (d)
92. When the metal mass is less than 1.0 g, the metal is the limiting reactant. When the metal mass is greater than 1.0 g, the bromine is the limiting reactant.

Chapter 5

11. All soluble. (a) Fe^{2+} and ClO_4^- (b) Na^+ and SO_4^{2-}
    (c) K^+ and Br^- (d) Na^+ and CO_3^{2-}
14. (a) $MnCl_2(aq) + Na_2S(aq) \longrightarrow MnS(s) + 2\,NaCl(aq)$
    (b) No precipitate   (c) No precipitate
    (d) $Hg(NO_3)_2(aq) + Na_2S(aq) \longrightarrow HgS(s) + 2\,NaNO_3(aq)$
    (e) $Pb(NO_3)_2(aq) + 2\,HCl(aq) \longrightarrow$
$$PbCl_2(s) + 2\,HNO_3(aq)$$
    (f) $BaCl_2(aq) + H_2SO_4(aq) \longrightarrow BaSO_4(s) + 2\,HCl(aq)$
15. $2\,NO_3^-(aq), 2H^+(aq) + Mg(OH)_2(s) \longrightarrow$
$$2\,H_2O(\ell) + Mg^{2+}(aq), \text{ exchange reaction}$$
17. (a) $Cu^{2+}(aq) + H_2S(aq) \longrightarrow CuS(s) + 2\,H^+(aq)$
    (b) $Ca^{2+}(aq) + CO_3^{2-}(aq) \longrightarrow CaCO_3(s)$
    (c) $Ag^+(aq) + I^-(aq) \longrightarrow AgI(s)$
18. (a) $Zn(s) + 2\,HCl(aq) \longrightarrow H_2(g) + ZnCl_2(aq)$
$Zn(s) + 2\,H^+(aq) + 2\,Cl^-(aq) \longrightarrow$
$$H_2(g) + Zn^{2+}(aq) + 2\,Cl^-(aq)$$
$Zn(s) + 2\,H^+(aq) \longrightarrow H_2(g) + Zn^{2+}(aq)$
    (b) $Mg(OH)_2(s) + 2\,HCl(aq) \longrightarrow MgCl_2(aq) + 2\,H_2O(\ell)$
$Mg(OH)_2(s) + 2\,H^+(aq) + 2\,Cl^-(aq) \longrightarrow$
$$Mg^{2+}(aq) + 2\,Cl^-(aq) + 2\,H_2O(\ell)$$
$Mg(OH)_2(s) + 2\,H^+(aq) \longrightarrow Mg^{2+}(aq) + 2\,H_2O(\ell)$
    (c) $2\,HNO_3(aq) + CaCO_3(s) \longrightarrow$
$$Ca(NO_3)_2(aq) + H_2O(\ell) + CO_2(g)$$
$2\,H^+(aq) + 2\,NO_3^-(aq) + CaCO_3(s) \longrightarrow$
$$Ca^{2+}(aq) + 2\,NO_3^-(aq) + H_2O(\ell) + CO_2(g)$$

$2\,H^+(aq) + CaCO_3(s) \longrightarrow Ca^{2+}(aq) + H_2O(\ell) + CO_2(g)$
    (d) $4\,HCl(aq) + MnO_2(s) \longrightarrow$
$$MnCl_2(aq) + Cl_2(g) + 2\,H_2O(\ell)$$
$4\,H^+(aq) + 4\,Cl^-(aq) + MnO_2(s) \longrightarrow$
$$Mn^{2+}(aq) + 2\,Cl^-(aq) + Cl_2(g) + 2\,H_2O(\ell)$$
$4\,H^+(aq) + 2\,Cl^-(aq) + MnO_2(s) \longrightarrow$
$$Mn^{2+}(aq) + Cl_2(g) + 2\,H_2O(\ell)$$
20. (a) $Ca(OH)_2(s) + 2\,HNO_3(aq) \longrightarrow$
$$Ca(NO_3)_2(aq) + 2\,H_2O(\ell)$$
$Ca(OH)_2(s) + 2\,H^+(aq) + 2\,NO_3^-(aq) \longrightarrow$
$$Ca^{2+}(aq) + 2\,NO_3^-(aq) + 2\,H_2O(\ell)$$
$Ca(OH)_2(s) + 2\,H^+(aq) \longrightarrow Ca^{2+}(aq) + 2\,H_2O(\ell)$
    (b) $BaCl_2(aq) + Na_2CO_3(aq) \longrightarrow BaCO_3(s) + 2\,NaCl(aq)$
$Ba^{2+}(aq) + 2\,Cl^-(aq) + 2\,Na^+(aq) + CO_3^{2-}(aq) \longrightarrow$
$$BaCO_3(s) + 2\,Na^+(aq) + 2\,Cl^-(aq)$$
$Ba^{2+}(aq) + CO_3^{2-}(aq) \longrightarrow BaCO_3(s)$
    (c) $2\,Na_3PO_4(aq) + 3\,Ni(NO_3)_2(aq) \longrightarrow$
$$Ni_3(PO_4)_2(s) + 6\,NaNO_3(aq)$$
$6\,Na^+(aq) + 2\,PO_4^{3-}(aq) + 3\,Ni^{2+}(aq) + 6\,NO_3^-(aq) \longrightarrow$
$$Ni_3(PO_4)_2(s) + 6\,Na^+(aq) + 6\,NO_3^-(aq)$$
$2\,PO_4^{3-}(aq) + 3\,Ni^{2+}(aq) \longrightarrow Ni_3(PO_4)_2(s)$
22. $Ba(OH)_2(aq) + 2\,HNO_3(aq) \longrightarrow$
$$Ba(NO_3)_2(aq) + 2\,H_2O(\ell)$$
24. $CdCl_2(aq) + 2\,NaOH(aq) \longrightarrow Cd(OH)_2(s) + 2\,NaCl(aq)$
$Cd^{2+}(aq) + 2\,Cl^-(aq) + 2\,Na^+(aq) + 2\,OH^-(aq) \longrightarrow$
$$Cd(OH)_2(s) + 2\,Na^+(aq) + 2\,Cl^-(aq)$$
$Cd^{2+}(aq) + 2\,OH^-(aq) \longrightarrow Cd(OH)_2(s)$
26. $Pb(NO_3)_2(aq) + 2\,KCl(aq) \longrightarrow PbCl_2(s) + 2\,KNO_3(aq)$,
    products: lead(II) chloride and potassium nitrate
28. $MgCO_3(s) + 2\,HCl(aq) \longrightarrow$
    manganese(II) hydrochloric
    carbonate   acid
$$MnCl_2(aq) + H_2O(\ell) + CO_2(g)$$
    manganese(II)   water   carbon
       chloride   dioxide
29. (a) Base, K^+ and OH^- (b) Base, Mg^{2+} and OH^-
    (c) acid, small amounts of H^+ and ClO^-
    (d) Acid, H^+ and Br^- (e) Base, Li^+ and OH^-
    (f) Acid, small amounts of H^+, HSO_3^-, and SO_3^{2-}
30. (a) Strong
    (b) Strong but insoluble, so solution will have a small number of OH^- ions
    (c) Weak (d) Strong (e) Strong (f) Weak
32. (a) $NaOH(aq) + HNO_2(aq) \longrightarrow NaNO_2(aq) + H_2O(\ell)$
$Na^+(aq) + OH^-(aq) + HNO_2(aq) \longrightarrow$
$$Na^+(aq) + NO_2^-(aq) + H_2O(\ell)$$
$OH^-(aq) + HNO_2(aq) \longrightarrow NO_2^-(aq) + H_2O(\ell)$
    (b) $Ca(OH)_2(s) + H_2SO_4(aq) \longrightarrow CaSO_4(aq) + 2\,H_2O(\ell)$
$Ca(OH)_2(s) + H^+(aq) + HSO_4^-(aq) \longrightarrow$
$$Ca^{2+}(aq) + SO_4^{2-}(aq) + 2\,H_2O(\ell)$$
$Ca(OH)_2(s) + H^+(aq) + HSO_4^-(aq) \longrightarrow$
$$Ca^{2+}(aq) + SO_4^{2-}(aq) + 2\,H_2O(\ell)$$
    (c) $NaOH(aq) + HI(aq) \longrightarrow NaI(aq) + H_2O(\ell)$
$Na^+(aq) + OH^-(aq) + H^+(aq) + I^-(aq) \longrightarrow$
$$Na^+(aq) + I^-(aq) + H_2O(\ell)$$
$OH^-(aq) + H^+(aq) \longrightarrow H_2O(\ell)$
    (d) $3\,Mg(OH)_2(s) + 2\,H_3PO_4(aq) \longrightarrow$
$$Mg_3(PO_4)_2(s) + 6\,H_2O(\ell)$$
(equation, complete ionic equation, and net ionic equation)
    (e) $NaOH(aq) + CH_3COOH(aq) \longrightarrow$
$$NaCH_3COO(aq) + H_2O(\ell)$$

$Na^+(aq) + OH^-(aq) + CH_3COOH(aq) \longrightarrow$
$$Na^+(aq) + CH_3COO^-(aq) + H_2O(\ell)$$
$OH^-(aq) + CH_3COOH(aq) \longrightarrow CH_3COO^-(aq) + H_2O(\ell)$

34. (a) Acid-base, $Fe(OH)_3(s) + 3\,HNO_3(aq) \longrightarrow$
$$Fe(NO_3)_3(aq) + 3\,H_2O(\ell)$$
 (b) Gas-forming; $FeCO_3(s) + H_2SO_4(aq) \longrightarrow$
$$FeSO_4(aq) + H_2O(\ell) + CO_2(g)$$
 (c) Precipitation, $FeCl_2(aq) + (NH_4)_2S(aq) \longrightarrow$
$$FeS(s) + 2\,NH_4Cl(aq)$$

36. (a) Fe: +3, O: −2, H: +1   (b) H: +1, Cl: +5, O: −2
 (c) Cu: +2, Cl: −1   (d) K: +1, Cr: +6, O: −2
 (e) Ni: +2, O: −2, H: +1   (f) H: +1, N: −2

37. (a) S: +6, O: −2   (b) N: +5, O: −2
 (c) Mn: +7, O: −2   (d) Cr: +3, O: −2, H: +1
 (e) P: +5, O: −2, H: +1   (f) S: +2, O: −2

40. Left side has best reducing agents. Right side has best oxidizing agents (except noble gases).

42. (a), (d), and (f)

44. (a) $2\,K + 2\,H_2O \longrightarrow 2\,KOH + H_2$
 (b) $Mg + 2\,HBr \longrightarrow MgBr_2 + H_2$
 (c) $2\,NaBr + Cl_2 \longrightarrow 2\,NaCl + Br_2$
 (d) $WO_3 + 3\,H_2 \longrightarrow 3\,H_2O + W$
 (e) $8\,H_2S + 8\,Cl_2 \longrightarrow 16\,HCl + S_8$

46. (a) Most reactive are in Groups 1A and 2A; least reactive are in Group 1B.
 (b) Ag will not react with Al^{3+}.

(c) Al will react with Pb^{2+} and Pb will react with Ag^+.
 (d) Al > Pb > Ag

48. (a) No   (b) No   (c) No   (d) Yes
 (e) Yes   (f) Yes   (g) No

50. (a) N.R.   (b) $Br_2 + 2\,NaI \longrightarrow 2\,NaBr + I_2$
 (c) $F_2 + 2\,NaCl \longrightarrow 2\,NaF + Cl_2$
 (d) $Cl_2 + 2\,NaBr \longrightarrow 2\,NaCl + Br_2$
 (e) N.R.   (f) N.R.

52. (a) $Br_2 + 2\,NaI \longrightarrow 2\,NaBr + I_2$
 (b) $Cl_2 + 2\,NaBr \longrightarrow 2\,NaCl + Br_2$
 (c) $F_2 + 2\,NaCl \longrightarrow 2\,NaF + Cl_2$

53. 0.12 M Ba^{2+}, 0.24 M Cl^-

55. (a) 0.254 M Na_2CO_3 (b) 0.508 M Na^+, 0.254 M CO_3^{2-}

57. 0.493 g $KMnO_4$

59. 5.08×10^3 mL

61. 0.0150 M $CuSO_4$

63. Method (b)

65. 0.205 g Na_2CO_3

67. 121 mL

71. 0.179 g AgCl, NaCl, 0.00833 M NaCl

73. (a) Step (ii) is wrong, and Step (iii) is wrong because Step (ii) is wrong
 (b) 0.00394 g citric acid

75. 1.192 M HCl

76. 0.167 M NaOH

77. 96.8%

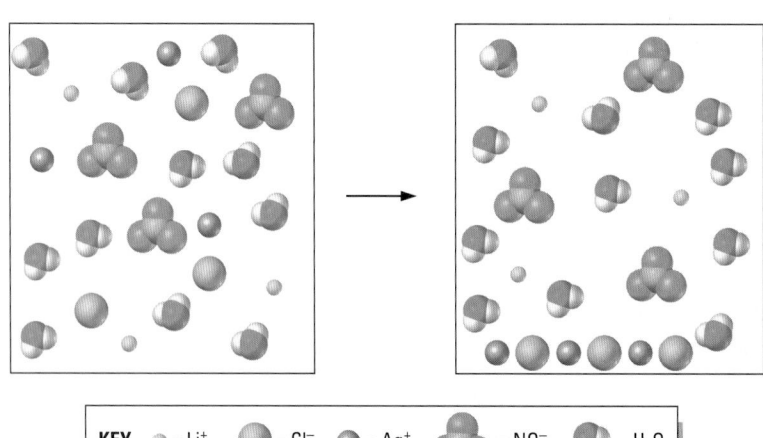

KEY ⬤ = Li^+ ⬤ = Cl^- ⬤ = Ag^+ ⬤ = NO_3^- ⬤ = H_2O

(Question 92, Case 1 (b))

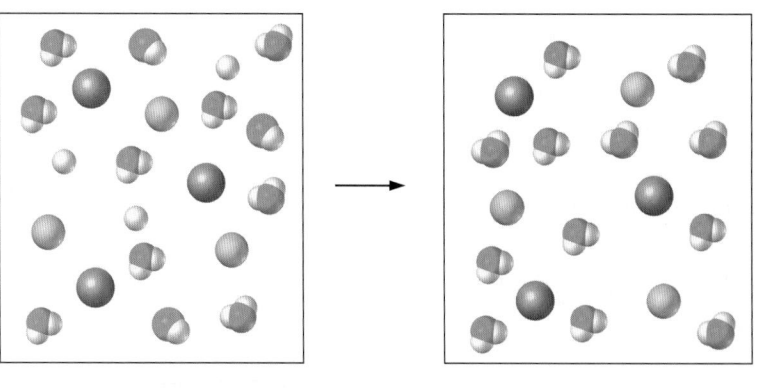

KEY ⬤ = Na^+ ⬤ = OH^- ⬤ = Cl^- ⬤ = H^+ ⬤ = H_2O

(Question 92, Case 2 (b))

79. 104 g/mol

82. (a) $Mg(s) + 4\,HNO_3(aq) \longrightarrow$

$$Mg(NO_3)_2(aq) + 2\,NO_2(g) + 2\,H_2O(\ell)$$

(b) Magnesium, nitric acid, magnesium nitrate, nitrogen dioxide, water

(c) $Mg(s) + 4\,H^+ + 2\,NO_3^-(aq) \longrightarrow$

$$Mg^{2+} + 2\,NO_2(g) + 2\,H_2O(\ell)$$

(d) Redox plus gas-forming reaction

84. (a) $2\,LiOH + H_2$, displacement

(b) $4\,Ag + O_2$, decomposition (c) $2\,LiOH$, combination

(d) N.R. (e) N.R. (f) $BaO + CO_2$, decomposition

86. $Cu_3(CO_3)_2(OH)_2 + 6\,HCl \longrightarrow 3\,CuCl_2 + 4\,H_2O + 2\,CO_2$

88. (a) No (b) Yes (c) No (d) Yes

90. (c) Mg is oxidized; $TiCl_4$ is reduced. Reducing agent is Mg; oxidizing agent is $TiCl_4$. Change in oxidation number during reduction of Ti is -4. Change in oxidation number during oxidation of Mg is $+2$.

92. Case 1 (a) Before: clear colorless solution; after: solid at the bottom of beaker with colorless solution above it

(b) See previous page.

(c) $Li^+ + Cl^- + Ag^+ + NO_3^- \longrightarrow Li^+ + AgCl(s) + NO_3^-$

Case 2 (a) Before and after: clear colorless solutions

(b) See previous page.

(c) $Na^+ + OH^- + H^+ + Cl^- \longrightarrow Na^+ + H_2O(\ell) + Cl^-$

94. (a) $Ba(OH)_2(aq) + H_2SO_4(aq) \longrightarrow BaSO_4(s) + 2\,H_2O(\ell)$

(b) $Na_2SO_4(aq) + Ba(NO_3)_2(aq) \longrightarrow$

$$BaSO_4(s) + 2\,NaNO_3(aq)$$

(c) $BaCO_3(s) + H_2SO_4(aq) \longrightarrow$

$$BaSO_4(s) + H_2O(\ell) + CO_2(g)$$

95. Students 1 and 3 will be successful. Student 2 will be unsuccessful.

97. Either $NaOH$ or H_2S, due to variable solubility of the cations with the anions.

99. Oxidizing and reducing agents are always reactants.

100. (d)

102. (a) and (d)

104. (a) $MgBr_2$, magnesium bromide; $CaBr_2$, calcium bromide; $SrBr_2$, strontium bromide

(b) $Mg(s) + Br_2(\ell) \longrightarrow MgBr_2(s)$

$Ca(s) + Br_2(\ell) \longrightarrow CaBr_2$

$Sr(s) + Br_2(\ell) \longrightarrow SrBr_2$

(c) Oxidation-reduction

(d) The point at which increase stops gives $\dfrac{\text{g metal}}{\text{g product}}$.

The different metals have different molar masses, so the ratios will be different.

106. (a) $Ag^+ + Cl^- \longrightarrow AgCl(s)$, $Ag^+ + Br^- \longrightarrow AgBr(s)$

(b) C and D and A and B (c) Bromide is heavier than chloride.

Chapter 6

13. (a) 399 Cal (b) 5.0×10^6 J

15. (a) 2.97×10^5 J (b) 7.10×10^4 cal (c) 71.0 kcal

17. 5×10^6 J, \$0.13

19. (a) The system: the plant (stem, leaves, roots, etc.); the surroundings: anything not the plant (air, soil, water, sun, etc.)

(b) To study the plant growing, we must isolate it and see how it interacts with its surroundings.

(c) Light energy and carbon dioxide are absorbed by the leaves and are converted to other molecules storing the energy as chemical energy and used to increase the size of the plant. Nutrients (minerals and water) are absorbed through the soil to assist in the chemical processes. The plant expels oxygen and other waste materials into the surroundings.

21. (a) The system: NH_4Cl; the surroundings: anything not NH_4Cl, including the water.

(b) To study the absorption of energy during the phase change of this chemical, we must isolate it and see how it interacts with the surroundings.

(c) The system's interaction with the surroundings causes heat to be transferred into the system and out of surroundings. There is no material transfer in this process, but there is a change in the specific interaction between the water and system.

(d) Endothermic

23. $\Delta E + 32$ J

25.

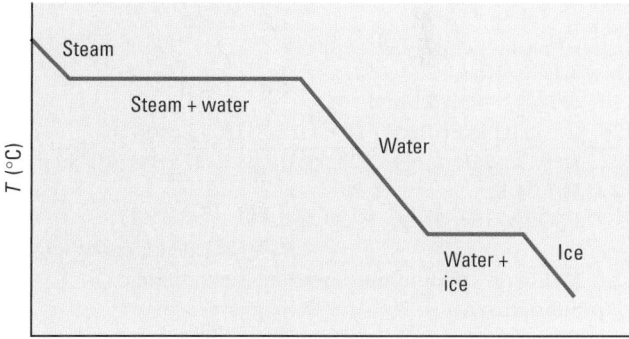

$\Delta E = 715.6$ kJ

28. Cu will get to body temperature faster.

30. 700 MJ

31. (a) H_2O takes more energy.

(b) More energy (1.48×10^6 J) is absorbed by the water than by the ethylene glycol (9.56×10^5 J).

32. 1.35 kJ

35. (a) $0.45\,Jg^{-1}\,^{\circ}C^{-1}$ (b) 25 J/mol·$^{\circ}$C

37. Gold

39. 170 °C

41. (a) Positive (b) Negative

43. 4.13×10^5 J

45. 4.99×10^5 J

47. -270 J

49.

Quantity of heat transferred out of the system

51. (a) X(s) (b) Heat of fusion

(c) Enthalpy of vaporization is positive.

53. Endothermic

55. Endothermic

57. 6.0 kJ of energy is used to convert ice into liquid water.

59. (a) 210 kJ (b) 33 kJ

61. (a) $\frac{1}{2}\,C_8H_{18}(\ell) + \frac{25}{4}\,O_2(g) \longrightarrow 4\,CO_2(g) + \frac{9}{2}\,H_2O(\ell)$

$$\Delta H^\circ = -2748.0 \text{ kJ}$$

(b) $100\,C_8H_{18}(\ell) + 1250\,O_2(g) \longrightarrow$

$800\,CO_2(g) + 900\,H_2O(\ell) \qquad \Delta H^\circ = -5.4960 \times 10^5$ kJ

(c) $C_8H_{18}(\ell) + \frac{25}{2}\,O_2(g) \longrightarrow 8\,CO_2(g) + 9\,H_2O(\ell)$

$$\Delta H^\circ = -5496.0 \text{ kJ}$$

63. $\dfrac{464.8 \text{ kJ}}{1 \text{ mol CaO}}$, $\dfrac{464.8 \text{ kJ}}{3 \text{ mol C}}$, $\dfrac{464.8 \text{ kJ}}{1 \text{ mol CaC}_2}$, $\dfrac{464.8 \text{ kJ}}{1 \text{ mol CO}}$,

$\dfrac{1 \text{ mol CaO}}{464.8 \text{ kJ}}$, $\dfrac{3 \text{ mol C}}{464.8 \text{ kJ}}$, $\dfrac{1 \text{ mol CaC}_2}{464.8 \text{ kJ}}$, $\dfrac{1 \text{ mol CO}}{464.8 \text{ kJ}}$

65. (a) 3×10^5 kJ (b) 3.63×10^4 kJ (c) 139 kJ produced
67. $\Delta H = -1450$ kJ/mol
69. 35.5 kJ
71. 6×10^4 kJ released
73. HF
75. For $H_2 + F_2 \longrightarrow 2\,HF$ reaction:
    (a) 594 kJ (b) -1132 kJ (c) -538 kJ;
    For $H_2 + Cl_2 \longrightarrow 2\,HCl$ reaction:
    (a) 678 kJ (b) -862 kJ (c) -184 kJ
    (d) ΔH of HF is most exothermic.
77. 20 °C
79. (a) 1.4×10^4 J (b) -42 kJ/mol
81. 6.7 kJ
83. 394 kJ/mol evolved
85. $\Delta H^\circ_{f,\text{PbO}} = -1220.$ kJ/mol
87. $\Delta H^\circ_{f,\text{PbO}} = -217.3$ kJ/mol, 2.6×10^2 kJ evolved
89. $Ag(s) + \frac{1}{2}Cl_2(g) \longrightarrow AgCl(s) \qquad \Delta H^\circ = -127.1$ kJ
91. (a) $2\,Al(s) + \frac{3}{2}O_2(g) \longrightarrow Al_2O_3(s) \quad \Delta H^\circ = -1675.7$ kJ
    (b) $Ti(s) + 2\,Cl_2(g) \longrightarrow TiCl_4(\ell) \qquad \Delta H^\circ = -804.2$ kJ
    (c) $N_2(g) + 2\,H_2(g) + \frac{3}{2}O_2(g) \longrightarrow NH_4NO_3(s)$
    $\qquad\qquad\qquad\qquad\qquad\qquad \Delta H^\circ = -365.56$ kJ
94. (a) $\Delta H = 2801.6$ kJ (b) Endothermic
96. $\Delta H = -69.14$ kJ
98. $\Delta H^\circ_{f,\text{OF}_2} = 18$ kJ/mol
100. 83.8 kJ evolved
101. 0.781 g propane
105. CH_4, -50.013 kJ/g; C_2H_6, -47.4832 kJ/g; C_3H_8, -46.353 kJ/g; C_4H_{10}, -32.605 kJ/g; $CH_4 > C_2H_6 > C_3H_8 > C_4H_{10}$
107. 720 kJ
109. 2.2 hours walking
111. Au reaches 100 °C first.
113. 75.4 g ice melted
114. $\Delta H^\circ_{f,\text{B}_2\text{H}_6} = 36$ kJ/mol
116. 2.19×10^7 kJ
118. $\Delta H^\circ_{f,\text{N}_2\text{H}_4} = 50.$ kJ/mol
120. (a) 36.03 kJ evolved (b) 1.18×10^4 kJ evolved
122. Step 1: $\Delta H^\circ = -137.23$ kJ; Step 2: 275.341 kJ; Step 3: 103.71 kJ
    $H_2O(g) \longrightarrow H_2(g) + \frac{1}{2}O_2(g)$
    $\qquad\qquad\qquad\qquad \Delta H^\circ = 241.82$ kJ, endothermic
124. Melting is endothermic. Freezing is exothermic.
126. Substance A
128. Greater. A larger mass of water can contain a larger amount of heat at a given temperature.
130. Each describes the thermal energy changes at constant pressure. They represent different chemical reactions.
132. The given reaction produces 2 mol of SO_3. Formation enthalpy from Table 6.2 is for the production of 1 mol of SO_3.
133. (a) 26.6 °C (Above $C_6H_8O_6$ masses of 8.81g, NaOH is the limiting reactant.)
    (b) $C_6H_8O_6$ limits in Experiments 1–3 and NaOH limits in Experiments 4 and 5.
    (c) Ascorbic acid has one hydrogen ion.
135. $w = 0$ J, $\Delta E = 310$ J

Chapter 7

2. A massless "particle" of light, $E = h\nu$
4. No more than two electrons can be assigned to the same orbital in an atom, and these electrons must have opposite spin.
6. Principle quantum number = 3, in the third shell of the atom's electrons. The orbital has a dumbbell-shaped probability profile (representative of p orbitals) aligned along the x-axis.
8. Atomic sizes increase and ionization energy decreases going down any column and from right to left on any period of the periodic table.
10. $1s^2 2s^2 2p^6 3s^2 3p^6 4s^1 \longrightarrow 1s^2 2s^2 2p^6 3s^2 3p^6 + e^-$
    First ionization
    $1s^2 2s^2 2p^6 3s^2 3p^2 \longrightarrow 1s^2 2s^2 2p^6 3s^2 3p^5 + e^-$
    Second ionization
    $4s$ valence electron takes much less energy to remove than the $3p$ core electron. $\text{IE}_2 \gg \text{IE}_1$
12. Noble gas notation is a shorthand way of representing an atom's core electrons by citing the noble gas with the same electron configuration in brackets. For example, the electron configuration of Mg is $1s^2 2s^2 2p^6 3s^2$ and can be represented using noble gas notation as $[\text{Ne}]3s^2$.
14. When an atom absorbs a photon in the visible region, the energy of the electrons changes, but the particles in the nucleus are not affected.
16. Short wavelength and high frequency
18. (a) Radio waves have less energy than infrared.
    (b) Microwaves have a higher frequency than radio waves.
20. (a) 3.00×10^{-3} m (b) 6.63×10^{-23} J/photon
    (c) 39.9 J/mol
22. (d) < (c) < (a) < (b)
24. 6.06×10^{14} Hz
26. 4.4×10^{-19} J/photon
28. Many types of electromagnetic radiation, including visible, ultraviolet, infrared, microwaves
30. 1.1×10^{15} Hz, 7.4×10^{-19} J/photon
32. Orange light is lower in energy (3.18×10^{-19} J/photon) than this x-ray (8.42×10^{-17} J/photon).
34. 1.2×10^8 J/mol
36. Photons of this light are too low in energy. Increasing the intensity only increases the number of photons, not their individual energy.
38. No
40. Line emission spectra are mostly dark, with discrete bands of light. Sunlight is a continuous rainbow of color.
42. Higher to lower; difference
44. (a) Absorbed (b) Emitted (c) Absorbed (d) Emitted
46. (a) Less (b) Less (c) More (d) Less
48. 3.577×10^{-19} J
50. 3.369×10^{-19} J
52. 4.576×10^{-19} J absorbed, 434.0 nm
54. 0.05 nm
56. (a) First electron: $n = 1$, $\ell = 0$, $m_\ell = 0$, $m_s = +\frac{1}{2}$
    Second electron: $n = 1$, $\ell = 0$, $m_\ell = 0$, $m_s = -\frac{1}{2}$
    Third electron: $n = 2$, $\ell = 0$, $m_\ell = 0$, $m_s = +\frac{1}{2}$
    Fourth electron: $n = 2$, $\ell = 0$, $m_\ell = 0$, $m_s = -\frac{1}{2}$
    Fifth electron: $n = 2$, $\ell = 1$, $m_\ell = 1$, $m_s = +\frac{1}{2}$
    (the fifth electron could have different m_ℓ and m_s values).

(b) $n = 3, \ell = 0, m_\ell = 0, m_s = +\frac{1}{2}$ and $n = 3, \ell = 0,$ $m_\ell = 0, m_s = -\frac{1}{2}$

(c) $n = 3, \ell = 2, m_\ell = 2, m_s = +\frac{1}{2}$ (Different m_ℓ and m_s values are also possible.)

58. (a) Cannot occur   (b) Can occur   (c) Cannot occur
    (d) Cannot occur   (e) Can occur

60. (a) $n = 4, \ell = 0, m_\ell = 0, m_s = +\frac{1}{2}$
    (b) $n = 3, \ell = 1, m_\ell = 1, m_s = -\frac{1}{2}$
    (c) $n = 3, \ell = 2, m_\ell = 0, m_s = +\frac{1}{2}$

62. Four subshells

64. Electrons do not follow simple paths as planets do.

66. Orbits have predetermined paths—position and momentum are both exactly known at all times. Heisenberg's uncertainty principle says that we can not know both simultaneously.

68. $d, p,$ and s orbitals; nine orbitals total

70. ^{13}Al: $1s^2 2s^2 2p^6 3s^2 3p^1$; ^{16}S: $1s^2 2s^2 2p^6 3s^2 3p^4$

72. ^{32}Ge: $1s^2 2s^2 2p^6 3s^2 3p^6 3d^{10} 4s^2 4p^2$

74. Oxygen, Group 6A, has 6 valence electrons with valence electron configuration of $ns^2 np^4$.

76. (a) $4s$ orbital must be full.
    (b) Orbital labels must be 3, not 2; electrons in $3p$ subshell should be in separate orbitals with parallel spin.
    (c) $4d$ orbitals must be completely filled before $5p$ orbitals start filling.

78. (a) V

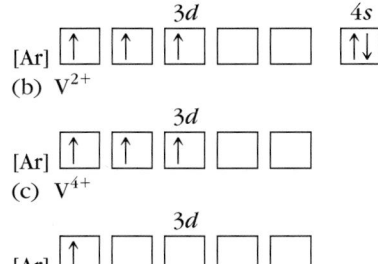

(b) V^{2+}

(c) V^{4+}

80. 18 elements; all possible orbital electron combinations are already used.

82. Mn: 5 unpaired electrons; Mn^{2+}: 5 unpaired electrons; Mn^{3+}: 4 unpaired electrons

84. $[Ar]3d^3 4s^2$

86. (a) ^{63}Eu: $[Xe]4f^7 6s^2$ (b) ^{70}Yb: $[Xe]4f^{14} 6s^2$

88. (a) $\cdot$Sr$\cdot$ (b) $:\ddot{\text{Br}}\cdot$ (c) $\cdot\dot{\text{Ga}}\cdot$ (d) $\cdot\ddot{\text{Sb}}\cdot$

90. (a) [Ar]   (b) [Ar]   (c) [Ne]; Ca^{2+} and K$^+$ are isoelectronic.

92. ^{50}Sn: $[Kr]4d^{10} 5s^2 5p^2$
    ^{50}Sn^{2+}: $[Kr]4d^{10} 5s^2$
    ^{50}Sn^{4+}: $[Kr]4d^{10}$

94. Ferromagnetism is a property of permanent magnets. It occurs when the spins of unpaired electrons in a cluster of atoms (called a domain) in the solid are all aligned in the same direction. Only metals in the Fe, Co, Ni subgroup (Group 8B) exhibit this property.

96. In both paramagnetic and ferromagnetic substances, atoms have unpaired spins and so are attracted to magnets. Ferromagnetic substances retain their aligned spins after an external magnetic field has been removed, so they can function as magnets. Paramagnetic substances lose their aligned spins after a time, and therefore cannot be used as permanent magnets.

98. number

100. P < Ge < Ca < Sr < Rb

102. (a) Rb smaller   (b) O smaller   (c) Br smaller
     (d) Ba^{2+} smaller   (e) Ca^{2+} smaller

104. Al < Mg < P < F

106. (c)

108. Na

110. (a) Al   (b) Al   (c) Al < B < C

112. (a) H$^-$ (b) N^{3-} (c) F$^-$

114. Red < yellow < green < violet

116. Seven pairs

118. (a) He   (b) Sc   (c) Na

120. (a) F < O < S   (b) S

122. (a) S   (b) Ra   (c) N   (d) Ru   (e) Cu

124. (a) Z   (b) Z

126. In^{4+}, Fe^{6+}, and Sn^{5+}

128. (a) False; shorter, not longer   (b) True   (c) True
     (d) False; inversely, not directly

130. (a) Directly related, not inversely related
     (b) Inversely proportional to *the square of* the principle quantum number, not inversely proportional to the principle quantum number
     (c) Before, not as soon as
     (d) Wavelength, not frequency

132. Ultraviolet, 91 nm or shorter wavelength needed to ionize electron

134. (a) $[Rn]5f^{14} 6d^{10} 7s^2 7p^1$ (b) Boron   (c) Et$_2$O$_3$, EtCl$_3$

136. $1s^2 2s^2 2p^4$

138. XCl

140. (a) Ground state
     (b) Could be ground state or excited state
     (c) Excited state   (d) Impossible
     (e) Excited state   (f) Excited state

142. $[Rn]5f^{14} 5g^{18} 6d^{10} 6f^{14} 7s^2 7p^6 7d^{10} 8s^2 8p^2$, Group 4A

144. (a) Increase, decrease   (b) Helium
     (c) 5 and 13   (d) He has only two electrons.
     (e) First electron is a valence electron, but the second electron is a core electron.
     (f) ^{12}Mg

Chapter 8

16. (a) $:\ddot{\text{Cl}}\!-\!\ddot{\text{F}}:$ (b) H$-\ddot{\text{Se}}-$H

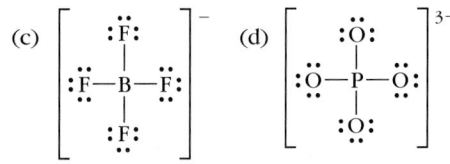

18.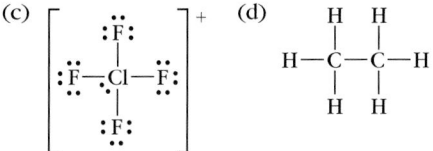

20. (a) 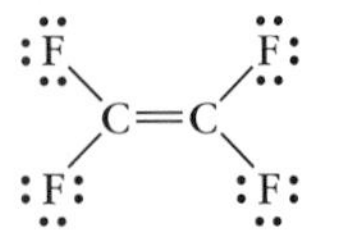 (b) 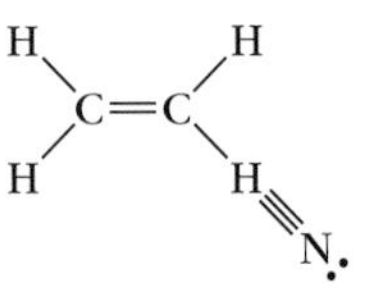

22. (a) Incorrect. F atom missing electrons.
    (b) Incorrect. Structure has 10 electrons, but needs 12 electrons.
    (c) Correct
    (d) Incorrect. Hydrogen atoms must have only two electrons.
    (e) Incorrect. Structure has 16 electrons, but needs 18 electrons. N atom doesn't follow the octet rule. It needs another pair of electrons in the form of a lone pair.
24. Four branched-chain compounds
26. (a) Alkyne (b) Alkane (c) Alkene
28. *cis*-2-pentene *trans*-2-pentene

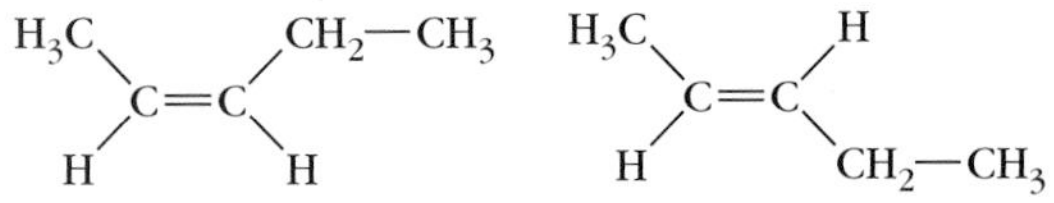

30. (a) 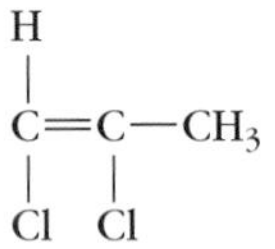 (b) 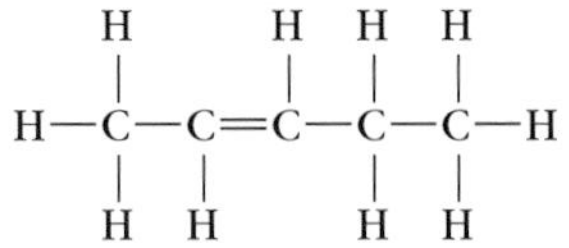
    (c) 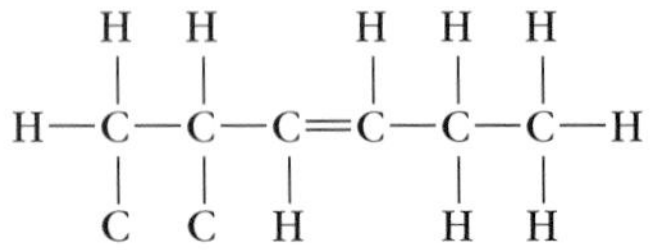
    (d) 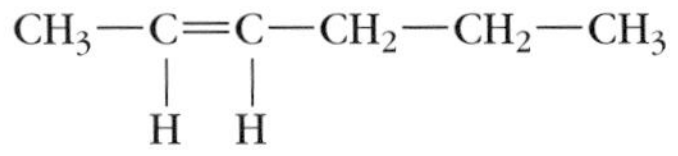
32. (a) B—Cl (b) C—O (c) P—O (d) C≡N
34. (a)
36. CO
38. HCO_2^-
40. −92 kJ
42. HF strongest bond; −538. kJ; −184 kJ; −97 kJ; −11 kJ. Reaction of H_2 with F_2 is most exothermic.
44. (a) C—N (δ^+ δ^-) H—C (δ^+ δ^-) C—Br (δ^+ δ^-) S—O (δ^+ δ^-)
    (b) O—S is most polar
46. (a) All bonds in urea are somewhat polar.
    (b) Most polar C═O; O end is partial negative.
48. Total formal charge on a molecule is always zero. Total formal charge on an ion is the ionic charge.
50. (a) 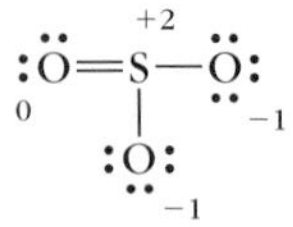 (b) 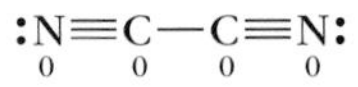
    (c) 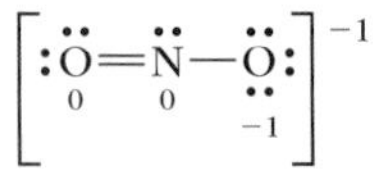
52. (a) 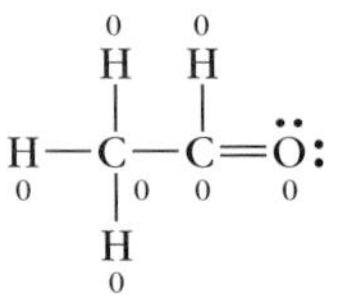 (b) 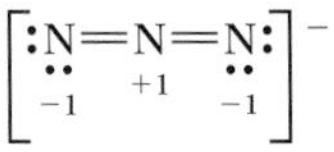
    (c) 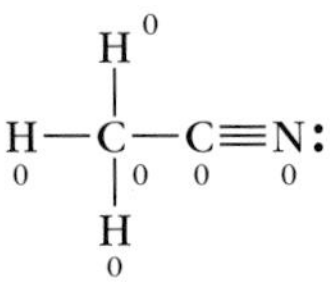
54. (a)

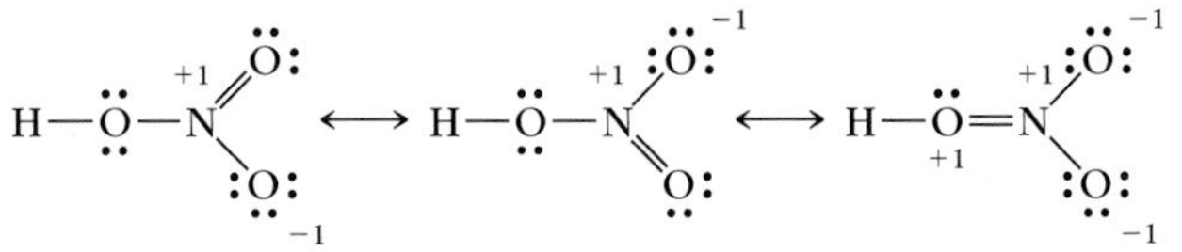

(b)

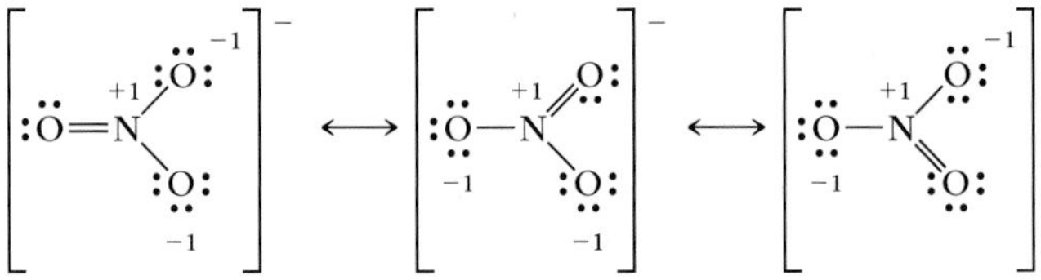

56. 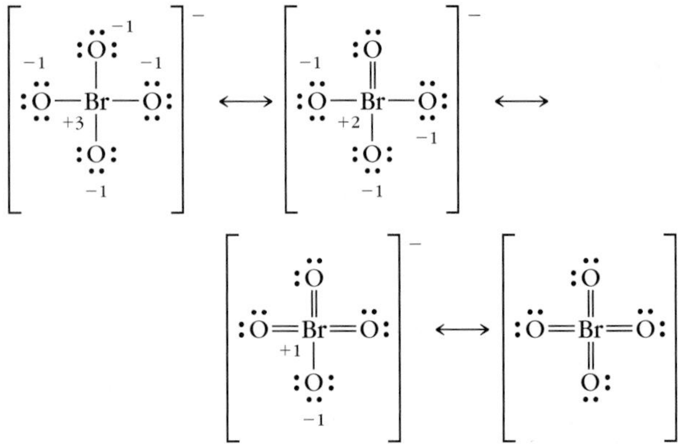

Best

59. (a) 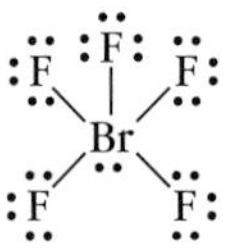 (b) 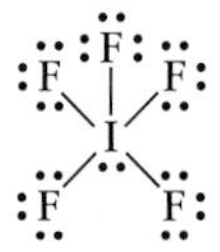 (c) 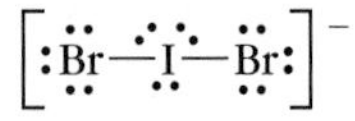
61. P, Cl, Se, Sn
63. Against
65. $C_{14}H_{10}$
67. Si—F
69. Yes. Close means similar electronegativities, therefore covalent bonds; far apart means different electronegativities, therefore ionic bonds.
71. (a) The C═C is shortest.
    (b) The C═C is strongest.
    (c) C≡N (δ^+ δ^-)
73. (a) 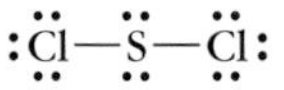 (b) 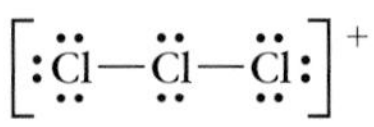
    (c) 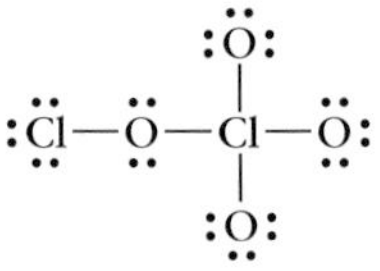 (d) 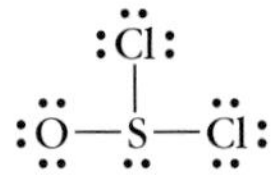
75. O—O < Cl—O < O—H < O═O < O═C
77. (a) Aromatic (b) Aromatic (c) Neither category
    (d) Alkane (e) Alkane (f) Neither category

79. Forgot to subtract one electron for the positive charge.

81. Atoms are not bonded to the same atoms.

83. (a)
H
:Ö:
H   H
H

(b)
:F:
H : : H

85. Cl: 3.0, S: 2.5, Br: 2.5, Se: 2.4, As: 2.1

87. NF_5; can't expand octet

Chapter 9

16. (a) H—Be—H Linear   (b)
H
:Cl—C—H
:Cl:
Tetrahedral

(c)
H
H—B
H
Triangular

(d)
:Cl: :
:Cl   Cl:
Se
:Cl   Cl:
:Cl:
Octahedral

(e)
:F:
:F   F:
P
:F   :F:
Trigonal bipyramid

18. (a) Tetrahedral, angular (120°)
(b) Tetrahedral, triangular pyramid (c) Both tetrahedral
(d) Both linear

20. (a) Both triangular planar (b) Both triangular planar
(c) Tetrahedral, triangular pyramid
(d) Both triangular planar. Three atoms bonded to the central atom gives triangular shape; however, structures with 26 electrons have a shape of triangular pyramid, and structures with 24 electrons have a shape of triangular planar.

22. (a) Both octahedral (b) Trigonal bipyramid and seesaw
(c) Both trigonal bipyramid
(d) Octahedral and square planar

24. (a) 120° (b) 120°
(c) H—O—N angle = 109.5°, O—N—O angle = 120°
(d) H—C—H angle = 120°, C—C—N angle = 180°

26. (a) 90° (b) 90° (c) 90°

28. The O—N—O bond angle in NO_2 is probably slightly smaller than that of NO_2^-.

30. Tetrahedral, sp^3 hybridized carbon atom

32.
:F:
:F—Ge—F:
:F:

Tetrahedral, sp^3 hybridized Ge atom

:F:
:F
Se:
:F   :F:

Seesaw, sp^3d hybridized Se atom

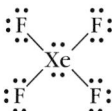

Square planar, sp^3d^2 hybridized Xe atom

34. (a) sp^3 (b) sp^3d^2 (c) sp^2

36. (a) 109.5° (b) 90° (c) 120°

38. (a) First carbon {CH_3 . . . } is sp^3 hybridized. Second carbon {. . . COOH} is sp^2 hybridized. (b) HCH and HCC angles (about the first carbon) are each 109.5°. CCO and OCO angles (at the second carbon) are each 120°.

40. N atom is sp^3 hybridized with 109.5° angles. The first two carbons are sp^3 hybridized with 109.5° angles. The third carbon is sp^2 hybridized with 120° angles. The single-bonded oxygen is sp^3 hybridized with 109.5° angles. The double-bonded O is not hybridized.

42. (a) The first two carbon atoms are sp^3 hybridized with 109.5° angles. The third and fourth carbon atoms are sp hybridized with 180° angles.
(b) C≡C (c) C≡C

44. (a)
:O=C=S:

(b)
H—N—O:
H   H

(c)
H—C=C—C=O
H   H   H

(d)
H   H   O
H—C—C—C—O—H
H   :O:
H

46. (a) 15 (b) 3 (c) 120° (d) 109.5° (e) 109.5°

48. (a) H_2O (b) CO_2 and CCl_4 (c) F

50. (a) Nonpolar
(b) Polar; H side of the molecule is more positive.
(c) Polar; H side of the molecule is more positive.
(d) Nonpolar

52. (a) The Br—F bond has a larger electronegativity difference.
(b) The H—O bond has a larger electronegativity difference.

53. Water and ethanol both interact via similar hydrogen-bonding forces. Cyclohexane interacts primarily with London forces, which are dissimilar to hydrogen-bonding forces found in water.

55. Boiling points increase as the strength of the interactive forces increases. Strongest interactive force is hydrogen bonding, then dipole-dipole, then London forces. London forces increase with increasing number of electrons (found in larger atoms).
(a) Group (IV) hydrides all interact by London forces. The larger the Group (IV) atom, the stronger the London forces and the higher the boiling point.
(b) NH_3 molecules interact via hydrogen bonding, so its boiling point is quite high compared with those of the other Group (V) hydrides. The other Group (V) hydrides interact mostly by London forces. Dipole forces are not very different between one and the next; however, the larger the Group (V) atom, the stronger the London forces and the higher the boiling point.

(c) H_2O molecules interact via hydrogen bonding, so its boiling point is quite high compared with those of the other Group (VI) hydrides. The other Group (VI) hydrides interact mostly by London forces. Dipole forces are not very different between one and the next; however, the larger the Group (VI) atom, the stronger the London forces, the higher the boiling point.

(d) HF molecules interact via hydrogen bonding, so its boiling point is quite high compared with those of the other Group (VII) hydrides. The other Group (VII) hydrides interact mostly by dipole-dipole and London forces. Dipole forces are not very different between one and the next; however, the larger the Group (VII) atom, the stronger the London forces, the higher the boiling point.

57. (a) No   (b) No   (c) Yes   (d) Yes   (e) Yes

59. Vitamin C is capable of forming hydrogen bonds with water.

61. (a) London forces   (b) No interactive forces
   (c) Intramolecular (covalent) forces   (d) Hydrogen bonds

64. (a)

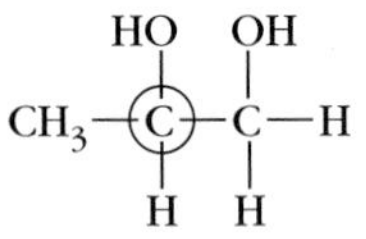

   (b) None

   (c)

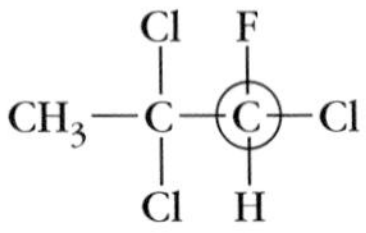

66. (a)

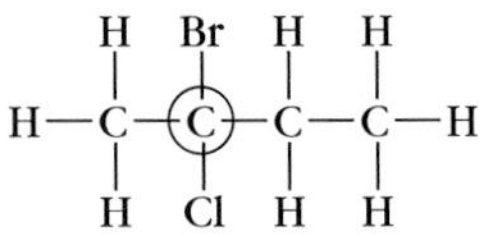

   (b) None

   (c)

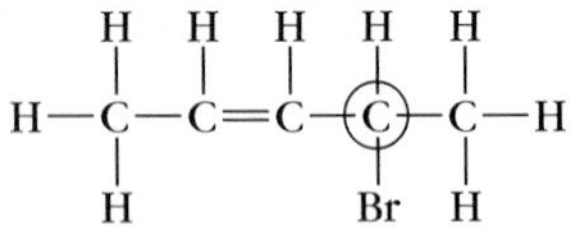

68. If a molecule has one or more chiral centers, the molecule is chiral.

70. (a) Pi electrons and electrons in *d* orbitals   (b) Frequencies of specific vibrational motions

72. G-C interacts with three hydrogen bonds, whereas A-T interacts with only two hydrogen bonds. Stronger attractive forces require higher melting temperature.

74.

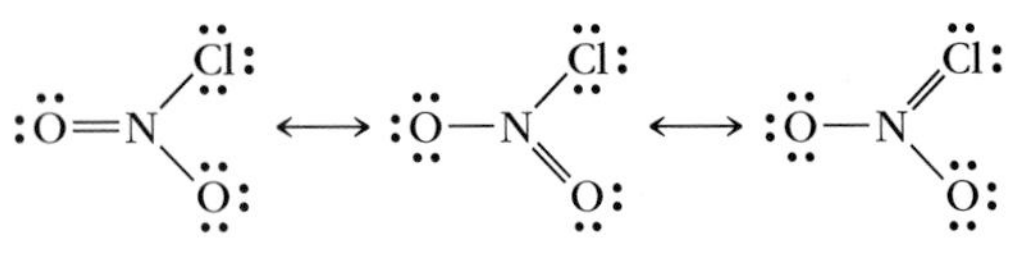

   Trigonal planar electron-pair and molecular geometries, 120° bond angles for O—N—O and and O—N—Cl

78. (a)

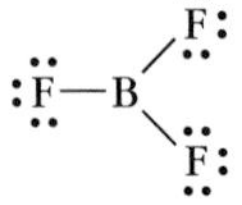

   Triangular planar electron and molecular geometries

(b)

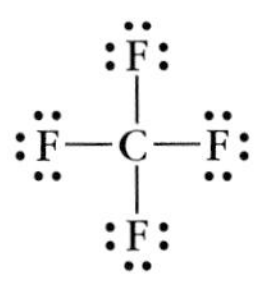

   Tetrahedral electron and molecular geometries

(c)

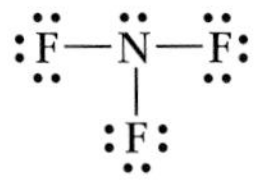

   Tetrahedral electron geometry; triangular pyramid molecular geometry

(d)

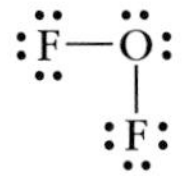

   Tetrahedral electron geometry; angular molecular geometry

(e)

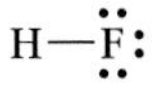

   Linear

80. (a)

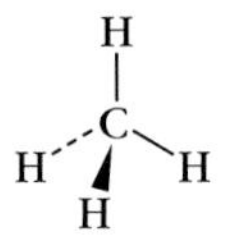

(b)

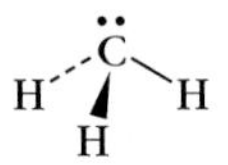

(c)

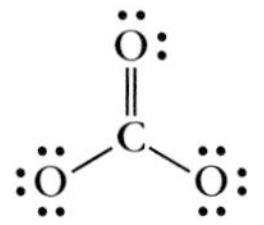

(d)

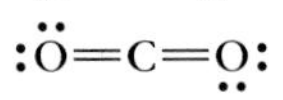

84. If the polarity of the bonds exactly cancel each other, a molecule will be nonpolar.

85.

Molecule or ion	Electron-pair geometry	Molecular geometry	Hybridization
ICl_2^+	Tetrahedral	Angular	sp^3
I_3^-	Triangular bipyramid	Linear	sp^3d
ICl_3	Triangular bipyramid	T-shaped	sp^3d
ICl_4^-	Octahedral	Square planar	sp^3d^2
IO_4^-	Tetrahedral	Tetrahedral	sp^3
IF_4^+	Triangular bipyramid	Seesaw	sp^3d
IF_5	Octahedral	Square pyramid	sp^3d^2
IF_6^+	Octahedral	Octahedral	sp^3d^2

87. (a) Nitrogen   (b) Boron   (c) Phosphorus   (d) Iodine

89. Five

91. Only (b) is correct.

Chapter 10

10. (a) 0.95 atm (b) 950. mm Hg (c) 542 torr
    (d) 99 kPa (e) 7 atm

12. 14 m

14. With a perfect vacuum at the top of the well, atmospheric pressure can push water only up to about 34 feet. So the well cannot be deeper than that, even if a high-quality vacuum pump is used.

16. Nitrogen serves to moderate the reactiveness of oxygen by diluting it. Oxygen sustains animal life as a reactant in the conversion of food to energy. Oxygen is produced by plants in the process of photosynthesis.

18.

Molecule	ppm	ppb	
N_2	780,840	780,840,000	↑
O_2	209,480	209,480,000	
Ar	9,340	9,340,000	>1 ppm
CO_2	330	330,000	
Ne	18.2	18,200	
H_2	10.	10,000	
He	5.2	5,200	
CH_4	2	2,000	
Kr	1	1,000	
CO	0.1	100	between 1 ppm and 1 ppb
Xe	0.08	80	
O_3	0.02	20	
NH_3	0.01	10	
NO_2	0.001	1	
SO_2	0.0002	0.2	≤1ppb

20. 1.5×10^8 metric tons SO_2, 2×10^6 metric tons SO_2

22. (a) Same (b) H_2 faster (c) More molecules of CO_2

24. $SOCl_2 < Cl_2O < Cl_2 < SO_2$

26. 4.2×10^{-5} mol CO

28. 154 mm Hg

30. 172 mm Hg

32. 26.5 mL

34. 176 K

38. 501 mL

40. 0.51 atm

42. (d) has the most molecules; (a) and (b) have the least molecules (both the same).

44. 1.9 L CO_2

46. 10. L O_2, 10. L H_2O

48. 21 mm Hg

50. 148 g/mol

52. (a) $2\ C_8H_{18}(\ell) + 25\ O_2(g) \longrightarrow 16\ CO_2(g) + 18\ H_2O(g)$
    (b) 1400 L CO_2

54. 64 K

56. 0.90 g

58. 3.7×10^{-4} g/L

60. (a) 154 mm Hg
    (b) $X_{N_2} = 0.777$, $X_{O_2} = 0.208$, $X_{Ar} = 0.0093$, $X_{H_2O} = 0.0054$, $X_{CO_2} = 0.0003$
    (c) This sample is wet. Table 1.3 gives percentages for dry air, so the percentages are slightly different due to the water in this sample.

62. (a) 29 g/mol (b) $X_{N_2} = 0.80$, $X_{O_2} = 0.20$

64. $X_{H_2O} = 0.033$; dry air has 0% water. This sample has 3.3% water.

66. 73.6%

68. 18. mL $H_2O(\ell)$, 22.4 L $H_2O(g)$, No. Vapor pressure of water at 0 °C is 4.6 mm Hg; we cannot achieve 1 atm pressure for this gas at this temperature.

70. Molecular attractions become larger at higher pressures. Molecules hitting the walls hit them with somewhat less force due to the opposing attractions of other molecules.

75. Ethane

77. (a) O_3 (b) $O + O_2$ (c) SO_3 (d) $H\cdot + \cdot OH$ (e) N_2O_4

79. (a) An example is $\cdot OH + CO \longrightarrow HCO_2\cdot$
    (b) An example is $6\ CO_2 + 6\ H_2O \longrightarrow C_6H_{12}O_6 + 6\ O_2$ (in plants)

81. C—F bond is stronger than C—Cl bond. It requires higher frequency, higher energy light to break it. Light that high in energy is effectively screened out by O_2 in the atmosphere.

83. $FCCl_3$, F_2CCl_3, F_3CCl

85. Yes

87. Primary pollutants (e.g., particle pollutants, including aerosols and particulates; sulfur dioxide; nitrogen oxides; hydrocarbons); secondary pollutants (e.g., ozone). See Section 10.12 for details.

89. Adsorption: firmly attaching to a surface. Absorption: drawing into the bulk of a solid or liquid.

91. 1.6×10^9 metric tons, 2×10^6 hours

93. $O_2 \xrightarrow{h\nu} O + O$; no.

95. SO_2, coal and oil, $2\ SO_2 + O_2 \longrightarrow 2\ SO_3$

97. $N_2 + O_2 \xrightarrow{\text{heat}} 2\ NO$; reaction takes elemental nitrogen and makes a compound of nitrogen.

98. Stratospheric ozone is beneficial as a sunscreen. Tropospheric ozone causes diminished respiratory capacity in humans.

101. 3×10^8 metric tons of H_2SO_4

103. (a)

105. See next page

107. (c)

109. (a) 41.9 g/mol (b) CH_2, C_3H_6
     (c)

111. No. Gas volume might have dropped due to low temperature, not necessarily because the balloon is defective.

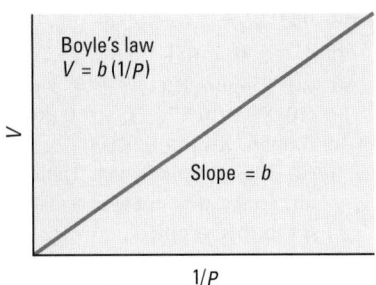

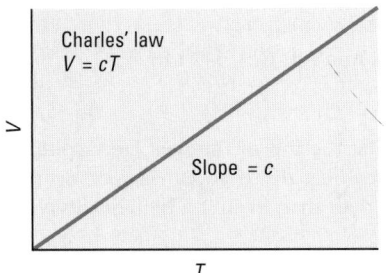

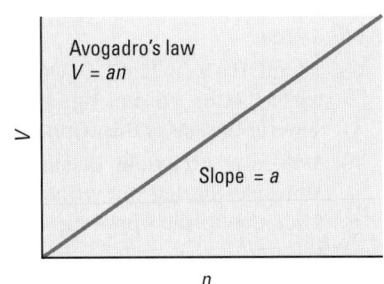

(Question 105)

Chapter 11

2. Molecules move around randomly in all directions. Molecules with higher temperature have higher average kinetic energy.

8. The unit cell is the smallest part of a crystal lattice that, when repeated along the directions defined by its edges, reproduces the entire crystal.

12. At higher temperatures, the molecules move around more. The increased random motion disrupts the intermolecular interactions responsible for surface tension.

14. Reduce the pressure.

16. The molecules of water in your sweat have a wide distribution of molecular speeds. The fastest of these molecules are more likely to escape the liquid state into the gas phase. The low-speed molecules left behind will have a lower average speed, and therefore a lower average kinetic energy, which, according to kinetic molecular theory, implies that the temperature becomes lower.

18. 1.5×10^6 kJ

20. 233 kJ

22. The 2.00 kJ required for Hg is more than the 1.13 kJ required for the H_2O sample.

24. NH_3 has a relatively high boiling point because the molecules interact using relatively strong hydrogen-bonding intermolecular forces. The increase in the boiling points of the series PH_3, AsH_3, and SbH_3 is related to the increasing London dispersion intermolecular forces experienced due to the larger central atom in the molecule (size: P < As < Sb).

26. Methanol molecules are capable of hydrogen bonding, whereas formaldehyde molecules use dipole-dipole forces to interact. Molecules experiencing stronger intermolecular forces (such as methanol here) will have higher boiling points and lower vapor pressures compared with those of molecules experiencing weaker intermolecular forces (such as formaldehyde here).

28. 0.21 atm, 40 °C

30. The interparticle forces in the solid are very strong.

32. The intermolecular forces between the molecules of H_2O in the solid are stronger than the intermolecular forces between the molecules of H_2S.

34. 27 kJ

36. LiF, because the ions are smaller, making the charges more localized and closer together, causing a higher coulombic interaction between the ions.

38. Highest melting point is (a). Extended network of covalent bonds in SiC are the strongest interparticle forces of the listed choices. Lowest melting point is (d). Both I_2 and $CH_3CH_2CH_2CH_3$ interact using the weakest intermolecular forces — London dispersion forces, but large I_2 has many more electrons, so its London forces are relatively stronger.

40. The freezer compartment of a frost-free refrigerator keeps the air so cold and dry that ice inside the freezer compartment sublimes [(s) $\rightarrow$ (g)]. The hailstones would eventually disappear.

42. An ideal gas would have a density of 130 g/L, which is much smaller than the density of CO_2 at its critical point (470 g/L).

44. (a) Molecular (b) Metallic (c) Network (d) Ionic

46. (a) Amorphous; it decomposes before melting and does not conduct electricity. (b) Molecular; low melting point and non-conducting. (c) Ionic; high melting point, and only liquid (molten ions) conduct electricity. (d) Metallic; both solid and liquid conduct electricity.

48. (a) Molecular (b) Ionic (c) Metallic (d) Amorphous

50. See Figure 11.19 and its description.

52. 0.219 nm

56. Diagonal is 0.700 nm; side length is 0.404 nm.

58. No. The ratio of ions in the unit cell must reflect the empirical formula of the compound.

60. (a) 152 pm
    (b) Radius of I^- = 212 pm and radius of Li^+ = 88 pm.
    (c) It is reasonable that the atom is larger than the cation.

The assumption that anions touch anions seems unreasonable; there would be some repulsion and probably a small gap or distortion.

62. (a) 3.51×10^{-16} J/photon (b) 2.11×10^{8} J/mol; X-ray

64. In a conductor, the valence band is only partially filled, whereas in an insulator, the valence band is completely full, the conduction band is empty, and there is a wide energy gap between the two. In a semiconductor, the gap between the valence band and the conduction band is very small, so that electrons are easily excited into the conduction band.

66. (c) Ag has the greatest electrical conductivity because it is a metal.

(d) P_4 has the smallest electrical conductivity because it is a nonmetal. (The other two are metalloids.)

68. A superconductor is a substance that is able to conduct electricity with no resistance. Two examples are $YBa_2Cu_3O_7$ and $HgBa_2Ca_2Cu_4O_8$.

70. $SiO_2(s) + 2\,C(s) \rightarrow Si(s) + 2\,CO(g)$; Si is being reduced and C is being oxidized.

$SiCl_2(\ell) + 2\,Mg(s) \rightarrow Si(s) + 2\,MgCl_2(s)$; Si is being reduced and Mg is being oxidized.

72. Doping is the intentional addition of small amounts of specific impurities into very pure silicon. Group III elements are used because they have one less electron per atom than the Group IV silicon. Group V elements are used because they have one more electron per atom.

74. Carbon atoms in diamond are sp^3 hybridized and are tetrahedrally bonded to four other carbon atoms. Carbon atoms in pure graphite are sp^2 hybridized and are bonded with a trianglular planar shape to other carbon atoms. These bonds are partially double-bonded, so they are shorter than the single bonds in diamond. However, the planar sheets of sp^2-hybridized carbon atoms are only weakly attracted by intermolecular forces to adjacent layers, so these interplanar distances are much longer than the C—C single bonds. The net result is that graphite is less dense than diamond.

76. Diamond is an electrical insulator because all the electrons are in single bonds that are shared between two specific atoms and cannot move around. However, graphite is a good conductor of electricity because its electrons are delocalized in conjugated double bonds that allow the electrons to move easily through the graphite sheets.

78. The amorphous solids known as glasses are different from NaCl because they lack symmetry or long-range order, whereas ionic solids such as NaCl are extremely symmetrical. NaCl must be heated to melting temperatures, then cooled very slowly, to make a glass.

80. Two examples of oxide ceramics are Al_2O_3 and MgO. Two examples of nonoxide ceramics are Si_3N_2 and SiC.

82. 1.46×10^4 kJ, 1.54×10^4 kJ

84. (a)

Triangular planar, angular.

(b) Dipole-dipole forces. (c) $CH_4 < SO_2 < NH_3 < H_2O$

86. (a) About 80 mm Hg (b) About 18 °C

(c) About 740 mm Hg (d) Diethyl ether and ethanol

(e) Diethyl ether would evaporate readily. The ethanol and water would evaporate more slowly.

(f) Water has the strongest intermolecular attractions.

88. The butane in the lighter is under great enough pressure that the vapor pressure of butane at room temperature is less than the pressure inside the light. Hence it exists as a liquid.

90. 1 and C, 2 and E, 3 and B, 4 and F, 5 and G, 6 and H, 7 and D, 8 and A

92. Each has the same fraction of filled space. The fraction of spaces filled by closest packed equal-sized spheres is the same, no matter what the size of the spheres.

Chapter 12

6. Pyrolysis is the process of heating coal at high temperatures in the absence of air. Three important compounds derived from coal tar, a product of pyrolysis, are benzene, toluene and xylene.

8. Reformulated gasolines are oxygenated gasolines with a lower percentage of aromatic hydrocarbons and lower volatility than ordinary gasoline. They are used to try to decrease the amount of hydrocarbons released into the air by spills and evaporation during filling, thereby decreasing the amount of urban air pollution.

10. The types of atoms, bonds, and degree of branching in various molecules cause variable ability for autoignition, the attribute described by the octane rating. The two molecules set as the standard for the octane rating are *n*-heptane (given a zero octane rating) and iso-octane (given the rating of 100). Compounds that autoignite even more readily than *n*-heptane get negative ratings, whereas compounds that autoignite even less readily than iso-octane have ratings over 100.

12. Ethylene glycol has two sites of hydrogen bonding, whereas ethanol has only one.

14. Crude oil is a liquid mixture of low molar mass alkane molecules, whereas coal is a complex and regular array of partially hydrogenated aromatic compounds.

16. Esters do not have the ability to hydrogen-bond like carboxylic acids do. The acids have a —COOH group that provides an acidic hydrogen. The esters are —COOR. Without the acidic hydrogen, the esters cannot undergo hydrogen bonding; therefore, they interact primarily with weaker dipole-dipole forces and London dispersion forces. Smaller intermolecular forces in the liquid will result in a lower boiling point.

18. Condensation polymers are made from linking two different functional groups on the monomer molecules. Both ester and amide linkages contain a carbonyl (C=O), since they are both made by reacting carboxylic acids. All condensation polymers have water as a byproduct.

20. The *trans* form is harder and more brittle than the *cis* form of rubber.

22. Addition polymers use monomers with C=C bonds and the polymerization process produces no additional product(s). Condensation polymers combine monomers that have carboxylic acid functional groups with alcohol or amine functional groups, and the process also produces water.

24. (a) 20–200 °C

(b) The octane rating of the straight-run gasoline fraction is 55.

(c) No. The octane rating is far lower than that of regular gasoline we buy at the pump (86–94); that means it would cause far more preignition than we expect from the gasoline. It would need to be reformulated to make it an acceptable motor fuel.

26. Gasolines contain molecules in the liquid phase that, at ambient temperatures, can easily overcome their intermolecular forces and escape into the vapor phase. That means all gasolines evaporate easily.

28. $CH_3—CH_2—OH$

30. The greenhouse effect is the trapping of heat by atmospheric gases. Global warming is the increase of the average global temperature. Global warming is related to an increase in the amount of greenhouse gases in the atmosphere.

32. CO_2 gets into the atmosphere by animal respiration, by the burning of fossil fuels and other plant materials, and by the decomposition of organic matter. CO_2 gets removed from the atmosphere by plants during photosynthesis, when it is dissolved in rain water, and when it is incorporated into carbonate and bicarbonate compounds in the oceans. Currently, atmospheric CO_2 production exceeds the CO_2 removal processes.

34. (a) $CH_3—CH_2—CH_2—OH$

(b)
$$CH_3—\overset{\overset{\displaystyle OH}{|}}{CH}—CH_3$$

(c)
$$CH_3—\overset{\overset{\displaystyle OH}{|}}{\underset{\underset{\displaystyle CH_3}{|}}{C}}—CH_3$$

36. (a)
$$CH_3—\overset{\overset{\displaystyle OH}{|}}{\underset{\underset{\displaystyle CH_3}{|}}{C}}—CH_2—CH_2—CH_3$$

(b)
$$CH_3—\overset{\overset{\displaystyle CH_3}{|}}{CH}—\overset{\overset{\displaystyle CH_3}{|}}{CH}—CH_2—OH$$

(c)
$$CH_3—\overset{\overset{\displaystyle OH}{|}}{CH}—CH_2—\overset{\overset{\displaystyle CH_3}{|}}{CH}—CH_3$$

(d)
$$CH_3—\overset{\overset{\displaystyle CH_3}{|}}{CH}—\underset{\underset{\displaystyle OH}{|}}{CH}—CH_2—CH_3$$

(e)
$$CH_3—\overset{\overset{\displaystyle CH_3}{|}}{\underset{\underset{\displaystyle OH}{|}}{C}}—CH_3$$

(f)
$$CH_3—\overset{\overset{\displaystyle OH}{|}}{CH}—CH_3$$

38. (a)
$$CH_3—\overset{\overset{\displaystyle O}{\|}}{C}—H$$
$$CH_3—\overset{\overset{\displaystyle O}{\|}}{C}—OH$$

(b)
$$CH_3—CH_2—CH_2—\overset{\overset{\displaystyle O}{\|}}{C}—H$$
$$CH_3—CH_2—CH_2—\overset{\overset{\displaystyle O}{\|}}{C}—OH$$

40. (a)
$$CH_3—\overset{\overset{\displaystyle }{\underset{\underset{\displaystyle CH_3}{|}}{CH}}—CH_2—CH_2—OH$$

(b)
$$CH_3—CH_2—\overset{\overset{\displaystyle OH}{|}}{CH}—CH_2—CH_3$$

(c)
$$CH_3—CH_2—\underset{\underset{\displaystyle CH}{|}}{CH}—CH_2—OH$$

42. Wood alcohol (methanol) is made by heating hardwoods such as beech, hickory, maple, or birch. Grain alcohol (ethanol) is made from the fermentation of plant materials, such as grains.

44. —OH groups are a common site of hydrogen-bonding intermolecular forces. Their presence would increase the solubility of the biological molecule in water and create specific interactions with other biological molecules.

46. (a) $CH_3COOCH_2CH_3$ (b) $CH_3CH_2COOCH_2CH_2CH_3$
    (c) $CH_3CH_2COOCH_3$

48. (a) $CH_3CH_2COOH + CH_3OH$
    (b) $HCOOH + CH_3CH_2OH$ (c) $CH_3COOH + CH_3CH_2OH$

50. Examples of thermoplastics are milk jugs (polyethylene), sunglasses and toys (polystyrene), and CD audio discs (polycarbonates). Thermoplastics soften and flow when heated.

52. (a)

$$\left(\begin{array}{cc} \overset{\displaystyle H}{|} & \overset{\displaystyle H}{|} \\ C & C \\ \underset{\displaystyle CH_2}{|} & \underset{\displaystyle H}{|} \\ \underset{\displaystyle CH_3}{|} & \end{array}\right)_n$$

(b)
$$\left(\begin{array}{cc} \overset{\displaystyle Cl}{|} & \overset{\displaystyle H}{|} \\ C & C \\ \underset{\displaystyle Cl}{|} & \underset{\displaystyle H}{|} \end{array}\right)_n$$

(c)
$$\left(\begin{array}{cc} \overset{\displaystyle H}{|} & \overset{\displaystyle H}{|} \\ C & C \\ \underset{\displaystyle O}{|} & \underset{\displaystyle H}{|} \\ C{=}O & \\ \underset{\displaystyle CH_3}{|} & \end{array}\right)_n$$

54.
$$\left(\begin{array}{cc} \overset{\displaystyle H}{|} & \overset{\displaystyle CH_3}{|} \\ C & C \\ \underset{\displaystyle H}{|} & C{=}O \\ & \underset{\displaystyle O}{|} \\ & \underset{\displaystyle CH_3}{|} \end{array}\right)_n$$

56. Isoprene, 2-methyl-1,3-butadiene; *cis* isomer

58. Carboxylic acid and alcohol

60. Carboxylic acid and amine, nylon

62. One thing is that the protein polymer's monomers are not all alike. Different side chains change the properties of the protein.

66.

68. Major end uses for recycled PET include fiberfill for ski jackets and sleeping bags, carpet fibers, and tennis balls. HDPE is converted into a fiber used for sportswear, insulating wrap for new buildings, and very durable shipping containers.

70. Protein, DNA

72.

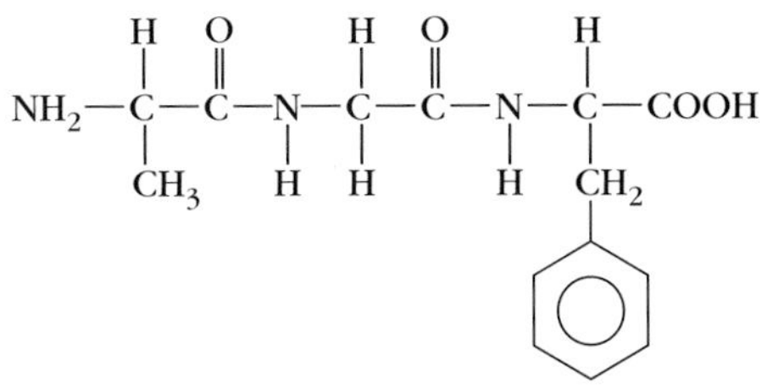

74. (a) A monosaccharide is a molecule composed of one simple sugar molecule, while disaccharides are molecules composed of two simple sugar molecules. (b) Disaccharides have only two simple sugar molecules, whereas polysaccharides have many.

76. Starch and cellulose

78. (a) Glycogen contains glucose linked together with the glycosidic linkages in *cis* positions, and cellulose contains glucose with the glycosidic linkages in *trans* positions. Humans do not have the enzyme required to break the *trans* linkage in cellulose. (b) Cows do have the enzymes for breaking the *trans* linkage of cellulose.

80. $CH_3CH_2CH_2CH_2CH_2CH_2CH_2CH_2CH_2CH_2OH$ is a larger molecule than CH_3CH_2OH. The polar alcohol group will interact well with the water; however, the nonpolar end of the molecule will not. The longer, nonpolar end of decanol will not be miscible in water, lowering the solubility compared with smaller, more polar ethanol.

82. Vulcanized rubber has short chains of sulfur atoms that bond together the polymer chains of natural rubber.

84.

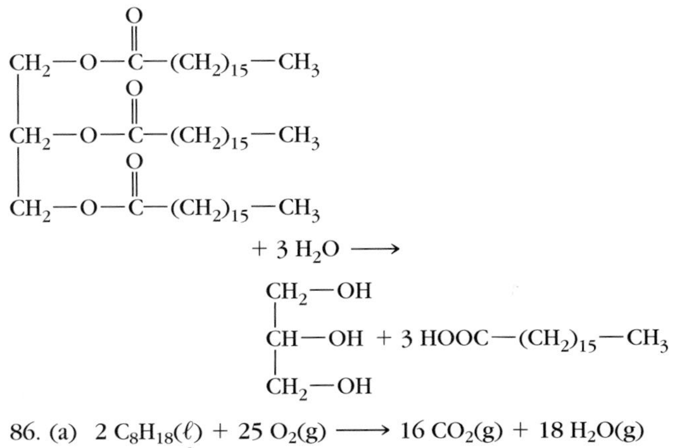

86. (a) $2\ C_8H_{18}(\ell) + 25\ O_2(g) \longrightarrow 16\ CO_2(g) + 18\ H_2O(g)$
(b) 1.4×10^3 L CO_2

88. Glycogen has the glycosidic linkages in *cis* positions, whereas cellulose has glycosidic linkages in *trans* positions.

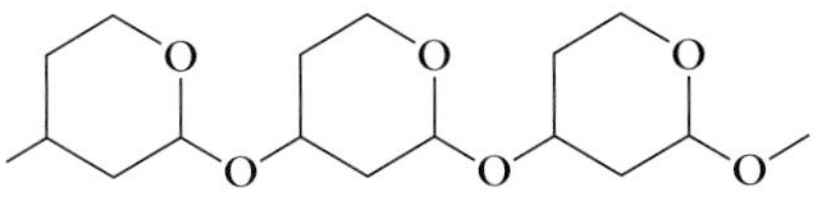

glycogen

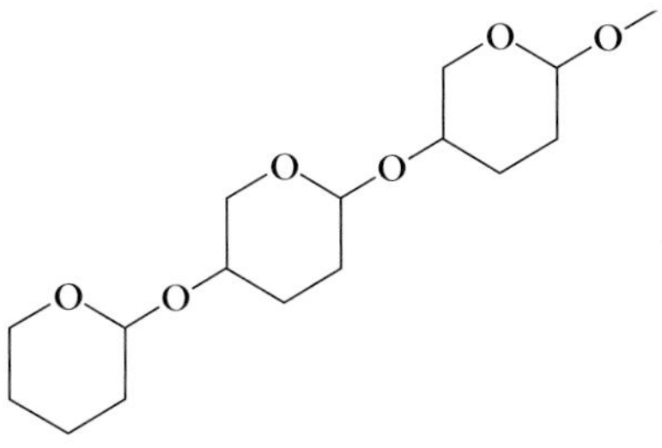

Cellulose

88. Butane in the lighter is under great enough pressure that the vapor pressure of butane at room temperature is less than the pressure inside the lighter. Hence it exists as a liquid.

90.

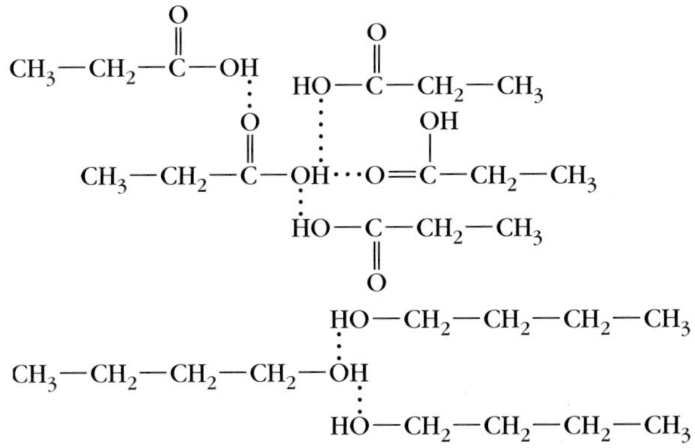

92. $CH_3—C{\equiv}C—H$

94.

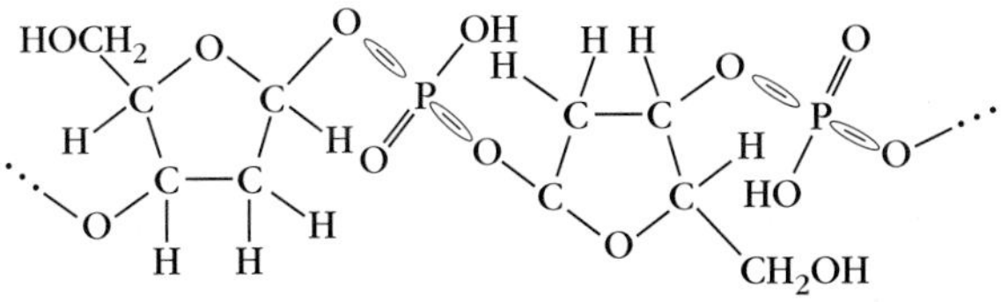

96. Some data that would be needed are: Sources and amounts of CO_2 generated over time to determine additional CO_2; photosynthesis rate of depletion per tree per year; average number of trees per acre; the number of acres of land in Australia that could support trees; the allowable tree density. Other data besides what is given may need to be contemplated, so consider it a challenge to think of other things you would need to know.

Chapter 13

14. (a) 1.3 times faster (b) Grind the metal into dust.

15. (a)

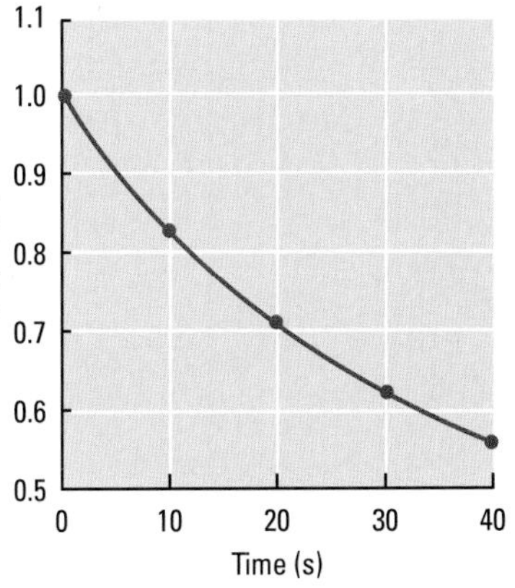

0.0167 M/s, 0.0119 M/s, 0.0089 M/s, 0.0070 M/s. Concentration of reactant is decreasing.

(b) Rate of change of [B] is twice as fast as the rate of change of [A].   (c) 0.0238 M/s

18. (a) The average rate is directly proportional to average $[N_2O_5]$ for each pair.   (b) $k = 0.29$ h^{-1}

22.

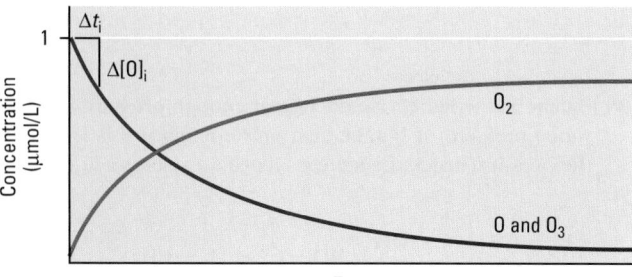

(a) At times close to zero, where the curve still looks linear, determine the change in concentration for a fixed change in time.

(b) The final rate can be determined to be zero. The change in concentration drops to zero after a fixed amount of time. This is seen late in the graph where the two lines become horizontal.

23. (a) Rate increases by a factor of nine.
    (b) Rate will be one fourth as fast.

25. (a) Rate = $k[NO_2]^2$ (b) Rate will be one fourth as fast.
    (c) Rate is unchanged.

27. (a) 9.0×10^{-4} M/h,   (b) 1.8×10^{-3} M/h,
    (c) 3.6×10^{-3} M/h
    (d) If initial concentration of $Pt(NH_3)_2Cl_2$ is high, rate of disappearance of $Pt(NH_3)_2Cl_2$ is high. If initial concentration of $Pt(NH_3)_2Cl_2$ is low, rate of disappearance of $Pt(NH_3)_2Cl_2$ is low. Rate of disappearance of $Pt(NH_3)_2Cl_2$ is directly proportional to $[Pt(NH_3)_2Cl_2]$.
    (e) Rate law shows direct proportionality between rate and $[Pt(NH_3)_2Cl_2]$.
    (f) When the initial $[Pt(NH_3)_2Cl_2]$ is high, rate of appearance of Cl$^-$ is high. When the initial concentration is low, rate of appearance of Cl$^-$ is low. Rate of appearance of Cl$^-$ is directly proportional to $[Pt(NH_3)_2Cl_2]$.

30. (a) Rate = $k[I][II]$ (b) $k = 1.04 \dfrac{L}{mol \cdot s}$

32. (a) Second order   (b) First order   (c) Third order
    (d) Rate = $k[A]^2[B]$ (e) $k = 3.3 \times 10^5 \dfrac{L^2}{mol^2 \cdot s}$
    (f) $\dfrac{\Delta[C]}{\Delta t} = 0.017 \dfrac{mol}{L \cdot s}, \dfrac{\Delta[D]}{\Delta t} = 0.051 \dfrac{mol}{L \cdot s}$

34. (a) NO is second order, O_2 is first order.
    (b) Rate = $k[NO]^2[O_2]$ (c) $25 \dfrac{L^2}{mol^2 \cdot s}$
    (d) $7.8 \times 10^{-4} \dfrac{mol}{L \cdot s}$
    (e) $-\dfrac{\Delta[NO]}{\Delta t} = 2.0 \times 10^{-4} \dfrac{mol}{L \cdot s}$
    $+\dfrac{\Delta[NO_2]}{\Delta t} = 2.0 \times 10^{-4} \dfrac{mol}{L \cdot s}$

36. (a) First order in A, and third order in B
    (b) First order in A, and first order in B

(c) First order in A, and zero order in B
(d) Third order in A, and first order in B

39. (a) Rate = k[phenylacetate]
    (b) First order in phenylacetate
    (c) $k = 1.3$ s^{-1} (d) $0.13 \dfrac{mol}{L \cdot s}$

42. (a) Rate = $k[NO_2][CO]$
    (b) First order in both NO_2 and CO   (c) $4.2 \times 10^8 \dfrac{L}{mol \cdot s}$

43. (a) Rate = $k[H_3O^+][CH_3COCH_3]$, first order in both H_3O^+ and CH_3COCH_3 and zeroth order in Br_2
    (b) $3.9 \times 10^{-3} \dfrac{L}{mol \cdot s}$ (c) $1.9 \times 10^{-5} \dfrac{mol}{L \cdot s}$

45.

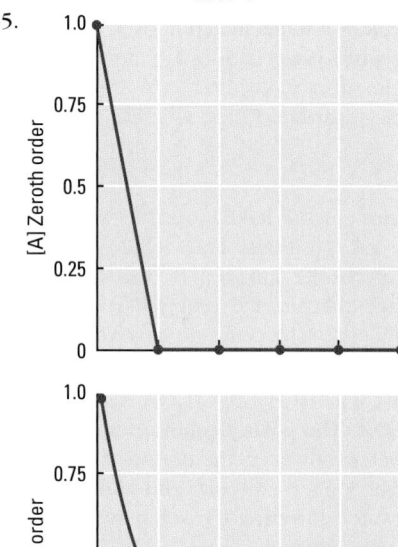

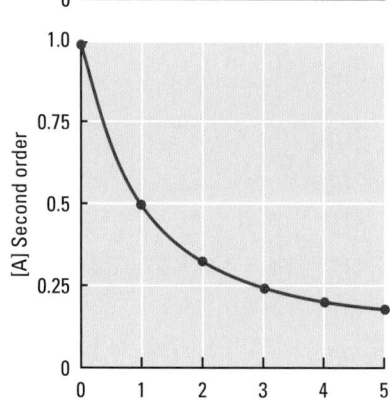

Only the second graph constructed here can be compared with one of the graphs in Figure 13.5—in particular, Figure 13.5(a). They both show a curved-downward functional dependence of [A] versus time, characteristic of first-order reactions.

47. (a) 0.16 mol/L   (b) 90 s   (c) 120 s

50. 260 s $[A]_o = 1.00 \times 10^{-3} \times$ mol/L

52. (a) 2.8×10^3 s (b) 1.4×10^4 s (c) 2.0×10^4 s
55. (a) Not elementary (b) Bimolecular and elementary
    (c) Not elementary (d) Unimolecular and elementary
57. $NO + O_3$; NO is an asymmetric molecule and Cl is a symmetric atom.
60. Ratio = 1.8
62. 10.7 times faster
64. (a) $E_a = 120$ kJ/mol, $A = 1.2 \times 10^{14}$ s^{-1}
    (b) $k = 1.7 \times 10^{-3}$ s^{-1}
66. (a) $E_a = 22.20$ kJ/mol, $A = 6.66 \times 10^7 \dfrac{L^2}{mol^2 \cdot s}$

    (b) $k = 8.39 \times 10^4 \dfrac{L^2}{mol^2 \cdot s}$
68. (a) 3×10^{-20} (b) 4×10^{-16}
    (c) 4×10^{-10} (d) 1.9×10^{-6}
70. (a) $1 \times 10^{-5} \dfrac{mol}{L \cdot s}$ (b) $25 \dfrac{mol}{L \cdot s}$
72. 1.0×10^2 kJ/mol
75. Exothermic
77. (a) 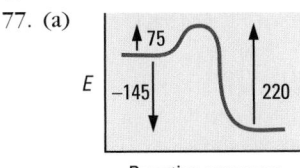 (b)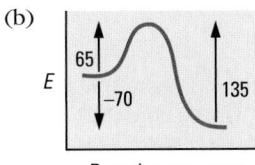

    Reaction progress   Reaction progress

    (c)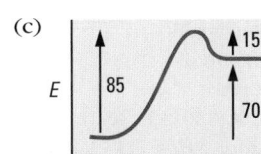

    Reaction progress
79. (a) Reaction (b) (b) Reaction (c)
81. (a) Reaction (c) (b) Reaction (a)
83. (a) Rate = $k[NO][NO_3]$ (b) Rate = $k[O][O_3]$
    (c) Rate = $k[(CH_3)_3CBr]$ (d) Rate = $k[HI]^2$
87. (a) Rate = $k[NO]^2[Cl_2]$
    (b) $NO + NO \rightleftharpoons N_2O_2$ Fast
        $N_2O_2 + Cl_2 \longrightarrow 2\,NOCl$ Slow
    (c) $NO + Cl_2 \longrightarrow NOCl + Cl$ Slow
        $NO + Cl \longrightarrow NOCl$ Fast
89. (a) $CH_3COOCH_3 + H_2O \longrightarrow CH_3COOH + HOCH_3$
    (b) Rate = $k[CH_3COOCH_3][H_3O^+][H_2O]$
    (c) Yes, H_3O^+ (d) $CH_3C(OH)OCH_3^+$, H_2O
92. (a) True (b) False (c) False (d) False
94. (a) Yes, homogeneous (b) No
    (c) Yes, heterogeneous (d) No
96. $\ln\left(\dfrac{k'}{k}\right) = \dfrac{E_a - E_a{}'}{RT}$
97. Approximately 26 times faster
100. 30. times faster
103. Catalysts make possible the production of vital products. Without them, many necessities and luxuries could not be made efficiently, if at all.
106. Rate = $k[H_2][NO]^2$
108. (a) First order in $HCrO_4^-$, first order in H_2O_2, and first order in H_3O^+

(b) Cancel intermediates, H_2CrO_4 and $H_2CrO(O_2)_2$, and add the three reactions.
    (c) Second step
109. $Pt(NH_3)_2Cl_2$ is a reactant, and Cl^- is a product.
110. Most mechanisms do not specify relative rates for the sub-processes. A bicycle gear-changing mechanism and the mechanism of an elevator are similar to a chemical mechanism in that they all describe how something is accomplished (production of products, gear changed, elevator lifted or lowered).
111. Catalysts make possible the efficient production of many modern materials and substances. Without them, the cost of producing many of these products would be much higher (because of higher energy requirements, for example), or we might not be able to make them quickly enough to meet the demand.
112. The catalytic role of the chlorine atom (produced from the light decomposition of chlorofluorocarbons) in the mechanism for the destruction of ozone indicates that even small amounts of CFCs released into the atmosphere pose a serious risk.
113. (a) cannot be correct.
114. Line A represents $[H_2O(g)]$ increase with time; Line B represents $[O_2(g)]$ increase with time; and Line C represents $[H_2O_2(g)]$ decrease with time.
116. (a)
118. Rate = $k[A]^2[B][C]^2$

Chapter 14

9. (a) 0 °C at 1 atm pressure (b) Dynamic
11.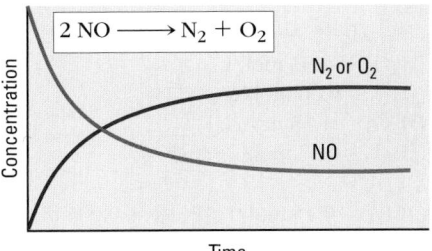

13. (a) $K_c = \dfrac{[H_2O]^2[O_2]}{[H_2O_2]^2}$ (b) $K_c = \dfrac{[PCl_5]}{[PCl_3][Cl_2]}$
    (c) $K_c = [CO]^2$ (d) $K_c = \dfrac{[H_2S]}{[H_2]}$

15. (a) $K_c = [Cl_2]$ (b) $K_c = \dfrac{[Cu^{2+}][Cl^-]^4}{[CuCl_4^{2-}]}$

(c) $K_c = \dfrac{[CO_2][H_2]}{[CO][H_2O]}$ (d) $K_c = \dfrac{[Mn^{2+}][Cl_2]}{[H_3O^+]^4[Cl^-]^2}$

17. (a) $K_c = [H_2O]^2$ (b) $K_c = \dfrac{[HF]^4}{[SiF_4][H_2O]^2}$

(c) $K_c = \dfrac{[HCl]^2}{[H_2O(g)]}$

19. Reaction has high activation energy.

21. (c), tripling reaction, cubes K

23. 0.40

24. (a) $K_P = \dfrac{P_{H_2O}^2 P_{O_2}}{P_{H_2O_2}^2}$ (b) $K_c = \dfrac{P_{PCl_5}}{P_{PCl_3}P_{Cl_2}}$

(c) $K_c = P_{CO}^2$ (d) $K_c = \dfrac{P_{H_2S}}{P_{H_2}}$

26. $K_p = 0.0161$

29. (a) 1.4 (b) 0.50 (c) 100 "If the [reactants] and [products] are equal *and their stoichiometric coefficients are the same,* then the equilibrium constant is always 1.0."

32. $K_p = 1.6$

33. $K_p = 2.0$

35. K_c 0.080

36. $K_c = 75$

38. (a) $K_c = 0.0168$ (b) 0.661 atm (c) 0.514

40. Reaction (b) is product-favored.
Most reactant-favored (c), then (a), then (b).

42. (a) $Ag_2SO_4(s) \rightleftharpoons 2\,Ag^+(aq) + SO_4^{2-}(aq)$

$\qquad\qquad\qquad\qquad\qquad\qquad K = 1.7 \times 10^{-5}$

$Ag_2S(s) \rightleftharpoons 2\,Ag^+(aq) + S^{2-}(aq) \qquad K = 6 \times 10^{-30}$

(b) $Ag_2SO_4(s)$ (c) $Ag_2S(s)$

44. (a)

	Butane $\rightleftharpoons$ 2-methylpropane	
Conc. initial	0.100 mol/L	0.100 mol/L
Change conc.	$-x$	$+x$
Equilibrium conc.	$0.100 - x$	$0.100 + x$

(b) $K_c = \dfrac{\text{[2-methylpropane]}}{\text{[butane]}} = 2.5 = \dfrac{0.100 + x}{0.100 - x}$

$\qquad\qquad\qquad\qquad\qquad\qquad\qquad x = 0.043$

(c) [2-methylpropane] = 0.024 mol/L,
[butane] = 0.010 mol/L

46. 3.36 g $C_6H_{12}(g)$

49. (a) 1.94 mol HI (b) 1.94 mol HI (c) 1.98 mol HI

51. (a) [CO] = [H$_2$O] = 0.33 mol/L,
[CO$_2$] = [H$_2$] = 0.66 mol/L
(b) [CO] = [H$_2$O] = 1.33 mol/L,
[CO$_2$] = [H$_2$] = 0.67 mol/L

53. 1.19%

55. (a) [PCl$_3$] = [Cl$_2$] = 0.15 mol/L, [PCl$_5$] = 0.005 mol/L
(b) [PCl$_3$] = 0.29 mol/L, [Cl$_2$] = 0.14 mol/L,
[PCl$_5$] = 0.01 mol/L

58. (b) because heat is a reactant, and increasing the temperature drives the reaction towards products

60. (a) Left (b) Left (c) Left

62. (a) Right (b) Left (c) Right

63. (a) Decrease
(b)

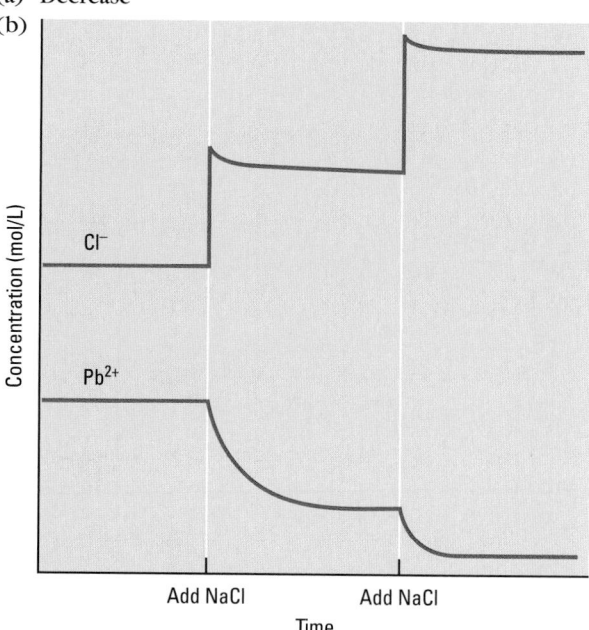

66. (a) (i) Right (ii) Right (iii) Left
(b) (i) Left (ii) No shift (iii) Right
(c) (i) Right (ii) Right (iii) No shift
(d) (i) Right (ii) Right (iii) Left

68. (a) Energy effect (b) Entropy effect (c) Neither

70. (a) Insufficient information is available.
(b) Greater than 1, products favored
(c) Less than 1, reactants favored

73. (a) First reaction: $\Delta H = -296.830$ kJ; second reaction: $\Delta H = -197.78$ kJ; third reaction: $\Delta H = -132.44$ kJ
(b) All three are exothermic; none are endothermic
(c) None of the reactions have entropy increase; the second and third reactions have entropy decrease; the first reaction entropy is about the same.
(d) Low temperature favors all three reactions.

75. $K_c = [H^+][OH^-]$, $K_c = \dfrac{[CH_3COO^-][H^+]}{[CH_3COOH]}$,

$K_c = \dfrac{[NH_3]^2}{[N_2][H_2]^3}$, $K_c = \dfrac{[O_2]^3}{[O_3]^2}$,

$K_c = \dfrac{[N_2O_4]}{[NO_2]^2}$, $K_c = \dfrac{[HCOOH]}{[HCOO^-][H^+]}$, $K_c = \dfrac{1}{[Ag^+][I^-]}$

First reaction:
(a) $(1.0 + x)(1.0 + x) = 1.0 \times 10^{-14}$
(b) Quadratic (c) N/A

Second reaction:
(a) $\dfrac{(1.0 + x)(1.0 + x)}{(1.0 - x)} = 1.8 \times 10^{-5}$
(b) Quadratic (c) N/A

Third reaction:
(a) $\dfrac{(1.0 + 2x)^2}{(1.0 - x)(1.0 - 3x)^3} = 3.5 \times 10^8$
(b) Not quadratic (c) Use approximation techniques

Fourth reaction:
(a) $\dfrac{(1.0 + \frac{3}{2}x)^3}{(1.0 - x)^2} = 7 \times 10^{56}$
(b) Not quadratic (c) Use approximation techniques

Fifth reaction:

(a) $\dfrac{(1.0 + \frac{1}{2}x)}{(1.0 - x)^2} = 1.7 \times 10^2$ (b) Quadratic   (c) N/A

Sixth reaction:

(a) $\dfrac{(1.0 + x)}{(1.0 - x)(1.0 - x)} = 5.6 \times 10^3$

(b) Quadratic   (c) N/A

Seventh reaction:

(a) $\dfrac{1}{(1.0 - x)(1.0 - x)} = 6.7 \times 10^{15}$

(b) Quadratic   (c) N/A

76. (a)

Species	Br_2	Cl_2	F_2	I_2
[E] (mol/L)	0.28	0.057	1.44	0.90

Species	H_2	N_2	O_2	
[E] (mol/L)	1.76×10^{-5}	4×10^{-14}	4.0×10^{-6}	

(b) I_2 (At this temperature, the lowest bond energy is predicted from the reaction that gives the most products.) Compare to Table 8.2; the product production decreases as the bond energy increases: 151 kJ I_2, 156 kJ F_2, 193 kJ Br_2, 242 kJ Cl_2, 436 kJ H_2, 498 kJ O_2, 946 kJ N_2. Lewis structures of I_2, F_2, Br_2, Cl_2, H_2 have a single bond, and more product atoms are produced than O_2 with double bond and N_2 with triple bonds.

78. (a) $K_c = 0.67$ (b) Same result

81. Cases (a) and (b) will have increased [HI] at equilibrium, Case (c) will have decreased [HI]. No significant change in Case (d) will be seen.

82. It is at equilibrium at 600 K. No more experiments are needed.

84. Exothermic

85. (a) (iii) (b) (i)

87. In the warmer sample, the molecules would be moving faster, and more NO_2 molecules would be seen. In the cooler sample, the molecules would be moving somewhat more slowly, and fewer NO_2 molecules would be seen. In both samples, the molecules are moving very fast. The average speed of gas molecules is commonly hundreds of miles per hour. In both samples, one would see a dynamic equilibrium, with some N_2O_4 molecules decomposing and some NO_2 molecules reacting with each other at equal rates.

90. Diagrams (b), (c), and (d)

92. (a) Diagram (d) (b) Diagram (b) (c) Diagram (c)

94. Dynamic equilibria introduce D^+ ions in place of H^+ ions for the acidic hydrogen.

$H_2O(\ell) \rightleftharpoons H^+(aq) + OH^-(aq)$

$D_2O(\ell) \rightleftharpoons D^+(aq) + OH^-(aq)$

$C_6H_5COOH(s) \rightleftharpoons C_6H_5COOH(aq) \rightleftharpoons$

$\qquad\qquad\qquad\qquad C_6H_5COO^-(aq) + H^+(aq)$

$C_6H_5COO^-(aq) + D^+(aq) \rightleftharpoons$

$\qquad\qquad\qquad C_6H_5COOD(aq) \rightleftharpoons C_6H_5COOD(s)$

96. Dynamic equilibria representing the decomposition of the dimer $N_2O_4(g)$ produce $NO_2(g)$ and $*NO_2(g)$, which will oc-casionally recombine into the mixed dimer, $O_2N^*{-}NO_2$ (g):

$O_2N{-}NO_2(g) \rightleftharpoons 2\ NO_2(g)$

$O_2^*{-}{^*}NO_2(g) \rightleftharpoons 2\ ^*NO_2(g)$

$O_2N^*{-}NO_2(g) \rightleftharpoons {^*}NO_2(g) + NO_2(g)$

Chapter 15

24. If the solid interacts with the solvent using similar (or stronger) intermolecular forces, it will dissolve readily. If the solute interacts with the solvent using different intermolecular forces than those experienced in the solvent, it will be almost insoluble. For example, consider dissolving an ionic solid in water and in oil. The interactions between the ions in the solid and water are very strong, since ions would be attracted to the highly polar water molecule; hence, the solid would have a high solubility. However, the ions in the solid interact with each other much more strongly than with the London dispersion forces experienced between the nonpolar hydrocarbons in the oil; hence, the solid would have a low solubility.

26. The dissolving process was endothermic, so the temperature dropped as more solute was added. The solubility of the solid at the lower temperature is lower, so some of the solid did not dissolve. As the solution warmed up, however, the solubility increased, dissolving the solid. The solution, saturated at the lower temperature, is no longer saturated at the current temperature.

28. When an organic acid has a large (nonpolar) section, it interacts primarily using London dispersion intermolecular forces. Since water interacts via hydrogen-bonding, it interacts with itself rather than with the acid. Hence, the solubility of the large organic acids decreases.

30. The positive H end of the very polar water molecule interacts with the negative ions. The negative end of the very polar water molecule interacts with the positive ions.

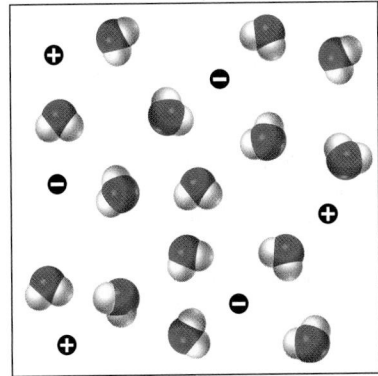

32. 1×10^{-3} M

34. 0.00732% by weight

36. $1\ \text{ppb} = \dfrac{1\ \text{g part}}{10^9\ \text{g whole}} \times \dfrac{1\ \mu\text{g part}}{10^{-6}\ \text{g part}} \times \dfrac{1000\ \text{g whole}}{1\ \text{kg whole}}$

$\qquad = \dfrac{1\ \mu\text{g part}}{1\ \text{kg whole}}$

38. 90. g

40. 1.6×10^{-6} g Pb

42. 96% H_2SO_4

44. 59 g

46. 1.2×10^{-7} M

48. 100.26 °C

50. (a) < (d) < (b) < (c)

52. $T_f = -1.65$ °C, $T_b = 100.46$ °C

54. $X_{H_2O} = 0.79999$, 7.1×10^2 g sucrose

56. 1.9×10^2 g/mol

58. 1.8×10^2 g/mol, $C_{14}H_{10}$

60. (a) 2.5×10^3 g (b) 104.2 °C

62. 30. atm

64. 1.2×10^4 g/mol

66. 5×10^{-4} g As

68. The lime-soda process relies on the precipitation of insoluble compounds to remove the hard water ions. The ion-exchange process relies on the high charge of the hard water ions to attract them to an ion-exchange resin, thereby removing them from the water.

70. Soap forms insoluble curds in the presence of hard water ions.

72. Risk: There is a small risk of byproduct formation (THMs) that may be linked with liver cancer. Benefit: Chlorine kills bacteria that pose a great risk to human health. Benefit outweighs the risk.

74. Fish take in oxygen by extracting it from the water. Plants take in carbon dioxide by extracting it from the air. The concentrations of these gases in calm water drop, unless they are replenished. The concentration of dissolved gases in the water is replenished by bubbling air through the water in the aquarium.

76. 4%

78. Water in the cells of the wood flowed out, since the osmotic pressure inside the cells was less than that of the sea water that the wood saturated.

80. 28%

82. 0.982 *m*, 10.2%

84. (a) = (c) < (d) < (b)

86. 1.77 g

88. Empirical and molecular formulas are both $C_{18}H_{24}Cr$.

90. (a) $AlCl_3(aq) + H_3PO_4(aq) \longrightarrow AlPO_4(s) + 3\ HCl(aq)$
    (b) 157 g $AlPO_4$ (c) 0.205 M Al^{3+} and 0.205 M PO_4^{3-}

92. (a)

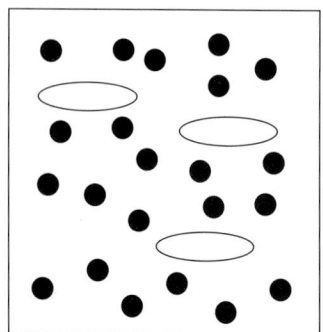

(b)

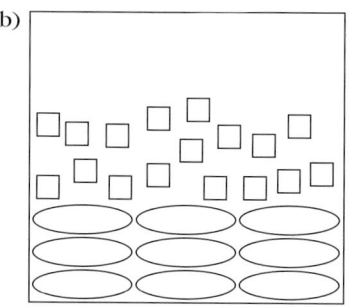

94. (a) saturated   (b) unsaturated   (c) super saturated
    (d) unsaturated

96.

Compound	Mass of compound	Mass of water	Mass fraction of solute
Lye	**75.0 g**	125 g	0.375
Glycerol	33 g	200. g	**0.14**
Acetylene	0.0015	**1.7×10^2 g**	0.000009

Compound	Weight percent of solute	Conc. of solute
Lye	37.5%	**3.75×10^5 ppm**
Glycerol	14%	**1.4×10^5 ppm**
Acetylene	0.0009%	**9 ppm**

98. Molecules slow down. The reduced motion prevents them from randomly moving as they had in the liquid state. As a result, the intermolecular forces between one molecule and the next begin to organize the solute molecules into a crystal. The presence of a nonvolatile solute disrupts the formation of the crystal. Its size and shape will be different from that of the solute. Intermolecular forces between the solute and solvent are also different from those of solvent molecules with each other. To form the crystalline solid, the solute has to be excluded. If the ice in an iceberg is in regular crystalline form, the water will be pure. So, melting an iceberg will produce relatively pure water.

100. (a) Sea water contains more dissolved solutes than fresh water. The presence of a solute lowers the freezing point. That means a lower temperature is required to freeze the sea water than to freeze fresh water.
    (b) Salt added to a mixture of ice and water will lower the freezing point of the water. If the ice cream is mixed at a lower temperature, it will freeze faster.

Chapter 16

12. (a) $HCO_3^- + H_2O \rightleftharpoons CO_3^{2-} + H_3O^+$
    (b) $HCl + H_2O \longrightarrow Cl^- + H_3O^+$

(c) $CH_3COOH + H_2O \rightleftharpoons CH_3COO^- + H_3O^+$
(d) $HCN + H_2O \rightleftharpoons CN^- + H_3O^+$

14. (a) $HIO + H_2O \rightleftharpoons IO^- + H_3O^+$
(b) $CH_3(CH_2)_4COOH + H_2O \rightleftharpoons$
$$CH_3(CH_2)_4COO^- + H_3O^+$$
(c) $CO_2COOH + H_2O \rightleftharpoons CO_2CO_2^{2-} + H_3O^+$
(d) $CH_3NH_3^+ + H_2O \rightleftharpoons CH_3NH_2 + H_3O^+$

16. (a) $HSO_4^- + H_2O \rightleftharpoons H_2SO_4 + OH^-$
(b) $CH_3NH_2 + H_2O \rightleftharpoons CH_3NH_3^+ + OH^-$
(c) $I^- + H_2O \rightleftharpoons HI + OH^-$
(d) $H_2PO_4^- + H_2O \rightleftharpoons H_3PO_4 + OH^-$

20. H_2SO_4 (b) HNO_3 (c) $HClO_4$ (d) $HClO_3$ (e) H_2SO_4

22. (a) I^-, iodide, conjugate base
(b) HNO_3, nitric acid, conjugate acid
(c) HCO_3^-, hydrogen carbonate ion, conjugate acid
(d) HCO_3^-, hydrogen carbonate ion, conjugate base
(e) SO_4^{2-}, sulfate ion, conjugate base; H_2SO_4, sulfuric acid, conjugate acid
(f) HSO_3^-, hydrogen sulfite ion, conjugate acid.

26. (a) reactant conjugate acid is H_2O, reactant conjugate base is HS^-, product conjugate acid is H_2S, product conjugate base is OH^-
(b) reactant conjugate acid is NH_4^+, reactant conjugate base is S^{2-}, product conjugate acid is HS^-, product conjugate base is NH_3
(c) reactant conjugate acid is HSO_4^-, reactant conjugate base is HCO_3^-, product conjugate acid is H_2CO_3, product conjugate base is SO_4^{2-}
(d) reactant conjugate acid is NH_3, reactant conjugate base is NH_2^-, product conjugate base is NH_2^-, product conjugate acid is NH_3

28. (a) reactant conjugate acid is CH_3COOH, reactant conjugate base is CN^-, product conjugate acid is HCN, product conjugate base is CH_3COO^-
(b) reactant conjugate acid is H_2O, reactant conjugate base is O^{2-}, product conjugate acid is OH^-, product conjugate base is OH^-
(c) reactant conjugate acid is H_2O, reactant conjugate base is HCO_2^-, product conjugate acid is $HCOOH$, product conjugate base is OH^-

30. (a) $CO_3^{2-} + H_3O^+ \rightleftharpoons HCO_3^- + H_2O$
$HCO_3^- + H_3O^+ \rightleftharpoons H_2CO_3 + H_2O$
(b) $H_3AsO_4 + H_2O \rightleftharpoons H_2AsO_4^- + H_3O^+$
$H_2AsO_4^- + H_2O \rightleftharpoons HAsO_4^{2-} + H_3O^+$
$HAsO_4^{2-} + H_2O \rightleftharpoons AsO_3^{3-} + H_3O^+$
(c) $NH_2CH_2COO^- + H_3O^+ \rightleftharpoons NH_2CH_2COOH + H_2O$
$NH_2CH_2COOH + H_3O^+ \rightleftharpoons NH_3CH_2COOH^+ + H_2O$

32. 3×10^{-11} M, basic
34. 3.6×10^{-3} M, acidic
36. pH = 12.40, pOH = 1.60
38. pOH = 12.51

42.

	pH	$[H_3O^+]$ (M)	$[OH^-]$ (M)	Acidic or basic
(a)	6.21	6.1×10^{-7}	6.1×10^{-8}	Acidic
(b)	5.34	4.5×10^{-6}	2.2×10^{-9}	Acidic
(c)	4.67	2.1×10^{-5}	4.7×10^{-10}	Acidic
(d)	1.60	2.5×10^{-2}	4.0×10^{-13}	Acidic
(e)	9.12	7.6×10^{-10}	1.3×10^{-5}	Basic

44. (a) 100 times more acidic (b) 10^4 times more basic
(c) 16 times more basic (d) 2×10^4 times more acidic

46. (a) $F^-(aq) + H_2O(\ell) \rightleftharpoons HF(aq) + OH^-(aq)$
$$K = \frac{[HF][OH^-]}{[F^-]}$$
(b) $NH_3(aq) + H_2O(\ell) \rightleftharpoons NH_4^+(aq) + OH^-(aq)$
$$K = \frac{[NH_4^+][OH^-]}{[NH_3]}$$
(c) $H_2CO_3(aq) + H_2O(\ell) \rightleftharpoons HCO_3^-(aq) + H_3O^+(aq)$
$$K = \frac{[HCO_3^-][H_3O^+]}{[H_2CO_3]}$$
(d) $H_3PO_4(aq) + H_2O(\ell) \rightleftharpoons H_2PO_4^-(aq) + H_3O^+(aq)$
$$K = \frac{[H_2PO_4^-][H_3O^+]}{[H_3PO_4]}$$
(e) $CH_3COO^-(aq) + H_2O(\ell) \rightleftharpoons$
$$CH_3COOH(aq) + OH^-(aq)$$
$$K = \frac{[CH_3COOH][OH^-]}{[CH_3COO^-]}$$
(f) $S^{2-}(aq) + H_2O(\ell) \rightleftharpoons HS^-(aq) + OH^-(aq)$
$$K = \frac{[HS^-][OH^-]}{[S^{2-}]}$$

48. (a) NH_3 (b) K_2S (c) CH_3COONa (d) KCN
50. (a) 2.19 (b) 5.13 (c) 8.07 (d) 9.02 (e) 13.30
(f) 1.54 (g) 9.69 (h) 7.00
52. $K_a = 1.6 \times 10^{-5}$
56. 8.85
58. 3.28
60. (a) CN^- (b) HS^- (c) $H_2(g)$
62. (a) $NH_3 + H_2PO_4^-$, reactant-favored
(b) $CH_3COO^- + H_2O$, product-favored
(c) $H_3PO_4 + SO_4^{2-}$, product-favored
(d) $CH_3COO^- + HF$, reactant-favored
64. (a) pH < 7 (b) pH > 7 (c) pH = 7
66. (a) pH > 7 (b) pH > 7 (c) pH > 7
68. All of them
70. $Na_2CO_3 + 2\,CH_3(CH_2)_{16}COOH \longrightarrow$
$$2\,CH_3(CH_2)_{16}COONa + H_2O + CO_2$$
72. Lemon juice
74. Dishwasher detergent is very basic and should not be used to wash anything by hand, including a car. If it gets into the engine area, it can also dissolve automobile grease and oil, which could prevent the engine from running correctly.
76. All three are Lewis bases; CO_2 is a Lewis acid.
78. Cr^{3+} and SO_3 are Lewis acids. CH_3NH_2 is a Lewis base.
80. (a) I_2 is a Lewis acid and I^- is a Lewis base.
(b) SO_2 is a Lewis acid and BF_3 is a Lewis base.
(c) Au is a Lewis acid and CN^- is a Lewis base.
(d) CO_2 is a Lewis acid and H_2O is a Lewis base.
82. T-shaped

It functions as a Lewis base.

Square planar

84. (a) Weak acid (b) Strong base (c) Strong acid
(d) Weak base (e) Strong base (f) Amphiprotic

86. (a) Less than 7 (b) Equal to 7 (c) Greater than 7
88. 2.82
90. (a) Increase (b) Stays the same (c) stays the same
92. 10 °C pH = 7.27, 25 °C pH = 7.00, and 50 °C pH = 6.631. At 10 °C, at 25 °C, and at 50 °C the solutions are neutral, since $[H_3O^+] = [OH^-]$
94. (a) $H_3O^+ = Cl^- \gg OH^-$
    (b) $Na^+ = ClO_4^- \gg H_3O^+ = OH^-$
    (c) $HNO_2 > H_3O^+ = NO_2^- \gg OH^-$
    (d) $Na^+ \cong ClO^- > OH^- = HClO \gg H_3O^+$
    (e) $NH_4^+ \cong Cl^- > H_3O^+ = NH_3 \gg OH^-$
    (f) $Na^+ = OH^- \gg H_3O^+$
96. Conjugates must differ by just one H^+.
98. Arrhenius theory: Electron pair on solvent or Arrhenius base (Lewis base) forms a bond with the hydrogen ion (Lewis acid). Brønsted theory: Proton (Lewis acid) is bonded to a Brønsted base using an electron pair (Lewis base).
100. Strongest HM > HQ > HZ weakest; $K_{a,HZ} = 1 \times 10^{-5}$, $K_{a,HQ} = 1 \times 10^{-3}$, $K_{a,HM} = 1 \times 10^{-1}$ or larger

Chapter 17

16. (c)
18. (a) pH = 2.12 (b) pH = 7.21 (c) pH = 12.44
20. (a) Lactic acid and lactate ion
    (b) Acetic acid and acetate ion (c) HClO and ClO⁻
    (d) HCO_3^- and CO_3^{2-}
22. 4.8 g
24. pH = 8.62
26. pH = 9.56, pH = 9.51
28. (a) No (b) Yes (c) No (d) No
30. (a) pH = 4.8 (b) pH = 8.52 (c) pH = 11.95
32. (a) pH = 5.03 (b) pH = 5.00 (c) pH = 4.1
34. At the equivalence point, the solution contains the conjugate base of the weak acid. Weaker acids have stronger conjugate bases. Stronger bases have higher pH, so the titration of a weaker acid will have a more basic equivalence point.
36. (a) Bromothymol blue (b) Phenolphthalein
    (c) Methyl red (d) Bromothymol blue, suitable-pH color change
38. 0.0253 M HCl
40. 93.6%
42. (a) 29.2 mL (b) 0.600 L (c) 1.23 L (d) 2.7 mL
44. (a) pH = 3.90 (b) pH = 8.45 (c) pH = 12.155
46. (a) pH = 0.824 (b) pH = 1.30 (c) pH = 3.82
    (d) pH = 7.00 (e) pH = 10.18 (f) pH = 12.477
    (See figure at top of next column.)
48. NO_2: $2 NO_2(g) + H_2O(g) \longrightarrow HNO_3(g) + HNO_2(g)$
    SO_3: $2 SO_2(g) + O_2(g) \longrightarrow 2 SO_3(g)$
    $SO_3(g) + H_2O(g) \longrightarrow H_2SO_4(g)$
50. $CaCO_3(s) + 2 H^+(aq) \longrightarrow Ca^{2+}(aq) + CO_2(g) + H_2O(\ell)$
52. (a) $FeCO_3(s) \rightleftharpoons Fe^{2+}(aq) + CO_3^{2-}(aq)$
    $$K_{sp} = [Fe^{2+}][CO_3^{2-}]$$
    (b) $Ag_2SO_4(s) \rightleftharpoons 2 Ag^+(aq) + SO_4^{2-}(aq)$
    $$K_{sp} = [Ag^+]^2[SO_4^{2-}]$$
    (c) $Ca_3(PO_4)_2(s) \rightleftharpoons 3 Ca^{2+}(aq) + 2 PO_4^{3-}(aq)$
    $$K_{sp} = [Ca^{2+}]^3[PO_4^{3-}]^2$$
    (d) $Mn(OH)_2(s) \rightleftharpoons Mn^{2+}(aq) + 2 OH^-(aq)$
    $$K_{sp} = [Mn^{2+}][OH^-]^2$$

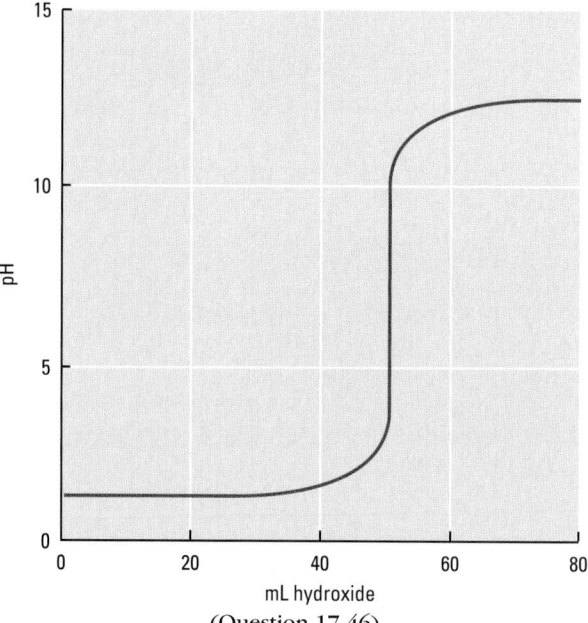

(Question 17.46)

54. $K_{sp} = 2.22 \times 10^{-4}$
56. $K_{sp} = 2.2 \times 10^{-12}$
58. $K_{sp} = 1.70 \times 10^{-5}$
60. 1×10^{-7} M Zn^{2+}
62. 3.1×10^{-5} mol/L
64. $K_{sp} = 1 \times 10^{-11}$ (b) $[OH^-]$ must be 0.008 M or higher.
66. 4.5×10^{-9} M
68. pH = 9.0
70. (a) $Ag^+(aq) + 2 CN^-(aq) \rightleftharpoons [Ag(CN)_2]^-(aq)$
    $$K_f = \frac{[[Ag(CN)_2]^-]}{[Ag^+][CN^-]^2}$$
    (b) $Cd^{2+}(aq) + 4 NH_3(aq) \rightleftharpoons [Cd(NH_3)_4]^{2+}(aq)$
    $$K_f = \frac{[[Cd(NH_3)_4]^{2+}]}{[Cd^{2+}][NH_3]^4}$$
72. 0.0078 M or more
74. (a) $Zn(OH)_2(s) + 2 H^+(aq) \longrightarrow Zn^{2+}(aq) + 2 H_2O(\ell)$
    $Zn(OH)_2(s) + 2 OH^-(aq) \longrightarrow [Zn(OH)_4]^{2-}(aq)$
    (b) $Sb(OH)_3(s) + 2 H^+(aq) \longrightarrow Sb^{3+}(aq) + 3 H_2O(\ell)$
    $Sb(OH)_3(s) + OH^-(aq) \longrightarrow [Sb(OH)_4]^-(aq)$
76. (a) H_2O, CH_3COO^-, Na^+, CH_3COOH, H^+, OH^-
    (b) pH = 4.95 (c) pH = 5.05
    (d) $CH_3COOH(aq) + OH^-(aq) \longrightarrow$
    $$CH_3COO^-(aq) + H_2O(\ell)$$
78. Ratio = 1.62
80. 0.020 M
82. (a) pH = 2.78 (b) pH = 5.40
84. $K_a = 2.4 \times 10^{-6}$
86. No change
90. The tiny amount of base (CH_3COO^-) present is insufficient to prevent the pH from changing dramatically if a strong acid is introduced into the solution.
92. $K_a = 2.3 \times 10^{-4}$
94. Blood pH decreases; acidosis
96. Sample A: $NaHCO_3$; Sample B: NaOH; Sample C: mixture of NaOH and Na_2CO_3; Sample D: Na_2CO_3

Chapter 18

16. (a) $H_2O(\ell) \longrightarrow 2 H_2(g) + O_2(g)$, reactant-favored
    (b) $C_8H_{18}(\ell) \longrightarrow C_8H_{18}(g)$, product-favored
    (c) $C_{12}H_{22}O_{11}(s) \longrightarrow C_{12}H_{22}O_{11}(aq)$, product-favored
18. (a) $\frac{1}{2}$ (b) $\frac{1}{2}$ (c) 50 of each
20. Six have two in each. Two have none in one. Most probably has two in each. Highest entropy has two in each. (See figure below.)
21. (a) Negative (b) Positive (c) Positive
23. (a) Increases (b) Increases (c) Increases
25. (a) Item 2 (b) Item 2 (c) Item 2
27. (a) $NaCl(s)$ (b) $P_4(g)$ (c) $NH_4NO_3(aq)$
29. (a) $Ga(\ell)$ (b) $AsH_3(g)$ (c) $NaF(s)$
31. (a) Negative (b) Positive (c) Negative (d) Positive
33. (a) Negative (b) Negative (c) Positive
35. $112\ J\ K^{-1}\ mol^{-1}$
37. (a) 2.63 J/K (b) 10.0 J/K (c) 2.45 J/K
39. (a) +113.0 J/K (b) +38.17 J/K
41. (a) −120.64 J/K (b) 156.9 J/K
    (c) −198.76 J/K (d) 160.6 J/K
46. (a) 58.78 J/K (b) −53.29 J/K (c) −173.93 J/K
    Adding more hydrogen makes the $\Delta S°$ more negative.
47. −247.7 J/K. Cannot tell without $\Delta H°$ also, since that is needed to calculate $\Delta G°$. $\Delta H° = -88.13$ kJ and $\Delta G° = -14.3$ kJ, so it is product-favored.
49. Product-favored at low temperatures. The exothermicity is sufficient to favor products if the temperature is low enough to overcome the decrease in entropy.
51. Exothermic reactions with an increase in disorder, exhibited by a larger number of gas phase products than gas phase reactants, never need help from the surroundings to favor products.
52. $CO_2(g)$ product means increased entropy, as in 51 above.

54. Entropy increase is insufficient to drive this highly endothermic reaction to form products without assistance from the surroundings at this temperature.
56. Product-favored. $\Delta H°$ is negative. Assuming the $S°$ of C_8H_{16} and C_8H_{18} are approximately the same value, $\Delta S°$ is negative, but not large enough to give $\Delta G°$ a different sign; hence, $\Delta G°$ is negative also.
57. (a) $\Delta S° = -37.52$ J/K, $\Delta H° = 851.5$ kJ, product-favored at low T.
    (b) $\Delta S° = -121.7$ J/K, $\Delta H° = 66.36$ kJ, never product-favored.
59. (a) $\Delta S_{univ} = -310.37$ J/K (b) -1559.83 kJ
    (c) Yes. Ethane is used as a fuel; hence, we might expect that its combustion reaction is product-favored.
62. $\Delta G° = \Delta H° - T\Delta S°$. Here, $\Delta H°$ is negative and $\Delta S°$ is positive, so $\Delta G° = -|\Delta H°| - |T\Delta S°| = -(|\Delta H°| + |T\Delta S°|) < 0$
65. $\Delta G° = 462.28$ kJ. Uncatalyzed, it would not be a good way to make Si.
68. (a) $\Delta H° = -174.47$ kJ, $S°_{C_2H_2} = 201\ J\ K^{-1}\ mol^{-1}$
    (b) $\Delta H° = 197.78$ kJ, $S°_{SO_3} = 257\ J\ K^{-1}\ mol^{-1}$
    (c) $\Delta H° = -905.47$ kJ, $S°_{NO} = 211\ J\ K^{-1}\ mol^{-1}$
69. (a) 385.7 K (b) 835.1 K
72. (a) $\Delta H° = 178.32$ kJ, $\Delta S° = 160.6$ J/K, $\Delta G° = 130.5$ kJ
    (b) Reactant-favored
    (c) No. It is only product-favored at high temperatures.
    (d) 1110. K
74. $\Delta G°_{f,Ca(OH)_2} = -867.6$ kJ
76. (a) 331 K (b) 340. K
78. (a) False (b) False (c) True (d) True (e) True
79. (a) $K = 4 \times 10^{-34}$ (b) $K = 5 \times 10^{-31}$
81. (a) $\Delta G° = -100.97$ kJ, product-favored.
    (b) $K = 5 \times 10^{17}$. When $\Delta G°$ is negative, K is larger than 1.

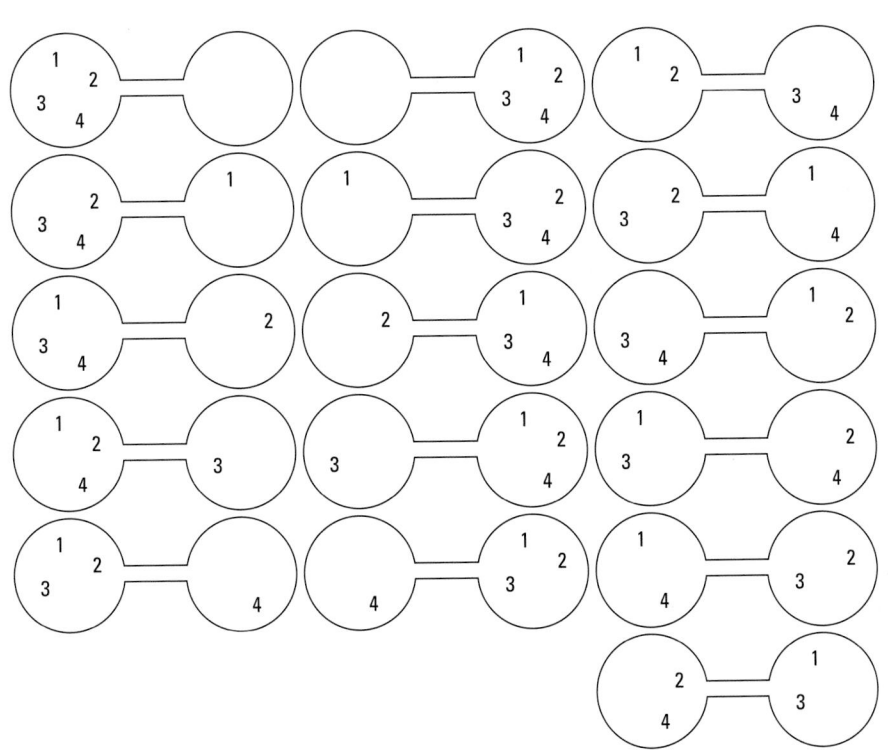

(Question 18.20)

83. (a) $K = 2 \times 10^{12}$ (b) $K = 2.0 \times 10^{-2}$
86. (a) $\Delta G° = -110.$ kJ (b) $\Delta G° = 8.55$ kJ
    (c) $\Delta G° = -33.8$ kJ
88. (a) $K_p = 1.49 \times 10^7$ (b) Product-favored
    (c) $K_c = 3.64 \times 10^8$
90. (a) can be used; (b) and (c) require work to be done.
93. (a) $7\,C(s) + 6\,H_2O(g) \longrightarrow 2\,C_2H_6(g) + 3\,CO_2(g)$
    $5\,C(s) + 4\,H_2O(g) \longrightarrow C_3H_8(g) + 2\,CO_2(g)$
    $3\,C(s) + 4\,H_2O(g) \longrightarrow 2\,CH_3OH(\ell) + CO_2(g)$
    (b) For C_2H_6, $\Delta H° = 101.02$ kJ, $\Delta G° = 122.72$ kJ,
    $\Delta S° = -72.71$ J/K
    For C_3H_8, $\Delta H° = 76.5$ kJ, $\Delta G° = 102.08$ kJ,
    $\Delta S° = -86.6$ J/K
    For CH_3OH, $\Delta H° = 96.44$ kJ, $\Delta G° = 187.39$ kJ,
    $\Delta S° = -305.2$ J/K
    None of these are feasible. $\Delta G°$ is positive. In addition, $\Delta H°$ is positive and $\Delta S°$ is negative, suggesting that there is no temperature at which the products would be favored.
95. (a) $2\,CuO(s) \longrightarrow 2\,Cu(s) + O_2(g)$
    $CuO(s) + C(s) \longrightarrow Cu(s) + CO(g)$
    $\Delta G° = -7.5$ kJ
    (b) $2\,Ag_2O(s) \longrightarrow 4\,Ag(s) + O_2(g)$
    $Ag_2O(s) + C(s) \longrightarrow 2\,As(s) + CO(g)$
    $\Delta G° = -125.97$ kJ
    (c) $2\,HgO(s) \longrightarrow 2\,Hg(s) + O_2(g)$
    $HgO(s) + C(s) \longrightarrow Hg(s) + CO(g)$
    $\Delta G° = -78.63$ kJ
    (d) $2\,MgO(s) \longrightarrow 2\,Mg(s) + O_2(g)$
    $MgO(s) + C(s) \longrightarrow Mg(s) + CO(g)$
    $\Delta G° = 432.26$ kJ
    (e) $2\,PbO(s) \longrightarrow 2\,Pb(s) + O_2(g)$
    $PbO(s) + C(s) \longrightarrow Pb(s) + CO(g)$
    $\Delta G° = 50.72$ kJ
    Cu, Ag, and Hg can be obtained by this method.
99. Five O—H bonds, seven C—O bonds, seven C—H bonds, five C—C bonds, and six O=O bonds are broken. Twelve C=O bonds and twelve O—H bonds are formed. $\Delta H° \cong -2873$ kJ. Interactive forces in condensed phases (solid glucose and liquid water) are being neglected in this calculation.
101. (a) 6.46 mol of ATP per mol of glucose (b) $\Delta G° = -106$ kJ (c) Product-favored
103. Combustion of coal, petroleum, and natural gas are the most common sources used to supply free energy. We also use solar and nuclear energy as well as the kinetic energy of wind and water.
105. (a) $\Delta G° = -86.5$ kJ, product-favored
    (b) $\Delta G° = -873.1$ kJ (c) No (d) Yes
107. Kinetic stability relates to the difficulties in the conversion of the reactants to products. Even very stable products may be difficult to form from some reactants. The kinetically stable materials usually require extensive reorganization of bonds, which likely involve very high-energy activated complexes and intermediates.
110. For $CH_4(g)$, $\Delta G° = -817.90$ kJ. For $C_6H_6(\ell)$, $\Delta G° = -3202.0$ kJ. For $CH_3OH(\ell)$, $\Delta G° = -702.35$ kJ.
    Organic compounds are complex molecular systems that require significant rearrangement of atoms and bonds to undergo combustion. This makes them likely candidates

for being kinetically stable. (See the answer to Question 107 from this chapter.)
111. (a) 5.5×10^{18} J/yr (b) 1.5×10^{16} J/day
    (c) 1.7×10^{11} J/s (d) 1.7×10^{11} W
    (e) 6×10^2 W/person
112. (a) 1.6×10^{11} kg (b) I must generate 1.7×10^{11} W, when a sprinter manages to muster up 900 W during a sprint. That means I must generate energy a billion times faster than a sprinter does while sprinting. I think I will not uphold this contract.
114. (a) Reaction 2 (b) Reactions 1, 4, and 5 (c) Reaction 2 (d) Reactions 2 and 3 (e) None of them
116. (a) $\Delta G° = 141.73$ kJ (b) No (c) Yes
    (d) $K_p = 9 \times 10^{-5}$
118. $\Delta G° = -RT \ln K = \Delta H° - T\Delta S°$
    $$\ln K = -\frac{\Delta H°}{RT} + \frac{\Delta S°}{R}$$
    If ln K is plotted against $1/T$, the slope would be $-\Delta H°/R$, and the y-intercept would be $\Delta S°/R$. The linear graph shows that $\Delta S°$ and $\Delta H°$ are independent of temperature.
120. (a) $\Delta G° = 31.8$ kJ (b) $K_p = P_{Hg(\ell)}$
    (c) $K_p = 3 \times 10^{-6}$ (d) $T = 450$ K
122. A scrambled egg is a very disordered state for an egg. The second law of thermodynamics says that the more disordered state is the more probable state. Putting the delicate tissues and fluids back where they were before the scrambling occurred would take a great deal of energy. Humpty Dumpty is a fictional character who was also an egg. He fell off a wall. A very probable result of that fall is for an egg to become scrambled. The story goes on to say that all the energy of the king's horses and men was not sufficient to put Humpty together again.
125. Many of the oxides have negative enthalpies of reaction, which means their oxidations are exothermic. These are probably product-favored reactions.
127. (a) (ii) (b) (i) (c) (ii)
129. NaCl, in an orderly crystal structure, and pure water, with only O—H hydrogen-bonding interactions in the liquid state, are far more ordered than the dispersed hydrated sodium and chloride ions interacting with the water molecules.
131. $\Delta G = 0$ means products are favored; however, the equilibrium state will always have some reactants present also. To get all the reactants to go away requires the removal of the products from the reactants, so the reaction continues forward.

Chapter 19

6. (a) Al oxidized; Cl in Cl_2 reduced; Cl_2 is the oxidizing agent; Al is the reducing agent; reactants' oxidation numbers, Al: 0, Cl: 0; products' oxidation numbers, Al: +3, Cl: −1.
   (b) Fe^{2+} oxidized; Mn in MnO_4^- reduced; MnO_4^- is the oxidizing agent; Fe^{2+} is the reducing agent; reactants' oxidation numbers, H: +1, O: −2, Mn: +7, Fe: +2; products' oxidation numbers, H: +1, O: −2, Mn: +2, Fe: +3.

(c) Fe and S in FeS are both oxidized; N in NO_3^- reduced; NO_3^- is the oxidizing agent; FeS is the reducing agent; reactants' oxidation numbers, Fe: $+2$, S: -2, N: $+5$, O: -2, H: $+1$; products' oxidation numbers, Fe: $+3$, S: $+6$, N: $+2$, O: -2, H: $+1$.

10. (a) $Zn(s) \longrightarrow Zn^{2+}(aq) + 2\,e^-$
(b) $2\,H_3O^+(aq) + 2\,e^- \longrightarrow 2\,H_2O(\ell) + H_2(g)$
(c) $Sn^{4+}(aq) + 2\,e^- \longrightarrow Sn^{2+}(aq)$
(d) $Cl_2(g) + 2\,e^- \longrightarrow 2\,Cl^-(aq)$
(e) $6\,H_2O(\ell) + SO_2(g) \longrightarrow$
$$SO_4^{2-}(aq) + 4\,H_3O^+(aq) + 2\,e^-$$

12. (a) $Al(s) \longrightarrow Al^{3+}(aq) + 3\,e^-$
$Cl_2(g) + 2\,e^- \longrightarrow 2\,Cl^-(aq)$
(b) $Fe^{2+}(aq) \longrightarrow Fe^{3+}(aq) + e^-$
$MnO_4^-(aq) + 8\,H_3O^+(aq) + 5\,e^- \longrightarrow$
$$Mn^{2+}(aq) + 12\,H_2O(\ell)$$
(c) $FeS(s) + 12\,H_2O(\ell) \longrightarrow$
$$Fe^{3+}(aq) + SO_4^{2-}(aq) + 8\,H_3O^+(aq) + 9\,e^-$$
$NO_3^-(aq) + 4\,H_3O^+(aq) + 3\,e^- \longrightarrow NO(g) + 6\,H_2O(\ell)$

14. (a) $3\,CO(g) + O_3(g) \longrightarrow 3\,CO_2(g)$. O_3 is the oxidizing agent; CO is the reducing agent.
(b) $H_2(g) + Cl_2(g) \longrightarrow 2\,HCl(g)$. Cl_2 is the oxidizing agent; H_2 is the reducing agent.
(c) $H_2O_2(aq) + Ti^{2+}(aq) + 2\,H_3O^+(aq) \longrightarrow$
$$4\,H_2O(\ell) + Ti^{4+}(aq)$$
H_2O_2 is the oxidizing agent; Ti^{2+} is the reducing agent.
(d) $2\,MnO_4^-(aq) + 6\,Cl^-(aq) + 8\,H_3O^+(aq) \longrightarrow$
$$2\,MnO_2(s) + 3\,Cl_2(g) + 12\,H_2O(\ell)$$
MnO_4^- is the oxidizing agent; Cl^- is the reducing agent.
(e) $2\,FeS_2(s) + 11\,O_2(g) \longrightarrow 2\,Fe_2O_3(s) + 8\,SO_3(g)$
O_2 is the oxidizing agent; FeS_2 is the reducing agent.
(f) $O_3(g) + NO(g) \longrightarrow O_2(g) + NO_2(g)$. O_3 is the oxidizing agent; NO is the reducing agent.
(g) $Zn(Hg)(amalgam) + HgO(s) \longrightarrow ZnO(s) + Hg(\ell)$
HgO is the oxidizing agent; Zn(Hg) is the reducing agent.

16. The generation of electricity occurs when electrons are transmitted through a wire from the metal to the cation. Here, the transfer of electrons would occur directly from the metal to the cation and the electrons would not flow through any wire.

18. Conventionally, in chemistry, they are written as reduction reactions.

20. (a) $Zn(s) + Pb^{2+}(aq) \longrightarrow Zn^{2+}(aq) + Pb(s)$
(b) Oxidation of zinc occurs at the anode. The reduction of lead occurs at the cathode. The anode is metallic zinc. The cathode is metallic lead.
(c)

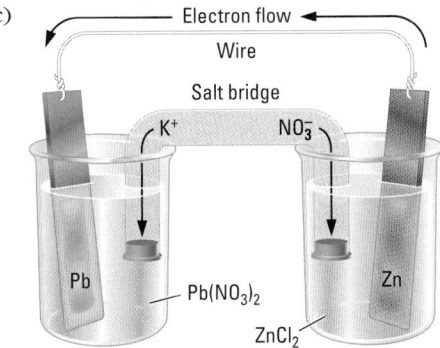

22. 7.5×10^3 C
24. (a) $Cu(s) \longrightarrow Cu^{2+}(aq) + 2\,e^-$
$Ag^+(aq) + e^- \longrightarrow Ag(s)$
(b) The copper half-reaction is oxidation, and it occurs in the anode compartment. The silver half-reaction is reduction, and it occurs in the cathode compartment.

26. Li is the strongest reducing agent, and Li^+ is the weakest oxidizing agent. F_2 is the strongest oxidizing agent, and F^- is the weakest reducing agent.

28. Worst oxidizing agent (d) $<$ (c) $<$ (a) $<$ (b) best oxidizing agent

30. (a) 2.91 V (b) -0.03 V (c) 0.65 V (d) 1.16 V
32. (a) Al^{3+} (b) Ce^{4+} (c) Al (d) Ce^{3+} (e) Yes (f) No
(g) Mercury(I) ion, silver ion, and cerium(IV) ion
(h) Hg, Sn, Ni, Al

34. $E°$ is greater than zero and $\Delta G°$ is less than zero.
36. (a) 1.55 V (b) -1196 kJ, 1.55 V
38. -409 kJ
40. 4×10^{-19}, 105 kJ
42. 3×10^9, -54 kJ
44. (a) 1.20 V (b) 1.16 V (c) 0.03 M
48. (a) $Ni^{2+}(aq) + Cd(s) \longrightarrow Ni(s) + Cd^{2+}(aq)$
(b) Cd is oxidized; Ni^{2+} is reduced; Ni^{2+} is the oxidizing agent; Cd is the reducing agent.
(c) Metallic Cd is the anode and metallic Ni is the cathode.
(d) 0.15 V
(e) Electrons flow from the Cd electrode to the Ni electrode.
(f) Toward the anode compartment.

50. A fuel cell has a continuous supply of reactants and will be useable for as long as the reactants are supplied. A battery contains all the reactants of the reaction. Once the reactants are gone, the battery is no longer useable.

52. (a) N_2H_4 half-reaction occurs at the anode and O_2 half-reaction occurs at the cathode.
(b) $N_2H_4(g) + O_2(g) \longrightarrow N_2(g) + 2H_2O(\ell)$
(c) 7.5 g N_2H_4 (d) 7.5 g O_2

54. Au^{3+}, Hg^{2+}, Ag^+, Hg_2^{2+}, Fe^{3+}, Sn^{4+}, Sn^{2+}, Ni^{2+}, Cd^{2+}, Fe^{2+}, Zn^{2+}

56. H_2 and Br_2 are produced. After the reaction is complete, the solution contains Na^+, OH^-, a small amount of dissolved Br_2 (though it has low solubility in water), and a very small amount of H_3O^+. H_2 is formed at the cathode. Br_2 is formed at the anode.

58. 0.16 g Ag
60. 5.93 g Cu
62. 3×10^5 g Al
64. 1.9×10^2 g Pb
66. 0.10 g Zn
68. (a) 0.04 g Li (b) 0.6 g Pb
70. (a) 1×10^6 g HF (b) 1.7×10^3 kWh
72. One is Au; there are others.
74. Galvanized iron has a coating of a more active metal. That metal corrodes instead of the iron.
76. 0.00689 g Cu, 0.00195 g Al
78. Worst reducing agent B $<$ D $<$ A $<$ C best reducing agent
80. (a) B is oxidized; A^{2+} is reduced.
(b) A^{2+} is the oxidizing agent; B is the reducing agent.

(c) B is the anode and A is the cathode.

(d) $A^{2+} + 2 e^- \longrightarrow A$

$\qquad B \longrightarrow B^{2+} + 2 e$

(e) A gains mass. (f) Electrons flow from B to A.

(g) K^+ ions will migrate toward the A^{2+} solution.

Chapter 20

11. (a) $^{238}_{92}U$ (b) $^{32}_{15}P$ (c) $^{10}_{5}B$ (d) $^{0}_{-1}e$ (e) $^{15}_{7}N$

13. (a) $^{28}_{12}Mg \longrightarrow ^{28}_{13}Al + ^{0}_{-1}e$

(b) $^{238}_{92}U + ^{12}_{6}C \longrightarrow 4\,^{1}_{0}n + ^{246}_{98}Cf$

(c) $^{2}_{1}H + ^{3}_{2}He \longrightarrow ^{4}_{2}He + ^{1}_{1}H$

(d) $^{38}_{19}K \longrightarrow ^{38}_{18}Al + ^{0}_{+1}e$

(e) $^{175}_{78}Pt \longrightarrow ^{4}_{2}He + ^{171}_{76}Os$

15. $^{231}_{90}Th, ^{231}_{91}Pa, ^{227}_{89}Ac, ^{227}_{90}Th, ^{223}_{88}Ra$

17. (a) $^{19}_{10}Ne \longrightarrow ^{19}_{9}Al + ^{0}_{+1}e$

(b) $^{230}_{90}Th \longrightarrow ^{0}_{-1}e + ^{230}_{91}Th$

(c) $^{82}_{35}Br \longrightarrow ^{0}_{-1}e + ^{82}_{36}Kr$

(d) $^{212}_{84}Po \longrightarrow ^{4}_{2}He + ^{208}_{82}Pb$

19. Binding energy per nucleon of ^{10}B is 1.038×10^{-15} kJ/nucleon. Binding energy per nucleon of ^{11}B is 1.111×10^{-15} kJ/nucleon. ^{11}B is more stable than ^{10}B because its binding energy is larger.

21. 5 mg

23. (a) $^{131}_{53}I \longrightarrow ^{131}_{54}Xe + ^{0}_{-1}e$

(b) 1.56 mg

25. 4.0 y

27. 35 d

29. 2.58×10^3 y

31. $^{239}_{94}Pu + 2\,^{1}_{0}n \longrightarrow ^{0}_{-1}e + ^{241}_{95}Am$

33. $^{12}_{6}C$

35. Cadmium rods (a neutron absorber to control the rate of the fission reaction), uranium rods (source of fuel, since uranium is a reactant in the nuclear equation), and water (used for cooling by removing excess heat and in steam/water cycle for the production of turning torque for the generator).

37. 1.6×10^3 tons of coal

39. Rad = measure of the amount of radiation absorbed. Rem includes a quality factor that better describes the biological impact of a radiation dose. The unit rem would be more appropriate when talking about effects of an atomic bomb on humans. The unit gray (Gy) is 100 rad.

41. Since most elements have some proportion of unstable isotopes that decay and we are composed of these elements (e.g., ^{14}C), our bodies emit radiation particles.

43. The gamma ray is a high-energy photon. Its interaction with matter is most likely just going to be imparting large quantities of energy. The alpha and beta particles are charged particles of matter that could interact and possibly react with the matter composing the food.

45. 0.13 L

47. (a) $^{0}_{-1}e$ (b) $^{4}_{2}He$ (c) $^{87}_{35}Br$ (d) $^{216}_{84}Po$ (e) $^{68}_{31}Ga$

49. 1.5 mg

51. 3.6×10^9 y

53. (a) $^{247}_{99}Es$ (b) $^{16}_{8}O$ (c) $^{4}_{2}He$ (d) $^{12}_{6}C$ (e) $^{10}_{5}B$

55. Alpha and beta radiation decay particles are charged ($^{4}_{2}He^{2+}$ and $_{-1}^{0}e^-$), so they are better able to interact with and ionize tissues, disrupting the function of the cancer cells. Gamma radiation, like x-rays, goes through soft tissue

without much being absorbed. This is less likely to interfere with the cancerous cells.

57. ^{20}Ne is stable. The ^{17}Ne is likely to decay by positron emission, to increase the ratio of neutrons to protons. The ^{23}Ne is likely to decay by beta emission to decrease the ratio of neutrons to protons.

59. A nuclear reaction occurred, making products. Therefore, some of the lost mass is found in the decay particles, if the decay is alpha or beta decay, and almost all of the rest is found in the element produced by the reaction.

Chapter 21

20. (a) Cathode (b) Chlorine gas

(c) 2.03×10^9 C (d) 1.6×10^3 kJ/mol

22. 1.1×10^3 kwh

24. 67.1 g Al

26. Answers in **bold.**

Formula	Name	Oxidation state of phosphorus
P_4	Phosphorus	0
$(NH_4)_2HPO_4$	**Ammonium hydrogen phosphate**	**+5**
H_3PO_4	Phosphoric acid	+5
P_4O_{10}	Tetraphosphorus decaoxide	+5
$Ca_3(PO_4)_2$	**Calcium phosphate**	**+5**
$Ca(H_2PO_4)_2$	Calcium dihydrogen phosphate	**+5**

28. 1.7×10^7 tons

30.

32. 962 K

34. Raw materials: sulfur, water, oxygen, and catalyst (Pt or V_2O_5)

$S_8(s) + 8\,O_2(g) \longrightarrow 8\,SO_2(g)$

$2\,SO_2(g) + O_2(g) \longrightarrow 2\,SO_3(g)$

$SO_3(g) + H_2SO_4(\ell) \longrightarrow H_2S_2O_7(\ell)$

$H_2S_2O_7(\ell) + H_2O(\ell) \longrightarrow 2\,H_2SO_4(aq)$

36.

38. $K_c = 2.7$

40. $K = 5.3 \times 10^5$. Putting sea water in the presence of $Ca(OH)_2$ will cause a precipitate of $Mg(OH)_2$ to form. The solid can be isolated after it settles.

42. 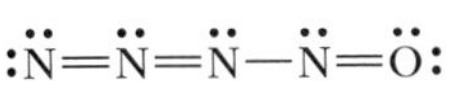 :N̈=N̈=N̈—N̈=Ö:

44. (a)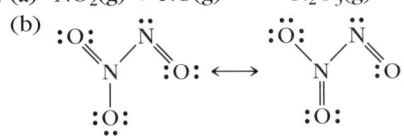

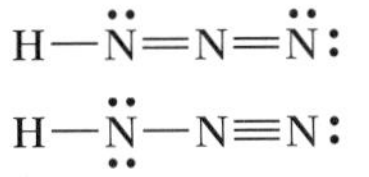

(b) +140. kJ with single and triple bond; +410. kJ with two double bonds.

46. (a) $NO_2(g) + NO(g) \longrightarrow N_2O_3(g)$

(b)

:Ö: N̈   :Ö: N̈
    \ / ..   \ / ..
     N   :O: ⟷   N   :O:
    | ‖
   :Ö: :Ö:

(c) O—N—O angle is 120°, N—N—O angle is a little less than 120°.

48. $S_2(g)$

50. (a) 6×10^2 kJ   (b) 3 V

Chapter 22

14. Ag: $[Kr]4d^{10}5s^1$ Au: $[Xe]4f^{14}5d^{10}6s^2$

16. Cr^{2+} and Cr^{3+}

18. $Cr_2O_7^{2-}$ (in acid) $> Cr^{3+} > Cr^{2+}$. E°_{red}, respectively, is $+1.33$ V > -0.74 V > -0.91 V, so the more positive the oxidation state, the better the oxidizing agent.

20. (a) $Fe_2O_3(s) + 3 CO(g) \longrightarrow 2 Fe(s) + 3 CO_2(g)$
(b) $Fe(s) + 2 H_3O^+(aq) \longrightarrow$
$$Fe^{2+}(aq) + 2 H_2O(\ell) + H_2(g)$$
or $2 Fe(s) + 6 H_3O^+(aq) \longrightarrow$
$$2 Fe^{3+}(aq) + 6 H_2O(\ell) + 3 H_2(g)$$

22. $6 H_3O^+(aq) + 3 NO_3^-(aq) + Fe(s) \longrightarrow$
$$Fe^{3+}(aq) + 3 NO_2(g) + 9 H_2O(\ell)$$

24. (a) $[Cr(NH_3)_2(H_2O)_3(OH)]^{2+}$ (b) The counter ion is an anion.

26. (a) $C_2O_4^{2-}$ with 2− charge and Cl^- with 1− charge.
(b) +3 charge   (c) $[Co(NH_3)_4Cl_2]^+$, with +1 charge

28. For $Na_3[IrCl_6]$: (a) Six Cl^- (b) Ir with +3 charge
(c) $[IrCl_6]^{3-}$ with 3− charge   (d) Na^+
For $[Mo(CO)_4Br_2]$: (a) Four CO and two Br^-
(b) Mo with 2+ charge   (c) $[Mo(CO)_4Br_2]$ (d) None

30. (a) Four   (b) Four

32. (a) $[Pt(NH_3)_2Br_2]$ (b) $[Pt(en)(NO_2)_2]$
(c) $[PtBrCl(NH_3)_2]$

34. (a) Four   (b) Four   (c) Six   (d) Six

36. (a) Monodentate   (b) Tetradentate   (c) Tridentate
(d) Monodentate

38. For example, $FeCl_3$

40. (a) Four chloride ligands coordinately-covalently bonded to a Mn^{2+} ion to make a complex anion with a 2− charge, called tetrachloromanganate(II).
(b) Three potassium counter ions combine with a complex anion composed of three oxalate anion ligands coordinately-covalently bonded to Fe^{3+} to form a neutral salt called potassium trioxalatoferrate(III).
(c) Two ammonia ligands (NH_3) and two cyanide ligands (CN^-) are coordinately-covalently bonded to a Pt^{2+} metal ion, forming a neutral compound called diamminedicyanoplatinum(II).

(d) Five water ligands and one hydroxide ligand are coordinately-covalently bonded to Fe^{3+} to form a complex cation with a 2+ charge called pentaquahydroxoiron(III).
(e) Hexacoordinated manganese(II) ion has two bidentate ethylenediamine ligands and two chloride ligands to form a neutral compound called diethylenediamminedichloromanganese(II).

42.

(a)   (b)
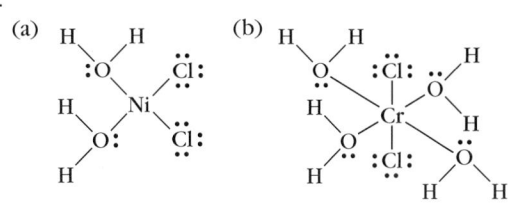

44.
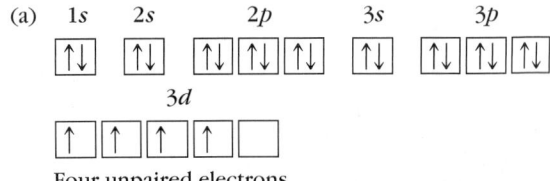

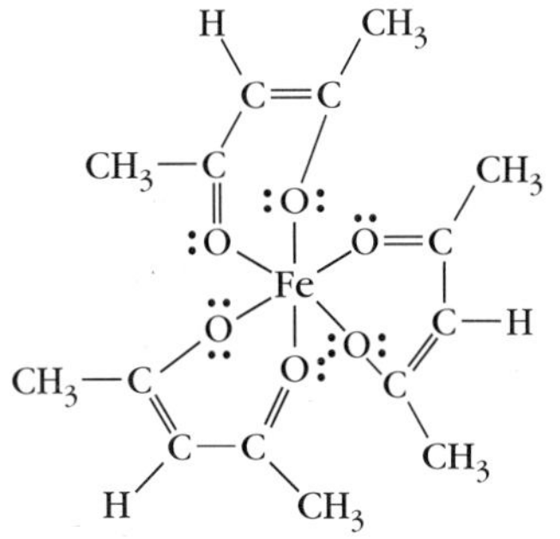

46. (a) and (b)

48. (a), (c), and (d)

50. (a) $[Ar]3d^4$ (b) $[Ar]3d^{10}$ (c) $[Ar]3d^7$ (d) $[Ar]3d^3$

52.
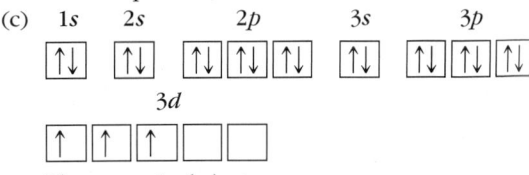

54. 14.8 g

56. 0.40 tons SO_2

58. (a) Four   (b) Six   (c) Four   (d) Six

60.

$$\left[\begin{array}{c} H_3N \overset{\textstyle :\overset{..}{\underset{..}{Cl}}:}{\underset{\textstyle :\overset{..}{\underset{..}{Cl}}:}{\underset{\displaystyle Co}{\big|}}} \begin{array}{c} NH_3 \\ NH_3 \end{array} \end{array}\right]^+ \; \left[\;:\overset{..}{\underset{..}{Br}}:\;\right]^-$$

$$\left[\begin{array}{c} H_3N \overset{\textstyle :\overset{..}{\underset{..}{Cl}}:}{\underset{\textstyle :\overset{..}{\underset{..}{Br}}:}{\underset{\displaystyle Co}{\big|}}} \begin{array}{c} NH_3 \\ NH_3 \end{array} \end{array}\right]^+ \; \left[\;:\overset{..}{\underset{..}{Cl}}:\;\right]^-$$

$$\left[\begin{array}{c} H_3N \overset{\textstyle NH_3}{\underset{\textstyle :\overset{..}{\underset{..}{Cl}}:}{\underset{\displaystyle Co}{}}} \begin{array}{c} NH_3 \\ \overset{..}{\underset{..}{Cl}}: \end{array} \end{array}\right]^+ \; \left[\;:\overset{..}{\underset{..}{Br}}:\;\right]^-$$

$$\left[\begin{array}{c} H_3N \overset{\textstyle NH_3}{\underset{\textstyle :\overset{..}{\underset{..}{Cl}}:}{\underset{\displaystyle Co}{}}} \begin{array}{c} NH_3 \\ \overset{..}{\underset{..}{Br}}: \end{array} \end{array}\right]^+ \; \left[\;:\overset{..}{\underset{..}{Cl}}:\;\right]^-$$

62. (a) False. The coordination number is <u>six</u>.
   (b) False. Cu^+ has <u>no</u> unpaired electrons.
   (c) False. The net charge is $3+$.

64. (a) $2+$

   (b)

$$:\overset{..}{\underset{..}{O}}: $$

(structure showing Pt bonded to two NH_3 groups and to an oxalate ligand with $C=O$ groups and $:\overset{..}{O}:$ atoms)

66. $Fe(s) + 2\,Fe^{3+}(aq) \longrightarrow 3\,Fe^{2+}(aq)$

68. Four

70. M^{2+}

72.

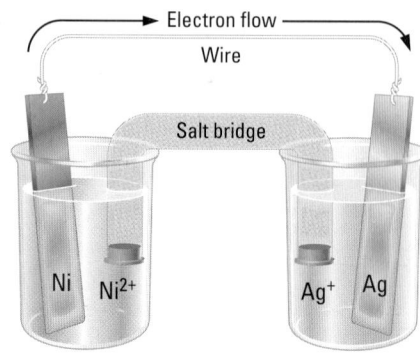

74. 15.8%

76. (a) $-1.38\ \text{V}$ (b) $1 \times 10^{-20}\ \text{M}\ Ag^+$

78. (a) No ions   (b) 2.00 mol ions   (c) 4.00 mol ions
   (d) 3.00 mol ions

80. (a) $[Co(NH_3)_6]Cl_3$
   (b) $[Co(NH_3)_6]Cl_3 \longrightarrow [Co(NH_3)_6]^{3+}(aq) + 3\ Cl^-(aq)$

82.

$$\left[\begin{array}{c} H_3N \overset{}{\underset{H_3N}{\underset{\displaystyle Pt}{}}} \begin{array}{c} NH_3 \\ NH_3 \end{array} \end{array}\right]^{2+} \; \left[\begin{array}{c} :\overset{..}{\underset{..}{Cl}} \overset{}{\underset{:\overset{..}{\underset{..}{Cl}}}{\underset{\displaystyle Pt}{}}} \begin{array}{c} \overset{..}{\underset{..}{Cl}}: \\ \overset{..}{\underset{..}{Cl}}: \end{array} \end{array}\right]^{2-}$$

GLOSSARY

absolute temperature scale (Kelvin temperature scale)   A temperature scale on which the zero is the lowest possible temperature and the degree is the same size as the Celsius degree.

absorb   To draw a substance into the bulk of a liquid or a solid (compare with adsorb).

achiral   A compound whose molecule is superimposable on its mirror image.

acid ionization constant (K_a)   The equilibrium constant for the reaction of a weak acid with water to produce hydronium ions and the conjugate base of the weak acid.

acid   A compound that ionizes in water to give hydronium ions, H_3O^+.

acid rain   Rain (or other precipitation) with a pH below about 5.6 (the pH of unpolluted rain water).

acid ionization constant expression   Mathematical expression in which the product of the equilibrium concentrations of hydronium ion and conjugate base is divided by the equilibrium concentration of the un-ionized conjugate acid.

acidic solution   A solution in which the concentration of hydronium ion is greater than the concentration of hydroxide ion.

actinides   The elements after actinium in the seventh period in which the $5f$ subshell is being filled.

activated complex   A molecular structure corresponding to the maximum of a plot of energy versus reaction progress; also known as the transition state.

activation energy (E_a)   The potential energy difference between reactants and activated complex; the minimum kinetic energy that reactant molecules must have to be converted to product molecules.

active site   The part of an enzyme molecule that binds the substrate to help it to react.

activity (A)   A measure of the rate of nuclear decay, given as disintegrations per unit time.

actual yield   The experimental quantity of product obtained from a chemical reaction.

addition polymer   A polymer made when monomer molecules join directly with one another, with no other products formed in the reaction.

adsorb   To attract and hold a substance on a surface (compare with absorb).

aerosols   Small particles (1 nm to about 10,000 nm in diameter) that remain suspended indefinitely in air.

air pollutant   A substance that degrades air quality.

alcohol   An organic compound containing a hydroxyl group (—OH) covalently bonded to a saturated carbon atom.

aldehyde   An organic compound characterized by a carbonyl group in which the carbon atom is bonded to a hydrogen atom; a molecule containing the —CHO functional group.

alkali metals   The Group 1A elements in the periodic table.

alkaline earth metals   The elements in Group 2A of the periodic table.

alkane   Any of a class of hydrocarbons characterized by the presence of only single carbon-carbon bonds.

alkene   Any of a class of hydrocarbons characterized by the presence of a carbon-carbon double bond.

alkyl group   A fragment of an alkane structure that results from the removal of a hydrogen atom from the alkane.

alkyne   Any of a class of hydrocarbons characterized by the presence of a carbon-carbon triple bond.

allotropes   Different forms of the same element that exist in the same physical state under the same conditions of temperature and pressure.

alpha (α) amino acids   Organic molecules containing a carboxyl group, an R group on the alpha carbon, and an amine group; building block monomers of proteins.

alpha carbon   The carbon adjacent to the acid group (—COOH) in an amino acid.

alpha (α) particles   Positively charged ($+2$) particles ejected at high speeds from certain radioactive nuclei; the nuclei of helium atoms.

alpha radiation   Radiation composed of alpha particles (helium nuclei).

amide   An organic compound characterized by the presence of a carbonyl group in which the carbon atom is bonded to a nitrogen atom (—CONH$_2$, —CONHR, —CONR$_2$); the product of the reaction of an amine with a carboxylic acid.

amide linkage   A linkage consisting of a carbonyl group bonded to a nitrogen atom that connects monomers in a polymer.

amine   An organic compound containing an —NH$_2$, —NHR, or —NR$_3$ functional group.

amorphous solid   A solid whose constituent nanoscale particles have no long-range order.

ampere   The unit of electrical current defined as the flow of one coulomb of charge per second.

amphoteric   Refers to a substance that can act as either an acid or a base.

anion   An ion with a negative electrical charge.

anode   The electrode of an electrochemical cell at which oxidation occurs.

anodic inhibition   The prevention of oxidation of an active metal by painting it, coating it with grease or oil, or allowing a thin film of metal oxide to form.

aqueous solution   A solution in which water is the solvent.

aromatic compound   Any of a class of hydrocarbons characterized by the presence of a benzene ring or related structure.

Arrhenius equation   Mathematical relation that gives the temperature dependence of the reaction rate constant; $k = Ae^{-E_a/RT}$.

atom   The smallest particle of an element that can be involved in chemical combination with another element.

atom economy   The fraction of atoms of starting materials incorporated into the desired final product in a chemical reaction.

atomic mass units (amu)   The unit of a scale of relative atomic masses of the elements; 1 amu = 1/12 the mass of a six-proton, six-neutron carbon atom.

atomic number   The number of protons in the nucleus of an atom of an element.

atomic radius   One-half the distance between the nuclei centers of two like atoms in a molecule.

atomic structure   The identity and arrangement of subatomic particles in an atom.

atomic weight   The average mass of an atom in a representative sample of atoms of an element.

autoionization   The equilibrium reaction in which water molecules react with each other to form hydronium ions and hydroxide ions.

average reaction rate   A reaction rate calculated from a change in concentration divided by a change in time.

Avogadro's law   The volume of a gas, at a given temperature and pressure, is directly proportional to the amount of gas.

Avogadro's number   The number of particles in a mole of any substance (6.022×10^{23}).

axial position(s)   Positions above and below the equatorial plane in a triangular bipyramidal structure.

background radiation   Radiation from natural and synthetic radioactive sources to which all members of a population are exposed.

balanced chemical equation A chemical equation that shows equal numbers of atoms of each kind in the products and the reactants.

bar A pressure unit equal to 100,000 Pa.

barometer An atmospheric pressure measuring device.

basal metabolic rate The energy required to maintain a body that is awake, at rest, and not digesting or metabolizing food.

base A compound that dissociates, ionizes, or reacts with water to give a hydroxide ion.

base ionization constant (K_b) The equilibrium constant for the reaction of a weak base with water to produce hydroxide ions and the conjugate acid of the weak base.

base ionization constant expression Mathematical expression in which the product of the equilibrium concentrations of hydroxide ion and conjugate acid is divided by the equilibrium concentration of the conjugate base.

basic solution A solution in which the concentration of hydroxide ion is greater than the concentration of hydronium ion.

battery (voltaic cell) An electrochemical cell (or group of voltaic cells) in which a product-favored oxidation-reduction reaction is used to produce an electric current.

becquerel A unit of radioactivity equal to 1 nuclear disintegration per second.

beta particles Electrons ejected from certain radioactive nuclei.

beta radiation Radiation composed of electrons.

bidentate ligand A ligand that has two atoms with lone pairs that can form coordinate covalent bonds to the same metal ion.

bimolecular reaction An elementary reaction in which two particles must collide for products to be formed.

binary molecular compound A molecular compound whose molecules contain atoms of only two elements.

binding energy The energy required to separate all nucleons in an atomic nucleus.

binding energy per nucleon The energy per nucleon required to separate all nucleons in an atomic nucleus.

biodegradable Capable of being decomposed by biological means, especially by bacterial action.

boiling The process whereby a liquid vaporizes when its vapor pressure equals atmospheric pressure.

boiling point The temperature at which the equilibrium vapor pressure of a liquid equals the external pressure on the liquid.

boiling-point elevation A colligative property; the difference between the normal boiling point of a pure solvent and the higher boiling point of a solution in which a nonvolatile solute is dissolved in that solvent.

bond Attractive force between two atoms holding them together, for example, as part of a molecule.

bond angle The angle between two bonds originating from the same atom in a molecule or polyatomic ion.

bond enthalpy (bond energy) The change in enthalpy when a mole of chemical bonds of a given type is broken, separating the bonded atoms; the atoms and molecules must be in the gas phase.

bond length The distance between the nuclei of two bonded atoms.

bonding pair A pair of valence electrons that are shared between two atoms.

boundary surface A surface within which there is a specified probability (often 90%) that an electron will be found.

Boyle's law The volume of a confined ideal gas varies inversely with the applied pressure, at constant temperature and amount of gas.

Brønsted-Lowry acid A hydrogen ion donor.

Brønsted-Lowry acid-base reaction A reaction in which an acid donates a hydrogen ion and a base accepts the hydrogen ion.

Brønsted-Lowry base A hydrogen ion acceptor.

buckyball Buckminsterfullerene; an allotrope of carbon consisting of molecules in which 60 carbon atoms are arranged in a cage-like structure consisting of five-membered rings linked to six-membered rings.

buffer See **buffer solution.**

buffer capacity The quantity of acid or base a buffer can accommodate without a significant pH change (more than one pH unit).

buffer solution A solution that resists changes in pH when limited amounts of acids or bases are added; it consists of a weak acid and a salt of its conjugate base, or a weak base and a salt of its conjugate acid.

caloric value The energy of complete combustion of a stated size sample of a food, usually reported in Calories, which are equal to kilocalories.

calorie (cal) (See also **kilocalorie.**) A unit of energy equal to 4.184 J. Approximately 1 cal is required to raise the temperature of 1 g of liquid water by 1 °C.

Calorie (Cal) A unit of energy equal to 4.184 kJ.

calorimeter A device for measuring the quantity of thermal energy transferred during a chemical reaction or some other process.

capillary action The process whereby a liquid rises in a small-diameter tube due to noncovalent interactions between the liquid and the tube's material.

carbohydrates Biochemical compounds

with the general formula $C_x(H_2O)_y$, in which x and y are whole numbers.

carbonyl group An organic functional group consisting of carbon double bonded to oxygen; $\verb|>|C{=}O$.

carboxylic acid An organic compound characterized by the presence of the carboxyl group (—COOH).

catalyst A substance that increases the rate of a reaction but is not consumed in the overall reaction.

catalytic cracking A petroleum refining process using a catalyst, heat, and pressure to break long-chain hydrocarbons into shorter-chain hydrocarbons, including both alkanes and alkenes suitable for gasoline.

catalytic re-forming A petroleum refining process in which straight-chain hydrocarbons are converted to branched-chain hydrocarbons and aromatics for use in the manufacture of other organic compounds and gasoline.

cathode The electrode of an electrochemical cell at which reduction occurs.

cathodic protection A process of protecting a metal from corrosion whereby it is made the cathode instead of the anode by connecting it electrically to a more reactive metal.

cation An ion with a positive electrical charge.

cell voltage The electromotive force of an electrochemical cell; the quantity of work a cell can produce per coulomb of charge that the chemical reaction produces.

Celsius temperature scale A scale defined by the freezing and boiling points of pure water, set at 0 °C and 100 °C, respectively.

cement A solid consisting of microscopic particles containing compounds of calcium, iron, aluminum, silicon, and oxygen in varying proportions and tightly bound to one another.

ceramics Materials fashioned from clay or other natural materials at room temperature and then hardened by heat.

CFCs See **chlorofluorocarbons.**

change of state A physical process in which one state of matter is changed into another (such as melting a solid to form a liquid).

Charles's law The volume of an ideal gas at constant pressure and amount of gas varies directly with its absolute temperature.

chelating ligand A ligand that uses more than one atom to bind to the same metal ion in a complex ion.

chemical change (chemical reaction) A process in which substances (reactants) change into other substances (products) by rearrangement, combination, or separation of atoms.

chemical compound A pure substance (e.g., sucrose or water) that can be decomposed into two or more different pure substances; homogeneous, constant-composition matter that consists of two or more chemically combined elements.

chemical element (element) A substance (e.g., carbon, hydrogen, or oxygen) that cannot be decomposed into two or more new substances by chemical or physical means.

chemical equilibrium A state in which the concentrations of reactants and products remain constant because the rates of forward and reverse reactions are equal.

chemical formula (formula) A notation combining element symbols and numerical subscripts that shows the relative numbers of each kind of atom in a molecule or formula unit of a substance.

chemical fuel A substance that reacts exothermically with atmospheric oxygen and is available at reasonable cost and in reasonable quantity.

chemical kinetics The study of the rates of chemical reactions and the nanoscale pathways or rearrangements by which atoms, ions, and molecules are converted from reactants to products.

chemical periodicity, law of Law stating that the properties of the elements are periodic functions of atomic number.

chemical property Describes the kinds of chemical reactions that chemical elements or compounds can undergo.

chemical reaction (chemical change) A process in which substances (reactants) change into other substances (products) by rearrangements, combination, or separation of atoms.

chemistry The study of matter and the changes it can undergo.

chemotrophs (See also **phototrophs.**) Organisms that must depend on phototrophs to create the chemical substances from which they obtain Gibbs free energy.

chiral A compound whose molecule is not superimposable on its mirror image.

chlor-alkali process Electrolysis process for producing chlorine and sodium hydroxide from aqueous sodium chloride.

chlorination Addition of chlorine or a chlorine compound, to kill bacteria in municipal water supplies; HOCl formed in water is the antibacterial agent.

chlorofluorocarbons (CFCs) Compounds of carbon, fluorine, and chlorine. CFCs have been implicated in stratospheric ozone depletion.

cis **isomer** The isomer in which two like substituents are on the same side of a carbon-carbon double bond, the same side of a ring of carbon atoms, or the same side of a complex ion.

cis-trans **isomerism** A form of stereoisomerism in which the isomers have the same molecular formula and the same atom-to-atom bonding sequence, but the atoms differ in the location of pairs of substituents

on the same side or on opposite sides of a molecule.

closest packing The arrangement of atoms so they are as close together as possible.

coagulate The process in which the protective charge layer on colloidal particles is overcome, causing them to aggregate into a soft, semisolid, or solid mass.

coefficients (stoichiometric coefficients) The multiplying numbers assigned to the species in a chemical equation in order to balance the equation.

cofactor An inorganic or organic molecule or an ion required by an enzyme to carry out its catalytic function.

colligative properties Properties of solutions that depend only on the concentration of solute particles in the solution, not on the nature of the solute particles.

colloid A state intermediate between a solution and a suspension, in which solute particles are large enough to scatter light, but too small to settle out; found in gas, liquid, and solid states.

combination reaction A reaction in which two reactants combine to give a single product.

combined gas law A form of the ideal gas law that relates the P, V, T of a given amount of gas before and after a change: $P_1V_1/T_1 = P_2V_2/T_2$.

combining volumes, law of At constant temperature and pressure, the volumes of reacting gases are always in the ratios of small whole numbers.

combustion analysis A quantitative method to obtain percent composition data for compounds that can burn in oxygen.

combustion reaction A reaction in which an element or compound burns in oxygen.

common ion effect Displacement of an equilibrium caused by introducing a reactant or product ion identical to an ion initially present from another source, such as a strong acid, strong base, or soluble salt.

complementary base pair Bases, each in a different DNA strand, that hydrogen-bond to each other: guanine with cytosine and adenine with thymine or uracil.

complex ion An ion with several molecules or ions connected to a central metal ion by coordinate covalent bonds.

compressibility The property of a gas that allows it to be compacted into a smaller volume by application of pressure.

concentration The relative quantities of solute and solvent in a solution.

concentration cell An electrochemical cell in which the voltage is generated because of a difference in concentrations of the same chemical species.

concrete A mixture of cement, sand, and aggregate (crushed stone or pebbles) in varying proportions that reacts with water and carbon dioxide to form a rock-hard solid.

condensation Process whereby a molecule in the gas phase enters the liquid phase.

condensation polymer A polymer made from the reaction of monomer molecules that contain two or more functional groups, with the formation of a small molecule such as water as a byproduct.

condensation reaction A chemical reaction in which two (or more) molecules combine to form a larger molecule, simultaneously producing a small molecule such as water.

condensed formula A chemical formula of an organic compound indicating how atoms are grouped together in a molecule so that chemically important atoms or groups of atoms are emphasized.

conduction band An energy band that contains electrons of higher energy than those in the valence band.

conductor A material that conducts electric current; has an overlapping valence band and conduction band.

conjugate acid-base pair A pair of molecules or ions related to one another by the loss and gain of a single hydrogen ion.

conjugated Refers to a system of alternating single and double bonds in a molecule.

conservation of energy, law of (first law of thermodynamics) Law stating that energy can be neither created nor destroyed—the total energy of the universe is constant.

conservation of mass, law of Law stating that there is no detectable change in mass in an ordinary chemical reaction.

constant composition, law of Law stating that a chemical compound always contains the same elements in the same proportions by mass.

constitutional isomers (structural isomers) Compounds with the same molecular formula that differ in the order in which their atoms are bonded together.

continuous phase The solvent-like dispersing medium in a colloid.

continuous spectrum A spectrum consisting of all possible wavelengths.

conversion factor (proportionality factor) A relationship between two measurement units derived from the proportionality of one quantity to another; e.g., density is the conversion factor between mass and volume.

coordinate covalent bond A chemical bond in which both of the two electrons forming the bond were originally associated with one of the two bonded atoms.

coordination compound A compound in which complex ions are combined with oppositely charged ions to form a neutral compound.

coordination number The number of coordinate covalent bonds between ligands and a central metal ion in a complex ion.

copolymer A polymer formed by combining two different types of monomers.

core electrons The electrons in the filled inner shells of an atom.

corrosion The deterioration of metals by oxidation-reduction reactions.

coulomb The unit of electrical charge equal to the quantity of charge that passes a fixed point in an electrical circuit when a current of one ampere flows for one second.

covalent bond Interatomic attraction resulting from the sharing of electrons between two atoms.

critical mass The minimum quantity of fissionable material needed to support a self-sustaining chain reaction.

critical pressure The pressure above which there is no distinction between liquid and vapor phases.

critical temperature The temperature above which there is no distinction between liquid and vapor phases.

cryogens Liquefied gases that have temperatures below $-150\ °C$.

crystal lattice The ordered, repeating arrangement of ions, molecules, or atoms in a crystalline solid.

crystalline solids Solids with an ordered arrangement of atoms, molecules, or ions that results in planar faces and sharp angles of the crystals.

crystallization The process in which mobile atoms, molecules, or ions in a liquid convert into a crystalline solid.

cubic close packing The three-dimensional structure that results when atoms or ions are stacked in the *abcabc* arrangement.

cubic unit cell A unit cell with equal-length edges that meet at 90° angles.

curie (Ci) A unit of radioactivity equal to 3.7×10^{10} disintegrations per second.

Dalton's law of partial pressures The total pressure exerted by a mixture of gases is the sum of the partial pressures of the individual gases in the mixture.

decomposition reaction A reaction in which a compound breaks down chemically to form two or more simpler substances.

delocalized electrons Electrons, such as in benzene, that are spread over several atoms in a molecule or polyatomic ion.

denaturation Disruption in protein secondary and tertiary structure brought on by high temperature, heavy metals, and other substances.

density The ratio of the mass of an object to its volume.

deoxyribonucleic acid (DNA) A double-stranded polymer of nucleotides; serves as the genetic information storage molecule.

deposition The process of a gas converting directly to a solid.

detergent(s) Molecules whose structure

contains a long hydrocarbon portion that is hydrophobic and a polar end that is hydrophilic.

diamagnetic Describes atoms or ions in which all the electrons are paired in filled shells so their magnetic fields effectively cancel each other.

diatomic molecule A molecule that contains two atoms.

dietary minerals Essential elements that are not carbon, hydrogen, oxygen, or nitrogen.

dimensional analysis A method of using units in calculations to check for correctness.

dimer A molecule made from two smaller units.

dipole moment The product of the magnitude of the partial charges ($\delta+$ and $\delta-$) of a molecule times the distance of separation between the charges.

dipole-dipole attraction The noncovalent force of attraction between any two polar molecules or polar regions in the same large molecule.

disaccharides Carbohydrates consisting of two monosaccharide units.

dispersed phase The larger than molecule-sized particles that are distributed uniformly throughout a colloid.

displacement reaction A reaction in which one element reacts with a compound to form a new compound and release a different element.

doping The addition of a tiny amount of some other element (a *dopant*) to improve the semiconducting properties of silicon.

double bond A bond formed by sharing two pairs of electrons between the same two atoms.

dynamic equilibrium A balance between opposing reactions occurring at equal rates.

electrochemical cell A combination of anode, cathode, and other materials arranged so that a product-favored oxidation-reduction reaction can cause a current to flow or an electric current can cause a reactant-favored redox reaction to occur.

electrochemistry The study of the relationship between electron flow and oxidation-reduction reactions.

electrode A device such as a metal plate or wire that conducts electrons into and out of a system.

electrolysis The use of electrical energy to produce a chemical change.

electrolyte A substance that ionizes or dissociates in water to form an electrically conducting solution.

electromagnetic radiation Radiation that consists of oscillating electric and magnetic fields that travel through space at the same rate (the speed of light: 186,000 miles/s or 2.998×10^8 m/s in a vacuum).

electromotive force (emf) The difference in electrical potential energy between the two electrodes in an electrochemical cell, measured in volts.

electron A negatively charged subatomic particle that occupies the space surrounding the nucleus.

electron affinity The energy change when an electron is added to a gaseous atom.

electron capture A radioactive decay process in which one of an atom's inner-shell electrons is captured by the nucleus, which decreases the atomic number by 1.

electron configuration The complete description of the orbitals occupied by all the electrons in an atom or ion.

electron-pair geometry The geometry around a central atom including the spatial positions of bonding and lone electron pairs.

electronegativity A measure of the ability of an atom in a molecule to attract bonding electrons to itself.

electronically excited molecule A molecule whose potential energy is greater than the minimum (ground-state) energy because of a change in its electronic structure.

electropositive elements Elements with electronegativities less than 1.3.

element (chemical element) A substance (e.g., carbon, hydrogen, and oxygen) that cannot be decomposed into two or more new substances by chemical or physical means.

elementary reaction A nanoscale reaction whose equation indicates exactly which atoms, ions, or molecules collide or change as the reaction occurs.

empirical formula A formula showing the simplest possible ratio of atoms of elements in a compound.

emulsion A colloid consisting of a liquid dispersed in a second liquid; formed by the presence of an *emulsifier* that coats and stabilizes dispersed-phase particles.

enantiomers A stereoisomeric pair consisting of a chiral molecule and its mirror-image isomer.

end point The point at which the indicator changes color during a titration.

endergonic Refers to a reaction that requires input of Gibbs free energy; applies to biochemical reactions that are reactant-favored at body temperature.

endothermic (process) A process in which thermal energy must be transferred into a thermodynamic system in order to maintain constant temperature.

energy The capacity to do work.

energy band In a solid, a large group of orbitals from neighboring atoms whose energies are closely spaced and whose average energy is the same as the energy of the corresponding orbital in an individual atom.

energy conservation The conservation of useful energy, that is, of Gibbs free energy.

energy density The quantity of energy released per unit volume of a fuel.

enthalpy change (ΔH) The quantity of thermal energy transferred when a process takes place at constant temperature and pressure.

enthalpy of solution The quantity of thermal energy transferred when a solution is formed.

entropy A measure of the probability of disorder in a system.

enzyme A highly efficient biochemical catalyst for one or more reactions in a living system.

enzyme-substrate complex The combination formed by the binding of an enzyme with a substrate.

equatorial position Position lying on the equator of an imaginary sphere around a triangular bipyramidal molecular or ionic structure.

equilibrium concentration The concentration of a substance (generally expressed as molarity) in a system that has reached the equilibrium state.

equilibrium constant (K) A quotient of equilibrium concentrations of product and reactant substances that has a constant value for a given reaction at a given temperature.

equilibrium constant expression The mathematical expression associated with an equilibrium constant.

equilibrium state A system in which opposite nanoscale processes are occurring at equal rates, with the result that there is no observable, macroscopic change.

equilibrium vapor pressure The pressure of a pure gas that is in equilibrium with its pure liquid phase.

equivalence point The point in a titration at which a stoichiometrically equivalent amount of a standard solution has been added to the substance whose concentration is to be determined.

ester An organic compound structurally related to carboxylic acids, but in which the hydrogen atom in the carboxyl group has been replaced by a hydrocarbon group R (—COOR).

evaporation The process of conversion of a liquid to a gas.

exchange reaction A reaction in which cations and anions that were partners in the reactants are interchanged in the products.

excited state The unstable state of an atom or molecule in which at least one electron does not have its lowest possible energy.

exergonic Refers to a reaction that releases Gibbs free energy; applies to biochemical reactions that are product-favored at body temperature.

exothermic Refers to a process in which thermal energy must be transferred out of a thermodynamic system in order to maintain constant temperature.

extent of reaction The fraction of reactants that has been converted to products.

Faraday constant (F) The quantity of electricity that corresponds to the charge on one mole of electrons, 9.6485×10^4 C/mol of electrons, commonly rounded to 96,500 C/mol of electrons.

fat A solid triester of fatty acids with glycerol.

ferromagnetic A substance that contains clusters of atoms with unpaired electrons whose magnetic spins become aligned, causing permanent magnetism.

first law of thermodynamics (law of conservation of energy) Energy can neither be created nor destroyed — the total energy of the universe is constant.

formal charge The charge a bonded atom would have if its electrons were shared equally.

formation constant (K_f) The equilibrium constant for the formation of a complex ion.

formula (chemical formula) A notation combining element symbols and numerical subscripts that shows the relative numbers of each kind of atom in a molecule or formula unit of a substance.

formula unit The simplest cation-anion grouping represented by the formula of an ionic compound; also the unit represented by any formula.

fractional distillation The process of refining petroleum (or other mixture) by distillation to separate it into groups (fractions) of compounds having distinctive boiling point ranges.

Frasch process Process for recovering sulfur from underground deposits by melting the sulfur with superheated water.

free radical A highly reactive atom, ion, or molecule that contains one or more unpaired electrons.

freezing-point lowering A colligative property; the difference between the freezing point of a pure solvent and the freezing point of a solution in which a nonvolatile solute is dissolved in the solvent.

frequency The number of complete waves passing a point in a given period of time (cycles per second).

frequency factor The factor (A) in the Arrhenius equation that depends on how often molecules collide when all concentrations are 1 mol/L and on whether the molecules are properly oriented when they collide.

fuel cell An electrochemical cell that converts the chemical energy of fuels directly into electricity.

fuel value The quantity of energy released when 1 g of a fuel is burned to form carbon dioxide and water.

functional group An atom or group of atoms that imparts characteristic properties and defines a given class of organic compounds (e.g., the —OH group is present in all alcohols).

galvanized Has a thin coating of zinc metal that forms an oxide coating impervious to oxygen, thereby protecting a less active metal, such as iron, from corrosion.

gamma radiation Radiation composed of energetic photons.

gas A phase or state of matter in which a substance has no definite shape and has a volume determined by the volume of its container.

gasohol A blended motor fuel consisting of 90% gasoline and 10% ethanol.

gene The unique sequence of bases in DNA that codes for the synthesis of a specific protein; carrier of a genetic trait.

Gibbs free energy A thermodynamic function that decreases for any product-favored system. For a process at constant temperature and pressure, $\Delta G = \Delta H - T\Delta S$.

glass Amorphous, clear solids formed from silicates and other oxides.

global warming Increase in temperature at earth's surface as a result of the greenhouse effect enhanced by increasing concentrations of carbon dioxide and other greenhouse gases.

glycogen A highly branched, high-molar-mass polymer of glucose found in animals.

glycosidic linkage The C—O—C bond that connects monosaccharides in disaccharides and polysaccharides; forms between carbons 1 and 4 or 1 and 6 of adjacent monosaccharides.

gram(s) The basic unit of mass in the metric system; equal to 1×10^{-3} kg.

gray The SI unit of absorbed radiation dose equal to the absorption of 1 joule per kilogram of material.

greenhouse effect Atmospheric warming caused by absorption of infrared radiation by molecules of carbon dioxide, water vapor, methane, ozone, and similar "greenhouse" gases in the atmosphere.

ground state The state of an atom or molecule in which all of the electrons are in their lowest possible energy levels.

groups The vertical columns of the periodic table of the elements.

Haber-Bosch process The process developed by Fritz Haber and Carl Bosch for the direct synthesis of ammonia from its elements.

half-cell One half of an electrochemical cell in which only the anode or cathode is located.

half-life, $t_{1/2}$ The time required for the concentration of one reactant to reach half its original value; radioactivity—the time required for the activity of a radioactive sample to reach half of its original value.

half-reaction   A reaction that represents either an oxidation or a reduction process.

halide ion   An ion $(1-)$ of a halogen element.

halogens   The elements in Group 7A of the periodic table.

heat (heating)   The energy-transfer process between two samples of matter at different temperatures.

heat capacity   The quantity of thermal energy that must be transferred to an object to raise its temperature by 1 °C.

heat of fusion   The enthalpy change when a substance melts; the quantity of thermal energy that must be transferred when a substance melts at constant temperature and pressure.

heat of sublimation   The enthalpy change when a solid sublimes; the quantity of thermal energy, at constant pressure, that must be transferred to cause a solid to vaporize.

heat of vaporization   The enthalpy change when a substance vaporizes; the quantity of thermal energy that must be transferred when a liquid vaporizes at constant temperature and pressure.

heating curve   A plot of the temperature of a substance versus the quantity of energy transfered to it by heating.

helium burning   The fusion of helium nuclei to form beryllium-8, as it occurs in stars.

Henderson-Hasselbalch equation   The equation describing the relationships among the pH of a buffer solution, the pK_a of the acid, and the concentrations of the acid and its conjugate base.

Henry's law   A mathematical expression for the relationship of gas pressure and solubility; $S_g = k_H P_g$.

Hess's law   If two or more chemical equations can be combined to give another equation, the enthalpy change for that equation will be the sum of the enthalpy changes for the equations that were combined.

heterogeneous catalyst   A catalyst that is in a different phase from that of the reaction mixture.

heterogeneous mixture   A mixture in which components remain separate and can be observed as individual substances, phases, or entities.

heterogeneous reaction   A reaction that takes place at an interface between two phases, solid and gas for example.

hexadentate   A ligand in which each of six different atoms donates an electron pair to a coordinated central metal ion.

hexagonal close packing   The three-dimensional structure that results when atoms in a solid are stacked in the *ababab* arrangement.

homogeneous catalyst   A catalyst that is in the same phase as that of the reaction mixture.

homogeneous mixture   A mixture in which the composition is the same throughout.

homogeneous reaction   A reaction in which the reactants and products are all in the same phase.

Hund's rule   Electrons pair only after each orbital in a subshell is occupied by a single electron.

hybrid orbital   Orbital formed by mixing atomic orbitals of appropriate energy and orientation.

hydration   The binding of one or more water molecules to an ion or molecule, or within a crystal lattice.

hydrocarbons   Compounds composed only of carbon and hydrogen.

hydrogen bond   Noncovalent interaction between a hydrogen atom and a very electronegative atom to produce an unusually strong dipole-dipole force.

hydrogen burning   The fusion of hydrogen nuclei (protons) to form helium, as it occurs in stars.

hydrogenation   An addition reaction in which hydrogen adds to the double bond of an alkene; the catalyzed reaction of H_2 with a liquid triglyceride to produce saturated fatty acid chains, which convert the triglyceride into a semisolid or solid.

hydrolysis   A reaction in which a bond is broken by reaction with a water molecule and the —H and —OH of the water add to the atoms of the broken bond.

hydronium ion H_3O^+; the simplest proton-water complex; responsible for acidity.

hydrophilic   "Water-loving," a term describing the polar part of a molecule that is strongly attracted to water molecules.

hydrophobic   "Water-fearing," a term describing the nonpolar part of a molecule that is not attracted to water molecules.

hydroxide ion OH^- ion; bases increase the concentration of hydroxide ions in solution.

hypertonic   Refers to a solution having a higher concentration of nanoscale particles and therefore a higher osmotic pressure than another solution.

hypothesis   A tentative explanation or prediction derived from experimental observations.

hypotonic   Refers to a solution having a lower solute concentration of nanoscale particles and therefore a lower osmotic pressure than another solution.

ideal gas   A gas that behaves exactly as described by Boyle's, Charles's, and Avogadro's laws.

ideal gas constant   The proportionality constant, R, in the equations $PV = nRT$; $R = 0.0821$ L atm mol^{-1} K^{-1} = 8.314 J K^{-1} mol^{-1}.

ideal gas law   A law that relates pressure, volume, number of moles, and temperature for an ideal gas; the relationships expressed by the equation $PV = nRT$.

immiscible   Describes two liquids that form two seperate phases when mixed.

induced dipole   A temporary dipole created by a momentary uneven distribution of electrons in a molecule or atom.

induced fit   The change in the shape of an enzyme, its substrate, or both when they bind.

inhibitor   A molecule or ion other than the substrate that occupies the active site of an enzyme causing a decrease in enzymatic reactivity.

initial rate   The instantaneous rate of a reaction determined at the very beginning of the reaction.

initiation   The breaking of a carbon-carbon double bond in a polymerization reaction to produce a molecule with highly reactive sites that react with other molecules to produce a polymer; first step in a chain reaction.

inorganic compound   A chemical compound that is not a hydrocarbon or derived from a hydrocarbon.

insoluble   Describes a solute, almost none of which dissolves in a solvent.

instantaneous reaction rate   The rate at a particular time after a reaction has begun.

insulator   A material that has a large energy gap between fully occupied and empty energy bands, and does not conduct electricity.

intermediate (reaction intermediate)   An atom, molecule, or ion produced in one step and used in a later step in a reaction mechanism; does not appear in the equation for the overall reaction.

intermolecular forces   Noncovalent attractions between separate molecules.

internal energy   The sum of the individual energies of all of the nanoscale particles (atoms, molecules, or ions) in a sample of matter.

ion   An atom or group of atoms that has lost or gained one or more electrons so that it is no longer electrically neutral.

ion product (Q)   A value found from an expression with the same mathematical form as the solubility product expression (K_{sp}) but using the actual concentrations rather than equilibrium concentrations of the species involved. (See **reaction quotient.**)

ionic compound   A compound that consists of positive and negative ions.

ionic hydrate   Ionic compounds that incorporate water molecules in the ionic crystal lattice.

ionic radius   Radius of an anion or cation in an ionic compound.

ionization constant for water (K_w)   The equilibrium constant that is the mathematical product of the hydronium ion concentration and the concentration of hydroxide ion in any aqueous solution; $K_w = 1 \times 10^{-14}$ at 25 °C.

ionization energy The energy needed to remove a mole of electrons from a mole of atoms in the gas phase.

isoelectronic Refers to atoms and ions that have identical electron configurations.

isomers Compounds that have the same molecular formula but different arrangements of atoms.

isotonic Refers to a solution having the same concentration of nanoscale particles and therefore the same osmotic pressure as another solution.

isotopes Forms of an element composed of atoms with the same atomic number but different mass numbers owing to a difference in the number of neutrons.

joule (J) A unit of energy equal to 1 kg m^2/s^2. The kinetic energy of a 2-kg object traveling at a speed of 1 m/s.

Kelvin temperature scale (See also **absolute temperature scale**) A temperature scale on which the zero is the lowest possible temperature and the degree is the same size as a Celsius degree.

ketone An organic compound characterized by the presence of a carbonyl group in which the carbon atom is bonded to two other carbon atoms ($R_2C{=}O$).

kilocalorie (kcal or Cal) (See also **calorie.**) A unit of energy equal to 4.184 kJ. Approximately 1 kcal (1 Cal) is required to raise the temperature of 1 kg of liquid water by 1 °C. The food Calorie.

kinetic energy Energy that an object has because of its motion. Equal to $\frac{1}{2}mv^2$, where m is the object's mass and v is its velocity.

kinetic-molecular theory The theory that matter consists of nanoscale particles that are in constant, random motion.

lanthanide contraction The decrease in atomic radii between the fourth-period and the fifth-period lanthanides due to the lack of electron screening by electrons in the $4f$ orbitals.

lanthanides The elements after lanthanum in the sixth period in which the $4f$ subshell is being filled.

lattice energy The net attractive and repulsive forces among ions in a crystal lattice.

law A concise verbal or mathematical statement that summarizes experimental observations.

law of chemical periodicity Law stating that the properties of the elements are periodic functions of atomic number.

law of combining volumes At constant temperature and pressure, the volumes of reacting gases are always in the ratios of small whole numbers.

law of conservation of energy (first law of thermodynamics) Law stating that energy can be neither created nor destroyed—the total energy of the universe is constant.

law of conservation of mass Law stating that there is no detectable change in mass in an ordinary chemical reaction.

law of constant composition Law stating that a chemical compound always contains the same elements in the same proportions by mass.

law of multiple proportions When two elements can combine in two or more ways, the mass ratio A : B in one compound is a small-whole-number multiple of the mass ratio A : B in the other compound.

Le Chatelier's principle If a system is at equilibrium and the conditions are changed so that it is no longer at equilibrium, the system will react to give a new equilibrium in a way that partially counteracts the change.

Lewis acid A molecule or ion that can accept an electron pair from another atom, molecule, or ion.

Lewis base A molecule or ion that can donate an electron pair to another molecule.

Lewis structure A structural formula for a molecule showing all bonding electrons and lone-pair electrons as dots.

Lewis dot symbol An atomic symbol with dots representing valence electrons.

ligands Atoms, molecules, or ions bonded to a central atom, such as the central metal ion in a coordination complex.

limiting reactant The reactant present in limited supply that controls the amount of product formed in a reaction.

line emission spectrum A spectrum produced by excited atoms and consisting of discrete wavelengths of light.

linear Molecular geometry in which there is a 180° angle between bonded atoms.

lipid bilayer The structure of cell membranes that are composed of two aligned layers of phospholipids with their hydrophobic regions within the bilayer.

liquid A phase of matter in which a substance has no definite shape but a definite volume.

London forces Forces resulting from the attraction between positive and negative regions of momentary dipoles in neighboring molecules.

lone-pair electrons Paired valence electrons unused in bond formation; also called nonbonding pairs.

major minerals Dietary minerals present in humans in quantities greater than 100 mg per kg of body weight.

macromolecule A very large polymer molecule made by chemically joining many small molecules (monomers).

macroscale Refers to samples of matter that can be observed by the unaided human senses; samples of matter large enough to be seen, measured, and handled.

main-group elements Elements in the eight A groups to the left and right of the transition elements in the periodic table; the s- and p- block elements.

mass A measure of an object's resistance to acceleration.

mass fraction The ratio of the mass of one component to the total mass of a sample.

mass number The number of protons plus neutrons in the nucleus of an atom of an element.

mass percent The mass fraction multiplied by 100%.

mass spectrometer An analytical instrument used to measure atomic and molecular masses directly.

mass spectrum A plot of ion abundance versus the mass of the ions produced by a mass spectrometer.

materials science The science of the relationships between the structure and the chemical and physical properties of materials.

matter Anything that has mass and occupies space.

melting point The temperature at which the structure of a solid collapses and the solid changes to a liquid.

meniscus A concave or convex surface that forms on a liquid as a result of the balance of noncovalent forces in a narrow container.

metabolism The series of chemical reactions that occurs as food nutrients are converted by an organism into constituents of living cells, to stored Gibbs free energy, or to thermal energy.

metal An element that is malleable, ductile, forms alloys, and conducts an electric current.

metal activity series A ranking of relative reactivity of metals in displacement and other kinds of reactions.

metallic bonding In solid metals, the nondirectional attraction between positive metal ions and the surrounding sea of negatively charged electrons.

metalloid An element that has some typically metallic properties and other properties that are more characteristic of nonmetals; in Groups 3A to 6A.

methyl group A —CH_3 group.

metric system A decimalized measurement system.

micelles Colloid-sized particles built up from many surfactant molecules; micelles can transport various materials within them.

microscale Refers to samples of matter so small that they have to be viewed with a microscope.

millimeters of mercury (mm Hg) A unit of pressure related to the height of a column of mercury in a mercury barometer (760 mm Hg = 1 atm = 101.3 kPa).

mineral A naturally occurring inorganic compound with a characteristic composition and crystal structure.

miscible Describes two liquids that will dissolve in each other in any proportion.

model A mechanical or mathematical way to make a theory more concrete, such as a molecular model.

molality (m) A concentration term equal to the moles of solute per kilogram of solvent.

molar heat capacity The quantity of thermal energy that must be transferred to 1 mol of a substance to increase its temperature by 1 °C.

molar heat of fusion The thermal energy transfer required to melt 1 mol of a pure solid.

molar mass The mass in grams of 1 mol of atoms, molecules, or formula units of one kind, numerically equal to the atomic or molecular weight in amu.

molar solubility The solubility of a solute in a solvent, expressed in moles per liter.

molarity (M) Solute concentration expressed as the moles of solute per liter of solution.

mole (mol) The amount of substance that contains as many elementary particles as there are atoms in exactly 0.0120 kg of carbon-12 isotope.

mole fraction (X) The ratio of moles of one component to the total number of moles in a mixture of substances.

mole ratio (stoichiometric factor) A mole-to-mole ratio relating moles of a reactant or product to moles of another reactant or product.

molecular compound A compound whose molecules contain atoms of two or more different elements.

molecular formula A formula that expresses the number of atoms of each type within one molecule of a compound.

molecular geometry The three-dimensional arrangement of atoms in a molecule.

molecular weight The sum of the atomic weights (in amu) of all the atoms in a compound's formula.

molecule The smallest particle of an element or compound that exists independently and retains the chemical properties of that element or compound.

momentum The product of the mass (m) times the velocity (v) of an object in motion.

monatomic ion An ion consisting of one atom bearing an electrical charge.

monodentate ligand A ligand that donates one electron pair to a coordinated metal ion.

monomer The small repeating unit from which a polymer is formed.

monoprotic acid An acid that can donate a single hydrogen ion per molecule.

monoprotic base A base that can accept only one hydrogen ion per molecule.

monosaccharides The simplest carbohydrates.

monounsaturated acid Refers to fatty acids, such as oleic acid, that contain only one carbon-carbon double bond.

mortar A mixture of cement, sand, and lime that reacts with water and carbon dioxide to form a rock-hard solid.

multiple proportions, law of When two elements can combine in two or more ways, the mass ratio A : B in one compound is a small-whole-number multiple of the mass ratio A : B in the other compound.

n-type semiconductor Semiconductor material made by doping silicon with a Group 5A element such as arsenic that leaves extra valence electrons.

nanoscale Refers to samples of matter (e.g., atoms and molecules) whose normal dimensions are in the nanometer range.

Nernst equation The equation relating the potential of an electrochemical cell to the concentrations of the chemical species involved in the oxidation-reduction reactions occurring in the cell.

net ionic equation A chemical equation in which only those ions undergoing chemical changes in the course of the reaction are represented.

network solid A solid consisting of one huge molecule in which all atoms are connected via a network of covalent bonds.

neurons Specialized cells that are part of animals' nervous systems and that function according to electrochemical principles.

neutral solution A solution containing equal concentrations of H_3O^+ and OH^-; a solution that is neither acidic nor basic.

neutron An electrically neutral subatomic particle found in the nucleus.

newton (N) The SI unit of force, equal to 1 kg times an acceleration of 1 m/s^2; 1 kg m/s^2.

nitrogen fixation The conversion of atmospheric nitrogen (N_2) to nitrogen compounds utilizable by plants.

noble gas notation An abbreviated electron configuration of an element in which filled inner shells are represented by the symbol of the preceding noble gas in brackets. For Al, this would be [Ne]$3s^23p^1$.

noble gases Gaseous elements in Group 8A; the least reactive elements.

nonbiodegradable Not capable of being decomposed by microorganisms.

nonbonding pair An unshared electron pair in a molecule or ion.

noncovalent interactions All forces of attraction other than covalent, ionic, or metallic bonding.

nonelectrolyte A substance that dissolves in water to form an electrically nonconducting solution.

nonmetal Element that generally does not conduct an electric current.

nonpolar covalent bond A bond in which the electron pair is shared equally by the bonded atoms.

nonpolar molecule A molecule that is not polar either because it has no polar bonds or because its polar bonds are oriented symmetrically so that they cancel each other.

NO$_x$ Oxides of nitrogen.

normal boiling point The temperature at which the vapor pressure of a liquid equals 1 atm.

nuclear burning The nuclear fusion reactions by which elements are formed in stars.

nuclear fission The highly exothermic process by which very heavy fissionable nuclei split to form lighter nuclei.

nuclear fusion The highly exothermic process by which comparatively light nuclei combine to form heavier nuclei.

nuclear magnetic resonance The process in which the nuclear spins of atoms align in a magnetic field and absorb radio frequency photons to become excited. These excited atoms then return to a lower energy state when they emit the absorbed radio frequency photons.

nuclear medicine The use of radioisotopes in medical diagnosis and therapy.

nuclear reaction A reaction involving one or more atomic nuclei and resulting in a change in the identities of the isotopes.

nuclear (atomic) reactor A container in which a controlled nuclear reaction takes place.

nucleon A nuclear particle, either a neutron or a proton.

nucleus (atomic) The tiny central core of an atom; contains protons and neutrons. (There are no neutrons in hydrogen-1.)

nutrients The chemical raw materials, eaten as food, that are needed for survival of an organism.

octahedral Molecular geometry of six groups around a central atom in which all groups are at angles of 90° to other groups.

octane number A measure of the ability of a gasoline to burn smoothly in an internal-combustion engine.

octet rule In forming bonds, main group elements (other than H) gain, lose, or share electrons to achieve a stable electron configuration characterized by eight valence electrons.

optical fiber A fiber made of glass constructed so that light can pass through it with little loss of intensity; used for transmission of information.

orbital A region of an atom or molecule within which there is a significant probability that an electron will be found.

order of reaction The reaction rate dependency on the concentration of a reactant or product, expressed as an exponent of a concentration term in the rate equation.

organic compound A compound of carbon with hydrogen, possibly also oxygen, nitrogen, sulfur, phosphorus, or other nonmetals.

osmosis The movement of a solvent (water) through a semipermeable membrane from a region of lower solute concentration to a region of higher solute concentration.

osmotic pressure (Π) The pressure that must be applied to a solution to stop osmosis from a sample of pure solvent.

overall reaction order The sum of the exponents for all concentration terms in the rate equation.

oxidation The loss of electrons by an atom, ion, or molecule, leading to an increase in oxidation number.

oxidation number (oxidation state) A comparison of the charge of an uncombined atom with its actual charge or its relative charge in a compound.

oxidation-reduction reaction (redox reaction) A reaction involving the transfer of one or more electrons from one species to another so that oxidation numbers change.

oxides Compounds of oxygen combined with another element.

oxidized The result when an atom, molecule, or ion loses one or more electrons.

oxidizing agent The substance that accepts electron(s) and is reduced in an oxidation-reduction reaction.

oxoacids Acids in which the acidic hydrogen is bonded directly to an oxygen atom.

oxoanion Polyatomic ion that contains oxygen.

oxygenated gasolines Blends of gasoline with oxygen-containing organic compounds such as methanol, ethanol, and *tertiary*-butyl alcohol.

ozone hole Region of ozone depletion in the stratosphere centered on the earth's poles, most significantly the South Pole.

ozone layer Region of maximum ozone concentration in the stratosphere.

p-block elements Main-group elements in Groups 3A through 8A whose valence electrons consist of outermost *s* and *p* electrons.

p-n junction An interface between *p*-type and *n*-type material that produces a rectifier that allows current to flow in only one direction.

p-type semiconductor Semiconductor material made by doping silicon with a Group 3A element such as boron that leaves a deficiency of valence electrons.

paramagnetic Refers to atoms or ions that are attracted to a magnetic field because they have unpaired electrons in unfilled electron shells.

partial hydrogenation Addition of hydrogen to some of the carbon-carbon double bonds in a triglyceride (a fat or oil).

partial pressure The pressure that one gas in a mixture of gases would exert if it occupied the same volume at the same temperature as the mixture.

particulate Atmospheric solid particles, generally larger than 10,000 nm in diameter.

parts per billion (ppb) One part in one billion (10^9) parts.

parts per million (ppm) One part in one million (10^6) parts.

parts per trillion (ppt) One part in one trillion (10^{12}) parts.

pascal (Pa) The SI unit of pressure; 1 Pa = 1 N/m^2.

Pauli exclusion principle An atomic principle that states that, at most, two electrons can be assigned to the same orbital in the same atom or molecule, and these two electrons must have opposite spins.

peptide linkage The amide linkage between two amino acid molecules; found in proteins.

percent abundance The percentage of atoms of a natural sample of the pure element that consists of a particular isotope.

percent composition by mass The percentage of the mass of a compound represented by each of its constituent elements.

percent yield The ratio of actual yield to theoretical yield, multiplied by 100%.

periodic table A table of elements arranged in order of increasing atomic number so that those with similar chemical and physical properties fall in the same vertical groups.

periods The horizontal rows of the periodic table of the elements.

petroleum fractions The mixtures of hundreds of hydrocarbons in the same boiling point range obtained from the fractional distillation of petroleum.

pH The negative logarithm of the hydronium ion concentration ($-\log[H_3O^+]$).

pH meter An instrument for measuring pH of solutions using electrochemical principles.

phase A state of matter; solid, liquid, or gas.

phase change A physical process in which one state or phase of matter is changed into another (such as melting a solid to form a liquid).

phase diagram A diagram showing the relationships among the three phases of a substance (solid, liquid, and gas), at different temperatures and pressures.

phospholipid Glycerol derivative with two long, nonpolar fatty-acid chains and a polar phosphate group; present in cell membranes.

photochemical reactions Chemical reactions that take place as a result of absorption of photons by reactant molecules.

photochemical smog Smog produced by strong oxidizing agents, such as ozone and oxides of nitrogen, NO_x, that undergo light-initiated reactions with hydrocarbons.

photodissociation The splitting of a molecule into two radicals by a photon of light.

photoelectric effect The emitting of electrons by some metals when illuminated by light of certain wavelengths.

photon A "massless" particle of light whose energy is given by $h\nu$, where ν is the frequency of the light and h is Planck's constant.

photosynthesis A series of reactions in a green plant that combines carbon dioxide with water to form carbohydrate and oxygen.

phototrophs (See also **chemotrophs.**) Organisms that can carry out photosynthesis and therefore can use sunlight to supply their free energy needs.

physical changes Changes in the physical properties of a substance, such as the transformation of a solid to a liquid.

physical properties Properties (e.g., melting point or density) that can be observed and measured without changing the composition of a substance.

pi (π) bond A bond formed by the sideways overlap of parallel *p* orbitals.

Planck's constant The proportionality constant, h, in the relationship $E = h\nu$. The value of h is 6.626×10^{-34} J · s.

plasma A state of matter consisting of unbound nuclei and electrons.

plastic A polymeric material that has a soft or liquid state in which it can be molded or otherwise shaped. See also **thermoplastic** and **thermosetting plastic.**

polar covalent bond A covalent bond between atoms with different electronegativities; bonding electrons are shared unequally between the atoms.

polarization The induction of a temporary dipole in a neighboring molecule or atom by momentary shifting of electron distribution.

polluted water Water that is unsuitable for an intended use, such as drinking, washing, irrigation, or industrial use.

polyamides Polymers in which the monomer units are connected by amide bonds; formed by reaction of diamines with diacids.

polyatomic ion An ion consisting of more than one atom.

polydentate Refers to ligands that can form two or more coordinate covalent bonds to the same metal ion.

polyester A polymer in which the monomer units are connected by ester bonds; formed by reaction of dicarboxylic acids with dialcohols.

polymer A large molecule composed of many smaller repeating units, usually arranged in a chain-like structure.

polypeptide A large polymer of amino acid residues joined by peptide linkages (amide bonds).

polyprotic acids Acids that can donate more than one hydrogen ion per molecule.

polyprotic bases Bases that can accept more than one hydrogen ion per molecule.

polysaccharides Carbohydrates that consist of many monosaccharide units.

polyunsaturated acid A carboxylic acid containing two or more carbon-carbon double bonds; commonly refers to a fatty acid.

positron A nuclear particle having the same mass as an electron, but a positive charge.

potential energy Energy that an object has because of its position.

precipitate An insoluble product of an exchange reaction in aqueous solution.

pressure The force exerted on an object divided by the area over which the force is exerted.

primary battery A voltaic cell (or battery of cells) in which the oxidation and reduction half-reactions cannot easily be reversed to restore the cell to its original state.

primary pollutants Pollutants that enter the environment directly from their sources.

primary structure of proteins The sequence of amino acids in the protein molecule.

principal energy level An energy level containing orbitals with the same quantum number ($n = 1, 2, 3 . . .$).

principal quantum number An integer assigned to each of the allowed main electron energy levels in an atom.

product A substance formed as a result of a chemical reaction.

product-favored system A system in which, when a reaction appears to be over, products predominate over reactants.

proportionality factor (conversion factor) A relationship between two measurement units derived from the proportionality of one quantity to another; e.g., density is the conversion factor between mass and volume.

proton A positively charged subatomic particle found in the nucleus.

pyrolysis The process of decomposing a compound by heating it to a high temperature.

pyrometallurgy The extraction of a metal from its ore using chemical reactions carried out at high temperatures.

qualitative In observations, nonnumerical experimental information, such as a description of color or texture.

quantitative Numerical information, such as the mass or volume of a substance, expressed in appropriate units.

quantum The smallest possible unit of a distinct quantity; for example, the smallest possible unit of energy for electromagnetic radiation of a given frequency.

quantum theory The theory that energy comes in very small packets (quanta); this is analogous to matter occurring in very small particles—atoms.

racemic mixture A mixture of equal amounts of enantiomers of a chiral compound.

rad A unit of radioactivity; a measure of the amount of radiation absorbed by a substance.

radioactive series A series of nuclear reactions in which a radioactive isotope undergoes successive nuclear transformations resulting ultimately in a stable, nonradioactive isotope.

radioactivity The spontaneous emission of energy and/or subatomic particles by unstable atomic nuclei; the energy or particles so emitted.

Raoult's law A mathematical expression for the vapor pressure of the solvent in a solution; $P_1 = X_1 P_1^0$.

rate The change in some measurable quantity per unit time.

rate constant (k) A proportionality constant relating reaction rate and reactant concentrations of reactants and other species that affect rate.

rate law (rate equation) A mathematical equation that summarizes the relationship between concentrations and reaction rate.

rate-limiting step The slowest step in a reaction mechanism.

reactant A starting substance in a chemical reaction.

reactant-favored system A system in which, when a reaction appears to be over, reactants predominate over products.

reaction intermediate (intermediate) An atom, molecule, or ion produced in one step and used in a later step in a reaction mechanism; does not appear in the equation for the overall reaction.

reaction mechanism A sequence of unimolecular and bimolecular elementary reactions by which an overall reaction may occur.

reaction quotient (Q) A value found from an expression with the same mathematical form as the equilibrium constant expression but with the actual concentrations in the mixture not at equilibrium.

reaction rate The change in concentration of a reactant or product per unit time.

redox reaction A reaction involving the transfer of one or more electrons from one species to another so that oxidation numbers change.

reduced The result when an atom, molecule, or ion gains an electron(s).

reducing agent The atom, molecule, or ion that donates electron(s) and is oxidized in an oxidation-reduction reaction.

reduction The gain of electrons by an atom, ion, or molecule, leading to a decrease in its oxidation number.

reformulated gasolines Oxygenated gasolines with lower volatility and containing a lower percentage of aromatic hydrocarbons than regular gasoline.

rem A unit of radioactivity; 1 rem has the effect of 1 röntgen of radiation.

replication The copying of DNA during regular cell division.

resonance hybrid The actual structure of a molecule that can be represented by more than one Lewis structure.

resonance structures The possible structures of a molecule for which more than one Lewis structure can be written, differing by the arrangement of electrons but having the same arrangement of atomic nuclei.

reverse osmosis Application of pressure greater than the osmotic pressure to cause solvent to flow through a semipermeable membrane from a concentrated solution to a solution of lower solute concentration.

reversible process A process for which a very small change in conditions will cause a reversal in direction.

röntgen (R) A unit of radioactivity; 1 R corresponds to deposition of 93.3×10^{-7} J per gram of tissue.

s-block elements Main-group elements in Groups 1A and 2A whose valence electrons are s electrons.

salt An ionic compound whose cation comes from a base and whose anion comes from an acid.

salt bridge A device for maintaining balance of ion charges in the compartments of an electrochemical cell.

saponification The hydrolysis of a triglyceride (a fat or oil) by reaction with NaOH to give sodium salts that are soaps.

saturated fats Fats (or oils) that contain only carbon-carbon single bonds in their hydrocarbon chains.

saturated hydrocarbon Hydrocarbon in which carbon atoms are bonded to a maximum number of hydrogen atoms.

saturated solution A stable solution in which concentration of solute is the concentration that would be in equilibrium with undissolved solute at a given temperature.

scanning tunneling microscope An analytical instrument that produces images of individual atoms or molecules on a surface.

second law of thermodynamics The total entropy of the universe (the system and surroundings) is continually increasing. In any product-favored system, the entropy of the universe is greater after a reaction than it was before.

secondary battery A voltaic cell (or battery of cells) in which the oxidation and reduction half-reactions can be reversed to restore the cell to its original state.

secondary pollutants Pollutants that are formed by chemical reactions of primary pollutants.

secondary structure of proteins Regular repeating patterns of molecular structure in proteins.

semiconductor A material exhibiting electrical conductivity intermediate between those of metals and insulators.

semipermeable membrane A thin layer of material through which only certain kinds of molecules can pass.

shell A collection of orbitals with the same value of the principal quantum number, n.

shifting an equilibrium Changing the conditions of an equilibrium system so that the system is no longer at equilibrium and there is a net reaction in either the forward or reverse direction until equilibrium is reestablished.

sievert The SI unit of effective dose of absorbed radiation.

sigma (σ) bond A bond formed by head-to-head orbital overlap along the bond axis.

simple sugars Monosaccharides and disaccharides.

single bond A bond formed by sharing one pair of electrons between the same two atoms.

smog A mixture of smoke (particulate matter), fog (an aerosol), and other substances that degrade air quality.

solar cell A device that converts solar photons into electricity.

solid A state of matter in which a substance has a definite shape and volume.

solubility The maximum amount of solute that will dissolve in a given volume of solvent at a given temperature when pure solute is in equilibrium with the solution.

solubility product constant (K_{sp}) An equilibrium constant that is the product of concentrations of ions in a solution in equilibrium with a solid ionic compound.

solubility rules General guidelines to predict the water solubilities of ionic compounds based on the ions they contain.

solute The material dissolved in a solution.

solution A homogeneous mixture of two or more substances in a single phase.

solvation The binding of a solute ion or molecule by one or more solvent molecules, especially in a solvent other than water.

solvent The medium in which a solute is dissolved to form a solution.

sp hybrid orbitals Orbitals formed by the combination of one s orbital and one p orbital.

sp^2 hybrid orbitals Orbitals formed by the combination of one s orbital and two p orbitals.

sp^3 hybrid orbitals Orbitals formed by the combination of one s orbital and three p orbitals.

sp^3d hybrid orbitals Orbitals formed by the combination of one s orbital, three p orbitals, and one d orbital.

sp^3d^2 hybrid orbitals Orbitals formed by the combination of one s orbital, three p orbitals, and two d orbitals.

specific heat capacity The quantity of thermal energy that must be transferred to 1 g of a substance to increase its temperature by 1 °C.

spectator ion An ion that is present in a solution in which a reaction takes place, but is not involved in the net process.

spectroscopy Use of electromagnetic radiation to study the nature of matter.

spectrum A plot of the intensity of light (photons per unit of time) as a function of the wavelength or frequency of light.

standard atmosphere (atm) A unit of pressure; 1 atm = 760 mm Hg exactly.

standard state conditions These are 1 bar pressure for all gases, 1 M concentration for all solutes, and a specified temperature.

standard enthalpy change The enthalpy change when a process occurs at a specified temperature and reactants and products are in their standard states.

standard equilibrium constant ($K°$) An equilibrium constant in which each concentration (or pressure) is divided by the standard-state concentration (or pressure); if concentrations are expressed in moles per liter (or pressures in bars) then the concentration (or pressure) equilibrium constant equals the standard equilibrium constant $\Delta G° = -RT \ln K°$.

standard hydrogen electrode The electrode against which standard potentials are measured, consisting of a platinum electrode at which 1 M hydronium ion is reduced to hydrogen gas at 1 bar.

standard molar enthalpy of formation The standard enthalpy change for forming 1 mol of a compound from its elements, with all substances in their standard states.

standard molar volume The volume occupied by exactly 1 mol of an ideal gas at standard temperature (0 °C) and pressure (1 atm), equal to 22.414 L.

standard reduction potential (E^0) The potential of an electrochemical cell when a given electrode is paired with a standard hydrogen electrode under standard conditions.

standard state The most stable form of a substance in the physical state in which it exists at 1 bar and a specified temperature.

standard solution A solution whose concentration is known accurately.

standard temperature and pressure (STP) Universally accepted experimental conditions for the study of gases, defined as a temperature of 0 °C and a pressure of 1 atm.

standard voltages Electrochemical cell voltages measured under standard conditions.

state function A property whose value is invariably the same if a system is in the same state.

steel A material made from iron with most P, S, and Si impurities removed, a low carbon content, and possibly other alloying metals.

steric factor A factor influencing the rates of chemical reactions that depends on the three-dimensional shapes of reactant molecules.

stoichiometric coefficients The multiplying numbers assigned to the species in a chemical equation in order to balance the equation.

stoichiometric factor (mole ratio) A factor relating moles of a reactant or product to moles of another reactant or product.

stoichiometry The study of the quantitative relations between amounts of reactants and products.

stratosphere The region of the atmosphere 12 to 50 km above sea level.

strong acid An acid that ionizes completely in aqueous solution.

strong base A base that ionizes completely in aqueous solution.

strong electrolyte An electrolyte that is completely converted to ions in aqueous solution.

structural formula Formulas written to show how atoms in a molecule are connected to each other.

structural isomers (constitutional isomers) Compounds with the same molecular formula that differ in the order in which their atoms are bonded together.

sublimation Conversion of a solid directly to a gas with no formation of liquid.

subshell An energy level consisting of an s orbital, or three p orbitals, or five d orbitals, and so on.

substance Matter of a particular kind; each substance, when pure, has a well-defined composition and a set of characteristic properties that differ from the properties of any other substance.

substrate A molecule or molecules whose reaction is catalyzed by an enzyme.

superconductor A substance that, below some temperature, offers no resistance to the flow of electric current.

supercritical fluid A substance above its critical temperature that has a density characteristic of a liquid, but the flow properties of a gas.

supersaturated solution   A solution that temporarily contains more solute per unit volume than the saturation concentration.

surface tension   The energy required to overcome the attractive forces between molecules at the surface of a liquid.

surfactant(s)   Natural and synthetic compounds that have both a hydrophobic part and a hydrophilic part.

surroundings   Everything that can exchange energy with a thermodynamic system.

system   In thermodynamics, that part of the universe that is singled out for observation and analysis. The region of primary concern.

temperature   A physical property that describes the direction of spontaneous transfer of thermal energy between objects.

tertiary structure of proteins   The overall three-dimensional folding of protein molecules.

tetrahedral   Molecular geometry of four atoms or groups of atoms around a central atom with bond angles of 109.5°.

theoretical yield   The quantity of product theoretically obtainable from a given quantity of reactant in a chemical reaction.

theory   A unifying principle that explains a body of facts and the laws based on them.

thermal equilibrium   The condition of equal temperatures achieved between two samples of matter that are in contact.

thermochemical equation   A balanced chemical equation, including specification of the states of matter of reactants and products, together with the corresponding value of the enthalpy change.

thermodynamics   The science of heat, work, and the transformations of one into the other.

thermoplastic   A plastic that can be repeatedly softened by heating and hardened by cooling.

thermosetting plastic   A polymer that melts upon initial heating and forms crosslinks so that it cannot be melted again without decomposition.

third law of thermodynamics   A perfect crystal of any substance at 0 K has minimum entropy.

titrant   The standard solution being added to the unknown solution in the course of carrying out a titration.

titration   A procedure whereby a substance in a standard solution reacts with a known stoichiometry with a substance whose concentration is to be determined.

titration curve   A plot of the progress of a titration as a function of the volume of titrant added.

torr   A unit of pressure equivalent to 1 mm Hg.

trace elements (See also **major minerals**)   The dietary minerals that are present in smaller amounts than the major minerals, sometimes far smaller amounts.

tracer   A radioisotope used to track the pathway of a chemical reaction, industrial process, or medical procedure.

trans isomer   The isomer in which two like substituents are on opposite sides of a carbon-carbon double bond, a ring of carbon atoms, or a complex ion.

transition elements   Elements that lie in rows 4 through 7 of the periodic table in which d or f subshells are being filled; comprising scandium through zinc, yttrium through cadmium, lanthanum through mercury, and actinium and elements of higher atomic number.

transition state   A molecular structure corresponding to the maximum of a plot of energy versus reaction progress; also known as the activated complex.

triangular bipyramidal   Molecular geometry of five groups around a central atom in which three groups are in equatorial positions and two are in axial positions.

triangular planar   Molecular geometry of three groups at the corners of an equilateral triangle around a central atom at the center of the triangle.

triple bond   A bond formed by sharing three pairs of electrons between the same two atoms.

triple point   The point on a temperature/pressure phase diagram of a substance where solid, liquid, and gas phases are all in equilibrium with each other.

troposphere   The lowest region of the atmosphere, extending from the Earth's surface to an altitude of about 12 km.

Tyndall effect   Scattering of visible light by a colloid.

uncertainty principle   The statement that it is impossible to determine simultaneously the exact position and the exact momentum of an electron.

unimolecular reaction   A reaction in which the rearrangement of the structure of a single molecule produces the product molecule or molecules.

unit cell   A portion of a crystal lattice defined as the smallest unit that can be replicated in each of three directions to generate the entire lattice.

unsaturated fats   Fats (or oils) that contain one or more carbon-carbon double bonds in their hydrocarbon chains.

unsaturated hydrocarbon   A hydrocarbon containing double or triple carbon-carbon bonds.

unsaturated solution   A solution that contains less dissolved solute per unit volume than the saturation concentration at a given temperature.

valence band   In a solid, an energy band (group of closely spaced orbitals) that contains valence electrons.

valence bond model   A theoretical model that describes a covalent bond as resulting from an overlap of orbitals on the bonded atoms.

valence electrons   Electrons in an atom's highest occupied principal shell and in partially filled subshells of lower principal shells.

valence-shell electron-pair repulsion model (VSEPR)   A simple model used to predict the shapes of molecules and polyatomic ions based on repulsions between bonding pairs and lone pairs around a central atom.

van der Waals equation   An equation of state for gases that takes into account noncovalent attractions between molecules and the sizes of gas molecules:

$$\left[P + a\left(\frac{n}{V}\right)^2\right][V - bn] = nRT.$$

vaporization   The change of a substance from the liquid to the gas phase.

vapor pressure   The pressure of the vapor of a substance in contact with its liquid or solid in a sealed container.

viscosity   The resistance of a liquid to flow.

volatility   The tendency of a liquid to vaporize.

volt (V)   Electrical potential energy difference defined so that 1 joule of work is performed when 1 coulomb of charge moves through 1 volt of potential difference.

voltaic cell   An electrochemical cell in which a product-favored oxidation-reduction reaction is used to produce an electric current.

water of hydration   The water molecules trapped within the crystal lattice of an ionic hydrate or coordinated to a metal ion in a crystal lattice or in solution.

wave functions   Solutions to the Schrödinger wave equation that describe the behavior of an electron in an atom.

wavelength   The distance between adjacent crests (or troughs) in a wave.

weak acid   An acid that is only partially ionized in aqueous solution.

weak base   A base that is only partially ionized in aqueous solution.

weak electrolyte An electrolyte that is only partially ionized in aqueous solution.

weight percent The mass fraction of a solute in a solution multiplied by 100%.

work A process that transfers energy to or from an object by a process other than heating.

x-ray crystallography The science of determining nanoscale crystal structure by measuring the diffraction of x-rays by a crystal.

zone refining A purification process in which a molten zone is moved through a sample being purified, causing the impurities to move along in the liquefied portion of the sample.

zwitterion A structure containing both a positive charge and a negative charge, commonly due to loss and gain of a hydrogen ion within the same molecule.

FIGURE CREDITS

Frontmatter—p. x, p. xi, p. xii, p. xiv, p. xvii, p. xviii, p. xix, p. xxi, p. xxviii: Charles D. Winters; **p. xv:** NY State Dept. of Economic Development, Albany, NY; **p. xx:** 1999 MBARI; **p. xxvi (top):** ©Lawrence Berkeley Laboratory/ Science Photo Library/Photo Researchers, Inc.; **p. xxvi (bottom):** Charles Steele; **p. xxvii:** Bernard Asset/Photo Researchers, Inc.; **p. xxix (top):** Oliver Strewe/Tony Stone Images; **p. xxix (bottom):** Photo Researchers, Inc; **p. xxx:** NASA.

Chapter 1—**Chapter opener, 1.2, 1.3, 1.4, 1.5, 1.6, 1.8, 1.10, 1.11, 1.12, 1.14, 1.16, 1.19a, 1.20, 1.21a&b, 1.24b&c, 1.25, unnumb. figs. pp. 2, 8, 9, 14, 15, 30 (top & bottom), Q1.14, Q1.17, Q1.32, Q1.43, Q1.44:** Charles D. Winters; **1.7:** George Semple; **1.9 (pricked finger):** © Martin Dohrn/Science Photo Library/Photo Researchers, Inc.; **1.9 (blood cells), Q1.45:** Ken Eward/Science Source/Photo Researchers; **1.17b:** Mehau Kulyk/Science Photo Library/Photo Researchers, Inc.; **1.19b:** X. Xu, S. M. Vesecky and D. W. Goodman; **1.21c:** Larry Cameron; **1.24a:** Grant Heilman; **unnumb. fig. p. 29:** Paul S. Howell, Rice University; **unnumb. fig. p. 22:** Oesper Collection in the History of Chemistry, University of Cincinnati; **unnumb. fig. p. 21:** Dr. Philippa Uwins, Centre for Microscopy and Microanalysis, The University of Queensland, Australia.

Chapter 2—**Chapter opener:** © Lawrence Berkeley Laboratory/Science Photo Library/Photo Researchers, Inc.; **2.4, unnumb. figs. pp. 50, 51, 55, 59 (middle), 63, 64:** Charles D. Winters; **unnumb. fig. p. 60:** Boeing; **unnumb. figs. p. 59 (bottom), p. 61:** George Semple; **unnumb. fig. p. 46:** Corbis-Bettmann; **unnumb. fig. p. 48:** Institut für Allgemeine Physik, TU Wien; **unnumb. fig. p. 49:** Courtesy of IBM Corporation, Almaden Laboratories; **unnumb. fig. p. 61:** E.F. Smith Collection, University of Pennsylvania; **unnumb. fig. p. 65:** Courtesy of Lawrence Berkeley Laboratory, CA.

Chapter 3—**Chapter opener:** George Kelvin; **3.1b, 3.4b, 3.5, 3.6, unnumb. figs. pp. 80, 86, 90 (middle), 92, 98, 100, 101, 108, 110, 111:** Charles D. Winters; **unnumb. fig. pp. 78, 100; unnumb. fig. p. 90 (top):** Marna G. Clarke; **unnumb. fig. p. 93:** Dane S. Johnson/Visuals Unlimited; **unnumb. figs. pp. 102, 107:** George Semple; **unnumb. fig. p. 78:** Sierra Antifreeze.

Chapter 4—**Chapter opener:** NASA; **4.1, 4.2, 4.4, 4.5, 4.6, unnumb. figs. pp. 122, 127, 128 (top), 129, 131, 137 (left), 144, 145, 146, Q4.44:** Charles D. Winters; **4.7:** Charles Steele; **unnumb. fig. p. 123:** North Wind Archives; **unnumb. fig. p. 126:** Krafft-Explorer/Science Source/Photo Researchers, Inc.; **unnumb. fig. p. 128 (middle):** The Granger Collection; **unnumb. fig. p. 130:** 1999 MBARI; **unnumb. fig. p. 137 (right):** Peter Skinner/Photo Researchers, Inc.

Chapter 5—**Chapter opener:** Charles Steele; **5.2, 5.3, 5.4, 5.6, 5.7, 5.8, 5.10, 5.11, 5.12, 5.13, 5.14, 5.15, 5.16, 5.18,** **5.19, unnumb. figs. pp. 169, 171, 177, 188:** Charles D. Winters; **unnumb. fig. p. 196:** Viatec/Recovery Systems, Inc., Richland, WA.

Chapter 6—**Chapter opener:** George Kelvin; **6.1:** Bernard Asset/Photo Researchers, Inc.; **6.2:** NY State Dept of Economic Development, Albany, NY; **6.3, 6.6, 6.7, 6.12, unnumb. figs. pp. 213, 231, 232, 234, 235, 242, 244:** Charles D. Winters; **6.17:** Jerrold J. Jacobsen; **6.12, Q6.99:** Bethlehem Steel; **unnumb. fig. p. 212:** P. H. Royer/Photo Researchers, Inc.; **unnumb. fig. p. 213 (bottom right):** Oesper Collection in the History of Chemistry, University of Cincinnati; **unnumb. fig. p. 219:** John Moore; **unnumb. fig. p. 224:** AFP/Corbis; **unnumb. fig. p. 225:** Warren Gretz/U.S. Department of Energy/National Renewable Energy Laboratory; **unnumb. fig. p. 228:** George Semple; **unnumb. fig. p. 245:** H. David Seawell/Corbis; **Q6.12:** Bethlehem Steel; **Q6.55:** E. R. Degginger/Photo Researchers, Inc.

Chapter 7—**Chapter opener:** Reuters New Media, Inc./Corbis; **7.3a:** Runk/Schoenberger/Grant Heilman Photography; **7.6:** David Parker/Science Photo Library/Photo Researchers, Inc.; **7.9:** Donald Potter, Department of Metallurgy, University of Connecticut; **unnumb. fig. pp. 269, 299 (right):** Alexander Tsiaris/Science Source/Photo Researchers, Inc.; **unnumb. figs. pp. 271, 300:** Charles D. Winters; **unnumb. fig. p. 277:** American Institute of Physics; **unnumb. fig. p. 279:** © Jerry Wachter/Photo Researchers, Inc.; **unnumb. fig. p. 280 (left):** Duomo/Corbis; **unnumb. fig. p. 280 (right):** Bill Beatty/Visuals Unlimited, Inc.; **unnumb. fig. p. 299 (left):** Mauro Fermariello/Science Photo Library/Photo Researchers, Inc.

Chapter 8—**Chapter opener:** Reuters New Media Inc./ Corbis; **8.8:** S. Ruren Smith; **unnumb. figs. pp. 316 (a and b), 325, 333 (top), 335, 338, 347:** Charles Winters; **unnumb. fig. p. 329:** Roger Ressmeyer/Corbis; **unnumb. fig. p. 333 (bottom):** George Semple; **unnumb. figs. pp. 317, 340:** Oesper Collection in the History of Chemistry/University of Cincinnati.

Chapter 9—**Chapter opener:** George Kelvin; **9.2:** Ken Eward/Biografx/Photo Researchers, Inc.; **9.3, 9.17, 9.28, unnumb. figs. p. 385, p. 380:** Charles D. Winters; **unnumb. fig. p. 395:** Vittorio Luzzatti; **unnumb. fig. p. 394:** DOE/Science Source/Photo Researchers, Inc.; **unnumb. figs. p. 391 (left), p. 391 (right):** Minden Pictures.

Chapter 10—**Chapter opener:** Jean Marc Barey/Agence Vandystadt/Photo Researchers, Inc.; **10.1.2, unnumb. figs. p. 411, p. 421, p. 427 (top) and (bottom):** Charles D. Winters; **10.3:** Leon Lewandowski; **10.17, unnumb. figs. p. 415, p. 431:** NASA; **10.19:** Bill Varie/Corbis; **unnumb. fig. p. 412:** Paul Schuelt/Eye Ubiquitous/Corbis; **unnumb. fig.**

p. 428: Donald Johnston/Tony Stone Images; **unnumb. fig. p. 433 (top):** Photo Researchers, Inc.; **unnumb. fig. p. 433 (bottom):** CC Studio/Science Photo Library/Photo Researchers, Inc.; **unnumb. fig. p. 447:** Jim Sugar/Corbis; **unnumb. fig. p. 445 (right):** Randy Faris/Corbis; **unnumb. fig. p. 445 (left):** Stephanie Maze/Corbis; **unnumb. fig. p. 442:** Hal Garb/AFP/Corbis-Bettmann; **unnumb. figs. p. 420, p. 443:** Corbis/Bettmann.

Chapter 11 — Chapter opener, 11.1, 11.3a.1, 11.11.2, 11.3b.1, 11.6.2, 11.9.1, 11.10, 11.17a, 11.17b, 11.27, 11.30a, 11.30b, 11.30.1, unnumb. figs. p. 460, p. 475, p. 504, p. 499 (right), p. 485: Charles D. Winters; **11.15:** Layne Kennedy/Corbis; **11.29:** IBM Research/Peter Arnold, Inc.; **11.36:** Deneve Feigh Bunde/Visuals Unlimited, Inc.; **unnumb. fig. p. 461:** Hermann Eisenbeiss/Photo Researchers, Inc.; **unnumb. fig. p. 477:** NCAR/TSADO/Tom Stack & Associates; **unnumb. fig. p. 484:** Wolfgang Kaehler/Corbis; **unnumb. fig. p. 490 (left):** Peter Arnold, Inc.; **unnumb. fig. p. 490 (middle):** Stewart Cohen/Tony Stone Images; **unnumb. fig. p. 490 (right):** Howard Kingsnorth/Tony Stone Images; **unnumb.fig. p. 498:** Courtesy of AT&T Bell Laboratories; **unnumb. fig. p. 499 (left):** National Renewable Energy Laboratory; **unnumb, fig. p. 505:** Novastock/PhotoResearchers, Inc.; **unnumb. fig. p. 487:** Archive Photos.

Chapter 12 — Chapter opener: Bill Ross/Corbis; **12.15, 12.16.2, unnumb. figs. p. 523, p. 533, p. 539, p. 540, p. 541, p. 543, p. 544, p. 553, p. 555, p. 535:** Charles D. Winters; **12.13a:** Chris Springmann/The Stock Market; **12.13b:** Dean Conger/Corbis; **12.13c:** William Whitehurst/The Stock Market; **unnumb. fig. p. 515:** Ashland Oil Company; **unnumb. fig. p. 516:** Leonard Lessin/Peter Arnold, Inc.; **unnumb. fig. p. 522:** Martin Bond/Science Photo Library/Photo Researchers, Inc.; **unnumb. fig. p. 529:** Courtesy of Statoil; **unnumb. fig. p. 530:** © Kim Fennema/Visuals Unlimited; **unnumb. fig. p. 533:** Patrick Grace/Photo Researchers, Inc.; **unnumb. fig. p. 550:** Custom Medical Stock Photo; **unnumb. fig. p. 549:** George Semple; **unnumb. fig. p. 555:** John Sohlden/Visuals Unlimited, Inc. **unnumb. fig. p. 560a:** CNRI/Science Photo Library/Photo Researchers, Inc.; **unnumb. fig. p. 560b:** Omikron/Science Source/Photo Researchers, Inc.; **unnumb. fig. p. 564:** Cabisco/Visuals Unlimited; **unnumb. fig. p. 538:** Courtesy of the National Academy of Sciences; **unnumb. fig. p. 554:** Du Pont Company; **unnumb fig. p. 532:** Neil Rabinowitz/Corbis.

Chapter 13 — 13.1a&b, 13.2 a,b&c, 13.4a&b: Charles D. Winters; **13.13a:** Larry Cameron; **13.13b:** Thomas Eisner with Daniel Aneshansley; **13.22:** Schmidt/University of Minnesota; **13.20:** AC/GM/Peter Arnold, Inc.; **unnumb. fig. p. 593:** AP Photo/Wideworld Photos.

Chapter 14 — Chapter opener: Thomas Hovland/Grant Heilman Photography; **14.3a, 14.3b, 14.8, unnumb. fig. p. 671:** Charles D.Winters; **14.7:** Richard Thom/Visuals Unlimited; **unnumb. fig. p. 636 (top), p. 636 (left):** Simon Fraser/MRC Unit, New Castle General Hospital/Science Photo Library/Photo Researchers, Inc.; **unnumb. fig. p. 636 (bottom):** Brent Nicastro; **unnumb. fig. p. 647:** Charles Pakek/Tom Stack & Associates; **unnumb. fig. p. 667:** Oesper Collection in the History of Chemistry/University of Cincinnati.

Chapter 15 — Chapter opener: Charles D. Winters; **15.1a.2, 15.5, 15.6.2, 15.20, 15.22, 15.24.2, unnumb. figs. p. 684, p. 692 (top), p. 692 (bottom), p. 694, p. 696, p. 703, p. 706, p. 713, p. 695:** Charles D. Winters; **15.19.2:** Science Source Photo Researchers; **15.13, 15.23:** NASA/Peter Arnold, Inc.; **unnumb. fig. p. 683:** AP Photo/John Gaps III/Wide World Photos; **unnumb. fig. p. 692:** AFP/Corbis; **unnumb. fig. p. 694:** D.R. & T.L. Schrichte/Tony Stone Images; **unnumb. fig. p. 695 (left):** George Semple; **unnumb. fig. p. 107:** Courtesy of Edgar Fahs Smith Memorial Collection, Department of Special Collections, University of Pennsylvania; **unnumb. fig. p. 719:** © Kim Fennema/Visuals Unlimited.

Chapter 16 — Chapter opener, 16.3, 16.5a, 16.5b, 16.5c, unnumb. figs. p. 733, p. 741, p. 758 (top), p. 758 (bottom), p. 760, p. 761, p. 763, p. 765, p. 769: Charles D. Winters; **16.4:** Marna G. Clarke; **unnumb. fig. p. 739:** Courtesy of the Arnold and Mabel Beckman Foundation, Photo by Yana Bridle; **unnumb. fig. p. 740:** Courtesy of D. Kirk Nordstrom and Charles N. Alpers, USGS.

Chapter 17 — Chapter opener: © Charlie Ott/Photo Researchers, Inc.; **17.1a,b&c, 17.12.2, 17.3, unnumb fig. p. 799:** Charles D. Winters; **17.4a,b&c:** Marna G. Clarke; **unnumb. fig. p. 779:** © Owen Franken/Corbis; **unnumb. fig. p. 787:** Charles Steele; **unnumb. fig. p. 798:** © Farrell Grehan/Teddy Roosevelt/Photo Researchers, Inc.; **unnumb. fig. p. 805:** SIU/Visuals Unlimited.

Chapter 18 — Chapter opener, 18.1a, 18.1b, 18.5a.2, 18.5b.2, 18.7, unnumb. figs. p. 824, p. 826, p. 847: Charles D. Winters; **18.15:** Courtesy of Phillips Petroleum, Co.; **unnumb. fig. p. 817:** Oesper Collection in the History of Chemistry, University of Cincinnati; **unnumb. fig. p. 828:** Rick Poley/Visuals Unlimited; **unnumb. fig. p. 838 (top):** George Semple; **unnumb. fig. p. 838 (bottom):** Bob Webster, Montana Power Company; **unnumb. fig. p. 839:** Charles Steele; **unnumb. fig. p. 841:** Corbis-Bettmann.

Chapter 19 — Chapter opener: Fisher/Thatcher/Stone; **19.1.2, 19.2, 19.4, 19.6, 19.20, 19.21, 19.24a, 19.24b.2, 19.26, 19.27.2; unnumb. fig. p. 909:** Charles D. Winters; **19.14.2:** Triach/Visuals Unlimited, Inc.; **unnumb. fig. p. 885:** Science Photo Library/Photo Researchers, Inc.; **unnumb. fig. p. 896:** Courtesy of Wilson Greatbatch Technologies, Inc; **unnumb. fig. p. 898:** © Kim Fennema/Visuals Unlimited; **unnumb. figs. p. 900 and 901:** John Helwig; **unnumb. fig. p. 907:** Don Smetzer/Tony Stone Images.

Chapter 20 — Chapter opener: U.S. Department of Energy/Photo Researchers, Inc.; **20.2:** Jack Fields/Photo Researchers, Inc.; **20.5.2, unnumb. fig. p. 945 (top):** Charles D. Winters; **20.8.2:** Phil Degginger/Tony Stone Images; **20.11:** SUNY Upstate Medical Center; **20.12:** CEA-ORSAY/CNRI/Science Photo Library/Photo Researchers, Inc.; **unnumb. fig. p. 934:** Bettmann/Corbis; **unnumb. fig. p. 936:** Courtesy of

Lawrence Berkeley Laboratory; **unnumb. fig. p. 938:** D.O.E./Science Source/Photo Researchers, Inc.; **unnnumb. fig. p. 939:** Science Photo Library/Photo Researchers, Inc.; **unnumb. fig. p. 941:** Naval Research Laboratory/Photo Researchers, Inc.; **unnumb. fig. p. 942:** Cliff Moore/Photo Researchers, Inc.; **unnumb. fig. p. 945 (bottom):** Nordion International; **unnumb. fig. p. 940:** Courtesy of Energy Technology Visuals Collection/Department of Energy; **unnumb. fig. p. 935:** Lawrence Berkeley Laboratory.

Chapter 21 — Chapter opener: NASA; **21.5.2, 21.12a.2, 21.12b.2, unnumb. figs. pp. 959, 964, 966, 970, 972 (left), 972 (right), 973, 975, 977, 978 (top right), 978 (left bottom), 978(a), 978(b), 978(c), 979:** Charles D. Winters; **21.7:** Reuters New Media/Corbis; **unnumb. fig. p. 966:** Richard Olivier/CORBIS; **unnumb. fig. p. 967:** Ottmar Bierwagen/Spectrum Stock; **unnumb. fig. p. 970:** Nada Pecnik/Visuals Unlimited; **unnumb. fig. p. 977:** Photo by Will Brown, Fisher Collection, Chemical Heritage Foundation, Philadelphia, PA; **unnumb. fig. p. 978 (top left):** Photo Researchers, Inc.; **unnumb. fig. p. 979:** Post Street Archives, Michigan; **unnumb. fig. p. 976:** Bettmann/Corbis.

Chapter 22 — Chapter opener: Oliver Strewe/Tony Stone Images; **22.6b:** Courtesy of Bethlehem Steel; **22.8, 22.12, unnumb. figs. p. 984, p. 991, p. 993, p. 996, p. 998 (a, b, bottom), p. 1007, p. 1008, p. 994 (bottom):** Charles D. Winters; **unnumb. fig. p. 994:** James Cowlin; **unnumb. fig. p. 995:** Chris Sharp/Photo Researchers, Inc.; **unnumb. fig. p. 996:** Asian Art & Archaeology/Corbis; **unnumb. fig. p. 996:** Jeff Greenberg/Visuals Unlimited, Inc.; **unnumb. fig. p. 997 (box):** U.S. Mint; **unnumb. fig. p. 999:** Joseph Nettis 1998/Photo Researchers, Inc.; **unnumb. fig. p. 1009:** Science Photo Library/Photo Researchers, Inc.

INDEX

Bold face refers to key term definitions. *Italics* refer to images or diagrams. A "t" indicates a table.

A

Absolute temperature scale, **419**-420. *See also* Celsius temperature scale; Fahrenheit temperature scale
 in Arrhenius equation, 594
 in thermodynamics, 817-818
Absorption, 445
Accuracy, *A.6*-**A.7**
Acetic acid, 171 *t*. *See also* Vinegar
 and buffer capacity, 783
 and buffer solutions, 777-778, 781*t*
 and conjugate acid-base pairs, *730*-731
 and oxidation of alcohols, *535*
 production of, 147, 613
 and production of aspirin, 540
 and titrations, 789-790
 as weak acid, 170-*172, 729, 734*
Acetyl coenzyme A, 842-*843*
Acetyl groups, 842-*843*
Acetylene, *329*
 production of, 519
 sigma and pi bonds in, *378*
 uses for, 408*t*
Acetylsalicylic acid, 2-4, 142-143, 353, 540, 767
Achiral molecules, **397**
Acid-base indicators, 740-741,786-792. *See also* pH
Acid-base pairs, conjugate, 730-732, 777-784
Acid-base reactions, **728**. *See also* Acids; Bases
 as exchange reactions, 170-178
 net ionic equations for, 175-177, 756
 and salts, 756-761
 and titrations, *199*
Acid-base titrations, **199**-200, 786-793
Acid ionization constants (K_a), **741**-742, 745*t*, 747-751
Acid ionization expressions, **741**
Acid rain, **793**-794
 and nitrogen compounds, 448, 488, 728, 793, 965
 pH of, 891
 and sulfur compounds, 440, 445-446, 466, 728, 793
Acidic solutions, **736**
Acidosis, 778
Acids, 170-172, 171*t*, 727-775. *See also* Acid-base reactions; Acid rain; *specific acids*
 amino, 397-398, 556-559, 607, 754-756, *843, 968*
 Brønsted-Lowry, **728**-734
 and buffer solutions, 777-786

Acids *(Continued)*
 fatty, 331-333, *335, 349, 391,* 540-541, *843*
 and formation of salts, 173-174, 756-757*t*
 Lewis, **766**-769
 monoprotic, 744
 organic, 174, 734, 744, 786-787
 polyprotic, 744, 746
 and solubility of salts, 797-798
 strengths of, 732-734, 751-756
 titrations of, 786-793
Acrylonitrile, 544, 546*t*, 549
Actinide elements, **62**-63, 292-293, 958
Action potentials, 894
Activated complexes, **589**-590, *592*
Activation energy (E_a), 589-**590**, *592*
 and catalysts, *606, 608, 611*
 determining from Arrhenius equation, 595-596
 and stability, 850
Active sites, **608**-*610*
Activity (*A*), **930**-931
Activity series, **186**-189, 187*t*, 884
Actual yields, **145**
Addition polymers, **543**-547
Addition reactions, 350
Adenine, *393*-395
Adenosine diphosphate (ADP), 215, 840-846, *848, 892,* 894
Adenosine triphosphate (ATP), 215, 840-846, *848, 892,* 894, *978*
ADP. *See* Adenosine diphosphate (ADP)
Adsorption, 445
Aerosols, **444**, 446, 714*t*
 and global warming, 528-*529*
Air. *See also* Atmosphere
 composition of, 408, *412*-413*t*, 432, 813, 957, 961-962*t*
Air bags, *428,* 966-967
Air pollutants, **444**-445. *See also* Pollution
Alcohol groups, 76, 81, 353, 682
Alcohols, **81**-83, *108,* 531-538.
 boiling points of, 81-82*t,* 532*t*
 hydrogen bonds in, 534
 nomenclature of, 534, A.31-A.32*t*
 oxidation of, 534-537
 solubilities of, 682-683, 685*t*
Aldehyde groups, *108,* 326, A.31-A.32*t*
Aldehydes, **535**-536
Alkali metals, *62*-63
Alkaline batteries, 895
Alkaline earth metals, *62*-63
Alkaline solutions. *See* Basic solutions
Alkalosis, 778

Alkanes, **80***t*-83
 cyclo-, 322-*323*
 isomers of, 83-86
 molecular geometry of, 367-*368*
 nomenclature of, A.28*t*-A.30
 in petroleum, 514, 519
 single covalent bonds in, 322
 straight-chain *vs.* branched-chain, 322
Alkenes, **328**-333, 514, A.30
Alkyl groups, **84***t*, 534, 551, A.28
Alkynes, **329**, A.30
Allotropes, **26**-28
 of carbon, 14, 25-*28, 243,* 500
 of iron, 991-993
 of oxygen, *26, 441*
 of phosphorus, 977-*978*
 and standard states, 243
 of sulfur, 967-*968*
Alloys. *See also* Brass; Bronze; Steel
Alnico, *991*
Alpha (α) carbons, **556**
Alpha (α)-helix, 559-*560, 562*
Alpha (α) particles, 42-46, *920*t.
 and alpha decay, 921-*923, 925*-926, 929, 944
 and artificial transmutations, 935
 and identification of radioactive elements, 65
 and nuclear fusion, 940-941
Alpha radiation, **920**
Aluminosilicates, 502, 959-960
Aluminum (Al), *25, 59*
 and activity series, 187*t*
 corrosion of, 907, 974-*975*
 in the earth's crust, *957, 959*-961*t*
 production of, 13, 838, 847-848, 899, 907, 961*t,* 974-976
 recycling of, 848, 907, 975
 uses for, 25, *60*
Alumininum oxide
Amalgam, dental, 883-884
Amide groups, A.32*t*
Amide linkages, **552**
Amides, **551**
Amine groups, 556, 755, A.32*t*
Amines, **551**, 734-735
Amino acids, **556**-559, 557*t*
 chirality and enantiomers of, 397-398
 in enzymes and proteins, 607
 essential, 557*t,* 968
 and zwitterions, 754-756
Ammonia, 171*t*
 and amines, 551
 as Brønsted-Lowry base, 728-*729*
 and buffer solutions, 781*t*

Ammonia *(Continued)*
in complex ions, 800–802*t*
and conjugate acid-base pairs, 731
in coordination compounds, 1003
hybrid orbitals in, *375*
molecular geometry of, *366–367*
production of, 132, 633–635, 640–641, 666–668, 989
sigma bonds in, *376*
and titrations, *792*
uses for, 408*t*, 965–966
as weak base, 172
Ammonium ions, 89*t*, 92
and Brønsted-Lowry bases, 728
and buffer solutions, 781*t*
and complex ion formation, 1004*t*–1006, *1009–1010*
and conjugate acid-base pairs, 731
Ammonium nitrate
as explosive, 184, 427
as fertilizer, 184
Amorphous solids, 478*t*–**480**, 501–505, 901. *See also* Glass
Amount of substance, units for, A.17*t*
Amperes, **875**, 904, A.17*t*
Amphiboles, *959*
Amphoterism, **768t,** 802–*803*
Amplitude, *267*
amu. *See* Atomic mass units (amu)
Amylopectin, 563–564
Amylose, 563–564
Analgesics, 3
Angular geometry, 368–*369*
Anions, **86**
nomenclature of, 92–93
Anodes, 43, 871–875, **872**
of batteries, 894–898
and corrosion, 907–909
and electrolysis, 899–904
Anodic inhibition, **909**
Antacids, 174–175, *177*, 761–762*t*
Antibonding molecular orbitals, A.24–A.25
Antifreeze. *See* Ethylene glycol; Propylene glycol
Antilogarithms. *See* Exponentials
Antimatter, 923, 947
Antimony (Sb), 64
Antioxidants, **349**
Aqua regia, 999
Aqueous solutions, **123,** 681, 776–811. *See also* Acids; Bases; Concentration; Solutions
electrolysis of, 901–904
reactions in, 195–199
titration of, 199–200
Aquifers, 716
Area, units for, A.18*t*
Argon (Ar)
in the atmosphere, *412*–413*t*, 432, 962*t*
Aromatic groups, *380*
Aromatic hydrocarbons, **349**–353
and catalytic reforming, 518
in coal, *522*
in natural gas, 522
and octane numbers, 516
in petroleum, *514*

Arrhenius, Svante, 527–528, 594, 728
Arrhenius equation, 593–596, **594**
Arsenic (As), *64*
in doped silicon, *497*–498
properties of, 26
Asbestos, *959*
Aspartame, *99, 109t,* 398, 554
Aspirin, *1–4, 98,* 540. *See also* Acetylsalicylic acid
digestion of, 767
production of, 142–143, 353, *540*
risks of, 31
Asymmetric molecules, **397**
Atmosphere, 408, 412–414, 432, 962*t. See also* Pollution
chemical reactions in, 439–440
elements obtained from, 962–967
free radicals in, 440, 442, *450*–451
ozone layer in, 441–444
thickness of, *415*
Atmosphere, standard (atm), **410**
and boiling, 463–464
Atom economy, **148**
Atomic mass units (amu), **51,** A.22*t*
Atomic numbers (Z), **51**–55, 921
Atomic orbitals, 283–284
Atomic radii, 46–51, **296**–300, *334,* 987–*988*
Atomic structure, **42**–46
Atomic theory, 22–24, 123
Atomic weight, 55–57, **56**
Atoms, **22**–24, 41–73
electron configurations for, 290*t*
sizes of, 46–51, **296–**302, *334,* 820, 987–*988*
ATP. *See* Adenosine triphosphate (ATP)
Autoionization of water, **736**–737
Average reaction rates, **576**–577
Avogadro, Amadeo, 58, 421
Avogadro's law, 420–422
and ideal gas law, 422–423, 425
Avogadro's number (N_A), **58**–59, 97, 123, 885, A.22*t*
Axial positions, **369**

B

Background radiation, *942*–944
Bacteria
Balanced chemical equations, **123,** 132–134. *See also* Chemical equations
Ball-and-stick models, *77, 322, 360*
Balmer series, *273, 275*
Band gaps, *494*
Band theory, 493–494
Barium sulfate, *166,* 186, 796*t,* 798–799
Barometers, *410–412*
Bars, 230, **410**
Bartlett, Neil, 304
Basal metabolic rates (BMRs), **251**
Base ionization constants (K_b), 742–743, 745*t,* 747–751
Base ionization expressions, **743**
Bases, 171*t*–173, **172,** 727–775. *See also* Acid-base reactions; *specific bases*
Brønsted-Lowry, **728**–734
and buffer solutions, 777–786

Bases *(Continued)*
Lewis, **766**–769
strengths of, 732–734
titrations of, 786–793
Basic oxygen process, *991*
Basic solutions, **736**
standard reduction potentials in ($E°$), A.40*t*
Batteries, *863,* 894–898. *See also* Electrochemical cells; Voltaic cells
alkaline, 895
dry cell, *876, 894*–895
lead-acid storage, *896*–897, 900
lithium, 896–898
mercury, *895*
nickel-cadmium, 870–871, 897
nickel-metal hydride, 900–901
primary, *894*–895
recharging of, *838*
recycling of, 897
secondary, *894*–898
Bauxite, 974
Beckman, Arnold, 739
Becquerel, Henri, 42, 919–920, 931
Becquerels (Bq), **931**
Bednorz, Georg, 496
Benefits *vs.* risks, 31*t*–32, 944
Benzene
as aromatic compound, 349
and catalytic reforming, 518
London forces in, *385*
molecular orbitals in, A.27
nomenclature of derivatives of, A.30–A.31
resonance structure of, *350*–351
Beryllium (Be)
formation of, 955
molecular orbitals in, A.25–A.26*t*
in nuclear reactions, 923, 935
Beta (β)-carotene, *380, 591*
Beta (β) particles, 42–44, 46, **920***t. See also* Electrons
and beta decay, 921–*923,* 925–926, 946, 956
Beta (β)-pleated sheets, 559–*560,* 562
Beta radiation, **920**
Bicarbonate ions, 89*t,* 93
and buffer solutions, 781*t*
and carbon cycle, 130–*131*
and hard water, 169, 657, 718–719
in the human body, 105
and Lewis acids, 769
Bidentate ligands, ***1006***
Big Bang, 955
Bimolecular reactions, **587,** 590, 592–593
rate laws for, 597–599
Binary molecular compounds, **78**–79
Binding energy (E_b), 53, **926**–928
Binding energy per nucleon, **927,** 955–956
Biochemistry, 63, 215
and biological catalysts, 607–613
biological periodic table for, *105*–107
and biological polymers, 542, 555–565
and biomolecules, 107–111, 392–396
and cells, 390–392, ***892–894***
and Gibbs free energy, 837–847
Biological periodic table, *105*–107
Biological polymers, 542, 555–565

Biomolecules, 107–111, 392–396
Bismuth (Bi), 64
Blast furnaces, 989–*990*
Blister copper, 994
Blood. *See also* Hemoglobin
   alcohol levels in, 533*t*
   as buffer solution, 778–*779*
   drug solubility in, 759
   gases dissolved in, 433, 694, 778–779,
      1011–1012
   as heterogeneous mixture, *13*
   osmotic pressure in, *710–711*
   pH of, 778–*779*
   and sickle cell anemia, 559–*560*
Blood sugar. *See* Glucose
Body-centered cubic unit cells, *482, 484*
Bohr, Niels, 272–273, 275–*277*, 340, 936
Bohr model of atoms, 271–277
   inadequacies of, 277, 280
Bohr radius (a_0), A.22*t*
Boiling point elevation, **702–704**
Boiling points, 225–226, **463–464**, 467*t*
   and intermolecular forces, 384–389
   and petroleum distillation, 514–516
   and vapor pressure, 463–464, 938
Boltzmann, Ludwig, 415, 817
Boltzmann constant (k), 817, A.22*t*
Boltzmann distributions, *415*
Bomb calorimeter, *238, 250*
Bond angles, **363**
Bond energies, 336*t*–338, A.24. *See also* Bond
      enthalpies
Bond enthalpies, **237–238**, 250, 318
   of covalent bonds, 318, 336*t*–338
Bond lengths, **333**
   covalent, 318, 333–334*t*
Bond orders, A.24
Bond polarity, 338–341
Bonding electrons, **318**
Bonding molecular orbitals, *A.24–A.25*
Bonds. *See* Chemical bonds
Boron (B)
   in doped silicon, *497*
   and materials science, 490–491
   molecular orbitals in, A.25–A.26*t*
Bosch, Carl, 666–667
Bose-Einstein condensates, 818
Boundary surfaces, *280*
Boyle, Robert, 16, 418
Boyle's law, 417–419
   and ideal gas law, 422–423, 425
Bragg, William and Lawrence, 488
Bragg equation, 488
Branched-chain alkanes, 83–85, *322*
   nomenclature of, 85–86, A.28–A.30
   and octane numbers, 516
Brass, 13, *996–998*
British thermal units (Btus), 523
Bromine (Br), *25, 64, 978*
   color of, 408
   and expanded octets, 347
   as oxidizing agent, 179–181*t*
   physical properties of, 25
   production of, 961*t*, 978–979
   properties of, 978–979
   uses for, 978–979

Bromthymol blue, *740–741, 786–788, 790*
Brønsted, J. N., 728, 766
Brønsted-Lowry acids and bases, **728–734**
   intermolecular, 755
Bronze, 681*t*, 993, *995–996*
Btus. *See* British thermal units (Btus)
Buckminsterfullerene, *28,* 243
Buckyballs, *28*
Buffer solutions, 777–786
   pH changes of, 784–786
   pH of, 779–782
Buffers, **777**
Buffers capacity, 782–784, **783**
Buoyancy, 431
Butane
   as fuel, 249
   isomers of, *83–84,* 1010
2-Butene, *cis-trans* isomerism in, *330*

C

Cadmium (Cd)
   in batteries, 870–871, 897
   as control rods, 937
   in electrochemical cells, 870–871,
      873–874
   and formation of the elements, 956
   as pollutant, 897
Calcium (Ca), *63*
   and activity series, 187*t*
   in fireworks, *265*
   in hard water, 169, 657, 715, 718–719, 764
   in neurons, 892–893
Calcium carbonate
   and carbon cycle, 130–*131*
   and gas-forming reactions, *177*
   and hard water, 718–719
Calcium chloride
   in Downs cells, 970
   in hot packs, 691–*692*
   and vapor pressure lowering, 707
Calcium hydroxide, 171*t*, 717, 719, 973
Calcium oxide
   in industrial processes, 973, 990
Calculators, electronic, A.10
Caloric values, **249**
Calories (cal), 213
Calorimeters, **238,** 240, 817
Calorimetry, 238–241
Cancer
   and chemotherapy, *573–579,* 585, 1003
   and free radicals, 349
   and radiation therapy, 942, 944, 946–947,
      949
   skin, 441
Candelas, A.17*t*
Capillary action, **462**
Carbohydrates, **108**–110
   as dietary fuels, 249–251*t, 843*
   fermentation of, 532
   as fuels, 249
   and photosynthesis, 844–845
Carbon (C), 16. *See also* Diamond; Graphite
   allotropes of, 14, 25–*28,* 243, 500
   and atomic mass units, 52
   in the human body, 105–106*t*

Carbon (C) (*Continued*)
   isotopes of, *58,* 634, 924, 933
   molecular orbitals in, A.25–A.26*t*
   and moles, 57
   and radioactive dating, 933–934
   standard state of, 243
   in steel, 991–992*t*
Carbon cycle, 130–*131,* 514, 933
Carbon dioxide
   as acid-producing compound, 728
   and acid rain, 793
   in the atmosphere, 413*t,* 432, 525–529, 962*t*
   and blood buffers, 778–779
   in car exhaust, 140
   and carbon cycle, 130–*131,* 933
   dipole moment of, *381*
   double covalent bonds in, 326–*327*
   and global warming, 248, 527–*529*
   as greenhouse gas, *525–527*
   phase diagram for, *474*
   storage of, in deep-ocean lakes, *131, 529*
   sublimation of, *469*
   supercritical, *474–475,* 530
Carbon monoxide
   in the atmosphere, 413*t*
   and automobile emissions, 449, 518, 522,
      615
   and bond energy, 336
   and hemoglobin, 1012
   in industrial processes, 140–141, 146–147,
      531, 613, *617*
   as monodentate ligand, *1006*
   as pollutant, 449, 451
Carbonate ions, 89*t*
   as bidentate ligands, 1004*t, 1006*
   and blood buffers, 778
   and buffer solutions, 781*t*
   and greenhouse effect, 526
   and hard water, 169
   in the human body, 105
   molecular geometry of, *366*
   as resonance hybrids, *344–345*
Carbonic acid, 171*t,* 177, 607–*608*
   and acid rain, 728, 793
   and blood buffers, 778
   solubility of, 798
Carbonic anhydrase, 607–*608*
Carbonyl groups, 326–*327, 536*
   as ligands, 1004*t*
   and ultraviolet-visible spectroscopy, 380
Carboxylic acid groups, **110,** A.32*t*–A.33
   in amino acids, 556, 755
   in carboxylic acid, *734, 753*
   in organic acids, 172, 556
Carboxylic acids, **535**–*536,* 538–539,
      734–735
   and polyamides, 551
   strengths of, 753–754
Carcinogens, 530
Carothers, Wallace, 551
Cast iron, 991
Catalysts, **517,** 572, 604–607. *See also*
     Enzymes
   biological, 607–613
   and chemical equilibrium, 635
   in fuel cells, *898*

Catalysts *(Continued)*
  heterogeneous, 613
  homogeneous, 607
  in industrial processes, 501, 613–617, *667*–*668*
  and partial hydrogenation, 541
  and polymerization, 543–544
Catalytic converters, 518–519, 572, 613, *616*, 661
Catalytic cracking, **517**, 519, 531
Catalytic reforming, 517–**518**
Cathode-ray tubes, 42–43
Cathode rays, 43
Cathodes, 43, 871–875, **872**
  in batteries, 894–898
  and corrosion, 908–909
  and electrolysis, 899–904
Cathodic protection, *909*
Cations, 86
  nomenclature of, 92
Cell voltages, **875**–879
  and concentration, 887–892
Cells, biological
  neurons, *892*–894
  noncovalent forces in, 390–392
Cellulose, 564–565
  as biopolymer (polysaccharide), 542, 555, 563
  as carbohydrate, 109–110
  cis-trans isomerism in, 564
Celsius temperature scale, **6**–7, *419*–*420*, A.19–A.21. *See also* Absolute temperature scale; Fahrenheit temperature scale
Cement, **501**–502, 655
Ceramics, 488–490, 496, **502**–504
Cesium, (Cs)
  and photoelectric effect, 270, 306
CFCs. *See* Chlorofluorocarbons (CFCs)
Chadwick, James, 45, 935
Chain reactions
  and nuclear fission, *936*–*937*
  and ozone depletion, 442
Changes. *See* Chemical changes; Physical changes
Changes of state. *See* Phase changes
Chargaff, Erwin, 394
Charges. *See* Electrical charges
Charles, Jacques Alexandre Cesar, *420*
Charles's law, 419–420
  and ideal gas law, 422–423, 425
Chelating ligands, **1006**, *1008*
Chemical bonds, **80**. *See also* Covalent bonds; Hydrogen bonds; Ionic bonds; Metallic bonds; Pi (π) bonds; Sigma (σ) bonds
Chemical changes, **11**–13
Chemical equations, 122–124
  balancing, 123, 132–134
Chemical equilibrium, 632–679, **633**
  as dynamic process, 172, 463, 634
  and entropy, 663–665
  and equilibrium constants, 635–655
  and Gibbs free energy, 828
  and Le Chatelier's principle, 655–663
Chemical formulas, **26**
  for ionic compounds, 90–91

Chemical fuels, **247**–249. *See also* Fuels
Chemical kinetics, 571–631, **572**
  and Arrhenius equation, 593–596
  and catalysts, 604–617
  and elementary reactions, 587–593, 597–599
  and enzymes, 607–613
  and equilibrium, 652
  and rate laws, 581–587, 597–599, 928, 930–931
  and reaction mechanisms, 599–604
  and reaction rates, 572–581
  and thermodynamics, 590, 813–814, 849–851
Chemical periodicity, law of, **61**
Chemical potential energy, 212, 847
Chemical properties, **12**–13
Chemical reactions, **11**, 162–209. *See also* Acid-base reactions; Combustion reactions; Oxidation-reduction reactions; Product-favored reactions; Reactant-favored reactions
  addition, 350
  in aqueous solution, 195–199
  in the atmosphere, 439–440
  bimolecular, 587, 590, 592–593
  chain, 442
  combination, *125*–*127*
  condensation, 540
  coupled, 838–839, 848–849
  decomposition, 16
  direction of, 650–652
  displacement, *125*, **128**–130
  disproportionation, *998*
  elementary, 587–593
  endergonic, 840–841
  and energy, 210–264
  and enthalpy changes, 231–236
  exchange, *125*, *130*–132, 163–178
  exergonic, 840–841
  extent of, 832
  first-order, *583*–585, 928, 930–931
  half-, 866–871
  heterogeneous *vs.* homogeneous, 572
  neutralization, 173–176
  orders of, 581–587
  photochemical, 440
  precipitation, 166–167, 802–805
  second-order, *583*–585
  substitution, 350
  unimolecular, 587–590
  zeroth-order, *583*–585
Chemical symbols, 24, 28–31
Chemistry, 1–40, **2**. *See also* Biochemistry; Nuclear chemistry
  analytical, 15–16, 139
  electro-, 863–917, **864**
  green, 148, *530*
  organic, 63
Chiral molecules, 396–399, **397**
Chlor-alkali process, 904, **971**–*972*
Chloride ions, 86
  in complex ions, 1003, *1009*–*1011*
  in the human body, 105
  in ionic compounds, 90
  as monodentate ligands, *1004t*, *1006*

Chlorination, 717–718
Chlorine (Cl), *64*, *972*
  color of, 408
  electronegativity of, 339–340
  in the human body, 106t
  isotopes of, *53*
  and ozone depletion, 64
  production of, 864, 904, 961t, 969, 971–*972*
  properties of, 971–972
  uses for, 62, 408t, 971–972
Chlorine monoxide radicals, 442
Chlorine nitrate, 443
Chlorofluorocarbons (CFCs), *441*
  disposal of, 185
  and elementary reactions, 588
  and ozone depletion, 441–444
Cholesterol, 192, 335, *537*
Chromium (Cr), *59*, *1000*–1002
  in batteries, 900–901
  complex ions of, *1011*
  corrosion of, 1000
  electroplating with, 864
  oxidation states of, *988*, *1000*
  paramagnetism of, 296
Chromosomes, 394
Chrysotile, 959
cis isomers, *329*
cis-trans isomers, **330**–333.
  in complex ions, *1010*–1011
  and ultraviolet-visible spectroscopy, 380
  and unimolecular reactions, 588–590
Cisplatin, *573*–*579*, 585, 1003
Citric acid cycle, 565, 842–*843*
Clean Air Act, 518–519
Cleaners, household, 763–766, *1008*
Climate change, 526. *See also* Global warming
Clocks, atomic, 278
Closest packing, **483**–**484**
Coal, 247–249, 522, 957
  and air pollution, 252, 445–446, 449, 522
  and carbon cycle, 130–*131*
  combustion of, 126, 514, 838, 938
  energy from, 523t–524
  as energy resource, 815, 842
Coal gas, 522
Coal tar, 522
Cobalt (Co)
  complex ions of, 802t, *1006*, *1008*–*1010*
  as essential dietary mineral, 1012
  ferromagnetism of, 296, 489
  in magnets, 991
  and nuclear medicine, 928, 947, 949
  oxidation states of, *988*
Coefficients, **123**
  stoichiometric, 124
Coenzymes, 842–*843*
Cofactors, **608**
Coinage, *55*, 188–189, *996*–*997*
Coke
  in coal, 522
Cold packs, 691–*692*
Colligative properties, **700**–707
Colloids, 444, **712**–714
Combination reactions, *125*–*127*

Combined gas law, **424**
Combustion analysis, **148**-*149*
Combustion reactions, *12*-*13*, *121*-124, **123.** *See also* Fuels
Common ion effect, 798-800, **799,** 805
Complementary base pairs, **394**
Complex ions, 800-802t, 908, 984, 999, 1002-1012. *See also* Coordination compounds
  isomerism in, 1009-1011
  and Lewis acids, 767
  nomenclature of, 1003-1005
Complexes. *See also* Complex ions; Coordination compounds
  chelated, *1008*
  enzyme-substrate, 610-*611*
Composites, 489-490
Compounds, **16**-*18*, 74-119. *See also* Coordination compounds; Ionic compounds; Molecular compounds
Concentration, 190-195
  and cell potentials, 887-892
  and reaction rates, 577-581
  of solutions, 694-700
Concentration cells, *889*
Concrete, 220t, *502*, 920
Condensation, 225-226, **465**-466. *See also* Phase changes
  and phase diagrams, *472*
Condensation polymers, **543,** 549-551, 556
Condensation reactions, **540**
Condensed formulas, **76**
Condensed matter. *See* Liquids; Solids
Conduction bands, **493**-494
Conductivity. *See* Electrical conductivity; Thermal conductivity
Conductors, **494**
Conformations, 819
Conjugate acid-base pairs, 730-732, **731**
  and buffers, 777-784
Conjugated pi (π) bonds, 380
Conservation of electrical charge, law of, 868, 904
Conservation of energy, law of, 214-218, **215**
  and Gibbs free energy, 847-849
  and Hess's law, 241
  and phase changes, 226-227
Conservation of mass, law of, **23,** 123, 168, 867
Conservation of matter, law of, **23,** 868, 904
Constant composition, law of, **23**
Constants, physical, A.22t
Constitutional isomers, **83**-86, 351-353, 1009
Constructive interference, **488**
Contact process, 968
Continuous phases, **712**
Continuous spectra, 272-273
Conversion factors. *See* Proportionality factors
Coordinate covalent bonds, **767,** *1002*-1012
  in complex ions, 743, *801*
Coordination compounds, 1002-**1003.** *See also* Complex ions
  isomerism in, 1009-1011
  and life, 1011-1012
  nomenclature of, 1003-1005

Coordination number, **1005**-1009
Copolymers, **548**-*549*
Copper (Cu), *6, 25, 59, 188, 993*-998
  and activity series, 187t
  in alloys with precious metals, 999
  in brass, 13, *996*-998
  in bronze, 681t, 993, 995-996
  in coins, *55,* 189, *996*-*997*
  complex ions of, 802t, 997-998, *1008*-*1009*
  in concentration cells, *889*
  in electrochemical cells, 872-875, 877-*878,* 885
  and electroplating, 904, 906
  in the human body, 1012
  oxidation of, *178, 181*-*182, 975*
  oxidation states of, *988*
  production of, 899, 961t, *994*-995
  smelting of, 152, 994
Copper(II) sulfate, 94, *998*
  pH of solutions of, *760*
Copper(II) sulfate pentahydrate, *100t,* 488, *997*-*998*
Core electrons, **291**
Corrosion, 864, **907**-908
  protection against, 909
Corundum. *See* Aluminum oxide
Cosmic radiation, 933, *942*-943
Coulomb-meters (C m), 380
Coulombs (C), 43, **875,** 904, A.18t
Counter ions, 1003
Coupled reactions, 838-839, 848-849
Covalent bonds, 315-358, **317**
  and bond energy, 336t-338, 388, 819
  and bond length, 333-334t
  and bond polarity, 338-341
  coordinate, 743, **767,** *801, 1002*-1012
  double, 108, 321, 325-333
  and electronegativity, 338-341
  and hybrid orbitals, 371-378
  and infrared spectroscopy, *362*
  and molecular geometry, 361-371
  nonpolar, 338-339
  polar, 339
  in proteins, *561*
  single, 318-325
  triple, 321, 325-333, 964
  and valence bond model, 371-379
COX enzymes, 3-5
Crick, Francis, 360, *394*-395, 489
Critical mass, **938**
Critical points, *472*
Critical pressure (P_c), *474*-475
Critical temperature (T_c)
  phase-change, *474*-475
  superconducting, 495t-496
Cross-linked polyethylene (CLPE), 544-*545*
Crude oil, 514. *See also* Petroleum
Crutzen, Paul, 442, 572
Cryogens, **962**
Crystal growing, 658
Crystal lattices, *95,* **481**
  entropy of, 818
Crystalline solids, 480-487
Crystallization, **467**
Cubic close packing, *483, 485*
Cubic crystal lattice sites, 486

Cubic unit cells, 481-*482*
Curie, Marie and Pierre, 25t, 42, 919-920, 931
Curies (Ci), **931**
Curium (Cu), 25t
Curl, Robert, 28
Cyclic hydrocarbons, 81
Cycloalkanes, 322-*323,* 514, 537
Cyclooxygenase (COX) enzymes, 3-5
Cyclopropane, *323*
Cytosine, 393-*395*

D

d-block elements, 292, *300,* 984. *See also* Transition elements
Dacron, 543, 550
Dalton, John, *22*-23, 123, 213, 421, 432
Dalton's law of partial pressures, **432**
Daughter products, 921-923
Davisson, Clinton, 279
Davy, Humphrey, 969
de Broglie, Louis, 277, 279-280
Debye (D), 380
Decay constants (k), 930-932. *See also* Half-lives ($t_{1/2}$); Rate constants (k)
Decomposition reactions, 16, *125,* **127**-129
Delocalized electrons, 344, 493. *See also* Resonance hybrids
Denaturation, 533, **611**-*612,* 766
Density, **7**-9t
  of alkenes, 330t
  of aromatic compounds, 352t
  of gases, 429-431
  of lead-acid storage batteries, 896
  of solids, 484, 500, 984, 998
  units for, A.18t
  of water and ice, 473, 477-478, 696
Dental amalgam, 883-884
Deoxyribonucleic acid (DNA), *41, 50,* **392**-396
  molecular model of, 360
  and x-ray crystallography, 489
Deoxyribose, *319, 393,* 397
Deposition, **465,** 469-470
  and phase diagrams, *472*
Derived units, A.18t
Desalinization, 17, 711
Destructive interference, **488**
Detergents, 462, **715**-716
  chlorine in, 971
  in household cleaners, 763-766
  and supercritical carbon dioxide, 530
Deuterium (D), *55,* 57t
  fusion of, 941
  and nuclear binding energy, 926-927
  as stable isotope, 924
Dextrins, 563
Dextrose. *See* Glucose
Diamagnetism, **296,** *986*

Diamond, 480–*481*
 as allotrope of carbon, *27, 243,* 500
 and Gibbs free energy, *847*
 as network solid, 500–*501*
 as pure substance, 14
 stability of, 850
Diatomic molecules, **26**
 molecular orbitals for, A.24–A.26
Dicarboxylic acids, 539, 551
1,2–Dichloroethene, *329*–330*t*, 1010
Dietary minerals, 105–107, **106**, *1012*
Dietary Reference Intakes (DRIs), 107
Diffraction, 488–*489*
 of electrons, *279*–280
 of light, 271–*272*
Diffraction gratings, *272, 274*
Digestion, *843*
Dihydrogen phosphate ions, 89*t*, 93
 and baking, 763
 and buffer solutions, 781*t*
 in the human body, 105
Dilution, 192–194
Dimensional analysis, *8*–10, 48–49, **A.4**–A.6
Dimers, 347
Dinitrogen oxide, 79*t*, 83, 408*t*, 444
Dinitrogen pentoxide, 79*t*, 488, 575–576
Dinitrogen tetraoxide, 440, *644*–646, 648,
 650–651, *664*–665
Dipeptides, *556*
Dipole-dipole interactions, 384, **386**–388. *See
 also* Noncovalent interactions
 and properties of solids, 478*t*–479*t*
 in solutions, 681, 686
Dipole moments (μ), **379**–382, 388
Disaccharides, **109**
Disorder. *See* Entropy (*S*)
Dispersed phases, **712**
Dispersion forces. *See* London forces
Displacement reactions, *125,* **128**–130
 as oxidation-reduction reactions, 186–189,
 864
Disproportionation reactions, *998*
Dissociation, 97
Distillation, 514–516, *962*–963
Disulfide linkages, 968
DNA. *See* Deoxyribonucleic acid (DNA)
Doping, **497**
Double bonds, 108, 321, **325**–333. *See also*
 Covalent bonds
Double-displacement reactions. *See* Exchange
 reactions
Dow, Herbert H. and Willard, 979
Dow process, *973*
Downs cells, *969*–970
Downs process, 969
Drain cleaners, *427*
Drugs. *See also* Aspirin; Diseases and condi-
 tions
 amines as, 735
 anesthetics, *735, 759*
 antacids, 174–175, *177,* 761–762*t*
 antihistamines, 759
 decongestants, 759–760
 enzyme inhibitors, 612–615
 pain relievers, *3*–5
 solubility in blood, 759

Drugs *(Continued)*
 sulfa, 612–613
Dry cell batteries, *876. See also* Primary batteries
Dry ice, *469*
Ductility, 491, 493
Dynamic equilibrium, **634**. *See also* Chemical
 equilibrium
 and ionization of acids and bases, 172,
 729–730
 in saturated solutions, 689
 and vapor pressure, 463

E
EDTA. *See* Ethylenediaminetetraacetate
 (EDTA) ions
Efficiency of energy conversions, 523,
 848–849
Einstein, Albert
 and nuclear chemistry, 919
 and photoelectric effect, 269, 271–272
 and special relativity, 927
Elastomers, 548
Electric automobiles, *900*–901
Electrical charges, 42, 875
 elementary (*e*), 43–*44,* A.22*t*
 units for, 875, 904, A.18*t*
Electrical conductivity, 478*t*–479*t*, 491
 of electrolytes, *97*
 of ionic compounds, 96
 and materials science, 489–490
 of network solids, 500
 and temperature, *494*
Electrical current, units for, 875, 904, A.17*t*
Electrical energy, 212, 847
Electrical potential difference, units for, A.18*t*
Electrical potential energy, 875
Electrical work, 885
Electricity generation
 from fossil fuels, 523–524, *838,* 938
 and greenhouse effect, 526
 hydroelectric, 247, *970,* 975
 from nuclear fuels, *918,* 929, *936*–939, 941
 from solar energy, *225,* 247, 499
Electrochemical cells, 864, **871**–875. *See also*
 Batteries
 and voltage, 875–879
Electrochemistry, 863–917, **864**
Electrodes, **872**. *See also* Anodes; Cathodes
Electrolysis, 864, **899**–904. *See also*
 Electroplating; Electrorefining
 of water, *29,* 128, 249, 837
Electrolytes
 acids and bases as, 728
 colligative properties of, 707
 strong *vs.* weak, 170
Electromagnetic induction, 885, 901
Electromagnetic radiation, 266–*268,* **267,** 920
Electromagnetic spectrum, 266–267*t*
Electromagnets, superconducting, 496
Electromotive force (emf), **875**. *See also*
 Standard reduction potentials (*E°*);
 Voltages

Electron affinities (EA), **305**–306*t*
Electron capture, **923,** *925*–926, 956
Electron configurations, **286**–293, 985*t*–987
Electron-density models, *322, 359*
Electron diffraction, *279*–280
Electron-pair geometry, **363**–364*t*. *See also*
 Valence-shell electron-pair repulsion
 (VSEPR)
Electron shells. *See* Shells
Electron spin, 284–286, 318
Electron transitions, 273–277
Electron-volts (eV), 875
Electronegativity, 338–341, ***339***
 and acid strength, **752**
Electronically excited molecules, **440**
Electronics, 490, 494, 496–500
Electrons, 42–*44,* **43,** 46, 52*t*
 bonding, 318
 charge on (*e*), 875, A.22*t*
 charge-to-mass ratio of (*e/m*), A.22*t*
 core, 291
 delocalized, 344, 493
 diffraction of, *279*–280
 and electrolysis, *904*–907
 lone pair, 318
 probability of finding, 280
 rest mass of (m_e), A.22*t*
 unpaired, 296, 968, 986
 valence, 289–291
Electroplating, 864, 899, *906*–907
Electrorefining, 960, 969–976, *994*–995, 999
Electrostatic energy, 212
Elementary charges, A.22*t*
Elementary reactions, 587–593, **588**
 rate laws for, 597–599
Elements, 16–*18,* 24–28, 41–73. *See also*
 Periodic table; Transition elements
 actinide, *62*–63, 292–293, 958
 d-block, 292, 984
 electron configurations for, 286–293,
 985*t*–987
 formation of, 955–956
 and human health, 105–106*t*
 lanthanide, *62*–63, 292–293, 958
 names of, 24–25*t*, 697, 999
 p-block, 291, *300,* 303, 984
 s-block, 291, *300,* 304, 984
 and standard enthalpies of formation, 244
 symbols for, 24, 28–31
 terrestrial, 957–961
 transuranium, 935
emf. *See* Electromotive force (emf)
Emissions, automobile, 447*t*
 and catalytic converters, 518–519, 572,
 613, *616*
 and natural gas, 522
 and oxygenated gasolines, 519–521
 and zero-emission vehicles, 901
Empirical formulas, **103**–105, 148–152
Emulsions, **713**–715
en. *See* Ethylenediamine (en)
Enantiomers, **397**–398, 769, *1011*
Endergonic reactions, **840**–841
Endothermic processes, **227,** 232, 237, **336**
 and elementary reactions, 590
 and Le Chatelier's principle, 661

Endothermic processes *(Continued)*
    and phase changes, 465, 467, 469
    and solutions, *687–688*
Endpoints, **786**
Energy (*E*), **12**–13, 210–264. *See also*
        Activation energy (*E*$_a$); Binding energy
        (*E*$_b$); Conservation of energy, law of;
        Enthalpy (*H*); Internal energy;
        Ionization energy (IE); Kinetic energy
        (*E*$_k$); Lattice energy; Potential energy
        (*E*$_p$); Quantum theory; Thermal energy
    and calorimetry, 238–241
    conservation of, 214–218, 847–849
    dispersal of, 815
    and enthalpy, 223–236, 243–247
    and entropy, 664–665
    of fuels, 247–252, *523t*–524
    and heat capacity, 218–223
    and Hess's law, 241–243
    nature of, 211–214
    resources of, *247*
    as stored in the human body, 110
    and thermochemical equations, 229–231
    units for, 212–214, 906, *A.17t, A.20t*
Energy bands, **493**–494
Energy conservation, 214–218, **847**–849
Energy conversions, efficiency of, 523,
        848–849
Energy density, **249**
Energy diagrams
    for catalyzed reactions, *606, 611*
    for elementary reactions, *589*
Energy distribution curves, *594*
Energy levels, 280–286
    principle, **281**
Energy resources, 815, 842, 847–848
Enthalpy (*H*), 223–229, **228**. *See also* Heat
        (*q*); Standard enthalpy changes (Δ*H*°)
    and chemical reactions, 231–236
    and elementary reactions, 590
    and product-favored reactions, 826*t*
    and solutions, 687–688, 690–*691*
Enthalpy of solution (Δ*H*$_{soln}$), **688**, 690–*691*
Entropy (*S*), 663–**665**, 817–821. *See also*
        Standard molar entropy (*S*°)
    changes in, 821–822
    and dispersal of matter, 816
    and product-favored reactions, 826*t*
    and solutions, 687–688, 691–692, 702,
        709, *820*
Environmental Protection Agency (EPA)
    and emissions testing, 449
    and NAAQS, 444, 447*t*, 451
    and oxygenated gasolines, 520
    and ozone depletion, 443
    and water supply, 717–718
Enzyme-substrate complexes, **610**–*611*
Enzymes, **607**–613
    as catalysts, 541, 562, *571, 604*, 607–613
    and molecular structure, *392*, 560–561
    and nitrogen fixation, 965
    and polymerization, 543
EPA. *See* Environmental Protection Agency
        (EPA)
Equations. *See* Chemical equations
Equatorial positions, **369**

Equilibrium. *See* Chemical equilibrium
Equilibrium concentrations, **644**–646,
        652–655
Equilibrium constant expressions, 638–639
Equilibrium constants (*K*$_c$), 635–643, **638,**
        647*t*. *See also* Formation constants
        (*K*$_f$); Ionization constants (*K*$_a$, *K*$_b$);
        Solubility product constants (*K*$_{sp}$);
        Standard equilibrium constants (*K*°)
    determining, 643–646
    and entropy, 663–665
    and Gibbs free energy, 831–836,
        886–887
    meaning of, 646–650
    and standard reduction potentials,
        886–887
    in terms of partial pressures (*K*$_p$), 643
    using, 650–655
Equilibrium vapor pressure. *See* Vapor pres-
        sure
Equivalence points, **199**, 786–*793*
Essential amino acids, *557t*, 968
Ester groups, A.32*t*
Esterification, 540
Esters, **540**–542. *See also* Polyesters
Ethane, *361*
    as alkane, *80t, 323–324, 328–329*
    in natural gas, 521
1,2-Ethanediol. *See* Ethylene glycol
Ethanol, *74*, 531–534
    boiling point of, 81–82*t, 388, 464, 532t,*
        534, 539
    dipole moment of, 388
    as enzyme inhibitor, 612
    and fermentation, 695
    hydrogen bonds in, 388, *683*
    and infrared spectroscopy, *362*
    and octane numbers, 517*t*, 519
    oxidation of, *535*
    production of, 519
Ethene. *See* Ethylene
Ether groups, A.32*t*
Ethyl alcohol. *See* Ethanol
Ethyl groups, *84t*
Ethylene
    as alkene, *328–329*
    polymerization of, *543–544*, 546*t*
    production of, 519, 848–849
Ethylene glycol, *77, 702*
    as antifreeze, *325, 328*, 702
    and boiling point elevation, 706
    and condensation polymerization, *549*
    and freezing point lowering, 705–706
    production of, 519
Ethylenediamine (en), *1004t, 1006, 1011*
Ethylenediaminetetraacetate (EDTA) ions,
        1006, 1008
Ethyne. *See* Acetylene
Evaporation. *See* Vaporization
Exchange reactions, *125*, **130**–132
    acid-base, 170–177
    gas-forming, 177–178
    precipitation and net ionic equations,
        163–169
Excited states, *274*
Exergonic reactions, 840–841

Exothermic processes, **227**, 232, 237, **336**
    and elementary reactions, 590
    and Le Chatelier's principle, 661
    and solutions, 687–688
Expanded octets, 347–348*t*, 369–371,
        378–379
Experimental error, A.7
Exponential notation. *See* Scientific notation
Exponentials, 594, A.14–A.15
Extent of reaction, **832**

F

Face-centered cubic unit cells, *482–485*
Fahrenheit temperature scale, 6–7,
        A.19–A.21. *See also* Absolute tempera-
        ture scale; Celsius temperature scale
Faraday, Michael, 476, *885*
Faraday's constant (*F*), **885**–888, 905–906,
        A.22*t*
Fats, 110–111
    cis-trans isomerism in, 331–333, 541
    as dietary fuels, 249–251*t, 843*
    and esterification, 540
    hydrogenated, 332–*333*
    partially hydrogenated, 335
    saturated *vs.* unsaturated, 332
Fatty acids, 331–333
    in cell membranes, *391*
    cis-trans isomerism in, *332, 335*, 541
    and digestion, *843*
    and esterification, 540–541
    and free radicals, 349
Femtosecond spectroscopy, 593
Fermentation, 532, 695, 763
Fermi, Enrico, 935
Ferromagnetism, **296**, 984
Fibrous proteins, 561
First law of thermodynamics, **215**. *See also*
        Conservation of energy, law of
First-order reactions, *583–585*, 928, 930–931
Fission. *See* Nuclear fission
Fluids, 474. *See also* Liquids; Solids;
        Supercritical fluids
Fluoridation of water, 883–884, 978
Fluoride ions, 86–87, 293
Fluorine (F)
    color of, 408
    electronegativity of, 339, 341
    and hydrogen bonds, 387–388
    in noble gas compounds, 66, 305, *964*
    and ozone depletion, 64
    production of, 904, 961*t*
Fluoroapatite, 978
Fog, 444, 446, 713–714*t*
Food. *See also* Nutrition
    and antioxidants, 349
    as fuel, 213, 249–252, *843*
    irradiation of, 945
    margarine, shortening, oils, 332, 335, 541
    preservatives for, 174, 945
Force, units for, A.18*t*, A.20*t*
Formal charges, 341–343
Formaldehyde, *108, 536*
    double covalent bonds in, *326*

Formaldehyde *(Continued)*
  sigma and pi bonds in, *376–377*
Formation constants (K_f), **801**–802*t*
Formula units, **95**
Formula weights, **98**
Fossil fuels, 247–249
  and air pollution, 447, 793
  and carbon cycle, 130–*131*
Fractional distillation, 514–516, 962–*963*
Franklin, Rosalind, 394–*395*, 489
Frasch process, **967**
Free radicals, **346**–347
  in the atmosphere, 440, 442, 450–451
  and polymerization, 543
  and reactive oxygen species, 349
Freezing point lowering, 703–707, **705,**
    970
Freezing points, 223–225, 465–469
  and phase diagrams, *472*
Frequency (ν), **267**
Frequency factors, **594**
Frisch, Otto, 936
Fuel cells, 841, **898**–899, 905
Fuel values, **249,** 523
Fuels, 11, *121. See also* Coal; Fossil Fuels;
    Hydrogen (H); Natural Gas; Nutrition;
    Petroleum
  biomass, 247–249
  chemical, 111, 247–249, 533
  food, 249–252
Fuller, R. Buckminster, 28
Fullerenes, *28*
Functional groups, **76,** 531
  acetyl, 842–*843*
  alcohol, 76, 81, 324, A.31–A.32*t*
  aldehyde, *108,* 326, 445, 451, A.31–A.32*t*
  alkyl, 84*t*, 534, 551, A.28
  amide, A.32*t*
  amine, 556, 755, A.32*t*
  aromatic, 380
  butyl, 84*t*
  carbonyl of ketone, *108,* 326–327, 380,
    536, 1004*t*, A.31–A.32*t*
  carboxylic acid, *110,* 172, 353, 556, *734,*
    753, 755, 868, A.32*t*–A.33
  ester, A.32*t*–A.33
  ether, A.32*t*
  ethyl, 84*t*
  halide, A.32*t*
  isopropyl, 84*t*
  ketone, *108,* 445, A.31–A32*t*
  methyl, 84*t*
  nomenclature of, A.28, A.31–A.33
  propyl, 84*t*
Fundamental particles, 44
Fundamental units, A.17*t*
Fusion, 223–225. *See also* Melting; Nuclear
    burning; Nuclear fusion

G

Galvani, Luigi, 872
Galvanic cells, 872. *See also* Electrochemical
    cells

Galvanized iron, *909*
Gamma (γ) radiation, 42–*43*, 267, **920***t*
  exposure to, 942
  and food irradiation, 945
  and nuclear medicine, 946–947
  and nuclear reactions, 922–923
Gas constant. *See* Ideal gas constant (R)
Gas laws, 417–426. *See also* Avogadro's law;
    Boyle's law; Charles's law; Combined
    gas law; Dalton's law of partial pres-
    sures; Ideal gas law; Law of combining
    volumes
Gases, **18,** 407–458. *See also* Gas laws
  in chemical equations, 123
  collection of, over water, *435–436*
  densities of, 429–431
  and dispersal of matter, 815–817
  entropy of, *819–821*
  and kinetic-molecular theory, *20, 409,*
    414–417, 437–439
  and Le Chatelier's principle, 659–660
  mixtures of, 432–436
  nonideal behavior of, *437–439*
  properties of, 408–412
  solubility of, 692–693
  uses for, 408*t*
Gasohol, 533
Gasoline, 351, *515,* 519–521, 813. *See also*
    Petroleum
Gay-Lussac, Joseph, 421
Geiger counters, *931,* 946
Genes, *394–395*
Genetic code, 392, 394–396
Genomes, 396
Geometric isomers. *See cis-trans* isomers
Germer, L. H., 279
Gibbs, J. Willard, 827
Gibbs free energy (G), **827**–831. *See also*
    Standard Gibbs free energy of forma-
    tion (ΔG_f°)
  and biological systems, 839–847
  conservation of, 847–849
  and equilibrium constants, 831–836,
    886–887
  and photosynthesis, 844–847
  and standard reduction potentials, 884–887
  and temperature, 829–831
Glass, **503**–504
  as amorphous solid, 478*t*–480
Global warming, **527**–*529*
  and burning of hydrocarbon fuels, 248
  and carbon cycle, 130–*131*
Globular proteins, 561, 607
Glucose, **108**–109*t*
  chirality and enantiomers of, 398
  as dietary fuel, 249–250
  enthalpy changes in, 229
  fermentation of, 532, 695
  metabolism of, 371, 839–845
  polymerization of, 555, *563*
  as product of carbohydrate breakdown,
    392, 563
  in starches, 563–564
  and urine tests, 998
Glycerol, *110*
  and digestion, *843*

Glycerol *(Continued)*
  and saponification, *714,* 765–766
  and triglycerides, 331, *540–542*
Glycogen, 563–564
Glycosidic linkages, **563**
Gold (Au), *25,* 188, 998–1000
  and activity series, 187*t*
  in coins, 188, 996–997
  complex ions of, 802*t*, 999, 1009
  density of, *9t,* 984, 998
  and electroplating, 906–907
  as precious metal, 974, 984
  properties of, 998–999
Graham, Thomas, 712
Grain alcohol. *See* Ethanol
Graphite
  as allotrope of carbon, 25, *27,* 243, 500
  as electrolysis anodes, *970, 974–975*
  and materials science, 490–491, 992
  as network solid, *500–501*
  stability of, 850
Gravitational energy, 212
Grays (Gy), **942**
Greatbatch, Wilson, *896*
Green chemistry, 148, *530*
Greenhouse effect, **525**–527
Greenhouse gases, 80, 413*t*, 525
  and car emissions, 140
  and global warming, 527–*529*
Ground states, **273**–*274*
Groundwater, 716, 740
Groups, functional. *See* Functional groups
Groups, in periodic table, **61**–*62*

H

Haber, Fritz, 633, 666–667
Haber-Bosch process, 633, 666–668, **667,**
    965, 989
Hahn, Otto, 936
Half-cells, **871**–875
Half-lives ($t_{1/2}$), **586**–587, 928–930, 932
Half-reactions, **866**–871
Halide ions, **92**
Hall, Charles Martin, 974, *976*
Hall-Héroult process, 974–*975*
Halogens, **62**–64, *63*
  as oxidizing agents, 179–182
Hard water, 169, 657, 715, 718–719, 764,
    1008
Hazardous wastes, 719–720*t*
HCFCs. *See* Hydrochlorofluorocarbons (HCFCs)
HDLs. *See* High-density lipoproteins (HDLs)
HDPE. *See* High-density polyethylene (HDPE)
Heat (q), 211, **216**–*218. See also* Enthalpy
    (*H*)
Heat capacity, 218–223, **219**
  and entropy measurements, 818
Heat exchangers, 938, 971
Heat of condensation, 466
Heat of crystallization, 467
Heat of deposition, 469
Heat of formation, 243. *See also under*
    Standard enthalpy changes ($\Delta H°$)
Heat of fusion, **224,** 467–468*t*, 477*t*,
    491*t*–492

Heat of sublimation, 469
Heat of vaporization, **225,** 465–467*t*, 477*t*
Heating curves, 470–*471*
Heisenberg uncertainty principle, 279–280
Helium (He). *See also* Alpha (α) particles
 in the atmosphere, 412–413*t*, *957, 962t*
 compounds of, lack of, 964
 and deep-sea diving, 433
 formation of, 955
 molecular orbitals in, A.24–A.25
 in natural gas, 521
Helium burning, **955–***956*
Hemoglobin, *361,* 559, 561, 778, 1011–*1012.*
  See also Blood
Henderson-Hasselbalch equation, 779–781,
  780, 784–785
Henry's Law, **693–**694
Héroult, Paul, 974
Hertz (Hz), 267
Hess's law, **241–**243
Heterogeneous catalysts, **613**
Heterogeneous mixtures, **13,***18*
Heterogeneous reactions, **572**
Hexadentate ligands, *1006,* **1008**
Hexagonal close packing, **483,** *485*
Hexamethylenediamine, *551–552*
Hexane, *80t*
 and catalytic reforming, 518
 and octane numbers, 517*t*
HFCs. *See* Hydrofluorocarbons (HFCs)
High-density lipoproteins (HDLs), 335
High-density polyethylene (HDPE), *544,*
 554–555t
Hodgkin, Dorothy Crowfoot, *487*
Hoffman, Darleane, *936*
Hoffmann, Felix, 3, *540*
Hoffmann, Roald, 5
Homogeneous mixtures, **13,** *18. See also*
  Solutions; Substances
Homogeneous reactions, **572**
Homogenous catalysts, **607**
Hormones, 537–*538, 562,* 735, *839*
Hot packs, 691–*692*
Human Genome Project, 396
Hund's rule, **288,** 292, A.23
Hybrid orbitals, *359,* 371–379, **373,** 441
Hydrated metal ions, 743–744. *See also*
  Complex ions; Coordination com-
  pounds
Hydration, *691*
Hydrocarbons, 79–83, **80.** *See also specific*
  hydrocarbon
 aromatic, 349–353
 and automobile emissions, 449
 and catalytic converters, 616
 combustion of, *122–123*
 multiple covalent bonds in, 328–333
 as pollutants, 445, 447–448, 451, 793
 saturated, 322–325, 367–*368*
 single covalent bonds in, 322–325
 unsaturated, 322–325
Hydrochloric acid, 171*t,* 728. *See also*
  Hydrogen chloride
 in aqua regia, 999
 in household cleaners, 765
 in industrial processes, 973

Hydrochloric acid *(Continued)*
 as monoprotic acid, 744
 preparation of, 192–194
 as stomach acid, 761–762
 as strong acid, 170–*171, 729,* 732–733,
  741, 752
 and titrations, 788–789, *792*
Hydrochlorofluorocarbons (HCFCs), 443–444
Hydroelectric power, 247, *970,* 975
Hydrofluorocarbons (HFCs), 444
Hydrogen (H), 16. *See also* Bohr model of
  atoms; Protons; Quantum mechanical
  model of atoms
 and activity series, 187*t*
 and bond enthalpies, 236–237, 336–337
 combustion of, *121,* 231–234, 247–249,
  649, 840–841
 and covalent bonds, 316–318
 and formation of the elements, 955–956
 as fuel, *121,* 231–234, 247–249, 649,
  840–841
 in fuel cells, **898–**899
 and hybridization, *372*
 and hydrogen bonds, 387–390
 in industrial processes, 132, 140–142, 146,
  531, 617, 633–*635,* 640–641, 666–668
 isotopes of, *55,* 57*t, 924–925*
 line emission spectrum of, *273, 276t*
 molecular orbitals in, A.24–A.25
 and nuclear binding energy, 926–927
 and nuclear magnetic resonance, 298–299
 production of, 248–249, 905–906, *961t*
 in standard hydrogen electrodes,
  876–878
Hydrogen bombs. *See* Nuclear weapons
Hydrogen bonds, 384, **387–**390. *See also*
  Noncovalent interactions
 in alcohols, 388, 534, *683*
 in complex ions, 998
 in DNA, 394–*395*
 and enzymes, *610*
 in nylon, 552–553
 and properties of solids, 478*t*–479*t*
 in proteins, 559–*561*
 in solutions, 681–683, 686–687
 in water, 388–*389,* 459, 461, 476–478,
  683
Hydrogen burning, **955–***956*
Hydrogen chloride, *78. See also* Hydrochloric
  acid
 and infrared spectroscopy, *362*
 and ozone depletion, 443
 as polar molecule, *379*
Hydrogen fluoride
 hydrogen bonds in, *388*
 as monoprotic acid, 744
 as weak acid, 729–730, 732–733
Hydrogen ions. *See* Protons
Hydrogen peroxide
 and catalysts, *604,* 608
 as oxidizing agent, 181*t*
 as reactive oxygen species, 349
Hydrogen phosphate ions, 89*t,* 93, 105, 745*t,*
  781*t*
Hydrogen sulfate ions, 89*t,* 93, 172, 745*t*–746,
  979

Hydrogen sulfide
 in natural gas, 521
 oxidation of, 180
 as pollutant, 444
Hydrogenation, 332–*333,* 335
Hydrolysis, 502, **541–***542,* 608, **757–***758*
Hydronium ions, *170*
 and acids, 728, 767
 and pH of solutions, 737–738
 in standard hydrogen electrodes, *876–878*
Hydrophilic molecules, **390**
 in cell membranes, 390–*391*
 in colloids, 713
 in proteins, 561, 636
 in solutions, 682
 in surfactants, 714–*716,* 764
Hydrophobic molecules, **390**
 in cell membranes, 390–*391*
 in colloids, 713
 in proteins, *561,* 636
 in solutions, 682
 in surfactants, 714–*716,* 764
Hydroxide ions, 89*t,* 172
 and bases, 728, 733
 and Lewis acids, 769
 as monodentate ligands, 1004*t, 1006*
Hydroxyapatite, 977–978
Hydroxyl radicals
 in the atmosphere, 440, 445, 448, 451
 as reactive oxygen species, 349
Hypertonic solutions, *710–***711**
Hypochlorous acid, 717, 751
 as oxoacid, 752–*753*
Hypotheses, 5. *See also* Laws, scientific;
  Theories
Hypotonic solutions, **710**

I

Ice, *459. See also* Water
 density of, 473, 477–478
 molecular shape of, 476–477
 pressure melting of, *475*
 sublimation of, *470,* 473
Ideal gas constant (*R*), 423*t,* 834, 887, A.22*t*
Ideal gas law, 20, 422–426, **423**
 deviations from, 437–439, 962
Ideal gases, **417–**422
Immiscibility, *682*
Impurities, 15–16, 990–991, 994
Indicators. *See* Acid-base indicators
Induced dipoles, **385**
Induced fits, **608–**610
Industrial processes
 basic oxygen, *991*
 chlor-alkali, 904, *971–972*
 contact, 968
 Dow, *973*
 Downs, 969
 Frasch, *967*
 Haber-Bosch, 633, 666–668, 965, 989
 Hall-Héroult, *974–975*
 lime-soda, 719
 Linde, *962–963*
 Ostwald, 966

Industrial Revolution, 526, 793
Industrial smog, 449. *See also* Smog
Inert gases. *See* Noble gases
Infrared radiation, 267
    and greenhouse effect, *525-526*
Inhibitors, **612-613**
    cyclooxygenase (COX), 3-5
    protease, 614-615
Initial rates, **579**
Inorganic compounds, **75**
Instantaneous reaction rates, **576-577**
Insulators, 491-496, **494**
Integrated circuits, 490, 498
Integrated rate laws, 582-585, 931
Interference, 488
Intermolecular forces, 383-392, **384**
    in gases, 437-439
    in liquids, 460, 478
    in solids, 460
Internal energy, **217-218**
International System of Units (SI), 18*t*, 47*t*,
    942, A.17-A.19
Intramolecular forces, 384
Iodine (I), *25, 64, 978-979*
    as catalyst, *605-607*
    color of, 408
    production of, 961*t, 978-979*
    properties of, 978-979
    sublimation of, *231*
    uses for, *978-979*
Ion exchange, 719
Ion products (Q), **803**
Ion pumps, *892-894*
Ionic bonds, *339, 478t-479t*
Ionic compounds, **86-92**
    aqueous solubility of, 163-166
    in combination reactions, 126
    crystal structures of, 485-487
    entropy of, 820
    nomenclature of, 92-94
    properties of, 95-97, 96*t*, 478*t*-479*t*
Ionic hydrates, **100***t*-101
Ionic radii, *301-302*, 820
Ionization, 97
    of acids and bases, 172, *729*
Ionization constants. *See also* Equilibrium con-
        stants (K_c)
    for acids (K_a), 741-742, 745*t*, 747-751,
        A.34*t*
    for bases (K_b), 742-743, 745*t*, 747-751,
        A.35*t*
    and pH, 781
    for water (K_w), 736, 751
Ionization energy (IE), **302-305***t*
Ionizing radiation. *See* Alpha radiation; Beta ra-
        diation; Gamma radiation
Ions, **44,** 86-89
    acid-base properties of, 761*t*
    complex, **800-802** *t*, 908, 984, 999,
        1002-1012
    electron configurations for, 293-296
    nomenclature of, 92-94
Iron (Fe), *25, 989-993*
    and activity series, 187*t*
    as catalyst, 501
    complex ions of, 802*t, 908, 1008*

Iron (Fe) (*Continued*)
    corrosion of, *907-909*
    in the earth's core, 957
    in the earth's crust, 957-958, 960-961*t*,
        989
    as essential dietary mineral, 106
    ferromagnetism of, 296, 489
    formation of, 955-956
    in hemoglobin, 1011-*1012*
    in industrial processes, *970,* 991-993
    and nuclear binding energy, *927-928*
    ores of, 989
    oxidation states of, *988-989*
    production of, 838, 851-852, 961*t,*
        989-991
    smelting of, 152
    stability of, *927-928*
    uses for, 25, 984, 989-991
Iron(III) oxide, *86, 162,* 989
    as catalyst, 668
    and corrosion, 908
    in industrial processes, 152, 851-852
Irradiation of food, 945
Island of stability, 65
Isoelectronic ions, **294***t,* 302
Isomers, **83-86,** 150, 531
    cis-trans, 329-333, 380, 541, *548,* 564,
        588-590, *1010*-1011 (*See also under*
        2-Butene)
    constitutional, 83-86, 351-353, 1009
    in coordination compounds and complex
        ions, 1009-1011
    linkage, 1010
    and melting points, 360
    optical, 707, 1010-*1011*
    stereo-, 1010
Isooctane. *See* 2,2,4-Trimethylpentane
Isotonic solutions, **710**
Isotopes, 23, **53-55**
    and atomic weight, 55-*58*
    disintegration rates of, 928-934
    stability of, 924-928

J

Joule, James P., *213*
Joule-Thompson effect, *962-963*
Joules (J), 212-213, A.18*t*
Julian, Percy Lavon, *538*

K

Kekulé, Friedrich A., 350
Kelvin temperature scale. *See* Absolute tem-
        perature scale
Kelvins (K), 419, 594, 817, A.17*t*
Kerosene, *9,* 514-515
Ketone groups, *108,* 445, A.31-A32*t*
Ketones, **536**
Kevlar, 552-554
Kidney stones, *805*
Kilocalories (kcal), 213
Kilograms (kg), A.17*t*
Kilowatt-hours, 906

Kinetic energy (E_k), **211-212.** *See also*
        Potential energy (E_p)
    and elementary reactions, 589
    and nuclear reactions, 935, 941
    and properties of gases, *414-417*
    and properties of liquids, 460, *463*
Kinetic-molecular theory, **19-22**
    and colligative properties of solutions, 705
    and elementary reactions, 587-593
    and properties of gases, 409, 414-417,
        437-439
    and properties of liquids, 460-462
    and temperature, 215
Knocking, 516, 518-519
Krebs, Hans, 842
Krebs cycle. *See* Citric acid cycle
Kroto, Harry, 28
Krypton (Kr)
    in the atmosphere, *412-413t,* 962*t*
Kwolek, Stephanie Louise, *554*

L

Lactase, 392
Lactic acid
    and baking, 763
    and buffer solutions, 781*t, 784-785*
    chirality and enantiomers of, *397,* 769
    oxidation of, *536*
    as weak organic acid, 734
Lactose, *109t,* 392, *392,* 713
Lanthanide contraction, **988**
Lanthanide elements, **62-63,** *292-293,* 958
Lasers, 490, 593
Latent heat of vaporization. *See* Heat of vapor-
        ization
Latex, 531, 547
Lattice energy, **690**
Lavoisier, Antoine, *123,* 728
Law of combining volumes, *421*
Laws, scientific, **5.** *See also* Avogadro's law;
        Boyle's law; Charles's law; Chemical
        periodicity, law of; Combined gas law;
        Conservation of electrical charge, law
        of; Conservation of energy, law of;
        Conservation of mass, law of;
        Conservation of matter, law of;
        Constant composition, law of; Dalton's
        law of partial pressures; First law of
        thermodynamics; Gas laws; Henry's
        Law; Hess's law; Hypotheses; Ideal gas
        law; Law of combining volumes;
        Multiple proportions, law of; Raoult's
        law; Second law of thermodynamics;
        Theories; Third law of thermodynamics
LDLs. *See* Low-density lipoproteins (LDLs)
LDPE. *See* Low-density polyethylene (LDPE)
Le Chatelier's principle, **655-663**
    and autoionization of water, 736
    and common ion effect, 799, 805
    and solubility, 692, 797
    and thermodynamics, 823, 836
Lead (Pb), *59*
    and activity series, 187*t*
    in batteries, *896-897*
    in chelated complexes, *1008*

Lead (Pb) *(Continued)*
  in nuclear reactions, 922–*923*, 935, 944
  as pollutant, 451, 519, 897
  and x-ray diffraction, 488
Lead-acid storage batteries, 896–897, 900
Leclanché dry cells, 894–895
Lemon juice, *171*
Length, units for, 18*t*, 46–49, A.17*t*, A.20*t*
Levi, Primo, 2
Lewis, Gilbert N., 290–291, *317,* 766
Lewis acids and bases, 766–769
Lewis dot symbols, **291***t*
Lewis structures, **318**–322
  and exceptions to the octet rule, 346–347*t*
  and resonance, 343–345, A.27
Libby, Willard, 933–*934*
Ligands, **1002**, 1004*t*
  and coordination numbers, 1005–1008
Light. *See also* Electromagnetic radiation
  dual nature of, 271
  speed of (*c*), 267, A.22*t*
Lime-soda process, 719
Limiting reactants, 140–145, **141,** 813
  enzymes as, 611
Linde process, 962–*963*
Line emission spectra, *273,* 276
Linear geometry, *361, 363*–365, 369, 376*t,*
  1009
Linkage isomers, 1010
Linkages. *See also* Polymers
  amide, 552
  disulfide, 968
  glycosidic, **563**
  peptide, **556,** *559*
  phosphate anhydride, 977
Lipid bilayers, **390**–*391*
Liquids, **18,** 459–479
  in chemical equations, 123
  entropy of, *819–820*
  equilibria involving, 639
  and kinetic-molecular theory, 19–*20*
  and phase changes, 465–475
  properties of, 460–462
  and vapor pressure, 463–464
Lithium batteries, 896–*898*
Litmus, 170–*171*
Logarithms, A.13–A.15
London forces, 384–386, **385,** 388. *See also*
  Noncovalent interactions
  and cells, 390–391
  and melting, 468–469
  and properties of solids, 478*t*–479*t,* 500
  in solutions, 681–683, *687*
Lone pair electrons, **318**
  and Brønsted-Lowry bases, 728–729
  in hybrid orbitals, 375
  in hydrated ions, 743
  and hydrogen bonds, 387–390
  and Lewis acids and bases, 766
  and molecular geometry, 366–369
Long-chain fatty acids, 331–333,
  540–541
Low-density lipoproteins (LDLs), 335
Low-density polyethylene (LDPE),
  544, 555*t*
Lowry, T. M., 728, 766

Luminous intensity, units for, A.17*t*
Lye. *See* Sodium hydroxide

M

Macromolecules, **543**
Macroscales, **18**–*19, 29–30,* 58
  and chemical reactions, 122, 134–140
Magnesium (Mg), *59, 63, 137, 973*
  and activity series, 187*t*
  alloys of, 973
  formation of, 955
  in hard water, 169, 657, 715, 718–719, 764
  in the human body, 106*t*
  production of, 972–*974*
  properties of, 973
  in sea water, 957, 972–973
  uses for, 973
Magnetic moments, *986*
Magnetic properties of materials, *296, 347,*
  496, 984, 986, A.23
Magnetic resonance imaging (MRI), *299,* 496
Main group elements, **61**–62
  electron configurations for, 286–289
Major minerals, dietary, 105–**107,** 1012
Malleability, 491
Mass, **46.** *See also* Conservation of mass, law
  of
  units for, 46–47*t,* 49, A.17*t,* A.20*t*
  vs. weight, 46, 57
Mass defect, 53
Mass fraction, **696,** 699*t*
Mass number (A), **51**–55, 921
Mass percent. *See* Percent composition by
  mass
Mass spectra, *54*
Mass spectrometers, 53–*54*
Materials science, 487–491, **490**
Mathematical operations, A.2–A.16
Matter, **2.** *See also* Conservation of matter,
  law of
  classification of, 17–*18*
  dispersal of, 815–817
  physical properties of, 6–10
  states of, *20*
Maximum work (w_{max}), 837–839
Mechanical energy, 211
Medicine. *See also* Drugs
  chemotherapy, *573–579, 585,* 1003
  imaging in, 946–948
  intravenous solutions, 199, *696*
  molecular, 2–5
  nuclear, 919, 928, 935, 946–949
  pacemakers, 896
  and protease inhibitors, 614–615
  radiation therapy, 942, 944, 946–947
  use of polyesters in, *550*
Meitner, Lise, 25*t,* 936
Meitnerium (Mt), 24–25*t*
Melting, 223–225, *465*–469
  and phase diagrams, *472*
  surface, *473, 476*
Melting point curves, *472*
Melting points, 7
  and intermolecular forces, 384, 388

Melting points *(Continued)*
  of ionic compounds, 95–96, 468*t*
  of metals, 491*t*–492, 984
  of molecular compounds, 468*t*
Membranes
  in cells, 390–*391, 710*–711, *892*–894
  semipermeable, *708,* 873, 971–*972*
Mendeleev, Dmitri, 6, 25*t, 61,* 266, 514
Mendeleevium (Md), 24–25*t,* 935
Meniscus, **462**
Mercury (Hg), *6, 462*
  in barometers, *410*
  in batteries, *895*
  complex ions of, 802*t*
  line emission spectrum of, *273*
  as pollutant, 15, 444, 895
  risks of, 31
Mercury batteries, 895, 972
Metabolism, 13, 109, **839**
  and Gibbs free energy, 839–844
Metal activity series. *See* Activity series
Metal ions, hydration of, 743–744
Metallic bonds, 384, 478*t*–479*t,* **493**–494
Metalloids, **26,** 61–62
  electronegativity of, 339
  in ionic and molecular compounds, 90
Metals, **24**–25, 61–*63,* 491–496, 984–989.
  See also Transition elements
  alkali, **62**–63
  alkaline earth, **62**–63
  in combination reactions, *126*
  and coordination compounds,
    1002–1003
  electronegativity of, 339
  hydrated ions of, as acids, 743–744,
    767–768
  in ionic compounds, 90
  melting points of, 491*t*–492, 984
  noble, *188,* 961
  obtaining from ores, 960–961, 974–976
  oxidation numbers of, 984, 988–989
  oxidation of, 178–182
  precious, 974, 984
  properties of, 478*t*–479*t,* 489, 491*t,*
    984–989
Metathesis reactions. *See* Exchange reactions
Meters (m), A.17*t*
Methane, *361*
  in the atmosphere, 413*t,* 444, 448, 962*t*
  combustion of, *12,* 134–137, *210*–211,
    338, 572
  as fuel, 248–249
  and Gibbs free energy, 828–829
  as greenhouse gas, 525
  hybrid orbitals in, 373
  in industrial processes, *617*
  intermolecular forces in, 383–384
  molecular geometry of, *363, 367*
  in natural gas, 521
Methanol, 531–534
  conversion of methane to, *617*
  as octane enhancer, 519
  and octane numbers, 517*t*
  production of, 140–*141,* 146, 531
Methyl alcohol. *See* Methanol
Methyl groups, *84t*

Metric system, 46–47t. *See also* International
    System of Units (SI)
Micas, 478t–479t, *959*
Micelles, 715–*716*
Microscales, **18**–*19*
Microwave radiation, 267, 382
Millikan, Robert, 43–44
Millimeters of mercury (mm Hg), **410**
Minerals, 480, **957**–961, 976–977
    dietary, 105–**107**, 1012
Miscibility, *682*
Mixtures. *See* Heterogeneous mixtures;
    Homogeneous mixtures; Solutions
Models, **5.** *See also* Molecular models
Molality, 696, **703**
    vs. molarity, *703*
Molar heat capacity (c_m), **221**–223
Molar mass, **59**, 97–98
    and gas densities, 429–431
Molar volume at STP (V_m), A.22t
Molarity, **190**–192, 698–700
    in reactions in aqueous solutions, 195–199
    vs. molality, *703*
Mole fractions, **433**, 700
Mole ratios, **135**–139
Molecular compounds, **75**–78
    nomenclature of binary, 78–79
    properties of, 96t, 478t–479t
Molecular formulas, **75**–78
    vs. empirical formulas, 103–105, 150–152
Molecular geometry, **363**–379. *See also*
    Valence-shell electron-pair repulsion
    (VSEPR)
Molecular models, 77, *322, 360*–361
Molecular orbitals, 373, A.23–A.27
Molecular polarity, **379**–383
Molecular shapes, **361**–371. *See also* Valence-
    shell electron-pair repulsion (VSEPR)
Molecular speeds, 20–*21*, 415–417
Molecular structures, **359**–406
    and acid strength, 751–756
    and biomolecules, 392–396
    and chirality, 396–399
    and entropy, 819
    and molecular models, 360–361
    and molecular polarity, 379–383
    and molecular shapes, 361–371
    and noncovalent interactions, 383–392
    and orbital hybridization, 371–379
Molecular weights, **97**
Molecules, **26**, 75
    single-element, 26–28
Moles (mol), **57**–59, A.17t
    in chemical reactions, 134–140
    of compounds, 97–101
    and gas laws, 420–422
Molina, Mario, 442, 572
Momentum, **279**
Monatomic ions, **87**–88
Monodentate ligands, *1006*
Monomers, **543**, 754
Monoprotic acids, **744**
Monosaccharides, **108**, 562–563
    and digestion, 843
Monounsaturated fatty acids, 332, **541**
Montreal Protocol, 443

Moseley, H. G. J., 61
MRI. *See* Magnetic resonance imaging (MRI)
Multiple covalent bonds, **325**–333. *See also*
    Covalent bonds
    and hybrid orbitals, 376–378
    and molecular geometry, 365–366
Multiple proportions, law of, **23**
Mylar, 550
Myoglobin, 1012

N

n-type semiconductors, **497**–498
NAD⁺. *See* Nicotinamide adenine dinucleotide
    ions (NAD^+)
Nanoscales, **18**–*19, 29*–*30*, 58
    and chemical reactions, 122, 134–140
    and equilibrium, 663–665
    and states of matter, 460, 463, 479
Nanotubes, 28
National ambient air quality standards
    (NAAQS), 444
Natural gas, 247–249, 519, 521–522. *See also*
    Methane
    and carbon cycle, 130–*131*
    combustion of, 514
    energy from, 523t
    as energy resource, 815, 842
Natural logarithms. *See* Logarithms
Nernst equation, 887–**888**, 893
Net ionic equations, **167**–169
    for acid-base reactions, 175–177, 756–757
    for precipitation reactions, 802–803
Network solids, 478t–479t, **500**–505, *684*
Neurons, 892–894
Neutral solutions, **736**
Neutralization reactions, 173–176
Neutrinos, 947
Neutron capture, 956
Neutrons, **45**–46, 52t. *See also* Nucleons
    and artificial transmutations, 935
    in nuclear reactions, 921–922, 936–938
Newtons (N), **410**, A.18t
Nicad batteries. *See* Nickel-cadmium batteries
Nickel (Ni)
    in batteries, 870–871, 897, 900–901
    as catalyst, 501
    in coins, 996–*997*
    complex ions of, 802t, *1002*–1003, *1008*
    ferromagnetism of, 296, 489
    in industrial processes, 970, *972*, 991–992t
    in magnets, 991
    oxidation states of, *988*
Nickel-cadmium batteries, 870–871, 897
Nickel-metal hydride batteries, 900–901
Nitric acid, 171t
    in acid rain, 448, 793, 965
    in aqua regia, 999
    as Brønsted-Lowry acid, 729
    as monoprotic acid, 744
    as oxidizing agent, *181t, 975, 997*
    as oxoacid, *752*
    production of, 966
    as strong acid, 741
Nitric oxide. *See* Nitrogen monoxide

Nitrite ions, 89t, *93*, 745t
Nitrogen (N), *964*
    in the atmosphere, 408, *412*–413t, 432,
      813, 957, 961–962t
    in blood, 433, 694
    and bond energy, 336
    and catalytic converters, *616*
    in fertilizers, *102*, 633, 661, 964–965
    in the human body, 105–106t
    and hydrogen bonds, 387–388
    in industrial processes, 132, 142, 633–*635*,
      640–641, 666–668
    molecular orbitals in, A.25–A.26t
    properties of, 25, 962–966
    triple covalent bonds in, *326*, 964
    uses for, 962–967
Nitrogen cycle, *965*
Nitrogen dioxide, *997*
    and acid rain, 728, 793
    in the atmosphere, 413t, 440, 813
    and automobile emissions, 518, 520, 522,
      661
    color of, 408, *997*
    as free radical, 346–*347*, 440
    in industrial processes, 966
    as pollutant, 445–451, 814
Nitrogen fixation, 661, 666, **964**–965
Nitrogen monoxide
    in the atmosphere, 440, 813–814
    in automobile emissions, 518, 520, 522,
      616, 661
    as free radical, 346–347, 440
    in industrial processes, 966
    and Le Chatelier's principle, 660–661
    natural formation of, *965*
    and ozone destruction, 647
    as pollutant, 445–451, 793, 814
NMR. *See* Nuclear magnetic resonance (NMR)
Nobel, Alfred, *128*
Nobel prizes, 128
Noble gas notation, **289**
Noble gases, *62*, **66**, 408
    in the atmosphere, 412–413t, 957,
      961–962
    compounds of, 66, 304–305, *364, 369*, 964
Noble gases electron configurations, 289,
    294t. *See also* Octet rule
    and covalent bonds, 316–317
Noble metals, *188*, 961
Nomenclature, A.28–A.33
    of alcohols, 534, A.31–A.32t
    of aldehydes, A.31–A.32t
    of alkanes, A.28t–A.30
    of alkenes, A.30
    of alkynes, A.30
    of amides, A.32t
    of amines, A.32t
    of anions, 92–93
    of benzene derivatives, A.30–A.31
    of branched-chain alkanes, 85–86,
      A.28–*A.30*
    of carboxylic acids, A.32t–A.33
    of cations, 92
    of complex ions, 1003–1005
    of coordination compounds, 1003–1005
    of elements, 24–25t, 697, 999

Nomenclature *(Continued)*
    of ethers, A.32*t*
    of functional groups, A.28, A.31–A.33
    of halides, A.32*t*
    of hydrocarbons, A.28–A.31
    of ionic compounds, 92–94
    of ions, 92–94
    of ketones, A.31–A.32*t*
    of molecular compounds, 78–79
    of organic compounds, A.28–A.33
    prefixes for, 78*t*
    Stock system for ionic compounds, 92
    of unbranched alkanes, A.28*t*
Noncovalent interactions, 383–392, **384.** See
    also Dipole-dipole interactions;
    Hydrogen bonds; London forces
    and biological activity, 390–392
    in enzymes, 607, 611
    in proteins, *561*
    in solutions, 681–687
Nonelectrolytes, **97**, 173
Nonmetals, **25**, 61–*62*
    in combination reactions, 126
    compounds of, 90
Nonpolar covalent bonds, **338–339**
Nonpolar molecules, **379**
    and acid strength, 751–752
Normal boiling points, **463**–*464, 472*
Normal freezing points, *472*
Novocain hydrochloride, *759*
NO$_x$, 449
Nuclear binding energy. *See* Binding energy
    (E_b); Binding energy per nucleon
Nuclear burning, **955**–*956*
Nuclear chemistry, 918–953
    and artificial transmutations, 935–936
    and disintegration rates, 928–934
    and nuclear reactions, 920–924
    and radiation, 942–945
    and radioactivity, 919–920, 945–949
    and stability of atomic nuclei, 924–928
Nuclear fission, 927, **936**–*940*
Nuclear fuels, 247
Nuclear fusion, 927, **940**–*941*
Nuclear magnetic resonance (NMR),
    298–*299*
Nuclear medicine, 919, 928, 935, **946**–949
Nuclear reactions, 920–924, **921**
    and artificial transmutations, 935–936
    chain, *936*–*937*
Nuclear reactors, *918*–919, 929, 936–**937**,
    971
Nuclear waste, *939*–*940*
Nuclei, 44–*45*, *47*
    stability of, 924–928
Nucleons, **921**
Nucleotides, *393*
Numbers, A.3–A.6
Nutrients, **839**, *843*
Nutrition, 106–111. *See also* Food
    and antioxidants, 349
    and dietary fuels, 213, 249–252, *843*
    and dietary minerals, 105–107, 1012
    energy units for, 213
    and essential amino acids, *557t*, 968
    and trace elements, 105–107

Nutrition *(Continued)*
    and vitamins, *105*, 148–150, 349,
        352–353, 487, 591, 685–*687*, 1012
Nylon, 478*t*–479*t*, 543, 551–*552. See also*
        Polyamides

O

Octahedral crystal lattice sites, 486
Octahedral geometry, *361, 363t, 369, 379t,*
    1009
Octane
    combustion of, 133, 140, 428, 849
    and octane numbers, 85, 517*t*
Octane enhancers, 518–519, 970, 979
Octane numbers, 351, **516**–517*t*
Octet rule, **318**
    exceptions to, 346–349
Octets, expanded, 347–348*t*, 369–371,
    378–379
Oil. *See* Crude oil; Petroleum
Oil-drop experiment, 43–*44*
Oils, 110–*111*
    cis-trans isomerism in, 331–333, 541
    and detergents, 715–*716, 764*–765
    and esterification, 540
Optical fibers, 490, **504**–*505*
Optical isomers, 707, *1010*–*1011*
Orbitals, **280**–286, *281*
    delocalized, 493
    and molecular geometry, 371–379
Order, nanoscale. *See* Entropy (*S*)
Orders of reactions, 581–587, **582**
Ores, 138–140, **960**–961, 974, 988–989, 993
Organic acids, 174, 734, 744
Organic chemistry, 63
Organic compounds, **75**, 529–531
    nomenclature of, A.28–A.33
Osmosis, **708**–*712*
    reverse, 711–*712*
Osmotic pressure (π), **708**–712
Ostwald process, 966
Overall reaction orders, 581–587, **582**
Oxidation, **179.** *See also* Oxidation-reduction
    reactions
Oxidation numbers, 182–186, **183**, 864
    of transition metals, 984, *988*–989
Oxidation-reduction reactions, **178**–182
    and breathing, 349
    and displacement reactions, 186–189
    and electrochemistry, 864–871
    obtaining pure elements using, 960,
        969–980
    and oxidation numbers, 182–186
Oxidative phosphorylation, 842
Oxidizing agents, **178**–182, 181*t*, 864
    and photochemical smog, 449
    and standard reduction potentials,
        880–881
Oxoacids, **752**–753
Oxoanions, **92**
Oxygen (O), 16
    allotropes of, *26*, 441
    in the atmosphere, 408, *412*–413*t*, 432,
        440, 813, 957, 961–962*t*

Oxygen (O) *(Continued)*
    biologically active, 349
    in blood, 433, 694, 1011–1012
    and combustion reactions, *121*–124,
        134–139, *413*
    discovery of, 127, 830
    in the earth's crust, *957*–959
    as free radical, 440, *450*–451
    in fuel cells, **898**–899
    isotopes of, 924
    molecular orbitals in, A.25–A.26*t*
    molecular speeds of, *415*–416
    as oxidizing agent, 178–179, 181*t*
    paramagnetism of liquid, *347*, A.23, A.26
    properties of, 25, **962**–963
    standard state of, 243
    uses for, 962–963
Oxygenated gasolines, 519–521
Ozone, 441–444
    as allotrope of oxygen, *26*, 441
    and automobile emissions, 449, 520
    destruction of stratospheric, 443, 588,
        592–593, 597–598, 647
    as greenhouse gas, 525
    as pollutant, 343, 347, 440, 445, 448–451
    as resonance hybrid, *343*–344
    stratospheric, 64, 343, 441–444
    tropospheric, 413*t*, 440, 445
    and water purification, 718
Ozone hole, *442*–443
Ozone layer, 441–*444*

P

p-block elements, **291**, 984
p-n junctions, **498**
p-type semiconductors, **497**
Paramagnetism, **296, 347**, 984, *986*,
    A.23
Parent nuclei, 921–923
Partial hydrogenation, 335, **541**
Partial pressures, **432**–436
    and equilibrium constants, 643
Particle accelerators, 496
Particulates, **444**
Parts per billion (ppb), 412, **697**, 699*t*
Parts per million (ppm), 412, **697**, 699*t*
Parts per trillion (ppt), **697**, 699*t*
Pascals (Pa), **410**, A.18*t*
Pauli exclusion principle, **285**, 288
Pauling, Linus, 277, 339–*340*
Pentane, 77
Peptide linkages, **556**, 559
Perc. *See* Tetrachloroethylene
Percent abundances, 55–*56*
Percent composition by mass, 101–102,
    148–152
Percent yields, 145–148, **146**
Perchloric acid, 171*t*, 741
    as oxoacid, *752, 753*
Periodic table, 60–66, **61**, *62*
    biological, *105*–107
    and electron configurations, 265–314
Periods, 61–*62*
PET. *See* Poly(ethylene terephthalate) (PET);
        Positron emission tomography (PET)

Petroleum, 247, 249, *513–521*
  in oil spills, *683,* 720
  and carbon cycle, 130–*131*
  and catalysts, 613
  and detergents, 715
  energy from, *523t*
  as energy resource, 815
  and smog, 449
  sulfur from, 967
Petroleum fractions, **514**–*515*
pH, *727,* **737**–741
  and buffer solutions, 779–782, 784–786
  measurement of, 890–891
  and zwitterions, 755
pH meters, *739*–740, 786, **891**
Phase changes, **224,** 465–475
  and enthalpy, 223–229, 232
Phase diagrams, **471**–474
Phenolphthalein, 170, 199, *786, 788, 790,*
  902, 908
Phosphate anhydride linkages, 977
Phosphate groups, *393, 842–843*
Phosphate minerals, 957, 976–977
Phospholipids, **390**–*391*
Phosphoric acid, 171*t, 393,* 744
  and fertilizers, 977
  in household cleaners, 765
  as polyprotic acid, 744, 746*t*
  production of, 976
Phosphorus (P), *25,* 978
  allotropes of, 977–*978*
  discovery of, 977
  in fertilizers, *102,* 964, 976–977
  in the human body, 106*t*
  production of, 976–978
  properties of, 25, 976–978
Photochemical reactions, **440**
Photochemical smog, **449**–451. *See also*
  Smog
Photodissociation, **440**–441, 443, 448,
  450–451
Photoelectric effect, 269–272, **270,** 306
Photography, 197–198, 801, 979
Photons, **271,** 362
  and chemical reactions in the atmosphere,
  440
Photosynthesis, 13, **844**
  and carbon cycle, 130, 514, 933
  and Gibbs free energy, 844–847
  and greenhouse effect, *525*–526
  and oxygen production, 963
Photovoltaic cells, *499*
Physical changes, **6**–7
Physical constants, A.22*t*
Physical properties, **6**
Pi (π) bonds, **376**–378, *A.25–A.27*
  conjugated, 380
Pig iron, 990
Planck, Max, 269, 271, 273
Planck's constant (*h*), **269,** 275, A.22*t*
Plasmas, **941**
Plastics. *See also* Polymers
  and catalytic cracking, 517
  polyethylene, *242*
  production of, 519, 848
  recycling of, *554–555t*

Plastics *(Continued)*
  types of, 543
Polar covalent bonds, **339**
Polar molecules, **379**–383
  and acid strength, 751–752
  in solutions, 682
Polarity. *See* Molecular polarity
Polarization, **385**
Pollution, 412, 444–451. *See also* Global
  warming; Greenhouse effect; *specific*
  pollutant
  and coal, 252, 445–446, 449, 522
  and heavy metals, 15, 444, 451, 519, 895,
  897
  and hydrocarbons, 445, 447–448, 451,
  793
  nitrogen compounds, 445–451, 793,
  814
  and ozone, 343, 347, 440, 445, 448–451
  and smog, *445,* 449–451, *642*
  sulfur compounds, 252, 444–447*t,* 451,
  793, 967–968
  thermal, *692*
Polyamides, **551**–554
  proteins as, 556
Polyatomic ions, **89***t*
Polyatomic molecules, molecular orbitals for,
  A.27
Polydentate ligands, **1006**
Polyesters, 549–551, **550**
Polyethylene, *242,* 328
  as polymer, 489, 543–546*t*
  production of, 519, *848*–849
  as thermoplastic, 543
Poly(ethylene terephthalate) (PET), *550,*
  554–555*t*
Poly(ethylene vinylacetate) (EVA), *549*
Polymers, **393,** 542
  addition, 543–547
  biological, 542, 555–565
  co-, 548–*549*
  condensation, 543, 549–551, 556
  of monosaccharides, 109
  properties of, 489
  synthetic organic, 542–555
Poly(methyl methacrylate), 546*t*
Polypeptides, **556**
Polypropylene, *242,* 328, 519, 546*t,* 555*t*
Polyprotic acids, **744,** 746*t*
  and titrations, *793*
Polyprotic bases, 746
Polysaccharides, **109,** 562–563
  and enzymes, 609
Polystyrene, 480, 543, 545–546*t,* 555*t*
Polytetrafluoroethylene, 543, 546*t*
Polyunsaturated fatty acids, 332, **541**
  and free radicals, 349
Poly(vinyl acetate) (PVA), 546*t*–547
Poly(vinyl chloride) (PVC), 546*t,* 555*t,* 972
Positron emission, **923,** *925–926,* 947
Positron emission tomography (PET), 923,
  947–948
Potassium (K), *188*
  and activity series, 187*t*
  in fertilizers, *102,* 964
  in the human body, 106*t*

Potassium (K) *(Continued)*
  in industrial processes, 974
  and ion pumps, *893–894*
  in neurons, *892*–894
  production of, 961*t*
Potential differences. *See* Standard reduction
  potentials (*E°*); Voltages
Potential energy (*Ep*), **212**
  chemical, 212, 847
  and covalent bonds, *317*
  electrical, 875
  and elementary reactions, 589
  and probability, 814–815
Precious metals, 974, 984
Precipitates, **166**
Precipitation reactions, 166–167
  prediction of, 802–804
  selective, 805
Precision, **A.6**
Prefixes, for SI units, **A.17**–A.18*t*
Preservatives, food, 174, 945
Pressure (*P*), **409**–412. *See also* Partial pressures; Vapor pressure
  atmospheric profile of, *413*
  critical, 474–475
  and gas laws, 417–422
  and Le Chatelier's principle, 659–660
  in liquids, 460
  in solids, 479
  and standard temperature and pressure
  (STP), 423, A.22*t*
  units for, 230, 410*t,* A.18*t,* A.20*t*
Priestley, Joseph, 127, 830
Primary alcohols, 534–*535*
Primary batteries, 894–*895*
Primary pollutants, **444**–445
Primary structures of proteins, **559**
Principle energy levels, **281**
Principle quantum numbers, **273**
Principles. *See* Hypotheses; Laws, scientific;
  Le Chatelier's principle; Pauli exclusion
  principle; Theories; Uncertainty principle
Probability. *See also* Entropy (*S*)
  and chemical reactions, 814–817
  of finding electrons, 280
  and nanoscale processes, 663–665,
  817
Problem solving, 8–10, A.2–A.16
  and dimensional analysis, 8–10, 48–49,
  App. A.2
  using estimation, 24, 58
Product-favored reactions, **633,** 813–814
  and coupling with reactant-favored reactions, 838–839, 848–849
  and electrochemistry, 864, 871
Products, **11,** 121–163
Propane
  combustion of, 234–*235*
  as fuel, 249
  in natural gas, 521
Properties. *See* Chemical properties; Physical
  properties
Proportionality factors, **9**–10, 47, 51*t,*
  A.4–A.6, A.19–A.21
Propylene, 328, 519, 544, 546*t*

Proteins
  amino acids in, 557*t*, 754
  as biocopolymers, 542, 555
  denaturation of, 766
  as dietary fuels, 249–251*t, 843*
  and enzymes, 607
  hydrogen bonds in, 559–*561*
  shapes of, *384,* 387
  toxic, 636
Protons, 44–45, 52*t. See also* Nucleons
  charge on (*e*), A.22*t*
  in nuclear reactions, 921–922, 935,
    940–941
  rest mass of (*m_p*), A.22*t*
Purification, 13–15. *See also* Electrorefining;
    Refining
  of elements, 200, *497,* 994–*995,*
    999–*1000,* 1009
  of water, 17, 180, 711, 716–718, 792
PVA. *See* Poly(vinyl acetate) (PVA)
PVC. *See* Poly(vinyl chloride) (PVC)
Pyrometallurgy, 989–*990*

Q

Quadratic equations, 654, **A.15**–A.16
Quadratic formula, 654–655, A.15
Qualitative data, **5**
Quanta, **269**, 362
Quantitative data, **5**
Quantities, A.3–A.6
Quantum mechanical model of atoms,
    277–280
Quantum numbers, 280–286, 283*t*
  fourth (*m_s*), 284–286
  principle (*n*), **273**, 281
  second (*l*), 281–283
  third (*m_l*), 283–284
Quantum theory, 268–271, **272**
Quartz, *18, 958. See also* Silicon dioxide
  in the earth's crust, 958
  as network solid, 478*t*–479*t*, 501, *684*
  and production of glass, 503–504

R

r processes, 956
Radiation, 942–945. *See also* Alpha (α) parti-
    cles; Beta (β) particles; Gamma (γ) ra-
    diation; Ultraviolet radiation;
    X-radiation
  alpha, 920
  background, *942*–944
  beta, 920
  cosmic, 933, *942*–943
  electromagnetic, 266–*268, 267,* 920
  infrared, 267, *525*–526
  microwave, 267, 382
  ultraviolet, 267, 271
Radiation therapy, 942, 944, 946–947
Radicals. *See* Free radicals
Radii
  atomic, 46–51, **296**–300, *334,* 987–*988*
  ionic, *301*–302, 820
Radioactive dating, 933–934
Radioactive decay
  and half-lives, 586, 928–930
  rates of, 930–933

Radioactive series, **922**, 944
Radioactivity, **42**, 919–920
  and alpha decay, 65, 921–*923*
  applications of, 945–949
  and beta decay, 921–*923*
Rads, **942**
Raoult's law, **700**
Rate constants (*k*), **578**. *See also* Half-lives
    (*t_{1/2}*)
  and decay constants, 930–932
  and temperature, 593–596
Rate laws, **578**–581
  for elementary reactions, 597–599
  for nuclear reactions, 928–934
  and orders of reaction, 581–587
Rate-limiting steps, **600**
Rates, **573**. *See also* Reaction rates
Reactant-favored reactions, **633**, 813–814
  and coupling with product-favored reac-
    tions, 838–839, 848–849
  and electrolysis, 899–904
Reactants, **11**, 121–163
  limiting, 140–145
Reaction intermediates, **600**
Reaction mechanisms, 599–604
Reaction quotients (*Q*), **650**–*651*
  and Gibbs free energy, *833*–834
  and Nernst equation, 888
Reaction rates, 572–577, **573**. *See also*
    Chemical kinetics
  and concentration, 577–581
  maximum, *611*
  and stoichiometry, 575–576
  and temperature, 593–596
Reactions. *See* Chemical reactions; Nuclear re-
    actions
Rechargeable batteries. *See* Secondary batter-
    ies
Recycling
  of aluminum, 848, 907, 975
  of auto batteries, 897
  of plastics, *554*–555*t*
Redox reactions. *See* Oxidation-reduction re-
    actions
Reducing agents, **178**–182, 181*t*, 864
  and industrial smog, 449
Reduction, **179**. *See also* Oxidation-reduction
    reactions
Refining
  of petroleum, 514–516, 613
  zone, *497*
Reformulated gasolines, 519–521
Refraction, 271
Refrigerants, 444
Relativity, special, 927
Rems, **942**
Replication of DNA, 394–*396, 395*
Resonance structures, 343–345, 350–351,
    441, A.23, A.27
  molecular geometry of, 365–*366*
Reverse osmosis, **711**–*712*
Reversible processes, **817**
Risks *vs.* benefits, 31*t*–32, 944
Roentgens (R), **942**
Röntgen, Wilhelm Conrad, 488, 942
Rounding, A.8–A.9

Rowland, F. Sherwood, *442,* 572
Rubber, 531, 543, 547–549, 968
Rust, 235, 813, 907–*908, 989*
Rutherford, Ernest, 44–46, 272, 277,
    920–921, 926, 935
Rydberg constant (*R_∞*), 275, A.22*t*

S

s-block elements, **291**, 984
  atomic radii of, *300*
  ionization energies of, 303
s processes, 956
sp hybrid orbitals, *373*–374
sp² hybrid orbitals, *374*
sp³ hybrid orbitals, *374*–375
sp³d hybrid orbitals, *378*
sp³d² hybrid orbitals, *378*
Sacrificial anodes, 909
Salicylic acid, 142–143, 353, 540, 767
Saline solutions, 696
Salt bridges, **872**–875, 884.
  in batteries, 894–897
  in chlor-alkali process, 972
  and corrosion, 908–909
Salts, **173**. *See also* Sodium chloride
  and acid-base reactions, 756–761
  and displacement reactions, *188*
Saponification, 542, *714*–715
Saturated fats, **332**, 335
Saturated hydrocarbons, **322**–325
  molecular geometry of, 367–*368*
Saturated solutions, **688**–*689,* 794. *See also*
    Solubility; Solubility product constants
    (*K_sp*)
SBR. *See* Styrene-butadiene rubber (SBR)
Scanning tunneling microscopy (STM), 22, *41,*
    48–**49**
Schrödinger, Erwin, 280–281, 286, 340
Scientific method, 5–6
Scientific notation, 43, A.10–A.12
Scuba diving, *694*
Sea water
  desalinization of, 17, 711
  ions in, 712*t*, 957, 972–973, 978
  vapor pressure of, 700–*701*
Seaborg, Glenn, 25*t*, 935
Seaborgium (Sg), 24–25*t*
Second law of thermodynamics, 822–827,
    823
Second-order reactions, *583*–585
Secondary alcohols, 534–536
Secondary batteries, **894**–898
Secondary pollutants, **449**
Secondary structures of proteins,
    559–560
Seconds (s), 904, A.17*t*
Seesaw geometry, *364t,* 369
Semiconductors, 26, 491–496, **494**
Semipermeable membranes, **708**, 873,
    971–*972*
Sex hormones, 537–*538*
Shared electron pairs. *See* Bonding electrons;
    Covalent bonds
Shells, **281**

Shifting an equilibrium, **655**. *See also* Le Chatelier's principle
SI units. *See* International System of Units (SI)
Sickle cell anemia, 559–*560*
Side chains, 556
Sieverts (Sv), **942**
Sigma (σ) bonds, **375–378**, *A.24*–*A.27*
Significant figures, 59–60, A.6–A.10
    and logarithms, A.13–A.14
Silica. *See* Silicon dioxide
Silicate minerals, 957–960
Silicates, 501
Silicon (Si), *14*
    in the earth's crust, *957–959*
    and electronics, 496–500
    production of, 14, 496–*497*, 961*t*
    in steel, 991–992*t*
Silicon chips, 490, 496–500
Silicon dioxide. *See also* Quartz
    in air bags, 967
    in ceramics, 502
    in the earth's crust, 958
    in optical fibers, 504–505
Silk, 560
Silver (Ag), *188*, 998–1000
    in coins, 188
    complex ions of, 768, 800–802*t*, *1009*
    in electrochemical cells, 872
    and electroplating, 904
    as precious metal, 974, 984
    properties of, 998–999
    purification of, 999–1000
    removal of tarnish from, 880
Silver bromide, 197–*198*, 794–796*t*, 801, 803, 979
Silver nitrate
    and displacement reactions, 129, *188*
    and photography, *198*
Simple cubic unit cells, *482*, 484–*485*
Single bonds, 108
Single covalent bonds, **318**–322. *See also* Covalent bonds
Slopes, 577, 583–584, App. A.8
Smalley, Richard E., *28*
Smog, *445*, 449–451, 642. *See also* Pollution
Soaps, 462, 542, 714–716
    and hard water, 718
    in household cleaners, 764–765
Soddy, Frederick, 920
Sodium (Na), *970*
    and activity series, 187*t*
    in the human body, 106*t*
    in industrial processes, 974, 983
    and ion pumps, 894
    in neurons, 892–894
    production of, 961*t*, 969–971
    properties, 969–971
Sodium azide, 428, 966–967
Sodium bicarbonate
    as antacid, 761–762*t*
Sodium chloride, 86, 90, 98
    as acid-base salt, 173–174, 756–757
    crystal structure of, *486–487*
    electrolysis of, 900–*901*, 904, 969–*972*

Sodium chloride (*Continued*)
    in intravenous solutions, 199
    as ionic compound, 86–*87*, 93–94*t*
    properties of, 95–96
    in sea water, 957
Sodium hydroxide, 62, 171*t*
    electrolysis of, 903–904
    in household cleaners, 764–766
    production of, 904, 969, 971–*972*
    and saponification, *542, 714*
    as strong base, 172
    and titrations, *788–793*
    uses for, 971–972
Soft water, 719
Solar cells, **225, 499**
Solar energy
    electricity generation from, *225, 247*, 499
    and fuel cells, 249, 899
    and nuclear fusion, *941*
    and photosynthesis, 844–*845*
Solidification. *See* Freezing points
Solids, **18**, 479–505
    amorphous, 480
    crystalline, 480–487
    electrical classes of, 491–496
    and materials science, 487–491
    network, 500–505
    and phase changes, 465–475
    properties of, 478*t*–481
Solomon, Susan, *443*
Solubility, 163–166, **688**–692
    and acids, 797–798
    and common ion effect, 798–800
    equilibria involving, 794–797
    and Le Chatelier's principle, 692, 797
    and noncovalent interactions, 681–687
    pressure dependence of, 693–694
    rules for, 164*t*
    and solubility product constants, 796–797
    and temperature, 692–693
Solubility product constants (K_{sp}), 794–797, **795**, A.36*t*–A.37*t*. *See also* Ion products (Q)
    and solubility, 796–797
Solutes, **190**, 681
Solutions, *13*–*14, 18*, 680–726, 681*t*. *See also* Acids; Aqueous solutions; Bases; Water
    acidic, 736
    basic, 736
    buffer, 777–786
    colligative properties of, 700–707
    concentration of, 190–195, 694–700
    enthalpy of, 688, 690–*691*
    entropy of, 687–688, 691–692, 702, 709, *820–821*
    equilibria involving, 639–640, 688–692
    hydrogen bonds in, 681–*683*, 686–687
    hypertonic, 710–711
    hypotonic, *710*
    isotonic, *710*
    neutral, 736
    noncovalent interactions in, 681–687
    and osmotic pressure, 708–712
    preparation of, 192–195
    saturated, 688–689, 794
    standard, 199, 786

Solutions (*Continued*)
    supersaturated, *688–689*
    and surfactants, 714–716
    and temperature, 692–693
    thermodynamics of, 687–688, 690–692
    unsaturated, *688–689*
Solvation. *See* Hydration
Solvents, **190**, 681
    freezing of, *705*
Space-filling models, 77, *322, 360*
Specific heat capacity (*c*), **219**–220*t*, 477*t*
Spectator ions, **167**
Spectra, 266–*268, 267*
    continuous, *272–273*
    electromagnetic, 266–267*t*
    line emission, **273**, 276
    mass, *54*
Spectroscopy, 276, **362–363**
    infrared, 362–363, *526, 964*
    mass, 53–**54**
    nuclear magnetic resonance (NMR), 298
    ultraviolet-visible, 380
Spin, electron, 284–286, 318
Spontaneous reactions. *See* Product-favored reactions
Square planar geometry, 364*t*, 369–370, *1009*
Square pyramidal geometry, 364*t*, 369–370
Stability
    of atomic nuclei, 924–928
    thermodynamic *vs.* kinetic, 813–814, 849–852
Standard atmospheres (atm), **410**
Standard conditions, **876**
Standard enthalpy changes (ΔH°), **230**
    and chemical reactions, 231–236, 245–246
    of formation ($\Delta H_f°$), 243–247, 245*t*, 818, A.41*t*–A.47*t*
    of fusion ($\Delta H_{fus}°$), 468 *t*
    of vaporization ($\Delta H_{vap}°$), 467*t*
Standard equilibrium constants (K°), **834**
Standard Gibbs free energy of formation ($\Delta G_f°$), 828, A.4 *t*–A.47*t*
Standard hydrogen electrodes, **876**
Standard molar entropy (S°), 818*t*, A.41*t*–A.47*t*
Standard molar volume, **423**
Standard reduction potentials (E°), **879**–884, 881*t*, A38*t*–A.40*t*
    and equilibrium constants, *886–887*
    and Gibbs free energy, 884–887
Standard solutions, **199**, 786
Standard states, **243**
Standard temperature and pressure (STP), **423**, A.22*t*
Standard voltages (E°), **876**
Starch, 563–564
    as biopolymer (polysaccharide), 542, 555, 563
State functions, **229**
Steel, 972, *974, 976, 984*, **991**–993
Stereoisomers, 1010. *See also Cis-trans* isomers; Optical isomers
Steric factors, **592**
Steroids, 537–538
Stock system, 92

Stoichiometric coefficients, **124**
Stoichiometry, **124**
   in aqueous reactions, *195*
   in gaseous reactions, 426–429
   and reaction rates, 575–576
   and reaction tables, 644–646
Stomach acid, 174–175.
Storage batteries. *See* Secondary batteries
STP. *See* Standard temperature and pressure
   (STP)
Straight-chain alkanes, 83–85, *322*
   and octane numbers, 516
Strassman, Fritz, 936
Stratosphere, *413*–414. *See also* Ozone hole;
   Ozone layer
Strong acids, **170**–171*t, 729, 732*–734
   ionization constants for, 741
   salts of, 756–757, 759–760
   and titrations, *788*–789, *792*
Strong bases, 171*t*–**172**, *732*–734
   salts of, 756–759
   and titrations, *788*–792
Strong electrolytes, **97**, 170–171*t*, 173
   strong acids as, 729
Structural formulas, **76**–77
Structural isomers. *See* Constitutional isomers
Styrene, 545–546 *t, 548*–549
Styrene-butadiene rubber (SBR), 548–*549*
Styrofoam, 489, 545. *See also* Polystyrene
Subatomic particles, 42–44
Sublimation, *465*, **469**–470, *979*
   and phase diagrams, *472*–473
Subshells, 281–283
Substances, **6**, *18*
Substitution reactions, 350
Substrates, **608**–*610*
Successive approximations, A.15–A.16
Sucrose, *16*
   chirality of, 397
   combustion of, *138*–139
   as nonelectrolyte, 97
Sugar, table. *See* Sucrose
Sulfa drugs, 612–613
Sulfur (S), *25, 59, 967*–968
   allotropes of, 967–968
   in coal, 522
   in the human body, 106*t*
   production of, 967–969
   properties of, 25, 967–969
   uses for, 967–969
   and vulcanized rubber, 547–*548*
Sulfur dioxide
   and acid rain, 440, 445–446, 728, 793
   in the atmosphere, 413*t*, 440
   and industrial smog, 449
   as pollutant, 252, 444–447*t*, 451,
     967–968
Sulfur trioxide
   and acid rain, 440, 446
   in the atmosphere, 440
   as pollutant, 444
   and production of sulfuric acid, 968
Sulfuric acid, 171*t*–172, **744**
   in acid rain, 466, 793
   in batteries, *896*–897
   as polyprotic acid, 744

Sulfuric acid *(Continued)*
   production of, 639, 968–969
   as strong acid, 741
Superconductors, 494–496, **495**
Supercritical fluids, *472, 474*–475, 530
Superphosphate, 976–977
Supersaturated solutions, *688*–**689**
Surface area, 572
Surface tension, **461**–462*t*, 477*t*
Surface water, 716
Surfactants, 462, **714**–716, 764
Surroundings, thermodynamic, *217,*814
Suspensions, 713
Synthetic organic polymers, 542–555
Système Internationale d'Unités. *See*
   International System of Units (SI)
Systems, thermodynamic, *217*, 814
Szilard, Leo, 936

T

T-shaped geometry, 364*t*, 369–370
Table salt. *See* Sodium chloride
Table sugar. *See* Sucrose
Teflon. *See* Polytetrafluoroethylene
Temperature (*T*), **6**–7. *See also* Temperature
   scales
   critical, *474*–475, 495*t*–496
   and denaturation, 611–*612*
   and entropy, 665, 823, 826
   and equilibrium constants, 638, 797
   and gas laws, 417–422
   and Gibbs free energy, 829–831
   and global warming, 527–*529*
   and kinetic-molecular theory, *20*–21
   and Le Chatelier's principle, 660–663
   and reaction direction, 829–831
   and reaction rates, 593–596
   and solubility, 692–693
   and standard temperature and pressure
     (STP), 423, A.22*t*
   and thermal energy, *215*–216, 815
   units for, A.17*t*, A.20*t*
Temperature scales, 6–7, *419*–420,
   A.19–A.21
Terephthalic acid, *539, 549, 552*
Terrestrial elements, 957–961
Tertiary alcohols, 534–536
Tertiary structures of proteins, **561**–**563**
Tetrachloroethylene, 530
Tetraethyllead, 518–519, 970
Tetrahedral geometry, *361*, 363*t*, 366–369,
   376*t, 1009*
Theoretical yields, **145**
   and chemical equilibrium, 633
Theories, **5**, 211. *See also* Atomic theory;
   Band theory; Hypotheses; Kinetic-mole-
   cular theory; Laws, scientific; Quantum
   theory
Thermal conductivity, 477*t*–479*t*, 489–491,
   500
Thermal decomposition, 127
Thermal energy, 212, *215*, 815, 847–848
Thermal equilibrium, **216**
Thermal pollution, *692*
Thermochemical equations, **229**–231

Thermodynamics, **211**, 812–862. *See also*
   First law of thermodynamics; Second
   law of thermodynamics; Third law of
   thermodynamics
   and chemical kinetics, 590, 813–814,
     849–851
   constants for, A.41*t*–A.47*t*
   and Le Chatelier's principle, 823, 836
   and nuclear binding energy, 927–*928*
   and systems, 217–218
Thermoplastics, **543**
Thermosetting plastics, **543**
Third law of thermodynamics, **818**
Thomson, Joseph John, 43, 277, 920
Three Mile Island, 939
Thymine (T), 393–*395*
Time, units for, 904, A.17*t*
Tin (Sn)
   in bronze, 681*t*, 993, 995–996
Titanium (Ti)
   production of, 961*t*, 970, 972, 983
   uses for, *60, 983*
Titrants, **786**
Titration curves, **787**–788, *790, 792*–*793*
Titrations. *See* Acid-base titrations
TNT. *See* Trinitrotoluene (TNT)
Torr, **410**
Trace elements, dietary, 105–**107**
Tracers, **945**–946*t*
trans Fatty acids, 332, *335*, 541
trans isomers, *329*
Transistors, 490, 494, 498
Transition elements, 61–**63**, *292*, 983–1017.
   electron configurations for, 292
   properties of, *492*, 984–989
Transition metals. *See* Transition elements
Transition states. *See* Activated complexes
Transmutations. *See* Nuclear reactions
Transuranium elements, 935
Triangular bipyramidal geometry, 361, *363t*,
   369, 379*t*
Triangular planar geometry, 361, 363*t*, 366,
   369, 376*t*
Triangular pyramidal geometry, 364*t*, 367,
   369
Tricarboxylic acid cycle. *See* Citric acid
   cycle
Triglycerides, *330*–333, 540–541, 685
2,2,4-Trimethylpentane, *85*, 516–517*t*, 616
Tripeptides, *558*
Triple bonds, 321, **325**–333, 964. *See also*
   Covalent bonds
Triple points, *472*
Tritium (T), *55*, 57*t*, 924, 929*t*, 941
Troposphere, *413*–414
   ozone in, 441–442
Tyndall effect, *713*

U

Ultraacidic water, 740
Ultraviolet radiation, 267, 271
   and ozone layer, 441, *443*, 448
   and water purification, 718

Ultraviolet-visible spectroscopy, 380
Uncertainty principle, 279–280
Unimolecular reactions, **587**–590
    rate laws for, 597
Unit cells, *481*–*482*
    and density, 484
    of ionic compounds, 485–487
Units, A.3–A.6. *See also* Dimensional analysis;
    International System of Units (SI)
Universal gas constant. *See* Ideal gas constant
    (*R*)
Unpaired electrons
    and color of sulfur, 968
    and paramagnetism, 296, 986
Unsaturated fats, **332,** 335
Unsaturated hydrocarbons, **328**–333
Unsaturated solutions, *688*–*689*
Unshared electron pairs. *See* Covalent bonds;
    Lone pair electrons
Uranium (U)
    fission of, *936*–*938*
    isotopes of, 55, 924
    and Manhattan Project, 919
    radioactive decay series of, 921–*923*,
        944
    radioactivity of, 42, 919–920, 931

V

Valence bands, 493–*494*
Valence bond model, **371,** A.23, A.27
Valence electrons, 289–291, **290**
Valence-shell electron-pair repulsion (VSEPR),
    359, **361**–371, A.23
van der Waals equation, **438**–439
van't Hoff, Jacobus Henricus, *707*
van't Hoff factors, 707
Vapor pressure, 436*t*, **463**–464
    lowering of, 700–702
Vapor pressure curves, *472*
Vaporization, 225–226, *463*, **465**–466. *See
    also* Phase changes
    and London forces, *385*
    and phase diagrams, *472*
Vinegar, 145, 170, 648, 763, 765. *See also*
    Acetic acid
Viscosity, *461*
Volatility, **463**
    and petroleum distillation, *515*
Voltages. *See also* Standard reduction poten-
    tials (*E°*); Standard voltages (*E°*)
    of electrochemical cells, 875–879

Voltaic cells, **871,** 894. *See also* Batteries;
    Electrochemical cells
Volts (V), **875,** A.18*t*
Volume (*V*)
    and enthalpy, 227–229
    and gas laws, 417–422
    and Le Chatelier's principle, 659–660
    molar, at STP (*V*$_m$), A.22*t*
    units for, 47*t*, 50–51, A.18*t*, A.20*t*
VSEPR. *See* Valence-shell electron-pair repul-
    sion (VSEPR)

W

Water, *75,98, 360,* 475–479. *See also* Acid
    rain; Ice; Sea water; Solutions
    in acid-base reactions, 79
    as acid or base, 730
    in the atmosphere, 412–414, 432
    autoionization of, 736–737
    boiling point elevation of, 704
    collection of gases over, *435*–436
    conductivity of, 316
    contaminants in drinking, 696–697
    decomposition of, 649
    in decomposition reactions, 128–*129*
    density of, *9t*, 473, 477–478, 696
    electrolysis of, *29*, 128, 249, 837
    fluoridation of, 883–884, 978
    freezing point lowering of, 705
    as greenhouse gas, 525
    hard, 169, 656–657, 715, 718–719, 764
    hydrogen bonds in, 388–*389*, 459, 461,
        476–478, *683*
    meniscus of, *462*
    molecular geometry of, *367*
    phase changes in, 223–227, 231
    phase diagram for, *472*
    physical states of, *7, 460*
    as polar molecule, 381–*382*
    purification of, 17, 180, 711, 716–718, 792
    sigma bonds in, *376*
    as solvent, *690*–691
    surface, 716
Water ionization constants (K_w), **736,** 751
Water of hydration, **100**
Watson, James, 360, *394*–395, 489
Wave equations, 280
Wave functions, 280
Wavelength (λ), **267**
Weak acids, **170**–171*t*, 729–730, 732–734
    and buffers, 777–778

Weak acids *(Continued)*
    ionization constants for, 741–742, A.34*t*
    salts of, 757–761
    and titrations, 789–792
Weak bases, 171*t*–**172,** 732–734
    and buffers, 777–778
    ionization constants for, A.35*t*
    salts of, 759–761
    and titrations, *792*
Weak electrolytes, **170**–171*t*, 173
    weak acids as, 741–742
Weight
    units for, A.20 *t*
    vs. mass, 46, 57
Weight percent, **696,** 699*t*
Werner, Alfred, *1009*
Wilkins, Maurice, *394*–395, 489
Wöhler, Friedrich, 529
Work (*w*), 211, **215**–*218*
    electrical, 885
    maximum (*w*$_{max}$), 837–839
    and volume change, *228,* 816

X

X-radiation, 267, 271, 798, 920
    exposure to, 942, 944
X-ray crystallography, **488**–489
Xenon difluoride, *369*
Xenon tetrafluoride, *369*

Y

y-intercepts, 577, 583–584, App. A.8
Yields, reaction, 145–148
Yucca Mountain, 939–*940*

Z

Zero-emission vehicles, 901
Zeroth-order reactions, *583*–585
Zewail, Ahmed H., *593*
Zinc (Zn)
    and activity series, 187*t*
    in batteries, *894*–895
    in brass, 13, *996*–998
    in coins, 55,189, *996*–997
    in electrochemical cells, 872–879, 885
    as essential dietary mineral, 1012
    and galvanized iron, *909*
    properties of, 984
Zone refining, **497**
Zwitterions, 754–756, **755**

Name	Symbol	Atomic Number	Atomic Weight	Name	Symbol	Atomic Number	Atomic Weight
Actinium*	Ac	89	(227)	Neodymium	Nd	60	144.24(3)
Aluminum	Al	13	26.981538(2)	Neon	Ne	10	20.1797(6)
Americium*	Am	95	(243)	Neptunium*	Np	93	(237)
Antimony	Sb	51	121.760(1)	Nickel	Ni	28	58.6934(2)
Argon	Ar	18	39.948(1)	Niobium	Nb	41	92.90638(2)
Arsenic	As	33	74.92160(2)	Nitrogen	N	7	14.00674(7)
Astatine*	At	85	(210)	Nobelium*	No	102	(259)
Barium	Ba	56	137.327(7)	Osmium	Os	76	190.23(3)
Berkelium*	Bk	97	(247)	Oxygen	O	8	15.9994(3)
Beryllium	Be	4	9.012182(3)	Palladium	Pd	46	106.42(1)
Bismuth	Bi	83	208.98038(2)	Phosphorus	P	15	30.973762(4)
Bohrium*	Bh	107	(264)	Platinum	Pt	78	195.078(2)
Boron	B	5	10.811(7)	Plutonium*	Pu	94	(244)
Bromine	Br	35	79.904(1)	Polonium*	Po	84	(210)
Cadmium	Cd	48	112.411(8)	Potassium	K	19	39.0983(1)
Cesium	Cs	55	132.90545(2)	Praseodymium	Pr	59	140.90765(2)
Calcium	Ca	20	40.078(4)	Promethium*	Pm	61	(145)
Californium*	Cf	98	(251)	Protactinium*	Pa	91	231.03588(2)
Carbon	C	6	12.0107(8)	Radium*	Ra	88	(226)
Cerium	Ce	58	140.116(1)	Radon*	Rn	86	(222)
Chlorine	Cl	17	35.4527(9)	Rhenium	Re	75	186.207(1)
Chromium	Cr	24	51.9961(6)	Rhodium	Rh	45	102.90550(2)
Cobalt	Co	27	58.933200(9)	Rubidium	Rb	37	85.4678(3)
Copper	Cu	29	63.546(3)	Ruthenium	Ru	44	101.07(2)
Curium*	Cm	96	(247)	Rutherfordium*	Rf	104	(261)
Dubnium*	Db	105	(262)	Samarium	Sm	62	150.36(3)
Dysprosium	Dy	66	162.50(3)	Scandium	Sc	21	44.955910(8)
Einsteinium*	Es	99	(252)	Seaborgium*	Sg	106	(266)
Erbium	Er	68	167.26(3)	Selenium	Se	34	78.96(3)
Europium	Eu	63	151.964(1)	Silicon	Si	14	28.0855(3)
Fermium*	Fm	100	(257)	Silver	Ag	47	107.8682(2)
Fluorine	F	9	18.9984032(5)	Sodium	Na	11	22.989770(2)
Francium*	Fr	87	(223)	Strontium	Sr	38	87.62(1)
Gadolinium	Gd	64	157.25(3)	Sulfur	S	16	32.066(6)
Gallium	Ga	31	69.723(1)	Tantalum	Ta	73	180.9479(1)
Germanium	Ge	32	72.61(2)	Technetium*	Tc	43	(98)
Gold	Au	79	196.96655(2)	Tellurium	Te	52	127.60(3)
Hafnium	Hf	72	178.49(2)	Terbium	Tb	65	158.92534(2)
Hassium*	Hs	108	(269)	Thallium	Tl	81	204.3833(2)
Helium	He	2	4.002602(2)	Thorium*	Th	90	232.0381(1)
Holmium	Ho	67	164.93032(2)	Thulium	Tm	69	168.93421(2)
Hydrogen	H	1	1.00794(7)	Tin	Sn	50	118.710(7)
Indium	In	49	114.818(3)	Titanium	Ti	22	47.867(1)
Iodine	I	53	126.90447(3)	Tungsten	W	74	183.84(1)
Iridium	Ir	77	192.217(3)	Uranium*	U	92	238.0289(1)
Iron	Fe	26	55.845(2)	Vanadium	V	23	50.9415(1)
Krypton	Kr	36	83.80(1)	Xenon	Xe	54	131.29(2)
Lanthanum	La	57	138.9055(2)	Ytterbium	Yb	70	173.04(3)
Lawrencium*	Lr	103	(262)	Yttrium	Y	39	88.90585(2)
Lead	Pb	82	207.2(1)	Zinc	Zn	30	65.39(2)
Lithium	Li	3	6.941(2)	Zirconium	Zr	40	91.224(2)
Lutetium	Lu	71	174.967(1)	—[‡]		110	(269)
Magnesium	Mg	12	24.3050(6)	—[‡]		111	(272)
Manganese	Mn	25	54.938049(9)	—[‡]		112	(277)
Meitnerium*	Mt	109	(268)	—[‡]		114	[285]
Mendelevium*	Md	101	(258)	—[‡]		116	[289]
Mercury	Hg	80	200.59(2)	—[‡]		118	[293]
Molybdenum	Mo	42	95.94(1)				

[†] The atomic weights of many elements vary depending on the origin and treatment of the sample. This is particularly true for Li; commericially available lithium-containing materials have Li atomic weights in the range of 6.96 and 6.99. Uncertainties are given in parentheses following the last significant figure to which they are attributed.

[*] Elements with no stable nuclide; the value given in parentheses is the atomic mass number of the isotope of longest known half-life. However, three such elements (Th, Pa, and U) have a characteristic terrestial isotopic composition, and the atomic weight is tabulated for these.

[‡] Not yet named.